Research Reports in Physics

Research Reports in Physics

S. M. Qaim (Ed.)

Nuclear Data

for Science and Technology

Proceedings of an International Conference,
held at the Forschungszentrum Jülich,
Fed. Rep. of Germany, 13–17 May 1991

Organized in Cooperation with
OECD-Nuclear Energy Agency
International Atomic Energy Agency

Springer-Verlag Berlin Heidelberg GmbH

Syed M. Qaim, Ph. D., D. Sc.
Forschungszentrum Jülich GmbH, Institut für Nuklearchemie, Postfach 1913,
W-5170 Jülich, Fed. Rep. of Germany

ISBN 978-3-642-63473-4   ISBN 978-3-642-58113-7 (eBook)
DOI 10.1007/978-3-642-58113-7

The use of general descriptive names, registered names, trademarks, etc. in this publication does not imply, even in the absence of a specific statement, that such names are exempt from the relevant protective laws and regulations and therefore free for general use.

Typesetting: Camera ready by authors

57/3140-543210 – Printed on acid-free paper

Foreword

This book describes the Proceedings of the International Conference on Nuclear Data for Science and Technology held at Jülich in May 1991. The conference was in a series of application oriented nuclear data conferences organized in the past under the auspices of the Nuclear Energy Agency-Nuclear Data Committee (NEANDC) and with the support of the Nuclear Energy Agency-Committee on Reactor Physics (NEACRP). It was the first international conference on nuclear data held in Germany, with the scientific responsibility entrusted to the Institute of Nuclear Chemistry of the Research Centre Jülich. The scientific programme was established by the International Programme Committee in consultation with the International Advisers, and the NEA and IAEA cooperated in the organization. A total of 328 persons from 37 countries and five international organizations participated.

The scope of these Proceedings extends to a wide range of interdisciplinary topics dealing with measurement, calculation, evaluation and application of nuclear data, with a major emphasis on numerical data. Both energy and non-energy related applications are considered and due attention is given to some fundamental aspects relevant to the understanding of nuclear data.

The energy related papers deal with both fission and fusion systems, with a special focus on future needs, particularly with respect to advanced reactor concepts, nuclear incineration, development of radiation resistant and low activation materials, etc. Methods of data processing, data validation via integral experiments and benchmark studies, and related areas like radiation shielding, neutron dosimetry, neutron standards, neutron diagnostics, decay data of relevance to decay heat calculations, etc. are treated in detail, both in review papers and original contributions. The increasing future nuclear data needs appear to be associated with transmutation and fusion related problems. Some scenarios regarding the energy systems in transition are authoritatively discussed.

The non-energy related papers cover areas of astrophysics, cosmic studies, geology and mining, nuclear medicine and medium energy research. The importance of neutron and charged particle induced reaction cross sections and the need of radioactive beams in nuclear astrophysics and related studies are elaborated. Similarly the role of nuclear analytical techniques in geological research and the existing data needs are critically analysed. Nuclear data relevant to cyclotron production of radioisotopes, in particular those used in emission tomography, as well as the data needs in neutron and charged particle therapy are discussed in detail. Medium energy research is an expanding and fast developing field with potential applications of data in areas like accelerator development, space exploration, cosmic investigations, spallation studies, etc. Some aspects of this research are addressed in several contributions.

Nuclear research today involves utilization of modern facilities and interdisciplinary techniques, as was substantiated in an opening discourse. As far as nuclear data research is concerned, in recent years some progress has been made in the production of high energy neutrons and development of charged particle accelerators. Remarkable refinements have also been achieved in background corrections and error analysis. Furthermore, interdisciplinary studies involving radiochemical techniques, accelerator mass spectrometry (AMS) etc. are gaining increased significance. The relevant papers give an overview of the state of art of experimental techniques used in nuclear data measurements.

Considerable efforts have been devoted in recent years to the evaluation of available experimental nuclear data, resulting in the updating of existing data files and contributing to further developments in evaluation methodology. These Proceedings describe the salient features of all the major data files available to date.

Nuclear theory is becoming an increasingly important tool in nuclear data research. With diminishing experimental facilities more reliance will be placed in the future on theory and model calculations. The energy region covering the transition from precompound to spallation processes deserves special attention. Several papers present new theoretical concepts and ideas valid for energies above 20 MeV.

These Proceedings reflect the changing trends in nuclear data research on a global scale. Cooperation among various institutions is increasing. In the low energy region evaluation and refinement of data continue. However, considerable experimental and theoretical work is now directed to higher energy regions and new applications. The reader thus gets an up-to-date information on the recent advances in the field.

The book contains invited and contributed papers presented during the conference but not the introductory lectures given during the Tutorial Session. All the contributed papers were subjected to peer review by a group of referees, consisting mostly of International Advisers and members of the International Programme Committee. The invited papers were not refereed. Each contribution is placed under a definite topic or subject heading according to the recommendation of the International Programme Committee, though in some cases the arrangement may appear to be rather arbitrary. A total of 285 papers contributed by about 850 authors are included. I would like to thank the authors and referees for their cooperation.

The preparation of these Proceedings demanded appreciable technical efforts and I am grateful to the Organizing Committee for the help extended to me.

Jülich, Germany
December 1991

Syed M. Qaim

Conference Committees

Conference Chairman

S.M. Qaim, KFA Jülich, FRG

International Programme Committee

K.-H. Böckhoff; CEC-CBNM Geel, Belgium · S.W. Cierjacks; KfK Karlsruhe, FRG
C. Coceva; ENEA Bologna, Italy
H. Condé; Univ. Uppsala, Sweden · H. Gruppelaar; ECN Petten, The Netherlands
G. Herrmann; Univ. Mainz, FRG · Y. Kikuchi; JAERI Tokai-mura, Japan
A. Michaudon; ILL Grenoble, France · C. Nordborg; NEA Data Bank , Saclay, France
S.M. Qaim; KFA Jülich, FRG (Chairman)
M. Salvatores; CEA Cadarache, France · J.J. Schmidt; IAEA-NDS, Vienna, Austria
O. Schult; KFA Jülich, FRG · A.B. Smith; ANL Argonne, USA
M.G. Sowerby; AERE Harwell, UK · G. Stöcklin; KFA Jülich, FRG
H. Vonach; IRK Vienna, Austria · P.G. Young; LANL Los Alamos, USA

Board of International Advisers

M. Berrada; Mohammed V Univ. Rabat, Morocco · J. Blachot; CEN Grenoble, France
J. Boldeman; ANSTO Lucas Heights, Australia
N.M. Butt; PINSTECH Islamabad, Pakistan · E.T. Cheng; TSI Res., Solona Beach, USA
P. Christmas; NPL Teddington, UK · F.E. Chukreev; IAEK Moscow, USSR
J. Csikai; Kossuth Univ. Debrecen, Hungary
A. Deruytter; CEC-CBNM Geel, Belgium · A.T.G. Ferguson; AERE Harwell, UK
E. Fort; CEA Cadarache, France · F. Fröhner; KfK Karlsruhe, FRG
R.C. Haight; LANL Los Alamos, USA
P.E. Hodgson; Univ. Oxford, UK · R. Jahr; PTB Braunschweig, FRG · Y. Kanda; Univ. Kyushu, Japan
S.S. Kapoor; BARC Bambay, India · I. Kimura; Univ. Kyoto, Japan
B. Kuzminov; FEI Obninsk, USSR · H. Küsters; KfK Karlsruhe, FRG
R.M. Lambrecht; ANSTO Lucas Heights, Australia
B. Leroux; Univ. Bordeaux, France · Liu Tingjin; IAE Beijing, P.R. China
M.A. Lone; AECL Chalk River, Canada
E. Menapace; ENEA Bologne, Italy · S. Pearlstein; NNDC Brookhaven, USA
R.W. Peelle; ORNL Oak Ridge, USA · F.G. Perey; ORNL Oak Ridge, USA
G.H. Ricabarra; CNEA Buenos Aires, Argentina
D. Seeliger; Tech. Univ. Dresden, FRG · K. Shirakata; PRNFDC Ibaraki-ken, Japan
D.L. Smith; ANL Argonne, USA · K. Sumita; Univ. Osaka, Japan
D.R. Weaver; Univ. Birmingham, UK · S.L. Whetstone; DOE Washington, USA

Organizing Committee

S.M. Qaim (Chairman) · G. Stöcklin · U. Herpers
B. Krahl-Urban (Coordinator) · Renate Mengels (Secretary)

Conference Sessions and Chairpersons

Opening Session
S.M. Qaim and G. Stöcklin; Jülich, Germany

Nuclear Data Relevant to Fission Reactors and some Fundamental Studies with Slow Neutrons
H. Küsters; Karlsruhe, Germany · S.S. Kapoor; Bombay, India
H. Weigmann; Geel, Belgium · M. Asghar; Algiers, Algeria

Data Testing and Validation for Reactors
J. Rowlands; Cadarache, France · N. Janeva; Sofia, Bulgaria · I. Kimura; Kyoto, Japan

Nuclear Data for Fusion Reactor Technology
R.C. Haight; Los Alamos, USA · F.M. Mann; Richland, USA
D.W. Muir; Vienna, Austria · A. Takahashi; Osaka, Japan
A.J. Deruytter; Geel, Belgium

Experimental Facilities and Techniques
H. Vonach; Vienna, Austria · Tang Hongqing; Beijing, P.R. China
G. Herrmann; Mainz, Germany · C. Coceva; Bologna, Italy

Nuclear Data Relevant to Standards
H. Condé; Uppsala, Sweden · A.D. Carlson; Gaithersburg, USA

Nuclear Structure and Decay Data
D.R. Weaver; Birmingham, UK · J. Blachot; Grenoble, France

Nuclear Data for Medical Applications
G.J. Beyer; Rossendorf, Germany · M.A. Chaudhri; Melbourne, Australia
F. Tárkányi; Debrecen, Hungary

Nuclear Data Relevant to Astrophysics, Geology, Neutron Dosimetry and some other Applications
H.V. Klapdor-Kleingrothaus; Heidelberg, Germany · M.A. Lone; Chalk River, Canada
C. Wagemans; Geel and Gent, Belgium · R. Jahr; Braunschweig, Germany

Medium Energy Data
Y. Uwamino; Tokyo, Japan · S.L. Whetstone; Washington, D.C., USA
N.G. Zaitseva; Dubna, USSR

Nuclear Models and Evaluation Methodology
J.J. Schmidt; Vienna, Austria · Y. Kikuchi; Tokai-mura, Japan
D. Seeliger; Dresden, Germany · E. Sheldon; Lowell, USA · F.H. Fröhner; Karlsruhe, Germany
Liu Tingjing; Beijing, P.R. China · H. Gruppelaar; Petten, The Netherlands
E. Fort; Cadarache, France · V.A. Konshin; Vienna, Austria

Closing Session
A.B. Smith; Argonne, USA

Tutorial Lectures

H. Liskien; Geel, Belgium
Experimental methods in the determination of nuclear reaction data

H.O. Denschlag; Mainz, Germany
Measurement of nuclear data using radiochemical methods

D. Seeliger; Dresden, Germany
Nuclear theory relevant to the understanding of nuclear data

J.J. Schmidt; Vienna, Austria
Importance of nuclear data in nuclear fission technology

D.L. Smith; Argonne, USA
Nuclear data for fusion technology

G. Stöcklin; Jülich, Germany
Nuclear data for biomedical sciences

Co-ordinators

J.J. Schmidt; Vienna, Austria · G. Stöcklin and S.M. Qaim; Jülich, Germany

Contents

Part II   Data Testing and Validation for Reactors

Part III   Nuclear Data for Fusion Reactor Technology

Part IV Experimental Facilities and Techniques

Part V   Nuclear Data Relevant to Standards

Part VI   Nuclear Structure and Decay Data

Part VII   Nuclear Data for Medical Applications

Part VIII	**Nuclear Data Relevant to Astrophysics, Geology, Neutron Dosimetry and some other Applications**

Part IX   Medium Energy Data

Part X   Nuclear Models and Evaluation Methodology

Closing Session

Opening Session

WELCOME AND GREETINGS

S.M. Qaim (KFA Jülich)

Ladies and Gentlemen,

It is an honour and a pleasure for me to welcome you all to the International Conference on Nuclear Data for Science and Technology 1991 in Jülich. As conference chairman it is my great privilege to greet the heads of several institutes, the chairman of the Board of Directors of the Research Centre Jülich, the officials from our Federal Ministry of Research and Technology in Bonn, as well as the representatives from the OECD-Nuclear Energy Agency in Paris and the International Atomic Energy Agency in Vienna. My special regards and compliments to the members of the International Programme Committee, the International Advisers and the fellow scientists, especially those from abroad, some of whom have travelled very long distances to reach Jülich.

This conference is in a series of meetings held previously at Harwell, Antwerp, Santa Fe und Mito at intervals of about 3 years. The number of registered participants so far has reached 320. Together with about 40 accompanying spouses and guests the total number is slightly larger than that at Santa Fe in 1985 and Mito in 1988. The present conference has, however, two distinct features as compared to the earlier ones: Firstly, the number of scientific papers is much larger. We received about 400 contributions. Due to time and space limitations the International Programme Committee accepted about 300, thus rejecting about 25 % of the submitted abstracts. Together with the invited papers the programme of the conference consists of about 350 presentations. The second feature concerns the geographical distribution of the participants. We have delegates from 37 countries. The majority of the participants originates from the OECD area. However, a significant number of 15 % comes this time from Eastern Europe including Soviet Union, and another 15 % from the P.R. China and developing countries. This has been possible partly through the financial supports provided by the IAEA, and especially by the KFA Jülich. My thanks and appreciation to both the organizations.

The field of nuclear data is very broad and involves interdisciplinary studies extending over many areas of chemistry, physics and engineering. It deals on the one hand with fundamental aspects of nuclear interactions and, on the other, with applications in various fields. This conference is application oriented but due consideration is given to the fundamental aspects relevant to the basic understanding of data. Whereas in the past the main impetus to nuclear data research was provided by fission energy related studies, in recent years the emphasis has been shifting to more scientifically oriented investigations and topics like fusion technology, nuclear incineration, accelerator development and medical applications have been gaining increasing attention. The International Programme Committee in consultation with the International Advisers has established the scientific programme of this conference which, we believe, reflects the changing trends in nuclear data research. My earnest hope and desire is that the conference will meet your expectations and that the deliberations will be in a friendly, constructive and fruitful atmosphere. I wish you all a pleasant stay in Jülich and its surroundings.

2

H.F. Wagner (BMFT Bonn)

Ladies and Gentlemen,

It is a great pleasure for me to welcome you to Germany and to KFA Jülich in the name of the Federal Ministry of Research and Technology.

Our connections to this conference are threefold: The Ministry is the 90 % shareholder of KFA and is, therefore, very pleased to have so many distinguished scientists in Jülich. Secondly, we financially support a great deal of scientific work in Germany pertinent to this conference. And thirdly, NEA, IAEA and the Research Ministry are, for some decades now, very close friends and it is, therefore, a particular pleasure to welcome these agencies as cooperators in organizing this conference.

It is greatly satisfying to us that the meeting is really international. In addition to the traditional participation from USA, Japan and Western Europe, a large number of delegates originate from Soviet Union, Eastern Europe, P.R. China and developing countries. For us it is also a happy occasion that several participants come from eastern part of Germany.

The field of nuclear data is very broad. Previously major emphasis was on data for energy applications. In recent years this emphasis has been shifting to other fields. The present conference is very interdisciplinary. The programme extends over many areas of nuclear chemistry and physics, reactor technology, nuclear medicine, nuclear models etc. Having worked myself in the field about 20 years ago, I know of the diversity involved. It is therefore befitting that it is held in a large research centre where large experimental facilities and interdisciplinary environment are available.

Scientific research in general, and large projects in particular, are becoming too expensive. The need for international cooperation cannot be overemphasized. It is hoped that conferences like this would foster international scientific cooperation.

On behalf of the Federal Ministry of Research and Technology I wish you a successful conference.

J. Rosen (NEA Paris)

Director General, Ladies and Gentlemen,

The OECD Nuclear Energy Agency is pleased to be cooperating with the Jülich Research Centre in the organisation of this latest in the series of international conferences on Nuclear Data, following previous conferences in Mito and Santa Fe. You may know that our Agency is composed of 23 countries: Canada, Japan, the United States, what we used to call Western Europe and Australia - and that within this Agency nuclear science is represented by the NEA Nuclear Data Committee and the NEA Committee on Reactor Physics; many of their members are also members of the International Programme Committee for this conference. Finally, many of you will be users of the NEA Data Bank in Saclay, which cooperates with other nuclear data and software centres worldwide, and supplies nuclear data to users in Western Europe and Japan, and computer programs in all IAEA member states except the United States and Canada.

Compared with previous conferences in the series, I myself have been struck by the increased time given to nuclear theory and evaluations, and to medium energy data. I see that fusion technology receives as much time as fission reactors, and that a respectable place is given to non-energy applications. Each of us will see different patterns in the programme. I myself see big progress in evaluation work: the new ENDF/B-VI standards and the re-analysis of resonance data notable for ^{238}U and ^{56}Fe have been important steps towards a magical C/E figure very close to unity for a wider range of reactor calculations based on the new generation of evaluated neutron data files: JENDL-3, ENDF/B-VI and JEF-2. NEACRP and NEANDC have set up an International Evaluation Cooperation linking these three projects, and we hope that the worksharing introduced between laboratories taking part will produce the further data needed for new generations of power reactors.

It is clear that for current reactors all the most important problems can be solved using the data and calculation methods now available, even if some data adjustment is needed. NEANDC and NEACRP have made an essential contribution to this development over the past 30 to 40 years, and on our estimates several billion dollars were well spent on nuclear data over that period. Improvements will continue, but the economic benefits from better data and analysis techniques lie in future applications rather than in modifying current reactor designs, so that new work on improving neutron data at thermal energies will represent only a small fraction of the earlier input.

The trend in current work towards higher particle energies also stands out from your programme.

The existing fund of nuclear data will maintain its value for many years, and it is the service work of the Data Bank and other data centres to make it available to users. However, the policy objective of the NEA must be to encourage scientific developments which will anticipate future needs in nuclear technology. A review of NEA's orientations for the 1990s has emphasized the need to modernise its nuclear science programme, and to clarify its committee structure. We have reached the conclusion that three distinct specialised committees for nuclear science is too many, and that the necessary overview of the whole field would be better achieved through a single committee which should identify and focus cooperation on the scientific issues of greatest impact for future technology.

We propose to set up a new NEA Nuclear Science Committee, whose scope would include the work now done through NEACRP and NEANDC but would be wider: it should as its first task concentrate on policy issues and on defining the new areas to be supported by NEA's work. Change would be evolutionary rather than revolutionary, and balance must be kept between supporting valued ongoing projects from the programmes of the existing committees, and encouraging new work.

Our consultations also showed a very high degree of appreciation for the services provided by the NEA Data Bank (we shall maintain them) and a serious shortage of new young people entering the nuclear science field. Studies carried out by NEANDC (following an earlier U.S. study) for nuclear physics, and by the NEA Nuclear Development Committee for the wider nuclear energy field show a clear demographic problem, and the scientists now retiring, who joined the field in the 1950s, will not be easy to replace. It is important that we present a clear development programme, and that we are able to justify our confidence in the need for nuclear energy in the new century. The best young scientists will join us only if we can offer them interesting work and a worthwhile future.

On behalf of my Director General, Dr. Uematsu, I wish you a very successful meeting.

4

V.A. Konshin (IAEA Vienna)

Ladies and Gentlemen,

It gives me great pleasure to convey the IAEA Director General's greetings and best wishes at the opening ceremony of the International Conference on Nuclear Data for Science and Technology. Let me thank the German Government for the hospitality extended to all of us and for making this conference possible.

Nuclear power and nuclear technology are facing serious public acceptance problems in several countries. Nevertheless, nuclear plants supplied about 17 % of the world's total electricity last year. At the end of 1990 there were 424 nuclear power plants in operation, and 83 under construction. They are a major element in the energy mix, and will remain so.

Considerable efforts are now underway in the IAEA to ensure that operational safety is maximized and that any accidents and incidents in nuclear power stations that might occur in the future will not have significant consequences.

Nuclear power stations of Germany are producing about 30 % of the electricity. There are 25 reactors in operation and 6 under construction. The German electricity industry is prepared to construct new nuclear reactors, for instance at the Greifswald site, using approved West German technology, and depending on the immediate granting of a construction licence.

The nuclear data base for fission reactor design calculations is now to a large extent satisfactorily established. However, development work in nuclear fission and fusion technology, nuclear safety and radiation protection and, to a lesser extent, in the nuclear applications area depends, among other things, on the availability of reliable data for nuclear, atomic and other characteristics of nuclear materials.

The present lack of material property data for plasma-facing walls in fusion reactors is a limiting factor for the whole fusion development programme. Research and feasibility studies on the burning of actinide waste in dedicated reactors or accelerators are also hampered by the absence of data or the inaccuracy of the data that are available. For an accurate assessment of the radiation hazards associated with the decommissioning of power and research reactors, better activation and decay data bases for reactor materials and activation products are needed. Also, advanced reactor designs may well need new or refined neutron fission and capture data.

Life in future will be computer-complex and will need much professionalism and group discipline. Constant updating of our information on very dynamic markets will cost us effort and time.

I think that most of us are thrilled by the challenges this implies.

So you have very big tasks to solve, and I close my remarks now by wishing you all success in your work.

NUCLEAR RESEARCH IN AN INTERDISCIPLINARY ENVIRONMENT

J. Treusch

Head of the Board of Directors
Research Center Jülich GmbH
5170 Jülich, Federal Republic of Germany

Abstract: The general structure of the Forschungszentrum Jülich (KFA) is described with particular emphasis on the interdisciplinary cross-fertilization between nuclear technology, nuclear physics, nuclear chemistry and other fields of KFA-activities as materials science, fusion technology and medicine.

The Research Centre Jülich was founded 34 years ago as the "Association for the Advancement of Nuclear Physics Research" (this name being changed to Nuclear Research Centre Jülich in 1967). This foundation was characterized by two basic motivations. On the one hand, the reestablishment of sovereignty for the Federal Republic of Germany permitted questions of nuclear technology to be dealt with. These questions, in particular the problem of energy conversion with modern reactor series, were to be the concern of Jülich - in addition to the Nuclear Research Centre in Karlsruhe. On the other hand, the government of the Federal State of North Rhine-Westphalia intended to create a research centre giving the universities in the Federal State the possibility of implementing large-scale experiments which would have been impossible for individual institutions.

The close relationship which the Research Centre Jülich enjoys both in cooperation with industry as well as with the surrounding universities resulting from these two objectives is most probably unique in the world and continues to have many positive consequences. One of which is that from the very beginning the KFA has never been a monolithic centre for nuclear technology but has always participated in the neighbouring research fields ranging from nuclear medicine and chemistry to materials research and solid-state physics.

This is not the occasion to enlarge upon the historical development of the Jülich Research Centre, nevertheless from the preceding remarks it is apparent that after the KFA had fulfilled its first important task, namely the development of the high-temperature reactor ready for industrial production, it did not need to search desperately for new tasks. On the contrary, the KFA has developed in a very organic process into a modern multidisciplinary research centre able to fulfil in an exemplary manner the demands of our present society for the far-sighted planning and implementation of scientific activities and their conversion into useful technology.

How then can these demands made by modern society be correlated with the corresponding multidisciplinary research at the KFA? The following quotation from the 1990 KFA Annual Report can clarify this relationship:

"As an essential element in the government's responsibilities, research must be oriented towards basic human needs. This is the principle guiding the tasks and objectives of the Research Centre Jülich (KFA), the largest of the 13 national research centres in the Federal Republic of Germany. With its staff of more than 4700 (including more than thousand guest researchers, postgraduate students and trainees), the Research Centre makes a contribution towards tackling the urgent basic problems in our society at an early stage with the aim of maintaining and improving our environment and living conditions, of offering solutions to technical problems, of enhancing the competitiveness of the Federal Republic in modern technology and extending knowledge of complex relationships in our world."

The major emphasis of the revised Jülich research programme is placed on mankind with the basic needs for energy, health and the environment, matter and material, as well as information and communication.

These four central topics are reflected in the four priority programmes characterizing the KFA in the nineties.

These programmes have the aim of indicating options for a long- term and environmentally friendly energy supply for a continuously growing world population, of recognizing at an early stage and preventing threats to the environment and health, of understanding the structure and properties of matter and using this knowledge to make new materials available for many fields of technology, and also to contribute towards developing technologies with which the ever increasing quantity of knowledge and information can be processed and exchanged.

The four priority programmes interrelate at a number of points; whether due to the joint utilization of the large-scale devices and effective infrastructure facilities of the KFA or due to the mutual encouragement of research areas as is for example apparent from the significance of materials research for fields ranging from energy to information technology, or whether due to joint nuclear engineering and nuclear physics expertise which, after all, has brought you all to Jülich this week.

Naturally, the solution of ever more complex technical and also social problems requires an interdisciplinary cooperation between many disciplines within the KFA and also outside its walls. The organization and practice of such interdisciplinary cooperation between physicists, chemists, biologists, physicians, engineers and technicians, even including social scientists, has a long tradition at the Research Centre Jülich and is a continuing challenge. This can be seen particularly clearly in the case of medical research; metabolic examinations of the heart and brain are only possible if the corresponding radiopharmaceuticals are provided by _chemists_ for the _physicians_, whereby the signals emanating from inside the body are detected by the methods of _physicists_ and processed in complicated _electronic_ devices. Cooperation between materials research and fusion physics is a topic of this conference and also figures in the KFA programme. The same is true of the scientific interactions ranging from basic research in nuclear physics and medium-energy nuclear physics up to high-energy physics and astrophysics.

These common exertions by a wide range of disciplines will continue to be the driving force in tackling the problems dealt with at the Research Centre Jülich. Indeed, this is an absolute prerequisite for providing scientific answers to tomorrow's

questions. It hardly needs to be said that this interdisciplinary cooperation is particularly fruitful since the four major research priorities of materials research, information technology, environmental research and energy technology are embedded in a broad spectrum of basic research ranging from medicine and biotechnology to solid-state research and the basic nuclear research already mentioned.

Within the framework of a programmatic reorientation of the _energy technology sector_, the fields of nuclear fission and nuclear fusion long represented at the KFA have been extended to include the topics of photovoltaics and hydrogen. These thematic priorities are supplemented by R&D work on advanced systems elements of the energy economy, such as fuel cell technology and electrolytic hydrogen production. Each of the projects is accompanied by systems analysis studies on the energy economy and the associated release of pollutants.

This combination of research work enables the options for the future energy supply to be further developed in breadth and depth both comparatively and interactively.

Fusion research at the KFA planned for the long term - with its major focus in the Institute of Plasma Physics - is a constituent part of the European Community fusion programme and is implemented within the framework of an association agreement concluded with EURATOM. Within this international cooperation, extending even beyond the European framework, the KFA's contributions concentrate on the systematic study of plasma-wall interaction. The TEXTOR tokamak is the central experimental facility. The KFA works together with its national and international partners in this programme towards creating the prerequisites for the next phase of fusion reactor development.

In the field of fission reactors, research work at the KFA in course of the past three decades has been mainly concerned with the development of the high-temperature reactor as an electricity generator and source of process heat for temperatures up to 950°C. The major successes of this work, carried out in close cooperation with industry, were the operation of the AVR and THTR 300. In accordance with the level of development now achieved, further work in this field of nuclear engineering now largely falls to industry. Problems of safety engineering and waste disposal are now the primary concerns of research and development in this field at the KFA.

The fast pace of development in _information technology_ to ever higher densities of information storage and to ever greater speeds of information processing and transport brings us ever closer to physical boundaries which must be thoroughly understood in order to permit further innovations. Although it is still the Coulomb force and not nuclear interaction that remains the decisive physical interaction, nevertheless the limits of conventional methods and materials are increasingly being reached. Basic research and also methods from nuclear physics (such as ion implantation) are required to open up novel techniques and materials. This is where the most recent KFA programme "Basic Research for Information Technology" appears on the scene. Its subject areas are

- semiconductor film systems and epitaxy
- superconductor film systems
- magnetic films
- new components and quantum structures
- metallization
- and crystal growth of compound semiconductors.

This unique cooperation - particularly as far as the combination of semiconductor and superconductor electronics is concerned - between a total of eight KFA institutes serves rather to prepare for tommorrow's technology than to provide industrial technology for today. Impressive and also internationally recognized successes in the preparation of high-temperature superconductor films, in growing gallium arsenide crystals, and in structural investigations using the scanning tunnelling microscope and electron microscope are just three examples of the speed with which programmatic reorientation can achieve success if it is able to make use of such a broad scientific and technical background as is available at the KFA.

In the same way as the programme on "Basic Research for Information Technology", the _environmental research programme_ at the KFA is also quite decisively characterized by interdisciplinary work. The realization, which has grown in the past few years, that our environment does not merely provide resources but is to the same extent the final repository for large, often polluting quantities of material has resulted in the observation of material flows from production and transport up to and including deposition becoming a central topic. This affects all sectors of the geosphere: soil, waters, atmosphere and stratosphere, as well as the biosphere. The scientific aspects dealt with by the KFA range from radioagronomy, biological waste water purification and hazardous waste incineration to air chemistry and the theoretical modelling of the primarily non-linear dynamics of all processes effective in this connection. It is quite obvious that the KFA makes wide use of its experience gathered in the field of nuclear engineering and waste disposal.

Materials research has long had an assured place in the Jülich research programme. This research concentrates on both metallic and also ceramic materials and composites. They are to be used wherever high temperatures for physical and chemical processes require high-temperature materials and structural ceramics - as indicated by the name of the programme. The higher the permitted temperatures the higher is the efficiency of the facility for energy extraction - either in the power station or car. High-temperature materials thus also represent a contribution towards efficient energy use.

In the past, the major interest of materials research was centred on the high-temperature reactor. Today the materials research programme covers a wide range of topics and problems starting with steels, iron and nickel alloys for long-term use at high temperatures in energy and chemical facilities or future fusion reactors. This programme also involves basic theoretical and experimental studies of new superalloys, of materials produced by powder metallurgy and high-temperature protective films. The Jülich boronated graphite protective films for the inner wall of fusion reactors have become a worldwide standard.

However, surfaces can also be altered and coated by using ion implantation, as already mentioned. This is a method which has progressed from nuclear physics via solid-state physics to materials research - and is notably also used in microelectronics. Irrespective of whether the result of this treatment is in the final analysis a moulding element for a machine tool, starting material for an artificial hip or microchip with buried conductor tracks - the road to success is characterized by cooperation between nuclear physicists, accelerator experts, solid-state physicists, materials researchers and engineers.

It was already noted at the beginning that the interdisciplinary and interinstitutional cooperation

required by the KFA must not become
an end in itself to the detriment of
scientific depth. A research atmosphere which encourages and promotes
scientific excellence, even if there
is no apparent application relevance,
is the most effective guarantee of
creativity. If there is also a readiness for communication and cooperation then completely unexpected results may arise which not only increase our knowledge but lead to concrete applications. In this sense,
the wide range of basic research carried out at the KFA is absolutely
indispensable for a number of
reasons. It creates the appropriate
scientific atmosphere for the development of research and application,
it ensures an uninterrupted dialogue
with university research - particularly by the joint utilization of the
large-scale devices DIDO, TEXTOR and
COSY - and it is always of assistance
in correctly assessing long-term
developments and any possible change
of paradigm. This also contributes
towards enhancing the reputation of
the Research Centre Jülich by leading
to a meeting of biology, medicine and
physics, for example in the utilization of the DIDO neutron source, by
integrating the KFA into the world-
wide community of medium-energy and
nuclear physics with the research and
development work at the COSY cooler
synchrotron, by opening up the boundaries between biology, biochemistry,
medicine and physics at the Institute
of Biological Structural Research
which is currently being established.
This all demonstrates that at the KFA
nuclear-oriented research profits
from its environment to the same extent as non-nuclear research fields
benefit from nuclear technology.

ENERGY SYSTEMS IN TRANSITION
UNDER THE CONDITIONS OF THE FUTURE

Wolf Häfele

Forschungszentrum Jülich GmbH

5170 Jülich

and

Zentralinstitut für Kernforschung Rossendorf

8051 Rossendorf

Federal Republic of Germany

Abstract: The energy situation continues to pose problems. An example is the question of the appropriate position of future nuclear energy uses. An answer to this question requires orientation. The design of scenarios is of help here if scenarios are not understood as predictions but as the identification of possible futures. The paper explores a few such scenarios that mostly envisage the time period till 2030. Special attention is given to the impact of abating CO_2 emissions.

(Nuclear energy today, future nuclear energy uses, energy scenarios, abating CO_2 emissions)

Introduction

Nuclear data and reactor development have been closely connected from the very beginning. Most nuclear scientists and engineers are aware of the fact that the capture cross-section of graphite was the big problem during the early stages of reactor development. The group around Enrico Fermi concluded that it was possible to have a configuration with natural uranium and graphite to become critical. While the group around Werner Heisenberg concluded that due to the absorption cross-section of graphite this would not be possible. That was a false conclusion. Nevertheless, it had tremendous consequences because the group around Werner Heisenberg had a critical assembly using natural uranium and heavy water, quite a different approach. Both groups were aware of the fact that one has to have a probability for resonance escape close enough to one to make a reactor configuration critical and both groups concluded that only a heterogeneous reactor can achieve this. So these were tremendously big boundary conditions for the evolution of reactor technology from the very beginning.

But one should also recall that the problem of making a reactor critical with natural uranium continued even after 1955. That was when I became involved in this field. The big question was to what extent a natural-uranium-fueled reactor can really have a high burn-up because when fission poisoning increases then the reactivity goes down and if there are only a few thousand MWd per tonne left, the reactor cannot become very economical. It was in particular the value of η, that is the fission neutrons produced per the absorption of one neutron, that was very much at stake. I remember that the British sources consistently provided low data, while others were more optimistic and this was the big question. One percent difference in that value made all the difference between an economical reactor and a non-economical reactor. That necessarily led to the consideration of the epithermal contribution to fission. This was eminently necessary and changes the value depending on the epithermal part of the neutron spectrum. The great challenge was to calculate the epithermal part of the neutron spectrum in a heterogeneous reactor. I tried very hard at that time to make a contribution and I recall that this led to the theory of neutron thermalization and this new theory of neutron thermalization raised the question of neutrons interacting with the molecules or the atoms of the moderator and the beginning of that aspect of solid-state physics which later had a tremendous impact for instance on the work here, in this research center. That means undertaking solid-state physics or condensed matter physics with the help of neutrons. This was the beginning when, among others, Karl Heinz Beckurts introduced this field as a new branch of scientific study. Using enriched uranium essentially dispensed with the issues of nuclear data in conjunction with nuclear reactors and, of course, we have to know the nuclear data, that is for sure. But there were no longer that tricky problems about criticality since we no longer used natural uranium but enriched uranium. The condition for criticality is that η be larger than one, that means the number of produced neutrons must be larger than the number of absorbed neutrons.

Breeding

The problem of neutron cross-sections comes up again when the condition is that η has to be larger than two when one not only wants to be critical but when one wants to breed. Again these strong interconnections between nuclear data and the design of nuclear reactors came to the forefront. I recall the early days of fast breeder development when the concern was that the neutron spectrum would be too soft.

At that time none of us knew the medium energy part of the value of η together with the calculation of the neutron spectrum. At that time this was an issue. It was not incidental that the first fast breeder reactors all used the metal form of the fuel, that means uranium and plutonium in metallic form without an oxide contribution, because that contributed an additional moderation and a softening of the neutron spectrum. EBR 1 and the Dounreay Fast Reactor were of that kind. It was a major step forward to consider plutonium oxide fuel and uranium oxide fuel in fast breeders. That was around 1960 and the years thereafter when the possibilities of calculating the neutron spectrum were better and when the nuclear data were at hand to make the calculations. Indeed it was possible to show and to demonstrate that even with an oxide-fueled reactor not only criticality but also breeding would be possible.

I do recall, however, great excitement in the year 1967 when the Schomberg data from Harwell [1] came to the forefront. Again they were very pessimistic in the sense of providing values for α between 100 eV and 10 keV twice as high as the assumptions so far used, the old data of Knoll's Atomic Power Laboratory. The Schomberg data were twice as bad as the Knoll's Atomic Power Laboratory data and later it was Gwin's data [2] which were somewhere in between and saved the option of breeding. Eventually you have to make a measurement of whether you can carry out fast breeding and it is then beyond the calculation and beyond the measurements in zero power critical facilities. Indeed the French have demonstrated it by closing the fuel cycle of the fast breeder technically, pragmatically technically twice, and demonstrating a breeding ratio of 1.15. So it is absolutely beyond doubt that you can breed with fast neutrons.

But, of course, in the past there was the issue of breeding with thermal neutrons by using ^{233}U and its value is sufficiently close to 2 so that whether one can breed or not continued for a long time to be an issue. The question was whether one can enhance the neutron economy by having some support of fast fission by using beryllium, for instance. The (n, 2n)-effect was at one time a big hope for thermal breeding. I myself during my year at Oak Ridge was given the task of making that calculation by Alvin Weinberg. What I had available was only a mechanical calculator and I had to work for three months to produce one figure, so that would no longer be conceivable today. But at that time I knew exactly what I was doing, I was aware of each part of the calculation. The result was an enhancement of the neutron multiplication effect by 1.06, a good figure, and it would indeed have helped thermal breeding significantly. However, there were the cross-section data again indicating that the threshold for (n, 2n) is only at 2 MeV so that only part of the fission neutrons would see the (n, 2n) possibility while the (n, α) cross-section has a maximum at two MeV leading among other things to ^{6}Li and ^{3}He. This is a build-up of poisoning which essentially destroys the whole benefit so the negative, the pessimistic value of the fast multiplication effect at that time was 0.99. One not only gains but also loses a percent due to that added poisoning, which essentially led to the conclusion that thermal breeding today is not readily possible, except if one removes the fission products.

I want to recall the molten salt reactor concept which is almost forgotten but it is a reactor concept where the fission products are continuously removed and thermal breeding is therefore essentially possible. I would like to add the remark that perhaps in the future a molten salt reactor might again become very important because it is the only reactor concept so far where the fission product inventory is minimal in the active core because it is constantly removed, and it is the fission product inventory which poses the fundamental safety problem of nuclear reactors.

I would like to say one last word about light water reactor breeding, pursued by Admiral Rickover and Professor Radkovski, who is now in Israel. Indeed it is a question of cross-section and a question of fission product poisoning whether thermal breeding by ^{239}Pu and perhaps even ^{235}U with very tight lattices is a reasonable way to go.

Nuclear Power Today

This has led altogether to the present situation of nuclear power. Nuclear power today is much larger than any of us expected it in 1960 or 1965. If someone had told us at that time that today we would be operating 320 GW$_e$ with nuclear reactors nobody would have believed them.

Table 1. Nuclear Power in Various Countries

	Reactors in Operation	Electricity Generated (MWe)	Percent of Electricity Generation	Reactors under Construction
North & Central America				
Canada	18	12,185	15.6	4
U.S.	110	98,331	19.1	4
South America				
Argentina	2	935	11.4	1
Europe				
Belgium	7	5,500	60.8	0
Bulgaria	5	2,585	32.9	2
Czechoslovakia	8	3,264	27.6	8
Finland	4	2,310	35.4	0
France	55	52,588	74.6	9
Hungary	4	1,645	49.8	0
Spain	10	7,544	38.4	0
Sweden	12	9,817	45.1	0
Switzerland	5	2,952	41.6	0
U.K.	39	11,242	21.7	1
West Germany	24	22,716	34.3	1
Asia				
Japan	39	29,300	27,8	12
South Korea	9	7,220	50.2	2
Taiwan	6	4,924	35.2	0
U.S.S.R.	46	34,230	12.3	26
Totals	426	318,271	16.0	96

Source: International Atomic Energy Agency, Vienna

To that extent it is a very great success. Table 1 gives an overview of nuclear reactors operating in various countries. The figures are for 1989, those for 1990 are only very slightly different. Indeed there are 426 power reactors in operation worldwide generating 318 GW$_e$ of electricity. This provides 16 % of all the electricity in the world. In various parts of the world this figure is different, here in the case of France it is as large as 74 %, while in other countries like for instance in the UK it is 21.7 %, in Argentina it is 11 % so the average is 16 %. 96 reactors are still under construction and perhaps not all of them will be completed, but assuming for a moment that they will be completed, as a scenario, one would arrive not at 318 GW$_e$, but almost at 400 GW$_e$.

The title of my presentation is "Nuclear Power in Transition under the Conditions of the Future". The important aspect is the comparison with today and the question which way is nuclear power going? Table 2 shows a few figures for 1989 that may be of interest giving the equivalent of the capacity with electric generation or translating that to $TW_{th}a$, which is the favorite unit for systems analysis comparisons, of a total of 11 TWa today of the overall energy consumption that is given here with the nuclear share of 16 % of world electricity capacity but only 5.7 % of world primary energy when one counts it in thermal units and when one takes everything into account.

Table 2. Nuclear Shares, 1989

Nuclear:	$318\ GW_e \hateq 1854\ TW_e h$
	$1854\ TW_e h \cong 0.63$ TWa, thermal
World's total:	11405 $TW_e h$, electricity generation
	11 TWa, primary energy
The nuclear share:	16 % of world's electricity
	5.7 % of world's primary energy

Source: IAEA, Ref. Data Series, 1990

Time Scales

Referring to the title of this presentation the question is raised "in transition" what does it possibly mean? If one says "in transition" one must have some vague idea about where to go in the long run. I find it important to present a rather simple picture which is kind of trivial but nevertheless it is important to reflect on. Namely that mankind is not doomed to failure by not having enough energy. That is a matter of principle but not necessarily of technology.

End use electricity will be environmentally benign and always necessary. But it is not possible to operate a transport system by this means. It is necessary to have a storable secondary energy and if this is hydrogen one also has practically a closed cycle as one has in the case of electricity between water splitting and water recombination. So that part of the system is in principle environmentally benign and not limited. If one now has nuclear energy or solar energy, which is nuclear energy in the sun, and if one applies breeding either by fast breeder reactors or thermal breeder reactors for that matter and/or fusion then the supply is unlimited in terms of primary energy and if one handles the fission products safely, in that case one has solved the problem even if there are 20 billion people or more in the world and this is only one alternative among many others, I have written many big books on this subject it so please accept this short statement; hard solar energy is required not soft solar energy. Not the absorber on the roof but the big receiver plants in the desert area, for instance, covering a hundred thousand or one million square kilometers. So if this is also a proof of existence, an existence theorem, if in principle such a solution exists, how far and how soon is it possible to go from here to there? This requires then a look at the possible time scale. One should follow the scheme shown in Fig. 1, which goes back to Marcchetti at IIASA where he considered the H/C ratio in the past for primary energy in the world by

considering wood, by considering coal, which was displacing wood, by considering oil, which was and still is displacing coal and gas.

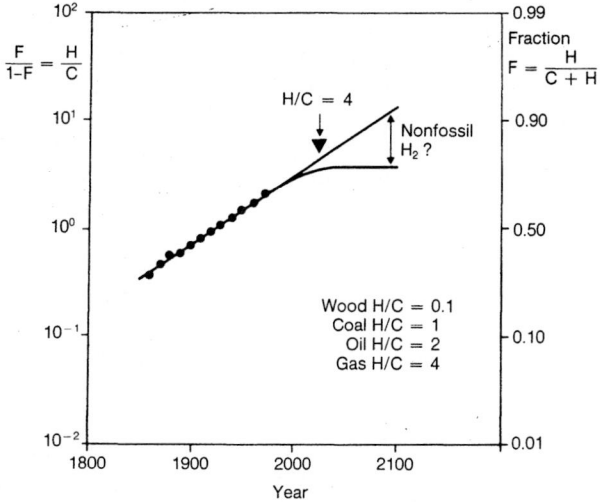

Source: C. Marchetti, IIASA

Fig. 1. Hydrogen to carbon ratio (H/C) of fossil fuels in the world, 1860 to 2100

There the H/C ratio gradually increased if in the case of wood, the water which is the ground state is not counted and therefore doesn't matter, then one sees a very strong regularity. The plotting is such that a linear curve here is a logistic curve. That means a s-shaped curve as the transition from the state zero to the state one and the message here is the time scale. In the past since 1860 the world has behaved as if it were hungry for hydrogen. Displacing the carbon atom by increasing the hydrogen share, then one can really ask when one has arrived at a large percentage of hydrogen and if that is 90 %. One can see this is at around the year 2100. This is a very crude but significant conclusion. It is not meant to be a precise prediction.

But we are talking about the next one hundred years therefore if we talk of these transitions and as a systems analyst I have always found it very useful to shape a problem into various time stages by considering time horizons and also in this case I will do that, referring to Table 3.

Table 3. Time Horizons*

Regional impacts of NO_x- und SO_2-emissions	now
Emergence of the CO_2-problem (doubling of the CO_2-content)	50 years
Supply from conventional resources (oil + gas)	50 years
Development of environmentally benign energy systems	100 – 140 years
Supply of non-conventional resources (oil, gas, coal)	250 years
Decay of additional CO_2-emissions in the atmosphere	500 – 1000 years
Decay of nuclear waste down to levels of deposits of natural resources	1000 years
Supply from nuclear resources	15000 years
Supply from solar radiation	billions of years

* rough indications only

The immediate problems with the energy systems we have today are the regional impacts of NO_x and SO_2 and this is now a problem. The emergence of the CO_2 problem, which is on everyones lips, I am not going to explain the CO_2 problem, has a time scale of 50 years. The supply from conventional, and I stress the notion conventional, resources of oil and gas also has a time scale of 50 years. If, however, one considers the development of environmentally benign energy systems, just the way I explained it, then one has a scale of 100 to 140 years while the supply of non-conventional resources of oil and gas and in particular coal referring here to tar sands and shale oil and coal has a time scale of 250 years.

Now looking at the waste problem, the decay of additional CO_2 emissions in the atmosphere, once realizing that one has gone too far, has a time scale of five hundred to one thousand years, that is the time constant of ocean mixing from the upper layers to the deep bottom. The decay of nuclear waste down to levels of deposits of natural resources has a thousand years, the supply from nuclear resources, fission breeding or fusion breeding 15000 years, and from the sun billions of years. I do think that the most urgent problem is around 2030 to 2040.

Scenarios and Features

Now we are dealing with the regional impacts, acid rain and the others, but the next 50 years are important. The question is, what will be the case for nuclear power? The case for nuclear power is highly controversial and I think it is helpful to consider two scenarios. I did this in an earlier article in Scientific American of last year in the September issue [3] For details reference should be made to that. One scenario is that we are staying at 400 GW_e and fortunately for some, and unfortunately for me, this is not unlikely. In such a case the supply of uranium is not an issue. There will be enough uranium for at least 100 years and this easily covers that time scale of 50 years here and because it is not a problem the reprocessing part of the fuel cycle and the necessity of breeding is not given. If we stay at the present capacity of 400 GW_e with nuclear power in that scenario, I want to emphasize if, then it is not necessary to provide additional sources by going through the fuel cycle and to go to fast breeding to provide plutonium. Instead we will have the other problem of having too much plutonium. Given today, 58 tonnes of plutonium are produced per year and we do not have the problem of breeding or providing plutonium but handling plutonium either by final disposal or by doing something with it. But these are so completely different scales 58 tonnes of plutonium on the one hand, while in the sixties I remember the great excitement when it was possible for us in Karlsruhe to buy 5 kilograms of plutonium at a price of $ 43 per gram. Today one will be given $ 20 in addition to the one gram of plutonium itself if one accepts it.

So the circumstances have certainly changed. However, if the other problems of energy transitions and energy problems come to the forefront and not only a single contribution by a particular kind of primary energy is being considered, but a whole system, and nuclear power is in demand, the picture changes drastically. In the past we had energy resources, conversion and utilization, now we have disposal of wastes, ecology and climate, risks and uncertainties, hypothetical risks, public acceptance, interfaces of technology, society and economy and in particular we have the CO_2 problem. Let's for a moment take the CO_2 issue seriously, and I do think we should do that, and have a look at Table 4. Today we have energy-related emissions of carbon

Table 4. CO_2 Emission Rates

		GtC/a as CO_2
1987/88	energy related	6
	non energy related	1
2005	the Toronto target	4.8
2030	the Jülich CO_2 reduction scenario	4.0
sustainable? 2080?		2.5

in the form of CO_2 which are essentially 6 gigatonnes of carbon per year, non-energy-related 1 gigatonne of carbon per year. The Toronto target of 1988 required a 20 % reduction for the year 2005 which is 4.8 gigatonnes and we do not know what to shoot for. Alvin Weinberg maintains that perhaps the most important aspect of studying energy systems today is to understand the acceptable emission of carbon to the atmosphere, or in other words that amount of carbon which can be taken from the upper surfaces of the ocean by natural mixing down to the deeper parts of the ocean. It may be somewhere in the region of 2.5 gigatonnes carbon per year. And this would be then in the midst of the transition to what will be happening in 2080. So these are the time scales that we had in mind.

There are two question marks: Is it sustainable? And do we have to have it by the year 2080? These remain open issues because there are no good answers. Nevertheless, in the spirit of the scenarios you can ask yourself what has to happen if by the year 2030 the carbon emissions would have to be reduced to 4 gigatonnes of carbon per year, which is a fair scenario question. Now in studying that one should realize that there are 5, some people say 6 possibilities, for the reduction of CO_2 emissions (see Table 5).

Table 5.   Five Possibilities for the Reduction of CO_2 Emissions

1.   changing the fossil fuel mix to higher hydrogen contents, less coal, more gas

2.   introducing biomass on a recycling basis

3.   introducing non carbon alternative sources: hydropower, solar (in whatever form)

4.   enhancing nuclear power

5.   introducing large scale energy conservation

Namely, changing the fuel mix, because the hydrogen content of natural gas is greater than that for coal and therefore it essentially means going from coal to gas primarily, then introducing biomass on a recycling basis. That means not burning more biomass than you plant; to have not a nuclear fuel cycle but a biomass fuel cycle, on balance you wouldn't produce any surplus amount of CO_2. Introducing non-carbon alternative sources like hydropower, solar, say in whatever form and, of course, enhancing nuclear power and inducing large-scale energy conservation. Now the message is, and that has been described elsewhere in my publications, that all of these steps are necessary.

I will say only a few words on each of these steps. In Table 6 you have the gigatonnes of carbon per TWa for the various fossil fuels.

Table 6. Carbon Content in Fossil Fuels; estimated global average, in [GtC/TWa]

Coal	0.751
Liquids	0.605
Gas	0.432

In the case of gas it is the lowest because the relative hydrogen content is the largest and in the case of coal it is the highest because coal is essentially CH_1 and not CH_4. The liquids as CH_2 are in between.

I will say a few words then on solar power and on biomass. The big thing, which is usually not taken into account, is the issue of energy densities as seen in Table 7.

Table 7. Energy Densities in [W/m²]

bio plantations	~ 0.2
solar photovoltaic in arid areas, overall	~ 10
consumption density in urban areas	~ 5

Bioplantations produce or give you 0.2 W/m² on a larger average. That means if one locally plants a garden or a particular field of limited size one can have more than 0.2 W/m², you can go as high as 0.8 W/m² or even 1 W/m² But if you go to large-scale averages it is very difficult to go beyond 0.2 W/m² While the consumption in urban areas is a rather unique figure throughout the world, be it developing countries or developed countries, it is practically always close to 5 W/m², which is then a factor of 5 times 5. The consumption density is 25 times larger than the supply density of biomass. If one wants to rely on biomass on a truly large scale one runs into logistic problems of collecting and managing these large amounts of biomass.

Solar-photovoltaics in arid areas has a better figure of 10 W/m² as a technical figure, not as a physical figure, and therefore is much more promising than the plantations.

A word on energy conservation, which is such a politicized issue. If you consider the primary energy consumption per gross domestic product (GDP) as shown in Table 8, or the wattyears per dollar, in this case the figures refer to the dollar of 1987, you see very big differences. Take the case of Japan, where only 0.22 wattyears of primary energy are necessary for a dollar of GDP while in China it is 2.66 wattyears. That is a factor of twelve. In the United States the figure is nicely above 0.22, in the Soviet Union even more, Germany is there somewhere in between, the world average is 0.6. The message is this: a gross national product is not a gross national product. A gross national

Table 8. Energy consumption per gross domestic product as of 1988

Country	PEC/GDP [Wyr/$ 1987]
FRG	0.32
India	1.43
Japan	0.22
PR China	2.66
USA	0.55
USSR	0.85
World	0.63

GDP: Gross Domestic Product
PEC: Primary Energy Consumption

Source: Commissariat a l'Energie Atomique: Memento sur l'energie, 1989

product is an interplay between capital investment, energy use, labor and know-how. And these four ingredients are completely different in the various parts of the world. So it is essentially an open economic and societal issue, not so very much a technical issue or a scientific issue, to what degree energy conservation is possible. It is one of the open questions today because it is not a technical question. You have to go into these considerations. In almost all cases strict energy conservation means substitution of energy consumption by capital. But capital is in such shortage. In this country, in Germany, we have a capital shortage in view of Eastern Germany, and the same statement is true of Western Europe with respect to Eastern Europe. So capital is a precious thing, and we must kind of optimize its use and cannot put it all into the displacement of energy.

Given these soft considerations which makes it all relative I want to tell you what I personally believe is the reasonable figure for the degree of energy conservation achievable in the foreseeable future. Here I concur very much with Alexei Makarov, with his recent study that was also published in Scientific American of September 1990 [4]. I do appreciate most of the studies of Alexei Makarov in Moscow. His result is that a reasonable figure for energy conservation, if you are willing to change the infrastructure and live up to the necessities of energy conservation, is 25 %, compared with today's situation. This is my personal opinion, it is not possible to prove it by scientific means because it is a political and socio-economical issue more than anything else.

Some Conclusive Figures of the 1989 Jülich

Reduction Scenario for CO_2

Asking then for the necessary contribution of nuclear power one can again go into scenarios. For the World Energy Conference in Montreal I presented a detailed scenario [5] taking into account the differences for the various world regions. I only want to mention the conclusive figures in Table 9, namely, that a roughly 25 % contribution not of

14

Table 9. The 1989 Jülich CO_2 Reduction Scenario
for 2030, the Case of Nuclear Power

- 2.2 TWa/a for electricity generation relates to a capacity of   1.25 TW$_e$

- 1.5 TWa/a for high temperature process heat generation relates to the equivalent of   ~0.75

- thus the total equivalent capacity is   ≈2 TW$_e$

- the OECD/IAEA high case for 2025* yielded a capacity of   2.16 TW$_e$

- the total number of 1 GW$_e$ units would be 2000 that is a factor of 6 compared with 330 such units in 1988

*Nuclear Energy and its Fuel Cycle, OECD/IAEA, Paris 1987

energy conservation but of nuclear power, to the energy demand of the year 2030, that means not taking all the burden of it but only 25 % of primary energy provision something like 2 TW$_e$ is necessary, which is a factor of five compared with the 400 GW$_e$, which was my first scenario. So the second scenario that I arrive at expects 2 TW$_e$ by the year 2030. 2.2 TWa/a, if one takes the Jülich scenario, more specifically relates to 1.25 TW$_e$ in that scenario. 1.5 TWa/a of high-temperature process heat relates to 0.75 TW$_e$ of these 2 TW$_e$ equivalent. So not all of it is for electricity, but the equivalent capacity is roughly 2 TW$_e$. I want to recall for reasons of comparison the OECD/IAEA high case for the year 2020 which expected, that was prior to Chernobyl, a similar figure of 2.16 TW$_e$. In terms of capital availability, in terms of construction capability, this is still a feasible figure, it requires some 40 plants per year, a figure that we have already achieved in the past, that means if we really go strong, then it is possible.

Conclusion

Now these are two scenarios and which way the future turns out to be is open. We should recall that nobody is able, and now is even no longer willing, to predict the future. But nevertheless it is necessary to provide an orientation, and I think that these two figures of 400 GW$_e$ or else 2000 GW$_e$ or 2 TW$_e$ provide the orientations for the year 2030 when we are in the midst of the transition to the ultimately sustainable energy futures.

REFERENCES

1. M. G. Schomberg, M. G. Sowerby, F. W. Evans: Proc. IAEA Conf. Fast Reactor Physics, Karlsruhe 1, 289 (1967)
2. R. Gwin et al.: ORNL-TM-2598 (1969)
3. W. Häfele: Scientific American, Special Issue, Vol. 263, No. 3, 136, (September 1990)
4. W. U. Chandler, A. Makarov, Z. Dadi: Scientific American, Special Issue, Vol. 263, No. 3, 120, (September 1990)
5. W. Häfele: Energy Systems under Stress, Position Paper held at the 14th Congress of the World Energy Conference, Montreal, Canada (1989)

Part I

Nuclear Data Relevant to Fission Reactors and some Fundamental Studies with Slow Neutrons

2.2. Nitrite fueled fast reactors

Reactivity coefficients in this type of core show different sensitivities with respect to oxide cores, at high energies, as is indicated in figs 9 and 10 for the Na void coefficient sensitivity in a PHENIX type of core to 10 % variations of the U-238 fission cross-section and the absorption cross-section of N-14. On the contrary, the sensitivity profiles for Pu-239 σ_f and U-238 σ_c are very similar (see for example fig. 11).

Moreover, the N-14 absorption plays an important role in the neutron absorption. For the PHENIX-type nitrite fueled reactor considered, the N-14 neutron absorption represents $\sim$ 3 % $\Delta K/K$ (to be compared to 0.2 % $\Delta K/K$ for oxygen in the corresponding oxide core).

2.3. Nitrite fueled thermal reactors

A few exploratory studies have been performed on the neutronics caracteristics of a nitrite-fueled PWR. N-14 absorptions modify significantly the neutron balance, since for a standard PWR lattice they represent $\sim$ 6 % of the total absorptions at begining of life and $\sim$ 4 % at 40 000 MWd/t burn-up.

Moreover for a plutonium nitrite-fueled PWR (with a MR = 0.6) the void coefficient sensitivities have been compared to those of the equivalent MOX-fueled PWR and some results are given in the following table :

Variation of isotope	σ_c		σ_f	
	Nitrite	Oxide	Nitrite	Oxide
U-238	- 0.3	+ 0.6	- 4.3	- 3.9
Pu-239	- 2.9	- 2.4	- 6.4	- 6.3
Pu-240	- 2.3	- 2.3	- 1.4	- 1.1
Pu-241	- 0.4	- 0.4	- 0.3	- 0.9

Percentage variation of total void effect for a variation of + 10 % of the respective σ in a nitrite or oxide Pu-fueled PWR.

2.4. Lead-cooled fast reactors

This type of reactor is presently studied in the USSR. The presence of lead modifies both the leakage and the neutron moderation caracteristics of the core. As an example we have compared a Na-cooled and a Pb-cooled PHENIX-type core. The coolant total void coefficients are compared in the following table (values in $10^{-5} \Delta K/K$) :

Void effect components	Na-cooled reactor	Pb-cooled reactor
Spectral component	+ 2700	+ 3850
Leakage component	- 3400	- 6250

which shows both the harder spectrum effect in the lead-cooled reactor, and the different density effect on leakage in the two systems.

2.5. Actinide burners

Transactinide data uncertainties will play a role in more advanced phases of any future design for this type of reactor, both in terms of criticality balance and in terms of reactivity coefficients. If actinides are added in large quantities to the fuel, the uncertainty level on K_{eff} can be of 2 $\div$ 3 % $\Delta K/K$ and for e.g. Na void coefficient in a fast reactor with actinide recycling, the uncertainty will be substantially higher than the present level of $\pm$ 20 %.

3. NEUTRON INTERACTION DATA NEEDS BY TYPE OF ISOTOPE

3.1. Major Actinides (Pu[239], Pu[240], Pu[241], Pu[242], U[235], U[238], U[233], Th[232])

3.1.1. Thermal data, related in particular to the moderator temperature coefficient

In this field, much progress has been made in the very recent past. However there still exists a controversy on the U[235] η shape. The evidence for a 1 $\div$ 2 % slope has still to be confirmed. However, even if the slope is confirmed only 30 % of the present discrepancy on moderator temperature coefficient in PWRs will be removed.

Different assumptions are made about the influence of solide state effects on Doppler breadening of U[238] resonances and the consequent effect on the temperature dependence of U[238] resonance capture. In particular, the choice of the appropriate Debye temperature (250° - 650°C for UO_2) has a relevant impact on the temperature coefficient, and it will play a different role at room temperature (at which experiments are performed) and at the power reactor operating temperature. This problem should also be studied for full MOX fueled LWRs and in particular for the case of a high moderating ratio MR (MR > 3).

3.1.2. Resonance (epithermal) range

The sensitivity study results presented previously for MOX fueled LWRs indicate that, if these concepts are considered, there is a clear incentive to improve the knowledge of the resonance data for Pu isotopes (Pu[239], Pu[240], Pu[241] and Pu[242]) in the resonance energy range. The NEACRP benchmark for HLWRs has indicated /1/ the order of magnitude of the dispersion of results for integral parameters, which cannot be accounted for only in terms of methods uncertainties.

3.1.3. Fission cross-section of Pu[239] in the 10 keV - 100 keV

This is a well known problem, on which is focused a sub-group of the International Cooperation on Evaluation /2/ and results of the present state of the art will be presented at this Meeting /3/.

It has to be recalled that the Pu[239] fission data have a large impact on the Na-void coefficient in LMFBRs, whatever the type of fuel (see previous paragraph). A large impact of Pu[239] fission data is also found in the assessment of the uncertainties in the calculation of control rod worth in LMFBRs of large size /4/. In fact, an uncertainty reduction from $\pm$ 6 to $\pm$ 2 % on these data, reduces by 25 % the total uncertainty on the control rod worth in this type of reactor.

3.1.4. U^{238} inelastic cross-section

Again in this case, the problem is well known, but despite much work in this field, discrepancies are not negligeable and they affect the criticality balance in most types of reactor. It is clear that for existing reactors, the large experimental evidence coming from operational experience allows us to accommodate this uncertainty. However the study of new reactor types, for which the critical balance results from different compensations, can suggest a further improvement of these data.

3.1.5. U^{233} and Th^{232}

Often neglected in high priority data requirement lists, these data are less will established, over the whole energy range, compared with Pu, U^{235} and U^{238} data. Th/U(233) cycle based reactors are still studied (India, Japan, Bresil), and the use of Th in place of U to alter long-lived minor actinide production, could become a future issue.

3.2. Minor Actinides

In this field the interest is obviously related to the studies of actinide-burner reactors, which are presently being made in many countries /5/.

For present exploratory studies the status of the data in the major data files is acceptable, since many integral experiments have been performed in the past fifteen years on power reactor irradiated fuel, separate isotope irradiation, critical experiments at zero power (see for example Refs 6-7) to validate them.

However, in more advanced phases of the design, the accuracy requirements will become much tighter and the present status of uncertainties for Np^{237}, Am^{241}, Am^{243}, Cm^{244} and Pu^{238} cannot be anymore satisfactory. In particular resonance data for Np^{237} and Am^{241} can become crucial, in particular if actinide fuel is considered, both in fast and in thermal spectra, in heterogeneous recycling mode.

Few integral experiments are presently available to check resonance data, and the analysis of the SUPERFACT experiment /8/, in which high content Np^{237} and Am^{241} fuel pins have been irradiated in the PHENIX reactor, will give first indications in this field.

A specific problem is represented by inelastic cross-sections (mainly for Np^{237} and Am^{241}), since most of the integral experiments mentioned above, give indications on capture, fission and, in some cases, (n,2n) cross-sections. A target accuracy of 20 % should be envisaged.

Finally, branching ratios for many reactions of interest in fuel cycle applications should be verified.

It is also to be remembered that the extended burn-up objective for present PWRs, enhances the need for an improved knowledge of higher actinide data.

3.3. Structural materials

The need for data improvement does not come from specific advanced reactor needs, but it is mostly related to the present state of these data, in particular data above 100 keV. In this case also a comparative study of the leading data files has been undertaken by one subgroup of the International Cooperation on Evaluation, and results will be reported at this meeting /9/. A typical example of the present dispersion among data files is given in fig. 12. The effort in this field should be kept, a further motivation being the large variety of advanced structural materials envisaged for future reactors both for core (e.g. the high Ni steel considered as a possible cladding candidate for the a European Fast Reactor) and for shielding.

3.4. Fission products

In this field, significant progress has been made in the past decade, both in terms of data evaluation and in terms of validation with integral measurements (see for example Ref. 10 and Ref. 11, presented at this meeting). A continuous effort is underway for criticality/safety purposes.

However studies on the envisaged possibility of transmutation of the most long-lived fission products (Tc^{99}, Cs^{135} and I^{129}) in fission reactors with a thermalized spectrum, indicate the need to carefully verify the resonance data for these isotopes. Moreover, inelastic cross sections for even-even nuclei have been pointed out as potentially discrepant data.

3.5. Other isotopes for specific applications

3.5.1. The use of nitride fuels, studied at present in the context of innovative reactor concepts (both for thermal and fast reactors), points out the need to improve the N^{14} and N^{15} data. The high (n, p) and (n, α) reactions in N^{14} have a significant effect on the critical balance (see paragraphs 2.2 and 2.3) and, if an enrichment in N^{15} is envisaged to reduce absorption, the data for this isotope should be better assessed, since very few data are presently available.

Finally, the tritium production reaction in N-14 should be considered, due to the potentially high production in this type of core.

3.5.2. The use of "inert" matrices instead of uranium for Pu fueled thermal reactors is envisaged as a possibility to enhance the capability to burn plutonium, without the effect of the build-up of new Pu from U^{238}. In that case the candidates can include elements like Ce /12/, for which data, including resonance data for Doppler effect calculation, will be needed.

3.5.3. Lead cooled fast reactors are studied at present in the USSR. We have indicated in a previous paragraph the impact of the moderating properties of lead on the core neutronics caracteristics, and this implies an assessment of Pb data (including inelastic cross-sections).

3.5.4. The transmutation of long-lived fission products (Tc^{99}, Cs^{135}, I^{129}) has already been mentioned. The possibility to perform this transmutation in special subassemblies located at the periphery of fast reactors is presently being studied, where the materials to be irradiated are surrounded by moderating materials (mostly hydrides), to thermalize the spectrum and to enhance the epithermal captures. Experiments of this type are underway both in PHENIX and FFTF (e.g. for the production of Co^{60} from Co^{59} targets). Even if the moderation is essentially due to hydrogen, the hydride compound (e.g. Calcium or yttrium hydride) characteristics play a role and the data for the materials involved should be known with sufficient accuracy.

Finally, special matrices have to be envisaged for the irradiation of some of these isotopes (as Ba for I-129) and the corresponding data should be verified in case of massive use in reactors.

Moreover the prediction of the amount of these isotopes produced in reactors will have to be as accurate as possible, and data related to their production will have to be verified against irradiated fuel analysis, and possibly improved if calculation/experiment discrepancies are found. As an example the uncertainty in Cs^{135} production in French PWRs is of the order of $\pm$ 30 %.

3.5.5. Needs for isotope production reactors

It has been pointed out that specific isotope production for medical application represents a field of potential future applications, and data needs will certainly arise for target isotopes.

4. DATA OTHER THAN NEUTRON INTERACTION

4.1. Delayed neutron data are still the object of validation. For major actinides (U^{235}, U^{238}, Pu^{239}), a subgroup of the International Cooperation for Evaluation has established a work plan for data improvement and validation, using a benchmark critical experiment to measure the effective fraction of delayed neutrons (β_{eff}) in the critical facility MASURCA (at Cadarache) under the sponsorship of the OECD NEA Committee for Reactor Physics. A report will be presented at this Meeting /13/.

In the case of dedicated actinide burners (i.e. fission reactors with minor actinide fuel), the need for reliable delayed neutron data for the concerned isotopes (Np^{237}, Am^{241}, Cm isotopes) is related to the fact that the β_{eff} for this type of reactor (which has impact on the kinetic performance) can be very low (preliminary studies have indicated /14/ values as low as 0.1 % $\Delta k/k$).

4.2. Decay heat data are at present validated in most major data files. Apart from the uncertainties related to very short cooling times, due to the models used to account for β decay, the present situation is fairly satisfactory. However, and again for the case of actinide burner reactors, extended data bases should be envisaged.

4.3. Photon-production data have become more widely available in the past few years for structural materials. The validation work and its implications, even for existing reactors, is still to be undertaken. This is an area of potential work in the future, of relevance to existing and advanced reactors.

4.4. Covariance data

The need for this type of data is not specific to innovative concepts. However their need has been recognized for the assessment of neutronics design caracteristics, data improvement with adjustment techniques, and target accuracy definition. The efforts in this field are still limited and future work in this field is an area for fruitful international cooperation, in particular in view of the convergence of the major existing data files. Reactor physics needs are mainly related to simplified covariance data and a general consensus on the features, formats, energy ranges etc, is still to be found.

5. CONCLUSIONS

Innovative reactor concepts for the medium and long term will require a substantial effort in reactor physics and the related nuclear data disciplines.

The spectacular progress both in the theoretical and in the experimental fields during the last thirty years should not hide the need to consolidate our knowledge and to avoid the loss of competence.

In the data field, even if it is perhaps premature to produce detailed compilations of requirements and target accuracies, it is already clear that many areas will need special attention, as has been briefly rewieved in this paper.

The well established link between reactor physics, by means of integral experiments, and nuclear data work, by means of data evaluation and differential measurements, will play an important role in support of studies on innovative reactor concepts. Improved safety and economics, a better perception by public opinion, will be based on a better understanding of fundamental principles and on reduced uncertainties in design calculations. To paraphrase N. BOHR, the complementary of "Klarheit" and "Warheit", in terms of complementarity of differential and integral experiment information, will be the unavoidable basis for progress towards a further development of nuclear energy.

6. ACKNOWLEDGMENTS

S. CATHALAU and R. SOULE performed most of the sensitivity studies mentioned in the present work.

7. REFERENCES

1. W. Bernnat et al "Advances in the analysis of the NEACRP HCLWR Benchmark Problems" Proc. PHYSOR'90 Conference, Marseille, April 1990

2. C. Dunford, Y. Kikuchi, M. Salvatores, paper presented at this meeting

3. E. Fort et al, paper presented at this meeting

4. M. Salvatores "CEA Data and Methods for Control Rod Calculations" Proc. IWGFR Meeting on Methods for Reactor Physics Calculations for Control Rods in Fast Reactors, Winfrith, Dec. 1988

5. See for examples the Proceedings of the First OECD/NEA Meeting on Information Exchange on Actinides and Fission Product Partitioning and Transmutation, Mito, November 1990

6. A. D'ANGELO et al., Nucl. Sci. Eng. 105, 244 (1990)

7. H. Kusters, Nucl. Technology 71 (1985)

8. L. Koch et al. J. of Less-Common Metals 122, 371 (1986)

9. H. Gruppelaar et al. paper presented at this meeting

10. A.J. Janssen et al. "Integral Data Test of JEF-1 Fission Product Cross Section" ECN Report, ECN-176 (1985)

11. J. Mondot et al., paper presented at this meeting

12. H. Küsters, private communication

13. A. Filip et al., paper presented at this meeting

14. D.F. Tsurickov et al., private communication

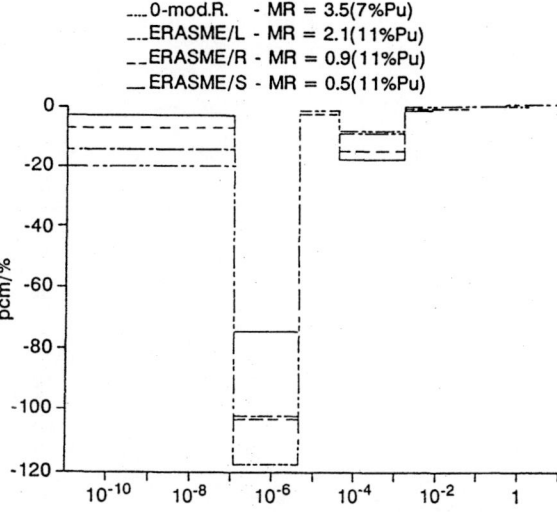

Figure 3 : Sensitivity to ^{240}Pu capture
(1% variation)

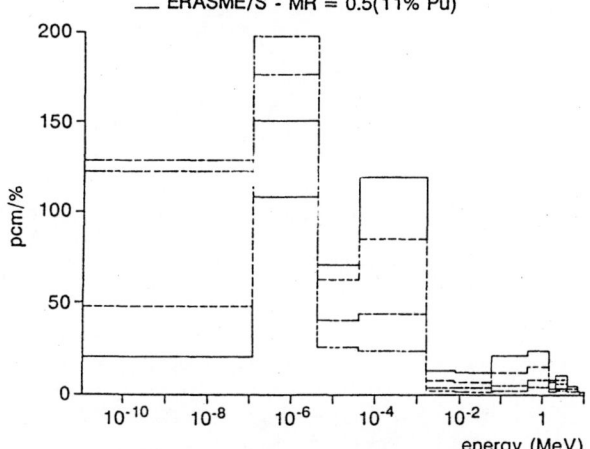

Figure 1 : Sensitivity to ^{239}Pu fission
(1% variation)

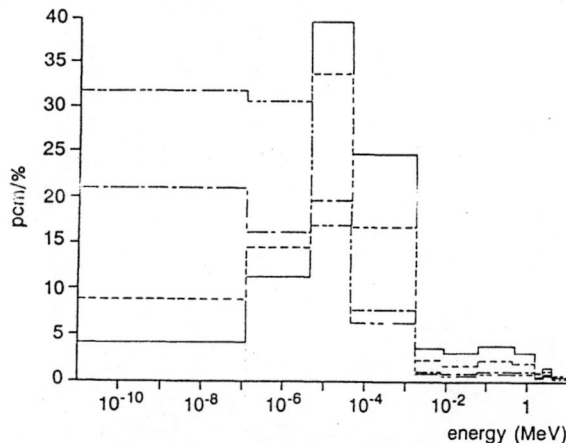

Figure 4 : Sensitivity to ^{241}Pu fission
(1% variation)

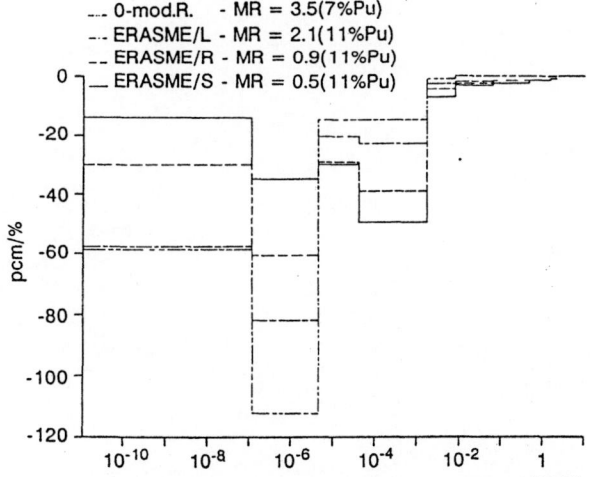

Figure 2 : Sensitivity to ^{239}Pu capture
(1% variation)

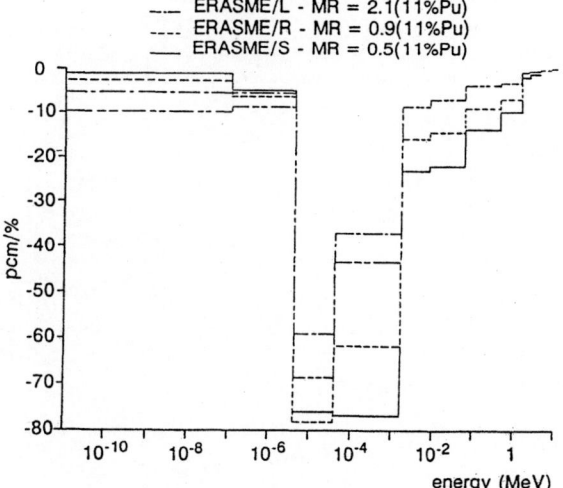

Figure 5 : Sensitivity to ^{238}U capture
(1% variation)

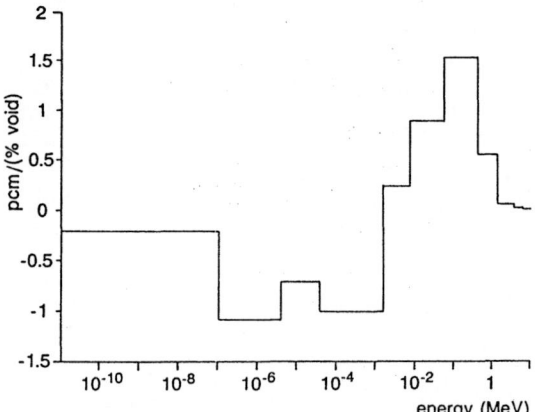

Figure 6 : Sensitivity of void coefficient to ^{239}Pu fission
1% variation - (MOX fuel LWR MR = 0.5)

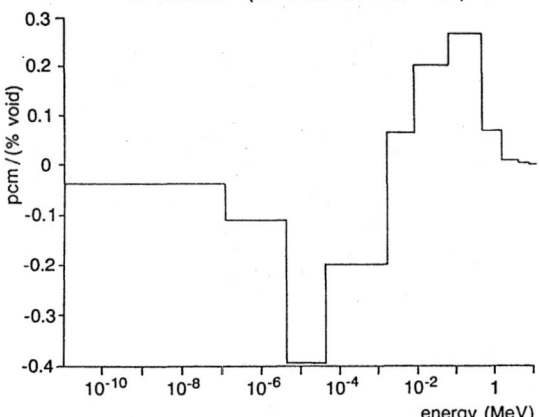

Figure 7 : Sensitivity of void coefficient to ^{241}Pu fission
1% variation - (MOX fuel LWR MR = 0.5)

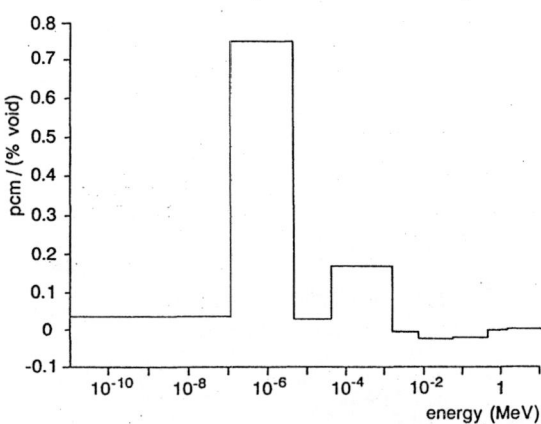

Figure 8 : Sensitivity of void coefficient to ^{240}Pu capture
1% variation - (MOX fuel LWR MR = 0.5)

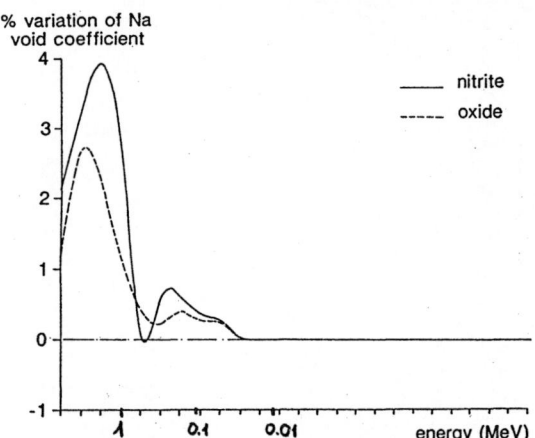

Figure 9 : Na void coefficient sensitivity to +10% σ_{in} ^{238}U
in PHENIX type cores(oxide and nitrite)

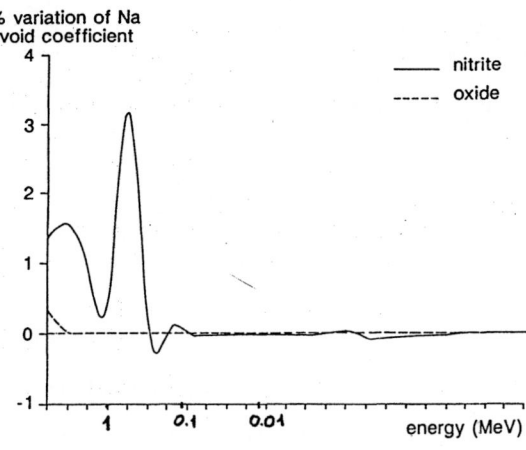

Figure 10 : Na void coefficient sensitivity to +10% σ_{abs} 0-16
and N-14 in PHENIX type cores(oxide and nitrite)

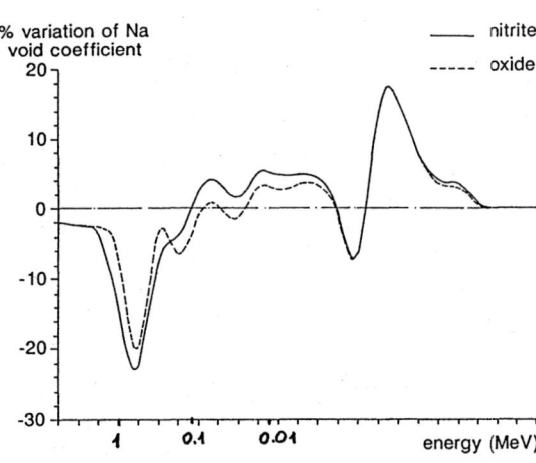

Figure 11 : Na void coefficient sensitivity to +10% σ_f ^{239}Pu
in PHENIX type cores(oxide and nitrite)

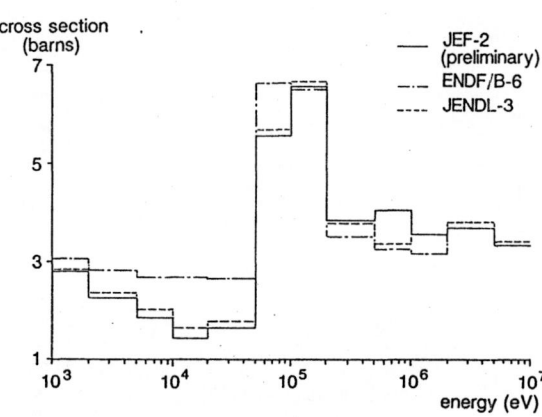

Figure 12 : Comparison of JEF-2, ENDF/B-6 and JENDL-3 libraries
Average total cross sections of CR-52

NUCLEAR AND NUCLEON DATA NEEDS FOR INCINERATION OF THE RADIOACTIVE WASTES FROM FISSION REACTORS WITH A PROTON ACCELERATOR

Y. Nakahara

Tokai Research Establishment
Japan Atomic Energy Research Institute
Tokai-mura, Ibaraki-ken, 319-11 Japan

Abstract: Brief summaries are given of the computer codes prepared at JAERI for the computer simulation calculations of the nuclear spallation reaction and for the nuclear design analyses of accelerator target/blanket systems for the trans-uranium waste incineration. Nuclear and nucleon data needs are reviewed for developing computer code systems and for making design analyses of the intermediate energy-high intensity beam proton accelerator and the incineration system.

A model incineration system and its performances are described. The proton energy has been taken to be 1.5 GeV. Two types of the systems have been investigated: the single composition type similar to the LMFBR assembly and the hybrid type. In the hybrid system the tungsten target is surrounded by the trans-uranium blanket. This system has been employed to make the power density distribution flatter. The neutronics and thermal-hydraulics analyses have been performed for Na cooled and Pb-Bi cooled assemblies. The allowable beam current is larger for the Na cooled assembly than for the Pb-Bi cooled one. The maximum incineration rate has been obtained for the tungsten loaded, Na cooled assembly, in which the proton beam of the energy 1.5 GeV and the current 22.6 mA bas been estimated to be able to incinerate the TRUW from 7.6 units of the 3,000 MWt, LWR.

Design studies of the engineering test accelerator (1.5 GeV, 10 mA) are being carried out at JAERI.

(nuclear data, nucleon data, nuclear spallation reaction, computer simulation, computer codes, incineration, trans-uranium waste (TRUW), target/blanket system, neutronics, thermal-hyraulics, proton, accelerator)

Introduction

Nuclear spallation reactions of the heavy metal are the strongest available neutron and meson sources. Persistent efforts to build the more and more intense spallation neutron sources have been made at several Laboratories, i.e., ISIS at RAL, SINQ at SIN, IPNS at ANL, LANSCE at LANL and KENS at KEK [1, 2].

The steadily increasing feasibility of the intense spallation neutron source has opened up the possibilities to introduce accelerators in the nuclear fuel cycle. At first, attention was paid to use the accelerator to produce the fissile material [3, 4]. Then the attension has been shifted to use it to incinerate the radioactive waste as the public concern about the waste increases.

In the design analyses of fissile breeding and trans-uranium (TRUW) waste incineration plants, nuclear data in the intermediate energy range 20 ∿2,000 MeV are very useful, if available. But the available nuclear data file is only one at present. That is the ENDF/B-VI High Energy Library for the Fe target in the energy range 1 ∿1,000 MeV [5].

Since there is no nuclear data file applicable to the analyses of the TRUW incineration with a proton linear accelerator, the analyses of nuclear spallation reactions of nuclei with protons and other nucleons have been performed with the use of Monte Carlo simulation techniques for all the elementary processes involved. The data used in the simulation are those for nuclear structures and nucleon-nucleon elastic and inelastic cross section, including the pion-nucleon cross section.

Several computer codes have been and have been being developed for specific purposes. Discussions are made on the theoretical models and methods used in the codes and also on the data needs for evaluating the computational schemes.

A model incineration system and its performances are described.

Computer Codes for the Nuclear Spallation and Incineration Analyses

A starting point of our efforts to develop computer codes for nuclear spallation and incineration studies was the famous NMTC code [6]. Several codes have been developed for specific purposes.

NUCLEUS [7]

This code is used for calculating the nuclear spallation reaction (intra-nuclear cascades + evaporations and/or high energy fissions) between a single target nucleus and a projectile without taking into consideration of the inter-nuclear nucleon transport processes, in order to make direct evaluations of physical and computational models.

The mass A of a target nucleus is limited to A=1, $6 \leq A \leq 250$. The upper limit of the applicable energy range is 3.5 GeV but 2.5 GeV for pions.

The nuclear data prepared in the code are those for the nculear structure. A nucleus is assumed to consist of three regions. To each region the following data are assigned, i.e., the radius, the proton and neutron density distributions, the Fermi energy distributions for protons and neutrons.

The nucleon-nucleon cross section data used in the code are those for (n, p), (p, p) elastic scattering, differential scattering, inelastic scattering with 1π and 2π productions, and also those for (π^-, p), (π^0, p), (π^+, p), (π^0, n) elastic scattering, charge exchange scattering, differential scattering, absorption and inelastic scattering with the 1π production.

NMTC/JAERI [8]

This code simulates the transport process in a heterogeneous bulk medium of nucleons in the intermediate energy range (≥ 15 MeV). The data used in the code are the same as those used in NUCLEUS except for the macroscopic geometry related data.

NMTA/JAERI [9]

This is the edit program of the output data from NMTC/JAERI, and gives the spallation product distribution and the heat deposition density distribution.

ACCEL [10]

This is a code system for nuclear design calculations of the accelerator target-blanket systems for the TRUW incineration and fissile breeding. This code system is designed to perform simulation calculations of all the reactions and nucleon transport processes in a heterogeneous medium through entire energy range from the incident particle energy down to the thermal neutron energy.

DCHAIN-SF [11]

This is an extended version of a one point depletion code DCHAIN2 [12]. DCHAIN-SF can treat build-up and decay processes of nuclides due to the reactions not only with neutrons but with other nucleons.

SPD [13]

In the TRUW incineration by spallation reactions, various kinds of nuclides are produced as the spallation (including fission) products. Most of the spallation products are neutron dificient, and far from the stability line. Since there are scarce data of them, it is necessary to make theoretical estimates of half lives and the decay heat.

SPD calculates the following quantities related with the β- and γ- decay. The Q- value, the half life, all the energy convertible to the thermal energy, kinetic energy of electrons, the energy of the γ-decay following the β-decay, γ-energy released by the position pair annihilation.

Theoretical Models used in the Nuclear Spallation Simulation

The nuclear spallation had been modelled as a two step process. The first step is the intra-nuclear cascades of nucleons initiated by an incident particle, during which several numbers of protons, neutrons and pions are knocked out of the nucleus. The second step is the competing decay of the residual nucleus by fission and/or particle evaporations. A comparative investigation of the high energy fission models of Nakahara, Takahashi, Atchison and Alsmiller, et al. was made by T.W. Armstrong, et al. [14].

In the two step model of the spallation two important reaction processes are not taken into consideration. One is the preequilibrium decay of the residual nucleus after the intra-nuclear

cascades. The preequilibrium decay process has been successfully analyzed by the exciton model proposed by Griffin [15] and improved extensively by Blann [16]. The other is the multi-fragmentation of a nucleus. Many models have been proposed to explain the mechanism of the fragmentation. But the actual fragmentation process is so complicated that none of them has succeeded in offering a convincible explanation.

Particle Emission from the Preequilium State

It is known that the spallation neutron spectra calculated by the two step model show remarkable underestimates in comparison with measured ones in the energy range above 20 MeV.

To improve the two step model, Nakahara and Nishida formulated the Monte Carlo algorithms for simulating particle emissions from the preequilibrium states [17], using the Kalbach's phenomenological formulation [18], which is an extension of the Griffin's exciton model [19]. These Monte Carlo algorithms have been incorporated in the HETC code [20] by Ishibashi, et al. [21, 22]. The program used was the RL Version [23], in which improvements were made by Atchison to incoporate the high energy fission. The flow of the three step model calculation is shown in Fig. 1.

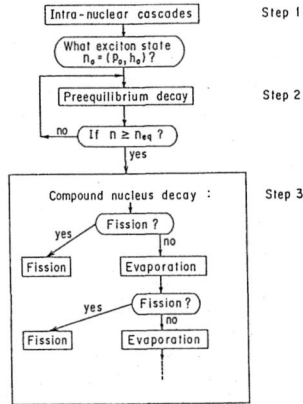

Fig. 1 Flow of the 3 step calculation.

The computational results are shown in Fig. 2 in comparison with the experimental data obtained by Cierjacks, et al. [24], for the 585 MeV protons incident on the lead target [25]. The dotted lines show the results of the standard HETC calculations. The solid lines indicates the results of the three step calculations, where in addition to the Monte Carlo exciton algorithm two important improvements have been made. For the momentum distribution of the intranuclear nucleons, the following probability function [26] is used instead of the degenerate Fermi distribution at the zero temperature,

$$W(p) = W_0\{\exp[-(p/p_0)^2] + \varepsilon_0\exp[-(p/q_0)^2] + \varepsilon_1\exp[-(p/q_1)^2]\}, \quad (1)$$

where p is the nucleon momentum, W_0 is a normalization factor, and

$$p_0 = (2/5)^{1/2}k_f, \quad \varepsilon_0 = 0.03, \quad q_0 = (6/5)^{1/2}k_f$$
$$\varepsilon_1 = 0.003, \quad q_1 = 0.5 \text{ GeV/c}$$

and k_f is the Fermi momentum. The other improvement is the use of a probability density function $f(E_c)$ for terminating the intranuclear cascade:

$$f(E_c) = 2 E_0^{-1} (1-E_c/E_0), \quad (2)$$

where E_0 is a parameter,

As seen from Fig. 2, the shoulders are well reproduced, especially in the backward direction.

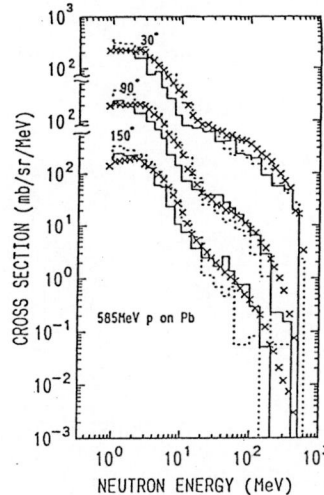

Fig. 2   Double differential cross section for the 585 MeV proton on Pb.
××× experimental data [24].
••• standard HETC calculation.
—— HETC with the exciton model.

Spallation Products

To make the assessment of the real feasibilities of the incineration of TRUW, it is necessary to estimate what kind of nuclides are produced as the spallation products. In Fig. 3 the mass yield cross section is shown for the case of a 3 GeV proton impinging on a silver nucleus [27]. The experimental values are due to Katcoff, et al. [28]. Our calculations show a good agreement with the measurements for A > 80, while they give considerably lower values for 40 < A < 80 and the increase for A < 30 is not reproduced in our calculations. A main cause of these discrepancies can be considered to be due to the absence of the fragmentation process in our computational scheme.

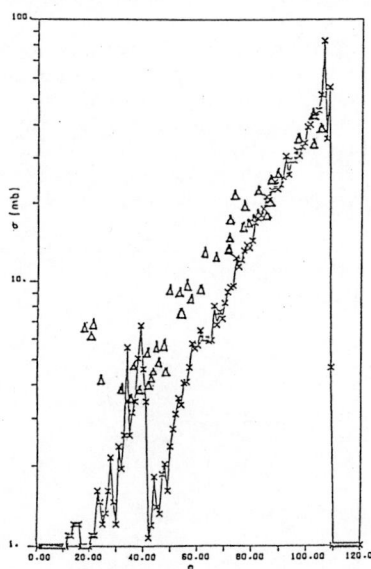

Fig. 3   Spallation product distribution for a silver nucleus at 3 GeV incident proton energy.
Δ experimental data [28].
× calculation.

Nuclear Multifragmentation

The distribution of positions of nucleon-nucleon collisions during intranuclear cascades is strongly localized, as shown by Baraschenkov, et al. [29]. The local hot regions are created temporarily. Characteristics of these hot regions can be considered to define the conditions for the subsequent multifragmentation. The hot region may be considered to correspond to the emitting source used in the light fragment emission analysis by Korteling, et al. [30].

On the other hand, Botvina, et al. proposed a cascade fragmentation evaporation (CEF) model [31]. Upon completion of the intranuclear cascade process, the space distribution of nucleons in the residual nucleus is shown to be inhomogeneous. Relaxation of such a unstable nucleus may cause collective oscillations of the nuclear matter and may lead to the fission or the fragmentation.

It seems that the models of Korteling, et al. and Botvina, et al. represent the two aspects of the fragmentation, i.e., the hot spot effect and the local density depletion effect, respectively. We are planning to investigate these effects extending editor functions of our NUCLEUS code [7].

Nuclear and Nucleon Data Needs in the Analyses of the TRUW Incineration

There are needs for four categories of data, i.e., nucleon-nucleon data, nucleon-nucleus data, the yield data of spallation and fission products and the decay constants.

As for the nucleon-nucleon cross sections, the data compiled by Bertini in early 1960 s are used widely even now [32]. Most of the data are those measured in 1950 s. Entering 1960 s and 1970 s, measurements became performed actively at several Laboratories, especially at SIN, TRIUMF and LAMPF. Bugg made a survey and evaluations of the data for the channels $pp \to pp$, $np \to np$, $pp \to pn\pi^+$ in the energy range up to 1 GeV [33]. Measurements on the channel $np \to pp\pi^-$ were planned at TRIUMF and LAMPF [34].

Nuclear and Nucleon Data

At present, the nuclear spallation process is simulated with the Monte Carlo method using the following data set, i.e.,
• nuclear radius,
• nucleon (p and n) density distribution in the nucleus,
• Fermi energy distribution in the nucleus,
• mass formula,
• level density parameter and the fission barrier height for the high energy fission,
• nucleon-nuleon reaction cross sections,
• pion-nucleon reaction cross sections.

For the level density parameter a_f for the high energy fission, use is made of an analytic interpolation-extrapolation expression [8],

$$a_f/a_n = aE^2 + bE + c, \text{ E in MeV.} \qquad (3)$$

The level density parameter for neutron emission has been chosen as $a_n = A/10$ from the numerical evaluations. Values of parameters in Eq. (3) have been determined by fitting Eq. (3) to the values $a_f/a_n = 1.07$, 1.02 and 1.01 at E = 150, 660 and 1,000 MeV, respectively, obtained by Iljinov, et al. [35].

The fission barrier heights for almost all nuclides have been calculated and evaluated by Iljinov, et al. [35] with the use of the liquid

drop model due to Meyers-Swiatecki [36] and
Nix [37], which permits the extrapolation to the
region of nuclides with A < 150, where no experimental information on E_f is available. For the
nuclides with A ≳ 225, we use the values obtained
by Kupriyanov, et al. [38].

These data are used in our Monte Carlo codes
[7, 8].

Nucleon-Nucleus Data

The thorough use of the Monte Carlo method
to simulate the nucleon transport and nuclear
spallation processes requires a very high
computation cost. If the cross sections for
nucleon (pion)-nucleus reactions are available,
it becomes possible to simulate the incineration
process efficiently without following the
intranuclear cascades.

As for the energy range of incident protons,
the upper limit as high as possible, at least
1.5 GeV is desired, because the neutron yields
increase almost linearly with the incident
proton energy up to that energy, as is shown for
a depleted uranium cylindrical target with a
diameter 10 cm [39], and also because it is not
made clear yet what is the optimum incident
proton energy for the TRUW incineration.

The structures of the nucleon-nucleus data
required in the incineration analyses are very
complicated, because they must include all the
channeles listed below, i.e.,
(p, elastic),
(p, nonelastic),
(p, $injpk\pi^+ l\pi^- m\pi^0$), where i, j, k, 1 and m are
integers, including zero,
(p, γ),
(p, d),
(p, t),
(p, α),
(p, complex),
where "complex" means spallation and fission
products, including γ, d, t and α. "Complex" is
even "more complex", since the processes such as
(p, $injpk\pi^+ l\pi^- m\pi^0$ complex)
can be considered to occur.

For neutrons, the same kind of data are
necessary in spallation and neutron transport
analyses, because not a small amount of neutrons
of the energy comparable to that of the incident
proton are produced in the spallation reaction.
Data of the energy-angle distributions of the
emitted particles are also required for almost
all the reactions listed above.

The nucleon-nucleus data are required for
the elements and nuclides, at least, listed below.

For the design analyses of an engineering
test accelerator (1.5 GeV, 10 mA), N, O, Ar (air
and water), O, Na, Mg, Al, Si, Ca, Mn, Fe
(ordinary concrete), Zn, Cu, (Mg), (Al) (aluminium
beam tube), Cr, Co, Ni, (Mn), (Fe) (stainless
steel), (Al), (Cu), (Fe) (accelerating cavity).

For the design analyses of an incineration
plant (a target/blanket system), Na, Fe, Zr, Y,
W, Pb, Bi, ^{238}U, ^{237}Np, ^{238}Pu, ^{239}Pu, ^{241}Am and
^{243}Am.

Decay Constants

It is important and necessary to make the
time evolution calculations of the build-up and
decay of all the nuclides produced in the target
and blanket, irradiated continuously by proton
beams and spallation neutrons, repectively, in
order to esimate the real feasibility of the
incineration strategy.

As the spallation products, a variety of
nuclides, especially many neutron-deficient
nuclides with the mass number greater than 180,
are produced [7]. Figure 4 shows the (N, Z)
distribution of nuclides generated in the
spallation of ^{237}Np and the decay constants of
which are undetermined yet [40]. The β decay
constants of some products were calculated by
using the β decay code developed by T. Yoshida
on the basis of the gross theory of β decay [41].

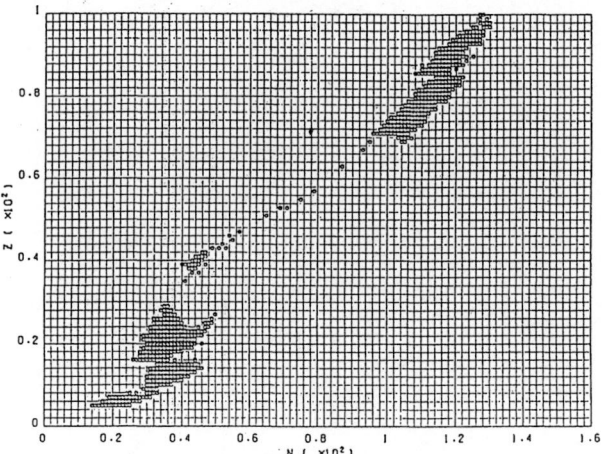

Fig. 4 Nuclides with unknown half life,
produced in the spallation of ^{237}Np.

A Model Incineration System

Target System

Design parameters of a model incineration
system are summerized in Table 1 and the strcuture
of the target system is shown in Fig. 5 [42].

A high intensity beam of protons hit the
target, which is a subcritical assembly with the
effective neutron multiplication factor less than
0.95. The height and the width of the target are
1 m, but the length is 2.6 m and 2 m, for the Na
cooled and the Pb-Bi cooled systems, respectively.
The protons are introduced into the system through
the beam window of the rectangular cross section
1 m × 0.1 m located at a depth of 0.5 ∿ 0.7 m
from the front surface.

The tungusten-loaded hybrid system has been
employed to make flatter the neutron flux distribution in the system, which consists of two
regions, i.e., the actinide region and the
tungsten region. The latter is located just
behind the beam window and has the same cross
section. The length is 60 cm.

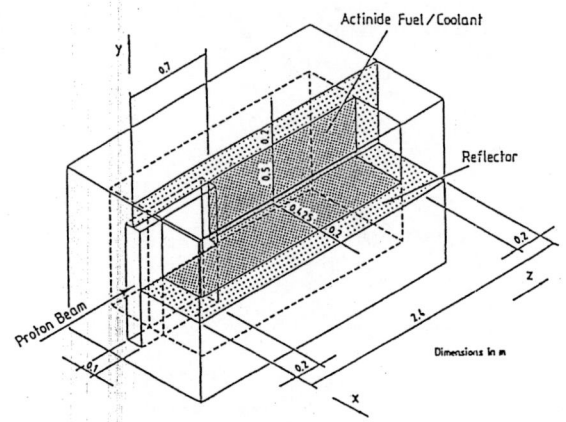

Fig. 5 Accelerator driven incineration
reactor.

Table 1 Design Parameters of a Model
Incineration System

Coolant	Na/Pb-Bi eutectic
Proton energy	1.5 GeV
Target	
Length	2.6 m/2 m
height	1 m
width	1 m
Tungsten region	
length	0.6 m
height	1 m
width	0.1 m
Reflector	
material	Stainless steel
thickness	0.2 m
Fuel	55Np-15Pu-30Zr*
	55Am Cm-35Pu-10Y
diameter/length	4 mm/1 m
clad	HT-9 steel
clad outer dimater	5.2 mm
clad thickness	0.3 mm
bond	Na

* Numbers indicate the wt%.
Pu is added in order to reduce the reactivity
swing.

Actinide Fuel

In our study, metallic alloys of actinides
are employed. The metallic alloy gives a harder
neutron spectrum than the oxide. The hard neutron
spectrum makes the incineration more effective.

The fuel consists of two kinds of alloys, as
described in Table 1 [42]. These alloys are
expected to have the sufficiently high phase
stability and have been proposed for an actinide
burner fast reactor.

Pu is added to keep the reactivity swing
within an acceptable range. The 30 wt% of Zr is
added to raise the melting point of Np from 640°C
to about 900°C.

The fuel assembly is similar to that commonly
employed in the Liquid Metal Fast Breeder Reactor.

Neutronics Calculations

The neutronics calculations were performed
by making use of the several codes.

The nucleon transport and nuclear reactions
in the energy range above 15 MeV were simulated
by making use of the Nucleon Meson Transport Code
NMTC/JAERI [8]. The spatial and energy distributions of the neutrons with the energy less than
15 MeV are obtained also in the NMTC/JAERI calculation. These distributions are used as the fixed
neutron source in the neutron transport calculation in the energy region below 15 MeV.

In that energy region the neutron transport
calculations were performed by using the three
dimensional Monte Carlo neutron transport code
MORSE-DD [43].

Burn-up calculations were performed with the
ORIGEN2 code [44]. This code is a point depletion
and decay code. The average total neutron flux
obtained from the transport calculation is used
as an input data. The neutron cross section and
the fission product yield library contains the
effective one energy group data. Due to the lack
of the lumped cross section data for the fission
product yield from the trans-uranium nuclides such
as ^{237}Np, the data for ^{239}Pu were used for them.
The ORIGEN2 code was used also to calculate the
nuclide concentration change as a function of the
burn-up.

Estimated parameters characterizing the
tungstem-loaded hybrid system are shown in Table
2 [42]. The burn-up rates are 7.0 % and 4.4 %
for the Na cooled and the Pb-Bi cooled systems,
respectively. They correspond to the incineration

of 202 kg and 89 kg, respectively. It means that
the Na cooled system can incinerate the actinides
from 7.6 units of 3,000 MWt PWR per year, where
the amount of the actinides per a PWR is taken as
26.5 kg for the burn-up of 33,000 MWd/t and the
99.5 % Pu recovery.

Results obtained with the point depletion
codes, however, should be considered as the
preliminary estimates. Problems associated with
the presence of the external neutron source, the
space dependent power peaking and depletion will
require careful analyses.

Table 2 Characteristics of the Tungusten-
loaded Hybrid System

Coolant	Na	Pb-Bi
Effective neutron		
multiplication factor	0.92	0.86
Actinide loading (kg)	2866	2013
Beam current (mA)	22.6	9.8
Neutrons per proton	38.1	52.8
Average neutron energy (keV)	739	629
Total neutron flux (n/cm^2·sec)	4.6×10^{15}	6.6×10^{15}
Operation time (days)	270	270
Burn-up rate (%)	7.0	4.4
(kg)	202	89

Thermal Hydraulics Analyses

The thermal hydraulics analyses were
performed to determine the maximum achievable
thermal power and proton beam [45].

The maximum center line temperature of the
actinide fuel must stay below the melting point
($\sim$ 900°C). The maximum temperature of the HT-9
cladding was limited to 650°C. The coolant
temperature at the inlet was set to 300°C.

The maximum clad and coolant outlet temperatures in the system cooled by Na are 492°C and
389°C, respectively. The maximum thermal power
is 691 MW, for which the proton beam current is
22.6 mA.

In the Pb-Bi cooled system, these values are
614°C, 441°C, 342 MW and 9.8 mA, respectively.

High Intensity Proton Linear Accelerator

The conceptual design study of the engineering test accelerator is being carried out [46].
Since the many technological innovations are
required, the two step development plan is considered. As the first step, the low energy part
(the Basic Technology Accelerator for the proton
beam with 10 mA and 10 MeV) will be studied,
because the beam quality is determined mainly at
the low energy stage.

Conclusion

The present status of the computer codes,
the nuclear and nucleon data has been reviewed
in relation with the TRUW incineration by the use
of a proton linear accelerator.

Owing to the scarcity of the nuclear data
covering the intermediate energy range, the Monte
Carlo algorithms have been developed to simulate
nuclear spallation processes, using the nuclear
strcuture and nucleon data. In order to evaluate
applicabilities of these methods also, various
types of nuclear data are indispensable.

It is important also that we can estimate
the spallation product distribution in a reasonable accuracy to show the actual feasibility of
the TRUW incineration.

Measurements and compilations of these data

28

up to the energy of GeV order will require a huge amount of financial and man power resources. International collaborations are crucial in the future.

Acknowledgements

Most of the work reported in this presentation has been done in collaboration with Dr. T. Nishida and the members of Spallation Research Group at JAERI. The collaboration with a group at Kyushu University (Prof. K. Ishibashi, et al.) must be acknowledged also.

Acknowledgements are also due to Dr. H. Takahashi (BNL, USA) for his year long encouragements.

REFERENCES

1. ICANS-IX: Proc. the 9th Meeting of the International Collaboration on Advanced Neutron Sources, Villigen, 1986 (edited by F. Atchison and W.E. Fischer)
2. ICANS-XI: KEK (Tsukuba), 1990, Proc., in preparation
3. C.M. Van Atta, J.D. Lee and W. Heckrotte, The Electronulcear Conversion of Fertile to Fissile Material, UCRL (1976)
4. P. Grand, H.J. Kouts, J.R. Powell, M. Steinberg and H. Takahashi, Conceptual Design and Economic Analysis of a Light Water Reactor Fuel Enricher/Regenerator, BNL-50838, Brookhaven National Laboratory (1978)
5. S. Pearlstein: Astrophys. J. 346, 1049 (1989)
6. W.A. Coleman and T.W. Armstrong, NMTC Monte Carlo Nucleon Meson Transport System, ORNL-4606, Oak Ridge National Laboratory (1970)
7. T. Nishida, Y. Nakahara and T. Tsutsui, Development of a Nuclear Spallation Simulation Code and Calculations of Primary Spallation Products, JAERI-M86-116, Japan Atomic Energy Research Institute (1986)
8. Y. Nakahara and T. Tsutsui, NMTC/JAERI A Simulation Code System for High Energy Nuclear Reactions and Nucleon-Meson Transport Processes, JAERI-M82-198 (1982)
9. T. Nishida and Y. Nakahara, Calculations of Heat Deposition in a Target Bombarded by High Energy Charged Particles, JAERI-M84-154 (1984)
10. Y. Nakahara, T. Tsutsui and Y. Taji, ACCEL Computer Code System for Nuclear Characteristics analyses of Accelerator Target/Blanket Systems, JAERI-memo 9502 (1981) (unpublished)
11. T. Nishida and Y. Nakahara, DCHAIN-SF A Computer Code for the Calculation of Build-up and Decay of Spallation and Fission Products, JAERI Internal memo (1988)
12. T. Tasaka, DCHAIN2 A Computer Code for Calculation of Transmutation of Nuclides, JAERI-M 8727 (1980)
13. T. Nishida and Y. Nakahara, SPD Spallation Product β- decay Program, JAERI Internal memo (1987)
14. T.W. Armstrong, P. Cloth, D. Filges and R.D. Neef, An Investigation of Fission Models for High-Energy Radiation Transport Calculations, Jül-1859, Kernforschungsanlage Jülich GmbH (1983)
15. J.J. Griffin: Phys. Rev. Lett. 17, 4 (1975)
16. M. Blann, Precompound Decay Models for Medium Energy Nuclear Reactions, UCRL-101262, Lawrence Livermore National Laboratory (1989)
17. Y. Nakahara and T. Nishida, Monte Carlo Algorithms for Simulating Particle Emissions from the Preequilibrium States during Nuclear Spallation Reaction, JAERI-M 86-074 (1986)
18. C. Kalbach: Phys. Rev. C23, 124 (1981) and C24, 819 (1981)
19. J.J. Griffin: Phys. Rev. Lett. 17, 4 (1966)
20. T.W. Armstrong and K.C. Chandler, HETC, Monte Carlo High Energy Nucleon Meson Transport Code, ORNL-4744 (1972)
21. K. Ishibashi, et al., in Proc. the 2nd Int. Symp. on Advanced Nuclear Energy Research-Evolution by Accelerators-, Jan. 1990, Mito, Japan, p. 704, Japan Atomic Energy Research Institute (1990)
22. H. Takada, Y. Nakahara and K. Ishibashi, in Reactor Engineering Department Annual Report (Apr. 1, 1987 - March 31, 1988), p. 39, JAERI-M88-221 (1988)
23. F. Atchison, A Theoretical Study of a Target Reflector and Moderator Assembly for SNS, RL-81-006, Rutherforrd and Appleton Laboratories (1981)
24. S. Cierjacks, et al.: Phys. Rev. C36, 1976 (1987)
25. K. Ishibashi, et al., in Proc. the 1989 Seminar on Nuclear Data, Nov. 1989, Tokai, Japan (Edited by Y. Nakajima and S. Igarashi), p. 362, JAERI-M90-025 (1990)
26. Y. Haneishi and T. Fujita: Phys. Rev. C33, 260 (1986)
27. T. Nishida and Y. Nakahara: Kerntechnik 50, 3 (1987)
28. Ref., R. Silberberg and C. Tsao: Astrophys. J. Suppl. Series No. 220, p. 396 (1973)
29. V.S. Baraschenkov, et al.: Sov. Phys. Usp. 16, 31 (1974)
30. R.G. Korteling, et al.: Phys. Rev. C41, 2571 (1990)
31. A.S. Botvina, A.S. Iljinov and I.N. Mishustin: Nucl. Phys. A507, 649 (1990)
32. H.W. Bertini, Monte Carlo Calculation on Intranuclear Cascades, ORNL-3383 (1963)
33. D.V. Bugg: Ann. Rev. Nucl. Part. Sci. 35, 295 (1985)
34. G.W. Hoffman, et al., Nuclear Structure and Reaction Studies at Medium Energies, DOE ER 40441-1 (1988)
35. A.S. Iljinov, E.A. Cherpanov and S.E. Chirginov: Sov. J. Nucl. Phys. 32, 166 (1980)
36. W.D. Meyers and W.J. Swiatecki: Ark. Fys. 36, 343 (1967)
37. J.R. Nix: Nucl. Phys. A130, 241 (1969)
38. V.M. Kupriyanov, et al.: Sov. J. Nucl. Phys. 32, 184 (1980)
39. T.W. Armstrong, P. Cloth, D. Filges and R.D. Neef, Theoretical Target Physics Studies for the SNQ Spallation Neutron Source, Jul-Spez-120 (1981)
40. T. Nishida, et al., in Proc. the 1989 Seminar on Nuclear Data, Nov. 1989, Tokai, Japan (Edited by Y. Nakajima and M. Igarashi), p. 343, JAERI-M 90-025 (1990)
41. M. Yoshida, GROSS-M and GROSS-P, Codes for Prediction of Beta-Decay Properties and Evaluation of their Applicability to Decay Heat Calculations, JAERI-M 6313 (1975)
42. H. Takada, et al., in Ref. 21, p. 375
43. M. Nakagawa and T. Mori, MORSE-DD A Monte Carlo Code using Multi-group Double Differential Form Cross Sections, JAERI-M 84-126 (1984)
44. A.G. Croff, ORIGEN2-A Revised and Updated Version of the Oak Ridge Isotope Generation and Depletion Code, ORNL-5621 (1980)
45. T. Takizuka, et al., in Ref. 21, p. 381
46. M. Mizumoto, et al., in Ref. 21, p. 219

VALIDATION OF FISSION PRODUCT CAPTURE CROSS SECTIONS BY THE ANALYSIS OF THERMAL AND EPITHERMAL INTEGRAL EXPERIMENTS

J. MONDOT, J.P. CHAUVIN, J.P. WEST[*]

DER/SPRC - CEA Cadarache nuclear Center
13108 - St Paul Lez-Durance, France

* Electricité de France, DER
1, Avenue du Général De Gaulle, 92141 Clamart, France

Abstract : Since 1984, a large collaborative effort has been made in France by CEA, EDF and FRAMATOME to get an experimental validation of the fission product cross sections in a wide range of spectral conditions. Three different programs have been realised in the MINERVE facility at Cadarache. These programs consist of measurements, using the oscillation technique, of the reactivity worths of the fission products contained in irradiated fuel samples cut from PWR fuel assemblies. The available experimental results have been analysed using the APOLLO code and multigroup cross section libraries processed from JEF1 and the recent JEF2 data base. The comparison between experimental and calculated results given in this paper is a contribution to the first stage of the JEF2 benchmarking on integral experiments.

(fission-products, JEF1, JEF2, benchmarking, integral-experiments thermal epithermal cross-sections)

I - INTRODUCTION

The oscillation experiments performed in MINERVE at Cadarache have already been analysed to validate the JEF1 fission product nuclear data. These results, presented in several publications [1, 2, 3] showed a good agreement for epithermal spectra, while indicating a slight underestimation of the poisoning effect for thermal lattices of about 7%.

In 1990, the JEF2 data [4] for the main fission product isotopes were processed at the CEA Saclay Nuclear Center [5], and made available to us for a first check against experimental results.

In order to get a first idea of the validity of this new set of multigroup cross sections, we have repeated the JEF1 benchmarking exercise, using exactly the same method and cross section set, except the JEF2/JEF1 substitution for the data concerning the 72 main fission products involved in the calculation.

This paper is aimed at presenting these new results which will contribute to the JEF2 benchmarking coordinated program [6].

II - DESCRIPTION OF THE EXPERIMENTAL TECHNIQUE

MINERVE is an experimental zero power pool reactor composed of a driver core made with high enriched MTR plates and a central square cavity containing a watertight cylindrical vessel.

Any kind of lattice of interest can be inserted in this central part to form the "Test-zone" in which different types of sample can be oscillated in order to measure their reactivity effects (fig. 1, 2).

The principle of the measurements is the following : the oscillator communicates cycling movements with a 2 mn period to an oscillation tube which contains the test samples.

The resulting reactivity variations are compensated by the motion of an automatic control absorber, made of rotating Cadmium sectors, placed in the external part of the driver core.

The signal delivered by this control device is recorded and analysed in real time by a computer in terms of reactivity changes. It has been demonstrated that the amplitude of the signal is proportional to the reactivity variation for small perturbations.

Fig. 1 : Radial cross section of Morgane experiment in Minerve

Fig. 2 : Axial cross section of the Morgane experiment in Minerve

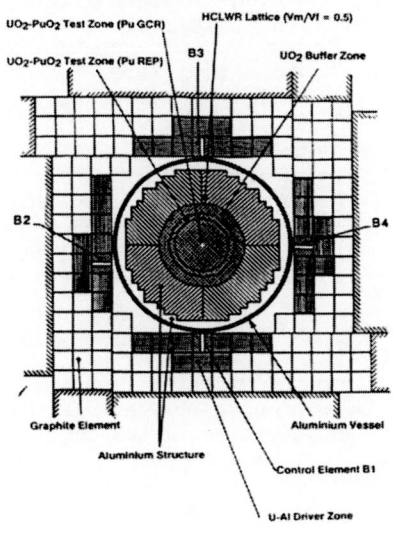

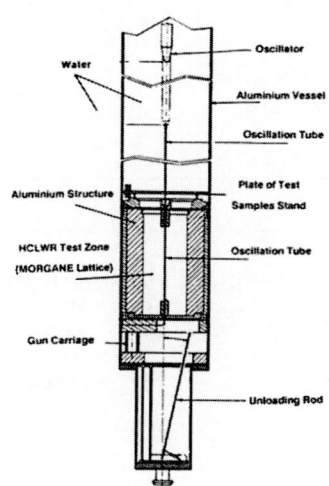

In fact, the oscillation tube is filled with the 10 cm high samples enclosed in a UO_2 (3% enriched) fissile column which remains the same for any type of sample.

Hence, the signal which is obtained represents the sum of several effects given by the alternative insertion of the different samples and also by the change of the position of the can.

In order to eliminate systematic errors coming from these last effects (edge effects for example), we only consider the differences of the signals relative to a reference case. This reference case has been chosen to be a 3.1% enriched UO_2 sample which corresponds to the beginning of life composition of the irradiated samples (origin of the burn-up scale).

The experimental uncertainties depend on the number of independent measurements and on the number of cycles for each measurement.

Each sample is oscillated several times and for 20 to 40 cycles.

Hence the standard deviation associated with the reactivity effect of a sample is derived as :

$$\sigma_i = \sqrt{\frac{(\sigma_s)^2}{N_{Ci}} + \frac{(\sigma_f)^2}{N_{Mi}}} \quad , \quad \text{where}$$

σ_s is the average standard deviation of the whole oscillation campaign for the sample i,

σ_r is the reproducibility deviation also averaged over the whole campaign,

N_{Ci} is the total number of cycles for the sample i,

N_{Mi} is the number of measurements for the sample i.

The above statistical uncertainties are taken into account together with those coming from the calculational scheme.

The different samples which have been oscillated during each experiment are described in table 1.

Table 1 : Experimental samples

Type	Number	Content
UO_2	5	0.2 % - 5 % of ^{235}U
UO_2/PuO_2	4	Simulating 4 burn-up levels 11 - 20 - 37 - 52 GWd/t
	2	5 % of Pu 10 % $\begin{cases} 69 \% \ ^{239}Pu \\ 20 \% \ ^{240}Pu \\ 8 \% \ ^{241}Pu \\ 3 \% \ ^{242}Pu \end{cases}$
Spent fuel	7	20 - 60 GWd/t

These three types of sample correspond to the following successive steps of measurements.

- Standard UO_2 samples with various 235U enrichments are used to calibrate the measured and calculated reactivity effects.

- Mox (UO_2-PuO_2) samples having each an isotopic content corresponding to a definite depletion have been especially constructed to give the reactivity effects of the heavy isotopes alone (235U depletion and Pu build-up).
- Finally the PWR irradiated fuel samples are oscillated to get the total reactivity effects.

Of course, short lived and gaseous FPs are not present in the samples.

The whole set of samples has been chemically analysed and the corresponding isotopic distributions are known with a very good accuracy.

Burn-up values were estimated from depletion calculations resulting in the same ^{148}Nd as the measurements. These values are consistent with those obtained from ^{235}U depletion to within 1%.

Reactivity effects of the fission products are deduced from the differences of the signals measured for irradiated fuel and mox samples corresponding to each burn-up.

III - EXPERIMENTAL RESULTS

The main characteristics of the three programmes performed between 1986 and 1989 are summarized in table 2 and figures 3, 4 and 5.

Table 2 Main characteristics of the experimental programmes

Programme	Lattice type
MORGANE/S	MOX fuel 11 % Pu tight hexagonal lattice Vm/Vu = 0.5
MORGANE/R	Same fuel but hexagonal lattice with Vm/Vu = 0.9
MELODIE	Standard UO2 3 % ^{235}U PWR lattice square pitch = 1.26 cm Vm/Vu = 1.7 (nominal conditions)

In each experiment, the same oscillation samples have been used. Due to the differences between the spectra, the fission product contribution to the total reactivity loss is different.

This proportion is always greater than 1/2 as one can see in table 3 for a 60 GWd/t burn-up :

Table 3 % proportion of the reactivity loss due to the fission products

	Heavy isotopes	Fission products
ERASME/S	48	52 (85%)*
ERASME/R	43	57 (85%)*
MELODIE	50	50 (81%)*

(*) In fact the major part (> 80%) of the fission product poisoning is due to 11 isotopes whose list is given later in detail.

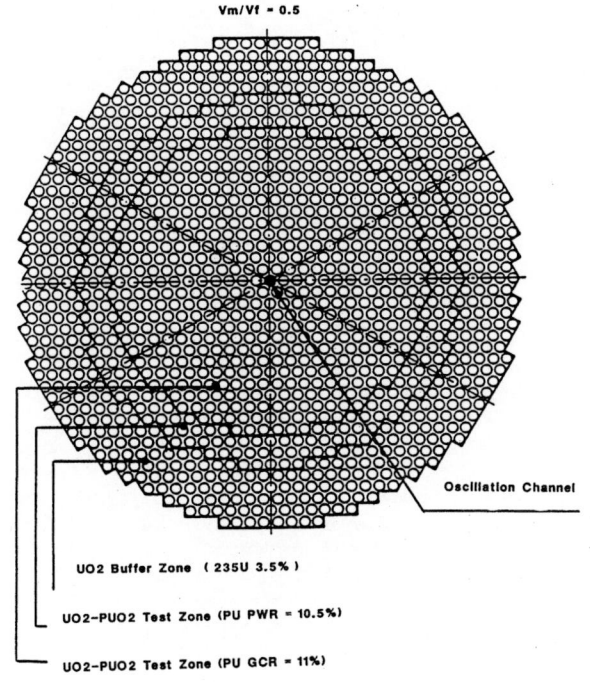

Vm/Vf = 0.5

Oscillation Channel

UO2 Buffer Zone (235U 3.5%)

UO2–PUO2 Test Zone (PU PWR = 10.5%)

UO2–PUO2 Test Zone (PU GCR = 11%)

Fig. 3 :   Morgane/s lattice

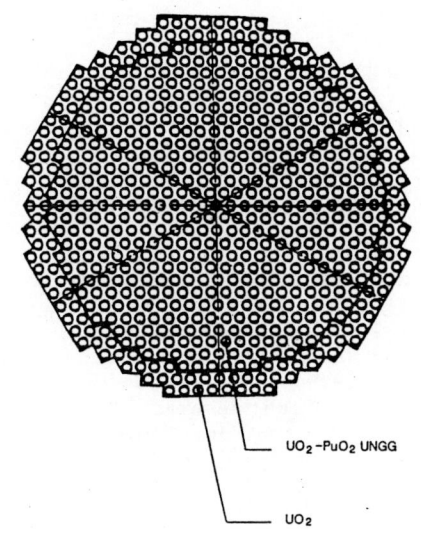

UO₂–PuO₂ UNGG

UO₂

Fig. 4 :   Morgane/r Vm/Vf = 0.9

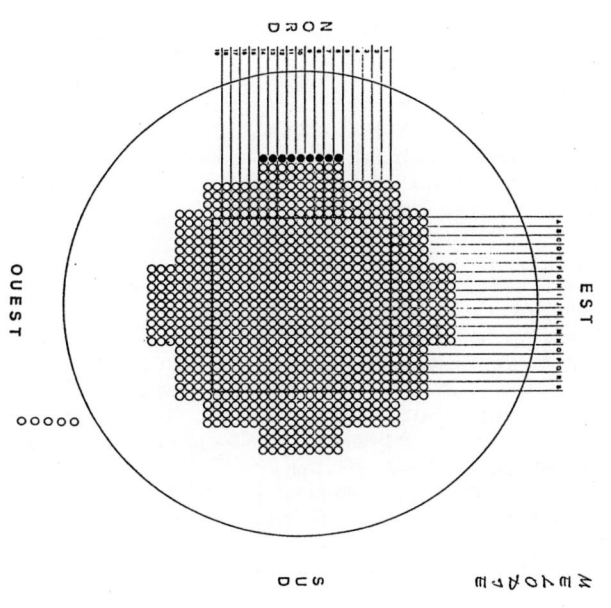

Fig. 5 :   Melodie lattice

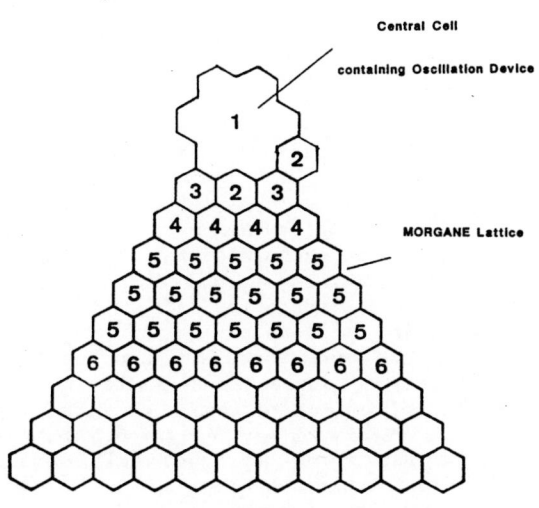

Central Cell

containing Oscillation Device

MORGANE Lattice

Fig. 6 :   Multicell geometry used by Apollo code

IV - CALCULATIONAL METHOD

The experiments have been analysed using the APOLLO code and the Apollib 86 multigroup library [7].

In a first stage the fission product cross sections were essentially derived from the JEF1 file, and recently we have substituted for these data the new JEF2 evaluations processed at Saclay [5] using the NJOY-THEMIS code system.

The Apollo-1 energy structure contains 99 groups (52 fast above 2.76 eV and 47 in the thermal range).

Concerning the calculation, we have used a first-collision probability method able to treat the energy-space dependence of the flux in the oscillation sample and the surrounding lattice.

This method is based on a multicell approximation which assumes that the neutrons are isotropic and uniformly distributed on each surface of the cells (U-DP0).

These multicell approximations of the integral transport equation are referred to as Roth x 6 (for hexagonal lattices) and Roth x 4 (for square lattices).

They have been tested for a large set of absorbing heterogeneities in PWR and HCLWR spectra in experimental programmes performed in the EOLE reactor [8, 9].

The results are in good agreement with two-dimensional reference calculations within 1-2% for reactivity worths.

An example of the geometry used in calculations for an hexagonal lattice is given in fig. 6.

These geometries represent the central oscillation region where all the details are explicitly described and a part of the surrounding zone where the external cells are asymptotic regarding the perturbation (fundamental mode).

Self-shielding effects for the heavy isotopes have been calculated for the sample and the fuel pins using a special Pic module which approximates the interaction effects between the two media. The self-shielding of the fission product has not been taken into account in this study. We intend to process self-shielding data in the form of effective resonance integrals to be used together with an equivalence principle to correct for fission product self shielding effects in a further stage.

The ability of this scheme to reproduce the experimental results has been checked by looking at :
- 235U reactivity effects for the calibration [2],
- Pu build-up effects using the results of reactivity changes between mox samples with increasing Pu content [2].

It must be observed, also, that the residual calculational errors for the heavy isotope effects are basically the same for the MOX and irradiated fuel samples and thus cancel when looking at the resulting fission products effects [2].

V - COMPARISON BETWEEN CALCULATED AND EXPERIMENTAL RESULTS

The results obtained for the global capture of the fission products are summarized in the table 4 - for the three experimental programmes and four burn-up levels from 25 to 60 GWd/t.

Table 4 : JEF2 calculation/experiment deviation for the global FP poisoning effect

Burn-up (GWd/t)	MOX 11% Vm/Vf=0.5	MOX 11% Vm/Vf=0.9	UO$_2$ 3% Vm/Vf=1.3
25.5	-7.8 ± 8.0	+5.6 ± 12.5	-4.2 ± 3.7
39.0	+2.2 ± 6.4	-3.3 ± 7.6	-7.8 ± 3.2
49.0	+1.0 ± 5.4	-2.5 ± 5.9	-8.9 ± 3.1
59.0	-3.7 ± 4.5	-7.7 ± 5.4	-7.8 ± 3.0
Mean	-0.9 ± 4.3	-4.9 ± 4.6	-8.2 ± 3.0

* 1 standard deviation including experimental and calculation uncertainties.

These results are the same within 0.1% as those previously obtained using the JEF1 cross sections.

One can observe that for hardened spectra the calculated results are in very good agreement with the measured values, within 1 sigma uncertainties.

However, for thermal spectra in the standard 3% UO$_2$ case it seems that there is a trend to a slight underestimation of the poisoning effect (-8%±3%).

This could be due to a corresponding underestimation of the fission product thermal capture cross sections.

Unfortunately, the analysis work including sensitivity calculations is still underway and a final conclusion cannot yet be drawn concerning the relation between the uncertainties in the cross sections, the evaluated variance-covariance matrices, and observed deviations.

In order to compare in more detail the JEF1 and JEF2 results we have analysed explicitly the 11 main isotopes regarding their poisoning effects.

These 11 isotopes which represent more than 80 % of the total effect are listed below in table 5.

Table 5 : Contribution of some F.P. isotopes to the global poisoning effect

F.P.	Morgane S (%)	Morgane R (%)	Melodie (%)
143Nd	4.2	4.8	12.3
145Nd	6.0	5.5	4.1
147Pm	6.2	6.1	4.2
149Sm	0.7	1.2	5.6
151Sm	1.4	1.8	6.9
152Sm	8.7	8.5	5.6
153Eu	7.8	7.3	5.7
99Tc	11.6	11.0	6.6
103Rh	12.3	13.7	14.0
131Xe	12.1	11.4	7.5
133Cs	14.3	13.5	8.6
Other F.P.	14.7	15.2	18.9

The individual contributions to the total poisoning effect are indicated for the three different spectra.

The differences between the JEF1 and JEF2 1 group capture cross sections (weighted by the spectrum) are plotted in the figure 7 for the 11 above mentioned isotopes and for the three experiments.

The abscissa is the slowing down density at the thermal cut-off, ie the proportion of the neutrons absorbed in the thermal range which is a good parameter to indicate the spectrum hardness.

VI - CONCLUSION AND FUTURE WORK

This first analysis of the Morgane and Melodie oscillation experiments has indicated the following trends :

a) The fission product poisoning is very correctly predicted using the JEF evaluation for epithermal lattices while a slight underestimation is observed for standard UO_2 PWR lattices (-8%).

This trend seems to be significant relative to the $\pm 3\%$ uncertainty of the measurement.

b) There is practically no difference between the total reactivity effect computed using JEF1 or JEF2 values and this is due to compensations of positive and negative effects for individual isotopic contributions.

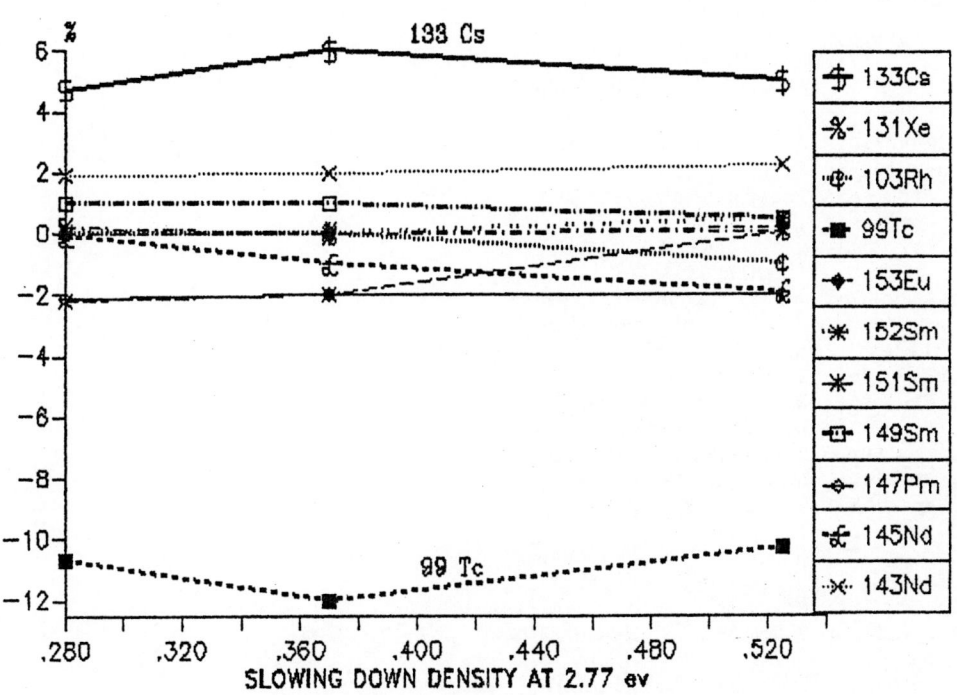

Fig. 7 :   JEF2/JEF1 averaged capture cross sections

One can observe that most of the differences are fairly evenly distributed roughly within $\pm 2\%$.

Only two isotopes seem to reveal rather large changes from JEF1 to JEF2 :

- Tc 99 : JEF2/JEF1 ~ 0.88
- Cs 133 : JEF2/JEF1 ~ 1.06.

Looking at the Tc 99 multigroup cross sections we have seen that at the energy of the second resonance (20.4 eV) the JEF2 infinite dilution average cross-section is 35% lower than in JEF1.

Concerning Cs133 several differences of opposite sign are observed but in the low energy range JEF2 values are higher than those of JEF1.

Finally, these compensations weighted by the importance of each isotope result in the very small difference that we have observed in the total fission product capture.

However, this analysis has to be completed using sensitivity coefficients and evaluations of the cross section uncertainties.

This work is underway and will be issued in the frame of the international JEF2 benchmarking.

The final conclusions will of course take into account other benchmarking studies and in particular those performed in the UK with another technique in the DIMPLE reactor [10]. A recent agreement between the AEA and CEA on collaboration in the field of LWR reactor physics will hopefully allow us to exchange experimental materials (samples) and calculational results. Thanks to this collaboration we will be able to get two independent validations of fission product reactivity effects in standard and advanced LWRs.

For what concerns the future, another experimental campaign is planned in 1992 in the EPICURE programme [11]. We will take advantage of this opportunity to extend the validation of the fission-product data by the study of over-moderated (Vm/Vf ~ 3 to 3.5) neutronic conditions. This will allow us to investigate the trends observed in this study of a growing underestimation of the fission product capture with the importance of the thermal range.

Possible deviations due to the calculational models will be investigated using the improved version of the Apollo code [12] (Apollo II) and also reference calculations regarding the self-shielding problems.

Finally, irradiation experiments recently completed in the MELUSINE reactor [13] will be analysed to get information on some individual fission product capture rates [95 Mo, 97 Mo, 103 Rh, 143 Nd, 145 Nd, 149 Sm, 153 Eu].

REFERENCES

[1] J. MONDOT (CEA), J.M. GOMIT, C. GARZENNE (EDF), P. CHAUCHEPRAT, A. SANTAMARINA (CEA) : "MORGANE/S : fission product capture measurements in a HCLWR tight lattice". Proc. ANS conf. JACKSON HOLE Sept. 88, Supplement : p. S-117.

[2] P. CHAUCHEPRAT (CEA), C. GARZENNE (EDF), J. MONDOT (CEA) : "SYNTHESIS of the experimental validation of global fission product capture for PWR's and advanced reactor lattices. Proc. PHYSOR conference Marseille 1990, Vol. I, p. IV.32.

[3] J.P. CHAUVIN, J.P. WEST, C. GARZENNE : Qualification of neutronic parameters in HCPWR Spectrum through ERASME, MORGANE and ICARE experiments. Conf. ENC'90 - Lyon (FRANCE) September 90, ENS/ANS Foratum, p. 857-867.

[4] C. NORDBORG, H. GRUPPELAAR, M. SALVATORES "Status of the JEF and EFF project", contribution to this meeting.

[5] H. TELLIER and A. CONSTANS : Personal Communication.

[6] E. FORT : Proposal for JEF2 validation. JEF DOC REPORT N° 296.

[7] A. KAVENOKY, "APOLLO : A general code for transport, slowing down and thermalization calculations in heterogeneous media", CONF. 73044 P1, Ann. Arbor, Michigan, April 1983.

[8] L. MARTIN-DEIDIER, A. SANTAMARINA, D. DOUTRIAUX, M. ROSHD, S. ZERO : CAMELEON : a Benchmark Experiment for Absorbers and burnable poisons in PWR assemblies. Conf. ANS - New-Orleans (USA) June 84. Vol. 46 - p. 755-756.

[9] L. MARTIN-DEIDIER, A. SANTAMARINA, S. CATHALAU, J.P. CHAUVIN, J.M. GOMIT : Undermoderated PWR neutronic qualification through the ERASME experiments. Topical Meeting on advances in Reactor physics, Mathematic and computation. PARIS (FRANCE) April 87. Vol. I - p. 97-106.

[10] J. MARSHALL, G. INGRAM, N.F. GULLIFORD, M. DARKE : Irradiated fuel Measurements in DIMPLE. Conf. ANS PHYSOR MARSEILLE (FRANCE) April 90. Vol. II - p. XI-45.

[11] J. MONDOT, J.P. CHAUVIN, J.C. LEFEBVRE, A. VALLEE et al : EPICURE : An experimental programme devoted to the validation of the calculationnal schemes for Plutonium recycling in PWR's. Conf. ANS PHYSOR - MARSEILLE (FRANCE) - April 90. Vol. I - p. VI-53.

[12] Richard SANCHEZ, J. MONDOT et al. : APOLLO II : A user-oriented, portable, modular code for multigroup transport assembly calculations. N.S.E. publication 100, 352-362 (1988).

[13] P. CHAUCHEPRAT, L. MARTIN-DEIDIER, J.M. GOMIT : ICARE experiments for the qualification of capture cross-section in the neutron spectrum of undermoderated reactors. International conference on nuclear data MITO (Japan) June 88. AA04-p. 61.

MEASUREMENT OF THE SUBTHERMAL NEUTRON INDUCED FISSION CROSS-SECTION OF 241PU

C. Wagemans*, P. Schillebeeckx**, A.J. Deruytter, R. Barthélémy

Commission of the European Communities, Joint Research Centre
Central Bureau for Nuclear Measurements, 2440 Geel, Belgium.

Abstract: The 241Pu(n,f) cross-section has been measured from 2 meV up to 20 eV. The experiments were performed at GELINA, using a dedicated set-up optimized for cross-section measurements below 1 eV. The present $\sigma_f(E)$-data below 40 meV are clearly compatible with a 1/v-shape, in contradiction to the measurements reported in the literature. Consequently, there is a probability of errors in the thermal normalization of these measurements. Furthermore, the Westcott g_f-factor calculated from the present $\sigma_f(E)$-data yields a value of 1.041 ± 0.003, which is lower than generally adopted.

(241Pu, fission cross-section, range 0.002-20 eV, Westcott g_f-factor)

Introduction

At the PHYSOR 90 Conference in Marseille we reported preliminary results[1] of 241Pu(n,f) cross-section measurements in the neutron energy region from 0.002 eV to 20 eV. These experiments were performed under optimized experimental conditions for low-energy measurements. Since the results were discrepant with all data previously reported in the literature, a verification measurement was urgently needed.

The results of this experiment are reported in the present paper.

Experimental conditions

In order to increase the neutron flux in the meV-region, a liquid nitrogen (77°K) cooled methane moderator has been installed at GELINA (Geel Electron LInear Accelerator). The measurements were performed at a 8.2 m flight-path, the accelerator being operated at a 40 Hz repetition frequency with 2 µs burst widths and an average electron current of 15 µA. An evaporated layer of 25 µg 6LiF/cm^2 used for the neutron flux determination and a 27.2 µg 241Pu/cm^2 deposit (prepared by suspension spraying of plutonium-acetate) were mounted back-to-back in the center of a large vacuum chamber. Both layers were prepared by the CBNM Sample Preparation Group. The 6Li(n,α)t reaction products and the fission fragments were detected in a low geometry with two 30 cm^2 large surface barrier detectors placed outside the neutron beam. The corresponding pulse-height versus time-of-flight spectra were stored in a HP 1000 - A 700 data acquisition system.

Both layer thicknesses were chosen in such a way that absorption and self-absorption effects are very small. Moreover, two separate experiments were performed with the neutron beam incident on a Li-Pu and on a Pu-Li sandwich resp.

For the preparation of the 241Pu-layer, a 241Pu solution was used which was freshly separated from the 241Am ingrowth. Its isotopic composition at that moment was: 238Pu (0.0343 weight %), 239Pu(1.7708 w.%), 240Pu(6.8355 w.%), 241Pu (86.2809 w.%) and 242Pu(5.0784 w.%).

* NFWO, Nuclear Physics Lab., University of Gent, Belgium
** CEC, JRC, IST, Ispra, Italy

The background was determined using the black resonances of Cd, Rh, Au and W. The background contribution due to neutrons from overlapping bursts was checked by operating GELINA at a 20 Hz repetition frequency. The signal-to-background ratio as a function of the incident neutron energy is shown in fig. 1.

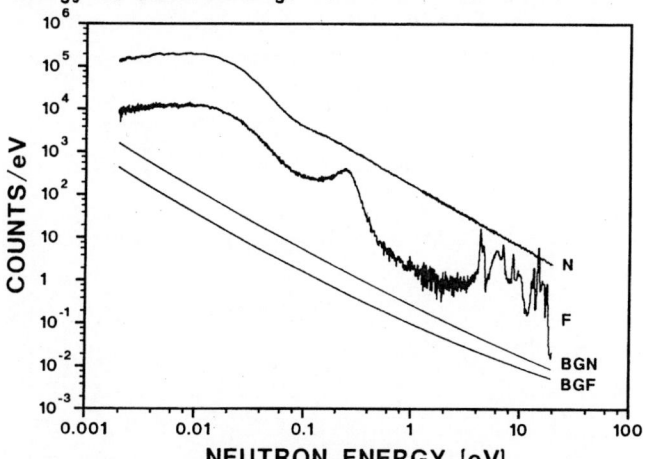

Fig. 1 Counting-rate spectra vs neutron energy with the corresponding background curves.

Results and discussion

Fission cross-section measurements

The measurements were performed relative to the 6Li(n,α)t reaction, for which a 1/v-shape was adopted. Hence the ratio of the background-corrected fission and (α+t) counting-rates yields the $\sigma_f(E)\sqrt{E}$ shape, which still needs to be normalized and corrected for the 239Pu(n,f) contribution. For this correction, the $\sigma_f(E)$-data of Wagemans et al.[1] were used.

The normalization was done in the thermal region via a linear least-squares fit of $\sigma_f(E)\sqrt{E}$ from 20 to 30 meV and adopting a σ_f^0-value of (1013±7) barn (ENDF-B6).

Fig. 2 shows the 241Pu(n,f) cross-section obtained in this way (upper curve) and the 239Pu(n,f) correction applied to the raw data (lower curve). The present 241Pu(n,f) data confirm the presence of a dip in the cross-section at 2 eV as observed by Derrien and de Saussure[2] during the evaluation of the Weston and Todd[3] fission data. They interpreted the dip to be due to a strong interference effect between resonances at 1.735 eV and 5.81 eV.

Table 1: ^{241}Pu fission integrals renormalized to $\int_{0.10\,eV}^{0.50\,eV} \sigma_f(E)dE = 264.58\ barn.\ eV$

E_1(eV)	E_2(eV)	This work [a]			Wagemans 1983 ref. 5	Weston 1978 ref. 3	Wagemans 1976 ref. 4	ENDF-B6
0.01	0.02	13.42	±	0.04	13.24	12.14		13.00
0.02	0.03	10.24	±	0.04	10.12	9.80	10.14	10.01
0.03	0.10	48.7	±	0.2	48.3	47.4	48.3	49.0
0.1	0.5	264.58	±	1	264.58	264.58	264.58	264.58
0.5	1.0	17.6	±	0.4		17.6		17.3
1	3	54.4	±	0.9		55.5		54.4
3	8	1183	±	14	1197	1201	1192	1170
8	12	540	±	11	537	538	533	514
12	20	1284 [b]	±	20	1339	1321	1323	1310
0.02	0.45	319.9	±	1	319.0	319.2	319.6	320.3
Renormalization factor		1			0.9710	0.9596	0.9706	0.9771

a) Errors excluding normalization uncertainty
b) Value obtained with a poor energy resolution for $\sigma_f(E)$

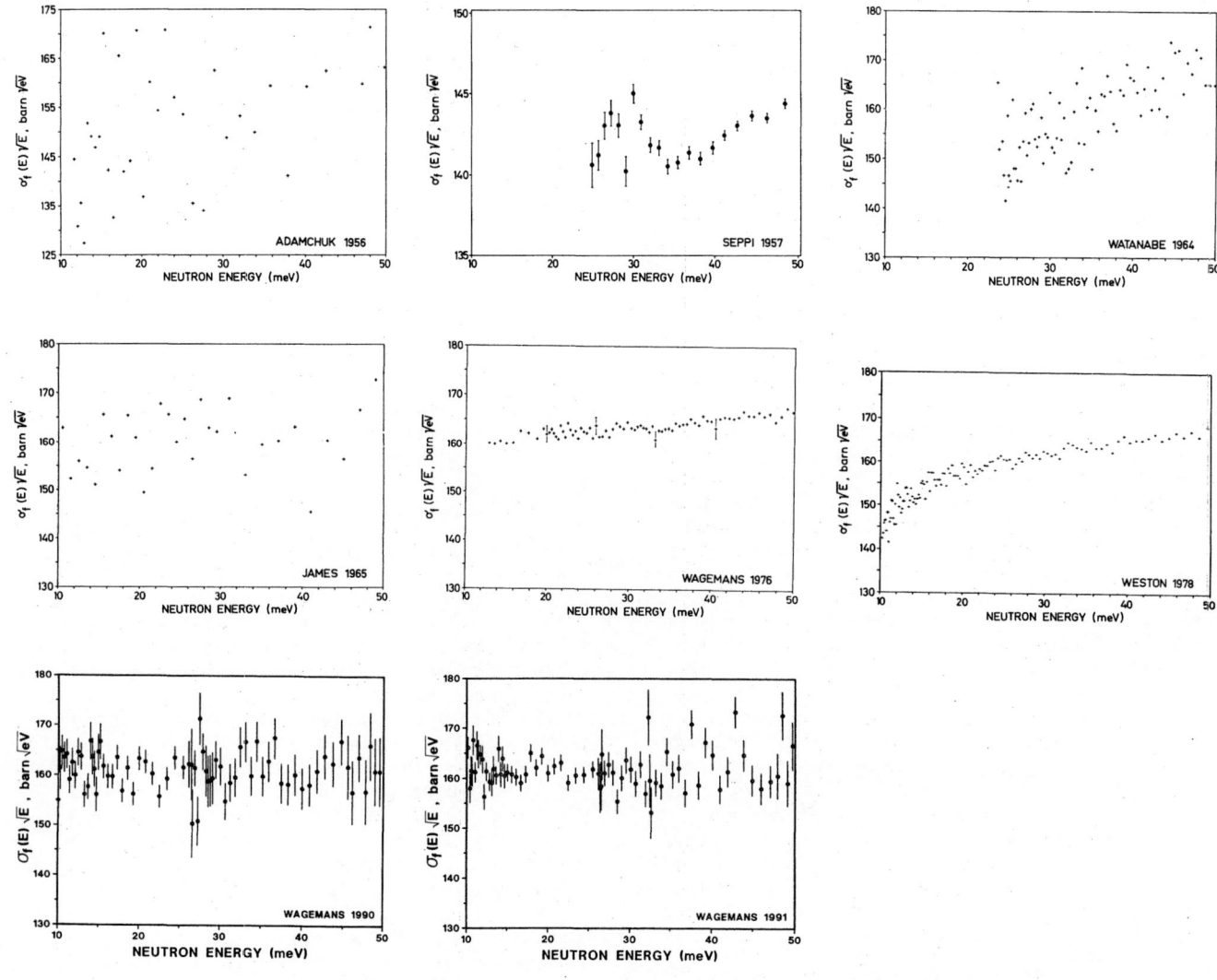

Fig. 4 Overview of the available $^{241}Pu(n,f)$ cross-section data below 50 meV.

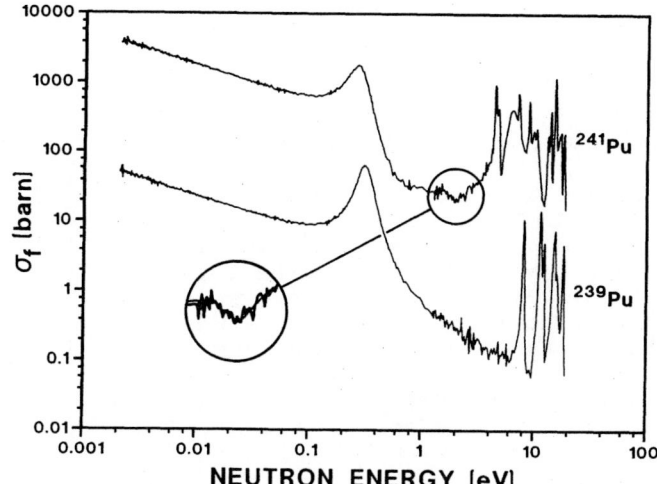

Fig. 2 ^{241}Pu(n,f) cross-section and ^{239}Pu(n,f) correction applied to the raw data.

Fig. 3 shows the $\sigma_f(E)\sqrt{E}$-curve from 0.002 eV to 1 eV. This figure illustrates the constancy of $\sigma_f(E)\sqrt{E}$ below 40 meV, demonstrating the almost perfect 1/v-shape of the ^{241}Pu(n,f) cross-section in this energy region. This result confirms our preliminary data[1] but strongly disagrees with all measurements previously reported. These older measurements suffer from a low statistical accuracy resulting in a large scatter of the data, or are subject to solid state effects, or suffer from absorption and/or self-absorption effects or uncertain background corrections.

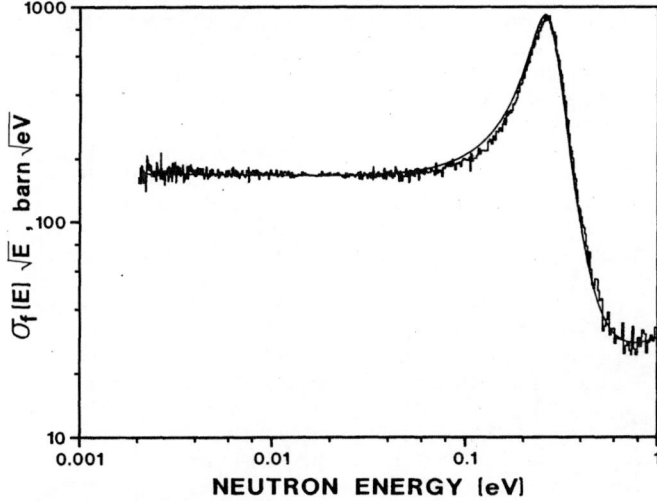

Fig. 3 $\sigma_f(E)\sqrt{E}$ curve for ^{241}Pu. The full line represents the ENDF-B6 evaluation.

This is illustrated in fig. 4, which gives an overview of the ^{241}Pu σ_f-data below 50 meV. From this figure it is obvious that the thermal normalizations of these older measurements are likely not to be very accurate. The discrepancy for the 0.26 eV resonance reported by Derrien and de Saussure[2] is likely to be a consequence of an erroneous thermal normalization.

Fission integrals

In table 1, selected fission integrals calculated from the present measurements are compared with the corresponding integrals obtained in our previous measurements [4,5], in those of Weston and Todd[3] and evaluated[2] for ENDF-B6. To avoid the effect of erroneous thermal normalizations, all data were renormalized to

$$\int_{0.10\,eV}^{0.50\,eV} \sigma_f(E)dE = 264.58\ b.eV$$

calculated from the present measurements.

An examination of this table shows that the present data and the renormalized experimental values [3-5] are in good agreement above 0.1 eV. Below this energy, a cross-section reduction due to absorption effects etc. can be observed. Moreover, the four experiments mentioned above lead to an arithmetic average

$$\int_{12\,eV}^{20\,eV} \sigma_f(E)dE = 1317\ b.eV.$$

This value is based upon a thermal value

$$\int_{0.02\,eV}^{0.03\,eV} \sigma_f(E)dE = 10.24\ b.eV.$$

and a value

$$\int_{0.02\,eV}^{0.45\,eV} \sigma_f(E)dE = 319.9\ b.eV.$$

as obtained from the present work. These three integrals were used as references for the ENDF-B6 evaluation, in which the following values were adopted: 1350, 10.24 resp. 326.03 b.eV. This implies that a modification of ENDF-B6 is likely to be needed below 300 eV.

Westcott g_f-factor

The value of g_f is determined by the shape of $\sigma_f(E)$ below 1 eV (see e.g. ref.1). The σ_f-data below 20 meV even contribute with roughly 30% to g_f. From fig. 4 it is clear that the cross-section data in this region are very scarce and discrepant. Nevertheless, the evaluated g_f-value had to be based on these σ_f-data.

New g_f-calculations were done using the $\sigma_f(E)$-values obtained in this work. The extrapolation towards zero energy was done via a least-squares fit $\sigma_f(E)\sqrt{E} = a + bE + cE^2$ which was applied to the data points in the energy region from 0.002 eV to 0.06 eV. The extrapolated part contributes only with 1.5% to the g_f-value at T = 20.44°C.

In this way, a g_f(T = 20.44°C)-value of 1.041±0.003 was obtained which is somewhat lower and more accurate than the evaluated values. ENDF-B6 e.g. reports a value of 1.0450 ± 0.0053.

Conclusion

The present experiments yield new $\sigma_f(E)$-data for ^{241}Pu from 0.002 eV up to 20 eV. These results point to erroneous thermal normalizations in some older measurements, which definitely had an impact on ENDF-B6. For the Westcott g_f-factor, a new and more accurate value was obtained, which is slightly lower than the evaluated values.

References

1. C. Wagemans, P. Schillebeeckx, A. Deruytter, R. Barthélémy, Proc. Int. Conf. PHYSOR 90, Marseille, France, Vol. 1, III-9 (1990)
2. H. Derrien and G. de Saussure, Nucl. Sci. Engn. 106, 415 (1990)
3. L. Weston and J. Todd, Nucl. Sci. Engn. 65, 454 (1978)
4. C. Wagemans and A. Deruytter, Nucl. Sci. Engn. 60, 44 (1976)
5. C. Wagemans and A. Deruytter, Proc. Int. Conf. on Nuclear Data for Science and Technology, Antwerp, Belgium, 69 (1983)

ON ALPHA OF 235U FOR SUB-THERMAL NEUTRON ENERGIES

H. Weigmann, J.A. Wartena and C. Bürkholz

Commission of the European Communities, Joint Research Centre
Central Bureau for Nuclear Measurements, 2440 Geel, Belgium.

Abstract: An experiment to measure the capture to fission ratio α of 235U in the sub-thermal neutron energy region, is described. The experiment is based on the measurement of specific low energy capture γ-rays and prompt fission γ-rays with a Ge detector. The liquid methane moderator of the Geel linac is used as a pulsed source of sub-thermal neutrons, the neutron energy being determined by time-of-flight. The method requires that the relative yields of the measured γ-rays per capture and fission event, respectively, do not vary over the energy range considered. Resultant α-values show an increase by about 9% in the subthermal region as compared to thermal energy, corresponding to a 1.2% decrease in η.

(capture to fission ratio, α, of 235U; sub-thermal neutron energy; temperature coefficient of reactivity)

Introduction

The detailed energy dependence of neutron cross sections and related parameters of fissile nuclei for sub-thermal neutron energies has recently found considerable attention because of its effect on the temperature coefficient of reactivity of thermal reactors [1]. One of the quantities thus considered is η, the number of fission neutrons emitted per neutron absorbed, of 235U. Recent measurements [2] of η have shown a small decrease by about 2% with decreasing neutron energy in the subthermal region, in qualitative agreement with conclusions drawn from integral experiments, while other measurements [3] of the same quantity seem to show a smaller (≤1%) variation in this energy region.

The measured [2] effect on η of about 2%, if present at all, is only about twice the experimental uncertainties. Furthermore, it has been argued [3], that the apparent variation of η might have been produced by a solid state effect - coherent backscattering of the incident neutrons from the metallic sample used. Therefore, an independent verification was highly desirable.

Under the assumption that the average number of neutrons per fission, $\bar{\nu}$, is constant, the decrease of η implies an increase of the capture to fission ratio α by a much larger fraction. A measurement of α is experimentally more difficult, but as less accuracy is required, may be an equally sensitive test of the effect under investigation. We have therefore performed an experiment to measure the neutron energy dependence of α of 235U in the sub-thermal energy region.

Experimental Method

The method is based on the measurement of specific low energy capture γ-rays and prompt fission γ-rays with a Ge-detector. It thereby rests on the assumption that the relative yields of the measured γ-rays per capture and fission event, respectively, do not vary as a function of neutron energy. This will only be fulfilled as long as the relative contributions of different resonances to the cross sections do not vary strongly, i.e. only for an energy interval which is smaller than the typical resonance width. The method is thus limited to the sub-thermal and near-thermal energy region. Apart from this limitation, systematic uncertainties are small because capture and fission γ-rays are measured simultaneously and under very much the same condition.

The liquid methane moderator of the Geel linac is used as a pulsed neutron source. After suitable collimation the neutron beam hits a 0.33 g/cm2 metallic U sample (93% 235U), which is positioned at 8.07 m distance from the neutron source. The neutron energy is determined by time-of-flight, the linac being operated with 40 Hz repetition rate. In order to protect the U sample against out of time neutrons, it is inserted into a cylindrical 6Li sleave. Also, a 2 cm thick Be filter is placed in the neutron beam in order to reduce the neutron flux at high energies (which may contribute to the background at later times, i.e. lower energies) by about a factor of 4 while having only a small effect on the flux below 6.8 meV neutron energy.

Capture and fission γ-rays emitted from the sample are detected in a well shielded HP Ge-detector of 80% relative efficiency. Data are stored in a ND-6600 data acquisition system in the form of two-dimensional spectra of 48 time-of-flight channels covering the time interval from 445 μs (corresponding to a neutron energy of 1.72 eV) to 25 ms, times 1024 amplitude channels. The time-of-flight channel width varies from 73.5 μs to 1.176 ms.

The 1024 amplitude channels cover the γ-energy interval from about 400 keV to 1.4 MeV.

An example of a γ-ray spectrum for the neutron energy interval from 4.5 meV to 6.1 meV is shown in Fig.1. Some γ-rays which will be refered to later on, are marked by their energy (in keV) and by "C" for capture, "F" for prompt fission, and "D" for delayed fission. Here "prompt" refers to a half-life which is short as compared to the smallest time-of-flight channel width, and "delayed" refers to a half-life much longer than the linac repetition period of 25ms. The distinction between these three categories is based on the literature [4-7]. Prompt (capture or fission) and delayed γ-rays are of course also distinguished by the fact that delayed γ-rays are present also at the longest neutron flight times of 22 to 25 ms where practically no neutron flux persists any more, or, if the half-life is sufficiently long, are present still after shut down of the linac.

Data analysis

Assuming that the relative yields of the individual γ-rays per capture and fission event are constant, the capture to fission ratio α is simply given by the ratio of areas under capture and prompt fission γ-ray lines. However, the analysis is complicated by the fact that in some cases γ-rays of different origin are so close in energy that they are not resolved in the γ-ray spectrum. As an example, the strongest "capture" γ-ray at 642.2 keV coincides with a delayed fission γ-ray ($T_{1/2}=93$min) of ^{142}La at 641.2 keV. In order to deal with this kind of interference, first the "background" spectrum of delayed fission γ-rays, as obtained from the last two time-of-flight channels (neutron flight-times t > 22 ms), after suitable normalization, is subtracted from the γ-ray spectra for all time-of-flight intervals. The normalization of the "background" γ-ray spectra is done with the aid of the strong delayed γ-ray lines at 1383.8 keV (^{92}Sr) and 1435.6 keV (^{138}Cs), which are well isolated in the γ-ray spectra for all values of t. In this way the fact that the dead-time of the data acquisition system is t-dependent is automatically taken into account in the background subtraction.

After subtraction of the "background" from delayed fission γ-rays, the areas under well defined capture and prompt fission γ-ray lines are determined for each time-of-flight interval. The contributions representing capture and fission are separately summed, and their ratio is formed. The following individual γ-rays have been considered in the analysis:

capture: 642.2keV and 909.7+912.4+914.6 keV (triplett).
fission: 706.8 keV, 814.6 keV, (1178+1181 keV), (1222+1224 keV), and 1278.9 keV.

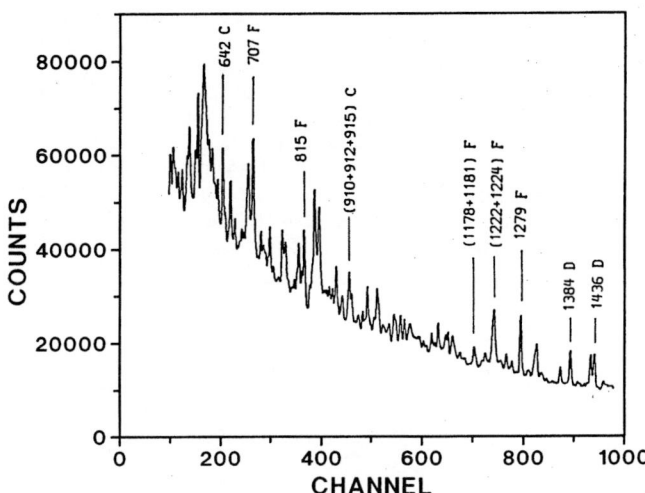

Fig. 1: Measured γ-ray spectrum for the neutron energy interval from 4.5 to 6.1 meV

As mentioned above, a basic uncertainty of the method rests in the possibility that the relative yields of individual γ-rays per capture or fission event might vary in the energy interval of interest. Therefore, a number of separate analysis runs have been performed which differ in the number of individual γ-rays included and their relative weights. In particular, the γ-lines given in parentheses above have been included only in part of the analysis runs. Also, the strongest capture γ-ray at 642.2 keV and the well separated prompt fission γ-ray at 1278.9 keV have been given extra weight in some runs or even used exclusively. The differences observed were generally smaller than the statistical errors, and are included in the systematic uncertainties assigned to the final results. Only for neutron energies above about 0.5 eV larger differences occur, and we conclude that the applicability of the method is limited to lower energies.

The measured capture and fission γ-ray intensities still have to be corrected for different γ-ray self-absorption in the U-sample. This difference depends on the neutron energy because the range of interaction points and thus the average amount of sample material which the γ-rays have to penetrate, does. The relevant correction on the final capture to fission ratio never exceeds 3%.

40

Results and discussion

The results obtained from the present experiment are shown in Figs.2 and 3. In Fig.2 the resultant α-values are shown, where the data have been normalized in the interval from 17 to 70 meV to a value of $\bar{a} = 0.1687$, which is shown as the horizontal line in the figure. Assuming a value of $\bar{v} = 2.4251$, the corresponding η-values are shown in Fig.3 together with the flat reference shape and the shape proposed by Santamarina et.al.[1]. The error bars in the figures are the linear sum of the statistical and systematic uncertainties. The numerical values may be obtained from the NEA data bank.

The measured increase of α for sub-thermal energies indicates that a bound state not far below neutron separation energy contributes to the cross sections in this region, and that its relative contribution varies over the energy region of interest. Thus the assumption made in the analysis that the yields of the measured γ-rays per capture and fission event are constant, is not necessarily fulfilled (a corresponding argument also holds for the 100 to 400 meV region). Different relative γ-ray yields for the assumed bound state as compared to the resonances predominating in the thermal region, may have an effect on the measured shapes of α and η in the sub-thermal region. However, in the alternative scenario where we assume the first bound state to be sufficiently far away such that α- and η-values stay constant in the sub-thermal region, there is no reason to expect varying relative γ-ray yields, and also the present measurements should produce constant values. Such a scenario is in contradiction with the present experimental results.

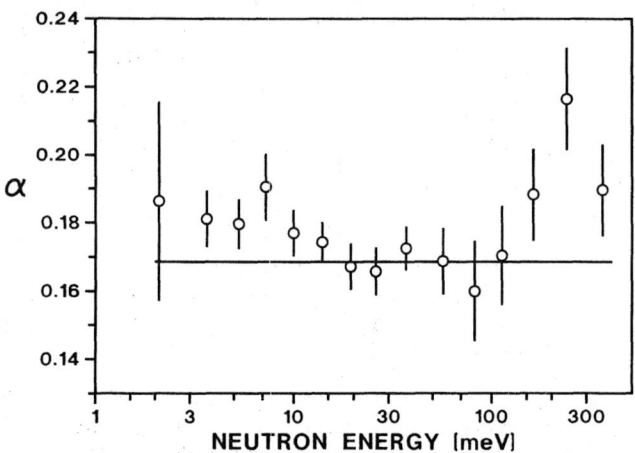

Fig. 2: Capture to fission ratio α of 235U for sub-thermal neutron energies

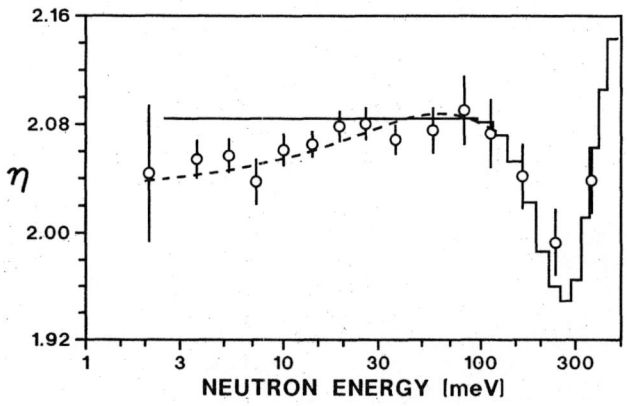

Fig. 3: η of 235U deduced from the α – values of Fig. 2.

References

1. A. Santamarina and H. Tellier, Proc. Intern. Conf. on Nuclear Data for Science and Techno-logy, Mito 1988, p.47; JAERI, Tokyo 1988.
2. H. Weigmann, P. Geltenbort, B. Keck, K. Schreckenbach and J.A. Wartena, Proc. Intern. Conf. on the Physics of Reactors, Marseille 1990.
3. M.C. Moxon, J.B. Brisland and D.S. Bond, Proc. Intern. Conf. on Nuclear Data for Science and Technology, Mito 1988, p.75; JAERI, Tokyo 1988; M.C. Moxon, private communication.
4. R.G. Graves, R.E. Chrien, D.I. Garber, G.W. Cole and O.A. Wasson, Phys. Rev. C8 (1973) 781.
5. T.A. Khan, D. Hofmann and F. Horsch, Nucl. Phys. A205 (1973) 488.
6. W.J. Schindler and C.M. Fleck, Nucl. Phys. A206 (1973) 374.
7. L.V. East and G.R. Keepin, Proc. Intern. Sympos. on Physics and Chemistry of Fission, Vienna 1969, p.647; IAEA, Vienna 1969.

^{56}Fe AND ^{60}Ni RESONANCE PARAMETERS

C. M. Perey, F. G. Perey, J. A. Harvey
N. W. Hill, and N. M. Larson
Oak Ridge National Laboratory
Oak Ridge, TN 37831-6356 USA

Abstract: High-resolution neutron transmission and differential elastic-scattering measurements were made for a ^{56}Fe-enriched iron target at the Oak Ridge Electron Linear Accelerator (ORELA). A natural iron target was used for transmission measurements below 160 keV. The data were analyzed from 5 to 850 keV. Parameters were obtained for 33 $\ell = 0$ and 242 $\ell > 0$ resonances. New ^{6}Li-glass transmission data were acquired for two ^{60}Ni-enriched sample thicknesses. The neutron width for the 2.253-keV resonance was determined to be 59.3 ± 0.6 meV and the radiation width 553 ± 50 meV.

(^{56}Fe and ^{60}Ni transmission measurements, ^{56}Fe differential elastic-scattering measurements, R-matrix analyses below 850 keV, resonance parameters)

Introduction

The ENDF/B-V evaluation for the resonance parameters of ^{56}Fe covered the energy region below 400 keV. New ORELA 200-m transmission and differential elastic-scattering measurements were used to update the resonance region and extend it to 850 keV [1]. In this note we present some of the results of this work. We also present precise resonance parameters for the 2.253-keV resonance of ^{60}Ni based upon new ^{6}Li-glass 80-m transmission data which supersede our earlier results [2] for this resonance.

Data Acquisition and Data Processing for ^{56}Fe

Transmission and elastic-scattering measurements were made at the 200-m flight path station by the time-of-flight technique using neutron pulses from the ORELA water-moderated tantalum target.

Two transmission measurements were made: one with unmoderated neutrons directly from the tantalum and the other with moderated neutrons. Two sample thicknesses and burst widths were used (see Table 1).

The measurement with moderated neutrons was made on a natural iron sample. Transmission data from 2 to 163 keV were obtained using a 1-cm-thick NE-110 scintillator epoxy-coupled to two 12.5-cm-diameter RCA 8854 photomultipliers. The measurement on a ^{56}Fe sample was made with an "effective" sample enrichment of 99.92% ^{56}Fe and covered the energy region from ≈100 keV to 20 MeV. Unmoderated neutrons from the tantalum target were used. A 2.5-cm-thick NE-110 scintillator also mounted between two RCA 8854 photomultipliers was used as the detector.

The data were corrected for the deadtime (1104 ns) of the digitizer and for various backgrounds.

The scattering measurements, covering the ≈10-keV to 5-MeV energy region, were made with filters and collimators which allowed both unmoderated and moderated neutrons to reach the sample made of 123.4 g of iron enriched to 99.87% in ^{56}Fe. The sample was a hollow cylinder suspended at the center of the scattering chamber. Six neutron detectors were located 19.1 cm from the center of the chamber at various angles from the direction of the incident neutron beam. Later one was placed in the direct beam to measure the product of the flux and the detector efficiency. Each neutron detector consisted of a NE-110 cylinder which was viewed at each end by RCA 8850 photomultiplier tubes.

All spectra were normalized by means of a neutron monitor detector and corrected for deadtime, constant room background, and geometrical factors to deduce a relative differential scattering cross section with an uncertainty of ≈5%.

Analysis of the ^{56}Fe Data

The transmission data were analyzed with the multilevel R-matrix (Reich-Moore) code SAMMY [3]. SAMMY is a constrained least-squares code which uses Bayes' theorem for the fitting process. Two main data sets were analyzed between 5 and 850 keV (see Table 1). Fit to the data was achieved with channel radii of 5.437 fm for the s- and d-wave resonances and 4.896 fm for the p-waves.

In order to insure that the resonance parameters based upon the analysis of the 200-m data would produce acceptable thermal scattering cross sections, data based upon transmission measurements made with a ^{6}Li-glass detector at the 17- and 80-m flight path stations were also used in the fitting process below 2 keV.

A sample of the fit to the total cross section data is shown in Fig. 1.

The differential elastic-scattering data were used as the principal tool to determine the spin and parity of the $\ell > 0$ resonances. The theoretical cross sections were calculated as a function of the incident neutron energy with the R-function code RFUNC [4] and compared to the experimental data at the six scattering angles. For a given resonance, various combinations of spins and parities were tested. The one which yielded the best agreement with the data was adopted.

The differential elastic-scattering data and theoretical curves for three of the six scattering angles are plotted in Fig. 1 from 550 to 600 keV. Figure 1 illustrates the usefulness of the differential data in assigning the spin and parity of resonances. The resonance at 561.4 keV (on top of an s-wave resonance) is fitted as a $p_{3/2}$ resonance in our analysis (underlined in Fig. 1, left). From the analysis of their 400-m transmission data Cornelis et al. [5] recommended a $d_{3/2}$ assignment for this resonance. With a $d_{3/2}$ assignment the fit to the transmission data is just slightly degraded but the agreement with the differential elastic-scattering data is lost as shown in Fig. 1, right. This demonstrates that the transmission data alone cannot always be depended upon to provide determination of spins and parities of $\ell > 0$ resonances.

Table 1. Summary of the 200-m data analyzed

Sample	Energy range (keV)	Burst width (ns)	Average sample thickness (at/b)
Transmission data			
^{nat}Fe	5 to 120	7	0.2144 ± 0.0005
^{56}Fe	120 to 850	4.5	0.2227 ± 0.0005
Differential elastic-scattering data			
^{56}Fe	40 to 850	6	0.0677 ± 0.0020

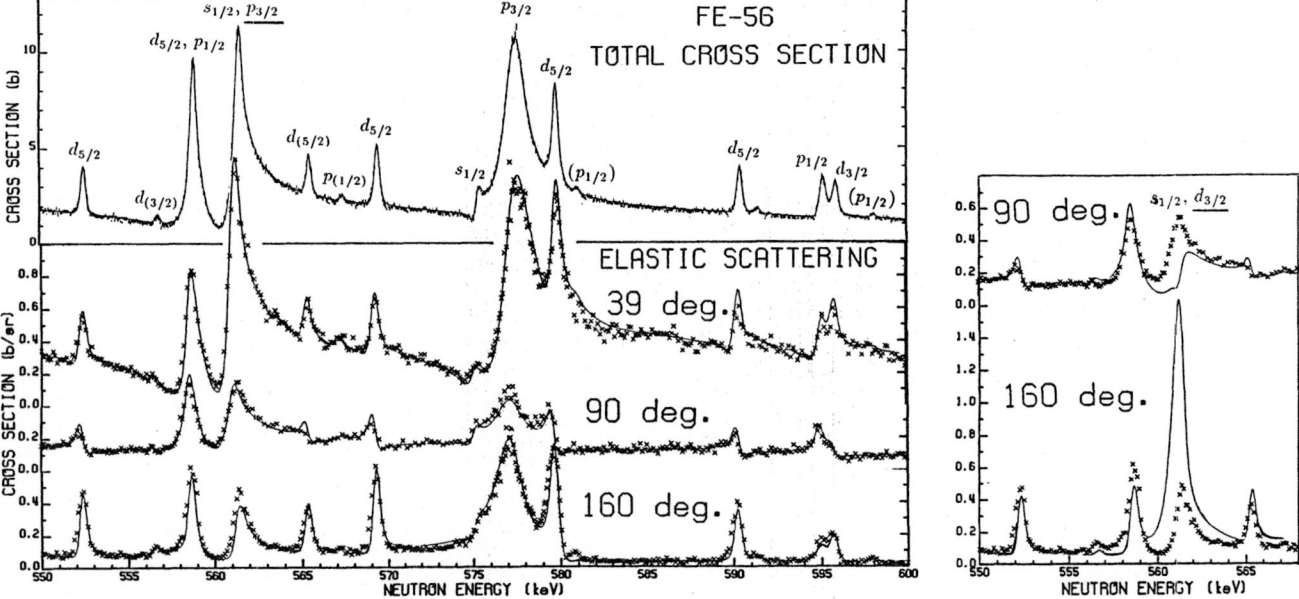

Fig. 1. <u>Left</u>: Sample of the fit to the total cross-section data and comparison of the elastic-scattering data at three angles with theoretical calculations. Parentheses indicate uncertain spin and parity assignments. <u>Right</u>: Comparison with the data when the $p_{3/2}$ resonance, at left, underlined, is changed to a $d_{3/2}$ resonance as recommended by Cornelis et al. [5]

Results for ^{56}Fe

Results of the simultaneous analyses of the transmission and differential elastic-scattering data were combined with the results of the ^{56}Fe capture data analysis of Corvi et al. [6] which extends to 350 keV. We reported parameters for 302 resonances in the 1- to 850-keV energy range [1], of which 26 were seen only in the capture data.

Thirty-three resonances were assigned as s-wave resonances. From the analysis of differential scattering data the orbital angular momentum, ℓ, of 184 resonances were definitely assigned, i.e., 76% of the $\ell > 0$ resonances analyzed in the transmission data. Eighty-five are p-wave and 99 are d-wave resonances. The spin, J, of 81% of the 85 p-wave resonances can be assigned with some degree of confidence but only 63% of the 99 d-wave resonances could be given a definite spin assignment.

The average radiation widths for the $\ell = 0$, 1, 2 resonances below 350 keV are 0.92 ± 0.41 eV, 0.45 ± 0.23 eV, and 0.75 ± 0.27 eV, respectively.

From the comparison of the distribution of the normalized reduced neutron widths of the 33 observed s-wave resonances with a Porter-Thomas distribution we conclude that up to four narrow s-wave resonances could have been missed. The average level spacing for the s-wave resonances, D_o, is given in Table 2 and compared with values obtained from three earlier analyses [5,7,8]. The normalized distribution of the s-wave level spacings is in good agreement with a Wigner distribution.

The plot of the cumulative sum of the s-wave reduced neutron widths as a function of energy, given in Fig. 2, reveals that almost 50% of the observed strength lies in two small energy intervals that span less than 10% of the energy range analyzed. Consequently, the s-wave strength function, S_o, based upon the total strength observed up to 850 keV is larger than the one based upon the strength observed only up to 360 keV (see Table 2). The two conspicuous large steps in the staircase plot could indicate the presence of particle vibration doorway states.

The final resonance parameter set generates the thermal capture and total cross sections recommended by Mughabghab et al. [9].

Table 2. Resonance parameter statistics for s-wave resonances compared with results of three earlier analyses

Source	Energy range (keV)	D_o (keV)	S_o (10^{-4})
Present work	5-850	25.4 ± 2.2	2.3 ± 0.6
Present work	5-360		1.7 ± 0.7
Cornelis et al. [5]	40-850	25.5	
Cornelis et al. [5]	240-850		2.6 ± 0.8
Cierjacks et al. [7]	450-900	19.6 ± 1.8	2.6 ± 0.8
Pandey et al. [8]	10-500	$25. \pm 5.$	2.6 ± 0.9
Pandey et al. [8]	10-200		1.9 ± 0.9

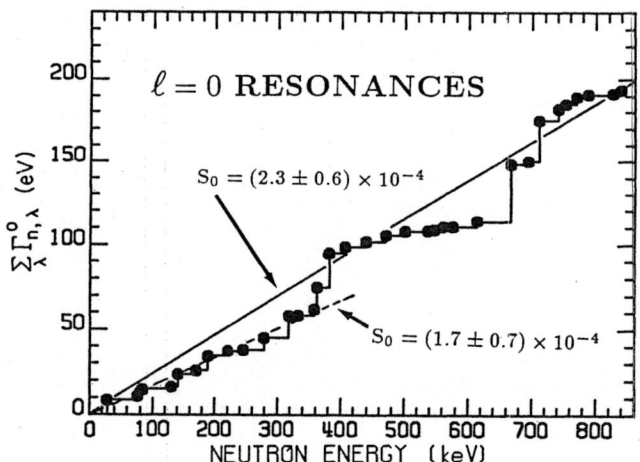

Fig. 2. Sum of the reduced neutron widths for s-wave resonances as function of incident neutron energy.

Data Acquisition and Results for ^{60}Ni

Transmission measurements have been made on two ^{60}Ni metal samples, enriched to 99.68%, using an improved two phototubes ^{6}Li-glass scintillation detector located at the 80-m flight-path station. This ^{6}Li-glass detector has a much better resolution function than the single photomultiplier detector used in the earlier measurements [2]. Also the metal samples were cast in a reducing atmosphere, whereas the samples in the earlier measurements were pressed powder which may have contained some oxide or water. The diameters of the samples were 2.54 cm with thicknesses of 0.0293 and 0.0837 atoms/barn respectively. The electron beam burst was 20 ns wide producing a beam power of 18 kw at 400 Hz. The earlier measurements had used 40-ns-wide bursts at 1000 Hz.

Since the sample thicknesses were different from those in the previous measurements, the two sets of transmission data cannot be directly compared. In the energy range from 1 to 5 keV the new data yield total cross sections that are about 0.1 barn smaller than those based upon the old data. The ORELA target moderation length and the detector resolution used in the earlier analysis is now realized to have been too small. Although the neutron width of the 2.25-keV resonance is insensitive to this resolution, it resulted in too large a value for its total width. The new transmission data for the two sample thicknesses are very consistent with each other and yield a substantially larger value for the neutron width of this resonance. Furthermore, the resonance parameter analysis of the new transmission data down to 500 eV yields the known thermal scattering cross section.

Our earlier resonance parameter analysis was updated for ENDF/B-VI using the new ^{6}Li-glass transmission data up to 115 keV and the old NE-110 transmission data above this energy. This analysis yields for the 2.25-keV resonance, with a $p_{1/2}$ assignment, a radiation width of 553 ± 50 meV, and a neutron width of 59.3 ± 0.6 meV (see Fig. 3). We have therefore obtained from a transmission experiment the capture area in this resonance to an accuracy of about 1%. This result should be useful in testing the consistency between transmission and capture experiments.

Conclusions

The purpose of this work was to extend the resolved resonance energy region for the ^{56}Fe ENDF/B-VI evaluation and to improve upon the low energy resonance

region for ^{60}Ni, in particular the parameters of the 2.25-keV resonance.

The ^{56}Fe report [1] on which part of this publication is based gives resonance parameters for an energy range twice as large as the one covered in the previous evaluation. These resonance parameters provide a complete and accurate description of the scattering cross section from thermal to 850 keV and are consistent with the accepted values for the thermal total and capture cross sections. Our parameters were compared with those obtained at other laboratories. In general, agreement is good between our parameters and those of Cornelis et al. [5] for the resonances reported in both analyses but it is not clear why they missed 40% of the $\ell > 0$ resonances we observed below 240 keV and 20% above that energy. The agreement with the parameters of Cierjacks et al. [7] is very poor.

Our improved knowledge of the resonance parameters for neutron interaction with ^{56}Fe and the extension of the energy region described by those parameters are of significant importance in reactor calculations since it eliminates the need to deal with the approximate unresolved resonance formalism.

Acknowledgments

The authors thank E. M. Cornelis, L. Mewissen, and F. Poortmans for sending us, before publication, the results of their ^{56}Fe transmission data analysis below 240 keV, and D. C. Larson for many useful discussions. We also thank R. E. Chrien from Brookhaven National Laboratory for the loan of the ^{56}Fe sample.

This research was sponsored by the Office of Energy Research, Division of Nuclear Physics, U.S. Department of Energy, under contract DE-AC05-84OR21400 with Martin Marietta Energy Systems, Inc.

References

1. C. M. Perey, F. G. Perey, J. A. Harvey, N. W. Hill, and N. M. Larson, ^{56}Fe Resonance Parameters for Neutron Energies Up to 850 keV, ORNL/TM-11742, Oak Ridge National Laboratory, 1990.
2. C. M. Perey, J. A. Harvey, R. L. Macklin, F. G. Perey, and R. R. Winters, Phys. Rev. C **27**, 2556 (1983).
3. N. M. Larson and F. G. Perey, Users Guide for SAMMY: A Computer Model for Multilevel R-Matrix Fits to Neutron Data Using Bayes' Equations, ORNL/TM-7485 (1980), ORNL/TM-9179 (1984), ORNL/TM-9179/R1 (1985), and ORNL/TM-9179/R2 (1990), Oak Ridge National Laboratory.
4. F. G. Perey, RFUNC - A Code to Analyze Differential Elastic-Scattering Data, ORNL/TM-1112, Oak Ridge National Laboratory, 1989.
5. E. M. Cornelis, L. Mewissen, and F. Poortmans, Proc. Intern. Conf. on Nuclear Data for Science and Technology, Antwerp, p. 135 (1983), and private communication (1985).
6. F. Corvi, A. Brusegan, R. Buyl, and G. Rohr, Proc. Consultants Meeting on Nuclear Data for Structural Materials, Vienna, Austria, 1983, International Nuclear Data Committee Report INDC(NDS)-152L, 1984.
7. S. Cierjacks and I. Schouky, Proc. Intern. Conf. on Neutron Physics and Nuclear Data for Reactors, Harwell, 1978 (OECD, Paris, 1978), p. 187.
8. M. S. Pandey, J. B. Garg, J. A. Harvey, and W. M. Good, Proc. Intern. Conf. on Nuclear Cross Sections and Technology, NBS Special Publication 425 (1975) vol. II, p. 748.
9. S. F. Mughabghab, M. Divadeenam, and N. E. Holden, Neutron Cross Sections, Vol. 1: Neutron Resonance Parameters and Thermal Cross Sections (Academic Press, New York, 1981).

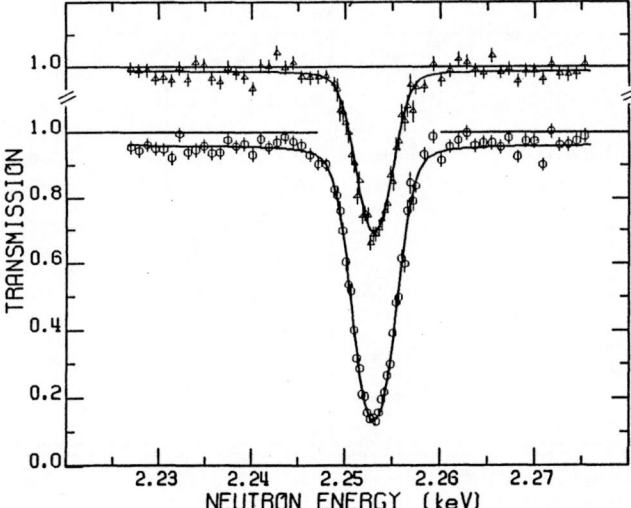

Fig. 3. Fits to the transmission data in the neighborhood of the 2.25-keV resonance.

RESONANCE NEUTRON CAPTURE IN STRUCTURAL MATERIALS

F. Corvi, G. Fioni*, A. Mauri** and K. Athanassopulos

Commission of the European Communities, Joint Research Centre
Central Bureau for Nuclear Measurements, B- 2440 Geel, Belgium.

Abstract: In view of the improvements recently introduced in γ-ray detection with the pulse-height weighting method, it was decided to revisit the field of data on neutron capture in structural materials. First, the [56]Fe capture results obtained in Geel some years ago were revised by applying to the raw data base the correct weighting. Secondly, [60]Ni capture measurements were performed at GELINA with a renewed detection setup. Resonance parameters are given for both isotopes in the range from 1 to 300 keV.

([56]Fe, [60]Ni, capture cross section, range 1-300 keV, capture areas, resonance parameters)

* Fellow of the CEC, now at ILL, Grenoble   ** seconded from ENEA, Italy

Introduction

Advances in the accuracy of the measurements performed with total energy detectors were made in recent years by replacing the previously used calculated weighting function with one determined thanks to an original experimental method. In this way the long-standing discrepancy between capture and transmission results for the neutron width of the 1.15 keV resonance was successfully solved [1,2]. Since it is very likely that the same kind of systematic errors exists in data concerning other resonances of this isotope or of other nuclei in the same mass region, it was recommended at the 1988 Mito Conference [3] to revisit the field of capture experiments performed with total energy detectors with special reference to resonance capture in structural materials of fast reactors. As a first step, it was decided to revise the [56]Fe capture results obtained in Geel some years ago [4,5] by applying to the raw data base the correct weighting function. In this way, resonance parameters and capture areas were re-calculated for 97 resonances in the range 1-300 keV.

In a second time, high resolution neutron capture measurements of [60]Ni were carried out at the Geel linac and the data were analysed up to 300 keV.

Capture in [56]Fe

A point not sufficiently clarified in our previous works [1,2] was the dependence of the weighting function on sample thickness. To investigate this effect, we performed a series of measurements of the 1.15 keV [56]Fe resonance normalized to capture in gold for a variety of iron samples and alloys. Two different detector systems were studied: the first one, which was employed in the old [56]Fe measurements, consisted of two cylindrical C_6D_6 scintillators viewing the sample at a right angle with respect to the neutron beam. The second one, used in the recent [60]Ni experiment, consisted of four detectors placed at an angle θ = 125° with respect to the neutron direction. This last arrangement was intended to minimize anisotropic effects of radiation emitted from resonances with spin J > 1/2.

The results, uncorrected for γ-ray and electron self-absorption, are listed in Table 1: one may notice that, for the θ = 90° case, the measured neutron width stays approximately constant and agrees with the transmission [6] value Γ_n = 61.7 ± 0.9 meV up to a thickness of 0.8 g/cm² corresponding to a 1 mm metallic sample. Then it decreases steadily.

In particular for the iron oxide sample, which has the same thickness as the one used in the original capture measurements [4], this decrease is about 13%. In this case the difficulty of gold calibration was avoided by normalizing the whole capture data set to the area of the 1.15 keV resonance whose parameters were determined from transmission [6]. The question left open is whether the weighting function used is still able to correctly deal with the degraded γ-ray spectral shapes emerging from a thick sample. This point was investigated using a mixed sample of Fe and Au: the neutron width obtained by normalizing to capture in the gold contained in the sample was only 5% larger than the thin-sample value. This shift is considered acceptable since the differences in spectrum shapes amongst the various [56]Fe resonances are much less than that between [56]Fe and Au spectra. The tests for θ = 125° reflect the same situation as for θ = 90°. This finding is very important because it shows that the weighting function, which was experimentally determined for the θ = 90° geometry [2], can be successfully applied to the θ = 125° case.

Table 1:   Values of the neutron width of the 1.15 keV resonance in [56]Fe obtained from capture measurements normalized to the 4.9 eV gold resonance

Sample		Γ_n (meV)	
Thickn. (g/cm²)	Chem.Isot. Composit.	θ = 90°	θ = 125°
0.3874	[56]Fe	60.1 ± 1.7	62.0 ± 1.6
0.7973	[56]Fe	61.3 ± 1.8	62.7 ± 1.8
1.185	[56]Fe	56.2 ± 1.7	
1.987	Fe_2O_3	53.8 ± 1.8	
1.970	22.7% Fe 77.3% Ni	52.3 ± 1.7	
1.826 [a]	93.5% Fe 6.5% Au	63.6 ± 2.0	65.3 ± 2.0
ORNL transmission		61.7 ± 0.9	

a) Self-calibration sample

Table 2: Capture areas and resonance parameters for ^{56}Fe + n from 1 to 300 keV.

E_o (keV)	l	2J	$g\Gamma_n$ (eV)	$g\Gamma_\gamma$ (eV)	$g\Gamma_n\Gamma_\gamma/\Gamma$ (eV)	E_o (keV)	l	2J	$g\Gamma_n$ (eV)	$g\Gamma_\gamma$ (eV)	$g\Gamma_n\Gamma_\gamma/\Gamma$ (eV)
1.15	1	1	0.0617	0.574	0.0557	179.92	1	3	13.0	0.53 ± 0.07	0.51 ± 0.07
2.35			(0.44 ± 0.10)x10^{-3}	0.84	(0.44 ± 0.10)x10^{-3}	181.34 b					2.68 ± 0.23
12.46			(2.7 ± 0.7)x10^{-3}	0.54	(2.7 ± 0.7)x10^{-3}	187.25	2		3.4	0.64 ± 0.16	0.54 ± 0.13
17.77			0.014 ± 0.002	0.54	0.014 ± 0.002	187.91	0	1	3600	1.05 ± 0.30	1.05 ± 0.30
20.19			(8.1 ± 1.6)x10^{-3}	0.84	(8.1 ± 1.6)x10^{-3}	188.32			0.13 ± 0.05	0.54	0.09 ± 0.03
22.81			0.257 ± 0.017	0.54	0.174 ± 0.008	190.14 b					0.91 ± 0.10
27.74	0	1	1520	1.00 ± 0.13	1.00 ± 0.13	193.17	2	5	40	1.22 ± 0.10	1.19 ± 0.10
34.26 a	1	1	1.58	0.41 ± 0.02	0.54 ± 0.03	195.95	1	1	66	0.55 ± 0.08	0.55 ± 0.08
36.75			0.39 ± 0.02	0.84	0.26 ± 0.02	201.78	2	3	48	1.62 ± 0.14	1.57 ± 0.14
38.45			0.49 ± 0.04	0.54	0.336 ± 0.018	203.91			0.08 ± 0.04	0.84	0.08 ± 0.04
46.09 a	1	1	10	0.58 ± 0.03	0.55 ± 0.03	206.14	2		3.42	1.52 ± 0.18	1.05 ± 0.10
52.18	1	3	24	0.79 ± 0.04	0.77 ± 0.04	208.20	1	3	22	0.73 ± 0.09	0.71 ± 0.08
53.60			1.0	0.64 ± 0.05	0.39 ± 0.02	209.00			0.28 ± 0.08	0.54	0.22 ± 0.04
53.73			0.034 ± 0.009	0.54	0.032 ± 0.008	210.01			0.13 ± 0.04	0.84	0.12 ± 0.04
59.28 a	1	1	8.0	0.86 ± 0.06	0.78 ± 0.05	210.91	2	5	18.0	1.27 ± 0.14	1.18 ± 0.12
63.52			1.6	1.15 ± 0.10	0.67 ± 0.04	216.04			0.35 ± 0.09	0.54	0.21 ± 0.04
73.04 a		3	20	0.69 ± 0.05	0.67 ± 0.04	220.90	0	1	1150	1.50 ± 0.21	1.50 ± 0.21
74.06	0	1	535	0.52 ± 0.08	0.52 ± 0.08	222.11	1		16.0	0.40 ± 0.08	0.39 ± 0.08
77.14	1	1	3.6	0.30 ± 0.03	0.27 ± 0.02	223.89			20.0	1.05 ± 0.12	1.00 ± 0.10
80.91	2	5	21	2.28 ± 0.12	2.05 ± 0.11	226.07 a	1	3	56.0	0.82 ± 0.11	0.81 ± 0.11
83.64	0	1	1250	0.52 ± 0.08	0.52 ± 0.08	230.13			15.0	0.56 ± 0.08	0.54 ± 0.08
90.42	1		28	0.80 ± 0.06	0.77 ± 0.05	232.76	2	3	40.0	2.14 ± 0.20	2.03 ± 0.17
92.77	2		3.2	1.48 ± 0.14	1.01 ± 0.06	235.09	2	5	60.0	2.38 ± 0.17	2.29 ± 0.16
92.99			1.06 ± 0.18	0.54	0.53 ± 0.05	241.86 a	2	3	30.0	3.48 ± 0.25	3.16 ± 0.22
96.27			0.48 ± 0.12	0.84	0.40 ± 0.08	243.61			0.22 ± 0.08	0.54	0.18 ± 0.06
96.44 a	1	1	1.3	0.97 ± 0.17	0.55 ± 0.06	245.25	0	1	435	0.69 ± 0.08	0.69 ± 0.08
96.69	2		5.0	1.69 ± 0.10	1.26 ± 0.12	246.60			0.16 ± 0.10	0.54	0.12 ± 0.07
102.78	1	3	42	0.64 ± 0.06	0.63 ± 0.06	252.79			0.13 ± 0.07	0.54	0.12 ± 0.06
103.16			1.52	1.44 ± 0.18	0.74 ± 0.06	253.83	2	5	84	1.66 ± 0.18	1.62 ± 0.18
106.02	2		5.6	1.62 ± 0.12	1.26 ± 0.09	256.40 a	2	3	12.0	1.05 ± 0.13	0.97 ± 0.11
112.81	2	3	11.0	1.10 ± 0.08	1.00 ± 0.07	260.18	1	3	29.0	0.61 ± 0.09	0.60 ± 0.09
121.13			0.055 ± 0.018	0.54	0.050 ± 0.016	261.14			0.11 ± 0.05	0.84	0.10 ± 0.05
122.69			0.15 ± 0.05	0.54	0.13 ± 0.04	264.02	1	1	140	0.43 ± 0.09	0.43 ± 0.09
122.92 a	1	1	92	0.59 ± 0.07	0.59 ± 0.07	264.71			0.34 ± 0.13	0.54	0.31 ± 0.12
124.29	1	1	7.5	0.61 ± 0.06	0.57 ± 0.06	267.31	2	5	165	0.91 ± 0.18	0.90 ± 0.18
125.28	2	3	20	1.21 ± 0.08	1.14 ± 0.08	267.89	1		1.29 ± 0.48	0.84	0.73 ± 0.28
130.05	0	1	500	0.62 ± 0.09	0.62 ± 0.09	270.16	1	1	144	0.42 ± 0.11	0.42 ± 0.11
130.37			1.58	0.81 ± 0.11	0.80 ± 0.11	275.06			0.51 ± 0.26	0.54	0.26 ± 0.14
140.26	0	1	2700	1.61 ± 0.28	1.61 ± 0.28	276.52	1	3	170	0.54 ± 0.12	0.54 ± 0.12
141.38			1.18 ± 0.27	0.84	0.69 ± 0.17	278.29	0	1	4000	1.15 ± 0.35	1.15 ± 0.35
142.55			1.50 ± 0.22	0.54	0.63 ± 0.11	281.32 a	2	3	16	1.84 ± 0.20	1.65 ± 0.18
150.03			0.39 ± 0.06	0.54	0.23 ± 0.03	283.62	1		1.04 ± 0.50	0.54	0.53 ± 0.27
154.08	1		0.90 ± 0.10	0.84	0.59 ± 0.05	284.37	2	5	24.0	1.38 ± 0.19	1.31 ± 0.18
161.91	2		13.0	1.20 ± 0.10	1.09 ± 0.09	285.83	2	5	30.0	2.05 ± 0.21	1.92 ± 0.19
169.28	2	3	12.0	1.83 ± 0.30	1.59 ± 0.23	289.00	2	3	24.0	1.27 ± 0.16	1.20 ± 0.15
169.30	0	1	1000	0.94 ± 0.14	0.94 ± 0.14	290.86	1	3	26.0	1.50 ± 0.17	1.41 ± 0.16
173.30			0.29 ± 0.06	0.54	0.23 ± 0.05	293.54	1	1	130	0.56 ± 0.09	0.56 ± 0.09
173.82	1	3	84	0.49 ± 0.06	0.49 ± 0.06	296.07			0.35 ± 0.12	0.54	0.27 ± 0.10
175.94			0.14 ± 0.06	0.54	0.13 ± 0.05						

a)   Spin taken from ref. [11]
b)   Peak consisting of a doublet according to ref. [10]

After these validation tests, we proceeded to the re-analysis of the ^{56}Fe results by applying the weighting determined in ref. [2] to the old raw data base, consisting of both time-of-flight and amplitude information. For any detail of experiment and analysis, the reader is referred to refs. [4,5]. The new corrected list of resonance parameters obtained using the area code TACASI [7] or the shape fitting code FANAC [8] is given in Table 2. The neutron widths were taken still from ref [9] while spin and parity values are from ref. [10], except when otherwise indicated. Only the certain J and l assignments were considered. Whenever available, the neutron width was kept as a fixed input parameter in the analysis and values of the radiation width Γ_γ and of the capture area $g\Gamma_n\Gamma_\gamma/\Gamma$ were derived. When a resonance was not seen in transmission, its neutron width was obtained by fixing $\Gamma_\gamma = 0.54$ eV or $\Gamma_\gamma = 0.84$ eV according to whether they were assigned as p-waves or d-waves, respectively, in ref. [10]. The values of the widths kept fixed are those listed without any error in Table 2. All resonances for which the spin is not given in Table 2, were analyzed assuming J = 1/2.

By comparing the present results to those obtained with the same data set but with the old weighting, one may note that the capture areas have varied by a relative amount going from -10% to +5%. For each resonance the amount and the sign of this variation should be related to the shape of its capture spectrum. Finally, from the data of Table 2 the following average radiation widths and their standard deviations were calculated for each l value:

$<\Gamma_\gamma>$ = 0.96 ± 0.39 eV for s-waves
$<\Gamma_\gamma>$ = 0.49 ± 0.21 eV for p-waves
$<\Gamma_\gamma>$ = 0.72 ± 0.33 eV for d--waves

Capture in ^{60}Ni

Neutron capture measurements of ^{60}Ni were performed at GELINA in the energy range 1-700 keV using two metallic nickel samples of 8 cm diameter enriched to 99.07% ^{60}Ni, on loan from ORNL: a sample of thickness N = 0.0180 at/b was used for the high resolution run at a 58.58m flight path in conjunction with the $\theta = 125°$ detector setup, while a sample of N = 0.0041 at/b was used at 28.41 m distance with the 90° setup, mainly in order to cover the s-wave resonances below 100 keV.

The linac was operated to provide 1 ns wide bursts of 100 MeV electrons at a repetition frequency of 800 Hz, the average beam power being 7 kWatt. The relative neutron flux was measured as a function of energy at the same time as capture with an ionization chamber containing three back-to-back deposits of ^{10}B of thickness 40 µg/cm^2; the chamber was placed in the beam about 0.8 m before the capture sample.

The dependence on energy of the prompt neutron scattering sensitivity relative to gold capture $\varepsilon_n/\varepsilon_c$ was measured for the new setup by comparing the counting rate from a graphite disk to that from a gold sample. Values of $\varepsilon_n/\varepsilon_c$ about 2 to 3 times larger than those of the $\theta = 90°$ setup [5] were found. These values are considered accurate to within ± 50%.

The data were normalized to the capture area of the 2.25 keV resonance which was measured relative to gold capture at 4.9 eV, using the thin sample. Since preliminary γ-spectroscopy measurements performed at Geel indicated a spin J = 3/2 for the 2.25 keV resonance, this calibration measurement was also performed with the $\theta = 125°$ setup. The data analysed with the TACASI area code by fixing Γ_γ = 0.583 eV [12] give rise to a value $g\Gamma_n = 62.9 \pm 2.0$ meV which should be compared with the value $g\Gamma_n = 59.3 \pm 0.8$ meV obtained in a recent accurate transmission measurement performed at ORELA [12]. The whole ^{60}Ni capture data set was normalized to the area of the 2.25 keV resonance asssuming $g\Gamma_n = 61.1$ meV, i.e. equal to the average of the two previous results.

Table 3: Comparison of the values of the capture widths of the ^{60}Ni s-wave resonances below 100 keV from the thick and thin sample run, respectively.

E_0 (keV)	Γ_n(eV)	Γ_γ(eV)	
		thick sample	thin sample
12.42	2358	2.27 ± 0.21	2.30 ± 0.14
28.67	698	0.36 ± 0.12	0.38 ± 0.04
43.13	104	0.19 ± 0.03	0.21 ± 0.02
65.23	443	0.73 ± 0.07	0.79 ± 0.04
86.91	398	0.43 ± 0.04	0.39 ± 0.02
98.05	1002	0.50 ± 0.11	0.46 ± 0.06

The preliminary result of a recent transmission experiment performed at Geel [13] gives $g\Gamma_n = 61.8 \pm 4$ meV in agreement with the previous values. The data were analysed with the shape code FANAC except for a few isolated low-energy resonances whose parameters were determined with the area code TACASI. An accurate determination of the radiation widths of s-wave resonances, and particularly of the important one at $E_0 = 12.4$ keV, was hindered, in the case of the thick sample, by the large contributions of the prompt neutron scattering background and of the multiple scattering. The thin-sample run was carried out mainly in order to check the thick-sample data for s-waves below 100 keV. The results, listed in Table 3, are very satisfactory: all values agree within the errors and their weighted average was taken as best estimate of the width. The ^{60}Ni resonance parameters are listed in Table 4 for the energy range 1-300 keV. The neutron widths $g\Gamma_n$ as well as the assignments for $l = 0$ resonances were taken from the results of the Oak Ridge transmission measurements [14]. When a resonance was not visible in transmission, it was analysed by fixing $g\Gamma_\gamma = 0.5$ eV. In absence of J and l assignments, all non s-waves resonances were considered as p 1/2 in the analysis.

References

1. F. Corvi, A. Prevignano, H. Liskien and P.B. Smith: Nucl. Instr. Meth. in Phys. Res. A265, 475 (1988)
2. F. Corvi, G. Fioni, F. Gasperini and P.B. Smith: Nucl. Sci. Eng. 107, 272 (1991)
3. M.G. Sowerby and F. Corvi, Proc. Int. Conf. on Nuclear Data for Science and Technology, 1988, Mito, Japan, p. 37, Saikon Tokyo (1988)
4. F. Corvi, A. Brusegan, R. Buyl, G. Rohr, R. Shelley and T. van der Veen, Proc. Int. Conf. on Nuclear Data for Science and Technology, 1982, Antwerp, Belgium, p. 131, Reidel (1983)
5. F. Corvi, A. Brusegan, R. Buyl and G. Rohr, Proc. Consult. Meet. on Nuclear Data for Structural Materials, 1983, Vienna, Report INDC (NDS) -152L (1984)
6. F.G. Perey, Proc. Int. Conf. on Nuclear Data for Basic and Applied Science, 1985, Santa Fe, p. 1523, Gordon and Breach, New York (1986)
7. F.H. Fröhner, Report GA-6906 (1966)
8. F.H. Fröhner, Report KfK-2145 (1976)
9. F.G. Perey, G.T. Chapman, W.E. Kenney, and C.M. Perey, Proc. Intern. Conf. on Neutron Data of Structural Materials for Fast Reactors, Geel, Belgium, 1977, p. 530, Pergamon Press (1979)
10. C.M. Perey, F.G. Perey, J.A. Harvey, N.W. Hill and N.M. Larson, Report ORNL/TM-11742 (1990)
11. C. Coceva, Y.K. Ho, M. Magnani, A. Mauri and P. Bartolomei, Proc. Int. Conf. on Capture Gamma Ray Spectroscopy, 1987, Leuven, p. 679, IOP Publ., Bristol (1988)
12. F.G. Perey, C.M. Perey, J.A. Harvey, N.W. Hill and N.M. Larson, this conference.
13. A. Brusegan, C.B.N.M. Geel, private communication (1991)
14. C.M. Perey, J.A. Harvey, R.L. Macklin, F.G. Perey, and R.R. Winters: Phys. Rev. C27, 2556 (1983)

Table 4: Capture areas and Resonance Parameters for ^{60}Ni + n from 1 to 285 keV.

E_0 (keV)	l	$g\Gamma_n$ (eV)	$g\Gamma_\gamma$ (eV)	$g\Gamma_n\Gamma_\gamma/\Gamma$ (eV)
1.33		0.00025 ±0.00003	0.50	0.00025 ±0.00003
2.25		0.0611 ±0.0018	0.583	0.0581 ±0.0015
5.53		0.041	0.50	0.038
12.21		0.24 ±0.04	0.50	0.16 ±0.02
12.42	0	2358	2.29 ±0.17	2.29 ±0.17
13.61		1.20	0.36 ±0.04	0.28 ±0.03
21.29		0.021 ±0.02	0.50	0.02 ±0.02
23.80		4.50	0.24 ±0.02	0.23 ±0.02
23.91		1.17	0.88 ±0.10	0.50 ±0.03
28.47		0.034 ±0.04	0.50	0.032 ±0.04
28.52		0.106 ±0.11	0.50	0.087 ±0.07
28.67	0	698	0.37 ±0.03	0.37 ±0.03
29.50		0.044 ±0.003	0.50	0.040 ±0.03
30.28	1.00		0.61 ±0.05	0.38 ±0.02
33.42		10.0	0.32 ±0.02	0.31 ±0.02
33.04		7.70	0.56 ±0.04	0.52 ±0.04
39.56		2.30	0.73 ±0.06	0.56 ±0.04
42.74		1.60	0.37 ±0.03	0.30 ±0.02
43.01		2.00	0.44 ±0.04	0.36 ±0.03
43.13	0	104.4	0.20 ±0.04	0.20 ±0.04
47.66		1.74	1.93 ±0.15	0.92 ±0.04
49.84		0.60	0.76 ±0.08	0.34 ±0.02
50.97		0.27 ±0.02	0.50	0.17 ±0.01
51.63		0.85	1.47 ±0.12	0.54 ±0.02
52.70		0.14 ±0.02	0.50	0.11 ±0.01
56.33		0.69	0.48 ±0.06	0.28 ±0.02
56.92		0.62	1.54 ±0.25	0.44 ±0.02
65.09		0.26 ±0.03	0.50	0.18 ±0.02
65.23	0	443	0.77 ±0.06	0.77 ±0.06
65.63		2.40	1.47 ±0.12	0.91 ±0.06
71.43		0.75	0.78 ±0.07	0.38 ±0.03
73.27		2.00	0.90 ±0.07	0.62 ±0.03
78.28		0.80	0.28 ±0.03	0.21 ±0.02
80.12		0.90	0.88 ±0.07	0.44 ±0.03
82.05		0.45	0.48 ±0.05	0.23 ±0.02
85.06		1.00	0.71 ±0.06	0.41 ±0.03
86.25		3.00	1.98 ±0.17	1.19 ±0.06
86.91	0	398	0.40 ±0.04	0.41 ±0.04
87.96		11.6	0.81 ±0.07	0.76 ±0.06
89.83		0.50	0.39 ±0.04	0.22 ±0.02
91.73		6.50	0.37 ±0.03	0.35 ±0.03
93.86		2.30	0.75 ±0.06	0.59 ±0.05
95.58		0.21 ±0.02	0.50	0.15 ±0.02
97.13		2.90	0.65 ±0.06	0.53 ±0.04
98.05	0	1002	0.47 ±0.05	0.47 ±0.05
99.48		7.40	1.01 ±0.08	0.86 ±0.06
102.16		0.40 ±0.04	0.50	0.22 ±0.01
107.94	0	649	0.64 ±0.06	0.64 ±0.06
108.54		3.30	1.03 ±0.08	0.78 ±0.04
111.58		4.00	3.11 ±0.27	1.73 ±0.08
112.09		2.50	0.45 ±0.04	0.38 ±0.03
112.35		3.50	0.54 ±0.05	0.47 ±0.03
113.41		3.00	1.00 ±0.09	0.74 ±0.05
113.97		0.03 ±0.004	0.50	0.03 ±0.01
120.50		0.17 ±0.05	0.50	0.13 ±0.04
120.65		7.50	0.84 ±0.10	0.76 ±0.09
120.97		2.60	1.76 ±0.16	1.05 ±0.06
123.74		31.5	0.72 ±0.06	0.71 ±0.05
127.44		0.023 ±0.006	0.50	0.022 ±0.005
127.79		67.4	0.46 ±0.04	0.46 ±0.04
129.90		2.50	1.16 ±0.10	0.79 ±0.05
133.64		20.9	0.52 ±0.04	0.51 ±0.04
135.65		0.18 ±0.05	0.50	0.13 ±0.03
136.15		15.5	1.11 ±0.10	1.04 ±0.09
136.42		6.70	2.04 ±0.18	1.56 ±0.10
137.36		0.091 ±0.02	0.50	0.077 ±0.015
137.58		5.60	0.48 ±0.04	0.44 ±0.04
138.69		0.032 ±0.01	0.50	0.03 ±0.01
139.18	0	30.7	0.80 ±0.07	0.80 ±0.07
139.69		26.2	1.12 ±0.10	1.078 ±0.09
140.13		31.6	1.83 ±0.16	1.73 ±0.14
145.89		1.00	0.59 ±0.06	0.37 ±0.02
147.65		0.052 ±0.009	0.50	0.047 ±0.007
148.94		8.50	1.09 ±0.09	0.97 ±0.07
151.54		14.6	0.37 ±0.03	0.36 ±0.03
154.49		162.6	0.47 ±0.04	0.47 ±0.04
156.46	0	472	0.25 ±0.03	0.25 ±0.03
156.80		0.031 ±0.01	0.50	0.029 ±0.01
160.36		20.4	0.52 ±0.05	0.51 ±0.04
161.41	0	1531	0.15 ±0.03	0.15 ±0.03
162.32		10.0	1.11 ±0.10	1.00 ±0.08
166.37		3.00	1.51 ±0.14	1.00 ±0.06
167.36		7.40	2.37 ±0.21	1.79 ±0.12
167.78		1.23 ±0.19	0.50	0.35 ±0.01
170.76		4.00	1.92 ±0.17	1.29 ±0.08
172.29		0.27 ±0.03	0.50	0.17 ±0.01
172.79		30.6 ±2.72	0.50	0.48 ±0.001
174.72		12.3	0.31 ±0.03	0.31 ±0.03
175.13		2.00	1.40 ±0.13	0.81 ±0.05
180.10		1.60	0.68 ±0.07	0.47 ±0.03
183.12		77.2	0.91 ±0.08	0.90 ±0.08
183.70		12.7	0.68 ±0.06	0.64 ±0.06
185.97	0	5237	≤0.70	≤0.70
186.83		2.00	1.42 ±0.14	0.82 ±0.05
191.13		0.016 ±0.007	0.25	0.016 ±0.006
192.81		62.2	0.61 ±0.06	0.61 ±0.06
194.36		19.9	0.93 ±0.09	0.89 ±0.08
194.62		3.00 ±0.30	0.50	0.42 ±0.006
196.05		0.16 ±0.02	0.50	0.12 ±0.01
196.80		5.00	1.33 ±0.12	1.05 ±0.08
198.97	0	3025	0.69 ±0.08	0.69 ±0.08
199.94		0.022 ±0.010	0.50	0.022 ±0.010
200.88	0	9.80	0.41 ±0.04	0.46 ±0.04
201.18		7.40	0.06 ±0.18	0.058 ±0.18
201.71		156	1.36 ±0.12	1.35 ±0.12
202.74		3.00	0.31 ±0.04	0.28 ±0.036
206.17		36.0	0.37 ±0.04	0.37 ±0.038
206.74		141	0.42 ±0.04	0.42 ±0.04
209.13		0.27 ±0.04	0.50	0.17 ±0.02
209.93		0.19 ±0.03	0.50	0.13 ±0.01
213.10		0.15 ±0.03	0.50	0.11 ±0.02
214.36		120	0.77 ±0.07	0.76 ±0.07
214.95		4.00	1.25 ±0.12	0.95 ±0.07
220.31		76.0	0.44 ±0.04	0.44 ±0.04
220.84		33.0	1.40 ±0.13	1.34 ±0.12
221.20		65.0	0.58 ±0.06	0.58 ±0.06
224.09		0.10 ±0.03	0.50	0.09 ±0.02
226.52		2.00	0.61 ±0.06	0.46 ±0.04
229.47		16.0	2.32 ±0.21	2.021 ±0.16
230.40		106	2.06 ±0.18	2.023 ±0.18
233.77		15.0	0.76 ±0.07	0.72 ±0.07
234.15		9.00	0.77 ±0.08	0.70 ±0.06
235.32		2.00	0.70 ±0.07	0.51 ±0.04
236.93		12.0	1.50 ±0.14	1.32 ±0.11
238.25		2.00	2.10 ±0.25	1.007 ±0.06
238.50		16.0	0.40 ±0.05	0.39 ±0.05
244.34		3.00	0.83 ±0.09	0.65 ±0.05
246.91		0.19 ±0.03	0.50	0.14 ±0.02
249.07		75.0	1.73 ±0.16	1.69 ±0.15
252.10	0	536	0.83 ±0.09	0.83 ±0.09
252.65		5.00	0.75 ±0.08	0.65 ±0.06
253.33		264	0.75 ±0.08	0.75 ±0.08
254.66		32.0	0.26 ±0.04	0.25 ±0.04
256.17	0	870	1.22 ±0.20	1.22 ±0.20
256.41		11.0	1.02 ±0.18	0.93 ±0.15
257.45	0	1826	0.44 ±0.09	0.44 ±0.09
257.91		23.0	0.07 ±0.03	0.07 ±0.03
258.74		13.0	0.78 ±0.10	0.74 ±0.09
260.07		0.16 ±0.04	0.50	0.12 ±0.02
261.09		52.0	0.49 ±0.05	0.48 ±0.05
262.99		0.84 ±0.08	0.50	0.31 ±0.01
263.40		44.0	1.24 ±0.12	1.20 ±0.11
265.83		0.22 ±0.05	0.50	0.15 ±0.02
266.26		56.0	0.54 ±0.06	0.54 ±0.06
268.82		1.50	0.62 ±0.08	0.44 ±0.04
269.28		98.0	0.65 ±0.07	0.64 ±0.07
269.58		4.20 ±0.38	1.00	0.78 ±0.01
273.94		0.26 ±0.09	0.50	0.17 ±0.04
277.05		1.50 ±0.14	1.00	0.58 ±0.02
277.38		17.2	1.89 ±0.19	1.70 ±0.15
278.48		367	0.77 ±0.08	0.77 ±0.08
279.39	0	225	0.27 ±0.04	0.27 ±0.04
280.40		145	0.87 ±0.09	0.87 ±0.09
282.04		15.1	0.27 ±0.04	0.26 ±0.04
282.95		175	1.91 ±0.19	1.89 ±0.18
283.40		108	1.31 ±0.14	1.29 ±0.13

MEASUREMENTS OF keV-NEUTRON CAPTURE GAMMA RAYS FROM SOME STRUCTURAL AND SHIELDING MATERIALS

M. Igashira, Y. Dozono, F. Uesawa, M. Shimizu, and H. Kitazawa

Tokyo Institute of Technology,
2-12-1 O-okayama, Meguro-ku, Tokyo 152, Japan

Abstract: We have measured capture gamma rays from the structural and shielding materials, Mg, Al, Si, Ca, Cr, Co and Zr, with an anti-Compton NaI(Tl) detector, at several neutron energies between 10 and 600 keV, employing a time-of-flight technique. Observed pulse-height spectra were unfolded using a response matrix of the detector in order to obtain capture gamma-ray spectra. Those spectra from Al, Si, Ca and Cr were compared with statistical model calculations. In general the calculations reproduce fairly well the observed spectra at neutron energies of several hundreds keV.

(capture gamma-ray spectrum, keV neutron, Mg, Al, Si, Ca, Cr, Co, Zr, statistical model calculation)

Introduction

Neutron-induced-photon production nuclear data are indispensable for shielding design calculation, for radiation damage estimate, and for radiation heating calculation. However, the data are scanty, especially in the keV-neutron region. For technological applications, the photon-production data contained in nuclear data libraries, e.g. ENDF/B-VI and JENDL-3, are not always satisfactory both in quantity and in quality. More accurate data are requested for a revision of photon-production data libraries. Therefore, we have measured keV-neutron capture gamma-ray spectra from structural and shielding materials to provide those nuclear data, and also to investigate the characteristics of resonance average spectra. The present paper reports on the results for Mg, Al, Si, Ca, Cr, Co and Zr.

Experimental Procedure and Data Processing

Experiments were carried out, employing a time-of-flight (TOF) technique. The experimental and data-processing methods have been described in detail elsewhere[1-3], and so are summarized briefly in the present paper. A pulsed proton beam (width: 1.5 ns at FWHM, repetition rate: 2 MHz, average beam current: 7-11 µA) from the 3.2-MV Pelletron accelerator of the Research Laboratory for Nuclear Reactors in Tokyo Institute of Technology produced keV neutrons via the $^7Li(p,n)^7Be$ reaction. Capture samples were natural metallic disks of 60 mm in diameter by 5-27 mm in thickness, and were located at a distance of 81 or 156 mm from the neutron source. Capture gamma rays from the sample were detected by a 76 mm in diameter by 152mm NaI(Tl) detector centered in an annular NaI(Tl) detector. The detector assembly operated as an anti-Compton gamma-ray spectrometer and was placed in a heavy shield consisting of borated paraffin, lead and cadmium. A ^6LiH shield, which absorbs effectively the neutrons scattered into the detector from the capture sample, was inserted into the collimeter of the detector shield. The spectrometer was located at about 80 cm from the capture sample and at an angle of 125° with respect to the proton beam direction. Two ^6Li-glass scintillation detectors were used as neutron monitors. One detector, located at 7° or 14° to the proton beam direction, was used to measure the incident neutron spectrum on the capture sample. The other detector located at 45° was used to monitor

the number of neutrons emitted from the source. Capture gamma-ray measurements were performed at several neutron energies between 10 and 600 keV with all samples. The neutron energy spread was 9-120 keV.

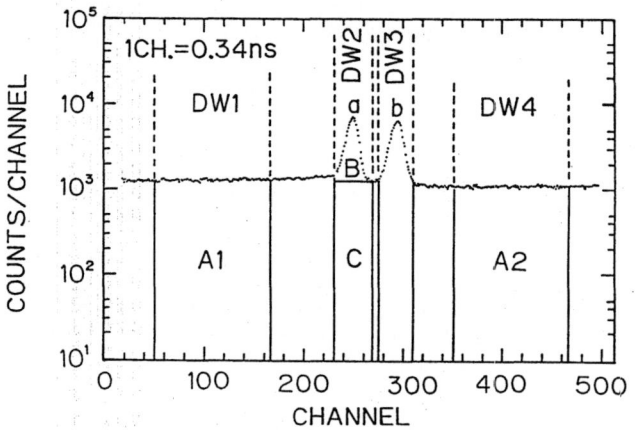

Fig. 1 Time-of-flight spectrum for the Co sample at the neutron energy of 540 keV.

Signals from the gamma-ray detector were taken in a mini-computer as two-dimensional data (pulse-height and TOF). A typical TOF spectrum for the Co sample is shown in fig. 1. The peak 'a' shows the capture gamma rays produced by the direct neutrons from the neutron source to the sample, and the peak 'b' shows the gamma rays produced by the $^7Li(p,\gamma)^8Be$ and $^7Li(p,p'\gamma)^7Li$ reactions. The part 'A1' or 'A2' can be regarded as the nearly time-independent background, which mainly consists of the natural background and the gamma rays due to the $^1H(n_{th},\gamma)^2H$ and $^{56}Fe(n_{th},\gamma)^{57}Fe$ reactions in the detector shield. Four digital windows (DW1-DW4) were set on the TOF spectrum as shown in fig. 1, and the gamma-ray pulse-height spectrum corresponding to each digital window was stored. DW2 was set to measure the capture gamma rays from the sample, and the other three windows were set to determine the background (C). The net capture gamma-ray pulse-height spectrum was obtained by subtracting the background (C) from the foreground (B+C).

The net capture gamma-ray pulse-height spectrum was unfolded by the computer program FERDOR[4], using a response matrix of the detector which was constructed by an interpolation method from experimentally determined response functions[3]. The unfolded capture gamma-ray spectrum $v(E_\gamma)$ (gamma rays/MeV/capture) was normalized as

$$\int E_\gamma \cdot v(E_\gamma)\, dE_\gamma = \overline{Bn} + En$$

$$\overline{Bn} \equiv \sum_i A_i \frac{<\Gamma_{\gamma 0}>_i}{<D_0>_i} Bn_i \Big/ \sum_i A_i \frac{<\Gamma_{\gamma 0}>_i}{<D_0>_i}$$

where A_i is the natural abundance of the ith isotope, Bn_i the neutron binding energy, $<D_0>_i$ the s-wave average level spacing, $<\Gamma_{\gamma 0}>_i$ the average radiative width for s-wave resonances, En the incident neutron energy, and E_γ the gamma-ray energy[5]. Correction for the gamma-ray attenuation in the sample was made by a Monte-Carlo calculation.

Results and Discussion

A part of measured capture gamma-ray spectra are shown in figs. 2 - 8 with incident energy En. Structure observed in the unfolded spectra can be attributed to the intrinsic one of spectra, never to an oscillation produced by the unfolding process used in the present analysis. The error shown in the figures contains the statistical error, the error of the gamma-ray detection efficiency, and the error in the spectrum unfolding. The error in the spectrum normalization is not included. However, it was estimated to be less than about 5 %.

The spectrum of Mg shown in fig. 2 mainly consists of gamma rays from the 431-keV $p_{3/2}$-wave resonance of ^{24}Mg. The strong transitions from the capture state to the ground (0.0 MeV, 5/2+), first-excited (0.59 MeV, 1/2+) and fifth-excited (2.56 MeV, 1/2+) states are observed. Those strong transitions are also observed in spectra at lower neutron energies. Gamma rays above 8 MeV are due to the neutron capture in ^{25}Mg.

The spectrum of Al in fig. 3 is composed of gamma rays from about 10 resonances. The strong transitions to the ground (0.0 MeV, 3+) and first-

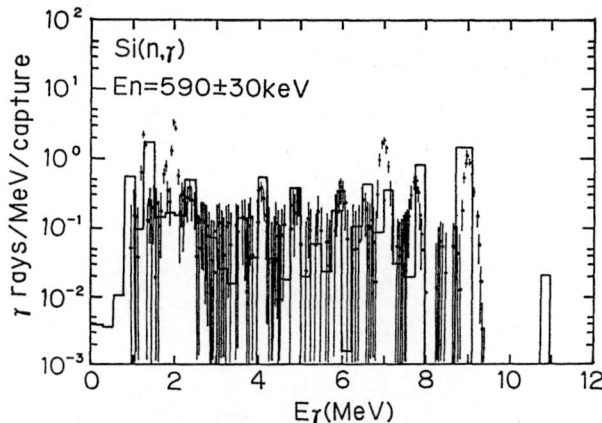

Fig. 4 Unfolded capture gamma-ray spectrum of Si at the neutron energy of 590 keV. The histogram shows the calculated spectrum.

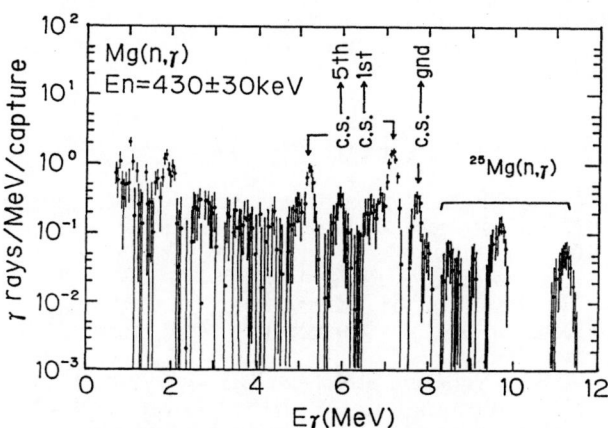

Fig. 2 Unfolded capture gamma-ray spectrum of Mg at the neutron energy of 430 keV.

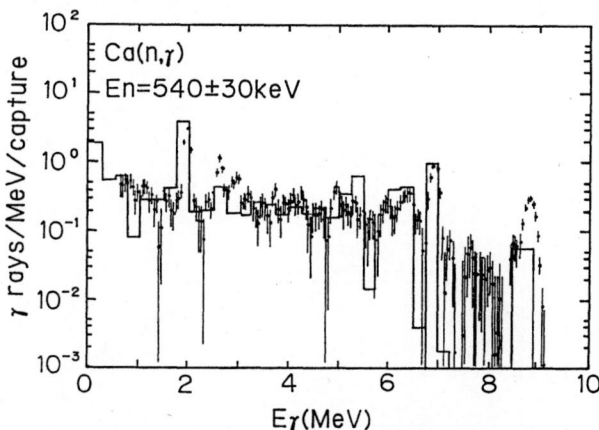

Fig. 5 Unfolded capture gamma-ray spectrum of Ca at the neutron energy of 540 keV. The histogram shows the calculated spectrum.

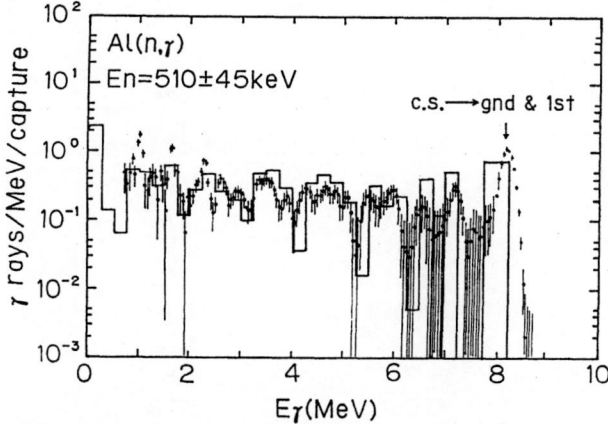

Fig. 3 Unfolded capture gamma-ray spectrum of Al at the neutron energy of 510 keV. The histogram shows the calculated spectrum.

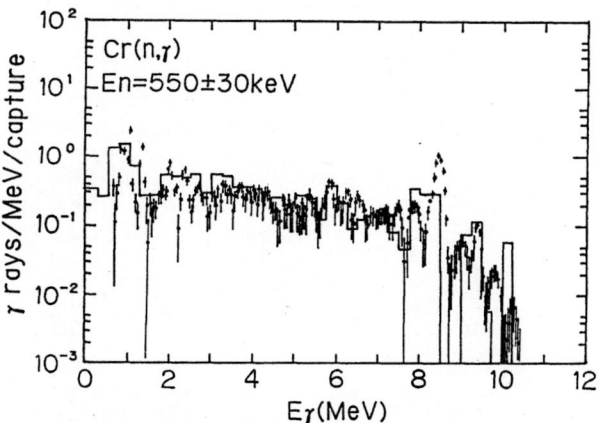

Fig. 6 Unfolded capture gamma-ray spectrum of Cr at the neutron energy of 550 keV. The histogram shows the calculated spectrum.

excited (0.031 MeV, 2+) states are observed not only in the spectrum but also in spectra at lower neutron energies.

The spectra of Si, Ca and Cr in figs. 4 - 6 are composed of gamma rays from several resonances. Strong transitions to low-lying states in each residual nucleus are observed in those spectra. Similar spectra are also observed at lower neutron energies.

The spectra of Co and Zr in figs. 7 and 8 are composed of gamma rays from many resonances. In both spectra, a characteristic hard spectrum is noticed.

The capture gamma-ray spectra of Al, Si, Ca, and Cr were calculated by the computer program

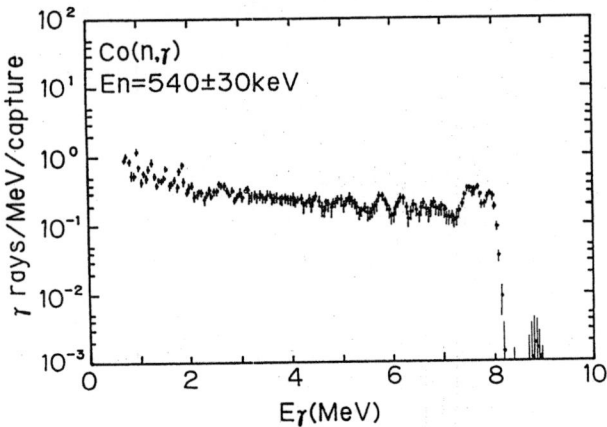

Fig. 7 Unfolded capture gamma-ray spectrum of Co at the neutron energy of 540 keV.

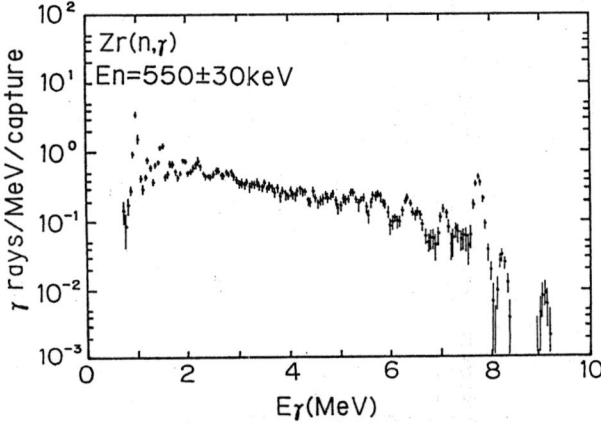

Fig. 8 Unfolded capture gamma-ray spectrum of Zr at the neutron energy of 550 keV.

CASTHY[6] based on the Hauser-Feshbach statistical model. The calculation includes electric dipole (E1), magnetic dipole (M1), and electric quadrupole (E2) transition. The conventional Brink-Axel gamma-ray strength function was used for the E1 transition, and the parameters of the giant E1 resonance were taken from the compilation of Berman[7]. For the M1 and E2 transitions, the single-particle transitions of Weisskopf's model were assumed. The nuclear level density distribution was obtained from the composite formula proposed by Gilbert and Cameron[8]. The level-density parameters were determined from a cumulative plot of the number of excited levels and from the condition that the constant-temperature-model formula smoothly connects to the Fermi-gas-model formula at an energy of Ex. Those derived parameters for the residual nuclei are shown in table 1. In the calculation, the capture gamma-ray spectra of natural silicon and chromium were obtained by summing up the weighted spectrum of each isotope in those elements, which was calculated using the natural abundance, the average radiative width for s-wave resonances, and the s-wave average level spacing[5].

The calculated spectra of Al, Si, Ca, and Cr are shown by the histograms in figs. 3 - 6. In general the calculations reproduce fairly well the observed spectra at neutron energies of several hundreds keV, although some peaks are not reproduced. The comparison between observed and calculated spectra were also performed at lower neutron energies, but the reproduction is unsatisfactory. Probably, it results from the fact that only several resonances contribute to each spectrum at lower neutron energies.

REFERENCES

1. M. Igashira, H. Kitazawa and N. Yamamuro: Nucl. Instr. and Meth. A245, 432 (1986).
2. M. Shimizu, M. Igashira, K. Terazu and H. Kitazawa: Nucl. Phys. A452, 205 (1986).
3. M. Igashira, H. Kitazawa, M. Shimizu, H. Komano and N. Yamamuro: Nucl. Phys. A457, 301 (1986).
4. H. Kendrick and S.M. Sperling: GA-9882 (1970).
5. S.F. Mughabghab, M. Divadeenum and N. E. Holden: Neutron Cross Sections, vol. 1, part A (Academic Press, New York, 1981).
6. S. Igarasi: J. Nucl. Sci. Technol. 12, 67(1975).
7. B.L. Berman: Atom. Data and Nucl. Data Table, 15, 319 (1975).
8. A. Gilbert and A.G.W. Cameron: Can. J. Phys. 43, 1446 (1965).

Table 1. Nuclear level-density parameters used for the statistical model calculation

Nuclei	a (MeV^{-1})	Ex (MeV)	T (MeV)	$\sigma^2(0)$	Δ(MeV)
28Al	3.94	9.05	2.024	5.06	0.0
29Si	3.95	11.84	2.024	6.68	2.75
30Si	3.96	12.49	2.024	4.46	3.35
31Si	3.97	11.43	2.024	3.15	2.25
41Ca	5.63	8.35	1.477	7.59	1.15
51Cr	8.34	9.30	1.099	8.43	1.53
53Cr	8.36	8.45	1.099	7.80	1.88
54Cr	8.40	9.35	1.099	3.91	2.73

The quantities a are the level-density parameter in the Fermi-gas-model formula, T are the nuclear temperature in the constant-temperature-model formula, Ex are the energy connecting both formulas, $\sigma^2(0)$ are the spin-cutoff factor in the vicinity of the ground state, and Δ are the pairing energy.

MEASUREMENTS OF ENERGY DEPENDENCE OF AVERAGE NUMBER OF PROMPT NEUTRONS FROM NEUTRON-INDUCED FISSION OF ^{235}U, ^{241}Am AND ^{243}Am FROM 0.5 TO 12 MeV

Yu. A. Khokhlov, I.A. Ivanin, Yu.I. Vinogradov, V.I. In'kov, L.D. Danilin,
V.I. Panin, V.N. Polynov
All-Union Scientific Research Institute of Experimental Physics

607200, Arzamas-16, Nizhnij Novgorod region, USSR

Abstract: The energy dependence measurements results of average number of prompt neutrons from neutron-induced fission of ^{235}U, ^{241}Am, ^{243}Am and total gamma-rays energy for ^{235}U and ^{243}Am from 0.5 to 12 MeV are presented. The measurements were carried out with neutron beam from uranium target of electron linac of All-Union Scientific Research Institute of Experimental Physics using time-of-flight technique on 28.5m flight-path. The neutrons and gamma-rays from fission were detected by a big liquid scintillator detector (BLSD) loaded with gadolinium, events of fission - by parallel plate avalanche detector for fission fragments. Measurement of energy dependence of $\bar{\nu}_P$, $E\gamma_{tot}$ and determination of BLSD efficiency (relative to $\bar{\nu}_P$ and $E\gamma_{tot}$ for ^{252}Cf) were carried out simultaneously. Least squares fitting results give $\bar{\nu}_P = 3.05 + 0.14 \circ En$ for ^{241}Am and $\bar{\nu}_P = 3.20 + 0.15 \circ En$ for ^{243}Am.

(Electron linac, fast neutrons, time-of-flight method, fission, prompt neutrons, parallel plate avalanche detector, liquid scintillator detector, ^{235}U, ^{241}Am, ^{243}Am)

Introduction

At the present time vast information about the energy dependence of prompt neutrons ($\bar{\nu}_P(En)$) for principal fissile isotopes is accumulated. These data for heavier isotopes, interesting for the problem of incineration of actinide nuclei, which pile up in industrial reactors, are small or absent.

This work is dedicated to investigation of $\bar{\nu}_P(En)$ for ^{241}Am and ^{243}Am nuclei at neutron energy range from 0.7 to 12 MeV.

Measurements for ^{235}U were carried out simultaneously to rise reliability of results because $\bar{\nu}_P(En)$ dependence shape was measured rather in detail [1].

In addition, in one of the experiments total gamma-ray energy of fission was measured to study the possible correlations between the $\bar{\nu}_P(En)$ and $E\gamma_{tot}(En)$.

Experimental procedure

Geometry of experiment is presented in fig.1.
Measurements were carried out with an accelerator electron beam having the following parameters:

    average electron energy - 50 MeV;
    average electron current - 220 μA;
    pulsed frequency   - 2400 Hz;
    pulse width   - 12 ns.

Collimated neutron beam from uranium target of linac fell on a combination of parallel plate avalanche detectors (PPAD) with samples of fissile isotopes placed in the centre of a big liquid scintillator detector (BLSD) of 400

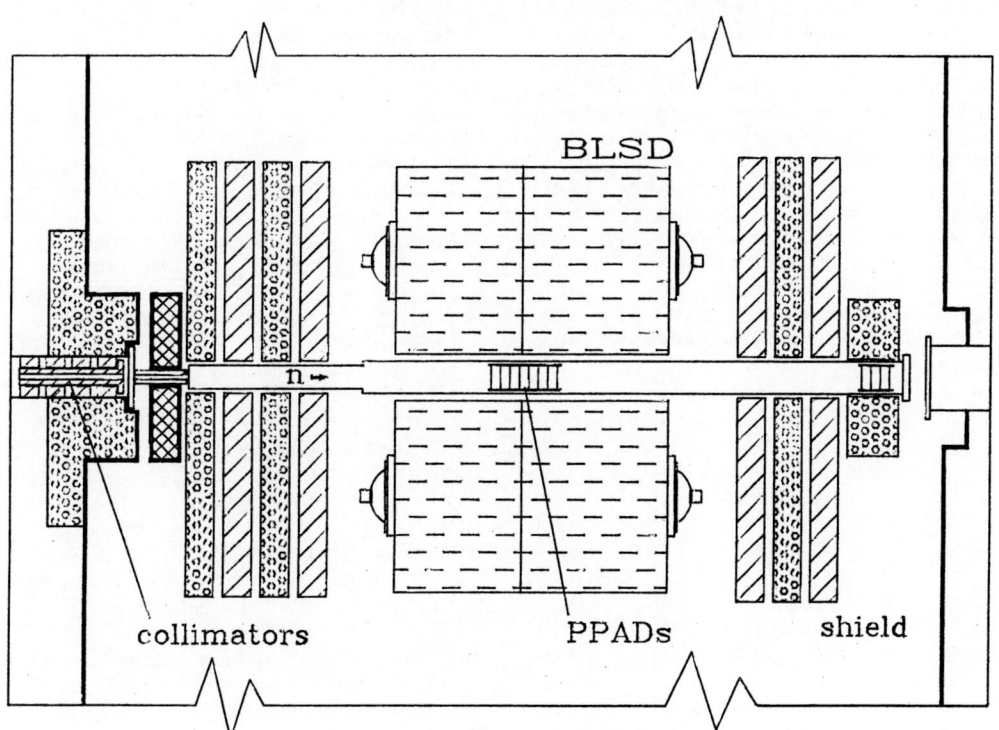

BLSD

n →

collimators   PPADs   shield

Fig. 1. Geometry of experiment

litres volume.

BLSD was filled with liquid scintillator on C_7H_8 base, containing gadolinium with 0.9 % weight concentration. Average life-time of neutron in detector was 10 μs. BLSD was installed behind shielding collimator 1.5 meters long, being a cylindrical system with alternating layers of iron shot with boron carbide and paraffin with boron carbide, which shielded the detector against gamma-ray and neutrons emitted from the linac target. Moreover, collimator formed neutron beam 20 mm in diameter.

To lower background of scattering neutrons on PPAD materials special steps were undertaken: fissile deposits were layered on 50-mm-diam and 2-mg/cm^2-thick silver foils (diameter of deposits was 20 mm), working gas (pentan) pressure in this experiment was 2 Torr. Vacuum frame of PPAD was a stainless steel tube four meters long, going from centre of formed collimator to other side of the shield. There were vacuum windows of PPAD frame shielded outside the liquid scintillator.

Selection of induced fission neutrons was done by a time-of-flight method. Start signal of time-of-flight measurement system was a fast scintillator detector pulse from a gamma-flash of the linac target, stop signal – a pulse from PPAD. Flight-path in this experiment was 28.5 m.

In a PPAD frame four detectors in centre of BLSD (three – with measuring isotopes and one – with ^{252}Cf) and two – behind rear shield (^{235}U and ^{252}Cf) were installed. Weights of samples in this measurement were : ^{235}U – 3.82 mg, ^{241}Am – 0.819 mg, ^{243}Am – 2.59 mg. Every isotope was layered on two foils. Quantity of impurity fissile isotopes in samples didn't exceed 0.1%. Clearance between PPAD electrodes was 3.2 mm.

Installed behind the BLSD, PPAD with ^{235}U layer was used for continuous measurement of correlated with initial neutron flow background in neutron registration channel. Act of fission in these PPADs allowed detection of pulses from BLSD. Additional shield absorbed neutrons, which were emitted during fission in the latter.

Pulse from PPAD opened a 30 μs time-gate for pulses counting from fission neutrons detected by BLSD with 0.8 μs delay to exclude the prompt fission γ-quanta and recoil protons from registration. Gamma-rays were detected in 50 ns-time-window triggered by a corresponding PPAD pulse. Registration of acts was forbidden if pulses from any PPAD coincided during 40 μs-interval.

BLSD neutron efficiency (67 % at 2.5-MeV gamma-ray threshold) was determined continuously during the experiment.

To provide such operating regime the period between linac pulses (416.6 μs) was divided in two time intervals: 11 μs after a control linac pulse for time-of-flight measurement and after 130 μs, 200 μs-"calibration window". During the first 11 μs four (three – for investigated isotopes and one – for background) two-dimensional spectra neutron multiplicity-time 4x(16x1024) and four charge-time spectra 4x(32x1024) (gamma-rays energy per fission) were registered. During the calibration interval two spectra of neutron multiplicity for determination of BLSD neutron detection efficiency (standard and background) were formed.

Moreover, pulse-height distribution from all PPADs were detected for efficiency control.

Measurement results' processing

Procedure of measurement results' $\bar{\nu}_p$ processing is stated in detail in a number of works, for example in ref. [2]. List of corrections, accounted during the measurement results processing is given in table 1.

Table 1. List of accounted correction

correction	value %	error %
false fission	+0.2 – +0.7	0.1
neutron spectrum	–	0.1
displacement of samples	–	0.03
dead time	-0.2 – +0.4	0.04
background time dependence	–	0.03
fragments anisotropy	–	0.2
delay gamma-rays	–	0.06
layers thickness	–	0.06

Measurement errors due to statistic fluctuations and equipment instability were calculated from the dispersion of separate measurement series.

Errors made during correction determination (0.3%) were summarized with the previous ones. The results presented have a total measurement error.

Detection bias when total gamma-rays energy being determined was 1.5 MeV. Errors were calculated from fluctuation of separate parts of experiment.

Results and discussion

Results of energy dependence measurements of $\bar{\nu}_p$ for uranium-235, americium-241 and -243 are presented in fig. 2. Data for ^{235}U obtained in present work agree within experiment errors with the last results of Gwin et al. [2]. Experimental data for ^{241}Am in energy interval 0.7-10 MeV are approximated by linear dependence $\bar{\nu}_p(En) = (3.047 \mp 0.023) + (0.139 \mp 0.007) \circ En$. Other measurements of $\bar{\nu}_p$ energy dependence are not published. Results of $\bar{\nu}_p(En)$ evaluation for this isotope, obtained in work [3] $\bar{\nu}_p(En) = 3.219 + 0.15 \circ En$ are higher than our data, because they are based on a value measured in a thermal point [4], and a tendency was estimated from the neutron binding energy. Results of experiment for ^{243}Am are approximated by $\bar{\nu}_p(En) = (3.199 \mp 0.012) + (0.154 \mp 0.05) \circ En$, that agreed well with evaluation $\bar{\nu}_p(En) = 3.2 + 0.16 \circ En$) [3], which was obtained from $\bar{\nu}_p$ systematic. However, measured $\bar{\nu}_p$ energy dependence has vivid deviation from a linear one in 6-7 MeV energy interval, that is stimulated, apparently, by a channel effect (n,n'f) [5]. The only data for ^{243}Am obtained by a French group in 6-15 MeV energy region are evaluated by dependence $\bar{\nu}_p(En) = 3.28 + 0.139 \circ En$ [6].

In fig. 3 the energy dependence path for a relation of total gamma-ray fission energy for ^{235}U, ^{243}Am to $E\gamma_{tot}$ for ^{252}Cf(sf) is given. Authors regard these results as preliminary, because they were obtained at the stage of experimental method perfection. Nevertheless measurement results reveal same path of $E\gamma_{tot}(En)$ dependence for ^{235}U, as in work [7],

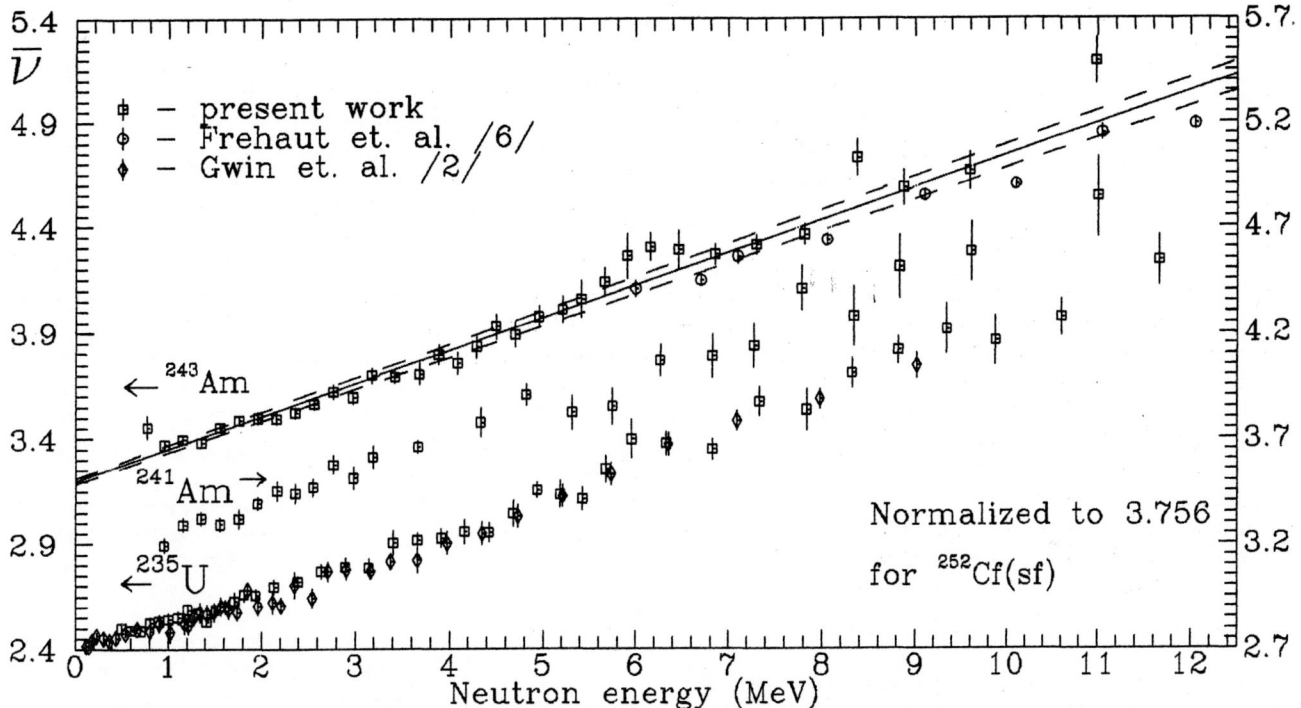

Fig. 2. Results of measurement of $\bar{\nu}_p(En)$ for ^{235}U, ^{241}Am and ^{243}Am.

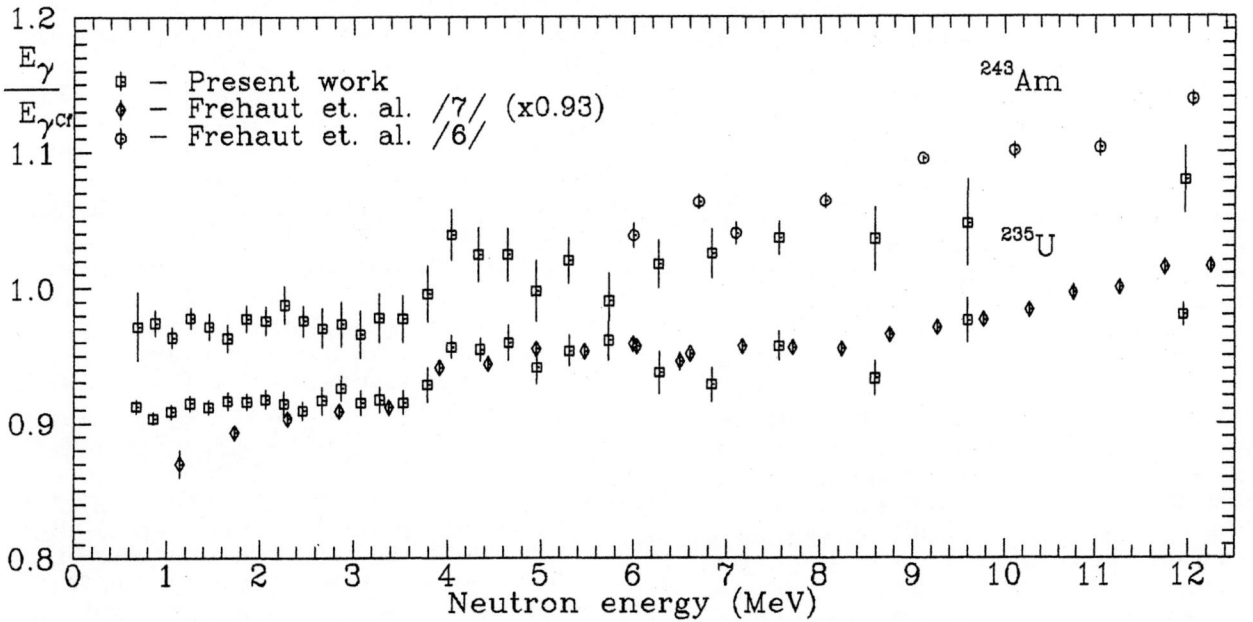

Fig.3. Energy dependence of $E_{\gamma tot}(En)$ for ^{235}U and ^{243}Am relative to ^{252}Cf $E_{\gamma tot}(sf)$.

but the absolute value of $E_{\gamma tot}$ in our work is less to 7-8 %. Absolute value $E_{\gamma tot}$ for ^{243}Am tallied not bad with data of the French group in the range $En > 6$ MeV [6]. For all at present time measured isotopes (^{232}Th, ^{235}U, ^{237}Np [7]) has a common regularity: $E_{\gamma tot}$ increase spasmodically near 4 MeV induced fission neutrons. For physical interpretation of this phenomena additional data of total kinetic energy dependence of fragments on excitation energy of fissile nuclei are needed.

REFERENCES

1. V.V. Malinovskiy, V.G. Vorobyova and B.D. Kusminov: Vopr. At. Nauki i Tekhn., Seriya Yadernye konstanty, 5(54), 19, (1983)
2. R. Gwin, R.R. Spenser and R.W. Ingle: Nucl. Sci. Eng. 94, 365 (1986)
3. Y. Kikuchi, Evaluation of neutron data for ^{241}Am and ^{243}Am, JAERI-M 82-096 and NEANDC(J)-86/U (1982), Japan Atomic Energy Research Institute.
4. A.H. Jaffey and J.L. Lerner: Nucl. Phys. A145, 1(1970)
5. Yu.A. Vasil'ev, Yu.S. Zamyatnin, Yu.I. Il'in, E.I. Sirotinin, P.V. Toropov and E.F. Fomushkin: Zh. eksp. teor. Fiz. 38, 671 (1960); Soviet Phis JETP 11, 483 (1960)
6. J.Frehaut, A.Bertin and R.Bois, in Compter rendu d'activite du Service de Phique et Nucleaire pour l'annee 1983. CEA-N-2396, Centre d'Etudes de Bruyeres-le-Chatel, 92 - Montroge (France), p.69 (1984)
7. J.Frehaut, A.Bertin and R.Bois, in Proc. Int. Conf. on Nuclear Data for science and technology, 6-10 September 1982, Antwerp, Belgium (Edited by K.H.Bockhoff), p.78 (1983)

Radiative Neutron-Capture in Some Low Lying Resonances of ^{52}Cr

L.C. Pratt
Queen Mary and Westfield College
Mile End Road
London
E1 4NS, UK

S. Croft
AEA Technology
Harwell Laboratory
Oxfordshire
OX11 0RA, UK

Abstract: Resonant radiative neutron-capture in the fast reactor structural materials such as Fe, Ni and Cr is of considerable technological importance. Nuclei in this mass region are also known to exhibit non-statistical single-particle capture effects. These are recognised through, 1) anomalously large partial radiation widths compared with the predictions of the statistical model, 2) significant correlations between the reduced γ-ray widths and the reduced (d,p) stripping widths (spectroscopic factors) leading to the same final state, and 3) significant correlations between reduced neutron widths and partial radiative widths. Explanation of these correlations and numerical calculations of the radiative widths have generally been made within the framework of the channel capture formalism of Lane and Lynn, and Lynn's valence model specialisation of this formalism.

In this work the absolute primary γ-ray spectra from some low-lying neutron resonances in ^{52}Cr have been measured using the Harwell electron linear accelerator HELIOS as a pulsed neutron source. A comparison of the absolute partial widths with the predictions of the valence model has been made.

INTRODUCTION

The elements in the mass region A=45 − 65 are major constituents of the steels used in fast breeder reactors. Accurate radiative capture cross-sections are required for these materials in order to calculate sufficiently well the neutronic characteristics of the core under temperature excursions and to predict neutron activation inventories. Knowledge of the capture γ-ray spectra are also needed so that suitable shielding may be designed and γ-ray heating distributions calculated. Improved capture cross-section data are required in the resolved resonance region so that better calculations of the unresolved and higher energy region can be carried out. An important part of these calculations is the need to understand the capture reaction mechanism.

The aim of this work was to investigate the neutron capture mechanism in ^{52}Cr. For closed shell nuclei such as ^{52}Cr, the neutron capture process is expected to be dominated by non-statistical effects [1,2]. Direct single particle E1 transitions from the $3s_{1/2}$ shell just above the neutron binding energy to low lying $2p_{1/2}$ and $2p_{3/2}$ states with large spectroscopic factors are expected to take place without exciting the target nucleus, this is in stark contrast to the expectations of the statistical and semi-direct (or multi-step) models. The simple direct valence capture process outlined above is embodied in the so called valence model of Lane and Lynn [3-5] which describes the transition probability of an orbital neutron in the field of a spectator core nucleus in terms of the final state spectroscopic factor, the reduced neutron width of the initial state and the radial integral of the neutron wavefunctions taken over all space. The s-wave resonances in ^{52}Cr at 50.19 keV and 96.23 keV are predicted to have large valence widths and should thus provide a promising testing ground for the valence model. The best test of the model is of course to compare the measured and calculated partial radiative widths on an absolute scale rather than to merely test for correlations or to compare total radiative widths. No absolute partial radiative widths have previously been made for ^{52}Cr.

EXPERIMENTAL METHOD

The neutron capture γ-ray measurements were made using a pair of HPGe detectors and the time-of-flight (TOF) technique. The experimental arrangement is shown schematically in Fig. 1. The Fast Neutron Target of the Harwell electron linear accelerator, HELIOS, was used as a pulsed source of neutrons. The accelerator was operated at a pulse repetition rate of 1 kHz and an electron-pulse width of 8 − 10 ns. Data was collected for a period equivalent to approximately 10 days of beam time at a beam power of 3 kW. The ^{52}Cr sample, in the form of highly enriched (99.83%) ^{52}Cr$_2$O$_3$, was contained in a thin Al can and had an n-value of 6.93 x 10^{-3} ^{52}Cr atoms/barn. It was positioned 12.45m from the neutron source in the carefully collimated F2 flight tube [6]. The neutron fluence incident on the sample was monitored by an in-line transmission ^{235}U fission chamber 5.82m from the source. The energy fluence rate incident on the sample in terms of the monitor rate was determined by replacing the sample by an identically canned sample of highly enriched (92.5% by weight) ^{10}B-powder with an n-value of 6.95 x 10^{-3} ^{10}B atoms/barn.

The capture γ-rays were detected by a pair of 110 cm^3 HPGe spectrometers placed on opposite sides of the thin walled evacuated carbon-fibre flight tube. For each detected event the neutron TOF and γ-ray pulse height (PH) were recorded enabling comprehensive off-line analysis to be carried out. For each detector 4096 amplitude channels and 2048 TOF channels were used. This represents an improvement on the system established and described by Mason [6 − 8] otherwise the two arrangements were the same.

The TOF histogram extended from 0.5 keV to 130 keV, the upper limit being imposed by the recovery time of the detectors following the γ-flash. The overall timing resolution at the highest energies was approximately 18 ns.

DATA ANALYSIS

The data were analysed as described in detail by Mason [6 − 8]. Briefly, the raw data was first corrected for background, losses due to the single shot operation of the detector control electronics and the neutron flux energy shape. On resonance and off resonance TOF windows were set about the resonances of interest. Exclusion windows were used to discard the narrow p-wave resonances that overlapped with the broad s-wave resonances. The relative strengths of the primary γ-ray transitions appearing in the PH spectra corresponding to the selected TOF windows were obtained by dividing the peak areas by the measured detection efficiency. The results from the full energy, single escape and double escape peaks were combined to give the best overall accuracy.

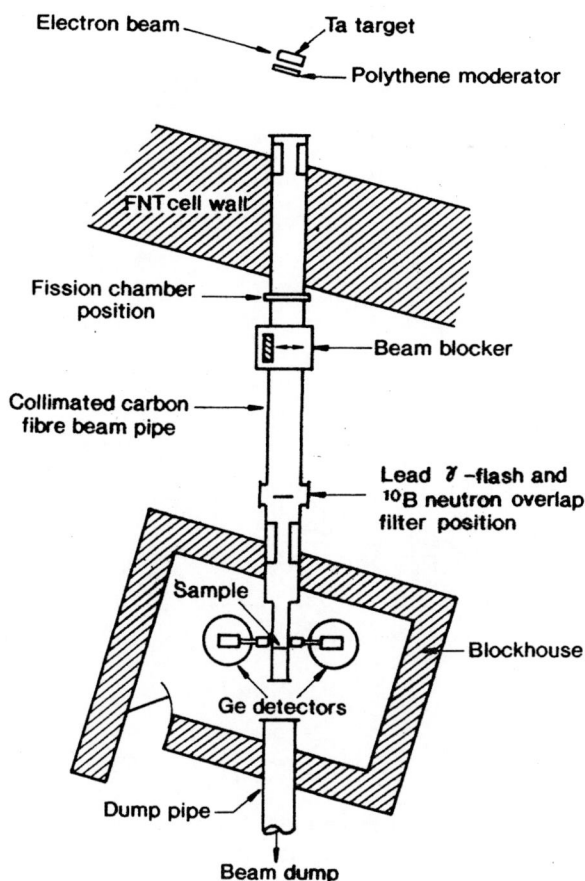

Figure. 1. Schematic representation of the various components of the F2 flight path on the FNT of HELIOS.

of the real, central potential well being varied within limits (~40 – 60 MeV) so that the eigenvalue energy of the solution matches the binding energy of the state. Solutions for the initial (scattering) and final (bound) state wavefunctions were found independently, giving rise to different values for the initial and final potential well depths V_λ and V_μ.

RESULTS

Only one previous measurement has been made of the γ-ray spectrum from a neutron resonance in ^{52}Cr. This was a relative measurement for the 1.626 keV p-wave. From Table. 1. it can be seen that our results compare favourably with it. The experimental partial radiative widths for the 1.626 keV p-wave and the 50.19 and 96.26 keV s-waves are given in Table. 2. The uncertainty values indicated refer only to the peak area counting statistics. To these must be added uncertainties of about 10% shape fitting analysis; 5% relative detector efficiency curve; 3% attenuation and geometry corrections and 7% run to run normallisation uncertainties.

Energy of final	Relative Intensity	
state,E_k/keV	This Work	Kopecky et al [14]
0	100.0±7.4	100.0±5.2
564	35.8±3.3	46.5±2.9
1006	48.4±4.7	54.1±2.9
1974	54.9±3.6	60.5±2.9
2320	8.1±2.8	22.7±2.3
2670	6.7±3.2	20.3±4.1
2708	17.4±2.3	37.8±4.7
3180	38.4±3.3	45.3±2.9
3262	76.1±4.6	82.6±3.5
3616	5.0±4.6	18.0±4.7
4187	9.3±2.3	37.2±6.4
4294	26.6±5.7	18.0±3.5

Table. 1. Relative intensities of the primary γ-rays from the 1.626 keV ^{52}Cr(n,γ)^{53}Cr resonance, normalised to 100 for the ground state transition.

Absolute values of the partial radiative widths were obtained by comparing, for each of the resonances analysed, the yield curve of the strong ground state transitions to the yield of the 478 keV γ-ray from the standard ^{10}B sample using the R-matrix shape fitting code REFIT [9]. In addition to the natural resonance shape, REFIT also makes allowance for, neutron self shielding, neutron multiple scattering and neutron resolution function distortions.

VALENCE MODEL CALCULATIONS

The absolute radiative partial width predicted by the valence model, $\Gamma^V_{\gamma\lambda\mu}$, was calculated from the product, $\Gamma^V_{\gamma\lambda\mu} = S_{dp}.\Gamma_n.Q^V_{\lambda\mu}$ where: S_{dp} is the (d,p) spectroscopic factor for the final state, μ; Γ_n is the neutron width of the initial state, λ, and $Q^V_{\lambda\mu}$ is a transition matrix overlap integral. For the evaluation of the $Q^V_{\lambda\mu}$ we have used a version of Eric Lynn's code VALNCP extended to include M1 transitions as described by the renormallised M1 operator. The basic assumptions and formalism embodied in VALNCP have been described extensively elsewhere [10 – 13].

The potential well used comprised a Saxon-Woods real potential with both spin-orbit and imaginary surface terms (nuclear radius = 4.930 fm; diffuseness = 0.62 fm; imaginary surface potential = –15 MeV; imaginary surface spreading width = 0.7 fm; spin orbit coupling parameter 0.436 x 10^{-2}; potential cut off radius = 8.650 fm). R-matrix solutions were found for the wavefunctions. This involves considering the wavefunction in two regions, internal and external to the nucleus, and matching them at the boundary. The depth

Energy of final state, E_k/(keV)	1.626 keV Resonance $\Gamma_{\gamma\lambda\mu}$ /eV	50.19 keV Resonance $\Gamma_{\gamma\lambda\mu}$ /eV	96.26 keV Resonance $\Gamma_{\gamma\lambda\mu}$ /eV
0	0.1141 ±0.0084	0.496 ±0.055	2.17 ±0.16
564	0.0409 ±0.0038	0.127 ±0.037	0.18 ±0.06
1006	0.0552 ±0.0054	0.043 ±0.047	0.31 ±0.19
1974	0.0627 ±0.0041		
2320	0.0092 ±0.0032	0.171 ±0.043	0.44 ±0.14
2670	0.0076 ±0.0037	<0.025	<0.15
2708	0.0199 ±0.0026	<0.025	<0.13
3180	0.0438 ±0.0038		
3262	0.0869 ±0.0053		
3616	0.0057 ±0.0053	0.092 ±0.023	0.49 ±0.13
4187	0.0106 ±0.0026		
4294	0.0304 ±0.0065		
Total	0.487 ±0.017	0.929 ±0.095	3.59 ±0.32

Table. 2. Experimental partial radiative widths.

Energy of final state, E_k keV	Neutron binding energy keV	lJ^π_μ	V_μ MeV	S_{dp}	50.19 keV $s_{1/2}+$ Resonance $\Gamma_n = 1704 \pm 18$ eV $V_\lambda = -45.0$ MeV		96.26 keV $s_{1/2}+$ Resonance $\Gamma_n = 6410 \pm 66$ eV $V_\lambda = -44.4$ MeV		1.626 keV $p_{3/2}-$ Resonance $\Gamma_n = 0.03135 \pm 0.00090$ eV $V_\lambda = -65.3$ MeV	
					Polarity	Width (eV)	Polarity	Width (eV)	Polarity	Width (eV)
0	7939	$p_{3/2}-$	-46.3	0.670	E1	5.478×10^{-1}	E1	1.279×10^{0}	E2	1.267×10^{-5}
									M1	5.314×10^{-6}
564	7376	$p_{1/2}-$	-49.5	0.430	E1	2.501×10^{-1}	E1	1.279×10^{0}	E2	1.267×10^{-6}
									M1	1.826×10^{-6}
1006	6933	$f_{5/2}-$	-48.9	0.346	E3	1.139×10^{-8}	E3	2.839×10^{-8}	E2	7.077×10^{-7}
									M1	2.735×10^{-7}
1289	6650	$f_{7/2}-$	-38.4	0.0625	E3	3.238×10^{-9}	E3	1.051×10^{-8}	E2	2.336×10^{-7}
1536	6403	$f_{7/2}-$	-38.0	0.0231	E3	9.475×10^{-10}	E3	3.178×10^{-9}	E2	7.474×10^{-8}
2320	5619	$p_{3/2}-$	-42.6	0.3205	E1	1.479×10^{-1}	E1	3.538×10^{-1}	E2	1.714×10^{-6}
									M1	1.613×10^{-6}
2670	5270	$p_{1/2}-$	-45.9	0.07	E1	2.423×10^{-2}	E1	6.350×10^{-7}	E2	2.813×10^{-7}
									M1	2.173×10^{-7}
2708	5231	$p_{3/2}-$	-45.8	0.01162	E1	4.742×10^{-3}	E1	1.140×10^{-2}	E2	4.765×10^{-8}
									M1	5.213×10^{-8}
3180	4759	$p_{3/2}-$	-41.2	0.0108	E1	4.236×10^{-3}	E1	1.527×10^{-8}	E2	3.108×10^{-8}
									M1	4.127×10^{-8}
3262	4677	$d_{5/2}+$	-55.7	4.83×10^{-3}	E2	6.048×10^{-7}	E2	1.526×10^{-6}	E1	4.073×10^{-5}
3616	4322	$p_{1/2}-$	-44.2	0.77	E1	1.132×10^{-1}	E1	2.759×10^{-1}	E2	7.866×10^{-7}
									M1	1.059×10^{-6}
					Totals	1.092 eV		2.505		7.279×10^{-5}

Table. 3. Partial radiative widths calculated using VALNCP.

The calculated partial radiative widths for the same three resonances are given in Table. 3. The uncertainty contributions in this case are 1 – 3% from Γ_n; ~ 15% from S_{dp} and ~ 10% from $Q^V_{\lambda\mu}$.

SUMMARY

The absolute partial radiative widths of the strong transitions to low-lying states in ^{53}Cr following resonant neutron capture by ^{52}Cr, have been measured. For the 50.19 keV and 96.26 keV resonances the valence capture model appears to reproduce the partial widths reasonably well. However, for the 1.626 keV p-wave the model fails completely. This may be because core-excitation, neglected in the valence model calculations, is very important in this instance.

ACKNOWLEDGEMENTS

The work described in this paper was undertaken as part of the Underlying Research Programme of the UKAEA. L.C. Pratt, wishes to thank the SERC and the UKAEA for financial support. We thank J.P. Mason for the invaluable help and advice throughout this work and are grateful to M.C. Moxon for his guidance in the use of his code REFIT. Our thanks also go to J.E. Lynn for making available his valence model code VALNCP and for many helpful discussions.

REFERENCES

1. B.J. Allen and A.R. de L. Musgrove: Advances in Nuclear Physics 10(1978)129–195.
2. B.J. Allen: Nuclear Data for Science and Technology, K.H. Böckhoff (ed), D. Reidel Publ. Co. (1983)707–718.
3. A.M. Lane and J.E. Lynn: Nucl. Phys. 17(1960)563–585 and 586–608.
4. J.E. Lynn: The theory of neutron resonance reactions, Clarendon Press (Oxford, 1968).
5. A.M. Lane and S.F. Mughabghab: Phys. Rev. C10(1)(1974)412–414.
6. J.P. Mason: UKAEA Report AERE–R–12078(1986).
7. J.P. Mason: Capture Gamma-ray Spectroscopy 1987. K. Abrahams and P. Van Assche (eds), IOP Conf. Ser N°88(1988) IOP(Bristol)pp S622–S624.
8. J.P. Mason: Nucl. Phys. A465(1987)413–428.
9. M.C. Moxon: UKAEA Report AERE–R 11986(1985).
10. S. Raman, R.F. Carlton, J.C. Wells, E.T. Jurney and J.E. Lynn: Phys. Rev. C32(1)(1985)18–69.
11. J.E. Lynn, S. Kahane and S. Raman: Phys. Rev. C35(1)(1987)26–35.
12. S. Kahane, J.E. Lynn and S. Raman: Phys. Rev. C36(2)(1987)533–542.
13. S. Raman, M. Igashira, Y. Donzono, H. Kitazawa, M.Mizumoto and J.E. Lynn: Phys. Rev. C41(2)(1990)458–471.
14. J. Kopecky, R.E. Chrien and H.I. Liou: Nucl. Phys. A334(1980)35–44.

Measurement of the Thermal Neutron Cross Section and Resonance Integral of the Reaction ^{137}Cs(n,γ)^{138}Cs

T. Sekine, Y. Hatsukawa and K. Kobayashi
Department of Radioisotopes,
Japan Atomic Energy Research Institute,
Tokai-mura, Ibaraki-ken 319-11, Japan

H. Harada and H. Watanabe
Nuclear Fuel Technology Development Division,
Power Reactor and Nuclear Fuel Development Corp.,
Tokai-mura, Ibaraki-ken 319-11, Japan

T. Katoh
Department of Nuclear Engineering, Nagoya University,
Furo-cho, Chikusa-ku, Nagoya 464-01, Japan

The reactor-neutron cross section of the reaction ^{137}Cs(n,γ)^{138}Cs has been measured by means of a radiochemical method. First, the thermal neutron cross section, including a contribution of epithermal neutrons, was determined and found to be twice as large as that of the previous work by D.C. Stupegia. Second, a modified Cd-ratio experiment was carried out, using a movable cadimium shield in a reactor. The thermal neutron cross section (for 2200 m·s^{-1} neutrons) was 0.25±0.02 b and the resonance integral 0.36±0.07 b. In the analysis of the experimental results, a possibility of the production of the ^{138}Cs isomer ($T_{1/2}$ =2.9 min) was disregarded. The model calculation for the isomeric yield ratio in s-wave neutron capture of ^{137}Cs showed that the cross-section values obtained have a probable error of +1.2% or +3.9%.

(neutron capture, cesium 137, cesium 138, cross sections, isomeric yield ratio, thermal neutron, resonance integral, nuclear transmutation, chemical separation, gamma-ray spectra)

Introduction

Data of neutron capture cross sections are fundamental for the investigation of nuclear transmutation of radioactive fission products, since the neutron is considered to be available in high intensity in fission reactors and in expected accelerator-based neutron sources. Experimental data of this kind, however, are scarce in literature.

The nuclide ^{137}Cs is one of the most important fission products in nuclear waste management. For the (n,γ) reaction of this nucleus, few reports have been published since Stupegia[1] measured first in 1960 the thermal neutron cross section. The present authors remeasured the cross section of the same reaction by taking advantage of a γ-ray spectrometer with high efficiency and high resolving power as well as a chemical process of purification of cesium to enhance the sensitivity of radioactivity measurement. In addition, the resonance integral of the reaction was determined.

In the present paper, the experimental results are described and a possible source of error in experimental cross sections is discussed. A part of the present work has been published previously.[2]

Thermal Neutron Cross Section

The CsCl containig about 0.4 MBq of ^{137}Cs was irradiated with reactor neutrons for 10 min together with a Co/Al-alloy flux monitor; the thermal neutron flux was typically 4×10^{13} n/cm^2/sec. After irradiation, the cesium was purified using zeolite; most of the ^{24}Na and ^{38}Cl radioactivity was eliminated. The ^{137}Cs sample was measured with a HPGe detector of 90% efficiency relative to a 3" × 3" NaI(Tl) detector and 2.3 keV energy resolution at 1.3 MeV. In a γ-ray spectrum, γ-rays associated with the decay of ^{138}Cs ($T_{1/2}$ = 33.4 min) were observed at 1010, 1436 and 2218 keV together with the 662-keV γ-ray of ^{137}Cs. The reaction cross section was determined from the intensity ratio between γ-rays of ^{137}Cs and ^{138}Cs.

On the basis of the cross-section value of 37.18 b for the ^{59}Co(n,γ)^{60}Co reaction, the thermal neutron cross section of the ^{137}Cs(n,γ)^{138}Cs was found to be 0.250 ± 0.013 b.

The value obtained is twice as large as that of the previous work by Stupegia.[1] Since both the cross-section values include a contribution of epithermal neutrons, one possibility for this discrepancy is a difference in neutron spectrum, if the resonance integral of the reaction is very large; the neutron spectrum in our experiment, which is characterized with an epithermal index in the Westcott convention[3] of 0.02, had a fraction of epithermal neutrons ten times more than that in Stupegia's.

2200-m/sec Neutron Cross section and Resonance Integral

A Cd-ratio experiment was carried out for the ^{137}Cs (n,γ)^{138}Cs reaction. The Cd shield used was a cylinder of 0.6-mm thick cadmium with a diameter of 7 cm and a length of 30 cm. Neither end of the shield was covered with cadmium and a target was placed at the center of the shield; thermal neutrons coming from the ends of the shield were considered in the analysis of experimental results. The ^{137}Cs target identical to the previous experiment was irradiated together with flux-monitor wires of Co/Al and Au/Al alloys and molybdenum, which differ from each other in sensitivity to epithermal neutrons. Irradiation was carried out with and without the Cd shield during a period of 10 or 2 min. The fluxes of thermal and epithermal neutrons were determined from the radioactivities produced in the flux monitors for each irradiation. It was found that the Cd shield reduced the thermal neutron flux to one-thirteenth without changing the epithermal neutron flux. Figure 1 shows γ-ray spectra of the ^{137}Cs irradiated with and without the Cd shield and purified as before. One can notice much smaller peaks of ^{138}Cs γ-rays in the spectrum of the ^{137}Cs irradiated with the Cd shield.

From the neutron fluxes and the reaction rates of ^{137}Cs(n,γ)^{138}Cs in irradiations with and without a Cd

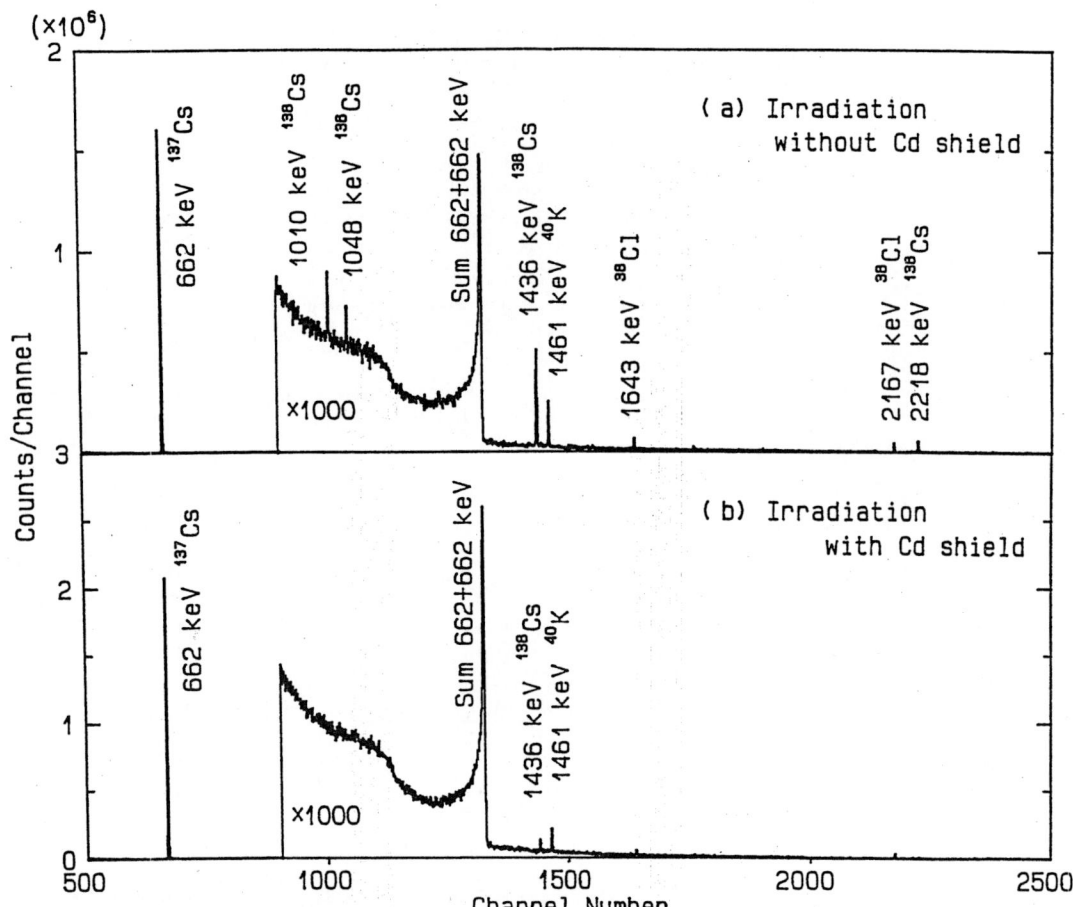

Fig.1 γ-ray spectra of ^{137}Cs samples irradiated with and without a Cd shield, and purifed chemically. Measurements were started 30 min after irradiation, and their counting periods were 30 min.

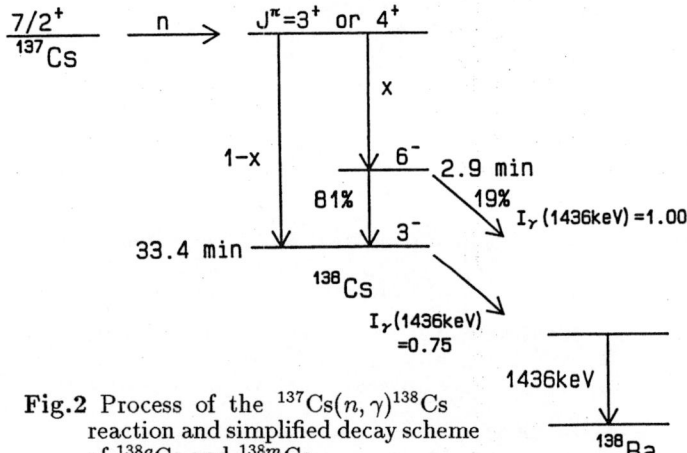

Fig.2 Process of the ^{137}Cs$(n, \gamma)^{138}$Cs reaction and simplified decay scheme of ^{138g}Cs and ^{138m}Cs.

Table 1 Comparison of resonance integrals for the ^{137}Cs$(n, \gamma)^{138}$Cs reaction

	Resonance integral (b)	Reference
Present work	0.36 ± 0.07	
ENDF/B-V	0.499	[4]
JENDL-3	0.680	[5][6]

shield, it was found that the thermal neutron cross section (for 2200 m·s^{-1} neutrons) is 0.25±0.02 b and the resonance integral for neutrons with energies over 0.5 eV 0.36±0.07 b. This result proved that the discrepancy mentioned in Section 1 was not caused by a difference in neutron spectrum.

Discussion

As indicated in Fig.2, there exists an isomer of ^{138}Cs with a half-life of 2.9 min. In the present work, however, it was impossible to measure the production cross section of the isomer. In the foregoing analyses, a possibility of the production of the isomer was disregarded. Even if the ground-state ^{138}Cs was produced only through the internal transformation of the isomer, the error of the above cross-section values is estimated to be +12%. The model calculation, presented by Huizega and Vandenbosch,[3] for the isomeric yield ratio in s-wave neutron capture showed that the cross-section values obtained have a probable error of +1.2% or +3.9% for a compound spin of 3 or 4.

In Table 1, the experimental resonance integral is compared with recent evaluations. Although the value of ENDF/B-V [Ref. 4] is close to that of the present work, the value of JENDL-3 [Ref. 5,6] is substantially larger.

REFERENCES

1. D.C. Stupegia: J. Nucl. Energy, A12, 16 (1960)
2. H. Harada, H. Watanabe, T. Sekine, Y. Hatsukawa, K. Kobayashi, T. Katoh: J. Nucl. Sci. Tech., 27, 577 (1990)
3. J.R. Huizenga and R.V. Vandenbosch: Phys. Rev., 120, 1305 (1960)
4. T.R. England, W.B. Wilson, R.E. Schenter and F.M. Mann: NP-2345 (1984)
5. M. Kawai, T. Watanabe, T. Nakagawa, S. Iijima, H. Matsunobu, T. Nishigori, M. Sasaki, T. Sugi, and A. Zukeran: Proc. Int. Conf. on Nucl. Data for Science and Technology, May 30, 1988, Mito, Japan, p.569
6. T. Nakagawa: private communication

NEUTRON SCATTERING IN ^{239}Pu FROM 0.2 TO 1.0 MeV

J.J. Egan, G.H.R. Kegel, G. Yue, A. Mittler, P.A. Staples, D.J. DeSimone
and M.L. Woodring

Department of Physics and Applied Physics,
University of Lowell
Lowell, Massachusetts 01854 USA

Abstract: Neutron elastic and inelastic scattering in ^{239}Pu has been studied via the time-of-flight technique. Neutrons were produced via the ^{7}Li(p,n)^{7}Be reaction at the University of Lowell 5.5 MV Van de Graaff Accelerator Laboratory. An angular distribution was measured at 570 keV for two level groups, ground plus first excited state and second plus third excited state. The scattering sample was a circular disk of 3.5-cm diameter and 0.2-cm thickness containing 28.7 g of plutonium. Extensive modifications were made to our accelerator-target-sample-detector configuration in order to accommodate the small scatterer. Elastic scattering from a vanadium scatterer provided line shapes for yield extraction from the time-of-flight spectra. Results are compared to ENDF/B-VI.

(Neutrons, differential cross sections, angular distributions, time-of-flight spectra)

Introduction

In this paper we report on the initial phases of our neutron scattering measurements using small size scatterers. Often it is difficult to obtain scatterers of sufficient mass with which to make neutron scattering cross section measurements using conventional experimental geometries. In earlier work on the actinides [1-5] we used ^{238}U and ^{232}Th scatterers whose masses were in excess of 100g. In this work a metallic plutonium scatterer with a mass of only 28.7g was used. We have commenced a series of angular distribution and excitation function measurements in the 0.2 to 1 MeV range concentrating on elastic scattering and inelastic scattering for states below 400keV in excitation. The low-lying levels of ^{239}Pu are illustrated in Fig.1.

To date inelastic scattering measurements on ^{239}Pu have been sparse, with the 1981 work of Haouat et al. [6] being the most significant in that they obtained angular distributions for the first four levels at 700keV incident neutron energy and angular distributions of three two level clusters at 3.4 MeV.

In this work the pulsed beam time-of-flight (TOF) technique was used in conjunction with Mobley magnetic bunching to obtain an

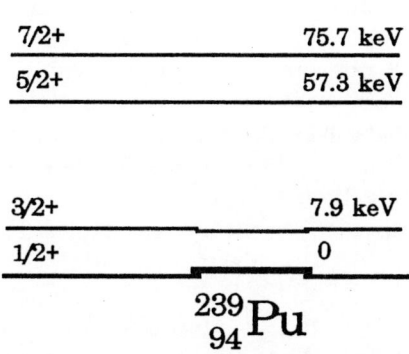

7/2+	75.7 keV
5/2+	57.3 keV
3/2+	7.9 keV
1/2+	0

$^{239}_{94}$Pu

Fig. 1. The low lying levels of ^{239}Pu.

angular distribution at 570 keV for two level clusters (1/2+, 3/2+) and (5/2+, 7/2+) in ^{239}Pu. The neutron energy was chosen because it is near a peak in the ^{7}Li(p,n)^{7}Be cross section giving a maximum neutron yield for these exploratory measurements.

Experimental Method

Neutrons were generated via the ^{7}Li(p,n)^{7}Be reaction at the University of Lowell 5.5 MV Van de Graaff accelerator laboratory. Thin lithium metal targets were produced in situ by a resistive heating evaporator incorporated into the target assembly. This target assembly along with the entire portion of the beam line beyond the Mobley bunching magnet has been completely redesigned since the work reported at the Mito Conference [5]. The evaporator deposits lithium onto a 0.25-mm thick tantalum planchet which is held in place by an indium vacuum seal on the end of the beam line. The tantalum target backing is cooled by a stream of water directed against it. The new beam line was designed to improve the vacuum by minimizing virtual leaks in the feed-throughs for the evaporator, beam-pulse pickups, beam-collimator current leads and other incorporated hardware, and by eliminating rubber O-rings in the vicinity of the target. We routinely achieve pressures of 10^{-7} torr in the assembly despite the fact that it is opened to atmosphere every other day to replace the tantalum target blanks in preparation for the evaporation of a new target. No significant target deterioration is apparent after 48 hours of running with proton beams of 8-10 μA .

The scatterer was a disk of metallic plutonium with a diameter of 3.46 cm and thickness of 2mm containing 28.7g of ^{239}Pu. It was clad in aluminum with a mass of 1.33g. Background measurements using an aluminum scatterer of appropriate mass were made after each plutonium run. The scatterer was placed 6.0 cm from the lithium target, as opposed to 10 cm in our previous measurements [1-5], in order to compensate for the lower yields of scattered neutrons from the small sample.

Scattered neutrons were detected with a 1-cm thick BC418 plastic scintillator mounted on an RCA 8850 photomultiplier tube with a tube base constructed according to the design of Kerns [7]. The detector was housed in a massive shield consisting of tungsten, lead, and lithium-carbonate loaded paraffin. Flight paths varying from 87cm to 116cm were employed. Direct target neutrons were shielded from the detector by a shadow bar consisting of a combination of borated polyethylene, copper and lead.

A second time-of-flight detector consisting of a 0.56-cm thick Pilot U plastic scintillator mounted on an RCA C31024 photomultiplier was used for normalization of scattered neutron runs to incident fluence runs made at zero degrees with the scatterer removed.

Data Analysis

Fig. 2. shows a time-of-flight spectrum at 125° for 570-keV neutrons. The aluminum background and a constant background due to radioactivity in the plutonium sample have been subtracted out. For the present analysis

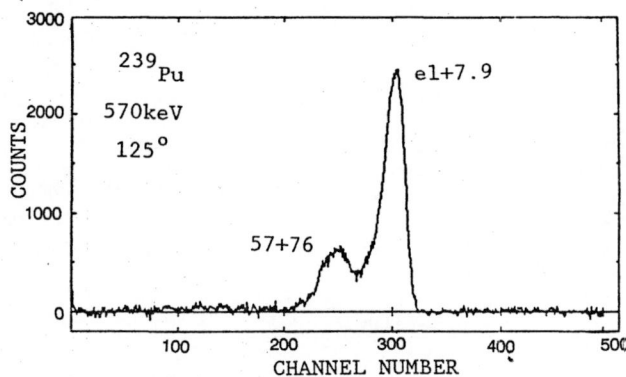

Fig. 2 Background subtracted time-of-flight spectrum for ^{239}Pu at 125° and E_n=570keV.

we have unfolded the spectrum into two components, ground plus 7.9-keV state and 57- plus 76-keV state. In order to extract the yields we used the peak shape from elastic scattering in vanadium obtained using a scatterer whose dimensions were identical to those of the plutonium scatterer. This scatterer contained 11 grams or 0.22 moles of vanadium. A vanadium spectrum for elastic scattering is shown in Fig. 3. For comparison we also show in Fig. 4 a TOF spectrum for ^{238}U at 570 keV for 125°. The uranium scatterer

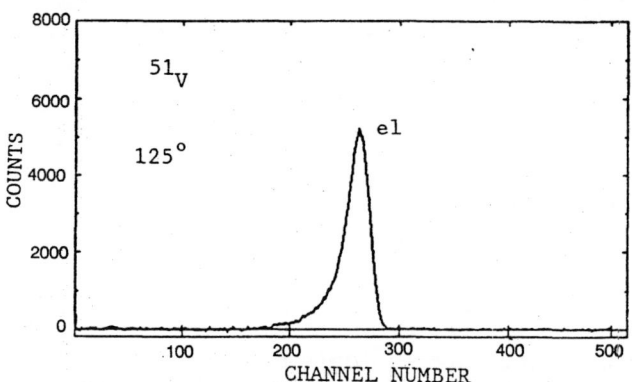

Fig. 3. Time-of-flight spectrum of neutrons elastically scattered from ^{51}V used to obtain the elastic peak shape.

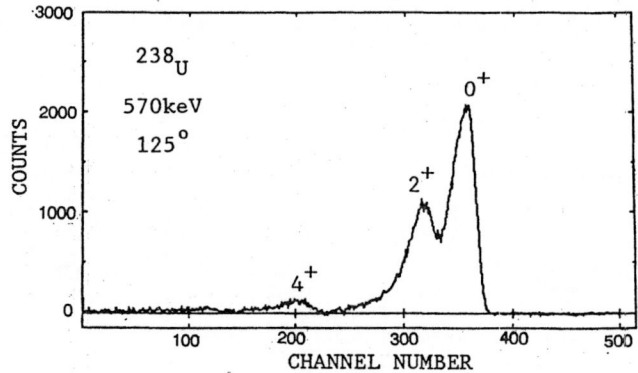

Fig. 4. Background subtracted time-of-flight spectrum of neutrons scattered from ^{238}U. The ground, 45-keV first excited and 148-keV second excited states are labled 0+, 2+, 4+ respectively.

has the same dimensions as the plutonium sample and contains 38.39g of ^{238}U.

The efficiency of the neutron detector was determined by the method described by Kegel et al. [8] at this conference. The data have been corrected for finite scatterer size effects using the code IMBUI [9].

Figures 5 and 6 show the angular distributions for the 1/2$^+$, 3/2$^+$ ground and first excited state combination and the 5/2$^+$, 7/2$^+$ second and third excited state combination respectively, compared to the ENDF/B-VI [10] evaluation. The elastic plus 7.9-keV angular distributions compare favorably with ENDF/B-VI. The combined inelastic results are generally lower than the evaluation.

Conclusion

The data shown here are the results of exploratory measurements with which we are

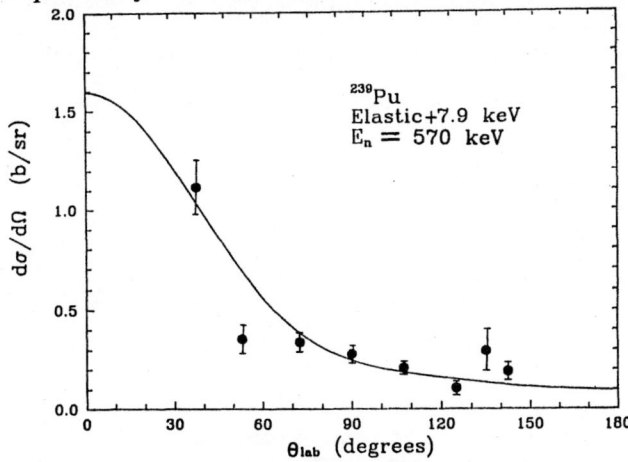

Fig. 5. Angular distribution at 570-keV for the ^{239}Pu ground-first excited state combination. The solid line represents ENDF/B-VI.

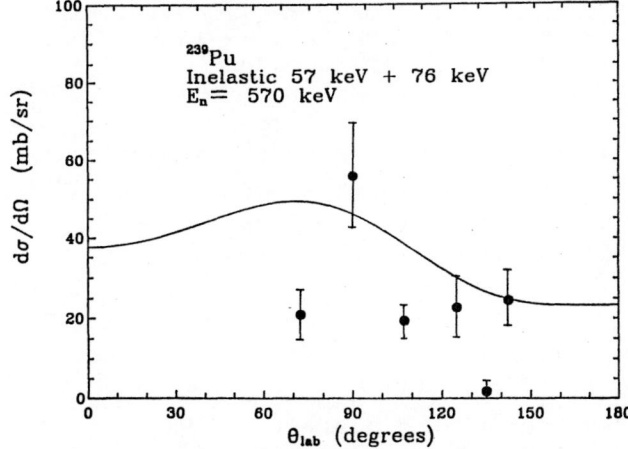

Fig. 6. Angular distribution at 570-keV for the ^{239}Pu second-third excited state combination. The solid line represents ENDF/B-VI.

assessing techniques for making neutron scattering measurements on small samples with reasonable run durations. Typical acquisition times for these measurements were about 10 hours. Further data analysis is in progress and we are measuring a 125° excitation function over the 0.2 to 1.0-MeV range.

Acknowledgments

This work was supported in part by the United States Department of Energy. The authors would like to thank C. Jen, M. O'Connor, C.A. Horton, W.L. Chang, C. Narayan, D.A. Shaw, D. Rondeau and J. Dumont for their assistance in hardware preparation, accelerator operation, and data acquisition.

References

1. L.E. Beghian, G.H.R. Kegel, T.V. Marcella, B.K. Barnes, G.P. Couchell, J.J. Egan, A. Mittler, D.J. Pullen and W.A. Schier, Nucl. Sci. Eng. 69, 191(1979).
2. G.C. Goswami, J.J. Egan, G.H.R. Kegel, A. Mittler and E. Sheldon, Nucl. Sci. Eng., 100, 48(1988).
3. C.A. Ciarcia, G.P. Couchell, J.J. Egan, G.H.R. Kegel, S.Q. Li, A. Mittler, D.J. Pullen, W. Schier and J.Q. Shao, Nucl. Sci. Eng. 91, 428(1985).
4. J.Q. Shao, G.P. Couchell, J.J. Egan, G.H.R. Kegel, S.Q. Li, A. Mittler, D.J. Pullen, W.A. Schier and E. D. Arthur, Nucl. Sci. Eng. 92, 350(1986).
5. J.J. Egan, A. Aliyar, C.A. Horton, G.H.R. Kegel and A. Mittler, Proceedings of the International Conference on Nuclear Data for Science and Technology, Mito, Japan 1988, S. Igarasi ed., Japan Atomic Energy Research Institute, Saikon Publishing Co., Tokyo (1988).
6. G. Haouat, J. Lachkar, Ch. Lagrange, J. Jary, J. Siguad and Y. Patin, Nucl. Sci. Eng. 81, 491(1982).
7. C.R. Kerns, IEEE Trans. on Nucl. Sci., NS-24, 353(1977).
8. G.H.R. Kegel, P.F. Dugan, J.J. Egan, C.K.C. Jen, A. Mittler and P.A. Staples (paper O9, this conference).
9. G.H.R. Kegel, IMBUI, Comput. Phys. Comm. 24, 205(1981).
10. P.G. Young, L.W. Weston and W.P. Poenitz, Evaluated Nuclear Data File B Version VI, ENDF/B-VI, Mat. 9437 National Nuclear Data Center, Brookhaven National Laboratory, Upton, NY, (1989).

AVERAGE FAST NEUTRON RADIATIVE CAPTURE CROSS-SECTIONS FOR FISSION PRODUCTS AND FOR ISOTOPES OF RARE EARTH ELEMENTS

M.V.Bokhovko, V.N.Kononov, E.D.Poletaev,
N.S.Rabotnov, V.M.Timokhov

Institute of Physics and Power Engineering,
Obninsk, USSR

Abstract: Average neutron capture cross sections from 5 to 400 keV for isotopes ^{103}Rh, ^{133}Cs, ^{147,152}Sm, ^{151,153}Eu, ^{nat}Eu and from 10 to 400 kev for isotopes ^{160,161,162,163,164}Dy, ^{nat}Dy, ^{170,171,172,173,176}Yb, ^{nat}Yb, ^{176,177,178,179,180}Hf, ^{nat}Hf, ^{158,160}Gd, ^{159}Tb, ^{165}Ho, ^{175}Lu, ^{181}Ta, ^{187}Os have been measured by time-of-flight method at the pulsed electrostatic accelerator EG-1 of Institute of Physics and Power Engineering. The experimental data were analyzed in the framework of statistical theory of nuclear reactions and new results were obtained for neutron and radiative strenght functions for s-, p- and d-neutrons.

The capture cross sections σ_c were measured by time of flight method at a pulsed electrostatic accelerator EG-1 FEI. A source of neutrons was the reaction ^{7}Li(p,n)^{7}Be. Promt γ-rays were detected using a liquid scintillator tank of the volume ∞ 17 l filled by standard toluene scintillator and ∞ 60% trimethylborate for the discrimination of the scattered neutrons. The neutron flux was measured by ^{6}Li glass detector NE-912 ∞ 0.5 mm thick placed before the sample and by detector composed of a ^{10}B plate and two NaI(Tl) cristals positioned bihind the sample.

Experiment was made in several steps. The measurements for the energy range 20-400 keV were performed on the flight path from a source neutrons to sample 2.4 m (fig. 1). Measurments from 2 to 20 eV for the normalization capture cross sections by the saturated resonance method were made on the same flight path. The flight path ∞ 0.72 m was used to measure capture cross sections from ∞ 4 to ∞ 140 keV.

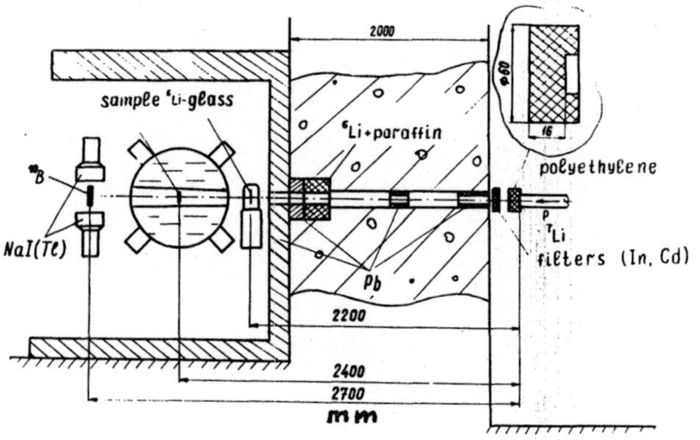

To obtain the values of capture cross sections weigting technique was employed and the capture cross section of ^{197}Au /1/ as well as the saturated resonances of ^{109}Ag(5.19 eV), ^{145}Nd(4.35 eV), ^{181}Ta(4.28 eV), ^{182}W(4.16 eV) and ^{197}Au(4.906 eV) were used for the normalization. The weigt function W(V) was found for our capture detector in the form /2/:
$$W(V) = C\, U^{-0.331}\, V^{1.75}$$
, where C - const., $U = B_n + E_n$, V - pulse hight of capture detector.

The errors of the experimental results were 4.6% at most part of neutron energy range and increase up to ∞ 15% at $E_n \infty$ 5 keV.

Experimental results for the capture cross sections of ^{170,172,174,176}Yb together with the experimental data of autohrs /3-7/ are shown in fig.2 as an example. The results of our measurements of σ_c agree with the corresponding data of other works mostly within the error limits. The results of the measurments of capture cross-sections for ^{103}Rh, ^{133}Cs, ^{147,152}Sm, ^{151,153}Eu, ^{nat}Eu, ^{159}Tb, ^{160,161,162,163,164}Dy, ^{nat}Dy, ^{171,173}Yb, ^{nat}Yb, ^{176,177,178,179,180}Hf, ^{nat}Hf, ^{165}Ho, ^{175}Lu, ^{181}Ta, ^{187}Os and detailed analysis of all experimental data have been publisched elswhere /8,9/.

The resulting experimental data were analyzed in terms of the statistical theory of nuclear reactions to obtain the neutron S_1 and radiative $S_{\gamma 1}$ strength

Fig.1. Block-diagram of the experiment on the flight path 2.4 m.

functions for s-, p- and d-neutrons.
The code EVPAR /10/ was used. The parameters were optimized by the maximum like-
lyhood method. S_o was assumed constant
and the data from ref. /11/ were applied.
The experimental data analysis has
shown the neutron strength functions for
s- and d- neutrons to be close, so we
asumed $S_2 = S_0$. The resulting optimal
parameters are presented in the table.
The values $< \sigma_c >$ averaged over the
Maxwell spectrum at kT = 30 keV are
also tabulated. They are useful in

investigating models of the processes
of nucleosynthesis.
The results of calculation of σ_c with
optimal parameters are shown in fig. 2 by
solid line. They give a fair description
of our experimental data over the whole
energy range. Radiative strength functions for p- and d- neutrons are not equal
to $S_{\gamma 0}$. If they are taken equal adequate
description of the capture cross-sections
holds only up to ∞ 100 keV (dashed line
in fig. 2).

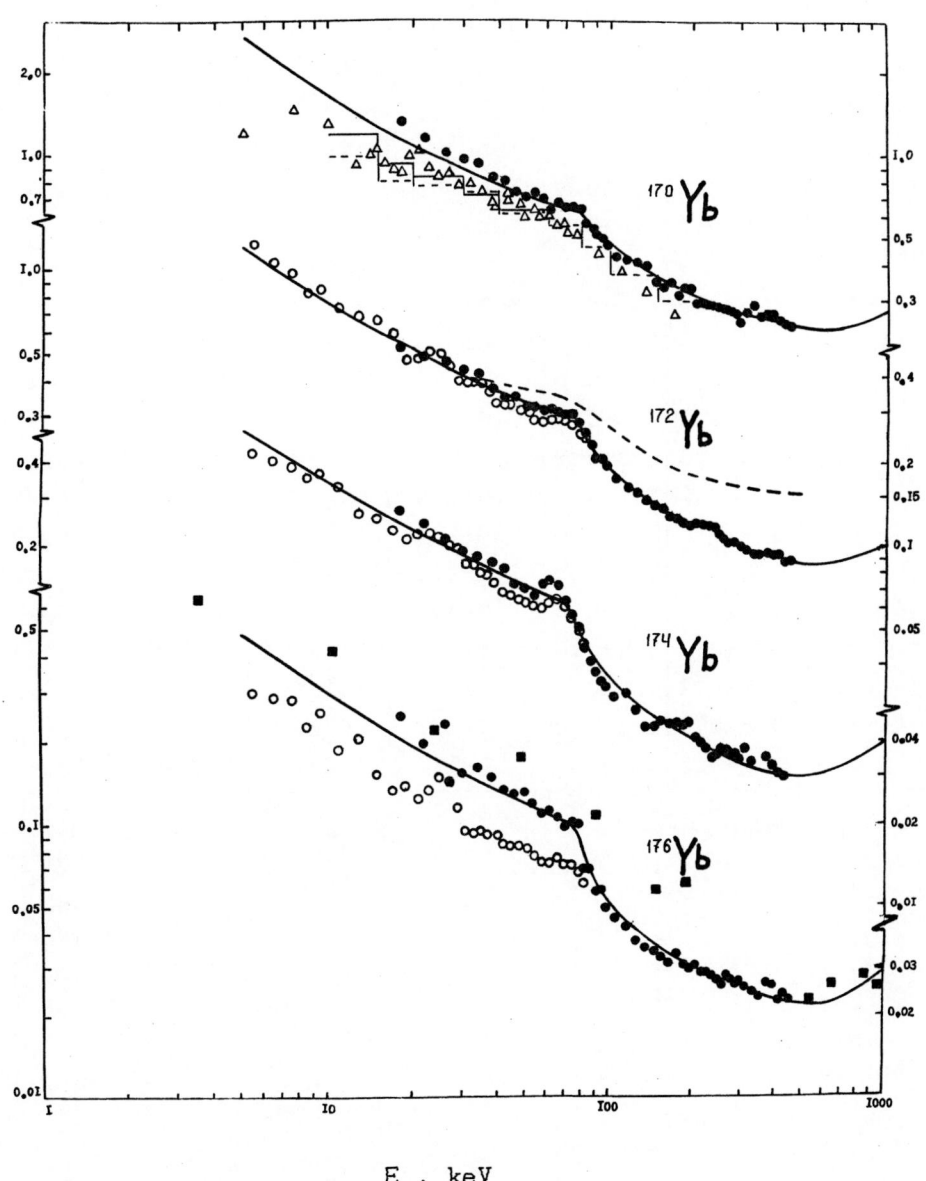

E_n, keV

Fig.2. Neutron-capture cross-sections for ^{170,172,174,176}Yb. ●/9/; ⌐⌐/3/; △ -/4/; ○ -/5/; ⌐ - /6/; ■ -/7/. Dashed line - the results of the calculations of σ_c with $S_{\gamma 0}=S_{\gamma 1}=S_{\gamma 2}$. Solid line-$\sigma_c$ calculated with the optimal set of parameters from Table 1.

Table 1. Results of the analysis of the experimental data
on the fast neutron capture cross sections.

Nuclide	$S_0 \times 10^4$ /11/	$S_1 \times 10^4$	$S_2 \times 10^4$	$S_{\gamma 0} \times 10^4$	$S_{\gamma 1} \times 10^4$	$S_{\gamma 2} \times 10^4$	$<\sigma_c>$, mb кT=30keV
^{103}Rh	0.53	4.3±1.4	0.5±0.1	51±17	51±17	51±17	741±37
^{133}Cs	0.72	1.3±0.4	0.7±0.2	77±20	42±15	47±15	507±28
^{147}Sm	4.8	0.8±0.2	4.8±1.	154±48	16±5	86±29	953±47
^{152}Sm	2.2	0.9±0.3	2.2±0.5	21±8	9±3	8±2	445±25
^{151}Eu	3.7	0.5±0.2	3.7±0.8	1739±678	909±300	626±206	3458±102
^{153}Eu	2.5	1.2±0.4	2.5±0.6	745±268	640±205	503±186	2447±73
^{159}Tb	1.55	0.8±0.2	1.6±0.5	417±83	249±75	78±24	1471±66
^{158}Gd	1.5	1.3±0.4	1.5±0.4	13±2.6	5.5±1.7	5.3±1.6	319±21
^{160}Gd	1.6	0.7±0.2	1.6±0.5	10.8±2.2	1.7±0.5	1.7±0.5	200±13
^{160}Dy	2.0	1.5±0.4	2.2±0.6	39.6±7.9	31.4±9.4	13.1±3.9	806±40
^{161}Dy	1.75	1.3±0.4	1.8±0.5	650±130	422 ±127	322±97	1836±92
^{162}Dy	1.8	1.3±0.4	2.0±0.6	17.3±3.5	10.4±3.1	7.5±2.3	427±21
^{163}Dy	1.9	1.1±0.3	2.1±0.6	165±33	129±39	61±18	1026±51
^{164}Dy	1.7	1.1±0.3	1.6±0.4	7.8±1.6	3.1±0.9	2.8±0.8	209±15
^{165}Ho	1.8	1.0±0.3	1.8±0.5	200±40	154±46	76±23	1134±60
^{170}Yb	2.4	1.1±0.3	2.5±0.8	56.4±11.3	32.7±9.8	16.9±5.1	989±50
^{171}Yb	1.7	1.0±0.3	1.7±0.5	266±53	121±36	69.7±20	1371±69
^{172}Yb	1.7	1.0±0.3	1.7±0.5	18.2±3.6	9.2±2.9	6.0±1.8	420±25
^{173}Yb	1.7	1.0±0.3	1.7±0.5	137±27	72.4±22	72±21.6	868±43
^{174}Yb	1.6	1.0±0.3	1.6±0.5	5.4±1.1	2.8±0.8	2.1±0.6	173±14
^{176}Yb	2.3	1.0±0.3	2.3±0.7	4.6±0.9	1.7±0.5	1.4±0.4	136±11
^{176}Hf	1.4	0.8±0.2	1.4±0.4	18.8±3.8	15.7±4.7	20.2±6.1	449±27
^{177}Hf	2.5	0.9±0.2	2.5±0.8	364±73	147±44	154±46	1663±83
^{178}Hf	2.2	0.8±0.2	2.2±0.6	8.7±1.7	10.9±3.3	7.5±2.3	327±20
^{179}Hf	1.7	0.8±0.2	1.7±0.5	150±30	67.4±20	138±41	858±43
^{180}Hf	1.9	0.8±0.2	1.9±0.6	4.4±0.9	2.8±0.8	3.7±1.1	169±14
^{175}Lu	1.82	0.5±0.2	1.8±0.5	187±37	161±48	108±32	992±50
^{181}Ta	1.7	0.6±0.2	1.7±0.5	150±30	96±29	70.3±21	761±38
^{187}Os	3.1	0.6±0.2	3.1±0.9	159±32	86±26	129±39	922±46

REFERENCES

1. L.E.Kazakov,V.N.Kononov,E.D.Poletaev, M.V. Bokhovko, A.A.Voyevodski, V.M.Timokhov: VANT. Ser. Yad. Const. 2, 44 (1985)
2. V.N.Kononov,E.D.Poletaev,V.M.Timokhov, P.A.Androsenko,G.V.Bolonkina: Preprint FEI -1589,Obninsk (1984).
3. H. Beer, G. Walter, R.L. Macklin, P.J.Patchett: Phys. Rev. C 30, 464 (1984)
4. H. Beer, F. Kappeler, and K. Wisshak and A. Ward Pichard: The Astrophys. Journ. Suppl. Ser. 46, 295 (1981)
5. V.S. Schorin, V.N. Kononov, E.d. Poletaev: Yad. Phys. 20, 1092 (1974)
6. B.J. Allen, D.D. Cohen: EXFOR 30543001
7. D.C. Stupegia, M. Schmidt, C.R. Keedy and A.A. Madson: Jour. of Nucl. Ener. 22, 267, (1968)
8. M.V.Bokhovko,A.A.Voyevodskiy,V.N.Kononov,G.N.Manturov,E.D.Poletaev,V.M.Timokhov:Preprint FEI-2168,Obninsk (1991)
9. M.V.Bokhovko,A.A.Voyevodskiy,V.N.Kononov,E.D.Poletaev,V.M.Timokhov: Preprint FEI-2169, Obninsk (1991)
10. G.N. Manturov, V.P. Lunev, L.V. Gorbacheva: VANT. Ser.: Yad. Const. 1, 50, (1983)
11. S.F.Mughabghab.:Neutron Cross Sections, vol.1,Part B, Acad.Press,New York(1984)

MEASUREMENT OF CAPTURE CROSS SECTIONS FOR ^{238}U, Au AND Sb

Katsuhei Kobayashi, Shuji Yamamoto and Yoshiaki Fujita

Research Reactor Institute, Kyoto University
Kumatori-cho, Sennan-gun, Osaka   590-04   Japan

Abstract: Capture cross sections for the ^{238}U(n,γ) reaction were measured relative to the standard cross section for the ^{10}B(n,$\alpha\gamma$) reaction with Fe and/or Si filtered beam neutrons of 24, 55 and 146 keV, and absolute capture cross sections for Au and Sb were measured between 0.01 and 1.0 eV, using the 46 MeV electron linear accelerator at the Research Reactor Institute, Kyoto University.   Present data at 55 and 146 keV for ^{238}U are in good agreement with the evaluations in ENDF/B-VI and JENDL-3 and with the recent measurements by Kazakov et al.   The results for Au showed a good agreement with the existing measurements and the ENDF/B-V data.   New data have been obtained for the Sb(n,γ) reaction whose data were not always enough in quality and quantity for practical application.

(capture cross sections, measurements, ^{238}U, Au, Sb, linac time-of-flight, Fe and Si filtered beams neutrons, 24, 55 and 146 keV, BGO scintillators, total absorption γ-ray detector, 0.01-1.0 eV)

Introduction

The new evaluated data for the ^{238}U (n,γ) reaction cross section appeared in the recent works[1-4] have been revised to be smaller by more than 5 to 10 % than old evaluated data and/or most of the experimental data in the energy region from 10 to 300 keV.   The revised data are in good agreement with the experimental data which have recently measured by Kazakov et al.[5].   The evaluated data also give a good agreement with the ^{238}U capture-related integral measurements[2]. In such a situation, much interest has been paid to the capture cross section, especially to the energy dependency for the ^{238}U data in the ten to hundreds keV energy region.   Then, the ^{238}U neutron capture cross sections were measured with Fe and/or Si filtered neutrons of 24, 55 and 146 keV using the linac time-of-flight method.

For absolute measurement of the Au (n,γ) and Sb(n,γ) reaction cross sections, determination of the capture detection efficiency and the absolute measurement of neutron flux incident upon the sample were carried out with a total absorption gamma-ray detector composed of BGO scintillators[6].   The capture cross sections were measured between 0.01 and 1.0 eV and compared with the existing data.

Experimental Methods

The capture cross section measure-

ments were made by the time-of-flight (TOF) method using the 46 MeV electron linear accelerator at the Research Reactor Institute, Kyoto University (KURRI). The experimental arrangement is shown in Fig. 1.   For the neutron filtered beam experiments with a pair of C_6D_6 scintillators, thick layers of Fe and/or Si were put at the entrance of the flight tube and in the middle position of the flight path[7].   A BF_3 counter was inserted in the neutron TOF beam to monitor the neutron intensity between the experimental runs.

Bursts of fast neutrons were produced from the water-cooled Ta photoneutron target with an octagonal water tank moderator.   Experimental parameters and conditions for the present measurements are summarized in Tables 1 and 2. Background measurements for the neutron filtered beams experiments were made with

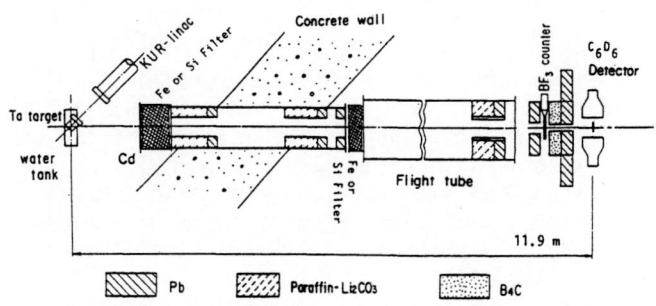

Fig. 1   Experimental arrangement for the capture cross section measurements.

Table 1 Experimental parameters and conditions for the U-238 capture measurement

Parameter	Fe filtered beam	Si filtered beam
Samples	^{10}B: 0.369 g/cm^2	^{10}B: 0.369 g/cm^2
	^{238}U: 1.5mm plate	^{238}U: 1.5mm plate
Filters	15* + 5 cm	30* + 46 + 16 cm
Neutron beam	24 keV	146, 55 keV
Detectors	C_6D_6 scinti.	C_6D_6 scinti.
Linac :		
repetition	250 pps	250 pps
pulse width	0.22 μs	0.22 μs
ave. current	140 μA	140 μA
energy	~30 MeV	~30 MeV
Analyzer:		
width	20 ns/channel	20 ns/channel
No. of channel	1024	1024

* Placed at the entrance of the flight tube.

Table 2 Typical experimental parameters and conditions for the Au and Sb capture measurements

Au sample	0.7165 g/cm^2
Sb sample	1.843 g/cm^2
^{10}B sample	0.996 g/cm^2
Detectors	BGO scinti. composed of 12 bricks
Linac :	
pulse repetition rate	50 Hz
pulse width	30 μs
ave. current	~60 μA
electron energy	~30 MeV
Analyzer:	
channel width	0.25 ~ 4 μs/channel
No. of channel	1024

a 3 mm thick plate of Pb instead of the ^{238}U sample, to investigate the effect of scattered neutrons experimentally. Background level for the Au and Sb measurements was determined by using a 0.5 mm thick Cd filter and notch filters of In at 1.46 eV and Au at 4.92 or Sb at 6.24 eV.

Data Taking and Reduction

Capture signals from the scintillators were fed into a fast amplifier and discriminator. A coincidence circuit was employed between the scintillators to improve the signal-to-noise ratio. A time digitizer was initiated by the KURRI linac burst and the TOF data were stored in the data acquisition system linked to a micro-computer.

The ^{10}B(n,$\alpha\gamma$) reaction was used as a well known reference cross section[8] for the measurement of incident neutron flux. Capture events for ^{10}B and ^{238}U samples with Fe and/or Si filtered beams are counted respectively, as the follow-

ing procedures:
$$C_B(24) = \eta_B(24) \, Y_B(24) \, F(24),$$
$$C_U(24) = \eta_U(24) \, Y_U(24) \, F(24),$$
for 24 keV, and
$$C_B(x) = \eta_B(x) \, Y_B(x) \, F(x) \, ,$$
$$C_U(x) = \eta_U(x) \, Y_U(x) \, F(x) \, ,$$
for x (146 and/or 55) keV, where C, η, Y and F are count, efficiency, yield and neutron flux, and the subscripts B and U indicate ^{10}B and ^{238}U samples, and 24 and x mean the relevant filter energies, respectively. By assuming the relation $\eta_U(24) = \eta_U(x)$, the capture cross section can be obtained as

$$Y_U(x) = \frac{C_B(24)}{C_U(24)} \cdot \frac{C_U(x)}{C_b(x)} \cdot \frac{Y_B(x)}{Y_B(24)} \, Y_U(24).$$

where $Y_U(24)$ is the reference yield of the ^{238}U capture cross section at 24 keV.

In order to make absolute measurements of Au and Sb capture events, the detection efficiencies for ^{10}B, Au and Sb samples were experimentally determined by the black-out method using a very large capture cross section at thermal neutron energy or big resonances[6]. Knowing the detection efficiencies with ^{10}B and Au measurements, the capture yield $Y_{Au}(E)$ can be derived as

$$Y_{Au}(E) = \frac{C_{Au}(E)}{C_B(E)} \cdot \frac{\varepsilon_B}{\varepsilon_{Au}} \, , \, Y_B(E) \, .$$

$Y_{Sb}(E)$ can be obtained as similar procedure as that for Au.

Results and Discussion

Present results for the ^{238}U(n,γ) reaction are given in Fig. 2, whose original drawings were taken from Ref. 3). When we normalized our data to 0.47 b at 24 keV in ENDF/B-VI, the cross sections at 55 and 146 keV are 0.291 and 0.147 b, respectively, with the experimental uncertainties of 6 to 8 %. Recent evaluations and Kazakov data are in reasonably agreement with the present measurements. Old measurements are rather higher than the present data and the new evaluations, especially in the energy region above 100 keV.

The results of Au and Sb capture cross sections between 0.01 and 1.0 eV are given in Figs. 3 and 4, and the experimental uncertainties are 3 to 4 % and 4 to 6 %, respectively. The present measurement for the Au(n,γ) reaction is in very good agreement with the existing measured and the evaluated data. The evaluated curve of JENDL-3 for the Sb(n,γ) reaction shows a general agreement with the present measurement, although the present values seem to show a little deviation, whose tendency is observed in the neutron total cross section of Sb measured by the authors very recently[9].

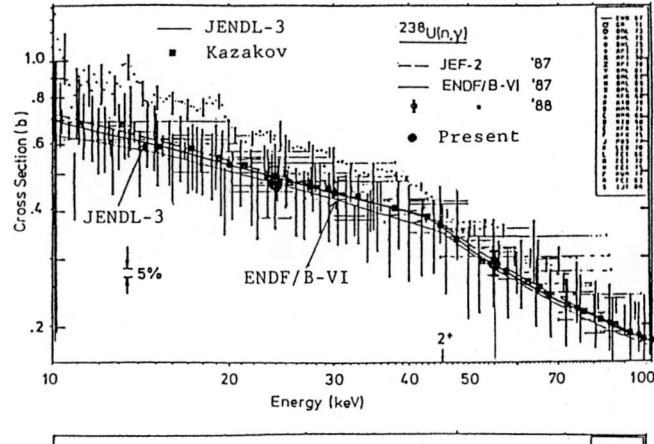

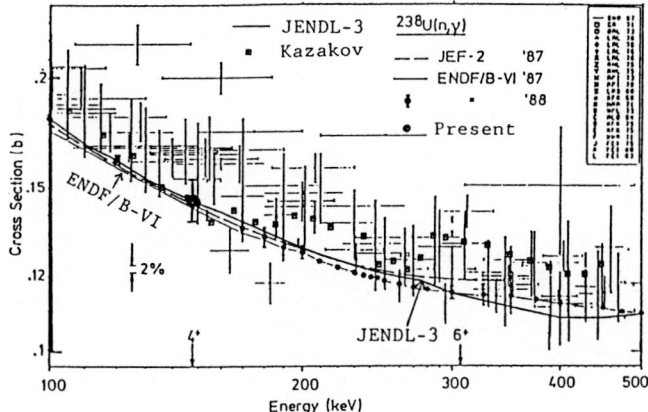

Fig. 2 Comparison of the present results, the previous measurements and recent evaluations for the ^{238}(n,γ) reaction.

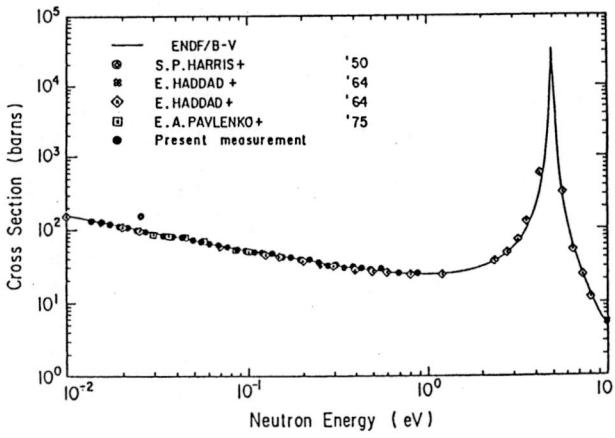

Fig. 3 Present result for the Au(n,γ) reaction.

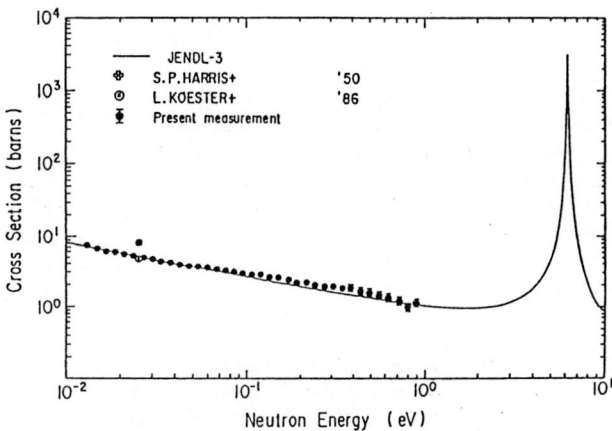

Fig. 4 Present result for the Sb(n,γ) reaction.

REFERENCES

1. F.H. Fröhner: Proc. Int. Reactor Physics Conf., Jackson Hole, p.III-171 (1988).
2. A.Hasegawa: JAERI-M 89-026, p.90 (1989).
3. F.H. Fröhner: Nucl. Sci. Eng., 103, 119 (1989).
4. Japanese Evaluated Nuclear Data Library, Version-3, -JENDL-3-, JAERI 1319 (1990).
5. L.E. Kazakov, et al.: Yad. Const., 3, 37 (1986).
6. K. Kobayashi, et al.: JAERI-M 91-032, p.189 (1991).
7. K. Kobayashi, et al.: JAERI-M 90-025, p.374 (1990).
8. IAEA Technical Reports Series No.227, "Nuclear Data Standards for Nuclear Measurements", IAEA, Vienna (1983).
9. K. Kobayashi, et al.: Annu. Rep. Res. Reactor Inst., Kyoto University, Vol.23, 1 (1990).

CROSS SECTIONS OF THE REACTIONS ^{55}Mn(n,2n)^{54}Mn, ^{58}Ni(n,2n)^{57}Ni AND ^{58}Ni(n,np)^{57}Co AVERAGED OVER THE U-235 FISSION NEUTRON SPECTRUM

O. Horibe[1] and H. Chatani[2]

1 Department of Reactor Engineering, Kinki University, Higashiosaka
577, Japan

2 Research Reactor Institute, Kyoto University, Kumatoricho
Osaka, 590-04, Japan

Abstract: Cross sections of ^{55}Mn(n,2n)^{54}Mn, ^{58}Ni(n,2n)^{57}Ni and ^{58}Ni(n,np)^{57}Co reactions are presented together with the 34 kinds of threshold reaction cross sections which have been measured earlier. The ^{55}Mn(n,2n)^{54}Mn reaction cross section values are also used to give specific variance-covariance matrix of the 35 kinds of cross sections. The ^{58}Ni(n,np)^{57}Co reaction includes the (n,pn) and (n,d) reactions. The obtained experimental values are compared with those of experimental values of other authors, or with the values calculated by the Horibe's formula, or with the integral values calculated with the cross section data in ENDF/B-V and JENDL-2, assuming the fission neutron spectrum shapes to be of Maxwell, or Madland-Nix, or Watt types.

(U-235 fission neutron, average cross section, (n,2n) and (n,np) reaction, ^{55}Mn, ^{58}Ni)

Introduction

Cross sections of ^{55}Mn(n,2n), ^{58}(n,2n) and ^{58}Ni(n,np) reactions were newly obtained by analyzing the gamma spectral data accumulated in our laboratory. This work is a continuation of the work reported in the Proceedings of Int. Conf. on 50 Years with Nuclear Fission [1].

In our measurements, attention was paid to minimize systematic errors. (1), irradiation neutron spectrum is to be that of the fission neutrons, so a U-235 fission plate was used, (2), cascade coincidence summing pulses increase or decrease the aimed photo peak pulses, then we must correct the total gamma ray counts with correction factors which are obtained by probability calculations and (3), it should be kept in mind to lessen the random coincidence summing pulses. Accordingly, counting rate is limited below some amounts; this means, the experiments were repeated many times to achieve the statistical accuracy.

The cross section of the ^{58}Ni(n,2n)^{57}Ni reaction was obtained from the photo peak area measured for 127 keV gamma of ^{57}Ni. Usually, the cross section is obtained from the measurement of 1379 keV gamma rays. We selected 127 keV peak, because the 1379 keV gamma rays are affected by ^{214}Bi natural background.

Experimentals

^{55}Mn(n,2n)^{54}Mn. The cross section was derived by a way similar to the procedure in our earlier measurements [1]. The obtained result is 0.150±0.017 mb. Table 1 shows the value together with our earlier measured values of the 34 kinds

Table 1(a). Cross sections averaged over the U-235 fission neutron spectrum, $\bar{\sigma}$.

No.	Reaction	$\bar{\sigma}$(mb)	$\pm\Delta\bar{\sigma}$(mb)	No.	Reaction	$\bar{\sigma}$(mb)	$\pm\Delta\bar{\sigma}$(mb)
1	^{27}Al(n,α)^{24}Na	0.705	0.044	19	^{59}Co(n,p)^{59}Fe	1.39	0.10
2	^{24}Mg(n,p)^{24}Na	1.38	0.094	20	^{115}In(n,p)^{115}Cd	0.00939	0.00104
3	^{46}Ti(n,p)^{46}Sc	11.6	0.8	21	^{115}In(n,p)^{115m}Cd	0.0486	0.0080
4	^{47}Ti(n,p)^{47}Sc	20.1	1.4	22	^{59}Co(n,α)^{56}Mn	0.170	0.012
5	^{48}Ti(n,p)^{48}Sc	0.305	0.020	23	^{113}In(n,n')^{113m}In	172	11
6	^{54}Fe(n,p)^{54}Mn	81.7	5.2	24	^{115}In(n,n')^{115m}In	198	12
7	^{56}Fe(n,p)^{56}Mn	1.13	0.07	25	^{55}Mn(n,2n)^{54}Mn	0.150	0.017
8	^{58}Ni(n,p)^{58}Co	106	7	26	^{59}Co(n,2n)^{58}Co	0.303	0.021
9	^{60}Ni(n,p)^{60}Co	2.19	0.17	27	^{27}Al(n,p)^{27}Mg	4.09	0.26
10	^{61}Ni(n,p)^{61}Co	1.33	0.10	28	^{42}Ca(n,p)^{42}K	2.92	0.22
11	^{64}Zn(n,p)^{64}Cu	29.4	7.5	29	^{43}Ca(n,p)^{43}K	2.53	0.19
12	^{67}Zn(n,p)^{67}Cu	1.01	0.09	30	^{84}Sr(n,p)^{84}Rb	5.36	0.45
13	^{103}Rh(n,p)^{103}Rh	0.119	0.013	31	^{86}Sr(n,p)^{86}Rb	0.664	0.067
14	^{140}Ce(n,p)^{140}La	0.0045	0.00029	32	^{92}Mo(n,p)^{92m}Nb	6.78	0.41
15	^{50}Ti(n,α)^{47}Ca	0.0210	0.0021	33	^{95}Mo(n,p)^{95}Nb	0.142	0.011
16	^{54}Fe(n,α)^{51}Cr	0.860	0.079	34	^{96}Mo(n,p)^{96}Nb	0.0256	0.0025
17	^{68}Zn(n,α)^{65}Ni	0.0771	0.0237	35	^{45}Sc(n,α)^{42}K	0.254	0.019
18	^{103}Rh(n,n')^{103m}Rh	955	101				

Table 1(b). Covariant matrix for variants in Table 1(a).

CORRELATION MATRIX (PRINTED VALUE HAS BEEN MULTIPLIED BY 100.)

ROW/COL	1	2	3	4	5	6	7	8	9	10	11	12	13	14	15	16	17	18	19	20	21	22	23	24	25	26	27	28	29	30	31	32	33	34	35
1	100																																		
2	96	100																																	
3	99	96	100																																
4	94	91	95	100																															
5	99	96	99	95	100																														
6	99	96	99	95	99	100																													
7	99	96	99	96	99	99	100																												
8	99	96	99	96	100	100	100	100																											
9	88	85	88	84	88	88	88	88	100																										
10	89	87	90	88	90	90	90	91	79	100																									
11	26	25	26	25	26	26	26	26	23	23	100																								
12	70	68	71	69	71	71	71	72	63	66	18	100																							
13	54	52	55	53	55	55	55	55	48	51	15	39	100																						
14	83	80	82	78	83	82	83	74	74	22	63	45	100																						
15	71	69	71	68	71	71	71	71	63	64	19	51	39	60	100																				
16	69	68	70	68	70	70	70	70	64	18	51	39	57	50	100																				
17	21	21	21	20	21	21	21	21	18	19	6	15	12	17	15	14	100																		
18	60	59	60	58	60	60	60	60	53	55	16	43	33	48	42	41	13	100																	
19	87	85	87	84	88	87	88	88	78	79	23	63	48	73	63	62	18	53	100																
20	52	51	53	51	53	53	53	53	47	49	14	34	30	44	38	38	11	32	47	100															
21	38	37	38	37	38	38	38	38	34	35	10	27	21	31	27	27	8	23	34	20	100														
22	94	91	95	91	95	95	95	95	84	86	25	68	52	79	68	67	20	57	84	51	36	100													
23	95	92	95	93	96	96	96	96	85	85	25	70	53	79	68	68	20	58	84	52	37	92	100												
24	97	95	98	96	99	99	99	99	87	91	25	72	56	81	70	70	20	60	87	53	38	95	97	100											
25	63	61	63	60	63	63	63	63	56	57	16	45	35	44	13	44	13	36	52	45	34	84	85	88	100										
26	87	85	88	85	88	88	88	88	78	80	23	63	49	73	63	62	18	53	78	47	34	84	85	88	56	100										
27	95	92	95	91	95	95	95	95	84	87	25	68	53	79	68	67	20	58	84	51	36	91	92	95	60	84	100									
28	81	78	80	74	80	81	81	81	72	72	21	57	44	68	58	56	17	48	71	43	31	77	77	78	96	85	84	100								
29	92	89	92	89	93	93	93	93	82	84	24	67	52	77	66	66	19	56	82	50	35	88	90	92	59	82	89	75	100							
30	75	73	75	72	75	75	75	75	67	68	19	54	42	62	54	53	16	46	67	40	29	72	72	75	47	67	72	61	70	100						
31	60	58	60	58	60	60	60	61	54	55	16	43	33	50	43	43	13	36	53	32	23	58	58	60	38	53	58	49	56	46	100					
32	99	96	99	95	99	99	99	99	88	90	26	71	55	82	70	70	21	60	88	53	38	95	96	99	63	88	95	81	93	75	60	100				
33	77	75	77	75	77	77	77	78	78	69	71	20	56	43	64	55	55	16	47	68	42	30	74	75	77	49	69	74	63	72	59	47	77	100	
34	66	64	66	63	66	66	66	66	59	60	17	47	36	53	43	64	65	42	59	63	54	62	50	40	66	52	100									
35	82	79	81	77	81	81	81	81	72	73	21	58	44	68	59	57	17	49	72	43	31	77	78	80	52	72	78	76	62	50	81	63	54	100	

Table 2. Nuclear constants used.

Reaction	Isotope abundance (%)	Half-life	γ-energy (KeV) & intensity (%)
^{55}Mn(n,2n)^{54}Mn	100	312 d	834.8 (100)
^{58}Ni(n,np)^{57}Co	68.3	271.65 d	122 (85.6)
^{58}Ni(n,2n)^{57}Ni	68.3	35.99 h	127 (12.88)

These constants were quoted from TABLE OF ISOTOPES, 7-th edition.

Table 3. Correction factors and ε_p used.

γ-energy, keV (nuclide)	ε_p (Error,4%) G-1☆	G-2☆	C# G-1	G-2	Γs^*
122 (^{57}Co)	.0894	.0295	1.0	1.0	.971
127 (^{57}Ni)	.0859	.0294	1.196	1.062	.971

#, correction factor for cascade coincidence summing.

*, self-shielding factor.

☆, G-1,G-2; source-detector distance: 8 and 30 mm, respectively.

Correction factor of ε_p due to sample sizes: 0.984.

of reaction cross sections and a covariant matrix correlated with other errors. Table 2 shows nuclear data used in the calculation.

^{58}Ni(n,2n)^{57}Ni. Reaction rates were obtained from photo peak areas of the 127 keV gamma using nuclear data and correction factors shown in Table 3. The cross section is calculated relative to the ^{27}Al(n,α)^{24}Na reaction cross section of 0.705 mb as reference. Table 4 shows results thus obtained.

Table 4. ^{58}Ni(n,2n)^{57}Ni reaction cross sections measured.

Source-detector (dead-time %)	Cross section (μb)	Weighted mean (μb)
30 mm (2.6)	4.92± 0.75	
30 mm (2.5)	5.10± 0.71	4.98± 0.38
8 mm (6.0)	4.90± 0.55	

^{58}Ni(n,np), ^{58}Ni(n,pn) and ^{58}Ni(n,d)^{57}Co total cross section. ^{57}Co nuclei are also produced by β^+ and EC decays of ^{57}Ni nuclei yielded by the ^{58}Ni(n,2n)^{57}Ni reaction. Thus, gammas from ^{57}Co produced from ^{57}Ni interfere with photo peak area of ^{57}Co which is the product of the above reaction, so that to obtain the cross section, the cross section modified for this interference must be subtracted from the cross section calculated from the observed photo peak area. The modified cross section is given by the ^{58}Ni(n,2n)^{57}Ni cross section multiplied by a numerical factor. The modification means to multiply the cross section of ^{58}Ni(n,2n)^{57}Ni reaction by a numerical factor. The numerical factor includes the effects of irradiation time, cooling time, time needed to get sufficient counts, and also of half-lives of ^{57}Co and ^{57}Ni. The values of the factor can be calculated and were about 0.5 to 1.005 (maximum) in our cases. The main term of the cross section is calculated from the photo peak area of 122 keV gammas of ^{57}Co. The modified term is estimated from the reaction cross section of ^{58}Ni(n,2n)^{57}Ni. We take this reaction cross section to be 4.98 μb as mentioned above. Then

Table 5. Cross sections published so far for the ^{55}Mn(n,2n)^{54}Mn and ^{58}Ni(n,2n)^{57}Ni reactions.

^{55}Mn(n,2n)^{54}Mn (mb)		^{58}Ni(n,2n)^{57}Ni (μb)	
0.150± 0.017	(ours)	4.98± 0.39	(Ours)
0.202± 0.010	(Kobayashi	5.77± 0.31	(Zijp)
0.202± 0.018	(Nasyrov)	4.9 ± 1.4	(Calamand)
0.26 ± 0.02	(Steinnes)	4.19± 0.22	(Mannhart)*
0.258± 0.013	(Calamand)	4.06± 0.21	(Sekine)▲
0.244± 0.015	(Fabry)	4 ± 0.9	(Braun)
0.227	(Horibe) #	3.8 ± 0.5	(Sekine)
0.19	(Roy)	3.60± 0.24	(Kobayashi)
0.21	(Francois)	2	(Pearlstein)
0.13	(Pearlstein)	6	(Roy)
		7.67	(Horibe)#

These cross sections without marks were reproduced from ref.[2].

Predicted by Horibe's empirical rule, to be published elsewhere.

* 5th ASTM-Euratom Symp. on Reactor Dosimetry, (1984).

▲ J. inorg. nucl. Chem. 43, 1981. p.1424

the cross section of the ^{58}Ni(n,np)^{57}Co reaction is 0.232±0.005 mb. Tables 2 and 3 show nuclear data, correction factors and ε_p used in the calculations. Γ_s value is calculated in a way similar to that in ref.[3].

Discussion

Table 5 shows the ^{55}Mn(n,2n)^{54}Mn cross sections so far published. Our value is close to 0.13 mb predicted by Pearlstein. We doubted such a low value and then deemed it to be the result of our gamma peak area estimation, since the peak was interfered with the natual background of the complex peaks of ^{228}Ac at 836 keV. Reliability of the measred area depends on that of the measured area of the complex peaks, so we calculated the ratio of the estimated area of the complex peaks to the area estimated for the natural background of 1460 keV gamma peak of ^{40}K in the measured gamma spectrum. We obtained a similar ratio in the measured natural background spectrum. The thus obtained two ratios were almost equal. Then, our area estimation is reliable.

For the ^{58}Ni(n,2n)^{57}Ni reaction, we obtained three cross section values as shown in Table 4. Each result is calculated from three spectra measured for different samples. These agree well with each other, in spite of the fact that the cascade correction factors, C, dispersed very much, as shown in Table 3. The errors cited for the results are fairly large, because counting statistic errors of the peak areas were 7~10%, and the background counts in the measured gamma spectra were as high as 2500~4000 counts/channel, so that slightly different selections of the background affect very much the estimations of peak areas. Table 5 shows cross sections published so far for this reaction. These measured values scatter around 4 μb. The scattering seems to stem from cascade summings due to annihilation gammas of positive betas and 1379 keV cascade gamma in decay of the parent nuclei ^{57}Ni as well as random coincidences due to intense gammas from the ^{58}Co produced by the ^{58}Ni(n,p) reaction, if samples are placed close to the detectors.

Table 6. Differences between ratios of calculated to measured cross sections, C/E and unity

Reaction	Spectrum Type / E(MeV) / Cross section Measured(mb)	JENDL-2 ENDF/B-IV Maxwell 1.985		JENDL-3T Madland-Nix 2.039		ENDF/B-V Watt 2.032	
		ENDF/V (C/E-1)%	JENDL-2 (C/E-1)%	ENDF/V (C/E-1)%	JENDL-2 (C/E-1)%	ENDF/V (C/E-1)%	JENDL-2 (C/E-1)%
^{115}In(n,n)^{115}In^m	198 ± 12	-12.5		-8.33		-9.44	
^{47}Ti(n,p)^{47}Sc	20.2 ± 1.4	+7.28		+11.5		+11.3	
^{58}Ni(n,p)^{58}Co	106 ± 7	-3.21	-5.94	+0.311	-2.45	+0.434	-2.34
^{54}Fe(n,p)^{54}Mn	81.7 ± 5.2	-3.88	-12.0	-0.40	-9.61	-0.16	-8.95
^{46}Ti(n,p)^{46}Sc	11.6 ± 0.8	-3.79		-6.03		-3.45	
^{27}Al(n,p)^{27}Mg	4.09 ± 0.26	+4.25	-7.78	+1.71	-9.80	+4.43	-7.46
^{56}Fe(n,p)^{56}Mn	1.13 ± 0.10	-0.974	-2.83	-12.1	-8.76	-7.35	-3.81
^{59}Co(n,α)^{56}Mn	0.170 ± 0.012	+0.647	+8.00	-14.7	-7.82	-9.24	-2.06
^{27}Al(n,α)^{24}Na	*0.705 ± 0.044	+16.2	+17.9	-3.63	-1.72	+3.08	+4.99
^{48}Ti(n,p)^{48}Sc	0.305 ± 0.350	+4.26		-12.5		-6.69	
^{55}Mn(n,2n)^{54}Mn	0.150 ± 0.017	+118	+122	+25.4	+28.0	+43.2	+46.1
^{58}Ni(n,2n)^{57}Ni	#4.98 ± 0.39	+6.83	+4.42	-48.6	-49.6	-40.6	-41.8
^{59}Co(n,2n)^{58}Co	#0.303 ± 0.021	-1.75	-8.78	-44.4	-48.0	-36.4	-40.6

*, this cross section value was used as the reference of the measured values.
#, unit is in μb.

Our measured cross section of 4.90 μb was obtained from the spectrum measured at 6% dead-time of our PHA, so that counting loss due to random coincidence amounts to about 8% and also counting losses for the others amount to about 4% from the dead times of 2.5 and 2.6% as shown in Table 4. If cascade and random coincidence probabilities are additive, then a resultant correction factor is given by the product of each correction factor. The cross sections corrected on trial with our estimated factors are 5.33 (6.0%), 5.33 (2.5%) and 5.14 μb (2.6%).

The ^{58}Ni(n,np)^{57}Co reaction cross section measured by Bruggeman et al. [3] is 0.216±0.005 mb. This value is somewhat lower than ours. However, our reference cross section used is 0.705 mb for the ^{27}Al(n,α)^{24}Na reaction, while they used 0.64 mb. When their value is renormalized, the two values become closer.

Our measured cross sections were compared with those evaluated by Ohsawa[4] for confirmation of the U-235 fission neutron spectrum. Table 6 shows deviations of the ratios of the calculated to our measured cross sections, denoted with C/E, from unit. Also shown C/E for our earlier measured cross sections. From these results, those calculated by assuming cross section data of ENDF/B-V for Maxwellian spectrum are better than the others and agreement between measured and calculated values are rather good over wide range of the effective threshold energies from 1.2 MeV for the ^{115}In(n,n')^{115m}In to about 13.5 MeV for the ^{58}Ni(n,2n)^{57}Ni [5], except the ^{55}Mn(n,2n) reaction.

ACKNOWLEDGEMENT

The authors are greatly indebted to Professor T. Yanabu (Professor Emeritus at the Institute for Chemical Research of Kyoto University) for critical reading of the manuscript and for encouragement throughout this work.

REFERENCES

1. O. Horibe, H. Chatani, Y. Mizumoto and T. Kusakabe: Proceedings of the International Conf. on 50 Years with Nuclear Fission 2, 927, published by the American Nuclear Society Inc. (1989)
2. K. Kobayashi and I. Kimura: NEANDC(J)-67/U, 42 (1980)
3. A. Bruggeman, W. Maenhaut and J. Hoste: Radiochem.Radioanal.Letters 18, 87 (1974)
4. T. Ohosawa: On the Fission Neutron Spectrum for U-235 in JENDL-3T, JAERI-M-89-026,114 (1989)
5. W. L. Zijp: ECN-70 (1979)

RESONANCE PARAMETERS OF 58Ni+n AND 60Ni+n FROM VERY HIGH RESOLUTION TRANSMISSION MEASUREMENTS.

A. Brusegan, G. Rohr, R. Shelley, C. Van der Vorst, C. Baracca* and Zhou Enchen**
Commission of the European Communities,
Joint Research Centre,
Central Bureau for Nuclear Measurements, B-2440 Geel, Belgium.

F. Poortmans and L. Mewissen
Studiecentrum voor Kernenergie, Mol, Belgium.

G. Vanpraet
Rijksuniversitair Centrum Antwerpen, Belgium.

Abstract: very high resolution neutron transmission measurements of enriched ^{58}Ni and ^{60}Ni samples were performed in the energy ranges 14eV up to 7.5MeV and 40eV up to 30MeV respectively. Time-of-flight and 1ns linac bursts were used to measure neutron energies. Total cross section data from 3 sample thicknesses have been analysed with an R-matrix multi-level code to deduce parameters of 72 resonances in ^{58}Ni and 208 in ^{60}Ni up to 305 and 550keV respectively, and the s-wave strength functions and average level spacings.

(Transmission ^{58}Ni and ^{60}Ni; $E_0, \Gamma_n, J, l, R, S_0, D_0$; energy ranges up to 305 and 550keV resp.)

Introduction

The series of measurements presented here was a continuation of our investigation into the neutron cross sections of structural materials required for reactor design; the total cross section is, moreover, needed in order to correct the capture data for self shielding and multiple scattering. These Ni measurements were one of the first to be performed at the Geel Electron Linear Accelerator (GELINA) after it was substantially upgraded by the addition of a compression magnet to produce a 1ns burst width, enabling very high energy resolution.

Experimental Method

With the linac operating at a frequency of 800Hz and a uranium target producing bursts containing a spectrum of white neutrons, several measurements were performed along collimated flight paths ranging from 50 to 400m. The neutron beam was moderated, except for the 400m case, immediately after the target and 3 different samples (on loan from ORNL) were used as shown in Table 1, which details the measurements performed.
It is of note that for the ^{6}Li detector of run 4 the glass was placed orthogonally to the neutron beam

and to the photomultiplier photocathode plane and was contained in an unshielded aluminium sphere (0.5mm thick). A ^{10}B$_4$C anti-overlap filter was permanently in the beam to absorb low energy neutrons from the previous burst and a 9mm thick lead or a 0.37g/cm^2 uranium slab was added to reduce the gamma flash in the measurements at 100, 200 and 400m.

During the measuring time a BF$_3$ detector was used as reference and the data acquisition system controlled the cycling of the samples in and out of the beam. For the 50 and 100m measurements the amplitude pulses from the PM's were recorded together with the time-of-flight spectra and for each sample position. The maximum gain variation was:
a)1.3% and 0.4% respectively for ^{58}Ni and ^{60}Ni with the boron detector, and,
b)less than 0.8% for the ^{6}Li glass.

The background was determined in the black resonance minima in the energy range 18.4eV up to 102keV. For the proton recoil detector it was measured at long flight times and immediately prior to the arrival of the next burst. Each detected neutron event was analysed in time by a 1ns Multistop Time Coder.

Data Analysis and Results

After correcting the measured data for background and deadtime, the calcutated neutron transmission spectra were analysed with the Reich-Moore multilevel routine MULTI [1]. The total cross sections of ^{58}Ni and ^{60}Ni have been measured by several groups; see for example refs.[2,3,4] and [5,6,7] respectively. In the actual analysis the radiation widths given in refs.[2] and [5] have been used, with the exception of the 2.25keV resonance in ^{60}Ni, for which the resonance parameters result from the simultaneous analyses of the 50 and the 100m data with the code REFIT [8] and the help of M.C.Moxon from Harwell.

The extension of the measurements down to very low energies has enabled the determination of

Table 1: Experimental Details.

Run	Isotope	Sample (at/b)	Purity (%)	F/path (m)	En.Range (keV)
1	^{58}Ni	0.0440	99.93	48.191	0.014 - 600
2	^{58}Ni	0.0440	99.93	198.056	47 - 7500
3	^{60}Ni	0.0746	99.07	48.191	0.04 - 600
4	^{60}Ni	0.0185	99.07	97.494	0.096 - 600
5	^{60}Ni	0.0746	99.07	198.056	47 - 7500
6	^{60}Ni	0.0746	99.07	387.739	180-30000
Detector Systems					
1&3	5mm ^{10}B$_4$C slab + four C$_6$D$_6$ (4"x2") scintillators				
4	1/4" x 6" Li glass (NE912) + 5" EMI 9823KQB PM				
2,5&6	1" NE110 plastic scintillator + 4 RCA 4516 PM's				

* CBNM Fellow   ** CBNM Visiting Scientist

Table 2 : ^{60}Ni Resonance Parameters ($p_{1/2}$ assumed unless quoted)

E_0 (eV)	J	l *	Γ_n (eV) #	E_0 (eV)	J	l	Γ_n (eV)	E_0 (eV)	J	l	Γ_n (eV)
-50000	1/2	0	12800.	233660			16 (2)	389380			52 (5)
-150	1/2	0	860.	236800			11 (3)	393420	(3/2)	1	168 (20)
2255	3/2	1	0.031 (2)	238440			18 (3)	398430			448 (22)
5532			0.045 (5)	248850			91.0 (36)	399330			35 (3)
12510	1/2	0	2337 (60)	252140	1/2	0	595 (30)	403240	3/2	1	200 (16)
13623			0.9 (4)	252500			5 (1)	403460			20 (2)
23789			5.23 (45)	253220	(3/2	1)	130 (10)	408550			25 (3)
23898			1.3 (4)	254500			40 (4)	410380	3/2	2	83 (9)
28699	1/2	0	663 (22)	256260	1/2	0	903 (65)	413860	1/2	0	342 (27)
30263			0.7 (3)	256290			44 (3)	415740			248 (20)
33021			8.8 (6)	257750	1/2	0	1878 (55)	415960	3/2	1	45 (3)
33396			9.4 (7)	260950			49 (4)	419050			30 (3)
39540			2.3 (5)	263290			34 (3)	422500			75 (5)
42725			0.7 (3)	266100			59.6 (48)	423100	1/2	0	1280 (52)
43105	1/2	0	98.3 (30)	269140			58.6 (60)	425090			78 (5)
47630			2.0 (3)	269410			28.4 (30)	429580			467 (28)
51594			0.9 (2)	277150			44.4 (40)	429750	3/2	1	17.5 (15)
56290			0.6 (1)	278340	3/2	1	185 (11)	432580	3/2	1	171 (11)
65225	1/2	0	425 (9)	279100	1/2	0	282 (25)	435810			322 (29)
73206			2 (1)	280260	3/2	1	71 (4)	435930	3/2	1	31 (3)
86170			3 (1)	282800			124 (12)	438170	1/2	0	880 (55)
86844	1/2	0	361 (11)	283200	3/2	1	99 (9)	438900			23 (3)
87940			11.6 (20)	290900			39 (3)	439220			32 (3)
91700			8.8 (18)	292120	1/2	0	151.1 (75)	440000			38 (3)
93814			4.4 (9)	292860	3/2	1	81.3 (57)	441390	3/2	2	58.7 (35)
97100			6.9 (7)	294750			25.6 (23)	444610			92 (7)
98097	1/2	0	947 (19)	295770			186 (11)	445750			105 (15)
99435			10 (5)	297440			180 (10)	448510	1/2	0	2022 (85)
107870	1/2	0	610 (15)	300525			67.1 (40)	449010			40 (4)
108490			3 (2)	302420			18.8 (19)	451560			131 (8)
111550			7 (2)	306650			24.4 (21)	453860			26 (4)
112050			4 (2)	306950			458 (27)	454810			220 (15)
112320			4 (2)	307660	(3/2	1)	83 (10)	455440	3/2	1	300 (25)
113370			6 (3)	311640			25.8 (26)	456470	1/2	0	978 (78)
120610			8 (2)	313810			17.3 (15)	458770			41 (4)
123690			34.8 (25)	317340			22.1 (22)	462950			40 (6)
127240			72.5 (35)	317390	1/2	0	3100 (93)	464180	1/2	0	1565 (63)
133590			26.0 (13)	321830			22 (3)	475840			184 (11)
136100			21.5 (11)	325830	1/2	0	7465 (225)	476130	3/2	1	100 (6)
136360			8.4 (9)	327670			32.9 (17)	476680			48 (5)
137510			6.6 (16)	329400			30 (6)	476840	3/2	1	70 (7)
139110	1/2	0	33.3 (17)	334320			44 (4)	480800			20 (5)
139640			19 (2)	334400	3/2	1	50.2 (41)	482200			15 (5)
140100			37 (2)	334770			136 (12)	486600	1/2	0	3500 (210)
148870			10 (1)	335950			27.4 (3)	488080	(3/2	1)	186 (19)
151480			16.8 (12)	336390			74.7 (60)	491720			148 (12)
154445	3/2	1	87 (6)	336490	3/2	1	40.0 (25)	495800			70 (15)
156460	1/2	0	470 (14)	338840	1/2	0	3550 (106)	499300	3/2	2	362 (18)
160310			22.4 (16)	341720			87.9 (35)	500170			65 (8)
161840	1/2	0	1356 (27)	343310			140 (10)	500230	1/2	0	4800 (390)
166280			6 (3)	343390			90 (7)	504080			190 (19)
167290			12.7 (25)	348480			93.1 (63)	512580			15 (3)
174640			11.8 (12)	349130			61.9 (50)	513370	3/2	2	473 (29)
175030			8.6 (20)	350220			51.8 (36)	514080	3/2	1	180 (16)
183040			77.5 (25)	354580	3/2	1	234.5 (94)	516670	3/2	2	550 (35)
183620			13.1 (12)	355780			16 (2)	518310			120 (12)
186660	1/2	0	5710 (270)	357960	1/2	0	1135 (57)	522890	1/2	0	5580 (225)
192700			64.3 (30)	359050			57 (5)	525370			60 (5)
194300			16.2 (24)	359650	3/2	1	639 (37)	526130			30 (4)
197730	1/2	0	3022 (160)	366170			85 (6)	527780	1/2	0	3580 (290)
200800	1/2	0	23.8 (12)	370850			53 (3)	528700	3/2	2	482 (30)
201630	3/2	1	71.3 (36)	371810			10 (5)	530600			30 (5)
206070			33.4 (20)	373800			25.5 (15)	531830			75 (11)
206640	(1/2)	1	142.7 (70)	377300	1/2	0	3764 (150)	536160			174 (12)
214220	(1/2)	1	129.5 (65)	379750			70 (4)	538500	3/2	2	117 (7)
220170			76.7 (46)	379930	3/2	1	140.5 (85)	539400			94 (7)
220730			30 (2)	381530			50 (5)	539880			20 (3)
221090	3/2	1	44.5 (25)	383520	1/2	0	385 (27)	541870	3/2	2	121 (7)
229360			19.5 (20)	388640	3/2	1	95.5 (57)	(1050000)	(1/2	0)	(45000)
230280	3/2	1	56.5 (28)	389150			73 (7)	(1200000)	(1/2	0)	(25000)

* Neutron widths uncertainties (see text) are quoted in parenthesis, which for spin and/or parity indicate assignment is uncertain.

Table 3 : ^{58}Ni Resonance Parameters ($p_{1/2}$ assumed unless quoted)

E_o (eV)	J	l *	Γ_n (eV) #	E_o (eV)	J	l	Γ_n (eV)	E_o (eV)	J	l	Γ_n (eV)
-68000	1/2	0	15000	108380	1/2	0	1010 (50)	a)184600	3/2	1	101 (5)
-14700	1/2	0	7420	110755	(3/2	1)	5.3 (16)	185990	3/2	1	24.7 (13)
a)13310	(1/2	1)	7.71 (85)	117875	3/2	1	8.0 (13)	191460	(3/2)	2	8 (4)
a)13630	(3/2	1)	1.08 (20)	119800	(5/2)	2	3.5 (17)	191900	1/2	0	2300 (100)
15305	1/2	0	1310 (20)	121400	(3/2)	2	6.5 (20)	196360	3/2	1	6.4 (20)
20016	(1/2	1)	1.71 (16)	123990	1/2	0	424 (30)	198180	(5/2)	2	7.0 (19)
21138	(3/2	1)	1.75 (71)	129960	3/2	1	12.3 (11)	201470	(3/2	1)	2.8 (15)
26638	(3/2	1)	1.72 (78)	137550	1/2	0	2010 (90)	202515	(1/2)	1	14.5 (20)
32387	1/2	1	16.6 (15)	140350	1/2	0	3240 (120)	206600	1/2	0	6920 (120)
34230	(3/2)	1	1.2 (8)	141860			30.9 (14)	a)207245	3/2	1	139 (12)
36125	1/2	0	18 (2)	145230	3/2	1	73.3 (13)	211050	1/2	0	43 (4)
47891	3/2	1	4.4 (4)	a)148780	3/2	1	76.4 (30)	a)216680			186 (9)
60135	3/2	1	13.6 (19)	151350			29.5 (13)	218120	3/2	1	10.9 (30)
61791			19.0 (21)	156980	3/2	1	21.2 (14)	a)231095	(1/2)	1	100.0 (85)
63340	1/2	0	3650 (90)	158050	1/2	0	5200 (180)	232330	(3/2)	2	4 (2)
69915	(1/2)	1	6.8 (11)	159830	(3/2)	2	7.0 (15)	233100	1/2	0	5350 (160)
82870	3/2	1	32.7 (21)	161255	(5/2)	2	9.7 (20)	234150	(3/2)	1	17.4 (42)
83410	1/2	0	12 (3)	166060			40.4 (20)	242770	3/2	1	18.2 (34)
a83835			27 (4)	167090	(3/2)	2	11.0 (26)	243950	5/2	2	6.1 (35)
89995	(3/2)	1	4.2 (21)	168590	3/2	1	6.1 (21)	245550	1/2	0	226 (10)
97615			19.9 (10)	168930	1/2	0	410 (20)	a)249510	(3/2)	1	160.4 (82)
101425	(5/2)	2	2.7 (8)	175200	3/2	1	38.7 (23)	272600	1/2	0	5010 (200)
105450	(3/2)	2	10.1 (11)	180660	(3/2)	2	14.9 (19)	281500	1/2	0	2000 (120)
107770	(3/2)	2	4.5 (20)	181305			20.1 (24)	305450	1/2	0	710 (25)

\# Neutron widths uncertainties (see text) are quoted in parenthesis.   a) Difficulties in fitting the measured shape.

* Parenthesis for spin and/or parity indicate that assignment is uncertain.

the potential scattering length (for s-wave resonances and both isotopes R = 6.0 ± 0.2 fm) with the parameters of the negative energy resonances given in Tables 2 and 3. Spins and parities were assigned by detailed examination of the resonance shape observed with the exceptionally high resolution of the 200 and 400m measurements and, for ^{58}Ni, taken from the results quoted in ref.[2]. The errors for the neutron widths are given in Table 2 and Table 3 thus: 123(11) = 123 ± 11 and 67.5(46) = 67.5 ± 4.6, etc. They include the uncertainties due to the statistics, to the background and to the error quoted for R. In the case of measurements performed in the moderated neutron beam, the width of the (highly asymmetric) resolution is known to within 20% and this uncertainty is also included in the errors.

In Table 3, for ^{58}Ni, the parameters are given only for those resonances which have been observed, i.e. for 18 s-wave (below 305keV) and 52 $l>0$ resonances (for energies below 250keV). When compared with ref.[2] our data show no indication of 5 very small s-wave resonances, 4 of which had uncertain assignment.

For ^{60}Ni, ref.[5] quotes 29 s-wave resonances below 450keV, 3 of which are re-assigned here. The 2.25keV p-wave resonance, similarly to the 1.15keV resonance of ^{56}Fe, has a neutron width much smaller than the capture width, i.e. (30.9 ± 2.0) meV and (533 ± 100) meV respectively, assuming $J^{\pi} = 3/2^{-}$. The capture area of this resonance provides a useful normalization value for the capture measurements. The s-wave strength functions and

the average level spacings are listed in Table 4; no correction for missed levels is applied and the quoted errors are $(2/N)^{0.5}.S_0$ and $(0.273N)^{0.5}.D_0$, respectively, where N is the number of resonances.

Regular level spacings of ^{60}Ni are observed and are discussed in ref.[9].

Table 4 : Average Level Spacings and Strength Functions for s-wave Resonances in ^{58}Ni and ^{60}Ni.

Isotope	D_0 (keV)	$S_0 \times 10^4$
^{58}Ni	17.1 ± 2.1	3.40 ± 1.13
^{60}Ni	16.1 ± 1.5	2.50 ± 0.62

References

1 G.F.Auchampaugh, Report CA-5473-MS (1974)
2 C.M.Perey, F.G.Perey, J.A.Harvey, N.W.Hill, M.M.Larson, R.L.Macklin, Report ORNL/TM-10841, ENDF-347
3 D.B.Syme, P.H.Bowen, Neutron Physics and Nuclear Data for Reactors, p.319, Harwell (1978)
4 F.Fröhner, Neutron Data of Structural Materials for Fast Reactors, p.138, Geel (1977)
5 C.M.Perey, J.A.Harvey, R.R.Macklin, F.G.Perey, R.R. Winters, Phys. Rev. C27-6, 2556 (1983)
6 D.B.Syme, P.H.Bowen, A.D.Gadd, in ref[4], p.703
7 F.Fröhner, in ref[4], p.268
8 M.C.Moxon, AEA-InTec-0470.
9 G.Rohr, contribution to this conference.

HIGH RESOLUTION NEUTRON TRANSMISSION MEASUREMENTS FOR U AND Sc THICK COMPOSITE FILTERS

A. Brusegan, C. Baracca* and Ch. Van der Vorst
Commission of the European Communities, Joint Research Centre
Central Bureau for Nuclear Measurements, B-2440 Geel, Belgium.

W.G. Alberts, M. Matzke
Physikalische- Technische Bundesanstalt, Braunschweig, Germany

Abstract: The spectral distribution of the neutron current behind U, U+W, U+Ge+Mn, U+Ge+Mn+W, Sc and Sc+Mn filter combinations has been investigated with high resolution transmission measurements in the energy range from 15 eV up to 500 keV. The time of flight technique and the 50 m total cross section beam facility of GELINA, running with 14 ns burst width, were used. The transmission, the energy positions, the widths and the areas of the observed "windows" have been calculated below 102 keV.

(Sc, U filters, neutron transmission, range 18.4 keV - 102 keV)

Introduction

Due to the resonance-potential interference term, the toal cross section for large s-wave resonances drops to near zero at energies below the peak. From these minima strong quasi-monochromatic neutron beams are obtained when a thick sample is put in the beam of a reactor and the transmitted flux is observed.

For applications to dosimetry, health physics, neutron physics and radio-biology, the development of specific filters [1], i.e. for different neutron energies, may meet the stringent request for quasi-monochromatic beams having high signal to background ratio, low γ-ray contribution and variable intensities (with the filter thickness).

The presence of more than one s-wave resonance produces other minima in the total cross section, which appear as unwanted contaminations of the beam: the addition of selected isotopes, removing the high order contaminations with little attenuation at the main line , has lead to the development of composite filters like U+Ge+Mn or Fe+Al+S [2]. A way to correct for the contaminations is provided by the addition of a 'difference' filter, which essentially screens the main transmitted line and weakly affects the rest of the neutron spectrum .

By subtracting two measurements, done without and with the 'difference' filter, an improved energy definition may be attained. The actual measurements, performed with the time of flight technique, provide the accurate transmission either of the composite and/or of the 'difference' filters.

Experimental method and analysis

The Sc filter consisted of a 33.5 cm thick metal sample and its "difference" filter was a 8.1 g/cm² powdered Mn specimen, canned in a thin (2 x 0.1) mm Al box. The 2.3 keV s-wave resonance of Mn attenuates the 2 keV line transmitted by the Sc filter.

The U filter was composed of 16.4 cm of metal uranium (with 0.3% ^{235}U), 0.3 cm Ge (it attenuates the 100 eV line) and of the same Mn sample described above.

A W sample, 0.15 cm thick, is the difference filter for U: it reduces the main 185 eV line. All the samples for the composite filters were supplied by PTB (Braunschweig, Germany).

The transmission measurements were carried out at the 100 MeV linac GELINA, running with a burst width of 14 ns, f=400Hz and i=20 μA or f=100 Hz and i=10 μA (for the Sc and the U measurement resp.). The neutrons, produced by a rotating U target, were moderated in a Be canned water moderator and detected after 48.2 m flight by a 0.5 cm thick sintered ^{10}B$_4$C slab viewed by 4 x (4"x3") NaI(Tl) scintillators; the assembly was screened with lead and borated paraffin. A barium-133 gamma ray source, placed in the vicinity of the scintillators, proved empirically that the gain variation of the PM's could be kept within 3% during all the measuring time. To reduce the detection of those slow neutrons which arrive at a time overlapping with the beginnig of the next burst, a ^{10}B$_4$C filter was placed permanently in the beam.

The acquisition system controlled the sample changer, providing a cyclic sequence during the measuring time.

Table 1: Lines in the transmission spectra of the Sc and of the Sc + Mn filters

Sc filter				Sc + Mn filter				Difference spectrum				Area 3/Area 1	
Av.energy (eV)	Width (eV)	Area 1	Error	Av.energy (eV)	Width (eV)	Area 2	Error	Av.energy (eV)	Width (eV)	Area 3	Error	Ratio	Error
1725.4	455.7	299.8	0.39	1292.6	184.0	35.69	0.22	1783.4	450.6	264.2	0.45	0.881	0.041
7419.3	141.0	7.577	0.040	7614.1		0.052	0.017	7418.	141.3	7.525	0.043	0.993	0.088
10465.	143.3	0.805	0.015	10477.	105.4	0.255	0.011	10459.	153.1	0.550	0.018	0.68	0.13
10988.	38.0	0.103	0.014	10981.	79.4	0.056	0.012	10995.		0.047	0.019	0.46	0.30
14302.	198.9	1.320	0.024	14294.	224.7	1.014	0.021	14326.	198.1	0.306	0.032	0.232	0.075
15383.	109.5	3.890	0.035	15387.	101.1	3.183	0.031	15363.	157.6	0.707	0.046	0.182	0.047
17265.	589.1	2.049	0.030	17197.	559.8	1.715	0.027	17613.	645.6	0.334	0.041	0.163	0.057
19310.	91.2	1.170	0.018	19305.	91.5	1.009	0.017	19344.	95.1	0.161	0.025	0.138	0.054
23653.	214.2	2.277	0.024	23410.	326.1	0.226	0.011	23680.	188.3	2.051	0.026	0.90	0.12
26117.	473.5	6.998	0.039	25958.	453.5	4.208	0.030	26358.	461.7	2.791	0.050	0.399	0.054
27453.	160.8	0.357	0.010	27503.	162.2	0.102	0.007	27433.	148.2	0.254	0.012	0.71	0.17
28990.	292.8	6.212	0.036	29007.	312.7	4.415	0.030	28948.	282.9	1.797	0.047	0.289	0.047
30528.	326.5	1.638	0.019	30533.	308.9	1.371	0.017	30500.	408.8	0.267	0.025	0.163	0.051
33335.	102.6	1.701	0.018	33333.	115.3	1.060	0.014	33339.	132.6	0.641	0.023	0.377	0.073
38679.	764.2	10.288	0.044	38876.	672.3	3.289	0.024	38587.	786.9	7.000	0.051	0.680	0.062
42245.	383.3	3.290	0.024	42313.	315.9	1.057	0.014	42213.	419.6	2.233	0.028	0.679	0.081
44008.	413.8	1.524	0.016	44035.	423.6	0.758	0.012	43982.	404.2	0.766	0.020	0.503	0.084
46823.	163.7	1.137	0.014	46814.	208.4	0.649	0.010	46835.	187.4	0.488	0.017	0.430	0.083
48042.	273.6	0.780	0.011	48044.	276.5	0.516	0.009	48038.	278.4	0.264	0.015	0.339	0.081
49697.	366.4	2.869	0.022	49693.	377.0	2.018	0.018	49706.	362.6	0.851	0.028	0.297	0.054
53863.	529.9	0.633	0.011	53908.	493.4	0.447	0.010	53755.	606.5	0.186	0.015	0.294	0.084
55858.	468.8	1.398	0.015	55834.	420.4	1.137	0.013	55964.	535.1	0.261	0.020	0.187	0.052
59917.	157.3	0.197	0.006	57889.	137.4	0.016	0.004	57919.	159.8	0.180	0.007	0.92	0.21
60926.	306.7	0.495	0.009	60939.	336.9	0.203	0.007	60917.	304.6	0.292	0.011	0.59	0.12
64515.	452.6	3.699	0.025	64463.	564.3	0.765	0.011	64528.	422.7	2.934	0.028	0.793	0.085
68032.	418.9	0.722	0.011	68071.	427.8	0.485	0.009	67952.	380.7	0.237	0.015	0.328	0.083
74653.	259.2	0.354	0.008	74689.	254.4	0.111	0.005	74637.	246.9	0.244	0.009	0.69	0.14
75962.	378.0	2.809	0.023	75987.	376.4	1.561	0.016	75931.	396.9	1.248	0.028	0.444	0.068
78333.	242.3	0.353	0.008	78324.	233.9	0.225	0.006	78349.	218.6	0.129	0.010	0.36	0.10
79201.	149.7	0.057	0.004	79183.	156.7	0.042	0.003	79250.	139.8	0.015	0.005	0.26	0.16
84466.	415.4	2.398	0.021	84402.	396.0	1.097	0.014	84520.	419.9	1.301	0.025	0.543	0.079
87914.	246.2	0.068	0.005	87929.	242.5	0.039	0.004	87893.	258.0	0.028	0.007	0.42	0.21
90777.	543.7	0.228	0.008	90773.	563.0	0.182	0.007	90792.	471.8	0.046	0.010	0.202	0.096
93329.	442.6	0.983	0.013	93332.	428.8	0.802	0.012	93318.	372.6	0.181	0.018	0.184	0.058
97917.	812.6	1.873	0.019	97889.	817.0	1.638	0.018	98117.	812.2	0.235	0.026	0.125	0.042
101710.	295.9	0.088	0.006	101740.	358.0	0.082	0.005	101260.		0.006	0.008	0.067	0.078

Table 2 : Lines in the transmission spectra of the U + Ge + Mn and of the U + Ge + Mn + W filters.

U + Mn + Ge filter				U + Mn + Ge + W filter				Difference spectrum				Area 3/Area 1	
Av.energy (eV)	Width (eV)	Area 1	Error	Av.energy (eV)	Width (eV)	Area 2	Error	Av.energy (eV)	Width (eV)	Area 3	Error	Ratio	Error
33.59	0.91	1.63	0.16	33.55	0.91	1.33	0.16	33.73	0.88	0.28	0.05	0.173	0.078
64.21	0.39	0.418	0.076	64.42	0.55	0.478	0.074						
98.53	1.49	1.583	0.098	98.75	1.03	1.279	0.094	97.60		0.299	0.043	0.189	0.075
185.02	1.75	10.05	0.20	182.47		0.47	0.14	185.15	1.49	9.48	0.14	0.94	0.15
592.90	1.04	0.032	0.008										
658.46	1.19	0.105	0.011	658.56	0.58	0.094	0.011	657.58	1.45	0.011	0.009	0.107	0.094
819.39	0.71	0.016	0.007										
848.92	1.68	0.028	0.007	848.50	1.68	0.023	0.007			0.005	0.004	0.17	0.16
933.67	1.56	0.066	0.008	933.74	1.42	0.052	0.007			0.014	0.006	0.20	0.14
955.20	0.97	0.067	0.008	954.35		0.018	0.006			0.048	0.005	0.72	0.286
987.52	0.95	0.069	0.008	987.08	0.95	0.069	0.008						
1136.80	2.13	0.043	0.006	1137.30	1.68	0.028	0.006	1135.90	1.75	0.015	0.005	0.35	0.20
1469.20	0.97												
1518.00	2.14	0.041	0.005	1518.60		0.024	0.005	1517.10		0.017	0.004	0.41	0.2
1593.20	1.51	0.055	0.006	1593.10		0.052	0.006			0.003	0.005	0.045	0.065
3428.00	2.44												
3847.40	6.60	0.014	0.003			0.002	0.003	3846.80		0.012	0.002	0.83	0.42
4503.60	6.19	0.020	0.003	4501.00	5.63	0.012	0.003			0.008	0.002	0.39	0.22
5627.20	9.37	0.007	0.002										
6025.40	2.04	0.015	0.003	6020.50		0.008	0.002			0.007	0.002	0.46	0.25
6155.30	11.79	0.086	0.004	6156.40	11.10	0.066	0.004	6151.80		0.020	0.005	0.23	0.11
6446.40	10.38	0.010	0.002	6446.20	5.89	0.007	0.002			0.003	0.002	0.26	0.21
14751.00	69.43	0.022	0.002	14745.00		0.019	0.002			0.003	0.002	0.11	0.10
17202.00	42.17	0.021	0.002	17202.00		0.017	0.002			0.004	0.002	0.20	0.13

Table 3: Relative weights of the main lines

FILTER	INTEGRAL (504 eV - 3169 eV)	INTEGRAL (18.4 eV - 102 keV)	RATIO
SC	299.79 ± 0.39	372.0 ± 1.1	0.806 ± 0.003
DIFFERENCE	264.17 ± 0.45	300.0 ± 1.4	0.881 ± 0.004
	INTEGRAL (177.1 eV - 188.6 eV)	INTEGRAL (18.4 eV - 102 keV)	
U + GE + MN	10.05 ± 0.20	17.75 ± 1.53	0.566 ± 0.050
DIFFERENCE	9.48 ± 0.14	12.11 ± 0.20	0.783 ± 0.017

For Sc the following combinations have been measured:
a) 'flux'; b) Sc; c) Sc+Mn; d) black resonance filters; e) Sc+Mn+ black resonance filters; f) Na (black at 2.85 keV).
For U:
a) 'flux'; b) U; c) U+Ge+Mn; d) U+Ge+Mn+W; e) U+W; f) black resonance filters; g) Co (black at 132 eV).

The black resonances have been used to define the background for the 'flux' in the energy range from 18.4 eV up to about 102 KeV and in order to monitor the minima in the Sc+Mn spectrum.

After dead time correction (160 ns), the background was fitted either in the minima of the spectra of the filters or in those of the black resonances. In this last case, the calculated functions were renormalized at the Na and Co (for Sc and U resp.) minima and subtracted from the 'flux' spectrum, after correction for the attenuation caused by the Na or Co filters. Due to the limited energy range covered by the black resonances, the transmission was calculated only from 18.4 eV up to 102 keV, but the measured data may be still used up to 500 keV to inspect the higher order contaminations. Our previous mesurement on the Fe+Al+S filter [3] has proved that the black resonance method for the background determination induces a systematic error of 2% on average for the actual detection system based on a $^{10}B_4C$ slab and NaI(Tl) scintillators. This error plus that due to the gain variation of the PM's raises the total average systematic error of the final transmission values to 5%.

Results

Tables 1 and 2 quote the results, i.e. the average energies, their widths (one standard deviation), the areas and their statistical standard deviation, only for those lines having a peak transmission larger than 0.01 and for neutron energies below 102 keV. The ratio Area3/Area1 gives the attenuation due to the difference filter. In a few cases only the peak may be observed, and its energy and width are given. In Table 1 the results for Sc, Sc+Mn and for their difference spectrum are presented.

In the case of the U measurement, only the U+Ge+Mn and the U+Ge+Mn+W filters are quoted in Table 2 with their difference. Table 3 summarizes the relative weights of the main lines of the Sc and U filters when compared to the total integral from 18.4 eV up to 102 keV.

References

1. R.C. Block and Brugger, Filtered Neutron Beams, in Neutron Physics and Nuclear Data in Science and Technology, (S. Cierjacks, Ed.), Neutron Sources for Basic Physics and Applications', Pergamon Press, Vol. 2 (1983).
2. E.Kondiaiah,R.P.Anand,D.Bhattacharya: "25 keV Neutron Beam Facility at the Reactor "APSARA", Nucl. Inst. & Meth. 111 (1973), 337-343.
3. A.Brusegan,H.G. Priesmeyer, M.Matzke, 'Investigation on an Optimized 24 keV Neutron Beam Filter',to be published.

EXPERIMENTS ON SEARCH AND INVESTIGATION
OF UNUSUAL NEUTRON RESONANCES

G.V.Muradyan, V.A.Stepanov, M.A.Voskanyan

I.V.Kurchatov Atomic Energy Inctitute, Moscow, USSR

Abstract: The measurements of neutron capture and scattering for ^{238}U have been performed in order to search and investigate neutron resonances with unusual properties. Resonance widths have been determined. For 721 eV resonance from capture data an anomalously low value of radiation width of Γ_γ=3.2±0.5 meV was obtained, which qualitatively conforms with the result from [1] and is indicative of pronounce of this resonance. It appears from scattering data that Γ_γ for this resonance seems to be larger than the mentioned value. Γ_γ amounts to 23±3 meV for 1211 eV resonance which similar to 721 eV resonance has a large fission width. This value coincides with the ^{238}U average radiation width within measurement errors and is significantly larger than that of 6.6 meV calculated in [1]. At the same time the results of capture γ-quanta spectrum measurement showed that this spectrum for both resonances mentioned is the same within errors and is not different from that for other resonances. The results obtained indicate that there is no reason to assert that 721 and 1211 eV resonances are associated with the states of the second well of fission barrier [1]. Probably, they correspond to unusual states of various nature.

(^{238}U, neutron resonances, radiative capture, scattering, radiation width, multiplicity spectrometry, capture γ-quanta spectrum)

The neutron resonances with unusual properties (unusual resonances) have been considered by Muradyan and a number of experiments for their search has been offered by him [2]. Unusual resonances correspond to non-statistical branches of exitation and can become apparent as weak resonances i.e. resonances with a small Γ_n neutron width. If decay widths of such resonances through their own channels differed from usual compound-state channels are small, the unusual resonances regardless of their nature, should have small radiation and total widths.

In order to reveal unusual neutron resonances as well as to study very interesting anomalies in the radiation width of ^{238}U resonances at 721 eV and 1211 eV [1], measurements of ^{238}U resonance widths have been carried out in the present paper. 721 eV and 1211 eV resonances have extremely large fission width, which makes it possible to consider them as states of the second well of the fission barrier [1]. These resonances can be regarded as unusual from a wider scope [2].

In a general case the measurement of weak resonance widths is still under investigation. That explains the lack of data for such resonances both on a Γ total width and on a Γ_γ radiation width. The reason lies in the fact that weak resonances can be revealed in total and capture cross-sections which are practically insensitive to Γ and Γ_γ. Sensitive to them are resonance areas in the scattering cross-section, measurement of which is not taken due to an insufficient effect and a significant potential scattering background, strongly hindering detection of weak resonances. However, a different situation is observed with unusual resonances. For the latter, weak as they are, a small value of neutron width in contrast to that for usual weak resonances is not associated with a small Γ_n/Γ ratio due to a small total width. Therefore, the area of unusual resonance in scattering cross section is much greater than that of usual weak resonance and, besides, the area of unusual resonances in capture cross section is more sensitive to Γ and Γ_γ [2]. Thus, if two resonances – usual and unusual – are equally weak in capture cross section, the unusual resonance becomes more apparent in scattering cross section.

To reveal unusual resonances, a sensitive technique to measure Γ and Γ_γ based on multiplicity spectrometry [3] has been created. This technique ensures the possibility to reveal unusual resonances in scattering cross-section measurement due to its high efficiency, which is very important in determining Γ when not all decay channels are known. It makes it possible to measure resonance areas in scattering, capture and fission cross sections simultaneously and under the same conditions. That enables Γ_γ to be obtained by different ways and a number of systematic errors to be eliminated.

The measurements have been performed with a 48 section two-layer NaI(Tl) crystal 4π-detector. The thickness and length of a scintillator in a 16 section internal layer is of 130 mm and 900 mm, respectively. The internal layer is designed to reduce the background by triggering the detector electronics by a total layer's pulse, and to increase the scintillator's thickness in order to diminish admixture of capture events in the scattering channel. The external layer consists of 32 sections, 130 mm of thickness and 600 mm of length. Scattering events are recorded by 480-keV γ-quanta emitted at capture of neutrons scattered on the sample in a small ^{10}B layer surrounding the sample under investigation. To shield the scintillator from scattered neutrons not captured in the ^{10}B layer, between the latter and crystals there is inserted a layer of ^6LiH / paraffin mixture. The measurements were performed at a 26-m flight path of the Electron Linear Accelerator of the I.V.Kurchatov Atomic Energy Institute with a resolution of 3 ns/m. A metal ^{238}U sample of a high impurity (99.996%), 31.92 g of weight, 6.42 10^{-3} at./b of thickness was used in capture and scattering measurements. The time-of-flight scattering spectrum was formed by selection of events recorded within γ-quantum energies of 0.35 < E < 0.60 MeV at the coincidence multiplicity k=1. The time-of-flight capture spectrum was formed by selection of events corresponding to k=2÷6 recorded within the range of 1.8 $\leqslant E_\gamma \leqslant$ 5.5 MeV.

Figure shows the dependence of capture events on a time-of-flight channel around 1211 eV resonance. Experimental data are marked with asterisks connected by a dotted line. Points with a solid line correspond to calculation in which the Doppler broadening, multiple neutron interaction in ^{238}U sample and neutron spectrometer resolution were taken into account. The calculated 1, 2, 3 curves were obtained at Γ_γ of 23 meV, 12 meV and 6.6 meV, respectively. It is seen that Γ_γ of

1211.4 eV resonance corresponds to 23 meV. The same value of Γ_γ is obtained from scattering data. Thus, the results obtained show that no anomalies in 1211 eV resonance radiation width of 23+3 meV are observed. The value of Γ_γ = 6.6 meV obtained in [1] in terms of equality in radiation widths of appropriate levels of the second well, is unacceptable.

The value of Γ_γ =3.2+0.5 meV was obtained for 721 eV resonance from the capture data. That confirms the result from [1] and indicate the pronounce of this resonance. The higher Γ_γ in [1] seems to be explained by the fact that fission effect in the capture channel was taken into account incompletely (that was noted in [1]) and that sensitivity of results to changes in γ-quanta spectrum is significantly higher. The latter results from a higher discrimination level of γ-quantum energy in [1] (4 Mev) in comparison with the present measurements (1.8 MeV).

An unexpected effect was revealed from the scattering data for 721 eV resonance. The experimental data for this resonance do not agree with the calculated data obtained at Γ_γ=3.2 meV, as well as with those obtained at Γ_γ=23.5 meV, and indicate that 3.2 < < 23.5 meV. In other words, Γ_γ obtained from scattering data is higher than that from capture data. It should not be attributed only to measurement errors, especially as sources of these errors are not seen so far, except for the statistics. The point is that such discrepancy can correspond to a certain physical situation different from a traditional one. From the analysis of widths determination it follows that Γ-

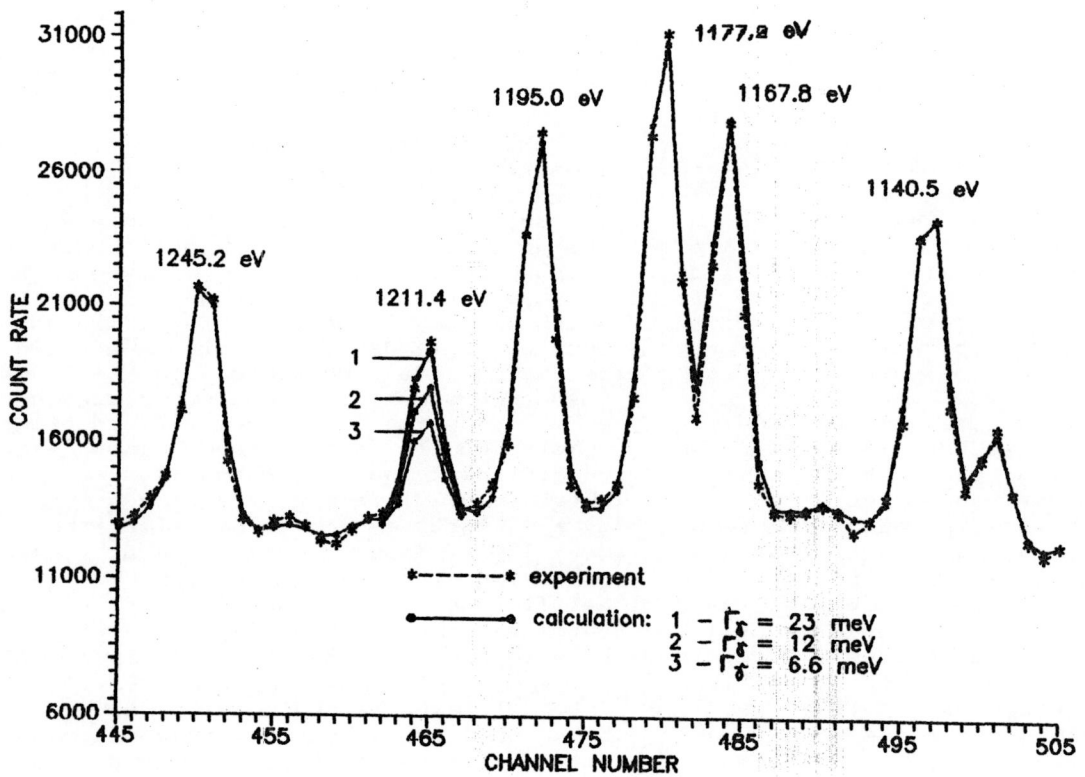

Γ_n but not Γ_γ is measured in scattering. The measurement of capture provides Γ_γ in the case when only scattering and capture channels are opened. Otherwise, Γ_γ from the capture data will be underestimated. The discrepancy obtained in the present paper can be explained with an assumption, in particular, that there exists a channel for 721 eV resonance which has not yet been revealed. In this case the decay through this channel is not recorded by the γ-detector with the given discrimination level.

If the 721 eV resonance has indeed a small Γ_γ, that can be attributed to its unusual properties. According to [2] low Γ_γ can be related to two situations. First, the unusual state decays, mainly, through own channels. In this case the measured Γ_γ corresponds to own radiation width of unusual state and the spectrum of capture γ-quanta for unusual resonance can be different from that for other resonances. Second, the decay takes place, mainly, through channels of neighbour compound-states. In this case the own radiation width should be much less than the measured value (3.2 meV) and the γ-quanta spectrum should be the same as for other resonances. In order to make this question clear the γ-quanta spectrum was measured at 2÷4 - fold coincidences picking out capture events. These measurements were performed with a metal ^{238}U sample, 2.6 g of weight, 9.73 10^{-4} at./b of thickness, enriched above 99.999%. The results showed that the γ-quantum spectra for 721 and 1211 eV

resonances are the same within the measurement accuracy and does not differ from those for other resonances with sufficiently good statistics. This indicates that if the 721 eV resonance is specified by unusual state, the width of this state γ-decay is significantly less than 3.2 meV. Note, that the radiation width of levels of the second well is estimated as ~ 4 meV [2].

To make the situation with 721 eV resonance clear it is necessary to improve statistical accuracy both in scattering and γ-quanta spectrum measurements. That can be realized at a more intensive neutron source. But regardless of future results one can say that properties of 721 and 1211 eV resonances are different. Therefore, there is no sufficient ground to state that both resonances are associated with the states of the second well. Probably, they correspond to unusual states of various nature.

REFERENCES

1. S.F.Auchampaugh, G. de Saussure, D.K.Olsen, R.W.Ingle, R.B.Perez and R.E.Macklin: Phys. Rev. C 31, 125 (1986)
2. G.V.Muradyan: Yadernaya Fizika 53, 883 (1991)
3. G.V.Muradyan, Yu.V.Adamchuk, Yu.G.Shchepkin and M.A.Voskanyan: Nucl. Sci. Engin. 90, 80 (1985)

ON THE STUDY OF NEUTRON RESONANCES IN ^{147}Sm

G.P.Georgiev[1], Yu.V.Grigoriev[2], Yu.S.Zamyatnin[1], V.I.Ivanov[1],
G.V.Muradian[3], L.B.Pikelner[1], I.A.Sirakov[1], N.B.Yaneva[4]

[1] Joint Institute for Nuclear Research, 141980 Dubna, USSR
[2] Physics-Power Engineering Institute, Obninsk, USSR
[3] Institute of Atomic Energy, Moscow, USSR
[4] Institute of Nuclear Research and Nuclear Power, Sofia, Bulgaria

Abstract: The pulsed booster IBR-30 (JINR, Dubna) is used as a neutrons source for the experiments on the measurement of multiplicities of γ-rays following the radiative neutron capture in ^{147}Sm in the energy range from 15 eV to 900 eV. The experiments were performed by the time-of-flight method using the multisectional 4π -detector of γ-rays, which allows one to obtain multiplicity spectra in individual resonances. The correlation was investigated between the multiplicity spectrum and spin of the resonance. Spins were assigned to ~90 resonances and radiative widths of about 30 resonances were measured.

(neutron capture, neutron scattering, capture γ-ray multiplicity, spin, radiative widths, ^{147}Sm, time-of-flight method, pulsed neutron source, multisectional detector)

Introduction

Radiative neutron capture reactions are successfully applied to the study of the structure and characteristics of nuclear levels at excitation energies close to the neutron binding energy. They also make a valuable contribution to the development of the model representation of the mechanism of the γ -decay of compound nuclear states. In particular, the registration of capture γ -rays was successfully used to make resonance spin assignments /1,2/.

A wider circle of problems can be solved with the recently developed method of γ -multiplicity spectrometry /3/.

This paper describes the multiplicity spectrometer built on the basis of the high intensity pulsed neutron source at the Laboratory of Neutron Physics,JINR,and reports on the first results obtained with this method in the ^{147}Sm experiment.

Experiment

In the experiment the multisectional scintillation γ -detector, Romashka - the given name, /4/ was installed on the 500 m flight path of the pulsed neutron booster IBR-30. Mean booster power is 10 kW and the multiplication coefficient - 200. At that the neutron pulse generated has the duration of about 4 μsec and a resolution of 8 ns/m. The pulse repetition rate is 100p.p.s.

The detector consists of 16 independent sections, crystals of NaI(Tl), measuring 122x122x x152 mm^3, each viewed with a photomultiplier. The detector has a through vacuum channel for the collimated beam to pass and a sample be positioned inside the detector (Fig.1).

Samples were made from the ^{147}Sm oxide enriched to 96.4%. They were prepared either 3.58x10^{-4} atom/barn or 8.94x10^{-5} atom/barn thick and put into thin-wall aluminium containers 100mm in diameter.

The neutron time-of-flight and γ -quanta coincidence multiplicity were measured for every neutron capture event. The events will be stored in the memory of the measuring module, if the sum energy of registered γ -quanta lies in the interval from 2 MeV to 8 MeV. The pulse detection threshold of each section is 100 keV.

This registration scheme allows us to simultaneously measure 16 time-of-flight spectra corresponding to coincidences of different multipli-

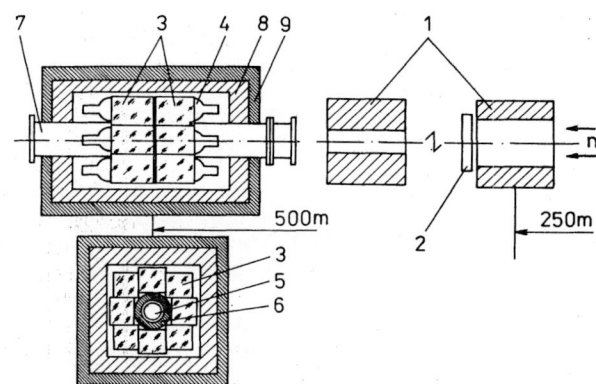

Fig.1. Longitudinal and transversial section view of the multisectional scintillation γ-detector "Romashka" (not in scale): 1 - collimator; 2 - filter; 3 - NaI(Tl) crystals; 4 - photomultipliers FEU-110; 5 - sample; 6 - converter; 7 - vacuum tube for sample positioning; 8 - lead shielding; 9 - B$_4$C shielding.

cities and to obtain γ -multiplicity distributions for every resonance.

In addition, the boron n-γ converter, positioned inside the detector, gave the opportunity to detect neutron scattering events by measuring monoenergy γ -quanta from the ^{10}B(n,α) reaction. Moreover, it served to decrease the detector sensitivity to scattered neutrons.

All the information from the detector came to the measuring module and was written into the computer memory as 16x4K files of 2 μsec per channel.

Results

a) Multiplicity spectra and spins

In result of the determination of areas under resonance peaks in spectra of different multiplicities, K, at K varying from 1 to 7 we have found the multiplicity distribution and average multiplicity value, $\langle K \rangle$, for every resonance, (Fig.2). As is seen in the figure the resonances form two clearly distinguishable groups with respect to the $\langle K \rangle$ value. These groups correspond to 3$^-$ and 4$^-$ spin values for the resonances with earlier known

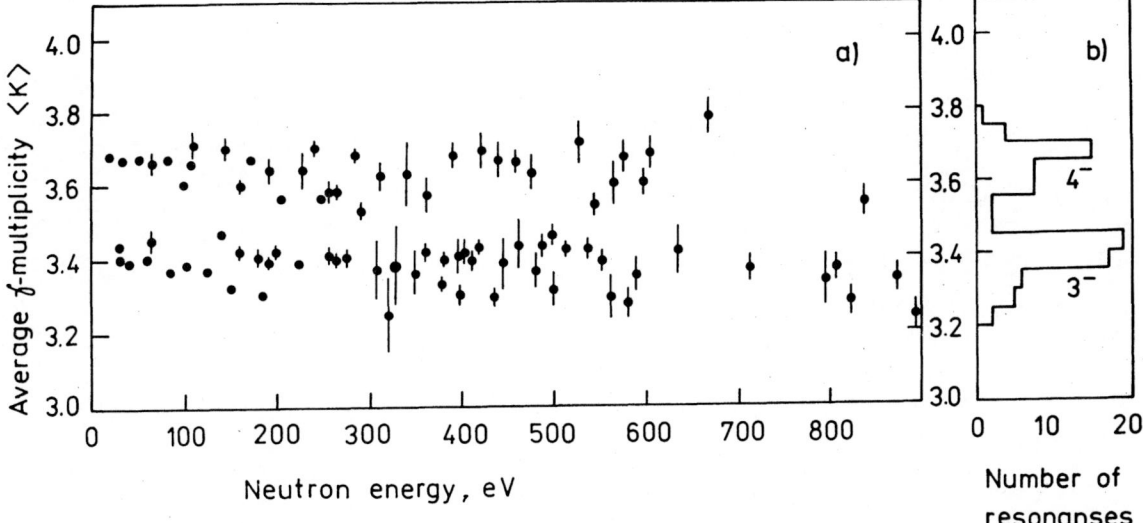

Fig.2. Experimental average values for γ-ray multiplicity $\langle K \rangle$: a) $\langle K \rangle$ values of resonances at energies from 15 eV to 900 eV; b) number of resonances as a function of $\langle K \rangle$.

spins (at energies up to ∼400 eV) /5,6/. Besides this fact such interpretation of the results is favoured by the good agreement between the spectra experimentally observed and calculated in the γ-cascade model with account for the efficiency of the detector for typical resonances with spins 3^- and 4^- /7/.

This made it possible to assign spins to about 90 resonances in ^{147}Sm in the energy interval from 15 eV to 900 eV and confirm most of the earlier made spin assignments. It is suggested to revise spin assignments for five resonances (161.0; 161.8; 359.2; 362.3 and 412.0 eV). Some of the resonances (65.1 and 257.0 eV) appeared to have intermediate $\langle K \rangle$ values. A per-channel examination of $\langle K \rangle$ values has shown that these resonances are the earlier unresolved double resonances with various spins (Fig.3). Additional measurements conducted with the ^{148}Sm isotope have prooved the resonance at 94.9 eV to belong to this isotope and not to ^{147}Sm.

b) Radiative widths

Simultaneous measurement of the time-of-flight spectra of neutron capture and neutron scattering has allowed us to make estimates on radiative neutron widths for a considerable number of resonances. These estimates were derived from the comparison of under-peak areas in neutron capture, S_γ, and neutron scattering, S_n, spectra with account for the relative effectiveness of detection of these two processes, $\varepsilon_\gamma / \varepsilon_n$, and known neutron width values Γ_n /5/:

$$\Gamma_\gamma = \frac{S_\gamma}{S_n} \cdot \frac{\varepsilon_n}{\varepsilon_\gamma} \, \Gamma_n.$$

The S_γ determination is based on the use of sum neutron capture spectra with coincidence multiplicities ranging from 2 to 8. The ratio $\varepsilon_\gamma / \varepsilon_n$ was evaluated by using some low energy resonances with known Γ_γ and Γ_n values /5/. The S_γ and S_n values obtained were corrected for the contribution of capture events to the scattering "channel" (4.2%) and of scattering events to the capture "channel" (from 2% to 7% in dependence on neutron energy). The corrections to be introduced were found in special experiments.

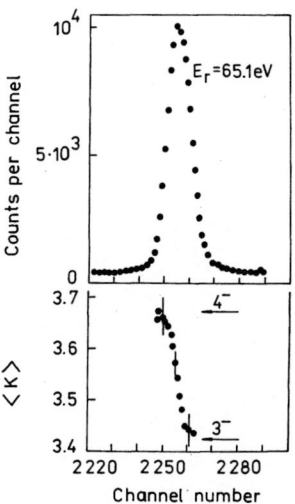

Fig.3. Per-channel determination of the average γ-ray multiplicity value in the vicinity of the 65.1 eV resonance. A sharp deviation in $\langle K \rangle$ value from the value characteristic of resonances with spin 4^- towards that corresponding to spin 3^- evidences the presence of two unresolved resonances with different spins.

In this way we have obtained the preliminary Γ_γ values for some 30 resonances at energies from 15 eV to 900 eV. Just a small deviation in the $\langle \Gamma_\gamma \rangle$ values for the groups of 3^- and 4^- resonances has been observed, but it was within experimental errors.

In the future we plan to undertake detailed data processing aimed at Γ_γ fluctuations value determination and investigation of possible correlations of Γ_γ with spin and neutron width of resonances.

References

1. C.Coceva, F.Corvi, P.Jiacoble, J.Carrado:

Nucl.Phys., A117, 586 (1968)

2. E.N.Karzhavina, Kim Sek Su, A.B.Popov: JINR
   Preprint P3-6092, Dubna, 1971

3. G.V.Muradian: Sov.J.Atomic Energy, 50, 394
   (1981); Nucl.Sci.Eng., 90, 60 (1985)

4. G.P.Georgiev, Yu.V.Grigoriev, Yu.S.Zamyatnin,
   G. V.Muradian et al.: JINR Communications,
   P3-88-555, Dubna (1988)

5. S.F.Mughabghab: Neutron Cross Section, v.1,
   Part B, Academic Press, 1984

6. A.B.Popov, Trzeciak K., Hvan Cher Gu: Sov.
   J.Nucl.Phys., 32, 603 (1980)

7. G.P.Georgiev et al.: Bulgarian Physical
   Journal, 18, No.1, 1991

MEASUREMENT AND ANALYSIS OF RESONANCE STRUCTURE
FOR U-238 TOTAL AND RADIATIVE CAPTURE CROSS SECTIONS
IN ENERGY RANGE 0.465-200 KEV

Yu.V.Grigoriev,V.N.Kostsheev,G.N.Manturov,V.V.Sinitsa
Institute of Physics and Power Engineering,Obninsk,USSR

H.V.Muradyan
(Kurchatov Atomic Energy Institute,Moscow,USSR)

G.P.Giorgiev, Yu.S.Zamyatnin, B.I.Ivanov
Joint Institute for Nuclear Research,Dubna,USSR

I.A.Sirakov, N.B.Yaneva
Institute for Nuclear Research and Nuclear Energy,Sofia,Bulgaria

Abstract: An analysis of measured transmissions and self-indications in the radiative capture cross section for U-238 was made. Average group constants for the BNAB system (total cross sections and self-shielding factors) are obtained by the code GRUCON. Also average resonance parameters were obtained by means of optimization of calculational to experimental deviations with the code EVPAR.

(U-238, experiment, transmission, self-indication in radiative capture, calculation, average cross sections, self-shielding factors, average resonance parameters, codes: GRUCON,EVPAR)

Introduction

Measured average cross sections, transmissions and self-indication ratios are usually used as source of information about cross section resonance structure in the unresolved resonance region. Transmissions are used to prepare self-shielding factors in group constant systems for practical calculations of fission reactors and shielding .

Much experimental information on transmission and self-indication in radiative capture and its temperature dependences are accumulated for U-238 by now. These data are available for thin as well as for thick filter samples. Analysis, for example in /1,2/ showed discrepancies in the interpretation of experimental results.

In this work new measurements of transmission and self-indication in radiative capture for U-238 in the energy range 0.465-200 keV are presented. The average resonance parameters were estimated from these data and group averaged total and capture self-shielding factors were obtained.

Experimental method

Transmissions and self-indication ratios of U-238 have been measured on the pulsed neutron source IBR-30 facility /3/. Transmissions were measured by the TOF method on a 1006-m flight path with low background detectors in the form of 16 He-3 counters with the 4 ns/m spectrometer resolution. Self-indication in radiative capture was measured by 16 NaJ(Tl) multisection scintillators. A thin sample of U-238 (0.00239 n/b) was used as radiator in the gamma ray detector. The gamma ray detector was placed on a 502-m flight path. The spectrometer resolution was 8 ns/m. This provided an energy resolution of about 10% at the energy 200 keV and about 1% at the 0.465 keV. The recording efficiency for gamma rays was about 80%. The neutron detector had a factually constant recording efficiency for neutrons of about 6%. This provided a higher energy resolution than for the gamma ray detector by more than a factor of 2. Transmission and self-indication ratios were measured simultaneously for 7 metallic U-238 filters with thicknesses from 0.0047 n/b to 0.306 n/b.

The present experimental data were measured under better conditions concerning thickness of filters and radiator samples and under approximatly analogous conditions concerning background and energy resolution as compared to older experiments /1,4/.

Results and discussions

The resonance self-shielding effects in the U-238 total cross section are significant over the whole energy range and even for thin samples. In the energy range above 50 keV the Tt(n) and Tc(n) curves converge and above 100 keV they coincide factually. This indicates that resonance self-shielding effects are absent in the radiative capture cross section above 50 keV .

It should be noted that the extrapolation of observed cross sections to zero sample thickness in order to define the average cross section presents a highly undefined procedure because of large experimental errors and bad theoretical foundation for the behaviour of observed cross sections at small thicknesses. It is more convenient to present average cross sections and transmissions and self-shielding factors by means of the subgroup method/5/. Here the experimental transmissions were aproximated by a sum of exponential functions, and two exponents have been enough for treatment of these experimental data. Optimal subgroup parameters were found by the least-square method. The total and capture shelf-shielding factors were calculated from the obtained subgroup parameters.

The code EVPAR/6/ was used for estimating average resonance parameters for the calculation of cross sections and their functionals in the unresolved region. The code EVPAR takes into account in the evaluation proces various experimental information: average total and different partial cross sections. The last modification of the code EVPAR accepts for analysis also transmissions. The estimated average resonance parameters are presented in tab.1 together with results of other authors. It should be noted that the obtained strength functions Sn0 and Sn1 and also Γc0 and D0 agree with averaged resolved resonance parameters . The scattering radius was taken with the energy dependence R1=R1'*(1-0.2*E) in order to agree with the data at low and high energies.

Table 1. Average resonance parameters
of U-238

value		p.w.	/4/	/6/	/7/
Sn1	l=0	1.10	0.89	0.93	1.11
		±0.05	±0.01	±0.03	±0.11
	l=1	1.7	1.87	2.30	2.20
		±0.2	±0.03	±0.07	±0.2
Γc(meV)	l=0	23.5	24.8*	22.9	22.9
		±0.07		±0.7	±0.9
D1(eV)	l=0	20.8*	20.8*	20.8*	20.8*
	l=1	6.9*	6.9*	4.4*	6.9*
R1(fm)	l=0	9.35	9.01	9.35	
	l=1	8.00	9.01	6.70	

* - fixed parameter at optimization

The self-shielding factors deduced from our experimental data and other data are presented in tab.2. Calculated values were produced by the code GRUCON /8/ from the estimated resonance parameters presented here.

The transmission and self-indication functions for U-238 in the energy range 4.65-10keV are shown in fig.1 as example. Here the experimental capture self-indication data are not described well by the calculations. As can be seen the calculated Tc(n) values are smaller than the experimental data. It seems that in this energy range resonance cross section peculiarities are observed. It would be interesting to treat jointly these measurements and the data of Moxon et al./11/ obtained with high resolution.

The energy dependence of the U-238 capture cross section calculated with the estimated average resonance parameters together with some experimental data are shown in fig.2.

Table 2. Capture self-shielding factors
of U-238

Elow,keV	exp	calc	/1/	/9/	/10/
100.	1.0	0.986	–	0.986	–
46.5	0.96	0.962	0.980	0.958	
	±0.02		±0.019		
21.5	0.94	0.914	0.931	0.910	0.879
	±0.02		±0.024		
10.0	0.84	0.765	0.868	0.830	0.781
	±0.03		±0.035		
4.65	0.78	0.666	–	0.719	0.659
	±0.04				
2.15	0.51	–	–	0.501	–
	±0.03				
1.00	0.36	–	–	0.304	–
	±0.02				
0.465	0.24	–	–	0.183	–
	±0.02				

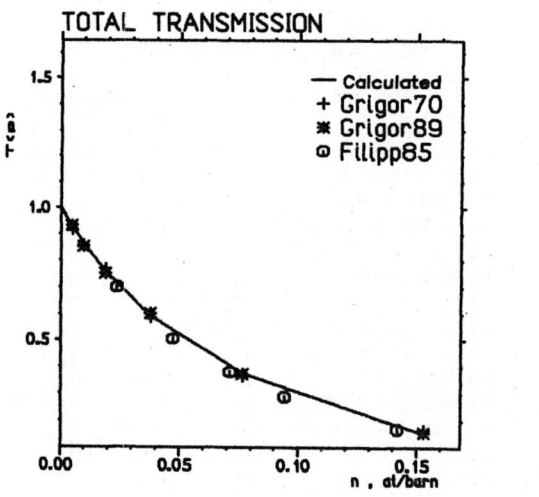

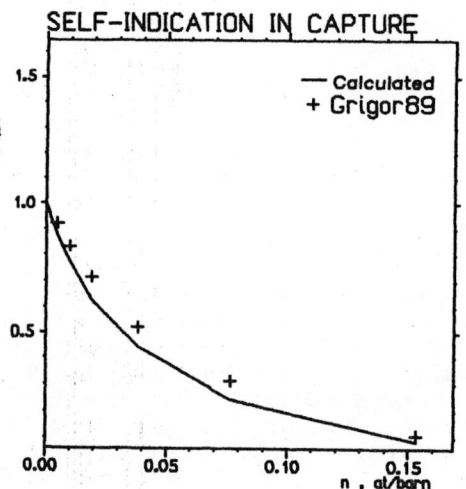

Fig.1. Transmission and self-indication functions
of U-238 in energy range 4.65-10 keV

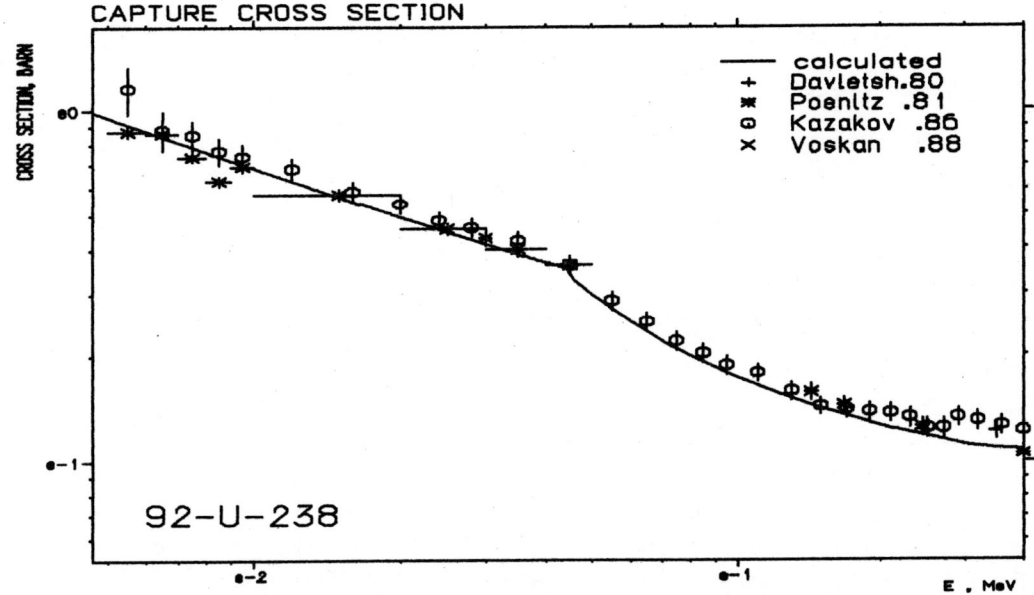

Fig.2. Capture cross section of U-238

References

1. M.V.Bokhovko,V.N.Kononov,G.N.Manturov
   et al.:VANT.Ser.:Nucl.Constants,3,
   11(1988)
2. V.N.Kostcheev,G.N.Manturov,
   V.V.Sinitsa: VANT.Ser.:Nucl.Constants,
   3,15(1989)
3. G.N.Georgiev,Yu.V.Grigoriev,
   Yu.S.Zamiatnin et al.,Measurement and
   analyses of transmission and self-
   indication radiative cross section
   U-238 in energy range 0.465-200 keV,
   P3-89-611, JINR, Dubna(1989)
4. A.A.Vankov,Yu.V.Grigoriev,M.N.Nikolaev
   et al.,In Proc.Conf. on Nucl. Data for
   Reactors. 1970, Helsinki, v.1, p.559,
   IAEA.Vienna(1970)
5. M.N.Nikolaev,V.V.Filippov: Nucl.
   Energy, 15(6), 493(1963)
6. G.N.Manturov,V.P.Lunev,L.V.Gorbacheva,
   In Proc.6-th Conf. on Neutron Physics,
   Oct.1983, Kiev.USSR, v.2, p.321
7. L.E.Kazakov,V.N.Kononov,G.N.Manturov
   et al.: VANT.Ser.: Nucl.Constants, 3,
   37(1986)
8. V.V.Sinitsa:VANT.Ser.:Nucl.Constants,
   5(59), 34(1984)
9. L.P.Abagian,N.O.Bazaziants,
   M.N.Nikolaev,A.M.Tsibulya,
   Group Constants for Calculation of
   Reactors and Shielding. Moscow,
   Energoizdat, 1981.
10. A.A.Vankov,V.F.Ukraintsev: VANT.Ser.:
    Nucl.Constants, 4, 58(1987)
11. M.E.Moxon,M.G.Sowerby,Y.Nakajima,
    C.Nordborg,In Proc.of Int.Reactor
    Phys. Conf. Sep.18-22,1988, Jackson
    Hole, v.1, p.281(1988)

STUDY OF THE DELAYED NEUTRON YIELD AND ITS TIME DEPENDENCE BY THE SUMMATION METHOD, AND THE SENSITIVITY OF THE YIELD TO PARAMETERS OF THE INDEPENDENT YIELD MODEL AND THE DECAY DATA

R.W.Mills, BNF Plc Sellafield works, attached to UKAEA Winfrith Technology Centre,

M.F.James, UKAEA Technology, Winfrith Technology Centre, Dorchester, Dorset, DT2 8HD. ENGLAND.

D.R.Weaver, School of Physics and Space Research, The University of Birmingham, Edgbaston, Birmingham. B15 2TT. ENGLAND.

ABSTRACT

This paper describes summation calculations of $\bar{v}_d$ and the time dependence of delayed neutron emission using a preliminary version of JEF2. The sensitivity of $\bar{v}_d$ to fission yields and P_n values, and the effects of the models used to generate complete yields and P_n sets are studied.

INTRODUCTION

The total number of delayed neutrons per fission $\bar{v}_d$ and the time dependence of the delayed neutron emission rate are important parameters for reactor design and safety studies, as they determine the kinetic response and behaviour of the reactor. There exist three ways of determining $\bar{v}_d$; firstly experimentally from integral measurements eg Keepin [1], secondly from summation calculations eg Liaw et al [2] using cumulative fission yields and P_n branching ratios; and thirdly the more empirical method, proposed by Pai et al [3] and modified by Tuttle [4], based upon systematics of the delayed neutron production with mass and charge of the fission nuclide. The time dependence of delayed neutron emission can be determined by experiment or by summation calculations using the P_n branching ratios, half-lives and inventories of the fission products following an irradiation eg Brady and England [5].

The proposed use of reprocessed fuel containing significant quantities of higher actinides has led to requests for $\bar{v}_d$ for these fuels. As experiments with these materials are often difficult, the summation method may be the most reliable way for these $\bar{v}_d$'s to be estimated if it can be shown to be more accurate than the empirical extrapolation method of Pai [3] and Tuttle [4]. However the uncertainties in the yields and branching ratios of the delayed neutron emitters must be reviewed in order to decide whether the summation method is significantly accurate for practical use.

The delayed neutron emitters exist on the extremely neutron rich side of the independent fission yield distribution, where few fission product yield measurements have been made except for the more common actinides such as ^{235}U and ^{239}Pu. Thus the models used to predict the charge distribution of the fission yields will have a significant effect on the $\bar{v}_d$. The different yield distributions for the fission of the higher actinides, particularly the movement of the light mass chain yield curve towards higher mass as the mass of the fissioning nuclide increases, mean that some precursors, relatively unimportant for ^{235}U fission, become much more significant. However measurements of the P_n values have been based mainly upon ^{235}U fission so theoretical estimates of these P_n branching ratios become much more important when considering higher actinides.

DELAYED NEUTRON EMISSION

The neutron emission is a result of β^- decay producing a daughter which has sufficient energy to throw off a neutron. The probability of a nuclide emitting a neutron as a result of a β^- decay is referred to as P_n. In a reactor or irradiated fuel sample the fission products present determine the delayed neutron emission rate, n_{emit}, which is thus given by:

$$n_{emit}(t) = \sum_i P_{ni}\, \lambda_i\, N_i(t) \qquad (1)$$

where P_{ni} is the P_n for nuclide i, λ_i is the decay constant of i, and $N_i(t)$ is the number density of i. N_i is determined by the fuel composition and the irradiation it receives (burnup, irradiation time and neutron spectra).

Therefore to generate the delayed neutron emission rate the irradiation must be specified and a calculation made of the inventory at each time t.

However, the total delayed neutron emission rate per fission, $\bar{v}_d$ can be calculated by integrating over all time for a single fission. Thus

$$\bar{v}_d = \sum_i P_{ni} \cdot R_i = \sum_i P_{ni} \cdot Yc_i \qquad (2)$$

The total decays of nuclide i per fission, R_i, is equal to the cumulative fission product yield of i; thus, for a pure sample of an actinide, if the cumulative yields, Yc_i are known the $\bar{v}_d$ can be calculated. The uncertainty in the $\bar{v}_d$ can be estimated from equation 2, by partial differentiation assuming Yc and P_n are independent as:

$$\sigma^2_{\bar{v}_d} = \left(\sum_i P_{ni} \cdot \sigma_{Yc_i}\right)^2 + \left(\sum_i \sigma_{P_{ni}} \cdot Yc_i\right)^2 \qquad (3)$$

THE KEEPIN 6 GROUP MODEL

The 6 group representation of the delayed neutron activity following a single fission pulse and long irradiation of 1 fission per second are given [1] by:

$$n_{emit}(pulse) = \bar{v}_d \sum_{k=1}^{6} a_k \lambda_k e^{-\lambda_k t} \qquad (4)$$

$$n_{emit}(long) = \bar{v}_d \sum_{k=1}^{6} a_k e^{-\lambda_k t} \qquad (5)$$

where t is the time after irradiation, a_k is the normalised group strength and the λ_k are the decay constants for the amalgamated delayed neutron groups. Note that a pulse is considered too short for any precursor to decay significantly during irradiation and the long irradiation allows all precursors to reach equilibrium.

FISSION PRODUCT YIELDS

The fission product yield set used for this work is UKFY2, submitted to and included in JEF2. This evaluation, described by the authors in 3 reports [6,7,8], extends and modifies the preliminary work on UKFY2 [9]. The evaluation has 4 stages: collection of experimental data, compilation and analysis of the data, using models to fill required gaps in the data, and adjusting the yields to conform to the physical conservation laws governing the fission process.

The independent yields are used to generate, with decay data, the cumulative yields and delayed neutron emission data. The independent yields are based upon the fractional independent yields and the chain yields. On the whole chain yields are well defined by measurements and empirical models; most systems of interest have in excess of 90% of chain yields measured. However the fractional independent yield distributions have very few measurements; even the most completely measured system, $^{235}U(Th)$, has 145 measured fractional independent yields (some measurements being discrepant) out of approximately 1000 required for UKFY2. UKFY2 uses Wahl's Zp model [10] fitted with data from the UKFY2 database to generate complete fractional independent yield sets, only $^{233}U(Th)$, $^{235}U(Th)$, $^{239}Pu(Th)$ and $^{235}U(F)$ having sufficient data to fit to the model. The model parameters for other systems are based upon extrapolation. Uncertainties remain about the effect of neutron spectra on the parameters and on extrapolation beyond U to Pu. Details of the fitting and data are given elsewhere

[6,7,8].

The decay data used for this work are based upon a preliminary version of JEF2, with the P_n values extended with the work of Lund [11] and Klapdor [12]. The half-lives were extended using the Japanese Chart of the Nuclides [13].

SUMMATION CALCULATION

From equations 2 and 3 values of $\bar{\nu}_d$ have been calculated from the UKFY2 datafile. Table 1 shows the values with the calculated uncertainty, the uncertainty components from the P_n and Yc values, values taken either from experiment or evaluations, and the ratio (calculation / experimentally based estimate).

TABLE 1.
The number of delayed neutrons per 100 fissions
a comparison of calculated & experimental values

NUCLIDE &ENERGY (*)	Calculated delayed neutrons	Standard Total	Deviation(%) From Pn's	From Yields	Evaluated	Ref	Calc/Eval
Th232 F	5.61039	0.390(7.0)	0.172	0.350	5.47 +/-0.12	T	1.026+/-0.074
Th232 H	2.83841	0.245(8.6)	0.084	0.230	2.85 +/-0.13	V	0.996+/-0.097
U233 Th	0.85440	0.083(9.7)	0.031	0.077	0.664+/-0.018	T	1.287+/-0.129
U233 F	0.92347	0.088(9.5)	0.034	0.081	0.729+/-0.019	T	1.267+/-0.124
U233 H	0.33727	0.065(19.3)	0.020	0.062	0.422+/-0.025	V	0.799+/-0.161
U234 F	1.15861	0.156(13.5)	0.043	0.150	1.06 +/-0.12	T	1.093+/-0.181
U235 Th	1.64484	0.114(6.9)	0.055	0.100	1.654+/-0.042	T	0.994+/-0.073
U235 F	1.85414	0.129(7.0)	0.047	0.120	1.714+/-0.022	T	1.082+/-0.077
U235 H	0.78682	0.082(10.4)	0.031	0.076	0.927+/-0.029	V	0.849+/-0.092
U236 F	2.24125	0.182(8.1)	0.065	0.170	2.31 +/-0.26	T	0.970+/-0.134
U238 F	4.05641	0.195(4.8)	0.111	0.160	4.510+/-0.061	T	0.899+/-0.046
U238 H	2.30629	0.196(8.5)	0.078	0.180	2.73 +/-0.08	V	0.845+/-0.077
Np237 Th	1.17786	0.149(12.7)	0.073	0.130	1.07 +/-0.10	W	1.101+/-0.171
Np237 F	1.18115	0.086(7.3)	0.040	0.076	1.22 +/-0.03	B	0.968+/-0.075
Np238 Th	1.38655	0.168(12.1)	0.051	0.160			
Np238 F	1.47849	0.172(11.6)	0.063	0.160			
Pu238 Th	0.33699	0.051(15.1)	0.026	0.044	0.456+/-0.051	T	0.739+/-0.139
Pu238 F	0.45152	0.072(15.9)	0.033	0.064	0.456+/-0.051	T	1.001+/-0.194
Pu239 Th	0.58948	0.054(9.2)	0.028	0.046	0.624+/-0.024	T	0.945+/-0.094
Pu239 F	0.66093	0.057(8.6)	0.029	0.049	0.664+/-0.013	T	0.995+/-0.088
Pu240 F	0.89203	0.107(12.0)	0.049	0.095	0.96 +/-0.11	T	0.929+/-0.154
Pu241 Th	1.30026	0.129(9.9)	0.047	0.120	1.56 +/-0.16	T	0.834+/-0.118
Pu241 F	1.37337	0.093(6.8)	0.046	0.081	1.63 +/-0.16	T	0.843+/-0.102
Pu242 F	1.81825	0.137(7.5)	0.066	0.120	2.28 +/-0.25	T	0.797+/-0.107
Am241 Th	0.38675	0.063(16.3)	0.032	0.054	0.44 +/-0.05	W	0.879+/-0.175
Am241 F	0.39153	0.074(18.9)	0.035	0.065	0.394+/-0.024	B	0.994+/-0.194
Am242mTh	0.62609	0.084(13.7)	0.038	0.075	0.69 +/-0.05	W	0.887+/-0.137
Am242mF	0.56542	0.094(16.6)	0.040	0.085			
Am243 Th	0.87007	0.151(17.4)	0.057	0.140			
Am243 F	0.86606	0.123(14.2)	0.055	0.110			
Cm242 Sp	0.11313	0.036(31.8)	0.021	0.029			
Cm243 Th	0.22969	0.066(28.7)	0.030	0.059			
Cm243 F	0.20907	0.044(21.0)	0.020	0.039			
Cm244 Sp	0.31890	0.079(24.8)	0.035	0.071			
Cm244 Th	0.34173	0.072(21.1)	0.031	0.065			
Cm244 F	0.33966	0.073(21.5)	0.033	0.065			
Cm245 Th	0.47658	0.084(17.6)	0.031	0.078	0.59 +/-0.04	W	0.808+/-0.151
Cm245 F	0.48874	0.112(22.9)	0.050	0.100			
Cf252 Sp	0.68913	0.154(22.3)	0.035	0.150	0.86 +/-0.10	M	0.801+/-0.200

Weighted Mean (Calc/Eval) = 0.959+/-0.022

*Th=Thermal, F=Fast, H=High(~14MeV), Sp=Spontaneous Fission

References:
B Benedetti(1982)   M Manero(1972)   T Tuttle(1975)
V Tuttle(1979)   W Waldo(1981)

Using equation 1 the inventory code FISPIN [14] was used to generate the n_{emit} at 204 time steps for all 39 fission systems in UKFY2 for both a single fission pulse (10^6 fission/s for 10^{-6}s) and an 'infinite' (1 fission/s for 10^{13}s) irradiation. Then Keepin's 6 group model was fitted to the pulse and infinite irradiation data concurrently (ie 408 data points). As well as fitting the 12 a_k and λ_k parameters, an attempt was made to use the set of λ_k values from Keepin [1] (Table 4-9, page 91), and only vary a_k. Table 2 gives the results for ^{235}U(Th) for the 12 and 6 parameter fits. Figure 1 compares the 12 parameter fits to n_{emit} for ^{235}U(Th) with other published sets of parameters ($\bar{\nu}_d$, a_k, λ_k).

SENSITIVITY OF $\bar{\nu}_d$ TO Zp PARAMETERS

The sensitivity of the $\bar{\nu}_d$ to the Zp parameters was studied by considering the fractional change in $\bar{\nu}_d$ following a small change in each Zp parameter used to generate the 'unadjusted' yields, i.e. those determined before the adjustment procedure to fit physical constraints. Each parameter x was varied in turn by + and - 1%, and the sensitivity S(x) of $\bar{\nu}_d$ to x found from

$$S(x) = \frac{\bar{\nu}_d(x+1\%) - \bar{\nu}_d(x-1\%)}{2\,\bar{\nu}_d(x)}.100 \qquad (6)$$

Variations of + and - 10% were also made, but the calculated sensitivities were not significantly different. Table 3 gives 1% sensitivities for ^{235}U(Th) and ^{235}U(F).

TABLE 2.
RESULTS OF FIT TO 6 GROUP PARAMETERS USING BOTH PULSE AND INFINITE IRRADIATIONS FOR ^{235}U(TH).

group no.	half-life s	decay constant s^{-1}	standard deviation	%stan. dev.	relative abundance	standard deviation	%stan. dev.
Fitting all parameters							
1	55.291	1.254E-02	2.250E-05	0.180	3.430E-02	2.909E-04	0.848
2	22.805	3.040E-02	1.563E-04	0.514	1.974E-01	2.048E-03	1.038
3	7.679	9.027E-02	4.451E-03	4.931	1.193E-01	9.091E-03	7.618
4	2.772	2.501E-01	1.016E-02	4.064	4.002E-01	1.407E-02	3.515
5	1.074	6.455E-01	5.122E-02	7.934	1.745E-01	1.748E-02	10.013
6	0.282	2.460E+00	8.903E-02	3.619	7.420E-02	5.143E-03	6.931
Reduced chi-squared= 0.268. Maximum difference of fit and data 0.9%							
Fitting abundances only:							
1	54.670	1.268E-02	0.000E+00	0.000	3.645E-02	4.364E-05	0.120
2	21.660	3.200E-02	0.000E+00	0.000	2.155E-01	3.057E-04	0.142
3	5.420	1.279E-01	0.000E+00	0.000	1.710E-01	1.406E-03	0.822
4	2.280	3.040E-01	0.000E+00	0.000	4.242E-01	1.952E-03	0.460
5	0.514	1.349E+00	0.000E+00	0.000	1.309E-01	1.504E-03	1.149
6	0.191	3.629E+00	0.000E+00	0.000	2.132E-02	8.042E-04	3.773
Reduced chi-squared= 0.879. Maximum difference of fit and data 2.2%							

TABLE 3
Sensitivity of $\bar{\nu}_d$ to input Zp model parameters.

System	$\Delta Z(A'=140)$	$d\Delta Z/dA'$	σ_z	σ_{50}	F_z	F_n	$\Delta A'_z$	ΔZ_{max}
U235T	-0.44	-0.046	1.2	1.0×10^{-3}	-0.99	0.10	-1.2×10^{-3}	-1.0×10^{-3}
U235F	-0.47	-0.10	1.0	4.5×10^{-4}	-0.95	0.037	-2.1×10^{-5}	2.3×10^{-5}

SENSITIVITY OF $\bar{\nu}_d$ TO DIFFERENT P_n SETS

To study the effect of different P_n sets upon $\bar{\nu}_d$, calculations of equation 2 were carried out using the UKFY2 cumulative yields and different P_n datasets. The results for ^{235}U(Th) and ^{235}U(F) are shown in table 4.

TABLE 4
$\bar{\nu}_d$ calculated using various P_n datasets.

Dataset:	no. of P_n's	$\bar{\nu}_d$ ^{235}U(Th)	$\bar{\nu}_d$ ^{235}U(F)
JEF2(prelim)	94 E	1.6448	1.8492
Lund(1986)	83 E	1.4455	1.5963
Mann(1986)	88 E	1.5665	1.7629
Brady(1988)	271 EM	1.6995	1.9092
Klapdor(1989)	209 M	1.2572	1.4044
JEF2(prelim) + Klapdor(1989)	251 EM	1.6447	1.8541
JEF2(prelim) + Klapdor(1989) + Lund(1990) + Brady(1988)	286 EM	1.6540	1.8625

E=Experimental and M=modelled P_n's.

SENSITIVITY OF REACTIVITY TO 6 GROUP DATA SETS

The reactivity of a critical system corresponding to a flux transient period is given by the 'inhour equation':

$$\rho = \frac{\Lambda}{T} + \sum_{k,m} \frac{(\beta_k{}^m)_{eff}}{(1+\lambda_k{}^m T)} \qquad (7)$$

where ρ = reactivity, Λ = neutron generation time, T = period and λ = decay constant. The superscript m refers to the material and the subscript k refers to the group number. The $(\beta_k{}^m)_{eff}$ term can be approximated [15] by $(\beta^m)_{eff}.a^m{}_k$. Thus for a given $(\beta^m)_{eff}$, if the prompt neutron term is negligible, the effect of different delayed neutron datasets can be compared by plotting $\sum_k a_k{}^m/(1+\lambda_k{}^m T)$ against T. This is the fractional change in delayed neutron emission rate due to a fission rate increasing exponentially with period T. Figure 2 compares this work and Brady and England's (1989) 6 group sets for ^{235}U(Th) with Keepin's experimentally derived values [1].

FIGURE 1

Comparison of $n_{emit}(t)$ calculated using published 6 group parameters(a_k, λ_k and $\bar{\nu}_d$) to the 12 parameter fits for ^{235}U(Th).

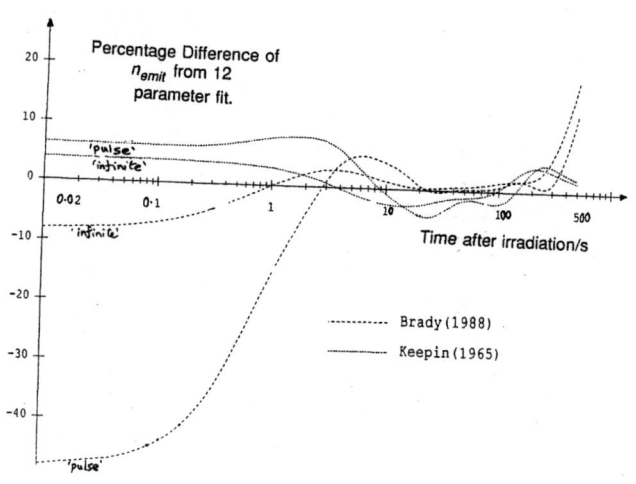

FIGURE 2

Comparison of $\sum\limits_{k} a_k^m/(1+\lambda_k^m T)$ against

period, T, for Brady and England, and the 6 and 12 parameter fits relative to the experimental Keepin data.

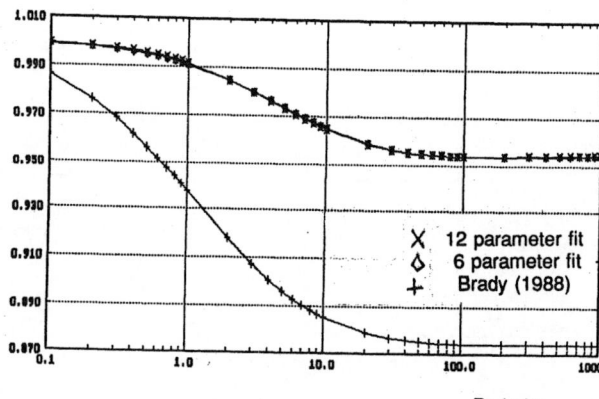

Period/s

CONCLUSIONS

$\bar{\nu}_d$

The $\bar{\nu}_d$ calculations show reasonable agreement with the evaluated data. Between the masses 235 and 241 for both fast and thermal there is good agreement, suggesting that the calculated uncertainties are reasonable. However there is a trend for calculated / evaluated $\bar{\nu}_d$ to decrease with increasing mass. Now, as none of the Zp parameters used in UKFY2 are mass dependent this might suggest that the change in the chain yield distribution on moving to higher fissioning mass makes more significant those delayed neutron precursors in the ^{235}U(Th)

valley region which are either underestimated or not currently included. An alternative hypothesis is that the fitted data range of U and Pu was too small to see a mass dependent effect on the Zp parameters.

Tuttle [4] quotes an estimated uncertainty of 10.7% overall for his empirical data, based upon the difference in the prediction and experimental values. This compares with about 8% overall for the $\bar{\nu}_d$ calculated above, and about 5% for the mass range 235-241. Thus the summation method would appear to give a better estimate of $\bar{\nu}_d$ globally; however some systems show differences with the evaluated values by up to 30%, although in each case this is within 2 standard deviations.

SENSITIVITY OF $\bar{\nu}_d$ TO DATA

Table 3 shows that the most important Zp parameters for the $\bar{\nu}_d$ determination are $\bar{\sigma}_z$, F_z, and $\Delta Z(A'=140)$. The two parameters $\bar{\sigma}_z$ and $\Delta Z(A'=140)$ largely determine the shape and positions of the Gaussian fractional independent fission yield distributions, and hence the yields of the neutron-rich precursors. The dependence on F_z reflects the preponderance of odd-Z precursors.

Table 4 shows that the majority of the delayed neutron emission comes from precursors whose P_n values have been measured and that the remaining component consists of a large number of precursors, ~190, with little overall effect.

6 GROUP MODEL

The summation method is the only alternative to the experimental measurement of n_{emit}. The results for ^{235}U(Th) give reasonable agreement with Keepin's experimental data considering the 4% standard deviation on the Keepin's $\bar{\nu}_d$. The Brady and England data also give good agreement between 2 and 300s, (figure 1).

SENSITIVITY OF REACTIVITY TO 6 GROUP DATA SETS

The reactivity parameter plot of $\sum\limits_{k} a_k^m/(1+\lambda_k^m T)$ against T

gives a more realistic comparison of the different 6 group parameters than that for n_{emit}. From Figure 2 one can see the very good agreement between the experimental results of Keepin and the derived values from this work when using both 12 and 6 parameters.

REFERENCES

1) G.R.Keepin, Physics of Nuclear Kinetics, Addison-Wesley(1965).
2) J.R.Liaw and T.R.England: Trans. Amer. Nucl. Soc. 28, 750(1978).
3) H.L.Pai: Ann. Nucl. Energy, 3, 125(1976).
4) R.J.Tuttle in Proc. consultants' meeting on delayed neutrons properties, Vienna, 26-30 March(1979).
5) M.C.Brady and T.R.England: Nucl. Sci Eng. 103 129(1989).
6) M.F.James, R.W.Mills and D.R.Weaver, UKAEA report AEA-TRS-1015(1991).
7) M.F.James, R.W.Mills and D.R.Weaver, UKAEA report AEA-TRS-1018(1991).
8) M.F.James, R.W.Mills and D.R.Weaver, UKAEA report AEA-TRS-1019(1991).
9) R.W.Mills, M.F.James and D.R.Weaver, in 50 Years With Nuclear Fission Meeting, April 1989, Washington DC, USA 1989 vol 1 p471(1989).
10) A.C.Wahl: Atomic Data and Nucl. Data Tables 39, 1(1988).
11) E.Lund, G.Rudstam, K.Aleklett, B.Ekstrom, B.Fogelberg and L.Jabobsson, in Proc. Specialists' meeting on Delayed Neutron Properties, Sept. 1986, Birmingham University, England(1986).
12) H.V.Klapdor, private communication, March 1989.
13) Y.Yoshizawa, T.Horiguchi and M.Yamada, The Chart of the Nuclides, INDC(JPN)99/L, Vienna(1984).
14) R.F.Burstall, UKAEA report ND-R-328(R) (1979).
15) J.Stevenson, in Proc. Specialists' meeting on Delayed Neutron Properties, Sept. 1986, Birmingham University, England(1986).
16) E.Lund, Report NFL-60, Unversity of Uppsala(1990).
17) F.M.Mann, in Proc. Specialists' meeting on Delayed Neutron Properties, Sept. 1986, Birmingham University, England(1986)
18) G.Benedetti, A.Cesana, V.Sangiust, M.Terrani and G.Sandrelli: Nucl. Sci. Eng. 80, 379-387(1982).
19) R.J.Tuttle: Nucl. Sci. Eng. 56, 37(1975).
20) F.Manero and V.A.Konshin: Atomic Eng. Rev. 10, 637(1972).
21) R.W.Waldo R.A.Karam and R.A.Meyer: Phys. Rev. C. 23, 1113(1981).
22) M.C.Brady, Thesis University of Texas, USA(1988).

DELAYED NEUTRON YIELDS FOR 39 FISSILE SYSTEMS

Jean Blachot
CEA ISN
Departement de Recherches Fondamentales
F38041 GRENOBLE cedex, France

Abstract: Accurate knowledge of absolute delayed neutron yields (νd) need t be improved.A method to derive the $\bar{\nu}d$ is to do a summation calculation usin cumulative fission yields (CU) and delayed neutron branching (Pn) for al precursors. For the "CU" we have taken the data from the JEF-2 file for 3 fissile systems.We have also done the calculations with the "CU" distributed b Wahl (3) for U233T, U235T and F , U238F, PU239T, PU241T.

In JEF-2 we have "Pn" data for 156 precursors ,almost 100 coming fro experimental results and the others from a calculation by Klapdor (8).

The results are compared with the last evaluations for the systems alread measured . The relative uncertainty is three times higher for the "CU" than fo the "Pn".

Introduction

Accurate knowledge in absolute Delayed Neutron Yields ($\bar{\nu}d$) need to be improved[1].These data for instance, play a key role in the reactor physics analysis of safety related parameter[2].A new version of JEF-2 was in development over the last few years.This version includes new decay and fission yields data.

This paper gives the last results obtained in using microscopic method with the JEF-2 file.

This method is based on a summation calculation,with delayed neutron emission probabilities "Pn" and cumulative fission yields "CU" taken from JEF-2.Problems not yet solved will also be discussed.

Method and Files

The calculation of '$\bar{\nu}d$' at the equilibrium using the microscopic method , need the 'Pn' and 'CU' before delayed neutron emission for each precursor.The method is well described by A. C. Wahl [3].We have to do a summation on all precursors.

$$\bar{\nu}d = \sum Cu(I) \times Pn(I)$$

The errors are derived from:

$$d\bar{\nu}d = \sqrt{(\sum (dCU(i)*Pn(i))^2 + (dPn(i)*CU)^2}$$

'Pn' Data

a) JEF-2 'Pn'

JEF-2 contain 'Pn' for 156 precursors , with 94 measured and 62 unmeasured :

-For measured 'Pn' :

97 are from a recent evaluation of Rudstam [4].

2 from Pfeiffer[5]

1 from England[6]

-For unmeasured 'Pn' :

62 Values are from a calculation of Klapdor et al.[8]

b) Other 'PN' file :

A file has been prepared by England [6] .This file has 'Pn' for 271 precurors with 85 measured

and the other 186 other calculated by the systematics of Kratz and Herrmann[7].

Comparison between JEF-2 and U.S Pn.

- measured

The first 10 precursors (U235T) are given below. Only 85As presents large difference between the 2 files.

PRECURSOR	Pn(%) (England)	Pn(%) (Rudstam)	CONTRIB(%)
137I	6.97	7.0	14
89Br	14	13.8	12
94Rb	10	10.0	9.4
90Br	24.6	25.2	7.9
88Br	6.26	6.58	7.15
138I	5.38	5.46	4.76
85As	70.9	59.4	4.72
95Rb	8.256	8.27	3.7
139I	9.81	10.0	3.63
87Br	2.54	2.52	3.21

- unmeasured

These Pn don't contribute too much to $\bar{\nu}d$. But it is important to see the differences between some Pn values calculated by Klapdor [8] for the JEF-2 and the corresponding U.S values calculated using the model of Kratz [7] .

PRECURSOR	Pn% (England)	Pn% (Klapdor)
Ga84	28	99.9
Ge85	16	99.7
Ge86	15	99.9
As88	19	88
Sn135	9.2	97.3
Sb137	18	99.5
Sb138	22	98.7
I142	13.8	97.3

'CU' Data

a) JEF-2 'CU' file :

This file has neutron induced fission yields for 21 isotopes which represent 39 fissile systems (Thermal, 1 MeV, 14MeV, Spontaneous).This file has been produced by James et al [9] .Details about the evaluation and the parameters of the model can be found in the original paper.The 'CU' had been calculated by the Zp model .

b) Other 'CU' files [3]
With his review paper on fission yields, Wahl has distributed data on tape for 6 fissile systems. Each fissile system have one 'CU' set calculated by the Zp model and an other one calculated by the A'p model .

Calculations

Table 1. shows an example of calculated delayed neutron yields(DNY) .The percentage contribution to the total $\bar{v}d$ is given in column (CONT%).

Table 1.

Delayed-Neutron yields per precursor for U235T
Sorted by importance(the 29 FIRST)

U5TY

Precursor	DNY	ER.Cu%	ER.Pn%	ER.DNY%	Cont%
53 137	2.29E-1	5.4E-3	4.0E-1	3.8E-2	14.0
35 89	1.96E-1	3.3E-3	5.0E-1	4.7E-2	12.0
37 94	1.54E-1	4.2E-3	2.2E-1	4.2E-2	9.4
35 90	1.29E-1	1.7E-3	9.0E-1	4.4E-2	7.9
35 88	1.17E-1	3.0E-3	2.6E-1	1.9E-2	7.2
39 98M	9.04E-2	5.0E-3	9.6E-1	3.0E-2	5.5
53 138	7.77E-2	4.2E-3	3.0E-1	2.2E-2	4.8
33 85	7.71E-2	4.8E-4	2.1E+0	2.5E-2	4.7
37 95	6.04E-2	2.3E-3	2.3E-1	2.0E-2	3.7
53 139	5.93E-2	2.1E-3	4.0E-1	2.0E-2	3.6
35 87	5.24E-2	2.1E-3	6.0E-2	5.4E-3	3.2
37 93	4.75E-2	5.4E-3	4.0E-2	7.3E-3	2.9
39 99	3.58E-2	4.8E-3	3.2E-1	1.0E-2	2.2
35 91	3.24E-2	5.9E-4	3.3E+0	1.2E-2	2.0
51 135	2.69E-2	6.3E-4	2.0E+0	1.0E-2	1.7
55 143	2.69E-2	4.2E-3	8.0E-2	7.0E-3	1.7
37 96	2.03E-2	5.4E-4	4.0E-1	7.2E-3	1.2
33 86	1.53E-2	1.7E-4	3.5E+0	5.7E-3	.94
55 144	1.42E-2	1.5E-3	3.2E-1	5.1E-3	.87
55 145	1.42E-2	3.4E-4	9.0E-1	5.0E-3	.86
52 136	1.35E-2	4.6E-3	4.0E-1	8.3E-3	.83
52 137	1.19E-2	1.6E-3	5.0E-1	4.6E-3	.73
53 140	1.18E-2	4.5E-4	1.0E+0	4.3E-3	.72
36 93	1.03E-2	1.7E-3	1.1E-1	3.4E-3	.63
35 92	7.82E-3	8.2E-5	3.0E+0	2.8E-3	.48
37 97	7.78E-3	1.0E-4	8.0E-1	2.7E-3	.48
51 136	7.27E-3	7.7E-5	4.0E+1	9.1E-3	.44
39 100	6.42E-3	2.2E-3	1.9E-1	2.6E-3	.39
36 94	6.41E-3	4.1E-4	2.2E+0	3.4E-3	.39

Table 2. gives the $\bar{v}d$ for 39 systems with the contribution of standard deviation from Pn and from CU.

Discussion

a) Relative error between Pn and CU.
We can see in Table 2 that the contribution of standard deviation for CU is three or four times greater than the contribution of Pn. So for improving accuracy on vd, we have first to improve the CU.

The SD_CU (Table 2.) for U238F is 3.94% and 6.25% for U235T.We have more than 200 experimental data in 235UT and less than 20 in U238F.The assignment of errors need to be reconsidered in the "CU".

Table 2. DELAYED NEUTRON YIELDS PER 100 FISSIONS

Nuclide	$\bar{v}d$	+/-	SD_Pn	SD_CU
Th232 F	5.87	0.33	2.2	5.42
Th232 H	2.88	0.21	1.9	7.08
U233 T	0.88	0.08	2.0	9
U233 F	0.95	0.08	2.3	8.75
U233 H	0.36	0.06	2.4	18.35
U234 F	1.19	0.14	2.3	12.36
U235 T	1.69	0.11	2.1	6.26
U235 F	1.89	0.12	2.4	6.23
U235 H	0.80	0.08	2.5	9.5
U236 F	2.31	0.17	2.3	7.2
U238 F	4.28	0.24	4.2	3.94
U238 H	2.37	0.19	3.1	7.48
Np237 T	1.22	0.14	3.2	11.32
Np237 F	1.22	0.08	2.9	6.49
Np238 T	1.44	0.15	2.8	10.75
Np238 F	1.53	0.16	2.9	10.37
Pu238 T	0.35	0.05	4.46	14.48
Pu238 F	0.47	0.07	3.94	15.32
Pu239 T	0.61	0.05	3.98	8.24
Pu239 F	0.68	0.06	3.76	7.78
Pu240 F	0.9	0.1	3.68	10.88
Pu241 T	1.37	0.12	3.13	8.96
Pu241 F	1.44	0.09	3.06	5.9
Pu242 F	1.92	0.14	3.54	6.64
Am241 T	0.41	0.06	4.84	14.7
Am241 F	0.41	0.07	4.06	17.19
Am242 T	0.64	0.08	3.76	12.71
Am242 F	0.59	0.08	3.67	15.28
Am243 T	0.91	0.14	3.29	15.49
Am243 F	0.91	0.11	3.24	12.78
Cm242 S	0.12	0.04	6.69	31.37
Cm243 T	0.25	0.06	4.49	26.43
Cm243 F	0.25	0.04	4.57	19.88
Cm244 S	0.34	0.07	4.14	22.72
Cm244 T	0.36	0.07	3.9	19.31
Cm244 F	0.36	0.07	4.01	19.81
Cm245 T	0.50	0.08	3.39	16.3
Cm245 F	0.51	0.11	3.59	21.1
Cf252 SF	0.73	0.15	3.1	21.19

b.) The Y98M :
Y98M has a contribution of 5.5% in U235T ,with a 'Pn' of 3.4% This value is the result of only one measurement at SOLIS81 [10], 'CU' has to take into account the isomeric ratio between isomeric and ground state.The ratio calculated by model [11] and used in JEF-2 is 84%. One recent measurement from ref[12] gives only 7%. The calculation with 7% would give $\overline{\nu d}$ = 1.62 in comparison with 1.69.

c.) Comparison between the Zp and the A'p models :
Using Pn of JEF2 anf CU of Wahl, we obtain the following $\overline{\nu d}$ with errors.

	Zp Model:		A'p Model:	
U233T	0.72	0.02	0.74	0.02
U235T	1.60	0.05	1.63	0.05
U238F	3.85	0.13	4.59	0.23
Pu239T	0.65	0.03	0.69	0.03
PU241T	1.29	0.05	1.51	0.06

The two models give similar results when they are derived from many experimental data.

d.) Comparison
Ronen [13] has presented $\overline{\nu d}$ derived from correlations with respect to the 2Z-N value of the nucleus.
The comparison between most of these values is given below

Fig. 1. ABSOLUTE DELAYED NEUTRON YIELDS AS A FUNCTION OF ELEMENT MASS

Element	Ronen	Present calc.
237NPF	1.10	1.22
238PUF	0.53	0.47
241AMF	0.53	0.41
242AMF	0.77	0.59
242AMT	0.648	0.64
243AMF	1.10	0.91
243CMT	0.267	0.25
244CMF	0.53	0.36
245CMT	0.648	0.50
245CMF	0.77	0.51

Some $\overline{\nu d}$ not calculated by us could be easily extrapolated on the Fig.1 which shows our $\overline{\nu d}$ versus A

REFERENCES
1. J.BLACHOT,M.C. BRADY,A. FILIP,R.W. FILIP,R.W.MILLS,D.R. WEAVER, NEACRP-L-323 (1990)
2. A. BENJELLOUN, Private Communication
3. A. C. WAHL, ADNDT 39 1 (1989)
4. G.RUDSTAM NFL-60 (1991).
5. B. PFEIFFER,K.L. KRATZ,H. GABELMANN,P. MOLLER, An. Report Mainz iKMZ 89-1, p 14
6. T. R. ENGLAND,M.C. BRADY,E.D. ARTHUR,R.J. LABAUVE,F.M. MANN, LA-UR-86 2693
7. K.L. KRATZ,G. HERRMANN, Z. PHYSIK 363 ,435
8. K. V. KLAPDOR,A.STAUD communication to JEF-2
9. M.F JAMES,R.W. Mills,D.R. WEAVER AEA-TRS-1099
10. G. ENGLER,E. NEEMAN NUC. PHYS. A367 29 (1981)
11.D. G. MADLAND,T.R. ENGLAND LA-6430-MS 1976
12. H.O. DENSCHLAG NEANDC(E)-262 (1985) 13. Y. RONEN J. PHYS. G. NUC. 16, 1891 (1990)

ABSOLUTE DELAYED NEUTRON YIELDS %

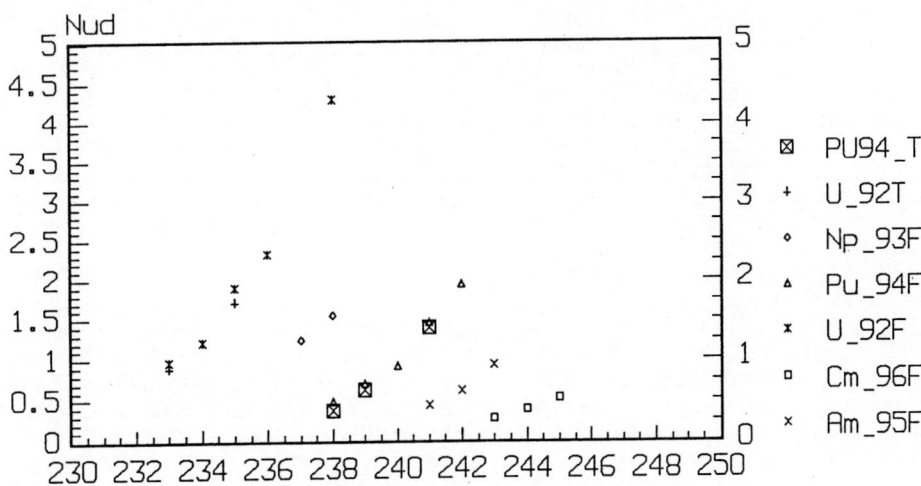

ELEMENT MASS

NUCLEAR PHYSICS INFORMATION NEEDED FOR ACCELERATOR DRIVEN TRANSMUTATION OF NUCLEAR WASTE

P. W. Lisowski, C. D. Bowman, E. D. Arthur, and P. G. Young

Los Alamos National Laboratory
Los Alamos, NM 87545 USA

Abstract: There is renewed interest in using accelerator driven neutron sources to address the problem of high-level long-lived nuclear waste. Several laboratories have developed systems that may have a significant impact on the future use of nuclear power, adding options for dealing with long-lived actinide wastes and fission products, and for power production. This paper describes a new Los Alamos concept using thermal neutrons and examines the nuclear data requirements.

(Transmutation, radioactive waste, spallation, fission energy, nuclear data)

Introduction

One of the most difficult problems associated with the worldwide use of fission energy is the question of management of the resulting high-level long-lived radioactive waste. That issue has been dealt with in most countries by considering some kind of long-term geologic repository, either by storing spent fuel material directly or with some combination of partitioning and recycling. In the United States, the repository solution has met with considerable public resistance, partly because of the difficulty of certifying container integrity for periods determined by regulatory agencies to be as long as 10,000 years and partly because of public concerns about storing high-level radioactive waste. In addition, the United States has the nearly unique problem of dealing with the accumulation of forty-five years of high-level waste associated with defense material production. At Los Alamos, Brookhaven National Laboratory (BNL), and the Japanese Atomic Energy Research Institute (JAERI), and elsewhere [1,2,3], studies are underway to investigate the role of spallation neutron sources in high-level nuclear waste disposal issues, either as a supplement to geologic storage or to produce energy from the waste. The work at Los Alamos began by considering defense applications; however, the great flexibility provided by an accelerator neutron source has led to the consideration of a highly efficient new energy production concept that has the potential to destroy its own high-level long-lived waste to the extent that surface disposal can be considered after managed storage for a period of time comparable with a human lifetime [4].

The concept of using a high power accelerator to produce fissile material or to transmute radioactive materials is not new [5]. It is the recent development of advanced high-power, high-efficiency accelerator technology which holds the key to the potential success of these systems. All three concepts discussed above use a medium-energy high current accelerator to produce neutrons through spallation. Both the JAERI and BNL designs use fast neutrons to fission actinides. Those systems may be an important component of the overall strategy needed to deal with high-level waste disposal of the high toxicity actinides that comprise the bulk of nuclear waste. They are inefficient when it comes to fission product destruction because there the principal means of transmutation is through neutron capture, a process which has a small cross section for fast neutrons. As shown in Fig. 1, because of their migratory nature, ^{99}Tc, ^{129}I, and ^{135}Cs dominate long-term risk scenarios [6] and need to be addressed as well. The Los Alamos concept discussed here provides a system based on thermal neutron interactions that provides a system based on thermal neutron interactions that can efficiently address both types of materials.

System Description

In the Los Alamos concept a high-power medium energy accelerator operating in the 600-1600 MeV range with beam currents from 50-140 mA is used to generate a high flux of neutrons through proton-induced spallation of a lead target. Those neutrons are moderated in surrounding D_2O, and interact with material circulating in piping within the D_2O. A schematic representation of the concept is presented in Fig. 2. The box labelled energy extraction is associated with extraction of energy resulting from fission of actinides

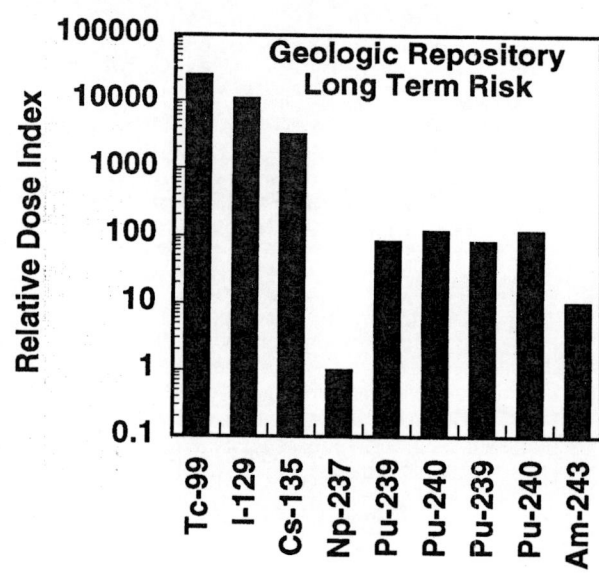

Fig. 1. Relative Radiation Dose Index for Some Fission Products and Actinides during geologic storage. From Ref. [6].

or other fuel. To retain high efficiency and to remove stable or short-lived transmutation endproducts, it is necessary to perform a chemical separation either as part of a slipstream or by processing the material in batches. Calculations using the Los Alamos High Energy Transport (LAHET) code system show that thermal neutron intensities in the D_2O blanket can reach levels in the range of 10^{15} - 10^{16} n/cm^2/s. Those intensities, typically one to two orders of magnitude larger that present in a PWR reactor and extending over a large volume, allow rapid transmutation of materials with small capture cross sections.

A key feature of the Los Alamos system is the generation of high thermal neutron fluxes to destroy actinide waste as well as fission products. Isotopes such as ^{237}Np and ^{241}Am are present in quantity in spent fuel or defense waste and are threshold fissioners; materials that have been traditionally believed impossible to destroy in a thermal system. Our calculations indicate that at sufficiently high thermal fluxes it is possible to efficiently burn those isotopes through a two-step process consisting of a neutron capture leading to a short-lived fissile isotope, followed by a second neutron interaction prior to the decay of the fissile isotope which usually leads to fission. This feature is illustrated in Fig. 3 for the case of ^{237}Np. The upper branch leads to fission of ^{239}Pu, the most probable outcome in a flux typical

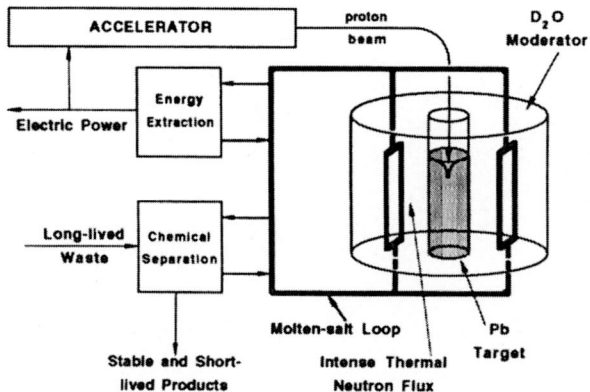

Fig. 2. Schematic diagram illustrating the Los Alamos concept for energy production and waste transmutation. The proton beam generates neutrons from a lead target which are moderated in the heavy water blanket. Molten salt circulating through the blanket carries fissile material for heat generation and power production. Nuclear waste is also circulated through the blanket and transmuted to stable and short-lived nuclides which are then chemically extracted.

of that in a thermal reactor system, and requires 3 neutrons with a release of only about 2.7, indicating that the material acts as a net neutron absorber or poison. The lower branch dominates in a high thermal flux and leads directly to fission of ^{237}Np with the use of 2 neutrons but release of 2.7; in that case ^{237}Np is a fuel. To our knowledge this is the first time thermal neutrons have been considered as a viable option for destroying threshold fissioning actinides. These results need confirmation, either experimentally or calculationally after measured differential reaction data are available for the short-lived nuclei. The potential for destruction of both fission and actinide waste coupled with in-situ fissile material production is the feature that makes the Los Alamos thermal system a candidate both for waste transmutation applications and as an advanced energy source with the possibility of destroying its associated long-term high-level waste.

Using the concepts discussed above, together with a system for in-situ fissile material production and utilization, we have developed a system to burn ^{232}Th. Applying methods implemented in the Molten Salt Breeder Reactor to the accelerator system, it appears that one could develop a practical energy producing system which operates far from criticality, has a fissile loading during operation of less than 100 kg, and has a cost and efficiency competitive with reactor systems [4].

Data Requirements

Much of the nuclear data currently needed for the Los Alamos system is available in sufficient accuracy, either through measurement, modelling, or nuclear systematics, to allow an initial positive assessment of the concept. As the design becomes more sophisticated, a considerable amount of new more accurate data will be required. Because of the vast range in both proton and neutron energy encountered in the accelerator and the target-blanket shield, it appears impractical at the present time to develop a comprehensive data library based on evaluated nuclear data and measurements such as was done for reactor systems. It may be necessary to rely on an improved suite of nuclear models benchmarked by measure-ments to provide the necessary information in all but a few critical areas.

It appears that the accelerator can be designed and operated at currents of up to 250 mA with beam losses small enough to allow hands-on maintenance for most of the major components. For sections where beam transitions take place,

for accident analysis, and for regions near the neutron production target, it will be important to determine yield data for thick target (p,xn) and (p,xγ) reactions and to carefully chose materials to be sure that induced activities are not unexpectedly high. The issue of shielding data needs associated with accelerators was recently addressed, [7] we note however that at minimum accurate total cross section measurements extending to the maximum neutron energy expected, together with enhanced modeling, will be required to provide needed transport information.

For the near term, the most important nuclear data needs for confirmation of the Los Alamos concept are for neutron production target yield, for spallation product mass yield, and for capture, fission, and total cross sections of short-lived actinides and fission products.

Verification of the calculated neutron yield, spectrum, and thermal neutron flux in the D$_2$O blanket resulting from medium energy proton interactions with a massive lead target is needed both to answer questions about overall efficiency, and to be able to accurately address radiation damage problems. Neutron leakage from thick targets depends heavily on effects associated with transport. Neutron capture, multiplication in the lead, and absorption in the spallation products which will build up as a poison over time will all be important to investigate as details of the system evolve.

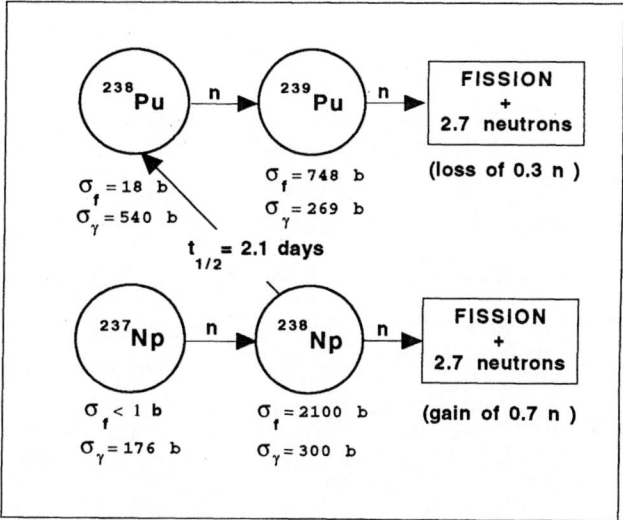

Fig. 3. Diagram illustrating two-step process for thermal burning of threshold fissioning actinides. The upper branch dominates at low fluxes.

Good resolution measurements of the fission and total cross sections of 16-hr ^{242}Am, 2.1-day ^{238}Np, and 1.3 d-^{232}Pa (produced in the conversion of ^{232}Th to ^{233}U in the energy concept) over the range from thermal to 1000 eV are important to verification of effective burning of higher actinides using low-energy neutrons and to transport calculations. For the fast neutron concepts of JAERI and Brookhaven, fission and capture data need to be extended to the MeV range. At present there are no direct measurements available, although some fission cross section information has been obtained from fission probabilities calculated from charged particle measurements and from integral tests.

Finally, better information is needed for differential neutron total and capture cross sections of fission products with half lives in excess of 11 years. The list of fission products under consideration is given in Table 1. Of those, ^{99}Tc and ^{129}I are most important from a risk standpoint, and each need better information over the range of interest.

Conclusions

Because accelerators provide neutrons in addition to those present when fissionable material is destroyed, it appears possible to achieve the goal of energy production and concurrent waste destruction that can not be reached in a system based on a fission neutron economy alone. It will be important, both for a system designed to destroy waste through transmutation alone, and for one using actinide fuel, to improve the existing nuclear data in order to show that such systems are scientifically practical and economically feasible.

Table 1. Long-Lived Fission Product Transmutation Candidates

Nuclide	Half-Life Yrs	Thermal Cross Section (barns)
^{79}Se	6.5×10^4	?(10)
^{90}Sr	2.9×10^1	0.9
^{93}Zr	1.5×10^6	2.5
^{99}Tc	2.1×10^5	20
^{107}Pd	6.5×10^6	1.8
^{126}Sn	1.0×10^5	0.14
^{129}I	1.6×10^7	27
^{135}Cs	3.0×10^6	8.7
^{137}Cs	3.1×10^1	0.25
^{151}Sm	9.0×10^1	152000

References

1. Y. Kaneko, T. Mukaiyama, and T. Nishida, Proc. of The 2nd International Symposium on Advanced Nuclear Energy Research - Evolution by Accelerators, Mito, Japan, 1990, Publ. by the Japanese Atomic Energy Research Institute, p. 369

2. M. Steinberg et al., Nucl. Tech. 48, 437, (1982)

3. J. A. Rawlins, Westinghouse Hanford Company, WHC-EP-0268 (1990)

4. C. D. Bowman et al., "Nuclear Energy Generation and Waste Transmutation Using an Accelerator-driven Intense Thermal Neutron Source", Los Alamos National Laboratory, to be published

5. C. Van Atta, J. Lee, and W. Heckrott, Lawrence Livermore Laboratory, UCRL-52144 (1976)

6. T. Pigford, Univ. Calif. Berkeley, UCB-NE-4176 (1990)

7. T. Nakamura, Proc. of The 2nd International Symposium on Advanced Nuclear Energy Research - Evolution by Accelerators, Mito, Japan, 1990, Publ. by the Japanese Atomic Energy Research Institute, p. 719

CROSS SECTIONS FOR ACTINIDE BURNER REACTORS

Felix C. Difilippo
Oak Ridge National Laboratory
P.O. Box 2008
Oak Ridge, TN 37831-6363, USA

Abstract: Recent studies have shown the feasibility of burning higher actinides (i.e., transuranium (TRU) elements excluding plutonium) in ad hoc designed reactors (Actinide Burner Reactors: ABR) which, because of their hard neutron spectra, enhance the fission of TRU. The transmutation of long-lived radionuclides into stable or short-lived isotopes reduces considerably the burden of handling high-level waste from either LWR or Fast Breeder Reactors (FBR) fuels. Because of the large concentrations of higher actinides in these novel reactor designs the Doppler effect due to TRU materials is the most important temperature coefficient from the point of view of reactor safety. Here we report calculations of energy group-averaged capture and fission cross sections as function of temperature and dilution for higher actinides in the resolved and unresolved resonance regions. The calculations were done with the codes SAMMY in the resolved region and URR in the unresolved regions and compared with an independent calculation.

Introduction

Neutronic and thermalhydraulic calculations have shown that it is possible to burn higher actinides in especially designed fast reactors which because of their particular hard neutron spectra enhance the fission of transuranic element (TRU). The analysis of a proposed Actinide Burner Reactor (ABR) plant (400 MWe) shows that it would be possible to burn in one year the amount of reprocessed TRU from eleven 1000 MWe Light Water Reactors (LWR) that had been operated for one year. The transmutation of long-lived radionuclides into stable or short-lived isotopes reduces considerably the burden of handling high-level waste from either LWR or Fast Breeder Reactors (FBR) fuels.

Detailed calculations of the capture and fission rates are essential to the aforementioned analysis whose accuracy depends on the careful preparation of the cross sections sets. Because of the large concentrations of higher actinides in these novel reactor designs, the Doppler effect due to TRU materials is the most important temperature coefficient from the point of view of reactor safety. Here we report calculations of energy group-averaged capture and fission cross sections as function of temperature and dilution for actinides in the resolved and unresolved resonance regions. The calculations were done with the codes SAMMY[2] in the resolved region, URR[3] in the unresolved regions, and the ENDF/B-V evaluation of resonance parameters (single-level Breit Wigner formalism).

In order to define dilutions of higher actinides in realistic systems we had chosen, as an example, one of the cases analyzed in Ref. 1. Details of the unit fuel cell are summarized in Table 1 with information on the TRU isotopic composition.

Table 1. Parameters of a Typical ABR Fuel Cell

Fuel:	Np–Pu–Zr alloy (Inner Core, IC) Am/Cm–Pu–Y alloy (Outer Core, OC) (Volumetric Fraction α_F=0.4031) Density: 9.788 g/cm³ (IC), 11.700 g/cm³ (OC) Isotopics[a]: IC: ^{237}Np, 58%; Pu, 22%; Zr, 20% OC: Am/Cm, 60%; Pu, 35%; Y, 5%
Coolant:	Na, α_c=0.3137
Fuel Pin Radius:	2mm

[a] Weight percent.

The dilution of the absorbing atom is parameterized in different ways according to the approximations used to compute the average cross section

$$\bar{\sigma}_{p,i}(T) = \frac{\int_{E_{i+1}}^{E_i} \phi(E)\sigma_p(E,T)dE}{\int_{E_{i+1}}^{E_i} \phi(E)dE} \qquad (1)$$

for the process p of the resonant absorber. In the Bondarenko treatment ϕ is approximated as proportional to $1/\sum_t(E)$ where the total cross section $\sum_t = n(\sigma_t + \sigma_o)$, n is the density of the resonant absorber with total cross section σ_t, and σ_o is the scattering cross section of the rest of the atoms of the mixture per absorbing atom. In other, more accurate, approximations (like the treatment of the resolved region in the AMPX system) $\phi(E)$ is computed in terms of the scattering cross sections of specific atoms per absorbing atom and their respective masses. Table 2 shows values of σ_o for the cell of Table 1.

Table 2. Scattering Cross Section Per Absorbing Atom for the Cells of Table 1

Absorbing Atom	Homogeneous Cell σ_o (barn)		Heterogeneous Cell[c] σ_e (barn)	
	All Isotopes	Sodium	Isolated Lump	Interactive Lumps
^{237}Np	30.2	4.0	173.0	52.0
^{241}Am[a]	48.1	5.4	237.0	71.0
^{243}Am	118.0	11.8	520.0	156.0
^{243}Cm[b]	52500.0	4836.0	210,000.0	63300.0
^{244}Cm	295.0	28.2	1230.0	372.0
^{245}Cm	5587.0	515.0	22600.0	6800.0

[a] 87.63% Am in Am/Cm alloy; 68.7% ^{241}Am; 31.3% ^{243}Am, (from PWR).
[b] 0.55% ^{243}Cm; 94.28% ^{244}Cm; 5.17% ^{245}Cm, (from PWR).
[c] Mean chord: 0.4cm; Dancoff Coefficient: 0.7. σ_e denotes the escape cross section in the Wigner's rational approximation.

Calculations of Cross Sections and Bondarenko Factors

The codes SAMMY[2] (for the resolved region) and URR[3] (for the unresolved region) were used to compute energy group-averaged cross sections as functions of dilution and temperature. The Bondarenko approximation for the energy dependence of the flux was used in both regions, so a universal parameter σ_o (potential scattering cross section of the rest of the mixture of the isotopes per absorbing atom) appears without any reference to the masses of the isotopes or the moderation process within the width of the resonances.

With the code SAMMY, we computed the point cross sections as functions of temperature with parameters from the ENDF/B-V evaluation which provides single-level Breit-Wigner resonance parameters and assumes g=0.5 for all resonances. Because SAMMY is based on an R-matrix theory description of the cross sections that require the spin of each resonance, we arbitrarily divided the set of resonances into a large number of spin families in order to eliminate the interference between resonances and to be consistent with the single-level evaluation.

The next step was to average the temperature-dependent cross section with the Bondarenko prescription:

$$\bar{\sigma}_{p,i}(\sigma_o, T) = \frac{\int_{E_{i+1}}^{E_i} \sigma_p(E,T)w(E,T)dE}{\int_{E_{i+1}}^{E_i} w(E,T)dE} \qquad (2)$$

where the weighting function w is

$$w(E,T) = \frac{1}{\sigma_t(E,T) + \sigma_o} \quad . \qquad (3)$$

The Bondarenko factor for process p, group i, temperature T and dilution σ_o was also computed according to its definition

$$B_{p,i}(T, \sigma_o) = \frac{\bar{\sigma}_{p,i}(\sigma_o, T)}{\bar{\sigma}_{p,i}(\infty, T)} \quad . \qquad (4)$$

We calculated the cross sections in the unresolved region with the code URR[3] which computes the probability distribution of the cross section at an input energy while playing Monte Carlo in the surrounding energy region with the average resonance parameters (also from the ENDF/B-V evaluation). After the calculation of the probability table, the code evaluates the shielded averaged cross sections and the Bondarenko factors. Using the ergodic hypothesis, these parameters are considered average values for the energy region around the input energy; this way of calculating the parameters is faster than the calculation of a ladder of resonances in a broad energy group. The use of ENDF/B-V data allowed us to compare the numerical approach presented here with similar calculations performed with the system of codes AMPX (used by the nuclear industry) which in its present version do not include a R-matrix description of the cross sections. To compare both approaches we select the unresolved region, because it is more important than the resolved region for the design described in Table 1, and the actinides ^{237}Np, ^{241}Am, ^{244}Cm, because they are the most shielded of each actinide element.

Figure 1 compares the AMPX results with the URR calculations for infinite diluted capture and fission cross sections of ^{237}Np, ^{241}Am, ^{244}Cm, in their respective unresolved region: 130eV–40keV; 50eV–10keV and 525eV–10keV.

The comparison is very good; ^{237}Np exhibits the structure of the subthreshold fission so a suitable energy grid would have been necessary to fully describe it which was not intended in the calculations of Fig. 1.

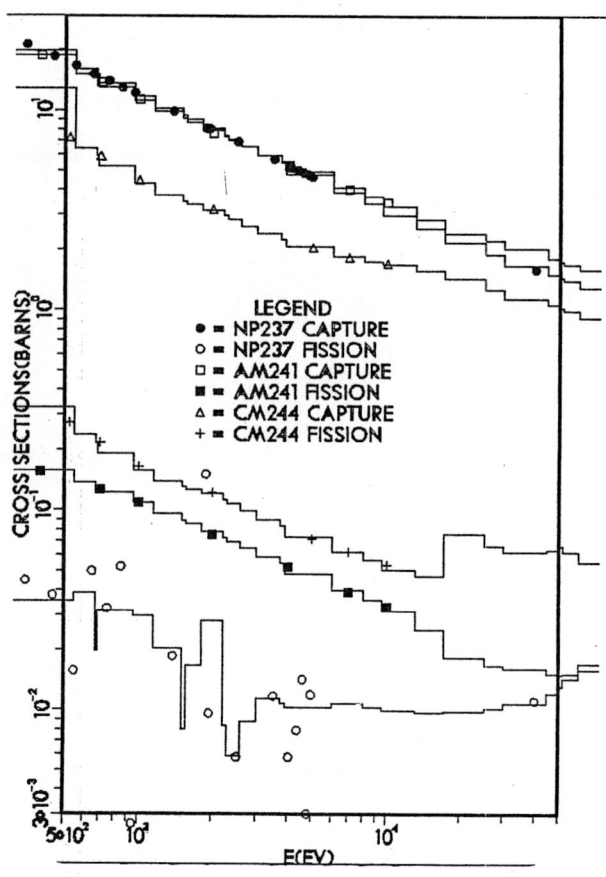

Fig. 1. Infinite diluted actinide cross sections in the unresolved region. Histogram calculated with the AMPX system, points calculated with the URR code.

The Bondarenko factors for the capture process at T=293°K and σ_o=100 barns are shown in Fig. 2 computed with the AMPX system and the URR code. Again the comparison shows consistency.

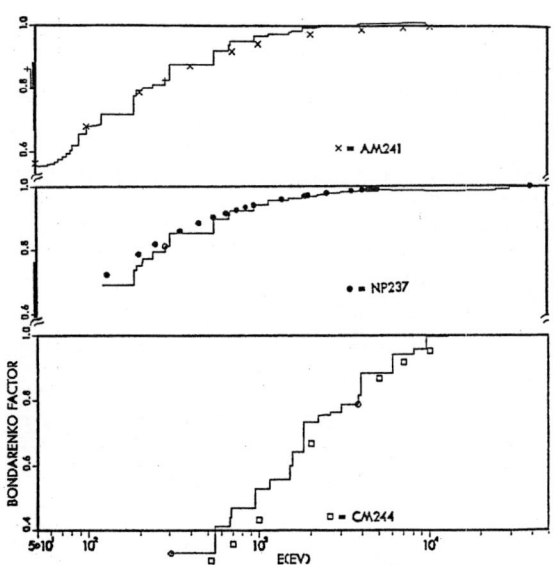

Fig. 2. Bondarenko factors for the capture process. Histogram calculated with the AMPX system; points calculated with the URR code.

Conclusions

Because of the major presence of higher actinides in ABRs of innovate design and their consequent impact on the Doppler coefficient of reactivity, effective capture and fission cross sections and Bondarenko factors were calculated for ^{237}Np, ^{241}Am, ^{243}Am, ^{243}Cm, ^{244}Cm, and ^{245}Cm in the resolved and unresolved resonance regions. The calculations were done with the computer codes SAMMY and URR, which are codes mainly used for the analysis of experimental data. This new approach has the following advantages: (1) It uses the same numerical tool already used in the analysis of experimental data, i.e., the method is more transparent to the original data; (2) improvements in the analysis of experimental data (e.g., the R matrix analysis performed by SAMMY) can be introduced directly in the preparation of the cross sections; and (3) the ergodic hypothesis implemented in the code URR is faster than explicit calculations of ladders of resonances in a broad energy group. Extensive calculations as functions of dilution factors and temperature were made with resonance parameters from ENDF/B-V (single Breit-Wigner resonance parameters) that allowed direct comparison of the results with an independent calculation with the AMPX system.

REFERENCES

1. T. Mukaiyama, H. Tanako, T. Hakizuka, T. Ogawa, Y. Gunjian, and S. Okajima, "Higher Actinides Transmutation Using Higher Actinides Burner Reactors," International Conference on the Physics of Reactors: Operation, Design and Computation, Marseilles, April 1990, Vol. 1, p. I-97.

2. N. M. Larson, "Updated Users' Guide for SAMMY: Multilevel R-Matrix Fits to Neutron Data Using Bayes' Equations," ORNL/TM-9179/R2 (1989).

3. L. C. Leal, G. deSaussure and R. B. Perez, "URR Computer Code: A Code to Calculate Resonance Neutrons Cross-Section Probability Tables, Bondarenko Self-Shielding Factors, and Self-Indication Ratios for Fissile and Fertile Nuclides," ORNL/TM-11297 (1989).

4. N. M. Greene, J. L. Lucius, L. M. Petrie, W. E. Ford, J. E. White and R. Q. Wright, "AMPX: A Modular Code System for Generating Coupled Multigroup Neutron-Gamma Libraries from ENDF/B," ORNL/TM-3706 (1976).

EVALUATION OF TOTAL FISSION CHARACTERISTICS FOR ^{235}U IN THE LOW ENERGY REGION

U. Gohs

Joint Institute for Nuclear Research, 141980, Dubna, USSR

Abstract: An energy conservation consistent evaluation of total fission characteristics for $J^\pi=4^-$ neutron resonances of ^{235}U has been performed on the basis of a phenomenological multimode model. Including the (n,γf) process, the total kinetic energy, average number of promptly emitted neutrons, multiplicity of γ-rays, and total energy of γ-rays for different resonances of uranium were calculated and compared with experimental data.

(4$^-$ neutron resonances of ^{235}U, fission modes, (n,γf) process)

Introduction

Hambsch et al. [1] measured total kinetic energy $\overline{TKE}$ and fission fragment mass distributions for low energy resonances of ^{235}U. The observed fluctuations in $\overline{TKE}$ ($\Delta\overline{TKE} \simeq 0.1$ %) and the fluctuations in the relative number of promptly emitted neutrons $\overline{\nu}_n$ ($\Delta\overline{\nu}_n \simeq 0.5$ %) [2] were discussed in a multimode model [3] and related to fluctuations in fission mode yields. In low energy reactions of neutrons with actinide nuclei, a two step reaction called (n,γf) process, where after emission of one or more γ-rays fission occurs, is possible [4]. The width $\overline{\Gamma}_{\gamma f}$ and the average γ-ray energy $\overline{E}_{\gamma f}$ of this process deduced from experiments are different [5]. We suppose that the difficulties arise from an overlap of different processes in the measurements of $\overline{\nu}_n$, the multiplicity of γ-rays $\overline{N}_\gamma$, and the total energy of γ-rays $\overline{E}_\gamma$.

Model

According to Furman et al. [6], the relative population $W_{d\lambda}$ of the fission mode d for resonances numbered by λ (with total fission width $\Gamma_{f\lambda}$) is related to the relative contribution $P_{k\lambda}$ of fission channel k as

$$W_{d\lambda} = \sum_k P_{k\lambda} W_d^k, \qquad (1)$$

where W_d^k is the relative population of mode d for fission channel k. All population values were deduced from experimental data [1]. Information about $P_{k\lambda}$ is available from the analysis of Moore et al. [7]. Following this idea and including the (n,γf) process, the total characteristic $\overline{X}_\lambda$ can be expressed by

$$\overline{X}_\lambda = \frac{\overline{\Gamma}_{\gamma f}}{\Gamma_{f\lambda}} \sum_d W_{d\lambda} \overline{X}_d^\gamma + (1 - \frac{\overline{\Gamma}_{\gamma f}}{\Gamma_{f\lambda}}) \sum_d \sum_k P_{k\lambda} W_d^k \overline{X}_d \quad (2)$$

with

$$\overline{X}_d^\gamma = \sum_A Y_d(A) \overline{X}_d^\gamma(A) \text{ and } \overline{X}_d = \sum_A Y_d(A) \overline{X}_d(A). \quad (3)$$

$\overline{X}$ stands for $\overline{TKE}$, $\overline{\nu}_n$, $\overline{N}_\gamma$ or $\overline{E}_\gamma$. The mass yield of fission mode d, i.e. $Y_d(A)$, is represented by a gaussian [1]. The index γ of the characteristic $\overline{X}$ is related to the (n,γf) process.

(n,γf)-process

In previous works, the (n,γf) process was discussed on the basis of following relations:

$$\overline{N}_\gamma = \overline{N}_\gamma^0 + \overline{\Gamma}_{\gamma f} \overline{N}_{\gamma f}/\Gamma_{f\lambda} \qquad (4)$$

$$\overline{E}_\gamma = \overline{E}_\gamma^0 + \overline{\Gamma}_{\gamma f} \overline{E}_{\gamma f}/\Gamma_{f\lambda} \qquad (5)$$

$$\overline{\nu}_n = \overline{\nu}_n^0 - \overline{\Gamma}_{\gamma f} \overline{E}_{\gamma f} (\partial\nu_n/\partial E^*)/\Gamma_{f\lambda} \qquad (6)$$

where $\overline{N}_{\gamma f}$ and $\overline{E}_{\gamma f}$ are the average multiplicity and average energy of prefission γ-rays, respectively. $\partial\nu_n/\partial E^*$ represents the average dependence of neutron multiplicity on excitation energy E^*. In this representation the γ-ray multiplicity $\overline{N}_\gamma^0$, the energy of γ-rays $\overline{E}_\gamma^0$, and the number of promptly emitted neutrons $\overline{\nu}_n^0$ are average values of the (n,f) reaction.

In the framework of the present model we assume that the excitation energy at the scission point decreases by $\overline{E}_{\gamma f}$ and the multiplicity of γ-rays increases by $\overline{N}_{\gamma f}$. Further we neglect channel effects in describing the (n,γf) process because of unknown transition states after emission of prefission γ-rays. So we use the same mass distribution as in the (n,f) reaction. In order to evaluate $\overline{\Gamma}_{\gamma f}$ and $\overline{E}_{\gamma f}$ for resonances of ^{235}U we include experimental results of Howe et al. [2], Scherbakov [5], and Shackleton [8].

Calculation of model parameters

The general energy balance in fission reads

$$\overline{Q}(A_L/A_H) + B_n + E_n = \overline{TKE}_d(A_L/A_H) + \overline{TXE}_d(A_L/A_H), \qquad (7)$$

where $\overline{Q}$ is the average energy release in fission for a given mass split A_L/A_H. B_n and E_n are the binding energy and the kinetic energy of the incidence neutron, respectively. $\overline{TXE}_d$ denotes the total fragment excitation energy. Both terms in the right-hand side of equation (7) can be expressed with reference to scission point:

$$\overline{TKE}_d(A_L/A_H) = \overline{E}_{c,d}(A_L/A_H) + \overline{E}_{pre,d}, \qquad (8)$$

$$\overline{TXE}_d(A_L/A_H) = \overline{E}_{def,d}(A_L) + \overline{E}_{def,d}(A_H) + \overline{E}_{dis,d}. \qquad (9)$$

The pre-scission kinetic energy $\overline{E}_{pre,d}$ and dissipative energy $\overline{E}_{dis,d}$ were calculated from Grossmann et al. [9]. Using experimental data of kinetic energy [1] and neutron multiplicity [2] as function of mass number we fitted the parameters $\bar{E}_{c,d}(A_L/A_H)$, $\bar{E}_{def,d}(A_L)$, and $\bar{E}_{def,d}(A_H)$ of the multimode model which reproduce these data within the experimental errors (Figs.1, 2).

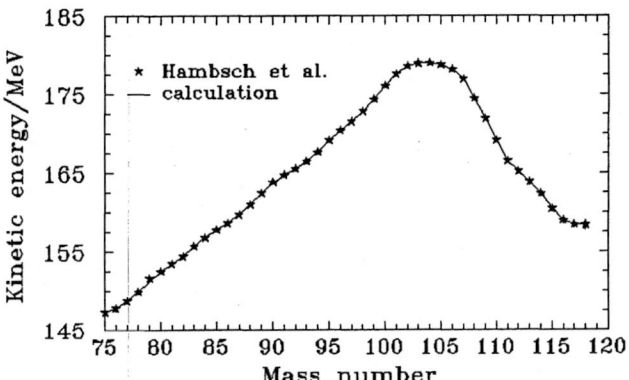

Fig. 1 Kinetic energy as function of mass number for thermal-neutron induced fission.

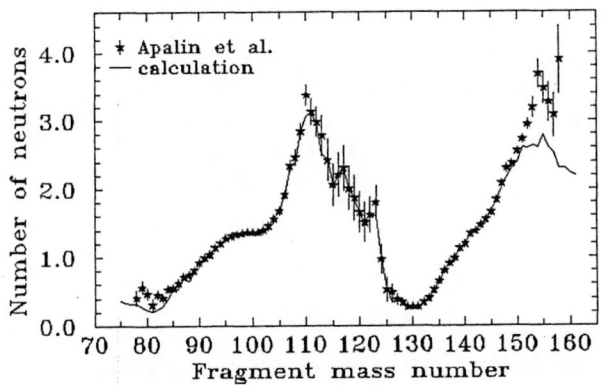

Fig. 2 Neutron number as function of mass number for thermal-neutron induced fission.

The average excitation energy of a fragment $\overline{XE}_d(A)$ is assumed to be given by

$$\overline{XE}_d(A) = \bar{\nu}_{n,d}(A)\,[\bar{B}_n(A) + \bar{\epsilon}_{n,d}(A)] +$$
$$\bar{N}^E_{\gamma,d}(A)\,\bar{E}^E_{\gamma,d}(A) + \bar{N}^S_{\gamma,d}(A)\,\bar{E}^S_{\gamma,d}(A), \quad (10)$$

where $\bar{B}_n$ is the average neutron binding energy and $\bar{\epsilon}_{n,d}(A)$ is the average neutron emission energy in the centre-of-mass system. $\bar{N}^E_{\gamma,d}(A)$ represents the average number of γ-rays of E1/E2-transitions. $\bar{N}^S_{\gamma,d}(A)$ includes statistical γ-rays [10], γ-rays of the (n,γf) process, and so called contraction γ-rays of the superlong fission path [10].
The most probable spin of a given fragment were calculated according to [11]. We assume that all neutrons and statistical γ-rays emitted from the fragments are s-neutrons and E1 γ-rays, respectively. The average spin of E1-, M1-, and E2-transition γ-rays was determined from experimental data [12], [13]. On the basis of energy- and angular momentum conservation relations we calculated the values $\bar{X}_d$ (Table 1).
For comparison we represent the deduced data of Brosa et al.

Table 1 Multimode model parameters for thermal neutron induced fission of ^{235}U

fission mode	Brosa	calculation			
	$\overline{TKE}^d$/MeV	$\overline{TKE}^d$/MeV	$\bar{\nu}^d_n$	$\bar{N}^d_\gamma$	$\bar{E}^d_\gamma$/MeV
stand I	183 ± 6	179.3	1.78	7.03	8.02
stand II	151 ± 4	168.6	2.56	6.79	6.29
superl.	159 ± 4	161.2	3.90	5.69	6.80

Total fission characteristics

Table 2 includes calculational and experimental values of $\overline{TKE}$, $\bar{\nu}_n$, $\bar{N}_\gamma$, and $\bar{E}_\gamma$, which show reasonable agreement within the error margins.

Table 2 Averages for thermal fission of ^{235}U

value	calculation	experiment
$\overline{TKE}$ [MeV]	170.61 ± 0.02	170.604 ± 0.005 [1]
$\bar{\nu}_n$	2.418 ± 0.010	2.4251 ± 0.0034 [14]
$\bar{N}_\gamma$	6.75 ± 0.25	6.51 ± 0.30 [15]
$\bar{E}_\gamma$ [MeV]	6.61 ± 0.15	6.43 ± 0.30 [15]

In Figs. 3 - 6 a comparison between experimental and calculational results of $\overline{TKE}$, $\bar{\nu}_n$, $\bar{N}_\gamma$, and $\bar{E}_\gamma$ for several 4$^-$ resonances is made. According to the present evaluation, the (n,γf) process for 4$^-$ resonances of ^{235}U is characterized by the following averages

$$\bar{\Gamma}_{\gamma f}\,\bar{N}_{\gamma f} = (0.42 \pm 0.10) \text{ meV}$$
$$\bar{\Gamma}_{\gamma f}\,\bar{E}_{\gamma f} = (410 \pm 120) \text{ eV}^2.$$

The difference of calculational average neutron number as function of fragment mass at 19.3 eV with respect to the thermal values is shown in Fig. 7. It is observed that this distribution changes in the mass range of fission mode standard I.
In order to show how the fission modes and the (n,γf) process influence on total fission characteristics we calculated $\bar{\nu}_n$ with and without account for (n,γf) process, respectively (Fig.8). The structures in the left-hand side arise mainly from fluctuations in fission modes whereas the decrease of $\bar{\nu}_n$ for small resonances is due to the (n,γf) process.

Summary

It has been shown that the observed experimental fluctuations in total fission characteristics can be explained within a multimode fission model with account for the (n,γf) process. Including the influence of fission modes we can get information about the (n,γf) process ($\bar{\Gamma}_{\gamma f}$ and $\bar{E}_{\gamma f}$) from measurements of $\bar{\nu}_n$, $\bar{E}_\gamma$, and $\bar{N}_\gamma$.
On the basis of the present evaluation one may fairly conclude that the neutron number as function of fragment mass depends on the contribution of fission modes which influences the calculation of preneutron emission mass distribution from measured fragment kinetic energies.

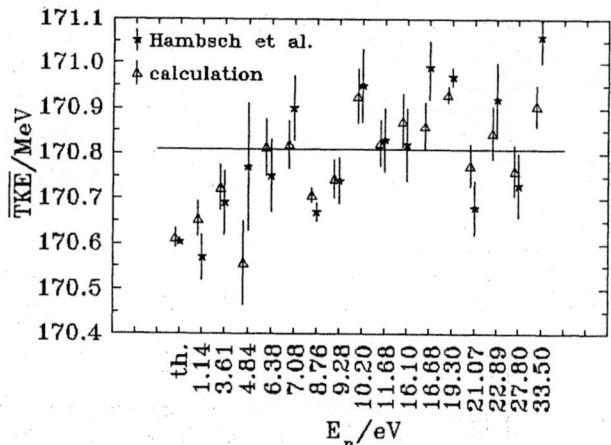

Fig. 3 $\overline{\text{TKE}}$ as function of resonance energy.

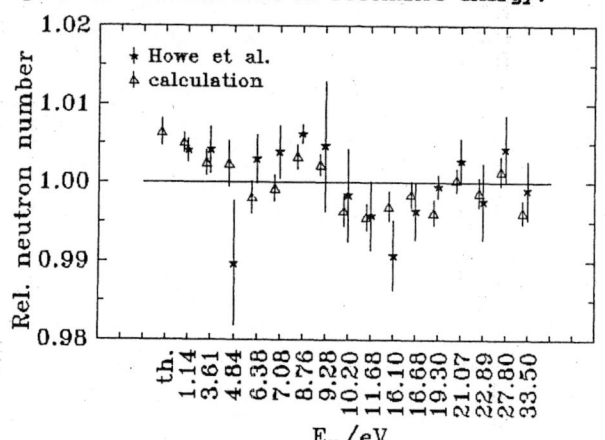

Fig. 4 Rel. $\bar{\nu}_n$ as function of resonance energy.

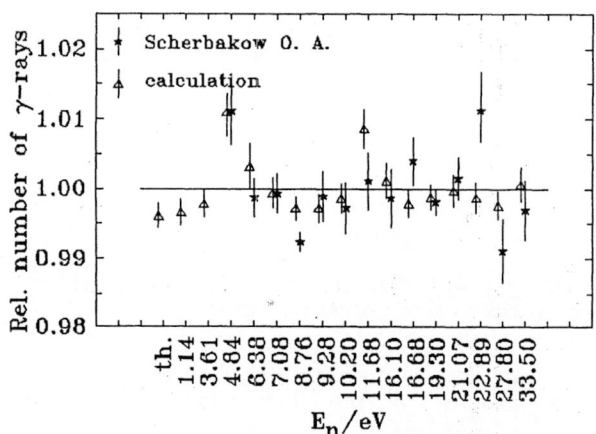

Fig. 5 Rel. $\overline{N}_\gamma$ as function of resonance energy.

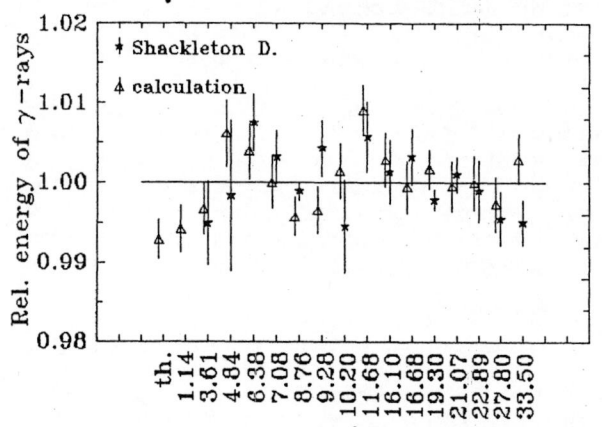

Fig. 6 Rel. $\overline{E}_\gamma$ as function of resonance energy.

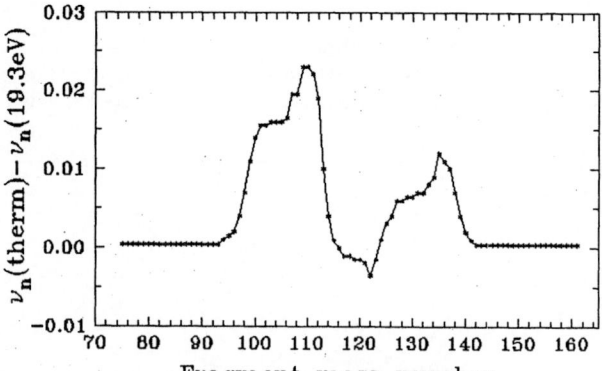

Fig. 7 Difference of $\bar{\nu}_n$ as function of mass number.

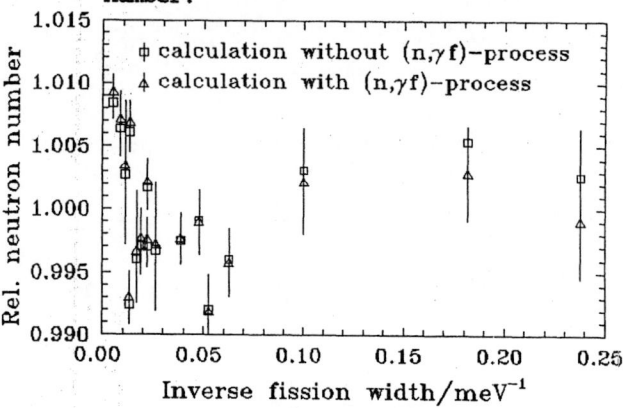

Fig. 8 Rel. $\bar{\nu}_n$ as function of inverse fission width.

REFERENCES

1) F.-J. Hambsch, H.-H. Knitter, C. Budtz-Joergensen, and J. P. Theobald, Nucl. Phys. A491, 56 (1989)

2) R. E. Howe, T. W. Phillips, and C.D. Bowman, Phys. Rev. C13, 195 (1976)

3) U. Brosa, S. Grossmann, and A. Mueller, Z. Naturforsch. 41A, 1341 (1986)

4) J. E. Lynn, Phys. Lett 18, 31 (1965)

5) O. A. Scherbakov, Phys. of Elem. Part. and Atom. Nucl. 21, 419 (1990)

6) V. I. Furman and J. Kliman, in Proc. of the XVII$_{th}$ Int. Symp. on Nucl. Phys, Nov. 1987, Gaussig, ZfK-646, 142 (1988)

7) M. S. Moore, G. de Saussure, and J.R. Smith, in Proc. IAEA Cons. Meet., Sept./Oct. 1981, Vienna, IAEA 1981 INDC (NDS)-129/GI, p.74

8) D. Shackleton, thesis, Paris, 1974

9) S. Grossmann, U. Brosa, and A. Mueller, Nucl. Phys., A481, 340 (1988)

10) R. Schmidt-Fabian, thesis, Heidelberg, 1989

11) P. Fong, in Proc. of a Symp. on Phys. and Chem. of Fission, May 1979, Juelich, vol.2, p. 373

12) W. R. Phillips, I. Ahmad, H. Emling, R. Holtzmann, R. V. F. Janssens, T. L. Khoo, and M. W. Dregert, Phys. Rev. Lett. 57, 3257 (1986)

13) M. A. C. Hotchkis, J. L. Durell, J. B. Fitzgerald, A. S. Mowbray, W. R. Phillips, I. Ahmad, M. P. Carpenter, R. V. F. Janssens, T. L. Khoo, E. F. Moore, L. R. Morss, Ph. Benet, and D. Ye, Phys. Rev. Lett. 64, 3123 (1990)

14) Neutron Cross Section, BNL, 1984

15) F. Pleasonton, R. L. Ferguson, and H. W. Schmitt, Phys. Rev. C6, 1023 (1972)

SIMULTANEOUS EVALUATION FOR CORRELATED DATA OF ^{239}Pu, ^{238}U, ^{235}U(n,f) AND ^{238}U(n,γ) CROSS SECTION

Liu Tingjin , Deng Jingshan

Chinese Nuclear Data Center ,
Institute of Atomic Energy ,
P. O. Box 275(41) , Beijing , P. R. China

Abstract: A way to construct the covariance matrices of the multi–set experimental data has been studied. The simultaneous evaluation for correlated data proposed in this paper has been applied to evaluate the cross sections of ^{239}Pu(n,f), ^{238}U(n,f), ^{238}U(n,γ) and ^{235}U(n,f) and the ratios $R[\sigma_f(^{239}$Pu$) / \sigma_f(^{235}$U$)]$, $R[\sigma_f(^{238}$U$) / \sigma_f(^{235}$U$)]$ and $R[\sigma_\gamma(^{238}$U$) / \sigma_f(^{235}$U$)]$, and their covariance matrices were constructed. The consistent relations of different reaction cross sections and the correlations of experimental data have been taken into account in the evaluations.

(neutron data evaluation,simultaneous evaluation,spline fitting,covariance matix, correlative data processing, fission cross section, capture cross section, ^{235}U, ^{238}U, ^{239}Pu)

Introduction

Among the most important cross sections required for nuclear engineering are the fission cross sections of ^{239}Pu ^{238}U, ^{235}U and the capture cross section of ^{238}U. Now there are many measurements of these quantities and their ratios to ^{235}U(n,f) cross section over the energy range 30 keV~ 20 MeV. It is necessary to critically collect and evaluate the available experimental data and to produce recommended cross sections with high accuracy and reliability.

The covariance matrices of experimental data are very important in nuclear engineering at present time. We have tried to find a way to construct the covariance matrix from the statistical property of experimental data. The covariance matrix includes all information of experimental errors, not only uncertainties but also the correlations.

Many evaluations for the cross sections of ^{239}Pu(n,f), ^{238}U(n,f), ^{238}U(n,γ) and ^{235}U(n,f) had been done. Most of them are isolated or individual evaluations, not considering the relations of different reactions and the correlations of experimental data. The present method deals wilh the simultaneous evaluation for correlated data, in which the consistent relations of different reactions are taken into account. With this developed method the cross sections of ^{239}Pu(n,f), ^{238}U(n,f), ^{238}U(n,γ) and ^{235}U(n,f) have been evaluated and their correlation covariance matrices have been given.

1. Evaluation for Experimental Data

The experimental data for $\sigma_f(^{239}$Pu$)$, $\sigma_f(^{238}$U$)$, $\sigma_\gamma(^{238}$U$)$, $\sigma_f(^{235}$U$)$, $R[\sigma_f(^{239}$Pu$) / \sigma_f(^{235}$U$)]$, $R[\sigma_f(^{238}$U$) / \sigma_f(^{235}$U$)]$ and $R[\sigma_\gamma(^{238}$U$) / \sigma_f(^{235}$U$)]$ (altogether 169 sets)were evaluated [1] .

2. Data Processing Method

2.1 Spline Curve Fitting for Individual Reaction With Knot Optimization [2]

With the program SPF of the multi–set data spline fitting, the above seven evaluated groups of experimental data for $\sigma_f(^{239}$Pu$)$, $\sigma_f(^{238}$U$)$, $\sigma_\gamma(^{238}$U$)$, $\sigma_f(^{235}$U$)$, $R[\sigma_f(^{239}$Pu$ / \sigma_f(^{235}U)]$, $R[\sigma_f(^{238}$U$) / \sigma_f(^{235}$U$)]$ and $R[\sigma_\gamma(^{238}$U$) / \sigma_f(^{235}$U$)]$ are fitted respectively by 3–order spline function through adjusting initial knots and Σ_k and optimizing the knots automatically to make χ^2 smaller. In the fitting, the width for each set of data, Σ_k , were taken very small ($10^{-4}\sim 10^{-7}$B) so that the errors $\Delta\sigma_i$ of fitted values could be considered as only statistical ones [3]. The fitting result of $R[\sigma_f(^{238}$U$) / \sigma_f(^{235}$U$)]$ is shown as an example in Fig. 1.

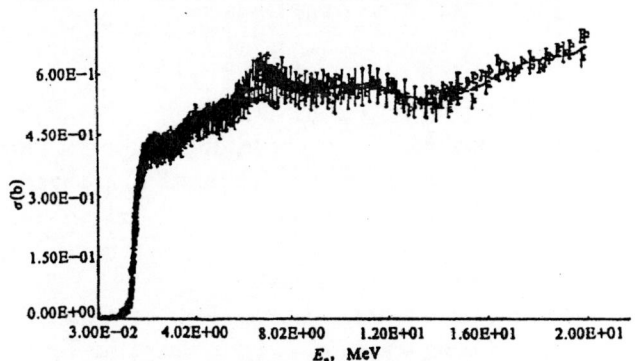

Fig. 1 The individual fitting curve of $R[\sigma_f(^{238}$U$) / \sigma_f(^{235}$U$)]$

2.2 Construction of Covariance Matrices for Experimental Data

Drawing a picture of experimental data, according to the discrepancy situation of the data, the systematic errors Sm can be estimated [1]. Fig. 2 is an example to obtain S_{11} of the 11-th interval (3~5 MeV) of $\sigma_f(^{235}U)$.

From the calculated statistical errors $\Delta\sigma_j$ $(j=1,2,...,N_E)$ and estimated systematic errors S_m in above way for each fitted curve, the covariance matrices for $\sigma_f(^{239}Pu)$, $\sigma_f(^{238}U)$, $\sigma_\gamma(^{238}U)$, $\sigma_f(^{235}U)$, $R[\sigma_f(^{239}Pu)/\sigma_f(^{235}U)]$, $R[\sigma_f(^{238}U)/\sigma_f(^{235}U)]$ and $R[\sigma_\gamma(^{238}U)/\sigma_f(^{235}U)]$ were constructed (Suppose no correlation exists between curves).

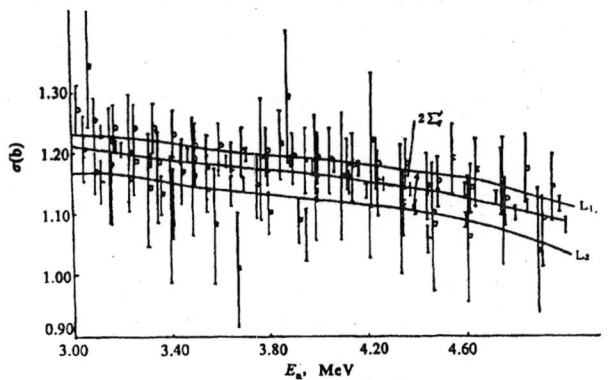

Fig. 2 An example to get systematic error :
S_{11}, the 11-th interval of $\sigma_f(^{235}U)$

2.3 Simultaneous Fitting for Multi-Curves

With the program of the simultaneous evaluation for correlative data [4], seven individual fitted curves were fitted simultaneously . The spline knots were basically taken as the same as those of individual fitting, but slightly adjusted if it is needed. The weights of each curves were changed according to their reliability. In fact, the weights for ratios were adjusted larger because of their higher reliability.

In order to analyze and compare the results, the simultaneous evaluation for independent data was also done, for which the covariance matrices are fully diagonal.

3. Results and Discussion

Using the simultaneous evaluation method for correlative data mentioned above, four cross sections and their covariance matrix were given, which are consistent and correlative with each other.

3.1 Comparision of Results

As an example, the comparison of the results obtained in different ways for ^{239}Pu is shown in Figs. 3. The simultaneous fitted curves for independent data disagree with individual fitted curves at high energy inter-

vals. These are due to adjusting of multi-curves to make them consistent.

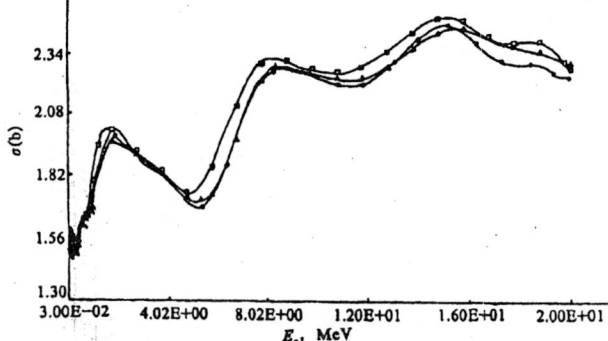

Fig. 3 The fitted cross section curve for $\sigma_f(^{239}Pu)$

-•- individual fitting curve
-▲- simultaneous evaluation for correlative data
-■- simultaneous evaluation for independent data

The simultaneous fitted curves for correlated data move roughly parallel to one side compared with the simultaneous fitted curves for independent data. This may be caused by the data correlation.

Fig. 4 shows the difference between the present evaluation curve and ENDF/B-V for $\sigma_f(^{235}U)$ at high energy range. The reason is that the $^{235}U(n,f)$ cross section curves recommended by ENDF/B-V are the results of individual evaluation, not adjusting for consistency of different measurements. Fig. 4 also shows the parallel shift of our present curve relative to ENDF/B-V curve, which is caused by data correlation introduced in our evaluation.

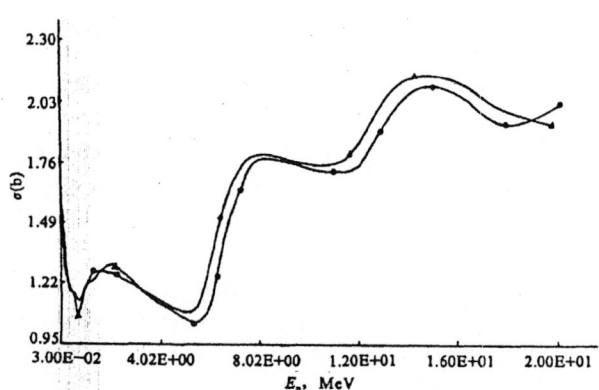

Fig. 4 The comparison between the present evaluation and ENDF/B-V for $\sigma_f(^{235}U)$

-▲- present evaluation   -•- ENDF/B-V

Fig. 5 also shows the parallel shift of our present fitted curve relative to Uenohara's (Japan) [5] curve for $^{235}U(n,f)$ cross section, which is the result of the simultaneous evaluation for independent data by Y. Uenohara. This is caused by data correlation introduced in our evaluation.

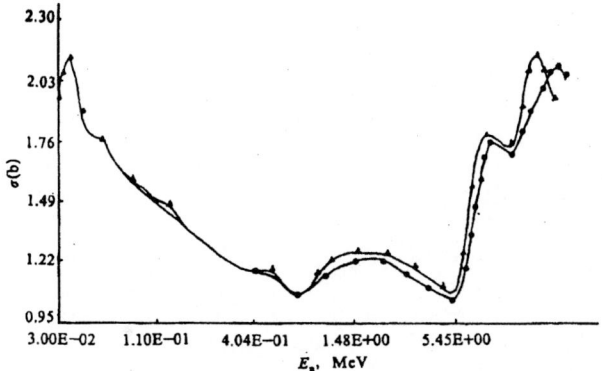

Fig. 5 The comparison between present evaluation
and Y. Uenohara's for $\sigma_f(^{235}U)$.

—▲— present evaluation, —●— Y. Uenohara(Japan)

3.2 The Covariance Matrix for Simultaneously Evaluated Data

30 output energy points were chosen to decrease the order of the covariance matrix because the number of the output points does not influence the correlated coefficients [6]. The output matrix can be divided as follows :

$$\begin{bmatrix} A & B & C & D \\ & E & F & G \\ & & H & I \\ & & & J \end{bmatrix}$$

where the diagonal matrices A, E, H and J are the covariance matrices of four evaluated cross sections themselves for $^{239}Pu(n,f)$, $^{238}U(n,f)$, $^{238}U(n,\gamma)$ and $^{235}U(n,f)$. Each of nondiagonal matrices B, C, D, F, G and I is covariance matrix between $\sigma_f(^{239}Pu)$ and $\sigma_f(^{238}U)$, $\sigma_f(^{239}Pu)$ and $\sigma_\gamma(^{238}U)$, $\sigma_f(^{239}Pu)$ and $\sigma_f(^{235}U)$, $\sigma_f(^{238}U)$ and $\sigma_\gamma(^{238}U)$, $\sigma_f(^{238}U)$ and $\sigma_f(^{235}U)$, $\sigma_\gamma(^{238}U)$ and $\sigma_f(^{235}U)$, respectively.

(1) The Evaluated Data Covariance Matrices for Each Reaction, A, E, H and J.

As an example Fig. 6 shows the two points correlation with others for $\sigma_f(^{239}Pu)$. From Fig. 6 it can be seen that the correlation coefficients at the same point of a reaction are 1, this is the inevitable result in physics. The correlation coefficients with other points are roughly the same as the input ones except nearby points, for which they are larger.

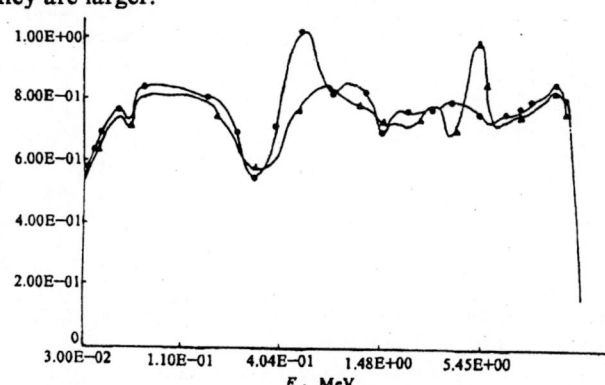

Fig. 6 The correlation for evaluated $^{239}Pu(n,f)$ cross section

—●— 0.5 MeV, —▲— 5.2 MeV

(2) The Correlation Coefficients at Same Point for Different Reactions

It was found that the correlation coefficients with $\sigma_f(^{235}U)$ are larger than others (the reason is that $^{235}U(n,f)$ cross section is used as "standard"), and the correlation coefficients between two reactions become larger if their cross sections are larger. Possibly the bigger cross sections contribute more to simultaneous fitting in adjusting of cross sections and ratios.

(3) The Correlation Covariance Matrices at Different Points for Different Reactions, B, C, D, F, G and I

The correlation coefficients at different energy points are smaller than that at the same energy point. There are positive correlation and negative correlation, which can be explained as follows : If the fitted data $R = \sigma_i / \sigma_j$ increases, either σ_i increases, or σ_j decrease, then correlation between σ_i and σ_j is negative; or, if σ_i increase faster than σ_j, then correlation between σ_i and σ_j is positive. It can be explained in the same way if the fitted data $R = \sigma_i / \sigma_j$ decreases.

4. Conclusion

The methods and programs have been developed for construction of the covariance matrices of experimental data and for simultaneous evaluation of correlated data. The simultaneous evaluation for correlation data is an advanced method, it introduces the consistent relations of different reaction cross sections by ratios and the correlation of experimental data.

As the newly measured experimental data and more advanced methods have been used in the present evaluation, so the recommended cross sections for $^{239}Pu(n,f)$, $^{238}U(n,f)$ $^{238}U(n,\gamma)$, and $^{235}U(n,f)$ with their correlation covariance matrices are more reasonable and reliable.

The authors express their appreciation to Drs. Liang Qichang (Chinese Nuclear Data Center, IAE), Tang Guoyou (Peking Univ.) for their discusion on the evaluations of $\sigma_f(^{239}Pu, ^{238}U)$ and $\sigma_{n\gamma}(^{238}U)$, Chen Baoqian, Li Shubing (Nankai Univ.) for their help in computer program and Liu Jianfeng, Zhang Xizhi, Lu Zuhui (Zhengzhou Univ.) for their beneficial advices.

REFERENCES

1. Liu Tingjin, Deng Jingshan : Comm. of Nucl. Data Progress 4, 49(1991)
2. Liu Tingjin, Zhou Hongmo, Liu Renqiu : Comm. of Nucl. Data Progress 2, 58(1989)
3. Zhou Hongmo, Liu Renqiu, Liu Tingjin : Atom. Ener. Scic. & Tech. 21, 4, 389(1987)
4. Liu Tingjin, Chen Baoqian : Mathematical Treatment and Program for Simul. Evaluation (to be published)
5. Y. Uenohara, Y. Kanda: Nuclear Data for Sci. and Techn., p639(1982)
6. Liu Tingjin, Chen Baiqian, Zhou Hongmo, Li Shubing : Atom. Ener. Scic. & Tech. 24(1), 15(1990)

EVALUATION AND TESTING OF n + ^{239}Pu DATA FOR ENDF/B-VI IN THE keV AND MeV ENERGY REGIONS

P. G. Young and R. E. MacFarlane

Theoretical Division, MS B243
Los Alamos National Laboratory
Los Alamos, New Mexico 87545, USA

Abstract: A new analysis of neutron-induced reactions on ^{239}Pu above E_n = 20 keV was performed as part of the evaluation activity for Version VI of ENDF/B. This study merges results from a new theoretical analysis of n + ^{239}Pu reactions with covariance analyses of experimental data for the ^{239}Pu total and (n,f) cross sections and for the prompt fission neutron multiplicity. The results of this study are combined with a new analysis of the resolved and unresolved resonance regions of ^{239}Pu to produce a new evaluation spanning the incident neutron energy range 10^{-5} eV to 20 MeV. Preliminary data testing calculations for several fast critical assemblies were performed in conjunction with the evaluation.

(Keywords: ^{239}Pu, neutron reactions, data evaluation, covariance analysis, benchmark data testing)

Introduction

Significant new experimental data [1] for ^{239}Pu became available in the keV and MeV neutron energy ranges since the previous update of the ENDF/B data base (ENDF/B-V.2). Especially important are new measurements of the neutron total cross section, the (n,f) cross section, and the prompt fission neutron multiplicity $\bar{\nu}_p(E_n)$. Additionally, as part of the ENDF/B-VI standards evaluation, the Cross Section Evaluation Working Group's Standards Subcommittee released in 1987 a comprehensive covariance analysis [2] that incorporated a large body of experimental data in a simultaneous analysis of standards and related cross section data, including measurements of the ^{239}Pu(n,f) cross section. To consolidate and incorporate the new ^{239}Pu experimental results as well as to check and update results for the ^{239}Pu(n,f) cross section from the simultaneous standards analysis, we performed new covariance analyses of the σ_{tot}, σ_{nf}, and $\bar{\nu}_p(E_n)$ data, including consideration of uncertainties and correlations within and among a variety of experiments. The results were combined with a recent analysis of the resolved and unresolved resonance parameters for ^{239}Pu [3] and with calculations of (n,n), (n,n'), and (n,xn) reactions from a new theoretical analysis.[4]

The assembled evaluation was tested by calculating several critical assemblies, including the Los Alamos fast criticals JEZEBEL, JEZEBEL-PU, and FLATTOP-PU, and the Argonne fast reactor critical ZPR-6/7.

Covariance Analyses and Results

For the ^{239}Pu(n,f) cross section, we initially planned to simply adopt results from the simultaneous covariance analysis of standards reactions for ENDF/B-VI. In addition to the actual standards cross section measurements, this thorough analysis included ^{238}U(n,γ), ^{238}U(n,f), and ^{239}Pu(n,f) cross sections together with interconnecting ratio data. However, the standards analysis was completed before results from two new measurements of the ^{239}Pu(n,f)/^{235}U(n,f) cross section ratio were available.[5,6] The new measurements and the ^{239}Pu(n,f) cross section from the standards analysis were found to disagree above E_n ~ 6 MeV, as is shown in Fig. 1. Furthermore, it was discovered that use of the standards ^{239}Pu(n,f) cross section in a preliminary version of the ENDF/B-VI evaluation led to an underprediction of k_{eff} by ~1% in the JEZEBEL ^{239}Pu bare sphere critical assembly.[7] Accordingly, it was decided to perform a new covariance analysis of the ^{239}Pu(n,f) ratio and absolute cross section data, along with new analyses of $\bar{\nu}_p(E_n)$ and the neutron total cross section of ^{239}Pu.

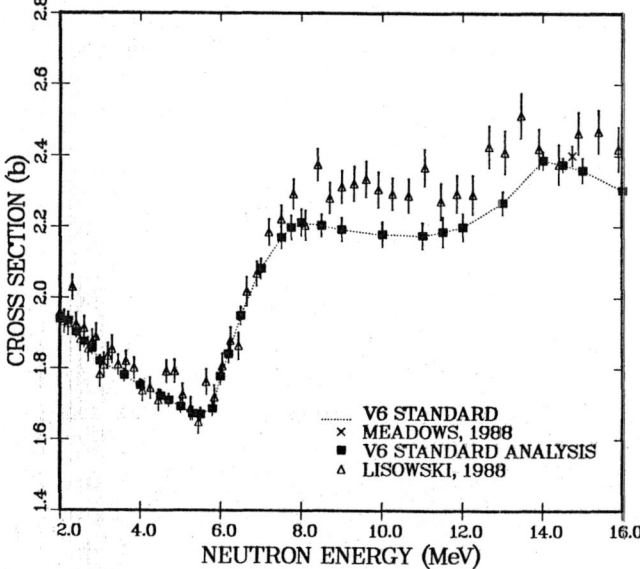

Fig. 1. Comparison of ENDF/B-VI standards evaluation of the ^{239}Pu(n,f) cross section with recent ratio measurements by Lisowski *et al.* [5] and Meadows, [6] converted to absolute cross sections. The dotted curve connects the solid squares, which are the covariance analysis results from the standards analysis.

All available experimental data and standard deviations were included in the analyses.[1] Where necessary, the $\bar{\nu}_p(E_n)$ and σ_{nf} data were converted to ENDF/B-VI standards prior to the analyses. Approximate correlation matrices were constructed based on the details of the various experiments. In the case of ^{239}Pu(n,f) cross section, separate analyses were performed of the absolute cross section data and of the cross section ratio data relative to ^{235}U fission. All covariance analyses were performed using the Baysian analysis code GLUCS.[8]

Results from the covariance analysis of the ^{239}Pu(n,f)/^{235}U(n,f) ratio data (labeled ENDF/B-VI) are compared to a sampling of experimental data in Fig. 2, including the newer measurements.[5,6] In Fig. 3 the final ENDF/B-VI results for the ^{239}Pu(n,f) cross section are compared to ENDF/B-V, together with results of the present and the standards covariance analyses. Note that the two covariance analyses and their uncertainties are quite consistent at neutron energies below ~6 MeV. The evaluation closely follows the standards analysis at lower energies, increased by approximately one standard deviation. At higher energies the new analysis lies above the standards values due to influence of the new measurements. [5,6]

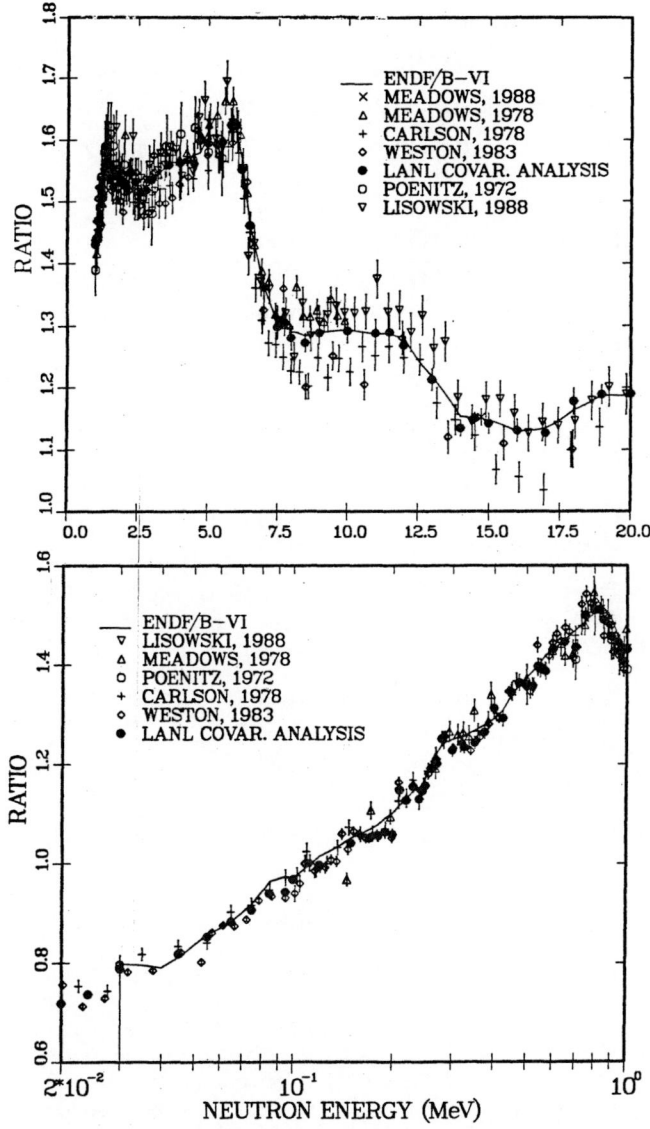

Fig. 2. Evaluated and measured neutron-induced ^{239}Pu(n,f)/^{235}U(n,f) fission cross-section ratio from 20 keV to 20 MeV. The solid circles are the results of the LANL covariance analysis, and the curve is the ENDF/B-VI evaluation. A small selection of experimental data is compared.

The ^{239}Pu neutron total cross section that resulted from the present covariance analysis is compared to ENDF/B-V.2 in Fig. 4. Similarly, results from the $\bar{\nu}_p(E_n)$ covariance analysis are given in Fig. 5 along with ENDF/B-V.2 and assorted experimental data. Much of the available data was measured relative to $\bar{\nu}$ for ^{252}Cf, and the data shown have been corrected for the ENDF/B-VI standard.

Integral Data Comparisons and Conclusions

To begin testing the new evaluations in ENDF/B-VI, preliminary calculations were made of several fast critical experiments that are ENDF/B benchmarks. [7] Results are presented here for the bare uranium and plutonium spheres GODIVA and JEZEBEL, the "dirty" plutonium sphere JEZEBEL-PU, the uranium-reflected plutonium sphere FLATTOP-PU, and liquid-metal fast breeder reactor benchmark ZPR-6/7. Homogeneous, one-dimensional, spherical calculations were made using P_3S_{16} transport theory for the assemblies and P_0S_8 theory for ZPR-6/7. Eighty-group cross-section libraries were generated for these calculations using the NJOY nuclear data processing system. [9]

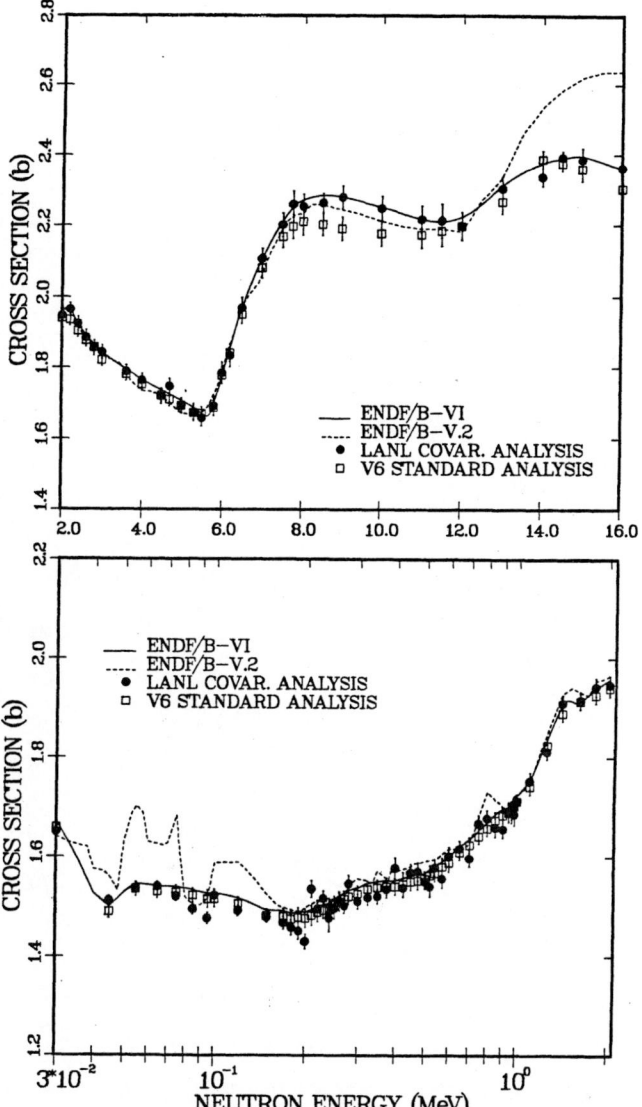

Fig. 3. Comparison of the LANL covariance analysis of the ^{239}Pu(n,f) cross section with the Version VI standards analysis and with the ENDF/B-VI and ENDF/B-V evaluations.

Results of these preliminary calculations of the integral assemblies are given in Table 1. Dramatic changes are not seen in the calculated k_{eff} eigenvalues and reaction rate ratios, although some improvements over ENDF/B-V can be noted. k_{eff} is slightly improved for two of the four ^{239}Pu assemblies shown, and there appears to be a small improvement in the reaction ratios generally. A more noticeable improvement is seen in comparing the ^{238}U/^{235}U and ^{237}Np/^{235}U fission reaction ratios for GODIVA and JEZEBEL. In particular, the asymmetry or bias between the ^{235}U and ^{239}Pu assemblies that was apparent with ENDF/B-V libraries appears to be reduced. Overall, the calculations indicate somewhat greater consistency among the integral calculations for the small criticals with ENDF/B-VI libraries than was the case with ENDF/B-V. The ZPR-6/7 results do not include heterogeneity, which would be expected to increase R_{eff} by about 1%. Calculations with more detailed models will be needed.

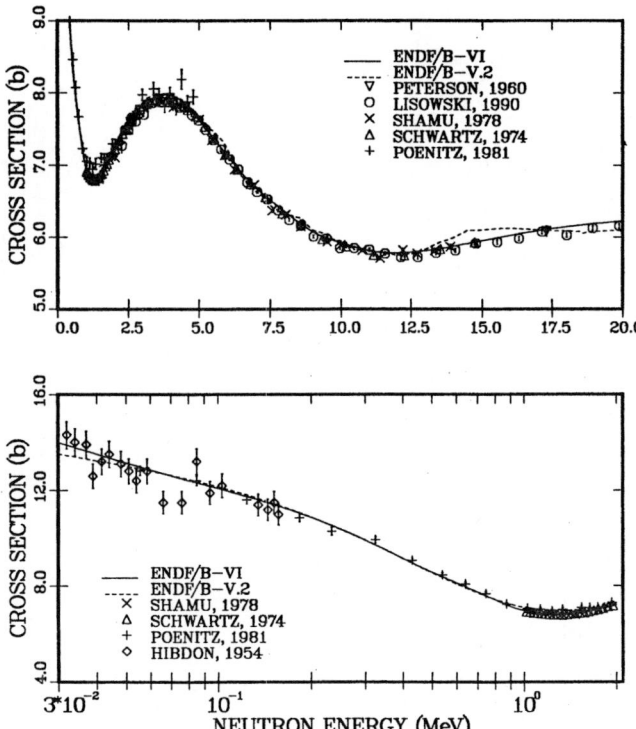

Fig. 4. Measured and evaluated neutron total cross section of ^{239}Pu from 30 keV to 20 MeV. The solid curve is ENDF/B-VI, based on the present covariance analysis, and the dashed curve is the ENDF/B-V evaluation. Only a selection of experimental data are shown.

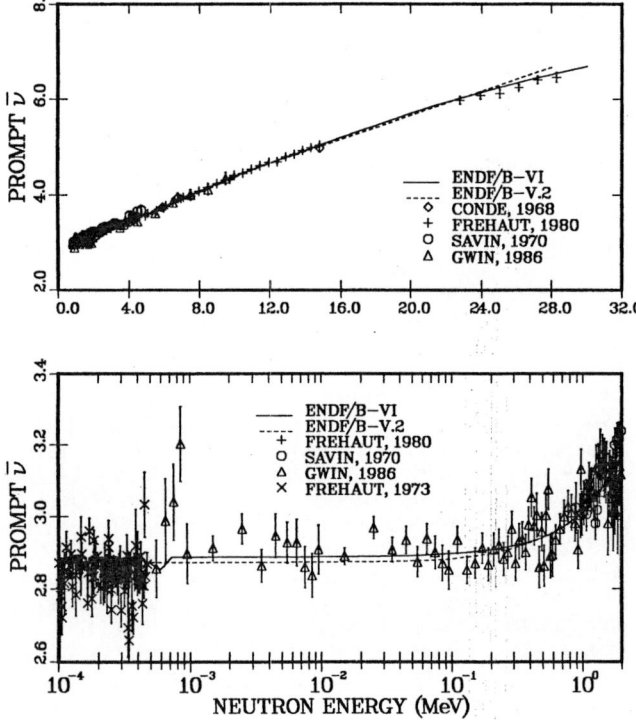

Fig. 5. Comparison of evaluated $\bar{\nu}_p(E_n)$ for n + ^{239}Pu reactions between 100 eV and 30 MeV with a selection of experimental data.

Table 1 Preliminary Critical Assembly Performance Parameters for ^{239}Pu.

Parameter[a]	ENDF/B-V.2	ENDF/B-VI	Assembly
k_{eff}	0.9990	0.9979	GODIVA
f^{28}/f^{25}	1.037	0.970	
f^{37}/f^{25}	1.043	0.961	
f^{49}/f^{25}	0.985	0.977	
f^{23}/f^{25}	0.986	1.000	
k_{eff}	0.9982	0.9988	JEZEBEL
f^{28}/f^{25}	0.961	0.974	
f^{37}/f^{25}	0.979	0.970	
f^{49}/f^{25}	0.966	0.975	
f^{23}/f^{25}	0.985	1.000	
k_{eff}	0.9918	0.9912	JEZEBEL-PU
f^{28}/f^{25}	0.954	0.965	
f^{37}/f^{25}	1.009	0.998	
k_{eff}	1.0064	1.0049	FLATTOP-PU
f^{28}/f^{25}	0.973	0.984	
f^{37}/f^{25}	1.000	0.987	
k_{eff}	0.9857	0.9890	ZPR-6/7
f^{28}/f^{49}	1.045	1.046	
f^{25}/f^{49}	1.009	1.037	
c^{28}/f^{49}	1.106	1.088	

[a]Ratios are calculated result divided by experimental result.

References

1. Experimental data available from the CSISRS compilation by the National Nuclear Data Center, Brookhaven National Laboratory.
2. A. D. Carlson, W. P. Poenitz, G. M. Hale, and R. W. Peelle, in Proc. *Int. Conf. Nucl. Data for Basic and Applied Science*, Santa Fe, N.M. (Gordon and Breach Sci. Publ., 1986) v.2, p. 1429; R. Peelle and H. Condé, in Proc. *Int. Conf. on Nucl. Data for Science and Tech.*, Mito, Japan, 30 May-3 June 1988 [Ed. S. Igarasi, Saikon Publishing. Co., Ltd., Toyko, 1988], p. 1005.
3. H. Derrien and G. de Saussure, Oak Ridge National Laboratory reports ORNL/TM-10986 (1989) and ORNL/TM-11490 (1990).
4. P. G. Young and E. D. Arthur, Proc. Int. Conf. *Nuclear Data For Science and Technology*, Jülich, Germany, 13-17 May 1991 (to be issued).
5. P. W. Lisowski, J. L. Ullmann, S. J. Balestrini, A. D. Carlson, O. A. Wasson, and N. W. Hill, "Neutron-Induced Fission Cross-Section Ratios for ^{232}Th, ^{235,238}U, ^{237}Np, and ^{239}Pu from 1 to 400 MeV," Proc. Int. Conf. *Nucl. Data for Science and Technology*, Mito, Japan, May 30 - June 3, 1988 (Ed. S. Igarasi, Saikon Publ. Co., Ltd., 1988) p. 97.
6. J. W. Meadows, *Nucl. Sci. Eng.* **85**, 271 (1983); J. W. Meadows, *Ann. Nucl. En.* **15**,421 (1988).
7. H. Alter, R. G. Kidman, R. J. LaBauve, R. Protsik, and B. A. Zolotar, "ENDF-202:Cross Section Evaluation Working Group Benchmark Specifications," Brookhaven Nat. Lab. report BNL-19302 (ENDF-202).
8. D. M. Hetrick and C. Y. Fu, "GLUCS: A Generalized Least-Squares Program for Updating Cross-Section Evaluations with Correlated Data Sets," Oak Ridge National Laboratory report ORNL/TM-7341 (1980).
9. R. E. MacFarlane, D. W. Muir, and R. M. Boicourt, "The NJOY Nuclear Data Processing System, Volume I: User's Manual," Los Alamos National Laboratory report LA-9393-M, Vol. I (ENDF-324) (1982).

MULTILEVEL S-MATRIX ANALYSIS OF ^{235}U NEUTRON CROSS-SECTIONS UP TO 93 EV USING THE COLLISION MATRIX UNITARITY METHOD

V.V.Kolesov, A.A.Lukyanov

Obninsk Institute of Nuclear Power Engineering
P.O.B. 249020, IATE, Obninsk, Kaluga reg., USSR.

Abstract: A new multilevel fit to ^{235}U total and fission cross-section data up to 93 eV has been obtained in S-matrix formalism. The collision matrix unitarity method has been used to construct elastic scattering and neutron capture resonance cross-sections from the total and fission resonance parameters. So, interference effects have been taken into account for all partial cross-sections. Resonance parameters have been written in ENDF/B format.
Group averaged cross-sections and neutron transmission have been calculated and compared with other evaluations and experimental results.

Introduction

With the increasing interest to HCLWR safety problems, the accuracy of neutron cross-section representation becomes important again. In the frame of our work for multilevel S-matrix parametrisation of neutron cross-sections for fissile nuclei in Adler-Adler approach [1], we have generated a set of adjusted multilevel parameters of ^{239}Pu in the energy region from thermal up to 465 ev [2].
In this work we present the results of ^{235}U neutron cross-section parametrisation, using the collision matrix unitarity method, described in [2].

Method

Our method is based on the Adler-Adler approach. The main feature of the method consists of using the unitarity of the collision matrix as additional condition in the traditional S-matrix formalism. In this case, the absorption cross-section σ_a can be calculated using only the total resonance parameters G_r^t and H_r^t in the same form, as for σ_t:

$$\sigma_a = \frac{\pi}{k^2} \sqrt{E} \sum_J \sum_{r=1}^{N(J)} \frac{G_r^a}{\nu_r} \psi - \frac{H_r^a}{\nu_r} \chi \, ,$$

where $N(J)$ is the number of resonances with a specified J, and

$$G_r^a = G_r^t \cos(2\phi_0) + H_r^t \sin(2\phi_0) - G_r \sqrt{E} \, ,$$

$$H_r^a = H_r^t \cos(2\phi_0) - G_r^t \sin(2\phi_0) - H_r \sqrt{E} \, ,$$

$$G_r = \frac{1}{2g(J)} \sum_{r'(J)} \frac{(\nu_{r'}+\nu_r)\,\xi_{r'} + (\mu_{r'}-\mu_r)\,\zeta_{r'}}{(\nu_{r'}+\nu_r)^2 + (\mu_{r'}-\mu_r)^2} \, ,$$

$$H_r = \frac{1}{2g(J)} \sum_{r'(J)} \frac{(\mu_{r'}-\mu_r)\,\xi_{r'} - (\nu_{r'}+\nu_r)\,\zeta_{r'}}{(\nu_{r'}+\nu_r)^2 + (\mu_{r'}-\mu_r)^2} \, ,$$

$$\xi_{r'} = G_r^t G_{r'}^t + H_r^t H_{r'}^t \,; \quad \zeta_{r'} = G_{r'}^t H_r^t - G_r^t H_{r'}^t \,;$$

where the sum over $r'(J)$ includes all resonances with a specified J. This form can be derived from the unitarity of the S-matrix. So, if we have the set of total and fission resonances with parameters μ_r, ν_r, G_r^t, H_r^t, G_r^f, H_r^f, we can calculate all partial cross-sections with resonance-resonance interference terms.

The total cross-sections, measurements of Michaudon et al. [3] and fission cross-sections measurements of Moore et al. [4], Wagemans et al. [5], Blons [6] have been used for our analysis. The spin assignments have been made using all earlier estimations, mainly [4,7].

We have used a least-square minimisation program, based on the Nelder-Mead simplex method [8]. This method is very convenient in our case, because some parameters (G_r^t, H_r^t, G_r^f, and H_r^f) can be exactly calculated from linear regression. So, only a part of the resonance parameters (μ_r and ν_r) must be determined by the simplex method.

Results

Neutron cross-sections in the energy range 3.5 to 93 eV are described by 145 resonances with R'= 0.936 fm. Table 1 shows some energy integrals, calculated from the resonance parameters with the NJOY code, compared with ENDF/B-IV data and some experimental results. Neutron transmission values are compared with corresponding experimental data from [9] in Table 2.

After the thermal region will be finished the file of resonance parameters can be organised in ENDF/B format, because the proposed method can be easily realised in the NJOY/RECONR module by a small change of the subroutine CSAA.

Table 1. Energy Integrals of the ^{235}U Neutron Cross-Sections.

Energy, eV	σ_f			σ_a			σ_c		
	calcu-lated	ENDF/B-IV	Gwin [10]	calcu-lated	ENDF/B-IV	Gwin [10]	calcu-lated	ENDF/B-IV	Gwin [10]
10 -15	42.17	45.40	42.60	95.70	92.80	97.80	53.53	47.40	55.20
15 -20.5	58.17	59.69	56.55	100.4	94.99	97.45	42.19	35.30	40.90
20.5-33	34.24	37.26	35.12	64.18	61.28	62.64	29.94	24.02	27.52
33 -41	63.19	64.50	60.38	100.8	98.91	100.0	37.62	34.41	39.62
41 -60	49.66	49.95	47.42	74.84	73.81	73.84	25.18	23.86	26.42

Table 2. Neutron Transmission Values of ^{235}U as Function of Sample Thickness.

n, $10^{-2} \frac{nucl}{barn}$		Energy, eV		
		10-21.5	21.5-46.5	46.5-93
0.257	1	.806	.834	.850
	2	.819$\pm$.017	.835$\pm$.010	.859$\pm$.010
0.386	1	.735	.722	.789
	2	.739$\pm$.017	.752$\pm$.017	.788$\pm$.014
1.02	1	.529	.561	.570
	2	.550$\pm$.016	.566$\pm$.016	.582$\pm$.017
2.14	1	.322	.358	.356
	2	.356$\pm$.009	.369$\pm$.011	.379$\pm$.011
4.14	1	.152	.186	.180
	2	.164$\pm$.007	.186$\pm$.007	.192$\pm$.009

Notes:
  1. Calculations with our resonance parameters.
  2. Results from [9] (4-th group: 46.5-100 eV).

References

1. F.T.Adler, D.B.Adler: in Proc. Conf. on Breeding, Econ. and Safety in Fast Reactors, ANL-6792, 695(1963).
2. A.A.Lukyanov, V.V.Kolesov et al.: in Proc. Int. Conf. on Nuclear Data for Basic and Applied Science (Gordon and Breach Publ.), v.2, 1691(1986).
3. A. Michaudon et al.: Nucl. Phys., 69, 545(1965).
4. M.S. Moor, J.D. Moses, G.A. Keyworth: Phys. Rev. C.18, 1328(1978).
5. C. Wagemans, A.J.Deruytter: Ann. Nucl. Energy, 3, 437(1976).
6. J.Blons: Nucl.Sci.Eng., 51, 130(1973).
7. G.A.Keyworth, C.E.Olsen, F.T.Seibel et al.: Phys. Rev. Lett., 17, 1077(1973).
8. J.A.Nelder, R.Mead: The Comp. Journal, 7, 308(1965).
9. A.A. Vankov, V.F. Ukraintsev et al.: Nucl. Sci. Eng., 96, 122(1987).
10. R.Gwin, E.G.Silver, R.W.Ingle, H.Weaver: Nucl. Sci. Eng., 59, 79(1976).

MULTILEVEL RESONANCE ANALYSIS OF ^{59}Co NEUTRON TRANSMISSION MEASUREMENTS

G. de Saussure and N. M. Larson

Oak Ridge National Laboratory
Oak Ridge, Tennessee 37831-6356 USA

Abstract: Large discrepancies exist between the high-resolution ^{59}Co neutron transmission data of Harvey et al. and the resolved resonance parameters of ENDF/B-VI. In order to provide new resonance parameters consistent with these data, the high-resolution transmission measurements have been analyzed with the computer code SAMMY. Results of that analysis are reported here.

(^{59}Co transmission measurements, R-matrix analysis, resonance parameters)

Introduction

The resonance parameters describing the low-energy neutron cross sections of ^{59}Co are of considerable interest, not only because ^{59}Co is used as a structural material in nuclear technology but also because the thermal capture cross section and the capture resonance integral of ^{59}Co are widely used as standards and for dosimetry [1].

High-resolution neutron transmission measurements through several thicknesses of ^{59}Co were performed in 1986 at the Oak Ridge Electron Linear Accelerator (ORELA), in conjunction with the ENDF-B-VI evaluation of ^{59}Co done at Argonne National Laboratory. These measurements have much better resolution than previously published measurements [2,3]. The results of one of these measurements were used in the ENDF/B-VI evaluation of the ^{59}Co total cross section above 100 keV [4,5], but the data were not used below 100 keV where the ENDF/B-VI cross sections are represented by resolved resonance parameters based on a recent compilation of Mughabghab et al. [6].

The ENDF/B-VI resolved resonance parameters were used to compute values of transmission on the same energy scale as the ORELA transmission measurements; large discrepancies were observed between the computed and measured transmissions, indicating that the ENDF/B-VI resolved resonance parameters were inadequate. Therefore, all six transmission measurements were analyzed using the Reich-Moore approximation option in SAMMY [7], and new resonance parameters were obtained which are consistent with both the high resolution ORELA transmission measurements and the thermal cross sections.

In the next section of this report the experimental conditions of the ORELA transmission measurements are given, and it is shown that the ENDF/B-VI resolved resonance parameters are inconsistent with the results of these measurements. In the last section the resonance analysis with the code SAMMY is described and the new resolved resonance parameters are discussed.

The High Resolution Transmission Measurements

In 1986 a series of high-resolution transmission measurements was performed at ORELA through several thicknesses of ^{59}Co and under various resolution conditions, partly at the request of A. B. Smith for the ENDF/B-VI evaluation [4,5]. Only one of these measurements was processed at the time: Run 10454TS. This run was done with a very thick sample, 0.3 atoms/barn, at a flight path of 200 m and with the collimation selecting the unmoderated photoneutrons produced in the Ta target. This run is mostly useful at high energy and indeed was used only to define the total cross section above 100 keV in ENDF/B-VI.

All six runs have now been reduced to transmission versus energy. The experimental parameters for these runs are summarized in Table 1. Two of the runs with sample thicknesses of 0.3 and 0.075 atoms/b used a flight path of 200 m with the collimation looking at the unmoderated neutrons from the Ta target; the remaining four runs used a flight path of 80 m looking at the water moderator, and samples varying in thickness from 0.005 to 0.3 atoms/barn. The samples were kept at the physical temperature of 293 K; for a Debye temperature of ^{59}Co of 445 K, the effective temperature is computed as 330 K. All these runs have considerably better time resolution than the previously published transmission measurements [2,3,8].

Table 1. Experimental Parameters of the Transmission Measurements

Run No.	Flight Path (m)	Sample Thickness (atoms/b)	Detector	Collimation	Linac burst width
10454TS	201.575	0.3008	NE-110	Unmoderated Ta target	5-6 ns
10460TS	201.575	0.0750		Unmoderated Ta target	5-6 ns
10432T1	80.394	0.3008	^{6}Li glass	Water moderator	12 ns
10432T2	80.394	0.0750		Water moderator	
10428T1	80.394	0.0210		Water moderator	
10428T2	80.394	0.0050		Water moderator	

In order to test the validity of the ENDF/B-VI resonance parameters, the transmission for the sample thicknesses used in the measurements were computed with these resonance parameters and with the code SAMMY, using the "no fit" option of the code. Comparisons of these computed transmission ratios with the measurements showed rather large discrepancies, indicating that the ENDF/B-VI resonance parameters were inadequate.

Figure 1 shows a comparison between the computed transmission ratio (dashed line) and the measurement of run 10432T1 (see Table 1) in the neutron energy range 2 to 3 keV. The small resonance at 2.28 keV is not in ENDF/B-VI, probably because it is too weak to have been detected in previous transmission measurements, which had poorer resolution. The resonance at 2.85 keV has clearly wrong parameters, perhaps the result of a typographical error. The resonance parameters of ENDF/B-VI were taken somewhat literally from Ref. 6, and indeed the parameters of the resonance at 2.86 keV are as given in that reference (neutron width of 3.497 eV, capture width of 0.11 eV). However, in ENDF/B-V as well as in previous evaluations [9-11] or in previous editions of the neutron cross-section compilation BNL-325 [12], the neutron width of that resonance is always given as approximately 0.1 eV. The source of the error in Ref. 6 and in ENDF/B-VI was not traced further.

Figure 2 shows a comparison between the transmission ratio computed with ENDF/B-VI (dashed line) and

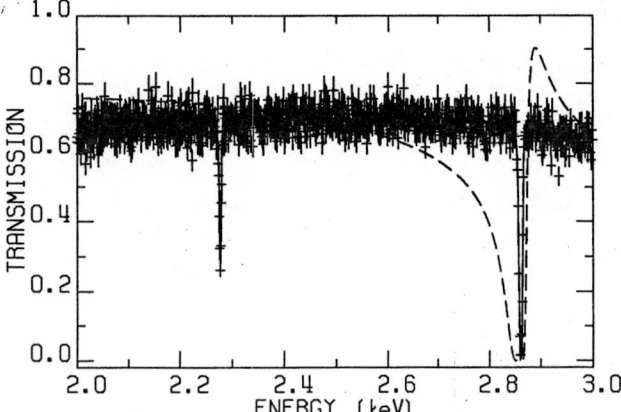

Fig. 1. ORELA data (crosses) vs transmission computed from ENDF/B-VI resonance parameters (dashed curve) and from parameters found in this work (solid curve), from 2 to 3 keV.

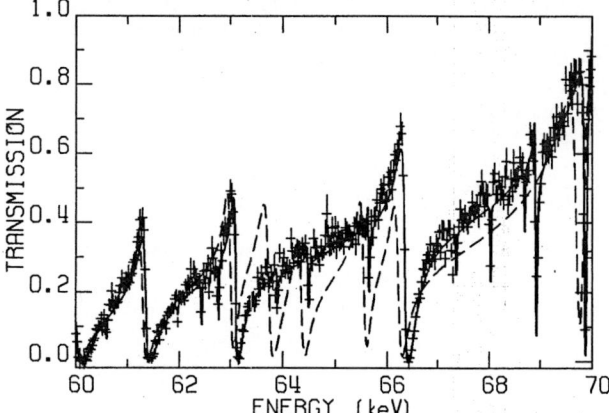

Fig. 2. ORELA data (crosses) vs transmission computed from ENDF/B-VI resonance parameters (dashed curve) and from parameters found in this work (solid curve), from 60 to 70 keV.

the measurement of run 10454TS over the neutron energy range from 60 to 70 keV. The figure illustrates the large discrepancies typical above 60 keV. The origin of these discrepancies could not be completely explained; in that energy region the resonance parameters published in Refs. 2 and 3 are inconsistent and ENDF/B-VI is not consistent with either of these references.

Because of these discrepancies between the experimental measurements and the calculations based on ENDF/B-VI, it was decided to analyze the six ORELA transmission measurements, using the Reich-Moore approximation and the resonance analysis code SAMMY. The assumptions of this analysis and its results are the subject of the next section.

Resonance Analysis of ^{59}Co Transmission Data

The spin J of most resonances and the orbital angular momentum ℓ for the smaller resonances cannot be determined from the data and were assigned somewhat arbitrarily as follows: (1) All the observed resonances were assumed to be due either to s-wave or to p-wave neutrons, a reasonable assumption. (2) Those resonances for which the neutron width divided by the square root of the neutron energy was larger than $0.1\sqrt{(eV)}$ were taken to be s-wave resonances; in many cases, the characteristic s-wave interference pattern between resonance and potential scattering can be observed. (3) Most s-wave levels were assigned a spin consistent with the assignment of Ref. 6, and therefore with ENDF/B-VI. (4) Some smaller resonances were also taken to be s-wave; these were selected so as to satisfy the long-range level-spacing correlation implied by the Dyson-Metha Δ_3 statistics [13].

(5) Since ^{59}Co has a spin of 7/2, the interaction with p-wave neutrons can lead to the resonance spin $J = 2$, 3, 4, or 5; however, for simplicity, all p-wave resonances were arbitrarily assigned spins of 3 or 4, as was also done in ENDF/B-V and ENDF/B-VI. All p-wave spin states were included in generating the potential scattering.

The neutron widths of ten large resonances above 100 keV were adjusted to account for the contribution below 100 keV of all the levels above that energy. Similarly, the parameters of four fictitious bound levels were adjusted to account for the contribution of the bound levels and to reproduce the evaluated cross sections in the thermal region. The parameters of the first resonance, at 132 eV, were not searched, because these parameters are well known and because the ORELA transmission measurements were not optimized for that low-energy range.

The scattering length was determined from the fit to the transmission and a value of 6.67 fm was obtained, in good agreement with the value 6.80 ± 0.70 fm given in Ref. 6. The set of resonance parameters obtained is consistent with the ORELA transmission measurements, as illustrated by the solid curves in Figs. 1 and 2, which show that the resonance parameters obtained here are more consistent with the measurements than are the ENDF/B-VI parameters. The value of χ^2 per degree of freedom for each of the data sets analyzed is close to unity, also indicating consistency between the resonance parameters and the measurements.

Figures 3 and 4 illustrate graphically the agreement between the measured transmission ratios and the calculations with the resonance parameters obtained in this

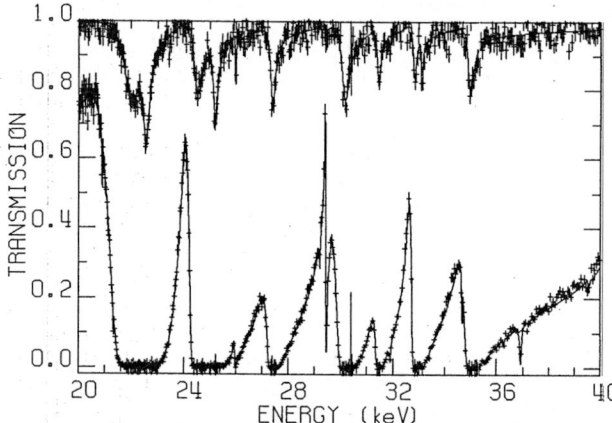

Fig. 3. Measured and computed transmission through 0.005 atom/barn (upper curve) and 0.3 atom/barn (lower curve) of ^{59}Co between 20 and 40 keV. The solid lines are computed with the resonance parameters obtained in this analysis.

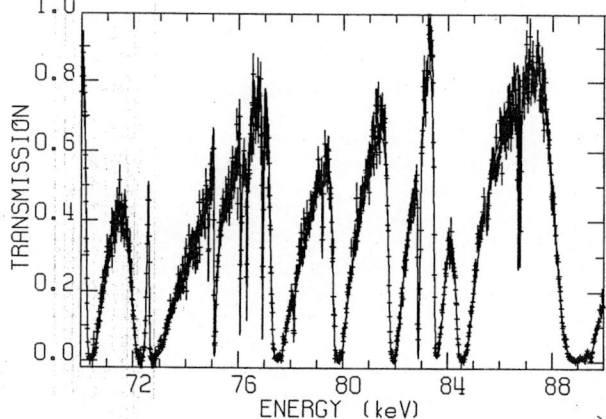

Fig. 4. Measured and computed transmission through 0.3 atom/barn of ^{59}Co between 70 and 90 keV. The solid line is computed with the resonance parameters obtained in this study.

study. In Table 2 the statistical properties and thermal parameters of the present evaluation are compared to the ENDF/B-VI values.

Table 2. Comparison of the present evaluation with ENDF/B-VI

	This study	ENDF/B-VI
Number of s-wave levels	80	117
$<\Gamma_n^0>$ (eV)	0.941	0.8226
$<\Gamma_\gamma>$ (eV) for $\ell=0$	0.4488	0.4445
Average s-wave level spacing (eV)	1299	1034
s-wave strength function	3.83×10^{-4}	3.93×10^{-4}
Number of p-wave levels	65	1
$<\Gamma_n^1>$ (eV)	0.237	2.806
$<\Gamma_\gamma>$ (eV) for $\ell=1$	0.36	0.37
Average p-wave level spacing (eV)	1562.5	
p-wave strength function	0.25×10^{-4}	
Cross sections at 0.0253 eV		
σ_t (b)	43.150	43.143
$\sigma_{n\gamma}$ (b)	37.170	37.190
σ_s (b)	5.980	5.953
Westcott g_γ factor	1.0000	1.0019
Channel radius (10^{-12} cm)	0.6672	0.620
Capture resonance integral (b)	75.74	75.72

The transmission measurements are not very sensitive to the value of the capture widths; therefore, the capture widths obtained in this analysis have large uncertainties. An attempt was made to adjust the capture widths to give the capture areas reported by Spencer et al. [14,15]; however, the attempt was not successful because the correspondence between the resonances observed in transmission and those reported with the capture measurements are not unambiguous; many multiplets are not resolved in the capture measurements. The problem is complicated by the large (20 to 40%) multiple scattering corrections that were made by Spencer et al. on the capture areas which had to assume prior values of the resonance parameters. A direct analysis of capture data with the code SAMMY cannot be done until SAMMY is modified to handle multiple scattering effects, an effort which is currently underway.

The capture cross section averaged over intervals from 1–100 keV is given in Table 3 to facilitate future evaluation work. The average capture cross section at 30 keV (averaged from 23–37 keV) is 36.9 mb, compared with 88 ± 30 mb of Gibbons et al. [16] and 48 ± 5 mb from Stroud et al. [17]. The average capture from 85–100 keV is 15.3 mb, compared with 18.0 ± 0.6 mb given by Spencer and Macklin [15] over that energy interval.

Table 3. Averaged capture cross section

Energy interval (keV)	Cross section (mb)
1 – 10	88.8
10 – 20	32.1
20 – 30	35.9
30 – 40	27.3
40 – 50	16.7
50 – 60	19.9
60 – 70	19.6
70 – 80	18.5
80 – 90	14.7
90 – 100	15.8

Conclusions

The investigations described in this report indicate that the representation of the neutron interactions with ^{59}Co in ENDF/B-VI is unsatisfactory below 100 keV, although the thermal cross sections are correctly reproduced. Indeed ENDF/B-VI fails to describe properly the most recent ORELA transmission ratio measurements. These transmission measurements were analyzed with the resonance analysis code SAMMY and a set of resonance parameters was obtained that is consistent with those transmission measurements. These resonance parameters are a dramatic improvement over ENDF/B-VI. An additional improvement could be achieved by the analysis of relevant capture measurements. Such an analysis would require a treatment of multiple scattering not yet available in the resonance analysis code SAMMY. The resonance parameters obtained from this work, together with the previously determined parameters of the 132-eV resonance, provide an accurate total cross section from 10^{-5} eV to 100 keV. The capture cross section is reproduced at thermal. Further analysis of the capture data from 132 eV to 100 keV is warranted.

Acknowledgements

The authors thank R. Q. Wright for help with the comparison of the ENDF/B evaluations and D. C. Larson for helpful reviews and suggestions. Thanks go also to J. A. Harvey for providing us with otherwise unobtainable data and for helpful discussions concerning the transmission measurements. This research was sponsored by the Office of Energy Research, Division of Nuclear Physics, U.S. Department of Energy, under contract DE-AC05-84OR21400 with Martin Marietta Energy Systems, Inc., and under Intragency Agreement No. 0046-C083-A1 with the Defense Nuclear Agency.

References

1. N. E. Holden, BNL-NCS-51388 (1981). Neutron Capture Cross-Section Standards for BNL-325 Fourth Edition.
2. J. B. Garg, Nucl. Sci. and Eng. 65, 76 (1978).
3. J. Morgenstern et al., Nucl. Phys. A102, 602 (1963).
4. P. Guenther, R. Lawson, J. Meadows, M. Sugimoto, A. Smith, D. Smith, and R. Howerton, ANL/NDM-107 (1988).
5. J. A. Harvey, private communication (1986). Data available at the NNDC.
6. S. Mughabghab, M. Divadeenam, and N. E. Holden, Neutron Cross Sections, Vol. 1, Part-A, Academic Press, New York (1981).
7. N. M. Larson, Updated Users' Guide for SAMMY, ORNL/TM-9179, R1 (1985) and R2 (1989). See also N. M. Larson and F. G. Perey, "Resonance Parameter Analysis with SAMMY," Proc. Int. Conf. Nuclear Data for Sci. and Tech., Mito, Japan, p. 573 (1988).
8. J. B. Garg, J. Rainwater, and W. W. Havens, CR-1860 (1964), "High Resolution Neutron Spectroscopy in the keV Region."
9. S. F. Mughabghab and T. J. Krieger, BNL-NCS-50468 (1975), Neutron Cross Sections of ^{59}Co below 100 keV.
10. B. A. Magurno, Ed., BNL-NCS-50446 (1975), ENDF/B-IV Dosimetry File.
11. R. Kinsey, Ed., BNL-NCS-17541 (1979), ENDF/B Summary Documentation (Section 27-Co-59 for MAT 1327, ^{59}Co, by S. F. Mughabghab).
12. S. F. Mughabghab and D. I. Garber, Neutron Cross Sections. Vol. 1. Resonance Parameters. BNL-325, Brookhaven National Laboratory (1973).
13. F. J. Dyson and M. L. Metha, J. Math. Phys. 4, 107 (1963).
14. R. R. Spencer and H. Beer, Nucl. Sci. and Eng. 60, 390 (1976).
15. R. R. Spencer and R. L. Macklin, Nucl. Sci. and Eng. 61, 346 (1976).
16. J. H. Gibbons, R. L. Macklin, P. D. Miller, and J. H. Neiler, Phys. Rev. 122, 182 (1961).
17. B. J. Allen, quoted from thesis of Stroud, CSISRS AN/SAN 30506.007, available from the National Nuclear Data Center, Brookhaven National Laboratory.

Averaged U238 Fission Cross Section Measurement
in Cf-252 Neutron Spectrum

A. Pazirandeh and Y. Ahmedi-zadeh
Physics Department, Teheran University
P.O. Box 19395-1943, Teheran IRAN
Nuclear Research Center, AEOI

ABSTRACT: With regard to the standardization of Cf-252 neutron spectrum, the averaged U-238 fission cross section was measured over the Cf-252 neutron spectrum using foil activation technique. The gamma spectra of the fission products were obtained on a pure germanium detector at different decay times in order to identify the gamma rays of short, medium and long half-life radioisotopes. Analyzing the gamma spectra of the irradiated urnium foils several distinct photopeaks of several fission products were identified. The pure activity of Te-132 was determined from the net area under 228.2 kev photopeak after correcting it for neighbouring photopeaks of Np-239 and Pa-234. The averaged U238 fission cross section was calculated from the corrected counts of Te-132 photopeak to be 329 ± 10 mb. The main sources of errors in the measurements are mainly due to inaccuracy of location of foils, neutron flux, counting statistics, weight of foils, impurity of foils and interferences from neighbouring photopeaks. The contirbution of fission neutrons from spontaneous fission of Cm-248, daughter of Cf-252, to the total flux density was estimated at 0.13 %.

INTRODUCTION

The accurate measurements of some basic reaction rates, such as capture and fission and the measurement of some basic parameters in reactors such as fission ratio and capture to fission ratio for different fissile and fissionable materials and neutron flux measurement require standard cross sections to be easily available and referenceable. These standard reactions are also needed for dosimetry purpose and detector calibration.

The use of these basic neutron monitor reactions, namely U-238 fission for the purpose of neutron fluence and flux density determination, require prior establishment of a measuring technique and the reliable nuclear data.

The Cf-252 spontaneous fission neutrons spectrum is rather well konwn [1-4] as a standard fission spectrum, with a small pertubation in the spectrum arising from scattered neutrons by the source material nuclei, is widely used for the assessment of some monitor reaction rates.

The uncertainty involved in the reaction rate measurements are attributed to variety of sources such as, fast neutron flux determination, detector calibration, U-foil gamma ray self absorption, fission yield, foil positioning, foil impurities, weight of foils, counting statistics, detector efficiency foil size. There is a discrepancy between measured fission reaction rate of U238 in Cf-252 spontaneous fission neutron spectrum and the calculated one using Madland-Nix, NBS and Maxwellian distributions and ENDF/B-V differential cross section values. The prime objective of the measurement is to establish a measuring technique to be referenced as a standard method.

EXPERIMENTAL ARRANGEMENT AND MEASUREMENTS

The irradiation facility was a Cf-252 neutron source of about 10 µg strength secured at the end of a thin metal rod located in a borated paraffin cask covered with a boral sheet. The position of the source was remotely controlled and moved up and down through a hole in the boral sheet and placed in the middle of a concrete room of 7 x 4 x 3.5 m whenever needed. The natural and depleted uranium metal foils of 9 mm diameter and 0.1 to 0.32 mm thick were fixed at a distance of 11.25 mm from the source center line. The irradiated foils were counted on a highly pure germanium detector and spectra were analyzed by an IBM-PC.

The foils were irradiated long enough, about 100 hours, to yield sufficient statistics for counting purpose. In analyzing the gamma-ray spectra of irradiated foils many fission products such as Nb-95, Mo-99, Te-132, I-133, Xe-135, Cs-137 and Ba-140 were identified which only one fourth of the gamma energy spectrum is shown in Fig. (1).

To calculate the averaged U-238 fission cross section one of the prominent photopeak of Te-132 i.e., 228.2 kev was chosen. The net area under the 228.2 kev photopeak of Te-132 was determined after correcting it for interferences from 228 kev photo peaks of Np-239 and 228.0 kev of Pa-234. The relative contributions from neighbouring gamma-rays of these two actinides to 228.20 kev of Te-132 were determined by obtaining the net area under 278.0 kev of Np-239 and under 1001.0 kev of Pa-234. Since the ratio of the net area under the photopeaks of Np-239 and/or Pa-234 remain constant the absolute contribution of each nuclide was determined.

In order to be ascertain of net count of 228.20 kev photopeak of Te-132, the half life of Te-132 was determined to be exact within 5 % of the published values. The purity of the uranium foils was determined by two techniques, neutron activation using standard uranium foil and chemical analysis which both showed purity to be 92.5 %.

Since the original source strength at the time of manufacturing was 100 µg, at the time of measurement more than 90 µg of Cf-252 has been decayed, by alpha emission, to Cm-248. The contribution of spontaneous fission neutrons of Cm-248 with an effective half life of 8.6×10^5 years [5] to the total neutron flux density was estimated at .13%.

The contribution of induced fissions in uranium foils, due to scattered neutrons from the walls, to the total spontaneous fissions was found to be negligible.

The gamma ray self absorption in the uranium foils for low energy gamma rays was taken into account. It is not needed to elaborate on the method of calculation here, description of the method is thoroughly discussed in most reactor shielding books including ref.[6]. To ascertain of mass attenuation coefficient of uranium the coefficient was measured by irradiating a very thin layer of natural uranium deposit as a gamma source of fission product gamma rays and then counting 228.20 kev photopeak of Te-132 with a cover of a uranium foil and without any cover. Then using normal attenuation equation the coefficient was determined.

The net area under 228.20 kev gamma ray was also corrected for branching ratio, fission yield, total efficiency of the HPGe detector, irradiation time, decay time, counting time. The averaged U238 fission cross section was determined from absolute activity of Te-132 to be 329 $\pm$ 10 mb. The neutron flux was determined from the information provided by the manufacturer.

The result of the averaged fission cross section measurements agrees quite well with the other published results [2] and has 3% discrepancy with the calculated value of 312 mb using ENDF/B-V and Los-Alamos Cf-252 neutron spectrum [2].

REFERENCES

1. Drapchinsky, L.V., Absolute measurement of fission cross section averaged over the fission spectrum. Neutron Physics & Nuclear Data measurements with accelerator and reactors. IAEA-TECDOC-345, 1985.

2. Handbook on Nuclear Activation Data, Technical Report No. 273, IAEA, 1987.

3. Nuclear Standard Reference Data, IAEA-TECDOC-335, 1985.

4. Mehta, M.K., and Schmidt, J.J., Workshop on "Applied Nuclear Theory & Nuclear Model Calculations for Nuclear Technology Applications", 15 Feb - 19 Mar, 1988, Trieste, Italy.

5. Enge, H.A., Nuclear Physics, Addison-Wesley, 1974.

6. T. Rockwell III, Reactor Shielding Design Manual D. Van Nostrand Company.

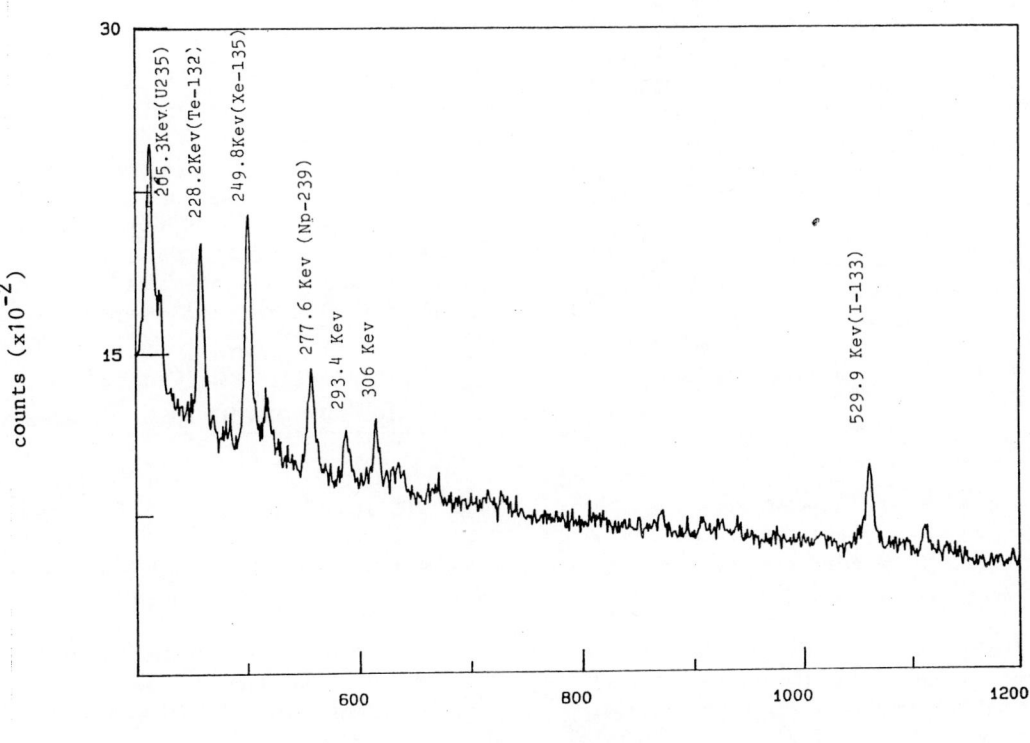

Fig. 1   Gamma ray spectrum of fast fission products

DOUBLE-DIFFERENTIAL CROSS-SECTION OF NEUTRON EMISSION FROM ^{235}U AT 4.9 MEV NEUTRON INCIDENCE ENERGY

G.N. Lovchikova, S.E. Sukhikh, A.V. Polyakov, A.M. Trufanov
Institute of Physics and Power Engineering, Obninsk, USSR

H. Märten, A. Ruben
Technische Universität Dresden, Germany

Abstract: The energy spectra of secondary neutrons from ^{235}U were measured at 4.9 MeV incident neutron energy. The experiment was performed by means of the time-of-flight technique at a pulsed-beam tandem accelerator with a tritium gas target as neutron source via the reaction T(p,n). Neutron emission spectra were measured at 30°, 45°, 60°, 90°, 120°, and 150° with reference to incident neutron beam direction. The double differential cross sections of elastic and inelastic scattering were obtained taking into account the contributions of prompt fission neutrons on the basis of a well-established theoretical model.

(double-differential neutron cross-section, inelastic scattering, neutron emission spectra, prompt fission neutron spectra, ^{235}U, time-of-flight technique, tritium gas target)

Introduction

The measurement of differential and integral characteristics of fast-neutron interaction with fissile nuclei is important for both applied and scientific purposes. Requirements for this data were repeadetly notized, e.g. for studies of fast critical assemblies as well as reactor calculations. Experimental investigations of neutron emission from fissile nuclei at incident neutron energies in the MeV region and beyond are also the basis for the study of reaction mechanisms. In this case, the analysis of secondary neutron spectra must include the decomposition of the different components due to neutron scattering and post-fission neutron emission. The experimental data basis is still insufficient. In the present work, a measurement of double-differential neutron spectra from ^{235}U at 4.9 MeV incident neutron energy is described. Fission neutron theory is applied in the data analysis in order to account for the post-fission neutron component. In this way, the cross section of neutron scattering can be explicitely deduced.

Experiment

Standard neutron time-of-flight spectroscopy was applied to measure the ^{235}U neutron emission spectra. The spectrometer is described in Ref. [1]. Monoenergetic neutron production was realized applying a tritium gas target bombarded by a pulsed proton beam from the EPG-10M tandem accelerator. The design and specifications of the gas target are given in Ref. [2]. The repetition rate of the proton pulses with about 1 ns width was 5 MHz. The energy of neutrons escaping from the target was 4.90 ± 0.06 MeV. Based on FWHM of the γ-peak, the time resolution was found to be 3 ns yielding an energy resolution of 40, 200, and 440 KeV at the neutron energies 1, 3, and 5 Mev, respectively. The ^{235}U sample was prepared as a hollow cylinder with 45 mm outer diameter, 40 mm inner diameter, and 54.8 mm in height. The sample mass corresponds to 1.093 mole. The ^{235}U sample was placed at an angle 0° with refernce to the proton beam direction at a distance of 17 cm from the centre of tritium gas target.

The time-of-flight spectroscopy of secondary neutrons was performed by the use of a scintillation detector at 200 cm flight path. The neutron detection with a cylindrical stilbene scintillator (6.3 cm in diameter and 3.9 cm height) viewed by FEU-30 photomultiplier was combined with the electronic pulse shape discrimination in order to suppress the γ-ray background efficiently. The detector was located in a heavy collimator and shielding system consisting of lithium hydride, paraffin, and polyethylene [1] for the suppression of the background due to neutrons scattered by the walls of the experimental hall and those escaping directly from the T target. The neutron detection threshold was 0.5 MeV. A further stilbene crystal (4.0 cm in diameter and 2.0 cm in height) detector was used for monitoring the incident neutron flux. In addition, a long counter at 300 cm distance from the T target at 90° and a current integrator for monitoring the total charge (number) of protons bombarding the T target were applied.

Measurement and Processing

Neutron emission spectra from ^{235}U were measured at the angles 30°, 45°, 60°, 90°, 120°, and 150° at the incident neutron energy 4.9 MeV. The stability of the accelerator and the electronic devices was monitored by continuous analysis of the shape and position of the neutron and γ-ray time-of-flight peaks in the monitor and main detector spectra, e.g. the direct peak of primary neutrons from the T target was considered. The experimental procedure consisted in the alternative measurement with and without ^{235}U sample in order to measure effect and main background spectrum. The direct spectrum of target neutrons detected by the monitor was simultaneously collected. Further, gas-in/gas-out runs were performed. Fig. 1 shows an example of raw experimental spectra at 120° in order to illustrate the effect/background ratio.

For the absolute normalization of the differential neutron cross sections, an additional measurement of the spectrum of neutrons scattered by hydrogen was carried out at the angle

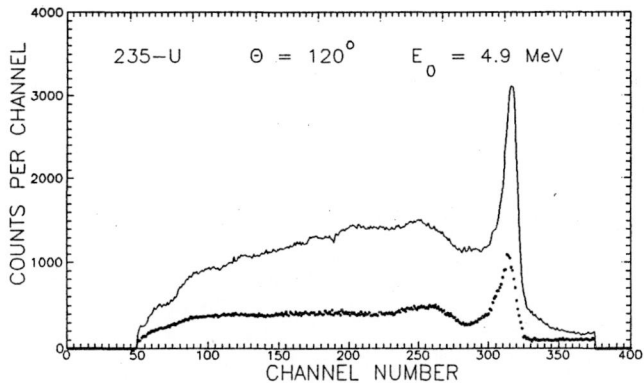

Fig. 1. Raw experimental time-of-flight spectra at 120° (effect versus background)

45° with refernce to the incident neutron baem. For this purpose, a polyethylene (CH_2) cylinder (10 mm in diameter and 50 mm in height) was used.

The neutron detector efficiency was determined by measuring the (standard) ^{252}Cf(sf) neutron spectrum under identical experimental conditions, i.e. the ^{252}Cf sample within a fast (ΔE) ionization chamber was located at the former ^{235}U sample position.

The analysis of differential neutron cross sections was performed in the standard way including the correction for systematic influences like secondary scattering effects and energy resolution. The total uncertainties were estimated considering statistical ones as well as systematic uncertainties due to

- time and energy calibration (detection threshold, time scale determination accounting for the γ-peak position),
- dimensions of the sample and detector,
- correction for neutron flux attenuation and multiple scattering in the samples (^{235}U and CH_2),
- spectrum normalization based on the (n,p) cross section measurement.

As already estimated in Ref. [3], the influence of flux attenuation and multiple scattering in the ^{235}U sample with the above mentioned geometry is very small in the energy range covered in the experiment.

Data Analysis and Results

The neutron emission spectra obtained for the angles 30°, 45°, 60°, 90°, 120°, and 150° include the contributions by inelastically scattered neutrons and prompt fission neutrons in the energy range below 4.9 MeV. Above 4.9 MeV, a pure post-fission neutron spectrum is measured. Around 4.9 MeV, the peak of elastically scattered neutrons appears. In order to obtain the double-differential cross-section of inelastic and elastic scattering, the spectrum of prompt fission neutrons has to be subtracted. There are no experimental informations about this spectrum at the given incident energy. However, recent experimental data are available for ^{235}U, ^{232}Th, and ^{239}Pu at 1.5 MeV incident energy [4]. These spectra were successfully reproduced by a complex statistical model approach to prompt fission neutron emission [5,6]. Fig. 2 represents the measured prompt fission neutron spectrum of ^{235}U at 1.5 MeV neutron incident energy in comparison with the calculation. The model includes a detailed energy balance in nuclear fission as function of fragment mass asymmetry. Its predic-

tive power was recently tested by systematic calculations in the neutron incident energy range up to 20 MeV for several actinide nuclei [7]. The code for predicting fission neutron distributions includes also the calculation of angular distributions with reference to both fission axis and incident beam direction. In the latter case, the angular distribution of fission fragments has to be accounted for. At 1.5 MeV incident neutron energy, the measured 90°/150° anisotropy could be well described by the model [4]. The good agreement between the calculated and measured fission neutron spectra confirms that the model applied allows a fairly complete consideration of the basic features of neutron evaporation from fission fragments.

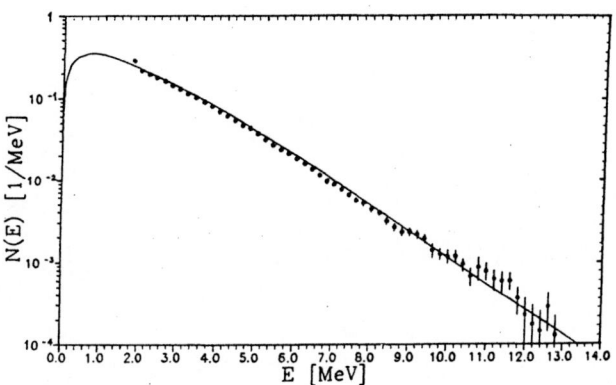

Fig. 2. The spectrum of prompt neutrons from the fission of ^{235}U induced by 1.5 MeV neutrons (dots - Ref. [4], solid line - calculation)

Therefore, the double-differential emission cross section of prompt fission neutrons from ^{235}U at 4.9 MeV neutron incident energy was calculated on the basis of the above statistical evaporation model. Based on the fission cross section at this incident energy point, the distribution was absolutely normalized. The spectrum normalization was confirmed by comparing the calculated with the measured spectrum at energies above 4.9 MeV. The difference between the measured total emission cross section and the calculated one for prompt fission neutrons is obviously the double-differential emission cross section of (elastically and inelastically) scattered neutrons. The final results are represented in the Figs. 3,4. Whereas the emission cross section is almost isotropic at energies below 3 MeV, elastical and direct inelastic processes yield the typical forward-peaking at 4.9 MeV neutron energy and just below.

Summary

The present work describes the measurement of double-differential neutron cross section of ^{235}U at 4.9 MeV incidence energy. The analysis of the data was performed accounting for theoretical emission distributions of prompt fission neutrons. It was shown that these calculations yield reliable results enabling the analysis procedure applied.

116

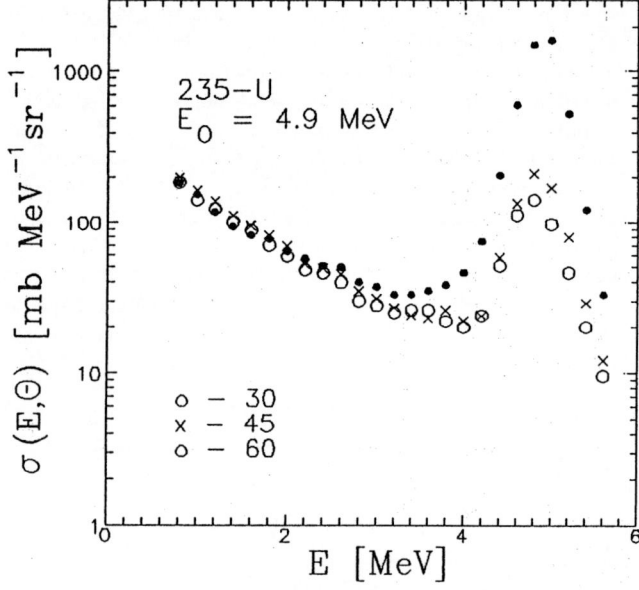

Fig. 3. Differential cross section of elastic and inelastic neutron scattering on ^{235}U at three emission angles

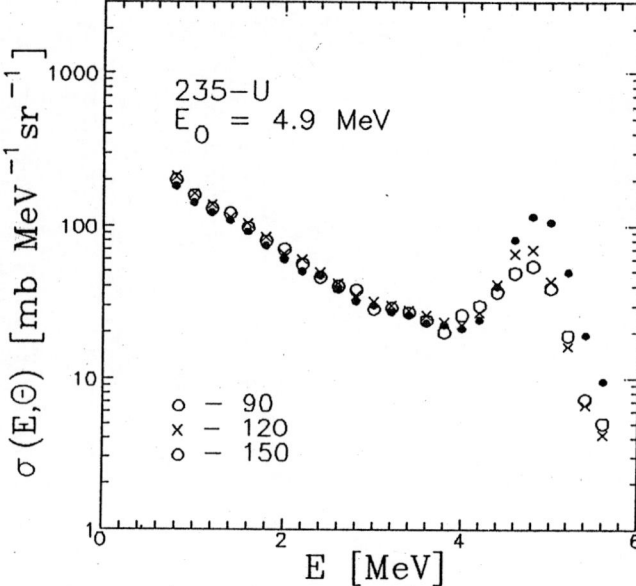

Fig. 4. Differential cross section of elastic and inelastic neutron scattering on ^{235}U at three emission angles

References

[1] A.M. Trufanov et al., "Instrumentation and Technique of an Experiment", No. **2**, 50 (1979)

[2] N.I. Fetisov et al., ibid. No. **6**, 22 (1980)

[3] O.A. Salnikov et al., Izv. AN USSR, Ser. Fiz. **XXX11**, 653 (1969)

[4] G.N. Lovchikova et al., Proc. Int. Conf. on the 50th Anniversary of Nuclear Fission, Leningrad, 16 - 20 October, 1989, in print

[5] A. Ruben et al., Z. Phys. **A338**,67 (1991)

[6] H. Märten et al., Proc. Int. Conf. on 50 Years with Nuclear Fission, 25-28 April, 1989, Gaithersburg, ed. by J.W. Behrens and A.D. Carlson (American Nuclear Society, 1989) Vol. II, p. 743

[7] H. Märten et al., these proceedings

COMPARATIVE ANALYSIS OF THE ENERGY SPECTRA OF NEUTRONS EMITTED BY FISSION FRAGMENTS AND EXCITED NUCLEI

O.I.Batenkov, M.V.Blinov

V.G.Khlopin Radium Institute, Shvernik Av. 28,
194021 Leningrad USSR

Abstract A comparison of the energy spectra of neutrons emitted in spontaneous fission and by nuclei excited in various nuclear reactions (equilibrium component) was carried out . A good agreement for fragments and nuclei far from closed shells was found and essential difference near to double-magic region (A=132) was observed.
(fission, nuclear reactions, neutron energy spectra, average neutron energy).

Experimental study of the energy spectra of neutrons emitted in various nuclear reactions of the type - (n,n'), (n,2n), (p,n), (α,n) in the mass region of fission fragments were carried out in various papers /e.g. 1-6 /. From analysis of the equilibrium parts of these spectra the information on tne level density parameters (a) and the nuclear temperature (T) was obtained for a number of nuclei at several excitation energies.

As soon as the majority of fission neutrons are emitted by excited fragments /7,8/, the values (a) and (T) were determined in this case too /9,10/. The results of the analysis showed, that as in the case of excited nuclei formed in nuclear reactions there is a smooth change of the parameter (a) with the nucleus mass (A) and the deviation from this course in the region of the closed shell A = 132. Besides, the values of the level density parameters (a) agree well with theoretical predictions / 11,12 / in the mass region far from closed shells and considerably differ in the closed shell region (A = 132, N = 82, Z = 50) /9/.

It is interesting to carry out a comparison of the characteristics of neutron emitted by excited fragments and nuclei. At the first stage it is desirable to carry out the direct comparison of the energy spectra, because on the determination, for example, of the level density parameters (a) there may be the influence of various approaches to determination of value (a).

One may expect some difference in the energy spectra, connected with the isospin part of the optical potential, as fission fragments are rather far from the β - stability band. It is possible however, that the specific character of fragment excitation influences considerably the neutron energy characteristics. In the latter case the dynamic effects in fission will play an important role in the formation of the neutron energy spectra.

In the data analysis we used the information on neutron energy spectra from paper /13/. The accuracy of mass determination was about 2 a.m.u. The average excitation energy of fragments was determined by the average number and energy of emitted neutrons and gamma quanta. The values of neutron average energies from excited nuclei for equilibrium component were determined from the reactions with 14MeV neutrons.

In Fig.1 the comparison of the average energies of fission neutrons and neutrons emitted by excited nuclei is presented for average excitation energies

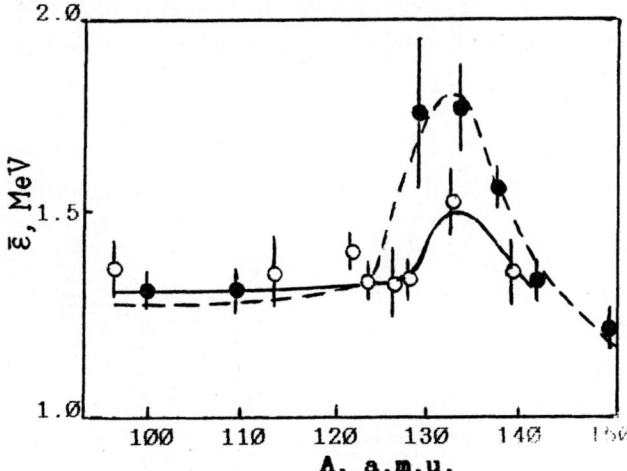

Fig. 1. Average energy of neutron (c.m.s. as a function of the nucleus mass.
(●) data for fission neutron /9/
(O) data for neutron emission at interaction of 14MeV neutrons with nuclei (equilibrium component) /1,2/.

about 20-22 MeV. It is evident from Fig.1 that the data agree well with each other in the mass region 90-120 and 140-150a.m.u. This confirms our conclusion made earlier for these excitation energies / 9 /. This

118

indicates the sufficiently fast dissipation of collective energy of fragment in comparison with neutron time emission. In the mass region 120-140a.m.u. the data differ rather considerably though there is very little information, and experimental errors are comparatively large.

Nevertheless, the results of the comparison, in our opinion showed that if in the mass region far from the double--magic region(Z=50, N=82) neutron energy spectra are close to each other in the two cases considered, then in the region near to A=132 they differ rather considerably.

It is the essential contribution of the nonequilibrium neutrons that might lead to this effect in this fragment mass region. However we did not find any presence of hard non-equilibrium component for the data /9/ which might effectively influence the value of the average energies of the fission neutrons. Therefore, one may assume that such a behaviour of the average energies of spectra is connected with the specific character of fragment excitation. This effect may be is determined by the fact that at the collective type of excitation of fragments near closed shells the excitation energy is distributed on a smaller number of degrees of freedom, than at the one-particle excitation in the reactions of (n,xn) type.

The obtained results of comparative analysis of neutron energy spectra showed the necessity of further precision experimental study of both fission neutrons and neutrons emitted by excited nuclei.

References

1. O.A.Salnikov, N.I.Fetisov, G.N.Lovchikova: Yadernaja Fizika 4, 1154(1966)(Rus)
2. V.B. Anufrienko, B.V.Devkin, G.V.Kotelnikova: Yadernaja Fizika 2, 826(1965)(Rus)
3. O.A.Salnikov, G.N.Lovchikova, G.V.Kotelnikova, N.I.Fetisov, A.M.Trufanov: Yadernaja Fizika 12, 1132(1970) (Rus)
4. G.N.Lovchikova, A.M.Trufanov, O.A.Salnikov,G.V.Kotelnikova, S.P.Simakov: Yadernaja Fizika 1, 41(1981) (Rus)
5. B.V. Zhuravlev, A.Titarenko, B.I.Trykova: Neitronnaya Fizika "Proc. of the Conf." part 3, 267 (1984)(Rus)
6. N.S. Birukov, B.V.Zuravlev, A.P.Rudenko, O.A.Salnikov, B.I.Trykova: Ibid p. 291
7. H.R.Bowman, S.C.Thompson, J.C.Milton, W.J.Swiatecki: Phys.Rev.126,2121(1962)
8. O.I.Batenkov, M.V.Blinov, V.A.Vitenko: Phys. and Chem. of Fission (" Proc. the Intern. Conf." 1979 Julich, IAEA, Vienna, v.2, 267, (1980)
9. O.I.Batenkov, A.B.Blinov, M.V.Blinov, S.N.Smirnov: Nuclear Data for Science and Technology "Proc. of the Intern. Conf " 1988 Mito , H.D.Lemmel(ed.), Saikon, Tokyo, 207, (1988).
10. C.Budtz-Jorgensen, H.H.Knitter: Nucl.Phys. A 490, 307, (1988)
11. A.V.Ignatyk, G.N.Smirenkin, A.S.Tishin: Yadernaja Fizika 22, 485(1975) (Rus)
12. A.V. Ignatyk, K.K.Istekov, G.N.Smirenkin: Yadernaja Fizika 29, 825(1979) (Rus)
13. O.I.Batenkov, A.B.Blinov, M.V.Blinov, S.N.Smirnov: "Proc. of the X Intern. Symp. on Nucl. Phys. Nucl. Fission" (Gaussig 1985) ZFK-592,1986.

GAMMA-RAYS OF SOME PRINCIPAL FUEL NUCLEI EXCITED BY MONOCHROMATIC NEUTRONS

A.A.Filatenkov,   M.V.Blinov,   S.V.Chuvaev,   V.M.Saidgareev

V.G.Khlopin Radium Institute,
Shvernik Av. 28, Leningrad, 194021, USSR

Abstract: Gamma-ray   spectra   in the interactions of ^{232}Th, ^{235}U, and ^{238}U with neutrons of energies 0.7-2.0 MeV, 3.0 MeV, and 14.9 MeV have been measured. Above 300 gamma-transitions have been revealed. The neutron flux was determined using small fission chambers, activation foils, and scintillation counters. Gamma-ray excitation cross sections of the ^{232}Th(n, n´γ)-, and ^{238}U(n, n´γ)- reactions obtained at neutron energies of 0.7-2.0 MeV and 3.0 MeV were transformed to the partial neutron scattering cross sections. The data indicate an appreciable level population by unobserved gamma-transitions from unknown high lying levels. Fission fragment gamma-ray intensities were determined at neutron energies of 3.0 MeV and 14.9 MeV in a time gate of 25 ns. The coincidence within experimental errors was found between 2^+ - 0^+ transition yields and independent yields of corresponding even-even fission fragments.

(Gamma-ray spectra, ^{232}Th, ^{235}U, ^{238}U, fission, inelastic scattering, cross section, excitation function, level population, independent fission fragment yield.)

Introduction

Gamma-radiation arising in the interaction of fast neutrons with nuclei furnish rather rich information. Thus, by means of high resolution gamma-spectrometers, the most detailed data can be obtained about partial neutron inelastic scattering cross sections /1-3/. Besides, gamma-spectra of fissile nuclei contain information about fission process (e.g. fission fragment yields, and angular momenta /4/). However, due to high complexity of the studied gamma-spectra and insufficient completeness and accuracy of level schemes of target nuclei and of fission fragments, the additional uncertainties appear in results obtained, sometimes rather remarkable ones. The simultaneous study of gamma-rays of three fissile nuclei, namely, ^{232}Th, ^{235}U, and ^{238}U, performed in a wide range of neutron energy, allowed remarkable increase in the reliability of information extracted.

Experimental Method

The measurements were carried out at the electrostatical accelerators EG-5 and NG-400 of Leningrad Radium Institute. Monochromatic neutrons were generated by means of ^{3}H(p, n)^{3}He, ^{2}H(d, n)^{3}He, and ^{3}H(d, n)^{4}He reactions. The NG-400 was operated in a pulse regime. Gamma-rays were registered with a Ge(Li)-detector, that was placed in a shield made of lead, and hydrogen- and boron-containing materials.

For neutron flux determination different methods were applied. Most carefully it was measured in the neutron energy interval of 0.7-2.0 MeV, where multidetector system was used which included three ionization chambers, two scintillation counters and several activation foils. The thin-wall chambers containing the ^{232}Th-, ^{238}U-, and ^{237}Np- layers were placed close to the sample. Two meter distant plastic scintillator registered target neutrons, passed through the sample and "undistorted", at the angle of 0 and 22 degrees, respectively. In some runs the foils of ^{197}Au and ^{115}In (enriched) were installed between fission chambers. Unfortunaly, the lack of place does not allow to present here the detailed results of neutron field measurements. Only it should be noted, that different methods gave the concordant results, excluding the ^{197}Au monitors, the activity of which always was higher than expected one. That revealed contamination from slow neutrons, the main source of which appear to be the water cooling of the target.

In the experiment, four cylindrical samples were used (^{232}Th, ^{235}U, ^{238}U, and $H_2C_2O_4$ - "background"), which were attached to a changing device managed by an on-line computer. This allowed to obtain under identical conditions all the main spectra as well as background. The weight of the used samples was 116.5 g, 100.2 g, 186,7 g, and 9.8 g, respectively.

The corrections for neutron field distortion and for gamma-rays self-attenuation were computed for the real geometry of experiment. The angular and energy distribution of neutrons flying out of target were calculated using the recommended c.m.s. data /5/.

The absolute efficiency of Ge(Li)- detector was determined by means of standard gamma-ray sources. During the experiment for this purpose the natural radioactivity of the ²³²Th-sample was used, which was beforehand measured with an accuracy of 2.5%.

Results

We already reported results of our investigation of these nuclei gamma spectra performed at neutron energy 3.0 MeV /3/, where both inelastic scattering gamma-rays and fission fragment gamma-rays were well observed. In order to increase the reliability of the conclusions drawn then, we supplemented our study by the 14.9 MeV-measurement (1989, fission fragment gamma-rays) and by the 0.7-2.0 MeV-measurement (inelastic scattering, 1990-1991).

Excitation Functions of the ²³²Th(n,n´γ)- and ²³⁸U(n, n´γ)-Reactions

These measurements have just been completed and only the first results may be reported here.

The data were measured in neutron energy interval of 0.7-2.0 MeV with steps of 20-50 keV. For the gamma-peak identification, five gamma-spectra measured under identical conditions were analysed simultaneously (²³²Th, ²³⁸U - inelastic scattering; $H_2C_2O_4$ and "empty" - background; and ²³⁵U - fission gamma-rays). As a result, only about 40 gamma-peaks from approximately 100 were interpreted as undoubtedly belonging to the inelastic scattering process.

In Fig. 1, several level excitation functions are shown. The experimental resuts agree within errors with experimental data of other authors. The agreement of these data with the statistical model calculations is not so well. However, the calculations made in this neutron energy region should be regarded as qualitative only, since they were carried out with the same parameters that were obtained at neutron energy 3.0 MeV. We consider that some of the parameters must be varied (e.g. at neutron energy 3 MeV the corrections for width fluctuations are far not so important as in the studied region, where only few channels are opened). Formely, at neutron energy 3.0 MeV we had shown that consideration of level population by unobserved gamma-transitions from unknown higher lying states leads to a reasonable agreement with the statistical model predictions. An effect of an additional level population appears at lower energies, too.

Fission Fragment Gamma-Rays

The remarkable contamination of gamma-rays emitted by fission fragments is a significant pecularity of gamma-spectra of fissile nuclei interacting with neutrons. Its share of the total gamma-ray production cross section exceeds often 50%. In spectra, the fission gamma-rays

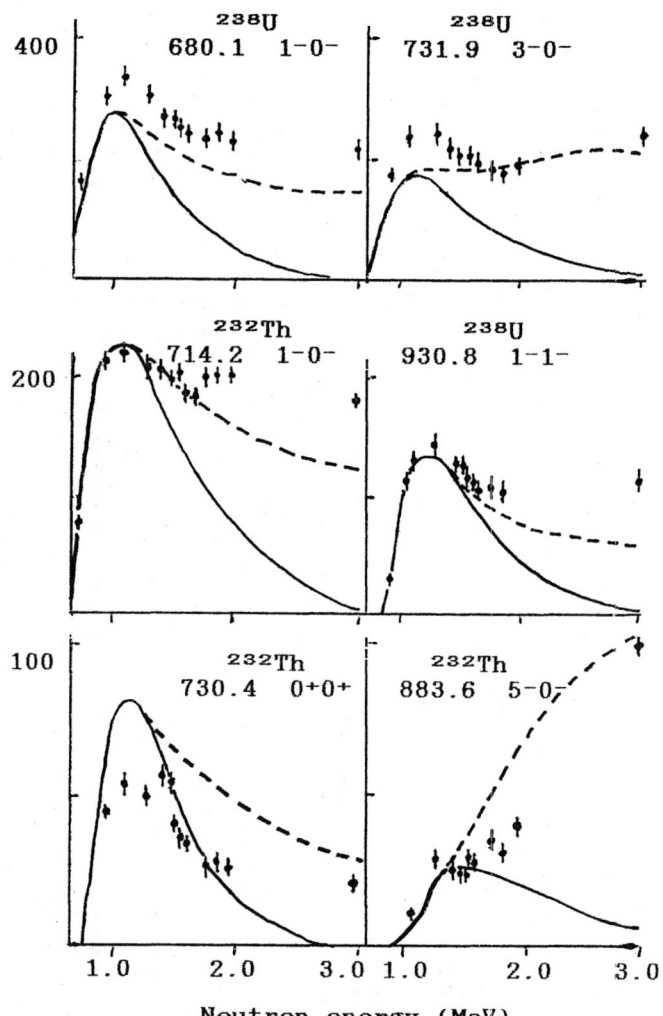

Fig. 1. The excitation functions of the ²³²Th(n, n´γ)- and ²³⁸U(n, n´γ)-reactions. Points – this experiment; curves – calculation: solid line – inelastic neutron scattering only; dotted line – the level population by -transitions from "continuum" (E* > 1.2 MeV) are added.

appear mainly in the form of an intense component with a continuous, exponentially tailed energy distribution. At the same time, it is also possible to observe some individual gamma-peaks from the fission fragments having high yield.

Formerly, during the analysis of the ²³²Th, ²³⁵U, and ²³⁸U gamma-spectra measured at a neutron energy 3 Mev, about 30 gamma-transitions were interpreted as belonging to the fission fragments /3/. The identification was done by a comparison of the experimental energy values and excitation cross sections of the gamma-transitions with the corresponding values obtained from avaiable data on fission cross sections /7/, on fission fragment yields /8/, and on level schemes of nuclei /9/. In this work the results obtained at 3 MeV are considered together with results obtained at a neutron energy

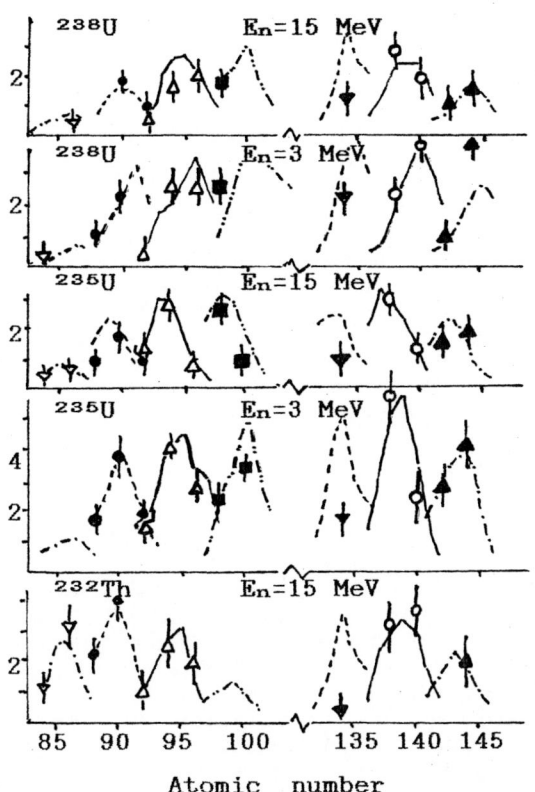

Fig. 2. 2⁺ - 0⁺ γ-transition yields of
even-even fission fragments.
⊽ - isotopes of Se, ◆- Kr, ◮ - Sr,
▪ - Zr, ▼- Te, ⌀- Xe, ▲- Ba.

of about 15 MeV, where the fission gamma-rays become predominant.

Since in both cases the time gating was used, the spectra of prompt gamma-rays were obtained, i.e. the gamma-transitions were observed that occur in fission fragments just after neutron emission but before beta-decays. Among them, the gamma-transitions 2⁺ - 0⁺ of even-even fission fragments manifest themselves most clearly. In Fig. 2, their measured yields are compared with the evaluated independent yields of the corresponding fission fragments. Generally, the agreement is quite satisfactory.

The only remarkable exception is the gamma-transition 1279.1 keV of ¹³⁴Te whose yield is consistently lower. It may be explained by a peculiarity of the level scheme of this nucleus. The level 1691.3 keV 6⁺ is isomeric, its lifetime is 160 ns. Since in the experiment the gamma-rays were measured that were emitted in a time interval of 25 ns, then only 10% of gamma-cascades going through this level could be registered. Thus, the observed low intensity of the gamma-transition

1279.1 keV is evidence for a large contribution of high spin states during the formation of fission fragments.

Another confirmation of high angular momenta of fragments is observation of the gamma-transitions 4⁺ - 2⁺ and 6⁺ - 4⁺ for some nuclides (e.g. ¹³⁸Xe, ¹⁴⁰Xe, ¹⁴²Ba, ¹⁴⁴Ba). The average intensities of these transitions were, in units of independent fragment yield, 0.8 and 0.6, respectively, that correspond to the average fragment spin value of 5.5 ħ.

Conclusion

The performed analysis exhibits the various possibilities of the high-resolution gamma-spectroscopy to study both neutron inelastic scattering and nuclear fission. In the inelastic scattering, many level excitation functions can be obtained. However, the level population by unobserved gamma-transitions is an important problem that should be taken into account. In the fission, the gamma-transitions between lowest states of fission fragments can be revealed even when the fission act is not registered.

For a further increase of experimental accuracy, an improvement in our knowlege about level schemes of actinide nuclei and fission fragments is requested.

REFERENCES

1. W.R.McMurray, I.J.Van Heerden: Zeitschrift fur Physik 253, 289 (1972)
2. D.W.S.Chan, J.J.Egan, A.Mitler, and E.Sheldon: Phys. Rev. C26, 841 (1982)
3. A.A.Philatenkov, M.V.Blinov, S.A.Egorov, V.A.Rubchenja, and S.V.Chuvaev, in Proc.Int.Conf. on Nuclear Data for Science and Technology, May/June 1988, Mito, Japan (Edited by S.Igarasi), p.79, Saikon, Tokyo (1988)
4. A.A.Filatenkov and S.V.Chuvaev, in Abstr.Int.Conf. "Fiftieth Anniversary of Nuclear Fission", Oct.1989, Leningrad, USSR, (Edited by L.Drapchinsky) p.83 , LNPI, Gatchina, 1989
5. M.Drosg and O.Schwerer: Handbook on Nuclear Activation Data, Technical reports ser.273, 83, IAEA, Vienna (1987)
6. E.Sheldon, L.E.Begian, D.W.S.Chan, A. Chang, G.P.Couchell, J.J.Egan, G.Goswami, G.H.R.Kegel, S.Q.Li, A.Mittler, D.J.Pullen, W.A.Shier, J.Q.Shao, and A.Wang: J.Phys.G: Nucl. Phys. 121, 443 (1986)
7. Z.Body: Handbook on Nuclear Activation Data, p.29, IAEA, Vienna (1987)
8. E.A.C.Crouch: Atom. Data and Nuclear Data Tables 19, 500 (1977)
9. M.Sakai: Atomic Data and Nucear Data Tables 31, 399 (1984)

EFFECTS OF THE TWO-FISSION-NEUTRON DISTRIBUTIONS IN REACTIVITY MEASUREMENTS WITH THE ^{252}Cf SOURCE

Felix C. Difilippo
Oak Ridge National Laboratory
P.O. Box 2008
Oak Ridge, TN 37831-6363, USA

Abstract: Stochastic descriptors based on the first and second moments of the probability distribution of the neutrons in phase space $\bar{\mu} = (\bar{r}, \bar{v})$ are extensively used in the nuclear industry for the non-intrusive measurement of nuclear parameters. One of these methods is based on the ^{252}Cf source which has been used in an unprecedented range of situations and reactivities; the interpretation of this method is sensitive to the distribution of two neutrons coming from the same fission, $\gamma(\bar{v}, \bar{v}')$. The few direct measurements of this $n - n$ correlation have shown that γ depends on $\cos\theta'' = \bar{v}.\bar{v}'/vv'$ and the neutron energies, the available experimental data was used to quantify the sensitivity of the ^{252}Cf method to the anisotropies of the $n - n$ correlation. It was found that it affects the interpretation of the reactivity measurement for low values of the multiplication constant.

Introduction

In a typical experiment the subcritical system to be measured is driven by a ^{252}Cf neutron source located inside a fission chamber (labeled Detector 1) and the system is monitored by two sets of neutron detectors (labeled Detector 2 and 3). The Fourier transforms, G_{i_ℓ}, of the correlation functions between the fluctuations of the signals of detectors i and ℓ define the ratio R,

$$R \equiv \frac{G_{12}^* G_{13}}{G_{11} G_{23}} \qquad (1)$$

which can be related to the reactivity ρ by the equation

$$\rho = \frac{R^* Y_1}{1 - Y_2 R^*} \qquad (2)$$

where $\rho = (1 - k)/k$. For convenient locations of the source and the detectors,[1] R* is almost equal to the value of R at low frequency; Y_1 and Y_2 are ratios of functionals and assumed known from calculations of the neutron distributions. Because by far the main source of correlation between the signals is the fission process (induced in the fissile material or spontaneous in the ^{252}Cf) and because we are analyzing second moments of the distribution the coefficients Y_1 and Y_2 are sensitive to the distributions of two neutrons coming from the same fission. More explicitly

$$Y_1 = Y_{10} E_f \qquad (3)$$

$$Y_2 = Y_{20} E_c \qquad (4)$$

where Y_{10} and Y_{20} depend on nuclear parameters and the first moment of the neutron and source distributions, respectively. E_f and E_c are the factors sensitive to the distribution of two neutrons coming from the same fission, $\gamma(\bar{v}, \bar{v}')$.

Effects of the n-n Correlations

For the important case of a "point" ^{252}Cf source located at $\bar{r}_s$

$$E_c = \frac{\int \chi_1^+(\bar{r}_s, \bar{v})\xi_c(\bar{v})d\bar{v} \int \chi_1^+(\bar{r}_s, \bar{v}')\xi_c(\bar{v}')C_c(\bar{v}, \bar{v}')d\bar{v}'}{[\int \chi_1^+(\bar{r}_s, \bar{v})\xi_c(\bar{v})d\bar{v}][\int \chi_1^+(\bar{r}_s, \bar{v}')\xi_c(\bar{v}')d\bar{r}']} \qquad (5)$$

where $\xi_c(\bar{v}) = S_c(E)/4\pi$ is the one-fission-neutron distribution for the ^{252}Cf and we have introduced the correlation associated with the two-fission-neutron distribution, $\gamma_c(\bar{v}, \bar{v}') \equiv \xi_c(\bar{v})\xi_c(\bar{v}')C_c(\bar{v}, \bar{v}')$; χ_1^+ is a function in space $\bar{\mu}$ evaluated where the source neutrons are born and related to the importance of these neutrons to maintain a chain reaction in the multiplicative assembly. A similar expression is obtained for E_f with subscript f substituting subscript c and an additional integral in real space $\bar{r}$ in the numerator and denominator of Eq. (5) to account for the spatial distribution of induced fission. Note that if the two fission neutrons are uncorrelated in $\bar{v}$ space, E_c would be equal to 1, a result that it is not compatible with the reactivity measurement of Ref. 2; of course, there are also measurements of the n-n correlation[3] that very directly show that C_c at least depends on $\cos\theta'' = \bar{v}.\bar{v}'/vv'$. A considerable simplification in the evaluation of Eq. (5) can be obtained if we assume that the correlation depends only on the angle between the neutrons and not on the energy, although Ref. 4 shows that the angular correlation might be also a function of the energy of the neutrons. Because of the lack of detailed experimental data and the simplification, the assumption is adopted in what follows. Equation (5) for E_c is then approximated by

$$E_c = \frac{\int w_c(\bar{r}_s, \bar{\Omega})d\bar{\Omega} \int w_c(\bar{r}_s, \bar{\Omega}')C_c(\cos\theta'')d\bar{\Omega}'}{[\int w_c(\bar{r}_s, \bar{\Omega})d\bar{\Omega}]^2} \qquad (6)$$

where θ'' is the angle between directions $\vec{\Omega}$ and $\vec{\Omega}'$ and w_c is the function χ_1^+ averaged with the fission spectra

$$w_c(\bar{r}, \bar{\Omega}) \equiv \int \chi_1^+(\bar{r}, E, \bar{\Omega})S_c(E)dE \qquad (7)$$

w_c is then the average importance of the fission neutrons born at $\vec{r}$ and moving in direction $\bar{\Omega}$; a similar expression is obtained for E_f.

Evaluation of the Effects of the n-n Correlation

The numerical analysis of Eq. (6) requires the specification of the kinetic adjoint distribution χ_1^+. We had approximated the kinetic adjoint with the static adjoint in the calculations of spherical tanks containing typical solutions of highly enriched U.

Transport calculations were performed with a 39 energy group library based on ENDF/B-V, the values of w_c are shown in Fig. 1 as a function of $\cos\theta$ and r (azimuthal symmetry) for the case of a tank of radius of 12.5 cm with k=0.785. Because of the point nature of the ^{252}Cf source, E_c picks the asymmetries of w_c and C_c at the location of the source. For example: if the source is located at 5cm from the tank the contributions to the integrals of Eq. (6) come, as Fig. 1 shows, from neutrons moving toward the tank; because $C_c > 1$ for these neutrons the value for E_c=1.20, i.e., larger than 1. Because E_f is computed as an integral over the induced fission distribution inside the tank, where the anisotropy is smaller, the sensitivity of E_f to the $n-n$ correlation is smaller. In general, E_f depends on the degree of skewness of the fission distribution, so E_f is sensitive to the location of the source and the subcriticality.

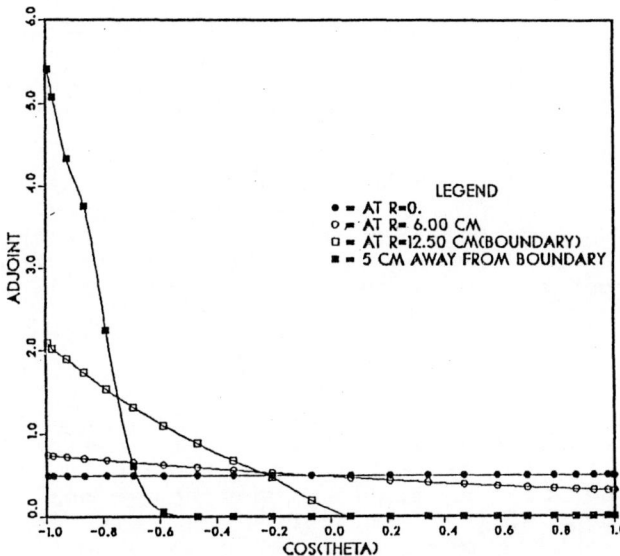

Fig. 1. Average importance of the fission neutrons born at r and moving in direction θ. Anisotropies of the two-fission-neutron distributions are picked at the peak of the importance.

The sensitivity of the reactivity measurement to the biases of Y_1 and Y_2 can be calculated from Eq. (2) as $\Delta\rho/\rho = \Delta Y_1/Y_1 + \rho(Y_2/Y_1)(\Delta Y_2/Y_2)$, because $Y_2 \sim Y_1$ any bias in Y_2 is amplified by a factor ρ which is larger than 1 for $k < 0.5$. In our example, a bias of 20% would appear in the calculation of Y_2 by not considering the anisotropy of the $n-n$ correlation; for $k = 0.2$ this bias would have produced an error of 80% in the measurement of ρ.

Conclusions

The complete specification of the two-fission-neutron distribution is very important for the application of the ^{252}Cf method to criticality safety and there is a need for more experimental data or correlations that summarize theoretical results for this application.

References

1. F. C. Difilippo, "Harmonic Analysis of Stochastic Descriptors and the Interpretation of ^{252}Cf Neutron Source Experiments," *Nuclear Science and Engineering*, 104, 123 (1990).

2. J. T. Mihalczo, E. D. Blakeman, G. E. Ragan, E. B. Johnson and Y. Hochiye, "Dynamics Subcriticality Measurement Using the ^{252}Cf Source-Drive Noise Analysis Method," *Nuclear Science and Engineering*, 104, 314 (1990).

3. J. S. Pringle and T. D. Brooks, "Angular Correlation of Neutrons from Spontaneous Fission of ^{252}Cf," *Phys. Rev. Letters*, 35, 1563 (1975).

4. C. B. Franklyn, C. Hofmeyer and D. W. Mingay, "Angular Correlation of Neutrons from Thermal-Neutron Fission of ^{235}U," *Phys. Letters*, 78B, 564 (1978).

COLD FRAGMENTATION PROPERTIES OF 252Cf (SF)

F.- J. Hambsch, H.- H. Knitter[1] and C. Budtz-Jørgensen[2]

Commission of the European Communities
Central Bureau for Nuclear Measurements, 2440 Geel, Belgium

[1] present address : Gerststraat 8, B-2490 Balen
[2] present address : Danish Space Research Institute, DK-2800 Lyngby

Abstract : The spontaneous fission of 252Cf was investigated experimentally in the cold fission region. The fission fragment mass- and nuclear charge distributions were determined as a function of mass in total kinetic energy (TKE) bins of 2 MeV width in the TKE-range from 1 to 15 MeV and parallel to the average Q_{max}-value. There is no odd-even effect in the mass yield, whereas proton and neutron odd-even effects δ_Z and δ_N show linear dependence on (Q_{max}-TKE), however with different slopes. The present results lead to the conclusion that δ_Z and δ_N cannot be interpreted as indicators of the intrinsic excitation energy at scission, and that the structure of five to six mass units observed in many fission parameters finds its explanation in the shape of the Q-value surface in fission as a function of mass and nuclear charge in even-even fissioning systems.

(252Cf(SF), cold fragmentation, fission fragment mass- and charge distributions, odd-even effects)

Introduction

Nuclear mass and charge distributions close to the reaction Q-values were measured for the spontaneous fission of 252Cf. In these high energy outskirts of the fission fragment distribution the yield has decreased by several orders of magnitude and therefore an efficient detection system with a high energy resolution is needed.

The interesting fact for such fragmentations is that those fragments carry nearly no excitation energy, which makes neutron emission unlikely. The fragments are close to their ground state. Investigations close to the reaction Q-values, therefore, give unique information about the roles of nuclear pairing, shell and liquid drop effects as well as on the possible ground state deformations of nuclei in the large nuclide range covered by the fission fragments.

The Experiment

The fission fragment detection is done with a Frisch-gridded twin ionization chamber [1], with energy resolution of <500 keV.

The nuclear charge information has been obtained from the experimental double ratio

$$R(Z_L) = \frac{(P_{anode,L} - P_{sum,L})/P_{anode,L}}{(P_{anode,H} - P_{sum,H})/P_{anode,H}} = \frac{X_L(E_L, A_L, Z_L)}{X_H(E_H, A_H, Z_H)}$$

X is the distance between origin and centre of gravity of the charge distribution of the fragments' ion trace in the detector gas. X(E) can be determined independently from the experiment [1]. P_{anode} is the anode signal and P_{sum} the sum signal of anode and grid [1].

For the whole fragment distribution $1.4 \cdot 10^8$ fission events were measured. Fragment energies were corrected for the energy loss of the fragments in the source carrier and for the pulse height mass defect.

Evaluation and Results

The scheme of the data analysis is illustrated by Fig. 1. This figure shows to the left an energy scale in MeV and to the right a percentage scale. Both scales are correlated with the light fragment mass scale as abscissa. The open and full circles represent the maximum Q-values as a function of mass split as calculated using the mass tables of Möller and Nix [2] and of Möller et al. [3], respectively.

The thick full line is a kind of average Q_{max}-value between the odd and even mass splits (open points). Parallel to this line eight fragment TKE bins are defined in steps of 2 MeV as indicated by the thin lines. For each TKE bin the nuclide yields were evaluated. As example the relative elemental yields for each light fragment mass as measured for bin 6 are plotted in the lower part of Fig.1. Thus, the nuclide yields as a function of the average total excitation energy available to both fragments are obtained [4].

Other cuts through the two-dimensional yield Y(A,TKE), like they are often used, e.g. for constant E_K values, are arbitrary, because they are not related to the same excitation energy available to both fragments as shown in Fig.1 by the dash-dotted line. Therefore comparisons between different fission nuclides must be handled with care.

As an example the mass resolution achieved for (Q_{max}-TKE) = 7 MeV is shown in Fig. 2.

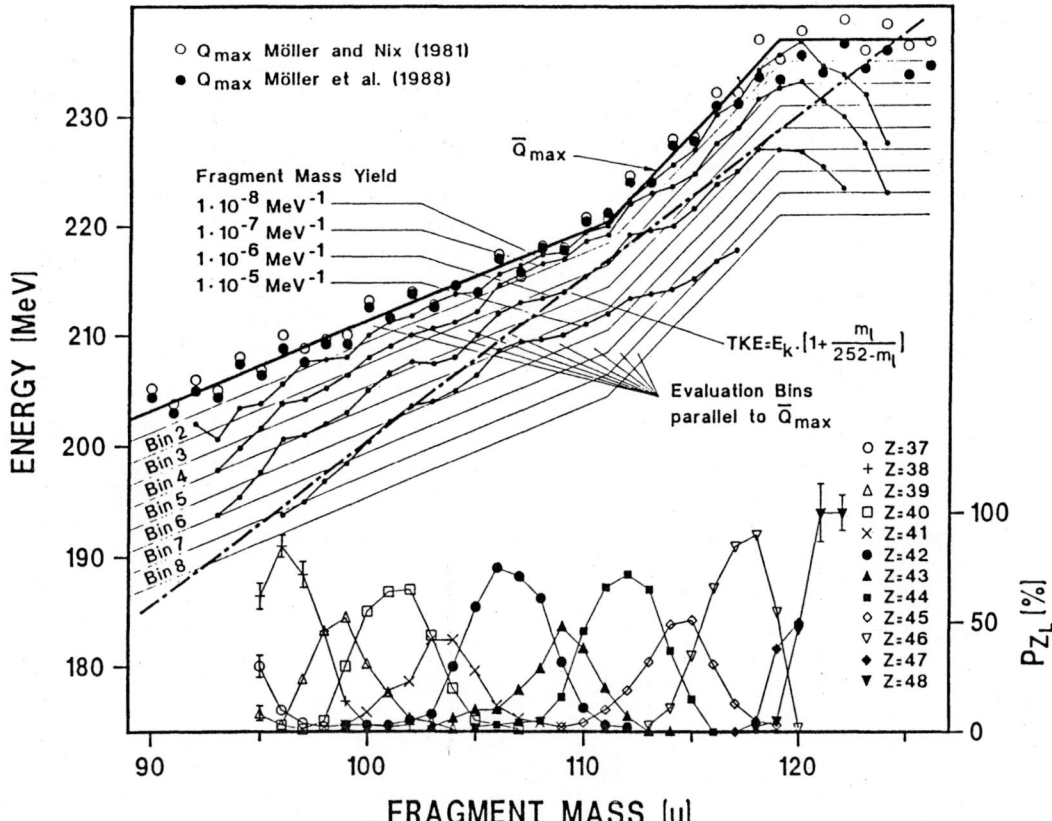

Fig.1 Maximum Q-values of ^{252}Cf(SF). TKE-evaluation bins. Lower part: Nuclear charge distributions versus light fragment mass.

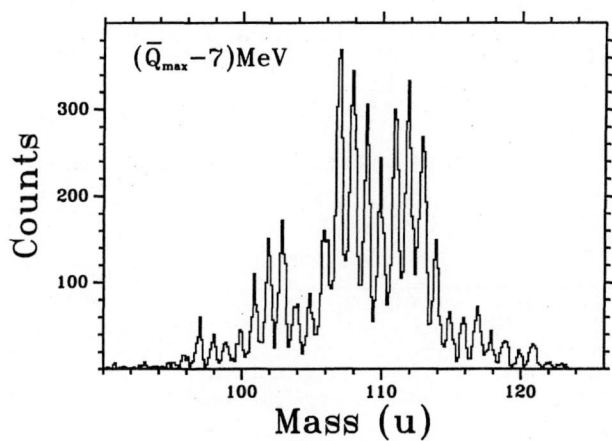

Fig.2 Light fragment mass distribution for $(Q_{max}-TKE)$ = 7 MeV.

The experimental nuclear charge distribution as measured for $(Q_{max}-TKE)$ =11 MeV is shown in Fig. 3 for two selected mass splits A_L/A_H=106/146 and 109/143. The whole elemental yield distributions as obtained from bin 6 are plotted in the lower part of Fig. 1

with its scale in percent on the right hand side.

Since the nuclide yields for each bin are measured, it is possible to sum up the nuclides with even-even, odd-odd, even-odd and odd-even proton and neutron numbers, respectively. These numerical values are given in Table 1. It is evident that also the odd-even effects for the mass, δ_A, nuclear charge δ_Z and neutron number, δ_N can be obtained. They are given in Table 1 also.

δ_A, calculated from the experimental yield, is essentially zero which gives evidence for the randomness of the neck-rupture at scission even at very small excitation energy. Whereas δ_Z and δ_N, calculated from the fitted distributions (yield > 100 counts), reveal linear dependence with $(Q_{max}-TKE)$. However the magnitude of δ_Z is about five times larger than of δ_N. The slopes for δ_Z and δ_N are $-(0.031\pm0.002)\text{MeV}^{-1}$ and $-(0.0045\pm0.0010)\text{MeV}^{-1}$, respectively. In Table 1 the relative yields for even-even (EE), odd-odd (OO), even-odd (EO) and odd-even (OE) fragmentations for the different evaluation bins are given too. From Table 1 it is clear that even fragmentation is favoured and odd fragmentation is dying out with decreasing exitation energy TXE. This is a well known fact found also for other fissioning systems. However this is due to the rather artificial cut through the fission fragment nuclide yield as will be shown later.

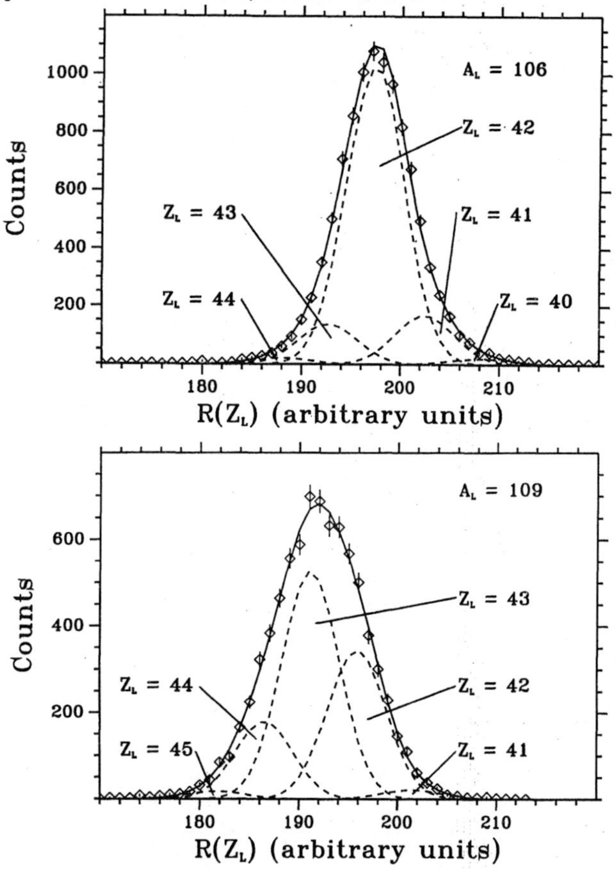

Up to now only results of the odd-even effect integrated over all fragments were presented. However the measurement permitted to evaluate proton and neutron odd-even effects also as function of mass split. This is done in the same way as for the integrated values and shown as an example in Fig. 4.

Fig.3 Experimental nuclear charge distribution for $(Q_{max}-TKE)=11$ MeV

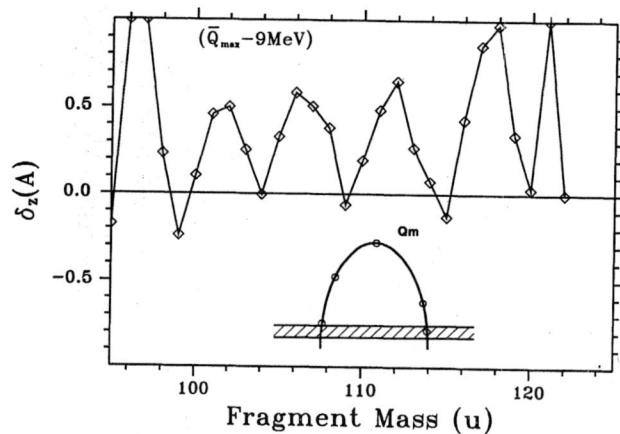

Fig.4 Proton odd-even effect as function of mass for $Q_{max}-TKE = 9$ MeV.

A clear undulatory structure with a period of five to six mass units is seen in $\delta_Z(A)$.

The experimentally measured nuclide yields are a direct picture of the structures visible in the Q-value energy surface $Q(A_L,Z_L)$ shown in Fig. 5.

This figure explains the behaviour of $\delta_Z(A)$ and other parameters showing structures of five to six mass units[4].

	BIN 1 (0-2)MeV	BIN 2 (2-4)MeV	BIN 3 (4-6)MeV	BIN 4 (6-8)MeV	BIN 5 (8-10)MeV	BIN 6 (10-12)MeV	BIN 7 (12-14)MeV	BIN 8 (14-16)MeV
δ_A	0.19±0.06	0.06±0.03	0.014±0.014	0.003±0.008	0.000±0.005	0.009±0.003	0.000±0.002	-0.003±0.002
δ_Z	-	0.48±0.04	0.45±0.01	0.473±0.007	0.370±0.005	0.304±0.003	0.238±0.002	0.183±0.002
δ_N	-	0.13±0.05	0.06±0.02	0.022±0.010	0.045±0.005	0.033±0.003	0.018±0.002	0.013±0.002
$<\sigma_Z^2>$	-	0.16±0.09	0.32±0.11	0.33 ±0.66	0.40 ±0.04	0.45 ±0.03	0.48 ±0.02	0.50 ±0.02
EE[%]	(59)	37.6 ±4	36.9 ±1.2	37.5 ±0.6	35.1 ±0.3	33.6 ±0.2	31.4 ±0.2	29.7 ±0.2
OO[%]	(0)	6.7 ±2	11.4 ±0.5	12.6 ±0.3	14.8 ±0.2	16.8 ±0.1	18.6 ±0.1	19.8 ±0.1
EO[%]	(33)	36.7 ±3	35.6 ±1.0	36.5 ±0.5	32.9 ±0.3	31.6 ±0.2	30.5 ±0.2	29.5 ±0.1
OE[%]	(7)	19.8 ±1	16.1 ±0.6	13.4 ±0.3	17.1 ±0.2	18.0 ±0.2	19.5 ±0.1	21.0 ±0.1

Table 1 :The odd-even effects for fragment mass, nuclear charge and neutron number as well as the nuclear charge variance, are given for the different TKE-evaluation bins. The four lower lines give the relative yields for even-even, odd-odd, even-odd and odd-even nuclear charge and neutron number, respectively.

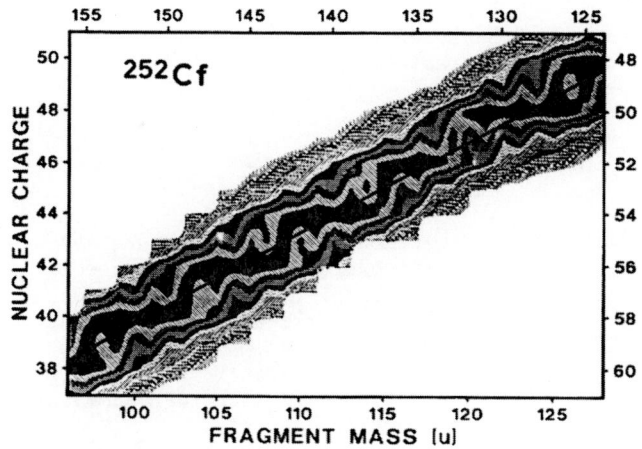

Fig.5 The Q-value energy surface Q(A,Z) in a grey-shaded representation.

Also the dying out of the odd-fragmentations and the decrease of the variance $\langle\sigma^2_Z\rangle$ of the charge distributions (see Table 1) can be understood due to the fact that mostly the even- charge fragmentations have the highest Q-value and the evaluation bins follow parallel to this Q-value and not parallel to individual Q-values of single charges. However, the measured data allow also the evaluation for the same excitation energy of even or odd- charge fragmentations. Fig. 6 shows the result.

The undulatory structure seen before (Fig. 4) essentially disappears when $\delta_Z(A)$ is evaluated for constant TXE(A,Z). Summing over all masses, gives slightly negative values for δ_A, δ_Z and δ_N. Also the abundancy of OO, EE, OE, and EO changes drastically giving now higher probability for odd-odd (OO) than for EE fragmentations.

considerations. The level densities close to the ground state are larger for OO-fragments than for EE-fragments and therefore favour fragmentations with broken nucleon pair.

The only physical cut through the landscape Y(A,Z,TKE) is the one evaluating the yields for constant Q(A,Z)-TKE. Doing so, the evaluation is performed for constant TXE(A,Z). The models proposed in the past, e.g. [5], linking odd-even effects to pair breaking and excitation energy (TXE ~ ln δ) are no longer valid, because δ_Z and δ_N close to zero would imply in these models intrinsic excitation energy close to infinity. Negative values for δ_Z and δ_N may not occur either. So δ_Z or δ_N cannot be interpreted as being a measure of the excitation energy of the fissioning nucleus at the moment of scission.

References

[1] C. Budtz-Jørgensen, H.- H. Knitter, Ch. Straede, F.- J. Hambsch and R. Vogt,
Nucl. Instr. Meth. A258, 209 (1987)
[2] P. Möller and J.R. Nix,
Atomic Data and Nucl. Data Tables 26, 165 (1981)
[3] P. Möller, W. D. Myers, J. W. Swiatecki and J. Treiner,
Atomic Data and Nucl. Data Tables 39, 225 (1988)
[4] H.- H. Knitter, F.- J. Hambsch and C. Budtz-Jørgensen
submitted to Nucl. Physics.
[5] H. Nifenecker, G. Mariolopoulos, J.P. Bocquet, R. Brissot, Ch. Hamelin, J. Crançon and Ch. Ristori
Z. Phys. A308, 39 (1982).

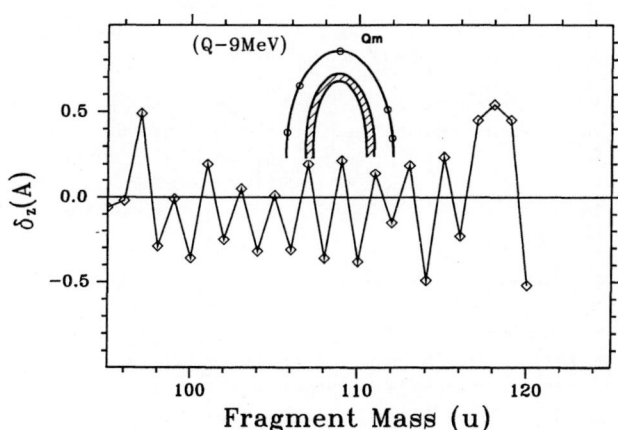

Fig.6 Proton odd-even effect at (Q-9 MeV) for cuts with constant distance from Q(A,Z).

The structure left over in Fig. 6 can be understood by level density

NUCLIDE YIELDS OF LIGHT FISSION PRODUCTS FOR ^{241}Am(2n$_{th}$,f) AT DIFFERENT KINETIC ENERGIES

P. Siegler*, T. Schunk, M. Mutterer, W. Schwab, J.P. Theobald
Institut für Kernphysik, Technische Hochschule Darmstadt, FR Germany
U. Quade⁺
Sektion Physik, Universität München, FR Germany
H. Faust
Institute Laue-Langevin, Grenoble, France

Abstract: The yield of light fission products from the reaction ^{241}Am(2n$_{th}$,f) were measured as function of mass A, atomic number Z, and kinetic energy E, using the recoil mass separator LOHENGRIN of the Institute Laue-Langevin at Grenoble. Mass distributions summed over ionic charge states were determined within the mass range $85 \leq A \leq 110$ at seven kinetic energies between 86.9 and 117.0 MeV. For the same range of masses, nuclear charge distributions were measured for five energy values between 98.0 and 116.2 MeV. From the data, we have also evaluated the energy integrated nuclear charge and mass yields, as well as isotonic and isotopic yields and their odd-even effects.

(^{241}Am(2n$_{th}$,f), ^{242}Am, mass separator LOHENGRIN, light fission products, mass and energy distribution, fractional independent yields, odd-even effects of elements and isotones)

Introduction

Precise data on fission fragment yields are of considerable interest both for the exploration of the fission process and for applied work. For ^{242}Am(n$_{th}$,f), experimental results on the mass and charge distributions are still scarce up to now. Only for the relatively long living (T$_{½}$ = 151a) isomeric state ^{242m}Am, an incomplete set of data on cumulative and independent yields of fission products has been obtained by radiochemical methods [1], mass spectrometry [2] and gamma spectroscopy [3]. A powerful instrument to measure complete sets of fission fragment yields, at least for the light group of fission fragments, is the recoil mass spectrometer LOHENGRIN [4] installed at the high-flux reactor of the Institute Laue-Langevin at Grenoble, France. In recent years, this instrument has provided accurate data on various even-even fissioning systems ranging from ^{233}U(n$_{th}$,f) [5] to ^{249}Cf(n$_{th}$,f) [6]. In some cases, the high thermal neutron flux of 5·10^{14} cm^{-2}s^{-1} makes it feasible to use double neutron capture to produce odd-Z even-N fissioning systems with a large fission cross section, e.g. to study the reaction ^{238}Np(n$_{th}$,f) starting from a ^{237}Np target [7]. We have applied this method to investigate the similar case of ^{242}Am(n$_{th}$,f) starting from a ^{241}Am target. Here, fission occurs with about equal rates from the 16h ground state of ^{242}Am and from the 151a isomeric state ^{242m}Am (Fig.1). In this contribution, we can only briefly describe the experiment and discuss the most prominent results. A detailed analysis of the data and complete listings of the measured yields will be contained in a subsequent paper.

Experiment

With the LOHENGRIN separator the recoiling fission fragments leaving the target are analysed at appropriate electric and magnetic field settings according to the ratios of the mass over ionic charge A/q and energy over ionic charge

---

* Present address: JRC-CBNM, Geel, Belgium
⁺ Present address: Gesellschaft für Reaktorsicherheit, Garching, FR Germany

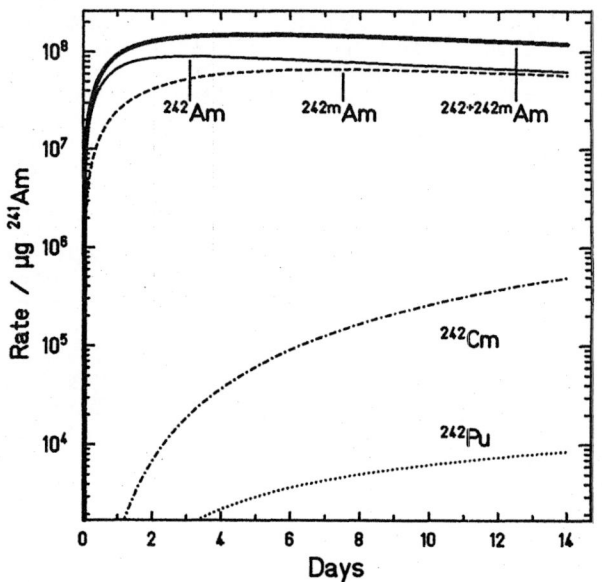

Fig. 1. Production rates of the ^{241}Am target versus time of exposure.

E/q, respectively. The individual mass yields are then determined from the fragment energies measured at the focal plane of the spectrometer, e.g. with a high-resolution ionization chamber [8], and by summing over the relevant ionic charge states.

In our experiment, we have used for measuring the mass distributions only the central part of the focal plane yielding an energy resolution $\Delta E/E$ of $\leq 1\%$ fwhm. Targets were thin layers of ^{241}Am (100 to 140 µg/cm^2 Am-oxide) electrodeposited onto titanium backings and covered with very thin carbon foils (20 µg/cm^2) which minimize energy loss corrections to values of ≈ 1 MeV. The nuclear charge measurements were performed by measuring the residual energy behind a homogenous absorber consisting of a stack of Parylene-C foils of a total thickness of 900 µg/cm^2 which reduces the initial fragment energy by about 70%. By this method described in detail in [5] nuclear charges can be separated with a re-

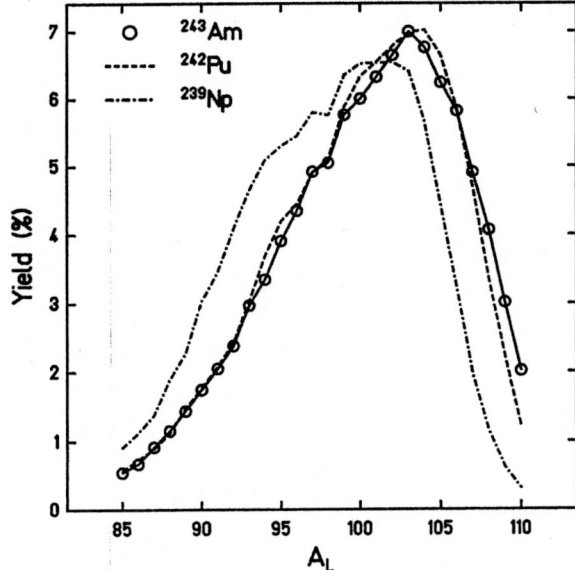

Fig. 2. Energy integrated mass distribution of light fission fragments from the reaction ^{241}Am$(2n_{th},f)$, compared to ^{237}Np $(2n_{th},f)$ [7] and ^{241}Pu(n_{th},f) [9].

solving power of $Z/\Delta Z$ of about 50 provided that pure mass over ionic charge states A/q are selected which carry the smallest possible contaminations of other masses. In our experiment we have controlled possible mass contaminations by measuring initial fragment energies with other segments of the ionization chamber located on both sides of the absorber stack along the focal plane. For the rather time consuming nuclear charge measurements we have covered the ^{241}Am targets with stabler nickel foils of 0.25 µm thickness. Here, energy corrections are in the range of 6 MeV. Since for both types of targets the distribution of the ionic charge states were different, the integrations over q were performed individually before normalizing the nuclear charge measurement to the measured mass yields.

Results and Discussion

Mass Distribution: The energy integrated mass distribution is plotted in Fig. 2 and compared with that of ^{241}Pu(n_{th},f) [9] for which the fragments share the same number of neutrons (N = 148) and with ^{237}Np$(2n_{th},f)$ [7] (N = 146). In the given mass range, the mean masses $\langle A_L \rangle$ are 100.3, 100.0, and 98.0 for ^{243}Am*, ^{242}Pu*, and ^{239}Np*, respectively. It is predominantly the neutron number which determines the mass distribution. Figure 3 shows the mass distributions measured at different light fragment kinetic energies E. As known from other mass distributions (e.g. ^{239}Np* [7]) the asymmetry of the spectral shape increases with kinetic energy. There is a structure in the mass distribution at the lowest energy measured (E = 86.9 MeV), i.e. in the region of cold deformed fission, with small minima at the mass numbers 96 and 100. At the highest energy of E = 117.0 MeV, i.e. close to cold compact fission, there is an enhancement of the yield around A = 106 as is also the case in the odd-Z nucleus ^{239}Np* [7] and in the even-Z nuclei ^{240}Pu* [10] and ^{242}Pu* [9]. The mean energy $\langle E_L \rangle$ of the light group of fission products is determined to be 102.2 MeV.

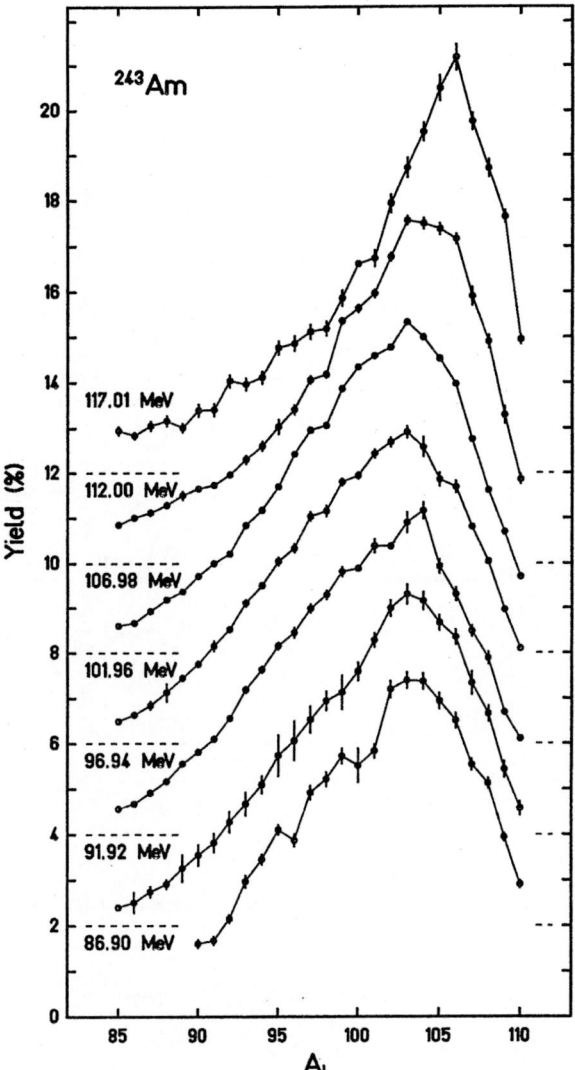

Fig. 3. Mass distribution for different kinetic energies. The ordinate is shifted for subsequent spectra.

Charge and Neutron Yield Distributions: Nuclide yields $Y(A_L,Z_L)$ have been determined for 129 nuclei, compared to the yields of only 30 nuclei given in the literature for the light group of fission fragments [1-3]. Figures 4 and 5 show the element and neutron yield distributions at different light fragment kinetic energies. Both sets of distributions closely resemble those found earlier for the odd-Z even-N system ^{238}Np(n_{th},f) [7]. With increasing kinetic energy, the element yields move progressively towards the atomic number Z_L = 42, the neutron yields towards N_L = 64, both numbers representing deformed shells with a deformation parameter ß ≈ .55 according to the model of Wilkins et al. [11]. The neutron yields show a pronounced odd-even staggering at all energies; from the energy integrated neutron distribution we derive a global odd-even effect of 10.4 ± 1.6 %, compared to 8.1 ± 1.5 % in the case of Np, and a similar enhancement of the local odd-even effect around N_L = 60 as observed before in the fission of other actinides [6,7]. For the element yields, there is a weak odd-even staggering observed (analysing the global odd-even effect in the standard manner leads to a value of

130

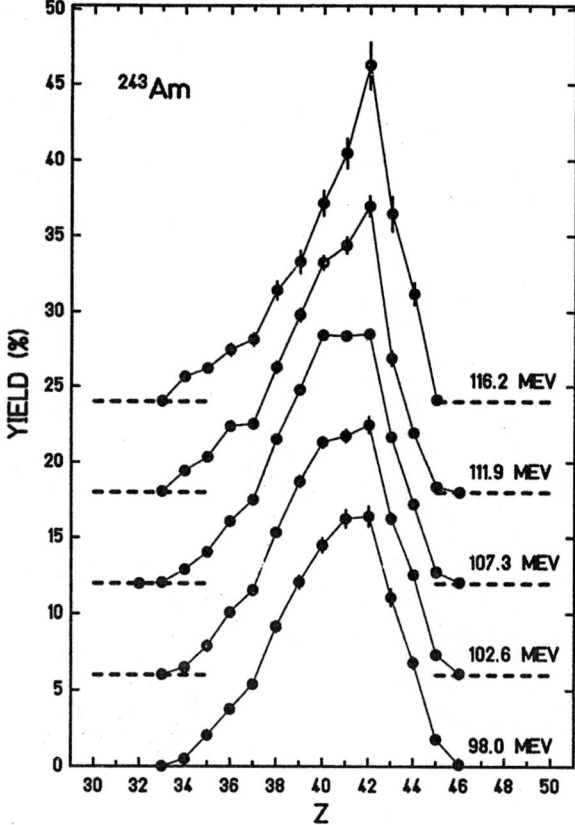

Fig. 4. Element yields at different kinetic
energies

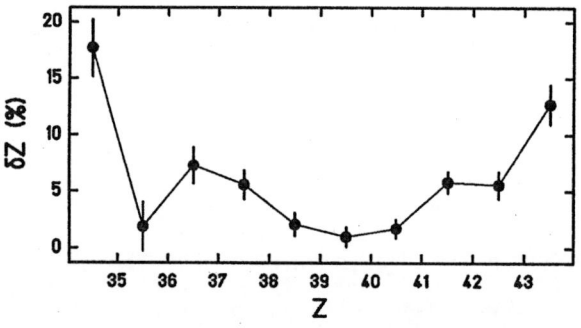

Fig. 6. Local proton odd-even effect, calculated
following Tracy et al. [12]

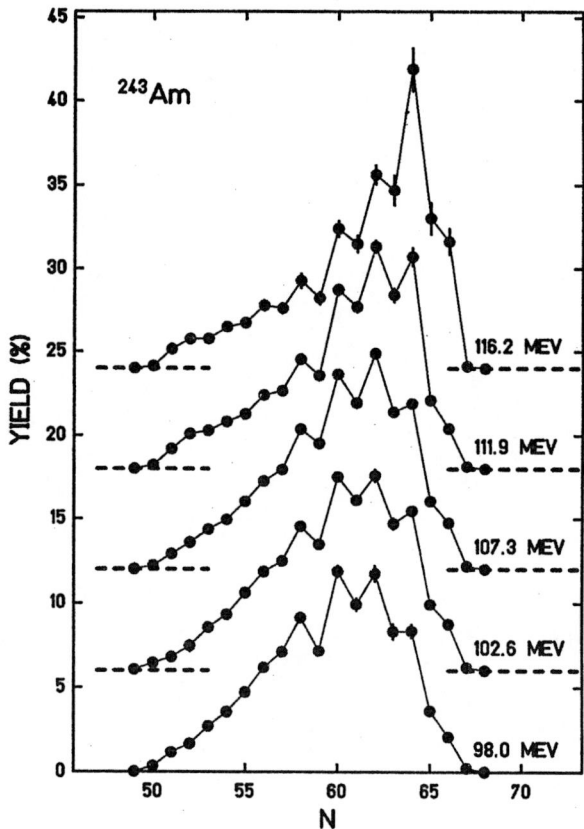

Fig. 5. Isotonic yields at different kinetic
energies.

Acknowledgement

The authors would like to thank the ILL Grenoble
for generously supporting the experiment and the
GSI Darmstadt for giving computer assistance.
This work has been funded by the German Federal
Minister for Research and Technology (BMFT) un-
der the contract number 06 DA 453.

4.5 ± 1.6 %) in spite of the odd atomic number Z
= 95 of americium. The local proton odd-even effect, as depicted in Fig. 6, indicates that at
least part of this effect may be due to the deformed proton shells located at Z = 38 und Z =
44 [11]. The surprisingly high value below Z =
35 is consistent with the result of Stumpf et
al. [13].

It is interesting to note that we observe also
structures in the mass dependence of the displacement of the mean isobaric nuclear charges from
the unchanged charge density values, in the
variances of the isobaric nuclear charge distributions, and a weak odd-even staggering in the
mean energy as function of atomic number Z.

REFERENCES

1. K. Wolfsberg and G.P. Ford, Phys. Rev.
   C3, 1333 (1971).
2. B.L. Tracy, J.E. Pleva, H.G. Thode, J.
   Inorg. Nucl. Chem. 35, 2639 (1973).
3. A.N. Gudkov et al., Sov. J. Nucl. Phys.
   41, 365 (1985).
4. E. Moll et al., Kerntechnik 19 , 374 (1977).
5. U. Quade et al., Nucl. Phys. A487, 1 (1988).
6. M. Djebara et al., Nucl. Phys. A496, 346
   (1989).
7. G. Martinez et al., Nucl. Phys. A515, 433
   (1990).
8. J.P. Bocquet, R. Brissot, H.R. Faust, Nucl.
   Instr. Meth. in Phys. Res. A267, 466 (1988).
9. P. Schillebeecks, PHD-Thesis, Gent (1989).
10. C. Schmitt et al., Nucl. Phys. A430,
    21 (1984).
11. B.D. Wilkins, E.P. Steinberg, R.R. Chaseman,
    Phys. Rev. C14, 1832 (1976).
12. B.L. Tracy et al., Phys. Rev. C5, 222 (1972)
13. P. Stumpf, U. Güttler, H.O. Denschlag, con-
    tribution to this conference.

FRAGMENT YIELDS FROM THE FISSION REACTIONS ^{232}U(n, f) AND ^{239}Pu (n, f)

J. Kaufmann[1], N. Boucheneb[2,3], G. Medkour[2], M. Asghar[2], F. Gönnenwein[1], W. Mollenkopf[1], G. Barreau[3], T.P. Doan[3], B. Leroux[3], A. Sicre[3] and P. Geltenbort[4]

1 Physikalisches Institut, Univ. Tübingen, Tübingen, Germany

2 Institut de Physique, USTHB, Algiers, Algeria

3 Centre d'Etudes Nuclearies, Bordeaux - Gradignan, France

4 Institut Laue - Langevin, Grenoble, France

Abstract: The distribution of fission fragment masses, charges and kinetic energies from the reactions ^{232}U(n,f) and ^{239}Pu(n,f) induced by thermal neutrons has been investigated with the detector based spectrometer Cosi fan tutte of the ILL in Grenoble. While ^{239}Pu(n,f) is an already well studied reaction, only few measurements on the thermal neutron reaction ^{232}U(n,f) have been done in the past. An especially interesting quantity is the charge odd-even effect which is found to be δ_p= (9.8 +- 1.0) % for the reaction ^{239}Pu(n,f) confirming data obtained at the Lohengrin spectrometer [1]. For ^{232}U(n,f) we have measured a pronounced odd - even effect amounting to δ_p = (20.3 +- 1.5) % which is in good agreement with former experiments by Djebara et al. [2]. In addition to the new results from the reaction ^{232}U(n,f), in both reactions a quite new effect called "cold deformed fission" [3] has been confirmed.

Introduction

The Cosi fan tutte spectrometer is based on high resolution axial ionization chambers and time of flight devices. It allows to study the fragment masses, charges and kinetic energies from one single set of multiparameter data. The performance of the instrument is outstanding for the light fragment group. Therefore, only results for this group are presented. For the light fragments one finds a mass resolution $\Delta M = 0.6$ u. This means that the energy resolution of the ionization chamber has to be $\Delta E = 400$ keV. For the measurements of nuclear charges we use a variant of Bragg Curve spectroscopy [4] yielding a charge resolution of 0.9 charge units for the group of the light fragments.

Fragment masses

The first result of the measurements to be presented is the global mass distribution of the ^{239}Pu(n,f) fragments. Fig.1 shows the present results compared to the data obtained on Lohengrin [1]. The results of the

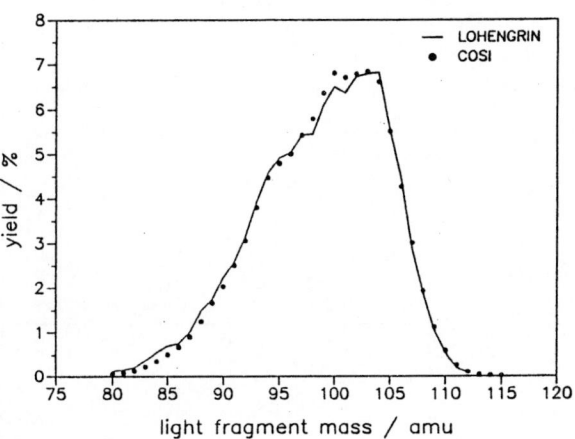

Fig.1 Fragment mass distribution of ^{239}Pu(n,f).

^{239}Pu(n,f) reaction have been taken to check the calibration of the spectrometer as the statistics of the ^{232}U(n,f) experiment (approx. $4.5 \cdot 10^5$ fission events) is poorer than for the ^{239}Pu(n,f) experiment (approx. $2 \cdot 10^6$ events). With this calibration the mass distribution of the ^{232}U(n,f) fragments (Fig. 2) is evaluated. This distribution was almost unknown before our measurements. In Fig. 2 some radiochemical data [5] are

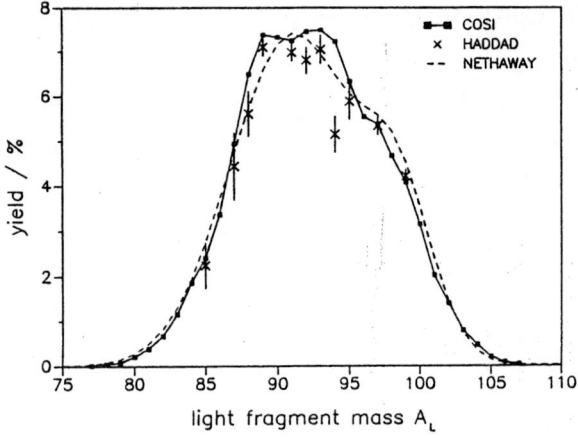

Fig. 2 Fragment mass distribution of ^{232}U(n, f).

shown describing the distribution only partially. In addition to the measurements in Fig. 2 there is also given a distribution calculated by D.R. Nethaway [6] by extrapolation of data from other uranium isotopes.

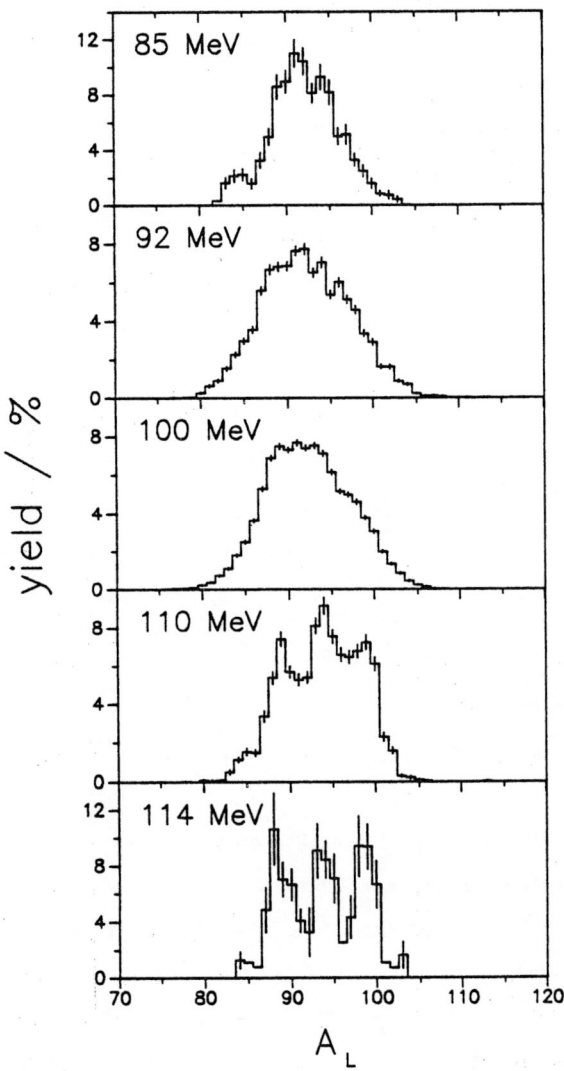

Fig. 3 Mass distribution of ^{232}U(n, f) at different fragment kinetic energies.

The mass distribution of the ^{232}U(n, f) fragments exhibits a more pronounced fine structure than the mass distribution of the ^{239}Pu(n, f) fragments. For the ^{232}U(n, f) reaction Fig. 3 displays mass distributions for given windows in the kinetic energy of the light fragment. Suprisingly, the structure in the mass distributions is marked both at high and low kinetic energies, while for average energies the fine structures appear to be washed out.

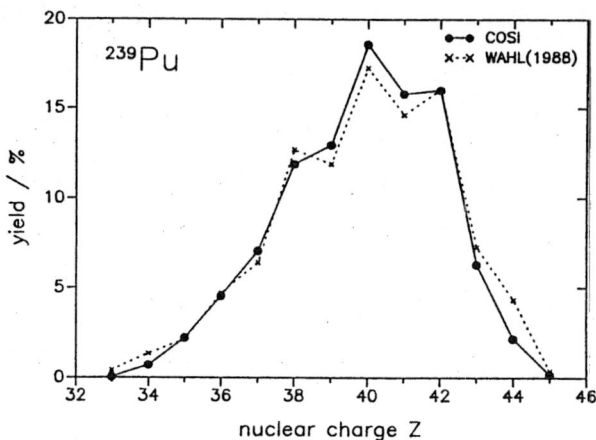

Fig. 4 Charge distribution of the ^{239}Pu(n, f) fragments.

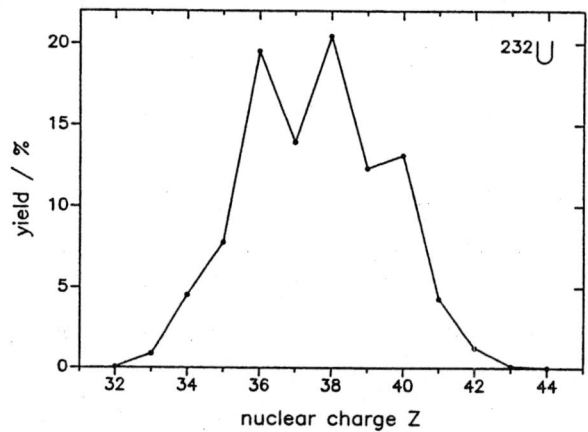

Fig. 5 Charge distribution of the ^{232}U(n, f) fragments.

Nuclear charges

Fig. 4 and Fig. 5 show the charge distributions of the fragments for ^{239}Pu(n, f) and ^{232}U(n, f), respectively. The results for ^{239}Pu(n, f) are compared with data compiled by A.C. Wahl [7]. In both reactions the yield of fragments with even proton number Z are clearly enhanced. Defining an odd-even effect

$$\delta_p = \frac{Y(\text{even } Z) - Y(\text{odd } Z)}{Y(\text{even } Z) + Y(\text{odd } Z)}$$

one obtains for ^{239}Pu(n, f) $\delta_p = (9.8 +- 1.0)$ % and for ^{232}U(n, f) $\delta_p = (20.3 +- 1.5)$ %. These values are in quite good agreement with former measurements [1,2].

A very interesting feature is the dependence of odd-even effects on fragment kinetic energy. The odd-even effect may be correlated to the probability of breaking proton pairs in the fissioning nucleus and may, hence, be used as a "nuclear thermometer" indicating the intrinsic excitation energy of the fissioning nucleus. As a consequence, one expects that at high fragment kinetic energies and, therefore, low excitation energies, the probability of breaking proton pairs is low, resulting in a high proton odd-even effect.

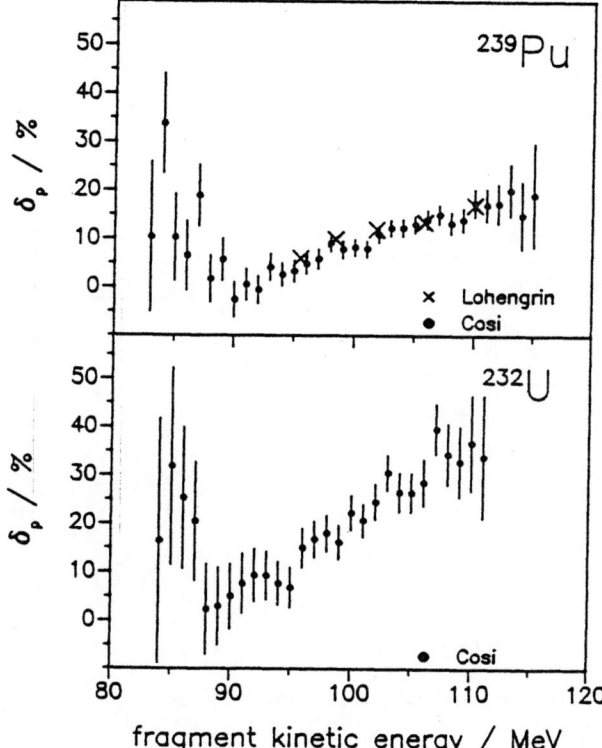

Fig. 6 Proton odd-even effect as a function of kinetic energy.

The present measurements are given in Fig. 6 together with the Lohengrin data for ^{239}Pu(n,f). From 90 to 110 MeV the proton odd-even effect rises continuously as expected. At high kinetic energies one reaches the limiting case of "cold compact fission" where no intrinsic excitation is left (the fragments are cold) and the whole energy available is converted into kinetic energy.

Suprisingly at very low kinetic energies (E < 90 MeV) the proton odd-even effect rises again (Fig. 6). This can be understood in a static scissionpoint model where

the nascent fragments, though heavily deformed, carry no intrinsic excitation energy. This limiting case is called "cold deformed fission". The discovery of a large odd-even effect at low kinetic energies puts into question the standard interpretation of odd-even effects indicated above.

Comparing Fig. 6 with Fig. 3 it becomes evident that the pronounced odd-even effect at high and low kinetic energies is correlated to the fine structure observed in the mass distributions. The fact that, in spite of neutron evaporation from the fragments, the fine structure in mass reappears at low kinetic energies points to low variances of the neutron emission number at these energies.

Conclusion

For both fission reactions, even for the well studied ^{239}Pu(n,f) reaction, a wealth of new data has been obtained for the light fragment group. Probably the most remarkable resuls are the discovery of fine structure in the mass distributions and strong odd-even effect in the charge distributions for kinetic energies of the fragments well below average. These features of the fragment distribution are considered to be evidence for the existence of cold deformed fission.

References

1. C. Schmitt, A. Guessous, J.P. Bocquet, H.G. Clerc, R. Brissot, D. Engelhardt, H.R. Faust, F. Gönnenwein, M. Mutterer, H. Nifenecker, J. Pannicke, C. Ristori and J.P. Theobald, Nucl. Phys. A 430, 21 (1984).

2. M. Djebara, M. Asghar, J.P. Bocquet, R. Brissot, M. Maurel, H. Nifenecker and C. Ristori, Nucl. Phys. A 425, 120 (1984).

3. W. Mollenkopf, J. Kaufmann, F. Gönnenwein, A. Oed, and P. Geltenbort, Nucl. Phys., to be published.

4. A. Oed, P. Geltenbort and F. Gönnenwein, Nucl. Instruments and Methods 205, 451 (1983).

5. M. Haddad, J. Crançon, G. Lhospice and M. Asghar, Radiochimica Acta 46, 23 (1989).

6. D.R. Nethaway, Lawrence Livermore Nat. Lab., private communication (1990).

7. A. C. Wahl, Atomic Data and Nuclear Data Tables 39, 1 (1988).

INVESTIGATION OF THE P-RESONANCE PROPERTIES
IN SLOW NEUTRON FISSION OF U-235

A.M. Gagarsky, S.P. Golosovskaja, A.B. Laptev, A.K. Petukhov
G.A. Petrov, Yu.S. Pleva, V.E. Sokolov and O.A. Shcherbakov

Leningrad Nuclear Physics Institute
Academy of Sciences of the USSR
188 350 Gatchina, Leningrad district, USSR

Abstract: A new method to obtain information about the properties of low energy p-resonances in heavy fissile nuclei is described. It is based on a study of neutron energy dependence of the forward-backward asymmetry of angular distribution of fission fragments, which is the result of s- and p-wave interference in neutron capture. The investigations of this effect for U-235 have been carried out using the neutron TOF-spectrometer GNEIS in Gatchina. The irregular energy behavior of the asymmetry coefficient α_{nf}^{fb} was observed in accordance with theory predictions. For some strong p-resonances the positions and effective widths have been determined at first.

(U-235, neutron, fission, time-of-flight, forward-backward asymmetry, parity violation, resonance, effective width)

Introduction

Nowadays, the information about the low-lying weak neutron p-resonances of heavy fissile nuclei is practically absent. These resonances are not observed neither in neutron total cross sections nor in partial ones.

On the other hand, in recent years there was a growth of interest in resonances of this kind inspired mainly by the studies of P- and T- violation effects in the compound nuclear systems and reactions. In a few theoretical and experimental works [1-3] the existence of the resonance enhancement in the vicinity of admixtured p-resonance was predicted and experimentally (in a case of P-violation) confirmed.

In the present work we have used the energy behavior of P-even forward - backward asymmetry of U-235 fission fragments relatively the direction of neutron beam to study the properties of p-wave resonances. An asymmetry of the following type $(\vec{p}_n \cdot \vec{p}_f)$, where $\vec{p}_n$ and $\vec{p}_f$ refer, respectively, to the neutron and fragment momenta, is a result of s- and p-wave interference in a slow neutron capture. The forward - backward asymmetry coefficients α_{nf}^{fb} in a simpliest two-level approximation may be written [2,3]:

$$\alpha_{nf}^{fb} = Q_{sp} \cdot \sqrt{\frac{\Gamma_p^n \cdot \Gamma_p^f}{\Gamma_s^n \Gamma_s^f}} \times$$

$$\times \text{Re} \left\{ \frac{E-E_s + \dfrac{i\Gamma_s}{2}}{E-E_p + \dfrac{i\Gamma_p}{2}} \cdot \exp(i\Delta\varphi_{sp}) \right\} \qquad (1)$$

The commonly used notations are employed in this equation and $\Delta\varphi_{sp}$ is a phase factor for interfering s- and p-resonances.

It can be seen from Eq.(1) that the asymmetry coefficient is enhanced in the vicinity of weak unobserved p-resonances and has a specific energy behavior. Thus, the observation of the energy behavior of $\alpha_{nf}^{fb}(E_n)$ is a new method to search for weak p-resonances and to study their properties. It ought to mention that the studies of this type have been carried out for nuclear fission for the first time.

Experimental set-up and results

This measurement have been done using the neutron TOF-facility GNEIS [4] of LNPI. Fission fragments were detected by means of a fast ($\tau \sim 0.1$ μs) multiplate (32 anodes, 16 double-side U-235 targets with ~ 22 cm and thickness ~ 150 μg/cm²) ionization chamber [5] (see Fig.1) filled with isobutan(C_4H_{10}) to pressure of ~ 1 atm.

A high voltage of ~ 3 kV is applied to the cathode (target). With an aim to change mutual orientation of $\vec{p}_f$ and $\vec{p}_n$ vectors, the chamber was turned periodically on 180° relatively the direction of the neutron beam. By means of appropriate discrimination of the kinetic energy spectra, U-235 fission fragments can be divided into light (N_1) and heavy (N_h) groups characterized by an opposite signs of the effect because of a two-particle nature of the fission reaction. In the course of an experiment, for every anode (i=1,2,... 32) TOF-spectra $N_{1,h}(E_n)$ for fission events were accumulated at two (j=1,2) opposite positions, so that :

$$\text{sign}(\vec{p}_n \cdot \vec{p}_f) = (-1)^{1+j} \qquad (2)$$

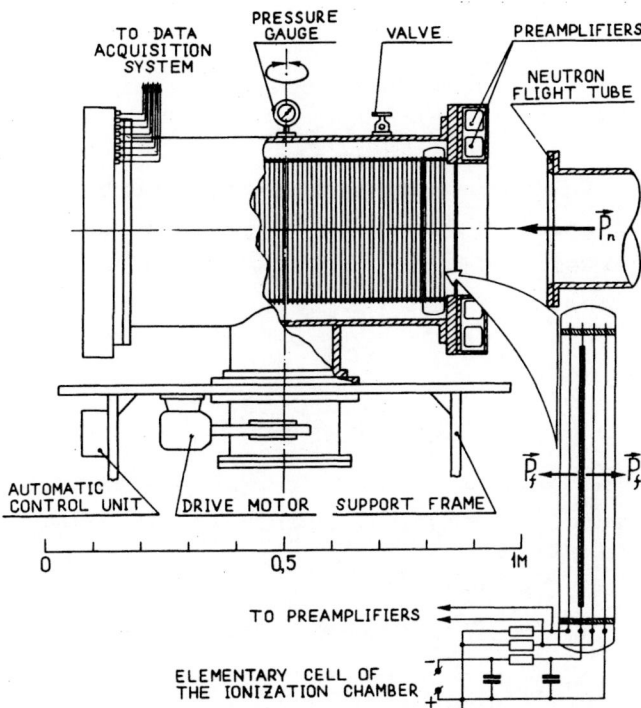

Fig. 1. Experimental set-up for forward-
backward asymmetry measurements.

An observed effect was calculated using
the following equation:

$$(\alpha_{nf}^{fb})_i = \frac{1}{2} \frac{(N_l/N_h)_{i,j=1} - (N_l/N_h)_{i,j=2}}{(N_l/N_h)_{i,j=1} + (N_l/N_h)_{i,j=2}} \qquad (3)$$

Having renormalized 32 TOF-spectra obtained to the unique value of flight path,
after the averaging, an effect $\alpha_{nf}^{fb}(E_n)$
and a false instrumental asymmetry $\alpha(E_n)$
were calculated simultaneously. The last
one is caused by inhomogeneities of the
targets, the neutron beam and a turn of
the chamber. A search for interference
effects caused by p-wave resonances of
U-235 was carried out in the energy range
from 1 eV to 135 eV divided into 6 intervals with variable width of the time
channel. A pure time for data accumulation was ~ 350 hours.

Some results of our measurements are
shown in Fig.2. Evidently, the unstatis-
tical irregularities are seen in the en-
ergy behavior of $\alpha_{nf}^{fb}(E_n)$. A shape of these irregularities varies from bipolar to
bell-shaped one. Apparently, the existence of the irregularities of this nature
at the 10^{-2} level is caused mainly by a
presence of the strongest $(\Gamma_p^n > \bar{\Gamma}_p^n)$ p-
wave resonances in the explored energy
range. At the same time, the analysis of
the instrumental asymmetry $\alpha(E_n)$ shows
an absence of the asymmetry statistically
significant at the level 10^{-3} ($\chi_{red}^2 =$
1.05 ± 0.05). A weighted integral value
of the experimental effect averaged over
the whole energy interval is

$$\overline{\alpha_{nf}^{fb}} = -(0.2 \pm 0.8) \ 10^{-4}, \ \chi_{red}^2 = 2.55 \pm 0.05$$

The effective parameters of p-resonances

With a view to find the parameters of
p-wave resonances we have chose from
$\alpha_{nf}^{fb}(E_n)$ 16 typical irregularities, for
which the value of the maximum effect was
~(3-7) standard deviations from an average value. On the average, 20-30 experimental points per one irregularity were
used. The further fitting procedure was
performed using the least squares code.

After some simplifications has been
done, based on an assumption about the
presence of one p-resonance and a few
nearest s-resonances, the fitting formula
was obtained which connects experimental
effect and the resonance parameters:

$$(\alpha_{nf}^{fb})_{exp}(E_n) = \frac{Const \cdot Q_{sp} \cdot \sqrt{\Gamma_p^n}}{E_n \cdot \sigma_{nf,exp}^{norm}(E_n) \cdot \Delta(E_n)} \ \times$$

$$\times \sum_k \left[g_{sk} \cdot \sqrt{\Gamma_{sk}^n \cdot \Gamma_{sk}^f \cdot (\Gamma_p - \bar{\Gamma}_\gamma)} \ \times \right.$$

$$\left. \times \int_{E-\Delta/2}^{E+\Delta/2} Re\left\{ \frac{\exp(i\varphi_{sp})}{\left(E-E_{sk}-\frac{i\Gamma_{sk}}{2}\right) \cdot \left(E-E_{sk}-\frac{i\Gamma_p}{2}\right)} \right\} \cdot dE \right]$$

(4)

where Q_{sp} is the spin factor, exact value
is unknown $(|Q_{sp}| \simeq 1)$; $g = (2J+1)/2(2I+1)$
is the statistical factor; Δ is the instrumental resolution (a rectangular reso-

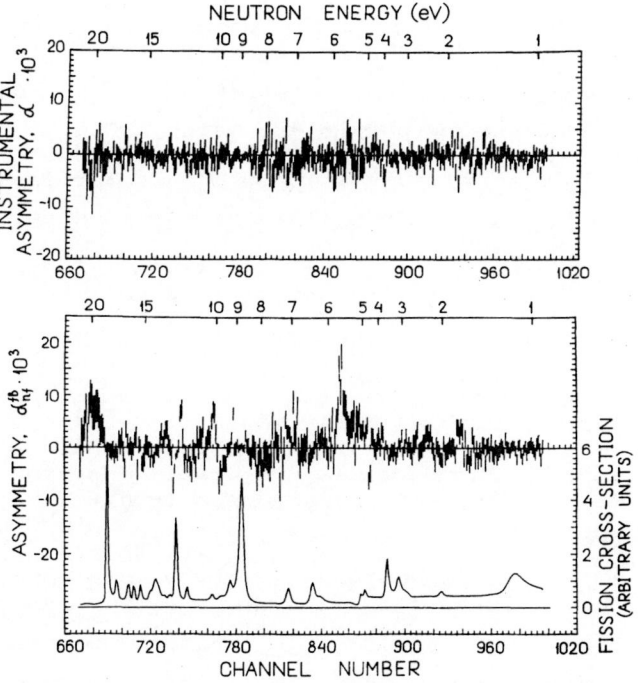

Fig. 2. Energy dependence of the forward-
backward asymmetry coefficient and
instrumental asymmetry.

Table 1. Effective parameters of some
p-resonances for U-235

E_p eV	Γ_p meV	$Q_{sp}^2 \cdot \Gamma_p^{n1}$ *) meV	$\Delta\varphi_{sp}$ rad
1.70±0.05	50±30	21.7±31.9	0.8±0.2
5.70±0.05	160±30	6.4±1.1	2.9±0.1
9.90±0.05	130±50	1.3±0.4	2.4±0.2
12.9±0.1	240±50	3.0±0.9	1.9±0.2
20.1±0.1	190±50	1.2±0.5	1.9±0.2
28.5±0.2	300±100	0.7±0.3	1.5±0.3
32.05±0.05	90±30	17.1±6.9	1.6±0.3
36.4±0.1	180±80	1.1±0.6	2.2±0.3
45.8±0.1	190±50	3.2±1.3	1.6±0.3
52.8±0.2	800±200	2.2±1.0	2.3±0.3
70.75±0.15	550±150	7.5±3.4	2.0±0.2
79.6±0.1	200±90	2.0±1.1	0.9±0.3
87.35±0.05	180±120	1.8±1.3	0.2±0.3
115.7±0.1	130±60	1.9±1.2	0.2±0.5
121.8±0.1	130±30	5.8±4.7	0.7±0.5
126.8±0.2	260±170	1.4±0.8	0.8±0.6

*) Γ_p^{n1} is reduced neutron width in
this transformed amplitude factor.

lution function is supposed); σ is the
observed fission cross section of U-235
normalized to resonance integral over the
range from 1 eV to 136 eV, deduced from
the data of [6]. With a view to simplify
the fitting formula, it was assumed that
effective values of Q_{sp} and Δ are equal
for all s-resonances involved, while the
fission width of the unknown p-resonance
is equal to the difference of total and
average radiation widths ($\Gamma_p^f = \Gamma_p - \overline{\Gamma}_\gamma$) [6].
The following parameters were varied:

$Q_{sp}\sqrt{\Gamma_p^n}$ -effective amplitude factor inclusive the neutron width of p-resonance;

E_p - energy of p-resonance;

Γ_p -effective total width of p-resonance;

$\Delta\varphi_{sp}$ -effective phase factor.

The results of the data analysis are
shown in Table 1. The most steady parameters of the fitting procedure are E_p
and Γ_p. Their values are weakly dependent
on a number of s-resonances involved. At
the same time, for effective $Q_{sp}^2 \Gamma_p^{n1}$ and
$\Delta\varphi_{sp}$ this dependence is very significant.
The values of $\Delta\varphi_{sp}$ are within the range
from 0 to π whilst $Q_{sp}^2 \cdot \Gamma_p^{n1}$ varies more than
order in value. As a rule, the quality of
fit is good.
The average values of the effective
parameters of p-resonances deduced using
the normal distribution are as follows:

$$\overline{Q^2 \cdot \Gamma_p^{n1}} = (5.0 \pm 1.5) \text{ meV}$$

$$\overline{\Gamma_p} = (240 \pm 50) \text{ meV}$$

$$\overline{\Delta\varphi_{sp}} = (1.5 \pm 0.2) \text{ rad}$$

An evaluation of effective number of
of fission channels ν_p^f for 16 p-resonances of U-235 was done using formula of
Wilets [7] and $\Gamma_p^f = \Gamma_p - \overline{\Gamma}_\gamma$:

$$\nu_p^f = \frac{2 \, (\overline{\Gamma_p^f})^2}{\overline{(\Gamma_p^f)^2} - (\overline{\Gamma_p^f})^2} = 2.4 \pm 1.6$$

Analogous estimate for reduced neutron
widths Γ_p^{n1} (using an assumption that $|Q_{sp}| = 1$) gives :

$$\nu_p^{n1} = 1.4 \pm 1.1$$

These estimations are in agreement
with the corresponding quantities for
known s-resonances [6] within the same
energy range:

$$\nu_s^f = 2.1 \pm 0.4 \quad , \quad \nu_s^{n0} = 1.5 \pm 0.4$$

Conclusion

The information obtained seems to be
very important both for nuclear spectroscopy of weak p-resonances and for
fundamental investigations of P- and T-
violation effects. These effects are expected to be resonancely enhanced in the
vicinity of p-resonances.
Nowadays, an analogous measurements
are in progress at the GNEIS-facility for
U-233 and the next are planned in a
nearest future for Pu-239. The experiment
described in this article was performed
on a limit of accessible statistical
accuracy. That is why a further development of such investigations is planned
with the use of more intensive neutron
source IBR-30 (Dubna).

References

1. G.V. Danilyan: Uspekhi Fiz. Nauk 131,
320 (1980)
2. O.P. Sushkov and V.V. Flambaum:
Uspechi Fiz. Nauk 136, 4 (1982)
3. V.T. Bunakov and V.P. Gudkov: Nucl.
Phys. A401, 93 (1983)
4. N.K. Abrosimov, G.Z. Borukhovich,
A.B. Laptev, V.V. Marchenkov, G.A.
Petrov, O.A. Shcherbakov, Yu.V. Tuboltzev and V.I. Yurchenko: Nucl.
Instr. Methods A242, 121 (1985)
5. A.M. Gagarsky, S.P. Golosovskaja, A.B.
Laptev, G.A. Petrov, A.K. Petukhov,
Yu.S. Pleva, V.E. Sokolov, G.V. Stab-
nikova and O.A. Shcherbakov, LNPI
preprint-1634, Leningrad (1990)
6. S.F. Mughabghab, Neutron Cross Sections: Vol.1, Part B, Academic Press,
London (1984)
7. L. Wilets: Phys. Rev. Lett. 9, 430
(1962)

ISOTONIC YIELDS AND NEUTRON EVAPORATION IN FISSION

M. Djebara and M. Asghar
Institut de physique
USTHB,   BP32   El Alia, Algiers
Algeria

Abstract: A Monte-Carlo simulation has been set up to treat the neutron evaporation from fission fragments. The influence of the neutron evaporation on different distribution and fission parameters has been studied.

It has been observed experimentally that the isotonic distribution of fragments in (n_{th}, f) shows an enhancement in yield for even neutron numbers. This odd-even effect is defined as: $\delta n = (Ye-Yo)/(Ye+Yo)$, where Ye and Yo represent the total yield for even and odd neutron numbers. Fig.1 shows the variation of n with the light fragment kinetic energy EL for $236U^*$, $240Pu^*$, $250Cf^*$ and $230Th^*$ [1-4]. Moreover, the local odd-even effect $\delta n(N)$ depends strongly on the neutron number N as shown in Fig.2. Since the measured isotonic distributions correspond to those obtained after the prompt neutron evaporation, one wonders if there is any relationship between these distributions and the primary ones. In order to understand this phenomenon a Monte-Carlo simulation of prompt neutron evaporation was carried out.

Monte-Carlo Simulation

For each primary mass division AL'/AH' the light fragment nuclear charge ZL and the light fragment kinetic energy EL were selected randomly according to a Gaussian distribution. The reaction Q-values and the neutron separation energies Bn were obtained from the Wapstra mass tables [5] completed, when needed with values from the Liran and Zelde mass tables [6]. The total fission kinetic energies EK were calculated with the help of energy and momentum conservation laws. The total excitation energy $E^*=Q-EK$ was distribued between the two fragments according to the mean number of neutrons evaporated by each fragment: $EL=E^*\gamma(AL')/(\gamma(AL') + \gamma(AH'))$. The neutron kinetic energy distribution εn was taken to be Maxwellian in shape with a nuclear temperature corresponding to EL. The evaporation was continued up to $EL^* < Bn + \varepsilon_n$.

Results and Discussion

i) Mass Distribution.

The post-mass distributions are reproduced well by the simulation. The heavy wing of the initial (primary) mass distribution shifts towards the light masses, while the light wing remains almost fixed; this enhances the maximum mass yield and reduces the width of the mass distribution. This is due to the increase of the number of emitted neutrons with the fragment mass number.

ii) Isotonic Distribution

The simulation shows that starting with a smooth isotonic yield distribution for the light group, simulated isotonic yield, after neutron evaporation, is well structured and is not sensitive to the initial distribution: the neutron evaporation destroys completely any initial structure due to neutron odd-even effect. The mean calculated neutron odd-even effect is $\delta n = 15$ % and it is the same for different fissioning nuclei. Fig.3 shows on for different fissioning nuclei, deduced from the work on Lohengrin mass separator [1-4]. Apart from $250Cf^*$ it seems that the δn is independent of the fissioning nuclei and its value is 6%. The simulated δn value can be shifted towards the experimental value if one acts on the difference of the Bn-values of fragments of different parity with respect to the neutron number N. This is not unreasonable because:
a) the Bn-values used for evaporation and calculated from the mass tables are correct only for fragments in their fundamental state;
b) towards the end of evaporation cascade, the gamma emission becomes very competetive. The higher value of δn for $250Cf^*$ is due to the fact that the range of N number is bigger and it includes two peaks situated at $N \simeq 60$ and $N \simeq 66$ in the $\delta n(N)$, Fig.2. Only the first peak is common to all the fissioning systems. The two peaks behave very differently as a function of EL. In the case of $250Cf^*$ while the first peak keeps its value of 15% for all EL values, the second one has a value that increases by a factor 3, when one goes from EL= 95 MeV to EL=

138

120 MeV, Fig.4. This second peak is responsible for the increase in on for 250Cf when EL goes up,Fig.1. We think that this peak results from a spherical shell effect (N = 82 and Z = 52), whose influence increases as EL goes up. The peak at N=60 is due to a deformed shell effect (N = 88). Contrary to 250Cf, this peak, for the other fissioning nuclei is again due to the same spherical shell effect. This explains the increase in δn with EL for these nuclei [7].

iii) Charge Distribution

The simulation shows that starting with a smooth and constant initial charge polarisation $\Delta Z(A')$= -0.5, the final $\Delta Z(A')$, after neutron evaporation, is structured as observed experimentally. We can reproduce the experimental data, if we introduce an initial proton odd-even effect whose value corresponds to the experimental one. Hence, the oscillations in $\Delta Z(A')$ results from proton odd-even effect and neutron evaporation. Furthermore, the results show that neutron evaporation does not affect the behaviour of $\langle \sigma z^2 \rangle$ as a function of EL; this is what is observed experimentally.

The conclusion is that the simulation of neutron evaporation helps to understand its influence on the different fission obsrevables.

REFERENCES

1. W. Lang, H. G. Clerc, H. Wohlfarth, H. Schrader and K. Schmitt; Nucl. Phys. A435,34 (1980).
2. C. Schmitt, A. Guessous, J. P. Bocquet, H. Clerc, R. Brissot, D. Engelhardt, H. R. Faust, F. Gonnenwein, M. Mutterer, H. Nifenecker, J. Pannicke, Ch Ristori and J.P. Theobald; Nucl. Phys. A430,21(1984).
3. M. Djebara, M. Asghar, J.P. Bocquet, R. Brissot, J. Crançon, Ch Ristori, E. Aker, D. Engelhardt, B. D. Wilkins, U. Quade and K. Rudolph; Nucl. Phys. A496,346(1989).
4. J.P. Bocquet, R. Brissot, H. R. Faust, M. Fowler, J. Wilhelmy, M. Asghar and M. Djebara; Z. Phys. A335,41(1990).
5. A.H. Wapstra and G. Andi; Nucl. Phys. A432,1(1985).6. Liran and Zelde; Atomic Data and Nuclear Data Tables 17,474(1976).
7. U.Quade, PhD Thesis, Munich, 1983.

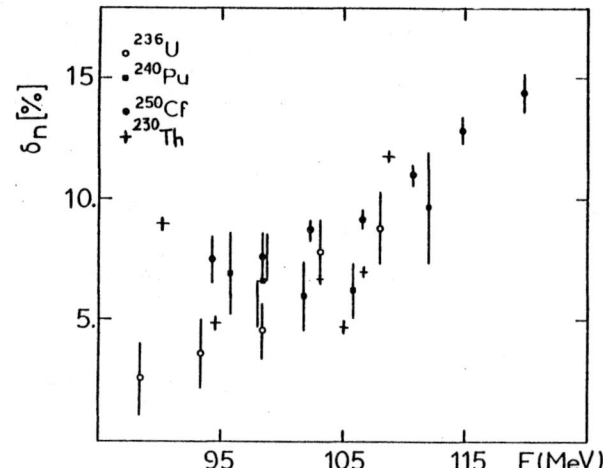

Fig.1 Neutron odd-even effect on as a function of light fragment kinetic energy EL.

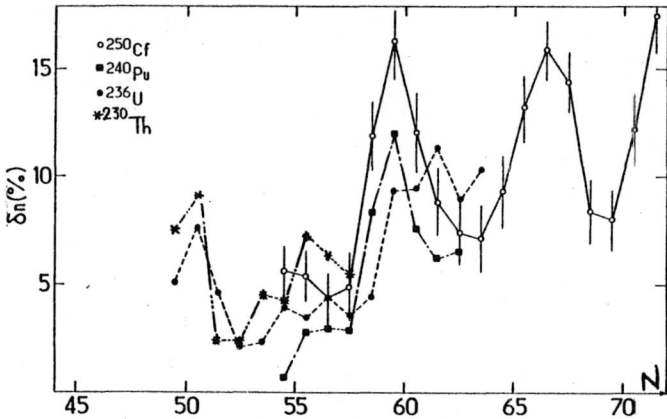

Fig.2 Neutron local odd-even effect on as a function of neutron number N.

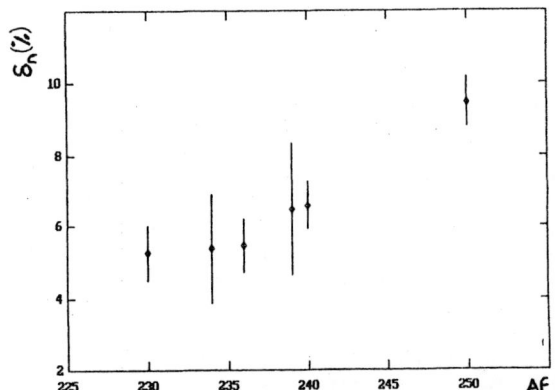

Fig.3 Neutron odd-even effect on as a function of fissioning neuclus AF.

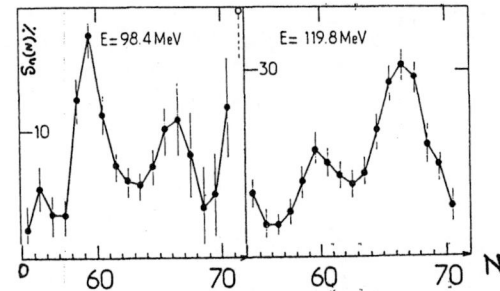

Fig.4 Neutron local odd-even effect on for 250Cf as a function of neutron number N.

URANIUM COLD FRAGMENTATION BY THERMAL AND FAST NEUTRONS

V.A. Khryachkov, A.A. Goverdovsky, B.D. Kuzminov, V.F. Mitrofanov,
N.N. Semyonova, A.I. Sergachov, A.I. Slyusarenko

Institute of Physics and Power Engineering, Obninsk, USSR

Abstract: Mass-energy spectra of ^{235}U cold fragmentation induced by thermal and fast (En= 1MeV) neutrons were obtained. Effect of the conservation of fissile nucleus excitation energy collected in the saddle point up to scission point was observed. It means that for Z-even target irradiated by thermal neutrons it is impossible to reach the region of true-cold-fragmentation.

(^{235}U, twin ionization chamber, cold fragmentation, fast neutrons)

INTRODUCTION

The binary nuclear fission is specified by a wide spectrum of fragments mass-energy, that considerably makes difficult the study of this complex and interesting phenomenon. The reaction energy Q consists of the Ek- fragments kinetic energy and their excitation E0, removed by the neutron and γ-emission after the nuclear rupture. The particular interest is generating by the study of so called cold fragmentation for which Ek is near or equil to Q. In this case the neutron emission energetically forbidden, and mass spectrum shows a bright structure due to shell and odd-even effects.

The detailed investigations of the cold fission have been performed in the recent ten years using combined detector, which allowe to obtained not only mass-energy, but and charge spectra [1]. The obtained data reflected all the essential properties of cold compact fission. However, one fundamental question has been unclear- fate of excitation energy collected by nucleus in the saddle point of fission barrier. In this work it is proposed to compare the cold fragmentation properties (yields, spectra) of U-235 irradiated by thermal and fast (En= 1 MeV) neutrons.

EXPERIMENTAL METHOD

The work has been realized on the accelerator KG-2.5 IPPE. As a fast neutron source the reaction $T(p,n)^3He$ was used. The thickness target (T-Ti) was 1mg/cm2 (water cooled copper backing). Thermal neutrons were received by fast neutron moderation in a polyethylene unit (20 cm thick).

Multidimensional fission fragment spectrometer has been built using twin ionization chamber with Frish-grids [2]. The uranium target presented a thin uranium tetrafluoride layer (50 mkg/cm2) with 3 cm in diameter. To obtain the cathode conductivity the backing and the layer were sprayed with gold (0,03mg/cm2) As an operating gas the mixture of 90%Ar+ 10% CH4 was used. The ionization chamber operated in gas passage conditions (with the intensity of 3 L/H). A special device maintained the gas mass in the chamber volume with the stability of 0.3% under the pressure of 1.05*E5 Pa. The bias voltage (-4.7 kV) was applied to the cathode (uranium target) and through resistor deviders to the Frish grids [2].

Experimental data evaluation was realized using the method analogue to [3].

RESULTS

Fig.1 gives a typical fragment mass spectra for different values of light fragment kinetic energy EL (in MeV). Fission events have been selected from all the array. They conform to a fragment yield angle within the limits $\leq 16^\circ$ relative to target normal (for good resolution). Primarily the attention is drawn to the fact that for EL $\geq$ 112 MeV the spectrum is formed by four essential components, grouping near ML = 102,96,90,84 a.m.u.(ZL = 40,38,36 and 34 respectively [1]). Every component is defined by its values of the reaction energy Q and by total kinetic energy TKE, that is why the mass spectra in question are modulated by individual excitation energy in the scission point E0. With the increase of EL (or TKE) E0 is decreasing giving rise to cold fragments spectrum changes. So, the fraction ML = 84 a.m.u. insreases progressively and for EL = 116 MeV it dominates in the spectrum (see fig.2 on the left). The situation is opposite for the group ML = 96 a.m.u.; with EL = 115

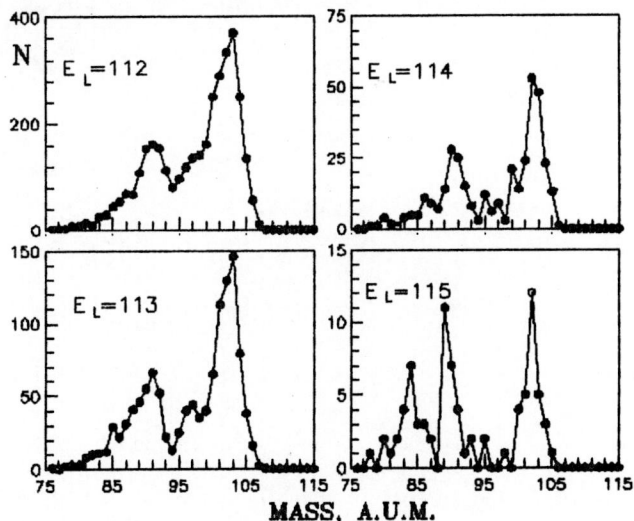

Fig.1. CF mass spectra for U-235 fission by thermal neutrons. $\vartheta \leq 16^\circ$.

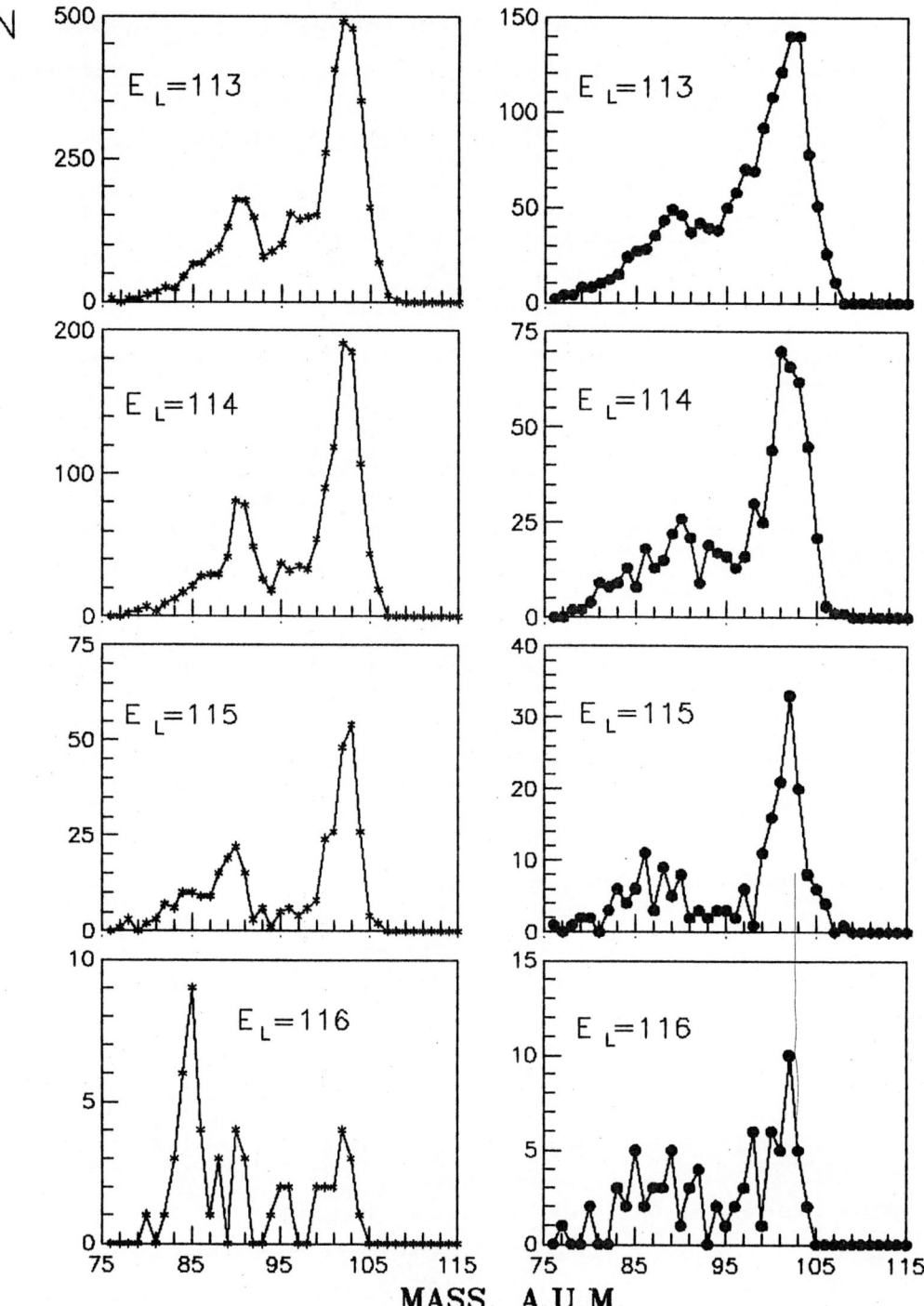

N

Fig.2. U-235 CF mass spectra for thermal (left) and fast En = 1 MeV
(right) neutrons.   θ ≤ 40°.

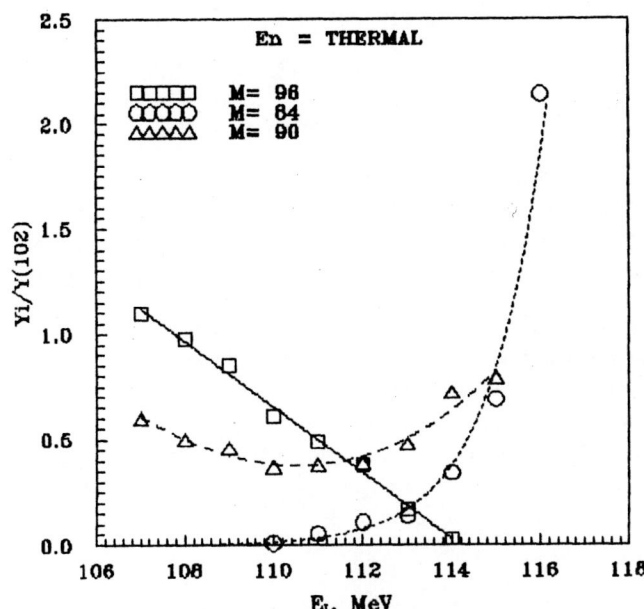

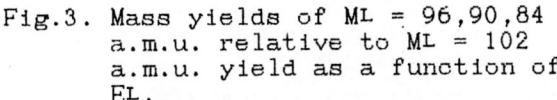

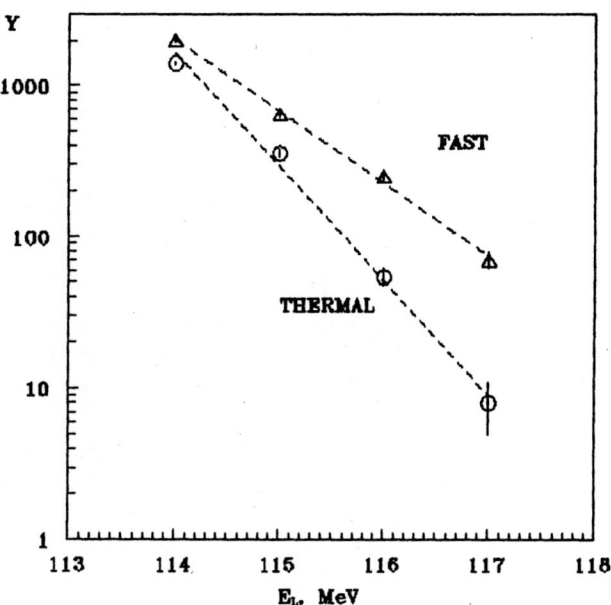

Fig.3. Mass yields of ML = 96,90,84
a.m.u. relative to ML = 102
a.m.u. yield as a function of
EL.

MeV already its contribution is statistically negligible. Figure 3 demonstrates spectrum peculiarities more visually. This figure presents relative group yields for ML = 96,90 and 84 a.m.u. as a function of EL. As a bench mark the component ML = 102 a.m.u. has been taken. Mass spectra were described by four Gaussian curves. First moments were given precisely (102,96,90,84) and were not variated.

On the background of a sufficiently smooth linear reduction of ML=96 relative contribution and of a approximetive constancy for ML=90 a.m.u. in all region of EL studied, the fragment addition around the mass 84 a.m.u. increases exponentionally. The explanation of this effect can be found from Q-values calculation in comparison with the TKE [2].

We can see from fig.4, which represents CF-yields for thermal and fast neutrons, that additional excitation contributed to the nucleus by 1-MeV-neutron conserves up to scission point. CF-spectra (fig.2 on the right) confirm this. Side by side with already known four-component spectra exists their qualitative and essential difference from the considered above ones for a fission by thermal netrons: for EL = 116 MeV the condition of the 102-mode prohibition have not yet been reached, that is one can observe some delay of a CF spectrum transformation in favour of the component ML = 84 a.m.u.

On returning to the fig.4 it is necessary to make some important remarks. First, during the U-235 fission by fast neutrons the fragments mean total kinetic energy, averaged over all fission modes increases relative to TKE in the thermal point only by 0,2 MeV [3] and can not influence the effects in question in a

Fig.4. Integral U-235 CF-yields for
thermal and fast neutrons.

cold fraction region. Second, the δEL-difference on the EL axis of Y yields increases in the region of EL $\geq$ 112 MeV. Taking into account that the minimum neutron binding energy in light fragments group is approximately 3-4 MeV [4] we can explain the δEL behaviour as a neutron emission manifestation and fix for the uranium the limit of the fragmentation without neutron emission (in all the region of masses from 80 to 105 a.m.u.), where EL = 115 MeV. This limit refers first of all to the fraction ML = 84 a.m.u., naturally. Unfortunately, in the frame of statistics obtained during the uranium fission by fast neutrons one can not realize a spectra decomposition to components and trace the δEL behaviour for each fragment group.

The data evaluation of cold uranium fragmentation by neutrons with the energy of 1 MeV permits to draw a conclusion that a nucleus exitation energy in a saddle point conserves up to the moment of fragment rupture. It means that it is impossible to observe a true cold (TKE=Q) uranium-235 fragmentation process neither with fast neutrons (E0=2 MeV) nor with thermal neutrons (E0=1 MeV). In order to make this it is necessary to obtain the condition of fission lower the barrier, which is connected with significant methodical difficulties.

REFERENCES

1. H.-G. Clerc, W. Lang, M. Mutterer
   et.al.: Nucl. Phys. A452, 277 (1986)
2. V.A. Khryachkov , A.A. Goverdovsky,
   B.D. Kuz'minov et.al.: Nucl.Phys.(Sov)
   53, 621 (1991)
3. C.A. Streade, Neutron induced fission
   of U-235, Thesis, Geel (1985)
4. P. Moller and J.R. Nix : At.Nucl.Data
   Tables 26, 165 (1981)

MASS YIELDS IN THE VERY ASYMMETRIC FISSION OF ^{249}Cf(n_{th},f) *,**

R. Hentzschel[1], H. R. Faust[2], H. O. Denschlag[1],
B. D. Wilkins[3], J. Gindler[3]

[1]Institut für Kernchemie, Universität Mainz, Germany
[2]Institut Laue-Langevin, Grenoble, France
[3]Argonne National Laboratory, Argonne, Illinois, USA

Abstract: Independent fission yields and fragment kinetic energies in the very asymmetric (A=87-72) fission of ^{249}Cf have been measured using the mass separator LOHENGRIN at the Institut Laue-Langevin in Grenoble. The mass yields were found to be up to a factor of 100 higher than predicted in the US ENDF-V file. A preliminary evaluation of the isotopic yields shows a strong increase of the proton odd-even effect from 7 % to 15 %. This can be explained by a decreasing excitation energy in this very asymmetric region.

Introduction

Yields in the fission of ^{249}Cf have been measured radiochemically for values > 0.15 % [1] and using physical separation techniques [2] in the mass region A=88-120. We have extended the mass yield curve to A=87-69 to investigate odd-even effects at extreme mass-asymmetry values of A_L/A_H=69/181 and to check the chain yield predictions of the ENDF-V file [3].

Target Preparation

The californium-targets were produced by electrodepositing ^{249}Cf (99.7 % isotopic enrichment) as Cf_2O_3 onto Ti-backings of 50 μm thickness. The electrodeposition was successful only after a careful purification of the californium by ion-exchange chromatography using α-hydroxyisobutyrate as elution agent. The sources with a layer of 25 μg of ^{249}Cf (area of deposit 3 x 20 mm) were covered with a Ni-foil of 0.25 μm thickness against loss of fissile material through sputtering.

Measurements

The experiments were performed in the mass separator for unslowed fission fragments LOHENGRIN at the high-flux reactor of the Institut Laue-Langevin in Grenoble. The targets in the source position were exposed to a neutron flux of 5.4·10^{14} cm^{-2}s^{-1}. The fission fragments were separated according to their ratios of mass to ionic charge state (A/q) and kinetic energy to ionic charge state (E/q). The mass resolving power A/ΔA was >500 and the energy dispersion was about 0.5 MeV. A big ionization chamber [4] with a divided anode allowed the simultaneous determination of the total energy E_{tot} and the specific energy loss ΔE of the fragments. ΔE is a function of the nuclear charge of the isotopes. The charge resolving power Z/ΔZ was about 40 for the light fragments.

---

* Dedicated to Prof. Dr. P. Armbruster on the occasion of his 60th birthday.

** This work forms part of the doctoral thesis of R. Hentzschel.

The energy distributions of the fission products were scanned at the mean ionic charge state from E_{kin}=99 to 119 MeV (A=87 to A=73). These distributions can be described as Gaussians with typical mean values of 107 MeV, and widths of 16 MeV FWHM. The ionic charge distributions were measured at mean kinetic energy (A=87 to A=74) for 5 or 6 ionic charge states and at E_{kin}=103 and 115 MeV (A=87 to A=78) for 3 or 4 ionic charge states. Because of the very low counting rates (7 counts per hour at A=70) the yields of A=72 to A=69 were measured only at the estimated mean kinetic energy (E=105 MeV) and ionic charge state (q=18).

The burn-up of the targets was monitored by scanning the energy distribution of fission fragments with A/q=4 once a day. After one to three days of increased burn-up the targets became stabilized. The half-life of the targets of 6.5 days is in good agreement with the neutron flux and the total cross section of 2125·10^{-24} cm^2. The mean kinetic energy of the fission fragments was reduced by 0.7 MeV during two weeks. This indicates a slight diffusion of californium into the Ti-backing.

Data Evaluation

The measured counting rates were corrected for burn-up of the fissile material, dead-time losses and for the energy dispersion of the spectrometer. After integration over the kinetic energy and the ionic charge states these relative yields were normalized to the absolute yield of the mass chain with A=88 [2]. The kinetic energies were corrected for the calculated energy loss in the target and the Ni-foil ($\approx$7 MeV). The nuclear charges were obtained by fitting the corresponding ΔE-spectra. Usually yields of three or four nuclides were obtained per mass chain.

Results

The mass yields integrated over kinetic energies and ionic charge states are shown in Fig.1 together with the prediction of Rider [3]. It appears that the predicted values are considerably lower than our experimental ones (e.g. A=74: 2.46·10^{-5} % (Rider) and 5.32·10^{-4} % (this work)).

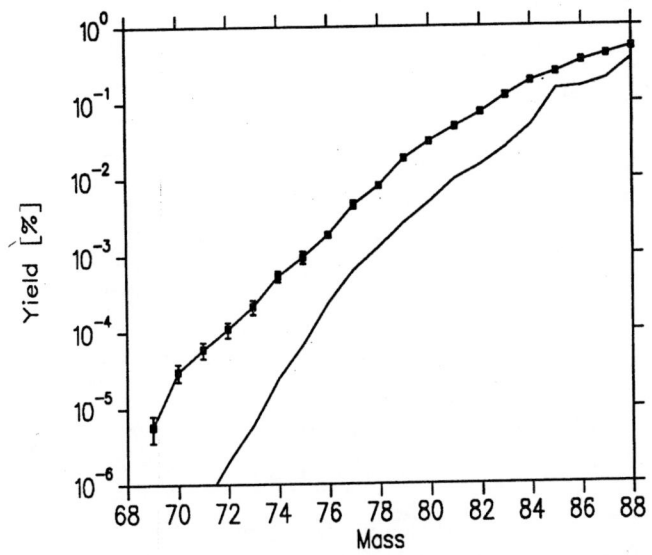

Figure 1: Isobaric chain yields in the thermal neutron indu-ced fission of ^{249}Cf. Squares are experimental values from this work. The drawn out lower line are predicted values from [3].

On the other hand, for the reaction ^{251}Cf(n_{th},f) Rider predicts chain yields (e.g. $1.22 \cdot 10^{-3}$ % for mass chain A=74) that are considerably higher than both his predictions and our experimental values for ^{249}Cf-fission. Such high yields in the fission of ^{251}Cf could bias our present results, since one can calculate that ^{251}Cf is growing in our target by a double (n,γ)-reaction to form a contamination of about 10 % after two weeks exposure time. Such a rather small contamina-tion could make a large bias in a mass region where the chain yield of the contamination would be 50 times larger than that of the studied reaction. In order to exclude such a bias, we have measured the relative intensity of A=74 at E_{kin}=105 MeV and q=19 as a function of exposure time. The results are shown in Fig.2 together with a calculated curve that would indicate the increase in counting rate due to breeding of ^{251}Cf, if the yield values for the fission of ^{249}Cf and ^{251}Cf of Rider were true. The figure shows no increase in counting rate, thus excluding a major bias of the present mass yield values. Obviously the yield values predicted for the very light fission products in the reaction ^{251}Cf(n_{th},f) should be checked.

Table 1: Experimental isobaric chain yields of ^{249}Cf(n_{th},f) summed over ionic charge states and integrated over the kinetic energy of the fission fragments

Mass	Yield [%]	Δ Yield [%]
88†	$5.20 \cdot 10^{-1}$	$1.2 \cdot 10^{-2}$
87	$4.21 \cdot 10^{-1}$	$8 \cdot 10^{-3}$
86	$3.43 \cdot 10^{-1}$	$1.9 \cdot 10^{-2}$
85	$2.47 \cdot 10^{-1}$	$6 \cdot 10^{-3}$
84	$1.89 \cdot 10^{-1}$	$4 \cdot 10^{-3}$
83	$1.22 \cdot 10^{-1}$	$3 \cdot 10^{-3}$
82	$7.37 \cdot 10^{-2}$	$1.6 \cdot 10^{-3}$
81	$4.79 \cdot 10^{-2}$	$1.3 \cdot 10^{-3}$
80	$3.01 \cdot 10^{-2}$	$1.5 \cdot 10^{-3}$
79	$1.84 \cdot 10^{-2}$	$1.4 \cdot 10^{-3}$
78	$8.15 \cdot 10^{-3}$	$2.3 \cdot 10^{-4}$
77	$4.61 \cdot 10^{-3}$	$5.6 \cdot 10^{-4}$
76	$1.86 \cdot 10^{-3}$	$1.5 \cdot 10^{-4}$
75	$9.54 \cdot 10^{-4}$	$1.7 \cdot 10^{-4}$
74	$5.32 \cdot 10^{-4}$	$8.9 \cdot 10^{-5}$
73	$2.13 \cdot 10^{-4}$	$5.7 \cdot 10^{-5}$
72	$1.09 \cdot 10^{-4}$	$2.4 \cdot 10^{-5}$
71	$5.35 \cdot 10^{-5}$	$1.3 \cdot 10^{-5}$
70	$3.02 \cdot 10^{-5}$	$7.6 \cdot 10^{-6}$
69	$5.7 \cdot 10^{-6}$	$2.2 \cdot 10^{-6}$

$\dagger$ Value from Djebara et al.[2]

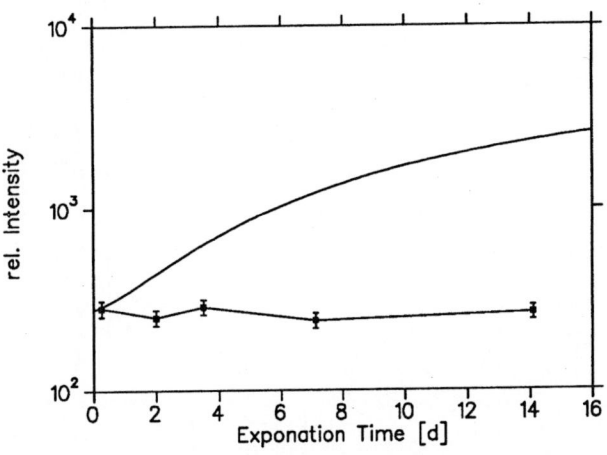

Figure 2: Relative experimental yield of A=74 at E_{kin}=105 MeV and q=19 as a function of the exposure time of the ^{249}Cf-target in the neutron flux.
The yield is corrected for the burn-up of the target. The drawn out upper curve is calculated with the predicted yields of ^{249}Cf(n_{th},f) and ^{251}Cf(n_{th},f) and the cross sections of the (n,f)- and (n,γ)-reactions.

A preliminary analysis of the individual isotopic yields with the method of the third difference [5] shows an increa-sing proton odd-even effect up to 15 %. (Fig.3). This is three times more than reported in the mass region 88-120 [2]. The same trend of increasing odd-even effects is also reported by Sida et al. [6] in reaction ^{235}U(n_{th},f) and Ditz in the reac-tion ^{239}Pu(n_{th},f) [7]. This behavior can be explained by a decreasing amount of energy available for excitation in the fission process in the very asymmetric region.

144

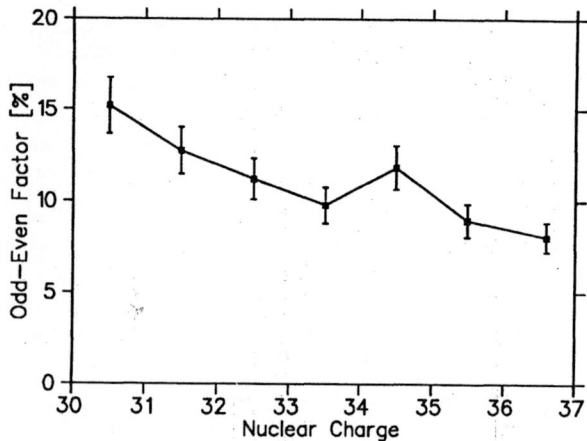

Figure 3: Proton odd-even effect as a function of the nuclear charge of the light fragments

This energy (Q-value of the mass split [8] minus the experimental total kinetic energy of the fragments) decreases from 45 MeV to 35 MeV as shown in Fig.4. Here the TKE was calculated from the kinetic energies of the light fission products under the condition of conservation of linear momentum.

A comprehensive data-evaluation is in progress and will be published.

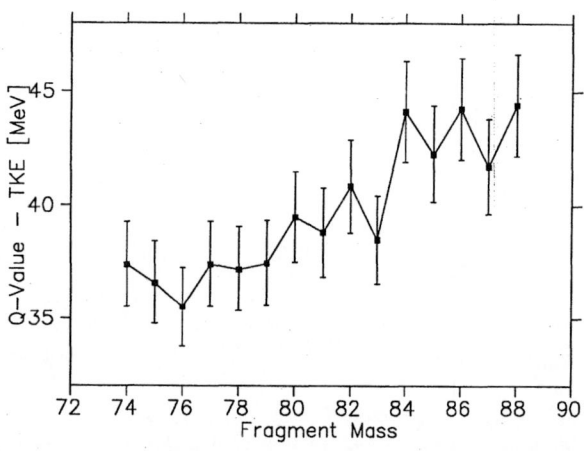

Figure 4: Difference Q-value [8] minus TKE as a function of the light fragment mass

Acknowledgement

We would like to thank Dr. N. Trautmann and B. Görtz for helping in the target preparation. This work has been supported by the German Federal Minister of Research and Technology (BMFT) under contract number 06-MZ-106.

References

[1] J.E. Gindler, L.E. Glendenin, D.J. Henderson, J.inorg.nucl.Chem.   Vol 43, 1743-1749 (1981)

[2] M. Djebara, M. Asghar, J.P. Bocquet, R. Brissot, J. Crançon, Ch. Ristori, E. Aker, D. Engelhardt, J. Gindler, B.D. Wilkins, U. Quade, K. Rudolph, Nucl.Phys. A496, 346 (1986)

[3] B.F. Rider, Report NEDO-12154-3(C), ENDF-322 (1981)

[4] J.P. Bocquet, R. Brissot, H.R.Faust, Nucl. Inst. Meth. A267 466-472 (1988)
(for a sketch of the set-up see fig. 1 in the contribution Stumpf et al., this proceedings)

[5] B.L. Tracy, J. Chaumont, R. Klapisch, J.M. Nitschke, A.M. Poskanzer, E. Roeckl C. Thibault, Phys. Rev. C5, 222 (1972)

[6] J.L. Sida, P. Armbruster, M. Bernas, J.P. Bocquet, R. Brissot, H.R. Faust, Nucl.Phys. A502 (1989) 233c-242c

[7] W. Ditz, doctoral thesis, Mainz (1991)

[8] P. Möller W.D. Myers, W.J. Swiatecki, J. Treiner, Atom. and Nucl. Data Tables Vol 39 Nr 1 (1988)

ODD-EVEN EFFECTS IN THE REACTION ^{241}Am(2n,f)*,**

P. Stumpf, U. Güttler, H. O. Denschlag
(Institut für Kernchemie, Universität Mainz, Germany)
H. R. Faust (ILL, Grenoble, France)

Abstract: Fission yields of the light fission products in the reaction ^{241}Am(2n,f) have been measured. The results indicate an odd-even proton effect in the yield distribution of this odd nucleus with Z=95.

(Fission, ^{241}Am, ^{242}Am, yields, odd-even effect, mass separator, LOHENGRIN)

Introduction

Odd-Even effects, i.e. a preferential formation of fission fragments with even proton (and/or neutron) number are of interest for fission theory as they are an indication for the temperature of nuclear matter at the scission point. They are also of interest from a practical standpoint for the prediction of yet unmeasured fission yields. Up to now odd-even proton effects in fissioning systems with an odd number of protons were not considered as, necessarily, in such a reaction a simultaneous formation of an even and an odd fragment takes place. We want to show in the following, however, that fragments of the light mass range, are preferentially even, whereas those of the heavy mass range are correspondingly odd. The consequences of this effect on fission theory and on fission yield systematics are the same as discussed above.

Experimental

The measurements were performed at the recoil mass separator LOHENGRIN [1,2] of the Institut Laue-Langevin in Grenoble (France). Targets of thin layers of ^{241}Am (100 μg AmO_2/cm^2) electrodeposited on some tantalum backings and covered by layers of 222 μg/cm^2 nickel were irradiated in the target position of the mass separator with thermal neutrons (flux: $5 \cdot 10^{14}$ s$^{-1} \cdot$cm^{-2}). The fissioning nucleus, a mixture of the two isomers $^{242m+g}$Am, was produced in a breeding reaction from the target material in the source position. The actual measurements of fission products started after two days of breeding.

In LOHENGRIN the separation of the fission products according to their mass (A), initial kinetic energy E_{kin} (to be corrected for an energy loss of about 7 MeV in target and Ni cover) and according to their ionic charge (q) takes place.

At the collector site, the beam of separated fission products entered the ionization chamber (BIC) [3] shown in Fig. 1. Due to a divided anode the ionization chamber allows one to measure the

----

* Dedicated to Prof. Dr. P. Armbruster on the occasion of his 60th birthday.
**This work comprises a part of the doctoral thesis of P. Stumpf, to be submitted to Johannes Gutenberg Universität, Mainz.

specific energy loss (dE) and the total energy of the fission fragments. This provides information on the mass and atomic number (Z) of each fission fragment measured.

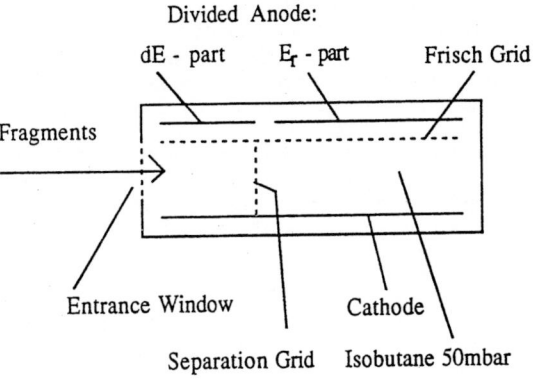

Fig. 1: Sketch of the Big Ionization Chamber (BIC) attached to the LOHENGRIN Mass separator.

The mass separator LOHENGRIN combined with the ionization chamber provides relative count rates of fission fragments of a specific mass, atomic number, kinetic energy and ionic charge state. In order to compare these numbers with radiochemical yields in every mass chain an integration over the distributions of kinetic energies and ionic charges is required.

Concerning the energy distribution, generally, measurements were carried out at five different values. The results were fitted to a Gaussian distribution with a small asymmetric low energy tail.

Due to the limitation in beam time and target material, the ionic charge distribution could not be measured completely for all mass chains. Therefore the evaluation is based essentially on the theory of Betz [4] which was checked experimentally only in a few mass chains. The parameters derived were used to correct the other mass chains in which only one or two ionic charge states near the maximum of the distribution were measured.

146

RESULTS

Relative count rates of fission fragments obtained at different values of their atomic number (Z) are shown in Fig. 2. Qualitatively one observes that points with even Z (30, 32, 34) tend to be higher than those with odd Z (31, 33). This staggering is, however, optically depressed by the use of a logarithmic scale required by the steep increase of the yields in this region of the mass yield curve. When we use the method of Tracy et al. [5] to calculate a local odd-even effect for protons (EOZ), we obtain the values shown in the inset of Fig. 2 for the mean numbers of Z indicated. A corresponding analysis for an odd-even effect for neutrons gives values of EON = 8.4 ± 2.2 and EON = 12,1 ± 2.4 for the mean neutron numbers N = 48.5 and 49.5, respectively. These values are, however, less significant, as they are affected by the prompt neutron emission. Therefore the following discussion will focus on EOZ.

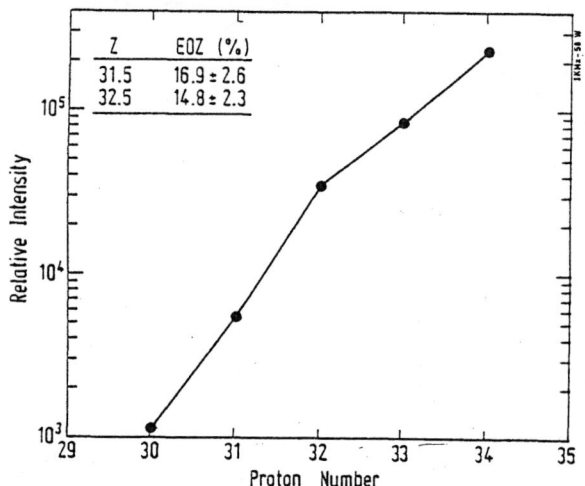

Fig. 2: Relative count rate of fission fragments of a particular atomic number (Z) (summed over the contributing masses and the corresponding distributions of kinetic energy and ionic charge states as discussed in the text)

DISCUSSION

The preferential formation of light fragments with even Z (and of heavy fragments with odd Z) may be due to two possible reasons:

1. The pairing term in the droplet model is δ/A ($\delta \approx 135$ MeV, A = mass number of atom). Consequently, the division of an odd nucleus into a light even fragment and a heavy odd fragment should be energetically more favourable (by about 1 MeV) than the division into a light odd fragment and a heavy even fragment.

2. The second argument is statistical. If we look at the Americium nucleus (Z = 95) like a bag with 47 pairs of protons and one single proton and we imagine that we would take out 33 species (either pairs of protons or the single proton, indiscrimately) to form the heavy fragment and 15 species to form the light fragment we would have a statistical probability

of 15/48 = 0.31 (or 33/48 = 0.69) for the single proton to end up in the light (or heavy) fragment. This would correspond to a value of EOZ = (0.69-0.31)·100 = 38 % for the light fragment with Z ≈ 30. This value is an upper limit because it assumes that no proton pair is broken on the way to scission. If we assume one additional proton pair to break, a process which would give 3 unpaired protons, using a binomial distribution we can calculate a statistical EOZ of 1.04. A rough estimate of the situation of ^{243}Am brings us to the conclusion that in this case one additional proton pair is broken in every second fission reaction. This would be in agreement with systematical estimates (see e.g. Gönnenwein [6]).

ACKNOWLEDGEMENT

We gratefully acknowledge helpful correspondence with Ray Nix and U. Brosa. This work has been funded by the German Federal Minister of Research and Technology (BMFT) under contract number 06-MZ-106.

REFERENCES

1. E. Moll, H. Schrader, G. Siegert, M. Asghar, J. P. Bocquet, G. Bailleul, J. P. Gautheron, J. Greif, G. I. Crawford, C. Chauvin, H. Ewald, H. Wollnik, P. Armbruster, G. Fiebig, H. Lawin, K. Sistemich, Analysis of ^{236}U-Fission Products by the Recoil Mass Separator "Lohengrin" Nucl. Instr. and Methods, 123, 615-617 (1975)

2. E. Moll, H. Schrader, G. Siegert, H. Hammers, M. Asghar, J. P. Bocquet, P. Armbruster, H. Ewald, H. Wollnik: Aufbau und Arbeitsweise des Spaltprodukt-Massenseparators Lohengrin am Hochflußreaktor in Grenoble (Design and Working Principles of the Lohengrin Mass Separator for Fission Products at the High Flux Reactor in Grenoble); Kerntechnik 19, 374-381 (1977)

3. J. P. Bocquet, R. Brissot, H. R. Faust: A Large Ionization Chamber For Fission Fragment Nuclear Charge Identification at the LOHENGRIN Spectrometer; Nucl. Instr. Methods in Phys. Res. A267, 466-472 (1988)

4. H. D. Betz: Charge States and Charge-Changing Cross Sections of Fast Heavy Ions Penetrating Through Gaseous and Solid Media; Rev. Mod. Phys. 44, 465-539 (1972)

5. B. L. Tracy, J. Chaumont, R. Klapisch, J. M. Nitschke, A. M. Poskanzer, E. Roeckl, C. Thibault: Rb and Cs Isotopic Cross-Sections from 40-60 MeV Proton Fission of ^{238}U, ^{232}Th, and ^{235}U; Phys. Rev. C5, 222 (1972)

6. F. Gönnenwein, J. P. Bocquet, R. Brissot: Energy Dissipation in Fission Close to the Barrier; in D. Seeliger, H. Kalka (Eds.), Proc. XVIIth Int. Symp. Nucl. Phys., Gaussig 1987; Report ZfK-646 (1988), p. 129-141

FISSION GAMMA-RAY MULTIPLICITY MEASUREMENTS IN ^{233}U, ^{235}U, ^{237}Np AND ^{239}Pu LOW ENERGY FISSION RESONANCES

E.Dermendjiev[1], A.A.Goverdovski[2], V.I.Furman[1], L.B.Pikelner[1], I.N.Ruskov[1], P.Siegler[3], Yu.S.Zamyatnin[1]

[1] Joint Institute for Nuclear Research, 141980, Dubna, USSR
[2] Institute for Physics and Power Engineering, Obninsk, USSR
[3] Institute for Nuclear Physics, Technical University, Darmstadt, FRG

Abstract: Fission gamma-ray multiplicity measurements in low energy ^{233}U, ^{235}U, ^{237}Np and ^{239}Pu fission resonances were made by using the Dubna pulsed reactor IBR-30 as a resonance neutron source. A large 6-section liquid scintillation detector and fission chambers with a combination of (^{233}U-^{235}U), (^{237}Np-^{235}U) and (^{239}Pu-^{235}U) targets were used for detecting 3 or more fission gamma-quanta in coincidence with a fission fragment pulse. The fission gamma-ray multiplicity data from these measurements are discussed relative to the prompt fission neutron multiplicity data available for low energy ^{233}U, ^{235}U and ^{239}Pu fission resonances. Fission gamma-ray yields from ^{237}Np fission resonances were measured over the neutron energy interval around 40 eV. Correlation between the fission gamma-ray multiplicities and the fission widths Γ_f of measured ^{237}Np resonances was observed, which might be explained by the $(n, \gamma f)$ reaction.

(resonance neutrons, ^{233}U, ^{235}U, ^{237}Np, ^{239}Pu, liquid scintillation detector, fission chamber, fission gamma-ray multiplicity)

Introduction

Recently Boldeman [1] has noted that there is a certain need for some important characteristics of heavy isotope fission like the total kinetic energy of both fragments TKE, the prompt neutron and the gamma-ray multiplicities ν_n and ν_γ to be measured in the neutron resonance energy region. Study of the interaction between S-wave neutrons and fissioning nuclei allows two spin states of the compound nucleus to be distinguished and possible correlations between TKE, ν_n and ν_γ to be established. Also, precise measurements of these values and their variation from resonance to resonance have an important practical meaning [2]. Some results on the variation of ν_γ-dependent relative fission gamma-ray yields R in fission resonances of ^{233}U, ^{235}U, ^{237}Np and ^{239}Pu are reported below. Our preliminary results on variation of R values in fission resonances of ^{235}U and ^{237}Np were reported in [3].

Experiment

All measurements of the R-values in fission resonances of U, Np and Pu isotopes were carried out on the Dubna IBR-30 pulsed reactor based TOF spectrometer. The repetition rate and the neutron pulse width were 100 s^{-1} and 4 µs, respectively. The neutron flight-path was equal to 58.5 m. The experimental set-up consisted of a large 6-section liquid scintillation detector (LSD) with a central hole and a multisection fission chamber (FC) inserted into the LSD as it is shown in Fig.1. The LSD is described in more detail in [4]. Fission gamma-quanta were detected by the LSD with a gamma-ray registration threshold that varied from $\sim$0.2 MeV up to $\sim$0.6 MeV depending on the measurement requirements. One should note that due to an extremely low subbarrier neutron fission cross section of ^{237}Np it was very difficult to extract the fission gamma-quanta from a much larger amount of ^{237}Np capture gamma-rays by detecting single gamma-rays [5] or double gamma-ray coincidences [6]. Therefore, unlikely the authors of [5] and [6], we have recorded only coincidences between 3 or more fission gamma-quanta. Below, this procedure is denoted with "3γ". The fission fragments were detected by FC. As it is shown in Fig.1, a fast-slow coincidence technique provided a selection of fission gamma-rays by using a slow coincidence between the fast "3γ" and slow FC pulses. Both the fast majority coincidence and the slow coincidence had a resolution time $\tau = 5 \cdot 10^{-8}$s and 0.5-2.10^{-6}s, respectively. The FC pulses and the "3γ-fission" coincidence pulses were recorded yielding the fission and coincidence TOF spectra.

Three identical FC containing 0.2 g of ^{237}Np, 2.1 mg of ^{239}Pu and 9.5 mg of ^{233}U, respectively, were used in our experiments. Each FC contained also a small ^{235}U target, which allowed the R values in ^{233}U, ^{237}Np and ^{239}Pu resonances to be measured relative to the ^{235}U resonance R-values. Due to reasonably chosen amounts of fissile isotopes in each FC no crossover interference was observed in any coincidence TOF spectra. Since the Dubna pulsed reactor based TOF spectrometer has a moderate energy resolution the resonance area method was used for the evaluation of the measured TOF spectra. The evaluation consisted in the calculation of the resonance area ratio $R_i = (\sum N_{co})_i / (\sum N_f)_i$ for each resonance, where the resonance areas $(\sum N_{co})_i$ and $(\sum N_f)_i$ were taken from coincidence and fission TOF spectra, respectively. Backgrounds in both coincidence and fission TOF spectra were determined by the black resonance method using Mn, Co, Rh and Cd filters.

Results

The relative yields R of fission gamma-rays in ^{233}U, ^{235}U and ^{239}Pu fission resonances are shown in Fig.2. The R values for each fissioning isotope were normalized to the $<R>$ value, which is the average over the neutron energy interval from 1 eV up to about 50 eV. This procedure is similar to the one used by Howe et al. [7]. For ^{235}U, particularily, it allows the R values to be compared with the ν_n resonance values and some other fission characteristics [7,8].

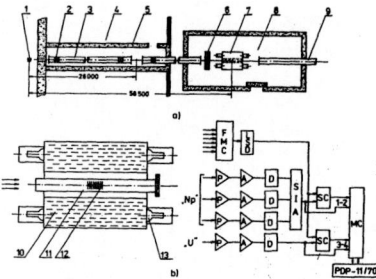

Fig.1. Experimental facility for relative fission gamma-ray yield measurements. 1 - IBR-30 pulsed reactor core; 2,6 - collimators; 3,9 - TOF pipes; 4,5 - shielding; 7,10 - liquid scintillation detector; 8 - experimental hall; 11 - multiplate fission chamber; 12 - fissile targets; FMC - fast majority coincidence; LSD - linear shaper and delay; SIA - summator and invertor-amplifier; SC - slow coincidence; MC - computerized measurement module.

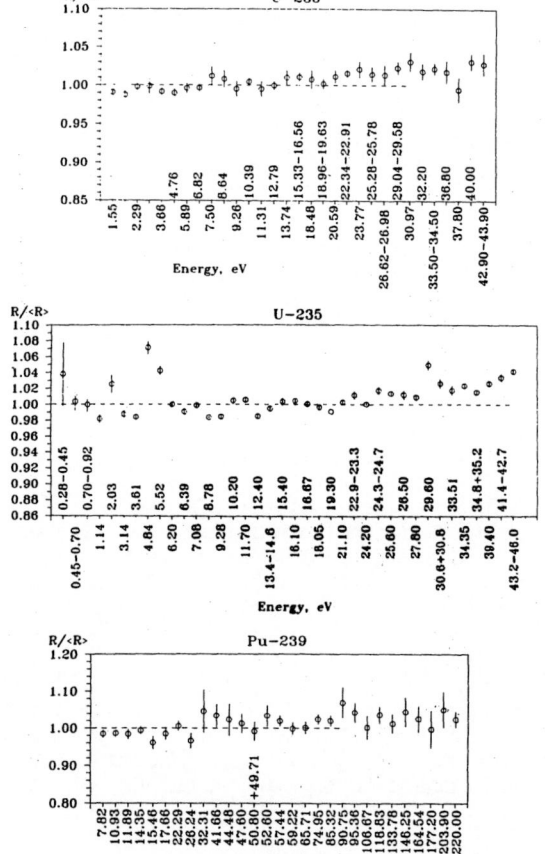

Fig.2. Normalized R/<R> values for ^{233}U, ^{235}U and ^{239}Pu isotopes. The <R> values are averaged over the energy ranges indicated with dashed lines.

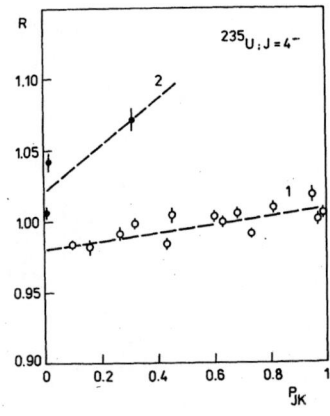

Fig.3. The R values of ^{235}U spin J^{π} = 4^{-} resonance group versus the branching P.

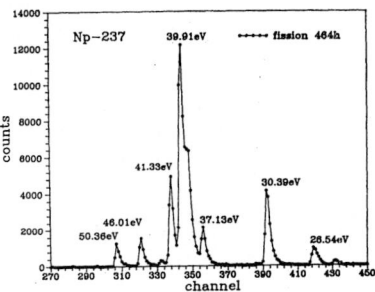

Fig.4. ^{237}Np fission TOF spectrum. Channel width of 2 μs.

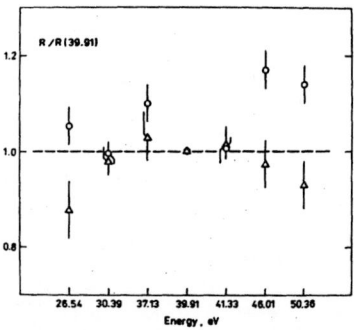

Fig.5. Resonance R values of ^{237}Np. o - low gamma-ray detection threshold of 0.2-0.3 MeV, Δ - high gamma-ray detection threshold of ∼0.6 MeV.

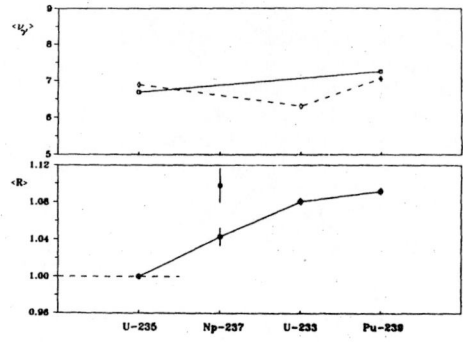

Fig.6. Comparison between the resonance <R> values and the thermal fission ν_{γ} values of Hoffman /12/ (dashed line) and of Verbinsky et al. /13/ (full line).

^{235}U

The ^{235}U resonance R values shown in Fig.2 represent five independent measurement series with a total measurement time of 860 hours and special background measurements with "black" Mn, Co, Rh and Cd filters. The weighted average R values for both $J^\pi = 3^-$ and $J^\pi = 4^-$ spin resonance groups were found to be 1.002±0.004 and 0.998±0.003, respectively. Due to their experimental errors these values are statistically indistinguishable and no spin dependence of R values exists. For $J^\pi = 4^-$ group of resonances a clear correlation between the R values and branching values P(4$^-$,1) /9/ was found. Here the branching P(4$^-$,1) = $\Gamma_f(4^-,1)/\Gamma_f(4^-)$, where the fission width of $\Gamma_f(4^-) = \Gamma_f(4^-,1) + \Gamma_f(4^-,2)$ corresponds to the fission channels with quantum numbers of $J^\pi K = (4^-,1)$ and $(4^-,2)$. The dependence of R = f(P), which is shown in Fig.3, surprisingly splits into two branches. For the main branch the correlation coefficient or r(R,P) = = 0.79±0.11 is statistically significant and confirms correlation between R and P values. Unfortunately, the second branch consists of only 3 weak resonances - 4.84 eV, 5.52 eV and 11.7 eV with small Γ_f and P values and this finding needs to be confirmed. We considered the dependence of the W_1 values /8/ versus the branching P. Here W_1 values are the weights of mass yield gaussian that is associated with Brosa's "Standard I" mode of fission /10/. A positive correlation between the R and W_1 values with a correlation coefficient of r(R,W_1) = 0.64±0.16 for those resonances that belong to the main branch, was found. One should note a significant positive correlation between the R and TKE /8/ values, which is characterized with a correlation coefficient of r(R,TKE) = 0.61±0.18. Comparing the ν_n values, taken from /7/ with the R values of the main branch resonances we found a negative correlation with a coefficient of r(R,ν_n) = = -0.57+0.19.

^{237}Np

The ^{237}Np resonance R value measurements were undertaken in order to compare the fission gamma-ray yields from even-even ^{236}U and odd-odd ^{238}Np compound nuclei. The ^{237}Np R-values were measured for a group of resonances that are lying in the energy range from 26 eV up to 50 eV, i.e. for the first resonance cluster in the ^{237}Np fission cross section. One should note that the resonances on both sides of the central group of unresolved resonances at $\sim$39 eV are very weak and e.g. have values of Γ_f = 22.5±3.2 ueV and 23.2±5.9 ueV /11/ for the resonances at 26.5 eV and 50.36 eV, respectively. Besides that the IBR-30 pulsed reactor based TOF spectrometer has a significant (in the ^{237}Np case) fast neutron background between neutron bursts, due to neutron multiplication and delayed neutrons from the reactor core. That is why a special attention was given to the background determination in both coincidence and fission TOF spectra. In addition to the Mn, Co, Rh and Cd "black" filter measurement data, another 15 energy intervals (in the energy range from 1 eV to 100 eV) with very weak resonances that have $\sigma_0\Gamma_f$ values less than 0.001 b.eV /11/ were chosen and used for more accurate background determination. The ^{237}Np fission TOF spectrum and the resonance R values are shown in Figs.4,5. Open points in Fig.5 represent 3 independent low gamma-ray detection threshold

(0.2-0.3 MeV) measurements, triangles correspond to the measurements with higher threshold ($\sim$0.6 MeV). No difference between the R values of resonances at 30.39 eV, 41.44 eV and resonance group at $\sim$39 eV was found. The R values of resonances at 26.54 eV, 37.13 eV, 46.01 eV and 50.36 eV were found to be larger, however the difference between the R values of both resonance groups disappears at a higher gamma-ray detection threshold. However, this surprising result needs to be confirmed in better background conditions.

Comparison between the ^{233}U, ^{235}U, ^{237}Np and ^{239}Pu average resonance values

To compare the average fission gamma-ray yields $<$R$>$ three similar FC with different combinations of fissile isotope targets as ^{233}U-^{235}U, ^{237}Np-^{235}U, ^{239}Pu-^{235}U and the same gamma-ray threshold of $\sim$0.2 MeV were used in these measurements. The $<$R$>$ values shown in Fig.6 are averaged over the neutron energy interval from 1 eV to 50 eV and normalized to the $<$R$>$ value of ^{235}U. For ^{237}Np the energy interval was from 26 to 50 eV. The open point in Fig.6 presents the R value for those ^{237}Np resonances that have higher fission gamma-ray yields than the other resonances belonging to the first cluster. Since no resonance ν_γ values are available, the $<$R$>$ values are compared with the thermal fission ν_γ values obtained by Hoffman and Hoffman /12/ and by Verbinsky et al. /13/. One can conclude that for ^{233}U both $<$R$>$ and Hoffman's ν_γ data disagree.

References

1. J.W.Boldeman: Proc. IAEA Consult.Meeting, Mito City, Japan, 24-27 May 1988. INDC(NDS)-220, 1989, p.21, IAEA, Vienna
2. Recommendations, Ibid., p.9
3. E.Dermendjiev, Yu.S.Zamyatnin, L.B.Pikelner, W.I.Furman: Int.Conf. "50th Anniversary of Nuclear Fission", Leningrad, USSR, October 16-20, 1989, Abstracts, p.116
4. K.Malecky, L.B.Pikelner, K.G.Rodionov, I.M.Salamatin, E.I.Sharapov: Preprint JINR 13-6609 (1972), Dubna
5. Yu.Ryabov, J.Trochon, D.Shackleton, J.Frehaut: Nucl.Phys. A216,395 (1973)
6. G.Z.Boruhovich, G.A.Petrov, E.N.Teterev, Tz.Panteleev, Yu.V.Ryabov, Tyan San Hak: Yad. Fizika, 14, 689 (1971)
7. R.E.Howe, T.V.Phillips, C.D.Bowman: Phys.Rev. C13, 195 (1976)
8. F.-J.Hambsch, H.-H.Knitter, C.Budtz-Jørgensen, J.P.Theobald: Nucl.Phys. A491, 56 (1989)
9. W.I.Furman, J.Kliman: Proc. 17th Symp. on Nucl.Phys. November 9-23, 1987, Gaussig; Report ZfK-646, 1988, p.142
10. S.Grossmann, U.Brosa, A.Möller: Nucl.Phys. A481,440 (1988)
11. S.Plattard, J.Blons, D.Paya: Nucl.Sci.Eng. 61,477 (1976)
12. D.C.Hoffman, M.H.Hoffman: Ann.Rev.Nucl.Sci., 24, 151 (1974)
13. V.Verbinski, H.Weber, R.Sund: Phys.Rev. C7, 1173 (1973)

PECULIARITY OF THE FISSION OF ^{239}Pu BY RESONANCE NEUTRONS

A.A.Bogdzel,N.A.Gundorin,U.Gohs,A.Duka-Zolyomi,J.Kliman,
V.Polgorski,A.B.Popov,Dao Anh Minh

Joint Institute for Nuclear Research,141980,Dubna,USSR

Abstract: Gamma-ray spectra from resonance neutron induced fission of ^{239}Pu at 0.2 eV to 230 eV were measured with the high resolution gamma-spectrometer at the pulsed reactor $IBR-30$ as a resonance neutron source.The fast 19-section ionization chamber with the ^{239}Pu targets was used for detecting fission gamma-rays in coincidence with fission fragments pulse inside the time window of 290 ns.On the ground of the observed gamma-peaks in the (0.1 - 1.6)MeV interval the fragment identification was made and the independent yields of some fragments were obtained in both the full interval energy and individual resonances.The results are compared with the thermal neutron data. (nuclear reaction ^{239}Pu(n,f);E_n=0.2-230.eV;measured independent fission fragment yields)

Introduction

Interest in peculiarities of resonance neutron fission is connected with the possibility of developments in the multi-fission-mode model on the ground of new experimental data. John et al. [1] and Cheifetz et al. [2] started to apply discrete promt gamma-ray spectroscopy of the fission process to determine independent yields of fragments from spontaneous fission of the ^{252}Cf nucleus. The aim of this work is to measure the yields of fragments from neutron resonance fission of ^{239}Pu by means of the gamma-ray spectroscopy method.

Experiment

The measurements of gamma-ray spectra were made with the fission gamma-ray arrangement /FIGARA/. This arrangement occupies beam No.5 of the IBR-30 booster at 57 m from the core,where the neutron beam parameters are:

Energy resolution	$1.9*10^{-3}E^{3/2}$
Neutron flux	$8.3*10^3E^{-1}$ n/cm^2 s eV
Recycling neutrons energy	0.17 eV

The basic units in the arrangement are the fast multilayer ionization fission chamber with ^{239}Pu targets /IFC/, Ge(Li)-detector and electronic system. The fission chamber was employed as the target and as the fast detector of fission fragments.

Ionization fission chamber details:

Fission material (g)	1.6
Number of targets	19
Density (mg/cm^2)	1.0
Target diameter (cm)	7.5
Enriched (%)	99.9
Efficiency (%)	60
Time resolution (nsec)	2.6
Number of sections	19

The semiconducting coaxial Ge(Li)-detector is used to measure gamma-ray spectra.

Ge(Li)-detector details:

Working volume (cm^3)	100
Relative efficiency (%)	10
Energy resolution at E =1333 keV	2.4
Distance from IFC to Ge(Li)-detector (cm)	28
Absolute efficiency at E=100 keV	7.10^{-4}

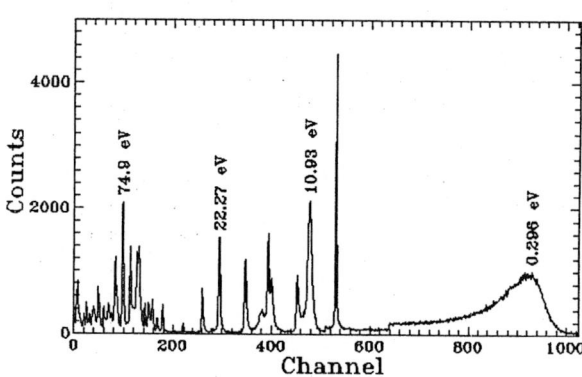

Fig. 1 Time of flight spectrum – ^{239}Pu–IFC

Hardware selects events in coincidence from the fission chamber and detector within the controlled nanosecond time window and measures the following three parameters for each event: the neutron time-of-flight,the time between fission and gamma-quanta detection event,the pulse amplitude in the gamma-detector.By analyzing these data one may obtain a kind of "sections" of respective incident neutron energy or gamma-quanta energy for different time windows of coincidences between fission and gamma-ray detection events.

The time-of-flight spectrum /TOF/ from the fission chamber for neutron energy at 0.2 eV to 230 eV is given in Fig.1.

The fission gamma-ray spectrum from 100 keV to 1.6 MeV /Fig.2/ is obtained after background subtraction from the measuring spectrum and contains more than 100 photo-peaks.

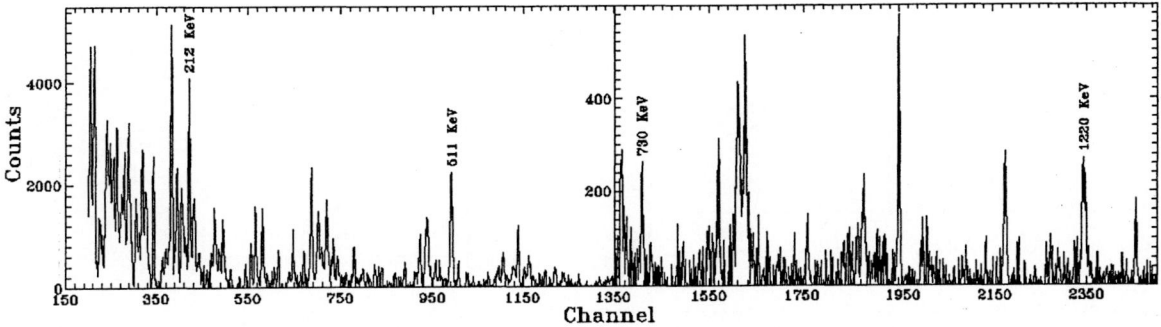

Fig. 2 Gamma spectrum of ^{239}Pu(n,f) reaction

These photo-peaks are used for identification of the fission fragments.The time-of-flight spectrum for one of gamma-lines with E =212 keV /Fig.3/corresponds to the 2^+-0^+ gamma-transition for 40-Zr-100 and characterizes the contribution of this fragment to individual resonances.

Determination of the yields of fragments consists of the calculation of ratio:

$$Y_i^{exp} = (\sum_k N_{co}^k)_i / (\sum_k N_f^k)_i * (K^{eff})^{-1}$$

for each resonance,where the resonance areas $(\sum_k N_f^k)_i$ and $(\sum_k N_{co}^k)_i$ were taken from fission /Fig.1/ and coincidence /Fig.3/ TOF spectra respectively, K^{eff} is the absolute efficiency of FIGARA which was determined using standard calibration set.

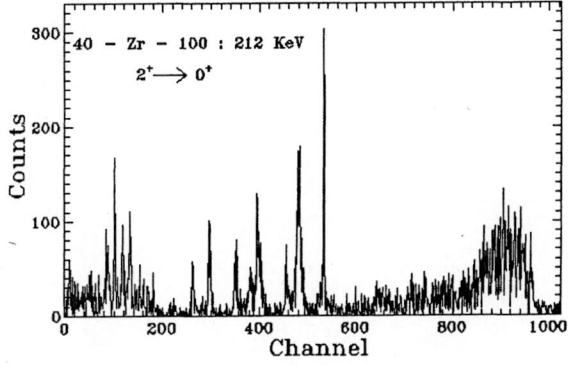

Fig. 3 Time of flight spectrum for 40-Zr-100

Results

1. Independent yields of some fragments from the fission of ^{239}Pu by resonance neutrons with energy from 0.2eV to 230eV were measured /Table 1/.
2. The experimental data obtained for thermal/Y_t/and resonance/Y_{res}/neutrons were compared with the recommended independent yields from thermal neutron fission [3].Resonance neutron fission has revealed the peculiarity: the yield decreases with mass number near standard I and increases - near standard II as demonstrated in Fig.4.Opposite peculiarity was found for resonance neutron-induced fission of ^{235}U by Hambsch et al.[4].
3. Contributions of independent yields to individual resonance from 0.2 eV to 75 eV were measured. Some of the preliminary results are shown in Figs.5-8.

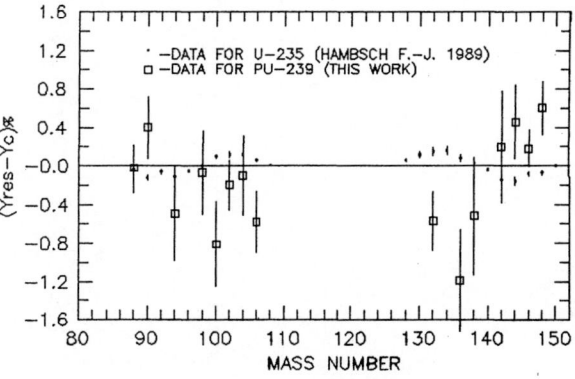

Fig.4. Difference vrs fragment mass.

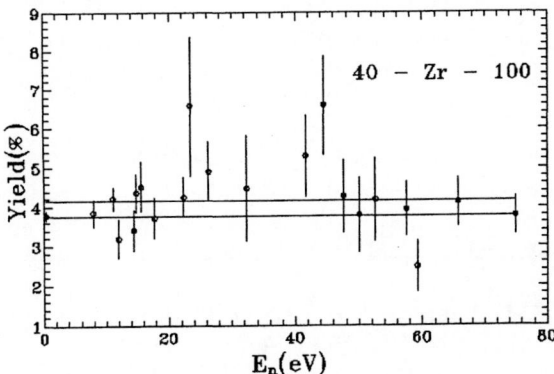

Fig. 5 Yields for 40-Zr-100 at some resonances

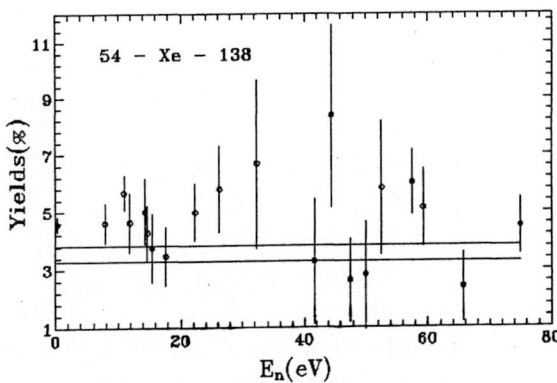

Fig. 6 Yields for 54-Xe-138 at some resonances

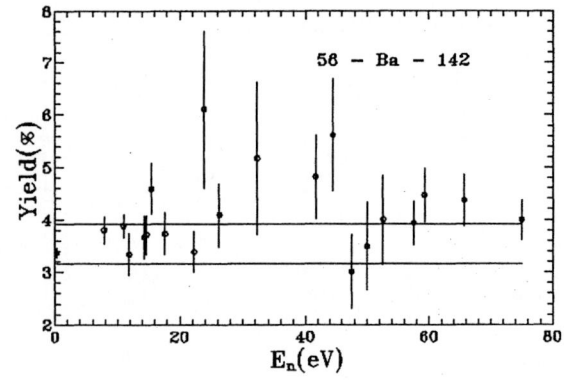

Fig. 7 Yields for 56-Ba-142 at some resonances

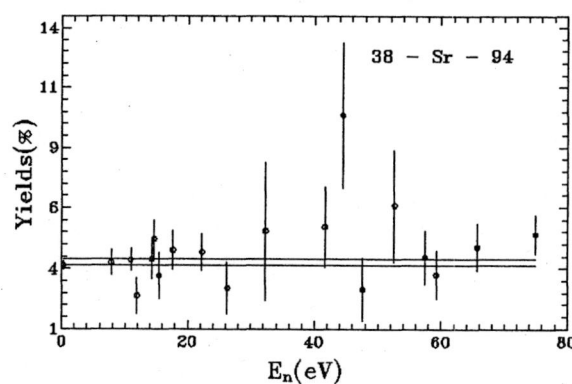

Fig. 8 Yields for 38-Sr-94 at some resonances

TABLE 1 INDEPENDENT YIELDS OF FISSION FRAGMENTS
FROM RESONANCE NEUTRON—INDUCED FISSION
OF 239-Pu

FRAGMENT Z-FR	MASS A	TRANSITION ENERGY E(KeV)	TRANSITION SPIN J^π	RECOMMENDED INDEPENDENT YIELD (THERMAL NEUTRON) $Yc(\%)$	EXPERIMENTAL INDEPENDENT YIELD (THERMAL NEUTRON) $Yt(\%)$	EXPERIMENTAL INDEPENDENT YIELD (RESONANCE NEUTRON) $Yres(\%)$
36-Kr	88	775	2^+-0^+	$0.79+0.03$	$0.80+0.21$	$0.77+0.23$
	90	707	2^+-0^+	1.18 ± 0.05	1.10 ± 0.19	1.58 ± 0.27
38-Sr	94	837	2^+-0^+	$3.14+0.16$	$2.92+0.32$	$2.65+0.33$
40-Zr	98	1223	2^+-0^+	$2.85+0.14$	$2.75+0.28$	$2.78+0.29$
	100	212	2^+-0^+	4.76 ± 0.24	4.50 ± 0.23	3.95 ± 0.20
	102	152	2^+-0^+	1.19 ± 0.12	1.19 ± 0.15	0.99 ± 0.14
42-Mo	104	192	2^+-0^+	$4.12+0.21$	$3.99+0.20$	$4.02+0.20$
	106	171	2^+-0^+	2.06 ± 0.10	1.80 ± 0.20	1.48 ± 0.21
52-Te	132	975	2^+-0^+	$2.36+0.07$	$2.32+0.28$	$1.79+0.23$
	134	1279	2^+-0^+	4.70 ± 0.51	$3.42\pm0.35*$	$2.70\pm0.28*$
54-Xe	136	1313	2^+-0^+	$3.02+0.36$	$2.86+0.30$	$1.83+0.17$
	138	589	2^+-0^+	4.08 ± 0.33	3.89 ± 0.27	3.56 ± 0.28
56-Ba	142	359	2^+-0^+	$3.27+0.26$	$3.09+0.25$	$3.49+0.32$
	144	199	2^+-0^+	2.05 ± 0.23	2.39 ± 0.14	2.51 ± 0.15
58-Ce	146	259	2^+-0^+	$0.95+0.01$	$0.79+0.13$	$1.13+0.18$
	148	159	2^+-0^+	1.09 ± 0.12	1.29 ± 0.12	1.70 ± 0.16

* — Relative Values

REFERENCES

1. Walter John,Frank W. Guy,J.J. Wesolowski, Phys. Rev. C2, 1451 (1970)
2. E. Cheifets,J.B. Wilhelmy,R.C. Jared,S.G. Thompson,Phys. Rev. C4, 1913 (1971)
3. Arthur C. Wahl,Atomic data and nuclear data tables 39, 118 (1988)
4. F.-J. Hambsch,H.-H. Knitter, C. Budtz-Jorgensen and J.P. Theobald, Nucl.Phys. A491, 56 (1989)

PAIRING EFFECTS IN NEUTRON FISSION CROSS SECTION FOR ACTINIDES

A.V. Ignatjuk[1], V.M. Maslov[2]

[1]) Physics and Power Engineering Institute, Obninsk, USSR
[2]) Institute of Nuclear Engineering, Minsk, USSR

Abstract: The step-like structure around 1 MeV in neutron-induced U-235 fission cross section is analyzed within a statistical theory. The key role in analysis plays the intrinsic state density w(U). The discrete character of quasi-particle excitations leads to step-like structure in w(U). Taking into account the collective excitations we were able to correlate the observed irregularity with the threshold behaviour of two quasi-particle state density at the outer U-236 saddle. Pairing gap and barrier parameters are consistent with the threshold.

(U-235, fission cross section, two-quasi-particle state density)

Introduction

Pairing correlation effects have been extensively studied using the angular anisotropy of fission fragments /1/. Step-like structure in K_0^2 parameter is interpreted as a manifestation of few-quasi-particle excitations of fissioning nucleus. Much less attention has been given to the analogous structure in fission cross sections. The well-known structure in the neutron-induced fission cross section of U-235 in the energy range 0.8 - 1.2 MeV may be considered as direct experimental evidence of such excitations. We present the example of taking into account the discrete character of quasi-particle excitations of even-even nuclei alongside with collective excitations. As a result we get a fair description of the experimental energy dependence and an independent estimate of the energy gap at saddle point.

Statistical description of neutron-induced fission cross section

In case of an even-even fissioning nucleus neutron emission and fission are the main competing reactions. In the double-humped fission barrier model /2/ neutron-induced fission of N-odd actinide target nuclei proceeds as an independent tunneling through inner (A) and outer (B) barriers. The transmission coefficients of both barriers are governed by density of levels $\rho(U,J,\pi)$ at the barrier deformations. The main factor, defining energy dependence of fission and inelastic scattering transmission coefficients is the level density. In the adiabatic approximation quasi-particle and collective state contributions to the total level density factorize /3/, that is

$$\rho(U,J,\pi) = Krot(U,J)Kvib(U)\rho_{qp}(U,J,\pi) \quad (1)$$

Here, $\rho_{qp}(U,J,\pi)$ is the quasi-particle level density, Krot(U,J) and Kvib(U) - factors of rotational and vibrational enchancement of level density. In accordance with shell correction calculations /4,5/ we assume that the U-236 inner saddle configuration is axially symmetric, as is ground state configuration. In that case $Krot(U,J) \approx \sqrt{U}$. The outer saddle mass asymmetry leads to doubling of the

level density against the level density of mass-symmetric shapes. In all cases $Kvib \approx \exp(U^{2/3})/3/$.

Now let's proceed with the density of quasi-particle levels $\rho_{qp}(U,J,\pi)$:

$$\rho_{qp}(U,J,\pi) = \frac{w(U)}{4\sqrt{2\pi}\sigma_{\parallel}\sigma_{\perp}^2} \exp\left[-\frac{J(J+1)}{2\sigma_{\perp}^2}\right] \quad (2)$$

Here, w(U) is the intrinsic quasi-particle state density, $\sigma_{\parallel}$ and $\sigma_{\perp}$ are angular momentum distribution parameters. The w(U) can be represented as the sum of n-quasi-particle state densities $w_n(U)$. The $w_n(U)$ are frequently represented by the Boltzman gas model:

$$w_n(U) = \frac{g^n(U-U_n)^{n-1}}{((n/2)!)^2(n-1)!} \quad (3)$$

where g - single particle state density, n- number of quasi-particles, U_n - threshold energy for excitation of the n-quasi-particle configuration, n = 2,4... for even-even nuclei and n = 1,3... for odd nuclei. The σ^2 can be represented as:

$$\sigma_{\parallel}^2 = \frac{\sum_n n\langle m^2\rangle w}{\sum_n w} \quad (4)$$

where $\langle m^2\rangle = 0.24 A^{2/3}$ is the average value of the squared projection of the angular momentum of single particle states.

As shown in /6/ the relation (3), with $U = n\Delta$, where Δ is the correlation function, overestimates threshold values and distorts an energy dependence of $w_n(U)$ as compared with a superfluid model with fixed number of quasi-particles. However, if the U_n estimate takes into account the energy dependence of the gap $\Delta_n(U)$ and a modified Pauli correction as in /7/, the results of the uniform pairing model /8/ could be reproduced with a factor-of-two difference for low n. The shell effects in $w_n(U)$ could be represented with the energy dependence of the a-parameter,

$$a = \begin{cases} a(1+\delta W f(U-E_K)/(U-E_K)), U>U_K \\ a(U)=a_K, U<U_K=0.47a\Delta^2 -m\Delta \end{cases} \quad (5)$$

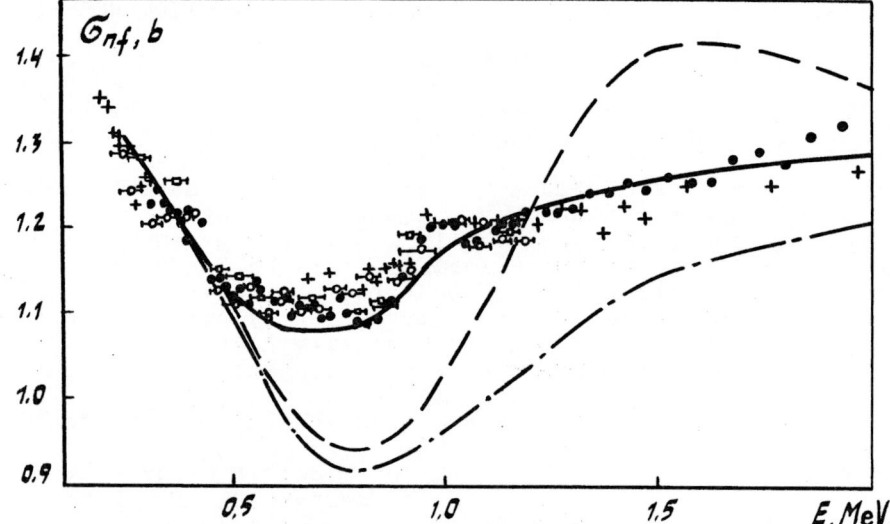

Fig. 1. Neutron fission cross section of U-235. Curves represent
calculations with different intrinsic state densities, shown in
fig. 2; data points: o ,□ -/15/, ● -/16/, + -/17/.

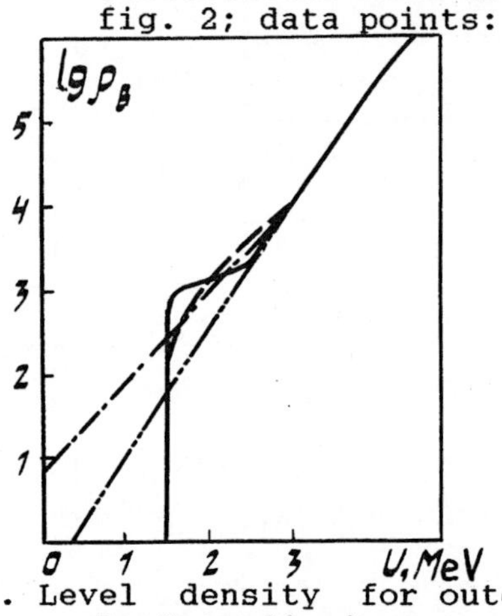

Fig. 2. Level density for outer saddle
used to calculate fission cross section.
Full curve corresponds to fission data
fit (see fig.1) broken curve- Boltzman
gas model /7/; dot-dashed curve -cons-
tant temperature model with parameters fixed in /14/; double- dot-dashed-
curve- pairing model /3/.

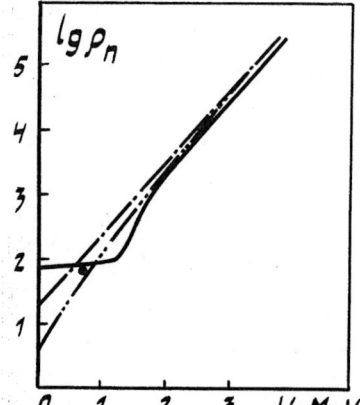

Fig. 4. Level density for residual U-235
nucleus. See curves on fig.2; data point
is defined as N(U)/U.

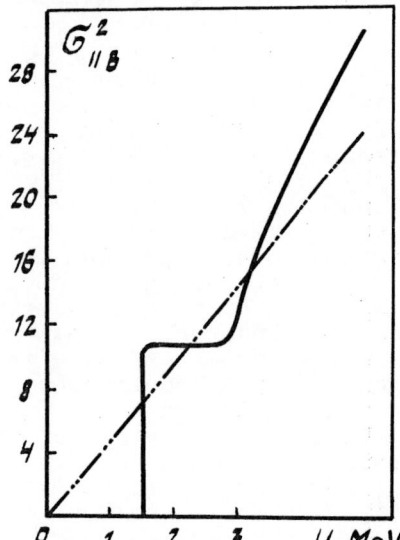

Fig. 3. Spin parameter $\sigma_{\|}^2$ energy depen-
dence for outer saddle. Curves represent
calculations with models, identified in
(see fig.2)

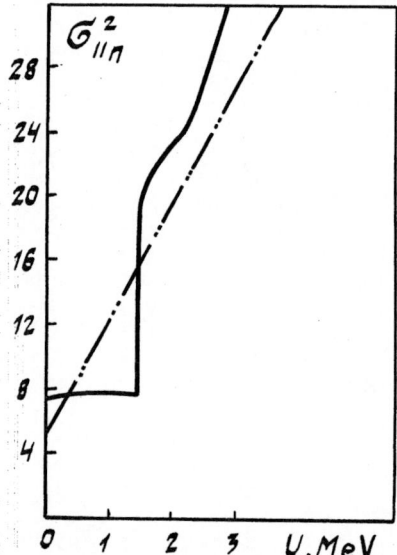

Fig. 5. Spin parameter $\sigma_{\|}^2$ energy de-
pendence for residual nucleus U-235.
See curves on fig. 2.

Here, $E_K = 0.15 a_K \Delta^2$, $f(x) = 1-\exp(-\int x)$, $m = 0,1,2$ in case of e-e, odd and o-o nuclei, respectively. An asymptotic value of the parameter a for fissioning nuclei, inner and outer saddle shell corrections δW_{fi} and gap parameter are the model parameters, as described in /9/. Supposing, that there is no error in threshold values, we would try to infer $w_n(U)$ by fitting fission cross section data. Before starting the discussion on that point it is necessary to state the following.

When the neutron energy E is less than $(U_2 + E_{fi} + B)$, where B – neutron binding energy, the fission cross section is governed by collective levels, lying within the gap. In case of the compound nucleus U-236 the cross section value is governed by the outer barrier, because $E_{fA} < E_{fB}$. Since the outer saddle is mass asymmetric, the bandheads with negative parity $K^\pi = 0^-$, 1^- ... are lying lower than in case of ground state deformations. As an initial guess for bandhead positions we use values from /12, 13/, later on they are slightly varied. Rotational bands constants are quoted in /9/.

Fission cross section of U-235

The data on the U-235 fission cross section are shown on fig.1. Suppose, that fission barrier parameters E_{fi} , $\hbar w_{fi}$ and gap parameter Δ are fixed by fitting fission data for $E \geq 2$ MeV /14/. Then $E = U_2 - E_{fB} - B \approx 0.8$ MeV corresponds to two-quasi-particle state excitation threshold U_2. That is the very point around which the step in fission cross section occurs. The 4-quasi-particle state excitation threshold occurs at $E_4 \approx 2.15$ MeV. So, the behaviour of the fission cross section in the neutron energy range $E = 0.8 - 2.15$ MeV is governed by $w_2(U)$ density. If you try to calculate $w_2(U)$ with the relation (3), the discrepancy of experimental data and calculated curve would correspond to the $w_2(U)$ approximation deficiency with a Boltzman gas model. That is, even with equal thresholds U the energy dependences of $w_2(U)$ in the Boltzman gas model and pairing model are different. In the former case $w_2 \approx (U - U_2)$, while in the latter $w_2(U)$ is almost constant with energy above threshold. One should keep in mind the uncertainty, involved in assessing the $\langle m^2 \rangle$ for saddle point deformations. To fit σ_f in the vicinity of the 2-quasi-particle state threshold one needs a functional dependence of w_2 differing from $w_2(U) \approx (U - U_2)$. Fig. 2 shows level densities for outer saddle. The curves shown correspond to densities, 1) fitting cross section data, 2) Boltzman gas model, 3) calculated with constant temperature model and 4) pairing model /3/. The $\sigma_{||}^2$ parameter (see (4)) is compared with results of superfluid model /3/ in fig. 3. The $\sigma_{||}^2$ values for densities labelled above 1) and 3) virtually coincide. So, by K_0^2 analysis alone one can't infer the deta-

iled shape of the level density. For the present work we calculate the residual nucleus level density with relations (1)-(5), introducing an o-e excitation energy shift: $\tilde{U} = U + \Delta$. Fig. 4, 5 show the comparison of $\rho(U)$, and $\sigma_{||}^2(U)$ for the models mentioned in the text above. All curves are normalized to the U-235 neutron resonance spacing.

The fig. 2 indicate, that for actinides the partial contributions of n- quasi-particle intrinsic state densities into $\rho(U)$ produce jump only for states with lowest number $n = 2$.

Summary

The U-235 fission data are unique in the sense that they allow to fix simultaneously the fission barrier height and the two-quasi-particle state excitation threshold. The step-like structure is most probably due to the jump-like excitation of the two-quasi-particle configurations for outer saddle. The excitation threshold value is consistent with fissioning nucleus pairing gap and barrier parameters, fitting cross section between .1 and 5 MeV. Similar irregularities do exist in other actinide fission cross sections : 233-U(n, f)(0.5 MeV), 239-Pu(n,f) (1.25 MeV), 241-Pu (n,f) (1.0 MeV). In parentheses are shown excitation threshold E values infered in similar analyses.

References

1. D.L. Shpak, Y.B. Ostapenko, G.N. Smirenkin: Sov.J.Nucl.Phys.13,950 (1971)
2. V.M. Strutinsky, S. Bjørnholm, in Nuclear Structure, p.431,Vienna(1968)
3. A.V. Ignatjuk, K.K. Istekov, G.N. Smirenkin: Sov.J.Nucl.Phys. 29,875 (1979)
4. P. Møller: Nucl.Phys. A192, 529 (1972)
5. W.M Howard, P. Møller: Atom. Data and Nucl. Data Tables 15, 219 (1980)
6. A.V. Ignatjuk, Y.V. Sokolov: Sov.J. Nucl.Phys. 11, 1229 (1974)
7. C.Y. Fu: Nucl.Sci.Eng. 86, 344 (1984)
8. L.G. Moretto:Nucl.Phys. A243, 77(1975)
9. A.V. Ignatjuk, A.B. Klepatskij, V.M. Maslov, E.Sh. Sukhovitskij: Sov.J. Nucl.Phys. 42, 569 (1985)
10. C.Y. Fu: Rad. Effects 95, 1115 (1986)
11. F.C. Williams: Nucl.Phys. A166, 231 (1971)
12. J.E. Lynn: AERE-R7468 (1974)
13. B.B. Back, O. Hansen H.C. Britt, J.D. Garrett: Phys. Rev. 5, 1924 (1974)
14. A.V. Ignatjuk, V.M. Maslov: Sov.J. Nucl.Phys. 51, 1227 (1990)
15. O.A. Wasson, M.M. Maier, K.C. Duval: Nucl.Sci.Eng. 81, 196 (1982)
16. A.D. Carlson, in Proc. IAEA Consult. Meeting on the 235-U Fast Neutron Fission Cross Section, 1983, Smolenice, p.61, IAEA, Vienna (1983)
17. W.P. Poenitz: Nucl.Sci.Eng. 64, 89, (1977)
18. G.V. Antsipov, V.A. Konshin, V.M. Maslov: INDC(CCP)-182 (1982)

PACKED CLUSTER MODEL FOR PHOTOFISSION ASPECTS OF THE GIANT RESONANCE

O.M. El-Bakly and L.S.El-Mekkawi

Atomic Energy Authority, Cairo, Egypt

Abstract: This work shows that when a gamma-ray resolves the **order** of a packed cluster,/14/ the higher energy peak of giant resonance results. Nuclear scission at forming a packed cluster in each fissioning side, confirms that it is essential for nuclear formation and locates the giant resonance peak at about 14 MeV. For resolved order of subclusters, the lower energy peak of giant resonance results. At final state, the order of packed cluster is lowered by a shell release whose collective quantum order gives the fission states at packed cluster resolution. The substates of the equally rotating shell yield two resolved fission states for the two subcluster resolutions.

Scission **level** , at subcluster resolution yields Th-232 peak at 6.5 MeV. By packed cluster model, the elongation of the fissioning nuclear shape occurs at the energy level of nuclear temperature and the Coulomb energy from scission is close to kinetic energy of fragments.

Fission from packed cluster, gives fragment rotational angular momentum normal to fragment motion confirming the average fragment angular momentum from neutron fission and expecting lower photofission fragment angular momentum at photon energy less than 11 MeV.

Introduction

The general shape of photofission cross sections /1-4/ shows a giant resonance at about 14 MeV.and at least a lower energy peak. Explanations were approached by the hydrodynamic model for strong long range correlation and the shell model approach assuming no correlation/9-12/. Brink/13/, Goldhaber and Teller/5/tried to explain the giant resonance by the vibration of the bulk of protons against all neutrons. In this paper, the packed cluster model/14/ displays its validity for details of photofission giant resonance and its theoretical cross sections, in Actinide region.

The Packed Cluster (P.C.) Model

The packed cluster is assumed to have a quantum order for each individual subcluster and a collective quantum order for the p.c. as a whole. The order of each cluster is its top quantum level. The simplest p.c. is called the core. It has quantum order 1 for a subcluster and 2 for the other and collective order 3 for p.c. Two nucleons can occupy the core yielding deutron photofission/14/. Subclusters with sublevels require subcluster order 2 and 3 and order 5 for p.c. As each level in a p.c. has an individual subcluster quantum order and a collective quantum order, the number of states per level is independent of its order. The 18 final substates of separated shell of order 3 are fully resolved when the gamma-ray resolves subcluster order 3 and 2/3 of these final states (12) are resolved, when the ray resolves subcluster order 2. At ray resolution of subclusters, the ray angular momentum perturbs the p.c. mass at the quantum order of the resolved subcluster charged level. The presence of two fissioning sides, at stage of removal of p.c. during fission suggests that a super packed cluster (S.P.C.) having individual

order 3,4 and a collective order 7 modiffies both subcluster energy. When the angular momentum of either subcluster reaches the quantum order of final function relative to a centre of possible p.c., two P.C. are formed ending the S.P.C. through scission.

Initial solution relative to ray

The Hamiltonian of two subclusters of a p.c. yields Schrödinger equation for relative motion of subclusters. For incident ray propagating along φ =0, magnetic vector potential $\bar{A}$ along θ= 0,the polarisation direction of initial solution is inclined at (α,β) relative to ray system. As A_φ =0, the initial solution responding to photon is a plane solution part of initial function. Its value relative to incident gamma ray is ψ_i.

$$\psi_i = \left[4\pi/(2\ell+1)\right]^{1/2} N_\ell \, r^\ell e^{-ar^2} \left[\sum_{m=-\ell}^{m=\ell} Y_\ell^m(\alpha,\beta)\right] Y_\ell^m(\theta,\varnothing)\right] \quad ; \quad a= \mu\, w_0/2\hbar \quad (1)$$

where μ is the reduced mass of subclusters.

Absorbtion of gamma-ray by P.C.

The perturbation Hamiltonian H' from the ray equivalent magnetic vector potential $\bar{A}$ interacts/15/ with resolved dual P.C. level of initial state, where r_1, r_2 are the separations of charge centres of subclusters from c.m. of p.c.

$$H'=\bar{A}_0 \left((z_1/m_1)e^{-ikr_1} - (z_2/m_2)e^{+ikr_2}\right)(e\hbar/c)\nabla$$
$$= S\bar{A}_0 (e\hbar/c).\nabla \quad (2)$$

This interaction yields a final state k through the matrix element H_{ik} and the time coefficient of change of states C(t).

$$H_{ik}=\psi_k^* H'\psi_i \, dT=\hbar\, e^{-i(w_{ik}-w)t} \, dC(t)/dt \quad (3)$$

The final state is considered to be the normalised final function for free motion ψ_k.

$$\Psi_k = k^{3/2} (i)^{\ell_f} y_f^m(\hat{k}) \, y_f^m(\hat{r}) \, J_{\ell_f + \frac{1}{2}}(kr)/(kr)^{\frac{1}{2}}$$

where $\ell_f = \ell - 1$ (4)

At end of p.c. perturbation the reduced mass of equal subclusters μ' reaches a kinetic energy due to photon energy excess from threshold resolution energy.

$$\hbar^2 k^2 / 2\mu' = \hbar w - E_{th,\ell}$$ (5)

Hence the parameter $u_\ell = k^2/2a$ can be defined for each fission mode. The final energy of subclusters yields the density of final states $\rho(E)$ which gives the probability per unit time W for the time varying magnetic vector potential.

$$W = \int \left[(4/t) H_{ik}^* H_{ik} \, \rho(E) (\sin^2(w_{ki} - w) t/2)/ \hbar^2 (w_{ki} - w)^2 \right] \, dE_{ik}$$ (6)

For e.m. wave, the number of events per area per time yields the cross section σ for the modes of p.c. fission.

$$\sigma = |S|^2 (\mu'/E) \{ 1.08726 (n_2^3 \, u_2^{3.5} e^{-u_2} + n_3^3 \, u_3^{3.5} e^{-u_3}) + 271.77 (3^3) u_5^{\frac{1}{2}} [(1 - u_5/3)^2/1.5 + u_5^2 ((4.6285(1 - 2u_5/21)^2 + 0.42857) + 0.174191 \, u_5^4 (1 - 3u_5/26)^2 + .001595 \, u_5^6)] e^{-u_5} \}$$ (7)

where u_2 is the value of u at $E_{th,2}$ for

order 2 resolution of subclusters, u_3 is at $E_{th,3}$ for order 3 subcluster and u_5 at $E_{th,5}$ for p.c. resolution. The value $(r_1 + r_2) = r_c$ in S of eq(3) is the separation of centres of charges of the two subclusters which is found greater than their c.g. separation, yielding antiphase resonance of subclusters at 327 MeV of the Δ-resonance region.

Energy transitions for scission

At p.c. removal, due to two nuclear scission energy levels of S.P.C., the fissioning nucleus modifies the subclusters' energy to $(E_q - E_s)$. The equal subclusters move with equal and opposite velocity v_o till they yield the angular momentum order of final function relative to possible p.c. centre at $0.6 r_c$, where two p.c are formed ending S.P.C. by scission.

$$m_s (v_o - v_{ri}) \times 0.6 \, r_c = J_i \hbar$$ (8)

where m_s = mass of final state subcluster = 36 nucleons, v_{ri} is the velocity of fissioning part at scission relative to its elongation velocity due to subclusters' vibration by photon magnetic vector potential. For forming a P.C. in 2 sides, there is rigid interaction between quantum frames giving rotational angular momentum to fission fragments. The values of fission fragment angular momentum J_L and J_H are perpendicular to the plane of fission fragment motion and subcluster motion, as assumed by Strutinski. Equating momentum before and after scission.

$$m_s v_o \cos \xi + (m_{fi} - m_s) v_{fi} = m_{fi} V_i$$ (9)

where m_{fi} is light or heavy fragment mass

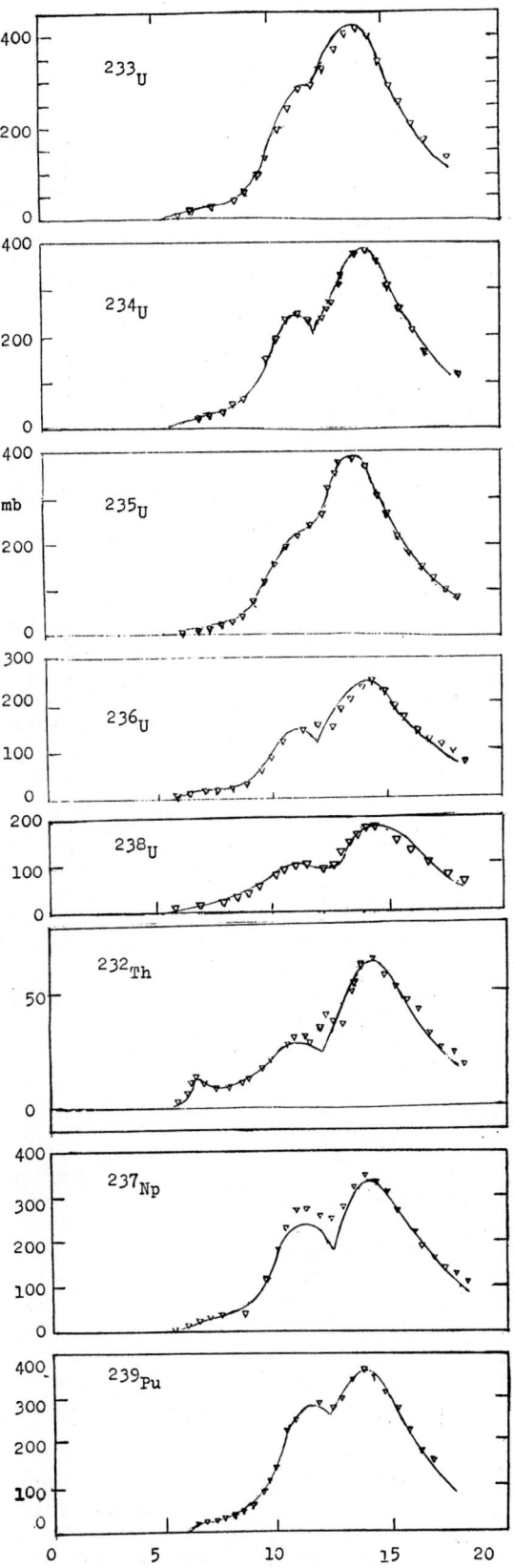

Figure 1: Photofission cross sections in mb for uranium-233, 234, 235, 236, 238, Th-232, Np-237 and Pu-239, versus incident photon energy in MeV. —is theoretical by P.C. model, ∇ is Livermore data.

The energy E_s is the excess energy given by S.P.C. to the fissioning parts to modify their separation energy from its value due to vibration of subclusters.

$$E_s = \frac{1}{2}(m_{fH} - m_s)v_{rH}^2 + \frac{1}{2}(m_{fL} - m_s)v_{rL}^2 \qquad (10)$$

The total kinetic energy of fragments at scission is:

$$E_k - E_c = \frac{1}{2}(m_{fL} \mathbf{V}_L^2 + m_{fH} V_H^2) \qquad (11)$$

where E_k is the total kinetic energy of fragments, E_c is Coulomb repulsion after scission.

$$E_q - E_s = 2(\frac{1}{2} m_s v_0^2) \qquad (12)$$

where E_q = energy released at scission less the energy E_x which goes to random energy of nucleons.

$$E_q = E_f - E_c - E_x \qquad (13)$$

where E_f is the total energy released in fission. Solving eq(9),...,(12), where $R = (E_k - E_c)/E_q$; $E_s' = E_s/E_q$; $\mu' = (m_{fH} - m_s)(m_{fL} - m_s)/(m_{fH} + m_{fL} - 2m_s)$. Hence,

$$0 = (R - m_s/2\mu)^2 + [2(m_s/2\mu - R)(\mu'/\mu - m_s/2\mu) - 2\mu' m_s - 2\mu' m_s/\mu^2]E_s' + [(\mu'/\mu - m_s/2\mu)^2 + 2\mu' m_s/\mu^2]E_s'^2 \qquad (14)$$

For eq(10) and eq(12), eq(8) gives:

$$J_i = 1.0577\left[(E_q - E_s)^{\frac{1}{2}} - 6(2E_s u')^{\frac{1}{2}}/(m_i - 36)\right] \qquad (15)$$

where m_i is fragment mass, E_q and E_s are In MeV; r_c makes antiphase resonance at 327 MeV. The original nuclear spin J_0 gives J_i equal parity. Due to high values of J_0 and J_i for neutron fission /16/, the emitted fission neutrons $\mathring{V}$ carry opposite spin. Hence for neutron induced fission, the average fragment angular momentum J_t is,

$$J_t = J_0/2 + J_i + 0.25 \mathring{V} \qquad (16)$$

For p.c. resolution, J_i gets the final function order 4, at nuclear scission, from either subcluster. Hence $E_s = 0$ and $E_q = 14.3$, so that the fissioning nucleus modifies subclusters' level to scission level 14.3 MeV ,but it does not alter directly the energy of fissioning nuclear parts from its value at fission of p.c. At subcluster resolution: J_i from either subcluster gets the final function order 1, at scission. Hence $E_s = 0$, $E_q = 1$ MeV.

Effect of nuclear scission level

At p.c. removal, the photon energy part for resolving p.c. quantum order is given to subclusters through S.P.C. Hence if the photon energy is equal to nuclear scission level(for p.c. resolution), the time coefficient for change of states C(t) eq(4) is maximum. Hence W of eq(5) is maximum and σ_5 (part of σ eq(6) of $u = u_5$) is maximum at 14.3 MeV,but σ_2 and σ_3 of σ may shift the giant resonance peak of σ eq(6) to slightly lower values. For subcluster resolution, when the photon energy less the subcluster resolution energy is equal to final subcluster energy for scission (1 MeV),the cross section is maximum as in case of Th-232 resonance

at 6.5 MeV where the resolved fission states increase to maximum value of 18 states.

Results and Discussion

From eq(8) and (15),scission levels are proportional to the square of isotope Δ-resonance frequency. For scission at p.c. resolution,the Coulomb energy from scission E_{c4} can be obtained from eq(14) and E_{x4} from eq(13). For fission at subcluster resolution, the Coulomb repulsion energy E_{c1} and the excitation energy E_{x1} can also be obtained for E_f and E_k (kinetic energy of fragment)/17/. The value J_t from (16) is close to experimental data /18/ as in table 1.

Table 1: Coulomb repulsion energy E_{c4}, E_{c1}, additional excitation energy E_{x1}, E_{x4} (all in MeV) and J_{t4} from p.c.model compared to experimental data.

Isotope	E_k	E_{c1}	E_{c4}	E_{x1}	E_{x4}	J_{t4}	J_{exp}
^{239}Pu	159.8	159.5	155.5	14.5	5.1	7.25	6.9
^{235}U	154.6	154.3	150.3	13	3.7	6.6	6.4
^{233}U	149.6	149.3	145.2	17.3	8	6.15	6.3

Table 2: Subcluster charges, their energy thresholds and p.c. resolution threshold with actual number of fission states n_2, n_3, n_5 for fission from P.C. resolution.

Isotope	Z_1	Z_2	E_{th2}	E_{th3}	E_{th5}	n_2	n_3	n_5
^{232}Th	9	12	5.4	7.5	12	12	15	3
^{233}U	17	22	5	8.6	11.85	11	22	3
^{234}U	19	25	5.4	8.4	12.3	11	20	3
^{235}U	17	22	5.3	8.6	12	12	20	3
^{236}U	16	21	5.3	8.2	12.2	11	18	3
^{237}Np	17	22	5.2	7.9	12.2	12	20	3
^{238}U	15	20	5.3	8	12.2	12	18	3
^{239}Pu	17	22	5.7	8.2	12.1	12	18	3

The experimental cross sections follow the final state 3 of collective order of separated shell and may deviate slightly from theoretical values for substates at subcluster resolutions. The peak widths agree with theoretical value of p.c. potential.

The thorium peak at 6.5 MeV is due to nuclear scission level for subcluster resolution and increases the fission states from 12 to 18.

The unsaturated p.c. with one subcluster from original nucleus may cause aspects for fission product instability till it reaches stable unsaturated state.

Summary and conclusion

The packed cluster model introduces nuclear scission level which locates the giant resonance peak at about 14 MeV, corresponding to nuclear scission level from p.c. collective resolution. Nuclear scission

level due to subcluster resolution is
close to its fission threshold causing
a small peak as in Th-232 peak at 6.4 MeV.
The final order 4 causes a minimum fission mass from 18 substates = 72 mass units.
There are 90 fission channel (possible)
from p.c. order 5 through 18 substates.
The cross sections presented confirm the
theoretical p.c. final fission states
and scission levels. They also confirm
p.c. peak widths as obtained from p.c.
theoretical potential. The release of p.c.
collective resolution energy at p.c. removal causes a large time coefficient of
change of states yielding the giant resonance. Scission at formation of 2 p.c.
shows that a p.c. is an essential unique
quantum structure in the nucleus. The average of Coulomb energy levels released
in fission is closest to kinetic energy
of fragments at photon energy below 11 MeV
and is smaller at high photon energy.

The rotational angular momentum from
p.c. fission confirms the average fragment
angular momentum normal to fission direction for neutron fission and high energy
photofission, but it expects 3 units lower
values of fragment angular momentum for
low photon energy (below 11 MeV).

References

1. A.Veyssiere,H.Bell, R.Bergere, P.Carlos
   and A. Lepretre,and K. Kernbath, Nucl.
   Phys. A 199 (1973) 45.
2. J.T.Caldwell, E.L.Bergman,R.A.Alvarez
   P.Meyer,Phys.Rev.C 21,1215(1958)
3. J.T. Caldwell, E.J. Dowdy, R.A.Alvarez,
   B.L.Bergman and P.Meyer,Nucl.Sc.Eng.
   73,153(1980)
4. B.L.Bergman, J.T.Caldwell ,E.J.Dowdy,
   S.S.Dietrich, P.Meyer and R.A. Alvarez
   Phys. Rev.c34,2201 (1986)
5. M Goldhaber and E.Teller,Phys.Rev 74,
   1046.(1948)
6. J.H.Jensen and P.Jensen,Z.Naturf. 5a
   343 (1950)
7. J.H.Jensen and H.Steinwedel, Z. Naturf
   5 a, 413 (1950)
8. J.M. Araujo, Nuovo Cimento 12,780(1954)
9.   J.I.Burhardt,Phys.Rev. 12,780
   (1954)
10. A.Reifman, Z.Naturf 8a , 502 (1953)
11. U.Khokhlow, Dokl,Akad.Nauk USSR 97
   (1954),239.
12. J.S. Levinger and D.C. Kent,Phys.Rev
   95 (1945),418
13. D.Brink,Nuc.Phys. 4,215(1957)
14. L.S.El-Mekkawi and O.M.El bakly,Int.
   Conf.on Nucl. Data for Science and
   Technology,p.1225 (1988 MITO)
15. A.Bohr,Proc. Int. Conf. on PUAE,
   Geneva, 1955 (United Nations,N.Y.1956)
   Vol2,P151
16. O.M.El Bakly,L.S. El-Mekkawi,Theoret-
   ical Low neutron energy induced fission cross sections (under pub.)
17. Hyde E.K., UCRL-9036 (1960)
18. Pleasonton, R.Ferguoson and H.Schmitt,
   Phys.Rev.16 ,1023 (1972)

MEASUREMENT OF THE NEUTRON TOTAL CROSS SECTIONS FOR Bi AND Pb: ESTIMATE OF THE ELECTRIC POLARIZABILITY OF THE NEUTRON

Yu.A. Alexandrov and V.G. Nikolenko

Joint Institute for Nuclear Research
101 000, Head Post Office, P.O. Box 79, Moscow, USSR

I.S. Guseva, A.B. Laptev,
G.A. Petrov and O.A. Shcherbakov

Leningrad Nuclear Physics Institute
Academy of Sciences of the USSR
188 350 Gatchina, Leningrad district, USSR

Abstract: The neutron total cross sections are measured with an accuracy $\Delta\sigma/\sigma \sim 10^{-3}$ for Bi and Si in energy range from 1 eV to 100 eV and for ^{208}Pb and C in the range from 1 eV to 10 keV using the time-of-flight facility GNEIS in Gatchina. An estimated value of the neutron electric polarizability is $\alpha_n = (2.6 \pm 1.4) \cdot 10^{-3}$ fm^3.

(time-of-flight measurement, total cross section, Bi, Si, ^{208}Pb, C, neutron electric polarizability)

Introduction

The electric polarizability α_n is one of the characteristics of the neutron as an elementary particle and determines the induced electric dipole moment in an external electric field : $\vec{D} = \alpha_n \cdot \vec{E}$. It is known [1] that information about polarizability of the neutron can be obtained from the neutron total cross section measurements for heavy nuclei. In Coulomb field, an addendum caused by polarizability is added to the neutron-nucleus interaction potential:

$$V = -\frac{1}{2}\vec{D}\vec{E} = -\frac{1}{2}\alpha_n Z^2 e^2/r^4 ,$$

which in total scattering cross section parametrization [2]

$$\sigma(k) = \sigma_0 + \sigma_1 k + \sigma_2 k^2 + O(k^4)$$

leads to the addendum linear depended on k, where k is the wave number. The parameter σ_1 depends only on the neutron polarizability:

$$\sigma_1 = \frac{8\pi^2}{3}\alpha_n b_{coh} m_n \frac{(Ze)^2}{\hbar^2} .$$

Here b_{coh} is coherent scattering length.

To date, several attempts have been performed to measure of α_n:

in Dubna [3], $\alpha_n < 5 \cdot 10^{-3}$ fm^3;

in Garching [4], $\alpha_n = (0.8 \pm 1.0) \cdot 10^{-3}$ fm^3;

in Harwell [2], $\alpha_n = (1.2 \pm 1.0) \cdot 10^{-3}$ fm^3.

The results of the total cross section measurements for Bi in the energy range from 1 eV to 100 eV and that of ^{208}Pb from 1 eV to 10 keV performed at the GNEIS facility for the purpose of estimating a value of α_n are given in this article. The first results of this measurement were published earlier in a review [5]. Simultaneously the total cross section measurements for silicon and carbon were carried out as a zero-test of the experiment. For silicon and carbon the polarizability influence should be negligibly small in comparison with that of bismuth and ^{208}Pb because the polarizability contribution to the cross section is proportional to the square of nuclear charge Z^2.

Experimental Setup

The total neutron cross section measurements were carried out at the neutron time-of-flight spectrometer GNEIS [6]. The 40 m flight path was used. The schematic layout of the experiment is shown in Fig. 1. The ionization ^{3}He-chamber was used as neutron detector. With a view to increasing the running speed and to decreasing the dead time of the whole detection system the ^{3}He-chamber was divided in 3 sections. To decrease experimental hall background the detector is located in a composite shielding (see Fig. 1). Resonance filters of Co(132 eV, 4.3 keV, 5.0 keV), W(18.8 and 4.16 eV), In(1.46 eV) and Al(34 keV) were placed in the beam for permanent background monitoring. The counts in a flat bottom of depths in time-of-flight spectra caused by strong resonances were taken as background. In other channels the background was obtained by means of interpolation. The neutron background weakly altered in

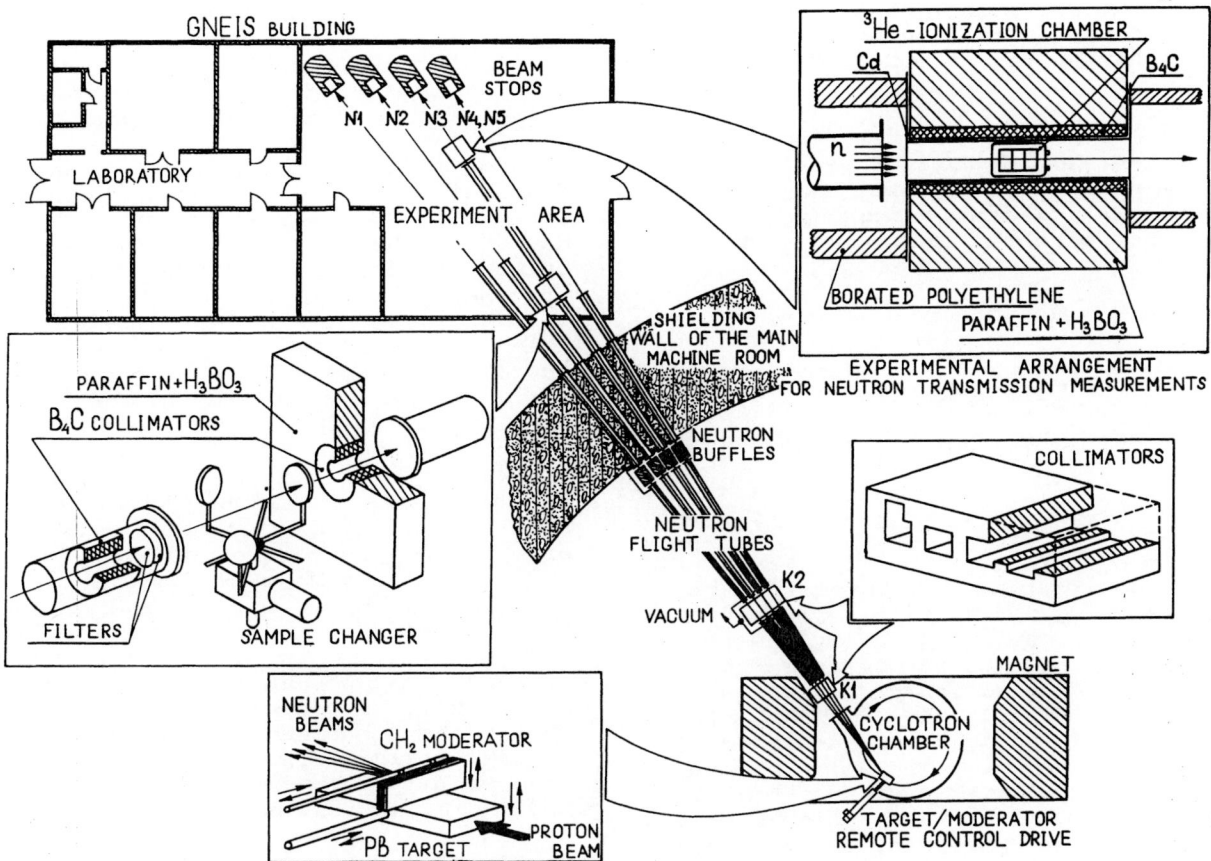

Fig. 1. Schematic layout of the GNEIS facility and experimental arrangement for transmission measurements.

the energy range from 1 eV to 100 eV and on the average was about 0.5%, but with increasing energy it grew and for E_n = 5 keV was about 10%. In order to suppress the overlap neutrons cadmium filter was placed in the beam. To compensate for neutron flux fluctuations the "sample in beam" and "sample out of beam" measurements were carried out in short (10 min.) runs with consecutive recording of the measured spectra on a magnetic tape. An experimental transmission spectrum for Bi measured during 10 minutes is shown in Fig. 2. For a distortion of the spectrum

shape caused by nonzero dead time to be equal for bismuth-silicon and ^{208}Pb-carbon pairs the samples thicknesses were chosen for their transmissions to be about equal in energy range of interest. The dead time of the whole detection system was measured by "two sources" method and was found to be equal to τ_d=0.660±0.012 μs.

Results

The measured total cross sections of ^{209}Bi and Si are shown in Fig. 3 and that of ^{208}Pb and C in Fig. 4. The given errors are statistical ones and reflect uncertainties of cross section energy dependence. Only the energy dependence of cross section is used for α_n-value estimation. The systematic error of σ^{tot} absolute value is about 1% and represents uncertainty addendum to σ^{tot}. In order to evaluate a value of polarizability it is necessary to subtract from total cross section the corrections for Schwinger and (n,e)-interactions, solid state correction, absorption cross section and neutron resonances contribution.

Bismuth

Analysis of the data for Bi using the method given in [3] with simultaneous variation of two parameters - polarizability and resonances contribution (parameter P_2 from [3]) - leads to α_n=(25±11)·

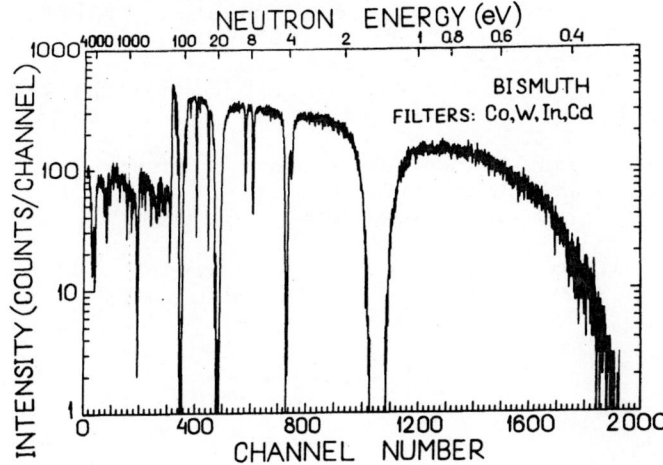

Fig. 2. Transmission spectrum for Bi obtained for 10 min of running time.

162

$\cdot 10^{-3}$ fm^3. One cannot obtain better precision of α_n for Bi data because of the large contribution of 800 eV resonance.

Lead-208

After the subtraction of all corrections mentioned above and suggesting a full account of the resonances contribution, the polynomial fit for ^{208}Pb leads to

$$\sigma(k) = (12.475 \pm 0.003) + (1.60 \pm 0.88)k$$

and value $\alpha_n = (2.6 \pm 1.4) \cdot 10^{-3}$ fm^3.
The fit for carbon leads to

$$\sigma(k) = (4.8015 \pm 0.0012) + (0.50 \pm 0.35)k$$

Discussion

In our opinion, the estimated value of α_n is in agreement with zero value. A further progress in increasing the precision of α_n-measurements could be connected with a substitution of neutron detector for a more effective one and with a more detailed background analysis and therefore increasing the high-energy boundary of measured cross section. That leads to an increase of polarizability contribution to the total cross section because it depends on energy as $\sqrt{E_n}$.

Quite recently, the two communications appeared about the analogous measurements performed at Garching [8] with a value of $\alpha_n = (0.4 \pm 1.5) \cdot 10^{-3}$ fm^3 or $\alpha_n = (-1.1 \pm 1.5) \cdot 10^{-3}$ fm^3 depending on value chosen for (n,e)-interaction amplitude, and at Oak Ridge [9] where a nonzero value was obtained for polarizability $\alpha_n = (1.20 \pm 0.15 \pm 0.20) \cdot 10^{-3}$ fm^3.

We think that Oak Ridge value of α_n does not close the problem of neutron polarizability. Analysis of our data for Bi has shown that α_n-value and its uncertainty significantly depend on the data processing method involved. For example, as mentioned above two parameters fit leads to $\alpha_n = (25 \pm 11) \cdot 10^{-3}$ fm^3. Proposed in [3] method when in fact only one parameter α_n is varied gives an error 7 times smaller but we think it would be very underestimated.

Authors of ref. [7] have pointed at the analogous problems. Thus, they have come to a conclusion that statistical error for α_n in references [3] and [4] (σ^{tot} of Bi and Pb measurements in low-energy range up to $E_n = 2$ keV) needs to be increased by 5 - 10 times. In the first place, it is due to a significant correlation between the fitted value of α_n and nucleus radius. Secondly, it is due to the fact that fitted α_n scatter, significantly larger than statistical errors, depending on variation of widths and a number of resonances taken into account at positive energy as well as on a peak value and position of hypothetical resonance at negative energy.

An obtained value of α_n and an analysis of both our and other results do not have final character; a continuation of the investigations is intended.

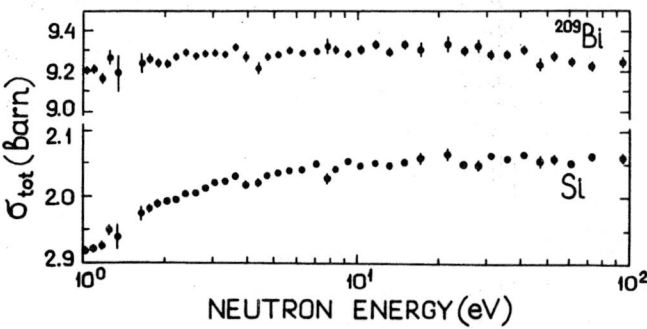

Fig. 3. Measured total cross sections of Bi and Si.

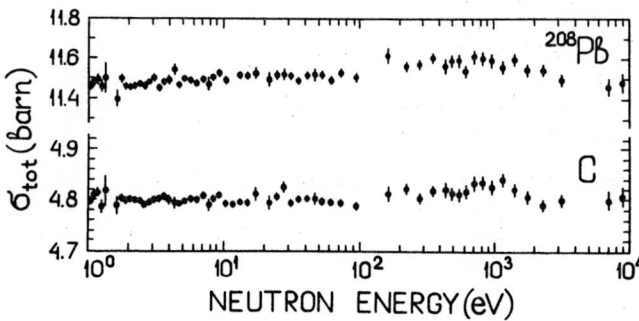

Fig. 4. Measured total cross sections of ^{208}Pb and C.

References

1. Yu.A. Alexandrov: Yadern. Fiz. 37, 253 (1983); Yadern. Fiz. 38, 1100 (1983)
2. J. Schmiedmayer, H. Rauch and P. Riehs: Phys.Rev.Lett. 61, 1065 (1988)
3. Yu.A. Alexandrov, M. Vrana, J. Manrique Garcia, T.A. Machekhina and L.N. Sedlakova: Yadern. Fiz. 44, 1384 (1986)
4. L. Koester, W. Waschkowsky and J. Meier: Z. Phys. A 329, 229 (1988)
5. O.A. Scherbakov, in Proc. VI Int. School on Neutron Physics, Alushta, 8 - 18 October, 1990, in press; LNPI Preprint 1664, Leningrad, 1990
6. N.K. Abrosimov, G.Z. Borukhovich, A.B. Laptev, V.V. Marchenkov, G.A. Petrov, O.A. Shcherbakov, Yu.V. Tuboltsev and V.I. Yurchenko: Nucl. Instr. Meth. A242, 121 (1985)
7. V.G. Nikolenko and A.B. Popov, JINR, P3-90-568, Dubna (1990)
8. Yu.A. Alexandrov, L. Koester, G.G. Samosvat and W. Waschkowski, JINR Rapid Communications, № 6[45]-90, p. 48-50, Dubna (1990)
9. J. Schmiedmayer, P. Riehs, J.A. Harvey and N.W. Hill: Phys. Rev. Lett. 66, 1015 (1991)

ELECTRIC PROPERTIES OF THE NEUTRON FROM
PRECISION CROSS SECTION MEASUREMENTS

Jörg Schmiedmayer* and Peter Riehs
Institut für Kernphysik der TU Wien**, A-1040 Wien, Austria

John A. Harvey and Nat W. Hill
Oak Ridge National Laboratory***, Oak Ridge, TN 37831 USA

Abstract: The total neutron scattering cross section σ_s contains non vanishing contributions from the electric polarizability α_n and from the n-e⁻ interaction because of their interference terms with coherent nuclear amplitudes. Both contributions can be separated only by an accurate understanding of nuclear amplitudes (for α_n and for the n-e⁻ interaction) as well as solid state amplitudes (for the n-e⁻ interaction). Because of the smallness of effects, the needed accuracy for the shapes of σ_s is given in orders of 10^{-3} to 10^{-4}, if we ask for basic electric properties of the neutron. Transmission measurements have been made at the Oak Ridge Electron Linear Accelerator ORELA with low background from the machine and a good n-TOF resolution using the neutron time of flight (n-TOF) technique. To minimize counting errors, detector signal heights (flash-ADC, 8 bit, 100 Mhz) and on line calculated dead time corrections were recorded on tape as a function of n-TOF.

The electric polarizability was determined to be $\alpha_n = (1.20 \pm 0.15 \pm 0.20)10^{-3}$ fm³. This is the first value different from zero and has been found with radiogenic ²⁰⁸Pb- and highly enriched ²⁰⁸Pb samples for the neutron energy range from 50 eV to 40 keV. α_n is deduced from the linear term in k of the shape of $\sigma_{s,corr}(k) = 11.508(5) + 0.69(9)k - 448(3)k^2 + 9500(400)k^4$ as a function of the wave number k in fm⁻¹, where $\sigma_{s,corr}$ is σ_s, corrected for Schwinger, n-e⁻ scattering and resonance contributions. The calculation of resonance contributions were made with our new set of nuclear parameters of ²⁰⁸Pb and other Pb isotopes.

To investigate the n-e⁻ interaction, solid state corrections below about 1 eV are of particular interest. As the presently used ones are almost without experimental verifications, work to improve solid state corrections is an important subject of our measurement program. Measured will be liquid Pb, liquid Bi and noble gases in transmission arrangement. The aim is to separate the shape of n-e⁻ scattering coming from the well known atomic form factor. First transmission measurements have been made with solid, powder and liquid samples in the energy range of orders of magnitude around eV.

Keywords: Nuclear Reactions: Neutron Scattering, Pb, ²⁰⁸Pb, Resonance Paramters,
           Particle Properties: Proton, Neutron

Introduction

In the neutron nuclear interaction the electric charge structure of the neutron (first inferred by its high magnetic moment and now established by the quark picture of hadrons) interacts with the Coulomb field of atomic nuclei. Contributions are expected for the electric polarizability α_n because of an induced electric dipole moment and for the n-e⁻ scattering. The latter is given by the sum of the FOLDY interaction coming from the magnetic moment of the neutron and from the interaction with the internal charge structure of the neutron, which is described by the mean-squared charge radius $<\rho r^2>$; It is a lowest order term of the electric form factor of the neutron.

The quantity α_n changes the scattering cross sections σ_s, in the order of 0.2 %. The n-e⁻ interaction shows negative contributions in orders of per cent in the eV region.

As it is not possible to switch off and on the effects of α_n or $<\rho r^2>$, a difference measurement is not possible. In addition, accurate calculations of absolute cross sections are too difficult. Therefore, the characteristic changes of σ_s as a funtion of the n-energy are asked, or the "shape of σ_s" has to be investigated.

The shapes of σ_s were obtained by transmission measurements with n-TOF at the neutron source ORELA at flight path lengths of 80 m for α_n and 18 m for the $<\rho r^2>$. On the other hand, the shape has to be calculated. In these calculations the parameters for α_n and $<\rho r^2>$ are fitted to get best agreement with the measured shapes of σ_s. For this purpose, accurate parameter sets of nuclear data are necessary for the used sample nuclides.

* Present address: Department of Physics, Harvard University, Cambridge, MA 02138, USA
** Supported by the Austrian Fonds zur Förderung der Wissenschaftlichen Forschung, nros 6849, 8489
*** Managed by the Office of Energy Research, Division of Nuclear Physics, U. S. Department of Energy, under contract DE-AC05- 84OR21400 with Martin Marietta Energy Systems, Inc.

The electric polarizability α_n

To investigate α_n, transmission measurements with ^{208}Pb (>99% enriched) and radiogenic ^{208}Pb (72.40% enrichment) compensated for ^{206}Pb and ^{207}Pb were made to determine the total scattering cross sections σ_s with a ^{6}Li glass-scintillator and a ^{10}B-loaded liquid scintillator. The shape of σ_s is

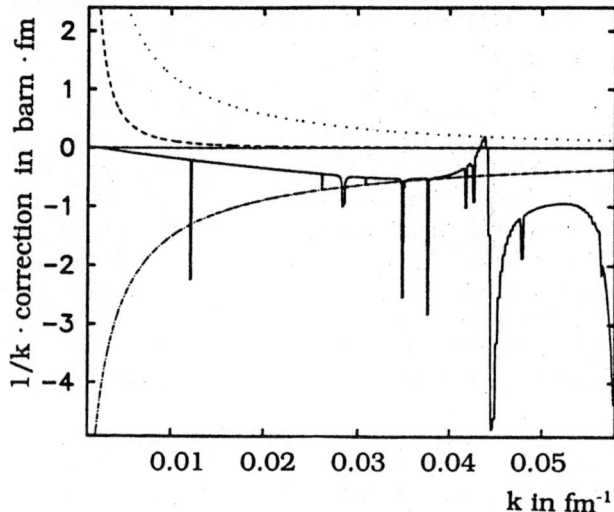

Fig. 1. Corrections of σ_s divided by k, where k(fm^{-1}) = $2.1986 \times 10^{-4} \sqrt{E(eV)}$. Shown are the corrections for resonances (·········), Schwinger (·········), n-e$^-$ scattering (·········) and the air (·········).

corrected for resonances, Schwinger- and n-e$^-$-scattering, as shown in Fig. 1, and is expressed as a function of the wave number k by [1]:
$$\sigma_{s,corr}(k) = 11.508(5) + 0.69(9)k - 448(3)k^2 + 9500(400)k^4,$$

α_n is deduced by the value of 0.69(9) of $\sigma_{s,corr}$ depending only on α_n. We obtained a value of

$$\alpha_n = (1.2 \pm 0.15 \pm 0.20) \times 10^{-3} \text{ fm}^3 \text{ [1]}. \quad (1)$$

The error of 0.15 is the statistical one and 0.2 stands for an absolute systematic error, which comes mainly from the background subtraction. Errors by the resonance corrections are negligible, because of the smallness of corrections and the good understanding of the ^{208}Pb sample cross sections after the development of the parameter sets within the framework of the REFIT cross section fit program [2], see Fig. 2. For the first time a value definitely different from zero is obtained; the accuracy is significantly better than in the previous results of $\alpha_n = (1.2 \pm 1.0) \times 10^{-3}$ fm^3 [3], $\alpha_n = (0.8 \pm 1.0) \times 10^{-3}$ fm^3 [4] and $\alpha_n = (1.17 + 0.43/-1.17) \times 10^{-3}$ fm^3 [5].

The mean-squared charge radius $\langle \rho r^2 \rangle$

The determination of $\langle \rho r^2 \rangle$ is deduced from the n-e$^-$-interaction, which is given in first Born approximation [6] as a function of σ_s by

$$\sigma_s = 4\pi(b_{Nc}^2 - 2b_{Nc}b_e Z(1-f(k)+...). \quad (2)$$

The contributions by α_n and by the Schwinger scattering are not written down. Z is the nuclear charge number and f(k) the atomic form factor. b_{Nc} and b_e are the scattering lengths of nuclear- and n-e$^-$-interaction. The quantity $\langle \rho r^2 \rangle$ can be calculated from b_e and the scattering length of the FOLDY-

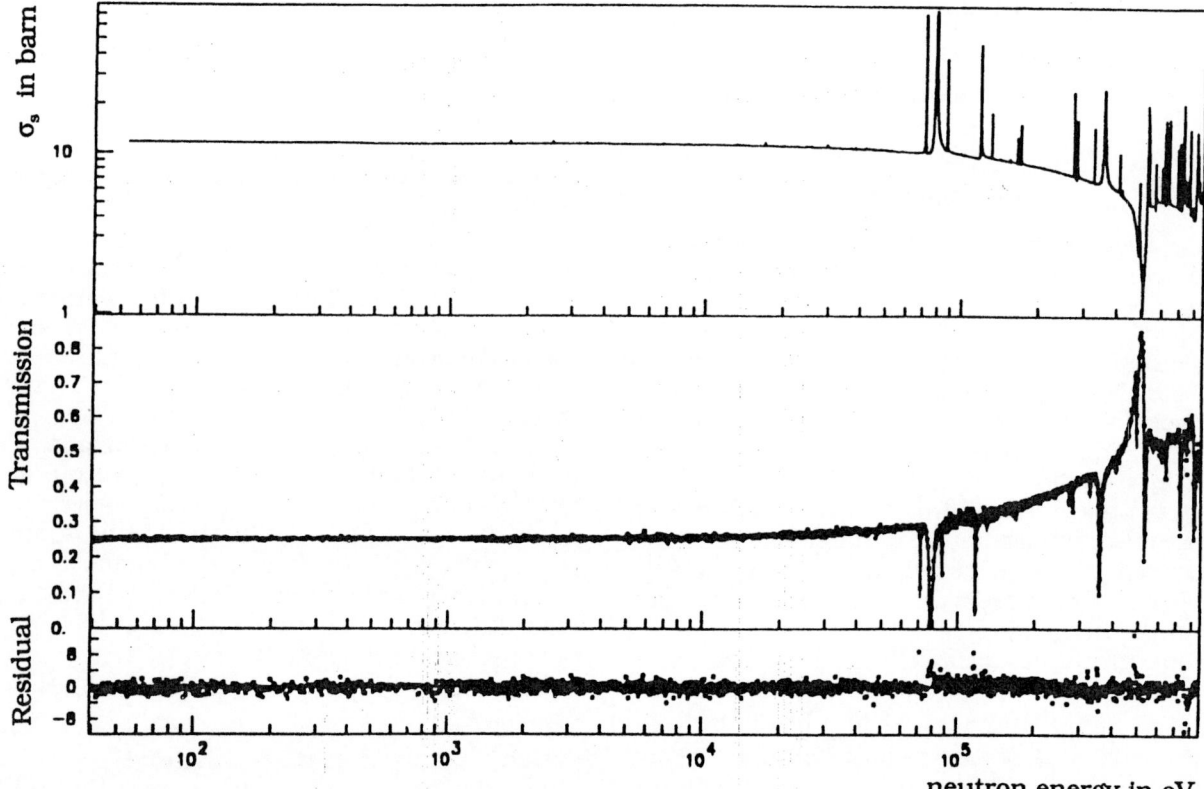

Fig. 2. Calculated cross section, transmission and residuals for ^{208}Pb, thickness of 0.1196 a/barn, measured at 80 m, ORELA flight path no 1, with a ^{6}Li glass scintillator, 110 x 12 mm.

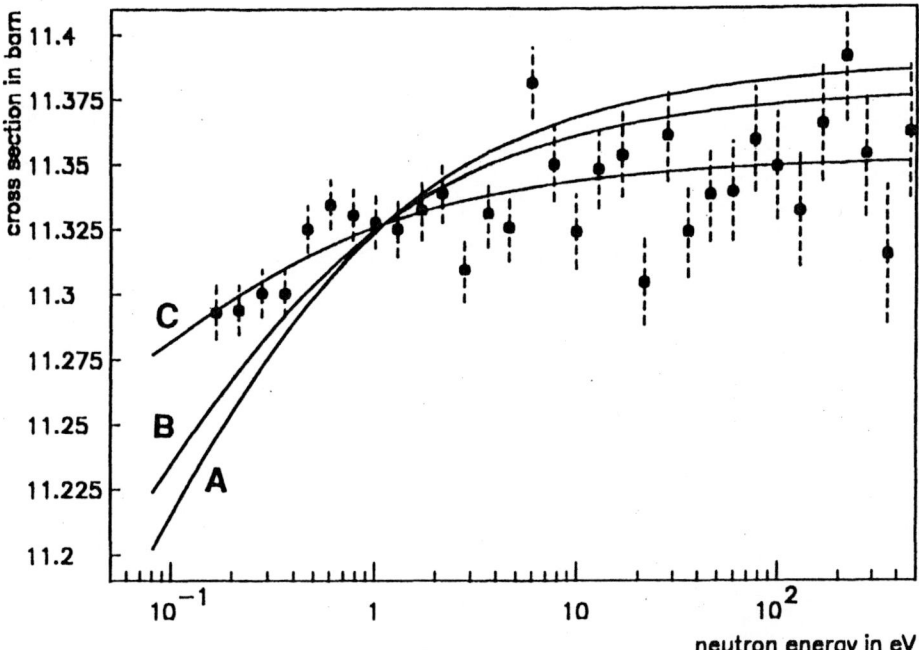

Fig. 3. Cross sections of liquid Pb at 350°C, 0.1559 a/barn, measured at ORELA, 18 m flight path, with a 1 mm ^{6}Li glass scintillator. Curve A, B and C are drawn for values of $\langle \rho r^2 \rangle$ of - 0.012 [10], + 0.014 [6] and + 0.075 fm^2. Points are corrected for solid state effects [11].

interaction $b_F = -1.467971 \times 10^{-3}$ fm by:

$$\langle \rho r^2 \rangle = 3 m_e a_0 (b_e - b_F)/m \qquad (3)$$

with the Bohr-radius a_0. The mass of neutron and electron is m and m_e.

The first experiment for b_e was made by FERMI and MARSHALL in 1947 [7]; spread over the last 40 years, for b_e about 15 different values can be found in the literature. The most accurate value came from two experiments (absolute cross section measurement at 1.26, 5.19, 18.8, 132 eV and refraction at 0 eV of Ref. [8]; angle correlation at 45°, 135°, 225°, 315° of Ref. [9]) and were evaluated [6] to give $\langle \rho r^2 \rangle$ = 0.014(3) fm^2. In disagreement to this a new value of $\langle \rho r^2 \rangle$ = - 0.012(4) fm^2 has been reported [10], where two previous measurements (transmission with n-TOF measurement above 1 eV; scattering with W-single crystals, nuclear scattering length compensated, but containing paramagnetic impurities) were reconsidered.

Using the n-TOF method and recording neutrons down to energies below 0.1 eV, we made preliminary transmission measurements with solid and powder samples as well as, later on, short time runs with a liquid lead sample at ORELA. The cross section data points, taken for liquid Pb, are shown in Fig. 3 and are corrected for: a) α_n of Ref. [1], b) the effective range of Ref. [3], c) the inelastic cross section of 1.3 mb, d) the Schwinger scattering, Ref. [6] and e) the solid state effects, Refs. [11]. So far, no corrections were made for the interference of resonances with the nuclear potential well.

The preliminary result is $\langle \rho r^2 \rangle$ = 0.075(8) fm^2. This high value might be mainly an effect by the solid state correction [11], because we obain without this correction a value of $\langle \rho r^2 \rangle$ = 0.020(8)

fm^2. As the continous shape of the cross section has not been measured with similar accuracy before and as the solid state corrections have been experimentally proofed in part only [11], the result of the present measurement is $\langle \rho r^2 \rangle$ = 0.020 ± 0.008 ± 0.050 fm^2. The statistical error is 0.008, the systematic one is 0.050 fm^2.

As a next step, we intend to improve the statistics of this measurement to have enough information for a simultaneous fit with 3 parameters: normalization, $\langle \rho r^2 \rangle$ and $c\lambda^2$. λ is the wavelength of the neutron.

References

1. Jörg Schmiedmayer, Peter Riehs, John A. Harvey Nat W. Hill, Phys. Rev. Lett. **66**, 1015(1991)

2. M. C. Moxon, in Neutron Data of Structural Materials for Fast Reactors, Geel 1977 (Edited by K. H. Böckhoff), p. 644, Pergamon Press

3. J. Schmiedmayer, H. Rauch, P. Riehs, Phys. Rev. Lett. **61**, 1065(1988)

4. L. Koester, W. Waschkowsky, J. Meier, Z. Phys. **A239**, 229(1988)

5. K. W. Rose, B. Zurmühl, P. Rullhusen, M. Ludwig, A. Baumann, M. Schuhmacher, J. Ahrens, A. Ziegler, D. Christmann, B. Ziegler, B. Schoch, Phys. Lett. **B234**, 460(1990)

6. V. F. Sears, Physics Reports **141**, 282(1986)

7. E. Fermi and L. Marshall, Phys. Rev. **72**, 1139(1947)

8. L. Koester, W. Waschowski and A. Klüver, Physica **137B**, 282(1986)

9. V. E. Krohn and R. G. Ringo, Phys. Rev. **D8**, 1305(1973)

10. Yu. A. Alexandrov, Nucl. Instr. and Meth. **A284**, 134(1989)

11. G. Placzek, B. R. A. Nijboer and L. Van Hove, Phys. Rev. **82**, 392(1951); K. Binder, Phys. Stat. Sol. **41**, 767(1970); J. R. Granada, Z. Naturforsch. **39a**, 1160(1984)

Part II

Data Testing and Validation for Reactors

THERMAL AND EPITHERMAL DATA ASSESSMENT
FOR FISSION REACTORS

Henry TELLIER

Service d'Etudes de Réacteurs et de Mathématiques Appliquées
Centre d'Etudes Nucléaires de Saclay - 91191 Gif sur Yvette - FRANCE

Abstract: Sensitivity calculations indicated that the wrong computation of the reactivity temperature coefficient of a multiplying medium could be explained by a change of the neutron cross section shapes for the major actinides in the thermal energy range. Consequently, new measurements of these shapes were recently performed in several laboratories. Nu-bar, the fission and total cross sections, the eta parameter and some resonance widths were measured again. The isotopes of interest are uranium 238, uranium 235, plutonium 239 and plutonium 240. Now the results of these new measurements are almost all available. The purpose of the present paper is the review of the new results and a proposal, if it is necessary, of a modification of some cross sections shapes which were previously used in the evaluated files. The new cross section values are validated against a wide set of integral experiments which includes buckling and temperature coefficient of reactivity measurements. It seems that we observe a satisfactory agreement between the new microcospic data and the values which are deduced from the integral experiments.

Introduction

For obvious safety reasons, the temperature coefficient of reactivity must be predicted with a very good accuracy for nuclear power reactors. This quantity is measured at the opportunity of the start-up experiments of the power reactors, but the result is not in agreement with the computed value. The later is too negative. As a power reactor is geometricaly very complicated, the reactor physicists used also specific critical facilities to measure this temperature coefficient. As the geometry of these facilitiesissimpler than the one of a power reactor we can be more confident in the computed value. Even in this favourable case, the measure value is different from the calculated one and the order of magnitude of the discrepancy is the same as for the power reactors. Several phenomena could cause a wrong calculation of the temperature coefficient. They are the spectrum shift, the thermalization and the Doppler effect. Each of these effects are related to different physical quantities.

The spectrum shift is sensitive to the shape of the neutron cross sections in the thermal energy range, the thermalization is sensitive to the model and the parameters which are used to represent the vibrations of the moderator molecule, and the Doppler effect is affected by the crystalline binding. These three ways were separately studied since several years. Sensitivity calculations showed that the discrepancy between the measured value and the computed value of the temperature coefficient can be explained by a modification of the shape of the cross section used in the

evaluated files. The required modifications can be obtained by introducing an extra resonance near the binding energy for both uranium 238 and uranium 235 [1]. As these resonances were not previously observed, reactor physicists required new measurement of some cross sections in the thermal and subthermal range for the major actinides. The results of these new and accurate measurements are now available. The purpose of this paper is to make a review of these new results and to see if they can explain or not the temperature coefficient discrepancy. In the following section, we will review the new neutron data for uranium 238 and 235, plutonium 239 and 240 and the crystalline binding effect. Some informations about the validation of these data with integral experiments will also be given.

Uranium 238

The uranium 238 case was very simple because the only cross section involved in the temperature coefficient is the capture cross section in the thermal and subthermal energy range. At low energy, the known resonances of this nuclei are the 4.4 eV "p" wave and the 6.67 eV "s" wave resonances. These two resonances induce for capture cross section a 1/v behaviour in the thermal range. The temperature coefficient measurements suggested a non 1/v dependence, the cross section must decrease with the energy faster than the 1/v shape. This can be obtained by adding and extra resonance at -0.005 eV as it was proposed by Erradi [2]. This assumption was supported by the measurement of the fission cross section at 0.025 eV which was performed in Grenoble [3]. The value obtained in this experiment cannot be explained by the contribution of the nearby resonances. This fission data suggests the existence of a bound level which can be more or less far from the zero energy. If it is close to the zero energy as in the Erradi assumption, this level must have an influence on the shape of all the neutron cross section in the thermal range. Consequently an accurate measurement of the capture cross section was needful. This measurement was performed at the Geel Laboratory and the results are now available [4]. As it is shown on the figure 1 which represents the variation of $\sigma_\gamma \sqrt{E}$ as a function of the neutron energy, and a comparison with the JEF evaluation, these experimental results confirm a 1/v behaviour for the capture cross section. The assumption of a resonance in the immediate vicinity of the zero energy and the explanation of a part of the temperature coefficient discrepancy by such a resonance must be dropped. Reasonably, we must consider that the uranium 238 capture cross section problem is solved in so far as the thermal energy range is concerned.

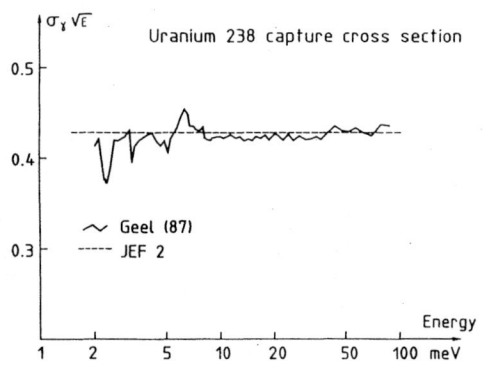

Figure 1 - Uranium 238 capture cross section of a metal sample and comparison with JEF 2.0.

Uranium 235

The case of uranium 235 is more complicated than the one of uranium 238 because the quantity which is involved is eta : a function of nu-bar, the fission and the capture cross sections. Several authors thought that the temperature coefficient calculation could be improved with a modification of the ^{235}U eta behaviour in the thermal range. Instead of a constant value, they proposed an increase with the energy as in the WIMS library [5] or the APOLLO library [6]. As it is shown on figure 2, these modification were compatible with the eta measurements which were available at that time. These eta measurements were performed a long time ago in Brookhaven [7] and in Harwell [8] and were not accurate enough. Obviously reactor physicists asked for new measurements. During the last years, numerous experiments were performed in the thermal energy range and in the cold neutron range. They are relative to nu-bar, the fission cross section, the total cross section and eta. Table 1 gives the list of these measurements and the energy range of interest.

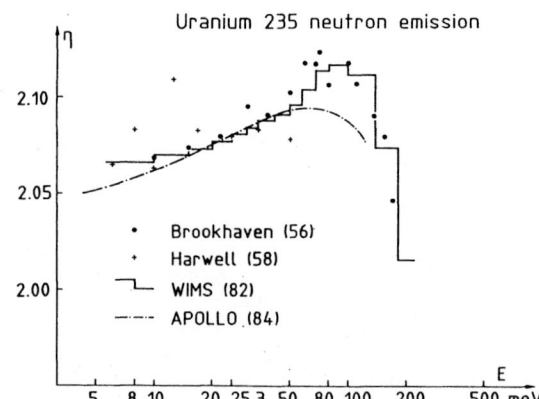

Figure 2 - Old measurements of uranium 235 eta and comparison with the multigroup libraries of the reactor calculation codes WIMS and APOLLO.

Table 1 : Recent Measurements of Uranium 235 Thermal Data.

Quantity	Energy range (meV)	Laboratory
ν-bar	5 - 1000	Oak Ridge (84) [9]
σ_f	10 - 500	Oak Ridge(84) [10]
σ_f	2 - 1000	Geel (85) [11]
σ_f	0.006 - 0.055	Grenoble (88) [12]
σ_f	20 - 1000	Gaithersburg (88)[13]
σ_t	0.0004 - 0.034	Moscow (86) [14]
η	2 - 450	Geel (87) [15]
η	3 - 400	Harwell (88) [16]
η	2 - 150	Grenoble(89) [17]
η	3 - 1000	Oak Ridge(90) [18]
α	2 - 400	Geel (91) [19]

Important conclusions can be deduced from the examination of these microscopic data.

a) <u>Nu-bar</u>

In the preceding years, only one measurement of nu-bar was performed in the energy range which is interesting for us.

It is the Oak Ridge experiment [9] which gives the ratio of uranium 235 ν_p over Californium 252 ν_{sp} for the neutron energies between 5 meV and 1 eV. As it can be seen on figure 3, nu-bar can be considered as constant below 1 eV. Nothing appears in the vicinity of the 0.290 eV resonance. As all the evaluated files adopted a flat shape for nu-bar, we can keep this assumption which is in agreement with the experimental results.

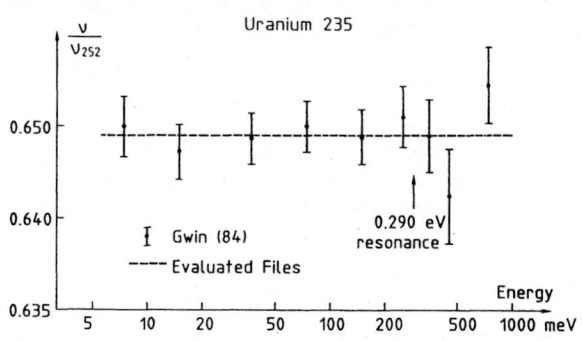

Figure 3 - Ratio of uranium 235 nu-bar to the one of californium 252 in the thermal energy range.

b) Fission cross section

Since 1984 several measurements of the uranium 235 fission cross section were performed in the thermal energy range [10, 11, 13]. The results are now available. Above 20 meV and in particular in the 0.290 eV resonance all these results are in agreement between them and with the most recent evaluated files ENDF/B5 and JEF2. In the subthermal energy range, below 5 meV, only the Geel experiment gives informations. According to this measurement the uranium 235 fission cross section reaches an 1/v shape for energies higher than it was assumed in ENDF/B5. Consequently is seems reasonable to recommend the JEF 2.0 evaluation which takes into account the Geel experiment and which is in agreement with the other results above 20 meV as it is shown on figure 4.

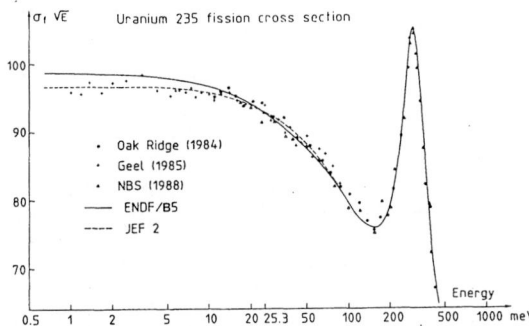

Figure 4 – Recent measurements of the uranium 235 fission cross section in the thermal and epithermal energy range.

c) Cold neutron cross sections

The knowledge of the cross sections for the very low energy neutrons or cold neutrons is not essential for the reactor physicists, because there is almost no neutron in that part of the spectrum in a nuclear reactor. But this knowledge is important for the evaluator for whom the cold neutron cross section is an extrapolation point. It is particulary important for uranium 235 because it can confirm or not the 1/v dependence of the cross sections. Two recent experiments were performed for uranium 235 : the absolute measurement of the total cross section made in Moscow [14] and the relative measurement of the fission cross section performed in Grenoble [12]. The results which are displayed an figure 5 confirm the 1/v behaviour of these cross sections for the neutron energy below 0.1 meV. Thus, it does not exist an extra resonance in the vicinity of the zero energy. On the same figure we can see the comparison with the JEF 2.0 recommendations. For the fission cross section the experimental results are normalized to the average value of $\sigma_f \sqrt{E}$ between 2 and 5 meV where the cross section has already a 1/v behaviour.

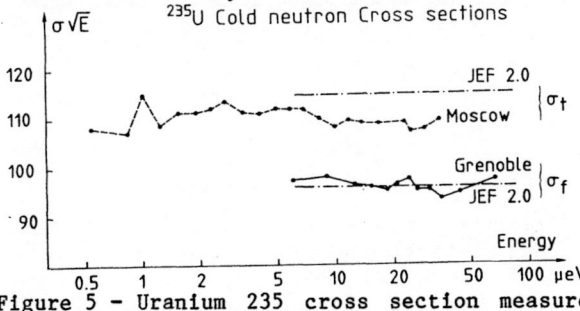

Figure 5 – Uranium 235 cross section measurements in the cold neutron energy range.

d) Eta measurements

To explain the temperature coefficient discrepancy it is $\eta = \nu\sigma_f/\sigma_a$ which is the most sensitive parameter. In all the evaluated files eta was assumed to have a constant value below 0.1 eV and the reactor physicists proposed an increase of eta between 5 and 100 meV. The number of neutrons emitted in a fission and the fission cross section were separately measured. A measurement of eta, or of the absorption cross section, was necessary. Chronologically four measurements of eta were performed in the range of interest

The first one is the Geel experiment [15]. This experiment was performed with a linac and a liquid methane moderator to inhance the importance of the low energy neutrons. It covered the neutron energy between 2 and 450 meV. As it is shown in figure 6 the results suggest an increase of eta between 2 and 80 meV. It is in the good direction for the reactor physics.

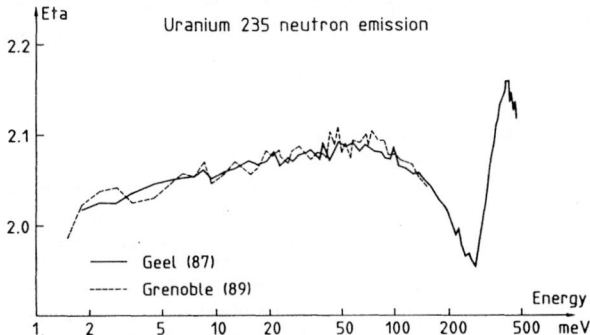

Figure 6 – Experimental results of the Geel and Grenoble measurements in the thermal energy range.

A second measurement was carried out with the Harwell linac [16]. Unfortunately the number of low energy neutron was not very high. Consequently the accuracy was not good enough. Nevertheless this experiment did not show a significant shape of eta versus the neutron energy as it is displayed on figure 7. It is contradictory with the Geel results.

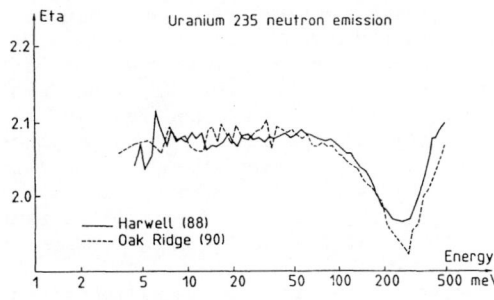

Figure 7 – Experimental results of the Harwell and Oak Ridge eta measurement in the low energy range.

In order to explain this disagreement, a third experiment was performed in Grenoble [17]. Instead of a linear accelerator, like in Geel, the neutron source was constituted by a cold neutron beam of the high flux reactor and more neutron of low energy were obtained. The accuracy was then expected to be better because the background would be lower. In the energy range between 2 and 150 meV, this experiment perfectly confirms the Geel results and the shape of eta versus energy as it is shown on figure 6.

Finally a fourth experiment was carried out with the Oak Ridge linac [18]. The preliminary results of the last experiment are compared with the Harwell results on figure 7. These two series of results seem more or less in agreement and do not show a significant shape of eta.

With these four experiments we have two sets of results, one which indicates a slope (The Geel and Grenoble data) and the second which does not (The Harwell and Oak Ridge data). But, as the low energy neutron flux was the highest in the Grenoble experiment, some evaluators adapted a higher weight for this experiment and concluded to an increase of eta between 2 and 80 meV. This shape is in good agreement with the tendency deduced from the temperature coefficient measurement. The preliminary point-wise version of the JEF 2.0 evaluation takes this change of eta into account. Independently it was also shown that this slope can be reproduced by the contribution of the resonances below 4 eV if we use an R-matrix formalism instead of the Breit and Wigner one which was formerly employed [20]. Nevertheless all the evaluators do not accept the slope and ENDF/B6, in particular, does not reproduce the Geel and Grenoble results. Actually we cannot consider the problem of the uranium 235 eta as definitely solved even if the slope seems more likely than in the past. Works are underway to try to understand why the four recent measurements of eta are not in satisfactory agreement.

e) Capture cross section

No capture cross section measurement was carried out for uranium 235 in the subthermal energy range. We have only indirect informations deduced from the total and fission cross sections measurements with cold neutron. According to the results of these experiments we can assume that below 100 μeV the capture cross section is very close to a 1/v shape. Between 100 μeV and the first resonance, the shape of the capture cross section must be evaluated. Each evaluated file has its own phylosophy. For example in JEF 2.0 the capture cross section was deduced from the experimental results of nu, eta and fission cross section in the subthermal energy range. In ENDF/B the capture cross section is considered as the contributions of the resolved resonance and bound level. Consequently, the result is not the same for the two evaluations as it can be seen on figure 8. The discrepancy below 20 meV is not very important for the reactor physicist because in a reactor the neutron flux is very low below 20 meV. But for the nuclear physics point of view it would be very interesting to have a measurement of the capture cross section in the thermal and subthermal range. In the last months an alpha measurement were performed at Geel in the energy range of interest [19]. From the preliminary results of this measurement we can deduced capture cross section values assuming that the fission cross section is known. If we adopt the fission cross section of JEF 2.0, the values which are obtained for the capture cross section are reported on figure 8.

They seem in satisfactory agreement with the values of JEF 2.0 which were evaluated before the publication of the alpha measurement. They are also coherent with the slope of eta.

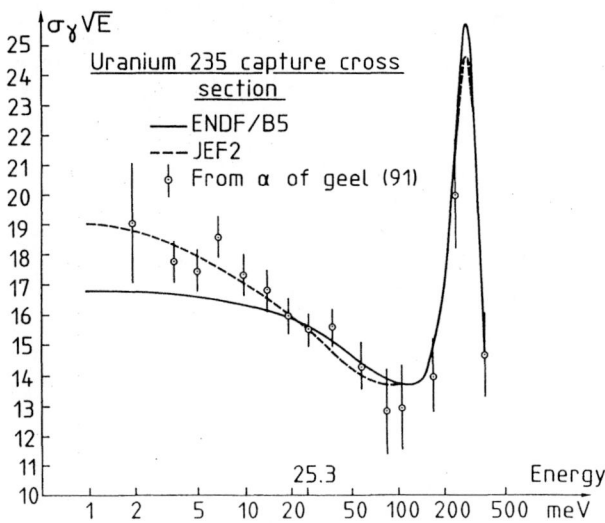

Figure 8 - Uranium 235 capture cross sections used in the ENFD/B5 and JEF.2 files and comparison with the values deduced from the Geel measurements of alpha.

Plutonium 239

All the old evaluations of the plutonium 239 neutron data, including ENDF/B5 are considered as not satisfactory by the reactor physicists. As a matter of fact, in all these files it was adopted a flat behaviour of nu-bar in the low energy range, the spin of the resonances was not considered and it was used a Breit and Wigner formalism to compute the cross sections. With the high burn up fuels and the recycling the plutonium became more and more important in the thermal neutron reactors. An updating of the plutonium 239 neutron data was strongly required. This updating was performed by Derrien et al. [21] who take into account new experimental results (except the Geel results which were not available in that time yet) and used a Reich and Moore formalism which is more convenient for the fissile nuclei. The list of these new measurements and the energy range of interest is given in table 2.

Table 2 : Recent measurements of ^{239}Pu low energy data

Quantity	Energy range (meV)	Laboratory
ν-bar	5 - 1000	Oak Ridge (84) [9]
σ_f	10 - 500	Oak Ridge(84) [10]
σ_f	2 - 1000	Geel (88) [22]
σ_t	700 - 1000	Oak Ridge(88) [23]
σ_t	10 - 1000	Oak Ridge(89) [24]

All the new evaluation of ^{239}Pu neutron data, and in particular JEF 2, are based on the resonance parameter set derived from the Derrien's analysis. In addition, in JEF 2, it was used a shape of nu-bar which accounts for the strong dip which was experimentally observed in the vicinity of the 0.3 eV resonance This structure is well reproduced by the Fort's theoretical calculation of nu-bar which includes the spin effect and the (n, γ_f) effect of the J = 1 resonances [25]. As it can be seen on figure 9, the agreement between the experimental points and the computed curve is quite good.

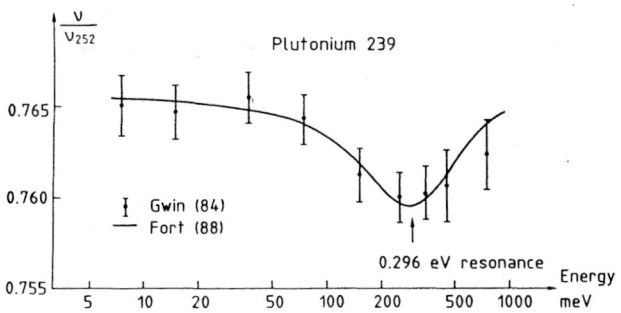

Figure 9 - Ratio of Plutonium 289 nu-bar to the one of the Californium 252 and comparison with the Fort's evaluation.

As the Geel fission data was released after the evaluation it was interesting to compare the new experimental results with the fission cross section obtained with the Derrien's parameters. This comparison is displayed on figure 10 which indicated a very good agreement. Up to now, it does not seem exist any problem of shape for the plutonium 239 data in the low energy range, below one electron volt.

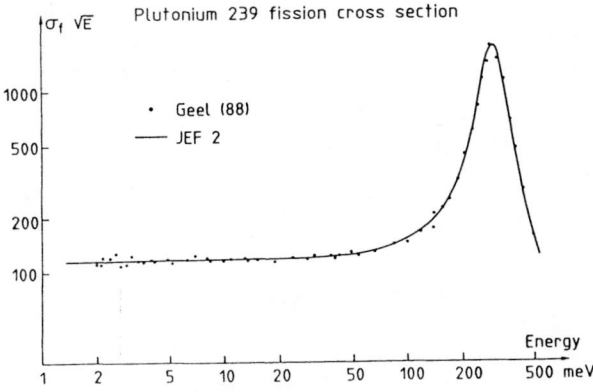

Figure 10 - Comparison of the last fission cross section of Geel and the recommended value of JEF.2 for Plutonium 239.

Plutonium 240

For this isotope, the cross section behaviour in the thermal energy range is mainly governed by the 1.056 eV resonance. Consequently it is necessary to have a very accurate knowledge of the parameters of this resonance. Only two measurements of the 1.056 eV resonance parameters were recently carried out, the experiment of Brookhaven [26] and the one of Oak Ridge [24]. In the Brookhaven experiment, total and capture cross sections with room temperature and cooled samples were used. In the Oak Ridge measurement, transmission measurements with seven thicknesses of sample were performed. Thus it was expected that the

results could be very satisfactory. Unfortunately, as it can be seen in table 3, the two sets of results are significantly discrepant.

Table 3 : Parameters of the ^{240}Pu 1.056 eV resonance

Γ_n (meV)	Γ_γ (meV)	Laboratory
2.32 ± 0.06	32.4 ± 0.6	Brookhaven (81) [26]
2.45 ± 0.02	30.3 ± 0.3	Oak Ridge (87) [24]

The difference between the two sets of resonance parameters induced a change of 1.2 percent or the contribution of the 1.056 eV resonance to the capture cross section at 0.025 eV and a change of 4.5 per cent to the contribution of the same resonance to capture integral. These effects are important every time that the plutonium 240 is significantly involved. This situation occurs in plutonium recycling in light water reactors or irradiated fuel analysis for instance. This discrepancy must be clarified.

Crystalline binding effects

Although, stricly speaking, the Doppler broadening of the resonances is not a nuclear data problem, this phenomenon introduces an extra parameter which has a non negligible impact in reactor physics. For a motionless target, the resonance cross section shape is well represented by a Breit and Wigner or a Reich and Moore formalism. But because of the thermal energy the target nuclei are not motionless. For a free gas, the velocity spectrum is a Maxwell Spectrum. Bethe and Placzek showed that in this case the cross section for a given temperature can be written as a convolution of the motionless cross section by a Gauss law. In practice we cannot consider the target nuclei in a nuclear reactor as a free gas because the fuel is generaly in a solid state and it exists bindings between the atoms of the crystal. When these bindings are weak, and it is fortunately the case for the uranium crystal, we always can use a free gas model. But, according to Lamb's results, instead of the physical temperature, we must use an effective temperature which is higher. This effective temperature can be computed with a crystalline vibration model. Generally the Debye model is considered as good enough. This model depends on a parameter Θ which is called the Debye temperature. Thus, the reactor physicist needs an accurate knowledge of the Debye temperature for uranium crystal. In reactor physics the importance of this parameter is greater for room temperature or cooled experiments than for the power reactor temperature. In the case of uranium metal fuel, it seems that the Debye temperature is well known and is equal to 260 K. But, very often the fuel is uranium dioxide and in this later case, the Debye temperature seems not to be well known [27]. Several methods were used to measure or to compute the Debye temperature and the various results are not in a very good agreement as it can be seen in table 4.

Table 4 : Experimental and theoretical determinations of the UO_2 Debye temperature

	METHOD	Θ K	REFERENCE
"EXPERIMENTAL RESULTS"	DOPPLER EFFECT	620	GOLINELLI (1980)[28]
	SPECIFIC HEAT	>300	BELLE (1961)[29]
	NEUTRON DIFFRACTION	377	WILLIS (1963)[30]
	ELASTIC PROPERTIES	875	MARLOWE (1969)[31]
	SPECIFIC HEAT	590	HOCH(1970) [31]
	ELASTIC PROPERTIES	385	FRITZ (1976)[32]
	TRANSMISSION	590	BRUGGER (1981)[33]
THEORETICAL COMPUTATIONS	DOLLING MODEL	620	DOLLING (1965)[34]
	BORN-VON KARMAN MODEL	550	ADKINS (1966)[35]
	THORSEN-JARVIS MODEL	630	BUTLAND (1974)[36]

Although, the reactor physicists must be more confident in a determination of the Debye temperature which uses reactor physics technics, for instance the direct Doppler measurements in a critical facility which indicate a Debye temperature equal to 620 K for the uranium dioxyde, it does not exist a general agreement for such a value [37]. The use of 620 K instead of 260 K gives for a room temperature reactor a decrease of 100.10^{-5} of the criticality factor. As we will see in the next section this effect can have a non negligible impact for the integral validation of the nuclear data. Consequently the Debye temperature of the uranium dioxide remains a problem which must be clarified.

Integral validation of the neutron data

All the informations provided by the measurements and the theoretical computation of the various neutron cross sections as a function of the incoming neutron energy are used by the reactor physicists through the evaluated files. As the multigroup libraries needed by the reactor core computation codes come from the processing of the evaluated files, all the incertainties or discrepancies of the original files have important consequences of the calculation of the reactor core parameters. Thus the reactor physicists want to check the validity of the evaluated files. For this purpose they perform specific experiments which can be easily computed. The result of this calculation depends on the microscopic data as a whole. These experiments are usually called integral experiments. For instance we can measure the critical size of a given multiplying medium or reaction rates and compare the result with the computed values. If the geometry of the multiplying medium is not too complicated and if the neutron spectrum can be considered as space independent, it is possible with the recent reactor physics codes, to make calculation without numerical approximations. Such necessary conditions can be obtained with uniform lattices or homogeneous media which constitute the main part of the integral experiments used for qualification purposes. For these clean experiments the difference ΔY_i between the measured value E_i and the completed value F_i can be considered as a consequence of the inaccuracy of some neutron data x_k. If we modify these neutron parameters which become $x_k + \Delta x_k$ the computed value becomes $F_i + \Delta F_i$. In practice, the modifications Δx_k are expected to be small and it is possible to make a first order expansion of ΔF_i as a function of Δx_k

$$\Delta F_i = \sum_k S_{ik} \Delta x_k$$

and to replace the partial derivatives by the sensitivity coefficients S_{ik}. If we choose a large set of integral experiments with different sensitivity coefficients, the minimization by the least square method of the quantity

$$Q = \sum_i (\Delta Y_i - \sum_k S_{ik} \Delta x_k)^2$$

leads to the required modifications Δx_k. It is the tendency research method.

In the case of the thermal neutrons reactors the sensitive parameters x_k are mainly the magnitudes of the cross sections at the 0.025eV energy, the shapes of the cross section are assumed to be the one of the microscopic data as they appear in the evaluated files. This method was applied to the qualification of one of the recent evaluated files, JEF 2.0. Sixty six critical size measurements were used for this work which allowed to obtain an accurate determination of the thermal data for uranium 235 and plutonium 239 [38]. The table 5 shows the comparison between the results of the tendency research, the original values of JEF 2.0 and the last recommendations of Divadeenam [39] and Axton [40]. The convergence between the integral determination and the differential measurements is very satisfactory.

Table 5
Actinide thermal neutron data

		JEF2.0	Divadeenam	Axton	Tendency Research
U 235	ν	2.437	2.425±0.003	2.426±0.005	2.435±0.004
	σ_f	582.5	582.6±1.1	585.1±1.6	582.3±1.1
	σ_c	98.8	98.3±0.8	96.1±1.7	98.0±2.0
U 235	ν	2.877	2.877±0.006	2.879±0.006	2.873±0.007
	σ_f	747.2	748.1±2.2	748.5±2.6	745.8±2.0
	σ_c	270.2	269.3±2.2	270.4±3.2	269.8±3.9

According to this study, the thermal data of the major fissile isotopes seem satisfactory for the reactor physist point of view. But we must not to forget that these new data do not completely solve the important problem of the temperature coefficient of reactivity which initiated the new interest for the neutron data in the low energy range. Only one third of the discrepancy can be explained by the use of the recent data. The calculated value of this coefficient remains to negative by $2.9.10^{-5}/°C$.

Conclusion

The status of the thermal and epithermal data of the major actinides was greatly improved since several years. The experimental measurements of the microscopic data which were required by the reactor physicists led to a better knowledge of the cross section in the low energy range. The magnitude given by the differential are in good agreement with the values deduced from the integral experiments. Nevertheless, it remains some problems which have to be further investigated. Let us mention, for instance, the slight discrepancy which still exist between the various measurements of the uranium 235 eta, the capture width of the first plutonium 240 resonance and the Debye temperature value which must be used for uranium dioxide.

Acknoledgements

The author is indebted to Dr. Moxon and Dr. Weigmann for having communicated their preliminary results and authorized the quotation before publication.

References

1. J.Bouchard, C. Golinelli et H. Tellier : Nuclear Data for Science and Technology-Anvers", 21 (1982)
2. L. Erradi : These Doctorat es Sciences. Paris Orsay (1982)
3. P. D'Hont, C. Wagemans, A. Emsallem and R. Brissot : Annals of Nuclear Energy 11, 485 (1984)
4. F. Corvi and G. Fioni : "Nuclear Data for Science and Technology- Mito" 127. (1988)
5. M.J. Halsall : "Thermal Reactor Benchmark Calculations. Brookhaven", EPRI-NP2855,21 (1983)
6. A. Santamarina; C. Golinelli and L. Erradi : ANS meeting "Reactor Physics and Shielding - Chicago", I, 48 (1984)
7. H. Palevsky, D.I. Hughes, R.L. Zimmerman and R.M. Eisberg : Journal of Nuclear Energy, 3, 177 (1956)
8. H.M. Skarsgard, and C.J. Kenward : Journal of Nuclear Energy, 6, 212 (1958)
9. R. Gwin, R.R. Spencer and R.W. Ingle : Nuclear Science and Engineering 87, 381 (1984)
10. R. Gwin, R.R. Spencer, R.W. Ingle, J.H. Todd and S.W. Scoles : Nuclear Science and Engineering 88, 37 (1984)
11. G Wagemans and A.J. Deruytter : "Nuclear Data for Basic and Applied Science - Santa Fe", I, 499 (1985)
12. C. Wagemans, A.J. Deruyter, R. Barthelemy, W. Mampe, P. Ageron and A. Michaudon : "Nuclear Data for Science and Technology - Mito", 131 (1988)
13. R.A. Schrack : "Nuclear Data for Science and Technology - Mito", 101 (1988)
14. A.V. Antonov, A.S. Voronin, A.I. Isakov, S.P. Kuznetsov, I.V. Meshkov, A.D. Perekrestenko, S.M. Solovev and A.V. Shelagin : Atommays Energiya, 61, 201 (1986)
15. J.A. Wartena, H. Weigmann and C. Burkholz : Report IEAA, Tecdoc 491, 123 (1987)
16. M.C. Moxon, J.B. Brisland and D.S. Bond : "Nuclear Data for Science and Technology - Mito", 75 (1988)
17. J.A. Wartena : Report NEANDC (E) 312, III (1989)
18. M.C. Moxon and J.A. Harvey : Private communication to the NEA Data Bank (1990)
19. H Weigmann (These proceedings)
20. G. de Saussure, L.C. Leal and R.B. Perez : "Physics of Reactors : Operation, design and computation. Marseille, April 23-27" Vol. I, III, 19 (1990)
21. H. Derrien, G. de Saussure and R.B. Perez : Nuclear Science and Engineering, 106, 434 (1990)
22. C.Wagemans, P.Schillebeeckx, A.J. Deruyter and R. Barthelemy : "Nuclear Data for Science and Technology - Mito", 91 (1988)
23. J.A. Harvey, N.W. Hill and F.G. Perey : "Nuclear Data for Science and Technology - Mito", 115 (1988)
24. R.R. Spencer, J.A. Harvey, N.W. Hill and L.W. Weston : Nuclear Science and Engineering, 96, 318 (1989)
25. E. Fort, J. Frehaut, H. Tellier and P. Long : Nuclear Science and Engineering, 99, 375 (1988)
26. H.J. Liou and R.E. Chrien : "Uranium and plutonium resonance parameters - Vienna", IAEA, INDC.129, 438 (1981)
27. H. Tellier : "Thermal Reactor Benchmark Calculations - Brookhaven" EPRI - NP 2855, 6-1 (1983)
28. C. Golinelli, J. Cray, M. Darouzet, L. Erradi, G. Granget, A. Santamarina and G. Gambiep : "Advances in Reactor Physics and Shielding - Sun Valley" 243 (1980)
29. J. Belle : "Uranium dioxide Properties and Nuclear Applications" US-AEC. 190 (1961)
30. B. Willis : Proceeding of the Royal Society, 274A, 134 (1963)
31. M.O. Marlowe and A.I. Kaznoff : International Symposium on Nuclear Fuels. American Ceramic Society. Colombus, 90 (1969)
32. I.J. Fritz : Journal of Applied Physics, 47, 4353 (1976)
33. R.M. Brugger and H. Aminfar : "Uranium and Plutonium resonance Parameters-Vienna" IAEA-INDC 129, 271 (1981)
34. G. Dolling, R.A. Cowley and A.D. Woods : Canadian Journal of Physics, 43, 1597 (1965)
35. C.R. Adkins, P.J. Persiani, R.N. Hwang and J.J. Kaganove : "Neutron cross-section Technology. Washington" CONF-660303, 134 (1966)
36. A. Butland : Annals of Nuclear Science and Engineering, 1, 575 (1974)
37. J. Rowlands : "Nuclear Data for the calculation of Thermal Reactor Reactivity coefficients - Vienna" IAEA-TECDOC 491, 7(1987)
38. H. Tellier, C. Van Der Gucht and J. Vanuxeem : "Advances in Mathematics, Computations and Reactor Physics - Pittsburgh" (To be publised) (1991)
39. M. Divadeenam and J.R. Sthen : Annals of Nuclear Energy, 11, 375 (1984)
40. E.J. Axton : Geel Report GE/PH/01/86 (1986)

NEW PERSPECTIVE IN COVARIANCE EVALUATION FOR NUCLEAR DATA

Y. Kanda
Department of Energy Conversion Engineering,
Kyushu University
Kasuga, Fukuoka 816, Japan

Abstract: Methods of nuclear data evaluation have been highly developed during the past decade, especially after introducing the concept of covariance. This makes it utmost important how to evaluate covariance matrices for nuclear data. It can be said that covariance evaluation is just the nuclear data evaluation, because the covariance matrix has quantitatively decisive function in current evaluation methods. The covariance primarily represents experimental uncertainties. However, correlation of individual uncertainties between different data must be taken into account and it can not be conducted without detailed physical considerations on experimental conditions. This procedure depends on the evaluator and the estimated covariance does also. The mathematical properties of the covariance have been intensively discussed. Their physical properties should be studied to apply it to the nuclear data evaluation, and then, in this report, are reviewed to give the base for further development of the covariance application.

(nuclear data evaluation, covariance, correlation, least squares method, sensitivity, design matrix, optical model parameter)

Introduction

Nuclear data are one of physical data bases which must be determined on the basis of measurements, which have always naturally uncertainties. In particular, nuclear data measurements seem to have large discrepancy among individual experiments. It can be understood as coming from systematic errors. This situation results in necessitating an artificial procedure to estimate the physical data base i.e. the evaluation in nuclear data field. In addition there is another inevitable reason that essential quantities as nuclear data base for application can not always be measured so that they should be deduced by the ways without experiments. Therefore, the uncertainties for the evaluated values must be estimated taking account of the errors in both the experiments and the evaluating procedures. In early stage, we had no suitable method to realize such comprehensive treatment of nuclear data while the error propagation law and simple weighted averaging were used to estimate the uncertainty in a locally limited region.

The nuclear data evaluation has been greatly sophisticated by introducing a concept of covariance [1]. Its ability makes it possible that the evaluated values and their uncertainties are assumed to have quantitative relation with existing measurements and that the uncertainty for shapes of evaluated excitation functions are given a numerical expression. Moreover, the introduction of the covariance matrices in evaluated data files accelerates the study of sensitivity and uncertainty analysis of reactor performance parameters in reactor physics [2]. Such potentialities of the covariance in the nuclear data field stimulate a large number of evaluators to study the characteristics of the covariance. The results of their efforts are summarized in a book by Smith [3], and you can also find a large number of literatures there.

The covariance is a concept developed in statistics. Therefore, utilization of the covariance involves estimation of the nuclear data by the statistical method. In early works, statistical procedures of experiments were also applied but they were based on a simple theory without covariance matrices. In order to adapt the statistical method to the nuclear data evaluation, understanding on both mathematical and physical attributes of the covariance should be separately and integrally deepened. The mathematical formalism has been developed to apply it for estimating the nuclear data and associated uncertainties. The covariance matrices in these formulae play an important part in the evaluation. The potentialities of the formulae can be displayed with association of the most suitable covariance formation. It is possible only by physical consideration for the covariances. However, so far mathematical properties of the covariance seem to be intensively discussed rather than physical ones. The latter have not been studied enough to estimate the right covariance at given conditions. It is easy when experiments are uniformly distributed in the interesting region. On the contrary it is puzzling when they are extremely limited to a specific region because it is doubtful whether the evaluated covariance from this condition is suitable in its application. A few manners have been proposed to evaluate covariances for nuclear data.

In this report, they are reviewed and compared, emphasizing the viewpoint mentioned above in order to develop the covariance evaluation manner fitting to the current nuclear data evaluation method.

Covariance Matrices

Mathematical Description

In the generalized least-squares method, the equations relating to the covariances discussed in this paper are given as

$$y \doteq \Phi\,(p-p_0) \tag{1}$$
$$p = p_0+(\Phi^tV^{-1}\Phi+M_0^{-1})^{-1}\Phi^tV^{-1}\{y-\Phi(p-p_0)\} \tag{2}$$
$$M = (\Phi^tV^{-1}\Phi+M_0^{-1})^{-1} \tag{3}$$

for the posterior parameter vector p and its covariance matrix M to be estimated in specific issues, respectively. The experimental data vector y can be approximately expressed as eq. (1) where Φ is called a design matrix or sensitivity matrix. The matrix V is a covariance matrix for the experimental data y. The vector p_0 and matrix M_0 are a prior vector for p and its covariance matrix for M, respectively. All the matrices are symmetric. The correlation matrix is defined as follows. The diagonal element u_{ii} of a covariance matrix U is called the variance which is physically equivalent to a square of the standard deviation at the data point i and the off-diagonal elements $u_{ij}(i \neq j)$ implies information of the uncertainties for the data i and j and of correlation between both the uncertainties. For convenience to compare distinct covariances and elements in an identical matrix each other, the correlation matrix C of the U is defined such that the diagonal element c_{ii} is 1.0 and the off-diagonal element $c_{ij}(i \neq j) = u_{ij}/(u_{ii}u_{jj})^{1/2}$.

Physical Implication

In common practice, the components of **y** are experimental cross section data which can be one-point or excitation-function data and absolute or relative ones. The parameter vector **p** dose not always consist of cross sections and may be other physical quantities which are properly defined depending on an evaluation model as well as Φ, as described in the later sections.

The objective of the nuclear data evaluation is to produce reliable physical constants on nuclear phenomena. The evaluated covariance matrix M is a mathematical expression for the representation of physical uncertainty of the evaluated nuclear data and implies the present-day achievement of the evaluation reliability.

A schematic correlation matrix is shown in Fig. 1 imaging the covariance of neutron-induced multi-reaction excitation functions for multi-target system. The partial correlation matrices C_{ii} and C_{ij} $(i \neq j)$ are the self-correlation one and mutual-correlation one between reactions i and j, respectively. To consider features of the covariances, extreme conditions are intentionally posed. If all the excitation curves are completely decided, all $C_{ij}=0$. Except this case, the diagonal elements of all C_{ii} are 1.0. The assumption that all the excitation-function shapes and their relative values are fixed leaves the results in which the diagonal elements of all the partial matrices $C_{ij}(i=j$ and $i \neq j)$ are 1.0 and the off-diagonal ones are 0.0. In this case, a normalization constant absolutely determines all the excitation curves. If the diagonal elements of all the partial matrices C_{ii} are 1.0 and the other elements are 0.0, all the distinct excitation-function shapes are fixed and they are normalized to the absolute values with the independently proper factors, respectively. In practical cases, the diagonal elements of C_{ii} are 1.0 and the others in C_{ii} are non-zero values allowing positive or negative sign: a characteristic trend of the absolute values of the non-diagonal elements in C_{ij} $(i \neq j)$ is nearly equal to 0.0, i.e. the correlation between the distinct reactions is very weak.

An example of the resultant correlation matrix for the optical model parameters is shown in Fig. 2: this has been estimated with the generalized least squares method in our laboratory [4]. More detailed description can be found later. The abbreviations V_0, r_0, a_0, W_s, r_s, and a_s and the real potential depth, radius, diffuseness, imaginary surface potential depth, radius, and diffuseness, respectively. The whole correlations are strong or moderate: this suggests that all the parameters must be simultaneously searched though only part of them are often done keeping the others constant in usual optical model parameter search.

Formation of Prior Covariances

Experimental Uncertainties

Experimenters always estimate the errors associated with their measurements and then attach a so-called total error to their objective result. They are generally interested in the experimental result itself and consider the error as something to guarantee the result so that they involuntarily lean toward estimating smaller error.

On the other hand, evaluators understand and expect that the error presented in the measurement report is computed from the whole uncertainties regarding all the experimental conditions and assures the reliability of the measurement. The error which ought to be a physical quantity is substantially charged such with the confused function by the distinct users of the word "error". If the values resulted by individual experimenters agree within the errors given by themselves there is no need to mention it at length. However, the discrepancies beyond them are most usual in the nuclear data field. The source of them is clearly the systematic errors which had not been identified by the experimenters. The evaluators must make efforts towards solving discrepancies by examining available experiments in order to find their sources unknown to the individual experimenters. This has been discussed in the nuclear data community over and over for a long time, but this is a fundamentally difficult problem and, naturally, any effective manners have not been established. One of the reasons is that there are experimental reports having no detailed documentations about the error sources and experimental conditions. It is, however, a difficult and nonproductive work because identification of the systematic errors is almost impossible owing to their latency, even if the thorough documentation is presented by the experimenters concerned.

On the basis of the reason discussed above, it is clear that the uncertainties given by experimenters must not be adopted as the elements of the matrix V without examining their validity in detail. The author is quite agreeable to Peelle's opinion [5]: "A particular discrepancy problem arises when a very small uncertainty is claimed for one experiment in a minimum document, while a thorough document asserts more modest claims for a competing experiment that appears to be just as accurate. Though opinion varies, this author believes that evaluators must be free to expand published uncertainties when documentation and common knowledge seem not to support claimed small uncertainties." Such uncertainties can be adopted as the elements of V. This is the weighted averaging of the available measurements taking account of the judgement by the evaluator.

To estimate the off-diagonal elements of V, the evaluator demands detailed information about the partial errors in the experiment concerned. The element is the summation of those products respective to the specific partial errors, which are the products of three factors of the partial errors at two data points and their correlation coefficient. A practical example can be found in Smith [3] and is partially quoted in Tables 1.

178

Prior Covariances V and Mo

The covariance matrices in the eqs. (1), (2) and (3) are V, Mo and M. The former two matrices of them are given by the evaluator and the last one is the objective covariance which is to be estimated by him and is arithmetically computed by using given V and Mo. Therefore, the most effort of the evaluator is required to construct both the covariances. In the evaluation emphasizing newly adopted measurements, V is most decisive for the result and should be reasonably prepared. On the other hand, Mo is an essential quantity when prior data po is considered to be most probable and the measurements taken into account in the evaluation are used to correct the prior data po. The relation between V and Mo is relative and competitive: the evaluators must judge what data are most and more correct and express quantitatively it through V and Mo. Such duties of the evaluators are unchanging and everlasting from the primitive nuclear data evaluations in which they intuitively selected the data which they believed most correct and drawn a curve with an eye-guide.

The prior covariance matrix Mo is representative of uncertainty for the prior parameter po and responsible for modification of po in processing with eqs. (2) and (3). An example is presented here to see the roles of Mo and po. An evaluated cross section curve and its covariance are expressed as po and Mo, respectively. New experimental data sets given as y and V are used to revise the previous evaluation po and Mo. Briefly speaking, this is a weighted averaging of the new data and the previous evaluation. The outcome of this procedure, p and M, is equal to the one which is evaluated with all the experimental information about the cross sections and their uncertainties of both the new data and the old data which were used in the previous evaluation. This is assured with sufficiency of estimation in the statistical theory. Muir presented a mathematical consideration [6].

Evaluation Designing and Posterior Covariances

The equations of the generalized least squares method presented in the previous section can be applied to various types of nuclear data evaluations by designing the "design matrix" Φ suitable to the combination of y and p. If an appropriate curve can be postulated it is fitted to available experimental data by adjusting parameters representative to the curve so as to reproduce experiments on average. In this sense, Φ is the "sensitivity matrix", which is the differential coefficient of the function expressing y with respect to p. Therefore, Φ depends on what model is contrived to parameterize the nuclear data. A few examples are presented here.

Average of Data

Average of existing experimental data has been the most popular evaluation method. It is easy to apply it to the data at an energy but not to the excitation-function data. The latter can be easily performed by the generalized least squares method. This was applied in the simultaneous evaluations of heavy nuclide reactions.

Simple average: In the case of the averaging of cross sections measured at the same energy, Φ is the unit matrix and V is the unit matrix for the simple averaging and is the diagonal matrix having weighting factors for the weighting averaging.

Curve fitting simultaneous evaluation: Main reactions for heavy nuclides were evaluated with simultaneous evaluations for the new versions of the major three evaluated files, i.e. ENDF/B-VI, JEF-2 and JENDL-3. Since their methodologies were already reported in detail, they are briefly summarized here though they are the early and typical applications of the generalized least squares method.

Poenitz [7] adopted as the fitting curves step functions which were the first terms in the Taylor-expansion of the excitation functions concerned. His group [8] applied the generalized least squares method, R-matrix evaluations and a procedure for combining the results of these evaluations to simultaneously evaluate the ENDF/B-VI standards.

The author's group [9] used second-order B-spline functions as the fitting curves in the simultaneous evaluation method developed in their laboratory [9, 10]. The fission cross sections of ^{235}U, ^{238}U, ^{239}Pu, ^{240}Pu, ^{241}Pu, and the capture cross section of ^{197}Au and ^{238}U were evaluated for JENDL-3 in the energy range from 50 keV to 20 MeV base on their absolute measurements and the seven kinds of relative ones for combinations of the nuclides. The prior experimental covariances were estimated from the partial errors and assumed specific correlation factors. The correlation only in neutron energies was taken into account. The resultant cross sections agree very well with those of ENDF/B-VI.

Fitting of Nuclear Model Formulas

Application of model formulas is the best if they can fairly well reproduce the experimental data. At present we can not always have such formulas so that we should come up with various ideas. In fitting a nuclear-reaction-model formula to an experimental excitation function, the matrix Φ is a partial derivative of the model formula with respect to the model parameters.

The optical-model-parameter-search method: developed by the author's group [4, 11] by the generalized least squares method can produce the covariance matrices for angular distribution data and optical model parameters. The data used in the fitting in the experiment of $^{56}Fe(n, n)$ at 7.96 MeV neutron energy [12]. In Figs. 3,4, and 5, the resultant optical model parameter correlation matrix, angular distribution compare with experimental data used in fitting, and angular distribution correlation matrix, respectively.

Hauser-Feshbach model: The author and his group [13] have studied the method to estimate the physical parameters in nuclear reaction model formulae and the developed method has been applied to estimate the covariance matrices with the nuclear reaction model calculation [14]. The Hauser-Feshbach model is used and the parameters to be estimated are the level density parameters for the relevant nuclides and the main optical model parameters for the neutron, proton and α-particle. In the latest studies about Fe [15] and Co and Ni [4], the estimation of the proper neutron optical parameters for the regarded nuclides is performed to decrease the ambiguity depending on the optical parameters. They are estimated for the neutron by the simultaneous evaluation of the total cross section and precise elastic angular distribution data. Naturally, such data are not always available for all the nuclides concerned because of absence of suitable data base. The parameters of the nuclides nearby are used instead of the proper ones. The optical model parameters for the proton and α-particles cannot be estimated by the same ways because of lack of appropriate data base so that there is nothing else other than the global parameters.

As an example, the results for ^{56}Fe are shown in Fig. 7 [15] . Since the Hauser-Feshbach model is used in this work, the calculations of the individual reactions are competitive with each other so that strong correlation between the reactions resulted.

The correlation matrix in Fig. 7 resulted from the uncertainties of the experiments, model formulae and parameters. An object of the latest studies estimating the proper optical model parameters for the specific nuclide is to decrease their uncertainties as far as possible.

Legendre polynomial fitting: To illustrate the effect of different fitting functions, the optical model fitting in the same experiment used above was analyzed with Legendre polynomial functions by the generalized least squares method. The obtained angular distribution correlation matrix is shown in Fig. 6. In Fig. 4, both the fitted curves are compared with the experiment. The resultant correlation matrices of Figs. 5 and 6 and their difference are beyond physical comprehension. Our attention has to be paid to the correlation matrix for the coefficients of the Legendre polynomials in Fig. 6. They show extremely strong correlation at all the elements, and then strong long-range correlation. If such strong-correlating functions are adopted to approximate excitation functions, the evaluated correlations are too strong to apply in practice. The step function used by Poenitz and second-order B-spline function used by the author have no overall correlation and very short-range correlation, respectively. This is a reason that the simultaneous evaluations are successful. Smith [3, 16] discussed the long-range correlation particularly from a view point of applying evaluated nuclear data and showed that there is weak long-range correlation. Evaluators must postulate an appropriate function taking account of such feature of correlation matrix.

Integral Data

The simultaneous evaluation of differential cross sections with integral data developed by the author's group [17, 18] can be usefully utilized to estimate the covariance matrices for the nuclear data. For this object, appropriate integral data are the average cross sections measured in the neutron source whose spectrum is well known, like the fission neutron source and accelerated-deuteron-thick-target source. The design matrix Φ is formed as expressing the sensitivities of the differential cross sections to the integral data. In this case, an integral data is equivalent to a large number of differential data points. At present, major ones are limited to the data with the neutron-induced ^{235}U fission and ^{252}Cf spontaneous fission sources. It is, however, expected to be displayed potentially also for the accelerator neutron sources to cover the energy regions lacking in the monochromatic neutron data.

Implication of Posterior Covariance

There is an interesting demonstration presented by Peelle [5] and quoted here in Fig. 8, where the upper part is a primitive but essential covariance matrix derived from the definition and the lower part is the covariance estimated by a least squares method, for a common data base, respectively. They were estimated by the SUR program [19] for the upper and the SURP program [20] for the lower, using the data sets expressed as the ratio of ^{239}Pu to ^{235}U fission cross sections. Twenty-six group uncertainty matrices were based on the scatter of 22 weighted data sets and then collapsed using flat weighting to the five-group structure indicated.

The SUR program provides uncertainty matrices for smooth cross sections by deriving matrix components from the scatter of sample data sets around the pre-existing evaluation using a formula obtained from the definition of the covariance matrix. An arbitrary energy grid with energies E_j is established, and cross section values σ_{jn} at each of these energies are obtained by interpolation for each of the input data sets. The elements of the uncertainty matrix to be associated with average cross section are obtained from the formula

$$V_{ij} = (1/W_{ij})\sum_n W_n(\sigma_{in}-\sigma_i)(\sigma_{jn}-\sigma_j). \quad (4)$$

In this formula, σ_i is the evaluated cross section at E_i, W_n is the relative weight assigned by the uncertainty evaluator to the n-th data set, and W_{ij} is the appropriate combined weight. If each experiment yielded an observed value at each data point, $W_{ij}=\sum W_n$. The SURP program is an extension of SUR and retains the fundamental assumption of SUR that independent experimental data sets are sampled from an underlying population except for a weighting factor, but tries to correct the misidentification of an experimental uncertainty matrix as the uncertainty matrix of the evaluation. The idea of SURP is to perform the least-squares average of the input data sets, using eq. (4) to give the input uncertainty matrix V_{ij} for each experiment.

The remarkable difference between both the results is found out in the values of the standard deviation while the correlation matrices are not so different. The least squares method usually results in the very small standard deviations. This effect can be commonly seen at the covariance matrices obtained with the (generalized) least squares methods because of the synergism of two reasons; the function of the methods are substantially, simply mentioning, averaging of the data concerned and they necessitated even distribution of a large number of the data points in order to be effectively applied. There is a question whether such a extremely small deviation can be acceptable. The arithmetic outcome from the statistical equations is based on the assumption that the experimental uncertainties are wholly statistical while actually existing data are possibly biased and scattered with systematic errors. Therefore, the Eq. (4)-type calculation for covariance matrix estimation is not to be unconditionally superseded.

Smith and Guenther [21] estimated the covariance matrix of the neutron optical parameters for the elements Z=39-51 using a straightforward statistical analysis of the collection of local parameter sets, which are deduced by parameter fitting simultaneously to the total cross section and angular distribution of elastically scattered neutrons for the specific elements. Regional optical model parameter formulae can be obtained by fitting appropriate functions to the local parameter sets. The deviations of the individual local parameters from the regional parameters calculated with the formulae obtained fitting to the local parameters are processed to construct the covariance matrix. These procedures are similar to the estimation with eq. (4). The result is shown in Fig. 9.

The consistency between this result in Fig. 9 and the covariance shown in Fig. 2 and 3 is very surprising. The latter is estimated from the data at an energy point. This intensively shows that the uncertainties of the parameters are very large while the optical model formula is appropriate to represent the angular distribution of the elastically scattered neutron.

Conclusion

The roles of the covariance matrices in the nuclear data evaluation have been reviewed, emphasizing their physical implication. Fundamentally, the covariances are the uncertainties depending mainly on the experimental errors. However, the uncertainties of the model formulae and the model parameters can be discussed from the estimated covariance through the generalized least squares method by appropriate design and parameterization of the evaluation. Although physical understanding of the posterior covariances is often difficult it is expected that potentialities of the covariances make it possible to develop new evaluation method.

The author is indebted to Dr. T. Kawano, and Mr. H. Tanaka and Mr. K. Kamitsubo for their contributions to the study.

References

1. F.G. Perey, G. de Saussure and R.B. Perez, in Proc. Int. Conf. on Advanced Reactors: Physics, Design and Economics, Sept. 1974, Atlanta, USA, (Edited by J.M. Kallfelz and R.A. Karam), p.578, Pergamon Press (1975)
2. Sensitivity and Uncertainty Analysis of Reactor Performance Parameters, Advances in Nuclear Science and Technology Vol. 14 (Edited by J. Lewins and M. Becker), Plenum Press, New york and London (1982)
3. D.L. Smith, Probability, Statistics and Data Uncertainties in Nuclear Science and Technology, To be published as An OECD Nuclear Energy Agency Nuclear Data Committee Series, Vol. 4.
4. T. Kawano, H. Tanaka, K. Kamitsubo and Y. Kanda, Estimation of Nuclear Reaction Model Parameters for ^{59}Co, ^{58}Ni, and ^{60}Ni, May 1991, Juelich, Germany (This Conference)
5. R.W. Peelle, Ref. 2, p.11 (1982)
6. D.W. Muir: Nucl. Sci. Eng. 101, 88 (1989)
7. W.P. Poenitz, in Proc. Conf. on Nuclear Data Evaluation Methods and Procedures, Sept. 1980, Brookhaven, USA, BNL-NCS-51363 (Edited by B.A. Magurno and S. Pearlstein) Vol. 1, p.249, Brookhaven National Laboratory (1981)
8. A.D. Carlson, W.P. Poenitz, G.M. Hale and P.W. Peelle, in Proc. Int. Conf. on Nuclear Data for Basic and Applied Science, May 1985, Santa Fe, USA (Edited by P.G. Young, R.E. Brown, G.F. Auchampaugh, P.W. Lisowski and L.Stewart) Vol. 2, p.1429, Gordon and Breach, New York (1986)
9. Y. Kanda, Y. Uenohara, T. Murata, M. Kawai, H. Matsunobu, T.Nakagawa, Y. Kikuchi and Y. Nakajima, in Ref. 8, Vol. 2, p.1567 (1986)
10. Y. Uenohara, M. Tsukamoto and Y. Kanda: J. Nucl. Sci. Tech., 20, 787 (1983)
11. T. Kawano and Y. Kanda: J. Nucl. Sci. Tech., 28, 156 (1991)
12. S.M,El-Kadi, C.E. Nelson, F.O. Purser and R.L. Walter: Nucl. Phys., A390, 509 (1982)
13. Y. Uenohara, H. Tsuji and Y. Kanda, in Proc. Int. Conf. on Nuclear Data for Science an Technology, May/June 1988, Mito, Japan (Edited by S. Igarasi), p.481, Saikon, Tokyo (1988)
14. Y.Kanda and Y. Uenohara, in Ref. 13, p.1041 (1988)
15. K. Kamitsubo, T. Kawano, H. Tanaka and Y. Kanda (Private Communication, 1991)
16. D.L. Smith, Some Comments on the Effects of Long-rang Correlations in Covariance Matrices for Nuclear Data, ANL/NDM-99, Argonne National Laboratory (1987)
17. Y. Kanda and Y. Uenohara, in Proc. Int. Conf. on Neutron Physics, Sept. 1987, Kiev, USSR (1987)
18. Y. Kanda, Y. Uenohara, D.L. Smith and J.W. Meadows, in Ref. 13, p.541 (1988)
19. F.C. Difilippo, SUR, A Program to Generation Error Covariance Files, ORNL/TM- 5223, Oak Ridge National Laboratory, March (1976)
20. C.R. Weisbin and R.W. Peelle, in Proc. Int. Specialists Symposium on Neutron Standards and Application, March 28-31,1977, NBS-493, Gaithersburg, USA, (Edited by C.D. Bowman, A.D. Carlson, H.O. Liskien and L. Stewart), p.269, National Bureau of Standards (1977)
21. D.L. Smith and P.T. Guenther, Covariance for Neutron Cross sections Calculated Using a Regional Model Based on Local-Model Fits to Experimental Data, ANL/NDM-81, Argonne National Laboratory (1988)

Table 1 An example of experimental covariance matrix and their data base. Error sources for a typical experiment[3].

Source of Error	Magnitude(%)[a]
E1 Event statistics	0.1-4.6
E2 Background	N[b]-0.1
E3 Event determination procedures	1.7-2.0
E4 Event determination calibration standards	0.6-2.1
E5 Sample assay	1.4-1.7
E6 Activity half-life	N -0.5
E7 Activity decay factors	<1.0[c]
E8 Isotopic abundance	<0.5
E9 Geometry	0.3-1.0
E10 Neutron source	N -0.2
E11 Neutron fluence	4.0
E12 Neutron absorption	0.1-0.5
E13 Neutron scattering	0.1-1.5

a Range of errors encountered for the measurements
b N = Negligible(<0.1%)
c <1.0 signifies that no error for this categtory is as large as 1%.

Total errors and the correlation matrix estimated from the data base presented above.

Reaction[a]	Total Error(%)	Correlation Matrix[b] 1	2	3	4	5	6	7	8
1 ^{7}Li(n,n't)^{4}He	4.6	100							
2 ^{27}Al(n,p)^{27}Mg	5.3	76	100						
3 ^{27}Al(n,α)^{24}Na	5.5	74	64	100					
4 Si(n,X)^{28}Al	4.8	85	74	71	100				
5 Ti(n,X)^{46}Sc	5.0	83	72	70	81	100			
6 Ti(n,X)^{47}Sc	6.2	65	56	54	63	61	100		
7 Ti(n,X)^{48}Sc	5.8	70	60	59	67	66	51	100	
8 ^{51}V(n,p)^{51}Ti	5.1	79	68	67	77	75	58	63	100

a Cross section nomalized to ^{235}U neutron fussion.
b Correlation parameters multiplied by 100.

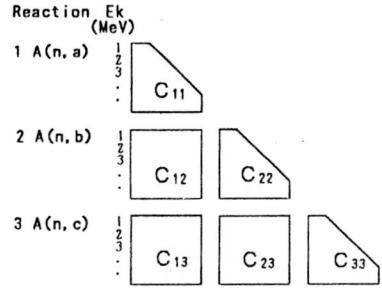

Fig. 1　Correlation matrix.
　　　　C_{ij} is the partial correlation matrix.

^{60}Ni　En=5.05 MeV
　　　error(%)
Vo	1.44	100				
Ro	1.10	-94	100			
Ao	3.43	68	-84	100		
Ws	4.11	-25	26	-21	100	
Rs	4.59	-32	20	16	-20	100
As	5.51	19	-18	9	-66	-39	100

Fig.2　An example of the correlation matrix for
　　　　the optical model parameters, ^{60}Ni(n,n),
　　　　En=5.05 MeV.

^{56}Fe　En=7.69 MeV
　　　error(%)
Vo	2.57	100				
Ro	1.19	-98	100			
Ao	1.37	28	-43	100		
Ws	2.74	-14	14	28	100	
Rs	2.03	-18	1	64	21	100
As	3.14	13	-5	-58	-82	-71	100

Fig.3　Correlation matrix of the optical model
　　　　parameters estimated from ^{56}Fe(n,n)
　　　　angular distribution data at 7.96
　　　　MeV[11].

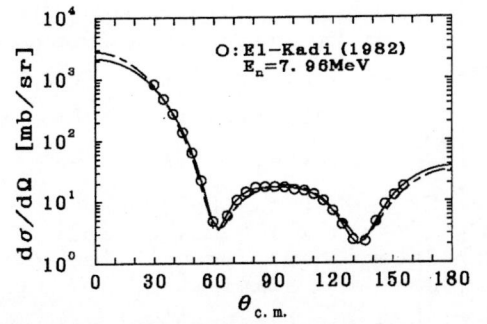

Fig.4　Comparison of the fitted curves with
　　　　^{56}Fe(n,n) angular distribution data at
　　　　7.96 MeV [11]. The dot-dashed line is
　　　　estimated by the optical model using the
　　　　parameters deduced by the generalized
　　　　least squares. The solid line is calcu-
　　　　lated with Legendre polynomials
　　　　(l_{max}=8).

Θ	dσ/dΩ	error(%)	correlation(%)					
0	2834.	1.24	100					
30	766.8	0.55	76	100				
60	3.370	4.60	60	95	100			
90	17.69	1.87	-10	44	39	100		
120	7.435	1.99	-25	-20	-13	2	100	
150	9.738	2.03	56	42	29	7	17	100
180	31.89	1.90	12	41	46	39	18	50	100

Fig.5　Correlation matrix for the optical model
　　　　curve shows in Fig.4 (dot-dashed line).

Θ	dσ/dΩ	error(%)	correlation(%)					
0	2177.	3.81	100					
30	714.3	2.73	96	100				
60	4.004	8.28	-6	-12	100			
90	17.09	2.58	-15	-14	-11	100		
120	6.869	1.96	-32	-21	13	-33	100	
150	12.20	1.61	-12	-9	31	5	27	100
180	36.73	11.9	44	30	12	14	-31	49	100

Fig.6　Correlation matrix for Legendre polynom-
　　　　inal curve shown in Fig.4 (solid line).

Group Boundary
En(MeV)　S.D.　Correlation Matrix

a. Results from SUR
0.04	3.0	100				
0.18	2.4	53	100			
0.50	3.2	-2	43	100		
1.35	2.5	14	9	53	100	
3.68	2.1	3	-28	-36	-10	100
20.0						

b. Results from SURP
0.04	0.6	100				
0.18	0.4	39	100			
0.50	0.6	-1	27	100		
1.35	0.4	4	7	31	100	
3.68	0.6	1	-2	-1	-4	100
20.0						

Fig.8　Examples of fission-ratio correlation
　　　　matrices estimated by SUR and SURP(see
　　　　text). S.D.: Standard Deviation.

Parameter	Error	Correlation Matrix					
		P1	P2	P3	P4	P5	P6
P1 = Vo	1.9%	1					
P2 = rv	1.1%	-0.88	1				
P3 = av	4.5%	0.17	-0.41	1			
P4 = W	11.5%	-0.47	0.50	0.20	1		
P5 = rw	3.4%	0.08	-0.14	0.64	0.53	1	
P6 = aw	15.0%	0.23	-0.27	-0.39	-0.89	-0.81	1

Fig.9　Correlation matrix for the optical model
　　　　parameters: This is formed from the
　　　　deviations of local parameters to re-
　　　　gional parameters.(see text)

^{56}Fe(n,p)
[MeV]
5.0	100									
6.0	40	100								
8.0	8	94	100							
10.0	11	82	90	100						
12.0	18	34	40	74	100					
14.0	14	16	16	35	73	100				
15.0	9	33	32	28	39	87	100			
16.0	4	40	37	20	13	66	95	100		
18.0	4	37	33	6	-12	46	84	97	100	
20.0	1	24	18	-10	-24	39	78	92	97	100

^{56}Fe(n,α)
5.0	99	40	8	12	18	14	9	4	1	1	100							
6.0	35	99	95	81	30	13	33	42	40	27	35	100						
8.0	3	93	99	84	29	8	29	39	37	23	3	95	100					
10.0	2	68	78	89	55	-6	-18	-23	-31	-45	3	67	74	100				
12.0	3	16	26	55	55	-16	-53	-69	-80	-87	3	13	18	79	100			
14.0	3	5	14	45	53	-17	-57	-74	-86	-92	3	2	5	71	99	100		
15.0	4	3	11	43	52	-17	-58	-76	-87	-92	3	-1	4	69	99	99	100	
16.0	4	1	10	42	52	-17	-58	-76	-88	-93	3	-2	2	68	99	99	99	100
18.0	4	0	8	41	51	-18	-59	-78	-88	-93	3	-4	1	67	98	99	99	99	100
20.0	4	0	8	40	51	-18	-59	-77	-88	-93	4	-4	0	67	98	99	99	99	99	100

Fig.7 Correlation matrix (%) for ^{56}Fe(n,p) and (n,α) reactions.

NUCLEAR DATA PROCESSING, ANALYSIS, TRANSFORMATION AND STORAGE WITH PADE-APPROXIMANTS

S.A.Badikov, E.V.Gay, M.A.Guseynov, N.S.Rabotnov

Institute of Physics and Power Engineering, Obninsk,USSR

Abstract: A method is described to generate rational approximants of high order with applications to neutron data handling. The problems consi-dered are: the approximations of neutron cross-sections in resonance regi-on producing the parameters   for Adler-Adler type formulae; calculations of resulting rational approximants' errors given in analytical form allo-wing to compute the error at   any energy point inside the interval of approximation; calculations of the correlation coefficient of error values in two arbitrary points provided that experimental errors are independent and normaly distributed; a method of simultaneous generation of a few rational approximants with identical set of poles; functionals other than LSM; two-dimensional approximation.

(Nuclear data, Pade-approximation, resonance analyses, covariational mat-rix, simultaneous fitting of several cross-sections, two-dimensional rational fitting)

Introduction

The main way to extract and compress useful information from experimental cur-ves is the analytical approximation. The polynomial approximation is the simplest, best developed and most popular one.

But the physics of a problem often dictates the necessity to use more comp-licated functions with special analytical features. Rational functions (Pade-appro-ximants) are extensively used in mathema-tical physics (see [1] for references) and offer wide opportunities especially in the case of resonance processes parti-cularly nuclear reactions. But their ap-plication was hindered so far by two obstacles. First: rational approximants unlike polynomial ones lead to nonlinear systems of equations in the least squares method (LSM). Second is a special form of approximant's instability – real pole-zero couples. We have proposed a method to circumvent both difficulties, outlined below briefly. The method is based on the recursive calculation of many approxi-mants differing by the choice of interpo-lation knots with their subsequent stati-stical optimization by discrete sorting. The best fitting approximants are anali-zed and noise poles removed.

A rational function $f(z) = P(z)/Q(z)$ may be parametrized:

1. By the coefficients p^n and q^m of the polynomials in the numerator and denomi-nator

$$f(z) = \sum_{n=0}^{N} p_N^n \, z^n / \sum_{m=0}^{M} q_M^m \, z^m. \qquad (1)$$

2. By the values of the polinomials' zeros

$$F(z) = P_N^N \prod_{n=1}^{N} (z - z_n) / Q_M^M \prod_{m=1}^{M} (z - z_m) \qquad (2)$$

3. By the parameters of polar expansion

$$f(z) = A_{N-M}(z) + \sum_{\beta} a_\beta / (z - z_\beta) =$$

$$= A_{N-M}(z) + \sum_{i=1}^{l_1} \frac{a_i}{z - p_i} + \sum_{k=1}^{l_2} \frac{\alpha_k (z - \varepsilon_k) + \beta_k}{(z - \varepsilon_k)^2 + \gamma_k^2}, \quad (3)$$

4. By the very values of the function for $N+M+1$ argument values

$$f(z) = f(z; f(z_1), f(z_2), \ldots f(z_{N+M+1}) \qquad (4)$$

Unlike polynomial case there is no simple Lagrange-type formula for this "suppor-ting ordinates" parametrization.

Pade-2 approximant for a function $f(z)$ is defined as a rational function $f^{[N,M]}(z)$ coinciding with $f(z)$ in $L=N+M+1$ points:

$$f^{[N,M]}(z_i) = f(z_i), \quad i=1,2,\ldots,L \qquad (5)$$

Eqs (5) result in the linear sistem for coefficients which may be solved either with determinants or with the recurrent formulae. The simplest of the latter is

$$f^L(\gamma; z) = \frac{P_{(L-1)}(z) + \gamma(z - z_{L-1}) P_{L-2}(z)}{Q_{(L-1)}(z) + \gamma(z - z_{L-1}) Q_{L-2}(z)} \qquad (6)$$

Annotated references for corresponding algorithms may be found in [2].

Discrete optimization

An attempt to optimize a rational ap-proximant by standart least squares me-thods results in very complicated set of equations within any parametrization. At the same time a rational interpolant with L supporting points may be calculated by recursive relations (6) without inversing high order matrices. We have prorosed and realised in computer codes a method of rational approximation of single variable functions based on discrete optimisation (sorting) and described in [3].Its essen-tial stages are: for $N_{ex} \gg L$ (N_{ex} is the number of experimental points) we choose an initial set of L supporting points

z_k, F_k; $1 \leq k \leq N_{ex}$ and using recursive algorithms interpolate them with a rational function. Then the values $f^L(z_i)$ for $i=1, 2, \ldots N_{ex}$ and the functional to be minimized, usually

$$S = \sum_{i=1}^{N_{ex}} [f^L(z_i) - F_i]^2 / 2\sigma_i^2 , \qquad (7)$$

are computed defining a starting interpolant. Then iteration process begins. On its l-th stage one of the supporting points is replaced by another experimental point, new approximant and its functional's value S^l are calculated.
If $S^l < S_{min}^{l-1} = min\{S^1, S^2, \ldots S^{l-1}\}$, then l-th set of supporting points is stored as current optimum and so on.
One of the advantages of discrete optimization as compared with continuous LSM is the possibility to use a variety of functionals. Theoretical estimates show that mean quadratic deviation of the approximant found by sorting from LSM solution for $N_{ex} \gg L$ is $\cong \sqrt{N_{ex}/L}$ times less than deviation of the last from exact curve, and the approximant is statistically equivalent to LSM solution. Note that fitting of experimental data with Pade-2 approximant (without sorting!) followed by statistical optimization with supporting ordinates for variables is known as Pade-3 [4,5].

Noise doublets

Let's increase the rank L in approximating a rational function $P_{N_0}(z)/Q_{M_0}(z)$.
Coming to $N=N_0$, $M=M_0$, we can reproduce the function exactly with any L_0 interpolation knots. Increasing L we must nevertheless get $f^L(z)=P_{No}(z)/Q_{Mo}(z)$, i.e. an approximant of incomplete rank. With finite accuracy overranking L by one leads to a noise singularity in remote point, by two - results in the approximant $(Az+B)P_{N_0}(z)/(Az+B')Q_{M_0}(z)$, where A and A', B and B' are close pairs. That's called "noise doublet":zero at $z=-B/A$ and a pole at $z=-B'/A'$. In polar expansion (3) the pole appears as a term $a/(z-b)$ where a is small and $b=-B'/A'$. Cancellation of these almost identical binoms in the numerator and denominator corresponds to omitting the term in polar expansion. So in spite of singular instability its local character (the term $a/(z-b)$ differs from zero considerably for z close to b only) allows to liquidate it relatively simply. These arguments are valid also when a noise poles result from random errors of input data.The poles appear mostly after overranking and indicate together

with the statistical criteria that analytical information in the approximated set is exhausted.
The method was tested and applied in many practical problems. The most significant examples are:
1. Convertion to analytical form the evaluated neutron cross-section data (BOSPOR lirbrary). 144 curves were approximated with resulting 20 -fold reduction of the numerical information to store. Fig 1. shows relatively complex case of approximation. The statistical properties are discussed in the next section.
2. Complete neutron data files of the isotopes $^{14,15}N$ and ^{19}F were prepared for BROND-library with the extensive use of Pade-approximation.

Information matrix of approximant

Supporting ordinates as parameters allow to simplify and clarify the statistical analysis of resulting approximant. For the covariational matrix of supporting ordinates to be diagonal the following equations are necessary and sufficient

$$A_{\mu\nu} = \sum_{i=1}^{N_{ex}} \frac{\partial f(z_i)}{\partial f_\mu} \times \frac{\partial f(z_i)}{\partial f_\nu} / \sigma_i^2 = \lambda_\mu \delta\mu\nu , \qquad (8)$$

Then supporting ordinates are statistically independent and their dispersions are $\overline{\Delta f_\nu^2} = 1/\lambda_\nu$. The derivatives in (8) are

$$\frac{\partial f(z)}{\partial f_\mu} = \frac{Q_M^2(z_\mu)}{Q_M^2(z)} \prod_{\nu \neq \mu} \frac{(z-z_\nu)}{(z_\mu - z_\nu)} , \qquad (9)$$

and (8) is equivalent to orthogonality of the polinomials in the numerator of (9) with the weight $[\sigma_i^2 Q_M^4(z_i)]^{-1}$ on the same set of arguments $z_1, z_2, \ldots z_{N_{ex}}$.
The method to construct the needed set of orthogonal polinomials is well known. If the covariational matrix is diagonal i.e. $\overline{\Delta f_\mu \Delta f_\nu} = \overline{\Delta f_\nu^2}\delta_{\mu\nu}$ then assuming the dispersion $\Delta^2(z)$ small we get the mean quadratic error of the approximant's value in an arbitrary point

$$\overline{\Delta^2(z)} = \sum_\nu [\partial f^L(z)/\partial f_\nu]^2 \overline{\Delta f_\nu^2} , \qquad (10)$$

and for a correlation coefficient of two values $f^L(z_1)$ и $f^L(z_2)$ in two arbitrary points z_1, z_2

$$\rho(z_1, z_2) = [\overline{\Delta^2(z_1)}\,\overline{\Delta^2(z_2)}]^{-1/2} \times$$

$$\sum_{\mu=1}^{L} \overline{\Delta f_\mu^2} \frac{\partial f^L(z_1)}{\partial f_\mu} \times \frac{\partial f^L(z_2)}{\partial f_\mu} , \qquad (11)$$

where $\overline{\Delta^2(z)}$ is defined by (10).

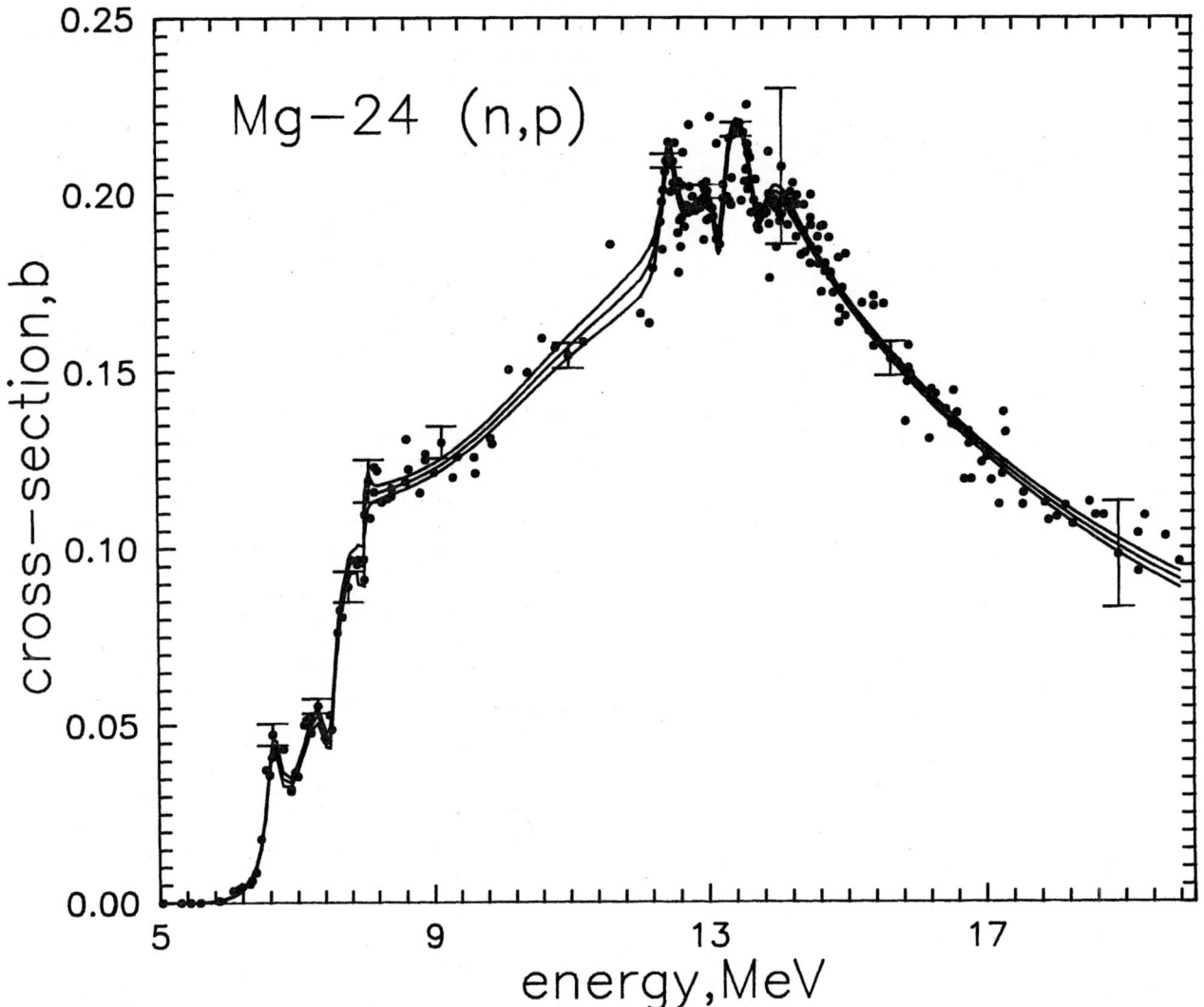

Fig.1. The results of Pade-approximation of a set of experimental data on the energy dependence of ^{24}Mg(n,p)-reaction cross-section. The data were taken from some 20 papers. The points are unspecified and unreferenced here because the figure is purely illustrative. Error-bars are indicated for a few points only but all of them were taken into account both in computing the approximant (central solid line) and in estimating "resulting error corridor" (two side lines) using Eqs.(9-10).All experimental errors were assumed independent and normally distributed. The number of points N_{ex} = 269, the number of parameters L = 37.

Pade-approximation with nondiagonal covariational matrix.

There exists a well known anomaly in the behavior of reduced neutron widths of ^{235}U resonances - first levels are about ten times weaker than the rest. We tried to determine short range energy dependence of $\bar{\Gamma}_n^o$ if any by Pade-approximating the cumulative reduced width as a function of resonance' number N

$$G(N)=\sum_{i=1}^{N} 2g_i\Gamma_{ni}^o=2g\overline{\Gamma_n^o}\sum_{i=1}^{N} x_i=2g\overline{\Gamma_n^o} X(N) \quad (12)$$

x_i are supposed to be x^2-distributed with one degree of freedom. Then covariational matrix is easily calculated

$$R_{mn} \overline{-(X(m)-m)(X(n)-n)} = \min (m,n) , \quad (13)$$

and it's inverse, the informational matrix with following non-zero elements

$$A_{nn} = 2 - \delta_{nN}; \quad A_{n\ n\pm1} = -I \quad (14)$$

Then the functional to be minimized is

$$S = \Delta_1^2 + \sum_{i=2}^{N} (\Delta_i - \Delta_{i-1})^2 , \quad (15)$$

where Δ_i are the differences between values to be approximated and the approximant. By calculating the finite differences of the approximant we get "smoothed" dependence $2g\Gamma_n^o(N)$. The results for ^{235}U resonances with spin J=3 are shown in Fig.2. We are not discussing now possible physical meaning of the results but they seem to be a good demonstration of the analytical potential of Pade-approximation.

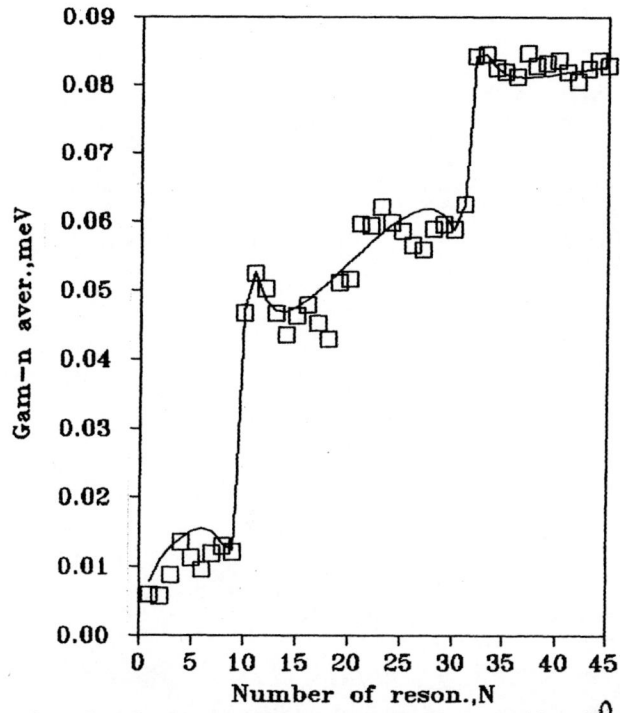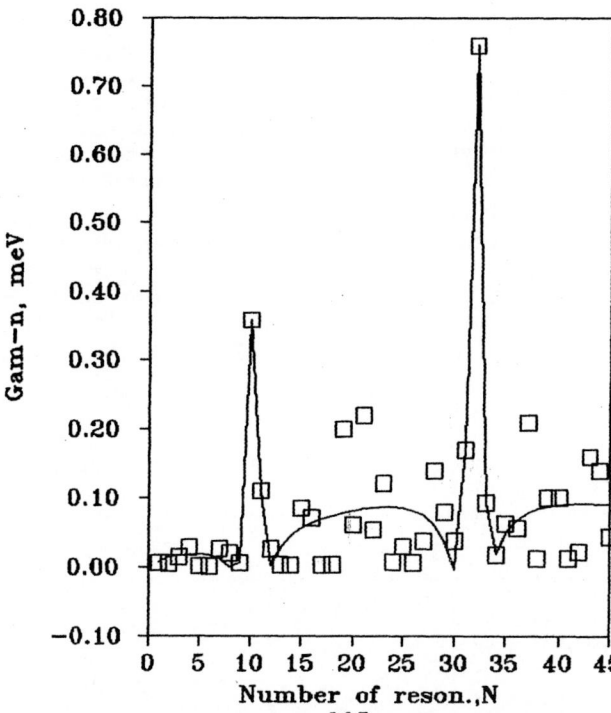

Fig.2. Left: reduced neutron widths $\Gamma_n^0(N)$ of the resonances of ^{235}U with spin J=3 averaged over first N levels as a function of N (squares) and the Pade-approximant (line). Right: the same for differential curve $\Gamma_n^0(N)$.

Simultaneous fitting of a few curves

We consider a problem of cross-sections fitting for different nuclear reactions going through the same compound nucleus. For the mininmization we choose the standart LSM functional

$$S = \sum_i S_i = \sum_i \sum_l (\sigma_{i,exp}(E_l) - \sigma_i^{(L)}(E_l, \vec{p})) w_i^l, \quad (16)$$

where w_i^l – is the statistical weight of $\sigma_i(E)$ at the energy $E = E_l$ determined by the experimental error. We use the algorithm of interpolatig a few rational curves with the same denominator optimizing the set of supporting points by sorting.

The analysis of the numerators and denominators allows to select the solutions satisfying physical conditions. If the total cross-section is fitted together with all the partial ones, the approximants' numerators are taken as

$$P^{(k)}(E) = P^{(t)}(E) - \sum_{i \neq k} P^{(i)}(E) \text{ or}$$

$$P^{(t)}(E) = \sum_i P^{(i)}(E)$$

Let's consider multyband (subgroup) parameters generation. Our way to calculate them is simultaneous approximation of some cross-sections' functionals $f(\sigma_o)$, $f_x(\sigma_o)$, x = el,f,c as functions of "background cross-section" σ_o by rational functions with identical denominators

$$f(\sigma_o) \equiv \langle \frac{1}{\sigma + \sigma_o} \rangle \cong \sum_{k=1}^{M} \frac{a_k}{\sigma_{tot,k} + \sigma_o}$$

$$f_x(\sigma_o) \equiv \langle \frac{\sigma_x}{\sigma + \sigma_o} \rangle \cong \sum_{k=1}^{M} \frac{a \sigma_{x,k}}{\sigma_{tot,k} + \sigma_o} \quad (17)$$

$$x = el,f,c$$

Here σ_{el}, σ_f, σ_c are elastic, fission and capture cross-sections respectively, σ – total cross-section, σ_o – background cross-section, $a_k, \sigma_{xk}, \sigma_{tk}$ multyband parameters, M – the number of subgroups, brackets $\langle...\rangle$ mean averaging over the group interval with standart neutron spectrum.

Table 1 contains the calculated multyband parameters compared with the results of the same calculations by momentum method for ^{239}Pu ($\Delta E = 40.84$-43.54 eV), ^{238}U ($\Delta E = 202.1$-215.5 eV), natural iron ($\Delta E = 1136.5$-1211.5 eV). The functions $f(\sigma_o)$, $f_x(\sigma_o)$, x=el,f,c calculated for 23 values of σ_o in $[0,10^6 b]$ interval with BROND resonance parameters were used as input information. In the Table Δ_x is maximum relative deviation (%) of the approximant of $\Sigma_{tot}(\sigma_o)$ and $\Sigma_x(\sigma_o)$ - self-shielded total and partial cross-sections

$$\Sigma_{tot}(\sigma_o) \equiv \langle \frac{\sigma}{(\sigma + \sigma_o)^2} \rangle / \langle \frac{1}{(\sigma + \sigma_o)^2} \rangle$$

$$\Sigma_x(\sigma_o) \equiv f_x(\sigma_o)/f(\sigma_o), \quad x=el,f,c. \text{ The needed}$$

accuracy of approximation is about 1.5%. The results in Table 1 show that the method of simultaneous approximation ensures either new quality — less subgroup number (2 instead of 3) or substantial quantitative improvement as compared to momentum method. Moreover the momentum method failed completely in case of ^{238}U while simultaneous approximation produced satisfactory accuracy (1.2%).

It must be stressed that multyband parameters corresponding to the results in Tables satisfy physical restrictions $\Sigma a_k = 1$,

$$\sum_k a_k \sigma_{tot,k} = \langle \sigma_{tot} \rangle, \quad \sum_x \sigma_{x,k} = \sigma_{tot,k}.$$

Table 1. The accuracies (%) of multyband calculations.

N	Method	Δ_{el}	Δ_f	Δ_c	Δ_{tot}	Nuclide
2	I	9.5	33	31	28	
	II	3.1	14	7.3	13	
3	I	1.9	0.7	4.1	3.9	^{239}Pu
	II	1.3	3.3	2.4	5.8	
4	I	0.32	1.2	0.76	1.1	
	II	0.04	0.4	0.48	0.78	
2	I	38	–	74	53	
	II	14	–	18	31	
3	I	12	–	10	34	^{238}U
	II	3.2	–	6.4	7.3	
4	I	6.2	–	9.1	19	
	II	0.33	–	1.2	1.1	
2	I	0.41	–	23	3.1	^{nat}Fe
	II	0.26	–	1.6	0.69	
3	I	0.02	–	1.1	0.28	
	II	0.008	–	0.13	0.14	

Two-dimensional approximation

A straightforward generalisation of discrete optimisation method for the case of two variables is relatively simple. We used the approximants

$$R_{[N,M]} = P_N(x,y;\vec{p})/Q_M(x,y;\vec{q}) \qquad (18)$$

where $\vec{p} = (p^{(0)}, \ldots p^{(N)}), \vec{q} = (q^{(0)}, \ldots, q^{(m)})$,

$$P_N = \sum_{k=0}^{N} p^{(k)} \phi_k , \quad q_M = \sum_{k=0}^{M} q^{(k)} \phi_k \qquad (19)$$

with the ordered products of variables' degrees for basic functions ϕ_k

$\phi_k(x,y) = x^{i_k} y^{j_k}$, $i_k \geq 0$, $j_k \geq 0$, $k=0,1,2,\ldots$

The intermediate problem of two dimensional rational interpolation of increasing order may be solved either by using recursive relations or by matrix bordering. The latter method was used to make a computer code and solve a few test problems.

A function of the type

$$F(x,y) = \sum_{k=1}^{2} \frac{h_k}{1 + \left(\dfrac{x - E_{1,k}}{r_{1,k}}\right)^2 + \left(\dfrac{y - E_{2,k}}{r_{2,k}}\right)^2} \qquad (20)$$

was calculated on a grid 13x10 in $[0,1]^2$, randomized with 5% relative error and approximated. The results are illustrated by Fig.3.

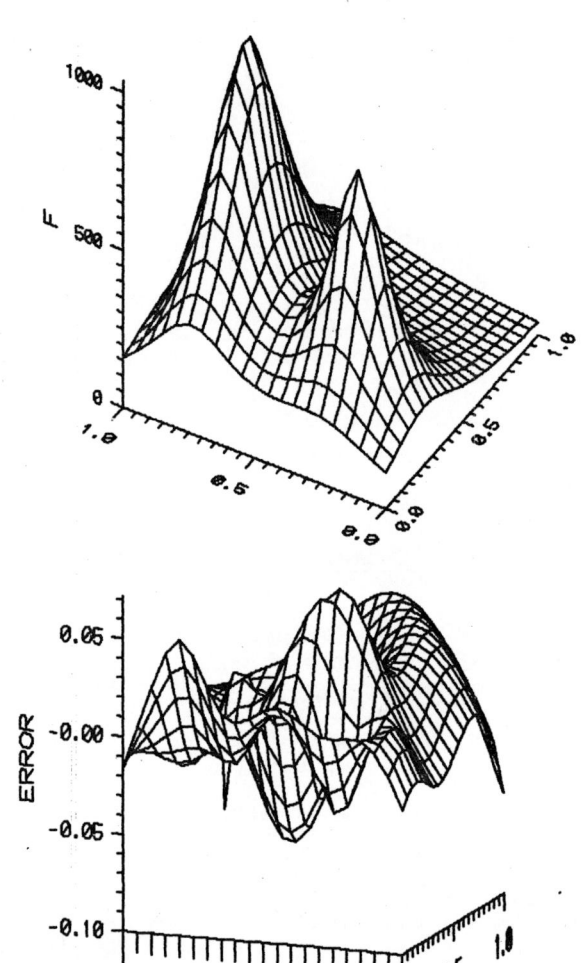

Fig.3. Top: the initial two-dimensional function F(x,y) (Eq(20)); $h_1 = 800$;

$E_{1,1} = E_{2,1} = 0.25$; $r_{1,1} = r_{2,1} = 0.125$; $h_2 = 1000$; $E_{1,2} = 0.375$; $E_{2,2} = 0.75$; $r_{1,2} = 0.25$; $r_{2,2} = 0.125$. Bottom: the relative error of the approximation $\Delta F/F = (F - F_{appr})/F$.

The number of parameters L=28, $\varkappa^2 = 0.981$.

Further possibilities

1. Trigonometric (Pade-Fourrier) approximants may be generated by the methods almost identical to those described above. The simplest case was successfully tested: the description of complicated angular distributions with strong forward peaks by change of the variable $x = \cos\Theta$.

2. Laplace- and Z-transforms of a sum of exponents are rational functions which opens prospects of using Pade-approximations in general exponential analysis including complex periods. Some examples may be found in [3].

3. Fredholm integral equations of the second type with a rational function in the right-hand side may be solved analytically for some special forms of equation's nucleus. For examples of using this property to correct resolution and Doppler broadening in resonance analysis see again [3].

4. In the sorting method described above a single supporting point is optimized in a small iteration so that $nL(N_{ex}-L)$ sets of supporting points are considred insteed of $N_{ex}!/[L!(N_{ex}-L)!]$ possible in principle (n is the number of big iterations, a few are usualy enough). For every set considered the functional S is calculated and the computer time is roughly proportional to $L(N_{ex}-L)^2$. This sets practical restrictions $L \leq 40$, $N_{ex} \leq 500$. A method of directed instead of random search was proposed which allows to reduce the number of the functional's computations to no more than $2\log_2 N_{ex} + 1$ in every small iteration with conciderable gain of speed. Some details are in [5].

Conclusions

Pade-approximants, unlike polynomial ones, provide an instrument of analytical continuation of a function under investigation which makes drastic diffe-
rence. The main mathematical advantage of rational functions is of course the existence of the poles which reproduce the poles of approximated function. If energy is the argument the real part of the pole's value more often than not may be interpreted as resonance energy and the imaginary part as a total half-width.

The supporting ordinates as a parameters to be evaluated by optimization are very convenient to visualize the statistical properties of the final approximant (error corridor, covariational matrix).

The method of discrete sorting does not require any starting set of parameters and the approximation process may be made highly automatic. If the information on experimental errors is incomplete the appearance of noise doublets is - together with and instead of x^2-criteria - a raliable signal to stop increaing the number of parameters.

So the rational functions (Pade-approximants) seem to provide most natural language in dealing with various problems in nuclear data domain as well as in some areas of neutron and nuclear physics in general.

REFERENCES

1. G.Baker, P.Graves-Morris: Pade-Approximants. Addison-Wesley (1981)
2. L.Wuytack, ed. Lecture Notes on Mathematics 765. Pade Approximation and its Applications.Berlin-Heidelberg-N.Y., Springer-Verlag (1979)
3. V.N.Vinogradov, E.V.Gai, N.S.Rabotnov: Analytical approximation of the data in nuclear and neutron physics. Energoatomizdat, Moscow (1987) (in Russian).
4. F.Nichitiu: Phase analysis in the nuclear interaction physics.Moscow, Mir (1988) (in Russian).
5. S.A.Badikov,E.V.Gai,M.A.Guseinov, N.S.Rabotnov: Proc. of the Third IMSL User Group Conference, Bologna,B11, (1990).

NUCLEAR DATA IMPROVEMENTS IN THE DECADE WITH SPECIAL EMPHASIS ON VERY RECENT DATA EVALUATIONS, AND THEIR APPLICATION TO THERMAL AND FAST REACTOR ANALYSIS

A. Mateeva-Küsters*) and E. Kiefhaber
Kernforschungszentrum Karlsruhe
Institut für Neutronenphysik und Reaktortechnik
P.O. Box 3640, W-7500 Karlsruhe

Abstract

The present analysis describes mainly the application of JEF-2 and JENDL-3 to thermal and fast reactor systems. For thermal reactors the burnup behaviour of PWR-fuel up to about 30 GWd/to is investigated; nuclide concentrations are compared to experimental results from postirradiation analyses. For fast reactors, criticality for a variety of critical assemblies is evaluated and the results are compared, both among each other and to experiment.

The results with the new data sets are discussed in some detail. Some intercomparisons of group cross sections are presented, as those for inelastic scattering (U 238, Pu 239 to 241), fission and capture data for U 235 and Pu 239. Also some results from the ENDF/B-VI and BROND-2 files are included, as far as available at present.

Introduction

The application of the European Joint Evaluated File JEF-1 to describe the physics behaviour of PWR power reactors [1] showed satisfactory agreement with experimental results of postirradiation analyses with the exception of the Cm 242 concentration at 30 GWd/t burnup. The good agreement of calculational results using JEF-1 data was also confirmed by an independent analysis on PWR-fuel cycle investigations at the University of Stuttgart [2]. No major surprises therefore are expected in using JEF-2 data for the analysis of PWRs, but expecting better results now for the Cm 242 concentration at 30 GWd/t burnup. For fast reactors, some calculations showed larger discrepancies with experimental results [3]: JEF-1 data could not have been used for fast reactor analysis without adjustments. The aim to develop the files JENDL-3 in Japan, of ENDF/B-VI in USA, JEF-2 in Europe and other files as BROND-2 in the USSR, was to calculate fast reactors with the basic group constant sets based on the various evaluated data files without major adjustments to experiments in critical facilities. The new versions of the data files have been distributed recently. This paper will concentrate on the application of JEF-2 and JENDL-3 to fast reactor systems to see whether the goal of using these unadjusted data sets for a reliable fast reactor analysis could be reached.

The processing of basic data to group constants, using the Karlsruhe version of NJOY, is carefully analysed to investigate whether major differences in C/E values for integral reactor quantities may result from the processing procedure. These investigations will be discussed in [7] and are not presented in this contribution.

*)On leave from the Institute for Nuclear Research and Nuclear Energy, Sofia, Bulgaria. Work done in the frame of a Governmental Agreement between Bulgaria and the FRG.

Test of the Data Files JENDL-3 and JEF-2 on the Burnup Behaviour of Nuclide Concentrations in PWRs

There have been published already many results on the test of the JENDL-3 data file for PWR application [4]. These concern criticality for benchmark cores, for HCLWR cores of the PROTEUS experiments, including the voided configurations, and a burnup benchmark. The prediction of compositions for urania, transurania and fission products was felt to be in satisfactory agreement with the results from ORIGEN2. In this chapter, a comparison of the calculated isotopic compositions of JENDL-3 and JEF-2 as well as with experimental results from post-irradiation analyses will briefly be presented.

Comparison of nuclide concentrations after irradiation of PWR-fuel, using JEF-2 and JENDL-3 data

In Table 1 the nuclide concentrations after about 30 GWd/tHM burnup in the Obrigheim power plant KWO are compared, using data sets derived from the JEF-1, JEF-2, and JENDL-3 data files. The irradiation history is simulated accurately. Only marginal differences can be observed between the results of the 3 data sets. The largest difference of about 9 % occurs for Np 237: JEF-2 gives a result lower by 9 % than JENDL-3. A similar deviation is observed for U236: The result, obtained on JEF-2 basis, is smaller by about 8 % than that using JENDL-3 (the JEF-1 result is in between JEF-2 and JENDL-3). For Pu 238 JEF-1 and JENDL-3 give identical results, JEF-2 is by about 6 % lower. All other results differ by no more than about 2 %.

Comparison of calculated nuclide concentrations with experimental results

In [1] JEF-1 data had been applied for this analysis. It was observed that, with the exception of the Cm 242 concentration, all other actinide concentrations at about 30 MWd/tHM are in satisfactory agreement with experimental results, which are not of very high precision, but allow to reveal major discrepancies. The same agreement is found again with the JEF-2

Table 1: Comparison of Nuclide Concentrations at 30 GWd/tHM burnup for PWR fuel (KWO), calculated with different data sets

Data Sets / Nuclide	JEF-1	JEF-2	JENDL-3
Pu 239	5.14 E-3 *)	5.13 E-3	5.14 E-3
Pu 240	2.00 E-3	2.02 E-3	2.01 E-3
Pu 241	1.07 E-3	1.08 E-3	1.07 E-3
Pu 242	3.87 E-4	3.91 E-4	3.89 E-4
U 234	2.97 E-6	2.92 E-6	2.99 E-6
U 235	8.63 E-3	8.67 E-3	8.53 E-3
U 236	3.78 E-3	3.66 E-3	3.94 E-3
U 238	9.47 E-1	9.47 E-1	9.47 E-1
U 233	3.16 E-9	3.15 E-9	3.15 E-9
Np 237	4.02 E-4	3.74 E-4	4.07 E-4
Pu 238	1.35 E-4	1.27 E-4	1.35 E-4
Am 241	9.75 E-5	9.84 E-5	9.94 E-5
Am 243	6.02 E-5	6.11 E-5	6.02 E-5
Cm 242	3.69 E-6	3.73 E-6	3.60 E-6
Cm 244	1.50 E-5	1.53 E-5	1.53 E-5

*) E-n always means 10^{-n}

Table 2: k_{eff} for Several Fast Critical Assemblies

Assembly	Fiss Fuel	Spectrum	k_{eff}		Comments
			JEF-2	JENDL-3	
GODIVA	U	Hard	1.0025	1.0124	1-dim. S_{16}, Transp. Appr.
JEZEBEL	Pu	Hard	0.9960	1.0014	1-dim. S_{16}, Transp. Appr.
ZPR III-10	U	↓ Softening ↓	1.0062	1.0033	2-dim. Diff. +Corr.*)
ZPR III-25	U		1.0025	0.9935	2-dim. Diff. +Corr.*)
SNEAK-3A1	U		1.0010	0.9888	2-dim. Diff. +Corr.*)
SNEAK-3A2	U	Soft	1.0014	0.9865	2-dim. Diff. +Corr.*)
ZPR III-48	Pu	FBR Prot.	0.9899	0.9972	2-dim. Diff. +Corr.*)
ZEBRA-6A	Pu	FBR Prot.	0.9819	0.9903	2-dim. Diff. +Corr.*)

*)The corrections are assumed to be the same as those derived originally for the KFKINR set; they include mainly transport and heterogeneity.

and the JENDL-3 data bases. The Cm 242 concentrations for all three data files are overpredicted by about a factor of 2 compared to experimental results. Already in [1] it was suspected that very probably the experimental results, which are difficult to obtain because of the very short half-life of Cm 242 ($T_{1/2}$ = 163 days), are in error. This aspect is still under investigation.

Check on Fission Product Nuclear Data

Fission product nuclear data have not been changed from JEF-1 to JEF-2. Tests of JEF-1 fission product data have been reported in [5] and, for JENDL-3, in [4]. Results with JEF-1 data are within the experimental error bars of 10 %. JENDL-3 gives a satisfactory good prediction for Sb-125 and Eu 154 concentrations.

Test of Recently Established Nuclear Data Files for Fast Reactor Applications

The fast energy range up to now was not appropriately described with cross sections for the unresolved resonance region, the capture and fission data especially for U 238 and Pu 239, and inelastic scattering for almost all heavy nuclides. Therefore, design calculations for fast reactors very frequently were performed by using adjusted data sets or using bias or "fudge factors" to integral quantities. Both schemes are unsatisfactory from the point of view of describing the important physical features of fast reactor systems of very different designs which are under discussion presently, i.e. with different fuel (oxide, metal, nitride), of special actinide burner reactors to reduce the long term hazard of nuclear waste, of modular and heterogeneous cores, and of other futuristic characteristica. The neutron spectrum usually varies from relative soft to relative hard spectrum systems. Therefore, a check of the quality of a modern data set should cover the whole range of neutron spectra above about 100 eV up to some MeV.

Observations from Tests of Data Sets based on JEF-2 and JENDL-3 for Fast Assemblies

Only a selection of the obtained results can be presented here. A full documentation of all investigations will be published as a KfK-report [7]. Table 2 shows the results for k_{eff} for a variety of fast reactor critical assemblies with uranium and plutonium fuel, based on the JEF-2 and JENDL-3 nuclear data libraries. The spectra of these systems range from very hard (GODIVA, JEZEBEL) to fairly soft neutron spectra. For the high leakage systems a high transport - S_{16} - approximation was used, the follow-up assemblies were calculated in 2-dimensional diffusion theory with transport corrections and corrections for heterogeneity and improved neutron slowing down.

In general, both data sets describe fairly well the criticality of both uranium and plutonium assemblies: With JEF-2 only for the assembly ZEBRA-6A (characteristic for a prototype fast reactor) k_{eff} is underpredicted by about 2 %.

This first assessment with non-adjusted data sets is encouraging, especially for the JEF-2 basis. The results were carefully analysed and lead to following more detailed conclusions. The criticality values given in Table 2 for GODIVA and JEZEBEL may be slightly overpredicted due to the application of the transport approximation; using instead the higher moments of the scattering matrices up to P_3 may lead to a reduction by about 0.3%. However, both data sets underestimate the criticality of JEZEBEL by about 1% relative to that derived for GODIVA. In addition, JENDL-3 gives higher criticality values by 0.9% for GODIVA and 0.5% for JEZEBEL. A better agreement would be obtained if $\nu\sigma_f$ would be slightly increased for Pu 239 in JEF-2 (which is confirmed by the fact that k_∞ for pure

Pu 239 is 1% lower for JEF-2 relative to JENDL-3), and slightly reduced for U 235 in JENDL-3 (assuming that σ_t of both isotopes is known reasonably well).

With increasing spectrum softness (the assemblies in Table 2 are arranged in that order) the criticality of U-fueled assemblies tends to an underprediction with JENDL-3 whereas it remains practically unchanged with JEF-2. This favorable tendency which gives confidence in the neutron production cross section of JEF-2, does, unfortunately, not continue when going to assemblies with even softer spectra. There exists a tendency to an overprediction as we found when considering e.g. the so-called steam density coefficient as measured in the SNEAK-3A series of experiments where the hydrogen concentration was increased stepwise. In this case $\Delta k(\Delta H)$ was in much better agreement with experiment when using JENDL-3 instead of JEF-2. This leads to the conclusion that in JENDL-3 the energy dependence of $\nu\sigma_f(E)$ and/or of $\alpha(E)$ for U 235 in the epithermal range up to a few keV may be more suitable than that of JEF-2 (having in mind that σ_c for U 238 of both files is in rather good agreement).

The already mentioned underprediction of criticality for the Pu-fueled critical JEZEBEL by JEF-2 becomes even more significant for the assemblies ZPR III-48 and ZEBRA-6A with softer neutron spectra (JENDL-3 shows a similar but slightly less pronounced tendency). A discrepancy of about 0.7% in k_{eff} resulting between both files is not very encouraging because both assemblies are considered to be representative benchmarks for medium-sized prototypes of LMFBR power reactors.

Our intercomparisons clearly demonstrate that for hard spectrum k_∞ experiments ,which are sensitive to the U238 inelastic scattering, such as ZEBRA-8H, SNEAK-8, ZPR IX-25, which have U-fuel, JEF-2 may lead to an underprediction by about 1%, but this underprediction increases considerably (to more than 2%) with JENDL-3 which clearly indicates that the energy loss caused by neutron inelastic scattering on U 238 is too large in JENDL-3 and should probably also be slightly reduced in JEF-2. For the Pu-fueled k_∞-experiment ZPR III-55 the difference between JEF-2 and JENDL-3 exceeds 1%, but both k_{eff}-values seem acceptable with no clear preference for one of the two files.

A rather striking discrepancy between both datasets was observed when comparing k_{eff} for similar compositions but differing in the fissile material. A typical example is ZPR-6-6A and ZPR-6-7 or other combinations, like SNEAK-2A-R1 and SNEAK-6A-Z1 or the inner core configurations of SNEAK 9A0 and SNEAK-9B. In all these cases we observed that k_{eff} (JENDL-3) - k_{eff} (JEF2) was roughly +1% for Pu-fueled compositions and -1% for U-fueled ones. In our opinion this surprising feature deserves further investigations, although the k_{eff}-values will probably stay in a ±1% uncertainty range, but in our opinion advanced nuclear data files should lead to better results because uncertainty bands of that amount were typical of adjusted group constant sets one or two decades ago.

Intercomparison of Important Nuclear Data from Different Modern Data Files

In this contribution it is possible to give only some selected examples for present day differences in the fundamental data files. In the high energy range the inelastic scattering of neutrons on heavy nuclides is of high importance. Fig. 1 shows a comparison between JEF-2 data and those from the adjusted KFKINR set for U 238. The data differ appreciably; although adjustment not necessarily improves the cross sections themselves, it might be concluded that JEF-2 data might be changed into the direction of JEF-1 data, which almost coincide with those of KFKINR. JENDL-3 data also differ from JEF-2 data. This is more pronounced for Pu 239 (Fig. 2) where a difference of up to 40% around 100 keV is seen. In Fig. 3 the inelastic scattering cross section for Pu241 is depicted for the data files JEF-2, JENDL-3, ENDF/B-VI and BROND-2: This situation, although not so important as for U 238, is highly unsatisfactory. Moreover, for JEF-2 and JENDL-3 the differences in the scattering matrices are remarkable for all heavy nuclides: Here the recently created international Task Force for re-evaluation of the inelastic scattering processes might bring some clarification. Fig. 4 shows the differences in the capture cross section of U 235 between JEF 2 and JENDL-3. The diffeences in the range from 100 eV to 1keV were already mentioned before. Special emphasis should be given to the energy dependence of σ_c in this range. Fig. 5 shows the comparison of the fission cross section of Pu 240 between JEF-2 and ENDF/B-VI. These large differences (up to about 80%) should be investigated. Fig. 6 gives the differences of the capture cross section for Pu 241 between the BROND-2 and JEF-2 libraries. Again, these differences should be clarified and removed.

Conclusion

In a first assessment, the recently evaluated data files JEF-2 and JENDL-3 have been tested for thermal reactors (PWR's) and for a some fast critical assemblies with uranium and plutonium fuel of varying neutron spectra (from very hard to soft). For PWR's nuclide concentrations after a burnup of about 30 GWd/t agree well with experimental results with the exception of Cm 242. For fast critical assemblies criticality values obtained with JEF-2 data are mostly very near to experimental ones. The discussion of these results show that further improvement seems to be necessary, especially for inelastic scattering of neutrons on heavy materials, the values of $\sigma_f(E)$ in the fast energy range, $\alpha(E)$ values for U 235 in the resonance range and the fisison cross section of Pu 240 and also of Am 243, and Cu 244, which are not discussed in this contribution.

Acknowledgement

The authors are very grateful to Dr. H. Küsters for his continuous interest in this work and for many fruitful discussions.

References

1. A. Mateeva, Qualification of the JEF-1 Nuclear Data Library for Pressurized Water Reactor Burnup Analysis, KfK-4461, Kernforschungszentrum Karlsruhe (1988) and A. Mateeva, in Proc. KTG, May 1989, Düsseldorf, Germany, p. 67.

2. D. Lutz, W. Bernnat, M. Mattes, in Proc. KTG, May 1989, Düsseldorf, Germany, p. 71.

3. J. Rowlands, N. Tubbs, in Proc. Int. Conf. on Nucl. Data for Basic and Applied Science, Santa Fe, NM (1985), p. 1493.

4. H. Takano, K. Kaneko, T. Nakagawa, in Proc. Int. Conf. on the Physics of Reactors: Operation, Design and Computation, April 1990, Marseille, France, V3, p. PI.21.

5. P. Chaucheprat, M. Mondot, C. Garzenne, in Proc. Int. Conf. on the Physics of Reactors: Operation, Design and Computation, April 1990, Marseille, France, V1, p. IV.32.

6. E. Kiefhaber (Comp.), The KFK-INR-Set of Group Constants; Nuclear Data Basis and First Results of its Appliction to the Recalculation of Fast Zero Power Reactors, KfK-1572, Kernforschungszentrum Karlsruhe (1972).

7. A. Mateeva-Küsters, E. Kiefhaber et al., KfK report, to be published 1991.

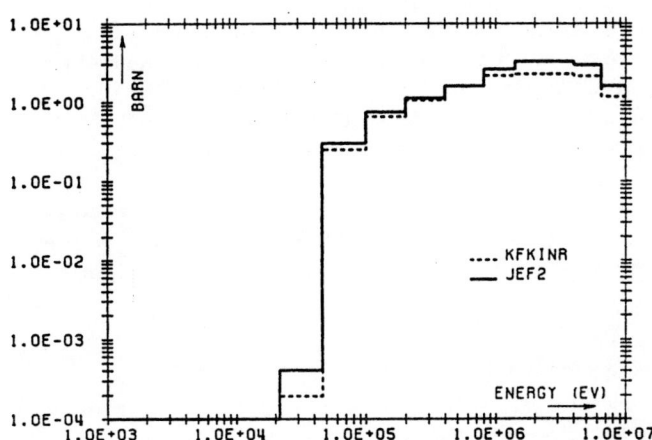

Fig. 1. Comparison of σ_{inel} for U 238

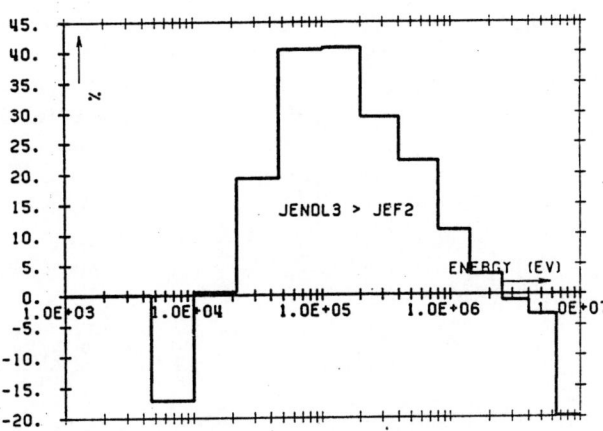

Fig. 2. Difference in % for σ_{inel} of Pu 239

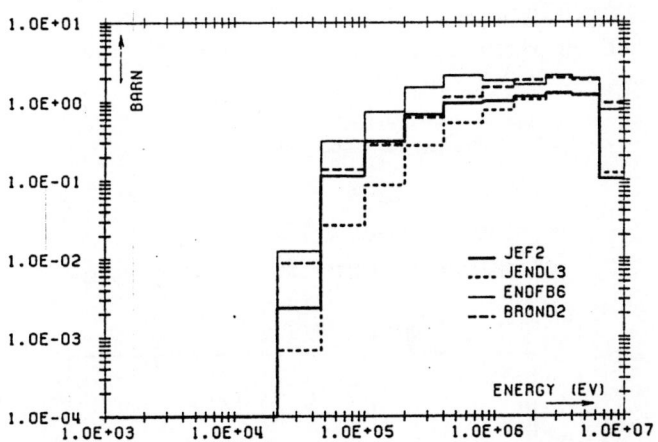

Fig. 3. Comparison of σ_{inel} for Pu 241

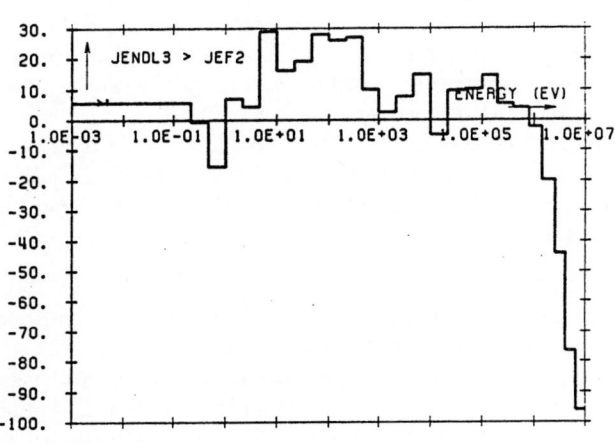

Fig. 4. Difference in % for σ_c of U 235

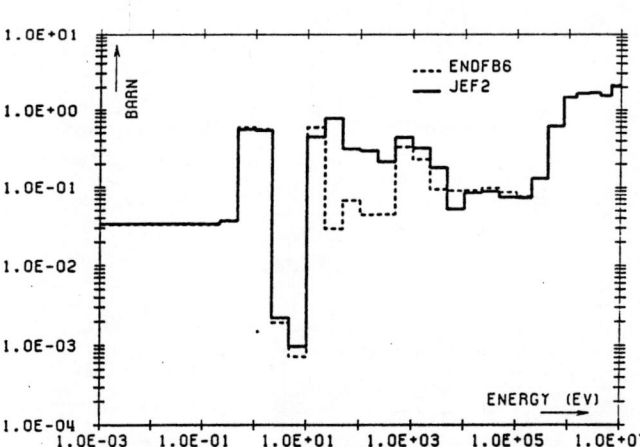

Fig. 5. Comparison of σ_f for Pu 240

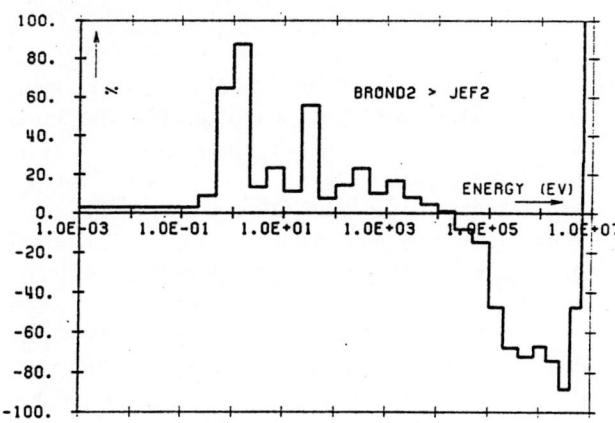

Fig. 6. Difference in % for σ_c of Pu 241

Analysis of Thermal Benchmarks Based on the Evaluated Nuclear Data Files JEF-2 and ENDF/B-VI [1]

W. Bernnat, M. Mattes, J. Keinert, R. Seidel

Institut für Kernenergetik und Energiesysteme (IKE),
Universität Stuttgart, Pfaffenwaldring 31, D-7000 Stuttgart 80

Abstract: For the validation of the recently released evaluated nuclear data files JEF-2 and ENDF/B-VI a first set of benchmark calculations for thermal systems important for criticality safety and LWR-neutronics were performed and compared with corresponding calculations based on JEF-1 and the MCNP-3B library (ENDF/B-IV, ENDL-85). The processing of the JEF-2 and ENDF/B-VI data which are given in the ENDF-6 format was performed by an updated version of the nuclear data processing system NJOY. Both multigroup libraries with 316 groups in MATXS-format and MCNP-libraries (continuous data with a very fine energy resolution) based on JEF-2 and ENDF/B-VI were generated. The ORNL uranyl nitrate benchmarks as well as the PNL Pu nitrate benchmarks were analysed by deterministic (ANISN/ONEDANT-S_N-method) and Monte Carlo methods (MCNP-3B). The group constants for the S_N-calculations were generated by a combination of TRAMIX and RSYST/CGM.

(Thermal criticality benchmarks, evaluated data, criticality safety, nuclear data processing)

Introduction

The evaluated nuclear data files JEF-2 (preliminary version 2.1) and ENDF/B-VI show several major changes relative to the previous available data files JEF-1, ENDF/B-IV and V. Especially in the resolved resonance range completely new resonance analysis based on the Reich-Moore formalism have been performed for major nuclides as ^{235}U, ^{238}U, ^{239}Pu and ^{241}Pu taking into account much more resolved resonances as in the previous data. In the thermal range significant modifications have been made for the thermal fissionable nuclides ^{235}U and ^{239}Pu.

It is necessary to validate these new evaluations for application in neutronic calculations of thermal systems. Therefore, as we have done in earlier studies based on JEF-1 [1], benchmark calculations with JEF-2 and ENDF/B-VI data have been carried out for selected thermal systems like homogeneous U and Pu-nitrate solutions [2]. This benchmarking was accomplished by comparisons of calculated and measured thermal spectra to check mainly the scattering law of H in H_2O [3]. The calculations were done with both a deterministic code for spectrum calculation with very detailed resolved resonance treatment by the calculation of the neutron slowing down equation for 8000 groups and the continuous Monte Carlo code MCNP-3B. The data libraries for these codes were processed by an updated version of the nuclear data processing system NJOY-89.62 [4]. The main scheme of data processing is shown in Fig. 1. The multigroup libraries for the deterministic spectrum calculations were generated in the MATXS format. 316 groups represent the total energy range. The group structure for the fast/epithermal range is VITAMIN-J (165 of 175 groups). For the thermal energy range we used our approved 151 group structure. The resonance range was covered by an superfine library with lethargy width of

$\Delta u = 0.001$. Libraries for MCNP were generated without or with minor thinning of the pointwise data (0.1 % accuracy) in the complete energy range (e.g. Fig. 2 shows a comparison of the total ^{239}Pu cross section as in the MCNP data set and from the pointwise (PENDF) file). All generated cross sections were carefully proved and

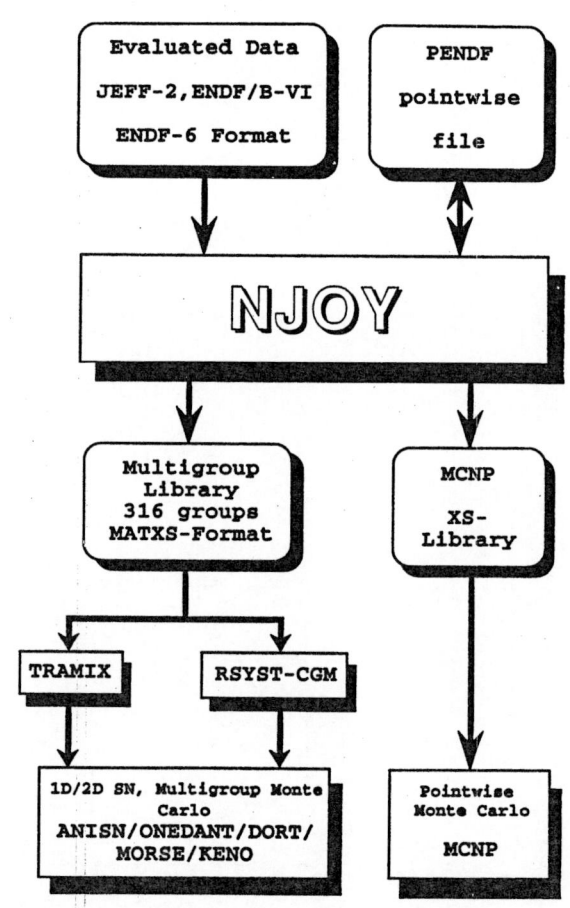

Fig.1: Generation of Multigroup Cross Section for Transport Calculations

[1]supported by Kernforschungszentrum Karlsruhe

compared with alternative evaluations by extensively use of graphical representations and comparisons: Many problems had to be solved in processing the ENDF-6 format by the present NJOY version. Due to a new representation of the energy-angle correlation of secondary neutrons used for some nuclides an extension of the MCNP-3B code is necessary. Since this extension is not yet implemented in our MCNP version assemblies with ^{235}U and ^{238}U could not be anaylised by MCNP with the generated data sets from JEF-2 and ENDF/B-VI. The scattering probability tables for the thermal energy range based on scattering law data were generated for a relatively large number of secondary energies ($\geq$ 96). The data were tested by calculation of infinite media spectra by MCNP and comparison with measurements or spectra calculated by our spectrum code CGM [5].

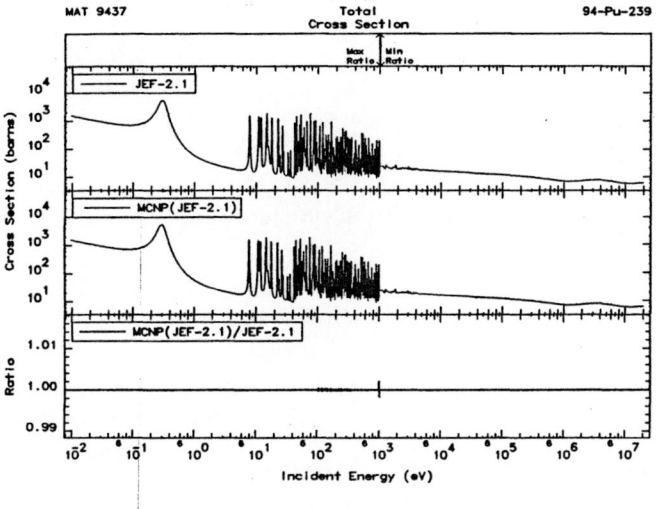

Fig.2: Comparison of the Total Cross Section of ^{239}Pu Represented in MCNP Library and in PENDF File

Benchmark calculations

The calculation of interesting integral parameters e.g. the effective multiplication factors of different configurations was performed by a deterministic calculation system based on integral transport theory and the discrete ordinate method (S_N) and by the continuous Monte Carlo code MCNP-3B. The deterministic system used the modul TRAMIX [6] for the aquisition of data in MATXS format and the shielding of resonance data (intermediate resonance (IR) method) and the modul RSYST/CGM for weighting the multigroup data in the resonance range (solution of the slowing down equation for a superfine energy grid) and calculation of cell spectra for 316 groups and final collapsing/homogenisation to 60 energy groups. The determination of flux spectra and k_{eff} was performed by the one-dimensional S_N codes ANISN and ONEDANT. Calculations were performed for different benchmarks using JEF-2 and partly ENDF/B-VI (e.g. for the main plutonium isotopes). The results were compared with values derived from JEF-1.

The MCNP-3B code was used to calculate the Pu nitrate systems PNL 6 through 12 on the basis of the MCNP-libraries (included in the code package distributed by NEA DATA Bank), the JEF-2 library and for ^{240}Pu, ^{241}Pu and ^{242}Pu also for the ENDF/B-VI library. The uranium systems could not yet be calculated by MCNP-3B due to the lack of treatment of the new representation of the energy-angle correlations of secondary neutrons in ENDF-6 format.

Results

Fig. 3 shows the thermal spectrum of a borated H_2O system calculated by MCNP-3B using JEF-2 and MCNP thermal scattering probabilites tables/libraries. The calculated spectra agree sufficiently well with measurements in the energy range important for reactor applications. Some discrepancies below 10 meV were seen, probably due to the method of selecting the secondary energies in MCNP-3B.

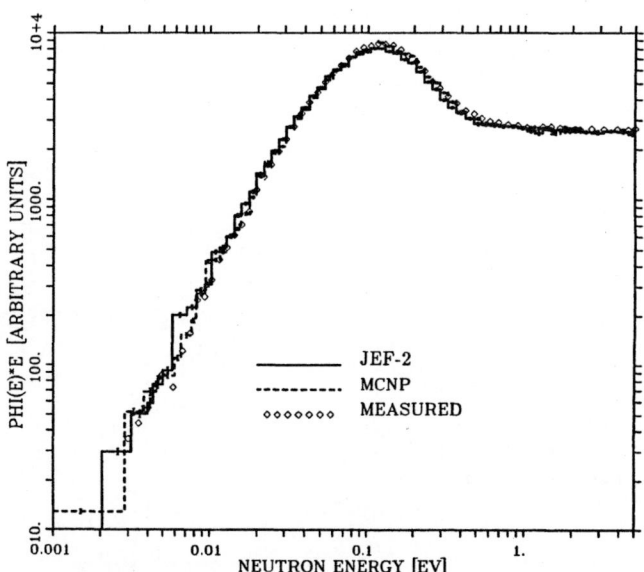

Fig.3: Comparison of Calculated and Measured Neutron Flux Spectra for a Infinite Borated H_2O System. The Calculations were Performed by MCNP Based on JEF-2 and MCNP Libraries

The analysis of an infinite ^{239}Pu-H_2O critical solution showed a sufficiently agreement for the JEF-1, JEF-2 and ENDF/B-VI libraries (Tab. 1).

	JEF - 1	JEF - 2	ENDF/B-VI (Pu-239)	JEF - 2 MCNP-3B
k_∞	0.9957	0.9963	0.9953	0.9960 ±0.0004

Table 1: Multiplication factor for an infinite homogeneous critical solution of Pu-239 atoms in water

To test the ^{235}U cross-sections the ORNL benchmarks 1 through 4 and 10 were analysed. The comparison of results given in Tab. 2 show that JEF-2 gives slightly lower values for the multiplication factors than JEF-1. The last column contains the continuos Monte Carlo results.

For the plutonium nitrate systems calculations based on JEF-1, JEF-2 and ENDF/B-VI were performed by the CGM/TRAMIX code system and MCNP-3B. The CGM results are shown in Tab. 3. Most of the results agree better than 1 % with experiment. Using ENDF/B-VI data for the Pu isotopes the results agree better than 0.8 % with the measured values. The results of MCNP calculations for the plutonium nitrate benchmarks are shown in Tab. 4. Most of the calculated values agree in the range of about 1 % with the experiments. The MCNP results agree well with the deterministic CGM/TRAMIX values based on the 316 group library.

The results shown are a first step in the validation procedure for thermal systems. The processing of JEF-2 and ENDF/B-VI data in ENDF-6 format could be demonstrated. Further analysis for heterogeneous systems (UO$_2$- and UO$_2$-PuO$_2$-lattices) will be performed.

Table 2: Multiplication factors for spherical uranyl nitrate criticals

Assembly	H/U-235 Ratio	JEF - 1 k_∞	JEF - 1 k_{eff}	JEF - 2 k_∞	JEF - 2 k_{eff}	MCNP-3B k_{eff}
ORNL - 1	1378	1.2145	1.0013	1.2130	0.9989	0.9920 ± 0.0038
ORNL - 2	1177	1.2098	1.0015	1.2081	0.9981	0.9993 ± 0.0035
ORNL - 3	1033	1.2029	0.9987	1.2010	0.9944	0.9925 ± 0.0041
ORNL - 4	972	1.2033	0.9997	1.2013	0.9959	0.9937 ± 0.0041
ORNL - 10	1835	1.0737	0.9986	1.0724	0.9981	0.9965 ± 0.0030

Table 3: Multiplication factors for spherical plutonium nitrate solution criticals

Assembly	H/Pu Ratio	JEF - 1 k_∞	JEF - 1 k_{eff}	JEF - 2 k_∞	JEF - 2 k_{eff}	ENDF/B-VI °) k_∞	ENDF/B-VI °) k_{eff}
PNL-6B *)	125	-	-	1.6543	1.0073	–	–
PNL-7A +)	980	1.5592	1.0078	1.5588	1.0075	1.5536	1.0042
PNL-8A *)	758	1.5838	1.0115	1.5831	1.0120	1.5759	1.0077
PNL-12A +)	1067	-	-	1.4889	1.0086	1.4841	1.0051

*) unreflected
+) H$_2$O reflected

°) Pu isotopes from ENDF/B-VI, others from JEF-2

Table 4: Multiplication factors calculated by MCNP-3B for plutonium nitrate solution benchmarks PNL 6 through 12 (CSEWG definition)

Assembly	H/Pu Ratio	MCNP-libraries k_{eff}	MCNP-libraries (1σ)	JEF - 1 k_{eff}	JEF - 1 (1σ)	JEF - 2 k_{eff}	JEF - 2 (1σ)	ENDF/B-VI °) k_{eff}	ENDF/B-VI °) (1σ)
PNL-6B *)	125	1.0182	0.0020	1.0162	0.0009	1.0106	0.0007		
PNL-7A +)	980	1.0080	0.0026			1.0104	0.0007		
PNL-8A *)	758	1.0101	0.0019	1.0124	0.0012	1.0092	0.0008		
PNL-9 **)	910	1.0110	0.0017			1.0173	0.0017		
PNL-10 ++)	210	0.9950	0.0027			1.0039	0.0025		
PNL-11 ++)	623					1.0064	0.0019	1.0038	0.0021
PNL-12A +)	1067	1.0039	0.0028			1.0104	0.0028		

*) unreflected sphere
+) H$_2$O reflected sphere
°) ^{240}Pu, ^{241}Pu and ^{242}Pu
**) unreflected cylinder
++) H$_2$O reflected cylinder

References

[1] Bernnat, W.; Mattes, M.; et al.: Analysis of Selected Thermal Reactor Benchmark Experiments Based on the JEF-1 Evaluated Nuclear Data Files.
IKE 6-157 (JEF-Report 7) (1986)

[2] ENDF-202 Cross Section Evaluation Working Group Benchmark Specifications.
BNL-19302 (ENDF-202) (1974)

[3] Young, J. C.; Huffmann, D.: Experimental and Theoretical Neutron Spectra.
GA-5319 (1964)

[4] MacFarlane, R. E.; Muir D. W.; Boicourt, R. M: The NJOY Nuclear Data Processing System, Volume I: User's Manual.
LA-9303-M (ENDF-324) (1982)

[5] Arshad, M.: Entwicklung und Verifikation eines Programmsystems zur Berechnung von Spektren und gewichteten Gruppenkonstanten für thermische und epithermische Spaltstoffsysteme. IKE 6-156 (1986)

[6] MacFarlane, R. E.; Stepanek, J.: TRAMIX, A General Purpose Code for Interfacing MATXS Cross-Section Libraries to Nuclear Transport Codes. PSI-report in preparation

THE MEASUREMENT OF FAST REACTOR IRRADIATED MINOR ACTINIDES AND COMPARISON WITH NEUTRON PHYSICS CALCULATIONS

L. Koch, K. Kammerichs, R. Wellum
Commission of the European Communities
Joint Research Centre, Institute for Transuranium Elements
Postfach 2340, 7500 Karlsruhe, Federal Republic of Germany

I. Broeders, B. Krieg
Kernforschungszentrum Karlsruhe (KfK)
Institute für Neutronenphysik und Reaktortechnik
Postfach 3640, 7500 Karlsruhe, Federal Republic of Germany

T. Matsumura
Komae Research Laboratory
Central Research Institute of Electric Power Industry (CRIEPI)
11-1,Iwato Kita 2-Chome, Komae-Shi, Tokyo, 201 Japan

Abstract: Pellets of ten different mixtures of actinide oxides including thorium, neptunium, uranium, plutonium of different isotopic compositions and americium were irradiated in the Material Testing Element 2 (MTE 2) in the reflector of the KNKII/2 reactor, Karlsruhe from August 1983 to October 1985. The nuclide compositions of the pellets after irradiation have been measured and calculated. Results of the calculations with different cross-sections and different approximations for the neutron flux density spectra are presented and compared with the measurement values.

Introduction

At the beginning of the operation of KNKII/2 in August 1983, two actinide irradiation experiments were started in the reactor: in an empty fuel pin of a core element various actinides (^{232}Th...^{244}Cm) in amounts of about 1 mg and some fission products (^{137}Cs, ^{106}Ru, ^{95}Zr/Nb, ^{139}La) in amounts of about 1 μg were inserted for irradiation. These samples will be available for radiochemical analysis in 1992. In a second experiment pellets of 10 different mixed actinide oxides were irradiated. These samples were unloaded in October 1985 after about 360 full power days. The radiochemical analyses of the products have been carried out and first results of the neutron physics calculations are now available and are reported here. Fuller details of the irradiation have been given previously [1].

Radiochemical Analysis

The products resulting from the irradiation were measured where possible by isotope dilution mass-spectrometry (IDMS), augmented by alpha- and gamma- spectrometry and for neptunium by ICP-MS. For each pellet a complete analysis was made to allow the individual nuclide concentrations to be normalised to the sum of all products.

Neutron Physics Analysis

Neutron physics analyses were carried out independently at CRIEPI, Tokyo and at KfK, Karlsruhe. In both evaluations the changes of isotope concentrations in the actinide oxide samples were calculated using burnup codes based on the depletion formalism of ORIGEN-2 [2].

Cross-sections

The 26 energy group ABBN structure was used for the standard calculations. For the CRIEPI analyses, group constants for the actinides were calculated from the Japanese Nuclear Data library, JENDL-3, whereby ^{239}Pu was calculated from ENDF/B-IV and for another group constant set (called ENDF/B-V(IV) in the following) the cross-sections for ^{235}U, ^{237}Np, ^{238}Pu, ^{241}Am, ^{242m}Am, ^{243}Am, ^{243}Cm, ^{244}Cm were calculated from ENDF/B-V, ^{242}Cm from JENDL-3 and all other cross-sections from ENDF/B-IV. For the calculation of the resonance self- shielding factors the Monte Carlo code VIM-2 was used [3].

At KfK modifications of the adjusted fast reactor group constant set KFKINR were used [4]. In the KFKINR/JEF-2 set the cross-sections for ^{237}Np, ^{238}Pu, ^{241}Am, ^{242}Am, ^{243}Am, ^{242}Cm and ^{243}Cm were calculated from the Joint Evaluated File JEF-2 whereas those for ^{242m}Am and ^{244}Cm were calculated from the JEF-1.1 point- file. All other group constants were taken from the KFKINR set. In another set named JEF-2, the data of ^{234}U, ^{235}U, ^{236}U, ^{238}U, ^{239}Pu, ^{240}Pu, ^{241}Pu and ^{242}Pu were taken from JEF-2 and also a set KFKINR/JENDL-3 was used in which the data of ^{241}Am, ^{242}Am, ^{242m}Am, ^{243}Am, ^{242}Cm, ^{243}Cm and ^{244}Cm were replaced with values from JENDL-3. In special cases a group constant set in the 69 group WIMS structure based on the Karlsruhe Nuclear Data library KEDAK [5] was used. The Fast Reactor cell code KAPER 4 [6] was applied to take into account energetic and spacial self-shielding in the pellets.

Neutron Flux Density Measurements

Level III of the MTE-2 where the pellets were irradiated is shown in Fig. 1. The vertical position is about 50 cm below core midplane, corresponding to the height of the lower edge of the lower axial blanket. The calculation of the neutron spectra is very difficult in this position especially in a very compact and hetrogeneous reactor like KNKII/2. To support the calculation of local neutron flux density spectra, multiple foil monitors, indicated by DK1, DK2, DK7, DK8 and DK9 in Fig. 1 were made available by

SCK/CEN Mol, who also measured the monitors and derived the reaction rates after irradiation. The unfolding of the measured reaction rates was carried out at CRIEPI using the code SAND-2 [7] with nuclear data from JENDL-3 and ENDF/B-V(IV). Initial flux approximations necessary for the unfolding were obtained from calculations with VIM-2.

In the subsequent analyses carried out at CRIEPI the spectra obtained from the SAND-2 calculations with JENDL-3 or ENDF/B-V(IV) were used. In the calculations done at KfK, spectra from the SAND-2 analysis with JENDL-3 were used and also spectra from 3 dimensional diffusion calculations with the code D3E [8] - although it is recognised that diffusion theory alone is not adequate for the calculations of local flux densities in highly hetrogeneous situations like the reflector zone of KNKII.

The strong local variation of the neutron flux within the irradiation space is demonstrated in Fig. 2 where neutron flux density spectra are shown for a core element of KNKII/2 and at the positions of the monitors DK1 and DK2.

Comparison of Neutron Physics Calculations with Measurements

The ratios of the calculated-to-experimental values (c/e) are given in Tabs. 1-3 for three representative samples. The results are not yet final and are still being reviewed. The samples include a standard fast reactor MOX mixture, a U/Pu mix containing second generation plutonium (Pu2nd), i.e. recycled from PWR fuel, and a standard fast reactor fuel containing 2% ^{241}Am. The first column in each case shows the CRIEPI results obtained with JENDL-3 and the second with ENDF/B-V(IV). 'SAND(DKx)' indicates that the spectrum measured by monitor DKx was used. The KfK results are given in columns 3 and 4.

Pellet 4 $(U_{0.75}Pu_{0.25})O_2$ (Tab. 1)

This pellet shows good agreement between calculated and experimental results in the majority of the isotopes of U and Pu. ^{234}U is consistently calculated higher than measured, ^{237}Np on the contrary is found in much higher amounts than calculated: this could be due to Np impurities existing in the starting material before irradiation. The amount of ^{241}Am found is also higher than that calculated and this is likely to be due to decay of ^{241}Pu before irradiation. In the last column of Tab. 1 it was assumed that 7% of the ^{241}Pu in the fresh pellet had decayed into ^{241}Am, corresponding to a decay time of 1.5 years.

Pellet 7 $(Pu2nd_{0.45}, U_{0.55})O_2$ (Tab. 2)

The agreement between calculated and experimental results is not so good in this pellet as in the previous one. ^{234}U shows again the poorest agreement for the uranium isotopes. The calculated results for ^{243}Am (except that using the JENDL-3 data set) are much higher than the measured values and in contrast the calculated ^{244}Cm values are much lower than those measured. The last column shows the results of the assumption that 9 % ^{241}Pu had decayed to ^{241}Am in the pellet before irradiation.

Pellet 10 $(U_{0.73}Pu_{0.25}Am_{0.02})O_2$ (Tab. 3)

Apart from ^{234}U, the calculated uranium and plutonium isotopes agree well with the measured values. ^{241}Am also shows good agreement but ^{243}Am and ^{244}Cm are underestimated in the calculations. Increasing the amount of ^{243}Am at the start of irradiation however was not found to improve the

agreement between measured and calculated values for these nuclides.

Discussion

At the present state of the evaluation it is difficult to draw conclusions on the cross-sections of minor actinides taken from the different libraries used in the calculations. That good agreement can be obtained is shown by sample 4 whereby the interference of contaminants in the starting material (Np in Pu for example or ^{241}Am grown in from ^{241}Pu) remains a problem to be eradicated by subsequent measurements on the starting material.

The most difficult part of the neutron physics analysis is the determination of the local fluxes. The neutron flux density spectrum in the irradiation region has considerable contributions below 1 eV (Fig. 2). It may therefore be useful to carry out the unfolding of the reaction rates measured by the monitors with an energy group structure that has a good resolution at low energies, i.e. the 69 group WIMS structure.

REFERENCES

1. R. de Meester, L. Koch, R. Wellum, I. Broeders, L. Schmidt, 'An experiment to measure neutron Cross-sections of Milligram Amounts of Actinides in a Fast Reactor', Int. Conf. on Nuclear Data for Basic and Applied Science, May 13-17, Santa Fe, 1985

2. M. J. Bell, 'ORIGEN- The Oak Ridge Isotope Generation and Depletion Code', ORNL-4628, 1973

3. L. B. Levitt and R. C. Lewis, 'VIM-1, A Non-multigroup Monte Carlo Code for Analysis of Fast Critical Assemblies', Al-AEC-12951, Atomic International, 1970

4. E. Kiefhaber, 'The KFKINR Set of Group Constants: Nuclear Data BASIS and First Results of its Application to the Recalculation of Fast Zero-Power Reactors', KFK 1572, 1972

5. B. Goel, B Krieg, ' Status of the Nuclear Data Library KEDAK-4', KFK 3838, 1985

6. R. Böhme, E.A. Fischer, 'The Fast Reactor Cell Code KAPER4', KfK 4435, 1988

7. W. N. McElroy et al, 'A computer Automated Iterative Method for Neutron Spectra Determination by Foil Activation', AFWL-TR-67-41, Vol I-IV, 1967

8. B. Stehle, 'D3D und D3E, Zweige eines FORTRAN-Programms zur Lösung der stationären Multigruppendiffusionsgleichungen in Rechteck-, Zylinder- und Dreieckgeometrie', KFK 4764, 1991

Table 1: c/e-values for sample 4: $(U_{0.75},Pu_{0.25})O_2$

	CRIEPI Results		KfK Results		
Isotope	JENDL-3 SAND (DK2)	ENDF/B-V(IV) SAND (DK2)	KFKINR/JEF-2 SAND (DK2)	KFKINR/JEF-2 D3E/D3D2FL	KFKINR/JEF-2 SAND (DK2) (with Am-241)
U-234	1.3924	1.4007	1.4233	1.4101	1.4245
U-235	0.9926	0.9958	1.0022	1.0008	1.0022
U-236	1.0824	1.0620	1.0596	1.1111	1.0596
U-238	1.0060	1.0107	1.0097	1.0062	1.0097
Np-237	6.239E-3	7.678E-3	5.855E-3	5.971E-3	7.808E-3
Pu-238	0.9658	0.9736	0.9763	0.9694	0.9996
Pu-239	0.9740	0.9570	0.9697	0.9820	0.9697
Pu-240	1.0552	1.0333	1.0194	1.0091	1.0193
Pu-241	1.2064	1.5396	1.1513	1.1044	1.1152
Pu-242	1.0614	1.0753	1.0798	1.0495	1.0755
Am-241	0.6007	0.7044	0.7113	0.6977	0.9604
Am-242m	0.4745	0.5514	0.4097	0.4037	0.6521
Am-243	1.1325	1.0914	0.9158	1.1045	0.9148
Cm-242	0.6607	0.7868	0.5357	0.5093	0.7756
Cm-244	1.3458	1.2306	0.8019	0.9852	0.8014

Table 2: c/e-values for sample 7 $(Pu^{2nd}_{0.45}U_{0.55})O_2$

	CRIEPI Results		KfK Results		
Isotope	JENDL-3 SAND (DK2)	ENDF/B-V(IV) SAND (DK2)	KFKINR/JEF-2 SAND (DK2)	KFKINR/JEF-2 D3E/D3D2FL	KFKINR/JEF-2 SAND (DK2) (with Am-241)
U-234	0.6753	0.6419	0.8275	0.8203	0.8359
U-235	0.9022	0.7759	0.9061	0.9009	0.9059
U-236	0.7650	1.3462	0.8576	0.8974	0.8573
U-238	0.9380	0.9848	0.9344	0.9335	0.9342
Pu-238	1.0497	1.0848	1.0563	1.0538	1.0931
Pu-239	1.0785	0.7551	1.0920	1.0936	1.0918
Pu-240	0.9495	1.2146	1.1215	1.1242	1.1210
Pu-241	1.7232	1.4348	1.1828	1.1787	1.0944
Pu-242	1.2012	1.2652	1.1667	1.1645	1.1635
Am-241	0.5202	0.4149	0.5418	0.5384	0.8660
Am-242m	0.5759	0.8735	0.4609	0.5002	0.8847
Am-243	1.1125	1.4208E+1	2.2833	2.9012	2.2814
Cm-244	0.06274	2.9419	0.09596	0.13362	0.09589

Table 3: c/e-values for sample 10 $(U_{0.73} Pu_{0.25} Am_{0.02})O_2$

	CRIEPI Results		KfK Results		
Isotope	JENDL-3 SAND (DK8)	ENDF/B-V(IV) SAND (DK8)	KFKINR/JEF-2 SAND (DK8)	KFKINR/JEF-2 D3E/D3D2FL	KFKINR/JEF-2 SAND (DK2)
U-234	0.6397	0.6316	0.6132	0.5851	0.6571
U-235	0.9726	0.9787	0.9825	0.9855	0.9657
U-236	1.1231	1.0921	1.1157	1.1203	1.1790
U-238	0.9982	0.9960	0.9973	0.9931	1.0022
Pu-238	1.2593	1.2137	0.9762	0.8345	1.2397
Pu-239	0.9865	1.0023	1.0029	1.0280	0.9718
Pu-240	1.0371	1.0343	1.0024	0.9933	1.0169
Pu-241	1.1843	1.0467	1.1168	1.0058	1.2602
Pu-242	1.1107	1.0742	1.0604	0.9878	1.1821
Am-241	1.0133	1.0176	1.0529	1.0639	1.0294
Am-242m	1.4532	1.3674	1.0404	0.9078	1.2718
Am-243	0.2149	0.2081	0.1536	0.1636	0.2162
Cm-242	1.2623	1.2136	0.9196	0.7539	1.2306
Cm-244	0.8650	0.8236	0.4699	0.4519	0.8370

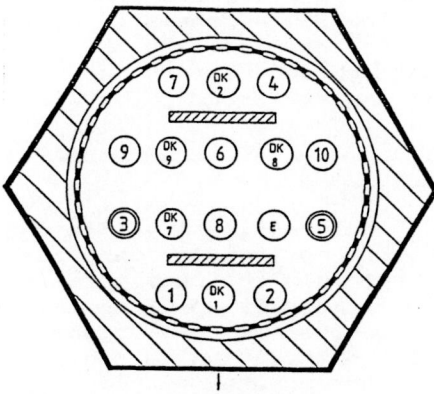

Core Center

Fig. 1. Actinide samples and monitors in the Material Testing Element MTE 2

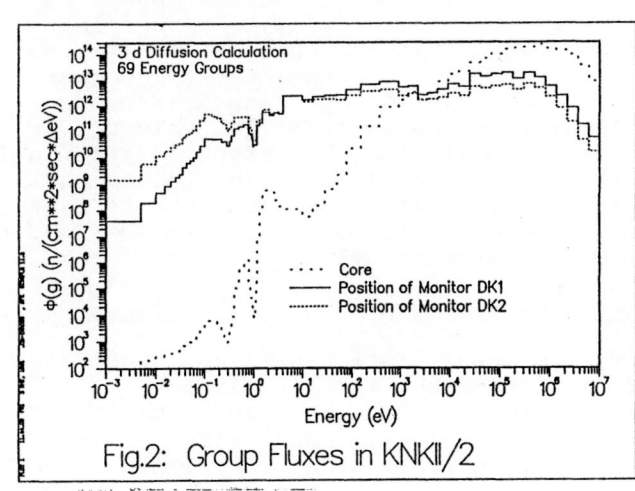

Fig.2: Group Fluxes in KNKII/2

NEUTRON DATA CHECK BY SAMPLE REACTIVITY MEASUREMENTS IN CRITICAL ASSEMBLIES WITH PREDETERMINED ADJOINT SPECTRA

K.Dietze, W.Hansen, G.Hüttel, H.Kumpf, E.Lehmann[*] and D.Richter

Zentralinstitut für Kernforschung Rossendorf
8051 Dresden, PF 19, Federal Republic of Germany

[*]now Paul Scherrer Institut, CH-5232 Würenlingen

Abstract: Up to now discrepancies between different evaluations of neutron data for structural materials of FBR as well as for fission products persist. Reactivity measurements as integral data check are usually difficult to interpret because of the simultaneous dependence of sample reactivity worth on absorption and scattering cross-sections. In order to avoid or at least to minimize this disadvantage the method of reactivity measurement has been modified with the goal to separate absorption and scattering effects by designing of special reactor configurations.

(structural materials, fission product nuclides, sample reactivity worths, cross-sections, f-factors, importance function)

Introduction

Nuclear data used for calculations of neutronic processes in fission reactors, mainly applied in form of group cross-sections, have been evaluated and adjusted constantly. Two ways for improving the prediction of such essential reactor parameters as keff, breeding rate or power distribution are usually applied:
1. Enhancing the knowledge about the fundamental nuclear data by means of differential experiments and by model calculations in order to refine established data files which can be transformed into group cross-section sets.
2. Adjusting existing group data by means of integral experiments.
The variety of differential and integral experiments led to a high level of accuracy for neutron data. Remaining discrepancies can be resolved with the help of specific integral tests. An interesting possibility was realized at the critical fast-thermal coupled facility RRR of the Central Institute for Nuclear Research Rossendorf. By arranging special fast core configurations the adjoint spectrum in the centre of the core could be adjusted so that a separation of absorption and slowing-down contribution has been obtained in reactivity measurements. These configurations are characterized by a nearly energy-independent or, contrary to that, a strong energy-dependent adjoint spectrum in order to avoid or to raise strongly the scattering effect in the sample reactivity. In this way, conclusions could be drawn about partial cross-sections in the fast region and resonance data as well.

Fast insertion zones in the reactor RRR

The zero power reactor RRR is a facility of the Argonaut type equipped with a fast insertion lattice (SEG) instead of the central reflector [1]. With a precisely defined combination of pellets made from enriched uranium, natural uranium, structural and other materials which were arranged in the channels of a cylindrical matrix the designed neutronic properties could be realized.

Configurations with energy-independent adjoint flux

The problem for the investigation of the absorption properties is that in usual fast systems the contribution of absorption to the measured reactivity value is accompanied by a scattering term in the same order of magnitude. These scattering contributions are influenced mainly by the energy dependence of the adjoint flux function. In order to suppress the scattering term fuel and other materials were combined in the reactor core so that a nearly energy-independent central importance function results. Consequently, the measured reactivity value can be interpreted as the absorption cross-section of the investigated material averaged over the neutron flux spectrum, whereas scattering processes provide small corrections only.
Two of such fast lattices (named SEG-4 and SEG-5) were arranged. Their central neutronic properties are shown in Fig 1. Whereas the SEG-4 consists of uranium, graphite and cadmium, the SEG-5 unit cell comprised boron carbide as absorbing material instead of Cd [2,3].

Insertion zones with a large slope of the adjoint spectrum

In order to investigate scattering cross-sections the principle described above was inverted in the configuration SEG-6 [4]. If the adjoint spectrum has a strong and monotonous energy dependence the dominant contribution to the reactivity value is the moderation term influenced by scattering processes.
This goal could be realized by an arrangement shown in Fig.2. A central neutron

absorbing insertion zone filled with boron carbide is surrounded by channels containing enriched (36%) and natural uranium. The central neutronic properties are shown in Fig.2 also.

Because of the small contributions to the neutron flux spectrum in the energy range below 10 keV the absorption part to the measured reactivity values is not very important for the majority of the investigated materials.

Configurations with soft neutron spectrum

Investigations in the configurations SEG-4 and SEG-5 have shown that resonance data used for corrections of the self-shielding effect in macroscopic samples contain obvious uncertainties. Fig. 3 shows an example for tungsten. The corrected curves for zero chord length offer considerable discrepancies for different data sets. Therefore, a fast facility was realized with important spectral components of neutrons in the resonance energy region. The comparison of measured and calculated dependence of specific reactivity on sample size give hints for the best suited resonance data.

Also in this configuration (named SEG-7) the energy dependence of the adjoint flux was minimized in order to avoid scattering contributions. Additionally, the experiments in this arrangement should contribute to the solution of the well-known discrepancy in the C/E-ratios of reactivity worths of the standards U-235 and B-10 reported in the literature and also found in all our SEG-configurations.

Experiments

Sample reactivity measurements were performed by the pile oscillation technique at the central position of the mentioned systems with an accuracy of dk/k $\geq$ E-08. The reactivity worths of the following structural and reactor materials as well as fission product nuclides could be studied over a wide range of sample sizes:

Fe,Cr,Ni,Mo,Mn,Zr,Ti,Cu,Ta,Nb,Cd,Pb,Be, Bi,Al,W,Co,V,Au,U-235,238,Na,H,B-10,C, Mo-95,97,98,100,Rh-103,Pd-105,Ag-109, Cs-133,Nd-143,145,Sm-149,Eu-153.

The predicted properties of the configurations have been verified experimentally as well. The real behavior of the energy dependence of the adjoint neutron flux was checked by neutron importance measurements using different neutron sources. The neutron spectrum as important weighing function was measured over a wide range of energy by proton recoil spectrometry applying hydrogen filled proportional counters and stilbene scintillation detectors.

The measurements were compared with calculations using the reactor code system RHEIN [5] developed in the CINR Rossendorf. Sample reactivity values were derived by perturbation theory taking into account the calculated neutron spectra and its adjoint functions.

The f-factors could be checked if the measured dependence of specific reactivity on sample size is corrected to zero chord length using different data sets in calculations of the self-shielding effect (e.g. see Fig. 3). The corrected curves should be horizontally.

The extrapolation of all curves to zero leads to the central reactivity worth (CRW) which was compared with calculated values (C/E-ratios). Remaining small scattering effects in systems with energy-independent adjoint flux have been corrected.

Results

Comprehensive experimental results and an analysis of the available data sets have been already published [6...11]. They can be used together with the description of the facilities for further data checks also in future. A repetition of all results in detail would exceed the frame of this report. As a selection, Fig. 4 shows results for FP nuclides in SEG-5. A comparison of reactivity values of selected materials in SEG-6 is given in Table 1 in order to illustrate the data situation of scattering cross-sections.

Summarizing our experiences only some general remarks shall be given here:

- The absorption cross-sections of the main structural materials Fe, Cr, Ni tend to be overestimated in most of the group data sets by 10...20 per cent, especially for Ni.

- For some other reactor materials the deviations in different data sets are even larger. In the cases of Cd, W and Sm-149 the situation is very unsatisfactory for several data sets.

- Most discrepancies exist in the capture self-shielding data. A considerable part of our present investigations in the configuration SEG-7 is directed to check fc-factors.

REFERENCES

[1] K.Fährmann,G.Hüttel and H.Krause: Kernenergie 17, 70 (1974)

[2] K.Fährmann and E.Lehmann: Kernenergie 24, 431 (1981)

[3] E.Lehmann,K.Fährmann,G.Hüttel, H.Krause and H.Kumpf: Kernenergie 29, 30 (1986)

[4] E.Lehmann,G.Hüttel,H.Krause and H.Kumpf: Kernenergie 34, 9 (1991)

[5] Chr.Reiche,H.-U.Barz,B.Kunzmann, E.Seifert,H.Wand: Reactor-Code-System RHEIN for ESER Computer,ZfK-668(1989)

[6] K.Dietze,K.Fährmann,G.Hüttel, E.Lehmann: Kernenergie 29, 401 (1986)

[7] K.Dietze,K.Fährmann,W.Hansen, G.Hüttel,H.Kumpf and E.Lehmann: Kerntechnik 53/2, 143 (1988)

[8] E.Lehmann,K.Dietze,K.Fährmann, G.Hüttel and H.Kumpf: ZfK-656 (1988)

[9] E.Lehmann,W.Hansen,G.Hüttel,H.Kumpf, D.Richter: ZfK-729 (1990)

[10]S.M.Bednyakov, M.V.Bokhovko, G.N.Manturov and K.Dietze: Sov.At.En. 67, 675 (1990)

[11]K.Dietze and H.Kumpf: Kernenergie 34, 1 (1991)

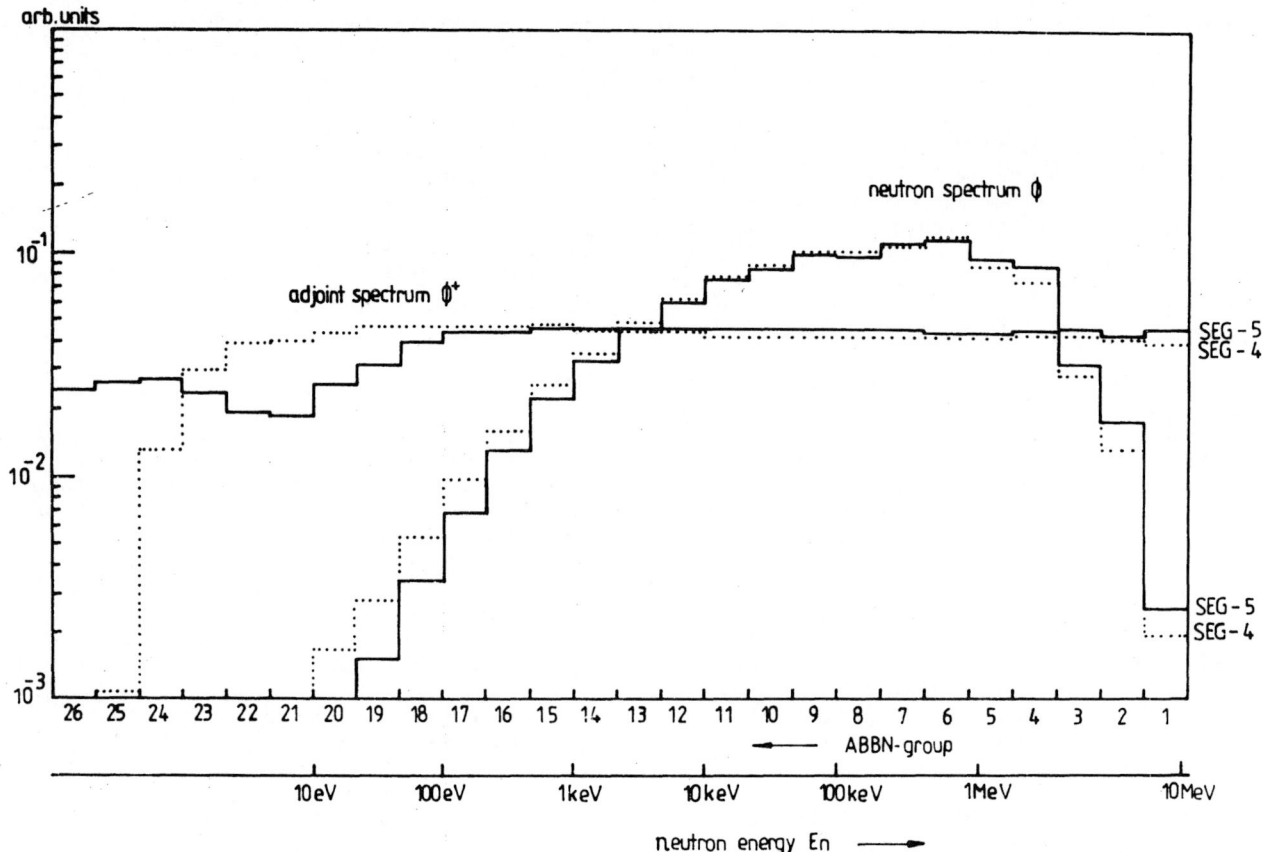

Fig.1
Neutron group spectra Ø and adjoint spectra Ø⁺ in the central channel of the SEG-4
and SEG-5 in ABBN-group representation.

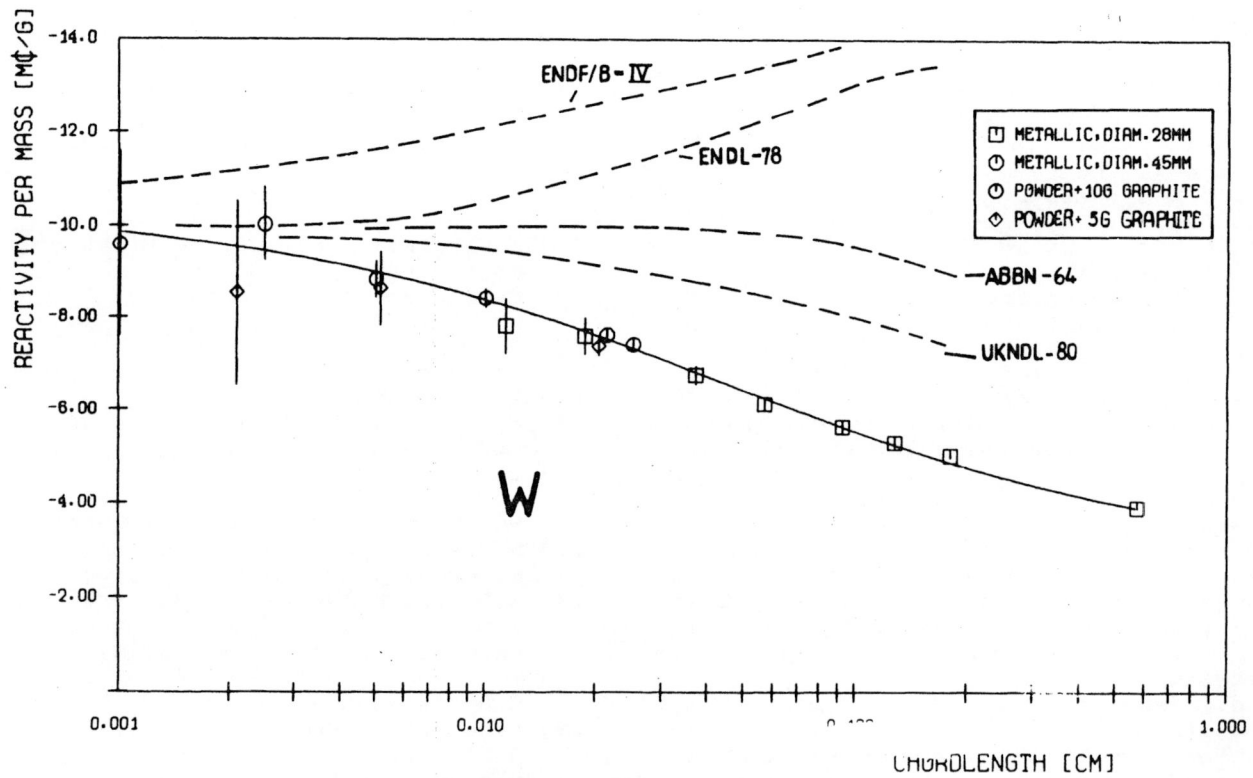

Fig.3
Measured specific reactivity of tungsten as a function of sample size in the SEG-5.
The dashed lines are corrected for zero chord length.

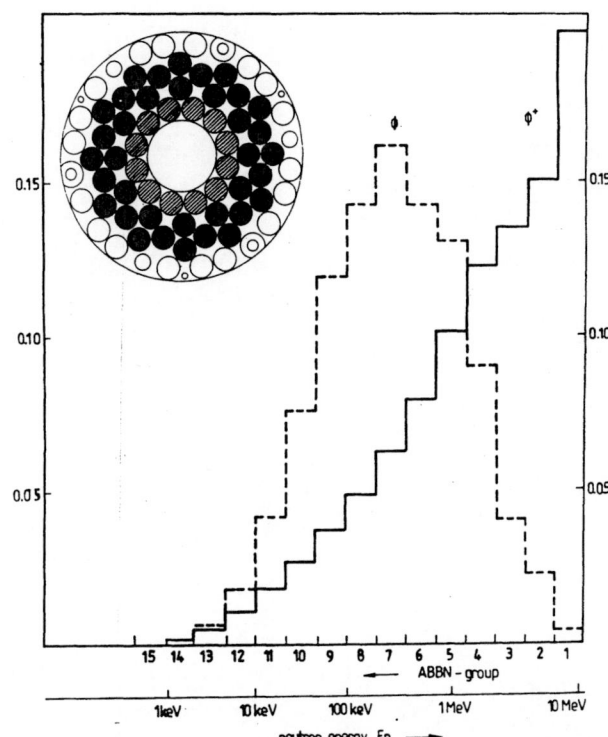

Fig.2
Neutron group spectrum Ø and adjoint spectrum Ø⁺ of the SEG-6. The configuration consists of boron carbide (white central part), enriched uranium (dashed) and natural uranium (black).

Table 1
Comparison of calculated and measured sample reactivities, per gram, for selected materials in the facility SEG-6, related to carbon.

	CALCULATIONS				EXPERIMENT
	ABBN-64	ABBN-78	JFS-II	KEDAK-3	
Al	.319	.306	.288	.320	.272+-.010
Fe	.157	.158	.157	.160	.166+-.006
Ni	.199	.219	.183	.195	.211+-.008
Mo	.232		.202		.231+-.007
Cd		.253		.253	.257+-.008
Cu	.170		.193		.198+-.004
Pb	.037	.051			.044+-.003
Bi	.033				.041+-.003
Th	.197		.168		.184+-.007
U-nat.	.076	.110	.079		.084+-.005
H₂O	17.76	17.86	17.84		17.47 +-.07

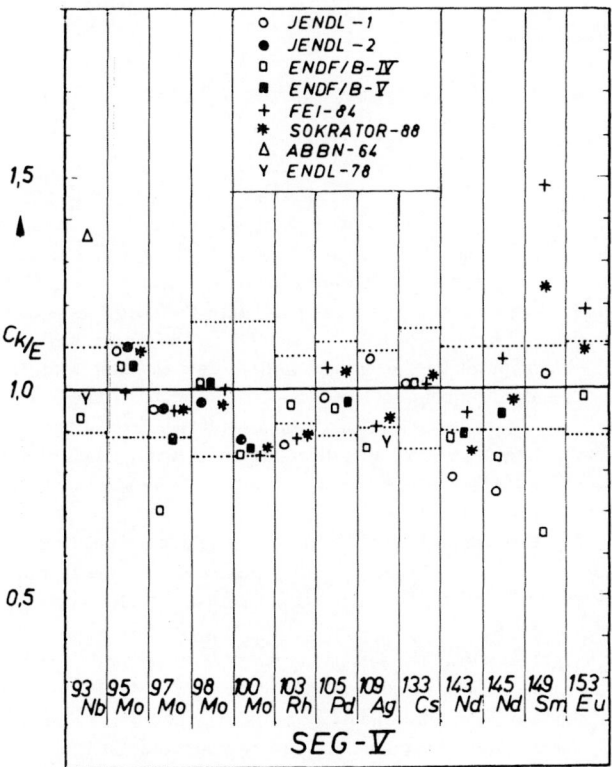

Fig.4
Ratios of calculated to experimental effective absorption cross-sections for fission product nuclides and different data sets k in the SEG-5. The experimental inaccuracies are marked by dotted lines.

VALIDATION OF RECENT DOSIMETRY FILES
IN Li(d,n) NEUTRON FIELD

J.R. Dumais, S. Iwasaki, N. Odano, M. Sakuma and K. Sugiyama

Department of Nuclear Engineering, Tohoku University,
Aramaki-Aza-Aoba, Aoba-ku, Sendai 980, JAPAN

Abstract: Validation of dosimetry cross sections has been carried out in a Li(d,n) neutron field for the four dosimetry files, i.e., the precedent IRDF-85, and the three recently compiled files, JENDL dosimetry file, IRDF-90 and ENDF/B-VI. The average cross sections for fourteen threshold reactions were measured, and compared with calculations. The tested reactions were (n,p) on ^{24}Mg, ^{46}Ti, ^{48}Ti, ^{54}Fe, ^{58}Ni and ^{60}Ni; (n,α) on ^{27}Al and ^{63}Cu; and ($n,2n$) on ^{23}Na, ^{55}Mn, ^{58}Ni, ^{59}Co, ^{93}Nb and ^{197}Au, respectively. Overall, good agreement was obtained for the recent files except for a few reactions which showed large discrepancies.

(validation, JENDL dosimetry file, IRDF-85, IRDF-90, ENDF/B-VI, Li(d,n) neutron source, thick target, activation reactions, (n,p) reactions, (n,α) reactions, ($n,2n$) reactions, average cross sections)

Introduction

Validation of the threshold dosimeter cross sections is usually performed by the integral experiments using fission neutrons from reactors or ^{252}Cf sources. Because of the inherent characteristics of the fission spectrum, however, the neutron intensity rapidly decreases with increasing the energy, and response ranges of the cross sections in the spectrum are mostly the threshold regions[1]. Therefore such integral experiments are not so appropriate for the test of top regions or high energy regions of the cross sections, which are especially important for the fusion energy development.

Another type of neutron fields which possess to some degree flat spectrum with relatively high intensity up to above 10 MeV can be obtained using accelerators[2]. Recently we have proposed utilization of a neutron field employing Li(d,n) reaction on a thick target and demonstrated its feasibility for the integral test using a 4.5 MV Dynamitron accelerator at Fast Neutron Laboratory of Tohoku University[3].

In the present work, the Li(d,n) neutron field was applied to validate the four files, the precedent dosimetry file, IRDF-85[4] and the three recently compiled or released files, ENDF/B-VI[5], IRDF-90[6] and JENDL dosimetry file[7]. In the validation, the experimental average cross sections in the field were compared with calculated ones based on the above four files for selected fourteen dosimetry reaction cross sections.

Characteristics of Li(d,n) Neutron Field

The neutron target was a metal disk of natural lithium of 8 mm in diameter and about 1 mm in thickness. The target was mounted in a low mass chamber made of copper with an air cooling system as shown in Fig. 1. During irradiation, the target was bombarded by a DC deuteron beam of 10 μA with energy 2 MeV accelerated in the Dynamitron. Since the Q-value of the ^{7}Li(d,n) reaction is rather high (+15.02 MeV), the source produce neutrons with energies from a few hundred keV to about 17 MeV.

The angular spectra of the neutron source were measured using a time of flight spectrometer with a goniometer in the angular range from 0° to 90° by 5° or 10° step. In the measurement, a 0.25-MHz pulsed beam of 0.2 μA and 2-nsec pulse width was used. Emitted neutrons were detected by an NE-213 scintillation counter (2″ϕx2″t), which was placed at a distance of 5.3 meter from the neutron target in a heavy shield collimator system on the goniometer.

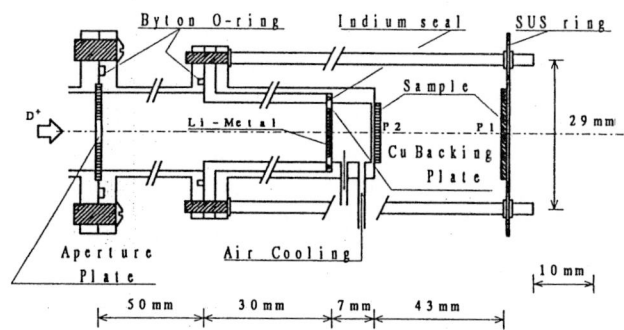

Figure 1: Chamber for the thick Li target. Irradiation positions (P1,P2) are also shown.

Relative efficiency curve of the neutron detector in the energy range up to 6 MeV was obtained by the measurement of the ^{252}Cf spontaneous fission spectrum[8], and that above 6 MeV was given by a Monte-Carlo calculation using O5S code[9]. The measured neutron spectrum (ϕ(E)) at 0° is shown in Fig. 2. Three arrows (A,B, and C) in the figure indicate peaks correspond to the ground and the broad excited states of the residual nucleus of ^{8}Be, respectively, and the lowest one (D) to neutrons from the (d,n) reaction on the contaminated carbon at the target surface. Because of the energy loss of the incident deuterons in the target, the shape of the peaks further broadened.

Measured angular distribution of the source neutrons in some energy intervals showed fairly flat below 15 MeV, while forward angular dependence was observed in the neutrons above 15 MeV.

Figure 3 shows the sensitivity functions (S(E)=ϕ(E)σ(E), where σ(E) is the cross section taken from IRDF-85) of the four representative reactions having various threshold energies for both the ^{252}Cf spontaneous fission neutrons and Li(d,n) neutrons spectra, respectively. In the case of the Li(d,n) source spectrum, the peaks of the respective functions shift appreciably up to the higher energy regions from the cases of the ^{252}Cf fission spectrum and is large even for the highest threshold reaction ^{58}Ni($n,2n$).

From above discussions and figures, the advantages of the Li(d,n) neutron source can be pointed out as follows: i) large

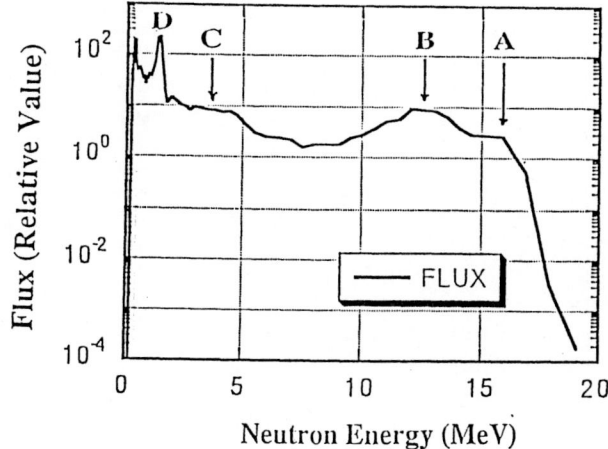

Figure 2: Measured neutron spectrum of the Li(d, n) source at 0°. Three arrows A,B, and C indicate peaks correspond to the ground and the broad excited states of the residual nucleus of ^{8}Be, respectively, and D to neutrons from the ^{12}C(d, n) reaction on the contaminated carbon.

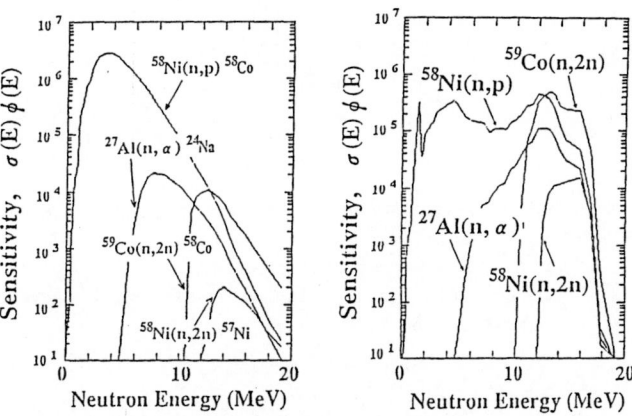

Figure 3: Response functions of the ^{58}Ni(n, p), ^{27}Al(n, α), ^{59}Co($n, 2n$) and ^{58}Ni($n, 2n$) reactions to the spectra of the ^{252}Cf (left) and Li(d, n) source (right), respectively.

response to the top of the cross sections especially for the high threshold reactions; ii) readiness of the spectrum measurement with less ambiguity by means of the TOF technique; iii) easy availability of the source only by a small accelerator.

On the other hand, the disadvantages of the source are also addressed: i) existence of the rather distinct structure and slight angular dependence in the spectrum; ii) less stability of the spectrum compared with the fission-driven neutron sources. However, these were not serious in the present experiment.

Benchmark Experiment

The average cross sections of the fourteen reactions (shown in Fig.4) have been measured in the neutron field relative to that of the standard reaction ^{27}Al(n, α)^{24}Na calculated using IRDF-85.

Two packets of the activation samples were prepared. Samples of each packet are arranged in a sandwich geometry as shown in Table 1 with their irradiation positions (see also Fig.1). Both packets were simultaneously irradiated for 48 hours. Measurement of gamma rays from the activated samples was performed with a calibrated high pure germanium detector to obtained the reaction rate, R.

At P2, the irradiated samples spanned over an angular

Table 1: List of the samples contained in the foil packets.

packets No./Pos.	foils (ordered from the source side)
#1/P1	Al,Nb,Ni,Ti,Zr,Fe,Mg,Al,Au,Nb,Al,Mn
#2/P2	Al,Nb,Cu,Ni,Fe,Nb,Al,NaCl

Note: Na(NaCl) and Mn(powder) were contained in thin plastic envelopes shaped in about 15 mmϕ x 1.5 mm in thick, Cu (15 mmϕx 1 mm in thickness), the rest are metal disks of 15 mmϕx0.1-0.2 in thickness; Pos.: irrad. position, see Fig. 1.

range from 0° to 20°, and the anisotropy effect of the neutron source on the samples were exactly corrected using the data of the angular spectra.

Average cross section of the reaction i, E^i is obtained from the following equation, relative to the average cross section $< \sigma^{Al} >_{Calc.}$ of the ^{27}Al(n, α) reaction,

$$E^i = (R^i/R^{Al}) < \sigma^{Al} >_{Calc.} F^i \qquad (1)$$

where, R^i is the reaction rate for i, R^{Al} is the reaction rate of the ^{27}Al(n, α) and F^i the correction factor of the flux difference between the foil positions for i and Al, which can be estimated from the reaction rates of the pair of the aluminum foils.

Comparison with Calculation

These data for the E^i were compared with corresponding calculated values, C^i from the following formula,

$$C^i = < \sigma^i >_{Calc.} = (\sum_g \phi_g^{ir} \sigma_g^i \Delta E_g)/(\sum_g \phi_g^{ir} \Delta E_g) \qquad (2)$$

where, ϕ_g^{ir} relative group flux of the Li(d, n) neutrons in a bin width of the group g ΔE_g, σ_g^i is i's group cross section taken from each dosimetry file.

In the error estimation of both the experimental and calculated average cross sections (E^i's and C^i's) and their ratios (C^i/E^i's), the correlations between the cross sections and those between the neutron flux were taken into account. The former covariance data were taken from the files (note: the covariances of JENDL[7] are tentatively the same as that of IRDF-85 if available). The later covariance data were estimated from the analysis of the spectrum data. Part of the flux covariances originated from the ^{252}Cf standard spectrum[8] and the Monte-Carlo calculations used for the estimation of the detector efficiency. Major experimental error sources in the R^i were the gamma counting statistics, the gamma detector efficiency calibration, and the flux corrections.

The results of the present benchmark experiment are summarized for each dosimetry file in Fig.4, where the C^i/E^i values with the ranges of the uncertainties are schematically compared for the four files. Large discrepancies are found for ^{23}Na($n, 2n$) (50 percent) in ENDF/B-VI and ^{63}Cu(n, α) (25 percent) in IRDF-85 compared with those of other reactions. Furthermore, all the libraries underestimated the average cross sections for the ^{54}Fe(n, p), and overestimated for the ^{55}Mn($n, 2n$) over 12 percent. Particularly, it is worth to note that the former reaction has been considered as one of the most significant and established reactor dosimetry reactions because its cross sections have already shown the excellent results in the fission neutron field benchmark tests [10,11]. In addition to the above reactions, the C/E values beyond each uncertainty are found in the following reactions; ^{58}Ni(n, p) and ^{59}Co(n,2n) of JENDL ; ^{23}Na($n, 2n$),^{58}Ni($n, 2n$) and ^{59}Co($n, 2n$)

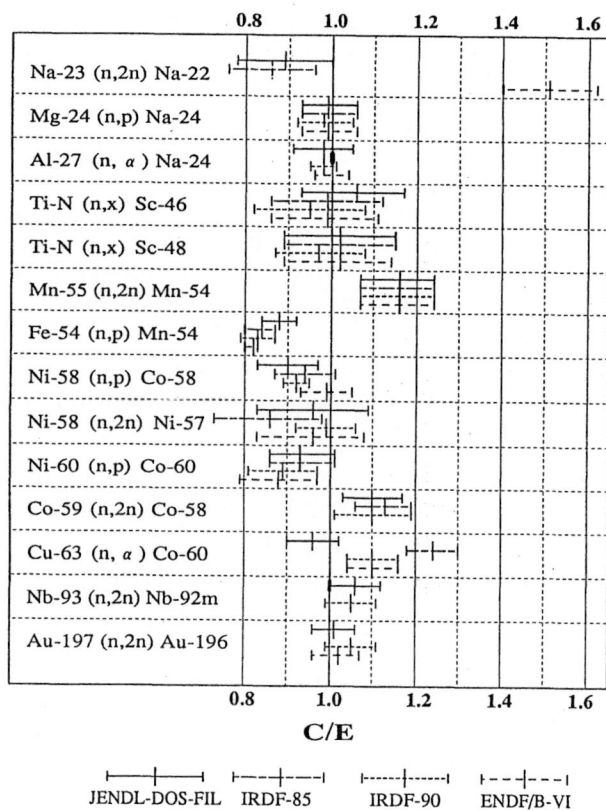

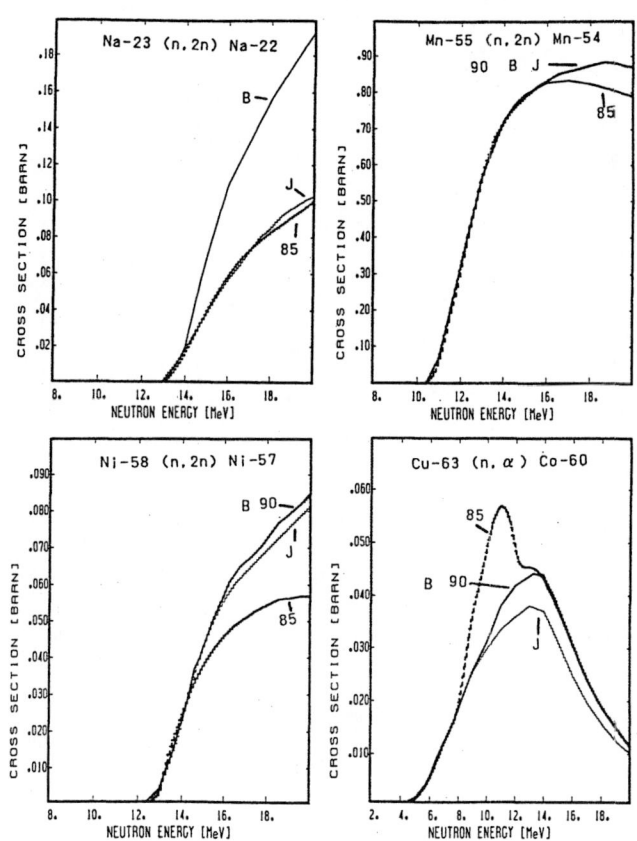

Figure 4: The C/E values of the $\mathrm{Li}(d, n)$ neutron spectrum average cross sections of the tested fourteen reactions in the four dosimetry files.

of IRDF-85; $^{60}\mathrm{Ni}(n,p)$ and $^{59}\mathrm{Co}(n,2n)$ of IRDF-90; $^{60}\mathrm{Ni}(n,p)$ and $^{63}\mathrm{Cu}(n,\alpha)$ of ENDF/B-VI, respectively.

The cross section curves of the four reactions which are picked up from among the reactions mentioned above are compared with each other in Fig. 5 a, b, c and d. From these figures, origins of the above discrepancies in the average cross sections between the files are obvious. Difference in the cross section above 16 MeV of the $^{55}\mathrm{Mn}(n,2n)$ among the files result in no effect for the average cross sections, because the region above 16 MeV is out of the sensitive range of the $\mathrm{Li}(d, n)$ neutron field.

Conclusions

The fourteen threshold dosimetry cross sections in the four dosimetry files have been tested in the thick target $\mathrm{Li}(d, n)$ neutron field. It has been demonstrated that the neutron source is suitable for the test of the cross sections, especially for the high threshold reactions. The differences in the cross sections between the files below 16 MeV were actually identified. There have been some reactions which showed rather large discrepancies in the four files; such as $^{23}\mathrm{Na}(n,2n)$ was serious for ENDF/B-VI and also not so good for other files; $^{54}\mathrm{Fe}(n,p)$ and $^{55}\mathrm{Mn}(n,2n)$ for all files. Further differential and integral studies on these reactions are needed.

With above exceptions, the performance of the recent dosimetry files JENDL and IRDF-90 were generally good. The IRDF-file have been improved by the revision up from 85 to 90.

References

1. D.L. Smith, Nuclear Cross Sections for Technology, Ed.

Figure 5: Comparison of cross section curves for picked up reactions from among ones which showed rather large discrepancies among the files or with the experiments. (a)$^{23}\mathrm{Na}(n,2n)$, (b)$^{55}\mathrm{Mn}(n,2n)$, (c)$^{58}\mathrm{Ni}(n,2n)$, and (d)$^{63}\mathrm{Cu}(n,\alpha)$. J: JENDL dos. file, 85: IRDF-85, 90: IRDF-90; and B: ENDF/B-VI.

by J.L. Fowler, C.H. Johnson, and C.D. Bowman, NBS Special Publication 594, USDOC/NBS, p285 (1980).

2. L.R. Greenwood, R.R. Heinrich, R.J. Kennerly and R. Medrzychowsky, Nuclear Technology 41. p109 (1978).

3. J.R. Dumais, S. Iwasaki, S. Tanaka, N. Odano and K. Sugiyama, Proc. of the 7th. ASTM-EURATOM Symposium on Reactor Dosimetry, Strasbourg, France, 27-31 August (to be published).

4. D.E. Cullen, N. Kocherov, and P.M. Mclaughlin, IAEA-NDS-41/R, rev.0 (1982). IRDF-85 is a modified version with additional cross section data.

5. BNL/ National Nuclear Data Center, ENDF/B-VI (1990),

6. N.P. Kocherov and H. Vonach, Proc. of the 7th. ASTM-EURATOM Symposium on Reactor Dosimetry, Strasbourg, France, 27-31 August (to be published).

7. S. Iijima, M. Nakazawa, K. Kobayasi, S. Iwasaki, T. Iguchi, Y. Ikeda, K. Sakurai, and T. Nakagawa, to be published in JAERI-M (1991).

8. W. Mannhart, IAEA-TECDOC-410, p158 (1987).

9. R.E. Textor and V.V. Verbinski, ORNL-4160, Oak Ridge National Laboratory (1968).

10. K. Sakurai, Reactor Dosimetry Method, Applications, and Standardization, STP 1001, Harry Farrar IV and E.P. Lippincott, editors, p277 ASTM (1989).

11. W. Mannhart, Handbook on Nuclear Activation Data, Technical Reports Series No. 273, p413, IAEA (Vienna, 1987).

RECENT ADVANCES IN UTILIZATION OF THE R-MATRIX
PARAMETERS FOR REACTOR APPLICATIONS*

R. N. Hwang

Argonne National Laboratory
9700 S. Cass Ave
Argonne, IL 60439, USA

Abstract: A simplified version of the rigorous pole representation of cross sections has been developed to facilitate utilization of the newly released ENDF/B VI resonance data based on the Reich-Moore parameters for reactor applications. The procedure is equivalent to the extraction of the Humblet-Rosenfeld-type parameters and the associated 'background' term explicitly from the rigorous pole and residue parameters which, in turn, are converted from a given set of the Reich-Moore parameters. The computational efficiency and its amenability to the existing reactor codes are significantly enhanced by the introduction of the pertinent analytic continuation in place of the smooth 'background' term via the non-linear least square fitting. The method has been successfully applied to all major nuclides examined and the results are presented.

Introduction

Two special features of the newly released ENDF/B VI resonance data that will have profound impacts on the cross section processing codes for reactor applications are the dramatic expansion of the resolved energy regions and the availability of the R-Matrix parameters based on the Reich-Moore formalism[2] for most major nuclides of practical interest. While the availability of the new resonance data has all but eliminated the long-standing difficulties attributed to the lack of sufficient resolved resonance information and the inadequacy of the Breit-Wigner single level representation for the closely spaced resonances, one obvious question of great practical concern is the compatibility of the new resonance representation with the existing reactor methodologies based almost exclusively on the traditional formalisms.

To facilitate the efficient utilization of the new R-matrix data, a simplified version of the rigorous pole representation described previously[3] has been developed so that the rigor of the Reich-Moore cross sections and the traditional feature of Doppler broadening via the Voigt profile essential to many existing codes can be preserved simultaneously. Extensive calculations have been carried out to demonstrate the viability of the proposed method and the results will be presented.

Natural extension of Rigorous Pole Representation

The theoretical justification of the rigorous pole representation is based on the rationale that the collision matrix must be single-valued and meromorphic in the momentum space. Any function that exhibits such properties must be a rational function according to the well known theorem in complex analysis. Thus, one obtains via partial fraction[3],

$$\sigma_x = \frac{1}{E} \sum_{\ell,J} \sum_{\lambda=1}^{N} \sum_{j=1}^{2(\ell+1)} \Re e \left\{ R_{\ell,J,j,\lambda}^{(x)} \cdot \frac{(-i)}{P_\lambda^{(j)*} - \sqrt{E}} \right\} \quad (1)$$

where x denotes the type of reaction under consideration and N is the total number of known R-matrix resonances. The genuine energy independent pole and residue parameters can be obtained from any given set of the Reich-Moore parameters using the WHOPPER code.[3] The expression leads immediately to the well-known Voigt profiles when subject to Doppler-broadening. However, the presence of $2(\ell+1)$ terms for each resonance to be evaluated in momentum domain is obviously undesirable from the point-of-view of computing efficiency, storage requirement and its amenability to existing codes. In the following discussion, it will be shown how the problem can be alleviated.

The $2(\ell+1)N$ poles for a given ℓ and J defined in Eq. 1 can be divided into two distinct classes. There are 2N s-wave-like poles with distinct spacings while the remaining $2\ell N$ poles are closely spaced independent of how well separated the input Reich-Moore resonances are. Such characteristics can be best visualized by examining the denominator of the rational function leading to Eq. 1.

$$P_{2(\ell+1)N}(\sqrt{E}) = P_{2N}^{(1)}(\sqrt{E}) \cdot P_{2\ell N}^{(2)}(\sqrt{E}) \quad (2)$$

where $P_{2N}^{(1)}(\sqrt{E})$ is a polynomial with its complex coefficients strongly dependent on the input resonance parameters whereas the coefficients of $P_{2\ell N}^{(2)}(\sqrt{E})$ are dominated by terms involving $k_o a$ with a = channel radius and $k_o = 2.197 \times 10^{-3}$ $[A/1+A)]$.

Thus, $P_{2\ell N}^{(2)}(\sqrt{E})$ reflects the higher order energy dependence of the penetration and shift factors and its roots are barely distinguishable. In addition, the magnitude of their imaginary components becomes of the order of $1/k_o a$ in the limit of small neutron widths. Such characteristics had already been illustrated analytically for the cases of a single level in Ref. 3. The sum of these pole terms involving poles with exceedingly large 'width' (or imaginary part) can be considered as the background contribution to the cross section with weak $\sqrt{E}$ dependence.

*Work performed under the auspices of the U.S. Department of Energy, Nuclear Energy Program.

In contrast, the roots of $P_{2N}^{(1)}(\sqrt{E})$ are distinct and always appear in pairs with their real components opposite in signs but not necessarily equal in magnitude spanning over $(\sqrt{E_{min}} \leq \sqrt{E} \leq \sqrt{E_{max}}$ and $-\sqrt{E_{max}} \leq \sqrt{E} \leq -\sqrt{E_{min}})$ on the real axis of the momentum domain. Let $p_\lambda^{(1)}$ and $p_\lambda^{(2)}$ be poles with positive and negative real components, respectively. It is important to note that the actual interval of interest in computing cross sections is half of the interval taken to be $(\sqrt{E_\ell} \leq \sqrt{E} \leq \sqrt{E_u})$ where E_u $(< E_{max})$ and E_ℓ $(> E_{min})$ are the upper and lower resolved region boundaries, respectively. Because of their short range nature of fluctuation, all terms involving $p_\lambda^{(2)}$ constitute another smooth component to σ_x that reflects the contributions from the tails of outlying poles outside the interval of practical interest.

Let $q_\ell^{(x)}(\sqrt{E})$ denote the contributions from those $2\ell N$ terms involving poles with giant 'width'. By taking advantage of the characteristics of poles described above, Eq. 1 can be cast into the same form as that of Humblet-Rosenfeld.[4]

$$\sigma_x = \frac{1}{E} \sum_\ell \Re e \left\{ \sum_J \sum_{\lambda=1}^{N} \left[R_{\ell,J,1,\lambda}^{(x)} \cdot \frac{2(-i)\sqrt{E}}{(p_\lambda^{(1)*})^2 - E} \right] \right.$$
$$\left. + s_\ell^{(x)}(\sqrt{E}) + q_\ell^{(x)}(\sqrt{E}) \cdot \delta_\ell \right\} \qquad (3)$$

where

$$s_\ell^{(x)}(\sqrt{E}) = \sum_J \sum_{\lambda=1}^{N} \left[\frac{R_{\ell,J,2,\lambda}^{(x)}(-i)}{p_\lambda^{(2)*} - \sqrt{E}} - \frac{R_{\ell,J,1,\lambda}^{(x)}(-i)}{(-p_\lambda^{(1)*}) - \sqrt{E}} \right] ; \qquad (4)$$

$$\delta_o = 0, \ \delta_\ell = 1 \ for \ \ell > o$$

Hence, for a given range of practical interest, the rigorous pole representation can be viewed as a combination of a 'fluctuating' term consisting of N poles with $\Re e \ p_\lambda^{(1)} > 0$ expressed in the energy domain consistent with the traditional formalisms and two 'non-fluctuating' terms attributed to the tails of outlying poles with negative real component and the poles with extremely large 'width' (or $\left| \Im m \ p_\lambda^{(j)} \right|$) for $\ell > 0$ states respectively. The striking behavior of the 'fluctuating' and 'non-fluctuating' components have been confirmed in recent calculations for all major nuclides specified by the Reich-Moore parameters in the ENDF/B VI files.

Eq. 3 represents no more than an empty form unless the background terms can be evaluated efficiently. The smooth behavior of these terms clearly suggests that their energy dependence can obviously be reproduced by other simpler functions within the finite interval of practical interest. It is well known in numerical analysis that the rational functions are best suited to approximate a well behaved function within a finite range. Hence, the obvious choice is to set the approximate functions $\hat{s}_\ell^{(x)}(\sqrt{E})$ and $\hat{q}_\ell^{(x)}(\sqrt{E})$ to be rational functions of arbitrary order. If the order of polynomials in the numerator and the denominator are taken to be MM and NN respectively, a total of (MM + NN + 2) complex coefficients (twice as many entries) can be used as the variables to be determined in the non-linear fitting procedure. Because of the extremely well behaved nature of $s_\ell^{(x)}(\sqrt{E})$ and $q_\ell^{(x)}(\sqrt{E})$, the MM and NN required are expected to be very small. One attractive feature of the proposed method is that the rational functions so obtained can be again expressed in the form of pole expansion via partial fraction, i.e.,

$$\hat{s}_\ell^{(x)}(\sqrt{E}) = \frac{P_{MM}^{\rightarrow}(\sqrt{E})}{Q_{NN}(\sqrt{E})} = \sum_{\lambda=1}^{NN} \frac{r_\lambda^{(x)}(-i)}{\alpha_\lambda^* - \sqrt{E}} \qquad (5)$$

$$\hat{q}_\ell^{(x)}(\sqrt{E}) = \sum_{\lambda=1}^{NN} \frac{b_\lambda^{(x)}(-i)}{\xi_\lambda^* - \sqrt{E}} \qquad (6)$$

if NN > MM.

Conceptually, the procedure is equivalent to determining the analytic continuations of $s_\ell^{(x)}(\sqrt{E})$ and $q_\ell^{(x)}(\sqrt{E})$ for $\sqrt{E_\ell} \leq \sqrt{E} \leq \sqrt{E_u}$. The pole and residue parameters so obtained can be viewed as 'pseudo' pole parameters. A code (WHOPJR) based on the MINPACK-package[5] has been developed to compute these 'pseudo' pole parameters. To provide sufficient accuracy to cross sections, NN of no greater than 3 is required.

The Doppler-broadening of Eq. 3 and Eq. 5 immediately leads to the similar form defined by the traditional formalisms:

$$\sigma_x(E,T) = \frac{1}{E} \sum_\ell \Re e \left\{ \sum_J \sum_{\lambda=1}^{N} 2\sqrt{E} R_{\ell,J,1,\lambda}^{(x)} \frac{\sqrt{\pi}}{\Delta_E} W\left(\frac{E-e_\lambda}{\Delta_E} \right) \right.$$
$$\left. + \hat{s}_\ell^{(x)}(\sqrt{E},T) + \hat{q}_\ell^{(x)}(E) \cdot \delta_\ell \right\} \qquad (7)$$

where

$$\hat{s}_\ell^{(x)}(\sqrt{E},T) = \sum_{k=1}^{NN} \frac{r_k^{(x)}}{\Delta_m} \left[\sqrt{\pi} \ W\left(\frac{\sqrt{E}-\alpha_k^*}{\Delta_m} \right) \right] \qquad (8)$$

$$e_\lambda = [p_\lambda^{(1)*}]^2 \qquad (9)$$

$$\psi(x,y) + i\chi(x,y) = \sqrt{\pi} \ y \ W(z) \qquad (10)$$

and $\Delta_E = 2\sqrt{E} \ \Delta_m$ denotes the Doppler width in energy and momentum domains respectively.

Thus, the only difference between Eq. 7 and the traditional formalisms is the presence of the 'background' terms explicitly defined by a handful of 'pseudo' poles with weak energy and temperature dependence. Its compatibility to all ENDF/B format based codes is quite apparent.

Results and Conclusions

Extensive calculations have been carried out in order to demonstrate the viability of the proposed method. The studies include all major actinides and two structural isotopes.

The viability of the method can be best illustrated by examining the relative errors in the resulting cross sections with respect to the directly computed Reich-Moore cross sections. Representative results corresponding to the absorption cross sections of ^{239}Pu and ^{238}U are shown in Fig. 1 and Fig. 2 respectively. These figures show the behavior of the absolute values of relative errors in great detail as a function of $\sqrt{E}$ over the $(\sqrt{E_\ell} \le \sqrt{E} \le \sqrt{E_u})$. By and large,

the relative errors and the cross section values are anti-correlated. The peaks of the former correspond to the valleys of the latter and vice versa. The lower curves represent the inevitable errors attributed to the computations of parameters when all poles defined by Eq. 1 are included. The errors are generally less than 10^{-9} and can reach 10^{-7} as E approaches zero. Thus, Eq. 1 is not only analytically rigorous but also numerically exact for practical purposes. The upper curves show the corresponding errors when the simplified version is used. It is important to note that the results are based on the tolerance limit of 10^{-5} and NN=3 in computing the background terms. The maximum errors are of the order of 10^{-4} near the valleys between resonances (much smaller for fissionable nuclides) and are clearly tolerable in practical applications.

It is, therefore, reasonable to conclude that the proposed method is not only readily amenable to methodologies based on the traditional formalisms but also preserves the rigor of the Reich-Moore cross sections. The procedure is being implemented in the MC^2-2 code.[6]

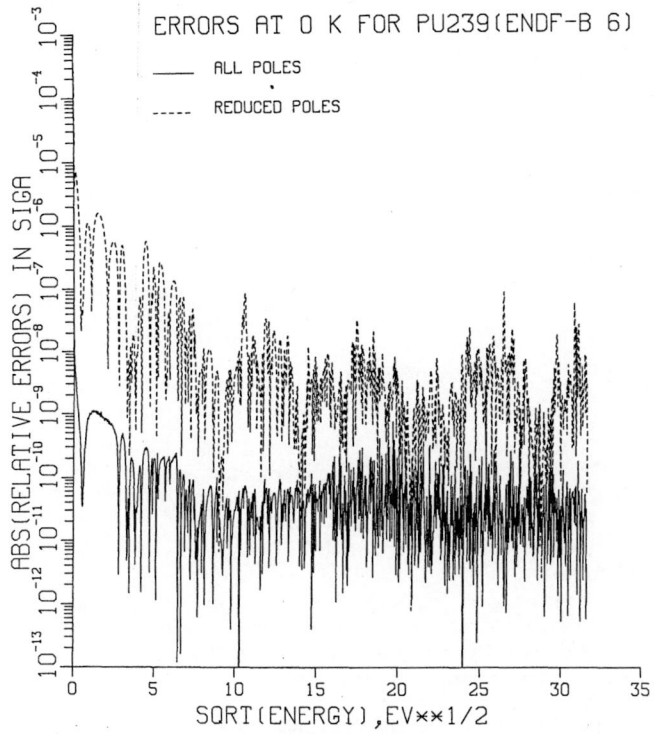

Fig. 1.   Relative errors in σ_a of ^{239}U based on the parameters converted from ENDF/B VI data.

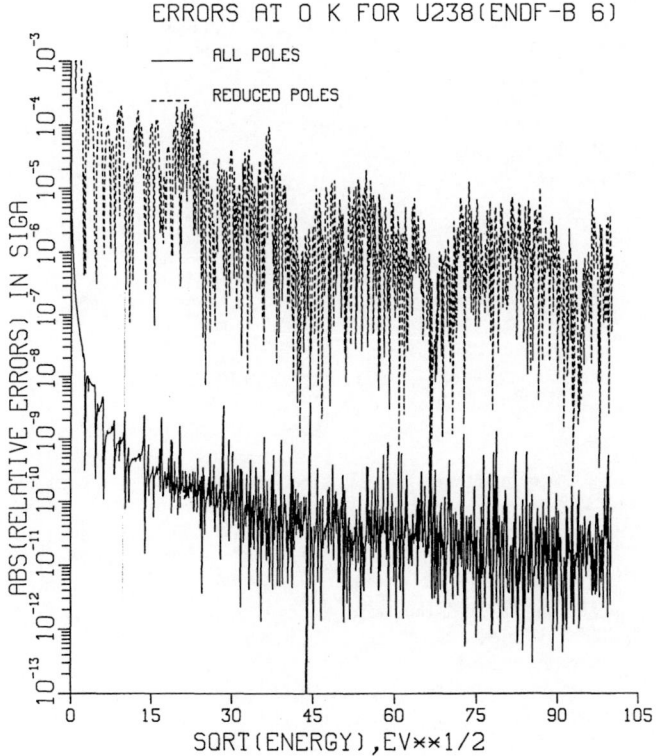

Fig. 2   Relative errors in σ_a of ^{238}U based on the parameters converted from ENDF/B VI data.

References

1.   P. F. Rose and C. L. Dunford, Ed., 'ENDF-102 Data Formats and Procedures for Evaluated Nuclear Data File ENDF-6,' BNL-NCS-44945 (1990).

2.   C. W. Reich and M. S. Moore, Phys. Rev. 111, 929 (1958).

3.   R. N. Hwang, Nucl. Sci. Engr. 96, 192 (1987).

4.   J. Humblet and L. Rosenfeld, Nucl. Phys. 26, 529 (1961).

5.   J. J. Moré, B. S. Garbow and K. E. H. Millstrom, 'User Guide for MINPACK-1,' ANL-80-74 (1980).

6.   H. Henryson, B. J. Toppel and C. G. Stenberg, 'Mc^2-2: A Code to Calculate Fast Neutron Spectra and Multigroup Cross Sections,' ANL-8144 (1976).

REPRESENTATION OF THE FISSION SPECTRUM
AND MULTIPLICATION CALCULATIONS

R. L. Perel, J. J. Wagschal

Racah Institute of Physics
The Hebrew University of Jerusalem
91904 Jerusalem, ISRAEL

Abstract: We investigated the different representations of the fission process as total fission on one hand and as the sum of direct fission, first chance fission, second, third and fourth chance fission on the other hand (MT = 18 versus MT = 19, 20, 21 and 38 in ENDF terminology).

The representations of these cross sections (MF = 3) in ENDF/B libraries are generally consistent, but the fission spectrum (MF=5) of the total fission given in libraries is generally not equivalent to the fission spectrum obtained from all partial spectra.

We calculated the effective multiplication factor k_{eff} of several simple near critical systems and the neutron leakage from a central 14 MeV neutrons source embedded in simple subcritical systems. We found that in certain cases there is a non-negligible difference in the results, depending on the representation of the fission spectrum that was used; in other cases the differences were small. These results are explained in the paper.

As a conclusion we propose improvements in the methodology of the representation of fission in ENDF-6 format and also suggest changes in the data of some materials in ENDF/B-VI.

(Fission spectrum, ENDF/B-VI, direct fission, first chance fission)

Introduction

The treatment of the fission spectrum in transport calculations has undergone many refinements over the last two decades. In early times [1] it was generally accepted to use the same, incident neutron independent, fission spectrum for all materials in a single calculation. Later [2], it became evident that the fission spectrum is not only material dependent, but it also depends on the incident neutron energy. Since the results of calculations (criticality or shielding) are sensitive to this effect, an increasing number of calculations are done with the full fission matrix. Another refinement of the description of the fission process, that is well known, is the more detailed description of the fission process as the sum of direct fission, first chance fission (E > ~ 6 MeV), second chance fission (E > ~ 12 MeV) and so on; in ENDF [3] terminology the fission process is differentiated as reactions MT=19, MT=20, MT=21 and MT=38 respectively, in addition to their sum, the total fission MT=18.

We were interested to see the sensitivity of simple neutron transport calculations to the representation of the fission process. Moreover, while the evaluators of ENDF/B libraries generally made sure that the detailed fission cross sections (MF=3) sum up to the total fission cross section, the fission spectrum (MF=5) of the total fission given in libraries is generally not equivalent to the effective fission spectrum obtained from all partial spectra. Typically (in ENDF/B-IV or for certain materials in ENDF/B-VI) the total fission spectrum is given as either a Maxwellian or as a Watt spectrum, while the partial spectra are given as mixture of a fission spectrum (Maxwellian or Watt) for the fission part of the reaction and of an evaporation spectrum for the inelastic part of the reaction. In other cases the data given for the detailed fission is either incomplete or completely missing. In any case, we would like to know how results of transport calculations are affected.

Representation of Fission

The fission process in ENDF/B format is described as total fission on one hand and as direct fission, first chance fission, second, third and fourth chance fission on the other hand (MT = 18 versus MT = 19, 20, 21 and 38 in

ENDF terminology). For all these reactions the ENDF/B libraries should define cross sections, $\sigma(E)$, (MF = 3), angular distributions - normally isotropic - (MF = 4) and energy distributions, $f(E,E')$, (MF = 5) of the secondary neutrons.

In addition, ENDF/B libraries define the number of secondary neutrons per fission, ν, in file MF = 1. MT = 452, 455 and 456 are given to describe the number of total, delayed and prompt neutrons per fission, respectively. No specific definition is given whether these data apply to MT = 18, 19, 20, 21 or 38. The tacit assumption is that the data surely apply to the total fission MT = 18, and - lacking any other way of assigning neutron multiplicities in the ENDF formalism - also to the other MT numbers. This might be in contradiction to physical reality, where the expected number of secondary neutrons from first chance fission for instance in general is not equal to the number of secondary neutrons from direct fission triggered by a neutron of the same initial energy [4].

In order to be consistent, the total and detailed fission cross section, multiplicity and spectrum should satisfy the following sum rules:

$$\sigma_{MT=18}(E) = \Sigma_{MT} \, \sigma_{MT}(E) \qquad (1)$$

$$\nu_{MT=18}(E) \, \sigma_{MT=18}(E) = \Sigma_{MT} \, \nu_{MT}(E) \, \sigma_{MT}(E) \qquad (2)$$

$$\nu_{MT=18}(E) \, \sigma_{MT=18}(E) \, f_{MT=18}(E,E') = $$
$$\Sigma_{MT} \, \nu_{MT}(E) \, \sigma_{MT}(E) \, f_{MT}(E,E') \,, \qquad (3)$$

where the summation over MT is performed on MT = 19, 20, 21 and 38.

The first sum rule is usually followed by all evaluations of cross sections that have been published in ENDF/B libraries. Currently there is no way to represent ν correctly for MT other than 18, and therefore there is no way for a user of ENDF libraries to check consistency, unless the implicit assumption is made that ν is independent of MT. With this assumption, equations (2) and (3) can be simpified as follows:

$$\nu(E) \; \sigma_{MT=18}(E) \; = \; \nu(E) \; \Sigma_{MT} \; \sigma_{MT}(E) \tag{2'}$$

$$\nu(E) \; \sigma_{MT=18}(E) \; f_{MT=18}(E,E') \; = \; \nu(E) \; \Sigma_{MT} \; \sigma_{MT}(E) \; f_{MT}(E,E') \; . \tag{3'}$$

Equation (2') was followed by the evaluators of ENDF/B-IV, however, in practice we have seen that the sum rule (3') is generally violated, i.e. the fission spectrum calculated directly from data given in MF=5, MT=18, is generally different from the spectrum calculated according to (3') based on data from MF=5, MT=19,20,21,38.

In general, evaluators of cross sections in ENDF/B libraries have always included all needed data for the total fission (MT=18). As to the more detailed fission description (MT=19, 20, 21, 38), one of the following options has been selected:
1. All data are given.
2. Only the partial cross sections (MF=3) are given, but no energy spectra (MF=5).
3. No data are given.
In ENDF/B-VI evaluators chose option 1 for U-238 and Pu-240, option 2 for U-235 and Pu-239, and option 3 for U-233. In ENDF/B-IV evaluators chose option 1 for U-235 and Pu-239.

Unless all data are given explicitly (option 1), one cannot check the impact of the more detailed representation of the fission process on the results of transport calculations.

Sensitivity to the Representation of the Fission Spectrum

In order not to obstruct the transport calculations with irrelevant data, we demonstrate the sensitivity of the results for "synthetical" bare spherical systems composed of a single isotope only. On one hand, we checked the change in the effective multiplication factor, k_{eff}, of nearly critical assemblies with suitable radius and density. On the other hand, we analyzed the leakage out of (somewhat smaller) subcritical spheres pulsed by a central 14-15 MeV neutron source. The definition of these systems is displayed in Table 1.

Table 1: Definition of the "synthetical" systems

Isotope	Density [g/cm^3]	Radius (k-calculation) [cm]	Radius (for Leakage) [cm]
U -235	18.5	8.6	7.8
U -238	18.5	-	5.0
Pu-239	15.44	6.385	5.8
Pu-240	15.44	9.7	9.0

The transport computations were done according to the CSEWG [5] recommendations:
- 34 energy groups (26 energy groups of half lethargy width with an upper energy of 10 MeV, augmented and refined by groups of 1 MeV width between 6 Mev and 15 MeV in order to be able to detect the sought effects)
- S$_n$ order of 16
- 40 radial intervals in the sphere.
Three anisotropic terms have been kept in the flux Legendre expansion. In all cases fission matrices, rather than incident energy independent fission spectra, have been used.

The group cross sections were generated using the NJOY.89 [6] code with data from ENDF/B-IV or -VI. This code can either generate the effective fission matrix as the sum of the detailed matrices of MT=19, 20, 21, 38 or compute the fission matrix with the total fission cross section and with the total fission spectrum (MT=18 only), depending on the input and on the availability of the relevant data in the ENDF/B file.

Inconsistent Spectra

The cases where full data were given (option 1 above) include U-235 and Pu-239 in ENDF/B-IV, but not in the recently released ENDF/B-VI, and U-238 and Pu-240 in ENDF/B-VI. In all these cases the fission spectra (MT=18, 19 and the fission part of MT=20, 21, 38) were described as a Watt- or Maxwell-spectrum, and the inelastic part of MT=20, 21 and 38 was described as an evaporation spectrum. In contrast to the other materials, where the numerical parameters (e.g. "temperature") describing the fission spectra are the same for total fission and for the fission part of the other fission reactions, for U-238 in ENDF/B-VI the parameters are different. As a result, with the exception of U-238, the spectrum of the total fission reaction is harder than the effective spectrum derived (according to formula (3')) from the partial spectra. For U-238 in ENDF/B-VI the evaluators chose the parameters of the total fission reaction in such a way that the average energy of the total fission spectrum is the same as the average energy of the corresponding sum of the partial spectra.

The results of the k_{eff} and leakage calculations for the different systems that were defined in Table 1 are summarized in Table 2.

Table 2: k and leakage calculated with different fission spectra

Iso-tope	MT	k 19+20+21	k 18	leakage* (neut/s) 19+20+21	leakage* (neut/s) 18
U -235[†]		1.00384	1.00385	28.330	28.453
U -238[†]		-	-	2.218	2.218
Pu-239[†]		1.00545	1.00589	27.966	28.283
Pu-240[‡]		1.00013	1.00217	45.670	48.420

* Source normalized to one 14-MeV neutron per second at the center of the sphere
[†] ENDF/B-IV data
[‡] ENDF/B-VI data

For Pu-239 and Pu-240 we found that k_{eff} computed with the total fission spectrum is higher than when computed with the spectrum obtained from the sum of partial fission spectra. For example a nearly critical sphere of pure Pu-239 has a k_{eff} of 1.00589 when calculated with ENDF/B-IV data and MT=18 fission spectrum. When the fission spectrum results from the sum of the partial fission processes, k_{eff} is only 1.00545, so that the result of MT=18 is higher by 0.4 mk (1 mk = a change of 0.001 in k_{eff}). The leakage out of a somewhat smaller sphere computed using MT=18 is higher by 1% than the one computed the other way (leakage of 28.28 vs. leakage of 27.97). The leakage is higher in all energy groups. Calculations of similar Pu-240 systems using ENDF/B-VI data gave even more pronounced differences: with the MT=18 spectrum k_{eff} is higher by 2 mk, and the leakage is higher by 6%. Remembering that the experimental ' uncertainties in k_{eff} of assemblies such as Godiva or Jezebel are 0.8 mk and 1.7 mk respectively, then the discrepancies found are not negligible at all.

On the other hand, for U-235 in ENDF/B-IV the differences in k_{eff} and in leakage between the two modes of calculation are quite small (0.01 mk and 0.4% respectively). For U-238 in ENDF/B-VI the difference in leakage is practically zero.

Even for the Uranium assemblies where the calculated differences in k_{eff} or in leakage are small, a more detailed analysis of the energy spectrum of the leakage shows much larger discrepancies. In the case of leakage from a subcritical U-235 assembly differences of 2% were found in the leakage for certain energy groups, and differences of 5% were found for the U-238 system (see Figure 1).

210

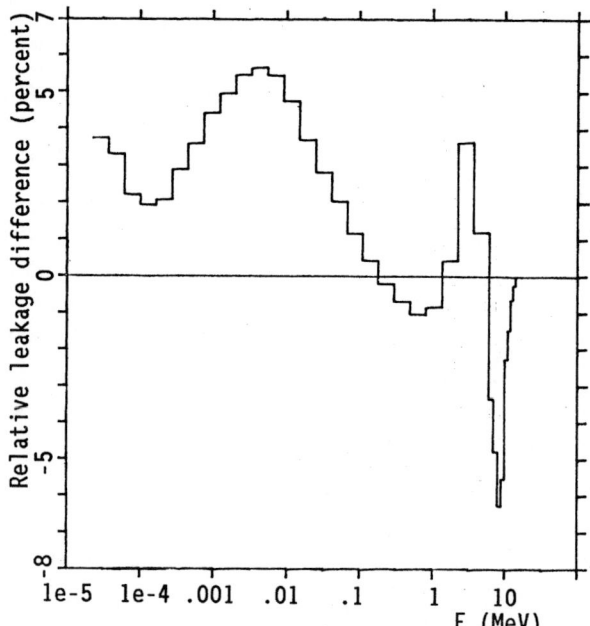

Figure 1. Leakage(MT=18) / Leakage(MT=19-21) - 1
as calculated for a centrally 14 MeV pulsed sphere
(U-238, R=5cm, ρ=18.5 g/cm^3) as function of energy.

Missing Spectra

The ENDF/B-VI evaluators of U-235 and Pu-239 chose not to give any spectral data (MF = 4 or 5) for the detailed fission reactions. According to the built-in documentation in the ENDF/B-VI file itself, the spectral data given for the total fission take into account the effects of first chance fission, second chance fission and so on. The energy distribution of the total fission is not given by any analytical formula, but as a set of discrete tables with an interpolation scheme. (The interpolation scheme for Pu-239 is LF=1, i.e. constant, contradicting the official recommendation for interpolation schemes for secondary energy spectra which should be linear-linear [3]).

By choosing not to give any spectral data for the detailed fission description, the contradictions shown above might have been avoided. On the other hand, the user has no possibility to do his own cross check for consistency. Without addition to the ENDF formalism that allows different values of ν for partial fission, there will probably be no way to display full and consistent data for partial fission spectra. The display of the total spectrum in discrete form might be more accurate, but the elegance of the analytical formula is lost: Actually the ENDF/B-VI U-235 and Pu-239 fission spectra are results of calculations which used analytical spectra of the Madland - Nix formalism.

Discussion

As we have pointed out, in U-235 and Pu-239 of ENDF/B-IV and in Pu-240 of ENDF/B-VI the fission part of all reactions (MT=18-21) is defined as a Maxwellian or a Watt spectrum, the temperature of it being the same for all reactions. The inelastic part of MT=20 and MT=21 is described as an evaporation spectrum, which is softer than the Maxwellian/Watt spectrum. Therefore the spectrum of total fission is harder than the effective spectrum of the sum of all partial reactions. Since normally in the fast energy range $\nu\sigma_{fis}$ increases with energy, it is clear that k_{eff} and the multiplication, as reflected in the leakage, in the inhomogenous problem, are higher for the single total fission representation. The increase in $\nu\sigma_{fis}$ in the region around 1 MeV - which is the most important energy region in the fission spectrum - is specially pronounced in the Pu isotopes, and therefore transport calculations are quite sensitive to the differences in spectra. In U-235 the change in $\nu\sigma_{fis}$

around 1 MeV is less pronounced, therefore transport calculations are less sensitive to the fission representation.

The energy-dependent Watt spectrum parameters of the U-238 total fission in ENDF/B-VI were chosen, so that the resulting spectrum has the same average energy as the effective fission spectrum obtained from the sum of the partial fission reactions. This choice of parameters led to a calculated leakage that is not sensitive to the way the fission matrix is computed. But even this clever way of choosing parameters in order to improve consistency, still leaves us with discrepancies of 5% in the energy spectrum of the leakage in certain energy groups.

Conclusion and Recommendations

While <u>evaluating and editing ENDF/B</u> cross section files, care should be taken to define the spectra of the total and of the detailed fission processes representations consistently. A change in ENDF methodology is required in order to enable definition of the average number of neutrons per fission (prompt+delayed, prompt, delayed) for each fission-MT: 18, 19, 20, 21, 38. As long as such a change has not been adopted, the consistent way is giving the spectral data (MF=4 and MF=5) for MT=18 only. For elegance and physical insight, an analytic representation (or sum of analytic representations) is preferred over discrete tables of numbers. In any case, the ENDF/B recommended linear-linear interpolation scheme should be used.

While <u>using cross section data</u> (for creation of group averaged cross sections) one should know whether the data file contains inconsistent information. Once again the rule is confirmed: Do not rely blindly on your cross section processing code, be sure to use the better data as appropriate in each case.

References

1. W. W. Engle, Jr., ANISN, A One-Dimensional Discrete Ordinates Transport Code with Anisotropic Scattering, K-1693, Oak Ridge Gaseous Diffusion Plant (1967).

2. J. J. Wagschal and A. Yaari, A Systematic Test of the ENDF/B-III Evaluated Cross-Section Library on 'Clean' Critical Assemblies, in Nuclear Data in Science and Technology II, IAEA-SM-170/19, International Atomic Energy Agency (1973).

3. ENDF-102, Data Formats and Procedures for the Evaluated Nuclear Data File, ENDF, Revised by R. Kinsey, National Nuclear Data Center, Brookhaven National Laboratory (1979).
See also:
ENDF-102, Data Formats and Procedures for the Evaluated Nuclear Data File ENDF-6, Edited by P. F. Rose and C. L. Dunford, National Nuclear Data Center, Brookhaven National Laboratory, BNL-NCS-44945, Informal Report (July 1990).

4. E. D. Arthur, P. G. Young, D. G. Madland, and R. E. MacFarlane, Evaluation and Testing of n + Pu-239 Nuclear Data for Revision 2 of ENDF/B-V, Nuc. Sci. Eng., **88**, 56-70 (1984).

5. ENDF-202, Cross Section Evaluation Working Group Benchmarks Specifications, National Neutron Cross Section Center, Brookhaven National Laboratory, (1974 with later revisions and additions).

6. R. E. MacFarlane, D. W. Muir, and R. M. Boicourt, The NJOY Nuclear Data Processing System, Vols 1 to 4, LA-9303-M, (ENDF-324), Los Alamos National Laboratory, (May 1982 to December 1985).

SN VALIDATION OF JEF-1 DATA ON THE PCA-REPLICA SHIELDING BENCHMARK

Massimo Pescarini

ENEA-INN
Viale G.B. Ercolani 8, 40138 Bologna, Italy

Abstract: The results of a DOT 3.5-E/JEF-1 validation on the (H2O/Fe) PCA-REPLICA (UKAEA-Winfrith) low-flux shielding benchmark are presented. The PCA-REPLICA experiment reproduces the ex-core radial geometry of a PWR and is closely related to LWR safety since it is dedicated to test the accuracy of the calculated neutron exposure parameters (fast fluence and iron displacement rates) in a pressure vessel simulator. The SN 1-D ANISN code is used to collapse cross sections from the VITAMIN-J (175 n) shielding library, based on the JEF-1 data, to a 28 group working library for 2-D calculations. A 3-D-equivalent synthesis (X,Y,Z) of 2-D and 1-D DOT 3.5-E SN calculations in plane geometry, gives the integral and spectral results for comparison with the respective experimental data. The underprediction of the in-vessel dosimeter measured activities depends probably on an overestimation of the Fe-56 inelastic scattering cross section in the JEF-1 natural iron file.

Introduction

This work represents a summary of the ENEA report /1/ containing the complete results of neutron transport calculations on the (H2O/Fe) PCA-REPLICA shielding benchmark. The preliminary results /2/ have been already presented at the 7th (UKAEA/OECD-NEA) International Conference on Radiation Shielding (Bournemouth, U.K., 1988). In particular, the PCA-REPLICA experiment was carried out in UKAEA-Winfrith (U.K.) in the ASPIS facility of the NESTOR low-flux experimental reactor. Moreover, the PCA-REPLICA experiment was the preliminary step of the U.K. NESTOR Shielding and Dosimetry Improvement Programme (NESDIP) strictly connected with the U.K. Sizewell B PWR project. It is included in the experimental shielding benchmarks for nuclear data testing, proposed by the Nuclear Energy Agency Committee on Reactor Physics (OECD/NEACRP). The experiment is addressed to the study of the neutron penetration across alternating layers of water and steel simulating the ex-core radial geometry of a typical PWR. Integral and spectral results from SN transport calculations are presented. The cross section data used are exclusively based on the JEF-1 Euro-Japanese nuclear data file. Fruitful advice and valuable suggestions were obtained from informal cooperation with the Institut fur Kernenergie (IKE-University of Stuttgart).

PCA-REPLICA Experimental Details

The PCA-REPLICA /3/ experimental facility duplicated exactly the 12/13 geometrical configuration of the ORNL PCA (Pool Critical Assembly) experiment /1*/. Unlike the PCA, where the neutron source was a low-power reactor, in PCA-REPLICA the source was a fission plate with a cross-sectional area identical to that of the PCA reactor source. The thin fission plate (whose rectangular dimensions were 63.5 x 40.2 x 0.6 cm) was made of highly enriched uranium (93.0 w % in U-235) alloyed with aluminium. During the experiment, it was irradiated by the NESTOR reactor through a graphite thermal column, in the ASPIS shielding facility. Beyond the fission plate, the REPLICA shielding array was arranged in a large parallelepiped steel tank filled with water and surrounded by a thick concrete shield. After the first water gap (12.1 cm), there was the stainless steel thermal shield simulator (5.9 cm), the second water gap (12.7 cm), the mild steel vessel (RPV) simulator (thickness T=22.5 cm) and finally a void box (29.58 cm) made of a thin layer of aluminium, simulating the cavity between the RPV and the biological shield in a real PWR. All these elements were perfectly orthogonally aligned and centred along the imaginary line (horizontal axis Z) passing through the centre of the fission plate. Along the Z axis, three types of threshold detectors gave the integral measurements: Rh-103 (n,n') Rh-103m, In-115 (n,n') In-115m and S-32 (n,p) P-32. Their effective threshold energies, in the U-235 fission spectrum, are 0.69, 1.30 and 2.8

MeV. Moreover, in the same spectrum, the Rh-103, In-115 and S-32 median energies are respectively 1.9, 2.6 and 4.0 MeV and the energy ranges corresponding to 90% of the response are respectively 0.72-5.8 MeV, 1.1-5.9 MeV and 2.3-7.3 MeV. Spectral measurements were performed in two positions: in the mild steel, at a quarter thickness (position T/4) of the RPV simulator (at 6.4 cm from the RPV inner surface: fission plate side) and in the void box, 3.5 cm beyond the RPV. Two kinds of spectrometer were used. In particular spherical hydrogen-filled proportional counters were employed (type SP-2; internal diameter 40.0 mm). Individual counters with gas fillings of approximately 0.5, 1.0, 3.0 and 10.0 atmospheres were used in combination, to cover the energy range from 50.0 keV to 1.2 MeV. Neutron fluxes between 1.0 and 10.0 MeV were determined with a spherical 3.5 ml organic liquid (NE213) scintillator. Complete experimental details are reported in ref. /3/.

Nuclear Data and Processing

The NEA VITAMIN-J /1*/ neutron shielding library, based on the JEF-1 /1*/ nuclear data file, is the primary cross section source for the transport calculations. It covers the energy range 1.0E-5 eV - 19.64 MeV with 175 groups and was obtained through the NJOY /1*/ data processing system (June 1983 version) based on the Bondarenko self-shielding factor approach for resonance absorption. The detector cross sections for Rh-103, In-115 and S-32, processed into the VITAMIN-J group structure by using the GROUPIE /1*/ code, were obtained from OECD/NEA-Saclay, satisfying in this way a need for standardization with a previous JEF-1 NEA-UKAEA validation /1*/ for the JEF-1 iron file on the pure iron ASPIS shielding benchmark. The Rh-103 cross sections were taken from the ACTL-82 dosimetry file, while those for In-115 and S-32 were taken from the ENDF/B V file. The preparation of a binary group-independent working library for the ANISN /1*/ code is carried out using the TAPEMAKER program.

Transport Calculations

The SN 1-D ANISN code is used to obtain a macroscopic problem-dependent working library, collapsing cross sections from the VITAMIN-J group structure to a 28 group structure for the calculations with the DOT 3.5-E /1*/ code. This structure contains 26 groups above 0.1 MeV, with an average lethargy width of 0.2. Secondly, a collapsed library of microscopic detector cross sections is obtained through ANISN. Plane geometry is used for both ANISN and DOT 3.5-E. The approximation S8-P3 is used for all the ANISN and DOT runs. In DOT, the fully symmetric S8 quadrature set is employed throughout. The convergence criterion for the pointwise scalar flux error is 1.0E-4 in ANISN and 1.0E-3 in DOT. The 1.0E-4 value is tested in additional DOT

212

calculations. Concerning the spatial
discretization methods, the weighted difference
model is used in ANISN whilst in DOT 3.5-E, in
addition, the exponential model is employed as
well. No leakage treatment is used in ANISN to
collapse the cross sections, while a
3-D-equivalent flux synthesis approach, implying
the separability of the variables for geometry and
source distribution, is adopted in DOT
calculations. In particular the results on the Z
axis (where detectors are located) are obtained
with the the following relationship for space,
energy and angular dependent flux: $\Phi(X,Y,Z) =$
$\Phi(X,Z) \times \Phi(Y,Z) / \Phi(Z)$. The use of DOT to
calculate $\Phi(Z)$, by inserting appropriate
reflection boundary conditions, is preferable to
the use of ANISN for this purpose, since the same
angular quadrature sets, consistent with the other
2-D calculations, can be employed. Only the
ex-NESTOR-core geometry of the PCA-REPLICA
experiment is described in the DOT calculations
and as Z is a symmetry axis for the (X,Z) and
(Y,Z) planes of the geometric model, only half of
these planes are considered with proper boundary
conditions. The reflection option is used on the Z
axis and the void option on the other three sides,
in the 2-D DOT calculations. Only a unit thickness
(1.0 cm) in a direction orthogonal to the Z axis
is described in the 1-D-equivalent DOT
calculation. Differently from 2-D cases, the
reflection option is assumed in this calculation
also for the boundary on the opposite side of the
Z axis. The thickness of the mesh intervals in
the Z and in the orthogonal X and Y axes is taken
not larger than 1.0 cm in the water and steel
zones. The mesh intervals are chosen such that the
detectors are found at the midpoint of the
corresponding interval. The fixed source
calculations with ANISN and DOT are performed
using volumetric sources derived from a pure
fission spectrum distribution obtained at NEA,
with the NJOY data processing system, from the
U-235 JEF-1 file. Since the calculations refer
only to the ex-NESTOR-core geometry, the
volumetric sources are exclusively assigned to the
fission plate mesh interval. The macroscopic cross
sections used in the steel zones are derived from
the totally self-shielded natural iron file
(JEF-1/MAT 4260; background cross section = 1.0E-6
barns).

Discussion of the Results

A general underprediction of the integral
experimental results is found in the water gaps,
in the mild steel RPV and in the void box. The
calculations employing the exponential model
always give higher C/E values with respect to
those using the weighted difference model.
Nevertheless, the computation time is reduced by
about 30% using the exponential method and the
results are safety-conservative with respect to
the weighted difference results. In particular the
weighted difference calculations give the
following results. The Rh-103 calculated
activities in water are below experimental values
in the range between 10% and 20%. The in-vessel
results for Rh-103 (see TAB. 1) seem insensitive
to the neutron penetration depth and the
underprediction remains below 10%. On the
contrary, for the higher threshold detectors
In-115 and S-32, the underprediction of the
calculated activities tend to increase with the
neutron penetration depth (see TAB. 1) in the RPV.
At the crucial T/4 position the calculated
activities reach the required target accuracy,
maintaining the underprediction below about 15%
but at the 3T/4 position the underprediction
increases to about 25%. In TAB. 1, the Winfrith
integral results /3/, using the 3-D Monte Carlo
code McBEND, are in better agreement than the DOT
3.5-E results with the experimental data.
Secondly, a different trend of the calculated S-32
activities is observed. Contrary to the DOT/JEF-1
results, in the McBEND/UKNDL calculations, the C/E
values for the S-32 activities do not decrease

with the neutron penetration depth in the RPV.
This is probably partially due to the fact that
McBEND used a pointwise UKNDL library including
the ENDF/B III-based DFN 908A /3/ iron file. In
fact, consistently with the previous results, this
UKNDL iron file behaved surprisingly better than
the more recent JEF-1 file (DFN 4260D), in a
Winfrith McBEND sensitivity analysis /1*/ on the
stainless steel Winfrith experiment for the JANUS
programme. Finally in the void box, a still
unexplained strong underprediction (about 25%) is
confirmed for all the detector calculated
activities. Similar strong underpredictions in the
void box (cavity) were obtained not only in the
ENDF/B IV-based transport calculations on the PCA
experiment /1*/, but also in the more recent ORNL
calculations /1*/ for the NESDIP2 Winfrith
experiment, using iron cross sections derived from
the ENDF/B V Mod-4 file (with reduced inelastic
scattering cross section). Considering all the
measurement positions, the spread of the C/E
values is contained within 25%, as for the results
of the calculations using the ENDF/B IV file in
the PCA international blind test /1*/. This seems
consistent with the fact that the JEF-1 natural
iron evaluation is very similar in the MeV region
to that of the ENDF/B IV file and, consequently,
similar results for fast flux above 1.0 MeV should
be reached in transport calculations. In FIG. 1,
the DOT spectral results (exponential model) are
compared with the experimental data and with the
Winfrith McBEND results /3/. The flux
underprediction in the MeV energy region is
confirmed and is particularly evident above about
3.0 MeV. Analyzing the spectral results at
different positions in the mild steel RPV,
appreciable spectral changes are found and this
confirms the need of dpa calculations to take into
account properly the in-vessel spectral dependence
of the neutron damage. Complete details on this
DOT 3.5-E/JEF-1 validation are reported in ref.
/1/ together with a dpa calculation.

Conclusion

The higher threshold detectors, In-115 and S-32,
show the greatest activity underpredictions. Since
their response energy ranges are entirely in the
MeV region, the underpredictions are probably
attributable to a too high Fe-56 inelastic
scattering cross section (energy threshold = 0.860
MeV) in the JEF-1 natural iron file. SN
validations /1*/ of the JEF-1 iron file on pure
iron benchmarks (ASPIS/NEA-UKAEA and
EURACOS/IKE-EURATOM) confirmed this conclusion.
Anyway, more precise information on PCA-REPLICA
can come only from a detailed sensitivity
analysis. Further SN validations on this benchmark
are in progress at ENEA-Bologna to test the new
iron files JEF 2.1, ENDF/B VI and JENDL-3. In
particular, an improvement in the MeV region
results is expected in the RPV when the JEF-2.1
and ENDF/B VI iron files are used: in both, the
Fe-56 inelastic scattering cross section has been
reduced in comparison with the JEF-1 file.

Acknowledgements

The author wishes to thank Dr. F. Fabbri, Mrs.
M.G. Borgia, Mr. G.C. Panini from ENEA-Bologna
together with Dr. G. Hehn (IKE-Stuttgart) and Dr.
E. Sartori (OECD/NEA-Saclay).

References

1. M. Pescarini, DOT 3.5-E/JEF-1 ANALYSIS OF THE
PCA-REPLICA (H2O/FE) SHIELDING BENCHMARK FOR
THE LWR-PV DAMAGE - PREDICTION, ENEA
RT/INN/90/21, 1990.

1*. see reference cited in ref. /1/.

2. M. Pescarini TRANSPORT CALCULATIONS FOR THE
PCA-REPLICA SHIELDING BENCHMARK: A VALIDATION

OF JEF-1 DATA AND SN CODES FOR THE PREDICTION OF LWR'S PRESSURE VESSEL NEUTRON DAMAGE, Proc. of the UKAEA/OECD-NEA 7th Int. Conf. on Radiation Shielding, Bournemouth (UK), 1988, VOL. I, pp. 219-225.

3. J. Butler, M.D. Carter, I.J. Curl, M.R. March, A.K. McCracken, M.F. Murphy, A. Packwood, The PCA Replica Experiment PART I, Winfrith Measurements and Calculations. AEEW-R 1736, 1984.

Detec. Pos.	Dist. from Fiss. Plate (cm)	Detector	C/E DOT 3.5-E/ JEF-1 (X,Y,Z) Synthesis Weighted Difference Model	C/E DOT 3.5-E/ JEF-1 (X,Y,Z) Synthesis Exponential Model	(*) C/E McBEND/ UKNDL (X,Y,Z) Monte Carlo	Ref. Location
8	39.01	Rhodium	0.90	0.97	1.01	RPV
		Indium	0.84	0.91	1.01	
		Sulphur	0.84	0.92	1.03	(1/4 T)
9	49.61	Rhodium	0.92	0.99	0.97	RPV
		Indium	0.75	0.81	0.88	
		Sulphur	0.79	0.87	1.02	(3/4 T)
10	58.61	Rhodium	0.75	0.82	0.85	Void Box
		Indium	0.68	0.74	0.83	
		Sulphur	0.76	0.85	1.00	

TAB. 1 Comparison of Calculated/Experimental (C/E) Ratios between SN DOT 3.5-E and Monte Carlo McBEND Calculations along the Z Horizontal Axis for the PCA-REPLICA Slab Geometry Benchmark

(*) Results reported from ref. /3/.

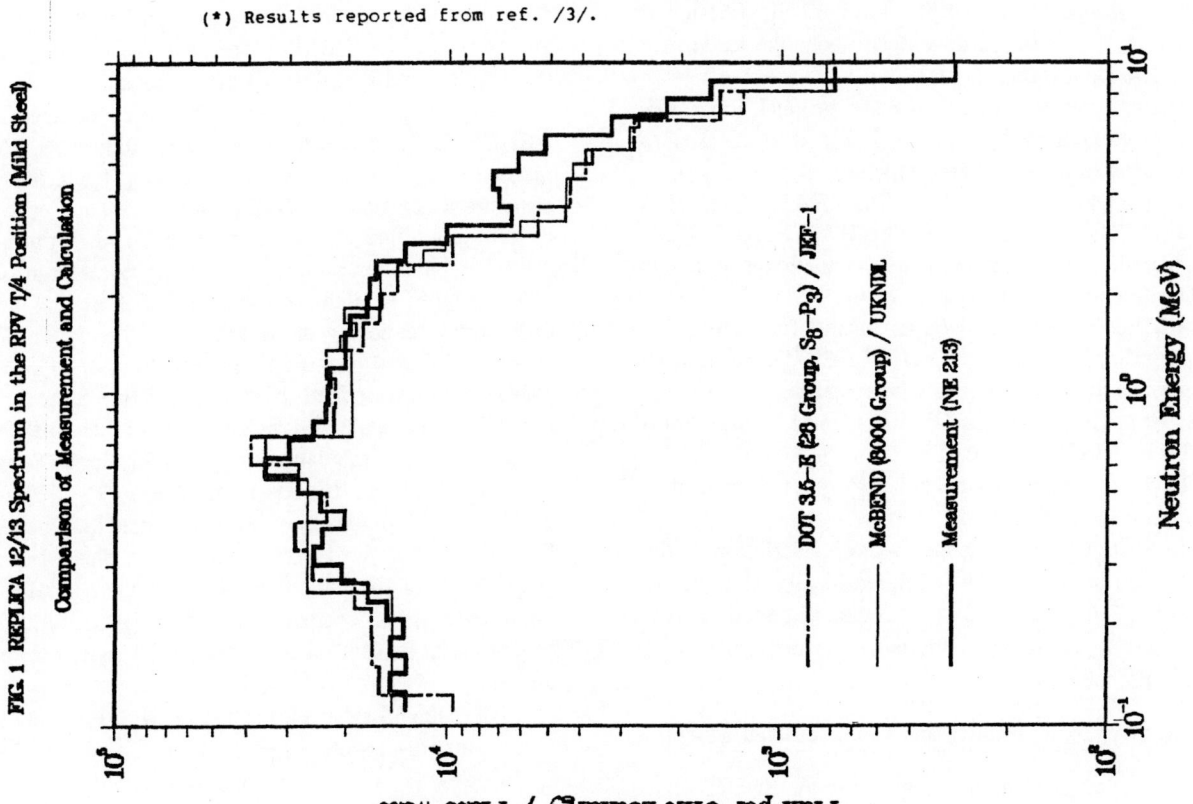

FIG. 1 REPLICA 12/13 Spectrum in the RPV T/4 Position (Mild Steel)
Comparison of Measurement and Calculation

DOT 3.5-E (28 Group, S_8-P_3) / JEF-1

McBEND (8000 Group) / UKNDL

Measurement (NE 213)

Flux per Unit Lethargy / Plate Watt

Neutron Energy (MeV)

BENCHMARK TEST OF JEF-2 AND ENDF/B-VI EVALUATIONS BY CALCULATING SOME CRITICALITIES

M. Caro   S. Pelloni

Paul Scherrer Institute

CH-5232 Villigen/PSI

Switzerland

Tel. 056 99 28 90

Abstract: The newer release of the Joint European File JEF-2.1 (first revision of JEF-2) and the American evaluation ENDF/B-VI have been tested in the fast energy range. A series of fast criticality benchmarks have been recalculated using the discrete-ordinates transport code ONEDANT from Los Alamos and 175 neutron group cross sections in the VITAMIN-J structure generated with the newest release of NJOY (Version 89.62).

It is found that both evaluations and particularly ENDF/B-VI lead globally to a better prediction of the experimental values compared to ENDF/B-V and JEF-1.

The eigenvalues k_{eff} calculated with ENDF/B-VI agree well generally within 0.5 % (the agreement was 1 % for JEF-1 and ENDF/B-V) and most spectral indices measured at the core centre deviate by up to 5 % from the experiment (this deviation was up to 10 % for JEF-1 and ENDF/B-V). The available fast data for ^{233}U and ^{232}Th now seem to be suitable for use in the neutronics calculations of fast systems.

Introduction

During the last years a large effort has been devoted to the improvement of resonance data, high energy cross sections, unresolved resonance parameters, and scattering laws [1]. Recently, a new format option for the representation of double-differential scattering cross sections (File 6 in ENDF/B terminology) [2] has been approved for use in the new evaluations. The main results of this cooperative, international work are Version 2 of the Joint European File (JEF-2) in Europe and ENDF/B-VI in the United States of America. JEF-2 has been developed under the leadership of the NEA Data Bank.

Since a revised version of the original file (JEF-2.1) was released in January 1991, we decided first to recalculate a series of 9 representative, simple, fast criticalities, namely JEZEBEL, GODIVA, JEZEBEL-23, BIG TEN, FLATTOP-25, FLATTOP-23, THOR, FLATTOP-Pu, and JEZEBEL-Pu, using the same calculational model as adopted in a previous study [3].

JEZEBEL and GODIVA are useful for testing fast data of ^{239}Pu, ^{235}U, and ^{238}U. JEZEBEL-23, JEZEBEL-Pu, and THOR are suitable for investigating ^{233}U, ^{240}Pu, ^{232}Th fast cross sections and fission spectrum data. Additionally, FLATTOP-Pu, FLATTOP-25, FLATTOP-23, and BIG TEN are particularly sensitive to the scattering data of the major actinides.

In Section II of this paper a short description of the cross section preparation and computational model used to obtain the results presented and discussed in Section III is given. Finally, Section IV summarizes some conclusions and recommendations.

Multigroup Cross Sections and Computational Model

First, pointwise neutron cross sections in the ENDF/B like PENDF format have been prepared for the nuclides re- quired in the 9 benchmarks using the modules RECONR, BROADR, and UNRESR of the newest release of NJOY (Version 89.62), which includes all updates of Version 89.0 up to June 1990 [4]. The RECONR module reconstructs pointwise (energy-dependent) cross sections using ENDF/B resonance parameters and interpolation schemes. BROADR Doppler-broadens and thins the pointwise cross sections. UNRESR computes effective self-shielded point- wise cross sections in the unresolved resonance region. RECONR and UNRESR can now use the more sophisti- cated Reich Moore formalism for treating the unresolved energy range. The PENDF files have been produced using the JEF-2.1 and ENDF/B-VI evaluations.

Second, multigroup neutron cross sections in the GENDF format have been generated with the GROUPR module. The 175 group VITAMIN-J [5] neutron group structure was used. The GROUPR format data sets consist of infinitely dilute vector cross sections and P_0 through P_4 scattering matrices processed at room temperature for all reaction types available on the PENDF files plus double- differential scattering cross sections (File 6 in ENDF/B terminology) and promt and delayed fission, when appro- priate. We have noticed that the use of File 6 for the main actinides is more frequent in ENDF/B-VI than in JEF-2. In the processing of the data with GROUPR the VITAMIN-E weighting spectrum [5] has been used.

The GENDF files have been converted into the MATXS file suitable for further processing with the reformatting code TRANSX [6] using the NJOY module MATXSR. Us- ing TRANSX, we corrected total and self-scattering cross sections with the Bell-Hansen-Sandmeier transport approx- imation.

All calculations were performed with the one-dimen- sional neutral-particle discrete-ordinates transport code ONEDANT [7] from Los Alamos using the $P_3\ S_{32}$ ap- proximations. The model choosen is analogous to that described in a previous work [3].

Value	JZBL	GDV	JZBL-23	BG TN	FLT-25	FLT-23	THR	FLT-Pu	JZBL-Pu
k_{eff}	0.9952	0.9934	0.9756	0.9928	0.9898	0.9836	0.9797	0.9887	0.9898
	1.0095	0.9995	0.9620	1.0016	0.9984	0.9766	0.9923	1.0054	1.0024
	0.9960	0.9954	0.9929	1.0063	1.0007	1.0026	1.0056	1.0025	0.9893
	1.0111	1.0028	-	-	1.0149	-	-	1.0207	-
C02/C28	-	-	-	-	-	-	1.2119	-	-
	-	-	-	-	-	-	1.2262	-	-
	-	-	-	-	-	-	1.0818	-	-
	-	-	-	-	-	-	-	-	-
F02/F28	-	-	-	-	-	-	0.9022	-	-
	-	-	-	-	-	-	0.8777	-	-
	-	-	-	-	-	-	0.9472	-	-
	-	-	-	-	-	-	-	-	-
N02/N28	-	-	-	-	-	-	1.3250	-	-
	-	-	-	-	-	-	1.0744	-	-
	-	-	-	-	-	-	1.0413	-	-
	-	-	-	-	-	-	-	-	-
F23/F25	0.9659	0.9676	-	0.9773	0.9621	-	-	-	-
	0.9185	0.9515	-	0.9691	0.9476	-	-	-	-
	1.0005	1.0016	-	0.9972	0.9949	-	-	-	-
	0.9867	0.9855	-	-	-	-	-	-	-
C28/F25	-	-	-	0.9998	-	-	0.8301	-	-
	-	-	-	1.0145	-	-	0.8048	-	-
	-	-	-	0.9836	-	-	0.8500	-	-
	-	-	-	-	-	-	-	-	-
F28/F25	0.9528	0.9535	0.9348	1.0142	0.9708	0.9384	0.9781	0.9734	0.9651
	0.9757	1.0595	0.9193	1.0322	1.0617	0.9178	0.9990	0.9844	0.9811
	0.9600	0.9541	1.0081	1.0519	0.9655	1.0030	0.9559	0.9730	0.9675
	0.9167	1.0364	-	-	1.0383	-	-	0.9406	-
N28/F28	-	-	-	-	-	-	0.7814	-	-
	-	-	-	-	-	-	1.2245	-	-
	-	-	-	-	-	-	0.9376	-	-
	-	-	-	-	-	-	-	-	-
F37/F25	0.9624	0.9609	0.9393	0.9720	0.9868	0.9483	0.9633	0.9766	0.9903
	0.9634	1.0167	0.9184	0.9487	1.0305	0.9224	0.9635	0.9600	0.9859
	0.9889	0.9742	1.0079	1.0724	1.0016	1.0184	0.9580	1.0042	1.0169
	0.9891	1.0640	-	-	1.0868	-	-	1.0143	-
F49/F25	0.9893	0.9922	-	0.9872	0.9983	-	-	-	-
	0.9878	1.0076	-	1.0032	1.0126	-	-	-	-
	0.9836	0.9860	-	0.9985	0.9936	-	-	-	-
	0.9724	0.9943	-	-	1.0007	-	-	-	-

Table 1: Calculated over Experimental (C/E) Ratios for the Eigenvalue and Spectral Indices Using JEF-2.1, JEF-1, ENDF/B-VI, and ENDF/B-V Data

Results and Discussion

Tables 1-2 give the results from the transport calculations with ONEDANT and the new data, and summarizes previously published results for the same 9 benchmarks, obtained using an older version of NJOY (Version 83.0) and data from JEF-1 and ENDF/B-V [3]. The old JEF-1 values presented in Tables 1-2 could be reproduced well within 0.1% using Version 89.62 of NJOY.

The new results include ratios (C/E) between calculated and experimental multiplication factors k_{eff}, and spectral indices measured at the core centre for the main actinides (Table 1) and for other nuclides (Table 2).

In Tables 1-2 "F" means fission, "C" neutron capture, and "N" (n,2n) reaction rates. The benchmark names were shortened.

The first line of results refers to JEF-2.1, the second line to JEF-1, the third line to ENDF/B-VI and the fourth line to ENDF/B-V.

Tables 1-2 show that both new evaluations, and particularly ENDF/B-VI lead globally to a more accurate prediction of the experimental values.

The eigenvalues k_{eff} calculated with ENDF/B-VI generally agree with the experimental eigenvalues to within 0.5 % (the agreement was 1 % for JEF-1 and ENDF/B-V)

whereas most spectral indices measured at the core centre deviate by up to 5 % from the experiment (this deviation was up to 10 % for JEF-1 and ENDF/B- V [3]). The eigenvalues k_{eff} obtained using JEF-2.1 data are consistently less than 1 and smaller, except for JEZEBEL-Pu, than those calculated with ENDF/B-VI. A similar trend is shown by JEF-1 and ENDF/B-V. The eigenvalue k_{eff} for GODIVA decreases by almost 0.7 % if the new evaluations are used, and its prediction is now less accurate.

The new data for ^{233}U and ^{232}Th now seem to be suitable for use in the calculations of fast spectra. ENDF/B-VI underestimates k_{eff} for JEZEBEL-23 by only 0.7 % (3.8 % and 2.4 % are the underpredictions of JEF-1 and JEF-2.1) and predicts to within 0.5 %, the spectral index ratio of fission in ^{233}U to fission in ^{235}U for JEZEBEL, GODIVA, BIG TEN, and FLATTOP-25 (the accuracies of JEF-2.1, JEF-1, ENDF/B-V are 3.8 %, 8.1 %, and 1.5 %). ENDF/B-VI predicts the eigenvalue for THOR to within 0.6 % (the deviation from JEF-2.1 and JEF-1 are 2.0 % and 0.7 %), improves the prediction of all measured spectral indices, and particularly that of the ^{232}Th (n,2n) reaction (the relative error is reduced to 4.1 % compared to 32.5 % for JEF-2.1 and 7.4 % for JEF-1).

The ^{238}U (n,2n) and fission cross sections have been improved in ENDF/B-VI.

The fast fission cross sections of ^{239}Pu and capture cross sections of ^{238}U have been improved in JEF-2.1. The spectral index ratio of fission in ^{249}Pu to fission in ^{235}U is predicted to within 1.2 % by JEF-2.1 (the corresponding relative errors for ENDF/B-VI, JEF-1, and ENDF/B-V are 1.6 %, 1.3 %, and 2.8 %). The ratio of capture in ^{238}U to fission in ^{235}U for BIG TEN is exact (the corresponding relative errors for ENDF/B-VI and JEF-1 are 1.6 % and 1.4 %).

Conclusions and Recommendations

Nine representative fast critical benchmarks have been calculated using the one-dimensional discrete-ordinates transport code ONEDANT and 175 neutron group NJOY cross sections in the VITAMIN-J energy structure from JEF-2.1 and ENDF/B-VI. Eigenvalues k_{eff} and spectral indices at the core centre have been compared with experimental values. From the analysis presented above, some conclusions and recommendations can be reached.

1. The new evaluations, and particularly ENDF/B-VI lead globally to a better prediction of the experimental values than ENDF/B-V and JEF-1, although some shortcomings (such as the resulting k_{eff} underprediction for GODIVA) still remain.

2. Fast data for ^{233}U and ^{232}Th have been revised, and seem now to be more suitable for calculations of fast spectra.

3. It is recommended to perform similar studies for more thermalized systems, and to quantify the effect induced by the explicit treatment of double-differential scattering cross sections in energy and angle (File 6).

Value	JZBL	GDV	BG TN
CV/F25	0.7166	-	-
	0.6957	-	-
	0.8452	-	-
	0.7826	-	-
CMn55/F25	1.4424	1.5614	-
	1.3750	1.3333	-
	1.2226	1.2839	-
	1.1667	1.1111	-
CCo59/F25	-	0.1625	0.8654
	-	0.1553	1.0000
	-	0.1622	0.8488
	-	0.1553	-
CNb93/F25	1.1959	1.1143	-
	1.1565	1.0000	-
	1.2396	1.1536	-
	1.1783	0.9867	-
CAu197/F25	0.9624	0.9488	0.9674
	0.9386	0.8670	0.9641
	0.9492	0.9483	0.9510
	0.9530	0.8550	-
CCu63/F25	0.8505*	0.8241*	0.9223*
	0.8300	0.8462	1.0244
*) ^{nat}Cu	1.0180	0.9921	1.0617
	0.8400	0.7521	-

Table 2: Calculated over Experimental (C/E) Ratios for Some Spectral Indices Using JEF-2.1, JEF-1, ENDF/B-VI, and ENDF/B-V Data

References

1. J. Rowlands, N. Tubbs, "The Joint Evaluated File: A New Data Resource for Reactor Calculations", Proc. Int. Conf. on Nuclear Data for Basic and Applied Science, Santa Fe (1985)

2. P. F. Rose, C. L. Dunford, "ENDF-102 Data Formats and Procedures for the Evaluated Nuclear Data File ENDF-6", BNL-NCS-44945 (July 1990)

3. S. Pelloni, "Benchmark Test of JEF-1 Evaluation by Calculating Fast Criticalities", EIR-Report 584 (JEF-Report 6) (June 1986)

4. R. E. MacFarlane, D. W. Muir, R. M. Boicourt, "The NJOY Nuclear Data Processing System, Volume I: User's Manual, LA-9303-M (May 1982)

5. E. Sartori, "VITAMIN-J, a 175 Group Neutron Cross Section Library Based on JEF-1 for Shielding Benchmark Calculations", JEF-1/Doc-100 (Oct. 1985)

6. R. E. MacFarlane," TRANSX-CTR: A Code for Interfacing MATXS Cross Section Libraries to Nuclear Transport Codes for Fusion System Analysis", LA-9863-MS (Febr. 1984)

7. R. Douglas O'Dell, F. W. Brinkley, D. R. Marr, "User's Manual for ONEDANT: A Code Package for One-Dimensional Diffusion Accelerated Neutral Particle Transport", LA-7396-M (Sept. 1986)

CORRECTION OF NEUTRON CONSTANTS OF LEAD BASED ON RESULTS OF INTEGRAL EXPERIMENTS

A.P.Vasilyev, Ya.Z.Kandiev, V.D.Lutov, A.I.Saukov, B.I.Suhanov

Institute of Technical Physics
454070, Chelyabinsk-70, USSR

Abstract: Results of measurements and calculations of neutron leakage spectra from a lead sphere 5 cm-thick and reflected from a hemisphere with a central DT-neutron source are presented. A comparison with results of similar experiments for lead spheres of different thicknesses made in USSR, USA and Japan has been carried out using 3-dimensional Monte-Carlo calculations (code PRIZMA and file of neutron data BAS).
The results of the comparison were used to adjust the lead neutron data.

(14 MeV neutrons, Pb, integral experiment, code PRIZMA, Monte-Carlo, nuclear data library BAS).

Introduction

Neutron leakage spectra from lead spherical samples up to 12 cm thick with a central DT-neutron source have been measured previously in many works to verify lead neutron data. These results /1-6/ are contradictory in some details.

In this work the neutron leakage spectra from 5 cm-thick lead sphere with a central 14 MeV neutron source as well as the spectra reflected from the back hemisphere have been measured using the time-of-flight (TOF) method.

Our results are compared with those from /1-6/ by means of three-dimensional (3-D) Monte-Carlo calculations using the PRIZMA /7/ code and nuclear data for lead from the libraries BAS /8/, ENDF/B-IV and ENDL-78. This comparison allows to locate the neutron energy ranges, where further corrections of the lead neutron data are required.

Measurements of the neutron leakage and reflected spectra

The measurements have been carried out using the TOF-method for a flight path 8.5 m (Fig.1). The neutron generator pulse width was ~ 15 ns, repetition frequency – 500 kHz. The average source yield was $3 \cdot 10^8$ n/s. The neutron generator target was positioned in the centre of a sphere through a hole 3 cm in diameter. The main detector was stilbene (7 cm x 7 cm) with a FEU-110 photoelectric multiplier, which was placed in a lead shield just against the collimator hole. In some measurements a steel rod, 3 cm in diameter and 40 cm long, was positioned between the target and the detector in order to protect the detector from direct 14-MeV neutrons.

The lead samples were a hollow sphere or a hemisphere with outer radius Ro=10 cm, inner radius Ri=5 cm and density 11.3 g/cm³. The measurements of neutron spectra reflected from the hemisphere allow to verify neutron angle distributions at large scattering angles, where the error of differential experiments is especially large. The measurements with hemispheres of different materials have been used to verify and to correct the neutron data when compiling the nuclear data library BAS.

Neutrons with energies between 0.2-15 MeV were detected in the experiment. Neutron yield monitoring during the measurements was performed by two pulse monitors. A high energy threshold of the monitors (~ 10 MeV) made this detectors to be insensitive to scattered radiation. The neutron yield was also monitored by two long counters and by measuring of the target ion current.

To minimize the γ-background the TOF method was completed by a device of n/γ-separation using the pulse form that allowed to make the accidental coincidence background 6-8 times lower.

The results given are normalized per one source neutron from the DT-peak (En>10 MeV, the experimental resolution being taken into account). Special measurements were conducted on the neutron generator operating in a stable mode to normalize the TOF spectra. Normalization accuracy is estimated to be in the range of 2-3 %.

The stilbene detector efficiency calculated is made to be more precise through comparing experimental and calculated spectra of U-238 and CH₂ – the materials with well-known cross-sections. Besides, for two neutron energies, i.e. 1.5 and 2.5 MeV, the efficiency was determined from the measurements using a CH₂ ring scatterer. The error in the efficiency determination is 5-7 %, which is the main part of the measurement total error for E<10 MeV.

Table 1. The stilbene relative efficiency ε(E) in this work.

E, MeV	ε(E)	E, MeV	ε(E)
0.25	0	1.5	1.74
0.35	1.38	3	1.45
0.5	1.70	5	1.2
0.7	1.8	7	1.05
0.9	1.80	9	1.00
		15	1.00

Other authors' experiments with the DT-neutron source

When comparing data with each other and with calculations, experimental peculiarities and measurements normalization were taken into account.

Errors of TOF measurements /1-4/ turn out to be comparable. The number of neutrons in DT-peak is usually measured with an error of 1-3 %, the main component of the error being the result of the normalization process. In the range of lower energy the main component in the error of the energy spectra is caused by inaccuracy in determining the efficiency of detector (5-7%).

The neutron leakage spectra from lead spheres depend only on the sample thickness and is independent of their radii what was tested in comparative calculations. This fact makes the comparison easier. It should be noted that for the samples of U-238 this is not true.

Calculations

The code PRIZMA (Monte-Carlo method) enables to describe experiments adequately, using:
- 3-D geometry in the calculations;
- the possibility to describe anisotropy and time dependence of the source intensity;
- the possibility to obtain high statistical accuracy on detectors of real size.

Calculations were performed in accordance with the normalization used in every experiment.

Calculations for our measurements are carried out in 3-D geometry taking into account the tube design. This enables to describe the spectrum and the neutron angular distribution emitted from the tube. Neutron removal out of the collimated beam was also accounted for.

The geometry description of the neutron reflection from the back hemisphere differs only in the absence of the front hemisphere. 3-D calculations for a sphere and hemisphere with the rod or without it were performed within one calculation, since the PRIZMA code allowed to calculate variants partially different in geometry or areas' composition concurrently and therefore in correlated manner.

Comparative calculations were performed for our measurements and for those in /1/ and /2/, using the lead neutron data from BAS (the initial version and the last one corrected on the basis of our results for the lead), ENDF/B-IV (mat 1288) и ENDL-78 (mat 7862). Direct results of calculations are normalized per one source neutron averaged in all directions. To perform comparison with measurements calculations were renormalized occasionally. Statistical uncertainty of Monte-Carlo calculations was about 0.1 % for the total neutron number in the detector volume.

Discussion of results

Calculations using the corrected version of BAS are in good agreement with our measurements. Measurements on the sphere without the rod verify the cross sections and the secondary neutron spectra about En=14 MeV. Measurements on the sphere with the rod verify the elastic scattering anisotropy at small angles, which measurements in the differential experiments is rather difficult. Measurements on the back hemisphere (with the rod) verify anisotropy of back scattering accepted in the BAS, but it requires to be defined more exactly.

Table 2. Neutron leakage spectra for lead samples as measured in this work and calculated.

E1, MeV	E2, MeV	Experiment sphere rod out	Experiment sphere rod in	Experiment hemi sphere rod in	BAS sphere	ENDF/B-IV sphere	ENDF/B-IV hemi-sphere	ENDL-78 sphere	ENDL-78 hemi-sphere
						hemi-sphere			
0.4	1.0	0.166	0.151	0.087	1.081	1.156	1.163	1.278	1.079	1.162
1.0	2.0	0.217	0.198	0.111	1.006	1.015	0.893	0.978	0.910	1.052
2.0	3.0	0.112	0.102	0.059	1.000	0.894	0.726	0.813	0.870	1.032
3.0	4.0	0.041	0.037	0.021	1.099	0.946	0.707	0.876	1.074	1.299
4.0	5.0	0.018	0.015	0.008	1.013	0.938	0.791	0.926	1.053	1.321
5.0	6.0	0.010	0.008	0.004	0.986	0.937	0.901	1.197	1.114	1.509
6.0	8.0	0.013	0.010	0.005	1.046	0.927	1.109	1.531	1.014	1.294
8.0	10.0	0.012	0.010	0.004	1.020	0.802	0.888	1.524	0.800	1.417
10.0	15.0	0.658	0.147	0.015	0.987	1.201	1.008	1.404	0.998	1.363

In the range 1-5 MeV, calculations using ENDF/B-IV lie lower than measurements and in the range En<1 MeV calculations using ENDF/B-IV are higher than all measurements besides /1/.

In Fig. 3 the results of /1-6/ and those of the authors are compared with calculations for the leakage of neutrons as a function of sphere thickness.

From the comparison of the measurements and the calculations we can conclude the following:

1. The measured number of neutron in the DT-peak and neutron removal from the range En>6 MeV agree in all experiments and are well described by calculations using all neutron data.

2. The total leakage given in /1/ is higher than the calculated one, and in /5/ it agrees with calculations. Earlier in /9/ we have shown inner contradiction to be present in /1/: for all thicknesses the measured neutron number with En<6 MeV is considerably larger if compared with the number given the (n,2n) reaction on all the neutrons escaping the range En>6 MeV according to the same experimental data /1/.

3. Calculations using BAS and ENDL-78 well describe the experiments in the range 1-5 MeV except the part of data in /1/ that are above the other measurements. Calculations using ENDF/B-IV are lower than other experiments and calculations in the range. So, the data BAS and ENDL-78 properly describe neutron leakage in the range 1-5 MeV, and ENDF/B-IV underestimates it.

4. Thus, in the range En<1 MeV calculations using BAS and ENDL-78 give correct number of neutron, and calculations using ENDF/B-IV overestimate it as well because all these libraries describe the total neutron number correctly.

5. The (n,2n) reaction cross sections determining the neutrons escaping from the DT-peak into the range En<6 MeV are practically the same in all the three libraries, but the secondary neutron distributions differ considerably, i.e. it turned out that a far smaller number of neutrons get into the range En=1-5 MeV in the calculations using ENDF/B-IV, if compared with BAS and ENDL-78, and a correspondingly larger number of neutrons get into the range En<1 MeV. Making compared all the measurements we can conclude that the distribution of the secondary neutrons from the reaction (n,2n) in BAS and ENDL-78 is more correct than in ENDF/B-IV.

Conclusion

Measurements of neutron spectra reflected from a hemisphere, made it possible to adjust neutron angular and energy distributions for back scattering, where the error in the differential measurements is especially high.

Comparison of all experimental results by means of 3-D calculations using the PRIZMA code allowed to make a conclusion about systematic overestimation of the neutron leakage for En<6 MeV and especially for En<0.4 MeV in /1/, where the discrepancy between calculations and the experiment /1/ is about 2-3 times larger. The energy distribution of neutrons in the reaction (n,2n) in ENDF/B-IV describes experimental results with less accuracy than those in ENDL-78 and BAS.

To increase the accuracy of nuclear data evaluation it is desirable to present results of measurements for the real geometry since spherization introduces an additional error depending on data used in such calculations.

REFERENCES

1. Y.Yanagi, A.Takahashi: Oktavian report A-84-02, Osaka, May (1984)

2. L.F.Hansen, H.M.Blann, R.J.Howerton, T.T.Comoto, B.Pohl: Nucl. Sci. Eng.92, 382 (1986)

3. A.A. Androsenko, B.V.Devkin, B.V.Zhuravlev, V.A.Zagryadskij, M.G.Kobazev, A.V.Markovskij O.A.Sal'nikov, S.P.Simakov, D.Yu.Chuvilin: Neutr. Phys. 3, 194, Moscow (1988)

4. S.Iwasaki, N.Odano, S.Tanaka, J.R.Dumais, K.Sugiyama: Nucl. Data for Sci. and Techn., MITO, 229-232 (1988)

5. S.Antonov, V.A.Zagryadskij: Neutr.Phys., 4, 218, Moscow (1988)

6. M.Kralic, Ya.Pulpan, M.Tihy, V.A.Zagryadskij, M.I.Krainev, D.Yu.Chuvilin: Voprosy atomnoy nauki i techniki, series: Termonuclear Fusion, 2, 46 (1990) in Russian

7. Ya.Z.Kandiev, E.S.Kuropatenko, A.I.Orlov, V.M.Shmakov: The 3d All-Union Sci. Conf. on the Shield. from Ionizing Irradiations of Nucl. Technical Installations. 27-29 October 1981, Tbilisi, 24 (1981)

8. A.P.Vasilyev, Ya.Z.Kandiev, V.I.Chitaikin: Neutron Physics, Moscow., 2, 119 (1984)

9. A.P.Vasilyev, Ya.Z.Kandiev, E.S.Kuropatenko, V.D.Lutov, A.I.Saukov, B.I.Suhanov: Neutron Physics, 3, 292, Moscow (1988)

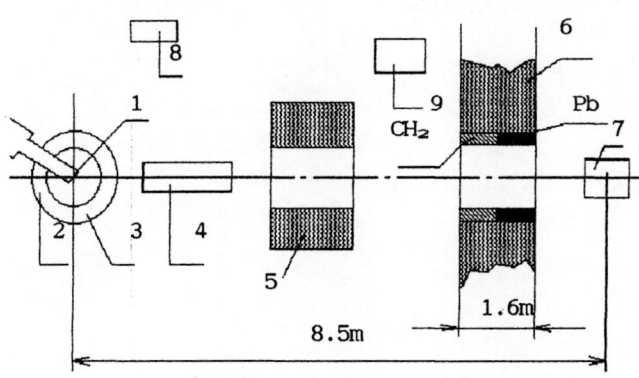

Fig. 1. Geometry
1 - neutron source, 2 - back hemisphere of sample, 3 - front hemisphere, 4 - steel rod, 5 - CH₂ - collimator, 6 - wall with collimator, 7 - detector, 8 - pulse monitor, 9 - counter

dN/dt, count/(ns*sourse count)

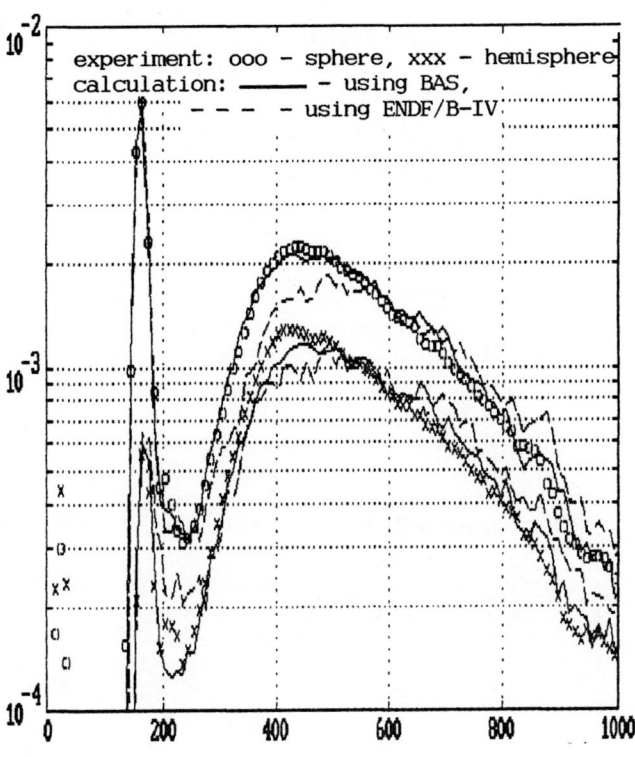

flight time, ns

Fig. 2. TOF spectra for Pb samples in this work

dN, neutron/sourse neutron

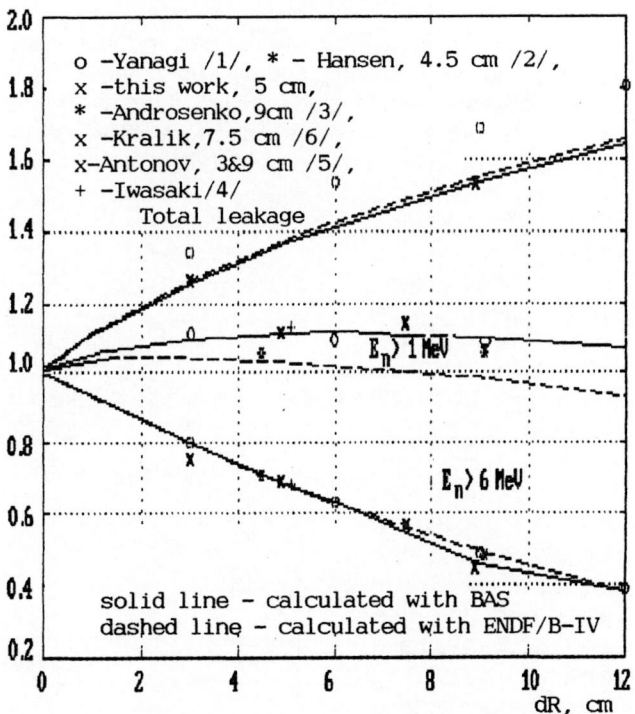

Fig. 3. Measured and calculated neutron leakage from Pb sphere

INTEGRAL EXPERIMENTS WITH REGARD TO THE THORIUM-BASED HYBRID FUSION BLANKET

Shu A. Hayashi, Chihiro Ichihara[*], Keiji Kobayashi[*]
Satoshi Kanazawa[**], Itsuro Kimura[**] and Junji Yamamoto[***]

Institute for Atomic Energy, Rikkyo University, Yokosuka 240-01, Japan
[*] Research Reactor Institute, Kyoto University, Kumatori Osaka 590-04, Japan
[**] Dept. of Nuclear Eng., Fac. of Eng., Kyoto University, Kyoto 606, Japan
[***] Dept. of Nuclear Eng., Fac. of Eng., Osaka University, Suita 565, Japan

ABSTRACT: In order to examine the existing nuclear data for thorium, which is the dominant material for the fusion-fission hybrid reactors, two kinds of integral experiments with thorium metal piles have been performed. In the reaction rate distribution measurement, we obtained the spatial distribution of the ^{197}Au(n,γ) reaction rates in three types of thorium metal piles. The measured values were compared with the theoretical calculations using MCNP Monte-Carlo code with JENDL-3 and ENDL-75 evaluated nuclear data files. It turned out that the JENDL-3 evaluation gave prediction superior to the ENDL-75 one. In the time-of-flight measurement, we obtained the neutron spectrum from a thorium scatterer (26.7 cm × 35.6 cm × 35.6 cm) by utilizing the signals from the associated α-particles as the trigger signals. The measured spectrum was compared with the MCNP calculation with JENDL-3 and ENDL-75 data files.

(integral experiment, D-T neutron source, hybrid fusion blanket, thorium pile, beryllium, stainless steel, reaction rate distribution, ^{197}Au(n,γ) reaction, TOF technique, associated particle method, Monte Carlo calculation, nuclear data file, JENDL-3, ENDF/B-IV, comparative evaluations)

INTRODUCTION

Fusion reactor is expected to be one of the dominant energy resources for the coming century. However, in spite of the rapid development of fusion related technologies in the past few decades, it seems difficult to develop commercial fusion reactor plants by early 21st century. Fusion-fission hybrid reactors, which consist of the fusion core surrounded by various kinds of "hybrid" blankets can take part in the nuclear energy cycle by producing energy and/or the nuclear fuel for both fission and fusion reactors, even before the fusion reactors become economically self-sustaining[1,2]. Thorium is one of the most promising hybrid blanket materials for breeding nuclear fuel for the fission reactors. Several groups have carried out experiments by combining 14 MeV neutrons and thorium piles[3-5]. However, the agreement between experiment and calculation in their works are not always satisfactory for the design work. We have been conducting integral experiments to examine the nuclear data of thorium for further investigation of

hybrid blanket designs[6]. Two kinds of experiments have been performed so far. One is to measure reaction rate distribution of gold foils in three kinds of thorium piles and the other is to obtain neutron spectra from a thorium scatterer by a time-of-flight (TOF) technique. In each case, the D-T neutron generator which utilizes a 300 kV insulated-transformer type accelerator with a duoplasmatron ion source installed in the Kyoto University Critical Assembly (KUCA) was used. General characteristics of this neutron generator are described in another report[7]. The bird's-eye view of the KUCA neutron generator is shown in **Fig. 1.**

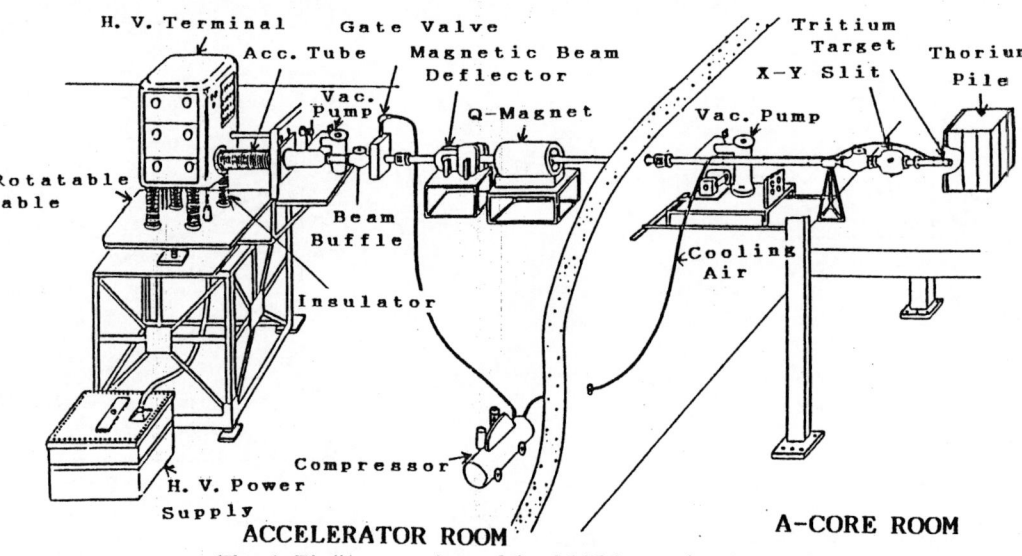

Fig. 1 Bird's-eye view of the KUCA accelerator system.

REACTION RATE MEASUREMENT IN THE THORIUM METAL PILES

About 70% of the reaction rate of ^{197}Au(n,γ) in the thorium pile (Type I) is induced by the neutrons between 0.1 and 10 MeV, where the calculated spectra using JENDL-3[8] and ENDL-75[9] in the $(n,2n)$ cross section for thorium are inconsistent. Therefore, the difference between these two data files can be evaluated by measuring the ^{197}Au(n,γ) reaction rate.

(1) Experiments

We have measured the reaction rate distribution of gold foils in three kinds of thorium piles, Type I (thorium pile, 32.8 cm × 25.5 cm × 25.5 cm), Type II (beryllium: 2.5 cm + thorium: 32.8 cm × 25.5 cm × 25.5 cm) and Type III (lead 2 cm + thorium: 32.8 cm × 25.5 cm × 25.5 cm). A gold foil of about 70 mg was set on the front surface of the thorium section of each pile and four other foils were set 5.5 cm separated from each other in the thorium region. **Fig. 2** shows the typical experimental arrangement. Neutron irradiation was done for about 10 hours with the D-T neutron generator. The γ-rays from the ^{198}Au decay were measured with a pure Ge detector, the relative detecting efficiency of which was about 30 % of 3"diam. × 3" thick NaI(Tℓ) scintillator. and the reaction rates were deduced after several kinds of corrections required.

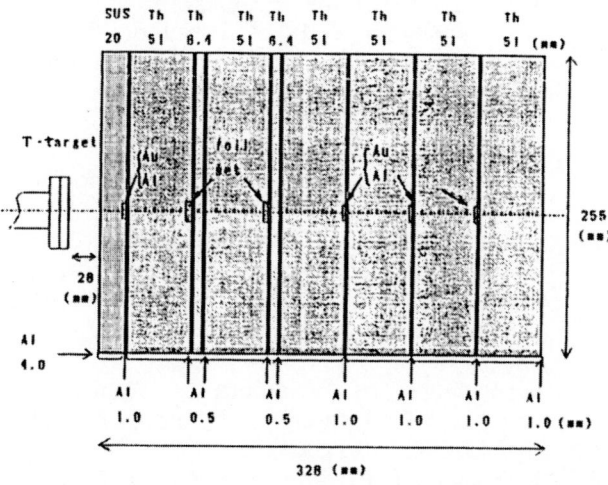

Fig. 2 Experimental arrangement of the thorium pile, Type II

(2) Calculations

The theoretical calculations of the neutron energy spectra were performed with MCNP Monte Carlo transport code[10] using the continuous energy neutron cross section libraries, FSXLIB[11] processed from JENDL-3 and BMCCS from ENDL-75. From the calculated neutron spectra and the activation cross section of ^{197}Au(n,γ) reaction derived from ENDF/B-V dosimetry file[12], the reaction rate of each position in the pile was calculated. The calculated and experimental values are shown in **Fig. 3**. Generally speaking, the prediction from JENDL-3 data gives a good agreement with the experiment, while the prediction from ENDL-75 gives slightly higher value for all piles. In Type I (thorium) pile, the prediction

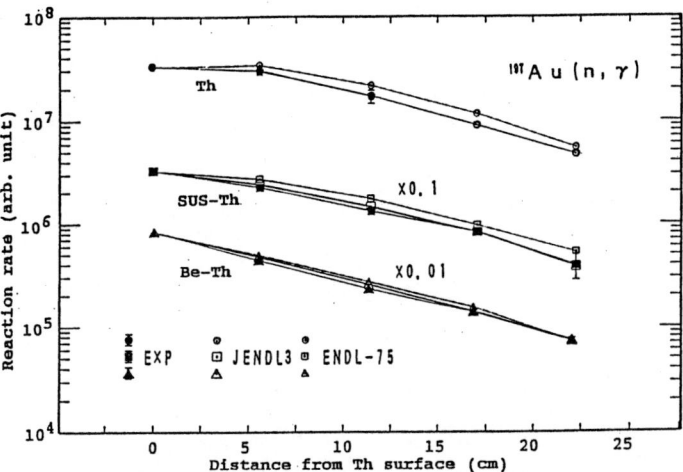

Fig. 3 Calculated and Experimental values of the ^{197}Au(n,γ) reaction rates (normalized at the thorium surface)

from JENDL-3 agrees with the experiment within experimental errors, while one from ENDL-75 overestimates by 10 to 20 %. In Type II (SS + thorium) and Type III (Be + thorium) piles, the prediction from JENDL-3 agrees within 5 to 10 % with the experiment. The prediction for Type II pile using ENDL-75 overestimates by about 20 %. However, the prediction of ENDL-75 for Type III pile is almost as good as JENDL-3 prediction, because the spectrum softening is occured by the beryllium region in the pile.

TIME-OF-FLIGHT MEASUREMENT

(1) Experimental Arrangement

The thorium scatterer, 20.3 cm × 35.6 cm × 26.7 cm rectangular prism, was formed by stacking thorium plates, 5.08 cm square x 1.27 cm and 0.32 cm thick. The angular neutron spectrum was measured by the TOF technique using the associated particle method. The experimental arrangement is shown in **Fig. 4**. The associated α-particles emitted by the D-T reaction are detected with a thin plastic scintillator and the neutrons scattered from the thorium pile are detected with a liquid scintillator at the 7 m flight path station. These signals of alpha and neutron detections work as the start and stop signals for a time analyzer, respectively. The electronic block diagram of the TOF experiment is shown in **Fig. 5**.

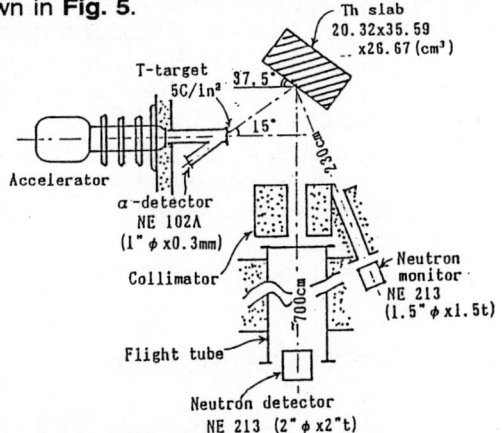

Fig. 4 Experimental arrangements of TOF technique using associated particle method

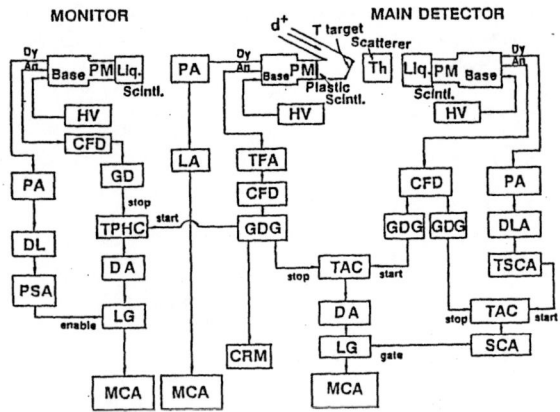

Fig. 5 Block diagram of TOF technique using associated particle method

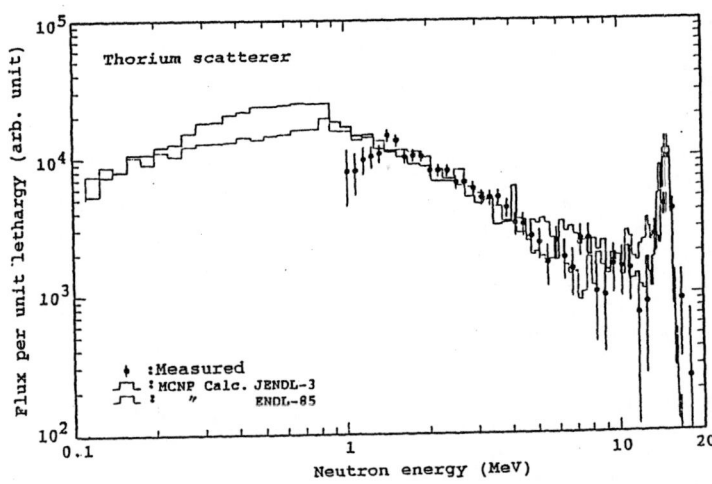

Fig. 6 Comparison of calculated and experimental neutron spectra from thorium scatterer

a) Alpha detection   A plastic scintillator, NE102A 2.54 cm diam. x 0.03 cm thick, was selected for the detection of α-particle, because of the consideration on its fast time response, large effective area, and radiation durability. The α-signals from the plastic scintillator were used to trigger the time analyzer.

b) Neutron detection   Two neutron detection systems were adopted. One is a main neutron detector of 5.08 cm diam. x 5.08 cm thick liquid scintillation counter, NE 213, which was set at the 7 m station and the other is a neutron monitor of 3.8 cm diam. x 3.8 cm thick liquid scintillation counter, NE 213, set at 230 cm position from the scatterer. When 14 MeV neutrons entered directly into the main neutron detector, the n-γ discrimination level was adjusted so as to observe neutrons whose energy was higher than 1 MeV.

(2) Results and discussion
The time spectrum obtained with the time analyzer is converted to the neutron spectrum taking account of the detector efficiency and other corrections. The detector efficiency was determined by the combination of the measured neutron spectrum from a lead scatterer with the same detection technique and the theoretically calculated one used with O5S Monte Carlo code. The neutron spectrum calculation was performed by a continuous energy Monte Carlo code, MCNP using the data contained in JENDL-3 file. The experimental and calculated neutron spectra from the thorium scatterer are shown in **Fig. 6**. General shape of the spectra agreed in the measured and calculated results. There are some discrepancies in the several resonances between calculated and experimental spectra. However, due to large statistical errors of experiment, it is difficult to examine the data file so far. The improved measurement is scheduled with larger detectors and fast detection circuits.

ACKNOWLEDGEMENT

The present work has been partially supported by Grant-in-Aid, Ministry of Education, Science and Culture, Japan. The authors wish to thank the staff of the KUCA for their assistance in the experiment and analysis.

REFERENCES

1. H. A. Bethe: *"The Fusion Hybrid"*, Nuclear News, **21**, p.41, (May 1978)
2. E. Greenspan: *"Fusion-Fission Hybrid Reactors"*, Advances in Nuclear Science and Technology, Vol. 16, p.289, Plenum Press (1984)
3. L. F. Hansen, C. Wong, T. T. Komoto, B. A. Pohl, E. Goldberg, R. J. Howerton and W. M. Webster: Nucl. Sci. Eng., **72**, 35 (1979)
4. V. R. Nargundkar, T. K. Basu and O. P. Joneja: Fusion Tech., **14**, 354 (1988)
5. A. Kumar, W. R. Leo, L. Green and G. L. Woodruff: Fusion Tech., **15**, 1430 (1989)
6. I. Kimura (*ed.*): *"The Fundamental Study of Thorium-Loaded Hybrid Reactor"*, Research Result Report of Grant-in-Aid. Ministry of Education, Science and Culture, Japan (May 1990)
7. C. Ichihara, H. Nakamura, K. Kobayashi, R. Ikegawa, T. Tsuruta, K. Kanda, T. Shibata: KURRI-TR-240 (1983)
8. K. Shibata, T. Nakagawa, T. Asami, T. Fukahori, T. Narita, S. Chiba, M. Mizumoto, A. Hasegawa, Y. Kikuchi, Y. Nakajima and S. Igarashi: *"Japanese Evaluated Nuclear Data Library, Version-3 -JENDL-3-"*, JAERI 1319, (1990)
9. *"LLNL Evaluated Nuclear Data Library"*, UCRL-50400
10. F. Briesmeister (*ed.*): *"MCNP-A General Monte Carlo Code for Neutron and Photon Transport"*, LA-7396-M, Rev. 2, (1986)
11. K. Kosako: *private communication*
12. *"ENDF/B-V; Evaluated nuclear data file, Version-V,"* BNL-17541 (ENDF-201), (1979)

MEASUREMENT OF LEAKAGE NEUTRON SPECTRA FROM VARIOUS SPHERICAL PILES OF FIVE ELEMENTS WITH 14 MEV NEUTRONS

Chihiro Ichihara[1], Shu A. Hayashi[2], Katsuhei Kobayashi[1], Junji Yamamoto[3], Akito Takahashi[3], and Itsuro Kimura[4]

1) Research Reactor Institute, Kyoto University, Osaka 590-04, Japan
2) Institute for Atomic Energy, Rikkyo University, Yokosuka 240-01, Japan
3) Dep. Nuclear Engineering, Fac. Engineering, Osaka University, Suita 565, Japan
4) Dep. Nuclear Enginering, Fac. Engineering, Kyoto University, Kyoto 606, Japan

Abstract: In order to examine the existing nuclear data files related to the fusion reactor materials, neutron leakage current from the sphere sample piles of five elements have been measured using the intense pulsed 14 MeV neutron source, OKTAVIAN and a time-of-flight technique. Measured samples were Aℓ, Ti, As, Se, and Zr. The thicknesses of the piles in unit of mean free path for 14 MeV neutrons for these elements were 0.5, 0.7, 0.8, 0.6, and 2.0, respectively. The neutron energy spectra from the above piles were obtained in the energy range from 0.1 to 14 MeV. The obtained data were compared with the theoretical calculations using MCNP Monte Carlo transport code and the evaluated data files, JENDL-3 and others. For each pile, some discussions and comments are presented for the integral assessment.

(neutron leakage spectrum, D–T neutrons, liquid scintillator, TOF technique, aluminum, titanium, arsenic, selenium, zirconium, nuclear data file, JENDL-3, ENDF/B-IV, Monte Carlo calculation, integral experiment, comparative evaluations)

INTRODUCTION

An integral experiment is useful for the examination of nuclear data files and calculation methods. We have measured the angular flux in large number of fission reactor candidate materials using a time-of-flight technique and photoneutrons from the electron linac at Kyoto University Research Reactor Institute[1].

The design study of the fusion reactors or fusion-fission hybrid reactors needs the evaluated nuclear data in higher energy region. The new version of Japanese Evaluated Nuclear data (JENDL-3)[2] was established and expected to respond these requirements. The present paper describes the results of the integral experiment performed at the OKTAVIAN facility and the calculation obtained by using MCNP[3] Monte Carlo neutron transport code with JENDL-3 and other evaluated nuclear data files for the comparison. For the selenium sample, however, no calculation was performed because its nuclear data were not available.

EXPERIMENT

Sample piles

Powdered sample of aluminum, titanium, arsenic, selenium, and zirconium were packed in spherical stainless steel shells so as to place a tritium neutron producing target at their centers. The details of the samples are listed in **Table 1**, where pile diameters and sample thicknesses (both in centimeters and in mean free paths for 14 MeV neutrons) are given.

Experimental setup

The experiment has been performed by the time-of-flight (TOF) technique using the intense 14 MeV neutron source facility OKTAVIAN[4] at Osaka University. A tritium target was placed at the center of the pile. The energy of the deuterons was about 250 keV. A cylindrical liquid organic scintillator NE-218 (12.7 cm-diam. × 5.1 cm-thick) was used as a neutron detector, which was located about 11 m from the tritium target, and 55° with respect to the deuteron beam axis. A pre-collimator made of 1.1 m thick polyethylene-iron multi-layers was set between the pile and the detector in order to reduce the background neutrons. The aperture size of this collimator was determined so that the whole surface of the pile facing the detector could be viewed.

Table 1 Characteristic parameters of sample piles

Pile	Diameter (cm)	Thickness	
		cm	MFPs
Aℓ	39.9	9.8	0.5
Ti	39.9	9.8	0.7
As	39.9	9.8	0.8
Se	39.9	9.8	0.6
Zr	61	27.5	2.0

Data Processing

The detector efficiency was determined by combining, 1) the Monte Carlo calculation, 2) the measured efficiency derived from the TOF measurement of ^{252}Cf spontaneous fission spectrum, and 3) the measured efficiency from the leakage spectrum from a graphite sphere, 30 cm in diameter with the similar detection system.

To monitor the absolute number of the source neutrons during each run, a cylindrical niobium foil was set in front of the tritium target and irradiated during the TOF experiment. From the γ-ray intensity of the induced activity, ^{92m}Nb and the integrated counts of the source neutron spectrum, the absolute neutron leakage spectrum can be obtained. This procedure is stated by Takahashi *et al.* in the **Ref.5**.

CALCULATION

Theoretical calculations were performed using MCNP, which is a three–dimensional continuous energy Monte Carlo transport code. The continuous energy library for the calculation of Aℓ, Ti, and Zr piles were FSXLIB[6] derived from JENDL–3. In addition to these JENDL–3 based libraries, BMCCS, and ENDL–85 attached with MCNP code were also used for the reference.

RESULTS

Measured and calculated spectra are shown from Figs.3 to 7. Brief discussions are given for each pile.

Aluminum

Though general shape of the experimental spectrum can be predicted with both JENDL–3 and ENDF/B–IV libraries, the discrepancies in the range from 4 MeV to the elastic scattering peak are observed for the Aℓ 40 cm pile (**Fig.1**).

Underestimation between 8 and 13 MeV The inelasic level scattering cross section mainly form this range. In JENDL–3, the energy levels are included up to 5.433 MeV. Based on the kinematics calculations, secondary neutrons should be scattered above about 8 MeV. The experimental spectrum actually contains this

contribution, but the spectrum in this area is much higher than the calculations with both JENDL–3 and ENDF/B–IV. This fact implies that the inelastic scttering cross sections to the discrete levels seems too low.

Underestimation between 3 and 8MeV This region is formed mainly from inelastcally scatterd neutrons and (*n,2n*) neutrons. Threfore, the evaluated nuclear data especially JENDL–3 might have improper cross sections. It is possible that the threshold energy for continuum levels is too high.

Overestimation aronud 0.5 MeV The calculated spectra with both libraries give overestimation around 0.5 MeV. Since the spectra in this energy range are formed by the neutrons scatterd so many times, it is considered that the difference in the total cross section has been mutiplied.

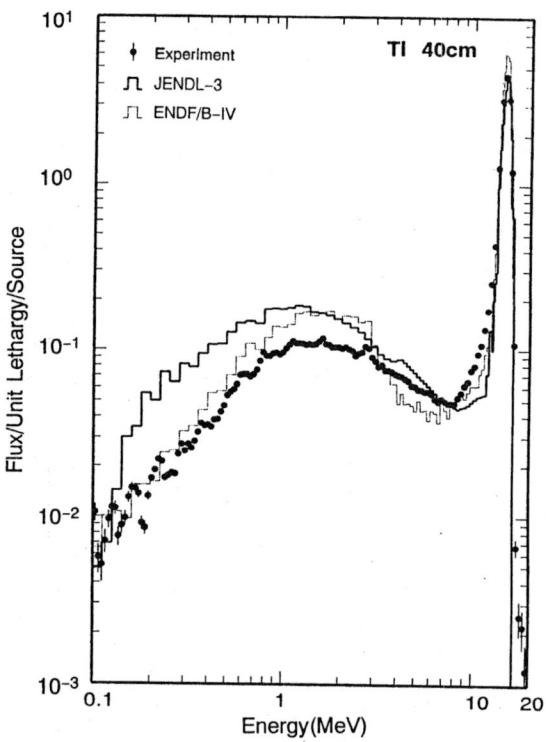

Fig.2 Experimental and Calculated spectra from Ti 40 cm pile

Titanium

For a Ti 40 cm pile, the calculated spectra with both JENDL–3 and ENDF/B–IV well predict the elastic scattering peak. However, the other part of the spectra differs considerably from the experiment (**Fig.4**).

Underestimation between 8 and 13 MeV Since the inelastically scattered neutrons are dominant in this energy range, both JENDL–3 and ENDF/B–IV are likely to have evaluated the inelastic level scattering cross sections too low.

Discrepancy between 3 and 8 MeV The calculated spectra with JENDL–3 and ENDF/B–IV give opposite results. Inelastic continuum and (*n,2n*) cross sections may be suspicious.

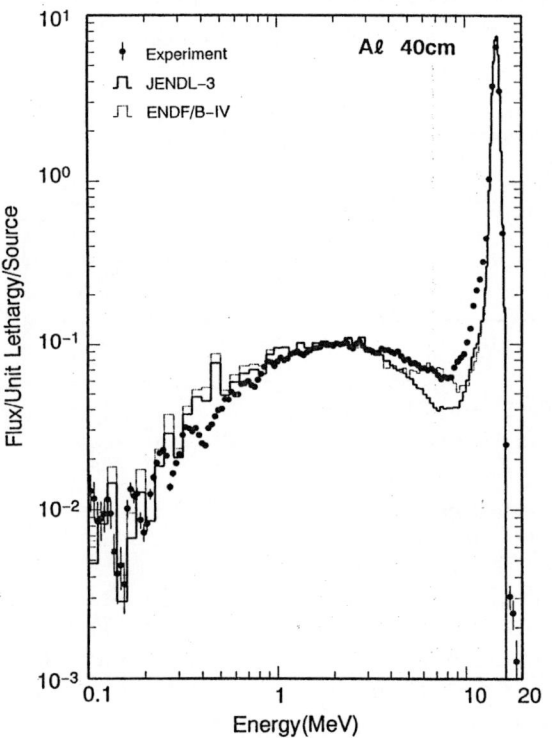

Fig.1 Experimental and calculated spectra from Aℓ 40 cm pile

Overestimation below 3 MeV Both JENDL-3 and ENDF/B-IV give large discrepancy from the experiment. Neutrons after (n,2n) reactions fall in this region and total cross section might have been checked.

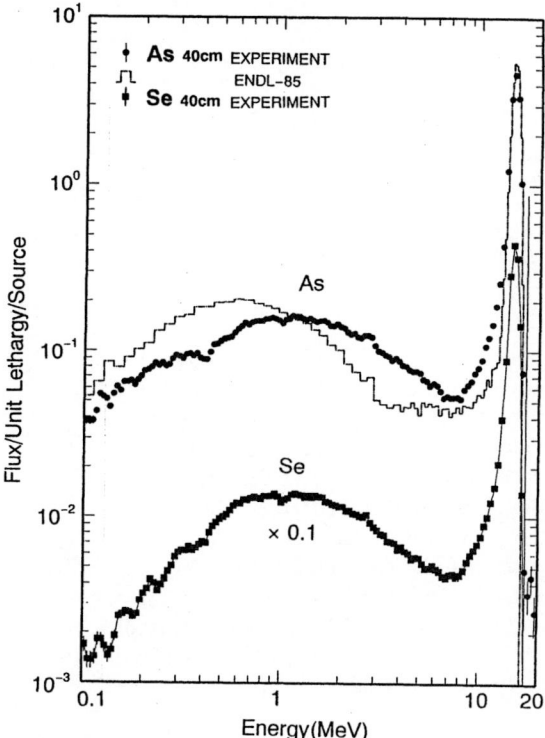

Fig.3 Experimental and calculated spectra from As 40 cm and experimental spectrum from Se 40 cm pile

Arsenic

Quite different spectrum from the experiment was obtained by the calculation using ENDL-85 data (**Fig.3**). Re-evaluation should be necessary for this element.

Selenium

Since no evaluated data were available, only measured spectrum is given in the **Fig.3** for the selenium sample.

Zirconium

For the Zr 61 cm pile, the calculated spectrum with JENDL-3 gives an improved prediction compared with the ENDL-85 spectrum (**Fig.5**).

Underestimation between 10 and 13 MeV Both calculation severely underestimate the spectrum in this range. This is possibly caused by too small evaluation the inelastic level scattering cross sections.

Overestimation between 1 and 10 MeV The JENDL-3 calculation gives much better prediction compared with the ENDL-85 one. There still exisits a slight overestimation (10 to 30 % of the spectral values) throughout the whole energy range below the elastic peak. This is probably due to the improper values for the continuum inelastic scattering and/or (n,2n) cross sections.

Overestimation with JENDL-3 under 1 MeV Though ENDL-85 predicts the experiment under 1 MeV fairly well, the calculation with JENDL-3 overestimates the experiment. This cause is not clear. However, total cross sections would have to be checked thoroughly.

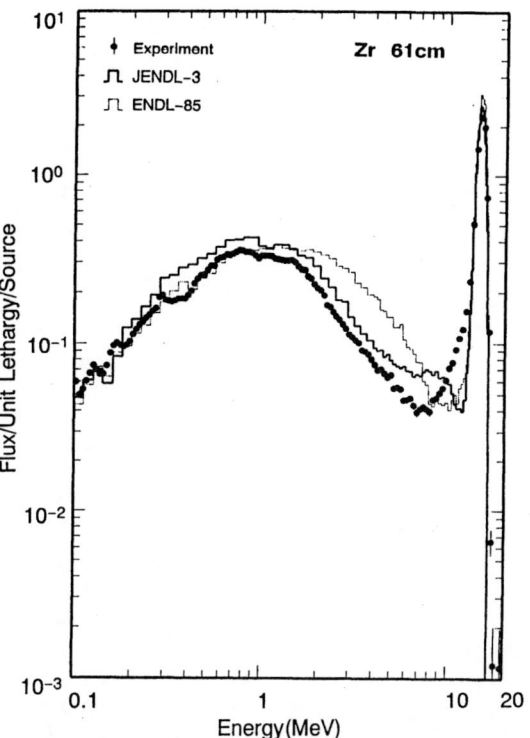

Fig.4 Experimental and calculated spectra from Zr 61 cm pile

ACKNOWLEDGEMENT

Part of this work has been supported by a Grant-in-Aid for Scientific Research from the Ministry of Education, Science and Culture, Japan (#60050040, #63050016, etc). This is also undertaken with the framework of the co-operative research program of the OKTAVIAN facility. We are very much grateful to Prof. Kenji Sumita of Osaka University for his continuous support for this work.

REFERENCES

1. S.A.Hayashi, I.Kimura, S.Yamamoto, H.Nishihara, S.Kanazawa, T.Mori, and M.Nakagawa: *J. Nucl. Sci. Technol.*, 24[9], 702(1987)

2. K.Shibata, T.Nakagawa, T.Asami, T.Fukahori, T.Narita, S.Chiba, M.Mizumoto, A.Hasegawa, Y.Kikuchi, Y. Nakajima, and S.Igarashi:*Japanese Evaluated Nuclear Data Library, Version-3 – JENDL-3 –*, JAERI 1319(1990)

3. J. F. Briesmeister ed.: *MCNP-A General Monte Carlo code for Neutron and Photon Transport, LA-73966-M*, Rev.2, (1986)

4. K.Sumita, A.Takahashi, T.Iida, J.Yamamoto, S.Imoto, and K.Matsushita: *Proc. 12th Symposium on Fusion Technology*, Sept., 1982, Jülich, FRG, Vol.1, 675(1982)

5. A.Takahashi, J.Yamamoto, K.Oshima, M.Ueda, M.Fukazawa, Y.Yanagi, J.Miyaguchi, and K.Sumita: *OKTAVIAN Report*, C-83-03(1983)

6. K. Kosako: *private communication*

CALCULATION AND MEASUREMENT OF NEUTRON AND GAMMA-RAY FLUXES IN LOW POWER REACTORS

Itsuro Kimura[*], *Yoshinori Sakurai*[*], *Satoshi Kanazawa*[*],
Katsuhei Kobayashi[**], *Seiji Shiroya*[**] *and Keiji Kanda*[**]

[*] *Department of Nuclear Engineering, Kyoto University,
Yoshida, Sakyo-ku, Kyoto 606-01, Japan*
[**] *Research Reactor Institute, Kyoto University,
Kumatori, Sennan-gun, Osaka 590-04, Japan*

Abstract: Neutron and gamma-ray flux distributions in a critical assembly (KUCA) were measured with gold wires and small thermoluminescence dosimeters (TLD) of magnesium orthosilicate, $Mg_2SiO_4(Tb)$ (MSO), respectively, and the results have been compared with those calculated by ANISN-JR with combined cross sections of neutrons and gamma-rays generated by RADHEAT-V3. For the comparison, we normalized the measured activity of the gold wire at the core center with the calculated one. The measured neutron flux distribution in the KUCA B-core with highly enriched uranium fuel and polyethylene moderator/reflector agree well with that calculated, however the absorbed doses of MSO in the core are about 10% higher than the calculated ones.

A similar comparison was carried out in four different irradiation fields (in void, and in three filter boxes) in an internal graphite reflector between two divided cores of a low power reactor (UTR-KINKI). Although the measured neutron flux distributions agree well with the calculated ones by TWOTRAN, the measured absorbed doses of MSO exceed considerably (about 35% in average) than those calculated.

(thermoluminescence dosimetry, magnesium orthosilicate, neutron kerma, RADHEAT-V3, neutron-gamma coupled calculation, cross sections, critical assembly, low power reactor, KUCA, UTR-KINKI)

Introduction

The importance of characterization of gamma-ray flux distribution in a reactor has risen recently, especially for the estimation of gamma-ray flux at experimental and therapeutic facilities in research and medical reactors. However there is hardly any established method to measure in-core gamma-flux distribution in a thermal reactor. Very recently, Abderrahim et al. have attempted to use three thermoluminescence dosimeters (TLD), namely ^{7}LiF, Al_2O_3 and BeO, for gamma-ray dosimetry in a reactor [1]. Integrated gamma dose in a power reactor has been measured with LiF dosimeters by King and Gilliam [2] recently. Hashizume et al. [3] showed the properties of magnesium orthosilicate activated with terbium, $Mg_2SiO_4(Tb)$ (MSO) as an excellent TLD and applied it to wide range gamma-ray dosimetry. Sato and Ono [4] tried to use MSO for gamma-ray dosimetry in therapeutic and other experimental facilities of the Kyoto University Reactor, KUR. In this work we have adopted MSO to measure gamma-ray distribution in the Kyoto University Critical Assemblies, KUCA and the results have been compared with those calculated by ANISN-JR with combined cross sections of neutrons and gamma-rays generated by RADHEAT-V3.

A similar comparison was carried out for gamma-ray distributions in four different irradiation fields at the Kinki University Reactor, UTR-KINKI. In this case, neutron and gamma-ray distributions were calculated by a two dimensional transport code TWOTRAN.

Experimental Procedures

MSO Detectors: The MSO powder of about 10 mg manufactured by Kasei Optonics Co. Ltd. was weighed and packed into aluminum tubes of 1 mm in inner diameter, 0.5 mm in thickness and 1 cm in length. The MSO detectors attached on an aluminum holder were irradiated in reactors. A few days after the irradiation for cooling, the MSO powder was taken out from its tube and was scattered on a sample plate of a TLD reader. An integral value of the thermoluminescence glow curve was digitally printed out for each MSO powder sample. The relation between this value and absorbed dose in Gy had been calibrated by using a ^{60}Co source. As an experimental error of absorbed dose determination, we took 10% from our preliminary experimental data.

The response of the MSO detector to slow neutrons were experimentally obtained using a supermirror neutron guide tube of the KUR. As a result, the obtained value is about 4.9×10^{-14} Gy·cm^2/n. Therefore, we can neglect the contribution of thermal neutrons.

Configuration of Reactors: We measured neutron and gamma-ray fluxes in the B-core of the KUCA and in four irradiation fields at the UTR-KINKI. The horizontal and vertical views of the KUCA B-core with highly enriched uranium fuel and polyethylene moderator/reflector are shown in Fig.1. The gold wires of 0.5 mm in diameter and the MSO detectors were irradiated along the small vertical gap surrounded by four aluminum sheaths of fuel elements, of which position is shown in the same figure.

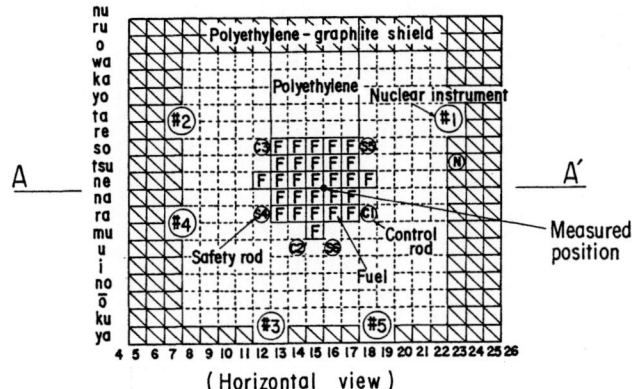

(Horizontal view)

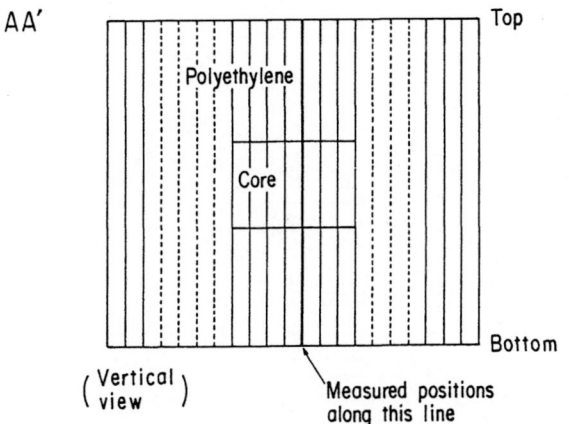

$\begin{pmatrix} \text{Vertical} \\ \text{view} \end{pmatrix}$

Fig. 1 Horizontal and vertical views of the B–core of KUCA.

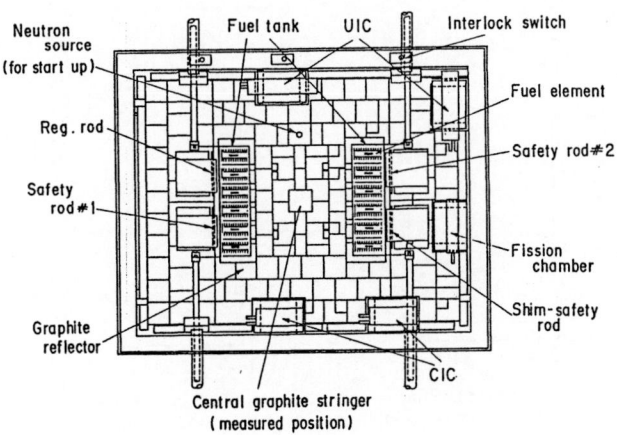

Fig. 2 Horizontal view of UTR–KINKI.

The UTR-KINKI has two divided cores with highly enriched uranium fuel and water moderator surrounded by a large graphite reflector. In the middle of the divided cores, a part of inner reflector (graphite blocks) can be withdrawn and one of three neutron filter boxes (Bi, Fe and polyethylene) can be inserted into this position to make irradiation fields for mainly biological samples. The horizontal view of UTR-KINKI is shown in Fig. 2.

Calculation

For neutron-gamma coupled calculations, we assumed an infinite slab sandwiched by two polyethylene reflectors for the KUCA B-core and a finite cylinder for the UTR-KINKI. Neutron and gamma-ray combined cross sections for each region were generated by RADHEAT-V3 [5] on the basis of the ENDF/B-IV and POPOP libraries. We took 100 and 18 groups for neutron and gamma-ray energies, respectively. For the transport calculation of neutrons and gamma-rays in the KUCA and the UTR-KINKI, we used ANISN-JR and TWOTRAN, respectively. For both cases, calculations were performed by P_3S_8 approximation.

Absorbed dose of gamma-rays was calculated from the obtained gamma-ray energy spectra and the mass absorption coefficients of MSO, and that of fast neutrons from the neutron spectra and kerma factors. The mass absorption coefficients and the kerma factors were taken from the references [6,7].

The contribution of delayed gamma-rays from fission products was estimated using the classical Way-Wigner's formula. We assumed that the delayed gamma-ray spectrum was similar to that of the fission gamma-rays. No cavity theory correction was considered, because the equivalent atomic number (11.1) of MSO is close to that of aluminum.

Results and Discussion

The measured and calculated distributions of the neutron flux (gold wire activation) and those of absorbed dose of MSO in the KUCA B-core are shown in Figs. 3 and 4, respectively. All data are normalized by the gold wire activation at the core center. The measured neutron flux distribution agrees well with the calculated one, except the positions close to the core-reflector boundary. The absorbed doses of MSO in the core are slightly (about 10%) higher than the calculated values.

As seen in Figs. 5 and 6, all of the measured neutron flux distributions in the UTR-KINKI satisfactorily agree with the calculated ones, however the measured absorbed doses of MSO are considerably higher (about 35% in average) than the calculated ones for all cases under the similar normalization to the KUCA. In this case, the contributions of fast neutrons and delayed gamma-rays were not taken into account. The former can be thought negligible, but the latter might cause the above discrepancy.

Acknowledgement

The authors are deeply thankful to Mr. T. Sato for his guidance to use MSO detectors. The present work was achieved by the Cooperative Use Programs of both the KUCA of Kyoto University and the UTR-KINKI of Kinki University.

REFERENCES

1. H.A. Abderrahim, R. Menil and H. Geens : Proc. 7th ASTM/EURATOM Symp. on Reactor Dosimetry, Strasbourg, August, 1990, in press
2. S.Q. King and D.M.Gilliam : ibid.
3. T. Hashizume, Y. Kato, T. Nakajima, T. Toryu, H. Sakamoto, N. Kotera and S. Eguchi : Proc. IAEA Symp. on Advanced Radiation Detectors, SM-143/11, p. 91 (1971)
4. K. Sato and K. Ono : UNTL-R-0073, p. 46 (1979)
5. K. Koyama, K. Minami, Y. Taji and S. Miyasaka : JAERI-M 7155 (1977)
6. J.H. Hubbell : NSRDS-NBS 29 (1969)
7. ICRP Report 26 (1977)

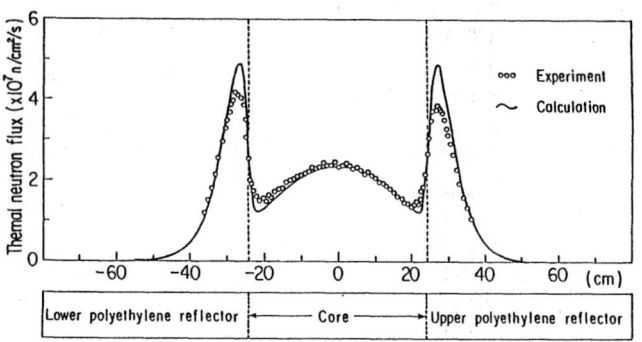

Fig. 3 Measured and calculated neutron flux distribution of MSO in the B-core of KUCA.

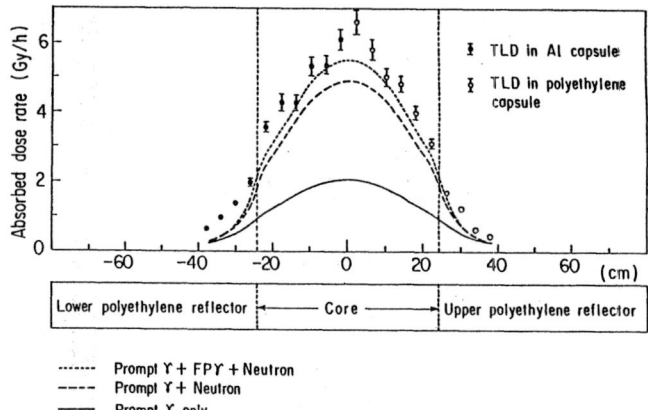

Fig. 4 Measured and calculated absorbed dose of MSO in the B-core of KUCA.

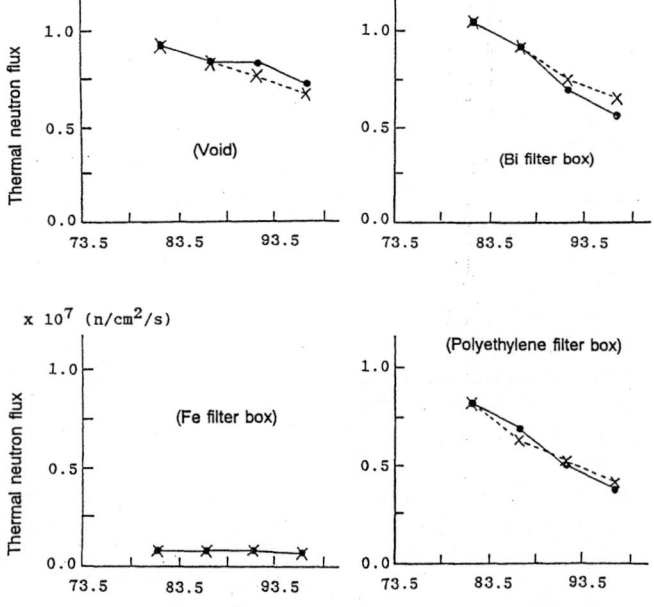

Fig. 5 Measured and calculated distributions of neutron fluxes in void and in three filter boxes located in the inner reflector between divided cores of UTR-KINKI.

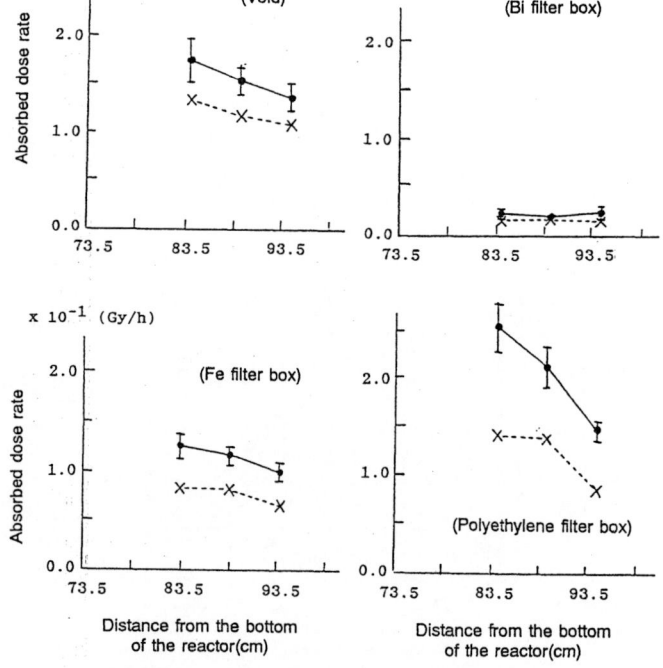

Fig. 6 Measured and calculated distributions of absorbed doses of MSO in void and in three filter boxes located in the inner reflector between divided cores of UTR-KINKI.

PREPARATION OF A CROSS SECTION LIBRARY FOR A 3 MW TRIGA REACTOR AND ITS VALIDATION THROUGH SOME APPLICATIONS

S.I. Bhuiyan, A.R. Khan, M.M. Sarker, M. Rahman
Z. Gulshan Ara, M. Musa, M.A. Mannan and I. Mele

Atomic Energy Research Establishment
P.O. Box 3787, Dhaka, Bangladesh

Abstract: A data base for the TRIGAP code has been generated for the 3 MW TRIGA Mark II reactor, Bangladesh, using the WIMS-D/4 code. Cross sections were calculated from zero burnup to 37% of initial ^{235}U in 20 burnup steps. The created TRIGAP library was tested and compared with experimental values or with values in the Safety Analysis Report (SAR). Excellent agreement of calculated values of temperature defect, flux distribution, temperature coefficient and peaking factor with experimental values establishes the validity of the library.

Introduction

The principal objectives of initiating this study were: a) Preparation of a Library for the Bangladesh 3 MW TRIGA reactor, and b) Application of the TRIGAP/1/ and WIMS/D-4/2/ codes for a few practical reactor calculations. The first step was to obtain nuclear cross section/group constants for homogenized fuel and moderator mixtures, and for nonfuel elements. This process of the analysis was performed with the standard IBM version of the WIMS/D-4 code distributed by OECD NEA Data Bank. The extended IJS version of a WIMS-D/4 library for isotopes specific to TRIGA fuel (Er-166, Er-167, Sm, H in ZrH) was used. Among them scattering cross section of hydrogen in zirconium hydride is especially important to determine the fuel temperature coefficient. WIMS with this library was used to generate a resonance shielded two group cross section library to support calculations with the TRIGAP code.

TRIGAP Library

The library is generated for LEU fuel element, LEU fuel follower and nonfuel elements. It also provides corrections of the cross section for temperature and for xenon and samarium. For operating conditions other than full power (3 MW) and equilibrium Xe and Sm, these coefficients are used to correct the cross sections. The library contains the unit-cell cross sections as a function of burnup. For the LEU fuel and LEU fuel followers the cross sections are tabulated in dependence of twenty burnup steps from zero burnup to 36.89% of initial ^{235}U. Detailed calculations show that all the three corrections are functions of burnup (τ). To handle this along with the coefficients for temperature, corrections at (τ_0) (12.83% burnup) the slope between burnup 0.0 and (τ_0), and the slope between (τ_1) (26.174% burnup) and (τ_0) are provided. For Xe, the corrections (delta) at (τ_0) are supplied with the slope at (τ_0). Similarly, the Sm corrections (delta) at (τ_0) are supplied with the slope at burnup 0.0 and (τ_1).

Cross sections of the fuel unit cells were calculated at nominal power 30 KW per element. The corresponding temperature was defined on the basis of the data obtained from the Safety Analysis Report (SAR)/4/ and also from experimental data. At nominal power of 30 KW/element, 330°C was used as average fuel temperature, 121°C as average cladding temperature and 40°C as bulk water temperature.

WIMS was run in a pin-cell option with energy condensation. The WIMS run in burnup mode covered at least 40% burnup of initial ^{235}U which was considered enough to cover the entire life-time of the reactor. This run in full power (3MW) included the expansion of the components of the fuel rod. The respective group constants at different burnup steps (1-20 in the library) for the main part of the TRIGAP library were actually generated from these WIMS outputs. Several WIMS runs at burnups (τ) = 0.0%, 12.864%, 26.149% of initial ^{235}U produced the required data base to generate the constants for temperature, Xe and Sm corrections.

To generate cross sections for the nonfuel elements, WIMS in the cluster option was run with the respective nonfuel element cell in the center surrounded by six fuel elements to provide for the fission neutron source.

Validation of TRIGAP Library

At several different burnups the following calculations were performed with the newly generated library for power levels of 50 w, 1 Mw and 3 Mw with or without equilibrium Xe. By choosing burnup steps: 0, 50 MWh, 350 MWh, and 750 MWh, the operating history of the reactor core under consideration was covered. The total energy produced upto 31 December 1989 was 606.41 MWh. A second burnup step was deliberately kept very short (50 MWh) to take into account Xe build-up. At each burnup step the temperature defect and Xe-value were calculated. The results, compared with those from the Log book and summarized in Table 1 show a very good agreement and confirms that TRIGAP library was successfully generated.

Calculation of Flux Distribution

Fast and thermal fluxes distributions were calculated and the results, compared with the projected values from SAR, showed good agreement (Fig.1).

Temperature Coefficient of LEU Fuel

The temperature coefficient (α_f) of LEU fuel was calculated for three different burnups (BU=0, BU=1000 MWd, BU=2000 MWd). The results of the temperature dependence of (α_f) compared with those of SAR for the LEU fuel are shown in Fig.2.

(α_f) of LEU fuel as a function of temperature and burnup can be explained as follows: The temperature coefficient of the LEU fueled core increases as a function of fuel temperature

Table 1. Comparison of calculated values of temperature defect and Xe-value (at 50 MWh) to projected values and experimental values from LOG BOOK

P		Calculated		SAR	Experimental
		$\Delta\rho$ ($)	ΔK ($)	$\Delta\rho$ ($)	ΔK ($)
1 MW	Temperature defect	1.15	1.32	–	1.02
3 MW	Temperature defect	3.59	4.06	3.43 – 5.14	3.64
1 MW	Xe-value	1.97	2.21	–	–
3 MW	Xe-value	2.97	3.20	3.57	–

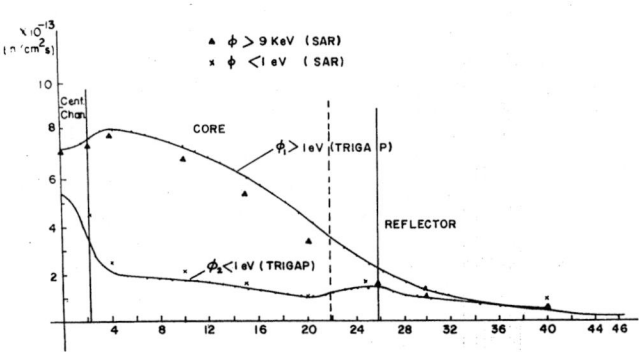

Fig.1. Comparison of calculated flux distribution and their projected values.

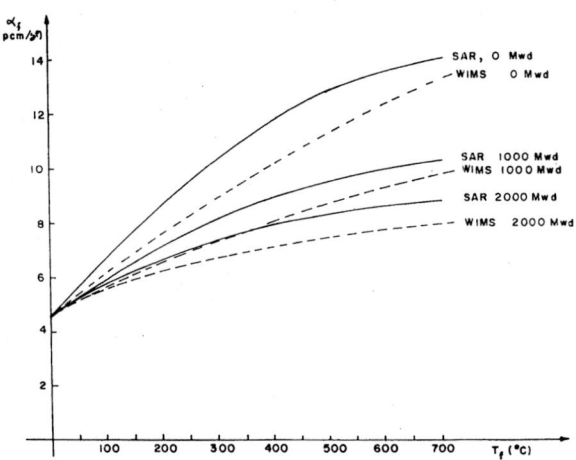

Fig.2. Temperature coefficient for LEU fuel.

because of the steadily increasing number of thermal neutrons being absorbed in the Er-167 resonance. It is also seen that after 1,000 and 2,000 MWd of burnup the coefficient is less temperature dependent and smaller in magnitude than that for the initial clean core because of the sizeable burnup of Er-167 and the resulting increased transparency of the resonance region around 0.5 eV to thermal neutrons.

Power Peaking Analysis

For the power peaking studies four core configurations were analyzed:
- Compact core, completely loaded with fuel elements with a) graphite, and b) water reflectors;
- Core with central thimble, c) graphite and d) water reflectors.

Radial Peaking Factor

The radial power density distribution was calculated using the program WIMS/D-4 in unit cell geometry. It is convenient to split this into two sub factors:

$$f_R = f_R^R \times f_R^G \qquad (1)$$

where, f_R^R is the radial rod power peaking factor, f_R^G is the power gradient peaking factor (due to irregularities like water gaps, irradiation channels etc.) f_R, calculated in unit cell geometry, is maximal at the outer radius where $f_R^R = 1.37$

for our LEU fuel.

Now, the gradient peaking factor, defined as the ratio between the maximum and the asymptotic power density ($r \rightarrow \infty$) around the irregularity in the core with the homogenized fuel unit cells, was calculated with WIMS-D/4 for water, graphite and void in the central region surrounded by 18 homogenized fuel unit cells. In case of water, $f_R^G = 1.85$, for graphite central region, $f_R^G = 1.32$, for the void central cell $f_R^G = 1.27$.

Hot Rod Factor

The hot rod peaking factor is defined as the ratio between the maximum power released by one fuel rod (P_{ROD}), and the average power per element in the core P_{CORE},

$$f_{HR} = (P_{ROD})_{MAX}/P_{CORE} \qquad (2)$$

$$P_{CORE} = P/N_{el} \qquad (3)$$

where P is the total reactor power which is 3 MW in our case. N_{EL} is the number of fuel elements which is considered to be 100 (95 LEU + 5 FF) for our TRIGA. Taking into account that all fuel elements have the same volume of fissionable material V_{EL}, the definition of f_{HR} in eq. (2) also applies to the ratio between average power density P_{ROD} of the hot rod and the core average power density P_{CORE} i.e.

$$((P_{ROD})_{MAX}/V)/(P_{CORE})/V = (P_{ROD})_{MAX}/P_{CORE} \qquad (4)$$

For the calculation of the hot rod factor, the 1-D code TRIGAP with the generated library was considered sufficient as the core has only one type of fuel. The calculated hot rod factors for the compact core for graphite and water reflectors was 1.45 and 1.56, respectively. But for the core with central thimble and graphite and water reflectors it was 1.58 and 1.67, respectively. In SAR for the compact core the hot rod factor f_{HR} = 1.60 and for core with central thimble f_{HR} = 1.70, thus showing their good agreement.

Total Peaking Factor

The total peaking factor defined as a product of hot rod factor, axial peaking factor and radial peaking factor in the SAR for the compact core is 3.5 and for core with central thimble is 5.63. The calculated values for compact core with graphite and water reflectors is 3.15 and 3.39, respectively. In case of the core with the central thimble with graphite and water reflectors the total peaking factors are 5.01 and 5.29, respectively. It can be observed that results of our calculation are slightly lower than that of the projected values. Probably the results can be improved if a better energy group structure is used (more than two energy groups) and even better if a 2-D diffusion code can be used.

Conclusion

A 3 MW TRIGAP Library for LEU fuel and LEU fuel-follower was generated and successfully tested. Since no 3 MW TRIGA cross section library is available currently, this library can be released and distributed internationally to fill the existing gaps. Calculations like fuel temperature coefficient and power peaking analysis using TRIGAP and WIMSD/4 was used for determining limiting conditions of normal operation.

Acknowledgement

This work was partially supported by the International Atomic Energy Agency (IAEA) through a Technical Cooperation Project, BGD/4/012.

REFERENCES

1. I. Mele, M. Ravnik, "TRIGAP - A Computer Programme for Research Reactor Calculations", IJS-DP-4238, Josef Stefan Institute, Ljubljana, Yugoslavia (1985).
2. WIMS-D/4 Programme Manual NEA Data Bank, Bat 45, 911191, Gif-sur-Yvette, CEDEX, France.
3. S.I. Bhuiyan, Z.G. Ara, M.M. Sarker, A.R. Khan, M. Musa and M.A. Mannan, "Analysis and Core-Life Calculation of 3 MW TRIGA Mark II Research Reactor Including Effects of Central Thimble Modifications", International Conference on the Physics of Reactors: Operation, Design and Computation, MERSEILLE-FRANCE, Vol. 4, p: 114-127, Proceedings of Physor '90, April 23 - 27, 1990.
4. Safety Analysis Report for Bangladesh (3 MW TRIGA Mark II) Research Reactor, G.A. Technologies Inc., USA (1981).

DEVELOPMENT OF A COMMON NUCLEAR GROUP CONSTANTS LIBRARY SYSTEM: JSSTDL-295n-104γ BASED ON JENDL-3 NUCLEAR DATA LIBRARY

Akira HASEGAWA

Japan Atomic Energy Research Institute,
Tokai-mura, Naka-gun, Ibaraki-ken, 319-11 Japan

abstract: JSSTDL 295n-104γ : a common group cross-section library system has been developed in JAERI to be used in fairly wide range of applications in nuclear industry. This system is composed of a common 295n-104γ group cross-section library based on JENDL-3 nuclear data file and its utility codes. Target of this system is focused to the criticality or shielding calculations in fast and fusion reactors using ANISN, DOT, or MORSE code. Specifications of the common group constants were decided responding to the request from various nuclear data users, particularly from nuclear design group in Japan. Group structure is decided so as to cover almost all group structures currently used in our country. This library includes self-shielding factor tables for primary reactions. A routine for generating macro-scopic cross-section using the self-shielding factor table is also provided. Neutron cross-sections and photon production cross-sections are processed by Prof. GROUCH-G/B code system and γ ray transport cross-sections are generated by GAMLEG-JR. In this paper, outline and present status of the JSSTDL library system is described along with two examples adopted in JENDL-3 benchmark test. One is for shielding calculation, where effects of self-shielding factor (f-table) is shown in conjunction with the analysis of the ASPIS natural iron deep penetration experiment. Without considering resonance self-shielding effect in resonance energy region for resonant nuclides like iron, the results is completely missled in the attenuation profile calculation in the shields. The other example is fast rector criticality calculations of very small critical assemblies with very high enrichment fuel materials where some basic characteristics of this library is presented.

(group constants library, JSSTDL system, JENDL-3, fast reactor, fusion reactor, shielding, self-shielding factor, iron, deep penetration, attenuation profile, ASPIS, criticality calculation.)

1. Introduction

A system called "JSSTDL-295n-104γ (neutron:295 gamma:104) group constants library system" has been developed, which is composed of a common 295n-104γ group cross-section library based on JENDL-3 nuclear data file and its utility codes for fast and fusion reactor applications.

Such a system has been requested for many years by various nuclear data users, particularly nuclear design group. Usually in Japan, shielding and criticality calculations have been performed in different calculation paths using the different group constants libraries, i.e., in different group structures and data sources due to the characteristics of the calculations involved. Finer group structure is inevitably necessary for the shielding calculation but relatively coarse group structure is sufficient for the criticality calculations. Hence there have been many requests for supplying a common library applicable both for the criticality and shielding calculations. Users are also very keen for using the latest nuclear data.

Responding to these requests, specifications of the common group constants were decided. Scope and detailed specification of this library is given in the next chapter. Used results of this library for the shielding calculation is given in the third chapter, where merits of using self-shielding factor tables are shown. Another example applied to criticality calculation is given succeedingly to exhibit some characteristics of this library.

2. Out line of JSSTDL-295n-104γ library system

This library is a common group cross-section library recently developed in JAERI by the author under the cooperation of Working Group on Standard Group Constants affiliated by the Committee of Group Constants of JNDC (Japanese Nuclear Data Committee). This work was promoted by Nakazawa Committee's recommendation /1/ deserted to the JNDC (Proposal on Post-JENDL-3 Activity Programme for JNDC, 1986) defining the working frame of JNDC after JENDL-3 project was completed. In this recommendation, it was stated that JNDC should supply commonly usable group cross-section library for primary data users like fast or fusion reactor designers in no delay of JENDL-3 release.

Responding to the recommendation and requests from reactor design groups, specifications of the common group constants were decided. In Table 1.1 specification of JSSTDL system is shown. Most users insist to maintain their own group structure which has been so far used. Therefore a universal group structure was decided to cover almost all group structures used in Japan, as seen in Table 1.2 ~ 1.3, so as to produce their own required group structure library from a master library. It was also decided to prepare a library for design codes most frequently used such as ANISN, DOT or MORSE. Resonance self-shielding factors were also considered for primary reactions. Scattering matrices were calculated up to P5 components and were stored independently for elastic and inelastic scatterings so as to use the different self-shielding factors. For secondary γ production cross sections, data are stored for the following 4 reactions, i.e., total, capture((n,γ) MT:102 only), fission(MT:18), other than capture and fission ((n,n'),(n,p),(n,α),...), to reflect self-shielding factor for capture and fission reactions in neutrons. Gamma transport cross-section are generated by GAMLEG-JR /2/.

A utility routine was developed to enhance portability of this system to other sites or machines. Routines for group collapsing and for generating region dependent macro-scopic cross sections were also developed and released with the library. In the Table 1.4, all of the routines developed are shown.

Up to now, 63 nuclides were processed and stored for JSSTDL library from JENDL-3 general purpose file /3/. Gamma data are furnished for 32 nuclides out of 63 processed nuclides. Almost all nuclides available of γ production data in JENDL-3 were processed. For the processing of JENDL-3 neutron and γ production cross-section, Prof.GROUCH-G/B /4/ system and its utility codes were fully used.

Table 1.1 Group cross-section library processing specification

Group structure	neutron:295 gamma:104
Weighting spectrum	Maxwellian from 1.0E-5 to 0.3224 eV the rest is 1/E.
Resonance reconstruction tolerance	0.1%
Self-shielding factor Temperature grid	300 600 900 2100 Kelvin
σ_0 grid	0 0.1778 1 10 10^2 10^3 10^4 10^5 10^6 barn
f-tab reactions	total, elastic, capture, fission.
Anisotropic Pl order	5

Table 1.2 Neutron group boundaries considered in JSSTDL system

Library name	groups
JSD-100	100
JSD-1000	100
BERMUDA-121	121
FNS-125	125
VITAMIN-C	171
VITAMIN-J(E+C)	175
GICX-42	42
ABBN-25	25
JFS-New	70
GAM-123*	92
MGCL-137*	91
WIMS-69*	28

Table 1.3 Gamma group boundaries considered in JSSTDL system

Library name	groups
CSEWG-94	94
LANL-12	12
STEINER-21	21
STRAKER-22	22
LANL-48	48
LANL-24	24
BERMUDA-36	36
HONEYCOMB -15	15

<== n.b. *: fast only

3. An example of shielding calculation

To demonstrate the effect of self-shielding factor for resonant nuclide, we performed an analysis of axial attenuation profile for ASPIS deep penetration experiment measured for natural iron block.

Experimental configuration is as follows, a low-power natural uranium converter plate, driven by the source reactor NESTOR, provided a large thin disc source of fission neutrons is placed at the interface of a graphite moderator and extensive iron shield (= 140 cm thickness). Analyses are made for the axial attenuation measurements with three threshold detectors and one low-energy activation detector, and for the spectrometer measurements at four selected positions in the iron shield.

Calculational model and analysis: As for the calculational model, we closely follow the model in the report of M. D. Carter et al. /5/, except some modifications.

For Fe natural cross-sections in the test region, which is thought as nearly pure material, following three methods are used to check self-shielding effects for the resonance cross-sections.

CASE 1: exact weighted cross-sections, this means that group cross-sections are generated by $1/\sigma_t$ (Fe) weighting, i.e. fully shielded cross-sections including higher Pl matrices (using special option of PROF-GROUCH-G/B). For

the macro cross-section generation for DOT code, GLIB.MAKE -MACROX utilities were used.

CASE 2: self-shielding effect is automatically considered by JSSTDL system.

CASE 3: no self-shielding effect is considered.

Anisotropy were included up to P5. Calculations were performed by DOT3.5 with S48 P5 R-Z: 53x92 meshes.

Results and discussions: The results are shown in Fig.3.1 for transmitted fluxes. And in Fig.3.2 axial attenuation profile of Rh-103 (n,n') reaction rate is shown. Circle marks in figures show experimental results. Triangle, plus, cross marks show CASE 1(exact weighting), CASE 2(self-shielding), CASE 3(no self-shielding) results respectively. From Fig.3.1, CASE 1 is the best, the next is CASE 2, the worst is the CASE 3. In this depth position, i.e., about one meter depth, the results of CASE3 (no self-shielding effect) is completely deviate from experiments. The difference is nearly factor 10, i.e., magnitude of order is different. The same tendency is seen in the axial attenuation profile for

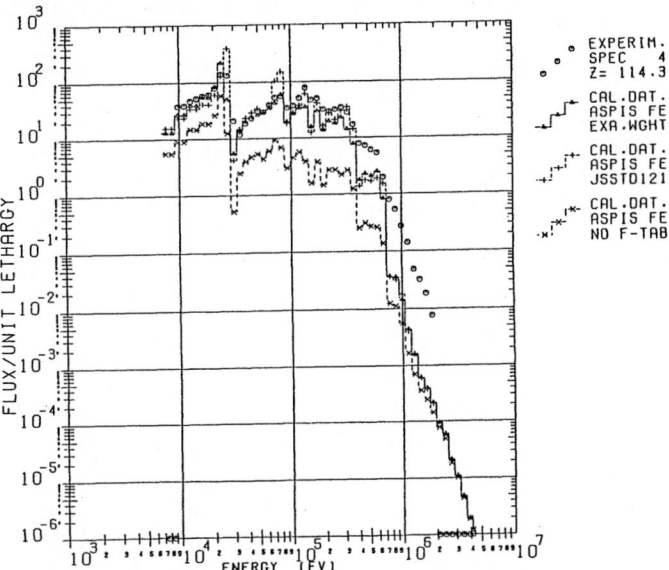

Fig. 3.1 Transmitted Spectrum in natural iron block at 114.3 cm position

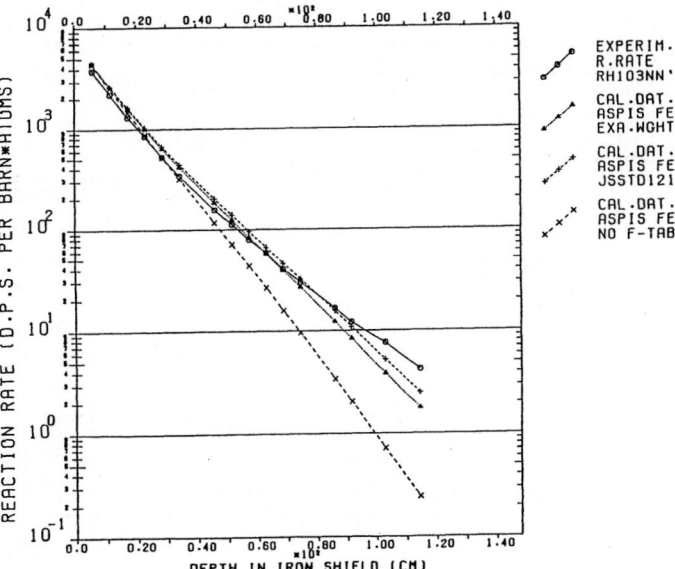

Fig. 3.2 Rh-103 (n,n') reaction rate axial attenuation profile

Rh-103 (n,n') given in Fig. 3.2. From these results, it is concluded that the best is using exact weighted cross-sections. But this method is very expensive, because every time we must generate group cross-sections using the quite time-consuming processing code for each nuclides and for each region. From the results shown above we conclude that self-shielding factor method is rather convenient even though it is not superior to the exact weighting method. No self-shielding factor results are completely bad. We cannot use the library without self-shielding factor for the resonant nuclides.

4. An example of criticality calculation

To demonstrate the applicability of this library to the criticality calculation, benchmark test results which have been performed by our group/6/ are shown here. This work was performed for the assessment of neutron cross sections in JENDL-3 for Pu-239 to the simple geometry, high enrichment, very small fast breeder critical assemblies. The discussions for the applicability of the neutron cross sections of JENDL-3 are given elsewhere/7/, where the results by JENDL-3 are concluded as to be quite satisfactory, here we restrict our discussion to the characteristics of the library.

In order to finalize the results in real calculation steps, many calculation conditions should be checked. Among them, group number effects or higher Pl truncation effects are typical one. Calculations for checking these conditions are fairly easily performed by using this library system. Examples for these two effects are shown in Table 4.1 ~ 2.

Table 4.1 Group number effects on some integral qunatities for Pu-239 core

assembly	group	Keff	spectral indices at core center
			$\sigma_f(U8) / \sigma_f(U5)$ (cal., C/E)
JEZEBEL	175	1.0010	0.2121 , 0.993
	100	1.0011	0.2121 , 0.993
	70	1.0004	0.2115 , 0.990
	25	1.0026	0.2108 , 0.986
JEZEBEL-Pu	175	0.9961	0.2075 , 1.007
	100	0.9961	0.2074 , 1.007
	70	0.9953	0.2069 , 1.004
	25	0.9973	0.2062 , 1.001
FLATTOP-Pu	175	0.9982	0.1818 , 1.010
	100	0.9982	0.1818 , 1.010
	70	0.9977	0.1813 , 1.007
	25	1.0011	0.1807 , 1.004

Table 4.2. Pl truncation effects for several integral parameters

	assembly	GODIVA			
	PL order	$L=1$	$L=2$	$L=3$	$L=5$
C/E	Keff	1.0013	1.0066	1.0066	1.0066
	$\sigma_f(U8) / \sigma_f(U5)$	0.9930	1.0012	1.0004	1.0004
	$\sigma_f(Np7) / \sigma_f(U5)$	0.9951	0.9992	0.9989	0.9989

Discussions of the results: Four different group structure libraries are generated from the JSSTDL-295n-104γ library. Generated group libraries are

1) VITAMIN-J 175 group   3) JFS-70   70 group
2) JSD-100   100 group   4) ABBN-25   25 group.

Criticality and central spectral indices for σ_f U-238 / σ_f U-235 are shown. From this table as for as these two integral data are concerned, one hundred group is sufficient for the calculation. But twenty five group library produce at most 0.3 % difference from reference data(175 group data)

for k_{eff}. This tendency comes partially from adopted weighting spectrum in the JSSTDL system i.e., 1/E instead of fission spectrum. These assemblies are more sensitive to the higher energy part around fission spectrum peak. Twenty five group structure is too coarse for these sensitive energy range.

From Table 4.2, Pl truncation effects are not so large for this assembly. $P2$ is sufficient. For the criticality calculation, anisotropy of neutron data for these energy region is not so important for the analysis.

5. Conclusion

Newly developed JSSTDL library system was introduced. This system is composed of a common 295n-104γ group cross-section library based on JENDL-3 nuclear data file and its utility codes. The target of this system is focused to the criticality or shielding calculations in fast and fusion reactors. Group structure is defined so as to cover almost all group structures currently used in Japan. Neutron cross-sections and photon production cross-sections were processed by Prof. GROUCH-G/B code system and γ ray transport cross-sections are generated by GAMLEG-JR.

As seen from ASPIS deep penetration analysis, self-shielding effects from the iron nuclide are very large. At about one meter penetration depth, the spectrum is underestimated by factor 10 in the resonance energy range if no self shielding effect is considered. This means that group constants library without self-shielding factor table can not be applicable to the shielding applications as for as resonance nuclides are concerned. In the criticality calculation, this is a very common knowledge.

This library system has been fully applied for the benchmark test of JENDL-3. We confirmed the applicability of this system to the fast reactor calculations both for criticality and shielding and also to the fusion neutronics calculations. This system: 295n-104γ JSSTDL library is now available through JNDC Nuclear Data Center with the associated utility codes in FORTRAN source. The package includes also 100n-40γ library for the users of shielding fields.

References

1. M. Nakazawa : "Proposal on Post-JENDL-3 Activity Programme for Japanese Nuclear Data Committee", JAERI-M 87-025 pp9 (1987).

2. K. Koyama, N. Yamano, K. Minami and S. Miyasaka:"RADHEAT-V3",JAERI-M 7155 (1977).

3. K. Shibata, T. Nakagawa, T. Asami :"Japanese Evaluated Nuclear Data Library, Version-3 -JENDL-3-",JAERI-1319 (1990).

4. A. Hasegawa :"Development of a Processing Code System Prof. GROUCH-G/B", unpublished work (1986).

5. M. D. Carter, A.K. McCracken and A. Packwood : 'The Winfrith Iron Benchmark Experiment, A Compilation of Previously Published Results for use in the International Comparison of Shielding Data-Sets Sponsored by NEA,' AEE Winfrith (1982).

6. D. Tian, A. Hasegawa, T. Nakagawa and Y. Kikuchi : in Proc. 1990 Symp. Nuclear Data, Nov. 1990, JAERI, p.148, JAERI-M 91-032 (1991).

7. Y. Kikuchi and Members of JNDC: 'JAPANESE NUCLAER DATA LIBRARY, VERSION-3' Int. Conf. on Nuclear Data for Science and Technology, May 1991 Julich,FRG, IP13 (1991).

CRISY: A CRITICAL ASSEMBLIES DOCUMENTATION SYSTEM

U. Salmi, R. L. Perel and J. J. Wagschal

Racah Institute of Physics
The Hebrew University of Jerusalem
91904 Jerusalem, Israel

Abstract: A database named CRISY (CRItical assemblies documentation SYstem) was set up. Data on experimental critical assemblies were collected, analysed and checked. Missing information was supplemented after a careful analysis of the data. Inconsistent data were improved as far as possible. Corrections, either specifically stated or implied, were added to the data. The processed data of various sources and formats were arranged in an electronic spreadsheet, which enables quick and easy deduction of the input necessary for transport codes. The CRISY spreadsheet is accompanied by detailed documentation of the data used and of the considerations leading to the final form. This database can be used for data testing validation calculations and for cross-section adjustment.

(CRISY, critical assemblies, database, integral measurements)

Introduction

The exact specifications of the dimensions and materials composition of any benchmark system are a necessary condition for any validation calculation or cross-section adjustment attempt. Experience has shown that currently available compilations of critical assemblies specifications [1] sometimes have a few major shortcomings. These include:
a. Lack of some information on the dimensions or material composition of an assembly;
b. Seemingly redundant data that indicate inconsistencies in the description;
c. Incomplete description of "corrections" applied to experimental data in order to compensate for experimental difficulties.

Validation and adjustment calculations should be carried out with great precision. Both the experimental and the calculational uncertainties should be minimal, therefore "one-dimensional" assemblies are preferred. We collected data on spherical assemblies, which can be calculated with a one dimensional transport code.

CRISY, a CRItical assemblies documentation SYstem, reflects the experience of generating input data for reactivity calculations of critical assemblies.

In this paper we are first going to outline the conceptual constitution of CRISY. This will be followed by an example illustrating how data on ^{233}U assemblies appear in the literature and how they are treated in the process of becoming part of CRISY. This detailed example will be followed by some additional general considerations encountered during the formation of CRISY and by some concluding remarks.

CRISY, General Outline

The CRISY system consists of two major parts. The first is an electronic spreadsheet that has input entries for the data given in the literature, output entries to be used as the necessary input for transport codes and the linking formulae used. Any change in the input data, will be reflected automatically in any quantity based on the modified data.

The other part consists of documentation of the data available for each critical assembly. The documentation includes:
1. A replica of the data used, a discussion of lacking or redundant data and a detailed explanation of the considerations that led to the final input data for the transport calculations.
2. A "completeness index" indicating the level of completeness, consistency and reliability of the information on each assembly.
3. A discussion of quoted experimental uncertainties and the corresponding uncertainties in calculated reactivities.

The necessary input data for transport codes, cited or computed in CRISY, are the radii and densities of all zones in each assembly, and the atomic composition of each zone.

The completeness index ratings are:

Index A was given to assemblies for which all the necessary data appear in the literature or can be calculated without any further assumptions, and all the data are consistent.

Index B was given to assemblies for which there is incompleteness or ambiguity in the meaning of data, but we can find a reasonable interpretation based on our knowledge and on comparison to other assemblies.

Index C was given to all other assemblies included in CRISY.

In order to enable each potential user of CRISY to verify our considerations or to form his own opinion, we first "copy" the description of each assembly to be included in CRISY, as it appears in the respective references.

^{233}U Assemblies

The information on assemblies with a ^{233}U core is presented next. Each assembly gets a name indicating in a schematic way the material content of consecutive radial zones. Similar assemblies are differentiated by a sequential number.
1. U23−1 (Jezebel) [2]:

    16.53(±0.4%) kg U, density 18.42$_4$ gr/cm^3, 98.13 at% ^{233}U, 1.24 at% ^{234}U, 0.03 at% ^{235}U, 0.60 at% ^{238}U.

2. U23-Be−1 [3]:

    7.601 kg U, density 18.62 gr/cm^3, diameter 3.622 in., 98.25 wt% ^{233}U; Be reflector, density 1.83 gr/cm^3, 1.652(±1%) in. thick.

3. U23-Be−2 [3]:

    10.012 kg U, density 18.62 gr/cm^3, diameter 3.972 in., 98.25 wt% ^{233}U; Be reflector, density 1.83 gr/cm^3, 0.805(±1%) in. thick.

The information on the next assembly is based on two different sources.

4a. U23-U−1 [3]:

    Core same as that of (2); U(93) reflector, density 18.80 gr/cm^3, 0.780(±1%) in. thick.

4b. U23-U-1 [2]:

7.601 kg U, density 18.64_4 gr/cm^3, 98.2 wt% ^{233}U, 1.1 wt% ^{234}U, 0.7 wt% ^{238}U; U(93.2) reflector, density 18.80 gr/cm^3, 0.783(±1%) in. thick.

Here again the information stems from two different references:

5a. U23-U-2 [3]:

core same as that of (3); U(93) reflector, density 18.80 gr/cm^3, 0.478(±1%) in. thick.

5b. U23-U-2 [2]:

10.012 kg U, density 18.62_1 gr/cm^3, 98.2 wt% ^{233}U, 1.1 wt% ^{234}U, 0.7 wt% ^{238}U; U(93.2) reflector, density 18.80 gr/cm^3, 0.481(±1%) in. thick.

6. U23-U-3 [4]:

2.37_1 kg ^{233}U isotope, density(U) 17.78 gr/cm^3, diameter 2.504 in., 97.9 wt% ^{233}U, 0.9 wt% ^{234}U, 0.2 wt% ^{238}U, 0.95 wt% W; U(93.17) reflector, density 18.8 gr/cm^3, 34.8±0.1 kg of ^{235}U, 1.886 in. thick.

7. U23-Un-1 [3]:

Same core as that of (2); natural U reflector, density 18.92 gr/cm^3, 2.090(±1%) in. thick.

8. U23-Un-2 [3]:

Same core as that of (3); natural U reflector, density 18.92 gr/cm^3, 0.906 in. thick.

9. U23-Un-3 (Flattop23) [2]:

5.74±0.03 kg U, density 18.42 gr/cm^3, 98.13 wt% ^{233}U, 1.24 wt% ^{234}U, 0.03 wt% ^{235}U, 0.60 wt% ^{238}U; natural U reflector, density 19.0 gr/cm^3, 7.84 in. thick (total diameter 19.00 in.).

Only the description of assemblies (1) and (9) is complete and consistent. For all other assemblies the data had to be supplemented by comparative considerations, knowledge and experience. This will be shown next.

In assembly (6) the sum of the wt% in the core is 99.95%, and therefore should be normalized. There is ambiguity to the meaning of the "(U) density" in the core description. The only interpretation consistent with the given weight and diameter is that the density is that of all the U isotopes, *excluding* the W. There is also an inconsistency of about 0.01 kg in the mass of the reflector in the redundant data. However, this can be neglected when compared to the 0.1 kg experimental error.

Assemblies (3), (5a), (8) are listed in Plassmann and Wood [3] as having a common core and different reflectors. An additional assembly with the same core was not included in CRISY, because there is not enough information about the composition of the W alloy reflector. The core diameter, weight and density are consistent. The description of assembly (5), an enriched U reflected assembly, appears also in Hansen and Paxton [2] with a more complete description of the core composition (5b). Since the three assemblies have the same core, this composition was also adopted for the other two. The thickness of the U(93) reflector was increased in [2] by 0.003 in., in order to take into account additional reflections by the machine and Kiva. We also added 1.02 wt% ^{234}U to the U(93) reflector's composition, as will be explained in the next section.

Assemblies (2), (4a), (7) are also listed in Plassmann and Wood [3] as having a common core, and the W alloy reflector assembly is once again not included in CRISY. This time however, the diameter, weight and density entries are inconsistent. The U(93) reflected assembly (4) appears also in Hansen and Paxton [2] with a slightly different density (4b). When we calculate the core radius from the weight and density of [2], we arrive at the same diameter as in [3]. Therefore we adopted the core data of the U(93) reflected assembly from [2] for *all three* assemblies. In addition, the U(93) reflector of (4b) was also modified in the same way as the U(93) reflector of assembly (5b) of the other core, in order to take the additional reflections into account.

The input specifications for neutronic calculations, derived from these original data according to the previous considerations, are listed in Table 1. One should note that only two assemblies get the completeness index A as explained earlier. One should also note that all compositions were translated to atomic percents.

General Considerations

In this section we discuss examples of general problems encountered while preparing the input for CRISY, and the way we solved them.

A problem common to many assemblies is the lack of information on the content of certain isotopes, namely ^{234}U and ^{241}Pu.

In natural U the ^{234}U wt% is negligible (only 0.005%). During the enrichment process (either by a centrifugue or by diffusion) the lighter isotopes are enriched at a higher rate. So, when the ^{235}U concentration reaches about 93%, the ^{234}U

Table 1. Dimensions and composition of critical assemblies containing ^{233}U cores

No.	NAME	INDEX	CORE R cm	DENS g/cc	U 233	U 234	U 235	U 238	W	REFLECTOR R cm	DENS g/cc	U 234	U 235	U 238	Be
1	U23_1	A	5.983	18.424	98.130%	1.240%	0.030%	0.600%							
2	U23-Be_1	B	4.600	18.644	98.219%	1.096%		0.685%		8.796	1.830				100.0%
3	U23-Be_2	B	5.044	18.621	98.219%	1.096%		0.685%		7.089	1.830				100.0%
4	U23-U_1	B	4.600	18.644	98.219%	1.096%		0.685%		6.589	18.800	1.025%	93.264%	5.711%	
5	U23-U_2	B	5.044	18.621	98.219%	1.096%		0.685%		6.266	18.800	1.025%	93.264%	5.711%	
6	U23-U_3	B	3.180	17.951	97.708%	0.894%		0.195%	1.202%	7.971	18.800	1.025%	93.234%	5.741%	
7	U23-Un_1	B	4.600	18.644	98.219%	1.096%		0.685%		9.909	18.920		0.720%	99.280%	
8	U23-Un_2	B	5.044	18.621	98.219%	1.096%		0.685%		7.346	18.920		0.720%	99.280%	
9	U23-Un_3	A	4.206	18.420	98.148%	1.235%	0.030%	0.587%		24.119	19.000		0.720%	99.280%	

concentration reaches about 1%. The description of "Godiva" (a bare enriched U assembly) in Hansen and Paxton [2] includes specifically 1.02 wt% of ^{234}U, and it is stated that "This value is insensitive to few percent change in the enrichment of ^{235}U". Nevertheless, the description of many assemblies containing enriched U does not mention any ^{234}U. Sinces 1.0 wt% of ^{234}U makes a difference in reactivity calculations, and since highly enriched U(~90%) does contain about 1 wt% of ^{234}U, we added it to all assemblies containing enriched U.

Plutonium is obtained by irradiation of ^{238}U by neutrons in a reactor, leading to the ^{239}Pu isotope. By further irradiation of the Pu, heavier isotopes are created. The longer the irradiation exposure, the higher the percentage of heavier Pu isotopes. For each percent of ^{240}Pu created, there is a typical percentage of ^{241}Pu. Yet the data about some Pu assemblies contain information about both ^{240}Pu and ^{241}Pu percentage, while the data about many other assemblies only contain information about a similar percent of ^{240}Pu and not on ^{241}Pu. Thus, we followed Pazy et al [5] and added a typical percentage of ^{241}Pu. This value has a large uncertainty, because of the short half life of ^{241}Pu (14.4 y).

Another example of missing data is found in a group of assemblies, all of which have a common core of Pu and different reflectors [3]. The Pu composition is given exactly, and it is stated that "the Pu contains a minor amount of inert diluent". In a later work [2], some of the assemblies in this group are described, and it is written that they contain 1.0 wt% of Ga. We conclude that the common core of all the assemblies in this group, in the work of Plassmann and Wood [3], contains also 1.0 wt% of Ga.

Another example of incomplete data that had to be complemented is the Ni coating of Pu and ^{233}U. The core parts consisting of Pu or ^{233}U are coated with a thin layer of Ni or similar metal for protection. The information about some of these assemblies includes remarks that the data were corrected empirically to compensate for Ni coating [4]. For some other assemblies there is a remark that the data were not corrected for the effect of Ni coating, therefore these assemblies should be calculated with the Ni taken into account explicitly. There is also a third group of Pu or ^{233}U assemblies with no indication at all if any correction was applied to the data to the effect of Ni coating, and we had to make a decision according to our best judgement.

Summary

We have demonstrated in this paper how data about critical assemblies were collected and processed in CRISY, in order to have complete and consistent input for neutronic calculations.

CRISY will be used for consistency tests of the new ENDF/B-VI data and for cross-section adjustment that can substantially reduce the uncertainty in calculated reactor performance and safety data [6].

References

1. See e.g. H.C. Paxton and N.L. Pruvost, Critical Dimensions of Systems Containing ^{235}U, ^{239}Pu and ^{233}U, LA-10860-MS (1986 Revision)

2. G.E. Hansen and H.C. Paxton, Re-Evaluated Critical Specifications of Some Los Alamos Fast-Neutron Systems, LA-4208 (TID-4500), Los Alamos Scientific Laboratory (1969)

3. E.A. Plassmann and D.P. Wood, Critical Reflector Thicknesses for Spherical ^{233}U and ^{239}Pu Systems, Nucl. Sci. Eng. 8, 615 (1960)

4. H.C. Paxton, Los Alamos Critical-Mass Data, LA-3067-MS, Rev. (UC-46), 1975

5. A. Pazy, G. Rakavy, I. Reiss, J.J. Wagschal, A. Ya'ari and Y. Yeivin, The Role of Integral Data in Neutron Cross-Section Evaluation, Nucl. Sci. Eng. 55, 280 (1974)

6. See e.g. J.R. White and T.F. DeLorey, Data Uncertainty Reduction in High Converter Reactor Design Using PROTEUS Phase II Integral Experiments, PHYSOR90 International Conference on the Physics of Reactors: Operation, Design and Computation, Vol 3. p. III-107 (1990)

MODELLING NEUTRON TRANSMISSION EXPERIMENTS AND DEMONSTRATION OF RESONANCE SHIELDING EFFECTS BY USING EVALUATED NUCLEAR DATA FILES

P. Vértes

Central Research Institute for Physics, Budapest, Hungary

Abstract A calculation model for neutron transmission and self-indication experiments with samples composed from one or more isotopes is presented. On the basis of this model a computer program – **FEDMIX** [1],[2] – has been developed. The data for calculation can be taken from any well known evaluated nuclear data file as ENDF/B, JENDL, BROND etc. Some typical calculated neutron transmission spectra and their dependence on experimental conditions are demonstrated.

The **FEDMIX** can also be applied to the exact calculation of lumped mutual screening of two resonance elements. The results of such calculations are compared with those of Bondarenko's f-factor method and it is found that the later one is inadequate in energy intervals where both isotopes have large resonances. It is also shown that the lumped cross-sections often cannot be calculated from measured transmissions and self-indications.

1. Formulae modeling the transmission experiments

The average transmission probability and self-indications are expressed by the following formulae

$$T(\Sigma h) = \frac{\int\limits_{\Delta E} dE \int\limits_{E-\epsilon}^{E+E_\circ} \Phi(E')R(E,E')e^{-\Sigma(E')h}dE'}{\overline{\Phi}} \quad (1a)$$

$$T_x(\Sigma_x, \Sigma h) =$$

$$= \frac{\int\limits_{\Delta E} dE \int\limits_{E-\epsilon}^{E+E_\circ} \Sigma_x(E')\Phi(E')R(E,E')e^{-\Sigma(E')h}dE'}{\overline{\Sigma_x\Phi}} \quad (1b)$$

where

$$\overline{\Sigma_x\Phi} = \int\limits_{\Delta E} \Sigma_x(E)\Phi(E)dE$$

$$\overline{\Phi} = \int\limits_{\Delta E} \Phi(E)dE, \qquad \overline{\Sigma_x} = \frac{\overline{\Sigma_x\Phi}}{\overline{\Phi}}$$

$\Sigma(E)$ is the total macro cross-section, ΔE is the energy interval for averaging, $\Phi(E)$ is the neutron flux (generally taken as 1/E). $\Sigma_x(E)$ is the macroscopic cross-section for the reaction x, and h - is the thickness of the filter. $R(E,E')$ is the resolution function of the neutron spectrometer. Let the mixtures of the filter and the self-indicating sample be composed of N_f and N_x isotopes, respectively, then

$$\Sigma(E) = \sum_{N_f} \rho_i^f \sigma^i(E), \quad \Sigma_x(E) = \sum_{N_x} \rho_i^x \sigma_x^i(E),$$

where ρ_i are the abundances of the isotope, σ^i and σ_x^i are the microscopic cross-sections for the isotope.

The total transmission probability through a mixtured sample is not a simple product of those for the composing isotopes unless their cross-sections are weekly depending on energy or they are in the unresolved resonance region.

2. The effect of resolution function on transmission spectra

The demonstrative calculations to be presented here were performed by the PC version of **FEDMIX** [2]. The data of JENDL-2 library were used. The curves in Figs.1-3 belong to different mixtures of uran isotopes U-235 and U-238. They show transmission spectra calculated at 300K taking and not taking into account the instrumental resolution. Its smoothing effect is well displayed in Figs.1 and 2. In the latter figure the effect of base length can also be seen. Fig.3 shows that the resolution effect may be negligible for lower energy region.

3. Mutual screening of resonance elements and the investigation of the validity of Bondarenko's f-factor method

The mixtures of U-235 and U-238 are investigated by means of a slightly modified form of the FEDMIX program. The resonance parameters of these isotopes are taken from JENDL-2. In this library between 1 and 100 eV the neutron cross-sections of both isotopes are determined by resolved resonance parameters. This corresponds to the 18-23 groups of Bondarenko's system. The U-238 has rare but large resonances while the resonances of U-235 are more dense but smaller. Different compositions of these two isotopes are taken and Bondarenko's factors $f_x^5(z_8)$ and $f_x^8(z_5)$ are calculated where the first are for (n,γ), elastic and fission cross-section of U-235, and the latter for those of the U-238. The parameters z_8 and z_5 are the ratio of U-238 to U-235 and that of U-235 to U-238, respectively. The functions $f_x^5(\sigma_0)$ and $f_x^8(\sigma_0)$ are also calculated, where σ_0 is the dilution cross-section as it is used in the Bondarenko's method. By comparing these two functions the $\sigma_0(z)$ function is determined for each type of cross-sections. Obviously, Bondarenko's f-factor method is only applicable if this function is a linear one and it is matched for all types of cross-sections. In Fig.4 the function $\sigma_0(z)$ is shown for U-235. The f-factor method is OK only for the 22-23 groups. The largest deviations are in groups 19-21. For U-238 (Fig.5) only the groups 18-21 are displayed because the f-factors for (n,γ) and fission cross-sections in group 22 become larger than 1.0 if U-235 is added to U-238. In group 23 these will decrease when more U-235 is added. The reason of this effect is that there is only one small $l=1$ resonance for U-238 in this group interval and its cross-sections are mainly determined by the negative energy resonances. On the other hand here there are many resonances for U-235. This also explains the success for the U-235 in these groups: the cross-sections of U-238 are smooth here.

4. Self-shielding calculated from self-indication and transmission probability

The results of transmission experiments are often used for the determination of f-factors characterizing the lumped reactions as

$$f_x = \frac{<\frac{\Sigma_x}{\Sigma}>}{<\frac{1}{\Sigma}>}\frac{1}{\overline{\overline{\Sigma}}_x} = \frac{\int\limits_0^\infty dh\,T_x(\Sigma_x, \Sigma h)}{\int\limits_0^\infty dh\,T(\Sigma h)} \qquad (2)$$

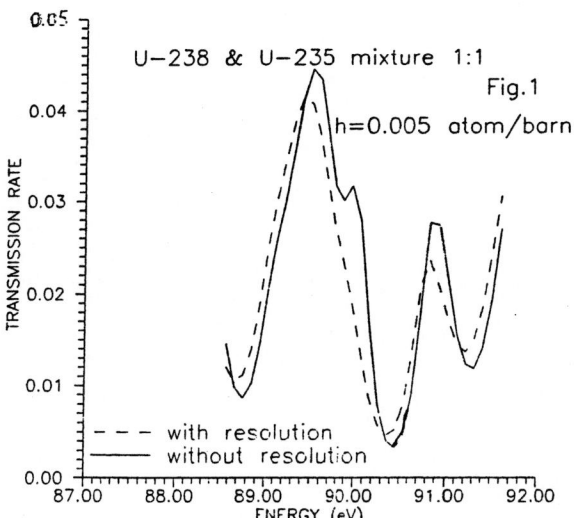

Fig.1
U−238 & U−235 mixture 1:1
h=0.005 atom/barn
- - - with resolution
——— without resolution

This formula may be important in the unresolved region where no unambigous way exists for the calculation of resonance self-shielding and therefore experimental determination is required.

This method can be checked in the resolved region. The f-factors in $\sigma_0 = 0$ dilution are calculated in groups 18-23 of the Bondarenko system from the resonance parameters of U-235 and U-238. Results are displayed in the Table below, where those calculated directly by integrating flux-weighted cross-sections are denoted by A and those obtained from the self-indication by formulae (2) are denoted by B.

Self-shielding factors $(T = 300K^o)$

group		U-235		U-238	
	f_s	f_γ	f_f	f_s	f_γ
18 A	0.9613	0.4851	0.5273	0.3405	0.0455
B	0.9651	0.5376	0.5828	0.3661	0.0524
19 A	0.9354	0.3839	0.4401	0.1043	0.0297
B	0.9441	0.4387	0.5083	0.1393	0.0481
20 A	0.8967	0.3006	0.4420	0.1577	0.0170
B	0.9056	0.3499	0.4962	0.1709	0.0333
21 A	0.9576	0.2770	0.3259	0.4319	0.0258
B	0.9614	0.3332	0.3850	0.4575	0.0578
22 A	0.9964	0.5809	0.6074	1.000	0.9967
B	0.9977	0.5982	0.6351	1.000	0.9969
23 A	0.9924	0.7106	0.6751	1.000	0.9998
B	0.9930	0.7321	0.6927	1.000	0.9998

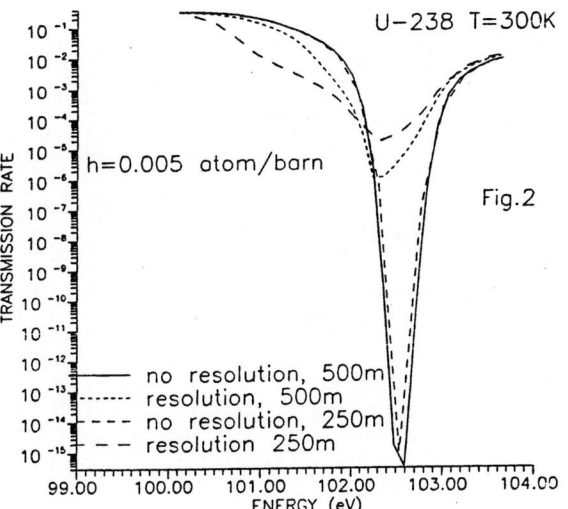

U−238 T=300K
h=0.005 atom/barn
Fig.2
——— no resolution, 500m
········· resolution, 500m
—·—· no resolution, 250m
- - - resolution 250m

The deviations are dramatic but the reason is quite straightforward as it can be shown that for successful integration in (2) $\Sigma(E)\Delta h \ll 1$ is required all over in ΔE which is not fulfilled in our case (and is often not fulfilled in experiments, too, as far as the lower part of the unresolved region is concerned).

References

[1] P. Vértes, FEDMIX: Neutron Transmission Functions and Lumped Averaged Cross-Sections from Standardized Evaluated Neutron Data, Computer Physics Communications, 56. (1989), 199-229.

[2] P. Vértes, Calculation of Transmission and other Functionals from Evaluated Data in ENDF Format by means of Personal Computers, KFKI-1991-10/G, Budapest, Hungary.

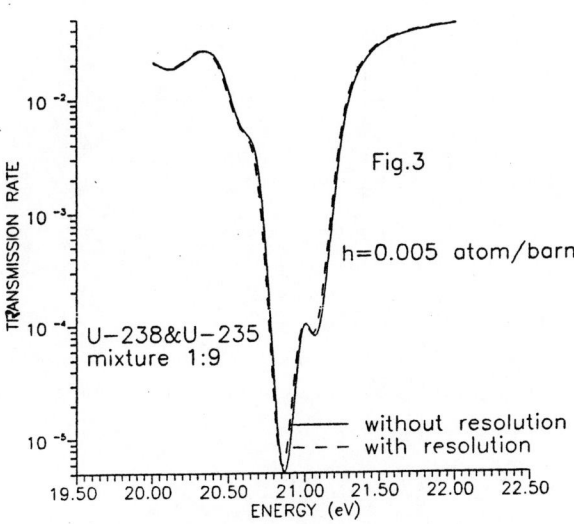

Fig.3
h=0.005 atom/barn
U−238&U−235 mixture 1:9
——— without resolution
- - - with resolution

240

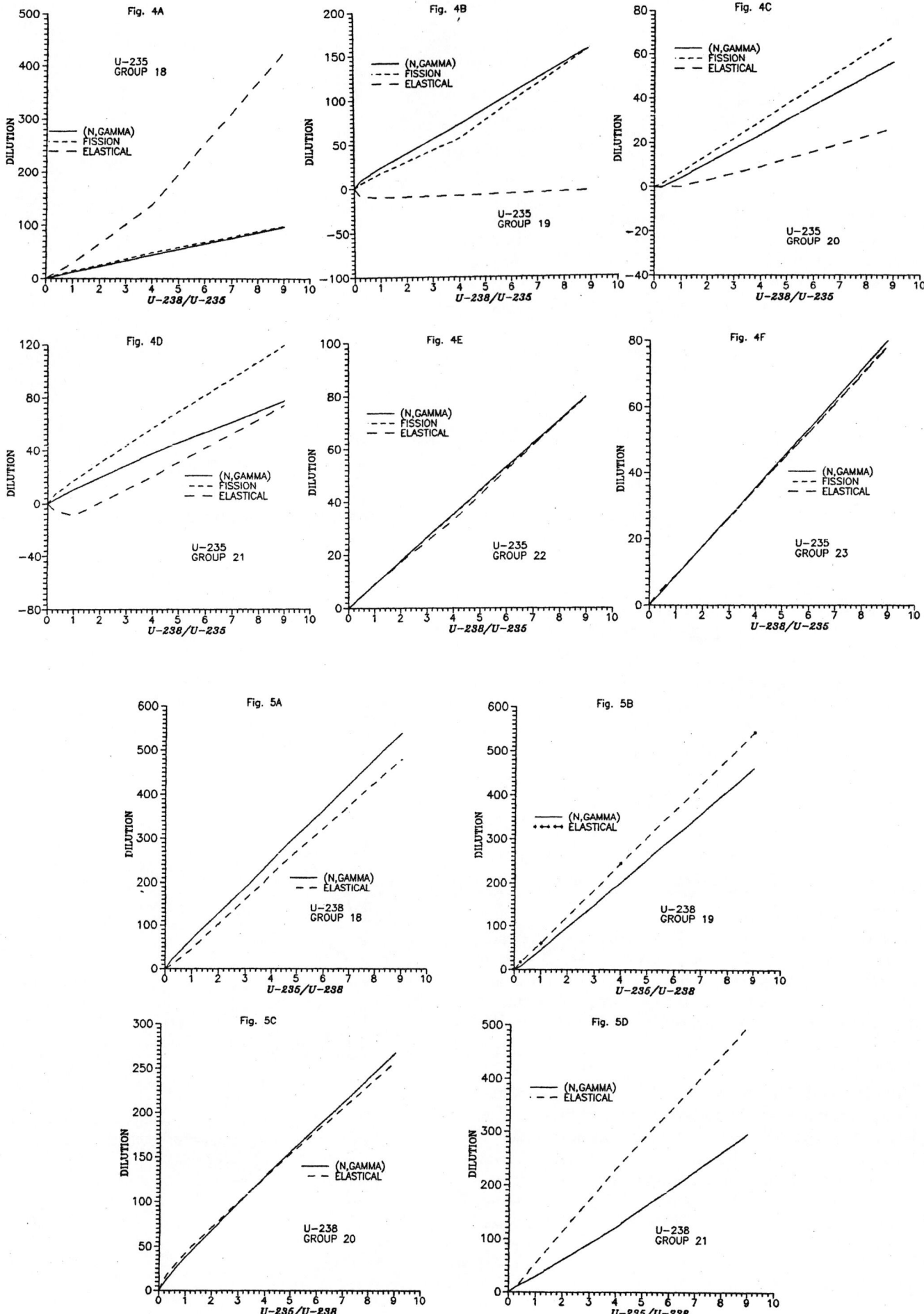

IMPACT OF DIFFERENT LIBRARIES ON THE PERFORMANCE CALCULATION
OF A MODUL-TYPE PEBBLE BED HTR

U. Ohlig, H. Brockmann, K.A. Haas, E. Teuchert
Institut für Sicherheitsforschung
und Reaktortechnik - ISR
Forschungszentrum Jülich GmbH
Postfach 1913, 5170 Jülich, Federal Republic of Germany

Abstract: A new multigroup library for the VSOP code system has been compiled from the ENDF/B-V and JEF-1 data files. The progress in comparison to the older standard library has been studied for one specific reactor design of the Modular High Temperature Reactor. The study covers various aspects of the performance of the reactor both for the initial core and for the equilibrium cycle.

Introduction

A new multigroup library has been generated from the ENDF/B-V /1/ and JEF-1 /2/ data files. It is intended for the GAM-THERMOS /3,4/ spectrum codes which are parts of the VSOP /5/ code system. This VSOP code is applied to the physics calculation of a reactor from its startup to the equilibrium cycle, to control and safety assessment, and to thermal evaluations. Up to now an older nuclear library based on the ENDF/B-II /6/ data files has been used.

The comparison of these two libraries is made for a very specific reactor: a MODUL-type Pebble Bed HTR and performed in three steps:

1. for the initial core with few nuclides,
2. for the equilibrium cycle containing the produced actinides and fission products,
3. for reactivity effects under control and accidental situations.

Nuclear Libraries

The basic nuclear library of the VSOP code system consists of two source libraries: an epithermal one given in a 68 group GAM-I structure, and a thermal one containing the data in a 30 group THERMOS structure. The older libraries are based on ENDF/B-II and BNL-325 /7/. They contain 160 absorber nuclides and 5 moderator nuclides with scattering kernels of different temperatures.

The new library is made of 175 nuclides. It has been prepared from the ENDF/B-V and the JEF-1 data files using the AMPX /8/ modules XLACS-2 and NPTXS. The weighting spectrum was a fission -1/E-Maxwellian. In the thermal energy range a Maxwell spectrum of 900 K was applied. The Doppler-broadening was calculated for a temperature of 900 K. The thermal cross sections of the moderator nuclides remained unchanged.

The ^{232}Th and ^{238}U resonance cross sections were calculated separately for different temperatures and for the specific configurations of the fuel elements by the code ZUT /9/.

Initial Core

As to Table 1 the difference of k-eff between the two libraries is strong for the initial core and small for the equilibrium cycle.

Table 1: Impact of Libraries on k-eff

Library	old	new	Δk-eff (new-old)
Initial Core	0.9987	1.0151	+ 0.0164
Equil. Cycle	1.0000	1.0017	+ 0.0017

In order to explain the reason for this finding the cross sections for only one nuclide have been taken from the new library: either for ^{235}U, or ^{238}U, or ^{239}Pu. The data sets of all other nuclides remained unchanged.

In this comparison it appears that the greatest difference results from ^{235}U. The k-eff-value increases by Δk-eff = + 0.0124. This increase is due to the increase of the new $\bar{\nu}$- and $\bar{\sigma}_f$-value by 0.3 % and 2.1 %, respectively in the thermal energy range. This is so important because 98.2 % of the produced neutrons in the thermal region are produced by ^{235}U. The increase in the thermal absorption cross section, which is mostly due to the increase of σ_f, does not compensate this effect.

If only ^{238}U is taken from the new library k-eff increases only by 0.0022. This difference is mostly due to the decrease of 1.1 % in the absorption cross section in the epithermal energy region, and this decrease results from the reduced Γ_γ-widths of the first three resonances of ^{238}U. The resulting resonance integral is about 1.2 % lower than the resonance integral calculated with the older resonance parameters.

In order to check the influence of ^{239}Pu we computed the same initial core as before but loaded with ^{238}U and 1.16 % ^{239}Pu. We got a decrease of - 0.0201 in k-eff when taking the ^{239}Pu data from the new library. This decrease is due to a decrease of 0.6 % in $\bar{\nu}$ and to a decrease of 1.1 % in $\bar{\sigma}_f$ in the thermal region. 99 % of the produced neutrons appear in the thermal range, therefore the differences in thermal $\bar{\nu}$ and $\bar{\sigma}_f$ have such a great influence.

Equilibrium cycle

The reduction in Δk-eff between the initial core and the equilibrium cycle is partly due to the nuclides which emerge during the burnup, and partly to the reduced importance of ^{235}U. The burnup calculation includes 60 different nuclides. For some of them the change in the cross sections is strong, e.g. ^{131}Xe, ^{105}Rh, ^{145}Nd, and ^{155}Eu. But their influence on k-eff is small, because their fractional absorption rate is in the order of 0.1 %.

For the few nuclides of higher importance the change in the cross sections is only moderate. Just like for the initial core, the new ^{235}U and ^{238}U data make the tendency for an increase of k-eff. A tendency of decrease results from the new cross sections of ^{239}Pu, ^{240}Xe, ^{135}Xe, and ^{149}Sm. Thus, the influence on the equilibrium cycle is no more than Δk-eff = 0.0017. This difference in k-eff can be compensated by slight reduction of the loading and unloading rate of fuel elements per day. Thus the transition from the old library to the new one can be balanced by an increase of the fuel residence time from 916.0 to 921.1 days achieving the same value of k-eff = 1.0 at the equilibrium cycle.

Table 2 illustrates the impact on reactor performance characteristics as caused by that transition in the loading rate. In the global data a slight decrease is observed in the conversion ratio. The power peaking increases by 4 %.

In the balance of neutrons the change in the neutron production rate from fissions in ^{238}U is relatively high which results from the improvement of $\nu\sigma_f$. But the importance of that effect on the total fission rate is very low. The changes in the production rates of the Pu-isotopes almost cancel out each other. A slight increase is observed in the fission products absorption. This is mainly due to the higher thermal absorption cross sections of ^{135}Xe, ^{149}Sm and to the higher epithermal absorption in ^{147}Pm.

In the breakdown of the fissile material balances the changes are of the order of 6 %. This illustrates that library changes can effect the isotopic mass balances to some degree.

Aspects of Control and Safety

For the two equilibrium cycles we investigated some effects which are basic for the assessment of the reactor. The temperature coefficients of the two cases differ slightly (Table 3). They have been obtained by changing the temperature in the fuel, moderator, and reflector by about 50°C. It appears that the Doppler coefficient remains almost unchanged, but the moderator and the reflector coefficient are different in the order of 20 %. Since the scattering matrix of the graphite has not been altered the reactivity changes must be assigned to the thermal cross sections of the absorbers and to their impact on the leakage.

The largest change of temperature occurs at down cooling of the reactor after shutdown. k-eff increases by 4.3 or 4.7 % when applying the old or new library, respectively. In Table 3 the shutdown requirement includes another 4 % as due to the ^{135}Xe decay. The contribution of Xenon is very similar for the two libraries.

In total the shutdown requirement is well covered by the shutdown system, which consists of little absorber spheres (KLAK) being poured into channels in the reflector. The capability of the KLAK system is about 15.5 %, and it is little dependent on the libraries.

Conclusion

When calculating with the new nuclear data library the multiplication factor k-eff increases about 0.0164 at the startup of the reactor, which is mainly due to changes in the cross sections of ^{235}U. At the turn to the equilibrium cycle the difference reduces to $\Delta\bar{k}$-eff = 0.0017 as due to various opposite tendencies in the data of the many involved nuclides. The change in the mass balance of the individual fissile isotopes is about 6 %. The impact on the temperature related properties is in the order of 4 %, and the influence on other safety related properties of the reactor is lower than 1 %. These results considerably confirm the validity of formerly received reactor calculations.

References

/1/ R. Kinsey, ENDF/B Summary Documentation ENDF-201, 3rd Edition, BNL-NCS-17541, Brookhaven National Laboratory (1979)

/2/ Index to the JEF-1 Nuclear Data Library, Volume 1: General Purpose File OECD, NEA Data Bank (1985)

/3/ C.D. Ioanou, J.S. Dudek, GAM-A Consistent P1 Multigroup Code for the Calculation of Fast Neutron Spectra and Multigroup Constants, GA-1850, General Atomics (1961)

/4/ H.C. Honeck, THERMOS - A Thermalization Transport Theory Code for Reactor Lattice Calculation, BNL-5826, Brookhaven National Laboratory (1961)

/5/ E. Teuchert, U. Hansen, K.A. Haas, VSOP-Computer Code System for Reactor Physics and Fuel Cycle Simulation, JÜL-1649, Forschungszentrum Jülich (1980)

/6/ H.C. Honeck, ENDF/B-Specifications for an Evaluated Data File for Reactor Applications, BNL-50066, Brookhaven National Laboratory (1966)

/7/ M.D. Goldberg, S.F. Mughabghab, B.A. Magurno, V.M. May, Neutron Cross Sections, BNL-325, Brookhaven National Laboratory (1966)

/8/ N.M. Greene, J.L. Lucius, L.M. Petrie, AMPX: A Modular Code System for Generating Coupled Multigroup Neutron-Gamma Libraries from ENDF/B, ORNL/TM-3706 Revised, Oak Ridge National Laboratory (1978)

/9/ E. Teuchert, K.A. Haas, ZUT-DGL-VSOP Programmzyklus für die Resonanzabsorption in heterogenen Anordnungen, KFA-IRE-70-1, Forschungszentrum Jülich (1970)

Table 2: Equilibrium Cycle

Library		old	new	relative change %
Global characteristics:				
Fuel residence time	d	916.0	921.1	+ 0.56
Conversion ratio		0.442	0.437	- 1.1
Power peaking	kW/ball	1.70	1.77	+ 4.1
Balance of neutrons:				
Production of neutrons ^{235}U	%	62.63	63.47	+ 1.3
^{238}U	%	0.25	0.29	+ 16.0
^{239}Pu	%	29.32	28.17	- 3.9
^{241}Pu	%	7.78	8.03	+ 3.2
Neutron losses: fission products	%	7.57	7.84	+ 3.6
Balance of fissile materials:				
Discharged fuel ^{235}U	kg/GWd_{th}	0.1398	0.1310	- 6.3
^{239}Pu	kg/GWd_{th}	0.0532	0.0511	- 4.0
^{241}Pu	kg/GWd_{th}	0.0228	0.0242	+ 6.1

Table 3: Reactivity Effects

Library		old	new	relative change %
Temperature coefficient:				
at operating condition	$\Delta k\text{-eff}/\Delta T/10^{-5}$	- 4.46	- 4.66	4.5
breakdown: fuel (Doppler)	"	- 3.68	- 3.78	2.7
moderator	"	- 1.68	- 2.00	19.0
reflector	"	+ 0.90	+ 1.12	24.4
at shutdown, cold	"	- 13.00	- 13.82	6.3
Shutdown requirement:				
full power -> zero power (50°C, Xe-decay)	$\Delta k\text{-eff}$	+ 0.083	+ 0.087	4.8
Capability of shutdown system				
KLAK	$\Delta k\text{-eff}$	- 0.155	- 0.156	0.6

PU-239 AND U-238 RESONANCE CROSS-SECTION REEVALUATION: INFLUENCE ON HCLWR CELL CHARACTERISTICS.

V.V. Kolesov, V.F. Ukraintsev, V.E.Kolesov*, N.A. Soloviev*.

Obninsk Institute of Nuclear Power Engineering,
P.O.B. 249020, IATE, Obninsk, Kaluga Region, USSR
* Institute of Physics and Power Engineering,
P.O.B. 249020, FEI, Obninsk, Kaluga Region, USSR.

Abstract: Reevaluations of resonance cross-sections for ^{239}Pu in the energy region 10^{-5}eV – 25keV and ^{238}U for 4-50keV were perfomed. The verification of these reevaluations on the HCLWR test cell model with Pu fuel is presented. The K∞, average cross-sections and spectral indexes for ENDF/B-IV data are compared with our corresponding results. Substitution of our reevaluations results in a K∞ change of 0.0067.

Introduction

During 1986-1990 a number of papers concerning resonance cross-section evaluation and cross-section reconstruction were published. In the resolved resonance region a modification of the S-matrix Adler-Adler scheme was proposed. In this method the unitarity of the collision matrix is used as additional condition for absorption and elastic scattering cross-sections reconstruction from total resonance parameters [1].

In the unresolved resonance region the analysis of average resonance parameters of ^{239}Pu, ^{235}U and ^{238}U on the basis of cross-section data and transmission data was perfomed by Monte-Carlo pseudoresonance ladder simulations[2].

A logical continuation of this work consists of the verification of our results on various benchmarks and cells. Such work is also useful for our calculational analysis of BFS critical assembly HCLWR core results.

So, the aim of this work consists of the estimation of HCLWR test cell characteristics variations, caused by our resonance cross-section reevaluations.

Main features of our ^{239}Pu and ^{238}U reevaluations.

Resolved resonance parameters evaluation have been perfomed in the energy region 10^{-5}eV – 500 eV for Pu-239 [1]. A comparison of energy-averiged cross-sections [1] with other author's data is presented in Table 1.

Simultaneous evaluation of averige cross-sections [3] and transmission data [2] was perfomed in the unresolved resonance region. The transmission data affect mainly the scattering radius R, it decreases from 9.5 fm. in the resolved range to 9.2 fm in the unresolved range of ^{239}Pu. So as the new high resolution data [4] are not available for us yet, we have no reason to change the average cross-section [3] These data we placed in a ENDF/B-IV library file.

Evaluated cross-sections of ^{238}U radiative capture in the unresolved resonance region is not adequate yet. Measurements of Kasakov et al.[5] agree well enough with Fröhner's [6] curve, estimated for JEF-2 and ENDF/B-VI. But experimental data of Muradyan [7], measured by a multisectional (n,γ) detector

Table 1. Comparison of Average ^{239}Pu Neutron Cross-Sections.

Energy eV	$\bar{\sigma}_f$					$\bar{\sigma}_a$			
	Calcu-lated	ENDF/B-IV	ENDF/B-IV*	Blons [15]	Data [17]	Calcu-lated	ENDF/B-IV	Gwin [16]	Data [17]
50-100	55.7	56.9	——	59.29	——	92.4	93.4	92.84	91.25
100-200	18.7	18.4	18.66±0.13	18.90	18.135	35.0	35.5	33.66	33.88
200-300	17.9	17.7	17.88±0.12	17.76	17.312	34.0	35.2	34.69	33.32
300-400	9.1	8.4	8.43±0.06	8.91	8.080	18.6	17.8	18.31	17.73
400-500	9.7	9.6	9.57±0/07	9.72	9.389	14.7	13.9	13.56	13.28

* ENDF/B-VI Standard Committee. Data from [17].

Table 2. Muradyan [7] Experimental ^{238}U(n,γ) Cross-Sections (barn $*10^4$) Data Compared with ENDF/B-V Data and its Descreapancies (%).

E_n, keV	4.5	5.5	6.5	7.5	8.5	9.5	15.	25.	35.	45.
A [7]	8910	8710	8610	7360	6280	6920	5740	4540	3960	3600
B ENDF/B-V	9185	9055	8196	7690	6939	7049	6289	4852	4360	3739
C (A-B)/B%	-3.0	-3.8	+5.1	-4.3	-9.4	-1.7	-8.7	-6.4	-9.2	-3.7

and an absolute method, is 5-7% lower in the 4-50keV region than the ENDF/B-V evaluated curve. Average resonance parameters derived from [7] (only S_1 parameter varied) are placed into the ENDF/B-IV file for cell calculations.

Benchmark cells.

The most popular HCLWR benchmark today is PROTEUS, but we have no reliable information about its equivalent simple cell model. So, we took the well documented test HCLWR cell model, proposed by Ishiguro [8] with $V_m/V_f = 0.6$ with 8% Pu fuel. We chose also the uranium benchmark BAPL-UO2-1 [9].

Data preparations and calculations.

The ENDF/B-IV files of U,Pu,H,O,Fe, Cr,Ni,Mn and Al isotopes together with our U-238 and Pu-239 files were processed by the NJOY code [10] to an output file in DTFR format. For calculations of σ_0 dilution and σ_e escape cross-sections and also of Dankov factors, we used our code, based on first-collision probability formulas. We calculated the effect of cell heterogeneity for Pu-239 and U-238 only.

The energy scale was divided into groups in a special way. Groups are generated, using moduls RECONR and BROADR of NJOY so that self-shielded cross-sections do not depend on σ_0 in all resonances of Pu-239 and U-238 up to 500 eV (for uranium fuel up to 30 eV resonances of U-235). In the thermal range the WIMS library grouping was used and the LANL-80 grouping for the fast range. The total group number was about 1350.

Calculations were mainly performed by the ANISN code [11] in P1/S2 approximation. Only one basical variant for each benchmark was calculated by P1/S4.

Results.

A preliminary estimation of the U-238 reevaluation's influence on cell characteristics was made on the SCHERZO [9] facility. The ABBN energy grouping and P3/S16 was used. ENDF/B-IV gave K∞=0.9881± 0.0002 and our U-238 K∞ =0.9925. A δK variation of about 0.005 was seen.

Three series of calculations for the HCLWR Ishiguro test cell were perfomed:
A) all data were taken from ENDF/B-IV;
B) only Pu-239 data were changed by our file;
C) Pu-239 and U-238 data were changed by our files.

The K∞ values for these variants are compared in Table 3.

Table 3. Calculated values of K∞.

	A	B	C
K∞	1.0802 ±0.0003	1.0845 ±0.0007	1.0869 ±0.0006
$\dfrac{(K∞ - K∞\,A)}{K∞\,A}$ %	——	0.43	0.67

Obviously our K∞ A is rather low, compared to the majority of other author's results, where K∞ is about 1.092 [14]. The main reason of this situation is a high value of the U-238 capture cross-section (see Tabl. 4, var. A). The situation becomes better after substitution of our results, but essential disagreement in this cross-section remains (see [13] data).

Comparison of reaction cross-section ratios shows a decrease of the important indexes C8/F9, C9/F9.

Summary.

The influence of Pu-239 and U-238 resonance cross-sections reevaluations on a HCLWR test cell model was estimated. The effect of cross-section changes is essential for K∞ and for cross-section ratios. The main effect is seen on Pu-239, where increasing of fission and decreasing of capture cross-sections leads to K∞ increasing.

It is reasonable to continue this work on a wide range of benchmarks with different neutron spectrum.

Table 4. Comparisons of Average Cross-Sections and Spectral Indexes.

Data	^{238}U			^{239}Pu			C9/F9	C8/F9	F8/F9	F5/F9
	σ_c	σ_f	σ_a	σ_c	σ_f	σ_a				
A	0.6922	0.0889	0.7811	3.5428	6.1634	9.7062	0.5748	0.1123	0.0144	0.9761
B (B-A)/A, %	0.6864 -0.84	0.0888 0.0	0.7753 -0.74	3.5020 -1.15	6.1966 0.54	9.6986 -0.08	0.5651 -1.69	0.1108 -1.34	0.0143 -0.55	0.9726 -0.36
C (C-A)/A, %	0.6880 -0.61	0.0888 -0.11	0.7768 -0.56	3.5087 -0.96	6.2054 0.68	9.7142 0.08	0.5654 -1.64	0.1109 -1.25	0.0143 -0.76	0.9727 -0.35
[12]	——	——	——	——	——	——	0.5471	0.0923	0.0141	1.0576
[13] MAX MIN	0.614 0.581	0.1 0.088	0.714 0.668	4.28 3.50	7.12 6.20	11.2 9.7	0.601 0.565	0.0862 0.0937	0.014 0.0142	0.8973 0.936

246

REFERENCES

1. A.A.Lukyanov, V.V.Kolesov et al.: in
   Proc.Int.Conf on Nucl Data for Basic
   and Appl.Sci.,$\underline{2}$,1619(1986).
2. A.A.Vankov, N.Janeva, V.F.Ukraintsev
   et al.: in Nucl.Sci.Eng.,$\underline{96}$,122,(1987)
3. V.A.Konshin Nuclear Constants for fissile nuclei, Energoatomizdat, (1984)
   (in Russian).
4. J.A. Harvey et al.:in Proc. Int.Conf.
   on Nucl.Data for Sci. and Tech.,Mito,
   $\underline{1}$, 115,(1988), JAERY.
5. L.E.Kazakov et al.: in Yad.Const.,$\underline{3}$,
   37,(1986),(in Russian).
6. F.H.Fröhner :Nucl.Sci.Eng.,$\underline{103}$,119,
   (1989).
7. G.V.Muradyan et al. :in Proc. of All-
   Union Conf. on Neut. Phys, $\underline{2}$, 242,
   (1988),(in Russian).
8. Y. Ishiguro, H. Akie et al.:NEACRP-A-
   798, NEA Data Bank(1986).
9. Cross Section Evaluation Working Group Benchmark Specifications, ENDF-202
   (Nov 1974).
10. R.E. MacFarlane et al.: The NJOY
    Nuclear Data Processing System,$\underline{1}$,
    LA-9309-M,ENDF-324, (1982).
11. W.W. Engle : A User's Manual for
    ANISN, K-1693,(1967).
12. J. Stepanek, P. Vontobel :NEACRP-A-
    851,report.
13. L.V. Majorov, M.S.Yudkevich : Neutr.
    Constants in Thermal Reactor Calcul.,
    Moskow,(1988),(in Russian).
14. W. Bernnat, Y.Ishiguro et.al.:in Proc
    of Int.Conf. on the Phys. of Reactors
    $\underline{1}$, I-54,(1990).
15. J. Blons : Nucl.Sci.Eng.,$\underline{51}$,130,(1973)
16. R. Gwin, E.G. Silver et al.: Nucl.Sci.
    Eng.,$\underline{59}$,79(1976).
17. H. Derrien, G. de Saussure : ORNL/TM
    -10986, Oak Ridge Nat. Lab. (1989).

APPLICATION OF THE ROMB - ASSEMBLY AND THE CYLINDRICAL HETEROGENEOUS CRITICAL ASSEMBLIES FOR TESTING OF NEUTRON DATA

E.N.Avrorin, A.P.Vasilyev, Ya.Z.Kandiev, E.S.Kuropatenko, A.I.Orlov, V.D.Perezhogin, Yu.A.Sokolov, V.A.Teryohin, E.Ya.Filippova

Institute of Technical Physics
454070, Chelyabinsk-70, USSR

Abstract: The detachable experimental blanket model (ROMB) is a cylinder of depleted uranium (weight = 6.5 t) with central zone disks 20 cm in diameter. There is the possibility to use simultaneously both the 14 MeV neutron source and critical or close to critical multiplying assembly in the central zone. Fission rates of U-238, U-235, Np-237, Pu-239 were measured at the distances up to 60 cm from the source and compared with results obtained by Weal et al. Parameters are presented for heterogeneous critical assemblies investigated by authors. The assemblies consist of the disk sets of fissionable materials interchanging with the disks of moderating or structural materials. The results of measurements are compared with Monte-Carlo calculations using nuclear data library BAS and the codes KLAN and PRIZMA.

(hybrid fusion blanket, ROMB, fission number measuring, 14 MeV neutron, heterogeneous critical assembly, codes KLAN, PRIZMA, Monte-Carlo, nuclear data library BAS)

Introduction

Computer codes and neutron data libraries used in designing hybrid fusion reactors can be tested by comparing calculated results with experimental ones obtained on simple models of blankets. Thus, for the blanket of natural uranium Weal et al /1/ measured (n,f) reaction rates of U-238, U-235, Pu-239 and (n,γ) reaction rates of U-238 in the cylindrical assembly H x D = 106.6 cm x 99 cm made of natural uranium and using a 14 MeV neutron source. The assembly consisted of rods. The clearances between the rods influenced the neutron transmission and made the comparison of calculations and the experiment difficult.

Isotopes of plutonium are accumulated in the blanket as uranium is burning up. At the initial stage U-235 and other fissile materials could be used in the blanket to increase the multiplication factor. That is why, it is a certain interest to investigate the neutron transmission from the 14 MeV source in the combined system, containing isotopes of various fissile elements, structural materials and a coolant, as well as heterogeneous critical (or close to critical) assemblies (CA) containing such layers.

ROMB - assembly

The detachable experimental blanket model (ROMB) was designed in the Institute of Technical Physics to perform these investigations. This model is a cylinder (H x D=90 cm x 70 cm, weight is 6.5 ton) of depleted uranium (0.5 % U-235). The mean density of the assembly is 18.75 g/cm³.

The ROMB-assembly consists of outer-zone rings 70/40 cm in diameter, of intermediate-zone rings 40/20 cm in diameter and of central-zone disks 20 cm in diameter. Besides, there are additional sets of such disks of U-235, Pu-239, various structural materials and moderators are available. These disks are of different thickness from 0.5 cm to 5 cm, to widen our possibilities in obtaining numerous variants of assemblies.

The ROMB-assembly has vertical holes where various neutron detectors are placed during the measurements; the holes are filled with cylindrical plugs of uranium.

The ROMB-assembly is installed in a hall (8m x 12 m x 18m), the walls of the hall are of borated concrete. The outer surface of the ROMB is covered by a cadmium sheet 1mm thick.

The 14 MeV neutron source was a generator NG-150M. Its tritium-scandium target, 2.8 cm in diameter, is introduced through a special horizontal hole in the ROMB and is placed at the assembly axis at a distance of 25 cm from its bottom.

The peculiarity of the ROMB is the possibility to operate with the 14 MeV source and the Critical Experiments Facility simultaneously. To do this the assembly bottom part, 20 cm in diameter, is introduced into the ROMB central channel. The top part is mounted in the channel beforehand and placed on a thin (2 mm steel) ring diaphragm. Then, after moving the 14 MeV source inwards, we lift the assembly bottom part reaching the desired degree of subcriticality (usually a central source neutron multiplication M=1-50). As a result, neutrons with E_0=14 MeV breed in the inner assembly, and this allows to investigate the distribution of neutrons having spectra characteristic for blankets of different compositions and concentrations of fissile elements both in this part of the assembly, and in the outer uranium blanket.

Measurements and calculations of neutron transport in the uranium blanket

Measurements similar to those described in /1/, were performed on the ROMB in the first series of the experiments. Fission rates of U-238, U-235, Np-237 and Pu-239, and also ^{19}F(n,2n) and ^{56}Fe(n,p) reaction rates were measured. Fissions were registered by semiconductor detectors and track mica detectors. Values of the fission cross-sections at En=14 MeV from the neutron data library BAS /2/ (see the table 1) were used for the absolute calibration of the detectors. Reactions ^{19}F(n,2n) and ^{56}Fe(n,p) were detected by means of activation detectors, made of teflon and iron.

Results of the measurements in the axial channel are presented in the figure below as relationship $\mathcal{H}_r(R)=4\pi R^2 n_r(R)$, where $n_r(R)$ is the number of reactions per one source neutron in the unit of volume and R is the distance between the source and the detector. The total experimental error doesn't exceed 3 % in the main range of the measurements. Results of Monte-Carlo calculations using the code PRIZMA /3/ and the neutron data library BAS /2/ are also presented in the figure. The source design was also taken into account. Detectors were replaced by the disks, 0.05 cm thick and 2.4 cm in diameter, placed on the axis.

On the whole, the agreement between experimental and theoretical results is good, especially for the threshold detectors. For detectors of U-235 and Pu-239 the calculations systematically underestimate experimental results by 5 to 10 % at R=30-50 cm. This is likely to be a result of the overestimated capture on U-238 in the resonance region. The results of K_{eff} calculations for heterogeneous CA, consisting of alternate layers of U-235, U-238 and CH_2 also indicate this.

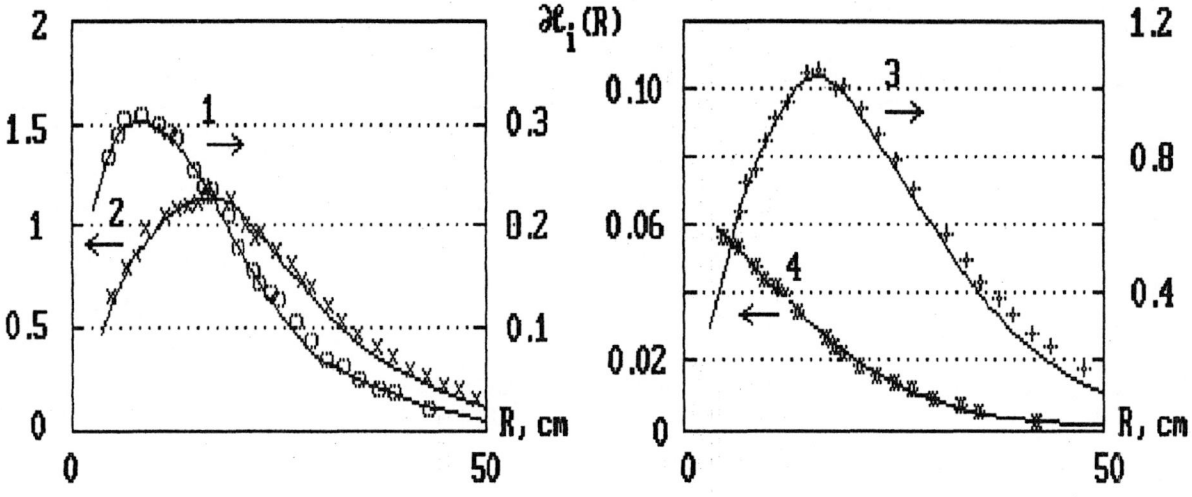

Fig. $\mathcal{H}_f(R)$ relationship for detectors: 1 – Np-237; 2 – Pu-239; 3 – U-235(90 %); 4 – depleted uranium (0.5 % U-235); (solid lines – calculation; o, x, +, * – experiment).

The integral values $\mathcal{H}_\infty = \int_3^7 \mathcal{H}_f(R)dR$, obtained by extrapolation of experimental and calculated curves $\mathcal{H}_f(R)$ for $R \to \infty$, are given in table 1; the extrapolation error is less than 1 %. The total error (2 6) is also given in table 1.

Table 1. Calculated and measured values

Detector material	σ_f, barn	experim. $\mathcal{H}_\infty$	calculated $\mathcal{H}_\infty$
depl.U(0,5%)	1.17	0.93±0.06	0.95±0.04
237Np	2.46	6.55±0.4	6.45±0.3
235U(90%)	2.00	29.9±1.8	27.9±1.4
239Pu(95,5%)	2.30	33.0±2.0	32.0±1.6

In the work /1/ the measured fission rate of U-238 is 1.18±0.06. Our two-dimensional calculation using the PRIZMA code and BAS has given for geometry of the experiment $\mathcal{H}_\infty$ =1.10. The calculated values given in the work /4/ are 0.95 using ENDF/B-IV and 1.11 using ENDL-75.

Cylindrical heterogeneous critical assemblies

Traditionally CA's were used to test and adjust neutron data used in calculations /5/. Only one-dimensional calculations were performed at that time, so the first assemblies approached the calculated ones as much as possible. Very often these assemblies were the sets of layers of fissile materials and reflectors, divided in two

hemispheres. The critical gap between them was measured in the experiment, i.e. the real CA was two-dimensional. Then corrections to the critical mass of fissile materials or to the reflector thickness, which would have to make this CA absolutely spherical, were determined in additional experiments, varying the mass of fissile material or the reflector. This procedure of "spherization" introduced additional errors in the K_{eff} value. Nowadays it is preferable to calculate the real experimental systems, this increase the accuracy of the comparison.

Spherical assemblies have another source of errors: crescent clearances between the layers that by our estimations give the error $\Delta K_{eff} = 0.1$ %. This error was unessential earlier, when the error in determining K_{eff} was ~ 0.5 %. As the calculation accuracy increased the requirements to the experimental accuracy increased too. Cylindrical assemblies, consisting of disks, eliminate this error. It should be noted, that application of disks allows to obtain a much greater variety of CA's than using spherical layers. All this forced us to investigate heterogeneous cylindrical CA's.

Every CA consists of two parts. The top part is mounted at the height of 2 m from the floor on a thin (2 mm thick) ring diaphragm.

The assembly bottom part is mounted on the cone duralumin cup fastened to a lifted rod. Both parts of the CA can consist of disks or hemispheres of fissile and other materials in different combinations, and can also have additional external reflectors. The top and bottom parts can be symmetric or can differ in geometry and composition as well.

In the initial state both parts are divided with some distance h^0, where h^0 is the gap between their neighbour surfaces. As the rod rises, the gap h decreases; and the multiplication factor M of neutrons emitted by miniature Cf-source nested in the assembly bottom part increases. The value of the gap h_{crit}, when the assembly becomes delayed critical, is determined by extrapolation of the inverse multiplication factor 1/M=f(h) to the zero (usually from values M=100-500). The error of determining h_{crit} is <0.01 cm.

The diameter of the disks is chosen to be 20 cm in order that they could enter the ROMB inner zone precisely.

As a rule, the disks' thickness is 1 cm. There are sets of disks made of U-235 (96 %), Be,

BeO, C, Al, Ti, Cd, Fe, CH_2, ^6LiH, ^7LiH, ^7LiD and so on, which allow to obtain different variants of heterogeneous CA's. The disks of Pu-239 are of smaller diameter (12 cm) and 0.5 cm thick. They are cladded with stainless steel.

A large series of critical measurements (138 cylindrical CA's with different reflectors, which are analogous to previously investigated spherical CA's, and 134 heterogeneous CA's) to verify the calculational accuracy in different parts of the neutron spectrum and especially in the energy range, where the influence of resonance structure of capture and fission cross-sections is significant, was performed. CA's with different combinations of a large number (up to 60) of layers from fissile, structural and moderating materials were investigated.

Data on some heterogeneous CA's are given in Table 2.

Table 2. Data on some heterogeneous cylinrdical CA's.

Bottom part	Top part	hcrit., cm	Keff. (calc)
$(1U+0.5CH_2)$x5	$(1U+0.5CH_2)$x5	0.10	0.999
$(1U+0.5Be)$x7	$(1U+0.5Be)$x6	0.22	1.002
$(0.5Fe+1U)$x9	$(0.5Fe+1U)$x9	0.47	0.996
$(0.5Fe+0.5CH_2+1U)$x9	$(0.5Fe+0.5CH_2+1U)$x7	0.64	0.991
$(0.5Fe+1CH_2+1U)$x9	$(0.5Fe+1CH_2+1U)$x7	0.30	0.987
$(0.5Fe+2CH_2+1U)$x5+1U	$(0.5Fe+2CH_2+1U)$x5	0.36	0.990
$(2U*+2U)$x5	$(2U*+2U)$x5	0.38	0.996
$(0.5Fe+2CH_2+0.5Fe+2U)$x4+0.5Fe+1CH_2	$1CH_2+(0.5Fe+2U+0.5Fe+2CH_2)$x4+0.5Fe	1.11	0.985
$(2CH_2+0.05Cd+2U+0.05Cd)$x4+1CH_2	$1CH_2+(0.05Cd+2U+0.05Cd+2CH_2)$x4	0.385	1.000
$(2CH_2+1U)$x4+1CH_2	$1CH_2+(1U+2CH_2)$x5	1.88	0.991
$10CH_2+3U$	$3U+10CH_2$	1.47	0.996
$10CH_2+0.05Cd+(1U*+1U)$x7	$(1U*+1U)$x5+1U*+0.05Cd+10CH_2	1.51	0.995
$10CH_2+(1Fe+1U)$x5+0.5Fe	$0.5Fe+(1U+1Fe)$x5+10CH_2	1.23	0.993
$10CH_2+(1.5Al+1U)$x5+0.5Al	$0.5Al+(1U+1.5Al)$x5+10CH_2	0.43	0.999

Designations: U = U-235(96 %); U* = depleted U (0.5% U-235); 1U means 1 cm thick layer of U; (...)x5 means the bracketed layers repeated 5 times.

Conclusion

Investigations, performed on the ROMB, proved it to be suitable to test the quality of data used in calculations.

Calculations for heterogeneous CA's with different combinations of fissile and structural materials allow to test the accuracy of neutron data and computer codes used, what is especially significant if one wants to investigate the uncertainties in predicting the influence of different effects in emergency situations, e.g. damage of nuclear reactors.

Investigations with Pu of various isotopic content including the cases with increased abundance of Pu-240 and Pu-241 might have been of especial interest both for nuclear reactors, and hybrid blankets of fusion reactors. Besides investigations for heterogeneous CA's, measurements of neutron transport from the 14 MeV source in these systems appeared to be necessary. Comparison of calculated and experimental results will make it possible to test the accuracy of neutron data for Pu-240 and Pu-241 in various energy ranges.

REFERENCES

1. J.W. Weale, H. Goodfellow, M.H. McTaggart and M.L. Mullender: J. Nucl. Energy, A & B14, 91 (1961)

2. A.P.Vasilyev, Ya.Z.Kandiev, V.I.Chitaikin: Neutron Physics, Moscow, 2, 119 (1984)

3. Ya.Z.Kandiev, E.S.Kuropatenko et al: The 3d All-Union Scientific Conference on the Shielding from Ionizing Irradiations of Nuclear Technical Installations. 27-29 October 1981. Theses of reports, Tbilisi, 24 (1981)

4. R.C. Haight, J.D. Lee: Nucl. Sci. and Eng. 61, 53 (1976)

5. H.C. Paxton: Los-Alamos critical-mass data. LAMS-3067 (1964)

Part III

Nuclear Data for
Fusion Reactor Technology

NEUTRON DIAGNOSTICS FOR TOKAMAK EXPERIMENTS

O.N. Jarvis

JET Joint Undertaking, Abingdon, Oxfordshire, OX14 3EA, England

Abstract

A brief account of the present status of tokamak performance is given, followed by a description of the properties of the ideal thermonuclear neutron source. The very comprehensive range of neutron measuring systems installed at a particular installation (The Joint European Torus) is then presented, along with some typical illustrative results. Some exceptional results are also presented to indicate the manner in which the properties of the actual neutron source deviate from the ideal, since the identification of such situations is essential to prevent the wrong interpretation being placed upon the most basic measurement, that of the total neutron yield. Areas in which the quality of either the nuclear data or the measuring techniques employed is deemed to be unsatisfactory will be highlighted in the course of the presentation.

(nuclear fusion, tokamak, neutron diagnostics, fission chambers, activation, tomography, neutron spectrometers, triton burn-up, gamma emission)

Introduction

The tokamak is presently the leading device in the attempt to obtain energy from controlled nuclear fusion. The most advanced operating tokamaks include JET (the European project), TFTR and DIII-D in the U.S.A. and JT-60 in Japan. The first three machines operate routinely with deuterium plasmas, whilst JT-60 intends to operate with deuterium during 1991.

The figure of merit used to describe plasma performance is the $n T\tau$ product, where n is the fuel ion density (m^{-3}), T is the ion temperature (keV) and τ is the energy confinement time (s). With d-t fuel, an $n T\tau$ product of above 5×10^{21} m^{-3} keV s should ensure ignition, when the plasma self-heating due to fusion reactions will be sufficient to maintain a constant plasma temperature. The highest $n T\tau$ value produced in deuterium so far [1] is about 8×10^{20} m^{-3} keVs, with T > 20 keV. The goal for the existing generation of tokamaks is energy break-even (Q = 1), with as much fusion power being produced from fusion reactions as heating power applied to the plasma. The best performance to date corresponds to Q of 0.8 if run in d-t plasmas. This discharge (in JET) was run with 4 MA plasma current and 17 MW of deuterium neutral beam injection heating power.

The generation of fusion power is assessed by recording the flux of neutrons emitted from the plasma. For the special case of the ^{3}He-deuterium mixture, the fusion reactions release protons rather than neutrons, so the fusion power is instead monitored by the emission of 16 MeV gamma-rays from the weak secondary reaction as these, unlike the protons, escape readily from the reaction vessel. ^{3}He-deuterium fuels are of interest at the present time because they do not lead to activation of the vessel but they have appreciably lower reactivities than deuterium-tritium and so offer scant prospect for exploitation as an energy source and will not be considered further.

Unfortunately, the observation of neutron emission is not uniquely related to the production of thermonuclear reactions; to demonstrate the latter, it is necessary to record their energy spectrum, which should be centred around 2.45 MeV in the case of d-d reactions and 14 MeV for d-t. Study of the neutron emission can also provide information on such plasma properties as position, extent, ion temperature and ion density. This introduces the need for measurements of spatial emission profiles, so that contours of neutron emission strength can be derived by tomographic means. Finally, a study of the 14 MeV neutron emission from deuterium plasmas permits the confinement of the 1.0 MeV tritons released from d-d reactions to be assessed; this has a direct bearing on the confinement of the 3.5 MeV alpha-particles from d-t fuels and the plasma self-heating capabilities.

Characteristics of neutron emission from tokamak plasmas

The following fusion reactions take place in a deuterium plasma, with nearly equal probability:

$$d+d \rightarrow {}^3He + n; \quad E_n = 2.45\ MeV,\ E_{He3} = 0.82\ MeV$$
$$\rightarrow T + p; \quad E_p = 3.00\ MeV,\ E_{T3} = 1.01\ MeV$$

The secondary reaction:

$$d+t \rightarrow {}^4He + n; \quad E_n = 14.06\ MeV,\ E_{He4} = 3.52\ MeV$$

proceeds as a by-product of the first. The probability that the 1.0 MeV triton undergoes a fusion reaction whilst slowing down in a deuterium plasma is about 1% for typical JET conditions; once thermalized, the further likelihood of fusion is negligible on the time-scale of the plasma duration (10 seconds) unless the ion and electron temperatures are unusually high.

The neutron production rate from deuterium plasmas is given by $r = 0.5\ n^2 <\sigma v>$, integrated over the plasma volume, where n is the deuterium ion density and $<\sigma v>$ is the velocity-averaged fusion cross-section. The fusion reactivity is given approximately by $<\sigma v> = a\ T^{-2/3} \exp(-b\ T^{-1/3})$ for modest ion temperatures. The strong temperature dependence results in neutron yields that can vary from 10^{13} n/s at 2 keV to 10^{16} n/s at 30 keV. The energy spectrum from thermal plasmas is a very nearly gaussian-shaped peak [2] at slightly above 2.45 MeV that has a width (fwhm) of $82.6 T^{1/2}$ keV for d-d reactions. For a plasma with spatially uniform ion and electron densities and temperatures, a measurement of the energy spread affords a measurement of the ion temperature. Once this has been obtained, the neutron yields can be used to determine the deuterium ion density.

The above discussion is appropriate for plasmas in thermal equilibrium, a condition that should apply with ohmically heated and ignited plasmas. However, in order to reach ignition conditions in present day machines it is necessary to use additional sources of plasma heating. The most successful involves the injection of fuel atoms (deuterium) at energies of up to 140 keV. The injected deuterium beam ions undergo fusion reactions with the thermal ions with a reactivity two orders of magnitude greater than for thermal fusion reactions in the original target plasma; however, the rapid rise in ion temperature resulting from beam heating powers of up to 20 MW rapidly raises the thermal contribution to a comparable level. The neutron energy spectrum is then largely determined by the beam energy and the directions of observation and injection relative to the magnetic field lines in the tokamak. Ion temperature information can still be extracted provided the ion temperature is neither too low nor too high. It should be noted that the energy spectrum due to 80 keV beam ions injected into a plasma with 20 keV ion temperature has a fwhm of nearly 25 keV; in general, the energy spectrum due to beam-thermal reactions is not gaussian for tokamak plasmas because directional information is retained as a consequence of the gyration of ions around the magnetic field lines [3].

The application of Ion Cyclotron Radio-Frequency (ICRF) heating can complicate matters considerably. Only in the case of ICRF tuned to resonate with ^{3}He minority ions in an otherwise pure deuterium plasma will the neutron energy spectrum reflect correctly the deuterium ion temperature. When the ICRF is tuned to minority hydrogen there is some second-harmonic acceleration of deuterons to very high energies (several MeV). These RF-accelerated deuterons will now undergo fusion reactions with the thermal ions, introducing an appreciable broadening of the neutron energy spectrum so that the desired ion temperature dependence is lost.

A further problem arises at JET due to the use of beryllium limiters, which are used for their low Z and beneficial effect of gettering oxygen and consequent reduction of the impurity ion concentration in the plasma. Regrettably, there is now a small concentration of beryllium present. Beryllium is the least tightly bound of all stable nuclei. ICRF-accelerated ions (protons, deuterons or ^{3}He ions) may undergo nuclear reactions with the beryllium impurity ions. For high RF power levels, the resulting neutron emission from nuclear reactions can exceed the emission from d-d fusion reactions. Neutron spectrometry is essential to diagnose this situation. Whether or not the beryllium contamination of the neutron emission is important in a particular discharge depends a variety of plasma parameters, including the bulk plasma density which, if sufficiently high, should prevent the accelerated ions from attaining sufficiently high energies for nuclear reactions to take place.

It is generally observed that the RF does not accelerate light ions to the highest energies if the plasma is undergoing frequent sawtooth oscillations. These oscillations take place when the magnetic energy, determined by the central current density, exceeds a limit set by magnetohydrodynamic considerations in a manner that is not fully understood. They are potentially beneficial in that they result in a mixing of the inner regions of the plasma, so expelling impurity ions from the centre, but limit the attainable central electron temperature so that one goal is to engineer long sawtooth-free periods. One way of achieving this is to apply high power neutral beam or RF heating, as these introduce fast particles whose presence has the effect of stabilizing the plasma against sawtooth oscillations. A natural consequence of sawtooth stabilization by means of high power RF heating is the generation of neutrons from nuclear reactions with impurity ions at a level which is competitive with the production from fusion reactions from d-d plasmas, as discussed above; this contamination will, however, be insignificant compared with the fusion neutron emission from d-t plasmas.

Many JET discharges employ ICRH and NBI simultaneously, in the hope of achieving a synergistic cooperation between the two heating techniques. The success of this approach is dictated by the general impurity level (which is frequently increased by the application of RF power). The neutron emission is usually highest with NBI heating alone. ICRH alone gives neutron yields intermediate between those from ohmically heated and NBI-heated plasmas. The most interesting aspect of synergism is the RF acceleration of fast beam ions. This has not so far been convincingly demonstrated owing to the difficulty of distinguishing it from the RF acceleration of thermal ions, which becomes more efficient at the higher temperatures obtained with beam heating.

Although JET power supplies are capable of sustaining full heating for ten seconds or longer, the duration of high neutron emission is short. The neutron emission rises to a peak one or two seconds after the application of full heating power, and then falls abruptly. Peak emission strengths of nearly 4×10^{16} n/s have been recorded at JET and slightly higher levels at TFTR. The fall is due to the ingress of plasma impurities (beryllium, carbon, nickel etc, depending on wall conditions) which are generated when the plasma temperature close to the wall exceeds about 10eV; the released impurities take about 1 second to penetrate to the plasma centre where they dilute the fuel ions and increase the radiative energy loss, cooling the plasma; furthermore, in the case of NBI heating, the increase in plasma density has the effect of reducing the penetration of the heating beams and diminishing their heating effect at the plasma centre. Some improvement can be achieved by sweeping the plasma vertically and radially to reduce the power loading on the limiting surfaces. Control of impurity production is now the major task for the large tokamaks and, at JET, a set of internal divertor coils will be installed for this purpose during 1992.

An example of the neutron emission from a JET discharge under conditions of combined heating (6MW RF and 13 MW D° -NBI) is shown in fig 1. The contributions from thermonuclear, beam-plasma and beam-beam reactions can be computed from the measured plasma parameters; that the calculations under-estimate the emission is attributed to the neglect of second-harmonic heating of deuterons by the RF.

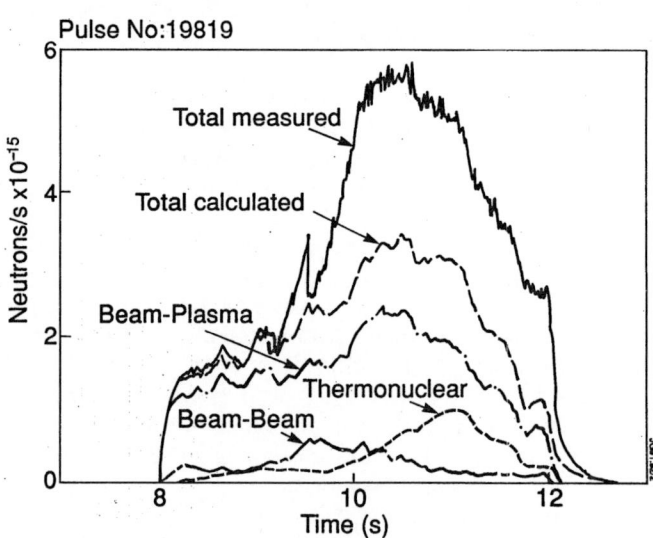

Fig 1. *Showing the neutron emission strength measured with the fission chambers and its calculated decomposition into beam-beam, beam-plasma and thermonuclear components; the excess between measurement and calculation is attributed to the second-harmonic acceleration of deuterons by the applied ICRF heating.*

This discharge is exceptional in that the d-d fusion reaction rate contribution to be attributed to RF-accelerated deuterons is large.

Neutron diagnostic systems

A comprehensive range of neutron diagnostic systems is installed on JET for use with deuterium plasmas and these will be described briefly in the following paragraphs. Less extensive ranges of neutron diagnostic systems have been installed at other tokamaks.

Time-resolved neutron yield monitors

These are three pairs of ^{235}U and ^{238}U fission chambers, embedded in polyethylene moderator and lead, respectively, and mounted on the vertical limbs of the transformer magnet at the horizontal mid-plane of the tokamak. Both pulse counting and current sampling modes of operation are used. Together, the two types of chamber allow neutron emission intensities from 10^{10} to 10^{20}n/s to be covered. So far, these chambers have operated very reliably. The major problem with their exploitation is that of obtaining an accurate absolute calibration. The direct method would be to move a point source of 2.5 MeV neutrons around the vacuum vessel so as to map out the volume occupied by typical plasmas. The use of accelerator-based sources is inconvenient but can be done; the accuracy achieved (at TFTR[4]) is at best 15%. The most practical alternative is to employ a ^{252}Cf radio-isotope neutron source. This is the customary approach and accuracies of 10 to 15% have been claimed (TFTR [4] and JET [5]). Unfortunately, the ^{252}Cf energy spectrum is different from the spectrum obtained from d-d reactions and detailed neutron transport calculations are needed to verify that the substitution is permissible. Such calculations indicate that it is permissible for machines that are of relatively open construction, such as TFTR, where the vacuum vessel is only partially surrounded by the copper field coils but that it may not be permissible when the construction is closed, as at JET where the spaces between the copper field coils are filled with high density concrete and spectral hardening may be important. Consequently, the activation method for calibrating the fission chambers is now favoured at JET[5], as will be described below.

Time-integrated neutron yield monitors

This diagnostic has two primary purposes: (i) to provide a reliable calibration of the time-resolved neutron yield monitors throughout the life-time of JET, and (ii) to measure the proportion of 14 MeV neutrons (from triton burnup) to 2.5 MeV neutrons (from d-d reactions) during the deuterium operating phase. The method requires the exposure of suitable materials to the neutron flux close to the vacuum vessel wall during a single discharge and their assay for induced radioactivity shortly thereafter. To obtain the time-integrated neutron flux (or fluence) from these measurements, the relation between emitted neutron yield and local neutron flux must first be established using neutron transport calculations [6].

A pneumatic transport system is used for moving samples between the irradiation positions at the surface of the vacuum vessel of the tokamak and the nuclear instrumentation used for the assessing the degree of activation. Useful reactions will have threshold energies somewhat below the energy of the fusion neutrons. Conventionally, the activation is determined from gamma-radiation measurements. For 2.5 MeV neutron studies, there are few candidate materials emitting suitable gamma-rays and their half-lives are much longer than the interval between discharges. Such materials can only be used once per several half-lives and the activity is measured long after the JET pulse has been completed. For 14 MeV studies, there is one interesting candidate [7] for which the half-life is short enough for the recycling of samples to be feasible. This is the ^{28}Si(n,p)^{28}Al reaction, emitting 1.8 MeV decay radiation and possessing a half-life of 2.2 minutes.

In addition to the conventional use of gamma-radiation measurements, we use delayed neutron counting of beta-delayed neutrons from fission events in fissionable materials (^{235}U, ^{238}U and ^{232}Th). For this, the relevant counting period is short (up to 2 minutes) and the samples can be recycled between discharges [8], a unique attribute for determining 2.5 MeV neutron fluences.

The accuracy of the yield measurements is believed to be about ±8% for 2.5 MeV neutrons using the ^{115}In(n,n')^{115m} reaction, ±6% using delayed neutrons and ±8% for 14 MeV neutrons using the usual ^{63}Cu(n,2n)^{62}Cu reaction.

The quality of the dosimetry reaction cross-section data for the materials used in these measurements is variable. It should be noted that activation data are needed not only for the neutron energies corresponding to the fusion reaction of interest, but also for lower energies since down-scattering from the vessel walls cannot be neglected. For d-d reactions (2.5 MeV neutrons), the delayed-neutron data are probably the most reliable, with indium nearly as good. The ^{64}Zn(n,p)^{64}Cu and ^{58}Ni(n,p)^{58}Co reactions have also been used; the nickel reaction data are apparently reliable but the zinc data are considered untrustworthy. For d-t reactions (14 MeV neutrons), the copper data are the best, the delayed-neutron data are barely adequate whilst data for silicon are not available for the entire energy range and reliance must be placed on model calculations, although the 14 MeV data-points are accurate. Other dosimetry reactions are available for d-t plasma measurements and are being investigated at the present time at JET. It is believed, but not demonstrated numerically, that the quality of the neutron transport cross-section data is adequate for these studies.

The Neutron Profile Monitor

The Neutron Profile Monitor consists of two multi-channel collimator assemblies, or cameras [10]. A 9-channel instrument is positioned to view downwards through the plasma, whilst a 10-channel instrument views horizontally through the pumping port door. The fan-like array of collimated lines-of-sight covers most of the plasma section. Each collimator channel is furnished with a NE213 liquid scintillator neutron spectrometer for use with 2.5 MeV (d-d) neutrons. In order to distinguish neutrons emitted directly from the plasma from neutrons which have scattered one or more times from the vessel walls, a 2.0 MeV energy discrimination is used. Gamma-rays are rejected by the use of a pulse-shape discrimination (PSD) circuit.

The absolute efficiency of each channel is calculated to about 10% accuracy; this entails performing neutron transport calculations to assess the effective solid angles and the attenuation of neutrons in passing through the walls of the vacuum vessel and profile monitor. A tomographic deconvolution of the line-integrated measurements can be carried out to obtain two-dimensional mappings of the neutron emission strength. Fig 2 illustrates the tomographic reconstruction method; this figure shows the neutron emission as being highly peaked on the plasma axis just before a sawtooth crash, whilst just after the crash the emission distribution flattens and may become hollow. Clearly, the fast ions have been ejected from the central region of the plasma by the action of the sawtooth crash. The change in emission profile is dramatic but the change in the total emission strength as recorded by the fission chambers is modest, which is in accord with the understanding that the fast ions are not lost from the plasma but are merely redistributed at a relatively large radius.

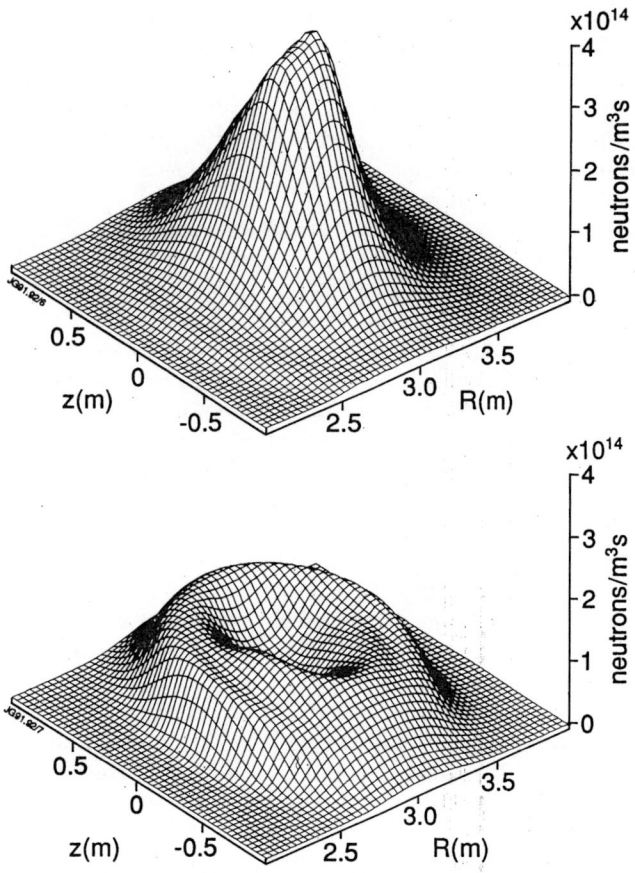

Fig 2. A tomographic reconstruction of the neutron emission profiles before (top) and after (lower) a sawtooth crash for a 4.7 MA discharge with 17 MW of applied D° - neutral beam heating.

Neutron Spectrometers

There is no single neutron spectrometer which can satisfy all the operational demands for deuterium plasma operation in JET. Consequently, several spectrometers of different design have been deployed [11] in the expectation that at least one of them will be within range for any given plasma operating condition. Because the neutron emission will usually be anisotropic for beam-heated plasmas, two orthogonal lines-of-sight have been provided so that energy spectra taken simultaneously can be compared in order to obtain information related to the trapped fast-ion fractions from the differences. The phenomenon of trapping arises because an ion tends to follow the magnetic field lines until its parallel motion has been converted entirely to perpendicular motion, according to the rule of conservation of magnetic moments, at which time its motion is reversed and the guiding-centre follows the so-called banana orbit.

Use of the "tangential" line-of-sight necessitates working in the torus hall, where a large radiation shield houses several small spectrometers, chief among which is a ³He ionization chamber, whilst the vertical line-of-sight leads to

the roof laboratory where there is a neutron time-of-flight spectrometer and another ³He ionization chamber.

³He ionization chamber spectrometers offer the best energy resolution, although this is marred by the low-energy tail due to recoil charged reaction products that reach the chamber wall and deposit a small amount of their energy there. These spectrometers have been optimized for use with neutrons with energies below 2 MeV; their geometrical properties could be improved for fusion applications. A neutron spectrum obtained with a ³He spectrometer by summing over a series of ohmically heated discharges is shown in fig 3.

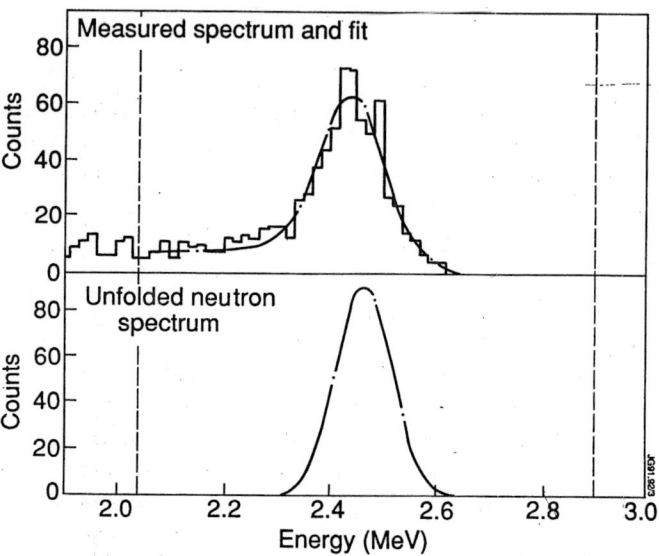

Fig 3. Pulse-height spectrum (top) obtained with a ³He neutron spectrometer for a series of ohmically heated discharges; the broadening of the unfolded neutron spectrum (lower) indicates an average temperature of 2.2 keV.

Fig 4 shows the spectrum obtained with the time-of-flight spectrometer for an ICRF-heated discharge with a considerable contamination from neutrons produced by ³He ions interacting with beryllium impurity ions.

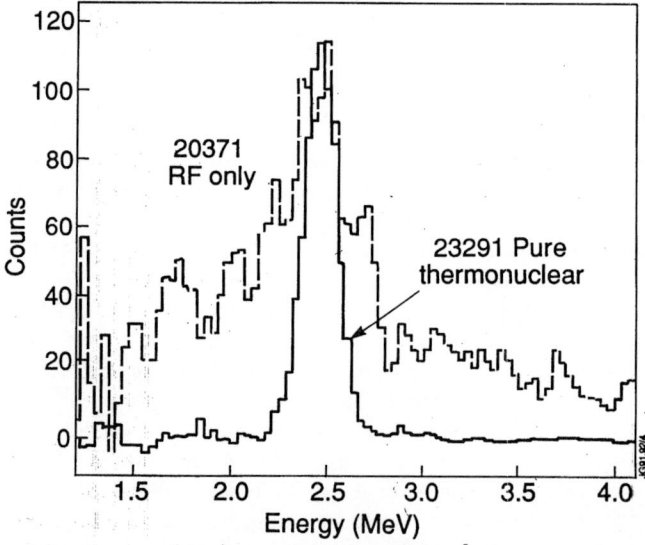

Fig 4. Neutron energy spectra recorded by the time- of-flight spectrometer during discharges with RF-heating tuned to hydrogen. The presence of beryllium gives rise to a very large contribution from neutrons produced by nuclear reactions with accelerated deuterons. A similar discharge obtained without RF heating shows only thermonuclear neutrons.

A major disadvantage of the ^{3}He ionization chamber spectrometer for fusion applications is that it cannot be used when there is a significant contribution from very energetic (>4MeV) neutrons, owing to the wall effect which reduces the signal amplitudes from all high energy events and superimposes them upon the signals from 2.5MeV neutrons.

It is worth noting that the study of the neutron emission during RF heating of plasmas containing significant levels of beryllium impurity is somewhat impeded by lack of cross-section data and neutron spectra for the relevant reactions-(^{3}He,n) and (d,n) on beryllium.

Time-Resolved 14 MeV Neutron Emission

The 1.0 MeV tritons emitted in d-d fusion reactions can undergo t-d fusion reactions whilst slowing down in the plasma; as a result there is a 14 MeV neutron emission accompanying the 2.5 MeV neutron emission.

Threshold reactions in a silicon diode can be used for detecting a low flux of 14 MeV neutrons in the presence of a much larger flux of 2.5MeV neutrons. Fig 5 shows an interesting example of the time-trace for the triton burnup obtained for a unique discharge in which the plasma maintained a low density, high temperature, condition for several seconds [12]. In this instance, the triton slowing down time was about 3 seconds.

A complication arises for this kind of measurement when H-minority ICRF heating is used (even in the absence of impurity ions), as it is possible for high energy deuterons to be generated and hence d-d neutrons may be produced which exceed the silicon threshold energy (around 5 MeV). When there is a significant contamination of the plasma from impurity ions, especially beryllium, RF heating may result in nuclear reactions which release neutrons with similar high energies, so that it is advisable that RF heating be not used when triton burnup is being studied.

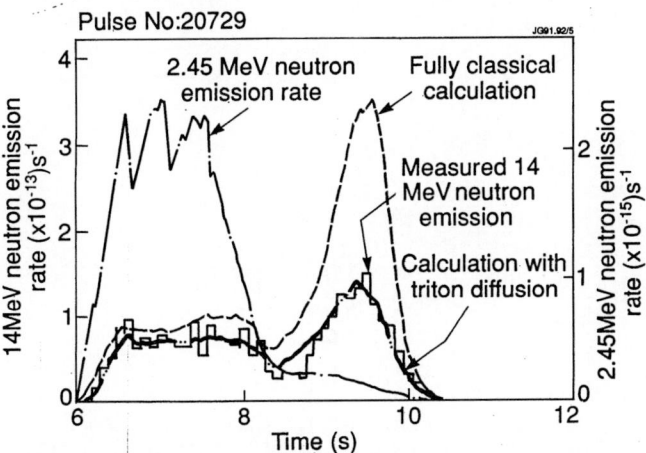

Fig 5. 14 MeV neutron emission from slowing down fusion reaction-product tritons; the 14 MeV emission, calculated from the observed 2.5 MeV emission, agrees well with the experimental measurements provided the effects of a loss mechanism such as charge-exchange are included.

Another complication arises from the possible contribution of neutrons from unfamiliar sources which may contribute when impurity ion concentrations are significant; one such reaction is that between fast tritons and beryllium, which results in the emission of energetic neutrons. In this instance ICRF heating is not involved. The available data on the ^{9}Be(t,n)^{11}B reaction are very limited.

Fusion Gamma-ray Studies

Whenever high power ICRF heating is applied, there is a significant possibility of accelerating light ions to MeV energies. These can then undergo fusion reactions with the bulk plasma ions and nuclear interactions with impurity ions. The signature of these reactions is the emission of unique gamma-ray lines. These gamma-rays can be studied with detectors (NaI, BGO, HPGe) placed in the roof laboratory; an example of a gamma-ray energy spectrum taken with a NaI detector is shown in fig 6.

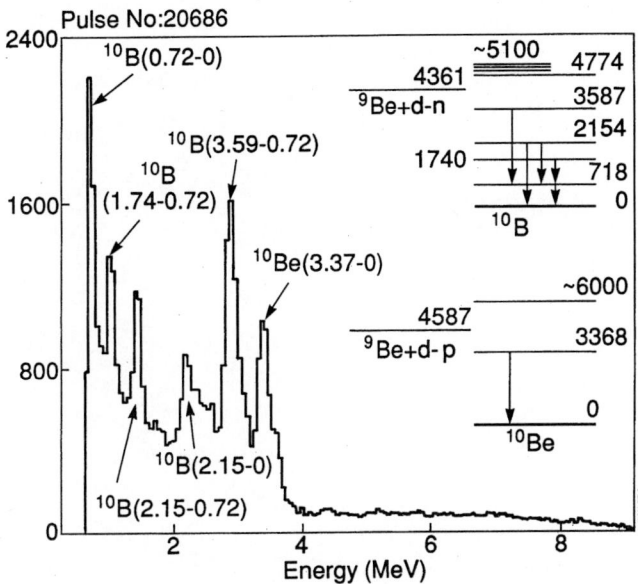

Fig 6. Gamma-radiation energy spectrum recorded during combined NBI and RF heating (hydrogen minority). The plasma is formed on the beryllium belt limiters. The presence of strong lines due to ^{9}Be(d,n) ^{10}B reactions explains the neutron spectrum of fig. 4.

The observation of certain gamma-rays provides useful information on the energy distribution of the fast ions and, hence, the effectiveness of the RF heating [13]. A complete analysis of the gamma-ray spectra is impeded by the lack of basic gamma-ray production cross-section data for light ions interacting with beryllium and carbon impurities.

Conclusions

A very comprehensive set of neutron diagnostics has been mounted on JET, and more modest sets on other tokamaks. The work at JET has demonstrated that all the proposed measurements can be made with a considerable degree of success, although the transient nature of the neutron emission imposes many problems. As the tokamak technology develops, the emission will tend more to the steady-state. This will make the measurements much easier to perform and the results should become even more reliable. As a result, the quality of the neutron production data involved in this work will be increasingly scrutinized. Potential areas of weakness have been pointed out in the text.

Acknowledgements

The author is indebted to the members of the Neutron Diagnostics Group and to the participating scientists from the Associated Laboratories, who have designed and operated the Neutron Diagnostics, and to the entire JET Team, without whom the experiment would not be possible.

258

References

1. The JET Team (presented by P-H Rebut), Recent JET Results and Future Prospects, 13th Int. Conf. on Plasma Physics and Controlled Nuclear Fusion Research, Washington, DC, USA, 1-6 October 1990. IAEA-CN-53/A-1-2.
2. H. Brysk, Plasma Physics 15 (1973) 611.
3. P. van Belle and G. Sadler, in Basic and Advanced Fusion Plasma Diagnostic Techniques (Varenna, 1986) Vol III, 767 EUR 10797 EN.
4. H.W.Hendel, R.W.Palladino, C.W.Barnes, et al, Rev. Sci. Instrum. 61 (1990) 1900.
5. O.N. Jarvis, G.Sadler, P. van Belle and T.Elevant, Rev. Sci. Instrum. 61 (1990)3172.
6. M.Pillon, O.N.Jarvis, J Kallne and M Martone, Fusion Technology 15 (1989) 1420.
7. G.Sadler, O.N.Jarvis and P.van Belle, Rev.Sci. Instrum. 61 (1990) 3175.
8. P. van Belle, O.N.Jarvis, G.Sadler, S.de Leeuw and P. D'Hondt, Rev. Sci. Instrum. 61(1990) 3178.
9. M. Pillon, O.N. Jarvis, J. Kallne and M. Martone, Fusion Technology 15 (1989) 1420.
10. J.M.Adams, A.Cheetham, S.Conroy, et al, in Proc. 16th Europ. Conf.on Controlled Fusion and Plasma Physics, (
    Venice,1989), Vol 13B, Part I, European Physical Society (1989), p.63.
11. O.N. Jarvis, G. Gorini, M. Hone, J. Kallne, G. Sadler, V. Merlo and P. van Belle, Rev.Sci. Instrum. 57 (1986) 1717.
12. S.Conroy, O.N. Jarvis, M. Pillon and G. Sadler, in Proc. 17th Europ. Conf. on Controlled Fusion and Plasma Heating, Amsterdam, 1990, Vol 14B, Part I, European Physical Society (1990), p.98.
13. G. Sadler, S. Conroy, O.N. Jarvis, P. van Belle, J.M.Adams and M.Hone, Fusion Technology 18 (1990) 556.

NUCLEAR DATA FOR FUSION MATERIALS RESEARCH

S. Cierjacks and K. Ehrlich

Kernforschungszentrum Karlsruhe, Association KfK - Euratom
Institut für Materialforschung
Postfach 3640, 7500 Karlsruhe, Germany

Abstract: Present problems in materials research for D-T fusion reactors are briefly reviewed. The significance of nuclear data for the determination of "standard" irradiation parameters is outlined and the status of available data is discussed. The neutron data base is reasonable in the energy range from thermal up to ~20 MeV for displacement cross sections, gas or solid transmutation production cross sections, and dosimetry reaction cross sections. Some improvements especially in the energy range from ~6-14 and above 15 MeV are, however, still needed. Since materials testing is also performed with light- and heavy-ion simulation methods, a large number of cross sections for various charged-particle-induced reactions is also required. Current comparative investigations of candidate high-intensity neutron sources for materials testing rely on various neutron data in excess of 20 MeV. For the latter two applications available results are rather sparce, so that more data are needed. The role of cross section systematics and nuclear model calculations in predicting unknown cross sections is adjointly discussed.

(**Key words:** Nuclear data for displacement damage, gas and solid foreign-element production, dosimetry and the investigation of novel high-intensity neutron sources)

1. Introduction

Presently, there are two main goals in fusion materials research which ought to be achieved simultaneously: (1) The development of materials which are resistant to large fluences of fusion neutrons with energies ≤ 14 MeV, and (2) the development of so-called "low-activation" materials for the first wall and other structural components which provide minimum activity during and at the end of a fusion reactor operation period. Thus, in addition to a large variety of engineering property specifications, a number of radiological requirements need to be fulfilled, not all of which may be satisfied simultaneously in an optimum manner. This presentation deals with the main aspects of fusion materials research, but excludes nuclear data requirements for materials optimization as "low-activation" versions. The latter data needs are covered in another survey of this conference.

Present achievements and problems in fusion materials research are briefly summarized in Section 2. The significance of nuclear data for the calculation of "standard" irradiation parameters is discussed in Section 3. Section 4 is devoted to nuclear data aspects involved in displacement damage calculations. The nuclear data needs for the determination of gas and solid transmution rates are outlined in Section 5. Dosimetry needs for specifying the irradiation environments in complex test cells are briefly described in Section 6. Data needs for the study of present candidate intense neutron sources for end-of-life fusion materials testing are discussed in Section 7.

2. Achievements and Current Problems

About 15 years of fusion materials research have led to a rather detailed understanding of many radiation damage effects and to first successes in the development of more-radiation-resistant materials. From these studies it is now well established that the macroscopically observed radiation damage effects are a consequence of two elementary interactions between the radiation and the atoms of the irradiated material (see Fig. 1):

1. Bombarding particles transfer recoil energy to the lattice atoms. If this energy exceeds the threshold energy for displacement, one or more vacancy-interstitial pairs (Frenkel defects) are produced. The number of defects and their spatial distribution depend on the kind and the energy of the radiation and on the composition and the structure of the material. Fusion neutrons, for instance, transfer higher energies to the lattice than fission neutrons, i.e. larger numbers of vacancies and interstitials are created per primary nuclear interaction in one or more localized zones (displacement cascades).

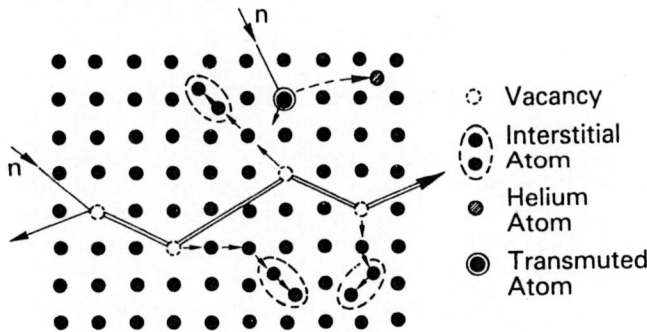

Fig. 1.   The interaction of high-energy neutrons with the nuclei of an ordered solid can lead to (1) displaced lattice atoms (interstitials) and vacancies and to (2) transmutation products (e.g. a He and a transmuted atom created by an (n,α)-reaction). From Ullmaier Ref. 1.

2. The incident radiation undergoes nuclear reactions which can produce considerable concentrations of foreign elements within the material due to nuclear trans-

mutations. Again, fusion neutrons are much more severe than fission neutrons, since the cross sections for transmutations are much higher. In particular the inert gas helium, that is produced by (n,α) and (n,n′α) reactions, plays an important role in the behavior of materials under fast neutron irradiations.

From previous work a large number of microscopical changes and macroscopic radiation damage effects have also been observed [2-4]. Regarding microstructural changes, small vacancy loops, threedimensional vacancy agglomerate voids, extended interstitial loops, helium bubbles etc. are characteristic defect structures that develop under irradiation; their creation depends largely on irradition temperature and fluence. The radiation-induced changes finally lead to changes in many macroscopic materials properties such as high- or low-temperature embrittlement under static (creep) or cyclic (fatigue) loads. Furthermore, there are various other deteriorations in special materials. A few examples are: (1) Decreases in the transition temperature and the critical current density of superconductors; (2) changes in the thermal expansion and decreases in the thermal and electrical conductivity of graphites; (3) degradations of the mechanical, dielectric and optical properties of ceramics and plastics envisaged for use as electrical and thermal insulators in fusion devices.

Despite this progress, fusion materials research is still faced with a large number of unsolved problems. Many of the observed radiation damage effects cannot yet be fully understood from first principles. The transfer of simulation data to fusion conditions is still afflicted by large uncertainties, especially for materials properties which are determined by more than one basic radiation damage effect. Thus there is a clear need for more basic research and for the generation of standard data sets which can serve as "calibration points" for engineering data. Finally, it should be mentioned that there is not yet any suitable first-wall candidate material that can survive the full life-time of envisaged fusion reactors.

3. Significance of Nuclear Data

It is now a well established method to quantify radiation damages by a few "standard" irradiation parameters such as: (1) Displacement damages in term of displacements per atom (dpa); (2) helium and hydrogen gas production rates in terms of atom parts per million (He(appm) or H(appm)); (3) solid foreign element production rates also in terms of atom parts per million (appm) and (4) involved flux levels. This allows to characterize many of the basic radiation damage effects in terms of "standard" parameter ratios, e.g. He(appm)/dpa or H(appm)/dpa etc. at given fluxes. The virtues of this characterization have been widely tested, and the limits of such attemps have also been identified [5]. It is mainly the calculation of the "standard" parameters which leads to the important nuclear data needs in fusion materials research. Two other tasks which are related to further nuclear data requirements are neutron dosimetry in various test en-

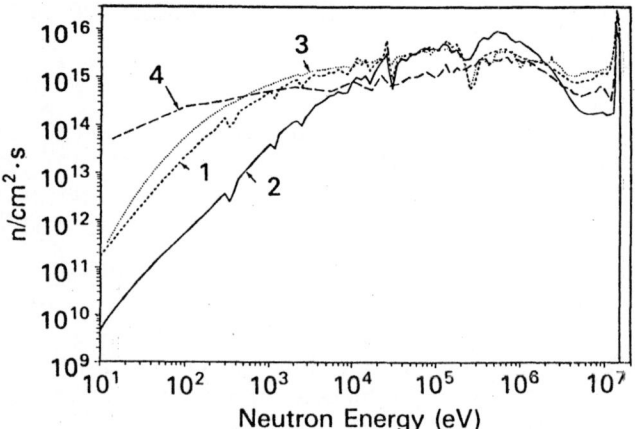

Fig. 2. Neutron flux spectra in the first wall of conceptual fusion reactor designs with (1) liquid lithium; (2) liquid lithium-lead eutectic; (3) solid lithium oxide; and (4) solid lithium-orthosilicate with beryllium-multiplication blanket concepts. A 14-MeV neutron wall loading of 5 MW/m² was commonly assumed.

vironments and the investigation of candidate high-intensity neutron sources for materials testing.

Unfortunately, high-intensity neutron sources are presently not available. Therefore, all previous and current fusion materials programs are based on simulation irradiation methods [6]. These involve e.g. fission neutrons from thermal and fast reactors [7,8], spallation neutrons from operating spallation sources [9] and light- and heavy-ion beams from several existing medium-energy accelerator facilities [10-13]. In all cases, a major task is the transfer of simulation irradiation results to fusion conditions. Thus "standard" damage parameters must be calculated for both, specific fusion reactor environments and the irradiation conditions applied in the various simulation measurements. This implies that the total nuclear data needs become correspondingly large: Regarding neutron data, the energy range cannot be restricted to the region of typical fusion reactor spectra (see Fig. 2) only. Sufficient data must be also available for the very low (thermal and subthermal for fission reactors) and the very high energies beyond 14 MeV (up to 50 MeV for d-Li neutron sources and up to a-few hundred MeV for spallation neutron sources). In addition, a large variety of nuclear-reaction and decay data is required for nuclear interactions of a number of light and heavy ions. The energies of previously applied light- and heavy-ion beams were mainly below ~50 MeV, but in more recent facilities also energies in excess of 100 MeV are utilized. The amounts and kinds of required nuclear data differ for the different tasks in fusion materials research, and thus are discussed seperately in the following chapters.

4. Displacement Damage

4.1 General aspects
Displacement damage is one of the most important irradiation effects in any type of solid material. All kinds of radiation transfer recoil energy to the lattice atoms and produce displacement damages, if this energy exceeds the threshold energy for displacement (typically some 10 eV). Even though displacement

damages are usually quantified by the number of displacements per atom (dpa), the more genuine quantity is the primary knock-on atom (pka) spectrum which determines all secondary events. The primary knock-on atom causes a large number of cascade displacements by secondary atom-atom interactions, and the number of secondary recoils depends on the primary recoil energy and the composition of lattice atoms in a material. There are presently recommended standard rules [14] and well established damage models and computer codes [15,16] for relating the pka spectrum to damage cross sections and "standard" dpa parameters. Such calculations treat all secondary displacements and rely on several atomic data (e.g. displacement energies, stopping power, recombination probability etc.) which are, however, out of the scope of this survey.

4.2 Special data needs

The prime quantity to be known for the determination of displacement damages is the pka spectrum. For its calculation all types of nuclear interactions of the incident particles with lattice atoms must be considered. Reliable estimates require elastic and inelastic scattering data and all important reaction cross sections, including radiative capture and multi-particle emission. For a proper treatment, complete angular distributions need also to be known, if the reaction mechanisms are substantially non-isotropic in the centre-of-mass system. Most complete present calculations [15] include also beta decay and the fraction of atom-recoil energy transfered back to nuclear interactions, in order to provide the net primary-recoil energy distributions.

In addition to neutron data, similar interaction data are, in principle, also required for light and heavy ions. Presently, however, pka spectra are often calculated from Coulomb-scattering cross sections only. This is reasonable for sufficiently low ion energies, since Rutherford scattering is dominant in this region. At higher energies (for protons and deuterons above ~ 10 MeV) non-elastic nuclear interactions increasingly become important and cannot be neglected any more. Improved displacement damage calculations for simulation irradiations with p- and d-beams of up to 50 MeV and α-beams of up to 100 MeV thus rely also on the availability of elastic and non-elastic charged-particle reaction cross sections [17].

4.3 Available data base

As discussed above the required nuclear data base depends largely on the irradiation simulation method and the type of involved radiation. Due to the most common irradiation methods, the discussion will concentrate on neutrons and light ions. Heavy ions are considered only marginally due to their restricted use for large-volume specimen applications.

Neutrons   As long as neutrons with energies ≤ 20 MeV are involved, a major fraction of the required neutron cross sections are available from existing evaluated data libraries. Already the various general-purpose files [18-22] contain a large number of the important total, differential, and partly even double-differential cross sections in the range from thermal to 20 MeV. Many additional cross sections are contained in some of the special purpose files such as EFF [18], EAF-1 [23] or REAC-2 [24]. A worldwide data library for fusion technology, finally containing tested data from various existing libraries, is presently under development at IAEA [25]. Also most of the required decay data are available from special decay data libraries [26,27]. Three deficiencies exist, however, in all available libraries: (1) A reduced accuracy for the data in the energy range from 6-14 MeV and above ~15 MeV, due to lack of sufficient experimental results. (2) The lack of data for energies above 20 MeV. (3) An insufficient coverage of double differential angular-dependent inelastic and reaction cross sections. The latter data are important, when the angular distributions are significantly non-isotropic, which often starts at energies of a couple of MeV, but becomes increasingly more pronounced at energies above ~10 MeV. This deficiency typically is not important for simulation irradiations in thermal and fast fission reactor spectra, but causes some problems for characterizing the situation under the various fusion conditions. In any case, such data are clearly necessary for materials testing with d-Li or spallation sources [9,28] which produce spectra with substantial "high-energy tails" (see Sect. 7).

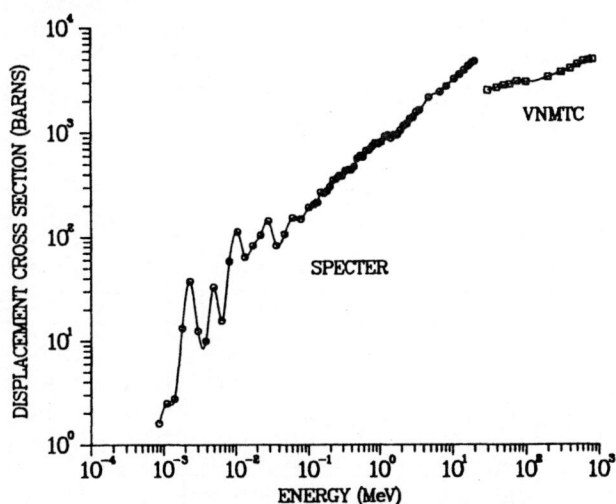

Fig. 3. The damage cross section for neutrons in iron vs neutron energy [29]. The two branches are calculated from ENDF/B-V (left side) and NMTC (right side) cross section data. It can be seen that the two branches do not match at the intersection point.

A pronounced discrepancy, which needs still to be solved [29], exists in the transition region from medium-to-high neutron energies for the displacement cross section of iron. This is shown in Fig. 3. It can be seen that there is a glaring discontinuity at 20 MeV. While cross section data for high energies were taken from intranuclear cascade model calculations, the corresponding results for low energies were obtained by using the well established neutron data from the evaluated general-purpose file ENDF/B-V.

Light ions.   Restricting to protons and deuterons which are most frequently used at energies below ~50 MeV [10,11], there are a few libraries which contain some of the required data in limited ranges of target nuclei, energies and reaction types [30-34]. Unfortunately, however, these libraries are by far not complete for more precise displacement damage calculations. Even though, there are a lot of new experimental data [35], most of these have not been adequately compiled nor incorporated in existing data libraries. On the other hand, the Russian library [34] compiled in the Kurchatov Institute contains a large number of measured thick target yields which cannot be translated into differential cross sections. Vice-versa, however, such experimental data can be favourably used for the verification of model predictions, if the latter data are suitably folded with differential particle ranges. In general, however, most of the proton- and deuteron-induced cross section data needed for the estimation of pka-spectra, and thus displacement damages, must presently come from model calculations.

Heavy ions.   Nuclear data needs are very similar as those for the light ions. Detailed differential data are, however, extremely sparce, and there exists, to our knowledge, little experimental information and no relevant compilation or evaluation. Suitable model codes for quick calculations are presently widely lacking. Fortunately, however, heavy ion irradiations have become less frequent in the past, due to the fact that their damage rates are orders of magnitude higher than those obtained under fusion conditions. In addition, mechanical tests are not possible, because of the small penetration of heavy ions, producing only damage layers of a few μm below the surface.

5. Gas and Foreign Element Production

5.1 General aspects

The damage due to transmutation products must be considered differently for gaseous and solid (foreign element) transmutation products. Regarding gas production, He and H play a particular role. The inert gas helium (with its isotopes ^{3}He and ^{4}He) has a severe effect under fast neutron irradiation. Helium is known to severely degrade the performance of materials by inducing high-temperature embrittlement, and by promoting cavity growth. Even for fission neutron spectra there exist already large differences in He production. In thermal reactors helium production in austenitic stainless steels is dominated by reactions of thermal neutrons with ^{10}B and ^{58}Ni, while in fast breeders the largest contributions come from reactions of fast neutrons with natural nickel. Therefore, further studies of this effect are continuing. Some of these, involving fission neutrons or light ions, however, fail to meet the proper He/dpa ratio obtained under fusion conditions. This does not apply for two techniques, one involving the Karslruhe Dual-Beam Facility [11] and the other employing specially designed deuteron- and proton-irradiation methods [10]. In the first case, proton and α-particle beams are used simultaneously to bombard suitably sized materials specimen: While a 0-100 MeV α-beam provides the necessary homogeneous He deposition and related displacement damage, additional adjusting displacement damage is provided by a 15-40 MeV proton beam delivering more than an order of magnitude higher current than that of the α-beam. In this way, the fusion-relevant He/dpa ratio can be matched by suitable beam current and particle energy

adjustments. For single proton or deuteron beams it has been shown by Jung [36] that the fusion-relevant He/dpa ratios in a test sample can occasionally be achieved by adjusting the incident particle energies.

The other gas transmutation product of concern in materials research is hydrogen (with its isotopes ^{1}H, ^{2}H and ^{3}H) from which ^{1}H is of most concern. This gas, which - in neutron irradiations - is produced predominantly by (n,p) and (n,n'p) and to a lesser extent by (n,d) and (n,t) rections, has an important effect on other materials properties. Hydrogen gas has been found to degrade various materials by inducing embrittlement at low (room) temperatures. For graphites and large carbon composites it has been pointed out by Conrads [1] that there may be a damage problem in lattices preloaded by large amounts of interstitially dissolved hydrogen.

Regarding solid transmutation products there is, in principle, a major concern that they might produce substantial concentrations of foreign elements in the lattice of a material. The critical level can be rather small depending on the material and its application. One major problem area refers to superconductors, stabilizers, insulators etc. for which properties such as electrical conductivity or resistivity are important. These properties are extremely sensitive on small impurities of solid foreign elements. A special case of this kind has been identified by Hodgson [37]. For presently considered insulator materials such as BeO, Al_2O_3, ALON or SiC etc. even ppm concentrations of e.g. Na and Mg cause already severe problems for both dc and r.f. applications. This is due to the fact that the mentioned foreign elements segrate to the surface, and thus destroy the initial insulation properties. Concerning metallic materials, solid transmutation products are of major concern, if they lead to nuclides or elements that (1) produce extremely high additional radioactivity (e.g Ag and Tb in sub-ppm quantities) and thus destroy the favourable properties of "low-activation" alloys, (2) are strong neutron absorbers (e.g. ^{10}B) and disturb the net neutron balance, (3) enhance, in small quantities, the embrittlement properties (e.g. H, O, N in zirkonium or P and S in ferritic steels).

5.2 Special data needs

For intercomparison and optimization purposes in simulation irradiations, it is necessary to know the hydrogen and helium gas production cross sections for a wide variety of technologically important elements and their isotopes. Furthermore, a broad data base is needed to estimate the solid element transmutations in various candidate fusion materials. For this kind of work the knowledge of energy-dependent reaction cross sections over the whole energy range from thermal or threshold to maximum energy is sufficient. There is no need for angular-dependent double-differential cross sections. For neutron irradiations or calculations of reference fusion conditions reaction cross sections for the elements are usually sufficient. In some cases, however, where isotopic cross sections are extremely large (e.g ^{6}Li, ^{10}B, ^{58}Ni), such data are required separately on an isotopic basis.

For simulation irradiations with light and heavy ions analogous data are required for all types of involved charged particles and energies. Since charged particles are continuously stopped during their passage through the material, all relevant excitation functions must be known from 0 to E(max), i.e. up to the incident beam energy. In several cases charged particle beams up to 100 MeV are used. This opens a large number of new reactions channels not possible at low energies.

5.3 Available data base

Neutrons. Nuclear data in the important range of the fusion spectra have been widely covered in the past. Sufficient results are currently available for energies from thermal to about 6 MeV and at 14 MeV in various evaluated neutron libraries.

For gas production cross sections Qaim et al. [38,39] measured systematically all types of gas production cross sections at 14.7 MeV. For the main gas producing reactions Forest [40] performed the most detailed systematics that now allows to predict the unknown 14-MeV cross sections with an accuracy of ~30-50%. An example from this work is shown in Fig. 4. Experimental data are, however, sparse in the intermediate energy range from 6 to 14 MeV and at energies above 15 MeV. For these energy ranges, IAEA's Nuclear Data Section is planning a Coordinated Research Program (CRP) that aims at new measurements, model calculations and data evaluations [41]. For the lacking data, in principle, also nuclear model calculations can be employed. There are a few model codes which allow quick calculations, and it is expected that these give already suitable results within an uncertainty of about a factor of 2. This is, in general, sufficient for fusion materials research programs.

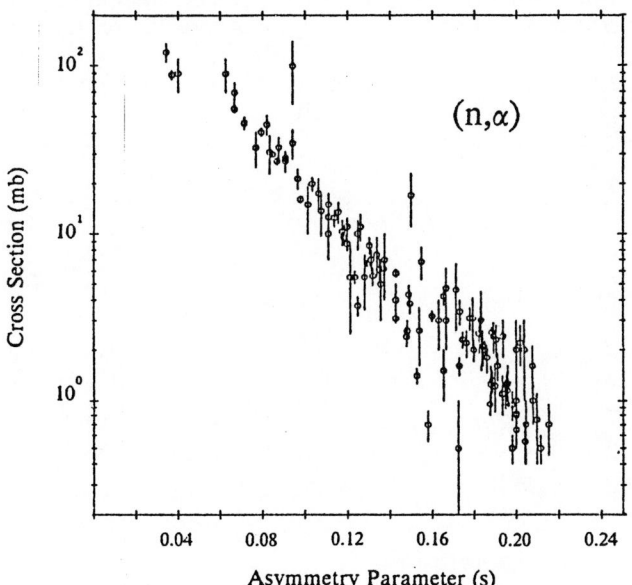

Fig. 4. Neutron-induced 14-MeV (n,α) cross-sections vs asymmetry parameter s = (N-Z)/A (redrawn from Ref. 40). A best 8-parameter fit in the two variables s and A (not shown) has been made separately for the two regions of Z≤50 and Z>50.

For solid transmutation products most of the necessary data for neutron irradiations below 20 MeV are available from the evaluated general-purpose data libraries [18-22] or from special-purpose starter libraries [23,24]. The latter two libraries, which are especially created for activation calculations, cover all kinematically possible reactions below 20 MeV (in REAC-2 often up to 50 MeV) for stable and radioactive nuclides with $T_{1/2} \geq 1$ day. A large fraction of the up to 10.000 reactions stem, however, from calculations with a simple semiempirical code (THRESH) [42]), and often no evaluation procedure has been made.

Light and heavy ions. Gas production cross sections for light ions like protons and deuterons are only available for a small number of elements or isotopes [30-34]. For optimization purposes in proton and deuteron simulation irradiations, which aimed at matching He/dpa ratios to fusion relevant conditions, Jung [43] has measured total He-production cross sections at some discrete proton and deuteron bombarding energies between 8 and 24 MeV. The majority of the required light-ion-induced gas-production cross sections must, however, still come from model calculations. Here e.g. the code ALICE [44] is capable to provide suitable predictions for most of the unknown cross sections. Complete special-purpose libraries that would be most desirable do not exist at present.

For solid transmutation products the situation is similar. Only very little experimental information is available for light ions. The two existing compilations from Oak Ridge and Karlsruhe are very old, and have been terminated in 1964 and 1979, respectively. The Brookhaven charged-particle library [32] is only a recovery of the Karlsruhe data base which originally had not been generated in form of a computer file. For light element between H and O White et al. [33] maintain a rather complete library. For (p,n) (d,n) and (α,n) reactions on all important isotopes of the elements from B to Mo Cierjacks et al. [45] have produced a starter library from model calculations employing an updated version of the ALICE code [46]. In the future, this library will be extended to include also the reactions (t,n), (d,2n) and (t,2n), and the mass range of target nuclei will be extended to all stable and radioactive nuclides with $T_{1/2} \geq 1$ day up to mass A = 209.

Regarding heavy ions, very little specific gas production and solid transmutation product data are available from experiments. In addition, no adequate nuclear model codes are available that can provide quick calculations.

6. Dosimetry

For fusion materials development dosimetry is an important means for determining the neutron invironment, when a material is irradiated in present test cells or in any future end-of-life fusion irradiation materials test facility. Employing multi-foil analysis techniques, various dosimetry reactions with different thresholds covering the whole range of the involved neutron spectra are necessary. In contrast to other nuclear data for materials research, dosimetry cross sections must be far more accurate (better than 5-10%) in order to be useful. For fission neutron irradiations the total number of reactions is comparatively small (~30 reactions are considered to be sufficient [47]). Required dosimetry data for covering the neutron energy range from thermal to 20 MeV are readily available from existing dosimetry files [48]. If d-Li or spallation sources are involved for bulk radiation damage studies (see Sect. 7) the number of dosimetry reactions would increase considerably. Even though some work had been started [49], no sufficiently extended dosimetry reaction files are available. For light- and heavy-ion simulation irradiations the dosimetry problem is not very pronounced. In most cases, the irradiation environment is well determined from suitable time-dependent current measurements of the irradiating particle beams.

7. High-Intensity Neutron Sources

7.1 Current activities and candidate source concepts

Presently, it is well agreed that a worldwide intense source of fusion neutrons is needed to evaluate suitable candidate fusion materials [50]. In the past, several concepts for an International Fusion Materials Irradiation Facility (IFMIF) have been studied and their results were jointly published [51,52]. On the request of the "IEA Executive Committee on an Intense Neutron Source", it has been recommended that four neutron source concepts - a beam-plasma source, a d-Li source, a t-H source and a spallation source concept - should be developed further [53]. The beam-plasma concept proposed by the Lawrence Livermore National Laboratory [54] is based on a mirror-type plasma machine fueled with tritium and subjected to some intense neutral deuteron beams injected perpendicular to its axis. The concept of a t-H source (referred to also as t-H₂O source) propagated by the Kernforschungszentrum Karlsruhe [55] is based on one or more intense triton beams impinging on one or more free water-jet targets. A tritium energy of 21 MeV produces a continuous neutron spectrum with a sharp cutoff energy at 14.6 MeV. The d-Li concept refers to a proposal from the Los Alamos National Laboratory [56], and represents a remarkable upgrade of the not realized FMIT design [57] in which neutrons are produced by one or more 35-MeV deuteron beams incident on one or more liquid lithium jet targets. The conceptual spallation source EURAC (European Accelerator) [58] adopted by the Institute of Nuclear Fusion of the Polytechnical University Madrid produces neutrons with a 600-MeV proton beam incident on a liquid lead target. While the first two concepts (beam-plasma and t-H) provide only neutrons with energies ≤14.6 MeV, the d-Li and the spallation source involve largely extended spectra with substantial "high-energy tails". These extend up to ~50 MeV for the d-Li and up to a-few hundred MeV for the spallation source EURAC. A recently established IEA IFMIF Working Group has been charged to develop detailed descriptions of the radiation environments of all four sources which aim at a critical examination of how well each source can simulate the environment of a fusion reactor. At present, source characterizations are made on the basis of neutron spectra, resulting pka spectra, displacement rates and transmutation species and rates.

7.2 Special data needs and existing libraries

Concerning beam-plasma and t-H sources additional data needs are relatively minor. Neutron producing source reactions are well known, and evaluated

data files have been produced [59]. The needs for displacement-damage and transmutation-rate calculations are the same as those discussed in Sects. 4,5. Since only neutron data for the energy range from thermal to 15 MeV are required, most of the necessary data are available from existing neutron libraries (exept again for angular-dependent non-elastic cross sections in the many-MeV region).

For the other two source concepts the situation is different. While for the d-Li source the source reaction cross sections are well known (except for energies below ~1.5 MeV [60]), the experimental source reaction cross sections for spallation sources are still discrepant, especially for energies above about 50 MeV [61,62]. For displacement-damage and transmutation-product calculations nuclear data above 20 MeV are widely lacking. There are, of course, a few libraries which contain point data for reactions cross sections above 20 MeV, but these libraries exists either only in their starter forms, containing data from simple model calculations, or only for a limited number of reactions on a few special target elements and isotopes. A library of the first type is REAC-2 [24], which covers point data for a-few thousend reactions, which are, however, largely based on simple THRESH-code [42] calculations. A data library based on sound evaluations is presently under development in Japan [63] for use with the ESNIT source (an intermediate-intensity, energy-variable source project for fusion materials research). The Japanese work finally aims at the inclusion of all neutron data up to 50 MeV needed for source construction and its utilization in basic and applied programs. At present, data evaluations for experimentally and theoretically determined cross sections are available for various isotopes of Fe, Cr, Cu, Al and Si needed for the prediction of displacement damages, gas, solid element and isotope transmutation rates. Fig. 5 shows a sample result from the this library.

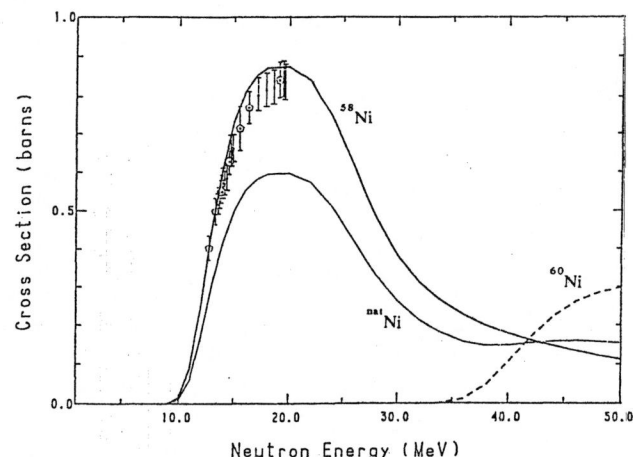

Fig. 5. Ni(n,xnp)^{57}Co production cross sections for the extended energy region from 1 to 50 MeV (from Ref. 63). This result is an example from the new medium-energy library under development at JAERI in support of materials research programs with ESNIT.

For spallation source applications nuclear data needs extent to energies of up to a-few hundred MeV.

For this energy range the American NNDC maintains some activities since several years. But the coverage of the excess energy region above 50 MeV needs further compilational and evaluational effort. Some work of this kind has recently also been proposed by IAEA's Nuclear Data Section [64]. Except for a few (p,xn) neutron production cross sections [61,62] and some transmutation product measurements for proton energies (mainly between 300 and 800 MeV [65]), other experimental information is rather sparse. Thus a large fraction of the necessary data must come from suitable model calculations. For this purpose, there are a few intranuclear cascade (INC) model codes with somewhat different versions such as [66,67]. In general, all of these have options for calculating the nescessary nuclear cross sections, but it is not yet clear to what extent of accuracy. One disadvantage of all INC codes is that they are Monte Carlo codes, and thus often require long computing times for cross section calculations. This does not apply for the ALICE code which had been recently generalized to allow for cross section calculations also in the a-few hundred MeV range [68]. Concerning specific nuclear libraries, a general deficiency is that at energies between 50 and a-few hundred MeV comprehensive evaluations of measured and calculated data are lacking.

8. Summary

The most urgent goal of current fusion materials research is the development of radiation-resistant materials that can survive large fluences of fusion neutrons over the lifetime of future power-producing reactors. Especially the first wall, but also other structural components, insulators, thermally and electrically conductive materials, and superconductive coils still cause major problems from the materials development point of view. Although 15 years of fusion materials research has led to a rather detailed understanding of many radiation damage effects and to first successes in the development of more-radiation-resistant materials, considerable further progress is needed. At present, high-intensity neutron sources for large-scale end-of-life testing of candidate fusion materials are still lacking. Therefore, current materials research programs still rely on various kinds of simulations irradiations e.g. with fission and spallation neutrons or with light and heavy ion beams. The transfer of such data to realistic fusion conditions is afflicted by large uncertainties, especially for materials properties that are governed by more than one basic radiation damage effect. For this reason, more recent irradiation studies attempt to optimize the simulation conditions, in order to match the important "standard" irradiation parameters to the relevant values in fusion reactor environments. Preparatory work for the selection of a suitable worldwide high-intensity high-energy neutron source currently concentrates on a critical examination of how well each of four selected candidates can simulate the environment of a fusion reactor.

For all of these tasks a large number of nuclear data is needed. The data needs range from total, differential elastic and inelastic scattering cross sections to elemental and isotopic transmutation production cross sections from thermal energies to a-few hundred MeV. Such data are not only needed for neutrons, but

also for several light and heavy ions and a large variety of technologically important elements and nuclei. The best situation exists presently for neutron data. Extended data libraries are well developed from previous fission and fusion reactor programs, and suitable upgrades for fusion applications are being made. Present shortcomings belong mainly to special elements and isotopes not sufficiently covered in the past. Concerning energy ranges and angular-distributions, there is still a need for improvements in the range from 6 to 14 MeV and the range from 20 to 50 MeV (d-Li) or even several 100 MeV (spallation). For light ions existing compilations are still poor; some of the existing libraries or compilations are even obsolete, since work had been finalized 15 to 20 ago. Fortunately, however, the required accuracy of the data for fusion materials research is modest (in general, an uncertainty by factor of 2 is acceptable), so that in most cases current model predictions of the unknown data are sufficient. For neutrons and light ions suitable nuclear model codes are readily available, and reasonable starter libraries could, in principle, be produced by quick calculations with these codes. This is, however, an unusual task for materials scientists. The production of comprehensive data libraries for fulfilling fusion materials research needs, thus seems to be a challenging task for the nuclear community in the future. Concerning heavy ions, the situation is much worse. There are, to our knowledge, no comprehensive experimental results, nor generalized model codes which can be used for broad-scale nuclear data predictions.

References

[1] S. Cierjacks, K. Ehrlich, E.T. Cheng, H. Conrads and H. Ullmaier, Nucl. Sci. Eng. 106, 99 (1990).

[2] For reviews see: H. Ullmaier and W. Schilling, Radiation Damage in Metallic Reactor Materials, in Physics of Modern Materials, Vol. 1, p. 301, IAEA, Vienna (1980); or J. Gittus, "Irradiation Effects in Crystalline Solids" Appl. Science Publ., London (1978)

[3] For review see: K. Ehrlich, J. Nucl. Mater. 100, 149 (1981).

[4] Proc. 3rd Intern. Conf. on Fusion Reactor Materials, Oct. 1987, Karlsruhe, FRG, J. Nucl. Mater. 155-157, (1988); Proc. 4th Intern. Conf. on Fusion Reactor Materials, Dec. 1989, Kyoto, Japan, J. Nucl. Mater. 179-181, (1991)

[5] P. Jung, Rad. Effects 113, 109 (1990)

[6] Proc. Int. Workshop on Radiation Damage Correlation for Fusion Conditions, Oct. 1989, Silkeborg, Danmark, J. Nucl. Mat. 174 (1990)

[7] F.A. Garner, J. Nucl. Mater. 174, 229 (1990)

[8] M.L. Grossbek, K. Ehrlich and C. Wassilew, J. Nucl. Mater. 174, 264 (1990)

[9] M. Victoria, W.V. Green, D. Gavillet, B.N. Singh and T. Leffers, J. Nucl. Mater. 155-157, 1075 (1988)

[10] P. Jung and H. Ullmaier, J. Nucl. Mater. 174, 253 (1990)

[11] A. Möslang, D. Kaletta and K. Ehrlich, Die Dual-Beam-Technik - Eine Einrichtung zur Simulation der Strahlenschädigung in Fusionsreaktorwerkstoffen, KfK Nachrichten, Jahrgang 19, 1/87, p. 37, (1987)

[12] D.J. Mazey, D.E.J. Bolster and W. Hanks, in Proc. Conf. on Dimensional Stability and Mechanical Behavior of Irradiated Metals and Alloys, April 1983, Brighton, UK, p. 9, British Nucl. Energy Soc., London (1983)

[13] D. Gulden and K. Ehrlich, in Proc. Conf. on Dimensional Stability and Mechanical Behavior of Irradiated Metals and Alloys, April 1983, Brighton, UK, p. 13, British Nucl. Energy Soc., London (1983)

[14] Standard Recommended Practice for Neutron Radiation Damage Simulation by Charged-Particle Irradiation, ASTM Committee E-10 on Nuclear Technology and Applications, ASTM E 521 - 77, June 1977

[15] L.A. Greenwood, in Proc. 14th Int. Symposium on Effects of Radiation in Materials, 1990, Philadelphia , USA, (Eds. N.H.

Packan, R.E. Stoller and A.S. Kumar), p. 663, ASTM, Philadelphia (1990)

[16] J.P. Biersack and L.G. Haggmark, Nucl. Instrum. and Meth. 174, 257 (1980)

[17] A. Möslang, Institut für Materialforschung, KfK, private communication, 1991

[18] C. Nordborg, H. Gruppelaar and M. Salvatores, Status of the JEF and EFF Projects, IP11, this Conference

[19] C.L. Dunford, Evaluated Nuclear Data File, ENDF/B-VI, IP12, ibid

[20] Y. Kikuchi and Members of JNDC, Japanese Evaluated Nuclear Data Library Version-3, JENDL-3, IP13, ibid

[21] I.A. Blokhin, A.V. Ignatyuk, B.D. Kuzminov, V.N. Manokhin, G.I. Manturov and M.N. Nikolaev, The Library of the Evaluated Neutron Data Files, BROND, IP14, ibid

[22] Liu Tingjing, Lian Qichang, and Cai Dunjiu, CENDL-2, IP15, ibid

[23] J. Kopecki, H. Gruppelaar and R. Forrest European Activation File for Fusion, IP19, ibid

[24] F.M. Mann, D.E. Lessor and J.S. Pintler, in Proc. Int. Conf. on Nuclear Data for Basic and Applied Science, May 1985, Santa Fe, USA, Gordon and Breach, p.207, (1986)

[25] A.B. Pashchenko and D.W. Muir, First Results of FENDL-1 Testing and Start of FENDL-2, INDC(NDS)-241, IAEA Nuclear Data Section, Nov. 1990

[26] M.R. Bhat, Evaluated Nuclear Structure Data File (ENSDF), IP17, this Conference

[27] R.A. Forrest, M.G. Sowerby, B.H. Patrick and D.A.J. Endacott, in Proc. Int. Conf. on Nuclear Data for Science and Technology, May/June 1988, Mito, Japan, (ed. S. Igarasi) p. 1061, Saikon, Tokyo (1988)

[28] T. Kondo, H. Ohno, M. Mizomoto and M. Odera, J. Fusion Energy 8 Nos. 3/4, 229 (1989)

[29] Proc. IAEA AG Meeting on Nuclear Data for Radiation Damage Assessment and Rel. Aspects, Sept. 1989, Vienna, IAEA-TECDOC-572, p. 179, and M.S. Wechsler, D.R. Davidson, L.R. Greenwood and W.F. Sommer, in Proc. 12th Int. Symp. on Effects of Radiation on Materials, (Eds. F.A. Garner and J.S. Perrin), p. 1199, ASTM, Philadelphia (1984)

[30] F.K. McGowan, W.T. Milner, and H.J. Kim. Nuclear Cross Sections for Charged Particle Induced Reactions, ORNL-CPX 1,2, Oak Ridge National Laboratory (1964)

[31] H. Münzel, H. Kleve-Nebenius, J. Lange, G. Pfennig, and K. Hemberle, ISSN 0344-8401, Kernforschungszentrum Karlsruhe (1979)

[32] S. Pearlstein, NNDC Evaluated Charged Particle Reaction Data Library, IAEA-NDS-59, Nuclear Data Section, IAEA, Vienna (1985)

[33] R.M. White, D.A. Resler and S.I Warshaw, Evaluations of Charged-Particle Reactions for Thermonuclear Applications, IP20, this Conference

[34] P.P. Dmietriev, Radionuclide Yield in Reactions with Protons, Deuterons, α-Particles and Helium-3, INDC(CCP)-263/G, International Nuclear Data Committee, IAEA, Vienna, (1986)

[35] S. Pearlstein, Data Requirements for Intermediate Energy Nuclear Applications, in Proc. of an IAEA AG Meeting on Intermediate Energy Nuclear Data for Applications, Vienna, Oct. 1990, IAEA-TECDOC, to be published

[36] P. Jung, J. Nucl. Mater. 144, 43 (1987)

[37] E.R. Hodgson, in Proc. Int. Panel on 14-MeV Intense Neutron Sources Based on Accelerators for Fusion Materials Study, Jan. 1991, Tokyo, (Eds. A. Miyahara and F.W. Wiffen), NIFS-WS-2, p. 41, (1991)

[38] S.M. Qaim, in Proc. IAEA AG Meeting on Nuclear Data for Fusion Reactor Technology, Dec. 1986, Gaussig, DDR, IAEA-TECDOC-457, p. 89, IAEA, Vienna (1988)

[39] S.M. Qaim, Nucl. Phys. A458, 237 (1986)

[40] R.A. Forrest, Systematics of Neutron-Induced Threshold Reactions with Charged Products at about 14.5 MeV, AERE R 12419, Harwell Laboratory (1986)

[41] A.B. Pashchenko, Proposal for an IAEA Coordinated Research Program on Helium Production Cross Sections, Invitational Letter for Participation, Dec. 1990

[42] S. Pearlstein, J. Nucl. Energy 27, 81 (1973), and S. Pearlstein, Programm THRESH (Dec.1975); code available from NEA Data Bank, Saclay, France

[43] P. Jung, J. Nucl. Mater. 174, 253 (1990)

[44] M. Blann, Code ALICE/85/300, UCID-20169, Lawrence Livermore National Laboratory (1984)

[45] S. Cierjacks, P. Oblozinsky and B. Rzehorz, Nuclear Data Libraries for the Treatment of Sequential (x,n) Reactions in Fusion Materials Activation Calculations, KfK-4867, Kernforschungszentrum Karlsruhe, April 1991

[46] M. Blann, Code ALICE, version 1990, private communication, January 1991

[47] V. Goulo, Report on the IAEA AG Meeting on Nuclear Data for Fusion Reactor Technology, Gaussig, GDR, Dec. 1986, INDC/P(87) - 3, IAEA/NDS, 1987

[48] F.M. Mann, Status of Dosimetry and Activation Data, Proc. Int. Conf. Nuclear Data for Science and Technology, Mito, June 1988, p. 1013, Japan Atomic Energy Research Institut, 1988

[49] D.J. Boerman, R. Dierckx, A Dupasquier, V. Sangiust and P. Tricherini, in Proc of an IAEA AG Meeting on Nuclear Data for Radiation Damage Assessment and Related Safety Aspects, Sept. 1989, Vienna, IAEA-TECDOC-572, p.91, IAEA, 1990

[50] Proc. Int. Fusion Materials Irradiation Facility (IFMIF) Workshop for the International Energy Agency, Febr. 1989, San Diego, USA, (Ed. R. Verbeek) Commission of the European Communities, DG XII, (1989)

[51] Special Issues of NS&E on High Intensity Fast Neutron Sources and Calibrated Neutron Fields for Fusion Technology and Fusion Materials Research, (Guest Ed. S. Cierjacks) Nucl. Sci. Engin. Volume 106 Nos. 2 and 3, Oct. and Nov. 1990

[52] Special Issues: IEA Workshop on an Int. Fusion Materials Irradiation Facility, (Ed. R.A. Krakowski), J. Fusion Energy 8, Nos. 3/4, Dec 1989

[53] D.G. Doran, F.M. Mann and L.R. Greenwood, J. Nucl. Mater. 174, 125 (1990)

[54] F. H. Coensgen, T.A. Casper, D.L. Correl, C.C. Damm, A.H. Futch, B.G. Logan and A.W. Molvik, Nucl. Sci. Eng. 106, 138 (1990)

[55] S. Cierjacks, Y. Hino and M. Drosg, Nucl. Sci. Eng. 106, 183 (1990).

[56] G.L. Versamis, G.P. Lawrence, T.S. Bhatia, B. Blind, F.W. Guy, R.A. Krakowski, G.H. Neuschaefer, N.M. Schnurr, S.O. Schriber, T.P. Wangler and M.T. Wilson Nucl. Sci. Eng. 106, 160 (1990)

[57] A.L. Trego et al., Fusion Materials Irradiation Test Facility: A Facility for Fusion Materials Qualification, in Proc. 5th Topical Meeting on Technology in Fusion Energy, April 1983, Knoxville, USA, (1983)

[58] J.M. Perlado, M. Piera and J. Sanz, J. Fusion Energy 8 Nos. 3/4, 181 (1989), and literature cited therein

[59] M. Drosg and O. Schwerer, Production of Monoenergetic Neutrons between O.1 and 23 MeV, in Handbook of Nuclear Activation Data, IAEA Report Series No 273, p. 83, IAEA (1987)

[60] F.M. Mann, F. Schmidtroth and L.L. Carter, HEDL-TC-1459, Westinghouse Hanford Company (1981)

[61] S. Cierjacks, Y. Hino, F. Raupp, L. Buth, D. Filges, P. Cloth and T.W. Armstrong, Phys. Rev C36, 1976 (1987)

[62] W.B. Amian et al., Validation Measurements of Neutrons from (p,xn) Reactions of Bombarding Energies at 597 and 800 MeV, KFA-IKP Annual Report 1990, Forschungszentrum Jülich, to be published

[63] K. Noda, Present Status and Future Plans of Nuclear Data for Neutron Energies up to 50 MeV and Development of Damage Parameter Calculation Codes, presented at a Meeting of the IEA Working Group on High-Energy, High-Flux Neutron Sources, April 1991, Culham, UK, 1991

[64] IAEA Advisory Group Meeting on Intermediate Energy Nuclear Data for Applications, Oct. 1990, Vienna, IAEA-TECDOC, to be published

[65] M. Victoria, Paul Scherrer Institute, Switzerland, private communication 1991

[66] T.W. Armstrong, P. Cloth, B.L. Colborn, D. Filges and G. Sterzenbach, in AECL-8488, p. 205, Atomic Energy of Canada Ltd. (1984)

[67] W.A. Coleman and T.W. Armstrong, Nucl. Sci. Eng. 43, 353 (1971)

[68] M. Blann, Needs for Experiment and Theory in Intermediate Energy Reactions, in Proc. of an IAEA AG Meeting on Intermediate Energy Nuclear Data for Applications, Vienna, Oct. 1990, IAEA-TECDOC, to be published

OBJECTIVES FOR LOW ACTIVATION MATERIALS AND THE DATA IMPLICATIONS

R.A. Forrest* and G.J. Butterworth

AEA Fusion, Culham Laboratory, Abingdon, Oxon., OX14 3DB, U.K.
(Euratom/UKAEA Fusion Association)

Abstract: The development of materials for the various regions of a fusion reactor such that the engineering property requirements, the nuclear constraints imposed by the need for adequate tritium breeding and the minimisation of residual induced radioactivity are simultaneously satisfied presents a technological challenge. Moreover, the long timescale for commercialisation of fusion energy offers an opportunity for developing materials specifically for fusion applications. In this paper the nuclear data that are needed to develop low activation materials are reviewed and some scenarios in which the use of these materials could be advantageous are considered. The nuclear data needs concentrate on the cross sections of neutron induced reactions for a wide range of target nuclides, although decay and radiological data are also discussed. The adequacy of the current data and strategies for improvement are reviewed.

Introduction

This paper considers the problem of designing low activation materials (LAMs) for future magnetic fusion reactors and the current activities to develop nuclear data libraries and inventory codes to predict how such materials will behave after irradiation. The need for and status of the data are reviewed and the considerations involved in defining criteria for LAMs are discussed.

It is planned that the initial demonstration and power producing fusion reactors will be based on the tokamak principle utilising the D-T reaction. One important distinguishing feature of fusion from fission is the fact that there is no radioactive product, and thus apart from the use of tritium as an intermediate fuel there is the potential for environmentally friendly energy production. However, as is well known the neutrons produced in the D-T reaction will interact with the reactor materials and the activation that they cause can be at a very high level. This is not inevitable, and by a careful choice of materials it may be possible to design a reactor which is environmentally benign, having the desired radiological properties for safety, routine operation, decommissioning and long-term disposal.

The long lead-time before the likely deployment of commercial fusion power gives ample time to develop the optimum materials for fusion systems, and it is essential that adequate effort be allotted to this task to ensure that the technology will be safe and environmentally acceptable and thus achieve its commercial as well as its scientific potential.

Low Activation Materials

Although it is easy to talk about low activation materials, it is in fact much more difficult to give a rigorous definition to the term. As this paper will discuss in more detail later, there can be conflicting radiological objectives that materials have to satisfy. For example, if a material is designed to have an activity and an associated γ-dose rate that is suitable for recycling at 100 years after service in the first wall then it may well be that its short-term properties relevant for maintenance are worse than those of a conventional material which would, however, be impossible to recycle.

Areas where LAMs may be of benefit

The areas where low activation materials may be advantageous cover safety, maintenance and waste disposal.

'Safety' refers to the need to design a fusion reactor so that the possibility of a serious accident is reduced to the minimum by ensuring that the amount of decay heat from materials is low, so that in the event of a loss of coolant accident there is little energy available to disrupt the layers of protection. In the event of a breach of containment, as few nuclides as possible with harmful biological impacts should be released.

'Maintenance' covers the protection of the staff and machines in the plant during operations to replace components e.g. the first wall. It is very design dependent and thus it is difficult at this stage to use it to define materials choice. Coolants will be required and the interaction of these with their containers must be understood to minimise radioactive corrosion products.

'Waste disposal' covers the longer term, when the reactor will require decommissioning. Sensible material choice in the design can make this operation much less costly and help in reducing the cooling time or the treatment of some of the materials before they can be disposed of or recycled. For many materials recycling will not be cost effective and it will be necessary to permanently dispose of them.

It is likely to be impossible to design a material for a particular application in the reactor that has optimum radiological properties for all the considerations above. However, using criteria similar to those outlined below it will be possible to optimise the material properties of various components in the reactor to satisfy some of the objectives. Before discussing the criteria for LAMs, it is necessary to consider the nuclear data needed for the calculation of activation. Calculations have to be relied on rather than integral measurements as there are at present no 14 MeV sources of sufficient intensity for the necessary experiments.

Data and calculational methods

In order to predict the isotopic composition of materials following irradiation in the various regions of a fusion reactor, it is necessary to use an inventory code which inputs data on cross sections and decay properties of all the nuclides involved and solves the following set of simultaneous differential equations:

---

*In Tec, Harwell Laboratory, Oxon., OX11 0RA, U.K.

$$\frac{dN_i}{dt} = -N_i (\lambda_i + \sigma_i \phi) + \sum_j (\lambda_{ij} + \sigma_{ij}\phi)N_j + \sum_k N_k \sigma^f_k \phi Y_{ik}$$

where N_i is the number of atoms of nuclide i, λ_i is the decay constant for nuclide i, σ_i is the effective total cross section (in the particular neutron spectrum) for removal of nuclide i, λ_{ij} is the decay constant for formation of nuclide i from nuclide j, σ_{ij} is the cross section for formation of nuclide i from nuclide j, σ^f_k is the fission cross section of nuclide k, Y_{ik} is the fission yield of nuclide i from nuclide k and ϕ is the neutron flux. The last term is only needed if fissionable nuclides are present. The equations have initial value boundary conditions and when solved give the values of N_i at some required time. For a realistic solution to this set of equations in the high fluxes of a fusion reactor, there must be cross sections for radioactive nuclides that are not initially present, but are formed by reactions.

The inventory code FISPACT

The inventory code used in the UK for fusion activation calculations is FISPACT, which is well documented [1]. In FISPACT the flux and spectrum defined in the equation above are fixed input quantities and do not change during the course of the period of irradiation. Strictly the set of equations are solved for a point in the material, in practice the reactor is divided into regions over which it is sensible to define a constant average flux.

FISPACT requires a large amount of data to be input before an inventory calculation can be done. The data required are: a cross section library, a neutron spectrum in the same energy group structure as the cross section data, a decay data library containing files for every nuclide that can be generated by reactions or decays and subsidiary libraries of biological hazard data.

Data libraries

In this section the libraries used by FISPACT are described. The number of nuclides which are considered in the data libraries is large. For UKDECAY3 [2], the currently available UK library, 1356 nuclides are present (of which 266 are stable) and for the radioactive nuclides, data are present on half-life, decay mode, average decay energies and γ-ray spectrum.

The cross section library contains data in a suitable energy group structure (100- or 175- energy groups) for all neutron induced reactions that are energetically possible, on target nuclides that are stable or have a half-life of longer than 0.5 days. In UKACT1.5 a total of 9650 reactions are given.

In addition to information on the number of atoms, FISPACT also generates radiological data for each nuclide, such as activity and the biological hazard due to ingestion and inhalation. The latter require data on the CEDE coefficients [3] that convert an activity to a dose inside the human body.

In the UK programme of work on assembling these data the view has been taken that completeness of the libraries is paramount. The fact that many of the early calculations of activation gave disparate results is explained by the exclusion of many of the nuclides or reactions from the data libraries.

Cross section data

For the European Activation File (EAF) [4] developed by ECN Petten and Harwell, the set of cross sections shown in Figure 1 is considered for each target nuclide. The treatment of isomers is extremely important in activation libraries. Many isomers have half-lives sufficiently long for the isomer to have to be considered as a target, while many isomers have different decay modes to the ground state nuclide. For these reasons, in all cases where the daughter nuclide of a reaction has an isomeric state (in general up to two isomers can be considered for each nuclide) the reaction cross section is split into a part to the ground state and the remainder to the isomer.

So far as the production of product nuclei is concerned both the reactions (n,t) and (n,nd) can be summed together as they yield the same nuclide. However, in the first reaction tritium is also produced while in the second the gas is deuterium. It is therefore necessary to list separately all reactions to the same daughter nuclide.

	N-2	N-1	N	N+1
Z	(n,3n)	(n,2n)	(n,n')	(n,γ)
Z-1	(n,nt)	(n,t) (n,nd)	(n,d) (n,np)	(n,p)
Z-2	(n,nα)	(n,α) (n,nh)	(n,h)	(n,2p)

Figure 1 Standard reactions for each target nuclide.

In the early cross section libraries little emphasis was placed on the capture reactions and many were neglected. The argument was that at neutron energies of about 14 MeV the capture cross sections were small ($\approx$ 1 mb) compared to the threshold reactions such as (n,2n). However, when realistic neutron spectra were calculated, even for near plasma components such as the first wall, the number of well-moderated neutrons was considerable and the capture reactions were seen to be much more important, especially for high atomic mass nuclides. Consequently a major effort to improve the capture data has been undertaken by the ECN group and these data will be available in the near future in the new EAF library.

Experimental data and calculational methods

A major deficiency in the early cross section libraries was the lack of data for radioactive targets. These are essential in the high neutron fluxes expected in fusion reactors ($> 10^{15}$ ncm^{-2}s^{-1} in the first wall).

The main problem with this philosophy of completeness is the lack of experimental data. With so many reactions to include in the library, the initial response was to employ simple model codes and systematics. By collecting the experimental data for a particular reaction and doing a least squares fit to an expression involving only Z and A, a good fit to the data can be achieved. In [5] this procedure is reported for (n,p), (n,α), (n,d), (n,t) and (n,h) reactions. In the (n,p) case data for 150 nuclides were used to give the fit to the 14.7 MeV cross section, using 4 parameters as shown below, where $s = (N-Z)/A$.

$$\sigma_{np} = 7.567 (A^{1/3} + 1)^2 \exp (-28.80s - 59.24s^2 + 0.2365A^{1/2})$$

By considering the spread of the data points it is possible to define an error factor φ such that the 1-standard deviation limits on σ_{np} are $\sigma_{np}\varphi$ and σ_{np}/φ, in this case $\varphi = 1.5$. Similar error factors were defined for the other reactions.

The accuracy of the data in the cross section libraries could be increased substantially if better calculational methods could be used for the cross sections for which there are no experimental data. To check how the various codes will perform when there are no experimental data, an intercomparison has been organised to calculate the $^{60}Co(n,p)$ cross section. The results from this study should show the spread of theoretical predictions and give some indication of the accuracy that could be expected for other unmeasurable nuclides. Another initiative to improve the data in libraries has been undertaken within the FENDL project. This involves the selection of 256 important reactions [6] and a graphical comparison of existing data and new calculations.

Current cross section data libraries for activation

REAC - This was the first of the libraries for fusion activation applications that included reactions on many radioactive targets. The current version (REAC3) described in [7] considers about 500 target nuclides and includes ENDF/B-VI evaluations.

EAF - This was originally based on the REAC library but has now been improved to make it complete. EAF-2 contains reactions on all nuclides with $T_{1/2} > 0.5$ days (667), and includes the best available data for capture reactions. More details about the status of the EAF libraries are given in [4].

UKACT - These libraries have the status of 'prototypes' for the EAF libraries. The current version (1.6) includes actinide targets (up to and including Cm).

The Japanese library [8] is designed for low flux applications where sequential reactions are not important (no radioactive targets are included). However, for the approximately 2000 reactions treated the data are extremely good, being based on detailed theoretical calculations.

FENDL - The activation part of FENDL-1 is not a complete library, but rather the best data that are available from all the other internationally available sources for a subset of 256 important reactions.

Status of classes of reactions

All the above libraries cover the same classes of reactions, and the status of each is given below.

(n,2n) There are many measurements at 14.5 MeV, and the systematic formulae available at this energy are reliable. Measurements have been made over the whole energy range and theory can reproduce the shape of the cross section well. For radioactive targets the predictions are probably accurate to about 25%.

(n,3n) There are fewer data available, and because of the generally steep shape of the excitation function at 14.5 MeV, the value is probably only accurate to a factor 2-3, but in cases where measurements have been made the accuracy is about 25%.

(n,γ) The new work means that data for this reaction are now more reliable. If however, the thermal part of the cross section is very important then for unmeasured targets the estimate could still be inaccurate. In other cases an overall accuracy of about 50% would be assigned unless a detailed evaluation or measurement were available.

(n,p) The systematic formulae for this reaction have already been mentioned, and on the basis of this an error of 50% is assigned. It should be noted that in cases where there have been measurements made at 14.5 MeV the accuracy is about 5-10%.

(n,α) This is also a well measured reaction, although it is more difficult to calculate and the shape of the cross section at low energies cannot always be reproduced well. The systematic formula is accurate to about 60% for unmeasured targets, although in other cases an accuracy of about 10-25% is possible.

(n,t) There is much less data available for (n,t) reactions and most of this is at 14.5 MeV (although the work by Wölfle et al [9] extends up to 19 MeV). The systematic formula is quoted as accurate to about 65%, although it might be safer to use a factor of 2.

(n,h) Only about 13 nuclides have been measured (up to 1985) and the data at 14.5 MeV are rather poor. The accuracy predicted by the systematic is about 90% (again safer to use a factor of 2).

(n,d), (n,np), (n,pn) All these reactions give the same daughter, and systematics for the sum of these have been given. The measured data at 14.5 MeV fall into two categories, depending on the relative magnitudes of the neutron and proton separation energies. The predictions for an unknown target are about a factor of 2.

(n,nα), (n,nt), (n,2p) The number of measurements on these two particle emission reactions is extremely limited and the systematics have not been investigated. Factors of 3-5 uncertainty would be expected; this would be an excellent area for some new measurements so that a systematic, however crude, could be constructed.

(n,n') The inelastic scattering cross section is only required for activation studies if there is an isomeric state. It is estimated that an accuracy of 50% can be assigned.

Branching ratios for the production of isomers have been measured for some reactions but systematics e.g. [4] must be used for others. For the case of the (n,2n) reaction and moderate isomer spin an accuracy of 60% is estimated, while for the other reactions it is probably a factor of 2. Even in the case of very high spin isomers the systematics are probably good to a factor of 3-5.

Recent experimental results on cross sections

The large number of cross sections in the current libraries (about 10,000 reactions) means that it is not possible to review the data in detail. It is useful, however, to note some areas in which current measurement programmes are contributing to the development of the libraries.

A particular problem already discussed is the formation of isomers. At present many of the branching ratios between the ground state and the isomer are determined by use of a systematic formula. This predicts the branching ratio for various reactions based on the spin of the isomer. In cases where the isomer spin is not too large then branching ratios [$b = \sigma_m/(\sigma_m+\sigma_g)$] are close to 0.5. However, as the spin gets larger than about 7 the branching ratio falls rapidly (e.g. $b = 0.004$ when $J^m = 12$). Tests of this systematic are therefore extremely valuable and two examples are discussed below.

An extreme case of a high spin long lived isomer is ^{178m}Hf [$J^\pi = 16^+$, $T_{1/2} = 31$ y]. Significant formation of this from W or Ta could restrict the use of these elements. The results at about 14.5 MeV of a measurement by Patrick et al [10] are shown in Table 1.

Table 1 Branching ratio data for some Hf isotopes

Reaction	Measured	Calculated	Systematic
^{179}Hf(n,2n)^{178m}Hf	2.8 10^{-3}	1.4 10^{-3}	1 10^{-3}
^{180}Hf(n,2n)^{179m}Hf	8.5 10^{-3}	3.7 10^{-3}	24 10^{-3}

The results in column 3 refer to GNASH calculations by Chadwick and Young [11]. Prior to this work there was considerable doubt that theory would be able to give a reasonable prediction for such high spin isomers. The general conclusions are that both the theory and systematics are quite good even for such high spins.

The study of production of isomeric states has been successfully performed by Qaim and co-workers. Many reactions such as ^{46}Ti(n,p)^{46m}Sc [12] have been measured over the energy range 4-11 MeV. Of particular interest is the ^{90}Zr(n,p)^{90m,g}Y [13] reaction since the isomer spin is quite large [$J^{\pi} = 7^{+}$] and thus can test the systematic formula. The measured isomer ratio is 0.25 and the prediction from the systematics is 0.22.

Improvement of cross section data

Since the present cross section libraries contain a large number of reactions it is important to be able to rank the reactions so that any improvements to the library are made efficiently. To do this, the dominant reactions have to be identified and a systematic approach is necessary for all but the simplest cases.

FISPACT contains an option to work out the important pathways in the production of a final nuclide. In this context a pathway is a set of nuclides, including the parent and final nuclide, linked by reaction and decays in a linear fashion. The code will identify all the possible pathways and calculate the percentage of the final nuclide formed via each pathway. An example is the production of ^{41}Ca from ^{48}Ti in first wall conditions which is dominated by the following pathway,

$$^{48}Ti(n,\alpha)^{45}Ca(\beta^{-})^{45}Sc(n,\alpha)^{42}K(\beta^{-})^{42}Ca(n,2n)^{41}Ca$$

which contributes 99.1% of all the ^{41}Ca produced from ^{48}Ti.

The result of such work is the list of 256 reactions mentioned above, and it is interesting to note that using the current data libraries a rare reaction type such as (n,h) or (n,2p) is occasionally shown to be important. It may be that the data for these rare reactions are overestimated and that more accurate data would remove them from the list but they may be genuinely important. In either case it is necessary to significantly improve the calculational methods or do careful experimental measurements on a few nuclides so that a systematic formula can be developed. This programme is iterative, as with better values for some of the poorly known reactions, the set of dominant reactions may change.

Additional reaction types

Recently the possible importance of sequential charged particle reactions has been investigated by Cierjacks and co-workers [14]. The effect being considered is the interaction of the charged particles, formed from the primary interaction of the neutrons, with the material. This two-step process can be represented by a pseudo cross section which can formally be used in the same fashion as a normal cross section.

In order to calculate this effect, data libraries are required for: the spectrum of charged particles (p, α and d have so far been considered) at a set of incident neutron energies, the cross section, e.g. (p,n), for the interaction of the charged particles with target nuclides and the slowing down of the charged particles in the material.

Preliminary calculations using a small set of 10 pseudo cross sections for a particular reactor region have shown that in some special cases (such as sodium) the effect can be extremely large, although in iron the effect is masked by the neutron induced reactions. Until realistic calculations have been carried out using a full set of pseudo cross sections, the practical importance of this effect cannot be judged.

The inclusion of actinides also increases the scope and complexity of the library. Uranium and thorium are present as impurities in materials that will be used in fusion reactors and the effect of fission of these impurities should be included. Work at Harwell is in hand to compile cross section data for about 50 actinides as targets and to increase the number of nuclides in the decay and biological hazard libraries. FISPACT also requires modification to read fission yield data and to include the fission channel within its calculational method.

It is believed that with the inclusion of the two types of additional reactions described above, the use of current data and codes should enable reliable calculations of inventories to be performed. There are not likely to be any 'surprises' left that radically change conclusions on materials choice.

Uncertainties

Currently none of the cross section files contain values for the uncertainties in the data. It would, however, be useful to be able to give estimates of the uncertainty, in, say, the activity of a material as a function of cooling time. At present within FISPACT it is possible to calculate sensitivity coefficients and use these with uncertainties of the cross section, to estimate the uncertainty in the number of atoms of a nuclide. The calculation of sensitivity coefficients is computationally time consuming and it would be unrealistic to attempt to calculate estimates for all nuclides. However, it should be possible to routinely calculate a small number of the sensitivity coefficients for the dominant nuclides, to estimate uncertainties on the total quantities such as activity.

Such a scheme demands a file of uncertainty data for the cross sections and it is planned that a preliminary version of this will be contained in EAF-3, which should be available at the start of 1992.

In conclusion, significant advances have been made in the cross section libraries, especially for capture reactions and isomer ratios and, although many deficiencies still need to be remedied, it is increasingly necessary to look at uncertainties outside the activation cross section database. There are large uncertainties in the composition of reactor materials, operational conditions, neutronic calculations and in assessment of the radiological consequences of activation. Radioactive inventories and their radiological implications are also dependent on reactor design and, until designs are more advanced, this feature remains a source of major uncertainty in LAM design.

Applications of LAMs

Strategies for design of LAMs

An important general method of improving the radiological properties of alloys is to substitute elements which have similar metallurgical properties but lower radiological impact for some of the alloying elements present in conventional materials. An alternative approach is represented by the development of novel alloys based on V or Cr rather than Fe, or materials such as ceramics, cermats and composites. The new materials will require extensive testing to ensure that their strength, ductility, joining properties and radiation resistance are adequate. Some examples of these approaches are noted below.

Austenitic stainless steels typically containing around 18% Cr and 12% Ni, with fcc structure, such as type 316 can be modified to give low activation versions by replacing the Ni by Mn. An example is OPTSTAB. However, the high void swelling in Cr-Mn steels appears to rule them out for first wall applications, though not elsewhere.

Martensitic steels such as FV448, which can be strengthened and hardened by heat treatment, can be changed to give low activation versions by replacing Mo and Nb strengthening additions with W and/or V and Ta. An example is LA12Ta.

The control of impurities in alloys is also extremely important, especially for the long-term dose rate. By setting limits on the γ-dose rate, acceptable concentrations of the elements in particular regions of the reactor can be calculated [15]. Table 2 shows the limits on impurities in a typical first wall and clearly for some elements such as Ag, Ho, Tb, Dy and Bi, the restriction is severe. It is therefore necessary to determine whether such low levels of impurities are feasible in the production of LAMs.

Table 2. Limits on impurities in first wall.

Permitted amount	'hands-on' limit $<25\ \mu Svh^{-1}$ at 100 y	'recycling' limit $<10\ mSvh^{-1}$ at 50 y
< 0.1ppm	Ag, Ho, Tb, Dy	
0.1 - 1 ppm	Bi, Nb, Pd, Gd, Sm, Eu	
1 - 10 ppm	Co, Ir, Er, Cd, Cs, Xe, Os	Co, Ag, Sm, Eu
10 - 100 ppm	Pt, Ba, Nd, Hf, Mo, Lu, Ar	Tb, Ho, Bi, Dy
100 - 1000 ppm	Al, K, Kr, Re, Ni, Tm, Cu	Nb, Pd, Cs, Gd, Nd, Ni, Cu, Ba, Ir
1000 ppm - 1%	Yb, Ca, Pb, Zr, Ru, Br, W, Cl, Sc, S, Rb, La, Ti	Xe, Er, Cd, Os, Pt, Hf, Kr, Lu
1 - 10%	Ta, Si, Sr, Sn, Y, Ce, Te	Mo, Rh, Ar, Al, K, Rb, Re, Zn
10 - 95%	In, Zn, P, Fe, Au, Rh, Hg	Fe, Yb, Na, Ca, Pb, Tm, Zr, Sc, Br, Sr, Ru, La, W, Sn, Sb, Cl
No limit	Remainder	Remainder

One requirement is the ability to measure the amount of impurity accurately at the ppb level. A particularly valuable technique is that of glow-discharge mass spectroscopy [16]. Another requirement is that steel makers can produce materials to the purity required. Considering the 100 years 'hands-on' limit, of the 17 most restricted nuclides, only Nb and Co are commonly assayed in bulk steels and even then only down to concentrations still well above the permitted 'hands-on' levels. More relevant experience of high purity material production is found in the 'superalloys' industry. In the production of these alloys the analysis would normally cover 44 elements, many at concentrations below 1 ppm. Many of the element concentrations attained in nickel-based superalloys are close to the limits required for LA steels and therefore by following similar production routes (advanced vacuum melting and refining) it should be possible to produce LA steels to the specification required.

A further technique that may hold promise for LAMs in certain cases is 'isotopic tailoring'. This would involve changing the proportions of the stable isotopes in an element from the naturally occurring values. Such a process will obviously only be applicable to elements with more than one stable isotope, which excludes elements as Al, V (^{50}V is only at 0.25%), Mn, Nb and Ta (^{180}Ta is only at 0.012%). In other elements it is possible to identify which of the stable isotopes is responsible for the production of undesirable nuclides. In many cases, however, nuclides responsible for long-term and short-term problems are produced from different stable isotopes and it might be practically difficult to satisfy both criteria. However, this is an area where a great deal of detailed work still remains to be done, although the likely cost of such an option would limit its use to special applications.

Importance of the neutron spectrum

Many studies of LAMs have concentrated on the components close to the plasma, such as the first wall and blanket. However, it must be remembered that while these will experience the most extreme conditions of neutron loading, their mass is only a small fraction of the whole reactor. In the case of the EEF reactor [17] about 0.6% of the mass (excluding concrete) is in the first wall while about 50% is in the shield. For regions away from the plasma the general trends in the neutron spectrum are: decrease in the 14 MeV fusion peak, increase in the moderated neutrons and an overall reduction in flux. Under these conditions the 1-group effective cross sections for threshold reactions are reduced, but the capture reactions are typically increased significantly. In [15] the calculations summarised in Table 2 are repeated for the region of the shield and while the general ranking of elements is similar in both cases there are detailed changes. The reduced value of the flux in the shield means that the levels are generally less restricted than in the first wall, although Eu is more tightly constrained in the shield compared to the first wall (0.03 ppm shield, 0.87 ppm first wall at 100 years cooling).

Some proposed criteria for LAMs

Some attempts to define objectives for LAMs have been made, as discussed below, although more work is required to understand the problem.

Accidents

Piet et al [18] have proposed four 'Figures of Merit' (FOM) for accidents. As an example, the prompt dose rating FOM requires that 100% release of inventory gives < 2 Sv prompt dose at the site boundary.

Rocco and Zucchetti [19] propose that the early dose at 1 km due to release of 1 kg of irradiated material < 50 mSv.

Maintenance

It is recognised that 'hand-on' maintenance of the fusion core is not a realistic goal for any LAM. However, hands-on maintenance for regions outside the blanket may be attainable, but will depend on careful choice of coolants so as to reduce activated corrosion products. Rocco and Zucchetti propose that the dose rate inside the plasma chamber $< 10^4$ Gyh^{-1} after 1 day cooling, based on considerations of dose to electronic components.

Recycling

A great deal of work on the possibility of recycling has been done in the UK e.g. [20], where issues of conservation of resources, reduction in volume of activated material and reduction of environmental impact have been addressed.

Two criteria have been proposed : the 'hands-on' limit of 25 μSvh^{-1} after 100 years cooling and the 'recycling' limit of 10 $mSvh^{-1}$ after 50 years cooling. The latter criterion refers to the ability to remotely remelt and refabricate components for reuse in a reactor.

Waste Disposal

Fetter et al [21] have studied the problem of waste disposal by considering the US regulations for shallow burial of low-level fission wastes (10CFR61). The methodology has been extended to calculate the specific activity limits for a range of nuclides relevant to fusion.

In the UK it is likely that much of the reactor when decommissioned will be placed in a geological repository. Using models of mobilisation and transport of nuclides back to the geosphere it is possible to calculate doses and consequently risks to the population.

In the EEF study on the environmental impact of fusion [17] comparison was made between the consequences of disposal of an entire reactor after decommissioning in a cementitious repository. Two reactors were considered, one using standard steels such as 316L and a 'low activation' version which was constructed from LAMs considered to date such as vanadium alloys. The broad conclusion of such a study is that in both cases disposal is 'safe' and that there is therefore no real advantage from the use of LAMs. However, two reservations on this conclusion must be made clear: firstly the LAMs used have not been optimised for post-disposal impact, and it seems reasonable to suspect that if this was done then LAMs could indeed give lower impact. Secondly, regulations change and public opinion would certainly demand that if LAMs could make geological disposal 'safer' at reasonable cost then these materials should be preferred. As a general conclusion the benefits of LAMs in this area are not proven, and a lower priority should be given to satisfying criteria at this timescale rather than the more important short- to mid-term timescale.

Conclusions

The nuclear data needed to develop LAMs are now well defined and data libraries offering completeness are available. Specific areas for improvement of the data are being identified and the next generation of libraries should enable inventory codes that can treat multistep reactions to make calculations of inventories to the reliability needed by the materials designer.

The status of the various classes of cross section data are reviewed and the main conclusion is that the available advanced calculational methods need to be applied in a 'mass production' fashion for radioactive targets where there are no experimental measurements. More measurements on the formation of isomeric states (especially with high spin) are needed to confirm the systematics that are used for the branching ratio to the isomer. Also, measurements on the rarer reactions such as (n,h) and (n,2p) would be valuable to establish reliable systematics.

Criteria for LAMs are often in competition and at the present time simplified criteria are being developed. It is clear that a range of LAMs will be needed in different regions of the reactor to exploit advantages over different timescales. The proposed criteria covering safety, maintenance, decommissioning, recycling and permanent disposal are considered and are found to be too narrowly focussed to provide clear guidance on material design. Though LAMs do not offer much benefit for geological disposal, current studies suggest that they will be most beneficial in situations where some design option, impossible with conventional materials, can be considered. Areas might include: achievement of passive safety designs, reduction of chronic effects following a major accident, easier maintenance, possibilities of recycling and the avoidance of large volumes of material requiring geological disposal.

References

1 R.A. Forrest and D.A.J. Endacott FISPACT-User Manual, AERE-M-3654, Harwell Report (1988)

2 R.A. Forrest, M.G. Sowerby, B.H. Patrick and D.A.J. Endacott, in Proc. Int. Conf. on Nuclear Data for Science and Technology, May/June 1988, Mito, Japan (Edited by S. Igarasi), p. 1061, Saikon, Tokyo (1988)

3 K.R. Smith, Dosimetric Data for FISPACT, CLM-R299, Culham Report (1990)

4 J. Kopecky, H. Gruppelaar and R.A. Forrest, European activation file for fusion, This Conference, IP19

5 R.A. Forrest, Systematics of neutron-induced threshold reactions with charged products at about 14.5 MeV, AERE-R-12419, Harwell Report (1986)

6 R.A. Forrest in Proc. Spec. Meeting on FENDL, May 1989, Vienna, Austria (Edited by V. Goulo), p.28, INDC(NDS)-223/GF (1989)

7 F.M. Mann and D.E. Lesser, REAC*3 nuclear data libraries, This Conference, C41

8 Y. Seki, H. Kawasaki, N. Yamamuro and S. Iijima, Revised graphs of activation data for fusion reactor applications, to be published as JAERI-M report (1991)

9 R. Wölfle, S.M. Qaim, H. Liskien and R. Widera: Radio. Acta 50, 5 (1990)

10 B.H. Patrick, M.G. Sowerby, C.G. Wilkins and L.C. Russen in Proc. Cons. Meeting on Neutron Activation, Sept. 1989, Argonne National Lab., USA (Edited by Wang DaHai), p. 69, INDC(NDS)-232/L (1990)

11 M. B. Chadwick and P.G. Young in Proc. Spec. Meeting on Neutron Activation Cross Sections for fission and fusion applications, Sept 1989, Argonne National Lab., USA (Edited by M. Wagner and H. Vonach), p. 241, NEANDC-259 'U'

12 N.I. Molla, S.M. Qaim and M. Uhl: Phys. Rev. C42, 1540 (1990)

13 S.M Qaim, M. Ibn Majah, R. Wölfle and B. Strohmaier: Phys. Rev. C42, 363 (1990)

14 S. Cierjacks and K. Ehrlich, Nuclear data for fusion materials research, This Conference, IT6

15 R.A. Forrest, M.G. Sowerby and D.A.J. Endacott, in Proc. 16th Symposium on Fusion Technology, Sept. 1990, London (1991)

16 G.J. Butterworth: Fusion Eng. and Des., 11, 231 (1989)

17 M.G. Sowerby and R.A. Forrest, A study of the environmental impact of fusion, AERE-R-13708, Harwell Report (1990)

18 S.J. Piet, E.T. Cheng and L.J. Porter: Fusion Tech., 17, 636 (1990)

19 P. Rocco and M. Zucchetti, Criteria for low activation materials in fusion applications, IEA Workshop - Low Activation Materials, April 1991, Culham

20 C.R. Gomer, D. Dulieu, K.W. Tupholme and G.J. Butterworth: Fusion Eng. and Des., 11, 423 (1990)

21 S. Fetter, E.T. Cheng and F.M. Mann: Fusion Eng. and Des., 6, 123 (1988)

NUCLEAR DATA NEEDS AND STATUS FOR FUSION REACTOR TECHNOLOGY

E.T. Cheng
TSI Research
225 Stevens Avenue, Suite 110
Solana Beach, California 92075, U.S.A.

D.L. Smith
Argonne National Laboratory
9700 South Cass Avenue
Argonne, Illinois 60439, U.S.A.

Abstract: A review was performed on the needs and status of nuclear data for fusion reactor technology. Generally, the status of nuclear data for fusion has been improved due to the dedicated effort of data developers in the past two decades. However, there are still deficiencies in the nuclear data base, particularly in the areas of activation and neutron scattering cross sections. Activation cross sections were found to be unsatisfactory in 83 of the 153 reactions reviewed. The scattering cross sections for fluorine and boron will need to be improved at energies above 1 MeV.

[fusion, data needs, status review, dosimetry, tritium production, activation, neutron multiplication, neutron emission cross sections]

Introduction

Fusion energy is a very important, inexhaustible energy source for future generations of mankind. The most promising near-term fusion fuel cycle is the deuterium-tritium (DT) cycle because of its large reaction cross sections at relatively lower plasma temperatures. Deuterium-deuterium (DD) and deuterium-helium-3 (DHe-3) cycles are in consideration for long-term applications. These fuel cycles produce neutrons as primary or secondary reaction products. Interaction of neutrons with engineering components such as the blanket and shield, which surround the fusion reacting plasma, is an important process as far as energy extraction and radiation shielding are concerned. Adequate knowledge of the nuclear data is necessary for the design of these engineering components and the assessment of their nuclear performance and radiological characteristics.

The types of needed nuclear data, according to the functional requirements for fusion reactor technology are, [1]: (1) Neutron flux determination - total cross sections, double-differential cross section data [(n,n), (n,n'), (n,2n), etc.], and dosimetry cross sections; (2) Tritium production cross sections - Li-6(n,alpha)T and Li-7(n,n'alpha)T. These cross sections are needed for the most promising DT fuel cycle; (3) Neutron multiplication - Be-9(n,2n) and Pb(n,2n); (4) Radiation hazards - activation cross sections and decay data; (5) Radiation damage - cross sections such as (n,2n), (n,n), (n,n'), and (n,charged particles) that lead to the calculation of atomic displacement, helium and hydrogen production cross sections, and transmutation related activation cross sections; and (6) Fusion reactions - DT, DD, and DHe-3 fusion reaction and scattering cross sections.

The nuclear data needs and status for fusion reactor development are under frequent review [1]. Significant progress in the development of nuclear data for fusion reactor technology has recently been achieved. An important effort undertaken by the IAEA Nuclear Data Section aims to produce an international fusion evaluated nuclear data library (IFENDL) [2]. This is possible primarily because the updated versions of cross section evaluations, such as ENDF/B-VI, JENDL-3, EFF-2, and BROND files, are available for international cooperation. Activation cross sections, particularly those leading to the production of long-lived radionuclides, are also considered for the 14 MeV energy range and lower energies [3]. In this paper, the results of a review on the nuclear data necessary for the development of fusion reactor technology are presented. Extensive review efforts are focusing primarily on the status of dosimetry cross sections, activation cross sections, and neutron emission cross sections such as (n,n), (n,n'), (n,2n), and (n,n'alpha). A brief update on the candidate fusion materials is also given. The nuclear data needs for neutronics in fusion reactors will become more stringent after data testing for the recently available evaluations, and after early experimental fusion reactor design stage.

Candidate Fusion Materials

Primarily, the candidate fusion

materials were determined due to considerations of high performance characteristics such as tritium breeding and nuclear heating, and low activation. The list of candidate materials is similar to that given in the last review [1]:

    a. Structure - Fe, Cr, Ni, Mn, V, Ti, W, Ta, Al, Si, and C. Note that aluminum is desirable only in the region behind the blanket to avoid production of a large quantity of long-lived Al-26 due to (n,2n) reactions.

    b. Tritium breeder/coolant - Li, H, O, Pb, F, Be, Si, and Zr.

    c. Neutron multiplier - Be and Pb.

    d. Magnet shielding - B, C, W, Pb, H, O, Fe, Cr, Ni, and Si.

    e. Magnet - Cu, Fe, Cr, Ni, Sn, Ti, Nb, Al, N, C, O, H, and He.

    f. Plasma facing materials - C, Cu, W, Hf, Ta, Re, Be, Li, and Pb.

    g. Biological shielding - Ca, Si, Ba, O, C, H, concrete, Fe, Cr, and Ni.

    h. Hybrid blanket - Th, U, and Pu.

Status of Nuclear Data

The review was carried out using standard reference works which provide a good sampling of the status of the neutron data endeavors relevant to fusion. The sources considered are:

1. Nuclear Wallet Cards, J.K. Tuli, Brookhaven National Laboratory, 1990.

2. NCRP Radioactivity Measurements Handbook, NCRP Report No. 58, 1989.

3. Table of Isotopes, C.M. Lederer et al., 1978; Table of Radioactive Isotopes, Browne and Firestone, 1986.

4. Neutron Cross Sections, Vol. II, V. McLane et al., 1988.

5. CINDA, IAEA, Vienna, 1935-1990.

6. Summary reports on recent IAEA IFENDL and CRP activities [2,3], and NEANDC Specialists' Meeting on Activation Cross Sections.

7. Proceedings of the last three meetings of the U.S. DOE fusion nuclear data program [4].

8. Major international conference proceedings (1979 Knoxville, 1982 Antwerp, 1985 Santa Fe, and 1988 Mito).

The present collection of information and subjective comments can in no way be comparable to the in-depth review accomplished by Jarvis in 1981 [5]. However, it identifies some of the major strong and weak points in the fusion nuclear data base.

Dosimetry Cross Sections

The results of a review of dosimetry cross sections compiled in Ref. 4 and used in Ref. 6 for fusion applications are given in Table 1. Out of the 34 candidate reactions, 15 were found to be acceptable for fusion dosimetry applications because of their adequate status on decay features and cross sections. Note that 3 of the 15 cross sections are employed in Ref. 6 and should be included in future considerations. All acceptable dosimetry cross sections except, K-39(n,2n)K-38, are also considered and maintained in the International Reactor Dosimetry File (IRDF-90) [7]. Future tests of these cross sections are required to confirm their adequacy in meeting fusion dosimetry needs.

Fusion Cross Sections

As given in the previous review, the status of experimental cross section data for the DT, DD, DHe-3 fuel cycles is adequate for fusion applications [1]. A new evaluation is underway at LLNL that will incorporate the recently measured data [8]. Fusion scientists should be encouraged to use the new evaluation when it becomes available.

Tritium Production Cross Sections

These cross sections include that for Li-6(n,alpha)T and Li-7(n,n'alpha)T reactions. Li-6 is a standard material for cross section evaluation. The Li-6 (n,alpha)T cross sections were satisfactory for fusion applications. The Li-7(n,n'alpha)T reaction was subject to intensive investigations that ended with a number of accurate measurements at energies around 14 MeV and from 7 to 10 MeV. However, there is still a weak point in understanding the Li-7(n,n'alpha) reaction in the threshold region (4-6 MeV), where the energy dependence is very sharp and there have been no new measurements. New evaluations with excellent accuracies at important energy ranges for tritium production are accessible for both Li-6 and Li-7 cross sections.

Neutron Multiplication Data

These cross sections include that for Be-9(n,2n) and Pb(n,2n) reactions at 14 MeV energy range. In general, there adequate experimental data for the above cross sections. Bismuth is less desirable as a neutron multiplier for fusion. However, it offers a satisfactory status of (n,2n) cross section data. New evaluations for these cross sections are available, with isotopic evaluations for

Table 1
Neutron Dosimetry Reactions Considered for Fusion Applications

Reaction	Decay Properties	Cross Sections	Satisfactory for Fusion Applications
O-16(n,alpha)C-13	A	M	No
Mg-24(n,p)Na-24	A	A	Yes(*)
Al-27(n,p)Mg-27	A	M	No(*,**)
Al-27(n,alpha)Na-24	A	A	Yes(*)
Si-28(n,p)Al-28	A	M/I	No(**)
P-31(n,p)Si-31	I	M/I	No(*)
Cl-35(n,2n)Cl-34m	I	I	No
K-39(n,2n)K-38	A	A	Yes
Ti-47(n,p)Sc-47	A/M	M/I	No(*)
Ti-48(n,p)Sc-48	A	A	Yes(*)
Fe-56(n,p)Mn-56	A	A	Yes(*)
Ni-58(n,2n)Ni-57	A	A	Yes(*)
Co-59(n,alpha)Mn-56	A	A	Yes(*)
Cu-63(n,gamma)Cu-64	A/M	I	No(*)
Cu-63(n,2n)Cu-62	I	A	No(*)
Zn-64(n,p)Cu-64	A/M	M	No(*,**)
Zn-64(n,2n)Zn-63	I	A/M	No
Rb-85(n,2n)Rb-84m	A	I	No
Zr-90(n,p)Y-90m	I	I	No
Zr-90(n,2n)Zr-89m	A/M	I	No(*)
Rh-103(n,n')Rh-103m	I	M	No(*)
In-115(n,gamma)In-116m	A/M	A	No(**)
In-115(n,n')In-115m	A	A	Yes(*)
Au-197(n,gamma)Au-198	A	A	Yes(*)
Au-197(n,2n)Au-196	A	A	Yes(*)
Hg-199(n,n')Hg-199m	A	I	No
U-235(n,f)	(+)	A	Yes(*)
Np-237(n,f)	(+)	A	Yes(*)
U-238(n,f)	(+)	A	Yes(*)
Ni-58(n,p)Co-58	A	A	Yes(*)
Nb-93(n,2n)Nb-92m	A	A	Yes(*)
F-19(n,2n)F-20	I	M	No(*)
Ti-46(n,p)Sc-46	A	I	No(*)
Mn-55(n,2n)Mn-54	A	A	No(*,++)

A - Adequate; A/M - Adequate/Marginal; M/I - Marginal/Inadequate;
I - Inadequate. [Note - For the cases of A/M and M/I, the status may vary
depending on intended use (i.e., the energy range of interest)].
* Also shown in the International Reactor Dosimetry File (IRDF-90).
** Improvement required if needed for fusion applications.
+ On-line detection of fission reaction rate is utilized.
++ Stay away because of contamination and physical form problems.

lead isotopes Pb-206, Pb-207, and Pb-208. Data testing activities are being conducted for Be and Pb neutron multiplication factors under the IAEA/NDS coordination [9]. A recent review on the neutron multiplication integral experiment for beryllium revealed that there are inconsistencies between experiments and calculations using the current beryllium evaluations [9].

Neutron Transport Cross Sections

The neutron reactions relevant to neutron transport calculations are total, elastic (n,n), inelastic (n,n'), and (n,2n). The (n,n'alpha) reactions also need to be considered in some cases with low z elements. These cross sections together with (n,p) and (n,alpha) are important for the determination of radiation damage (atomic displacement), heat generation

(kerma), and shielding (neutron flux attenuation) in candidate blanket and shielding materials. Table 2 presents the results of a review on important candidate fusion materials. Although most of these materials, excepting fluorine and boron, appear to have adequate status of nuclear data for their applications in fusion, at least for early design studies, it should be noted that a general deficiency exists in the double-differential scattering data area. The deficiency is due to the lack of experimental data in the energy range between 10 and 14 MeV. The scattering cross sections for fluorine and boron will need to be improved, particularly at energies above 1 MeV.

Activation Cross Sections

Necessary activation cross sections needed were recently identified, with low activation as an important goal for fusion reactor development [10]. These cross

Table 2
Status of Neutron Cross Sections for Transport Calculations

Material	Reactions	Application	Status	Comments
Vanadium	(n,n),(n,n'), (n,2n)	Structure	Adequate	1,2,3
Iron	(n,n),(n,n'), (n,2n)	Structure/ Shielding	Adequate	1,2,3
Nickel	(n,n),(n,n'), (n,2n)	Structure/ Shielding	Adequate	1,2,3
Chromium	(n,n),(n,n'), (n,2n)	Structure/ Shielding	Adequate	1,2,3
Titanium	(n,n),(n,n'), (n,2n)	Structure	Adequate/ Marginal	1,2,4
Copper	(n,n),(n,n'), (n,2n)	Magnet	Adequate	1,2,3
Zirconium	(n,n),(n,n'), (n,2n)	Breeder compound	Adequate	1,2,3,5
Silicon	(n,n),(n,n'), (n,2n)	Structure/ Shielding	Adequate/ Marginal	2,3,6
Carbon	(n,n),(n,n'), (n,alpha)	Structure/ Shielding	Adequate	1,2,3
Fluorine	(n,n),(n,n'), (n,p),(n,alpha)	Breeder compound	Marginal/ Inadequate	2,3,7
Oxygen	(n,n),(n,n'), (n,p),(n,alpha)	Breeder compound	Adequate/ Marginal	1,2,3,7
Boron	(n,n),(n,alpha) (n,t'alpha)	Shielding	Marginal/ Inadequate	1,2,3,8
Tungsten	(n,n),(n,n'), (n,2n)	Shielding	Adequate	1,2,3
Beryllium	(n,n),(n,n'), (n,2n)	Neutron multiplier	Adequate	1,2,3
Lead	(n,n),(n,n'), (n,2n)	Neutron multiplier	Adequate	1,2,3
Lithium	(n,n),(n,n'), (n,alpha), (n,n'alpha)	Tritium breeding	Adequate	1,2,3

Comments: 1. Measured data is available; 2. Model calculations available; 3. New evaluation available; 4. re-evaluation needed; 5. isotopic evaluations desirable; 6. more reliable (n,2n) data needed; 7. (n,n), and (n,n') data need improvement; 8. (n,n) and (n,alpha) data need improvement at energies above 1 MeV.

sections were reviewed, weighing the status of both the cross sections and decay features of their activation products. Table 3 shows a list of activation reactions whose cross sections are adequately known, while Table 4 gives the list of activation reactions whose cross sections are insufficiently known. About 83 of the 153 reactions given in Ref. 10 are rated as unsatisfactory. Note that the radionuclides with inadequate decay features, namely half-life, decay mode, and branching ratio (in the case involving isomers), are mainly long-lived, although deficiencies of decay information exist for some unimportant short-lived radionuclides. These long-lived radionuclides are Ag-108m, Mn-53, Hf-178m2, Pt-193, Mo-93, Tc-98, Be-10, Nb-92, Nb-94, Ni-59, Fe-60, Re-186m, Hf-182, and Zr-93.

Activation cross sections leading to the production of long-lived radionuclides are very important in determining the waste disposal and recycling aspects of fusion reactor materials. Out of the 21 activation cross sections recommended for new measurements [1], 13 were considered in the recent IAEA CRP program [3]. The status of these cross sections and the decay features of their long-lived reaction products are under review by the IAEA CRP effort. Currently, there is an increased concern for the low activation fusion reactor due to the sequential charged-particle induced activation reactions [11]. A review on the status of these sequential reaction processes should be conducted when adequate information is provided.

Summary and Conclusions

The nuclear data needs and status for the development of fusion reactor technology were reviewed. The nuclear data status has been improved due to the diligent effort of data developers during the past two decades. However, there are still deficiencies in the nuclear data base, particularly in the areas of activation and scattering cross sections. Activation cross sections are

Table 3
List of Neutron Activation Reactions
Whose Cross Sections are Considered Adequate

Al-27(n,alpha)Na-24; Al-27(n,p)Mg-27; Bi-209(n,2n)Bi-208;
Co-59(n,2n)Co-58; Cr-52(n,2n)Cr-51; Cu-63(n,g)Cu-64;
Cu-65(n,2n)Cu-64; Cu-63(n,alpha)Co-60; Cu-63(n,2n)Cu-62;
Cu-65(n,p)Ni-65; Cu-65(n,g)Cu-66; F-19(n,2n)F-18; F-19(n,g)F-20;
Fe-56(n,2n)Fe-55; Fe-54(n,p)Mn-54; Fe-56(n,p)Mn-56;
Hf-179(n,2n)Hf-178m2; Mg-24(n,p)Na-24; Mn-55(n,g)Mn-56;
Mn-55(n,2n)Mn-54; N-14(n,p)C-14; Na-23(n,2n)Na-22; Na-23(n,g)Na-24;
Nb-93(n,2n)Nb-92m; Nb-93(n,2n)Nb-92; Nb-93(n,g)Nb-94;
Nb-93(n,n')Nb-93m; Nb-93(n,n'alpha)Y-89; Y-89(n,2n)Y-88;
Ni-58(n,p)Co-58; Ni-58(n,n'p)Co-57; Ni-58(n,2n)Ni-57;
Ni-57(n,p)Co-57; Ni-60(n,p)Co-60; Ni-58(n,alpha)Fe-55;
O-16(n,p)N-16; Pb-204(n,2n)Pb-203; Pb-208(n,g)Pb-209;
Re-187(n,g)Re188; Re-187(n,g)Re-188m; Re-185(n,2n)Re-184;
Ta-181(n,g)Ta-182; Ti48(n,p)Sc-48; Ti-46(n,p)Sc-46;
Ti-47(n,n'p)Sc-46; V-51(n,alpha)Sc-48; V-50(n,2n)V-49;
V-51(n,n'alpha)Sc-47; W-186(n,2n)W-185; W-184(n,g)W-185;
W-182(n,p)Ta-182; Zr-90(n,2n)Zr-89; Zr-96(n,2n)Zr-95.

Table 4
List of Neutron Activation Reactions
Whose Cross Sections are Considered Inadequate

Ag-109(n,2n)Ag-108m; Ag-107(n,g)Ag-108m; Al-27(n,2n)Al-26;
Al-27(n,n'alpha)Na-23; Bi-209(n,g)Bi-210; Bi-209(n,n'alpha)Tl-205;
Tl-205(n,2n)Tl-204; Ca-44(n,g)Ca-45; Ca-42(n,alpha)Ar-39;
Ca-43(n,n'alpha)Ar-39; Ca-40(n,2p)Ar-39; Co-59(n,g)Co-60;
Cr-50(n,g)Cr-51; Cu-63(n,p)Ni-63; Cu-65(n,t)Ni-63; Fe-58(n,g)Fe-59;
Fe-59(n,g)Fe-60; Fe-54(n,np)Mn-53; Hf-178(n,2n)Hf-177m;
Hf-177(n,g)Hf-178m2; Hf-179(n,g)Hf-180m; Hf-180(n,g)Hf-181;
Hg-204(n,2n)Hg-203; Tl-203(n,g)Tl-204; Hg-198(n,2n)Hg-197;
Hg-198(n,2n)Hg-197m; Hg-200(n,2n)Hg-199m; Hg-198(n,g)Hg-199m;
Hg-200(n,p)Au-200m; Hg-196(n,p)Au-196; Hg-196(n,n'p)Au-195;
Hg-196(n,alpha)Pt-193; Mg-25(n,np)Na-23; Mg-24(n,t)Na-22;
Mg-24(n,n'p)Na-23; Mg-26(n,g)Mg-27; Mo-95(n,p)Nb-95;
Mo-96(n,n'p)Nb95; Mo-97(n,t)Nb-95; Mo-98(n,g)Mo-99; Tc-99(n,2n)Tc98;
Tc-98(n,2n)Tc-97; Mo-92(n,alpha)Y-88; Mo-92(n,g)Mo-93;
Mo-94(n,2n)Mo-93; N-14(n,n'alpha)B-10; Ni-60(n,t)Co-58;
Ni-58(n,g)Ni-59; Ni-60(n,2n)Ni-59; Ni-62(n,He-3)Fe-60;
Ni-64(n,n'alpha)Fe-60; O-17(n,alpha)C-14; O-18(n,n'alpha)C-14;
Bi-209(n,g)Bi-210; Pb-204(n,t)Tl-202; Pb-206(n,alpha)Hg-203;
Pb-207(n,n'alpha)Hg-203; Pb-204(n,p)Tl-204; Pb-206(n,t)Tl-204;
Re-185(n,g)Re186m; Re-187(n,2n)Re-186m; Re-187(n,p)W-187;
Re-187(n,alpha)Ta-184; Re-185(n,alpha)Ta-182; Si-28(n,n'p)Al-27;
Si-28(n,alpha)Mg-25; Si-28(n,n'alpha)Mg-24; Ta-180(n,t)Hf-178m2;
Ti-48(n,alpha)Ca-45; Ti-46(n,n'alpha)Ca-42; V-50(n,n'alpha)Sc-46;
V-51(n,t)Ca-45; Sc-45(n,n'alpha)K-41; K-41(n,t)Ar-39;
W-183(n,n'p)Ta-182; W-184(n,t)Ta-182; W-182(n,n'alpha)Hf-178m2;
W-186(n,n'alpha)Hf-182; Zn-64(n,g)Zn-65; Zr-90(n,t)Y-88;
Zr-94(n,g)Zr-95; Zr-92(n,g)Zr-93; Zr-94(n,2n)Zr-93.

not satisfactory in 83 of the 153 reactions reviewed. The scattering cross sections for fluorine and boron will need to be improved at energies above 1 MeV. There is a weak point in the understanding of the Li-7(n,n'alpha)T reaction in the threshold energy range (4-6 MeV).

Acknowledgement

This work was supported by the United States Department of Energy.

References

1. E.T. Cheng, "Review of the Nuclear Data Status and Requirements for Fusion Reactors," Proc. Intl. Conf. Nuclear Data for Science and Technology, June 1988, p. 187, Japan Atomic Energy Research Institute (1988).

2. A.B. Pashchenko, D.W. Muir, ed., "First Results of FENDL-1 Testing and Start of FENDL-2," INDC(NDS)-241, International Atomic Energy Agency, Nov. 1990.

3. Wang DaHai, ed., "Activation Cross Sections for the Generation of Long-lived Radionuclides," INDC(NDS)-232/L, Proc. IAEA Consultants' Meeting, ANL, Sept. 1989, International Atomic Energy Agency, January 1990; also see H. Vonach, this conference.

4. S. Grimes, ed., "Sixth Coordination Meeting of Nuclear Physics Program to Meet High Priority Nuclear Data Needs of Fusion Energy," CONF-8909333, U.S. Department of Energy, June 1990.

5. O.N. Jarvis, "Nuclear Data for Fusion Reactors," European Applied Research Reports, 3(1981)127.

6. I. Kimura, K. Kobayashi, "Calibrated Fission and Fusion Neutron Fields at the Kyoto University Reactor," Nucl. Sci. Eng., 106(1990)332.

7. N.P. Kocherov and H. Vonach, "International Reactor Dosimetry File (IRDF-90) Status and Testing," Proc. 7th ASTM/Euratom Symp. Reactor Dosimetry, 1990, Strasbourg, France.

8. R. M. White, et al., "Evaluations of Charged-particle Reactions for Thermonuclear Applications," this conference.

9. D.W. Muir, et al., ed., "Proc. IAEA AGM Nuclear Data for Neutron Multiplication in Fusion Reactor First Wall and Blanket Materials, Chengdu, China, Nov. 1990," IAEA report, 1991.

10. E.T. Cheng, "Activation Cross Sections for Safety and Environmental Assessment of Fusion Reactors," Proc. NEANDC Specialists' Meeting on Activation Cross Sections for Fission and Fusion Energy, Argonne, Sept. 1989, NEANDC-259 'U', OECD, Paris, 1990.

11. S. Cierjacks, "Nuclear Data Needs for 'Low Activation' Fusion Materials Development," Fusion Eng. Design, 13(1990)220.

REPORT ON THE IAEA COORDINATED RESEARCH PROGRAM ON ACTIVATION CROSS SECTIONS FOR THE GENERATION OF LONG-LIVED ACTIVITIES OF IMPORTANCE IN FUSION REACTOR TECHNOLOGY

H. Vonach

Institut für Radiumforschung und Kernphysik d. Univ. Wien,
A-1090 Wien, Boltzmanngasse 3, Austria

Abstract: A coordinated research program on the measurement of activation cross sections was initiated in 1988 by the IAEA in order to obtain reliable data for 16 reactions of special importance for fusion technology proposed by E. Cheng (^{27}Al(n,2n)^{26}Al, ^{63}Cu(n,p)^{63}Ni, ^{94}Mo(n,p)^{94}Nb, ^{109}Ag(n,2n)^{108m}Ag, ^{179}Hf(n,2n)^{178m2}Hf, ^{182}W(n,n'α)^{178m2}Hf, ^{151}Eu(n,2n)^{150m}Eu, ^{153}Eu(n,2n)^{152g}Eu, ^{159}Tb(n,2n)^{158g}Tb, ^{158}Dy(n,p)^{158g}Tb, ^{193}Ir(n,2n)^{192m2}Ir, ^{187}Re(n,2n)^{186m}Re, ^{62}Ni(n,γ)^{63}Ni, ^{98}Mo(n,γ)^{99}Mo, ^{165}Ho(n,γ)^{166m}Ho, ^{191}Ir(n,γ)^{192m2}Ir). 9 laboratories from 7 countries participate in the program.

First results of the CRP were reported at an IAEA consultants' meeting in Argonne in Sept. 1989. At this time cross sections for 8 of the requested reactions had already been obtained at $E_n \approx 14$ MeV either by measurements or evaluation of measurements performed outside the CRP, and measurements on 5 more reactions were in progress. It can be expected that adequate knowledge about all of the requested reactions - with one possible exception, ^{191}Ir(n,γ)^{192m2}Ir - will be obtained before the next CRP coordination-meeting in Nov. of this year.

Introduction

At the 1986 Gaussig Advisory Group Meeting on Nuclear Data for Fusion Reactor Technology [1] a list of important activation cross-sections for fusion reactor materials was presented by E.Cheng and adopted by Working Group I as high priority data requests for fusion technology. He pointed out, that present design studies demanded operation at a very low rate of production of long-lived radionuclides in order to reduce the problems of waste disposal and reactor maintenance. In order to meet these requirements, not only all activation cross-sections for the fusion materials themselves have to be known, but also for all possible impurities (practically all elements) if they lead to long-lived ($t_{1/2} \geq 5$ years) activities. This problem was furtheron discussed in more detail by Dr. Cheng in his contribution to the 1987 INDC Meeting [2]. He reported that present estimates of the admissable concentrations of various elements in fusion reactor materials varied by large factors due to insufficient knowledge of the relevant activation cross-sections, and the radionuclides probably most important for the radioactive waste problem have been identified.

For these reasons the INDC felt that a coordinated research program (CRP) of the IAEA (on the measurement and calculation of these cross-sections) could make an important contribution to fusion reactor technology and recommended that the Nuclear Data Section of the IAEA started a CRP with the following scope.

The CRP should concentrate on the determination of a relatively small number of important activation cross-sections, selecting those reactions, which as yet cannot be estimated reliably from systematics and/or theory and where no reliable measurements exist. The list of these reactions is given in column 1 of table 1.

Following this recommendation the Nuclear Data Section of the IAEA initiated the requested CRP in 1988 and succeeded in obtaining the participation of 9 laboratories from 7 countries (KFA, Jülich, Germany, Argonne National Lab. and Los Alamos National Lab., USA, IAE, Beijing and University Sichuan, P.R. China, KRI, Leningrad, USSR, JAERI, Japan, Univ. Debrecen, Hungary and IRK, Vienna, Austria) in this very specific task.

A first research coordination meeting (RCM) of this CRP was held in Argonne, U.S.A., in Sept. 1989 in close cooperation with the NEANDC specialist meeting on activation cross-sections and a second RCM will gather the participants in Vienna in Nov. of this year.

Results obtained up to the first RCM at Argonne

The results obtained at that time are summarized in table 1. Results at $E_n \approx 14$ MeV had been obtained in the first year of the CRP for 9 of the 12 requested threshold reactions. Cross-sections were determined by direct measurement or - in case of isomer production in (n,2n) reactions - as difference between the total (n,2n) cross-sections and the formation cross-sections for the short lived isomers (IRK Vienna results). The agreement between these two very different methods was satisfactory except for one striking discrepancy. For the ^{109}Ag(n,2n)^{108}Ag reaction three direct measurements gave consistently results lower by a factor of three than Vonach's value derived as $\sigma_{long} = \sigma_{n,2n\ total} - \sigma_{short}$. This discrepancy however has been removed in the mean-time as will be discussed later.

In addition to these results a number of measurements were reported both at 14 MeV, around 10 MeV and for a broad-spectrum of fast neutrons (from 7 MeV deuterons irradiating a thick Be-target) when the irradiations had been completed, but the sample activities had not yet been determined. For two of the capture cross-sections measurements had been started at the Sichuan university, P.R.C. No work had been started up to that time on only three of the requested reactions.

Furthermore in addition to the specific goal of the CRP, the reactions of table 1 show a considerable number of additional results on reactions for formation of long-lived activities in reactions induced by 14 MeV neutrons were reported by some of the participants [3].

Results obtained since the Argonne RCM meeting

Since the Argonne meeting considerable further progress has been reported.

a) The discrepancy concerning the ^{109}Ag(n,2n)^{108m}Ag cross-section has been resolved. A new measurement of the half-life of ^{109}Ag at PTB, Germany [4] resulted in a value of 418 ± 15 years, more than a factor of 3 larger than the presently adopted value of 125 years (s. table 1). Accordingly the directly measured values of this cross-section in table 1 have to be multiplied by a factor of 3.34. This results in values of 695 - 878 mb in reasonable agreement with the indirectly derived value 665 ± 73 mb.

b) The analysis of the cross-section measurements at KFA Jülich and ANL proceeds as anticipated and results for all "measurements in progress" - with exception of ^{63}Cu(n,p) results - will be presented at the next RCM meeting in november of this year.

c) Calculations of the excitation functions for the reactions ^{27}Al(n,2n), ^{62}Ni(n,γ), ^{63}Cu(n,p), ^{94}Mo(n,p), ^{109}Ag(n,2n)^{108m}Ag and ^{187}Re(n,2n)^{186m}Re have been performed by Prof. Yamamaro, which are in very good agreement with the 14 MeV measurement for the threshold reactions and thus allow a complete description of these cross-sections.

d) The need for the very difficult measurement of the reaction ^{158}Dy(n,p)^{158g}Tb was investigated in detail by E. Cheng and H. Vonach. Using Vonach's estimated upper limit of 50 mb, E. Cheng calculated the admissible levels of Dy impurities in fusion reactor materials and found out, that these do not pose any serious problem. Thus no further work is needed.

e) The situation concerning the ^{62}Ni(n,γ)^{63}Ni reactions was studied by H. Vonach. It was demonstrated that existing neutron capture γ-ray measurements [5] can be used to predict the ^{63}Ni production via this reaction with sufficient accuracy so that no further activation measurements are needed.

f) Measurements of the ^{98}Mo(n,γ)^{99}Mo cross-section for 30-110 keV neutrons were completed at the Sichuan University, measurements of the ^{165}Ho(n,γ)^{166m}Ho cross-section are in progress.

g) A new measurement of the ^{60}Ni(n,2n)^{59}Ni was performed [6] using accelerator mass-spectroscopy for determination of the ^{59}Ni production in order to investigate the large discrepancy between the measurements of Greenwood and Bowers [7] and theory. A value of 407 ± 102 mb at E_n = 14.8 MeV was obtained in complete agreement with theoretical expectations but about a factor of 4 larger than Greenwoods' result. In this case the discrepancy does not seem to be due to incorrect decay data. A measurement of the ^{59}Ni half-life [6] confirmed

the accepted value within the experimental errors and it must be assumed that Greenwoods results suffer from undetected systematic errors.

h) Measurements of the ^{98}Mo(n,γ)^{99}Mo cross-sections for En = 1-2 MeV were performed and completed at the Khlopin Radium Institute Leningrad, in addition the ^{192}W(nn',α) cross-section was measured at 14 MeV.

i) At JAERI [8] new cross-section measurements were performed and analyzed for 10 of the CRP reactions. Good agreement with the results reported in table 1 was found except for the case of ^{187}Re(n,2n)^{186m}Re, where the new result is smaller by about a factor of 4 than Vonach's result derived indirectly from the total and ground-state (n,2n) cross-sections.

Summary and conclusions

A) It can be expected that the objectives of the CRP can be achieved almost completely by the end of this year by combination of experimental, theoretical and evaluation work. Only one reaction ^{191}Ir(n,γ)^{192m}Ir will probably not be covered adequately because both measurements and theoretical estimates of the cross-section are very difficult.

B) Experience in two cases ^{109}Ag(n,2n)^{108m}Ag and ^{60}Ni(n,2n)^{59}Ni has shown that apparent large deviations from the theoretical expectations were either due to incorrect decay data or unknown systematic measurement errors. This means that at least for (n,2n) reactions any such deviations should be considered with great caution and only be accepted after confirmation by several independent measurements. Accordingly the cross-sections for the reactions ^{50}V(n,2n)^{49}V, ^{92}Mo(n,2n)^{91}Mo(β^+)^{91}Nb and ^{94}Mo(n,2n)^{93}Mo where the results of ref. [6] deviate strongly from theory definitely need further investigation.

REFERENCES

1. Proceedings of the IAEA Advisory Group Meeting on Nuclear Data for Fusion Reactor Technology, Gaussig, 1.-5. Dec. 1986, IAEA-TECDOC-457 (1988) p. 16
2. Minutes of the 16th INDC-Meeting, Report INDC/(89) (1989), A.J. Deruytter ed.
3. Proc. of the IAEA consultants' meeting activation cross-section for the generation of long-lived radionuclides of importance in fusion reactor technology, Report INDC/NDS/232/L (1990), Wang Da Hai ed.
4. U. Schützig, H. Schrader and K. Debertin: Contribution to this conference
5. H. Beer and R.R. Spencer: Nucl. Phys. A 240, 29 (1975)
6. D. Weselka et al.: Contribution to this conference
7. L.R. Greenwood and D.L. Bowers: Proc. NEANDC-Specialist meeting on neutron activation cross-sections for fusion and fission applications. Argonne 13.-15. Sept. 1989, NEANDC-259-U (1990) p. 85
8. Y. Ikeda, A. Kumar and C. Konnu: Contribution to this conference

Table 1: Data Status of activation cross-sections selected for the CRP on long-lived nuclides for fusion technology at the first RCM at Argonne Sept. 1989

Reaction	Half-life (years)	Status	Laboratory	Neutron Energy (MeV)	Cross-Section (mb)
^{27}Al(n,2n)^{26}Al	$7.2 \cdot 10^5$		several measurements available from outside CRP, data need satisfied, evaluation still needs to be done		
^{63}Cu(n,p)^{63}Ni	100	2)	KFA Jülich	7.6	
		2)	ANL/LANL/JAERI	10.,14., B.S.*	
			ANL	14.9	54 ± 4**
^{94}Mo(n,p)^{94}Nb	$2 \cdot 10^4$		ANL	14.8	53.1 ± 5.3**
^{109}Ag(n,2n)^{108m}Ag	125^{xxxx}	1)	IAE Beijing	14.77	230 ± 7
		1)	Debrecen	14.50	263 ± 20
		3)	IRK, Vienna	14.0 - 15.0	665 ± 73
		2)	ANL/LANL/JAERI	10.,14., B.S.*	
		1)	KRI Leningrad	14.9	208 ± 37
^{179}Hf(n,2n)^{178m2}Hf	31	2)	Harwell	14.8	5.9 ± 0.6***
		2)	IAE Beijing	4.2 - 14.8	
		4)	Oxford/LANL	14	2.9
		2)	ANL/LANL/JAERI	10.,14., B.S.*	
^{182}W(n,na')^{178m2}Hf	31	2)	Harwell	14.8	
^{151}Eu(n,2n)^{150m}Eu	36	1)	IAE Beijing	14.77	1219 ± 28
		3)	IRK, Vienna	14.0 - 15.0	1325 ± 94
		2)	KFA, Jülich	9.6 - 10.6	
		2)	ANL/LANL/JAERI	10.,14., B.S.*	
		1)	KRI Leningrad	14.9	1090 ± 84
^{153}Eu(n,2n)^{152g}Eu	13.6	2)	ANL/LANL/JAERI	10.,14., B.S.*	
		1)	IAE Beijing	14.77	1544 ± 42
		3)	IRK, Vienna	14.0 - 15.0	1442 ± 60
		1)	KRI Leningrad	14.9	1740 ± 145
^{159}Tb(n,2n)^{158g}Tb	150	2)	KFA Jülich	8.6 - 10.6	
		1)	IAE Beijing	14.77	1968 ± 56
		2)	ANL/LANL/JAERI	10.,14., B.S.	
		3)	IRK, Vienna	14.0 - 15.0	1930 ± 49
^{158}Dy(n,p)^{158g}Tb	150		no work done		
^{193}Ir(n,2n)^{192m2}Ir	241	3)	IRK, Vienna	14.0 - 15.0	184 ± 44
^{187}Re(n,2n)^{196m}Re	$2 \cdot 10^5$	3)	IRK, Vienna	14.0 - 15.0	591 ± 122
^{62}Ni(n,γ)^{63}Ni	100		no work done		
^{98}Mo(n,γ)^{99}Mo(β^-)^{99}Tc	$2,1 \cdot 10^5$		research contract given to Sichuan Univ.		
^{165}Ho(n,γ)^{166m}Ho	$1.2 \cdot 10^3$		research contract given to Sichuan Univ.		
^{191}Ir(n,γ)^{192m2}Ir	241		no work done		

1) measurement performed for CRP
2) measurement in progress for CRP
3) evaluation of existing data performed for CRP
4) calculation performed for CRP

* Broad neutron spectrum from 7 MeV deuterons on thick Be-target
** measurement performed outside of CRP by L. Greenwood, ANL
*** preliminary value
**** Status of 1989, since revised to 418 ± 12 year, see text

NEANDC WORKING GROUP ON ACTIVATION CROSS SECTIONS.
Comparison of activation cross section measurements performed with different neutron source
reactions in the 5-13 MeV range

D.L. Smith and J.W. Meadows
Argonne National Laboratory, Argonne, Ill. 60439, USA
H. Vonach and M. Wagner
Institut für Radiumforschung und Kernphysik d. Univ. Wien,
A-1090 Wien, Boltzmanngasse 3, Austria
R.C. Haight
Los Alamos National Laboratory, Los Alamos, NM 87545, USA
and
W. Mannhart
Physikalisch Technische Bundesanstalt,
D-3300 Braunschweig, Bundesallee 100, Germany

Abstract: On recommendation of the Nuclear Energy Agency (NEANDC) an international working group on activation cross-sections was formed in 1989, which, at its first meeting, initiated an intercomparison of activation cross-section measurements using the $D(d,n)^3He$ and $H(t,n)^3He$ reactions. For this purpose excitation functions for the $^{58}Ni(n,p)^{58}Co$ and $^{93}Nb(n,2n)^{92m}Nb$ reactions have been measured at the Argonne National Laboratory and the Physikalisch-Technische Bundesanstalt, Braunschweig (PTB) using neutrons produced by means of the $D(d,n)^3He$ reaction and in a collaboration Los Alamos National Laboratory - Institut für Radiumforschung, Vienna (IRK) by means of neutrons produced by the $H(t,n)^3He$ reaction in order to test the different corrections needed to account for the deficiencies of the respective source reactions. Measurements at Argonne were performed in the neutron energy range E_n = 5-10 MeV, at PTB in the range E_n = 8-14 MeV and at Los Alamos for 5-13 MeV neutrons. Reasonable agreement was obtained between the results of the three groups for most of the investigated neutron energy range, there are however also some small discrepancies, both in the gamma-efficiency calibration and some cross-section values which need further study. The data obtained in this intercomparison substantially improve our knowledge on the excitation function of the two important dosimetry reactions.

1. Introduction

The Nuclear Energy Agency Nuclear Data Committee (NEANDC) decided in 1988 to form a Working Group on Activation Cross Sections. This action was taken to explicitly stimulate international cooperative efforts directed toward the development of needed activation cross section information, largely to meet the growing demands for fusion energy applications. Progress in the development of a comprehensive data base of activation cross-section evolves from an intricate amalgamation of results from experimental measurements, nuclear theory, nuclear model calculations and evaluations. The most important purpose of this Working Group is to encourage and facilitate such communication, e.g., by means of the establishment of special joint research projects, the exchange of technical information, the organization of specialists' meetings, etc. About thirty individuals responded to a written invitation to join the Working Group. At the first meeting of the Working Group held at Argonne in September 1989, two formal projects were adopted. The first involves an intercomparison of activation cross-sections for the reactions $^{58}Ni(n,p)^{58}Co$ and $^{93}Nb(n,2n)^{92m}Nb$ using the $D(d,n)^3He$ and $H(t,n)^3He$ neutron source reactions. The second concerns a multi-laboratory intercomparison of nuclear model computational techniques, focused on examination of the $^{60}Co(n,p)^{60}Fe$ reaction.

This presentation will report mainly on the results of the first project which has been almost completed. Within this intercomparison of activation cross-section measurements the cross-sections for the reactions $^{58}Ni(n,p)^{58}Co$, and $^{93}Nb(n,2n)^{92m}Nb$ were measured in the energy range 5-14 MeV and threshold to 14 MeV respectively at 3 different laboratories (Argonne, PTB Braunschweig and Los Alamos/IRK Vienna cooperation) using both the $D(d,n)$ (Argonne, PTB) and the $H(t,n)$ (Los Alamos/Vienna) source reactions.

2. Experimental and data analysis procedures

2.1. Common features of the three experiments

All experiments use the same basic irradiation geometry. Neutrons produced at zero degree by means of either the $D(d,n)$ or $H(t,n)$ reaction are used to irradiate Ni and Nb discs back-to-back with the ^{238}U deposit of a thin-walled fission-chamber. Thus all measurements were performed relative to the ^{238}U fission cross-section and the ENDF/B-VI [1] values and their associated uncertainties were used for that cross-section. In all experiments similar procedures [2,3] were used for the extrapolation (1-2%), fragment absorption (< 1%) and dead-time corrections of the fission rates and for accounting for the position differences of the samples and the ^{238}U deposit.

The ^{58}Co and ^{92m}Nb-activities of the samples were measured by means of intrinsic Ge-detectors by evaluating the full energy peaks of the 810.76 and 934.4 keV γ-lines emitted in the decay of ^{58}Co and ^{92m}Nb using the branching ratios from the Nuclear Data Sheets [4,5]. All measurements of ^{58}Co were performed after complete decay of ^{58m}Co. The efficiency of these detectors had been determined by means of standard γ-sources to about 1.5% (see table 1) in all three laboratories and the calibrations of the three detectors (ANL, IRK, PTB) have been intercompared by exchange of irradiated samples (see section 2.2). Appropriate corrections for the finite dimensions of the samples (< 2%) and γ-attenuation in the samples (< 3%) were applied to all measurements.

2.2. Intercomparison of the efficiency calibrations of the Ge-detectors used at ANL, PTB and IRK

The efficiency calibrations of the Ge-detector used at the three laboratories were compared by exchange of irradiated samples. One Ni and one Nb sample irradiated at Los Alamos were measured both at IRK, Vienna and PTB, Braunschweig. Two Ni samples (one from the ANL and one from the Los Alamos/IRK experiment) were measured both at Argonne and IRK, Vienna. In addition, the efficiency calibration of the Vienna Ge-detector was compared with the Vienna 4 π-γ Na-detector [6] by measuring one of the Ni samples which had been irradiated with ≈ 4.6 MeV neutrons and thus could be considered as a pure ^{58}Co source with both detectors. The results of these comparisons are given in table 1. The uncertainties given were calculated from the estimated uncertainties of the respective detector calibration and the (much smaller) statistical uncertainties. As the table shows, there is very good agreement between all activity measurements performed at the two different γ-detectors at IRK and at the PTB Ge-detector. The two activity comparisons between IRK and ANL do show discrepancies (≈ 3-4%), somewhat outside the expected range. These differences, however, are opposite in sign for the two measurements. Thus they don't indicate a systematic difference in the efficiency calibration of the two detectors. Further measurements are needed to trace the origin of these differences between the IRK and ANL results.

2.3. Activation cross-section measurement performed at Argonne National Laboratory

Approximately monoenergetic neutrons of 5.2-9.8 MeV were produced by means of the D(d,n)^{3}He reactions in a gas target. The gas target had a length of 2 cm and was operated at 0.2 MPa. The entrance window to this cell was a 3.2 mg/cm^2 Mo foil. This foil lay against a Au grid so that it would sustain a higher beam current than otherwise possible with a self-supporting window of the same area (in excess of 10 microamperes average). The mean neutron energy and resolution were determined by calculations for each run. The analysis took into account the incident deuteron energy (determined by magnetic analysis), energy loss in the gas target entrance window and the deuterium gas, and kinematics of the source reaction. An uncertainty of 40 keV is assumed for each determined energy. This added an uncertainty estimated to be about 5% for the ^{93}Nb(n,2n) reaction near threshold. The samples were 2.5 cm in diameter. Two samples (Ni and Nb) were placed back-to-back against the fission chamber which contained the ^{238}U monitor deposit. The samples were located about 3.8 cm from the end of the gas cell. The data analysis correction procedures accounted for:
1) geometric effects (calculated as a few

percent, with < 1% error)
2) neutrons from the evacuated gas target (the measured correction to the fission rate amounted to more than 30% at 7 MeV deuteron energy, but the corresponding corrections were less for the other reactions, and the errors were < 3%),
3) breakup neutrons from the D(d,np)D reaction (calculated as no more than 10% net in the worst case, with 2% error contribution),
4) the fact that the primary neutrons were not truly monoenergetic (small calculated corrections except near threshold where they amounted to several percent, and the errors were < 1% and
5) neutron absorption and multiple scattering (net corrections were no more than 4%, with < 1.5% error). These corrections tended to cancel in the cross-sections, to varying degrees, but the cancellation was least effective for the ^{93}Nb(n,2n)^{92m}Nb measurements.

2.4. Activation cross-section measurement performed at PTB

Monoenergetic neutrons of 8.5-13.5 MeV were produced at the PTB cyclotron by means of the D(d,n)^{3}He reaction in a gas target (length 30 mm, pressure 0.2 MPa, deuteron current about 2 μA). Ni and Nb samples (10 mm diameter, 1 mm thick) were irradiated back to back with the IRK fission chamber (see section 2.5) at zero degree at a distance of 98 mm from the target. The neutron energy at zero degree was measured by time of flight with a NE213 detector at a distance of 12 m (time resolution about 1 nsec). Therefore the neutron energy averaged over the sample area was derived by means of the Monte-Carlo Code SINEMA [7]. The uncertainty of the average neutron energies derived in this way amounts to 20 keV. Contributions of neutrons from window and beam stop (determined by gas-out measurements) to activities and fission rates remained below 13% for all cases. The correction due to breakup neutrons from the D(d,np)D reaction was calculated using the new PTB precision measurements for the double-differential neutron emission cross-sections for this reaction [8]. An uncertainty of 3% was assigned to these corrections for the ^{58}Ni(n,p) and ^{238}U(n,f) reactions, for the ^{93}Nb(n,2n)^{92m}Nb reaction the breakup correction is negligible due to the high reaction threshold. Neutron attenuation between the centers of the Ni and Nb samples and the fission deposit were calculated using the total cross-sections from Neutron Cross-Sections Vol. 2 [9]. The contributions to the activities and fission rates due to neutrons scattered elastically and inelastically in the samples and fission chamber (≈ 3-9%) were derived from Monte Carlo calculations.

2.5. Activation cross-section measurements performed by the Los Alamos - IRK Vienna collaboration

Approximately monoenergetic neutrons of 4.6-13.2 MeV were produced by bombarding a pressure gas-cell (H$_2$ at 0.91 MPa) [10] with the triton beam of the Los Alamos Tandem accelerator. Nb and Ni discs, 0.5 mm thick, 10 mm diameter, 99.9% purity were used as samples. The ^{238}U deposit of the IRK fission chamber [11] had an Al-backing of 0.2 mm thickness and a steel wall 0.3 mm thick. The target-sample distance amounted to 30 mm. The neutron energy distributions averaged over the sample area were calculated by integration both over the length of the gas cell and area of the sample using the cross-sections of Drosg [12] for the source reaction. No corrections for neutrons scattered by the target assembly were needed [13] due to the kinematic focussing of the neutrons. Typical irradiation times were 30 min resulting in neutron fluences of 1-2 x 10^{12} neutrons cm^{-2}.

The influence of neutrons produced in the entrance foil and target backing was taken into account by means of gas-out measurements. These gas-out measurements have been the main problem of this experiment. At neutron energies above 11 MeV only about half of the ^{238}U fissions in the gas-in run are due to H(t,n) neutrons, the remaining being due to tritium breakup neutrons produced in the target backing [13], whereas for a high-threshold reaction like $^{93}Nb(n,2n)^{92m}Nb$ the fraction of the activity due to breakup neutrons is only about 15%. Thus cross-section measurements for such reactions are extremely sensitive to the accurate subtraction of the gas-out contributions. In principle this can be done exactly by normalizing both measurements to the same integral charge. For technical reasons discussed in another contribution to this meeting [14] this charge normalization could only be performed with very large uncertainties (≈ 30%) which did not allow accurate measurement of the $^{93}Nb(n,2n)^{92m}Nb$ cross-section. Thus the results of the Nb irradiations were only used to derive accurate normalization factors for the gas-out measurement, by demanding agreement between the Nb(n,2n) cross-sections derived from the Los Alamos/IRK experiment and the accurate PTB results. This allowed to derive the gas-out fractions with an accuracy of ≈ 10% much better than the direct measurements. The normalizations derived in this way were then used in the analysis of the $^{58}Ni(n,p)^{58}Co$ data. For this reaction - which has an excitation function very similar to $^{238}U(n,f)$ - the sensitivity to the correct gas-out fractions is much less, therefore accurate cross-sections could be obtained inspite of the described uncertainties.

For the analysis of the activity measurements the efficiency calibration derived from the IRK 4π-γ [6] detector (see section 2.2) was used as it is somewhat more precise for the activities measured in this work than the standard calibration based on calibrated single-line sources.

3. Results

3.1. $^{58}Ni(n,p)^{58}Co$ cross-sections

The cross-sections for all reactions are summarized in table 1. The uncertainties given are effective 1σ errors including contributions from all identified sources of error added in quadrature. Uncertainties in the $^{238}U(n,f)$ standard cross-sections, counting, deposit and sample-masses, γ-efficiencies, and uncertainties of all corrections applied in the fluence and activity determinations were taken into account including corrections for the uncertainty in the correction for gas-out neutrons. For the measurements using the $D(d,n)^3He$ source reaction also the uncertainty in the correction for the contribution of D(d,np)D break-up neutrons was taken into account.*)

In figure 1 the results of the three measurements are compared with each other, with other recent data [11,13,15-22] and the predictions of the four most recent nuclear data files [13, 23-25]. The result of this comparison can be summarized in the following way:

a) There is good agreement between the results of all three experiments over the entire neutron energy range investigated. Moreover, our data agree well with the accurate measurements reported in refs. 19 and 21 for the 14 MeV region.

b) In general the data obtained in this intercomparison support the results of ref. 13,18,20 and indicate that the cross-section of ref. 15 may have been somewhat low in the neutron energy range 6-10 MeV.

c) All recent evaluations [1, 23-25] are in reasonable agreement with all new measurements.

3.2. $^{93}Nb(n,2n)^{92m}Nb$

The cross-sections for this reaction are listed in table 2. Only results from the PTB and Argonne measurements are given. As discussed before the uncertainties in the normalization of the gas-out runs did not allow to derive sufficiently accurate cross sections from the Los Alamos/IRK experiment. The stated uncertainties include all contributions listed in detail for the $^{58}Ni(n,p)^{58}Co$ measurements. The results obtained in this work are shown in figure 2 together with other recent measurements [26-32] and the statistical model calculation by Strohmaier [33]. There is excellent agreement between the results of the new PTB measurements, the new measurements of Santry et al. [32] and the calculation of Strohmaier [33], whereas the recent measurements of Mannan and Qaim [30] seem to be definitely too low. Close to the threshold the ANL value is higher than the PTB data by a factor of ≈ 1.6; yet due to the strong energy dependence of the cross section in this region this difference may be explained by the uncertainties of the energy scale reported for both experiments.

4. Summary and conclusions

The measurements performed within the intercomparison project of the working group on activation cross-sections has demonstrated that the D(d,n) reaction can be used safely for accurate cross-section measurements up to 13 MeV. For the H(t,n) reaction this could be demonstrated only for the $^{58}Ni(n,p)^{58}Co$ reaction, that is for cross-section measurements in which the standard and the reaction to be measured have similar excitation functions. For reactions with excitation functions widely different from the standard the described problems with the normalization of the gas-out runs prevented a sufficiently accurate comparison of cross-section measurements with the different source reactions. For the $^{93}Nb(n,2n)^{92m}Nb$ reaction some new measurements with H(t,n) neutrons and improved accelerator performance are needed to

---

*) Detailed information on all uncertainty components for all measured cross-section values is available from the authors.

achieve the full objectives of the intercomparison exercise.

In addition the described experiments have substantially improved our knowledge on the excitation functions of both the reactions ^{58}Ni(n,p)^{58}Co and ^{93}Nb(n,2n)^{92m}Nb so that it appears worthwhile to upgrade the existing dosimetry evaluations for both reactions.

Finally it has to be noticed that - as demonstrated in the case of the ^{93}Nb(n,2n)^{92m}Nb excitation functions - in case of discrepancies between careful calculations and experimental data systematic errors of the measurements may be the reason in many cases.

References

1 P. Rose, ed. ENDF/B-VI Summary Documentation, Report ENDF-201, 4th edn., BNL, in press.
2 J.W. Meadows, Report ANL/NDM 39 (1978)
3 J.W. Meadows, Report ANL/NDM 97 (1986)
4 L.K. Peter, Nucl. Data Sheets 61, 189 (1990)
5 P. Luksch, Nucl. Data Sheets 30, 573 (1980)
6 G. Winkler and A. Pavlik, Int. J. Appl. Radiat. Isotopes 34, 1167 (1983)
7 B.R.L. Siebert et al., Report PTB-ND-23 (1982)
8 S. Cabral et al., Nucl. Sci. Eng. 106, 308 (1986)
9 V. McLane, C.L. Dunford and P.F. Rose, Neutron Cross Sections, Academic Press, N.Y. (1988)
10 R.C. Haight and J. Garibaldi, Nucl. Sci. Eng. 106, 296 (1990)
11 M. Wagner et al., Ann. Nucl. Energy 15, 363 (1987)
12 M. Drosg et al., Report LA 10444-MS (1985)
13 H. Vonach, M. Wagner and R.C. Haight, Proc. NEANDC Specialists' Meeting on Neutron Activation Cross Sections for Fission and Fusion Applications, Argonne 13.-15. Sept. 1989, NEANDC-259 U, 165 (1990)
14 M. Wagner, H. Vonach and R.C. Haight, Contribution to this conference
15 D.L. Smith and J.W. Meadows, Report ANL/NDM-10, Argonne National Laboratory (1975) and Nucl. Sci. Eng. 58, 314 (1975)
16 Mien-Win Wu and Jen-Chang Chou, Nucl. Sci. Eng. 63, 268 (1977)
17 H.A. Hussain and S.E. Hunt, Int. J. Appl. Radiat. Isot. 34, 731 (1983)
18 N.W. Kornilov, Atomnaya Energiya 59, 128 (1985)
19 A. Pavlik, G. Winkler, M. Uhl, A. Paulsen and H. Liskien, Nucl. Sci. Eng. 90, 186 (1985)
20 W. Mannhart, private communication
21 Y. Ikeda, Ch. Konno, K. Oishi, T. Nakamura, H. Miyade, K. Kawade, H. Yamamoto and T. Katoh, Report JAERI-1312, Japan Atomic Energy Research Inst., Tokai-mura (1988)
22 Zhao Wenrong, Lu Hanlin, Yu Weixiang and Yuan Xialin, Report CNDC-89014, Inst. of Atomic Energy, Beijing, People's Rep. of China (1989)
23 EFF-2: M.Uhl et al., Contribution to this conference
24 JENDL-3: K. Shibata et al., Report JAERI-1319, Japan Atomic Energy Research Inst., Tokai-mura (1990)
25 BROND: Y.N. Manokhin, ed., Report INDC (CCP)-283, IAEA, Vienna (1988)
26 A. Paulsen and R. Widera, Z.Physik 238, 23 (1970)
27 Lu Hanlin , Huang Jianzhou, Fan Peiguo, Cui Yunfeng and Zhao Wenrong, Nuclear Data for Science and Technology, Proc.Int.Conf. Antwerp, 6-10 Sept. 1982, p. 411, Reidel Publ. Comp., Dordrecht (1983)
28 S. Dároczy et al., Proc. 6th All Union Conf. on Neutron Physics, Kiev, Oct. 2-6, 1983, vol. 3, 191 (1983)
29 J. Csikai, Zs. Lantos and Cs.M. Buczkó, Proc. IAEA Advisory Group Meeting on Properties of Neutron Sources, Leningrad, USSR, June 9-13, 1986, Report IAEA-TECDOC-410, 296, IAEA, Vienna (1987)
30 A. Mannan and S.M. Qaim, Phys.Rev. C38, 630 (1988)
31 R. Wölfle, et al., Appl. Radiat. Isotopes (formerly Int. J. Radiat. Appl. Instrum. Part A) 39, 407 (1988)
32 D.C. Santry and R.D. Werner, Can. J. Physics 68, 582 (1990)
33 B. Strohmaier, Ann. Nucl. Energy 16, 461 (1989)

Table 1: Intercomparison of ^{58}Co and ^{92m}Nb activity measurements

Sample	Activity [Bq]			
	IRK (Ge)	IRK (4π-γ)	PTB	ANL
Ni 11 (IRK)[a]	218.5 ± 2.4	215.2 ± 0.9	218.42 ± 2.18	-
Ni 19 (IRK)[a]	387.6 ± 4.3	-	387.60 ± 3.88	-
Ni 6 (IRK)[a]	443.0 ± 6.0	-	-	459.86 ± 5.06
Ni 5 (ANL)[b]	97.9 ± 1.2	-	-	94.88 ± 1.50
Nb 2 (IRK)[a]	1010.0 ± 10.3	990.6 ± 5.1	1002.0 ± 10.0	-

Reference time: a) end of the irradiation of the foils
b) 25-04-1991, 0.00h Central European daylight saving time

Table 2: Cross Sections for the ^{58}Ni(n,p)^{58}Co reaction

PTB			ANL			IRK/LANL		
E_n[MeV]	ΔE_n (FWHM) [MeV]	σ [mb]	E_n[MeV]	ΔE_n (FWHM) [MeV]	σ [mb]	E_n[MeV]	ΔE_n (FWHM) [MeV]	σ [mb]
8.463 ± 0.02	0.117	637 ± 24	5.32 ± 0.04	0.27	531 ± 13	4.596 ± 0.020	0.612	422 ± 7
9.135 ± 0.02	0.110	636 ± 24	5.87 ± 0.04	0.25	583 ± 14	5.772 ± 0.024	0.544	557 ± 11
9.370 ± 0.02	0.108	649 ± 24	6.40 ± 0.04	0.24	600 ± 14	6.895 ± 0.029	0.506	646 ± 16
9.502 ± 0.02	0.106	647 ± 24	6.93 ± 0.04	0.22	637 ± 16	7.994 ± 0.032	0.512	658 ± 21
9.658 ± 0.02	0.106	629 ± 23	7.42 ± 0.04	0.21	624 ± 16	9.111 ± 0.036	0.520	641 ± 25
9.859 ± 0.02	0.103	634 ± 23	7.92 ± 0.04	0.20	642 ± 17	10.159 ± 0.040	0.492	630 ± 29
10.223 ± 0.02	0.099	638 ± 24	8.41 ± 0.04	0.18	619 ± 16	11.149 ± 0.044	0.480	618 ± 18
10.563 ± 0.02	0.096	615 ± 23	8.90 ± 0.04	0.16	638 ± 17	12.204 ± 0.047	0.464	563 ± 16
10.744 ± 0.02	0.096	610 ± 23	9.38 ± 0.04	0.17	609 ± 18	13.209 ± 0.051	0.470	499 ± 12
10.911 ± 0.02	0.096	605 ± 23	9.87 ± 0.04	0.20	593 ± 18			
11.392 ± 0.02	0.094	595 ± 23						
11.964 ± 0.02	0.092	564 ± 21						
12.485 ± 0.02	0.089	536 ± 20						
12.925 ± 0.02	0.087	497 ± 19						
13.469 ± 0.02	0.085	442 ± 19						

Table 3: Cross Sections for the ^{93}Nb(n,2n)^{92m}Nb reaction

PTB			ANL		
E_n[MeV]	ΔE_n (FWHM) [MeV]	σ [mb]	E_n[MeV]	ΔE_n (FWHM) [MeV]	σ [mb]
9.135 ± 0.02	0.110	1.165 ± 0.217	9.38 ± 0.04	0.17	23.2 ± 1.4
9.370 ± 0.02	0.108	13.27 ± 0.52	9.87 ± 0.04	0.20	97.0 ± 6.1
9.502 ± 0.02	0.106	29.00 ± 1.09			
9.658 ± 0.02	0.106	64.80 ± 2.29			
9.859 ± 0.02	0.103	97.07 ± 3.41			
10.223 ± 0.02	0.099	157.80 ± 5.50			
10.563 ± 0.02	0.096	215.90 ± 7.70			
10.744 ± 0.02	0.096	241.90 ± 8.60			
10.911 ± 0.02	0.096	266.70 ± 9.80			
11.392 ± 0.02	0.094	334.30 ± 12.50			
11.964 ± 0.02	0.092	388.00 ± 14.70			
12.485 ± 0.02	0.089	422.40 ± 16.70			
12.925 ± 0.02	0.087	435.20 ± 17.60			
13.469 ± 0.02	0.085	445.90 ± 18.80			

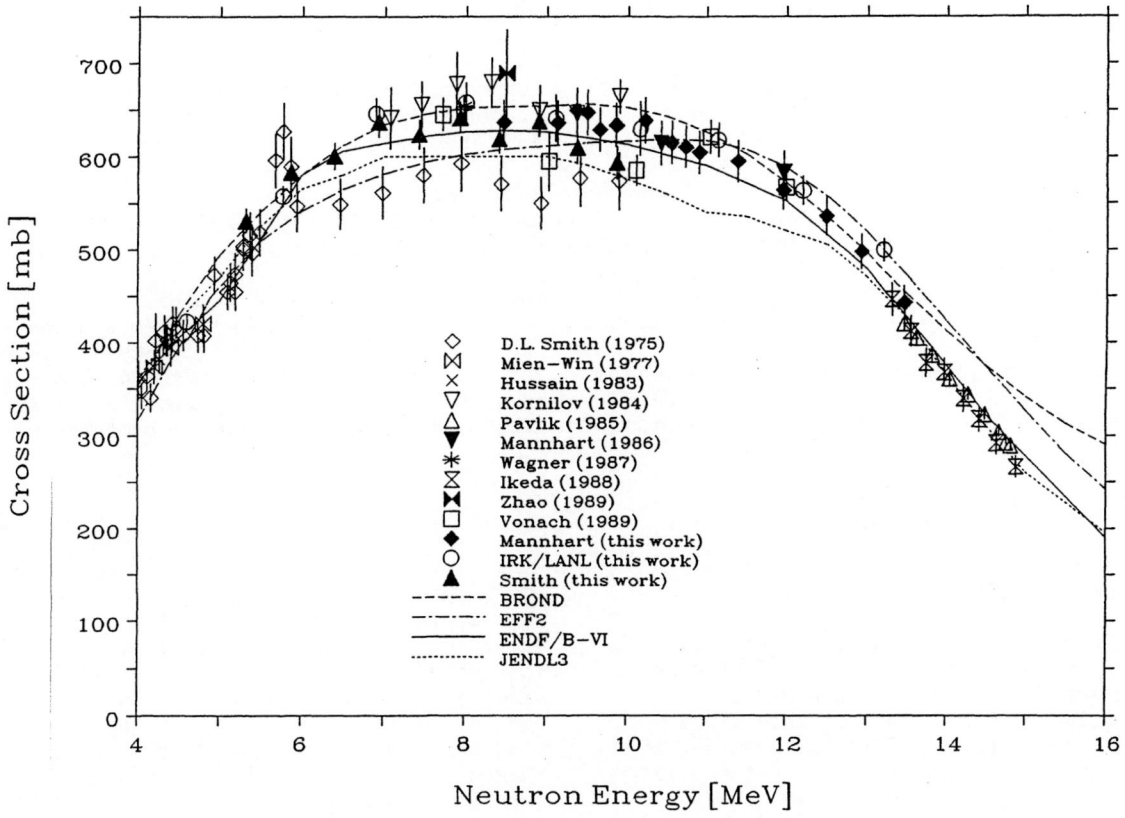

Fig. 1 Intercomparison of the ^{58}Ni(n,p)^{58}Co cross section measured within the working group and comparison to previous experiments and recent evaluations.

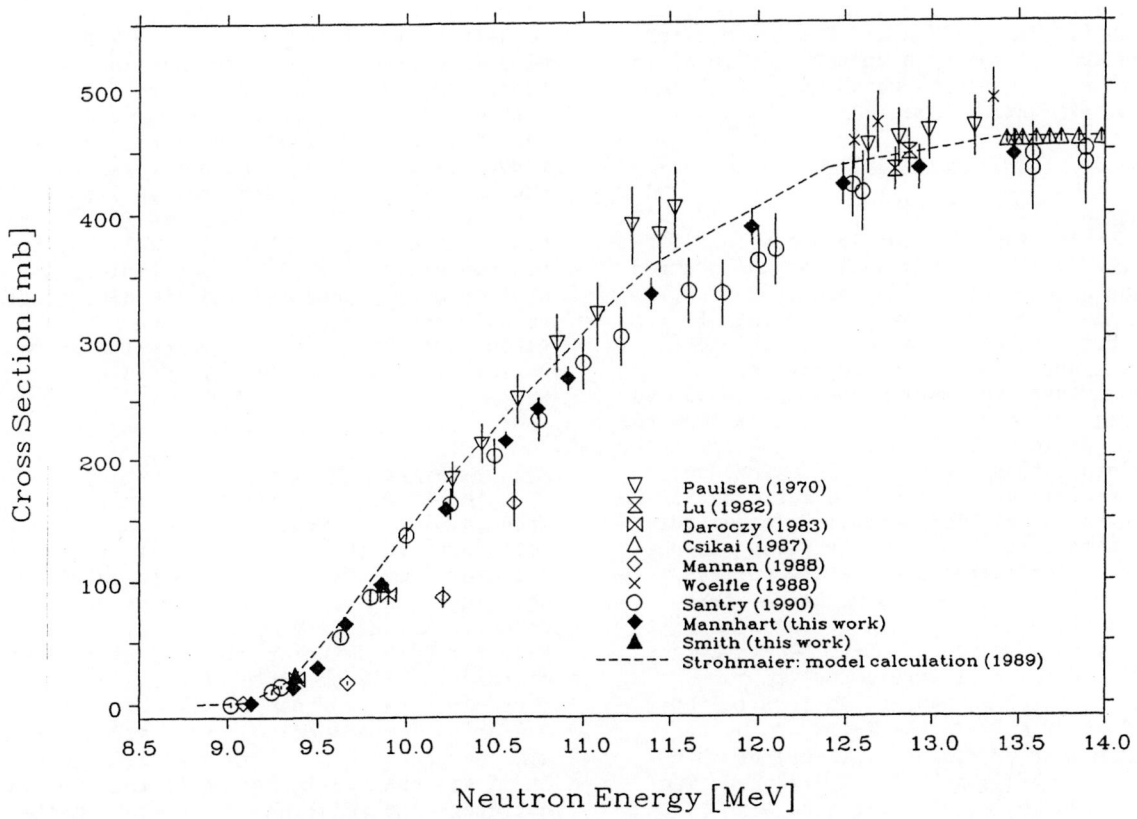

Fig. 2 Intercomparison of the ^{93}Nb(n,2n)^{92m}Nb cross sections measured within the working group and comparison with previous experiments and the statistical model calculations of Strohmaier.

MEASUREMENTS OF THE NEUTRON CROSS SECTION FOR Fe-54(n,alpha)Cr-51 BETWEEN 5.3 AND 14.6 MEV

J. W. Meadows, D. L. Smith, L. R. Greenwood*, L. P. Geraldo**
Argonne National Laboratory
Argonne, Ill. 60439, U.S.A.

W. Mannhart, G. Börker
Physikalisch-Technische Bundesanstalt
W-3300 Braunschweig, Federal Republic of Germany

Abstract: Measurements of the neutron cross sections of Fe-54(n,α)Cr-51 and of Fe-54(n,p)Mn-54 are reported. Two experiments were performed, one at ANL between 5.3 and 9.8 MeV and one at PTB between 9.1 and 14.6 MeV. In both experiments the D(d,n)He-3 reaction served as the neutron source reaction. The neutron fluence was monitored with the Al-27(n,α)Na-24 reaction in the PTB experiment and with a U-238 fission chamber in the ANL experiment. Details of the data corrections are discussed and the results are compared with ENDF/B-VI and data from the literature.

(Neutron cross section measurements, Fe-54(n,α)/U-238(n,f) and Fe-54(n,p)/U-238(n,f), E_n = 5.3 to 9.8 MeV; Fe-54(n,α)/Al-27(n,α) and Fe-54(n,p)/Al-27(n,α), E_n = 9.1 to 14.6 MeV)

Introduction

The Fe-54(n,α)Cr-51 reaction contributes to the helium accumulation in steel, and this is of some technical importance in intense, fast neutron fields. The data base of this cross section is sparse (in the threshold region, for example) and also contradictory, and there is a complete absence of data between 10 and 12.5 MeV. To improve our knowledge of this cross section, measurements have been made between 5.3 and 14.6 MeV.

The measured cross section ratios were normalized with cross section values taken from ENDF/B-VI for U-238(n,f) and from a recent evaluation [1] for Al-27(n,α). The two monitor reactions are related by a recent Al-27(n,α) to U-238(n,f) experiment [2] which is part of the Al-27(n,α) evaluation [1].

Experimental Methods

ANL Experiment

At the ANL Fast Neutron Generator Facility, a deuterium gas target was used for the neutron production. The water-cooled target, 2.54 cm in diameter and 2 cm long, was pressurized at 0.2 MPa. With deuterons betwen 2.5 MeV and 7 MeV, neutrons from 5.3 to 9.8 MeV were produced. A low-mass fission chamber was placed at zero degrees at a distance of 3.2 cm from the target. An enriched U-238 deposit (< 6 ppm of isotopes other than U-238) of 386 μg/cm^2 and 2.54 cm in diameter was employed for the fluence measurements. A detailed specification of this deposit, sample no. U-238-60, is given elsewhere [3]. Natural iron samples of 98.5 % purity, 2.54 cm in diameter and 2.9 mm thick, were irradiated back-to-back to the fission deposit in front of the fission chamber. Separate runs with an empty gas cell were conducted to determine the neutron background from the beam stop (1.9 mm of Au) and the walls of the cell. The neutron energy resolution (FWHM) of this experiment was between 0.16 and 0.27 MeV. The neutron energy scale was calibrated with protons by observing the thresholds for the Li-7(p,n) and B-11(p,n) reactions.

PTB Experiment

Deuterons between 6.3 MeV and 12.0 MeV, extracted from the PTB compact cyclotron CV-28, interacted with a deuterium gas target 1.1 cm in diameter and 3 cm 2 long. Neutrons between 9.1 and 14.6 MeV were produced via the D(d,n)He-3 reaction. The target cooled by streaming air worked with a gas pressure of 0.2 MPa. Disks of high-purity metallic foils 1 cm in diameter, of aluminum (99.999 %) and iron (99.99+ %) were irradiated at a distance of 3.9 cm from the target. Each iron foil (2 mm thick) was sandwiched between two aluminum foils 1 mm thick. The 'gas-out' corrections ranged from 0 to 3.6 % for Fe-54(n,α), from 1.5 % to 21 % for Fe-54(n,p) and from 0.2 % to 9.5 % for Al-27(n,α). Before each irradiation the neutron energy was measured by the time-of-flight method, operating the cyclotron in a pulsed mode (about 1 ns pulse width and 1 MHz repetition frequency). The flight path between the target and an NE213 neutron detector 10.2 cm in diameter and 2.54 cm thick, was 12 m. With the Monte Carlo code SINENA [4] simulating the neutron production in the gas target, the neutron energy measured and its distribution was transformed to that of the close sample-to-target geometry. The energy resolution was between 174 keV and 204 keV (FWHM).

Data Analysis

Radioactivity counting

The Cr-51 and Mn-54 radioactivity of the iron samples was measured with various Ge(Li) detectors. In the ANL experiment three different detectors were involved allowing fairly long counting times per sample. The sample-to-detector distances varied from 1 to 2 cm. At PTB, a single detector was used with a sample distance of 1.6 cm. The nominal efficiencies were corrected for sample size effects (radial and axial dependence) and for self-absorption within the samples. A PTB standard source of Cr-51 was used to intercompare the various detectors. In addition, a few iron samples were measured at ANL as well as at PTB. For the measured photopeaks of 320.1 keV (Cr-51) and 834.8 keV (Mn-54) a fair consistency between the various independently calibrated detectors was found which confirmed the calibration uncertainties of the ANL detectors (2 % for Cr-51 and 1.5 % for Mn-54) and of the PTB detector

---

*Present address: Battelle Pacific Northwest Laboratories, Richland, Wash. 99352, U.S.A.
**Present address: IPEN, 01000 Sao Paulo, Brazil

(1.5 %). The Na-24 activity of the aluminum foils of the PTB experiment was measured at a sample-to-detector distance of 5.3 cm, reducing the correction for summing coincidence losses from 9.7 % to 2.4 %. Other details are given in Ref. [2]. Recommended decay parameters [5] for the half-life ($T_{1/2}$) and the photon emission probability (h_γ) were used:

Cr-51 $T_{1/2}$ = 27.71(3) d h_γ = 0.0985(9)

Mn-54 $T_{1/2}$ = 315.5(5) d h_γ = 0.99975(3)

A value of (5.8 ± 0.1) % was adopted for the isotopic abudance of Fe-54 [6].

Influence of scattered neutrons

Neutrons scattered from the wall of the gas target, from support structures, the sample holder, the fission chamber material and from neighboured samples and also scattered within the samples, increase the reaction rates of the nuclear reactions investigated. To take this into account, Monte Carlo calculations were done which determined the individual contributions to each reaction. The code described in Ref. [7] simulates elastic and inelastic scattering processes. Collisions of a second and higher order were neglected, an approximation which is valid as long as the neutron transmission through the scatterer is large.

In the ANL experiment these corrections were between 12.7 % (5.3 MeV neutrons) and 8.9 % (9.8 MeV neutrons) for U-238(n,f), between 7.6 % and 5.6 % for Fe-54(n,p) and between 5.9 % and 4.1 % for Fe-54(n,α). In the PTB experiment, with a different geometry the corrections were between 3.8 % (9.1 MeV neutrons) and 3.3 % (14.6 MeV neutrons) for Al-27(n,α), between 5.7 % and 5.4 % for Fe-54(n,p) and between 4.1 % and 3.5 % for Fe-54(n,α). In ratio measurements only the difference between these values remains as a final correction factor.

Correction for breakup neutrons

A second effect which increases the reaction rates is due to breakup neutrons of the D(d,np) reaction. At the maximum neutron energy of 9.82 MeV used in the ANL experiment, these corrections were 10.5 % for U-238(n,f), 1.8 % for Fe-54(n,p) and negligible for Fe-54(n,α). In the PTB experiment the corrections at 14.64 MeV were 9.7 % for Al-27(n,α) and 21.9 % for Fe-54(n,α). For Fe-54(n,p) the correction is extreme, with a ratio of the contribution to the radioactivity of 2.628:1 for the breakup neutrons compared with the 14.64 MeV neutrons. These corrections were calculated on the basis of detailed measurements of breakup neutron spectra. The measurements covered the deuteron energy range between 5.3 and 13.3 MeV and were done for neutron emission angles up to 15 degrees [8]. Details and examples are given in the reference. The uncertainty of the corrections was estimated at 3 % of the amount of correction.

Final uncertainties

Further uncertainty components which have been included were the time variation of the empty gas cell correction (0.5 - 2.5 %), the non-uniform distribution of the activity in the samples (0.5 %), impurities of the sample material (0.0 - 0.5 %), fission fragment losses (1 %) and extrapolation to zero bias (1.1 %), the deposit mass (0.7 %), geometry factors

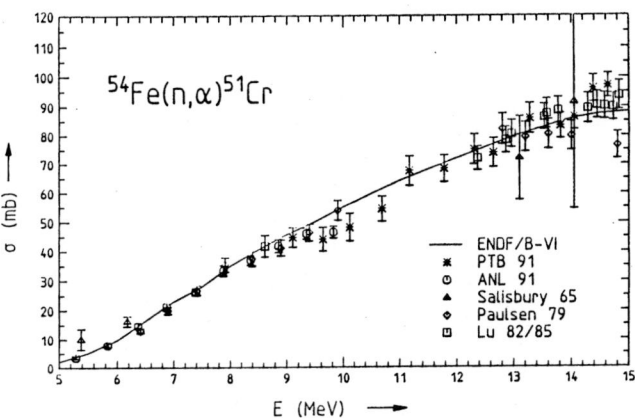

Fig. 1. Experimental data from the present work (ANL91 and PTB91) for the Fe-54(n,α) reaction

between the different sample positions (1.4 %) and the influence of the energy scale uncertainty. The last component is negligible for the relatively flat Fe-54(n,p) cross section but amounts to between 9.0 % and 0.1 % for the Fe-54(n,α) reaction in the energy range from 5.3 to 9.8 MeV. The uncertainties of the normalization cross sections were between 0.7 and 0.9 % for U-238(n,f) and between 0.5 and 3.2 % for Al-27(n,α).

The final uncertainties (one standard deviation) were obtained by combining the individual components quadratically. The range of the results was: 4.1 % - 13.3 % (0.8 % - 9.1 %) for Fe-54(n,α) and 3.2 % - 4.7 % (0.2 % - 3.4 %) for Fe-54(n,p) in the ANL experiment and 4.2 - 9.7 % (2.5 - 8.6 %) for Fe-54(n,α) and 3.8 - 8.6 % (0.8 - 1.4 %) for Fe-54(n,p) in the PTB experiment. The values of the counting statistics are separately listed in brackets. The large uncertainties of the Fe-54(n,p) data of the PTB experiment at high neutron energies reflect the amount of the breakup correction, which strongly increases with the energy.

Results and Conclusions

The data from the present work for the Fe-54(n,α) reaction between 5.3 and 14.6 MeV are plotted in Fig. 1. They are compared with the ENDF/B-VI evaluation and data from the few other experiments which have covered a similarly large energy range. The references were taken from the CINDA index [9]. A fair agreement between the ANL and the PTB experiment was found in the overlap region. With the exception of the radioactivity counting, in each of the two experiments slightly different experimental techniques and data analysis methods were used, which indicates a high degree of reliability for both experimental procedures. Below 9 MeV, the present data are in fair agreement with the ENDF/B-VI evaluation. Around 10 MeV, our data indicate a shoulder in the excitation function and are about 15 % lower compared with ENDF/B-VI. At high neutron energies, our data show a larger slope than the ENDF/B-VI evaluation. This trend is in agreement with another data set (Lu82/85). All other experiments were undertaken with 14 MeV neutrons. This is shown in Fig. 2 where all post-1975 experiments are plotted and compared with our data and with ENDF/B-VI. Above 14.3 MeV, our data are about 10 % higher than

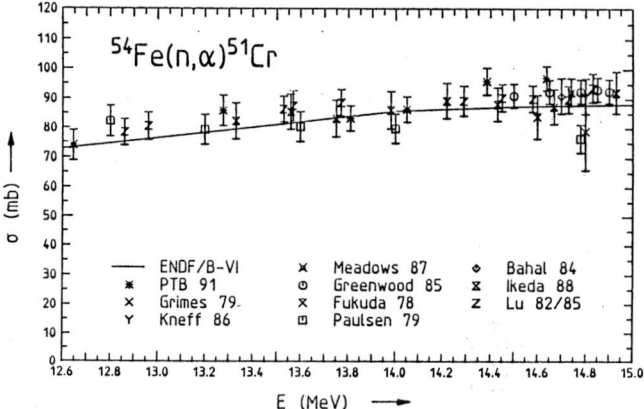

Fig. 2. The Fe-54(n,α) reaction between 12.5 and 15 MeV

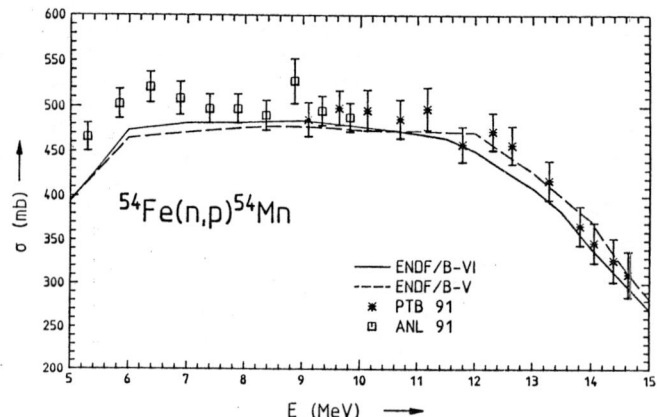

Fig. 3. Experimental data from the present work for the Fe-54(n,p) reaction

ENDF/B-VI and also higher than most of the data from other experiments.

In Fig. 3 our data on the Fe-54(n,p) reaction are plotted. For this reaction too, the agreement between the ANL and PTB experiment in the overlap region is very satisfactory. Above 7 MeV most of our data agree within the uncertainties with the ENDF/B-VI evaluation. Nevertheless, our data show a general tendency to be about 3 - 4 % higher than ENDF/B-VI. At the low energies in Fig. 3 this deviation amounts to 10 %. At high neutron energies substantial correction factors for breakup neutrons were applied to our data. A useful test of the validity of these corrections is a comparison with the 14 MeV data obtained with the T(d,n)He-4 neutron production reaction which is free of such corrections. There is usually a large amount of data for around 14 MeV (see Fig. 2, for example). A disadvantage is that these data very often show a large spread in their values and there is no guarantee that the average of these data represents a 'true' result. Two recent evalutions of 14 MeV data [10, 11] were used for a comparison. Both were performed at 14.70 MeV. Our 14.64 MeV data were extrapolated to this energy by using the individual slopes of the ENDF/B-VI evaluation for each reaction. The result is summarized in Table 1. The PTB experiment also comprised a measurement of the Fe-56(n,p) reaction which is not shown in the present work. The corresponding 14.64 MeV data point is included in the comparison in Table 1.

Within the uncertainties there is agreement between our data and the evaluations. For the Fe-54(n,α) reaction the agreement is at least within two standard deviations. However, our data on Fe-54(n,α) and Fe-54(n,p) are about 10 % and 8 % higher than the evaluated data. There is no clear correlation of these deviations with the magnitude of the breakup corrections. The net correction for Fe-54(n,α) is 12.2 %, the difference between a 21.9 % correction of this

reaction and a 9.7 % correction of the Al-27(n,α) monitor reaction. The Fe-56(n,p) reaction has a very similar net correction of 11.9 %. But for the Fe-54(n,p) reaction the breakup correction is an order of magnitude larger. Only 27.5 % of the measured reaction rate is from the 14.64 MeV neutrons.

The agreement shown in Table 1 is very encouraging. It indicates that also very large breakup corrections can be performed at a high level of confidence. Data obtained with the D(d,n)He-3 reaction are rarely superior to those measured with the T(d,n)He-4 reaction. This is due to the increasing uncertainties resulting from the breakup corrections. But in the intermediate region between 10 and 13 MeV neutron energy, the quality of data measured with the D(d,n)He-3 reaction does not seem to be seriously handicapped by corrections for breakup neutrons.

References

1. M. Wagner, H. Vonach, A. Pavlik, B. Strohmaier, S. Tagesen, J. Martinez-Rico, Physics Data 13-5 (1990)
2. G. Börker, H. Klein, W. Mannhart, M. Wagner, G. Winkler, in Proc. Int. Conf. on Nuclear Data for Science and Technology, May/June 1988, Mito, Japan, (Edited by S. Igarasi) p. 1025, Saikon, Tokyo (1988)
3. J.W. Meadows, Report ANL/NDM-118, Argonne National Laboratory (1990)
4. B.R.L. Siebert, H.J. Brede, H. Lesiecki, Report PTB-ND-23, Phys.-Tech. Bundesanstalt (1982)
5. U. Schötzig, H. Schrader, Report PTB-Ra-16/3, Phys.-Tech. Bundesanstalt (1989)
6. P. De Bièvre, I.L. Barnes, Int. Journal of Mass Spectrometry and Ion Processes 65, 211 (1985)
7. D.L. Smith, J.W. Meadows, Report ANL/NDM-37, Argonne National Laboratory (1977)
8. S. Cabral, G. Börker, H. Klein, W. Mannhart, Nucl. Sci. Eng. 106, 308 (1990)
9. CINDA-A (1935 - 1987) and CINDA90 (1988 - 1990), IAEA, Vienna (1990)
10. B.P. Evain, D.L. Smith, P. Lucchese, Report ANL/NDM-89, Argonne National Laboratory (1985)
11. T.B. Ryves, European App. Res. Rept. - Nucl. Sci. Technol. 7, 1241 (1989)

Table 1. Comparison with evaluated cross sections (in mb) at 14.70 MeV

Reaction	Present work*	Evaluation	Ref.
Fe-54(n,α)	97.2 ± 4.0	87.9 ± 2.4	[10]
Fe-54(n,p)	306.2 ± 26.7	284.4 ± 5.7	[10]
Fe-56(n,p)	109.0 ± 3.1	108.4 ± 0.5	[11]

*Extrapolated from 14.64 MeV to 14.70 MeV

NEUTRON ACTIVATION CROSS SECTIONS FOR ^{60}Ni(n,p)^{60m}Co
^{60}Ni(n,p)$^{60m+g}$Co AND ^{58}Ni(n,p)^{58m}Co REACTIONS
IN THE 5 TO 12 MeV ENERGY RANGE

S. Sudár and J. Csikai

Institute of Experimental Physics, Kossuth University,
H-4001 Debrecen, Pf.105, Hungary

S.M. Qaim and G. Stöcklin

Institut für Chemie 1 (Nuklearchemie),
Forschungszentrum Jülich GmbH, 5170 Jülich, FRG

Abstract: Excitation functions were measured from 5 to 12 MeV for ^{60}Ni(n,p)^{60m}Co, ^{60}Ni(n,p)$^{60m+g}$Co and ^{58}Ni(n,p)^{58m}Co reactions using dd gas targets at the variable energy cyclotrons at Jülich and Debrecen. The measured data for the ^{60}Ni(n,p)$^{60m+g}$Co reaction deviate considerably from the older data and do not show the earlier found structure. Our data, together with some other new measurements, indicate that for this reaction the cross sections in the libraries should be revised. The data for the ^{60}Ni(n,p)^{60m}Co and ^{58}Ni(n,p)^{58m}Co reactions give new information in this energy range.

(Cross section, ^{58}Ni,^{60}Ni, (n,p) reaction, excitation function, metastable states)

Introduction

Neutron cross sections in the 5 to 12 MeV neutron energy range are rather inaccurate or even lacking for many nuclear reactions, especially those leading to the formation of long-lived or rather short-lived products. The data for ^{60}Ni(n,p)$^{60m+g}$Co reaction in the neutron cross section libraries are based mainly on a measurement by Paulsen [1] which shows a questionable structure in the 8 to 12 energy range. Preliminary results of a new measurement [2] also indicated that this excitation function needs reinvestigation.

Our knowledge on the formation of metastable states in neutron induced reactions is rather scanty. Detailed studies can give new possibilities to check theoretical reaction models.

Experimental

Irradiation and neutron flux monitoring

The quasi-monoenergetic neutrons in the 5.3 MeV to 12 MeV neutron energy range were produced via ^{2}H(d,n)^{3}He reaction using deuteron gas targets at the variable energy Compact Cyclotron CV28 [3] at Jülich and the MGC-20 cyclotron at Debrecen. The nickel samples for ^{60}Ni(n,p)$^{60m+g}$Co reaction consisted of nickel disc (1.3 cm φ x 2 mm) pressed from 99.999 % pure nickel powder at Jülich or a stack of Ni foils (1.9 cm φ x 2 mm) or Ni disc (1.9 cm φ x 0.8 mm) from 99.995 % pure metal at Debrecen. The samples for the measurement of ^{58}Ni(n,p)^{58m}Co and ^{60}Ni(n,p)^{60m}Co reactions were Ni foils (1.9 cm φ 0.15 mm) of 99.9 % pure metal.

Al and in some cases Fe foils of the same size as the Ni sample were attached in front and at the back to monitor the neutron flux density. The samples and the monitor foils were placed at a distance between 1 cm and 3 cm from the end of the gas target in 0° direction to the deuteron beam for the irradiation. The mean neutron energy for each sample was calculated using a Monte Carlo program [4]. Two irradiations for the metastable states and one for ^{60}Ni(n,p)$^{60m+g}$Co were done at each deuteron energy. The calculated mean neutron energies for the foils and the Ni disc were slightly different because of the differences in distances. The typical irradiation times were 30 minutes for measurements on the cross sections of the metastable states and 4 to 12 hours for measurement on the ^{60}Ni(n,p)$^{60m+g}$Co reaction.

Measurement of radioactivity

The radioactivities of the ^{60}Ni(n,p)$^{60m+g}$Co and the monitor reactions products were determined using a 35 cm^{3} coaxial Ge(Li) detector at Jülich and an intrinsic germanium detector at Debrecen. The detectors were connected to computer based analyzers for the measurements and evaluation. The count rates were corrected for self absorption, pile-up, coincidence loss, geometry, efficiency of the detector and γ-ray abundance.

The measurement of the radioactivities of the ^{60}Ni(n,p)^{60m}Co and ^{58}Ni(n,p)^{58m}Co products presented some difficulty. These measurements were done at Jülich. The ^{60m}Co decays with a half-life of 10.47 min via IT (99.75 %) and β^{-} (0.25 %). The 1332 keV γ-ray associated with the latter process has low and uncertain intensity, therefore, it is not suitable for cross section measurement. In spite of the strong conversion of the IT, both the low energy γ-ray (58.6 keV) and the Co X-ray could be measured by a planar intrinsic germanium detector having Be window.

The ^{58}Ni(n,p)^{58m}Co product decays with a half-life of 9.15 hour and the IT is very strongly converted (I_{γ}=0.0363 %). Therefore, the cross section for the formation of the meta stable state can be measured either through the Co X-rays or via the activity of the ground state as a function of growth time. We chose the first method, because the X-rays were detected by the planar Ge detector used for the measurement of the activity of ^{60m}Co. Since the Co X-rays originated from different sources the intensity of the X-rays was determined as a function of time. Fe X-rays were also detected from the decay of ^{58g}Co and they partly overlapped the Co lines; therefore, they were evaluated together. The intensities of the components were determined by fitting a complex decay curve to the measured data. The relevant decay data of ^{58m}Co and ^{60m,g}Co are summarized in Table 1.

Table 1. Decay Data of Radioactive Products[*]

Nuclide	$T_{1/2}$	E_γ or E_x [keV]	I_γ or I_x [%]
^{58m}Co	9.15±0.10 h	6.915	7.8±0.4
		6.930	15.4±0.8
		7.649	2.79±0.15
^{60}Co	5.271±0.001 y	1173.2	99.90±0.02
		1332.5	99.9824±0.0005
^{60m}Co	10.47±0.04 min	6.915	8.8±0.5
		6.930	17.4±0.9
		7.649	3.16±0.17
		58.60	2.01±0.12

[*] All data taken from ref. 5

The efficiency calibration of the planar intrinsic germanium detector with Be window was performed by means of calibrated point like γ-ray and X-ray sources (^{241}Am, ^{55}Fe, ^{57}Co, ^{133}Ba) supplied by Amersham and PTB. To get a reliable efficiency curve, it was very important to take into account the coincidence summing of the gamma and the X-ray lines. Since our samples were not point like, the efficiency was measured as a function of the position, both in radial and longitudinal directions.

The efficiencies of the Ge(Li) detector at Jülich and the intrinsic Ge detector at Debrecen were determined by calibrated point sources (^{60}Co, ^{54}Mn, ^{137}Cs, ^{57}Co, ^{22}Na) and a ^{226}Ra extended source, supplied by Amersham, PTB and Hungarian National Bureau of Standards.

Table 2. Measured Cross Sections for ^{60}Ni(n,p)$^{60m+g}$Co reaction

E_n [MeV]	σ_{ref}[*] [mb]	^{60}Ni(n,p)$^{60m+g}$Co [mb]
Jülich		
5.230±0.10	482±34	13.9±2.6
6.300±0.10	597±42	37.6±3.7
6.800±0.11	612±43	52.2±4.9
7.290±0.11	620±44	63.4±5.9
7.770±0.12	625±38	68.5±6.4
8.240±0.12	627±44	79.1±10
8.710±0.12	627±44	94.2±8.7
9.170±0.13	624±44	92.5±9.5
9.630±0.13	618±44	96.9±9.0
10.080±0.14	611±46	105.5±9.6
Debrecen		
5.427±0.055	508±41	17.2±2.2
5.935±0.055	574±46	29.0±3.0
6.433±0.055	604±48	43.8±4.0
6.847±0.057	613±49	50.9±5.0
8.070±0.057	627±50	82.2±7.8
8.480±0.055	628±50	81.6±8.0
8.960±0.057	626±50	90.0±9.0
11.000±0.054	589±47	115.0±11
12.300±0.081	527±42	147.0±24

[*] Ref. cross sections data (^{58}Ni(n,p)$^{58m+g}$Co) were taken from IRDF-90

Results and Discussion

The results for the ^{60}Ni(n,p)$^{60m+g}$Co reaction are summarized in Table 2, which also gives the mean neutron energy, error of the mean neutron energy and the International Radiation Dosimetry File (IRDF-90) cross sections for the ^{58}Ni(n,p)$^{58m+g}$Co reaction used as an internal standard. The error of the mean neutron energy was estimated from the error in the deuteron energy, error in the thickness of the entrance window, stopping power of the window material and the D_2 gas. The neutron fluxes and the activities of ^{58m}Co,$^{60m+g}$Co, ^{60m}Co have been corrected both for the neutron production in the window and beamstop material, doing gas-in/out experiments, and for the break-up neutrons, using the results in ref. 7. The error contributions were added in quadrature.

The data from the present investigation, earlier measurements and the ENDF/B-V evaluation for ^{60}Ni(n,p)$^{60m+g}$Co reaction are shown in Fig. 1. It can be seen, that the structure in the cross section above 9 MeV, described by Paulsen [1], does not exist in our data. Our results are agreement with the data of Vonach et al [2] measured recently using a p-t neutron source.

The results for the ^{60}Ni(n,p)^{60m}Co and ^{58}Ni(n,p)^{58m}Co reactions are summarized in Table 3. In these two cases different monitor reactions were used. The errors in the neutron energies and cross sections were estimated as described above. Each cross section value for the ^{60}Ni(n,p)^{60m}Co reaction is the weighted mean of four data points, obtained from the two irradiations at each energy (each irradiation gave two values, one from X-ray and the other from γ-ray counting). The data for the ^{58}Ni(n,p)^{58m}Co reaction are the mean of the values from the two irradiations. The data are shown in Fig. 2, with a value for the ^{60}Ni(n,p)^{60m}Co reaction from Paulsen [6] at 8.09 MeV, which was the only data point below 14 MeV. It can be concluded that the cross sections for the formation of the metastable states constitute significant parts of the full activation cross sections for these reactions.

Table 3. Measured Cross Sections for ^{58}Ni(n,p)^{58m}Co and ^{60}Ni(n,p)^{60m}Co Reactions

E_n [MeV]	σ_{ref} [mb]	^{58}Ni(n,p)^{58m}Co [mb]	^{60}Ni(n,p)^{60m}Co [mb]
5.208±0.10	2.36±0.16[a]	135±12	8.8±1.0
6.273±0.10	3.94±0.31[b]	229±24	32.3±3.9
6.774±0.11	27.3±1.5 [a]	217±17	39.7±4.0
7.256±0.11	23.1±1.8 [b]	239±22	48.3±5.0
7.738±0.12	36.4±2.4 [b]	251±18	55.2±5.6
8.209±0.12	47.0±3.1 [b]	276±23	64.5±6.3
8.675±0.14	63.7±4.3 [b]	273±23	68.4±6.7
9.132±0.14	76.2±4.5 [b]	281±22	72.1±6.7
9.590±0.15	84.8±4.9 [b]	302±23	79.7±7.4
10.04±0.15	89.9±4.8 [b]	271±20	81.8±7.4

[a] σ_{ref} ^{56}Fe(n,p)^{56}Mn,

[b] σ_{ref} ^{27}Al(n,α)^{24}Na. All ref. data from IRDF-90

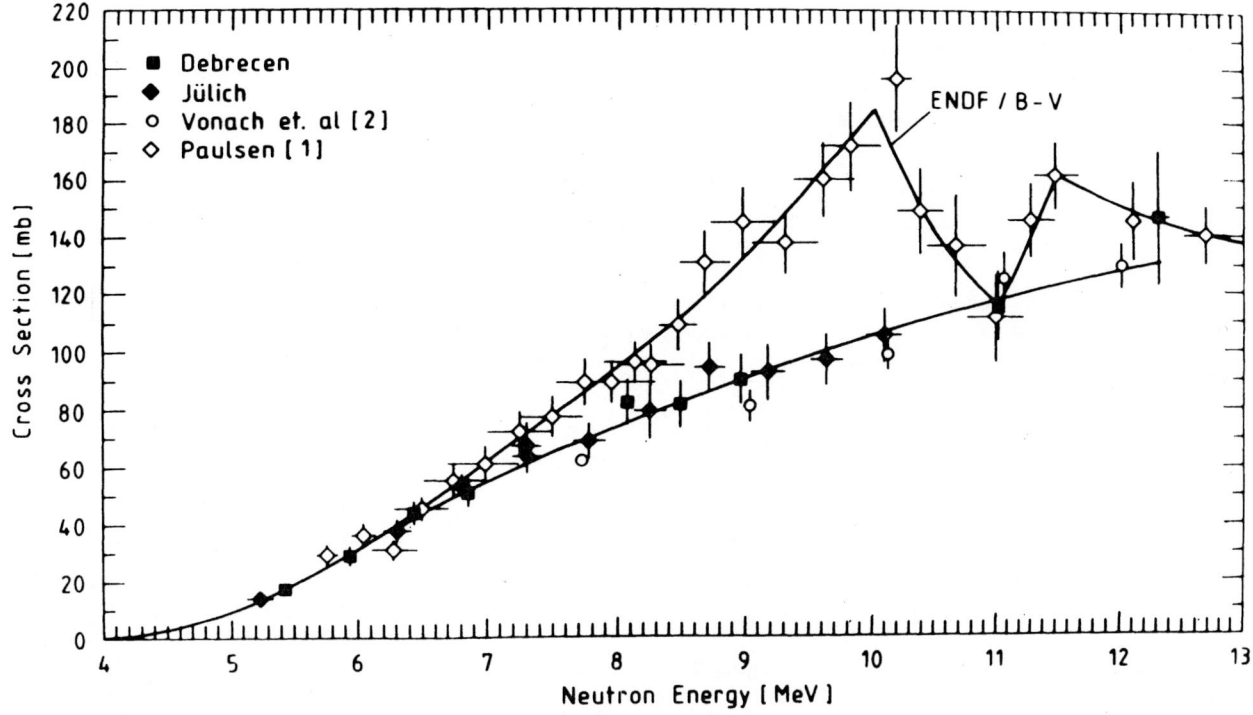

Fig. 1. Excitation function of $^{60}Ni(n,p)^{60m+g}Co$ reaction. The curve through our data is an eye-guide

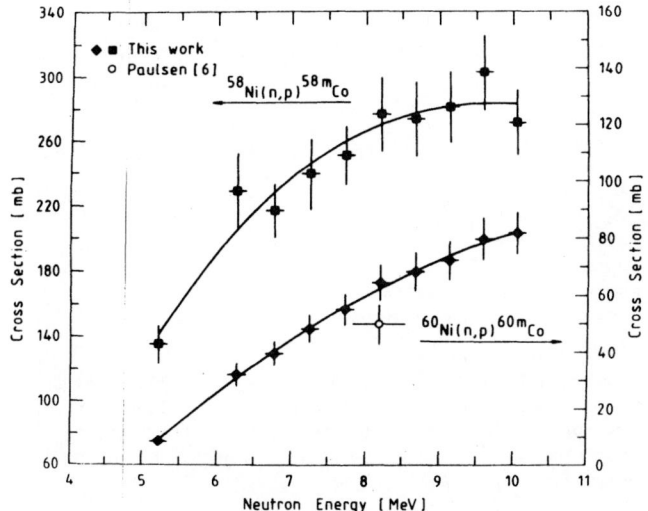

Fig. 2. Excitation function of $^{58}Ni(n,p)^{58m}Co$ and $^{60}Ni(n,p)^{60m}Co$ reactions

Acknowledgement

We would like to thank the staff of the cyclotrons at Jülich and Debrecen for providing deuteron beam. The Debrecen authors thank P. Raics, S. Nagy and S. Szegedi for their help in irradiations.

REFERENCES

1. A. Paulsen: Nukleonik 10, 91 (1967)

2. H. Vonach. M. Wagner and R.C. Haight, in Proc. NEANDC Specialists' Meeting on Neutron Activation Cross Section for Fission and Fusion Energy Applications, September 1989, Argonne, USA, (Edited by M. Wagner and H. Vonach) p. 165, OECD, Paris(1990)

3. S. M. Qaim, R. Wölfle, M.M. Rahman and H. Ollig: Nucl. Sci. Eng. 88, 143 (1984)

4. A. Suhaimi, Studies of (n,t) reactions on light nuclei, Jül-2169, KFA Jülich (1988)

5. E. Browne, R.B. Firestone and V.S. Shirley, Table of Radioactive Isotopes, J. Wiley & Sons, New York (1986)

6. A. Paulsen: Z. für Physik 205, 226 (1967)

7. S. Cabral, G. Börker, H. Klein and W. Mannhart: Nucl. Sci. Eng. 106, 308 (1990)

This work was supported by the Hungarian Research Foundation (Contract No. 1734/91) and the German-Hungarian bilaterial agreement (No. 13(X237.1)).

ACTIVATION CROSS SECTION MEASUREMENT AT NEUTRON ENERGIES OF 9.5, 11.0, 12.0 AND 13.2 MeV USING ^{1}H(^{11}B,n)^{11}C NEUTRON SOURCE AT JAERI

Y. Ikeda, C. Konno, M. Mizumoto, K. Hasegawa, S. Chiba,
Y. Yamanouchi and M. Sugimoto

Japan Atomic Energy Research Institute,
Tokai-mura, Ibaraki-ken 319-11, Japan

Abstract: Some activation cross sections at neutron energies of 9.5, 11.0, 12.0 and 13.2 MeV have been measured by the foil activation technique using a novel neutron source via ^{1}H(^{11}B,n)^{11}C reaction at JAERI. The experimental procedure is described and the data obtained are discussed by comparing with cross sections in JENDL-3 Dosimetry File. Applicability of the present neutron source to the foil activation technique is also noted.

(activation technique, TANDEM accelerator, dosimetry application, nuclear data, monoenergetic neutron source, ^{1}H(^{11}B,n)^{11}C reaction)

Introduction

The deficiency in experimental data of neutron cross sections in the energy region between 10 MeV and 13 MeV has been often stressed. This shortage has been mainly due to the lack of the appropriate monoenergetic neutron sources. Although a use of the reaction of ^{1}H(t,n)^{3}He for generating such monoenergetic neutrons has been reported at LANL[1], the availability is not sufficient to conduct a program to provide systematic data. So far, only a cross section measurement for the ^{58}Ni(n,p)^{58}Co and ^{60}Ni(n,p)^{60}Co has been reported by Vonach and Wagner[2]. Recently, use of ^{1}H(^{11}B,n)^{11}C reaction to generate monoenergetic neutrons from 10 to 13 MeV has been investigated in JAERI by using the TANDEM heavy ion accelerator[3]. This source with significant low background neutrons was applied success-fully in the secondary γ–ray production measurement[4].

Utilizing this particular neutron source, we have carried out activation cross section measurements to provide data at energies of 9.5, 11.0, 12.0 and 13.2 MeV for several important reactions in the dosimetry application.

Experiment

Reactions investigated

The reactions investigated in the present experiment are listed in Table 1 along with the associated decay data of the produced radioisotopes[5]. They are very important reac-tions for the dosimetry application where the accurate cross sections are needed. In particular, the reaction of ^{27}Al(n,α)^{24}Na is used as a standard for cross section determination experiments. The reactions ^{90}Zr(n,2n)^{89}Zr and ^{93}Nb(n,2n)^{92m}Nb are the most promising monitor reactions for the 14 MeV neutron flux. The reactions ^{47}Ti(n,p)^{47}Sc, ^{64}Zn(n,p)^{64}Cu and ^{115}In(n,n')^{115m}In are spectral indices for fast neutrons with relatively low threshold energies.

Neutron source and irradiation

Monoenergetic neutrons were generated by bombarding the hydrogen gas target with ^{11}B beam, the energy of which ranged from 45 to 68 MeV by using the TANDEM accelerator at JAERI. Four different irradiations were conducted changing the energy of the incident ^{11}B beam. To subtract contributions of the parasitic neutrons produced in the structural materials of the target assembly to the total reaction rate, irradiation run without ^{1}H gas in the target cell was carried out for each irradiation time. Irradiation usually lasted for 10 hours for both gas in and gas out runs. Typical neutron spectrum emitted from the target to the forward direction is illustrated in Fig. 1, which corresponds to the neutron peak energy of 9.5 MeV.

The size of the irradiation sample was 20 mm in diameter and 1 mm in thickness. Sample foils were stacked into a package in the

Table 1 Reactions and associated decay data

Reaction	a	T$_{1/2}$	Eγ	b
^{27}Al(n,α)^{24}Na	100	15.02 h	1368	100
^{47}Ti(n,p)^{47}Sc	7.32	3.35 d	159.4	68.0
^{48}Ti(n,p)^{48}Sc	73.99	43.7 h	983.5	100
^{64}Zn(n,p)^{64}Cu	48.89	12.70 h	511.0	35.8
^{90}Zr(n,2n)^{89}Zr	51.46	78.4 h	909.2	99.87
^{93}Nb(n,2n)^{92m}Nb	100	10.15 d	934.46	99.15
^{115}In(n,n')^{115m}In	95.77	4.3 h	336.2	46.7
^{197}Au(n,2n)^{196}Au *	100	6.183 d	355.65	87.7

* Flux monitor reaction
a; Abundance(%)
T$_{1/2}$; Half-life
Eγ; γ-ray energy(keV)
b; Branching ratio(%)

order; Al/Zr/Nb/In for the irradiation at 13.2 MeV, and Al/Ti/Nb/Zn/In for the irradiations at 9.5, 11.0 and 12.0 MeV. Two gold foils as neutron flux monitor were attached on rear and back sides of each package. The sample package was placed at a distance of about 100 mm from the center of the target cell.

Gamma-ray counting and cross section determination

After irradiation, reaction rates were derived from γ–ray counts measured with a Ge detector, performing necessary corrections. A correction factor for the neutron flux fluctuation during the irradiation was obtained from the output data of neutron flux monitor recorded in a chart. Net reaction rates were finally obtained by subtracting the data for irradiation with gas out from that with gas in, the beam current of which was adjusted to be equal to one with gas in.

As the flux monitor, the ^{197}Au(n,2n)^{196}Au reaction was employed by considering its large reaction cross section value and rather flat response around energy from 10 to 14 MeV. Because the neutron flux level was expected not to be strongly enough for using ^{27}Al(n,α)^{24}Na as the monitor. The cross section for the ^{197}Au(n,2n)^{196}Au reaction was taken from JENDL Dosimetry File[5].

Cross sections for the reactions of interest were obtained by using measured reaction rates and the neutron fluxes determined by the monitor reaction. A contribution of the low energy neutron component below the peak of neutron to the net reaction rate was estimated from a neutron spectrum which was obtained by adjusting a measured spectrum with TOF method by the SAND-II unfolding code, utilizing the measured

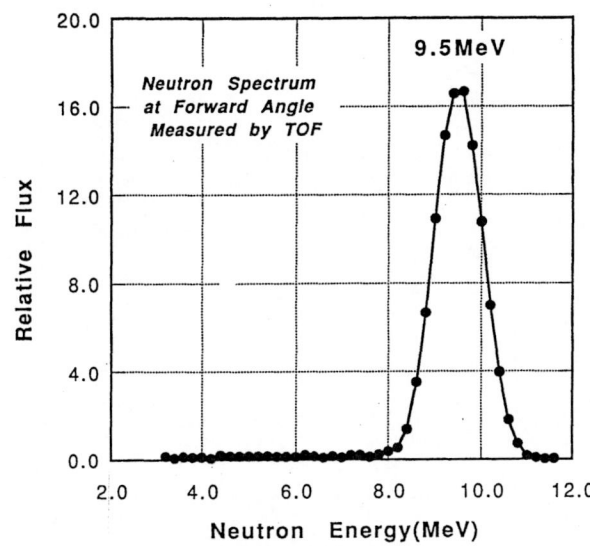

Fig. 1 Neutron spectrum with 9.5 MeV peak at forward direction measured by TOF technique.

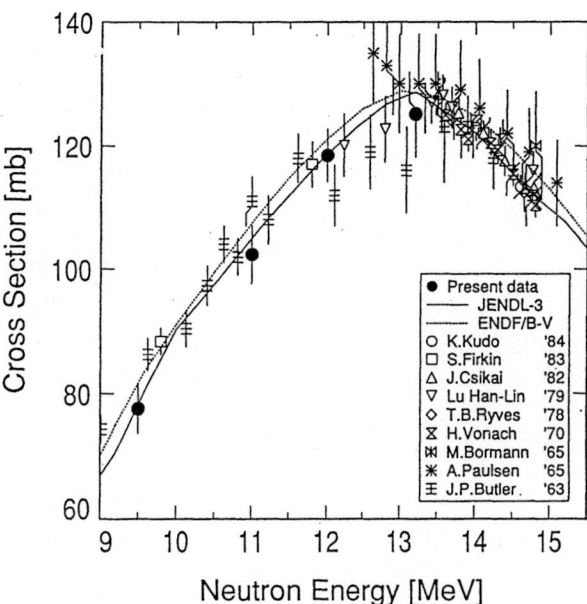

Fig. 2 Cross section data for ^{27}Al(n,α)^{24}Na reaction.

reaction rates. The uncertainty of mean neutron energy was estimated to be within ± 0.5 MeV. No correction for the shift in the mean energy was considered for the cross section of ^{93}Nb(n,2n)^{92m}Nb, which is steep in the energy range from 9 to 13 MeV.

Results and discussion

Cross sections obtained in the present experiment are summarized in Table 2.

^{27}Al(n,α)^{24}Na

The present data were in a good agreement with data in JENDL-3 within experimental errors. In particular, the data at 13.2 MeV gives a reasonably good consistency with values of the main body of data around 14 MeV region previously reported, as shown in Fig. 2. Since this reaction cross section at 14 MeV has been studied with satisfactory accuracy, the present result ensures us the reliability of the present experimental procedure.

^{47}Ti(n,p)^{47}Sc

Two data at 11.0 and 12.0 MeV measured in the present experiment gave slightly larger values than those in JENDL-3 and

ENDF/B-V. However, the measured data show a reasonably smooth connection with the data around 14 MeV measured at FNS[7]. Meanwhile, Mannhart[8] reported that the cross sections below 8 MeV to the threshold should be considerably lower by 10 -20 % than ENDF/B-V evaluation. Concerning the reverse trend of the cross section between the different energy ranges, data in the energy region from 8 to 10 MeV were still highly required.

^{48}Ti(n,p)^{48}Sc

Contrary to ^{47}Ti(n,p)^{47}, data for this reaction at 9.5, 11.0 and 12.0 MeV were lower than those in JENDL-3 and ENDF/B-V. Since this reaction is often used for dosimetry purpose, further examination is needed in the evaluation by taking the present data into account.

^{64}Zn(n,p)^{64}Cu

Two data at 9.5 and 11.0 MeV were measured in the present experiment. The data were in very good agreement with that in JENDL-3 Dosimetry File. However, the data around 14 MeV measured at FNS[7] were considerably lower than those in the JENDL-3 Dosimetry File. This result encourages us to measure the data at energy from 12 to 13 MeV using the present neutron source.

Table 2   Cross sections measured

Reactions	Cross sections (mb)			
	9.5MeV	11.0 MeV	12.0 MeV	13.2 MeV
^{27}Al(n,α)^{24}Na	77.6 ±3.9 **	102.4 ± 4.5	118.5 ± 4.3	125.1 ± 6.9
^{47}Ti(n,p)^{47}Sc	--------	142 ± 20	152 ± 14	------
^{48}Ti(n,p)^{48}Sc	24 ± 5	36.0 ± 2.2	44.3 ± 2.0	------
^{64}Zn(n,p)^{64}Cu	266 ± 38	275 ± 17	------	------
^{90}Zr(n,2n)^{89}Zr	---------	- - - - - -	------	305 ± 17
^{93}Nb(n,2n)^{92m}Nb	64 ± 8	219 ± 12	345 ± 13	444 ± 23
^{115}In(n,n')^{115m}In	274 ± 34	230 ± 16	164 ± 16	93 ± 6
^{197}Au(n,2n)^{196}Au *	748	1549	1795	1970

* Flux monitor reaction. Cross section values were taken from JENDL-3 Dosimetry File.
** The uncertainties in the monitor reaction cross sections were not considered.

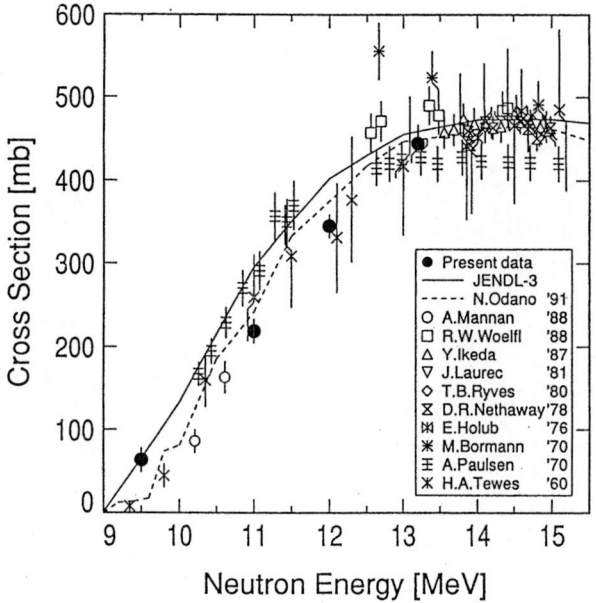

Fig. 3　Cross section data for ^{93}Nb(n,2n)^{92m}Nb reaction.

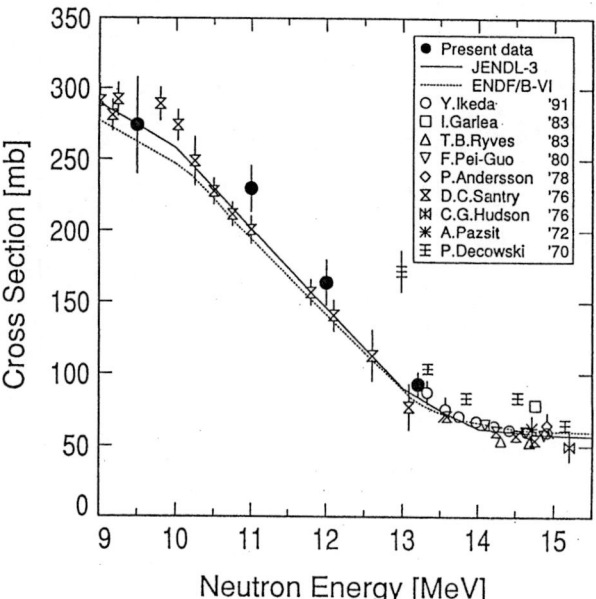

Fig. 4　Cross section data for ^{115}In(n,n')^{115m}In reaction.

^{90}Zr(n,2n)^{89}Zr

Data at one energy point at 13.2 MeV gave a slightly smaller value than that in JENDL-3. Nevertheless, good consistency with the data measured at FNS[7] was found. To confirm the validity of the evaluation, measurement of cross section at 12.5 MeV is required.

^{93}Nb(n,2n)^{92m}Nb

Since there were a large number of experimental data available around 14 MeV region, reasonable convergence could be expected in the evaluation. However, the present data at 11.0, 12.0 and 13.2 MeV were systematically lower than those of the JENDL-3 Dosimetry File as shown in Fig. 3. A reevaluation of this cross section including the present data is strongly recommended.

^{115}In(n,n')^{115m}In

The data measured are shown in Fig. 4. The data at 11, 12 and 13.2 MeV were systematically higher by about 5 % than those in JENDL-3 and ENDF/B-V evaluations. It was considered that the contribution of the low energy neutrons to the reaction rate should be significantly larger owing to the low threshold energy at 0.33 MeV for this particular reaction. Thus, further examination of the reaction sensitivity to the low energy part of the spectrum should be addressed to reduce the experimental error.

Summary

Though there are considerably large numbers of data in the energy region from 13.5 to 15 MeV, data below 13 MeV were very limited, as shown in Figs. 2 to 4. The present data supplemented significantly the data needed in this particular energy region.

In summary, it was demonstrated that the mono-energetic neutron source via ^{1}H(^{11}B,n)^{11}C was very powerful to provide experimental data in the deficient region. Further measurements on other important reactions are planned for the future.

Acknowledgements

Authors wish to express their sincere thank to the operation crew of the TANDEM accelerator at JAERI.

REFERENCES

1. R. C. Haight "Activation with an Intense Source of Monoenergetic Neutrons in the Range 8-14 MeV", Proc. of a Specialists' Meeting on "Neutron Activation Cross Section for Fission and Fusion Energy Applications", Argonne National Lab., USA, 13th-15th sept. 1989, OECD NEA, Paris (Editors M. Wagner and H. Vonach) pp95-97.

2. H. Vonach and M. Wagner,"Neutron Activation Cross-Section of ^{58}Ni and ^{60}Ni for 8-12 MeV Neutrons", ibid. pp165-176.

3. S. Chiba, M. Mizumoto, K. Hasegawa, Y. Yamanouchi, M. Sugimoto, Y. Watanabe and M. Drosg,"The ^{1}H(^{11}B,n)^{11}C Reaction as a Practical Low Background Monoenergetic Neutron Source in the 10 MeV Region," Nucl. Instr. & Methods, A281 581 (1989).

4. M. Mizumoto, K. Hasegawa, S. Chiba, Y. Yamanouchi, Y. Kawarasaki, M. Igashira, T. Uchiyama, H. Kitazawa and M. Drosg, "Gamma-ray Production Cross Sections of Some Structural and Shielding Materials," Proc. Int'l. Conf. on Nuclear Data for Sci. and Technol., May 30-June 3, 1988, Japan. (Editor S. Igarashi) Saikon Publishing Co. LTD., Tokyo (1988) pp 197-200.

5. C. M. Lederer and V. S. Shirley;"Table of Isotopes," 7th edition, John Wiley &Sons, Inc., New York (1978).

6. JENDL-3 Dosimetry File, to be published.

7. Y. Ikeda, C. Konno, K. Oishi, T. Nakamura, H. Miyade, K. Kawade, H. Yamamoto and T. Katoh, "Activation Cross Section Measurements for Fusion Reactor Structural Materials at Neutron Energy from 13.3 to 15.0 MeV Using FNS Facility," JAERI-1312 (1988).

8. W. Mannhart, D. L. Smith and J. W. Meadows, "Measurement of the ^{47}Ti(n,p)^{47}Sc Reaction Cross Section", Proc. of a Specialists' Meeting on "Neutron Activation Cross Section for Fission and Fusion Energy Applications", Argonne National Lab., USA, 13th-15th sept. 1989 OECD NEA, Paris (Editors M. Wagner and H. Vonach) pp95-97.pp121-134.

DIFFERENTIAL AND INTEGRAL CROSS SECTION MEASUREMENTS OF SOME
(n, charged particle) REACTIONS ON TITANIUM

S.M. Qaim, N.I. Molla, R. Wölfle and G. Stöcklin

Institut für Chemie 1 (Nuklearchemie)
Forschungszentrum Jülich GmbH
5170 Jülich, Federal Republic of Germany

Abstract: Cross sections were measured for the ^{47}Ti(n,p)^{47}Sc, ^{48}Ti(n,p)^{48}Sc, ^{48}Ti(n,α)^{45}Ca and ^{50}Ti(n,α)^{47}Ca reactions over the neutron energy range of 5.4 to 10.5 MeV using the activation technique. Of special interest was the (n,α) reaction on ^{48}Ti where the soft β^- emitting product (^{45}Ca) was separated radiochemically and measured via low-level counting. From the measured isotopic data and systematics the helium emission cross sections for neutrons on natural titanium were deduced. Integral cross sections were measured for ^{46,47,48}Ti(n,p)^{46,47,48}Sc processes with 20 and 25 MeV d(Be) breakup neutrons. The data are compared with spectrum averaged cross sections deduced from recently determined excitation functions and neutron spectral distributions.

((n, charged particle) reaction, helium emission, cross section, excitation function, integral data, neutron spectral distribution)

Introduction

Cross sections of (n, charged particle) reactions are needed for estimating nuclear transmutation, nuclear heating and radiation damage effects in fusion reactor technology (FRT). Furthermore, data on the formation of long-lived activities are of interest in connection with the development of low activation materials. Integral measurements provide useful validations of excitation functions.

Excitation Functions

Irradiations

Cross sections were measured by the activation technique for several neutron induced reactions on isotopes of titanium. About 5 g Ti metal powder (> 99.9 %) was pressed to a pellet ($\varnothing$ = 2 cm; thickness 0.5 cm). Monitor foils (Al or Fe, each 200 μm) were then attached in front and at the back of the sample, and irradiations done in the 0° direction with quasi-monoenergetic neutrons produced via the ^{2}H(d,n)^{3}He reaction on a D$_2$ gas target at the Jülich Compact Cyclotron CV28 [cf. 1].

Neutron flux monitoring

The neutron flux density was determined via the monitor reactions ^{56}Fe(n,p)^{56}Mn and ^{27}Al(n,α)^{24}Na [cf. 2]. The monitor cross sections were taken from the literature [cf. 3,4]. The mean neutron flux density was obtained by taking a mean of the flux values from the front and back monitor foils.

Radiochemical separation

For studies on the ^{48}Ti(n,α)^{45}Ca reaction it was mandatory to perform a chemical separation since the product is a long-lived soft β^- emitting radioisotope (T$_{1/2}$ = 163 d, E$_{\beta^-}$ = 258 keV). The irradiated Ti pellet was cut in three pieces, mixed with 0.25 g dry CaCl$_2$ carrier and about 0.5 g graphite powder. The mixture was heated in a quartz tube in a stream of Cl$_2$, whereby Ti reacted and was removed as TiCl$_4$. The quartz tube was then thoroughly rinsed with HCl. The mixture was filtered, and to the filtrate 60 mg Sc carrier was added. It was then precipitated as Sc(OH)$_3$ and centrifuged off. This cycle was repeated three times. Thereafter, the solution was mixed with oxalic acid, whereby calcium was precipitated as oxalate. The precipitate was converted to CaCO$_3$ by heating at 500°C. It was then weighed, spread on an Al planchet, fixed with glue and counted. The radiochemical yield was ~ 80 %.

Measurement of radioactivity

The radioactivity of the activation products ^{47}Sc (T$_{1/2}$ 3.42 d; E$_\gamma$ = 159 keV; I$_\gamma$ = 68.5 %), ^{48}Sc (T$_{1/2}$ = 1.82 d; E$_\gamma$ = 984 and 1312 keV; I$_\gamma$ = 100 % each) and ^{47}Ca (T$_{1/2}$ = 4.5 d; E$_\gamma$ = 1297 keV; I$_\gamma$ = 77.2 %) was measured via conventional γ-ray spectroscopy using a well calibrated Ge(Li) detector. The count rates were corrected for self-absorption, pile-up and coincidence effects. The coincidence loss correction was important for ^{48}Sc. The self-absorption correction was significant for the 159 keV γ-ray of ^{47}Sc. It was determined experimentally using an ^{123}I source ($E_\gamma = 159$ keV).

The activity of ^{45}Ca was measured via low-level anticoincidence ß$^-$ counting. The efficiency of the ß$^-$ detector was determined using ^{46}Sc($E_ß = 358$ keV) and ^{60}Co($E_ß = 313$ keV) sources of approximately the same thickness as the ^{45}Ca sample. An extrapolation of the efficiency curve to the lower energy of ^{45}Ca($E_ß = 258$ keV) was done exponentially.

Cross section data

The decay rates were corrected for contributions from background neutrons (gas out/gas in results and breakup of deuterons on D_2 gas). The cross sections were then calculated using the well known activation equation. The principal sources of errors in our activation measurements using γ-ray spectroscopy have been described earlier [cf. 1,5]. In the case of soft ß$^-$ counting the efficiency of the detector had an uncertainty of ± 20 %. The total error involved in the ^{48}Ti(n,α)^{45}Ca reaction cross section was therefore higher.

The results for the investigated reactions are given in Table 1. The cross sections for the ^{48}Ti(n,p)^{48}Sc reaction agree well with the literature values [cf. 3,6]. Rather extensive information existed on the ^{47}Ti(n,p)^{47}Sc reaction [cf. 3,6]. A recent detailed measurement in the 1 to 4.3 and 6.9

to 8 MeV energy regions, however, revealed [7] that the cross sections are appreciably lower than the ENDF/B-V values. We show those results as well as our results up to 8.5 MeV in Fig. 1. Our data agree well with the recently reported values [7]. The older data are thus somewhat discrepant. Above 9 MeV, however, our data agree with the ENDF/B-V values.

The excitation functions of the ^{48}Ti(n,α)^{45}Ca and ^{50}Ti(n,α)^{47}Ca reactions are shown in Figs. 2 and 3, respectively. Except for a few reports in the 14 to 15 MeV region [cf. 6,8], no information existed. Our measurements thus describe the first systematic studies on the excitation functions of these two important helium producing processes on titanium. Furthermore, the results on the ^{48}Ti(n,α)^{45}Ca reaction should satisfy the request for data on the formation of the long-lived activation product.

We obtained helium emission cross sections for elemental titanium from the activation (n,α) cross sections for ^{48}Ti and ^{50}Ti measured in this work, and from Hauser-Feshbach calculations and systematics for ^{46}Ti, ^{47}Ti and ^{49}Ti. In the region of 14 MeV the values were raised by 10 % to correct for the (n,n'α) contribution [cf. 9]. Since ^{48}Ti and ^{50}Ti constitute a summed abundance of 79.0 % in natural titanium, the error due to uncertainties in estimated (n,α) cross sections for ^{46,47,49}Ti should not be very critical. We estimated the total error in helium emission cross section to be about 30 %. The results are shown in Fig. 4 together with those obtained by charged particle detection [8] and mass spectrometry [10] at 14.8 MeV. Around 14 MeV the data obtained via the three techniques agree. Below 14 MeV helium emission cross

Table 1. Activation Cross Sections for some Neutron Induced Reactions on Isotopes of Titanium

E_n (MeV)	Reaction cross section (mb)			
	^{47}Ti(n,p)^{47}Sc	^{48}Ti(n,p)^{48}Sc	^{48}Ti(n,α)^{45}Ca	^{50}Ti(n,α)^{47}Ca
5.37±0.24	69.0± 7.4	0.30±0.05		
6.48±0.27	85.0± 9.3	3.9 ±0.05	1.3±0.3	
7.52±0.31	92.6± 9.4	10.3 ± 1.1	2.0±0.5	0.06±0.01
8.53±0.31	112.8±12.6	18.3 ± 2.0	4.1±1.1	0.20±0.04
9.49±0.39	121.0±18.4	26.2 ± 3.5	9.2±2.8	0.56±0.11
9.98±0.40	133.2±18.4	31.2 ± 3.6		0.90±0.13
10.47±0.39	147.5±23.1	37.0 ± 4.2	7.2±2.3	1.14±0.16

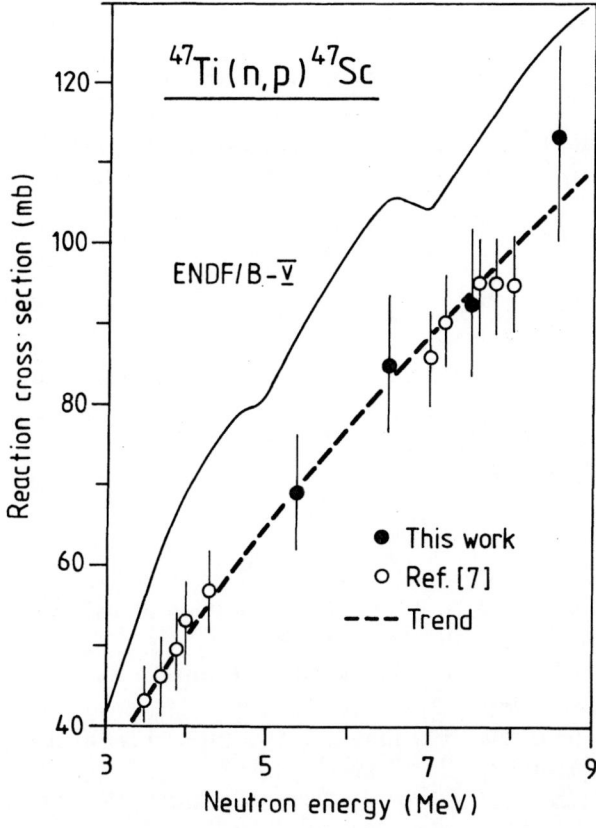

Fig. 1 Excitation function of the ^{47}Ti(n,p)^{47}Sc reaction.

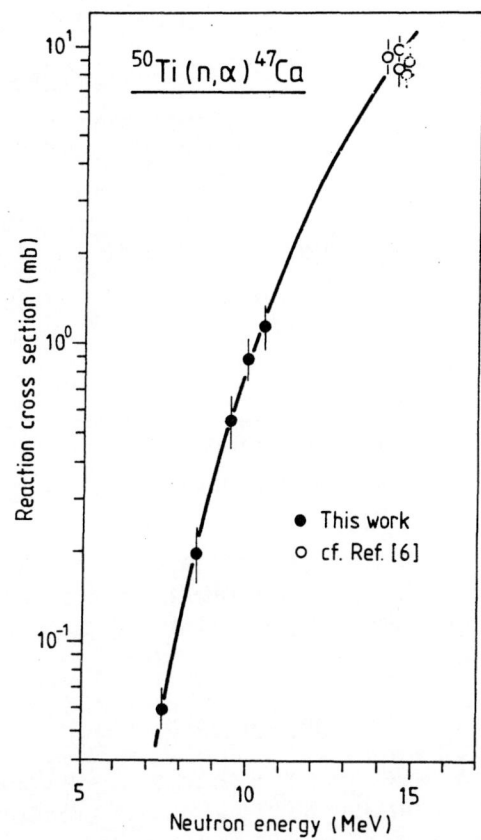

Fig. 3 Excitation function of the ^{50}Ti(n,α)^{47}Ca reaction.

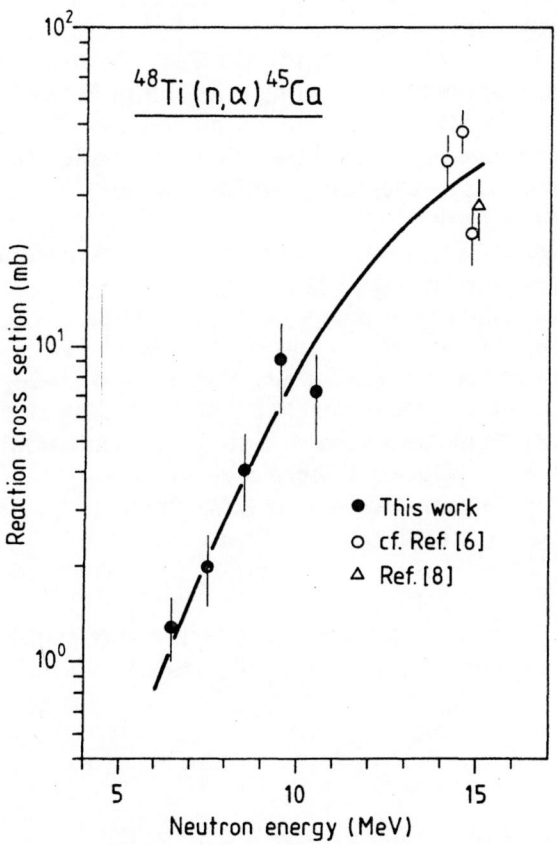

Fig. 2 Excitation function of the ^{48}Ti(n,α)^{45}Ca reaction.

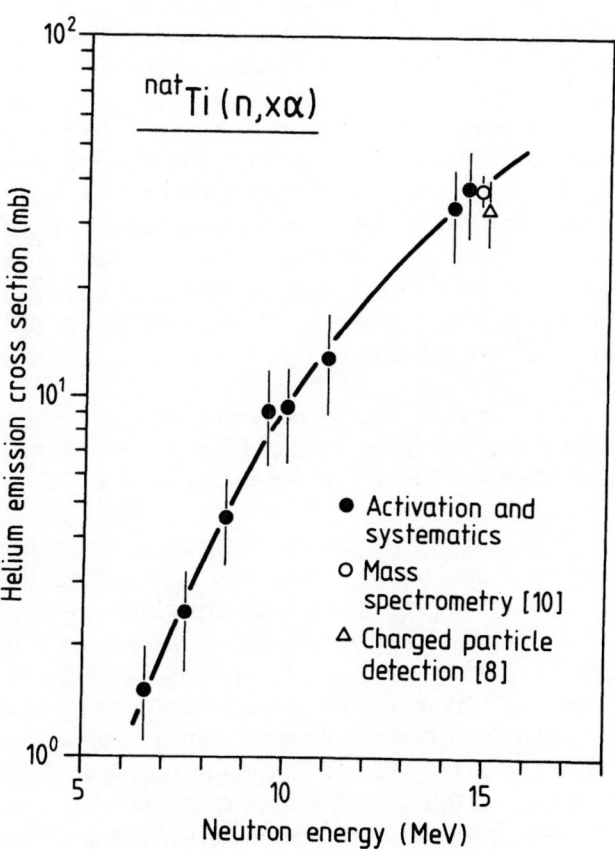

Fig. 4 Helium emission cross sections for elemental titanium as a function of neutron energy.

Table 2. d(Be) Breakup Neutron Spectrum Averaged Cross Sections

Nuclear reaction	20 MeV d(Be) neutrons			25 MeV d(Be) neutrons		
	measured σ (mb)	calculated σ (mb)*	deviation (%)	measured σ (mb)	calculated σ (mb)*	deviation (%)
^{46}Ti(n,p)^{46}Sc	164 ± 16	155	5.8	200 ± 18	185	8.1
^{47,48}Ti(n,x)^{47}Sc †	14.4 ± 1.5	13.4	7.5	23.5 ± 2.5	21.4	9.8
^{48}Ti(n,p)^{48}Sc	19 ± 2	18	5.6	25 ± 3	23.6	5.9

*Obtained by integration of the known excitation function over the neutron spectral distribution.
† Cross section normalized to ^{47}Ti + ^{48}Ti (summed abundance 81.2 %).

sections for titanium were based so far on model calculations [cf. 11]. Our results thus provide the first information based mainly on experimental studies.

Integral Data

We measured a few cross sections using two broad neutron spectral distributions produced via the breakup of 20 and 25 MeV deuterons from the Jülich Isochronous Cyclotron (JULIC) on a thick Be target. The neutron spectrum characterization was done via the multiple foil activation technique in combination with the iterative code SAND-II (for details cf. [12]). In all 15 reactions with thresholds between 0.3 and 19 MeV were used.

For cross section measurements high purity Ti foils were irradiated together with the monitor foils in the 0° direction at a distance of 4 cm from the Be converter. The reaction products were assayed by γ-ray spectroscopy as described above. The neutron flux density averaged over the spectrum (for $E_n \geq 2$ MeV) was determined via the ^{27}Al(n,α)^{24}Na reaction.

The experimentally obtained integral cross sections for the three investigated reactions are given in Table 2. The errors amount to ~ 12 %.

The d(Be) breakup neutron spectrum averaged cross section was also determined by integration of the known excitation function over the neutron spectral distribution. For this purpose the excitation functions measured recently by us up to 20 MeV [13, 14 and this work] were used. The differential data beyond 20 MeV were obtained by extrapolation of the experimental curves. The results are given in Table 2 and agree with the measured integral data within 5 to 10 %. The consistency in the results validates both the measured excitation functions and the unfolded neutron spectra.

REFERENCES

1. S.M. Qaim, R. Wölfle, M.M. Rahman and H. Ollig: Nucl. Sci. Eng. 88, 143 (1984)
2. M. Ibn Majah and S.M. Qaim: Nucl. Sci. Eng. 104, 271 (1990)
3. Evaluated Nuclear Data File (ENDF)/B-V, Dosimetry File (1979)
4. H. Vonach, Technical Report Series No. 227, p. 59, IAEA, Vienna (1983)
5. S.M. Qaim and R. Wölfle: Nucl. Sci. Eng. 96, 52 (1987)
6. V. McLane, C.L. Dunford and P.F. Rose, Neutron Cross Sections, Vol. 2, Academic Press, New York (1988)
7. W. Mannhart, D.L. Smith and J.W. Meadows, in Proc. Specialists' Meeting on Neutron Activation Cross Sections for Fission and Fusion Energy Applications, September 1989, Argonne, USA (Edited by M. Wagner and H. Vonach), p. 121, OECD, Paris (1990)
8. S.M. Grimes, R.C. Haight and J.D. Anderson: Nucl. Sci. Eng. 62, 187 (1977)
9. S.M. Qaim: Nucl. Phys. A458, 237 (1986)
10. D.W. Kneff, B.M. Oliver, H. Farrar IV and L.R. Greenwood: Nucl. Sci. Eng. 92, 491 (1986)
11. C. Philis, R. Howerton and A.B. Smith, Titanium-II: An Evaluated Nuclear Data File, ANL/NDM-28, Argonne National Laboratory, USA (1977)
12. R. Wölfle, S. Sudár and S.M. Qaim: Nucl. Sci. Eng. 91, 162 (1985)
13. N.I. Molla, S.M. Qaim and M. Uhl: Phys. Rev. C42, 1540 (1990)
14. N.I. Molla, S.M. Qaim, H. Liskien and R. Widera: Appl. Radiat. Isotopes 42, 337 (1991)

DOUBLE-DIFFERENTIAL NEUTRON EMISSION CROSS SECTIONS OF ^{51}V, ^{181}Ta, ^{nat}W AND ^{238}U AT 14 MEV NEUTRON INCIDENCE ENERGY

H. Al Obiesi, T. Elfruth, T. Hehl*), M. Toepfer, D. Seeliger,

K. Seidel and S. Unholzer

Technische Universität Dresden, Institut für Kern- und Atompysik,
Mommsenstr. 13, O-8027 Dresden, Federal Republic of Germany

Abstract: For elements used in fusion reactor blanket and shield designs double-differential neutron emission cross sections were measured by time-of-flight spectrometry at a pulsed d-T neutron generator. The obtained data are compared with evaluated library data. Discrepancies are discussed in context with a statistical model of multistep-compound and multistep-direct processes.

(neutron emission cross section, V, Ta, W, U, neutron nuclear data, evaluated data, statistical reaction model)

Introduction

V, Ta and W are in structure and shielding materials of fusion reactor designs. U is used in hybrid reactor projects for neutron multiplication, energy enhancement by fission and plutonium breeding. For neutron transport calculations of blanket, shield and other reactor components energy- and angle-differential neutron emission cross sections (DDX) at incident neutron energies $E_o \leq 14$ MeV are needed.

The DDX measured with a time-of-flight spectrometer [1] are compared with both previous experimental results and evaluated data of the libraries ENDF/B-VI [2] and ENDL-83 [3]. Independently to the measurement the DDX are calculated with the SMD/SMC-model [4] including direct collective and statistical single-particle excitations as well as multiple chance emissions from the compound nucleus cascade and (for U) from the fission fragments [5]. The calculated cross sections are used to discuss discrepancies between library and measured data. As far as the model describes the main components of the neutron emission at $E_o = 14$ MeV, it may be used for data evaluation at $E_o < 14$ MeV.

Experiment

The 14 MeV neutron generator was operated in pulsed mode with pulses of 2 ns f.w.h.m. and repetition rates of 2.5 and 5 MHz, respectively. The neutron flight-path was about 5 m. The neutron production was determined by counting the α-particles of the d-T reaction. Neutron emission spectra from ring samples were taken at emission angles $\vartheta = 15°$ ... $165°$ in steps of $15°$. The data corrected for differential nonlinearity of the spectrometer, dead time, uncorrelated background, source anisotropy, flux attenuation and multiple scattering in the sample were transformed from time to energy spectra. The angle-integrated DDX are shown in Fig. 1. The angular distributions at the emission energy range $E = (5.5 \pm 0.1)$ MeV (as an example) are presented in Fig. 2.

Calculation

The DDX are calculated as the sum of the neutron spectra $(n_1, n_2,...)$ of all possible emission chances of a nucleus with mass number A that decays after neutron bombardment in the chain $(A+1) \rightarrow A \rightarrow (A-1) ...$. At each stage the neutron emission competes with γ- and proton emission, and in the case of U also with fission additionally leading to fission neutrons (n_f). The α-emission can be neglected. The n_1 are assumed to be emitted in the pre-equilibrium and in the equilibrium phase of the reaction. Pre-equilibrium neutrons arise from multistep-direct and from multistep-compound processes. Their statistical treatment with code EXIFON [4] includes as two-step direct processes (SMD) both single-particle and collective excitations (phonons 2^+ and 3^-). The SMC-neutron emission is calculated by solving the master equations from 5-exciton states up to the equilibrium of excited particles and holes. The n_2 and n_3 are emitted from the equilibrated nuclei; the n_f are evaporated from the accelerated fragments. Global input parameters are used (surface-delta interaction $V_o = 19.4$ MeV, $r_o = 1.4$ fm, single-particle level density $g = A/13$ MeV^{-1}, phonon data from systematics). Angular distributions are assumed to be isotropic with the exception of the SMD-component which is described with the Kalbach-Mann-parametrization [6]. The calculated $\sigma(E)$ and angular distributions are shown in Figs. 3 and 2, respectively.

Results

^{51}V: The neutron emission is underestimated in the library ENDL-83, whereas the ENDF/B-VI evaluation follows up to about $E = 10$ MeV the experimental data represented by an uncertainty-weighted evaluation of the three experimental distributions. For $E > 10$ MeV SMD-neutron emissions are strongly underestimated in ENDF/B-VI. The forward-peaked angular distribution also caused by SMD-emissions is not represented by the library data.

---

*) Present address: Universität Tübingen, W-7400 Tübingen, Federal Republic of Germany

302

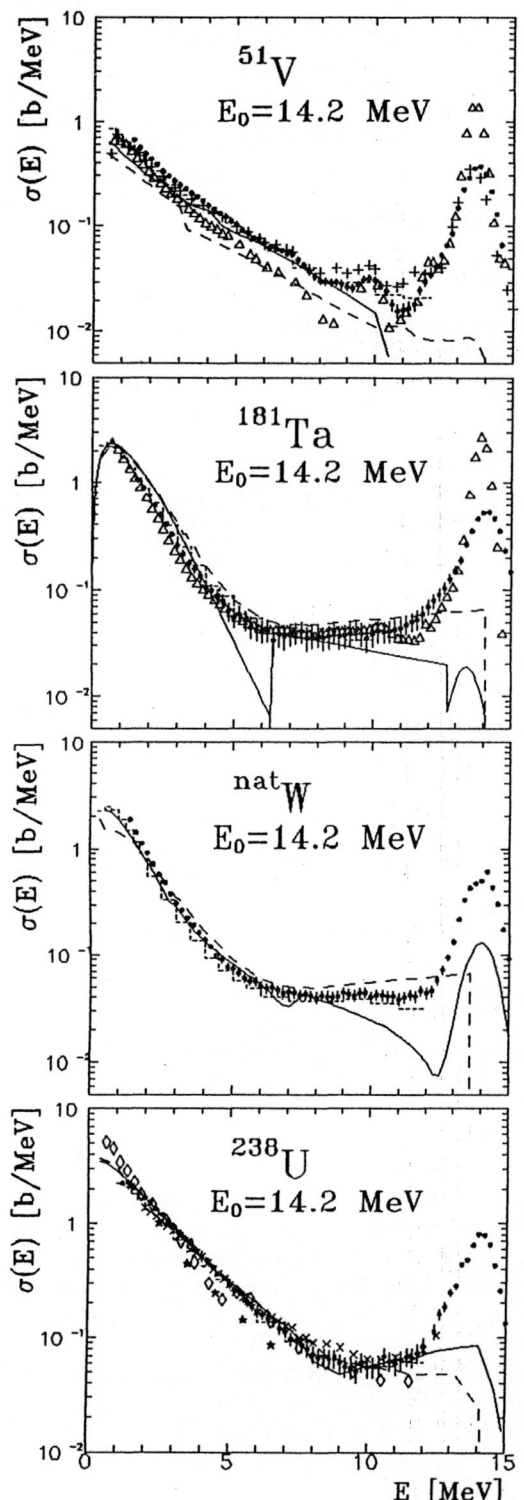

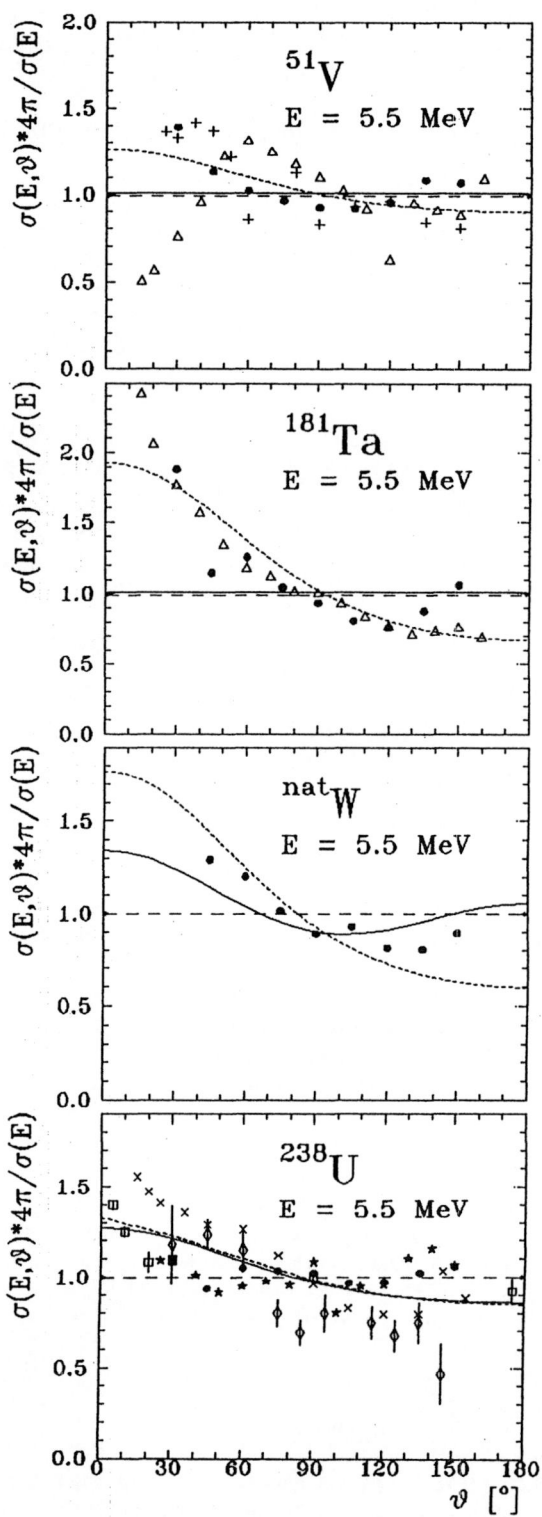

Fig. 1. Differential neutron emission cross sections. Experimental results (● present work, + [7], △ [8], ◇ [9], ⋆ [10], × [11]), evaluated data (dotted histogram) for Ta and W [12], present work for V and U) and data of the libraries ENDF/B-VI (solid line) and ENDL-83 (dashed line).

Fig. 2. Angular distributions of neutrons emitted at E=5.5 MeV. Experimental results (● present work, + [7], △ [8], ◇ [9], ⋆ [10], × [11], □ [13]), calculations with the above described model (dotted line) and data of the libraries ENDF/B-VI (solid line) and ENDL-83 (dashed line).

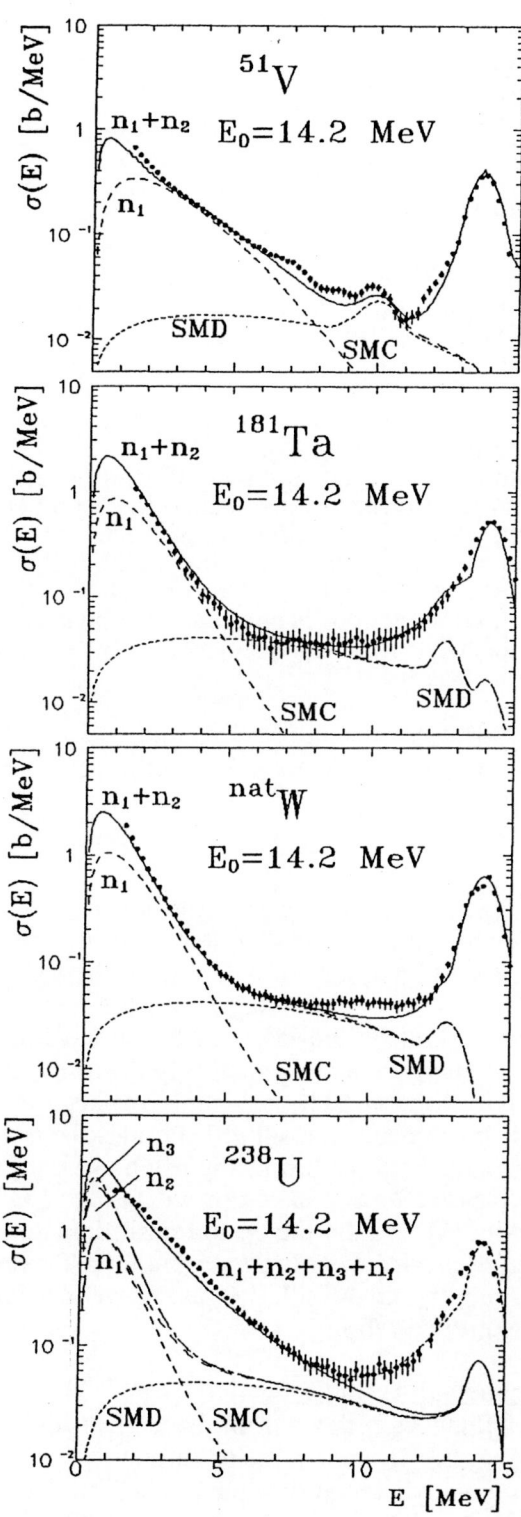

^{181}Ta : The library data overestimate the neutron emission for energies E < 5 MeV. For E > 5 MeV the energy distribution in ENDF/B-VI is artificially, not physically parametrized. The obvious anisotropy of the angular distribution is not described by the library data.

^{nat}W : Discrepancies between experimental and library data are observed for E > 10 MeV. The angular distribution is anisotropic, what is not evaluated in ENDL-83.

^{238}U : The energy distribution of the evaluated experimental spectrum is well described with the data of the libraries. The angular distribution has a smaller forward-peak due to the large number of n_f- compared to SMD-n_1-neutrons. It is described by ENDF/B-VI for the choosen emission energy.

The SMD/SMC-model calculations represent the experimental data at E_o = 14 MeV sufficiently. The main reaction components are adequately described. Thus, it may be used for evaluations at E_o < 14 MeV too.

References

[1] H. Al Obiesi, T.Elfruth, T. Hehl, M.Toepfer, D. Seeliger, K.Seidel and S. Unholzer, Neutron Emission Cross Section of Vanadium, Tantalum and Tungsten at 14 MeV Neutron Incidence Energy, INDC(GDR)-058/L, IAEA, Vienna (1990)

[2] H.D. Lemmel, ENDF/B-6 – The U.S. Evaluated Nuclear Data Library for Neutron Reaction Data, IAEA-NDS-100, IAEA, Vienna (1990)

[3] R.J. Howerton, R.E. Dye and S.T. Perkins, The Evaluated Nuclear Data Library ENDL-82, UCRL-50400, Vol.15, Lawrence Livermore National Laboratory (1983)

[4] H. Kalka, Statistical Multistep Reactions, INDC(GDR)-059/L, IAEA, Vienna (1990)

[5] H. Märten, in Proc. Int. Conf. on 50 Years with Nuclear Fisson, April 1989, Gaithersburg, USA (Edited by J.W. Behrens and A.D. Carlson), ANS(1989) Vol. II, p. 743, Illinois (1989)

[6] C. Kalbach and F.M.G. Mann: Phys. Rev. $\underline{C23}$, 112 (1981)

[7] M. Baba, M. Ishikawa, T. Kikuchi, H. Wakaboyashi and N. Hirakawa, in Proc. 1987 Seminar on Nuclear Data, November 1987, Tokai, Japan (Edited by T. Nakayama and A. Zukeran), JAERI-M 88-065, p. 365, Tokai (1988)

[8] A. Takahashi, E. Ichimura, Y. Sasaki and H. Sugimoto, Double and Single Differential Neutron Emission Cross Sections at 14.1 MeV, OKTAVIAN-Report A-87-03, Osaka University (1987)
A. Takahashi, Y. Sasaki and H. Sugimoto, Measurement and Analysis of Double Differential Neutron Emission Cross Sections at E_n=14.1 MeV for ^{93}Nb and ^{181}Ta, INDC(JPN)-118/L, IAEA, Vienna (1989)

[9] J.L. Kammerdiener, Thesis, UCRL-51232, Livermore (1972)

[10] J. Voignier, U-238 Elastic, U-238 Inelastic Neutron Scattering at 14 MeV, CEA-3503 (1968) and EXFOR 21094.004 (1980)

[11] S. Guanren, in Proc. Advisory Group Meeting on Nuclear Theory for Application, April 1987, Beijing (P.R. China)

[12] A. Pavlik and H. Vonach: Physics Data $\underline{13-4}$ (1988)

[13] D.J. Degtjarev, B.E. Leshchenko and V.A. Pljuko: Jadernaja Fizika $\underline{34}$, 299 (1981)

Fig.3. Differential neutron emission cross sections. Present experimental results (•) and calculations with the above described model (peak of elastically scattered neutrons with experimental resolution included).

DOUBLE-DIFFERENTIAL GAMMA-RAY PRODUCTION CROSS SECTIONS
OF Al, Fe AND Si FOR NEUTRONS BETWEEN 8.5 AND 14.2 MeV

M. Drosg
Inst. Exp. Phys., Univ. Vienna, Boltzmanngasse 5,
A-1090 Wien, Austria

D. M. Drake
SST-8 of LANL,
Los Alamos, NM87545, USA

K. Hasegawa, S. Chiba and M. Mizumoto
Physics Division, JAERI,
Tokai-mura, Naka-gun, Ibaraki-ken, Japan

Abstract: Gamma ray production cross sections have been measured for neutron energies of 8.5, 10.0, 12.2 and 14.2 MeV at 35°, 55°, 75° and 90° for gamma ray energies up to 5 MeV. Similar to experiments with pulsed white neutron sources the time-of-flight method was applied to select the monoenergetic portion of the ^{3}H(p,n)^{3}He neutron source spectrum so that data in the "gap" region could be measured. To make up for the reduced intensity a highly efficient NaI-detector was used. It was placed into an annulus detector to reduce the Compton contribution to the spectra by an anticoincidence condition. A measurement with a threefold smaller sample taken under otherwise identical conditions was used to check the computer codes correcting for neutron and gamma-ray interactions in the sample. This resulted in substantial improvements of these codes allowing the extraction of gamma-ray continua in the spectra.

(Fe, Si, ^{27}Al, cross section, gamma-ray production, NaI detector, tandem, Monte Carlo, neutron transport, gamma transport)

Introduction

Neutron induced gamma-ray production cross sections are needed for the controlled thermonuclear reactor design calculations. Despite recent efforts [1] data in the "gap" region between 8 and 14 MeV neutron energy are sparse. The present investigation covers the energy range between 8.52 MeV and 14.24 MeV for the materials iron, aluminum and silicon in an angular range between 35° and 90°. The corrections for the finite sample size effect were checked by using two samples of strongly different masses.

Experimental Procedure

Neutron Source

Before the recent upgrade of the tandem Van-de-Graaff of IBF at LANL which makes it possible to use the ^{1}H(t,n)^{3}He source over the energy range between 8.5 and 14.2 MeV there was no single monoenergetic neutron source available for making (n,x gamma) experiments. Therefore the neutron source reaction ^{3}H(p,n)^{3}He which we used in this experiment had to be combined with the "white neutron" technique to suppress events caused by background neutrons in the non-monoenergetic range of this reaction. Compared to a real white neutron source "monoenergetic" sources have the advantage of a much higher intensity and consequently a much smaller neutron energy width [2,3]. As was pointed out before [3,4] the p-T reaction has several advantages over the d-D reaction. Although the specific neutron yield in the required energy range is practically the same for these two reactions [4], the signal-to-background ratio is much better for p-T [5]. Besides the neutron background from the beam stop can be kept small in the p-T case by using materials with a high (p,n)-threshold [5]. Finally the beam transit time through the target is considerably shorter in the p-T case improving the time resolution and hence the signal-to-background ratio.

Experimental Arrangement

Bunched protons from the tandem accelerator of IBF at LANL were directed through a 5.3 mg/cm^2 molybdenum window into a 3 cm long gas cell filled with 2.9 atm of tritium and with a beam stop made of ^{58}Ni.

The samples (upright cylinders) with dimensions of 3 cm x 3 cm (Fe) and 1.6 cm x 3 cm (Al, Si) were placed at 0° at a distance of 96.1 cm from the neutron target. This flight path was sufficient to separate the events caused by the monoenergetic neutrons from those stemming from the tritium break-up (Q=-6.26 MeV) and from the ^{58}Ni beam stop (Q=-9.35 MeV) by time-of-flight tagging. Thus monoenergetic neutrons were provided over the required energy range.

The neutron flux was determined relative to the n-p cross section standard with a proton recoil counter telescope placed at 0° at a distance of 67.4 cm from the neutron target. This calibration was carried over to the production runs by means of a neutron monitor and the beam current integrator. The energy dependence of the calibrations by the counter telescope and that of the calculated neutron yields (using collected beam charge and known production cross sections [6]) had rms deviations of less than 1%.

The gamma-ray spectrometer was a 6cmx15cm NaI crystal centered in a NaI annulus inside a large shield-collimator at a distance of 98 cm from the center of the sample. To improve the peak-to-tail ratio of the response of the spectrometer the annulus was used in the anticoincidence mode suppressing Compton and escape events.

Recording both the arrival time and the pulse-height not only allows the selection of the "monoenergetic" events but also allows the subtraction of time-independent background and of background from neutrons scattered into the detector [7].

Data Reduction

To study the response functions and the efficiency of the spectrometer, spectra of discrete gamma-rays of known intensity in the energy range from 0.28 to 4.4 MeV were measured with the sources at the same position as the scattering sample. After correcting for the background the gamma-ray spectra were unfolded by using the response functions, and the full-energy-peak detection efficiency was applied to the spectra.

Corrections

To make systematic errors small, relatively small samples were used: 3.4 moles (Fe), 2.5 moles (Al) and 1.9 moles (Si). To check the quality of the corrections an additional measurement was done with a tubular cylinder of iron with the same outer dimensions but with a mass of only 1.2 moles.

For the neutron and gamma-ray transport corrections a Monte Carlo code (refining the procedure used in GAMSCAT [8]) was applied with the data bases ENDF/B-VI for Al and Si and ENDF/B-V for Fe. The correction was done in two steps. The code MNSCAT compares for each sample and each incoming neutron energy the distribution of interacting neutrons in each unit cell of the sample with that of a fictious very low density sample. This corrects for the attenuation and multiple scattering of the neutrons (and for the energy and yield anisotropy of the source if necessary).

Starting out with this distribution the code COMPCALC corrects at each angle for gamma attenuation and Compton scattering. Using the

measured spectrum as a first guess for the undisturbed gamma-ray spectrum in the unit cells of the sample it is adjusted until a simulated spectrum is found that gives at the location of the detector a spectrum that equals the measured gamma-ray spectrum. Because our pulse-height range was limited to energies below 5 MeV the missing high energy part of the spectrum had to be simulated [9].

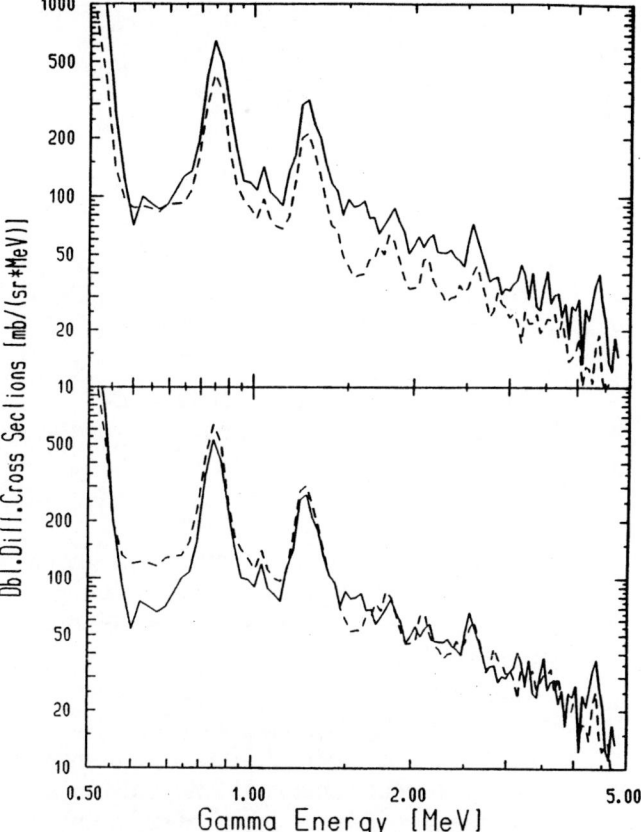

Fig. 1. Comparison of the results for the iron samples at 90° and 14.24 MeV. Upper part: uncorrected, lower part with corrections. Full line: full cylinder, dashed line: hollow cylinder

Fig. 1 demonstrates the capability (and the limits) of these corrections. Inspite of masses different by a factor of three the corrected spectra agree very well with regard to the intensities of the lines and the high energy continuum. However, at lower energies the continuum differs considerably. Part of this discrepancy is unsubtracted beam connected background. This is suggested by the good agreement of the uncorrected spectra in this energy range.

Results

Data

For all 3 samples Al, Fe and Si data at neutron energies of 8.52, 10.00, 12.24 and 14.24 MeV have been taken at 35° (8.52 and 10.00 MeV only), at 55°, at 75° and 90°. For 6 spectra (10 MeV at 35°, and 14 MeV at 55°) sample-out runs are available. In these cases reliable information on gamma-ray continua can be extracted. Fig. 2 demonstrates the effect of this background subtraction on the final answers.

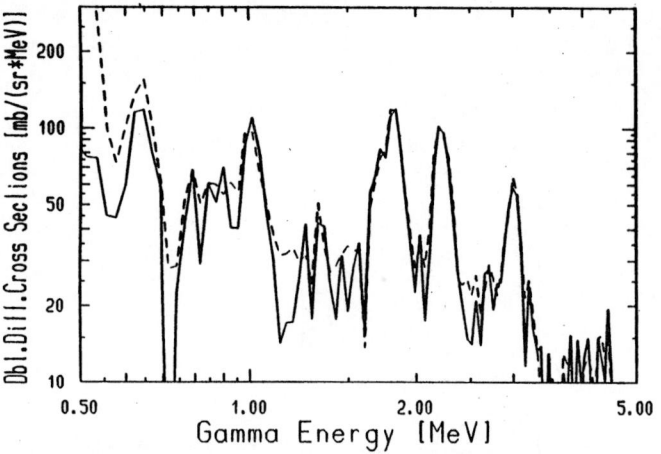

Fig. 2. Effect of background subtraction from the sample-out run. Al at 55° and 14.24 MeV.

The numerical results of all 42 double differential gamma spectra will be made available as an internal report.

Sources of Error

Uncertainties in the data come from several sources with magnitudes depending on the sample and neutron energy. However, some contributions can be generalized.

The uncertainty in the gamma-ray counting efficiency is less than 3% below 3 MeV, increasing to about 5% at 5 MeV. The absolute uncertainty of the neutron flux measurements is 3%. This scale uncertainty has a correlation factor of 0.94 among the four energies. The random error (from the foreground and background runs) ranges from about 1% for peak integrals to about 0.2 mb/(sr.MeV) for continua.

The **maximum** magnitude of the corrections was 2% for dead time, 5% for beam heating in the gas target, and 40% for the finite sample size. An error contribution of 10 % of each correction seems realistic. The uncertainty introduced by the unfolding procedure cannot be assessed quantitatively. It is negligible for intense peaks, and appreciable for the continuum.

Comparison

At 90° and 14 MeV there are abundant data available. Therefore these data were selected for the comparison in Table 1. Because of the relatively poor resolution of the NaI(Tl) spectrometer weak neighboring lines are included in the data.

Conclusion

The present experiment provides a consistent set of gamma-ray production cross sections for the materials Al, Fe and Si which brigdes the difficult "gap" region.

Table 1. Comparison of the cross sections for one gamma line of each sample with previous data.

Sample	Gamma-ray Energy [MeV]	Cross Sections [mb/sr]	[10]	[11]
Al	3.0	8.1±1.0	7.9±1.6	7.3±0.9
Fe	1.24	28.8±2.0	23.0±2.3	28.2±3.4
Si	1.78	32.8±2.8	29 ±7	27.5±3.3

References

1. M. Mizumoto, K. Hasegawa, S. Chiba, Y. Yamanouti, Y. Kawarasaki, M. Igashira, T. Uchiyama, H. Kitazawa, M. Drosg, Proc. Int. Conf. on Nuclear Data for Science and Technology, Mito, Japan, 1988 (S.Igarasi, Ed.), Saikon, Tokyo (1988)
2. J.K. Dickens, F.G. Perey, Nucl. Instr. Meth. 82, 301 (1970)
3. M. Drosg, Nucl. Sci. Eng. 106, 279 (1990)
4. M. Drosg, O. Schwerer, Production of Monoenergetic Neutrons Between 0.1 and 23 MeV: Neutron Energies and Cross Sections, in Handbook on Nuclear Activation Data, K. Okamoto, Ed., IAEA Tech. Rep. Ser. 273, Vienna 1987
5. M. Drosg, G.F. Auchampaugh and F. Gurule, Neutron Background Spectra and Signal-to-Background Ratio for Neutron Production Between 10 MeV and 14 MeV by the Reactions ^{3}H(p,n)^{3}He, ^{1}H(t,n)^{3}He and ^{2}H(d,n)^{3}He, LA-6459-MS, Los Alamos Scientific Laboratory (1976)
6. M. DROSG,"Angular Dependences of Neutron Energies and Cross Sections for 11 Monoenergetic Neutron Source Reactions, Computer-Code DROSG-87: Neutron Source Reactions (O. Schwerer, Ed.) Documentation series, IAEA, Nuclear Data Section, Vienna (1987), available costfree from IAEA, A-1400 Wien, Austria
7. D.M. Drake, L.R. Veeser, M. Drosg, G. Jensen, Proc.Conf. on Nuclear Cross Sections and Technology, (NBS-SP-425), p.813, Wash. D.C.(1975)
8. D.L. Smith, Sample size effects in fast-neutron gamma-ray production measurements in solid cylinder samples,ANL/NDM-17, Argonne National Laboratory (1975)
9. R.J. Howerton, E.F. Plechaty, Nucl. Sci. Eng. 32, 178 (1968)
10. F.C. Engesser, W.E. Thompson, J. Nucl. Energy, 21, 487 (1967)
11. D.M. Drake, E.D. Arthur, M.G. Silbert, Nucl. Sci. Eng. 65, 49 (1978)

EXPERIMENTAL AND CALCULATED EXCITATION FUNCTIONS FOR DISCRETE-LINE GAMMA-RAY PRODUCTION DUE TO 1–40 MeV NEUTRON INTERACTIONS WITH ^{56}Fe

J. K. Dickens, C. Y. Fu, D. M. Hetrick, D. C. Larson, and J. H. Todd

Oak Ridge National Laboratory
Oak Ridge, TN 37831-6356 USA

Abstract: Measuring cross sections for gamma-ray production from tertiary reactions is one of the ways to gain experimental information about these reactions. To this end, inelastic and other nonelastic neutron interactions with ^{56}Fe have been studied for incident neutron energies between 0.8 and 41 MeV. An iron sample isotopically enriched in the mass 56 isotope was used. Gamma rays representing 70 transitions among levels in residual nuclei were identified, and production cross sections were deduced. The reactions studied were ^{56}Fe$(n, n')^{56}$Fe, ^{56}Fe$(n, p)^{56}$Mn, ^{56}Fe$(n, 2n)^{55}$Fe, ^{56}Fe$(n, d+n, np)^{55}$Mn, ^{56}Fe$(n, t+n, nd+n, 2np)^{54}$Mn, ^{56}Fe$(n, \alpha)^{53}$Cr, ^{56}Fe$(n, n\alpha)^{52}$Cr, and ^{56}Fe$(n, 3n)^{54}$Fe. Experimental excitation functions have been compared with cross sections calculated using the nuclear reaction model code TNG, with generally favorable results.

(Nuclear Reactions ^{56}Fe$(n, x\gamma)$ $E = 0.8 - 41$ MeV; measured $\sigma(E_\gamma$, $\theta_\gamma = 125$ deg); ^{52,53}Cr, ^{54,55,56}Mn, ^{54,55,56}Fe γ-ray production. Enriched sample.)

Introduction

The U.S. National Research Council, in 1986, reporting [1] on desirable trends in physics research stated, "Experimental nuclear research is advancing toward the study of nuclei in states of higher excitation energy...." For research utilizing neutron beams, this trend has been manifest in greater interest in measurements for incident neutrons above 20 MeV, a trend already established by 1985 and discussed in reports by Rapaport [2] and by Lisowski et al. [3] at the Santa Fe Conference. We, at the Oak Ridge Electron Linear Accelerator (ORELA) had succeeded in extending the useful incident-neutron energy range of our in-beam, high-resolution gamma-ray spectrometric system [4] to obtain useful and interesting measurements for E_n to 40 MeV, and presented [5] some preliminary results at the same conference. Using this experimental system we reported [6] at the Mito conference on a systematic study of ^{10,11}B$(n, x\gamma)$ reactions for E_n up to 25 MeV which featured first measurements on the tertiary reaction ^{11}B$(n, n\alpha\gamma)^7$Li. The concept of using this system for study of previously unmeasured tertiary reactions was thus established.

The TNG statistical model code [7] has capabilities of computing all energetically available tertiary-reaction cross sections. Successful comparisons of calculated cross sections with experimental cross sections for E_n above 20 MeV should be a stringent test for applicability of the fundamental statistical-model concepts.

Experiment

The emphasis for this experiment was on the observation and measurement of tertiary reactions since there are in the literature a number of reports [8-16] on inelastic scattering gamma-ray production. The present data were also reduced for inelastic scattering gamma-ray production for incident neutron energies up to 20 MeV, primarily to serve as a measure of the reliability of the present measurements. Inelastic-scattering data for incident neutron energies between 22 and 41 MeV, however, are new, and they can be utilized to estimate total inelastic-scattering cross sections for neutron interactions with ^{56}Fe for these more energetic incident neutrons [5].

Measurements were made for incident neutron energies between 0.8 and 41 MeV. A 63-g iron sample isotopically enriched in the mass 56 isotope was used. Gamma rays representing 70 transitions among levels in residual nuclei were identified, and production cross sections were

deduced for the reactions given in the abstract. Values obtained for production cross sections as functions of incident neutron energy are given in an Oak Ridge National Laboratory Report [17] which also documents additional experimental details.

Calculations

The TNG code [7] is based on a unified Hauser-Feshbach [18] (H-F) and Pre-Compound (P-C) model [19]. The H-F part is multi-step, but angular distributions can be calculated only for the binary step. The P-C part is single step and angular distributions are calculated on the basis of partial wave interference (partial relaxation of the random phase approximation in the H-F formalism). Pairing corrections and spin cutoff factors as functions of excitation energy and exciton number, based on the BCS model, for the exciton level densities in the P-C component are included [20]. Gamma-ray production cross sections for each discrete gamma ray can be calculated. Other quantities calculated include total, elastic, nonelastic, (n, γ), (n, n'), (n, p), (n, α), (n, f), $(n, 2n)$, (n, np), $(n, n\alpha)$, (n, nf), ..., $(n, 4n)$, ..., associated secondary-particle and gamma-ray production spectra, the angular distributions of the first outgoing particles, and the production cross sections of isomeric states.

Besides the level density parameters which affect the individual reaction cross sections, the most important parameters in the TNG calculation for ^{56}Fe are the optical-model parameters which determine directly the nonelastic cross section for E_n above 8 MeV. The parameters given by Arthur and Young [21], with some adjustment in the imaginary part [22], gave the best overall results. All other parameters were either taken from standard sources (level energies, Q-values, etc.) or were taken from an earlier global analyses [7].

Results and Comparisons

Excitation functions, both experimental and calculated, are presented for twelve discrete-line gamma-ray transitions in Figs. 1 and 2. In Fig. 1 are exhibited data for neutron-only emission reactions, while in Fig. 2 are shown data for reactions involving charged particles. The inelastic scattering data shown in Fig. 1 agree reasonably well with most of the earlier data; for the $(n, 2n\gamma)$ data the agreement among the several data sets is somewhat poorer. The agreement of the present data with

ORNL-DWG 91M-7109

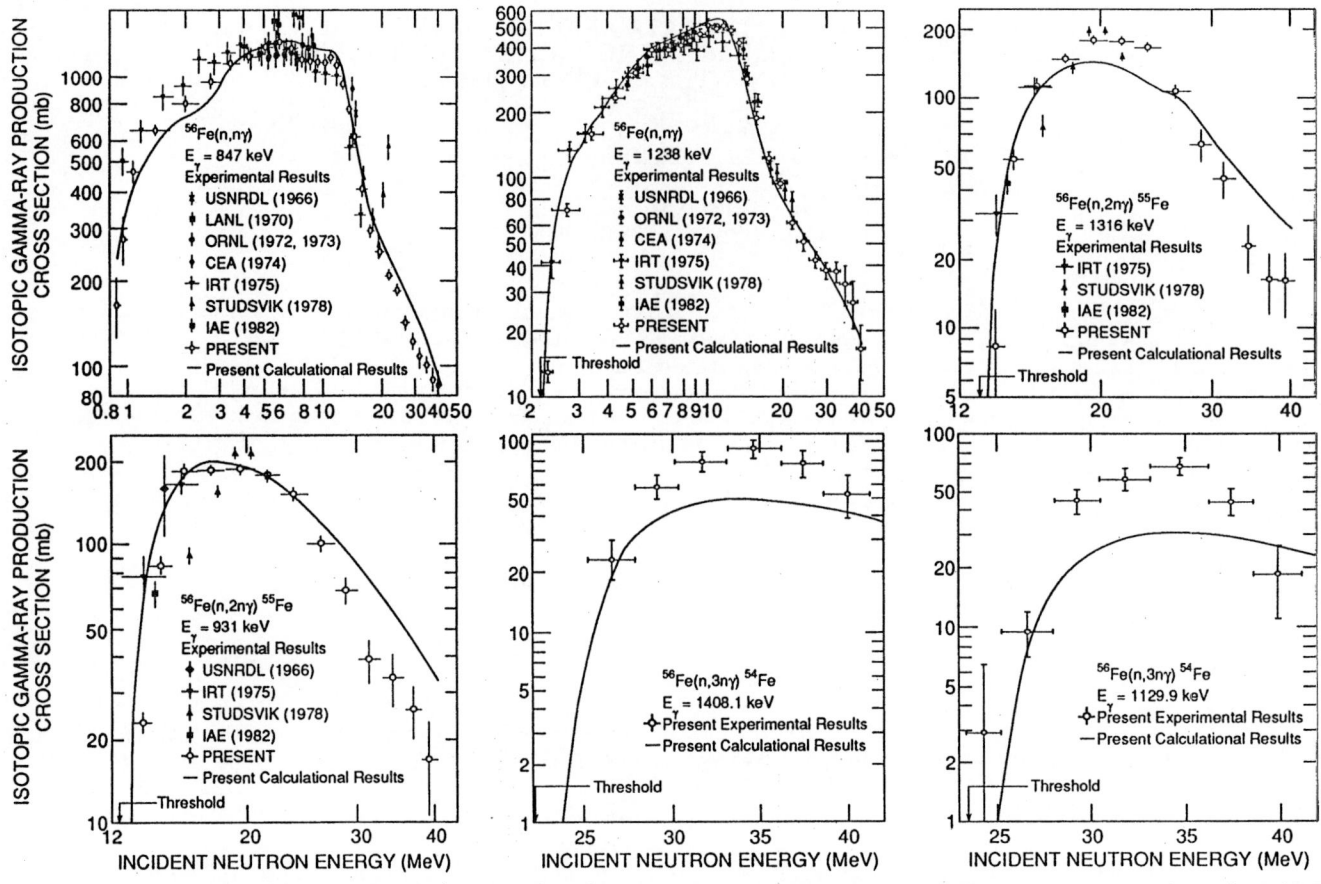

Fig. 1. Six experimental excitation functions compared with calculations. Prior data are: USNRDL (ref. 8), LANL (ref. 9); ORNL (ref. 10,11); CEA (ref. 12); IRT (ref. 13); Studsvik (ref. 14); and IAE (ref. 15).

calculation is quite good for $E_n < 20$ MeV, except for the resonance region [13,16] around $E_n \sim 2$ MeV, for $E_\gamma = 847$ keV. This good agreement may have been anticipated, however, since the parameters used in the TNG calculations were optimized to give a good accounting of data existing in 1986 [22] and were used in developing the ENDF/B-VI evaluation for iron [23].

The calculations shown for 20 MeV $< E_n <$ 40 MeV are the first to be done with this set of parameters developed for $E_n < 20$ MeV and represent an "extrapolation" of the theoretical estimates from the lower-energy calculations. Although the comparisons show somewhat larger variations for $E_n > 20$ MeV than for $E_n < 20$ MeV, this set of parameters is qualitatively correct in predicting large cross sections for reactions having large experimental values and predicting small cross sections for weakly-observed transitions. The model also does well in reproducing level-spin dependent differences observed in the measurements.

Future Work

Direct experimental knowledge of tertiary reaction cross sections is difficult to obtain, particularly if the residual nucleus is stable. Nuclear model calculations are used to obtain estimates of these important cross sections which are often the dominant generator of charged particles in the 15-MeV energy region. Gamma-ray production data provides a very useful window to these cross sections and can serve as benchmarks against which to test and improve nuclear models. The next task is to determine if improvements in calculated excitation functions for E_n above 20 MeV can be obtained through adjustments of parametric values. We plan, also, to study similar reactions for other sample nuclei; preliminary results [24] for ^{59}Co$(n,x\gamma)$, for example, exhibit large yields for

the (n,np) and $(n,2np)$ reaction channels for E_n above 20 MeV.

Acknowledgments

Research sponsored by the Office of Energy Research, Division of Nuclear Physics, U.S. Department of Energy, under contract DE-AC05-84OR21400 with Martin Marietta Energy Systems, Inc.

References

1. National Research Council, Commission on Physical Sciences, Mathematics and Resources, Board on Physics and Astronomy, Physics Survey Committee, William F. Brinkman, chairman, *Physics Through the 1990s An Overview*, National Academy Press, Washington (1986), p. 23.
2. J. Rapaport, "Physics With Monoenergetic Neutrons Below 100 MeV," in *Proc. Int. Conf. on Nuclear Data for Basic and Applied Science, May 1985, Santa Fe, New Mexico*, (Edited by P. G. Young, R. E. Brown, G. F. Auchampaugh, P. W. Lisowski, and L. Stewart), Vol. 2, p. 1229, Gordon and Breach, New York (1985).
3. P. W. Lisowski, S. A. Wender, and G. F. Auchampaugh, "The WNR/PSR Facility – Neutron Physics Capabilities from Sub-Thermal to 800 MeV," in *Proc. Int. Conf. on Nuclear Data for Basic and Applied Science, May 1985, Santa Fe, New Mexico*, (Edited by P. G. Young, R. E. Brown, G. F. Auchampaugh, P. W. Lisowski, and L. Stewart), Vol. 2, p. 1245, Gordon and Breach, New York (1985).
4. Z. W. Bell, J. K. Dickens, D. C. Larson, and J. H. Todd, *Nucl. Sci. Eng.* **84**, 12 (1983); D. C. Larson and J. K. Dickens, *Phys. Rev. C* **39**, 1736 (1989).

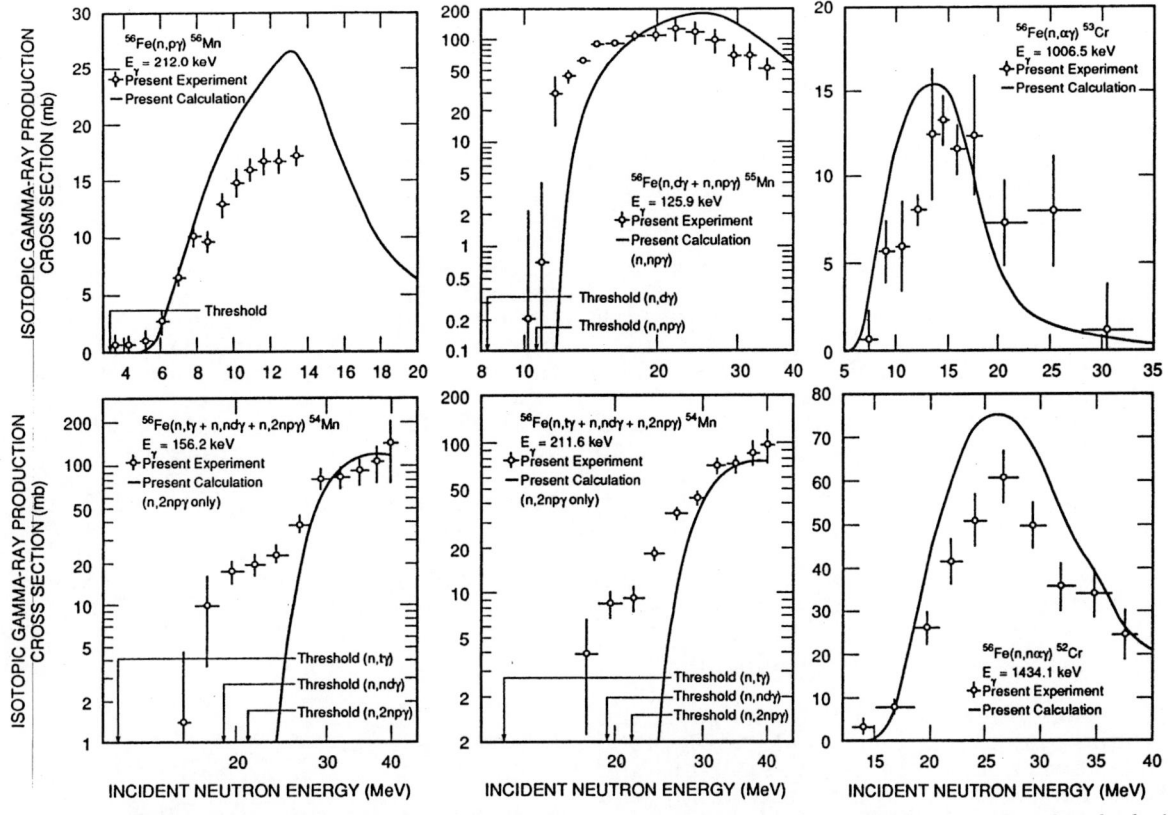

Fig. 2. Six experimental excitation functions involving secondary charged particles compared with calculations.

5. D. C. Larson, "High-Resolution Structural Material $(n, x\gamma)$ Production Cross Sections for $0.2 < E_n \leq 40$ MeV," in *Proc. Int. Conf. on Nuclear Data for Basic and Applied Science, May 1985, Santa Fe, New Mexico*, (Edited by P. G. Young, R. E. Brown, G. F. Auchampaugh, P. W. Lisowski, and L. Stewart), Vol. 1, p. 71, Gordon and Breach, New York (1985).

6. J. K. Dickens and D. C. Larson, "^{10,11}B$(n, x\gamma)$ Reactions for Incident Neutron Energies Between 0.1 and 25 MeV," in *Proc. Int. Conf. on Nuclear Data for Science and Technology, May/June 1988, Mito, Japan*, (Edited by S. Igarasi), p. 213, Saikon, Tokyo (1988).

7. C. Y. Fu, *Atomic Data and Nuclear Data Tables* **17**, 127 (1976); C. Y. Fu, "A Consistent Nuclear Model for Compound and Precompound Reactions with Conservation of Angular Momentum," in *Proc. Int. Conf. Nuclear Cross Sections for Technology, October 1979, Knoxville, Tenn.*, (Edited by J. L. Fowler, C. H. Johnson, and C. D. Bowman), p. 757, U.S. Government Printing Office, Washington; C. Y. Fu, *A Consistent Nuclear Model for Compound and Precompound Reactions with Conservation of Angular Momentum*, ORNL/TM-7042, Oak Ridge National Laboratory (1980).

8. F. C. Engesser and W. E. Thompson, *J. Nucl. Eng.* **21**, 487 (1967).

9. D. M. Drake, J. C. Hopkins, and C. S. Young, *Nucl. Sci. Eng.* **40**, 294 (1970).

10. J. K. Dickens, G. L. Morgan, and F. G. Perey, *Gamma-ray Production Due to Neutron Interactions with Iron for Incident Neutron Energies Between 0.8 and 20 MeV: Tabulated Differential Cross Sections*, ORNL-4798 (1972).

11. J. K. Dickens, G. L. Morgan, and F. G. Perey, *Nucl. Sci. Eng.* **50**, 311 (1973).

12. J. Lachkar, J. Sigaud, Y. Patin, and G. Haouat, *Nucl. Sci. Eng.* **55**, 168 (1974).

13. V. J. Orphan, C. G. Hoot, and V. C. Rogers, *Nucl. Sci. Eng.* **57**, 309 (1975).

14. V. Corcalciuc, B. Holmqvist, A. Marcinkowski, and G. A. Prokopets, *Nucl. Phys.* **A307**, 445 (1978).

15. Shi Xia-Min, Shen Rong-Lin, Xing Jin-Qiang, and Din Da-Zhao, *Chinese J. Nucl. Phys* **4**(2), 120 (1982).

16. F. Voss, "Messung und Fluktuationsanalyse von γ-Produtionsquer-schnitten nach inelastischer Neutronenstreuung an ^{56}Fe und ^{27}Al zwischen 0.8 und 13 Mev," Dissertation, Karlsruhe Univ. (1972).

17. J. K. Dickens, J. H. Todd, and D. C. Larson, *Cross Sections for Production of 70 Discrete-Energy Gamma Rays Created by Neutron Interactions with ^{56}Fe for E_n to 40 MeV: Tabulated Data*, ORNL/TM-11671, Oak Ridge National Laboratory (1990).

18. W. Hauser and H. Feshbach, *Phys. Rev.* **87**, 366 (1952).

19. C. Y. Fu, *Nucl. Sci. Eng.* **100**, 61 (1988); K. Shibata and C. Y. Fu, *Recent Improvements of the TNG Statistical Model Code*, ORNL/TM-10093, Oak Ridge National Laboratory (1986).

20. C. Y. Fu, *Nucl. Sci. Eng.* **86**, 344 (1984); C. Y. Fu, *Nucl. Sci. Eng.* **92**, 440 (1986).

21. E. D. Arthur and P. G. Young, *Evaluated Neutron-Induced Cross Sections for ^{54,56}Fe to 40 MeV*, LA-8626-MS (ENDF-304), Los Alamos National Laboratory (1980).

22. C. Y. Fu and D. M. Hetrick, *Update of ENDF/B-V Mod-3 Iron: Neutron-Producing Reaction Cross Sections and Energy-Angle Correlations*, ORNL/TM-9964 (ENDF-341), Oak Ridge National Laboratory (1986).

23. C. Y. Fu, D. M. Hetrick, F. G. Perey, and C. M. Perey (unpublished, 1990); C. Y. Fu, D. M. Hetrick, C. M. Perey, F. G. Perey, N. M. Larson, and D. C. Larson, "Improvements in ENDF/B-VI Iron and Possible Impacts on Pressure Vessel Surveillance Dosimetry," in *Seventh ASTM Euratom Symposium on Reactor Dosimetry, Strasbourg, France, August 1990*, (Proceedings to be published).

24. T. E. Slusarchyk, *Preliminary Cross Sections for Gamma Rays Produced by Interaction of 1 to 40 MeV Neutrons with ^{59}Co*, ORNL/TM-11404, Oak Ridge National Laboratory (1989).

MEASUREMENT OF DOUBLE DIFFERENTIAL (n,xα) CROSS SECTIONS OF natNi, 58Ni, 60Ni, natCu, 63Cu AND 65Cu IN THE 5 TO 14 MEV NEUTRON ENERGY RANGE

E.Wattecamps

Commission of the European Communities, Joint Research Centre,
Central Bureau for Nuclear Measurements, B-2440 Geel, Belgium

Abstract : Alpha yield data by neutron irradiation of natural Ni, natural Cu and enriched single isotopes were measured. The neutron energy was varied from 5 to 10 MeV and a single energy point at 14 MeV. The prompt α-particle production was measured simultaneously by five dE-dE-E telescopes. The α yield data of natural Ni and some of the natural Cu were measured relative to the elastic scattering cross section of H, whereas the yield data of the samples with enriched isotopes were measured relative to the yield of 58Ni. By integration of the double differential data over α-particle energy and over emission angle the total prompt α yield is obtained which is compared with the data from the ENDF/B-VI evaluation.

(Measurement, (n,xα), cross section, multitelescope, α-particle energy, emission angle, total α yield, neutron energy 5 to 14 MeV.)

The experimental set-up.

In the frame of the CBNM research programme on light ion production data in structural materials of fission and fusion reactors a multitelescope was used at the Van de Graaff laboratory. The multitelescope is made of five dE-dE-E telescopes. Each telescope provides the α-particle energy spectrum. By integration over all energies and all angles the total α-particle yield is deduced. By measuring under identical conditions with a hydrogeneous sample or a sample with known α-particle yield, the multitelescope provides double differential cross section data.

The multitelescope is drawn in Fig.1, and was described earlier [1,2]. The sample under investigation is a disc of 30mm diameter located in the middle of the telescope housing. Samples are mounted on a vertical sample changer which carries four samples, for instance: Ni, Ta, polyethylene and Am. All samples have the same cross sectional area and can be slided sequentially at the central location, which is located at 8.0 cm distance from the gaseous deuterium target or from the tritiated Ta target. Each telescope is made of two proportional counters dE and one surface barrier detector or energy detector E. A typical measurement provides five energy spectra, and two energy loss spectra. The telescope at 14° acquires in addition an energy spectrum with a known amount of pulses added from a pulse-generator, thus providing the means for a dead-time correction. The 241Am sample is used for

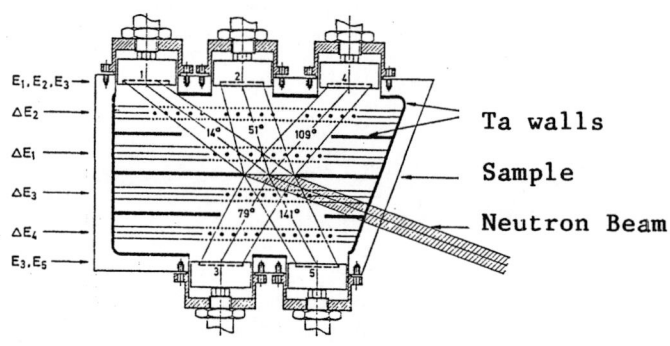

Fig.1 The multitelescope.

tests and for the calibration of gains and of the solid angles.

A summary of samples and a summary of measurements is given in Table 1 and 2.

Some of the measured differential data.

An α-particle energy spectrum for Ni at 8.0 MeV neutron energy and at 51° is shown in Fig.2. In the same drawing also the unfolded spectrum is shown, which is obtained from the measured spectrum after correction for energy loss in the sample and in the counter gas. To test the unfolding procedure, the unfolded spectrum was folded again and the resulting spectrum coincides well with the original spectrum.

The mean α-particle energy (mean over all α-particle energies) versus angle is shown in Fig.3 for the 58Ni (n,xα) reaction at 8.0 MeV neutron energy.

Table 1. Summary of samples

sample	weight mg/cm²	58Ni %	60Ni %	61Ni %	62Ni %	64Ni %	63Cu %	65Cu %
Ni	3.13±0.03	68.3	26.1	1.1	3.6	0.9	-	-
Cu	2.76±0.03	-	-	-	-	-	69.2	30.8
58Ni	4.06±0.03	99.9	-	-	-	-	-	-
60Ni	4.09±0.03	99.7	-	-	-	-	-	-
63Cu	4.09±0.03	-	-	-	-	-	9.,9	-
65Cu	4.22±0.03	-	-	-	-	-	0.3	99.7

Table 2. Summary of measurements

E_n[MeV]	Ni	Cu		60Ni	63Cu	65Cu
14.1	H			58	58	58
10.0	H	Ni		58	58	
9.5	H	Ni	H	58		
9.0	H	Ni		58	58	58
8.5	H	Ni				
8.0	H	Ni	H	58	58	58
7.5	H	Ni				
7.0	H	Ni		58	58	58
6.5	H	Ni	H	58		
6.0	H	Ni				
5.5	H	Ni				
5.0	H	Ni	H	58		

Mean α energies at the five angles were integrated over the entire angular yield of 4π sr with individual weights pertaining to each telescope. This integration yields a mean energy (mean by energy and angle) of α-particles versus neutron energy, shown in Fig.4.

The angular yield distribution of (n,xα) for 58Ni, is shown in Fig.5 for neutron energies ranging from 5.0 to 10 MeV. The data are given in α yield per unit solid angle per unit of deuteron charge on the target and are relative angular yield distributions. The error bars are deduced from the statistical accuracy of the measured foreground minus background and from the estimated error of the unknown yield below the experimental bias at approximately 3.5 MeV. Angular yield data of Ni and of the Cu data were normalized to the H(n,p) scattering yield at 14°. Some of the Cu data and those of single isotope samples were normalized to the total angular yield data of 58 Ni(n,xα).

α yield cross sections and ratios

The measured (n,xα) cross sections of Ni and Cu are shown and compared with the ENDF/B-VI data in Fig.6a,b,c. (Evaluation distributed April 90). The evaluated ratio for α yield by elements was calculated from the ENDF/B-VI values of 58Ni, 60Ni, 63Cu and 65Cu, assuming zero yield for 61Ni, 62Ni and 64Ni. All the measurements of Ni were made relative to

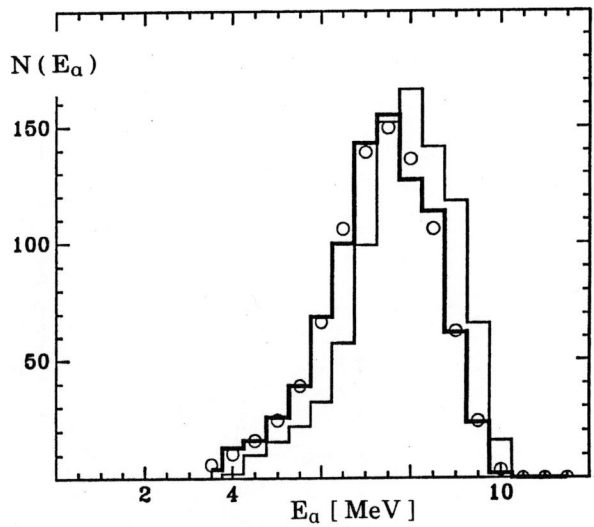

Fig.2. Measured (——), unfolded (——) and folded (O) α-particle energy spectra. Energy in the lab system.

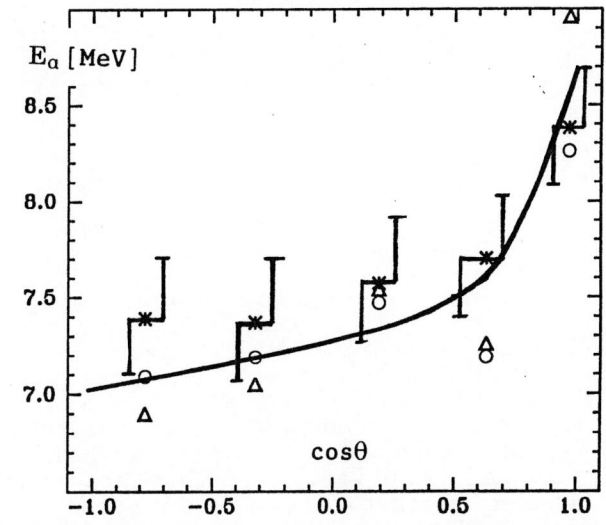

Fig.3. Mean α-particle energy versus angle. Three runs under comparable conditions. Typical error bars drawn for some points.

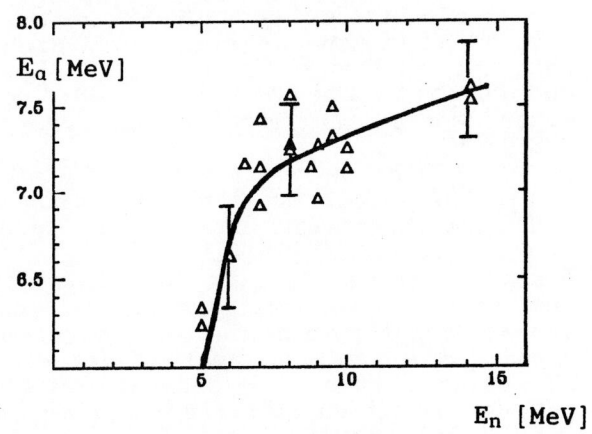

Fig.4. Mean α-particle energy for 58Ni(n,xα) versus neutron energy. Typical error bar amounts to 4%.

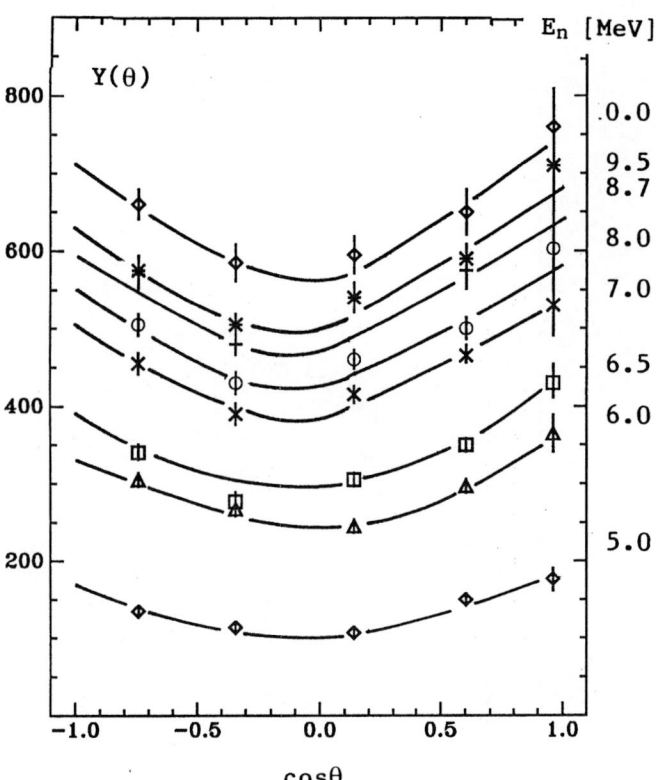

Fig.5. Measured angular (n,xα) yield.

elastic scattering of H, whereas part of the Cu measurements were made relative to H and some relative to Ni. Our measurements together with evaluations and measurements from other authors were compared earlier.[4]. The discrepancy between measurement and evaluation of our nickel data and the agreement for our copper data remains, but new data became available by our recent measurements of (n,xα) of single isotopes. Yield ratio measurements of ^{60}Ni, ^{63}Cu and ^{65}Cu with respect to ^{58}Ni are shown in Fig.6d,e,f, together with the ratio of the ENDF/B-VI data. Ratios for ^{60}Ni to ^{58}Ni agree well. Our measured ^{63}Cu to ^{58}Ni ratio is much larger than the evaluation which is coherent with our older measurements of Cu to Ni. The ^{65}Cu to ^{58}Ni ratio measurements are also larger than the ENDF/B-VI data.

In Fig.6c we compare Cu to Ni ratios:
. from measurements with natural samples
. from measurements with single isotopes
. from the ENDF/B-VI file.
Our two measured sets of Cu to Ni ratio's are larger than the evaluated data by almost equal amounts. The measurements with single isotope samples were done as the earlier ones with elemental samples, but in different periods of time, with independent and cautious preliminary calibrations. The numercal values of measred cross sections and cross section regios are listed in table 3.

Conclusions

. Our measurements indicate a relatively low yield for Ni(n,xα).
. Good agreement between measured and evaluated data for Cu(n,xα) and for the ratio of ^{60}Ni(n,xα) to ^{58}Ni(n,xα) is obtained.
. Our measured α yield data of natural and single isotope samples are coherent.
. Double differential α yield data can be made available for Ni, Cu, ^{58}Ni, ^{60}Ni, ^{63}Cu and ^{65}Cu.
. In view of the discrepancy between measured and evaluated yield data of Ni one wonders whether future efforts should go first to double differential data or to other means for measuring integrated α yield data.
. A known (n,xα) cross section for use as a reference cross section would circumvent the need for normalization by elastic scattering on H. The proton recoil technique needs a measurement at a small angle with quite large background and other gain settings than the (n,xα) measurement. The use of ^{58}Ni(n,xα) is tempting, but its cross section should be better known.

Aknowledgement

I am very thankful to Dr.H.Liskien for many stimulating discussions.

References:

[1] A.PAULSEN, H.LISKIEN, F.ARNOTTE, R. WIDERA, Nucl.Sci.Eng.,78,377 (1981)
[2] E.WATTECAMPS, H.LISKIEN, F.ARNOTTE, Proc.Int.Conf.Nuclear Data for Science and Technology, Antwerp, Belgium, (1982), p. 156.
[3] E.WATTECAMPS, F.ARNOTTE, CBNM, Annual Programme Progress Report, 62 (1984), 28 (1987), 38 (1988)
[4] E.WATTECAMPS and F.ARNOTTE, Proc.Int.Conf.,Fast Neutron Physics,Dubrovnik, Yugoslavia, (1986), p.258.

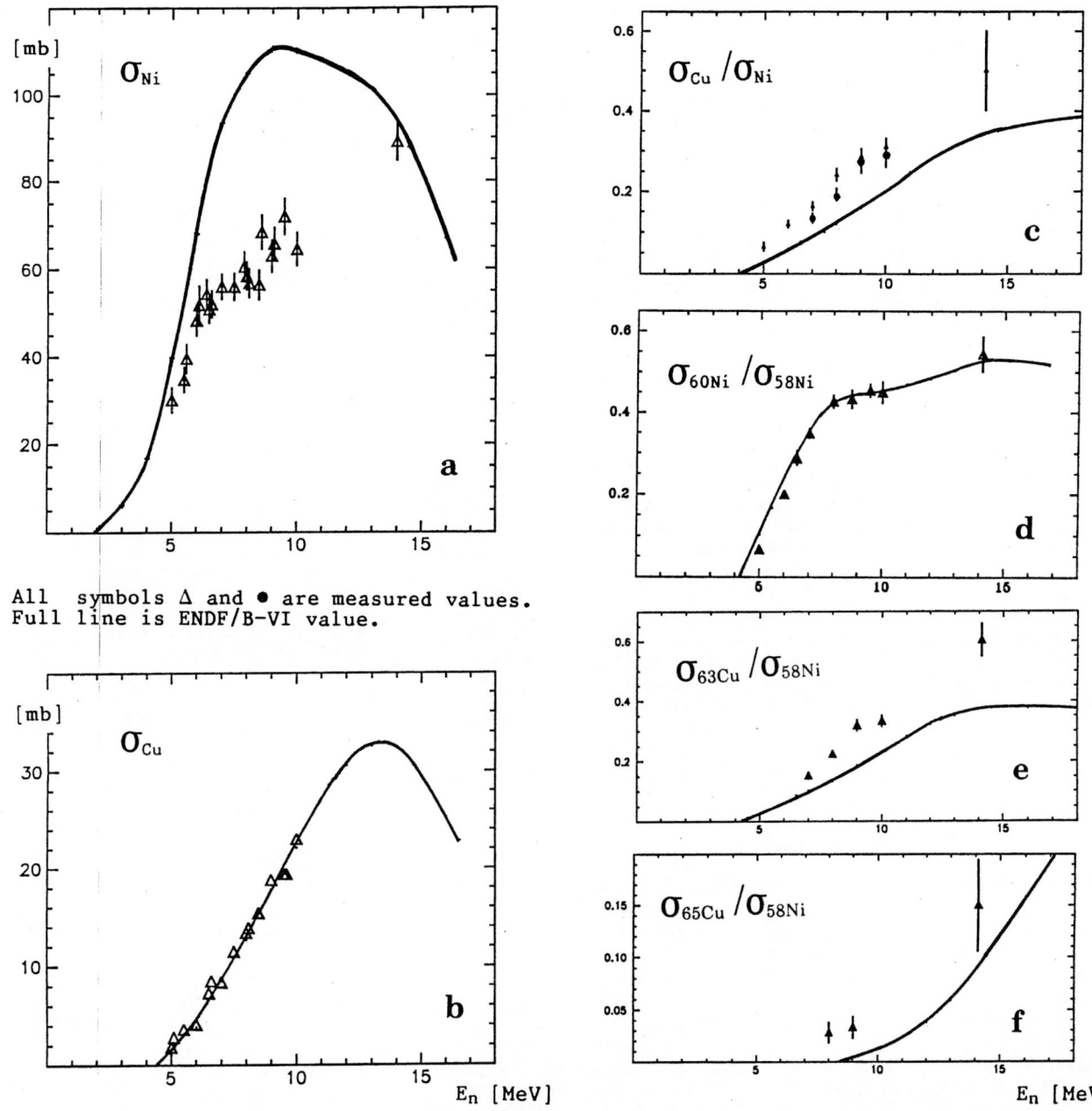

All symbols Δ and ● are measured values.
Full line is ENDF/B-VI value.

Fig.6a,b. Measured Ni(n,xα) and Cu(n,xα) cross sections compared with the ENDF/B-VI
   c.   Cross section ratios of Cu(n,xα) to Ni(n,xα), measured by elements (Δ)
         and by isotopes (●), compared with the ratio from ENDF/B-VI.
d,e,f. Measured cross section ratios of ^{60}Ni(n,α) to ^{58}Ni(n,α), ^{63}Cu(n,α) to
       ^{58}Ni(n,α) and ^{65}Cu(n,α) to ^{58}Ni(n,α) compared with ENDF/B-VI.

Table 3. Numerical values of cross sections (in mbarn) and cross section ratios

E_n[MeV]	σ_{Ni}	σ_{Cu}	σ_{Cu}/σ_{Ni}		$\sigma_{^{60}Ni}/\sigma_{^{58}Ni}$	$\sigma_{^{63}Cu}/\sigma_{^{58}Ni}$	$\sigma_{^{65}Cu}/\sigma_{^{58}Ni}$
14.0	89.0±4.5			.502±.100	.544±.044	.604±.057	.150±.045
10.0	64.5±3.9	23.0±1.0	.311±.022	.290±.030	.450±.027	.333±.020	
9.5	71.9±4.2	19.4±1.0			.454±.017		
9.0	64.4±3.8	18.8±0.8	.287±.020	.272±.027		.319±.020	.033±.011
8.5	62.4±3.5	15.4±0.8			.433±.023		
8.0	58.6±3.3	14.3±0.6	.242±.017	.192±.017	.427±.017	.224±.012	.028±.010
7.5	56.4±3.2	11.5±0.6					
7.0	56.0±3.0	8.4±0.4	.163±.012	.134±.013	.348±.013	.152±.014	
6.5	52.4±3.2	8.4±0.4			.287±.017		
6.0	50.5±3.6	4.1±0.4	.119±.010		.200±.010		
5.5	37.2±3.0	3.6±0.4					
5.0	30.1±3.0	2.4±0.4	.065±.012		.067±.010		

TRITIUM BREEDING RATIO IN Li/Pb-Li/Pb-Li-C SPHERES
MEASURED WITH Li$_2$CO$_3$ PELLETS AND/OR LiF TLDs

J. Cetnar, T. Iguchi, M. Nakazawa [U. of Tokyo, Bunkyo-ku, Tokyo 113],
K. Sugiyama [Tohoku U., Sendai 980], A. Takahashi, K. Sumita [Osaka U., Suita 565]
and Inter-university research group*, Japan

A series of fusion neutronics integral experiments have been made for Li, Pb-Li and Pb-Li-C combined spheres of 120cm diameter at the intense D-T neutron source facility OKTAVIAN, Osaka U.. The Tritium Production Rate (TPR) distribution inside the three kinds of spheres were measured with Li$_2$CO$_3$ pellets and/or LiF-TLDs, including some reference activation rates as neutron spectral indices. To analyze experimental data, neutron transport calculations were carried out using the ANISN and the MCNP codes with the nuclear data libraries based on JENDL-3 and ENDF/B-IV. Determination of the Tritium Breeding Ratio (TBR) was done using the least squares method employing covariance matrixes on experimental data. Good benchmark data on TBR are obtained in one-dimensional spherical geometry and are applied to integral check on the latest nuclear data file JENDL-3. The analysis for the Pb-Li-C sphere experiment shows that the TBR is 1.15($\pm$7%), experimentally determined from the LiF-TLD measurement.

(fusion neutronics, benchmark data, Li$_2$CO$_3$ pellets, LiF TLD, tritium breeding ratio, neutron transport codes, nuclear data files)

Introduction

One of the main goals of a fusion reactor system is to achieve self-sufficiency of tritium production in the blanket. A design of a fusion reactor blanket with neutron multipliers has been suggested in order to achieve the Tritium Breeding Ratio (TBR) greater than 1.0 at least. A series of integral experiments were carried out to measure TBR in lithium spherical assemblies. We investigated various types of the blanket starting from a bare large lithium sphere of 120cm diameter. On the progress of our research work a graphite reflector was added to the systems to improve TBR by neutron leakage reduction. For checking validity of neutron transport calculations for fusion reactor blankets the measured distribution of Tritium Production Rate (TPR) is compared with calculational ones obtained from codes; ANISN and MCNP with nuclear data libraries; JENDL-3 and ENDF/B-IV.

Outline of the experiment

The experimental setup consists of spherical layers covering about 99.9 % of total solid angle. A target of neutron source is placed in the center of the system. Spherical assemblies enable us the usage of one-dimensional transport calculations and simplify angular distribution problems on complex uncertainty of nuclear data. The experiments were held in the intense 14MeV neutron source facility "OKTAVIAN" at Osaka University. The assembly was placed at two kinds of beam transport lines of OKTAVIAN; one is the pulse beam line where pulsed DT neutrons of order 10^9 n/s can be produced from air-cooled Ti-T target mainly for a TOF experiment, and another is the D.C beam line using a water-cooled target which can increase the neutron yield up to 10^{11} n/s for a heavy irradiation experiment. For the absolute measurement of neutron yield the 'Large Solid Angle Activation Foil Method' has been applied. The method reduces an uncertainty from the position of detector and the neutron source angular and spatial distributions. The accuracy of the present method can be estimated about 2-3% through comparing with a recoil proton counter telescope calibrated in the Japanese 14MeV neutron standard of the Electro-Technical Laboratory [1]. We investigated the following spherical assemblies:
1. Li(50cm)
2. Pb(10cm)+Li(40cm)
3. Pb(10cm)+Li(40cm)+C(20cm)

Radial distributions of the TPR inside the Li and Pb-Li spheres irradiated at the D.C beam line were measured by the Dierckx's liquid scintillation counting method using lithium carbonate pellets Li$_2$CO$_3$ (enriched in ^{7}Li and ^{6}Li). They were placed between the cylindrical stainless steel canned Li metal plugs and inserted in the 5cm dia. radial experimental holes of the Li sphere of 0 and 90 deg. directions [2]. Measurements of activation rates were also made on the reference reactions of ^{115}In(n,n'), ^{27}Al(n,α) and ^{93}Nb(n,2n). For the Pb-Li-C assembly LiF TLDs, namely ^{6}Li enriched LiF (UD136N)

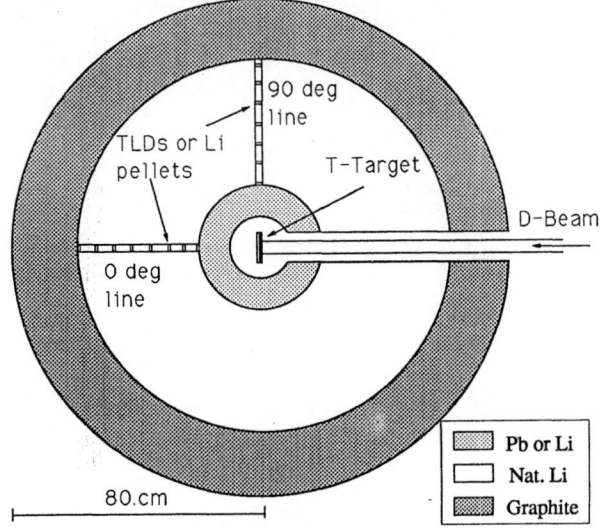

Fig. 1. Li/Pb-Li/Pb-Li-C combine spherical systems

combined with ^{7}Li enriched (UD137N), were used to measure TPR [3]. In order to obtain TPR from TLD measurements as well as total TBR from TPR, some corrections were made through neutron transport calculation using the ANISN code with the nuclear data library JENDL-3 (125 neutron energy groups, P$_5$ Legendre expansion order). An uncertainty analysis employing covariance matrixes was developed to obtain an error estimation of the TBR derived from TPR measurements.

Methodology of TBR estimation

Derivation of TBR by integrating the local TPR over the whole lithium zone was assisted by its theoretical distribution to overcome problems with step space dependencies of measured values. In case of the TLD application to TBR measurements, additional effort is necessary to obtain local TPR from TLD measurements. The present method is briefly described as follows;

Idea of the measurement is based on the fact that TLDs UD-136N and UD-137N consist of the same material apart from isotope of lithium. Detectors of ^{6}Li are sensitive to neutrons mainly due to reaction ^{6}Li(n,α)T, thus the response function to neutrons is similar to tritium production cross section. Both detectors has the same sensitivity for gamma radiation. If we take difference of responses from UD-136N and UD-137N, it is considered the response only for neutrons. In this case the presence of gamma radiation does not disturb the measurement. However, since the difference response function of TLDs (TLDD) differs from the tritium production cross section (fig. 2), an influence of neutron spectrum uncertainty on the TPR derivation from TLD responses must

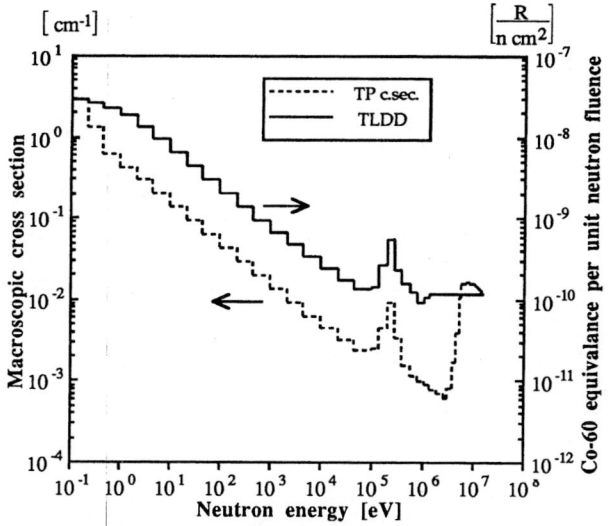

Fig. 2. UD-136N/UD-137N TLD difference response function (self-shielding corrected) and macroscopic tritium production cross section of natural lithium

be of concern. For this purpose the neutron spectrum is treated as a variable. When the neutron flux and cross sections are given in the group structure one can write equations:

$$TPR(r) = \sum \phi(r,E) \cdot \Sigma_{TP}(E) \qquad (1)$$

$$TLDD(r) = \sum \phi(r,E) \cdot f_{TLDD}(E) \qquad (2)$$

$$\phi(r,E) = \phi^*(r,E) \cdot \phi^{tot}(r) \qquad (3)$$

where Σ_{TP} is the macroscopic tritium production cross section and f_{TLDD} is the difference of TLDs response function, $\phi(r,E)$ is the position dependent neutron spectrum, $\phi^{tot}(r)$ is the total flux, and $\phi^*(r,E)$ is the normalized neutron spectrum.

To solve the equations (1) and (2) the general least squares method was used. Generally, it minimizes function (4) conserving side equations (1) and (2) expressing TPR and TLDD responses at each point of measurement.

$$M = [\vec{y} - \vec{y}^m]^T \cdot G_y \cdot [\vec{y} - \vec{y}^m] \qquad (4)$$

where $\vec{y} = [TLDD, f_{TLDD}, \Sigma_{TP}, \phi_1^*, \phi_2^*, .., \phi^{tot}]$ is a vector of the true values, while $\vec{y}^m$ is a vector of corresponding measured values. G_y is the weight matrix equal to reverse covariance matrix of vector y.

The TLD response functions were quoted from Ref. [4] and, to derive their covariance matrix, their accuracy was checked by comparing with the results of direct measurements on the TLD response. A self-shielding correction for thermal neutrons was also made. Macroscopic tritium breeding cross section and its covariance matrix were prepared from JENDL-3 based data including uncertainty evaluation of Ref. [5]. The neutron spectrum is treated as uncertain to take into

consideration the errors imposed by differences between the shapes of the TLDD response function and the tritium production cross section. Its initial guess is given from neutron transport calculation. For this the present initial total flux uncertainty (r.s.d. of 1σ) is set 100% while normalized neutron spectrum uncertainty was reasonably assumed around 20%, which can be estimated from analysis of the measured neutron spectrum in our other experiments. Calculation with a more pessimistic assumption of spectrum uncertainty of 80% were also performed. The correlation matrix of input spectrum was adequately assumed so that the final integral uncertainty might not be underestimated. Under the above assumptions the measured value of TPR and the final TBR uncertainty derived from TLD measurements is not strongly affected by additional a priori information concerning neutron spectrum.

Results and discussion

Table 1 summarizes calculated values of TBR from $^6Li(n,\alpha)T$ and $^7Li(n,n'\alpha)T$ reactions (T6, T7) using ANISN + JENDL-3, MCNP + JENDL-3 and MCNP+ENDF/B-IV together with experimental data. It can be noticed that tritium production by ^6Li isotope is strongly gained in the systems with the neutron multiplier and with the reflector. Findings from the experiments are as follows:

i. Fig. 3 presents C/E ratios of TBR measurements in the Li sphere. JENDL-3 based calculations agree well to experiment. In MCNP+ENDF/B-IV based calculations T7 is overestimated about 10%. There is slight underestimation in T6 distribution on the outer border due to neutrons scattered from concrete walls. $^{115}In(n,n')$ reaction rate is underestimated about 10-20% in all calculation. Other foil reaction rates show agreement within 1-5%.

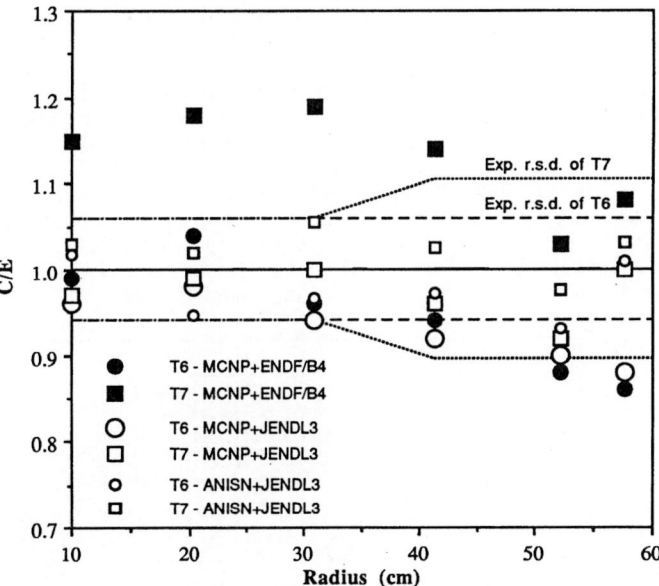

Fig. 3. C/E of TPR in Li Sphere

Table 1 Experimental and calculational values of TBR in the Li, Pb-Li, Pb-Li-C spheres

Spherical assembly		Exp. (r.s.d.)	MCNP + ENDF/B-IV Calc.	C/E	MCNP + JENDL-3 Calc.	C/E	ANISN+JENDL-3 Calc.	C/E
Li	T6	0.217(5.4%)	0.188	0.867	0.184	0.847	0.205	0.948
	T7	0.471(5.8%)	0.518	1.100	0.440	0.935	0.467	0.991
	Total	0.687(5.6%)	0.705	1.026	0.624	0.907	0.672	0.977
Pb-Li	T6	0.337(5.5%)	0.346	1.027	0.326	0.969	0.366	1.086
	T7	0.186(7.5%)	0.212	1.135	0.167	0.898	0.162	0.886
	Total	0.519(6.0%)	0.556	1.075	0.494	0.952	0.527	1.016
Pb-Li-C	T6		0.967		0.906		0.920	
	T7		0.234		0.187		0.190	
	Total	1.15(7%)	1.202	1.044	1.094	0.950	1.110	0.965

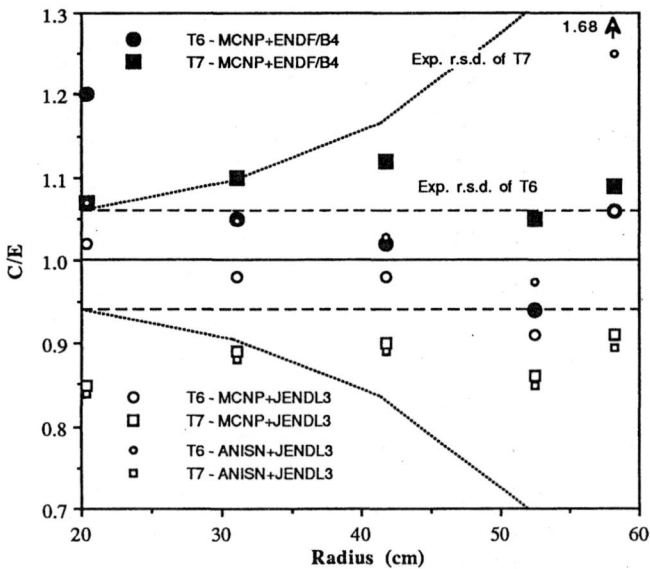

Fig. 4. C/E of TPR in Pb-Li Sphere

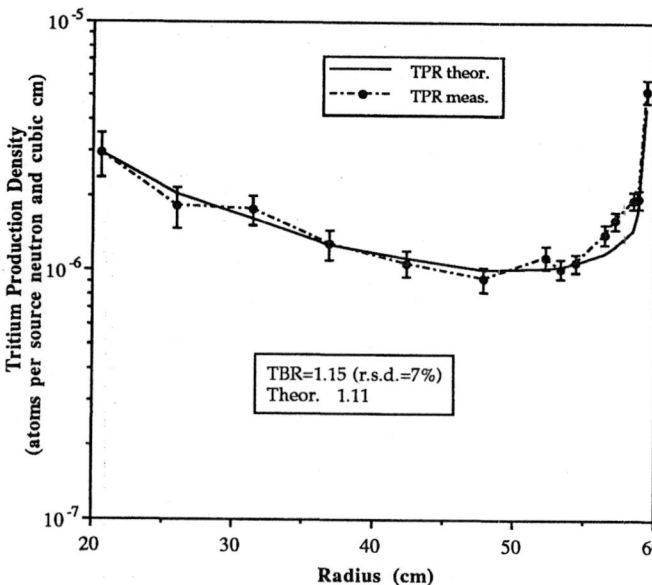

Fig. 6. Tritium Production Rate in the Pb-Li-C sphere

ii. In the Pb-Li sphere experiment all foils rates are underestimated about 10-20% but ^{115}In(n,n') rates go even 30% down. Calculation with ENDF/B-IV overestimates both T6 and T7 rates while in JENDL-3 results T7 is lower than experimental values (fig. 4). The ANISN calculation in this case fails to give a precise distribution of T6 on the outer border of the lithium sphere due to inadequate modeling of interaction with neutrons scattered from the concrete walls surrounding the assembly.

iii. Fig. 5 shows theoretical and measured responses of TLDs in the Pb-Li-C sphere. The distribution of TPR in the lithium zone of the Pb-Li-C system is presented on fig 6. The measured TBR is higher than JENDL-3 calculated value and lower than that from the ENDF/B-IV calculations. Spatial distribution of TPR shows gains in the vicinity of the graphite reflector which can be resulted from better reflection of neutrons than calculational predicted. The final value of TBR is estimated about 1.15 with relative standard deviation of 7.0%. In case of pessimistic assumptions on the neutron spectrum uncertainty the final r.s.d. rises to about 9.0%. This means that the real value of TBR is greater than 1.0 with probability about 95% at least, in the pessimistic case.

Table 2 presents the main sources of experimental errors. Concerning TLD application on TBR measurements one can raise its precision by improving data on TLD response function.

Table 2 Partial errors forming uncertainty of total TBR [%]

	Source strength	Detec. meas.	TPR c.sec.	TLD resp.func.	Neutron spectrum
Li	5.3	1.0			
PbLi	5.3	2.0			
PbLiC	3.0	2.0	1.0	5.0	2.0 (6.0*)

*data in case of pessimistic assumptions on neutron spectrum uncertainty

Conclusions

Good benchmark data on TBR in the Li/Pb-Li/Pb-Li-C spheres have been obtained to check the accuracy on the nuclear data library related to the DT fusion reactor blanket nuclear designs. In general, the present data on T6 show good agreement with JENDL-3 based calculations while T7 are slightly underestimated. In particular for the Pb-Li-C sphere experiment, the TBR has been verified to exceed 1.0 with good reliability. However, some inconsistences, in particular on spatial TPR distribution on the outer border of lithium in all systems, still remain.

The present study has been partly supported by Grant-in-Aid for Scientific Promotion of Fusion Research from the Ministry of Education, Science and Culture.

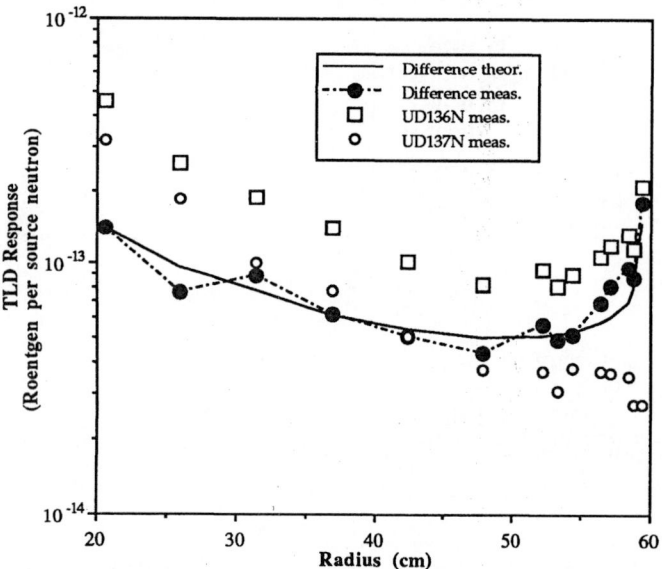

Fig. 5. Theoretical and measured responses of UD136N, UD137N and their difference in the lithium zone of Pb-Li-C sphere

References

1) T. Iguchi and M. Nakazawa: New Large Solid Angle (4π) Activation Foil Method.., J. of Nucl. Science and Tech.. Vol. 24, No. 11 (1987).
2) K. Sugiyama, T. Iguchi, et al: Neutronic Experiments in a Li-Pb Spherical assembly, Proc. 14th SOFT p 1243 (1987).
3) J. Cetnar, T. Iguchi, M. Nakazawa, K. Sugiyama, A. Takahashi, K, Sumita: Tritium Breeding Ratio in Li, Li-C, Pb-Li, Pb-Li-C, Be-Li, Be-Li-C Spheres, Proc. 2nd Spec. Meet. Nucl. Data for Fusion Reactor, JAERI-M 91-062, p71 (1991).
4) H. Hashikura, K. Haikawa, S. Tanaka and S. Kondo: Calculation of Neutron Response of Thermoluminescent Dosimeters, J. of the Fac. of Engineering, the U. of Tokyo (B) Vol. XXXIX, No.1 (1987).
5) K. Furuta, Y. Oka and S. Kondo: Evaluation of Covariance Data for ^{6}Li and ^{7}Li Cross Section in JENDL3-PR1, UTNL-R-0167 (1984).

*Inter university research group:
S.Iwasaki, N.Matsuyama, J.R.Dumais, M.Sakuma (Tohoku U.),
J.Kaneko, K.Harada, P.Strasser (U. of Tokyo), S.Itoh (Nagoya U.),
J.Yamamoto and K.Yamanaka (Osaka U.), Japan.

ELASTIC AND INELASTIC NEUTRON SCATTERING ON CARBON-12

G. Börker, W. Mannhart, B.R.L. Siebert
Physikalisch-Technische Bundesanstalt
W-3300 Braunschweig, Fed. Rep. of Germany

Abstract: Cross sections of elastic and inelastic neutron scattering on C-12 have been measured at 13.33, 14.06, 14.53 and 15.82 MeV. The measurements were performed with the PTB multi-angle neutron time-of-flight spectrometer. Neutrons were generated via the D(d,n)He-3 reaction in a deuterium gas target. The data analysis comprises a complete Monte Carlo simulation of the scattering geometry. The data were finally normalized to the n-p cross section by means of an additional scattering sample of polyethylene. The results are compared with ENDF/B-VI and data from the literature.

(Carbon-12, differential neutron cross sections of elastic and inelastic scattering, E_n = 13.33, 14.06, 14.53 and 15.82 MeV)

Introduction

The present work continues our series of precision measurements of scattering on light nuclei [1 - 3]. Differential cross sections of elastic and inelastic scattering on C-12 have been measured for neutron energies of 13.33, 14.06, 14.53 and 15.82 MeV. With a complete Monte Carlo simulation of the neutron time-of-flight (TOF) spectra measured and a subsequent normalization to the n-p scattering, many of the possible experimental error sources were excluded. The analysis of the inelastic data has been restricted to an analysis of scattering to the first excited level of C-12 of 4.439 MeV. The time-of-flight peaks of scattering to the higher levels of 7.65 and 9.64 MeV are masked behind a broad energy distribution of neutrons stemming from the D(d,np) breakup reaction. A precise analysis requires the inclusion of breakup data in the Monte Carlo simulation. This is in progress but has not yet been finalized. Detailed experimental data are available for this purpose [4].

Experimental Methods

The measurements were carried out at the PTB multi-angle neutron time-of-flight spectrometer facility. Neutrons were produced with the D(d,n)He-3 reaction in a deuterium gas target, 1.1 cm in diameter and 3.0 cm in length, with a gas pressure of 0.2 MPa. Our CV28 compact cyclotron was operated in pulsed mode with an average beam current of 1 µA. The pulse width was about 1.5 ns with a repetition frequency of 1 MHz. A graphite cylinder, 2.45 cm in diameter and 5.0 cm in height, acted as the scattering sample and was placed in the center of the facility. The sample-to-target distance was 17.5 cm. At a distance of 12 m from the center, five NE213 neutron detectors, each separated by an angle of 12.5 degrees, were located behind a fixed massive collimator system. The scattering angles were varied by turning the cyclotron with the gas target around the sample. Another NE213 detector acted as a neutron monitor and was always positioned at an angle of 60 degrees to the deuteron beam direction. Scattering angles between 12.5 and 160 degrees were covered by our arrangement. At low scattering angles an additional polyethylene sample (with a known hydrogen content) of the same size as the carbon sample was used to allow a normalization of our scattering data based on the well-known n-p scattering cross section. Each measurement was accompanied by a "gas out" run to determine the

neutron background from the structure materials of the gas target and a "sample out" run for the background of neutrons scattered in the air or penetrating the collimator system. At each detector pulse height, the pulse shape of the n-γ discrimination and the time-of-flight were recorded. The detector thresholds were adjusted to about 1.7 MeV neutron energy and the exact value (within ± 10 keV) was determined with a Cf-252 neutron source by comparing the measured spectrum with the predicted shape (Fig. 3 of Ref. [3]).

Data Analysis

Photons from the inelastic scattering to the 4.439 MeV level of C-12 served as a time reference. With two-dimensional cut conditions neutrons and photons were separated. The measured background spectra were subtracted from the remaining neutron time-of-flight spectra. Data from different angle sets were normalized with the monitor count rates. The corrected spectra were compared with those obtained from Monte Carlo simulations, taking into account multiple scattering effects, fluence attenuation and the finite geometry of the target, sample and detector. The STREUER code [5] simulates neutron production in the gas target, neutron transport between target, sample and detector and scattering in the sample. The code starts with evaluated ENDF/B-V cross sections of carbon as first trial functions. These data were modified iteratively until optimum agreement was achieved between experiment and simulation. At the same time, the mean energy of the incident neutrons and the scattering angles were varied by comparing the simulated neutron peak positions with the experimental ones. The code calculates the mean neutron energy, the mean scattering angle and the corresponding FWHM of the associated distributions. In the present experiment the FWHM of the initial energy and the angle was 120 - 135 keV and 6.3 degrees, respectively. The effective mean scattering angle, which usually deviates from the geometric one, can be determined very precisely due to the high sensitivity of the n-p scattering to the angle position. The corresponding uncertainties were 20 keV for the mean initial neutron energy and 0.2 degrees for the mean scattering angle.

Before comparing the neutron peak content of experiment and simulation, the simulated spectra were folded with the experimental time resolution and multiplied by the energy-dependent neutron detector efficiencies calculated with the NRESP code [6]. The spectra

318

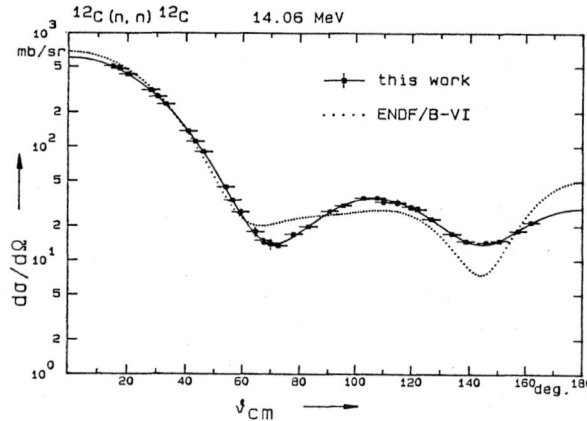

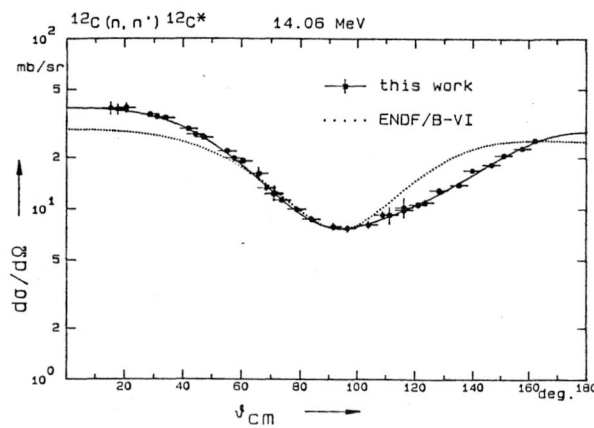

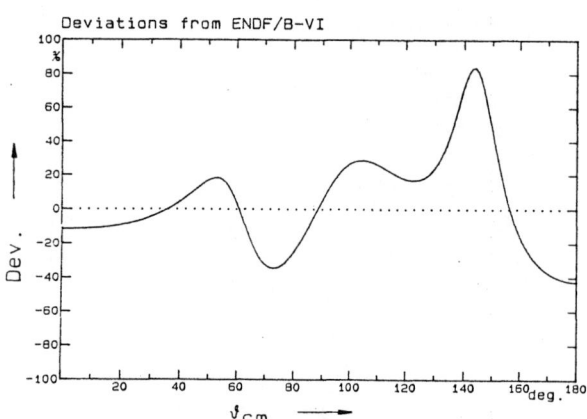

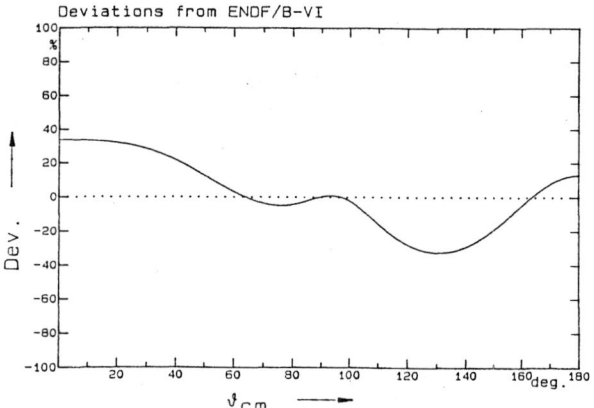

Fig. 1. Differential elastic scattering cross section at 14.06 MeV in the center-of-mass system. In the lower part the deviations between our data and ENDF/B-VI are plotted.

Fig. 2. Differential inelastic scattering cross section (4.436 MeV level) at 14.06 MeV. Same notation as in Fig. 1.

were normalized to the neutron fluence of the gas target which was measured with one of the NE213 detectors before the scattering experiment. The accuracy of the fluence determination was verified in previous experiments with a proton recoil telescope (Fig. 1 of Ref. [2]). The n-p scattering data of the polyethylene sample were analysed with an identical procedure. A comparison of the results with the evaluated reference values of ENDF/B-V was used in the final step to normalize the carbon data to the n-p cross sections. This procedure cancelled the previous fluence normalization. The differences between the two normalizations were of the order of 1 %. The mean scattering angles were transformed to the center-of-mass system, and center-of-mass differential cross sections were derived from the data.

Results

The statistical uncertainties of the differential cross section data ranged between 0.5 and 10 % for the elastic and between 2.0 and 20 % for the inelastic scattering data, with typical values of about 2 % and 4 % in both cases. Systematic contributions were the energy-dependence of the neutron detector efficiency (2 %), the normalization to the n-p scattering (1 %), the multiple scattering correction (1 %) and the angle uncertainty of 0.2 degrees. The magnitude of last component depends on the slope of the angular distribution and was up to 2 %

for the elastic data and substantially less for the inelastic. The components were added quadratically resulting in typical values of 3.3 % (one standard deviation) for the elastic and 4.7 % for the inelastic data. The differential scattering data obtained at 14.06 MeV are plotted in Figs. 1 and 2. The differential cross section for elastic scattering is shown in Fig. 1. The experimental data transformed to the center-of-mass system are given with error bars. The horizontal bars indicate the FWHM of the scattering angle distribution. The solid line through the data is the result of a Legendre polynomial fit to the data. The differential data from ENDF/B-VI are shown for comparison. In the lower part of the figure the deviations betweeen our data and ENDF/B-VI were scaled. These deviations were large and of the order of 30 % around the first minimum and between 40 and 80 % at large angles. In Fig. 2 the measured differential cross sections of inelastic scattering to the first excited level of 4.439 MeV are shown. Here too, the deviations between our data and ENDF/B-VI are large. It should be remembered that the shape of the elastic and inelastic angular distributions has not been changed from ENDF/B-V to ENDF/B-VI, i.e., the same structural deficiencies of the angular distributions as before are preserved in the recent version of the file. In contrast to this is the situation regarding the integrated scattering cross sections. Substantial modifications of the elastic and inelastic cross sections have been made between ENDF/B-V and ENDF/B-VI. This can be seen in Table 1, where our integrated data are compared with both evaluations. For the elastic and inelastic cross sections the agreement between our data and ENDF/B-VI is much better than with ENDF/B-V. Most of our data agree within their uncertainties with ENDF/B-VI. The maximum difference is 4.3 % for the elastic and 7.0 % for the inelastic scattering data.

Our 14.06 MeV data can be compared with an evaluation performed by Hansen [7]. This evaluation comprised six different experiments and was done at 14.1 MeV. In Fig. 3 our differential cross section data for elastic scattering are plotted relative to the recommended data set of Hansen. The solid line is the fit to our data. Our data agree within ± 10 % with the result of Hansen's evaluation. The uncertainty of Hansen's evaluation is 5 %. Compared with the large deviations from ENDF/B-VI, the agreement of our data with Hansen's evaluation is very satisfactory. The deviations between our data and those from individual experiments of Hansen's evaluation are much larger due to the fact that this evaluation averaged differential data with a spread of ± 30 % around the final result. The integrated elastic cross section of (821 ± 10) mb and the inelastic cross section of 206 mb of Hansen's evaluation agree very well with our 14.06 MeV data, shown in Table 1.

The differential cross section data of this experiment for elastic and inelastic scattering measured at 13.33, 14.53 and 15.82 MeV show in all cases deviations from ENDF/B-VI of a similar magnitude as those shown for the 14.06 MeV data in Figs. 1 and 2. In particular the shape of the ENDF angular distributions generally contradicts that of the experimental data. For example, a comparison of our elastic angular distribution measured at 15.82 MeV with that of Olsson et al. [8] at 16.5 MeV showed, despite the energy difference, a very similar pattern of the two angular distributions, the maximum deviations between the shapes being of the order of ± 20 %. This is in contrast to the ENDF/B-VI angular distribution which showed a totally different type of structure with deviations from the experimental data of between - 70 % and + 100 %.

Conclusions

Precise differential cross section data were obtained based on a complete Monte Carlo simulation of the scattering geometry with inclusion of multiple scattering processes and

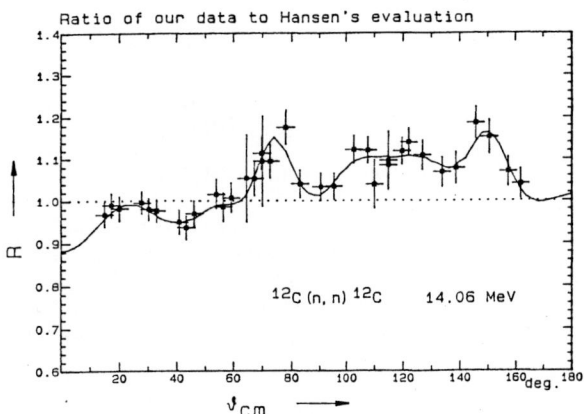

Fig. 3. Ratio of our elastic scattering data (14.06 MeV) to Hansen's evaluation (14.1 MeV).

with a final normalization to the well-known n-p scattering cross section. The simulation reduced the data analysis to the determination of ratios of experimental and simulated neutron time-of-flight peak content and avoided a complex separation between single and multiple scattering contributions. Due to a high degree of experimental redundancy, for example in the determination of the incident neutron energy from the scattering measurements and from a direct TOF measurement at zero degrees, and with the data normalization based on n-p scattering and a fluence measurement, possible error sources were minimized. Our data confirm the elastic and inelastic cross section data of the recent ENDF/B-VI evaluation but do also indicate an urgent need to revise the angular distributions of ENDF/B-VI which are highly inconsistent with our and other experimental data. A complete documentation of all numerical figures of the present work is in progress.

References

1. R. Böttger, H.J. Brede, H. Klein, H. Schölermann, B.R.L. Siebert, in Proc. Int. Conf. on Nuclear Data for Basic and Applied Science, May 1985, Santa Fe, N.M., USA, (Eds. P.G. Young et al.) p. 1455, Gordon and Breach, New York (1986)
2. G. Börker, R. Böttger, H.J. Brede, H. Klein, W. Mannhart, B.R.L. Siebert, in Proc. Int. Conf. on Nuclear Data for Science and Technology, May/June 1988, Mito, Japan, (Edited by S. Igarasi), p. 193, Saikon, Tokyo (1988)
3. G. Börker, R. Böttger, H.J. Brede, H. Klein, W. Mannhart, B.R.L. Siebert, Report PTB-N-1, Physikalisch-Technische Bundesanstalt (1989)
4. S. Cabral, G. Börker, H. Klein, W. Mannhart, Nucl. Sci. Eng. 106, 308 (1990)
5. B.R.L. Siebert, H.J. Brede, H. Klein, D. Schlegel-Bickmann, H. Schölermann, Report PTB-ND-20, Physikalisch-Technische Bundesanstalt (1981)
6. G. Dietze, H. Klein, Report PTB-ND-22, Physikalisch-Technische Bundesanstalt (1982)
7. L.F. Hansen, Report UCRL-95890 (preprint), Lawrence Livermore National Laboratory (1986)
8. N. Olsson, B. Trostell, E. Ramström, Nucl. Phys. A496, 505 (1989)

Table 1. Integral scattering cross sections (in mb)

E_n MeV	$\Delta E^*/2$ MeV	This work	ENDF/B-VI	ENDF/B-V
		Elastic		
13.325	0.060	896.3 ± 18.7	914.3	853.8
14.060	0.062	812.7 ± 17.1	825.3	798.5
14.529	0.063	814.1 ± 17.5	849.1	803.0
15.823	0.067	1020.4 ± 22.5	1013.8	886.3
		Inelastic (4.439 MeV level)		
13.325	0.060	212.6 ± 4.4	227.4	204.4
14.060	0.062	207.2 ± 4.3	210.4	187.6
14.529	0.063	181.1 ± 3.9	184.6	169.4
15.823	0.067	140.9 ± 3.3	146.6	137.9

*ΔE is the FWHM energy resolution

SMALL ANGLE SCATTERING CROSS SECTIONS
OF 14.8 MEV NEUTRONS FROM Fe, Ni and Cr*

Qi Huiquan, Chen Hongbin, Chen Yingtang
Chen Qiankun, Chen Zhenpeng, Chen Zemin

Department of Physics, Tsinghua University, Beijing, P. R. China

Abstract: A position sensitive neutron detector was used to measure the differential cross sections of 14.8 MeV neutrons scattered in the angular range of $2.5°$ to $16°$ from Fe, Ni and Cr. The data were corrected for flux attenuation, multiple neutron scattering and finite geometry by Monte Carlo method. The angular distributions were calculated with coupled channels code ECIS79. The measured and calculated results were compared and discussed.

(small angle scattering, 14.8 MeV neutron, Fe, Ni, Cr)

Introduction

There is a forward peak in the angular distribution of elastic scattering for ~ 14 MeV neutrons, so the precise knowledge of the differential cross section at small angle is important. But the data for the angles smaller than $15°$ are very scarce because of the difficulties in geometrical arrangement of most measurements. By use of the optical model, the differential cross sections at small angles can be calculated. But most of the optical model parameters published in the past were obtained without experimental data in the small angle region. They cannot fit the small angle data very well. Fe, Ni and Cr are important structural materials for nuclear fusion equipment. But only a few experimental data of small angle scattering were measured for Fe (Ref.[1-3]) and Ni(Ref.[2]) and no data has been published yet for Cr. So in view of the importance in application and for obtaining the optimum optical model parameters, the measurements of the differential cross sections of neutron scattering at small angles are needed.

Experimental Arrangement

The measurements of the small angle scattering cross sections in the past several years in Tsinghua University have been reported in Refs.[4-6]. Only a brief description about the experimental arrangement is given here.

The measurements were carried out on the Cockcroft Walton accelerator with a rotational target in Tsinghua University. The neutrons are generated from T(d,n)He-4 reaction. The angle between the associated alpha particles and the deuteron beam is $170°$, as close as possible to $180°$, to reduce the kinematic shifts of the associated neutron beam caused by the change of the effective energy of the incident particles. The neutron beam is collimated in $\sim \pm 0.5°$ by a square double truncated conical collimator located in a shielding which is 70 cm in thickness of iron, 60 cm of paraffin and 10 cm of lead. The T-Ti target is shielded all around by iron, lead and paraffin. The main detector is a position sensitive neutron detector which consists of a long cylindrical liquid scintillator tube (LLS) of 100 cm in length and 5.3 cm in inner diameter and two photomultipliers at both ends of the tube. The position information of the incident neutrons is extracted from the time difference between the output signals of the two photomultipliers. The samples are small discs, their parameters are shown in Table 1. All samples are natural elements metals. The Cr sample is made by pressing powder into the final shape.

Table 1. The Parameters of the Samples

sample	thickness (cm)	density g/cm**3	trans. (%)	purity (%)
Fe	2.0	7.8	64.5	99.99
Ni	1.6	8.9	68.2	99.5
Cr	2.1	4.8	70.0	99.5

The associated alpha particle time of flight technique and n-gamma discrimination were used to reduce the background. The block diagram of the electronic system is similar to that in Ref.[6].

Experimental Procedure

The differential scattering cross section is obtained from the following formula

$$\sigma(\theta) = \frac{N(\theta) \, k}{N_0 \, \varepsilon(\theta) \, n \, t \, d\Omega(\theta)}$$

where $N(\theta)$ is the net counts of the scattering neutrons at the angle θ in the element $A(\theta)$ of LLS. N_0 is the net counts of perpendicular incident neutrons on the element $B(0)$ at the middle point of LLS. Both $N(\theta)$ and N_0 are normalized to the alpha counting produced by T(d,n)He-4 reaction. $\varepsilon(\theta)$ is the intrinsic efficiency ratio of the element $A(\theta)$ to the element $B(0)$, k is the ratio of the solid angle subtended by the element $B(0)$ to that by the scattering sample at the neutron source. $d\Omega(\theta)$ is the solid angle subtended by the element $A(\theta)$ at the scattering sample. n is the number of nuclei in a unit volume of the sample, and t is the thickness of the sample.

*Supported in part by the International Atomic Energy Agency, the National Natural Science Foundation of China and the General Company of Nuclear Industry of China.

When N_0 is to be measured, the LLS is put at the center of the collimated neutron beam and perpendicular to it. When $N(\theta)$ is to be measured the LLS is put just over the collimated beam. When the background is measured, the arrangement is the same as that in the scattering measurement except that the sample is put at the other end of the collimator. The ratios of the elastic scattering neutron counts to the background are 10:1 to 4:1 for various samples and angles. When the LLS is risen 9 cm from the center of the neutron beam, the scattering angles that can be measured are between 2.5° and $12.^\circ$ When the LLS is displaced 14 cm horizontally and 9 cm vertically then the angle range is between 4° and $16.^\circ$

The angular uncertainty is $\pm 0.8.^\circ$ The energy resolution of the measuring system is 1 MeV. It cannot separate the neutrons scattered elastically from those scattered inelastically from low-lying excited levels. So the results

cancelled through the N_0 counting. The uncertainty of the scattering cross section caused by this is about 3%.

(3). The sum of errors caused by the Monte Carlo corrections, the correction for dead time of alpha counting, and the uncertainty of the number of nuclei in the samples is less than 3%.

Hence the total errors are 7%--10% in various samples and angles. The comparisons of experimental and calculated data of the angular distributions at small angles given by present work and others (Refs.[1-3]) for Fe, Ni, and Cr are shown in Fig.1 and those between 0° and 180° are shown in Fig.2. As the figures show, the consistence between present data and others is excellent. The curves represent the differential cross sections of elastic scattering (the inelastic scattering cross sections from the low-lying excited levels are negligible) calculated by the

Table 2. The Differential Cross Sections for Fe, Ni and Cr

θ (deg)		2.5	4.0	6.0	8.0	10.0	12.0	14.0	16.0
$\sigma(\theta)$	Fe	3.31±0.13	3.19±0.10	2.68±0.08	2.36±0.07	2.38±0.07	2.11±0.06	1.62±0.11	1.30±0.10
	Ni	3.43±0.14	3.29±0.13	3.02±0.11	2.81±0.10	2.76±0.10	2.18±0.08	2.25±0.16	1.32±0.11
(b/sr)	Cr	2.88±0.14	2.77±0.11	2.66±0.09	2.38±0.08	2.36±0.08	2.10±0.07	1.89±0.13	1.44±0.12

of present measurement include the contributtion of such inelastic scattering neutrons.

In order to minimize the counts of the neutrons scattered by the objects other than the sample, the materials in the forward direction of the sample were reduced as much as possible. The wall thickness of the aluminium box, used for shielding the liquid scintillation detector from ambient room light, is reduced to 0.5 mm. The supporter of LLS is as thin as possible. The distance between LLS and the cross member of the supporter is 50 cm. The distance between LLS and the floor is 183 cm. The walls of the measuring hall are far enough from the LLS (larger than 5 m)

Results and Discussion

The results of the measurements from Fe, Ni and Cr are show in Table 2. The results were corrected for the neutron flux attenuation, multiple neutron scattering and finite geometry by Monte Carlo method. According to our calculations the Schwinger scatterings can be neglected for Fe, Ni and Cr.

The total uncertainty of the average data in six runs is about 5% except those at $2.5.^\circ$ 14° and 16°, which were measured one or two times only. The errors in Table 2 are statistical errors only. The other major sources of error are:

(1).The uneven distribution of the collimated neutron beam and the uncertainty of its center determination (about 0.5 cm). The error of N_0 caused by this is smaller than or equal to 5%.

(2). The edge of the sample hit by the collimated neutron beam is not very clear-cut and some small angle scattering neutrons at the surface of the collimator would hit the sample. A part of this uncertainty can be

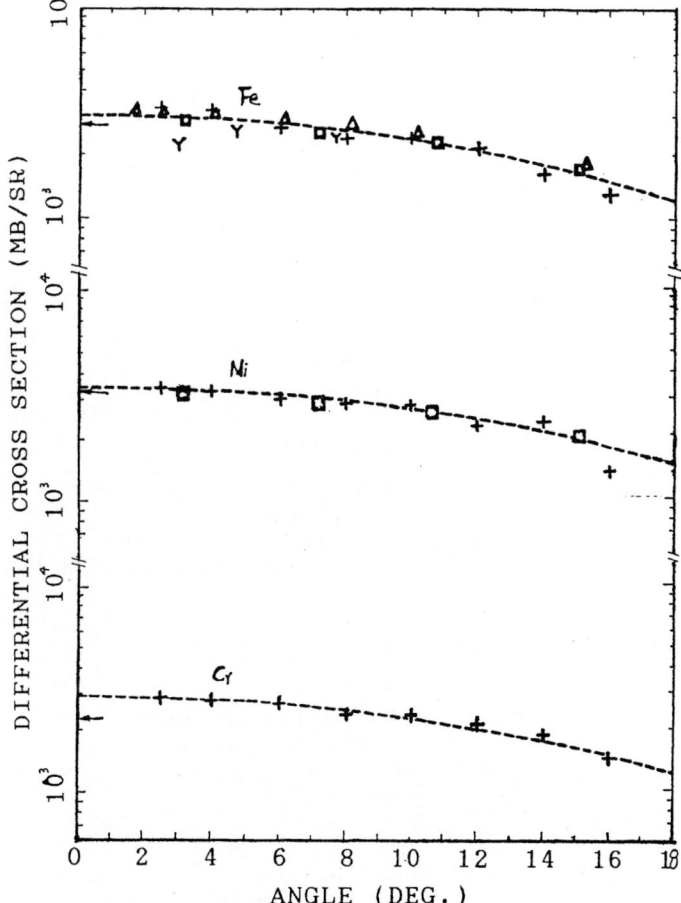

Fig.1 Differential cross sections at small angles

$\triangle$ Ref.1 $\square$ Ref.2 Y Ref.3
$+$ present work $---$ calculated results

322

coupled channel (CC) code ECIS79. Arrows are the Wick limits. All of the scattering cross sections of Fe, Ni and Cr at zero degree are larger than their Wick limits, repectively. As shown in the figures, the small angle scattering cross section for Fe, Ni and Cr at 14.8 MeV can be well described by the coupled channel model. The optimum parameters used in the coupled channel model are as follows:

The geometrical parameters (given in fm) are:
$\gamma_o = 1.15$ $\gamma_v = 1.19$ $\gamma_s = 1.32$ $\gamma_{so} = 1.20$
$a_o = 0.73$ $a_v = 0.80$ $a_s = 0.56$ $a_{so} = 0.74$

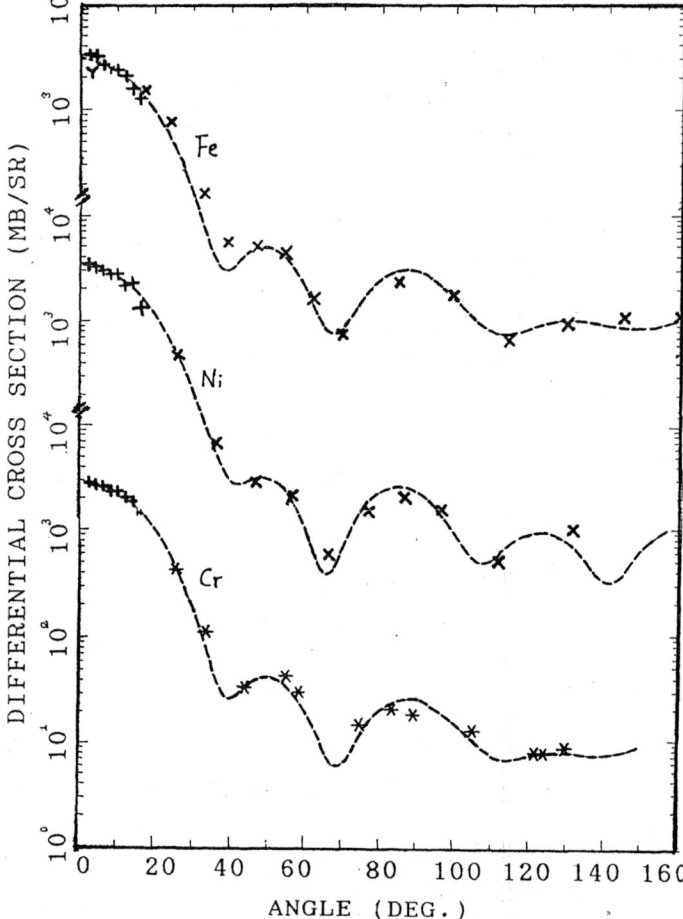

Fig.2   Differential   cross   sections
between 0° and 180°
Y Ref.3   X Ref.7   ✶ Ref.8
+ present work --- calculated results

The depths of the potentials (given in MeV) for Fe, Ni and Cr are shown in Table 3.

Table 3.   The Depths of the Potentials (given in MeV) for Fe, Ni and Cr

	Vo	Wv	Ws	Vso	β_2
Fe	49.32	1.23	5.58	4.91	0.23
Ni	50.50	1.59	7.18	9.07	0.14
Cr	50.77	1.43	6.53	5.67	0.16

The subscripts o, v, s and so represent real, volume imaginary, surface imaginary and spin-orbit, respectively. β_2 is the second order deformation parameter.

REFERENCES

1. Li Jingde, Wang Shiming, Xie Daquan, Ma Gonggui, Zou Yiming and Chen Shuying: Chinese J. of Nucl. Phys. 11, 19 (1989)
2. W. P. Bucher, and C. E. Hollandsworth: Phys. Lett. 58B 277(1975)
3. G. Deconninck: Annales de la Societe Scient de Bruxelles, Ser.I. 75, 102 (1961) (EXFOR 21858.003)
4. Qi Huiquan, Liu Yunchang, Chen Zhenpeng, Wu Xuechao, Wang Wanhong and Chang Jing: in Proc. of the Int. Conf. on Nuclear Data for Basic and Applied Science, May 1985 Santa Fe, U.S.A. (Edited by P. G. Young) Vol.2, p.1355, Gordon and Breach Science (1986)
5. Qi Huiquan, Liu Yunchang, Chen Zhenpeng, Wu Xuechao, Wang Wanhong, Chen Qiankun, and Xu Guang: INDC(CPR)-005/G (1985); INDC(CPR)-C06/L (1985)
6. Qi Huiquan, Chen Yingtang, Chen Zemin, Chen Zhenpeng, Chen Qiankun, Li Hucheng: in Proc. of the Int. Conf. on Nuclear Data for Science and Technology, May/June 1988, Mito, Japan, (Edited by S. Igarasi), p. 799, Saikon, Tokyo (1988)
7. L. F. Hansen, F. S. Dielrich, B. A. Pohl, C. H. Poppe and C. Wong: Phys. Rev., C31, 111 (1985)
8. P. H. Stelson, R. L. Robinson, H. J. Kim, J. Rapaport and G.R. Satchler: Nucl. Phys., 68, 97 (1965)

MEASUREMENT OF 7Li(n, n' γ) (478keV) INELASTIC ANGULAR DISTRIBUTION DERIVED VIA THE SHAPE ANALYSIS OF THE DOPPLER SHIFTED γ-RAY SPECTRUM (DSM) AT 14.9 MeV*

S. L. BAO, F. Z. HUANG, J. Q. LIU, G. Y. TANG

Institute of Heavy Ion Physics,
Peking University,
Beijing 100871, P. R. China

H. Y. ZHOU, G. Y. FAN, C. L. WEN, M. HUA

Institute of Low Energy Nuclear Physics,
Beijing Normal University,
Beijing 1000871, P. R. China

Abstract: The data were determined via shape and position analysis of γ-ray spectra measured in 5 lab angles, which were also simulated with Monte Carlo method. the results were expressed by the coefficients of Legendre polynomial in CM. The data were calculated with DWUCK4 code. Comparisons between the experimental and theoretical results are presented.

(7Li(n,n'γ)(478keV), Inelastic Angular Distribution, 14.9 MeV, DSM, DWUCK4)

1. THE PRINCIPLE AND METHOD

The experimental principle and method in detail can be found in an earlier report [2]. Because 73 fs, the half life of the first excited state of 7Li, is so short that the decay γ-ray actually is emitted from the moving 7Li*. According to the kinematics of the nuclear reaction the recoil 7Li* is emitted in a forward cone in Lab. Because of the Doppler effect the shape of the measured γ-ray spectrum strongly depends on the observation angle in lab. Then the angular distribution could be derived by the shape analysis of the measured γ-ray spectra. According to the reaction kinematics if the γ-ray was observed in angle α, the Doppler shift ΔE is the function of $\cos(\theta_{cm})$:

$$\Delta E_\gamma = \frac{E_\gamma}{C} * v_{lab}[\cos\theta_{lab}*\cos\alpha + \sin\theta_{lab} * \sin\Phi*\sin\alpha],$$

by the relation between the speed of the recoil in cm and in lab system, ΔE_γ can be expressed only by the function of the cm angle θ_{cm}, which can be expressed in Legendre polynomial coefficient in cm system. In the formula, E_γ is the level(478keV), v_{lab} and θ_{lab} are the speed and the direction of the recoil in lab system, Φ is the azimuthal angle, α is the γ-ray observation angle and C is the speed of light.

2. THE EXPERIMENT

We measured the γ-ray spectra at the pulsed 400 keV neutron generator in the Institute of Low Energy Nuclear Physics of Beijing Normal University using thick solid target of T(d, n) source (TTi). The 300 keV D beam with the average beam intensity 8 - 10 μA impinged on a pure metallic lithium sample, which was enriched to 99.98% and was sealed in Ar gas environment with a 0.2 mm Al shell. The size of the sample was 36.8 mm in diameter and 28.6mm in height. The sample was positioned at 0° and was 13.4 cm far from the neutron target. The average neutron energy is 14.9 MeV with 0.5 MeV energy spread. A post pulsed system was used in the measurement with repeat frequency 3.16 MHz and 1 ns time resolution, which was used to reduce the scattering neutron background. A plastic scintillator NE211 was used as a neutron monitor of TOF system. The electronics used in the experiment was the same as in reference[10]. The γ-ray spectra were measured in 40°, 55°, 90°, 125° and 140° in lab. The measuring system was checked regularly by a 226Ra -ray

---

*: This work is supported by China Splendid Youth Teacher Foundation and by the the IAEA under the contract No. 5430/RB.

324

source in order to estimate the system stability.

3. DATA ANALYSIS

A. MONTE CARLO SIMULATION

A Monte Carlo simulation of the γ-ray spectrum was done for the experiment [4]. Some corrections were included in the calculation, which were the primary neutron energy spread, solid angle correction for the sample to neutron target, energy resolution of the pure Ge detector and the speed reducing effect of recoil $^7Li^*$ before emitting the γ-ray.

B. THE BEST FITTING

In order to compare the experimental data with the Monte Carlo calculation the original experimental data need to be compressed to 1 keV per channel. Then the measured and simulated γ-ray spectra were fitted with the least squares method using Legendre polynomial. Then the angular distribution of $^7Li^*$ was expressed by the coefficients of 5 term Legendre polynomials in CM [2]. The ADI data were derived from the angular distribution of $^7Li^*$.

4. DWBA CALCULATION

range approximation DWBA. During the calculation the ^7Li ground state(3/2-), the first level (0.478keV,1/2-) and the second state(4.63MeV,7/2-) were treated as rotational band with K=1/2 [5]. In this work the spherical optical potential parameters and the residual interaction potential parameters together with deformation factor β_2 were automatically searched by a complex shape method [13]. In that case some subroutines in original DWUCK4 were changed slightly to fit the parameter searching, during which, the 7 energy points of the measured elastic plus inelastic scattering ADX data from Hugue [11] in the energy range from 8 to 14 MeV, the evaluated total cross section data from ENDF/B-VI in the same energy range and our earlier results of the ADI data [1, 2] in the energy points 8, 8.25, 10 MeV were used for the parameter adjusting with the best fitting method. The detailed calculation can be seen in a separate paper [3].

5. RESULT AND CONCLUSION

The data for ADI of the reaction $^7Li(n,n'\gamma)$ at the first level were compared with the coefficients of 5 term Legendre polynomials (have been normalized to Ao) and are shown in Fig. 1. In the fig. the experimental data at 8 and 8.5 MeV from reference 2, at 10MeV from reference 1 and at 14.9 MeV from this work together with the theoretical calculation in the energy range from 8 to 20 MeV, in step of 1 MeV were shown. In the fig. the earlier calculated data by Young [7] using same code for the same reaction channel are also shown. The optical potential parameters in the calculation were treated in different ways by us. The uncertainty bars in fig. 1 for experimental data are the fitting uncertainties. The ADI data from evaluated ENDF/B-VI file are also compared.

From the result, one sees that the DWBA calculation can basically be used for the description of the reaction and the inelastic angular distribution of the reaction at the first level has the similar trend as the experimental data. From above fact, we can conclude that the measured data at 14.9 MeV fill the gap and the DWBA theory can be used as a energy range. But the more experimental data will be useful for theoretical calculation and the coupled channel model may give the best fitting of the experimental data.

REFERENCE

1. Bao Shanglian, Huang Feizeng: Inter. Report IHIP/FN9002, Institute of Heavy Ion Physics, Peking Univ., Beijing 100871, P.R. China, Feb., 1990.
2. H. Liskien, S. L. Bao: Proc. of the Inter. Conf. on Fast Neutron Physics, Dubrovnik, Yugoslavia, May 1986, P275
3. Huang Feizeng, Bao Shanglian: to be published in AEA/NDS (tech. report) & in Chinese J. of Nucl. Phys.
4. Bao Shanglian, Xu Dongwu: Internal Report IHIP/FN8802, Institute of Heavy Ion Physics, Peking Univ., Beijing 100871, P. R. China, Feb., 1988.

5. A. J. Oastler, PhD thesis, 1977,
   Dept.of Physics, Birmingham Univ.,
   United Kingdom.
6. S. Chiba, M.Baba, H.Nakasima:Double-
   Japn. J. of Nuclear Science
   Tech., 22,771(1987).
7. P. G. Young, LANL, USA, private
   communication(1988).
8. H. D. Knox, R. M. White and R. O.
   Lane Nucl. Sci. Eng., 223, (1979),
   p69.
9. H. D. Knox and R. O. Lane, Nucl Phys.
   A359(1981), 131.
10. Zhou Hongyu,Fan Guoying, Wen Chenlin,
    Hua Ming, Institute of Low Energy
    Nuclear Physics and Bao Shanglian,
    Institute of Heavy Ion Physics,
    "Differential Cross sections at
    125° for γ-rays Produced from
    Lithium, Fluorine, Sodium and
    Chlorine under 14.9MeV Neutron
    Bombardments", report to IAEA under
    contract No. 5357/RB, Sep., 1989.
11. H. H. Hogue: Nucl. Sci. and Eng.,
    69, 22(1979).
12. P. D. Kunz, Documentation for PSR-
    235/DWUCK Code Package(1986).
13. Zhou Delin,private communication.
    (1986).

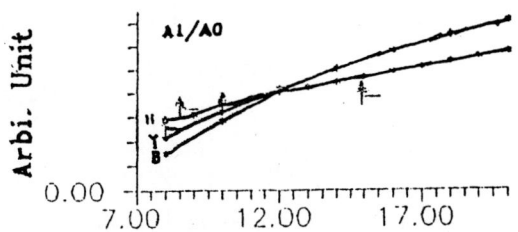

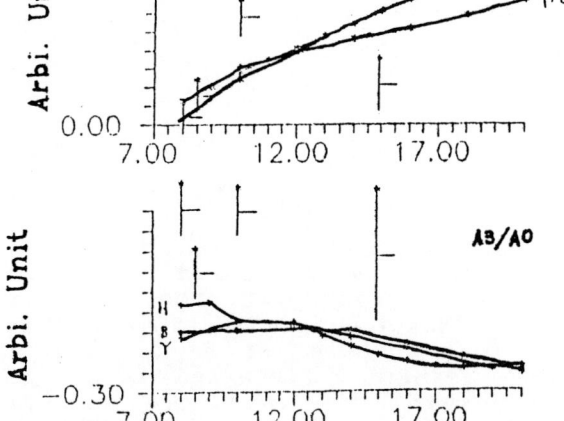

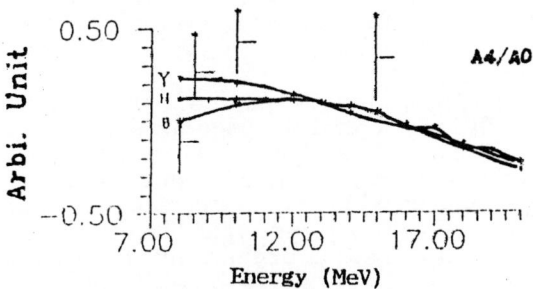

Energy (MeV)

Fig1. The Comparison of Coefficients
of Legendre polynomials

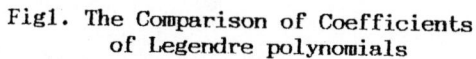

$\uparrow$: Ex. data
H : Our calculated Data
Y : P. G. Young Calculated
B : ENDF/B-VI Data

Table 1. The Searched Optical Parameters and Chi-square

V_{RO}	R_{RO}	a_{RO}	C_{RO}	R_{IO}	a_{IO}	C_{IO}	V_{LS}	o
-40.66	1.31	0.74	0.43	1.18	0.47	0.39	28.4	1.79

Table 2. The Searched Interaction Potential Parameters, β_2 and chi-square

$\beta_2{}^2$	V_R	R_{R1}	a_{R1}	C_{R1}	V_{I1}	R_{I1}	a_{I1}	C_{I1}	
1.50	-40.66	1.39	0.99	0.34	49.36	1.34	0.35	0.39	6.79

DOUBLE-DIFFERENTIAL NEUTRON EMISSION CROSS SECTION OF ^{9}Be

E. Dekempeneer, F. Poortmans, H. Weigmann, J.A. Wartena and
C. Bürkholz

Commission of the European Communities, Joint Research Centre
Central Bureau for Nuclear Measurements, 2440 Geel, Belgium.

Abstract: Double-differential neutron emission cross sections of ^{9}Be have been measured for incident neutron energies from about 2 to 11 MeV. Time-of-flight was used to measure the energy of incident neutrons. The energy spectra of neutrons emitted from the Be sample were obtained by unfolding the pulse-height spectra observed in eight NE-213 liquid scintillator detectors viewing the sample under different angles with respect to the incident beam. The data were normalized to the ^{12}C elastic scattering cross section at about 2 MeV.

(^{9}Be(n,2n), double-differential neutron emission cross section)

Introduction

Because of its large (n,2n) cross section, Be is a preferred candidate material for future fusion reactor blankets. Neutron transport calculations for the design of Be containing blankets require a detailed knowledge of the neutron emission cross section of Be, including the secondary neutron energy and angular distributions. In spite of its importance, experimental data on neutron emission from Be are not numerous.

Experimental investigations until now include fairly complete measurements at 14 MeV incident energy [1,2], including an investigation of the associated α-particle spectra[3]. At lower energies, neutron emission spectra have been measured at a few energies by Drake et al. and Baba et al.[4,5]. Extensive angular distribution measurements at many incident energies have been done for elastic scattering and for inelastic scattering to the 2.43 MeV level of ^{9}Be by Hogue et al.[6] and more recently by Sugimoto et al.[7].

Evaluations of the ^{9}Be(n,2n) cross section are involved because of the many different reaction mechanisms contributing. The total (n,2n) cross section is fairly well decribed in most evaluations. However, with respect to its sub-division into the different reaction channels, the earlier evaluations show large discrepancies with recent experimental data [7], and also a more recent evaluation [8] meets considerable difficulties with this sub-division, mainly due to contradictory basic data. In an attempt to improve the experimental data base, and also to aid in current evaluation efforts [9], we have measured double differential neutron emission cross sections of ^{9}Be for incident neutron energies from 1.6 to 11 MeV.

Experimental Method

The experimental methods used to measure double-differential neutron emission cross sections at the Geel linac have been described in detail earlier [10], and only a brief outline will be given here.
The uranium target of the linac, which is operated with 800 Hz repetition rate, provides a pulsed white neutron source. A well collimated neutron beam hits the ^{9}Be sample (0.0616 atoms/b) which is positioned within an evacuated tube at about 60 m from the neutron source. Neutrons emitted from the sample are detected by 8 NE-213 liquid scintillation detectors which are positioned 20 cm from the sample under different angles with respect to the incident beam. Pulse shape discrimination is used to distinguish neutrons from γ-rays. For each detected event the following information is stored in a HP 1000F data aquisition computer: the detection time giving the time-of-flight and thus the energy of the incident neutron, pulse height, pulse shape information, and detector identification. Only one event per linac cycle is accepted during data aquisition.
The shape of the incident neutron flux is measured with a ^{235}U fission chamber placed in the neutron beam at 30 m distance from the linac target.

After dead-time correction and background subtraction, secondary neutron energy spectra are obtained by unfolding the registered pulse height spectra with the aid of the FORIST-code [11] and the detector response matrix determined earlier [10].These secondary neutron energy spectra are then corrected for self-screening and multiple scattering in the sample and for attenuation of the emitted neutrons by the sample and by the 1.5 mm thick wall of the Al vacuum tube.

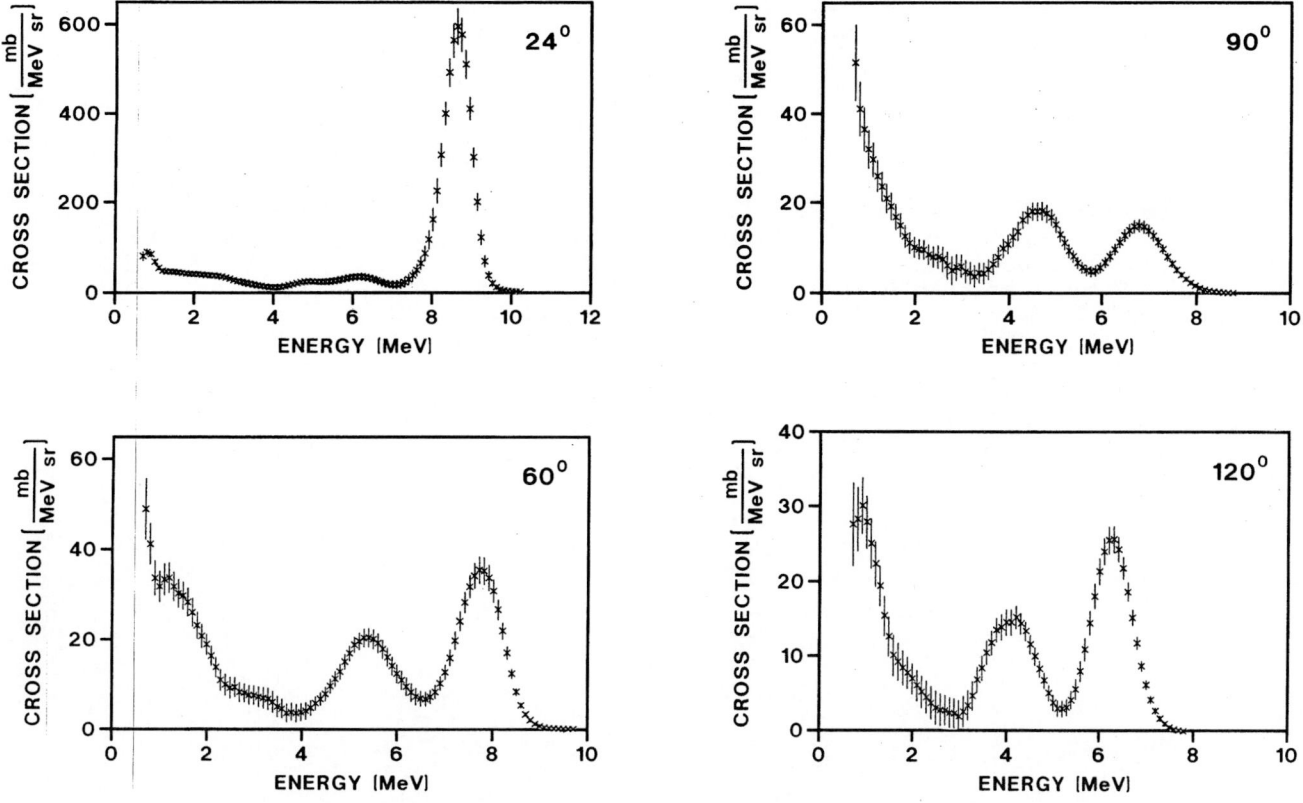

Fig. 1 :   Double-differential neutron emission cross sections
           of 9Be for the incident neutron energy interval
           from 8.37 to 8.75 MeV.

The Monte Carlo code written earlier [10] has been extended to include the (n,2n) reaction in the calculation of the multiple scattering corrections; a very approximate description of the secondary neutron energy distributions associated with these processes is used as an input to the code. The corrected secondary neutron energy spectra are converted to double-differential neutron emission cross sections with the aid of the neutron flux distribution measured with the 235U fission chamber, and a normalization factor which takes into account absolute detector solid angles and flux normalization. The latter factor is obtained from a measurement, under identical experimental conditions, of the 12C differential elastic scattering yield below 2 MeV neutron energy.

Results

As an example of the results obtained, Fig.1 shows double-differential neutron emission cross sections of 9Be for the incident neutron energy interval from 8.37 to 8.75 MeV and four emission angles. The emission angles indicated in the figures are in the lab. system, and are nominal, i.e. no correction has been applied for the opening angle of the detectors. The error bars indicated in Fig.1 represent the quadratic sum of

statistical errors (counting statistics) and of systematic uncertainties (all strongly correlated between different data points) associated with the measurement of the shape of the incident neutron flux, with the unfolding procedure and with the applied corrections for dead-times, backgrounds, multiple scattering and attenuation of the emitted neutrons. They do not include an additional normalization uncertainty of ~ 4% common to all data points. In total, double differential cross sections have been obtained for 8 emission angles and 23 incident energy intervals between 1.6 and 11.1 MeV (at the lowest energies only elastic scattering occurs). The data have been transferred to the NEA data bank and to the University of Birmingham for inclusion in the current evaluation [9]. An example of a comparison between the latter evaluation and the present experimental data is shown in ref.[9].

References

1. A. Takahashi, J. Yamamoto, T. Murakami, K. Oshima, H. Oda, K. Fujimoto and K. Sumita, Proc. Intern. Conf. Nuclear Data for Science and Technology, Antwerp 1982; D. Reidel, Dordrecht 1983.

2. M. Baba, M. Ishikawa, T. Kikuchi, H. Wakabayashi and N. Hirakawa, Proc. Intern. Conf. Nuclear Data for Science and Technology, Mito 1988; JAERI, Tokyo 1988.
3. D. Ferenc, B. Antolkovic, G. Paic, M. Zadro and S. Blagus, Nucl. Sci. Eng. 101 (1989) 1.
4. D.M. Drake, G.F. Auchampaugh, E.D. Arthur, C.E. Ragan and P.G. Young, Nucl. Sci. Eng. 63 (1977) 401.
5. M. Baba, T. Sakase, T. Nishitani, T. Yamada and T. Momota, Proc. Intern. Conf. Neutron Physics and Nuclear Data, Harwell 1978; OECD 1978.
6. H.H. Hogue, P.L. von Behren, D.H. Epperson, S.G. Glendinning, P.W. Lisowski, C.E. Nelson, H.W. Newson, F.O. Purser and T. Tornow, Nucl. Sci. Eng. 68 (1978) 38.
7. M. Sugimoto, P.T. Guenther, J.E. Lynn, A.B. Smith and J.F. Whalen, Nucl. Sci. Eng. 103 (1989) 37.
8. S.T. Perkins, E.F. Plechaty and R.J. Howerton, Nucl.Sci.Eng. 90 (1985) 83.
9. G.M. Field, T.D. Beynon and H. Gruppelaar, these proceedings.
10. E. Dekempeneer, H. Liskien, L. Mewissen and F. Poortmans, Nucl.Instr. Meth. in Phys. Res. A256 (1987) 489.
11. R.H. Johnson, FORIST, Neutron Spectrum Unfolding Code, Radiat. Shield. Int. Center, Computer Code Collection PSR-92, ORNL 1975.

GAMMA-RAY PRODUCTION CROSS SECTION MEASUREMENTS OF SOME STRUCTURAL MATERIALS BETWEEN 7.8 AND 13.0 MeV

Kazuo Hasegawa, Motoharu Mizumoto, Satoshi Chiba,
Masayoshi Sugimoto and Yoshimaro Yamanouti
Japan Atomic Energy Research Institute, Tokai-Mura, Ibaraki-Ken, Japan

Masayuki Igashira and Hideo Kitazawa
Tokyo Institute of Technology, O-okayama, Tokyo, Japan

Abstract: Gamma-ray production cross sections have been measured for Al, Si, Fe, Ni, Cu, Pb and Bi at incident neutron energies from 7.8 to 13.0 MeV using monoenergetic neutron sources. Neutrons were produced at the JAERI Tandem accelerator by the $^2H(d,n)^3He$ reaction for 7.8 and 10.0 MeV and $^1H(^{11}B,n)^{11}C$ reaction for 11.5 and 13.0 MeV. Emitted gamma-rays were measured with a 7.6 cm dia. x 15.2 cm long NaI(Tl) detector surrounded by a 25.4 cm dia. x 25.4 cm long annular NaI(Tl) detector operated in an anti-coincidence mode. Time-of-flight technique was used to reduce backgrounds. To obtain gamma-ray energy spectra, measured pulse height data were unfolded by the FERDOR code. Several correction factors were calculated by the Monte Carlo method such as angular distribution of source neutrons, neutron multiple scattering, outgoing gamma-ray attenuation and Compton scattering in the sample. Differential cross sections are given as a function of gamma-ray energy over the region from 0.7 to 12 MeV. Discrete gamma-ray and total gamma-ray production cross sections are also presented.

(gamma-ray production cross section, Al, Si, Fe, Ni, Cu, Pb, Bi, 7.8 – 13.0 MeV neutron, monoenergetic neutron source, Tandem accelerator, NaI, anti-coincidence, unfolding)

Introduction

Fast neutron induced gamma-ray production cross sections for whole gamma-ray energy spectra are required for fusion reactor design calculations. These cross section measurements were carried out mainly with the white neutron source using electron linear accelerator. Their neutron energy resolution, however, was generally broader than that for the monoenergetic neutron source due to the wide energy bin in order to compensate for the low counting statistics especially in higher neutron energy region.

Most of the previous experiments with monoenergetic sources were below 7 MeV or greater than 14 MeV, because of the availability of the neutron production reactions and/or acceleration energy for most of the Van de Graaff accelerators. In the present study, $^2H(d,n)^3He$ reaction was used in the energy region up to 10 MeV. Above that energy, break-up reaction cross section corresponding to $^2H(d,np)d$ reaction increases rapidly with incident deuteron energy, therefore, the source with $^1H(^{11}B,n)^{11}C$ reaction had been newly developed for neutron production in 10 – 13 MeV[1].

Our aims are to provide gamma-ray production cross section data from 8 to 13 MeV with higher accuracy using monoenergetic neutron sources.

Experimental Procedure

The experiments were carried out using the Japan Atomic Energy Research Institute (JAERI) Tandem accelerator. The experimental arrangement was described previously[2]. A positive in-terminal ion source and a negative ion source were used for the $^2H(d,n)^3He$ reaction and for $^1H(^{11}B,n)$ reaction, respectively. The beam was chopped and bunched to a 2 MHz and 4 MHz repetition rate for the $^2H(d,n)$ reaction and $^1H(^{11}B,n)$ reaction, respectively. Beam duration was about 4–9 ns at FWHM depending on the beam condition.

The same gas cell target was used both for $^2H(d,n)$ and $^1H(^{11}B,n)$ neutron source. The deuterium or hydrogen gas was contained in a cylindrical stainless steel cell with a thickness of 0.5 mm and a length of 3 cm. The entrance window was a 5-μm thick Mo foil, and the beam stopper was a 1.5-mm thick Au plate. The D_2 or H_2 gas pressure was 0.2 MPa and monitored by a semiconductor vacuum gauge and DC amplifier

system. This target assembly was electrically isolated so that the beam current could be read directly using the digital current integrator. The electron suppression was made with a bias of −300 V to repel secondary electrons.

Gamma-ray spectra were measured with a 7.6 cm dia. x 15.2 cm long NaI(Tl) detector, which was set in a 25.4 cm dia. x 25.4 cm long NaI(Tl) annular detector. These two detectors were operated in an anti-coincidence mode to suppress the Compton backgrounds and annihilation components of the pulse height spectra. The gamma-ray detector was located at about 80 cm from the sample. Time-of-flight technique of the gamma-ray detection was used to improve the signal to background ratio. The whole gamma-ray detector system with a heavy shield can be rotated around the sample to measure the angular distributions. To reduce the background, the detector shield was composed of lead and borated paraffin and an iron shadow bar was prepared between the gas target and the spectrometer.

Samples were used for some major fusion reactor structural materials such as Al, Si, Fe, Ni, Cu, Pb and Bi. The samples had a cylindrical shape and typical size was 3.0cm dia. x 3.0cm long except for Ni and Cu, which size was 2.0cm dia. x 3.0cm long. To check the correction of the finite sample size effects, 2.0cm dia. x 3.0 cm long samples were also used for Si and Pb at En = 10 MeV. The sample was placed at 10 cm from the neutron target. The measured angle was ordinary 90°, but in order to make a comparison with the different measurements and the investigation for the angular dependence, another angle of 125° was measured at 11.5MeV.

The neutron spectra were measured employing a 5 cm dia. x 1.27 cm long NE213 detector. A conventional TOF technique with a pulse shape n–γ discrimination was used. The bias of the discriminator was set using the ^{22}Na pulse height spectrum and detection efficiency was calculated by O5S code[3]. Fig.1 shows a neutron spectrum measured at the angle of 0° for the incident ^{11}B energy of 62 MeV(56.4 MeV at the center of the gas cell). The energy spread in the spectrum is mainly due to the TOF resolution.

Neutron flux was calculated using the evaluated values of $^2H(d,n)^3He$ reaction cross section[4], beam current, gas pressure and temperature. For $^1H(^{11}B,n)$ reaction, activation method using $^{27}Al(n,\alpha)^{24}Na$ reaction was employed. The NE213 detector system was also used to check the calculated flux.

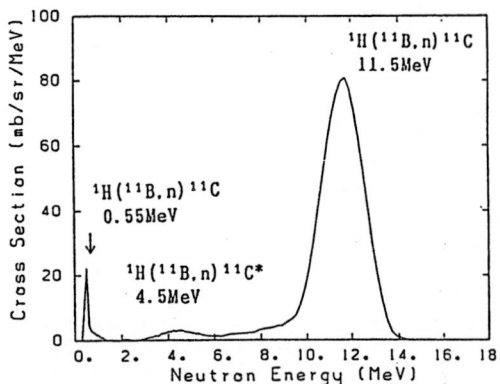

Fig. 1 Neutron energy spectrum at 11.5 MeV produced by $^1H(^{11}B,n)^{11}C$ reaction (E_B = 62.0 MeV, 0°)

Data Reduction

The procedure of data reduction was described in detail elsewhere[5]. After background subtraction, FERDOR code[6] was used for unfolding the pulse height spectra.

The energy spectra are distorted by the finite sample size effects, i.e., attenuation of the incident neutron flux and emitted gamma-rays, multiple scattering of neutrons, and production of the secondary gamma-rays due to Compton scattering in the sample. Because measurement of the continuum gamma-ray spectra is the main objective of this work as well as discrete ones, correction for Compton scattering inside the sample is essential. In the present work, these effects were corrected by Monte Carlo method.

The $^2H(d,n)$ reaction produces low energy background neutrons by the $^2H(d,np)d$ break-up reaction in the energy range of the present study. Similarly, the $^1H(^{11}B,n)^{11}C$ source also produces low energy neutrons which decay to the first excited state of residual ^{11}C above the incident ^{11}B energy of 56.8 MeV(neutron energy of 11.9 MeV). Gamma-rays produced by these low energy background neutrons contribute mainly to low energy part of the spectra and have to be corrected. Contributions from these low energy neutrons were estimated by using the data measured at ORELA[7] and correction was made by taking a ratio of calculated gamma-ray spectra with and without this background. This correction was applied for 10MeV data with $^2H(d,n)$ neutrons and 13 MeV data for $^1H(^{11}B,n)^{11}C$ source. The 7.8MeV data taken with $^2H(d,n)$ source and the 11.5 MeV data with $^1H(^{11}B,n)$ source were not corrected because contribution from these low energy neutrons were negligibly small.

Source of uncertainties and their typical magnitudes contained in the present results are summarized in Table 1. As a total, uncertainty for Al sample was estimated to be about 13%. For heavier samples, correction for gamma-ray attenuation is larger and statistics is generally poorer.

Results

Differential cross sections measured in this experiment for Fe at 11.5 MeV is shown in Fig. 2 along with the ORELA experimental data of Chapman et al.[8] and the evaluations of JENDL-3 and ENDF/B-IV. Present results tend to show smaller values than the data of Chapman for $E\gamma < 4$ MeV.

Gamma-ray production cross sections for discrete components were obtained by integration of the differential cross section over the energy of the peak component. An example of discrete gamma-ray production cross section is shown in Fig. 3.

Total gamma-ray production cross sections were obtained by integration of the differential cross sections over energy. To eliminate the discrimination of the noise and influence of 0.511 MeV annihilation gamma-rays, gamma-ray energy spectra were integrated from 0.7 MeV. Obtained cross sections are summarized in Table 2 and are shown in Figs. 4-7 for Al, Fe, Cu and Pb, respectively, compared with other experiments and evaluated data. For Al, the experimental results and the evaluated ones are in good agreement within uncertainty except for the data of Morgan[9]. For Fe, the two data sets of Chapman and Dickens[10] diverge above 5 MeV. Present results approach the results of Dickens near 13 MeV. JENDL-3 follows the data of Chapman, but ENDF/B-IV follows the experimental data using monoenergetic neutron sources.

Summary

Gamma-ray production cross sections of Al, Si, Fe, Ni, Cu, Pb and Bi at neutron energies from 7.8 to 13.0 MeV were obtained with higher accuracy using monoenergetic neutron sources. Differential, discrete, and energy integrated gamma-ray production cross sections were presented and compared with other experimental results and evaluated data.

Acknowledgement

The authors wish to thank JAERI Tandem accelerator staff for machine operation. We would like to acknowledge the useful discussions of Profs. K. Sugiyama and K. Kotajima at Tohoku University, and Prof. M. Drosg at University of Vienna.

Table 1 Source of Errors

Source	Error
Statistics	1–50%
Detection Efficiency	3% ($E\gamma$<2MeV) 8% ($E\gamma$=11MeV)
Response Matrix	5%
Sample Atomic Density	< 1%
Neutron Flux	6% for $^2H(d,n)$ neutrons 7% for $^1H(^{11}B,n)$ with NE213 6–9% for $^1H(^{11}B,n)$ with $^{27}Al(n,\alpha)^{24}Na$ activation
Correction Factor	5% for neutron transport 10% of correction for gamma-ray transport 2% for low energy neutron
Dead Time Correction	2%

Table 2 Total gamma-ray production cross sections (b/sr)

	$^2H(d,n)$			$^1H(^{11}B,n)^{11}C$		
En(MeV)	7.8	7.8	10.0	11.5	11.5	13.0
ΔEn(MeV)	0.1	0.1	0.1	0.3	0.3	0.3
angle	90°	125°	90°	90°	125°	90°
Al	0.104 ±0.013	0.115 ±0.015	0.116 ±0.016	0.132 ±0.021	0.124 ±0.018	0.127 ±0.019
Si	0.079 ±0.010		0.095 ±0.016	0.117 ±0.020	0.112 ±0.017	0.122 ±0.019
Fe	0.275 ±0.032		0.306 ±0.037	0.319 ±0.044	0.336 ±0.040	0.285 ±0.036
Ni				0.274 ±0.034		0.224 ±0.034
Cu				0.326 ±0.040	0.305 ±0.040	0.267 ±0.038
Pb	0.383 ±0.054		0.274 ±0.039	0.180 ±0.034	0.194 ±0.036	0.231 ±0.046
Bi	0.375 ±0.054		0.221 ±0.039	0.170 ±0.034		0.205 ±0.038

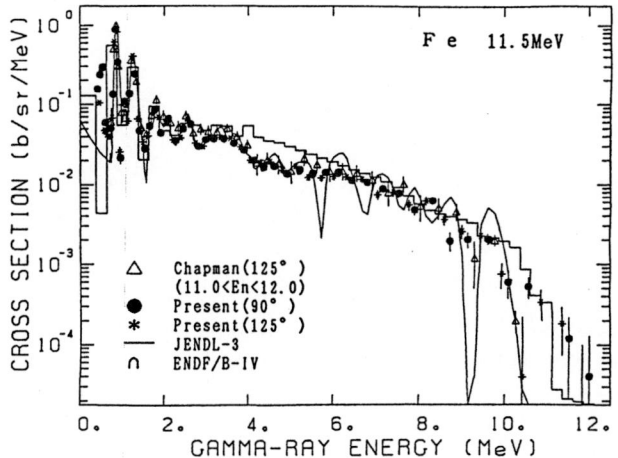

Fig. 2 Differential cross section for Fe at En = 11.5 MeV

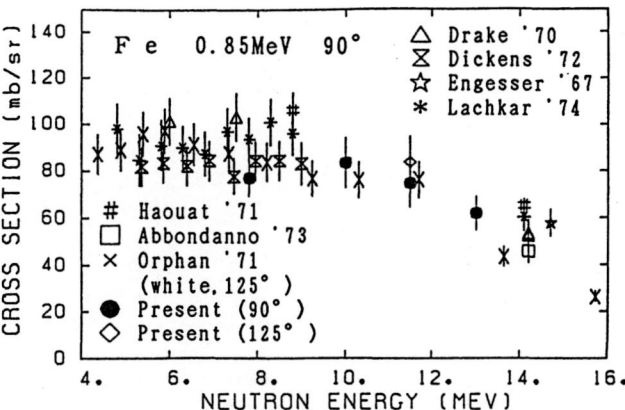

Fig. 3 Discrete gamma–ray production cross section for Eγ = 0.85 MeV from Fe

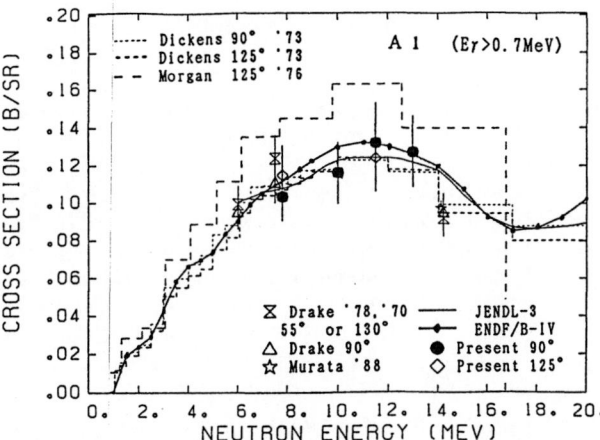

Fig. 4 Total gamma–ray production cross section for Al

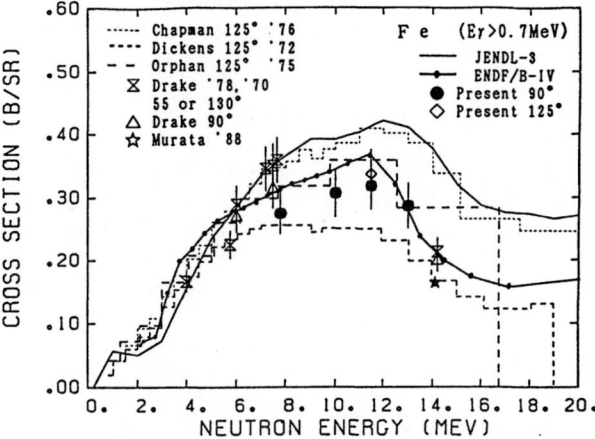

Fig. 5 Total gamma–ray production cross section for Fe

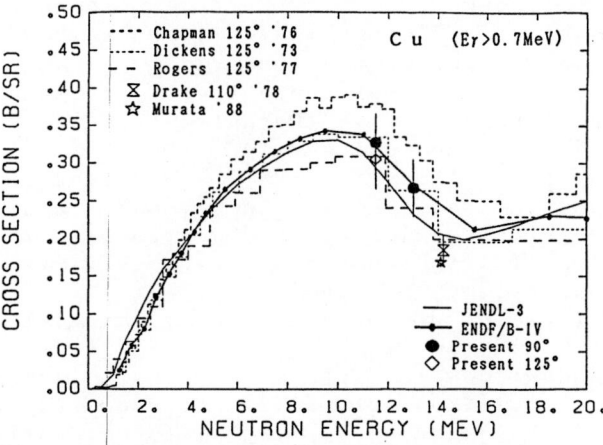
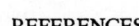

Fig. 6 Total gamma–ray production cross section for Cu

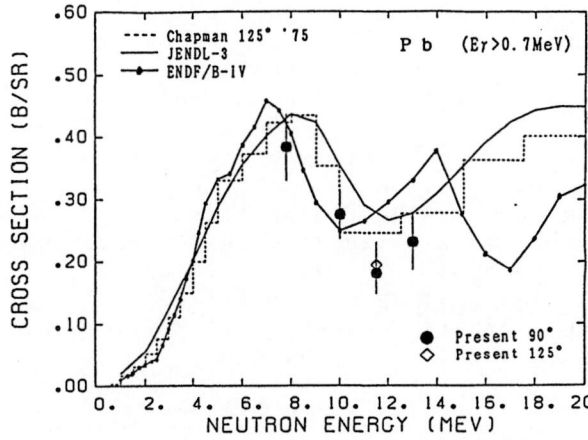

Fig. 7 Total gamma–ray production cross section for Pb

REFERENCES

1. S.Chiba, M.Mizumoto, K.Hasegawa, Y.Yamanouti, M.Sugimoto, Y.Watanabe and M.Drosg: Nucl. Instr. and Meth. A281, 581(1989)
2. M.Mizumoto, K.Hasegawa, S.Chiba, M.Igashira, T.Uchiyama, H.Kitazawa and M.Drosg in Proc. Int. Conf. on Nuclear Data for Science and Technology, May/June 1988, Mito,Japan (Edited by S.Igarasi), p.197, Saikon, Tokyo(1988)
3. R.E.Textor and V.V.Verbinski,O5S, ORNL–4160, Oak Ridge National Laboratory(1968)
4. H.Liskien and A.Paulsen, Nucl. Data Tables 27, 147(1973)
5. K. Hasegawa and M.Mizumoto, Detector System for Gamma-Ray Production Cross Section Measurements and Data Analysis (in Japanese), JAERI–M 89–042, Japan Atomic Energy Research Institute(1989)
6. L.Harris Jr., H.Kendrick, S.M.Sperling, An Introduction to the Principles and Use of the FERDOR Unfolding Code, GA–9882, Gulf Radiation Technology(1970)
7. J.K.Dickens, G.L.Morgan, G.T.Chapman, T.A.Love, E.Newman and F.G.Perey, Nucl. Sci. Eng. 62, 515(1977)
8. G.T.Chapman, G.L.Morgan and F.G.Perey, ORNL–TM–5416, Oak Ridge National Laboratory(1976)
9. G.L.Morgan and F.G.Perey, Nucl. Sci. Eng. 61, 337(1976)
10.J.K.Dickens, G.L.Morgan and F.G.Perey, ORNL–4798, Oak Ridge National Laboratory(1972)

PROGRESS IN THE MEASUREMENT OF GAMMA-RAY PRODUCTION CROSS SECTIONS INDUCED BY 14.9 MEV NEUTRONS

Fan Guoying, Zhou Hongyu, Zhu Xiaoge, Hua Ming
Wen Shenlin, Lan Liqiao, Liu Shuzhen

Institute of Low Energy Nuclear Physics
Beijing Normal University
100875 Beijing, China

Abstract: The energy spectra and the production cross sections of gamma rays from the reactions induced by 14.9 MeV neutrons have been measured at five different angles. The experiments were carried out with a pulsed fast neutron time-of-flight facility and a high resolution Ge(Li) spectrometer. A Monte Carlo program has been developed for calculating the corrections of the neutron flux attenuation and gamma ray self-absorption in the large hollow cylindrical sample with a position correlation. The lifetimes of the levels extracted from the Doppler broadening of gamma ray spectra have been studied.

(Gamma ray, cross section, Monte Carlo simulation, Doppler broadening)

1. Introduction

The production cross sections of gamma ray induced by 14 MeV neutrons are basic nuclear data for the design of nuclear plants and fusion reactors. Discrete gamma ray spectra and production cross sections are able to provide valuable information for understanding the structure of energy levels and the mechanism of nuclear reactions for various nuclei. In recent years , we have done systematic measurements of neutron induced gamma-ray production cross sections for Li, Be, C, O, F, Na, Al, Si, Cl, V, Fe, Co, Cu, Nb, and Pb at 14.9 MeV.

In order to improve the measurements for the low energy gamma rays, large hollow cylindrical samples have been proposed. A Monte Carlo program for calculating the corrections of the neutron flux attenuation and gamma ray self-absorption was developed. A preliminary study of Doppler effect in gamma spectra induced by 14 MeV neutron will be reported at the end.

2. Measurements of γ – Production Cross Sections

The experiments were performed at the fast neutron time-of-flight facility in the Institute of Low Energy Nuclear Physics [1]. The neutrons via the T(d, n)⁴He reaction are produced by bombarding a tritiated titanium target with a beam of 300KeV deuterons accelerated by the 400 kV Cockcroft – Walton accelerator.

The beam pulsing system consists of a high frequency and high voltage chopper and a spiral loaded wave guide buncher. The repetition frequency is 3.16 MHz. The FWHM of neutron pulses is 1—1.5 ns, the avarage neutron yield is about 5×10^{8} n/s. Absolute neutron flux was determined by the associated particle technique with a Si(Au) detector at $135°$ with respect to the deutron beam direction . A Ge(Li) detector was placed inside the large 4π shield with a sensitive volume of 110.7 cm^3 and a resolution of 2 KeV for 1332.5 KeV γ — peak of ^{60}Co. A careful absolute efficiency calibration was made using a series of standard γ — photon sources. The samples are 3 cm in diameter and 3 cm high with the natrual abundance and high chemical purity.

The time-of-flight technique was used for reducing the background caused by primary and scattered neutrons. In the time-of-flight spectra, the overall time resolution for gamma rays was 7 nsec while the window placed over the gamma peak was as large as $\sim$ 35 ns. Another time window was set for taking the gamma spectrum from activation of the sample and the surrounding. Two different ratio amplifiers were used for getting the gamma ray energy spectra.

The energy spectra of gamma rays from Si samples at 90° are shown in Fig 1 . The experimental data were analysed by a fitting program which is capable of resolving multiplets. The neutron flux attenuation and multiple scattering, gamma ray

*: This work is suported by Nuclear Data Centre of China and IAEA under the Contract No. 5147 / R1 / RB.

self-absorption in the sample, absorption of neutron by target backing and cooling water, the anistropy and size of neutron source were taken into account. The calculations of corrections and differential cross section were completed by GAM FOR program. The error of neutron flux determination was not more than 2.5% . The total uncertainty of differential cross section was about 5% for the main gamma lines. The production cross sections of main gamma lines for the interactions of 14.9 MeV neutrons with Si are presented in Table 1 .

3.Monte Carlo Simulation

In order to improve the data of low energy gamma – ray production cross section, a large hollow cylindrical sample was considered for the experiments. Therefore, the normal analytical method [2] for calculating the corrections of the neutron flux attenuation and γ self-absorption. with position correlation faced complicated and difficult problems. A Monte Carlo program was therefore developed. In order to simplify, we assume that the neutron beams are shot into the sample parallel to one another. In the beam direction, the position coordinate of a neutron is a random variable X_n , it deduces $e^{-\mu_n X_n}$ exponential density function. The range X_γ of gamma-ray in the samples is also a random variable, the density function is $e^{-\mu_\gamma X_\gamma}$. We reckoned the two effects together, the estimation η is defined as follows:

$$\eta = \begin{cases} 1 \text{ neutron and gamma ray event occur} \\ 0 \text{ neutron and gamma ray event avoided} \end{cases}$$

The total successful trial is equal to $\sum_{i=1}^{Nn} \eta i$. The even value is

$$\overline{\eta} = \frac{1}{Nn} \sum_{i=1}^{Nn} \eta i$$

The correct factor for two effects is

$$f = \frac{1}{\overline{\eta}}$$

When $\mu_n \rightarrow 0$, the correction factor f_γ of gamma ray self-absorption can be obtained; when $\mu_\gamma \rightarrow 0$ we get the correction factor f_n of neutron flux attenuation.

Finally, the spherical attenuation of the neutron flux has to be considered into the calculation. A weighting factor Kb related to the distance ρb from the neutron source to the point of the random event has been included.

$$Kb = \frac{\rho^2}{\rho_b^2}$$

Where the ρ^2 is the square of the average distance from the neutron source to the sample. Now:

$$\eta = \frac{1}{Nn} \sum_{i=1}^{Nn} \eta i \ Ki$$

The formula and the flow diagram of the Monte Carlo program are reported in reference 3 . The program was written by FORTRAN—77 in an IBM—PC microcomputer. A comparison of this program with the analytical method has been made. The calculated results for the production cross sections of 847 KeV and 1238.2 KeV gamma-rays of Fe sample at 90° are listed in Table 2 . For three Cu samples with different sizes, the production cross sections of three low energy gamma rays were calculated by the Monte Carlo program. The results are agree with each other . From a comparison with the other methods, we can conclude that the calculated results of the Monte Carlo program are satisfactory. The program is suitable not only for the hollow samples but also for the real ones.

4.Application of Doppler Shift Method

Doppler shift method of γ — spectroscopy is widely used in the lifetime measurement of nuclear states populated in reactions induced by charged particles. We attempted to introduce it into nuclear data produced by fast neutrons. In gamma-ray spectra from light and medium nuclei at 14 MeV neutron bombarding energy, the Doppler effects have been observed. The gamma energy spectra of 1808.7 KeV line produced in the interaction of 14.9 MeV neutrons with Al at 55° , 90° , 140° are shown in Fig 2 . The shapes of the gamma energy spectra present an angular dependence.

In order to extract the lifetime of an excited state from the Doppler broadening of gamma ray spectra, first question we must make clear was the mechanism of reactions. We chose $^{27}Al(n, pn\gamma)^{26}Mg$ reaction as the first object, in which only two levels are populated (2^+, 1808.7 KeV; 2^+, 2938.4 KeV) and the cascade transitions from high levels are weak. On the other side, the line of 1808.7 KeV was one of the most intensive lines in the spectrum of reaction $n+^{27}Al$, and has express Doppler effects. Analysis of the γ-line forms was provided by the Monte Carlo program which simulated the processes of slowing-down, multiple scattering of recoils, γ−ray emission and γ−absorption. About 10000 histories of recoil were simulated. The last step was compared with the experimental spectrum. Program permited one to take into account the cascade and side feelings for schemata of any degree of complexity. It was easy to calculate γ line form for each τ(lifetime) and E (energy of gamma ray) by spline interpolation and compare calculated and experimental forms of γ−line by x^2−method. The primary calculation was done with approximate values of stopping power parameters for Mg in Al, which were known from systematics of electronic stopping power coefficients. The values of τ obtained from the γ−line shape analysis at 55° and 140° were in a good agreement. After taking the values of x into account, the average values of τ were 0.70±0.03 ns for 1808.7 KeV state, and 0.20±0.03 ns for 2938.4 KeV.

5. Conclusion

Study of the γ−radiation arising from the reactions induced by fast neutrons is going forward in Low Energy Nuclear Physics Institute. A number of structural and shielding materials has been measured. The energy spectra and the differential production cross sections of gamma rays have been obtained. By means of improvements of experimental technique and data analysis method, the accuracy of data is getting ahead. A preliminary study of the lifetimes extracted from the Doppler broadening of gamma ray spectra showed that the DSAM, which is widely used in the reactions induced by charged particles, can be taken in the reactions induced by fast neutrons with some restrictions.

REFERENCE

1. Yan Yiming, Tang Lin, Zhou Hongyu, Wen Chenlin, Chin.J.Nucl.Phys. Vol. 10, No. 2, 166(1988)
2. Zhou Hongyu, Wang Wanhong, Yan Yiming, Fan Guoying, Wen Chenlin and Tang Lin, Proc. Internation Conference on nuclear Data for Science and Technology, Mito(1988), P311
3. Zhu Xiaoge, Master thesis(1990).

Table 1. γ Production Cross Sections for 14.9 MeV neutron interaction with Si

E (KeV)	Probable reaction	σ(mb/sr, 90)
389.7	^{28}Si(n, α γ)^{25}Mg	2.06±0.10
585.1	^{28}Si(n, α γ)^{25}Mg	4.59±0.20
947.6	^{28}Si(n, α γ)^{25}Mg	2.85±0.14
983.4	^{28}Si(n, p γ)^{28}Al	2.09±0.10
1778.8	^{28}Si(n, n γ)^{28}Si	38.8 ±2.3
2838.6	^{28}Si(n, n γ)^{28}Si	5.27±0.50
5099.7	^{28}Si(n, n γ)^{28}Si	2.50±0.45
6878.6	^{28}Si(n, n γ)^{28}Si	2.92±0.40

Table 2. A Comparison of two Methods for σ(E, 90°) with Fe Sample

Method	σ(847KeV, 90°)	σ(1238KeV, 90°)
Previous	64.95±2.98mb	28.79±0.85mb
Monte Carlo	65.28±2.00mb	28.34±1.40mb

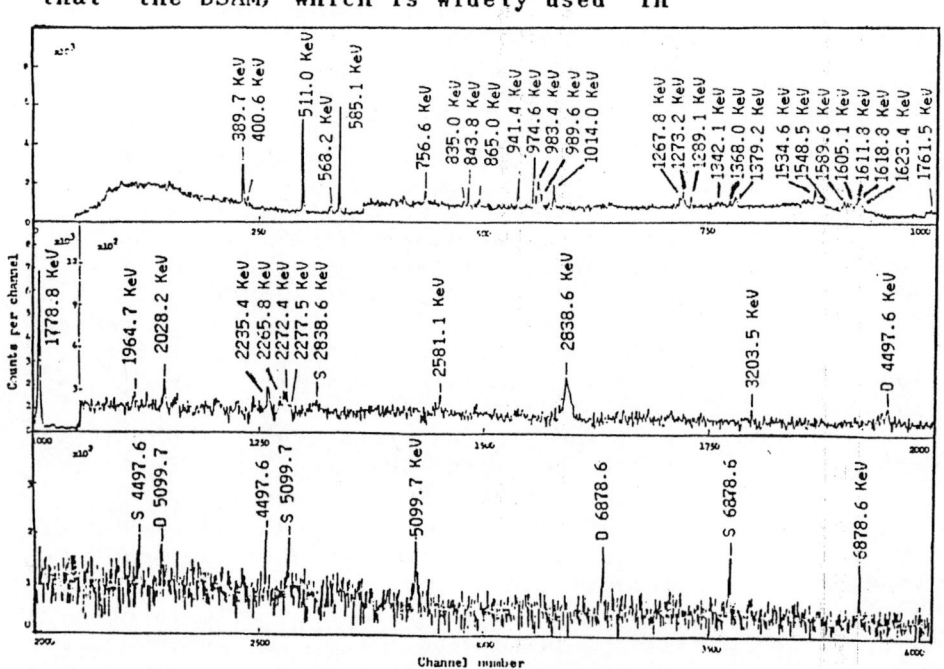

Fig 1. The energy spectra of gamma rays from Si samples

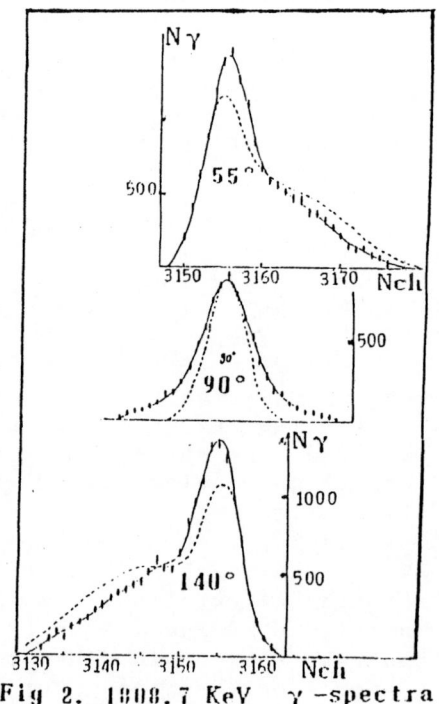

Fig 2. 1808.7 KeV γ−spectra at 55°, 90°, 140°

NEUTRON LEAKAGE SPECTRA AND TRITIUM PRODUCTION RATE OF A HYBRID-REACTOR BLANKET-ASSEMBLY

J. Brose, T. Elfruth, J. Klose, M. Toepfer, D. Seeliger,
K. Seidel and S. Unholzer
Technische Universität Dresden, O-8027 Dresden, Federal Republic of Germany
D. Hebert and W. Stolz
Bergakademie Freiberg, O-9200 Freiberg, Federal Republic of Germany
D. Markovskij and G. Shatalov
I.V. Kurchatov Institute of Atomic Energy, 123182 Moscow, U.S.S.R.

Abstract: The neutronics of a fusion reactor blanket is investigated at a mock-up consisting of steel, uranium, lithium-aluminium, polyethylene and steel in slab geometry. Beside functionals of the uranium part the lithium production rate of the LiAl-slabs per one 14 MeV neutron bombarding the assembly is measured with both Li_2CO_3-probes and liquid scintillator technique and LiF-thermoluminescence probes. The neutron spectra leaking several slab assemblies per one 14 MeV incidence neutron are measured with time-of-flight technique. The measured quantities are compared with calculated ones using the three-dimensional Monte Carlo codes MORSE, BLANK and MCNP and data of the libraries ENDF/B and ENDL.

(fusion reactor, blanket, tritium production, neutron leakage spectrum, thermoluminescence probe, liquid scintillator, time-of-flight measurement, neutron transport code, nuclear data library)

Introduction

The investigation of the neutronic properties of the blanket at a mock-up is one phase in the fusion reactor design. Evaluated nuclear data and transport codes tested at benchmarks are used to calculate the important functionals of a material assembly corresponding to the blanket design. Comparing the results with the measured values, just those parts of data, codes and their interplay are tested which are most sensitive for the functionals.

A central functional of each breeding blanket is the tritium production coefficient. For a hybrid-reactor blanket with uranium the U-fission and the Pu-breeding coefficient are of interest. The differential neutron flux density is the connecting link between these functionals.

As parts of a more complex investigation of a hybrid-blanket tritium production rates within the mock-up and neutron spectra leaking the assembly and their components are measured and calculated.

Mock-up

The geometrical arrangement is shown in Fig. 1. The 14 MeV neutrons are produced with a d-T neutron generator. The source strength is measured by the associated α-particles. It is about $3*10^9$ neutrons per second in pulsed mode and about 10^{10} neutrons per second with steady beam.

The slab assembly is 100 cm high, 100 cm wide and consists of the homogeneous materials

I – steel (thickness 2 cm) – simulating the first wall
II – uranium (10 cm) depleted in ^{235}U
  • for neutron multiplication
  • fission energy production
  • ^{239}Pu breeding

III – LiAl (4cm) with Li-content of 5 p.c.mass for ^{3}H breeding
IV – polyethylene (2 cm) as moderator
V – steel (11 cm) – simulating a shield

The quantities measured per one 14 MeV neutron are

k_t – tritium production rate

k_f – fission rate of ^{238}U

k_b – breeding rate of ^{239}Pu

$\Phi(t)$ – time-differential neutron fluence

$\Phi(E)$ – energy-differential neutron fluence.

Calculations are carried out with the three-dimensional Monte Carlo codes MORSE[1], BLANK[2] and MCNP[3] and with data of the libraries ENDF/B and ENDL.

Tritium Production Rate

The tritium production is measured by two independent methods.
  • The ^{3}H induced in Li_2CO_3 probes is determined by β-counting in liquid scintillator following Fritscher et al. [4].
  • LiF thermoluminescence probes are irradiated and glow-curve maxima are recorded.

The distribution of k_t is measured along the z-axis (Fig. 1.) as well as in y-direction. Results are plotted in Fig. 2 together with the results of two calculations.

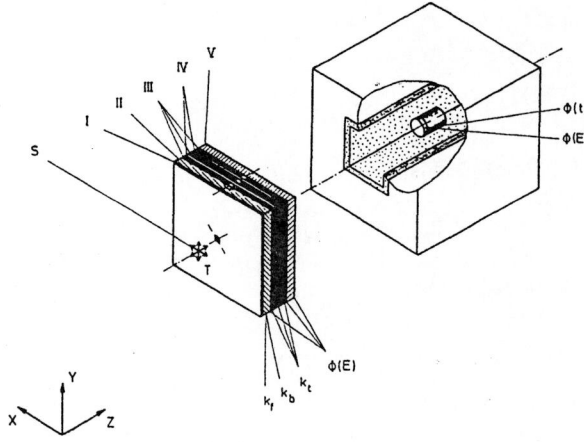

Fig. 1. The mock-up. S – neutron source; I ... V – slabs of different materials; k_f, k_b, k_t, $\Phi(E)$, $\Phi(t)$ – coefficients and spectra measured within the slabs or at the position of the neutron detector surrounded by a collimator.

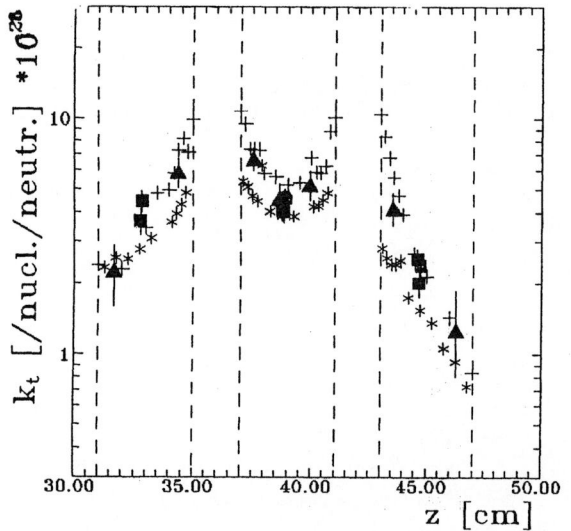

Fig. 2. Tritium production per Li-nucleus and 14 MeV neutron as function of the distance to the neutron source measured with Li_2CO_3-probes (▲) and LiF-probes (■) and calculated with BLANK [ENDL-78] (∗) and MORSE [ENDF/B-IV] (+).

The relative distribution strongly influenced by the moderator is described by the calculated results, but there are differences concerning the absolute values.

Neutron Leakage Spectra

Neutron leakage spectra are measured at distances source-detector of z = 2.76 m and 4.50 m for the slab assemblies I + II, I + ... + IV and I + ... + V. A typical example of the results is shown in Fig. 3 together with the calculated spectrum.

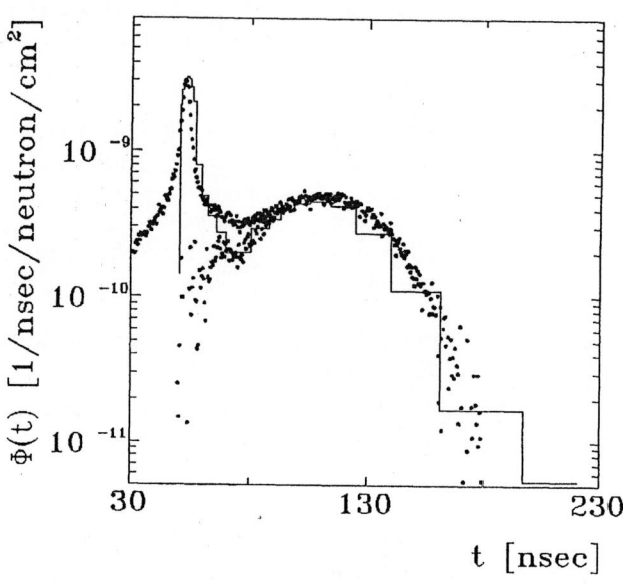

Fig. 3. Time-differential neutron fluence per one source neutron for the assembly I + II, both measured (•) and calculated with MCNP[ENDF/B-IV] at z = 2.76 m.

Differences between the measured and calculated spectrum at t ≈ 70 ns are caused by different source-neutron peak-shapes. If a measured peak-shape is fitted to the height of the leakage peak and separated, the agreement with the calculated spectrum is sufficient. The influence of the collimator had carefully to be taken into account.

References

[1] S.K. Fraley, Users Guide to MORSE-SGL, ORNL (1976)

[2] S.V. Marin, Programma rascheta prostranstvenno-energeticheskogo raspredelenija nejtronov v odnomernoj geometrii (BLANK), report IAE-2832, Moscow (1977)

[3] J.F. Briesmeister, MCNP – A General Monte Carlo Code for Neutron and Photon Transport – Version A, LA-7396, Rev. 2, 1986

[4] U. Fritscher, S. Kappler and D. Rusch: Nucl.Instr. Methods <u>153</u>, 563 (1978)

MEASUREMENTS AND ANALYSES OF ANGULAR NEUTRON FLUX SPECTRA ON LIQUID NITROGEN, LIQUID OXYGEN AND IRON SLABS

Yukio Oyama, Hiroshi Maekawa and Kazuaki Kosako

Japan Atomic Energy Research Institute
Tokai-mura, Naka-gun, Ibaraki-ken 319-11, Japan

Abstract: Angular neutron flux spectra leaking from slab samples of liquid nitrogen and oxygen, and large diameter slabs of iron have been measured with deuterium-tritium fusion reaction neutrons. The measuring energy range is 0.05 to 15 MeV by use of the time-of-flight method and the leaking angles are varied in 0 to 66.8 degrees. The results were analyzed by the transport codes of MCNP and DOT3.5 to test the newly released JENDL-3 nuclear data file. The JENDL-3 results showed good agreements with the present experiments.

(Fusion, Liquid Nitrogen, Liquid Oxygen, Iron, Angular Neutron Flux, TOF, JENDL-3, MCNP, DOT3.5)

Introduction

In a D-T burning fusion reactor, nitrogen and oxygen are important elements in air for skyshine problem. Especially oxygen is contained in ceramic materials for tritium breeding blanket and insulators. The experimental data for these materials are very few[1] because it is difficult to make liquid sample. Iron is also a key material as a shielding component for protection of super conducting magnet from radiation heating. In this case, the radiation dose after deep penetration of neutrons in material is of interest. The characteristics of deep penetration of neutrons are sensitive to total and elastic cross sections and also to group structure of cross section library for shielding calculations.

To examine nuclear data, calculational code and library, neutron angular flux spectra leaking from slabs of liquid nitrogen, liquid oxygen and iron have been measured at the Fusion Neutronics Source (FNS) facility by the time-of-flight method.[2-4] In the present paper, MCNP[5] analyses are performed with JENDL-3[6] and compared to the experiments. For the iron, the deterministic code DOT3.5 with groupwise cross section is also applied to examine the problems in the deep penetration.

Experimental

Slab Sample

Slab assembly of liquid gas was made in a cylinder with a double wall for thermal insulation. Figure 1 shows the container for liquid gas samples. The void between the walls was evacuated. Aluminum

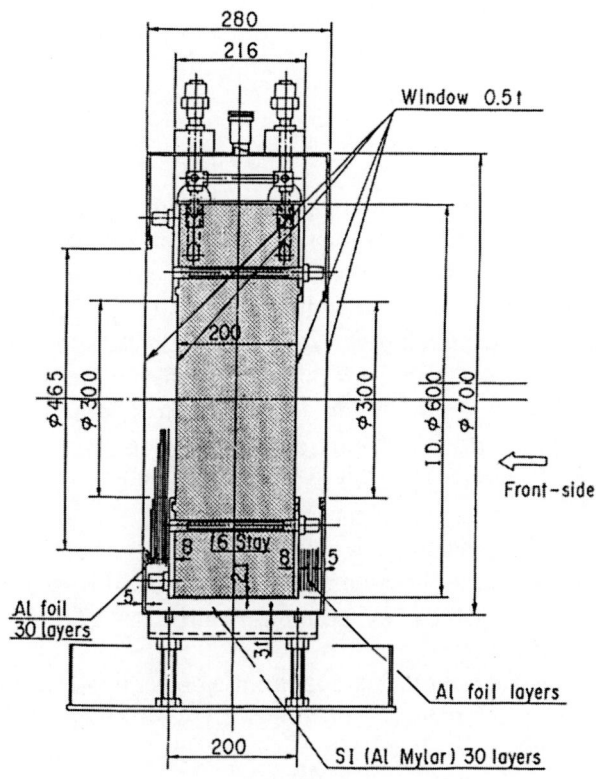

Fig. 1 Container of liquid nitrogen and oxygen

foil layers between the walls were inserted to prevent the inner container from heat loss due to thermal radiation. The central area of the container walls was made of very thin windows of 300 and 465 mm in diameter to reduce neutron scattering for measuring angles. The thickness of the inner cylinder is 200 mm and the diameter 600 mm.

Iron slab was made by combination of 50 and 100 mm-thick cylindrical plates of 1 m in diameter. The slab thickness for the measurement was changed

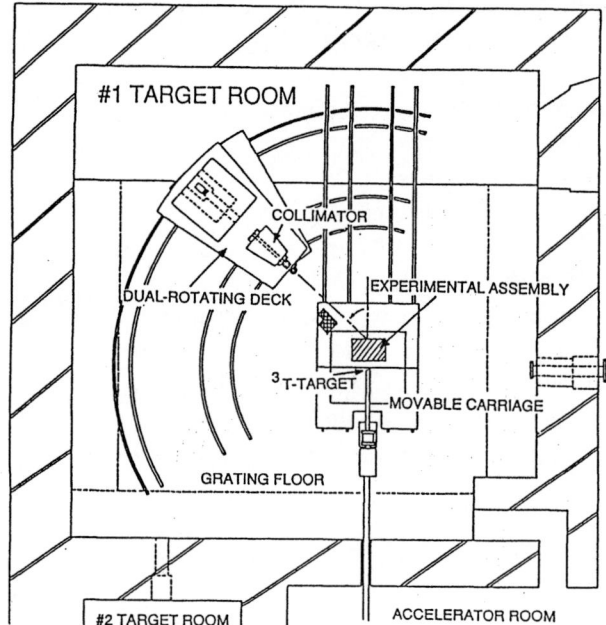

Fig. 2 Experimental arrangement for TOF

from 50 to 600 mm.

Measurement

Figure 2 shows the experimental arrangement. All sample materials were set at the distance of 200 mm from the D-T neutron source. The D-T neutrons were generated at the tritium-titanium target bombarded with 350 keV deuterons. An NE213 scintillator of 50.8 mm diameter and 50.8 mm long was used for the neutron detector of the time-of-flight system. The scintillator detected neutrons leaking only from the central area on the slab surface behind the source by the collimator. The flight path was about 7 m and the measuring angles were selected in 0, 12.2, 24.9, 41.8, 66.8 degrees. The measuring energy range was 0.05 MeV to 15 MeV by combination of the two bias scheme.[2]

The measured flux spectrum was reduced to the following quantity.

$$\phi(\Omega,E) = \frac{C(E)}{\varepsilon(E)\cdot\Delta\Omega\cdot A_s\cdot S_n\cdot T(E)}$$

[n/sr/m^2/lethargy/source neutrons],

where $C(E)$ is detected neutron counts, $\varepsilon(E)$ the efficiency, $\Delta\Omega$ the solid angle subtended by the detector to the point on the center of slab surface, A_s the effective measured area defined by the detector-collimator system on the plane perpendicular to its axis and $T(E)$ the attenuation correction due to scattering by air.

Calculation

The Monte Carlo calculations were performed by

MCNP[5] with the pointwise cross section library FSXLIB[7] based on the latest Japanese evaluated nuclear data file (JENDL-3)[6]. The calculational model is shown in Fig. 3. For the liquid samples, the container was fully simulated including small curvature of the window due to evacuation of the gap of the double walls. Five point estimators were located at the same distance as the measurement. The collimator was simulated by use of no-importance region where the neutron tracing was immediately terminated if neutron entered this area.

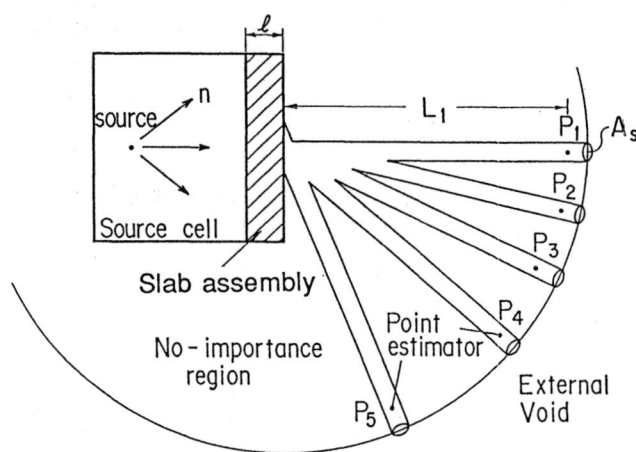

Fig.3 Calculational model for MCNP

For the iron case, the two dimensional discrete ordinate code DOT3.5[8] was also applied with the 125 group cross section set FUSION-J3[9] of P5-S16 approximation. Since the measuring angles are corresponding to the angles of S16 quadrature set, the measured flux can be estimated by the following average.

$$<\phi(\eta)>_r = \frac{\sum_r 2\pi r\cdot<\phi(\eta,r)>_\mu\cdot\Delta r}{\sum_r 2\pi r\cdot\Delta r},$$

where

$$<\phi(\eta,r)>_\mu = \frac{\sum_\mu \omega_{\mu\eta}\cdot\phi(\mu,\eta,r)}{\sum_\mu \omega_{\mu\eta}},$$

and $\omega_{\mu\eta}$ is the angular weight for S16 quadrature set, μ the azimuthal angle, η the polar angle (measuring angle) and r the radial position of the calculated angular flux.

Results and Discussions

Nitrogen

Figure 4 shows the results at 24.9 degree with the MCNP calculations. The nitrogen cross section in JENDL-3 is modified as JENDL-3M because a trivial mistake was found in charged particle emission reactions of (n,d), (n,t), etc. In the figure, the elastic peak of the calculations is lower than the experiment at large angles. For inelastic peaks between 1 and 10 MeV, the JENDL-3M shows fairly good agreement, in contrast to the ENDF/B-IV. From these facts, the angular distribution of elastic cross section should be checked.

Oxygen

The comparison between the MCNP calculations are shown in Fig. 5. The JENDL-3 agrees very well with the experiment in the whole energy region. The ENDF/B-IV is lower below 2 MeV. However, the calculations tend to underestimate the elastic peak at large angle by 20% or more. This suggests to check the elastic cross section as well as nitrogen.

Iron

The iron experiment was analyzed by the both transport codes of MCNP and DOT3.5 for four thicknesses of slab. The results for the 400 mm-thick slab is shown in Fig. 6 at 24.9 degree. The MCNP calculation shows fairly good agreement with the experiment. The agreement is within 20% for the whole angle and all thicknesses. But the JENDL-3 is worse below 400 keV. For the flux above 10 MeV, the C/E deviates with angle and thickness, because the signal-to-background ratio is also worse for them.

On the other hand, the DOT3.5 calculation underestimates the flux below 2 MeV by 30-40%. The trend of underestimation increases with the slab thickness. This is mainly caused by use of groupwise cross section for the infinite dilution without self-shielding correction.

Conclusions

From the comparison with the MCNP calculations, it is seen that:
1) For nitrogen, the modified JENDL-3 agrees well
   except angular distribution of elastic reaction peak.
2) For oxygen, the JENDL-3 agrees well, but the elastic
   peak shows the same trends as nitrogen.
3) For iron, the JENDL-3 agrees up to 600 mm-thick slab, except for the energy range below 0.4 MeV.

References

1. L. F. Hansen, J.D. Anderson, E. Goldberg, J. Kammerdiener, E. Plechaty, C. Wong, Nucl. Sci. Eng., 40, 262 (1970)
2. Y. Oyama and H. Maekawa, Nucl. Instr. Method, A245, 173 (1986)
3. idem., Nucl. Sci. Eng., 97, 220 (1987)
4. Y. Oyama, S. Yamaguchi, H. Maekawa, J. Nucl. Sci. Technol., 25[5], 419 (1988)
5. Los Alamos Radiation Transport Group (X-6), LA-7396-M (1981)
6. K. Shibata, T. Nakagawa, T. Asami, T. Fukahori, T. Narita, S. Chiba, M. Mizumoto, A. Hasegawa, Y. Kikuchi, Y. Nakajima, S. Igarashi, JAERI 1319 (1990)

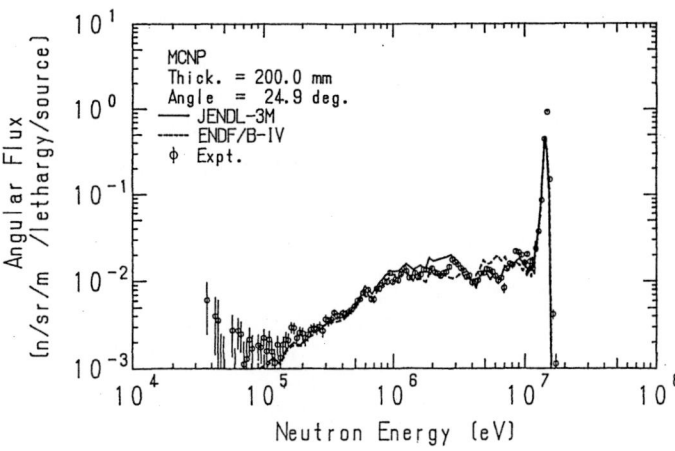

Fig. 4 Measured result with the MCNP calculation for 200 mm-thick nitrogen slab

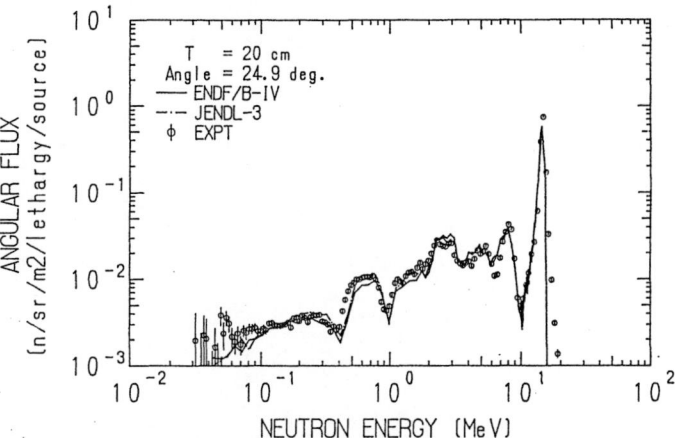

Fig.5 Measured result with the MCNP calculation for 200 mm-thick oxygen slab

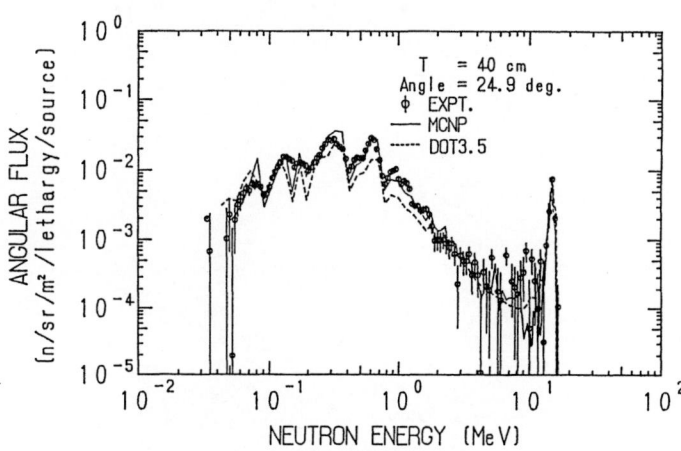

Fig. 6 Measured results with the MCNP and the DOT calculations for 400 mm-thick iron slab

7. K. Kosako, private communication (1990)

8. W. A. Rhodes and F. R. Mynatt, ORNL-TM-4280 (1973)

9. K. Maki, K. Kosako, Y.Seki, H. Kawasaki, JAERI-M 91-072 (1991) (In Japanese)

METHODOLOGY FOR THE GENERATION OF NEUTRON KERMA FACTORS FROM JEF2 LITHIUM-7 DATA

M. Konieczny, N.P. Taylor † and T.D. Beynon

School of Physics and Space Research,
University of Birmingham, UK

† AEA Technology, Culham, UK

Abstract : The nuclear data processing code NJOY89 is designed to generate neutron kerma factors using double-differential charged-particle data from JEF-2 and ENDF/B-VI. No such data is given for ^7Li in JEF-2, file 6 containing double-differential neutron distributions instead. An alternative approach, involving the multigroup neutron kerma code ENBAL2 and some pre-processing using NJOY89, is presented. At present, this is the only method of utilizing the double-differential neutron data for the generation of neutron kerma factors. An outline of the methodology involved is given, together with a description of errors discovered in NJOY89, and their solutions.

(Neutron kerma , ^7Li , JEF-2 , double-differential , energy-balance , NJOY89)

Introduction

Nuclear evaluated data processing codes are essential tools for the interpretation of basic data stored in files and its application to real problems. The NJOY nuclear data processing system [1] has taken up an eminent rôle because of its comprehensive set of features and has been adopted as one of the standard processing codes for the JEF project. The newest version, NJOY89, includes changes designed to support the new ENDF-6 format [2], which will be used for both the ENDF/B-VI and JEF-2 evaluated nuclear data libraries.

NJOY/HEATR

Of interest in this work, is the ability of NJOY to calculate neutron kerma factors via its HEATR module. Until now, HEATR has used neutron and photon files to calculate heating through an energy-balance approach [3], where the energy allocated to neutrons and photons is simply subtracted from the available energy to obtain the energy deposited locally (ie) the local heating

$$k_j(E) = (E + Q_j - \overline{E}_{jn} - \overline{E}_{j\gamma}) \, \sigma_j(E)$$

where Q_j is the mass-difference Q-value for reaction j, $\overline{E}_n$ is the total energy of secondary neutrons including multiplicity, and $\overline{E}_\gamma$ is the secondary photons including photon yield.

However, it has always been recognised that such an approach may give unsatisfactory results if the neutron and photon data used are not consistent. A better approach would be to sum over the total kinetic energy carried away by each species of secondary charged particle, including the recoil nucleus.

$$k_j(E) = \sum_l \overline{E}_{jl}(E) \, \sigma_j(E)$$

Previously, the ENDF format has not supported a charged particle capability, a deficiency rectified by the ENDF-6 format. Consequently, NJOY89 has been modified to make use of these changes. HEATR now uses the new energy-angle correlated charged-particle and recoil-nucleus data stored in double-differential format in file 6 of ENDF-6, although the facility to calculate heating using secondary neutron distributions given in files 4 and 5, and photon data in files 12, 13 and 15, has also been retained.

Unfortunately, no charged-particle data is available for ^7Li in JEF-2. Instead, accurate neutron distributions are given in the double-differential format of file 6. Consequently, HEATR fails to recognise this data as either ENDF-5 or ENDF-6 format, and is not equipped to process it. Thus, an alternative method is required to generate neutron kerma factors from JEF-2 ^7Li data.

ENBAL2

Neutron kerma factors may be generated from JEF-2 ^7Li data in multigroup form using the code ENBAL2 [4]. This is possible because input to ENBAL2 is in the form of group-to-group transfer matrices, these being generated from JEF-2 data using the GROUPR module of NJOY89 [5], which unlike HEATR, is capable of processing the file 6 neutron distributions in addition to neutron distributions as presented in ENDF-5 format.

ENBAL2 employs the traditional energy-balance methodology, subtracting the secondary neutron and photon energy from the available energy. However, ENBAL2 does not use photon data to calculate the photon energy transferred. The photon energy in question is the energy released by deexcitation of the residual nucleus following a reaction. Using kinematics, the average excitation energy of the residual nucleus may be calculated from the incident neutron energy, the atomic weight of the target nucleus, and the mean kinetic energy of the secondary neutrons, thus ensuring that energy-balance errors are avoided. One possible drawback of this approach is the assumption that residual nuclei always deexcite to the ground-state. However , for ^7Li, this is true. Using a multigroup structure removes the need of calculating the average energy of outgoing neutrons for each incident energy. For any group, g, in the multigroup structure, the mean energy of neutrons in that group, E_g, is easily calculated from a knowledge of the group structure and a weighting function representing the neutron spectrum; hence knowing the energy group to which neutrons are scattered defines their average energy. Naturally, using a multigroup approach has an inherent drawback – that of averaging behaviour over a broad energy range. However, the degree of approximation depends on the number of energy groups employed. Hence, by using a reasonably large number of groups, accurate results are assured.

Methodology

Using a pre-release version of the JEF-2 ^7Li data, trial runs were conducted to demonstrate the feasibility of this method.

Pointwise data was read into NJOY89 at 0 K. Using BROADR, the data was Doppler broadened to 850 K, to simulate typical conditions in the blanket of a fusion system.

A multigroup scattering matrix was generated for each reaction type using the module GROUPR . The Vitamin-J 175 energy-group group-structure was used together with a *thermal + 1/E + fission + fusion* weighting function. Cross section data from file 3 was incorporated into the scattering matrix via an extra edit position, giving a complete matrix of size 175x179 .

Each scattering matrix was output from NJOY in DTF format generated by the DTFR module.

ENBAL2 has been substantially modified, allowing it to read DTFR output directly, and store it in compact FIDO format. This data is then used by ENBAL2 to generate partial neutron kerma factors for each reaction type.

Validation

The methodology outlined above has been tested using a selection of JEF1 nuclides. This processing differed only in the respect that scattering matrices were generated by GROUPR using file 4 and 5 neutron distributions, (as is the norm for data in ENDF-5 format), rather than file 6, as for ENDF-6. Since ENDF-5 format data can be processed by NJOY89/HEATR, the accuracy of our kermas could be evaluated through comparison with kermas generated using HEATR. In all cases, excellent agreement was observed. Figure 1. shows the two sets of results for ^7Li.

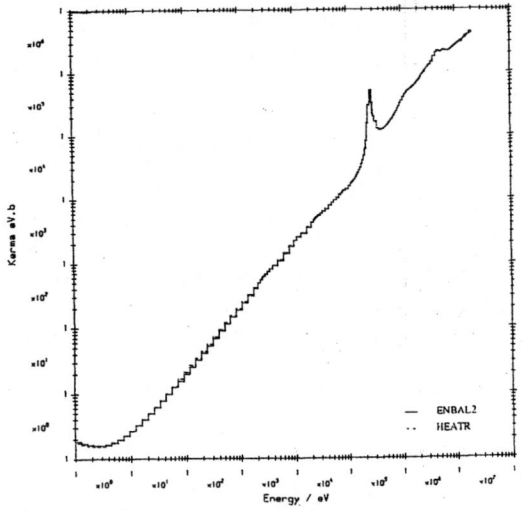

JEF-1 Lithium-7 Kermas

Fig. 1

Modifications to NJOY89

During the course of our evaluation procedures several coding errors in NJOY89 were detected and rectified.

According to ENDF formatting procedures regarding file 6, each section describing a particular reaction type (MT), may have NK subsections, each subsection describing one reaction product or reaction channel (LIP). However, in subroutine GETSED of NJOY89, the restriction is made that only one tabulated subsection is allowed, with the consequence that only the first tabulated subsection is ever processed, all subsequent subsections being ignored. At present, this problem is best avoided by lumping all reaction mechanisms into a single section (LIP=0), although in the longer term this may prove to be an unsatisfactory solution.

Errors were also detected within the GROUPR module, GROUPR failing to produce any transfer matrices from our tabular double-differential data, whilst giving no indication of any errors having occured. This was traced to a coding error in subroutine F6LAB which is responsible for converting tabular data into Legendre coefficients. Further modifications to this subroutine were needed to ensure that the transfer matrices subsequently produced were self-consistent.

Conclusions

At present we are unaware of any other codes capable of utitlizing tabular double-differential neutron data to generate neutron kerma factors. The methodology outlined here involves no significant assumptions or approximations in the case of ^7Li, although its accuracy for heavier nuclides should be considered carefully.

References

1. R.E. MacFarlane, D.W. Muir, and R.M. Boircourt, " The NJOY Nuclear Data Processing System, Volume I: User's Manual " LA-9303-M, (ENDF-324) (1982)

2. P.F. Rose and C.L. Dunford, Eds., " ENDF-102, Data Formats and Procedures for the Evaluated Nuclear Data File, ENDF " preliminary version for the ENDF-6 format available from the National Nuclear Data Centre, Brookhaven National Laboratory, Upton, NY (1988)

3. R.E. MacFarlane, D.W. Muir, and R.M. Boircourt, " The NJOY Nuclear Data Processing System, Volume II: The NJOY, RECONR, BROADR, HEATR, and THERMR Modules " LA-9303-M, LA-9303-M, Vol. II (ENDF-324) (1982)

4. N.P. Taylor " ENBAL2, A Program to Generate Multigroup Kerma Factors " University of Birmingham, Dept. of Physics report 81/02 (1981)

5. R.E. MacFarlane, D.W. Muir, and R.M. Boircourt, " The NJOY Nuclear Data Processing System, Volume III: The GROUPR, GAMINR, and MODER Modules, " LA-9303-M, Vol. III (ENDF/–324) (1987)

RE-ANALYSIS OF SWINPC BERYLLIUM INTEGRAL EXPERIMENT

Liu Lian-yan

Institute of Applied Physics and Computational Mathematics
P.O. Box 8009, Beijing, P.R. China

Abstract: ANISN-MORSE coupled method is developed to investigate the perturbation of detecting system to experimental results, in SWINPC beryllium multiplication integral experiment. Calculations indicate that the in-system measurement may result in considerably large deviation to the measured values. Correction is made to the experimental multiplication, which is still less than ENDF/B-VI and Los Alamos evaluation calculations by 7% for 2.5/17.35 Be sphere.

(beryllium, integral experiment, perturbation, ANISN, MORSE)

Introduction

It is expected that beryllium will play the part of neutron multiplier for fusion reactor blanket. The beryllium nuclear data are essential to determine the neutron multiplication capability. In order to clarify some doubts about Be-9 evaluation, efforts have been devoted to beryllium integral experiments [1-3]. Especially, a new round of beryllium multiplication integral experiments are underway in Osaka University (OU), Japan, Idaho National Engineering Laboratory (INEL), U.S.A. and South West Institute of Nuclear Physics and Chemistry (SWINPC), China. However, at the Chengdu IAEA Advisory Group Meeting, INEL [4,5] and SWINPC [6] gave apparently inconsistent experimental results. It is necessary and urgent to achieve a definite conclusion on these two experiments. Problems about SWINPC experiment, including D-T neutron source characteristic measurement scheme, impurity of Be sample and non-spherical geometry bias, have been analyzed in detail [7], which are shown to cause little deviations to experimental results. In this paper, emphasis is laid on a troublesome point, perturbation of in-system detecting method of SWINPC experiment.

SWINPC Experiment and Detecting System

Total absorption method was used in SWINPC experiment to measure the neutron multiplication of beryllium sphere (Ri/Ro: 12.8/17.35, 6.9/17.35, 2.5/17.35). Central D-T source neutrons were multiplied in beryllium via Be-9(n,2n) reaction. Surrounding the Be sample is a large volume of spherical polyethylene of three layers, in which, most of the neutrons leaking out of Be sphere were moderated and absorbed. By measuring the H-1(n,r) reaction rates, the beryllium leakage multiplication could be obtained with a minor calculation correction by the following expression:

$$M_L = M_{app} \cdot F_c$$

where M_{app} is the ratio of total H-1(n,r) rates in polyethylene sphere for with-Be sample system and for no-Be sample system. F_c represents the calculational correction factor dealing with non-hydrogen (carbon) absorption in polyethylene sphere and neutron leakage out of the system.

Measurement of the H-1(n,r) was carried out using U-235 fission chamber detector, which was placed into polyethylene sphere (Fig. 1). In this case, a serious problem may be introduced. The neutron flux around the detector may be perturbed away from the actual values, because of the occurrence of detecting duct. Therefore, the measured M_{app} could be questionable.

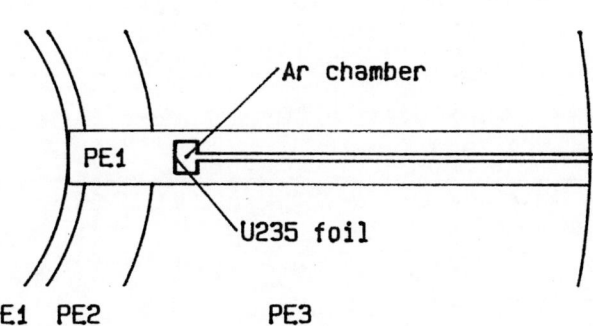

(PE1,PE2,PE3 denote polyethylene of weight density 0.92,0.56,0.94)

Fig. 1. Detecting system of SWINPC Be integral experiment

Calculational Method

In order to ascertain the perturbation of neutron flux around the detector, obviously one-dimensional calculation is not enough. However, for such a large experimental system (138cm in diameter), it is not economical and efficient to arrange three-dimensional Monte Carlo modeling, even impossible to reach satisfactory accuracy. In consideration of the small size of disturbed region, ANISN coupling MORSE method is resorted to. Unperturbed (no detector duct) neutron flux is determined by 1-D Sn transport code ANISN [8]. MORSE [9], 3-D Monte Carlo transport code adopting the same format nuclear constants as ANISN, is a judicious choice to be used to study the flux behavior in the perturbed region, while the boundary conditions of MORSE calculation are provided by ANISN.

• Nuclear constants

25-group, P3 constants, processed by NJOY [10] from ENDF / B-VI.

• MORSE boundary

MORSE boundary is a hypothetical cylinder, enclosing the detector duct. Determination of the cylinder dimension is a tradeoff between calculation accuracy and CPU time. The cylinder should be large enough, so the boundary flux can be regarded as unperturbed, on the other hand, as small as possible so as to save calculation costs.

• MORSE boundary source

The surface source intensity on the MORSE boundary is defined by:

$$dI = \varphi_g(r,\mu) \, |\vec{\Omega} \cdot \vec{n}| \, dS \, d\vec{\Omega}$$

$\varphi_g(r,\mu)$ is the neutron flux by ANISN, $\vec{\Omega}$ representing the neutron direction vector, $\vec{n}$ the normal vector for surface cell dS. After discretation, the source neutron probability distribution versus group number, position and direction is attained. The subroutine SOURCE in MORSE is modified to accomplish the source particle sampling from the probability distribution. Thus, ANISN and MORSE are coupled.

Results and Conclusions

ANISN calculations were conducted for no-Be sample system and for with-Be sample (2.5 / 17.35) system. ANISN flux was stored to prepare source neutron probability distribution for SOURCE in MORSE. Similar to the experiment measurement procedure, at every detected point in the polyethylene sphere, MORSE was run to compute the perturbed detector responses, i.e. U-235 fission rates. Unperturbed and perturbed detector responses at various radii are given in Table 1. It is noted that this in-system measurement method suffers large response perturbation, especially for with-Be sample system, over 25% at some positions.

Table 1. Detector response perturbation by MORSE

radius (cm)	response for no-Be system			response for with-Be system		
	unperturbed	perturbed	ratio	unperturbed	perturbed	ratio
20.0	5.088−1	4.856−1	1.048	1.436	1.318	1.089
21.0	5.071−1	4.803−1	1.056	1.310	1.179	1.112
23.0	4.979−1	4.692−1	1.061	1.059	9.387−1	1.128
25.0	4.856−1	4.777−1	1.017	8.740−1	7.482−1	1.168
27.0	4.710−1	4.263−1	1.105	7.207−1	5.939−1	1.214
28.0	4.619−1	4.241−1	1.089	6.480−1	5.209−1	1.244
30.0	4.302−1	3.772−1	1.140	4.720−1	3.843−1	1.228
35.0	3.300−1	2.958−1	1.116	2.070−1	1.770−1	1.169
39.0	2.513−1	2.215−1	1.135	1.090−1	9.127−2	1.194
44.0	1.699−1	1.514−1	1.122	5.140−2	4.557−2	1.128
50.0	1.011−1	8.777−2	1.152	2.200−2	1.963−2	1.121
56.0	5.670−2	5.265−2	1.077	9.800−3	8.254−3	1.187
63.0	2.430−2	2.225−2	1.092	3.500−3	2.562−3	1.366

Table 2. Perturbation of integral quantities

	unpertuebed	perturbed	relative
A_H^0	1.325+3	1.192+3	1.112
A_H^M	2.854+3	2.423+3	1.178
M_{app}	2.154	2.033	1.060

Table 3. Comparison of multiplications by experiment and calculation

	experimental		calculated	
	uncorrected	corrected	B-VI	LAN
M_{app}	1.832	1.942	2.138	2.084
M_L	1.616	1.713	1.851	1.840

Based on Table 1, perturbations of integral quantities are calculated and listed in Table 2. The total hydrogen absorption rates (relative values) in the polyethylene is obtained by weightedly integrating the detector response R(r) over the whole polyethylene region with hydrogen atomic density as the weighting factor, i.e.

$$A_H = \int n_H(r) \, R(r) \, 4\pi r^2 dr$$

The apparent multiplication Mapp is the ratio of total H(n,r) absorptions for with−Be sample system and for no−Be sample system. Table 2 reveals that astonishing deviations occur to H(n,r) rates. Thanks to the relative method employed, apparent multiplication Mapp, however, is only lowered down 6%. The experimental results must be corrected, though. A reasonable approach is to multiply the measured apparent multiplication by relative perturbation coefficient (1.06). Table 3 shows the comparison between corrected experimental results and calculated multiplications.

From the above analysis, the following conclusions are drawn:
1. The SWINPC experiment detecting method is of fatal defect. To avoid such a detector perturbation, it is suggested to adopt small−sized detector system of the same material as or similar to its surroundings.
2. After making perturbation correction, SWINPC beryllium multiplication integral experiment still differs from the INEL experiment, which agrees with Los Alamos evaluation calculation very well.

Acknowledgement

The work is partially supported by IAEA under research contract No. 5740 / RB.
Thanks are due to Prof. Cai Shaohui and Dr. Deng Li for beneficial discussions with the author.

REFERENCES

1. T.K. Basu, V.R. Nagrundkar, P. Cloth, D. Filges, and S. Taczanowski, Nucl. Sci. Eng., 70, 309 (1979)
2. V.A. Zagryadskij, M.I. Krajnev, D.V. Markovskij, V.M. Novikov, D.Yu Chuvilin, G.E. Shatelov, Measurement of Neutron Leakage from U, Th and Be Spherical Assemblies with a Central 14−MeV Source, INDC(CCP)−272/G, IAEA, Vienna, Austria (1987)
3. R.S. Hartley, N.E. Hertel, and J.W. Davidson, Fusion Eng. and Design, 10, (1989)
4. J.R. Smith and J.J. King, Maganese Bath Measurement of the Neutron Multiplication in Bulk Beryllium, Proc. IAEA AGM, Chengdu, China, Nov. 19−21, 1990
5. J.W. Davidson and M.E. Battat, Calculation of the INEL Beryllium Multiplication Experiment, Proc. IAEA AGM, Chengdu, China, Nov. 19−21, 1990
6. Yuan Chen, Gang Chen, Rong Liu, Haiping Guo, Wenjiang Chen, Wenmian Jiang, and Jian Shen, Experiments of Neutron Multiplication in Beryllium, Proc. IAEA AGM, Chengdu, China, Nov. 19−21, 1990
7. Liu Lianyan and Zhang Yuquan, Analysis of SWINPC Beryllium Multiplication Integral Experiment, Proc. IAEA AGM, Chengdu, China, Nov. 19−21, 1990
8. W.W. Engle, Jr. A User's Manual for ANISN, A One−Dimensional Discrete Ordinate Transport Code With Anisotropic Scattering, K−1693, ORNL (1957)
9. M.B. Emmett, The MORSE Monte Carlo Radiation Transport Code System, ORNL−4972 (1975)
10. R.E. MacFarlane, D.W. Muir, and R.M. Boicourt, The NJOY Nuclear Data Processing System, Vol.I: User's Manual, LA−9303−M, LANL(1982)

MEASUREMENT OF NEUTRON MULTIPLICATIONS IN BERYLLIUM ASSEMBLIES

Yuan Chen, Gang Chen, Rong Liu, Haiping Guo,
Wenjiang Chen, Wenmian Jiang and Jian Shen

Southwest Institute of Nuclear Physics and Chemistry
P.O.Box 525-74, Chengdu 610003, P.R. China

Abstract: The measurements of 14 MeV neutron multiplication in beryllium have been carried out using the total absorption method. A pure water sphere and a polyethylene sphere were used as the neutron moderators and total absorbers. The results measured by ^{235}U fission chambers and ^{6}Li glass detectors were compared. The effects of the structure of ^{235}U fission chambers on the neutron multiplications have been studied in detail. The light guides used in ^{6}Li glass detectors were also analyzed. The measured results obtained from the relative and the efficiency-determined methods have been compared with the ANISN calculations. It is shown that the measured results are 3% − 15% and 1% − 10% lower than the calculations using ENDF/B-IV and ENDF/B-VI respectively.

Introduction

Beryllium is a promising candidate for the material of the fusion reactor blanket due to its high cross-section of the (n,2n) reaction. Several previous experiments had been carried out to measure the neutron multiplications. But the experimental errors were large and the agreements between the experiments and the calculations were not satisfactory.[1-6] To get the better results from the experiments we have undertaken a project to measure the neutron multiplications in beryllium. The thickness of beryllium was up to 14.85 cm and some good results were obtained.

Methods of the Experiment

The method used in the experiment is "Total Absorption Detector (TAD)" method. The relative and the efficiency-determined methods were used and introduced as follows.

Relative Method

It has been discussed in the previous works that the apparent neutron multiplication and leakage neutron multiplication in the beryllium assemblies are

$$M_a = {}^{m}H_A / {}^{0}H_A \tag{1}$$

and

$$M_L = M_a (1 - {}^{0}X_A - {}^{0}L) + {}^{m}X_A + {}^{m}L \tag{2}$$

where M_a is the apparent multiplication and M_L is the leakage multiplication.[7] We denote the quantities with superscripts "0" and "m" to represent the case of the absence of beryllium from the system and the case of the locating the beryllium in the system. ${}^{m}H_A$ and ${}^{0}H_A$ are the neutron absorption rates by hydrogen in TAD and normalized with the source neutrons. The ratio of ${}^{m}H_A$ and ${}^{0}H_A$ is also equal to the ratio of the integral neutron counts of

1/V detector in the TAD. X_A are the neutron absorptions by non-hydrogen absorbers in the moderator and L the neutrons leaking out of the moderator. X_A and L are obtained by calculations.[8]

Efficiency-determined Method

The leakage neutron multiplication can also be obtained using the efficiency-determined method if the neutrons leaking from Be are separated into two groups. namely the neutrons directly leaking from the D–T source and the secondary, nonelastic-reaction neutrons, as

$$M_L = N_T + (M_a - N_T) \, \varepsilon_{14MeV} / \varepsilon_{Cf} \tag{3}$$

where

$$N_T = e^{-\Sigma \cdot \lambda} \tag{4}$$

is the part of neutron leaking directly from the D–T source. Σ is the macroscopic cross section of the nonelastic scattering, and λ is the thickness of Be. ε_{14MeV} is the detective efficiency of the neutron detector in TAD to the D–T neutrons and ε_{Cf} is the detective efficiency of the detector in TAD to the ^{252}Cf neutrons. These two efficiencies are proportional to the detective efficiencies of TAD to the two groups of neutrons.

Experimental Set-up

Beryllium Assembly

The beryllium assembly consists of six spherical shells. A hole of 3 cm in diameter was made on the sphere as the channel to put the neutron source system. The thicknesses of Be used in the experiment are 4.55, 8.40, 10.45 and 14.85cm. The purity of Be was carefully analyzed and it was found the neutron absorptions by impurities in the Be are negligible according to the requirement of the experiment.

Total Absorption Detector

A pure water sphere and a polyethylene sphere were used as the total absorption detectors. The diameter of the water sphere is 1.5 m. The beryllium sphere embodied in a outer polyethylene spherical shell is located in the center of the water sphere. The inner radius of the shell is 20 cm and the thickness is 1.7 cm. The spherical shell and the target chamber were sheathed with a very thin polyethylene bag. So the beryllium assembly was isolated from water.

The polyethylene sphere used as the TAD consists of eight blocks without frame in it. The diameter of the sphere is 1.38m.

Target Chamber

The target chamber consists of a TiT target and a beam drift tube which is made of Al. A Si(Au) surface-barrier detector is set in the target chamber at the angle of 178.2° to the beam line and used to monitor the yield of source neutrons. The average energy of the deuteron is 100 keV and so the anisotropic factor of the associated alpha particle emission at this direction is 1.220. The part of D-D neutrons in the source neutrons was controlled to be smaller than 1%. The detection error of the yield of the neutron source is found to be smaller than 2%.

Neutron Detectors

Both ^{235}U fission chambers and ^{6}Li glass detectors were used in the experiment. The structure of the fission chamber has been discussed in the previous reports. The ^{6}Li glass detector consists of a ^{6}Li glass, a light guide which is made of perspex and a xp-1115 photomultiplier. The property of the light guide has been studied and will be discussed later in this article. Fig.1 shows the pulse-height spectra of the associated alpha detector, the fission chamber and two ^{6}Li glass detectors.

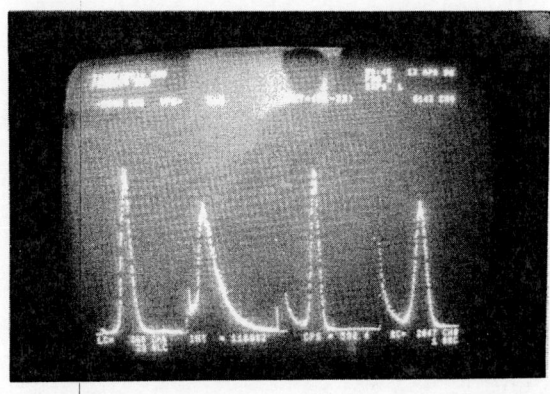

Fig.1. The Pulse-height Spectra of the Associated Alpha Detector, the Fission Chamber and Two Lithium Glass Detectors (from left to right)

Measurements

Both relative and efficiency-determined methods were used. For the relative method, the neutron distributions in the moderator were measured at 5 angles of 0°, 40°, 80°, 120° and 150° to the beam line, when the beryllium sphere was located in and removed from the center of the moderator respectively. Measured counts were normalized with the associated alpha particle counts and integrated over the region of the moderator.

To determine the detector efficiencies for the two groups of neutrons, beryllium sphere was removed from the moderator and put the D-T source or ^{252}Cf neutron source into the center of the system. The efficiencies were normalized to one source neutron.

Results and Discussion

The leakage neutron multiplications were obtained and tabulated in Table 1. The Neutron reflecting from the moderator into beryllium and the neutrons streaming through the beam duct were concerned and corrected in the final results. It can be found that on the same moderator the results got from two measurement models are nicely consistent and the deviations are about 1%. This fact demonstrates that it is suitable to use the ^{252}Cf source to simulate the secondary, nonelastic reaction neutrons from Be sphere. The agreements between the results from two moderators are also consistent. It is found the measured results are 3%-15% lower than the ANISN calculation using ENDF/B-IV and 1%-10% lower than the calculated ones using ENDF/B-VI.[8]

The effect of the outer case of the fission chamber on the neutron multiplication has been studied. The chamber is a copper cylinder with 2 cm of height and 3 cm in diameter. The thickness of the wall is 1 mm and the thickness of the bottom is 0.5 mm. We prepared several copper caps with different wall thicknesses. Some of the measurements were performed when the caps put onto the detectors. The results are shown in Table 2. The first row lists the thicknesses of the caps(wall/bottom). It can be found the neutron absorptions by copper caps are large but the changes of the multiplications are small and negligible.

On the water sphere system, we estimated the perturbation of the neutron distribution attributed to the Ar chamber of the ^{235}U fission chamber. The measurements were performed after the water behind the chamber was removed. Only a very small deviation was observed, and the changes of detector response caused by the Ar chamber is smaller than 1%.

It is found if the efficiency-determined method is used and ^{6}Li glass directly connects to the photomultiplier, the measured results by ^{6}Li glass detector are satisfactorily consistent with the measured results by ^{235}U fission chambers. But when the relative method is used, a sufficient length of the perspex light guide must be added between the glass and

Table 1. The Leakage Neutron Multiplications in Beryllium

Detectors	Methods	Moderators	Thicknesses of Be(cm) 4.55 8.40 10.45 14.85			
^{235}U	relative	water sphere	1.284		1.629	1.749
		PE sphere	1.333	1.548	1.635	1.774
	efficiency determined	water sphere	1.275		1.612	1.733
		PE sphere	1.323	1.531	1.625	1.751
^6Li	efficiency determined	PE sphere		1.497		1.815

Table 2. The Effect of the Copper Cases of the Fission Chambers on the Neutron Multiplication

Copper Caps	With Be		Without Be		Ma
0	56831	1	31810	1	1.787
0.5/0.5	53894	0.948	30216	0.950	1.784
1.0/0.5	52681	0.927	29192	0.918	1.805
2.0/1.0	49684	0.874	27475	0.864	1.808
3.0/1.5	46374	0.816	25580	0.804	1.813

the photomultiplier. This is caused by the vacuum chamber of the photomultiplier that disturbs the neutron distribution in the TAD. It was also shown by the experiments that the neutron count rate of the ^6Li glass detector increase sand approaches a constant with the length of the light guide increases. The results are shown in Fig. 2. When a proper light guide is used, the ^6Li glass detector can do the same work as the ^{235}U fission chambers do. The similar results have been obtained in the experiments of neutron multiplication in depleted uranium assembly. The measured results obtained using the ^6Li glass detector with 6 cm length of the light guide are consistent nicely with the results obtained using the fission chamber.[9]

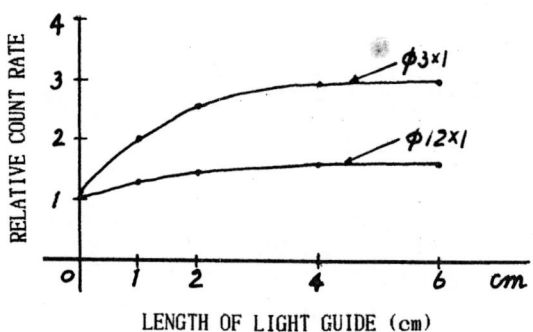

Fig. 2. Relative Count Rates of ^6Li Detectors Change with the Lengths of the Light Guide. The Numbers in the Figure Are the Sizes of ^6Li Glass.

Conclusions

The experiments of neutron multiplication in beryllium have been carried out. The overall experimental error on the water sphere is 3.7% and on the polyethylene sphere is 2.8%. We obtained good consistency among the methods of measurements. Neutron detectors used in the experiments have been studied in detail. The measured results are lower than the calculated ones. Further studies of the integral experiments are needed.

Acknowledgements

Our thanks are due to IAEA for the partly support under the research contract No. 5828/RB. We thank Liu Lian-yan, Zhang Yu-quan and Liu Cheng-an from IAPCM of China for their sincerely cooperations with us.

Reference

1. T. K. Basu, V. R. Nargundkar, P. Cloth, D. Filges and S. Taczanowski, Nucl. Sci. Eng. 70, 309 (1979)
2. V. A. Zargryadskij, M. I. Krajnev, D. V. Markovskij, V. M. Novikov, D. Yu chuvilin and G. E. Shatalov, INDC(CCP)-272/G, IAEA, Vienna, Austria (1987)
3. R. S. Hartley, "Neutron Multiplication in Beryllium." PhD Dissertation, Univ. of Texas, Austin, Texas (1987)
4. N. E. Hertel, J. W. Davidson and R. S. Hartley, 8th Topical Meeting on the Tech. Fusion Eng. paper O-NUL-5, Salt Lake City, Utah, Oct. 9-13, 1988
5. C. Wong, E. F. Plechaty, R. W. Bauer, R. C. Haight, L. F. Hansen, R. J. Howerton, T. T. Komoto, J. D. Lee, S. T. Perkins and B. A. Pohl, Fusion Technology, 8, 1165 (1985)
6. E. T. Cheng, "Review of the Nuclear Data Status and Requirements for Fusion Reactor," Proc. Int. Conf. on Nuclear Data for Sci. Tech. Mito, Japan, 30 May-3 June, 1988
7. Y. Chen, G. Chen, R. Liu, H. Guo, W. Chen, W. Jiang and J. Shen, Fusion Technology, to be published.
8. L. Liu, Y. Zhang, "Analysis of SWINPC Beryllium Neutron Multiplication Integral Experiment," presented to the Conference.
9. Y. Chen, et al, "Neutron Multiplication in Depleted Uranium Assembly," private communication.

DOUBLE-DIFFERENTIAL NEUTRON EMISSION CROSS SECTION OF U-238 AND TH-232 FOR 18 MEV INCIDENT NEUTRONS

M.Baba, S.Matsuyama, T.Ito, N.Ito, K.Maeda and N.Hirakawa

Department of Nuclear Engineering, Tohoku University
Aoba, Aramaki, Aoba-ku, Sendai 980, Japan

Abstract: Energy-angular double-differential neutron emission cross sections of U-238 and Th-232 were measured for 18 MeV incident neutrons using the TOF technique. The measured neutron emission spectra of both nuclides show marked disagreement with current evaluated nuclear data. The angular distributions of continuum neutrons are described fairly well by Kalbach-Mann systematics.

(Double-differential Neutron Emission Cross section, U-238, Th-232, 18 MeV incident energy, Evaluated Data, Angular Distribution, Kalbach-Mann Systematics)

Introduction

Neutron emission spectra for fast neutron interaction with U-238 and Th-232 are basic nuclear data for the design of accelerator based reactors since they dominate neutron energy-spatial distributions in reactors. They are also of interest as the reference data for fission neutron spectrum produced by fast neutrons. However, experimental data are very few especially at incident neutron energy above 14 MeV.

In the present study, we have measured double-differential neutron emission cross sections (DDXs) of U-238 and Th-232 for incident neutrons of 18 MeV using Tohoku University Dynamitron T-O-F spectrometer, and deduced angle-dependent emission spectra and neutron angular distributions. This paper presents data comparison with evaluated nuclear data, and semi-empirical analyses of neutron emission spectra and angular distributions.

Experiments and Data Analyses

Experiments were carried out using the pulsed neutron TOF technique with a 4.5 MV Dynamitron accelerator as a pulsed neutron generator. The methods of experiment and data analyses were almost identical with our previous studies /1-3/ except that a post-acceleration beam chopper /4/ was newly employed in the present study to improve energy resolution.

Experiment

Primary neutrons were obtained via the d-T reaction at 0-deg using a tritium-loaded titanium target and pulsed deuteron beam provided by the Dynamitron, about 1.5 ns in FWHM and 2 MHz repetition rate. The energy spread of primary neutrons was about 400 keV. In the source neutron spectrum, parasitic components were not negligible because of higher deuteron beam energy. Their influences were taken into consideration with a care in data correction.

Scattering samples were metallic cylinders of elemental uranium and thorium, 2-cm diam and 5-cm long; they were encased in thin aluminum cans and placed 10-12 cm from the neutron target on a remotely-controlled sample changer.

The absolute cross sections were determined relative to the n-p scattering cross section.

Secondary neutron detector was a NE213 scintillator, 14-cm diam and 10-cm thick, equipped with 2-bias pulse-shape discriminators /1-3/ for effective rejection of gamma-ray back-grounds over neutron energy range from 0.8 to 18 MeV. The detector was mounted in a massive shield on a turning table. The flight path length was around 4-m. Overall timing resolution was about 2.0 ns. Energy dependence of detector efficiency was determined by TOF measurement of Cf-252 fission neutrons and calculation using a Monte-Carlo code /1-3/.

Another NE213 neutron detector, 5-cm diam and 5-cm thick, was used for flux normalization and inspection of primary neutron spectrum.

Data were taken for the samples and empty can at 6 angles between 30 and 120-deg.

Data analyses

The raw data were corrected for the effects of 1)detector efficiency, 2)sample-out backgrounds, 3)sample activity, 4)finite sample-size, and 5)parasitic neutrons.

Backgrounds due to sample-activity were subtracted as flat background in TOF distributions /1/. The influences of 4) and 5) were estimated concurrently by a Monte Carlo simulation code /1-3/. The data used in the calculation were prepared by combining JENDL-3 and experimental ones; neutron emission spectra were made using experimental data and Kalbach-Mann systematics as described in the next section since the evaluated spectra were largely different from measured ones. Spectra and intensity of parasitic neutrons required for the correction were determined from the source spectrum measurements considering their time dependence caused by build-up of contaminant elements.

Results and Discussion

Data comparison

The results of neutron emission spectra of U-238 and Th-232 are shown in Fig.1, compared with the corresponding data derived from JENDL-3 and ENDF/B-VI. In this neutron energy, multiple-chance fissions are energetically possible and will have large contribution to neutron emission as well as scattering and cascade reactions.

The experimental spectra of U-238 are almost angle-independent in low energy but show strong forward rise above 5 MeV. They are generally closer to ENDF/B-VI providing angle-dependent spectra than JENDL-3 which lacks angular-dependence and pre-compound components. It is noted that experimental data show stronger angle-dependence than ENDF/B-VI which expresses the angular dependence of pre-compound neutrons by Kalbach systematics /5/; this observation is

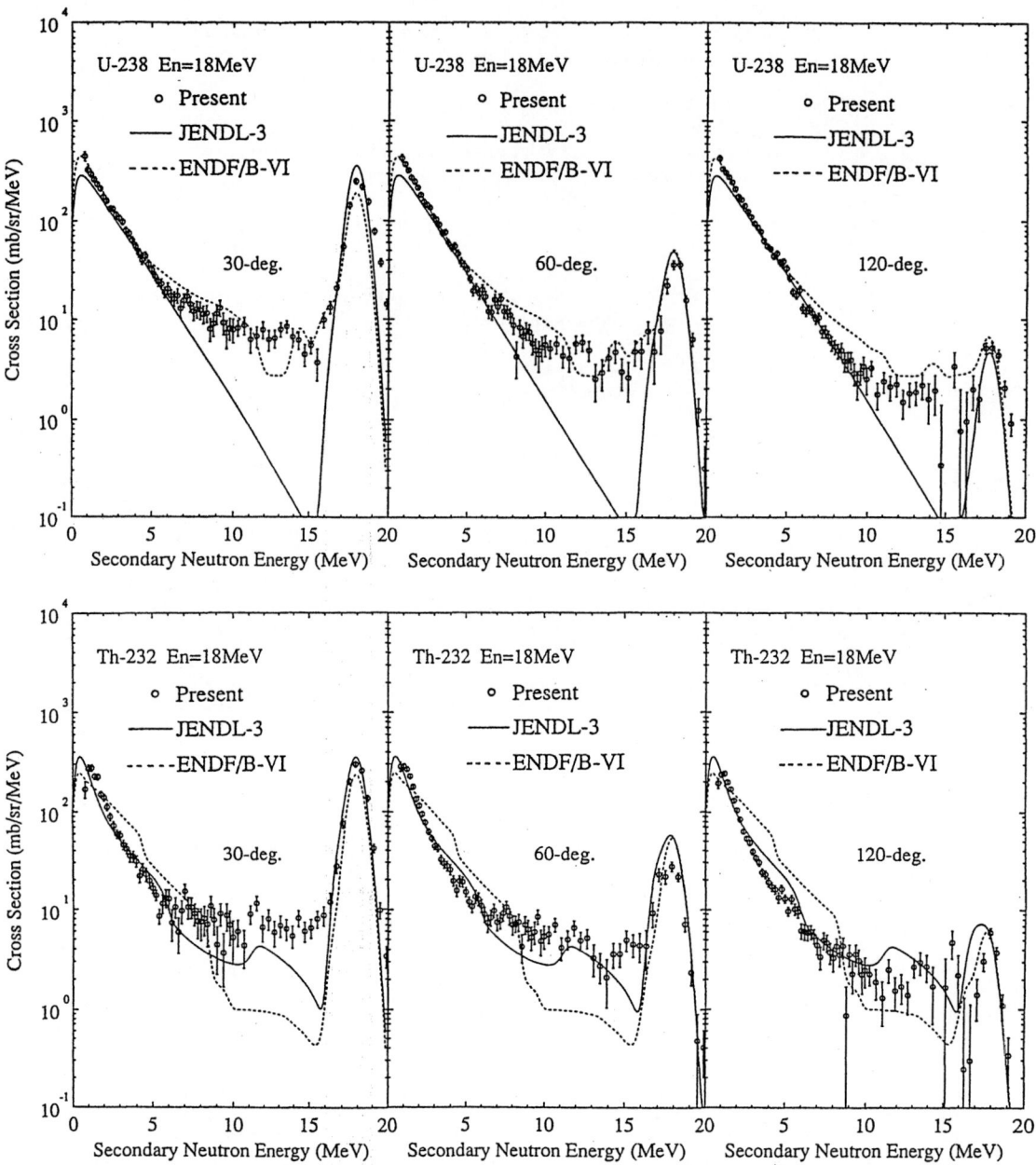

Fig.1: DDXs of U-238 (upper) and of Th-232 (lower), compared with evaluated data.

consistent with the results of angular distributions described in the following.

The neutron emission cross sections of Th-232 are, compared with those of U-238, similar in shape but generally smaller in magnitude probably because of much less fission cross section of Th-232. The present data are traced better by JENDL-3 than ENDF/B-VI but still discrepant in the energy region above 8 MeV since JENDL-3 does not provide angle-dependence of emission spectra.

Analyses of Spectra and Angular distributions

The experimentally observed spectra and angular distributions were analyzed semi-empirically using conventional descriptions /5-8/.

Firstly, angle-integrated spectra were fitted with combination of the cascade neutron spectrum by Le Couter /7/, the exciton spectrum by Blann /8/ and fission neutron spectrum. The results of fitting is shown in Fig.2(a)(b). The spectrum of U-238 is not reproduced consistently using the fission spectrum in JENDL-3 while

good fits are obtained for Th-232 and other elements /3/. This tendency was observed also in the case of 14 MeV data /1/. Then, we changed the fission spectrum to be softer than JENDL-3. We found that, as shown in Fig.2(c), overall fit was possible by assuming fission spectrum to be the Maxwellian with temperature around 1.33 MeV; it is interesting to note that the assumed spectrum is very close to that in ENDF/B-VI. This result suggests that the fission spectrum in JENDL-3 is too hard for 18 MeV incident energy.

Then the angular distributions of secondary neutrons from U-238 and Th-232 are compared with systematics by Kalbach & Mann (K-M) /6/ and by Kalbach /5/ which are useful for description of angular distributions /1,2/. In the calculation, we assumed that fission neutrons were isotropic and the fraction of Multi-step direct (MSD) process, required for the calculation, can be replaced with that of pre-compound process obtained from the spectrum fits. The results are not sensitive to the MSD fraction and energy-

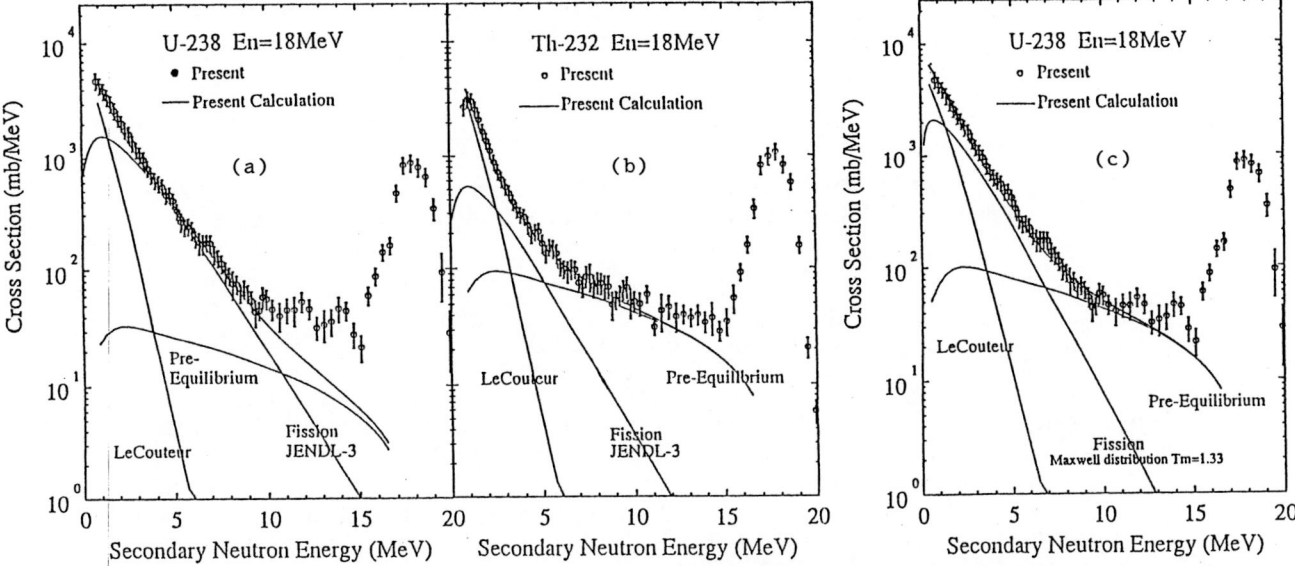

Fig.2(a)(b)(c): Analyses of neutron emission spectra of U-238 and Th-232.

angular distribution of fission neutrons. In fact, the results of U-238 are similar with those of Th-232 which emits much less fission neutrons.

The results are illustrated in Fig.3. The K-M systematics reproduces the experimental results better than the Kalbach systematics for both U-238 and Th-232. This results are consistent with the case for 14 MeV neutrons /1/, but in contrast with the cases for lighter and nonfissionable elements, for which the K-M systematics generally overemphasize the forward rise /3/.

The present results suggest the mass-dependence of scattered neutron angular distribution. Further studies will be interesting.

Summary

We have measured double-differential neutron emission cross sections of U-238 and Th-232 for incident neutrons of 18 MeV. The experimental emission spectra show marked disagreement with JENDL-3 and ENDF/B-VI in spectral shape and angular-dependence. The spectrum analyses suggest fission spectrum might be softer than in JENDL-3. The angular distributions of secondary neutrons are were found to be traced better by K-M systematics than Kalbach systematics for both nuclides.

Acknowledgement

This work was supported by Japan Atomic Energy Research Institute, and Grant-in-Aid for Scientific Research, Ministry of Education and Culture. The authors wish to thank Messrs. T. Iwasaki, M.Fujisawa, R.Sakamoto and H.Nakamura for their collaboration in accelerator operation and experiments.

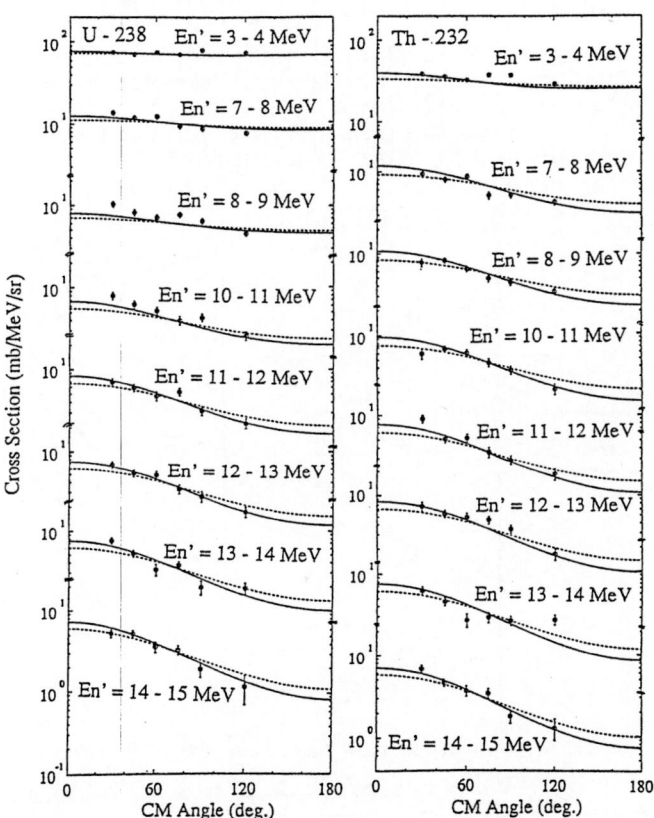

Fig.3: Secondary neutron angular distributions, circle; experimental values, solid line; Kalbach-Mann systematics, dash line; Kalbach systematics.

REFERENCES

1. M.Baba,H.Wakabayashi,N.Ito,K.Maeda and N.Hirakawa; J.Nucl.Sci.Technol.,27(7) 601(1990)
2. M.Baba,M.Ishikawa,T.Kikuchi,H.Wakabayashi, and N.Hirakawa;Proc.Int.Conf.Nucl.Data for Sci.Technol.,May/June 1988,Mito Japan (Edited by S.Igarashi)p.209, Saikon 1988
3. M.Baba,M.Ishikawa,N.Yabuta,T.Kikuchi, H.Wakabayashi and N.Hirakawa;ibid.,p.291
4. S.Matsuyama,M.Fujisawa,M.Baba,T.Iwasaki, S.Iwasaki,R.Sakamoto,N.Hirakawa and K.Sugiyama; This Conference, B51
5. C.Kalbach; ibid., C37(6) 2350(1988)
6. C.Kalbacn and F.M.Mann; Phys. Rev., C23(1) 12(1982)
7. K.J.Le Couteur; Proc.Phys.Soc.,A65 718(1952)
8. M.Blann and F.M.Lanzafame; Nucl.Phys. A142 559(1959)

CROSS SECTIONS FOR THE PRODUCTION OF HELIUM AND LONG–LIVING RADIOACTIVE ISOTOPES BY PROTONS AND DEUTERONS

P. Jung

Institut für Festkörperforschung, Forschungszentrum Jülich,
Postfach 1913, D–5170 Jülich, Germany, Association EURATOM–KFA

Abstract: Cross sections of protons of 8,16 and 24 MeV and of deuterons of 9 and 14 MeV for the production of helium and of radioactive isotopes with half–lifes exceeding 5 days were measured in foils of Al, Ti, V, Fe, Ni, Cu and type 316 stainless steel. The helium production cross section are compared to data for 14.8 MeV neutrons and 750 MeV protons. Helium production cross sections of protons and deuterons in the 14–16 MeV energy range closely match those of 14.8 MeV fusion neutrons. Nuclear reaction cross sections which can serve for dosimetry are indicated.

Introduction

Nuclear reactions can change mechanical properties of materials by changing the chemical composition and by adding new constituents, e.g. hydrogen and helium. Helium is virtually insoluble in metals and causes dimensional changes and embrittlement by the formation of bubbles. Therefore helium is considered one of the most detrimental elements in terms of mechanical property deterioration.

As long as strong 14 MeV neutron sources are not available, the development of structural materials for future fusion reactors largely depends on simulation experiments employing other irradiation sources. Among other things, these sources are valued by their ability to match the helium production in a fusion reactor. Irradiations with light ions are extensively used to study mechanical properties under irradiation [1]. It is the purpose of the present work to determine their cross section for the production of helium and other transmutation products in metals which are under consideration for fusion application, and to compare them to results of 14 MeV neutrons.

Experimental Details

Metal	Purity [wt%]	d [μm]	r	n
Al	99.999	11–80	12.3	1.71
Ti	99.97	50	9.7	1.66
V	99.95	51	6.7	1.69
Fe	99.997	37	5.8	1.66
Ni	99.9999	5–33	5.0	1.68
Cu	> 99.999	38	5.3	1.64
316 L	–	36	≈5.5	≈1.66

Table 1: Material purities (natural composition), thicknesses and parameters for proton range determination. The main constituents (wt%) of 316 L are 65.1% Fe, 17.7% Cr, 12.3% Ni and 2.3% Mo.

Specimen parameters are given in Table 1. Suppliers and pretreatment were reported in Ref. [2]. The specimens were soldered with indium to a copper bar which was mounted in a vacuum chamber. Irradiation with protons of 8,16 and 24 MeV and of deuterons of 9 and 14 MeV was performed at the Jülich compact cyclotron at temperatures around 280 K. Beam current was determined by deviding the heating power deposited in the copper bar by the particle energy as derived from the cyclotron frequency. Errors of the determination of heating power and energy were 5% and about 7%, respectively. This gives an estimated total uncertainty in particle dose of about 10%. The demand of a well defined particle energy for cross section determination is opposed

by energy degradation in foils of finite thickness. The projected range of protons [3] above 2 MeV can be approximated by an exponential law:

$$R_P = r \, E_0^n \qquad (1)$$

The parameters for protons are included in Table 1. For deuterons r will be smaller by about a factor of $\sqrt{2}$. The relative spread of energy in the foil is then given by:

$$\frac{\Delta E}{E_0} = \frac{E_0 - E}{E_0} = 1 - (1 - \frac{d}{R_P})^{1/n} \qquad (2)$$

While thin foils are favourable in terms of well defined energy, they bring out the problem of surface losses. As α–particles from nuclear reactions have energies of about 5–10 MeV they will have ranges R_α in the order of 10–40 μm. Some α–particles which are produced within this distance from the surface will leave the specimen, while on the other hand some α–particles which are produced within this distance in the specimen holder are injected into the specimen as depicted in Fig. 1.

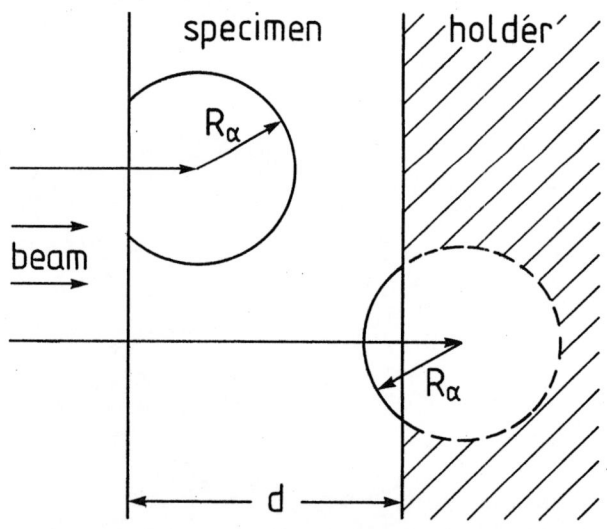

Fig. 1: Schematic view of helium (range R_α) losses to the surfaces and injection from the holder into a specimen of thickness d.

The error introduced by these surface effects depends on the relative size of the production cross sections in the specimen σ_s and the holder σ_h and on the ratio of thickness d and range R_α. A straightforward geometrical consideration gives for the helium concentration c deposited in a foil of thickness d compared to a infinitely thick foil:

for $\sigma_h \ll \sigma_s$:

$$\frac{c(d)}{c(\infty)} = d/2R_\alpha \qquad \text{for } d < R_\alpha \qquad (3a)$$

$$\frac{c(d)}{c(\infty)} = 1 - (R_\alpha/2d) \qquad \text{for } d > R_\alpha \qquad (3b)$$

for $\sigma_h = \sigma_s$:

$$\frac{c(d)}{c(\infty)} = \frac{1}{2} + (d/4R_\alpha) \qquad \text{for } d < R_\alpha \qquad (4a)$$

$$\frac{c(d)}{c(\infty)} = 1 - (R_\alpha/4d) \qquad \text{for } d > R_\alpha \qquad (4b)$$

for $\sigma_h = 2\sigma_s$:

$$c(d) = c(\infty) \qquad (5)$$

The above estimate is based on the assumption that specimen and holder differ only in their σ values but have identical R_α's. It can be seen that in the most unfavourable case ($\sigma_h \ll \sigma_s$), the surface effects would become less than 20% only for $d > 2.5 \, R_\alpha$. Such thicknesses are intolerable from the above point of view of energy degradation.

It was experimentally verified in Ref. [2] by varying the thickness of aluminum and nickel foils on the copper bloc that Al on the holder approximates the above relations for $\sigma_h \ll \sigma_s$, while for Ni a relation $\sigma_h \gtrsim \sigma_s$ is more appropriate. The latter relation was also adopted for the other metals. For the given specimen thicknesses the corrections for surface effect according to equations (3) to (5) amounted to less than 10%, while the spread of energy according to equation (2) was typically ± 0.3 MeV.

Helium content of the specimens was determined by melting them in a vacuum furnace and recording the helium by mass spectrometry. In all specimens only ^{4}He was observed, while the ^{3}He signal was negligible.

Parts of the specimens were also analyzed in a γ-ray spectrometer to record long-term activation [2].

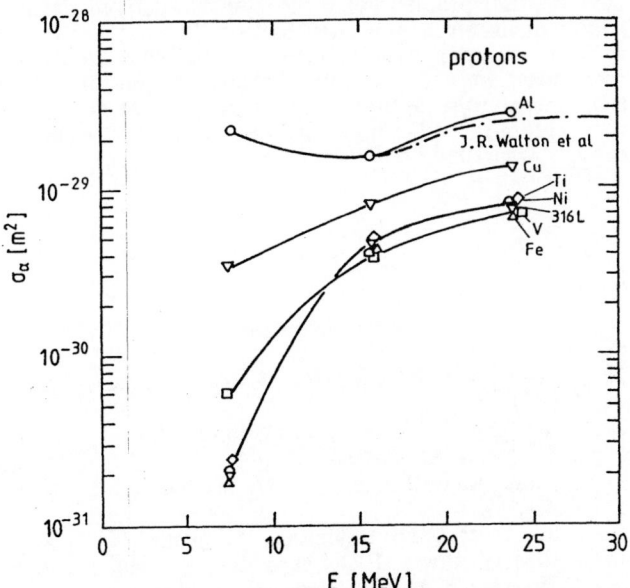

Fig. 2: Helium production cross section in Al, Ti, V, Fe, Ni, Cu and type 316 L stainless steel as function of proton energy. The dash-dotted line gives literature data of Ref. [4].

Results

Helium production cross sections by protons and deuterons are given in Figs. 2 and 3 and in Table 2 as function of energy.

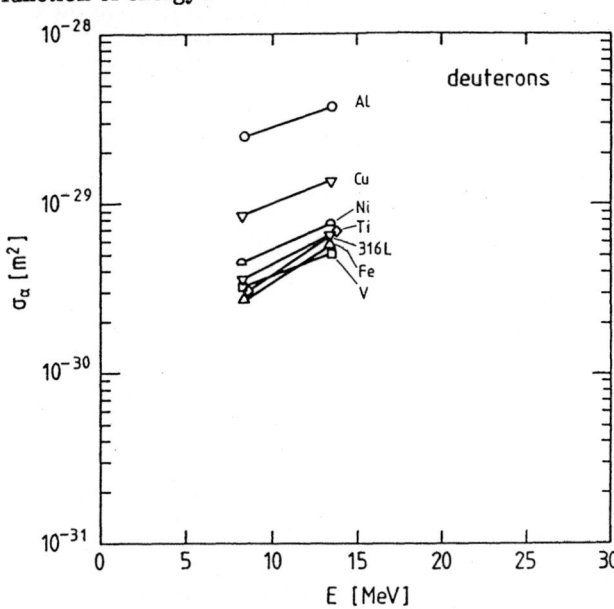

Fig. 3: Helium production cross sections in Al, Ti, V, Fe, Ni, Cu and type 316 L stainless steel as function of deuteron energy.

Table 2: Production cross sections [10^{-28} m^2] of ^{4}He by proton and deuterium irradiation. The energy ranges [MeV] indicate the degradation of the particle energies in the specimens.

Target	Protons		Deuterons	
	E	σ_α	E	σ_α
Al	7.6±0.4	0.22$_5$	8.3±0.7	0.25$_2$
	15.7±0.3	0.15$_8$	13.5±0.5	0.36$_5$
	23.8±0.2	0.28$_0$		
Ti	7.6±0.4	0.0024$_0$	8.4±0.6	0.032
	15.8±0.2	0.050	13.6±0.4	0.069
	23.8±0.2	0.084		
V	7.4±0.6	0.0060	8.2±0.8	0.032
	15.7±0.3	0.039$_5$	13.4±0.6	0.050
	23.7±0.3	0.072		
Fe	7.5±0.5	0.0019	8.3±0.7	0.028$_5$
	15.7±0.3	0.045	13.5±0.5	0.059
	23.8±0.2	0.065		
Ni	7.6±0.4	0.0020	8.3±0.7	0.045$_5$
	15.7±0.3	0.041$_5$	13.5±0.5	0.077
	23.7±0.3	0.084		
Cu	7.5±0.5	0.035	8.2±0.8	0.086
	15.7±0.3	0.080	13.4±0.6	0.0135
	23.8±0.2	0.135		
316 L	7.5±0.5	0.0020	8.3±0.7	0.036
	15.7±0.3	0.047	13.5±0.5	0.066
	23.8±0.2	0.074		

354

For both particles Al shows the highest cross sections, followed by Cu, while the other metals fall close together. Direct comparison of the present data to literatur is only possible for Al: Proton cross sections from Ref. [4] are slightly below the present data, probably because surface losses have not been taken into account [4].

The helium production cross sections of 16 MeV protons and 14 MeV deuterons are compared to the cross sections of 14.8 MeV neutrons [5] and of 759 MeV protons [6] in Fig. 4. Obviously protons as well as deuterons in the 14—16 MeV energy range match the 14.8 MeV cross sections quite well, while the cross sections of the high energy protons exceed the neutron data, especially for heavier elements significantly.

Doses of charged particle irradiation are commonly determined by charge accumulation. This is only possible in vacuum but not when cooling by flowing gas is applied. In this case or when beam inhomogeneities must be expected, activation analysis can be used for dose measurements. In Figs. 5 and 6 cross sections for the production of long living isotopes by proton and deuterons in the present materials are given which are best suited for this purpose. For details of γ—spectroscopy measurements see Ref. [2].

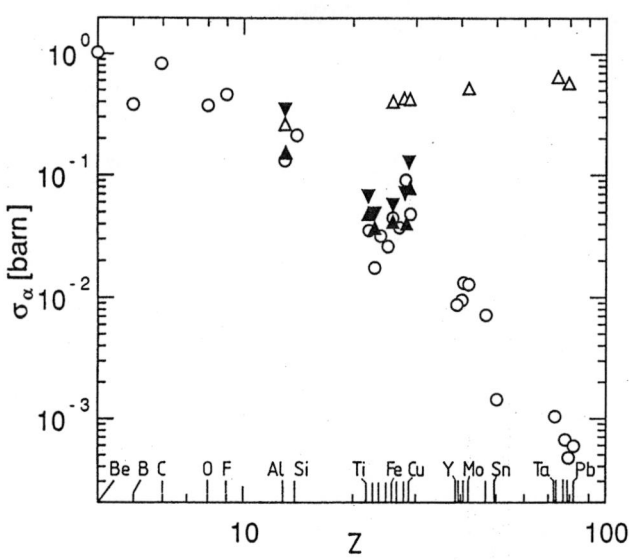

Fig. 4: Helium production cross sections of 16 MeV protons (▲) and 14 MeV deuterons (▼), compared to 14.8 MeV neutrons (o) from Ref. [5] and 750 MeV protons (△) from Ref. [6].

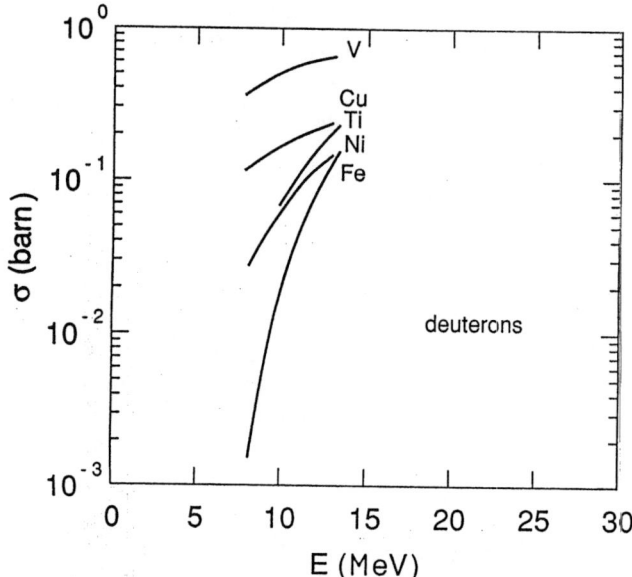

Fig. 6: Cross sections of deuterons for the production of long—living radioactive isotopes. Isotopes and half—lifes are for Ti: ^{48}V (1.391 Ms), V: ^{51}Cr (2.39 Ms), Fe: ^{56}Co (6.68 Ms), Ni: ^{58}Co (6.125 Ms) and Cu: ^{65}Zn (21.06 Ms).

Cross sections for other isotopes have been reported in Ref. [2]. Only in deuteron irradiated iron, σ_α was found to be closely matched by the cross section of the ^{56}Fe $(d,\alpha)^{54}$Mn reaction. This means that only in this case helium production rates can be derived from a known nuclear reaction cross section. In all other metals various isotopes and reactions contribute to helium production in such a way that estimation of helium production from α—producing reaction cross sections is very complex and the above determination of helium content is virtually indispensable.

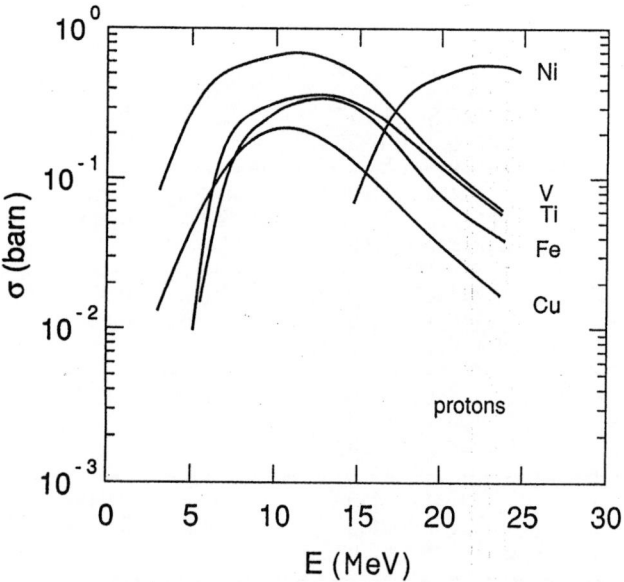

Fig. 5: Cross sections of protons for the production of long living radioactive isotopes. Isotopes and half—lifes are for Ti: ^{48}V (1.391 Ms), V: ^{51}Cr (2.39 Ms), Fe: ^{56}Co (6.68 Ms), Ni: ^{57}Co (23.3 Ms) and Cu: ^{65}Zn (21.06 Ms).

References

1. P. Jung and H. Ullmaier: J. Nucl. Mater. 174, 253 (1990)
2. P. Jung: J. Nucl. Mater. 144, 43 (1987)
3. H.H. Andersen and J.F. Ziegler: Hydrogen, Stopping Powers and Ranges in All Elements, Pergamon Press, 1977
4. J.R. Walton, D. Heymann, A. Yanis, D. Edgerley and M.W. Rowe: J. Geophys. Res. 81, 5689 (1976)
5. D.W. Kneff, B.M. Oliver, H. Farrar IV and L.R. Greenwood: Nucl. Sci. Eng. 92, 491 (1986)
6. S.L. Green, W.V. Green, F.H. Hegedus, M. Victoria, W.F. Sommer and B.M. Oliver: J. Nucl. Mater. 155—157, 1350 (1988)

EXCITATION FUNCTIONS OF SOME (n,p),(n,2n) AND (n,α) REACTIONS ON NICKEL, ZIRCONIUM AND NIOBIUM ISOTOPES IN THE ENERGY RANGE 13.64 - 14.83 MeV

N.I. Molla, R.U. Miah, M. Rahman
and Aysha Akhter

Institute of Nuclear Science and Technology
Atomic Energy Research Establishment
P.O. Box 3787, Savar, Dhaka
Bangladesh

Abstract: Cross section of ^{58,61,62}Ni(n,p), ^{90,96}Zr(n,2n), ^{90,92,94}Zr(n,p), ^{90,94}Zr(n,α), ^{93}Nb(n,2n) and ^{93}Nb(n,α) reactions in the energy range 13.64 to 14.83 MeV have been measured by activation technique in combination with high resolution HPGe detector gamma ray spectroscopy. Sample irradiations have been done in a ring geometry arrangement at J-25 neutron generator. Cross section data of this work along with the recent literature values plotted against neutron energy provide excitation functions for some of the reactions.

(Cross section, (n,2n), (n,p) and (n,α) reactions, excitation function, 13.64 - 14.83 MeV, activation)

Introduction

Measurements of cross section data for fast neutron induced reactions as a function of energy are important for validation support of nuclear model calculations and applications in fission and fusion reactor technology. Neutron cross section data in the energy range from thermal to about 16 MeV for FRT-relevant structural materials are needed for design calculations and the engineering requirements are diverse. To satisfy them extensive measurements and evaluations have been done in recent years with desired accuracy at 14-15 MeV and some cases from threshold upto 19 MeV by many laboratories. The progress and status of neutron cross section data has been reviewed recently by Qaim /1/. We report here excitation functions of some common reactions on the isotopes of nickel, zirconium and niobium in the energy range 13.64 - 14.83 MeV. The present work aims at reconfirming those cross section data already exist in this energy range and presenting some newer points in between.

Experimental

Cross sections were measured by using activation method. High purity nickel, zirconium and niobium foils (1 cm^2, 0.4 - 1.02 gm), each sandwiched between two aluminium foils, were irradiated at 0°, 30°, 70°, 90°, 110° and 150° with respect to beam direction in a ring geometry arrangement. Neutrons were produced at J-25 neutron generator using dt reaction with 120 keV deuteron and 400 μA beam current. The neutron energy versus emission angle was determined by measuring the ratio of the ^{89}Zr to ^{92m}Nb specific activities produced both in Zr and Nb foils by (n,2n) reactions /2/. The irradiation time was 3-4 hours. Neutron flux densities were determined via monitor reaction ^{27}Al(n,α)^{24}Na; cross section data for this reaction was taken from Vonach /3/.

After irradiations the radioactivity of the reaction products was measured by using a high resolution HPGe detector which was previously calibrated with a set of standard gamma sources. The gamma spectra were analysed in a IBM PC-XT system. The count rates at the end of irradiations were corrected for dead time loss, coincidence effects, detector efficiency and gamma transition intensities. The principal sources of errors and their magnitudes considered in the cross section measurements were reported earlier /4/. By combining those errors in quadrature the total error for each cross section value was estimated.

Results and Discussions

The cross section data measured in the energy range 13.64 - 14.83 MeV for fourteen reactions on the isotopes of niobium, zirconium and nickel are given in table 1-3. The total uncertainty in the data lies between 8% and 17%. Cross section data for ^{93}Nb(n,2n)^{92m}Nb, ^{93}Nb(n,α)^{90m}Y, ^{90}Zr(n,2n)^{89}Zr, ^{90}Zr(n,p)^{90m}Y, ^{92}Zr(n,p)^{92}Y and ^{96}Zr(n,2n)^{95}Zr reactions from the present work together with the recent literature values /4-18/ are plotted as a function of neutron energy in fig. 1-6. In most of the cases they agree fairly well with the literature data.

Table 1. Activation cross sections for fast neutron induced reactions on Niobium

Neutron energy (MeV)	^{93}Nb(n,2n)^{92m}Nb Cross section (mb)	^{93}Nb(n,α)^{90m}Y Cross section (mb)
14.83	501±46	6.68±0.54
14.58	483±45	6.48±1.11
14.30	–	6.34±0.52
14.10	496±46	6.49±0.53
13.88	451±42	6.42±0.54
13.64	487±45	5.94±0.49

Table 2. Activation cross sections for fast neutron induced reactions on some isotopes of Zirconium

Neutron energy (MeV)	^{90}Zr(n,2n)^{89m}Zr Cross section (mb)	^{90}Zr(n,2n)^{89}Zr Cross section (mb)
14.83	115.2±16.2	810±72
14.58	108.1±14.2	690±62
14.10	68.1±12.3	640±58
13.88	73.9±11.3	560±51
13.64	56.4± 9.0	460±43

Neutron energy (MeV)	^{90}Zr(n,p)^{90m}Y Cross section (mb)	^{90}Zr(n,α)^{87m}Sr Cross section (mb)
14.83	14.7±1.3	3.9±0.32
14.58	15.4±2.0	3.8±0.31
14.30	14.1±1.1	-
14.10	14.0±1.0	3.7±0.30
13.88	12.9±0.9	2.6±0.21
13.64	12.9±0.0	2.6±0.21

Neutron energy (MeV)	^{92}Zr(n,p)^{92}Y Cross section (mb)	^{94}Zr(n,p)^{94}Y Cross section (mb)
14.83	20.1±1.6	10.4±1.0
14.58	20.5±1.0	11.2±1.0
14.30	-	10.1±1.1
14.10	18.2±1.5	-
13.88	16.6±1.4	-
13.64	16.3±1.3	8.3±0.8

Neutron energy (MeV)	^{94}Zr(n,α)^{91}Sr Cross section (mb)	^{96}Zr(n,2n)^{95}Zr Cross section (mb)
14.83	5.2±0.5	1604±137
14.58	5.8±0.5	1539±132
14.10	5.7±0.6	1478±127
13.88	5.1±0.4	1459±125
13.64	4.1±0.4	1443±124

Table 3. Activation cross sections for fast neutron induced reactions on some isotopes of Nickel

Neutron energy (MeV)	^{58}Ni(n,2n)^{57}Ni Cross section (mb)	^{58}Ni(n,p)^{58}Co Cross section (mb)
14.83	28.4±4.0	310±31
14.30	34.2±3.0	300±30

Neutron energy (MeV)	^{58}Ni(n,2n)^{57}Ni Cross section (mb)	^{58}Ni(n,p)^{58}Co Cross section (mb)
14.10	23.3±2.1	168±19
13.88	19.4±1.8	152±18
13.64	19.9±1.7	222±23

Neutron energy (MeV)	^{61}Ni(n,p)^{61}Co Cross section (mb)	^{62}Ni(n,p)^{62}Co Cross section * (mb)
14.83	119.4±14.5	24.7±2.9
14.30	113.6±13.5	36.4±3.3
14.10	70.6±9.8	24.0±2.3
13.88	61.2±7.7	24.2±2.2
13.64	75.1±10.6	28.5±2.3

*Partial cross section i.e. cross section for the formation of only one ($T_{\frac{1}{2}}$: 14 min) of the isomeric states.

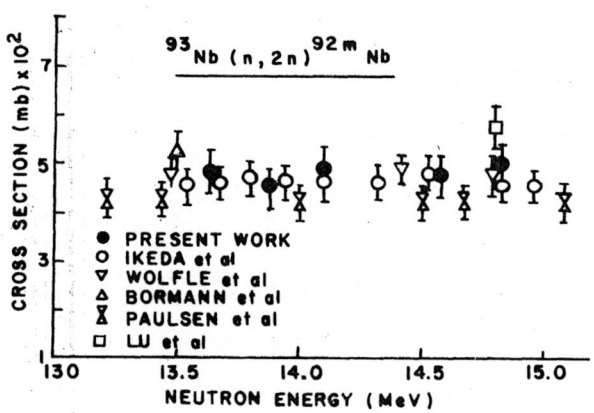

Fig.1 Excitation function of ^{93}Nb(n,2n)^{92m}Nb reaction.

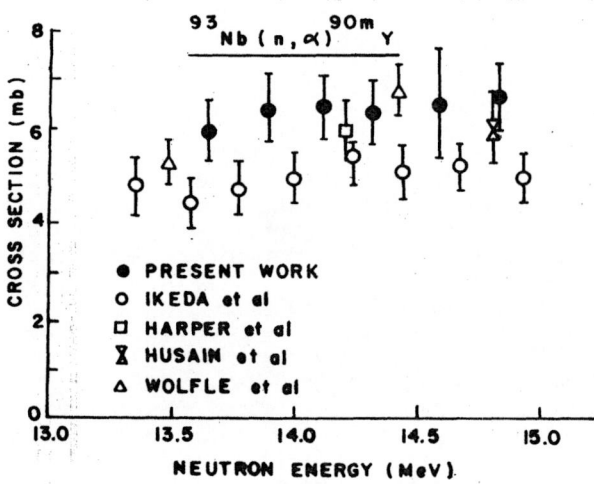

Fig.2 Excitation function of ^{93}Nb(n,α)^{90m}Y reaction.

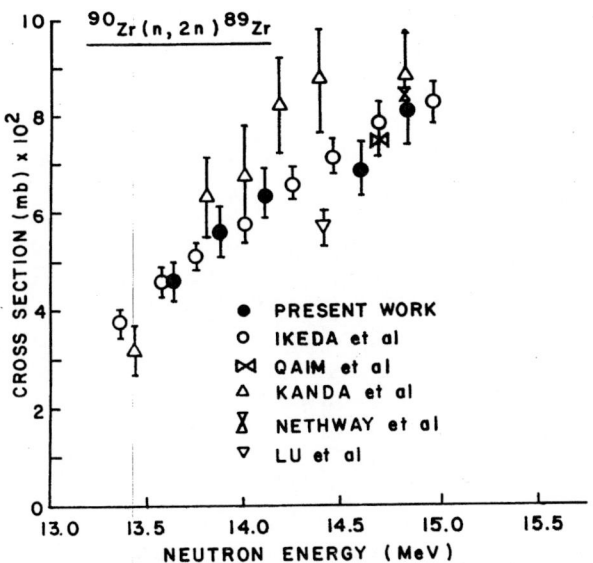

Fig.3 Excitation function of $^{90}Zr(n,2n)^{89}Zr$ reaction.

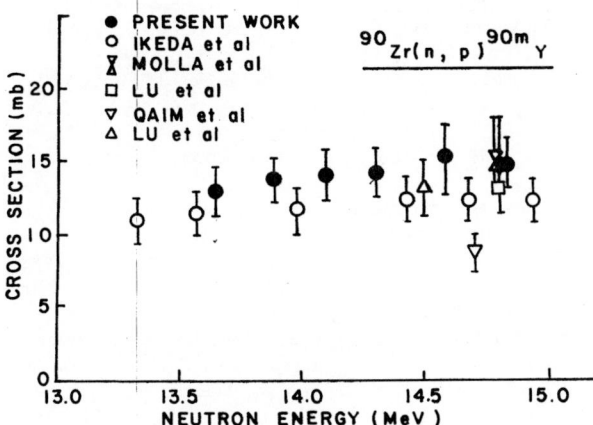

Fig.4 Excitation function of $^{90}Zr(n,p)^{90m}Y$ reaction.

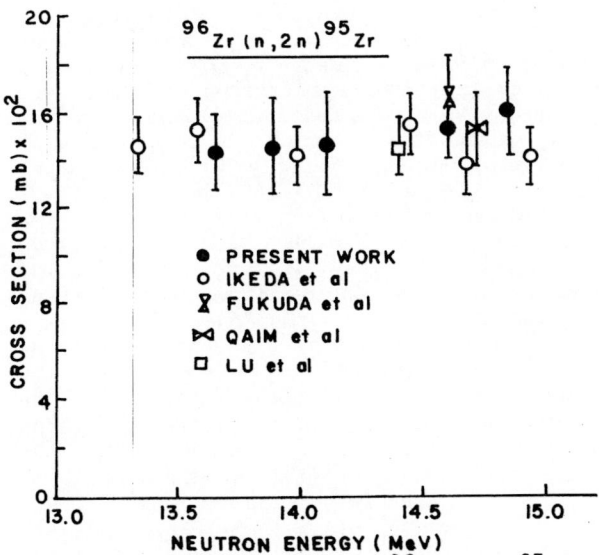

Fig.5 Excitation function of $^{96}Zr(n,2n)^{95}Zr$ reaction.

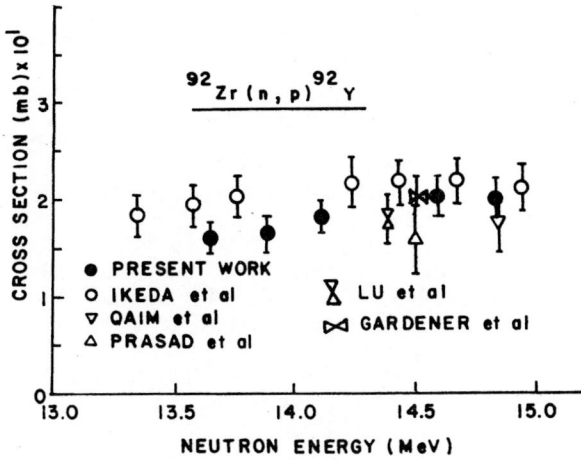

Fig.6 Excitation function of $^{92}Zr(n,p)^{92}Y$ reaction.

Acknowledgement: We thank Dr. M.A. Mannan, Dr. S. M.M.R. Chowdhury and Dr. F.R. Al-Siddique for their active support of the neutron nuclear data measurements programme in the Institute. The technical assistance of IAEA is gratefully acknowledged.

REFERENCES

1. S.M. Qaim: Proc.Int.Conf. on Nuclear Data for Science and Technology, Mito, Japan, May/June (1988)
2. J. Csikai: Handbook of fast neutron generators, CRC Press, Vol.1,3 (1987)
3. H. Vonach: Nuclear Data Standards for Nuclear Measurements, Technical Report Series 227, 59, IAEA (1983)
4. N.I. Molla, M.M. Rahman, S. Khatun, A.K.M. Fazlul Hoque, R.U. Miah and Aysha Akhter: INDC (BAN)-003 GI, INT (86)-8
5. Y. Ikeda, C. Konno, K. Oishi, T. Nakamura, H. Miyade, K. Kawade, H. Yamamoto and T. Katoh: Japan Atomic Energy Research Institute Report, JAERI-1312(1988)
6. S.M. Qaim and G. Stocklin: Proc. of the 8th Symp. on Fusion Technology, Noordijkerhout, the Netherland (1974), CEC, Luxemburg, EUR 5182e, 939 (1974)
7. R. Wolfle, A. Mannan, S.M. Qaim, H. Liskien and R. Widera: Int. J. Appl. Radiat. Isot. 39, 407(1988)
8. M. Bormann, H.H. Bissem, E. Magiera and R. Warnemunde: Nucl. Phys. A 157, 481 (1970)
9. A. Paulsen and R. Widera: Z. Phys. 238, 23(1970)
10. W. Lu, N. Ranakumar and R.W. Fink. Phys.Rev. C1, 358 (1970)
11. R.C. Harper and W.L. Alford: J.Phys. G8, 153 (1982)
12. L. Husain, A. Bari and P.K. Kuroda: Phys.Rev. C1, 1233(1970)
13. Y. Kanda   Nucl. Phys. A185, 177(1972)
14. D.R. Nethway: Nucl. Phys. A190, 635(1972)
15. W. Lu, N. Ranakumar and R.W. Fink: Phys. Rev. C1, 350(1970)
16. R. Prasad and D.C. Sarkar: IL Nuovo Cimento, 3A, No.3, 467(1971)
17. D.G. Gardener and S. Rossenblum: Nucl. Phys. A96, 121 (1967)
18. F. Fukuda, F. Matsuo, S. Shirahama and Ikumabe: Report INDC (JAP)-420(1978)

MEASUREMENT OF THE ^{60}Ni(n,p)^{60}Co CROSS SECTION IN THE NEUTRON ENERGY RANGE 6 - 13 MeV

M. Wagner and H. Vonach

Institut für Radiumforschung und Kernphysik der Universität Wien,
A - 1090 Vienna, Austria

R.C. Haight

Los Alamos National Laboratory, Physics Division, Los Alamos,
New Mexico, USA

Abstract: The cross section for the reaction ^{60}Ni(n,p)^{60}Co was measured by foil activation in the energy range 5.9 - 12.6 MeV relative to the ^{58}Ni(n,p)^{58}Co cross section, which in turn had been determined in a parallel experiment relative to the ^{238}U(n,f) cross section. Quasi-monoenergetic neutrons were produced via the H(t,n)^{3}He source reaction. The induced ^{58}Co and ^{60}Co activities were measured by standard γ-ray counting with a calibrated high-purity Ge γ-ray datector. The results for the ^{60}Ni(n,p)^{60}Co cross section are compared to the outcome of previous experiments and to recent evaluations.

(^{60}Ni(n,p)^{60}Co, activation cross section, E_n = 5.93 - 12.64 MeV)

Introduction

Until recently the measurement of cross sections in the 8 - 12 MeV neutron energy range has been difficult owing to the deficiencies of the most common source reactions such as D(d,n), T(p,n) and Li(p,n), which produce secondary neutrons from three-body break-up processes and provide only modest neutron fluxes. Thus cross sections for many nuclear reactions of technical interest, particularly those for the formation of long-lived nuclides, are rather inaccurate or even lacking in this energy region. In the case of the ^{60}Ni(n,p)^{60}Co reaction the authors previously had carried out activation cross section measurements in the neutron energy range 7.7 - 12.0 MeV [1] making use of the advantages of the H(t,n)^{3}He reaction. The results obtained for the ^{60}Ni(n,p)^{60}Co cross section considerably deviated from literature data [2] over the entire energy range investigated. In order to examine whether this trend extends to lower neutron energies, too, new measurements in a broader energy range were carried out.

Experimental and data analysis procedures

Irradiation and neutron fluence measurement

Approximately monoenergetic neutrons from ≈ 5.9 to 12.6 MeV were produced in the high-pressure (9.1 x 10^5 Pa) hydrogen gas cell of the Los Alamos FN Tandem Van-de-Graaff accelerator [3] by means of a triton (t$^+$) beam with incident energies from 11.5 to 19.9 MeV. Nickel samples of natural isotopic composition, 20 mm in diameter and 1.0 mm thick, with a purity of 99.98 % were positioned at zero degree relative to the incident triton beam at a distance of 4.0 cm from the end of the gas cell. At each energy two samples were irradiated with the gas cell filled with H$_2$ gas ("gas-in" irradiation runs) and evacuated ("gas-out" irradiation runs) in order to correct for parasitic neutrons produced in the gas cell structure. In addition, a ^{238}U fission chamber was placed behind the samples and the fission yield was measured both for the "gas-in" and the "gas-out" runs.

The neutron energy distributions averaged over the sample area were calculated by integration both over the length of the gas target and the area of the samples using the cross sections evaluated by M. Drosg et al. [4] for the source reaction. The widths of the energy distributions varied from ≈ 0.75 MeV FWHM at the lowest neutron energy (5.93 MeV) to 0.54 MeV at 12.64 MeV. No corrections for neutrons scattered by the target assembly were needed as discussed in ref. [1] due to the kinematic focusing of the neutrons emitted in the source reaction. Typical irradiation times were ≈ three hours resulting in a neutron fluence of ≈ 2.8 x 10^{12} to 1.0 x 10^{13} neutrons cm^{-2}.

The cross sections for the ^{60}Ni(n,p)^{60}Co reaction were measured relative to the ^{58}Ni(n,p)^{58}Co cross section, which had been determined in a parallel experiment relative to the ^{238}U(n,f) cross section (see ref. [5]). In the regions where the ^{58}Ni(n,p)^{58}Co cross section exhibits a nonlinear dependence on the energy, the respective neutron energy distributions were folded with the excitation function of the reference reaction using 20 keV wide steps.

The effect of secondary neutrons produced by (t,n) reactions and triton break-up particularly in the 0.5 mm thick Au target backing can be corrected in principle if the activities of both ^{58}Co and ^{60}Co produced in the "gas-out" irradiation runs are subtracted from the activities produced in the "gas-in" runs after normalization of the measured activities produced in both runs to the same integrated charge. In practice, this could not be done with sufficient accuracy because the ratio of the neutron production (as measured with the fission chamber) to the integrated triton charge showed large (up to 30 %) fluctuations depending on the focusing of the triton beam. Therefore we used the charge normalization only in the range of neutron energies up to 10 MeV where the "gas-out" fraction and its influence on the cross section is fairly small; for the neutron energies above 10 MeV we used the fission chamber results for the relative normalization of the reaction rates in the "gas-in" and "gas-out" irradiation runs. This was done by choosing the "gas-in"/"gas-out" ratio for the ^{238}U fission rate according to the condition that the normalization should reproduce the precision measurements carried out by Mannhart for the ^{93}Nb(n,2n)^{92m}Nb reaction [5]. As the ratios of the "gas-out"/"gas-in" reaction rates for ^{93}Nb(n,2n)^{92m}Nb and ^{238}U(n,f) are very different in the energy region 10 - 13 MeV, the cross-section ratio quite sensitively depends on the relative normalization of the "gas-out" and "gas-in" runs. Thus it was possible to derive the "gas-out"/"gas-in" ratio for ^{238}U(n,f) with an accuracy of about 10 % from the known

^{93}Nb(n,2n)^{92m}Nb to ^{238}U(n,f) cross section ratio. Accordingly, an uncertainty of ± 20 % in the neutron energy range from 5.93 to 10 MeV, ± 15 % between 10 and 11 MeV and ±10% above 11 MeV was assigned to the normalization factors for the reaction rates in the respective "gas-out"- and "gas-in" irradiation runs. A small correction was applied to account for the fact that the tritons hitting the Au beam-stop foil in the "gas-out" runs had a slightly higher energy than in the respective "gas-in" runs due to the energy loss in the H$_2$ gas.

Measurement of the induced ^{58}Co and ^{60}Co activities

The measurement of the γ-ray activities from the decay of the reaction products ^{58}Co and ^{60}Co was carried out at the Institut für Radiumforschung und Kernphysik, Vienna, Austria, by means of standard γ-ray counting with a calibrated intrinsic Ge γ-ray detector (for details see ref. 1). The measurements were performed in exactly the same way, the same efficiency calibration was used and all corrections discussed in ref. 1 were applied. The size of the corrections was almost the same as in our previous measurement [1], their uncertainties are given in table 2.

Results and discussion

The results of this measurement are listed in table 1 and shown in Fig. 1 together with cross sections measured in previous experiments [1 - 2, 6 - 9], the most recent evaluations of the ^{60}Ni(n,p)^{60}Co excitation function from the threshold up to 20 MeV [10 - 13] and the result of the evaluation of experimental cross sections in the 14 MeV range [14]. Table 2 states the principal sources of uncertainty, the resulting uncertainties and their estimated correlations within this work and with the former measurement [1]. The general trend found in ref. 1 is confirmed, the results obtained there agree with the new ones within the limits of the uncorrelated uncertainties except at neutron energies from $\approx$ 9 to 10 MeV and above 11.9 MeV. The ^{58}Ni(n,p)^{58}Co cross section measured in the 10 MeV region in ref. 1 is obviously too low indicating a hidden problem, so that the values given in this work appear to be more reliable. In the 12 MeV range there might have existed an unexplored difficulty in the normalization of the ratios of the gas-out"-to-"gas-in" reaction rates for ^{60}Ni(n,p)^{60}Co in our previous experiment. Yet, both the results given in ref.1 and obtained in this work confirm that in the 6 - 10 MeV energy range the cross sections measured by Paulsen [2] are definitely too high. The outcome of this work is generally consistent with the new evaluations EFF-2 [11] and ENDF/B-VI [12] but not with BROND [10] and JENDL-3 [13] up to an energy of $\approx$ 10.5 MeV, since these evaluations have obviously been fitted to the data published by Paulsen.

In order to check our data, we compared the integral ^{60}Ni(n,p)^{60}Co cross section measured in the ^{252}Cf fission neutron spectrum, $<\sigma>$ = 2.39 ± 0.13 mb [15], with the spectrum-averaged cross sections calculated from the excitation functions given in the above mentioned evaluations. For this purpose, the Cf neutron spectrum evaluated by Mannhart [16] was used. The results obtained by taking the excitation function as recommended in ENDF/B-VI ($<\sigma>$ = 2.510 ± 0.035 mb) and EFF-2 ($<\sigma>$ = 2.443 ± 0.034 mb) agree well

with the experimental $<\sigma>$ value, whereas BROND ($<\sigma>$ = 3.450 ± 0.048 mb) and JENDL-3 ($<\sigma>$ = 3.373 ± 0.047 mb) overestimate it considerably.

REFERENCES

[1] H. Vonach, M. Wagner and R.C. Haight, in Proc. Specialists' Meeting on Neutron Activation Cross Sections for Fission and Fusion Energy Applications, 13-15 Sept. 1989, Argonne, Illinois (Ed. by M. Wagner and H. Vonach), p.165, OECD, Paris (1990).

[2] A. Paulsen, Nukleonik 10, 91 (1967).

[3] R.C. Haight and J. Garibaldi, Nucl. Sci. Eng. 106, 296 (1990).

[4] M. Drosg, G. Haouat, W. Stoeffel and D.M. Drake, Report LA - 10444-MS, Los Alamos National Laboratory (1985).

[5] D.L. Smith, J.W. Meadows, H.Vonach, M. Wagner, R.C. Haight and W. Mannhart, Comparison of Activation Cross Section Measurements Performed with Different Neutron Source Reactions in the 5 - 13 MeV Range, Proc. of this Conference.

[6] B.M. Bahal and R. Pepelnik, Report GKSS-84-E, Germany (1984).

[7] L.R. Greenwood, Proc. 13th Int. Symp. on Influence of Radiation on Material Properties, Part II, ASTM STP 956 (ed. by F.A. Garner et al.) p.743, American Society for Testing and Materials, Philadelphia (1987).

[8] Y. Ikeda et al., Report JAERI-1312, Japan Atomic Energy Res. Inst., Tokai-mura (1988).

[9] Zhao Wenrong et al., Report CNDC - 89014, Institute of Atomic Energy, Beijing, People's Republic of China (1989).

[10] BROND: Y.N. Manokhin ed., Report INDC(CCP)-283, IAEA, Vienna (1988).

[11] EFF-2: M. Uhl et al., Proc. of this Conference.

[12] ENDF/B-VI: H.D. Lemmel, Report IAEA-NDS-100, Rev. 3, IAEA, Vienna (1990).

[13] JENDL-3: K. Shibata et al., Report JAERI 1319, Japan Atomic Energy Research Institute, Tokai-mura (1990).

[14] H. Vonach, A. Pavlik, S.Tagesen and M.Wagner, Proceedings of this Conference.

[15] W. Mannhart, in: Reactor Dosimetry, Proc.4th ASTM-EURATOM Symp., Gaithersburg, MD, 1982), Vol. 2, p. 637.

[16] W. Mannhart, in: Proc. AGM on Properties of Neutron Sources, Leningrad 9 - 13 June 1986, IAEA-TECDOC 410, p.158, and Documentation Series of the IAEA Nuclear Data Section, IAEA-NDS-98.

Table 1: Experimental results

E$_n$[MeV]$^{a)}$	ΔE_n [MeV]$^{b)}$	^{58}Ni(n,p)^{58}Co reference c.s.$^{c)}$	^{60}Ni(n,p)^{60}Co
5.931 ± 0.025	0.746	567.2 ± 11.8	20.3± 1.0
6.970 ± 0.029	0.694	641.6 ± 16.6	45.9± 1.8
7.944 ± 0.033	0.648	655.0 ± 20.8	71.3± 3.0
8.900 ± 0.036	0.620	644.1 ± 24.3	88.0± 4.4
9.954 ± 0.040	0.582	631.4 ± 28.2	112.7± 7.3
10.953 ± 0.043	0.574	619.5 ± 20.4	136.7± 8.3
11.858 ± 0.046	0.542	582.9 ± 16.6	153.5± 7.9
12.640 ± 0.048	0.540	534.6 ± 14.1	155.5± 8.3

a) centroid of the neutron energy distribution
b) half-width (FWHM) of the neutron energy distribution
c) taken from a recent measurement by the authors [5] relative to the ^{238}U(n,f) cross section; the ^{58}Ni(n,p)^{58}Co excitation function as obtained in [5] was folded with the respective calculated neutron energy distributions.

360

Table 2: Sources of uncertainty and resulting errors in the measured $^{60}Ni(n,p)^{60}Co$ cross sections (all values are given in percent)

	Neutron energy [MeV]								Estimated correlations to the values given in ref. 1 (in percent)
	5.93	6.97	7.94	8.90	9.95	10.95	11.86	12.64	
Random uncertainties[a]:									
γ-ray counting ^{58}Co statistics for the net count-rates	≤0.1 ————————————————					≤0.1	0.13	0.15	-
^{60}Co	1.54	1.07	1.07	0.84	0.83	1.06	1.02	1.09	
Subtraction of the activities induced by secondary (="gas-out") neutrons (net-effect)	2.42	1.29	1.57	2.67	4.20	4.63	3.84	4.20	-
Systematics uncertainties[b]:									
Efficiency of γ-ray ^{58}Co counting:	1.02 ———————————————————————————— 1.02								-
^{60}Co	1.02 ———————————————————————————— 1.02								100
γ-ray self-attenuation ^{58}Co in the sample:	0.44 ———————————————————————————— 0.44								-
^{60}Co	0.36 ———————————————————————————— 0.36								100
Correction for γ-ray summation (for ^{58}Co)	0.4 ———————————————————————————— 0.4								100
Reference cross-section [c]	2.08	2.58	3.18	3.77	4.46	3.30	2.84	2.64	≈ 20 [d]
Contribution of the neutron energy uncertainty due to the slope of the respective excitation functions [e]: ^{58}Co	0.35	<0.1 ——————————— <0.1					0.41	0.57	-
^{60}Co	3.00	1.58	1.03	0.86	0.90	0.70	0.36	0.0	-
Total	4.93	3.82	4.17	5.04	6.45	6.04	5.17	5.36	

The uncertainties for the half-lives and the survival of ^{58}Co and ^{60}Co, the emission probability of the 811 keV γ-rays from the ^{58}Co decay and the isotopic abundance of ^{58}Ni and ^{60}Ni were smaller than 0.1% and thus negligible.

a) uncorrelated uncertainties
b) 100% correlated uncertainties within this work unless otherwise specified
c) ≈ 30% correlated
d) correlation via the $^{238}U(n,f)$ reference cross section
e) ≈ 50% correlated via the uncertainty caused by a 0.35 cm displacement of the t^+-beam from the axis of the experimental set-up.

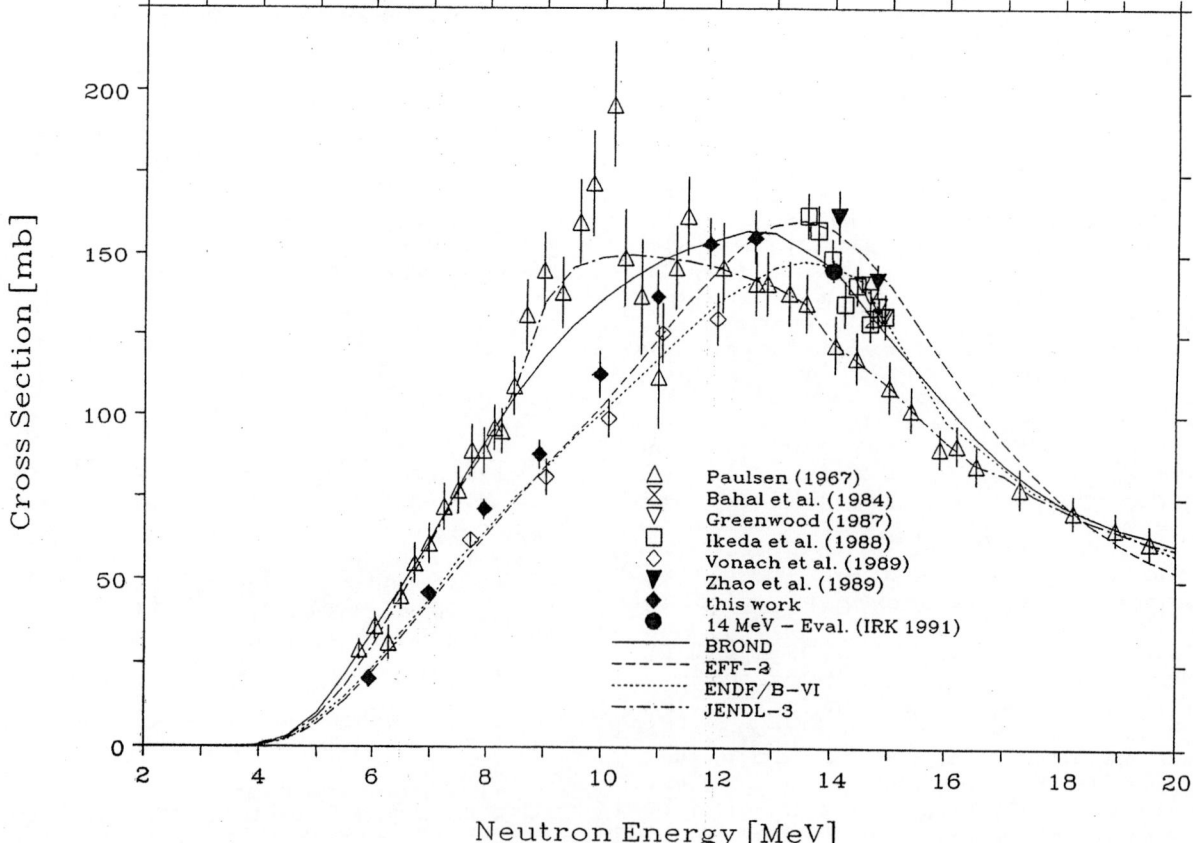

Fig. 1: Comparison of the measured $^{60}Ni(n,p)^{60}Co$ cross section with previous experiments and recent evaluations.

MEASUREMENT OF FORMATION CROSS SECTIONS OF
SHORT-LIVED NUCLEI BY 14 MEV NEUTRONS

K. Kawade, H. Yamamoto, T. Katoh, A. Taniguchi, T. Ikuta and Y. Kasugai

Department of Nuclear Engineering, Nagoya University,
Furo-cho, Chikusa-ku, Nagoya, 464-01 Japan

T. Iida and A. Takahashi

Department of Nuclear Engineering, Osaka University,
Yamadaoka, Suita-shi, Osaka, 565 Japan

Abstract: Activation cross sections for the reactions leading to short-lived nuclei with half-lives between 0.5 and 20 min were measured systematically at neutron energies of 13.4 to 14.9 MeV for various structural materials of fusion reactor by the activation method. Up to now, 53 cross sections have been obtained by using the intense 14 MeV neutron source facility of Osaka University (OKTAVIAN).

(Activation cross section, (n,2n), (n,p), (n,n'p), (n,t), (n,α), 13.4-14.9 MeV neutron, short-lived nuclei, activation method, structural material, fusion reactor)

Introduction

Neutron activation cross section data around 14 MeV have become important from the viewpoint of fusion reactor technology, especially for calculations on radiation damage, nuclear transmutation, induced activity and so on. Recently, cross sections were measured systematically and with good accuracy for 110 reactions on 26 elements by Ikeda et al.[1] and the quality of data base for activation and dosimetry files has been greatly improved. However, formation cross sections of short-lived nuclei have often not been measured with a reasonable accuracy, or there are no available data on some reactions, because of difficulty in measuring short-lived nuclei. This work has been done to provide more reliable activation cross sections leading to short-lived nuclei with half-lives between 0.5 and 20 min.

Experiments

The d-T neutrons were generated by the intense 14 MeV neutron source facility of Osaka University (OKTAVIAN). A pneumatic sample transport system as shown in Fig. 1 was used for the irradiation of samples. The T-target-to-sample distance was 15 cm and a typical neutron flux was about 5×10^7 n/cm$^2 \cdot$s at the irradiation position. The irradiation angles were 0°, 50°, 75°, 105°, 125° and 155°, which covered neutron energies ranging from 14.9 to 13.4 MeV. The neutron flux at the sample position was measured with use of the ^{27}Al(n,p)^{27}Mg (T$_{1/2}$=9.46min) reaction, whose cross sections were determined by referring to the standard ^{27}Al(n,α)^{24}Na reaction (ENDF/B-V). The samples were sandwiched between two aluminum foils of 10 mm x 10 mm x 0.2 mm^t. Induced radioactivities were determined by measuring γ-rays emitted from the irradiated sample and Al monitor foils with 12% and 16% HpGe detectors at 5 cm. Measured reactions and target isotopes are listed in Table.1. In deducing cross sections, corrections were made for fluctuation of the neutron flux, contribution of low energy neutrons below 10 MeV, sum effect of cascade γ-rays, thickness of sample, self-absorption of γ-ray in the sample material and interfering

reaction. Details of the experimental procedure are given elsewhere[2].

Table 1. Measured reactions and isotopes

(n,2n):	^{14}N, ^{31}P, ^{54}Fe^g, ^{63}Cu, ^{87}Rb^m, ^{90}Zr^m, ^{92}Mo^m ^{92}Mo^g, ^{113}In^{m,g}
(n,p) :	^{19}F, ^{25}Mg, ^{27}Al, ^{28}Si, ^{29}Si, ^{37}Cl, ^{50}Ti^{m+g} ^{51}V, ^{52}Cr, ^{53}Cr, ^{54}Cr, ^{60}Ni^m, ^{62}Ni^{m,g}, ^{66}Zn, ^{68}Zn^m, ^{86}Sr^m, ^{88}Sr, ^{94}Zr, ^{97}Mo^m, ^{107}Ag^m
(n,np):	^{26}Mg, ^{29}Si, ^{30}Si, ^{53}Cr, ^{54}Cr, ^{67}Zn, ^{87}Sr, ^{98}Mo^m
(n,t) :	^{32}S
(n,α):	^{26}Mg, ^{30}Si, ^{31}P, ^{54}Cr, ^{55}Mn, ^{64}Ni, ^{63}Cu^m, ^{65}Cu^m, ^{69}Ga, ^{71}Ga^m, ^{87}Rb^m, ^{89}Y^m, ^{92}Mo^m

* m and g mean the metastable and ground states of the reaction products, respectively.
* (n,np) means [(n,d)+(n,n'p)+(n,pn)].

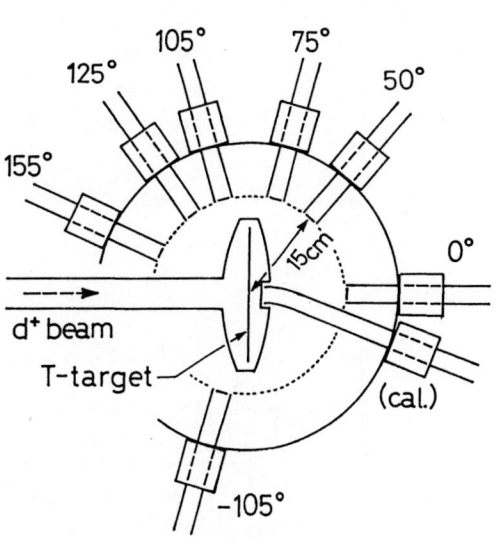

Fig. 1 Pneumatic sample transport system.

362

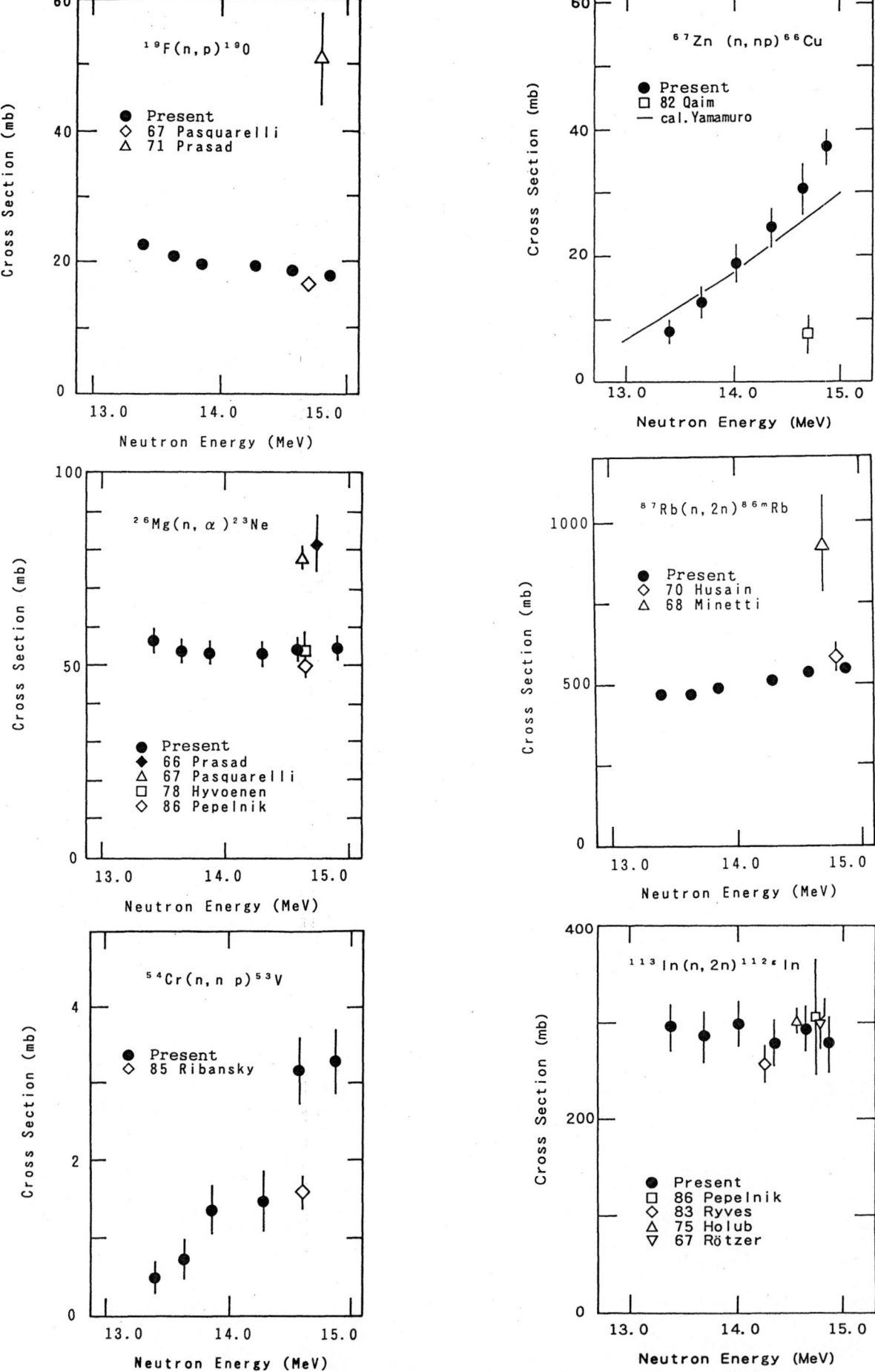

Fig. 2 Some of the cross sections of short-lived nuclei measured at OKTAVIAN.

Results and Discussion

Up to now 53 cross sections have been measured. Some of the results are shown in Fig.2. Accuracies of the obtained cross sections were around 3.5% in case of good statistics. Numerical data for each reaction are given in Ref. [2-4]. In Fig.3, systematic trends for (n,2n), (n,p), (n,n'p) and (n,α) reactions at 14.9 MeV are given as a function of the parameter (N-Z)/A, where N, Z and A are neutron, proton and mass numbers of the target nuclei[5-7]. No definite difference between data of present short-lived nuclei ($T_{1/2}$=0.5-20 min) and long-lived nuclei (longer than 1 h) is seen. Now the measurements are in progress to provide cross sections for more than 100 reactions.

The authors thank Prof. K. Sumita for his support to this work and Messrs. H. Sugimoto, M. Datemichi and S. Yoshida for the operation of the OKTAVIAN accelerator. We also thank Dr. T.Nakagawa, Dr. T. Asami and Dr. S. Igarashi of the JAERI NuclearData Center.

REFERENCES

1. Y. Ikeda, C. Konno, K. Oishi, T. Nakamura, H. Miyade, K. Kawade, H. Yamamoto and T. Katoh: JAERI 1312 (1988).
2. K. Kawade, H. Yamamoto, T. Yamada, T. Katoh, T. Iida and A. Takahashi: JAERI-M 90-171(1990).
3. T. Katoh, K. Kawade and H. Yamamoto: JAERI-M 89-083 (in Japanese).
4. K. Kawade, H. Yamamoto, T. Yamada, T. Katoh, T. Iida and A. Takahashi: to be published.
5. S. M. Qaim : Nucl. Phys. A382, 255(1982).
6. Y. Ikeda, C. Konno, T. Nakamura, K. Oishi, K. Kawade, H. Yamamoto and T. Katoh: Proc. Int. Conf. Nucl. Data for Sci. and Tech., Mito, ed. S. Igarashi, Saikon, Tokyo, 1988, p. 257.
7. O. Horibe and H. Chatani: Proc. Int. Conf. Nucl. Data for Sci. and Tech., Mito, ed. S. Igarashi, Saikon, Tokyo, 1988, p. 493.

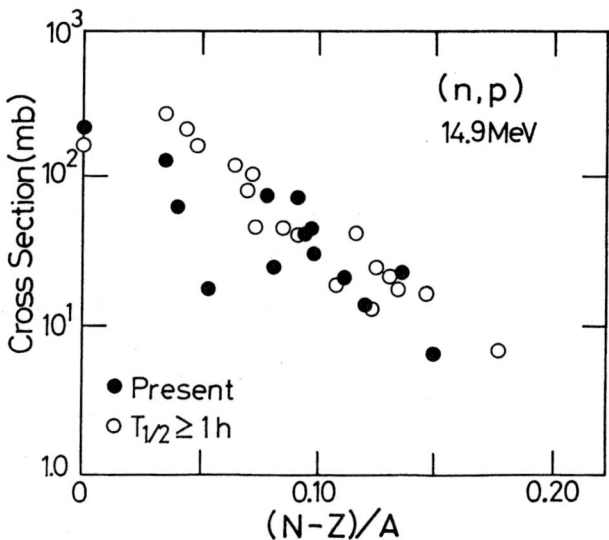

Fig. 3-2 Systematics of (n,p) reaction cross sections at 14.9 MeV. $\bigcirc$;$T_{1/2}$>1h, $\bullet$;$T_{1/2}$<20min.

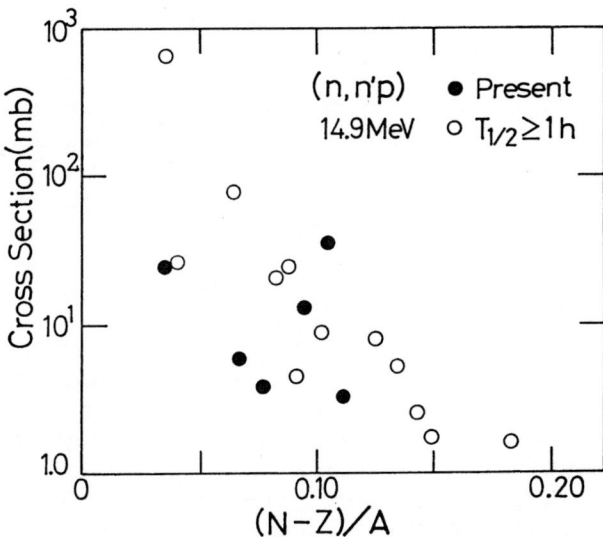

Fig. 3-3 Systematics of (n,n'p) reaction cross sections at 14.9 MeV. $\bigcirc$;$T_{1/2}$>1h, $\bullet$;$T_{1/2}$<20min.

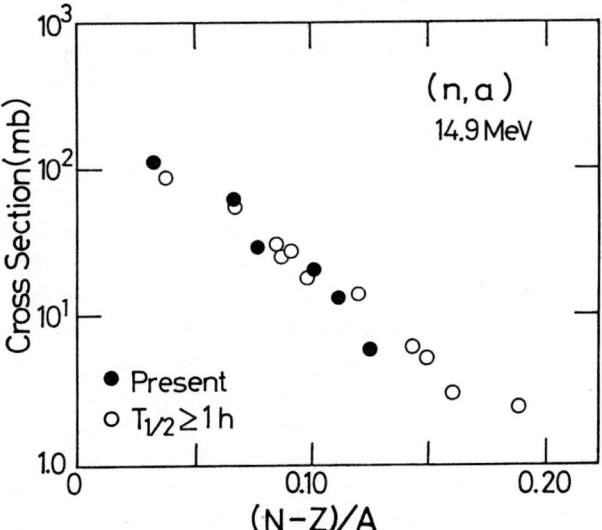

Fig. 3-4 Systematics of (n,α) reaction cross sections at 14.9 MeV. $\bigcirc$;$T_{1/2}$>1h, $\bullet$;$T_{1/2}$<20min.

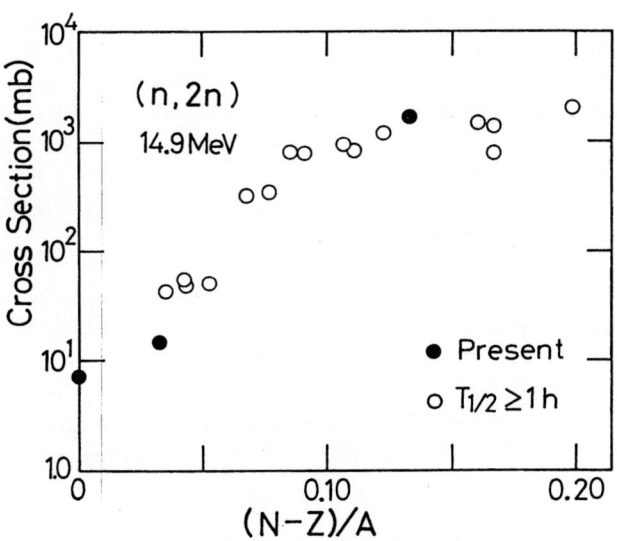

Fig. 3-1 Systematics of (n,2n) reaction cross sections at 14.9 MeV. Open circles show cross sections[1] with half-lives longer than 1 h . Closed circles show present results of short-lived nuclei.

MEASUREMENTS OF LONG-LIVED ACTIVATION CROSS SECTIONS BY 14 MEV NEUTRONS AT FNS

Y. Ikeda, A. Kumar* and C. Konno

Japan Atomic Energy Research Institute
Tokai-mura, Ibaraki-ken 319-11 Japan
* University of California, Los Angeles
Los Angeles, CA 90024-1597 U. S. A.

Abstract: Fourteen long-lived activation cross sections at 14.8 MeV have been measured using the FNS facility in order to provide experimental data meeting the requirement in the radioactive wastes disposal assessment in the D-T fusion reactor. The measurements were conducted under the JAERI/USDOE collaborative program on fusion neutronics. Measured data were compared with data available in the literature and discussed in terms of accuracy requirements.

(long-lived activation cross section, D-T fusion reactor, radioactive waste disposal, 14 MeV neutron, gamma-ray, Ge detector, fusion neutronics source(FNS))

Introduction

The production cross sections of long-lived radioactivi-ties in structural materials for 14 MeV neutrons are key of importance from the view point of waste disposal assess-ment in the D-T fusion reactor. An endeavor to measure long-lived activation cross sections initiated a few years ago as an IAEA-CRP[1] to respond to the nuclear data requirement, and significant efforts have been devoted to experimental measurements, theoretical predictions and evaluations. However, still there is need to measure more systematic data from the fusion reactor application point of view. In this context, an experiment to measure long-lived activation cross sections at 14 MeV has been conducted in the framework of the JAERI/USDOE collaborative program on fusion neutronics.[2] The major objective of this experiment was to verify systematically the induced radioactivity calculation codes and activation cross section data libraries currently available[3-5] in simulated D-T fusion neutron environments. This paper describes the experiments at FNS[6] in some detail and gives results for fourteen reaction cross sections at 14 MeV.

Experiment

Reactions, half-lives of products, natural abundance, γ-ray energies and their branching ratios used in the measurement are given in Table 1. Decay data were taken from the Table of Isotopes.[7] The long-lived radioactivities with half-lives from 5 to more than 10^6 years were studied in this experiment. And only the radioactivities associated with γ-ray emissions were chosen as the objects.

Samples of Al, ^{61}Ni, Cu, Nb, Mo, Ag, ^{151}Eu, ^{153}Eu, Tb, Dy, Hf, W, Re and Bi were irradiated with D-T neutrons for 4 days (8 hours irradiation per day), which were generated at FNS[6] by bombarding ^{3}T target with 350 keV deuterium (d^+) beam. It resulted in 1.7×10^{17} neutrons of total neutron yield at the target. Four sets of foil packages were placed at angles of $0°$, $45°$, $90°$ and $135°$ with respect to the incident d^+ beam, the distances of which were about 50 mm from the D-T source. This configuration was assumed to provide cross section data in the energy range from 13.5 to 15.0 MeV. Since the stacked sample package was considerably thick and significant neutron flux depression was expected in the sample, multiple thin Nb foils were inserted between samples for the neutron flux monitor. Neutron flux at each sample was determined with the ^{93}Nb(n,2n)^{92m}Nb reaction rate. The neutron fluence at the samples was estimated to be $0.8 - 2.0 \times 10^{15}$/cm^2.

More than 1.7 year after irradiation, γ-ray counting was started using a high detection efficiency Ge detector (115 % relative to 3" x 3" NaI(Tl)). Gamma-ray spectrum was analysed by a GENIE system provided by CANBERRA. The detector efficiency was calibrated by several standard γ-ray sources, the uncertainty of which was estimated to be about 3 % .

Results

The reaction rates of concerned radioisotopes were derived from γ-ray peak counts with necessary corrections, i. e. decay constant, cooling time, collection time, detector efficiency, natural abundance of the target material, γ-ray branching ratio, sample weight, self-absorption of γ-ray, neutron flux fluctuation during irradiation, and so forth.

The cross sections were obtained from the reaction rates divided by neutron flux determined by using a cross section value of 455 mb for ^{93}Nb(n,2n)^{92m}Nb reaction around 14 MeV. In Table 2, present data are summarized along with data available in the literature and IAEA-CRP[1].

^{27}Al(n,2n)^{26}Al

The present data were close to the data reported by Iwasaki[8] and slightly higher than data reported by Sasao[9].

^{61}Ni(n,np)$^{60m+g}$Co

Present measurement gave considerably larger cross section for this reaction than data in JENDL-3[10] and REAC-ECN3[11]. However, from the reaction systematics, our data seemed reasonable.

^{63}Cu(n,a)$^{60m+g}$Co

For the common reaction, many data have been reported. The present data were in a good agreement with the recent measurement by Greenwood[12].

^{93}Nb(n,n')^{93m}Nb

Due to the low threshold energy below 0.1 MeV, a precise evaluation of the reaction cross section has been required. The present measurement is in excellent agreement with a recent evaluation by Odano[13], but is slightly larger than the data by Ryves[14], which were so far the only available experimental data in the 14 MeV region.

^{94}Mo(n,p)$^{94m+g}$Nb

Table 1   List of Reactions Measured and Associated Decay Data of Products

	Reaction	Half-Life	Abundance	γ-ray Energy(keV)	γ-ray Branching	Final Spin State
1	^{27}Al(n,2n)^{26}Al	(7.16±0.32)×10^{5}y	100	1808.65	99.76±0.04	5$^+$
2	^{61}Ni(n,np)$^{60m+g}$Co	5.217 y	88.84[a]	1332.5	99.98	5$^+$(g),2$^+$(m)
3	^{63}Cu(n,α)$^{60m+g}$Co	5.217 y	69.1	1332.5	99.98	5$^+$(g),2$^+$(m)
4	^{93}Nb(n,n')^{93m}Nb	13.6±0.3y	100	30.4	(K/L+M+...0.12)	1/2$^-$
5	^{94}Mo(n,p)^{94}Nb	(2.03±0.16)×10^{5}y	9.3	871.10	100.0	6$^+$
6	^{109}Ag(n,2n)^{108m}Ag	127±7y	48.17	434.0	90.5	6$^+$
7	^{151}Eu(n,2n)^{150m}Eu	35.8±1.0y	97.7[a]	333.96	94±3	(4,5$^-$)
8	^{153}Eu(n,2n)$^{152m2+g}$Eu	13.2±0.3y	99.92[a]	344.3	27.2±0.4	8$^-$
9	^{159}Tb(n,2n)$^{158m+g}$Tb	150±30y	100	944.2	43±3	3$^-$
10	^{158}Dy(n,p)$^{158m+g}$Tb	150±30y	0.1	944.2	43±3	3$^-$
11	^{179}Hf(n,2n)^{178m2}Hf	31±1y	13.7	325.56	94.1±0.3	16$^+$
12	^{182}W(n,n'α)^{178m2}Hf	31±1y	26.3	325.56	94.1±0.3	16$^+$
13	^{187}Re(n,2n)^{186m}Re	2×10^{5}y	62.6	137.0 [b]	10.0	(8$^+$)
14	^{209}Bi(n,2n)^{208}Bi	(3.68±0.04)×10^{5}y	100	2614.47 [c]	100.0	(5$^+$)

a) Enriched ^{61}Ni (88.84%), ^{151}Eu(97.70%) and ^{153}Eu(99.92%)sample were used.

b) Natural background subtraction was needed.

c) Decay of ground state of ^{186}Re to ^{186}Os.

Although the present data were obtained from the activities produced in the natural molybdenum, subtracting the contribution of the ^{95}Mo(n,np)$^{94m+g}$Nb derived on the basis of the reaction systematics[15,16], they were in a good agreement with the data by Greenwood[17], which was the only cross section reported so far.

^{109}Ag(n,2n)^{108m}Ag

No experimental data for the cross section ^{109}Ag(n,2n)^{108m}Ag (T1/2=127y) was reported before IAEA-CRP report[1], data in which are tabulated in Table 2.

As long as the half-life of 127 ± 7 year was used, the present experiment gave cross section of 212 ± 11 mb. This value was very close to the other experimental values, whereas it was three times smaller than that of an evaluation by IRK. This large difference between experimental data and the evaluation should be attributed to the wrong half-life data for the deduction of experimental data.

^{151}Eu(n,2n)^{150m}Eu

Only two cross section data have been reported previously by Qaim[18] and Nethaway[19]. Along with data reported in IAEA-CRP[1], the cross section seemed somewhat convergent around 1.2 b. The present data were also in a good agreement within experimental error with those reported .

^{153}Eu(n,2n)$^{152m+g}$Eu

Since the isomeric state of 96 min de-excites with 100 % isomeric transitions to the ground state, the cross section was obtained for the ^{153}Eu(n,2n)$^{152m1+g}$Eu. Only one data was available for this production (T$_{1/2}$=13.2y) by Qaim[17] before IAEA-CRP report appeared. In the report of IAEA-CRP, we found the data of 1,544 mb at 14.77 MeV by Beijing and 1,740 mb at 14.7 MeV by KRI Leningrad. The IRK Vienna gave an evaluation of 1,442 mb at 14 - 15 MeV. These were rather scattered around 1,500 mb.

The present measurement gave cross section about 1.6 b being the middle value among the other data. However,the status of the cross section seemed a little bit controversial, because of, as mentioned, rather scattered data different with each other exceeding the experimental errors.

^{159}Tb(n,2n)$^{158m+g}$Tb

Terbium-158 has two states, one isomeric state deexciting 100 % isomeric transition with half-life of 10.5 sec and the ground state decaying with 150 year half-life. The isomer is so short-lived that the cross section is sum of ^{159}Tb(n,2n)$^{158m+g}$Tb.

Two experimental data have been reported by Qaim[18] and Prestwood[20] before IAEA-CRP report. In the IAEA-CRP report, one experimental data and one evaluation were given by IAE Beijing and IRK Vienna, respectively. Those data seemed very close each other to be around 1950 mb. The measurement at FNS through the inter-laboratory collaboration among ANL, LANL and JAERI gave a cross section of 1,600 mb at 14.8 MeV. The value was considerably lower than those previously reported even though the experimental errors were taken into account.

A cross section of 1775 mb was obtained from the present measurement, which located in the middle of the data mentioned above.

^{158}Dy(n,p)$^{158m+g}$Tb

For the first time, an experimental data was given by the present measurement. However, the error was so large due to poor statistics in γ-ray counting that present data provided only the lower and upper limit in this reaction cross section. One data available to be referred was found in the REAC-ECN3[11], which gave the cross section of 10 mb from the reaction systematics. This value seemed too low in comparison with the range of cross section value measured.

^{179}Hf(n,2n)^{178m2}Hf

No data has been reported so far before IAEA-CRP summary. In the inter-laboratory collaboration work, the cross section of 6.6 ± 0.3 mb was given. Harwell[1] reported an experimental data of 5.9 ± 0.6 mb at 14.8 MeV. One calculation has been given by Oxford/LANL to be 2.9 mb at 14 MeV.

Present measurement showed the value of 6.3 ± 0.6 mb at 14.8 MeV, which was very close to those of the Inter-laboratory collaboration work and Harwell[1].

^{182}W(n,n'a)^{178m2}Hf

The present measurement gave the cross section of 14 ± 8 mb for the first time. The present data was lower by one order of magnitude than that in REAC-ECN3[11].

^{187}Re(n,2n)^{186m}Re

For the first time, the present measurement gave an experimental cross section at 14.8 MeV. Due to poor counting statistics, experimental error was quite large. The data showed a range of upper and lower limit as in the case of ^{158}Dy(n,p) reaction cross section. However, the evaluations in IRK[1] and REAC-ECN3[11] were much larger than the present data. Further confirmation of the experimental data should be

Table 2 Cross Sections Measured for Long-Lived Activation Products

Reaction	Present Work		References		
	E_n(MeV)	Cross Section (mb)	E_n(MeV)	Cross Section(mb)	
^{27}Al(n,2n)^{26}Al	14.8	47 ± 5	14.7	40.6 ±4.4	8)
	14.5	30 ± 7	14.8	35 ±7	9)
	14.1	19 ±10	14.5	16.2	11)
			14.1	19.0	10)
^{61}Ni(n,np)$^{60m+g}$Co	14.8	50.5 ± 8.2	14.5	18	11)
			14.1	17.4	10)
^{63}Cu(n,α)$^{60m+g}$Co	14.8	40.4 ± 2.3	14.8	40.1±1.2	12)
	14.5	43.8 ± 2.5			
^{93}Nb(n,n')^{93m}Nb	14.8	43 ± 9	14.1	36 ± 4	14)
	14.5	44 ± 9			
^{94}Mo(n,p)^{94}Nb	14.8	58 ± 17	14.8	53.1±5.3	16)
	14.5	48 ± 20			
	14.1	44 ± 18			
^{109}Ag(n,2n)^{108m}Ag	14.8	212 ± 11	14.77	230±7	*-1)
			14.56	263±20	*-2)
			14 - 15	665±73	*-3)
			14.9	191±7	*-4)
			14.9	208±37	*-5)
^{151}Eu(n,2n)^{150m}Eu	14.8	1276 ± 64	14.77	1219±28	*-1)
	14.5	1170 ± 59	14 - 15	1325±94	*-3)
	14.1	1215 ± 53	14.8	1127±55	*-4)
			14.9	1090±84	*-5)
			14.7	1270±149	17)
			14.8	1180±150	18)
^{153}Eu(n,2n)$^{152m+g}$Eu	14.8	1659 ± 83	14.8	1712±155	*-4)
	14.5	1533 ± 77	14.77	1544±42	*-1)
	14.1	1326 ± 75	14 - 15	1442±60	*-3
			14.9	1740±145	*-5)
			14.7	1542±138	17)
^{159}Tb(n,2n)$^{158m+g}$Tb	14.8	1775 ± 88	14.77	1968±56	*-1)
			14.8	1600+-88	*-4)
			14 - 15	1930±49	*-3)
			14.7	1801	17)
			14.8	1930	19)
^{158}Dy(n,p)$^{158m+g}$Tb	14.8	100 ± 80	14.5	10.7	11)
^{179}Hf(n,2n)^{178m2}Hf	14.8	6.3 ± 0.6	14.8	5.9±0.6	*-6)
			14	2.9	*-7)
			14.9	6.0+-0.3	*-4)
^{182}W(n,n'α)^{178m2}Hf	14.8	(1.4±0.8)x10^{-2}	14.5	0.176	11)
^{187}Re(n,2n)^{186m}Re	14.8	135 ± 65	14.1	693	10)
			14.5	605±258	11)
			14-15	591±122	*-3)
^{209}Bi(n,2n)^{208}Bi	14.8	2450 ± 260	14.1	2176	10)
			14.5	2340	11)

* Measurement, evaluation or calculation performed for CRP:
  *-1); IAE Beijing   *-5); KRI Leningrad
  *-2); Debrecen   *-6); Harwell
  *-3); IRK, Vienna   *-7); Oxford/LANL
  *-4); ANL/LANL/JAERI

required.

^{209}Bi(n,2n)^{208}Bi

Although the error was large, the present measurement gave a data reasonably close to those in REAC-ECN3[11] within the error.

Acknowledgements

Authors would like to express their sincere thank to the members of the FNS facility for their operation of accelerator. The counting of low energy photon of 93mNb was performed by Dr. Kawade of Nagoya University. The U.S activity was supported by USDOE.

REFERENCES

1. INDC(NDS)-232/L, "Activation Cross Sections for the Generation of Long-Lived Radionuclides of Importance in Fusion Reactor Technology", Proc. of an IAEA Consultants' Meeting, Argonne National Laboratory, 11th - 12th. Sept 1989.

2. T. Nakamura, M. A. Abdou, "Overview of JAERI/USDOE Collaborative Program on Fusion Blanket Neutronics Experiments", Tokyo, Japan 1988, Fusion Eng. Design, 9, 303 (1989).

3. Y. Ikeda, A. Kumar, C. Konno, T. Nakamura and M. A. Abdou, "Experiment on Induced Activities and Decay-Heat in Simulated D-T Neutron Fields:JAERI/USDOE Collaborative Program on Fusion Neutronics", Proc. of 9th Topical Meeting on Technology of Fusion Energy, Oct. 7-11, 1990, Chicago, U.S.A.

4. A. Kumar, Y. Ikeda, M. A. Abdou and T. Nakamura,"Analysis of Induced Activities Measurements Related to Decayheat in Phase-IIC Experimental Assembly: JAERI/USDOE Collaborative Program," ibid.

5. A. Kumar, Y. Ikeda and C. Konno,"Experiment and Analysis for Measurements of Decay Heat Related Induced Activities in Simulated Line Source Driven D-T Neutron Fields of Phase IIIA: JAERI/USDOE Collaborative Program on Fusion Neutronics", ibid.

6. T. Nakamura, H. Maekawa, J. Kusano, Y. Oyama, Y. Ikeda, C. Kutsukake, Shi. Tanaka and Shu. Tanaka," Present Status of the Fusion Neutron Source(FNS)," Proc. 4th Symp. on Accelerator Sci. Technol., RIKEN, Saitama, 24 - 26 November 1982, pp 155-156.

7. C. M. Lederer and V. S. Shirley;"Table of Isotopes," 7th edition, John Wiley &Sons, Inc., New York (1978).

8. S. Iwasaki, J. R. Dumais and K. Sugiyama, Proc. Int'l. Conf. on Nuclear Data for Sci. and Technol., May 30-June 3, 1988, Japan. (Editor S. Igarashi) Saikon Publishing Co. LTD., Tokyo (1988) pp 295-297.

9. M. Sasao, T. Hayashi, K. Taniguchi, A. Takahashi and T. Iida, IPPJ-805, Inst. Plasma Phys., Nagoya Univ.(1987), and Phys. Rev. C 35, 2327 (1987).

10. K. Shibata, T. Nakagawa, T. Asami, T. Fukahori, T. Narita, S. Chiba, M. Mizumoto, A. Hasegawa, Y. Kikuchi, Y. Nakajima and S. Igarashi, "Japanese Evaluated Nuclear Data Library, Version-3 -JENDL-3-", JAERI-1319 (1990).

11. H. Gruppelaar, H. A. J. Van Der Kamp, J. Kopecky and D. Nierop, "The REAC-ECN-3 Data Library with Neutron Activation and Transmutation Cross Sections for Use in Fusion Reactor Technology," ECN-207 (1988).

12. L. R. Greenwood, "Recent Research in Neutron Dosimetry and Damage Analysis for Materials Irradiations", Proc. Symp. on the Effects of Radiation on Materials, June 23-25, 1986 Seattle, Washington, U.S.A.

13. N. Odano, S. Iwasaki and K. Sugiyama, "Evaluation of Cross Sections for the Dosimetry Reactions of Niobium," Presented in this conference. Paper No. C32

14. T. B. Ryves and T. Kolkowski, J. Phys. G: Nucl. Phys. 7, 52 (1981).

15. S. M. Qaim, Nucl. Phys. A382, 255 (1982).

16. Y. Ikeda, C. Konno, K. Oishi, T. Nakamura, H. Miyade, K. Kawade, H. Yamamoto and T. Katoh, "Activation Cross Section Measurements for Fusion Reactor Structural Materials at Neutron Energy from 13.3 to 15.0 MeV Using FNS Facility," JAERI-1312 (1988).

17. L. R. Greenwood, D. G. Doran and H. L. Heinisch, Phys. Rev. C 35, 76 (1987).

18. S. M. Qaim, Nucl. Phys. A 224, 319(1974).

19. D. R. Nethaway, Nucl. Phys. A 190, 635(1972).

20. R. J. Prestwood, D. B. Curtis, D. J. Rokop, D. R. Nethaway and N. L. Smith, Phys. Rev. C 30, 823(1984).

CAPTURE CROSS SECTIONS OF TERBIUM, THULIUM, HAFNIUM AND TUNGSTEN

Xu Haishan, Xiang Zhengyu, Mu Yunshan

Li Yexiang and Wang Shiming

Institute of Nuclear Science and Technology Sichuan University

P. O. Box 390—1 Chengdu, 610064, P. R. of China

and

Liu Jianfeng

Zhengzhou University, Physics Department

Zhengzhou, 450002, P. R of China

Abstract: Radiative capture cross sections of terbium, thulium, hafnium and tungsten were measured relative to that of gold over the neutron energy range of 0.3—1.6 MeV, using a large liquid scintillation counter and the time—of—flight technique.

The capture cross sections were calculated from 0.3—2.0 MeV for four nuclides by the Hauser—Feshbach statistical theory with width fluctuation correction.

(Capture cross section, terbium, thulium, hafnium tungsten, large liguid scintillation, time—of—flight)

1. Introduction

With the progress of nuclear energy development neutron data, including (n. γ) cross sections of fission products, are getting very important to satisfy needs of reactor design, safe operation of reactors, etc. In the neutron energy range 0.3 to 2.0MeV, there are six available experimental data [1—6] for rare earth nuclei. A comparison of such data at a few energy points shows obvious differences. The calculations of neutron capture cross sections for natural thulium were carried out by three authors [7—9]. But we couldn't find any calculations for the other three nuclides in the publications.

2. Experiments

2A. Experimenta method A detaile descriptio of the experimental method and data evaluation was given in ref. 10 and 11. The schematic drawing of the experimental setup and the simplified block diagram of the electronics used in the measurement are also given in ref 12.

The 2.5—MV Van de Graaff accelerator provided 10 ns proton pulses. Neutrons were produced by the T[p, n]³ He reaction. A large liquid scintillator, 100cm in diameter, served for the detection of prompt capture gamma ray. The neutron capture cross sections of natural Tb, Tm, Hf and W were measured relative to that of gold. The gold sample weighed 118.70g with a purity higher than 99.9%. The natural Hf and W samples are metal, other samples are oxide. Their purity is better than 99.9%.

2B. Experimental results: The standard neutron capture cross sections of gold were taken from ref. 13.

The measured neutron capture cross sections of natural Tb.

Tm. Hf and W are given in Fig. 1—4.

The corrections for multiple neutron scattering in the samples were calculated by the Monte Carlo method.

The resulting total error for the measurements is 10 to 12%. The following uncertainties are included: 5 to 8% uncertainty in extrapolation to zero pulse height, 7% uncertainty in standard cross sections, 2% uncertainty in the correction of multiple scattering. 2% uncertainty in the detection efficiency, 1 to 2% in the background subtraction, and 2 to 4% statistical error.

3. Calculations

3A. Theoretical Model: The capture cross sections were calculated from 0.3 to 2.0 MeV for natural terbium, thulium, hafnium and tungsten by the optical model and the Hauser—Feshbach statistical theory with width fluctuation correction. The nonstatistical effects such as potential capture and radiative capture in elastic and inelastic channels of a compound nucleus were included in the calculations. The calculated results show that the nonstatistical contribution to the capture cross sections is negligible compared with that of the statistical effects. According to the statistical theory [14], the capture cross section is given by

$$\sigma n, \gamma = \frac{\pi}{k^2} \sum_{l, J\pi} \frac{2J+1}{2(2I+1)} \frac{T_{alj} T_\gamma^{J\pi}}{T^{J\pi}} W_{alj}^{J\pi}$$

According to Axel's model [16], the photo—absorption cross section $\sigma_a(E_r)$ can be deduced from the giant dipole resonance shape of a statically deformed nucleus:

$$\sigma_a(E_r) = \sum_{i=1}^{2} \sigma_{gi} E_r^2 \Gamma_{gi}^2 / [(E_{gi}^2 + E_r^2)^2 + E_r^2 \Gamma_{gi}^2]$$

3B. **Model Parameters**: Becchetti—Greenleas [17] optical potential was used to calculate the neutron transmission coefficients and the parameters were determined according to the agreement of the calculated neutron total cross sections, elastic and inelastic scattering cross sections with experimental values. Since these cross sections calculated by using the universal parameters already better agree with experimental values before the continuous inelastic channels are open, the parameters were not changed. When the continuous inelastic channels are open, these cross sections were fitted to experimental values by adjusting the energy level density paramenters of the residual nucleus. Gilbert — Cameron s formula [18] was used to calculate the energy level density and the parameters were roughly adjusted first according to the known low—lying levels then, together with the giant dipole resonance parameter σ_g, were adjusted carefully. according to the agreements of the calculated (n,γ) cross section with the experimental value. The adjusted level density parameters of the compound nuclei and the dipole resonance parameters for the calculation of the absorption cross sections.

Table 1. Level Density and Giant Dipole Resonance Parameters

	E_x (MeV)	T (MeV)	E_0 (MeV)	U_0 (MeV)	a (MeV^{-1})	σ_{g1} (b)	σ_{g2} (b)	Γ_{g1} (MeV)	Γ_{g2} (MeV)	E_{g1} (MeV)	E_{g2} (MeV)
^{159}Tb	4.0	0.465	−0.73	0.73	20.54	0.2150	0.2330	2.80	5.30	12.23	15.96
^{160}Tb	3.44	0.480	−1.755	0.00	21.58	0.196	0.248	2.90	5.10	12.07	15.88
^{169}Tm	3.6	0.47	−0.98	0.61	21.96	0.242	0.259	2.76	4.7	12.12	15.50
^{170}Tm	3.38	0.4955	−1.41	0.00	21.49	0.242	0.259	2.76	4.7	12.12	15.50
^{176}Hf	4.54	0.474	−0.052	1.19	21.91						
^{177}Hf	3.30	0.430	−0.150	0.64	21.51	0.211	0.334	2.290	5.180	12.59	14.88
^{178}Hf	3.90	0.485	−0.160	1.04	19.24	0.211	0.334	2.290	5.180	12.59	14.88
^{179}Hf	3.80	0.495	−0.64	0.64	19.26	0.211	0.334	2.290	5.180	12.59	14.88
^{180}Hf	4.4	0.485	−0.11	1.37	19.80	0.211	0.334	2.290	5.180	12.59	14.88
^{181}Hf	3.97	0.490	−0.54	0.64	20.66	0.211	0.334	2.290	5.180	12.59	14.88
^{180}W	4.5	0.4733	−0.1053	1.12	21.88						
^{181}W	3.5	0.445	−0.22	0.72	21.70	0.211	0.334	2.290	5.180	12.59	14.88
^{182}W	4.3	0.46	0.44	1.45	21.30						
^{183}W	4.3	0.505	−0.64	0.72	20.01	0.211	0.334	2.290	5.180	12.59	14.88
^{184}W	4.6	0.495	0.50	1.30	20.45	0.192	0.304	2.290	5.180	12.59	14.88
^{185}W	3.4	0.45	−0.75	0.72	20.46	0.211	0.334	2.290	5.180	12.59	14.88
^{186}W	4.886	0.4614	−0.199	1.58	22.18						
^{187}W	4.022	0.482	−1.106	0.72	19.32	0.211	0.334	2.290	5.180	12.59	14.88

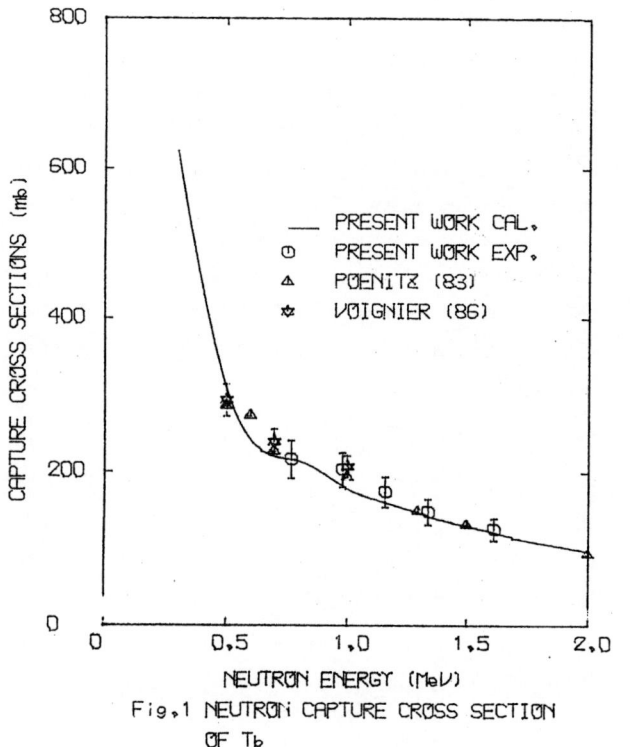

Fig.1 NEUTRON CAPTURE CROSS SECTION OF Tb

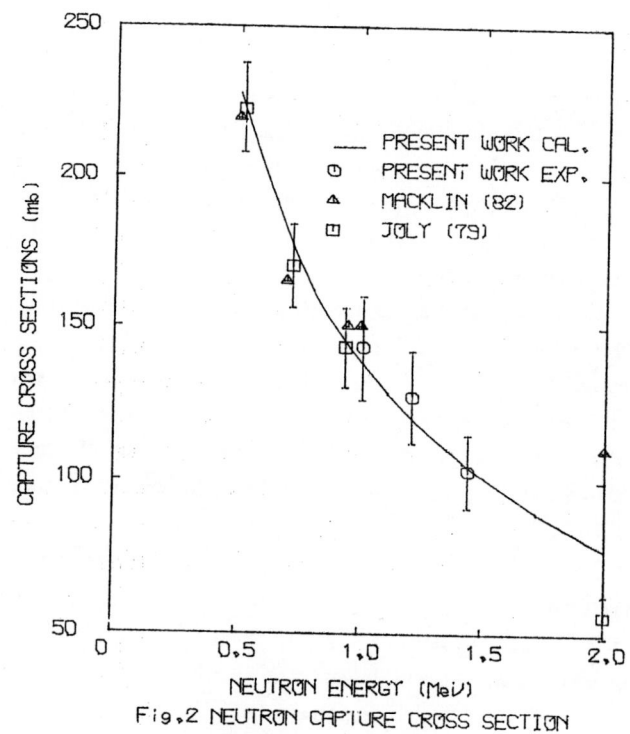

Fig.2 NEUTRON CAPTURE CROSS SECTION OF Tm

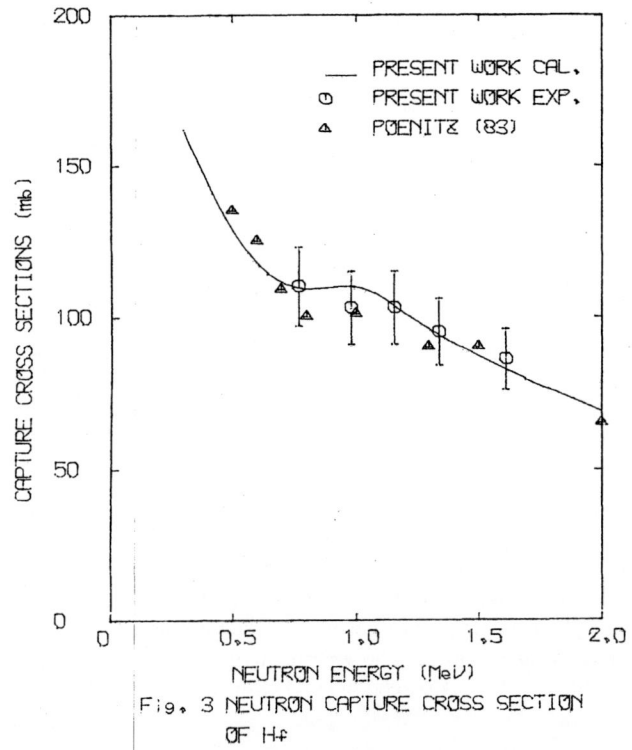

Fig. 3 NEUTRON CAPTURE CROSS SECTION OF Hf

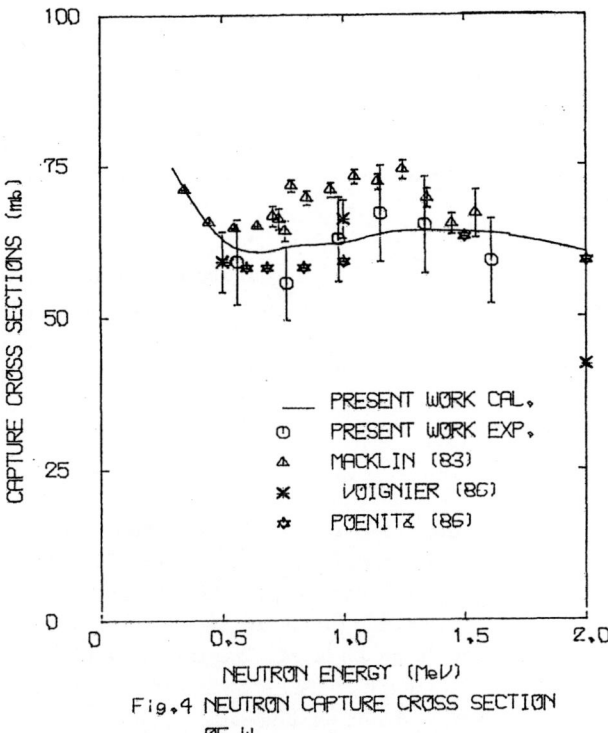

Fig.4 NEUTRON CAPTURE CROSS SECTION OF W

3C. Calculated Results The calculated capture cross sections of terbium, thulium, hafnium and tungsten are given in Table 2.

Table 2. Calculated Capture Cross Sections (mb)

E_n (MeV)	0.3	0.5	0.8	1.0	1.2	1.4	1.6	2.0
T_b	623.5	310.1	212.5	176.9	153.9	136.6	120.8	95.2
T_m	406.2	228.1	162.9	138.5	120.8	106.9	95.7	77.2
H_f	161.8	128.6	108.8	109.8	100.9	91.0	82.7	68.5
W	74.6	62.6	61.7	62.3	63.9	63.9	63.5	60.3

Discussion

A comparison between the present results of capture cross sections with other works is shown in Fig. 1—4, respectively.

T_b: the present experimental results are consistent with the present calculated values, and are also in good agreement with the experimental data of Poenitz [1], but lower than the results of Voignier [2].

T_m: The present experimental results are in good agreement with the experimental data of Joly [3] but higher than the results of Jiang [5]. Our experimental results in the lower than 1MeV energy range are in agreement with the resultsof Macklin [4], but for energy higher than 1MeV our experimental data are lower than his data.

H_f: Our experimental results are in good agreement With the expermental data of Poenitz [1].

W: The present experimental data are about 10% lower than the experimental data of Macklin [6], but are in good agreement with the data of Poenitz [1] and Vioginier [2].

References

1. W. P. Poenitz, P. T. Guenther and A. B. Smith, ANL—83—4, 239(1983)
2. J. Voignier, S. Joly and G. Grenier, Nucl. Sci. Eng. ,93, 43(1986)
3. S. Joly and G. Grenier, Nucl. Sci. Eng. , 70, 53 (1979)
4. R. L. Macklin, Nucl. Sci. Eng. ,82, 143(1982)
5. S. S. Jiang, Chin. Nucl. Phys. ,24, 136(1982)
6. R. L. Macklin, Nucl. Sci. Eng. ,84, 98(1983)
7. H. Qi, Chin. Nucl. Phys. ,3, 1, 1(1981)
8. P. G. Young, 82 Antwerp 792(1982)
9. D. G. Madland, LA—9841, 38(1983)
10. Xu Haishan, Xian Zhengyu, Mu Yunshan Che Yaoshuand Liu Jinrong, Chin. J. Nucl. Tech. , 9,5 (1986)
11. Xu Haishan, Xian Zhengyu Mu Yunshan Che Yaoshun Liu Jinrong and Li Yexiang, Chin. J. Nucl. Phys. ,9, 127(1987)
12. Yunshan Mu, Haisha Xu, Zhengy Xiang, Yexian Li, Shimin Wang and Jianfeng Liu, Nucl. Sci. Eng. , (1991) to be published.
13. ENDF/ B—V
14. E. Vogt, Advances in Nucl. Phys. ,1, 161(1978)
15. Z. D. Su, Chin. J. High. En. Phys. and Nucl. Phys. ,3,80(1979)
16. P. Axel, Phys. Rev. ,126, 671(1962)
17. F. D. Becchetti and G. W. Greenlees, Phy. Rev. , 182, 1190(1969)
18. A. Gilbert and A. Cameron, Can. J. Phys. , 43, 1446(1965)

MEASUREMENT AND ANALYSIS OF ^{98}Mo(n,γ)^{99}Mo REACTION CROSS SECTION

Wang Chunhao, Xia Yijun, Long Xianguan, He Fuqing
Yang Jingfu, Yang Zhihua, Peng Xiufeng, Liu Mantian
Luo Xiaobing

Institute of Nuclear Science and Technology, Sichuan University
Chengdu 610064 P. R. of China

Lu Hanlin

Institute of Atomic Energy, P. O. Box 275(3), Beijing

Abstract: Neutron capture cross sections of ^{98}Mo have been measured relative to that of ^{197}Au for neutron energy from 29 to 1100 KeV, using the activation method. Neutrons were generated via the ^{7}Li(p,n)^{7}Be and T(p,n)^{3}He reactions with a 2. 5 MV Van de Graaff accelerator at Sichuan University. Neutron energy selection was achieved by choice of angle of neutron emission with respect to the incident proton beam. The activities after irradiation were measured with a calibrated high resolution HPGe detector. The accuracy of the measurement is 7. 1 to 8. 7%. The experimental results were compared with existing data.

(Nuclear reaction ^{98}Mo(n,γ)^{99}Mo, $E_n = 29 - 1100$ KeV, activation method, neutron capture cross sections, radioactive waste)

Introduction

Neutron capture cross sections are important data for nuclear reaction theory, astrophysics and design of nuclear reactors. Along with the development of fusion research, more capture cross section data for the generation of long — lived radionuclides are needed for estimation of radioactive waste and selection of materials leading to low — level activities. ^{98}Mo(n,γ)^{99}Mo reaction is one of the important reactions. Therefore a detailed investigation of the ^{98}Mo(n,γ)^{99}Mo reaction is quite useful.

The experimental method was the activation technique, it is completely selective for a given nuclide in a natural sample, and the experiments are relatively simple to carry out. However, severe uncertainties may arise, all steps must be performed with extreme care. Our measurements are carried out with a ^{197}Au foil as a standard which is activated simultaneously. In this way most systematic uncertainties can be avoided.

The irradiations were carried out by neutrons produced via the ^{7}Li(p,n)^{7}Be and T(p,n)^{3}He reactions at 2. 5 MV Van de Graaff accelerator. The induced γ — ray activity of the samples and the corresponding gold reference foils were measured with a 100 cm^3 HPGe detector.

Experiment

The neutron activation technique consists of two steps: neutron irradiation of a suitable sample and absolute counting of the induced activity. The proton beam collimated to a diameter of $<$8 mm was incident on the target producing neutron. The target cooling was achieved by compressed air. The sample of Mo was a natural metal disk of 0. 8 mm thickness. The diameter of Mo and Au samples were 20 mm. The purities were 99. 9% for Mo and 99. 999% for Au. The samples were sandwiched between two gold foils with thickness 0. 1 mm and were wrapped in 0. 3 mm thickness cadmium foils. A group of samples mounted on the surface of an Al ring (88 mm diameter) centered at the neutron source were irradiated simultaneously at 0°, 30°, 60°, 90° and 120° degrees with respect to the incident proton beam. Therefore neutrons between 29 and 230 KeV were produced by the ^{7}Li(p,n)^{7}Be reaction, and between 215 and 1100 KeV by the T(p,n)^{3}He reaction. Two irradiation runs for each proton energy were performed. The proton beam currents were generally 8 to 14 μA and the duration of irradiation was about 20 to 70 hours in each run.

The accuracy of activation measurements depends critically on the mass of the irradiated samples. In principle the samples should be as thin as possible because then scattering and self — absorption effects are small during the irradiation and during the activity determination afterwards as well. To minimize neutron scattering of the target pipe, wall of pipes should be as thin as possible.

Since the measured molybdenum cross sections are normalized to that of gold in this experiment, we required only a knowledge of the relative flux at the site of the sample. The

Table 1. Sources of errors

Source of error	Error(%)
Au standard cross section	5. 0—7. 0
Error in decay scheme	4. 0
HPGe detector efficiency	≤1. 5
Statistics	1. 0
Irradiation history	<2. 0
Correction for neutron scattering and attenuation	1. 0
Background	1. 0
Gamma self absorption in the sample	0. 5—1. 0
Weighing error of sample	0. 5
Total error	7. 1—8. 7

neutron flux was monitored with a long counter at 0° at a distance of 1. 9 m from the source. In order to record the neutron flux as a function of time during the irradiation, the integral count rate of the long counter per 2 minutes (or 5 minutes) was recored continuously by nicrocomputer multiscaler and stored on magnetic disk for calculating the correction of nonuniform irradiation history.

The last step of the experiment is the absolute determination of the induced activity in the investigated sample and in the gold reference sample. The most accurate way for this measurement is high resolution gammaray spectroscopy with a 100 cm³ HPGe detector in a low background environment. The absolute efficiency was calibrated by ^{241}Am, ^{109}Cd, ^{57}Co, ^{54}Mn, ^{65}Zr, ^{152}Eu, ^{60}Co and ^{22}Na sources at different source—to—detector distances. The coincidence sum effect at the short source to detector distance was considered for ^{152}Eu, ^{60}Co and ^{22}Na. This method guarantees clean background corrections and allows us to sort out the studied nucleus via a characteristic gamma ray line.

Analysis and Result

Because of ^{99}Mo beta decay to the isomeric state of ^{99}Tc ($T_{1/2}=6. 02$h), the activation cross section for ^{98}Mo(n,γ) ^{99}Mo reaction can be determined either by measuring the number of ^{99}Mo or ^{99m}Tc produced. The decay of the isomeric state of ^{99}Tc can easily be detected via the 140 KeV gamma transition in ^{99}Tc. Therefore, ^{98}Mo(n,γ)^{99}Mo reaction cross section was obtained by measuring the number of ^{99m}Tc produced.

Spurious activations were produced by room scattered neutrons, In all irradiations, a similar sample group was placed at a distance of 2 metres to measure the background of the experimental room. To minimize spurious neutron activations each sample in every irradiation was shielded with cadmium. The other source of neutrons of spurious energies arises as a result of scattering (including elastic, inelastic and multiple scattering) from the beam pipe, target materials, sample holder and sample itself, since there was no simple

way for measuring such effects or experimentally distinguishing from the activation by the primary source neutrons. These corrections and neutron flux attenuation due to target and sample were estimated using a Monte—Carlo program.

The main uncertainties came from the Au standard cross section, the decay scheme, absolute activity determination, statistics of the activity measurement, various corrections etc. The main errors are listed in Table 1.

The capture cross sections of ^{98}Mo obtained with the values of the ^{197}Au(n,γ) reaction recommended by ENDF/B-v [1] are listed in Table 2 and plotted in Fig. 1 together with experimental error. For comparison, the neutron capture cross sections of other authors are also plotted in Fig. 1.

Table 2. Neutron capture cross sections of ^{98}Mo

Neutron energy (KeV)	Cross section (mb)
29± 8. 9	74. 2±6. 4
59±20. 6	43. 6±3. 8
121±32. 9	33. 8±2. 9
196±28. 1	29. 8±2. 6
215±43. 1	30. 6±2. 6
230±11. 7	29. 4±2. 5
376±100. 6	28. 2±2. 4
655±144. 0	30. 1±2. 1
962±121. 0	20. 2±1. 5
1100±68. 2	16. 3±1. 2

Discussion

It can be seen from Fig. 1 that our cross section curves are similar in shape with the results of Stupegia et al [2]and Trofimov and Nemilov [3]. Our results are in good agreement with measurements reported by Lyon and Macklin [4] and Anand et al[5]. In the energy range of 300—1100KeV,

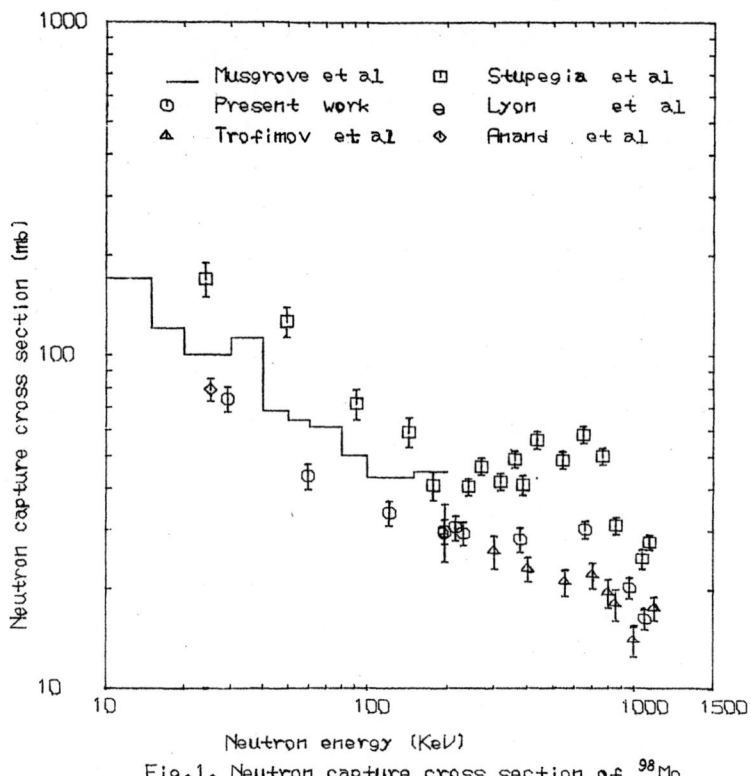

Fig.1. Neutron capture cross section of ^{98}Mo

The present resutlts are agreement with the data of Trofimov and Nemilov [3] within the experimental error. However, our result are lower than the values of Musgrov et al [6]in the energy range from 10 to 200 KeV. The discrepancy between Musgrove's and our results obviously exceeds the given experimental error. In the work reported by Stupegia et al[2] neutron capture cross sections of ^{98}Mo were measured relative to the ^{235}U fission cross sections as reported by White [7]. Since the result of White are systematically higher than the ENDF/B – V evaluation, Stupegia's results are systematically higher than the existing data. For comparison, the results of Stupegia et al [2]have been normalized to the ^{235}U fission cross sections recommended by ENDF/B—V.

The authors thank the International Atomic Energy Agency and Chinese Nuclear Data Center for encouragement and financial support.

REFERENCES

1. "Evaluated Neutron Data File For ^{197}Au. ENDF/B—V", compiled by R. Kinsey (1979)

2. D. C. Stupegia, Marcia Schmidt, C. R. Keedy and A. A. Madson: Journal of Nuclear Energy 22, 267(1968)

3. U. N. Trofimov and U. A. Nemilov: YK 3/57, 15 (1984)

4. W. S. Lyon and R. L. Macklin: Phys. Rev. 114, 1619 (1959)

5. R. P. Anand, M. L. Jhingan, D. Bhattacharya and E. Kondaiah: Nuovo Cimento A50, 247(1979)

6. A. R. DEL. Musgrove, B. J. Allen and J. W. Boldeman, in Proc. Int. Conf. on Neutron Physics and Nuclear Data for Reactors and other Applied purposes, AERE Harwell, Sept 1978, OECD, Paris(1979) p. 449

7. P. White: J. Nucl. Energy (Parts A/B Reactor Sci. Technol) 19, 325(1965)

MEASUREMENTS OF ACTIVATION CROSS SECTIONS OF IRON, NICKEL, COBALT, AND ZIRCONIUM ISOTOPES AT NEUTRON ENERGIES 13.6 - 14.9 MeV

M.V.Blinov, A.A.Filatenkov, S.V.Chuvaev, V.M.Saidgareev

V.G.Khlopin Radium Institute,
Shvernik Av. 28, Leningrad, 194021, USSR

Abstract: Measurements of a number of neutron activation cross sections are reported, which were performed under identical conditions. The main attention was paid to production of undistorted neutron field. For this purpose the scattering masses of the target chamber and sample holders were minimized. The ^{93}Nb(n, 2n)^{92m}Nb reaction was used as a standard for neutron fluence determination.

(Activation cross sections, 14 MeV neutrons, niobium, iron, nickel cobalt, zirconium)

Introduction

In the recent years an increased interest was renewed in nuclear data obtained at neutron energy about 14 MeV. These data are important for different nuclear physics applications, especially for fusion reactor technology, neutron dosymetry, neutron activation analysis and for development of nuclear models. At present, there is a bulk of experimental data on nuclear reaction cross sections in this neutron energy region. However, it is quite a typical situation, where the data scattering exceeds remarkable the errors cited by authors that often excluded a possibility to obtain an evaluation of high accuracy. To a large extent, this situation is, probably, due to the fragmentary character of many experiments.

One of the most promising way to improve this state appear to be a systematic measurements of a large number of various reaction cross sections in the identical and well known conditions. From this point of view, the studies conducted in Japan /1/ are very useful. In a series of measurements started by us, above 100 cross sections will be determined also. The conditions of our experiment differ from those of /1/ mainly by the characteristics of the used neutron source. It is of significantly less intensity, but it has a better monochromaty and more accurately correspond to the approach of an infinite small source.

Experiment

The measurements were carried out at neutron generator NG-400. Neutrons of the ^{3}H(d, n)^{4}He-reaction were used. The ion beam diameter was about 5 mm. The thickness of the titan-thritium target backing was 0.3 mm, the target chamber has the side walls of 1.2 mm in thick and was cooled by the air jet.

In order to reveal the role of scattered neutrons, and additional measurements were carried out, in which the water cooling of the target was used, and besides, the polyethilene "head" with walls of 10 mm was put on the target chamber. In a such manner, the cross sections of the ^{197}Au(n, γ)^{198}Au- and ^{115}In(n, n')^{115m}In-reactions were deduced, which have a high sensitivity to the small contamination of neutrons with $E_n < 1$ MeV and $E_n = (1-6)$ MeV, respectively. In Fig. 1 they are compared with the corresponding cross sections determined in the normal conditions.

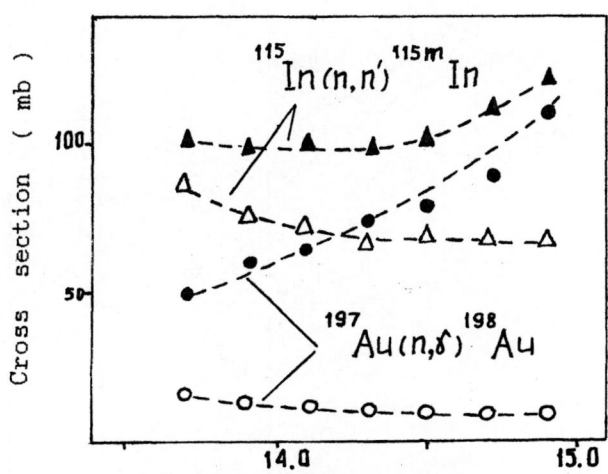

Neutron energy (MeV)

Fig. 1. Cross sections of the ^{115}In(n, n')^{115m}In and ^{197}Au(n, γ)^{198}Au reactions. $\triangle\,O$ - measured in normal conditions; $\blacktriangle\,\bullet$ - measured with scatterer.

374

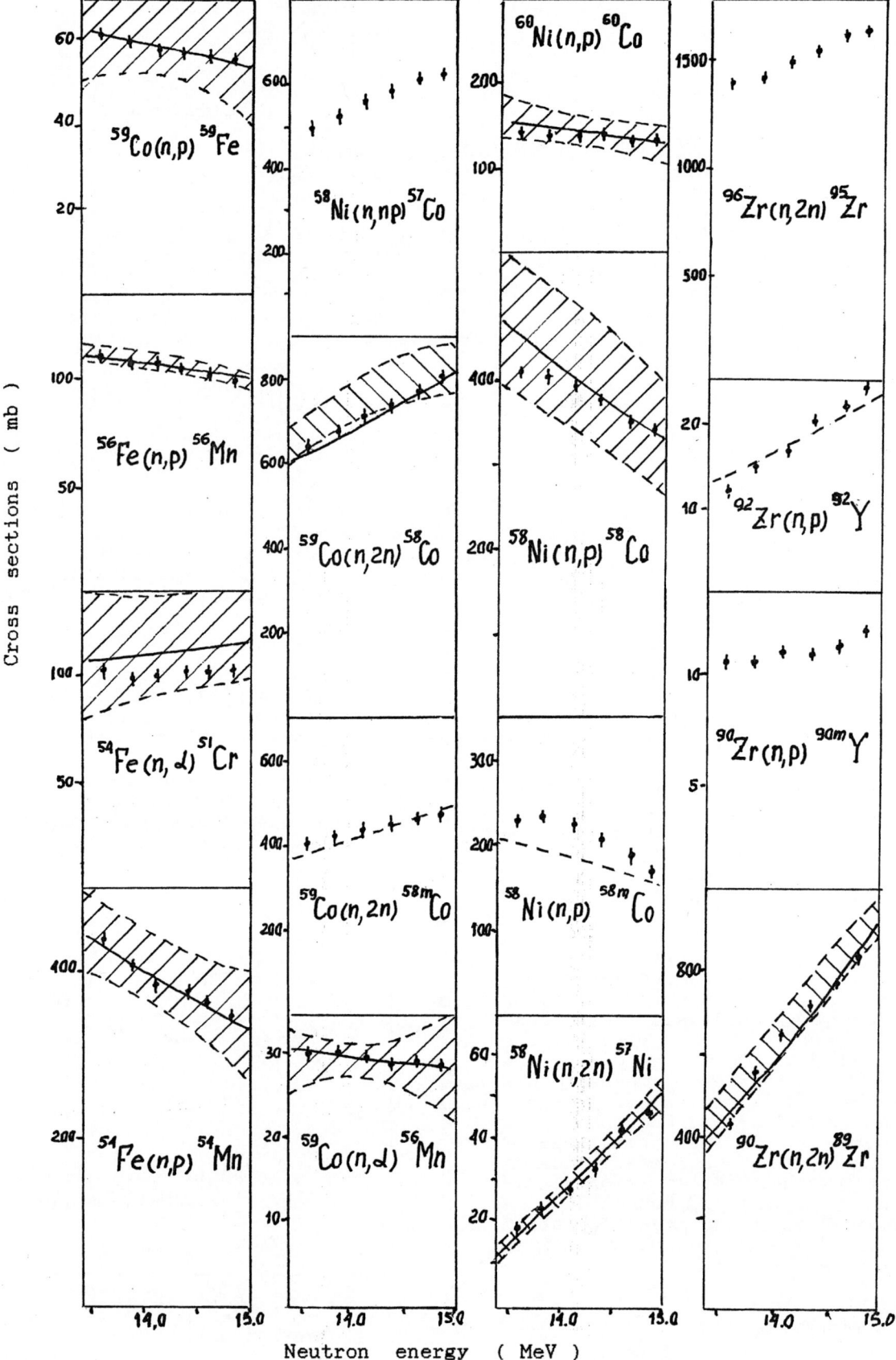

Fig. 2.   Cross   sections   of   the   studied   reactions.
⬤ - this work;
── - evaluation /5/.   Error   limits   of
evaluation   are   shown.

In all cases, the irradiated samples were placed in seven thin-wall holders arranged at the angles of 0 - 140 degrees (relative to the beam direction). The sample thickness was 0.08 - 0.4 mm, the total thickness of the sample assembly did not exceed 2 mm; the distant from the target center to the assembly center was 4 cm; the first and the last foils in each assembly were foils of niobium used for neutron fluence determination. The cross section of this reaction was shown to change no more than +1% in the neutron energy range 14.1 - 14.8 MeV /2/ and the uncertainty of the evalluation of this cross section at 14.7 MeV is 1.6% /3/. The neutron flux variations during irradiation were detected with a scintillator counter placed at the angle of 95 degrees.

The angular and energy distribution of neutrons was calculated for the real geometry using the recommended data on the $^3H(d, n)^4He$ reaction in c.m.s. /4/.

The induced gamma-ray activity was measured with a Ge(Li)-detector. Its energy resolution was 2.7 keV and the peak efficiency was 7.9% at 1332 keV gamma-ray energy. The spectrometer was provided with various programs for spectra treatment.

Results

Some cross sections measured in neutron energy region of 13.6 - 14.9 MeV are presented in Fig. 2. The evaluated data are taken from Ref. 5. As a rule, evaluation uncertainties are much more than experimental errors, that underlines an enormous situation which must be improved.

The results reported here were obtained on the first stage of measurements, in which an especial attention has been paid to the exact parametrization of neutron field of the standard D-T - source. The authors consider that the cross sections of above 100 reactions will be measured in these strictly determined conditions.

REFERENCES

1. Y.Ikeda, Ch.Konno, T.Nakamura, K.Oishi, K.Kawade, H.Yamamoto, and T.Katoh, in Proc.Int.Conf. on Nuclear Data for Science and Technology, May/ June 1988, Mito, Japan (Edited by S.Igarasi), p.257, Saikon, Tokyo (1988).
2. H.Liskien, and V.E.Levis: Nuclear Standard Reference Data.-IAEA-TECDOC-335, p. 324 (1985).
3. T.B.Ryves: Ibid. - p. 431.
4. M.Drosg and O.Schwerer: Handbook on Nucler Activation Data, Technical reports ser.273, 83, IAEA, Vienna (1987)
5. V.N.Manokhin, A.B.Pashchenko, V.I.Plyaskin, V.M.Bychkov, V.G.Pronyaev: Ibid. - p.305.

14.6 MeV NEUTRON INDUCED REACTION CROSS-SECTION MEASUREMENTS

A.Ercan, M.N.Erduran, M.Subaşı, E.Gültekin,
G.Tarcan, A.Baykal, M.Bostan

Çekmece Nuclear Research and Training Centre
Physics Department
P.K.1 , 34831 Havaalani, Istanbul-Türkiye

Abstract : Neutron induced reaction cross-sections at 14.6 MeV were measured for Na, Mg, Al, K, P, Si, Cr, Mn, Fe, Co, Ni, Cu and Zn. High purity chemicals in powder form were irradiated at SAMES T-400 low energy ion accelerator with neutron yield of about 10^{10}/sec.

Introduction

Accurate and precise cross-section data for fast neutron induced reactions are required in the field of biology, agriculture, fusion reactor technology and also needed for testing the nuclear models. The discrepancies in the available data arise mainly from the incorrect determination of the effective neutron flux or the mean neutron energy. Prior to this work, the mean neutron energy from ^{3}H(d,n)^{4}He reaction in the irradiation position was carefully determined as 14.6(1) MeV using the Zr/Nb(n,2n) reaction cross-section ratio method[1]. Terminal voltage of the neutron generator was 200 kV and 500 μA beam current was applied. The neutron yield of 10^{10}/sec from a 0.4 Ci/cm^2 TiT target provides an average neutron flux of 10^9 n/sec.cm^2 at a distance of 1 cm. To monitor the neutron flux during the measurements, a BF$_3$ long counter was calibrated using the well-known ^{27}Al(n,α)^{24}Na reaction cross-section (114 mbarn at 14.6 MeV) [2].

Experimental

The samples in powder form were irradiated in polyethylene containers (13.2 mm diameter, 8 mm height) placed 11 mm away from TiT target and transferred from/to the irradiation position by a pneumatic system; the transfer time to place the irradiated samples to the counting position was about 15 seconds. The mean neutron flux within the sample was determined by measuring the activity of Al foils which were irradiated simultaneously within the sample carriers.

In order to determine the effective neutron flux, the BF$_3$ long counter was employed and flux was sampled with 1 second dwell time. The FEP efficiency of the gamma spectrometer consisting of a 82cc HPGe detector and 8K MCA was determined with 2-3% overall accuracy in the energy range from 80 keV to 2 MeV [3]. Detector-sample distance was chosen 6 cm, so that coincidence summing effects were negligible. The gamma spectra were taken with good statistic and corrected for the dead-time .

Data Reduction and Results

The cross-sections are calculated from the well-known activation formula:

$$\sigma = \frac{P}{G.C.(N/A).h.\Phi.z.f.\Omega.\epsilon}$$

where
P: Full energy peak area
G: Weight of the sample(g)
C: Concentration of the element
N: Avogadro number(6.022 x 10^{23}/mol)
A: Atomic weight of the element(g/mol)[4]
h: Isotopic abundance [4]
f: Relative gamma-ray intensity [5]
Ω: Counting solid-angle (photon absorption and corrections due to the inhomogeneous activity distribution in the sample are taken into account).
ϵ: intrinsic efficiency of the gamma detector [3]

The effective flux Φ depends on the irradiation time t_i and on the decay constant λ of the residual nucleus:

---

1 This work supported by the IAEA under the Research Contract TUR-5670/RB

$$\Phi = \frac{\int_0^{t_i} \Phi(t).e^{\lambda.t}.dt}{\int_0^{t_i} e^{\lambda.t}.dt}$$

If the neutron flux is sampled with Δt dwell time, the numerical evaluation can be carried out exactly ;

$$\Phi = <\Phi(t)> \frac{(1-e^{-\lambda.\Delta t}).n.\sum_{k=1}^n M_k.e^{\lambda.k.\Delta t}}{(e^{\lambda.t_i}-1).\sum_{k=1}^n M_k}$$

here M_k = const. Φ_k denotes the monitor counts in the k-th time interval (t_i=n. Δt) and $<\Phi(t)>$ is the time averaged flux during t_i, obtained from the Al internal flux monitor.

$$z = (1-e^{-\lambda.t_i}).e^{-\lambda.t_c}.(1-e^{-\lambda.t_m})/\lambda$$

where

t_c: Cooling time
t_m: Measuring time.

The results are given in Table 1 with the compiled values from Pashchenco [6] for the reaction cross-sections induced by 14.5 MeV neutrons.

References

[1] M.N.Erduran, M.Subaşı, M.Bostan, S.Dökmen, H.Atasoy, C.Özbayli and A.Ercan,ÇNAEM Report AR-252,(1988)

[2] M.Subaşı, M.N.Erduran, G.Tarcan, Y.Ozbir, A.Baykal, E.Gültekin and A.Ercan,ÇNAEM Report AR-249,(1988)

[3] A.Ercan, M.Bostan, E.Gültekin and M.N.Erduran, II. Balkan Conference on Neutron Activation Analysis, Bled - Yugoslavia(1989)

[4] J. K. Tuli, Nuclear Wallet Cards, BNL (1990)

[5] Table of radioactive isotopes,1986 Authors E.Browne and R.B.Firestone Wiley-Interscience(Ed.V.S.Shirly)

[6] A.B.Pashchenco,INDC(CCP)-323(1991), Reaction Cross-sections induced by 14.5 MeV and by Cf-252 and U-235 Fission Spectrum Neutrons.

Table 1. 14.6 MeV Neutron Induced Reaction Cross-sections measured at Çekmece Nuclear Research and Training Centre (ÇNAEM) and the compiled values given by Pashchenco[6] at 14.5 MeV

Reaction	Cross-section (mbarn) ÇNAEM	Ref.[6]
59Co(n,p)59Fe	48(4)	47(10)
59Co(n,α)56Mn	33(3)	33(2)
58Ni(n,p)58mCo	218(60)	175(20)
58Ni(n,p)58Co	159(21)	310(25)
58Ni$(n,2n)$57Ni	46(2)	34(2)
60Ni(n,p)60mCo	114(8)	95(10)
61Ni(n,p)61Co	144(10)	95(10)
62Ni(n,p)62mCo	18(2)	17(3)
62Ni(n,p)62Co	24(4)	37(5)
64Ni(n,α)61Fe	6(1)	5(1)
63Cu(n,α)60mCo	12(5)	–
63Cu$(n,2n)$62Cu	316(53)	551(30)
65Cu(n,p)65Ni	14(5)	20(3)
65Cu(n,α)62mCo	4.9(4)	2(1)
65Cu$(n,2n)$64Cu	971(200)	968(20)
64Zn(n,p)64Cu	140(40)	176(20)
64Zn$(n,2n)$63Zn	167(10)	178(15)
66Zn(n,p)66Cu	69(10)	66(5)
66Zn$(n,2n)$65Zn	786(28)	690(60)
67Zn(n,p)67Cu	68(7)	45(5)
68Zn(n,p)68mCu	5.0(5)	5(1)
68Zn(n,α)65Ni	11.0(7)	12(2)
70Zn(n,p)70mCu	5.7(19)	8.7
70Zn$(n,2n)$69mZn	698(48)	448(29)

Reaction	Cross-section (mbarn) ÇNAEM	Ref.[6]
23Na(n,p)23Ne	41(8)	44(4)
23Na(n,α)20F	100(20)	150(10)
24Mg(n,p)24Na	169(17)	186(15)
25Mg(n,p)25Na	72(13)	55(5)
27Al(n,p)27Mg	66(3)	74(5)
31P(n,α)28Al	138(11)	110(10)
28Si(n,p)28Al	226(27)	260(25)
29Si(n,p)29Al	135(14)	115(15)
30Si(n,α)27Mg	62(5)	87(20)
41K(n,p)41Ar	52(3)	43(5)
41K(n,α)38Cl	36(3)	34(1)
50Cr$(n,2n)$49Cr	30(2)	23(4)
52Cr(n,p)52V	85(6)	78(10)
52Cr$(n,2n)$51Cr	400(34)	315(30)
53Cr(n,p)53V	42(5)	40(3)
54Cr(n,α)51Ti	12(2)	14(2)
55Mn(n,p)55Cr	63(9)	45(10)
55Mn(n,α)52V	24(12)	29(2)
55Mn$(n,2n)$54Mn	890(65)	809(35)
54Fe(n,p)54Mn	340(30)	295(30)
54Fe(n,α)51Cr	98(10)	85(15)
54Fe$(n,2n)$53Fe	9(2)	7(2)
56Fe(n,p)56Mn	123(6)	110(10)
57Fe(n,p)57Mn	135(10)	80(10)

Part IV

**Experimental Facilities
and Techniques**

Accelerator mass spectrometry and its impact on nuclear data studies

M. Suter

Paul Scherrer Institut c/o HPK ETH Hönggerberg, 8093 Zürich, Switzerland

The present status of accelerator mass spectrometry (AMS) is summarized in terms of sensitivity, efficiency and accuracy. AMS studies and applications related to nuclear data are reviewed. These include cross section determinations and half-life measurements.

Introduction

Accelerator mass spectrometry (AMS) is an analytical method which combines mass spectrometry, accelerator technology, and nuclear detector techniques. This new technique has improved selectivity and sensitivity by several orders of magnitude over conventional mass spectrometry. These improvements permit the detection of long-lived cosmogenic radioisotopes such as ^{10}Be, ^{14}C, ^{26}Al, ^{36}Cl, ^{41}Ca and ^{129}I in mg size samples at isotopic ratios in the range of 10^{-10} to 10^{-15}. The main applications so far have been geophysical studies in which these radioisotopes are used as natural tracers to study transport and exchange processes, or for dating. These radioisotopes have also been inestigated in extraterrestrial material such as meteorites, moon rocks and cosmic dust.

To interpret the measured radioisotope concentrations, production processes and the corresponding cross sections have to be known. Also, accurate values of the half-lives are needed.

The application of AMS to nuclear sciences has been reviewed previously by Kutschera [1, 2]. The production of long-lived radioisotopes by cosmic rays has been discussed in other review papers [3, 4, 5].

Here, the principles of AMS are reviewed and the present capabilities and limits of the technique are discussed. Specific problems related to absolute measurements as needed for the determination of half-lives and production cross sections are also briefly mentioned. Recent half-live and cross section measurements performed with AMS or related to this technique are summarized.

Accelerator mass spectrometry

The principle of tandem accelerator mass spectrometry is depictured in Figure 1. The purified and pretreated material under investigation is loaded into the ion source and bombarded with positive Cs ions to produce a negative ion beam. Because in some cases interfering elements do not form stable negative ions, the source can act as a **first filter**. The mass(es) of interest are usually selected with a first magnetic mass spectrometer (**second filter**). The ions which pass the mass filter are accelerated to the high voltage terminal of the tandem van de Graaff accelerator. There, they pass through a thin foil or a narrow canal filled with gas. This causes electrons to be stripped away and most ions end up in a positive, multiply charged state. Most multiple charged molecules ($q \geq 3$) are are unstable, the stripping process constitutes therefore a **third filter**. The positive ions are then accelerated back to ground potential and analyzed in a high energy mass spectrometer (**fourth filter**). Finally, the ions are identified in a detector system (**fifth filter**). With this multiple filtering system the high sensitivity required can be obtained while at the same time the detection efficiency can be kept high.

For precise measurements, an adequate normalization procedure is needed. In most cases the stable isotopes are measured sequentially in a slow or a fast sequence. In some systems, all isotopes of interest are measured simultaniously.

Additional technical information and references can be found in review articles presented at previous AMS conferences [6, 7, 8].

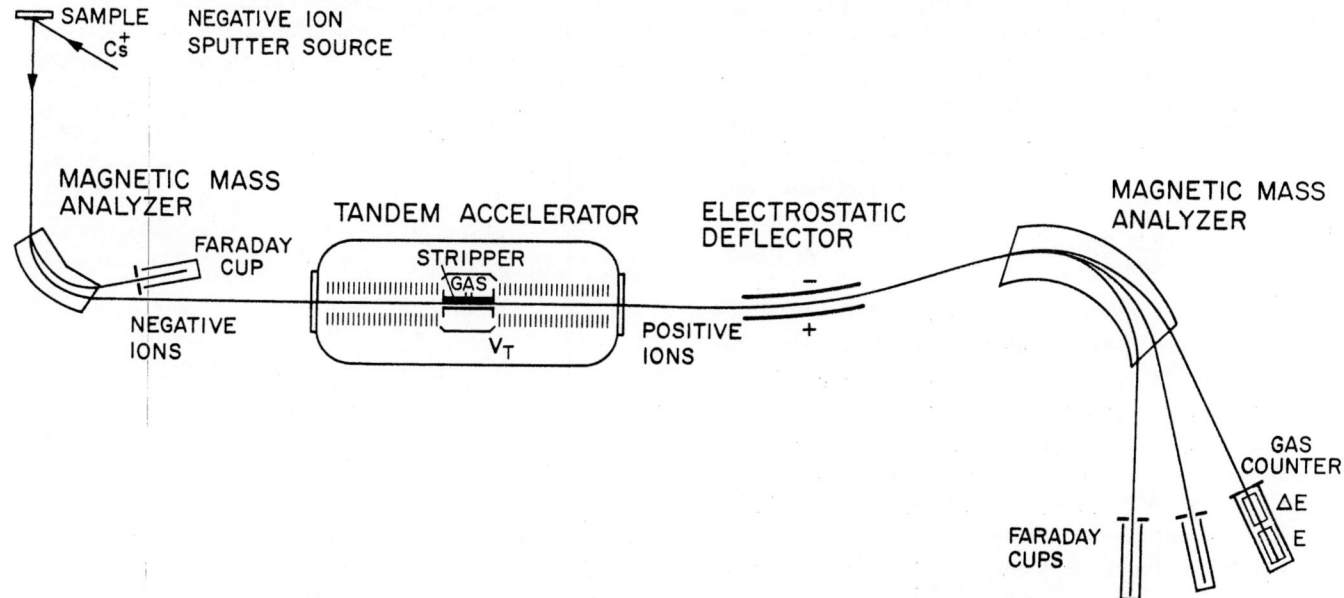

Figure 1. : Principle of AMS: The high sensitivity is reached by applying several filtering step including a tandem accelerator and nuclear detection techniques for the final identification.

Performance

Sensitivity: AMS systems based on small tandem accelerators operating at terminal voltages around 2 MV are commercially available. They are capable of analyzing ^{14}C, ^{10}Be, ^{26}Al and ^{129}I with isotopic ratios as low as 10^{-14} or 10^{-15}. Radioisotopes with severe isobaric interferences such as ^{36}Cl (^{36}S) and ^{41}Ca (^{41}K) require larger accelerators with terminal voltages in the range of 6 to 10 MV in order to reach similar sensitivities. Other radioisotopes (e.g. ^{53}Mn, ^{59}Ni and ^{60}Fe) have only been detected with the largest tandem accelerators available, sometimes in combination with post accelerators.

Efficiency: The detection efficiency varies from system to system. The output of the ion source depends on the samples, the design and operating conditions. For some elements (e.g. C and Cl), the yields of negative ion formation can be as high as 10 %, for others (Al, CaH$_3$) it might be as low as 10^{-3}. Stripping efficiencies are a function of terminal voltage and density of the stripper material and are typically in the range of 10 to 70 %. Additional losses have to be considered in the beam transport system and the final detector system, but they can be kept small with appropriate designs. Under favourable conditions the overall efficiency can be as high as a few percent.

Precision and accuracy: 5 - 10 % precision can be reached relatively easily. Higher precisions as required for ^{14}C dating need special care in the design and the measuring procedures. In recent years, several laboratories have shown that ^{14}C/^{12}C ratios can be measured with 0.2- 0.3 % precision.

In general, AMS measurements are performed relative to standards, which are measured periodically. In this case, many systematic errors cancel and mostly variations of systematic errors have to be considered. Standards are usually calibrated by decay counting (specific activity determination). This requires the precise knowledge of the corresponding half-life. Other standards are based on concentration measurement of highly enriched samples by conventional mass spectrometry which are then further diluted to obtain the low concentrations needed in AMS. For ^{14}C it has been agreed to use an oxalic acid standard distributed and certified by NIST (former NBS) [9]. For other radioisotopes equivalent standards are in preparation [10, 11]. When discussing accurary of AMS data, one has to consider uncertainties in the nominal values of the standards, especially when comparing results obtained with different sets of standards.

When no standards are available (e.g. for half-life measurements), absolute isotope ratios have to be measured directly with AMS. In this case, mass fractionation effects have to be considered. These can take place during sample preparation (chemistry), in the ion source (sputtering, ionization), in the beam transport system (mass dependent losses due to residual magnetic fields), in the stripping process, and finally in the detector system. If the element under investigation has several stable isotopes, mass fractionation effects can be studied by measuring the stable isotope ratios. Most of these effects can be assumed to be linear with mass and the appropriate corrections can be applied.

Cross section measurements

Early estimates on the production of long-lived radioisotopes were based on natural concentrations produced by the cosmic ray flux. When high energy accelerators became available, measurements of production cross sections were undertaken. The produced radioisotope contents were usually determined by decay counting, requiring long irradiation times and complex radiochemistry. The development of AMS initiated new and more detailed studies of long-lived radioisotope production mechanisms. The motivation for these new studies is primarily related to the availability of radioisotope data extraterrestrial material, especially depth profiles in meteorites or lunar samples. The interpretation requires a precise knowledge of the production processes. Production cross sections as function of energy are derived from irradiations of stacks of thin targets. With 4-π irradiations of thick targets, profiles are measured and compared to calculations with various simulation codes which also include the creation of secondary particles and their contribution to the radioisotope production.

In this summary, only thin target results are discussed. Cross section data obtained with AMS are listed in Table I. Most data are available for proton induced reactions, which usually cover the energy range from threshold up to several GeV. The targets irradiated and analyzed reflect practical needs: O, Mg, Al, Si, Ca and Fe make up dominant part of the total mass in terrestrial and extraterrestrial rocks.

It has been realized that radioisotope production induced by alpha particles can be significant in some cases although only about 12 % of the galactic cosmic ray flux are alpha particles. This is due to the fact that the corresponding production cross sections are considerably larger than the ones for p induced reactions (see Figure 2). Neutron induced reactions are important becauce the secondary particle flux in some depth consists primarily of neutrons. However, experiments are more difficult to perform due to the lack of the intense monoenergetic neutron sources in the desired energy range. Recently, a few neutron induced cross sections have been measured with neutrons from proton irradiated Be- or Li-targets

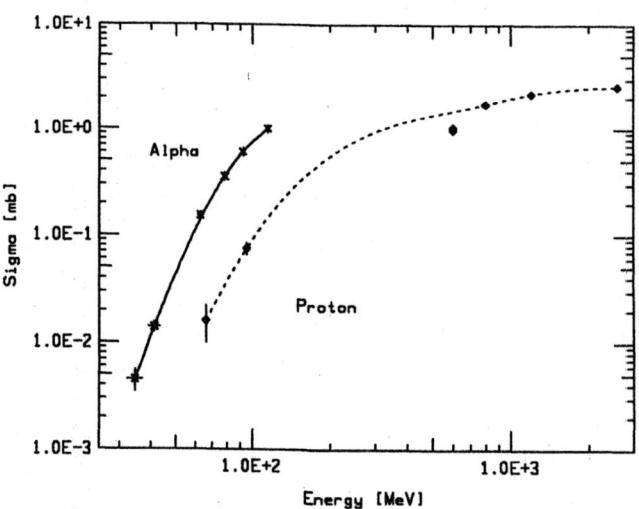

Figure 2. : Comparison of proton and alpha induced cross sections for the production of ^{10}Be in Mg [14, 15]

Table 1: Production cross section of ling-lived radioisotopes measured with AMS

BEAM Isotope	Target	Energy range	Ref.	BEAM Isotope	Target	Energy range	Ref.
PROTON				ALPHA			
[10]Be				[10]Be			
	O	52 - 2600	[14]		C	79 - 165	[15]
	Mg	66 - 2600	[14]		N	70 - 170	[15]
	Al	67 - 2600	[14]		O	70 - 170	[15]
	Si	50 - 2600	[14]		Mg	70 - 170	[15]
	Mn	63 - 2600	[14]		Al	70 - 170	[15]
	Fe	62 - 2600	[14, 16]		Si	70 - 170	[15]
	Ni	57 - 2600	[14, 16]		Fe	61 - 171	[17]
[26]Al					Ni	67 - 171	[17]
	Mg	25 - 2600	[18]	[26]Al			
	Al	67 - 2600	[14, 19]		Mg	30 - 120	[15]
	Si	50 - 2600	[18]		Al	30 - 80	[15]
	Mn	171 - 2600	[14]		Si	60 - 120	[15]
	Fe	600 - 2600	[14]				
	Ni	600 - 2600	[14, 16]				
[41]Ca				NEUTRON			
	Ti	35 - 149	[20, 21]	[10]Be			
	Fe	108 - 600	[20, 21]		B	Thermal	[22]
	Ni	98 - 600	[20, 21]		N	10 - 40	[13]
[60]Fe				[14]C			
	Ni	150	[23]		O	22 - 37	[12, 13]
				[26]Al			
					Al	10 - 40	[13]
					Si	30 - 40	[12, 13]
				[36]Cl			
					[36]Ar	Thermal	[24]

[12, 13]. The [26]Al production from Si agrees in this case quite well with the corresponding proton induced cross sections (see Figure 3). Thermal neutron induced reactions are studied using irradiations in nuclear reactors.

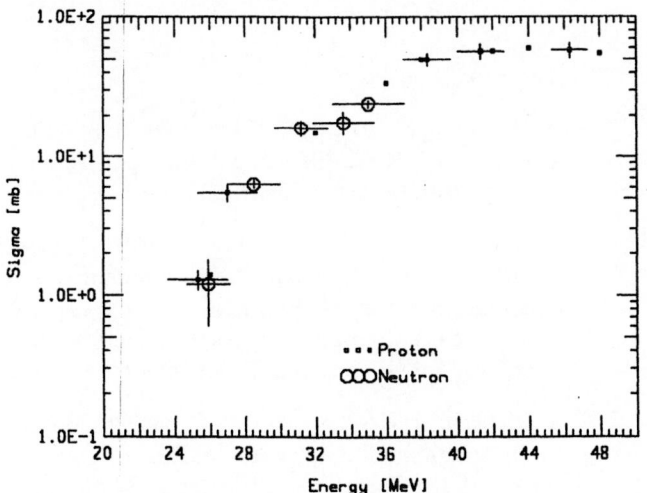

Figure 3. : Comparison of proton and neutron induced cross sections for the production of [26]Al in Si [12, 42]

Half-life determinations

Dating techniques using radioisotopes require the precise knowledge of the corresponding half-lives. This is also true, when a specific activity is compared with concentration data obtained by (accelerator) mass spectrometry. Some recent half-life measurements, which have been determined by AMS or which are interesting for comparison are shown in Table II. In general the radioisotope contents are determined by (accelerator) mass spectrometry and specific activities by decay counting. The half-life is then calculated from these two measurements.

In 1981 two AMS values (108 and 101 y) for the [32]Si half-life were published [25, 26], which differed significantely from previous estimates based on geochemical data [27, 28] or on activity measurements of irradiated samples [1]. A newer result (172 y) published a few years later and obtained by following the decay rate over 4 years disagreed significantly from the AMS data [29]. In a following AMS measurement performed at Zurich, systematic errors were studied extensively by analyzing all 3 stable Si isotopes [30]. This AMS result (133 ± 9 y) is consistent with the previous AMS data, but is significantly smaller than the decay counting result. At present, the cause for this large discrepancy is not clear.

Table 2: Half-lives measued by AMS and their comparison with other results

Isotope	Half-life	Technique	Ref.
^{32}Si	108 ± 18 y	AMS	[25]
^{32}Si	101 ± 18 y	AMS	[26]
^{32}Si	133 ± 9 y	AMS	[30]
^{32}Si	172 ± 4 y	DECAY	[29]
^{41}Ca	113 ± 12 ky	MS	[35]
^{41}Ca	103 ± 7 ky	AMS	[34]
^{44}Ti	46.4 ± 1.7 y	MS	[39]
^{44}Ti	48.2 ± 0.9 y	MS	[40]
^{44}Ti	54.2 ± 2.1 y	AMS	[38]
^{44}Ti	66.6 ± 1.6 y	DECAY	[41]
^{60}Fe	1.5 ± 0.3 My	AMS	[23]

Further investigations on the ^{32}Si half-life are in progress at various places. A further AMS study has been made at Aarhus, in which ^{32}Si is compared to ^{31}Si both beeing produded in the same irradiation. This measurement leads to a half-life value which is close to the value decay counting result [31, 32].

In recent years, it became possible to detect ^{41}Ca with AMS on a routine basis [33]. A novel technique for half-life determination was used in this case :^{36}Cl and ^{41}Ca concentrations were measured in several meteorites with different terrestrial ages [34]. Based on the known half-life of ^{36}Cl, the ^{41}Ca half-life was be determined. The new value (103×10^3)is consistent with a previously published half-life of Mabuchi et al [35] but in disagreement with the earlier results [36]. A new determination recently performed at Argonne National Laboratory [37] seems to confirm the value of Fink et al and Mabuchi et al.

The ^{44}Ti half-life measurements based on AMS [38] and conventional mass spectrometry [39, 40] are consistent but significantly smaller than the result obtained by following the decay rate over several years [41], a situation which is very similar to the one for ^{32}Si. The half-life of ^{60}Fe was only poorly known before an AMS measurement was performed [23].

Conclusions

During the last 10 years AMS has become the superior method for the detection of long-lived radioisotopes at natural abundances. By optimizing the various processes associated with this technique, performances has significantly improved. The through-put was increased primarily through ion source developements and proper stripper design. Backgrounds were reduced in many cases to a level in which the contamination during sample preparation or the radioisotope content of carrier material are the limit of the sensitivity. Accuracy was pushed to the 0.2 % level for ^{14}C dating. Based on present experience, further improvements can be expected in the next few years.

The fact that some AMS facilities are already overloaded with requests for analysis, shows the strong interest in this technique and its versatility. The large variety of AMS applications in dating, tracer experiments and cosmic ray flux studies stresses the need for more detailed knowlegde of nuclear processes such as production processes and decay rates.

As shown here, a large number of production cross sections have already been measured with AMS and new half-life studies were initiated. There are still open questions and also some significant discrepancies in existing data (cross sections and half-lives), More data are needed especially for the production due to secondary particles such as neutrons and muons.

References

[1] W. Kutschera, Nucl. Instr. and Meth. B17, 377 (1986)

[2] W. Kutschera and M. Paul, Ann. Rev. Nucl. Part. Sci. 40, 411 (1990)

[3] D. Lal and B. Peters, in S. Flügge, ed., Handbuch der Physik 4612, 551 (1963)

[4] R.C. Reedy, J.R. Arnold and D. Lal, Ann. Rev. Nucl. Part. Sci. 33, 505 (1983)

[5] D. Lal, Ann. Rev. Earth Planet. Sci. 16, 355 (1988)

[6] A.E. Litherland, Nucl. Instr. and Meth. B5, 100 (1984)

[7] W. Wölfli, Nucl. Instr. and Meth. B29, 1 (1987)

[8] M. Suter, Nucl. Instr. and Meth. B52, 211 (1990)

[9] M. Stuiver and H.A. Polach, Radiocarbon 19, 355 (1977)

[10] K. Inn, S. Raman, B.M. Coursey, J.D. Fassett and R.L. Walker, Nucl. Instr. and Meth. B29, 27 (1987)

[11] P. Sharma, P.W. Kubik, U. Fehn, H.E. Gove, K. Nishiizumi and D. Elmore, Nucl. Instr. and Meth. B52, 410 (1990)

[12] M. Imamura, H. Nagai, M. Takabatake, S. Shibata, K. Kobayashi, K. Yoshida, H. Ohashi, Y. Uwamino and T. Nakamura, Nucl. Instr. and Meth. B52, 595 (1990)

[13] T. Nakamura, H. Sugita, M. Imamura, Y. Uwamino, Shibata, H. Nagai, M. Takabatake and K. Kobayashi, these proceedings, Int. Conf. on Nuclear Data for Science and Technology (1991)

[14] B. Dittrich, U. Herpers, H. J. Hofmann, W. Wölfli, R. Bodemann, M. Lüpke, R. Michel, P. Dragovitsch and D. Filges, Nucl. Instr. and Meth. B52, 588 (1990)

[15] H.-J. Lange, T. Hahn, R. Michel, B. Dittrich, U. Herpers, R. Rösel, H.J. Hofmann and W. Wölfli, these proceedings, Int. Conf. on Nuclear Data for Science and Technology (1991)

[16] M. Lüpke, R. Michel, B. Dittrich, U. Herpers, P. Dragovitsch, D. Filges, H.J. Hofmann and W. Wölfli, these proceedings, Int. Conf. on Nuclear Data for Science and Technologie (1991)

[17] B. Dittrich, Ph. D. Thesis, Köln (1990)

[18] R. Bodemann, R. Michel, B. Dittrich, U. Herpers, R. Rösel, H.J. Hofmann, W. Wölfli, B. Holmqvist, H. Conde and P. Malmborg, these proceedings, Int. Conf. on Nuclear Data for Science and Technology (1991)

[19] R.J. Schneider, J.M. Sisterson, A.M. Koehler, J. Klein and R. Middleton, Nucl. Instr. and Meth. B29, 271 (1987)

[20] D. Fink, J. Sisterson, S. Vogt, G. Herzog, J. Klein, R. Middleton, A. Koehler and A. Magliss, Nucl. Instr. and Meth. B52, 601 (1990)

[21] D. Fink, M. Paul, G. Hollos, S. Theis, S. Vogt, R. Stueck, P. Englert and R. Michel, Nucl. Instr. and Meth. B29, 275 (1987)

[22] D. Lal, K. Nishiizumi, R.C. Reedy, M. Suter and W. Wölfli, Nucl. Phys. A468, 189 (1987)

[23] W. Kutschera, P.J. Billquist, D. Frekers, W. Henning, K.J. Jensen, M. Xiuzeng, R. Pardo, M. Paul, K.E. Rehm, R.K. Smither and J. L. Yntema, Nucl. Instr. and Meth. B5, 430 (1984)

[24] S.S. Jiang, T.K. Hemmick, P.W. Kubik, D. Elmore, H.E. Gove and S. Tullai-Fitzpatrick, Nucl. Instr. and Meth. B52, 608 (1990)

[25] D. Elmore, N. Anantaraman, H.W. Fulbright, H.E. Gove, H.S. Hans, K. Nishiizumi, M.T. Murell and M. Honda, Phys. Rev. Lett. 45, 589 (1980)

[26] W. Kutschera, W. Henning, M. Paul, R.K. Smither, E.J. Stephenson, J.L. Yntema, D.E. Alburger, J.B. Cumming and G. Harbottle, Phys. Rev. Lett. 45, 592 (1980)

[27] H.B. Clausen, J. Glaciol. 12, No.66, 411 (1973)

[28] D.J. Demaster, Earth & Planet. Sci. Lett. 48, 209 (1980)

[29] D.E. Alburger, G. Harbottle and E.F. Norton, Earth & Planet. Sci. Lett. 78, 168 (1986)

[30] H. J. Hofmann, G. Bonani, M. Suter, W. Wölfli, D. Zimmermann and H.R. von Gunten, Nucl. Instr. and Meth. B52, 544 (1990)

[31] M.S. Thomsen, J. Heinemeier, P.Hornshoj, H.L. Nielson and N. Rud, Nucl. Instr. and Meth. B31, 425 (1988)

[32] M.S. Thomsen, J. Heinemeier, P.Hornshoi, H.L. Nielson and N. Rud, Nucl. Physics A, submitted (1991)

[33] D. Fink, R. Middleton, J. Klein and P. Sharma, Nucl. Instr. and Meth. B47, 79 (1990)

[34] D. Fink, J. Klein and R. Middleton, Nucl. Instr. and Meth. B52, 572 (1990)

[35] H. Mabuchi, H. Takahashi, Y. Nakamura, K. Notsu and H. Hamaguchi, J. Inorg. Nucl. Chem. 36, 1687 (1974)

[36] C.M. Lederer and V.S. Shirley, Table of Isotopes, 7th edition (Wiley, New York) (1978)

[37] M. Paul, I. Ahmad and W. Kutschera, Z. Phys. A, submitted (1991)

[38] D. Frekers, W. Henning, W. Kutschera, K.E. Rehm, R.K. Smither, J.L. Yntema, R. Santo, B. Stievano and N. Trautmann, Phys. Rev. C 28, No.4, 1756 (1983)

[39] J. Wing, M.A. Wahlgren, C.M. Stevens and K.A. Orlandini, J. Inorg. Nucl. Chem. 27, 487 (1965)

[40] P.E. Moreland Jr. and D. Heymann, J. Inorg. Nucl. Chem. 27, 493 (1965)

[41] D.E. Alburger and G. Harbottle, Phys. Rev. C41, 2320 (1990)

[42] M. Furukawa, K. Shizuri, K. Komura, K. Sakamoto, and S. Tanaka, Nucl. Phys. A174, 539 (1971)

NEW NEUTRON FACILITIES FOR NUCLEAR DATA MEASUREMENTS AT $E_n > 10$ MeV

H. Condé[a], R. Haight[b], H. Klein[c], and P. Lisowski[b]

(a) Department of Neutron Research, Uppsala University, Uppsala,
Sweden
(b) Los Alamos National Laboratory (LANL), Los Alamos,
U.S.A.
(c) Physikalisch-Technische Bundesanstalt (PTB), Braunschweig,
Germany

Abstract: A variety of neutron sources producing "monoenergetic" neutrons of up to 200 MeV in energy and "white" spectral distributions up to 800 MeV is described in this paper. The $H(t,n)^3He$ reaction can be employed at LANL as the ideal monoenergetic neutron source for energies $E_n < 14$ MeV because a triton beam with energies of up to 22 MeV is available on an FN tandem. A rotating, high-pressure hydrogen gas target allows high-intensity neutron fields to be produced which are well-suited for studying the neutron-induced activation of long-lived isotopes. At PTB, the $D(d,n)^3He$ reaction must still be used for the same purpose because any radiation hazard which may be caused by the handling of tritium must be avoided. However, the unavoidable contribution of low-energy neutrons due to deuteron breakup reactions must be corrected for. For this purpose, the spectral fluence has been determined for projectile energies of up to 13.3 MeV and neutron emission angles of up to 15 degrees, which are occasionally taken into account in close-geometry irradiations. At the The Svedberg Laboratory in Uppsala the upgraded cyclotron is now equipped with a thin ^7Li target in order to produce a "monoenergetic" neutron beam with energies from 50 to 200 MeV, well collimated to a solid angle of about 10^{-4} sr. A highly efficient recoil proton spectrometer allows the primary spectral neutron fluence to be reconstructed and other (n,p) reactions to be investigated for isovector excitations. At LANL a spallation neutron source is now available at Target-4 of the WNR facility. Part of the 800 MeV proton beam from LAMPF is chopped and bunched to provide 40 macropulses/s and a total of 32000 micropulses/s with a time width of 150 ps and a separation of > 1 microsecond. Seven beam lines with flight paths between 7 and 90 m (350 m is aimed at) are available at angles of 15 to 90 degrees to the incident proton beam. The spectral distributions differ for these production angles and allow low or high energy ranges to be emphasized. Neutron-induced charged particle and photon production and fission have already been investigated.

($H(t,n)^3He$ for $E_n \leq 14$ MeV; $D(d,n)^3He$ and $D(d,np)D$ for $E_n \leq 14$ MeV; $^7Li(p,n)^3Be$ for 50 MeV $\leq E_n \leq 200$ MeV; spallations neutron source with 30 MeV $\leq E_n \leq 800$ MeV)

Introduction

Neutron-induced reactions can be investigated by means of monoenergetic neutron sources for energies up to ~ 20 MeV only [1]. But practical problems already arise in the energy region below 14 MeV which is of particular interest in fusion reactor technology, since really monoenergetic neutron sources are not easily realized. Two different attempts to overcome the lack of neutron cross section data in this energy region are discussed. First, the ideal $H(t,n)^3He$ reaction [2] is employed at LANL to investigate long-lived, neutron-induced activation. Advantage is taken of a high-energy triton beam and a rotating high-pressure hydrogen gas target in order to produce an intense neutron field for energies up to 14 MeV [3]. Secondly, the $D(d,n)^3He$ reaction is still the "workhorse" although the low-energy neutrons from the deuteron breakup reaction $D(d,np)D$ cannot be avoided. In order to overcome this drawback the angular spectral fluence of these breakup neutrons were carefully determined at PTB [4] to allow corrections to be made for their influence.

In the neutron energy range above 30 MeV (p,n) reactions on light nuclei are preferably employed to produce a spectral neutron fluence which is strongly peaked at the desired energy [5]. At the The Svedberg Laboratory of the Uppsala University, neutrons from the $^7Li(p,n)$ reaction are now available in the energy region

50 MeV $\leq E_n \leq$ 200 MeV chiefly to investigate isovector excitations in (n,p) reactions [6].

Spallation neutron sources make it possible to determine the energy-dependent cross section of neutron-induced reactions simultaneously over a wide energy range. At target-4 of WNR there is a unique installation with several neutron beam lines offering different white neutron spectra with energies up to 800 MeV [7].

The main features of these four facilities, particularly the specification of their neutron sources and their application in recent experiments, will be discussed.

An Intense Monoenergetic Neutron Source using the H(t,n) Reaction for $E_n \leq 14$ MeV

The neutron source reaction with the highest specific yield at zero degrees is the $H(t,n)$ reaction [1,2]. At the Los Alamos Ion Beam Facility (IBF), this reaction is being used with intense triton beams and a high-pressure hydrogen gas target [3].

The tritons are accelerated by an FN tandem Van de Graaff accelerator to energies up to 22 MeV with beam currents of 6 to 10 microamperes. At the upper beam energy and taking into account the energy loss in the entrance foil to the gas cell, the maximum mean neutron energy at zero degrees from the H(t,n) reaction is 14.4 MeV. For maximum intensity, the neutron production target is placed at the "straight-through" port of the analyzing magnet, that is without any magnetic analysis. The beam passes through a series of collimators and focusing magnets that serve to remove off-energy components of the beam.

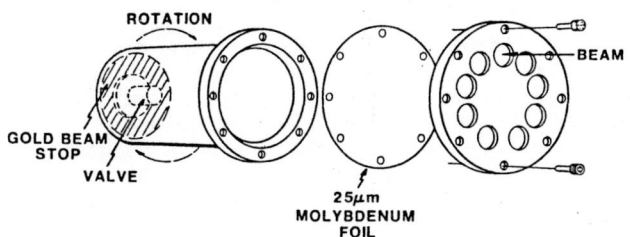

Fig. 1. Expanded view of the rotating gas cell.

The neutron-producing target is the high-pressure hydrogen gas cell previously described [3]. Briefly, the cell consists of a molybdenum entrance foil which is supported by a nine-hole collimator, a stainless-steel body, and a gold beam stop. (Fig. 1) This cell is pressurized to 1.1 MPa with normal hydrogen. To allow high beam currents, the cell is rotated at up to 16 revolutions per second and the beam stop is cooled by an external water-air mist. In this way neutron fluence rates up to 10^{11} n/(sr·s) are achieved. To monitor the beam current, a graphite brush makes electrical contact with the rotating cell which is electrically isolated from the rest of the beam tube.

To ensure that the beam is positioned where it should be in the cell, we tape a piece of RAREX (gadolinium oxysulfide) to the rear surface of the rotating cell (and temporarily remove the air-water mist cooling). A closed-circuit television camera monitors the light from this fluorescent material. With beam currents of less than a microampere, ample light is produced by neutrons and gamma rays from the cell to indicate the position of the beam. For irradiations where the sample is close to the gas cell, an accurate determination of the beam position is very important for obtaining the maximum fluence and for determining the average energy of the neutrons (see Fig. 2).

This neutron source is characterized by a strong monoenergetic peak and a weaker continuum. The source reaction itself, H(t,n), is monoenergetic over this energy range with no breakup contributions. Because of energy loss in the gas and energy straggling both in the entrance foil and in the gas, the "monoenergetic peak" has a finite width. The finite angular range around zero degrees subtended by the irradiated sample increases the width of this peak.

The neutron energy spectrum as calculated is given in Fig. 2 for z = 2 cm and several y-positions. The shape of the flux spectrum is due to the finite thickness of the gas cell (46 mm) and the kinematics of the reaction. The need for good alignment of beam, cell, and sample is obvious from the figure.

Flux monitoring during irradiations is carried out by monitor activation reactions and by a ^{238}U fission chamber. Monitor reactions such as ^{58}Ni(n,p)^{58}Co and ^{27}Al(n,α)^{24}Na are useful in this energy region. A fission chamber can be placed just behind the sample position and, in fact, is often used to hold the sample.

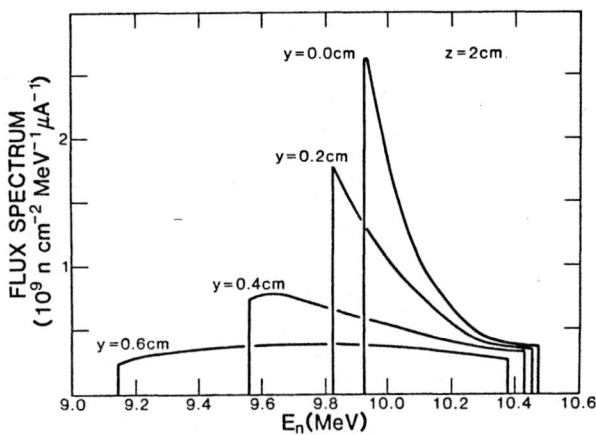

Fig. 2. Neutron energy spectrum 2 cm behind the gas cell for various lateral positions (y) from the beam axis. Again, the H(t,n) reaction is used with the cell pressurized to 1.0 MPa.

The best arrangement is to have a fissile (^{238}U) deposit of the same diameter as that of the irradiated sample. No geometry corrections need be made in this case.

There are also breakup neutrons from interactions of the tritons with collimators, the entrance foil, and the beam stop. To investigate these breakup neutrons, two approaches were taken. The first was to compare counting rates in the fission chamber with the cell "empty". This technique should be adequate for assessing breakup in the entrance foil. An additional component comes from breakup in the beam stop, however, since the triton energy is higher in the empty cell runs. Breakup in the collimators depends to some extent on beam tuning and therefore the runs with the cell full and the cell empty had to be made with no change in the beam tuning. The results are given in Table 1 for a 2 cm diameter ^{238}U fission deposit 4 cm from the end of the gas cell. The ratio here of peak neutrons to breakup neutrons is averaged over the ^{238}U fission cross section and should be considered to be a lower limit because of the additional component in the empty cell runs.

Table 1. Investigation of Breakup Neutrons with a 238U fission chamber

E_t (MeV)	$\langle E_n \rangle$ (MeV)	Peak/Breakup[a]
10.0	4.6	> 10.7
15.2	8.0	> 3.1
17.8	11.0	> 2.2
20.0	12.7	> 1.0

[a] averaged over the ^{238}U(n,f)

A second method was to measure the neutron spectrum by time-of-flight. A plastic scintillator, 5 cm in diameter and 1 cm thick, was placed 1.36 meters behind the gas cell. Runs with a shadow bar were made to estimate the contribution of neutrons scattered from the walls and floor of the room before reaching the detector. The detector efficiency was calculated by the SCINFUL program [8]. The resulting spectra for two different triton energies are shown in Fig. 3.

388

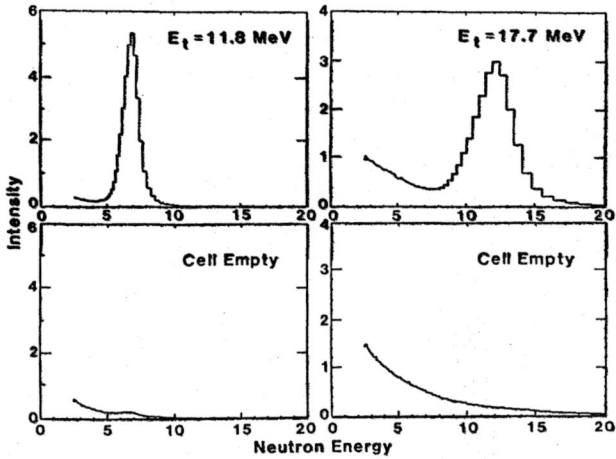

Fig. 3. Neutron spectrum showing the source and the breakup neutrons.

From these two sets of data, it is clear that the breakup neutrons are a larger fraction of the total neutrons at the higher triton energies. Their effect, of course, depends on the thresholds of the reactions studied and those of the monitor reactions.

This neutron source has been used for several studies of neutron activation cross sections [8,9]. An investigation of helium production by neutrons is in progress.

How to use the D(d,n)^{3}He reaction for neutron energies up to 14 MeV

The fast neutron facility at PTB with a ns-pulsed compact cyclotron and a multi-detector neutron time-of-flight spectrometer [11] was originally designed to investigate the neutron interaction with tissue and tissue-equivalent material in the energy range from 6 MeV to 14 MeV. As neither a triton beam nor a tritium gas target are practicable, due to strict radiation protection regulations, the D(d,n) reaction must be applied. High resolution ($\Delta E_n/E_n \leq 1$ %) and accurate ($\Delta\sigma/\sigma \approx 1 - 2$ %) scattering experiments [12] have been performed, but restricted to inelastic scattering with excitation energies up to 6 MeV due to the unavoidable low-energy neutrons chiefly stemming from the deuteron breakup reaction, D(d,np)D. Recently, the angular-dependent spectral neutron fluence of the breakup reaction has been carefully measured [4] in order to enable corrections to be made for the contributions from breakup neutrons, provided that the corresponding cross sections are reliably evaluated.

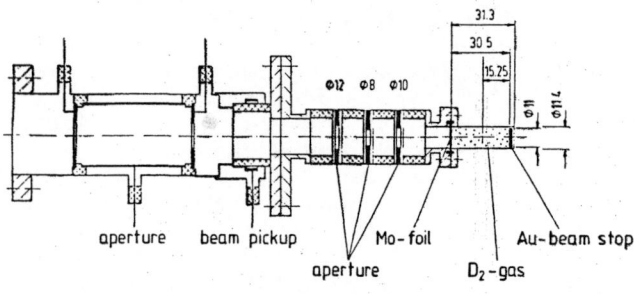

Fig. 4. Gas target assembly.

Deuterium Gas Target

The conventional deuterium gas target (Fig. 4) consists of a thin, 3 cm long steel cylinder, a 5 µm Mo entrance foil (sealed by In rings) and a 0.5 mm thick Au beam stop, which is replaceable as the projectile deuterons are implanted in the beam stop and increasingly produce D(d,n) neutrons due to this self-target build-up.

The target is pressurized with deuterium up to 0.25 MPa such, that the energy loss of the projectile in the gas, the energy straggling of the energy loss in the entrance foil and the beam energy width result in the desired neutron energy width [13]. The net neutron spectra obtained from gas-in and gas-out runs show a clean gap between the D(d,n) and the D(d,np) breakup neutrons (Fig. 5) provided that deuterium gas of the highest purity is employed (no cold trap necessary !). A fluence rate of $2 \cdot 10^8$ n/(sr·s) is achieved for D(d,n) neutrons at zero degrees and $\Delta E_n = 0.1$ MeV.

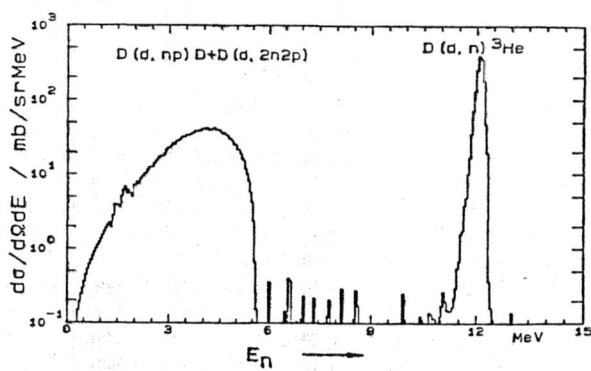

Fig. 5. Net spectral fluence of zero degrees neutrons from D + d at E_d = 9 MeV.

Deuteron Breakup Reaction

As the D(d,n) reaction is strongly anisotropic [1], irradiations should be performed in a small forward cone only ($\vartheta_n \leq 10(15)$ deg). The angular spectral fluence of breakup neutrons undergoes considerable changes, however, even in this small range of neutron emission angles (Fig. 6a).

The reaction was therefore measured in steps of 0.5 MeV from threshold to 13.5 MeV deuteron energy and for neutron emission angles up to 15° in steps of 2.5° taking advantage of a well-specified neutron time-of-flight spectrometer [4]. The energy-integrated cross section of 0° breakup neutrons increases steeply from threshold on and exceeds the D(d,n) yield at $E_d \sim 9$ MeV.

The most interesting result of these investigations, however, is that the neutron emission behaves very similarly for both reactions, at least for emission angles up to 10°. The spectral fluence is almost identical if drawn on a relative energy scale and renormalized for the same D(d,n) neutron yield (Fig. 6 b). The spectral neutron fluence of breakup neutrons applied in close geometry of target and sample can therefore easily be calculated on the basis of the measured 0° spectral fluence.

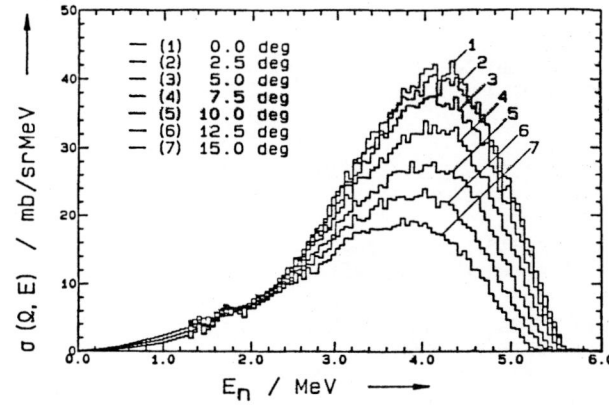

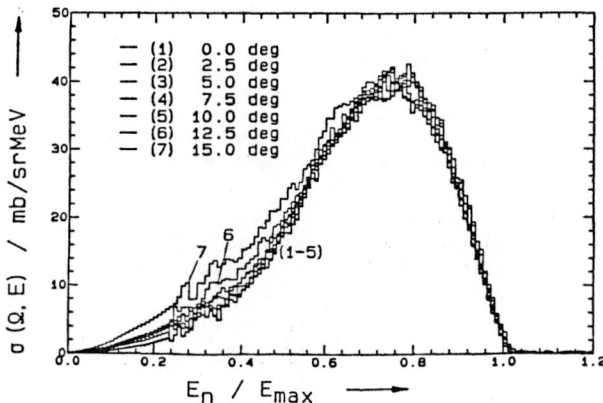

Fig. 6. (a) Spectral angular cross section of the D(d,np) reaction for E_d=9.02 MeV
(b) same as in (a) but on a relative energy scale (E_{max} denotes the maximum breakup neutron energy at the corresponding angle) and renormalized for the yield of D(d,n) neutrons at zero degrees (data from [4]).

Application

On this basis, neutron-induced activation was recently investigated for reactions with rather low thresholds [9,14]. Even if more than half of the neutron-induced events such as activation or fission events from the $^{238}U(n,f)$ reaction (which is commonly used for reference) are due to breakup neutrons, reliable cross sections can be deduced with an accuracy comparable to the data obtained with the "ideal" source [9].

Attempts to extend the analysis of recent scattering experiments [12,15] to excitations higher than 6 MeV by including the primary breakup neutrons in the complete simulation of the scattering experiments are in progress.

Neutron Energy Range 50 MeV $\leq E_n \leq$ 200 MeV

A facility for the study of neutron-induced reactions in the energy region from about 50 MeV to 200 MeV has been built at the upgraded Gustaf Werner cyclotron of the The Svedberg Laboratory [6]. Well-collimated, monoenergetic neutron beams are produced with a fairly long distance between the neutron source and the reaction target in order to reduce background radiation.

The facility is initially being used in studies of light-particle emission from neutron induced-reactions, with the emphasis on the (n,p) reaction to investigate isovector excitations. To that end a magnetic spectrometer with large angular and momentum acceptance has been constructed to allow measurements of energy and angular distributions of the protons produced in the (n,p) reactions in various nuclei.

Studies of charged-particle emission from neutron-induced reactions involve some experimental complications. For example, it is difficult to obtain high signal counting rate, good energy and angular resolution, and low background simultaneously. It was also found desirable to include particle identification and make it possible to observe particles emitted in the forward direction, including 0°.

Neutron Facility

The experimental equipment is shown in Fig. 7. Protons from the cyclotron approach the neutron target from the left. The neutrons are produced by the $^7Li(p,n)^7Be$ reaction, using 100 - 200 mg/cm^2 thick discs of lithium, enriched to 99.984 % in ^7Li and mounted in a water-cooled rig with four target holders, each 26 mm in diameter. The targets can be wobbled to avoid overheating. A liquid-nitrogen trap is put into the beam line to reduce carbon buildup on the targets. After its passage through the target, the proton beam is deflected and focussed onto a lead-shielded, water-cooled graphite block at the end of an 8 m long tunnel.

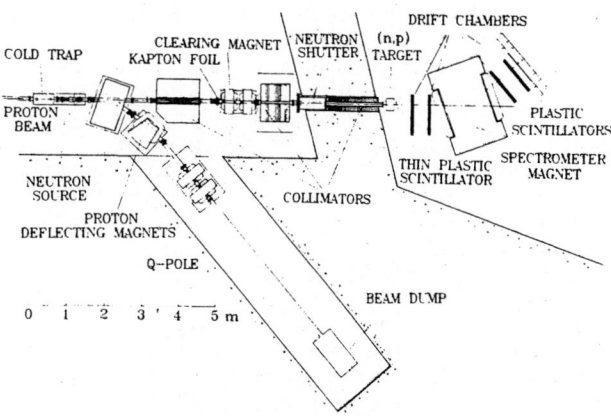

Fig. 7. Overview of the Uppsala (n,p) facility.

The neutron beam is defined by a system of three collimators. The first one consists of a 1.1 m long iron cylinder of the revolver type, with four axial holes put into the desired position by remote control. Three of these holes are used to collimate the beam to three solid angles (60, 80 and 100 μsr). The neutron channels are double-coned in shape with a central waist defining the solid angle. The second and third collimator serve as scrapers for the beam defined by the first collimator. The vacuum system is terminated after the first collimator by a thin Kapton foil. Charged particles produced in this foil and along the collimator channel are deflected by a clearing magnet placed behind the first collimator.

The most prominent features of the neutron-production facility are the very good shielding between the beam dump and the experimental area, and the long distance between the neutron source

Table 2. Characteristic Data for the TSL Neutron Facility

<u>Proton beam from TSL cyclotron:</u>
Proton current:	10 μA (max)
Proton energy:	50 - 200 MeV
Pulse width:	3 - 4 ns
Energy width:	400 keV

<u>Neutron beam:</u>
Neutron production reaction:	^{7}Li(p,n)^{7}Be
Solid angle of neutron beam:	60, 80 and 100 μsr
Neutron flux*:	1 x 10^6 s^{-1}
Neutron energy spread*:	0.7 MeV
Distance neutron production to reaction target:	≥ 8 m
Neutron/proton ratio in neutron beam:	1.2 x 10^{-5}

*(at 10 μA, 100 MeV, 100 mg/cm^2 of ^{7}Li, 60 μsr)

and the reaction target, allowing the use of TOF techniques to reject low-energy neutrons. Characteristic data on the facility are given in Table 2.

Detector System for (n,p) Reaction Studies

The (n,p) target should contain as much material as possible without impairing the energy resolution in view of the relatively low intensity of the neutron beam. A sandwiched target system was therefore constructed with a stack of up to 6 thin target layers interspaced by up to 9 multiwire chambers. The first two detectors produce veto signals for an efficient rejection of charged particles which contaminate the neutron beam.

The momentum analysis of protons from the (n,p) reaction is performed in a light ion spectrometer (LISA) consisting of a dipole magnet, four drift chambers and three scintillators (see Fig. 7). With one position and one magnetic field setting, the spectrometer accepts protons with a momentum bite of 22 % and a solid angle of detection of 14 msr.

The four drift chambers are used to determine the particle trajectories. The position resolution is 0.3 mm (FWHM) for each coordinate and the detection efficiency is 98 %. A trigger signal is generated by a triple coincidence between a thin plastic scintillator placed between the target box and the first drift chamber and two large scintillators placed behind the fourth drift chamber. The last two scintillators are also used for particle identification. Helium-filled bags are placed in the magnet gap and between the first two drift chambers to minimize the energy spread from small-angle scattering along the proton trajectories. Characteristic data on the spectrometer are shown in Table 3.

Neutron Fluence Measurements

Six targets, one of which is a CH$_2$ reference foil, are normally mounted in the target box. This foil allows (n,p) differential cross sections to be measured with reference to the n-p scattering cross section and serves as an internal flux monitor.

A separate neutron flux monitor, also based on n-p scattering, is placed in the neutron beam behind the spectrometer. This monitor records the neutron flux, independent of the position of the magnet. The neutron beam is dumped 10 m from

Table 3. Characteristic Data for the Light Ion Spectrometer (LISA)

<u>Acceptance:</u>
Momentum bite:	22 %
Solid angle:	14 msr

<u>Dispersion (E_p = 100 MeV, B = 0.8 T):</u>
11 mm/Δp/p% over last (4th) drift chamber

<u>Spectrometer efficiencies</u>
Total spectrometer efficiency	75 %

<u>Angular and energy resolution:</u>
Angular resolution:	0.3°
Energy resolution spectrometer (100 MeV):	
Spectrometer (small angle scattering and straggling):	2.0 MeV
(n,p)-target (85 mg/cm^2, carbon):	0.6 MeV
Energy resolution neutron facility (100 MeV):	
Proton energy spread:	0.5 MeV
Neutron beam (100 mg/cm^2, ^{7}Li)	0.7 MeV
Total energy spread (FWHM)	2.3 MeV

the spectrometer in a well-shielded cave (not shown in Fig. 7).

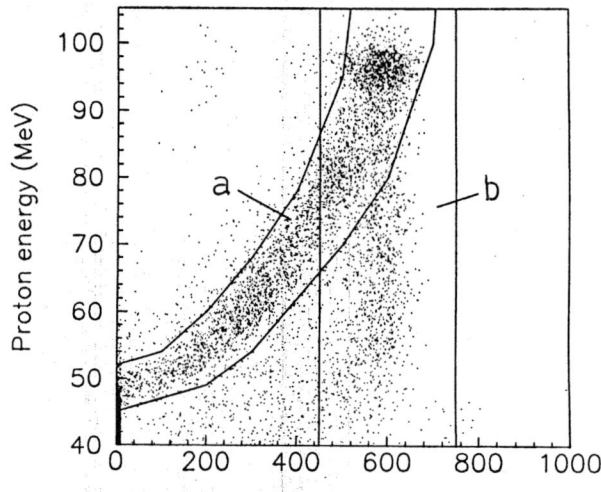

Fig. 8. Scatter plot (E_p vs n-tof) with two bands indicating (a) the n-p scattering events generated by neutrons in the tail up to maximum energy (diagonal) and (b) the n-tof window on the maximum primary neutron energy (vertical).

A scatter plot of the proton energy versus neutron time-of-flight is shown in Fig. 8, recorded using a CH$_2$ target. The n-p scattering events generated by neutrons in the tail up to full energy form a diagonal band (a) shown in the figure, whereas events induced by the maximum primary neutron energy correspond to the vertical band (b), which in turn corresponds to the setting of the time-of-flight window. The proton spectra from n-p scattering (see Fig. 8 band (a)) are obtained by subtracting carbon spectra from those of CH$_2$ after normalizing according to the carbon content of CH$_2$ in relation to that of the graphite sample and to the integrated neutron flux. From these spectra and n-p cross section values, which are calculated from phase shifts using the SAID

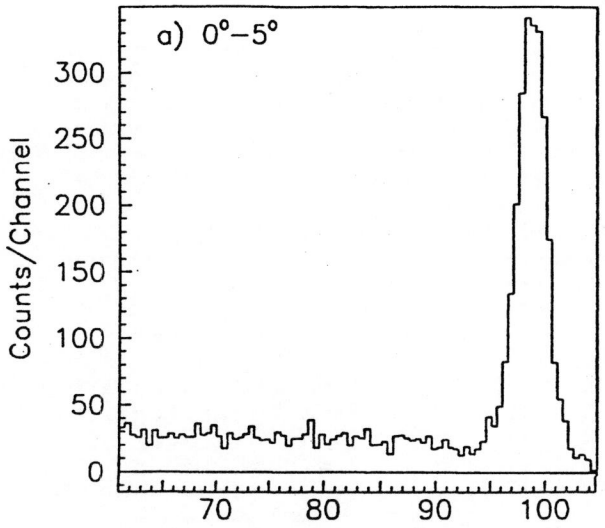

Fig. 9. Neutron spectrum corresponding to band
(a) in Fig. 3 after subtracting the
contribution from the $^{12}C(n,p)^{12}B$
reaction and normalizing on the basis of
the n-p scattering cross sections.

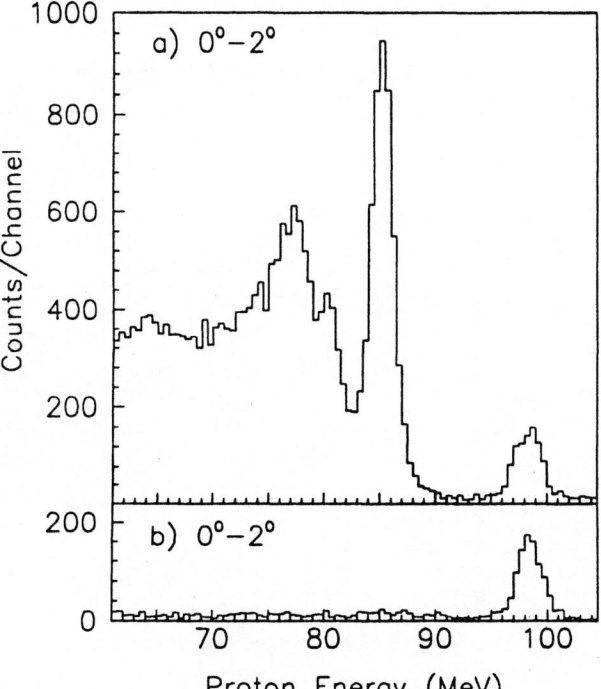

Fig. 10. (a) Uncorrected proton energy spectrum
for (n,p) reactions in ^{12}C induced
by 100 MeV neutrons. The peak at
98.5 MeV is due to n-p scattering
from the hydrogenous material of
the target box.
(b) The corresponding normalized background spectrum.

computer code of Arndt et al. [16], the neutron
spectra are reconstructed. Such a spectrum is
shown in Fig. 9. It can be seen that the full
energy peak is accompanied by a low-energy tail.

The accuracy of the relative n-p cross
section (in relation to the value calculated
with SAID) is 2 - 3 % statistically and about
5 % systematically.

Experimental Results

Measurements of double-differential (n,p)-
reaction scattering cross sections at 100 and
160 MeV for ^{12}C, $^{54,56}Fe$, ^{90}Zr and ^{208}Pb are in
progress with the aim of studying isovector
multipole excitations in nuclei.

An uncorrected proton spectrum from the
$^{12}C(n,p)$ reaction is shown in Fig. 10a with the
corresponding normalized background in 10b. The
cross sections obtained in this measurement
exhibit the same main features as those obtained
by Brady et al. [17] at a neutron energy of
65 MeV. The three peaks observed correspond to
transitions to the ground state of ^{12}B and to
groups of excited states at 4.4 and 7.7 MeV.

Further developments to the facility and
the spectrometer may include studies of (n,n),
(n,n'), (n,d) and (n,α) reactions, as well as
polarized protons and neutrons.

WNR Spallation Neutron Source

At Los Alamos National Laboratory, the WNR
target-4 facility provides a source [7] that is
useful for time-of-flight experiments in the
neutron energy range from 0.1 to above 750 MeV.
This broad range is made possible by using the
pulsed 800 MeV proton beam from LAMPF to produce
spallation neutrons from a water-cooled cylindrical tungsten target, 7.5-cm long and 3-cm in
diameter. The target is located in a vacuum
chamber, 1.8-m diameter x 1.2-m high, mounted
inside a massive shield of iron and concrete
with seven holes for neutron beamlines. Each
hole in the shield has a remote-controlled,
removable mechanical shutter mechanism to shut
off the neutron beam from its flight path and to
provide some initial neutron collimation.
Downstream of each shutter there is additional
collimation and permanent magnets to deflect
contaminant charged particles. All flight paths
have one or more detector stations allowing six
experiments to be carried out simultaneously.
Data is acquired with a CAMAC-based Micro-Vax
data acquisition system. A layout of the WNR
Facility, highlighting the features of the
target-4 spallation source is shown in Fig. 11.

Beam Structure

In normal operation the WNR beam is chopped
into a pulse approximately 20 ns long containing
3×10^8 protons and bunched into the phase
acceptance of the accelerator. The separation
between those pulses can be adjusted in 360 ns
increments and is typically set between 1 and
4 microseconds, depending on the frame-overlap
requirements of the WNR experiments. At the exit
of the linac the beam consists of a train of
micropulses approximately 60 ps wide contained
in a macropulse envelope extending over ~ 800
microseconds. As many as 70 macropulses/s are
produced from the linac (at an interval of 1 µs
this would amount to 56,000 micropulses/s and
2.5 µa beam on the target). Limitations in the

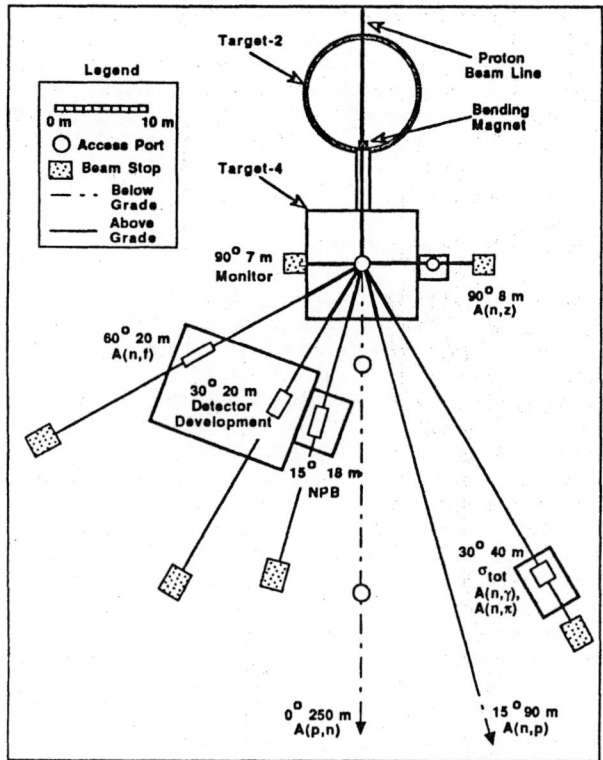

Fig. 11. Layout of the WNR target-4 facility. The proton beam enters from the top. Flight paths and detector stations associated with other WNR facilities were omitted from this drawing for clarity.

LAMPF beam transport system at present restrict the WNR to a maximum of 60 macropulses/s.

The LAMPF proton pulse disperses after acceleration to a time width of approximately 150 ps at the WNR target. There are two non-intercepting, wall-gap beam-pulse monitors [18] located along the proton beam line. The one chiefly used by experimenters is mounted about 23-m upstream of the center of target-4 and provides a precise fiducial point for time-of-flight experiments. The beam current is determined using a high-sensitivity toroid located near the center of target-2 (see Fig. 11). A monitor detector at a flight path of 7 meters which views the neutron production target at an angle of 90° is used by linac operators to adjust the beam position and size for optimum neutron output.

Experimental Facilities

In addition to the monitor flight path, there are six neutron beam lines, between 8 and 90 meters long, which view the target at neutron emission angles from 15 to 90 degrees. The beam lines are arranged to provide some tailoring of the neutron spectrum for various experiments. At 90 degrees, for example, the neutron flux above about 100 MeV drops rapidly, lower energy neutrons are less attenuated by the target, and the flight path uncertainty is the smallest, making these flight paths more suitable for measurements using the lower end of the energy range. In contrast, at 15 degrees the neutron spectrum extends to nearly 750 MeV and lower energy neutrons are more strongly absorbed by the target, resulting in a somewhat 'harder' spectrum.

Flight path lengths, experiment station locations and recent experiments are listed in Table 4. Information about facilities under construction is indicated in square brackets.

Over the past year several experiments have been performed on each flight path to measure the outgoing neutron spectrum. The measurements on 60R [19] have used neutron scattering from hydrogen as a standard, whereas all of the

Table 4. WNR Target-4 Flight Paths

Angle	Station Distance (m)	Recent Experiments
90L	8	0 to 30 MeV (n,p),(n,d),etc.
30L	[20] 40	3 to 750 MeV σ_t;(n,xγ);(n,π)
15L	90	30 to 350 (n,p); (n,d)
15R	11	(n,γ);(n,xγ);n-p Bremsstrahlung
30R	[20]	Defense program detector calibration
60R	20	$\sigma_f(E_n),\theta_f$, m_f);0.5-400 MeV (n,f)
90R	7	Facility neutron monitor

others (15L, 30L, and 90L) have used new fission cross section data [19] from WNR for either ^{238}U(n,f) or ^{235}U(n,f) to determine the neutron flux over the range from 3 - 250 MeV. Above 250 MeV the measurements assume that the fission cross section is independent of energy and has the same value as at 250 MeV. This assumption is based on the shape of high-energy neutron and proton-induced fission data. A comparison of measured and calculated [20] neutron spectra is shown in Fig. 12. Because of the difficulty in

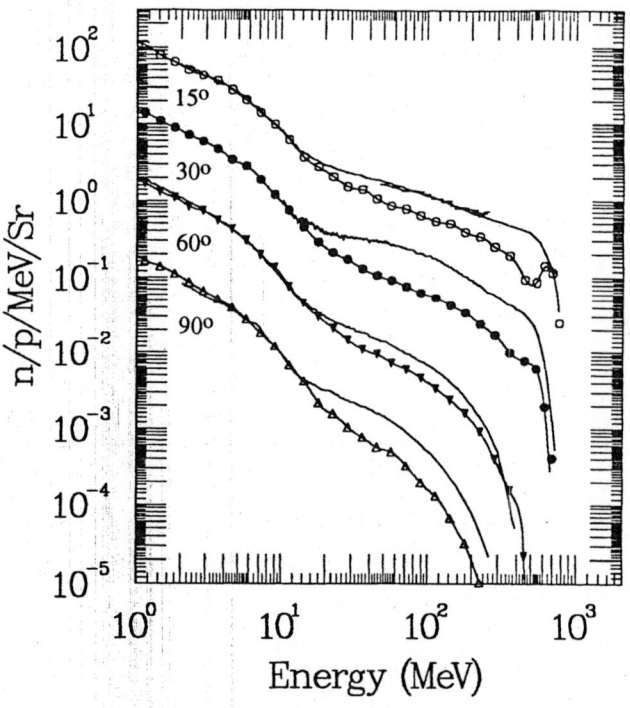

Fig. 12. Calculated and measured spectra from the WNR tungsten target. The curves with symbols show the results of INC calculations. The other lines are drawn through the measured data. To offset the curves, the 15° results have been multiplied by 1000, the 30° results by 100, and the 60° results by 10.

determining the proton beam current, the spectra were normalized to intra-nuclear cascade model predictions over the region near 10 MeV where previous measurements and calculations have shown good agreement. Two independent measurements performed on the 15L and 15R flight paths have the same relative normalization and are in excellent agreement with each other.

In addition to the target-4 'white' source, a monoenergetic neutron beam is also available at the WNR facility. Monoenergetic neutrons are produced with an energy spread of about 1 MeV at a reaction angle of 0° via the ^{7}Li(p,n) reaction and can be used simultaneously during target-4 operation. A detector station at a flight path of 250 m can use this neutron source for measurements near 800 MeV.

Recent Experimental Results and Conclusions

Listed in Table 4 are a few examples of experiments recently performed at WNR. The (n,xγ) work on the 30L flight path is re-presented at this conference by Ref. [21], and the fission work by Ref. [22,23]. The medium-energy (n,p) studies are described in two papers [24,25].

Facility improvements for the future include plans for upgrading the proton intensity by 75 % and investigating possibilities for producing polarized neutron beams using Mott-Schwinger or double scattering. We have begun a major new basic physics experimental program to measure the neutron-proton bremsstrahlung cross section [26] and are planning experiments to investigate cross sections for energetic neutrons incident on biological constituent elements and on shielding materials to provide data for the United States Space Exploration Initiative. Starting in 1992 we will be involved with experiments associated with the Los Alamos accelerator transmutation of radioactive waste program [27].

The WNR target-4 neutron source is proving to be a unique resource for neutron nuclear physics, one which is complementary to other sources, providing useful intensities for continuous coverage of an energy range practically untouched in the past.

REFERENCES

1. M. Drosg, O. Schwerer:
   "Production of Monoenergetic Neutrons Between 0.1 and 23 MeV",
   In: Handbook on Nuclear Activation Data, IAEA Technical Report Series No. 273, Vienna, p. 83 (1987)
2. M. Drosg:
   Nucl. Sci. Eng. 106, 279 (1990)
3. R.C. Haight, J. Garibaldi:
   Nucl. Sci. Eng. 106, 296 (1990)
4. S. Cabral, G. Börker, H. Klein, W. Mannhart:
   Nucl. Sci. Eng. 106, 308 (1990)
5. F.P. Brady, J.L. Romero:
   Nucl. Sci. Eng. 106, 318 (1990)
6. H. Condé, S. Hultqvist, N. Olsson, T. Rönneqvist, R. Zorro, J. Blomgren, G. Tibell, A. Hakansson, O. Jonsson, A. Lindholm, L. Nilsson, P.-U. Renberg, A. Brockstedt, P. Ekström, M. Österlund, F.P. Brady, Z. Szeflinski:
   Nucl. Instrum. Meth. A292, 121 (1990)
7. P.W. Lisowski, C.D. Bowman, G.J. Russell, S.A. Wender:
   Nucl. Sci. Eng., 106, 208 (1990)
8. J. K. Dickens:
   "RSIC Peripheral Shielding Routine Collection SCINFUL, Scintillator Full Response to Neutron Detection",
   Oak Ridge National Laboratory Report PSR-267 (1988).
9. D.L. Smith, J.W. Meadows, H. Vonach, M. Wagner, R.C. Haight, W. Mannhart:
   "Comparison of Activation Cross-Section Measurements Performed with Different Neutron Source Reactions in the 5 - 13 MeV Range,"
   these proceedings
10. M. Wagner, H. Vonach, R.C. Haight:
    "Measurement of the ^{60}Ni(n,p)^{60}Co Cross Section in the Neutron Energy Range 5 - 12 MeV",
    these proceedings
11. H.J. Brede, M. Cosack, G. Dietze, H. Gumpert, S. Guldbakke, R. Jahr, M. Kutscha, D. Schlegel-Bickmann, H. Schölermann:
    Nucl. Instrum. Meth. 169, 349 (1980)
12. G. Börker, R. Böttger, H.J. Brede, H. Klein, W. Mannhart, B.R.L. Siebert:
    "The Differential Neutron Scattering Cross Section of Oxygen Between 6 and 15 MeV",
    In: Nuclear Data for Science and Technology, Ed. S. Igarasi, Saikon Publ. Comp., Tokyo, 193 (1988)
13. H. Klein, H.J. Brede, B.R.L. Siebert:
    Nucl. Instrum. Meth. 193, 634 (1982)
14. J.W. Meadows, D.L. Smith, R.L. Greenwood, L.P. Geraldo, W. Mannhart, G. Börker:
    "Measurements of the Neutron Cross Section for ^{54}Fe(n,α)^{51}Cr between 5.3 and 14.6 MeV",
    these proceedings
15. G. Börker, W. Mannhart, B.R.L. Siebert:
    "Elastic and Inelastic Neutron Scattering on Carbon-12",
    these proceedings
16. R.A. Arndt, L.D. Roper, R.A. Bryan, R.B. Clark, B.J. Verwest, P. Signell.:
    Phys. Rev. D28, 97 (1983)
17. F.P. Brady, T.D. Ford, G.A. Needham, J.L. Romero, D.S. Sorenson, C.M. Castaneda, J.L. Drummond, E.L. Hjort, B. McFachern, N.S.P. King, D.J. Millener:
    UCD-CNL 246 (1987) and Phys. Rev. C, 43, 2284, (1991)
18. D.A. Lind:
    Can. J. Phys., 65, 637 (1987)
19. P.W. Lisowski, A. Gavron, W.E. Parker, J.L. Ullmann, S.J. Balestrini, A.D. Carlson, O.A. Wasson, N.W. Hill:
    "Fission Cross Sections in the Intermediate Energy Range",
    NEANDC Specialists's Meeting on Neutron Cross Section Standards for the Energy Region Above 20 MeV, Uppsala, Sweden, May 21 - 23, 1991
20: R.E. Prael and Henry Lichtenstein:
    "User Guide to LSC: The LAHET Code System", LA-UR-89-3014, Los Alamos National Laboratory (1989)
21. D.C. Larson, J.K. Dickens, R.O. Nelson, S.A. Wender:
    "White Source Gamma-Ray Production Spectral Measurement Facilities in the US",
    these proceedings

394

22. A.D. Carlson, O.A. Wasson, P.W. Lisowski,
    J.L. Ullmann, N.W. Hill:
    "Measurements of the ^{235}U(n,f) Cross Section
    in the 3 to 30 MeV Neutron Energy Region",
    these proceedings
23. P.W. Lisowski, A. Gavron, W.E. Parker,
    S.J. Balestrini, A.D. Carlson, O.A. Wasson,
    N.W. Hill:
    "Fission Cross Section Ratios for
    ^{233,234,236}U Relative to ^{235}U from 0.5 to
    400 MeV",
    these proceedings
24. A. Ling, A. Aslanoglou, F.P. Brady,
    R.W. Finlay, R.C. Haight, C.R. Howell,
    N.S.P. King, P.W. Lisowski, B.K. Park,
    J. Rapaport, J.L. Romero, D.S. Sorenson,
    W. Tornow, J.L. Ullmann:
    "Ground State Gamow-Teller Strength in
    ^{64}Ni(n,p)^{64}Co Cross Sections at
    90 - 240 MeV",
    submitted to Phys. Rev., (1991)
25. D.S. Sorenson, J.L. Ullmann, A. Ling,
    P.W. Lisowski, N.S.P. King, R.C. Haight,
    F.P. Brady, J.L. Romero, J.R. Drummond,
    C.R. Howell, W. Tornow, J. Rapaport,
    B.K. Parks, X. Aslanoglou:
    "The Energy Dependence of the Gamow-Teller
    Strength in p-shell Nuclei Observed in the
    n,p Reaction",
    submitted to Phys. Rev. Lett., (1991)
26. Lectures from the Workshop on Nucleon-
    Nucleon Bremsstrahlung,
    (Compiled by B.F. Gibson, M.E. Schillaci,
    and S.A. Wender), LA-11877-C, Los Alamos
    National Laboratory (1990)
27. P.W: Lisowski, C.D. Bowman, E.D. Arthur,
    P.G. Young:
    "Nuclear Physics Information Needed for
    Accelerator Driven Transmutation of Nuclear
    Waste",
    these proceedings

NEUTRON SOURCES FOR FUSION MATERIALS TESTING

F. H. Coensgen
Lawrence Livermore National Laboratory
Livermore, CA, USA

G. P. Lawrence
Los Alamos National Laboratory
Los Alamos, NM, USA

S. Cierjacks
Kernforschungszentrum Karlsruhe
7500 Karlsruhe, Germany

Abstract: Three neutron sources designed for evaluation of materials to be used in fusion reactors are described in this paper. In all three sources, neutrons are produced by injection of energetic particles into a target. They differ in the nuclear reaction utilized for neutron production and hence in the atomic species and energy of the injected particles and in the composition and construction of the targets. Each of these neutron sources, to a large extent, meets the evaluation criteria developed by the international fusion materials community at the request of the Executive Committee for Annex 2 of the International Energy Agency's Implementing Agreement on a Program of Research and Development of Radiation Effects in Fusion Materials. These criteria are[1]:

- Neutron flux ≥ 2 MW/m^2. The lower limit is that anticipated for a demonstration reactor. Accelerated materials testing requires flux rates several times this value. The upper limit is determined by the ability to maintain required temperatures in the materials specimens.
- Neutron spectrum should be as close as possible to that in the first wall of a fusion reactor. Spectrum tailoring should also be possible.
- Fluence. Irradiations of demo-relevant fluences (100 dpa, i.e. about 10 MW $\cdot$ yr $\cdot$ m^{-2}) require high source availability (> 70%).
- Irradiation volume. A volume of 10 liters is required in a region of flux ≥ 2 MW $\cdot$ m^2 ($0.9 \cdot 10^{18}$ n $\cdot$ m$^{-2} \cdot$ s^{-1} uncollided flux).
- Flux gradients. < 10% cm^{-1}.
- Accessibility. The proposed test volume must allow easy change-out of experimental assemblies.
- Time structure of neutron flux. A quasicontinuous operation is mandatory.

[Keywords: Fusion, Materials Testing, Nuclear Data, Neutron Source]

Beam Plasma Neutron Source

The Beam Plasma Neutron Source (BPNS)[2] is a plasma-based deuterium-tritium (D-T) system that is designed to provide end-of-life (≈ 100 displacements per atom) irradiation in one to two years. The source design is based on demonstrated plasma physics and technologies developed in the magnetic fusion energy program. However, because BPNS is a driven system rather than a power producer, it has been possible to design a simple system that is relatively small, has relatively low cost, and requires little, if any, development. By limiting the D-T reactions to a small volume, it is possible to provide the required neutron flux in sufficient volume with a minimum reaction rate, minimizing induced radioactivity and tritium consumption.

Neutron production is achieved by the injection of energetic (150 keV) deuterium atoms across a magnetic field into a fully ionized tritium plasma column that is confined radially by the magnetic field. The target plasma is sufficiently dense ($n_e \simeq 3 \cdot 10^{21}$ m^{-3}) to stop most of the injected D$^\circ$ atoms and is hot enough ($T_e \simeq 0.2$ keV) to increase the reaction rate significantly above that obtained with a solid target. A flux of 5 to 10 MW $\cdot$ m^{-2} of uncollided 14-MeV neutrons is achieved at the plasma surface for 50 MW of D$^\circ$ injection into a 0.16-m-diameter tritium plasma column that is confined in a 4-T linear magnetic field. For these parameters the total neutron production rate is $3.6 \cdot 10^{17}$ n $\cdot$ s^{-1} corresponding to approximately 1 MW of fusion power.

Expressed in dimensionless units, the plasma parameters of BPNS have been achieved in the 2XIIB experiment.[3] The injection geometry at the BPNS midplane is shown in Fig. 1, where it is seen that access for injection of deuterium atoms and sample placement is defined by the quadrupole magnetic. This magnet provides for magneto hydromagnetic (MHD) stability of the plasma system. An important feature of BPNS is that power deposited at the center flows along the plasma column to end regions where the expanding magnetic field allows it to be deposited on large area walls and to be removed by conventional water cooling. Density of the plasma column is maintained by tritium flow from the end chambers. The fueling rate is adjusted by controlling the pressure difference between end cells. A complete description of the system is given in Ref. 2.

Equal-intensity contours of the uncollided 14.1-MeV neutron flux are shown in Fig. 1 as solid lines. Close to the reacting plasma at the midplane (shaded area), the 14.1 MeV neutron flux decreases approximately as r^{-1}. At large radii the flux decreases more rapidly due to the finite length of the reacting plasma. The flux contours shown in the vertical plane can be elongated by re-aiming the

beams to lengthen the volume of neutron production. For constant beam power, the total neutron production rate remains constant. This rate scales approximately linearly with injection power, suggesting the feasibility of phased construction. In Fig. 2 the neutron irradiation capabilities of a number of BPNS operating points are compared with that of FMIT, an accelerator-based neutron source described by A. L. Trego et al.[4] BPNS operation at 50 MW provides an 8-liter sample volume in which the 14.1-MeV neutron flux exceeds $10^{18} n \cdot m^{-2} s^{-1}$ corresponding to approximately 2 MW $\cdot m^{-2}$. The test volume for wall loading greater than 5 MW $\cdot m^{-2}$ is 0.4 liter.

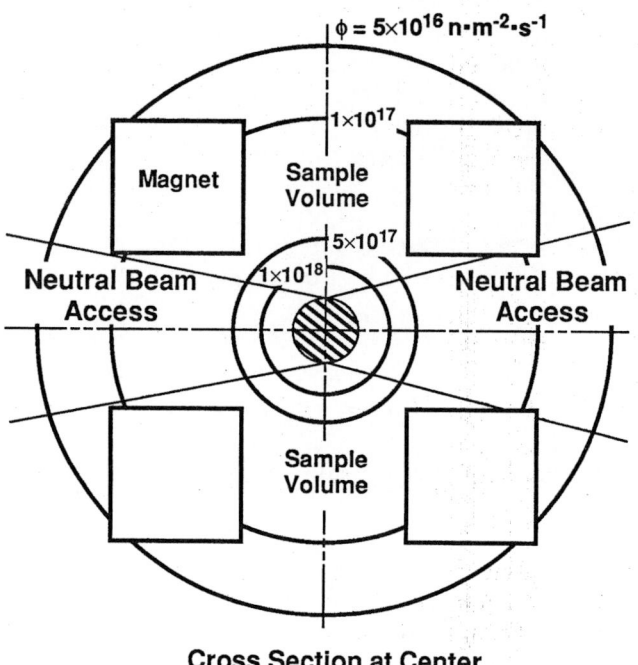

Cross Section at Center

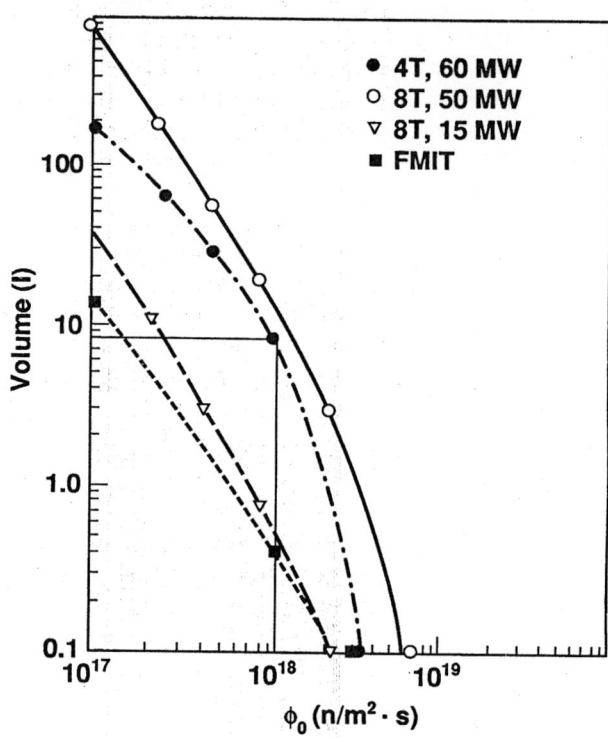

Fig. 2. BPNS sample volumes with $\phi > \phi_0$, 150-keV, D° injection.

In addition to the 14-MeV flux, a low-energy secondary neutron component arises from interaction of the 14-MeV primary neutrons with the surrounding structures. Intensity and energy spectrum of the secondary component depend on the composition and location of structures. Two neutron spectra at a radial position of 0.08 m in the midplane have been derived for the 50-MW, 4-T operation and are shown in Fig. 3. In one case, the structure was limited to a 0.08-m-thick vessel of 0.215 m internal radius. For the second case, a 0.30-m-thick aluminum cylinder (internal radius = 0.223 m) was placed outside of the vacuum vessel. The spectrum in a materials sample is further modified by interaction of the ambient neutron population with the sample holder, coolant, instrumentation, and the sample itself. The high-energy component of the neutron spectrum is essentially a narrow peak at 14.1 MeV independent of position in the test volume.

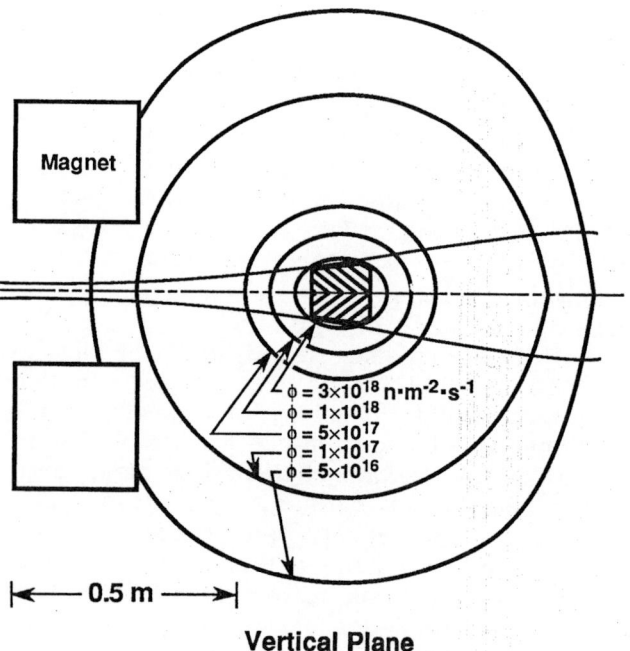

Vertical Plane

Fig. 1. BPNS cross-sections showing flux contours of uncollided 14-MeV neutrons.

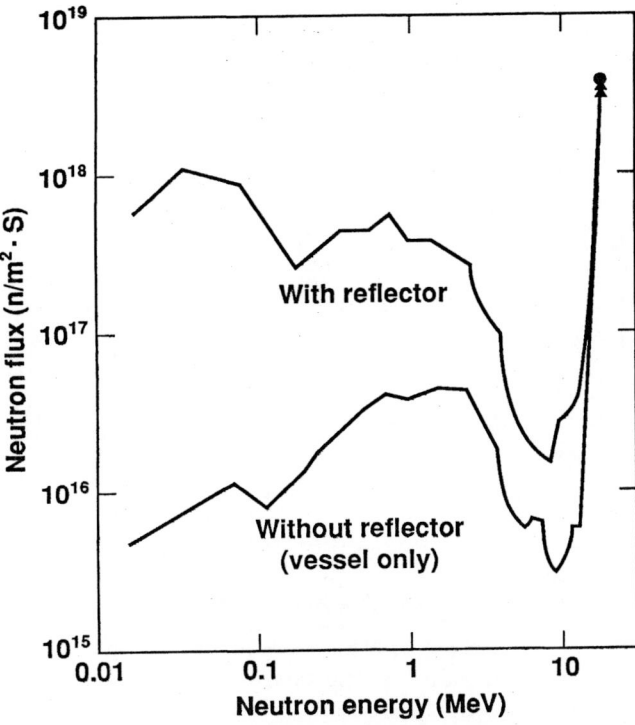

Fig. 3. BPNS neutron energy spectra at
z = 0, r = 0.08 m.

Deuterium-Lithium Neutron Source

In the Deuterium-Lithium (D-Li)[5] Neutron Source concept, 35-MeV deuterons incident on lithium generate a fusion-like spectrum from the Li(d,n) reactions. Much of the relevant technology was developed by the FMIT program in the early 1980s.[4] Although not implemented, FMIT was to provide a 100-mA D+ beam to a single lithium jet target, generating a 0.5-liter test volume exposed to a flux of 10^{18} n · m^{-2} · s^{-1}, and a 0.1-liter volume at 10^{19} n · m^{-2} · s^{-1}. Flux gradients were relatively steep because of the strong forward peaking of the neutron yield.

Recently an improved D-Li source concept[5] has been proposed that employs linac technology advances to supply higher beam currents than FMIT, and that uses a two-target/two-beam configuration to provide larger test volumes at a given neutron flux as well as lower flux gradients. The reference scheme, sketched in Fig. 4, consists of two D+ accelerator modules, each delivering a 250-mA continuous wave (cw) beam to two lithium jet targets oriented at 90°. An initial step in a staged deployment might consist of a single 250-mA (or lower current) module, with the output beam split and transported to two targets. As implied in the sketch, the total D+ current could ultimately be expanded to 1000 mA.

Figure 4 tabulates energy, current, frequency, and radio frequency power in the different accelerator sections of the D-Li source driver. Other parameters are listed elsewhere.[5] Preliminary beam simulations and accelerating structure designs were done to predict beam performance and power consumption. Structure frequencies are much higher than in FMIT, allowing a compact accelerator and improved beam quality. A 1-MW cw 350-MHz RF source is available to drive the drift tube linac (DTL). The high-energy beam transport (HEBT) in each module incorporates a cavity that increases the beam energy spread to 1.0-MeV (rms), and an octupole magnet that flattens and widens the transverse beam distribution. At the target the beam distribution is Gaussian (1 cm rms) along the jet flow, while in the cross-flow direction it has a 4-cm-wide uniform distribution with Gaussian edges (1 cm rms). These measures reduce the peak power deposition in the jet to 0.7 MW/cm^2. In the reference configuration, the two lithium targets are oriented at 90° and spaced 10 cm from a common vertex. The jets are formed by nozzles into 2-cm-thick ribbons flowing at 17.3 m/s along a concave steel back-wall. A free surface faces the beam. The 35-MeV D+ ions stop in the lithium jet, depositing most of their energy, but centrifugal pressure induced by the curved flow path prevents vapor formation. Thermal-hydraulic simulations were done to confirm jet behavior, showing that the output lithium temperature (672° C) is less than that estimated for FMIT because of the broad beam distribution.

Neutrons are generated in the lithium target primarily through the ^{6}Li(d,n)^{7}Be stripping reaction, with smaller contributions from other processes. The neutron angular distribution is forward peaked. Within a cone centered at 0°, the spectrum has a peak at an energy of 14 MeV. A small fraction of the flux is in a high energy tail (up to 50 MeV) arising from the ^{7}Li(d,n)^{8}Be reaction.

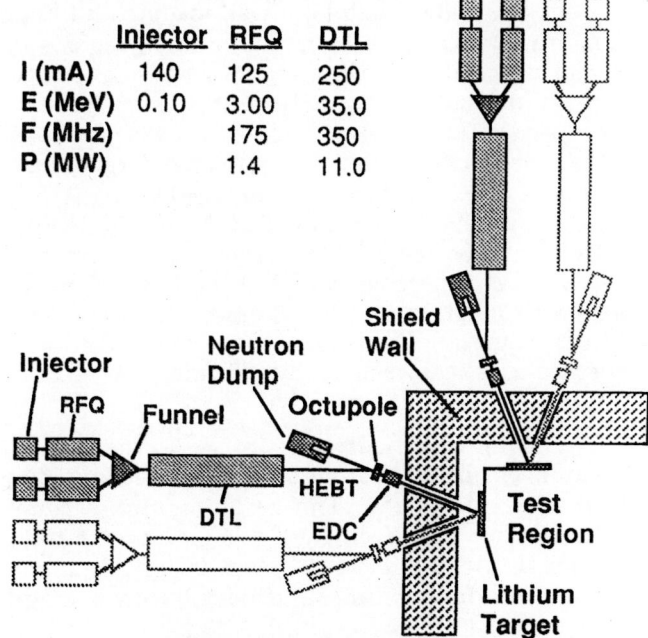

	Injector	RFQ	DTL
I (mA)	140	125	250
E (MeV)	0.10	3.00	35.0
F (MHz)		175	350
P (MW)		1.4	11.0

Fig. 4. Reference D-L neutron source: two 250-mA modules and two targets. Lightly-drawn modules indicate upgrade path.

Flux-volume and spectral properties in the test cell are determined by the beam distribution (in space and energy) within the target and by the dependence of the reaction cross-sections on D^+ energy and neutron emission angle. The uncollided neutron flux at a grid of points in the test cell was computed by generating a differential cross-section table from fits to existing Li(d,n) data, and then integrating over the D^+ energy-loss function in the target and summing contributions from all parts of the two beam-target interaction zone.[5] The calculated flux contours in the plane of the D^+ beams are shown in Fig. 5, using a conversion to equivalent 14.1-MeV neutron wall-loading power to allow comparison with fusion-reactor parameters (10 MW $\cdot$ m^{-2} = 4.42 $\cdot$ 10^{18} n $\cdot$ m^{-2} $\cdot$ s^{-1}).

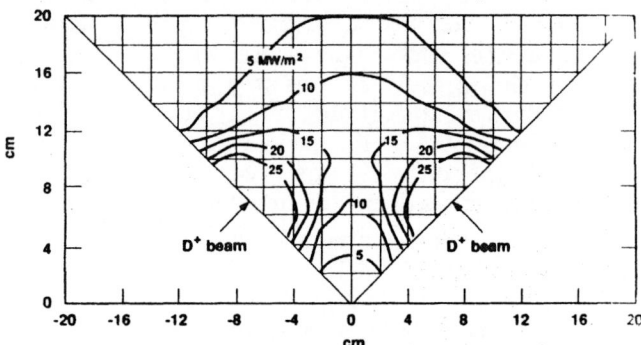

Fig. 5.   Neutron flux (wall-loading power) contours for two 250-mA beams: relative target orientation 90°.

Evident in the figure (reference case) are (1) large regions of low flux gradient and moderate flux in the center of the test cell, (2) small regions (≈ 0.1 liter) with high wall loading and high flux gradient close to the targets, and (3) regions in which small specimens can be exposed simultaneously to different flux levels. Comparison of this flux map with maps for other situations[5] reveals that as the relative target angle increases, flux gradients are reduced and the test volume at medium-to-low flux levels increases. Conversely, as the relative target angle decreases, flux gradients become steep but the test volume at high wall loadings greatly increases. Gradients and flux/volume relations can thus be tailored to suit specific user requirements by adjusting the relative target orientation and spacing.

Using the neutron flux grid, the useful volume enclosing regions of a specified average flux was calculated. The resulting flux-volume relations are displayed in Fig. 6 for several total currents. At the reference level (500 mA), a test volume of 10 liters can be achieved with average wall loadings up to 4 MW $\cdot$ m^{-2}. The flux-volume relations at fixed current and volume-current relations at a fixed flux can be represented by simple exponential expressions. For example, at 5 MW $\cdot$ m^{-2} wall loading, the dependence on current is approximately V (liters) $= 28\ I^{1.8}$ (A), which

explicitly reveals the strong advantage of scaling to higher D^+ currents.

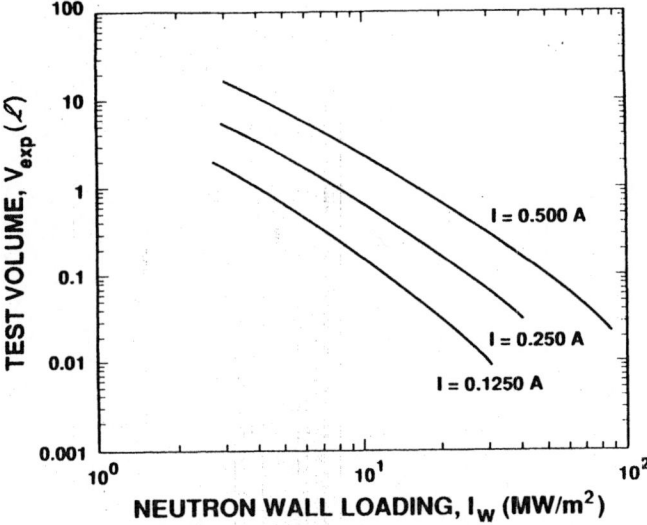

Fig. 6.   Test volume vs. uncollided neutron wall-loading power as a function of total beam current.

Neutrons produced in the lithium jet have a spectrum that depends on emission angle and distance from the target. At forward angles and away from the target, the stripping reaction dominates and there is a peak at half the beam energy. At large emission angles and closer to the target, the spectrum contains a higher proportion of lower-energy (few MeV) evaporation neutrons. Figure 7 shows the uncollided flux as a function of

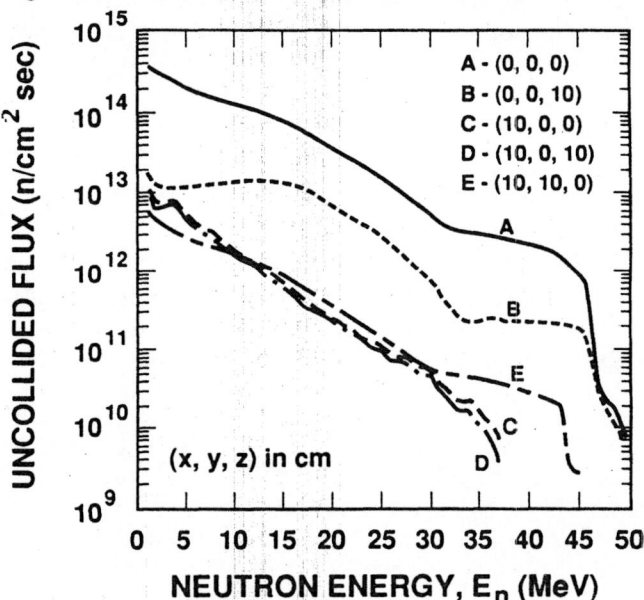

Fig. 7.   Uncollided neutron flux as a function of energy for several points within test region.

energy at various positions (x,y,z) within the test cell. The x-axis is along the beam, the y-axis along the jet flow, and the z-axis transverse to the flow. Position (0,0,0) is the center of the target back plate, the point of maximum flux. The high energy neutron tail is visible here, but even at this location only 15% of the neutrons have energies above

14 MeV. At other cell locations, the high-energy portion of the spectrum contains a fraction not larger than 20%. Preliminary analysis of damage data in copper[5] suggests that the high-energy tail will not significantly degrade the usefulness of the D-Li source as a fusion-damage simulator.

High-Intensity 14-MeV Cutoff Neutron Source

In the High-Intensity 14-MeV Cutoff Neutron Source described in Ref. 6, neutron production is based on the $^1H(t,n)^3He$ reaction. Energetic tritium ion beams are injected into two hydrogenic targets in a geometry similar to that described for the D-Li source. Tritium ions are accelerated in an RFQ/DTL accelerator similar to that described by Los Alamos National Laboratory for 250-mA deuterium beams.[5] The hydrogenic targets are provided by free water jets in which the beam-deposited power can be absorbed without disruption of the flow if a suitable beam-energy spread is provided. When the water jet targets are oriented at an angle of 120°, neutron fluxes $\geq 10^{19}$ n · m^{-2} · s^{-1} can be achieved in a volume of 0.16 liter, and fluxes $\geq 10^{18}$ n · m^{-2} · s^{-1} can be achieved in a volume of 4.2 liter. For a beam power of 17.5 MW, the corresponding values are 0.33 liter and 9.1 liter, respectively. The outstanding feature of this source is a high spectral intensity over the range from 1 to 14 MeV with a sharp cutoff at 14.6 MeV.

The global spectrum of this neutron source is shown in Fig. 8, where it is seen that the spectrum integrated over all angles from 0° to 67° is almost flat from ~1 to 14 MeV and exhibits a sharp energy cutoff level at 14.6 MeV.

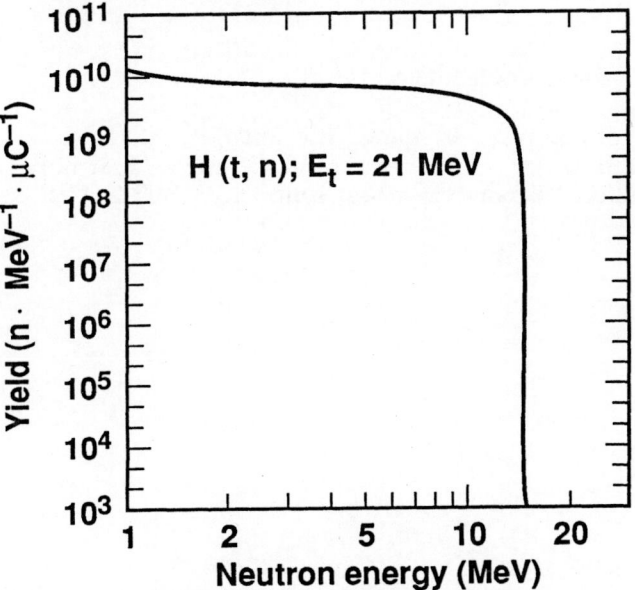

Fig. 8.   14-MeV cutoff neutron source global neutron spectrum (integral from 0° to 67°).

Figure 9 shows a flux-contour mapping for a configuration in which 250-mA beams are incident on two targets oriented at 120° and centered 8 cm from the vertex. The triton-beam diameters are

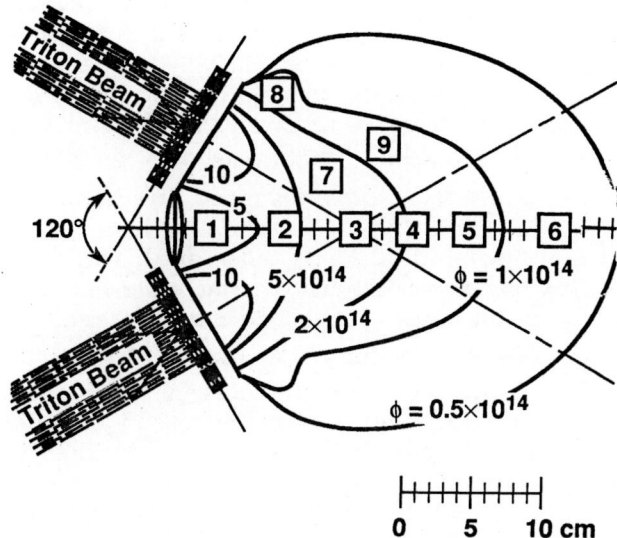

Fig. 9.   14-MeV cutoff neutron source neutron flux contours.

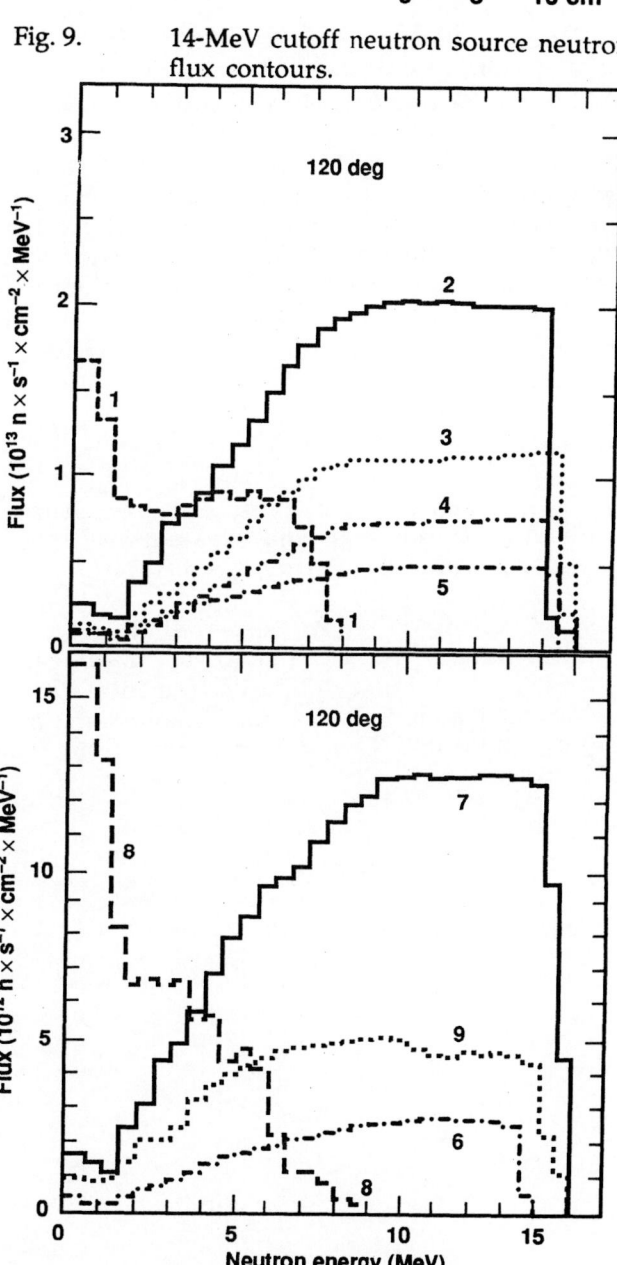

Fig. 10.   14-MeV cutoff neutron source primary neutron energy spectra at positions indicated in Fig. 9.

400

4 cm, and the contour lines refer to regions enclosing the indicated average neutron flux levels. It can be seen that there are relatively large regions enclosing the indicated average neutron flux levels and also that there are relatively large regions with low flux gradients (i.e., between $0.5 - 2 \cdot 10^{18}$ n $\cdot$ m$^{-2} \cdot$ s^{-1}). This provides good irradiation of sample specimens in the form of long bars or rods at intermediate flux levels when small variations of flux levels over the whole sample are required. Close to the water targets there are two isolated regions with very high flux levels ($>1 \cdot 10^{19}$n $\cdot$ m$^{-2} \cdot$ s^{-1}) but high flux gradients in which small materials specimens can be exposed. Similar calculations were performed for other relative target orientation angles and spacings.[7]

Uncollided neutron flux spectra at the positions indicated in Fig. 9 were estimated by simple superposition of contributions from both target volumes to obtain the spectra shown in Fig. 10. Spectra for positions not too close to the targets extend to cutoff levels close to 14.6 MeV and typically show a broad flat top over several MeV at the upper end. Below the flat-top region the spectral intensity usually decreases rapidly with decreasing energy. The most suitable test-cell regions to simulate first-wall conditions of a fusion reactor are at positions 3 and 4 where fairly large samples can be irradiated at flux levels in the range of $1-3 \cdot 10^{18}$ n $\cdot$ m$^{-2} \cdot$ s^{-1}.

References

1. D. G. Doran and J. E. Leiss, "Neutron Source Evaluation Process and Evaluation Panel Report," J. of Fusion Energy, **8**, 3/4 (1989).

2. F. H. Coensgen, T. A. Casper, D. L. Correll, C. C. Damm, A. H. Futch, B. G. Logan, and A. W. Molvik, "High-Performance Beam Plasma Neutron Sources for Fusion Materials Development," Nucl. Sci. and Eng., **106**, 138 (1990).

3. W. C. Turner, J. F. Clauser, F. H. Coensgen, D. L. Correll, W. F. Cummins, R. P. Freis, R. K. Goodman, A. L. Hunt, T. B. Kaiser, G. M. Melin, W. E. Nexsen, T. C. Simonen, and B. W. Stallard, "Field-Reversal Experiments in a Neutral-Beam-Injected Mirror Machine," Nucl. Fus., **19**, 1011 (1979).

4. A. L. Trego, J. W. Hagan, E. K. Opperman, and R. J. Burke, "Fusion Materials Irradiation Test Facility--A Facility for Fusion Materials Qualification," Nuclear Technology/Fusion, **4** (2) Part 2, 695 (1983).

5. G. L. Varsamis, G. P. Lawrence, T. S. Bhatia, B. Blind, F. W. Guy, R. A. Krakowski, G. H. Neuschaefer, N. M. Schnurr, S. O. Schriber, T. P. Wangler, and M. T. Wilson, "Conceptual Design of a High-Performance Deuterium-Lithium Neutron Source for Fusion Materials and Technology Testing," Nucl. Sci. and Eng., **106**, 160-82 (1990).

6. S. Cierjacks, Y. Hino, and M. Drosg, "Proposal for a High-Intensity 14 MeV Cut-off Neutron Source Based on the ^{1}H(t,n)^{3}He Source Reaction," Nucl. Sci. and Eng., **106**, 183 (1990).

7. S. Cierjacks and Y. Hino, "Additional Studies Related to the Proposal for a Novel High-Intensity 14 MeV Cutoff Neutron Source Based on the ^{1}H(t,n)^{3}He Source Reaction," Proceedings International Panel on 14 MeV Intense Neutron Sources Based on Accelerators for Fusion Material Study, Tokyo, Report NIFS-WS-2 (Jan. 1991).

Acknowledgements

*Work performed under the auspices of the U.S. Department of Energy by Lawrence Livermore National Laboratory under contract W-7405-ENG-48.

PREREQUISITES FOR ACCURATE ACTIVATION CROSS-SECTION MEASUREMENTS

G. Winkler

Institut für Radiumforschung und Kernphysik d. Univ. Wien,
A-1090 Wien, Boltzmanngasse 3, Austria

Abstract: Main emphasis is put on the measurement of activation cross-sections for fast-neutron induced nuclear reactions. The particulars that effect the reliability and accuracy of experimental energy-dependent cross-section data are discussed in the frame of relevant basic topics such as sample assay, source-field characterization, definition of the energy scale, optimization of irradiation conditions, modifications of the neutron field, fluence monitoring techniques and the implementation of references or standards, determination of the relevant activity induced in the samples, consideration of interferences, allocation of corrections, assignment and reporting of uncertainties. Examples are given and suggestions are made based on experiences and recent work in particular by the author and his co-workers.

[nuclear cross-section measurements, fast neutrons, activation measurements, accuracy, sample assay, neutron sources, neutron field characterization, energy scales, fluence monitoring, radioactivity measurements, interferences, corrections, uncertainties]

Introduction

As for applications the quantitative knowledge of nuclear activation cross sections is most important for the realization and control of nuclear fission and fusion energy devices. Although there already exists a vast number of data, the reliability and accuracy is still not satisfying for special purposes. A general view of the present status, data needs and ongoing efforts can be obtained form the very recent report on a specialists' meeting on Neutron Activation Cross Sections for Fission and Fusion Energy Applications [1]. On this occasion also an NEANDC working group on activation cross sections was constituted to deal with problems of common immediate interest and introduce collaborative research programs. Thoughts and remarks about requirements for obtaining good-quality cross section data are scattered throughout a huge amount of literature; review-like articles are, for instance, [2] and [3], where the latter particularly considers the impact of sample- and target properties on the accuracy of nuclear cross-section data.

The measurement of an energy-(E)-dependent microscopic reaction cross section $\sigma_R(E)$ is based on evaluation of the dependence of the number of reaction events R on the fluence [particles/cm^2] and on the number of sample nuclei which are exposed to this fluence. For the idealized case of a directed monoenergetic or quasi-monoenergetic fluence $\phi(E)$ of neutrons with energy (E), provided that the sample is sufficiently small and that a very small fraction of the incident neutrons experience interactions, this relation is in its simplest form given by

$$R = N_0 \, \sigma_R(E) \, \phi(E) \tag{1}$$

with N_0 the number of sample nuclei of a specific isotope, for which the microscopic reaction cross sections $\sigma_R(E)$ is studied. If the reaction product is a radionuclide, R can be determined by measuring at a time t after the end of the irradiation the decay rate A(t) from the number of nuclei produced by evaluating

$$A(t) = \lambda \, R \, F(\varphi(\bar{t}), \lambda, \bar{t}) \, e^{-\lambda t}, \tag{2}$$

where λ is the decay constant of the product nucleus and F is a function of the radiation history $\varphi(\bar{t})$ (i.e. the variation of the flux density [neutrons/cm^{2}s] with time $\bar{t}$ during the irradiation), which takes into account decay during irradiation.

Most practical experiments differ from this idealized situation. On the one hand (equ. (1)), reaction products of a specific species may be produced by a range of fluences with various energies (either due to the neutron source characteristics or due to interactions in the sample, target construction, or in the environment), with the spectral intensities even being different in different portions of the sample. On the other hand (equ. (2)), radioactive reaction products of various kinds may be produced by the effective neutron spectrum via different nuclear reactions or via different species in the sample (including impurities) leading to interferences in the measurement of the activity of interest. The quantitative estimate of the modifications to equations (1) and (2) is crucial for applying corrections or giving uncertainty statements in activation cross-section measurements.

Assay of the physical and chemical sample properties and the number of sample nuclei

For the majority of measurements the number N_0 of sample nuclei of a specific isotope can be deduced from the sample mass using information on the chemical and isotopic composition of the sample material. The mass may be determined by weighing (or via weight differences) which is in general the simplest way if applicable, or by other methods such as counting the emitted particles (or photons) in the case of unstable samples, or mass-spectrometric techniques (such as isotope dilution mass spectrometry) or other chemical or physical methods of quantitative analysis [4,5]. When relatively massy samples are used in an experiment, usually not the weighing errors are decisive, but the uncertainties in stoichiometry, impurities (including radiochemical purity) and isotopic abundance values (especially if enriched isotope samples are used). Samples, sample backings, or sample containers of low electrical conductivity may cause wrong balance readings due to the build-up of electric charges during handling. Discharging source backings or empty vials by means of an ultraviolet lamp before weighing and earthing the forceps may overcome such problems. For very low mass samples in particular it is possible to define the mass relative to a mass standard of the same or another material employing a transfer technique, for instance observing the relative responses in a proper radiation field. An example is the calibration of fissile deposits [3,6,7,8].

The physical and chemical stability of the sample material with respect to environmental conditions must be assured or investigated. Variations of the sample mass as a function of time are indications for sample instability or stoichiometric changes (e.g., [9,3]). They may result in mass values being too low or too high according as, for instance, sublimation takes place or oxidation or chemical reaction with CO_2 from the air. Thin-walled containers of organic materials (e.g. perspex or polyethylene), which may be profitable for avoiding interferences from activation of the encapsulation, may not provide a barrier good enough to exclude such interferences (e.g., [3,9]). Literature information on the most suitable form or compound to choose should be collected (e.g., [10]). Special environments (e.g., dry inert-gases or protective layers) may be necessary for handling (e.g., [3]). Whenever reasonable, alternative independent methods for sample assay should be applied to obtain a realistic estimate of the uncertainty of N_0. Essential specifications of the sample supplier should be checked when possible (for instance testing purity specifications by activation analysis). One should make certain that the analyzed part of the sample material is representative of that used in the experiment. Sample preparation and analyzing techniques may occasionally lead to an elemental or isotopic diversification.

Selection of the appropriate neutron source

Since activation experiments integrate over the time of the irradiation, no information can be immediately derived about the energy of the effective neutron as it is possible in combination with time-of-flight measurements. Therefore, pointwise measurements of energy-dependent cross sections require the usage of neutron sources well defined with respect to their neutron emission spectrum. The characteristics of the most important source reactions for the production of monoenergetic neutrons in the energy range 0.1-23 MeV are described in ref. [11] providing a lot of other useful references. The angular dependence of neutron energies and differential production cross sections are also available in the form of a computer code [12]. The ranges of monochromacy are discussed and possible interferences are dealt with. Most of the source reactions when used over a wide energy range suffer from the deficiency that the primary neutron group is contaminated by secondary neutron groups, either due to exited states in the exit channel or due to breakup of the projectiles or the target nuclei. More than 90% of all activation cross-section measurements have so far employed the reactions $T(p,n)^3He$, $T(d,n)^4He$, $D(d,n)^3He$, $^7Li(p,n)^7Be$ using electrostatic accelerators to produce incident charged particle beams of the proper energies (see, e.g. [2]). Discussions on other source reactions besides those treated in ref. [11] can be found, e.g., in refs. [13,14].

As long as there is a clear gap between primary and secondary neutrons and the yield of the latter is not dominating, depending on the slope and shape of the excitation function to be measured, corrections can well be applied for the contributions of parasitic neutron groups in a way of a bootstrap procedure starting with measurements at lower neutron energies and proceeding to higher ones, under the condition that the interfering neutron field can be characterized sufficiently well (e.g., [15], using the $D(d,n)^3He$ reaction up to 10 MeV). These corrections can be tested if parts of the excitation function to be measured can be covered by different source reactions (e.g., [15], using $^7Li(p,n)^7Be$ and $D(d,n)^3He$).

Contaminant neutrons may not only be inseparably linked to the source reaction itself, but can also be produced by reactions of the incident charged particles with the beam stop, other components of the target construction such as the entrance window or covering layers, or materials of the beam line. The influence of these structural components on the sample response (and possibly also on the fluence standard) can be investigated by means of a measurement with the primary neutron source medium being not present, e.g. in the gas-out condition in case of a gas target. But even so for a gas target an exact simulation of both the entrance-window and beam-stop effects may be difficult since the energy-loss due to the cell filling is missing. (For solid-state targets, see e.g. [8]).

A difficult region for neutron activation is from 8-14 MeV neutron energy, where the data base is still sparse. Neutrons down to about 12 MeV can be produced via the $T(d,n)^4He$ reaction using backward emission angles and deuterons in the few-MeV range (e.g., [16-18]). If backward angles are used, a contamination from higher-energy neutrons originally emitted at smaller angles and inscattered by the target backing or other structural material may be significant and has to be corrected for [16-18]. With the $D(d,n)^3He$ source, breakup neutrons seriously contaminate the neutron field above 8 MeV. To cover the critical region, reactions like $^{14}C(d,n)^{15g}N$ and $^{15}N(d,n)^{16g}O$ have been used (e.g., [19]), but without success, probably due to a deficiency in the corrections applied for contaminant lower-

energy neutrons associated with these source reactions (see ref. [15]). Two new developments have improved the situation for neutron cross section measurements in the region 8-14 MeV:

(a) Using time-of-flight techniques neutron emission spectra of the breakup neutrons were carefully measured [20] for the $D(d,n)^3He$ reaction for projectile energies up to 13.3 MeV relative to the yield for the main neutron peak for emission angles up to 15° (in order to permit irradiations at relatively short sample-source distances). This information has already been used to correct new activation data for $^{27}Al(n,\alpha)$ and $^{24}Mg(n,p)$ in the energy region 8-14 MeV [21].

(b) The $^1H(t,n)$ reaction was implemented at LANL employing a pressurized gas cell [22,23]. This reaction is in principle truly monoenergetic up to a neutron energy of 17.6 MeV, but contaminated by breakup neutrons produced by the triton beam mainly in the beam stop and entrance window. It has been shown that due to the high yield with the neutrons focused in the forward direction the lower-energy neutron background may be tolerable [22].

Within the NEANDC working group on activation cross sections a comparison of cross section measurements on the reactions $^{58}Ni(n,p)^{58}Co$ and $^{93}Nb(n,2n)^{92m}Nb$ is under way in the energy range 5-14 MeV using the $D(d,n)^3He$ source at PTB [20] and the $^1H(t,n)$ source at LANL [22] in order to check the validity of the different correction procedures needed to account for the deficiencies of these two source reactions [23].

The energy scale

Every real differential experiment is actually an integral experiment, with tighter limits on the experiment parameters. For example, a truly monochromatic source does not exist. Therefore, an experimental result has to be interpreted as an average of the microscopic cross section over the parameter space, which in activation experiments is the effective neutron energy profile. For nonlinear shapes of the excitation function the immediate average cross section extracted from the experimental data, σ_m, more or less differs from the cross section at the average energy $\sigma(E_m)$ (e.g., [15,24]). As a quick rule, for a smooth cross section and a symmetric energy profile the cross section at the mean energy E_m (averaged over the profile) will be larger or smaller than the uncorrected result σ_m dependent on the 2nd derivative of $\sigma(E)$ being negative or positive at E_m [15,24].

Paying attention to a reliable energy scale is particularly important when the excitation function is structured or shows a steep slope. Activation experiments at different energies can be performed simultaneously by exposing samples at different neutron emission angles relative to the incident charged-particle beam exploiting reaction kinematics. For beam energies in the few-MeV range the accurate knowledge of these angles is important. Energy scales can be transferred and checked by performing irradiations at different facilities at different beam energies, but with some samples exposed to about the same neutron energy at both facilities [16-18]. With lower beam energies the achievable span in neutron energy is much smaller but the experiment is less sensitive to the sample and beam spot position and to the beam energy. Shifts in the position of the beam spot centre can be detected by observing the response ratios (induced activities per unit mass) in samples being arranged around the beam spot symmetrically on the right and left side of the incident charged-particle beam axis [16-18]. Furthermore, the energy scale can be verified by observing the response ratio for a pair of monitor reactions with very different or opposite slopes of their excitation functions in the energy range in question (e.g., [9,16,25,26]). Shifts in the energy scale can also be detected by observing the spectra of charged particles produced by neutron reactions in a Si detector [27].

In any case, the uncertainty of the effective neutron energy will depend on the uncertainty of the effective beam energy initiating the source reaction. Therefore, uncertainties in the thicknesses and uniformities of the neutron producing target layer, entrance window, possible dead layers or other absorption layers (e.g., due to bad vacuum conditions or dirty targets [3]) have to be estimated, as well as uncertainties in the stopping-power data for the energy-loss calculations. An additional uncertainty in the average energy may arise from the dependence of the gas density on the cell temperature for a gas-target (dependent on the beam current), which is contingent on the energy deposition [15]. Therefore, it may make a difference whether the cell is operated at constant pressure or constant volume [28].

Neutron energy distribution profiles should be calculated (e.g., [29]) using relativistic kinematics, taking into account the relevant interaction processes (including small-angle scattering) and target parameters, to determine the effective average neutron energy and the energy spread (FWHM) (e.g., [29,7,8,18]). A critical target parameter with solid state targets is the concentration profile of the hydrogen isotope occluded in it (e.g., [7,8,29]).

Optimization of irradiation conditions and/or corrections for effects of sample-size, target-size and other environmental conditions

In practice an ideal experiment with (almost) massless samples on (almost) massless supports combined with an (almost) massless detector equipment and an extremely low-mass neutron producing target construction and beam line is not possible for various reasons. Therefore often corrections are needed (e.g., [15]) to take into account changes of the response of the samples (and of the fluence monitor, which may be affected differently) due to modifications of the incident neutron field by interactions in the samples themselves and in the material neighbouring the samples and the neutron source. The respective energy-dependent cross sections are required to consider the effects of scattering processes (in- and outscattering), attenuation due to nonelastic processes, production of secondary neutrons (by inelastic scattering or (n,2n) processes) or other radiation (see, e.g. refs. [8,15-18]). A proper choice of the materials for target backing, beam stop, entrance window, etc. may keep the corresponding corrections low (e.g., the use of Al instead of Cu or Fe). Impurities in the respective materials may occasionally play a role. Analytical calculations or Monte Carlo simulations (e.g., [15]) for the experiment may be useful to estimate corrections and their uncertainties. In particular, activation enhancement due to elastic scattering in the irradiated samples (leading to an increase of the average path length) has to be considered. One should take all efforts to design an experiment in a way that the corrections mentioned above are kept as small as possible and reasonable (compared to other sources of uncertainties). For instance, sample supports and target mounts should be of low-mass design (e.g., [16-18,30]).

Whenever possible the experiment should be performed in a low-scatter area. The effect of room-return neutron background can be studied by conducting irradiations for various source-sample distances (e.g.,[7]).

Neutron fluence measurement and the implementation of standards

The direct measurement of an activation cross section requires the knowledge of the fluence $\phi(E)$ according to equ. (1). Measurement of the fluence on an absolute scale (that means that the result is not proportional to a cross-section value) is possible by means of the associated-particle method (e.g., [31,32,8]) or using "black" detectors (e.g., [33]). The associated particle technique is capable of a very high accuracy for the determination of the neutron flux (e.g., [34, 8]), but is usually difficult to use. To exploit its potential requires defining the geometry with at least the same precision. Its practical applicability is restricted to certain neutron sources and the corresponding neutron energy bands, and for some source reactions, such as the $T(p,n)^3He$ reaction, demands the detection of low-energy associated particles in the presence of a huge background of Coulomb scattered charged particles from the incident beam with nearly self-supporting T-targets. The "black" neutron detector method cannot be applied in an open neutron field, but needs a collimated neutron beam.

Therefore, most nuclear cross-section measurements are in practice done relative to a cross section "standard", that means using a material and a specific reaction on it, the cross section of which is very well established [35-37]. The cross section in question is then derived from the ratio of the measured reaction rates in the investigated sample material and the reference material, being proportional to the reference cross section. Besides the good knowledge of the cross section value for reasons mentioned below the reference reaction should meet the requirements of

(a) a flat response (that means that the excitation function does not vary rapidly with energy),

(b) a proper reaction threshold (not too close to the energy region where it is used),

(c) sufficient sensitivity,

(d) in case that an activation fluence monitor is used, the half-life of the reaction product being sufficiently long compared to the irradiation time in order to enable proper integration over a time-dependent neutron flux.

In many cases the best choice is a reference which is least sensitive to contaminant lower-energy neutrons stemming either from the source or from reactions in surrounding materials (see conditions (b) and (c)). In most favourable cases the reference excitation function may exhibit a long flat plateau around its maximum (e.g., $^{93}Nb(n,2n)^{92m}Nb$ in the 14 MeV region [37,38]). The conditions (a) and (b) assure that the response of the reference is less sensitive to the correct energy scale of the experiment. In any case the actual value taken for the reference cross section should be stated to permit eventual renormalization of the data in case of new or better knowledge of the reference cross section involved.

Using reference reactions does not only involve the choice of a proper standard but also its implementation, that is employing a detection technique together with the reference material, which can in fact exploit the accuracy to which the standard is known. The most fundamental reference is certainly the hydrogen elastic scattering cross section, which is structureless and, also on theoretical grounds, currently the best-known cross section over a wide energy range, at least for $E_n < 10$ MeV [35,36]. But that does not mean that fluence monitoring based on this reference is preferable to any other technique, because the cross section is not the only quantity involved. The uncertainty of the efficiency of the (telescope) counter employed to detect the proton recoils and the uncertainty in the real density of the hydrogen atoms in the radiator foil (e.g. polyethylene) and its long-term stability are other limiting factors (e.g., [16,17]). Therefore, the overall uncertainty of recoil telescope measurements is still typically about 2.5% as international intercomparisons have shown [26]. Counting loss corrections have to be determined under actual irradiation conditions, for instance by looking for coincidence losses of pulses fed coincidently into all detector channels from an external pulser [16,17].

The fission reactions $^{235}U(n,f)$ and $^{238}U(n,f)$ are very reliably evaluated [39,40] over the whole energy range up to 20 MeV and can be implemented relatively easily using a well-defined fission deposit and prompt detection of fission fragments, for instance by means of a low-mass ionization chamber (e.g., [7,8,41]). Due to the quasi-threshold behaviour of its excitation function, the reaction $^{238}U(n,f)$ is most useful as a reference in the fast-neutron energy region ($\geq$ 2 MeV, when the excitation function reaches its first plateau) being less sensitive to parasitic low-energy neutrons. A discrepancy of 2-3% to generally accepted values of the $^{238}U(n,f)$ cross section has been recently pointed out for the 14 MeV range [8]. The problems that may arise with the usage of fission deposits, the corrections to be taken into account, the sources of uncertainties and suggestions for improvements have been extensively discussed in recent papers [3,42].

The geometrical set-up of an experiment often requires the transfer of the fluence value determined by the monitor to the position of the irradiated sample. Uncertainties in the knowledge of the respective positions relative to the source and in the fluence

404

gradient will affect the results. Problems will arise if sample and reference are of different shape and/or dimensions. It is up to the ideas of the experimentalist to implement possibilities that permit to better define the fluence conversion factors. Examples can be found, for instance in ref. 7, where due to a very low response different sizes of sample and reference had to be chosen and the irradiations performed close to the neutron source. In another experiment [18] the source-sample-reference geometry could be better established by using the ratio of activities induced in samples exposed symmetrically relative to the direction of the incident ion beam and a-priori information both on the source reaction kinematics and its angular distribution and on the shape of the excitation function to be studied, in this way getting more precise information on the position of the centre of the neutron source [18,43].

As can be gathered from the outline above, obviously the simplest way to monitor the neutron fluence is to employ as reference chemically stable metal monitor foils that can be obtained with sufficient purity and where the induced radioactivity of an activation product can be reliably measured without interference. These foils may be typically used in foil sandwich arrangements. In particular there exist several good choices for such references in the 14 MeV region (see [37,38,30]).

In any experiment the effect of possible contaminant neutrons has to be estimated or measured not only for the sample but for the reference as well.

Measurement of the relevant activity induced in the sample

Although, according to equ. (1), this task is a crucial one, it cannot be treated here in much detail. So much depends on the special circumstances.

Several comprehensive textbooks and review-articles are available on this subject (e.g., [44,45,46]). The majority of activity measurements is done comparatively to other sources of known activity or by means of detector devices pre-calibrated by reference sources [47]. In the following some remarks not commonly found in textbooks shall be given only.

In any case, a careful analysis of the induced activities may guard from large errors due to unexpected events. "Surprises may always occur". If γ-ray activities are to be measured, the radionuclei produced can best be identified by recording full γ-ray and x-ray spectra by means of a high-resolution Ge detector. Interferences may show up which one has not expected (see, e.g., [3, 18]) due to reasons already mentioned in the introduction. Adequate timing of the measurements can be very important to yield an optimum signal-to-background ratio. Commercial programs for data handling may work very well for routine tasks, but several critical cases have been found leading to errors in the spectrum evaluation. The fit to the background should always be checked by visual inspection. The efficiency calibration of the detector should be verified from time to time.

Decay scheme parameter uncertainties will often impede gaining accurate activation cross-section data. For the frequently used reference reaction $^{65}Cu(n,2n)^{64}Cu$, for instance, the accuracy could only be improved to about 1.5% in the 14 MeV range [48] when the decay scheme parameters of ^{64}Cu could be more reliably determined [49], which suddenly also eliminated discrepancies between previous γ-ray and β-ray measurements. In case that the half-life value $T_{1/2}$ is not very well known, the optimum choice for the time of the measurement - if other circumstances permit - will be around $t = 1/\lambda = 1.44\ T_{1/2}$ after the end of the irradiation (see equ. (2)). Special attention should be payed to corrections of the counting efficiency for sample-size effects (e.g., [7]). Samples irradiated close to the neutron source may be inhomogeneously activated. High-efficiency γ-ray counting with (almost) 4π geometry, e.g. with a large well-type NaI(Tl) detector [50], has proven to be a very powerful instrument both for the metrology of radionuclides with complex decay schemes (exploiting the effect of coincidence summing) being rather independent of a precise knowledge of the decay-scheme parameters,

and for measurements with volume sources [51,52]. Unfortunately this technique lacks in energy-discriminating capabilities. In several cases β-ray counting has proven to be very useful (e.g.,[8,53]). The problem of β^+ annihilation in γ-ray counting of β^+-ray emitters has been addressed in [37].

Interferences due to x-ray fluorescense may arise when characteristic x-rays are measured, either due to radiation from the sample itself [3] or induced by background [41,7]. A measurement with a blank (not activated) sample is always recommended.

For very thin foils the escape of recoiling radionuclides from a nuclear reaction may on the one hand cause a loss of radioactivity [54] and on the other hand contaminate adjoining material, e.g. in sandwich arrangements [3]. A thin separating organic catcher foil helps.

In order to properly correct for the radioactive decay during the irradiation (equ. (2)) every activation experiment requires to monitor the time dependence of the neutron yield in relative units.

Experiment reporting

To convey the full value of an experiment to the users of data and to enable data evaluators to generate full covariance matrix information the following requirements should be met [55,56].

All parameters that directly enter into the outcome of the experiment should be quoted, such as decay-scheme parameters, reference cross sections etc. to enable renormalization procedures. All corrections made and their estimated uncertainties should be mentioned. The allocated different sources of uncertainties should be stated in a detailed manner and the resulting uncertainty components given as standard deviations or equivalent standard deviations distinguishing between random and non-random uncertainties [55,56]. The outline should explain if uncertainty components are uncorrelated, correlated or partially correlated (containing random and non-random parts) over the energy range of an experiment. If an 'orthogonal' set of uncertainties has been defined such that there is no correlation between the different contributions for each single data point, the total uncertainties result from adding the uncertainty components in quadrature [57].

Ref. 15 was among the first ones that fully adhered to the recommended practice, refs. 7-9,16-18,30,41,48 provide other examples of proper experiment reporting.

REFERENCES

1. Proc. Specialists' Meeting on Neutron Activation Cross Sections for Fission and Fusion Energy Applications, 13-15 Sep. 1989, Argonne National Laboratory, USA (Edited by M. Wagner and H. Vonach), OECD, Paris (1990)

2. D.L. Smith, Remarks concerning the accurate measurement of differential cross sections for threshold reactions used in fast-neutron dosimetry for fission reactors, ANL/NDM-53, Argonne National Laboratory (1976)

3. G. Winkler, Nucl. Instr. and Meth. A282, 317 (1989)

4. Proc. 12th World Conf. of the Intern. Nuclear Target Development Society, 25-28 Sep. 1984, Antwerpen, Belgium (ed. by. J. van Audenhove and J. Pauwels), Nucl. Instr. and Meth. A236, 429-677 (1985), in particular: C. Wagemans; ibid., p. 429

5. C. Wagemans: Nucl. Instr. and Meth. A282, 4 (1989)

6. J.W. Meadows, D.L. Smith, G. Winkler, H. Vonach and M. Wagner, in ref. 1, p. 153

7. M. Wagner, G. Winkler,, H. Vonach and G. Petö: Annals of Nuclear Energy 15, 363 (1988)

8. G. Winkler, V.E. Lewis, T.B. Ryves and M. Wagner, The $^{238}U(n,f)$ cross section and its ratio to the $^{27}Al(n,\alpha)^{24}Na$ and $^{56}Fe(n,p)^{56}Mn$ cross section in the 14 MeV region, this conference

9. M. Wagner, G. Winkler, H. Vonach, Cs. M. Buczko and J. Csikai: Annals of Nuclear Energy 16, 623 (1989)

10. Gmelin, Handbook of Inorganic Chemistry; Springer, Berlin,

Heidelberg, New York (many volumes with updates for the various chemical elements)

11. M. Drosg and O. Schwerer, in Handbook on Nuclear Activation Cross Sections(ed. by K. Okamoto), IAEA, Techn. Rep. Ser. 273, Vienna (1987)

12. M. Drosg, Computer Code DROSG87: Neutron Source Reactions (ed. by O. Schwerer), available from the IAEA Nuclear Data Section, Vienna, Austria

13. Proc. Advisory Group Meeting on Properties of Neutron Sources, 9-13 June 1986, Leningrad, USSR, IAEA-TECDOC-410 (ed. by K. Okamoto), IAEA, Vienna (1987)

14. Proc. Consultants' Meeting on Neutron Source Properties, 17-21 March 1980, Debrecen, Hungary, Rep. INDC(NDS)-114/GT (ed. by K. Okamoto), IAEA, Vienna (1980)

15. G. Winkler, D.L. Smith and J.W. Meadows: Nucl. Sci. Eng. 76, 30 (1980)

16. A. Pavlik, G. Winkler, H. Vonach, A. Paulsen and H. Liskien: J. Phys. G: Nucl. Phys. 8, 1283 (1982)

17. A. Pavlik, G. Winkler, M. Uhl, A. Paulsen and H. Liskien: Nucl. Sci. Eng. 90, 186 (1985)

18. H. Liskien, M. Uhl, M. Wagner and G. Winkler: Annals of Nuclear Energy 16, 563 (1989)

19. A. Paulsen: Nukleonik 10, 91 (1967)

20. S. Cabral, G. Börker, H. Klein and W. Mannhart: Nucl. Sci. Eng. 106, 308 (1990)

21. G. Börker, H. Klein, W. Mannhart, M. Wagner and G. Winkler, in Proc. Int. Conf. on Nuclear Data for Science and Technology, 30 May - 3 June 1988, Mito, Japan (ed. by S. Igarasi), p. 1025, Saikon, Tokyo (1988)

22. D.L. Smith, J.W. Meadows, H. Vonach, M. Wagner, R.C. Haight and W. Mannhart, this conference

23. R.C. Haight, in ref. 1, p. 95 (1990); R.C. Haight and J. Garibaldi: Nucl. Sci. Eng. 106, 296 (1990)

24. D.L. Smith, Some comments on resolution and the analysis and interpretation of experimental results from differential neutron measurements, ANL/NDM-49, Argonne National Laboratory (1979)

25. V.E. Lewis and K.J. Zieba: Nucl. Instr. Meth. 174,141 (1980)

26. V.E. Lewis, Metrologia 20, 49(1984)

27. T.B. Ryves and K.J. Zieba: Nucl. Instr. and Meth. 167, 449 (1979)

28. J.W. Meadows, D.L. Smith and G. Winkler: Nucl. Instr. and Meth. 176,439 (1980)

29. A. Pavlik and G. Winkler, Calculation of the energy spread and the average neutron energy of 14 MeV neutrons produced via the T(d,n)^{4}He reaction in solid Ti-T targets, Report INDC(AUS)011/LI (INT(86)-6), IAEA Nuclear Data Section, Vienna (1986)

30. S.J. Hasan, A. Pavlik, G. Winkler, M. Uhl and M. Kaba: J. Phys. G: Nucl. Phys. 12, 397 (1986)

31. M.M. Meier, in Proc. Intern. Specialists' Symp. on Neutron Standards and Applications, 28-31 March 1977, Nat. Bureau of Standards, Gaithersburg, USA (ed. by C.D. Bowman et al., Special Publ. 493, p. 221, U.S. Government Printing Office, Washington (1977)

32. J.L. Fowler, J.A. Cookson, M. Hussain, R.B. Schwartz, M.T. Swinhoe, C. Wise and C.A. Uttley: Nucl. Instr. Meth. 175, 449 (1980)

33. F. Gabbard, same as ref. 31, p. 212 (1977)

34. H. Vonach, M. Hille, G. Stengl, W. Breunlich, and E. Werner: Z. Physik 237; 155 (1970)

35. Z. Bödy, in Handbook on Nuclear Activation Data, Technical Report Series No. 273, p. 29, IAEA, Vienna (1987)

36. Nuclear data standards for nuclear measurements, Technical Report Series No. 227, IAEA, Vienna (1983)

37. T.B. Ryves, A simultaneous evaluation of some important cross sections at 14.70 MeV, in ref. 1, p. 293; more details in: European Appl. Res. Rept.-Nucl. Sci. Technol. 7, pp. 1241-1334 (1989)

38. M. Wagner, H. Vonach, A. Pavlik, B. Strohmaier, S. Tagesen, and J. Martinez-Rico, Evaluation of cross sections for 14 important neutron-dosimetry reactions, Physics Data No. 13-5, Fachinformationszentrum Karlsruhe (1990)

39. N. Kocherov and H. Vonach, Intern. Reactor Dosimetry File (IRDF-90): Status and Testing, Proc. 7-th ASTM Symposium on Reactor Dosimetry, 27-31 Aug. 1990, Strasbourg, France (1990)

40. ENDF-201, 4th edition, ENDF/B-VI Summary Documentation, ed. by P. Rose, Brookhaven National Laboratory Report (in press)

41. M. Wagner, G. Winkler, H. Vonach and H. Liskien, in Proc. Int. Conf. on Nuclear Data for Science and Technology, 30 May - 3 June 1988, Mito, Japan (ed. by S. Igarasi), p. 1049, Saikon, Tokyo (1988)

42. V. Drapchinsky: Nucl. Instr. Meth. A282, 308 (1989)

43. H. Liskien, Improved neutron fluence accuracies in activation experiments, in ref. 1, p. 111 (1990)

44. A Handbook of Radioactivity Measurement Procedures, NCRP - Report No. 58, 2nd edition, Nat. Council on Radiation Protection and Measurements, Bethesda, MD., USA (1985)

45. W.B. Mann, A. Rytz and A. Spernol: Appl. Radiat. Isotopes 39,No. 8, 717-937 (1988)

46. G. Winkler, Measurement of Radioactivity, in Proc. Interreg. Training Course on Ensuring Measurement Accuracy, 11-21 Sep. 1984, Seibersdorf, Austria (ed. by Internat. Measurement Confederation -TC8), Vol. II, p.144, Budapest, Hungary (1984)

47. G. Grosse and W. Bambynek, International Directory of Certified Radioactive Sources, Physics Data 27-1, Fachinformationszentrum Karlsruhe (1983)

48. G. Winkler and T.B. Ryves: Annals of Nuclear Energy 10,601 (1983)

49. P. Christmas, S.M. Judge, T.B. Ryves, D. Smith and G. Winkler: Nucl. Instr. and Meth. 215,397 (1983)

50. W. Mannhart and H. Vonach: Nucl. Instr. Meth. 136,109 (1976)

51. A. Pavlik and G. Winkler, Int. J. Appl. Radiat. Isotopes 34, 1167 (1983)

52. G. Winkler and A. Pavlik, Int. J. Appl. Radiat. Isotopes 34, 547 (1983)

53. T.B. Ryves, P. Kolkowski and K.J. Zieba: Metrologia 14,127 (1978)

54. P.Kolkowski and T.B. Ryves: Nucl. Instr. Meth. A276,539 (1989)

55. P. Giacomo, Metrologia 17, 69 (1981)

56. NEANDC-A-134, Treatment of data uncertainties in the EXFOR international data exchange files and recommendations from NEANDC to experimentalists and evaluators, Nuclear Energy Agency, Nuclear Data Committee Document from 20 March 1981 (1981)

57. J.W. Müller, Nucl. Instr. and Meth. 163, 241 (1979)

RAPID RADIOCHEMICAL SEPARATION PROCEDURES FROM AQUEOUS SOLUTIONS

N. Trautmann and G. Herrmann

Institut für Kernchemie, Universität Mainz, D-6500 Mainz,
Federal Republic of Germany

Abstract: The state-of-the-art in the field of rapid radiochemical separation procedures is illustrated by two systems using aqueous solutions as input: ARCA, an apparatus for repetitive separations, and SISAK, a centrifuge system for continuous separations. Applications presented are: Solvent extraction of element 105 investigated with 34-s ^{262}105 using ARCA, and decay scheme studies of the fission product 1.0-s ^{110}Tc by means of SISAK.

(High performance solution chromatography, microprocessor controlled Automated Rapid Chemistry Apparatus ARCA, extraction behaviour of element 105; on-line multistage solvent extraction, centrifuge system SISAK, decay studies of 1.0-s ^{110}Tc)

Introduction

The investigation of the chemical properties of the heaviest elements produced in nuclear reactions on a one-atom-at-a-time scale requires chemical procedures with excellent selectivity, high yield and fast performance because in such reactions complex product mixtures are generated and the half-lives of the longest-lived isotopes of the transactinide elements are in the range of only seconds. Similar requirements exist for decay studies of very short-lived nuclides isolated from fission, spallation or other complex nuclear reactions. Rapid chemical methods offer one possibility for unravelling complex mixtures of radioactive species. There are two main approaches to fast radiochemical separations: the batchwise discontinuous and the continuous operation. In discontinuous procedures, a nuclide is produced, chemically separated, and measured sequentially whereas in continuous procedures the target is irradiated permanently and extraction of the species from the target, chemical separation, and counting are performed on-line.

Two systems for the fast separation of individual elements are de- scribed: i) The ARCA system for high performance solution chromatography controlled by a microprocessor, and ii) the SISAK centrifuge facility for multistage solvent extraction. In both systems a gas-jet is used for the transport of the activities from the target area to the chemistry set-up. Illustrative applications are presented.

Rapid Chemical Methods

ARCA - an Apparatus for fast, repetitive separations

ARCA (Automated Rapid Chemistry Apparatus) is mainly designed [1] for the study of chemical properties of transactinide elements, especially of element 105, in aqueous solutions. The separations are carried out in an automated high-performance liquid chromatography system with micro-columns. The apparatus is built symmetrically and allows operations in an alternating mode.

The main components of ARCA consist of i) a catcher/chemistry unit that collects the activity from the gas-jet and carries out the chemical separation, ii) the peripheral components for high performance liquid chromatography (pumps, valves etc.), and iii) the electropneumatic part together with a control unit coupled to a microprocessor. The solvents contact only inert materials.

The catcher/chemistry part of ARCA is shown in Fig. 1. The body of this unit consists of three teflon parts pressed against each other to obtain a leaktight seat between the stationary and movable components. Three inlet tubes are connected to the upper part by fittings and the solvents are fed into the system through 0.3 mm (i.d.) capillaries connected to the fittings on the left and right side. The middle fitting connects the gas-jet capillary to one of the two polyethylene frits pressed in a Kel-F slider. Below the frit in the central position the bore has a connection to a ventilation system to pump off the carrier gas during collection of the clusters with the attached reaction products on the frit. Simultaneously, previously collected reaction products are washed from the other frit and a chemical separation takes place on the right side in Fig. 1. At the same time a

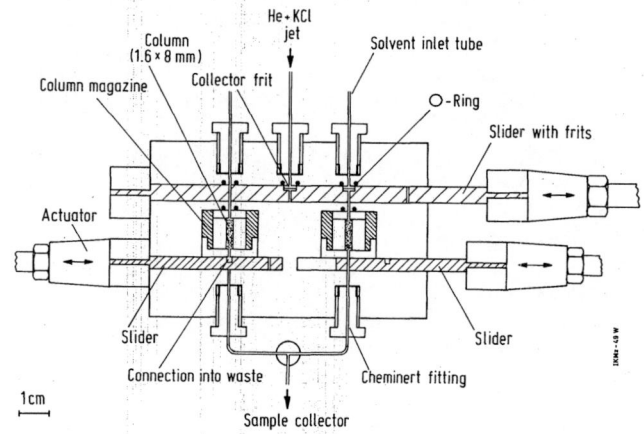

Fig. 1. View of the catcher/chemistry unit of the microprocessor controlled Automated Rapid Chemistry Apparatus ARCA [1].

magazine carrying 20 columns can be pushed forward to bring a new column into position (left side). After the separation is finished on the right column, the slide with the frits is moved to the left to start a new activity collection on the right frit, to bring a new column into position and to begin with a chemical separation on the left column. The columns are made by drilling 8 mm long holes with a diameter of 1.6 mm into Kel-F plugs pressed into holes in the stainless steel magazine. Pneumatically driven actuators push the magazine forward in a stepwise mode. Depending on the position the effluent from the column can be directed either to the sample collector or into the waste.

The sample preparation for α- and fission-fragment pulse height analysis occurs by fast evaporation of the effluent to dryness by means of intense infra-red light and hot helium-gas followed by a flaming and cooling step. Measurements could be started about 30 s after the end of collection.

Solvent extraction of element 105

The chemical properties of element 105 in aqueous solution are expected to be similar to those for the group VB elements and systematic trends between niobium and tantalum may continue for element 105. Recently it was shown that element 105 can be adsorbed on glass surfaces from concentrated nitric acid [2], a behaviour very characteristic for the VB elements. However it was not possible to extract fluoride complexes of 105 into methyl isobutyl ketone [2], under conditions where tantalum can be extracted quantitatively. To study the halide complex extraction of element 105, liquid extraction chromatography with triisooctylamine (TIOA) from hydrochloric acid solutions has been performed with ARCA. The data are compared with the properties of the VB elements niobium and tantalum, the IVB elements zirconium and hafnium and the "pseudo"-VB element protactinium. The extraction curves for Nb, Ta, Zr, Hf and Pa from hydrochloric acid into TIOA, as shown in Fig. 2, were obtained from batch experiments [3]. Investigations on element 105 have been performed with the longest-lived isotope 34-s $^{262}105$

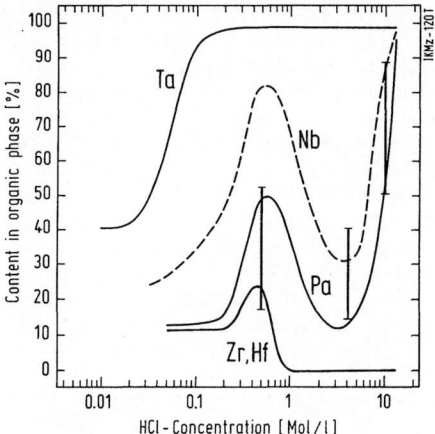

Fig. 2. Extraction of Ta, Nb, Pa, Zr, Hf and 105 into triisooctylamine in dependence on the HCl concentration. The behaviour of element 105 is indicated by the vertical bars [4].

produced with an average rate of about 0.5 atoms per minute in the $^{249}Bk(^{18}O,5n)$ reaction [2]. With ARCA automated separations of $^{262}105$ from HCl and HCl/HF solutions using micro-columns of TIOA adsorbed on an inert support were made [3] and it could be shown that either from 12 M HCl/0.02 M HF or 10 M HCl, 105 is adsorbed on the column more like Nb and Pa. In elutions with 0.5 M HCl/0.01 M HF and with 4 M HCl/0.02 M HF the positions of the 105-activity were found [4] to be not significantly different from that of Pa (Fig. 2). Tantalum is not eluted under these conditions. The results from these experiments indicate that the extraction behaviour of element 105 is different from that of its direct group VB homolog tantalum over a wide range of HCl concentrations but is similar to that of protactinium and niobium in the up and down of the extraction coefficients. This indicates the formation of oxo- or hydroxohalide complexes of element 105, like $[NbOCl_4]^-$ and $[PaOCl_4]^-$, in contrast to pure halide complexes, like $[TaCl_6]^-$. In total, 1700 separations had to be made to register 230 decay events from $^{263}105$.

Chemical studies of element 105 with diisobutylcarbinol as the extractant [5] and with α-hydroxy-iso-butyric acid solutions on cation exchange columns [4] have also been performed using the ARCA system. Furthermore, the new nuclide $^{263}105$ was discovered [6] in the bombardement of ^{249}Bk with 93-MeV ^{18}O ions and this activity was separated from the reaction product mixture with ARCA using cation exchange chromatography in 0.5 M α-hydroxy-iso-butyric acid. After chemical separation, ^{263}Ha was found to decay by spontaneous fission and by α-ray emission with a half-life of 27 s [6].

SISAK - a centrifuge system for fast, continuous separations

Fast continuous separations can be accomplished with multistage solvent extraction methods using high speed centrifuges for rapid phase separation, the so-called SISAK system (Short-lived Isotopes Studied by the AKUFVE technique). For fast performance small centrifuges were developed [7] and very recently, a mini-system, SISAK 3, has been put into operation [8] enabling to work with very low liquid flow-rates which makes it possible to feed the effluent directly into an α-spectrometric system of passivated ion-implanted planar Si-detectors (PIPS).

The SISAK set-up as outlined in Fig. 3 [9] consists of four main parts, namely a target system, a degassing unit, an array of four centrifuges for fast solvent extractions and a detection and data acquisition system. In the target area the reaction products recoil out of a thin target, are thermalized in a pressurized gas and attached to KCl clusters. In this form, they are transported through a capillary from the irradiation site to the degassing unit. There the reaction products together with the clusters are dissolved in a liquid phase and the mixture of gas and liquid is then fed into a degassing centrifuge designed for efficient gas/liquid separations [8]. The solution leaving the degassing unit is pumped to a centrifuge battery of one to four centrifuges, depending on

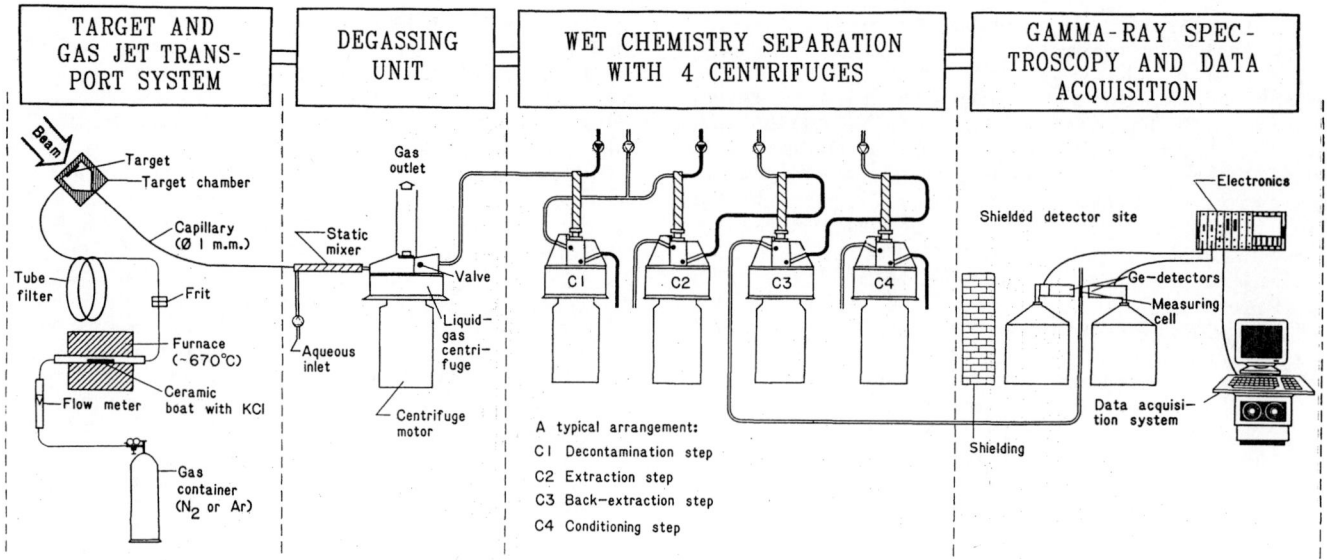

Fig. 3. Schematic drawing of the fast on-line separation system SISAK (not to scale) [9].

the number of extraction steps required to get a pure fraction of the element to be investigated. In the SISAK 3 version the centrifuges have only 0.3 ml internal volume. This yields at a maximum flow rate of 3 ml/s per phase a minimum hold-up time of only 0.1 s. The aqueous and organic phases are mixed in Y-formed inlet tubes of the centrifuge and separated at speeds of 25000-30000 revolutions per minute. The centrifuges are made of titanium stabilized with palladium, the other parts (pumps, tubes etc.) of titanium, teflon or Kel-F. The detection and data acquisition system comprises standard equipment for γ-ray spectroscopy. The measuring cell is either a thin-walled plastic flow cell or a small ion exchange column on which the isolated activity is retained for some time.

Decay studies of short-lived isotopes: 1.0-s ^{110}Tc

Thus far, SISAK has been applied for decay studies of short-lived nuclides of 13 elements [10]. It allows the investigation of nuclides with half-lives down to about one second. Especially the mass region of neutron-rich nuclei around mass number 110 is of interest where transitions between spherical and deformed nuclear shapes can occur. In order to extend the information on the nuclear properties towards the middle of the 50 and 82 neutron shells where maximum deformation is expected, ^{110}Ru has been studied after the β^--decay of ^{110}Tc by γ-ray singles and $\gamma\gamma$(t) coincidence measurements [11]. The ^{110}Tc activity was produced by thermal-neutron induced fission of ^{249}Cf and chemically isolated by a rapid solvent-extraction procedure as outlined in Fig. 4. The fission products are thermalized and transported from the target position by a KCl/N$_2$ gas-jet to a static mixer. There they are dissolved in diluted sulfuric acid containing tetravalent cerium to oxidize technetium to the pertechnetate. The gas-liquid mixture is pumped into a degassing centrifuge

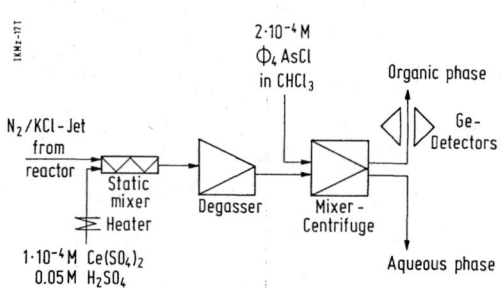

Fig. 4. Flow sheet for the rapid chemical separation of technetium from fission products [11].

where the nitrogen carrier gas and the fission noble gases are removed from the liquid. In the next step technetium is extracted into tetraphenylarsonium chloride dissolved in chloroform. After the extraction the organic phase is pumped continuously through a teflon tube to a measuring cell (2 cm^3) placed in front of γ-ray detectors. The overall hold-up time, i.e., the delay time from the target position to the detectors, has been determined to be 2.8 s.

The γ-ray singles spectra and the coincidences were measured simultaneously. The coincidence data were stored in list mode on magnetic tape as three parameter events and the spectra were created off-line by gating all strong γ-lines observed in each of the two detectors.

The level scheme for ^{110}Ru as fed by β^--decay of 1.0-s ^{110}Tc is presented in Fig. 5. For the neutron-rich ruthenium isotopes the transition between spherical and deformed nuclear shapes is very smooth and occurs over a coexistence between shapes of different deformation in the direction of strong asymmetric

REFERENCES

1. M. Schädel, W. Brüchle, E. Jäger, E. Schimpf, J.V. Kratz, U.W. Scherer and H.P. Zimmermann: Radiochim. Acta 48, 171 (1989)

2. K.E. Gregorich, R.A. Henderson, D.M. Lee, M.J. Nurmia, R.M. Chasteler, H.L. Hall, D.A. Bennett, C.M. Gannett, R.B. Chadwick, J.D. Leyba, D.C. Hoffman and G. Herrmann: Radiochim. Acta 43, 223 (1988)

3. J.V. Kratz, H.P. Zimmermann, U.W. Scherer, M. Schädel, W. Brüchle, K.E. Gregorich, C.M. Gannett, H.L. Hall, R.A. Henderson, D.M. Lee, J.D. Leyba, M.J. Nurmia, D.C. Hoffman, H. Gäggeler, D. Jost, U. Baltensperger, Ya Nai-Qi, A. Türler and Ch. Lienert: Radiochim. Acta 48, 121 (1989)

4. H.P. Zimmermann, J.V. Kratz, M.K. Gober, M. Schädel, W. Brüchle, E. Schimpf, H. Gäggeler, D. Jost, J. Kovacs, U.W. Scherer, A. Weber, K.E. Gregorich, A. Türler, B. Kadkhodayan, K.R. Czerwinski, N.J. Hannink, D.M. Lee, M.J. Nurmia and D.C. Hoffman: Annual Report, Inst. f. Kernchemie, Univ. Mainz (1991)

5. M.K. Gober, J.V. Kratz, H.P. Zimmermann, W. Brüchle, M. Schädel, E. Schimpf, H. Gäggeler, J. Kovacs, U.W. Scherer, A. Weber, K.E. Gregorich, A. Türler, B. Kadkhodayan, K.R. Czerwinski, R.A. Henderson, D.M. Lee, M.J. Nurmia and D.C. Hoffman: Annual Report, Inst. f. Kernchemie, Univ. Mainz (1991)

6. J.V. Kratz, M.K. Gober, H.P. Zimmermann, M. Schädel, W. Brüchle, E. Schimpf, K.E. Gregorich, A. Türler, N.J. Hannink, K.R. Czerwinski, B. Kadkhodayan, D.M. Lee, M.J. Nurmia, D.C. Hoffman, H. Gäggeler, D. Jost, J. Kovacs, U.W. Scherer and A. Weber: submitted to Z. f. Physik A (1991)

7. G. Skarnemark, P.O. Aronsson, K. Brodén, J. Rydberg, T. Björnstad, N. Kaffrell, E. Stender and N. Trautmann: Nucl. Instr. Meth. 171, 323 (1980)

8. H. Persson, G. Skarnemark, M. Skålberg, J. Alstad, J.O. Liljenzin, G. Bauer, F. Haberberger, N. Kaffrell, J. Rogowski and N. Trautmann: Radiochim. Acta 48, 177 (1989)

9. G. Skarnemark, M. Skålberg, J. Alstad and T. Björnstad: Physica Scripta 34, 597 (1986)

10. G. Skarnemark, J. Alstad, N. Kaffrell and N. Trautmann: J. Radioanal. Nucl. Chem. 142, 145 (1990)

11. T. Altzitzoglou, J. Rogowski, M. Skålberg, J. Alstad, G. Herrmann, N. Kaffrell, G. Skarnemark, W. Talbert and N. Trautmann: Radiochim. Acta 51, 145 (1990)

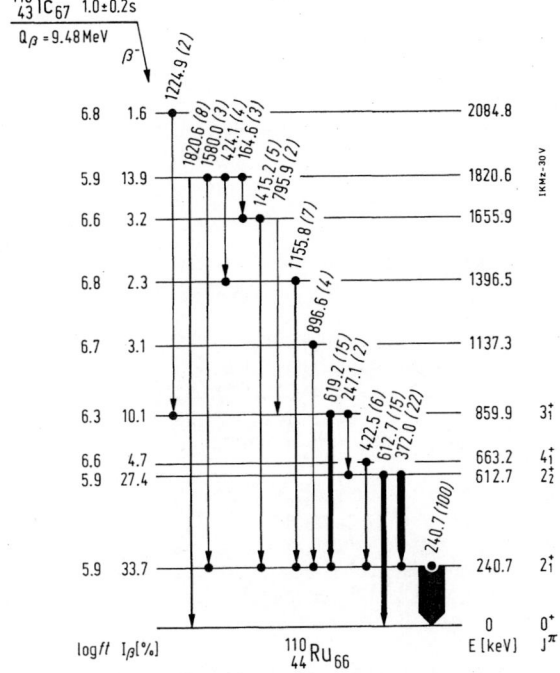

Fig. 5. Level scheme of ^{110}Ru as fed by β^--decay of 1.0-s ^{110}Tc [11].

deformation with pronounced softness towards triaxiality. Attempts have been made to describe the experimental data by different nuclear models.

Conclusions

The rapid chemical separation systems ARCA and SISAK are well suited for studies with and of exotic nuclei with short half-lives. So far they have mainly been used for investigations on fission products, actinide and transactinide elements. The time scale reached is close to the limit caused by diffusion of chemical species through boundary layers between two phases and by the velocity of phase separation. ARCA allows fast and repetitive separations within a few seconds whereas SISAK delivers a steady source of short-lived activity enabling detailed decay studies on nuclides down to a half-life of about one second and opening the possibility for on-line α-spectroscopic measurements.

FACILITIES FOR MEASURING (n,z) REACTIONS

S.M. Grimes
Ohio University, Athens, Ohio 45701 U.S.A.

and

R.C. Haight
Los Alamos National Laboratory, Los Alamos, NM 87545 U.S.A.

Abstract: Reactions of the type (n,z), where z denotes an outgoing charged particle, are quite difficult to study. If the charged particle is to be detected, a thin target is usually required, making the signal small compared to the background.

Two recently developed spectrometers take different approaches to this problem. An instrument developed at Ohio University uses a single detector system consisting of a ΔE detector and an E detector. The imposition of a coincidence condition and the requirement that the time–of–flight match that of a particular type of particle allow simultaneous measurements of both protons and alpha particles over the outgoing energy range reached by 6–11 MeV neutrons.

An alternative approach has been taken in designing a spectrometer at Los Alamos. Here, the source is also pulsed but is a white source. Thus, the charged–particle time–of–flight cannot be used as a separate constraint on the particle identification. A multi–detector array consisting of ΔE and E detectors has been assembled for studying (n,z) reactions between 4 and 30 MeV. The wide range of outgoing particle energies has made the initial measurements focus only on alpha particles, but future runs looking at protons are planned.

Results obtained with both systems will be presented, as will a comparison with other similar spectrometers.

I. Introduction

Reactions of the type (n,z) (a neutron–induced charged–particle–producing reaction) are of great importance in applied nuclear physics. Methods for treatment of cancer with neutrons involve ionization caused by the secondary charged reaction products of neutron–induced reactions. Fusion reactors will produce large neutron fluxes; both energy production and material damage considerations rest on (n,z) cross sections.

Many measurements of (n,z) cross sections have been based on measurement of residual radioactivity. These measurements are usually much simpler and require less equipment. The cross section is automatically integrated over energy and angle, which tends to improve counting statistics. Difficulties with this technique include the fact that some residual nuclei are stable, some reaction channels cannot be separated (for example (n,d) and (n,n'p)) and the fact that no emission energy or emission angle information is obtained.

Direct measurements of the outgoing particle spectra have some advantages but also some significant disadvantages. They do yield direct information on the angular and outgoing energy dependence of the charged–particle spectra. This information is especially important in determining kerma for medical purposes but is also of importance in reaction mechanism determinations. Direct particle spectrum measurement also is not affected by the stability or instability of the final nucleus. If the particles themselves are detected, a thin target must be used. This and the angular limits imposed by the need to get an angular distribution limit the counting rate. Finally, although there is no ambiguity concerning the type of particle emitted, there can be uncertainty as to what type of reaction was involved, that is, (n,2p) protons cannot be separated from (n,n'p) protons, for example. On the other hand, the availability of spectral shape information does allow a significant test of reaction models even if some

channels cannot be separated. The most favorable situation, of course, is where both activation and emission spectra information is available.

Measurements made for nuclear data purposes require a different approach than measurements carried out for spectroscopic purposes. The latter would likely focus on low lying levels of the residual nuclei and would likely concentrate on the reactions (n,p) and (n,d). These would measure charge–exchange and particle state strength distributions, respectively. Nuclear data measurements would normally be directed toward the decay channels with largest cross sections. For charged particle decay channels, these would be protons and alpha particles. Simple considerations of binding energy and Coulomb barrier explain the much smaller cross sections for production of deuterons, tritons and helium–3. The deuteron is only bound by 2.2 MeV, so picking up a proton costs the proton binding energy minus the 2.2 MeV. This, coupled with the Coulomb barrier which is the same as for protons, puts the deuteron at a disadvantage when competing with protons. Similarly, the triton and helium–3 are also weakly bound. Their approximate binding energy of 7 MeV is balanced by the need to pull two nucleons out of the target nucleus. This puts them at an even larger disadvantage than the deuteron. An alpha particle, on the other hand, is bound by about 28 MeV, more than compensating for the energy needed to pull three nucleons out of the nucleus. Surveys of (n,z) cross section values [1,2] support these predictions.

A spectrometer designed to measure cross sections for nuclear data applications would be optimized for the detection of protons or alpha particles. For most nuclei, equilibrium reactions comprise a large fraction of the reaction cross section at neutron energies below 20 MeV. These tend to produce many low energy particles. Light nuclei may have large cross sections for break–up and these modes may decay too rapidly for formation of a compound nucleus. Because of the low Coulomb barriers in these nuclei and because many of the break–up modes correspond to products of approximately equal mass, the charged particles produced in (n,z) reactions on light targets also tend to be at low energies.

II. Facilities

Two recently developed spectrometers for (n,z) reactions have attacked the problem of detecting charged particles over a wide range of energies in different ways. The first is a spectrometer [3] built and used at Ohio University. It has been used in a number of (n,xp) and (n,xα) measurements for monoenergetic neutrons in the neutron energy range 8 ≤ E_n ≤ 11 MeV. A diagram of the spectrometer is shown in Fig. 1. The spectrometer contains three detectors, all of which must be triggered for an event to be valid. A burst of neutrons approximately 1 ns in duration is produced by a short burst of deuterons incident on a gas target of deuterium. The charged particles produced in the target foil by the neutron beam then traverse two proportional counters in sequence and then are stopped after a 2.1 meter flight in a plastic scintillator. The pulse in the latter detector gives both pulse–height and timing information. By comparing the magnitude of the pulses in the two ΔE counters and that in the E detector and using the time–of–flight as well, the event is overdetermined. This allows identification of the particle and determination of its energy as well.

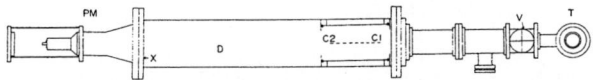

Fig. 1   The complete counter telescope showing the drift tube as well as the ΔE and E detectors [T = target ladder, V = isolation valve, (C_1, C_2) = two halves of proportional counter, D = drift tube, X = scintillator and PM = photomultiplier tube].

An unusual feature of this spectrometer is that it has been used to measure simultaneously protons and alpha particles over an energy range of about a factor of ten. This results in a range of pulse heights in the ΔE detector spanning about a factor of forty and about a factor of thirty in the E detector. The measurements with this spectrometer have been concentrated on targets for which A is about 55; these yield proton spectra with a range from 1 to 10 MeV and alpha particles with a range of 3.5 to 12 MeV when bombarded with neutrons of 11 MeV. Cross sections as small as 10 mb integrated over energy and angle have been measured with this spectrometer. Examples of spectra [4] obtained with this spectrometer are shown in Figs. 2 and 3.

It should be noted that, as the spectrometer has been used, it could detect deuterons, tritons and helium–3 particles as well. As indicated previously, these three channels have negative Q values for neutron–induced reactions. Small deuteron cross sections have caused the determination of these cross sections in many cases to yield no more than upper limits ($\stackrel{\sim}{<}$ 2 mb), while the (n,t) and (n,^{3}He) reaction channels are closed or nearly so for these targets at energies at or below 11 MeV.

The use of the D(d,n) reaction to produce the neutrons has the disadvantage that this source is not monoenergetic once the threshold for the reaction

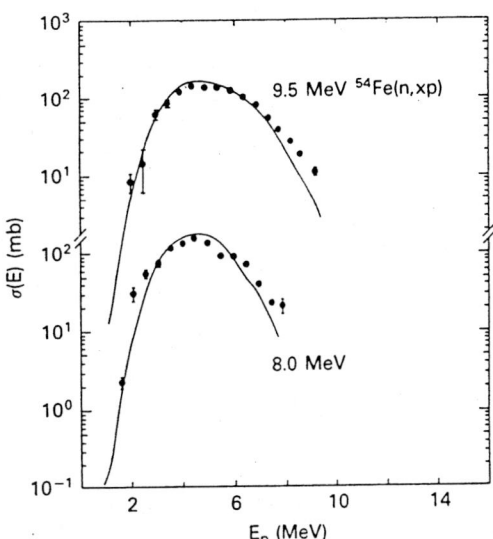

Fig. 2   Proton spectra from the ^{54}Fe(n,xp) reaction at 8– and 9.5–MeV bombarding energies. Circles represent data points and the line, the Hauser–Feshbach calculation.

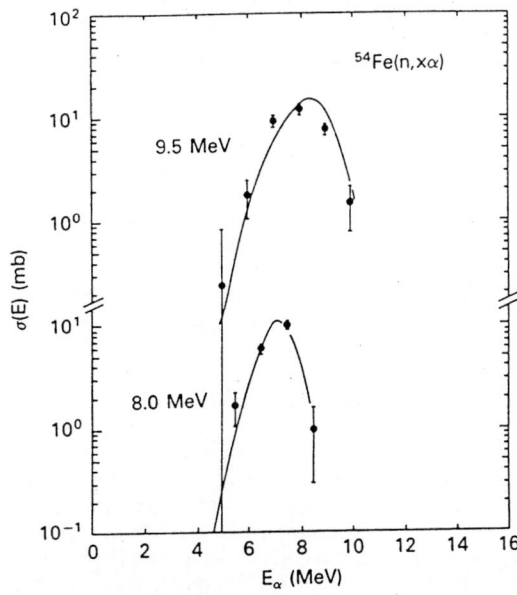

Fig. 3   Alpha spectra from the ^{54}Fe(n,xα) reactions at 8 and 9.5 MeV. Curves are Hauser–Feshbach calculations.

D(d,n+p)D is reached (E_d ≈ 4.4 MeV). As the bombarding energy is increased, the break–up neutrons increase in both number and energy. Particularly for the 11 MeV measurements, reactions induced by the break–up neutrons were appreciable. Corrections for such reactions were made using the ^{3}He(d,n+p)^{3}He reaction. As has been shown [5], the spectrum of neutrons produced by this reaction is similar to that of the break–up of deuterons by deuterons. By measuring the charged particle yield from bombardment with neutrons produced by this reaction, a correction to the measured charged particle spectra from neutrons produced by a deuterium target can be made.

412

A summary of the results from measurements on six targets in the A ~ 55 is that for neutrons between 8 and 11 MeV most of the charged particles are produced in compound nuclear reactions. Proton production is largest, followed by alpha production. The cross sections for other charged particles are so small they are compared to or less than the errors in the cross section determinations for protons and alphas. Second—stage decay channels open up over this range, but do not play the dominant role [6] that they do at 14 MeV. Cross sections have been measured for stainless steel, ^{58}Ni, ^{60}Ni, ^{63}Cu, ^{65}Cu, ^{54}Fe and ^{56}Fe.

Further measurements with this spectrometer are planned or underway. The cross section for n—d elastic scattering is being measured and the (n,α) cross section for ^{13}C will be measured soon.

A very different approach was used in designing the spectrometer used for (n,z) reaction studies at Los Alamos (WNR). This neutron source is produced by allowing 800 MeV protons to bombard a W target. Short (< 1 ns) bursts of neutrons spanning the energy range from kilovolts to 800 MeV are produced. These are allowed to traverse a 10 m flight path and then strike the (n,z) target. The short slight path and the angle at which the neutrons are observed (90°) means that neutrons above 50 MeV are few in number and would not be characterized by good resolution. Thus, the spectrometer is most useful for neutrons from 5 to 50 MeV.

The detection system must meet even more stringent requirements than that for the Athens spectrometer. A much larger range of charged particles can be produced at this facility. A coincidence between two detectors is highly desirable, to facilitate particle identification and to define the direction of travel of the particle. Once the particle is identified and the energy determined, the flight time from the (n,z) target can be evaluated and a correction to the neutron—time—of—flight can be calculated.

The wide range of energies which must be covered has made it impossible to detect both protons and alpha particles simultaneously. Data taken to date have been for $(n,x\alpha)$ reactions.

Fig. 4 shows the experimental set—up. Note that the neutron flight path allows not only an energy determination but collimation of the beam. In the chamber, counter telescopes consisting of proportional counters (ΔE counters) and solid state counters allow measurements to be made of cross sections at four angles at once. Fig. 5 shows the telescopes and scattering chamber.

Initial measurements have been made of the excitation function for $(n,x\alpha)$ reactions on ^{10}B and ^{27}Al. Cross sections for (n,α) on ^{10}B to the ground state of ^{7}Li are related through detailed balance to the cross section for ^{7}Li(α,n_0) of interest in astrophysics. A plot of the data showing the correlation between alpha particle energy and time—of—flight is shown in Fig. 6. The line corresponding to the most energetic alphas from ^{10}B(n,α) is especially evident.

Corresponding measurements for ^{27}Al$(n,x\alpha)$ have also been made. Based on the systematics observed near 14 MeV, it would be expected that the opening of energetically allowed channels would give increased production of alphas as the energy increases and this is found to be the case. A spectrum integrated over angle in a narrow bombarding energy range near 14 Mev is shown in Fig. 7.

The data for ^{10}B(n,α) appear to be usable over the range .9 MeV $\leq E_n \leq$ 6 MeV. This range results from the large cross section for low neutron energy and the very low cross section for higher neutron energies.

WNR facility LAMPF

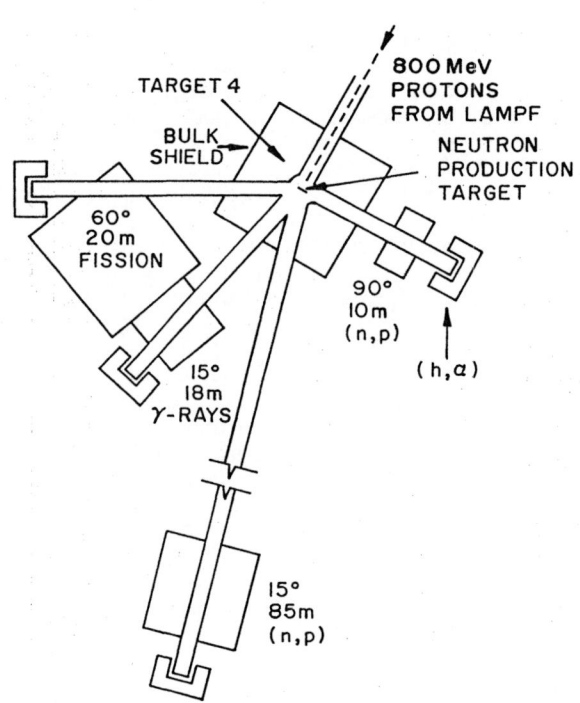

Fig. 4 Overview of the Target 4 area at WNR. The (n,z) measurements described here were performed on the 10 m leg at 90°.

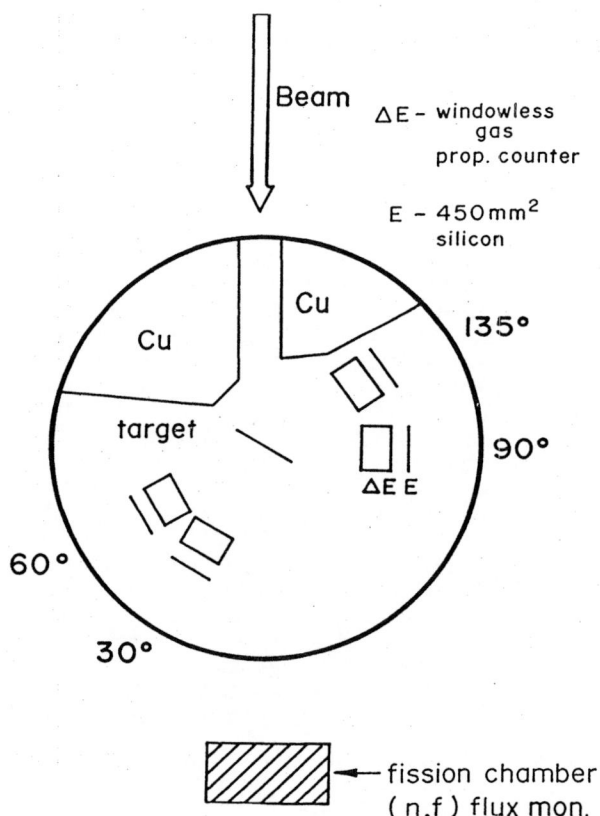

Fig. 5 Scattering chamber, counter telescopes and shielding used in the (n,z) measurements. The chamber is 10 m from the neutron target.

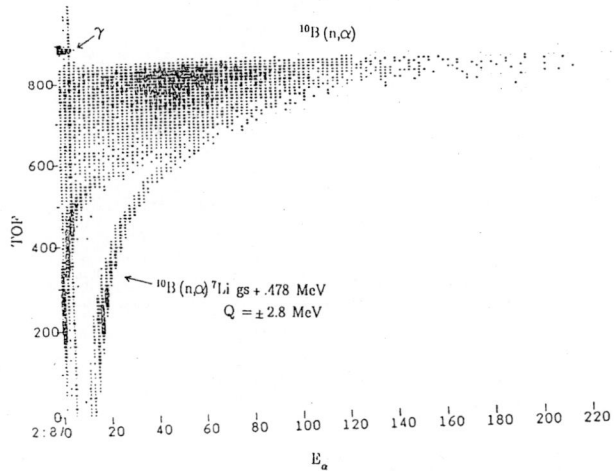

Fig. 6   Two–dimensional plot of alpha energy versus time of flight for the $^{10}B(n,x\alpha)$ reaction.

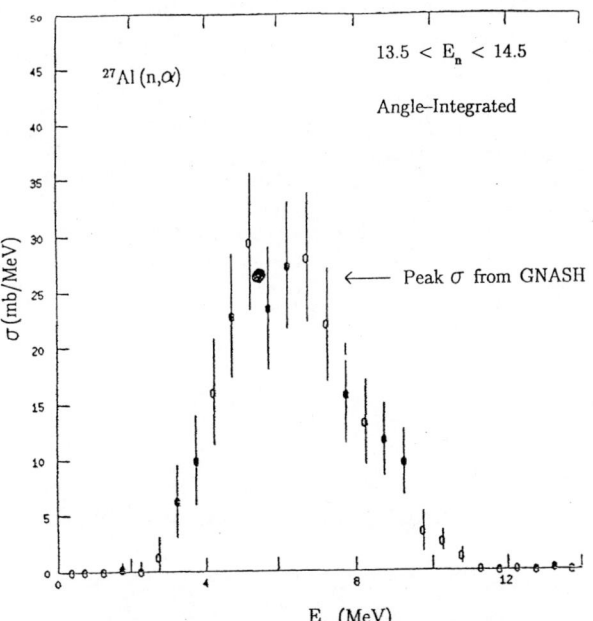

Fig. 7   Angle integrated cross section for $^{27}Al(n,x\alpha)$ at about 14 MeV neutron energy. The calculated cross section is the result of a multi–step Hauser–Feshbach calculation.

For the $Al(n,x\alpha)$ reaction the range of usable data is from 5 to about 35 MeV. Other measurements have been made for the $^{12}C(n,x\alpha)$ reaction.

Future plans include measurements of the $(n,x\alpha)$ cross sections for ^{14}N, ^{16}O, ^{56}Fe and ^{59}Co. These experiments should allow an improved understanding of reaction systematics and level densities from 10 to 30 MeV. Changing detectors will allow protons to be studied, and a measurement of the n–p elastic angular distribution is also planned.

III. Summary

Two recently developed charged particle spectrometers have been described. Both have been designed to look at a wide range of outgoing charged particle energies. They, therefore, have more in common with spectrometers in use at Vienna [8] and Geel [9] than with the other spectrometer [10] at WNR designed to focus on direct reactions for (n,p) reactions

above 100 MeV. The spectrometer [11] at UC Davis spans a somewhat wider range of energy, but has not been used to obtain cross sections for particles at very low outgoing energies.

References

[1] S.M. Qaim, Nucl. Phys. A382, 255 (1982)
[2] S.M. Qaim, R. Wölfli and H. Liskien, Phys. Rev. C34, 489 (1986)
[3] M. Ahmad, S.L. Graham, S.M. Grimes, R. Longfellow, H. Satyanarayana and G. Randers–Pehrson, Nucl. Inst. and Meth. 228, 349 (1985)
[4] S.K. Saraf, C.E. Brient, P.M. Egun, S.M. Grimes, V. Mishra and R.S. Pedroni, Nucl. Sci. Eng. 107, 365 (1991)
[5] S.M. Grimes, P. Grabmayr, R.W. Finlay, S.L. Graham, G. Randers–Pehrson and J. Rapaport, Nucl. Inst. and Meth. 203, 269 (1982)
[6] S.M. Grimes, R.C. Haight, K.R. Alvar, H.H. Barschall and R.R. Borchers, Phys. Rev. C19, 2127 (1979)
[7] R.S. Pedroni, N. Boukharouba, S.M. Grimes, V. Mishra and R.C. Haight, Bull. Amer. Phys. Soc. 33, 1577 (1988)
[8] C. Derndorfer, R. Fisher, P. Hille, G. Stengl and H. Vonach, Nucl. Inst. and Meth. 187, 423 (1981)
[9] A. Paulsen, H. Liskien, F. Arnotte and R. Widera, Nucl. Sci. Eng. 78, 377 (1981)
[10] R.C. Haight et al., Proceedings of the Conference on Pre–Equilibrium Reactions, Semmering, Austria, Feb. 1988, ed. by B. Strohmaier, p. 298
[11] F.P. Brady, Proceedings of the Conference on Neutron–Nucleus Collisions, Burr Oak, Ohio, September 1984 (American Institute of Physics), ed. J. Rapaport, R.W. Finlay, S.M. Grimes and F.S. Dietrich, p. 382

NEUTRON SCATTERING: TECHNOLOGICAL ACHIEVEMENTS AND ILLUSTRATIVE RESULTS

S. Chiba[a], A. Takahashi[b], H. Klein[c] and A. Smith[d]

ABSTRACT

Contemporary neutron scattering endeavors (energies $\leq$ = 25 MeV), using monoenergetic sources and the time-of-flight technique, are reviewed. Facilities and techniques are described, with attention to the optimization of measurement systems. Discrete scattering results are illustrated in fundamental and applied contexts. Techniques for and results from continuum neutron emission studies are discussed, with the implications on physical models and on neutron applications in energy systems.

a) Japan Atomic Energy Research Institute; b) Osaka University; c) Physikalisch-Technische Bundesanstalt Braunschweig, and d) Argonne National Laboratory.

PREFACE

Good knowledge of fast-neutron scattering is essential for a diversity of nuclear applications, and the fast neutron remains a unique hadronic probe for fundamental studies.

TECHNIQUES AND METHODS

Neutron scattering experiments are a complex combination of: 1) the production of monoenergetic neutrons, 2) the scattering process, and 3) the detection of scattered neutrons, and spectral fluence is a concern. Many fast-neutron spectrometers suffer from poor resolution and/or efficiency. These shortcomings are mitigated with the time-of-flight technique, as powerful neutron sources are available, good velocity resolutions are realized, and organic scintillators provide efficient neutron detection from $\approx$ 0.1 - 25 MeV.

The optimization of the experiments requires consideration of the effect of geometrical and material characteristics on the scattered-neutron response-rate, R, and on the angular and energy resolution. R may be estimated by

$$R \approx (N)_{beam} \times \left[n_t \cdot \ell_t \cdot \frac{d\sigma_t}{d\Omega_n}\right]_t \times \left[D_s^{-2} \cdot F_s \cdot \ell_s \cdot n_s \cdot \frac{d\sigma_s}{d\Omega_n}\right]_s$$
$$\times \left[D_d^{-2}, F_d, \ell_d \; n_d, \sigma_{np}\right]_d,$$

where N is the number of projectiles/sec, n_i the specific particle densities/cm**3, ℓ_i sizes in cm, F_i areas in cm**2, D_i distances in cm, and $d\sigma_i/d\Omega$ differential cross sections; and the index "i" refers to the target (t), scatterer (s) and detector (d). Obviously, R increases with the dimension ℓ_t, and the volume ($F_i \cdot \ell_i$) and particle densities (n_i) of the scatterer and detector. On the other hand, the energy spread of the primary neutrons increases with the energy loss of the projectiles and hence with areal density $n_t \cdot \ell_t$.

The resolution of the spectrometer improves with the length of the flight path (D_d) and decreasing linear dimensions (ℓ_i). Finally, the angular resolution improves with the smaller solid angles (F_i/D_i**2) subtended by the scatterer and detector. A compromise between these factors must be found, consistent with the energy and angular resolution goals, and with intensities requisite for sufficient statistical accuracies. In addition, technical (e.g., shielding) and methodological (e.g., corrections) considerations influence the optimization procedure. Nevertheless, scattering experiments can be designed to provide reasonable energy ($\approx$ 1%) and angular (e.g, $\leq$ 2 deg.) resolutions, and multiple-detector systems greatly increase efficiency.

Neutron Sources:- Reactions commonly used to produce monoenergetic neutrons are outlined in Fig. 1. Elemental metallic lithium is evaporated

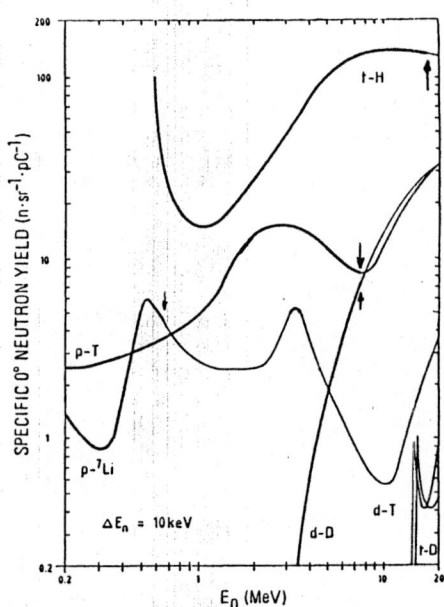

Fig. 1. Energy dependence of the 0-deg. yields of common neutron sources (1). Arrows indicate the upper and lower limits of monoenergetic neutrons.

----

onto thin backings and hydrogen gas contained in cells (1,2). The 0 deg. yield is optimal, and the neutrons are not polarized if unpolarized projectiles are used. Energy gaps are encountered (particularly from 10 - 13 MeV), and attention must be given to corrections for the contributions from secondary source-neutron groups (e.g., Li-7(p,n)Be-7* and D(d,np)D). The region above ≈ 15 MeV is practically accessible only with tritium targets. Additional "innovative" source reactions have recently been examined. For example, the H-1(B-11,n)C-11 reaction produces kinematically collimated monoenergetic neutrons over the range 7 - 12 MeV (filling the 11 - 13 MeV "gap") (3). The reaction has been used in activation (4) and (n,n',γ) (5) measurements, and may have a wider potential if problems of intensity and moderate energy resolution can be overcome. The H-1(t,n) reaction is, in principle, attractive but it requires the acceleration of relatively high energy tritium beams, and thus will probably remain of very limited use (6).

It is obvious that the peak pulsed current (N) should be as high as possible as it only enhances the response rate and improves the signal/background ratio. Proper ion bunching and effective cooling of the entrance foil and/or backing of the target are essential. Alternatively, a heavy-metal grid can be used to support the gas-cell window (7), concurrently providing physical support of the thin metal foil and efficient heat conduction at high currents (e.g., ≤ 20 μA).

Scattering Samples and Irradiation Geometry:-
Scattering samples should be as compact as possible (e.g., spheres or solid cylinders). With these the angular resolution will not significantly change over the experimental angular range (≈ 10 - 170 deg.). Hollow samples of equivalent mass somewhat reduce multiple-event effects but at the expense of incident-neutron and angular resolution (see Fig. 2-b and -c). Sample-size effects (e.g. flux attenuation and multiple scattering) should be addressed by means of Monte-Carlo simulations (8). Corrections of ≥ 10% can be reliably handled using an iterative cross section determination. Special care has to be taken with light elements as large energy ranges are involved and resonance structure can be a serious concern.

The target-sample distance cannot be too short as the detectors must be shielded from the primary source, and that becomes increasingly difficult as the scattering angle decreases. A small offset of the source centroid from the reference axis can cause a large shift in the scattering angle (e.g., Δθ ≈ 0.5 deg. for a 1 mm source shift 12 cm from the sample). In order to detect and correct for such angular shifts, measurements must be made symmetrical with respect to the reference axis, or alternatively, advantage can be taken of the kinematics of the C(n,n) or H(n,n) scattering (9) to determine the true scattering angles to ≤ 0.2 deg.

Neutron Time-of-Flight Detectors:- Organic scintillators are used as fast and efficient detectors with the thickness selected for the flight paths and energy resolution goals. The efficiencies are ≈ 10% - 50%, depending on

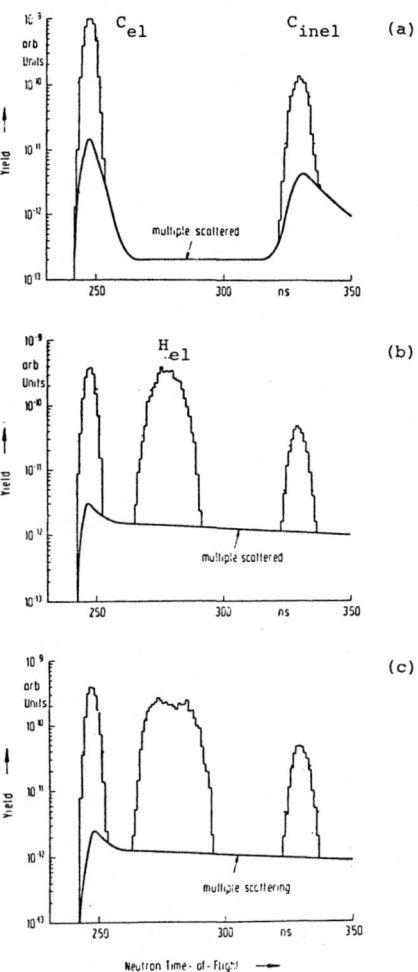

Fig. 2. Simulated TOF spectra for scattering of 10 MeV neutrons from carbon (a) and polyethylene (b) solid samples, and from a hollow polyethylene sample (c) of the same mass as (b).

neutron energy and detector thresholds. Constant-fraction timing of the phototube signals provides sub-nsec resolutions with a large dynamic range. Walk corrections can be made if multi-parameter data acquisition is used, including pulse-height information (10). The time resolution is determined not only by the thickness of the detector (neutron transit time) but also by its volume. Monte-Carlo calculations show that multiple scattering from the carbon of the scintillator, before the interaction with the hydrogen producing sufficient light for detection, can significantly increase the neutron time within the scintillator. Thus it is suggested that several detector sizes be used concurrently (11). For energies above ≈ 300 keV, pulse-shape analysis can be applied to several types of scintillators in order to discriminate against photon induced events, and in practice reduce backgrounds by orders of magnitude.

Multi-Detector Collimation:- Multi-detector arrangements require a compact collimator-shield system. The shape of the entrance throats must be carefully designed (see Fig. 3), and the materials, particularly the liners of the channels, well selected considering energy range and geometries. Monte-Carlo simulations are employed for optimization (12). Long flight paths (> 5 m) can be obtained with a variable collimator (10) or a movable accelerator (11). Consideration must be given to resonance outscattering from air when flight paths of ≥ 10 m are used.

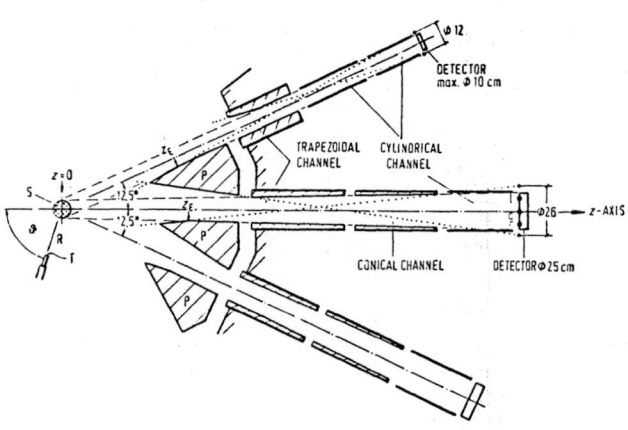

Fig. 3. Design of a multi-detector collimator system for the scattering of 10 MeV neutrons (12).

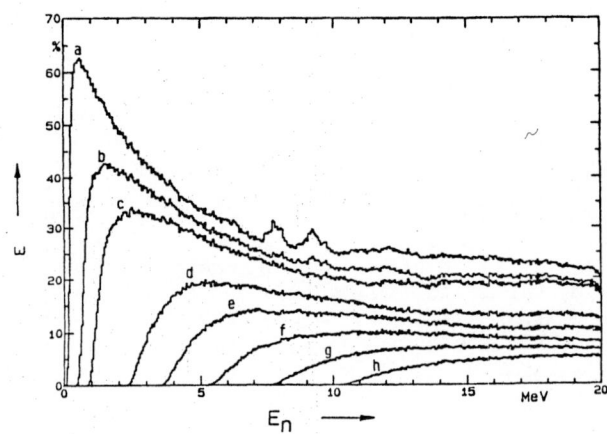

Fig. 4. Calculated neutron detection efficiencies of a NE213 detector (15).

Neutron Detection Efficiency and Normalization:- The energy dependence of the detection efficiency must be well known. Several common methods determine this efficiency to better than ± 2%. With appropriate targets and a versatile accelerator, the angular distributions of various neutron-producing reactions can be used, providing the respective reaction cross sections are well known (1). A quasi-absolute calibration, independent of the cross section data, can be established if the target and projectile can be interchanged so that the kinematic conditions are the same in the CM system (13). Nevertheless, consideration must be given to the effective emission angle and to target scattering effects if uncertainties of < 2% are to be obtained. The use of the time correlated associated particle method over a wide energy range involves a much larger effort (14). However, this method determines the efficiency on an absolute scale independent of evaluated reference data. Below 20 MeV, response functions can be reliably calculated from evaluated data and light output functions of the particular system (15). It has been shown that the calculated neutron fluence can agree with response functions measured with a proton-recoil telescope to ± 1.5% (9). Calculated efficiencies (11) display pronounced structure for low thresholds (see Fig. 4), which is supported by experimental observation (16) and can be attributed to resonances in the neutron reactions with carbon, particularly the (n,α) reaction (see Fig. 5).

It is common to use H(n,n) scattering for detector calibration, and polyethylene samples are well suited for monoenergetic work. However, the C(n,n) contribution is not easily subtracted using an "equivalent" carbon sample due to background and multiple-event effects (see Fig. 2). Thus a complete simulation of the experiment is needed to achieve the desired accuracy (8).

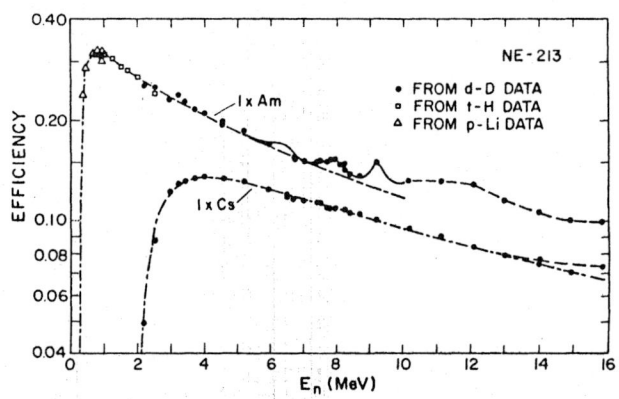

Fig. 5. Experimentally determined detection efficiencies of a NE213 scintillation detector (16).

Detectors can also be calibrated by using a Cf-252 source within a fast fission detector (17). A gas scintillator or ionization chamber with high fission rates (≥ 10**5/sec) can provide time resolutions of ≈ 1 nsec with a fission detection efficiency of ≥ 95%. With a careful analysis of the TOF spectrum (18), the energy dependent efficiency is obtained to ≈ 10 MeV by comparison with standard evaluations of the Cf-252 fission-neutron spectrum.

Data Analysis and Uncertainties:- Differential cross sections are commonly parameterized in terms of Legendre polynomial expansions. The spectral distribution of the scattered neutrons, corrections for sample size effects and detector efficiencies, and the density distribution of scattering angles must be considered in such parameterizations. Correlated uncertainties, such as those due to a common angle offset and the detector efficiency, must be addressed. The energy distribution of the primary neutrons should be provided for further evaluation. Many of the relevant parameters can only be estimated, but a complex scattering experiment, including source, scatterer and detector, can be completely simulated using the cross sections extracted from the iterative analysis. In addition to the sample size corrections, the mean values of the associated moments of the energy distributions and mean scattering angles are obtained from the simulations.

Finally, uncertainties of $\geq$ 2% and $\approx$ 1% can be achieved for differential and angle integrated cross sections, respectively. The major contributors to these uncertainties are the detection efficiency, the iterative analysis including correction procedures, and the finite energy and angular resolution.

DISCRETE SCATTERING

Neutron elastic and inelastic scattering from a light multiplier, Be-9, is illustrated in Fig. 6. The angle-integrated elastic- and inelastic-scattering cross sections are obtained to accuracies of $\approx$ 2.5% and < 10%, respectively. These scattering results imply (n,2n) cross sections that lead to major changes for ENDF/B-VI. The Be-9 scattering is due to negative parity levels, while low-lying positive-parity levels are not excited. This

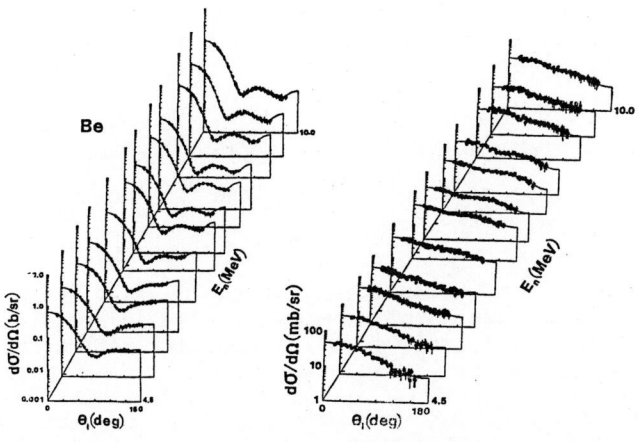

Fig. 6. Elastic- (left) and inelastic- (Excitation = 2.429 MeV, right) scattering from Be-9 (19).

----

suggests a possible "shape coexistence" in the Be-9 nucleus, with two rotational bands based on the g.s.(3/2-) and 1.69 MeV (1/2+) states (19). In Fig. 6 an extensive sample of modest accuracy was used to determine the overall result to good precision. An alternative is a precise "standard" measurement sequence involving fewer measurements, each with much greater care. Illustrative of such an approach is the elastic-scattering cross section of oxygen of ref. 20. In this case the differential values are known to 3 - 8% and the angle integrated cross section to $\approx$ 2.5%. Uncommon attention was given to identifying and quantifying the various measurement uncertainties, and the correlation matrix was obtained. Scattering of <$\approx$ 20 MeV neutrons from light nuclei is dominated by the resonance behavior. The interpretation commonly

uses the R-Matrix formalism, with the parameters of the matrix determined by fitting the experimental data. These parameters reflect nuclear structure information which can be obtained from the shell model and subsequently used in the R-Matrix calculations. The procedure, pursued in an iterative manner, has been successfully employed by the Ohio University Group (21), as illustrated by the C-13 results of Fig. 7. This correlated shell-model/R-Matrix technique has been exploited by the Livermore Group to predict a wide range of neutron-induced reactions, as cited elsewhere in these proceedings. Scattering from light targets at energies >$\approx$ 20 MeV has a relatively energy-smooth behavior, as illustrated in ref. 22. One could hope to describe such results with a conventional optical model, but the success is qualified. The shortcomings have been attributed to inappropriate potential forms, resonance fluctuations, coupling with the giant resonances, and/or ℓ-dependent potentials. Some of these alternatives have been examined without great success.

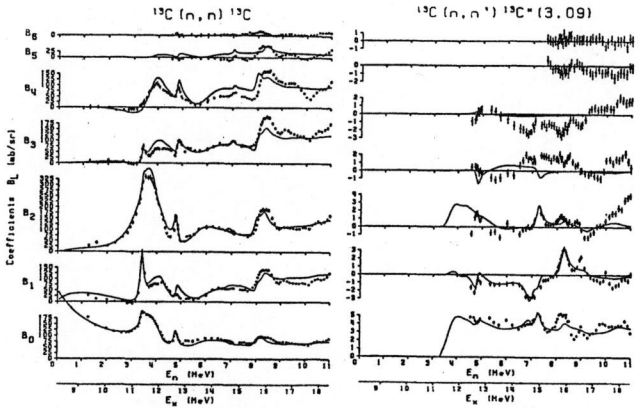

Fig. 7. Measured (symbols) and calculated (curves) elastic- and inelastic-scattering (21) from C-13.

Medium and heavy nuclei encompass the majority of the structural materials, fission products, and actinides. Structural materials in the A $\approx$ 50 - 65 region are largely dynamic vibrators, and an example of neutron scattering (23,24) from them is Ni-58. Spectrometeric capability of the measurements is approaching that of the best charged-particle spectrometers as illustrated in Fig. 8. Differential cross sections were obtained for many of these neutron groups, and the elastic scattering is the basis for comprehensive vibrational-model interpretations (23). The real and imaginary potential strengths display physically reasonable energy dependencies (see Fig. 9), that could only be achieved with the vibrational model. The geometries of the model are energy dependent, a characteristic that is only partially alleviated when the dispersion relationship is considered. The model nicely describes the elastic scattering over a large energy range (see Fig. 10). Nearly as good agreement is obtained for the excitation of the yrast (2+) vibrational level (23). The important factors are; i) the provision of a detailed data base extending over a wide energy range makes possible a rigorous interpretation, ii) the requirement of a vibrational model, and iii) the

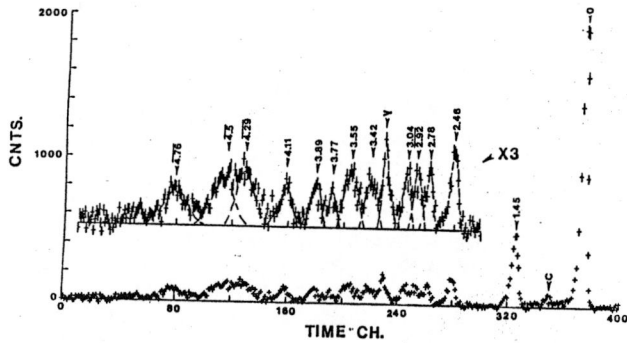

Fig. 8. Velocity spectrum (23) obtained by scattering 8 MeV neutrons from Ni-58 at 80 deg. Excitation energies are noted numerically.

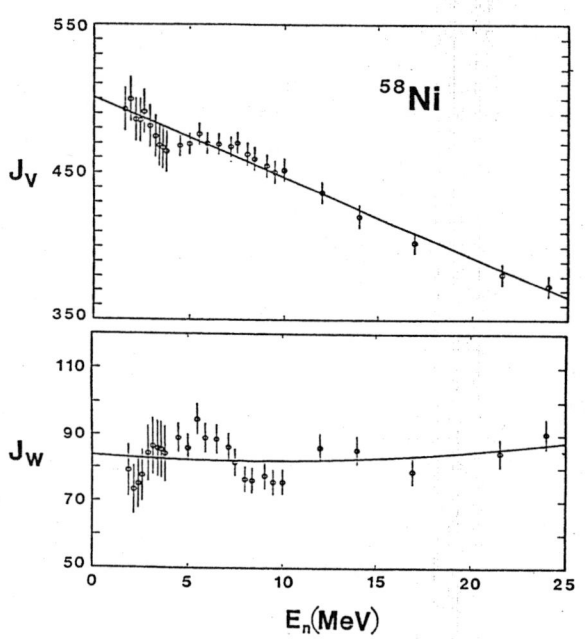

Fig. 9. Strengths of the real (J_v) and imaginary (J_w) Ni-58 vibrational potential expressed as volume integrals (23).

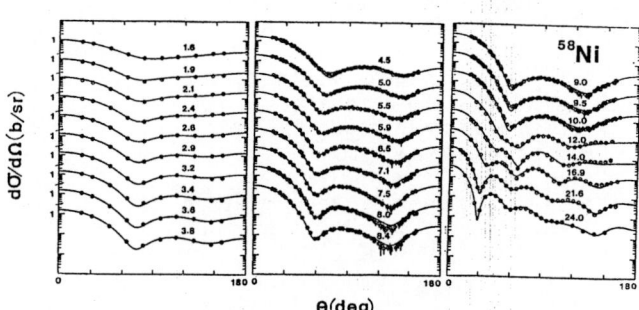

Fig. 10. Measured (symbols) and calculated (curves) elastic-scattering cross sections of Ni-58.

energy dependence of the model geometries.

Remaining structural materials are largely in the $A \approx 90 - 100$ region, with changing scattering properties as the target mass moves away from the $N = 50$ closed shell and collective processes become a consideration. Zirconium scattering is illustrative of the results in this mass region.

The elastic scattering is reasonably known to ≈ 10 MeV, and at 24 MeV (25,26), but from 10 - 24 MeV there is nothing (an all too typical situation). Near the shell closure neutron scattering is described using the spherical optical model, including the dispersion relationship implying a coupling of real and imaginary potentials, resulting in an energy dependent surface term of the real potential and a small contribution from volume absorption. The energy dependencies of these two contributions for zirconium are shown in Fig. 11. The parameters of the dispersive optical model remain energy dependent and can be projected to the bound energy region using the method of moments, where the neutron-based potential is equivalent to the shell-model potential and should describe the bound particle- and hole-states. The quality of such an extrapolation varies with the target but is particularly good near shell closures, as illustrated by the Pb-208 results of ref. 27. A characteristic of the dispersive model is the

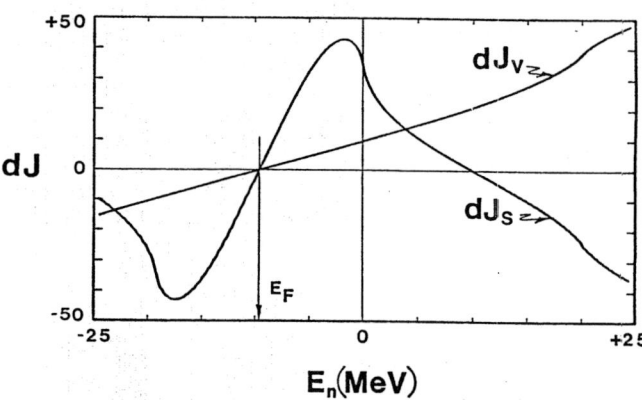

Fig. 11. Volume (dJ_v) and surface (dJ_s) dispersion contributions to the zirconium real potential (25).

"Fermi Surface Anomaly", which is significant only in the bound regime. The dispersion relationship implies energy dependent strengths and geometries of the real potential, but it is not the only source of such dependencies. The behavior of the imaginary potential is isotope dependent, with minimum strengths near the shell closures. Indeed, the isovector imaginary strength near shell closures may have an opposite sign from that commonly used in global models.

The sensitivity of the typical fast reactor to fission-product inelastic scattering is $\approx 25\%$ that for fast neutron capture. Provision of the full scope of data will require a coordinated measurement and calculational program using theory to extrapolate a few precise measurements. Near shell closures the knowledge of elastic and inelastic fission-product scattering can be similar to that outlined above for $A \approx 100$ structural materials. In the strong fission-product vibrators recent measurements confirm a large direct-reaction component of the cross sections for the excitation of the yrast 2+ levels of the even isotopes (28). However, at the lower energies of primary concern to the fast reactor, the inelastic-scattering processes are primarily of a compound-nucleus nature. Their magnitudes are large (e.g., ≈ 1 b at ≈ 1 MeV), and generally exceed those commonly encountered in major evaluated file systems. Many of the heavier fission products are ridged rotors with large deformations, leading to low-lying excited states expected to have large inelastic cross sections at low incident-neutron energies. Even the stable elements are generally multi-isotopic and the requisite measurement samples difficult to obtain. Remarkably little is known of scattering from these heavy rotors, and the situation is not likely to change.

Most actinides are rotors with characteristic low-level rotational band structure, and similar bands based upon vibrational configurations at higher energies. These structures remain uncertain above ≈ 1 MeV. Experimental knowledge of scattering from the common odd actinides (e.g., U-235 and Pu-239) is meager due to the complexity of the low-lying level structure and large fission cross sections. There is little technological hope of grossly improving this shortfall in the foreseeable future. Low-energy (<≈ 1 MeV) scattering from the common even actinides (e.g., U-238 and Th-232) has been been carefully studied for many years, and accuracy goals of ≈ 3% for the elastic- and inelastic-scattering cross sections are being approached. Above the maximum energy of reasonably understood level structure there are considerable uncertainties in both elastic- and inelastic-scattering cross sections. The latter can not be reasonably measured in the presence of the large fission-neutron contribution, and primary reliance must be placed upon the precise determination of the elastic-scattering cross sections which, together with the well known total and fission cross sections, imply the inelastic-scattering cross section until the (n,2n) reaction becomes significant. It is now possible to measure common even actinide elastic-scattering cross sections with 2 - 3% accuracies (see Fig. 12). These results include inelastic contributions due to the first several levels of the g.s. rotational band, but that is of no applied concern as the relevant energy transfer is far smaller than considered in macroscopic engineering calculations at these energies.

CONTINUUM NEUTRON EMISSION AND 14 MeV PROCESSES

Energy-angle double differential neutron emission cross sections (DDX, $d^2\sigma/(dE \cdot d\Omega)$) induced by $\leq$ 14 MeV neutrons are of special importance in fusion-reactor neutronic design as they determine neutron economy and energy loss, tritium production, nuclear heating, radiation damage and induced activities in the first wall, blanket and surrounding structures. Moreover, they are a challenge for contemporary reaction theories.

Experimental Techniques:- DDX measurements are particularly demanding as a very large range of neutron velocity spectra must be dealt with. A severe source of background is the source itself (particularly the T(d,n) reaction), thus the shielding must be particularly effective. A massive shadow bar consisting of copper, tungsten and paraffin, several tens of cm in length, is commonly used to protect the detector from the primary neutron source. The optimum position,

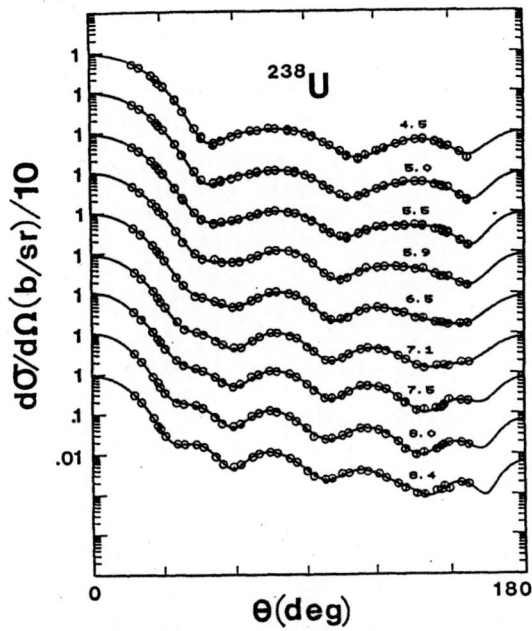

Fig. 12. Differential "pseudo" elastic scattering cross sections (29) of U-238. Symbols indicate measured values, curves the results of Legendre fitting, and numerical values incident energies.

----

shape and combination of materials varies with incident energy and measurement angle. Inscattering of the neutrons from the collimator wall is significantly reduced with careful collimator design (30). The shielding must also carefully control room- and air-scattered neutrons.

When the D(d,n) source is used at deuteron energies above ≈ 5 MeV there is additional background due to the D(d,np) break up and X(d,n) reactions (X → C, O, N, etc.). The latter contributions are determined by making measurements without the deuterium in the target cell at considerable expense in experimental time. The background from the D(d,np) breakup reaction can not be exactly subtracted by any means, thus the D(d,n) reaction is useful in DDX measurements only below E_d ≈ 7 MeV. The breakup component can be simulated by the He-3(d,np) reaction (31), but in a detailed measurement the difference between D(d,np) and He-3(d,np) spectra still presents a serious problem. Similarly, use of the T(d,n) source will lead to addition background neutrons from the D(d,n) reaction due to the accumulation of deuterium in the target assembly, which will vary with the target history.

In DDX measurements the neutron detector bias is set as low as possible in order to see the "whole" velocity spectrum. As a result, the detector system becomes sensitive to low-energy neutrons with a deterioration of the signal/background ratio, reduced effectiveness of pulse-shape discrimination of neutron and γ-ray signals, and poorer time resolution, in conflict with the experimental objectives. In addition, the detection system becomes more sensitive to instabilities over the requisite long measurement periods, placing demands on the stability of a "fragile" system. Some of these problems can be mitigated by using two or more detector biases and a multi-parameter data-acquisition system. A "clean" neutron pulse is important if the elastic component is to be well resolved and a formidable background in the (n,n') continuum region due to the "tail" on the burst is to be avoided. A pulsing system consisting of a 8 MHz post-acceleration chopper and associated components has been developed at Tohoku University (32) to meet these stringent pulse requirements. With it the experimental phase space is enlarged, particularly at higher energies and forward angles, and the background in the continuum region significantly reduced.

Data Correction Methods:- In principle, the correction of DDX data for multi-events and flux attenuation is similar to that of a conventional scattering experiment. The difficulty is that the DDX result must be accurately known from the incident to ≈ zero energy, and thus the data base for the correction procedures must be of a very wide energy scope. The data in even the best evaluated files is seldom well enough known for good corrections, and thus iterative estimates must be used. These are frequently based upon model calculations such as multi-step statistical, precompound, and direct-reaction models, and on systematics such as suggested by Kalbach and Mann (33,34). In this way the accuracy associated with DDX corrections has greatly improved.

Typical Results and Interpretations:- Light Element DDX results are characterized by discrete level structures that persist up to several MeV excitations, and a wide continuum component due to few-body breakup reactions, as illustrated in Fig. 13. α-cluster structure plays an important

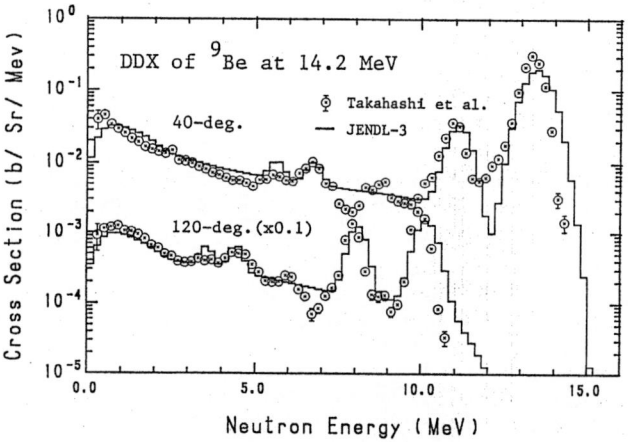

Fig. 13. DDX spectra of Be-9 at 14.2 MeV. Measured values are indicated by symbols, and the JENDL-3 evaluation by the curves. Angles are numerically noted.

---

role in understanding DDX in these cases. Except for the D(n,2n)p reaction, where a rigorous Faddeev calculation applies, the understanding of the few-body breakup is largely phenomenological, using a simple phase-space model or the theory of the Final State Interaction (FSI) (35). For example, the neutron spectrum from the Li-6(n,n'd)α reaction at 18 MeV can be fairly well understood by assuming a strong interaction

of d- and α-particles in the final state (36). The neutron scattering to the 2.2 and 4.3 MeV levels in Li-6 do not result in sharp peaks, but rather broad continuum-like spectra due to the fact that these levels are not bound, but the resonance states between d and α and the characteristics of the d-α scattering favors broad spectra. This can be incorporated into the calculations by taking into account the d-α phase shift. In contrast, the spectra from the C-12(p,p')3α can be explained by the four-body simultaneous breakup model (37). If the transition matrix element is properly converted from the proton to the neutron, the same model can potentially explain the C-12(n,n')3α neutron spectra (37). Knowledge of reaction mechanisms through few-body breakup process in light elements will have to proceed by accumulating the experimental data by target and projectile unless some innovative progress in few-body reaction theory is forthcoming. The Coupled-Discretized Continuum-Channels (CDCC) method, given attention in the past few years (38), provides a quantitative method to calculate light-mass breakup cross sections with relatively few parameters.

DDX spectra of Li-7 and Be-9 are of special importance in fusion applications. For the JENDL-3 evaluation, Shibata calculated the DDX of Be-9 using the sequential decay model, with the intermediate states expressed by the delta-function, plus the phase-space model for simultaneous few-body breakup (39). This is an approximation of FSI theory, but for Be-9 it reproduces the measured DDX results quite well (see Fig. 13). It is a practical approach for calculating the breakup spectra that will be employed in higher-energy (> 20 MeV) evaluations.

The reaction mechanisms in the medium-weight elements leading to the continuum spectra are thought to be fairly well understood (i.e., direct low-lying excitations via doorways, the precompound process with a limited number of degrees of freedom, and the compound process from a system in equilibrium). Neutrons are emitted at every stage of the reaction. In contemporary calculations the DWBA, coupled-channels, multi-step precompound, and compound theories are used to give a fairly good description of the measured spectra. The principal issues in this area now are; level density formulae, level density parameters (energy dependent and independent), Kalbach constants, and exit-channel optical potentials. The capability is illustrated by the Nb-93(n,n') spectrum of Fig. 14, measured by Takahashi et al (40). The calculated results were obtained using the code SINCROS-II (41) and Kalbach systematics. In this case, theory (+ systematics) can reproduce the data very well. The quantum mechanical approach (42) may bring new understanding of reaction mechanisms in this region.

In the actinide elements the situation is complicated by fission-neutron emission. Fission can be taken into account by statistical calculations (e.g.,the double humped barrier model), or phenomenologically as a competing process. Baba et al. (43) have shown that the DDXs of U-238 and Th-232 at 14 and 18 MeV are explained by the statistical model and Kalbach-Mann systematics, if it is assume that

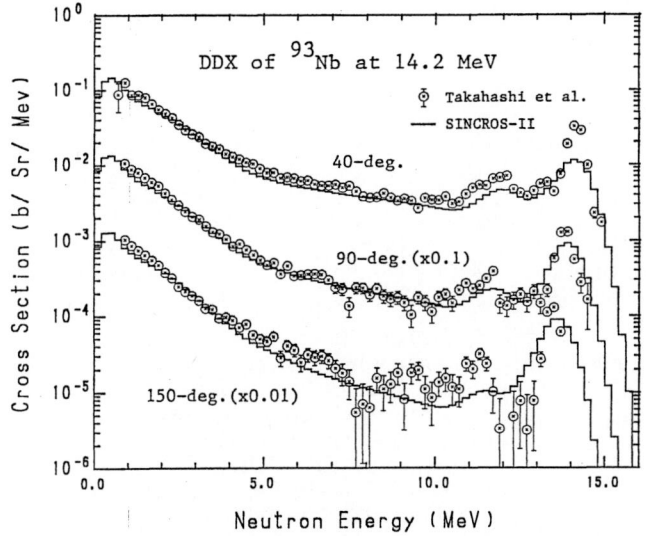

Fig. 14. Measured (symbols) (40) and calculated (curves) (41) DDX spectra of Nb-93 at 14.2 MeV.

the angular distribution of the fission neutrons is isotropic in the laboratory system. In contrast to the medium weight region, Kalbach-Mann systematics better describe the (n,n') angular distributions than the Kalbach systematics alone. This suggests that the mechanisms determining precompound angular distributions may be different in the medium- and heavy-element regions, and thus the (n,n') results can potentially contribute to the understanding of nuclear reaction mechanisms in the MeV region.

References

1. M. Drosg and O. Shwerer, IAEA Tech. Report 273 (1987).
2. C. A. Uttley, Neutron Sources for Basic Physics and Applications, Pergamon (1983).
3. S. Chiba et al., Nucl. Instr. Methods A281 581 (1989).
4. Y. Ikeda et al., JAERI Report, JAERI-M 91-032 (1991).
5. K. Hasegawa et al., elsewhere in these proceedings.
6. R. Haight, NEA Report, NEANDC-259U (1989).
7. J. Meadows et al., Nucl. Instr. Methods 176 439 (1980).
8. H. Klein et al., Proc. Conf. Nucl. Data for Sci. and Tech., Reidel (1983).
9. G. Boerker et al., Proc. Conf. Nucl. Data for Sci. and Tech., Saikon (1989).
10. A. Smith et al., Nucl. Instr. Methods 50 277 (1967).
11. R. Boettger et al., Proc. Conf. on Nucl. Data for Sci. and Tech., Reidel (1983).
12. D. Schlegel-Bickmann et al., Nucl. Instr. Methods. 169 517 (1980).
13. M. Drosg, Nucl. Instr. Methods 105 573 (1972).
14. J. L. Fowler et al., Nucl. Instr. Methods 175 449 (1980).
15. G. Dietze and H. Klein, PTB-Report, ND-22 (1982).
16. M. Drosg et al., Los Alamos Report, LA-7987-MS (1980).
17. A. Smith et al., Nucl. Instr, Methods 140 397 (1977).
18. R. Boettger et al., Nucl. Sci. Eng. 106 377 (1990).
19. M. Sugimoto et al., Nucl. Sci. Eng. 103 37 (1989).
20. G. Boerker et al., PTB Report, PTB-N-1 (1989).
21. D. A. Resler, Thesis, Ohio University, DOE Report, ER/02490-5 (1987).
22. J. Petler et al. Phys. Rev. C32 673 (1985).
23. A. B. Smith et al., ANL Report, ANL/NDM-120 (1991), and references cited therein.
24. P. P. Guss et al., Nucl. Phys. A438 187 (1985).
25. S. Chiba et al., ANL Report, ANL/NDM-119 (1991), and references cited therein.
26. Y. Wang and J. Rapaport, Nucl. Phys. A517 301 (1990).
27. R. Finlay et al. Phys. Rev. C39 804 (1989).
28. S. Chiba et al., Ann. Nucl. Energy, 16 637 (1989).
29. A. Smith, et al., to be published (1991).
30. D. Glasgow et al., Nucl. Instr. Methods. 114 521 (1975).
31. S. Grimes et al., Nucl. Instr. Methods 203 269 (1982).
32. S. Matsuyama et al., elsewhere in this conference.
33. C. Kalbach and F. Mann, Phys. Rev. C23 112 (1981).
34. C. Kalbach, Phys. Rev. C37 112 (1981).
35. J. Gillespie, Final State Interactions, Holden-Day (1964).
36. S. Chiba et al., Proc. Conf. on Nucl. Data for Sci. and Tech., Saikon (1988).
37. Y. Watanabe et al., JAERI Report, JAERI-M 91-032 (1991).
38. e.g., in Prog. Theor. Phys., Supplement 89 (1986).
39. K. Shibata et al., JAERI Report, JAERI-1319 (1990).
40. A. Takahashi et al., IAEA Report, INDC(JPN)-118/L (1989).
41. N Yamamuro, JAERI Report, JAERI-M 90-006 (1990).
42. e.g., Y. L. Luo and M. Kawai, Phys. Lett. B235 211 (1990).
43. M. Baba et al., JAERI Report, JAERI-M 91-059 (1991).

WHITE SOURCE GAMMA-RAY PRODUCTION SPECTRAL MEASUREMENT FACILITIES IN THE U.S.A.

D.C. Larson and J.K. Dickens
Oak Ridge National Laboratory
Oak Ridge, Tennessee 37831-6356 USA

and

R.O. Nelson and S.A. Wender
Los Alamos National Laboratory
Los Alamos, New Mexico 87545 USA

Abstract: The two primary neutron sources for measuring gamma-ray production (GRP) cross sections for basic and applied work in the U.S.A. are the Oak Ridge Electron Linear Accelerator (ORELA) located at the Oak Ridge National Laboratory (ORNL) and the Weapons Neutron Research (WNR) facility located at the Los Alamos National Laboratory (LANL). ORELA is based on a 180-MeV electron linear accelerator, while the WNR facility uses the Los Alamos Meson Physics Facility 800 MeV proton beam to produce neutrons. The facilities collectively cover the neutron-energy range from thermal to over 700 MeV. The paper describes the present capabilities for GRP measurements at each facility.

(gamma-ray production measurements, neutron white sources, gamma-ray detectors, ORELA, WNR)

Introduction

Gamma-ray-production (GRP) cross sections play an important role in applied neutronics work. Nuclear heating studies, transport calculations and radiation shielding design all require GRP cross-section input. Radiation damage studies requiring cross sections for reactions which are difficult to measure directly also benefit from GRP work since GRP measurements for these reactions, when coupled with nuclear model calculations, provide unique information on these difficult cross sections.

Pulsed high-intensity white neutron sources have the advantage of providing neutrons over a wide energy range thus data may be acquired simultaneously at all incident neutron energies. Because the energy dependence of GRP cross sections is of primary interest, white sources have a definite advantage over monoenergetic sources.

Historically, much of the early GRP data was obtained at ORNL and LANL. Dickens et al. [1] at ORELA reported GRP measurements for $0.1 < E_n < 20$ MeV for 22 materials at 125 deg, using a NaI detector system. Drake et al. [2] at LANL measured GRP cross sections at 14.2 MeV for 20 samples at three angles between 90 and 130 deg, also using a NaI detector system. While these measurements formed the basis for GRP cross sections in ENDF/B-IV, the gamma-ray energy resolution was generally inadequate to extract cross sections for individual gamma-ray lines. Both facilities have continued to improve their GRP measurement systems, and in the following sections we outline GRP measurement systems currently in use.

ORELA GRP Measurement Systems

General

The ORELA facility is a pulsed neutron source which is used over 4000 h/yr for neutron experiments, providing data for both basic and applied research. It has two neutron-producing targets and ten evacuated flight paths with 18 underground detector stations ranging in length from 9 to 200 m. The ORELA facility is described in detail most recently by Bockhoff et al. [3]. GRP measurement facilities are currently located at seven of the stations, with flight paths ranging from 18 to 155 m. Properties of these seven systems form the basis of the ORELA section of this paper. Since 1970, over 150 papers and reports have been published describing GRP and radiative capture measurements performed at ORELA [4]. In addition, these data have been heavily utilized by evaluators for the ENDF/B nuclear data evaluation system. Priorities in the ORELA GRP measurement program are strongly influenced by user requests contained in the U.S. DOE Data Request List [5].

High Resolution Germanium Detector System

A high-resolution gamma-ray detection system is located on Flight Path (FP) 8 at the 20-m station. It uses a well-shielded intrinsic germanium detector with a nominal volume of 150 cm^3, located 40-cm from the center of the beam and at 125 deg with respect to the incident neutron direction. Since gamma-ray angular distributions of bound states are symmetric about 90 deg and since the second term of the Legendre expansion polynomial $P_2(125) = 0.0$, multiplying the differential measurement results by 4π gives the total GRP cross section to a good approximation (for decaying states with $J = 0, 1/2, 1$, or $3/2$ the results are exact). A response function has been measured using sources in situ for the detector, giving an absolute efficiency of 8×10^{-5} at 1332 keV. The detector has an intrinsic efficiency of 26% at $E\gamma = 1332$ keV relative to the standard 7.5 cm $\times$ 7.5 cm NaI detector 25 cm from the source. The measured energy resolution is nominally 2 keV for a 1-MeV gamma ray, while the timing resolution is measured to be 7 ns for a 511-keV gamma ray.

The beryllium block ORELA target is normally used to provide neutrons for experiments with this system, covering the energy range from $0.1 < E_n < 40$ MeV. The gamma-ray energy range is $0.1 < E_\gamma < 8$ MeV, using several amplifier gain settings as needed to study higher-energy gamma rays created in different samples.

The electronics provide pulse-height (ph) analysis from the detector, and correlate the detector event with the time-of-flight (tof) of the neutron inducing the event. About thirty tof-vs-4096 channel ph spectra are generally acquired, covering the energy range of interest. The data-acquisition system [6], based on an IBM PS/2 personal computer, provides up to 15 tags for looking at backgrounds, coincidence events, etc. The ph information is digitized at the 20-m experimental station with the 12 bits of information transferred to the counting area to essentially eliminate noise contributions from the ORELA rf to the detector resolution.

The incident neutron flux is measured with a 0.6-cm diam $\times$ 0.6-cm long NE-110 plastic scintillator placed in the beam and coupled to a photomultiplier tube via a 5-cm light pipe. Pulse height vs tof information is sent through the electronics and stored in the PS/2 system for later analysis. Response functions for flux analysis are provided by the SCINFUL code [7].

Data reduction consists of removing any sample-out background (< 1% of sample-in counting rate), performing deadtime corrections appropriate for a one stop/start system, obtaining the photopeak areas (generally done with a version of TPASS [8]), and folding in the flux to obtain the GRP cross sections and associated uncertainties. The system routinely extracts integrated cross sections as low as 1 mb for $E_\gamma = 1$ MeV. Measurements have been completed for ^{10,nat}B, V, ^{nat,53,54}Cr, ^{56,57}Fe, Co, and ^{58,60,nat}Ni. A paper by Dickens et al. at this conference presents results from the ^{56}Fe measurement. Measurements are planned for Ti and Cu.

A second germanium detector system, very similar to this one, is currently located on FP-6 at the 20-m station. It is being used in a collaborative experiment with the National Institute for Science and Technology to measure the ^{10}B$(n, \alpha\gamma)$ cross section. A paper by Schrack et al. at this conference describes this measurement.

C_6D_6 Detector System

This low gamma-ray energy resolution system, located on FP-7 at the 40-m station, is a modification of an earlier C_6F_6 system. Its primary use is to measure pulse-height spectra from capture gamma rays, and from these data obtain GRP cross sections and the neutron capture cross section. The neutron energy range normally covered is the resonance region from 1 keV to 1 MeV, while the photon energy range is normally $0.1 < E_\gamma < 10$ MeV. The flight path is 40 m from the neutron source to the sample and 8 cm from the sample to the C_6D_6 detectors.

Two diametrically opposed 10-cm diameter C_6D_6 detectors are located at 90 deg to the incident neutron direction. They each have a volume of 550 cm^3 and subtend a solid angle of about 2.4 sr. The measured energy resolution is 9% for a 1-MeV photon, and the timing resolution is <1 ns for a ^{60}Co source. The efficiency for $E_\gamma > 1.5$ MeV is essentially flat and is about 4%. The region around the sample and detectors is constructed from low-Z materials and has all excess mass removed.

The neutron flux is monitored by a Li-glass detector viewed by two phototubes located 50-cm before the sample. Absolute normalization is done via the saturated gold resonance technique.

The data-acquisition system is based on a PS/2 personal computer, which stores ph vs tof information for the two C_6D_6 detectors, the summed signal, and detector coincidences. Similar information is retained for the two phototubes viewing the lithium-glass monitor detector. Tags are used for storing information useful for data reduction.

Data reduction consists of the usual background and deadtime corrections. Detector responses are calculated with the electron gamma-ray transport code EGS [9], incorporating a careful modeling of material in the vicinity of the sample and detectors. Using these responses, ph spectra can be unfolded and corrected for attenuation and multiple scattering and normalized to obtain GRP cross sections in the resonance region. The calculated responses can also be used to form weighting functions which, when applied to the ph spectra, provide capture yields for resonances. Prior to replacement of the C_6F_6 detectors by the present ones, an earlier version of this system [10] was used by R. L. Macklin to measure capture cross sections for over 170 isotopes. An earlier version of the present system was used to measure the capture cross section for the 1.15 keV resonance in ^{56}Fe, reported at the Mito Conference [11]. Future work includes measurement of capture cross sections and spectra for the structural materials.

BaF$_2$ Multicrystal Spectrometer/Multiplicity Detector

The initial use of this new system, located on FP-4 at the 20-m station, will be to measure the neutron energy dependence of capture and fission cross sections by recording capture and fission gamma-ray multiplicities. In addition, it will be used to measure capture GRP spectra and cross sections for structural materials. The detector system consists of 12 hexagonal BaF$_2$ crystals, each with volume of 826 cm^3. The measured energy resolution of the system is 14% for a ^{137}Cs source with a system timing resolution of < 1 ns. The system efficiency is about 80% for a ^{60}Co source. Data acquisition is done via a Multibus/Camac based system which uses a PS/2 based system for storage. Information is sorted according to pulse height, tof, and number of detectors fired (multiplicity). If desired, the system can also do event-by-event recording to Exabyte high-density tape. Measurement of the incident neutron flux is via a parallel plate BF$_3$ counter, located 1 m ahead of the sample with the tof spectra stored in the PS/2. At present, data reduction is in progress for the first measurement which is the ratio of capture to fission for ^{235}U from 0.005 eV to 100 eV.

NE-213 Based Neutron and Gamma-Ray Emission Measurement System

A three-detector system to measure neutron and gamma-ray production cross sections has been implemented on FP-9 in the Shield Test Station. Neutrons are provided by the beryllium block ORELA target. Detector events are recorded for $1 < E_n < 25$ MeV incident neutron energy, and $1 < E_\gamma < 10$ MeV gamma-ray energy. The source to sample flight path is 47 m, with the sample to detector distances ranging from 5 to 20 cm. Three NE-213 detectors are currently in use; 5.1-cm-diameter detectors are located at 30 and 64 deg with respect to the incident beam direction, and a 12.7-cm-diameter detector is located at 125 deg.

Pulse shape discrimination (PSD) is used to separate the detected neutrons and photons, and 2048 channel ph spectra are stored as a function of incident neutron tof in a PS/2 based data-acquisition system. This system and its associated software also provide for storing PSD window information including the capability to adjust the bias channel for each PSD window. Tags can be used to store results from different detectors, coincidences, etc.

Flux determination is done via a ^{238}U fission chamber located at 27 m and a small NE-110 plastic scintillator located at 45 m which has been intercalibrated with the fission chamber.

Data reduction for the first measurement with this system (on a sample of iron) is in progress. Future work will focus on measurement of (n, xn) and $(n, x\gamma)$ cross sections for the structural materials for $1 \leq E_n \leq 25$ MeV.

Other Systems

Two other systems at ORELA which measure gamma-ray spectra, but do not provide GRP cross sections directly, are available. The first of these is a large liquid scintillator tank, located on FP-6 at the 150-m station, used to measure capture cross sections for fissile and fertile isotopes. The neutron-energy range covered is typically 0.01 eV $< E_n <$ 200 keV, with photons of energy $0.2 < E_\gamma < 15$ MeV being registered. The detector, located 155 m from the neutron source, is filled with 3000 liters of NE-224 loaded with trimethylborate and surrounded by 32 12.7-cm photomultiplier tubes. To reduce backgrounds, the tank is optically divided in half so any event must produce two gamma rays to be counted. The system energy resolution is 40% for a ^{60}Co source, has a system timing resolution of 6 ns and an efficiency of 85% for $E_\gamma > 1.5$ MeV. Flux measurement is via a 1-mm-thick ^{6}Li glass detector located 15 cm ahead of the tank. Pulse heights from the detectors are summed and stored in a PS/2 data acquisition system, along with the corresponding tof. Results obtained with this system include capture measurements for ^{238}U, and older capture data for ^{235}U and ^{239}Pu.

The second system is a C_6F_6 NE-213 detector system located on FP-5 at the 20-m station. It is used primarily to measure capture and fission cross sections in the actinide region. The neutron energy range covered is 0.01 eV $< E_n <$ 400 keV, where photons of energy $0.1 < E_\gamma < 10$ MeV are registered. Two diametrically opposed, 10-cm diameter, 400-cm^3 C_6F_6 detectors are located at 90 deg to the incident beam direction. The system has 9% energy resolution at 1 MeV with a timing resolution of 3 ns and an efficiency of 4% for $E_\gamma > 1.5$ MeV. Two NE-213 detectors with PSD are also located in the plane of the C_6F_6 detectors to detect fission neutrons, if desired. Neutron flux is measured by a parallel plate ^{10}B ionization chamber located 50 cm ahead of the sample. Data acquisition is done via a PS/2 system storing ph, tof, and PSD information. Calculated detector responses provide weighting functions which, when applied to the ph spectra, give the desired capture yields for observed resonances. This system has been used to measure capture cross sections of ^{237}Np, ^{239,240,241}Pu, and ^{241,243}Am.

WNR GRP Measurement Systems

General

Two spallation neutron sources are available at Los Alamos National Laboratory. The Weapons Neutron Research (WNR) facility provides high-energy neutron beams spanning the energy range from ~1 to over 700 MeV, and the Los Alamos Neutron Scattering Center (LANSCE), a moderated neutron source, produces intense beams of thermal and epithermal neutrons. These neutron sources are described in detail by Lisowski et al. [12].

Two flight paths at the WNR facility are used for gamma-ray measurements. One is located at 15 degrees to the right with respect to the incident proton beam (15R) and has a detector station at 18 m. The second flight path is located at 30 degrees to the left (30L) with a detector station at 40 m. A second detector station will be installed this year at 22 m on the 30L flight path. At LANSCE, flight paths of 8 m and 21 m are used for gamma-ray measurements of capture on radioactive samples and for the investigation of parity non-conservation in neutron resonances.

The neutron flux is measured during experiments by thin, fission ionization chambers using foils of both ^{235}U and ^{238}U for all of the measurements performed at the WNR facility [13]. The ^{235,238}U(n, f) cross sections have been measured with good precision up to 100 MeV at our laboratory relative to the H(n, p) reaction [14].

The data are acquired using VAXstation/CAMAC based data acquisition systems using the Indiana University Cyclotron Facility's version of the XSYS code. The data acquisition systems provide the capability of storing large 2-dimensional arrays of neutron time of flight (tof) versus gamma-ray pulse height as well as storing data event by event on 8mm tape with a capacity of 1.5 GB per cassette. To reduce the deadtime associated with CAMAC readout the data are buffered in the CAMAC crate and read out between macro pulses.

High-Resolution Germanium Detector System

High gamma-ray energy resolution measurements similar to those performed at ORELA have been performed at the WNR facility for incident neutron energies up to 400 MeV, and with samples of interest to a wide variety of programs, including the Mars Observer mission to map the elemental composition of the Martian

surface. Results of these experiments will be used to improve both the models used in calculating and predicting reaction cross sections, and the resulting evaluated data bases.

The Ge detectors are located at 90 and 125 degrees with respect to the neutron beam. A typical timing resolution is 5 ns for $E_\gamma = 1$ MeV. The gamma-ray energy resolution is typically about 2.5 keV. The detector efficiencies have been measured with calibrated radioactive sources at energies from 122 keV to 2.6 MeV. From 2.6 to 8 MeV the efficiencies are calculated with the Monte Carlo code CYLTRAN [15]. The measured absolute efficiencies (times the solid angle subtended) for $E_\gamma = 1332$ keV range from 7 to 35×10^{-4}.

The data are stored in two-dimensional arrays, typically having 512 channels of neutron tof versus 4096 channels of gamma-ray pulse height. The raw event datum with the full ADC resolution of 8192 channels is usually written to a disk and then automatically transferred to 8 mm magnetic tape. This allows re-sorting the data with better dispersion if needed.

The data analysis includes: correcting for dead times, binning both the fission chamber beam monitor tof spectra and the Ge detector tof spectra into suitable neutron energy bins, extracting the areas under the peaks, and calculating the cross section from the extracted yields, taking into account the flux measured with the fission chamber, the sample size and thickness, and the measured detector efficiency. Finally corrections are made for neutron multiple scattering and gamma-ray attenuation in the sample.

Measurements of the strong transitions in natural B, C, N, O, Mg, Al, Si, S, Ca, Ti, Cr, Fe, and Mn have been made for the Mars Observer mission. Scoping studies have looked at the neutron-induced gamma-ray yields in Y, ^{238}U, and ^{232}Th. More detailed measurements have been performed for ^{14}N, ^{56}Fe, and ^{204,206,207,208}Pb.

BGO Five-Crystal Spectrometer

This detector system has been used to measure gamma-ray production cross sections for ^{nat}C, ^{10,11}B, ^{14,15}N, ^{nat}Eu, ^{nat}Gd, and ^{181}Ta. BGO was chosen because the ratio of gamma-ray to neutron sensitivity for BGO is generally greater than for other scintillators. The cylindrical BGO crystals are 7.6 cm diameter and 7.6 cm long. A typical gamma-ray energy resolution for these detectors is 5% at $E_\gamma = 15$ MeV. These detectors are usually used in the gamma-ray energy range, $0.5 < E_\gamma < 16$ MeV, and have a timing resolution of about 1.5 ns. This setup has been used on the 15R, 18 m flight path. The neutron energy range covered is normally $\sim 1 < E_n < 400$ MeV. The detectors span the angular range from 40 to 145 degrees.

The data acquisition setup is similar to that for the Ge detectors. Due to the lower gamma-ray energy resolution of the detectors the two-dimensional spectra have a typical size of 1024 channels of tof with 512 channels in pulse height.

Response functions for these detectors have been calculated with the CYLTRAN code. The calculated response functions have been verified with measurements at the tagged photon facility at the University of Illinois.

The analysis procedure involves the following steps: (1) Subtraction of the time-random background and the background from slower neutrons produced by a previous beam pulse, (2) Subtraction of blank and sample-out backgrounds, (3) Binning the data into suitable neutron and gamma-ray energy bins, (4) Unfolding the detector response using the deconvolution code, FERD [16], and (5) Calculating the cross sections using the flux (determined from the fission chamber data), the sample thickness, the detector efficiency, and corrections for neutron multiple scattering and gamma-ray attenuation in the sample. The cross-section data are fit to the obtain the coefficients of the usual Legendre polynomial expansion as a function of the incident neutron energy.

BGO Fast Neutron Capture Spectrometer

This system was designed to extend our capabilities to measure higher energy gamma rays. It has been used in the energy range $4 < E_\gamma < 50$ MeV. The spectrometer consists of a 10×15 cm BGO crystal surrounded by a plastic anti-coincidence shield and a large amount of Pb and polyethylene shielding. The entire system is mounted on a carriage which rotates about the sample to allow the measurement of angular distributions. This spectrometer is located at 19 m on the 15R flight path. The BGO crystal front face is 61 cm from the sample. The timing resolution is nominally 2.0 ns. The energy resolution is 8% at 4.44 MeV, and the efficiency calculated using the EGS code for a 1.2 MeV wide window around the full energy peak is 11% at $E_\gamma = 40$ MeV. The cosmic-ray rejection provided by the shield is ∼99%.

This system has been used to measure the $^{40}Ca(n, \gamma)$ cross section and angular asymmetries to extend measurements beyond the region of the isovector giant quadrupole resonance. The analysis consists of determining the cross sections from the peak corresponding to the ground state capture gamma ray. Data were measured at 55, 90 and 125 degrees. From the 55 and 125 degree data the asymmetry is calculated.

High-Energy BaF$_2$-NaI Gamma-Ray Telescope

Because the background due to neutrons scattered from the sample into the detector is often the most difficult background to properly subtract in gamma-ray experiments at a spallation neutron source, a neutron insensitive detector is especially desirable. In the gamma-ray energy range from 15 to 200 MeV the pair production mechanism can be exploited to reduce the neutron sensitivity of a detector. We are currently constructing such a detector, similar to the one described by Bertholet et al. [17].

This detector will be used this year on the 15R, 18-m flight path to measure high-energy gamma rays from the neutron-proton Bremsstrahlung process. Two identical detectors will be located at +90 and −90 degrees with respect to the beam. Each detector will be composed of segmented 1-cm-thick BaF$_2$ pair converters in front, followed by 2 thin, segmented plastic "delta-E" detectors backed by a calorimeter of 16 rectangular, $10 \times 10 \times 40$ cm NaI crystals. By measuring the energy loss in the delta-E detectors the photon events can be selected. A gamma-ray energy resolution of 15% and an absolute efficiency of 30% at $E_\gamma = 150$ MeV were calculated for this detector using the CYLTRAN code.

These data will be acquired event by event. Events corresponding to photon-electron showers will be selected by setting conditions on the various spectra. We plan to measure the inclusive gamma-ray production cross section using a liquid hydrogen target this year.

BaF$_2$ Capture Spectrometer

The construction of a 4π solid angle, BaF$_2$ spectrometer system is currently nearing completion [18]. The spectrometer consists of 8 cubic crystals ($15 \times 15 \times 15$ cm) with a square 4-cm aperture for the neutron beam and a square 4-cm entrance for a sample holder. The planned use for this system is to measure capture cross sections on small quantities of radioactive samples in the resonance region using an 8-m flight path at LANSCE. The measurements will span the neutron energy range, 0.025 eV $< E_n < 30$ keV, and the gamma-ray energy range from 6 to 10 MeV. Tests have demonstrated a system timing resolution of 250 ps. The efficiency is ∼100% for gamma-ray energies less than 10 MeV. The system was designed with the aid of Monte Carlo codes.

The flux for these experiments will be measured using a ^{6}Li sample located 22.5 cm behind the radioactive sample by measuring the known $^6Li(n, \alpha)$ reaction using a silicon surface barrier detector. These data have applications to nuclear astrophysics, reactor physics (both fission and fusion), and accelerator transmutation of waste. Two of the BaF$_2$ crystals have also been used in a pilot study to examine parity non-conservation in neutron resonances through the capture reaction.

Summary

GRP measurement systems based on the ORELA white neutron source cover the complete neutron-energy range from thermal to 40 MeV and measure gamma rays from 0.1 to 15 MeV. GRP cross sections for $(n, x\gamma)$ reactions are measured with a high-resolution germanium detector system, and low-resolution spectra at three angles are obtained as a byproduct of the NE-213 based (n, xn) system. Capture gamma-ray spectra in the resonance region can be measured with the new C$_6$D$_6$ detector system or the BaF$_2$ based multicrystal spectrometer. Papers by Dickens et al. and Schrack et al. at this conference provide examples of GRP measurements at ORELA.

The WNR facility is presently used for GRP measurements in the neutron energy range from ∼1 MeV to over 400 MeV, but neutrons with energies of over 700 MeV are produced. Instruments are available, or are being constructed to measure gamma rays in the energy range from 0.1 to 200 MeV. Spectrometers available include: high-resolution Ge detectors, BGO crystals for angular distribution and fast neutron capture measurements, and a gamma-ray telescope to study high-energy gamma-ray production. Unique measurements of capture gamma rays from radioactive samples and parity violating resonances are planned using a large volume BaF$_2$ calorimeter at LANSCE. The neutron energy range covered is from thermal to 100 keV.

Acknowledgements

Research at Oak Ridge National Laboratory is sponsored by the Office of Energy Research, Division of Nuclear Physics, U.S. Department of Energy, under contract DE-AC05-84OR21400 with Martin Marietta Energy Systems, Inc. Research at the Los Alamos National Laboratory is under the auspices of the U.S. Department of Energy under contract W-7405-ENG-36.

References

1. J. K. Dickens, G. L. Morgan, G. T. Chapman, T. A. Love, E. Newman, and F. G. Perey, Nucl. Sci. Eng. 62, 515 (1977).
2. D. M. Drake, E. D. Arthur, and M. G. Silbert, Nucl. Sci. Eng. 65, 49 (1978).
3. K. H. Bockhoff, A. D. Carlson, O. A. Wasson, J. A. Harvey, and D. C. Larson, Nucl. Sci. Eng. 106, 192 (1990).
4. R. W. Peelle, J. A. Harvey, F. C. Maienschein, L. W. Weston, D. K. Olsen, D. C. Larson, and R. L. Macklin, Neutron Research and Facility Development at the Oak Ridge Electron Linear Accelerator 1970-1995, ORNL/TM-8225 (July 1982), and report in preparation for the period 1982-1991.
5. P. F. Rose and A. Daly, editors, U.S. DOE Nuclear Data Committee "Compilation of Requests for Nuclear Data," BNL-NCS-52028 (January 1987), current under update.
6. B. D. Rooney, J. H. Todd, R. R. Spencer, and L. W. Weston, A Data Acquisition Work Station for ORELA, ORNL/TM-11454 (September 1990).
7. J. K. Dickens, SCINFUL: A Monte Carlo Based Computer Program to Determine a Scintillator Full Energy Response to Neutron Detection for E_n Between 0.1 and 80 MeV: User's Manual and FORTRAN Listing, ORNL-6462 (March 1988).

426

8. J. K. Dickens, *TPASS, A Gamma-Ray Spectrum Analysis and Isotope Identification Computer Code*, ORNL-5732 (March 1981).

9. W. R. Nelson et al., Report SLAC-265, Stanford Linear Accelerator Center (December 1985).

10. R. L. Macklin and B. J. Allen, *Nucl. Inst. Methods* **91**, 565 (1971).

11. F. G. Perey, J. O. Johnson, T. A. Gabriel, R. L. Macklin, R. R. Winters, J. H. Todd, and N. W. Hill, *Proc. Int. Conf. on Nuclear Data for Sci. and Tech., May/June 1988, Mito, Japan*, p. 379, Saikon, Tokyo (1988).

12. P. W. Lisowski, C. D. Bowman, G. J. Russell, and S. A. Wender, *Nucl. Sci. Eng.*, **106**, 208 (1990).

13. S. A. Wender, R. C. Haight, R. O. Nelson, C. M. Laymon, A. Brown, S. Balestrini, W. McCorkle, T. Lee, and N. W. Hill, *A Fission Ionization Detector for Neutron Flux Measurements at a Spallation Source*, LA-UR-90-3399 (1990).

14. P. W. Lisowski, "Fission Cross Sections in the Intermediate Energy Range", *Proc. NEANDC Specialists' Meeting on Neutron Cross Section Standards for the Energy Region above 20 MeV, Uppsala, Sweden, May 21-23, 1991*, to be published.

15. J. A. Halblieb, Sr., and W. H. Vandevender, *CYLTRAN*, Sandia Laboratories, SAND 74-0030 (1974).

16. B. W. Rust, D. T. Ingersoll, and W. R. Burrus, *A User's Manual for the FERDO and FERD Unfolding Codes*, ORNL/TM-8270 (1983).

17. R. Bertholet, M. Kwato Njock, M. Maurel, E. Monnand, H. Nifenecker, P. Perrin, J. A. Pinston, F. Schussler, D. Barneoud, C. Guet, and Y. Schutz, *Nucl. Phys.* **A474**, 541 (1987).

18. P. E. Koehler and H. A. O'Brien, *Cross Section Measurements on Radioactive Targets*, LA-UR-90-3491 (1990).

The Absolute Response Function of the Harwell Deuterated Benzene Total Energy Detectors to 6.13 MeV γ-rays

S. Croft   and   M. Bailey

AEA Technology, Harwell Laboratory, Oxfordshire
OX11 0RA.

Abstract: The absolute γ-ray response function of the Harwell deuterated benzene liquid scintillator detectors has been measured at 6.13 MeV and compared with spectra calculated using the state-of-the-art electron-gamma tracking code EGS4. The experiment was performed in a nearly monoenergetic photon field generated using the $^{19}F(p,\alpha\gamma)^{16}O$ reaction at the Harwell 5 MV Van de Graaff facility. A particularly open irradiation geometry was used to permit accurate modelling and ease of replication by other groups. The absolute yield of photons was measured using a calibrated 113 cm³ HPGe spectrometer. However, as the absolute yield, angular distribution and resonance widths of the 6.13 MeV, 6.92 MeV and 7.12 MeV photons from the 340.5 keV resonance of the $^{19}F(p,\alpha\gamma)^{16}O$ reaction were also measured using the same HPGe detector, the source reaction itself can be used as a transfer standard for other workers.

The scintillation detectors studied have been used by the Harwell electron linear accelerator group in the determination of the radiative neutron capture cross-section of ^{56}Fe in the keV neutron energy range. As an aid to understanding the response of the scintillators under these circumstances we have also measured the primary capture γ-ray spectrum from the 1.15 keV neutron resonance in ^{56}Fe. This was done using a pair of HPGe detectors on the 12.5 m flight path at HELIOS, the Harwell electron linear accelerator.

INTRODUCTION

The total energy pulse-height weighting function (PHWF) technique proposed originally by Maier-Leibnitz and first implemented by Macklin and Gibbons[1] has been widely applied to small non-hydrogenous liquid scintillators (C_6D_6, C_6D_{12}, C_6F_6) for many years now for the determination of fast neutron total capture cross-sections. In the early 1980's a troublesome discrepancy emerged between the parameters obtained for the 1.15 keV resonance in ^{56}Fe using the PHWF technique and other methods[2]. This resonance is of major importance in reactor calculations since it determines to a large extent the Doppler coefficient of reactivity on stainless steel. The discrepancy was also of great concern, because this resonance is an almost pure capture resonance, making it a near ideal case to test both capture measurement techniques and resonance analysis codes. Hence systematic discrepancies between different experimental approaches when applied to the 1.15 keV ^{56}Fe resonance also casts doubt on all other measurements.

In 1982 an NEANDC task force was set up to address this problem, and it soon became clear that the PHWF technique was not working as expected for hard capture γ-ray spectra. As a result of this, Harwell undertook the work reported below. A fuller account of the response function measurement at 6.13 MeV is to be published elsewhere[3].

RESPONSE FUNCTION MEASUREMENTS

The Harwell Maier-Leibnitz detectors, shown schematically in Figure 1, contain 377 cm³ of NE-230 scintillator[4] housed in a beryllium cell of internal radius 50 mm and height 50 mm. The inside surfaces of the cell are coated with reflective paint and the cell volume is made up by a nitrogen bubble to accommodate thermal expansion of the liquid. The cell is sealed by a quartz plate and optically coupled directly to the quartz window of a Thorn-EMI 9832QB photomultiplier tube[5].

The experiment was performed using a nearly monoenergetic and isotropic photon field produced using the $^{19}F(p,\alpha\gamma)^{16}O$ reaction by bombarding a 150 μg cm^{-2} CaF$_2$/Ta target with 4.9 μA beams of 375 keV protons. A particularly open irradiation geometry, Figure 2, was used so as to minimise the amount of scattering taking place and also to permit the geometry to be accurately represented in computer simulations using the EGS4 electron-gamma transport code[6].

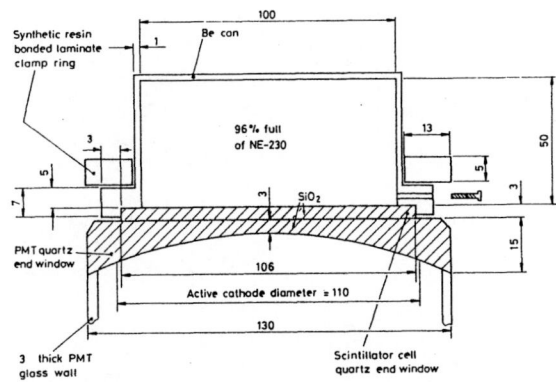

Figure 1: Schematic diagram showing the layout of material in the immediate vicinity of the C_6D_6 liquid scintillator.

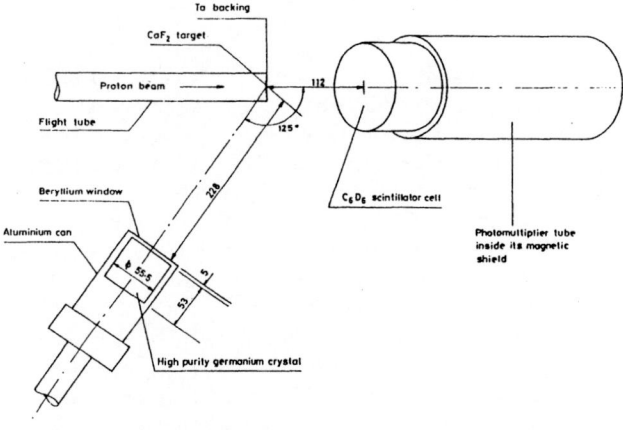

Figure 2: Plan view of the open experimental arrangement used to determine the absolute response function of the Harwell C_6D_6 total energy detector at 6.13 MeV. All dimensions are in mm.

Even so, the 0.265 mm Ta target backing was found to act as a significant electron radiator in its own right and to strongly influence the pulse height distribution in a way that depended on the angular position of the detector with respect to the target and the proton beam. An aluminium plate 9mm thick placed in front of the detector was found to dramatically reduce the low pulse height continuum caused by these electrons.

Figure 3 shows a comparison between the measured and calculated pulse height distributions (PHDs) with the detector axis tilted at 45° to the horizontal and mounted 112 mm from the target in the straight ahead position with respect to the proton beam. Referred to an energy threshold of 200 keV the ratio of the measured to the calculated detection efficiency obtained was 1.19. At distances of 150 mm and 170 mm the ratio was a little higher at 1.23. The energy deposition spectrum predicted by EGS4 was resolution-broadened according to the method described in [3] before making these comparisons.

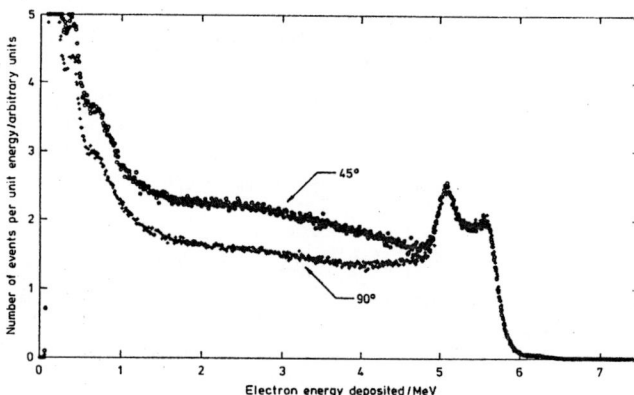

Figure 4: A comparison of the pulse-height spectra obtained at polar angles of 45° and 90° about the proton beam axis with the detector 112mm from the source.

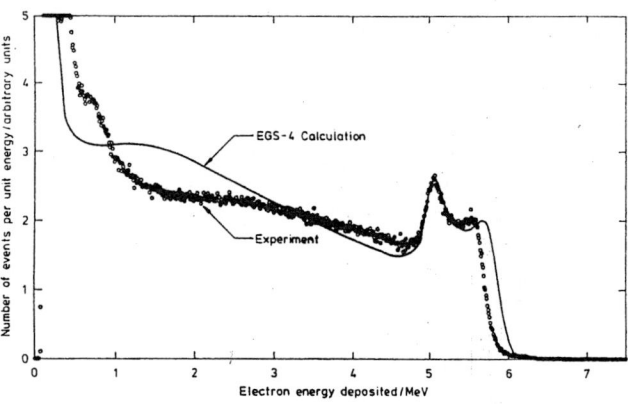

Figure 3: An absolute comparison between the measured and calculated PHD from the Harwell C_6D_6 total energy detector to 6.13 MeV photons. The detector was mouted at 45° with respect to the proton beam 112mm from the target. The target backing was 0.265mm of tantalum.

The sensitivity of the detection efficiency of the experimental arrangement is emphasised in a different way in Figure 4 which shows the effect of rotating the detector around from 45° to 90° with respect to the proton beam direction. At the 90° position the effective thickness of tantalum is larger. However the result is that there appear to be fewer low pulse-height events.

Figure 5 illustrates how, by introducing 9 mm of aluminium in front of the scintillator, the contribution of scattered electrons to the PHD can be dramatically reduced. In this figure the detector was mounted at 45° in the vertical plane with respect to the proton beam at a distance of 150 mm from the target.

The overall uncertainty associated with the experimental efficiency values is estimated to be ±5.4% at the 1σ level, and is dominated by the uncertainty in the absolute efficiency of the HPGe detector (±5.2%). The uncertainty in the EGS4 calculations is difficult to assess but is believed to be approximately ±5%.

The large influence which the Ta target backing exerts on the measured efficiency was clearly demonstrated by EGS4 calculation by performing runs as a function of Ta thickness[3]. For the geometry described above the effect of the Ta is to increase the apparent efficiency by about 80% over that which would be obtained with an ideal, transparent target backing. Clearly electron generation and transport in the materials in the vicinity of the source and the detector can have a major influence on the observed response and therefore represent a severe test of any electron transport code used to model it. This implies that great care must be taken in the design and modelling of neutron capture cross-section measuring facilities.

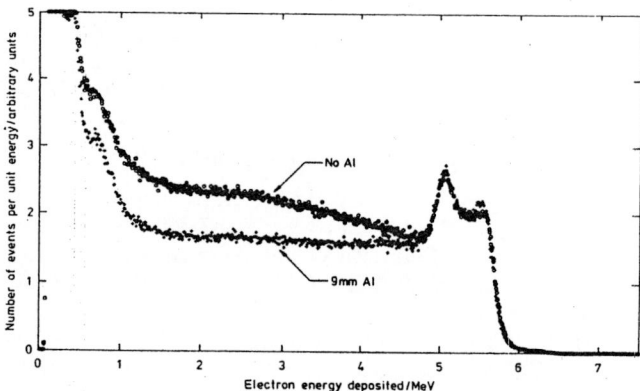

Figure 5: Comparison of spectra taken at a polar angle of 45°, 150mm from the source, with and without a 9mm thick aluminium disc on the front face of the detector. The effect of the aluminium is to dramatically reduce the relative importance of the low pulse-height continuum.

CAPTURE GAMMA-RAY SPECTRUM MEASUREMENT

Gayther *et al* [7] have used the deuterated benzene detectors studied here in a measurement of the radiative neutron capture cross-section of ^{56}Fe in the keV neutron energy range. Because the pulse-height weighting technique introduces a dependency on the spectrum of capture γ-rays emitted, we undertook a measurement of the primary capture γ-ray spectrum from the 1.15 keV resonance in ^{56}Fe. This was done using a pair of HPGe detectors on the 12.5 m flight path from the Fast Neutron Target at HELIOS, the Harwell electron linear accelerator. The experimental configuration and analysis procedures were as described in detail by Mason[8]. The sample was a disc of natural iron 2 mm thick.

The spectrum of primary γ-rays emitted following neutron capture by ^{56}Fe in the 1.15 keV resonance has only been measured twice before. Our results are broadly consistent with these as can be seen from Table 1.

E_γ keV	$I_\gamma(E_\gamma)$ [9]	$I_\gamma(E_\gamma)$ [10]	$I_\gamma(E_\gamma)$ This work
7645.5	51.0 ±3.4	41.0 ±4.	43.5 ±3.9
7631.4	100.0 ±4.3	100.0 ±4.	100.0 ±4.5
7510.1	7.0 ±1.4	5.2 ±0.8	5.71±0.62
7279.4	2.21±0.82	1.9 ±0.7	1.69±0.58
6507.0	2.64±0.48	< 1.0	1.6 ±1.0
6382.0	36.5 ±1.2	35.0 ±5.	35.4 ±2.1
6018.9	3.61±0.77	1.1 ±0.5	0.83±0.71
5921.4	7.8 ±1.1	5.6 ±1.3	4.9 ±1.5
5192.9	—	< 1.2	< 1.0
4950.5	5.9 ±1.0	5.0 ±1.4	4.9 ±1.3
4810.9	< 0.48	1.5 ±1.0	< 1.0
4589.3	—	< 1.0	< 1.0
4463.3	5.8 ±1.1	5.5 ±1.0	5.16±1.30
4408.0	1.73±0.67	1.3 ±0.8	1.66±0.85
4399.9	—	< 1.0	1.06±0.87
4276.7	1.35±0.67	1.7 ±0.8	1.92±0.93
4219.0	1.35±0.67	1.2 ±0.8	2.0 ±1.0
3855.5	1.25±0.91	1.9 ±1.1	1.7 ±1.1
3794.6	< 0.48	—	< 1.0
3745.3	< 0.48	—	< 1.0
3721.3	1.9 ±1.0	—	1.30±0.95
3508.4	< 0.48	—	< 1.0
3439.0	< 0.48	—	0.86±0.80
3267.0	1.9 ±1.2	—	3.0 ±1.8

Table 1: Relative spectrum of primary γ-rays emitted following the resonant capture of 1.15 keV neutrons by ^{56}Fe.

CONCLUSIONS

The absolute γ-ray response function of the Harwell Maier-Leibnitz detectors has been measured at 6.13 MeV in a simple low-scatter geometry. The state-of-the-art Monte-Carlo code EGS4 was used to calculate the response function expected assuming that the light output function for electrons, L, as a function of electron energy, E, is simply proportional to the electron energy, i.e. $L(E)=\alpha.E$ where α is the constant of proportionality (set equal to unity in this work). A shape discrepancy (see Figure 3) which could not be accounted for was observed. The linearity of the detection system was confirmed however by making measurements with a variety of γ-ray sources. The effect of various resolution broadening functions, including asymmetric forms was investigated but found not to improve the fit significantly. Even allowing for a modest light response off-set $[L(E)=\alpha.(E-E_0)$, $E_0 \lesssim 50$ keV] failed to fully account for the shape mis-match at high pulse heights.

The measured-to-calculated detection efficiency ratio was approximately 1.2 compared with an overall uncertainty of only about 7%. The thin Ta target backing was shown to have a remarkably strong influence on the observed pulse-height distribution. Aluminium filters in front of the detectors appeared to decouple the detector from the target-generated electrons. These findings have important consequences when translated to realistic capture cross-section measuring geometries.

The relative spectrum of primary γ-rays from the resonant capture of 1.15 keV neutrons in ^{56}Fe has been measured. Good agreement is found with the work of Chrien *et al* and Wells *et al*.

ACKNOWLEDGEMENTS

It is a pleasure to acknowledge the help of Dr. J.P. Mason in performing the capture spectrum measurement and of Mr. J.E. Jolly in setting up the response function measurement. The support and encouragement of Dr. D.B. Gayther throughout the work was greatly appreciated. Our thanks also go to Mr. R.B. Thom and Mr. J. Tilbury for their guidance in running the EGS4 code.

This work was funded through the UKAEA's Underlying Research Programme.

REFERENCES

1. R.L. Macklin and J.H. Gibbons: *Phys Rev* 159 (1967) 1007–1012.

2. M.S. Coates, D.B. Gayther, G.D. James, M.C. Moxon, B.H. Patrick, M.G. Sowerby and D.B. Syme: *Conference on Nuclear Data for Science and Technology 1983* K.H. Böckhoff (ed), D. Reidel Publ. Co. pp 977–986.

3. S. Croft and M. Bailey: *Nucl Instr Meth in Phys Res A* In press.

4. Nuclear Enterprises Limited, Sighthill, Edinburgh EH11 4BY, Scotland.

5. Thorn EMI Electron Tubes Limited, Bury Street, Ruislip, Middlesex HA4 7TA, England.

6. W.R. Nelson, A. Hirayama and D.W.O. Rogers: *Stanford Linear Accelerator Report* SLAC-265 (1985)

7. D.B. Gayther, J.E. Jolly and R.B. Thom: *Conference on Nuclear Data for Science and Technology 1988* S. Igarasi (ed), Saigon Publ. Co. pp 157–159.

8. J.P. Mason: *Nucl Phys* A465 (1987) 413–428.

9. R.E. Chrien, M.R. Bhat and O.R. Wasson: *Phys Rev* C1(3) (1970) 973–975.

10. J.C. Wells Jr, S. Raman and G.G. Slaughter: *Phys Rev* C18(2) 1978) 707–713.

MEASUREMENTS OF GAMMA-RAY SPECTRA FROM NEUTRON CAPTURE IN SINGLE RESONANCES

C. Coceva*, A. Spits, G. Fioni and A. Mauri*

CEC-JRC, Central Bureau for Nuclear Measurements, B-2440 Geel, Belgium
* ENEA, C.R.E. "E. Clementel", Viale G.B. Ercolani 8, I-40138 Bologna, Italy

Abstract: A measurement is described of partial radiative widths of neutron resonances of Cr-53.   Neutron energy is selected by time-of-flight and gamma energy is measured by a Ge-crystal with anticoincidence shielding. Partial widths are measured for single s-wave resonances from 4 to 30 keV.   An improved Average Resonance Capture method is applied to p-wave resonances from 13 to 70 keV. Transition probabilities are obtained to final states with excitations up to 4.3 MeV.   A new method is described to deduce absolute values of $\Gamma_{\gamma i}/\Gamma_\gamma$.   Estimates are obtained of E1 and M1 strengths and of their energy dependence.

(^{53}Cr(n,γ), E = 2.5-70 keV; measured E_γ, I_γ. Deduced E1 and M1 strengths)

Introduction

Experiments on the gamma decay of nuclear states formed by neutron capture provide valuable information on the properties of radiative transitions between specific nuclear states, below the electric giant dipole resonance. In such measurements a pulsed white neutron source is used. The initial state of a transition is singled out by the neutron energy, while the final state is determined by the energy of the emitted photon.   Gamma rays emitted after neutron capture are detected by a Ge crystal and its output pulses are processed by a bi-dimensional time and amplitude analyzer from which the neutron time-of-flight and the gamma-ray energy are obtained.

In this way, many possible transitions can simultaneously be measured between a number of initial states (neutron resonances) and final states (ground state and excited states of the compound nucleus).   For each neutron capture event, the detection probability for a given transition is in general extremely small, of the order of 10^{-6}, because of the low efficiency of Ge crystals (in the full-energy peak), and because partial radiative widths may be a very small fraction of the total width. For this reason, high neutron fluxes are required at the sample position. Therefore, in the past, most experiments were carried out at low neutron energy on nuclei having a high level density.

In particular, rather scarce data are available for nuclei in the mass region of structural materials, in spite of their practical importance.

Experiment

We describe here an experiment on neutron capture in ^{53}Cr, where partial gamma widths to final states with excitation up to 4.2 MeV could be measured for single resonances with energy up to 30 keV. This was obtained at the CBNM, Geel, by using the high-intensity pulsed neutron source GELINA, and a gamma-ray detector system with a very low Compton and escape-peak background.   Such a detector, shown in Fig. 1, consists of a 5.4 cm diameter, 6 cm long Ge crystal (NGe in Fig. 1) operated in anticoincidence with an annulus of NaI(Tl) and two other Ge crystals (LEGe and PGe in Fig. 1), completely surrounding the central detector.

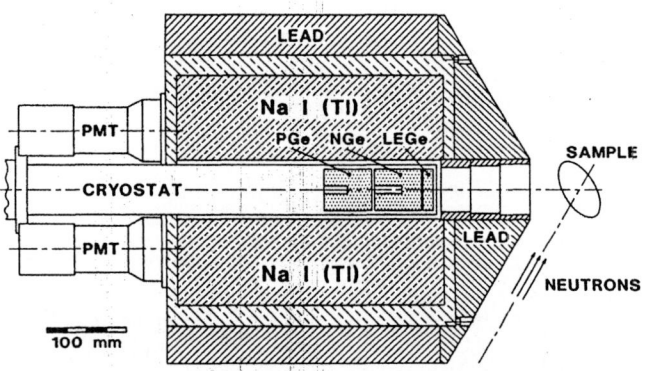

Fig. 1   Gamma detector with anticoincidence shielding

The three Ge crystals are mounted in the same cryostat arm and are inserted in the throughhole along the axis of a 30 cm diameter, 30 cm long NaI(Tl) scintillator. By this device, the Compton background of a 6 MeV gamma peak is reduced by a factor 10 and the single-escape peak by a factor 7.   Such a reduced background results in a much improved statistical accuracy in the determination of full-energy peak areas. The 96% enriched ^{53}Cr sample, consisting of 46.5 g of Chromium oxide, was placed at a flight distance of 25 m. The 130 MeV linear accelerator GELINA was operated at a beam power of 8 kw, electron burst width 1 ns, repetition frequency 800 Hz.

With the main amplifier of the central Ge detector operating with a formation time

2 μs, the pulse pile-up rejection circuit caused an approximately constant dead time of 10% from 3 keV up to about 70 keV. Above that energy, the dead time is rapidly increasing due to the effect of the gamma-flash from the linac.

The t.o.f. spectrum of capture gamma-rays between 2.5 and 70 keV is shown in Fig. 2. From the raw data, 30 amplitude spectra were built, corresponding to the t.o.f. intervals indicated by the shadowed areas. Each spectrum of 8192 channels covered the gamma energy interval from 0.5 to 9.8 MeV, with a resolution of 4.2 keV at 6 MeV. The count rate of gamma-rays with energy $E_{\gamma i}$, originated from neutron capture in the interval δE_λ, expressed in terms of the transition intensity $I_{\lambda i} = (\Gamma_{\gamma i}/\Gamma_\gamma)$, is

$$G_{\lambda i}(\delta E_\lambda) = N(E_\lambda)\,\delta E_\lambda\,C(\delta E_\lambda)\,I_{\lambda i}\,\epsilon(E_{\gamma i}) \qquad (1)$$

where $N(E_\lambda)$ is the rate of neutrons with energy E_λ incident on the sample, per unit energy; $C(\delta E_\lambda)$ is the fraction of captured neutron per incident neutron within δE_λ, and $\epsilon(E_{\gamma i})$ is the overall efficiency for detection of gamma rays with energy $E_{\gamma i}$, in the full energy peak.

If we label with index j all transitions leading to the ground state, we have

$$\Sigma_j [G_{\lambda j}(\delta E_\lambda)\,/\,\epsilon(E_{\gamma j})] = N(E_\lambda)\,\delta E_\lambda\,C(\delta E_\lambda) \qquad (2)$$

owing to the fact that $\Sigma_j\,I_{\lambda j} = 1$.

Therefore, from the measured count rates, one obtains the intensity:

$$I_{\lambda i} = [G_{\lambda i}(\delta E_\lambda)/\epsilon(E_{\gamma i})]\,/\,[\Sigma_j G_{\lambda j}(\delta E_\lambda)/\epsilon(E_{\gamma j})] \qquad (3)$$

This expression is independent of
a) energy width δE_λ,
b) neutron flux,
c) sample thickness,
d) any corrections to the count rates affecting primary and secondary transitions in the same way.

To get $I_{\lambda i}$ one needs to determine just the relative efficiency of the Ge detector.

From (2) we notice that the capture rate (and then the capture cross-section) could be obtained through an absolute determination of the efficiency ϵ and of the neutron flux N.

In our experiment the energy dependence of the efficiency between 0.34 and 10.8 MeV was obtained from radioactive sources of ^{152}Eu, ^{60}Co and ^{24}Na, and a measurement of thermal neutron capture gamma-rays in ^{14}N, having well known [1] intensities. In a first analysis 10 gamma-ray spectra were sorted, 8 of which correspond to capture in single s-wave resonances at 4.2, 5.7, 6.2, 8.2, 19.7, 25.9, 27.2 and 29.5 keV. One spectrum corresponds to a single p-wave resonance at 12.1 keV and one to the sum of the remaining 21 intervals containing 27 known [2] p-wave resonances. From this last spectrum, average intensities can be extracted as in the Average Resonance Capture (ARC) method; but we have here a double advantage:
i) one can select intervals containing only a given kind of resonances,
ii) background intervals can be removed.

In order to evaluate the relative capture rate (2), gamma-peak areas at the following energies (keV) were evaluated: 834.8, 2619.5, 3393.1, 3719.7, 4265.0, 4872.3, 5291.6 and 9718.8+E, E being the neutron energy. These make up all the observed

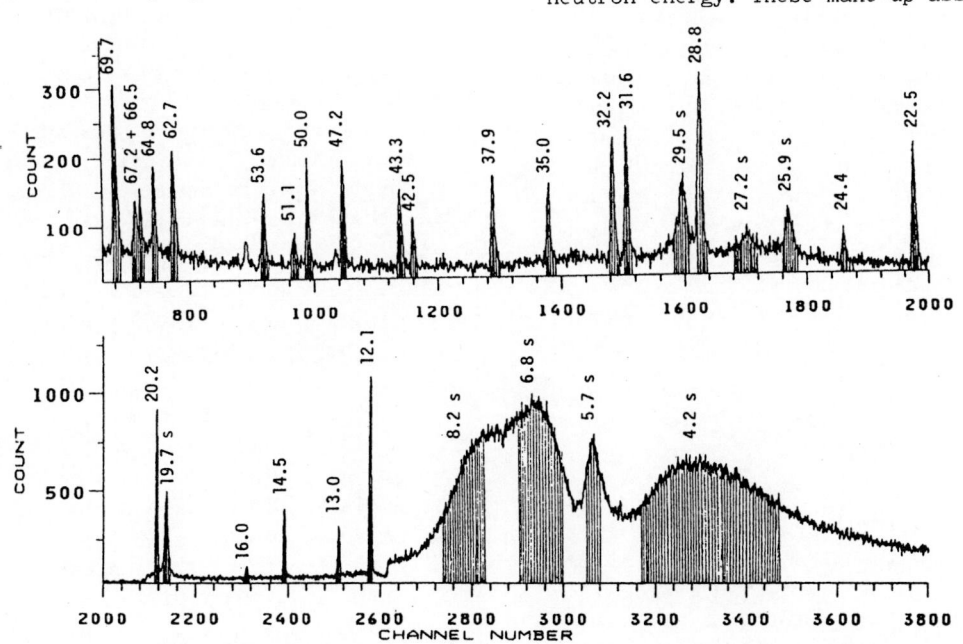

Fig. 2   Time-of-flight spectrum of neutron capture in ^{53}Cr.
Peaks are labelled with resonance energies in keV units

Table 1. Measured values of $\Gamma_{\gamma i}(E_\gamma)/\Gamma_\gamma$

E_γ(keV)	E_f(keV)	I^π_f	$J^\pi = 1^-$			$J^\pi = 2^-$					$J^\pi=2^+$	$J^\pi=0,1,2,3^+$
			E=4.2	E=6.8	E=27.2	E=5.7	E=8.2	E=19.7	E=25.9	E=29.5	E=12.1	13<E<70
5463.3	4256.2	$2^+(3^-)$	0.3	0.8		5.7	0.5	0.2	1.3	1.8	0.3	0.5
5480.2	4239.3	$2^+(3^-)$	0.7	0.2		1.2	0.9				1.5	1.1
5521.0	4197.5	2^+	0.4	1.4	0.6	0.5	0.4	0.4	0.4	0.9		0.3
5592.2	4127.6	3^-				1.1	1.9				1.3	2.1
5792.1	3927.4	2^+	0.5	1.9	2.1	2.6	7.0	3.3	5.6	4.3	1.5	4.1
5858.5	3860.9	2^+	1.2	0.5	3.0	5.1	8.1	0.2	4.3	4.3	2.7	1.4
5920.9	3798.6	4^+										4.9
5999.4	3720.0	$1^+(2^+)$	5.5	0.6	21.3	4.0	9.5	12.5	12.7	5.2		2.0
6064.3	3655.1	4^+										3.3
6559.9	3159.4	4^+										5.6
6645.6	3073.8	2^+	10.2	3.4	5.7		4.7	13.5	1.1	23.9		4.4
6889.7	2829.5	0^+	4.2	0.4	8.7							1.2
7099.7	2619.6	2^+	10.0	2.6	3.2	35.7	54.3		1.2	1.8	1.5	2.3
7895.3	1823.9	4^+										6.7
8884.3	834.8	2^+	37.2	48.7	20.1		39.6	10.3	12.5		59.0	16.9
9718.8	0	0^+	16.0	22.8								36.7

transitions ending to the ground state of ^{54}Cr. The soundness of this assumption is confirmed by the agreement of the deduced intensities for the 4.2 keV resonance (see Table 1) with the intensity standards determined at thermal energy by Loper and Thomas [3] in an independent way.

Results

From a preliminary analysis carried out on a part of the collected data, we give here, in Tab. 1, the transition intensities to 16 final states of ^{54}Cr. The first three columns give the transition energy (leaving out the neutron energy E), the energy of the final state and its spin and parity, respectively. The remaining 10 columns give, for each neutron energy E(keV), the values of $\Gamma_{\gamma i}/\Gamma_\gamma$ in units of transitions per 100 captures. Missing entries correspond to unobserved intensities. To be noted the strong intensities to the ground state and to the two first 2^+ states. Partial radiative widths are then inferred using the total radiative widths given by Brusegan et al. [2]. From the level spacings D_λ given in the same work [2], the average reduced widths

$$k(E_\gamma) = \langle \Gamma_{\gamma i \lambda}/E^3_{\gamma i} \, D_\lambda \rangle_\lambda$$

are obtained for each $E_{\gamma i}$, separately for E1 and M1 radiations. By fitting such values with a power law $k = \alpha E^x$, we find that the minimum $\chi^2(x)$ corresponds to a power $x = 2$ for E1 radiation, in rough agreement with the Giant Dipole Resonance (GDR) hypothesis of Brink-Axel. For M1 radiation, the energy behaviour is not in contrast with a constant value of $k(E_\gamma)$.

The overall electric dipole strength, deduced from 59 observed values of $I_{\lambda i}$, is

$$\langle k(E_\gamma)/A^{2/3}\rangle_{E1} = 1.5 \times 10^{-9} \ \text{MeV}^{-3} = 0.02 \ \text{W.u./MeV}$$

where W.u. stands for Weisskopf unit. This result is in agreement with the evaluation of Krusche and Lieb |4| in this mass region.

The magnetic dipole strength was obtained by averaging over 17 observed values of $I_{\lambda i}$, most of which are from the ARC spectrum:

$$\langle k(E_\gamma)\rangle_{M1} = 8.9 \times 10^{-9} \ \text{MeV}^{-3} = 0.4 \ \text{W.u./MeV}.$$

This result is in excellent agreement with the semiempirical estimate proposed by Kopecky [5], while the value deduced in ref. [4] from thermal neutron capture in the range 20<A<80 is 0.2 W.u./MeV.

The error in the dipole strengths should be smaller than 20%, including the systematic calibration error and statistical errors in the measured intensities.

REFERENCES

1. T.J. Kennett, W.V. Prestwich, J.S. Tsai: Nucl. Inst. Meth. in Phys. Res. A249, 366(1986)

2. A. Brusegan, R. Buyl, F. Corvi, L. Mewissen, F. Poortmans, G. Rohr, R. Shelley, T. Van der Veen, I. Van Marcke, in Proc. Santa Fe Conference on Nuclear Data for Basic and Applied Science, p. 633 (1985)

3. G.D. Loper, G.E. Thomas: Nucl. Instr. Meth. 105, 453 (1972)

4. B. Krusche, K.P. Lieb: Phys. Rev. C34, 2103 (1986)

5. J. Kopecky, in Proc. 4th Int. Symp. on Neutron Capture γ-Ray Spectroscopy, Grenoble, p. 423 (1981)

NEUTRON ENERGY SPECTRA FROM PROTON IRRADIATED THICK Li TARGETS

Gunter H.R. Kegel, Patrick F. Dugan*, James J. Egan, Causon K.C. Jen,
Arthur Mittler, and Parish A. Staples

Department of Physics and Applied Physics, University of Lowell
Lowell, Massachusetts 01854-9985, USA

Abstract: At our laboratory thick metallic Li targets are used to generate pseudo-white neutron spectra. We use these spectra for a quick determination of the efficiency of our time-of-flight neutron detectors. This efficiency is obtained as the ratio of measured and calculated spectra. We describe our procedure.

(Neutrons, time-of-flight, pseudo-white neutron spectra, detection efficiency.)

1. Introduction

The accuracy of neutron scattering data obtained in time-of-flight (TOF) experiments is, at best, equal to the accuracy with which the neutron detector efficiency is known. The detector efficiency depends on the detected neutron energy; it can be determined by measuring the neutron energy spectrum of a source which produces an accurately known neutron spectrum covering a broad range of neutron energies. Fission neutrons from the spontaneous fission of Cf-252 are commonly used for this purpose. Neutron TOF work requires "start" and "stop" signals; fission fragments of Cf-252 are usually detected to generate "start" signals. These fragments have a short range, hence a thin Cf source must be used so that the number of neutrons generated is small. It follows that the efficiency determinations with Cf-252 are prolonged processes, which require substantial time for their completion. The desirability of frequent efficiency checks has led us to examine the use of the p-n reaction on a thick lithium target to generate a broad energy neutron spectrum suitable for efficiency determinations. We view this technique as a supplement to, rather than a substitute for, the use of fission neutron spectra. Its main merit is the speed with which the desired data are obtained. We review our results in this paper.

2. Target Preparation

The preparation of lithium targets is similar to that described earlier [1]. A piece of about 0.5g cut from a metallic lithium rod [2] is etched in methanol until lustrous, then quenched in acetone. A stainless steel backing plate is cleaned; a drop of cyanoacrylate ester cement [3] is placed in its center and is covered by the cleaned metallic lithium. Next the lithium is compressed against the backing plate causing it to spread forming a disc with a 2cm diameter. Prior to use the target is cleaned by vigorous rubbing with tissue paper moistened with acetone. The target is water cooled when irradiated.

Since the cyanoacrylate ester cement is a poor heat conductor, a large temperature differential may exist between lithium target and its backing possibly causing the target to melt. To avoid this situation we use a lithium disc substantially larger than the proton beam diameter. The attendant increase in the lithium-backing contact area reduces the thermal resistance of the layer.

3. Thick target yield

The number of neutrons $\Delta^2 N$ emitted into the solid angle $\Delta\Omega$ when n protons with energy E traverse a layer of a metallic Li target where they lose energy ΔE is given by

$$\Delta^2 N = \frac{n}{\varepsilon} \frac{d\sigma}{d\Omega} \Delta\Omega \, \Delta E = \frac{Q}{e\varepsilon} \frac{d\sigma}{d\Omega} \Delta\Omega \, \Delta E. \quad (1)$$

Here $d\sigma/d\Omega$ is the differential cross section of the Li-7(p,n)Be-7 reaction. This reaction and also the Li-7(p,n)Be-7* reaction have been reviewed by Liskien and Paulsen[4]. Here ε denotes the the Li atomic stopping power, see e.g. Whaling[5] or Andersen and Ziegler[6], Q is the total proton charge, and e is the charge of the electron. A calculated [7] neutron energy spectrum for Li-7(p,n)Be-7 is shown

*Deceased.

in Fig. 1 for a proton energy of 5.5 MeV and for other parameters as specified. The spectrum for the weak, second neutron group is of similar shape. The calculated spectrum is based on the data by Liskien and Paulsen [4] and on an approximation for proton energies under 1950 keV as described in [7]. For dosimetry purposes it is necessary to relate the total number of neutrons produced (angle and energy integrated) to the proton charge Q for different proton energies E_p. Fig. 2 shows this relation, including both the first and the second neutron groups.

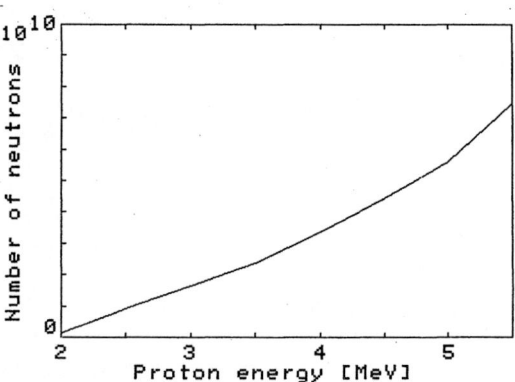

Fig. 2. Total number of neutrons generated by irradiating a thick metallic Li-7 target with 10^{-6}C of protons.

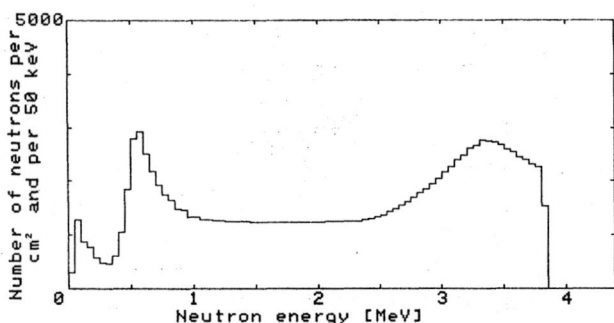

Fig. 1. First neutron group fluence from a thick lithium target at a 1-m distance for 10^{-6}C of 5.5-MeV incident protons.

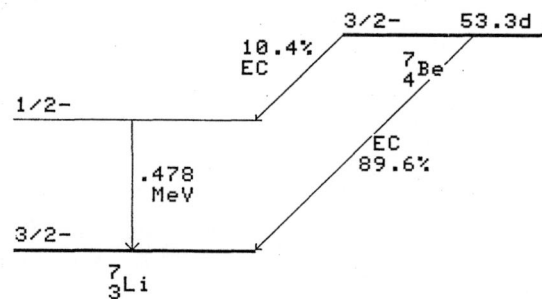

Fig. 3. Be-7 decay scheme.

4. Absolute neutron yield

In some applications the absolute efficiency of a detector must be known. There are two procedures to obtain this datum: the first utilizes proton current integration to obtain Q and hence $\Delta^2 N$, Eq. (1); the other procedure involves the Be-7 activity of the target. Since each neutron leaves one radioactive Be-7 nucleus behind the total number of neutrons produced, N, is readily obtained by assaying the Be-7 content of the target after irradiation. The decay scheme of Be-7 is simple; it is shown in Fig. 3. The half life of Be-7 (53.3d) is sufficiently long to permit assaying without haste.

In recent years we have frequently used thick lithium targets with 25μA / 4MeV proton beams to generate intense neutron fluences. In each case Q was measured and N was found by assaying. The ratio N/Q averaged over 25 runs was $3.108 \times 10^{+15}$ neutrons/C with a standard deviation of 4.8%. The calculated ratio [7] refers to a monoisotopic target of 100% Li^7. Since fluctuations in the isotopic composition of Li have been reported [8] we have established a procedure for measuring the Li-6 to Li-7 ratio [9] and we found that our lithium metal contains 92.5 at% of Li-7. With this datum we can compute a N/Q ratio and we obtain

$3.110 \times 10^{+15}$ neutrons/C in good agreement with our measurement.

5. Proton multiple scattering

As protons penetrate the target they undergo many small deflections due to small angle Rutherford scattering. Eventually these reflections will lead to a measurable angular spread of the beam. Since both, the number of neutrons emitted and their energy are functions of the emission angle the effect of an angular spread must be examined. An analysis (see, e.g. Segré [10]) shows that for a Li target with an incident proton energy of 5.5 MeV the RMS angular spread does not exceed 5° while the proton energy remains above the p-n threshold. The data by Liskien and Paulsen [4] show that the Li differential cross section changes at most by 2% for a 5° change in the beam direction and that the energy shift of the neutron generated does not exceed 3 keV. Hence in most cases the proton angular spread can be neglected. The finite solid angle subtended by the neutron detector also introduces a small change in the number and the energy of the neutrons detected and in most cases this effect is negligible. A computer code under development in our laboratory will include these two corrections for demanding applications.

6. Results

A typical thick target neutron spectrum is shown in Fig. 4. In this measurement the detector response was optimized for inelastically scattered neutrons. The

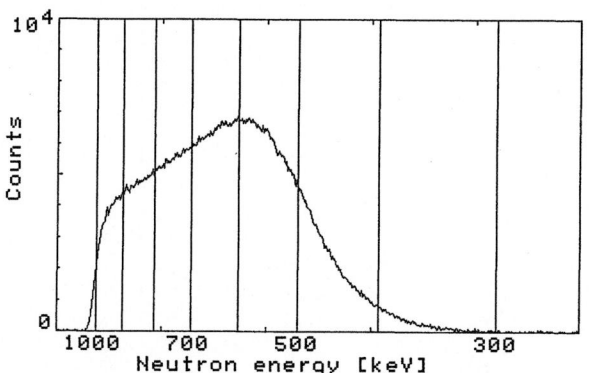

Fig. 4. A thick Li target neutron spectrum. The incident proton energy was 2.807 MeV.

background has been subtracted and the low-energy trail has been removed (see, e.g. [11]). Finite resolution effects have been neglected. The number of neutrons detected in each 100-keV wide interval has been divided by the corresponding thick-target yield figures. The result is the relative detector efficiency; it is shown in Fig. 5.

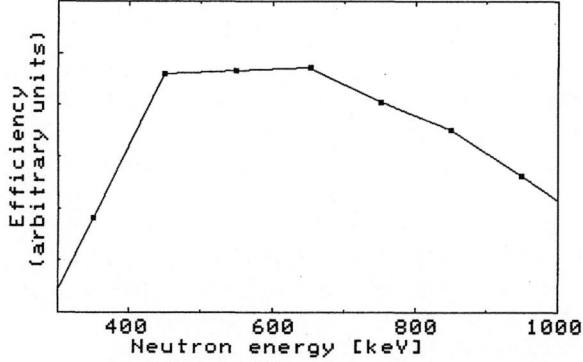

Fig. 5. The relative detector efficiency for neutron from 300 to 1000 keV.

7. Acknowledgements

The authors appreciate the help received from their collegues in the Neutron Scattering Group at the University of Lowell. Work reported in this paper has been supported by the US Department of Energy.

References

1. G.H.R. Kegel, Nucl. Inst. and Meth. in Phys. Research <u>B40/41</u>, 1165(1989).
2. Lithium Corporation of America, Bessemer City, NC, USA.
3. Sold under the name of "Super Glue", Loctite Corp. Cleveland, OH, USA.
4. H. Liskien and A. Paulsen, At. Data Nucl. Data Tables <u>15</u>, 57(1975).
5. W. Whaling in Handbuch der Physik, Vol XXXIV, ed. S. Flügge (Springer, Berlin, 1958).
6. H.H. Andersen and J.F. Ziegler, Hydrogen Stopping Powers and Ranges in all Elements, Pergamon, New York, (1977).
7. G.H.R. Kegel, Comput. Phys. Commun. <u>36</u>, 321(1985).
8. C. Lederer and V. Shirley, Table of Isotopes, pg. 3, Wiley, New York(1978).
9. C. Jen and G.H.R. Kegel, to be published.
10. E. Segré, Nuclei and Particles, pg. 39, Benjamin, New York, 1965.
11. G.H.R. Kegel, C. Ciarcia, G.P. Couchell, and J. Shao in Proc. Int. Conf. on Nucl. Data for Science and Technology Sept. 1982, Antwerp, Belgium (Edited by V.H. Böckhoff), pg. 897, Reidel, Boston, (1983).

A NEW APPROACH TO SOLVE THE PROBLEM OF SOURCE BREAK-UP NEUTRON INTERFERENCE IN SECONDARY NEUTRON SPECTRUM MEASUREMENT

Qi Bujia, Tang Hongqing, Zhou Zuying, Sa Jun

Ke Zunjian, Sui Qingchang and Xia Haihong

China Institute of Atomic Energy

P.O.Box 275-46, Beijing 102413

People's Republic of China

Abstract: To measure secondary neutron double differential cross sections in the incident neutron energy region of 9-13 Mev, an abnormal fast neutron TOF spectrometer was worked out and tested. Double differential cross sections of ^{238}U measured by both normal and abnormal TOF spectrometers were compared and a reasonable agreement was achieved.

(fast neutron, time of flight, spectrometer, secondary neutron spectrum, ^{238}U, double differential cross section)

I. Introduction

There are scarce measurements of secondary neutron double differential cross sections in the incident neutron energy range 9 to 13 MeV upto now. A major difficulty for this is lack of an applicable monoenergetic neutron source. When monoenergetic neutron energy reaches 8 MeV, the break-up neutrons from the d+D or p+T reaction start to become significant. It is difficult to get a pure secondary neutron spectrum induced only by monoenergetic neutrons in this incident energy region. To solve this problem, an abnormal fast neutron TOF facility was built and tested.

II. Principle of the Method

The idea of the method (the abnormal fast neutron TOF spectroscopy) is to make two groups of secondary neutrons, which are induced by monoenergetic and break-up neutrons of the d+D reaction neutron source respectively, separate enough in their flight time. For this purpose, an unusual arrangement was made. For a normal TOF spectrometer, the distance between neutron source and sample is about 15-20 cm and that between sample and neutron detector is several meters. However, for the abnormal TOF facility the distance between neutron source and sample is much longer than that used in the normal TOF spectrometer. In our case, it is 2.2 meters. So the difference of the flight time between monoenergetic and break-up neutrons of the D+d reaction over this distance is much larger than usual(usually the difference is 3 ns, but for the abnormal TOF spectrometer it is 35.0 ns assuming that the monoenergetic neutron energy is 10 MeV and break-up neutron energy 3.5MeV) . Meanwhile, the distance between sample and neutron detector is shortened. In our case it is 70 cm . As a result, the flight time of secondary neutrons with the lowest energy (say 1.5 MeV, cut by neutron detector bias) induced by monoenergetic neutrons is less than that with the highest energy(about 3.5 MeV) induced by break-up neutrons. Therefore two groups of neutrons which are induced respectively by monoenergetic and break-up neutrons of the D+d reaction, are well separated in the time of flight spectrum. The contamination of break-up neutrons from the source to the wanted secondary neutron spectrum can be fully eliminated and a pure secondary neutron spectrum induced only by monoenergetic neutrons can be obtained.

An ideal TOF spectrum of secondary neutrons induced by 10 Mev and 3.5 MeV neutrons is shown in Fig.1.

III. Experimental Facility

The abnormal fast neutron TOF facility is shown in Fig.2. The beam pick-off, gas target, neutron detector, electronics and data acquisition system are the same as in ref.1.

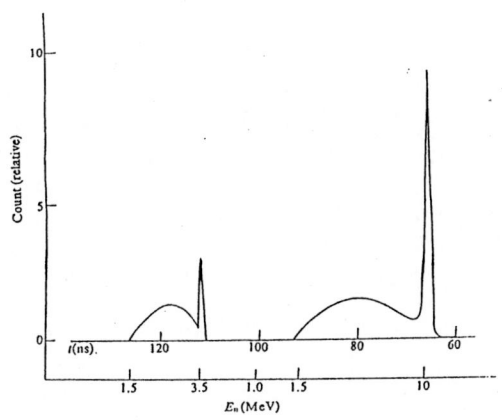

Fig.1 Ideal time of flight spectrum of secondary neutrons induced
by 10 MeV and 3.5 MeV neutrons.
Distance between neutron source and sample is 220 cm.
Distance between sample and neutron detector is 70 cm.

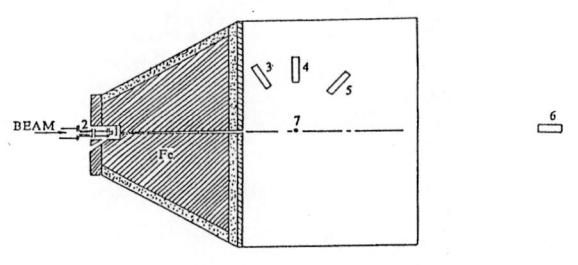

Fig.2 Schematic diagram of the abnormal fast neutron TOF
spectrometer
1 - neutron source   2 - beam pick-off
3, 4 and 5 - neutron detectors
▨ - paraffin + lithium carbonate
▧ - lead
▥ - iron

IV. Performance of the Spectrometer

The incident neutron TOF spectrum of the D+d reaction at the bombarding energy Ed=7.4 MeV measured by the normal spectrometer is shown in Fig.3. The figure shows that the break-up neutrons with energy below 3.5 MeV are significant.

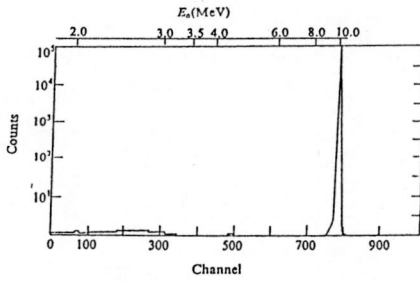

Fig.3 Time of flight spectrum of neutrons produced
through the D(d,n) reaction in the deuterium
gas target at 0 degree Ed = 7 MeV

It is the main trouble in secondary neutron spectrum measurement. Because of its interference, the lower energy part below 3.5 MeV of the spectrum is contaminated. It is difficult to obtain a pure secondary neutron spectrum of ^{238}U or ^{209}Bi for 10 MeV incident monoenergetic neutrons. The secondary neutron TOF spectra of ^{12}C, ^{238}U and ^{209}Bi induced by 10 or 12 MeV monoenergetic and break-up neutrons of the D+d reaction measured by means of the abnormal TOF spectrometer are shown in Figs.4, 5 and 6. It is obvious that two groups of secondary neutrons induced respectively by monoenergetic and break-up neutrons are well separated in the TOF spectrum. Therefore the contamination of break-up neutrons from the source to the secondary neutron spectrum to be measured is fully eliminated and a pure secondary

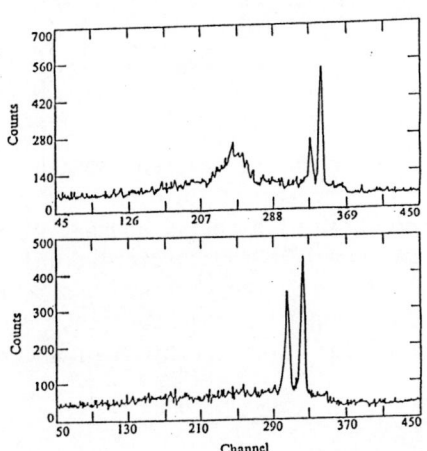

Fig.4 TOF spectra of d-D reaction neutrons with
monoenergetic energies 12 and 10 MeV
scattered from carbon at 90 degrees
Top: En=12MeV,   bottom: En=10MeV

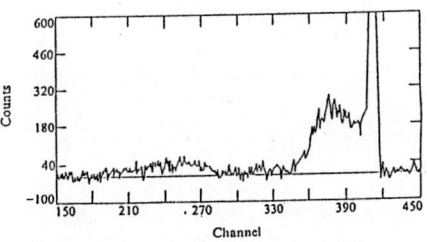

Fig.5 Net neutron TOF spectrum of ^{238}U induced by
the D+d reaction neutrons (Ed = 7 MeV)
Random background subtracted
45 degrees

neutron spectrum induced only by 10 MeV monoenergetic neutrons is obtained. The full width at the half maximum of the elastic peak in the spectrum is 1.6 ns. For 4 MeV neutrons the energy resolusion is 12% .

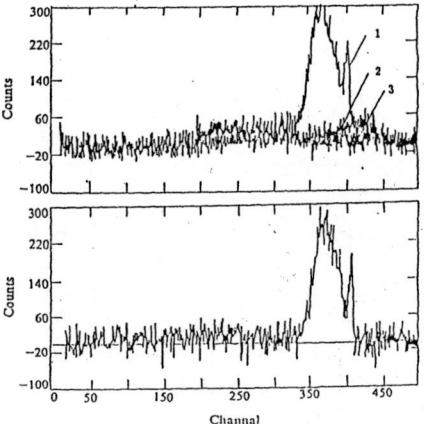

Fig.6 Neutron TOF spectra of ²³⁸U induced
by 10 MeV neutrons at 120 degrees
Random background subtracted
Top: 1- sample in, target gas (deuterium) in
2- sample out, target gas in
3- sample in, target gas out
4- sample out, target gas out
(not shown in the figure)
Bottom: [1] - [2] -[3] + [4]

The secondary neutron TOF spectrum of ²³⁸U is converted into energy spectrum and compared with that measured by means of the normal TOF spectrometer, which is shown in Fig.7. The crosses and circles in the figure indicate the double differential cross sections of secondary

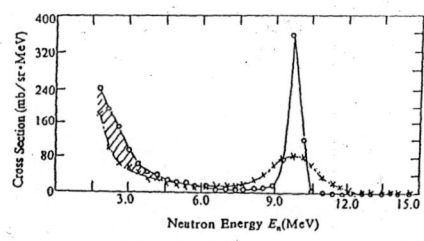

Fig.7 Double differential neutron emission spectra
of ²³⁸U induced by 10 MeV neutrons
at 45 degrees
-O- using the normal TOF spectrometer
-x- using the abnormal TOF spectrometer

neutrons of ²³⁸U at 45° for the incident energy of 10 MeV measured by the abnormal and normal TOF spectrometers respectively. The shadowed area is the source break-up neutron contribution which is moved away in the spectrum measured by the abnormal TOF spectrometer. The source break-up neutron contribution to the lower energy part of

the secondary neutron spectrum is quite minor for backward angles as seen in Fig.8.

In summary, although the energy resolution of the abnormal spectrometer is worse than that with normal TOF technique, the D+d break-up neutron interference in secondary neutron spectrum measurement can be fully

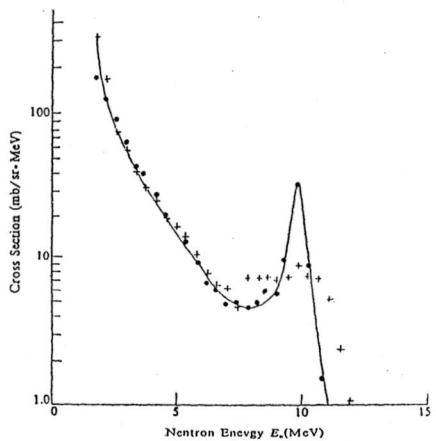

Fig.8 Double differential neutron emission spectrum
of ²³⁸U induced by 10 MeV neutrons
at 120 degrees
-•- using the normal TOF spectrometer
-x- using the abnormal TOF spectromete

eliminated . Therefore with conventional neutron sources (D+d, T+p reactions), secondary neutron spectrum measurement can be realized in the 8-13 Mev incident neutron energy region.

The error, time resolution and angular resolution for the abnormal spectrometer are roughly the same as for the normal TOF system.

References

1. Sa Jun, Tang Hongqing, Zhou Zuying, Sui Qingchang, Qi Bujia, Li Zhongfang, Yu Chunying and Shen Guanren: A multi-detector fast neutron TOF spectrometer, to be published.

MEASUREMENT OF HIGHLY ACTIVE ISOTOPE FISSION CROSS-SECTIONS WITH NUCLEAR EXPLOSION NEUTRONS

E. F. Fomushkin, G. F. Novoselov, V. V. Gavrilov, Ju. I. Vinogradov

All-Union Scientific Research Institute of Experimental Physics
607200, Arzamas, Nizhnij Novgorod region, USSR

Abstract: Unique properties of nuclear explosion as a pulsed source of neutrons give the possibility to make physical measurements on low weight samples and under high background conditions. A special technique for such measurements was developed. The fission fragments are registered by dielectric track detectors. Policarbonate films are glued onto the cylindrical surface of a wheel, driven by electromotor. Fissile isotope layers are separated from the detector surface by slit collimators. ^{235}U samples are used, as a standard. Weight of the investigated isotope layer is 2-10 mkg, and of ^{235}U 10-100 mkg. After the test films are chemically treated; the distributions of the fragment tracks are scanned with an optic microscope. The fission cross-section curves are normalized over the integral measurement results for the neutrons of a fast reactor. The cross-sections of ^{240}Pu, $^{242m,243}Am$, $^{243,244,245,246,247,248}Cm$ are measured. In some cases nuclear explosion neutron spectra contain the group of $T(d,n)$ reaction neutrons. Fission cross-sections of six curium isotopes by neutrons with average 14.1 MeV energy are given.

(Nuclear explosion neutrons, fission cross-section, dielectric track detector, enriched fissile samples, elimination of background and seismic effects, pulsed reactor, absolutizing of cross-section curves)

Introduction

Nuclear explosion as a pulsed neutron source used in time-of-flight measurement possesses unique peculiarities. A large neutron yield $(N=10^{23-24})$, generated in a single $(f=1 \ s^{-1})$ relatively short pulse $(\Delta t \approx 100 \ ns)$ allows to conduct nuclear-physical measurements with samples of small weight and high specific activity.

Comparative characteristics of nuclear explosion and the best laboratory pulsed neutron sources

Pulsed neutron sources may be compared by two generalized parameters: source quality $M=N/\Delta t^2$ and effect concentration in time $T=M/(\Delta t \cdot f)$. The first one characterizes information accumulation rate about the effect investigated with a given number of atomic nuclei in a sample and at a given energy resolution; the second – the relation of registered effect to background, which is not bound with a source, ex. , induced by investigated sample activity [1].

When compared with the two parameters of the best laboratory neutron sources, the nuclea explosion substantially exceeds them. Thus, the isochronous cyclotron of KfK, Karlsruhe, ($N=2 \cdot 10^{14} s^{-1}$, $\Delta t=0.8 \ ns$, $f=5 \cdot 10^4 s^{-1}$, $M=3 \cdot 10^{32}$, $T= 10^{37}$) and linear accelerator ORELA of ORNL, Oak Ridge,($N=10^{14} s^{-1}$, $\Delta t=3 \ ns$, $f=700 \ s^{-1}$, $M=1.1 \cdot 10^{31}$, $T=5 \cdot 10^{37}$) are 6-8 orders smaller than a nuclear explosion ($M=10^{38}$, $T=10^{45}$); i.e. in explosion measurements an absolute value of a permissible background may be many orders higher, than in the analogous measurements with the laboratory neutron sources. These relations explain, why the explosion measurements with rare isotopes of comparatively short periods of α-decay and of spontaneous fission, to a certain degree, are valid and expedient.

Alongside with the advantures a nuclear explosion has some principal drawbacks that sharply limits the volume of scientific problems, which can be solved with its help. In particular, if classical electronic methods of particles' detection, information transfer and recording are used, than the possibility of individual nuclear reaction' products' recording is completely excluded. An integral current regime of recording with data registering in an analogue form is to be used. It complicates the identification of background components and the comparison of data obtained at the explosion with the results of calibration measurements in standard neutron fields on the laboratory facilities and so on.

In their nuclear explosion measurements to register fission fragments the american collegues used semiconductor detectors; an analogue signal was recorded with an optical system and high speed camera [2].

Device for Fission Fragment Registration

We use the dielectric track detectors of fission fragments (polymer films) in the similar measurements. This completely eliminates α-particle background effect, electromagnetic induction effects on the apparatus, the effect of neutron and γ-radiation on the cables of information transmission lines and so on. Time-base scanning is carried out via high velocity mechanical displacement of a polymer film against a slit collimator with fissile substance layer. In fig. 1 the block-diagram of the experimental facility is shown. The main parameters are: film displacement velocity v=120-190 m/s; collimator slit width Δx=0.1-0.2 mm; time-of-flight distance L=250-350 m. Time resolution is Δx/(v·L)=1.5-3 ns/m, and energy resolution (full width at half-height) increases from 1-2 % at neutron energy E=0.1 MeV up to 10-25 % at E=5 MeV.

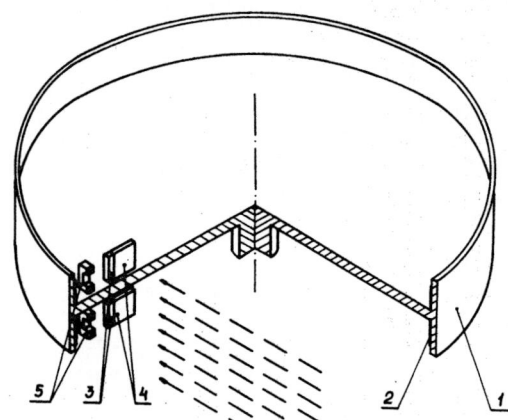

Fig. 1. Scematic view of the measurement device. 1-light alloy wheel, 2-polymer films, 3-fissile isotope layers, 4-metallic substates, 5-slit collimators.

A general view of the facility is presented in fig. 2, the full name −"Device for Fission Fragments Registration", the abbreviated one (in Russian) − UROD ("monster" − in English). As it is evident from fig. 2, special attention is paid to protection of the register unit (a drum with a film and collimators with layers) against seismic effect of a nuclear explosion. Spring (2) and rubber (7) shock-absorbers, air piston dumper (1) are used. When in the working position, each collimator is 0.5 mm (not more) from a film surface. In the interval between neutron pulse passage and seismic wave arrival all collimators are moved away from a film surface by a special device to a safe position.

As a rule, high enriched samples of transuranium isotopes are used in these measurements. In each facility three collimators are mounted − one with an investigated isotope layer, one with ^{235}U layer as a standard; one collimator with a layer, placed outside a straight neutron flux and used to measure scattered neutron background. This background does not exceed 2-3 %. Layer weight is 2-10 mkg for an investigated isotope and 10-100 mkg for a standard sample.

Fig. 2. Device for Fission Fragment Registration. 1-air piston dumper, 2-spring shock-absorbers, 3-posts of frame, 4-device body, 5-positioning table, 6-posts of table, 7-foundation with ribbon absorbers.

Experimental results processing

For neutron radiation output the vacuum channels are used in time-of-flight measurements in underground nuclear explosions. After the experiment the measuring devices are extracted from the working positions, the films are chemically treated. The films are scanned along the length and, consequently over a neutron time of flight, with an optical microscope. Depending on various factors each film registers $10^4 - 5·10^5$ fission fragment tracks.

To make time and consequently energy reference in track distribution along a film length, a neutron-spectrometric approach is usually used. At a flight distance the neutrons pass through comparatively thin steel plugs in the vacuum channels of radiation output. In total cross-section of neutron interaction with iron nuclei there are some noticeable resonances, which form typical minima in track distributions. It allows to put marks on the track distributions obtained, to make a time scale with a 0.2 - 0.3 % uncertainty, not more.

Energy dependence of an investigated isotope fission cross-section is calculated according to a standard formula of ratio measurements:

$\sigma_{fx}(E_i)=K\cdot\sigma_{fu}(E_i)\cdot\Delta n_i(X)/\Delta n_i(U)$, where

$\Delta n_i(X)$ and $\Delta n_i(U)$ — numbers of tracks, registered in certain energy intervals, from an investigated isotope layer and an uranium sample;
E_i — neutron energy, corresponding to a center of gravity of i-th interval;
σ_{fx} and σ_{fu} — fission cross-sections of investigated and standard isotopes;
K - a normalizing constant, depending on a number of fissile nuclei in the layers, on a probability of fission fragment passage through a collimator, but it does not depend on neutron energy.

As basis curve $\sigma_{fu}(E)$, V.Kon'shin evaluation [4] of ^{235}U fission cross-section is usually used with the account of ^{238}U nuclei fission, which are in a layer.
To absolutize the measurement results, i.e. to calculate normalizing constant K the effective fission cross-sections are measured for investigated and standard isotopes with the pulsed reactor neutrons on 90%-enriched metallic ^{235}U [5,6]. In this case thoroughly calibrated layers are placed in a special assembly with glass track detectors of fragments. The assembly with layers and detectors is placed in a central reactor cavity, where the angular distribution of neutrons is isotropic. A subsequent comparison of the effective cross-section value measured with the integration result for $\sigma_{fx}(E_i)$ curve over a neutron spectrum of the reactor allows to calculate K-constant and, consequently, to absolutize the fission cross-section curve of the investigated isotope. Thus, in our measurements a nuclear explosion is used, as a rule, only to determine the relative dependence of the fission cross-section curve. Data are absolutized in the laboratory with the help of calibrated layers and certified neutron sources. Besides, there are no analogous methods of data registration and presentation in our technique; all initial results and, consequently, processing methods are of discrete nature. As it was already mentioned, the dielectric track detectors used almost completely eliminated the background problem.

Some results and conclusion

While working with nuclear explosion neutrons our group measured the fission cross-sections of ^{240}Pu, ^{242m}Am, ^{243}Am and six curium isotopes. The measurement results on the nuclides mentioned are published [7,8]; the comparison with the data of other authors can be found in some compilations and evaluations [9].
In some cases nuclear explosion neutron spectrum comprises a group of neutrons , generated in T(d,n) reaction. Average energy of this quasimonochromatic line is 14.1 MeV. In this energy range the time resolution of our technique makes impossible the spectrometric analysis of the fission section studied. However, dT-neutron group is well enough

separated from a prolonged spectrum of lower energy neutrons, as a rule. It allows to estimate a fission cross-section of the isotope investigated by quasimonochromatic neutrons of 14.1 MeV average energy. These measurement results for curium isotopes are presented in the table.

Table 1. Curium isotopes fission cross-sections by dT-neutrons

Isotope	E_n, MeV	σ_f, b
^{243}Cm	14.8	2.98±0.105
^{244}Cm	14.1	2.93±0.20
^{245}Cm	14.1	2.75±0.17
^{246}Cm	14.1	2.73±0.19
^{247}Cm	14.1	2.48±0.18
^{248}Cm	14.1	2.26±0.17

The result for ^{243}Cm was obtained on an accelerator, it agrees with the data from explosion measuremehts.
In conclusion, we should mention that the measurements with nuclear explosion neutrons have their peculiarities. This report aim is to show the advantages and difficulties typical of such neutron source, the probable ways to cope with the difficulties, while solving a certain physical problem.

REFERENCES

1. G.V.Muradjan, in Proc. I-st Int. Conf. on Neutron Physics, 1987, Kiev, USSR, v.4,p.47, ZNIIatominform, Moscow(1988)
2. B.C.Diven: Anual Review of Nuclear Science 20, 79 (1970)
3. A.Hemmendinger: American Scientist 58, 622 (1970)
4. V.A.Kon'shin, Yaderno-fizicheskie konstanty delyashchikhsya yader, Moscow (1984)
5. Yu.B.Khariton, A.M.Voinov, V.F.Kolesov, M.I.Kuvshinov, A.A.Malinkin, A.I.Pavlovskii, in "Voprosy sovremennoj ehksperimentalnoj i teoreticheskoj fiziki", p.103, "Nauka", Leningrad (1984)
6. E.F.Fomushkin, E.K.Gutnikova, B.K.Maslennikov, A.M.Korochkin: Nucl. Phys. 17, 24 (1973). In Russian
7. E.F.Fomushkin, E.K.Gutnikova, G.F.Novoselov, V.I.Panin: Atomic Energy 39, 295 (1975). In Russian
8. E.F.Fomushkin, G.F.Novoselov, Yu.I.Vinogradov, V.V.Gavrilov et al: Nucl. Phys. 31, 39 (1980); 33, 620 (1981); 36, 582 (1982). In Russian Atomic Energy 62, 278 (1987); 62, 279 (1987); 63, 242 (1987); 69, 258 (1990). In Russian Voprosy atomnoj nauki i tekhniki. Ser.: Yadernye konstanty 35, 11 (1979); 57, 17 (1984) in Proc. V-th All-Union Conf. on Neutron Physics,1980, Kiev,USSR,v.3,p.25, ZNIIatominform, Moscow(1980). In Russ.
9. T.Nakagawa, Evaluation of Neutron Nuclear Data for Curium Isotopes, JAERI-M 90-101, Japan Atomic Energy Research Institute(1990)

The measurement of the ^{10}B content of enriched samples of boron using neutron time-of-flight techniques

M.C. Moxon, J.B. Brisland, S. Croft and D.S. Bond

AEA Technology
Harwell Laboratory
Oxfordshire
OX11 0RA, UK.

<u>Abstract:</u> The combined 8.8 m and 14.7 m neutron transmission spectrometer of the condensed matter target (CMT) of HELIOS, the Harwell electron linear accelerator, has been used to measure the attenuation of the neutron flux as a function of neutron energy of six aluminium clad samples of boron, enriched in ^{10}B. Five of these samples are to be used to determine the energy dependence of the neutron flux in an external beam from a nuclear reactor by measuring the response of various neutron detectors following the introduction of the samples into the beam. The sixth sample has been used to measure the energy dependence of the neutron flux in capture measurements performed at HELIOS.

Assuming that over the neutron energy region below 1 keV the total ^{10}B cross-section can be represented by a constant plus a 1/v term, it was possible to determine the ^{10}B content of four of the five filter samples from these measurements. The ^{10}B content of the thickest filter sample, sample number five, could not be measured as the sample had too great an attenuation over the neutron energy range studied, (0.1 eV to 200 eV). The measurements confirmed that the ^{10}B cross-section in the neutron energy region between 100 and 0.5 eV can be represented by a constant plus a 1/v term to an accuracy of better than 1%.

After the measurement the uniformity of the sixth sample was measured by X-radiography and it was destructively assayed to determine its ^{10}B content. From these measurements the sample thickness was determined to an accuracy of $\pm$ 0.5%. Using this sample thickness the value of the 1/v term at 1 eV was determined to be 609.45 $\pm$ 3.5 (0.57%) b.(eV)$^{-1/2}$. The ^{10}B contents obtained were in good agreement with those measured by chemical assay.

INTRODUCTION

For neutron energies below about 100 keV the ^{10}B(n,α) cross-section is a recognised standard. The original aim of this work was to exploit this fact for the determination of the ^{10}B - content of five boron filters. The filters were to be used to determine the energy spectrum of an external reactor neutron beam by a convolution technique [1]. The main constituent of these samples was powdered boron metal, enriched in the ^{10}B isotope. To determine the amount of ^{10}B in each sample from their masses, both a chemical and isotopic analysis is required, and, like most powdered metals, boron tends to oxidise, consequently an estimate based on the sample mass may yield an overestimate of the amount of boron. It was, therefore, felt to be prudent to determine the ^{10}B content of the filters directly as a check on the chemical and isotopic assays. This was achieved using an experimental set-up designed for high-precision total cross-section measurements. The approach involved measuring the transmission of each filter as a function of neutron energy and comparing it with the expected variation.

By including a sixth sample, which was destructively assayed immediately after the measurements on it were complete, the whole series of measurements were placed on a common absolute scale. This sample had previously been used to determine the energy fluence-rate of neutrons from the fast neutron target (FNT) at HELIOS, the Harwell electron linear accelerator [2]. As a result of the work reported here, the FNT energy fluence-rate measurements can be refined, and the absolute total ^{10}B cross-section over the neutron energy interval 0.5 eV to 100 eV determined to good accuracy.

MEASUREMENT

The layout of the flight path used in the measurements is shown in Figure 1. The flight path views the lower moderator of the condensed matter target (CMT) of HELIOS. The neutron beam passes through an in-line ^{235}U beam monitor and is collimated down to a diameter of 20mm at the sample position, 6m from the source. The beam is then allowed to diverge, passing through the 8.8m detector and the 14.7m detector, to the beam dump at 20m.

The samples under investigation, together with the samples used to provide the energy calibration of the flight-path and the filters used to determine the background, are mounted on computer controlled sample changers at 6m from the neutron source. The operating conditions of the linac (pulse width and neutron production rate) were continuously monitored, and data was only collected when these were within tight tolerance bounds. The various samples and filters were cycled periodically into the beam to ensure that all measurements were taken under as near identical conditions as possible. As an empirical check on the sample uniformity, the transmission of each sample was measured several times with the neutron beam incident on a different portion of it. Satisfactory agreement was obtained on all occasions.

The neutron detector at the 8.8m station was a 1mm thick by 76mm diameter lithium-glass scintillator, viewed edge-on by a 127mm EMI 9823 QB photomultiplier tube. The 14.7m detector is similar but it incorporates a 3mm thick by 127mm diameter lithium glass scintillator. The signals from each detector are fed via separate gated integrators into a time of flight scaler, enabling a histogram of events versus time-of-flight (TOF) and pulse-height (PH) to be built-up in the memory of an on-line PDP 11/34 data acquisition and experiment control computer. Recording both TOF and PH information enables very accurate background determination.

The neutron energy calibration of the TOF scale was obtained by measuring the centroid positions, in the TOF histogram, of the resonance dips produced by samples of Co, Ag, Gd, and Hf. The resonances used are listed in Table 1.

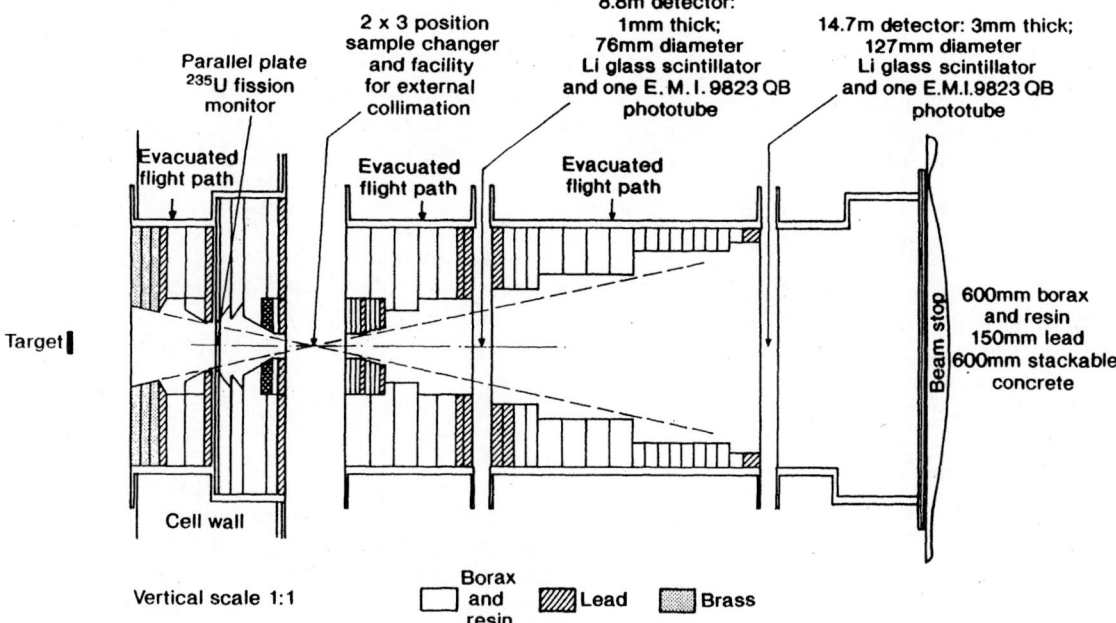

Figure 1: Schematic diagram illustrating the collimation system and the deployment of detectors for the transmission spectrometer used in these measurements.

Nuclide	Resonance Energy /eV
^{177}Hf	1.0964(15)
^{177}Hf	2.38370(20)
^{155}Gd	2.568(13)
^{109}Ag	5.19(1)
^{179}Hf $\}$	5.79(10) *
^{177}Hf	"
^{155}Gd	6.30(2)
^{177}Hf	6.5691(14)
^{178}Hf	7.7718(17)
^{177}Hf	8.83880(82)
^{177}Hf	13.9707(25)
^{107}Ag	16.30(5)
^{109}Ag	30.4(1)
^{109}Ag $\}$	40.8(5) **
^{107}Ag	"
^{59}Co	132.0(5)

Table 1. List of resonance neutron energies used in the energy calibration of the TOF scale.

* Unresolved doublet 5.68620(98)eV and 5.89370(90)eV.

** Unresolved doublet 40.1(1)eV and 41.5(1)eV.

Resonance energies for the Co, Ag, and Gd nuclides were taken from Mughabghab et al [3] while for the Hf nuclides the data of Moxon et al was adopted [4]. The reciprocal root neutron energy $y = E^{-1/2}$ was assumed to vary linearly with TOF, t, and a weighted linear least squares fit to the form y=a.t+b was performed in order to determine the calibration coefficients a and b. The weighting function took into account the uncertainty in the resonance energies, the uncertainty in locating the resonance dip centroids and the inherent uncertainty in the source plane position due to fast neutron moderation processes.

The background was measured using the black resonance technique with allowance for the attenuation of off-resonance neutrons by the resonance filters. This called for no-filter, single thickness and double thickness filter runs to be made on each of open-beam and sample-present runs. Filters of Ag, Cd, Hf and Co were used for this purpose.

SAMPLES

The assay list of the main constituents present in each of the two batches of powdered boron metal investigated is given in Table.2. Also given in Table 2. are the values of the scattering and reaction cross-sections at 1 eV assumed for each of the constituents in the analysis stage. Aluminium has been included in the list because each of the powder samples was canned in aluminium. In the case of samples 1 – 5 the total thickness of aluminium presented to the neutron beam was nominally 4.0mm, for sample number 6 it was (0.767±0.007)mm.

Constituent	Abundance in weight. %		Scattering cross-section at 1 eV σ_s/b	Reaction cross-section at 1 eV σ_R/b
	Samples 1 – 5	Sample 6		
^{10}B	77.39	92.46	2.23	610.27
^{11}B	6.36	5.13	3.60	0.87 x10^{-3}
C	2.66		4.75	0.54 x10^{-3}
O	10.22	0.84	3.764	0.043
Na	0.18		3.20	0.064
Mg	–	1.30	3.34	0.010
Al	–	–	1.43	0.037
S	0.20		0.98	0.083
K	0.56		1.50	0.334
Ca	0.09		2.90	0.0684
Ti	0.04		4.00	0.970
Cr	0.87		3.80	0.493
Mn	0.10		2.10	2.115
Fe	0.87	0.27	10.9	0.407
Ni	0.42		17.3	0.704
Cu	0.03		5.70	0.716
Zn	0.01		4.20	0.175
	100.00	100.00		

Table 2. The assay of the constituents present in each sample together with the assumed values for their scattering and reaction cross-sections at a neutron energy of 1 eV.

444

ANALYSIS

After correction for dead time effects, the TOF histograms from each run were normalised to the ^{235}U beam monitor. The transmission, T(t), at TOF, t, was calculated from the ratio of the corrected count rates, C, minus the background count rates, B, with (s) and without (o) the sample in the neutron beam.

$$T(t) = [C_s(t) - B_s(t)]/[C_o(t) - B_o(t)] \qquad (1)$$

Assuming that the total microscopic cross-section for each of the individual constituents identified in Table.2 can be represented as a sum of a scattering cross-section, σ_{si}, which is independant of neutron energy, and a reaction cross-section, $\sigma_{Ri}(E)$, which follows the $1/v$ – law we may write

$$\ln\left[\frac{1}{T(t)}\right] = \alpha + \beta/\sqrt{E(t)}$$
$$= \alpha + \beta.(a.t + b) \qquad (2)$$

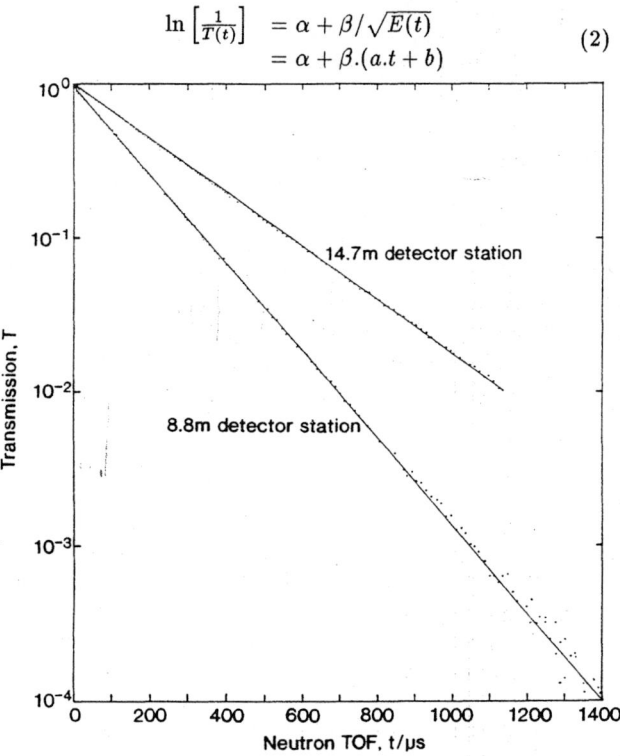

Figure 2: A plot of the measured neutron transmission function for sample number 6. The solid lines are best fit linear fits to the data as described in the text.

Sample number	Measured ^{10}B areal number density (atoms/b)	Measured to calculated ^{10}B-content
1	0.03598($\pm$0.91%)	1.005
2	0.009429($\pm$0.083%)	0.966
3	0.02049($\pm$0.22%)	1.037
4	0.006774($\pm$0.082%)	1.179
5	Too thick to determine	——
6	0.006941($\pm$0.28%)	0.9987

Table 3. Comparison of measured (by neutron transmission) and calculated (from constituent assay using the cross-section data given in Table. 2) ^{10}B contents of the samples studied. The percentage uncertainties quoted are at the one standard deviation level and are calculated from the consistency in the set of β-values obtained for each sample. No allowance has been made for the uncertainty in the value of the ^{10}B cross-section assumed.

where a and b are the TOF calibration coefficients discussed earlier, $\alpha = \sum n_i.\sigma_{si}, \beta = \sum n_i.\sigma_{Ri}(E=1eV)$, n_i is the areal number density (atoms per barn) of the ith constituent and the summation extends over all the constituents. The parameters α and β are determined by a linear least squares fit to the data.

RESULTS

For neutron energies above about 100 eV the background was poorly determined while for transmissions less than about 0.01% the counting statistics were poor. Thus least squares fits to Equation 2. to determine the parameters α and β were restricted to the energy range 0.5 – 100 eV. The parameter α is relatively poorly determined from these measurements as it constitutes only a small part (less than 15%) of the interaction probability. By contrast the parameter, β, which describes the energy dependency of the interaction probability and is determined overwhelmingly by the ^{10}B content of the samples, can be obtained very accurately ($< 1\%$).

The accuracy of the fits can be judged from Figure 2. which shows the curve calculated from the weighted mean parameters and the transmission data for sample number 6.

In Table 3. a comparison is made between the measured and predicted ^{10}B contents. In all cases where a measurement was feasible the agreement is good, except for sample 4 whose uniformity should be questioned.

After the measurements sample 6 was X-radiographed to confirm its uniformity (already empirically assessed in part during the transmission measurements by moving the sample in the beam) and assayed. From the results obtained with this sample we obtain a value for the $1/v$ component of the ^{10}B cross-section of 609.45$\pm$3.5 (0.57%) b.eV$^{-1/2}$ at 1 eV. The uncertainty, estimate at the one standard deviation level, is calculated from the quadrature sum of the following terms: β -parameter, $\pm$0.28% (this includes an allowance for the sample uniformity as discussed above); weight fraction of ^{10}B, $\pm$0.15%; sample can dimensions, 0.47%, and sample mass, 0.03%.

CONCLUSION

Neutron transmission has been used to determine the ^{10}B content of several enriched boron samples. Over the neutron energy region between 0.5 eV and 100 eV the ^{10}B cross-section can be represented by a constant plus a $1/v$ term to an accuracy better than 1%. The $1/v$ term of the ^{10}B cross-section was measured to be 609.45 $\pm$3.5 b.eV$^{-1/2}$. This is in excellent agreement with the value of 609.92 $\pm$1.91 taken from ENDF/B-V [5].

REFERENCES

1. J. Constantine, Harwell Laboratory, Private Communication (1988)

2. J.P. Mason : Nucl. Phys. A465(1987)413–428

3. S.F. Mughabghab, M. Divadeenam and N.E. Holben: Neutron Cross-sections. Vol. 1. Academic Press, London, Part A(1983), Part B(1985).

4. M.C. Moxon, D.A.J. Endacott, T.J. Haste, J.E. Jolly, J.E. Lynn and M.G. Sowerby: Unpublished UKAEA Harwell Report AERE–R7864(1974)

5. E. Wattecamps, The ^{10}B(n,α) cross-section: Nuclear data standards for nuclear measurements, IAEA Technical Reports Series 227(1983)19–23

THE NEUTRON LEAKAGE SPECTRA FROM BERYLLIUM AND ALUMINIUM
SPHERES FOR 14 MEV NEUTRON SOURCE WITHIN THE ENERGY
INTERVAL 6-15 MEV

B.E.Leshchenko,Yu.N.Onishchuk,G.A.Prokopets,D.Yu.Chuvilin

Kiev State University, Kiev, USSR

Abstract: The leakage neutron spectra with the 14 MeV neutron source from the 5 cm thick beryllium and 7.5 cm aluminium spherical shells were measured by the time-of-flight technique in the energy interval 6-15 MeV. The experimental results are in good agreement with other experimental data that have been obtained by the time-of-flight and total absorption methods.The experimental results have been compared with the calculations by the BLANK code based on the ENDF/B-IV, ENDL-78, ENDL-83 and JENDL-2 library data.

(neutron leakage spectra, time-of-fligt technique, beryllium, aluminium, experiment, calculations)

Introduction

The research on 14 MeV neutron interactions with different materials plays an important role in thermonuclear reactor blanket studies.In spite of the fact that in the past years many differential and integral experiments have been performed[1-4], still there is a need to have more reliable data.

Now some nuclear data files and corresponding codes are used for calculations of thermonuclear blanket design. These files contain the values of working constants for calculation of the functions in a format corresponding to the flux formats: kerma factors, γ-ray sources, activation cross sections,radiation damage cross sections, cross sections of reactions with secondary neutrons and charged particles yield, cross sections and yields of fission products and elements in the chain of nuclear transformations of the fissile material.

Experimental results with spherical samples are particularly useful to test the accuracy of transport codes, recommended sets of cross sections and physical assumptions in the neutronic calculations. Spherical geometry, single materials and simple compounds facilitate the interpretation of the results, when discrepancies between measurements and calculations are found. In these experiments the neutron source is placed in the centre of sphere and neutron spectra are measured both inside the sample and outside (neutron leakage spectra). Complete emission spectra, normalized to one source neutron,commonly characterize the constants for incident neutron energy. They are the sum of energy spectra of all interaction processes, each of them having its partial weight.

In this paper the results of the neutron leakage spectra from the 5 cm thick beryllium and 7.5 cm aluminium spheres for 14 MeV neutron source within the energy interval 6-15 MeV are reported.

Experimental Procedure

Measurements of neutron leakage spectra were done with neutron spectrometer based on a pulsed low voltage neutron generator. The 14 MeV initial neutron flux was generated by the T(d,n)^{4}He reaction at the 130 KeV deuteron beam energy. The parameters of spectrometer enable to measure neutron time-of-flight spectra with good enough energy resolution (neutron burst duration <2ns, repetition frequency 7.25 MHz).

The secondary neutron time-of-flight spectra have been measured in the energy interval of 6-15 MeV with a 81-mm-diam and 81-mm- long plastic scintillator, located at a distance 10.15 m from the tritium target at 0° to the deuteron beam direction. The detector was biased to 5.7 MeV of the recoil proton energy. The bias level was determined by special calibration procedure with γ-ray sources and has been controlled during the measurements.The detecting efficiency of the plastic scintillator as a function of energy has been calculated. Details on the geometry and electronics can be found in [5].

A 30-mm-diam and 40-mm-long plastic scintillator was used as a neutron monitor. It was located at a distance of 3.3 m from the tritium target. The monitor bias was set to 4.8 MeV.

To determine the absolute value of neutron yield the original monitoring procedure was used. The coefficient of neutron beam reduction by spheres has been measured. The whole procedure has been automatized and a remote control system has been used to move the samples to and away from neutron source. For beryllium and aluminium spheres the values of coefficient were 1.54±0.02 and 1.60± 0.02, respectively.

Particular attention was paid to the correct background estimation.

The statistical uncertainty of most of the experimental points within the interval 6-15 MeV did not exceed 8-10 %.

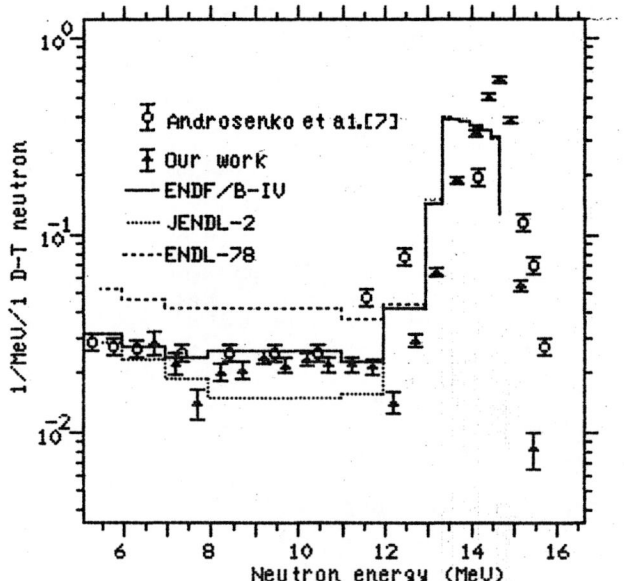

Fig. 1. Neutron leakage spectra from beryllium sphere.

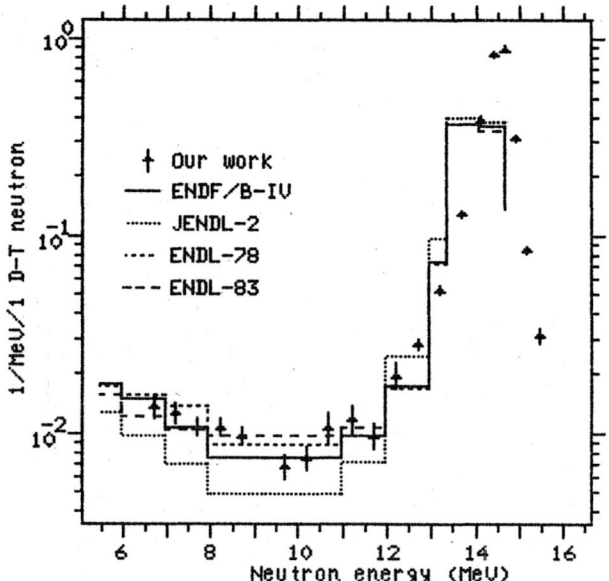

Fig. 2. Neutron leakage spectra from aluminium sphere.

Results and Discussion

The measured leakage spectra for 14 MeV neutrons from beryllium and aluminium spheres are shown in Fig.1 and Fig.2, respectively. The comparison of experimental results with calculations has been made.

For calculations of neutron leakage spectra from spherical samples one-dimensional BLANK code has been used [6]. To test the sensibility of secondary neutron spectrum to the shape of primary neutron beam two models of neutron source spectrum were compared: 1)mono-line with effective energy 14.69 MeV corresponded to the neutron yield from target at 0° to the deuteron beam (deuteron beam slowing-down in the target was taken into account); 2) equal distribution of neutrons within the energy interval 13.4 – 14.8 MeV (mean energy 14.1 MeV). It corresponded to the kinematical distribution of DT-neutrons resulting from 130 KeV deuterons with the accuracy of a few percents. The discrepancy in spectra shape calculation within the energy region below 12 MeV did not exceed the accuracy of experiment and majority of experimental points for beryllium lie in calculated corridor with the data of ENDF/B-IV file.

The calculations with the BLANK code based on the ENDL-78, ENDF/B-IV and JENDL-2 library data have been compared with experimental data for beryllium in Fig.1. The ENDL-78 and JENDL-2 data gave sufficient disagreement within the energy region 7-12 MeV both in our data and the experimental results by Androsenko et al. [7], and for the ENDF/B-IV a good agreement with both experiments was obtained. Some discrepancy of results for neutron source area is probably connected with different energy resolution.The measurements by Zagryadskii et al. [8]

where absorption method was used for the same sphere indirectly, satisfy both the time-of-flight experiments.

Exerimental results for aluminium are shown in Fig.2 in comparison with calculations.The tested files ENDF/B-IV, ENDL-78 and ENDL-83 produce the value of neutron flux in the energy interval 7-12 MeV well enough. The use of JENDL-2 data results in the reduced neutron flux by over 40% against the experimental data. This may be explained by incorrect set of data in a file for neutron emission in inelastic scattering.

REFERENCES

1. H.Maekawa,JAERI-M 90-025, p.69(1990)
2. C.Ichihara, S.A.Hayashi, K.Kobayashi, I.Kimura, J.Yamamoto, A.Takahashi, JAERI-M 91-062, p.255(1991)
3. L.F.Hansen, UCRL-97188, LLNL(1987)
4. T.Elfruth,T.Hehl,M.Toepfer,D.Seeliger, K.Seidel: Kerntechnik,55,N3,156(1990)
5. B.E.Leshchenko,Yu.N.Onishchuk,A.B.Me-lnichenko,A.A.Borisov,D.V.Markovskii, D.Yu.Chuvilin, IAE Report-5230/8,Moscow(1990)
6. S.V.Marin,D.V.Markovskii,G.E.Shatalov, IAE Report-2832,Moscow(1977)
7. A.A.Androsenko,P.A.Androsenko,B.V.De-vkin,B.V.Juravlev,V.A.Zagryadskii,M.G. Kobozev,D.V.Markovskii,O.A.Salnikov, D.Yu.Chuvilin,in Proc.1st Int.Conf. on Neutr.Phys,Sept.1987,Kiev,USSR,vol.3, p.194,CNIIatominform,Moscow(1988)
8. V.A.Zagryadskii, D.V.Markovskii, V.M. Novikov, D.Yu.Chuvilin, G.E.Shatalov: Fus.Eng.Design,9,353(1989)

NEW TYPE OF EXPERIMENT FOR AVERAGE RESONANCE PARAMETER ESTIMATION IN UNRESOLVED NEUTRON RESONANCE REGION

W.K.Basenko, S.N.Ezhov, G.A.Prokopets

Kiev State University, USSR

Abstract: A method of experimental study of neutron cross section resonance structure effects in the unresolved energy region is described. The usefulness and range of validity of this resonance self-indication method is discussed.

(neutron unresolved resonances, total and scattering cross section cross-correlation, average resonance parameters)

Introduction

It is not necessary to emphasize the importance of detailed knowledge of the resonance structure of neutron cross sections: from the viewpoint of nuclear power engineering the needs are the same as from the viewpoint of nuclear reaction theory. The only straightforward way of the experimental investigation of the problem is to measure the excitation functions using neutron spectrometers with as good resolution ΔE as possible, so

that $\Delta E < \overline{\Gamma} < \overline{D}$, where $\overline{\Gamma}$ and $\overline{D}$ are the average resonance width and resonance spacing. But in the energy region of unresolved or overlapping $(\overline{\Gamma} \gtrsim \overline{D})$ resonances it is convenient to make use of transmission or self-indication measurements on neutron beams with an energy spread ΔE, that allows to extract only energy averaged values connected with the resonance structure such as mean $\overline{\Gamma/D}$ ratios, a self-shielding factors etc [1]. Further new version of the self-indication method is presented and its possibility concerning statistical characteristics of resonance structure is demonstrated.

Foundation of the method

It is proposed to change the usual self-indication experimental arrangement in such a way that a parallel neutron beam with an intensity distribution over energies p(E) would be scattered first by a thin sample throgh a definite laboratory angle θ whereupon the transmission of this scatterd beam through a second sample of the same material must be measured. The spectral density of the scattered beam $\Phi(E',\theta)$ under this condition may be written as:

$$\Phi(E',\theta) = \frac{p(E'/f(\theta)) \, \sigma_s(E'/f(\theta),\theta) \cdot l_s}{f(\theta)} \quad (1)$$

The introduced notation is as follows:
l_s - the scatterer thickness in inverse

barns;
$E' = f(\theta) \cdot E$ - scattered neutron energy;
$f(\theta) = [m/(M+m)]^2 \{[(M/m)^2 - \sin^2(\theta)]^{1/2} + \cos(\theta)\}^2$, where m and M are neutron and target nucleus masses - kinematical factor for elastic scattering;
$\sigma_s(E,\theta)$ - differential scattering cross section under initial neutron energy E.
Hence the transmission of this scattered neutron beam by the screen with thickness L:

$$T(L,\theta) = \frac{\displaystyle\int_{\theta_o - \frac{1}{2}\Delta\theta}^{\theta_o + \frac{1}{2}\Delta\theta} \int_0^\infty \Phi(E',\theta) \cdot \exp\left[-\frac{L\,\sigma_t(E')}{\cos(\theta - \theta_o)}\right] \sin\theta \, d\theta \, dE'}{\displaystyle \Phi(E',\theta) \, \sin\theta \, d\theta \, dE'} \quad (2)$$

$\sigma_t(E')$ -total neutron cross section and $\Delta\theta$ - experimental angular resolution at mean scattering angle θ_o.
Under some limitation the expression (2) may be reduced to the form:

$$T(L,\theta) \approx T(L) - L\frac{\exp[-<\sigma_t> \cdot L]}{<\sigma_s(\theta)>} \mathrm{cov}_\theta\{\sigma_s,\sigma_t\} \quad (3)$$

Here T(L) - the transmission in the initial neutron beam,

$$<\sigma_i> - \int p(E)\sigma_i(E)dE \Big/ \int p(E)dE$$

is a mean cross section value over the energy region $\Delta E \ll <E>$ which is determined by the neutron source line shape p(E); $\mathrm{cov}_\theta\{\sigma_s,\sigma_t\} = r_{st}$ - covariance between $\sigma_s(E,\theta)$ and $\sigma_t(E')$ for energy shift $<\delta E> = <E \cdot [1-f(\theta)]>$.

The model calculations

Let us first evaluate the expected energy resolution of the measurements taking into account that the realistic attainable minimal scattering angles could amount to several degrees. Then one may hope to obtain $<\delta E>/<E> \approx 10^{-4}$ for moderate-mass nuclei and $<\delta E>/<E> \approx 10^{-8}$ for heavy nuclei. Therefore in principle the method is powerful enough to investigate rather fine resonance structure effects in this integral type measurement.
To test the sensitivity to the average resonance parameters computer-simu-

lated experiments were carried out.

Model cross section excitation curves were generated as a series of Breit-Wigner s- and p-resonance terms with Porter-Thomas distribution for neutron reduced widths Γ_n and Wigner distribution for resonance spacings D_1. Radiative widths were adopted to be of constant value. The code could calculate also the cross section which is Doppler-broadened.

Extensive calculations have been performed. In Fig. 1 one can see an example of these calculations that is related to the hypothetic case of a nucleus with A = 240 and a Gaussian neutron source line shape which is centered at E = (45.0 ± 1.25) keV. The angular resolution has been chosen to be $\Delta\theta = 3°$. The transmission dependence on the mean energy shift $<\delta E(\theta)>$ is shown with (curve b) and without (curve a) taking into account of Doppler broadening. The energy interval of $\Delta E = 2.5$ keV includes about N ≈ 500 resonances with an average width that is slightly less than 1 eV when the Doppler width is about 4 eV.

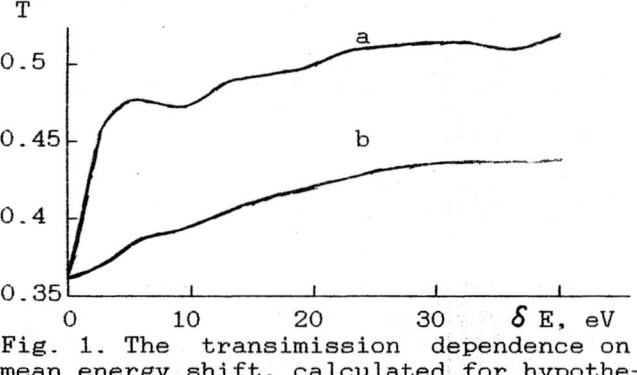

Fig. 1. The transimission dependence on mean energy shift, calculated for hypothetic nucleus with A = 240.

The main conclusions from analysis of similar calculations are as follows. The influence on the results of the source line shape is insignificant if the condition $\Delta E \ll E$ is fulfilled and if the number of resonances on this energy interval is N ⩾ 50. On the contrary, the average energy shift dependence of the neutron transmission $T[<\delta E(\theta)>]$ is rather sensitive to the angular resolution $\Delta\theta$, which must not exceed about 4° in order to prevent smoothing away the important peculiarities. The slope of the initial part of the transmission curves (small δE) is quite sensitive to the resonance widths and corresponds to the transition from overlapping firstly narrow and then broad resonances. When $<\delta E(\theta)>$ is raised, the transmissions approach the values T(L) for the initial neutron beam because of the cross section correlation loss ($r_{ot} \to 0$). Doppler broadening results in smoothing out the $T(<\delta E(\theta)>)$ dependence and the sensitivity to resonance widths distribution will be kept if the mean reduced neutron widths satisfy the approximate condition $\Gamma_n^{(0)} \gtrsim 0.3/(A+1)^{1/2}$ eV.

As an application of this method one may regard experiments on reactor filtered neutron beams with mean energy variation $2 \leqslant E \leqslant 144$ keV. Especially favourable condititons in this energy region exist for nuclei with A ≈ 80 where the information on resonance parameters is scarce.

Preliminary results of a control-type experiment [2] on ^{59}Co nuclei at the Kiev reactor filtered neutron beam with E = (59.0 ± 1.15) keV are shown in Fig. 2 (points) together with results of calculations (solid line) on the basis of the information about ^{59}Co resolved resonance parameters [3].

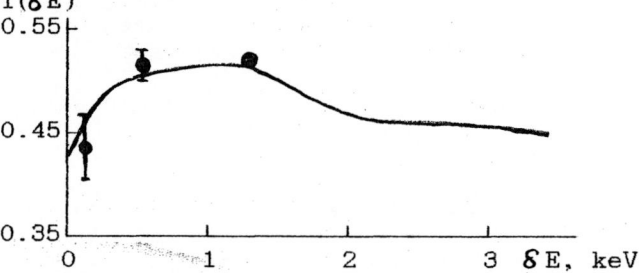

Fig. 2. The transmission dependence on mean energy shift for ^{59}Co nucleus at neutron energy E = (59.0 ± 1.15) keV.

The experimental values are consistent within the errors with the calculated values that demonstrates an agreement with the concept described here.

In spite of more complicated data interpretation, the method is useful as a complementary one to the thick-sample transmission and self-indication methods.

This circumstance may be elucidated if we make use of an equidistant resonance spacings approach together with a single-channel approximation [1]. For the sake of simplicity let us neglect also potential scattering phase shift.

Then for moderate screen thickness we obtain from eq.(3)

$$T(L,\theta) = 1 - 2L\sigma_{max} + 1/<\sigma_t> \cdot \qquad (4)$$
$$\cdot L \cdot s \sigma_{max} [\frac{1}{(1+s)^2} + \frac{\beta^2(1+s^2)}{[(1+\beta^4)s^2+\beta^2(1+s^4)]}];$$

where σ_{max} - a cross section in a resonance maximum;

s - $\pi\Gamma/2D$;

β - $tg(\pi\delta E(\theta)/2D)$.

Only parameter of the model depends explicitly on resonance spacing D but not on Γ/D ratio and may be determined by the proposed method. The rest parameters of eq. 4 may be fixed in usual type measurements that makes it possible to determine Γ and D separately.

Instead of resonance parameters the knowledge of S-matrix statistical properties is of great importance at higher neutron energies. Hence the method would be particularly useful also for time-of-flight measurements on accelerator based quasimonochromatic neutron sources with mean energy up to some MeV. The subject of interest here is the energy-correlation between modulus and real part of the diagonal S-matrix elements , as follows from eq. 3. This information may be used, for example, to evaluate compound nuclei mean lifetimes.

REFERENCES

1. A.A.Lukyanov, Neutron Cross Sections Structure., Moscow, Atomizdat, 1978.
2. W.K.Basenko, L.L.Litvinsky, G.A.Prokopets, O.A.Purtov, Preprint KIYaI-89-30, Kiev, 1989.
3. Mughabhab S.F. et al. Neutron Cross Sections 1, Academic Press INC, NY,1981

HIGHLY-ENRICHED SAMPLES OF URANIUM AND TRANSURANIC ELEMENT ISOTOPES FOR SCIENTIFIC INVESTIGATIONS

S.M.Abramychev, N.V. Balashev, V.N. Vjachin, A.A. Druzhinin,

A.M. Korochkin, V.G. Lapin, E.A. Nikitin, V.N.Polynov,

S.P. Vesnovskii, E.F. Fomushkin.

All-Union Scientific Research Institute of Experimental Physics

607200, Arzamas-16, Nizhnij Novgorod region, USSR

Abstract: This paper considers the production of highly-enriched samples of uranium, plutonium, americium and curium isotopes by electromagnetic separation method. The equipment used enabled production of isotopes in amount sufficient for setting up nuclear-physical and other experiments. For some isotopes, unique characteristics of isotope enrichment have been obtained.

To maintain research and investigation of transuranic element nuclear properties, in particular the laboratory of isotope electromagnetic separation was established in VNIIEF.

This laboratory is located in a special building, equipped with all the necessary machinery for operation with radioactive substances. The basic facility of the laboratory is electromagnetic mass-separator S-2, designed specially for high-efficiency separation of isotopes of heavy radioactive elements with a small relative difference of mass numbers.

Sector-type separator has a magnetic field of a hyperbolic form $H = H_0 \cdot r_0/r$ with a mean trajectory radius $r_0 = 1000$ mm and ion beam deflection angle of $114.6°$. Magnetic field intensity at mean trajectory is $H_0 = 4500$ Oe.

With total trajectory length of 6000 mm, dispersion per 1% of the relative difference of mass numbers is 20 mm. We have used the belt geometry of ion beam, ion current up to 10 mA, accelerating voltage - up to 40 kV. The ion source of a magnetron type, designed for temperatures up to $1000°$ C, is commonly used in experiments.

Waterless trichlorides of actinoid elements are used as a working medium in this source. Single loading of the starting material is ~1g. Utilization factor of the ion source material is ~5% in average. Isotopic beam receiver enables simultaneous trapping of all isotopes of the separating elements.

Mass-separator S-2 and auxilliary equipment construction provide systematic operation with transuranic elements, possessing high alpha-activity level, without their contact with the room air environment. The possibility of deactivation of the basic separator units with subsequent regeneration of the separating material, aimed at reuse for separation, is provided. Regeneration factor is 50-80%.

Mas-separator S-2 is placed in isolated room, its operation and parameters are remote-controlled. Radioactive element handling is carried out in two production lines, connected by locks of boxes; separating material being in one line and enriched isotopes - in another. All operation with ion source and isotopes receiver are performed in the main line boxes, which are sealed with a separation chamber on both sides of mass-separator. Supply and control circuits of the source and receiver are laid into these boxes through hermetic insulators. Boxes are filled with dry inert gas (nitrogen).

To prevent isotopic impurities, enriched isotopes are handled in different boxes and operator areas of the two box lines are located in separate rooms.

Enriched isotopes are stored in ion receiver boxes, made super-pure copper and aluminium. Extraction of the stored isotopes was performed by hot solutions of 6 M nitric (for copper box) or 6 M hydrochloric (for aluminium box) acids.

Purification of the enriched isotopes from inert and radioactive impurities was performed by co - precipitation method, extraction and ion exchange. These extraction methods provided sufficient purification of enriched isotopes from the material matrix and radioactive impurities.

If necessary, an additional, more perfect purification of the sepaawted isotopes from microimpurities was carried out using cromatography and extraction methods.

Isotopic content of the separated fractions of uranium and transuranic elements was determined by mass-spectrometer MI-3304. This device is a two-cascade double-focusing mass-spectrometer. The first cascade is a mass analyzer with a uniform magnetic field and angle of beam rotation of 90°; the second one is energy-analyzer (cylindrical capacitor with 90° angle of rotation).

Mass-spectrometer is equipped with tristrip ion source with surface ionization. Ion recording was performed by counting method. Dead time of a counting channel ~15 ns. Threshold of isotopic sensitivity at $\Delta M/M = 1/238 \sim 5*10^{-7}$, uranium sensitivity - 1500 at/ion.

Measurement accuracy of isotopic ratio as a function of its value is:

~0.1% rel.(2σ) - for ratio > 10^{-2}

~1÷3% rel.(2σ) - for ratio < 10^{-4}

To certify the products for determination inventory and content of such isotopes as uranium-232, plutonium-236, 238, curium-242 we used alpha-radiometry and alpha-spectrometry methods.

For nuclear-physical and other investigations special layers were made of highly-enriched actinide isotopes, using traditional methods: electrochemical deposition of transuranic elements from water-organic or organic media and vacuum spraying.

Separations of uranium and transuranic element isotopes have been carried out since 1969. Highly-enriched isotopes of uranium (233, 234, 235, 236, 238), plutonium (238-242, 244), americium (241-243) and curium (243-248) have been obtained for nuclear-physical, applied research and technical applications.

Majority of the separated isotopes have unique isotopic purity (data are presented in the table). Isotopes constitute from units of milligrams to grams, depending on the content in the starting material. ^{241}Pu, ^{243}Cm, and ^{245}Cm

isotopes of superhigh enrichment have been obtained by double separation, that allowed us to determine experimentally half-life of these nuclides relative to spontaneous fission.

Highly-enriched actinide isotopes are used in nuclear-physical and applied research for nuclear constant measurements in fission physics, mass-spectrometry, radiochemical experiments etc.

Using the obtained actinoid isotopes, we have carried out some measurements: neutron measurement of energy dependences of americium and curium isotope fission cross-section, v(En) measurement of plutonium isotopes, v measurement at spontaneous fission of curium isotopes and other measurement.

Isotopes are used in many scientific research institutes of our country.

Isotopic products are produced as layers, targets, solutions and other products; distributed by v/o "Izotop" within our country and by association of foreign trade "Tekhsnabeksport" can be delivered abroad on a contract basis.

Table. Content of the main isotope in prepared samples, obtained on VNIIEF electromagnetic separator.

Isotope	Content in prepared sample %
Uranium-233	99.83 - 99.97
Uranium-235	99.97
Uranium-236	98.0
Uranium-238	99.997
Plutonium-238	99.6
Plutonium-239	99.5
Plutonium-239	99.9977*)
Plutonium-240	99.0
Plutonium-240	100.0**)
Plutonium-241	99.6
Plutonium-241	99.998*)
Plutonium-242	99.96
Plutonium-244	84.60 - 98.97
Americium-241	>99.99
Americium-242	55.1 - 85.6
Americium-243	>99.99

Curium-243	93.4	
Curium-243	99.99$^{*)}$	
Curium-244	99.33 - 99.82	
Curium-245	97.86 - 98.64	
Curium-245	99.998$^{*)}$	
Curium-246	99.50 - 99.83	
Curium-247	70.3 - 90.2	
Curium-248	95.8 - 96.82	

$^{*)}$ Obtained by double enrichment

$^{**)}$ Obtained due to the decay of enriched ^{244}Cm

REFERENCES

1. L.N Prokhorova, V.G. Nesterov, G.N. Smirenkin, G.V. Grishin, E.A. Nikitin, V.N. Polynov and V.V. Ratchev
Vykhod nejtronov pri spontannom delenii chetno-chetnykh izotopov kjurija. AE, t.33, p.767-770, (1972).

2. A.A Druzhinin, V.N. Polynov, S.P. Vesnovskij, A.M. Korochkin, A.A Lbov and E.A. Nikitin
Period poluraspada ^{245}Cm po otnosheniju k spontannomu deleniju. DAN SSSR, t.280, v.6, p.1351-1352, (1982).

3. A.A Druzhinin, V.N. Polynov, A.M. Korochkin, E.A. Nikitin and L.J. Lagutina
Period spontannogo delenija ^{239}Pu i ^{241}Pu. AE, t.59, v.1, p.68-69, (1985).

4. V.N. Polynov, A.A Druzhinin, A.M. Korochkin, E.A. Nikitin, V.A. Bochkarev, V.N. Vyachin, V.G. Lapin and M.Yu. Maximov
Period spontannogo delenija ^{243}Cm. AE, t.62, v.4 p.277, (1987).

5. E.F. Fomushkin, G.F. Novoselov, Ju.I. Vinogradov, V.V. Gavrilov, B.K. Maslennikov, V.N. Polynov, V.M. Surin and A.M. Shvetsov
Energeticheskaja zavisimost´ sechenija delenija ^{243}Am bystrymi nejtronami. VANT, ser."Jadernye konstanty", v.3(57), p.17-19, (1984).

6. E.F. Fomushkin, G.F. Novoselov, Ju.I. Vinogradov, V.N. Vyachin, V.V. Gavrilov, A.S. Kochelev, V.N. Polynov, V.M. Surin and A.M. Shvetsov
Sechenie delenija ^{243}Cm bystrymi nejtronami

. AE, t.62, v.4, p.278-279,(1987).

7. E.F. Fomushkin, G.F. Novoselov, Ju.I. Vinogradov, V.V. Gavrilov and V.A. Zherebtsov
Izmerenie energeticheskoj zavisimosti sechenija delenija ^{247}Cm nejtronami v diapazone 0.02 - 3.0 MeV. AE, t.62, v.4, p.279-280, (1981).

8. E.F. Fomushkin, G.F. Novoselov, Ju.I. Vinogradov, V.V. Gavrilov, V.J. Jaikov, B.K. Maslennikov, V.N. Polynov, V.M. Surin and A.M. Shvetsov
Izmerenie energeticheskoj zavisimosti sechenija delenija ^{242}Am nejtronami. Ja.F, t.33, v.3, p.620-624, (1981).

9. A.A Druzhinin, V.K. Grigor´ev, A.A Lbov, S.P. Vesnovskii, N.G. Krylov and V.N. Polynov
Sechenie radiacionnogo zakhvata bystrykh nejtronov ^{242}Pu. AE, t.42, v.4, p.314-315, (1977).

10. E.F. Fomushkin, E.K. Gutnikova, A.M. Korochkin, B.K. Maslennikov
Izmerenie effektivnykh sechenij delenija izotopov Cm$^{244-248}$ nejtronami bystrogo reaktora. Ja.F, t.17, p.24-26, (1973).

11. V.V. Golushko, K.D. Zhuravlev, Ju.S. Zamjatin, A.M. Kroshkin and V.N. Nefedov
Srednee chislo nejtronov, ispuskaemykh pri spontannom delenii ^{244}Cm, ^{246}Cm i ^{248}Cm. AE, t.34, p.135-137, (1973).

12. V.A. Drunin, Ju.V. Lobanov, D.M. Nadkarni, Ju.P. Kharitonov, Ju.S. Korotkin, S.P. Tretyakova and V.I. Krashonkin
K voprosu o nenabljudenii v Berkli spontannogo delenija kurchatovija. AE,t.35, p.279-281, (1973).

MEASURMENTS OF BREMSSTRAHLUNG SPECTRA FROM 50 MEV ELECTRONS ON Ta
AND PHOTONEUNTRONS ENERGY DISTRIBUTION FROM LIGHT, MEDIUM AND HEAVY NUCLEI

N.V. Zav'yalov, V.I. In'kov, M.S. Dudorov, E.N. Donskoj,
M.V. Savin, M.K. Saraeva and Yu.A. Khokhlov

All-Union Scientific Research Institute of Experimental Physics,
607200 Arzamas, Nizhnij Novgorod region, USSR

Abstract: The results and methods of measuring the spectra of bremsstrahlung radiation from thick tantalum targets (H = 2 mm and H = 6 mm), produced by 50 MeV electrons are described as well as the measurement results for photoneutron spectra and yields from small sampels of C, Al, Fe, Ni, Cu, Ta, Pb and ^{238}U mounted in measured field of bremsstrahlung radiation. The theoretical analysis of the data was done.

(linac, bremsstrahlung radiation, photoneutrons, time-of-flight method)

Introduction

In work [1] it was shown that in a theoretical description of experimental data on photoneutron spectra the real excitation function of a compound nucleus would be taken into account. However, as there are no experimental data on bremsstrahlung radiation spectra from thick targets the excitation function is obtained via spectra of single-shot electron interaction (Schiff-spectrum) in the majority of works.

In the present work measurements on 50 MeV electron bremsstrahlung radiation spectra were done. The same geometry was used to measure the energy distribution and photoneutron yields from the sampels of C, Al, Fe, Ni, Cu, Ta, Pb and ^{238}U mounted in the measured field of bremsstrahlung radiation.

Technique

Linac LU-50 was used to conduct the measurements. Bremsstrahlung radiation characteristics were measured for two tantalum targets (H = 2 mm and H = 6 mm) using D(γ,n) reaction with heavy water. A thin-walled caprolon container filled with heavy water was mounted at 1645 mm from a tantalum target. The rectangular container (inner dimensions 1x7x7 cm^3) was so oriented that the bremsstrahlung beam entered its narrow side. The magnetic field and aluminium filters 20 mm thick were used to absorb electrons passing through the targets. Bremsstrahlung radiation spectra were measured at 0°, 2° and 4° to the electron beam axis. The background was measured with the distilled ordinary water instead of D$_2$O sample. The calibrated resistive current detector was used to measure the total charge of the electrons falling onto the target. The energy distribution of neutron produced by the D(γ,n) reaction was measured by a time-of-flight method with 0.2 ns/m resolution. The calibrated

scintillation detector, mounted at 53.5 m from the sample at the end of evacuated neutron guide, was used to register the neutrons at 90° to the electron beam axis. To reduce the overall background and to eliminate the neutron production on the walls and frames, the gamma-quanta beam was collimated in the target room, and the neutron beam was collimated with the help of steel iris, placed along the neutron guide. For the detector an additional boron-water shield with a set of removable collimators was applied at 50 m flight path. At the neutron guide input (10 m point) the boron carbide filter (H = 21 mm) was mounted. At the end of the flight path an effective trap for neutrons and γ-quanta was mounted. To eliminate the overloadings of apparatus and photomultiplier of scintillation detector, caused by the bremsstrahlung radiation scattered on the sample, and, what is more important, to reduce the afterpulses the photomultiplier operated in a controlled gain coefficient mode.

A multistop time-to-digital convertor registered the time distribution. The energy calibration was checked by placing a thick graphite absorber in the flight path on a 25 m mark and observing the neutron absorbtion resonances of ^{12}C. Neutron registration threshold of the scintillation detector was determined and periodically controlled by a peak of total absorption ^{241}Am γ-quanta.

Small samples of the elements investigated (40-150 g) were used to measure photoneutron spectra. Samples of natural isotopic composition were mounted instead of heavy water in the same collimated flux of the bremsstrahlung radiation emitting from 2 mm thick tantalum target.

The accelerator was not switched off while the samples were remotely moved.

Data Reduction and Results

The photoneutron spectra were determined

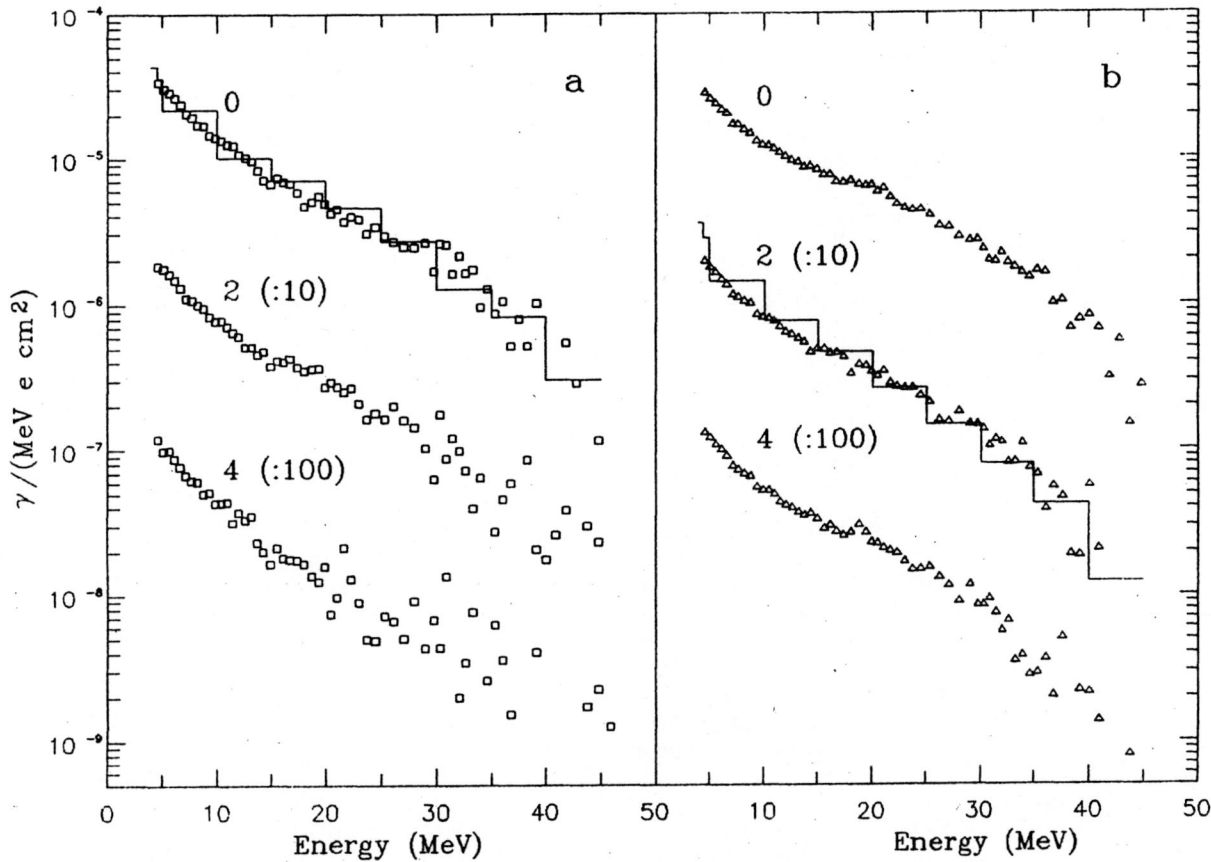

Fig. 1. Measured and calculated (-) 50 MeV bremsstrahlung intensity spectra from 2 mm (a) 6 mm (b) Ta radiator as a function of photon energy at three angles.

taking account of the relativistic correction and the attenuation in the sample and the plugs and filters of the neutron guide.

The energy of J-quanta was found according to the formulas in ref. [2]. The cross sections for the D(J,n) reaction were taken from work [3]. The experimental bremsstrahlung radiation spectra obtained and the calculated data by Monte Carlo code are presented in Fig. 1. Real geometry Monte Carlo calculations accounted all main interactions of charged particles and J-radiation with a substance. The calculational results were averaged over the measured spectrum of accelerated electron incident on the target.

The photoneutron spectra measured are given in Fig. 2. Here, a solid line shows the results of theoretical estimations for Ta and Pb. The energy distribution of evaporation neutron was calculated on the basis of formalism from ref. [4], where level density is determined in Fermi gas model approximation. The preequilibrium neutron spectrum is described by formula, obtained based on approach [5,6]:

$$N(E_n) = E_n \int_{Q+E}^{E_{max}} E^{-3/2} \sum_{n_0}^{M} n(n-1)(n+1) \; X$$

$$X \; ((E-Q-En)/E)^{n-2} X(E) dE \qquad (1)$$

Where Q,E_n - energy coupling of neutron in

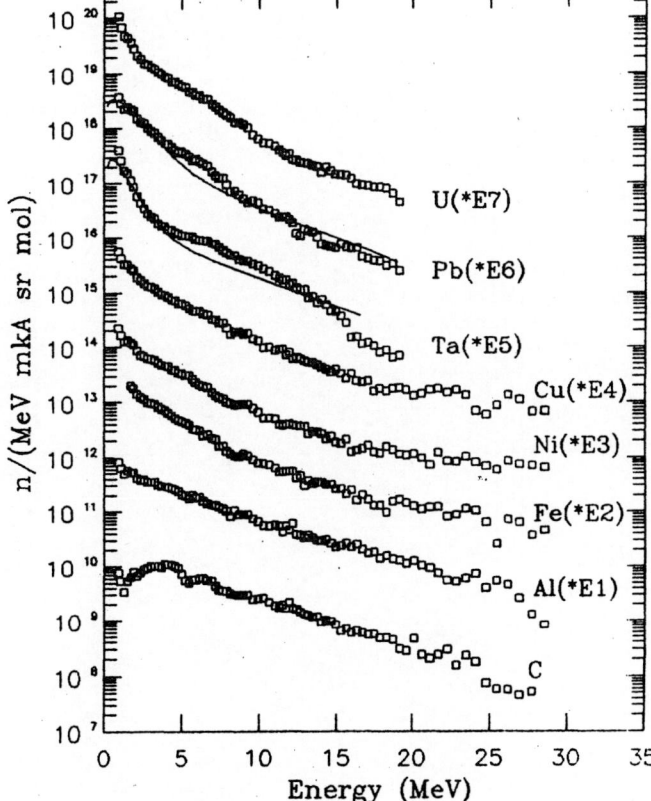

Fig. 2. Photoneutron spectra. Lines give calculation with the parameters: [181]Ta a = 23.8, b = 0.15 (fraction of the preequilibrium); [206]Pb a = 8.8, [207]Pb a = 8.9, [208]Pb a = 6.55, b = 0.2.

a nucleus and incident γ-rays, X(E) - nucleus excitation function, determined according to the measured spectrum of the bremsstrahlung radiation, $M = 1.1(aE)^{1/2}$.

Discussion

The data obtained may be the basis for the following conclusions.

The bremsstrahlung radiation spectra at various angles differ slightly in their forms, that does not contradict the theoretical considerations and agrees with the experimental data [7,8]. The energy distribution of the bremsstrahlung radiation from thick targets significantly differ from Schiff spectrum.

The photoneutron spectra formed under 50 MeV electron bremsstrahlung radiation have continuous disribution and even for light nuclei the resonances are not revealed which are caused by the transitions to the excited states. The preequilibrium group is obvious in the spectra of photoneutrons.

Statistical model calculations, averaged over the excitation spectrum and taking into account the preequilibrium processes for Ta and Pb together with the ordinary parameter values of level densities, reproduce the experimental results qualitatively. However, the applied simplified model of preequilibrium decay does not allow us to describe the results in the regions of $E_n = (5-8)$ MeV and $E_n \geq 12$ MeV.

REFERENCES

1. YU.N. Shubin, V.S. Stavinskij, in Proc. 3rd All-Union Conf. on Neutron Physics, June 1975, Kiev, USSR, vol.3, p.133, ZNIIatominform, Moscow (1976). In Russ.
2. N.K. Sherman: Nucl. Inst. and Meth. 79, 197 (1970)
3. S.I. Nagornyj, Yu.A. Kasatkin, E.V. Ikonin, I.K. Kirichenko: Yadernaya Fisika 44, vyp.5(11), 1171 (1986). In Russ.
4. A.E. Savel'ev: Byulleten' ZYaD. vyp.7, Prilozhenie 1, Atomizdat, (1977). In Russ.
5. M. Blann: Phys. Rev. Lett. 21, 1375 (1968)
6. V.K. Luk'yanov, V.A. Seliverstov, V.D. Toneev: Yadernaya Fizika 21, 992 (1975). In Russ.
7. A.A. O'Dell, Jr., C.W. Sandifer, R.B. Knowlen, W.D. Georg: Nucl. Inst. and Meth. 61, 340 (1968)
8. R.P. Lambert, J.W. Jury, N.K. Sherman: Nucl. Inst. and Meth. 214, 349 (1983)

APPLICATION OF THIN-FILM BREAKDOWN COUNTERS
FOR REGISTRATION OF FISSION FRAGMENTS
IN PHOTOFISSION AND NUCLEAR REACTOR PHYSICS EXPERIMENTS

A. N. Smirnov, O. I. Osetrov, I. Yu. Gorshkov, V. D. Dmitriev,
V. P. Eismont, S. V. Khlebnikov, G. P. Tyurin

V. G. Khlopin Radium Institute, 197022, Leningrad, USSR

V. G. Baty, V. Ya. Golovnya, O. G. Savchuk

Institute of Physics and Technology, 310108, Kharkov, USSR

S. I. Gulnic, I. V. Juce, E. M. Lomonosova, S. E. Tchigrinov

Institute of Nuclear Energetics, 220100, Minsk-Sosny, USSR

Abstract: Specific examples of arrangements based on thin-film breakdown counters are described. The set-ups were used for investigation of γ-induced fission of spontaneously fissioning nuclei and in experiments at the fast-thermal critical assembly.

(Thin-film breakdown counters, fragments registration, photofission, linac, bremsstrahlung, spontaneously fissioning nuclei, critical assembly)

Introduction

Thin-film breakdown counters (TFBC) for fission fragments /1/ combine threshold properties of solid state nuclear track detectors with high timing properties of surface barrier detectors. The outstanding characteristics of TFBC together with the low price render the TFBC as a very suitable detector for experiments on fundamental and applied nuclear physics where the commonly used techniques for fission fragment registration are not effective enough.

Photofission Experiments

At present new information on fission phenomena may be obtained from investigation of induced fission of even-even spontaneously fissioning nuclei (e.g. ^{248}Cm, ^{252}Cf) in the fission threshold region. However, the available intensities of the monochromatic particle sources are rather low for the measurements of fission cross sections and fragment angular distributions. Induced fission studies of that type can be effectively performed with the bremsstrahlung photons produced by linacs with high peak values of pulsed current.

In Fig. 1 a detector set-up which has been used at the Kharkov Institute of Physics and Technology Linac for measurements of photofission cross sections and fragment angular distributions of ^{248}Cm is shown. The TFBC system could be operated at a distance of 3 cm from the fissile targets at electron beam peak currents Ip 1 A , beam pulse duration τ= 2 μs. This peak current was about 2 orders of magnitude higher than usually used. The set up consisted of 8 channels for registration of curium fission fragments in the angular range 0°- 90° relative to the beam axis. The two channels were used to monitor γ-flux by 237-Np(γ,f)-reaction. The total efficiency ($\Omega/2\pi$) of the multiple TFBC arrangement was 3%. Time selection of the induced fission events from the spontaneous ones was based on utilization of multichannel time-to-digital converters. The time zero signals were given by linac's synchronisation system. The stop signals were taken from each of TFBC. To reduce the number of spontaneous fission fragments registered by TFBC the pulsed bias voltage gated by the linac's synchrosystem was used. One of the time spectra of TFBC pulses corresponding to the ^{248}Cm(γ,f)-fragments registration at 0° relative to the beam axis at E$_e$= 10 MeV is shown in Fig. 2. Experiments performed in the electron energy region 7 - 20 MeV demonstrate that the TFBC arrangement can be effectively used for γ-induced experiments on the subbarrier (E$_\gamma$ ≥ 5 MeV) fission of ^{242,244,246,248}Cm and near-barrier (E$_\gamma$ ≥ 7 MeV) fission of ^{252}Cf.

Reactor Physics Experiments

Another problem of interest is the investigation of the parameters of the active zone of critical assemblies. It is, for instance, the fission rate distributions along different directions relative to the zone axis.

Compactness of TFBCs and their insensitivity to high doses of γ - radiation (up to 10^6 Gy) make them competitive with SSNTD

measurements, which are inoperative, laborious and don't give a sufficient statistical accuracy. For the measurements of the characteristics of active zone of the fast-thermal critical assembly (FTCA-5) of the Institute of Nuclear Energetics the sandwich consisting of TFBC and the fissile material convertor is used. By remote control the sandwich is moved within the channel of 20 mm in diameter, placed along the vertical axis of the assembly. During the tests of the sandwich in the center of the active zone the pulse counting rate versus the TFBC applied voltage (counting characteristics) and the pulse counting rate versus the value of the assembly power were measured (see Fig. 3). The fisson rate distributions along the assembly axis for converters with ^{232}Th, ^{235}U, ^{236}U, ^{238}U, ^{237}Np and ^{239}Pu were measured with the statistical accuracy of about 2 %. The result for ^{235}U is presented in Fig. 4. The obtained results show that the detection properties of TFBCs don't change under the conditions of the critical assembly operated in wide range of power (two orders of magnitude), TFBCs remain operational at counting rates up to 800 pulses per second, the operational resource of TFBC is 10^7 breakdowns, the reduction of the detection efficiency due to reduction of their sensitive area after the first 10^5 counts is no more than 1 %. A comparison of the distributions for ^{235}U and ^{238}U obtained with TFBC and SSNTD shows that the results are equal within the experimental accuracy. The distributions obtained for the above six nuclides after the calibration measurements for the corresponding spectra of neutrons will be used for the determination of the fission cross section relations.

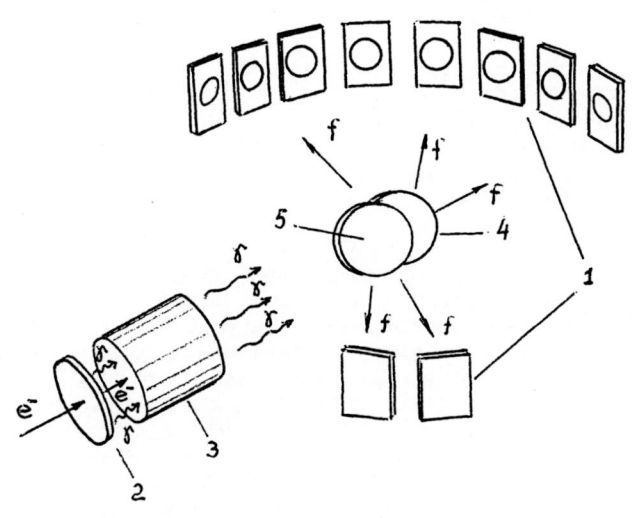

Fig. 1. Schematic drawing of the experimental set-up for photofission studies.

1 - TFBC, 2 - tungsten radiator, 3 - aluminium absorber, 4 - curium target, 5 - neptunium target.

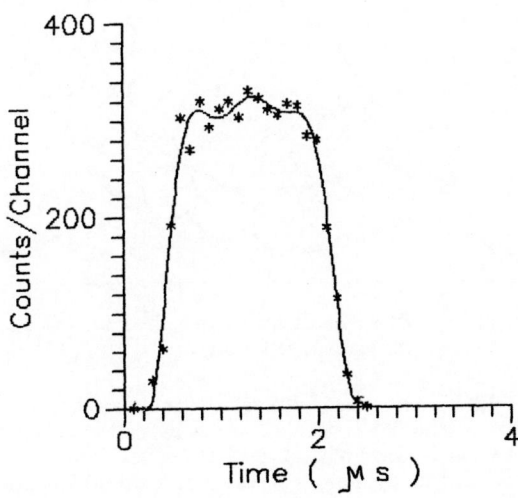

Fig.2. Time distribution of TFBC signals corresponding to fragments registration of ^{248}Cm photofission induced by pulsed bremsstrahlung photons (I_p – 1 A, E_{e^-} = 10 MeV, = 2 μ s).

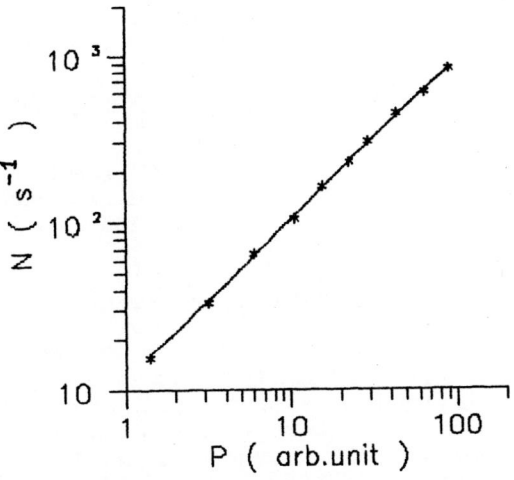

Fig.3. The dependence of TFBC with ^{238}U target count rate on the power of critical assembly FTCA–5.

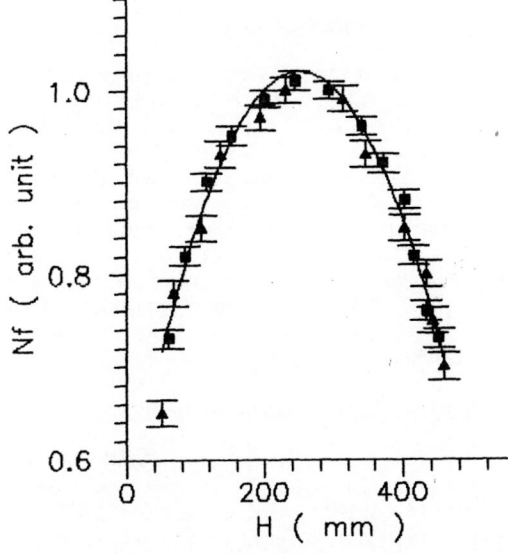

Fig.4. Axial distribution of ^{238}U fission density in the critical assembly FTCA–5.
▲ – measured using solid state detectors
■ – measured using TFBC

REFERENCE

1. A. N. Smirnov, V. P. Eismont, Pribory i technika eksperimenta, no. 6, p. 5, 1983 (in Russian).

458

VALIDATION EXPERIMENTS OF NUCLEAR CHARACTERISTICS OF THE FAST-THERMAL SYSTEM HERBE

M. Pešić, N. Zavaljevski, *P. Marinković,
M. Milošević, D. Stefanović, D. Nikolić, S. Avdić

'Boris Kidrič' Institute of Nuclear Sciences - IBK - Vinča
*Faculty of Electrical Engineering - ETF - Beograd
P.O.Box 522, 11001 Beograd, Yugoslavia

Abstract: In 1988/90 a coupled fast-thermal system HERBE at RB reactor, based on similar facilities, is designed and realized. Fast core of HERBE is built of natural U fuel in RB reactor center surrounded by the neutron filter and neutron converter located in an independent Al tank. Fast zone is surrounded by thermal neutron core driver. Designed nuclear characteristics of HERBE core are validated in the experiments described in the paper. HERBE cell parameters were calculated with developed computer codes: VESNA and DENEB. HERBE system criticality calculation are performed with 4G 2D RZ computer codes GALER and TWENTY GRAND, 1D multi-group AVERY code and 3D XYZ few-group TRITON computer code. The experiments for determination of critical level, dρ/dH, and reactivity of safety rods are accomplished in order to validate calculation results. Specific safety experiment is performed in aim to determine reactivity of flooded fast zone in possible accident. A very good agreements with calculation results are obtained and the validation procedures are presented. It is expected that HERBE will offer qualitative new opportunities for work with fast neutrons at RB reactor including nuclear data determination.

(RB reactor, HERBE, coupled fast-thermal core, validation experiments, computer codes)

Introduction

The investigation of fast neutrons at the RB reactor was initiated in late seventies. The external neutron converter [1], the experimental fuel channel [2] and the first version of coupled fast-thermal system [3] were designed up to now. Determination of fast neutron spectra and other characteristics of the realized neutron fields were dominant experiments performed at the RB reactor in few past years. New computer programs for the reactor calculations and experimental data evaluations were developed in Nuclear Engineering Laboratory (NET).

Coupled Fast-Thermal Core HERBE

The feasibility study of the HERBE system at RB reactor began in 1988 [4]. The goal was to design a coupled fast-thermal core flexible enough to simulate neutron spectra in diverse fast reactors. The existing nuclear fuel should be used with minimum reactor system modification.

HERBE fast core was designed of RB reactor natural uranium fuel elements placed as tight as possible in the reactor center. The fast core was surrounded with a neutron filter and a neutron converter, forming so called the 'fast zone'. A vertical experimental channel (VCH) was inserted in center of the fast core.

The exact composition and width of the neutron filter zone depends of desired neutron spectrum in the fast core. The neutron Cd filter zone and natural uranium buffer zone are designed in the same cylindrical aluminum tank.

RB 80% enriched UO_2 fuel elements were used as the neutron converter surrounding the neutron filter zone in the external cylindrical aluminum tank. Each fuel element consists of 9 hollow fuel segments supported by bottom spacers made of cylindrical Al tubes.

The horizontal cross section of HERBE fast zone designed at RB reactor is shown in Fig. 1.

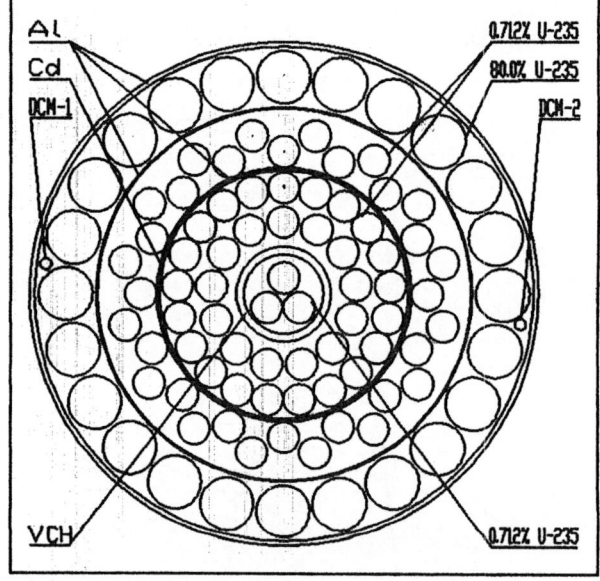

Fig. 1. HERBE fast zone horizontal cross section

HERBE thermal (driver) core was based on RB reactor standard core (12 cm square lattice pitch) with 44 fuel elements of 80% enriched UO_2 moderated and reflected by heavy water. The coupling zone is heavy water.

Two moderator leak sensors (DCM) were placed in the neutron converter zone and connected with the RB reactor safety system. RB control rod is replaced with new safety rod with higher reactivity. The fuel elements in neutron converter zone are sealed so that moderator can not enter the fast zone during postulated flooding accident.

HERBE calculation

The coupled fast-thermal system HERBE is used for the validation of assembly spectrum codes and global reactor calculations. The critical experiments, as well as control rod calibrations, were performed.

Several computer codes for determination of neutronic parameters of coupled fast-thermal system HERBE are developed in NET Laboratory: VESNA [5], DENEB [6], VEGA [7] AVERY [8] and GALER [9]. Programs TWENTY GRAND [10] and TRITON [11] are obtained from NEA Data Bank and installed at NET computer VAX 8250.

The fast zone of system HERBE in normal operating conditions does not contain moderator. Thus there is a considerable leakage in axial direction, due to the significant void volume fraction. This leakage is the most accurately described by Benoist theory, implemented in program DENEB. The 25-group macroscopic cross sections are calculated using programs VESNA (for thermal core) and DENEB (for the fast zone). Since the neutron converter zone is axially inhomogeneous, two sets of group constants are generated with program DENEB. The group constants of the filter zone containing natural uranium and cadmium, are homogenized, prior to the diffusion calculations.

Reactor is first calculated using one dimensional transport code AVERY (with the geometric axial buckling) in two cases: with converter zone consisting of enriched fuel only and with converter consisting of Al supporters only. Two sets of few group constants, generated by AVERY code, for two or three dimensional diffusion calculations are used. The effective multiplication factor is calculated using four group two dimensional diffusion program GALER which gives geometric buckling in prescribed regions, necessary for the new iteration of group constants in program AVERY.

Program AVERY is used after that twice with the same materials, but with geometric buckling from program GALER, generating few group constants for the final diffusion calculations. Global HERBE system diffusion calculations are performed with so prepared macroscopic cross sections using programs GALER, TWENTY GRAND and TRITON.

After series of trial runs critical height is estimated around 138 cm and all calculations are repeated for that height. Characteristics of coupled fast-thermal system HERBE described by Avery's kinetics parameters are shown in Table 1.

Table 1. Nuclear Characteristics of HERBE

Parameter	Value (AVERY code)
k_{11}	0.4153
k_{12}	0.1921
k_{21}	0.5651
k_{22}	0.8168
ℓ_{11} (ms)	0.207
ℓ_{12} (ms)	0.555
ℓ_{21} (ms)	0.576
ℓ_{22} (ms)	0.728
ℓ (ms)	0.622
β_{eff} (delay neutrons)	6.857 10^{-3}
β_{eff} (photoneutrons)	1.008 10^{-3}

The results of four group diffusion calculations and HERBE critical experiment are presented in Table 2.

The calculations of reactivity of system HERBE when heavy water penetrates through outer container of fast zone are performed using programs TWENTY GRAND and GALER for several water levels. The experiments for validation of these calculation are presented in the next section.

For the RB reactor safety rods cross section generation program VEGA is used. Supercell consists of three fuel elements and a safety rod (rod position in) or moderator (rod position out). The results of global diffusion calculations performed by TRITON code and experiments are presented in Table 3. The agreement is very good, indicating that safety rod cross sections are properly generated.

Experimental Validation

Results of criticality calculation of HERBE core by TWENTY GRAND and GALER computer codes compared with experimental result, show very good agreement.

Approach to the critical level of HERBE system is performed by increasing moderator level in thermal core of the RB reactor. Neutron flux is monitored by 3 independent BF_3 counters connected in the measuring system controlled by VAX-8250 computer. On-line data evaluation and extrapolation of critical height at the actual subcritical level of moderator is done by computer code INVMULX [12]. Extrapolation to the critical height is performed either using all measured data by the least square method or by straight line drown through the last two points of measurement.

Determination of reactivity gradient near heavy water critical level ($d\rho/dH$) is realized as another experiment performed for comparison with calculation results.

Using two BF_3 counters covered with Cd foil a neutron flux time distribution, in supercritical reactor, was recorded in 2K channel memory of MCA operated in MCS mode. The asymptotic reactor period was determined from true exponential part of neutron flux time dependence by the least square method and reactivity was calculated using inhour Nordheim relation. The kinetics parameters of HERBE system were determined using computer code AVERY.

Comparison of experimental (0.249±0.002 $/cm) and calculation results (0.248 $/cm) for the $d\rho/dH$ shows very good agreement.

Table 2. Results of HERBE critical height determination

Program/ Experiment	Moderator Height (cm)	$k_{eff} - 1$
TWENTY GRAND	138.0	− 0.00163
GALER	138.0	− 0.00237
TRITON	138.0	0.00326
EXPERIMENT:	H_c(cm) = 137.94 ± 0.02	

Reactivity worth of safety rods of HERBE was determined in series of measurements in which each safety rod was inserted in the RB reactor thermal core separately or in combination with the others. Time neutron distribution after shut-down, using one BF_3 counter, was recorded in 1K channel memory of MCA operated in MCS mode with 50 ms dwell time.

After data smoothing, the IM [13] computer code based on inverse method for reactivity determination was used. A good agreement of calculation and experimental results can be observed from Table 3., except for the SR3 whose material composition is not known with high reliability.

The analysis of possible accidents at HERBE system has shown that there is finite probability for the fast core outer tank cracking which results in fast and high reactivity increase. In order to verify the computer codes used for that reactivity calculation a special safety experiment using controlled flooding of HERBE neutron converter zone is performed. The decanting device is designed for semiautomatic transfer of heavy water moderator from HERBE thermal core into the neutron converter zone. Due to safety precautions, this operation is performed at the RB reactor high subcriticality.

For each heavy water level in neutron converter zone the critical level of the system is measured by standard criticality approach procedure using three independent neutron counting channels with BF_3 detectors. Control of the measuring system and on-line data evaluation is performed by computer code INVMULX from VAX-8250 computer.

Results of the experiments and calculations are presented in Table 4. showing acceptable agreement. The calculation by both computer codes overestimate the inserted reactivity, almost in all cases, confirming possibility to predict reactivity value during postulated flooding accident including failure of both moderator leakage sensors (DCM).

Table 3. Results of measurement and calculation of reactivity of HERBE safety rods

| Safety rod | Reactivity ρ_{SR}/β_{eff} ($) | |
	Calculation (TRITON)	Measurement (IM)
SR1	− 2.20	− (2.26 ± 0.08)
SR2	− 2.20	− (2.20 ± 0.08)
SR3	− 1.44	− (1.64 ± 0.04)
WLM	− 2.72	− (2.75 ± 0.04)
SR1+SR2	− 4.92	− (4.74 ± 0.13)
SR1+SR2+SR3	− 6.56	− (7.01 ± 0.19)
SR1+SR2+SR3+WLM	− 10.30	− (10.13 ± 0.25)

Table 4. Critical height of HERBE with flooded neutron converter zone by moderator

| Moderator height (cm) | HERBE Critical Height (cm) | | |
| | Experiment | Calculation | |
		GALER	TWENTY GRAND
0.0±0.2	138.86±0.02	138.23	137.84
20.0±0.2	137.64±0.02	−	136.97
40.0±0.2	136.44±0.02	137.67	135.94
60.0±0.2	135.11±0.02	135.81	134.26
80.0±0.2	133.50±0.02	133.31	132.29
100.0±0.2	131.88±0.02	131.36	130.61
120.0±0.2	130.46±0.02	130.22	129.64

Conclusion

Coupled fast-thermal system HERBE, designed in 1990, is successfully utilized for experimental research of fast neutron field and verification of models, data and computer codes used in reactor calculations [14]. New experiments for neutron spectrum determination and kinetic parameters validation are planed in near future.

Acknowledgements

This work was performed by support of Science Foundation of Republic Serbia.

The authors wish to acknowledge professor dr D.Popović for consultation and the RB reactor operation staff for the assistance in HERBE core realization and performing the experiments.

References

1. P.Strugar, O.Šotić, M.Ninković, M.Pešić, D.Altiparmakov: Kernenergie 24, 101 (1981)
2. M.Pešić, H.Marković, M.Šokčić, I.Mirić, M.Prokić, P.Strugar: Kernenergie 27, 461 (1984)
3. M.Pešić: Kernenergie 30, 142 (1987)
4. M.Pešić, P.Marinković, in Proc. Int. Conf. '50 Years with Nuclear Fission', Vol.II, p.784, Gaithersborg MD/Washington DC (1989)
5. M.Milošević, Program VESNA for LWR and HWR Cell Calculation, IBK-NET-26, Vinča (1989)
6. M.Milošević, Program DENEB for Complex Reactor Cell Calculation, NET Computer Code Library, Vinča (1990)
7. M.Milošević, Program VEGA for Thermal Reactor's Cassette Calculation in XY Geometry by Collision Probability Method, IBK-NET-30, Vinča (1989)
8. M.Milošević, M.Pešić, Computer Code AVERY for Calculation of Coupled Fast-Thermal Core Kinetics Parameters, Report IBK-NET-27, Vinča (1989)
9. M.Milošević, Multi-Group Diffusion Computer Code GALER for Reactor Calculation in RZ Geometry Based on Galerkin's Method, IBK-NET-29, Vinča (1989)
10. M.L.Tobias, T.B.Fowler, The TWENTY GRAND Program for the Numerical Solution of the Few Group Neutron Diffusion Equations in Two-Dimensions, ORNL-3200 (1964)
11. A.Daneri, G.Maggini, E.Salina, TRITON − A Multi-Group Diffusion Depletion Program in Three-Dimensions, EURATOM-Fiat-ARS, FN-E-97 (1967)
12. M.Pešić, N.Hadžimahmutović, N.Zavaljevski, Computer Code INVMULX for On-line Reactivity Measurement by Inverse Multiplication Method, NET Computer Code Library, Vinča (1989)
13. M.Milošević, Computer Code IM for Reactivity Determination by Inverse Method, NET Computer Code Library, Vinča (1988)
14. M.Pešić, N.Zavaljevski, M.Milošević, D.Stefanović, D.Popović, D.Nikolić, P.Marinković, S.Avdić: Annals of Nuclear Energy, in press (1991)

INVESTIGATION OF TOTAL CROSS SECTION CHARACTERISTICS IN THE REGION OF UNRESOLVED RESONANCES

E. F. Fomushkin, Ju. I. Vinogradov, V. V. Gavrilov, G. F. Novoselov, A. M. Shvetsov

All-Union Scientific Research Institute of Experimental Physics
607200, Arzamas, Nizhnij Novgorod region, USSR

Abstract: Neutron transmission through layers of different materials is measured. The source of neutrons is the uranium target of the electron linac LU-50. As a detector a sectioned fission chamber, containing 6-7 g of ^{235}U, is used. The chamber is placed at 52.18 m of flight path, the investigated sample at 25 m. The neutron pulse duration is 10-13 ns, the total resolution time 18 ns. CAMAC standardized equipment registers the time-of-flight neutron spectra, controls the background components and amplitude spectra of fragments for each chamber section. The whole measurement procedure is automatic, including the periodic sample movements. The transmission measurements are made for 7-8 values of thickness. The results are presented in $T(t, \Delta E_i)$-tables for 35-40 energy intervals of ΔE_i. The transmission $T(t)$ for ΔE intervals of neutron energy may be presented as Laplace transform of the total cross-section probability density $p(\sigma)$ in a given energy interval with the sample thickness (t) as transformation parameter. An algorithm for estimation of probability density semiinvariants and restoration of the $p(\sigma)$-function was created to treat measured transmission functions $T(t)$. The measurements of the transmission functions for Al, Pb, Fe and others are carried out. The results of the approximation by the semiinvariants of the obtained experimental data are used in the practical work. The experimental data and the results of their approximation for Al and Al$_2$O$_3$ are presented.

(neutron transmission, electron linear accelerator, ionization chamber, scintillation detector, time-of-flight method, CAMAC standard, total cross-section, automation of experiment, distribution density, moments, semiinvariants, resonance structure)

Introduction

To calculate the direct penetration of neutrons through the shieldings, one should account for the resonance structure of the total cross-sections. However, the total neutron cross sections are measured not for all nuclides with precise enough energy resolution. Alternatively, this problem can be solved by measuring the neutron transmission, $T(t)$, vs the investigated sample thickness, t, for a given energetic interval $(E, E+\Delta E)$. In this case, the nesessary energy resolution or interval width, ΔE, is determined by the technical requirements, as a rule.

Experimental equipment

In the Institute of Experimental Physics, the electron accelerator LU-50 [1] is used to measure time-of-flight data neutron transmission. The neutron source is a ^{238}U target. As neutron detectors, ionization chambers (IC) are used with 90%-enriched ^{235}U layers, filled with a mixture of argon (90%) and methane (10%) up to 1000 mm Hg pressure. To mo-

nitor the primary neutron flux at 25 m flight distance, an IC containing 1.5 g of ^{235}U is used. The main three-sectional IC with more than 6 g of ^{235}U is mounted at a flight path of L=52.182±0.045 m. To minimize electromagnetic induction effects during the linear accelerator operation, some actions were undertaken: the pre-amplifiers are mounted directly on the chambers' housings, the main IC with anode voltage blocks and pre-amplifiers' supplies are placed in the box with a brass screening grid, all circuits use transformer decouplings and so on.

The sample investigated is placed at 28 m; to automatically move this sample (into the neutron flux or out of it) a remote-controlled platform (RCP) is used.

The larger part of the neutron flight path is evacuated; to attenuate scattered neutron background at 26 m and 50 m collimators of 40 mm and 60 mm in diameter are mounted. The chambers were positioned correctly in the neutron beam with the help of radiographical dielectric trac detectors and films.

In the time-of-flight measurements

the start signal is the bremstrahlung radiation pulse, registered by a scintillation detector (SD),based on an anhydrogenous liquid scintillator, ZhS-52. The detector is placed at 50 m flight distance in a parallel channel of neutron output. Coaxial communication lines are used to transfer the analogous signals from IC and SD to the computer center.RCP control lines have their outputs here as well. A microcomputer of "Electronics-60" type connected to a central computer local net is the base of the measurement/control system in this experiment [2]. NIM and CAMAC standard apparatus selects pulses, codes the parameters registered, controls the sample motion.

Processing of experimental results

All IC amplitude pre-selection and pulse timing are obtained with "Constant Fraction" discriminators. The pulses from 0-20 μs (effect) and 200-220 μs (background) "windows" only are processed further. Two parameters are coded: the pulse amplitude (with 50μs amplitude numerical converter) and time of its arrival according to the start pulse (with a 10 ns time-code converter). The amplitude and time distributions are accumulated in programs in an external memory device (EMD) "Electronics-256" of 0.5 Mbytes capacity joined to the microcomputer line with an adapter. A frequent enough motion of the sample (into the flux and out of it), the results storing in the corresponding EMD zones, allows to reduce the inaccuracies due to the measuring channels' characteristic drift.

The total volume of the information stored is the sum of amplitude and time-of-flight spectra:
A_{cpw} and T_{cpw} (c=0,1,2,3; p=1,2; w=1,2),
    where
c- a measuring channel number;
    (c=0 - a monitoring IC, c=1,2,3 - the
        main IC sections),
p- a sample location (p=1 - in the flux,
    p=2 - out of flux),
w- a time "window" (w=1 - a "window" 1 -
    effect, w=2 -a "window" 2- background).

Measurement procedure and results

The length of each amplitude spectrum is 256 channels, and of each time-of-flight spectrum-2048 channels 10 ns long. After the subsequent series of measurements is completed, the information stored is recorded in file through the net channel on the central computer discs.

For each sample thickness there are several runs of two hours duration. Hardware and software of the experiment [3] allow to automaticaly conduct all runs with the sample mounted. Besides sample motion monitoring, the obtained information storage and accumulation, in the course of the experiment the preliminary proceeding of data stored is done, the measurement protocol is registered. It

allows to analyse the experimental data. Resulting spectra are shown in fig. 1.

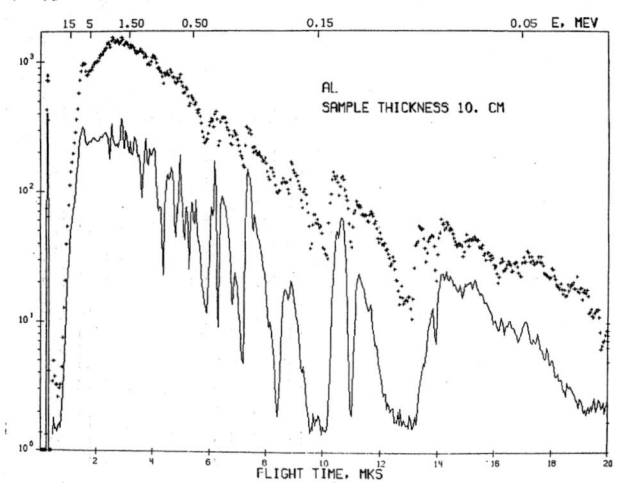

Fig. 1. Time-of-flight neutron spectra, ++++ -without sample,──── -with sample.

The processing of the experimental information comprises the following operations: the background subtraction, the computation of transmission values $T(E_i, E_i+\Delta E_i)$ in given energy intervals for each measurement series, normalized to the monitoring camera testimony, the statistic processing of all series for a given sample thickness with calculation of average values $\langle T(E_i, E_i+\Delta E_i)\rangle$ and statistical uncertainties.

Usually, for each investigated material the results are presented in the form of a table of $T(t, \Delta E_i)$ for 35-40 energy intervals $\Delta E_i=E_i-E_{i-1}$ and for 7-10 values of material thickness t. The measurements are conducted in the energy range 0.10 - 15 MeV. In fig. 2,3 the results of transmission function measurements for Al and Al₂O₃ are shown.

Interpretation of the results

Mathematical interpretation of the data obtained is based on the following considerations [4]. If the neutron beam, distributed with equal probability in $E_{i-1} \leq E \leq E_i$ energy interval, penetrates through the material layer of thickness t, the transmission is described by

$$T(t) = \frac{1}{(E_i - E_{i-1})} \int_{E_{i-1}}^{E_i} \exp(-\sigma(E)\cdot t)dE , \quad (1)$$

where $\sigma(E)>0$ is the total cross-section. Substituting the Riemann integral by a Lebesgue integral, the same function may be written in the form

$$T(t) = \int_0^\infty p(\sigma)\cdot \exp(-\sigma\cdot t)d\sigma , \quad (2)$$

where $p(\sigma)\geq0$ is the probability density of the total cross-section $\sigma(E)$ in interval $\Delta E_i=E_i-E_{i-1}$.

In presentation (2), the transmission function $T(t)$ is the Laplace transformation of the non-negative function $p(\sigma)$, defined on the positive axis, $\sigma \geq0$; in

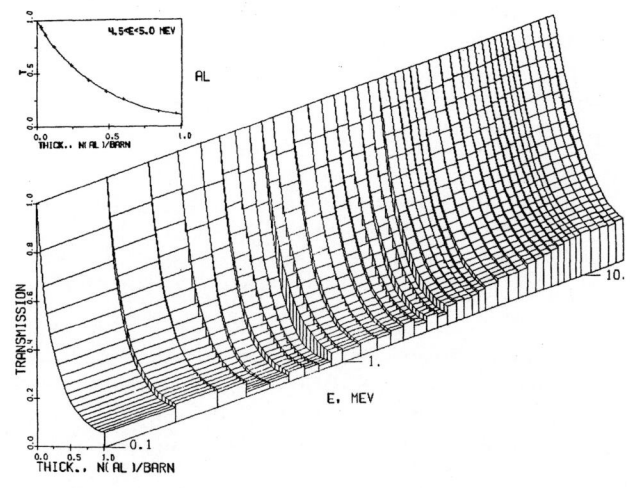

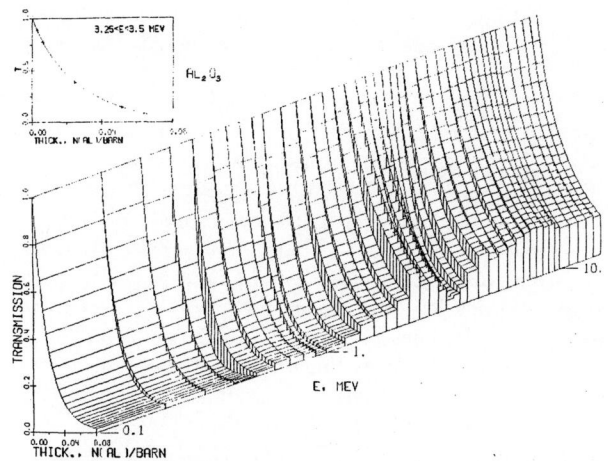

Fig. 2,3. Neutron transmission dependences on sample thickness and energy for Al and Al2O3.
In insertions - curves for defined energy groups;
+ - experimental points,
—— - approximation results by the first four semiinvariants.

this case, the parameter of the transformation, t, is a real positive number. Laplace transformation may be considered as a particular case of a generating function of moments [5]. So, the expansion of the transmission into a power series determines the initial moments of the probability density, $p(\sigma)$, for the total neutron section in the interval ΔE_i, i.e.

$$T(t) = \sum_{k=0}^{\infty} M_k \cdot (-t)^k / k! \quad , \qquad \text{where} \quad (3)$$

$$M_k = \int_0^{\infty} \sigma^k \cdot p(\sigma) d\sigma \quad , \quad (k=1,2,3,\ldots).$$

Expansion of the logarithm of the transformation or the curve of observed cross-section $S(t) = -\ln T(t)/t$ into a power series allows to calculate semiinvariants of the function $p(\sigma)$:

$$S(t) = \sum_{k=1}^{\infty} X_k \cdot (-t)^k / k! \quad , \qquad \text{where} \quad (4)$$

$X_1 = M_1 = \langle \sigma \rangle$ — an average value of total section in ΔE_i -interval;

$X_2 = \mu_2 = M_2 - M_1^2$ — the second central moment (variance) of the probability density, $p(\sigma)$;

$X_3 = \mu_3$ — the third central moment of the function $p(\sigma)$;

$X_4 = \mu_4 - 3\mu_2^2$ — and so on.

To estimate the values of semiinvariants and their uncertainties according to the results of $T(t, \Delta E_i)$ measurements, a program packet was composed. To restore the density function $p(\sigma)$ on the known values from the first few semiinvariants (moments) one of the Karl Pearson distribution types is usually used: beta-, gamma-distributions, Gram-Charlier series et al [5]. If the Gram-Charlier series are used, the probability density distribution of the total cross-section in the energy interval investigated may be written as

$$p(\sigma) = G(z) \left\{ 1 + \frac{X_3}{3!} \cdot H_3(z) + \frac{X_4}{4!} \cdot H_4(z) + \ldots \right\}, \quad (5)$$

where $z = (\sigma - \langle \sigma \rangle) / X_2^{1/2}$ — standardized variate;
$G(z) = (2\pi)^{-1/2} \cdot \exp(-z^2/2)$ — Gaussian distribution;
$H_j(z)$ — Chebyshev-Hermite polynomials.

Conclusion

The obtained data and the described method of the approximation represent the interest for the practical calculations of the transmission coefficients through the materials of construction. The work on the comparative analysis of the proposed method of the calculations and of the experimental data with one described in the scientific literature is started.

REFERENCES

1. Ju. A. Khokhlov, N. V. Zav'jalov, I. A. Ivanin, V. I. In'kov, N. P. Sitnikov, A. V. Tel'nov, in Proc. of this Conference
2. N. N. Zaljalov, A. L. Luk'janov, V. A. Skotnikov, S. V. Khlystov, G. V. Akamsina: Voprosy atomnoj nauki i tekhniki. Ser.: Metodiki i programmy chislennogo reshenija zadach matematicheskoj fiziki 2, 71 (1987)
3. N. N. Zaljalov, V. E. Zezjulin, A. L. Luk'janov, S. V. Khlystov: Voprosy atomnoj nauki i techniki. Ser.: Matematicheskoe modelirovanie fizicheskikh processov 1, 86 (1991)
4. E. F. Fomushkin: Voprosy atomnoj nauki i techniki. Ser.: Yadernye konstanty 1, 22 (1990)
5. M. G. Kendall, A. Stuart, Distribution Theory, London (1962)

An empirical look at the calculation of the unassociated particle anisotropy factors for the determination of neutron production from the D(d,n)³He reaction at bombarding energies below 500 keV

S. Croft   and   D.S. Bond

AEA Technology
Harwell Laboratory
Oxfordshire
OX11 0RA
UK

Abstract: The neutron field from the D(d,n)³He reaction can be inferred from a measurement, at some convenient angle, of the yield of protons from the competing D(d,p)T reaction. The number of protons detected is related to the differential neutron yield by the so called anisotropy factor function. The anisotropy factor may be calculated if the differential charged particle cross-sections are known. In principle these may be measured to high accuracy and so it should be possible to use the D(d,p)T channel as an accurate and absolute measure of the D(d,n)³He strength.

In this paper we review some aspects of the present state of the art of neutron fluence determinations by the unassociated particle technique for the D(d,n)³He reaction using the D(d,p)T reaction. The approach adopted has been an empirical one involving the use of differential cross-section data from as many sources as possible in order to estimate how the data base itself limits the accuracy of the method.

INTRODUCTION

The D(d,n)³He reaction is widely used to produce pseudomonoenergetic neutron beams in the range 2 MeV to 3.5 MeV. Harwell Laboratory for instance, operates a 500 kV Van de Graaff machine [2] which over recent years has been extensively used for the absolute calibration of a variety of neutron detection systems designed as diagnostics for JET, the Joint European Torus. JET is a tokomak fusion experiment currently operating with D-D plasmas and so the D(d,n)³He reaction is the natural choice for calibration purposes

JET is a copious producer of neutrons and because of this many of the instruments have inherently low efficiency. The calibration experiments are therefore necesserily long and are most conveniently carried out using moderatively thick solid targets, so as to achieve a reasonably high and stable yield. To maintain good neutron energy resolution the irradiations are performed, wherever possible, at an angle of about 105° to the incident beam direction, as at around this angle the neutron energy is insensitive to the deuteron energy and hence to the target layer thickness and the deuterium depth distribution within the target. Neutron production at 0° is also advantageous on occasions because the differential neutron cross-sections are forward peaked and only slightly dependent on angle. The neutron energy is also at its highest at 0° and often the ratio of direct to background neutron fluence can be at its maximum.

The targets used at Harwell [3] comprise deuterium gas occluded in a titanium layer of specified thickness deposited on a thin copper disc. Tests [4] have shown that the gas may be taken to be uniformly distributed throughout the titanium layer and that the neutron yield from thick targets is consistent with a deuterium to titanium atom ratio of between 1.6 and 1.7. The targets are inclined at 45° to the incident ion beam and viewed by a collimated silicon surface barrier detector mounted at 83°.

As with all absolute neutron calibrations a major obstacle is the determination of the absolute neutron fluence. Experimental differential cross-section data for the D(d,n)³He reaction have been evaluated and compilations produced [5,6] and in principle could be used together with mass stopping power tables [7] and current monitoring to estimate the differential neutron production rate from the target. However, below 500 keV the mass stopping power of deuterons in titanium-deuteride is subject to a scaling uncertainty of at least 7%, the differential production cross-section is uncertain to several percent and the deuterium target loading distribution is unknown. Thus the absolute production rates obtained by this route would be inaccurate. Furthermore current integration alone cannot be used to monitor for signs of target deterioration such as burn-off, carbon deposition, changing deuterium depth profile or to infer the neutron emission spectrum. In principle these deficiencies can be overcome by measuring the yield and spectrum of charged particles eminating from the target [8,9].

The associated particle technique, (APT) of producing a reference pseudomonoenergetic neutron flux from the D(d,n)³He reaction [10,11] exploits the fact that each neutron produced is accompanied by a ³He particle with kinetic properties uniquely determined by the interaction energy and the angle of emission of the neutron. As the angular distribution of the reaction can be accurately determined by measuring the charged particle, it is also possible to infer the differential neutron distribution by reference to a fixed associated particle (time uncorrelated) detector.

Experimentally, however, it is often more convenient to infer the neutron yield from the D(d,n)³He reaction by monitoring, at a fixed angle, the yield of protons from the competing D(d,p)T reaction [12]. The use of the proton branch in this way is referred to as the unassociated particle technique (UPT). It relies on knowing accurately the relative differential cross-sections of the two reaction branches.

Since the last anisotropy factor calculations were published [13] results from apparently very carefully performed differential cross-section measurements of the D(d,n)³He and D(d,p)T reactions below 500 keV have appeared in the open literature [14-19] extending and improving the established data base [20-25]. Furthermore a new evaluation of mass stopping powers of hydrogen ions has become available [7]. It is therefore timely to reassess the calculation of anisotropy factors for the D(d,n)³He reaction. In this paper we take an empirical look at several aspects of the problem.

ANISOTROPY FACTORS

The anisotropy factor, R, is defined as the ratio of the number of neutrons per unit solid angle emitted about the angle, θ_n, to the number of protons per unit solid angle emitted about the angle θ_p. For target layers of uniform composition, R depends on the angle at which the unassociated protons are observed, θ_p, the neutron emission angle, θ_n, The incident deuteron energy, E, the composition of the target, x, (as defined by the composition descriptor TiD_x), and the energy-loss, ΔE, of the incident deuterons in the target. Neglecting multiple scattering effects, anisotropy factors may be calculated from Equation 1.

$$R(x, \Delta E, \theta_p, \theta_n) = \frac{\int_{E-\Delta E}^{E} \frac{1}{S(t)} \cdot \frac{d\sigma_{d,n}}{d\Omega}(\theta_n, t).dt}{\int_{E-\Delta E}^{E} \frac{1}{S(t)} \cdot \frac{d\sigma_{d,p}}{d\Omega}(\theta_p, t).dt} \qquad (1)$$

where $S(t)$ is the mass stopping power for deuterons of energy t in the target material and $d\sigma/d\Omega$ is the microscopic angular differential cross-section of the specified reaction branch at the specified angle as a function of deuteron energy.

In the experiments designed to measure the differential cross-sections, the charged particles from the two reactions are detected simultaneously under identical target conditions and geometry. The measurements of the two branches are therefore correlated. The ratio of the two branches should consequently be determined more accurately than their individual absolute values. Thus the D(d,p)T reaction should serve as a faithful monitor for the D(d,n)³He branch, particularly when using thin targets. With this in mind we choose to work with relative shape factors rather than absolute values. We write the differential cross-section $d\sigma(\theta, t)/d\Omega$ as the product of the total microscopic cross-section $\sigma(t)$ and an angular factor $g(\theta, t)$ and introduce the D(d,n)³He to D(d,p)T branching ratio $b(t) = \sigma_{d,n}(t)/\sigma_{d,p}(t)$. In this notation the anisotropy factor for thin targets becomes

$$R(\theta_p, \theta_n) = b(t).f(\theta_p, \theta_n, t) \qquad (2)$$

where the angular factor $f(\theta_p, \theta_n, t) = g_{d,n}(\theta_n, t)/g_{d,p}(\theta_p, t)$. In a preliminary empirical attempt to assess the quality of the data base, we have calculated $b(t)$, $f(83°, 0°, t)$ and $f(83°, 105°, t)$ using data from as many sources as possible.

RESULTS

The branching ratio data base below 350 keV is shown in Figure 1. Over the energy interval from 350 keV to 500 keV only three determinations have been made. The scatter in the experimental data is larger than expected from the uncertainties quoted by the measuring authors.

As a convenient way of representing the disperate data we tried fitting various subsets of it by a first order Padé approximation using the code DEMING [26]. The results of this exercise are summarised in Figure 1. The Padé approximation has the functional form:

$$b(t) = a_1(1 + a_2.t)/(1 + a_3.t) \qquad (3)$$

The best fit parameters together with the various covariance terms, for the case when all the data were included in the fit were as follows:

$a_1 = 9.6940 \times 10^{-1}$
$a_2 = 5.1736 \times 10^{-3}$
$a_3 = 3.4601 \times 10^{-3}$
$cov(1,1) = 2.897 \times 10^{-4} \qquad cov(2,2) = 3.127 \times 10^{-6}$
$cov(1,2) = -2.735 \times 10^{-5} \qquad cov(2,3) = 2.350 \times 10^{-6}$
$cov(1,3) = -2.001 \times 10^{-5} \qquad cov(3,3) = 1.773 \times 10^{-6}$

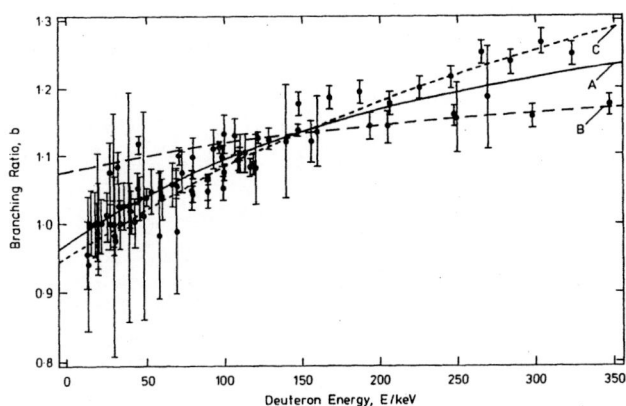

Figure 1: Plot of b, the $(\sigma_{d,n}/\sigma_{d,p})$ branching ratio as a function of deuteron incident energy. The data were drawn from references 14, 16–19, 21, 22, and 24 (88 data points in total). The data of Arnold et al [22] have been re-evaluated using the angular distribution data of Krauss et al [19]. The data from Theus et al [14] is based on the results given in their Table 1. as their Tables 2. and 3. are known to be in error [18]. Curve A is the best fit first order Padé approximation to all the data. Curves B and C are the corresponding fits when only the data from [14] and [16–19] are included respectively.

This fit predicts that the branching ratio is uncertain by about ±1.8% at low energies and ±9.0% at 500 keV. At practical running energies (upwards of about 100 keV) the uncertainties compare unfavourably with other techniques of fluence measurement such as foil activation and the proton recoil telescope method. The work of Theus et al [14], Brown and Jarmie [16–18] and Krauss et al [19] are considered to be the most accurate. However, they are inconsistent with each other, span only a limited energy range, and as can be seen from Figure 1. fits to them in isolation can be markedly different to the best global fit. In the absence of a more refined evaluation we favour the global fit which yields the parameter set listed above.

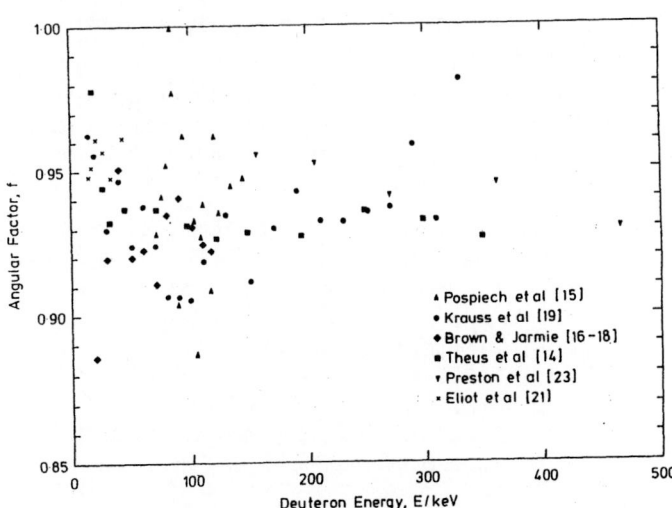

Figure 2: A plot of the angular factor, $f(\theta_p, \theta_n, E)$, as a function of incident deuteron energy, E, evaluated for $\theta_p = 83°$ and $\theta_n = 0°$ using the data from references 14–19, 21 and 23 (74 data points in total).

466

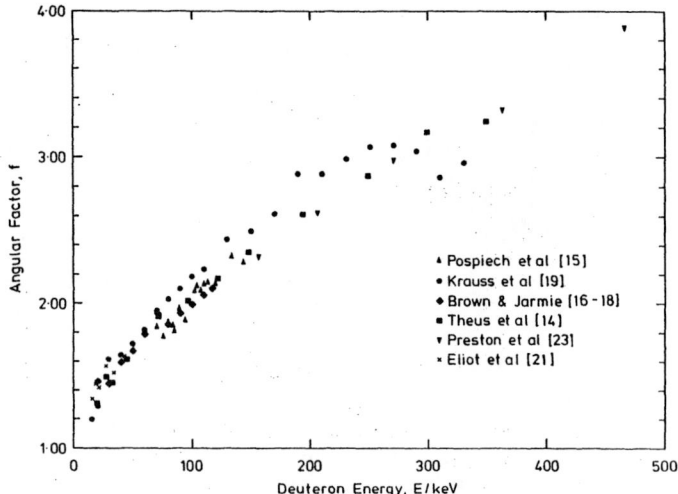

Figure 3: As Figure 4. but evaluated for $\theta_n = 105°$.

The f-factors calculated using the angular distribution coefficients drawn from a number of different sources are shown in Figures 2. and 3. for $\theta_p = 83°, \theta_n = 0°$ and $105°$ respectively. Lack of covariance information on the coefficients makes it difficult to estimate the uncertainties on the individual calculated values. The data of Theus et al [14] are favoured because they were collected under excellent experimental conditions and with high statistical accuracy. However, if the scatter in the data is indicative of how accurately the f-factors can be calculated in practice the situation is again somewhat discouraging, with deviations of the order of $\pm 3\%$ being typical in the $\theta_n = 0°$ case and somewhat larger in the $\theta_n = 105°$ case.

CONCLUSIONS

1. It is timely for a new evaluation of the data on the $D(d,n)^3He$ neutron source reaction to be undertaken. It is important that this should contain recommended results in numerical form. Ideally the data on the competing $D(d,p)T$ reaction should be evaluated simultaneously in a self-consistent fashion. This would considerably help workers wishing to apply the UPT while at the same time improving the standardisation and accuracy of the method.

2. Additional differential cross-section measurements of the $D(d,n)^3He$ and $D(d,p)T$ reactions are urgently needed, particularly in the sparsely populated interval between 300 keV and 500 keV.

3. It is noted that many authors report their angular distribution data in the form of Legendre Polynomial coefficients extracted from their data. Although these enable the angular distribution to be calculated at all angles, it is not possible to estimate the uncertainty on the result unless the full covariance matrix of the fitted parameters is also provided. Authors are therefore urged to present their data in numerical form and to provide covariance information on all derived parameters when appropriate.

4. The authors know of no cases in which the angular yield of $D(d,n)^3He$ neutrons from a thick titanium-deuteride target has been measured absolutely relative to the angular yield of protons from the competing $D(d,p)T$ reaction. As this is a situation of great prac-

tical importance it seems desirable that such a programme of measurements be carried out.

5. The unassociated particle anisotropy factors calculated by Ruby and Crawford [13] are due for revision.

ACKNOWLEDGEMENTS

We thank Dr N.P. Hawkes for rewarding discussions about this work.

REFERENCES

1. J. Csikai: Nucl. Instrum. and Meths. in Phys. Res. A280(1989)233-250.

2. High Voltage Ltd, 8 Kildare Close, Eastcote, Ruislip, Middlesex, HA4 9LH, UK.

3. Multivolt Ltd, 26 Loppets Road, Crawley, Sussex, RH10 5DW, UK.

4. C.A. Uttley: Harwell Laboratory, Priv. Comm.

5. H. Liskien and A. Paulsen: Nucl. Data. Tables 11(1973)569-619.

6. M. Drosg: Proc. of the IAEA Consultants' Meeting On Neutron Source Properties, Debrecen, Hungary, 17–21 March, 1980 Edited by K. Okamoto INDC (NDS) 114/GT(June, 1980) Paper R-8 pp 201-240

7. J.F. Janni: Atomic Data and Nuclear Data Tables 27(1982)147-339, 341-529.

8. P.B Johnson: Nucl. Instrum. and Meths. 114(1974)467-475

9. P.B. Johnson and T.R. Armstrong: Nucl. Instrum. and Meths. 148(1978)85-92

10. C.M. Bartle and P.A. Quin: Nucl. Instrum. and Meths. 121(1974)119-127

11. M.M. Meier: NBS SP 493(1977)221-226

12. T.B. Ryves and D. Sharma: Nucl. Instrum. and Meths. 128(1975)455-459

13. L. Ruby and R.B. Crawford: Nucl. Instrum. and Meths. 24(1963)413-417

14. R.B. Theus, W.I. McGarry and L.A. Beach: Nucl. Phys. 80(1966)273-288

15. G. Pospiech, H. Genz, E.H. Marlinghaus, A. Richter and G. Schrieder: Nucl. Phys. A239(1975)125-133

16. N. Jarmie and R.E. Brown: Nucl. Instrum. and Meths. B10/11(1985)405-410

17. R.E. Brown and N. Jarmie: Radiation Effects 92(1986)45-58

18. R.E Brown and N. Jarmie: Phys. Rev. C41(4)(1990)1391-1400

19. A. Krauss, H.W. Becker, H.P. Trautvetter, C. Rolfs and K. Brand: Nucl. Phys. A465(1987)150-172

20. W.A. Wenzel and W. Whaling: Phys. Rev. 88(1952)1149-1154

21. E.A. Eliot, D. Roaf and P.F.D. Shaw: Proc. Roy. Soc. A216(1953)57-65

22. W.R. Arnold, J.A. Phillips, G.A. Sawyer, E.J. Stovall Jr., and J.L. Tuck: Phys. Rev. 93(1954)483-497

23. G. Preston, P.F.D. Shaw and S.A. Young: Proc. Roy. Soc A226(1954)206-216

24. K.G. McNeill: Phil. Mag. (Series Seven) 46(1955)800-804

25. P.R. Chagnon and G.E. Owen: Phys. Rev. 101(1956)1798-1803

26. P.M. Rinard and A. Goldman: LA-11082-MS (Nov.,1987)

A SPECIFIC LIGHT ION " dE-E-T " TELESCOPE

E.Wattecamps, G.Rollin and M.Keters *

Commission of the European Communities, Joint Research Centre,
Central Bureau for Nuclear Measurements, B-2440 Geel, Belgium,
(*) Industriele Hogeschool van het Rijk, Hasselt, Belgium.

Abstract: A telescope was developed and tested. It is specific in its particle identification, and is used at the 7 MeV-300ps burst-width Van de Graaff accelerator. The telescope is made of two multi-wire-parallel-plate-avalanche-counters (MWPPAC), providing energy-loss and time, and a hexagonal pilot-U scintillator, viewed by two photomultipliers (PM), providing time and energy. By measuring the ratio of amplitudes on opposite sides of the scintillator, and by applying an on-line time- and pulse-height-correction procedure the time-resolution of the pilot-U scintillator reduces to 300 ps and the pulse-height resolution reduces from 35 to 28 %. Two-dimensional spectra of dE versus E, and E versus time-of-flight, are shown for ^{241}Am and for samples of Ni and Ta, irradiated by 8.0 MeV neutrons. By selecting areas in these two-dimensional spectra one deduces particle energy spectra with outstanding foreground to background ratio and reasonable statistical accuracy in short acquisition times.

(Telescope, light ion, neutron induced cross section, Ni(n,α), energy-loss, energy, time-of-flight.)

Introduction

To measure (n,p), (n,d), (n,t) or (n,^{4}He) cross section data in the neutron energy range from 5 to 15 MeV, as requested for fusion reactor design, a telescope was developed and tested for use at the 7 MeV Van de Graaff accelerator with its 300 ps burst-width. The requested cross sections are relatively small and many reaction channels become available in the neutron energy range of interest. Therefore the telescope should be quite efficient and separate well the different species of particles. A typical telescope [1,2,3] has an energy-detector at the end of the particle trajectory. Our new telescope provides a pulse-height and a time-of-flight, thus measuring contemporaneously the energy and the mass of the impinging particle.

To perform a time-of-flight measurement with reasonable timing accuracy and to achieve acceptable counting rate the detectors' time resolution should be of the order of 300 picoseconds and the flight path length should be about 40 cm. To define the trajectory of the particles on their way from the (n,x) sample to the energy detector and to identify the particle type, we have inserted two MWPPAC's in the flight path. These transmission detectors are known to be fast and efficient.

The experimental set up

The telescope in front of the deuterium target is shown in Fig.1.

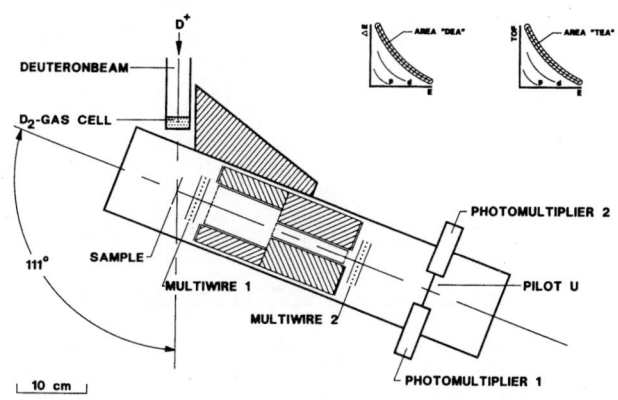

Fig.1 The " dE-E-T " telescope in front of the deuterium target

The sample, the two MWPPAC, the pilot-U scintillator and the two light guides are in a single vacuum tight housing filled with 4 torr isobutane counting gas. The sample is mounted on a computer controlled robot [4], which carries samples of 5 cm diameter, for instance: Am, Ni and Ta. The pressure of the counting gas is controlled to within three digits by a baratron and the purity of the gas is maintained by a controlled but small continuous isobutane flow.

Each MWPPAC is made of two aluminised (1.35 ng per cm^2) polymer foils (0.27 mg per cm^2) 6 mm apart. In between the cathode foils an anode is inserted (anode at 3 mm from the cathode foil) made of gilded tungsten wires of 20 micrometer diameter, every 1mm. The construction and the

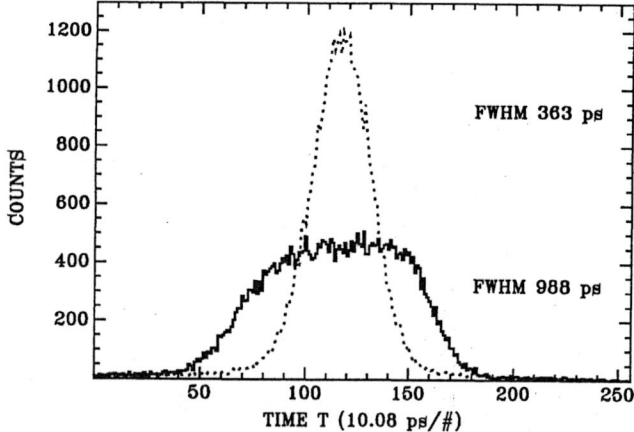

Fig.2a
The time spectrum T, PM 1 versus PM 2, without and with time correction (full and dotted line respectively).

performance of the MWPPAC is described in ref.[5]. The time resolution of a MWPPAC for a mono-energetic ^{241}Am alpha particle is 500 picoseconds and the pulse-height resolution is 45%.

The energy detector (by pulse-height) and time-of-flight detector at the end of the 38 cm flight-path is made of an hexagonal pilot-U scintillator of 6 cm sidelength and 1 mm thickness, observed on two opposite edges by Hamamatsu H2431 photomultipliers. The construction and performance of this device is described in refs.[6,7].

Test with alpha particles of ^{241}Am

The quality of the timing accuracy of the pilot-U scintillator with its two photomultipliers is shown in Fig.2.a. The full line gives the uncorrected start-stop time distribution which has a FWHM of 988 ps, quite steep slopes and a broad almost flat top. Since the ratio of amplitudes is measured for each event, and since the light attenuation across the scintillator was measured previously, the correction for the propagation time of the light across the scintillator is possible [6,7]. The corrected start-stop time disribution has a Gaussian shape with a FWHM of 363 ps for two identical photomultiplier chains , thus reducing to 240 ps for a single detector chain.

The pulse-height spectrum, seen by one chain, for a monoenergetic alpha particle of ^{241}Am, is shown in Fig.2.b. The FWHM amounts to 39% and the high energy side has an important wing. A 5.5 MeV alpha particle crossing the isobutane gas and the two MWPPAC's reaches the pilot-U scintiallator with 4.1 MeV energy and gives rise to a pulse height at channel number 44. The calibration of the energy scale in Fig. 2b, one dimensional spectrum of 128 channels, therefore is 92 KeV per channel. This value is a first order calibration applying for entire irradiation of the scintillator.

After correction the pulse-height spectrum of the pilot-U system will be composed of two halves, and the corrected contribution of one chain is shown by the dotted line. This time, the FWHM is 28% and there is no tail towards large pulse-heights. After correction for light attenuation across the scintillator the peak of the spectrum is shifted to a higher channel number and the callibration factor becomes 86 KeV per channel.

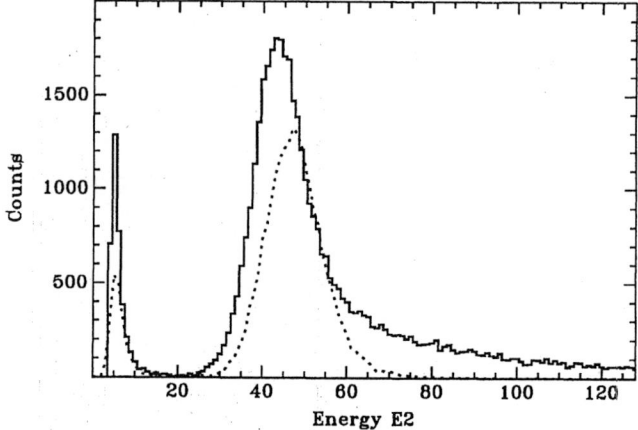

Fig.2b
The energy spectrum E2 (by pulse-height), observed by chain of PM 2, without (full line) and with correction (dotted line).

Test with Ni(n,x) and Ta(n,x) at 8.0 MeV neutron energy

The ^{241}Am sample was replaced by nickel (4.41 mg.cm^{-2}), or by tantalum (216 mg.cm^{-2}), and irradiated with monoenergetic neutrons of 8.0 MeV from the d,d neutron source of the pulsed Van de Graaff accelerator for equal fluences in a period of approximately 2.5 hours (accelerator parameters: pps 2.5x10^6 , mean current 1.8 microampere and pulse width 430 picoseconds). The energy by pulse-height and the time-of-flight are corrected for the finite size of the pilot-U scintillator as previously done for mono-energetic alpha particles of ^{241}Am, but now for a broad spectrum of energies. This is performed as for mono-energetic particles since the correction procedure relies on the ratio of measured amplitudes. The two-dimensional spectra are illustrated by isocontour lines at levels 2, 4, 8 and 16.

In Fig.3.a two zones are visible in the spectrum of energy-loss versus energy (by pulse- height). In the upper zone the spectra are quite broad in energy-loss and in energy, as expected, and slightly tilted from large loss at low energy to smaller loss at higher energy. The upper zone is not

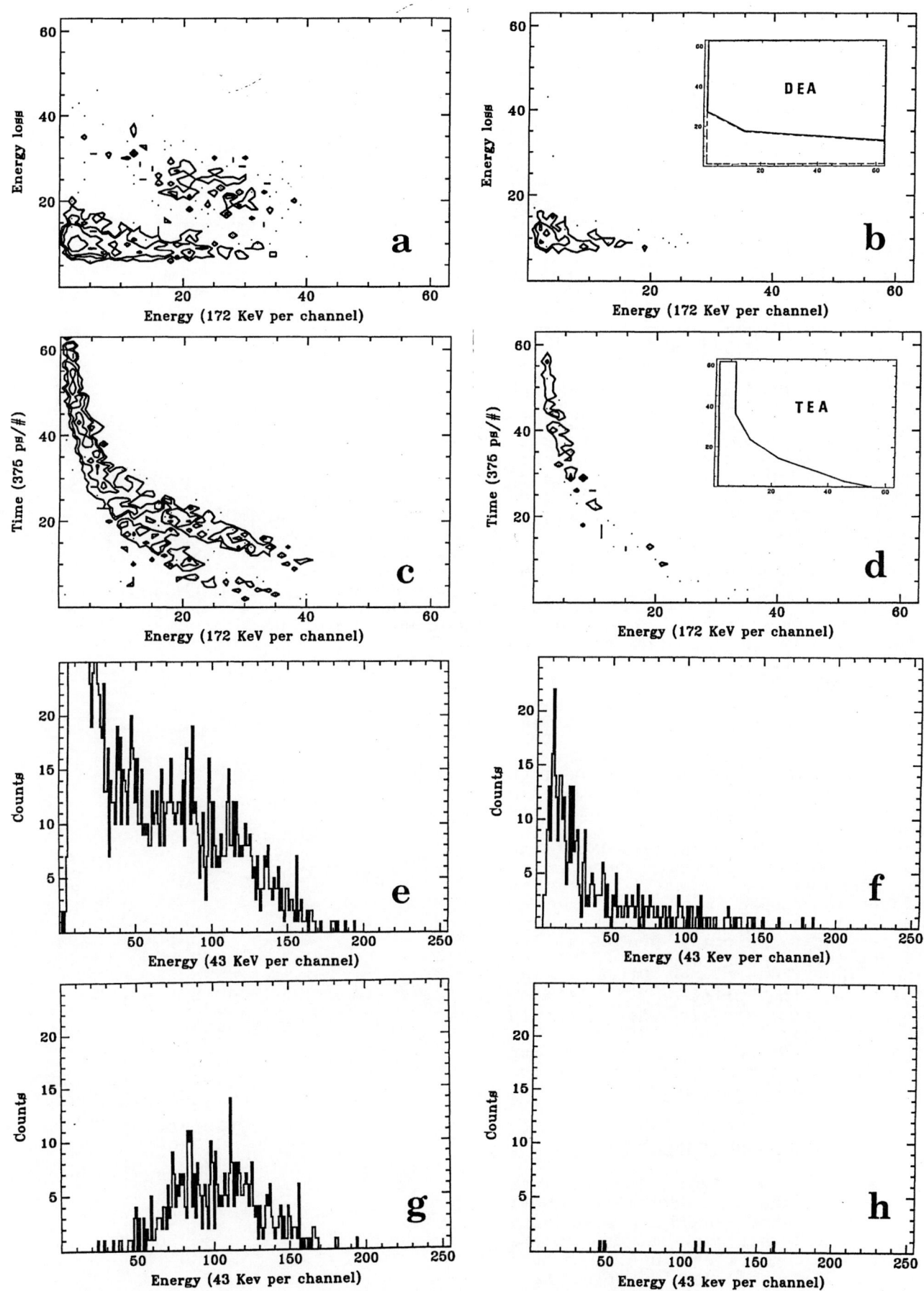

Fig.3. Spectra from Ni(n,x) (left) and from Ta(n,x) (right) at 8 MeV neutron energy.
a. and b.   Two-dimensional spectra of energy-loss versus energy (pulse-height).
c. and d.   Two-dimensional spectra of time-of-flight versus energy (pulse-height).
e. and f.   Energy spectra without conditions on energy-loss or time.
g. and h.   Energy spectra of alpha particles pertaining to energy loss area DEA and
             to the time-of-flight area TEA.

470

present in the corresponding spectrum of the tantalum run, shown in Fig. 3b.

The two-dimensional spectra of time-of-flight versus energy (by pulse-height) are shown in Fig.3.c and 3.d. A time channel is 375 ps wide. The time origin is shifted to allow, within the range of 64 channels, a detailed representation in the time range of interest. The spectrum with Ni (Fig.3.c) shows two well separated zones, and the spectrum with Ta (Fig.3.d) shows only one zone and even there, less events than in the Ni run.

One-dimensional spectra of energy (by pulse-height), shown in Fig.3.e and 3.f, are deduced from the previous more-dimensional spectra, by integration over all energy-losses and all times. Similarly in Fig.3.g and 3.h one-dimensional spectra of energy are obtained by integration over all energy-losses and all times, respectively pertaining to events in the upper zones which are known from the earlier calibrations with 241 Am to correspond with alpha particles. For definition of areas of interest see upper right corner in Fig. 3.b and 3.d.

Conclusion

The comparison of the final one-dimensional energy spectra, Fig. 3.g and 3.h, of Ni and Ta runs leads to the following conclusions:

. the signal left in the Ta run is 2 counts per hour and thus very small.
. the counting rate of alpha particles in the Ni run is 160 counts per hour.
. the small background and the reasonable fore-ground counting rate allow us to achieve the requested statistical accuracy in reasonable measuring times.
. additional analysis is needed to find out the lowest measurable energy.
. an energy- and time- correction procedure should be implemented to account for energy loss and corresponding velocity changes in the flight path.
. the linearity of the pulse height scale of the pilot-U scintillator versus alpha particle energy should be investigated.
. the actual telescope seems promising to identify other light ion reaction products and an irradiation of a Li sample is planned.

Aknowledgements

We would like to acknowledge Dr.F.J.Hambsch and Dr.B.Jäckel for their support in matters concerned with data handling. Our thanks are also due to the Van de Graaff accelerator staff for their excellent and never lasting help during the irradiations.

References

[1] S.M.Grimes, R.C.Haight, K.R.Alvar, H.H.Barschall and R.R.Borchers: Phys.Rev.C19, 2127 (1979).

[2] M.Ahmad, S.L.Graham, S.M.Grimes, R.Longfellow, H.Satyanrayana and G.Randers-Pehrson: Nucl.Instr. and Meth., 228, 349 (1985).

[3] A.Paulsen, H.Liskien, F.Arnotte and R.Widera, Nucl.Sci.Eng.,78, 377 (1981).

[4] G.Rollin and E.Wattecamps: Automation of a neutron cross section measurement system at the Van de Graaff accelerator. Internal report GE/R/VG/63/89

[5] H.Tulpinck: thesis (1988), Industriele Hogeschool van het Rijk, Hasselt, Belgium, unpublished.

[6] E.Wattecamps, G.Rollin and M.Keters: A 27 cm2 scintillator for alpha particle detection of some MeV with 28% energy- and 240 ps time-resolution, submitted Nucl.Instr. and Meth. (1991)

[7] M.Keters: thesis (1991), Industriele Hogeschool van het Rijk, Hasselt, Belgium, unpublished.

A NEW METHOD FOR MEASUREMENT OF (n,2n) CROSS SECTION

T. Iwasaki, H. Kimiyama, S. Meigo, S. Matsuyama,
M. Baba, K. Kanda and N. Hirakawa

Department of Nuclear Engineering, Tohoku University, Sendai, Japan 980

Abstract: A new method, where the intervals of detected times of neutrons are measured, has been developed for measuring the fast neutron induced (n,2n) cross section. By the method, (n,2n) events are easily distinguished on a measured time interval spectrum since two neutrons from one (n,2n) reaction produce one unique time interval. This method can be applicable to every nuclide. Demonstration experiments on various conditions were performed by developing a He-3 counter banked detector system (HCBD) with the characteristics of high efficiency and appropriate neutron capture-time. The time interval spectra expected by this method were observed in the experiments. The efficiency and the neutron capture-time of the HCBD were measured using a Cf-252 source and agreed very well with those calculated using Multi-KENO. A fluence monitor suitable for the HCBD was also developed.

(n,2n) cross section, measurement technique, interval of detected time, He-3 counter

I INTRODUCTION

Accurate measurements of the fast neutron induced (n,2n) cross section are required for designs of the blankets in the fusion reactors. The following techniques have been employed to measure the (n,2n) cross section; the multiplicity technique, the activation technique and the scattering spectrum technique. In these techniques, the multiplicity technique[1] has been applied for most of the nuclides since the technique has no limitation in principle. However, by the technique, the number of (n,2n) events can not be easily obtained from the measured spectrum and is derived by using the expressions including the terms of second, third, fourth and higher order power of efficiency. As a result, the cross section becomes very sensitive to the efficiency and the corrections for that.

We have developed a new method in which the intervals of detected times of neutrons are measured and (n,2n) events are easily distinguished on the measured time interval distribution. This method can be applicable to every nuclide.

We describe the principle of this method in Sec.II and the demonstration of this method using a He-3 counter banked detector (HCBD) in Sec.III Section IV explains characteristics (efficiency and neutron capture-time) of the HCBD. In Sec.V, a brief description of a fluence monitor suitable for the HCBD is presented.

II PRINCIPLE OF METHOD

In this method, the following experimental condition is assumed; the <u>time interval of reactions (Tr)</u> is long enough to neglect the <u>capture-time of detector (Tc)</u> which is defined as the life time of neutrons from their birth to the capture in the sensitive volume of the detector.

At first, consider the <u>interval of neutron detected time (Td)</u> when one neutron is emitted from one reaction. Under the condition mentioned above, the Td is equal to the Tr. The distribution of Tr, i.e. Td, is given as the following equation;

$$S(t)dt = So \cdot EXP(-So \cdot t)\ dt \qquad (1)$$

where So is the average reaction rate. Further, Td distributes widely in time scale since the average Tr (i.e., 1/So) is large.

Next, considering the (n,2n) reaction which produces two neutrons simultaneously, the distribution of Td differs from the distribution given by Eq.1. The Td between the first and second neutron from the (n,2n) reaction depends only on Tc. The Td due to the (n,2n) neutrons distributes only on a short time region.

Consequently, for the general cases accompanied with both (n,2n) and other reactions, the distribution of Td is expressed by the summation of the foreground distribution due to the (n,2n) reaction and the background distribution expressed by Eq.1 (see Fig.1). It should be noted that the Td between one of the (n,2n) neutrons and that from another reaction is approximately equal to Tr since the Tc is negligibly small.

Owing to the difference of the distributions, the (n,2n) events can be distinguished. The number of (n,2n) events can be accurately obtained by subtracting the background which is fitted to Eq.1 in the longer time region and is extrapolated in the shorter time region.

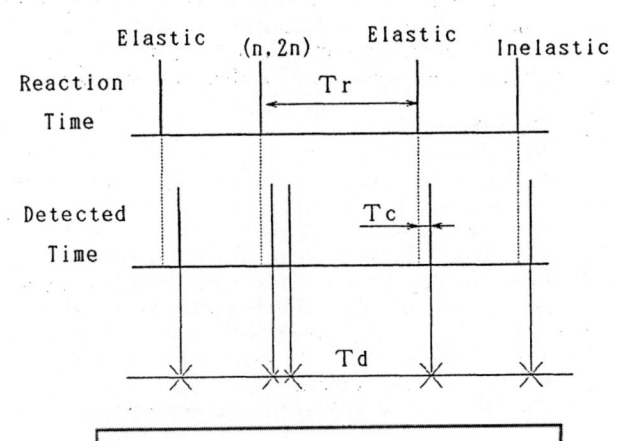

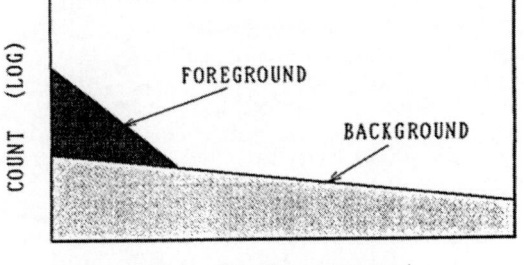

Fig.1. Expected spectrum by the present method.

III DEMONSTRATION EXPERIMENT

III.A He-3 counter banked detector (HCBD)

A detector system suitable for the present method was constructed with 28 He-3 counters (1" in diameter, 10" long) and 480 polyethylene moderator blocks (5cm×5cm×20cm) as shown in Fig.2. A square horizontal opening (5cm×5cm) penetrates the HCBD for neutron beam. A sample is placed in the center of the HCBD. The HCBD has the high efficiency because of the 4π-geometry and numbers of counters /2/. The neutron capture-time is appropriate since the detector captures thermalized neutrons with the optimized distribution of the counters.

III.B Experiment

All experiments mentioned here were performed using a 4.5MV Dynamitron accelerator in Tohoku University. Source neutrons were produced by T(d,n) reaction and were collimated with an iron collimator (50cm long). The neutron target was shielded by a water tank (1.2m cubic) to reduce background neutrons. The whole surface of the HCBD was covered with Cadmium plates against thermalized neutrons. The distance between the target and the sample was about 1.4m.

The He-3 counters were bunched to four groups. Each group was connected to a high voltage power supply, a preamplifier and an amplifier. The signals from the four amplifiers were summed, discriminated and transformed to the timing signals. The timing signals were split alternatively to the start or the stop connector of the Time-to-Amplitude Converter (TAC). Two TACs were utilized with the TAC range of 2msec and 50μsec. Using the spectrum measured by the shorter range TAC, the losses due to the dead time can be estimated.

Demonstration experiments were performed for various beam conditions, i.e., various counting rate of the HCBD (100~5000cps). Four samples were used; Beryllium, Lead, Graphite and Polyethylene. The samples have the diameter of 1~2cm and the length of 0.5~2cm. The runs without the sample or with a long shadow-bar were also carried out.

III.C Result

Figure 3 shows typical time interval distributions for the following samples with the deutron beam currents (the counting rates) shown below; (a) Beryllium, ~1.1μA (~1000cps) (b) Beryllium, ~0.1μA (~100cps) and (c) polyethylene, ~0.1μA (~100cps). In Fig.3, the first two spectra have a peak and a smooth distribution as expected above and shown in Fig.1. The peak and the smooth spectrum respectively correspond to the foreground and the background.

Further, the background spectrum of (a) shows a steep slope since the counting rate is high, i.e., So is large in Eq.1. That of (b) is flat because of the low counting rate. In the spectrum (c), only the flat distribution is found. This is reasonable since the polyethylene sample includes no nuclide to cause the (n,2n) reaction. It is found that the observed distributions behave as expected in all cases.

The fitted background to Eq.1 is shown in Fig.4 with the measured spectrum and the foreground spectrum which the background is subtracted. From this figure, it is clear that the background spectrum is easily fitted and estimated.

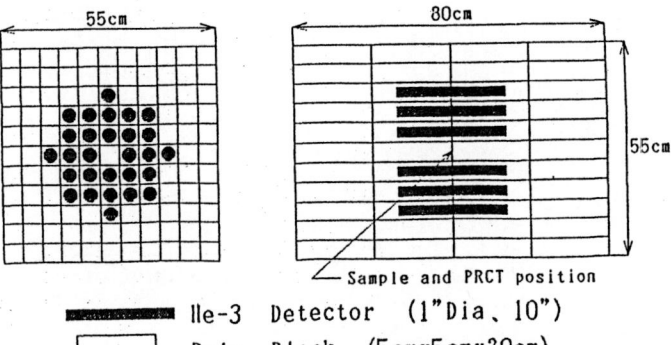

Fig.2. Schematic view of HCBD.

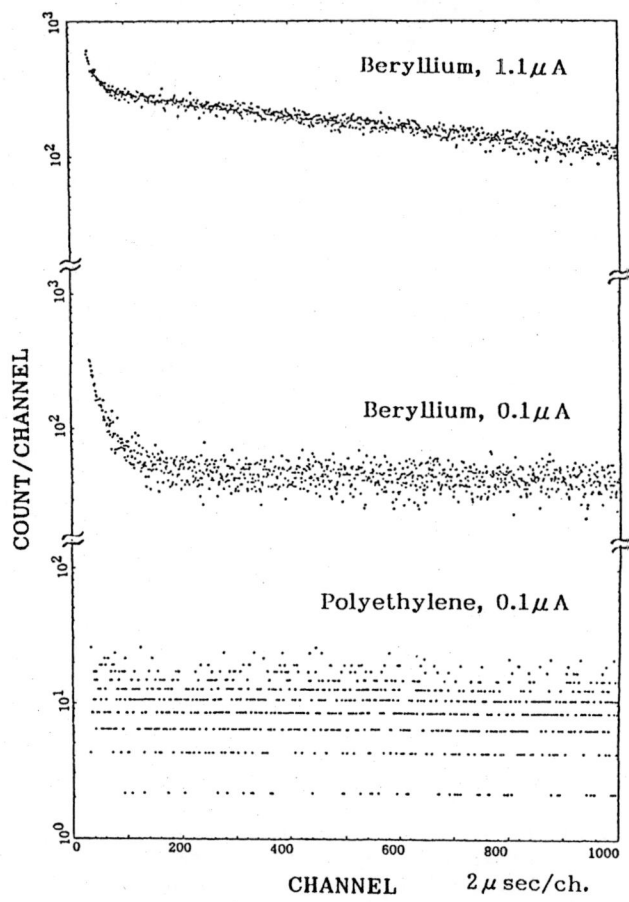

Fig.3. Measured time interval distributions.

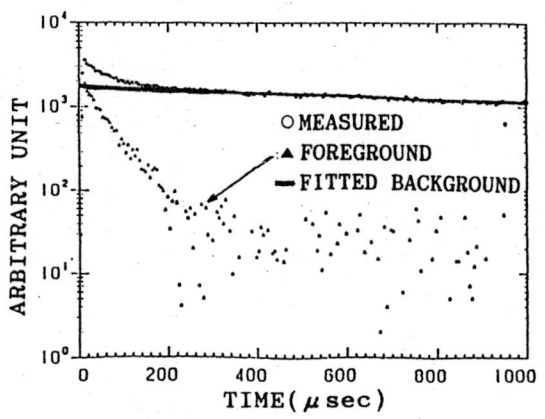

Fig.4. Fitted background and foreground obtained.

IV EFFICIENCY AND CAPTURE-TIME OF HCBD

The efficiency and the capture-time of each He-3 counter in the HCBD were measured using a Cf-252 neutron source (180 fissions per second) which was electro-deposited on a platinum plate. The plate was put in a stainless-steel chamber in front of a surface barrier solid-state-detector (SSD) to detect fission fragments. This Cf-252 chamber was placed at the sample position in the HCBD. Time intervals between the signals from the SSD and the He-3 counter were accumulated. The measured distribution directly corresponds to the capture-time distribution of each counter in the HCBD. The efficiency per neutron was derived by using the number distribution of emitted neutrons per fission/3/.

The efficiency and the capture-time of the HCBD were also calculated by Multi-KENO/4/ with the Multi Group Cross Section Library (MGCL). The Multi-KENO was slightly modified and coupled to a code to edit the detector characteristics.

Both the measured and the calculated efficiency for each counter position are presented in Table 1 where the numbers in the column of "Position Number" correspond to the numbers in Fig.5. A typical capture time distribution of the HCBD by the calculation is shown in Fig.6 with the measured distribution. The calculated results agree well with the experimental values.

Table 1 Efficiency of HCBD

Position Number	Efficiency ε (%) Measured (dε)	Calculated
1	2.896 (0.020)	2.89
2	2.301 (0.023)	2.25
3	1.405 (0.014)	1.38
4	1.150 (0.005)	1.11
5	0.631 (0.005)	0.61
6	0.589 (0.005)	0.55
Total	40.45 (0.13)	39.6

V FLUENCE MONITOR

The fluence at the sample position is one of the key quantities to derive the cross section. We developed a compact PRCT (proton recoil counter telescope), shown in Fig.7, which was small enough to be inserted into the HCBD and set at the sample position. The fluence at the sample position could be determined with the uncertainty of about 2% by using this PRCT.

VI CONCLUSION

In order to measure the (n,2n) cross section, we have developed a new method where the intervals of detected times of neutrons are measured. Using a detector consisted of 28 He-3 gas proportional counters and 480 polyethylene blocks, demonstration experiments were performed for various conditions. The expected distributions were observed in the experiments and the background is easily fitted by the least squares method. By this method, the number of (n,2n) events could be accurately obtained.

Further, since the calculated efficiency and capture-time agreed with the measured results for a Cf-252 source, the calculation method is verified to be applicable to estimate the characteristics of the HCBD for various energy spectra of (n,2n) neutrons from various nuclides.

Acknowledgment

The authors would like to thank Messrs. R.Sakamoto and M.Fujisawa for the accelerator operation. A part of this work was supported by Grand-in-Aid for Scientific Research Promotion from Ministry of Education, Science and Culture.

REFERENCE

1. J.Frehaut, Nucl. Instr. Meth., 135, 551(1976)
2. E.J.Dowdy, J.T.Caldwell and G.M.Worth, Nucl. Instr. Meth., 115, 573(1974)
3. R.R.Spencer, R.Gwin and R.Ingle, Nucl. Sci. Eng., 80, 603(1982)
4. Y.Naito, M.Yokota and K.Nakano, Multi-KENO: A Monte Carlo Code for Criticality Safety Analysis, JAERI-M-83-049, Japan Atomic Energy Research Institute(1983)

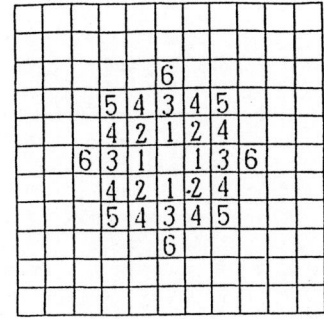

Fig.5. Counter position of Table 1

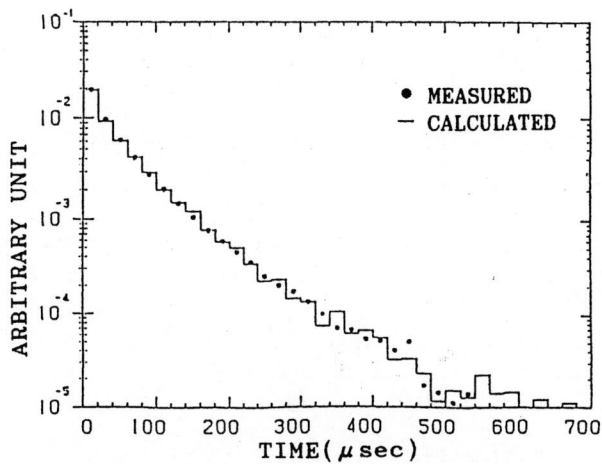

Fig.6. Capture time distributiom.

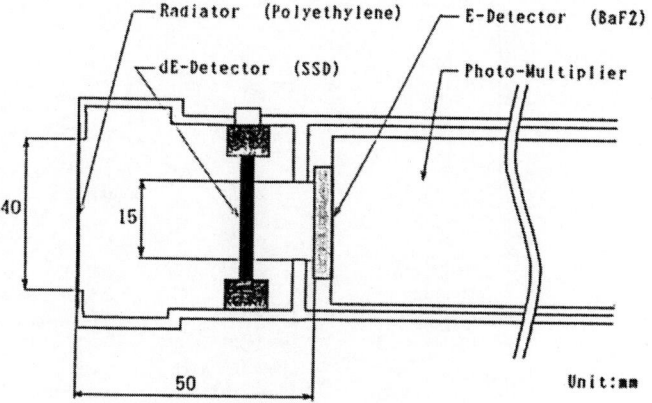

Fig.7. Schematic view of PRCT.

A POST-ACCELERATION BEAM CHOPPER FOR
4.5 MV DYNAMITRON PULSED NEUTRON GENERATOR

S.Matsuyama, M.Fujisawa, M.Baba, T.Iwasaki, S.Iwasaki,
R.Sakamoto, N.Hirakawa and K.Sugiyama

Department of Nuclear Engineering, Tohoku University,
Sendai 980, Japan

Abstract: A post-acceleration beam chopper has been installed for Tohoku University 4.5 MV Dynamitron pulsed neutron generator in order to improve the energy resolution in fast neutron time-of-flight experiments by reducing the pulse width. The post-chopper worked as expected and has been applied successfully for double-differential neutron emission cross section measurements; typical results are shown for lithium-6 and natural carbon.

(post-acceleration beam chopper, time-of-flight, neutron emission cross section)

Introduction

We have been conducting the double-differential neutron emission cross section (DDX) measurements using the time-of-flight (TOF) method with 4.5 MV Dynamitron accelerator as a pulsed neutron generator /1,2/. As is well known, the energy resolution in fast neutron TOF experiments is governed by the duration of beam pulses used to produce the neutrons as well as the time resolution of a neutron detector.

The terminal pulser of the Dynamitron, which consists of a lens-chopper and a klystron buncher assembly, can deliver 1-2 ns pulses in fwhm at 2,1,0.5,..MHz repetition rates. However, a "tail" component exists in both sides of the beam pulse and restricts overall energy resolution in the DDX measurements. Then, we have designed and installed a post-acceleration beam chopping system (PACS) synchronized with the accelerated beam pulses in order to eliminate the "tail" and even shorten the pulse width. PACS sweeps the accelerated beam pulses across a chopping slit and reduces the pulse duration by employing a set of deflector plates. For the design of PACS, a Monte-Carlo calculation was employed considering the beam emittance, optics and space-time correlation.

PACS proved to be very useful for the fast neutron TOF experiments. This paper describes PACS and the typical DDX results for lithium-6 and natural carbon.

Design Calculation

Initial design parameters for PACS were obtained by the simple equation by J.B.Marion and W.M.Good /3/. Then, we made detailed analyses of space-time behavior of chopped beam by using a Monte-Carlo calculation, since a single deflector chopper often degraded the beam quality /4/. In the calculation, we took into account the beam optics, divergence, space-time correlation and the bunching effects for realistic estimation of beam behavior. Parameters were determined from the performance tests of the Dynamitron and the geometric condition of the existing beam transport system, since PACS should be installed with minimal modification of the existing beam transport system.

The calculation showed that PACS would achieve desired beam duration (< 10 ns) without degrading the beam quality significantly.

The optimum design parameters of PACS were as follows; deflector spacing = 2.5 cm, deflector length = 50 cm, drift space = 2.5 m, slit opening = 5 mm and RF voltage = 10 kVp-p maximum at 8 MHz frequency /5/.

Construction of the Post-Chopper

The layout of PACS is shown in Fig. 1. The deflector is situated in upstream side of the switching magnet and the chopping slit is in downstream side of the image slit in +15 degree

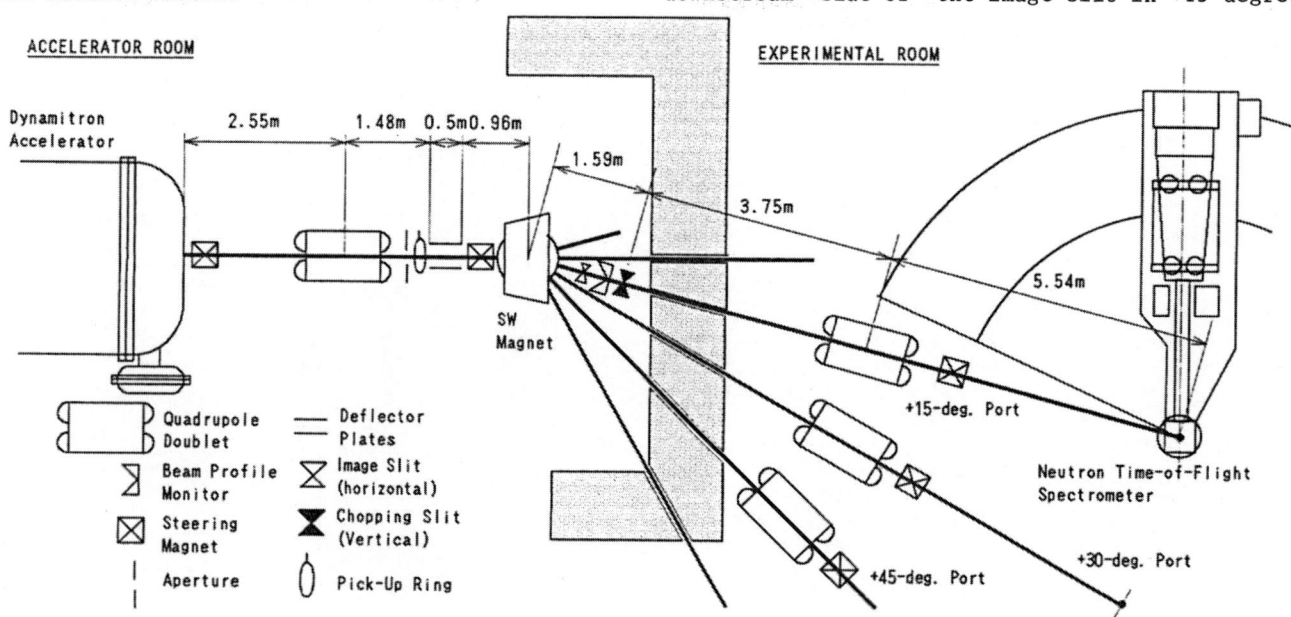

Fig. 1 Layout of PACS for Dynamitron Accelerator

port which led to a neutron TOF spectrometer. While it is preferable to place the chopping slit in upstream side of the switching magnet for common use in all beam lines, the space limitation between the accelerator and the switching magnet does not allow that installation. The distance between the deflector and the chopping slit is about 2.6 m. The deflector sweeps the beam vertically to avoid the interference with the energy stabilizing system. The beam is focused to less than 5 mm vertically on the chopping slit by a quadrupole doublet.

The electronic block diagram of PACS is shown in Fig. 2. The system is composed of three stages; 1) beam pick-up, frequency multiplier up to 2 MHz and phase shifter for synchronization with the accelerated beam pulses, 2) frequency quadrer for derivation of 8 MHz RF frequency and 3) high voltage resonator.

In order to synchronize PACS with the pulsed beam, we obtain the timing signal by frequency multiplication to 2 MHz and phase adjustment of the beam pick-off signal derived just before the deflector. The phase between the beam pulses and PACS is adjusted so as to maximize the target current. The phase delay showed satisfactory long term stability and needed only occasional readjustment. Therefore, no special phase stabilizing system is employed.

Resulting 2 MHz signal is then multiplied by a factor of four to derive 8 MHz RF for the deflector. The output amplitude in this stage is adjusted by a pin-diode attenuator after filtering. The frequency-, phase-, and amplitude-controlled signal is fed into a RF power-amplifier via a driver.

The high voltage RF is generated with a tank circuit (see Fig.2) driven by a commercial wide-band RF power-amplifier. This system has advantages of minimum number of high voltage items, simple tuning and easy maintenance. Optimum tuning point is found to minimize standing-wave-ratio (SWR) by adjusting the variable capacitors remotely from the control room. The inductor used in the tank circuit is cooled by two fans. The tank circuit and the deflector are wholly enclosed with an aluminum box to reduce RF noise into experimental devices.

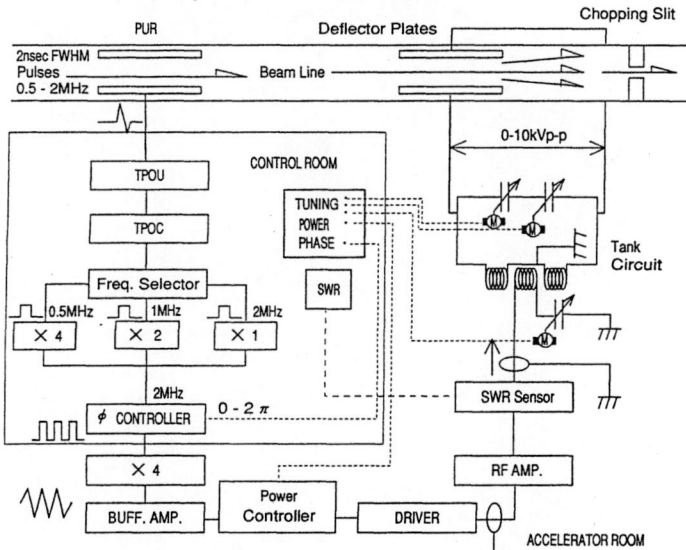

Fig. 2 Electronic block diagram of PACS

The performance of PACS has been tested by observing the TOF spectra of gamma-rays and neutrons. The measured gamma-ray TOF spectra

showed that PACS performed as expected by the simulation.

Figure 3(a) and 3(b) illustrate the measured neutron emission spectra for 14.1 MeV neutrons from polyethylene and nickel samples, respectively. These measurements were made at flight path length of 6.5 m using a NE213 liquid scintillator, 5" in dia. and 2" thick. As shown in Fig. 3(a), the separation between the peaks becomes clearer by operating PACS. The valleys between the peaks are reduced to about one tenth. In Fig. 3(b), the first levels of Ni-58 & 60 are separated satisfactorily from the ground state even at flight path length of 6.5 m. Then PACS proves to be very effective for studies of fast neutron scattering, especially for closely-spaced levels.

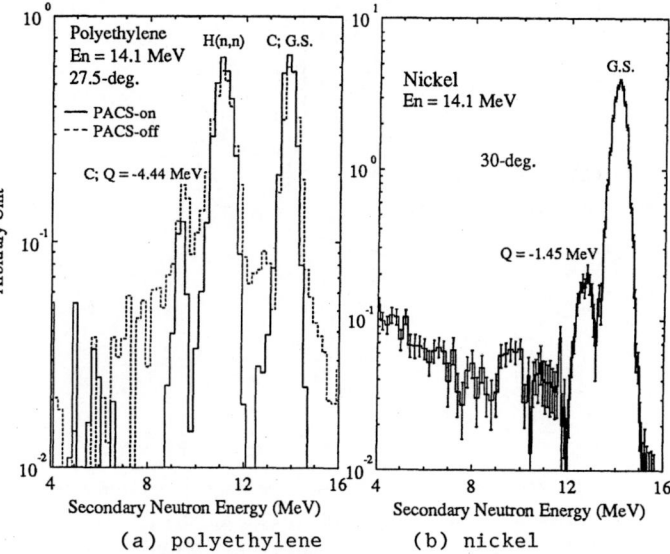

(a) polyethylene   (b) nickel

Fig. 3 Neutron emission spectra for 14.1 MeV incident neutrons measured using PACS; dashed line shows the results without operating PACS

Double-Differential Neutron Emission Cross Section of Lithium-6 and Carbon

PACS has been applied successfully for various DDX measurements; here typical results of lithium-6 and natural carbon are presented for 14.1 MeV incident neutrons.

The details of the experiments and data analysis have been described previously /1,2,6/. The neutron detector was a NE213 liquid scintillator, 14 cm in dia. and 10 cm thick, and placed about 6 m from the samples. Two separate neutron-gamma discriminators were used to get sufficient signal to background ratio over wide range of neutron energy. Samples were enriched lithium-6 (94.5 % Li-6 and 4.5 % Li-7) and natural carbon. By using PACS and an optimized "walk" of timing discriminator, we obtained markedly improved energy resolution. Neutron emission spectra were measured at 12 laboratory angles.

The raw data were corrected for the effects of sample-out background, parasitic and target scattered neutrons, finite sample size and impurity element Li-7 in the Li-6 sample /2,6/.

Lithium-6

Typical results of neutron emission spectra are shown in Fig. 4, compared with our previous data by Chiba /6/ and those derived from evaluated nuclear data /7,8/. The present results gener-

476

ally reproduce our previous data except that the discrete level cross sections become higher in forward angles owing to improved energy resolution /5/. Those results imply that the "tail" did not distort the previous data seriously. For derivation of partial cross sections, we analyzed the neutron emission spectra using least-squares method where the spectrum was assumed to consist of five discrete components and continuum parts due to the (n,2n) and the (n,n'd) reactions; the spectra of continuum neutrons were assumed to be evaporation for the (n,2n) and 3-body simultaneous break-up for the (n,n'd) reaction /6/. The experimental spectra were described very well by those assumptions /5/.

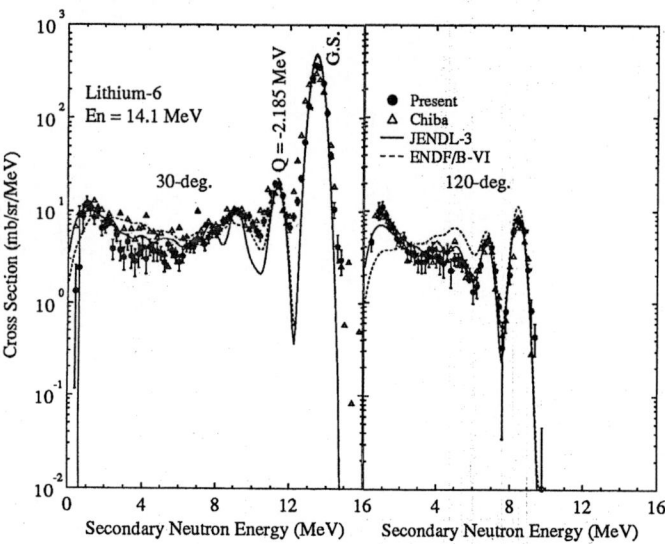

Fig. 4 Neutron emission spectra of lithium-6, compared with our previous data and evaluated ones

Carbon

The neutron emission spectra are illustrated in Fig. 5 together with our previous results /2/ and evaluated nuclear data /7,8/. While the present data are generally close to the previous ones for the ground and the third levels, the cross sections of the first and the second levels are markedly lower owing to greatly reduced beam pulse tail.

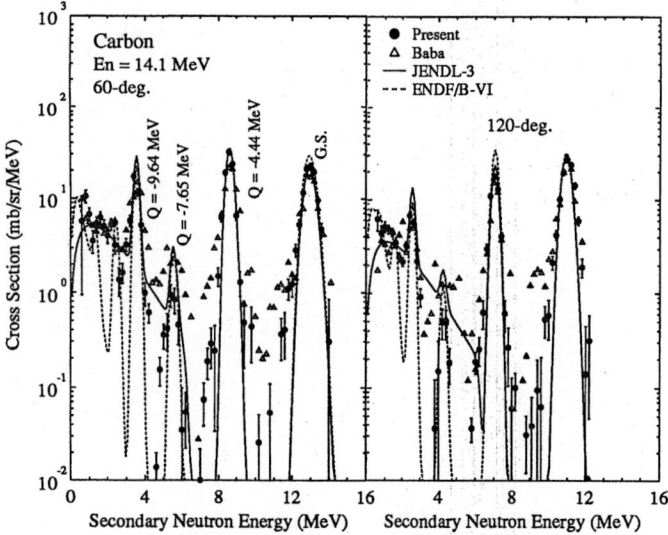

Fig. 5 Neutron emission spectra of carbon, compared with our previous data and evaluated ones

Summary

A post-acceleration beam chopper has been installed for the Tohoku University 4.5 MV Dynamitron accelerator and proved to be very effective for neutron TOF measurements. PACS has been applied for double-differential neutron emission cross section measurements. New results of neutron DDXs for lithium-6 and natural carbon were obtained with sufficient energy resolution.

Acknowledgement

The authors wish to thank Messrs T.Takahashi, K.Komatsu and T.Nagaya for there support in fabrication of PACS.

REFERENCES

1. M.Baba,M.Ono,N.Yabuta,T.Kikuchi,and N.Hirakawa: Radiation Effect,92-96,223(1986)
2. M.Baba,M.Ishikawa,T.Kikuchi,H.Wakabayashi and N.Hirakawa:Proc.Int.Conf. on Nuclear data for Science and Technology,May/June 1988, Mito,Japan,209(1988)
3. J.B.Marion & J.L.Fowler:Fast Neutron Physics Part I,Interscience Publishers,Inc.,New York (1960)
4. T.K.Fowler and W.M.Good:Nucl.Instr.Meth., 7,245(1960)
5. M.Baba,S.Matsuyama,M.Fujisawa,S.Iwasaki,and R.Sakamoto:Proceedings of the 1989 Seminar on Nuclear Data,JAERI-M 90-025, Japan Atomic Energy Research Institute(1990)
6. S.Chiba,M.Baba,H.Nakashima,M.Ono,N.Yabuta, S.Yukinobu and N.Hirakawa: J.Nucl.Sci.Technol.,22(10),771(1985)
7. JAERI Nuclear data center: "JENDL-3", JAERI 1319, Japan Atomic Energy Research Institute(1990)
8. BNL/National Nuclear Data Center: ENDF/B-VI(1990)

HIGH EFFICIENCY CHARGED-PARTICLE SPECTROMETER USING GRIDDED IONIZATION CHAMBER FOR FAST-NEUTRON INDUCED REACTIONS

M.Baba, N.Ito, S.Matsuyama and N.Hirakawa
Department of Nuclear Engineering, Tohoku University
Aoba Aramaki, Sendai 980, Japan

Abstract: A high efficiency charged particle spectrometer for fast neutron induced reactions has been developed using a gridded-ionization chamber taking advantage of its large solid angle and capability of energy-angle determination. It is characterized by high stopping-power and low background to be applicable for alpha-particles emitted by 15 MeV neutrons and protons for MeV incident neutrons. The spectrometer has been applied successfully for (n,alpha) and (n,p) reactions.

(Gridded-ionization chamber, charged particle emission, fast neutron induced reaction, energy-angular distribution, high efficiency, (n,alpha), (n,p) reactions)

Introduction

Energy and angular distribution data for secondary charged particles emitted by neutron induced reactions are of prime importance for assessment of radiation damage and nuclear heating in fusion and accelerator-based reactors. Various spectrometers have been developed for study of charged particle emission reactions /1-4/. However, experimental data of charged particle spectra are scanty because of experimental difficulties of low counting rate and serious background caused by limited sample thickness and low geometrical efficiency of spectrometers.

A gridded-ionization chamber (GIC) will be useful for secondary charged-particle measurements owing to its large geometrical efficiency close to 100 %, and capability of energy-angular determination and particle selection. GIC has been applied extensively for fission-fragment detection /5/; nevertheless, only very few applications have been done for light charged-particle detection /6/.

We have been developing GIC appropriate for studies of neutron induced charged particle emission reactions and applied for (n,alpha) and (n,p) reactions. This paper describes the feature and performance of GIC developed.

Gridded Ionization Chamber

Figure 1 illustrates the schematic view of GIC developed in the present study which is a twin chamber with a common cathode and sample. Cathode, grid and anode electrodes are stacked with three insulating rods (teflon) and are enclosed in a vessel made of stainless steel. High pressure gas of krypton or argon mixed with a few % carbon-dioxide is used as counting gas.

Provided that the charged-particles emitted from cathode stop between grid and cathode, anode signal Pa and cathode signal Pc are expressed by the following equations /5/:

$$Pa = Ca \cdot E(1-(\sigma \cdot \overline{x}/d) \cdot \cos\theta) \sim Ca \cdot E, \qquad (1)$$

$$Pc = Cc \cdot E(1-(\overline{x}/d) \cdot \cos\theta), \qquad (2)$$

where Ca and Cc are the gain factors of the electronics, E and θ are the energy and emission angle of particle, respectively, σ is the grid inefficiency, $\overline{x}$ is the distance between the cathode and the center-of-gravity of ionization distribution, d is the spacing between the cathode and grid (2.5 cm).

By combining the anode and cathode signals, we can derive the energy and emission angles of particles simultaneously, since grid-inefficiency could be less than a few %. In addition, we can select particles of concern by adjusting gas pressure since $\overline{X}$ differs largely depending on particle species.

For application of GIC to fast neutron induced reactions, achievement of high stopping power and background suppression are the key issues. Then, the present GIC has been designed with the following features.

a) High pressure operation: GIC withstands counting gas pressure up to 12 atom to stop alpha-particles emitted by 15 MeV neutrons; in this condition it can be applied to protons up to around 6.5 MeV using krypton. In addition, to ensure electric field required for charge collection under high pressure gas, high voltage can be applied to all electrodes up to around 5 kV.

b) Background suppression: All the electrodes are made of heavy elements, Ta or W, and chamber wall is lined with Pb plate. Neutron beam is defined within sample size using a collimator. Additional shield and ring electrode are placed to suppress backgrounds from the chamber windows and structures or those by particles with longer ranges.

c) Sample changer: Samples can be changed without opening GIC to keep gas purity.

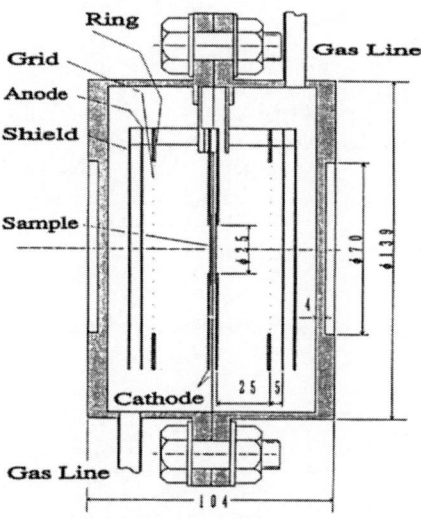

Fig.1 Schematic view of the present GIC.

478

A vacuum pumping system is equipped to evacuate GIC and gas feed line prior to gas filling for elimination of electronegative contaminants that deteriorate the performance of gas-filled counter. Similarly an oxygen-trap is employed in gas feed line to remove oxygen contaminant in the counting gas. In the course of test, we noted that reduction of out-gas from GIC and high gas purity are important for reliable performance of GIC, and that krypton is more susceptible to impurity while it is preferable because of much less background production.

Signals from anode and cathode are stored in two dimensional data array. The coincidence signal between anode and cathode gates the pulse height analyzer to reduce backgrounds and dead time. The shaping time of spectroscopy amplifier was set to 2 micro-second as a compromise between energy resolution and resolving time referring the test measurements mentioned below.

Backgrounds were measured by replacing samples with W foil. Flux normalization between foreground and background runs was made using a NE213 scintillator placed on the collimator axis.

Performance of GIC

Firstly, we tested the GIC performance to confirm proper operation under high pressure counting gas, using the Li-6(n,t) reaction for thermal neutrons and H(n,p) reaction for MeV neutrons whose energy and angular distributions are known well. Neutron beam was provided by Tohoku University 4.5 MV Dynamitron accelerator.

Figure 2 shows typical anode spectrum for the reactions; triton peak and proton edge are seen clearly. The intrinsic energy resolution of the present GIC was estimated to be better than 45 keV from the triton peak width and proton half-width; this value is good enough for the present purpose. The backgrounds in low energy region are due to alpha particles from oxygen and recoiled-gas atoms. Contribution of these backgrounds can be reduced significantly by processing two-parameter data.

Measurements of the triton peak and proton half-height as a function of grid-to-cathode field strength indicated that saturation of charge collection, i.e., complete charge collection, was achieved in the field strength higher than around 0.22kV/cm/(kg/cm**2) /7/.

Then, the angular distribution was derived for tritons from Li-6; the results reproduced expected isotropy of tritons by considering properly the sample geometry /7/. For the H(n,p) reaction, direct derivation of angular distribution was difficult because of strong energy-angular correlation; therefore, experimental anode and cathode spectra were compared with simulation calculation using the Monte Carlo technique considering energy loss of protons and grid inefficiency. Good agreement was observed except for lower energy region where large backgrounds distorted the spectra /7/.

Therefore, the present GIC performs properly even at high pressure over 10 atm by employing adequate electrode potential and gas purity.

Application of GIC and Discussion

The presently developed GIC has been applied for measurements of differential cross section of alpha-particles and protons from Ni, Cu and Al for incident neutrons of 4-6.5 MeV and 15 MeV. Typical examples are shown here.

Figure 3 shows two dimensional spectrum for Ni(n,p) reaction at 5.0 MeV neutron energy. The straight and curved lines indicate the 0- and 90-deg emission angles, respectively. Events from Ni(n,p) reaction distribute in the area between these two lines. The absolute cross section was determined relative to proton yields from the H(n,p) reaction except for 15 MeV data. The data were corrected for energy loss of charged particles within samples by an iteration method using the raw data as initial guess. Figure 4 shows double-differential Ni(n,p) cross section at En=5.1 MeV; peaks corresponding to low-lying levels of Co-58 are seen as well as evaporation-like continuum component. Energy- and angular-integrated cross sections are in good agreement with the established values based on many activation data /7/. This reveals the validity of the present experimental and data analyses methods.

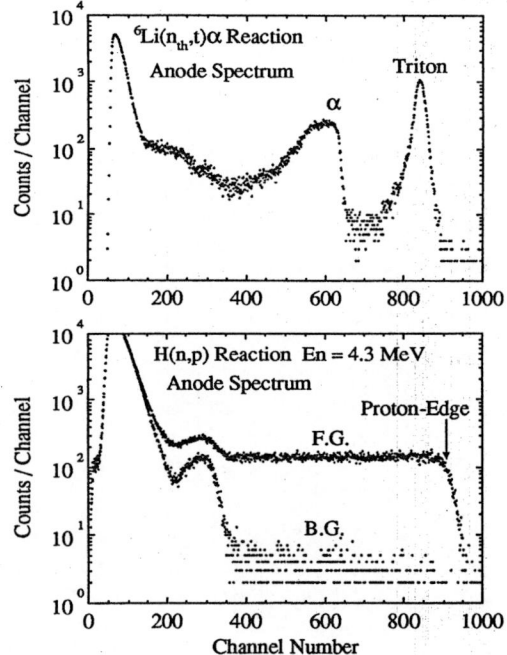

Fig.2 Anode spectra for Li-6(n,t) and H(n,p) reactions.

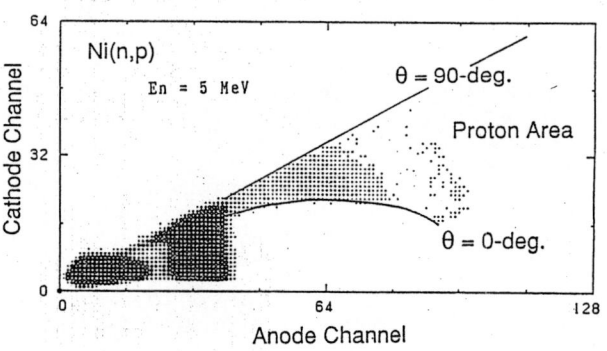

Fig.3 Two dimensional spectrum for Ni(n,p) reaction at 5 MeV neutron energy.

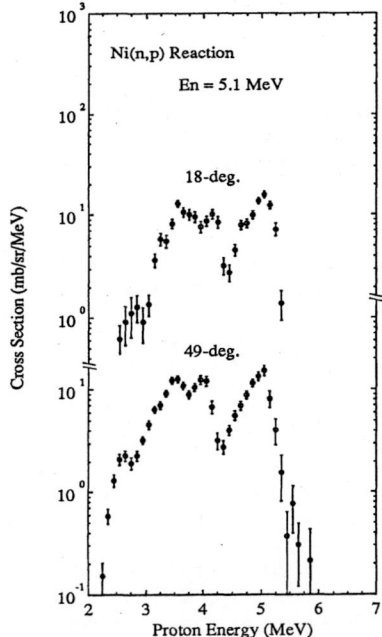

Fig.4 Proton emission spectra from Ni
for 5.1 MeV neutrons.

Figure 5 presents the alpha-particle spectra for natural nickel at 15 MeV incident neutron energy in comparison with the angle-integrated spectrum for Ni-58 by Grimes et al /2/. The present data at 60-deg are in good agreement with those by Grimes et al. The data at 26-deg show slightly harder spectrum and indicate anisotropic alpha emission. For MeV incident neutrons, average alpha-particle energies of the present data are in good agreement /7/ with those by Paulsen et al /3/. Figure 6 shows preliminary results for the Ni(n,alpha) reaction cross section; the present data are close to the evaluated values of ENDF/B-VI.

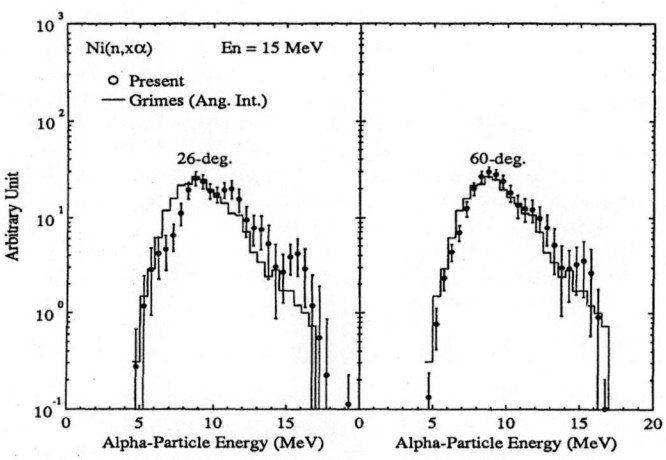

Fig.5 Alpha emission spectra from Ni
for 15 MeV neutrons.

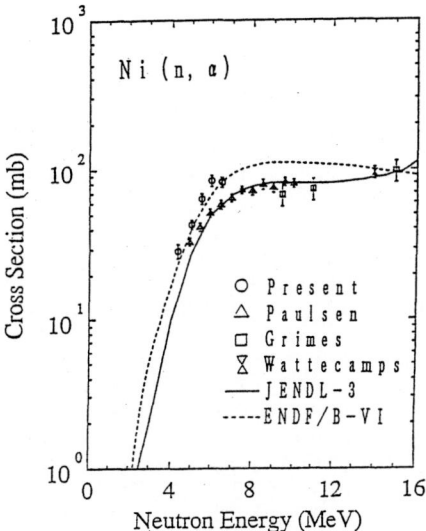

Fig.6 Ni(n,alpha) cross section.

Similarly, the present GIC was applied for (n,p) and (n,alpha) reactions on Cu and Al.

It will be noteworthy that, with the present GIC, data accumulation of only a few to several hours provides sufficient counting statistics to obtain the data shown above. Therefore, GIC is a useful means for studies of energy-angular distributions of secondary charged particles, especially of alpha-particles.

For further application of GIC, however, technical developments are required for reduction of backgrounds due to neutron interaction with counter gas since they mask low energy parts of emission spectrum significantly in the case of lighter elements like Al.

Acknowledgement

The present work was partly supported by Grant-in Aid of Scientific Research, Ministry of Education and Culture, and Japan Atomic Research Institute. The authors wish to thank Messrs. T.Iwasaki, R.Sakamoto and M.Fujisawa for their aids in accelerator operation and experiments.

REFERENCES:

1. M.Ahmad, S.L.Graham, S.M.Grimes, R,Longfellow, H.Satyanarayana and G.Randers-Pehrson; Nuc.Instr.Methods, 228 349(1985)
2. K.A.Alvar, H.H.Barschall, R.R.Borchers, S.M.Grimes and R.C.Haight; Nucl.Instr.Methods, 148 303(1987)
3. A.Paulsen, H.Liskien, F.Arnotte and R.Windera; Nucl.Sci.Engi.,78 377 (1981)
4. C.Derndorfer, R.Fisher, P.Hill, G.Stengl and H.Vonach; Nucl.Instr.Methods, 187 423(1981)
5. C.Butz-Jorgensen, H.H.Knitter, Ch.Strade, F.J,Hambsch and R.Vogt; Nucl.Instr.Methods, A258 209 (1987)
6. H.H.Knitter, C.Butz-Jorgensen, D.L.Smith and D.Marletta; Nucl.Sci.Eng., 83 229(1983)
7. N.Ito, M.Baba, S.Matsuyama and N.Hirakawa; JAERI-M 91-032 (JAERI 1991) p.302

INVESTIGATION OF PRESCISSION NEUTRON EMISSION IN ^{235}U(n_{th},f) USING A BACK-TO-BACK GRIDDED IONISATION CHAMBER

M.S.Samant, R.P.Anand, R.K.Choudhury, D.M.Nadkarni and S.S.Kapoor
Nuclear Physics Division
Bhabha Atomic Research Centre, Bombay-400 085

Abstract: Measurement of prompt neutron energy spectra and angular distributions from mass and kinetic energy selected fission fragments in ^{235}U(n_{th},f) have been carried out using a back-to-back gridded ionisation chamber. Neutron energy was determined by time of flight and the energy and angle of the fission fragments were measured using a back-to-back gridded ionisation chamber method developed earlier[1]. The data were analysed in a self consistent manner wherein the experimentally deduced emission spectra of neutrons from individual fragments were used to calculate theoretical angular distributions for the case of neutron emission from fully accelerated fragments. A comparison of these calculated neutron anisotropies with those measured have yielded information on the number of prescission neutrons emitted for different fragment kinetic energy bins and mass pairs. While the overall prescission contribution is found to be small ($\approx$ 5%), this fraction is shown to be very significant by looking at the cases of very large fragment kinetic energy where the contribution of neutron emission from moving fragments can be suppressed. The emission neutron spectra were also used to extract information on the fragment level density parameter and it was found that the nuclear temperatures can be explained for all fragment masses with the level density parameter $a \approx A/7$ except in the mass region of A = 128-144 amu where $a \approx A/10$ gives a closer agreement.

Introduction

Prompt neutron emission in thermal neutron induced fission of ^{235}U and spontaneous fission of ^{252}Cf has been studied in great detail[2] to understand the mechanism of emission of the prompt neutrons. While it is well established that fission neutrons are emitted primarily during deexcitation of the fully accelerated fragments, the question regarding a small fraction which may be emitted either during saddle to scission dynamics, or during the act of scission, or during the fragment acceleration phase is still not satisfactorily answered experimentally. More accurate information on this component (ν_{pre}) will not only lead to further refinement in the models to calculate fission neutron spectra, but will also provide a better insight on the fission process. Recent studies on prompt neutron emission in heavy ion induced fission reactions show that in many systems the number of prescission neutrons emitted is much higher than that expected from statistical models, thereby leading to the conclusion of prolonged saddle to scission transition times in the fission process[3]. This calls for a reexamination of the characterstics of neutron emission in the low energy fission process. However the analysis of recently measured anisotropy data[4,5] for the case of ^{252}Cf(sf) has lead to the conclusion that the yield of prescission component is less than 5%. The question of whether or not the prescission neutron component is present, is answered by comparison of the experimental neutron-fragment angular correlations with those calculated from kinematical considerations assuming neutron emission from fully accelerated fragments. The calculated neutron angular distribtion is, however, very sensitive to the emission spectrum of neutrons in the rest frame of fragments. In the earlier analysis, the emission spectra have been obtained from evaporation type calculations and any uncertainty in the calculation of the emission spectra would affect the conclusions on the prescission neutron emission. In the present work, the aim is to carry out a self consistent analysis by using the experimentally measured neutron multiplicities and neutron spectra in the rest frame of fragments as a function of fragment mass and kinetic energy for calculation of the neutron-fragment angular correlations to be able to reach model independent conclusions regarding prescission neutron emission for the case of ^{235}U(n_{th},f).

Also reported are results on fragment temperatures as a function of fragment mass deduced from the C.M. spectra of the prompt neutrons emitted from the fragments in ^{235}U(n_{th},f). The results are analysed with the statistical evaporation code ALICE-II using shell dependent level densities of the excited fission fragments. The level density parameter of the fragments is determined by comparing the measured T_{eff} with that obtained from the statistical cascade calculations which explicitly take into account multiple neutron emission.

Experimental Setup

The experiment was carried out at the CIRUS reactor at B.A.R.C.,Bombay. A back-to-back gridded ionisation chamber[6] was used in the first experiment to measure the energy and angle(θ) of both the fission fragments. The chamber consisted of a central cathode and two parallel plate ionisation chambers with frisch grids in a back-to-back geometry. The distance between the anode and the grid was 0.7 cm and between the cathode and grid was 3.0 cm. A ^{235}U source of 100 μg/cm^2 thickness on a thin VYNS backing was mounted in the centre of the cathode. The complete assembly was then housed in a brass chamber ,which was filled with P-10 gas at 1.1 atm pressure. The gas was then continuously purified by passing it over heated calcium filings. A 5cmx5cm NE213 liquid scintillation detector was used to detect neutrons. The detector was placed at a distance of 70 cm from the ^{235}U target along the field direction of the ion-chamber. The neutron energy was measured by time of flight (TOF), with the start and stop signals taken from the common cathode and the neutron detector respectively.Pulse shape discrimination (PSD) was used to differentiate neutrons from gammas. The pulse heights of the two collectors (V_{c1}, V_{c2}), grids (V_{a1}, V_{a2}), neutron TOF, PSD and pulse height signals of the neutron detector were recorded event by event on a magnetic tape. In a second set of experiments a pair of surface barrier detectors in 0° geometry was used to measure the neutron emission spectra.

A total of 3.6×10^6 coincident events were collected. The time resolution as seen from the prompt gamma peak is $\approx$ 2 ns. Singles binary events were also recorded for the calibration of the ion-chamber. The efficiency of the neutron detector as a function of neutron energy was determined by using a ^{252}Cf neutron source and was found to agree well with the results of Monte Carlo calculations.

Analysis and Results

The singles binary data from the ion-chamber was analysed to obtain the calibration of the grid pulses for event by event angle determination by using the expression[6],
$$V_a = V_c (1 - R*Cos\theta/D)$$
after incorporating energy loss correction for target thickness as described below. The ratio V_a/V_c is an index of the measure of the angle of the fragment with respect to electric field direction of the ion chamber. Fig.1 shows the collector pulse heights for most probable light and heavy fragments as a function of V_a/V_c, where the shift in the peaks due to energy loss in the target is clearly seen. The pulse heights of V_a and V_c were corrected for the energy loss by normalising the V_c pulse heights to the 0° values as shown by the dashed line in the Fig.1. For a given collector pulse height, V_a is expected to have a uniform distribution between the values corresponding to the emissions of fragments at 0° and 90° with respect to the field direction. For each mass and energy of the fragments, the grid distributions were plotted from which it was possible to accurately determine the grid pulse heights corresponding to 0° and 90° angles for any mass and energy of the fragment, thus providing calibration for determining the angles as a function of fragment mass and energy. By applying this correction procedure,the resolution in angle and energy showed significant improvements.For a fragment pair, the angle θ_L, was determined for the light mass from one chamber and the angle θ_H for the heavy mass was determined from the opposite chamber. Ideally the angle $\theta_L = \theta_H$ as the fragments move in opposite directions, the distribution of ($\theta_L - \theta_H$) should peak around 0°. The measured angular resolution is therefore given by (FWHM) of ($\theta_L - \theta_H$) for various fragment pairs, and was found to be around 4°- 5°.

The emission neutron spectra were determined event by event from the measured energy of neutrons after collimating the events to a cone of half angle of 18° with respect to the field direction. These spectra compared very well with those obtained from the 0° measurement. The emission neutron spectra were then fitted to the evaporation spectrum [7]
$$N(\eta) = Const. \; \eta^\lambda /T_{eff}.exp(-\eta / T_{eff})$$
where $T_{eff} = (11/12)T$,where T is the nucler temperature. Here $\lambda = 1$ for $\bar{\nu} \leq 1$ and $\lambda = 1/2$ for the cases when $\bar{\nu} > 1$ where $\bar{\nu}$ is the neutron multiplicity. Fig.2 shows the values of T_{eff} as a function of fragment mass. Statistical model evaporation cascade calculations show that after incorporating proper spread in the excitation energy of the fission fragments, the nuclear temperatures can be explained for all fragment masses with the level density parameter a $\approx$ A/7 except in the mass region of A= 128-144 amu where a $\approx$ A/10 gives a closer agreement to the experimental data as shown in Fig.2.

The neutron angular distributions for various fragment mass pairs were obtained by normalising the neutron coincident data with the singles data for various fragment total kinetic energy (TKE) windows. The neutron laboratory angular distributions for the typical mass pair of 96/140 are shown in Fig.3 for two different TKE bins. Fig.4 shows the measured laboratory anisotropies for two different mass pairs 96/140 and 98/138, as a function of TKE along with the results of the calculated anisotropies for these cases obtained by using a Monte Carlo procedure. The calculations were done by assuming that the neutrons are emitted from fully accelerated fragments using the measured ν_T (A,TKE) and T_{eff}(A,TKE). The contributions from the light and heavy fragments were added in proportion to the neutron multipllicities (ν_L and ν_H) for the different TKE bins. It is found that the measured and calculated anisotropies show significant deviations for cases of higher TKE values. This is so because for these cases the contribution of neutrons from individual fragments is suppressed and hence the prescission neutron contribution stands out. From the fraction of ν_{pre} obtained from the calculation it is possible to deduce the actual number of prescission neutrons. Fig.5 shows the measured fraction of ν_{pre} , the actual number of ν_{pre} and the ν_T for the two typical cases of 96/140 and 98/138. It appears that a considerabely large fraction of ν_{pre} at high TKE comes not only from the suppression of ν_T but also from the fact that the actual number of ν_{pre} also seems to increase with TKE.

This intresting information should be helpful in giving a better insight in the mechaninism of emission of scission neutrons.

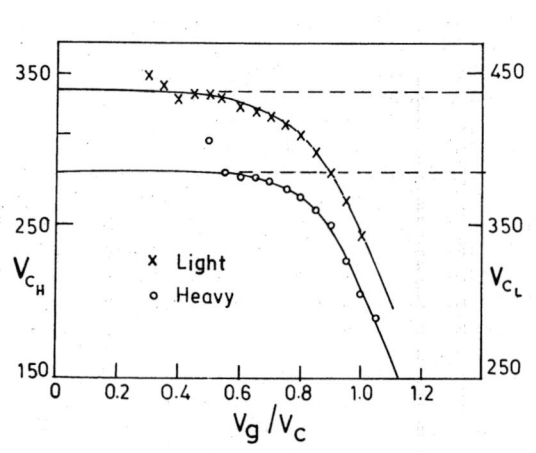

Fig.1 V_c Vs V_G/V_c

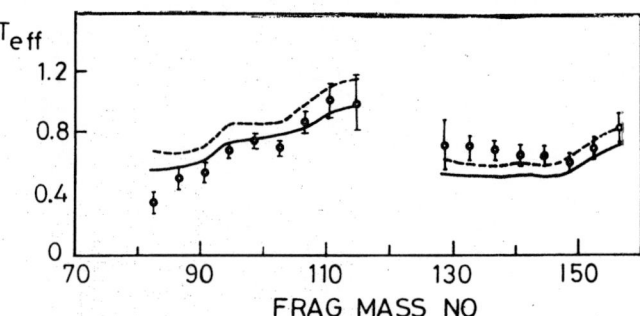

Fig.2 T_{eff} Vs Mass No.
---- calc. with a = A/10
———— calc. with a = A/7

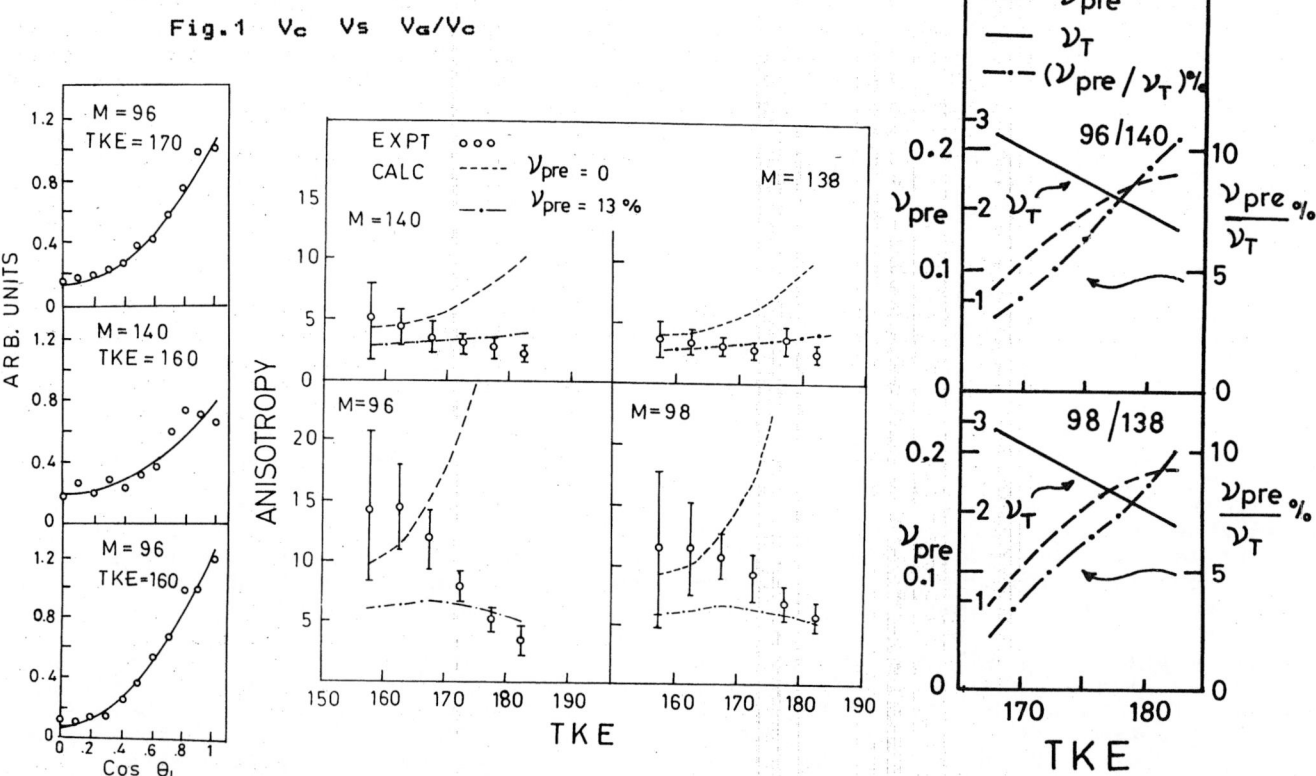

Fig.3 Ang. Distn.
       of Neutrons
o o o Expt.; ———— Fit

Fig.4 Anisotropy   Vs   TKE

Fig.5 $\nu_{pre}, \nu_T, \nu_{pre}/\nu_T$ Vs TKE

REFERENCES

(1) S.S.Kapoor, R.Ramanna and
    P.N.Rama Rao, Phys.Rev. 131 (1963)283

(2) H.Marten,D.Richter,D.Seeliger and
    W.Neubert, Proc.Int.Conf.on Nuclear
    Physics Nuclear   Fission,
    Gausing,1985; Zfk- 592 (1986)1;
    and references given therein.

(3) S.S.Kapoor, Proc. INDC Meeting on
    Physics of Neutron Emission in
    Fission, Mito City,   Japan, IAEA,
    June,1989,INDC-(NDS)-220,p221

(4) H.Marten and D.Seeliger, Proc. Int.
    Conf. on Nucl.Data for basic and
    applied science, Santa Fe, USA,
    May, 13-17, 1985

(5) C.Butz-Jorgensen and H.H.Knitter,
    Nucl. Phys. A490 (1988)307

(6) R.K.Choudhury,S.S.Kapoor,D.M.Nadkarni
    and P.N.Rama Rao,
    Nucl. Inst. & Meth. 164 (1979) 323

(7) K.J.Lecouter and D.W.Lang,
    Nucl. Phys. 13 (1959)32

A PULSED NEUTRON TOF FACILITY AT CHIANG MAI

T. Vilaithong, S. Singkarat, W. Pairsuwan, J.F. Kral,

D. Boonyawan, D. Suwannakachorn, S. Konklong,

P. Kanjanarat and G. G. Hoyes

Fast Neutron Research Facility, Department of Physics,

Faculty of Science, Chiang Mai University

Chiang Mai 50002, Thailand.

Abstract: A high-resolution neutron TOF facility has been developed and optimized for use at the Fast Neutron Research Facility (FNRF), Chiang Mai University. A pulsed neutron burst of 1.5 ns width was successfully obtained at 2-MHz maximum pulsing rate and average deuteron current of about 30 μA. Liquid scintillator 25.0 cm diameter by 10.0 cm thick is located at an extended fligt path of 10 m inside a well-shielded tunnel. A new-collimating and shielding system along the flight path has been constructed to reduce some inscattered backgrounds. A fixed detector station and a sample rotation technique provides the capability of measuring scattered neutrons over an angular range of 20 to 160 degrees. The overall timing resolution of the system is about 3.3 ns corresponding to an energy-resolution of 485-keV at 14-MeV neutron energy.

Typical results of elastic and inelastic scattering of 14.1-MeV neutrons incident upon carbon and tantalum are presented and discussed.

(High-resolution neutron spectrometer, pulsing system, time-of-flight, long flight path, natural carbon and tantalum, neutron-induced reactions).

Introduction

The Fast Neutron Research Facility (FNRF) at Chiang Mai University was established in 1982 with support from the International Atomic Energy Agency (IAEA). The facility operates a continuous beam of 150 kV at a 800 μA electrostatic accelerator producing 14-MeV neutrons for fast neutron activation analysis and study of neutron-induced reactions. An associated alpha-particle TOF spectrometer was successfully established /1/ with moderate energy resolution (1.2 MeV at 14 MeV neutron energy) and a well-behaved time-independent background. However, measurements with this method usually involve long hours of data accumulation, thereby requiring a high quality deuteron beam and good stability of associated electronics.

Since 1984 we have initiated a pulsed neutron facility at the FNRF by modifying an existing continuous beam generator to incorporate beam chopping and bunching devices. The design of the nanosecond pulsed beam line and related electronics system was adopted from the OKTAVIAN facility at the Osaka University in Japan/2-3/. The project is now fully operational and provides us with a good quality pulsed neutron beam and a higher neutron intensity. In addtion, the extended neutron flight path of 10 meters improves energy resolution to the order of a few hundred keV.

A preliminary test of the pulsed neutron facility has been performed with a neutron scattering experiment for carbon and tantalum. Double differential cross sections (DDX), emphasizing the energy range of 1 to 8-MeV, are presented and compared with existing data from other sources/4,5/.

Pulsed Neutron Production

Neutrons are produced from an AID J25 electrostatic accelerator by the $T(d,n)^4He$ reaction. Continuous deuterium ions (D_1^+) from a radio frequency plasma ion source are accelerated to 140-kV potential supplied by 9-kHz switching-frequency Cockcroft-Walton type d.c. high-voltage power supply. The deuteron beam which comprises 75 % atomic deuterons (D_1^+) and 25 % molecular deuterons (D_2^+) is analysed by an analysing magnet. The D_1^+ ions are bent 90° to a pulsed beam channel while the D_2^+ ions are excluded by momentum discrimination.

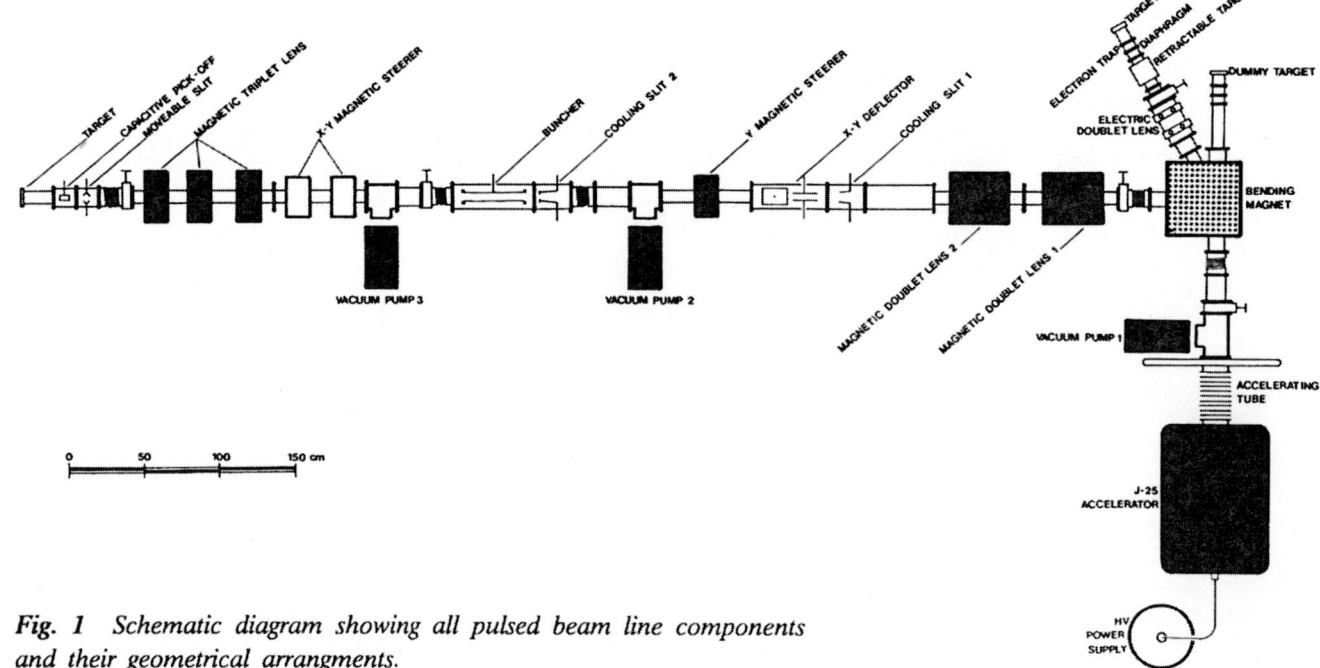

Fig. 1 *Schematic diagram showing all pulsed beam line components and their geometrical arrangments.*

Beyond the analysing magnet, the beam is transported through a series of collimating slits and quadrupole focussing magnets. The beam is chopped by a double-plate deflection system and then bunched to pulses with widths of 1.5 to 2.0 ns at the neutron production target by a double-gap klystron buncher. To ensure a minimum loss in transporting the beam to the target, series of beam transport calculations /6/ were carried out to predict and estimate beam sizes and beam divergences at various crucial beam components. Beam sizes both in horizontal and vertical axes were monitored by means of a crossed-wire beam profile monitor /7/. The schematic diagram shown in Fig. 1 is a layout of the beam line components from the ion source to the target.

A burst of deuterium ions incident on a tritium target produces a short pulse of neutrons from the D-T reaction. At a beam current of about 20 μA on target we can attain a neutron pulse of about 1.5 ns width at 2-MHz repetition rate. Figure 2 shows neutron pulse widths before and after bunching. The ratio of the number of neutrons at the peak to the background is approximately 2,000.

Experimental Facility

For the present high-resolution neutron time-of-flight measurement, a BC-501 liquid scintillator of diameter 25.0 cm and thickness 10.0 cm is used. This main scintillator is coupled directly to a RCA 8854 photo-multiplier tube. The PMT base is a modified version of the one suggested by Randers-Pehrson, et al. /8/. The neutron detector is located at an extented flight path of 10 m inside a well-shielded underground tunnel. Collimating and shielding systems were constructed from various absorbing materials to minimize unwanted in-scattered backgrounds. The schematic diagram for the shielding and collimating systems are shown in Fig.3. A sample rotation technique provides the capability of measuring scattered neutrons over an angular range of 20 to 160 degrees.

A data acquisition system is controlled by a MicroVAX II computer through a multiparameter-buffer system (MBS) unit. Each reaction event detected by the main detector is recorded sequentially in list mode on disk. Each event contains (1) the pulse-height, (2) the time-of-flight, and (3) the n-gamma pulse shape discrimination data. Our offline analysis software allows dynamic selections for each correlated parameter in contrast to a conventional hardware-resolution routine.

Time-of-Flight Measurements

The time-of-flight technique is employed for neutrons energy measurements. By utilizing a fast beam pick-off signal of 1-MHz rate with 2-3 ns rise time as a time reference signal, the neutron flight

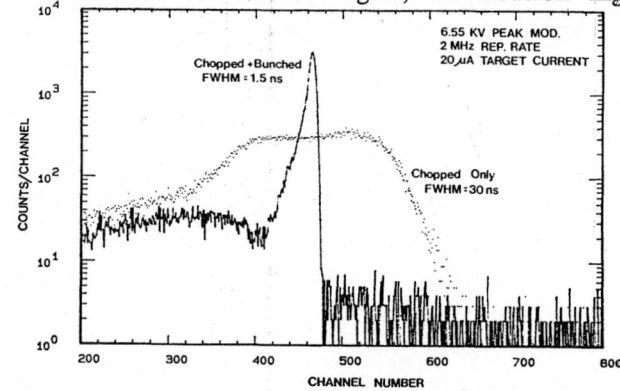

Fig. 2 *TOF spectrum from a neutron monitor showing beam pulse-widths before and after bunching.*

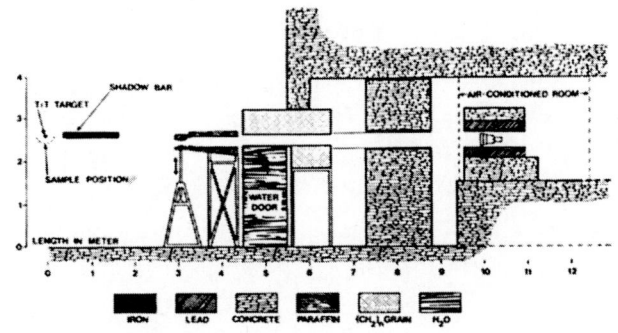

Fig. 3 *Geometrical arrangement of the time-of-flight spectrometer.*

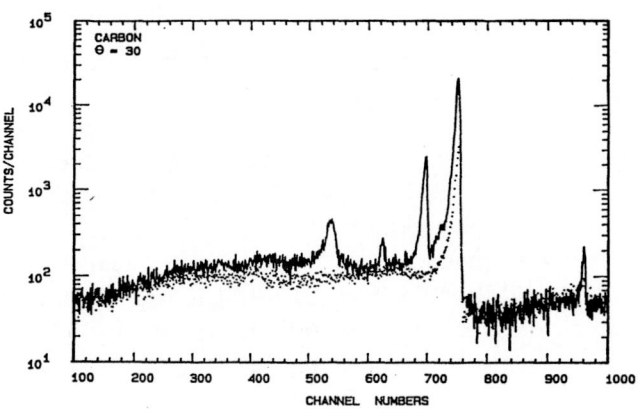

Fig. 4 *A TOF spectrum with background normalized to the flux monitor values for scattering of 14.1 MeV neutrons onto carbon.*

time (TOF) is measured with great accuracy. The n-gamma pulse shape discrimination (PSD) is also employed to reduce the gamma-ray background. The neutron pulse height information (PHA) is acquired directly from a dynode of the PMT base for a proper threshold selection for offline analysis. The TOF, PHA and PSD signals are synchronized and a coincidence signal is produced to trigger a master gate input of the MBS unit to record the three parameters corresponding to each neutron event.

Typical TOF spectrum for carbon is shown in Fig. 4. The time-dependent spectrum recorded without a sample, which is used for a background subtraction,

is also shown.

Conversion from TOF spectra to DDX spectra are accomplished by comparison with the n-p scattering cross sections obtained by using a polyethylene scattering sample. The neutron detection efficiency is calculated with the Monte Carlo program of Cecil et al./9/. Multiple scattering corrections are carried out with the MUSCC3 code/10/. The systematic error resulting from the detector

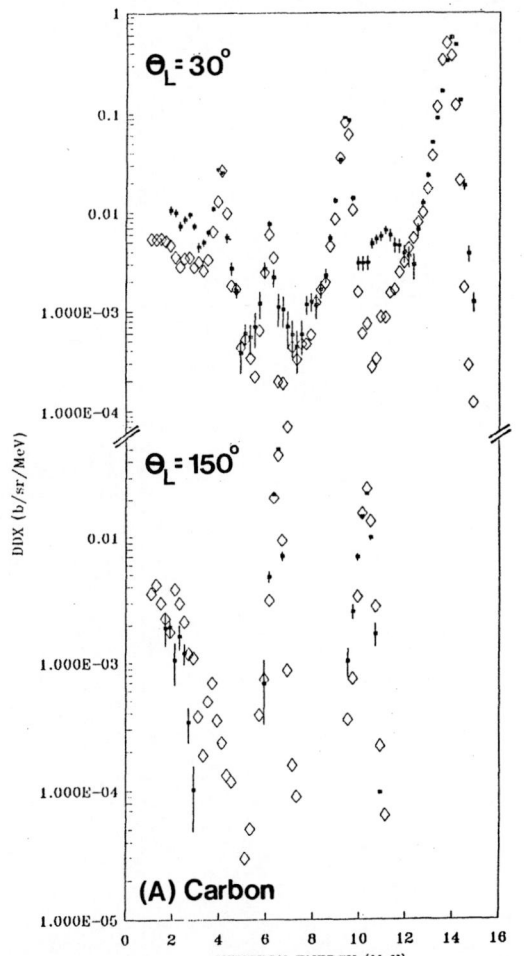

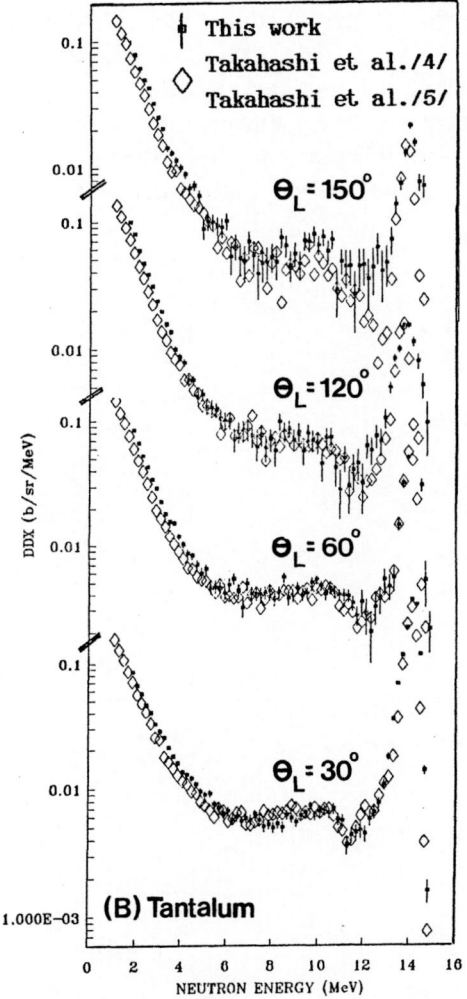

Fig. 5 *DDX spectra for (a) carbon and (b) tantalum. Multiple scattering corrections are applied only for carbon. The error bars respresent statistical errors.*

efficiencies, the background subtraction, and the solid-angle calculation is approximation 15 %. The multiple scattering correction raises this error to about 20 %. The statistical uncertainties for a 0.2-MeV bin vary from 5 percent to 15 percent for the DDX spectra at forward angle and increases up to 30 percent for backward angle spectra.

We employed a 25-cm diameter and 10-cm thick liquid scintillator at 10 m flight path to acquire an overall timing resolution of about 3.3 ns corresponding to an energy resolution of 485-keV at 14-MeV neutron energy.

Results and Discussion

Double differential cross section spectra for carbon and tantalum at various scattering angles are shown in Fig. 5. We compare our measurements with data of Takahashi et al./4,5/. Our preliminary results are in reasonable agreement with results from other sources. This would indicate that our pulsed neutron TOF facility works satisfactorily.

Acknowledgements

The authors wish to thank professor A. Takahashi for his assistance throughout the whole project. We would like to thank Dr. Alan George who worked on an initial beam line assembly. This work was supported in part by the International Atomic Energy Agency, Vienna, the International Program in the Physical Science (IPPS), Uppsala University, and the National Research Council of Thailand.

References

1. S. Singkarat, N. Chirapatpimol, D. Suwannakachorn, W. Pairsuwan, G.G. Hoyes and T. Vilaithong: Proc. Nucl. Data for Sci. and Tech., Mito, Japan, May 30-June 3, 1988, Edited by S. Igarasi.

2. K. Sumita, A. Takahashi, T. Iida, J. Yamamoto, S. Imoto and K. Matsuda: Proc. 12th SOFT, Julich, Germany, Permagon Press , 675, 1982.

3. A. Takahashi, J. Yamamoto, K. Oshima, M. Ueda, M. Fukazawa, Y. Yanagi, J. Miyaguchi and K. Sumita: J. Nucl. Sci. Tech. 21, 577, (1984).

4. A. Takahashi, H. Sugimoto and E. Ichimura: Proc. Conf. Fast Neutron Physics, Dubrovnik, Yugoslavia, May 26-31, 1986.

5. A. Takahashi, Y. Sasaki and H. Sugimoto: CRP report, NDC(JPN)-118/L, Osaka University, Osaka, Japan, 1989.

6. K.L. Brown, D.C. Carey, Ch. Iselin and F. Rothacker: CERN 80-04, CERN, Geneva, Switzerland, 1980.

7. L. Godfrey, G.G. Hoyes and W. Pairsuwan: Nucl. Instr. and Meth. B51, 294 (1990).

8. G. Randers-Pehrson, R.W. Finlay and D.E. Carter: Nucl. Instr. and Meth. 215, 433 (1983).

9. R.A. Cecil, B.D. Anderson and R. Madey: Nucl. Instr. and Meth. 161, 439 (1979).

10. E. Ichimura and A. Takahashi: OKTAVIAN report, A-87-02, Osaka University, (1987).

LINEAR ACCELERATOR OF ALL-UNION SCIENTIFIC RESEARCH INSTITUTE OF EXPERIMENTAL PHYSICS FOR NEUTRON SPECTROMETRY

Yu.A. Khokhlov, N.V. Zav'yalov, I.A. Ivanin, V.I. In'kov, N.P. Sitnikov, A.V. Tel'nov
All-Union Scientific Research Institute of Experimental Physics

607200, Arzamas, Nizhnij Novgorod region, USSR

Abstract: The electron linear accelerator intended for neutron spectrometry with a nominal energy of about 55 MeV at currents of 10 A in electron pulse repetition rate of 2400 Hz started in 1985 is described. The special features of linac construction, the measuring complex assembly, the results of spectrum measurements and neutron yield of uranium target are given. The possibilities of accelerator development for increasing the pulse current and pulse length shortening are considered.

(Linear accelerator, time-of-flight method, electron spectrum, neutron target, neutron spectrum, two-resonator injector, magnet compression system)

Introduction

An electron linear accelerator together with a time-of-flight method may be used as effective tool of neutron spectrometry in an energy range from few hundreds keV to few tens MeV. It was demonstrated in the end of 60-th and early 70-th [1-3]. Usage of a linear accelerator for research of the nuclear reactions under a neutron action was preferable to other sources of neutrons, because of comparatively low cost and simultaneous measurement of reaction characteristics in a wide range of neutron energy at good resolution.

The present work is dedicated to description of electron linear accelerator of All-Union Scientific Research Institute of Experimental Physics (LU-50), which was intended for the neutron spectrometry with time-of-flight method and was put into operation in 1985 with the help of Moscow Radio Engineering Institute [4].

Electron Linear Accelerator LU-50

The arrangement of linac, control and computer centre is shown in fig. 1. The beam guide system provides accelerated electron transport into two target rooms: the first is disposed on axis of the linac for photoneutron generation and the second - at the beam divergence angle of 43° for activation analysis and experiments with an electron pulse of picosecond length [5]. The neutrons from the water cooled uranium target get through a shield and vacuum neutron guides into the experiment rooms, which are stationed on flight-paths from 9 to 55 m.

The neutron detectors mounted in the experiment rooms behind the shaping collimators are connected with computer centre by signal cables.

Accelerated electron beam of the linac has the following parameters :

Electron energy (MeV)	55,
Peak current (A)	10,
Pulse length (ns)	10,
Pulse repetition rate (Hz)	2400.

The electron linear accelerator was created on the basis of implementing energy stored in an iris waveguide. Application of this acceleration principle and also high pulse repetition rate determined main characteristics of accelerating sections and RF-generator.

The iris waveguide consists of two accelerating sections with a variable geometry, each having a waveguide transformer on the input and a terminating load on the output. The main parameters of these accelerating sections are :

length (cm)	379.5,
accelerating wave length (cm)	16.5,
working mode	$2\pi/3$,
electric field at the onset (kV/cm)	100,
electric field at the end (kV/cm)	95,
RF power filling-up time (ns)	970.

An accelerating resonator with a diode gun is used as an injector in the linac LU-50. Used cylindrical resonator is excited in the E_{010} mode. An electric strength in the cavity is 300 kV/cm at a RF gap of 2.6 cm.

The electron diode gun delivers a pulse current about 100 A at a pulse length of 10 ns and at an injection voltage of 40...50 kV in a cathode-anode gap of 4.5 mm. The plane and spherical impregnated aluminato-barium cathodes with a 16 mm diameter are used.

A power supply of the accelerating sections

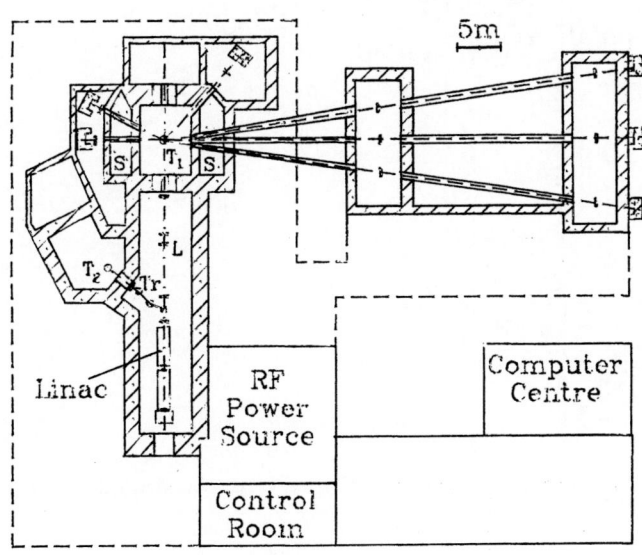

Fig. 1. The arrangement of linac LU-50

and the cavity is provided from a RF generator, which consists of a RF magnetron exciter and magnetron amplifier with the output pulse power of 30 MW and the average power of 100 kW, which is equally divided into two guides. One of them feeds the first accelerating section, a part of power (about 10 %) supplies the cavity via a circulator and a variable phase-shifter. The other RF generator output is connected with the second section via a circulator. The first and the second sections have 13.5 and 15 MW, respectively, at a pulse length of 1 μs and at maximum average power of 50 kW. A RF generator modulator of the LU-50 consists of five equal sections, which operate on a two-stage scheme with a partial discharge of a reservoir capacitance. Modulating pulses from the output cascade of one of the sections, come onto an exciter cathode, pulses from other cascades go to a RF power amplifier cathode. The main performances of this modulator are :

Peak voltage (kV)   50,
Peak current (A)   500,
Pulse length (μs)   1.2,
Pulse repetition rate (Hz) 2400.

The energy spectrum of accelerated electrons of LU-50 is presented in fig. 2. It has a marked flattop, which is typical for a stored energy operation linacs.

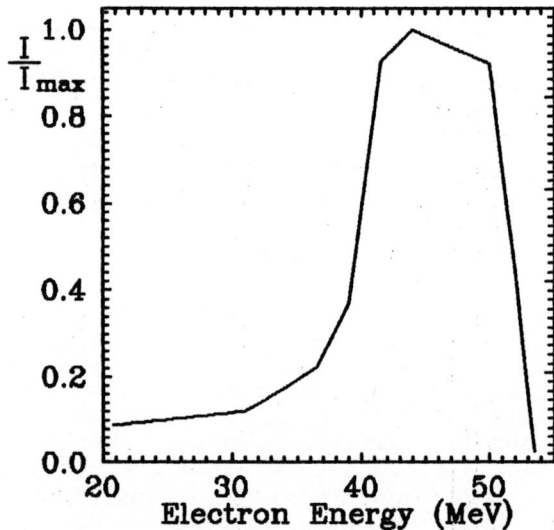

Fig. 2. Electron energy spectrum of LU-50.

LU-50 Neutron Target

With the switched off accelerator LU-50 the neutron target (fig. 3) is put in a special box in the underground part of a target room. During experiments the target is remotely installed in an operating position. The neutron target RM-100 [6] was designed in I.V. Kurchatov Institute of Atomic Energy. Its top and bottom parts are made of tantalum and uranium-238 dioxide, respectively. The tantalum rods and uranous oxide are positioned in cylindrical fuel elements 100 mm long and 5 mm in diameter. Cooling water from a closed circuit passes through a space between the fuel elements. The total weight of target uranium-238 is 3.5 kg. Energy distribution and neutron yield from the target is measured with parallel plate avalanche detector of fission fragments with an uranium-235 layer by a time-of-flight method. In fig. 4 the obtained energy

spectrum is plotted. The average energy of the spectrum is 1.69 MeV, the average neutron yield is $2 \cdot 10^{13}$ neutrons per second at a maximum beam power.

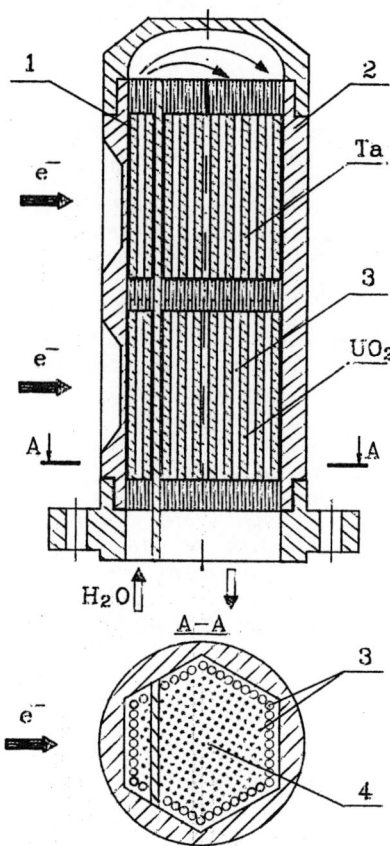

Fig. 3. Target RM-100 design:
1- target-radiator, 2- frame,
3- fuel element, 4- target-converter

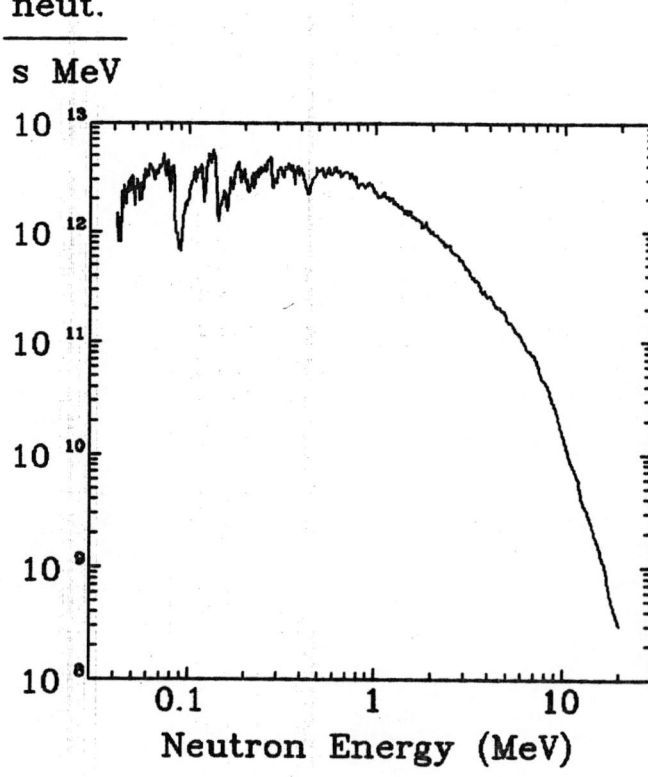

Fig. 4. Neutron spectrum of uranium target

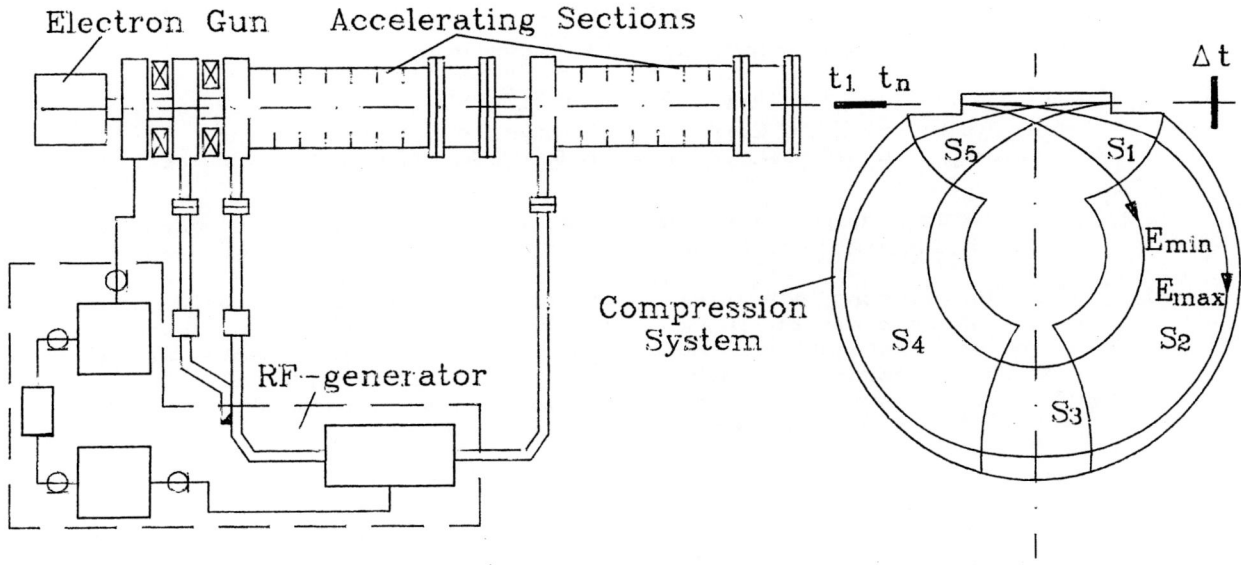

Fig. 5. Accelerator LU-50 with magnet compression system

LU-50 Development Trends

The injection system used in the linac LU-50 based on a cylindrical cavity limits a potential of beam current rise. Therefore, investigations of an optimum injection system are carried out, allowing to achieve a pulse electron current from 15 to 20 A with improved phase-energy characteristics of electron bunches and with preservation of a pulse length, a pulse repetition rate and an accelerated electron energy. It was demonstrated, that application of a klystron-type buncher base on two cavities, analogous to the used cavity, may produce a pulse current of about 20 A. An average energy divergence between bunches in every pulse will be from 2.0 to 2.5 MeV depending on a bunch number and an accelerated charge in it, that will allow us to calculate and to design a magnetic system for all electron bunches compression in a pulse (as in ref. [7]).

A compression system for LU-50 is designed as a five-sector magnet with circular faces between sectors and homogeneous magnetic fields. It will be posted in a vertical position, that is defined by the existing arrangement of the linac room. After a new buncher have been made, we suppose to obtain the following beam performance :

Electron energy (MeV)	55,
Peak current (A)	150,
Pulse length (ns)	0.5,
Pulse repetition rate (Hz)	2400.

That powerfully rises an efficiency of the time-of-flight neutron experiments. A scheme of the accelerator, the two-cavity injector and the compression system is shown in fig. 5.

Conclusion

At the present time the linac LU-50 is widely used for investigation of nuclear reactions, which are induced by fast neutrons. The results of some experiments are presented in the materials of this conference. Last years the linac operation was rather stable. The experience stored allowed to start the developments improving accelerator operation, which considerably increase the efficiency of neutron time-of-flight investigations.

References

1. M.V. Savin, Yu.A. Khokhlov, Yu.I. Il'in and Yu.V. Shain: Preebory i Tekhnika Experimenta 6, 27, (1969)
2. M.V. Savin, Yu.A. Khokhlov, Yu.S. Zamyatnin and I.N. Paramonova, in Proc. Int. Conf. on Nucl. Data for Reactors, June 1970, Helsinki, 2, 157, Vienna, IAEA (1970)
3. M.V. Savin, Yu.A. Khokhlov, I.N. Paramonova, V.A. Chirkin, V.N. Ludin and N.N. Zalyalov: Atomnaya Energiya 49(4), 236, (1980)
4. G.P. Antropov, N.A. Arkhangelov, B.V. Bekhtev, V.A. Boyko, R.M. Voronkov, Yu.Ya. Glazunov, V.A. Danilichev, V.I. Demchishin, V.V.Dobrokhotov, N.V. Zav'yalov, Yu.I. Il'in, V.I. In'kov, V.N. Kapalin, V.L. Noga, V.A. Pavlov, Yu.S. Pavlov, V.I. Panin, V.S. Saushkin, N.P. Sitnikov, V.G. Smekalin, Yu.V.Sokolov, Yu.A. Khokhlov and A.M. Shmygov: Voprosy atomnoy nauki i tekhniki /Tekhnika phizicheskogo experimenta/ 2(23), 3, (1985)
5. N.V. Zav'yalov, I.A. Ivanin, V.I. In'kov, N.P. Sitnikov, A.V. Tel'nov, Yu.A. Khokhlov, A.V. Galkin and A.V. Grigorenko: Preebory i Tekhnika Experimenta 3, 56, (1990)
6. G.P. Kiknadze, I.I. Kryuchkov and Yu.V. Chushkin, The Heat Exchange Crisis under Self-organizing of Whirlwind Vortex Structures in a Flux of Coolant, IAE-4841/3, I.V. Kurchatov Institute of Atomic Energy (1989)
7. D. Tronc, J.M. Salome and K.H. Bockhoff: Nucl. Instr. and Meth. in Phys. Research A228, 217, (1985)

490

COMPLEX OF INTENSIVE PULSE NEUTRON SOURCES OF MOSCOW MESON FACTORY FOR TIME-OF-FLIGHT EXPERIMENTS

M.I. Grachev, S.K. Esin, N.V. Kolmychkov, V.M. Lobashev,
V.A. Matveev, V.G. Miroshnichenko, S.F. Sidorkin, Yu.Ya. Stavissky

Institute for Nuclear Research of Academy of Sciences of USSR
60-th October Anniversary Prospect 7a, 117312 Moscow.

Abstract: The authors describe an assembly of neutron sources for physical research based on proton beams of the Moscow meson factory (MMF). The expected neutron intensity in the 4π solid angle in the regime of the resonance selector is $3 \cdot 10^{16}$ n/s (25 ns, 400 Hz) while the peak thermal neutron flux on the moderator surface is about $5 \cdot 10^{15}$ n/cm^{2}s and the pulse duration is about 50 ms.

(pulse neutron source, beam dump, meson factory)

The construction of the meson factory based on the high-current proton accelerator is in the final stage in the Institute for Nuclear Research of Academy of Sciences. The linear accelerator makes it possible to produce proton and H$^-$ beams with energy of 600 MeV and average current 0.5 to 1.0 mA. The beam time structure represents 100 ms pulses with the repetition rate of 100 Hz /1/.

On the base of the meson factory the complex of neutron sources with the compressing storage ring with the charge-exchange injection is being constructed. A one-turn ejection of protons from the storage ring onto the neutron target enables the neutron pulses with the duration 5 to 300 ns and average intensity up to $6 \cdot 10^{16}$ n/s in 4π to be produced (Fig. 1) /2,3/.

The neutron sources complex is constructed on the base of two proton beams that can be used simultaneously (Fig. 2). The first beam with the average current up to 1 mA is directed from the accelerator straight onto the neutron target of the quasistationary source of thermal and cold neutrons. The second one with the average current up to 0.5 mA first is compressed in the storage ring and then is directed onto the neutron target of pulse neutron source or into the lead slowing down time spectrometer /4/. The neutron complex also includes a number of proton beam stops located within the experimental hall. They have vertical

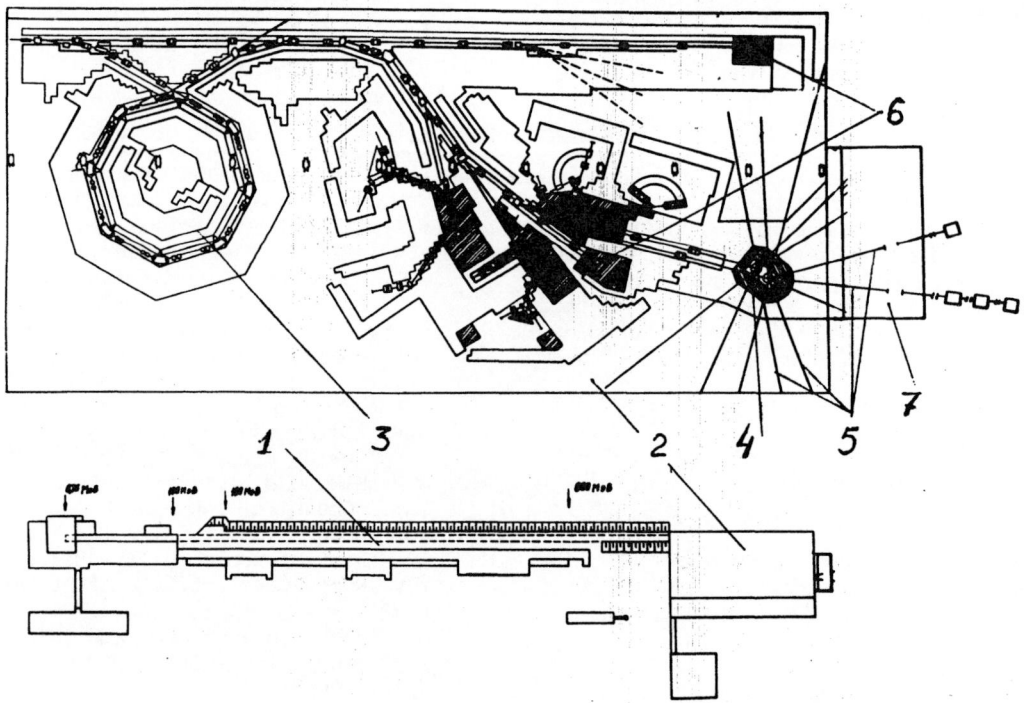

Fig.1. Moscow meson factory: 1-proton linac; 2-experimental hall; 3-proton storage ring; 4-complex of neutron sources; 5-neutron guides; 6-beam stops; 7-neutron laboratory.

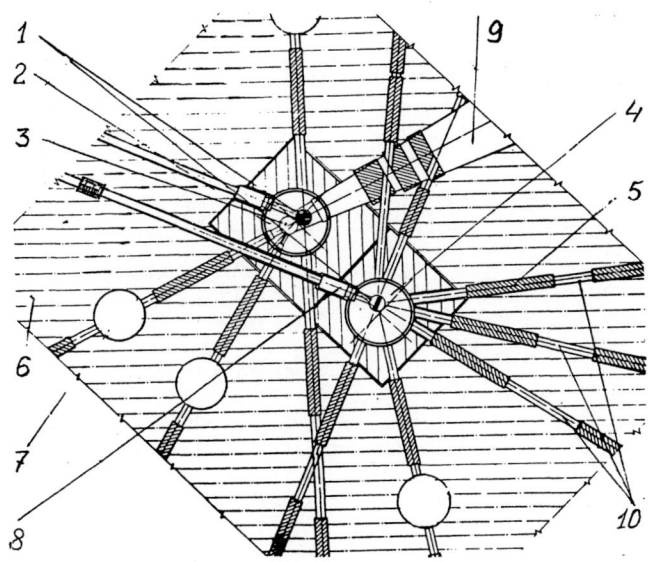

Fig.2. Complex of neutron sources: 1-neutron targets; 2-heavy water tank; 3-liquid deuterium moderator; 4-vacuum tank; 5-shut-off gate; 6-heavy concrete; 7-vertical wells 1 m in diameter; 8-remote-controlled vacuum seal; 9-wide aperture canal; 10-resonance neutrons canals.

canals intended for radiation physics experiments and for the neutron-deficient nuclides production /5/.

Pulse and quasistationary neutron sources are placed into the cylindrical boxes of radiation shield made of steel and heavy concrete. In the radiation shield there is a system of neutron guides 20 cm in diameter supplied with the guillotine shut-off gates. To install the additional experimental equipment neutron guides are supplied with the vertical wells about 1 m in diameter located in 3.5-4 m from the sources (Fig.2).

The pulse neutron source consists of the water-cooled target in the form of the assembly of natural uranium rods enclosed in stainless steel cans. On the first stage of the operation it is supposed to use the tungsten plates. The moderators are adjoined to upper and lower surfaces of the target. To reduce the background due to scattered neutrons and prevent the pulse lengthening the target and the moderators are placed into the vacuum tank (Fig. 3). The thin (~2cm) lower light-water moderator radiates to the neutron guides that go out of the experimental hall and are up to 500 m in length. These neutron guides are supplied with the intermediate pavilions. On top of the target there is a replaceable moderator intended for experiments with thermal and cold neutrons. It is supposed to use the light-water moderator of 4 cm thickness as well as the liquid hydrogen one. These moderators are supplied with the Be-reflectors of 15 cm thickness. Corresponding neutron guides go to the experimental hall.

The thermal neutrons peak flux density averaged across the water moderator radiating surface of area of 400 cm^2 is about $2\cdot10^{15}$ n/cm^{2}s at the pulse duration of 50 μs. With reflectors installed the peak flux density will increase up to $5\cdot10^{15}$ n/cm^{2}s on the radiating surface of 100 cm^2. Table 1 presents techni- in various operation conditions in comparison with a number of facilities for research in the resonance neutron energy range.

In case of using the tungsten target the average intensity and flux density of thermal and cold neutrons will be lower by a factor of 1.7 than that of the natural uranium target.

The time-of-flight spectrometer based on the pulse neutron source will allow one to develop neutron data measurements for advanced nuclear technologies (fast reactors, fusion and electrical breeding facilities) that require extensive optimizing computations of the following values: neutron balance, heat release, new isotopes production due to fission, (n,γ), (n,p), (n,α), $(n,2n)$, $(n,\lambda n)$-reactions, radiation shield dimensions, radiation damages rate. The high intensity of the pulse neutron

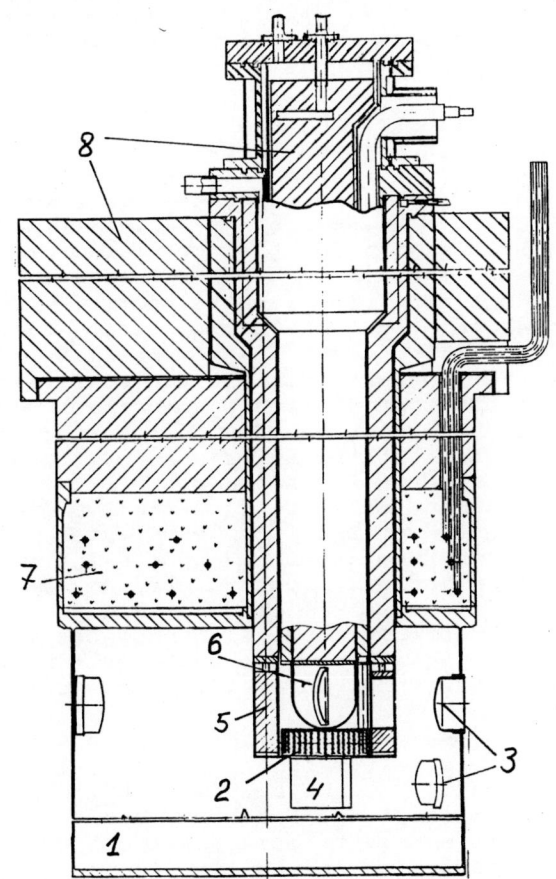

Fig.3. Pulse neutron source: 1 - vacuum 2 - assembly of ^{238}U rods (W-plates); 3-thin aluminum windows; 4-lower water moderator of 2 cm thickness; 5-reflector upper moderator; 6-water or liquid-hydrogen upper moderator; 7 - upper part of thermal shield; 8-removable steel shield.

Table 1

accelerator	average neutron intensity in the 4π solid angle (n/s)	pulse duration (ns)	repetition rate (Hz)	remarks
MMF	$3 \cdot 10^{16}$	25	400	p, 600 MeV
MMF	$7.2 \cdot 10^{15}$	50	100	–
MMF	$4.8 \cdot 10^{16}$	320	100	–
ORELA	$3 \cdot 10^{14}$	24	1500	e, 160 MeV
ISIS	$4 \cdot 10^{16}$	200	50	p, 800 MeV

source enables one to increase the accuracy of neutron data of key importance and to determine neutron cross-sections that cannot be measured up to now because of the insufficient intensity of existing facilities (fission products, transuranium isotopes). At the same time it is planned to investigate γ-spectra due to the neutron capture on isolated resonances and its correlations, to study in details (n,p) and (n,α)-reactions, oriented nuclei fission, under-threshold fission, P-parity non-conservation in neutron reactions on isolated resonances and, correspondingly, the mechanism of nuclear amplification of these effects. The corresponding experiments transmission for the design neutron intensity will be 10-100 times higher than that of the best existing facilities. The experimental program also includes condensed matter physics research using the cold and thermal neutrons beams with flight bases within the experimental hall.

The lead slowing down time spectrometer enables one to obtain the highest sensitivity for rare processes studying in the neutron energy range from thermal energies to 50 keV. It is expected that the sensitivity for the proton beam current of 0.15-0.2 mA, which is limited by cooling conditions, will be $\sim 10^8$ times higher than that of the electron linac based spectrometer. The possibility arises to measure, for instance, fission cross-sections of far transuranium elements using samples of 10^{-12} g.

The quasistationary source of thermal and cold neutrons includes the water-cooled target, consisting of uranium rods or tungsten plates, the heavy-water moderator in the aluminum tank ~ 1.5 m in diameter and the liquid-deuterium moderator of the volume of 20 l (Fig.4). The liquid-deuterium moderator radiates into the wide-aperture canal intended for the neutron-antineutron oscillations detection.

At the proton current of 0.5 mA the calculated values of the thermal neutron flux density at the bottom of the experimental canal and the flux density of cold neutrons with energy ≤ 5 meV on the cold moderator radiating surface are about $3 \cdot 10^{19}$ and $1 \cdot 10^{19}$ n/cm^{2}s, respectively. The efficient neutron temperature is about 40 K, the radiating surface of the cold moderator has the area of ~ 1000 cm^2.

With the removal of the liquid deuterium moderator the through tangential canal is formed. This canal can be used

to carry out the direct measuring of the neutron-neutron scattering length and other experiments in intensive thermal neutrons fields. The evaluation shows, that, using the scattering chamber cooled by liquid helium (vacuum of 10^{-10} Torr), the accuracy of $\sim 3\%$ in (n,n)-scattering length value can be obtained in a 2 months run /6/. After completing these programs it is planned to reequip the quasistationary source into the pulsed one.

Proton beam stops are located behind the meson targets. The beam stop core is the assembly of tungsten plates. The vertical canal which is 6 cm in diameter and is shifted to the first wall side gives the access to the beam stop

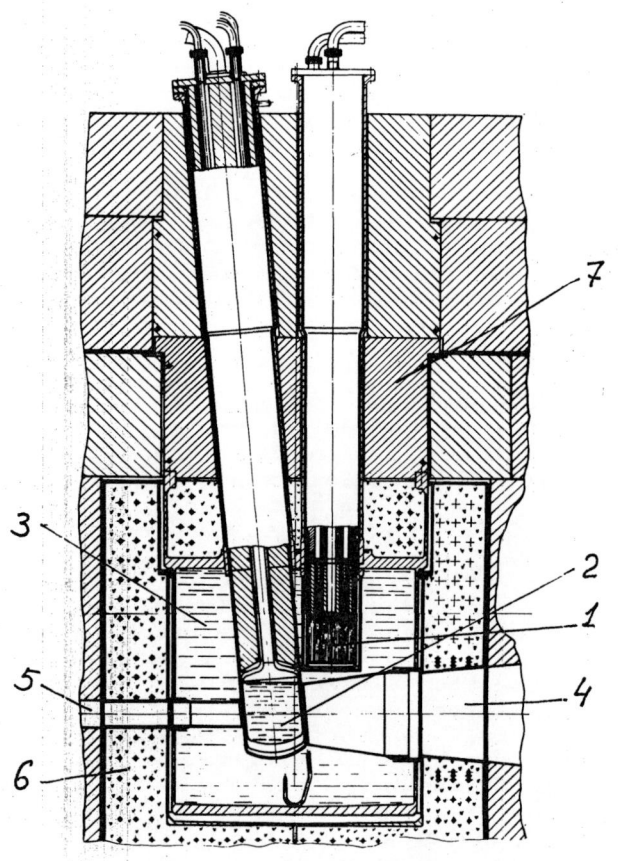

Fig. 4. Quasistationary source of thermal and cold neutrons: 1-neutron target; 2-liquid deuterium moderator; 3-heavy water moderator; 4-canals for n-ñ experiment; 5-extension of wide aperture canal; 6-thermal shield; 7-removable steel plug.

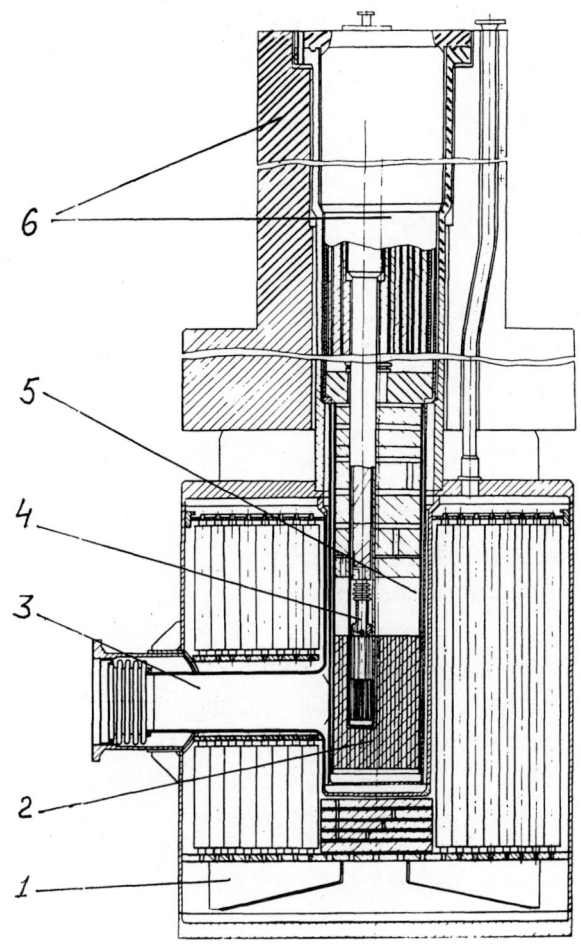

core. A sample being irradiated can be placed at the canal bottom or hanged up to the shielding plug. The average proton energy in the canal is about 450 MeV. The average fast neutrons flux density inside the beam stop is about $2 \cdot 10^{14}$ n/cm^{2}s for the beam current of 0.5 mA, while the ratio of the displacements number to the (n,ω)-reactions number, which determines the radiation effect on the medium, is close to that of the neutron spectrum on the first wall of fusion facilities /5/.

References.

1. V.M. Lobashev, A.N. Tavkhelidze, Vestnik Academii Nauk SSSR, 21 (1983) 115.
2. Yu.Ya. Stavissky, High-current proton accelerators of medium energy as neutron sources for physical research. In: Proc. of Int. Sem. on Intermediate Energy Physics, Moscow, 1989.
3. M.I. Gratchev, The Status of a Construction of an Experimental Complex of Moscow Meson Factory. In: Proc. of Int. Sem. on Intermediate Energy Physics, 1989.
4. V.N. Aseev, K.D. Ilieva, M.V. Kazarnovsky et al., Neutron Spectrometer based on time of neutron slowing-down in lead on the basis of MMF proton beam. In: Proc. of Int. Sem. on Intermediate Energy Physics, Moscow, 1989.
5. S.E. Boriskin et al., Inst. Nucl. Res. Preprint, P-0278, Moscow, 1983.
6. M.V. Kazarnovsky et al., Measurement of Neutron-Neutron Scattering Length at Moscow Meson Factory (Proposal of Exp.). Proc. of Int. Sem. Moscow, 1989.

Fig. 4. Proton beam dump: 1-thermal shield; 2-tangsten plates; 3-proton canal; 4-canal for sample irradiation; 5-ampoule of proton beam dump; 6-steel shield.

DOUBLE DIFFERENTIAL NEUTRON EMISSION CROSS SECTION OF (p,n) REACTION ON SILVER ISOTOPES USING 9, 11 AND 13 MeV PROTONS

S.M. Saleem and M. Ahmad

Nuclear Physics Division
Pakistan Institute of Nuclear Science & Technology
Nilore- Islamabad, Pakistan

Tang Hongqing, Zhou Zuying, Qi Bujia and Shen Guanren

China Institute of Atomic Energy
P.O. Box 275(46), Beijing 102413
Peoples Republic of China

Abstract: Neutron emission spectra from (p,n) reactions on ^{107}Ag and ^{109}Ag in the incident energy range from 9 to 13 MeV were measured by means of time-of-flight technique at the HI-13 tandem Van de Graaff accelerator of Institute of Atomic Energy, Beijing, China. The neutron emission spectra from (p,n) reaction were measured by TOF spectrometer with three liquid scintillation detectors in the angular range from 20 to 160 degree in steps of 10 degree. The emission spectra were converted into double differential neutron emission cross sections as a function of neutron energy in the center of mass system with the help of a computer code. The spectra were fitted using evaporation model of Weisskopf-Ewing. The comparison between experiment and theory is also presented.

(double differential neutron emission cross section, (p,n) reaction, Ag isotopes)

Introduction

The level density and its parameter are very important not only in nuclear structure but also in the statistical calculation of nuclear reaction. The information on level density obtained directly from experiment is so far limited to low excitation energy region and a small region above the neutron binding energy. To expand nuclear level density information to a wider excitation energy range, a method of extracting nuclear level density information is to study emission particle spectra from compound nucleus reactions. Vonach discussed the method in detail [1]. We measured the neutron emission spectra from (p,n) reactions on ^{107}Ag and ^{109}Ag in the incident energy range from 9 to 13 MeV. The emission spectra so obtained were analyzed using evaporation model of Weisskopf-Ewing.

Experimental set up

The experiment was performed around HI-13 tandem Van de Graaff accelerator of IAE Beijing having a terminal voltage of 13.4 MeV [2,3]. The beam current was measured with the Faraday Cup [4]. The time resolution of the spectrometer is about 1.2 ns. A schematic diagram of the tandem accelerator is given in Fig. 1. The target assembly used was designed in such a way that all the four targets were in a straight line on a single platform. In that arrangement one target was used for alignment with the help of quartz crystal. The other was blank for background purpose and the rest two were for isotopic ^{107}Ag and ^{109}Ag targets. The thickness of

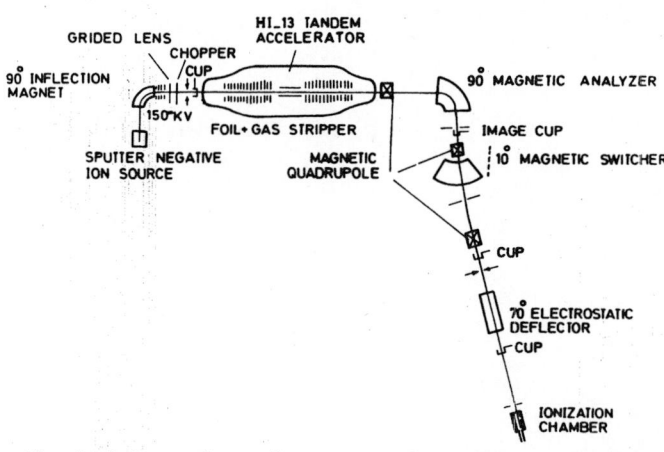

Fig. 1 Schematic diagram of 13MV tandem accelerator.

the samples was 5.109 and 5.011 mg/cm^2 respectively. With this arrangement, the targets could be changed from one to another position without breaking the vacuum. The spectrometer is described in detail in Ref. [5]. The experimental arrangement is shown in Fig. 2. The pulsed proton beam with a repetition rate of 2 MHz and beam width of about 1 ns was focused on to the target. The beam pick off signal was taken out right in the front of the target chamber. The cylindrical target chamber is made of aluminum having a diameter of 44 cm and 45 cm in height with thin wall of 3 mm. The target is so positioned that the proton beam impinges the target surface perpendicularly.

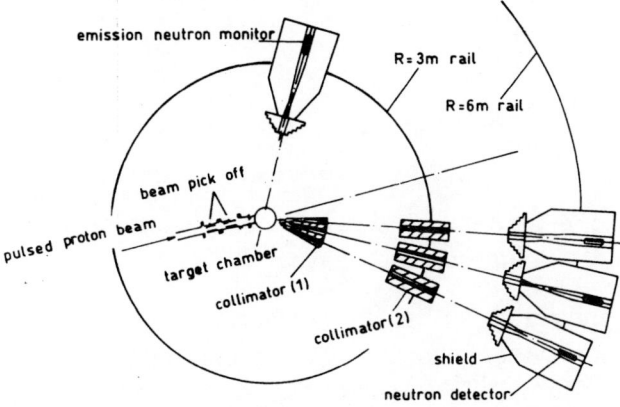

Fig. 2 Three detector fast neutron TOF spectrometer at the HI-13 tandem accelerator.

The neutron emitted from (p,n) reactions were detected by a three detector TOF spectrometer. The neutron detectors being used at present are made in China from ST-451 liquid scintillator having the same properties as that of NE-213 liquid scintillator from Nuclear Enterprise Ltd England and are of size 10.5 cm in diameter and 5.0 cm in length, coupled with properly shielded XP-2041 photomultiplier tubes. In front of the shielding there were collimators and a shadow bar. The whole system could be rotated on a rail and could be moved forward and backward freely. The angular distribution could be varied from 20° to 160° with respect to the incident beam. The accuracy of angle setting was better than 0.1 degree. The neutron flight path was 5.21 m, 6.55 m and 5.17 m respectively for detectors 1, 2 and 3 for this experiment. The detector bias was set at about 0.85 MeV. To reduce gamma-ray back ground, a n-gamma discriminator module (Canberra 2160 A) was used.

Measurement and Results

The efficiency of the neutron detector was determined by measuring n-p scattering cross section by means of an associated particle from fast neutron TOF spectrometer and Monte Carlo calculations. With neutron energy below 11 MeV, the experimental curve was used. Above 11 MeV, because the scattering angle of neutron was small and the energy of recoil proton was lower, calculated values were incorporated.

For each sample, the emission neutron TOF spectra were measured in the angular range from 20 degree to 160 degrees in steps of 10 degree. The

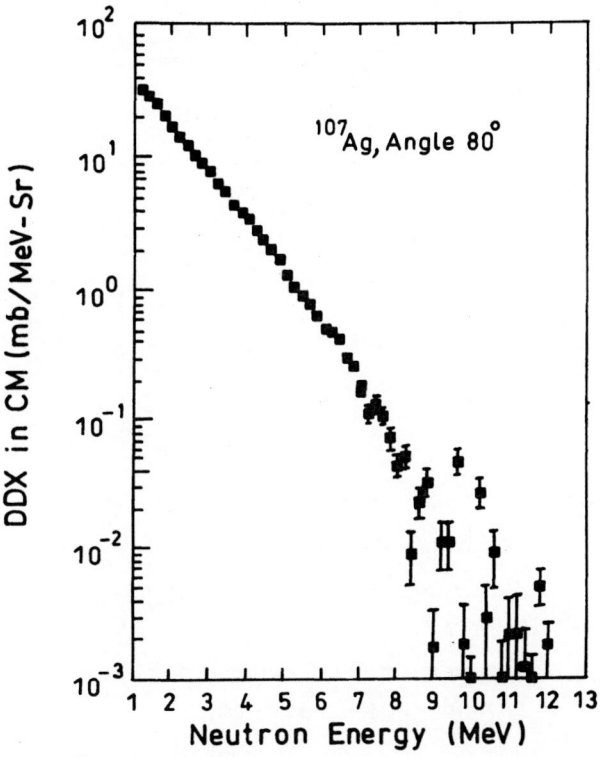

Fig. 3 Double differential neutron emission cross section of ^{107}Ag(p,n)^{107}Cd reaction at 80°.

background subtracted TOF spectra were converted into the double differential neutron emission cross sections. Fig. 3 shows the double differential neutron emission cross section of ^{107}Ag(p,n)^{107}Cd reaction at 80 degrees. Because of the wall of the target

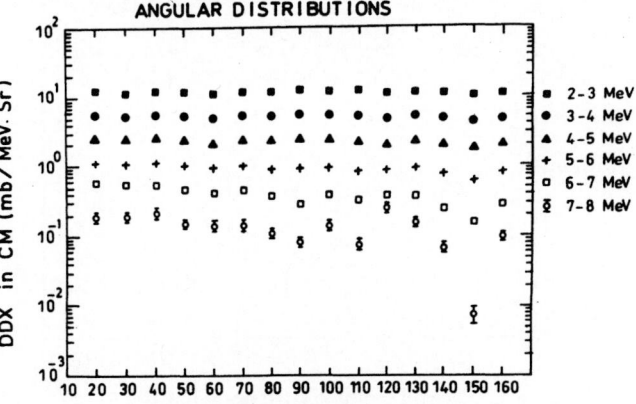

Fig. 4 Angular distribution of neutron emission from ^{107}Ag(p,n)^{107}Cd reaction.

chamber, the target frame and the samples themselves influence the emission neutrons for which the corrections were made in above conversions. The angular distribution of the emission neutron from the ^{107}Ag(p,n)^{107}Cd reaction is shown in Fig. 4. Using Legendre polynomials to fit the double differential neutron emission cross sections, we get the energy

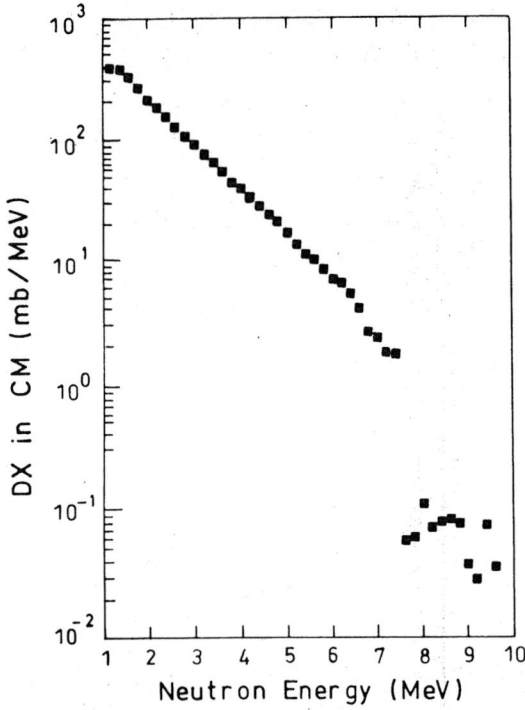

Fig. 5 Energy spectrum of the neutron emission from ^{107}Ag(p,n)^{107}Cd reaction.

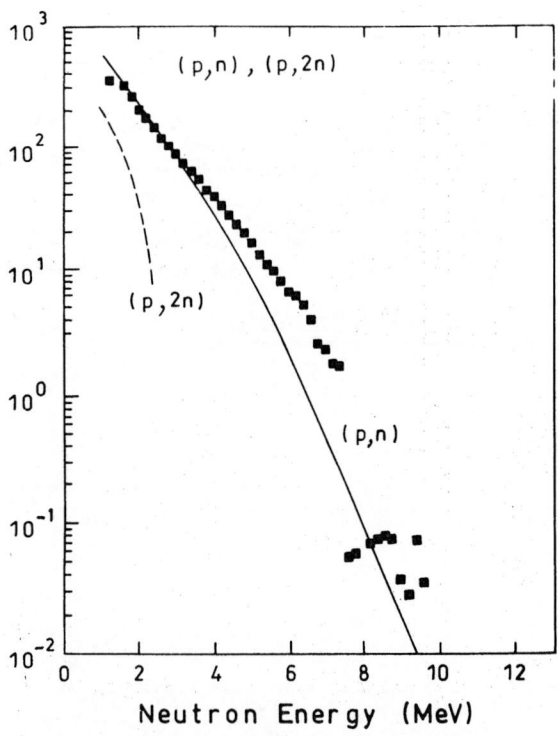

Fig. 6 Comparison between experimental data and theoretical curves.

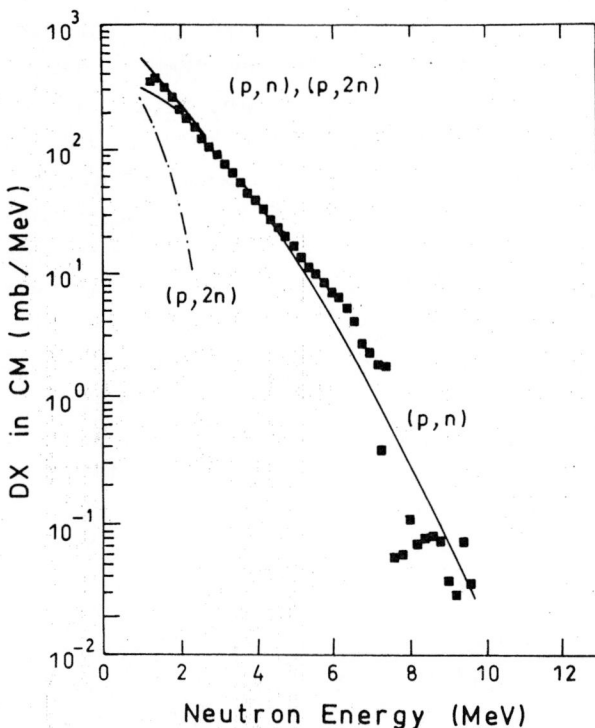

Fig. 7 Comparison between experimental data and theoretical fitted curves.

spectra (experimental spectra) in the 4- space. Fig. 5 is the energy spectra of the emission neutrons from ^{107}Ag(p,n)^{107}Cd reaction.

Discussion

From the angular distribution and the energy spectra, it is clear that the pre-equilibrium contribution

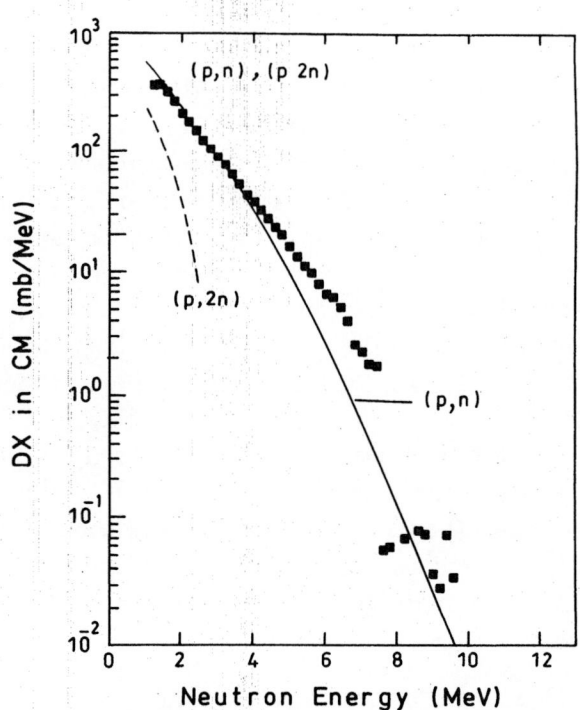

Fig. 8 Comparison between experimental data and theoretical curve fitted by the least squares method.

is not significant so the experimental spectra were fitted only by the evaporation model. Gilbert-Cameron [6] level density formula was used in the fitting and a set of the latest parameters 'a' and 'Δ' from reference [7] were obtained for the theoretical curves. The comparison between experimental data and theoretical curves which uses these parameters is shown in Figs. 6 and 7. Finally the least-squares fitting gives the best results and is shown in Fig. 8.

REFERENCES

1. H. Vonach, Extraction of level density information from non-resonant reactions, BNL - NCS-51694, page 247 (1983)

2. Yu Ju-Xian. Nucl. Instr. and Meth. **184**, 157 (1981)

3. Tian Ya-min, Yu-Xian, Liu Zhen- Ying, Yang Sui-Chun, Qin Jiu-Chang, Guan Xia-Ling, Du Xue-Ren, Ning Shu-Lan, Yang Bing-Fan, Yu-Yun Feng, Wang Ji-Dong, Han Rong-Ji, Yang Wei-Min, Bi De-Cai and Zheng Zhan Yi: Nucl. Instr. and Meth. **A244**, 39 (1986)

4. Qin Jiuchang, Yu Juexian, Yang Sui-Chun, Du Xue-Ren, Liu Zhen Ying, Wang Haipeng, Chen Wenkui, Hu Yaoming, Zhang Guilian, Yang Bing-Fan, Yu-Yun Feng, Guan Xialing, Ning Shulan, Yang Wei-Min, Ge Jiyun, Bi De-Cai and Zheng Zhan Yi: Nucl. Instr. and Meth. in Physics Research **A268**, 316 (1988)

5. Sa Jun, Tang Hongqing, Zhou Zuying, Sui Qingchang, Qi Bujia, Li Zhong fang, Yu Chunying and Shen Guanren, A multi-detector fast neutron TOF spectrometer at HI-13 tandem accelerator, Progress Report 1988-1989 page 119, Beijing National Tandem Accelerator Laboratory, China Institute of Atomic Energy, Edited by Sun Zuxun, 1990.

6. A. Gilbert and A.G.W. Cameron, Can. J. Phys. **43**, 1446 (1965)

7. Zhuang Youxiang, Chin. J. Nucl. Phys. **3**, 199 (1986)

TAILORING OF HEAFELY NEUTRON GENERATOR SPECTRA TO SUIT FUSION BLANKET PROGRAM

O.P.Joneja, J.-P.Schneeberger and M. Schaer

Institut de Génie Atomique, Ecole Polytechnique Fédérale de
Lausanne,CH-1015,Lausanne, Switzerland

Abstract: Fusion data validation requires experiments to be conducted with sources capable of delivering neutron spectra closer to that of a fusion machine. (D,T) driven tokamaks under fabrication, which are the forerunner of the near fusion machines, incorporate medium and medium heavy elements for their fabrication. The presence of these materials result in absorption, scattering and other nuclear reactions, which will alter the virgin neutron spectra. At the LOTUS facility, 14 MeV Haefely neutron generator is used for fusion blanket studies. This generator employs cooling water, silicon oil, polyethylene and several medium heavy elements. The presence of lighter elements may significantly change the neutron spectra, studies are therefore conducted by using Cd or B_4C absorbers held in front of the generator. It is experimentally found that by using 1 mm Cd or 2.6 mm thick B_4C, the reaction rates for higher threshold detectors are affected at the most by 11%, whereas the Au foil response is reduced by a factor of 1.4 and 3.5 respectively.

(tokamak, Haefely neutron generator, cross-section, absorber, threshold)

Introduction

At the LOTUS facility, Haefely neutron generator [1,2] is used for fusion blanket experiments. It measures 38 cm in dia and 58 cm in height and is housed in a 2 m thick concrete vault having inner dimensions $3.6 \times 2.4 \times 3.0$ m^3. It employs several materials consisting of 20 different elements [3] for its construction and operation. The neutron emission is found to be axially symmetric [4] due to symmetric location of the construction materials. The presence of low Z materials inside the generator would modify neutron spectra. In order to reduce the slow component of neutrons emerging from the generator, Cd sheet or B_4C [5] filled in an aluminum container is held in front of the generator. Although, it is almost impossible to remove completely the slow component of neutrons without affecting the higher energies, the present effort is therefore directed to reduce the proportion of slow neutrons such that it plays insignificant role in fusion studies. The inference about the flux reduction is drawn from the reaction rate measurements using several threshold detectors.

Experimental Results and Calculations

Fig.1 depicts the experimental arrangement where 1 mm Cd or 5mm thick B_4C absorber sandwiched between 1 mm thick aluminum sheets having cross-section 50×50 cm^2, is held infront of the generator. The B_4C powder is filled very slowly so as to get uniform thickness. The filling density obtained was 1.3 g/cm³ i.e 52% of the natural density. Zr, Al, and In activation foils of 17.8 mm dia and Au foils of 12 mm dia are employed to measure the reaction rates at the front and back surface of the absorber as shown in Fig.2. Monte Carlo calculations are made with

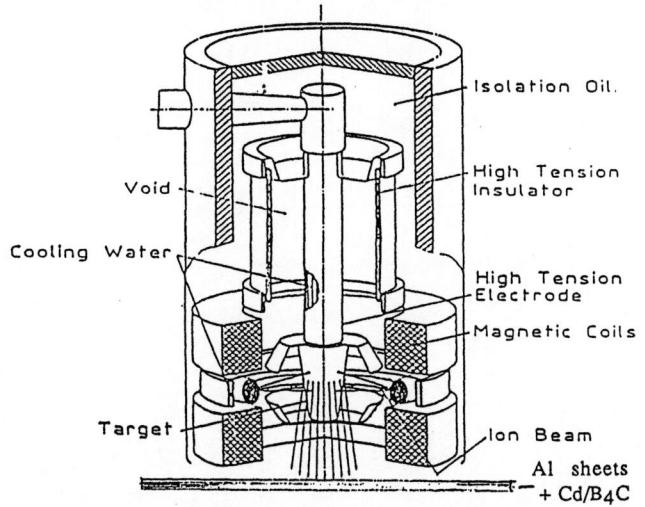

Fig.1 Schematic view of Haefely neutron generator along with the absorber

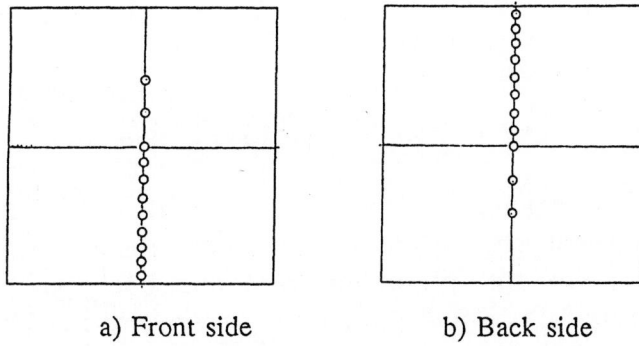

a) Front side   b) Back side

Fig.2 Location of activation detectors on either side of the absorber

[6] Morse combinatorial geometry package and Claw-IV cross-section library to determine the neutron fluxes inside the absorber regions for different thicknesses of the Cd or B₄C. Figs. 3 & 4 depict these results. It can be seen that the use of 5 mm thickness of B₄C results in considerable reduction of the low energy neutrons over a broad energy band in comparison to the Cd absorber.

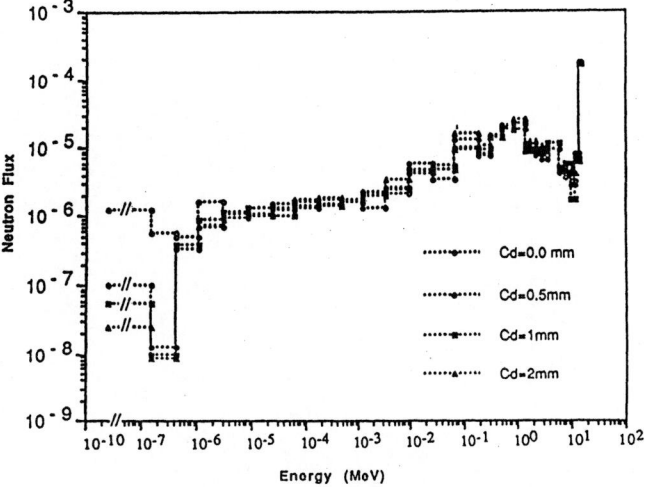

Fig.3 Neutron flux per source neutron in Cd sheets (50x50 cm2) of different thicknesses

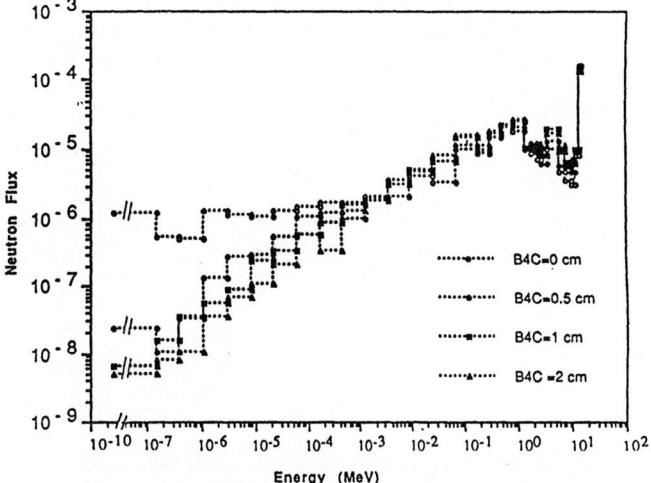

Fig.4 Neutron flux per source neutron in Boron Carbide (50x50 cm2) of different thicknesses

The measurements are carried out at 9 different radial locations covering a distance of 24.89 cm from the centre of the absorber plane. Figs. 5 & 6 show the results obtained with the Cd and B₄C respectively. Source nor-

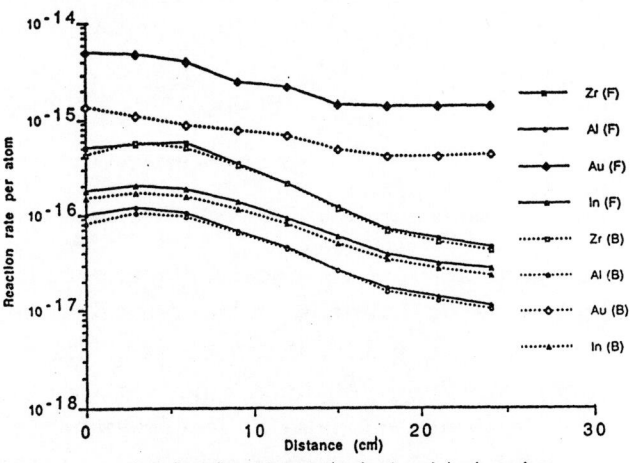

Fig.5 Reaction rates at the front and back surface of the Cd absorber for various radial distances

500

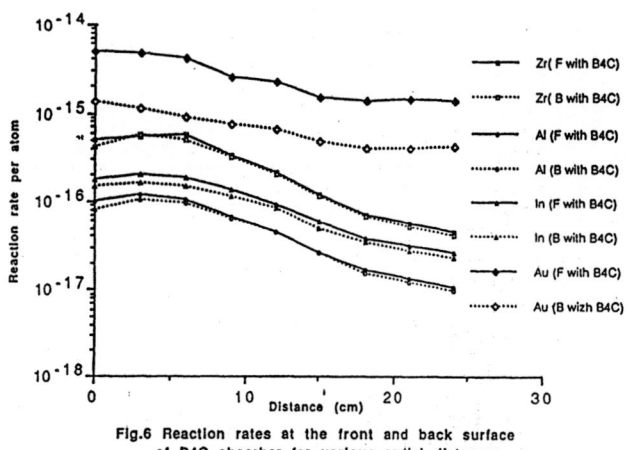

Fig.6 Reaction rates at the front and back surface
of B4C absorber for various radial distances

malized reaction rates of two different measurements for Zr and Al are depicted in Fig.7, whereas the corresponding results for In and Au are shown in Fig.8. The

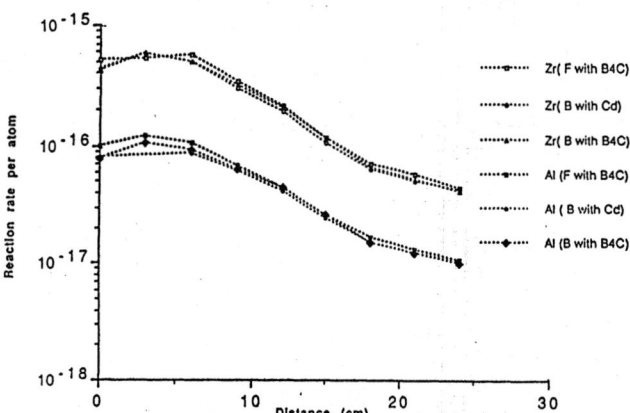

Fig.7 Zr and Al reaction rates at the front and back surface
of the Cd and B4C absorber for various radial distances

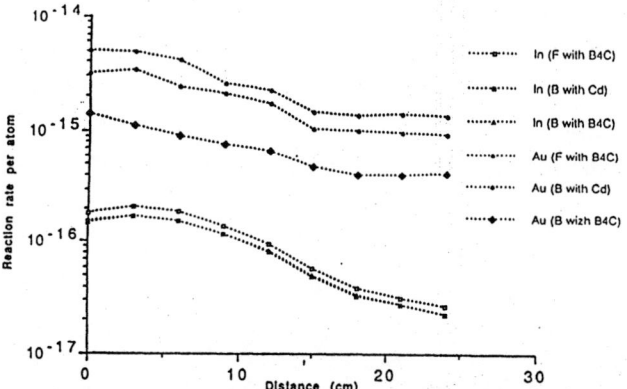

Fig.8 In and Au reaction rates at the front and back surface
of the Cd and B4C absorber for various radial distances

experimental front to back surface reaction rate ratios for various radial positions are shown in Figs. 9 &10. It can be seen that for higher threshold reactions, the variation in ratios is small and more or less identical for Cd and B₄C. On the otherhand, the ratios for Au varies considerably. Also, Au reaction rate ratios are found to strongly depend upon the location of the foils. Due to the difficulty

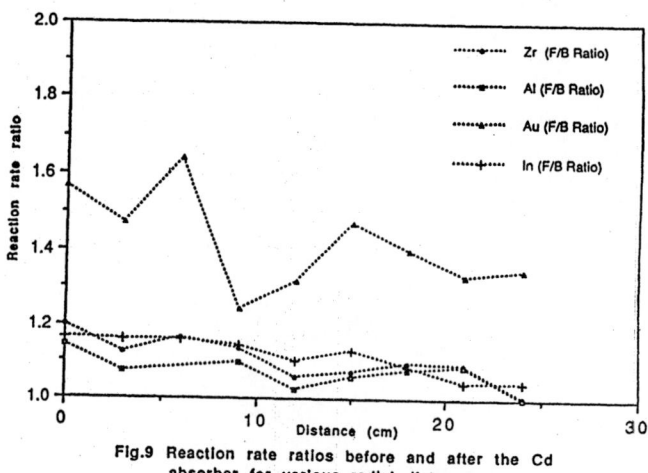

Fig.9 Reaction rate ratios before and after the Cd
absorber for various radial distances

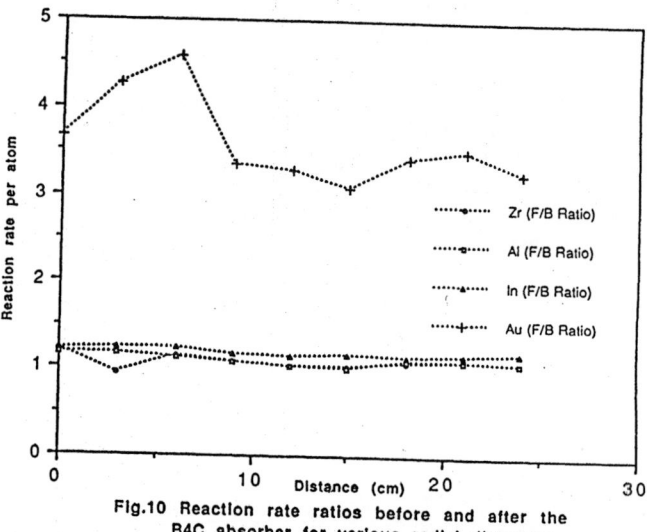

Fig.10 Reaction rate ratios before and after the
B4C absorber for various radial distances

in calculating the reaction rates for the small sizes of foils, aggregate reaction rate ratios for the entire front and back planes are calculated to determine the ratios. The calculations are performed by Monte-Carlo code MCNP using point cross-section data of ENDF-B-IV file. The code considers a complete 3-D description of the generator and the experimental arrangement. Track length and surface estimators are used to determine fluxes and thereby detector responses. For the purpose of comparison the experimental values are transformed by the programme RRPLN (Reaction Rate for the Plane). The experimental and calculated values are summarised in Table I. There is a good aggrement for all other reaction rates ratios, except that of Al in the case Cd absorber. Also, Au response is found to be strongly dependent on the composition of the concrete vault.

TABLE I Comparison of experimental & calculated reaction rate ratios

Absorber	Zr Foils		Al Foils		Au Foils	
	Expt.	Cal.	Expt.	Cal.	Expt.	Cal.
Cd 1 mm	1.11 ± 0.04	1.11 ± 0.02	1.03 ± 0.03	1.10 ± 0.02	1.40 ± 0.05	1.18 ± 0.17
B$_4$C 5 mm	1.08 ± 0.04	1.11 ± 0.02	1.08 ± 0.04	1.11 ± 0.02	3.50 ± 0.14	2.94 ± 0.40

Conclusions:

The Cd absorber is not found very effective in comparison to B$_4$C due to its sharp cutoff at 0.4 ev. The use of 2.6 mm thickness of B$_4$C reduces Au response by a factor of 3.5 and higher threshold reaction rate ratios by 8%. The slow component of neutrons can be further reduced if the B$_4$C density or thickness is increased. The good agreemant between the experimental and calculated ratios suggests that the generator description considered for the calculations and the cross-section set employed is quite adequate. The efforts are being made to achieve better accuracy on aggregate Au response by making calculations on Cray2 computer, with a large number of histories.

Acknowledgements:

We are thankful to Dr. Anil Kumar of University of California, Los Angeles for providing the Haefely generator source input description. Our thanks are due to Dr.V.R.Nargundkar (Invited Prof. from BARC, Bombay, India) who performed MCNP calculations by considering the experimental geometry and the Haefely neutron generator.

REFERENCES

1. K.A.Schmidt and G.Reinhold, "The HAEFELY-GFK Fast Neutron Generator", Int.Conf. on Particles and Radiation Therapy, Lawrence Berkley, California Sept.15-17, 1976.

2. K.A.Schmidt "Neutronengenerator" Deutsches Patent No.2112 215(1971);"Closed System Neutron Generator Tube",U.S.Patent No. 3786 258, 1972.

3. P.-A.Haldi,"Haefely Neutron Generator, 3-D Combinatorial Geometry Model-1",Internal Report Institute de genie atomique,EPFL.

4. O.P.Joneja et al,"Haefely Neutron Generator Characterisation from Neutronic Considerations",Internal Report,Institute de genie atomique,EPFL,LPR-176,june,1990.

5. A.Kumar and J.-P.Schneeberger, "A Proposal for Experimental Testing of Some Fast-Fission Hybrid Blankets at LOTUS Facility", Internal Report, Institute de genie atomique,EPFL, LPR-147, Jan.1986.

6. O.P Joneja and J.-P.Schneeberger,"Tailoring of Haefely Neutron Generator Spectra to Suit Fusion Blanket Program", Internal Report, Institute de genie atomique,EPFL, LPR-177, Aug. 1990.

ABSOLUTE DIFFERENTIAL CROSS SECTION OF
THE ^{14}N(D , α)^{12}C REACTIONS FROM 2 - 12 MeV

M.A. Chaudhri, Department of Medical Physics,

Austin Hospital, Melbourne 3084,

and the University of Melbourne, Australia.

Abstract: The absolute differential cross sections of the ^{14}N(d, α)^{12}C reactions, for the αo, α_1 and α_2 groups (going respectively to this ground, 4.433 MeV and 7.616 MeV states), have been measured at 90 degrees (lab. system) in the deuteron energy range of 2 -12 MeV in 50 - 100 KeV steps. Thin targets (150-300 ug/cm²) of Adenin ($c_5H_5N_5$), evaporated into 20ug/cm² thin self supporting carbon foils, were used for the measurements. The overall experimental error ranged from about 11% for the α_1, group till about 14% for the α_2 group.

Introduction

One of the most important applications of low energy accelerators is the analysis and depth profiling of elements and isotopes in/on material surfaces. This analytical technique is based on the observation of promptly emitted charged particles, neutrons and photons from the sample bombarded with charged particle beams. An accurate knowledge of the relevant nuclear cross section data is essential for exploiting the full potential of the nuclear analytical methods. This helps in selecting the appropriate experimental conditions, interface elimination, depth profiling, interlaboratory comparison etc..

Due to the high Q-value (13.575 MeV) the ^{14}N(d, α)^{12}C reaction is very well suited for nitrogen determination and depth profiling in/on material surfaces. However, the cross section of this reaction in the literature is available only up to about 3 MeV/1-3/, which has possibly hampered utilization of its full potential for analytical purposes. By using the Tandem Accelerator of the Max-Planck-Institute of Nuclear Physics, Heidelberg, we measured the exitation functions for the α_0, α_1, and α_2 groups (going respectively to the ground, 4.433 MeV and 7.616 MeV states in carbon) of this reaction at 90 degrees (lab. system). The energy range covered was from 2-12 MeV in energy steps of 50 kev up to 5 MeV and 100 keV (higher energies).

Experimental Set Up

The cross section measurements were carried out using the Tandem Van de Graff accelerator of the Max-Planck-Instituut, Heidelberg. The deuteron beam passed through 90 degree analysing and switching magnets and was focussed, at the target about 18m away, with the help of two quadropole lenses. The energy resolution of the beam was about 3-5 KeV and the absolute energy was reproducible to about ± 15KeV.

Scattering Chambers

A brass scattering chamber of 25cm internal diameter was used for these measurements. The top and bottom covers of the chamber could be rotated separately under vacuum. Therefore, two solid state detectors fixed to each of the covers could be rotated from 10⁰ - 170⁰ with respect to the incoming beam, independent of each other. The target holder, containing up to 4 targets could be introduced from the bottom cover. With an elaborate collimating system, which also had anti-scatter slits, the deuteron beam could be selected to any desired dimensions. For the present measurements the beam size was limited to 2mm x 4mm. The beam current was measured with an insulated Faraday Cup connected to an "Elcor" current integrator, with an accuracy of about ±1%.

Detectors and Electronics

ORTEC Solid State detectors, 50mm² active area and conventional electronics were used for measuring the alpha particles. Carefully prepared and accurately measured stainless steel collimators were placed on the detector, in order to define the solid angle and to ensure that only particles coming from the target were detected. By adjusting the bias on the detectors the width/depth of the sensitive zone was selected so as to stop the three alpha groups of interest completely, while only partially stopping the protons being produced simultaneously through the (d,b) reactions. In this way all three alpha groups could easily be identified at all the deuteron energies.

Targets

The nitrogen targets were prepared by vacuum evaporation of Adenin ($C_5H_5N_5$), which contains 51.85% nitrogen, onto 20ug/cm² thin self supporting carbon foils. By careful planning and execution of the evaporation, it was ensured that the targets were homogeneous at least in the centre. In this way a number of targets 150-300 ug/cm² thick and 12mm diameter were prepared.

The thickness of the targets was determined by a microbalance to an accuracy of about ± 10%. The homogeneity of the targets was estimated by determining the alpha particles coming out from different parts of the targets at the same deuteron energy. In this way it was determined that, at least in the centre, the targets were homogeneous to better than ± 2%. However, due to relatively low melting point of Adenin (365°C) the deuteron beam intensity had to be limited in order to avoid evaporation of the target during measurements. It was soon established that at lower deuteron energies a beam intensity of about 20mA and at higher energies of 40mA, did not cause appreciable evaporation of the target.

Method

The measurements were carried out with the target placed at an angle of 60 degrees with respect to the incoming beam. For up to 3 MeV deuteron energies, targets of about 150-200 ug/cm² were used, while for higher energies the targets used were 200-300 ug/cm² thick. For the excitation functions the deuteron energy was raised in steps of 50keV for 2 - 5MeV and in steps of 100 keV for 5 - 12 MeV. In order to minimize any evaporation/deterioration of the target during deuterons bombardment, beam intensities of around 20nA were used at lower energies, while for higher deuteron energies the beam current was increased to about 40 nA without noticing any appreciable target evaporation. However, after every 3 - 4 runs the deuteron energy was brought back to a reference energy and this measurement repeated. This was necessary in order to estimate the loss, if any, of the target material during irradiation and to compensate for it. However, it was observed that only in the energy range of 2 - 2.5 MeV some deterioration took place, which was taken into consideration when calculating the absolute cross section in that range. At all energies, various alpha groups were counted for an accumulated charge of 20 μ Coulomb as measured by the Faraday Cup.

Results and Discussion

A typical alpha particle spectrum at 10 MeV deuteron energy is shown in Fig. 1. As can be seen all three alpha groups are very well resolved and separated. Due to the high Q values, the α_0 and α_1 groups have virtually no background while the α_2 group has some "background" towards lower energy side of the peak, which has to be subtracted from the total α_2 peak in order to extract the number of α_2 particles. It was due to this reason that the errors in the estimation of α_2 counts were higher than for α_0 and α_1 groups. All the pulses, which were smaller than the α_2 pulses, were biased off with the help of a "bias amplifier" so that only the three alpha groups of interest were fed into the multichannel analyser and counted.

The excitation functions (differential cross sections at different energies) of the three alpha groups at 90° (lab.) are shown in fig. 2 (a), (b) and (c). No experimental errors are shown in the figures. The statistical errors as well as those due to the subtraction of background for the three alpha groups were 1.3 - 4% for α_0, less than 1% for α_1 and 2.5 to 8.5% for α_2 groups respectively. These errors when combined with other experimental errors, such as 10% in the estimation of the target thickness about 2% in the estimation of the solid angle, about 2% (max.) for the charge collections, gave total errors in the absolute differential crossection of the three groups to be no more than about 12%, 11% and 13.5% respectively.

Absolute excitation functions were also measured at 135° and 165° (lab.) in the same energy range. But, due to space limitations this data is not included in this paper. However, the detailed version of this paper with excitation functions at all three angles and numerical values with errors would be published elsewhere.

I would like to thank Prof. L. Lassen of the University of Heidelberg for his help with the measurements.

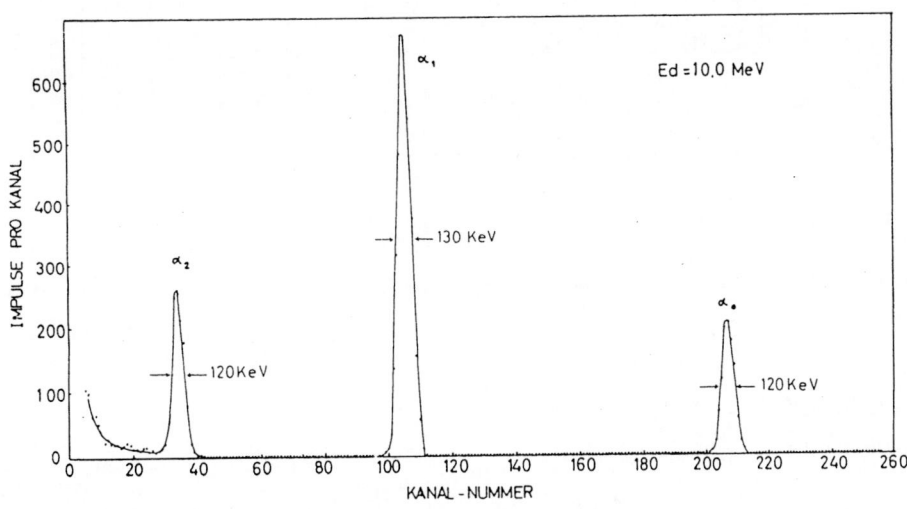

Fig. 1. A spectrum, of alpha particles for the ¹⁴N(dα)¹²C reaction at 10 MeV deuteron energy, clearly showing well-resolved α_0, α_1 and α_2 groups.

504

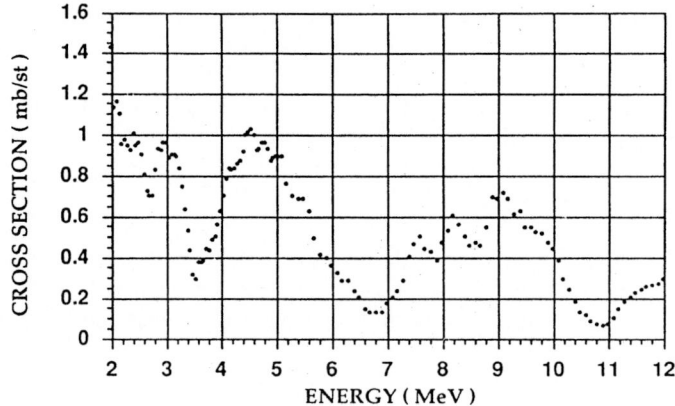

Fig. 2(a) α_0 group. Total experimental errors in the cross section of less than 11%.

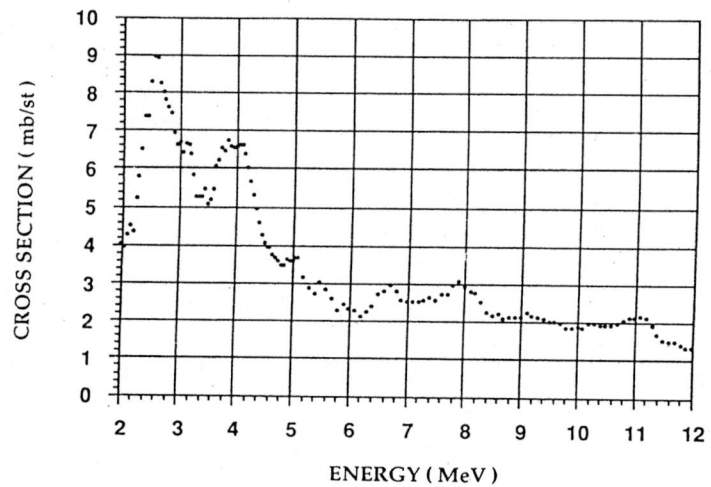

Fig. 2(b) α_1 group. Total experimental errors in the cross section of less than 12%.

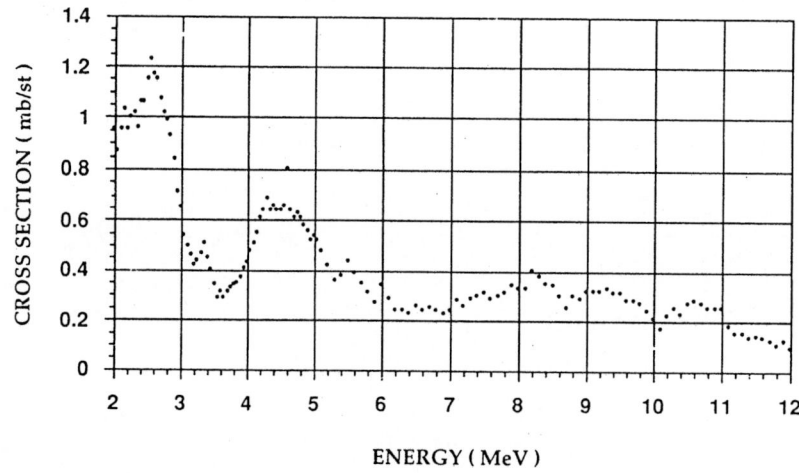

Fig. 2(c) α_2 group. Total experimetal errors in the cross section of less than 13.5%.

References:

1. G. Debras and G. Deconnick: J. Radio anal. Chem. 38, 193 (1977)

2. V. Gomes Porto, N. Ueta, R.D. Douglas, O. Sala, D. Wilmore, B.A. Robson and P.E. Hodgson: Nucl. Phys. A136, 385 (1969)

3. R.A. Jargis: Nuclear Cross Section Data for Surface Analysis Vol II, Department of Physics, University of Manchester (1979)

Part V

Nuclear Data
Relevant to Standards

MEASUREMENT OF THE ^{10}B(n,α$_1$γ)^{7}Li CROSS SECTION IN THE 0.3 TO 4 MeV NEUTRON ENERGY INTERVAL

R. A. Schrack and O. A. Wasson

National Institute of Standards and Technology,
Gaithersburg, MD 20899 USA, and

D. C. Larson, J. K. Dickens, and J. H. Todd

Oak Ridge National Laboratory,
Oak Ridge TN 37831 USA

Abstract: A relative cross section measurement of the ^{10}B(n,α$_1$γ)^{7}Li cross section was made using the ORELA neutron source. The reaction was measured by observing the 478 keV photon using a 30% efficiency germanium detector. The neutron flux was monitored with a high efficiency plastic scintillator. Data analysis required the use of extensive Monte Carlo calculations for corrections to the data. The measured cross section differs as much as 40% from the ENDF/B-VI evaluation for neutron energies greater than 1.5 MeV.

[^{10}B(n,α$_1$γ)^{7}Li cross section, germanium detector, neutron time-of-flight, black detector, Monte Carlo, neutron standards]

Introduction

The recently proposed cross section for the ^{10}B(n,α$_1$γ)^{7}Li reaction in ENDF/B-VI differs markedly from the previous evaluation (ENDF/B-V) in the neutron energy region above 1 MeV. The change was largely based on inverse reaction data, since no new data for the reaction was available. A measurement of the reaction was made at the National Institute of Standards and Technology (NIST) (then the National Bureau of Standards) in 1978 for neutron energy from 5 to 700 keV [1]. The NEANDC-endorsed Working Group on the ^{10}B(n,α) Standard Cross Sections recognized the need for a new high energy measurement in April of 1989. The NIST Neutron Interactions and Dosimetry group, in collaboration with personnel of the Oak Ridge Electron Linear Accelerator Laboratory (ORELA), undertook a new high energy measurement. This paper is a report on the results obtained in that measurement.

Experimental

Flight path 6 of ORELA was used with the reaction measurement made at 18.88 M. At that location a sintered boron disc, 6.35 cm in diameter and about 0.3-cm thick, was placed in the beam with the disc perpendicular to the neutron flux. The boron sample was enriched to 96.2% ^{10}B and was 0.023-atom/barn thick. The reaction rate was determined by detection of the 478 keV gamma ray in an N-type germanium detector placed approximately 48 cm from the boron sample at an angle of 125 degrees to the neutron beam. The detector efficiency is approximately 30% and has about 5 ns time resolution. The energy resolution was about 1 keV at 478 keV, which was more than adequate to define the 478 keV gamma ray emitted by the excited ^{7}Li nucleus remaining after the alpha particle emission. Since the gamma rays are emitted at all angles from the recoiling nucleus, the gamma ray is Doppler shifted and observed as a broad line which increases in width from about 18 keV to over 30 keV as the neutron

energy increases from 0.1 to 4.0 MeV. The relative neutron flux was monitored at 155.03 m with a large scintillation detector constructed of NE-110(12.7-cm diam by 17.8-cm long) termed the "black detector" because of its high efficiency [2]. Neutron energies were determined by time-of-flight.

The detector signals were processed in a system composed of NIM modules and the data collected on a personal computer based on the 80386 processor using software developed at ORELA [3]. The data were collected in 32 separate 24 hour runs and transmitted by Internet to NIST for analysis.

The electronics were configured such that a maximum of one detector signal was recorded for each linear accelerator pulse. The data were stored in two-parameter arrays, with pulse height and flight time determining where an event was stored. Bin widths varied as a function of flight time to conserve storage space while maintaining a maximum energy bin of 8% for the germanium detector and about 1.5% for the black detector.

In addition to the main two-parameter storage arrays for the desired neutron energy region, the gamma-flash and events occuring in a time region just before the next linac pulse were stored to determine the ambient background. As data checks and monitoring aids, one-parameter arrays of pulse height for all time and time-of-flight for all pulse height were also saved. Scalers were also used to record various event categories.

The linear accelerator was operated at 600 pulses per second with a pulse width of about 20 ns. An average of about 0.4 events were recorded per linac pulse. The data were dead-time corrected in the analysis with a maximum correction factor of about 1.6.

A representative germanium detector pulse height distribution in the region of interest is shown in figure 1. The bin width is about 1 keV. There is no apparent overlap of the 478 keV line with any other gamma ray for the neutron energies covered. No 478 keV gamma rays produced by neutrons scattering into the shielding materials

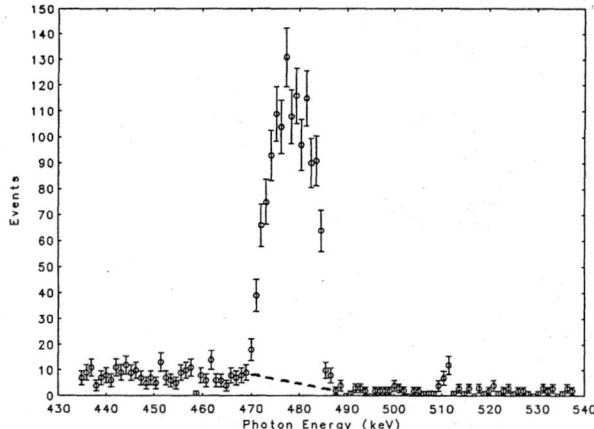

Fig. 1. The pulse height distribution of events in the germanium detector for 0.316 MeV neutrons is shown. The pulse heigh bins have been labelled in the graph with the associated photon energy. The dashed line shows the initial estimate of the background. The error bars indicate the statistical uncertainty.

were observed in extensive background measurements. We observed no 478 keV gamma rays produced by previous neutron bursts or other ambient neutron sources. The region on either side of the 478 keV line was determined to be free of gamma rays and was used to obtain Compton-event background values. Since part of the "background" for the pulse height region below the 478 keV line is caused by the line itself, a self-consistent technique was developed to determine the background to be subtracted, as shown by the dashed line in the figure.

Representative black detector pulse height distributions are shown in figure 2 for three neutron energies. Because of the large range (40:1) in pulse height it was necessary to take data with two different gain settings. The low gain setting was used to obtain data for neutron energies above 1.0 MeV. The high gain setting (about 2.6 times greater than the low gain) was used to obtain data for neutron energies below 2.5 MeV. Good agreement for the flux was found in the overlap region.

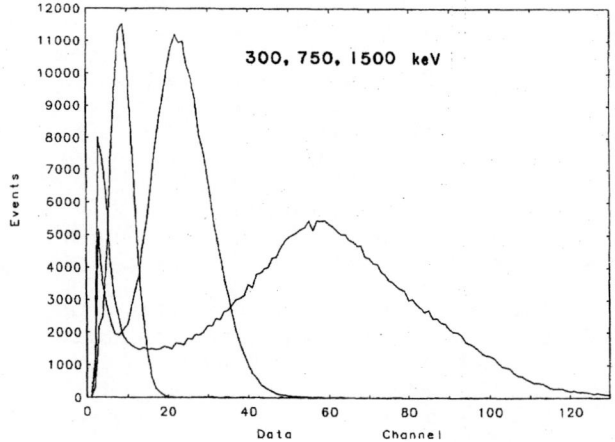

Fig. 2. Three representative pulse height distributions are shown for events recorded in the black detector. The three peaks, from left to right, show the pulse height distributions for neutrons of energies 300, 750, and 1500 keV.

Data Analysis

A series of programs was designed to divide the analysis into convenient steps. The dead-time correction required the combination of germanium and black detector data but all other programs handled the detectors separately until the final combination to produce the relative cross section.

Two major Monte Carlo simulation programs were used to provide input to the analysis. The first program determined the effect of multiple scattering and/or absorption of neutrons in the boron sample. This program [4] was developed at ORELA and was based on comprehensive neutron interaction codes. Figure 3 shows the correction factor determined by this program. The second program determined the efficiency of the black detector.

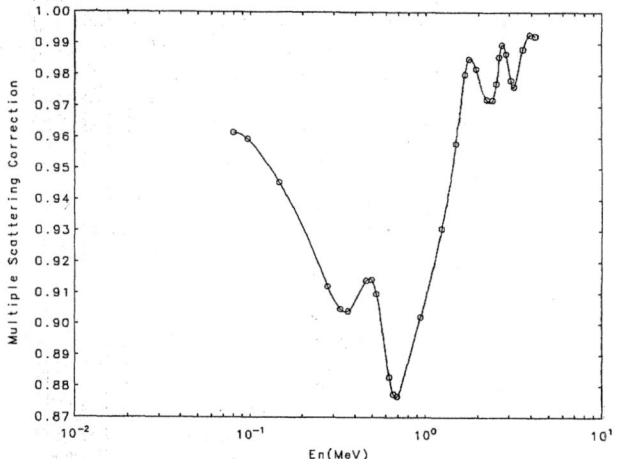

Fig. 3. The multiple scattering correction for neutron interactions in the boron sample is shown for the range 100 to 4500 keV. These results were determined by a Monte Carlo program and then fit with a least squares spline fitting program. The estimated error induced in the final cross section by this correction should be less than 0.2%.

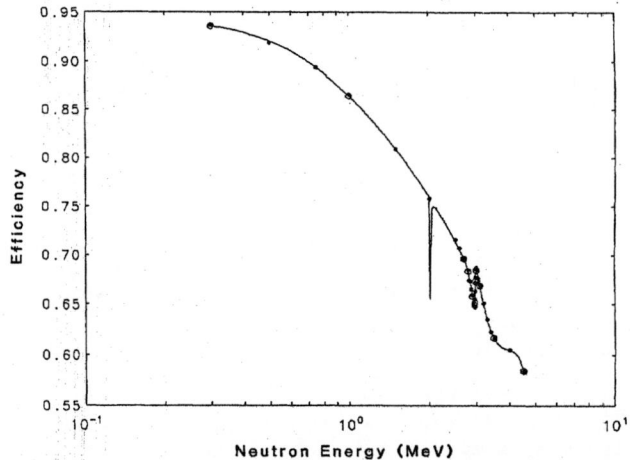

Fig. 4. The black detector efficiency as determined by a Monte Carlo program for the energies above 300 keV. At lower energies a different efficiency calculation is used. The greatest uncertainty is at the highest energies where it reaches 0.6% at 4.5 MeV. The uncertainty in this correction is a factor of ten smaller than the uncertainty in the result arising from statistical uncertainties.

This type of detector was originally devised by Poenitz [2] and further developed at NIST [5]. Figure 4 shows the results obtained by this program. Note that the accuracy of the reported cross section rests directly on the accuracy of these Monte Carlo results.

Because the energy (or time) bin structure of the two measurements differed, the black detector data were re-binned to match the bin structure of the germanium results. Since the time resolution of the black detector data is better than that of the germanium detector, it must be smeared slightly to match the detector resolutions before the two sets of data can be divided to produce the relative cross section. Finally, the relative cross section shape is given absolute values by normalizing the experimental results to the ENDF/B-VI values [6] in the region from 200 to 360 keV. The final results are shown in figure 5. The ratio of experimental results to ENDF/B-VI is shown in figure 6.

Measurement Uncertainties

The largest source of uncertainty is the statistical uncertainty associated with the count rate in the germanium detector. All other uncertainties are sufficiently smaller that the total uncertainty of the result would not be significantly reduced if they were reduced. The uncertainties associated with the Monte Carlo calculations affect the final results less than 0.2 %. (All measures of uncertainty are given in terms of one standard deviation.) The self consistent background correction factor for the germanium detector induces an uncertainty of less than 0.1 %. The procedure for calculating the black detector efficiency has an uncertainty of about 0.3% up to about 1.0 MeV. and rises to 0.6% at 4.5 MeV. The addition in quadrature of non-statistical errors to the statistical errors would have negligible effect. There are two data points below 1.5 MeV, where the fit is otherwise excellent, that differ significantly from the ENDF/B-VI values. It is tempting to associate the 5% difference at 420 keV to uncompensated air in the flight path between the two detectors but no evidence for this can be found. The 4% difference at 1.03 MeV may be statistical as the difference is only 2 standard deviations. Because of the observed major differences between our data and the ENDF/B-VI evaluation for neutron energies greater than 1.5 MeV, we completely and carefully reviewed all aspects of the experiment. No apparent systematic sources of error were identified.

Acknowledgment

We wish to thank the ORELA operating group for efficient accelerator operation. This work was sponsored in part by the U.S. Department of Energy.

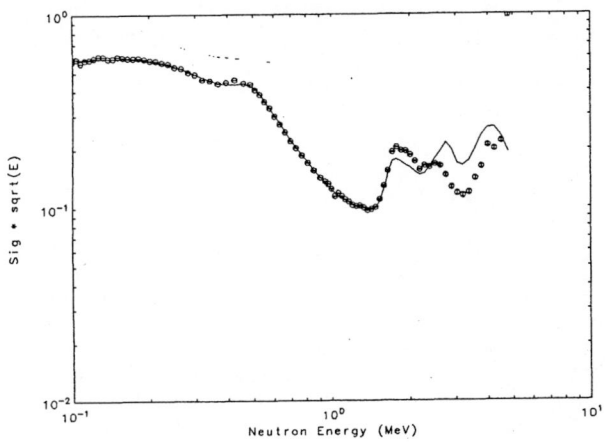

Fig. 5. The ^{10}B(n, $\alpha_1\gamma$)^{7}Li cross section results are shown as circles. The uncertainties ($\pm1\sigma$) is in all cases less than the circle diameter. The vertical scale is the cross section in barns/atom times the square root of the energy in MeV. The solid line shows the values given by the ENDF/B-VI evaluation. The experimental data are normalized to the ENDF/B-VI values in the neutron energy region from 200 to 360 keV.

References

1. R.A. Schrack, G.P. Lamaze and O.A. Wasson: Nucl. Sci. and Eng. 68, 189 (1978)

2. W.P. Poenitz: The Black Neutron Detector, ANL-7915 (1972)

3. B.D. Rooney, J.H. Todd, R.R. Spencer and L.W. Weston: A Data Acquisition Work Station for ORELA, ORNL/TM-11454 (1990)

4. J.K. Dickens: SCINFUL, ORNL-6462 (1988)

5. G.P. Lamaze, M.M. Meier and O.A. Wasson: Nuclear Cross Sections and Technology, R.A. Schrack and C.D. Bowman, eds, National Bureau of Standards Special Publication 425, p. 73-74 (1975)

6. A.D. Carlson, W.P. Poenitz, G.M. Hale, R.W. Peelle and C.Y. Fu: The ENDF/B-VI Neutron Cross Section Standards, ENDF-351, to be published

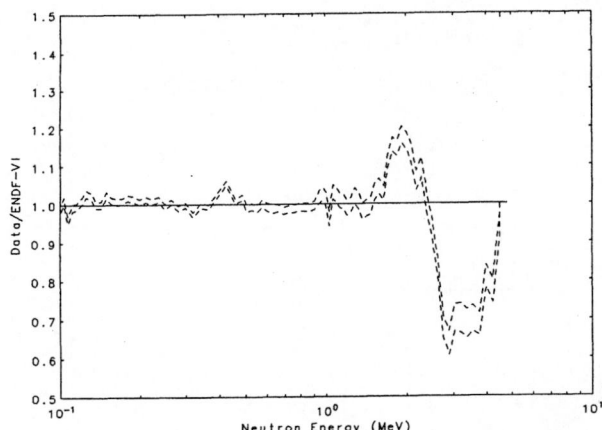

Fig. 6. The ratio of experimental data to ENDF/B-VI values is shown as a function of neutron energy. The dashed lines indicate the upper and lower values of one standard deviation as determined by the statistical errors.

ABSOLUTE MEASUREMENTS OF NEUTRON INDUCED FISSION CROSS-SECTIONS OF ^{235}U, ^{238}U, ^{237}Np AND ^{239}Pu USING THE TIME CORRELATED ASSOCIATED PARTICLE METHOD (TCAPM)

K. Merla, P. Hausch, C.M. Herbach, G. Musiol, G. Pausch, U. Todt
Technical University Dresden, Institute of Nuclear and Atomic Physics
D-O-8300 Pirna-4, Pratzschwitzer Str. 15, Germany

L.V. Drapchinsky, V.A. Kalinin, V.I. Shpakov
Khlopin Radium Institute, Roentgen Str. 1, 197022 Leningrad, USSR

<u>Abstract:</u> The comprehensive program of absolute fission cross-section measurements at the Technical University Dresden (TUD) in collaboration with the Khlopin Radium Institute Leningrad (KRI) was completed in 1990.

The Time Correlated Associated Particle Method (TCAPM) was used for neutron flux determination in all measurements. To achieve minimum uncertainties of the cross-section results the experimental setups were optimized individually for each specific neutron energy and the sources of all uncertainties were investigated carefully. The majority of all the small correction values are derived from experimental data only.

The investigations result in a set of 18 cross-section values for ^{235}U, ^{238}U, ^{237}Np and ^{239}Pu at 5 neutron energies in the region 2 to 20 MeV with single standard deviations of 1.10 to 3.17 %.
(^{235}U, ^{238}U, ^{237}Np, ^{239}Pu, fission cross-section, absolute measurements, TCAPM)

Experimental Method

The principle of our TCAPM measurements [1] is shown in Fig. 1. An AP detector registers the associated charged particles (AP) of the neutron production reaction within a fixed cone Ω_{AP}. The neutrons belonging to the registered AP form a cone Ω_n which has to be intercepted completely by the homogeneous fission-foils placed inside an ionization fission chamber (FC). Then the fission cross-section is given by the formula
$\sigma_f = N_f/(N_{AP} \cdot t)$ where are N_f the number of fission events registered in coincidence with an AP, N_{AP} the number of counted AP's and t the areal density of fissionable nuclei of the fission-foils.

All the necessary corrections which depend on the actual experimental conditions are determined from spectra which are collected during the whole measuring time by means of a CAMAC system [2].

Experimental System

The measurements at neutron energies of 2.6 MeV and 14.7 MeV were carried out at the neutron generator of the TUD and published in detail already in [1,3]. Main subject of this paper are measurements at neutron energies of about 4.5 MeV, 8.5 MeV and 18.8 MeV. Because the application of the TCAPM at these energies requires deuteron energies in the MeV region these measurements have been performed at the 5 MV tandem Van-de-Graaff accelerator of the CINR Rossendorf.

The kinematical and geometrical conditions are carefully optimized for each neutron energy point (Tab. 1).

The neutron production targets were mounted in a rotating target holder system. It allowed the use of uncovered Ti-T targets without noticeable T-escape out of the foil and a continuously work of more than two days with one $(CD_2)_n$-foil

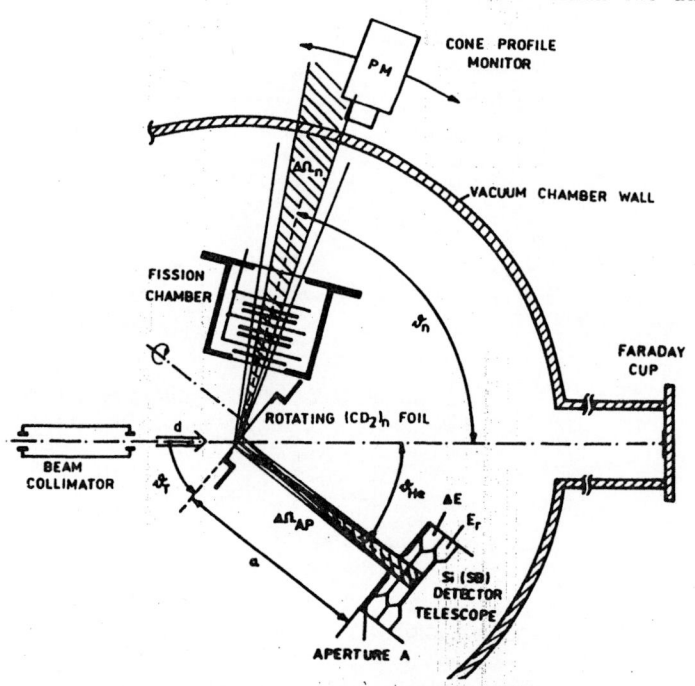

Fig.1. Experimental setup of the TUD/KRI TCAPM measurements

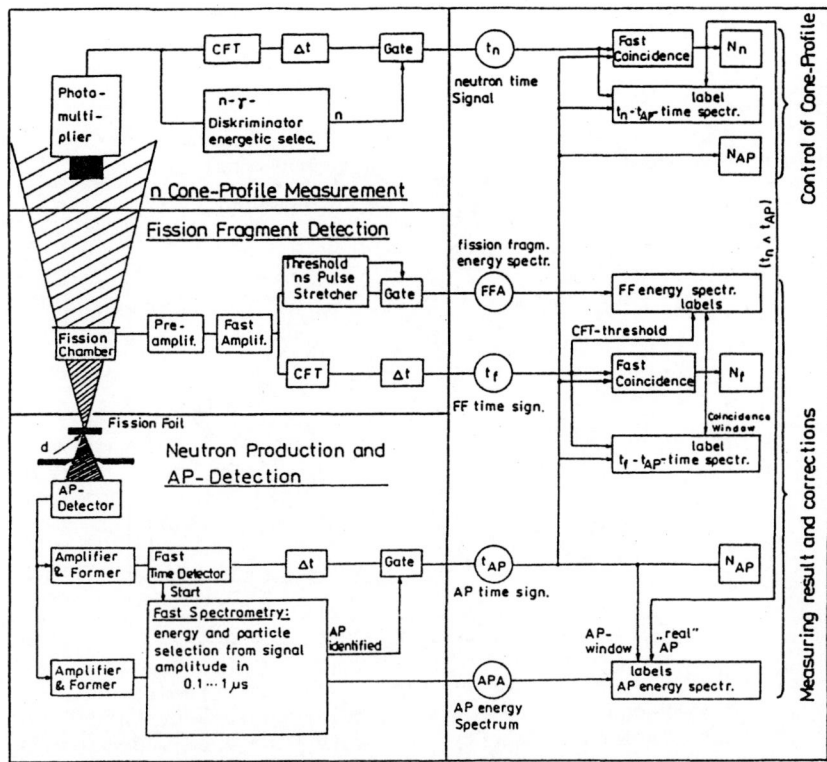

Fig. 2. Signal processing in TUD/KRI TCAPM measurements

Table 1. Main parameters of the experimental setup

Neutron energy E_n /MeV	$(4.45-4.9)\pm 0.2$	$(8.2-8.65)\pm 0.2$	18.8 ± 0.2
Neutron production reaction	$D(d,n)^3He$		$T(d,n)^4He$
Deuteron beam E_d /MeV	5.0- 5.8	9.50	6.00
I_d /nA	200 -600	300 -800	300 -650
Neutron producing target	self-supporting $(CD_2)_n$-foils		self-supp. Ti-T
t /mgcm^{-2}	0.5 - 1.8		3.1- 3.6
ϑ_n /°	52 - 53.5	75	40
AP detect. angle ϑ_{AP}/°	$(36.5-38) \pm 1$	$(40.8- 42.5)\pm 1$	68 ± 1
AP cone aperture ϕ /mm	4 - 5.5	5 - 6	5 - 9
a /mm	110 -119	118 -122	89 -122
Detect. thickness ΔE /µm	9 - 13	9 - 11	9 - 10
E_r /µm	37 -170	48 - 60	46 - 76
LAR outp.pulse lenght/µs	0.5	0.5	0.25
Mean AP counting rate/s^{-1}	2990-4800	3200-3900	2200-3500
Measured FWHM of the neutron cone profile /°	4.5- 4.8	4.8- 5.0	6.0- 6.3

(previously cross-linked by an intense electron beam) when using the T(d,n) and D(d,n) neutron production reactions, respectively.

The AP detection system [4,5] is based on a telescope of 2 Si(SB)-detectors for a ΔE and the rest energy E_r and a fast particle identification system (PI). The basic ideas of the system are:
- a fast timing signal (STROBE) with µs resolution is obtained as fast coincidence of the pulses formed by CFT of the ΔE and E_r detector current pulses
- amplitude information of the ΔE and E_r channels is obtained by integrating the clipped charge pulses using fast integrators (start of integration by the STROBE; suppression of deuteron pulses by a constant fraction timing (CFT) threshold in the ΔE channel)
- generation of the total energy ($E_g = \Delta E + E_r$) and particle amplitude (AP=aΔE+b·E_r) signals by the PI. The AP's are selceted by discriminator thresholds in the E_g and AP spectra. The analysis is done at the AP/E_g spectrum at measure-

ments of 4.5 and 8.5 MeV/18.8 MeV respectively (see Fig. 3)
- only if all threshold conditions are fulfilled and no pile-up is detected by the fast integrators, the delayed STROBE will pass the PI and represents the AP timing signal t_{AP} containing the complete AP information.

Because counting losses only occur due to the fixed dead-time of the PI circuit (~1 µs), which influences the total AP rate only but not the measured cross-section, no dead-time correction is neccessary.

The fission events are detected by a multiplate ionization FC filled with methane (p ~ 110 kPa), operated at a voltage of 400 V and containing up to 9 fission-foils [6]. The distance of 3 mm between electrodes and fission-foils represents a compromise to reach good separation of fission fragment (FF) pulses from α-events under the restriction of a compact chamber design. The fission-foils (t = 160-520 µg/cm^2) were manufactured at the KRI by HF-sputtering (^{235}U, ^{238}U,

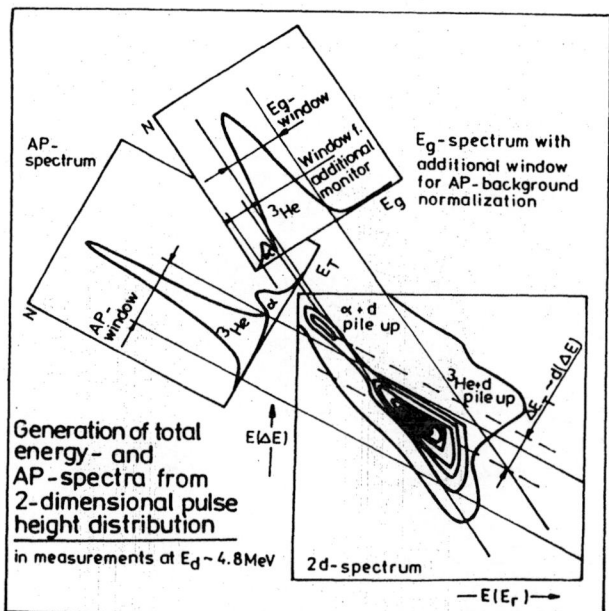

Fig. 3. Generation of E_g- and AP-spectra

^{237}Np), vacummevaporation (^{239}Pu) and a special "Metalloorganika" evaporation technique (^{238}U) from material of very high isotropic purity (>99.99 weight-% of fissionable material from main nuclide). The foil parameters areal density and layer inhomogeneity were determined at the TUD by low-geometry α-counting and α-scanning / RBS-measurements respectively [7]. The quoted t values refer to half-lives from [8].

The FC pulses were processed by a current sensitive preamplifier and a fast linear amplifier ($t_{int} \sim$ 10 ns). From its output pulses the FF energy loss spectrum and timing signal are obtained by stretching the pulse to a lenght of ~ 1 µs and by CFT, respectively (Fig. 2).

Corrections and Error Contributions

The numbers N_f of coincident fissions and N_{rc} of random coincidences were obtained from the coincidence timing spectrum. Because the TDC conversion range of ~140ns is smaller than the deadtime of the fast AP channel the random coincidence spectrum becomes "white" so that the background of random coincidences within the time window easily can be determined.

Fission fragment counting losses due to the CFT threshold (c_{ex}) were calculated by linear extrapolation of the plateau region, observed in the coincident FC spectrum, to pulse height zero.

An additional correction c_c was introduced to consider background events above the CFT-threshold in the FF spectrum which are correlated to AP's. At E_n= 18.8 MeV these events are caused mainly by the ^{12}C(n,n')3α reaction at the FC gas CH$_4$ [6], at lower neutron energies by (n,xα), (n,p) or (n,n') reactions at several FC materials. The correction guarantees the consequent linear extrapolation of the plateau region to zero pulse height in the case of low CFT-threshold.

The AP-background correction c_{AP} was determined by the normalization of background spectra to the effect spectra. In the measurements at neutron energies of 4.5 MeV and 8.5 MeV the background is caused by ^{12}C(n,α) reactions at the target material (CD$_2$)$_n$. Therefore background spectra were registred periodically during the runs by replacing the (CD$_2$)$_n$-foil by a (CH$_2$)$_n$-foil of comperable thickness. In measurements at

E_n = 18.8 MeV the sources of the background are (n,α)-reactions at the Ti, O (and Al if covered targets were used). Because it isn't possible to get a non-tritiated target of comperable parameters (identical thickness, O$_2$ depht-profil and thickness of the Al-layer) the different background portions resulting were considered individually by normalization of spectra measured with several non tritiated foils every of them comparable to the Ti-T-foil in one parameter .

The correction c_{sc} of neutron scattering and effective fission foil thickness due to the cone aperture was calculated by means of a Monte-Carlo simulation at the KRI and is the only one not based on experimental data.

Up to 1987 the correction c_{abs} of FF absorption was determined by analytical calculations based on a value for FF range of (7.5 $\pm$ 2.0) mg/cm^2 and considering the FF angular distribution. Then it became obvious that this correction is an systematical uncertainty of all fission cross-section measurements [7]. Caused by microscopic inhomogeneities and surface structures of the fissionable layers the fragment range should be essentially lower and a individual parameter of each fission-foil. Therefore the absorption was determined experimentally at every fission-foil by two different technologies. The first one, proposed by White [9], bases on the measurement of the value Q (fission rate divided by a measure of the layer thickness) at several foils of identical chemical composition but different thicknesses ς_m. The results give a relation between Q and ς_m, which is extrapolated to zero sample thickness. Assuming a detection efficiency

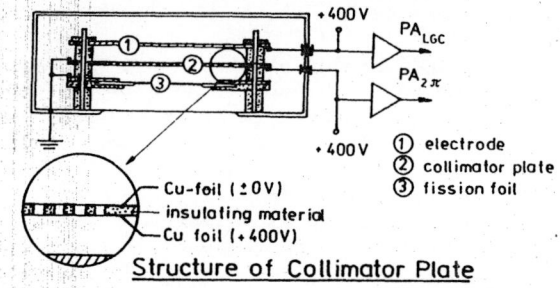

Structure of Collimator Plate

Fig. 4. Basic scheme of the collimator chamber

of 100 % for g_m= 0 the individual absorption for each foil can be determined [7].

The second method uses a collimator chamber (Fig. 4) which allows the determination of fission rates in 2π-geometry (like in the FC) and low geometry (as measure for the total number of fissions induced, because absorption processes are negligible) simultaneously. When the ratio of the geometry factors of 2π- and low geometry parts of the chamber is known the absorption for each fission-foil is calculable directly. The FF ranges determined by these measurements amount e.g. for ^{235}U-foils to (3.44 - 5.99) mg/cm^2. They are essentially lower than those formerly used and cause higher absorption corrections [7].

Results

The results of some measurements are based on several runs with different experimental conditions. To compare these separate runs an experimental ratio was introduced

$$r_{exp} = \frac{N_f(1-c_{rc})(1-c_c)(1-c_{ex})}{N_{AP}(1-c_{AP})}$$

It includes all corrections, which are obtained from the recorded spectra and may have changing values due to the actual threshold settings, detector and target properties. So it easily allows to control the confirmity of the results of the several runs. Based on this value the cross-section results (see Tab. 2) are derived by

$$\sigma_f = r_{exp} \frac{(1+c_{sc})}{t(1-c_{abs})(1-c_{nl})} .$$

The results of the measurements at neutron energies of 2.6 and 14.7 MeV were revised basing on the experimentally determined FF absorption and new values of the layer areal density and are: σ_f(2.6 MeV, ^{235}U) = $1.240\cdot10^{-28}$m^2 ± 1.94 %
σ_f(14.7 MeV, ^{235}U) = $2.096\cdot10^{-28}$m^2 ± 1.14 %
σ_f(14.7 MeV, ^{238}U) = $1.228\cdot10^{-28}$m^2 ± 2.11 %
σ_f(14.7 MeV, ^{237}Np)= $2.285\cdot10^{-28}$m^2 ± 1.93 %
σ_f(14.7 MeV, ^{239}Pu)= $2.449\cdot10^{-28}$m^2 ± 1.10 %.

Table 2. Corrections (c), error contributions (e) and final results of the measurements

Nuclide		^{235}U			^{238}U				^{237}Np			^{239}Pu		
Neutron energy/MeV		4.45	8.46	18.8	4.8	5.1	8.2	18.8	4.9	8.4	18.5	4.9	8.65	18.8
Statistics of effect	e	1.26	1.52	1.01	2.22	0.88	0.87	1.81	0.97	1.42	1.03	1.27	1.08	2.52
Random coincidences	c	1.40	7.46	2.82	0.48	0.68	1.19	1.07	0.94	3.00	1.32	0.64	1.75	4.55
	e	0.17	0.59	0.21	0.17	0.15	0.11	0.22	0.12	0.25	0.36	0.11	0.18	0.63
Correlated background	c	x	x	1.72	1.02	1.72	0.56	6.34	*	x	*	x	x	0.34
	e	x	x	0.04	0.07	0.61	0.04	0.06	*	x	*	x	x	0.13
FF spectr. extrapol.	c	1.18	2.25	1.67	0.87	1.27	1.24	1.20	1.56	1.95	1.25	1.50	1.00	2.57
	e	0.26	0.67	0.16	0.27	0.13	0.14	0.20	0.23	0.55	0.26	0.31	0.21	0.85
FF absorption	c	3.84	3.45	3.44	5.07	5.07	4.48	3.80	4.65	7.13	4.53	3.36	3.24	3.44
	e	0.73	0.74	0.76	1.59	1.59	1.42	1.57	1.44	2.07	1.38	0.53	0.54	0.58
AP-background	c	2.32	2.82	5.62	3.30	1.23	0.94	5.65	1.84	3.09	2.58	2.30	1.62	5.92
	e	0.67	0.23	1.35	0.60	0.58	0.29	0.60	0.40	0.85	0.69	0.36	0.31	1.74
Neutron scattering &	c	1.30	0.23	0.44	1.07	0.80	1.16	0.54	0.36	0.33	0.82	0.25	0.36	0.34
effect. foil thickn.	e	0.40	0.40	0.40	0.40	0.40	0.40	0.40	0.40	0.40	0.40	0.40	0.40	0.40
Cone neutr. outside ang. extent of foils	e	x	x	0.70	0.02	0.01	0.01	0.36	0.02	x	0.71	0.03	0.04	1.02
Fission foils • Areal density	e	0.72	0.82	0.82	1.15	1.15	1.15	1.00	0.71	0.90	0.71	0.60	0.60	0.60
• Inhomogeneity	e	1.02	1.00	1.02	0.54	0.54	0.54	1.59	0.41	0.50	0.41	0.88	0.86	0.88
σ_f /10^{-28}m^2		1.094	1.855	2.068	0.562	0.554	1.041	1.363	1.542	2.265	2.310	1.773	2.395	2.473
$\Delta\sigma_f/\sigma_f$ /%		2.15	2.39	2.43	3.11	2.39	3.17	3.17	2.03	2.91	2.25	1.85	1.68	2.39

x corr. negligible - not considered; * corr. backgr. events eliminated by new data storage technique

REFERENCES

[1] R. Arlt, W. Grimm, M. Josch, G. Musiol, H.G. Ortlepp, G. Pausch, R. Teichner, W. Wagner, I.D. Alkhazov, L.V. Drapchinsky, V.N. Dushin, S.S. Kovalenko, O.I. Kostochkin, K.A. Petrzhak, V.I. Shpakov, Proc. Conf. Nucl. Cross-Sect. and Techn., Knoxville, USA, 1979, NBS Special Publ., Vol. 594, p. 990

[2] G. Pausch, W.D. Fromm, C.M. Herbach, K. Merla, G. Musiol, H.G. Ortlepp, Proc. 15. Int. Symp. on Nucl. Phys., Gaussig, GDR, 1985, ZfK-592, p. 160 (1986)

[3] R. Arlt, W. Meiling, G. Musiol, H.G. Ortlepp, R. Teichner, W. Wagner, I.D. Alkhazov, O.I. Kostochkin, S.S. Kovalenko, K.A. Petrzhak, V.I. Shpakov, Kernenergie 24(1981), p. 48

[4] C.M. Herbach, M. Josch, K. Merla, G. Musiol, H.G. Ortlepp, G. Pausch, L.V. Drapchinsky, E.A. Ganza, S.S. Kovalenko, O.I. Kostochkin, V.I. Shpakov, INDC(GDR)-037/G(1985)

[5] H.G. Ortlepp, C.M. Herbach, K. Merla, G. Musiol, G. Pausch, see ref. [2], p. 156

[6] C.M. Herbach, K. Merla, G. Musiol, H.G. Ortlepp, G. Pausch, I.D. Alkhazov, L.V. Drapchinsky, E.A. Ganza, S.S. Kovalenko, O.I. Kostochkin, S.M. Solovjev, V.I. Shpakov, INDC(GDR)-036/G(1985)

[7] K. Merla, P. Hausch, C.M. Herbach, G. Musiol, G. Pausch, A. Rink, U. Todt, L.V. Drapchinsky, V.A. Kalinin, V.I. Shpakov, submitted to Nucl. Instr. and Meth.

[8] A. Lorenz (ed.), INDC(NDS)-149/NE(1985)

[9] P.H. White, Nucl. Instr. Meth. 79(1970), p. 1

[10] K. Merla, P. Hausch, C.M. Herbach, G. Musiol, H.G. Ortlepp, G. Pausch, U. Todt, I.D. Alkhazov, L.V. Drapchinsky, E.A. Ganza, V.A. Kalinin, O.I. Kostochkin, V.I. Shpakov, to be publ. in Nucl.Sci.Eng.

THE ^{238}U(n,f) CROSS SECTION AND ITS RATIO TO THE ^{27}Al(n,α)^{24}Na AND ^{56}Fe(n,p)^{56}Mn CROSS SECTION IN THE 14 MeV REGION

G. Winkler *, V.E. Lewis **, T.B. Ryves **, M. Wagner *

* Institut für Radiumforschung und Kernphysik d. Univ. Wien (IRK)
A-1090 Wien, Boltzmanngasse 3, Austria
** Division of Radiation Science and Acoustics, National Physical Laboratory (NPL)
Teddington, Middlesex TW11 0LW, UK

Abstract: The response of a ^{238}U fission chamber in an ^{3}H(d,n)^{4}He neutron field was measured against the activation of iron and aluminium foils and the neutron fluence as determined from measurement of the associated alpha particle fluence. Using a value for the mass of the ^{238}U deposit obtained from independent measurements, the ^{238}U(n,f) cross section was derived and also the ratio of this to the cross section for the ^{27}Al(n,α)^{24}Na and ^{56}Fe(n,p)^{56}Mn reactions was examined. The results were consistent, but indicated a discrepancy of 2 to 3% in the present ENDF/B-VI evaluation of the ^{238}U(n,f) cross section in the 14 MeV range (included in the International Reactor Dosimetry File).

[^{238}U(n,f), ^{27}Al(n,α)^{24}Na, ^{56}Fe(n,p)^{56}Mn, cross sections, 14 MeV range, neutron fluence measurements, associated alpha-particle counting, fission deposit mass assay]

Introduction

The reactions ^{238}U(n,f), ^{27}Al(n,α)^{24}Na, ^{56}Fe(n,p)^{56}Mn are most important and useful references in the fast-neutron energy region. To ensure the coherence and accuracy of neutron fluence measurements, the consistency and reliability of the respective cross section values is a prerequisite, which could be tested in the present work.

Therefore, the mass assay of a well-defined highly-enriched ^{238}U deposit, as gained previously by counting the emitted alpha activity and measuring fast-neutron fission yield ratios relative to other calibrated uranium deposits [1], was checked independently in irradiation experiments with 14.8 MeV and 14.5 MeV neutrons. The ^{238}U deposit was employed as a fluence monitor in a fission chamber, and its mass determined relative to the mass of an aluminium and an iron foil, which were activated simultaneously, using the known ^{238}U(n,f)/^{27}Al(n,α)^{24}Na, ^{238}U(n,f)/^{56}Fe(n,p)^{56}Mn cross section ratios. The reliability of the ^{24}Na and ^{56}Mn activity measurement was checked by intercomparing the results obtained at NPL and IRK, where different independent counting techniques were used. In the joint irradiation experiment at $\approx$ 14.5 MeV the T(d,n)^{4}He neutron field was also well defined with respect to fluence by counting the associated alpha particles from the source reaction, which permitted a value for the ^{238}U-deposit mass based on the adopted ^{238}U(n,f) cross section only to be derived.

Vice versa based on the neutron fluence values obtained from associated-particle counting and the activation foil technique, and the ^{238}U-deposit mass as determined independently by alpha-activity counting in previous work, evidence could be given for the value of the ^{238}U(n,f) cross section as compared to presently available evaluations.

Experimental procedure

Neutron irradiations

Quasimonoenergetic neutrons were produced via the T(d,n)^{4}He reaction by the action of a d^{+} beam of 150 keV deuterons (about 25 μA) from the NPL SAMES accelerator on a fresh target of titanium tritide in a low-scatter area. The beam diameter was limited to 10 mm by a tantalum collimator. The Ti layer was sufficiently thick to stop the deuterons completely and supported by a 0.5-mm aluminium backing inclined to the deuteron beam by 45°. It was glued to a thin aluminium plate at the edges and cooled by a blast of air. The mean deuteron interaction energy was estimated from previous studies [2,3] to be initially 85 $\pm$ 5 keV, falling to 80 $\pm$ 8 keV after one day's use.

Two irradiations (designated A and B in the following) were made on consecutive days. For these the neutron emission was measured by determining the associated alpha-particle fluence using a silicon surface barrier detector positioned about 900 mm from the target and at 90° to the deuteron beam. The assembly is schematically shown in Fig. 1.

The IRK fission ionization chamber [4] was mounted about 55 mm and 60 mm from the target face at 45° to the deuteron beam with the fissile deposit plane parallel to the Ti layer. The chamber was operated with a voltage gradient of about 450 Vcm^{-1} and a continuous gas flow of pure Ar as counting gas. The fissile material was about 160 μg ^{238}U enriched to 99.98%, in the chemical form UF$_4$, deposited as a 10-mm-dia layer on a 0.2 mm Al backing [1,4]. For each run a 10.0-mm-dia, 0.50 mm thick Al foil and a 9.56-mm-dia, 0.10 mm thick Fe foil (99.99% purity each) were attached to the fission counter in order to additionally monitor the fluence by the activation method. The Al foil was mounted on the inner face and the Fe foil on the outer face of the front wall at a mean distance from the ^{238}U layer of 0.45 $\pm$ 0.05 mm and 1.05 $\pm$ 0.15 mm, respectively. The foils were centered to each other and to the fission deposit (Fig. 2).

The effective mean neutron energy at 45° was calculated using relativistic reaction kinematics as 14.51 $\pm$ 0.02 MeV and 14.50 $\pm$ 0.02 MeV respectively for the above mean deuteron energies (compare [5,6]), with an energy spread (FWHM/2) of $\approx$0.12 MeV [5].

Finally an Fe and an Al foil were mounted close to the neutron-producing target and irradiated intensely. The induced activities were measured at both NPL and IRK in order to compare the different methods used for measuring activity by the two laboratories.

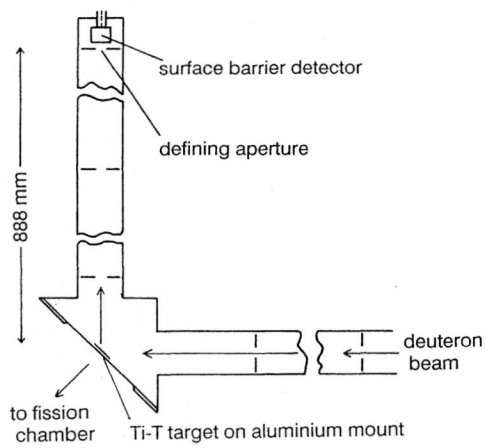

Fig. 1. Arrangement for measuring the d-T neutron fluence using associated-alpha-particle counting

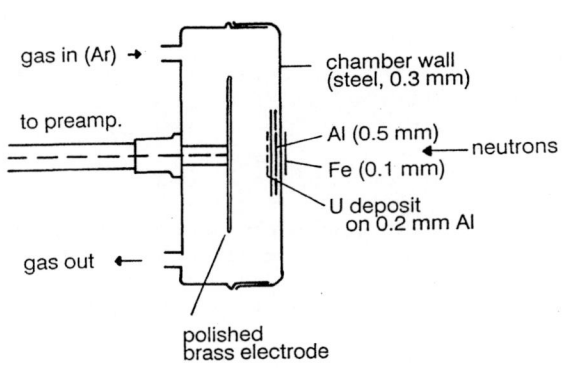

Fig. 2. Fission chamber with activation foils

The irradiation experiments at 14.8 MeV neutron energy as described in detail in ref.[4] involving the IRK fission counter and Al monitor foils were repeated at IRK using distances neutron source to ^{238}U layer between 105 mm and 33 mm. The contamination of the neutron field by neutrons from the D(d,n)^{3}He reaction was monitored by observing the relative intensities of associated alpha particles from the T(d,n)^{4}He reaction to those of associated protons from the D(d,p)T reaction at 150° relative to the deuteron beam, employing the differential particle emission cross sections [7,8]. A mylar foil ≈ 1 mg/cm^2 thick in front of a surface barrier detector served to separate the proton and the alpha-particle peaks. The contamination by D-d neutrons was found to be < 0.1%. Compared to [4] the operating voltage of the IRK fission counter was lower by ≈30% achieving the same good quality of the fission fragment pulse height spectra.

Measurement of neutron fluences
Associated alpha-particle counting:
The measured alpha-particle fluence was converted from the LAB to the centre-of-mass (CM) system yielding the overall alpha or neutron fluence into 4π, which was converted into the LAB neutron fluence at the chosen angle. Assuming the T(d,n)^{4}He reaction to be isotropic in the CM system this was done by calculating the appropriate solid angle conversion factors using relativistic kinematics. For the chosen arrangement (Fig.1) the alpha conversion factor was insensitive to the uncertainty in the effective deuteron energy due to the unknown tritium distribution in the target. Possible movements of the beam spot centroid on the inclined target caused a random uncertainty for the distance between source and defining aperture.

The counting threshold for the alpha spectrum was set at one third of the peak height. Previous work [9] had indicated that the net loss of genuine events from the tail of the alpha peak making allowance for the effect of other background, is (0.3 ± 0.3)%. The uncertainty due to the dead-time loss correction was estimated to be 5% of the loss.

Correction factors taking into account the extended size of the beam spot and the sizes of the ^{238}U layer and activation foils were applied for the positions used (e.g., [10]). Attenuation due to nonelastic processes in the target backing, the Al and Fe foils, the outer wall of the fission counter and the backing of the ^{238}U deposit was taken into account. The effect of neutron scattering by the target backing and target holder into the fission counter was estimated and corrected.

Table 1 summarizes the various sources of uncertainties taken into account quoting standard deviations or equivalent standard deviations. Adding the partial uncertainties in quadrature yields an overall random uncertainty of ≈ 0.55%, an overall non-random uncertainty of ≈ 0.75%, and a total uncertainty of 0.9 - 1.0 % for both runs. The results of the measurements are given in table 3.

Iron and aluminium foil activation:
The neutron emission rate was monitored by means of the associated particle detector and the fission chamber throughout the irradiations which also enabled corrections for decay of activity during irradiations. At NPL the induced activities of ^{56}Mn and ^{24}Na were measured using a 4πβ gas-flow proportional counter. The β-ray counting efficiencies of the foils were measured by the 4πβ-γ coincidence technique [11] in separate measurements using foils activated intensely.

For a set of these foils the induced ^{56}Mn and ^{24}Na activities were also measured at IRK within three days after irradiation employing a 12.7 x 12.7 cm NaI(Tl) well-type detector [12,13]. (Using a discrimination level of 22.1 keV and an additional Al well liner with a wall thickness of 1.5 mm the counting efficiency was calculated to be 0.794 ± 0.004 for ^{56}Mn and 0.8260 ± 0.0024 for ^{24}Na.) There was agreement between the IRK and NPL activity values within 0.4%, in good consistency with the respective uncertainty estimates. The half-life values taken were 14.960 ± 0.006 hours for ^{24}Na [14,15] and 2.5785 ± 0.006 hours for ^{56}Mn [14,15].

Table 1. Values of quantities and uncertainties used for the determination
of the neutron fluence using associated alpha-particle counting

Distance of the alpha detector aperture from the source		888.000 ± 1.5 mm*
		± 1.0 mm
Diameter of the alpha detector aperture		3.995 ± 0.010 mm
Factor converting alpha fluence in LAB to that in CM		1.004 --
Factor converting neutron fluence in CM to that in LAB,	Run A	1.031 ± 0.001
	Run B	1.030 ± 0.001
Alpha-particle detector efficiency		0.997 ± 0.003
Attenuation by aluminium target backing		0.997 ± 0.001
Attenuation by fission counter wall and iron foil		0.996 ± 0.001
Attenuation by aluminium foil and backing of ^{238}U layer		0.996 ± 0.001
Inscattering by target and target holder		1.008 ± 0.002
Distance between ^{238}U layer and Ti target layer,	Run A	60.150 ± 0.14 mm*
	Run B	54.900 ± 0.15 mm*
Correction for sizes of beam spot and ^{238}U layer or sample foils,	Run A	0.993 ± 0.001
	Run B	0.992 ± 0.001
Dead-time corrected alpha events,	Run A	23.06 * 10^6 ± 0.03%*
	Run B	19.84 * 10^6 ± 0.03%*
Uncertainty for dead-time loss correction		± 0.03%

Random uncertainties are indicated by asterisks (*)

The fluence at the position of the iron foil and the aluminium foil was calculated using a 14.50 MeV cross section value of 110.95 ± 0.55 mb [16] for the ^{56}Fe(n,p)^{56}Mn reaction and 115.55 ± 0.60 mb [16,17] for the ^{27}Al(n,α)^{24}Na reaction, respectively. The mean fluence over the ^{238}U layer was calculated applying an inverse square relationship and correcting for the attenuation due to the material between monitor foil and fission deposit. The uncertainty contributions are summarized in Table 2.

Fission chamber response

The extrapolation of the pulse-height spectrum from the discrimination level to zero pulse height [4] amounted to (0.82 ± 0.25)% of the total counts. The correction for selfabsorption of the fission fragments was calculated to be (0.3 ± 0.1)% [4], the increase of the total number of fission events due to lower energy neutrons produced by nonelastic processes was estimated as (0.3 ± 0.2)%. The total uncertainty of the number of fission counts was ± 0.75% for both runs A and B, governed by counting statistics. Separate test runs were performed, for example with the ^{238}U deposit replaced by a bare Al foil or the Ti-T target replaced by a tungsten beam stop. No significant interferences were found.

Table 2. Relative uncertainties (SDs) for
Fe and Al foil activation

Source of uncertainty	Fe:(± %)	Al:(± %)
Cross section value	0.5	0.5
Mean neutron energy	0.2	0.2
Beta-ray counting efficiency	0.3	0.3
Distance foil - ^{238}U layer	0.1	< 0.1
Half life, survival factor	0.1	< 0.1
Attenuation	< 0.1	< 0.1
Counting statistics	0.5	0.2
total uncertainty	0.8	0.7

Results

The neutron fluences experienced by the ^{238}U layer are listed in Table 3 persuant to the techniques that were used to measure them. Consistency within each run confirms both the adopted absolute values of ^{56}Fe(n,p)^{56}Mn and ^{27}Al(n,α)^{24}Na cross section and their ratios. With these fluence values, the fission counter response and the ^{238}U(n,f) cross section from IRDF ([18], data adopted from [19]), mass values for the ^{238}U deposit could be derived, which are inversely proportional to the fission cross section used. The results are averages over the two runs with the random or non-random character of the various uncertainty contributions having been considered. They are shown in the upper part of Fig.3 together with the results from the repeated (see text) and previous irradiation experiments [4] at IRK, which have been renormalized to the presently used cross section data [17,18]. A good consistency is indicated between NPL and IRK in the measurement of 14.5-14.8 MeV neutron fluence with the ^{238}U layer acting as transfer instrument. The variances of these data are all ≈50% correlated, either due to common cross sections or measuring techniques having been used. The lower part of Fig.3 shows values for the ^{238}U mass as obtained from α-activity counting using the ^{238}U half-life 4.468 * 10^9 (± 0.11 %) years [14] (which is essentially based on [22] only!). The fission foil intercomparison result done at ANL [1] can be considered to be traceable eventually to α-particle counting, too. A discrepancy of 2-3% is obvious between the results from direct α-particle counting and the irradiations experiments using the adopted ^{238}U cross sections ([18], 1.1593 b (±0.7%) at 14.5 MeV).

From the fluence values documented in Table 3, the corresponding fission chamber response, and the weighted average of the ^{238}U fission deposit mass from direct α-particle counting (159.6 ± 0.4 μg), one obtains for the ^{238}U cross section

1.196 ± 0.013 b (±1.1%) at 14.5 MeV.

This value agrees better with that obtained in a recent evaluation by Ryves [16] (done

Table 3. Fluences and their SDs in $10^{10}\mathrm{cm}^{-2}$ in the ^{238}U layer
averaged over its area after attenuation by target,
Fe and Al foils, counter wall and ^{238}U backing

Technique applied	Run A	Run B
Associated alpha-particles	4.140 ± 0.040	4.270 ± 0.040
Iron foil activation	4.175 ± 0.035	4.300 ± 0.035
Aluminium foil activation	4.140 ± 0.030	4.250 ± 0.030

simultaneously for various standard cross sections) than that from IRDF-90 [18,19]. In this context it may be noted that there is almost perfect agreement (within 0.2%) between the evaluation by Ryves [16] and the other recent evaluation [17] for the ^{27}Al(n,α)^{24}Na cross section.

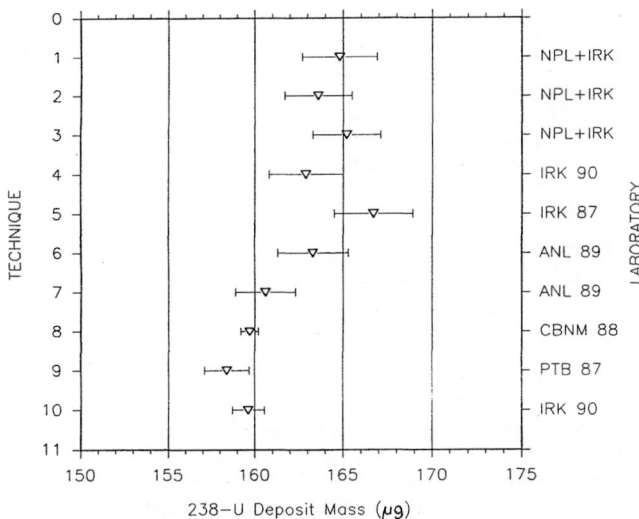

Fig. 3. ^{238}U deposit mass values and their uncertainties (SDs), based on:

1) $\sigma(^{238}\mathrm{U}(n,f)$ at 14.5 MeV) [18,19] and associated-particle counting, this work
2) $\sigma(^{238}\mathrm{U}(n,f))/\sigma(^{56}\mathrm{Fe}(n,p)^{56}\mathrm{Mn})$ at 14.5 MeV, this work
3) $\sigma(^{238}\mathrm{U}(n,f))/\sigma(^{27}\mathrm{Al}(n,\alpha)^{24}\mathrm{Na})$ at 14.5 MeV, this work
4) $\sigma(^{238}\mathrm{U}(n,f))/\sigma(^{27}\mathrm{Al}(n,\alpha)^{24}\mathrm{Na})$ at 14.8 MeV, this work
5) $\sigma(^{238}\mathrm{U}(n,f))/\sigma(^{27}\mathrm{Al}(n,\alpha)^{24}\mathrm{Na})$ at 14.8 MeV [4]
6) fast-neutron yield ratios relative to 5 calibrated uranium deposits [1]
7) α-particle counting, low solid angle [1]
8) α-particle counting, low solid angle [20]
9) α-particle counting, into 2π [21]
10) α-particle counting, into 2π [at IRK]

REFERENCES

1. J.M. Meadows, D.L. Smith, G. Winkler, H. Vonach and M. Wagner, in Proc. Specialists' Meeting on Neutron Activation Cross Sections for Fission and Fusion Energy Applications, 13-15 Sep. 1989, Argonne National Laboratory, USA (Edited by M. Wagner and H. Vonach), p. 153, OECD, Paris (1990)

2. V.E. Lewis and E.J. Axton: Nucl. Instr. and Meth. 167, 401 (1979)

3. T.B. Ryves and K.J. Zieba: Nucl. Instr. and Meth. 167, 449 (1979)

4. M. Wagner, G. Winkler, H. Vonach and G. Petö: Annals of Nuclear Energy 15, 363 (1988)

5. A. Pavlik and G. Winkler, Calculation of the energy spread and the average neutron energy of 14 MeV neutrons produced via the T(d,n)^{4}He reaction in solid Ti-T targets, INDC(AUS)-011/LI [INT/86-6], IAEA, Vienna (1986)

6. V.E. Lewis and K.J. Zieba, Nucl. Instr. and Meth. 174, 141 (1980)

7. R.B. Thens, W.I. McGarry and L.A. Beach: Nucl. Phys. 80, 273 (1966)

8. Handbook on Nuclear Activation Data, Technical Report Series No. 273, IAEA, Vienna (1987)

9. D.J. Thomas and V.E. Lewis: Nucl. Instr. and Meth. 179, 397 (1981)

10. H. Vonach, M. Hille, G. Stengl, W. Breunlich, and E. Werner: Z. Physik 237, 155 (1970)

11. T.B. Ryves, P. Kolkowski, and K.J. Zieba: Metrologia 14, 127 (1978)

12. G. Winkler and A. Pavlik, Int. J. Appl. Radiat. Isot. 34, 547 (1983)

13. A. Pavlik and G. Winkler, Int. J. Appl. Radiat. Isot. 34, 167 (1983)

14. Nuclear data standards for nuclear measurements, Technical Report Series No. 227, IAEA, Vienna (1983)

15. Table de Radionucléides, Laboratoire de Métrologie des Rayonnements Ionisants, CEA, Saclay (1982 and 1987)

16. T.B. Ryves, A simultaneous evaluation of some important cross sections at 14.70 MeV, same as ref. 1, p. 293; more details in: European Appl. Res. Rept. - Nucl. Sci. Technol. 7, pp. 1241-1334 (1989)

17. M. Wagner, H. Vonach, A. Pavlik, B. Strohmaier, S. Tagesen, and J. Martinez-Rico: Physics Data No. 13-5, Fachinformationszentrum Karlsruhe (1990)

18. N. Kocherov and H. Vonach, Intern. Reactor Dosimetry File (IRDF-90): Status and Testing, Proc. 7-th ASTM Symposium on Reactor Dosimetry, 27-31 Aug. 1990, Strasbourg, France (1990)

19. ENDF-201, 4-th edition, ENDF/B-VI Summary Documentation, ed. by P. Rose, Brookhaven National Laboratory Report (in press)

20. B. Denecke, Report GE/R/RN/02/88, Central Bureau of Nuclear Measurements, Geel, Belgium (1988)

21. K.F. Walz, Physikal. Techn. Bundesanstalt, Braunschweig, FRG, private communication (1987)

22. A.H. Jaffey, K.F. Flynn, L.E. Glendenin, W.C. Bently and A.M. Essling: Phys. Rev. C4, 1889 (1971)

MEASUREMENTS OF THE ^{235}U(n,f) CROSS SECTION
IN THE 3 TO 30 MeV NEUTRON ENERGY REGION*

A. D. Carlson and O. A. Wasson

National Institute of Standards and Technology
Gaithersburg, MD 20899, U.S.A.

P. W. Lisowski and J. L. Ullmann

Los Alamos National Laboratory
Los Alamos, NM 87545, U.S.A.

N. W. Hill

Oak Ridge National Laboratory
Oak Ridge, TN 37831, U.S.A.

Abstract: To improve the accuracy of the ^{235}U(n,f) cross section, measurements have been made of this standard cross section at the target 4 facility at Los Alamos National Laboratory (LANL). The data were obtained at the 20—meter flight path of that facility. The fission reaction rate was determined with a fast parallel plate ionization chamber and the neutron fluence was measured with an annular proton recoil telescope. The measurements provide the shape of the ^{235}U(n,f) cross section relative to the hydrogen scattering cross section for neutron energies from about 3 to 30 MeV neutron energy. The data have been normalized to the very accurately known value near 14 MeV. The results are in good agreement with the ENDF/B-VI evaluation up to about 15 MeV neutron energy. Above this energy differences as large as 5% are observed.

(^{235}U(n,f) standard cross section; H(n,n) standard cross section; annular proton recoil telescope; fission; fission chamber; fluence; neutron; standard)

Introduction

The ^{235}U neutron fission cross section is one of the most commonly used neutron cross section standards. In certain energy regions almost all fission cross section measurements have been made relative to this standard. It should be noted that any improvement in the ^{235}U(n,f) cross section, in the region where it is used as a standard, improves all cross section measurements which have been made relative to this standard. Though many measurements of this cross section have been made, significant differences in the data exist, particularly at high neutron energies. There is recent interest in neutron fluence standards in the upper MeV energy region and notably above 20 MeV as a result of applications in radio—therapy, fusion, accelerator shielding, radiation damage, etc. In response to this need the present shape measurements were made which are normalized to the very accurately known ^{235}U(n,f) cross section near 14 MeV neutron energy.

Experimental Details

The measurements were made at the 20 m station of the Weapons Neutron Research (WNR) target 4 neutron time—of—flight facility at LANL. The experiments were performed during several different running periods with different data acquisition systems and experimental conditions. The neutron fluence and fission reaction rates were determined with an annular proton telescope (APT) and a multiplate fission chamber, respectively. A white spectrum of neutrons was produced by 800 MeV protons from the proton linear accelerator of the Los Alamos Meson Physics Facility bombarding a tungsten target 7.5 cm long and 3 cm in diameter.

The 20 m flight path is at an angle of 60 degrees with respect to the incident proton beam. For this work, the spacing between the sub—nanosecond width microstructure pulses was about 4 μs. The experiment employed a collimated beam which passed through an evacuated flight path tube with a 2.54—cm thick polyethylene (CH_2) filter to reduce overlap effects due to previous micropulses and permanent magnets to deflect charged particles from the beam. The beam then entered the fission chamber, passed through 0.46 m of air, entered the APT, and was finally dumped in a concrete slab 5 m from the APT.

Neutron Fluence Detector

A proton telescope with an annular geometry was used to measure the energy dependence of the neutron fluence. Recoil protons, emitted from a thin CH_2 film placed in the neutron beam, were counted in a lithium drifted silicon (Si(Li)) detector which was shielded from the neutron beam by a carefully aligned copper shadow shield suspended in the center of the beam. The present detector is a design improved over that used in Ref. 1 since it has a larger evacuated containing vessel and a tapered copper shadow shield to reduce the background. This APT detector was used in measurements [2] of the ^{235}U(n,f) cross section on the National Institute of Standards and Technology linac. Significant improvements [3] in the performance of this detector system have been made compared to that given in Ref. 2. For the measurements a CH_2 film having a thickness of 2.08 mg/cm^2 and a diameter of 12.7 cm was used. The background was determined from a series of measurements made with and without the CH_2 film in place and with and without a tantalum cap over the Si(Li) detector. The tantalum cap is sufficiently thick to eliminate proton recoil events but was assumed to be transparent to neutron and gamma ray backgrounds.

*Research sponsored in part by the U.S. Department of Energy.

For each measurement, the ambient background was determined from a time window located just before the WNR micropulse. The time dependent background was determined from the quantity A−B+C. Where A is the measurement obtained with the CH_2 film and the Ta cap, B is the measurement obtained with no CH_2 film and the Ta cap, and C is the measurement obtained with no CH_2 film and no Ta cap. Normalization of the various background runs was obtained by using the fission chamber as a monitor. The background in the telescope was small, a maximum of a few percent. Measurements with CH_2 and CH films indicated there were no statistically significant contributions from neutron reactions with carbon.

Monte Carlo calculations were made of an additional background associated with neutrons which scatter from the shadow shield and then either strike the CH_2 film or scatter from the containing vessel and strike the CH_2 film. This background was found to be negligible.

Further information about the APT, and typical pulse height distributions for this detector and the fission chamber employed in this experiment are shown in Ref. 3. The bias channel used for the APT was neutron energy dependent and was set to include the proton recoil events and discriminate against background events. The intrinsic resolution of the 3−mm thick Si(Li) detector was better than 2% for ^{241}Am alpha particles. The pulse height resolution observed in this experiment was dominated by the angular spread of the proton recoils and their energy loss in the CH_2 film. The efficiency of the APT detector was calculated taking into account the angular distribution of the proton recoils and the finite size of the CH_2 film. Relativistic transformations from the center of mass to laboratory system were used in these calculations. The hydrogen scattering cross section from the ENDF/B−VI evaluation [4] was used in these calculations.

Fission Ionization Chamber

A fast multiplate fission ionization chamber [5] with 3 mm plate spacings was used to measure the ^{235}U fission reaction rate. It was used at room temperature and contains a 1.5 atmosphere gas mixture of 70% methane and 30% argon. A single pulse height bias was used for the entire neutron energy range for each of the ^{235}U deposits. The chamber contained 200 $\mu g/cm^2$ ^{235}U deposits of 10.2−cm diameter on stainless steel backings of 0.00127−cm thickness. A backing with no fission deposit for background estimation and one with a ^{252}Cf deposit for diagnostic measurements were included in the chamber. The ^{252}Cf deposit was also used to match the gains of the sets of electronics associated with each of the fission chamber plates so the background from neutron interactions in the backing material could be determined. Other deposits used for fission cross section ratio measurements [6] to the ^{235}U(n,f) cross section were also contained in the chamber. The neutron beam at the chamber is 12.7−cm diameter. The diameter of the supporting structure for the fission backings was 15.2 cm thus the detector could easily be aligned to ensure that the neutron beam would not strike this structure.

Corrections for fission events which do not escape from the deposit were made using the expression given by Carlson [7]. This expression takes into account both angular distribution and momentum effects. The deposits were all facing away from the neutron producing target. The maximum correction relative to the 14 MeV value where

the data are normalized is 0.5%. The background contribution in the fission chamber due to overlap neutrons from previous micropulses was calculated using the neutron spectra of Russell [8]. This contribution was less than 0.1%.

The ambient background for the fission chamber was measured by using a time gate located just before the WNR micropulse. The results of a run with the fission chamber out of the beam agreed with that of an ambient run. A determination of the background from neutron interactions in the backing material indicated a negligible effect for the energy range of these cross section measurements. A check was made to determine if neutrons scattering from the shadow shield in the annular proton telescope caused a background in the fission chamber. The test was performed by moving the telescope entirely out of the beam. No change in fission chamber count rate was observed. Further tests for background which related to both the fission chamber and the APT were performed. These tests included a run to check for high energy charged particles in the beam, a run to check for "cross talk" backgrounds related to the other flight paths, and a run with the target out of position. These runs indicated no significant background. Monte Carlo calculations were made of the background from neutrons which scatter from the beam defining collimators into the detectors. This contribution was negligible.

Data Acquisition

The APT and each fission foil employed essentially the same electronics. The electronic system permitted fast timing which is needed for the use of the time−of−flight technique and some integration of the pulse to provide reasonable pulse height resolution. A tagging method was used which allowed the timing information from all the detectors to be digitized in one analog to digital converter (ADC). The energy scale was established from transmission measurements of carbon resonances. Similarly all the pulse height signals were digitized in a single separate ADC. The data were taken and stored in an event by event mode. Each event is composed of three words: the tag which defines the detector in which the event occurred, the digitized time−of−flight and the digitized pulse height. Storing the data in this manner allows the experiment to be "replayed" so that shifts, etc. can be noted and handled appropriately. It does, however, require the storage of a significant amount of data.

Analysis and Results

The APT data were sorted with an energy dependent bias, divided by the efficiency, corrected for backgrounds and the transmission through the materials between the fission chamber deposits and the APT, and grouped into energy groups. The resulting data are shape measurements of the neutron fluence. The fission chamber data, which uses an energy independent bias, were sorted, corrected for backgrounds, grouped and divided by the fluence to produce a cross section shape. The statistical uncertainty, one standard deviation, yfor the measurements varies from about 1% at the lowest energy to about 2% at the highest energy. The dead time correction for the APT as a function of neutron energy is almost identical to that for each of the fission chamber deposits. Since the ratio of these quantities is used in the analysis, no correction was made for dead time effects.

520

A separate measurement [9] of the ^{235}U(n,f) cross section was made using the same fission chamber and at the 20 m station used for the present investigation, however, the neutron fluence was determined with two new proton telescopes. The low energy telescope (LET) uses a thin CH$_2$ film and a Si(Li) detector in a vacuum enclosure at ~15 degrees from the beam line. The geometry allows very good proton recoil pulse height resolution. This detector is located downstream from the fission chamber. The other telescope is the medium energy telescope (MET) which is also at an angle of 15 degrees from the beam but is operated in air downstream from the LET. The MET is composed of a CH$_2$ disk and three collinear proton detectors. The detectors are a 0.16-cm thick plastic scintillator, a 0.6-cm thick plastic scintillator and a 15.2-cm thick CsI scintillator. The MET is used in a coincidence mode in order to reduce the background. By combining the results from these two detectors, the fluence can be determined from 3 to greater than 250 MeV.

The results of the present investigation are shown in Fig. 1. They agree well with the ENDF/B-VI evaluation below 15 MeV. At higher energies differences as large as 5% exist. Also shown in this figure are the results obtained by Alkhazov [10] and those of the separate measurement [9] at this facility. These measurements are in generally good agreement with the present work and suggest that the ENDF/B-VI evaluation is low above 15 MeV.

References

1. G.S. Sidhu and J.B. Czirr: Nucl. Instr. and Methods, **120**, 251 (1974).

2. A.D. Carlson and B.H. Patrick: Proc. of an International Conference on Neutron Physics and Nuclear Data for Reactors and Other Applied Purposes, Harwell, U.K., 1978, p. 880.

3. A.D. Carlson, O.A. Wasson, P.W. Lisowski, J.L. Ullmann, and N.W. Hill, Proc. of the International Conf. on Nuclear Data for Science and Technology, Mito, Japan, May 30–June 3, 1988, p. 1029.

4. A.D. Carlson, W.P. Poenitz, G.M. Hale, R.W. Peelle and C.Y. Fu: ENDF-351, ENDF/B-VI Neutron Cross Section Standards, to be published.

5. R.C. Extermann, G.F. Auchampaugh, J.D. Moses, and C.E. Olsen, Nucl. Instr. and Meth. **189**, 477 (1981).

6. P.W. Lisowski, J.L. Ullmann, S.J. Balestrini, A.D. Carlson, O.A. Wasson, and N.W. Hill, Proc. of the Conf. on Fifty Years with Nuclear Fission, Gaithersburg, MD, p. 443 (1989).

7. G. Carlson, Nucl. Instr. and Meth. **119**, 97 (1974).

8. G.J. Russell, private communication (1989).

9. P.W. Lisowski, V. Gavron, W.E. Parker, J.L. Ullmann, S.J. Balestrini, A.D. Carlson, O.A. Wasson, and N.W. Hill, Proc. Specialists Meeting on Neutron Cross Section Standards above 20 MeV, May 21–23, 1991, Uppsala, Sweden to be published.

10. I.D. Alkhazov, L.V. Drapchinsky, V.A. Kalinin, S.S. Kovalenko, O.I. Kostochkin, V.N. Kuzmin, L.M. Solin, V.I. Shpakov, K. Merla, G. Musiol, G. Pausch, U. Todt, K.-M. Herbach, Proc. of Intl. Conf. on Nuclear Data for Science and Technology, Mito, Japan, May 30–June 3, 1988, p. 145.

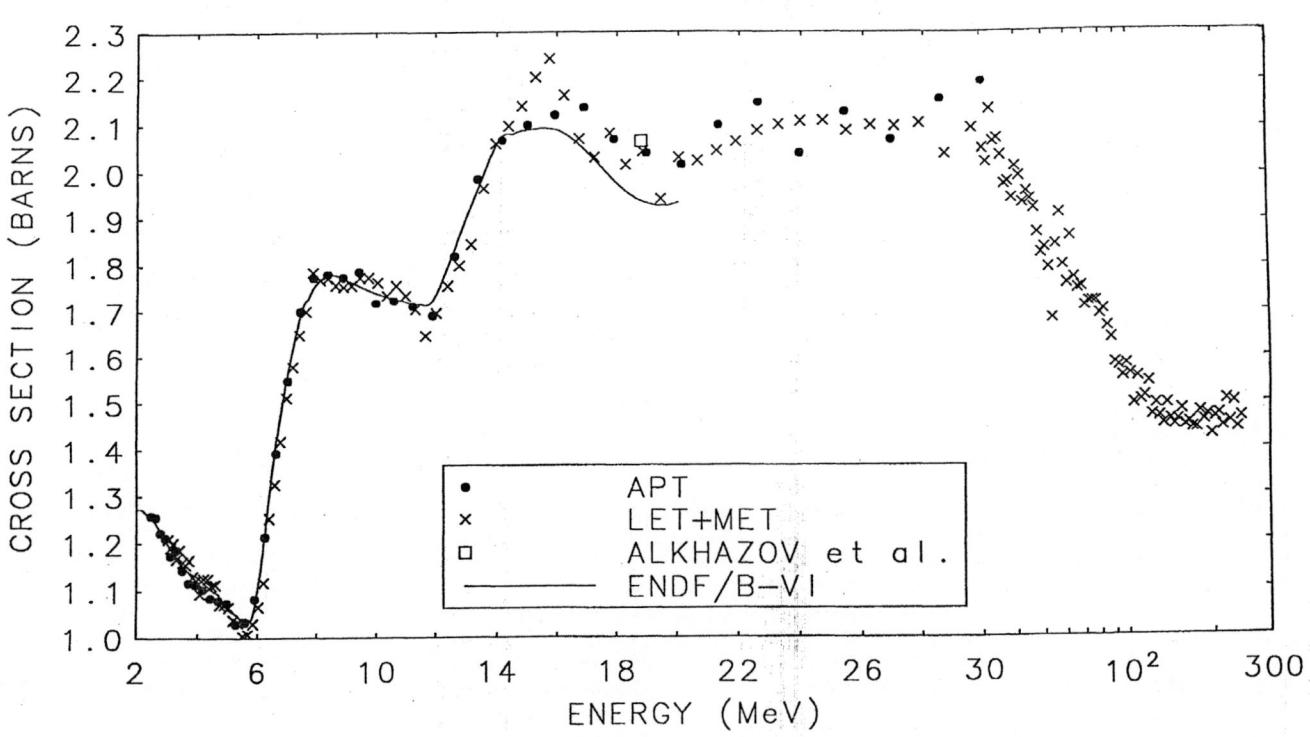

Fig. 1. Comparisons of the most recent measurements of the ^{235}U(n,f) cross section with the ENDF/B-VI evaluation. The data shown are the present measurements (labelled APT), those of Ref. 9 (labelled LET + MET) and Ref. 10. The statistical uncertainties, one standard deviation, on all these data are ~ 1–2%. The energy scale in the figure changes from linear to logarithmic at 30 MeV.

IAEA COORDINATED RESEARCH PROGRAMME ON X- AND GAMMA-RAY STANDARDS FOR DETECTOR CALIBRATION

A. L. Nichols

AEA Technology, Chemical Physics Department,
Winfrith Technology Centre,
Dorchester, Dorset, DT2 8DH, UK.

Abstract: A Coordinated Research Programme (CRP) has recently been completed under the auspices of the IAEA Nuclear Data Section on the Measurement and Evaluation of X- and Gamma-ray Standards for Detector Calibration. This exercise has resulted in the production of an agreed set of recommended half-lives and x- and gamma-ray emission probabilities for a selected number of radionuclides judged as important for detector efficiency calibrations.

(gamma-ray detectors, evaluated decay data, calibration, half-lives, gamma-ray emission probabilities, x-ray emission probabilities).

Introduction

A number of factors affect the feasibility of achieving accurate measurements of the gamma-ray emissions from radionuclides, including source preparation and counting geometry, sensitivity of the gamma-ray spectrometer, and derivation of the energy-efficiency curve for the detector system. The resulting quality of the decay data will depend to a considerable extent on the calibration curve, and hence the validity of the radionuclide standards used to establish the form of this energy dependence. Half-lives and x- and gamma-ray emission probabilities of suitable radionuclides need to be known to good accuracy. Furthermore, these data must be generally accepted by the scientific community so that inconsistencies do not occur between laboratories.

There has long been a need to formulate an internationally-accepted file of x- and gamma-ray decay data for detector calibration. Such a recommendation was proposed at the IAEA Transactinium Nuclear Data Advisory Group Meeting at Uppsala, May 1984 [1], and the International Nuclear Data Committee (INDC) responded by requesting the organisation of an experts' meeting associated with the International Committee for Radionuclide Metrology (ICRM). Hence, an IAEA consultants' meeting was held at the Centre d'Etudes Nucléaires de Grenoble on 30 and 31 May 1985 to establish an IAEA Coordinated Research Programme (CRP) on X- and Gamma-ray Standards for Detector Calibration [2]. Efforts within this IAEA-CRP focussed on the need to generate an acceptable data set of half-lives and x- and gamma-ray emission probabilities of radionuclide standards for the efficiency calibration of high-resolution detectors.

Coordinated Research Programme (CRP)

Nine groups experienced in decay data measurements and evaluations agreed to participate in the work of the IAEA-CRP. Representatives maintained contact through the IAEA Nuclear Data Section, and met for the first time at ENEA Headquarters in Rome from 11 to 13 June 1987 [3]. A final meeting was held at PTB, Braunschweig from 31 May to 2 June 1989 to report progress and formulate a technical data report [4]. Staff from the following laboratories participated formally in this CRP by performing the required measurements and evaluations:

- CEC-JRC, Central Bureau for Nuclear Measurements (CBNM), Geel, Belgium (represented by W Bambynek);
- Faculty of Science, Hiroshima University, Hiroshima-Shi, Japan (represented by Y Yoshizawa);
- Idaho National Engineering Laboratory (INEL), Idaho Falls, Idaho, USA (represented by R G Helmer);
- Laboratoire de Métrologie des Rayonnements Ionisants (LMRI), Gif-sur-Yvette, France (represented by N Coursol);
- National Institute of Standards and Technology (NIST), Gaithersburg, Maryland, USA (represented by F J Schima);
- National Office of Measures (OMH), Budapest, Hungary (represented by T Barta and R Jedlovsky);
- National Physical Laboratory (NPL), Teddington, Middlesex, UK (represented by P Christmas);
- Physikalisch-Technische Bundesanstalt (PTB), Braunschweig, FRG (represented by K Debertin);
- AEA Technology, Winfrith Technology Centre, Dorchester, Dorset, UK (represented by A L Nichols).

The programme was coordinated by A Lorenz of the IAEA Nuclear Data Section, with assistance from H D Lemmel.

Selection of Calibration Nuclides

The aims of the CRP were as follows:

a) select appropriate calibration nuclides;
b) assess the status of the existing data;
c) identify data discrepancies;
d) encourage measurements to meet the data needs;
e) evaluate and recommend improved calibration data.

These objectives were achieved by reviewing the existing data on an annual basis, coordinating measurements and initiating new studies. A recommended list of 36 nuclides evolved (Table 1), and emission probabilities were expressed as the absolute probability of the x- or gamma-ray photon per decay. The resulting data cover the energy range from approximately 5 keV to 3.55 MeV (^{56}Co), while a less rigorous assessment has also been made of suitable high-energy gamma-rays emitted by ^{66}Ga and specific neutron- and proton-capture reactions.

Table 1. Recommended Radionuclides for X- and
Gamma-ray Standards

Nuclide	Half-life [d]
^{22}Na	950.8 ± 0.9
^{24}Na	0.62356 ± 0.00017
^{46}Sc	83.79 ± 0.04
^{51}Cr	27.706 ± 0.007
^{54}Mn	312.3 ± 0.4
^{55}Fe	999 ± 8
^{56}Co	77.31 ± 0.19
^{57}Co	271.79 ± 0.09
^{58}Co	70.86 ± 0.07
^{60}Co	1925.5 ± 0.5
^{65}Zn	244.26 ± 0.26
^{75}Se	119.64 ± 0.24
^{85}Sr	64.849 ± 0.004
^{88}Y	106.630 ± 0.025
^{93m}Nb	(5.89 ± 0.05)x10^3
^{94}Nb	(7.3 ± 0.9)x10^6
^{95}Nb	34.975 ± 0.007
^{109}Cd	462.6 ± 0.7
^{111}In	2.8047 ± 0.0005
^{113}Sn	115.09 ± 0.04
^{125}Sb	1007.7 ± 0.6
^{125}I	59.43 ± 0.06
^{134}Cs	754.28 ± 0.22
^{137}Cs	(1.102 ± 0.006)x10^4
^{133}Ba	3862 ± 15
^{139}Ce	137.640 ± 0.023
^{152}Eu	4933 ± 11
^{154}Eu	3136.8 ± 2.9
^{155}Eu	(1.77 ± 0.05)x10^3
^{198}Au	2.6943 ± 0.0008
^{203}Hg	46.595 ± 0.013
^{207}Bi	(1.16 ± 0.07)x10^4
^{228}Th	698.2 ± 0.6
^{239}Np	2.350 ± 0.004
^{241}Am	(1.5785 ± 0.0024)x10^5
^{243}Am	(2.690 ± 0.008)x10^6

Data Measurements and Evaluations

All available measurements were taken into
account during an evaluation, including
experimental data from laboratory reports and
written private communications. Comprehensive
statements of the precise evaluation procedure
were prepared after each assessment, as well as
details of any changes made to the reported data.
Under normal circumstances, the evaluations
involved the determination of a weighted mean for
each parameter. It was agreed that no individual
measurement should contribute more than 50% to
the sum of weights when more than one value of
the same parameter was reported, and the
uncertainty of the datum was increased if
necessary. If the set of accepted experimental
data proved to be inconsistent, one of several
possibilities was adopted:

a) recommend the unweighted mean;
b) reject some measured values on the basis of
   objective or subjective judgements;
c) change the weights.

An appropriate method of changing weights was
preferred rather than outright rejection of data.

Specific radionuclides were identified as
requiring further half-life measurement to
resolve discrepancies and uncertainties. Various
inconsistencies in the relevant x- and gamma-ray
emission probabilities were also highlighted.
This exercise resulted in recommendations for
further studies to resolve specific aspects of
the more uncertain decay schemes. Most of these
measurements have been completed and incorporated
into the final evaluations to produce the
IAEA-CRP data set.

Conclusions

The accomplishments of the IAEA-CRP on X-ray and
Gamma-ray Standards for Detector Calibration
include:

a) assessments of the existing relevant data
   during 1986/87;
b) coordination of measurements within the
   existing programme;
c) evaluation of the data on the basis of all
   measurements;
d) preparation of a report which summarises the
   work of the CRP and includes the agreed data
   set [5].

A set of recommended half-life and emission
probability data has evolved from the resulting
measurement and evaluation exercise which
represents a significant improvement in the
quality of these specific decay parameters. It
is anticipated that the resulting data will be
acceptable on an international basis, and that
the tabulations will be used with confidence to
calibrate x- and gamma-ray detectors in
preparation for experimental studies of other
radionuclides.

REFERENCES

1. Transactinium Isotope Nuclear Data - 1984,
   Proc. IAEA Advisory Group Meeting, Uppsala,
   1984, IAEA-TECDOC-336 (International Atomic
   Energy Agency, Vienna, 1985).

2. A. Lorenz, Gamma-ray Standards for Detector
   Calibration, Summary Report of a Consultants'
   Meeting Held at the Centre d'Etudes Nucléaires
   de Grenoble, France, 30-31 May 1985,
   INDC(NDS)-171/GE (1985).

3. P. Christmas, A. L. Nichols and A. Lorenz,
   Gamma-ray Standards for Detector Calibration,
   Summary Report of a Research Coordination
   Meeting Held in Rome, Italy, 11-13 June 1987,
   INDC(NDS)-196/GE (1987).

4. P. Christmas, A. L. Nichols and H. D. Lemmel,
   X- and Gamma-ray Standards for Detector
   Efficiency Calibration, Summary Report of a
   Research Coordination Meeting Held in
   Braunschweig, Federal Republic of Germany,
   31 May - 2 June 1989, INDC(NDS)-221 (1989).

5. IAEA, X-ray and Gamma-ray Standards for
   Detector Calibration, to be issued as
   IAEA-TECDOC report (1991).

ENERGY SPECTRA AND ANGULAR CORRELATIONS OF NEUTRONS EMITTED IN SPONTANEOUS FISSION OF ^{248}Cm AND ^{252}Cf

O. I. Batenkov, A. B. Blinov, M. V. Blinov, A. S. Krivokhatski,
B. M. Alexandrov
V. G. Khlopin Radium Institute, 194021 Leningrad, USSR
Bao Shanglian
Institute of Heavy Ion Physics, Peking University,
100871 Beijing, P.R. China

Abstract: The ^{248}Cm spontaneous fission neutron spectrum has been measured in the energy range 0.1-10 MeV. An excess of neutrons in the low energy part of the spectrum over the Maxwell distribution has been observed. The angular correlations between the two neutrons emitted and the fragment motion direction was studied for ^{252}Cf spontaneous fission.

(Spontaneous fission, ^{248}Cm, ^{252}Cf, time-of-flight technique, neutron spectra, angular neutron correlations, neutron spectrum standard).

Introduction

A knowledge of the spontaneous fission neutron spectra is of interest both for studying the process of heavy nuclei fission, and for practical tasks in connection with the accumulation of heavy elements in nuclear reactors.

The available information is very limited [1-4]. For all the nuclides, except ^{252}Cf (the international standard of neutron spectrum), the spectrum shapes were measured with a low accuracy and in a comparatively narrow range of neutron energies. For ^{248}Cm there is information practically only about the average energy of the spectrum [4].

At the same time ^{248}Cm is of special interest because, on the one hand, its half-life ($\sim 4\times10^5$ years) is essentially greater than that of ^{252}Cf(2.7years) and on the other, the intensity of ^{248}Cm spontaneous fission is high enough ($\sim10^4$ fiss/mgs) which enables to use it in various scientific and practical purposes.

Information on neutron-neutron energy and angular correlations is very useful for a number of practical applications. However studies of those correlations were rather few [5-6]. We began systematic investigations of the correlations between the energies and emission angles of two neutrons and the fragment characteristics.

In this paper the results of measurement of ^{248}Cm spontaneous fission neutron spectrum are described and information on the initial stage of the correlation experiments is given for ^{252}Cf.

Method and Apparatus

A multidimensional time-of-flight neutron spectrometer was used for measurement of the ^{248}Cm spontaneous fission neutron spectrum. The ^{248}Cm neutron spectrum was measured relative to ^{252}Cf.

For this purpose a special source of fissions was manufactured, representing a thin platinum disc, on one side of which was Cf and on the other, Cm layer. The californium and curium used for preparation of the source were of high purity. The isotope composition of the curium source was as follows: ^{244}Cm-0.04%, ^{245}Cm - -0.2%, ^{246}Cm - 4.5%, ^{248}Cm - 95.2%. The share of spontaneous fission in the curium layer from ^{252}Cf impurities did not exceed 5%. The intensity of fissions in the curium layer was 2×10^4 fiss/min, and in the californium layer-10^5fiss/min. Both sources of fissions were covered with thin Al$_2$O$_3$ films (~ 40 μg/cm^2 thick). The electrons knocked out from the films at fragments passing, were registrated by means of detectors based on microchannel plates (MCP), due to which the registration and the time reference was done for all the fission events of Cf and Cm.

For the study of the neutron-neutron correlations the same set-up was used. The two fission detectors (semiconductor detectors or detectors based on MCP) were built into the chamber additionally for registration of the energies, velocities and directional motion of the fragments.

In the correlation experiment the registration of the fourfold coincidence events (neutron-neutron-fragment-fragment)

524

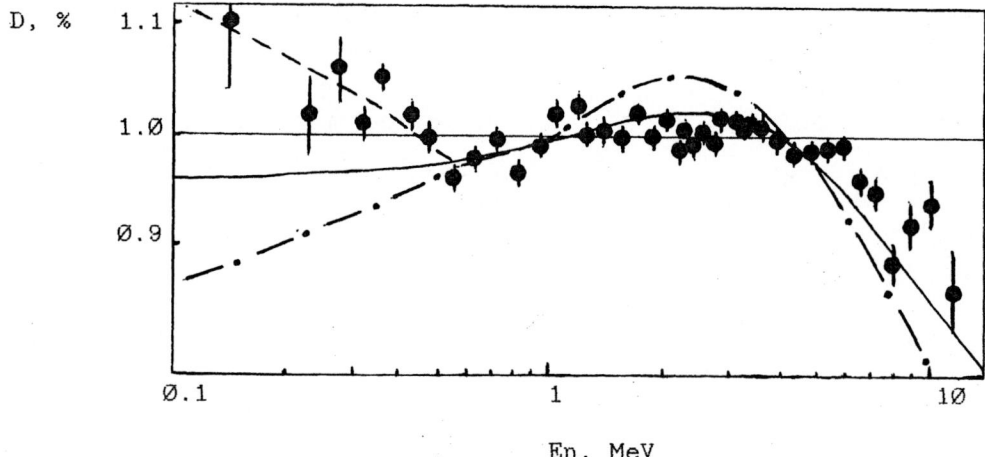

Fig. 1. The deviation of the integral neutron spectra for Cf and Cm
from Maxwell distributions.
   (●) - the data of this paper for ²⁴⁸Cm (T=1.38 MeV)
   (——) - the evaluated data [3] for ²⁵²Cf (T=1.42MeV).
   (-·-) - the results of the theoretical calculations for
       ²⁵²Cf [7,8]

was carried out. The neutron detectors were placed at different angles relative to the direction motion of fragments and to each other.

A stilbene crystal 50x30 mm with a photoelectron multiplier FEU-30 was used as the neutron detector. To decrease the number of background neutrons and to improve the time resolution, a two-threshold system of neutron registration was used. The values of the upper and low thresholds were a function of the neutron energy.

The time resolution for energies over 1 MeV was 0.6 ns, and for small energies, ~1ns. The separation of neutrons and gamma-quanta was done by means of the pulse shape. The suppression factor for neutron energies above 0.5 MeV was 10^4, below 0.5 MeV - 10^2. All this enabled to carry out measurements in the neutron energy range 0.1 - 10 MeV.

Cf and Cm neutron spectra were measured simultaneously, using the same neutron and fragment electronic channels. This reduced to minimum the effects of instability of the electronics and the detectors. The Cm spectrum measurements were done on three time-of-flight bases : 15, 30, 60 cm.

Control measurements for Cm spectrum were also done, turning by 180° and placing the neutron detector at an angle 70° relative to the normal to the layer. All the data agreed well within the measurement error limits (2-8 %).

Results and Discussion

The ²⁴⁸Cm spontaneous fission neutron spectrum is shown in Fig.1 in relation to the Maxwell distribution with T= 1.38 MeV. The value T=1.38 MeV was chosen from the best fitting of the energy range 0.75 - 6 MeV.

The errors in Fig.1 include statistical and systematic errors, as well as errors in determination of the standard spectrum shape [3]. As the measurements for Cm and Cf were carried out under strictly equal conditions of registration of fragments and neutrons, the systematic errors were reduced to minimum and slightly influenced the final results

Proceeding from average energy of neutron spectrum $\overline{E}$=2.069∓0.008 MeV, the average Maxwell temperature of the spectrum (T eff.= 2/3 $\overline{E}$) was determined. It turned out to be T=1.37 ∓0.005 MeV.

Studying the ²⁵²Cf spontaneous fission neutron energy spectrum [3] showed that there is a difference between the measured spectrum and various statistical calculations in the low-energy part of spectrum (En < 0.5 MeV) [7 - 8]. In the study [9], on the basis of analysis of differential measured data, it was treated as a manifestation of a new mechanism of neutron emission. This effect seems to be connected with the emission close to the instant of the nucleus scission. However, the discrepancy of the experimental and the calculated data did not yet give confidence in such an interpretation of results. Measurements done in this work for the first time point out to the reality of the given effect. As one can see from Fig.1, in the neutron energy range below 0.5 MeV quite a strong excess of the spectrum intensity over the Maxwell distribution with T=1.38 MeV is observed for ^{248}Cm. Theoretical calculations of the ^{248}Cm spontaneous fission neutron spectrum are absent. The course of the statistical calculated spectrum for ^{252}Cf [7-8] in the low energy range goes markedly lower than the corresponding Maxwell distribution. Calculations done for the thermal neutron fission of ^{239}Pu [10] (in these cases the neutron multiplicities (ν) for ^{248}Cm and ^{239}Pu are close) also show a lower intensity of the spectrum in comparison with the corresponding Maxwell distribution in the neutron energy range less than Ø.5 MeV. Therefore, the observed excess of neutrons in the low energy region for ^{248}Cm may be treated as a manifestation of the neutron emission mechanism connected with emission of neutrons at the earlier stages of spontaneous fission process. As the contribution of additional mechanism, connected with emission of low energy neutrons (En < Ø.5 MeV), considerably greater for ^{248}Cm than for ^{252}Cf, then it seems interesting to measure neutron spectra in this energy region in fission of various nuclides.

For determination of the low-energy neutron component characteristics we carried out measurement of the angular distribution of the neutrons of various energies for ^{252}Cf spontaneous fission. From the analysis of the experimental results based on the model of neutron evaporation from fully accelerated fragments a higher yield of low-energy neutron component was obtained mainly in the forward direction in relation to the fragment motion.

The study of neutron-neutron fragment angular correlations showed that there is a very strong correlation of emission angles for two neutrons. The ratio of the probabilities of this emission along the fission axis to that for the perpendicular direction is more than 25. This is connected mainly with the kinematic effect of the fragment motion. It is necessary to carry out a more detailed analysis of the results for final conclusion.

The experimental data showed that the average covariance of neutrons equals < cov(ν_L, ν_H)>=-Ø.28 $\mp$ Ø.Ø6 and the average variance of the distribution of neutrons emitted by one fragment < $\sigma^2(\nu_i)$ > = 1.22 $\pm$ Ø.11. These results on cov(ν_L, ν_H) and $\sigma^2(\nu_i)$ agree with the literature data [6,11,12].

Thus the study of the ^{248}Cm spontaneous fission neutron integral spectrum, diferential energy spectra and angular correlations in ^{252}Cf fission was carried out in this paper. The results obtained showed the expediency of the measurements of neutron spectra in low energy region in fission of various nuclides. The results demonstrate too that ^{248}Cm can be suggested for application as a source of standard spectrum for long-term measurements requiring a practically unchanged intensity of fission in a sample.

REFERENCES

1. Alexandrova S.A.,Bolshov V.I., Atomnaya Energia, 36, 282 (1974)(Rus)
2. Belov L.M.,Blinov M.V.,Kazarinov N.M., Krivokhatski A.S., Yadernaja Fizika, 9, 421 (1969)(Rus)
3. Mannhart W., Proc. IAEA Group Meeting on Neutron Sources, (1986 Leningrad, USSR) Vienna IAEA ,158 (1986)
4. Zhuravlev K.D.,Zamyatnin J.S., Kroshkin N.I., Neitronnaya Fizika, Kiev,4, 57 (1974)
5. Blinov M.V.,Vitenko V.A.,Krisjuk I.T., Neitronnaya Fizika Obninsk, 4, 171 (1974)
6. Gavron A., Fraenkel Z., Phys. Rev., 9, 632 (1974)
7. Madland D., Labauve R., Preprint LA-UR-84-129, (1984)
8. Gerasimenko B.F., Rubchenya V.A., Neitronnaya Fizika, Kiev,1, 349,(1984)
9. Batenkov O.I., Blinov A.B., Blinov M.V., Smirnov S.N., Atomnaya Energia, 64, 429 (1988)(Rus)
10. Marten H.,Sealiger D., Nucl. Sci. Eng. 93, 370, (1986)
11. Nifenecker H., C.Signarbieux C., Babinet R., Poitou J., Physics and Chemistry of Fission (Proc of the Intern. Conf. Rochester 1973) Vienna IAEA 2, 117 (1974)
12. Alchazov I.D., Dmitriev V.D., Kovalenko S.S., Kuznesov A.V., Malkin L.Z., Petrchzak K.A., Petrov B.F., Shpakov V.I., Yadernaja Fizika, 47, 1214(1988)(Rus).

ABSOLUTE MEASUREMENT OF CROSS SECTIONS OF
^{27}Al(n,α)^{24}Na AND ^{56}Fe(n,p)^{56}Mn AT En = 14.6 MeV

Bao Zongyu, Rong Chaofan, Zhang Shuping, Yang Xiaoyun, Yu Yiguang and Ding Shengyao

Radiometrology Division, Institute of Atomic Energy,
P.O.Box 275, Beijing 102413, P.R.China

ABSTRACT: The cross sections for the reactions ^{27}Al(n,α) ^{24}Na and ^{56}Fe(n,p)^{56}Mn were precisely measured via the activation technique at En=14.6MeV. The monoenergetic neutrons were produced through D-T reaction at the Cockcroft accelerator of CIAE. The neutron fluence was determined by the associated α-particle method. The absolute activities of the reaction products were measured by a $4\pi\beta$-γ coincidence device. Taking account of various corrections, the cross sections of ^{27}Al(n,α)^{24}Na and ^{56}Fe(n,p)^{56}Mn are obtained as 113.2 ± 1.5 mb and 111.4 ± 1.5 mb, respectively.

(Activation method, associated α-particle, $4\pi\beta$-γ coincidence)

INTRODUCTION

The cross sections of ^{27}Al(n,α)^{24}Na and ^{56}Fe(n,p)^{56}Mn are often used in determining fast neutron fluence rate. Although they have been measured and evaluated by plenty of authors, it is interesting to measure them precisely at the 14 MeV neutron standard field of the Radiometrology Division of CIAE via the activation technique. The neutron fluence was measured by the associated α-particle device, which has been successfully examined through the international comparison among several national standard laboratories organized by BIPM/1/. The absolute activities of the reaction products were measured by a $4\pi\beta$-γ coincidence counter. As one knows, in the absolute measurement of activity, the coincidence detector is superior to beta-particle detector or to gamma ray detector, because it avoids the difficulties in the efficiency calibration. Furthermore, it could be rather independent of nuclear parameters of samples, such as the branching ratio and the internal conversion coefficient. Also, the present $4\pi\beta$-γ coincidence device was perfectly tested by domestic intercomparisons.

EXPERIMENT

Measurement of the neutron fluence:

The neutrons were produced by the reaction T(d,n)^{4}He at the 600 keV Cockcroft accelerator of CIAE with an experimental room 20x10x7 m^3. The ϕ 8 mm Ti-T target on a ϕ 14x0.25 mm Mo backing was 0.5 mg/cm^2 thick and of $\sim$ 20 GBq activity. On the way of the incident deuteron beam there were two ϕ6 mm diaphragms to limit the size of its spot. The target was cooled by water or by air and placed at 45° to the incident deuteron beam. The deuteron energy was 200 keV. The accelerator terminal voltage was calibrated by the resonance reaction ^{19}F(p,$\alpha\gamma$)^{16}O. The neutron energy at the position of the samples, which were also at 45° direction to the deuteron beam, was found to be 14.57±0.02MeV, obtained with the Monte-Carlo calculation, considering the thickness and the size of both the target spot and the sample. This result was in good agreement with that measured by means of the Silicon detector method.

The neutron fluence was determined by counting its associated α-particle with an Au-Si surface barrier detector. It was positioned vertically to the deuteron beam and was 1004.2mm apart from the Ti-T target. Along the way in between there were several diaphragms to protect against the background. The smallest one was ϕ4 mm only, covered with 1μm Al foil, and was placed just in front of the Au-Si detector, so that the uncertainty of the solid angle of the α-detector could be minimized as well. In order to monitor the time-dependent variation of the neutron fluence rate during the irradiation period,instantly and continuously, the associated α-particle signals were recorded by a multichannel analyzer, which worked as a multi-scaler with the channel width of 1 minute.

Sample and irradiation:

The aluminium and iron samples with purity of 99.99% were ϕ 20x0.2 mm and were at the distance 80 - 130 mm from the neutron target. They were held by the circles made from very thin copper alloy wire of ϕ 0.5 mm. The Al and Fe foils were irradiated separately or sandwiched. 7 runs of experiment were carried out. Each of them varied in the geometric arrangement or in the irradiation time.

Measurement of activity:

The absolute activities of the samples after the irradiation were measured by a $4\pi\beta$-γ coincidence device: a methane gas flow proportional β-counter positioned between two ϕ7.6x7.6 cm NaI(Tl) scintillator γ-counters. The signals of these counters and the coincidence of them were recorded and processed by an on-line micro-computer. The window center of the single channel analyzer for the γ-detection was set at 2754 keV(^{24}Na) or at 846.7 keV(^{56}Mn). The samples of different thickness and diameter were irradiated at the same time and measured later on. In this way the beta efficiency(E_β) extrapolation factor (1 - E_β)/E_β was found to be as +0.25%(Al foil) and -0.9%(Fe foil). Meanwhile, little difference of the specific activities (less than 0.3% of the statistics) was observed between the samples of ϕ20 mm and ϕ5 mm foils.

CORRECTION, UNCERTAINTY AND RESULT

In addition to the routine decay corrections, where the half-lives were taken as 14.959 hr for ^{24}Na and 2.5785 hr for ^{56}Mn, the others were as follows:

1. The room background was determined by $1/R_e^2$ rule. Here, Re was the effective mean distance, deduced according to the real distance R between the target spot and the sample and the diameters of them as well. This correction varied 1.3% - 0.3% for R = 80 - 130 mm.

2. The influence of the charged particles ^{3}He, p and T from d+D reaction was measured and found to be less than 0.2% within 15 hr irradiation period.

3. The absorption and the scattering of the neutron target assembly were 1.1% and 0.5%, respectively, if it was cooled by water, and 0.5% and less than 0.1%, if cooled by air. The self-absorption of the samples was 0.1% (Al foil) and 0.2% (Fe foil).

4. The dead time correction of detection system of the associated α-particle was 0.4% - 0.7%, depending on the deuteron beam intensity.

5. The influence of other isotope reaction ^{57}Fe(n,np+d+pn)^{56}Mn was 0.2%.

The main uncertainties of the measurement are listed in Table 1. The result of each run differed from others among the 7 experimental runs was less than 3σ (σ= statistics). Thus, the weighted average cross sectiond from them were obtained as 113.2 ± 1.5 mb and 111.4 ± 1.5 mb for ^{27}Al(n,α)^{24}Na and ^{56}Fe(n,p)^{56}Mn and are shown as in Fig.1 and Fig.2, respectively.

The experimental and some evaluated cross sections for ^{27}Al(n,α)^{24}Na and ^{56}Fe(n,p)^{56}Mn, published in the last ten years are also shown in Fig.1 and Fig.2, respectively. For comparison, all the data have been normalized to the point of En= 14.6 MeV, according to the shapes of the evaluated cross section curves/2/. Except a few data of too great uncertainty(see the notes of the figures), the weighted average cross sections of the others are 113.5 ± 0.5 mb for ^{27}Al(n,α)^{24}Na and 110.7 ± 0.7 mb for ^{56}Fe(n,p)^{56}Mn. Meanwhile, the latest evaluation of Ryves/3/ gave 114.1 ± 0.5 mb and 109.9 ± 0.9 mb, respectively. As one sees, the present ones agree with them within the uncertainties, but are lower by 0.5% ^{27}Al(n,α)^{24}Na and higher around 1% for ^{56}Fe(n,p)^{56}Mn.

REFERENCES

1. V.E.Lewis, Metrologia, 20, 49(1984).
2. A.Pavlik and G.Winkler, INDC(AUS)-011/LI, INT(86)-6, 1986.
3. T.B.Ryves, Proc. Specialists' Meeting on Neutron Activation Cross Sections for Fission and Fusion Energy Applications, ANL, 1989 (OECD-Paris, 1990) p.293.

Table 1. Uncertainties of Measurement

Source of uncertainty	Uncertainty(%)
1. Neutron fluence	
Solid angle of α-particle detector	0.4
Background correction for α-particle	0.2
Anisotropy of neutron	0.3
Solid angle of sample	1.0
Room background correction	0.1
Absorption correction of target assembly	0.4
Scaterring correction of target assembly	0.2
Absorption and scaterring of the sample's frame	0.2
Self-absorption and scaterring of sample	0.1(Al) 0.2(Fe)
Neutron energy	0.3
2. Activity measurement	
Beta-efficiency extrapolation factor	0.1
Statistics	0.3
Interference of other isotope	0.1(Fe)
Total	1.33(Al) 1.34(Fe)

528

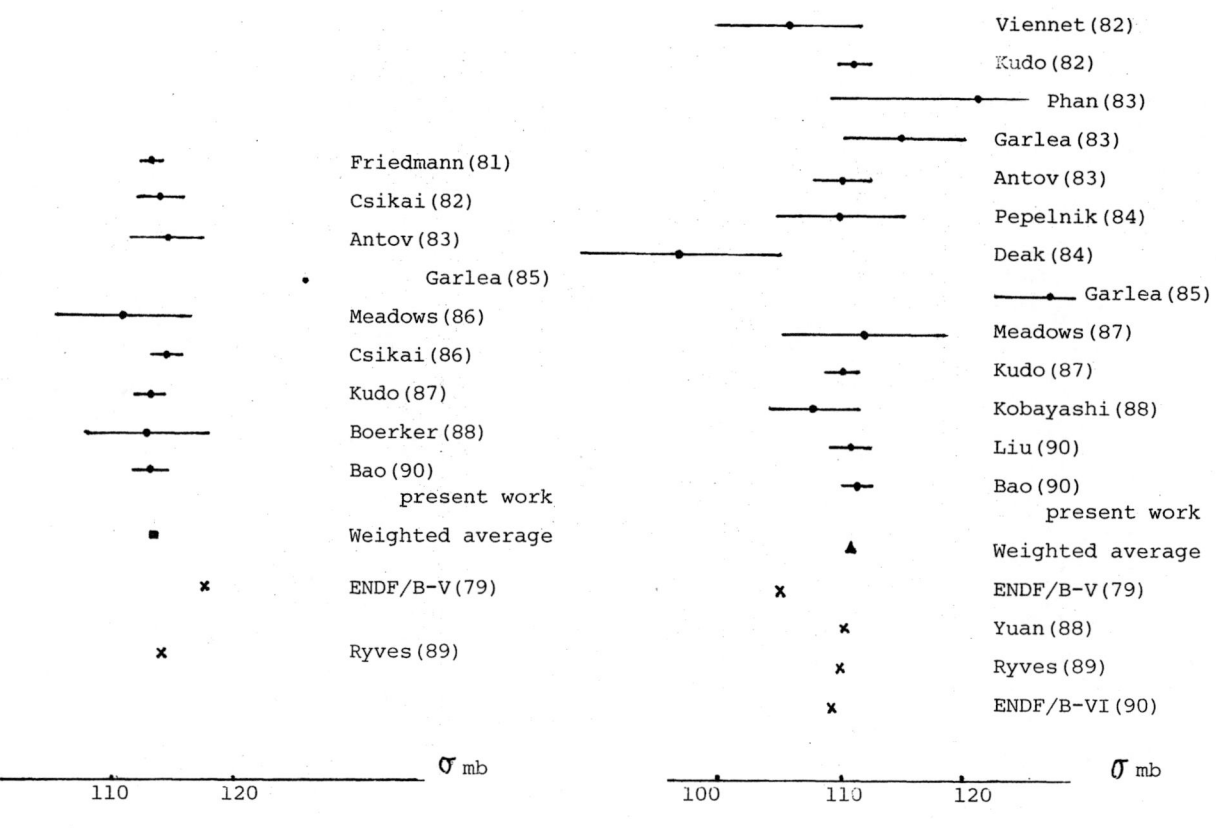

Fig.1 The cross section of ^{27}Al(n,α)^{24}Na
at En = 14.6 MeV.

Fig.2 The cross section of ^{56}Fe(n,p)^{56}Mn
at En = 14.6 MeV.

Figure note: • Experimental data.
 ✗ Evaluated data.
 ■ Weighted average from 8 experimental points in Fig.1, except Garlea(85).
 ▲ Weighted average from 10 experimental points in Fig.2, except Phan(83),Deak(84)
   and Garlea(85).

Antov(83), Bulgarian J. Phys. 10, 601.
Bao(90),Present work.
Boerker(88), Int. Conf. on Nucl. Data for Sci.
   & Technology, Mito, P.1025.
Csikai(82), Int. Conr. on Nucl. Data for Sci.
   & Technology, Antwerp, p.414.
Csikai(86), Z.Physik/A325, 69.
Deak(84), Nucleonika,29,87.
Friedmann(81), Z.Physik/A302, 271.
Garlea(83), INDC(ROM)-15/GI.
Garlea(85), Revue Roumaine de Physique,29,421.
Kobayashi(88), Int.Conf. on Nucl. Data for
   Sci. & Technology, Mito, p.261.
Kudo(82), NEANDC(J) 83/U,1.

Kudo(87), J. Nucl.Sci. & Technology, 24, 684.
Liu(90), Acta Sinica of Nanjing University (to
   be published).
Meadows(86), Radiation Effects 92, 123.
Meadows(87), Ann. Nucl. Energy 14, 459.
Pepelnik(84), NEANDC-E-252 U,p.30.
Phan(83), INDC(VN)-2/GI.
Ryves(89), NEANDC-259 'U', p.293.
Viennet(82), Int. Conf. on Nucl. Data for Sci.
   & Technology, Antwerp, p.407.
Yuan(88), Int. Conf. on Nucl. Data for Sci. &
   Technology, Mito, p.517.
ENDF/B-V(79), Evaluated Nuclear Data File.
ENDF/B-VI(90), Evaluated Nuclear Data File.

CROSS SECTION DATA FOR PROTON, ^{3}He AND α-PARTICLE INDUCED REACTIONS ON ^{nat}Ni, ^{nat}Cu AND ^{nat}Ti FOR MONITORING BEAM PERFORMANCE

F. Tárkányi, F. Szelecsényi
Institute of Nuclear Research of Hungarian Academy of Sciences
H-4001 Debrecen, Hungary

P. Kopecký
Nuclear Research Institute Department of Radiopharmaceuticals
250 68 Rez, CSFR

Abstract: For monitoring the beam performance excitation functions have been measured for ^{nat}Ni(p,x)^{57}Ni process up to 30 MeV, ^{nat}Ti(p,x)^{48}V reactions up to 18 MeV, as well as for several alpha and ^{3}He induced reactions on ^{nat}Cu and ^{nat}Ti up to about 40 MeV. The measured excitation functions are critically compared with the literature data and recent compilations.

(^{nat}Ni, ^{nat}Ti and ^{nat}Cu target; proton, alpha and ^{3}He induced nuclear reactions; charged particle monitor reactions)

Introduction

Monitor reactions play an important role in estimating and checking the beam performance not only for neutron fields but also for charged particle irradiations. Only very few laboratories have the possibilities to check the energy and the intensity of the incident charged particles by a magnetic field or a sophisticated time of flight method. The use of a "perfect", large Faraday-cup for beam intensity monitoring is also possible only at some laboratories and only in limited simple irradiation geometries.

Unfortunately, the existing literature data on the recommended standard nuclear reactions are, even in the most simple cases, contradicting and for many reactions rather scarce. A recent compilation by the IAEA shows that for real standardization of these reactions more experimental data with better accuracy are needed [1]. In the frame of this work we try to extend and to complete the existing data base for several important low energy monitor reactions used in every day practice in isotope production and in charged particle activation analysis.

Experimental Procedure

The excitation functions were measured by the stacked-foil technique. High purity metallic foils (Goodfellow Metals, Cambridge, England) were activated together as a stack for cross checking the investigated excitation functions on different targets (Ti+Cu, Ni+Cu).

Irradiations were carried out at the Rez U-120M cyclotron, the Jülich CV-28 cyclotron and the Debrecen MGC-20E cyclotron using different incident energies to get thin stacks and overlapping energy regions. The total integrated charge was detected by a Faraday-cup. The energy of the incident beam was controlled by magnetic deflection at the Rez and Debrecen cyclotrons and by normalization to other monitor reactions at the Jülich experiment. The beam energy degradation was calculated in accordance with Williamson et al. [2] and controlled by a method based on the principle of flux constancy along the stack [3].

The induced activities were determined via standard, high resolution gamma-ray spectrometry without chemical separation. Special attention was paid to the necessary corrections (dead time, pile-up, cascade losses, etc) to obtain absolute activities. All irradiated foils were measured several times in order to follow the decay and for calculation of cumulative cross sections.

The data evaluation and the error calculation were similar to those described in more detail in [4]. The cross sections are accurate to ±6-15 %. The uncertainties are lower near the maximum and higher near the threshold. The error of the corresponding energy values is small at the front foils of the different irradiations (±0.5 MeV), and the uncertainty varying in accordance with the straggling, the foil non-uniformities and the calculation method up to ±1.5 MeV.

Results and Discussions

The obtained results are summarized in Table 1.

Monitor reactions for protons

Chemically resistant and mechanically stable ^{nat}Ni and ^{nat}Ti foils could be used very effectively for the simultaneous energy and flux determination. The completed experimental data show good agreement and can serve as a good data base for application of these reactions as monitors, similar to proton induced reactions on natural Cu targets.

Table 1 Summary of investigated excitation functions

Particle	Target	Incident energies (MeV)	Energy range (MeV)	Final nucleus	Literature
p	^{nat}Ni	30, 22, 18	0 - 30	^{57}Ni, ^{57}Co ^{56}Co, ^{55}Co	5, 6, 7, 8 10
	^{nat}Ti	18 14.5	0 - 18 (18 - 30)	^{48}V, ^{47}Sc ^{44m}Sc, ^{44g}Sc ^{43}Sc	9
^{3}He	^{nat}Ti	36 28	0 - 36	^{48}V, ^{47}Sc ^{46}Sc, ^{44m}Sc ^{44g}Sc	11
	^{nat}Cu	36 28	0 - 36	^{61}Cu, ^{65}Zn ^{63}Zn, ^{62}Zn ^{67}Ga, ^{66}Ga,	12, 13, 14 15
α	^{nat}Ti	40 20	0 - 40	^{48}V, ^{47}Sc, ^{46}Sc, ^{44m}Sc ^{44g}Sc	11, 16
	^{nat}Cu	40 20	0 - 40	^{61}Cu, ^{65}Zn ^{63}Zn, ^{67}Ga, ^{66}Ga	10, 12, 17 18, 19, 20 21, 22, 23

In the Fig. 1 and 2 we present the obtained excitation functions for production of ^{57}Ni from ^{nat}Ni and for ^{48}V from ^{nat}Ti, respectively. For illustration of the agreement of the existing results the earlier data are also reproduced.

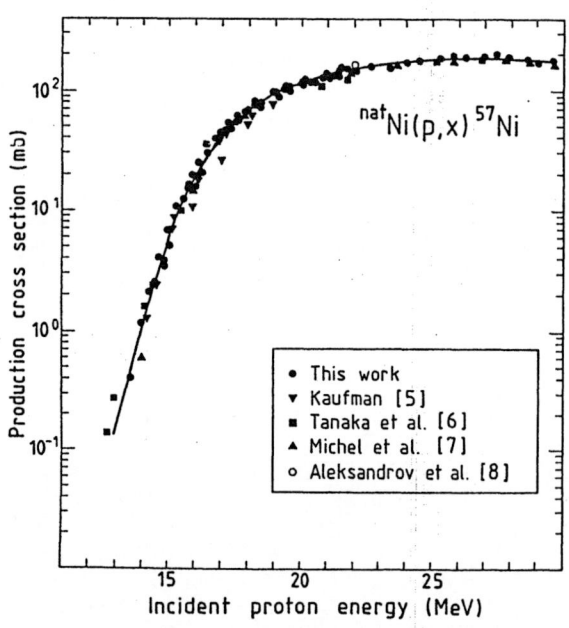

Fig.1. Excitation function for the production of ^{57}Ni from proton bombardment of ^{nat}Ni.

Monitor reactions for ^{3}He particles

When comparing the published results obtained for the proposed monitor reactions it could be concluded that:
- for the reactions on ^{nat}Ti only one work was done with some discrepancies in cross sections between two energy regions [11]
- many works have been done for the ^{3}He reactions on natural Cu (see Table 1) but the results are contradictory. The existing data could be separated roughly in two groups which disagree by a factor of about 2. The situation is similar in

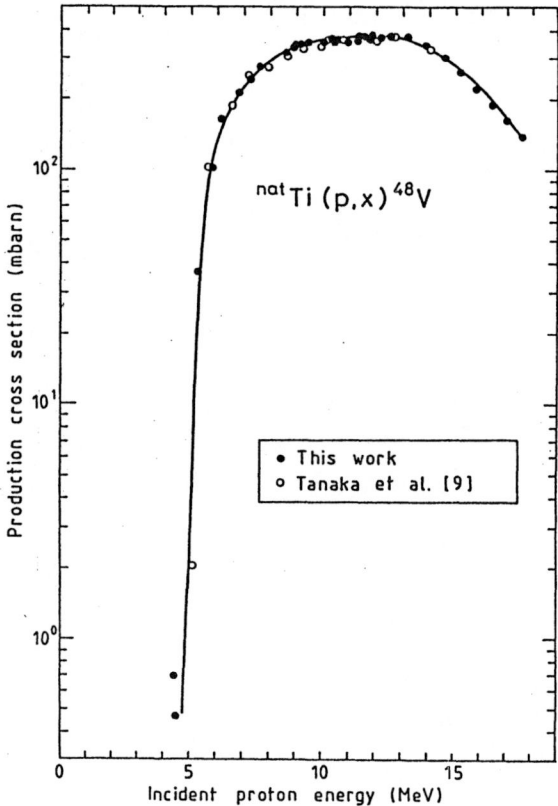

Fig.2. Excitation function for the ^{nat}Ti(p,x)^{48}V process.

alpha particle induced reactions on ^{nat}Cu as discussed later.

Our experimental results for ^{3}He on Cu leading to the formation of ^{66}Ga support the higher values [12,15], as illustrated in Fig.3.

In the case of the monitor reaction ^{nat}Ti(^{3}He,x)^{48}V the agreement is good with the results obtained in low energy irradiation by Weinreich et al. [11], (see Fig.4.).

Monitor reactions for alpha particles

Our values obtained on ^{nat}Cu both in absolute values and energy scale are similar to those in ^{3}He induced reactions and are in agreement with the data of [12,20,21,23] but are systematically higher than in [18,22]. Fig.5. shows the obtained cross sections for ^{66}Ga together with the results given by the authors mentioned above.

A good agreement with the published data could be observed on the excitation function of the ^{nat}Ti(α,x)^{51}Cr reaction, as represented in Fig.6.

Conclusion

On the basis of the obtained experimental results and the critical comparison with the literature data it could be concluded that:
- for monitoring the proton beam, from all the points of view (target material, cross section data, nuclear decay data) the reactions on ^{nat}Ni and ^{nat}Ti could be

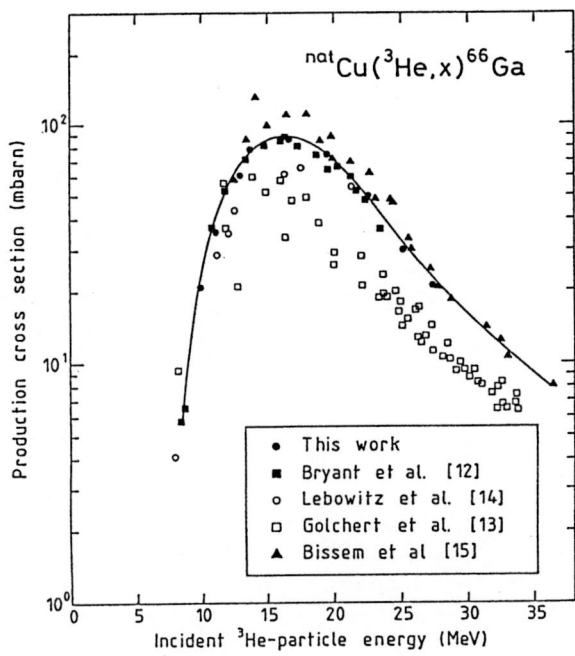

Fig.3. Excitation function for the production of ^{66}Ga from ^{3}He bombardment of ^{nat}Cu.

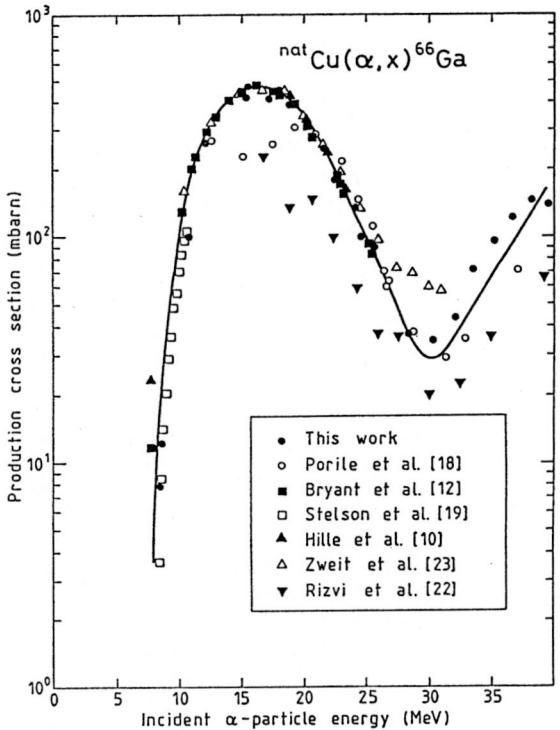

Fig.5. Excitation function for the ^{nat}Cu$(\alpha,x)^{66}$Ga process.

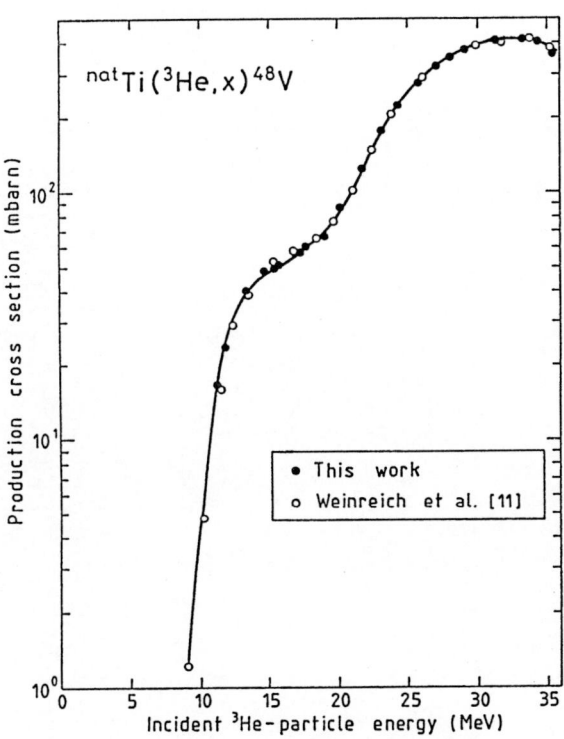

Fig.4. Excitation function for the production of ^{48}V from ^{3}He bombardment of ^{nat}Ti.

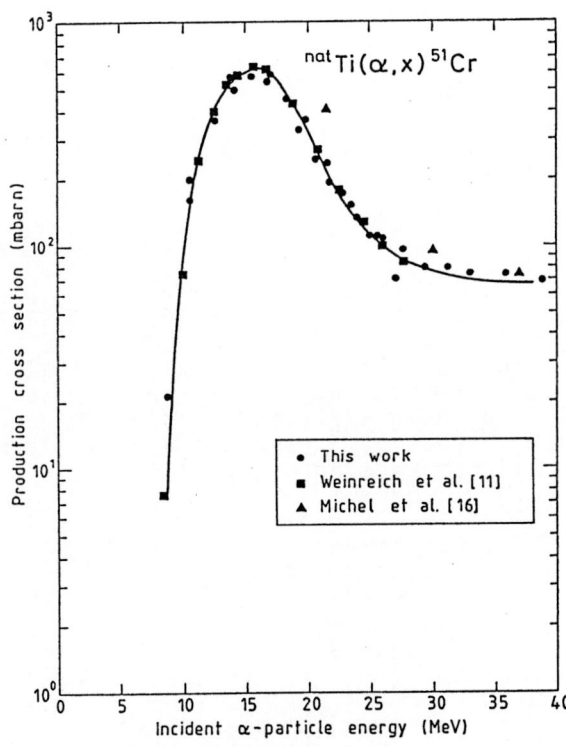

Fig.6. Excitation function for the production of ^{51}Cr from alpha particle bombardment of ^{nat}Ti.

used effectively, especially the ^{nat}Ni$(p,x)^{57}$Ni and ^{nat}Ti$(p,x)^{48}$V reactions
- for monitoring the ^{3}He and alpha particle beams the reactions induced on ^{nat}Ti are also very useful

- in spite of numerous measurements on ^{3}He and alpha particle induced reactions on the most commonly used natural Cu target, the literature data are surprisingly contradictory. Further detailed measurements are necessary; attention has to be paid especially to the beam current measurement which is perhaps the most important source of systematic error.

References

1. O. Schwerer and K. Okamoto Ed.: Status Report on Cross Sections of Monitor Reactions for Radioisotope Production INDC(NDS)-218/GZ+, 1989
2. C.F. Williamson, J.P. Boujot and J. Pickard: Report CEA-R-3042 (1966)
3. F. Tárkányi, F. Szelecsényi and S. Takács: Abstracts of the fifth Symp. on Medical Application of Cyclotrons (Edited by S.-J. Heselius) p.29 (1989)
4. F. Tárkányi , F. Szelecsényi and P. Kopecky: Int. J. Appl. Radiat. Isotopes 42, 513(1991)
5. S. Kaufman: Phys. Rev. 117, 1532(1960)
6. S. Tanaka, M. Furukawa and M. Chiba: J. Inorg. Nucl. Chem. 34, 2419(1972)
7. R. Michel, H. Weigel and W. Herr: Z. Phys. A286, 393(1978)
8. V.N. Aleksandrov, M.P. Semyonova and V.G. Semyonov: Atomnaya Energiya 62, 411(1987)
9. S. Tanaka and M. Furukawa: J. Phys. Soc. Jap. 14, 1269(1959)
10. M. Hille, P. Hille, M. Uhl and W. Weisz: Nucl. Phys. A198, 625(1972)
11. R. Weinreich, H.J. Probst and S.M. Qaim: Int. J. Appl. Radiat. Isotopes 31, 223(1980)
12. E.A. Bryant, D.R.F. Cohran and J.D. Knight: Phys. Rev. 130, 1512(1963)
13. N.W. Golchert, J. Sedlet and D.G. Gardner: Nucl. Phys. A152, 419(1970)
14. E. Lebowitz and M.W. Greene: Int. J. Appl. Radiat. Isotopes 21, 625(1970)
15. H.H. Bissem, R. Georgi and W. Scobel: Phys. Rev. C22, 1468(1980)
16. R. Michel, G. Brinkmann and R. Stück: Radiochimica Acta 32, 173(1983)
17. K.G. Porges: Phys. Rev. 101, 225(1956)
18. N.T. Porile and D.L. Morrison: Phys. Rev. 116, 1193(1959)
19. P.H. Stelson and F.K. McGowan: Phys. Rev. B133, 911(1964)
20. I.A. Watson, S.L. Waters, D.K. Bewley and D.J. Silvester: Nucl. Instr. Meth, 106, 231(1973)
21. H.P. Graf and H. Münzel: J. inorg. nucl. Chem 36, 3647(1974)
22. I.A. Rizvi, M. A. Ansari, R.P. Gautam, R.K.Y. Singh and A.K. Chaubey: J. Phys. Soc. Jap. 56, 3135(1987)
23. J. Zweit, H. Sharma and S. Downey: Int. J. Appl. Radiat. Isot. 38, 499(1987)

PRODUCTION OF A REFERENCE MONOENERGETIC NEUTRON FIELD USING D(d,n)^{3}He REACTION

K. Kudo, N. Takeda, Matiullah* and A. Fukuda

Quantum Radiation Division, Electrotechnical Laboratory,
1-1-4 Umezono, Tsukuba, Ibaraki 305, Japan

Abstract: A reference energy neutron field, produced by a D(d,n)^{3}He reaction using a Cockcroft-Walton type accelerator, has been developed. Calculated results showed that neutrons which are emitted at an angle of 100°, with respect to the incident deuteron beam direction, from a Ti-D solid target have a very sharp distribution around 2413 keV regardless of the target thickness, its age and the incident deuteron beam energy. A ^{3}He proportional counter was, first, calibrated in a reference energy field of 2413 keV, emitted at an angle of 100°, and then applied in measurements of the pulse height spectra at various emission angles to the deuteron beam direction. From the measured pulse height spectra, information concerning the mean energy and spread were derived.

(monoenergetic neutrons, NRESP code, ^{3}He(n,p)T reaction, ^{3}He proportional counter, D(d,n)^{3}He reaction, Ti-D target, reference energy, χ^2 fitting, fusion diagnostics, folded spectrum, energy calibration)

Introduction

Production of monoenergetic neutrons with least possible spread is highly desirable in certain situations like nuclear cross section measurements, response measurements of neutron detectors and energy calibration of neutron spectrometers. To do so, several most common nuclear reactions are used at the electrostatic accelerators such as a Cockcroft-Walton and a Van de Graaff accelerators. These include T(d,n)^{4}He, D(d,n)^{3}He, T(p,n)^{3}He and ^{7}Li(p,n)^{7}Be reactions and cover the range of neutron energies, depending upon the type and energy of the incident particles, from several tens of keV to 20 MeV except the energy gap between 7 MeV and 12 MeV. A major difficulty encountered in using such reactions is the existence of spread in the neutron energy. This spread in neutron energies corresponds to the energy spreads of bombarding charged particles which are slowed down in a target material and produce reactions at various depths in the target. In addition to the target thickness, neutron energy and its spread from such reactions also depend upon the energy of the incident particle beam and the depth distribution of the target atoms.

In the energy calibration of neutron spectrometers, reference energy points of monoenergetic neutrons which are independent of the bombarding particle energy, target thickness, age and composition are needed. In the case of a ^{3}He proportional counter, a thermal neutron peak, produced by a ^{3}He(n,p)T reaction is very often used as one of the invariable energy references for the calibration of the energy axis and determination of the detector resolution. However, in precise spectrometry of fast neutrons, for instance, required in fusion diagnostics of DD or DT plasma, the extrapolation of the energy calibration from thermal peak contributes towards uncertainties in the higher energy region of fast neutrons. Moreover, the energy resolution used for thermal peak can not be applied to the higher energy region. Establishment of easily reproducible reference energy fields of fast neutrons with limited widths even under conditions of different bombarding energy and target thickness, is therefore, highly desirable in order to improve the quality of neutron spectrometry.

Seagrave first suggested the idea of production of the monoenergetic DD and DT neutrons [1] and Ryves et al. [2,3] calibrated a Si-SSD and Zr/Nb activation foils with 14.00 MeV neutrons emitted at a judiciously chosen angle. In our earlier paper, we have reported the establishment of a standard irradiation field of DT neutrons using T(d,n)^{4}He reaction. An associated α-particle technique was employed for neutron spectra measurements. Neutrons emitted in the direction of 96° - 98° with respect to the direction of deuteron beam were found to have a sharp distributions around 13.98 MeV to 14.00 MeV with the energy spread less than 10 keV [4].

In this paper, we report the extension of the above technique to DD neutron field and establish a monoenergetic neutron field of 2413 keV, emitted at a special angle of 100°, regardless of the target thickness, its age and incident deuteron energy.

Characteristics of Reference Energy DD Neutrons

A 300 kV Cockcroft-Walton type accelerator was used for producing neutrons of energies ranging from 2.4 MeV to 3.2 MeV. All the conditions of accelerator, beam transport system and target assembly were similar as those used in DT neutron standards except the target [4].

Figure 1 (a) and (b) show calculated energy spectra of DD neutrons, emitted at angles of 85° to 115° with respect to the deuteron beam direction for beam energies of 130 keV and 260 keV, respectively. These spectra were calculated on the

---

* Guest Researcher: Centre for Nuclear Studies (CNS), P.A.E.C, Nilore, Islamabad, Pakistan.

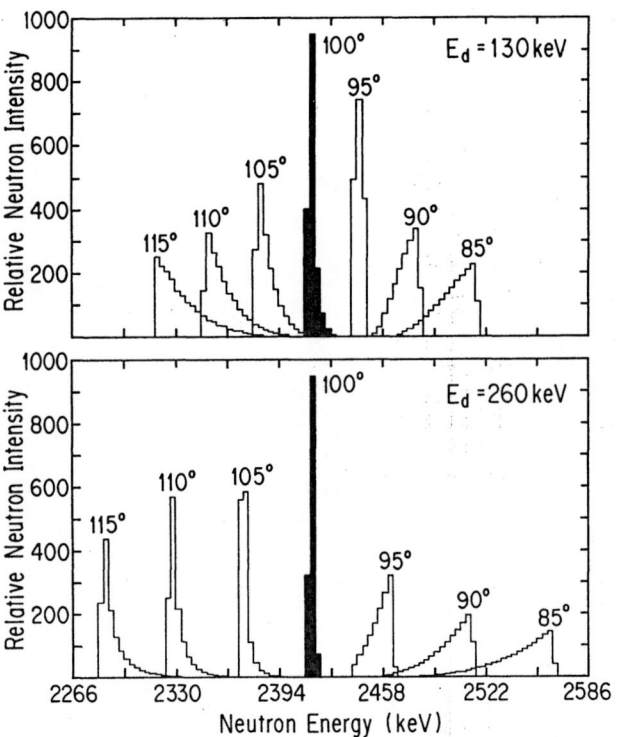

Fig. 1. Calculated neutron energy distributions for incident deuteron beam energies of (a) 130 keV and (b) 260 keV. Calculations were made for thick Ti-D target for the listed emission angles.

basis of relativistic kinematics of a two-body reaction, taking into account the angular differential cross section of D(d,n)^{3}He reaction [5] and the stopping power [6] of deuterons in a homogeneously distributed Ti-D target. It must be borne in mind that the above calculations were made for a thick target (as used in the experiments) ignoring the effect of angular straggling of deuterons inside the target material. As can be seen in Fig. 1, neutrons produced at the listed emission angles have energy ranging from 2320 keV - 2520 keV for a fixed incident deuteron beam energy of 130 keV. The range of energy becomes wider ranging from 2280 keV - 2570 keV with an increase in the incident deuteron energy from 130 keV to 260 keV. However, at an emission angle of 100°, the energy distribution has a very sharp peak around 2413 keV for both 130 keV and 260 keV incident beam energies. Full width at half maximum (FWHM) at this angle is less than 10 keV. On the basis of these calculated spectra which would be confirmed experimentally, we conclude that DD neutrons emitted at 100° to incident beam direction provides an excellent reference field for the calibrations of energy and resolution of neutron spectrometers.

Measurements of Pulse Height Spectrum at 2413 keV With a ^{3}He Proportional Counter

A ^{3}He proportional counter is known to have good characteristics for neutron spectrometry in a wide energy range of thermal to 20 MeV using ^{3}He(n,p)T and ^{3}He(n,n)^{3}He reactions. The latter reaction was used in determination of the mean

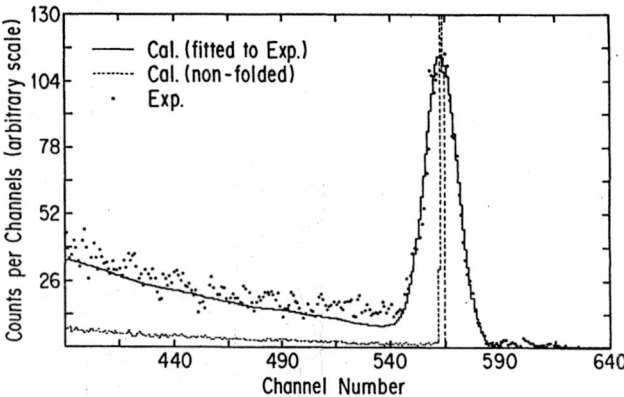

Fig. 2. The Monte-Carlo simulated spectrum (solid line) fitted to the experimental spectrum (individual points) along with non-folded spectrum (broken line).

neutron energy and spread in the DT neutron field [4]. However, in this present work concerning the development of a reference DD neutron field, we have used ^{3}H(n,p)T reaction for spectra measurements in order to avoid various complications involved in using ^{3}He(n,n)^{3}He reaction. A ^{3}He proportional counter, containing mixture of ^{3}He and Ar gases at 600 kPa (^{3}He at 400 kPa and Ar at 200 kPa), was used in the measurements of the energy distribution of DD neutrons. The detector had an active length of 15 cm with a diameter of 2.5 cm.

Precise pulse height spectrum, corresponding to the monoenergetic neutrons of energy 2413 keV, was calculated by the modified NRESP code [7]. Calculated pulse height spectrum was folded with a Gaussian distribution whose resolution is assumed to be proportional to the inverse of square root of the pulse height of a detector signal. Figure 2 illustrates non-folded pulse height spectrum (broken line) around the ^{3}He(n,p)T reaction peak, calculated by the modified NRESP code. Properly folded spectrum (solid line) fitted to the experimental spectrum (shown as individual points) is also shown in the above figure. The data was fitted to the experimental values measured at an emission angle of 100° for the incident deuteron energy of 260 keV.

Further investigations were made in order to compare the pulse height spectra at the reference angle of 100° by using deuteron energies of 75 keV, 110 keV, 130 keV, 220 keV and 260 keV. The deviation in the peak positions for the incident deuteron beam energies in question was found to be less than ±0.15% and the resolution of the detector system was (2.6 ± 0.2)% to the 2413 keV. Any significant shift has not been observed for five experimental spectra using different deuteron energies. These results suggest that neutron energy corresponding to 100° does not depend upon the different reaction energies of deuterons in a Ti-D target and hence confirm the calculated results. The spread in energy of neutrons which are emitted at this angle is negligible as compared to that of the detector system. To conclude, neutrons emitted at an angle of 100° with respect to the incident deuteron beam direction provides a reference point of 2413 keV with FWHM less than 10 keV in spite of the fact that energy spread of this reference energy can not be measured by the

[3]He proportional counter because of its poor resolution.

Comparison of Pulse Height Spectra at Different Angles Using a [3]He Proportional Counter

Figure 3 (a) and (b) show DD neutrons pulse height spectra at 0°, 45°, 75° and 100° for deuteron energies of 130 keV and 260 keV respectively. Individual points represent experimental

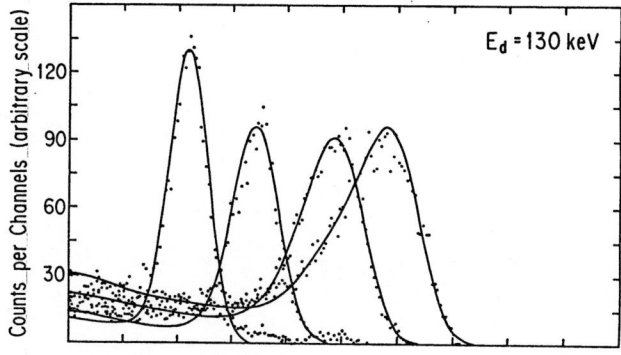

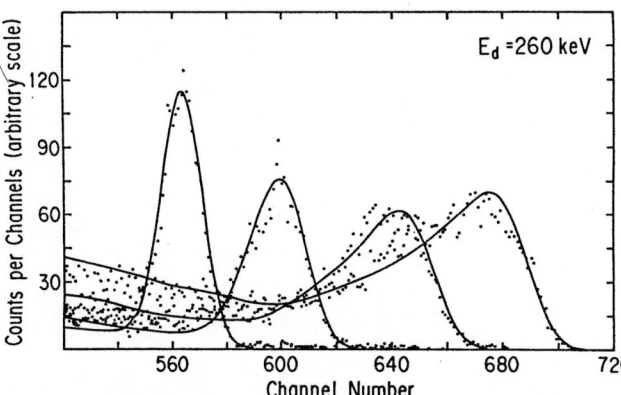

Fig. 3. Comparisons of calculated (solid line) and measured (individual points) pulse height spectra at the listed emission angles for deuteron energy of (a) 130 keV and (b) 260 keV (see, text for details).

values whilst solid curves correspond to the values calculated by the modified NRESP code assuming the spectrum at each angle calculated on the basis of relativistic kinematics for a thick

Table 1. Mean energy and spread in neutrons emitted at different angles. Upper values in the table correspond to 260 keV and lower values to 130 keV incident deuteron energies.

Angle (deg)	Mean Energy (keV)	Energy Spread FWHM (keV)
0	3040	113
0	2840	99
45	2870	79
45	2720	76
75	2630	43
75	2560	40
100	2413	6
100	2413	6

homogeneous target. The mean energy and the spread in DD neutrons emitted at each angle were determined after fitting the properly folded spectrum to an experimental spectrum using a χ^2 fitting procedure. Table 1 shows mean energy and spread in DD neutrons at the listed emission angles with respect to the incident beam direction. Differences in energies and spread at different angles of emission can clearly be seen in the table. The measured energy distribution at each angle is seen to be shifted towards the lower energy direction, for example, about 1.5% at an angle of 0° for both deuteron energies as compared with the calculated energy distribution.

Conclusion

In order to produce a reference monoenergetic field of DD neutrons, independent of the incident deuteron energy, target thickness and its composition, computer calculations were carried out. These calculations were based on relativistic kinematics of two-body reaction and indicated that neutrons which are produced as a result of the D(d,n)[3]He reaction and emitted at an angle of 100° to the incident deuteron beam direction have a very sharp distribution around 2413 keV with a FWHM less than 10 keV, ignoring the angular straggling of deuterons in the target material.

The pulse height spectra were measured with a [3]He proportional counter at the reference angle of 100° for different deuteron energies ranging from 75 keV to 260 keV. No change has been observed in five experimental spectra from the [3]He(n,p)T reaction for different incident energies. From the measured pulse height spectra using a calibrated [3]He detector in a reference field, information concerning the mean energy and spread were derived. Based on these results, we conclude that the energy of neutrons, emitted at an angle of 100° to the incident beam direction, does not depend upon the reaction energy of deuterons in a Ti-D target and hence neutrons emitted at this angle provides a reference energy point of 2413 keV. The establishment of the aforesaid standard field at our laboratory (ETL) would mainly be used for the calibration of the precise spectrometers, used in studying fusion diagnostics.

REFERENCES

1. J.D. Seagrave: LAMS-2162 (1958)
2. T.B. Ryves and K.J. Zieba: Nucl. Instrum. Methods 167, 449 (1979)
3. V.E. Lewis and K.J. Zieba: Nucl. Instrum. Methods 20, 49 (1984)
4. K. Kudo and T. Kinoshita: Nucl. Eng. & Design/Fusion 10, 145 (1989)
5. H. Liskien and A. Paulsen: Nucl. Data Tables 11, 569 (1973)
6. H.H. Anderson and J.F. Ziegler: Stopping Powers and Ranges in All Elements (Pergamon, New York, 1977)
7. G. Dietze and H. Klein: PTB report ND-22 (1982)

Part VI

Nuclear Structure and Decay Data

NEW CALCULATION OF BETA-DECAY DATA
OF NUCLEI FAR FROM STABILITY

M. Hirsch, A. Staudt, K. Muto[1] and H.V. Klapdor-Kleingrothaus

Max-Planck-Institut für Kernphysik, Heidelberg, Germany
[1] Tokyo Institute of Technology, Tokyo, Japan

Abstract: Beta-decay properties of nuclei far from stability are of decisive importance for the design and safety of nuclear reactors and for the understanding of the synthesis of heavy elements in astrophysical scenarios. Experimental data on β-decay are only partially available. Thus, theoretical predictions of β-decay observables are required. A second-generation microscopic calculation of both neutron-rich and neutron-deficient isotopes is presented. This approach uses the proton-neutron quasiparticle RPA.

1. Introduction

Considerable progress in the understanding of nuclear β-decay far from stability has been achieved in recent years [1], allowing for an improved second-generation microscopic calculation of β-decay properties of neutron-rich [2] and neutron-deficient isotopes [3]. In particular reliable predictions for neutron-rich isotopes are of importance for astrophysical conclusion on the element synthesis via the r-process and the age of the galaxy by means of cosmochronometers [1,4].

In addition β-decay observables such as half-lives, mean energies and β-delayed neutron emission probabilities are essential input parameters for decay heat calculations for nuclear fission reactors [5,6]. The decay heat power is an important quantity for the design and safety of nuclear reactors. A major source of heat after a shutdown of the reactor are the β- and γ-rays from the decay of the accumulated fission products.

Fig. 1 summarizes the current experimental knowledge of β-decay data in the region of fission products. Black dots denote stable isotopes, unstable isotopes are shown for two different cases. Nuclei for which only half-lives were experimentally determined but no detailed decay schemes are available are

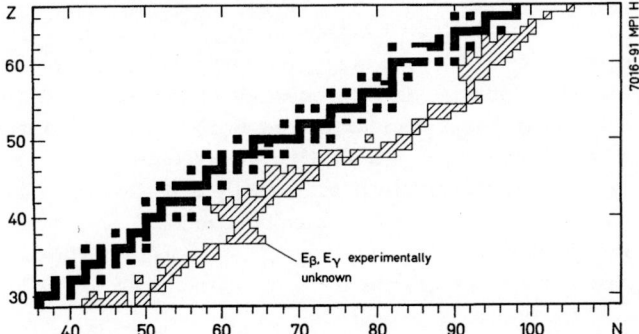

Fig.1: Schematic view of the current experimental knowledge on β-decay data in the region of fission products.

shown as grey-shaded area. For the white area between these two borderlines experimental mean energies can be found in ref. [7] and [8].

From Fig. 1 it can be seen that experimental mean β- and γ-energies are lacking for more than ∼ 150 isotopes. The situation is even worse for probabilities of β-delayed neutron emission. Thus, for an accurate calculation of the decay heat input from theoretical predictions is required.

There have been two earlier calculations of β-decay. The first one is the gross theory calculation of Takahashi et al. [9]. A more advanced TDA-calculation was presented by Klapdor et al. [10]. Though the TDA-approximation puts the predictions of β-decay rates on a more reliable basis than the older gross theory, the calculation was restricted to β⁻-decay due to some neglections. The second-generation microscopic model presented here can account for both β⁻- and β⁺-decay and in addition leads to a considerable improvement over the earlier approaches.

2. Model description

We use the proton-neutron quasiparticle random phase approximation (pn- QRPA) [2,11] to calculate the β-strength function of the individual isotopes. The model consists of three main steps. As a starting point single particle energies are calculated in the Nilsson model, which takes into account nuclear deformation. The pairing interaction is treated in the BCS-approximation. In the subsequent RPA-calculation proton-neutron residual interactions are accounted for by adding a schematic Gamow-Teller interaction to the Hamiltonian.

Extensive discussions of the details of the model and investigations of the influence of the model parameters on the results can be found in the literature [2,3,11].

Once the β-decay matrix elements have been determined the calculation of all interesting β-decay observables is straightforward.

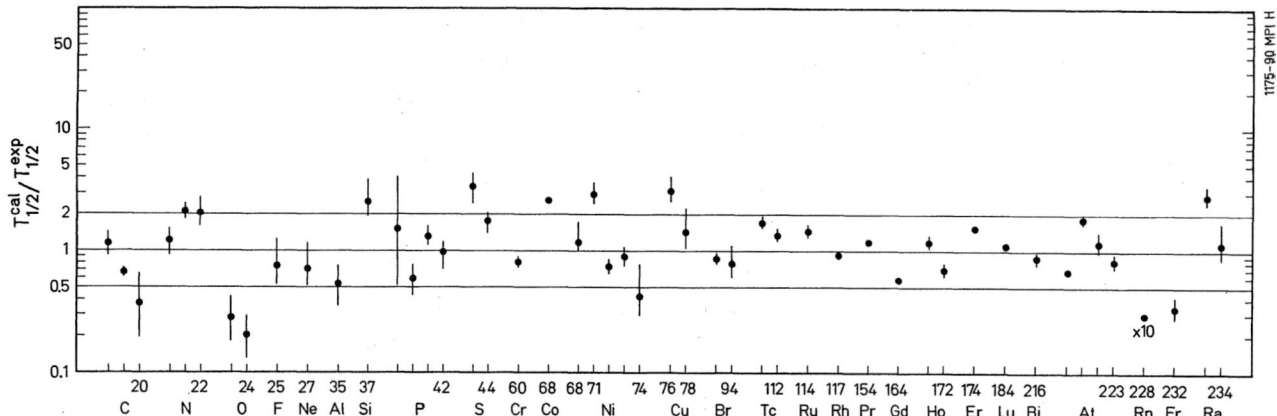

Fig. 2: Accuracy of the calculated β⁻-decay half-lives. Q_β was taken from reference [12]. Note that all experimental values have been measured after the completion of the calculation.

3. Results and discussion

With the model described above half-lives for both β⁻- and β⁺-decaying isotopes up to the particle drip lines have been calculated [2,3].

Fig. 2 demonstrates the reliability of the model for the prediction of β⁻-decay half-lives far from stability. In the Figure ratios of calculated to experimental values are shown for convenience. Note that all experimental values have been measured after the completion of the calculation.

Predictions of half-lives require an accurate knowledge of Q_β values. In order to calculate the half-lives of unknown nuclides we take Q_β values from the mass formulas of Hilf et al. [12], Groote et al. [13] and Möller and Nix [14], respectively. These mass formulae, especially the one of Hilf et al. [12], give reliable results for the prediction of decay rates. This can be seen from Fig. 2 and has already been pointed out in ref. [2]. Hence three sets of QRPA predictions are available [2].

Results on β⁺-calculations are shown in Fig. 3 for the example of $_{51}$Sb. Comparison with Fig. 2 indicates that the β⁺-calculation should be as reliable as the β⁻-calculation.

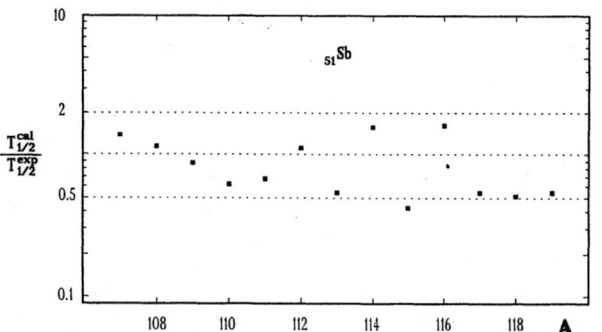

Fig. 3: Calculated β⁺-decay half-lives compared to experimental data

Besides half-lives mean β- and γ-energies are of interest in reactor physics, as discussed above. We applied the same β-strength distributions that were used for the half-life calculation to determine mean

β- and γ-energies [15].

The accuracy of the model is shown in Fig. 4, where some examples of calculated mean γ-energies are shown, compared to the most recent experimental data of Rudstam et al. [8].

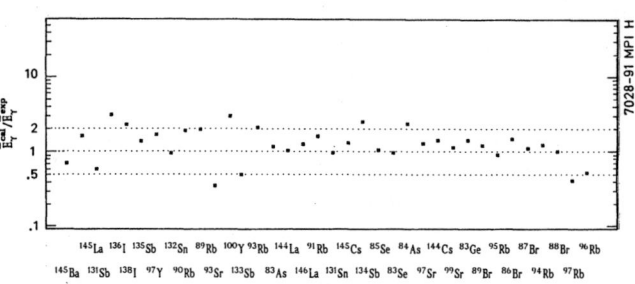

Fig. 4a: Degree of agreement between calculated and experimental mean γ-energies for some examples.

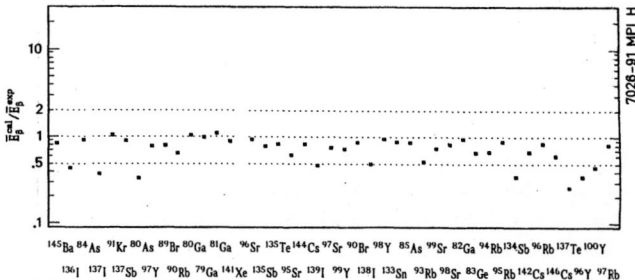

Fig. 4b: The same as Fig. 4a, but for mean β-energies.

Far from stability with increasing β-decay energy and decreasing neutron binding energy β-delayed neutron emission (βdn) is getting energetically possible. Since this decay mode acts as a competitive process to the more common γ- deexcitation βdn-values are important for the evaluation of mean γ-energies and thus have impact on decay heat calculations.

In addition the occurence of βdn-emission influences the abundances of the heavy elements synthesized in the astrophysical r-process.

Despite their importance β-delayed neutrons have been treated rather stepmotherly in physical applications in the past. For example, estimates on βdn-emission were made with the assumption of a

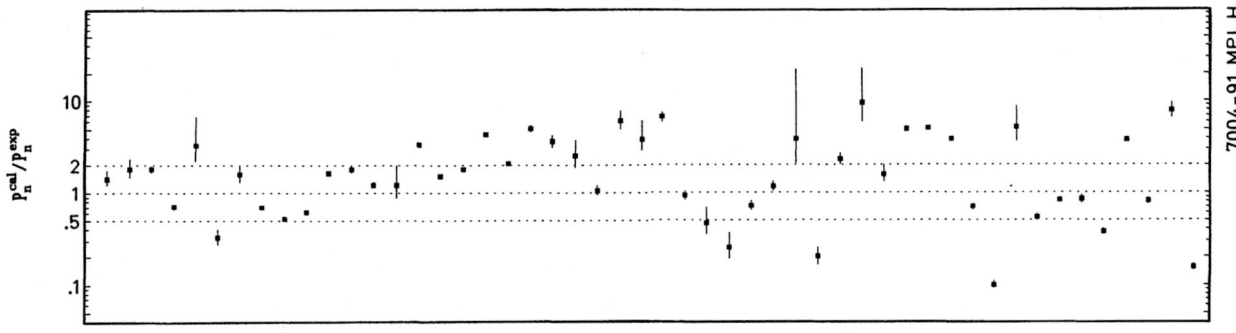

Fig. 5: Degree of consistency between calculated and experimental β-delayed neutron emission probabilities. The error bars indicate the experimental uncertainty only.

constant β-strength function [16].

With the new microscopic calculation presented here reliable predictions of probabilities of βdn-emission are possible, as demonstrated in Fig. 5. In the Figure we compare our calculation with experimental data of ref. [17]. The error bars indicate the experimental error only.

For the prediction of βdn-emission of unknown isotopes we use the same mass formulae [12-14] as in the calculation of half-lives. Thus, a consistent data set of β-decay properties is available now.

4. Summary

A new large scale microscopic calculation of β-decay properties of both neutron-rich and neutron-deficient isotopes has been presented. For β⁻-decay the model was shown to give reliable results for half-lives, mean energies and β-delayed neutron emission. An important advantage of this approach is that all interesting β-decay observables can be calculated simultaneously without any change of the incorporated model parameters.

In the case of β⁺-decay up to now only half-lives have been calculated and shown to be in good agreement with experimental data. These β-decay matrix elements will also allow the calculation of β-delayed proton emission in future.

References:

[1] H.V. Klapdor, Prog. Part. Nucl. Phys. **10** (1983) 131, and **17** (1986) 419

[2] A. Staudt, E. Bender, K. Muto and H.V. Klapdor-Kleingrothaus, At. Data Nucl. Data Tables **44** (1990) 79

[3] A. Staudt, M. Hirsch, K. Muto and H.V. Klapdor-Kleingrothaus, Phys. Rev. Lett. **65** (1990) 1543 and M. Hirsch, A. Staudt, K. Muto and H. V. Klapdor-Kleingrothaus, to be published

[4] F.-K. Thielemann, J. Metzinger and H.V. Klapdor, Z. Phys. **A 309** (1983) 301

[5] A. Tobias, Prog. Nucl. Energy **5** (1980) 1

[6] H.V. Klapdor and J. Metzinger, Proc. of Int. Conf. on Nucl. Data for Science and Technology, May 30 - June 3, 1988, Mito, Japan, ed. S. Igarasi, Saikon Publishing Co., Tokyo, Japan, p. 827

[7] E. Browne and R.B. Firestone, Table of Radioactive Isotopes, ed. V.S. Shirley, John Wiley & Sons, New York, 1986

[8] G. Rudstam, P.I. Johansson, O. Tengblad, P. Aagaard and J. Eriksen, At. Data Nucl. Data Tables **45** (1990) 239

[9] K. Takahashi, M. Yamada and T. Kondoh, At. Data Nucl Data Tables **12** (1973) 101

[10] H.V. Klapdor, J. Metzinger and T. Oda, At. Data Nucl. Data Tables **31** (1984) 81

[11] K. Muto, E. Bender and H.V. Klapdor, Z. Phys. **A 333** (1989) 125

[12] E.R. Hilf, H. v.Groote and K. Takahashi, CERN-Report 76-13 (1976) 142

[13] H.v. Groote, E.R. Hilf and K. Takahashi, At. Data Nucl. Data Tables **17** (1976) 418

[14] P. Möller and J.R. Nix, At. Data Nucl. Data Tables **26** (1981) 165

[15] M. Hirsch, A. Staudt and H.V. Klapdor-Kleingrothaus, to be published

[16] K.-L. Kratz and G. Herrmann, Z. Phys. **A 263** (1973) 435

[17] E. Lund and G. Rudstam, University of Uppsala, Research Report NFL-60 (1989)

Fission Product Decay Power--AESJ Recommendation

Shungo Iijima[+], Tadashi Yoshida
Nuclear Engineering Laboratory, Toshiba Corporation
Ukishima-cho, Kawasaki 210, Japan

Kanji Tasaka, Toshio Katoh
Nagoya University, Chikusa-ku, Nagoya 464-01, Japan

Jun-ichi Katakura
Japan Atomic Energy Research Institute
Tokai-mura, Ibaraki-ken 319-11, Japan

Ryuzo Nakasima
Hosei University, Fujimi-cho, Chiyoda-ku, Tokyo 102, Japan

<u>Abstract</u>: A recommendation regarding the FP decay power was issued from the Research Committee on Standardization of the Decay Heat Power in Nuclear Reactors of Atomic Energy Society of Japan (AESJ). The recommendation is primarily based on the results of summation calculations, obtained through using the JNDC FP Nuclear DataLibrary, Version 2. It consists of decay power values, their estimated accuracy, and their β- and γ-ray components for five fissionable nuclides; U-235, -238, Pu-239, -240 and -241. A set of 33-term exponential polynomials given therein help the users in calculating the decay-power curve for any irradiation history and cooling time. The time evolution for the γ-ray component energy spectra is also given.

Introduction

Fission-product (FP) decay power evaluation plays quite an important role in design, operation and safety precautions for nuclear facilities. In this respect, it was argued that a special activity was indispensable to study the decay power within Japan and to make efforts for improving its calculation accuracy. This argument led to the establishment of the Decay Heat Evaluation Working Group (DHEWG) by the Japanese Nuclear Data Committee in the early 1970's. This paper briefly reviews the activities on decay power study in Japan and, then, introduces the AESJ (Atomic Energy Society of Japan) Recommendation regarding the decay power data.

A Preceding History

In 1980, DHEWG completed the first version of the FP Nuclear Data Library for the summation calculation of the FP decay power[1]. This led to notable improvement in reliability in the calculated decay power, which was demonstrated by comparison with the integral measurements. This library was revised in 1989 after a long period of elaboration and then made open as the second version[2].

As well as the theoretical study mentioned above, a series of experimental studies were conducted in the 1980's by M. Akiyama et al. at University of Tokyo[3]. In their experiments, they measured the β-ray and γ-ray energies released from the aggregate FPs produced in uranium and plutonium isotope samples after neutron irradiations in YAYOI reactor. These high preci-

sion measurements provided excellent benchmarks for verifying DHEWG calculation reliability.

In 1987, the Research Committee on "Standardization of Decay Heat Power in Nuclear Reactors" was organized by the Atomic Energy Society of Japan. The Committee consisted of people from academies, utilities, industries, and national laboratories, who were interested in application of the decay power data in engineering problems, as well as the decay power study itself.

The Committee, after its three year study, submitted a report titled "Reactor Decay Power and Recommended Values" (in Japanese)[4]. A summary of this report is also published in English, carrying its essential parts[5].

Recommended Decay Heat Power

The Committee recommends the values for the total and its β- and γ-ray components of the FP decay powers. These recommendations are primarily based on summation calculations, making use of the JNDC FP Nuclear Data Library, Version 2. The good results obtained by use of this Library have already been described elsewhere[6]. The present Recommendations cover the following features.

① High precision; As an example, the error (on 1-σ level) associated with the ^{235}U decay power 1 second after an infinite irradiation is estimated to be 2.1 %.

② The Recommendations are based primarily on the summation calculation. This means that every quantity, such as the time evolution for the FP inventories, their activity and γ-ray spectra, are available strictly on the same basis as the recommended decay power itself.

+) Deceased November 14th, 1990

③ Not only the total decay power, but its β- and γ-ray components are provided. The time evolution for the γ-ray energy spectra are also given. Table 1 shows an example of the recommended spectra. The γ-ray spectrum from ^{134}Cs, which is produced by neutron capture in ^{133}Cs, is included as well.

④ Varieties of the fissile nuclides are extended from the existing standards or recommendations. Five fissiles, ^{235}U, ^{238}U, ^{239}Pu, ^{240}Pu and ^{241}Pu, are included in the present Recommendations.

⑤ This is applicable, even after quite a long cooling time, say, 10^{13} seconds (320,000 years).

⑥ The decay powers for any burn-up history and any cooling-time are easily reproduced with a set of 33-term exponential polynomials.

⑦ The effects of neutron capture in FPs for the ^{235}U and ^{239}Pu decay power are given in a tabular form. (For other fissiles, one of these should be substituted).

⑧ Actinide decay powers from ^{239}U and ^{239}Np are given in a simple analytical form.

Using Recommended Data

Table 2 is an excerpt from the tabulated recommendation for the decay powers after a fission pulse. The fission-pulse function $f_i(t)$, where t and i stand for the cooling time and the fissile index, can also be reproduced by a 33-term exponential polynomial,

$$f_i(t) = \sum_{j=1}^{33} \alpha_{ij} \exp(-\lambda_j t), \qquad (1)$$

with constants α_{ij} and λ_j given in Table 3. Furthermore, by integrating expression (1), we obtain the decay power at time t after the end of operating (or irradiation) time T as,

Table 1   Normalized γ-Ray Spectra for U-235 After a 3-Year Irradiation

Energy (MeV)	Time After Shutdown (s)			
	0.0	1.0E+1	1.0E+2	1.0E+3
0.0-1.0	3.989E-01	4.214E-01	4.685E-01	5.253E-01
1.0-2.0	3.287E-01	3.304E-01	3.307E-01	3.333E-01
2.0-3.0	1.713E-01	1.626E-01	1.414E-01	1.228E-01
3.0-4.0	7.043E-02	5.690E-02	4.010E-02	1.777E-02
4.0-5.0	2.426E-02	2.255E-02	1.599E-02	8.299E-04
5.0-6.0	5.752E-03	5.479E-03	3.045E-03	4.121E-05
6.0-7.0	6.871E-04	6.238E-04	3.270E-04	1.193E-07
7.0-8.0	1.168E-05	4.380E-07	2.732E-10	0.0
8.0-9.0	9.479E-07	6.985E-09	4.241E-12	0.0
9.0-10.0	5.761E-08	1.950E-11	0.0	0.0
(MeV)	1.0E+4	1.0E+5	1.0E+7	1.0E+9
0.0-1.0	6.159E-01	7.240E-01	9.755E-01	1.000E+00
1.0-2.0	3.081E-01	2.565E-01	1.065E-02	7.678E-07
2.0-3.0	7.006E-02	1.944E-02	1.384E-02	1.025E-07
3.0-4.0	5.688E-03	9.153E-05	5.359E-06	1.230E-08
4.0-5.0	2.046E-04	4.000E-07	8.678E-12	2.392E-10

In this Table 1.0E-5 reads 1.0×10^{-5}

Table 2   Recommended Decay Heat After a Pulsed Neutron Irradiation (MeV/S/fission)

t(s)	U-235	U-238	Pu-239	Pu-240	Pu241
0.0	1.370	2.647	8.17-01	1.035	1.479
1.0	7.41-01	1.308	5.09-01	6.24-01	8.42-01
3.0	4.04-01	6.59-01	2.91-01	3.53-01	4.58-01
1x10^1	1.45-01	2.00-01	1.09-01	1.29-01	1.57-01
3x10^1	4.50-02	5.45-02	3.63-02	4.14-02	4.79-02
1x10^2	1.24-02	1.43-02	1.05-02	1.15-02	1.29-02
3x10^2	3.21-03	3.51-03	2.85-03	2.97-03	3.18-03
1x10^3	9.21-04	9.39-04	8.80-04	8.99-04	9.22-04
3x10^3	2.87-04	2.80-04	2.62-04	2.70-04	2.72-04
1x10^4	6.03-05	5.47-05	4.86-05	4.94-05	4.90-05
3x10^4	1.36-05	1.21-05	1.10-05	1.07-05	1.02-05
1x10^5	2.64-06	2.54-06	2.50-06	2.43-06	2.32-06
3x10^5	5.61-07	5.85-07	6.04-07	5.74-07	5.57-07
1x10^6	1.61-07	1.62-07	1.59-07	1.49-07	1.57-07
3x10^6	4.60-08	4.34-08	4.12-08	3.89-08	4.20-08
1x10^7	9.68-09	8.69-09	8.37-09	8.13-09	8.17-09
3x10^7	1.68-09	1.76-09	1.91-09	2.04-09	2.28-09
1x10^8	2.28-10	2.72-10	3.20-10	3.53-10	4.10-10
1x10^9	4.10-11	3.00-11	2.73-11	2.57-11	2.57-11
1x10^{10}	5.04-14	4.12-14	4.07-14	3.93-14	3.88-14
1x10^{11}	9.79-16	1.03-15	2.40-15	2.46-15	1.11-15
1x10^{12}	8.64-16	9.07-16	2.02-15	2.08-15	9.69-16
1x10^{13}	3.01-16	3.04-16	4.56-16	4.59-16	3.07-16

In this Table 1.0-05 reads 1.0×10^{-5}

$$F(T,t) = \sum_{i=1}^{5} w_i \sum_{j=1}^{33} \alpha_{ij} \exp(-\lambda_j t)(1 - \exp(-\lambda_j T))/\lambda_j$$

$$(2)$$

where w_i stands for the contribution from the i-th fissile. As is exemplified in the ANS5.1 Standard[7], expression (2) can easily be extended to cases, where the operation history is more complicated. Figure 1 compares the present recommendation with the ANS5.1 for the pulsed-fission case, where the difference is larger than any other irradiation and easier to see.

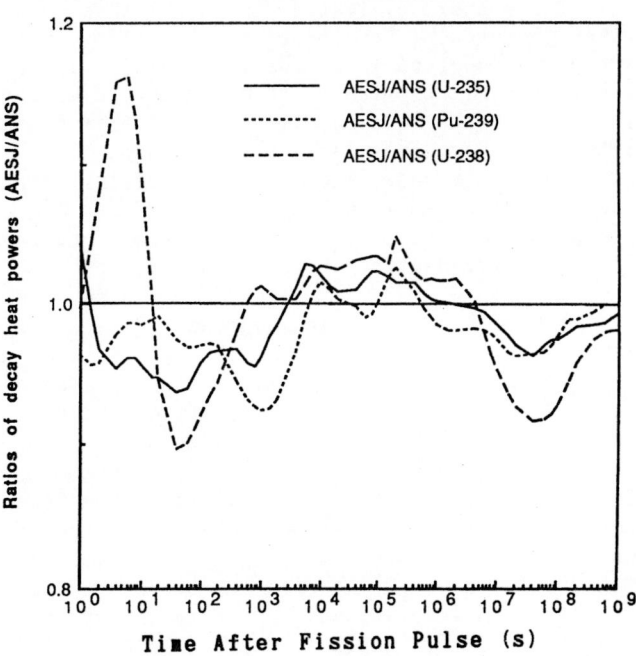

Fig. 1   Ratio of AESJ Recommendation to ANS 5.1

Table 3   Constants in the Exponential Formula Fittted to the Recommended FP Decay Heat Power

	λ	α :U-235	α :U-238	α :Pu-239	α :Pu-240	α :Pu-241
1	3.290E+00	1.994E-01	5.414E-01	5.245E-02	8.614E-02	1.717E-01
2	1.001E+00	4.664E-01	7.961E-01	2.144E-01	2.884E-01	4.332E-01
3	5.157E-01	5.771E-02	3.069E-01	1.221E-01	1.385E-01	1.875E-01
4	2.951E-01	3.465E-01	5.338E-01	1.928E-01	2.240E-01	3.079E-01
5	1.959E-01	5.639E-02	1.569E-01	5.537E-02	9.055E-02	1.382E-01
6	1.037E-01	1.692E-01	2.228E-01	1.190E-01	1.368E-01	1.582E-01
7	3.488E-02	4.445E-02	5.477E-02	3.561E-02	4.211E-02	5.116E-02
8	1.330E-02	2.283E-02	2.605E-02	1.955E-02	2.217E-02	2.483E-02
9	5.004E-03	3.954E-03	5.877E-03	2.537E-03	2.452E-03	2.832E-03
10	3.591E-03	1.061E-03	2.524E-04	1.291E-03	1.503E-03	1.605E-03
11	1.357E-03	1.166E-03	1.291E-03	1.097E-03	1.079E-03	1.148E-03
12	5.645E-04	6.399E-04	6.560E-04	6.645E-04	6.835E-04	6.816E-04
13	1.850E-04	1.940E-04	1.846E-04	1.647E-04	1.748E-04	1.794E-04
14	5.435E-05	3.005E-05	1.338E-05	6.469E-06	3.575E-06	5.659E-06
15	4.918E-05	1.487E-06	1.172E-05	1.444E-05	1.604E-05	1.268E-05
16	1.922E-05	8.346E-06	7.381E-06	6.447E-06	6.347E-06	6.225E-06
17	8.422E-06	1.648E-06	1.617E-06	1.832E-06	1.810E-06	1.667E-06
18	2.443E-06	4.679E-07	5.497E-07	5.746E-07	5.372E-07	4.880E-07
19	6.925E-07	9.897E-08	7.942E-08	8.668E-08	7.746E-08	7.199E-08
20	6.202E-07	8.368E-08	9.450E-08	8.078E-08	7.981E-08	1.082E-07
21	1.503E-07	-1.001E-08	1.107E-08	1.744E-08	1.774E-08	1.823E-08
22	1.277E-07	8.752E-08	-8.740E-08	-1.628E-07	-1.662E-07	-1.793E-07
23	1.262E-07	-5.134E-08	9.985E-08	1.672E-07	1.685E-07	1.794E-07
24	2.714E-08	2.337E-09	1.685E-09	1.129E-09	1.183E-09	1.153E-09
25	2.251E-08	-1.445E-10	8.759E-10	1.726E-09	2.029E-09	2.592E-09
26	8.985E-09	1.416E-11	2.096E-11	2.244E-11	2.360E-11	2.396E-11
27	4.366E-09	3.536E-12	2.345E-12	2.341E-12	2.735E-12	1.739E-12
28	7.707E-10	5.096E-11	2.792E-11	1.819E-11	1.622E-11	1.361E-11
29	7.280E-10	3.586E-11	3.512E-11	3.882E-11	3.755E-11	4.007E-11
30	2.430E-10	2.005E-14	3.827E-14	3.709E-14	4.060E-14	4.473E-14
31	2.198E-13	3.889E-16	4.424E-16	1.843E-15	1.928E-15	5.238E-16
32	1.026E-13	5.285E-16	5.361E-16	5.289E-16	5.143E-16	5.414E-16
33	9.550E-15	7.535E-17	6.912E-17	6.829E-17	6.644E-17	6.035E-17

The authors evaluated the error associated with the present Recomendations starting from the error inherent in the basic data[8]. In the publications[4,5], these estimated errors on the $1-\sigma$ confidence level are given in tables along with the recommended decay power values. In those publications, the correction factors for the effects of neutron capture in FPs are also given as the functions of reactor·type, neutron flux, and irradiation and cooling time.

Concluding Remarks

The present Recommendations are applicable to a wide range of problems, related to decay power in BWR, PWR and other nuclear facilities. The recommended values are reproduced, not only by the simple fitting formula given in the text, but also by summation calculation, through using the JNDC FP Nuclear Data Library, Version 2. The latter calculation can provide full information on generation, transmutation and activity of each FP nuclide in a manner consistent with the recommended decay power value. Part of the data presented is useful in a wider range of applications. For example, the γ-ray spectra, such as given in Table 1, can be used in the γ-ray source calculations in a operating reactor.

REFERENCES

1. K. Tasaka, H. Ihara, M. Akiyama, T. Yoshida, Z. Matumoto and R. Nakasima: JAERI 1287 (1982)
2. K. Tasaka, J. Katakura, H. Ihara, T. Yoshida, S. Iijima, R. Nakasima, T. Nakagawa and H. Takano: JAERI 1320 (1990)
3. M. Akiyama, K. Furuta, T. Ida, K. Sakata and S. An: J. At. Energy Soc. Japan, 24, 709 (1982) and 24, 803 (1982) (in Japanese)
4. "Reactor Decay Power and Recommended Values" At. Energy Soc. Japan (1989) (in Japanese)
5. K. Tasaka, J. Katakura, T. Yoshida, T. Katoh and R. Nakasima: JAERI-M 91-034 (1991)
6. T. Yoshida, H. Ihara, J. Katakura, K. Tasaka and R. Nakasima : Proc. Int. Conf. on Nucl. Data for Sci. and. Technol., Mito (1988) p.889
7. "American Nucl. Soc. Proposed ANS5.1 Standard", American Nuclear Society (1978)
8. J. Katakura and S. IIjima, J. Nucl. Sci. Technol., submitted for publication

Average Beta and Gamma Energies of Fission Products in the Mass Range 98-108

P-I. Johansson, G. Rudstam and J. Eriksen
The Studsvik Neutron Research Laboratory, University of Uppsala
611 82 Nykoping,Sweden
H.R. Faust
Institute Laue - Langevin, 156 X F-038042 Grenoble, France
J. Blachot
Centre d'Etudes Nucléaires, 85 X F-38041 Grenoble, France
J. Wulff
Technical University, D-3300 Braunschweig, Germany

Abstract: Continuous spectra of beta particles and gamma rays emitted in the decay of short-lived fission products in the mass range 98-108 have been measured at the isotope-separator LOHENGRIN at ILL [1]. The gamma rays were measured using a NaI(Tl) crystal with high efficiency and the beta particles with a telescope consisting of a thin plastic ΔE-detector in front of a high purity Si detector. The average beta and gamma energies (per decay) obtained from the spectra are important data for the calculation of the decay heat by the summation technique. The spectra can also be used for various applications in reactor technology. The measurement was carried out with the same equipment as the corresponding OSIRIS measurement [2]. The results of the present experiment have given beta spectra for the nuclides Zr-99, Nb-99m, Nb-99, Zr-100, Nb-100, Zr-101, Nb-101, Mo-101, Tc-101, Zr-101, Zr-102m, Nb-102, Tc-102, Nb-103, Mo-103, Tc-103, Mo-104, Tc-104, Mo-105, Tc-105, Nb-106, Mo-106, Tc-106, Tc-107, Ru-107, Mo-108, Tc-108, Ru-108, and Rh-108m. Gamma spectra have been obtained for Y-98, Zr-98, Y-99, Zr-99, Nb-99, Zr-101, Nb-101, Mo-102, Nb-102, Nb-103, Mo-103, Tc-103, Mo-104, Nb-104, Tc-104, Mo-105, Nb-105, Tc-105, Mo-106, and Tc-106. Experimental gamma spectra are compared with data derived from the ENSDF file [3], and beta results are compared with calculated spectra using decay schemes and Q_β values taken from Nuclear Data Sheets. All results given in this report are to be considered as preliminary.

(Beta spectra, gamma spectra, beta energy, gamma energy, fission products, mass range 98-108, on-line mass separator)

Introduction

The aim of the measurement has been to obtain continuous spectra of the beta particles and gamma rays emitted in the decay of short lived-fission products in the mass range 98 to 108. The measurement is motivated by discrepant results from summation calculations and integral measurements on the decay heat developed by decaying fission products in the fuel elements. The nuclides of interest were those with high beta disintegration energies and not too small fission yield. The mass ranges 79-98 and 130-147, the low and high fission yield peak from thermal fission of U-235, have been covered by the measurement at OSIRIS at Studsvik [2]. The data obtained at LOHENGRIN cover a part of the low mass peak of thermal fission of Pu-239 which is not available at OSIRIS.

Experimental method

At the OSIRIS experiment [2] the beta particles and gamma rays were measured in two independent experiments. The equipment used at OSIRIS was brought to ILL and used in the present experiment. Only basic information stressing important changes from the OSIRIS measurement are pointed out here. The technique used to measure the gamma and beta spectra from fission products was to collect a sample from the on-line isotope separator LOHENGRIN, to bring it to the detectors, and to measure simultaneously gamma and beta from the decaying fission products. A sample from the separator normally contains several isobars. By a proper choice of collection, waiting, and counting times, the composite pulse spectra obtained in the measurements can be decomposed into contributions from different components in the sample provided that the number of measurement is at least equal to the number of components. The pulse spectra are then converted into energy spectra by using the detector response functions. The procedure is fully described in Ref [2].

The beta and gamma measurement

Beta and gamma spectra were measured simultaneously with the experimental setup shown in Fig. 1. The beta particles were measured with a 1.5 cm thick high-purity germanium (HPGe) detector in coincidence with the 1 mm thick ΔE-detector, which was placed in between the sample and the HPGe-detector. For the gamma mesurement the same equipment was used for the OSIRIS and the ILL experiments but changes had to be made due to different experimental conditions. The main detector was a well shielded NaI(Tl) crystal, 12.5 cm in diameter and 12.5 cm in length. The collimator of length 15 cm was decreased to 5 cm. The background counts were reduced by requiring coincidences with beta pulses in the ΔE-detector (see Fig. 1 and next section). The response function for the NaI crystal has been measured with a collimator of length 15 cm. The impact on the response function by using a 5 cm collimator will be further studied.

The sample and monitor

Fission products were obtained using a Pu-239 target and collected on a movable tape [1]. For each mass the electric and magnetic field of the mass separator were chosen to optimize the yield of that mass and to minimize contamination from other masses. The length and width of the tape exposed to the beam was 100 cm and 0.7 cm respectively. For fission products with short half-lives (< 1 s) the tape was kept running continuously and the data collecting system was permanently open for counting. For fission products with longer half-lives the tape was first moved to the position for exposure. A sample was then collected during a selected collection time and then moved to the detector position where counting started after a predetermined waiting time. During counting a new sample was collected. To get enough counting statistics each cycle had to be repeated hundreds of times.
The number of decaying nuclides during each run was obtained by monitoring gamma spectra from the fission products with a Ge(Li) detector (see Fig. 1).

Treatment of data from beta measurement

A small gamma effect in the beta spectra was corrected for using results from runs with 15 mm Al in front of the beta detector, and without coincidence requirement. A Kurie-plot was carried out for each beta spectrum to yield the end-point energy. The beta spectrum was then extrapolated to the maximum beta energy. The loss of energy of beta particles in the ΔE detector was found to be 300+-50 keV from a Kurie-plot of a Ru-106 - Rh-106 spectrum. The low energy cut-off in the HPGe-spectrometer was about 200 keV. The energy range 0-500 kev is therefore not covered by the measurement. The fraction of beta particles below 500 keV was obtained from the ratio of the number of beta particles detected in the HPGe-detector and in the ΔE-detector. This fraction was then used in an extrapolation of the beta spectra to 0 energy.
Some measured beta spectra are shown in Figs. 2-5. The open circles with error bars are the results from the present measurement. Spectra calculated from decay schemes are shown as a solid lines. In some cases the spectra can be compared to earlier results from Studsvik [2]. Those are then shown as filled circles.

Treatment of data from gamma measurement

By an unfolding technique using measured response functions the pulse spectra were converted to energy spectra. The results obtained are relative and have to be normalized. The normalization is done using absolute gamma branching ratios and known spectrum of Sr-95 [2]. Some of the results are shown in Figs. 6-9. In some cases the gamma branching ratios are either unknown or have large uncertainties. The final result will then be based on the number of beta particles detected

in the ΔE-detector and the number of gamma rays detected in the NaI(Tl) detector.

Results

Several details have to be checked before final spectra, end-point energies, and average gamma and beta energies can be presented. Beta data will then be given for Zr-99, Nb-99m, Nb-99, Zr-100, Nb-100, Zr-101, Nb-101, Mo-101, Tc-101, Zr-101, Zr-102m, Nb-102, Tc-102, Nb-103, Mo-103, Tc-103, Mo-104, Tc-104, Mo-105, Tc-105, Nb-106, Mo-106, Tc-106, Tc-107, Ru-107, Mo-108, Tc-108, Ru-108, and Rh-108m. Gamma spectra will be given for Y-98, Zr-98, Y-99, Zr-99, Nb-99, Zr-101, Nb-101, Mo-102, Nb-102, Nb-103, Mo-103, Tc-103, Mo-104, Nb-104, Tc-104, Mo-105, Nb-105, Tc-105, Mo-106, and Tc-106.

REFERENCES

1. P. Armbruster, M Asghar, J.P. Bocquet, R. Decker, H. Ewald, J. Greif, E. Moll, B. Pfeiffer, H. Schrader, F. Schussler, G. Siegert and H. Wollnik, Nucl. Instr. and Meth. 139(1976)213

2. G. Rudstam, P. I. Johansson, O. Tengblad, P. Aagaard and J. Eriksen, Atomic Data and Nuclear Data Tables 45,(1990)239

3. Evaluated Nuclear Structure Data File (ENSDF), 1988 version, communicated through C. Nordborg

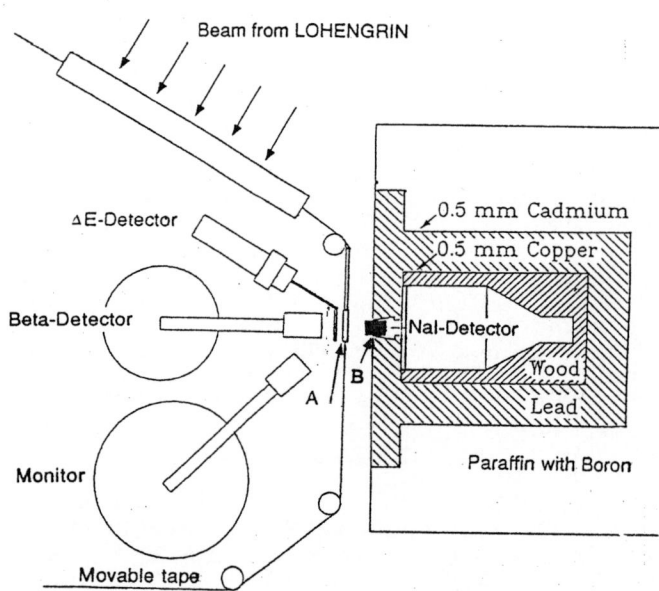

Figure 1. Experimental arrangement for measuring gamma and beta spectra.

A: Zig-zag pattern device to concentrate the activity
B: Beta absorber (5 cm polythene)

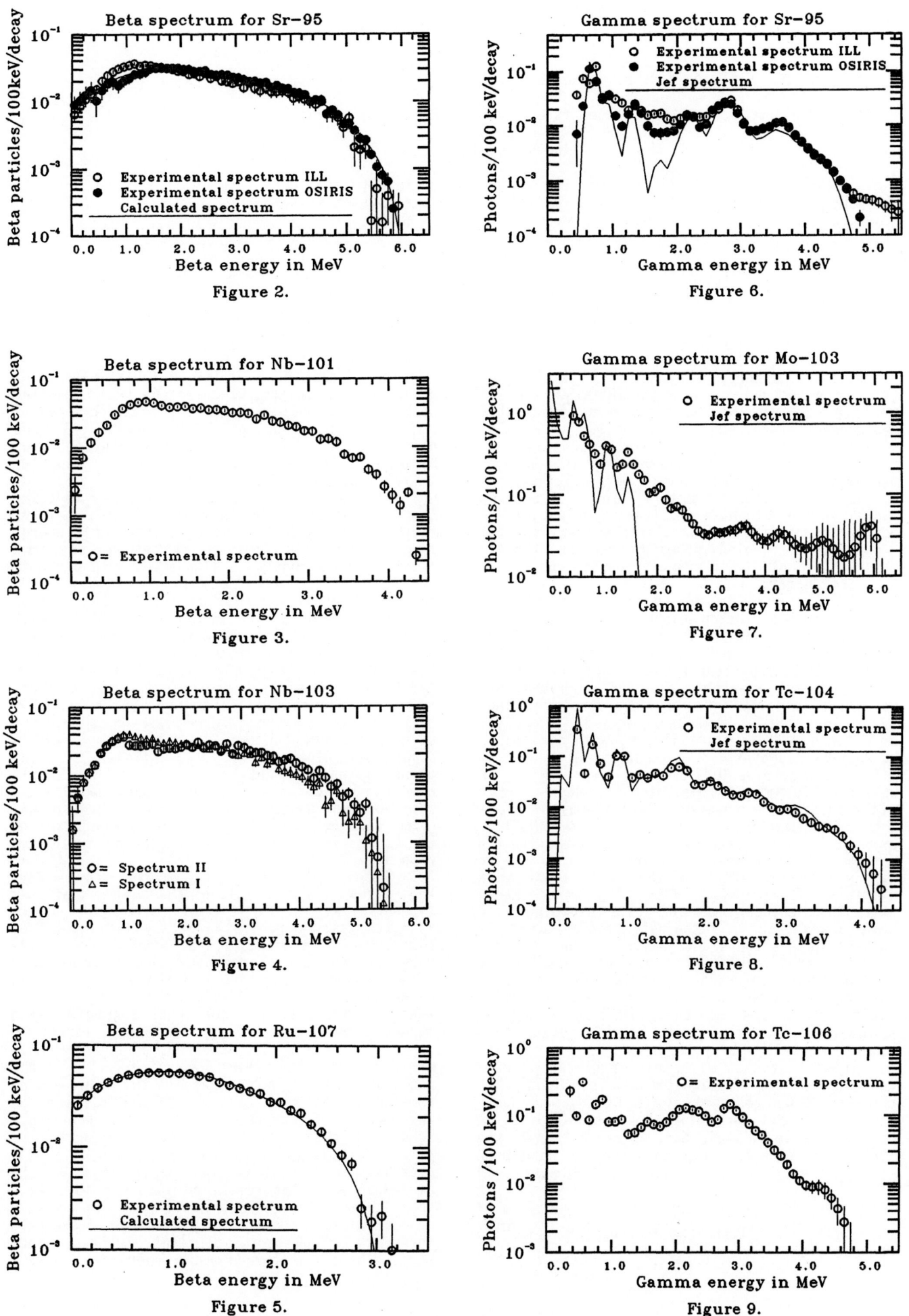

Beta spectrum for Sr-95

Figure 2.

Gamma spectrum for Sr-95

Figure 6.

Beta spectrum for Nb-101

Figure 3.

Gamma spectrum for Mo-103

Figure 7.

Beta spectrum for Nb-103

Figure 4.

Gamma spectrum for Tc-104

Figure 8.

Beta spectrum for Ru-107

Figure 5.

Gamma spectrum for Tc-106

Figure 9.

MEASUREMENT OF BETA-DECAY STRENGTH DISTRIBUTIONS OF FISSION-PRODUCT ISOTOPES USING A TOTAL ABSORPTION GAMMA-RAY SPECTROMETER

R. C. Greenwood, R. G. Helmer, M. A. Oates, M. H. Putnam and K. D. Watts

Idaho National Engineering Laboratory,
EG&G Idaho,
Idaho Falls, Idaho 83415, USA

Abstract: A total absorption gamma-ray spectrometer (TAGS), based on a large NaI(Tl) detector, has been developed at the INEL ISOL facility. With this system measurements of total absorption spectra for $^{138-144}$Cs, ^{141,142}Ba, $^{142-145}$La, and ^{145}Ce have been made. From these spectra we will deduce the beta-decay feeding distributions as a function of the daughter level energy.

(radioactivity, mass separation, beta-decay distributions, ^{141}Ba, ^{139}Cs)

Introduction

A total absorption gamma-ray spectrometer (TAGS) has been developed at the INEL ISOL facility [1,2] to undertake a program of systematic measurement of beta-decay feeding (or beta-strength) distributions of neutron-rich fission-product nuclei. An understanding of the beta-strength distributions can provide sensitive tests of models of nuclear structure. For neutron-rich nuclei, these data have important applications in both astrophysics, in calculations involving the r-process of nucleosynthesis, and in fission reactor technology, in calculations of the decay-heat source term.

The radionuclides to be studied consist of the spontaneous fission products from two ~100 µg electrodeposits of ^{252}Cf (fabricated at ORNL). These nuclides are transported by a He-jet system to an isotope separator and after separation are collected on a transport tape by which they are moved periodically into the detector. The TAGS system, previously described in Ref. [3], consists of a large NaI(Tl) scintillator, with dimensions of 25.4-cm diameter x 30.5-cm length and having a 5.1-cm diameter x 20.3-cm long axial well. A shielded 300-mm² x 1.0-mm thick Si charged-particle detector is located in a thin-walled tape-transport line inside the well of the NaI(Tl) crystal. The Si detector is oriented to view the bottom end of the well, with the source transport tape passing in front of the Si detector at a distance of ~20 mm. With this arrangement, extraneous coincidence effects, resulting from beta backscattering or the detection of internal-conversion electrons or gammas in the Si beta detector are estimated to be, at most, a few percent each. In actual operation, useful information on the distribution of beta-feeding intensities is obtained from the gamma singles spectra as well as from the gamma spectra measured in coincidence with the decay betas.

Initial studies with this spectrometer have concentrated on fission-product nuclides with complex, but reasonably well understood decay schemes, to serve as 'validation cases'. Specifically, because of their ease of separation in the INEL ISOL facility, TAGS measurements have been completed for $^{138-144}$Cs, ^{141,142}Ba, $^{142-145}$La, and ^{145}Ce. In this paper, techniques for analysis of these data are illustrated for the ^{141}Ba and ^{139}Cs radionuclides.

Results

The initial phase of the analysis consists of selecting the time multiscaled spectral bins (for both singles and coincidences) that best represents the radionuclide of interest and when necessary subtracting out the contributions from background (for singles) and parent or daughter activities.

In order to interpret such spectra in terms of a beta-feeding distribution, one must know the response of the NaI(Tl) detector to single-energy photons and to a sequence or cascade of gamma rays. We have chosen to simulate the response of the NaI(Tl) detector to a single gamma ray by means of a Monte Carlo code, CYLTRAN [4]. These calculated response functions are convoluted with a Gaussian function to give the full-energy peaks widths comparable to those of the observed peaks. The next step in simulating the observed sum spectrum is to compute the coincidence sum spectrum for each cascade in the decay scheme. For this purpose, a program was written to compute the coincidence sum spectrum for up to 7 gamma rays in a cascade. Given the detector efficiency for any gamma-ray energy (obtained from the Monte Carlo calculation of the response functions), this routine also computes the total spectrum from this gamma cascade, including all the components from the escape from the detector of one or more gamma rays.

The decay of ^{141}Ba [$T_{1/2}$ = 18 min, Q(β^-) = 3230 keV] gives a good example of the analysis of this type of data. In Fig. 1 the measured coincidence spectrum is compared with the simulated spectrum computed for the published decay scheme [5]. Here, this initial simulation uses the published beta feeding intensities and the specific gamma cascades that occur in this scheme. It is clear that in this case the published scheme has too much intensity to the highest energy level. To improve the fit to the measured spectrum, we have modified the decay scheme by adding levels in the gap near 2200 keV and by changing the beta-feeding intensities to some levels. It should be noted that when such changes are made in the decay scheme one must arbitrarily choose the multiplicity of the gamma cascades used as well as the energies of the individual gamma rays in the cascade. And, the resulting fits can vary with these choices. However, for ^{141}Ba this is not much a problem since a very high percentage of the beta intensity is related to portions of the original decay scheme that are correct; therefore, this

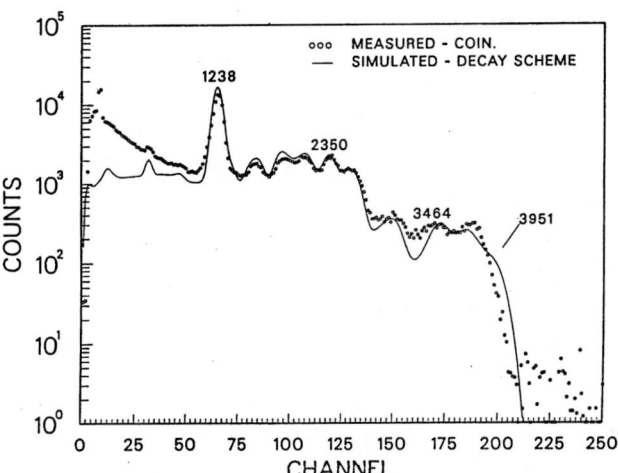

Fig. 3. TAGS coincidence spectrum for ^{139}Cs compared to the simulated spectrum for the published [4] decay scheme.

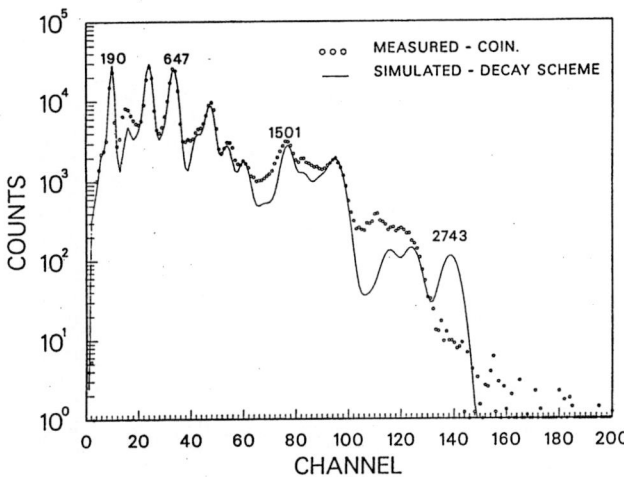

Fig. 1. TAGS coincidence spectrum for ^{141}Ba compared to the simulated spectrum for the published [4] decay scheme. This and the other spectra are shown at 20 keV per channel.

ambiguity will not have a significant effect. After adjustments were made to the decay scheme the fit shown in Fig. 2 was obtained. Similar results were obtained from the analysis of the ^{141}Ba singles spectrum.

This example shows the power of this method in testing a propose decay scheme.

Figure 3 shows the measured coincidence spectrum for the decay of ^{139}Cs [$T_{1/2}$ = 9 min, $Q(\beta^-)$ = 4213 keV]. This case is of special interest because it has a ground-state beta branch of ~80%. The simulated spectrum computed from the published decay scheme [5] again shows that, as expected, some levels are missing (see the 3200 keV region) and that some beta feeding branches are too strong. After modification of the decay scheme by the addition of levels and adjustment of the beta feedings, it was clear that the simulated spectrum still lacked a continuous low-energy component, and a little thought suggests an obvious candidate. For ~80% of the decays there are no gamma rays in coincidence with a beta gate, but these same betas do produce bremsstrahlung radiation either in or after having passed through the Si detector that can be detected in coincidence in

the NaI(Tl) detector. So, our next step in development of an analytical protocal was to write a bremsstrahlung simulation routine. For a given end-point energy and Z, this routine (1) calculates the beta distribution (for an allowed shape); (2) corrects for the electrons that do not produce a pulse in the Si detector that is above the gate cutoff energy; (3) generates the corresponding bremsstrahlung spectrum for the electrons in each 20-keV bin and adds these contributions; (4) attenuates the resulting photons for some absorbing material; and then (5) convolutes this photon spectrum with the NaI(Tl) detector response functions. Figure 4 shows that for ^{139}Cs the simulated spectrum including this bremsstrahlung component is in excellent agreement with the measured spectrum.

Although the addition of the bremsstrahlung component worked well in the ^{139}Cs case, this is not a very sensitive test of the quality of the simulated bremsstrahlung spectra. Figure 5 compares the measured coincidence spectrum for the two beta components from a source of ^{90}Sr - ^{90}Y with the corresponding simulated spectrum. This agreement is quite satisfactory, considering that the only free parameters used are the count rate scaling factors.

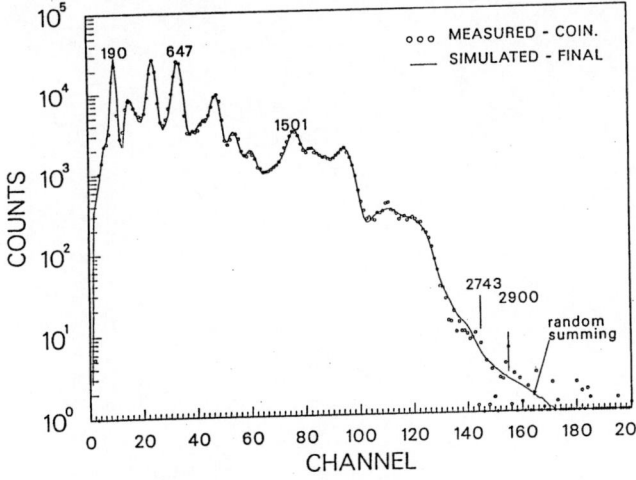

Fig. 2. TAGS coincidence spectrum for ^{141}Ba compared to the simulated spectrum for the modified decay scheme.

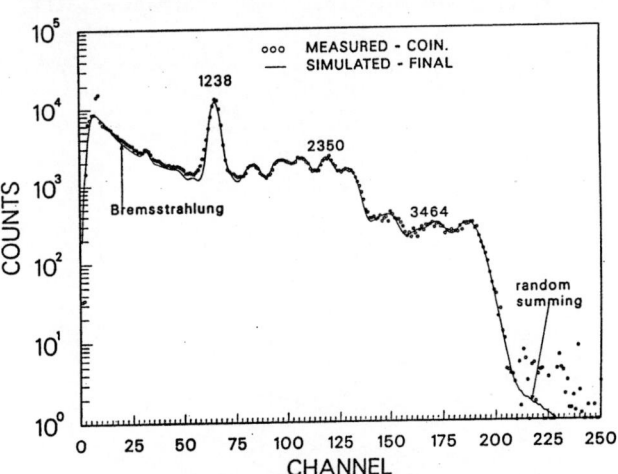

Fig. 4. TAGS coincidence spectrum for ^{139}Cs compared to the simulated spectrum for the modified decay scheme and including a bremsstrahlung component.

550

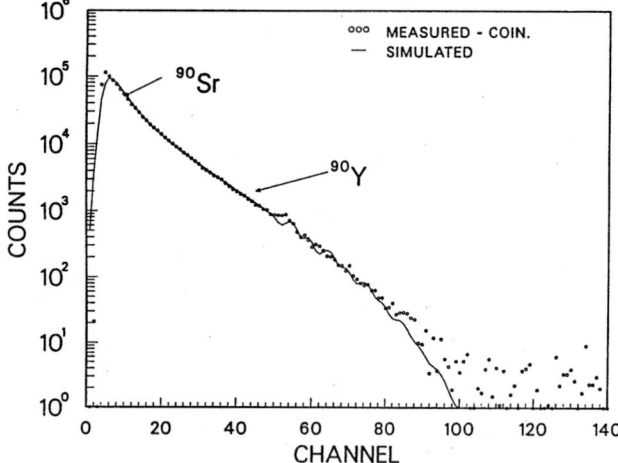

Fig. 5. TAGS coincidence spectrum for ^{90}Sr - ^{90}Y compared to the two simulated bremsstrahlung components.

We note that all of the above discussion has been in terms of the beta feeding of discrete levels or pseudo-levels. However, since the purpose of these measurements is to provide gross beta-strength distributions to be compared with theoretical calculations, it is useful to give the results in energy bins. In our case, since the energy of our pseudo-levels are defined no better than 50 keV, a bin width of 200 keV seems useful. As an example, Table 1 gives the binned results for the ^{141}Ba analysis, compared to the published decay data. As has been noted already, the major differences between the present results (coincidences and singles) and the published data are in the distribution of the beta intensity in the higher energy bins.

It is clear from these initial analyses that this TAGS system will be quite useful in defining the beta-feeding distributions for radionuclides that have complex decay scheme, or are produced in such small quantities or have such short half-lives, that the complete decay scheme can not be accurately determined.

This work was performed under the auspices of the U. S. Department of Energy under contract number DE-AC07-76IDO1570 with EG&G Idaho, Inc.

Table 1. Binned beta-feeding distribution for ^{141}Ba.

Bin energy	Beta-feeding (%)		
	Published [4]	INEL singles	INEL coin.
100	17	10.5	12.5
300	< 0.6	3.6	4.3
500	23	18.4	18.9
700	25	24.7	24.1
900	13.7	13.5	13.2
1100	5.3	5.2	5.3
1300	0.0	1.1	1.1
1500	5.0	6.5	6.1
1700	3.5	5.8	5.5
1900	5.7	6.3	5.7
2100	0.0	1.2	1.0
2300	0.4	1.9	1.6
2500	0.5	1.2	0.8
2700	0.5	0.18	0.14
2900	0.0	0.03	0.03

References

1. R. A. Anderl, J. D. Cole and R. C. Greenwood, Nucl. Instrum. Methods B26, 333 (1987).

2. R. C. Greenwood, R. A. Anderl, J. D. Cole and H. Willmes, Phys. Rev. C35, 1965 (1987).

3. R. C. Greenwood, R. A. Anderl, A. J. Caffrey, J. D. Cole, M. W. Drigert, R. G. Helmer, M. A. Lee, C. W. Reich and D. A. Struttmann, in Proc. Intern. Conf. on Nuclear Data for Science and Technology, May/June 1988, Mito, Japan (ed. by S. Igarasi) p. 359, Saikan, Tokyo (1988).

4. J. A. Halbleib and T. A. Mehlhorn, Nucl. Sci. Eng. 92, No. 2, 338 (1986); J. A. Halbleib and W. H. Vandevender, Nucl. Sci. Eng. 61, 288 (1976); and J. Halbleib in "Monte Carlo Transport of Electrons and Photons", eds. T. M. Jenkins, W. R. Nelson and A. Rindi (Plenum Press, New York, 1988) p. 246.

5. Evaluated Nuclear Structure Data File (ENSDF), ed. J. K. Tuli.

NUCLEAR DECAY SCHEME MEASUREMENTS AT THE UK NATIONAL PHYSICAL LABORATORY

S A Woods

Division of Radiation Science and Acoustics,
National Physical Laboratory,
Teddington, Middlesex, TW11 0LW, UK

Abstract: The National Physical Laboratory (NPL) maintains a wide variety of facilities appropriate to the measurement of nuclear decay data. An overview of these facilities is given.

(Radionuclide standardisation, nuclear spectrometry)

Introduction

NPL has the responsibility to develop and maintain the UK primary standards of radioactivity. The techniques of absolute counting, such as 4π beta-gamma coincidence counting, and alpha-, beta- and gamma-ray spectrometry used in this work are applied by NPL to the determination of selected nuclear decay data. Such data are of vital importance in many areas, such as the safe handling and storage of fuel in the nuclear power industry, use of radionuclides in medicine for diagnosis and therapy and environmental protection.

Standardisation of Radionuclides

The accuracy of measurement of nuclear data and the efficiency of their use are dependent on many factors. In particular, one principal factor is the ability to accurately determine the absolute disintegration rate of a radionuclide together with its half-life.

4π Beta-Gamma Coincidence Counting

The most powerful technique used for the standardisation of radionuclides is 4π beta-gamma coincidence counting [1]. This is the principal method used by NPL for the accurate determination of the activity concentration of its national standards of radioactivity.

The method relies on the radionuclide decaying by simultaneous emission of both electromagnetic and particulate-like radiations. Detection of these events in two separate counters and the recording of coincidences between them allows the activity to be determined absolutely. The overall uncertainty associated with such an absolute standardisation is typically of the order of 1%.

NPL has extended the 4π beta-gamma coincidence method to nuclides with complex decay schemes, through techniques such as "computer discrimination", involving energy-discriminating beta-ray detectors, "efficiency tracing", for pure beta-ray emitters and "correlation counting" for nuclides with delayed states.

Re-entrant Ionisation Chambers

The effort required to effect a standardisation using 4π beta-gamma coincidence counting is measured in days or even months. Given the transient nature of radioactivity, programmes of work on the determination of nuclear data may require the regular replenishment of standardised sources. Similar considerations apply in situations where the use of nuclear data depends on standardised sources, for example, in the calibration of gamma-ray detectors. Fortunately, for beta-gamma emitting nuclides, and some pure beta-emitters a convenient means exists for the effective retention of the results of a standardisation, namely the re-entrant ionisation chamber [2], which can far surpass any other secondary standard instrument in terms of long-term stability and reproducibility. Typically, it is not unreasonable to expect such an instrument to retain its calibration to better than one part in one thousand for a period of 10-20 years.

At NPL, several such ionisation chambers are maintained and calibrated against national standards of radioactivity for more than 50 radionuclides. In most cases the additional uncertainty introduced by using the ionisation chamber to determine activity, rather than resorting to absolute counting is small; with care, it can be kept to less than 0.1%.

Ionisation chambers are also particularly suitable for the precise measurement of half-lives. Provided the measurement of ionisation chamber response is in current-mode many of the complications of other techniques may be avoided, such as dead-time, pile-up, base-line shift and change of response/sensitivity with time. In addition, most ionisation chambers can operate over a very wide dynamic range without displaying any significant change in sensitivity. For measurements of half-lives ranging from a few days to tens of years, the ionisation chamber technique is considered superior to any other and almost all of NPL's half-life measurements are made by this method.

Nuclear Spectrometry

For the full understanding of the decay scheme of a radionuclide, it is essential to know the complete nature of the emitted radiations. To this end, NPL maintains facilities for use in alpha-, beta- and gamma-ray spectrometry.

Alpha-Ray Spectrometry

NPL has recently constructed a high-resolution alpha-spectrometer. The spectrometer is identical to that designed by Bortels at CEC-JRC, CBNM, Belgium [3]. Only the electronics differ in that the ADC and analyser are replaced with an EG&G Spectrum-Ace-8k plug-in MCA card coupled to an 80386 processor. Preliminary measurements indicate that with a selected 50 mm² PIPS detector (Canberra Semiconductor NV), a resolution of 9 keV will be achievable. Using such a spectrometry system, alpha-particle energies and emission probabilities may be accurately determined.

Beta-ray Spectrometry

NPL maintain a high-resolution, double focusing $\pi\sqrt{2}$ magnetic beta-ray spectrometer of 35 cm optic radius. This spectrometer has been described in detail by Christmas and Cross [4,5] and Woods et al [6]. The spectrometer is used to determine energies and intensities associated with beta-ray and internal conversion electron spectra. The latter are of particular importance in the measurement of decay data for actinide nuclei, the electron spectra of which are highly complex. Actinides decay predominantly by the emission of alpha-particles to low-lying energy levels of the daughter nuclei, the subsequent de-exiting transitions being highly internally converted. Using the beta-ray spectrometer, decay modes may be elucidated and conversion coefficients determined. Such data are extremely important for use in reactor calculations.

Gamma-ray Spectrometry

A number of high-purity germanium detectors are maintained by NPL. Each spectrometer is accurately calibrated for both energy and efficiency for samples in the form of either solutions in ampoules or solid gamma-reference sources. Thus, accurate information on gamma-ray energies and emission probabilities may be obtained; by the use of an annihilator, positron emission probabilities may also be determined. Additionally, measurement of radionuclidic contaminant levels may be made, a measurement of direct importance to the standardisation of radionuclides.

Other Facilities

One problem often faced in the study of nuclear decay data is the presence of radionuclidic contaminants in the sample under investigation. At NPL, an electromagnetic separator is employed to provide isotopically pure samples and sources.

The NPL Isotope Separator

The use of the isotope separator facilitates the more accurate standardisation of radionuclides which are not otherwise available in a sufficiently pure form, also the preparation of high efficiency sources for beta-ray spectrometry and other decay scheme studies. In order to produce pure samples of longer lived radionuclides the appropriate isotope may be separated prior to irradiation in a nuclear reactor to produce the material of interest.

A "Freeman" ion source is employed to produce a beam of ions at energies up to 60 keV and incorporates a 90 degree sector electromagnet with a maximum field strength of 0.6 Tesla, producing a path radius of one metre. The mass number of interest is selected with the help of an NPL designed mass meter which calculates the mass number of the focused beam from the measured magnetic field and the accelerating voltage.

Typical enrichment factors are of the order of 2000 with a separation of 10 mm between adjacent beams at mass number 100. The overall throughput varies by nearly four order of magnitude, depending on the chemical form of the material being separated. Collection can be by implantation at full beam energy or at a retarded energy to produce high efficiency sources on thin plastic foils. A beam stabiliser permits collection for extended periods without manual adjustment.

An Example of the Use of the NPL Facilities: ^{124}I

The increasing use of Positron Emission Tomography (PET) has highlighted the need to standardise and improve the quality of decay scheme data for ^{124}I. ^{124}I decays to ^{124}Te by both electron capture and positron emission followed in most cases by gamma emission. Measurements have been made on pure samples of ^{124}I [7]. These samples, prepared from solutions supplied by the University of Manchester, UK, using the NPL isotope separator, were used to determine the gamma-ray and positron emission probabilities, using the nuclear spectrometry facilities described above, and to accurately measure the half-life, using the secondary standard re-entrant ionisation chamber. An absolute standardisation was carried out for the determination of emission probabilities and also to provide a calibration figure for the commercially available NPL secondary standard radionuclide calibrator (ISOCAL IV, supplied by NE Technology, Reading, UK).

Conclusion

The facilities at NPL are ideal for the determination of radionuclide decay data. Further information on the availability and use of these facilities may be obtained from the author.

References

[1] NCRP, A Handbook of Radioactivity Measurements Procedures, 2nd Ed., Report No 58, National Council on Radiation Protection, Bethesda, MD, USA (1985).

[2] M J Woods, W J Callow and P Christmas: Int. J. Nucl. Med. Biol. 10, 127 (1983).

[3] G Bortels, D Mouchel, R Eykens, E Garcia-Torano, M L Acena, R A P Wiltshire, M King, A J Fudge and P Burger: Nucl. Instr. and Meth. A295, 199 (1990).

[4] P Christmas and P Cross: J. Phys. E 6, 533 (1973).

[5] P Christmas and P Cross: Metrologia 14, 147 (1978).

[6] S A Woods, P Christmas, P Cross, S M Judge and W Gelletly: Nucl. Instr. and Meth. A264, 333 (1988).

[7] D H Woods, S A Woods, M J Woods, J L Makepeace, C W A Downey, D Smith, A S Munster, S E M Lucas and H Sharma: The Standardisation and Measurement of Decay Scheme Data of ^{124}I, submitted to Int. J. Appl. Radiat. Isot.

PROJECT "INTERNATIONAL TABLE OF NUCLIDES PROPERTIES"
RELEVANT TO THE SHELL STRUCTURE OF THE ATOMIC NUCLEI

I. P. SELINOV

Lebedev Physical Institute of the USSR Academy of Sciences, Moscow, USSR

V. P. Chechev

Institute of Radium, Leningrad, USSR

Abstract: The Nuclear Physics Department of the USSR Academy of Sciences submits for discussion a project on international table of nuclides properties.

The proton-neutron system of the nuclides is suggested as one of possible versions. A stepwise form of the curve of binding energy of nucleon pairs (E_2) on the 2β-stability line, drawn on the p-n system, shows that alongwith the proton- and neutron shells there exist (in the general system of the lower n- and p-levels) the nucleon shells combining the constituent particles. The structure of the nucleonic shells is manifested, besides the E_2 values, in the regularities in the nuclide abundances. The deviation from these regularities - anomaly in the abundance of the Te and Xe isotopes - reveals the existence of symmetrical fission of the transuranium elements in the cosmic nucleosynthesis (nuclear structure, nuclear models, isotopes, diagrams)

Project "international table of nuclides properties"

The nuclide data given in the charts of nuclides published in the USA, FRG, USSR, and Japan for some of the isotopes are somewhat different. So, the Nuclear Physics Department of the USSR Academy of Sciences submits for consideration to the International Conference on Nuclear Data for Science and Technology a project on the international table of nuclides properties to be published regularly by analogy with the particle properties table edited by the International Particle Data Group [1].

Suggested as one of possible versions of this table of nuclides is the proton-neutron system of the nuclides [2].

Proton-neutron system of nuclides

Displayed in the coordinates of the system of nuclides, as in the charts of nuclides published in the FRG, USA and Japan, are the numbers of protons and neutrons in the nuclei. But, unlike these charts, the system of the nuclides has the 2β-stability line, drawn across the minimal mass isobars of the even mass numbers (M^{even}), which discriminates between the regions of β^- - and $\beta^+, \mathcal{E}$ - radioactive nuclides. As shown by the pointers of the circles, all the β- radioactive nuclides are transformed to the nuclides on the 2β-stability line which have energetically-optimal ratio of the numbers of neutrons and protons. Analogous to this, the ions with the excess or deficit of electrons, while becoming neutral, transform to the atoms with a normal number of electrons. Just as in the Mendeleev Periodic Table the atoms of the elements are arranged in the sequence of filling the lower energetic levels in the s,p,d,f-shells with electrons, the minimal mass isobars on the 2β-stability line are arranged in the sequence of alternating occupation of the **general** shell system of the lower n- and p-levels with protons and neutrons. As a consequence, the 2β-stability line has a stepwise form adequate to coupling on it of two groups of nucleons: 2p2n (α^4) and 2p4n (α^6). These groups of nucleons are identical with the composition of helium atoms ^{4}He and ^{6}He, and, therefore, such alternation of p and n can be called heterohelionic. Owing to this regularity there exist two types of pleiads of the β-stable isotopes for the stable and radioactive Z^{even} elements: in case of α^4-alternation of p and n- three 2β-stable isotopes, in case of α^6 - alternation - four or five.

Proton-neutron system of nuclides and shell structure of nuclei

The proton-neutron system of the nuclides represents the real shell structure of the atomic nuclei just as the periodic system of the elements represents the shell structure of the atoms. In 1934 Selinov [3] and, irrespective of him, Guggenheimer [4] discovered magic numbers of neutrons: N_m= 20, 50 and 82. In the same year Elsasser [5] discovered also magic numbers of protons of the filled shells. The stepwise form of the curve of binding energy of nucleon pairs (E_2) [2] on the 2β-stability line shows that there exists a third aspect of the shell nuclear structure - nucleon shells in the general system of the lower n- and p- levels manifested also in the periodicity in other nuclear properties. The nucleon shell is completed on the 2β-stable nuclides with a magic number of nuclides. However, these magic numbers, unlike the shell model of the nucleus, represent a sum of the numbers of neutrons in the α^4-kern* (14 n) and in n-shells:

$$N_m = \sum_{n=2}^{n} (n^2 + 3n + 4)_{n \geqslant 2} + 14n$$

The numbers of protons in nucleon shells are expressed by eq. $^nZ = 2(3n-2)$ and they are increasing by six in each consecutive shell: $^nZ = 6$.

From the number of neutrons and protons and a regular distribution of α^4 and α^6 in the nucleon shells pre-

dicted are hypothetical pleiads of β-stable isotopes of the elements with Z = 102 through 156. All the β-stable isotopes of the transuranium elements with Z = 92 through 100, predicted yet in 1950 [6,7] by similar regularities in the numbers of α^4 - and α^6 - alternations of p and n in the nucleon shells, have been discovered. Therefore, predictions of pleiads of β-stable isotopes for farther transmendelevium elements with Z $\geqslant$ 102 will probably be confirmed

* The nuclear core where nuclides, multiples of α-particle, $^{24}_{12}Mg_{12}$, $^{28}_{14}Si_{14}$, a.o. have high maxima of E_2 and maximal abundances

Nucleon shells and regularities in abundances of the nuclides

The regularities in alternation of protons and neutrons in the nucleon shells are also manifested in the abundances of the isotopes of the elements [2]. In pleiads of the isotopes of the Z^{even} elements the maximum abundance is characteristic of the isotopes on which the α^4- and α^6 - alternations of p and n on the 2β-stability line are completed. The M^{odd} isotopes are less abundant than the neighbouring M^{even} isotopes. Besides, a considerably larger abundance is characteristic of the isotopes with a magic number of neutrons. At the same time there are some enigmatic exclusions from these regularities. Yet in 1950 there was discovered a striking anomaly in the abundances of the isotopes of Te and Xe. For Te, the β-stable isotopes ^{128}Te and ^{130}Te have the maximal abundance. As to Xe, the isotopes of odd mass numbers: ^{129}Xe and ^{131}Xe have the maximal abundances and are more abundant than their intermediate M^{even} isotope ^{130}Xe. An anomalous high abundance of these nuclides was supposed to be explained by the fact that their normal abundance arising in the neutron nucleosynthesis [8] was replenished with nuclear fission fragments of the transuranium elements atoms. In [9] it was suggested that the fragments of symmetrical fission of the nobelium isotopes were added. More than 30 years had elapsed before this hypothesis was proved experimentally. Recently it has been discovered [10] that the nuclides $^{258}_{100}$Fm, $^{260}_{101}$Md, $^{262}_{102}$No, and their neighbours fission symmetrically mainly into fragments with M=128 through 131 with the numbers of protons and neutrons close to magic ones: Z_m=50, N_m=82. The β^- -decay of fragments generates isotopes with a maximal abundance, e.g.,

$$^{260}_{101}Md \longrightarrow {}^{130}_{50}Sn + {}^{130}_{51}Sb \text{ and }$$
$$^{130}_{50}Sn \xrightarrow{\beta^-} {}^{130}_{51}Sb \xrightarrow{\beta^-} {}^{130}_{52}Te \text{ (35%)};$$

here the 2β -stable $^{130}_{54}$Te screened by $^{130}_{52}$Te from this chain of β^- -decays has an anomalous small abundance (4.1%)

The curve of nuclides abundance in the Solar system exhibits one more analogous maximum of the abundances of nuclides with M=192 through 195 which also have an anomalous high abundance The heaviest β-stable ^{192}Os has a maximal abundance (41%) and is considerably more abundant than the 2β -stable M^{even} isotopes in the middle of the pleiad. The platinum isotope with an odd mass number ^{195}Pt has a maximal abundance (32.9%) and is more abundant than its other isotopes. One could suggest that this maximum in abundances of the nuclides also arises during a symmetrical fission of nuclei of the heaviest transuranium elements, For example, in fission of the finite nuclide $^{410}_{152}$of a nucleon shell with n=7 and its neighbouring nuclides. This nuclide, having a magic number of neutrons $N_m= 14n +\sum\limits_{2}^{7}(n^2+3n+1)=258n$, could be generated in the neutron star from which the Solar system

appeared.

It follows that the mass numbers and abundances of isotopes of the elements cannot be considered incidental, as in them manifested are the regularities of nucleosynthesis, "nuclear code" of alternation of p and n on the 2β -stability line and the laws of conservation in the symmetry of the real nucleon shells described in the monograph [2].

REFERENCES

1. Particle Data Group. Particle Properties Data Booklet:Review of Particle Properties, NORTH-HOLLAND, AMSTERDAM (1990)

2. I.P. Selinov: Stroenie i sistematika atomnykh yader. Prilozhenie: Tablitsy atomov, atomnykh yader i sub"yadernykh chastits. Moscow:Nauka (1990)

3. I.P. Selinov: The periodic system of stable isotopes and the relation of the mass and charge of the nuclei, Phys. Zs. d. Sowietunion, 7 (1) 82-98 (1935), ZhEhTF, 4 (7) 666 (1934)

4. Guggenheimer J.J: J.phys.radium, 5, 253 (1934)

5. Elsasser W:Sur le principe de Pauli dans Les noyaux, J.phys. radium, 5, 389, 457, 635 (1934)

6. I.P. Selinov: Sistema atomnykh yader i nekotorye zakonomernosti v svojstvakh izotopov: Frenkel' Ya I : Printsipy teorii atomnykh yader, 1st ed M.:L Izd-vo AN SSSR, Pril II, 273 (1950)

7. I.P. Selinov: Atomnye yadra i yadernye prevrashcheniya, M.:L Gostekhizdat (1951)

8. Ya. M.Kramarovskij, V P Chechev: Sintez ehlementov vo Vselennoj,Moscow, Nauka (1987)

9. I.P. Selinov: New Isotopes and Systematics of Nuclides, Proc. 3-d Int. Conf. on Peaceful uses of Atomic E Energy, Geneva, 15, 389 (1964)

10. E.K. Hulet: Spontaneous Fission Properties of the Heavy Elements:Bimodal Fission, UCRL-99956 Preprint, (Nov. 11 1988)

DECAY HEAT CALCULATION
An International Nuclear Code Comparison

B. Duchemin
DAMRI/LPRI, CE Saclay, B.P. 52, F-91193 Gif-sur-Yvette CEDEX, France

C. Nordborg
NEA Data Bank, Bât. 445, F-91191 Gif-sur-Yvette CEDEX, France

The results of an international code comparison on decay heat are presented and discussed. Participants from more than ten laboratories calculated, using the same input data, decay heat for thirteen cooling times between 1 and 10^{13} sec. Two irradiation cases were proposed: fission pulse and $3 \cdot 10^7$ seconds of irradiation of ^{235}U fuel. The results are analysed and compared. This inter-comparison shows that, if the same input data are used, almost all of the participating codes give very similar results for the decay heat and consequently also for the fission product contribution.

(deacy heat, computer codes, benchmark, fission pulse, irradiation, fission products)

Introduction

This paper contains the results of an international effort to compare the results obtained by codes which calculate decay heat either for pulse fission or after irradiation. These decay heat calculations are very important in many aspects of a reactor operation. As examples could be mentioned: reactor shut-down procedures and re-fueling. The comparison was initiated during the specialist meeting on data for decay heat predictions held at Studsvik, Sweden, in September 1987. The objective was to verify that the discrepancies appearing between different calculations are due to the data used (yields, capture cross-sections, nuclide mean energies, ...) and not to the method used to solve the generalised Bateman equations which describe the nuclide evolution during irradiation and cooling.

The participants were asked to calculate the decay heat, if possible with uncertainties, for two cases using a given set of input data. The results should contain, for many different cooling times, information on the total decay heat, the β and γ parts separately and, a list of the nuclides which contributed more than 1% to the total decay heat. The two cases agreed upon were:

A. a ^{235}U fission pulse (Benchmark 1),

B. a $3 \cdot 10^7$ seconds irradiation of ^{235}U with no burn-up (Benchmark 2).

The cooling times chosen ranged from 1 second to 10^{13} seconds. The input data contained information on: yields, spectroscopic data, and capture cross-sections. Only the decay heat due to fission products should be calculated. The decay heat due to actinide formation was omitted from this comparison.

Contributions Received

The contributions received could be divided into two classes: one for analytical solutions of the equations and one for numerical solutions. The name of the participating codes could be found in Table 1.

Table 1 Participating codes.

Code name (A=Analytical) (N=Numerical)	Name of participant (Only first name given)	Country
AFPA (A)	P.M. Rubtsov	USSR
CINDER-10 (A)	T. England	USA
CINDER (A)	Wang Dao	P.R. China
DCHAIN (A)	K. Tasaka	Japan
FISP6 (A)	A. Tobias	UK
FISPIN (N)	M.F. James	UK
INVENT (A)	G. Rudstam	Sweden
KORIGEN (N)	H.W. Wiese	Germany
MECCYCO (N)	G. Gillet	France
ORIGEN-S (N)	M.C. Brady	USA.
PEPIN (A)	B. Nimal	France

Input data

It was decided to use only a subset of the fission yield and decay data necessary for a realistic decay heat calculation, in order to facilitate the comparison of results. The heavy mass peak A=131 to 140 was selected and the data were compiled in the ENDF-5 format at the NEA Data Bank.

The objective of the pulse fission calculation was to test the ability of the codes to use the fission yields and to treat correctly the mass decay-chain calculation, whereas the objective of the long irradiation $(3 \cdot 10^7$ seconds) calculation was to test the ability of the codes to treat the capture (or absorption) problem.

Some inconsistencies were noted in the input data, but they had very little influence on the results:

- For some nuclides there was an inconsistency between independent and cumulative yields, but this affected the results less than 2 per-mil at 1 second cooling time and 1 per-mil after.

- The branching ratios did not, in all cases, sum up to 100 %, but this had no real influence on the results.

- The capture cross-section branching ratios to ground and isomeric states were not specified. This had ≤ 1 per-mil effect on the results, as was shown by the PEPIN contribution, where results had been obtained for the two hypothesis: 100 percent on ground-state, and 50 percent on ground state and 50 percent on isomeric state.

Results

Following the NEANDC/NEACRP Task Force meeting on Decay Heat Predictions held at the NEA Data Bank on 21st and 22nd September 1989, when preliminary results were presented, some contributions were revised with subsequent corrections to the codes:

- Double precision.
  In the case of INVENT and KORIGEN some calculational parameters were modified to double precision in order to better reproduce the results at long cooling times.

- Capture effects in benchmark no. 2.
  The first set of results from the CINDER code did not take into account the capture effects, but this was corrected in a second set of results.

- Wrong input data.
  The first results from benchmark no 2 from the code KORIGEN were discrepant and the reasons were two: misunderstanding of the interpretation of the one group capture cross section and a typing error resulting in an irradiation of $1\ 10^7$ instead of $3\ 10^7$ seconds.

- Burn-up effect.
  In the specification of benchmark no. 2 it was stated that no burn-up should take place, which in itself is unrealistic, but avoids further complications in a inter-comparison benchmark like this. One code, INVENT, which had a built-in burn-up calculation routine had to be modified to keep the amount of Uranium constant during the long irradiation. This resulted in an additional 0.04 % uncertainty in the results for this code.

- Delayed neutron effects.
  Following the discussions of the first benchmark results, the FISP6 code was revised to avoid any manual manipulation of the fission yields (to account for delayed neutron effects). This involved extension of the linearised decay chains to account for the delayed

neutron effects as well as some minor data manipulation amendments to the FISP6 code itself.

- Subdivision of the irradiation time.
  In Benchmark 2 a subdivision of the irradiation time was not prescribed in the specifications. Theoretically in case of constant coefficients, the solution does not depend on this subdivision. Numerically, however, a certain dependence is introduced by treating the chain equations with respect to the half-lives of the nuclides in the chains, as in the code KORIGEN. High total decay heat values were first noted for KORIGEN at 10^7 and 10^8 seconds cooling time but this dependence was since strongly reduced. The values are still slightly above the average but well within the quoted uncertainties. The same tendency was also found in the results of the ORIGEN-S code.

- Output precision.
  The somewhat scattered results for the ORIGEN-S code was due to the fact that the results were given with only two significant decimals. One effect of this was for example that the fission product contribution at 10^9 seconds cooling time for a fission pulse added up to over 100 percent. A second iteration with results containing more significant figures was not performed.

- Uncertainties.
  Only the INVENT and KORIGEN results were given with uncertainties. In the case of KORIGEN, the uncertainties resulted from energy release only. The INVENT code used, in the calculation, all uncertainties given in the input data. Typical values for the total decay heat uncertainties for Benchmark 1 varied from about 5 percent at very short and very long cooling times down to about 1.3 percent at medium cooling times. The corresponding values given for benchmark no 2 were: about 1 percent at short cooling times up to about 10 percent at very long cooling times.
  (It should be noted that these quoted uncertainties are relevant to this exercise only and not to a real case of decay heat calculation, as the input data did not correspond to a realistic case and did not included a complete set of uncertainties.)

Following these iterations, the results from the different codes agree very well (within about 1/2 %) and only minor differences are found. No systematic differences between the results of the Analytical and Numerical codes could be found.

Conclusion

The overall results of the calculation of decay heat and contributing fission products are in very good agreement for the codes. The results given by the CINDER code (contribution by Wang Dao) are on the average 1 to 3 percent lower that other results.

The objective of this exercise was to ensure that the discrepant results for decay heat calculation obtained in different laboratories were due to different input data and

558

not to the different solutions of the equations describing the fission product evolution. This objective has been reached.

This inter-comparison has permitted the participants to gain confidence in their codes and minor modifications to a few of the codes have been noted as a result of this exercise. Scientists who would like to test their codes against these benchmarks, can obtain the specifications (1), the input data, and the detailed results (2) from the NEA Data Bank.

REFERENCES

1. B. Duchemin, C. Nordborg: NEANDC-246 "U" and NEACRP-303"L"

2. B. Duchemin, C. Nordborg: NEACRP-319"L" and NEANDC-275"U"

Figure 1 Total decay heat for benchmark 1.

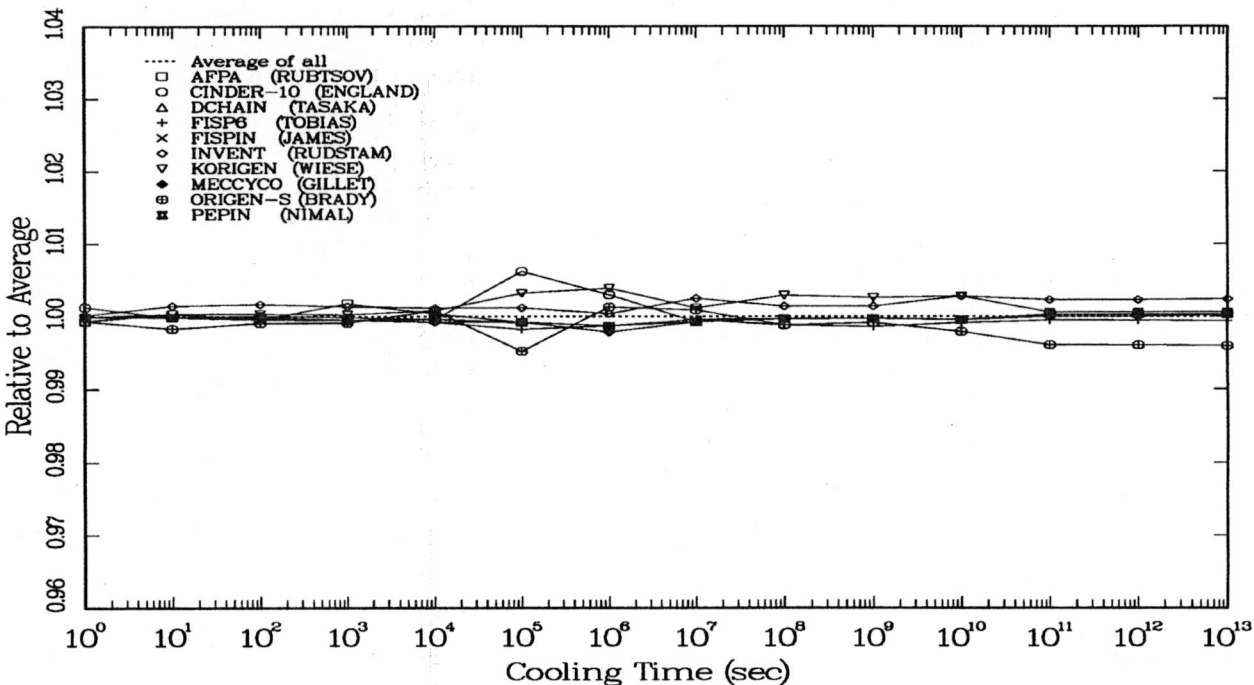

Figure 2 Total decay heat for benchmark 2.

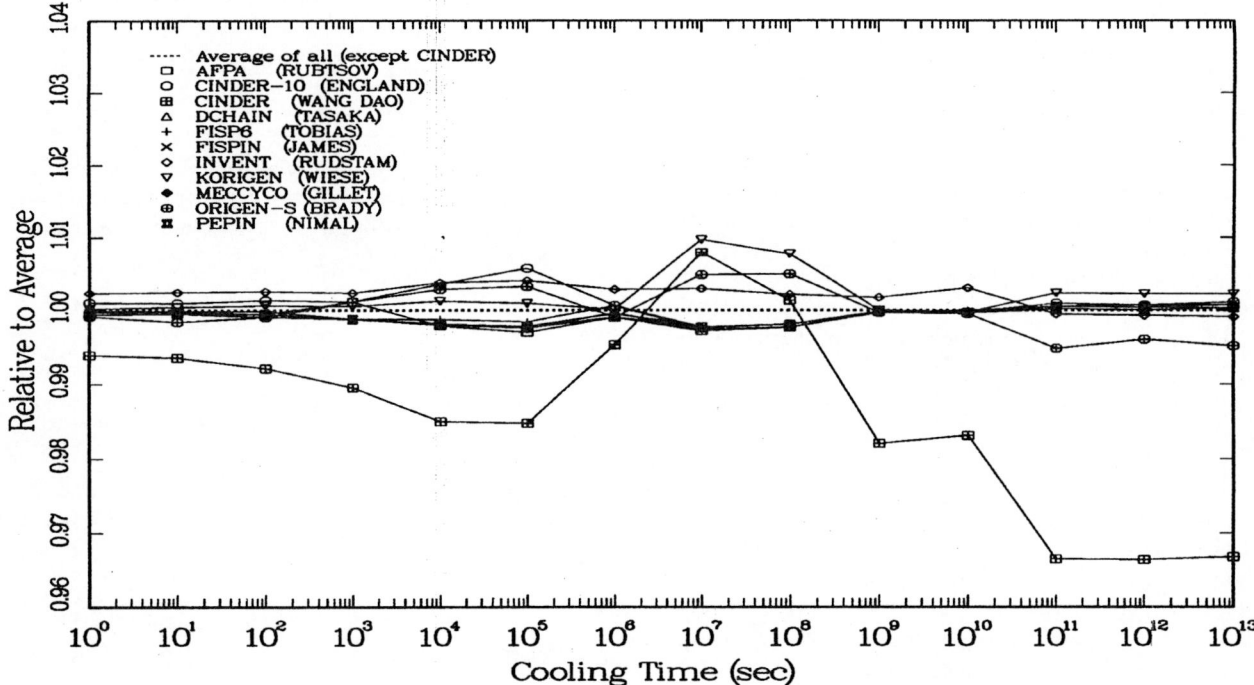

DETERMINATION OF THE HALF-LIFE OF ^{59}Ni
AND THE ^{60}Ni(n,2n) CROSS SECTION

D. Weselka[*], B. Schneck, W. Rühm, L. Zerle, G. Korschinek, and E. Nolte

Faculty of Physics, Technical University of Munich,
D-8046 Garching, Germany

H. Vonach

Institut für Radiumforschung u. Kernphysik der Universität Wien (IRK),
Boltzmanngasse 3, A-1090 Wien, Austria

Abstract: The Cross section of the reaction ^{60}Ni(n,2n)^{59}Ni for 14.8 MeV neutrons has been determined to be (410±120) mb by measuring the contents of ^{59}Ni in a fast neutron irradiated nickel foil via accelerator mass spectrometry (AMS). This new cross section is 4 times larger than the value of 100 mb as reported in literature. The half-life has been deduced to be (9.5±2.5)x10^4 y by using the result of a radioactivity measurement of ^{59}Ni produced by thermal neutrons in the ^{58}Ni(n,γ)^{59}Ni reaction.

^{59}Ni half-life, ^{58}Ni(n,γ)^{59}Ni, activation cross section, AMS, ^{60}Ni(n,2n) cross section at E_n=14.8 MeV

Introduction

Due to its relative long half-life ^{59}Ni is an interesting nuclide to determine the terrestrial age of meteorites [1] or to monitor the alpha component of solar wind in lunar samples using the ^{56}Fe(α,n)^{59}Ni reaction [2]. In cosmic samples, which often contain large amounts of ^{58}Ni, ^{59}Ni is mainly produced by neutron capture ^{58}Ni(n,γ). For the purpose of dating, the knowledge of its half-life is, of course, important. The presently accepted half-life of ^{59}Ni is (7.5±1.3)x10^4 y [3]. Yet other values, partly much higher, have been reported as well: 2.3x10^5 y [4,5], 7.5x10^5 y [6], (1.0±0.25)x10^5 y [7] and (7.6±0.5)x10^4 y [8]. All these results were obtained from radioactivity measurements.

An activation measurement of a nickel foil irradiated with 14.8 MeV neutrons yielded a cross section σ of (104±25) mb [9] for the reaction ^{60}Ni(n,2n)^{59}Ni under the assumption of the presently accepted half-life of $T_{1/2}$ = 7.5x10^4 y for ^{59}Ni [3]. In contrast, however, calculated values of this cross section are in the range of 400...500 mb [10-13]. Since the produced activity A is proportional to σ/τ (mean life τ = $T_{1/2}$/ln2), the discrepancy between the measured and the calculated values for σ could be explained by a half-life which is 4 to 5 times longer. In order to decide whether the presently accepted half-life or the (n,2n) cross section based on the activation of ref. 9 are correct, measurements of both quantities were repeated.

The ^{60}Ni(n,2n) cross section was measured by the AMS method, as this method does not require the knowledge of the half-life. Since N(^{59}Ni)/N(^{60}Ni) = σφt (φt fluence) the cross section is obtained from the measurement of the number N of ^{59}Ni nuclei. For the measurement of the half-life ^{59}Ni was produced by means of the ^{58}Ni(n,γ)^{59}Ni reaction in a thermal neutron field. The activity was measured relative to that of a ^{55}Fe reference source produced by irradiation of ^{54}Fe in the same neutron field.

Half-life measurement

30 mg ^{58}Ni powder packed in thin aluminium foil was irradiated along with a ^{54}Fe foil and Co foils with thermal neutrons in the Munich research reactor. The ^{58}Ni enrichment was 99.89% and the exposure time lasted 6 days. The nickel powder was dissolved and purified two times by an ion exchange column with 9n HCl and the NiCl$_2$ reduced in H$_2$. After this treatment the unwanted Co activities from the ^{58}Ni(n,p)^{58}Co reaction were below 10 Bq. At this level the ^{58}Co activity did not interfere with the ^{59}Ni measurement (see fig. 1). Finally the sample was melted and rolled out to foils of 4.8 and 0.95 mg/cm^2 thickness. The neutron fluence was determined from the ^{60}Co activity produced in the Co foils by the ^{59}Co(n,γ) reaction to be (3.9±0.4)x10^{17}/cm^2.

---

[*] Present address: Institut für Hochenergiephysik der Österr. Akad. d. Wiss., A-1050 Wien, Austria

The 6.93 keV ^{59}Co K$_\alpha$ radiation from the ^{59}Ni electron capture decay and the ^{55}Mn 5.89 keV radiation from ^{55}Fe were measured with a Si(Li) X-ray detector. The X-ray spectrum is shown in fig. 1. Using these measurements and the values from table 1 the half-life of ^{59}Ni can be calculated:

$$T_{1/2}(^{59}\text{Ni}) = \frac{A_{55} \cdot T_{1/2}(^{55}\text{Fe}) \cdot N_{58} \cdot \sigma_{58} \cdot \omega_{\text{Co}}}{A_{59} \cdot N_{54} \cdot \sigma_{54} \cdot \omega_{\text{Mn}}}$$

Table 1:	^{55}Fe	^{59}Ni
cross section $\sigma(n,\gamma)$	2.25 b	4.6 b
fluorescence yield ω	0.314	0.381
time	1000 s	70229 s
no. of nuclei $N \times 10^{17}$	8.23	503.1
counts A	7504	1937
half-life $T_{1/2}$	2.7 a	

The L/K capture ratios of both decays are almost equal. The possible correction is below 0.5% and was neglected. The detector efficiency is the same for both only slightly different X-ray energies. The influence of self absorption was determined by measuring the thin and the thick foil activities. Hence follows, including the values of table 1 and correction for self absorption, a ^{59}Ni half-life of $(9.5 \pm 2.5) \times 10^4$ y in fair agreement with the presently accepted value.

^{60}Ni(n,2n) cross section

The AMS experiment was performed at the Munich accelerator laboratory using the Tandem van de Graaff and the linear radiofrequency accelerator by detecting completely stripped ^{59}Ni ions in a Bragg ionization chamber. The system is described elsewhere [14,15]. The natural nickel foil had been irradiated by neutrons of (14.8±0.6) MeV at the Rotating Target Neutron Source-II (RTNS-II) at Lawrence Livermore National Laboratory with a fluence of $(5.18 \pm 0.72) \times 10^{17}$ cm^{-2} [16]. The irradiated foil was diluted by a factor of 10 with nonradioactive nickel and purified by the same method as described above. The negative nickel ions were generated in a high current ion source with typical currents of 5 μA. The ions were accelerated to 4.75 MeV/nucleon. The transmission from the ion source to the detector was about 10^{-7}. The detection limit, defined by one detected event per hour, was N(^{59}Ni)/N(^{58}Ni) = 2×10^{-10}. The calibration source was produced by thermal neutron irradiation in the Munich research reactor. The thermal and epithermal neutron fluences were monitored by the simultaneous irradiation of gold foils with and without 1 mm thick cadmium shielding. Using a cross section $\sigma(^{58}\text{Ni}(n_{th},\gamma)) = (4.6 \pm 0.3)$ b, the calibration source was calculated to have a concentration of N(^{59}Ni)/N(^{58}Ni) = $(1.05 \pm 0.21) \times 10^{-8}$. The concentration of the diluted nickel foil was measured to be

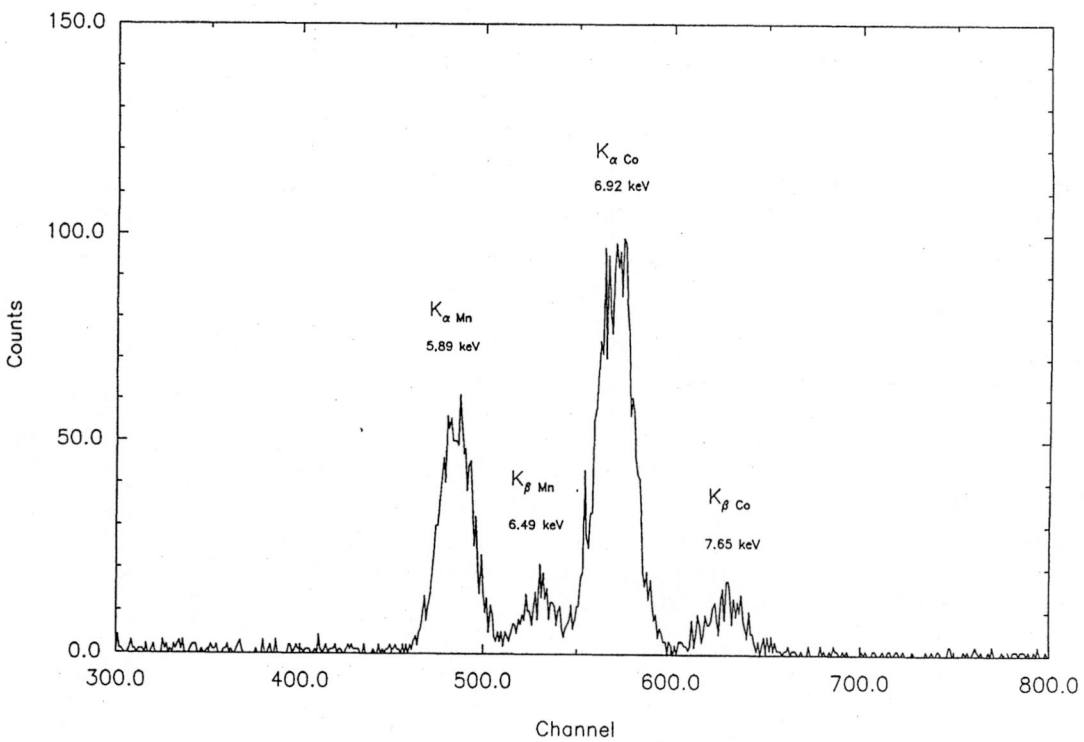

Fig.1. X-ray spectra of ^{59}Ni (Co K$_\alpha$ and K$_\beta$) and ^{55}Fe (Mn K$_\alpha$ and K$_\beta$) from the ^{58}Ni(n,α)^{55}Fe reaction induced by the fast neutron flux of the reactor.

$N(^{59}Ni)/N(^{58}Ni) = (0.81\pm0.20)\times10^{-8}$, corresponding to $N(^{59}Ni)/N(^{58}Ni)=(0.81\pm0.20)\times10^{-7}$ for the original irradiated foil. The uncertainty includes the errors of the ^{59}Ni concentration of the calibration source and the statistical error of the measurement between the ^{59}Ni concentration of the two sources.

With the isotopic abundances of ^{58}Ni and ^{60}Ni and using $N(^{59}Ni)/N(^{58}Ni) = \sigma(^{60}Ni(n,2n))\phi t(0.261/0.6827)$, the cross section $\sigma(^{60}Ni(n,2n)^{59}Ni)$ for 14.8 MeV neutrons is obtained to be (410 ± 120) mb. The experimental value for the cross section of 410 mb is in good agreement with calculated values between 400 and 500 mb [10-13] (for comparison see fig. 2).

Conclusions

Our results confirm both the presently accepted value of the half-life of ^{59}Ni and the theoretical predictions of the $^{60}Ni(n,2n)$ cross section at $E_n=14.8$ MeV. The result of ref. 9 is not compatible with our new data.

We would like to thank Dr. B. Patrick (Harwell) for providing us the neutron irradiated nickel foil and the neutron fluence measurements.

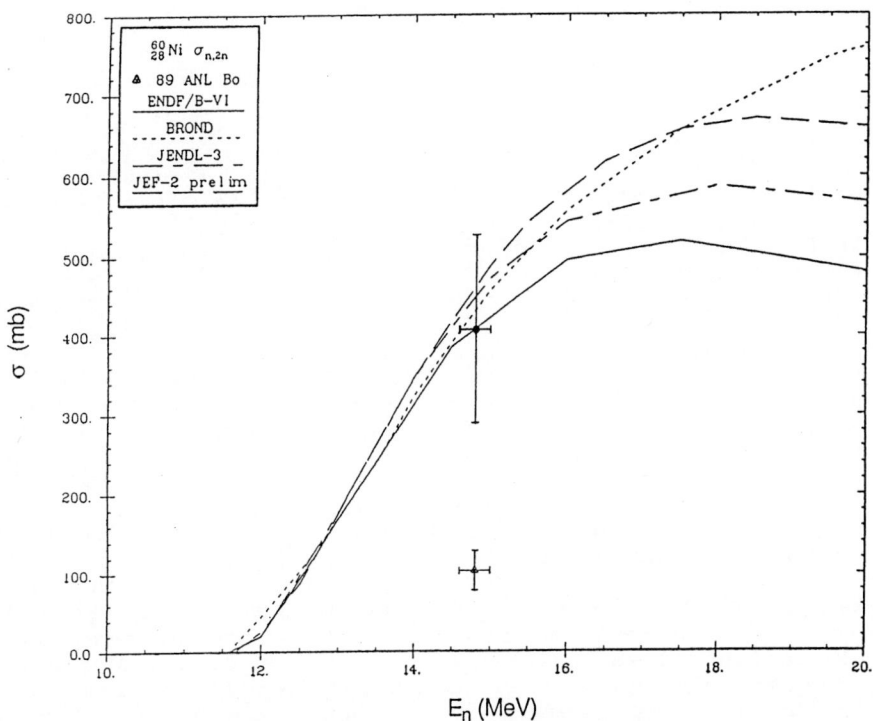

Fig.2. Calculated cross section for the $^{60}Ni(n,2n)^{59}Ni$ reaction [10-13] together with the experimental values of this work ($\bullet$) and of ref. 9 ($\triangle$).

References

[1] R. C. Reedy, J. R. Arnold and D. Lal, Science 231, 41 (1986).

[2] R. C. Reedy and K. Marti, in: The Sun in Time (Univ. Ariz. Press, 1990), Los Alamos Preprint LA-UR-89-1483.

[3] A. R. Brosi, C. J. Borkowski, E. E. Conn and J. C. Griess, Jr, Phys. Rev. 81, 391 (1951).

[4] G. Friedlander, Technical report, BNL-AS-2, 1949.

[5] J. M. Hollander, I. Perlman and G. T. Seaborg, Rev.Mod.Phys. 25, 497(1953).

[6] H. W. Wilson, Phys.Rev. 82, 548 (1951).

[7] B. Saraf, Phys. Rev. 102, 466 (1956).

[8] K. Nishiizumi, R. Gensho and M. Honda, Radiochim. Acta 29, 113 (1981).

[9] D. L. Bowers and L. R. Greenwood, J. Radioanalytical and Nuclear Chemistry 123, 461 (1988).

[10] ENDF-201, 4 edition, ENDF/B-VI Summary Documentation, edited by P. Rose, BNL-Report (in press).

[11] V. N. Manokhin ed., Report INDC(CCP)-283, IAEA, Vienna (1988).

[12] K. Shibata et al., Report JAERI 1319, Japan Atomic Energy Research Institute, Tokai-mura (1990).

[13] M. Uhl et al., Proc. Int. Conf. on Nuclear Data for Science and Technology, Jülich 12-17 May 1991 (to be published).

[14] P. W. Kubik, G. Korschinek and E. Nolte, Nucl.Instr.Meth. B1, 51(1984).

[15] G. Korschinek, H. Morinaga, E. Nolte, E. Preisenberger, U. Ratzinger et al. Nucl. Instr. Meth. B29, 67 (1987).

[16] B. Patrick, priv. comm.

PRECISION MEASUREMENTS OF RADIOACTIVE DECAY DATA

U. Schötzig, H. Schrader and K. Debertin
Physikalisch-Technische Bundesanstalt,
D-3300 Braunschweig, Germany

Abstract: Half-lives, gamma- and X-ray emission probabilities of radionuclides have been measured with high accuracy. Half-life values were obtained by following the radioactive decay with an ionization chamber and new measurement results are reported for ^{90}Sr and ^{108}Ag^m.

Gamma- and X-ray emission probabilities p per decay have been derived from emission rates measured with Ge and Si(Li) detectors, and activities measured with $4\pi\beta$-γ coincidence systems or a $4\pi\gamma$ NaI(Tl) counting system. Results of p are reported for ^{56}Co, ^{57}Co (14.4 keV), ^{109}Cd, ^{123}Te^m, ^{125}Sb and ^{125}I. Relative uncertainties of the order of 1% or less were achieved.

(Half-lives, gamma-ray emission probabilities)

Introduction

Accurate radioactive decay data are considered by the Physikalisch-Technische Bundesanstalt (PTB), the German national standards laboratory responsible for the realization of the activity unit and the dissemination of radioactivity standards to users, to be a prerequisite for performing high-quality activity measurements. Among these data, half-lives and photon (γ- and X-ray) emission probabilities, are most important. With these arguments in mind the International Atomic Energy Agency (IAEA) recently initiated a coordinated research programme on "Gamma-Ray Standards for Detector Calibration" [1] in which the PTB participated. The aim was to produce a list of well-evaluated recommended decay data for about 30 radionuclides.

The certified activity value for a standard refers to a fixed reference date, and the half-life must be known with high accuracy in order to calculate the activity for the date of the standard's use which may well be several half-lives later. Photon emission probabilities are needed both for the improvement of some basic activity measurement procedures in a standards laboratory and for the full-energy peak efficiency calibration of semiconductor γ- or X-ray spectrometers, or whenever an activity has to be derived from a photon emission rate measured with such a spectrometer. Although many data are available with a sufficiently low uncertainty for most practical purposes, there are other data for which discrepant values have been published and whose uncertainties quoted in data evaluations should be reduced [1].

Over the past 20 years, half-lives of about 40 radionuclides and photon emission probabilities of about 30 radionuclides have been measured at the PTB with high accuracy. These radionuclides are chiefly those used, produced or occurring in environmental radioactivity measurements, in the nuclear fuel cycle or in nuclear medicine. The present work reports on results recently obtained for ^{90}Sr, ^{108}Ag^m (half-lives), ^{56}Co, ^{57}Co, ^{109}Cd, ^{123}Te^m, ^{125}Sb and ^{125}I (photon emission probabilities).

Half-lives

The expert group involved in the above-mentioned IAEA programme [1] postulates that the relative uncertainty in any detector efficiency calibration arising from the uncertainty in the half-life of the sources used should not exceed 0.1%. Assuming that a source may remain in use for a period of 15 years or 5 half-lives, whichever is the shorter, means that a half-life of 15 years, for example, should have an uncertainty not exceeding 0.15%, while a half-life of 3 years should be known to 0.03%. This accuracy has been achieved, for example, for ^{24}Na, ^{51}Cr, ^{60}Co, ^{88}Y, ^{111}In, ^{134}Cs and ^{198}Au but we are nowhere near accomplishing this for ^{55}Fe, ^{75}Se, ^{109}Cd, ^{155}Eu and ^{207}Bi, for example. An evaluation of published half-life data has revealed that the reduced chi-square from the weighted-mean calculations is frequently appreciably higher than unity, which means that experimenters tend to underestimate the uncertainties of their half-life measurements.

In the 1983 status report, Walz et al. [2] described the basic procedures of half-life measurements at the PTB. The decay of the radioactive substance in a particular source, usually a sealed ampoule filled with a few ml of an aqueous solution, is followed by measurements with a well-type ionization chamber filled with argon at a pressure of 2 MPa over a period of several half-lives or, for long-lived radionuclides like ^{133}Ba, ^{152}Eu and ^{154}Eu, over up to 15 years. The long-term stability of the chamber is routinely checked by means of a radium reference source. The reproducibility of the ionization chamber measurements is of the order of $2 \cdot 10^{-4}$.

The initial activity of the sources was chosen sufficiently low to prevent any saturation effects in the chamber. γ-ray emitting impurities were identified by means of γ-ray spectrometry with germanium detectors, and their contributions in terms of current were subtracted from the total ionization current applying the relevant current-to-activity calibration factors.

From the measured data, half-lives were determined by regression analysis. Sub-groups within each data set were evaluated separately in order to discover possible systematic effects and to help in the estimation of realistic uncertainty values. In those cases in which the decay could be followed over several half-lives the relative standard deviations of the fit were of the order of 0.01%. However, this is not a sufficient indication that all systematic effects have been properly corrected for. For this reason, uncertainties derived from the sub-group evaluations were deemed more realistic.

Results obtained earlier at the PTB are tabulated in the references [2, 3]. The recent measurement results are summarized in table 1 and compared with those of 1983 [2]. In the case of ^{90}Sr new measurements were started in 1986 with a modified source avoiding plastic materials with varying absorption effects near the source. The source of ^{108}Ag^m was produced by an (n,γ) reaction containing an activity fraction of 0.5% of ^{110}Ag^m at the beginning of the half-life measurements in 1984. The result obtained here is considerably higher than that published by Harbottle [4] of 127(7) a, probably influenced by the reported instrument instabilities, but it agrees within the uncertainties with the value of 310(132) a from (n,2n) reaction data by Vonach et al. [5].

Table 1: Half-life measurements

Nuclide	n	$t/T_{1/2}$	$T_{1/2}$		Comments
^{85}Kr	638	1.2	3915(3)	d	
^{85}Kr	250	0.45	3909(8)	d	[2]
^{90}Sr	205	0.15	10513(14)	d	mod.source
^{90}Sr	123	0.1	10900(90)	d	[2]
^{108}Ag^m	233	0.02	418(15)	a	imp.^{110}Ag^m
^{133}Ba	787	1.5	3841(5)	d	
^{133}Ba	383	0.7	3842(18)	d	[2]
^{152}Eu	888	1.15	4936(2)	d	imp.^{154}Eu
^{152}Eu	557	0.5	4939(6)	d	[2]
^{154}Eu	671	1.5	3139(2)	d	
^{154}Eu	265	0.6	3136(4)	d	[2]

n: Number of measurements
t: Duration of measurements

Gamma- and X-ray emission probabilities

The quality of available data has been continually improved with the increasing use of germanium detectors, and uncertainties quoted in data evaluations are now of the order of 1% for most gamma und X-rays of practical importance.

In our studies we have concentrated on those cases where either discrepant data were published or where the accuracy achieved was not considered sufficient. Data obtained in the past have been published in [8-20]. In the following, data measured recently and not yet published are given.

The measurement procedures are described in detail, for example, in reference [9]. Photon emission probabilities, $p = R/A$, have been derived from source emission rates, R, measured with carefully calibrated germanium or Si(Li) detectors, and activities, A.

Activities have been determined by means of absolute methods being usually employed in national standards laboratories, like the $4\pi\beta$-γ coincidence method (in the case of ^{57}Co, ^{109}Cd, ^{123}Te^m and ^{125}I) or the $4\pi\gamma$ NaI(Tl) scintillation counting method (for ^{56}Co and ^{125}Sb, with 24.5 % of the decays running via the isomeric state in ^{125}Te^m). Typical uncertainties of such measurements are of the order of 0.3 %. This high degree of accuracy has been confirmed by numerous international comparisons under the auspices of the Bureau International des Poids et Mesures (BIPM).

The full-energy peak efficiency calibration curves for the semiconductor detectors were obtained with activity-calibrated standard sources of more than 20 radionuclides. Only those calibration nuclides were used for which the photon emission probabilities can be obtained very accurately from decay-scheme considerations, calculations or measurements with other types of detectors the efficiency of which was derived from other calibration principles. In energy regions where no adequate calibration standards were available, supporting Monte-Carlo calculations were performed and used for interpolation purposes (necessary for the high energy part of ^{56}Co). The calibration uncertainty is 0.8% (standard deviation) in the energy region between 100 and 1500 keV and up to 2% between 5 and 100 keV and between 1500 and 3000 keV.

Table 2 lists recent results for ^{56}Co, ^{57}Co (14.4 keV), ^{109}Cd, ^{123}Te^m, ^{125}Sb and ^{125}I.

For the emission probability of the 14.4-keV γ-ray of ^{57}Co frequently used for the efficiency calibration of Si(Li) detectors, there is a remarkable discrepancy of 4.4 % between the hitherto recommended value of 0.0954(12) [6] and our measurement value of 0.0916(15). The latter, however is in excellent agreement with the value that results from using the theoretical conversion coefficient quoted in reference [7].

The listed results of p for ^{125}Sb are valid for the radioactive equilibrium of ^{125}Sb with the 145 keV isomeric state of ^{125}Te^m ($T_{1/2}$ = 57.14 d). The agreement with the most recent work [21] is fairly good.

Table 2: Measured emission probabilities

Nuclide	Energy (keV)	p
[56]Co	733.7	0.00190(7)
	787.9	0.00315(10)
	846.8	0.99935(25) [1])
	896.6	0.00086(20)
	977.5	0.01449(15)
	1037.8	0.1417(13)
	1140.3	0.00137(5)
	1175.1	0.02288(21)
	1238.3	0.6692(60)
	1272.2	0.00024(10)
	1335.6	0.00118(6)
	1360.2	0.0429(4)
	1442.8	0.00185(7)
	1462.3	0.00065(8)
	1640.5	0.00072(12)
	1771.4	0.1547(14)
	1810.8	0.00638(8)
	1963.7	0.00724(10)
	2015.2	0.0304(5)
	2034.8	0.0789(13)
	2113.1	0.00376(10)
	2212.9	0.00395(14)
	2276.4	0.00128(19)
	2373.7	0.00082(22)
	2598.5	0.1725(28)
	3009.6	0.0116(3)
	3202.0	0.0332(7)
	3253.4	0.0812(17)
	3273.0	0.0193(4)
	3451.2	0.00972(20)
	3547.9	0.00200(5)
[57]Co	14.4	0.0916(15)
[109]Cd	88.0	0.0368(7)
[123]Te[m]	159.0	0.832(5)
[125]Sb [2])	27.4 Kα	0.640(11)
	31.1 Kβ	0.134(3)
	35.5	0.0593(10)
	117.0	0.00248(6)
	172.6	0.00206(6)
	176.2	0.0666(7)
	204.1	0.00332(7)
	208.1	0.00242(7)
	227.9	0.00128(7)
	321.0	0.00413(7)
	380.4	0.0149(2)
	408.0	0.00180(9)
	427.9	0.292(3)
	443.5	0.00294(9)
	463.4	0.1025(11)
	600.6	0.1759(18)
	606.6	0.0498(5)
	635.9	0.1110(11)
	671.4	0.0177(2)
[125]I	35.5	0.0655(13)

[1]) From decay scheme
[2]) In equilibrium with [125]Te[m]

Data compilation

A list of half-lives and photon emission probabilities [22] of 200 radionuclides has been compiled on the basis of existing data bases and on our own measurements. This list is recommended for use in German laboratories officially engaged in legally imposed activity measurements and in laboratories accredited by the German Calibration Service (DKD).

References

1. A. L. Nichols: Nucl. Instr. and Meth. in Phys. Res. A286, 467 (1990)
2. K. F. Walz, K. Debertin, H. Schrader: Appl. Radiat. Isot. 34, 1191 (1983)
3. H. Schrader: Appl. Radiat. Isot. 40, 381 (1989)
4. G. Harbottle: Radiochimica acta 13, 132 (1970)
5. H. Vonach, M. Hille, P. Hille: Z. Phys. 227, 381 (1969)
6. E. Browne and R. B. Firestone: Table of Radioactive Isotopes, ed. V. S. Shirley, Wiley, New York (1986)
7. H. H. Hansen: Europ. Appl. Res. Rept. Nucl. Sci. Technol. 6, 777 (1985)
8. K. Debertin, U. Schötzig, K.F. Walz, H.M. Weiß: Annals of Nuclear Energy 2, 27 (1975)
9. K. Debertin, U. Schötzig, K.F. Walz, H.M. Weiß: Proc. USERDA Symp. on X- and γ-Ray Sources and Applications, Ann Arbor, p. 59 (1976)
10. U. Schötzig, K. Debertin, K.F. Walz: Int. J. Appl. Radiat. Isot. 28, 503 (1977)
11. U. Schötzig, K. Debertin, K.F. Walz: Nucl. Sci. Eng. 64, 784 (1977)
12. U. Schötzig, K. Debertin, K.F. Walz: Gamma ray emission probabilities and half-lives of selected fission products, INDC(NDS)-87/Go+Sp., p. 261 (1978)
13. K. Debertin, W. Peßara, U. Schötzig, K.F. Walz: Int. J. Appl. Radiat. Isot. 30, 551 (1979)
14. K. Debertin: Nucl. Instr. and Meth. 165, 279 (1979)
15. U. Schötzig, K. Debertin, K.F. Walz: Nucl. Instr. and Meth. 169, 43 (1980)
16. U. Schötzig, K. Debertin: Int. J. Appl. Radiat. Isot. 34, 533 (1983)
17. U. Schötzig: Nucl. Instr. and Meth. 206, 441 (1983)
18. E. Funck, U. Schötzig, K.F.Walz: Int. J. Appl. Radiat. Isot. 34, 1215 (1983)
19. U. Schötzig: Nucl. Instr. and Meth. in Phys. Res. A286, 523 (1990)
20. B. M. Coursey, D. D. Hoppes, A. T. Hirschfeld, S. M. Judge, D. H. Woods, M. J. Woods, E. Funck, H. Schrader, A. G. Tuck: Appl. Radiat. Isot. 41, 289 (1990)
21. L. Longoria-Gandara, M. U. Rajput, T. D. MacMahon: Nucl. Instr. and Meth. in Phys. Res. A286, 529 (1990)
22. U. Schötzig, H. Schrader: Halbwertszeiten und Photonen-Emissionswahrscheinlichkeiten von häufig verwendeten Radionukliden, Report PTB-Ra 16/3, Physikalisch-Technische Bundesanstalt, Braunschweig, 1989

MEASUREMENT OF HALF-LIVES OF SHORT-LIVED NUCLEI

H. Yamamoto, K. Kawade, T. Katoh,
A. Hosoya, M. Shibata, A. Osa

Dept. of Nuclear Engineering, Nagoya University,
Nagoya, 464-01 Japan

and

T. Iida and A. Takahashi

Dept. of Nuclear Engineering, Osaka University,
Osaka, 565 Japan

Abstract: The half-lives of short-lived nuclei produced by 14 MeV or thermal neutron bombardments were measured with Ge detectors in the spectrum multi-scaling mode. The corrections for the pile-up loss and the dead time were performed by applying both source and pulser methods. Half-lives of short-lived nuclei, ^{51}Ti, ^{60m}Co, ^{89m}Zr, ^{91g}Mo, ^{91m}Mo, ^{97m}Nb, ^{104m}Rh, ^{108}Ag and ^{109m}Pd, were determined experimentally within 0.1 % accuracy.

(Half-life measurement, short-lived nuclei, neutron irradiation, Ge detector, spectrum multi-scaling)

Introduction

The half-life is one of the most fundamental constants of radioactive isotopes. In the procedure for cross section measurements of short-lived nuclei, an uncertainty affects the results strongly. It is required that the half-life values are precise and reliable. Most of the values previously published have been obtained with GM counter, ionization chamber, proportional counter and scintillation counter. The Ge detectors have been widely used for measuring the intensity and energy of γ-rays for about 20 years because of their excellent energy resolution.

In case of half-life measurement of short-lived nuclei, the counting rate changes remarkably during the measuring period. The purpose of the present study is to establish a proper method of correction for the dead time and pile-up by using a pulser method (signals from a pulser are counted simultaneously during the measurement) and a standard radiation source method (γ-rays from a standard source are counted at the time of measurement), and to measure precisely the half-lives of short-lived nuclei ($T_{1/2}$ =1-15min) produced by thermal or 14 MeV neutron irradiation.

Experimental

During the measuring time the counting rate greatly changes. For the correction of the pile-up loss and the dead time the source method seems most reliable. But statistical fluctuations of the reference source may occur and the peak area evaluation might be affected by the decaying Compton background. On the other hand the pulser method gives good statistics and no effect to the γ-ray

spectrum. However the peak shape of the pulser is different from that of the γ-radiation and the constant-pulser produces no random pile-up pulse by itself. Hence there might be some difference between the two methods at high counting rates. The variation of the peak intensity ratios of the source and the pulser was examined.

Sources of ^{51}Ti, ^{60m}Co, ^{89m}Zr, ^{91g}Mo, ^{91m}Mo and ^{97m}Nb were produced by 14 MeV neutron bombardments. The 14 MeV neutrons were generated by the intense neutron generator of Osaka University (OKTAVIAN). Sources of ^{51}Ti, ^{60m}Co, ^{104m}Rh, ^{108}Ag and ^{109m}Pd were produced by thermal neutron irradiation at the TRIGA-II reactor of Rikkyo University (100 kW). Pneumatic tubes were used for the sample transfer at the 14 MeV neutron facilities.

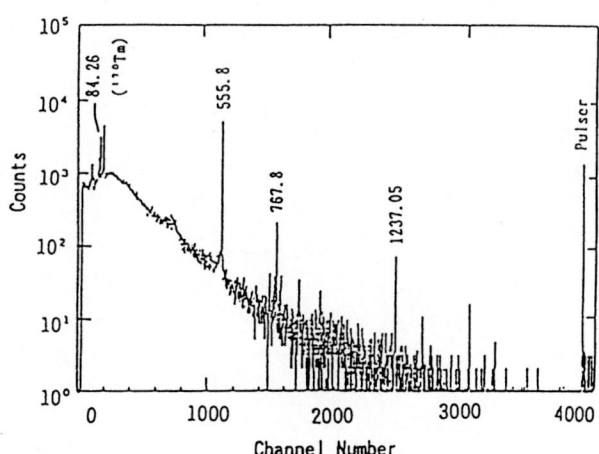

Fig. 1 γ-ray spectrum in the decay of ^{104m}Rh. γ-rays from ^{170}Tm and pulses were simultaneously measured for the correction of pile-up losses

The γ-rays were measured with the ORTEC 15 and 20 % Ge detectors in the spectrum multi-scaling mode. Decay was followed for about 10 times the half-life at equal intervals of 1/3 to 1/6 of the half-life. The standard source (^{133}Ba, ^{137}Cs or ^{170}Tm) was measured together with the short-lived activity for the correction of the pile-up loss and the dead time (source method). A constant-pulser with a rate of 60 cps was also connected to the preamplifier (pulser method). Peak areas are determined by summing all recorded counts in the peak channel. This summing method is similar to that by Debertin and Schötzig[1]. Half-life measurements were repeated 3 to 8 times for each short-lived nucleus.

Results

If both the methods are reasonable, the peak intensity ratios of the source and the pulser should be constant. In the high counting rate region they showed a clear deviation. When the initial counting rates were less than 9x10³ cps, no deviation was observed [2]. Data points in the constant ratio region were only used in the least-squares analysis. In order to estimate systematic errors and the effect of counting rate, the fitting points of the data were changed [2]. The fluctuation thus induced was included in the experimental error as a systematic error.

As an example, singles γ-ray spectrum and the decay curve in the decay of 4.3 min ^{104m}Rh are shown in Fig.1 and Fig.2, respectively. The results are summarized in Table 1 together with production reactions, γ-rays followed and previous works [3]. The results are shown in Figs.3-6 together with previous works taken from ref.4. This work with Ge detectors has improved the precision and the reliability by applying both source and pulser methods and changing the fitting data points. Most of the present results have shown agreement with previous works. Present results show somewhat shorter half-lives in general.

If the correction of the pile-up loss and the dead time is not sufficient, the obtained values should show longer half-lives. The difference of half-lives between previous and present works as a function of half-life is shown in Fig.7. The difference in the half-lives of short-lived nuclei under a few min is large. The present results are in good agreement with previous ones above 10 min in half-life. From these results it is concluded that the proper corrections of the pile-up loss and the dead time are very important in short-lived nuclei.

Table 1 Results of half-life measurement

Nuclide	Production reaction	Eγ (keV)	Half-life Present	Half-life Reference[a]
^{51}Ti	^{51}V(n,p) ^{50}Ti(n,γ)	320.1	5.759(9)m	5.76(1)m
^{60m}Co	^{60}Ni(n,p)^m ^{59}Co(n,γ)^m	58.6 1332.5	10.424(20)m	10.47(4)m
^{89m}Zr	^{90}Zr(n,2n)^m	587.7	4.145(9)m	4.18(1)m
^{91g}Mo	^{92}Mo(n,2n)^g	511.0	15.473(34)m	15.49(1)m
^{91m}Mo	^{92}Mo(n,2n)^m	653.0	62.2(10)s	65.2(8)s
^{97m}Nb	^{97}Mo(n,p)^m	743.3	58.7(18)s	60(8)s
^{104m}Rh	^{103}Rh(n,γ)^m	555.8	4.256(13)m	4.34(5)m
^{108}Ag	^{107}Ag(n,γ)	633.0	2.353(9)m	2.37(1)m
^{109m}Pd	^{108}Pd(n,γ)^m	188.9	4.663(11)m	4.69(1)m

a) taken from ref.3.

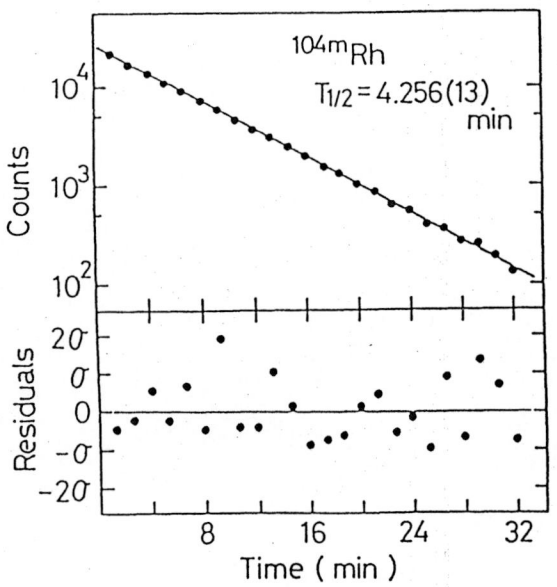

Fig. 2 Decay curve of 4.3 min ^{104m}Rh and residuals obtained from a least squares fitting analysis.

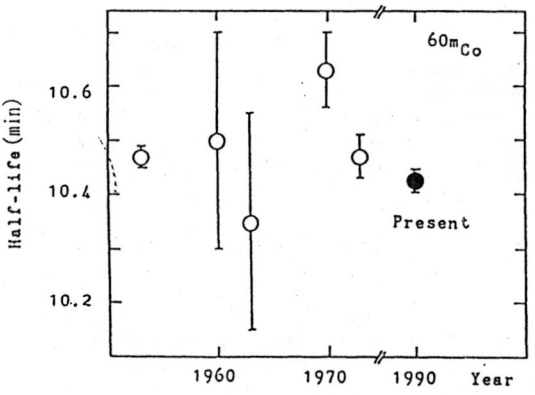

Fig. 3 Half-life of ^{60m}Co. Previous works are taken from ref. 4.

Summary

The half-lives of short-lived nuclei were measured with Ge detectors in the spectrum multi-scaling mode. The corrections for the pile-up loss and the dead time were performed by applying both source and pulser methods. Half-lives of short-lived nuclei, ^{51}Ti, ^{60m}Co, ^{89m}Zr, ^{91K}Mo, ^{91m}Mo, ^{97m}Nb, ^{104m}Rh, ^{108}Ag and ^{109m}Pd, were determined experimentally within 0.1 % accuracy.

REFERENCES

1. K. Debertin and U. Schötzig: Nucl. Instr. and Meth. <u>140</u>, 337(1977)
2. M. Miyachi, H. Ukon, M. Shibata, Y. Gotoh, H. Yamamoto, K. Kawade, T. Katoh, T. Iida and A. Takahashi: Proc. of Int. Conf. Nucl. Data for Sci. and Tech. 897(1988,Mito)
3. E. Browne and R. B. Firestone: "Table of Radioactive Isotopes", John Wiley & Sons, New York (1986)
4. C. M. Lederer and V. S. Shirley: "Table of Isotopes" , (7th Ed.), John Wiley & Sons, New York (1978)

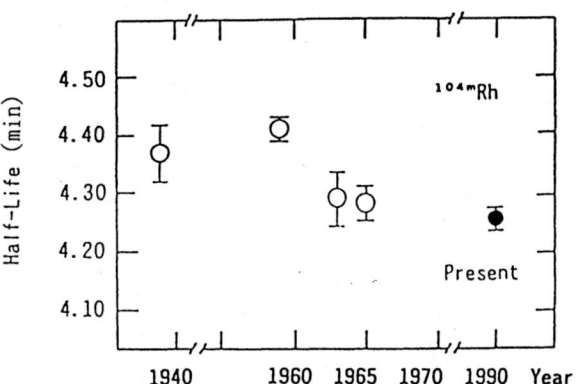

Fig. 5 Half-life of ^{104m}Rh.

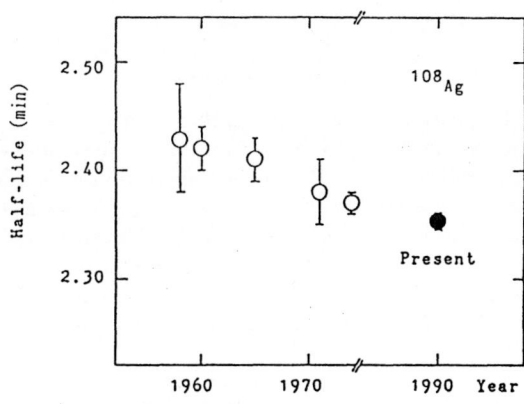

Fig. 6 Half-life of ^{108}Ag.

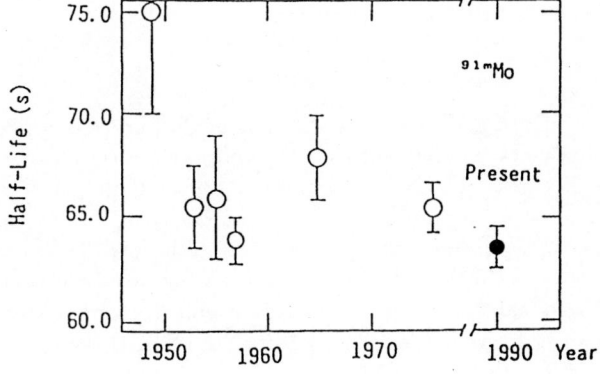

Fig. 4 Half-life of ^{91m}Mo.

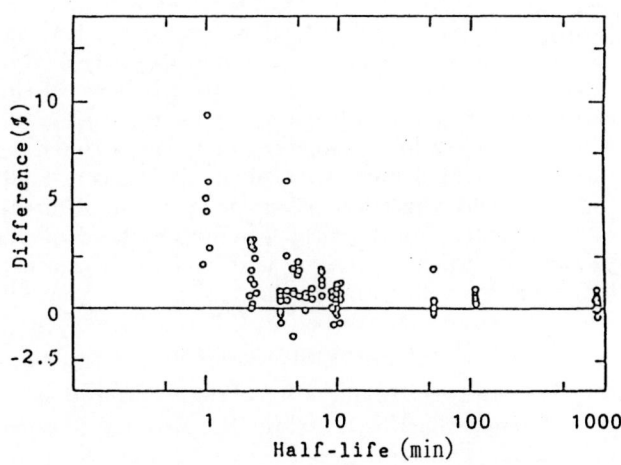

Fig. 7 Difference of half-lives between previous and present works in percent(%).

DECAY STUDIES OF ^{245}Cm

R.D. Daniels, J.N. Mo and W.R. Phillips,
Dept. of Physics, Manchester University, Brunswick Street, Manchester M13 9PL, U.K.

S.A. Woods and P. Christmas,
Division of Radiation Science and Acoustics, National Physical Laboratory, Teddington, Middlesex TW11 OLW, U.K.

L.R. Morss and I. Ahmad,
Argonne National Laboratory, Argonne, Illinois, U.S.A.

Abstract: Preliminary results are presented from internal conversion electron and gamma decay measurements following ^{245}Cm α–decay which are underway at the National Physical Laboratory, U.K. Multipole mixing ratios, from relative subshell intensities, have been established experimentally for the following transitions: 42, 54, 57, 79, 133, 136, 175 and 190 keV. This is the first time that mixing ratios have been determined experimentally for transitions following the parent decay. The measured values are compared with the previously adopted values. Relative gamma ray intensities are given for 12 transitions and these are also compared with the currently adopted values.

Radioactivity: ^{245}Cm(α), Relative Intensities (ce), δ^2, Relative Intensities (gamma), NPL $\pi\sqrt{2}$ β–ray spectrometer.

Introduction

The establishment of reliable actinide decay schemes is essential for both reactor design and decay heat calculations at the beginning of irradiated fuel handling and storage. Additionally, actinide decay data are also of interest for Nuclear Structure Physics. As the dominant decay mode of low energy levels in heavy elements is internal conversion, the measurement of conversion electron spectra is vital in constructing accurate decay schemes.

One nucleus identified in the current UK Chemical Nuclear Data request list [1] as requiring such decay measurements is ^{245}Cm. ($T_{\frac{1}{2}} = 8540(100)$ y)

Although the level scheme of ^{241}Pu has been determined previously by many workers [2], in particular using single particle transfer reactions, no measurements of γ-ray multipolarity mixing ratios have been published, either from $\gamma - \gamma$ correlation experiments or from the direct measurement of conversion electron subshell ratios. The published [2] values for the internal conversion coefficients, and hence multipolarity mixing ratios, are based solely on an intensity balance of the decay scheme and knowledge of neighbouring nuclides. Also, the values for gamma ray emission probabilites currently adopted are from a single measurement [3], which includes some transitions not yet placed in the decay scheme. The UKCND committee also requests [1] the redetermination of P_X and P_α for ^{245}Cm which this project is attempting to meet.

Experimental details

This study is a continuation of the collaboration between Manchester University and the National Physical Laboratory, who are involved with decay data measurements as requested by the UKCND committee. For this project, 63kBq of ^{245}Cm was chemically separated from a supply of ^{249}Cf at the Argonne National Laboratory, USA and deposited in the form of CmO_2 as a narrow line source (width 0.5mm, height 20mm) on an aluminium backing disc. Initial activity measurements by α and γ-spectrometry indicated that the source was > 95% ^{245}Cm.

Measurement of conversion electrons has been done using the iron–free $\pi\sqrt{2}$ double focussing β–ray spectrometer [4] at the National Physical Laboratory. It is an ideal tool for the measurement of conversion electrons following the decay of actinide nuclei due to its high resolution and the development of a 16 element gas flow proportional counter [5], increasing counting rates and thus enabling actinide nuclides of low specific activity to be studied. Additionally, the development of a windowless detector [6], comprising three Channel Electron Multipliers, now allows measurement of internal conversion lines below the window cut-off ($\sim$17keV).

Measurements of the gamma ray transitions have been made with a single germanium detector (Ortec n-type GMX 18190; active area= 84 cm^2), which is sited inside 100mm thick lead shields. Background radiation levels are routinely measured.

Results

Data collection and analysis of the internal conversion electron spectrum of ^{245}Cm for the energy range 0 - 300 keV are nearly complete. Relative shell intensities for 8 transitions following the decay of ^{245}Cm have been determined experimentally for the first time, and are given in table 1. The corresponding mixing ratio for these transitions are presented in table 2, along with the values from reference 2. As can be seen, there is good agreement for the 42, 54 and 57 keV transitions between the previously adopted values and these measurements. There appears to be less E2 admixture present than that deduced from an intensity balance determination for the 133, 136, 175 and 190 keV transitions. In addition, there is some experimental evidence for the postulated 95.69 keV transition from the electron conversion data but as yet no limit for this has been set.

The γ–spectrum measurement, to determine photon emission probabilities, is nearly complete and preliminary results are given in table 3. Relative intensities are presented for 12 transitions, normalized to the strongest line (175 keV), and compared with those previously determined in reference 3. As can be seen, there are some marked differences between the two determinations for some of the transitions.

The redetermination of α-particle emission probabilities using the newly commissioned α-spectometer at NPL is to commence in the near future.

Table 1
Relative Shell Intensities for conversion electrons

E_γ (keV)	Shell	Relative shell intensity
42	M1	74 ± 2.5
	M2	33 ± 1.4
	M3	21 ± 1.6
	N1	18.3 ± 0.6
	N2	8.8 ± 0.4
	N3	5.6 ± 0.3
54	L1	35 ± 7
	L2	14 ± 5
	M1	22 ± 9
	M2	16 ± 9
57	L1	8 ± 4
	L2	21 ± 2
	L3	18 ± 3
	M1	31 ± 17
	M2	37 ± 25
79	L1	25 ± 2
	L2	4.5 ± 1.7
133	L1	154 ± 7
	L2	25 ± 2
	L3	2.1 ± 0.1
	M1	33 ± 3
	M2	6.2 ± 0.5
	(N1+N2)	10 ± 2
	Higher Shells	3.3 ± 0.4
136	L1	4.8 ± 0.9
	L2	0.6 ± 0.2
	M1	1.3 ± 0.4
	(N1+N2)	0.3 ± 0.1
175	K	1000 ± 26
	L1	230 ± 12
	L2	38 ± 4
	L3	2.6 ± 0.3
	M1	54 ± 2.6
	M2	10 ± 1
	(N1+N2+N3)	17.6 ± 0.9
	Higher Shells	5.5 ± 0.5
190	K	23 ± 1.8
	L1	5.3 ± 0.4
	L2	1.9 ± 0.4

Table 2
Mixing ratio, δ^2, deduced from relative shell intensities

E_γ (keV)	δ^2 NDS[2]	δ^2 This study (Rosel et al ICCs)[7]
42	0.053 ± 0.040	0.042 ± 0.021
54	0.053 ± 0.040	0.06 ± 0.04
57	0.35 ± 0.07	0.75 ± 0.40
79	0.25 ± 0.35	0.028 ± 0.022
133	0.25 ± 0.35	0.0171 ± 0.0025
136	0.25 ± 0.35	≤ 0.059
175	0.25 ± 0.35	0.0208 ± 0.0118
190	0.25 ± 0.35	0.410 ± 0.121

Table 3
Relative gamma ray intensities

E_γ (keV)	Relative Intensity (Dickens & McConnell)[3]	Relative Intensity (This study)
42	3.68 ± 0.18	5.03 ± 0.15
54	0.70 ± 0.04	1.11 ± 0.07
57	0.38 ± 0.02	0.49 ± 0.05
65	0.12 ± 0.04	
69	0.07 ± 0.03	
79	1.58 ± 0.09	1.66 ± 0.13
90	0.23 ± 0.03	0.02 ± 0.06
94	0.38 ± 0.04	
133	29.2 ± 1.5	39.0 ± 1.8
136	1.18 ± 0.07	1.51 ± 0.07
140	0.06 ± 0.02	
162	0.09 ± 0.04	0.06 ± 0.03
165	0.09 ± 0.04	
175	100	100
186	0.11 ± 0.04	0.36 ± 0.03
190	2.03 ± 0.13	1.86 ± 0.06
211	0.07 ± 0.02	
233	0.16 ± 0.04	0.08 ± 0.01

No intensity balance of the level scheme has been done, as this is awaiting the exact determination of gamma ray emission probablities and a determination of the absolute acceptance factor the the beta spectrometer, which will enable absolute electron intensities to be calculated.

Conclusions

As can be seen, there are some major discrepancies between the currently adopted decay scheme of ^{245}Cm and the preliminary results presented here (in particular those of the gamma ray data). It is hoped that when absolute electron emission probabilities and absolute photon emission probabilities are determined, an intensity balance of the decay scheme may be made. To this end, it is important that the alpha particle emission probabilities are also measured, so that a consistent set of decay data is collected from the same source. Such measurements should establish the decay scheme with sufficient accuracy to fulfil the requirements of the UKCND Committee.

Acknowledgements: One of us (R.D.D.) would like to acknowledge the receipt of a CASE award from the Science and Engineering Research Council in conjunction with the National Physical Laboratory during the course of this work. The authors are also indebted for the use of ^{245}Cm to the Office of Energy Research, U.S. Dept. of Energy through the Chemistry Division at Argonne National Laboratory.

References

1. A.L. Nichols,
   AEEW–M2613 (1990)

2. Y.A. Ellis–Akovali,
   Nucl. Data Sheets 44 (1985) 407

3. J.K. Dickens and J.W. McConnell,
   Phys. Rev. C22 (1980) 1344

4. P. Christmas and P. Cross,
   J. Phys. E6 (1973) 533

5. S.A. Woods, P. Christmas, P. Cross, S.M. Judge and W. Gelletly,
   Nucl. Instr. and Meth. A264 (1987) 333

6. J. Pearcey, S.A. Woods and P. Christmas,
   Nucl. Instr. and Meth. A294 (1990) 516

7. F. Rosel, H.M. Fries, K. Alder and H.C. Pauli,
   Atom. Data Nucl. Data Tables 21 (1978) 441

Part VII

**Nuclear Data for
Medical Applications**

NUCLEAR DATA FOR MEDICAL RADIONUCLIDE PRODUCTION.
RADIOISOTOPES FOR EMISSION TOMOGRAPHY

G. Stöcklin

Institut für Chemie 1 (Nuklearchemie)
Forschungszentrum Jülich GmbH
5170 Jülich, Federal Republic of Germany

Abstract: Neutron deficient cyclotron-produced radionuclides play an increasing role in medicine, where they are used as radiopharmaceuticals for functional imaging with Positron Emission Tomography (PET) or Single Photon Emission Tomography (SPET). Nuclear data measurements are essential for new production routes since model calculations generally cannot predict unknown data with the required accuracy.

(Medical radionuclides, cyclotron radionuclide production, functional imaging, PET, SPET)

Introduction

Positron Emission Tomography (PET) and Single Photon Emission Tomography (SPET) are major tools in modern nuclear medicine for monitoring regional biochemical, physiological and pharmacological functions at a molecular level. These require physiological substrates or drugs labeled with suitable positron emitters and single photon emitters, respectively [1,2]. Short-lived neutron deficient radioisotopes of high specific activity and high radionuclidic purity are needed.

We can distinguish between two types of medical radionuclides according to their uses:
- Diagnostic radionuclides
  - for SPET requiring a dominant γ-line with high detection efficiency for present day crystal detectors, i.e. with main γ-energies between about 70 and 250 keV, such as ^{99m}Tc, ^{123}I, ^{201}Tl
  - for PET positron emitters such as ^{11}C, ^{13}N, ^{15}O, ^{18}F
- Therapeutic radionuclides for endotherapy
  - ß⁻-emitters, e.g. ^{67}Cu, ^{90}Y, ^{131}I, ^{153}Sm
  - α-emiters, e.g. ^{212}Bi, ^{211}At
  - Auger electron emitters, e.g. the EC nuclides ^{51}Cr, ^{75}Se, ^{77}Br, ^{125}I

In this review we shall consider some recent nuclear data for diagnostic radionuclides only.

Nuclear reaction cross section data are needed for
- determining the optimum energy range for production of a specific radioisotope
- calculating the yields of the desired radioisotope and of the impurities.

Several review articles have covered this topic up till about 1987 [3-6]. An IAEA consultants meeting on data requirements for medical radioisotope production was held in 1987 and the proceedings contain valuable information [6].
The present status and needs can be summarized as follows:

Up to 20 MeV the status of data is generally good. This includes the light positron emitters for PET. Data needs arise, however, while searching for alternative routes for producing radioisotopes which are generally produced at higher energies, e.g. iodine-123. Above 20 MeV data needs are more important and several examples of recent measurements will be given.

The use of nuclear theory for predicting cross section data is limited. Statistical and precompound model calculations predict the data up to 50 MeV generally only within 30 to 50%. Critical impurity data accuracy requirements are generally not met. Considerable discrepancies occur near the thresholds of reactions. Between 50 and 100 MeV predicted data have higher uncertainty.

The case of iodine-123

Among the iodine radioisotopes iodine-123 is the most important nuclide for diagnostic purposes, in particular as an analog tracer for SPET radiopharmaceuticals due to its suitable nuclear properties. Table 1 shows the decay properties of the major

Table 1.   Decay Properties of Major Iodines

Nuclide	$T_{1/2}$ (Days)	Decay Mode	Major γ-Energy MeV(%)	Dose Constant* (g rad/μCi h)
^{123}I	0.6	EC	159(83.3)	0.08
^{124}I	4.2	EC,β^+	603(81)	0.47
^{125}I	60.2	EC	35(6.7)	0.31
^{126}I	13.0	EC,β^+,β^-	389(34.1)	0.64
^{130}I	0.5	β^-	536(99)	0.10
^{131}I	8.1	β^-	364(81.2)	0.41

* Equilibrium dose constant for non-penetrating radiation

iodine isotopes. The 159 keV γ-line of iodine-123 together with the short half life makes it an ideal tracer with the lowest dose constant of only 0.08 rad g/μCi h. Iodine-131 as a therapeutic isotope has a significantly higher dose constant.

Iodine-123 can be produced by direct or by indirect processes as shown in Table 2. Inherently, the indirect processes via the xenon-123 precursor exhibit higher radionuclidic purities. This is particularly true for the iodine-124 impurity which due to its high dose constant [cf. Table 1] is unwanted for diagnostic patient studies. The relevant radionuclidic purity is that after labeling, synthesis and shipment, i.e. at the time of application. Therefore, in Table 2 the impurity is given 24 hours after the end of bombardment (EOB). If the iodine-123 is used on-site at earlier times the impurity is lower. At small cyclotrons the (p,n)- and (d,n)-reaction on ^{123}Te and ^{122}Te, respectively, can be used if highly enriched material is available.

The ^{122}Te(d,n)^{123}I reaction gives rise to low ^{124}I-impurity [7,8]. However, the low ^{123}I-yield restricts its use to local production. The ^{123}Te(p,n)^{123}I reaction on the other hand has higher yields and can even be used at small baby-cyclotrons [see ref. 9 and citations therein]. The excitation function is shown in Fig. 1.

It can be seen that the energy range from 14.5 to about 11 MeV can be used. The impurities are acceptable when at least 91% enriched ^{123}Te is used. Higher enrichments would be desirable.

Generally TeO$_2$ targets are irradiated and the iodine-123 is separated from the target by dry distillation [see for example 9].

For a medium size cyclotron (30-40 MeV) the ^{124}Xe(p,x)^{123}Xe-reaction is the most suitable one with respect to yield and purity. The production possibilities of iodine-123 via the three possible routes are shown in Fig. 2.

Table 2.   Radionuclidic Composition of Iodine-123 Produced via Commonly Used Processes (24 hours after EOB)

Reaction	Optimum particle energy range (MeV)	Target enrichment (%)	Thick target yield (mCi/μAh)	Radionuclidic content (%)					
				^{123}I	^{124}I	^{125}I	^{126}I	^{130}I	^{131}I
Indirect									
^{127}I(p,5n)^{123}Xe	65→50	Natural	15	99.5	-	0.5	-	-	-
^{124}Xe(p,2n) ^{123}Cs $\underline{6\ min}$ ^{123}Xe	29→23	99.9	~12	>99.9	-	<0.01	-	-	-
^{124}Xe(p,pn)^{123}Xe									
Direct									
^{124}Te(p,2n)	26→23	99.9	10.6	98.4	1.61	-	<0.01	0.038	-
^{123}Te(p,n)	14.5→11	91.0	4.0	97.3	1.64	-	0.17	0.9	-
^{122}Te(d,n)	16→8	96.5	2.0	97.4	0.26	-	0.20	1.4	0.7

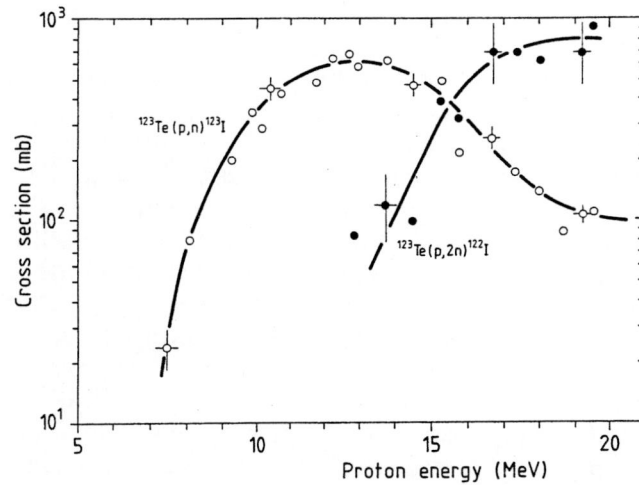

Fig. 1. Excitation functions of ^{123}Te(p,xn)^{122,123}I reactions measured of 85.4 % enriched ^{123}Te, extrapolated to 100% enrichment [9].

Several groups have measured yield data of this production process [10-15]. It has been shown that the major routes are the (p,2n)- and (p,pn)-processes, while the contribution of the (p,2p)-

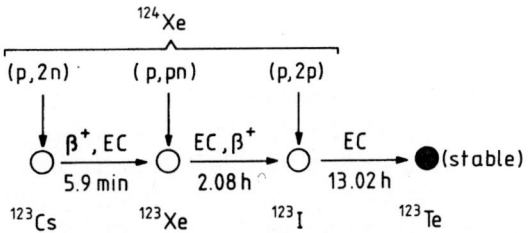

Fig. 2. Production possibilities of ^{123}I in proton induced reactions on ^{124}Xe [15].

reaction is negligible. The excitation functions as measured by the Jülich group [15] are shown in Fig. 3. A stacked gas cell technique described earlier [16,27] has been used. The results on the (p,2n)-reaction obtained with 20% enriched ^{124}Xe using radiochemical separations of the ^{123}Cs are in good agreement with those determined by direct counting of the irradiated cells. Since the (p,2n)-reaction is the major process contributing to the formation of ^{123}I, the optimum energy range for the production is from 29 → 23 MeV. The data reported by Kurenkov et al. [13] are shown for comparison.

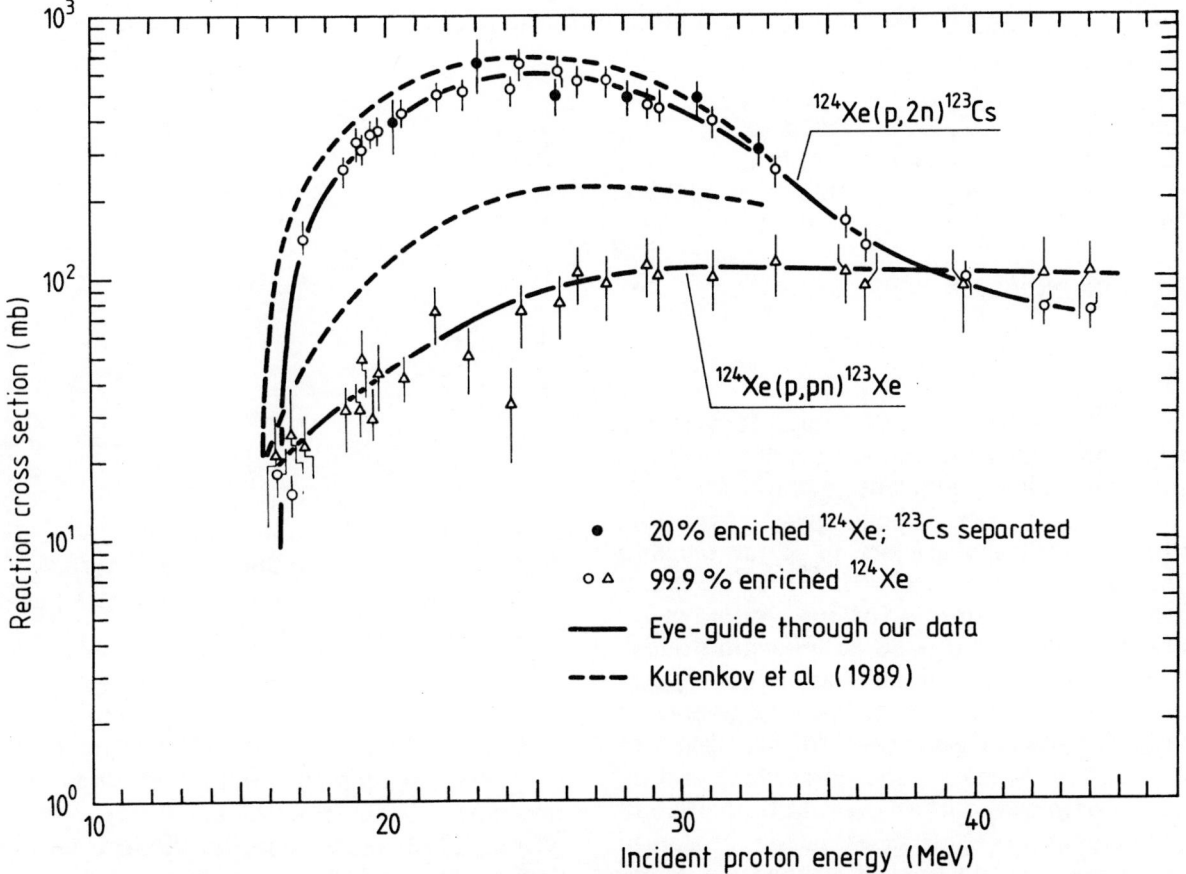

Fig. 3. Excitation functions of ^{124}Xe(p,2n)^{123}Cs and ^{124}Xe(p,pn)^{123}Xe reactions [15]. Data were obtained radiochemically (solid points) and via direct counting of irradiated cells (open circles and triangles). The results of Kurenkov et al. (1989) up to 33 MeV [13] are also shown.

While the results for the (p,2n)-reaction agree within the experimental error, significant deviations up to a factor of two occur between the (p,pn)-data. Excitation functions obtained from nuclear model calculations [17,18] deviate widely from each other and are significantly different from the experimental data. Thus, for reactions of such economic importance experimental measurements are mandatory.

The case of positron emitters

The positron emitters used in PET studies are listed in Table 3. The most important ones are ^{11}C,

Table 3.   PET Radionuclides

Radionuclide	Half-Life	% ß+ (MeV)
Commonly used nuclides		
Carbon-11	20.3 min	99.8 (0.96)
Nitrogen-13	9.97 min	100 (1.198)
Oxygen-15	2.03 min	99.9 (1.72)
Fluorine-18	1.83 h	96.9 (0.635)
Gallium-68*	1.13 h	90 (1.899)
Rubidium-82*	1.26 min	96 (3.4)
Less frequently used nuclides		
Bromine-75	1.62 h	76 (1.70)
Selenium-73	7.1 h	65 (1.32)
Copper-62*	9.73 min	97.8 (2.93)
Iodine-122*	3.6 min	77 (3.1)
Iodine-124	4.17 d	25 (1.55;2.15)
Potassium-38	7.63 min	100 (2.7)

*Generator-produced

^{15}O, ^{13}N, ^{18}F. Carbon-11 is the major isotope for labeling organic molecules such as metabolic substrates or drugs, since replacement of a ^{12}C by ^{11}C does not change its biochemical properties. Oxygen-15 and nitrogen-13 are used for labeling small molecules only, e.g. H$_2$^{15}O, ^{13}NH$_3$ for blood flow measurements. Fluorine-18 if not present in the original molecule can be used as an analog tracer substituting a hydrogen atom. This, however, can change the biochemical properties. The generator isotope gallium-68 is mainly used for calibration of the PET scanner (transmission correction) and as inorganic complexes. The generator produced rubidium-82 cation is used as potassium analog to measure regional blood flow in the heart. Nuclear data for the production of the commonly used positron emitters are well known (for a review cf. [3,4,6]).

Among the less frequently used positron emitters the sulfur analog selenium-73 is of interest, since there is no sulfur positron emitting isotope with a suitable half life. On the other hand, many biomolecules contain sulfur and can principally be designed as selenium analog using ^{73}Se with its relatively long half life of 7.1 h. In earlier times the reactor produced ^{75}Se (T$_{1/2}$ = 120 d) was used in conjunction with a simple γ-camera. This tracer is only available in low specific activity and is ill suited for in-vivo application due to its unfavorable decay characteristics. The cyclotron isotope ^{73}Se can be produced at a medium energy machine, and the most economic production route is the ^{75}As(p,3n)^{73}Se-reaction on the monoisotopic arsenic in the form of As$_2$O$_3$ (cf.[19] and citations therein).

The excitation function as measured with the stacked foil technique using electrolytic deposition of arsenic on Cu- or Al-backing is shown in Fig. 4. For comparison the excitation function as obtained from model calculations [20] is also shown. Except for the energy range near threshold the agreement is very good and the data could have been used for the estimation of production yields.

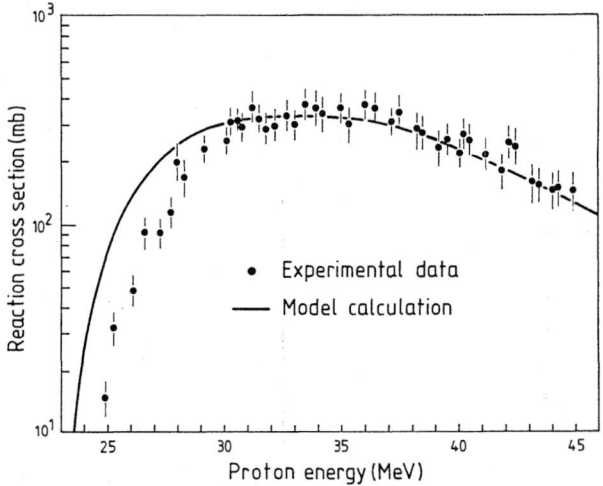

Fig. 4. Excitation function of the ^{75}As(p,3n)^{73}Se-reaction [19] and comparison with model calculation [20].

For production purposes the excitation functions of ^{72}Se (T$_{1/2}$ = 8.4 d) and ^{75}Se (T$_{1/2}$ = 120 d) must also be known in order to evaluate radionuclidic impurities at the optimum energy range. Fig. 5 shows that the optimum energy range is E$_p$ 40 → 30 MeV. The ^{72,75}Se impurities in this range amount to < 0.2%.

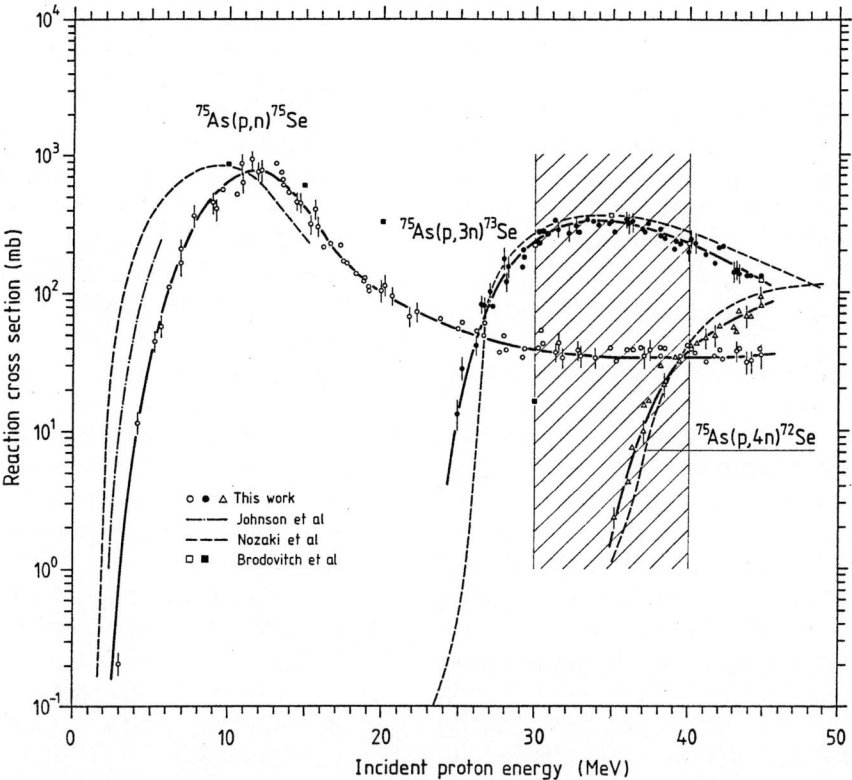

Fig. 5. Excitation functions of ^{75}As(p,xn)^{72,73,75}Se reactions. Typical error bars in cross sections are given. The optimum energy range for the production of ^{73}Se is $E_p = 40 \rightarrow 30$ MeV [19].

For clinical PET without an on-site cyclotron a satellite concept can provide tracers from a nearby accelerator. Thus fluorine-18 ($T_{1/2} = 110$ min) radiopharmaceuticals can be transported to the hospital if the time does not exceed about 2 hours. In addition, isotope generators are used. Unfortunately, the number of useful positron emitting isotope generators is very limited (for a review cf. [21-25]). One of the most important ones is the ^{82}Sr-^{82}Rb system providing short-lived ($T_{1/2} = 75$ sec) ^{82}Rb cations for regional myocardial blood flow from the ^{82}Sr ($T_{1/2} = 25.3$ d) parent. ^{82}Sr is produced via the ^{85}Rb(p,4n)^{82}Rb reaction [26] and particularly by spallation of Mo with 500-800 MeV protons [23,24]. Production at a medium size cyclotron is also possible [27,28] using natural krypton. The excitation functions for ^{3}He- and α-particle irradiation of ^{nat}Kr are shown in Fig. 6 [28].

Several krypton isotopes are contributing via ^{nat}Kr(^{3}He,xn)- and ^{nat}Kr(α,xn)-reactions to the formation of ^{82}Sr. However, large amounts of ^{83}Sr ($T_{1/2} = 31.4$ h) and ^{85}Sr ($T_{1/2} = 64.9$ d) as well as the shorter lived ^{81}Sr ($T_{1/2} = 22.2$ min) are also formed. This is also true for the commonly used spallation process. The presence of ^{83}Sr and ^{85}Sr as impurities is a disadvantage since the generator needs extra shielding. Furthermore, if ^{83}Sr has not

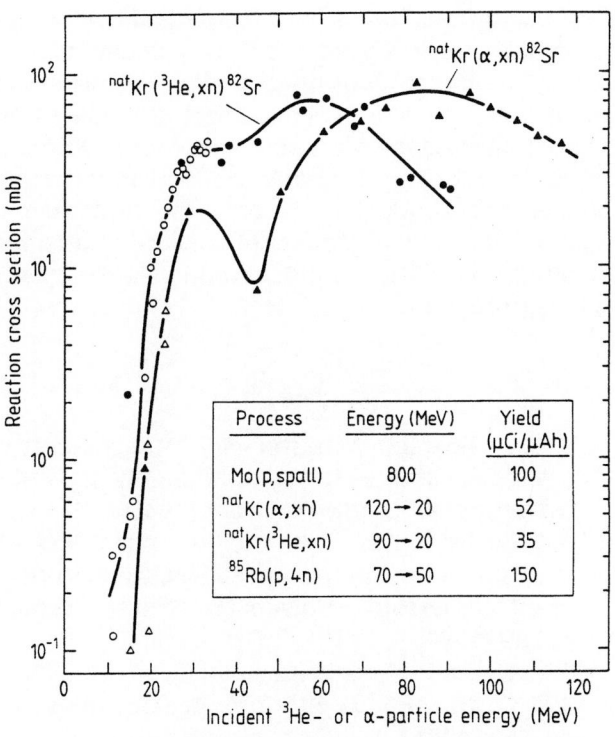

Fig. 6. Excitation functions for the formation of ^{82}Sr by high energy α- and ^{3}He-irradiation of ^{nat}Kr [27,28].

decayed out prior to loading of the generator, the daughter ^{83}Rb which is not a ß$^+$ emitter is also formed and would contribute to the patient dose. In any case the bombardment of natural krypton with 90 MeV ^{3}He or 120 MeV α-particles is a possibility to produce ^{82}Sr in an economic way, even though the yields are lower by a factor of 2 to 3 than those of the presently used reactions (see insert in Fig. 6).

Conclusion

At present there are no immediate nuclear data needs for medical radionuclide production. Most of the data for routine production at small and medium energy accelerators are well known. High energy accelerators may play an increasing role for some generator parent isotopes and less frequently used PET and SPET isotopes, particularly when a parasitic production mode is possible. Here some data needs may arise. High energy proton beam cancer therapy will also need accurate nuclear data with respect to the activation of body constituents and the formation of secondary particles. Finally, "new" radionuclides both diagnostic and therapeutic ones will appear and complete the spectrum of medical isotopes of the future, even though a certain saturation seems to be reached.

The problems in medical isotope production lie presently in the field of high beam current target development and target chemistry. Radiolytic and thermal processes drastically change the target composition and often lead to a significant deviation of the practical yields from the calculated ones. Finally, computer controlled and/or robotic handling, processing and quality control procedures are of great importance for a safe and economic radio-isotope production. It should be kept in mind that radioisotope production is only the first step in radiopharmaceutical preparation, involving also fast automated labeling syntheses and reliable quality control procedure.

REFERENCES

1. M.E. Phelps, J.C. Maziotta and H.R. Schelbert (eds.): Positron emission tomography and auto-radiography. Raven, New York (1986)
2. G. Stöcklin and G. Kloster in: P.J. Ell, B.L. Holman (eds.): Computed emission tomography. Oxford University Press, Oxford, p. 229 (1982)
3. S.M. Qaim: Radiochimica Acta 30, 147 (1982)
4. S.M. Qaim and G. Stöcklin: Radiochimica Acta 34, 25 (1983)
5. A. Hashizume in: Proc. Int. Conf. Nuclear Data for Science and Technology Mito, June 1988, (S. Igarasi, ed.), Saikon, Tokyo, p. 1067 (1988)
6. Proceedings of the IAEA Consultants' Meeting on Nuclear Data for Medical Radioisotope Production, Tokyo, April 1987, (K. Okamoto, ed.) INDC (NDS)-195/GZ
7. J.H. Zaidi, S.M. Qaim and G. Stöcklin: Int. J. Appl. Radiat. Isot. 34, 1425 (1983)
8. G.J. Beyer, G. Pimentel, D. Roeda, H. Virtanen and O. Solin: Proc. 3rd Symp. on Medical Applications of Cyclotrons, Turku, Finland, Ann. Univ. Turku D 17, 160 (1986)
9. B. Scholten, S.M. Qaim and G. Stöcklin: Appl. Radiat. Isot. 40, 127 (1989)
10. M.L. Firouzbakht, R.R. Teng, D.J. Schlyer and A.P. Wolf: Radiochimica Acta 41, 1 (1987)
11. L.A. Iljin, I.A. Germakow, V.I. Godov, A.S. Rostsin, B.V. Zabrodin, A.V. Motornij and B.A. Shustrov in: Current Questions in Medical Radiation Physics (Collection of Scientific Works) p. 97, Leningrad (1987)
12. F. Tárkányi, Z. Kovács, S.M. Qaim and G. Stöcklin: Radiochimica Acta 47, 25 (1989)
13. N.V. Kurenkov, A.B. Malinin, A.A. Sebyakin and N.I. Venikov: J. Radioanal. Nucl. Chem. Lett. 135, 39 (1989)
14. N.A. Konyahin, V.N. Mironov, N.N. Krasnov, P.P. Dmitriev, V.P. Lapin and M.V. Paparin: At. Energ. 67, 129 (1989)
15. F. Tárkányi, S.M. Qaim, G. Stöcklin, M. Sajjad, R.M. Lambrecht and H. Schweickert: Appl. Radiat. Isot. 42, 221 (1991)
16. F. Tárkányi, S.M. Qaim and G. Stöcklin: Radiochimica Acta 43, 185 (1988)
17. P. Grabmayr and P. Nowotny: Int. J. Appl. Radiat. Isot. 29, 261 (1978)
18. B.V. Zuravlev, S.P. Ivanoca, N.N. Krasnov and Yu.N. Subin: At. Energ. 60, 337 (1986)
19. A. Mushtaq, S.M. Qaim and G. Stöcklin: Appl. Radiat. Isot. 39, 1085 (1988)
20. S.M. Qaim, A. Mushtaq and M. Uhl: Phys. Rev. C38, 645 (1988)
21. R. Lambrecht: Radiochimica Acta 34, 9 (1983)
22. Y. Yano and H.O. Anger: J. Nucl. Med. 9, 412 (1968)
23. R. Robertson, D. Graham and I.C. Trevna: J. Label. Comp. Radiopharm. 19, 1368 (1982)
24. K. E. Thomas and J.W. Barnes in: Radionuclide Generators: New Systems for Nuclear Medicine Applications (F.F. Knapp and T.A. Butler eds.) p. 123, American Chemical Society, Washington D.C. 1984
25. S.M. Qaim: Radiochimica Acta 41, 111 (1987)
26. I. Huszár, He Youfeng, J. Jegge and R. Weinreich: J. Lab. Compd. Radiopharm. 26, 168 (1989)
27. F. Tárkányi, S.M. Qaim and G. Stöcklin: Appl. Radiat. Isot. 39, 135 (1988)
28. F. Tárkány, S.M. Qaim and G. Stöcklin: Appl. Radiat. Isot. 41, 91 (1990)

NUCLEAR DATA RELEVANT FOR PARTICLE RADIATION THERAPY TODAY AND TO MORROW

A. Wambersie and V. Grégoire

Université Catholique de Louvain
Unité de Radiothérapie, Neutron- et Curiethérapie
Cliniques Universitaires St-Luc
1200 Brussels, Belgium

Abstract: Control of the primary tumor remains one of the major goal in cancer treatment and, as far as radiotherapy is concerned, the replacement of conventional photon beams by other types of ionizing radiation is a very promising approach.

High-LET radiations, such as neutrons, have been introduced to take advantage of the higher sensitivity of the hypoxic cells to these radiations as compared to photons. Assuming that only the tumors are hypoxic, a therapeutic gain should be expected from this reduction in OER. Other biological differences have been demonstrated between high-LET and photons such as a decrease in the variation of sensitivity between different cell lines and between the different cell cycle phases, and a lower repair. In other words, after high-LET different cell populations tend to response on a similar way and this could bring a benefit when the tumor is more radioresistant than the normal tissues. Clinical advantages of high-LET have been demonstrated in salivary gland tumors, prostate adenocarcinomas and is assumed in paranasal sinuses sarcomas and fixed lymph nodes of the head and neck.

Low-LET particle beams like protons and helium ions have been introduced to take advantage of the higher physical selectivity (better dose distribution) and no advantage is expected from a differential effect between tumors and normal tissues. Clinical benefit have been demonstrated in uveal melanomas, chordomas and chondrosarcomas of the base of the skull, paraspinal tumors. Since increase in the dose distribution is always a benefit, lots of patients treated by photons could in theory, benefit from protons.

By using heavy ions, one aims at combining both advantages of the dose distribution and of the differential effect between tumors and normal tissues. All patients that would benefit from neutrons should be candidate for heavy ions. In addition, all patients whose tumor requires a high physical selectivity could also be treated by heavy ions since the potential therapeutic gain will be further increased by the biological advantages. The encouraging results from Berkeley with neon ions justify the different heavy ions programmes started in Japan (HIMA) or in Europe (GSI-Darmstadt and EULIMA) and stress the need for an international collaboration aiming at designing cooperative clinical studies.

Introduction

Control of the primary tumor remains one of the most important challenge in cancer treatment. The Commission of the European Communities recently published that out of 100 cancer patients, only 60% had a limited disease at the time of presentation and only 40% will be cured: 22% will benefit from surgery, 12% from radiotherapy and 6% from an association of surgery and radiotherapy. Although radiotherapy has a important curative role, the radiation oncologist has still to face a substantial amount of local recurrences. To circumvent this problem, the replacement of conventional photons beams by other type of ionizing radiations is one of the most promising approaches.

When discussing the place of non-conventional ionizing radiations, one has to distinguish :
-high-LET radiations which produce different types of biological effects, and which aim at improving the differential effect between tumor and normal tissues (e.g. : fast neutrons);
-particle beams which only improve the physical selectivity of the irradiation, i.e. the dose distribution (e.g. proton beams or helium ion beams);
-the two approaches can be combined and one could seek after a high physical selectivity with high-LET radiation (e.g.: heavy ions).

The potential advantage of high LET radiations

1:Radiobiological considerations:

Historically, neutrons were introduced in radiotherapy to take advantage of the higher sensitivity of hypoxic cells after high-LET irradiations as compared to photons. Considering that normal tissues are well oxygenated, a reduction in OER should lead to a therapeutic gain.

It's known that other differential properties between neutrons and photons could also bring a potential biological advantage **(Figure 1)**:

- a reduction -with or without inversion- of the difference in intrinsic radiosensitivity between cell lines [2,11].

- a reduction in the difference of sensitivity between the cell cycle phases [9].

- a lower repair phenomenon and in consequence a lower dependance of the biological effect on the fractionation scheme [28].

Since all cells population tend to respond to a similar way after neutrons, there will be a benefit from these particles or from high-LET particles in general, only when the cancer cell population is more resistant than the normal tissues (A. Wambersie in[21].

From these radiobiological considerations one can thus stress the importance of an accurate patient selection and a high physical selectivity in neutrontherapy or in high-LET therapy in general. As a matter of fact, a wrong selection of the patients suitable for neutrons will lead to a misestimation of the therapeutic gain. On the other hand, since the difference in response between tumors and normal tissues is lowered after high LET, it is of prime importance to reach an high selectivity in the dose delivery in order to minimize the complications resulting from irradiations of large volume of normal tissues.

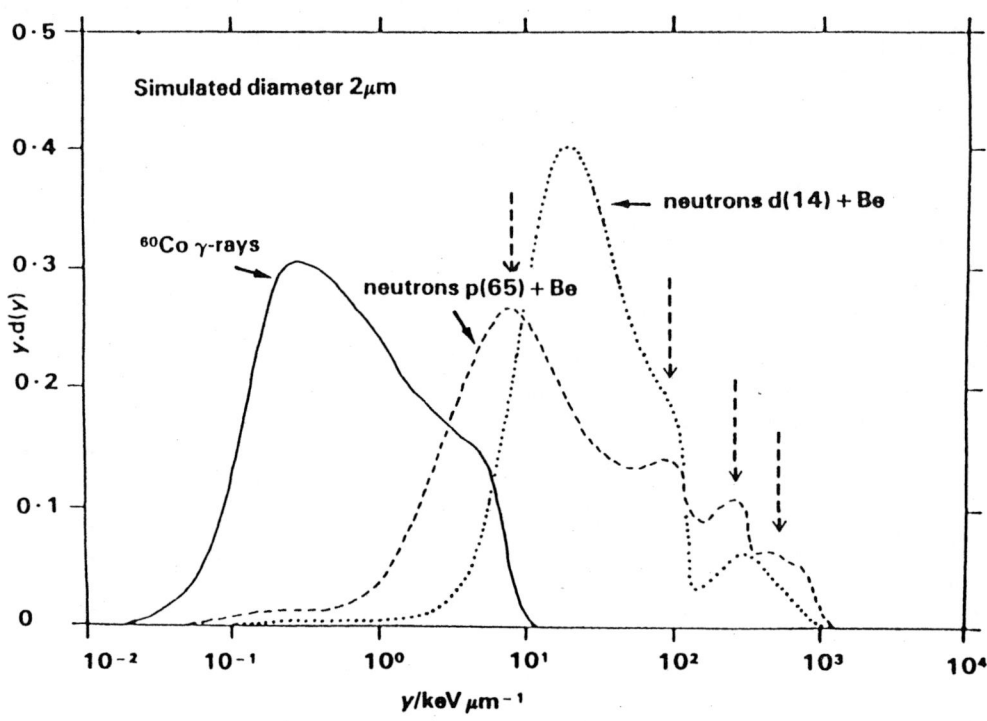

Figure 1: Comparison of energy deposition after irradiation with fast neutrons and gamma-rays. The 50-100 times higher energy deposition after neutrons explains the biological differences between low and high-LET. After Pihet in [28].

2:Survey of clinical data of neutrontherapy:

Fast neutron therapy is applied routinely in about twenty centers throughout the world (**Table I**). Over 15.000 patients have been treated so far with neutrons, either as the sole irradiation modality or in combination with other radiotherapy techniques, the longest follow-up exceeding 15 years.

The available clinical data now enable us to identify the tumor types and/or sites for which neutrons were shown to bring a benefit, and, on the other hand, tumors for which neutrons should not be used. In addition, there are tumor sites for which the available information is incomplete or for which the reported results are conflicting [25].

The tumors for which fast neutrons were found to be superior to conventional X-rays are in general, slowly growing and well differentiated tumors (**Table II**): salivary gland tumors [12] [16] [27], paranasal sinuses [10], some locally advanced tumors of the head and neck with fixed metastatic lymph nodes [13] [14], soft tissue sarcomas [17] [23], bone- and chondrosarcomas [17], prostatic adenocarcinomas [24] [23], melanomas [25]. By contrast, disappointing results were obtained for brain tumors [25]; these observations are in agreement with the radiobiological data and especially the high RBE value observed for CNS. However, a possible benefit for neutron boost in brain tumors should be further investigated [5]. In addition, neutrons should not be used for tumors showing an exquisite radiosensitivity to X-rays (e.g. seminomas, lymphomas, or in general poorly differentiated, rapidly growing tumors), and for which neutrons would then reduce a favorable differential effect

Table I

THE NEUTRONTHERAPY FACILITIES IN THE WORLD		
Center	Neutron Producing Reaction	Comments
EUROPE		
MRC-Clatterbridge, U.K.	p(62)+Be	rotational gantry variable collimator
Orléans, France	p(34)+Be	vertical beam
UCL- Louvain-la-Neuve, Belgium	p(65)+Be	vertical beam (multileaf collimator and horizontal beam in preparation)
Hamburg, Fed.Rep.Germany	(d + T)	rotational gantry
Heidelberg, Fed.Rep.Germany	(d + T)	"
Munster, Fed.Rep.Germany	(d + T)	"
Essen, Fed.Rep.Germany	d(14)+Be	rotational gantry
UNITED STATES		
M D Anderson- Houston, Texas	p(42)+Be	rotational gantry variable collimator
Cleveland, Ohio	p(43)+Be	horizontal beam
UCLA - Los Angeles	p(46)+Be	rotational gantry variable collimator
Seattle, Washington	p(50)+Be	rotational gantry multileaf collimator
Fermilab	p(66)+Be	horizontal beam
ASIA		
National Institute of Radiological Sciences (NIRS) - Chiba, Japan	d(30)+Be	vertical beam multileaf collimator
Institute for Medical Sciences (IMS) - Tokyo, Japan	d(14)+Be	horizontal beam
Korea Cancer Center Hospital (KCCH)-Seoul, Korea	p(50.5)+Be	rotational gantry
King Faisal Hospital - Riyadh, Saudi Arabia	p(26)+Be	rotational gantry
AFRICA		
National Accelerator Center (NAC) Faure, Rep.South Africa	p(66)+Be	rotational gantry variable collimator

From ICRU [15] and Tsunemoto et al. [27]

Potential advantages of precision radiotherapy: proton and helium ion beams

Historically, the major improvement in the efficiency of radiation therapy was the replacement of conventional X-rays (200 kV X-rays) by high-energy photons or electrons. The clinical benefit was rapidly evident for all, or for the majority of the patients.

We are now close to make a further step with the introduction of proton beams (**Table III.**) The higher physical selectivity makes them superior to high-energy photons. But no advantage has to be expected from the biological point of view, high energy protons being considered as low-LET radiations. For the present discussion, we can assume that helium ion beams are similar to proton beams.

The clinical benefit of proton beams has been demonstrated for several well selected tumor types or sites for which a physical selectivity is essential. The best example is the uveal melanoma, which has been treated since many years by proton beams at the Harvard cyclotron [1]. One thousand and six patients have been treated between 1975 and December 1986 : a 5 years local control rate of 96.3 ± 1.5% and a survival of 80 % were reported

582

(H. Suit in [21]. These results compare well with those reported from Berkeley, where the 5 years recurrence rate and survival for patients treated with helium ions were 3.2% and 81% respectively [19] .

An active program (OPTIS) is also carried out at the PSI-Villigen in Switzerland, where 462 patients were treated up to October 1988. Survivals at 2 and 4 years of 95 % and 88 % respectively were reported (L. Zografos in [20].

In addition, at least 4 proton therapy programmes for uveal melanoma are in preparation in Europe : in Uppsala - Sweden, Clatterbridge - United Kingdom, Nice - France, and Louvain-la-Neuve - Belgium. For the three last ones, proton therapy is planned with the cyclotrons currently used for fast neutron therapy [29].

Table II

SUMMARY OF THE CLINICAL RESULTS OBTAINED WITH HIGH-LET RADIATIONS (FAST NEUTRONS AND NEON IONS)		
Tumor site (or type)	Local control rates after:	
	Fast neutrons	Neon ions
- salivary gland tumors	67 % (24%)	80 % (28%)
- paranasal sinuses	67 %	63 % (21%)
- fixed lymph nodes	69 % (55%)	
- sarcomas	53 % (38%)	45 % (28%)
- prostatic adenocarcinomas	77 % (31%)	100 % (60-70%)

( ) for comparison, the local control rates currently obtained with conventional low-LET radiations.

Besides uveal melanoma, there are other localizations where the high physical selectivity of the proton beams can be fully exploited : radioresistant tumors close to critical organs such as chordomas or chondrosarcomas of the base of the skull, and paraspinal tumors. At the Harvard cyclotron, with protons, local control rates of 82 % and 63 % were reported at 5 and 10 years respectively (H.D. Suit in [20]). A local control rate of 70 % for chordomas, chondrosarcomas and meningiomas was reported from Berkeley after helium ion treatment (J.R. Castro in [21].

Table III

PROTONTHERAPY FACILITIES (first of July 1990)			
Site	Date of first treatment	Date of last treatment ¶	Total number of patients treated
Berkeley-184, California (USA)	1955	1957	30
Uppsala (Sweden)	1957	1976	73 (1)
Harvard, Mass. (USA)	1961	-	4994 (in June 1990)
Moscow (USSR)	1965	-	1945 (in May 1990)
Dubna (USSR)	1967	1977	80 (2)
Leningrad (USSR)	1975	-	508 (in December 1987)
Chiba (Japan)	1979	-	65 (in October 1989)
Tsukuba (Japan)	1983	-	178 (in April 1990)
PSI/SIN (Switzerland)	1984	-	719 (in October 1989)
Uppsala (Sweden)	1988	-	13 (in May 1990) (3)
Clatterbridge (U.K.)	1989	-	73 (in May 1990)
Louvain-La-Neuve (Belgium)	1991	-	4 (in May 1991)

¶ when the facility has been closed
(1) first protontherapy programme at Uppsala
From Sisterson [22]

(2) reopening of the facility scheduled
(3) new protontherapy programme at Uppsala

In addition, there is an increasing number of new projects which aim at treating with protons many other tumor types, and larger proportions of patients. As a matter of fact, since an improved physical selectivity is, in itself, always a benefit, all photon patients could be, in principle, potential candidates for proton treatment.

Prostatic adenocarcinomas, soft tissue sarcomas, some head and neck and rectal tumors are treated with protons at the Harvard cyclotron [1]. In Japan, 250 MeV protons are used at the University of Tsukuba for different localizations, including deep seated tumors (T. Kitagawa in [21]. The Swedish and Russian proton therapy programmes have been described [18]. An ambitious therapy programmes using 250 MeV protons is in preparation at he PSI in Villigen [4]. In Orsay also, a proton therapy programme using a 200 MeV synchrocyclotron is in preparation (J. Dutreix, personal communication).

However, one of the most impressive projects is probably the Loma Linda project at Los Angeles. A variable energy synchrotron (70-250 MeV), and 3 treatment rooms with isocentric rotating gantry, will be the "core" of a large oncology department. An additional horizontal fixed beam will be reserved for eye and brain irradiation. The facility is scheduled for completion by the end of 1989; once all treatment rooms will be fully operational, the center is expected to have a capacity of 1 000 new proton beam patients per year [26].

Potential advantages of heavy ions

Heavy ions combine the advantage of a high physical selectivity with the potential advantage of high-LET radiation for the treatment of some tumor types. As far as the physical selectivity is concerned, heavy ion beams are similar to proton or helium ion beams; they have even a smaller penumbra but it is questionable whether this factor could be of clinical relevance. More important is the fact that, with heavy ions, the higher RBE at the level of the spread-out Bragg peak acts in synergism with the advantage of the dose distribution. Furthermore, the high-LET at the level of the spread-out bragg peak alters the biological effect and this has to be taken into account when prescribing the irradiation modality (e.g. fractionation). The LET depends on the type of particles as well as on the width of the spread-out Bragg peak; factors such as RBE, OER, and repair capacity should then be determined.

The energies required to obtain a sufficient beam penetration are typically :
- for carbon ions : 400 MeV/amu;
- for neon ions : 620 MeV/amu
- for argon ions : 860 MeV/amu

Heavy ion therapy programs are justified by three sets of arguments :

1) the radiobiological and clinical data indicating that, for the treatment of some tumors types and/or sites, high-LET radiations could bring a benefit compared to low-LET radiations.

2) the importance of a high physical selectivity which has been clearly demonstrated with low-LET radiations. In addition, as discussed above, high physical selectivity is even more important for high-LET than for low-LET.

3) the encouraging clinical results reported from Berkeley are an additional argument, although they were obtained on limited, selected groups of patients [6], [7], [8]. It is however interesting to notice that the best results obtained with neon ions were obtained for tumors for which neutrons have been proven to be of interest (**Table II**). Great care has however to be done before drawing definite conclusions from these results since the series are quite small and non-randomized.

General principles for patients selection for heavy-ions therapy are presented on **Table IV**. They are based on the clinical experience from neutrontherapy -high LET advantage- and from protontherapy -high dose distribution advantage. Some potential additional indications have been also proposed including all tumors in children where avoiding irradiation of growing tissue is of prime importance.

Only a few heavy-ion therapy facilities are planned in the world : the facility at the NIRS in Japan which is under construction and in Europe the GSI project in Darmstadt-FRG and the EUropean Light Ions Medical Accelerator (EULIMA) project. Due to their high cost and complexity, an international cooperation is necessary in order to ensure a rapid exchange of clinical and radiobiological information and to design randomized cooperative trials.

Nuclear data needed for improving particles therapy

In photon therapy, an accuracy on dose delivery as high as 3.5 % (i.e. one standard deviation on the absorbed dose at the specification point) is required. This requirement is due to the steepness of the dose-effect relations for local tumor control and normal tissue complications. For high-LET radiations, the available clinical and radiobiological data indicate that the dose-effect relations are as steep as those observed for photons, and consequently the same degree of accuracy has to be achieved [5]. Furthermore, as discussed above, at least the same physical selectivity (dose distribution) is required due to a reduced differential effect with high-LET radiations.

Table IV

GENERAL PRINCIPLES OF PATIENT SELECTION FOR HEAVY-ION BEAM THERAPY To take advantage of the biological **AND** physical characteristics of the beams
A - The **radiobiological advantage** (high-LET) is thought to be the most important factor, followed by the physical selectivity of the beams

a. where high-LET radiation already demonstrated to be useful
    - prostatic adenocarcinomas
    - salivary gland tumors
    - paranasal sinuses
    - fixed lymph nodes
    - sarcomas, etc.
 	
b. where additional information is needed
    - pelvic tumors, digestive tract (rectum)

B .- The **physical selectivity (dose distribution)** of the beams is thought to be the most important factor followed by the radiobiological advantage of high-LET

Tumors in technically difficult situations, but where high-LET radiation may be better than low-LET radiation (e.g. slowly growing tumors)
    - <u>adjacent to CNS</u> : meningioma, pharyngioma, chordoma, optic nerve, glioma, AVM, paraspinal cord tumor, etc.
    - <u>root of neck disease</u> : upper oesophagus, post cricoïd carcinoma, etc
    - <u>thoracic disease</u> : mesothelioma, etc.
    - <u>others</u>: paraaortic lymph node, etc.

C.- Additional indications
    - tumors in children
    - very poor prognosis disease : unresectable hepatoma, pancreas, retroperitoneal sarcoma, recurrent after previous radiotherapy, etc. .

After G.R.H. Sealy in [21].

Optimization of neutrontherapy requires nuclear cross section data for: 1) the selection of source reaction for neutron production, 2) the design of collimators and shields, 3) the calculation of absorbed dose in the irradiated tissues, including heterogeneity corrections, 4) microdosimetry, and 5) studies of the influence of radiation quality on biological effects. Under the auspices of the International Atomic Energy Agency, a Coordinated Research Program has been underway since 1987 to assess the status of these nuclear data, to coordinate research efforts, to report recent progress, and to recommend acceptance of appropriate data and further research where necessary [30].

Modern neutron therapy facilities are usually based on monoenergetic proton beams bombarding beryllium targets to produce a spectrum of neutrons in the energy range below $\sim$ 70 MeV. Data for the $^9Be(p,n)^9B$ source reaction and for adjunct reactions producing background neutrons, i.e. $^{12}C(p,n)$ and $^{16}O(p,n)$ are available but additional measurements are needed. Collimators and shields can be optimized through neutron transport calculations, but, until recently, evaluated data bases extended only to 20 MeV, well below the energies of most of the source of neutrons. It is now possible to calculate the performance of conceptual designs with new experimental data that go well above 20 MeV for some materials. The extension of data libraries to at least 70 MeV for the materials of importance (H, Be, C, N, O, Ca, Fe, Ni, Cu, Zr, W, and Pb) is important, but in addition, Benchmark experiments in the 20-70 MeV range are critically needed to test these new data bases.

Significant progress in the determination of kerma factors has occurred in recent years, especially for carbon up to 30 MeV. Extension of the carbon measurements to 70 MeV is critical for neutron therapy and dosimetry applications. For selected other materials, and most importantly for oxygen, additional data are needed at all energies up to 70 MeV. These nuclear data and protocols designed for absorbed dose determinations are essential for neutrontherapy.

References

1 M. AUSTIN-SEYMOUR, M.URIE, J. MUZENRIDER, C. WILLETT, M. GOITEIN, L. VERHEY, R. GENTRY, P.McNULTY, A. KOEHLER, H.SUIT: Radiotherapy and Oncology 17, 29-35, 1990

2 G. W BARENDSEN, J.J. BROERSE. Int. J. Radiation. Oncology Biol. Phys. ,3, 211-214, 1977.

3 J.J. BATTERMANN, G.A.M. HART, AND K. BREUR Br. J. Radiol. 54, 899-904, 1981.

4 H. BLATTMANN Radiotherapy and Oncology, 17, 17-20, 1989

5 N. BRETEAU, B. DESTEMBERT, A. FAVRE, C. PHELINE, SCHLIENGER, Strahlentherapie und Onkologie 165, 320-323, 1989

6 J.R. CASTRO In G. Kraft and U. Grundinger (Ed.). Third workshop on heavy charged particles in biology and medicine, GSI - Darmstadt, Report 87-11, ISSN 0171-4546, 1987, KO I-5.

7 J.R. CASTRO, J.M. COLLIER, P.L. PETTI et al. Int. J. Radiation. Oncology Biol. Phys. 17, 477-484, 1989

8 J.R. CASTRO and M.M. REIMERS Int. J. Radiation. Oncology Biol. Phys. 14, 711-720, 1988

9 J.D. CHAPMAN In R.E. Meyn and H.R. Withers (Eds). Radiation Biology in Cancer Research, Raven Press, New-York, 21-32, 1988.

10 R.D. ERRINGTON Bull. Cancer (Paris), 73, 569-576, 1986.

11 B.FERTIL, P.J. DESCHAVANNE, J. GUEULETTE, A. POSSOZ, A. WAMBERSIE, and E.P. MALAISE Radiation Research, 90, 526-537, 1982.

12 B.R. GRIFFIN, G.E. LARAMORE, K.J. RUSSELL, T.W. GRIFFIN AND J. EENMAA Radiotherapy and Oncology, 12, 105-111, 1988.

13 T.W. GRIFFIN, R. DAVIS, F.R. HENDRICKSON, M.H. MAOR, G.E. LARAMORE, L. DAVIS Int. J. Radiat. Oncol. Biol. Phys., 1984, 10, 2217-2223.

14 T.W. GRIFFIN, R. DAVIS, G.E. LARAMORE, D.H. HUSSEY, F.R. HENDRICKSON, A. RODRIGUEZ-ANTUNEZ Int. J. Radiation. Oncology Biol. Phys. 9, 1267-1270, 1983

15 International Commission on Radiation Units and Measurements (ICRU), Clinical Neutron Dosimetry, Part I, Bethesda, Maryland, 20814, USA (in press).

16 W. KOH, G.E. LARAMORE, T.W. GRIFFIN et al. Am. J. Clin. Oncol. 12, 316-319, 1989

17 G.E. LARAMORE, J.T. GRIFFETH, M. BOESPFLUG et al. Am. J. Clin. Oncol. 12, 320-326, 1989

18 B. LARSSON Journal Européen de Radiothérapie, 3, 223-234, 1984.

19 D. LINSTADT, J.R. CASTRO, D. CHAR et al. Int. J. Radiation. Oncology Biol. Phys. 19, 613-618, 1990

20 Proceedings or the International Workshop on Particle Therapy; Meeting of the EORTC Heavy Particule Therapy Group, October 28-29, 1988, University of Essen, Fed. Rep. of Germany Strahlentherapie und Oncologie vol. 165, 1989

21 Proceedings of the EULIMA Workshop on the potential value of light ion beam therapy, November 3-5, 1988, Centre Antoine-Lacassagne (CAL), Nice,France

22 Proton Therapy Cooperative Group (PTCOG) Newsletter - Ed. Janet Sisterson, Harvard Cyclotron Laboratory, 44 Oxford Street, Cambridge MA 02138, n°6, June 1990

23 F. RICHARD, L. RENARD, A. WAMBERSIE Bull. Cancer (Paris), 73, 562-568, 1986

24 K.J. RUSSELL, G.E. LARAMORE, J.M. KRALL, F.J. THOMAS, M.H. MAOR, F.R. HENDRICKSON, J.N. KRIEGER, and T.W. GRIFFIN The Prostate 11: 183-193, 1987.

25 G. SCHMITT, A. WAMBERSIE Radiotherapy and Oncology 17, 47-56, 1990

26 J.M. SLATER and D.W. MILLER In G. Kraft and U. Grundinger (Ed.). Third workshop on heavy charged particles in biology and medicine, GSI - Darmstadt, Report 87-11, ISSN 0171-4546, 1987, KO K-2.

27 H. TSUNEMOTO, S. MORITA, S. SATHO, Y. IINO, S. YUL YOO Strahlentherapie und Onkologie, in press.

28 M. TUBIANA, J. DUTREIX, A. WAMBERSIE, Introduction to Radiobiology, Taylor & Francis Ed., London, 1990

29 S. VYNCKIER, J.P. MEULDERS, P. ROBERT, A. WAMBERSIE European Journal of Radiation Oncology, 5, 245-247, 1984.

30 R.M. WHITE, J.J. BROERSE, P.M. DELUCA et al. in Proceedings of the Seventh Symposium on Neutron Dosimetry, Berlin, 14-18 October 1991 (in preparation)

DETERMINATION OF KERMA FACTORS FOR A-150 PLASTIC AND CARBON FOR NEUTRON ENERGIES ABOVE 20 MeV

U.J. Schrewe[1], H.J. Brede[1], R. Henneck[2], S. Gerdung[3], A. Kunz[3], H.G. Menzel[4], J.P. Meulders[5], P. Pihet[3], H. Schuhmacher[1] and I. Slypen[5]

[1]Physikalisch-Technische Bundesanstalt, W-3300 Braunschweig, Federal Republic of Germany
[2]Physikinstitut der Universität Basel, CH-4056 Basel, Switzerland
[3]Fachrichtung Biophysik und Physikalische Grundlagen der Medizin, Universität des Saarlandes, W-6650 Homburg, Federal Republic of Germany
[4]Commission of the European Communities, B-1200 Bruxelles, Belgium
[5]Unité de Physique Nucléare, Université Catholique de Louvain, B-1348 Louvain-la-Neuve, Belgium

Abstract: Kerma factors for carbon and A-150 tissue-equivalent plastic were measured in nearly monoenergetic neutron beams with energies of 26 and 38 MeV, respectively. The kerma was measured with low-pressure proportional counters (PC) with walls made of graphite and A-150 plastic material. Corrections had to be made for significant fractions of low-energy neutrons which were present in the beams. The correction factors for kerma were determined from time-of-flight measurements using NE213 scintillation detectors and PCs. For the kerma factor weighted mean neutron energies of 26.3(29) and 37.8(25) MeV the kerma factors for carbon were determined to be 34.7(29) and 41.0(37) pGy cm^2 and for A-150 plastic to be 75.2(51) and 78.4(53) pGy cm^2.

(Kerma factors, monoenergetic neutrons, C and A-150 plastic, E_n = 26, 38 MeV)

Introduction

It is well established that fast neutron radiotherapy is superior to photon therapy for some kinds of oncologic diseases [1]. Modern therapy facilities use neutron beams of broad energy distributions between 20 and 70 MeV. From the present state of our radiobiological knowledge, it can be concluded that the demand for accuracy in dosimetry is almost identical for neutrons and photons, i.e. the uncertainties in neutron dosimetry (one standard deviation) should be less than 3.5 % as quoted recently [2]. However, to achieve such accuracy, our knowledge of various neutron interaction data must be considerably improved. For instance, although the absorbed dose in tissue is required, dosimetry is usually performed with dosemeters made of tissue substitutes, so that the dosemeter readings must be corrected for the difference in the composition of the elements. The calculation of corrections requires kerma factors, k_f, (kinetic energy released in matter, K, divided by the neutron fluence, Φ) of the individual elements, chiefly hydrogen, carbon and oxygen.

Kerma factors can be calculated from the cross sections σ_j for interactions of type j

$$k_f = K/\Phi = n \sum_j E_j(E_n) \cdot \sigma_j(E_n) \qquad (1)$$

where n denotes the number of target nuclides per unit mass, E_j the average energy transferred to charged particles and E_n the neutron energy. However, precise cross section data are scarcely above 20 MeV. In this work kerma factors were therefore carefully determined by directly measuring both the kerma and the corresponding fluence in approximately monoenergetic neutron fields. The feasibility of this method has been previously demonstrated [3,4,5]. Details of measurements, data analysis and considerations concerning the accuracy will be published elsewhere [6,7].

Kerma and Fluence Measurements

The experiments were performed at the monoenergetic neutron beam facility of the Paul-Scherrer Institute, (PSI), Switzerland [8]. Intensive neutron beams were produced by bombarding a 2 mm thick beryllium target with protons of energies of 31.9 MeV and 43.3 MeV. A collimator produced neutron beams with a solid angle of 2.5 10^{-5} sr and a fluence per unit proton beam charge of about 10 cm^{-2}nC^{-1} at a distance of 7.8 m from the target. Because of the proton energy loss in the target (5.8 and 4.4 MeV), the negative reaction Q-value (-1.85 MeV), scattering processes and the fact that reactions via excited states and break-up occur, complex spectral distributions were produced. They showed intensive peaks around 26 and 37 MeV (denoted as the nominal neutron energy) but also significant low-energy tails.

Since kerma and absorbed dose are numerically almost equal, provided that a charged secondary particle equilibrium in the detector cavity has been achieved, kerma can be deduced by measuring the absorbed dose. Low-pressure proportional counters (PC) with walls made of A-150 tissue substitute or graphite were used (types: LET1/2, Far West Technology, Goleta, USA, cavity diameters: 12.7 mm, wall thickness: 2.5 mm) in order to measure the absorbed dose. Build-up caps of the respective materials with thicknesses between 5 and 15 mm ensured the equilibrium of secondaries in the cavity. The PCs were filled with tissue-equivalent gas mixtures on the basis of propane or isobutane [9,10]. For the pressures used, the mass of the gas in the cavity per unit area corresponds to that of a sphere with a diameter of 1 μm and a density of 1 g cm^{-3}. The absorbed dose in the wall material can be determined from the pulse-height distribution of the PC, relying on the cavity chamber principle [11]. The calibration procedures and the corrections required will be described in more detail in other papers [6,7]. A proton recoil telescope (PRT) as first described by Bame et al. [12] was used to

determine the neutron fluences. Recoil protons
produced by elastic scattering in a polyethylene
(PE) radiator (mass per unit area of
144.7 mg cm^{-2}) were detected and separated from
other reaction products with four detectors in
series, two proportional counters and two silicon solid-state detectors (thicknesses of 1 mm
and 5 mm). For measurements with the greatest
neutron energy, a 0.5 mm thick aluminium absorber was inserted in front of the first solid-
state detector. Proton background produced in
the PE and outside the radiator foil was taken
into account by replacing the PE radiator with a
graphite disk. The fluence of neutrons with the
nominal neutron energy was determined from the
number of recoil protons by using:
1. the geometry of the detector assembly [12],
2. corrections taking into account the influence of the PRT components, e. g. scattering
   on apertures and other parts which were
   simulated by a Monte Carlo code [13],
3. n-p cross sections of Arndt et al. [14]
   slightly modified on the basis of the work
   of Fink et al. [15],
4. further corrections for deadtime losses,
   inelastic proton scattering in the silicon
   detectors and neutron fluence attenuation in
   various parts of the telescope.

Time-of-Flight Measurements

A significant fraction of the kerma due to
low-energy neutrons had to be separated and
subtracted. Time-of-flight (TOF) measurements
carried out with the PC and scintillation
detectors have been combined to take into
account disturbing influences over the entire
energy range.

The application of TOF techniques on the
PCs has been described earlier [10]. The method
has the potential advantage that it enables the
kerma to be directly determined as a function of
the neutron energy, thus making possible a
complete separation of kerma fractions below a
fixed energy. However, since the timing resolution capacity of the PCs, the length of the
flight paths and the neutron intensities were
limited, TOF signals could not be converted
unambiguously into neutron energies. The
conditions of the present experiment, with
flight paths up to 12 m and a PC timing
resolution of about 80 ns (FWHM), restricted
this separation method to the neutron energy
range of 0.1 to 10 MeV.

The energy range above 10 MeV and below the
nominal beam energy was made accessible by
scintillation detector TOF spectroscopy. Two
cylindrical scintillators (NE213, 5.1 cm diam.,
5.1 cm length; Stilbene, 2.5 cm diam., 2.5 cm
length) applied with pulse-shape techniques in
order to discriminate between photon and
neutron-induced events were used together with
TOF techniques to determine the spectral neutron
fluences.

Measurements were carried out for various
pulse-height thresholds with energies between
0.2 and 4 MeV. The corresponding detector efficiencies were calculated with a Monte Carlo code
[16]. The threshold values strongly influenced
the resulting absolute fluences; discrepancies
of up to 30 % were obtained for the thresholds
used. The absolute fluences extracted from the
scintillation measurements were therefore

discarded and in further analyses only the non-
discrepant relative spectral fluences were used.

Evaluation and Results

The total kerma, K_{tot}, measured with the PC
can be expressed as the sum of three parts: the
desired kerma, K_o, for neutrons of nominal
energy representing a neutron energy range E_2 to
E_{max}, and two background contributions, K_1 and
K_2, which were obtained by means of TOF measurements. K_1 was produced by neutrons of
energies from 0 to E_1 and K_2 from E_1 to E_2.

The ratio of nominal and total kerma can be
expressed by:

$$K_o/K_{tot} = (1-K_1/K_{tot})/(1+K_2/K_o) \qquad (2)$$

K_1/K_{tot} was obtained directly from the PC-
TOF measurements [10]; the small fraction of
kerma below the experimental energy threshold of
0.1 MeV was neglected. K_2/K_o was calculated from
the relative spectral neutron fluences measured
and kerma factors evaluated from the most recent
measurements [6] and theoretical data [17,18].
The values of E_2 were 20.0 MeV and 32.0 MeV for
the two respective measurements. The value of E_1
used was 10.0 MeV in the analysis. A variation
of E_1 between 5 MeV and 12 MeV does not significantly change the ratio K_o/K_{tot}.

Kerma and fluence were measured one after
another and normalized to the proton beam charge
on the neutron target. The experimental results
are compiled in Table 1. In Figs. 1 and 2,
theoretical [17 - 24] and experimental [6,25]
kerma factors are compared. The ratios of the
kerma in A-150 to that in carbon, which can be
determined with smaller uncertainties than the
individual kerma factors, are 2.15(13) for
26.3 MeV and 1.85(11) for 37.8 MeV.

One application in radiation therapy is to
use the ratio of kerma factors for tissue and
A-150 to determine the absorbed dose in tissue
from a A-150 dosemeter reading. By replacing the
older evaluated kerma factors of Caswell [17]
with the measured ones, the absorbed dose in
tissue deduced from the dosemeter reading is
increased by 8 %.

Acknowledgment
We gratefully acknowledge the support of
the PSI cyclotron staff, in particular Dr. P.A.
Schmelzbach. We thank Dr. B.R.L. Siebert for
providing his Monte Carlo code. Furthermore, we
appreciated the help of Mrs. E. Heinemann,
P. Dahmen and F. Langner. The work was partly
supported by the Commission of the European
Communities, contract B17-0030.

Table 1. Experimental results for A-150 (A) and
carbon (C). Numbers in brackets denote
the uncertainty (one standard
deviation) of the last digits of the
respective result.

En MeV	M	K_o/k_{tot}	K_o / Q pGy/nC	Φ_o / Q cm^{-2}nC^{-1}	k_f pGy cm^2
26.3(29)	A	0.562(11)	858(33)	11.44(64)	75.2(51)
	C	0.721(9)	395(24)	11.44(64)	34.7(29)
37.8(25)	A	0.493(11)	769(31)	9.81(55)	78.4(53)
	C	0.605(11)	402(29)	9.81(55)	41.0(37)

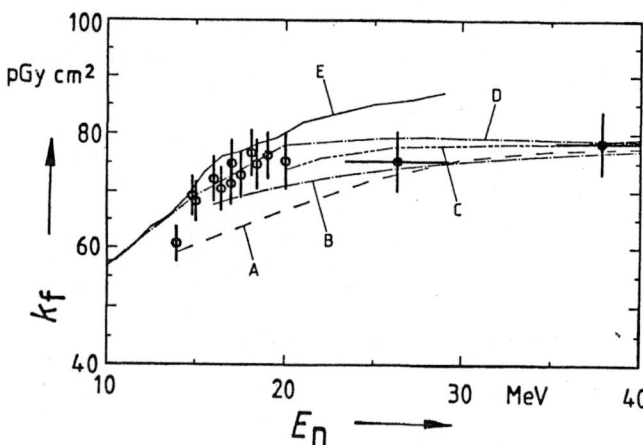

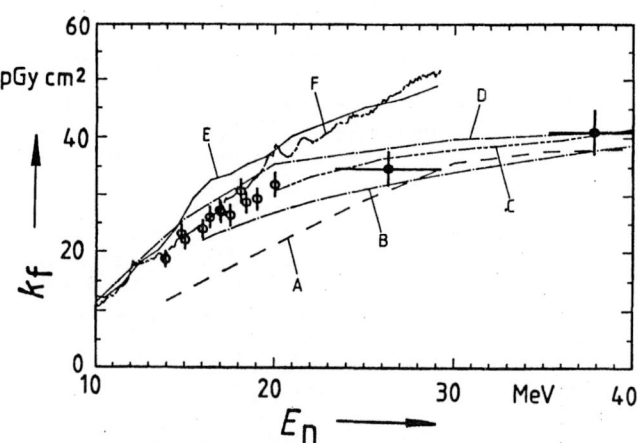

Fig. 1. Kerma factors, k_f, for A-150 plastic as a function of neutron energy. Measured values from this work (filled circles) and from Pihet et al. [6] (open circles). The curves are calculated kerma factors of (A) Behrooz [22], (B) Brenner [18], (C) Dimbylow [21], (D) Wells [20] and (E) Caswell [17].

Fig. 2. Kerma factors, k_f, for ^{12}C as a function of neutron energy. Measured values from this work (filled circles) and from Pihet et al. [6] (open circles). The curves are calculated kerma factors; (A) - (E) see Figure 1, (F) Caswell 1988 [24].

REFERENCES

1. A. Wambersie, J. van Dam, G. Hanks, B.J. Mijnheer, J.J. Battermann In: Proc.of Advisory Group Meeting of the IAEA: Atomic and Nucl. Data, Vienna (1988), p. 51 - 72; and A. Wambersie and F. Richard, IAEA, CRP meeting, Jan. 1989
2. B.J. Mijnheer, J.J. Battermann and A. Wambersie, Radiotherapy and Oncology 8, 237 (1987)
3. H.G. Menzel, G. Bühler, H. Schuhmacher, H. Muth, G. Dietze and S. Guldbakke, Phys. Med. Biol. 29, 1537 (1984)
4. P.M. DeLuca, H.H. Barschall, R.C. Haight and J.C. McDonald, Radiat. Res. 100, 78 (1984)
5. G. Bühler, H.G. Menzel, H. Schuhmacher and G. Dietze, Radiat. Prot. Dosim. 13, 13 (1985)
6. P. Pihet, H.G. Menzel, S. Guldbakke and H. Schuhmacher(submitted to Phys. Med. Biol.)
7. H. Schuhmacher, H.J. Brede, R. Henneck, A. Kunz, J.P. Meulders, P. Pihet and U.J. Schrewe (submitted to Phys. Med. Biol.)
8. R. Henneck, C. Gysin, M. Hammans, J. Lourdan, W. Lorenzon, M. A. Pickar I. Sick, S. Bursynski and T. Stammbach, Nucl. Instrum. and Meth. A259, 329 (1987)
9. Microdosimetry. ICRU-Report 36, Bethesda MD, International Commission on Radiation Units and Measurements (1983)
10. U.J. Schrewe, H.J. Brede, F. Langner and H. Schuhmacher, Nucl. Instrum. and Meth. A299, 226 (1990)

11. A. Rubach and H. Bichsel, Phys. Med. Biol. 27, 893 (1982)
12. S.J. Bame, E. Haddard, J.E. Perry and R.K. Smith, Rev. Sci. Instr. 28, 997 (1957)
13. B.R.L. Siebert, H.J. Brede and H. Lesiecki, Nucl. Instrum. and Meth. A235 (1985), and B.R.L. Siebert, private communication
14. R.A. Arndt, J.S. Hyslop and L.D. Roper, Phys. Rev. D35 128 (1987)
15. G. Fink, P. Doll, T. D. Ford, L. Garret, W. Heeringa, K. Hofmann, H. O. Klages and H. Krupp, Nucl. Phys. (in press)
16. R.A. Cecil, B.D. Anderson, R. Madey, Nucl. Instrum. and Meth. 161, 439 (1979)
17. R.S. Caswell, J.J. Coyne and M.L. Randolph, Radiat. Res. 83 217 (1980)
18. D.J. Brenner, Phys. Med. Biol. 26, 437 (1983)
19. R.G. Alsmiller and J. Barish, Health Phys. 33, 98 (1977)
20. A. H. Wells, Radiat. Res. 80, 1 (1979)
21. P.J. Dimbylow, Phys. Med. Biol. 25, 637 (1980)
22. M.A. Behrooz, E.J. Gillespie and D.E. Watt, Phys. Med. Biol. 26, 507 (1981)
23. P.J. Dimbylow, Phys. Med. Biol. 27, 989 (1982)
24. R.S. Caswell, J.J. Coyne, H.M. Gerstenberg and E.J. Axton, Radiat. Prot. Dosim. 23, 11 (1988)
25. J.L. Romero, F.P. Brady, T.S. Subramanien in: Proc. Int. Conf. on Nucl. Data for Basis and Applied Science, Santa Fe, 1985

MEASUREMENT OF C, Mg, AND Fe KERMA FACTORS AND THE ^{19}F(n,2n)^{18}F CROSS SECTION FOR 18 TO 27 MeV NEUTRONS

C. L. Hartmann, P. M. DeLuca, Jr., and D. W. Pearson

Department of Medical Physics
University of Wisconsin
Madison, WI 53706 USA

Abstract

C, Mg, and Fe kerma factors were measured at neutron energies of 18, 23, and 25 MeV using microdosimetric techniques. Small spherical proportional counters served to measure neutron kerma, while detection of the activity induced in $(C_2F_4)_n$ samples by the ^{19}F(n,2n)^{18}F reaction determined the fast neutron fluence. In a separate experiment, we measured the ^{19}F(n,2n)^{18}F reaction cross section for neutron energies of 18, 21, 23, 25, and 27 MeV by comparing activity induced in $(C_2F_4)_n$, Zr, and Au samples following simultaneous irradiation. For both experiments, deuteron bombardment of a tritium gas target produced monoenergetic fast neutrons accompanied by a secondary flux of low energy neutrons. These secondary neutrons made a significant contribution to the neutron kerma, but did not affect activation measurements. At each energy, a time-of-flight spectrometer determined the secondary neutron spectrum, which we used to correct our kerma measurements.

Introduction

Neutron kerma is the average energy transferred from neutrons to charged particles per unit mass of material. The kerma factor is the kerma per unit fluence. For energies below 20 MeV, neutron kerma factors can be determined from microscopic neutron cross section measurements. Above 20 MeV, cross section information is so sparse that kerma factors must be estimated primarily from nuclear model calculations. Although neutron radiotherapy and structural damage studies require accurate kerma factors, the results calculated by different nuclear models vary by up to almost a factor of two in the 20 to 60 MeV range.

Here we report kerma factor measurements for C, Mg, and Fe at 18, 23, and 25 MeV neutron energy. The dose measured in small spherical proportional counters with walls made of the respective materials determined neutron kerma, while the response of a liquid scintillator monitored the fast neutron fluence. For a fraction of each irradiation, the scintillator response was calibrated by exposing it to neutrons simultaneously with teflon* samples. Detection of fluorine activity served to determine the absolute fast neutron fluence. In a separate experiment, we also measured the ^{19}F(n,2n)^{18}F reaction cross section for neutron energies of 18, 21, 23, 25, and 27 MeV by comparing activity induced in $(C_2F_4)_n$, Zr, and Au samples following simultaneous irradiation. This data provided more accurate fluence measurement at neutron energies above 20 MeV [1].

Experimental Method

Neutrons from the ^{3}H(d,n)^{4}He reaction were produced by bombarding a tritium gas target with deuterons from the University of Wisconsin Tandem Accelerator. A 101.6 mm long and 6.4 mm inner diameter stainless steel cylinder, with a 2.5 μm Mo entrance foil and a 0.25 mm Au beam stop, contained the tritium gas at about 1 atm pressure. The neutron spectrum consisted of a monoenergetic group of neutrons at

the energy of interest, and a distribution of lower energy neutrons with a maximum energy about 19 MeV lower than the monoenergetic group.

For kerma factor determinations, we positioned proportional counters symmetrically about the deuteron beam axis at 50°, at distances of 50-60 cm from the center of the tritium target. Activation samples were also positioned at 50°, about 40 cm from the tritium target.

Three proportional counters, constructed with walls of graphite (CPC), magnesium (MgPC), and iron (FePC), measured the neutron dose. The counters have a 1.27-cm-inner-diameter, 1.3-mm-thick wall, and are of standard Far West Technology** design. We filled each counter with a mixture of Ar and CO_2 (10% CO_2 by weight). The CPC gas was at a pressure of 60 Torr, while the MgPC and FePC were filled to a pressure of 30 Torr.

The dose response calibration of the proportional counters was similar to that used in previous investigations [2,3,4]. The energy deposited in the spherical gas volume was calibrated for the MgPC and FePC with an internal ^{244}Cm alpha-particle source using tabulated values for the collision mass stopping power [5] and the known counter diameter. For the CPC, the energy deposited was calibrated either with the internal alpha-particle source or by the position of the maximum carbon recoil event size (carbon edge). A precision electronic pulser relates each gain to the alpha-particle or carbon edge calibration. By integrating over the energy deposition spectra, we deduced the dose to the proportional counter cavity gas.

At each energy, we measured the neutron spectrum with a time-of-flight spectrometer that consisted of a 5.1 × 5.1 cm NE 213 liquid scintillator positioned 2 to 2.5 m from the pulsed neutron source (a bunched, chopped deuteron beam incident on the tritium gas target). The Monte Carlo response code SCINFUL [6] was used to calculate the energy-dependent detector efficiency. We determined the number of lower energy neutrons relative to the ^{3}H(d,n)^{4}He neutron

---

* polytetrafluoromethane, $(C_2F_4)_n$

** Far West Technology, Goleta, Ca.

fluence by (1) by normalizing the breakup spectrum to the number of $^3H(d,n)^4He$ neutrons in the scintillator-recorded time-of-flight peak, and (2) normalizing the spectrum to the number of $^3H(d,n)^4He$ neutrons expected based on activation measurements. These two approaches resulted in corrections to the measured kerma that varied by less than 15 %. To correct the 23 and 25 MeV kerma factor measurements, we used the low energy neutron spectra determined by taking the average of results of methods (1) and (2).

For the kerma measurements, another liquid scintillation detector monitored the $^3H(d,n)^4He$ neutron fluence. We calibrated the scintillator during a portion of each proportional counter irradiation by comparing its response with the activation induced in 2.54 cm-diameter, 0.63 cm-thick teflon samples. A high purity intrinsic germanium spectrometer detected annihilation quanta resulting from fluorine activation. For the 25 MeV measurement activation results were corrected for ^{11}C activity.

To improve the accuracy in determining absolute neutron fluence above 20 MeV, we made additional measurements of the $^{19}F(n,2n)^{18}F$ cross section at 18, 21, 23, 25, and 27 MeV. Neutrons bombarded Zr, Au, and teflon samples. We corrected ^{18}F, ^{89}Zr, and ^{196}Au activities for variations in neutron flux, attenuation of neutrons during irradiation, self-absorption of sample gamma rays, and activity due to contaminating lower energy neutrons. $^{19}F(n,2n)^{18}F$ cross section values were deduced by comparing F activity with activity in Zr and Au due to the $^{nat}Zr(n,xn)^{89}Zr$ and $^{197}Au(n,2n)^{196}Au$ reactions. This experiment is described in more detail in ref. [1].

We corrected the proportional counter dose measurements for the background due to gamma rays as well as for neutrons due to deuteron interactions in the tritium containment cell by subtracting the signal acquired with an empty target. This resulted in the dose to the gas due to neutrons from deuteron interactions in the tritium gas.

For the 23 and 25 MeV measurements, we estimated the kerma contributed to C, Mg, and Fe by lower-energy d-T breakup neutrons by folding the breakup spectra with energy-dependent kerma factors tabulated by Caswell, et al. [7]. Multiplying the resulting dose per $^3H(d,n)^4He$ neutron fluence by the average ratio of stopping powers in the fill gas to the counter wall for the breakup neutron spectrum, we determined the dose in the gas due to d-T breakup neutrons per monoenergetic neutron fluence, and then subtracted it from the measured gas dose per fluence due to all d-T neutrons.

Finally, we converted the net dose to the proportional counter gas to dose in the wall (equivalent to wall kerma), by multiplying it by the average ratio of mass stopping powers in the wall to the gas. The ratios of stopping powers for charged particles resulting from 18, 23, and 25 MeV neutron interactions in the proportional counter walls were determined using Brenner and Prael's double-differential secondary-particle production cross section calculations [8] for the CPC and the ENDF-B/VI cross section tabulations [9] for MgPC and FePC. For 23 and 25 MeV neutron energies, we assumed that the Mg and Fe cross sections have the same relative magnitude as in the ENDF-B/VI evaluation at 20 MeV.

Perturbations in the charged particle spectrum caused by the proportional counter gas have been investigated by Bühler, et al. [10] and DeLuca, et al. [4]. DeLuca [4] found that for 14.7 MeV neutrons, the effect of Ar + CO_2 gas at 30 Torr was negligible for the MgPC but resulted in an increase of about 3% per 10 Torr for the FePC. For Fe, we scaled the relative dose contribution in the Fe counter measured by DeLuca [4] by the ratio of kerma factors in Ar + CO_2 and

Fe at 1.8 and 2.4 MeV in order to correct the estimated dose due to breakup and at 18, 23, and 25 MeV to correct the net dose due to fast neutrons.

Results

For neutron energies below 20 MeV, kerma factors have been calculated from measured microscopic cross sections. Above 20 MeV, the sparseness of cross section information has resulted in kerma factors derived almost completely from nuclear models, resulting in large uncertainties. It is difficult to accurately measure all the cross sections necessary to make kerma calculations similar to those made by Caswell below 20 MeV[7]. However, direct measurement of kerma factors can be used to normalize and verify model-based calculations.

The uncertainties in our measurements are caused primarily by the background due to d-T breakup neutrons from the measured dose. The average breakup spectrum is assumed to be uncertain by about 11%, while the measured dose to the proportional counter gas per monoenergetic neutron fluence is uncertain by about 8-10% due to uncertainty in the activation fluence measurement, error caused by assuming that W/e is constant for all charged particle types and energies, uncertainty in the stopping powers used to calibrate the proportional counters, and statistical error in the dose measurement. The dose caused by d-T breakup neutrons results in a 25-50% adjustment to the measured dose.

Relative neutron fluence was monitored by means of a liquid scintillation counter, while absolute fluence was determined from induced ^{18}F activity. Figure 1 summarizes the reaction cross sections measured and employed. Details of this determination are given in Ref. [1]. The cross section value at 18 MeV agrees with both evaluations while values at higher energies follow the trend indicated by Ref. [11].

The present carbon kerma factor determinations are plotted in Figure 2 (solid dots) along with previous microdosimetric determinations [3,12,13] and an ionization chamber measurement [14] as well as values deduced from measured [15], evaluated [7,16] and calculated [17] neutron cross sections. Our value at 18 MeV is in substantial agreement with previous measurements. Although the values measured at 23 and 25 MeV are consistent with those based on charged

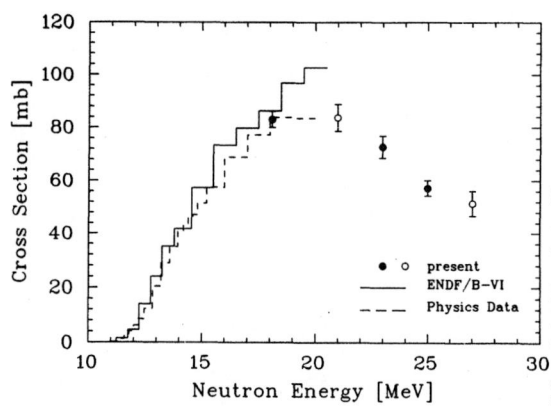

Figure 1: $^{19}F(n,2n)^{18}F$ cross section measurements and evaluations: closed circles indicate the cross sections used in our kerma factor determinations, while open circles indicate cross sections measured but not used in the kerma factor determinations.

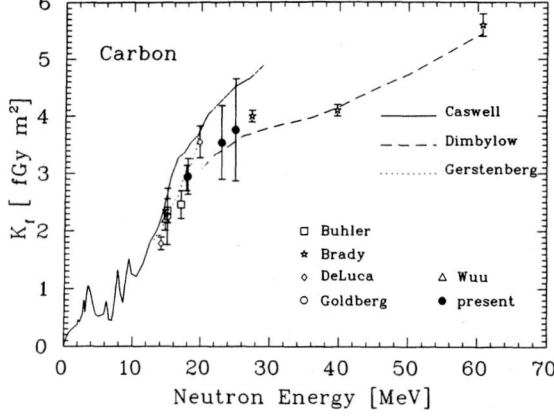

Figure 2: Carbon kerma factor measurements and calculations.

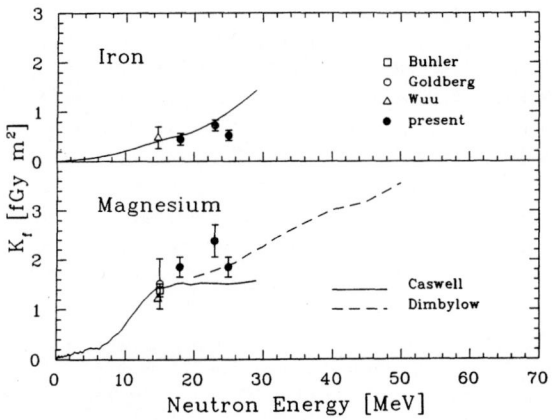

Figure 3: Iron and magnesium kerma factor measurements and calculations.

particle measurements [15] as well as Dimbylow's calculation [17], the uncertainty introduced by the correction for lower energy neutrons only allows the rejection of the evaluation based on limited microscopic data at 18 and 23 MeV [7].

Magnesium kerma factor values (Figure 3) are systematically larger than values deduced from evaluated cross sections [7], following more closely the trend indicated by Dimbylow [17]. In iron (also Figure 3), measured values are in agreement with the evaluation by Caswell, et al. [7] for lower neutron energies, but indicate that the iron kerma factor does not increase as rapidly as the evaluation predicts.

References

[1] C.H. Hartmann and P.M. DeLuca, Jr.: Nucl. Sci. Eng., in press

[2] P.M. DeLuca, Jr., H.H. Barschall, R.C. Haight and J. McDonald: Radiat. Res. 100, 78 (1984)

[3] P.M. DeLuca, Jr., H.H. Barschall, M. Burhoe and R.C. Haight: Nucl. Sci. Eng. 94, 192 (1986)

[4] P.M. DeLuca, Jr., H.H. Barschall, C.L. Hartmann and D.W. Pearson: Nucl. Inst. Meth. 40/41, 1279 (1989)

[5] H.H. Anderson and J.F. Ziegler, Hydrogen Stopping Powers and Ranges in All Elements 3 Pergamon Press, New York (1977)

[6] J.K. Dickins, Scintillator Full Response to Neutron Detection, ORNL-6462, Oak Ridge National Laboratory (1988)

[7] R.L. Caswell, J.J. Coyne and M.L. Randolph: Radiat. Res. 83, 217 (1980)

[8] D.J. Brenner and R.E. Prael: Atomic Data and Nuclear Data Tables 41 71 (1989)

[9] ENDF/B-VI, National Nuclear Data Center, Brookhaven National Laboratory, Upton, NY (1990)

[10] G. Bühler, H. Menzel, H. Schuhmacher, G. Dietze and S. Guldbakke: Phys. Med. Biol. 31 no. 6, 601 (1986)

[11] M. Wagner, H. Vonach, A. Pavic, B. Strohmaier, S. Tageson and J. Martinez-Rico: Physics Data 13-5 206 (1990)

[12] G. Bühler, H. Menzel, H. Schuhmacher and S. Guldbakke, in Proc. 5th Symposium on Neutron Dosimetry, Munich/Neuherberg, FRG, EUR-9762, p. 309, Commission of the European Communities, Luxemborg (1985)

[13] C. Wuu and L. Milavickas: Med. Phys. 14 No. 6., 1007 (1987)

[14] E. Goldberg, D.R. Slaughter and R.H. Howell, Experimental Determination of Kerma Factors at E_n =15 MeV, UCID-17789, Lawrence Livermore Laboratory (1978)

[15] F. Brady and J. Romero, Final Report of the National Cancer Institute, Grant No. 1R01 CA16261 University of California, CA (1979)

[16] H.M. Gerstenberg, R.L. Caswell and J.J. Coyne: Radiat. Prot. Dosim. 23 No. 1, 41 (1988)

[17] P. Dimbylow: Phys. Med. Biol. 27 no.8, 989 (1982)

INTERMEDIATE-STATE STRUCTURE IN THE ^{12}C(n,n3α) AND ^{16}O(n,n4α) BREAKUP INDUCED BY 22.5 AND 25.4 MeV NEUTRONS

B. Antolković

Ruđer Bošković Institute, POB 1016, 41001 Zagreb, Croatia, Yugoslavia

Abstract: The n3α and n4α final states induced by the interaction of 22.5 and 25.4 MeV neutrons with ^{12}C and ^{16}O are studied in a kinematically complete measurement, using nuclear emulsions as a target and a 4π detector. By determining the excitation energies of the correlated 2α, 3α and 4α particles in each event, the intermediate-state excitation-energy spectra of the reactions ^{12}C(n,n)^{12}C(α)^{8}Be(2α) and ^{16}O(n,n)^{16}O(α)^{12}C(α)^{8}Be(2α) are deduced. Partial cross sections for n+^{12}C inelastic scattering to various reaction bins of the ^{12}C excitation-energy spectrum are tabulated.

(Reaction: ^{12}C(n,n3α), ^{16}O(n,n4α); E_n=22.5 and 25.4 MeV. Deduced partial cross sections for various reaction chains in the ^{12}C(n,n)^{12}C(α)^{8}Be(2α) breakup. Determined intermediate-state structure of the reaction ^{16}O(n,n)^{16}O(α)^{12}C(α)^{8}Be(2α). Monte Carlo analysis)

Introduction

The increasing use of neutrons in clinical radiotherapy has created the need for neutron-induced charged-particle cross sections on tissue elements. Because of the considerable difference in the content of ^{12}C and ^{16}O in tissue and tissue-equivalent materials the information on the neutron interaction with ^{12}C and ^{16}O nuclei is of special interest. There are very few experimental data on the ^{12}C(n,n3α) breakup above 20 MeV and nearly total absence of information on the five-body breakup ^{16}O(n,n4α).

This paper presents results of the study of the ^{12}C(n,n3α) and ^{16}O(n,n4α) reactions induced by 22.5 and 25.4 MeV neutrons. The prime objective of the experiment is to extract the excitation-energy spectra of the ^{12}C and ^{16}O intermediate states of the multistep reactions ^{12}C(n,n)^{12}C(α)^{8}Be(2α) and ^{16}O(n,n)^{16}O(α)^{12}C(α) ^{8}Be(2α), respectively, and the branching ratios to the excitation-energy bins in successive intermediate-state spectra.

In the multiparticle breakup of light nuclei, the number of open channels increases rapidly with the number of outgoing particles and with the projectile energy. Nevertheless, for all reaction chains of a priori unknown presence, the nominal excitation energies of the 2α, 3α and 4α correlations can be computed and included implicitly in the excitation-energy spectra of interest.

A detailed study performed recently [1] has shown that the Monte Carlo simulated distributions computed for a specific reaction chain (defined by intermediate-state excitation-energy bins) are the same, regardless of the reaction mechanism assumed. It has also bee shown that the known ^{12}C excitation-energy spectra of given branching ratios to the transition via ^{8}Be$_{gs}$ permit a Monte Carlo modelling of the four-body n3α final state for any variable or correlation of variables and for any specific experimental condition [2,3].

Experimental procedure and results

Nuclear emulsions acting as a radiator and a 4π detector were exposed to neutrons of 22.5 and 25.4 MeV from the Ohio University neutron facility. The n3α and n4α final states induced by the neutron interaction with ^{12}C and ^{16}O were studied in a kinematically complete measurement. An extensive description of the processing and analysis of data, including corrections due to the loss of events during the scanning procedure, is given elsewhere [ref. 1 and references therein].

Reaction n+^{12}C→n+3α

The experimentally obtained ^{12}C excitation-energy spectra for E_n=22.5 and 25.4 MeV are shown in Fig. 1a and b, respectively. The spectra of the events pertaining to the transitions via the ^{8}Be$_{gs}$ (computed $E^*_{^8Be}$<0.2 MeV for at least one αα pair - see ref. [1]) are shown separately. The absolute

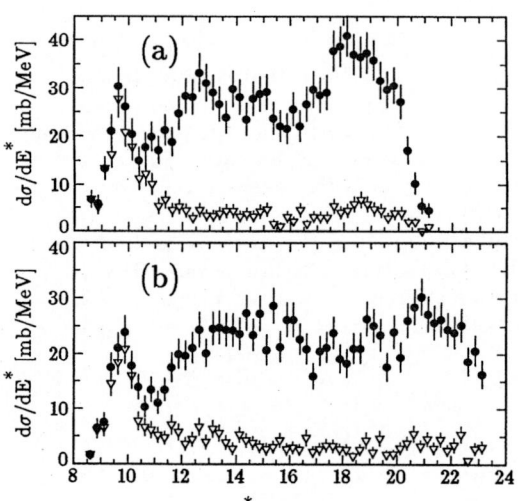

Fig. 1. ^{12}C excitation energy spectra of the reaction n+^{12}C-n+3α for (a) E_n=22.5 MeV and (b) E_n=25.4 MeV, respectively. Full points correspond to all events and triangles to only those events proceeding via the ^{8}Be ground state transition.

TABLE 1. Partial cross sections of neutron inelastic scattering to various bins of the ^{12}C excitation spectrum. The cross sections for the transition via the ^{8}Be$_{gs}$ are given in parentheses.

^{12}C Excitation Energy(MeV)		Neutron Energy (MeV)	
Boundaries	Mean Energy	22.5	25.4
<10.25	9.6	31.0(26.9)	24.0(20.9)
10.25 - 11.25	10.8	17.5(10.0)	12.2(6.4)
11.25 - 12.25	11.8	23.6(6.9)	17.7(5.3)
12.25 - 13.50	12.8	29.7(5.2)	23.0(5.3)
13.50 - 14.50	14.0	26.4(3.8)	24.9(4.0)
14.50 - 15.50	15.0	27.4(3.3)	25.0(3.1)
15.50 - 16.50	16.0	22.9(2.6)	24.1(3.1)
16.50 - 17.50	17.0	28.5(2.6)	19.6(3.2)
17.50 - 18.50	18.0	38.5(4.8)	20.5(2.4)
18.50 - 19.50	19.0	35.4(5.4)	24.0(3.4)
19.50 - 20.50	20.0	26.1(3.1)	21.8(2.5)
20.50 - 21.50	21.0	5.0(0.7)	28.0(4.0)
21.50 - 22.50	22.0		25.0(3.9)
22.50 - 23.50	23.0		13.9(1.6)
sigma $_{n+^{12}C\to n+3\alpha}$(mb)		319.1(76.6)	309.4(70.4)

cross sections were defined by normalizing the peaks of the 9.63 MeV state in ^{12}C at 22.5 and 25.4 MeV to 31 and 24 mb, respectively. The latter values were obtained by interpolating the data of Meigooni et al [4]. The partial cross sections for the ^{12}C excitation-energy bins of 1 MeV width were extracted from the ^{12}C excitation-energy spectra. In Table 1 the partial cross sections for the inelastic neutron scattering to various bins of ^{12}C excitations are listed for all events and for those involving the ^{8}Be$_{gs}$ transition (shown in parentheses). The uncertainties in the measured partial cross sections are in the range 7-12% and for total cross sections they amount to 15%. In Fig. 2 our data at 22.5 and 25.4 MeV are shown as an extension of the data in the region 12-19 MeV, measured recently [2]. Recent data on ^{12}C(n,n)^{12}C$_{9.63}$ inelastic scattering [4-7] are also shown in Fig. 2.

In Fig. 3 the Monte Carlo simulated α-particle distributions, based on the data extrapolated from the present results, are compared with experimental single α-particle spectra from the n+^{12}C interaction at 27.4 MeV [8]. The centre-of-mass isotropy has been assumed for inelastic scattering to all excitation-energy bins. The agreement is good, except at high energies. The present data do not

contain the α+^{9}Be$_{gs}$ final channel and thus the Monte Carlo simulated 3α distributions are kinematically limited to the lower energy region.

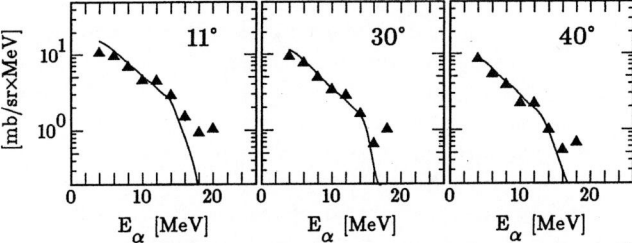

Fig.3   Alpha-particle spectra from neutron interaction with ^{12}C at 27.4 MeV /ref.8/, and Monte Carlo simulated distributions based on the data extracted from the present work. For details, see text.

Reaction n+^{16}O→n+4α

The complexity of the system is considerably increased with only one additional α-particle in the final channel. Since the four α-particles are indistinguishable, only the ^{16}O excitation energy is single-valued. Four ^{12}C excitation energies and six ^{8}Be excitation energies should be computed for all possible triads and pairs out of four α-particles. Only one triad/or pair represents a true correlation and the others form a continuum of spurious events. However, as will be shown later, if the reaction chain involves the ^{8}Be$_{gs}$ transition, i.e. if at least one αα pair (for example, $\alpha_i\alpha_j$) has the excitation energy less than 0.2 MeV, the situation is simplified. In this case, all pairs except $\alpha_i\alpha_j$ should be rejected (for explanation, see Fig. 5), as well as all triads except $(\alpha_i\alpha_j)_{gs}\alpha_k$ and $(\alpha_i\alpha_j)_{gs}\alpha_\ell$, one of the later two being spurious.

Fig. 4 shows the level scheme of the intermediate states involved in the reaction ^{16}O(n,n)^{16}O(α)^{12}C(α)^{8}Be(2α). The ^{16}O excitation-energy spectra are also inserted and are shown separately for all events and for only those proceeding via ^{8}Be$_{gs}$. The densely populated ^{16}O levels are not resolved and therefore the spectrum is bin averaged. The ^{16}O excitations of less than 16.25 MeV are missing since they decay into four α-particles whose energy is too low to be detected by the nuclear emulsion technique.

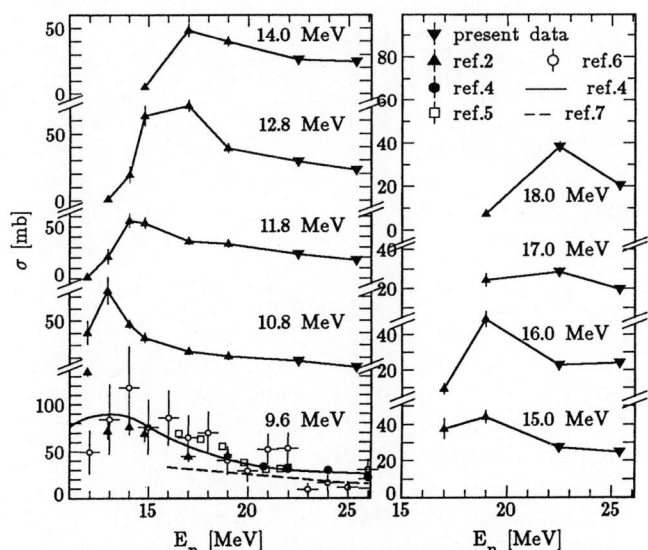

Fig. 2. Partial cross section for neutron inelastic scattering to various bins of the ^{12}C excitation spectrum. Lines are drawn through experimental points for better visibility.

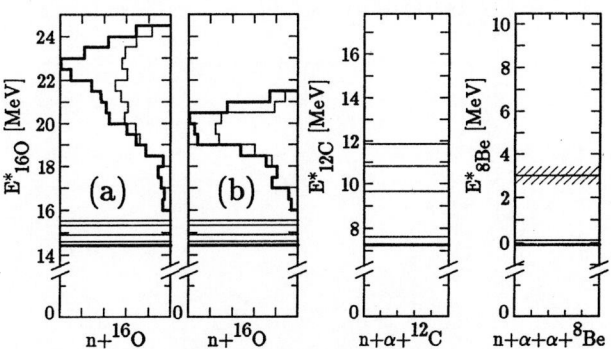

Fig. 4. Excitation energy level schemes of intermediate states in the n+^{16}O breakup, showing a few levels above the 4α breakup (Q=-14.4 MeV). ^{16}O excitation energy spectra at a) 25.4 and b) 22.5 MeV are given for all events and for those proceeding via the ^{8}Be$_{gs}$ (thick and thin line histograms).

Figs. 5a and 5b display the ^{8}Be excitation-energy spectrum for 22.5 and 25.4 MeV, respectively. Events proceeding via the ^{8}Be$_{gs}$ transition contribute to the enhanced peaks around zero excitation energy (pairs $(\alpha_i\alpha_j)_{gs}$ in area A) and to the well resolved distribution of five accompanyig spurious pairs (area 5A). The rest of events form area 6B, which contains six indistinguishable $\alpha\alpha$ pairs. The spectrum 6B is structureless due to 5/6 of spurious events and therefore there is no convincing evidence for the presence of the intermediate ^{8}Be nucleus in the 3.0 MeV state. The ratio of the events proceeding via the ^{8}Be$_{gs}$ to all events amounts to 72.5% and 61.8% for 22.5 and 25.4 MeV, respectively.

By analysing the ^{8}Be excitation-energy spectrum of the $\alpha_k\alpha_\ell$ pairs, complementary to the $\alpha_i\alpha_j$ pair in ^{8}Be$_{gs}$, a small (4%) but well resolved group of 2^8Be$_{gs}$ events due to the reaction ^{16}O(n,n)^{16}O(2^8Be$_{gs}$) is found. Because of the kinematical considerations (see ref. [2]), the enhancement of data in the distributions 6B below 1 MeV has to be assigned to the reaction chain pertaining to the transition via ^{9}Be$_{2.43}$.

In Fig. 6a we present the ^{12}C excitation-energy spectrum at 22.5 MeV for events proceeding via the ^{8}Be$_{gs}$ and containing the triads $(\alpha_i\alpha_j)_{gs}\alpha_k$ and $(\alpha_i\alpha_j)_{gs}\alpha_\ell$. The structure of the spectrum clearly indicates the presence of the 9.63 and 10.8 MeV states of ^{12}C, although half of the data are spurious. To unfold the spectrum, we used the Monte Carlo simulated ^{12}C excitation-energy distributions for the ^{16}O(n,n)^{16}O(α)^{12}C$_x$(α)^{8}Be$_{gs}$(2α) sequence (x indicating 9.63 and 10.8 MeV, alternately). The excitation energy of the ^{16}O interme-

diate state has been chosen at random from the experimentaly determined density distribution (see ref. [1]) of contributing ^{16}O excitation-energy bins. The computed ^{12}C excitation-energy distributions for the transition via ^{12}C$_{9.63}$ and ^{12}C$_{10.8}$ are shown in Fig. 6b and c, respectively, and their fit to experimental data is given in Fig. 6a. The absence of higher excited states (11.8, 12.8 MeV) is due to their unnatural parities, which forbid the 3α decay via a 0^+ ^{8}Be ground state. In decoding other ^{12}C excitation-energy spectra similar procedure was applied.

Conclusion

In the present paper we have determined partial cross sections of inelastic neutron scattering to various bins of ^{12}C excitation energies with given branching ratios to the transition via ^{8}Be$_{gs}$. As shown recently [refs. 2,3] and confirmed in the present paper (Fig. 3), these data permit a Monte Carlo modelling of the four-body n3α final state for any variable or correlation of variables, a feature required in various applications.

Due to the limited space, only the main features of the reaction n+^{16}O→n+4α have been given, marking the basic lines of the elaboration of data. Nevertheless, it is clearly seen that the sequential decay via ^{12}C and ^{8}Be intermediate states takes over the main part of the total cross section. Especially, the transition via the ^{8}Be$_{gs}$ is highly involved. More details will be published elsewhere.

Acknowledgement

The help of Prof. R.W. Finaly and the staff of the Ohio University neutron facility during the irradiation of nuclear plates is gratefully acknowledged.

REFERENCES

1. B. Antolković and M. Turk, Nucl. Phys. A524, 285 (1991)
2. B. Antolković, G. Dietze and H. Klein, Nucl. Sci. Eng. 107, 1 (1991)
3. B. Antolković and M. Turk, Radiat. Prot. Dosim. 23, 19 (1988)
4. Ali S. Meigooni, J.S. Petler and R.W. Finlay, Phys. Med. Biol. 29, 643 (1984)
5. N. Olsson, B. Trostell and E. Ramström, Nucl. Phys. A496, 505 (1989)
6. B. Antolković, I. Šlaus, D. Plenković, P. Macq and J.P. Meulders, Nucl. Phys. A394, 87 (1983)
7. D.J. Brenner and R.E. Prael, Nucl. Sci. Eng. 88, 97 (1984)
8. T.S. Subramanian, J.L. Romero, F.P. Brady, J.W. Watson, D.H. Fitzgerald, R. Garrett, G.A. Needham, J.L. Ullmann and C.I. Zanelli, Phys. Rev. C28, 521 (1983).

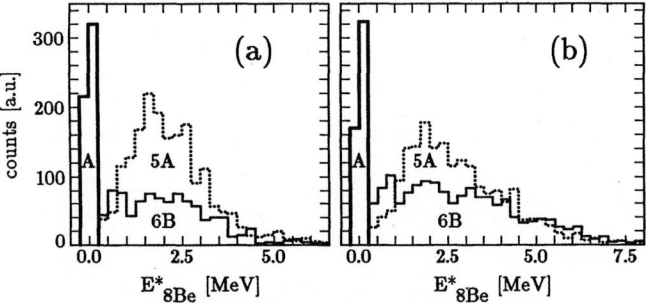

Fig.5 ^{8}Be excitation energy spectra for a) 22.5 MeV and b) 25.4 MeV. For details, see text.

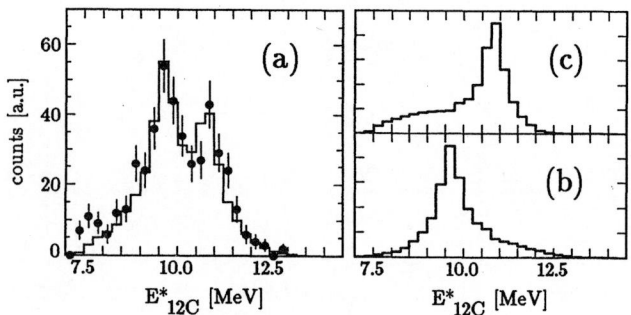

Fig.6 Experimental ^{12}C excitation energy spectrum at 22.5 MeV involving the ^{8}Be$_{gs}$ transition (Fig. 6a). Monte Carlo simulated distributions for neutron inelastic scattering to ^{12}C$_{9.6}$ and ^{12}C$_{10.8}$ states (both decaying via ^{8}Be$_{gs}$) and their fit to the experimental spectrum are given in Figs. 6b, c and a, respectively.

PRODUCTION OF TUNGSTEN-188 AND OSMIUM-194 IN A NUCLEAR REACTOR FOR NEW CLINICAL GENERATORS

S. Mirzadeh[*], F. F. Knapp, Jr. and A. P. Callahan

Nuclear Medicine Group, Health and Safety Research Division,
Oak Ridge National Laboratory, Oak Ridge TN 37831-6022

ABSTRACT

Rhenium-188 and iridium-194 are potential candidates for radioimmunotherapy with monoclonal antibodies directed against tumor-associated antigens. Both nuclei are short-lived and decay by high energy β^- emission. In addition, both nuclei emit γ-rays with energy suitable for imaging. An important characteristic is availability of ^{188}Re and ^{194}Ir from decay of reactor-produced parents (^{188}W and ^{194}Os, respectively) in convenient generator systems. The ^{188}W and ^{194}Os are produced by double neutron capture of ^{186}W and ^{192}Os, respectively. The large scale production yields of ^{188}W in several nuclear reactors will be presented. We also report a new measurement for the cross-section of ^{193}Os$(n,\gamma)^{194}$Os reaction and discuss the feasibility of producing sufficient quantities of ^{194}Os.

Key words: Radionuclide Generator, ^{188}W/^{188}Re, ^{194}Os/^{194}Ir, Nuclear Reaction Cross-section, Nuclear Reactor

INTRODUCTION

We are currently developing the ^{188}W($t_{1/2}$=70 d)/^{188}Re($t_{1/2}$=17 h)[1-6] and ^{194}Os($t_{1/2}$=6.0 y)/^{194}Ir($t_{1/2}$=20 h)[6,7] biomedical generator systems. The ^{188}Re and ^{194}Ir daughters both decay by emission of high energy β^- particles (E_β^{max} = 1.89 and 2.24 MeV, respectively) and are potential candidates for therapy.[7-9] The decay of ^{188}Re follows with emission of a predominant 155 KeV γ-ray with moderate intensity (15.8%) which can be efficiently detected with gamma cameras for *in vivo* biodistribution and kinetic studies. The decay of ^{194}Ir follows with emission of a 328 KeV γ-ray (13.0%) which is also within the detection range of gamma cameras. In addition to the emission of an abundant high-energy particle, an appropriate physical half-life and stable daughter, both nuclei have suitable characteristics for protein labeling through bifunctional chelates. Rhenium-188 is of special interest since it is an analogue of Tc and recent advances in the chemistry of Tc for biomedical applications of ^{99m}Tc could in principal be extended to Re.

The ^{188}W and ^{194}Os parent nuclei are produced in a fission nuclear reactor with double neutron capture on ^{186}W and ^{192}Os, respectively, according to the schemes shown in Figure 1a&b. In the case of ^{188}W, we report here large scale production yield of ^{188}W from ORNL-HFIR, BNL-HFBR, MURR and FFTF reactors[#] and compare the theoretical and experimental data. For ^{194}Os production, the thermal neutron capture cross-section of primary reaction, ^{192}Os(n,γ), is well known and is used extensively for neutron activation analysis of Os, however, the reported cross-section for the second reaction, ^{193}Os(n,γ), differs significantly.[10-14] An attempt has been made to resolve this discrepancy and our results are presented in this report.

---

[*]Author for correspondence
[#]HFIR: High Flux Isotope Reactor, Oak Ridge Nat. Lab.
HFBR: High Flux Beam Reactor, Brookhaven Nat. Lab.
MURR: Missouri University Research Reactor
FFTF: Fast Flux Test Facility, Westinghouse Hanford Co.

EXPERIMENTAL

Tungsten-188: Typically, 50 mg of enriched ^{186}W as WO_3 (95% enrichment) was encapsulated in a quartz ampule and irradiated for a period of 21 days (one reactor cycle) in the hydraulic tube of the ORNL-HFIR. Neutron fluxes and other irradiation conditions for various reactors are given in Table 1. After irradiation, the target was generally allowed to decay for a period of 10 days to reduce the ^{187}W activity to a level comparable to that of ^{188}W activity. Subsequently, the quartz ampule was crushed, WO_3 was dissolved in 0.1 $\underline{M}$ NaOH and a small aliquot ($\sim$10 μL) was taken for assay.

Osmium-194: An ^{192}Os target (27.1 mg with 99.3% enrichment) encapsulated in a quartz ampule was irradiated for one cycle (24 d) at position "Modified V-16" together with 2.0 and 5.5 mg of high purity Fe as flux monitors. The induced radioactivities in the ampules were measured directly without chemical separation. However, due to high levels of radioactivity, it was necessary to allow the samples to decay for a period of about 6 months before the first measurement.

Radioactivity Measurements: A calibrated 50-cm^3 high-purity Ge detector (EG&G ORTEC, Oak Ridge, TN) coupled to a AccuSpec PC-based multichannel analyzer (Nuclear Data/Canberra Inc., Meriden, CT) was used for radioactivity measurements. The activity of ^{187}W and ^{188}W were quantitated by measurement of the intensity of their predominate γ-rays at 685.5 keV (31.6%) and 155.0 (15.4%)[14] keV (from ^{188}Re daughter), respectively. For the determination of ^{194}Os, the 328.4 (13.0%) keV γ-ray (from ^{194}Ir) was used. All other relevant nuclear data were taken from references 15 and 16.

RESULTS AND DISCUSSION

The experimental yields of ^{188}W from four reactors (HFIR, HFBR, MURR, and FFTF) and available maximum neutron fluxes are given in Table 1. The large-scale production yield of ^{188}W at EOB from the HFIR hydraulic tube at 80 MW power and for one cycle irradiation ($\sim$ 21 d) at a thermal neutron flux of 2×10^{15} n.s^{-1}cm^{-2} is 4 mCi/mg of ^{186}W which is lower than the theoretical value by almost a factor of five. However, for a one day irradiation at the HFBR the yield of ^{188}W is lower than the theoretical value by a factor of two. The extent of discrepancy between the theoretical and experimental values cannot be totally attributed to the neutron self-shielding in the target material since the effect was found to be insignificant in the HFBR experiment in which two targets of 5 and 8 mg were irradiated together and the induced activities of ^{188}W were found to vary within 2% (Table 1). The higher yield of ^{188}W from the HFBR versus HFIR, in spite of lower thermal neutron flux, is indicative of some contribution from resonance neutron absorptions in the epithermal region. The neutron flux spectrum is harder in the HFBR core. The low yield of 0.3 mCi/mg at MURR reflects the lower neutron flux. As indicated in Table 1, the yield of ^{187}W at EOB is about 850 times higher than ^{188}W for a 21-d irradiation in the HFIR.

Relative to the cross-section of 1.28 mb for ^{58}Fe(n,γ)^{59}Fe flux monitoring reaction, a cross-section of $\sim$ 250 b was calculated for the ^{193}Os[n,γ]^{194}Os reaction. A summary of the reported cross-section for this reaction is given in Table 2. Based on our data 12 mCi of ^{194}Os can be produced by irradiating 25 mg of enriched ^{192}Os for 60 d at HFIR ($\phi_n = 2 \times 10^{15}$ n.cm^{-2}.s^{-1}) and the production of ^{194}Os at the HFIR is currently being explored.

ACKNOWLEDGEMENTS

Research supported by the Office of Health and Environmental Research, U.S. Department of Energy, under contract DE-AC05-84OR21400 with Martin Marietta Energy Systems, Inc. The authors thank Ms. L. S. Ailey for secretarial assistance.

REFERENCES

1. Blachot J., Hermment J. and Moussa, A., *Appl. Radiat. Isot.*, 20, 467 (1968).

2. Muxeeb H., Попobuy B., Pymep H., Cabcabcb Г., Boakoba H., *Isotopenpraxis*, 8, 248 (1972).

3. Ehrhardt G., Ketring A. P., Turpin T. A., Razavi M. S., Vanderheyden J.-L. and Fritzberg A. R., *J. Nucl. Med.* 28, 656 (1987).

4. Callahan A. P., Rice D. E. and Knapp, F. F., Jr., *NucCompact-Eur./Amer. Commun. Nucl. Med.* 20, 3 (1989).

5. Kodina G., Tulskaya T., Gureev E., Brodskaya G., Gapurova O. and Drosdovsky B., *Production and Investigation of Rhenium-188 Generator*, in Technetium and Rhenium in Chemistry and Nuclear Medicine 3 (M. Nicolini and G. Bandoli, editors), Corina International, (1990), pp 635-641.

6. Mani R. S., *Radiochimica Acta*, 41, 103 (1987).

7. Mirzadeh S., Callahan A. P., and Knapp, F. F., Jr., *Iridium-194 — A New Candidate for Radioimmunotherapy (RAIT) from an Osmium-194/Iridium-194 Generator System*, 38th Annual Meeting of the Society of Nuclear Medicine, Cincinnati, Ohio, June 11-14, 1991, *J. Nucl. Med.*, (1991), in press.

8. Venkatesan P. P., Shortkroff S., Zalutsky M. R., and Sledge C. B., *Nucl. Med. Biol.*, 17, 357 (1990).

9. Griffiths G. L., Knapp F. F., Jr., Callahan A. P., Chang C.-H., Hansen H. J., and Goldenberg D. M., *Cancer Research*, (1991) submitted.

10. Linder M., *Phys. Rev.*, 84, 240 (1951).

11. Williams D. C. and Naumann R. A., *Phys. Rev.*, 134 B289 (1964).

12. Report BNL-325 3rd ed. (Mughabgab S. F. and Carber D. J.) Vol. 1, U.S. Government Printing Office, Washington D.C. (1973).

13. Casten R. F., Namenson A. I., Davidson W. F., Warner P. D. and Borner H. G., *Physics Letters*, 76B, 280 (1978).

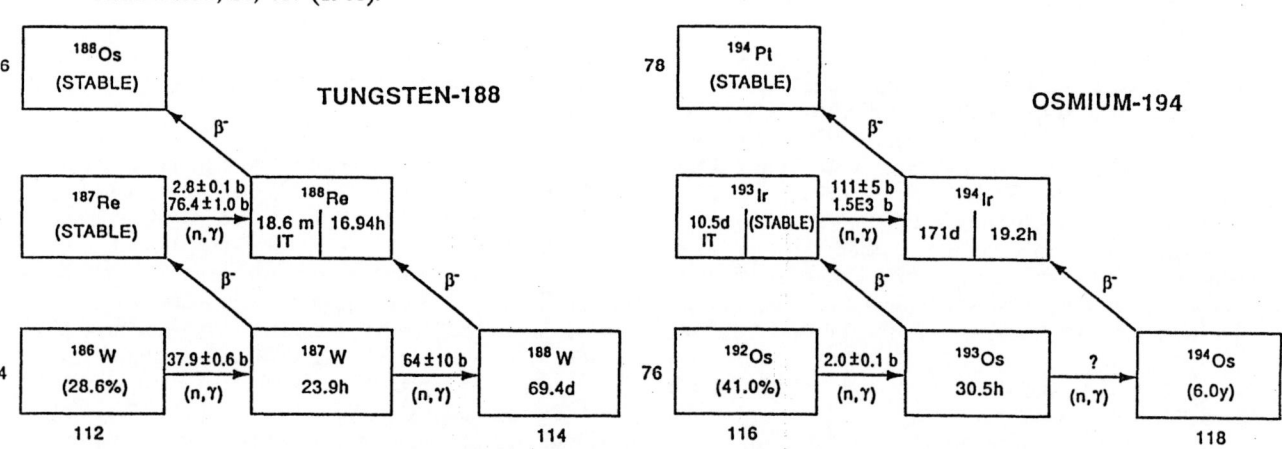

Figure 1a&b. Production Schemes of ^{188}W and ^{194}Os in a Nuclear Reactor

Table 1. Production of Tungsten-188

Reactor/ Irradiation position	Power MW	Neutron Flux, n.s⁻¹.cm⁻² Thermal (2200 ms⁻¹)	Fast (≥ 1 MeV)	Target mass, g as WO_3 (Enrichment, %)	Irradiation Period	Yield, mCi/mg at EOB ^{187}W	^{188}W
HFIR-ORNL HT position #5	100	2.5×10^{15}	1×10^{15}	10.3 (97.3)	21 d	NM	4.49
" #5	80	2×10^{15}	8×10^{14}	49.2 (96.07)	19.5 d	NM	3.94
" #3	80	2×10^{15}	8×10^{14}	50.2 (96.07)	21.1 d	3.25×10^3	3.88
" #5	80	2×10^{15}	8×10^{14}	50.2 (96.07)	~40 d [b]	NM	6.22
HFBR-BNL V-14(core-edge)	60	8.25×10^{14}	9.0×10^{13}	4.97 (97.06)	24.0 h	8.8	2.62×10^{-2}
				8.03 (97.06)	24.0 h	NM	2.57×10^{-2}
MURR Flux Trap	5	4.5×10^{14}	?	92.1 (96.07)	63.4 d	NM	~ 0.3
FFTF-Hanford	291	$(1-2) \times 10^{14}$ [a]	-	14.0 (96.07)	10 d [c]	-	3.89×10^{-2}
				12.7 (96.07)	10 d	-	3.81×10^{-2}

[a] Epithermal flux, [b] Several short and a 3-d shut down, [c] See reference 17
EOB = End of Bombardment, HT = Hydraulic Tube, NM = not measured

Table 2. Thermal Neutron Capture Cross-section of ^{193}Os

Reactor/ Irradiation position	Thermal Neutron Flux, (2200 ms⁻¹) n.s⁻¹cm⁻²	Target mass, g as Os metal (Enrichment, %)	Irradiation Period	Effective Cross-section, b	References
HFBR-BNL	4.5×10^{14} *	27.1 (99.3)	24 d	240[#]	this work
ORNL-Graphite	~1×10^{12}	-	-	200	Lindner (1951)[10]
MTR	3×10^{14}	50	150 d	8	Williams (1964)[11]
?	?	?	?	1540	Mughabghab (1973)[12]
ILL/GAMS1-3	8×10^{14}	84 (99.1)	on-line	38±10	Casten (1978)[13]

*Measured   [#]Relative to the cross-section of 1.28 mb for ^{58}Fe$(n,\gamma)^{59}$Fe flux monitoring reaction

14. Coursey B. M., Calhoun J. M., Cessna J., Hoppes D. D., Golas D. B., Schima F. J., Unterweger M. P., Callahan A. P., Mirzadeh S., and Knapp F. F., Jr., *Radioactivity and Radiochemistry*, 3, 38 (1990).

15. Table of Isotopes, (Lederer C. M. and Shirley V. S. Eds.), 7th ed, Wiley, New York (1978).

16. Neutron Cross Sections, (Mughabghab, S. F., Divadeenam M. and Holden N. E.), Vol. 1, part A&B, Academic Press, (1981).

17. Schenter R. E. *et al.* (1990) and Report OSU-NE-MED-9004 (Binney S. E. *et al.*), *Analysis of ^{188}W/^{188}Re from MIP Test*, (1990).

598

FISSION NEUTRON SPECTRUM AVERAGED CROSS SECTIONS OF SOME THRESHOLD REACTIONS ON MOLYBDENUM AND NICKEL: RELEVANCE TO THE PRODUCTION OF ^{99}Mo AND ^{57}Co

J.H. Zaidi, H.M.A. Karim, M. Arif and I.H. Qureshi

Nuclear Chemistry Division
Pakistan Institute of Nuclear Science and Technology
Islamabad, Pakistan

S.M. Qaim and G. Stöcklin

Institut für Chemie 1 (Nuklearchemie)
Forschungszentrum Jülich GmbH
5170 Jülich, Federal Republic of Germany

Abstract: Fission neutron spectrum averaged cross sections were measured by the activation technique in combination with radiochemical separations and high-resolution γ-ray spectroscopy for the low-yield reactions ^{95}Mo(n,p)^{95}Nb, ^{98}Mo(n,α)^{95}Zr, ^{58}Ni(n,2n)^{57}Ni and ^{58}Ni(n,n'p+pn+d)^{57}Co. The cross sections lie between 5 and 253 μb. The relevance of those data to the production of ^{99}Mo and ^{57}Co is discussed. The amounts of long-lived ^{95}Nb and ^{95}Zr in the (n,γ)-produced ^{99}Mo-^{99m}Tc generator are only of some significance from the viewpoint of waste disposal. A radiochemical separation of ^{57}Ni followed by removal of its decay product can lead to high purity ^{57}Co in small amounts.

(Fission neutron spectrum, threshold reaction, cross section, long-lived impurity, ^{99}Mo-^{99m}Tc generator, ^{57}Co activity)

Introduction

The radioisotope ^{99m}Tc (T$_{1/2}$ = 6.0 h; IT = 100 %; E$_\gamma$ = 141 keV; I$_\gamma$ = 89 %) is commonly used in diagnostic nuclear medicine. It is generally obtained via the ^{99}Mo-^{99m}Tc generator system. The parent radioisotope ^{99}Mo (T$_{1/2}$ = 66 h) is produced either in the fission of ^{235}U or via the ^{98}Mo(n,γ)^{99}Mo reaction on natural molybdenum. In the latter case, besides (n,γ) processes, fast neutrons induce ^{95}Mo(n,p)^{95}Nb and ^{98}Mo(n,α)^{95}Zr reactions which lead to the formation of long-lived products. An estimate of those impurities was highly desirable.

The radioisotope ^{57}Co (T$_{1/2}$ = 270 d; EC = 100 %; E$_\gamma$ = 122 keV; I$_\gamma$ = 85.6 %), tagged with vitamin B$_{12}$ is used in nuclear medicine for gastro-intestinal absorption studies, etc. Furthermore, it finds application in Mössbauer spectroscopy. It is produced presently via the ^{56}Fe(d,n)^{57}Co reaction at a cyclotron. Its production in a nuclear reactor was not considered. The nuclear processes to be investigated include

$$^{58}\text{Ni(n,2n)}^{57}\text{Ni}\xrightarrow[36\text{ h}]{\text{EC,}\beta^+}{}^{57}\text{Co, and}$$
$$^{58}\text{Ni(n,n'p+pn+d)}^{57}\text{Co.}$$

Experimental

Cross sections were measured by activation and identification of the radioactive products. The pertinent techniques have been described in several publications [cf. 1-3]. Here we give only some salient features.

Irradiations

High purity MoO$_3$ (99.999 % Puratronic, Johnson Mathey Chemicals Limited, England) or 99.66 % enriched ^{58}Ni powder (Isotec Inc., Ohio, USA), each weighing 25 mg to 100 mg, was filled in a quartz ampoule, which was sealed, studded with neutron flux monitor foils and then placed in an NRX type irradiation capsule. Irradiations were performed in the core of the 5 MW swimming pool type Pakistan Research Reactor (PARR-I) at PINSTECH.

The fast neutron flux densities effective at the ampoules were determined using the neutron threshold reactions ^{54}Fe(n,p)^{54}Mn, ^{56}Fe(n,p)^{56}Mn, ^{58}Ni(n,p)^{58}Co and ^{60}Ni(n,p)^{60}Co [cf. 4]. Through measurement of the ratio $\emptyset_f/\emptyset_{th}$ for monitor reactions having different thresholds it was concluded that the neutron spectrum used in these irradiations was approximately a fission spectrum. A similar conclusion was reached via multiple foil activation and spectrum unfolding technique [5].

Radiochemical separations

Niobium and zirconium were first separated from the matrix activity (containing large quantities of ^{99}Mo and ^{99m}Tc) by scavenging with Fe(OH)$_3$. The precipitate was then dissolved in conc. HCl and radioniobium extracted with di-isopropylketone. The aqueous phase was diluted to 2 M HCl and zirconium extracted with 2-thenoyltrifluoroacetone (TTA) solution in benzene (for details cf. [3]).

Cobalt and nickel were separated radiochemically from the irradiated ^{58}Ni. The target was dissolved in aqua regia, evaporated to dryness and redissolved in 9 M HCl. The mixture was loaded on a pretreated Dowex 1x8 resin column (Cl$^-$ form, 100-200 mesh). Nickel was eluted with 9 M HCl and cobalt with 4 M HCl.

Counting

The radioactivity of the reaction products and the monitor reactions was determined by γ-ray spectroscopy using a co-axial 30 cm^3 Ge(Li) detector in conjunction with a 4K Canberra Series 85 (model 8503) multichannel analyzer. The counting efficiency of the detector was determined using standard sources calibrated to ± 2 %.

Calculation of cross sections

The count rates of the chemically separated radionuclides were converted to decay rates by applying corrections for radiochemical yields, intensities of γ-rays, efficiency of the detector, geometry of counting, absorption, self-absorption, pile-up effects and coincidence losses. Cross sections were calculated using the well known activation formula. The main sources of errors have been described earlier [cf. 2,3].

Results and Discussion

Cross section data

The nuclear reactions investigated, their Q-values (calculated from the mass excesses given in [6]) and the measured cross sections are given in Table 1. The errors incorporate both statistical and systematic errors.

Our cross section for the ^{95}Mo(n,p)^{95}Nb reaction agrees with two previous values [4,7] but disagrees with a recent value [8]. In the case of ^{98}Mo(n,α)^{95}Zr reaction two values existed [4,7], which were, however, discrepant. Our value agrees with the result given in Ref. [4]. For the ^{58}Ni(n,2n)^{57}Ni reaction our cross section value agrees more or less with the previous data [4,9]. Two values existed on the ^{58}Ni(n,n'p)^{57}Co reaction [1,10]. Our measured value is somewhat higher.

Table 1.   Cross Sections of some Low Yield Nuclear Reactions Induced by Fission Neutrons on Molybdenum and Nickel

Nuclear reaction*	Q-value (MeV)	Cross section (μb)	
		This work	Literature [Ref.]
^{95}Mo(n,p)^{95}Nb †	-0.14	140 ± 10	140 ± 10 [4] 150 ± 10 [7] 200 ± 20 [8]
^{98}Mo(n,α)^{95}Zr	+3.20	14 ± 2	14 ± 2 [4] 8 ± 1 [7]
^{58}Ni(n,2n)^{57}Ni	-12.20	5.1 ± 0.3	4.9 ± 1.4 [4] 3.8 ± 0.5 [9]
^{58}Ni(n,n'p)^{57}Co**	-8.18	253 ± 15	216 ± 5 [10] 240 ± 35 [1]

*For studies on ^{58}Ni a highly enriched sample was used. Its % isotopic composition was: ^{58}Ni(99.66), ^{60}Ni(0.28), ^{61}Ni(0.01), ^{62}Ni(0.04) and ^{64}Ni(0.01).
† Cross section value for this reaction is the sum of ^{95}Mo(n,p)^{95m}Nb and ^{95}Mo(n,p)^{95g}Nb processes.
**Cross section value for this reaction is the sum of (n,n'p), (n,pn) and (n,d) cross sections. The Q-value for the (n,d) reaction is -5.95 MeV.

Table 2. Calculation of Radioactivity of ^{57}Co and Impurities under Production Conditions*

Target	Indirect production	Direct production [†]		
	^{57}Co activity KBq (μCi)**	^{57}Co activity KBq (μCi)	Radioactive impurities	
			^{58}Co MBq (mCi)	^{60}Co KBq(μCi)
^{nat}Ni	1.2(0.032)	150(4.0)	253(6.8)	81.6(2.2)
^{58}Ni	1.7(0.046)	221(6.0)	372(10)	0.86(0.02)

*Assuming weight of ^{nat}Ni-foil and enriched ^{58}Ni-powder = 1 g; $\emptyset_f$ = 6.7×10^{12} n cm^{-2}s^{-1}; σ-values as determined in this work and given in the literature; all values at end of bombardment.

**The indirectly produced ^{57}Co, i.e. via ^{58}Ni(n,2n)^{57}Ni $\xrightarrow[36h]{EC,\beta^+}$ ^{57}Co process, does not contain any impurities.

[†] The direct production route consists of ^{58}Ni(n,n'p+pn+d)^{57}Co processes.

Relevance of present data to the production of ^{99}Mo and ^{57}Co

The amounts of ^{99}Mo and the two major long-lived impurities, viz. ^{95}Nb (T$_{1/2}$ = 35.0 d) and ^{95}Zr (T$_{1/2}$ = 64.0 d), were calculated for an irradiation time of 120 h (for details cf. [3]). A nominal 2.78 GBq (75 mCi) ^{99}Mo-^{99m}Tc generator will contain 88 KBq (2.38 μCi) of ^{95}Nb and 7.4 KBq (0.20 μCi) of ^{95}Zr. Both radioniobium and radiozirconium remain adsorbed on the alumina generator column and are of no significance as far as elution of ^{99m}Tc is concerned. However, the activities need to be taken into account while disposing off the expired generator column.

The amounts of ^{57}Co and two long-lived impurities, viz. ^{58}Co (T$_{1/2}$ = 70.0 d) and ^{60}Co (T$_{1/2}$ = 5.26 y), were calculated for an irradiation time of 120 h. The results are given in Table 2. There are two routes for the formation of ^{57}Co activity: (i) indirect, i.e. through the decay of ^{57}Ni (T$_{1/2}$ = 36 h) which is formed via the ^{58}Ni(n,2n)-reaction, and (ii) direct, i.e. via the ^{58}Ni(n,n'p+pn+d)-process. The yield of ^{57}Co is much higher in the second case but the product contains appreciable amounts of ^{58}Co and ^{60}Co impurities. There is only about 40 % increase in the amount of ^{57}Co activity using enriched ^{58}Ni in comparison to natural target. The contamination from ^{60}Co, however, is almost 100 times reduced in case of enriched target.

Conclusion

The amounts of long-lived ^{95}Nb and ^{95}Zr are of some relevance only while disposing off the (n,γ) produced ^{99}Mo-^{99m}Tc generators. ^{57}Co of high purity cannot be easily produced using fission neutrons due to the competing ^{58}Ni(n,p)^{58}Co reaction on the same target nucleus. However, a radiochemical separation of the ^{58}Ni(n,2n)-reaction product ^{57}Ni, followed by removal of its decay product can lead to high purity ^{57}Co. At a high flux reactor amounts of about 100 KBq could be obtained.

REFERENCES

1. R. Wölfle and S.M. Qaim: Radiochim. Acta 27, 65 (1980)
2. J.H. Zaidi, H.M.A. Karim and S.M. Qaim: Radiochim. Acta 38, 123 (1985)
3. J.H. Zaidi, H.M.A. Karim, M. Arif, I.H. Qureshi, S.M. Qaim and G. Stöcklin: Radiochim. Acta 49, 107 (1990)
4. A. Calamand, in Handbook on Nuclear Activation Cross Sections, Technical Reports Series No. 156, p. 273, IAEA, Vienna, (1974)
5. A. Mannan, I.H. Qureshi, M.Z. Iqbal, R. Wölfle and S.M. Qaim: Radiochim. Acta 51, 49 (1990)
6. C.M. Lederer and V.S. Shirley: Table of Isotopes, 7th ed., Wiley, New York (1978)
7. I.M. Cohen, C. Magnavacca and G.B. Baro: Radiochim. Acta 34, 157 (1983)
8. O. Horibe: Radiat. Eff. 92, 517 (1986)
9. T. Sekine and H. Baba: J. Inorg. Nucl. Chem. 40, 1977 (1978)
10. A. Bruggeman, W. Maenhaut and J. Hoste: Radiochem. Radioanal. Letters 18, 87 (1974)

CROSS SECTIONS OF PROTON INDUCED NUCLEAR REACTIONS ON Kr GAS RELEVANT TO THE PRODUCTION OF MEDICALLY IMPORTANT RADIOISOTOPES ^{81}Rb AND ^{82m}Rb

Z. Kovács and F. Tárkányi

Institute of Nuclear Research of the Hungarian Academy of
Sciences, Debrecen, Hungary

S.M. Qaim and G. Stöcklin

Institut für Chemie 1 (Nuklearchemie),
Forschungszentrum Jülich GmbH, 5170 Jülich
Federal Republic of Germany

Abstract: Excitation functions were measured for the formation of medically important radioisotopes ^{81}Rb and ^{82m}Rb in proton induced nuclear reaction on Kr gas of different isotopic compositions. From the raw data absolute cross section values as well as differential and integral yields were calculated for the ^{82}Kr(p,n)^{82m}Rb, ^{82}Kr(p,2n)^{81}Rb, ^{83}Kr(p,n)^{83}Rb, and ^{83}Kr(p,2n)^{82m}Rb nuclear reactions. The optimum conditions for the production of ^{81}Rb and ^{82m}Rb are discussed.

(Proton induced reactions, cross section, excitation function, thick target yield, enriched ^{82}Kr and ^{83}Kr)

Introduction

The cyclotron produced radioisotopes ^{81}Rb ($T_{1/2}$=4.58 h) and ^{82}Rb ($T_{1/2}$= 1.2 m) are widely used in nuclear medicine, the former mainly through its short-lived daughter ^{81m}Kr ($T_{1/2}$=13.3 s) for lung ventilation and perfusion studies, and the latter for myocardial blood flow measurements. In both the cases generator systems are involved. Since ^{81}Rb is often contaminated with other rubidium isotopes, the ^{81}Rb/^{81m}Kr generator demands extra shielding. The ^{82}Sr/^{82}Rb generator, on the other hand, is relatively expensive. We considered it worthwhile to study the conditions for the production of ^{81}Rb in a pure state as well as to investigate the possibility of use of ^{82m}Rb, ($T_{1/2}$=6.5 h) as a longer-lived substitute for ^{82}Rb. In both the cases cross section measurements on enriched ^{82}Kr and ^{83}Kr were involved.

At the commencement of this work some yield data existed [1,2] for proton induced reactions only on ^{nat}Kr. In the meantime some cross sections have also been reported for ^{nat}Kr[3]. We report on cross section data measurements on natural and enriched Kr gas targets.

Experimental

Excitation functions were measured by the activation method using a "stacked gas cell" system. Stainless steel cells with 20 um titanium windows were filled with target gases ^{nat}Kr, 73 % enriched ^{82}Kr and 73 % enriched ^{83}Kr up to pressures of about 3 bars. 5 cells were irradiated in a row simultaneously for 30 min with protons of incident energies between 30.3 and 20 MeV at beam currents of about 200 nA. Cu foils were used for beam monitoring and energy degradation. The isotopic compositions of target gases as well as some other details have been described elsewhere [4].

After 3 hours the irradiated cells were evacuated and rinsed with RbCl solution. A 2 ml portion was taken and counted directly. The rest was precipitated as $Rb[B(C_6H_5)_4]$ and also measured by γ-ray spectrometry.

From the raw experimental data for the three sets of target gases of different isotopic compositions, effective cross sections were calculated for the formation of several rubidium isotopes. Therefrom the absolute cross sections were calculated by a mathematical analysis [4,5]. The errors of the absolute cross section values were estimated to be ±20 %.

Results

Detailed results have already been reported [4]. The absolute cross section values for the (p,n) and (p,2n) reactions on ^{82}Kr and ^{83}Kr are shown in Figs. 1 and 2, respectively. Since the data were deduced by calculation from results on 73 % enriched ^{82}Kr and ^{83}Kr, the excitation functions are depicted as smooth curves. These data were used to calculate the differential and integral yields of the radioisotopes concerned.

Optimum conditions for production of ^{81}Rb

The process involved in the production of ^{81}Rb is the ^{82}Kr(p,2n)^{81}Rb reaction. The optimum energy range is E_p=27 $\rightarrow$ 19 MeV which, for 100% enrichment of ^{82}Kr, gives a ^{81}Rb thick target yield of 48 mCi (1776 MBq)/μAh with only 7 % ^{82m}Rb impurity, originating from the (p,n) reaction. From the measured cross section data the yield of ^{81}Rb and the contamination level of other Rb isotopes can be estimated for any other energy range or isotopic composition of the target gas. Some yield values are given in Table 1.

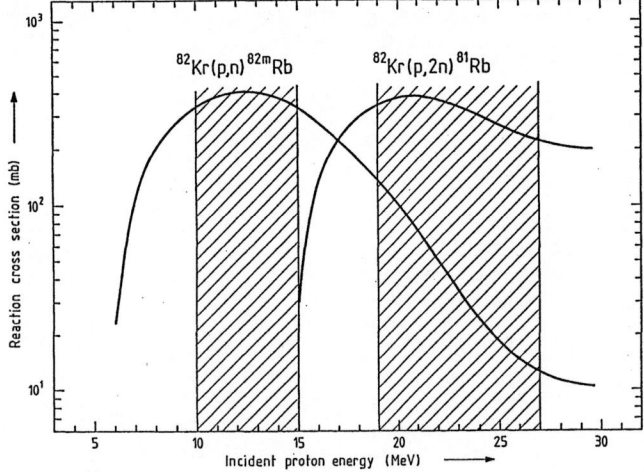

Fig.1. Absolute cross sections for (p,n) and (p,2n) reactions on ^{82}Kr. The optimum energy ranges for the production of ^{81}Rb and ^{82m}Rb are shown.

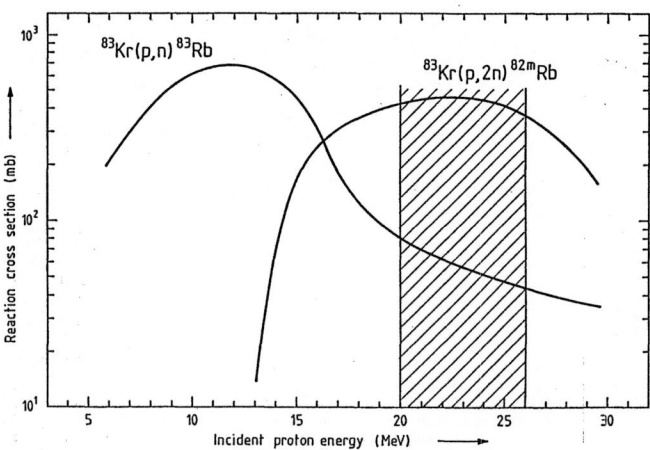

Fig.2. Absolute cross sections for (p,n) and (p,2n) reactions on ^{83}Kr. The optimum energy range for the production of ^{82m}Rb is shown.

Table 1. Nuclear process and energy ranges for the production of ^{81}Rb and ^{82m}Rb

Nuclear reaction	Energy range (MeV)	Product yield^x mCi(MBq)/μAh	Impurity (%)
^{82}Kr(p,2n)^{81}Rb	27 → 19	48 (1776)	^{82m}Rb (7)
	18 → 15	8 (296)	^{82m}Rb (68)
^{82}Kr(p,n)^{82m}Rb	15 → 10	10 (370)	^{81}Rb (0.01)
^{83}Kr(p,2n)^{82m}Rb	26 → 20	23 (850)	^{83}Rb (0.05)

^{x}Assuming 100 % enrichment of the target gas

Our detailed measurements on enriched ^{82}Kr show that ^{81}Rb can be produced also at a low energy cyclotron. From highly enriched ^{82}Kr gas the expected yield of ^{81}Rb over the energy range of 18 → 15 MeV is 8 mCi (296 MBq)/μAh. The level of contamination from ^{82m}Rb, however, is high. In comparison the expected yield via the ^{79}Br(α,2n)^{81}Rb process is only about 2 mCi (74 MBq)/μAh, even at high energy (E_α=30 MeV).

Optimum conditions for production of ^{82m}Rb

For the production of ^{82m}Rb at a low energy cyclotron ($E_p \leq 20$ MeV) the ^{82}Kr(p,n)-process is the method of choice; at a medium-sized cyclotron ($E_p \leq 30$ MeV), however, the ^{83}Kr(p,2n)-reaction is more suitable. The optimum energy ranges and yields are given in Table 1. We recommend the ^{82}Kr(p,n)^{82m}Rb process as the method of choice since the cost of highly enriched ^{82}Kr is lower in comparison to ^{83}Kr and since a low energy cyclotron is sufficient for production purposes.

Radiation dose estimates showed [6] that the dose from ^{82m}Rb to an individual organ is about 100 times higher than that from ^{82}Rb. On the other hand, the longer half-life of ^{82m}Rb may be of advantage with respect to the imaging time and

feasibility of extended studies. In absolute terms the dose from ^{82m}Rb is acceptable. It can therefore be used as a longer-lived substitute for the generator produced ^{82}Rb ($T_{1/2}$=1.2 min).

This work was supported by the German-Hungarian bilaterial agreement (No. 13(X237.1)).

REFERENCES

1. E. Acerbi, C. Birattari, M. Bonardi, C. Martinis and A. Solomone: Int J. Appl. Radiat. Isot. 32, 465(1981)

2. Mulders J.J.L. Int. J. Appl. Radiat Isot. 35, 475(1984)

3. G. F. Steyn, S. J. Mills, F. M. Nortier and F.J. Haasbroek: Appl. Radiat. Isot. 42, 361(1991)

4. Z. Kovács, F. Tárkányi, S. M. Qaim and G. Stöcklin Appl. Radiat. Isot. 42, 329(1991)

5. F. Tárkányi, S. M. Qaim and G. Stöcklin: Radiochimica Acta 43, 185(1988)

6. Z. Kovács, F. Tárkányi, S.M. Qaim and G. Stöcklin: Appl. Radiat. Isot. in press

EXCITATION FUNCTIONS OF PROTON INDUCED NUCLEAR REACTIONS ON [111]Cd AND[112]Cd. PRODUCTION OF [111]In.

F. Szelecsényi, F. Tárkányi, L. Andó, P. Mikecz
Institute of Nuclear Research of the Hungarian Academy of Sciences
H-4001 Debrecen, Hungary,

Gy. Tóth
Biomedical Cyclotron Laboratory, University Medical School
H-4032 Debrecen, Hungary,

P. Kopecký, A. Rýdl
Department of Radiopharmaceuticals, Nuclear Research Institute
250 68 Rez, CSFR

Abstract: Excitation functions were measured for the formation of ^{111}In in proton induced nuclear reactions on isotopically enriched [111]Cd and [112]Cd in the energy range up to 30 MeV. Absolute cross section values as well as differential and integral yields were deduced and compared with the earlier published data and with the experimental yield values obtained under production conditions. Routine production and chemical separation of ^{111}In are described.

(Enriched [111]Cd and [112]Cd target, proton induced reactions, excitation function, thick target yield, chemical separation)

Introduction

The radioisotope [111]In ($T_{\frac{1}{2}}$ =2.81 d) is one of the most widely used single photon emitters for medical application, mainly for antibody and cell labelling.

The excitation functions for its production have been studied previously mainly in terms of theory of nuclear reactions [1,2]. In the context of routine isotope production, cross section measurements have been carried out only on natural cadmium target [3].

The aim of this study was to complete and to extend the existing experimental data base from the point of view of large scale production (using enriched targets) sufficient for medical purposes.

Experimental

A detailed study of excitation functions was performed on enriched [111]Cd*([111]Cd/95.9 %/, [112]Cd/2.0 %/) and enriched [112]Cd* ([112]Cd/98.1 %/, [111]Cd/ 0.11 %/). Targets were prepared via electrolytic deposition on commercially available thin Ni** foils for cross section studies and on nickel plated thick Cu backing for production targets. Typical target thicknesses were 15 and 300 μm, respectively. The target preparation was similar to that described in detail in [4]. The excitation functions were measured using stacked foil technique. Irradiations were carried out with protons at the MGC-20E cyclotron in Debrecen (18 MeV incident energy) and at the U-120M cyclotron in Rez (22 and 30 MeV incident energies). The beam currents were monitored by a Faraday-cup

* Supplied by Technabexport, Moscow, USSR
**Supplied by Goodfellow Metals, Cambridge, England

and via monitor reactions induced in the Ni backings and [nat]Cu foils inserted into the stacks [5,6,7]. The energy of the incident beam was determined with magnetic deflection and the beam energy degradation was followed via monitor reactions as described in [8]. The activity of the targets and the monitor foils was measured via standard high resolution gamma-ray spectrometry. The cross sections were calculated using the well known activation formula. The overall uncertanties were estimated from the sums of the individual errors.

Excitation Functions

Excitation functions have been determined for the [111]Cd(p,n) [111]In, [112]Cd(p,2n)[111]In production reactions and for the [111]Cd(p,p')[111m]Cd, [111]Cd(p,2n)[110]In, [111]Cd(p,2n)[110m]In, [111]Cd(p,3n)[109]In, [112]Cd(p,pn)[111m]Cd, [112]Cd(p,3n)[110]In and [112]Cd(p,3n)[110m]In side reactions. Fig.1(a) gives the measured effective cross sections as a function of energy for the production of ^{111}In from the used (not fully enriched) targets. The absolute cross section for the [111]Cd(p,n)[111]In and [112]Cd(p,2n) [111]In reactions derived on the basis of target enrichments and the fitted effective cross sections are presented in Fig.1(b). The results shown in Fig.1(a) and Fig.1(b) have maximal errors of ±15 % and ±20 %, respectively. The obtained absolute cross section values in the overlapping energy region are in overall in good agreement with the data published earlier in [1,2], however, the energy scale of the results of Tanaka et al. [1] and Skakun et al. [2] seems to be shifted by about 1 MeV towards lower energy.

From the cross sections of the

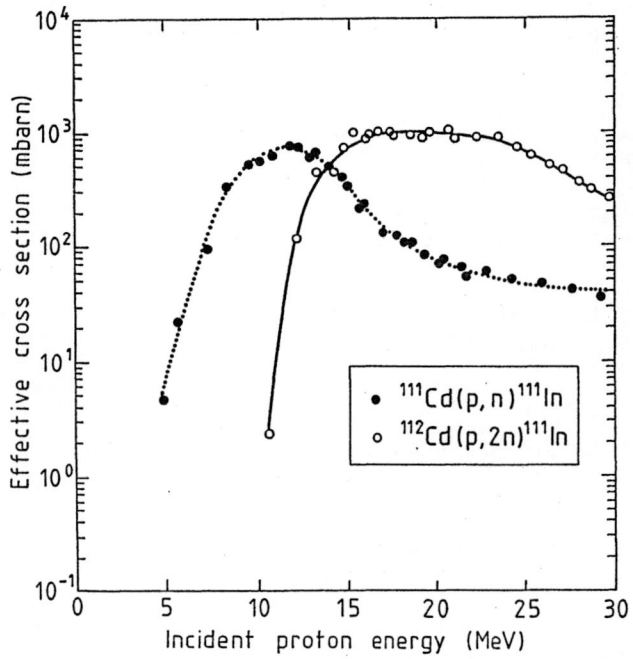

Fig.1(a). Excitation functions for the formation of ^{111}In in proton induced reactions on ^{111}Cd and ^{112}Cd of enrichments given in the text.

Fig.1(b). Absolute cross sections for the formation of ^{111}In in proton induced reactions on ^{111}Cd and ^{112}Cd derived from the effective cross sections.

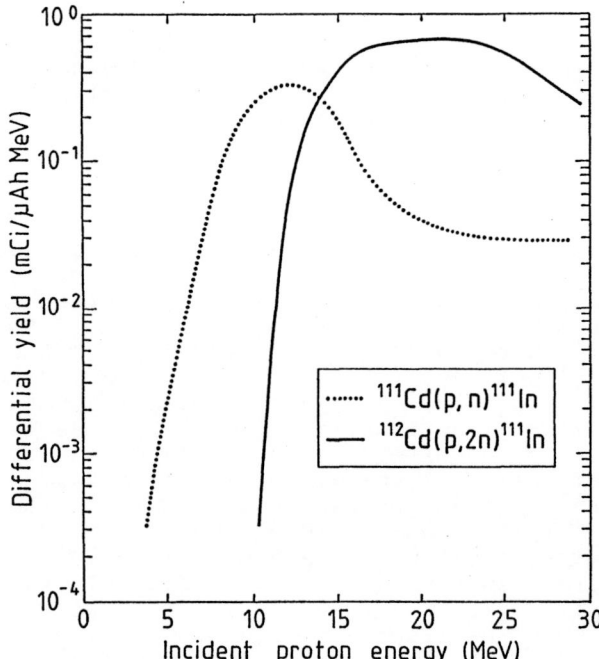

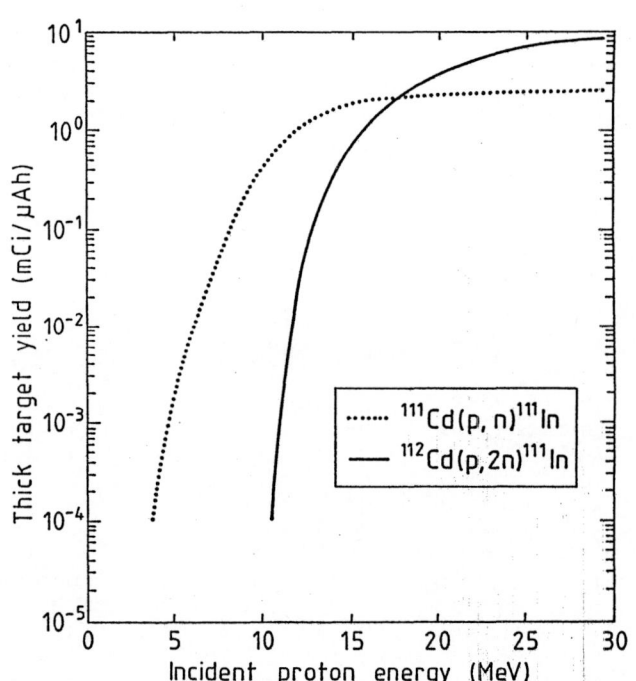

Fig.2(a). Differential yield of ^{111}In as a function of proton energy on ^{111}Cd and ^{112}Cd target nuclei.

Fig.2(b). Integral thick target yield of ^{111}In for protons on ^{111}Cd and ^{112}Cd.

measured reaction channels the impurity levels could be deduced for actual production conditions using thick targets. Isomeric ratios for $^{110m}In/^{110g}In$ could also be calculated for studies of reaction mechanism.

Thick Target Yields

The differential yields and the integral thick target yields of $^{111}Cd(p,n)$ ^{111}In and $^{112}Cd(p,2n)$ ^{111}In nuclear reactions calculated from the excitation functions as a function of the incident energy are shown in Fig.2(a) and Fig.2(b). The experimental thick target yields obtained with low and high beam intensities and the measured impurity levels are in good agreement with the calculated values from the excitation functions. For the radionuclidic purity of the final product, the content of ^{109}In ($T_{\frac{1}{2}}$ =4.2 h), $^{110m,g}In$ ($T_{\frac{1}{2}}$ =4.94 h, 69 m) and ^{114m}In ($T_{\frac{1}{2}}$ =49.5 d) are important. Using highly enriched ^{111}Cd or ^{112}Cd only the $^{111}Cd(p,3n)^{109}In$, $^{111}Cd(p,2n)^{110m,g}In$ or the $^{112}Cd(p,3n)^{110m,g}In$ reaction channels are dominant. With proper choice of energy range and using a non-significant cooling time these impurities could be decrease to the accepted levels (^{110m}In < 0.1 %, ^{110g}In < 0.1 %, ^{114}In < 0.1 %). The excitation functions of contaminating reaction channels and the detailed calculations on impurity levels will be published in detail elsewhere.

Routine Production and Chemical Separation

From the obtained excitation functions and from the literature data in other possible production routes it could be concluded that at low energy cyclotrons the method of choice for the production of ^{111}In is the $^{111}Cd(p,n)$ process while above 18 MeV the $^{112}Cd(p,2n)$ reaction results in higher yield of ^{111}In. At the Debrecen cyclotron, in principle in the 18 → 7 MeV (545 μm) energy range 2.133 mCi/μAh (78.921 MBq/μAh) thick target yield could be achieved and at the Rez U-120M cyclotron in the energy range 30 → 15 MeV (1335 μm) this value reaches 7.768 mCi/μAh (287.42 MBq/μAh). Taking into account the price of the enriched target materials and the technical difficulties in the preparation of the required very thick targets, the practical energy range and consequently the yields are usually lower. However, it is still sufficient for production even at a low energy cyclotron.

For routine production two chemical separation methods - an ion exchange and the other solvent extraction - were developed and compared. In both the cases the irradiated target material was dissolved from the backing in concentrated HBr. For ion exchange the

resin Dowex 50 WX2 was used [9] . The solution was passed through the column, and washed with 8 M HBr. The absorbed radioindium was eluted by conc. HCl. The extraction was performed in a continous flow extractor system developed by us. The indium isotopes were extracted from a 6 M HBr solution by isopropyl ether, and the organic phase was washed by 6 M HBr saturated with isopropyl ether. The radioindium was reextracted into injection water. The chemical yields in both the methods were about 90 %.

References

1. K. Otozai, S. Kume, A. Mito, H. Okamura and R. Tsujino: Nucl. Phys. 80, 335 (1966)
2. E.A. Skakun, A.P. Klyucharev, Yu.N. Rakivnenko and I.A. Romanii: Izv. Akad. Nauk. SSSR, Ser. Fiz. 39, 31 (1975)
3. F.M. Nortier, S.J. Mills and G.F. Steyn: Int. J. Appl. Radiat. Isotopes 41, 1201 (1990)
4. F. Tárkányi, F. Szelecsényi, Z. Kovács, and S. Sudár: Radiochimica Acta 50, 19 (1990)
5. F. Tárkányi, F. Szelecsényi and P. Kopecky: Int. J. Appl. Radiat. Isotopes 42, 513 (1991)
6. A. Gütter: Nucl. Phys. A383, 98 (1982)
7. P. Kopecky: Int. J. Appl. Radiat. Isotopes 36, 657 (1985)
8. F. Tárkányi, F. Szelecsényi and S. Takács: Abstracts of the fifth Symp. on the Medical Application of Cyclotrons, May 1989, Turku, Finland (Edited by S.-J. Heselius) p.29 (1989)
9. F. Nelson and D.C. Michelson: J. Chromatog. 25, 414 (1966)

CROSS SECTIONS FOR THE 100 MEV PROTON-INDUCED NUCLEAR REACTIONS
AND YIELDS OF SOME RADIONUCLIDES USED IN NUCLEAR MEDICINE

N.G.Zaitseva, C.Deptula, O.Knotek, Kim Sen Khan, S.Mikolajewski,
P.Mikeć, E.Rurarz, V.A.Khalkin, V.A.Konov*, L.M.Popinenkova*

Joint Institute for Nuclear Research,
Head Post Office, P.O.Box 79,
101000 Moscow, USSR

* Institute of High Energy Physics,
Protvino, Moscow Region, USSR

Abstract: Cross sections and yields have been determined for some radionuclides, frequently used in nuclear medicine (^{52}Fe, ^{77}Kr/^{77}Br, ^{82}Sr, ^{123}Xe/^{123}I, ^{128}Ba, ^{201}Tl), by bombardment of natural and enriched targets with protons of energy $\leq$ 100 MeV. The experimental cross sections are compared with the excitation functions calculated by the code Overlaid ALICE in the energy range from the thresholds up to 100 MeV. At the end of bombardment the thick target yields were as follows (mCi/μA·h, in the selected energy range): for ^{52}Fe from ^{55}Mn - 0.8 (36–100 MeV) and from ^{59}Co - 0.1 (64–100 MeV); ^{77}Kr from ^{79,81}Br - 313 (24–100 MeV); ^{82}Sr from ^{85,87}Rb - 0.43 (36–100 MeV); ^{123}Xe from ^{127}I - 270 (48–100 MeV); ^{128}Ba from ^{133}Cs - 8.4 (43–100 MeV); and for ^{201}Tl from: ^{206}Pb - 3.5 (48–57 MeV), ^{207}Pb - 3.0 (57–68 MeV), ^{208}Pb - 1.5 (68–76 MeV) 32 hours at the end of the bombardment.

(^{52}Fe/^{52m}Mn, ^{77}Kr/^{77}Br, ^{82}Sr/^{82}Rb, ^{123}Xe/^{123}I, ^{128}Ba/^{128}Cs, ^{201}Tl, stacked-foil technique, cross section, excitation function, isotopic enrichment, thick target yield, proton-induced nuclear reactions, radioactive impurity).

When investigating the conditions for production of radionuclides for nuclear medicine in reactions between target elements and accelerated ions, one usually concentrates on determining the dependence of the production cross section for nuclides on the energy of bombarding particles. The information on the excitation functions of the reactions is necessary for choosing both the target material and the energy intervals which allow the maximal yield of the required nuclides and the minimal unwanted isotope impurity. Knowing these parameters one can organize production of large amounts of radionuclides and radiopharmaceutical preparations on their basis |1|.

We determinated production cross sections and yields of some neutron-deficient radionuclides (Table 1) |2|. An analysis of the published data shows that the information on their production cross sections in the proton energy range from 40 to 100 MeV is scarce, if any.

Table 2 lists the reaction yield for the nuclides of interest. The excitation functions of these reactions were determined by the well-known stacked-foil method. The target materials were metals or their salts.

The targets were irradiated by a 100±0.5 MeV proton beam of intensity up to 2.8 μA, extracted from the linear accelerator LU–100 at the Institute of High Energy Physics (Serpukhov, Moscow Region). The exposure varied from 0.1 to 1 μA·h, depending on the task. The beam spot at the place of irradiation was about 6 cm², which allowed irradiation of 2–3 target assemblies at the same time. It should be mentioned that LU–100 is a reliable accelerator, which provides good irradiation conditions for different tasks including the study of nuclear reactions.

Our experimental data on the yields and maximum production cross sections of the radio-

Table 1. Decay data |3| of some nuclear reaction products used in cross section measurements

Nuclide ($T_{1/2}$)	Mode of decay branching (%)	Main γ-ray energies(keV) abundance (%)
^{52}Fe(8.3h)	β^+(56),EC(44)	168.7(99.2)
^{52m}Mn(21m)	β^+(98),EC(2)	1434.1(98.3)
^{77}Kr(1.2h)	EC(99.9),β^+(0.1)	129.7 (80.0) 146.4 (37.6)
^{77}Br(57h)	EC(99.3),β^+(0.7)	239(23.8) 520.7 (23.2)
^{82}Sr(25.5d)	EC(100)	–
^{82}Rb(1.27m)	β^+(95),EC(5)	776.5(13.4)
^{123}Xe(2.1h)	EC(87),β^+(13)	148.9(48.6)
^{123}I(13.2h)	EC(100)	159 (83.3)
^{128}Ba(2.4d)	EC(100)	273(14.5)
^{128}Cs(3.6m)	β^+(51),EC(49)	443 (25.8)
^{201}Tl(73.5h)	EC(100)	135.3(2.8) 167.4(10.6)

nuclides are given in Table 2. They are the arithmetical mean of several gamma-spectrometric measurements of nuclear activities in the irradiated targets from several independent trials. The standard deviations did not exceed 10%, the systematic error was estimated to be ±20%.

The cross sections, energies and yields determined are in good agreement with the results of other authors obtained under comparable conditions |4–12|. The theoretical excitation functions of the nuclear reactions were calculated by

Table 2. Targets, production cross section and yields of radionuclides in irradiation of targets with protons of $E_p \leqslant 100$ MeV

Chemical form of target and isotope abundance (%)	Nuclear reaction	Energy range (MeV)	σ_{max}, mb (E_p, MeV)	Thick target yield, mCi/μA·h
Mo_{met} $^{55}Mn(100)$	$^{55}Mn(p, 4n)^{52}Fe$	100–36	1.3(55)	0.8
Co_{met} $^{59}Co(100)$	$^{59}Co(p, 2p6n)^{52}Fe$	100–64	0.36(77)	0.1
KBr $^{79}Br(50.69)$ $^{81}Br(49.31)$	$^{79,81}Br(p, 3n) +$ $+ (p, 5n)^{77}Kr$	100–22	107(35)	313
RbCl $^{85}Rb(72.17)$ $^{87}Rb(27.89)$	$^{85,87}Rb(p, 4n) +$ $+ (p, 6n)^{82}Sr$	100–36	180(52)	0.43
NaI $^{127}I(100)$	$^{127}I(p, 5n)^{123}Xe$	100–48	350(57)	270
CsCl $^{133}Cs(100)$	$^{133}Cs(p, 6n)^{128}Ba$	100–43	322(63)	8.4
Pb_{met} $^{206}Pb(94)*$	$^{206}Pb(p, 6n) \rightarrow$	57–49	(60)	3.5
$^{207}Pb(89)**$	$^{207}Pb(p, 7n) \rightarrow$	68–57	(72)	3.0
$^{208}Pb(97.5)***$	$^{208}Pb(p, 8n) \rightarrow$	76–68	(83)	1.5
	$\rightarrow ^{201}Bi \rightarrow ^{201}Pb \rightarrow ^{201}Tl$			

Contents of other Pb isotopes (%):	^{204}Pb	^{206}Pb	^{207}Pb	^{208}Pb
*	0.01		4.04	1.96
**	0.01	2.06		8.24
***	0.08	0.82	1.6	

the programme ALICE |13| on the computer CYBER in the Institute of Nuclear Research (Swierk, Warsaw).

The experimental and calculated cross sections were in good agreement for the reactions $^{55}Mn(p,4n)^{52}Fe$, $^{85+87}Rb(p,4n+6n)^{82}Sr$ and $^{79+81}Br(p,3n+5n)^{77}Kr$ at proton energies above 60 MeV, 50 MeV and 30 MeV respectively (Figs 1–3). The cross sections calculated for the reaction $^{133}Cs(p,6n)^{128}Ba$ (Fig. 4) appeared to be 2–3 times larger than the experimental ones at the proton energy above 50 MeV, but both the excitation functions have practically the same shape.

The target for irradiation of separated lead isotopes ^{208}Pb, ^{207}Pb and ^{206}Pb foils was assembled as shown in Fig. 5a. Thus, in the same experiment we could find the dependence of the cumulative yields of ^{200}Tl, ^{201}Tl and ^{202}Tl radioisotopes in the reactions (p,xn), (p,pxn) and (p,αxn) in the 48–100 MeV range. The proton energies corresponding to the maxima of the curves shown in Fig. 5 b,c,d are 8–10 MeV higher than those calculated by the ALICE programme for the reactions like (p,xn).

Our measurements show that at the proton energy from 76 MeV to 50 MeV the yield of ^{201}Tl from the target of 95%–enriched lead-208, 207, 206 can be about 8 mCi·$\mu A^{-1}h^{-1}$ with the ^{200}Tl and ^{202}Tl impurities being not more than 4.5% and 0.5%, respectively.

The results of our experiments can be helpful in production of nuclear medical isotopes at proton accelerators of energy $\leqslant$ 100 MeV.

The excitation functions calculated for different nuclear reactions are also of interest. Their comparison with the experimental ones shows that they allow the yields of the radio-

nuclides with unknown production cross sections to be estimated to within an order of magnitude.

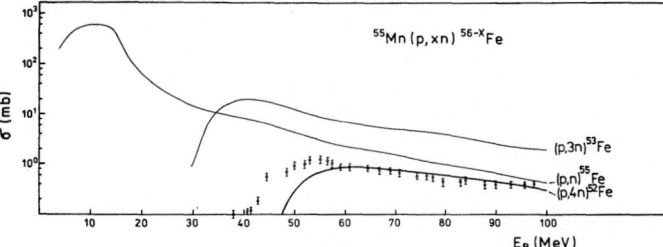

Fig. 1. Measured (points) and calculated (curves) excitation functions of the reactions $^{55}Mn(p,xn)^{56-x}Fe$, x = 1,3,4.

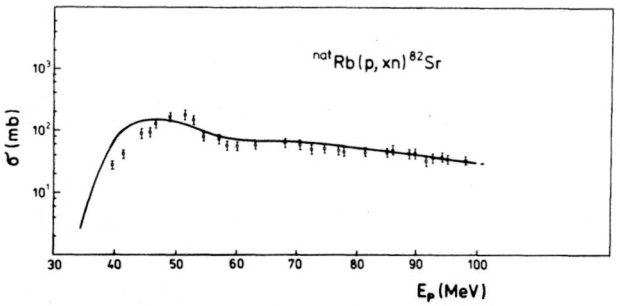

Fig. 2. Measured (points) and calculated (curves) excitation functions of the reactions $^{nat}Rb(p,xn)^{82}Sr$, x = 4 and 6.

608

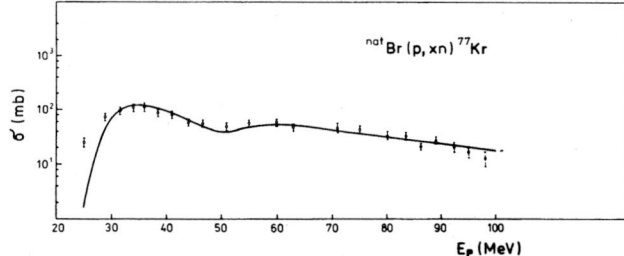

Fig. 3. Measured (points) and calculated (curves) excitation functions of the reactions $^{nat}Br(p,xn)^{77}Kr$, x = 3 and 5.

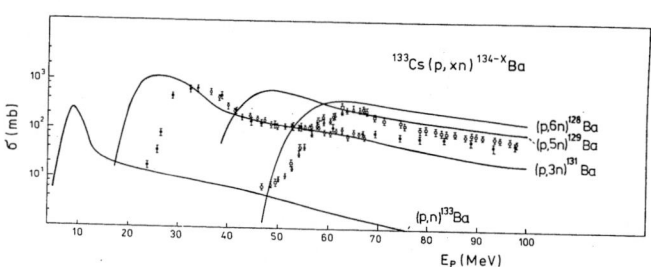

Fig. 4. Measured (points) and calculated (curves) excitation functions of the reactions $^{133}Cs(p,xn)^{134-x}Ba$, x = 1, 3,5,6; ▢ , ● — the present paper, △ , ○ — Ref. |11|.

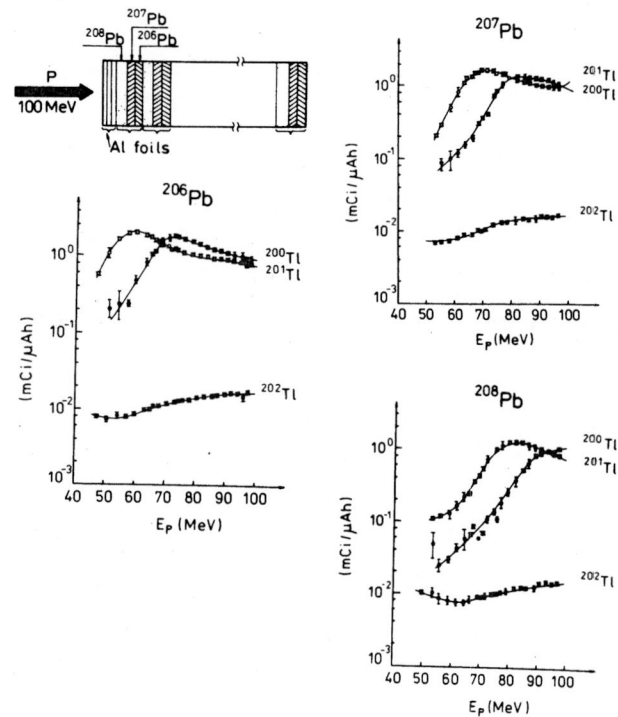

Fig. 5. a) Position of the Al and Pb foils in the target with respect to the proton beam direction; b),c),d) yields of $^{200,201,202}Tl$ from the enriched ^{206}Pb, ^{207}Pb and ^{208}Pb targets plotted against the proton energy.

References

1. Qaim, S.M.: Radiochim. Acta 30, 147 (1982)
2. Zaitseva, N.G., Deptula, C., Knotek, O., Kim Sen Khan, Mikolajewski, S., Mikeć, P., Rurarz, E., Khalkin, V.A., Konov, V.A. and Popinenkova, L.M.: Radiochim. Acta 54, 57 (1991)
3. Reus, U., Westmeier, W.: Gamma–Ray Catalog from Radioactive Decay. Atomic Data and Nuclear Data Tables 29, Parts I, II (1983)
4. Suzuki, K.: Radioisotopes 34, 537 (1985)
5. Sharp, R.A., Diamond, R.M., Wilkinson, G.: Phys. Rev. 101, 1493 (1956)
6. Weinreich, R., Knieper, J.: Int. J. Appl. Radiat. Isot. 34, 1335 (1983)
7. Qaim, S.M., Stöcklin, G., Weinreich, R.: Int. J. Appl. Rad. Isotopes 28, 947 (1977)
8. Dmitriev, P.P.: Radionuclide Yield in Reactions with Protons, Deuterons, Alpha Particles and Helium–3. (Handbook) INDC (CCP)–263/G+CN+SZ, p.23, IAEA: Vienna (1986)
9. Horiguchi, T., Noma, H., Toshizawa, Y., Takemi, H., Hasai, H., Kiso, Y.: Int. J. Appl. Rad. Isotopes 31, 141 (1980)
10. Lagunas–Solar, M.C., Carvacho, O.F., Liu, Bo–Li; Jin, Y., Sun Zhao Ziang: Appl. Radiat. Isot. 37, 823 (1986)
11. Lagunas–Solar, M.C., Little, F.S., Moore, H.A.: Int. J. Appl. Rad. Isotopes 33, 619 (1982)
12. Zaitseva, N.G., Kovalew, A., Knotek, O., Khalkin, V.A., Ageev, V.A., Kluchnikov, A.A., Linev, A.F.: Report JINR P6–85–254, Dubna (1985); Radiochem. (in Russian) 29, 247 (1987)
13. Blann, M.: Overlaid ALICE Code, CCO/3494–29, Rochester (1976)

AN ANALYSIS OF REACTION CROSS-SECTION CALCULATIONAL METHODS FOR THE PRODUCTION OF MEDICAL RADIOISOTOPES

V.P.Lunev, N.V.Kurenkov*, A.B.Malinin*,
V.S.Masterov, Yu.N.Shubin

Institute of Physics and Power Engineering, Obninsk, USSR
*Institute of Biophysics of USSR Ministry of Public Health,
Moscow 123182, USSR

Abstract: The results of calculations of charged particle induced reactions for the production of radioisotopes ^{123}I, ^{123}Cs and ^{123}Xe are discussed. The excitation functions for reactions ^{124}Xe(p,2n)^{123}Cs, ^{124}Xe(d,3n)^{123}Cs, ^{124}Xe(p,pn+np)^{123}Xe, ^{127}I(p,5n)^{123}Xe, ^{124}Xe(d,p2n)^{123}Xe, ^{123}Te(p,n)^{123}I, ^{124}Te(p,2n)^{123}I, and ^{121}Sb(α,2n)^{123}I are calculated on the basis of statistical model in energy range up to 80 MeV. The calculations and experiment are generally in agreement. It is shown, that consideration of the radiative channel is very important near threshold, particularly for neutron deficient nuclei and can drastically influence the results of calculations. The analysis performed showed that the calculations with code ALICE can server as a reasonable evaluation for the excitation functions of various reactions.

(Cross section, excitation function, ^{124}Xe, ^{123}Xe, ^{123}Te, ^{123}I, ^{123}Cs, ^{124}Te, ^{121}Sb, radioisotope production, statistical model, computer code)

1. Introduction

Owing to its suitable radioactive properties, ^{123}I is considered [1] as one of the best radionuclides for in vivo diagnostics. For its production more than twenty kinds of reactions have been proposed, but selection should be made from several essential points of view, that is, the cost of enriched isotope for a target, ease of recovery after irradiation, production yield rate, impurities of the other radioactive iodine isotopes in the final product. It should therefore be useful to evaluate the excitation functions for the reactions of interest. Such evaluations can be performed with the help of model calculations. One of the most widely used and convenient to operate computer code is ALICE [cf. 2-4], which enables to calculate the excitation functions of various reactions induced by light charged particles in energy range up to 200 MeV [4]. We used code ALICE-87 [5] to calculate the excitation functions for the production of ^{123}I, ^{123}Cs and ^{123}Xe in the reactions induced by protons, deuterons and α-particles and compared the results of calculations with experiment wherever available and with the calculations using other computer codes, based on the statistical model STAPRE [6] and GROGI-2 [7].

2. Results and Discussion

2.1 Reactions to produce ^{123}I

^{123}Te(p,n) reaction has been investigated since the beginning of 60s, but a detailed study has been performed only in 1988 [9]. The results of calculations of the excitation functions with code ALICE-87 give the maximum cross section value close to experimental data, although the whole curve is shifted to the smaller energy region. The calculated yield of ^{123}I in energy range 11-14.5 MeV is evaluated to be 3 mCi/μA·h, close to experimental value, also.

Fig.1 shows the results of excitation function calculations for ^{124}Te(p,n)^{124}I and ^{124}Te(p,2n)^{123}I reactions in energy range from 5 to 35 MeV, performed with code ALICE [5] (full line) and with STAPRE [6] (dashed line). The experimental data [10] are also given. It should be noted that ALICE calculations are in better agreement with the experimental data for both reactions, although there is large shift of the calculated curves to the small energies. The reaction on Sb have been the first ones, which have produced isotopes ^{123}I and ^{124}I. The results of calculations of excitation functions of ^{121}Sb(α,n)^{124}I, ^{121}Sb(α,2n)^{123}I, ^{123}Sb(α,3n)^{124}I and ^{123}Sb(α,4n)^{123}I - have been analyzed and compared with the available experimental data. The calculated excitation functions are in reasonable agreement with the experimental data on ^{121}Sb(α,n) reaction from [12], but differ considerably from experimental data from paper [11]. On the other hand, the experimental data on ^{121}Sb(α,n) reaction from [12] exceed considerably calculated results. For the ^{123}Sb(α,4n) reaction no experimental data exist.

2.2 Reactions to produce ^{123}Cs and ^{123}Xe

The excitation functions of ^{127}I(p,5n)^{123}Xe reaction have been measured repeatedly [cf.8]. There are considerable differences in the cross sections at maximum, which have been discussed in [13]. Fig. 2 shows the excitation functions calculated with code ALICE for ^{127}I(p,5n) reaction and also for ^{127}I(p,3n) and for ^{127}I(p,7n) reactions, which lead to admixtures of ^{125}I and ^{121}Te. We see that there is satisfactory agreement between calculation and experiment, especially if one shifts calculated curve for (p,5n) reaction to 3-4 MeV to larger energies. This observation was noted in [2]. However for ^{127}I(p,7n) reaction the shift is needed in the opposite direction. It should be noted that the calculation is in good agreement with

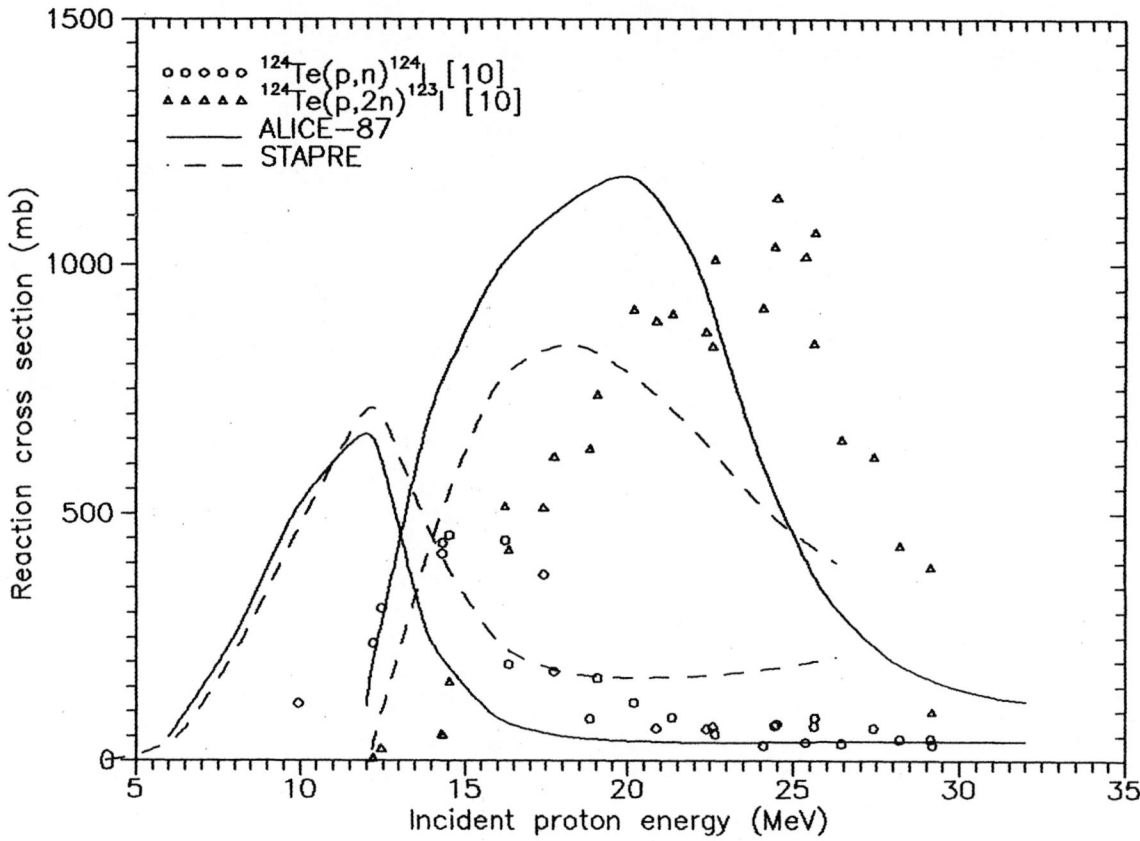

Fig. 1. Excitation functions of ^{124}Te(p,xn)^{123,124}I reactions [10].

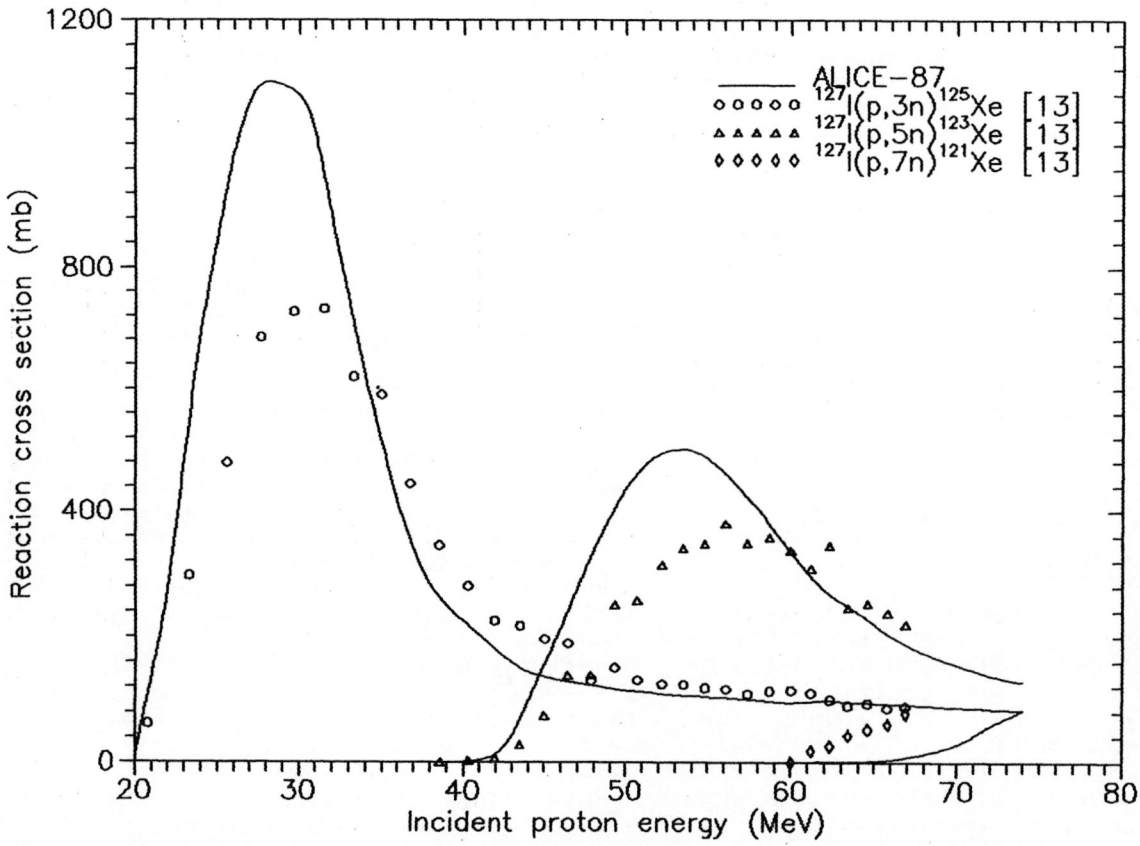

Fig. 2. Excitation functions for ^{127}I(p,xn)^{121,123,125}Xe reactions [13].

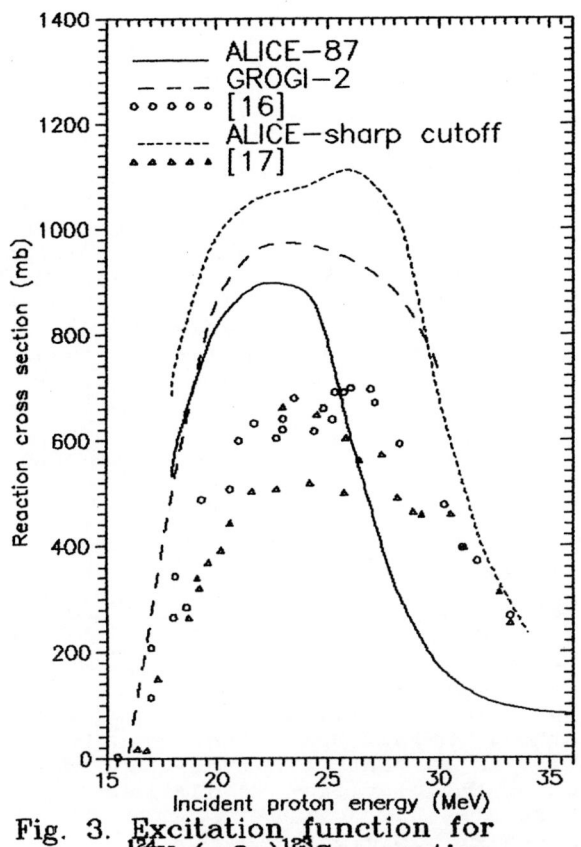

Fig. 3. Excitation function for ^{124}Xe(p,2n)^{123}Cs reaction.

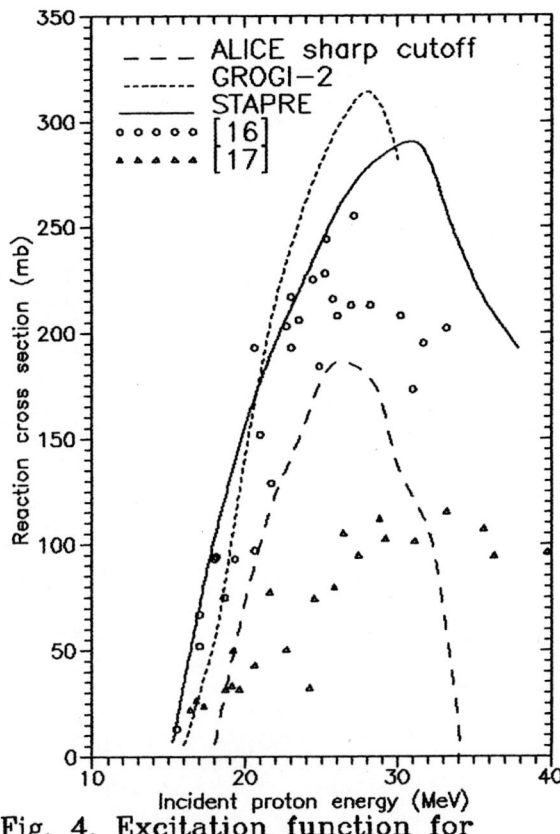

Fig. 4. Excitation function for ^{124}Xe(p,pn)^{123}Xe reaction.

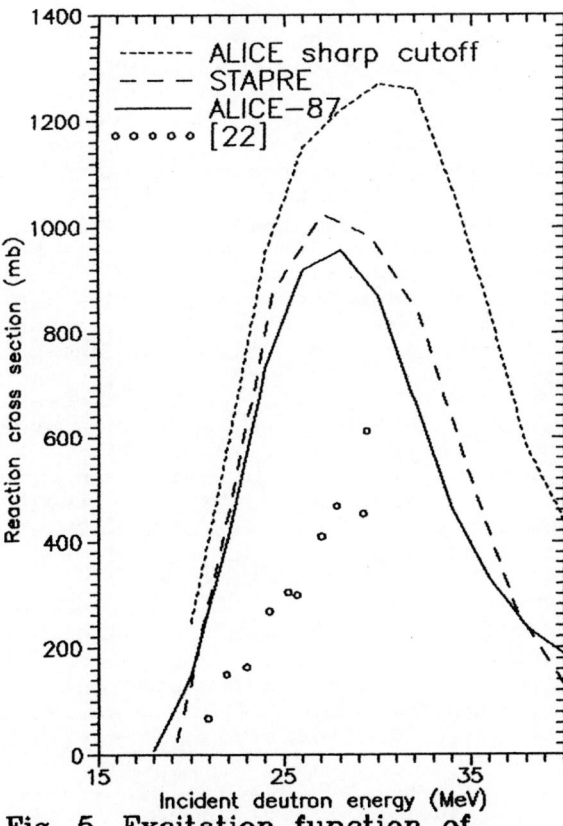

Fig. 5. Excitation function of ^{124}Xe(d,3n)^{123}Cs reaction.

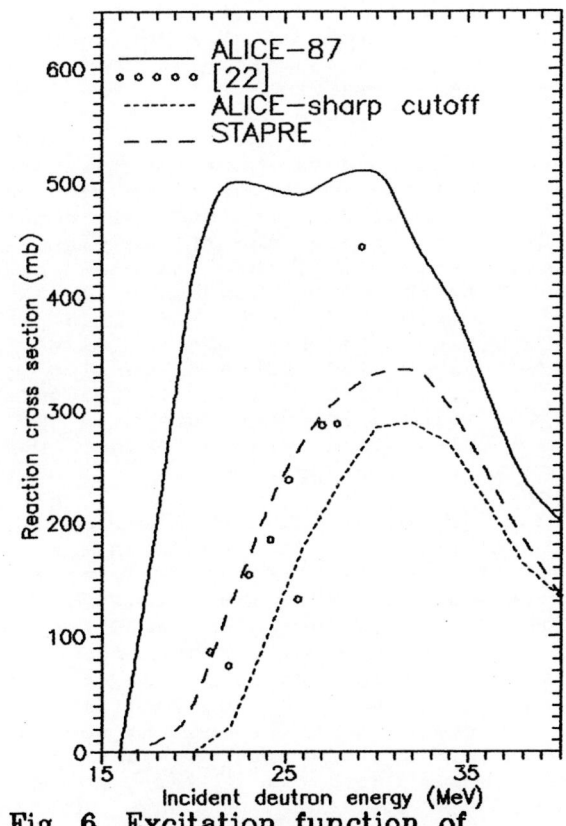

Fig. 6. Excitation function of ^{124}Xe(d,p2n)^{123}Xe reaction.

612

experimental data for (p,5n) reaction, obtained by Suzuki [14].

New possibilities arise to produce [129]I if the reactions on enriched [124]Xe are used [cf. 2]. The main advantages of these reactions - high production yield rate, ease of chemical process after irradiation, high isotope purity. Detailed calculations have been performed in [15] for a number of reactions on isotopes of Xe with the code GROGI-2. The production yield rates of [129]I and [129]Xe were determined, which have been confirmed later by experiment [16]. The results of calculations of [124]Xe(p,2n)[123]Cs reaction excitation functions with codes ALICE and GROGI-2 [15] are compared with the experimental data [16] in Fig.3. The results of another recent measurements [17] are also shown. We see better agreement between experimental data and ALICE calculations, although there is some difference of cross sections in maximum and shift of calculated curve to smaller energies.

However the situation is different if the results for [124]Xe(p,pn+np) reactions are considered. For this reaction the calculations with ALICE give a two-humped curve with maxima at energies near the threshold, which is hard to explain. We investigated the influence of the deuteron- and α-channel competition, the preequilibrium mechanism effect and other factors. The calculations were repeated with the codes GROGI-2 and STAPRE. After analysis we concluded that this problem is connected with the assumption about the competition between nucleons and γ-rays. This problem was discussed in paper [18] in detail. The calculations performed with GROGI-2 and STAPRE, taking into account γ-ray competition, show that for realistic radiative width values, normalized to the experimental neutron resonance radiative widths, the spurious peak near threshold disappears. It should be noted that for unreasonable low radiative width used for the calculations with STAPRE and GROGI-2 codes, results are analogous to ALICE results. Pearlstain showed [19] that the peak near threshold can be removed by using the classical cutoff option (INV=2) of ALICE. M. Blann [20] repeated this test in slightly different way, using standard version of ALICE (INV=0), but limiting the transmission coefficients in magnitude. In this case the calculations are in rather good agreement with the experiment, if $T_l \geq 10^{-2}$ (Fig.4).

The [124]Xe(d,3n)[123]Cs and [124]Xe(d,p2n)[123]Xe reactions have not been investigated much. Only recently differential yields of [129]I were measured on natural Xe [21]. The results of calculations with ALICE are compared with the experimental data for [124]Xe from paper [22] in Figs. 5 and 6. Considering the calculated and experimental data for these reactions it is easy to see almost complete analogy with the results of the comparison for the [124]Xe(p,2n) and [124]Xe(p,pn+np) reactions. The only difference is that the peak near threshold for the (d,p2n) reaction is not such sharp. However, it's origin reason is of the same

nature. The calculated production yield of [129]I in energy range 20-40 MeV for the (d,3n) reaction is about 30 mCu/μA·h.

3. Conclusion

The analysis performed shows that statistical model used for the excitation function description of the nuclear reactions, induced by light charged particles, gives satisfactory results. For the reactions such as (p,pxn) or (d,pxn) it is very important to take into account the radiative channel, in particular for the neutron deficient nuclei near threshold. It is considered that the calculations with the code ALICE can serve as a reasonable evaluation method for the excitation functions and production yield rates, especially if experimental data are available.

REFERENCES

1. Myers W.G., Anger H.O.:J.Nucl.Med., 3, 183 (1962)
2. Grabmayr P., Nowotny R. : Int.J.Appl. Rad.Isot., 29, 262 (1978)
3. Sueki K., Hashizume A., Tendow Y., Ohkubo Y., Kitao K.: IAEA Consultant's Meeting on "Data Requirements for Medical Radioisotope Production", 191. Tokyo (1978)
4. Blann M.: ibid., 115
5. Blann M.: ALICE-87 IAEA-NDS-93 (1988)
6. Uhl V., Stromaier B.: STAPRE Preprint IRK-76/01, Vienna, (1976)
7. Gilat J.: BNL-50246(T-580), (1970)
8. Hashizume A., Tendov Y., Kitao K. IAEA INDC(JPN)-144 (1990)
9. Scholten B., Qaim S.M., Stöcklin G. : Appl.Rad.Isot. 40, 127 (1989)
10. Kondo K., Lambrecht R.M.,Wolf A.: Int. J.Appl.Rad.Isot., 28, 395 (1977)
11. Watson I.A., Waters S.L., Silvester D. J.: J.Inorg.Nucl.Chem. 35, 3047 (1973)
12. Calboreanu A., Pencea C., Salagean O. Nucl.Phys., A383, 251 (1982)
13. Lagunas-Solar M.C., Carvacho O.F., Liu B.-L., Jin Y., Sun Z.X. : Appl.Rad. Isot., 37, 823 (1986)
14. Suzuki K. Radioisotopes 35, 235 (1986)
15. Juravlev B.V., Ivanova S.P., Krasnov N.N., Shubin Ju.N.: Atomnaja Energija 60, 337 (1986) In Russian
16. Kurenkov N.V., Malinin A.B., Sebyakin A.A., Venikov N.I.: J.Radioanal. Nucl. Chem., Letters, 135, 39 (1989)
17. Tarkanyi F., Qaim S.M., Stöcklin G. et. al.: Appl.Rad. Isot., 42, 221 (1991)
18. Shubin Yu.N.: Rep. in IAEA Advisory Group Meeting on Intermediate Energy Nuclear Data for Applications, Vienna, 1990
19. Pearlstain S.: Private communication, 1990
20. Blann M.,: Private communication, 1990
21. Tarkanyi F.,Kovacs Z.,Qaim S.M., Stöcklin G.: Radiochimica Acta 47, 25 (1989)
22. Malinin A.B., Sebyakin A.A., Venikov N.I. J.Radioanal.Nucl.Chem., Letters, in press.

EXCITATION FUNCTION AND YIELD FOR ^{97}Ru PRODUCTION
IN ^{99}Tc(p,3n)^{97}Ru REACTION IN 20 – 100 MeV PROTON ENERGY RANGE

N.G.Zaitseva *, E.Rurarz **, M.Vobecký *, Kim Hyn Hwan *, K.Nowak *,
T.Téthal *, V.A.Khalkin *, L.M.Popinenkova ***

* Joint Institute for Nuclear Research, Head Post Office, P.O.Box 79,
101000 Moscow, USSR

** Soltan Institute for Nuclear Studies, Swierk, Poland
*** Institute for High Energy Physics, Protvino, Moscow Region, USSR

Abstract: Proton-induced reactions on ^{99}Tc target were studied in order to evaluate the ^{99}Tc(p,3n)^{97}Ru reaction as a source of high radionuclidic purity ^{99}Ru.
Measurements of excitation functions for the main reaction ^{99}Tc(p,3n)^{97}Ru and
competing ^{99}Tc(p,pxn)^{99m,96,95}Tc reactions were performed in a stacked-foil
experiment using gamma spectrometry. The maximum cross section for ^{97}Ru
in the first reaction was found to be 440 mb ($\pm$15%) at 32 MeV. Results are compared
with calculations based on the hybrid model of nuclear reactions (Overlaid Alice
and Alice 85/300). For the proton energy range 99 – 20 MeV the cumulative ^{97}Ru
yield of 10.5 mCi/µA·h was estimated.

(^{97}Ru, ^{99m,96,95}Tc, ^{99}Tc target, stacked-foil technique, cross section, excitation function, thick target yield, proton-induced nuclear reactions).

Introduction

During the past 10 to 15 years attention
has been drawn to the ^{97}Ru nuclide. Its unusual
combination of excellent physical and chemical
properties (T$_{1/2}$ = 2.9 d, 100% EC, two gamma
lines 215.7 keV (85.8%) and 324.5 keV (10.2%),
no beta emissions to contribute to the radiation dosage, a few different valence states)
make this neutron deficient radioisotope a
desirable label for syntheses of radiopharmaceuticals for nuclear medicine applications.

It has not come into clinical use because
of the lack of efficient and economical production method of multicurie yields via particle accelerators. The reactor and accelerator methods suggested so far to produce ^{97}Ru
include: ^{96}Ru(n,γ)^{97}Ru |1|, ^{103}Rh(p,2p5n)^{97}Ru
|2-4|, ^{nat}Mo(^{4}He,xn)^{97}Ru, ^{nat}Mo(^{3}He,xn)^{97}Ru
|5-8| and ^{99}Tc(p,3n)^{97}Ru |9|. Reactor made
^{97}Ru does not yield a carrier free product
which limits its application in radiopharma-
cology. Cyclotron production with ^{3}He or ^{4}He
particle bombardment of natural molybdenum
targets produce comparatively low yields
($<$ 100 µCi/µAh) of ^{97}Ru. The proton spallation
reaction (p,xpyn) on Rh gives $\sim$1.4 mCi/µAh
and constitutes today the most useful procedure
for the production of no-carrier added (NCA)
^{97}Ru for use in radiopharmaceutical syntheses
|2,3|.

Very preliminary work presented in the
form of abstract |9| indicates for the ^{99}Tc(p,
3n)^{97}Ru reaction yield of $\sim$ 5 mCi/µAh. No
details, especially regarding the excitation
function and target thickness (the most convenient proton energy range) have been reported.
It has been pointed out by several authors that
the target material (Tc) is rather exotic and
radioactive (T$_{1/2}$ = 2.13·10^5 y) and therefore
is not very useful for practical applications.
However the content of ^{99}Tc in spent fuel
elements from atomic power station can be as
high as kilograms per 1 ton of uranium. The

extraction of Tc from solutions which are formed in the reprocessing of nuclear fuel elements
and preparation on its basis of metallic Tc
is not more difficult and costly than the
production of targets from highly enriched
isotopes used for production of medical radioisotopes.

All the above facts reinforce our optimism
regarding Tc as target material, because we
should rather expect the use of conventional
cyclotrons for the production of ^{97}Ru. We
could not find the excitation function for
^{99}Tc(p,3n)^{97}Ru reaction, so we have run the
computer code Alice 85/300 and its older ver-
sion Overlaid Alice |10| which gave the cross
section of 750 mb at proton energy around
31 MeV.

The objective of our experiment is to
measure the excitation functions of proton
induced nuclear reactions on Tc in the proton
energy range from thresholds up to 100 MeV.
These data will provide a basis for selecting
optimal irradiation condition for ^{99}Tc (p,
3n)^{97}Ru reaction and various by-product nuclides as well as an estimation of specific
radionuclide yields in thick targets.

Experimental Procedures

The excitation functions and yields were
measured applying the stacked-foil technique.
High purity metal foils of Tc were supplied
by the Institute of Physical Chemistry, Soviet
Academy of Sciences. The stack consisted
of 25 numbered Tc-foils (each of square shape
15$\pm$0.2 mm x 15$\pm$0.2 mm, 0.2$\pm$0.01 mm thickness,
weights varied between 0.4147 and 0.5079 g)
was exposed to an external proton beam current
generated from linear accelerator (LU-100)
at the Institute for High Energy Physics
in Serpukhov (USSR).

In order to cover the energy range from
16 to 99 MeV and to accomplish irradiations
at proton energies lower than 100 MeV the
lead degrader foils were employed. Experiments

614

were carried out at the proton beams with energies of 100, 73 and 60 MeV. The stack of foils for irradiation was placed in an Al target holder. The proton beam intensities varied from 50-100 nA. The duration of irradiations was 6-7 hours. Several aluminium foils were placed in front of the stack and used as monitor employing the $^{27}Al(p,3pn)^{24}Na$ reaction |11|. Degradation of the proton energy in the Tc, Al, Pb foils were calculated according to the range-energy tables |12|.

One additional measurement was carried out at the bombarding proton energy (35.4± 0.3) MeV at the cyclotron (U-120M) of the Nuclear Research Institute (NRI), Czechoslovak Academy of Sciences (Řež). The stack contained 10 Tc foils and Cu monitoring foils were applied. The cross sections for $^{63}Cu(p,2n)^{62}Zn$ and $^{65}Cu(p,n)^{65}Zn$ reactions are very well known |13|. Beam current of 500 nA was applied for irradiation time 15 hours.

The gamma-ray spectra of each target foil were measured over periods of several days. A gamma-ray spectrometer CICERO-8K (SILENA, Milan) equipped with HPGe detector (Canberra, USA) FWHM 1.72 keV for ^{60}Co, efficiency 20% and associated electronics was used to measure and identify the induced activities on targets and monitor foils. The spectrometer was calibrated for gamma-ray energy and efficiency using activity certified sources of ^{133}Ba and ^{152}Eu (Institute for Research, Production and Application of Radioisotopes, Prague, Czechoslovakia). The gamma-ray energies and half-lives of the various reaction products were different and easy to distinguish: ^{97}Ru ($T_{1/2}$ = 2.9 d) 215.7 keV (85.8%), 324.5 keV (10.2%); ^{95}Tc ($T_{1/2}$ = 20.0 h) 765.8 keV (93.9%), 1073.7 keV (3.75%); ^{96}Tc: 778.2 keV (100%), 718.5 keV (82.2%), 849.9 keV (97.8%), 1126.8 keV (15.2%); ^{99m}Tc ($T_{1/2}$ = 6.01 h) 140.5 keV (87.7%) |14,15|.

The number of counts in the relevant photopeak areas for the measured gamma-ray spectra and absolute activities were evaluated using the computer of SILENA spectrometer. These activities (with the known number of atoms in target foils and beam intensities) were then used to determine the cross section values (and production yields) by applying the well known activation formulas. The total error in the cross sections (and yields) is partly systematic and amounts to between 10-15%, we assumed for the cross sections and yields the error ±15%.

The data taken at 100, 73, 60 and 35 MeV join smoothly. Therefore the beam energies derived for the foils should be correct to ±1 MeV with an estimated energy straggling of the same magnitude |16|.

Results and Discussion

An analysis of the observed radioactivities from Tc target clearly demonstrates the absence of radionuclides other than Ru (^{97}Ru) and Tc (^{96}Tc, ^{95}Tc and ^{99m}Tc – isomer excitation in the target material by (p,p') process). Some of the short lived Ru radionuclides that might also be formed were not detected because they decayed during target transportation to Dubna. No peaks were noted which must have been due to impurities.

$^{99}Tc(p,3n)^{97}Ru$. The excitation function for the $^{99}Tc(p,3n)^{97}Ru$ reaction (E_{thr} = 18.3 MeV) was measured in the 99 – 20 MeV energy range. The results are given in Fig. 1. The maximum cross section is (438±66) mb at 32 MeV. The shape of the excitation function for $^{99}Tc(p,3n)$ reaction is quite well predicted by both Alice codes. However, agreement between experimental and theoretical values at the maxima of excitation function is rather poor. All three calculations (Overlaid Alice code with n_o = 4 (2,2,0) and Alice 85/300 code with n_o = 3 (1.2, 0.8, 1) and n_o = 4 (2,2,0)) overestimate the cross sections at the maximum in comparison with experimental one by a factor of 2.

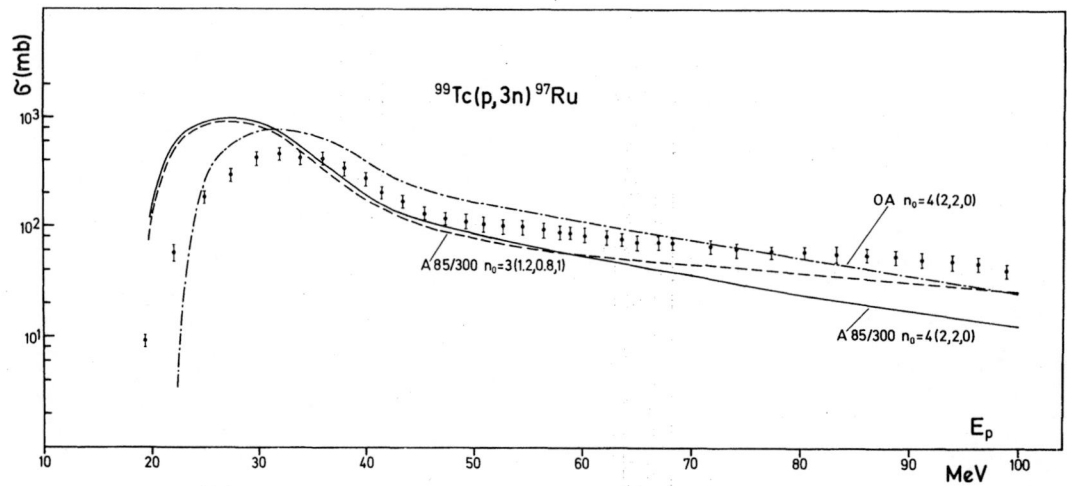

Fig. 1. Experimental (dots with error bars) and theoretical excitation functions for ^{99}Tc (p,3n)^{97}Ru reaction. The Overlaid Alice (OA) and Alice 85/300 codes (hybrid model of nuclear reactions implemented into both codes) were carried for estimation of the theoretical cross sections. The exciton numbers used are indicated.

^{99}Tc(p,p3n)^{96}Tc. The excitation function for the ^{99}Tc(p,p3n)^{96}Tc reaction (E_{thr} = 25.9 MeV) was measured in the energy range 99 – 29.8 MeV (Fig. 2). The maximum cross section of the excitation function is (197$\pm$30)mb and occurs at 47 MeV. The calculated excitation functions for this reaction with the Alice 85/300 code are nearly identical for two different exciton numbers (n_0 = 3 (1.2,0.8,1) and n_0 = 4 (2,2,0)) and are in excellent agreement with experimental data up to 70 MeV proton energy.

^{99}Tc(p,p'γ)^{99m}Tc. The measured excitation function in the 99–16 MeV proton energy range for excitation of ^{99m}Tc in (p,p') reaction is shown in Fig. 3. All the isomer cross sections are direct formation cross sections since the parent ^{99}Ru in the isobaric chain is stable and does not decay to ^{99m}Tc. The excellent agreement was achieved with Alice 85/300 code and exciton number n_0 = 3 (1.2, 0.8, 1). The calculated curve fits the data consistently in the energy range from 16 up to 70 MeV. This fact may be explained by the dominance of preequilibrium processes in (p,p') reaction, well described by the new version of the Alice code.

As we did not have thin enough Tc foils it was not possible to measure a very interesting part of the excitation function below 16 MeV.

The yields. A thick metallic Tc target (12.2 g/cm²) can be utilized to degrade the incident proton energy from 99 to 19 MeV. In this manner, ^{97}Ru yield of about 10.5 mCi/μAh may be achieved, allowing the production Ci quantities of ^{97}Ru if more than 5 μA of 100 MeV proton beams are available.

The most suitable energy region for the production of ^{97}Ru by proton bombardment of Tc extends from 19 MeV up to 40 MeV. Not only are the highest yields of ^{97}Ru achieved in this region, but its production can be performed with medium sized cyclotrons (for E_p = 40 MeV $\sim$ 6 mCi/μAh of ^{97}Ru may be produced).

Conclusions

The present work describes the basis for a new method to produce ^{97}Ru of high radionuclidic purity. Further study is clearly needed to fully evaluate the potential of this new method for producing ^{97}Ru and especially in the separation chemistry of Tc and Ru.

Acknowledgements. Our thanks are due to the Directors of the IHEP (Serpukhov) and NRI (Rež) for making available the irradiation facilities and to the staff of the accelerators for the irradiations. We are also grateful to Prof. Dr. M. Blann for making available to the Warsaw University the code Alice 85/300.

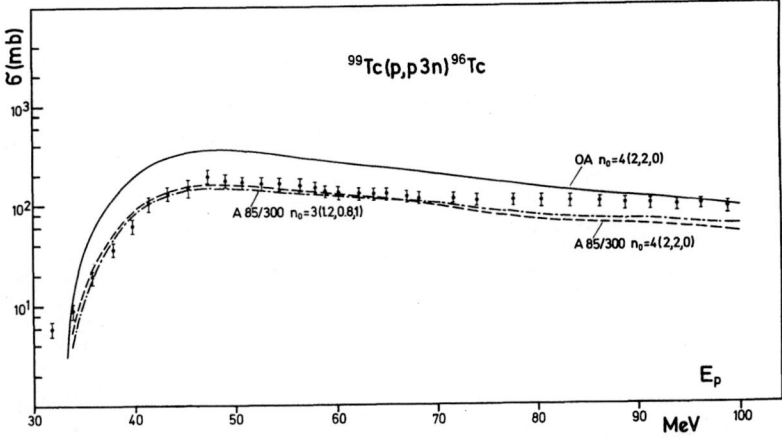

Fig. 2. Same as Fig. 1, for the reaction ^{99}Tc(p,p3n)^{96}Tc.

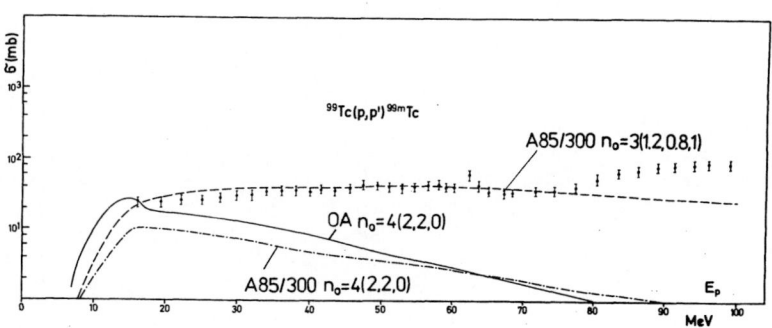

Fig. 3. Same as Fig. 1, for the reaction ^{99}Tc(p,p')^{99m}Tc.

References

1. G.Subramanian, I.G.Mc Afee and I.K.Poggenburg: J.Nucl.Med. 11, 365 (1970)
2. M.C.Lagunas-Solar, M.I.Avila, N.I.Navarro and P.C.Johnson: Int.J.Appl.Rad.Isot. 34, 915 (1983)
3. M.C.Lagunas-Solar, M.I.Avila and P.C.Johnson: Appl.Rad.Isot. 38, 151 (1987)
4. T.H.Ku, P.Richards, S.C.Srivastava, T.Prach, and L.G.Stang: "Production of ^{97}Ru for medical applications" Nuclear Medicine and Biology (C.Reynaud, Ed.) Pergamon Press, Paris (1982) 17
5. H.P.Graf and H.Münzel: J.Inorg.Nucl.Chem. 36, 3647 (1974)
6. D.Comar and C.Crouzel: Radiochem.Radioanalyt. Lett. 27, 307 (1976)
7. M.Gessner, S.Music, B.Babarovic and M.Vlatkovic: Int.J.Appl.Rad.Isot. 30, 578 (1979)
8. G.Comparetto and S.M.Qaim: Radiochimica Acta 27, 177 (1980)
9. E.Lebowitz, M.Kinsley, P.Klotz, C.Backsmith, A.Ansari, P.Richards and H.L.Atkins: J.Nucl. Med. 15, 511 (1974) (Abstract)
10. M.Blann: Overlaid Alice Code, COO-3494-29, Rochester (1976); Alice Livermore 82, Report UCID-19614 (1982); Code Alice 85/300, UCID-20169 (1984).
11. J.Tobailem, Ch.-H.de Lassus St-Genies and L.Leveque, Note CEA-N-1466 (1) Saclay (1971)
12. J.P.Janini: Atomic Data and Nuclear Data Tables, Proton range-energy tables, 1 keV – 10 GeV. Part 2, Elements, v.27, № 4/5 (1982)
13. P.Kopecky: Int.J.Appl.Radiat.Isot. 36, 657 (1985)
14. C.M.Lederer and V.S.Shirley (eds), Table of Isotopes, 7-th Ed. J.Wiley, New York (1978)
15. U.Reus and W.Westmeier: Gamma-Ray Catalog from Radioactive Decay. Atomic Data and Nucl. Data Tables 29, Parts 1 and 2 (1983)
16. C.Tschalar: Nucl.Instr.Meth. 61, 141 (1968)

616

OPTIMIZATION OF ISOTOPE PRODUCTION BY CROSS SECTION DETERMINATION

A. Hermanne, N. Walravens, O. Cicchelli

VUB Cyclotron , Laarbeeklaan 103,1090 Brussels, Belgium.

Abstract: In order to optimize the production of ^{201}Tl and ^{67}Ga using the (p,xn) reactions on enriched ^{203}Tl and ^{68}Zn targets we established a new set of cross section curves for the most relevant reactions. Target stacks made of up to 15 thin (50μm) natural Tl or Zn foils interleaved with brass degradation foils were irradiated with incident proton energies from 42 to 10 MeV in overlapping experiments. Yields for 200,201,202m,203,204mPb and 66,67,68Ga were computed from measured γ emission rates. Where possible, cross sections curves for the nuclear reactions involved were computed. Thick target yields and contamination ratios for enriched targets were derived and show good agreement with results from actual production runs.

(201Tl production, 67Ga production, stacked foil techniques, cross section, thick target yield, contamination ratio)

Introduction

In order to be able to produce GBq batches of commercially distributed radioisotopes for clinical use, yields and contamination ratios must be optimized. For the myocardial imaging nuclide ^{201}Tl the reaction of major interest at medium energy cyclotrons is ^{203}Tl(p,3n)^{201}Pb ($t_{1/2}$= 9h, EC > 99%) --> ^{201}Tl ($t_{1/2}$= 73h). Competing (p,xn) reactions on the two natural Tl nuclides (^{203}Tl : 29.5%; ^{205}Tl : 69.5%) are responsible for contamination with ^{200}Pb ($t_{1/2}$= 21.5 h, EC 100%) --> ^{200}Tl ($t_{1/2}$= 26.1h) and ^{202m}Pb ($t_{1/2}$= 3.6h, EC 11 %, IC 89%) --> ^{202}Tl ($t_{1/2}$= 293.5h). Standards of radionuclide purity require contamination ratios of less than 1% for ^{200}Tl and ^{202}Tl at time of calibration resulting in a strict limit of less than 1.5% of ^{200}Pb at the end of the first chemical separation, while up to 15% of ^{202m}Pb can be permitted.

The situation for obtaining large quantities of the tumor and infection imaging ^{67}Ga ($t_{1/2}$=78.3h, EC 100 %) is similar. Due to the stable nuclide composition of Zn (^{64}Zn: 48.9%, ^{66}Zn: 27.8%, ^{67}Zn: 4.1%, ^{68}Zn: 18.6%, ^{70}Zn: 0.6%) the most interesting production reactions are ^{67}Zn(p,n)^{67}Ga and ^{68}Zn(p,2n)^{67}Ga while ^{66}Ga ($t_{1/2}$=9,5h, β+ 56,5%, EC 43,5%) and ^{68}Ga ($t_{1/2}$=68.2min, β+ 90%, EC 10%) are the contaminants. Because of its short half life ^{68}Ga is not significantly contributing to contamination levels at end of chemistry (12-24h after end of bombardment) while up to 2% of ^{66}Ga is permitted.

As irradiation conditions and target geometry vary greatly between cyclotron centers an evaluation of relevant physical data and of the production conditions is best done at the actual irradiation facility. We used our CGR-560 cyclotron to establish a new set of cross section curves between 5 and 42 MeV for the most relevant reactions occuring in ^{201}Tl and ^{67}Ga production in order to optimize our automated chemistry systems [1],[2].

Materials and methods

1.Targets and irradiation conditions : target stacks of up to 15 foils with an exposed surface of 20mm diameter were irradiated for 6 min at 3μA in external collimated proton beams (Φ=18 or 12mm).
*^{201}Tl production: foils were made by electrodeposition of 59.2 mg/cm^2 (50μm) natural Tl metal on 50μm brass backings. Incident energy (calculated from accelerator settings) ranged from 42.6 to 25.2 MeV in 17 overlapping experiments.
*^{67}Ga production : commercially available pure 50μm Zn foils (>99.5%, 35.7mg/cm^2, Goodfellow) spaced by 50μm brass foils. Incident energy varied from 42.2 to 9.9 MeV (9 experiments).

2.Energy degradation : mean proton energy in each foil was computed using the stopping formulae coefficients for Tl and Ga of Andersen and Ziegler [3] and the tabulated stopping powers as published in the report of Janni [4] for brass.
3.Activity measurements : the induced activity of the different radionuclides was assessed on a coaxial Ge(Li) detector (10% relative efficiency, 2 keV FWHM resolution at 1332 keV). Efficiency calibration for different measuring geometries was performed using standard ^{57}Co, ^{133}Ba, ^{152}Eu and ^{22}Na sources. For Tl targets aliquots (1ml) were measured after complete dissolution of the target material (no traces of backing foil activity). The Zn foils were measured without chemical treatment after a cooling period of at least 4 hours for assessment of ^{68}Ga and at EOB+24h for the other nuclides. Dead time was less than 5% (excepted for the ^{68}Ga measurement : up to 10% correction).
4.Data analysis : yields of the nuclides of interest were calculated from the γ emission rates using the desintegration data from Erdtman and Soyka [5]. Corrections for decay during bombardment, waiting time and measuring time were made. Cross sections were constructed where possible from the smoothed yield curves taking into account the contributions of the nuclear reactions on the stable isotopes making up the target material .

The overall uncertainty in cross sections as derived from the quadratic summation of the uncertainties on target thickness (6%), integrated beam current (2%), counting statistics (2%), detector efficiency (5%), and physical decay data (5%) is 8.5%. Maximal error on energy values of data points is 0.6 to1.0 MeV due to uncertainties on incident energy and energy degradation.

Results and discussion

The cross section values for the reactions yielding Pb nuclides are given in Table 1. and Figures 1 to 4.

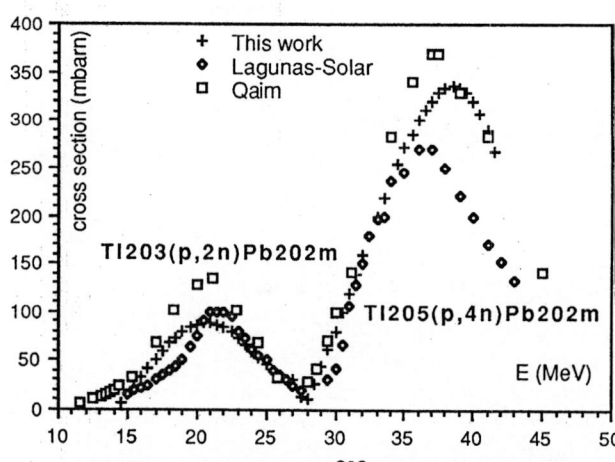

Fig 2. : cross sections for ^{202m}Pb yielding reactions

Comparison with data points extracted from the reaction cross section figures published by the group of Lagunas-Solar for natural targets [6] and ^{205}Tl enriched targets (above 35MeV) [7] and from the effective formation cross section graphs on natural Tl of Qaim and coworkers [8], shows good agreement for the ^{201}Pb (Fig.1) and ^{203}Pb (Fig. 4) yielding reactions. For the ^{202m}Pb yielding reactions a large scatter exists between the 3 data sets (Fig. 2) with the results of this work comparing well with the values of Lagunas-Solar [6] for the ^{203}Tl(p,2n) and with Qaim [8] for the ^{205}Tl(p,4n). In Fig. 3 the overall form of the cross sections agree , but our maximum value for the ^{205}Tl(p,2n)^{204m}Pb reaction is

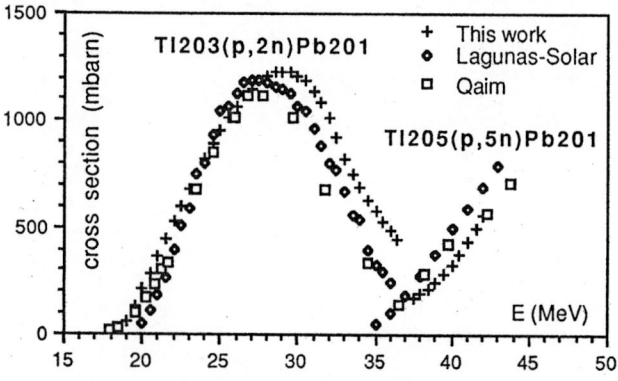

Fig 1. : cross section for ^{201}Pb yielding reactions.

Table1. : Cross sections for ²⁰³Tl(p,xn), ²⁰⁵Tl(p,xn), ⁶⁶Zn(p,n), ⁶⁸Zn(p,xn) reactions

E_p	$^{203}Tl(p,2n)$ ^{202m}Pb	$^{203}Tl(p,3n)$ ^{201}Pb	$^{203}Tl(p,4n)$ ^{200}Pb	$^{205}Tl(p,2n)$ ^{204m}Pb	$^{205}Tl(p,3n)$ ^{203}Pb	$^{205}Tl(p,4n)$ ^{202m}Pb	$^{205}Tl(p,5n)$ ^{201}Pb	$^{66}Zn(p,n)$ ^{66}Ga	$^{68}Zn(p,n)$ ^{68}Ga	$^{68}Zn(p,2n)$ ^{67}Ga	$^{68}Zn(p,3n)$ ^{66}Ga
12	-	-	-	6.1	-	-	-	495	536	269	-
14	-	-	-	27.2	-	-	-	475	356	438	-
15	14.6	-	-	47.3	-	-	-	423	286	528	-
16	32.6	-	-	69.3	-	-	-	352	229	617	-
17	51.2	-	-	89.8	77	-	-	276	183	700	-
18	67.5	-	-	106.1	240	-	-	214	146	772	-
19	80.1	58	-	116.2	418	-	-	138	118	830	-
20	86.9	209	-	118.0	599	-	-	120	97.5	879	-
21	88.0	365	-	112.3	771	-	-	104	83.3	883	-
22	83.2	522	-	99.3	926	-	-	90.7	73.4	861	-
23	73.5	675	-	80.5	1056	-	-	78.8	67.5	813	-
24	60.5	819	-	59.3	1155	-	-	68.4	64.1	740	22
25	46.7	950	-	40.1	1218	-	-	59.5	62.3	643	57
26	35.6	1062	-	29.2	1244	-	-	51.6	61.2	453	92
27	31.8	1150	-	28.8	1233	-	-	44.8	60.1	417	125
28	-	1208	-	27.3	1188	-	-	39.1	58.1	382	156
29	-	1230	57	25.8	1116	43.8	-	33.8	55.2	348	185
30	-	1210	126	24.4	1025	80.4	-	39.4	50.8	314	212
31	-	1139	252	23.1	820	119	-	25.5	35.0	283	236
32	-	1012	420	21.8	614	159	-	22.2	30.4	253	256
33	-	821	588	20.6	514	198	-	19.3	28.5	225	273
34	-	690	798	19.5	427	236	-	16.2	-	200	287
35	-	577	978	18.5	353	271	-	14.5	-	176	296
36	-	482	1155	17.5	291	300	-	12.6	-	156	300
37	-	403	1302	16.5	241	321	-	11.2	-	138	299
38	-	337	1368	15.6	203	334	191	9.5	-	125	293
39	-	281	1386	14.8	179	333	248	8.3	-	115	281
40	-	239	1365	14.1	166	321	328	7.2	-	108	264
42	-	165	1108	12.5	-	-	559	-	-	-	-

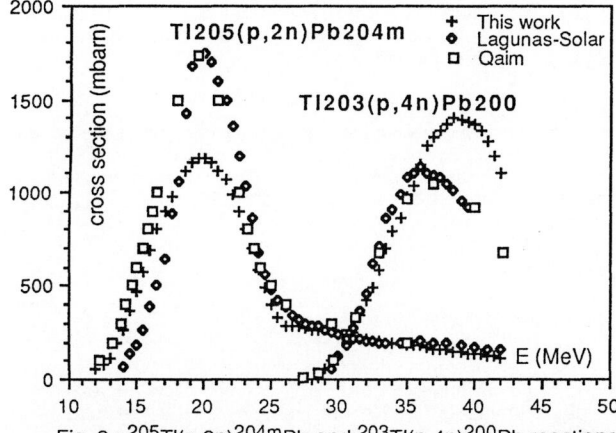

Fig. 3 : ²⁰⁵Tl(p,2n)²⁰⁴ᵐPb and ²⁰³Tl(p,4n)²⁰⁰Pb reactions

20% lower than those of the 2 other authors while the ²⁰³Tl(p,4n)²⁰⁰Pb reaction shows a 40% higher maximum . Formation of ¹⁹⁹Pb occured only above 39 MeV proton energy. The maxima of the cross sections measured by Lagunas-Solar show a slight shift to lower energy for most reactions.

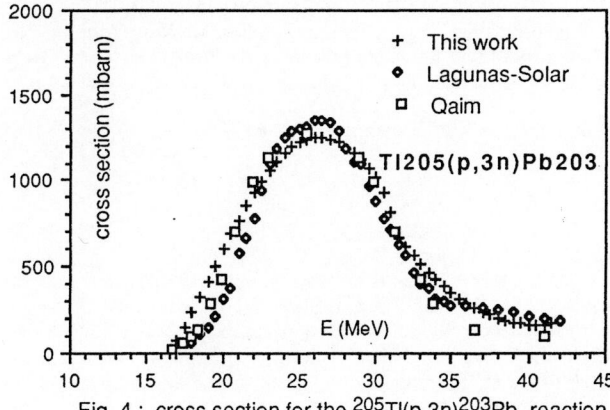

Fig. 4 : cross section for the ²⁰⁵Tl(p,3n)²⁰³Pb reaction.

Zn targets : the cross section values for some reactions yielding ⁶⁶,⁶⁷,⁶⁸Ga nuclides are given in Table1. and Figures 5 to 7.

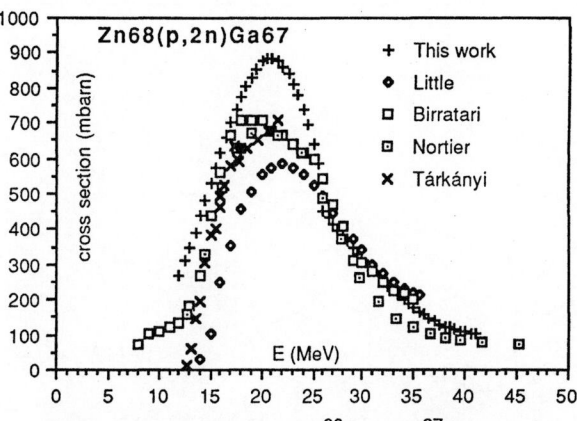

Fig. 5 : cross section for the ⁶⁸Zn(p,2n)⁶⁷Ga reaction

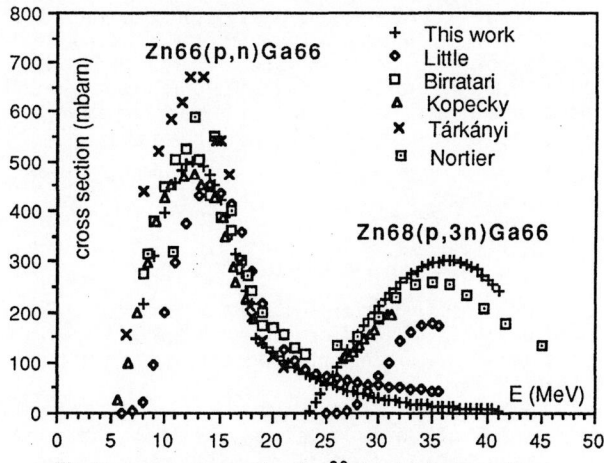

Fig. 6 : cross sections for the ⁶⁶Ga yielding reactions

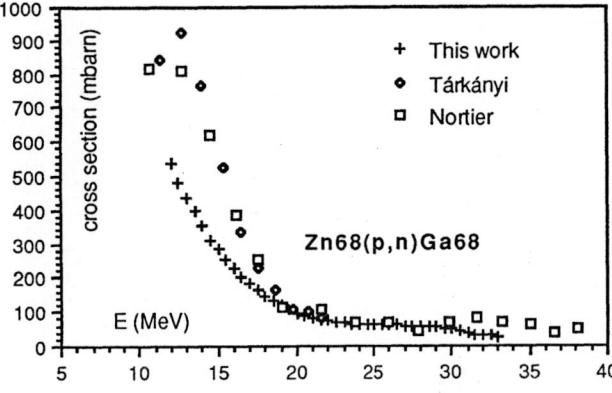

Fig. 7 : cross section for $^{68}Zn(p,n)^{68}Ga$ reaction

Our cross sections for the $^{68}Zn(p,2n)^{67}Ga$ reaction (Fig. 5) are significantly higher than those given by Little and Lagunas-Solar [10] and even than the highest values tabulated by Tárkányi et al. [9], Nortier et al. [11] and those taken from Birratari and Bonardi [13]. For the ^{66}Ga yielding reactions (Fig. 6) large scatter of cross section values exists between the data of all authors (this work, Kopecky [12], [9], [10], [11], [13]. A small shift to higher energy is observed for the results of Little and Lagunas-Solar [10]. In Fig. 7 significantly lower cross sections below 20 MeV is shown for this work compared to Tárkányi et al. [9] and Nortier et al. [11].

Thick target yields and production data

Yields at end of bombardment (EOB) for target thicknesses from 100μm to 700μm and for incident energies between 20 and 35 MeV are computed from the cross section curves. We considered natural Tl or Zn and highly enriched targets.

<u>^{203}Tl (95%) targets</u> : yield for ^{201}Pb and contamination ratio $^{202m}Pb/^{201}Pb$ as derived from our cross section data are shown in Fig. 8 .

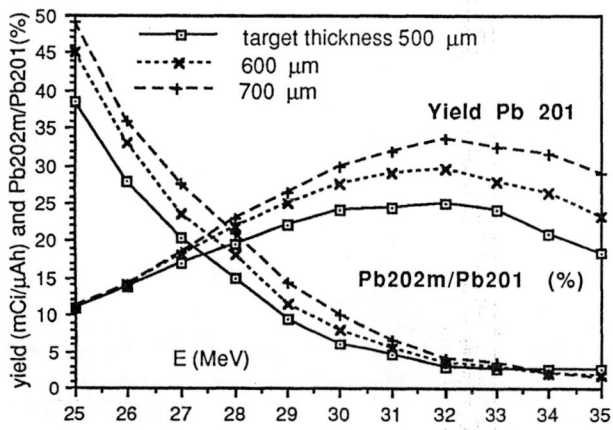

Fig. 8 : Yield and $^{202m}Pb/^{201}Pb$ ratio for thick ^{203}Tl(95%) target.

The $^{200}Pb/^{201}Pb$ ratio exceeds the imposed 1.5% limit (EOCS1) if the incident proton energy exceeds 30 MeV. Comparison with actual production results shows good agreement (8h irradiations at 80μA, target thickness 650μm, yields for all irradiations are converted to saturation activity A_s and are corrected for decay during chemistry procedure).

Ep(MeV)	A_s(mCi/μA)		$^{202m}Pb/^{201}Pb$(%)	
	cross section	production	cross section	production
30	371	286	8,5	7,8
29,2	345	279	11,5	9,8

Differences in yield are explained by losses during chemistry, decreasing yield at high incident current (80μA compared to 3μA) , uncertainty on beam centering and on target thickness.

<u>^{68} Zn (98%) targets</u> : yield for ^{67}Ga and $^{66}Ga/^{67}Ga$ contamination ratio as derived from our cross section data are shown in Fig. 8 Due to its short $t_{1/2}$ of 68min the ^{68}Ga contamination is not relevant in production conditions. Comparison with pre-production

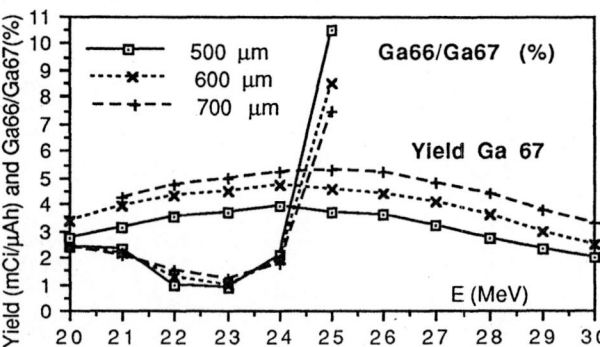

Fig. 9 : Yield and $^{66}Ga/^{67}Ga$ ratio for thick ^{68}Zn (98%) target

runs (irradiation of 30min at 5μA, targets 550μm thick, correction for decay during cooling period of 96h) shows good agreement.

Ep(MeV)	Y(mCi/μAh)		$^{66}Ga/^{67}Ga$(%)	
	cross section	production	cross section	production
22,4	3,8	3,4	1	4,1
24,75	4,2	3,65	8	9,1

Conclusion

Cross section data for the nuclear reactions involved in GBq production of ^{201}Tl and ^{67}Ga were derived from stacked foils irradiations on natural metallic Tl and Zn targets. Reasonable good overall agreement with several published data sets are obtained , but for some reactions large deviations in peak values were found. The thick target yields and contamination ratios derived from our cross section data for the most relevant nuclides in large scale production, are in good agreement with values measured on actual production batches.

Acknowledgment

We wish to thank our colleagues of the chemistry and mechanical groups at the VUB Cyclotron for the help in various aspects of experiment preparation, irradiation set ups and collection of production data. This reseach is part of an ongoing effort for cross section studies supported by a FGWO 4-0099-89 grant.

References

[1] Van den Winkel P., Waegeneer R., De Vis L., Terriere D., De Backer G., Hermanne A. and Vinchent R.: Proc.Symposium on Low Cost Automation, Milano, W165-W172, (1989).
[2] De Vis L. : Menu-gestuurde software en hardware voor het geautomatiseerd Tl-201 produktieproces van het VUB-Cyclotron, Vrije Universiteit Brussel (1990).
[3] Andersen H.H., Ziegler J.F. : Hydrogen, Stopping Powers and Ranges in all Elements, Vol. 3., Pergamon Press, (1977).
[4] Janni J.F. : Calculations of Energy Loss, Range, Path length,Straggling, Multiple Scattering, and the Probability of Inelastic Nuclear Collisions for 0.1 to 1000 MeV Protons, AFWL-TR-65-150, (1969).
[5] Erdtmann G., Soyka W. :The Gamma Rays of the Radionuclides, Verlag Chemie, Weinheim, New-York, (1979).
[6] Lagunas-Solar M.C., Jungerman J.A., Peek N.F., Theus R. : Int. J. Appl. Radiat. Isotopes, 29, 159, (1978).
[7] Lagunas-Solar M.C.,Jungerman J.A., Paulson D.W. : Int. J. Appl. Radiat. Isotopes, 31, 117, (1979).
[8] Qaim S.M., Weinreich R., Ollig H. : Int. J. Appl. Radiat. Isotopes, 30, 85, (1979).
[9] Tárkányi F., Szelecsényi F., Kovács Z. : Radiochemica Acta, 50, 19, (1990).
[10] Little F.E., Lagunas-Solar M.C. : Int. J. Appl. Radiat. Isotopes, 34, 631, (1983).
[11] Nortier F.M., Mills S.J. and Steyn G.F. : Int. J. Appl. Radiat. Isotopes, 42, 353, (1991).
[12] Kopecky P. : Int. J. Appl. Radiat. Isotopes, 41, 606, (1990)
[13] Birattari C., Bonardi M. : Ottimizzazione delle condizioni di irraggiamento per la produzione del radioisotopo ^{67}Ga con il ciclotrone di Milano. , INFN, TC 80/9, (1980).

PRODUCTION OF GALLIUM-66, A POSITRON EMITTING NUCLIDE FOR RADIOIMMUNOTHERAPY

Saed Mirzadeh[1]* and Yung Yee Chu[2]

[1]Nuclear Medicine Group, Health and Safety Research Division, Oak Ridge
National Laboratory, Oak Ridge, TN 37831-6022
[2]Chemistry Department, Brookhaven National Laboratory, Upton, NY 11973 USA.

ABSTRACT

Excitation functions for production of ^{66}Ga via α-induced nuclear reactions on enriched ^{66}Zn have been measured with $E_\alpha \leq 27.3$ MeV and $E_\alpha \leq 43.7$ MeV employing the stack-thin target technique. In addition, the induced activity of ^{67}Ga in the same sets of targets allowed an evaluation of the excitation functions of the corresponding nuclear reactions.

INTRODUCTION

Radiolabeled monoclonal antibodies for the purpose of radioimmunotherapy are currently of considerable interest. Several β^- and α-emitting nuclides have been identified for this application (e.g. ^{67}Cu[1], ^{90}Y[2], ^{188}Re[3], ^{211}At[4] and ^{212}Pb/^{212}Bi[5]). In principle, β^+-emitting nuclides should also be considered for therapeutic applications; not only is the radiation dose per decay comparable to those of β^--emitters, but the quantitative imaging ability of positron emission tomography (PET) would enhance dosimetry for β^+-emitters. Furthermore, β^+-emitters could be produced in medical cyclotrons. Among possible β^+-emitters, ^{66}Ga is the most interesting. The convenient 9.45 h half-life, the large β^+ branch (51.2%), with high end-point energy (4.15 MeV), and comparable EC branch (44%) with abundant short-range electrons make ^{66}Ga an attractive candidate for therapeutic applications.

The production routes of ^{66}Ga via peripheral interactions are summarized in Table 1. In this paper we report the preliminary data for production of ^{66}Ga via the following reactions:

I. ^{64}Zn[α,2n]^{66}Ge(2.3 h, EC)$\rightarrow$^{66}Ga
II. ^{64}Zn[α,np]^{66}Ga

The excitation functions were measured with $E_\alpha \leq 27.3$ MeV at the National Institutes of Health (NIH) cyclotron and with $E_\alpha \leq 43.7$ MeV at the 60" cyclotron at Brookhaven National Laboratory (BNL). In addition, the induced activity of ^{67}Ga in the same sets of targets allowed an evaluation of the excitation functions of the corresponding [α,n] and [α,p] reactions.

EXPERIMENTAL

For excitation function measurements, thin targets of ^{66}Zn ($\sim$16 μg.cm^{-2}) were prepared by vacuum evaporating 99.6% enriched ^{66}Zn as metal (ORNL) onto 7.664 mg.cm^{-2} high-purity Al support foils (99.999%, AESAR/Johnson Matthey, Seabrook, NH). These Al-supported Zn targets were covered with the same Al foils to avoid recoil losses of the product nuclides. The 15.33 mg.cm^{-2} Al in each target sample eliminated the need for additional degrader foils. Stacks of sealed targets (between 6-12 target foils in any one experiment) were mounted on water-cooled blocks and irradiated for durations ranging from 10 to 30 minutes with 1.0 μA of 27.3 at the NIH cyclotron or 43.7 MeV at the 60"

---

*Author for correspondence

cyclotron at BNL. After irradiation, the individual samples were mounted on counting cards for assay by γ-ray spectroscopy. For determination of the 19.0-m ^{67}Ge yield, attempts were made to count the samples quickly after irradiation, however, due to high levels of short-lived activities (predominantly 2.3-m ^{28}Al), it was necessary to allow the samples to decay for a period of about an hour before the first measurement. The same sets of samples were used for four independent experiments (three at NIH and one at BNL). Precise determination of the thickness of ^{66}Zn in the thin targets will be made at the conclusion of these studies.

A calibrated 50-cm^3 high-purity Ge detector, FWHM $\sim$1.8 at 1332 KeV, (EG&G ORTEC, Oak Ridge, TN) coupled to a AccuSpec PC-based multichannel analyzer (Nuclear Data/Canberra Inc., Meriden, CT) was used for radioactivity measurements. Typically, samples were counted at a distance of 10 cm from detector surface to eliminate the coincidental summings. The radioactivities in each sample were followed for several half-lives and the decay curve analysis was performed with the CLSQ code[6]. The relevant nuclear decay data, taken from reference 7, are summarized in Table 2. The energy of the cyclotron α-particles were deduced from the operating characteristics of the cyclotrons, and in the case of BNL 60" cyclotron, the beam energy was corrected by 1.5 MeV[8]. The Range Tables of Hubert et al.[9] was used to determined the energy of the degraded incident α-particles.

RESULTS AND DISCUSSION

The excitation function of ^{64}Zn[α,2n]^{66}Ge reaction which was measured in this work and that reported by Porile et al.[9] are shown in Figure 1. Above $E_\alpha = 20.0$ MeV the cross-section increases rapidly and reaches to a maximum of 94 mb at $E_\alpha = 33$ MeV. From the threshold up to 27 MeV (where our two measurements at the BNL and the NIH overlapped) our measured cross-sections agreed well but they were higher than reported values by almost a factor of 10. In the earlier study, quantitation of ^{66}Ga activity was made by the measurement of its annihilation radiation in a NaI detector. Activity of ^{66}Ga was most likely overcorrected while corrections were made for the contribution from long-lived β^+ emitters. At the higher-energy end of the excitation function, the agreement between our data and that of earlier measurements are generally good. The cumulative cross-section values for the production of ^{66}Ga from direct [α,pn] reaction and indirectly from the decay of ^{66}Ge are also shown in Figure 1. The cumulative cross-section reaches to a maximum of 900 mb at E_α=32.5 MeV with a threshold of about 19 MeV. At the maximum region of the excitation function, the agreement between our current and earlier measurements are surprisingly good. The subtraction of the excitation function of reaction I from the cumulative excitation function yields the excitation function for reaction II. The relative probability of reaction II to I, $\sigma_{(\alpha,pn)}/\sigma_{(\alpha,2n)}$, in the maximum region (30≤E_α≤40) remains rather constant at 6.5±1.0, indicative of substantially lower binding energy of protons in this mass region. The same set of α-activated targets yielded excitation functions for production of ^{67}Ga via the (α,p)and (α,n) reactions and the results are shown in Figures 2a and 2b, respectively. In this case, the $\sigma_{(\alpha,p)}/\sigma_{(\alpha,n)}$ is close to unity at the maximum of the excitation functions which occurs at E_α=20 MeV, about 13 MeV lower than that of (α,2n) or (α,pn) reactions. The peak of the excitation function of (α,n) reaction is larger than of (α,2n) by almost a factor of 10. However the situation is reversed in the case of (α,p) and (α,pn) reactions, where at the maximum of the excitation functions the ratio of $\sigma_{(\alpha,pn)}$ to $\sigma_{(\alpha,p)}$ is ~2.5. The errors of the cross-sections values are estimated at ~10% at the maximum of the excitation functions and at ~30% near the threshold. The incident α-particle energies are most accurate at the highest energy with a relative error of ~2%. This error increases to ~10% below 16 MeV due to straggling process.

Table 1. Gallium Isotopes of Interest in Nuclear Medicine

Isotope	$t_{1/2}$	Mode of Decay	E_β^{max}, MeV	E_γ, MeV (I_γ, %)
^{66}Ga	9.40 h	ß$^+$(56.5%), EC(44%)	0.367(0.82%) 0.747(0.97) 0.935(3.03%) 1.84(0.54%) 4.15(51.2%)	833.6(6.12%) 1039.0(38.4%) 2190.0(5.74%) 2752.1(23.5%)
^{67}Ga	3.26 d	EC(100%)	-	167.0(77.4%)
^{68}Ga	68.3 m	ß$^+$(90%), EC(10%)	~0.8(~2) 1.9(89%)	1077.4(2.93)

Table 2. Peripheral Reactions for Production of Carrier-free ^{66}Ga

Nuclear Reaction	References
^{64}Zn(α,2n)^{66}Ge(2.3 h)→	Porile et al. (1959)[10]
^{64}Zn(α,pn)	Porile et al. (1959)[10]
^{63}Cu(α,n)	Porile et al. (1959)[11]
^{65}Cu(α,3n)	Porile et al. (1959)[11]
$^{Nat.}$Cu(α,xn)	Goethals et al. (1990)[12]
^{66}Zn(^{3}He,3n)^{66}Ge(2.3 h)→	
^{66}Zn(^{3}He,p2n)	
^{65}Cu(^{3}He,3n)	
^{66}Zn(d,2n)	
^{68}Zn(d,4n)	
$^{Nat.}$Zn(p,xn)	Howe et al. (1958)[13]
$^{Nat.}$Zn(p,xn)	Hille et al. (1972)[14]
$^{Nat.}$Zn(p,xn)	Little, et al. (1983)[15]
$^{Nat.}$Zn(p,xn)	Kopecky et al. (1989)[16]
^{66,67,68}Zn(p,xn)	Tarkanyi et al. (1990)[17]

ACKNOWLEDGMENTS

Research supported by the Office of Health and Environmental Research, U.S. Department of Energy, under contract DE-AC05-84OR21400 with Martin Marietta Energy Systems, Inc.

REFERENCES

1. Mirzadeh S., Mausner L. F. and Srivastava S. C., *Appl. Radiat. Isot.*, 37, 29 (1986).
2. Kozak R. W., Raubitschek A., Mirzadeh S. *et al. Cancer Research*, 49, 2639 (1989).
3. Mirzadeh S. Rice D. E. and Knapp F. F., Jr., *Appl. Radiat. Isot.* (1991) in press.
4. Lambrecht R. M. and Mirzadeh S., *Appl. Radiat. Isot.* 36, 443 (1985).
5. Ruegg C. L., Anderson-Berg W. T., Brechbiel M. W., Mirzadeh S., Gansow O. A. and Strand M., *Cancer Research* 50, 4221 (1990).
6. Cummings J. B., National Academy of Sciences, National Research Council, Nuclear Science Series NAS-NS-3107 (1962).
7. Table of Isotopes (Lederer C. M. and Shirley V. S. Eds.) 7th ed. Wiley, New York (1978).

8. Mirzadch S. and Lambrecht R. M., <u>Measurement of the Cyclotron Beam Energy of the Alpha Particle by Rutherford Scattering in Gold Foil</u>, *IAEA INDC(NDC)-195/GZ* January 1988.

9. Hubert F., Fleury A. and Bimbot R., <u>Range and Stopping Power Tables</u>, *Ann. Phys.* (Suppl.) 5, 1-124 (1980).

10. Porile N. T., *Phys. Rev.* 115, 939 (1959).

11. Porile N. T. and Morrison, D., *Phys. Rev.* 116, 1195 (1959).

12. Goethals P., Coene M., Slegers G. *et al.*, *Eur. J. Nucl. Med.*, 16, 237 (1990).

13. Howe A. H., *Phys. Rev.* 109, 2083 (1958).

14. Hille M., Hille P., Uhl M., Weisz W., *Nucl. Phys.* A198, 625 (1972).

15. Little F. E. and Lagunas-Solar M. C., *Appl. Radiat. Isot.*, 34, 631 (1983).

16. Kopecky P, *Appl. Radiat. Isot.*, 41, 606 (1990).

17. Tarkanyi F., Szelecsenyi F., Kovacs Z., *Radiochimica Acta*, 50, 19 (1990).

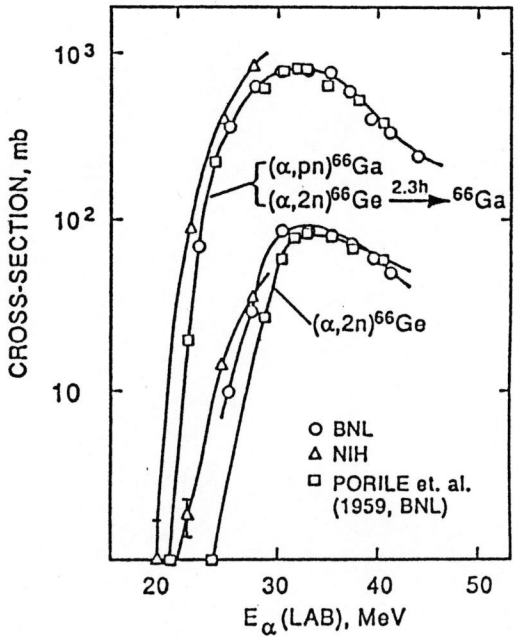

Figure 1. Excitation Functions for Production of ^{66}Ga via α-Induced Reaction on ^{64}Zn.

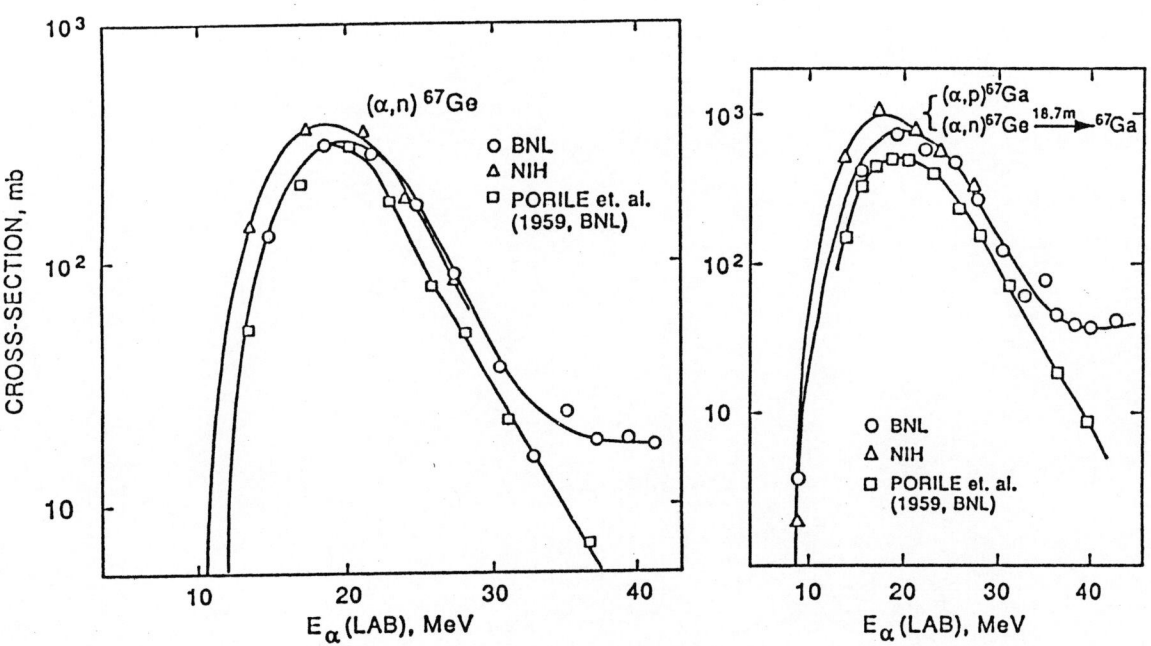

Figure 2a&b. Excitation Functions for Production of ^{67}Ga via α-Induced Reactions on ^{64}Zn

Part VIII

Nuclear Data Relevant to Astrophysics, Geology, Neutron Dosimetry, and Some Other Applications

DATA NEEDS IN ASTROPHYSICAL STUDIES

C. Rolfs

Ruhr-Universität Bochum, Experimentalphysik III, Bochum, Germany

Abstract: The general needs of improved and new nuclear physics data for various astrophysical scenarios are discussed. Particulat emphasis is given to the areas of charged-particle-induced reactions, electron screening effects, and radioactive ion beams.

1. Introduction

Investigations during the last 50 years have shown that we are connected with distant space and time not only by our imagination but also through a common cosmic heritage: the chemical elements that make up our bodies. These elements were created in the hot interiors of remote and long-vanished stars over many billions of years. Their fuels finally spent, these giant stars met death in cataclysmic explosions, scattering afar the atoms of heavy elements synthesized deep within their cores. Eventually this material, as well as material lost during the red-giant stages, collected into clouds of gas in interstellar space; these, in turn, slowly collapsed giving birth to new generations of stars, thus leading to a cyclic evolution still going on. The present picture is that all the elements from C to U (the "metals") have been produced entirely within stars during their fiery lifetimes and explosive deaths. A few of the lighter elements were formed before the stars even existed, during the birth of the Universe itself. In addition, a few of the lighter elements were synthesized in intergalactic space by cosmic rays. Thus, theories of nucleosynthesis have identified the most important sites of element formation and also the diverse nuclear processes involved in their synthesis.

The detailed understanding of our cosmic heritage combines astrophysics and nuclear physics and forms what is called nuclear astrophysics. In a sense, nuclear astrophysics involves much of astrophysics, because there are few important events in astronomy, cosmology and cosmogony that have not left nuclear clues. We may gather these clues, study the properties of the atomic nuclei in the laboratory and, if we are lucky, figure out what has happened. Impressive progress has been made here, in the theories of mucleosynthesis over the last 30 years and this has been rewarded[1]. However, there remain puzzles and problems, which challenge the basic ideas underlying nucleosynthesis in stars and elsewhere. Thus, the grand concept of elemental nucleosynthesis will not be truly established, until we attain a deeper and more precise understanding of the many nuclear processes operating in astrophysical environments. In fact, without such information, much of the work done so far will, at the best, be incomplete, and, at the worst, possibly incorrect. Therefore, new nuclear-physics data will not represent a "fine structure" information, it will not represent a type of "cleaning-up work" to be done, but represent the essential next step in this exciting scientific field. In a sense, nuclear astrophysics is simply nuclear physics but applied to the interesting problems of astrophysics. Although, initially, the intellectual motivation for doing such experiments lies in astrophysics, experience in the past has shown on many occasions that the evaluation of the collected data has provided unexpected intellectual rewards in nuclear physics itself.

There are several areas where new or improved nuclear data are needed (for details see refs. 2-5 and references therein), such as:

(i) nuclear reactions induced by charged particles, which are involved in the energy generation in stars and the associated nucleosynthesis during the various stellar burning phases (hydrogen burning, helium burning, etc, producing the metals up to the iron region) as well as in primordial nucleosynthesis;

(ii) the cross section of these charged-particle-induced processes are influenced by electron screening effects of the stellar plasma;

(iii) neutron-induced reactions play an important role in the early phase of the Universe as well as in stars, where they are involved in the nucleosynthesis of the transiron elements via the s- and r-processes;

(iv) the long-living radioactive nuclides of the transiron elements can be used as chronometers, barometers, and thermometers of the s- and r-processes;

(v) in hot and explosive astrophysical environments (such as the big bang or supernovae) short-living nuclides are involved in the nuclear burning processes;

(vi) spallation processes produce some light nuclides via cosmic rays in interstellar space;

(vii) a good understanding of neutrino properties is needed for the understanding of neutrino astronomy (e.g., the solar neutrino problem and the neutrino flash(es) from supernovae);

(viii) in the early Universe a quark-gluon plasma may have existed forming later hadrons (protons and neutrons), the details of which influence the primordial (inhomogeneous) nucleosynthesis.

In the following only the areas (i), (ii) and (v) will be discussed further.

2. Charged-particle-induced nuclear reactions

Owing to the effect of the Coulomb barrier E_C, the cross section $\sigma(E)$ drops rapidly (on an exponential scale) at low energies, and it becomes more difficult to measure the relevant cross sections. For low incident energies and for s partial waves, the penetration factor for the Coulomb barrier is approximately proportional to $\exp(-2\pi\eta)$, where η is the Sommerfeld parameter. It is convenient to factor out this energy dependence, as well as an additional factor $1/E$ arising from the squared de Broglie wavelength. The cross section can thus be expressed as

$$\sigma(E) = S(E)(1/E)\exp(-2\pi\eta), \qquad (1)$$

where the function $S(E)$ defined by this equation contains all the strictly nuclear effects. It is, therefore, referred to as the nuclear or astrophysical S factor. With improved experimental techniques direct measurements of $\sigma(E)$ can be extended toward lower energies, but in practice one hardly reaches the relevant stellar energy regions for quiescent burning ($E_O/E_C \simeq 0.01$, the "energies far below the Coulomb barrier" or the "sub-Coulomb energies"). The observed energy dependence of $\sigma(E)$, or equivalently of $S(E)$, must, therefore, be extrapolated to the stellar energy region (essentially to zero energy). However, these extrapolated data represent only lower limits; if there are resonances near the particle threshold at $-E_R$ or $+E_R$, they can completely dominate the $S(E)$ factor for low stellar energies.

As an illustrative example consider the $^2H(d,\gamma)^4He$ reaction, which is involved in both primordial and stellar nucleosynthesis. Its S-factor was shown in earlier work with $E \gtrsim 0.4$ MeV to decrease steeply with decreasing energy, and the angular distributions were of the form $\sin^2\theta\cos^2\theta$. Both observations indicated that the reaction mainly proceeds via a Direct Capture (DC) process, i.e. an E2 transition from a $^1D_2$ scattering state to the $^1S_0$ component of the ^4He ground state. This is consistent with the identical boson character of the entrance channel, which requires L+S to be even, and with the usual E1 and M1 isospin selection rules. Because of the D-wave centrifugal barrier in the entrance channel, the S factor has been expected to continue to decrease at low energies. In spite of challenging experimental problems, low energy measurement ($E \gtrsim 0.02$ MeV) were carried out recently[6,7]. The new results deviate from the trend at higher energies: the essentially constant S-factor at low energies demonstrates the dominance of an s-wave entrance channel, and therefore an E2 transition of the type $^5S_2 \to {}^5D_O$. This picture was confirmed by the isotropic angular distributions observed at low energies. The low energy isotropy and constant S factor are thus a clear demonstration of the existence of a D-state admixture in the ^{4}He ground state. Various analyses indicate a D-state admixture of the order of 1 to 10%. Thus, the strongly bound double magic nucleus ^{4}He is not spherical in shape, as might be expected, but has an intrinsic quadrupole moment. Note that these low energy studies appeared at first (from a purely nuclear physics point of view) to be of comparatively little interest. However, as shown by this example and many others, such data often provide new and unexpected intellectual rewards in nuclear physics. This work also demonstrates the general need for low energy data and once again warns of the dangers of extrapolating data over large energy ranges.

3. Electron screening

In equation 1 it is assumed that the Coulomb potential of the target nucleus and projectile is that resulting from bare nuclei, and thus the potential would extend to infinity. However, for nuclear reactions studied in the laboratory, the target nuclei are usually in the form of neutral atoms or molecules. The atomic (or molecular) electron cloud surrounding the target nucleus acts as a screening potential: an incoming charged projectile experiences no repulsive Coulomb force until it penetrates the electron cloud; thus, the projectile effectively sees a reduced Coulomb barrier. This in turn leads to a higher cross section, $\sigma_s(E)$, than would be the case for bare nuclei, $\sigma_b(E)$, with an enhancement factor $f(E) = \sigma_s(E)/\sigma_b(E) \simeq \exp(\pi\eta U_e/E)$, where U_e is the electron screening potential (e.g. $U_e \simeq Z_1 Z_2 e^2/R_a$, with R_a an atomic radius). Note that $f(E)$ increases exponentially with decreasing incident energy. For energy ratios $E/U_e \gtrsim 1000$, shielding effects are negligible, and laboratory experiments can be regarded as essentially measuring $\sigma_b(E)$. However, for $E/U_e \lesssim 100$, shielding effects cannot be neglected and become important for understanding low-energy data. Relatively small enhancements from electron screening at energy ratios $E/U_e \simeq 100$ can cause significant errors in the extrapolation of cross sections to lower energies, if the curve of the cross section is forced to follow the trend of the enhanced cross section, without correction for the screening[8]. Notice that for astrophysical and other applications (stellar and fusion plasmas) the value of $\sigma_b(E)$ must be known because the screening in these applications is quite different from that in laboratory nuclear reaction studies, and $\sigma_b(E)$ must be explicitly included for each situation. Recent low-energy studies[9] of ^{3}He(d,p)^{4}He clearly show such screening effects, as well as their dependence on the aggregate state of the target. Theoretical analyses suggest that a Born-Oppenheimer approximation cannot decribe the data. Clearly, a thorough understanding of screening effects requires additional efforts in theory as well as in experiment, where improved low-energy data for other reactions are needed.

4. Radioactive ion beams

In the hot and explosive burning phases of stars (or the early Universe) where E approaches E_C, nuclear burning times can be greatly reduced, even down to seconds. If the lifetime of a radioactive nucleus is longer than or of the same order as the burning time, that nucleus will be involved in the nuclear burning processes. A quantitative understanding of the observed nuclear ashes from these astrophysical scenarios requires a knowledge of the nuclear reaction rates, and this in turn requires measuring predominantly p- and α-induced reactions involving these radioactive nuclides. If the half-life of the nuclides is longer than a day or so, they may be made into a radioactive target[10]. However, in a great majority of interesting cases, the half-life is shorter (e.g. $T_{1/2} = 1$ sec for ^{8}Li). If the final nucleus is stable, as in ^{8}Li(α,n)^{11}B, the reaction can be studied experimentally via the inverse reaction[11]. If this possibility does not exist, the only direct method is to produce the radioactive nuclides in an accelerator, separate them, accelerate them in a second accelerator, and finally allow the radioactive ion beam to interact with a hydrogen or helium target. All of this must be achieved in a time shorter than the decay lifetime of the radioactive nuclides.

One of the most ambitious projects to develop radioactive ion beams is the TRIUMF proposal involving an isotope separator and linear accelerators[12]. A promising project is already well under way at Louvain-la-Neuve[13,14] to study the reaction $^{13}N(p,\gamma)^{14}O$. The ^{13}N nuclei are produced in the $^{13}C(p,n)^{13}N$ reaction by using an intense 30 MeV proton beam from one cyclotron. The ^{13}N are then ionized in an ion source, injected into a second cyclotron, accelerated to the appropriate energy, and finally directed onto a hydrogen target. After a number of test runs a successful measurement of this capture reaction has already been performed. After completion of this experiment, other radioactive ion beam studies with light nuclides are planned. Several other projects are in planning or development stages.

These projects are of particular value to nuclear astrophysics because such a facility would have the capability of systematically measuring $\sigma(E)$ for both light and medium mass nuclides. Further progress in some currently active areas of nuclear astrophysics will be greatly retarded if it is not possible to assist theoretical advances in estimating nuclear reaction rates with sound experimental data. It is clear that additional research and technical development are needed on all aspects of the problem (ion sources, accelerators, targets, detectors, and radiation safety), before nuclear reactions involving short-lived radioactive nuclides can be routinely studied in the laboratory.

References

1. W.A. Fowler, Rev. Mod. Phys. 56 (1984) 149
2. G.J. Mathews and R.A. Ward, Rep. Prog. Phys. 48 (1985) 1371
3. C. Rolfs, H.P. Trautvetter and W.S. Rodney, Rep. Prog. Phys. 50 (1987) 233
4. C. Rolfs and W.S. Rodney, Cauldrons in the Cosmos (University of Chicago Press, 1988)
5. C. Rolfs and C.A. Barnes, Ann. Rev. Nucl. Part. Sci. 40 (1990) 45
6. F.J. Wilkinson and F.E.Cecil, Phys. Rev. C31 (1985) 2036
7. C.A. Barnes, K.H. Chang, T.R. Donoghue, C. Rolfs and J. Kammeraad, Phys. Lett. B197 (1987) 315
8. H.J. Assenbaum, K. Langanke and C. Rolfs, Z. Phys. A327 (1987) 461
9. S. Engstler, A. Krauss, K. Neldner, C. Rolfs, U. Schröder and K. Langanke, Phys. Lett. B202 (1988) 179
10. S. Seuthe et al., Nucl. Phys. A514 (1990) 471
11. T. Paradellis et al., Z. Phys. A337 (1990) 211
12. L. Buchmann et al., Nucl. Instr. Meth. B26 (1987) 151
13. T. Delbar, M. Huyse and J. Vanhorenbeek, Belgian Interuniv. Report RIB 01 (1988)
14. W. Galster et al., Phys. Rev. (submitted)

MEASUREMENTS OF keV NEUTRON CAPTURE CROSS SECTIONS OF ^{122,123,124,125,126}Te WITH A 4π BARIUM FLUORIDE DETECTOR

K. Wisshak, F. Voß, F. Käppeler

Kernforschungszentrum Karlsruhe, Institut für Kernphysik,
POB 3640, W-7500 Karlsruhe, Germany

Abstract: The neutron capture cross sections of ^{122,123,124,125,126}Te have been measured in the neutron energy range from 10 to 200 keV using the Karlsruhe 4π Barium Fluoride Detector for the registration of capture gamma-ray cascades. Neutrons were produced via the ^{7}Li(p,n)^{7}Be reaction by bombarding metallic Li targets with the pulsed proton beam of a 3.75 MV Van de Graaff accelerator. The neutron energy was determined by time of flight. The flight path of the experiment was 78 cm, the time resolution 1 ns. The cross sections were determined relative to the standard cross section of gold. This new experimental setup allows to determine the cross section ratio with an overall uncertainty of ~1 % which is a significant improvement compared to older techniques. The data are used for studies of the nucleosynthesis of heavy elements in the so called s-process.

(neutron capture cross section, tellurium isotopes, 10<E_n<200 keV, barium fluoride scintillator, 4π gamma-ray detection, nucleosynthesis, s-process)

1. Introduction

The isotopes of tellurium play an important role in the studies of the synthesis of heavy elements in the so called s-process [1]. It is the only element with three isotopes, i.e. ^{122,123,124}Te, only produced by the s-process . The isotopic abundances of these isotopes are known with uncertainties of 0.1 %. Therefore, accurate stellar cross sections offer a unique possibility to check the "local approximation" predicted by the classical s-process model [2] that the product of s-abundance and stellar cross section is constant for a chain of three neighboring isotopes. In addition such data allow to set constraints on existing stellar models.

2. Experiment

In the present experiment the neutron capture cross sections of ^{122,123,124,125,126}Te have been measured in the neutron energy range from 10 to 200 keV using the Karlsruhe 4π Barium Fluoride Detector [3] for the registration of capture gamma-ray cascades. This detector consists of 42 hexagonal and pentagonal crystals forming a spherical shell of BaF$_2$ with 10 cm inner radius and 15 cm thickness. It is characterized by a resolution in gamma-ray energy of 7 % at 2.5 MeV, a time resolution of 500 ps, and a peak efficiency of 90 % at 1 MeV. Capture events are registered with ~95 % probability.

Neutrons were produced via the ^{7}Li(p,n)^{7}Be reaction by bombarding metallic Li targets with the pulsed proton beam of a 3.75 MV Van de Graaff accelerator. The neutron energy was determined by time of flight (TOF). The flight path of the experiment was 78 cm, the overall time resolution 1 ns. The repetition rate of the accelerator was 250 kHz, the average beam current 2 μA.

In order to optimize the signal to background ratio for different neutron energies three runs have been performed, selecting the energy of the proton beam 10, 30, and 100 keV above the reaction threshold at 1.881 MeV. This yields continuous neutron spectra in the energy ranges 10 - 70 keV, 3 - 100 keV and 3 - 200 keV , respectively. This is exactly the range of interest for s-process studies where the cross section has to be determined as an average over a Maxwellian distribution at kT=30 keV.

During the experiment a 64 bit word was recorded from each event on magnetic tape containing the sum-energy and TOF information together with 42 bits indicating those detector modules that have contributed to the sum-energy signal.

3. Samples

Metallic, isotopically enriched samples have been used ranging from 0.4 g (^{123}Te) to 4 g (^{126}Te) in weight. The weight was selected proportional to the inverse of the cross section in order to get similar capture yield per sample. An exact characterization of the sample is one of the severest problems for accurate cross section measurements. This was particularly difficult in the case of tellurium, since this element is known to absorb oxygen easily. Therefore, the material was carefully analysed for oxygen and possible hydrogen contaminations. In addition the isotopic analyses of the supplier were checked independently. The sample material was stored under argon atmosphere, the samples were welded into a thin polyethylene foil during the experiment, and their weight was controlled before and after the measurement to check for oxygen absorption. The metallic ^{123}Te powder supplied by ORNL proved to contain 12 % oxygen, while for the other isotopes which we had on loan from the USSR values below 3 % were measured. Hydrogen was not detected in the sample material.

4. Data Evaluation

The data evaluation is described in detail in Ref. 4. The events recorded on magnetic tape were sorted off line into 5 two dimensional spectra per sample according to multiplicities 1,2,3,4 and ≧5. This is very effective to separate capture events (multiplicity ≧3) from background (multiplicity ≦2). The shape of the cross section is determined by projecting the two dimensional spectra on the TOF axis in the region of the binding energy of the respective isotope. This is shown for the three different runs and the ^{122}Te sample in Fig. 1, where the background due to capture of sample scattered neutrons is also included. The ratio of total to capture cross section for this isotope is 23:1 at 30 keV. This yields a corresponding signal to background ratio at 30 keV, which increases from 4.0 to 5.8 and 7.1 for the runs with 200, 100 and 70 keV maximum neutron energy, respectively. Since the binding energy of ^{122}Te is 6.9 MeV, and since the background is dominated by capture events in ^{135,137}Ba, which are mainly located above 8 MeV, a large part of the background is eliminated by choosing appropriate limits in the projection. At low neutron energies the signal to background ratio is geting worse. In view of the comparably small cross section of the isotopes investigated here, it is not possible to extend the evaluation below 10 keV with sufficient statistical accuracy .

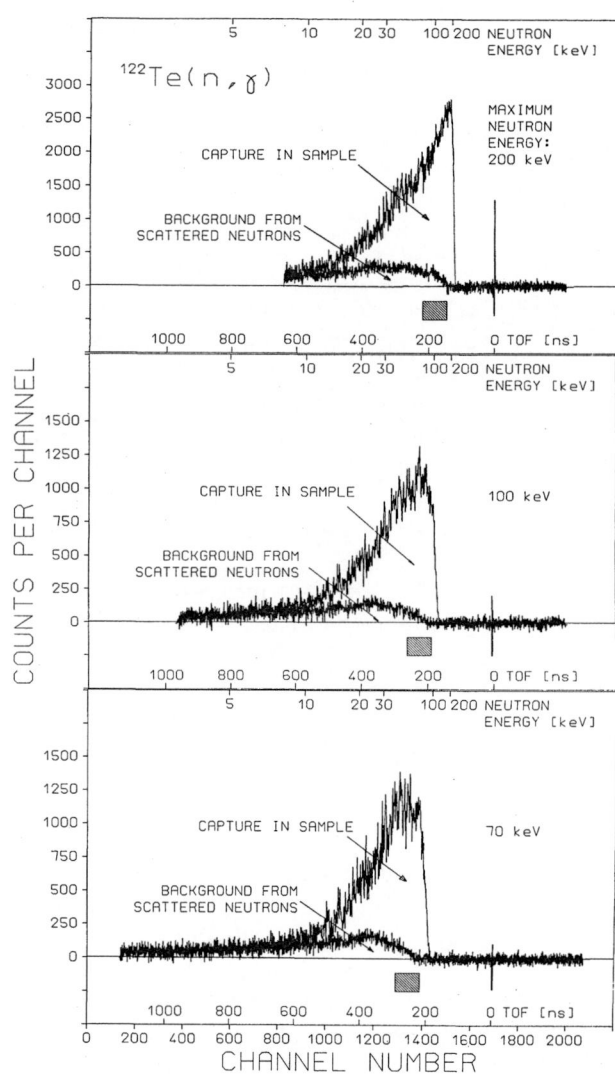

Fig.1 TOF spectra for neutron capture in the ^{122}Te sample demonstrating the different signal to background ratio in runs with 200, 100 and 70 keV maximum neutron energy (the dashed box indicates the TOF region used for normalization of the cross section shape (see Fig.2))

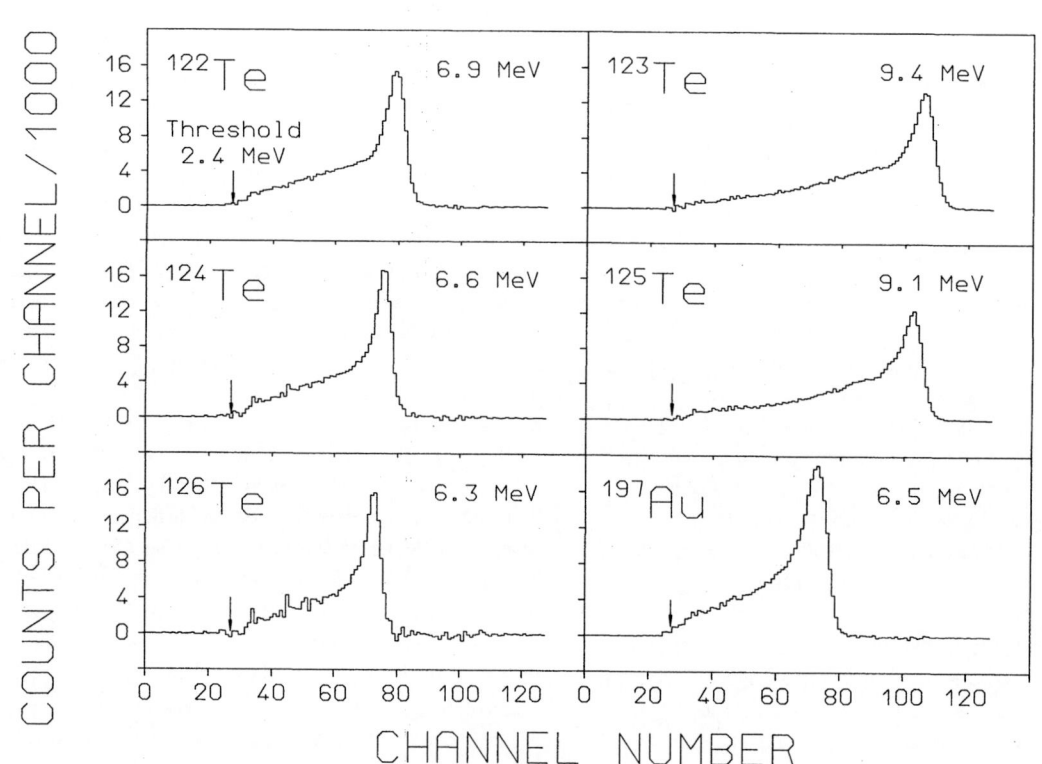

Fig.2 Sum-energy spectra used for normalization of the cross section shape (run with 200 keV maximum neutron energy). The experimental threshold at 2.4 MeV is indicated by arrows.

The normalization of the cross section is performed in the region with optimum signal-to-background ratio that is indicated by dashed boxes in Fig. 1. In this TOF region the two-dimensional spectra are projected on the sum-energy axis. The result is shown in Fig. 2. The fraction of events below the experimental threshold of 2.4 MeV indicated by arrows was determined using capture cascades calculated according to the statistical model and theoretical values for the detector efficiency. It is about 6 % for the even and 3 % for the odd tellurium isotopes. In the cross section ratio σ(Te)/σ(Au) only the ratio of this fraction for sample and standard has to be known, which proved to be very insensitive to the adopted detector efficiency. Thus the normalization could be performed with an uncertainty of ~0.7 %, which is the main systematic uncertainty of the present experimental method.

The correction for multiple scattering and self-shielding was calculated using the SESH code [5]. The small sample masses required corrections below 1 % for the odd and below 3 % for the even isotopes. In case of the ^{123}Te sample the oxygen content was properly taken into account.

In Fig. 3 the sum-energy spectra of the ^{122}Te sample are shown in dependence of the detector multiplicity. Multiplicities ≧5 are obtained for ~50 % of the events in the even isotopes and for ≧60 % in the odd isotopes. As can be seen, the background affects mainly the spectra with multiplicity 1 and 2 below ~3 MeV (channel number 40).

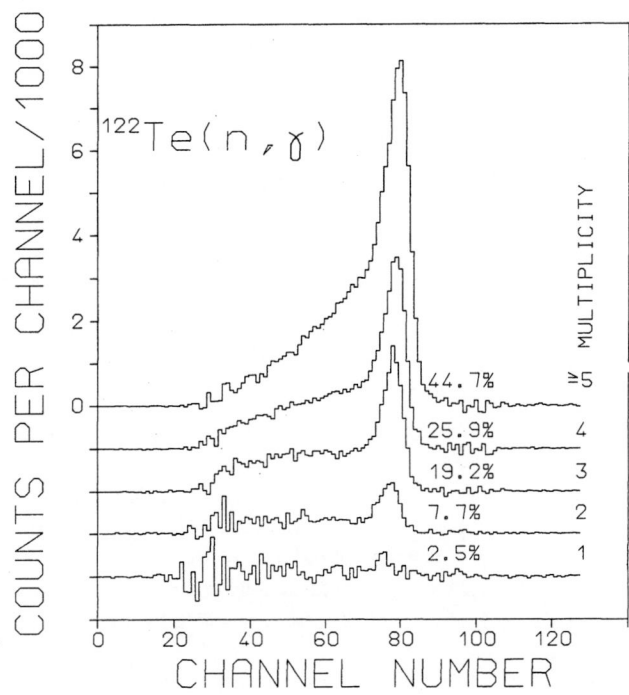

Fig.3 Sum-energy in dependence of the detector multiplicity as measured with the ^{122}Te sample.

5. Results

The experimental cross section ratios σ(Te)/σ(Au) have been converted into absolute values using the gold cross section from literature [6,7]. The result is shown in Fig. 4 for the three s—only isotopes. The statistical uncertainty in the quoted energy bins is less than 1 %. The systematic uncertainty is ~2 %. It is dominated by the 1.5 % uncertainty of the gold cross section. In s-process studies cross section ratios of two tellurium isotopes are sufficient in most cases. Then, the latter uncertainty cancels out and an overall uncertainty of ~1 % is finally obtained. In Fig. 4 the present results are compared to the data of Macklin and Winters [8]. Despite of the agreement between the two data sets, it was not before the present measurement that the s-abundances in the tellurium isotopes could be worked out in sufficient detail to constrain current stellar models.

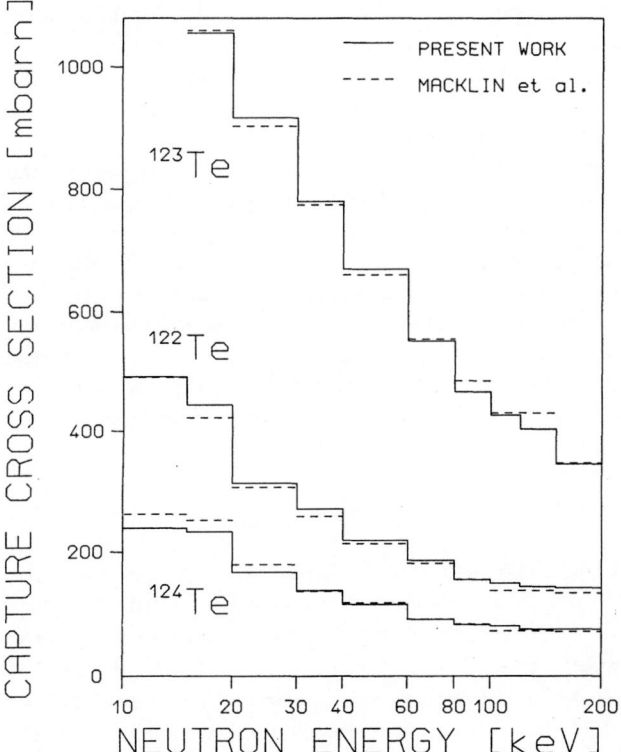

Fig.4 The neutron capture cross section of the three s-only isotopes ^{122,123,124}Te. A comparison is made to the data of Macklin and Winters [8].

References

1. F. Käppeler, H. Beer and K. Wisshak: Rep.Progr.Phys. **52**, 945 (1989)
2. F. Käppeler, R. Gallino, M. Busso, G. Picchio, and C.M. Raiteri: Ap.J. **354**, 630 (1990)
3. K. Wisshak, K. Guber, F. Käppeler, J. Krisch, H. Müller, G. Rupp, and F. Voß: Nucl.Instr.Meth. **A292**, 595 (1990)
4. K. Wisshak, F. Voß, F. Käppeler, and G. Reffo: Phys.Rev. **C42**, 1731 (1990)
5. F.H. Fröhner, SESH-A Fortran IV Code for Calculating the Self-Shielding and Multiple Scattering Effects for Neutron Cross Section Data Interpretation in the Unresolved Resonance Region, GA-8380, Gulf General Atomic (1968)
6. R.L. Macklin, private communications.
7. W. Ratynski and F. Käppeler: Phys.Rev. **C37**, 595 (1988)
8. R.L. Macklin and R.R. Winters, Neutron Capture of ^{122}Te, ^{123}Te, ^{124}Te, ^{125}Te, and ^{126}Te, ORNL-6561, Oak Ridge National Laboratory (1989)

^{147}Pm – AN EXPERIMENTAL STELLAR CROSS SECTION FOR A SHORT-LIVED s-PROCESS BRANCHING POINT

Th. W. Gerstenhöfer, F. Käppeler, K. Wisshak

Kernforschungszentrum Karlsruhe, Institut für Kernphysik,
W-7500 Karlsruhe, Federal Republic of Germany

G. Reffo

ENEA, Laboratorio Dati Nucleari, I-40138 Bologna

Abstract: Radioactive nuclei with $t_{1/2} < 10$ yr hold a key position in the synthesis of the heavy elements. In the s-process, neutron captures occur at rates of ~1yr and are, therefore, comparable to the beta decay rates of some of the involved isotopes, such as ^{147}Pm ($t_{1/2} = 2.6$ yr). The resulting branching of the neutron capture flow yields a characteristic abundance pattern for the neighboring isotopes that can be analyzed in terms of the physical conditions during the s-process. In case of ^{147}Pm, these studies yield an estimate for the stellar neutron density. For the quantitative discussion of this scenario, the neutron capture cross section of ^{147}Pm was measured for the first time in the keV range via two independent methods: (i) direct detection of the prompt capture gamma-rays by means of Moxon-Rae detectors, using the TOF technique at a neutron flight path of only 2 cm, and (ii) activation in a quasi-stellar neutron spectrum and detection of the induced ^{148}Pm activity. Both measurements were carried out relative to the standard cross section of ^{197}Au.

(^{147}Pm, neutron capture cross section, E_n = 25 keV, activation technique, TOF method, s-process, stellar nucleosynthesis)

Introduction

The studies of stellar nucleosynthesis in the s-process (s = slow neutron capture) are presently becoming sufficiently detailed that the observed abundances can be reproduced quantitatively. Particular emphasis is put on the interpretation of the abundance patterns in s-process branchings, which occur when the neutron capture chain encounters an unstable isotope with a half-life comparable to the typical neutron capture time of about 1 year. The strength of the resulting s-process branching is determined by the physical conditions at the s-process site, i.e. by the prevailing neutron density and temperature. In turn, this information can be deduced from such branchings by s-process analyses, and allows to constrain models of the Red Giant phase of stellar evolution.

The branchings at A = 147, 148, 149 are especially suited for such an investigation. In this case, the branching factor, which is defined by the rates for beta decay, $\lambda_\beta = \ln2/t_{1/2}$, and for neutron capture, $\lambda_n = \langle\sigma\rangle n_n v_T$ (v_T being the mean thermal velocity),

$$f_\beta = \frac{\lambda_\beta}{\lambda_\beta + \lambda_n} \, , \qquad (1)$$

can be inferred from the two s-only isotopes ^{148}Sm and ^{150}Sm. The solid lines connecting the various isotopes in Fig. 1 indicate the s-process neutron capture flow in the mass region 144 < A < 150 . The s-process branchings that occur at ^{147}Nd, and at the promethium isotopes 147, 148, and 149 merge again at ^{150}Sm. Since the r-process beta decay chains at A = 148 and 150 are shielded by the respective Nd isobars, ^{148}Sm and ^{150}Sm are of pure s-process origin. While ^{148}Sm is partly bypassed, ^{150}Sm experiences the total s-process flow. Hence, the abundance ratio of these nuclei – weighted with their stellar (n,γ) cross sections – represents a sensitive measure of the branching factor [1]. A precise determination of these cross sections is presently under way at Karlsruhe using the 4π BaF$_2$ detector [2, 3].

The last piece of information required to deduce the neutron density from these branchings are the neutron capture cross sections of the unstable branch point nuclei. So far, experimental data are not available for any of these. To our knowledge, this contribution describes the first measurements on ^{147}Pm at keV neutron energies. Apart from their immediate impact on the branching analysis, these results can also be used for improving statistical model calculations on the neighboring branch point nuclei.

Measurements

The measurements were carried out at the Karlsruhe 3.75 MV Van de Graaff accelerator. Neutrons were produced by means of the ^{7}Li(p,n)^{7}Be reaction. For proton energies of 1912 keV this reaction is known to yield a quasi-stellar neutron spectrum for a thermal energy of kT = 25 keV, very close to the conditions during the s-process [4, 5]. The ^{197}Au(n,γ) reaction was used as a cross section standard [5, 6].

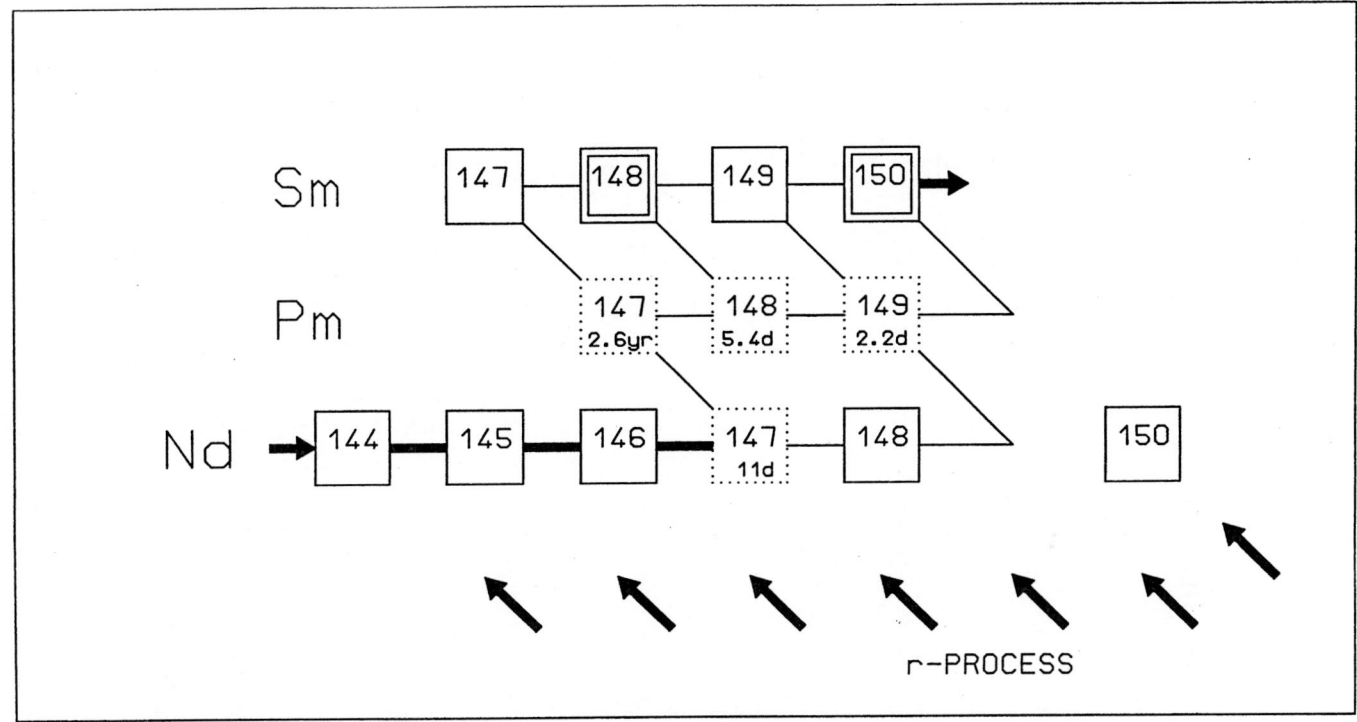

Fig. 1 The *s*-process flow in the mass region 144 < A < 150 (solid lines). Branch points are indicated by dotted boxes.

A. Sample preparation

A ^{147}Pm sample was prepared such that it could be used in both measurements. Freshly separated Pm_2O_3 powder with a total activity of about $1.5 \cdot 10^{11}$ Bq was sandwiched between two lead foils, and then pressed to a compact disk. The foils (10 mm in diameter and 0.1 mm thick) were of enriched ^{208}Pb (99.8 %) in order to minimize backgrounds in the TOF measurement.

This sample was mounted between two 5 μm thick polyethylene foils that were fixed to a glass fiber frame such that the sample was centered in the opening of 110 mm diameter. Since all neutrons from the ^{7}Li(p,n)^{7}Be reaction at $E_p = 1912$ keV are emitted in a forward cone of 120 deg, this geometry ensures that the frame was completely outside of the neutron beam, at least at the small target-sample distances used in the measurements. For transportation, the sample was enclosed in a suited steel container.

B. Activation measurement

The ^{147}Pm sample was irradiated between two gold foils in a flux of about 10^9 neutrons per second for 47.9 h. The induced ^{148}Pm activity was then counted by means of a HPGe detector. From the decay scheme of Fig. 2, gamma-ray lines associated with the decay of the short-lived ground state ($t_{1/2} = 5.4$ d) are expected at 611 and 1465 keV. The decay of the isomer ($t_{1/2} = 41.3$ d) yields lines at 630, 726, and 1014 keV, whereas the 550 and 915 keV transitions are common to both decay modes. The relative gamma-ray intensities of the ^{148}Pm transitions were taken from [7]. The gold activity was determined via the well-known intensity of the 411.8 keV transition [8].

Among the observed lines, those at high energies were most useful for the activity determination, because the low energy part of the spectrum was dominated by gamma-rays from a ^{146}Pm impurity and from sample-induced backgrounds. Fig. 3 shows the measured gamma-ray spectrum of the activated Pm sample, and the insets illustrate the relevant lines in detail. Note, that the isomer decay appears fainter due to its longer half-life.

The number of ^{148}Pm nuclei produced during the irradiation are

$$N_A = \Phi \cdot N_T \cdot \sigma \cdot f_b , \qquad (1)$$

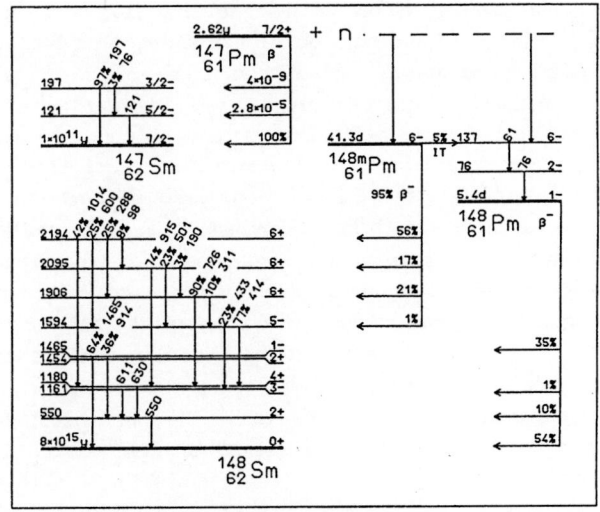

Fig. 2 The decay scheme of ^{148}Pm and the relevant transitions used in the activation measurement. Energies are in (keV), intensities in (%).

634

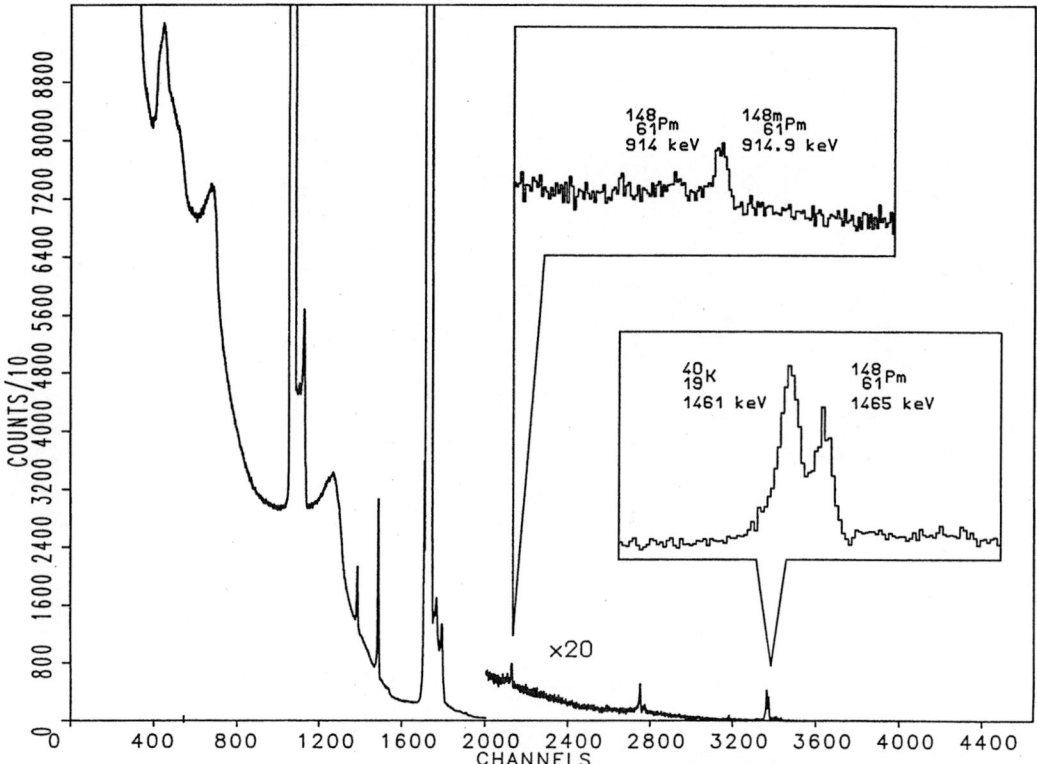

Fig. 3 The gamma-ray spectrum of the activated promethium sample. The relevant transitions for the cross section determination are presented in the insets.

where Φ is the integrated neutron flux, N_T the number of target atoms per cm^2, and σ the spectrum averaged cross section. The factor f_b corrects for fraction of produced nuclei that decay during the activation.

The number of activated nuclei can be deduced from the gamma-ray spectrum of Fig. 3,

$$N_A = \frac{C_A}{K_\gamma \cdot \varepsilon \cdot I_\gamma \cdot f_w \cdot f_m} \ , \qquad (2)$$

where C_A are the net counts of the ^{148}Pm lines. K_γ is the correction factor for absorption in the sample and in a 3 mm thick lead shielding between sample and detector, ε denotes the efficiency of the HPGe detector, and I_γ the relative intensity per decay. The time factors f_w and f_m account for the decays during the waiting time between irradiation and activity measurement [5].

Combining equations (1) and (2) with the corresponding expressions for the gold samples, and adopting the gold reference cross section for a thermal energy of kT = 25 keV (σ = 648 ± 10 mbarn [5]), yields eventually the stellar cross section for ^{147}Pm. However, the high specific activity of the promethium sample led to considerable dead time effects in the gamma-ray spectrum, which are difficult to quantify reliably. This problem can be circumvented by using the fact that the spectrum contains the 122 keV gamma-ray line of ^{147}Pm as well. Since mother and daughter activity are observed in the same spectrum, the ratio N_A/N_T can be expressed by the ratio of the respective count rates

$$\frac{N_A}{N_T} = \frac{C_A}{C_T} \frac{(K_\gamma \cdot \varepsilon \cdot I_\gamma \cdot f_w \cdot f_m)_T}{(K_\gamma \cdot \varepsilon \cdot I_\gamma \cdot f_w \cdot F_m)_A} \ . \qquad (3)$$

Application of this method to the transitions at 1014 and 915 keV yields a preliminary cross section in agreement with the calculated value of 1275 ± 260 mbarn [1]. Before

a quantitative result can be presented, additional studies of the dead time and pile-up effects must be performed. From the actual status of data analysis we estimate a 10 % uncertainty for the final cross section.

C. Time-of-flight experiment

Three Moxon-Rae detectors with different converter materials [9] were used to detect the prompt gamma-rays from the ^{147}Pm(n,γ)^{148}Pm reaction. The overall time resolution of the setup of ≤1 ns allowed the separation of the capture events from the prompt gamma-ray peak in the TOF spectrum even at a flight path of only 2 cm. Hence, the sample could be exposed to a high neutron flux of $\sim 10^8$ s^{-1}, resulting in a significant signal-to-background ratio. Though data analysis is still under way, it can be stated that this technique is well suited for investigating even small amounts of radioactive samples.

References

1. R. R. Winters, F. Käppeler, K. Wisshak, A. Mengoni and G. Reffo: Ap. J. 300, 41 (1986)
2. K. Wisshak, F. Voß and F. Käppeler, contribution to this conference
3. K. Guber, K. Wisshak, F. Voß and F. Käppeler (in preparation
4. H. Beer and F. Käppeler: Phys. Rev. C 21, 534 (1980)
5. W. Ratynski and F. Käppeler: Phys. Rev. C 37, 595
6. R. L. Macklin: private communication
7. C. M. Lederer and V. S. Shirley: *Table of Isotopes* (New York: Wiley 1978)
8. R. L. Auble: Nucl. Data Sheets 36, 411 (1983)
9. S. Jaag: Diplom thesis University of Karlsruhe (1990)

R-PROCESS ABUNDANCES AND NUCLEAR PROPERTIES FAR FROM STABILITY

K.-L. Kratz[1], H. Gabelmann[1] P. Möller[1,2] , B. Pfeiffer[1], A. Wöhr[1] and F.-K. Thielemann[3],

[1]Institut für Kernchemie, Universität Mainz, 6500 Mainz,
Federal Republic of Germany

[2]Lund Institute of Technology, 22100 Lund, Sweden

[3]Harvard-Smithsonian Center for Astrophysics, Cambridge, MA 02138, USA

Abstract: Recent measurements of β-decay properties of the 'waiting-point' nuclei ^{79}Cu, ^{80}Zn and ^{130}Cd, together with new QRPA shell-model predictions of so far unknown $N \simeq 50$ and $N \simeq 82$ isotopes in the r-process path, have allowed to explain the detailed isotopic composition in the $A \simeq 80$ and $A \simeq 130$ r-abundance peaks. The correlation between nuclear data far from stability and r-abundances suggests that the r-process involves a high-neutron-density β-flow equilibrium environment. Based on these results, the r-process components of nuclei in the $90 \leq A \leq 100$ mass range were predicted for freeze-out conditions ($n_n \simeq 10^{20}$, $T_9 \simeq 1$) and compared to the solar-system r-process abundances.

[$T_{1/2}$ and P_n of ^{79}Cu, ^{80}Zn and ^{130}Cd, QRPA-predictions; r-process 'waiting-point' concept, constraints on r-process conditions, prediction of r-abundances]

Introduction

Nucleosynthesis theory predicts that the formation of the nuclear species beyond iron occurs in nature as a consequence of neutron-capture processes (see, e.g., [1-3]). Strong support for this view is provided by the fact that the abundance features in the heavy element region are correlated with the neutron shell closures at $N = 50$, 82 and 126. Moreover, the splitting of the solar-system abundance peaks (see, e.g., Fig. 2.1 in [3]) reveals evidence for two distinct neutron-capture processes characterized by two quite different neutron fluxes. This has led historically to the definition of two nucleosynthesis processes, the s-process (slow neutron capture) and the r-process (rapid neutron capture), that are identified with quite different astrophysical environments.

The r-process is an important nucleosynthesis mechanism for a number of reasons: (1) it is crucial to an understanding of the $A > 60$ elemental composition of the Galaxy; (2) it is the mechanism that forms the long-lived Th-U-Pu nuclear chronometers which are used for cosmochronology; (3) it provides an important probe for the temperature - density conditions in explosive events; and last but not least (4) it serves to provide useful clues to and constraints upon the nuclear properties of very-neutron-rich heavy nuclei.

Despite its importance, the r-process is probably the most poorly understood of the stellar nucleosynthesis mechanisms. One reason for this lack of understanding is the fact that the r-abundances are so rare (only $\sim 10^{-6}$ $M_\odot$ per supernova event) that they have been difficult to observe astronomically. Another reason is that the extremely neutron-rich isotopes formed during the r-process have been impossible to study in terrestrial laboratories.

Therefore, with respect to the relevant physical quantities, it has been essential to find the most reliable predictions for nuclei far off stability using - as far as possible - a consistent set of nuclear models (see, e.g., [3,4]). On the other hand, it remains of utmost importance to continue experimental efforts to underline measure at least some of the crucial β-decay properties of isotopes lying in the r-process path. In the following, we present new nuclear-physics data, which - in combination with recently improved QRPA shell-model predictions - allow one to put constraints upon the stellar conditions under which the r-process has operated [5-7].

Beta-Decay Properties of 'Waiting-Point' Nuclei

A striking feature of the solar-system r-process abundances ($N_{r,\odot}$) is the presence of sharp peaks at $A = 130$ and $A = 195$ and to a lesser extent around $A = 80$ (see, e.g., Fig. 2.4 in [3]), which are correlated with the positions of the neutron closed shells at $N = 50$, 82 and 126. It is in the reproduction of the ($N_{r,\odot}$) that the nuclear-physics properties of the isotopes produced during the r-process are most crucial. At the magic neutron numbers, the r-process path bends and comes closest to β-stability. Hence, it is only in these regions where experiments in terrestrial laboratories may be possible at the extreme borderlines of production and detection. And in fact, within the last few years some of the important so-called 'waiting-point' nuclei at $A \simeq 80$ and $A \simeq 130$ could be measured.

Most of the experiments [10-12] were performed at the CERN- ISOLDE on-line mass separator by irradiating a ^{238}UC-graphite target with a beam of 600 MeV protons at a current of 2.2 μA from the Synchro-Cyclotron. The target was heated to $2,200 - 2,400°C$, and connected to a high-temperature plasma-discharge ion source via a heated transfer line. This line served as a kind of thermochromatographic column allowing preferential extraction of volatile species, whereas less- or non-volatile elements were retained. Growth-and-decay curves, with collection and cycle times optimized for each isotope of interest, were measured by delayed-neutron multiscaling. In this way, the β-decay half-lives ($T_{1/2}$) of the classical 'waiting-point' nuclei ^{130}Cd$_{82}$ [10] and ^{79}Cu$_{50}$ [11] could be measured for the first time. Furthermore, the β-delayed neutron emission probabilities (P_n) of the two $N = 50$ isotones ^{79}Cu and ^{80}Zn were determined [11]. In addition, the $T_{1/2}$ of $^{74-78}$Cu were measured with high accuracy; and two new isotopes lying 'beyond' the r-process path defined in [5,6], i.e. ^{81}Zn$_{51}$ and ^{84}Ga$_{54}$, were identified. Table 1 summarizes the $T_{1/2}$ and P_n-values of all $N \simeq 50$ and $N \simeq 82$ isotopes [8-13] lying in or close to the r-process path.

Constraints from r-Abundances and Nuclear Properties

Based on these experimental results, together with QRPA predictions [4] of β-decay properties of so far unknown $N = 50$

Table 1. $T_{1/2}$ and P_n-values of $N \simeq 50$ and 82 isotopes in or close to the r-process path.

Isotope	$T_{1/2}$ [ms]	P_n [%]	Reference
^{77}Cu	469 ± 8	not det.	[11]
^{78}Cu	342 ± 11	not det.	[11]
^{79}Cu	188 ± 25	55 ± 17	[11]
^{80}Zn	540 ± 10	1.0 ± 0.5	$T_{1/2}$[8,9,11];P_n[11]
^{81}Zn	290 ± 50	7.5 ± 3.0	[11]
^{81}Ga	1221 ± 5	12 ± 2	$T_{1/2}$[11,13];P_n[13]
^{82}Ga	599 ± 2	21 ± 1	$T_{1/2}$[11];P_n[13]
^{83}Ga	308.4 ± 1.3	55 ± 8	$T_{1/2}$[11];P_n[13]
^{84}Ga	85.4 ± 9.5	70 ± 15	[11]
^{84}Ge	984 ± 23	9.5 ± 2.0	[12]
^{85}Ge	535 ± 47	14 ± 3	[11]
^{85}As	2032 ± 12	53 ± 10	[11]
^{130}Cd	195 ± 35	~ 2.5(est.)	[10]
^{131}In	278 ± 3	1.7 ± 0.2	[13]
^{133}In	180 ± 20	87 ± 9	[12]

and 82 'waiting-point' nuclei, strong evidence has been obtained that the $N_{r,\odot}$ in the $A \simeq 80$ and 130 r-abundance peaks result from a simultaneous high-neutron-density (n,γ)-(γ,n) and a beta-flow equilibrium [5,6]. While for the $A \simeq 130$ peak a consistent picture was obtained, establishing - after correction for P_n-branching - a direct relation between the experimental $T_{1/2}$ of ^{131,133}In and ^{130}Cd and the $N_{r,\odot}$ of their stable isobars ^{131,132}Xe and ^{130}Te (see, e.g., Fig. 2 in [5]), initially some difficulties were encountered in the $A \simeq 80$ region. The r-process yields related to the $N = 50$ shell closure form an abundance plateau rather than a sharp peak which, in addition, exhibits a pronounced odd-even staggering. It could, however, be shown recently [14] that this behaviour originates from β-delayed neutron branching of $26 \leq Z \leq 34$ isotopes (see Fig. 1). It is worth to be mentioned in this context that our new data on ^{79}Cu$_{50}$ and the low P_n-value of ^{80}Zn$_{50}$ agree closely with the earlier QRPA predictions [5], thus strengthening our above arguments.

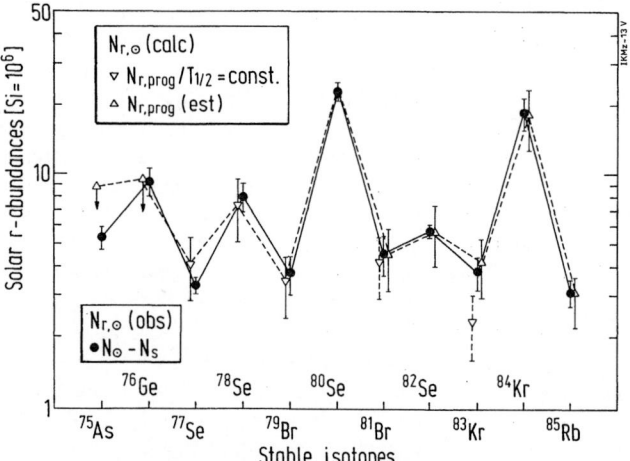

Fig. 1. Comparison of observed $N_{r,\odot}$ [2] with calculated abundances, $N_{r,\odot}$(calc), assuming approximate equality of progenitor abundance times β-decay rate (from [14]).

Given the result that the r-process occurs in a classical (n,γ) equilibrium environment, one can now ask whether the isotopic patterns found in the $N_{r,\odot}$ together with our knowledge of nuclear properties can give a clue to the stellar conditions under which the r-process has operated. In [5,6], the necessary time scale as

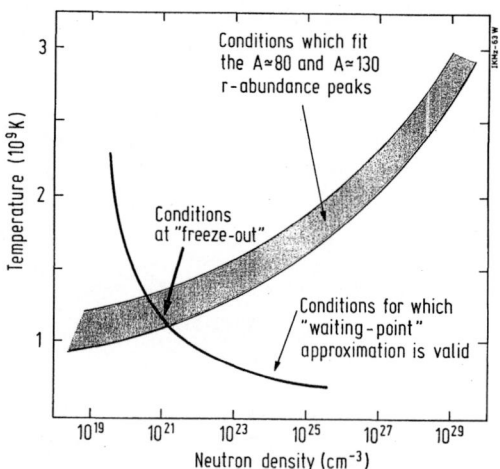

Fig. 2. Neutron densities (n_n) and temperatures (T) consistent with observed $N_{r,\odot}$-peaks and measured $T_{1/2}$ and P_n-values [5,6]. Also shown are the conditions for which the classical 'waiting-point' approximation is valid [15]. The thick arrow indicates the freeze-out conditions for a β-flow r-process.

well as stellar temperature (T) and neutron density (n_n) ranges have been determined which give the 'correct' steady-flow r-process path through the 'waiting-point' nuclei ^{76}Fe to ^{80}Zn and ^{127}Rh to ^{130}Cd. With the additional information from the $N_{r,\odot}$ of ^{81}Br and ^{83}Kr which originate from ^{81,83}Ga, and the $N_{r,\odot}$ of ^{131,132}Xe which were formed by the progenitors ^{131,133}In, rather tight constraints for the T - n_n conditions could be given (see dashed area in Fig. 2).

Recently, Mathews and Cowan [7] have deduced a critical neutron density of $n_n \simeq 10^{20}$cm^{-3} from our measured $T_{1/2}$ and the above assumption that the r-process does indeed flow through the closed-shell 'waiting-point' nuclei. For a classical r-process, this value corresponds to the minimum n_n required for the 'waiting-point' approximation [15]. Combining our T - n_n conditions [5,6] with the above critical value for n_n, one can determine the freeze-out temperature for the r-process (see Fig. 2). One may, however, argue that initially the r-process occurs at slightly higher temperature and significantly higher neutron density, at a different r-process path further away from β-stability involving isotopes with shorter $T_{1/2}$ than for the 'waiting-point' nuclei. What one probably observes as final solar-system abundances seems to reflect the T - n_n conditions just at freeze-out. This then could also explain that the $N_{r,\odot}$-peaks are so sharp. Such narrow peaks can only occur if most of the r-process material was produced in a single well defined environment, i.e. at the freeze-out conditions depicted in Fig. 2, rather than some combination of different environments with different T and n_n.

Prediction of R-Abundances in the $90 \leq A \leq 100$ Region

If we accept the implication of the above data that for the neutron closed shell nuclei at $A \simeq 80$ and $A \simeq 130$ there exists a β-flow equilibrium which implies that the ratio of progenitor abundance ($N_{r,prog}$) to progenitor half-life is constant, we then can use this requirement to also predict the $N_{r,prog}$ and - after correction for P_n-branching during β-decay back to stability - the $N_{r,\odot}$ of isotopes in between the magic neutron numbers, i.e. in between the r-abundance peaks.

As a first test, we have chosen the mass region $90 \leq A \leq 100$ for two reasons: (1) the stellar origin of ^{96}Zr is of special interest in connection with the recently observed correlated ^{50}Ti—^{96}Zr isotopic anomalies in several EK inclusions of the Allende me-

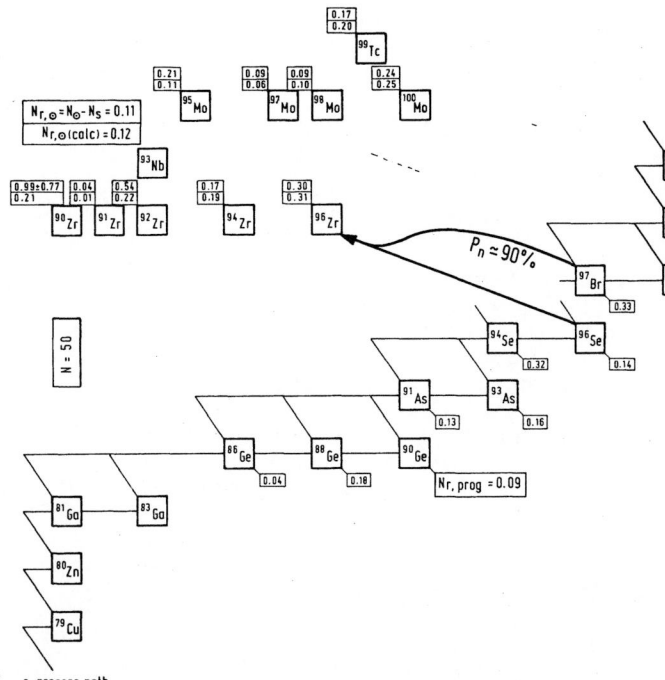

r-process path

Fig. 3. Schematic view of the r-process path at freeze-out conditions and r-abundance features in the $90 \leq A \leq 100$ mass region. From the progenitor abundances ($N_{r,prog}$), after β-decay and P_n-branching, the final abundances ($N_{r,\odot}$(calc)) of the stable isotopes ^{90}Zr to ^{100}Mo were derived. They are in fair agreement with the observed r-process residuals $N_{r,\odot}$ from [2]. Under these conditions, ^{96}Zr would be a 'r-only' isotope, mainly originating from the precursor ^{97}Br. For further discussion, see text.

teorite (see, e.g., [16,17]), and (2) ^{100}Mo is a 'r-only' isotope which can - if necessary - be used for an absolute normalization of the calculated r-abundances to the observed $N_{r,\odot}$ [2].

For freeze-out conditions (see Fig. 3.), we first have calculated relative r-abundances for each isotopic chain ($\sum N_{r,prog}(Z) = 1$; for details of the procedure, see [3,5]). These relative $N_{r,prog}$ were then weighted according to their predicted $T_{1/2}$ [18] and normalized to the neutron closed shell isotopes ^{80}Zn$_{50}$ and ^{130}Cd$_{82}$. In both cases, P_n-branching is so small that one can assume approximate equality between initial $N_{r,prog}$ and final $N_{r,\odot}$. At this point it turned out that the values of $N_{r,\odot}(^{80}\text{Se})/T_{1/2}(^{80}\text{Zn}) \simeq 42.5$ s^{-1} and $N_{r,\odot}(^{130}\text{Te})/T_{1/2}(^{130}\text{Cd}) \simeq 11.5$ s^{-1} were not equal, as initially expected from the global waiting-point concept. This is a potentially important result which may indicate that - similar to what is assumed for the s-process [2] - also the r-process has different components (each with a β-flow equilibrium) to build up the $N_{r,\odot}$ from $A \simeq 70$ to $A \simeq 240$.

A schematic view of the r-process path at freeze-out conditions and the r-abundance features in the $90 \leq A \leq 100$ region is shown in Fig. 3. The most interesting result is, that we obtain absolute r-abundances in fair agreement with the observed $N_{r,\odot}$ from [2], which need no further scaling, if we normalize to ^{130}Cd. This is most evident from the simultaneous reproduction of the abundance of the 'r-only' isotope ^{100}Mo and the $N_{r,\odot}$ of ^{99}Tc, which both originate from the progenitor ^{100}Kr. Another important result is, that under the above $T - n_n$ conditions ^{96}Zr would also be a 'r-only' isotope, mainly produced from β-delayed neutron decay of ^{97}Br. Although these results are rather sensitive to the neutron density assumed, even an order-of-magnitude agreement of our predictions with the (also rather uncertain) $N_{r,\odot}$-residuals [2] will put additional constraints on the r-process environment as well as on nuclear properties far off stability.

Conclusion

While continued improvements have been evident in nuclear physics, the astrophysical site of the r-process is still not clearly established. However, from the new nuclear input data and recent astronomical observations (for details, see [3,7]), one can draw the following conclusions on the r-process environment: (1) Shock-induced explosive He or C burning in supernovae (SN) or He-core flashes in low-mass stars seem to be ruled out. All present calculations show that for realistic stellar models, neither enough neutrons are produced by (α, n)-reactions nor is the available exposure time long enough to allow a β-flow process which can reproduce the observed $N_{r,\odot}$-pattern. (2) Only models assuming type II SN sites seem to fit the astronomical data on [Eu/Fe] and [Fe/H] ratios, and the evidence points to the inner neutronized cores. Hence, taking together all recent information [3,5,7], the current state of affairs indicates that the core ejections of highly neutronized matter from low-mass type II SN may be the most likely site for the r-process.

We would like to thank the staff of the ISOLDE separator at CERN for the excellent support given for the experiments. This work was supported by grants from BMFT (06 MZ 106), DFG (Kr 806/1), and from NSF (AST-8913799).

REFERENCES

1. G.R. Burbidge, E.M. Burbidge, W.A. Fowler and F. Hoyle: Rev. Mod. Phys. 29, 547 (1957)
2. F. Käppeler, H. Beer and K. Wisshak: Rep. Prog. Phys. 52, 945 (1989)
3. J.J. Cowan and F.-K. Thielemann: Phys. Rep. (1991), in print
4. P. Möller, J.R. Nix, K.-L. Kratz and W.M. Howard: Proc. Int. Symp. on Nuclear Astrophysics, MPA/P4, 266 (1990)
5. K.-L. Kratz, F.-K. Thielemann, W. Hillebrandt, P. Möller, V. Harms, A. Wöhr and J.W. Truran: Inst. Phys. Conf. Ser. 88, J. Phys. G Suppl. 14, 5331 (1988)
6. K.-L. Kratz: Revs. Mod. Astron. 1, 184 (1988)
7. G.J. Mathews and J.J. Cowan: Nature 345, 491 (1990)
8. B. Ekström B. Fogelberg, P. Hoff, E. Lund and A. Sangariya-vanish: Phys. Scr. 34, 614 (1986)
9. R.L. Gill, R.F. Casten, D.D. Warner, A. Piotrowski, H. Mach, J.C. Hill, F.K. Wohn, J.A. Winger and R. Moreh: Phys. Rev. Lett. 56, 1874 (1986)
10. K.-L. Kratz, H. Gabelmann, W. Hillebrandt, B. Pfeiffer, K. Schlösser, F.-K. Thielemann and the ISOLDE Collaboration: Z. Physik A325, 483 (1986)
11. K.-L. Kratz, H. Gabelmann, P. Möller, B. Pfeiffer, A. Wöhr and the ISOLDE Collaboration: Verhandl. DPG (VI) 26, 438 (1991); and subm. to Z. Physik A
12. G. Nyman: priv. communication
13. Proc. Specialists' Meeting on Delayed-Neutron Properties, Univ. of Birmingham Rep., ISBN 07044 03267 (1987)
14. K.-L. Kratz, V. Harms, W. Hillebrandt, B. Pfeiffer, F.-K. Thielemann and A. Wöhr: Z. Physik A336 357 (1990)
15. A.G.W. Cameron, J.J. Cowan and J.W. Truran: Ap. Space Sci. 91, 235 (1983)
16. C.L. Harper, L.E. Nyquist, C.Y. Shih and W. Wiesmann: Proc. Int. Symp. on Nuclear Astrophysics, MPA/P4, 138 (1990)
17. F.-K. Thielemann, K.-L. Kratz, B. Pfeiffer, P. Möller and W. Hillebrandt: ibid., p. 286 (1990)
18. P. Möller and K.-L. Kratz: Table of QRPA-Predictions of $T_{1/2}$ and P_n-values (1991), unpublished.

(n,p) AND (n,α) CROSS-SECTION MEASUREMENTS WITH ASTROPHYSICAL APPLICATIONS

C. Wagemans*
Nuclear Physics Laboratory, State University, B-9000 Gent

S. Druyts, H. Weigmann, R. Barthélémy
Commission of the European Communities, Joint Research Centre
Central Bureau for Nuclear Measurements, B-2440 Geel, Belgium.

P. Schillebeeckx**, P. Geltenbort
Institut Laue-Langevin, F-38042 Grenoble, France

Abstract: (n,p) and (n,α)-cross-sections with astrophysical importance have been measured for light nuclei with masses between 15 and 50 u. Thermal neutron induced measurements were performed at the high flux reactor of the I.L.L. (Grenoble); the resonance and keV neutron energy region was covered at the Geel Linear Accelerator. The importance of these data is illustrated by means of a typical s-process nucleosynthesis reaction network.

(^{14}N, ^{33}S, ^{35,36}Cl, ^{40}K, ^{41}Ca, ^{50}V, (n,p) and/or (n,α) cross-sections, range 0.020 eV - 1 MeV, Maxwellian averaged cross-sections)

Introduction

In his Noble Prize lecture W. Fowler [1] defines the goals of nuclear astrophysics as follows: "(i) Understand energy generation in the sun and other stars at all stages of stellar evolution (...). (ii) Understand the nuclear processes which produced under various astrophysical circumstances the relative abundances of the elements and their isotopes in nature".

The present results contribute to realize this second goal by providing previously unavailable nuclear data. The (n,p)- and (n,α)-cross-sections reported are indeed important for the neutron-dominated phase in the stellar evolution. As explained e.g. in the review paper by F. Käppeler et al. [2], the s-process nucleosynthesis takes place at stellar temperatures of a few 10^8 K, which corresponds to neutron energies of the order 10 to 30 keV. In the "two pulses model" the s-process nucleosynthesis is caused by two consecutive neutron bursts, the main one originating from the ^{13}C(α,n)^{16}O reaction (T ≈ 1.5 x 10^8 K or ≈ 12 keV neutron energy), the smaller one being due to the ^{22}Ne(α,n)^{25}Mg reaction (T ≈ 3 x 10^8 K or ≈ 25 keV neutron energy). Consequently, large (n,p)-and/or (n,α)-reaction cross-sections in this neutron energy region will have a direct impact on the abundances of the naturally occurring isotopes synthesized via explosive nucleosynthesis.

Experimental conditions

The (n,p)- and (n,α)-reactions have been studied under two different experimental conditions:
1. At the high flux reactor of the Institut Laue-Langevin (Grenoble) these reactions are investigated with thermal neutrons. The experimental conditions are excellent: at the end of the curved neutron guide H22D, the flux is 5 x 10^8 n/cm^2.s and the number of fast neutrons and γ's is reduced by a factor of 10^6 compared to the corresponding quantity at the entrance of the guide.
This high neutron flux permits a low detection geometry, so surface barrier detectors of various thicknesses are used here.
The neutron flux calibration is done via the well known ^{235}U(n$_{th}$,f) reaction. The thermal neutron induced reaction cross-sections determined in this way are used later on for the normalization of the measurements done at GELINA.

2. At the linear electron accelerator GELINA of the Central Bureau for Nuclear Measurements (Geel), the keV-neutron energy region is investigated with high resolution. Here large area surface barrier detectors as well as gridded ionization chambers are used. The neutron flux is determined via the ^{10}B(n,α)^{7}Li and/or the ^{235}U(n,f) reaction.

Results

Table 1 gives a survey of the thermal reaction cross-sections obtained. Great care was given to the sample definition [7,8]. In the case of ^{36}Cl(n,p)^{36}S this resulted in a cross-section value of

* National Fund for Scientific Research, Belgium
** Now at CEC, JRC, I.S.T., Ispra, Italy

(46±2) mbarn, i.e. one order of magnitude smaller than the only other available result [9]. As a typical example, fig. 1 shows the proton energy distribution from the $^{35}Cl(n_{th},p)$-reaction and fig. 2 the α-spectrum from the $^{41}Ca(n_{th},\alpha)$-reaction.

Table 1: Thermal neutron induced cross-sections

(n,p)-reactions	σ_p[mb]	ref.	(n,α)-reactions	σ_α[mb]	ref.
$^{14}N(n,p)^{14}C$	(1800)	3	$^{33}S(n,\alpha)^{30}Si$	115	6
$^{35}Cl(n,p)^{35}S$	440	4	$^{40}K(n,\alpha)^{37}Cl$	390	5
$^{36}Cl(n,p)^{36}S$	46	4	$^{41}Ca(n,\alpha_0)^{38}Ar$	40	7
$^{40}K(n,p)^{40}Ar$	4400	5	$^{41}Ca(n,\alpha_1)^{38}Ar$	140	7
$^{50}V(n,p)^{50}Ti$	(0.8)	3	$^{41}Ca(n,\gamma\alpha)^{38}Ar$	8	7

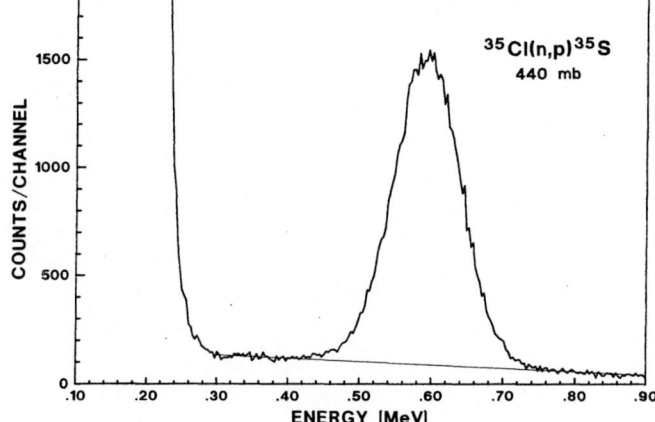

Fig. 1: Energy distribution of the protons emitted in the $^{35}Cl(n_{th},p)^{35}S$ reaction.

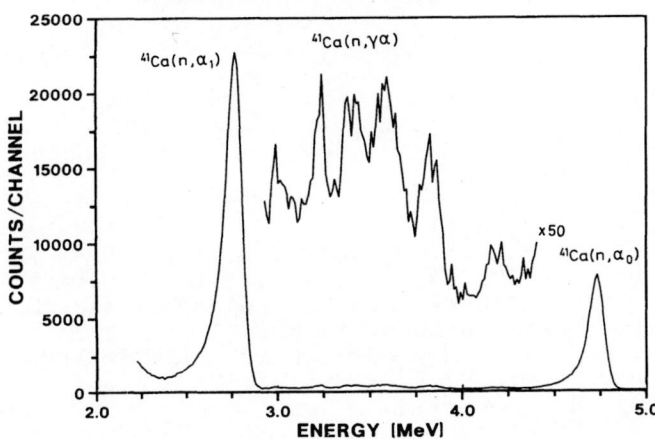

Fig. 2: Energy distributions of the α-particles from $^{41}Ca(n_{th},\alpha)^{38}Ar$.

Significant thermal cross-section values often indicate that resonances might be expected at higher neutron energies. This has been confirmed so far at GELINA for $^{35,36}Cl(n,p)$, $^{40}K(n,p)$ [10], $^{33}S(n,\alpha)$ [11], $^{40}K(n,\alpha)$ [10] and $^{41}Ca(n,\alpha)$. As typical examples, the $\sigma(E_n)$-results obtained for $^{35}Cl(n,p)^{35}S$ and $^{41}Ca(n,\alpha)^{38}Ar$ are shown in figs. 3 and 4, resp. Especially in the case of $^{41}Ca(n,\alpha)$, prominent resonances with cross-sections up to almost 100 barn can be observed.

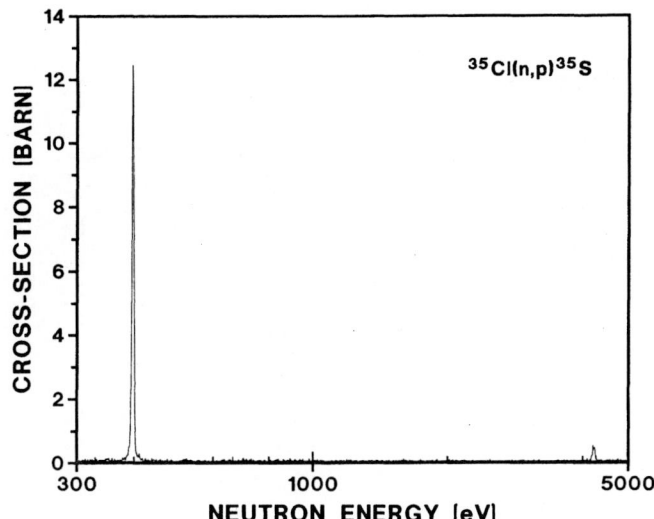

Fig. 3: Preliminary result for the $^{35}Cl(n,p)^{35}S$ reaction cross-section from 300 to 5000 eV neutron energy.

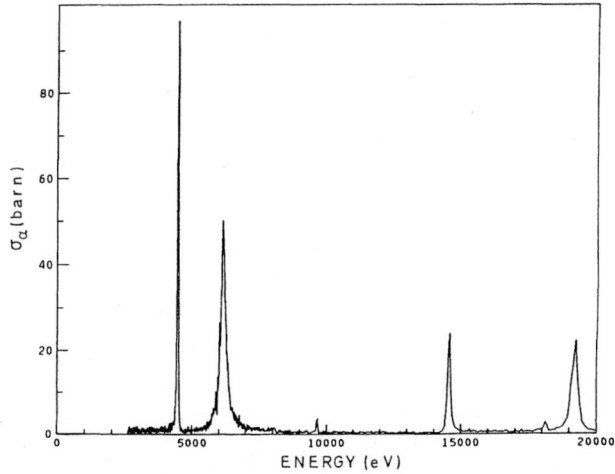

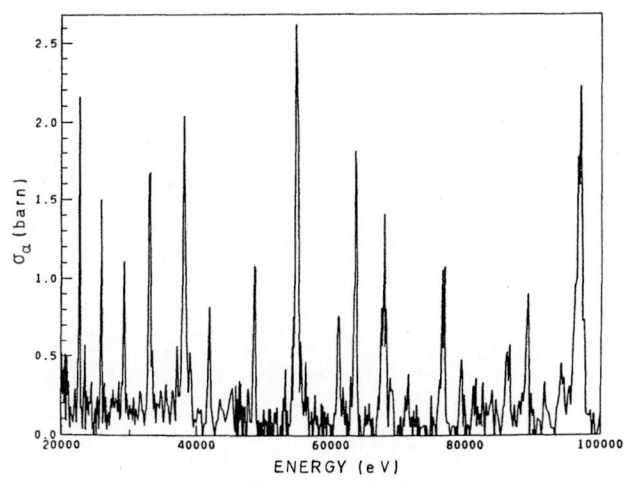

Fig. 4: Preliminary results for the $^{41}Ca(n,\alpha)^{38}Ar$ reaction cross-section in the neutron energy region from 3 to 100 keV.

Discussion

The need for experimental (n,p)- and (n,α)-cross-section data for s-process nucleosynthesis calculations is illustrated in Table III of the compilation by Bao and Käppeler [12], which shows that almost only model calculated cross-section values are available.

To compare these calculated values with our experimental data, the differential cross-section data $\sigma_p(E_n)$ and $\sigma_\alpha(E_n)$ have to be converted into Maxwellian averaged cross-sections $\langle\sigma\rangle$ defined as:

$$\langle\sigma\rangle = \frac{2}{\sqrt{\pi}}\frac{\int_o^\infty \sigma(E_n)\,E_n\,exp[-E_n/kT]\,dE_n}{\int_o^\infty E_n\,exp[-E_n/kT]\,dE_n},$$

kT being the stellar temperature. A typical example is given in fig. 5, which shows the Maxwellian averaged cross-section obtained for ^{33}S(n,α) [11].

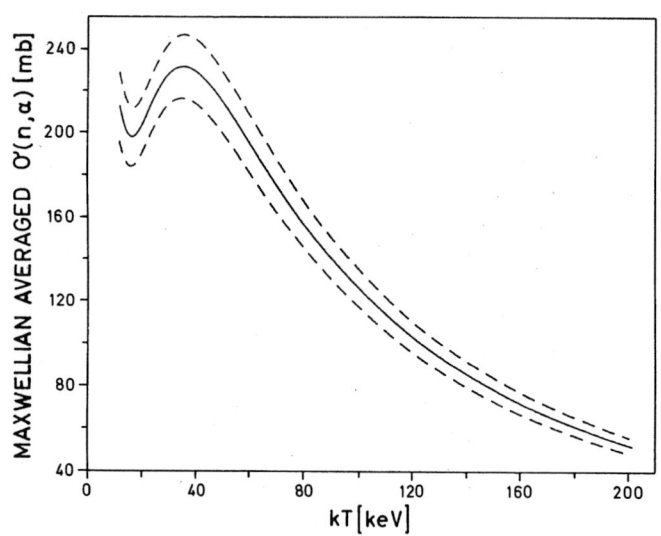

Fig. 5: Maxwellian averaged cross-section as a function of the stellar temperature (kT, in keV) for the ^{33}S(n,α)^{30}Si reaction.

Most of the reactions studied relate to the old ^{36}S discrepancy [11,13] (i.e. calculated ^{36}S abundances are considerably higher than the observed value of 0.015%) or to the potential chronometric pair ^{40}K/^{40}Ar [14]. Fig. 6 gives an overview of the nucleosynthesis reaction sequences in this region, unmarked arrows indicating (n,γ)-reactions. It is clear that the ^{36}S production will be strongly influenced by the ^{33}S(n,α) and the ^{35,36}Cl(n,p)-reactions. The ^{40}K/^{40}Ar ratio on the other hand will be strongly affected by the ^{40}K(n,p)^{40}Ar and the ^{40}K(n,α)- and ^{41}Ca(n,α)-reactions.

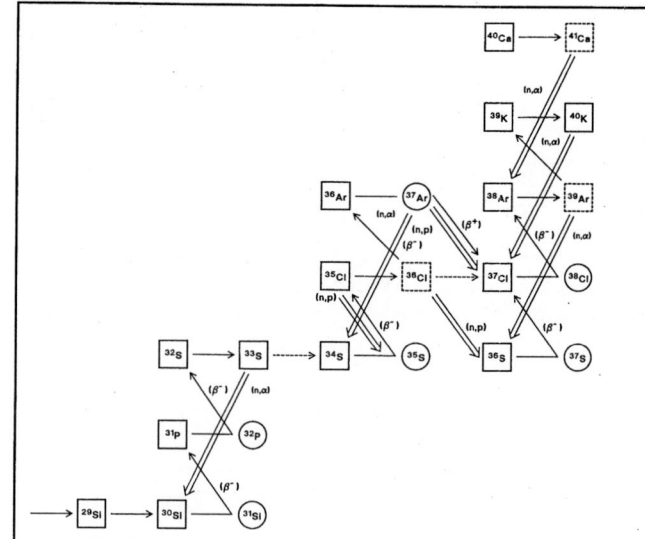

Fig. 6: s-process nucleosynthesis reaction network in the sulphur region. A full line stands for a (n,γ) reaction, a dotted line means a less probable transition due to competition with other reactions.

References

1. W. Fowler, Science <u>226</u>, 922 (1984).
2. F. Käppeler, H. Beer and K. Wisshak, Rep. Prog. Phys. <u>52</u>, 945 (1989).
3. preliminary results.
4. S. Druyts, C. Wagemans, P. Geltenbort and P. Schillebeeckx, Ann. Geophysicae, Suppl. to Vol. 9, C 366 (1991).
5. A. Emsallem, M. Asghar, C. Wagemans and H. Weigmann, Nucl. Phys. <u>A368</u>, 108 (1981).
6. C. Wagemans, Proc. Int. Symp. on Neutron Induced Reactions, Smolenice (CSFR), D. Reidel Publ. Cy. Boston (USA), 344 (1986).
7. C. Wagemans, Nucl. Instr. & Meth. <u>A282</u>, 4 (1989).
8. R. Eykens, A. Goetz, A. Lamberty, J. Van Gestel, J. Pauwels, C. Wagemans, S. Druyts and P. D'hondt, Nucl. Instr. & Meth. in print.
9. J. Andrzejewski et al., JINR Dubna Report P3-87-319 (1987).
10. H. Weigmann, C. Wagemans, A. Emsallem and M. Asghar, Nucl. Phys. <u>A368</u>, 117 (1981)
11. C. Wagemans, H. Weigmann and R. Barthélémy, Nucl. Phys. <u>A469</u>, 497 (1987).
12. Z. Bao and F. Käppeler, At. Data and Nuclear Data Tables, <u>36</u>, 411 (1987).
13. W. Howard, W. Arnett, D. Clayton and S. Woosley, The Astrophys. Journ. <u>175</u>, 201 (1972).
14. H. Beer and R. Penzhorn, Astronomy and Astrophysics <u>174</u>, 323 (1987).

THE STELLAR (n,γ) CROSS SECTIONS FOR [87]Rb AND [192]Pt

D. Neuberger, M. Tepe, F. Käppeler

Kernforschungszentrum Karlsruhe, Institut für Kernphysik,
W-7500 Karlsruhe, Federal Republic of Germany

Abstract: The stellar neutron capture cross sections of [87]Rb and [19]Pt have been measured by the activation technique relative to the gold standard cross section. The samples were sandwiched between gold foils and were irradiated in the kinematically collimated neutron field produced via the [7]Li(p,n)[7]Be reaction at a proton energy of 1912 keV. The energy distribution of the resulting neutron spectrum allows directly to determine the stellar cross section for a thermal energy of kT = 25 keV. For [87]Rb, previous measurements exhibit sizable discrepancies that were resolved in the present study. This isotope is one of the neutron magic nuclei with N = 50, which are characterized by particularly small cross sections and correspondingly large s-process abundances. The impact of the improved accuracy of the [87]Rb cross section on the s-process nucleosynthesis in the mass region 84 < A < 90 is discussed. The [192]Pt cross section has been investigated for the first time at keV energies. This isotope is of pure s-process origin and defines the branchings of the neutron capture chain at [191]Os and [192]Ir. Analysis of these branchings yields information on the s-process neutron density during helium shell burning in Red Giant stars.

([87]Rb, [192]Pt, neutron capture cross sections, E_n = 25 keV, activation technique, s-process, stellar nucleosynthesis)

Introduction

The stellar cross sections of [87]Rb and [192]Pt are of importance for two different aspects of s-process nucleosynthesis:

(i) The neutron magic nucleus [87]Rb is characterized by a comparably small (n,γ) cross section. Therefore, it builds up to a correspondingly large abundance in the low neutron density environment of the s-process. Together with the other N = 50 isotopes, [87]Rb governs the abundance pattern in the mass region around 85 < A < 90. This region is particularly important, since it contains significant abundance contributions from different nucleosynthesis sites. In the first place, the separation of these contributions requires reliable cross sections for the involved isotopes. In view of the rather large uncertainties and discrepancies resulting from previous studies of the [87]Rb cross section, a new measurement was, indeed, called for.

(ii) [192]Pt can be considered as of pure s-process origin. Its abundance defines the branchings of the s-process path at [191]Os and ^{192}Ir, and can, therefore, be used to derive an independent estimate for the s-process neutron density. This feature is illustrated in Fig. 1, which shows that the s-process flow is partly bypassing [192]Pt due to the competition between neutron capture and beta decay at A = 191, 192. The branching factor, which is defined by the rates for beta decay, λ_β = ln2/$t_{1/2}$, and for neutron capture, λ_n = <σ> n_n v_T (v_T being the mean thermal velocity),

$$f_\beta = \frac{\lambda_\beta}{\lambda_\beta + \lambda_n} \, , \qquad (1)$$

can be expressed by the ratio <σ>N_s([192]Pt)/[<σ>N_s([192]Pt) + <σ>N_s([192]Os)], as well. Since [192]Pt is of pure s-process origin, the product <σ>N_s([192]Pt) is determined by its stellar cross section <σ> and its observed abundance. Comparison with the overall <σ>N_s(A)-curve [1] yields eventually the s-process neutron density, n_n, via Eq. 1. However, estimates of the neutron density from this particular branching were obscured by large uncertainties due to the lack of experimental data for the [192]Pt capture cross section at keV neutron energies.

Measurements

All measurements were carried out via the activation technique at the Karlsruhe 3.75 MV Van de Graaff accelerator. Neutrons were produced by means of the [7]Li(p,n)[7]Be reaction that was shown to yield a quasi-stellar neutron spectrum for a thermal energy of kT = 25 keV, very close to the conditions during the s-process [2,3]. The [197]Au(n,γ) reaction was used as a cross section standard [3,4].

A. The [87]Rb cross section

Eight different Rb_2SO_4 samples with thicknesses between 0.7 and 67 mg/cm^2 were used in a series of activations in order to study the effect of systematic uncertainties experimentally. The samples were exposed to fluxes of about 10^9 neutrons per second for two to four half-lives of [88]Rb. The induced activity was then counted by means of a 4π electron spectrometer consisting of two Si(Li) detectors in close geometry. Detection of the beta decay

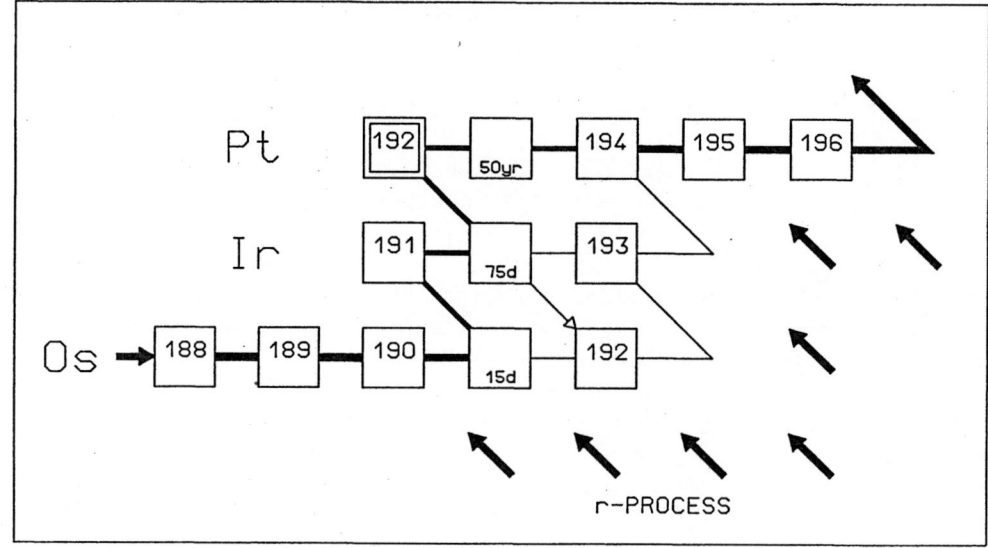

Fig. 1 The *s*-process branchings at ^{191}Os and ^{192}Ir.

electrons offered the advantage of a well defined efficiency. In this way, the uncertainties of the relative gamma-ray intensities in the ^{88}Rb decay could be avoided, which were found to be responsible for part of the previous discrepancies [5].

For the standard *s*-process temperature corresponding to a thermal energy of kT = 30 keV, the resulting stellar cross section is

$$\frac{\langle \sigma v \rangle}{v_T} = 18.0 \pm 0.5 \text{ mbarn.}$$

In Fig. 2, this result is compared to previous data [6,7,8, 9], which exhibit considerably larger uncertainties.

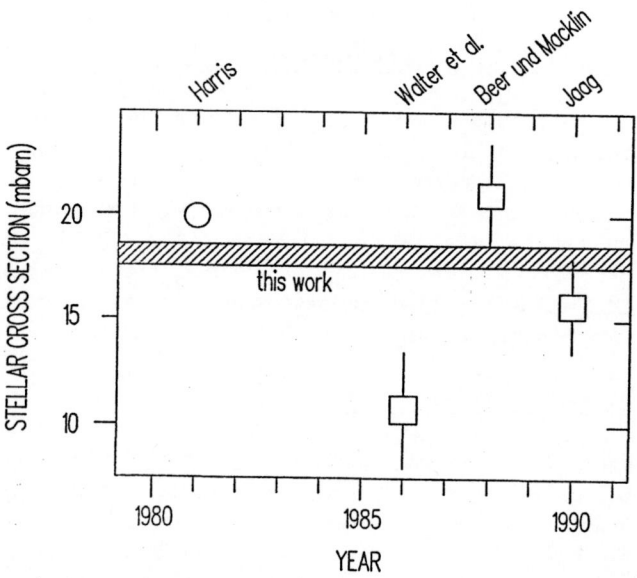

Fig. 2 The stellar ^{87}Rb cross section compared to previous work.

B. The ^{192}Pt cross section

This work represents the first experimental study of the neutron capture cross section of ^{192}Pt in the keV energy range. Because of its low natural abundance (0.79%), this isotope is not available in sufficient amounts and with sufficiently high enrichment to allow for a measurement with a direct detection technique. However, also the activation technique is hampered due to the large half-life of the product nucleus ^{193}Pt (~50 yr) and by the fact that ^{193}Pt decays only via electron capture from the L-shell. This implies a low specific activity of the irradiated samples and a severe limitation of the sample thickness.

Two samples of 165 and 356 mg/cm^2 thickness were prepared by electrodeposition of metallic platinum on ~0.1 mm thick carbon disks (50% enrichment in ^{192}Pt). After activation of these samples over 16 days, and with an optimized shielding for the above mentioned 4π Si(Li) spectrometer, the Ir LX-rays could be detected with an acceptable signal-to-background ratio (Fig. 3). The comparably large uncertainty of the resulting stellar cross section

$$\frac{\langle \sigma v \rangle}{v_T} (30 \text{ keV}) = 196 \pm 56 \text{ mbarn}$$

is determined by the uncertainties of the ^{193}Pt decay [10]. Any improvement in the half-life and in the LX-ray intensities of this isotope were highly desirable to improve the accuracy of this cross section. It is important to note that the present result is two times smaller than the average of previous theoretical estimates [6,11]. The correspondingly smaller empirical $\langle \sigma \rangle N_s$-value of ^{192}Pt yields a smaller branching factor f_β and hence a higher neutron density than was previously estimated from that branching [12].

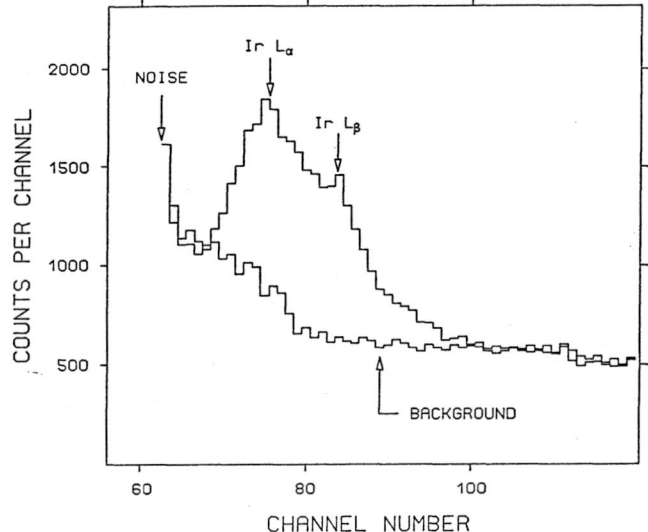

Fig. 3 LX-ray spectrum of an activated ^{192}Pt sample.

Astrophysics

This paragraph presents a brief summary of the most important astrophysical aspects. A detailed discussion of the involved astrophysics may be found in Ref. 5 for ^{87}Rb and in Ref. 10 for ^{192}Pt.

A. The s-process at N = 50

The new result for ^{87}Rb complements the set of improved stellar cross sections for the neutron magic nuclei with N = 50 [13]. With these data, detailed analyses of the isotopic abundance pattern in the mass region 85 < A < 90 were carried out, using the classical s-process approach [1, 8] as well as recent stellar models for helium burning in low mass stars [12, 14, 15] and for helium and carbon burning in massive stars [16]. With the classical approach, the various s-process contributions in the mass range 70 < A < 90 could be described satisfactorily without overproducing the observed abundances. Very similar results were obtained with the stellar models as well, except for the neutron magic nuclei ^{86}Kr and ^{87}Rb, which were overproduced by 40 % and 20 %, respectively. However, before this difference can be considered as significant, the stellar ^{84}Kr cross section needs further improvement in order to define the branching factor of ^{85}Kr with sufficient accuracy.

B. The s-process branchings at A = 191, 192

The new ^{192}Pt cross section was used for analyzing the s-process branchings of Fig. 1. In particular, it was of interest to deduce an improved estimate for the s-process neutron density for comparison with the results obtained from similar cases. With the classical model, one obtains a neutron density

$$n_n = (4.3 \pm_{2.5}^{3.4}) \cdot 10^8 \text{ cm}^{-3} ,$$

in good agreement with the mean value of $(3.4 \pm 1.1) \cdot 10^8$ cm^{-3} from the branchings at A = 95, A = 147, 148, and at A = 185, 186 [12]. As far as the abundance pattern in this mass region is concerned, it was found that the A = 191, 192 branching can be equally well described with the classical approach and with the s-process model of helium shell burning in low mass stars.

References

1. F. Käppeler, H. Beer and K. Wisshak: Rep. Prog. Phys. 52, 945 (1989)
2. H. Beer and F. Käppeler: Phys. Rev. C 21, 534 (1980)
3. W. Ratynski and F. Käppeler: Phys. Rev. C 37, 595 (1988)
4. R. L. Macklin: private communication
5. D. Neuberger: Diplom thesis University of Karlsruhe (1991)
6. M. J. Harris: Astrophys. Space Sci. 77, 357 (1981)
7. G. Walter, H. Beer, F. Käppeler and R.-D. Penzhorn: Astron. Astrophys. 155, 247 (1986)
8. H. Beer and R. L. Macklin: Ap. J. 339, 962 (1989)
9. S. Jaag: Diplom thesis University of Karlsruhe (1990)
10. M. Tepe: Diplom thesis University of Karlsruhe (1991)
11. J. A. Holmes, S. E. Woosley, W. A. Fowler and B. A. Zimmerman: Atomic Data and Nucl. Data Tables 18, 305 (1978)
12. F. Käppeler, R. Gallino, M. Busso, G. Picchio and C. M. Raiteri: Ap. J. 354, 630 (1990)
13. F. Käppeler, W. R. Zhao, H. Beer and U. Ratzel: Ap. 355, 348 (1990)
14. D. E. Hollowell and I. Iben, Jr.: Ap. J. 340, 966 (1989)
15. R. Gallino, C. M. Raiteri, M. Busso, G. Picchio and A. Renzini: Ap. J. 334, L45 (1988)
16. C. M. Raiteri, M. Busso, R. Gallino and G. Picchio: in preparation

NUCLEAR DATA FOR GEOLOGY AND MINING *

J. Csikai

Institute of Experimental Physics,
Kossuth University, H-4001 Debrecen,
Pf. 105, Hungary

Abstract: This paper briefly reviews the present status of nuclear techniques used in geology and mining and an attempt is made to identify nuclear (and atomic) data that are needed to improve these methods and to extend their possible applications. The following topics are discussed: role of nuclear techniques in chemical analysis of rocks and minerals, main fields of the geological applications, advances in the techniques and methods, status of required nuclear data.
(nuclear analytical techniques, activation and prompt radiation methods, nuclear data in the earth sciences)

Introduction

Nuclear techniques are widely used for fast, nondestructive, multielemental analysis of geological samples. The most abundant elements in atomic percent in the earth's crust taking into account the atmosphere, hydrosphere and 20 km thick surface layer of the lithosphere are as follows: O(52.3), H(17), Si(16.7), Al(5.5), Na(1.9), Fe(1.5), Ca(1.5), Mg(1.4), K(1.1), Ti(0.2), C(0.14), P(0.04), N(0.03), Mn(0.03), S(0.03), Cl(0.03), F(10^{-3}). The atomic percentages given in brackets indicate that 95 percent of the atoms in the range of interest for geology and mining consists of light elements Z ≤ 15. The major elements in most rock types are Si, O, Al, Fe, Mg, Ca, Na, K, Ti, and P. Concentrations of these elements in the well-known terrestrial rocks vary from 1 to 4 orders of magnitude, therefore, careful investigations are needed to develop the most suitable methods for the elemental analysis of minerals, standard geological materials, lunar soil samples, processing solutions or slurries and ocean bottom cores. These methods should be applicable for small and bulk samples using sampling, in-situ and on-stream procedures. The main difficulty in the utilization of atomic and nuclear methods for the chemical analysis of geological samples is in connection with the complexity of the matrices. An evaluation of more than 12000 research papers published in the last decade indicates that 36 different nuclear methods were applied among the about 140 instrumental analytical techniques used for the determination of 15 elements [1,2]. Considering the physical principles involved in the measuring process these techniques can be merged into a few groups, e.g. optical, nuclear, electrical, chromatography, miscellaneous, hyphenated (combined instrumental methods). Though the different variants of optical techniques are used most frequently in the elemental analysis, however, the expensive infrastructural tools as small accelerators and research reactors have kept their leading positions in solving special problems [3,4,5]. For example, in the analysis of geostandard materials, in the well-logging and bulk-media assay or in the nondestructive depth profiling. Recently a comprehensive review on trends in the instrumental analysis of geochemical materials was given by Braun and Zsindely [6]. The percentages in Table 1 show that optical and nuclear techniques are mainly used for elemental analysis of geological samples. The nuclear methods used for the analysis of different elements have the following ranking:
heavy metals > multielement determinations > alkali earth > metals > other non-metals > noble metals > alkali metals.

The frequencies of use of instrumental techniques for the analysis of different types of geological samples are demonstrated in Table 2. An evaluation of the data on the distribution of different techniques applied to the analysis of geochemical reference standards (GRS) during the 1959-1988 period indicates that the nuclear methods have the first ranking in this important field. Table 3 shows that between 1960 and 1985 the role of X-ray fluorescence technique has increased in addition to the neutron activation analysis.

Table 1. The weight of instrumental analytical techniques used in the determination of groups of elements in geological materials in 1979-1988.

Method	ALK	ALE	HME	OLM	NOM	REE	ACT	GAS	HAL	OHM	ANI	MUL
Optical	41.5	29.0	48.3	22.2	59.3	36.9	12.5	33.3	18.0	33.9	23.1	36.1
Nuclear	46.3	61.4	37.8	61.2	17.7	41.2	50.1	66.7	20.0	27.3	0.0	40.9
Electrical	0.0	0.0	2.6	0.0	5.3	0.7	8.3	0.0	22.0	3.3	7.7	1.0
Chromatography	7.3	3.2	2.0	0.0	7.1	3.5	8.3	0.0	34.0	10.7	38.4	4.9
Miscellaneous	0.0	1.6	1.9	0.0	0.0	0.0	0.0	0.0	2.0	5.0	30.8	0.5
Hyphenated	4.9	4.8	7.4	16.6	10.6	17.7	20.8	0.0	4.0	19.8	0.0	16.6

ALK - alkali metals; ALE - alkali earth metals; HME - heavy metals; OLM - other light metals; NOM - noble metals; REE - rare earth elements; ACT - actinides; GAS - gases; HAL - halogens; ONM - other non metals; ANI - anions; MUL - multielements

*This work was supported by the Hungarian Research Foundation (Contract No. 1734/91.)

Table 2. Distribution of instrumental analytical techniques in analysis of geological matrices in 1979-1988.

Matrix	OPT	NUCL	ELEC	CHROM	MISC	HYPH
Phosphates	2.7	3.1	0.0	13.0	0.0	5.4
Silicates	26.1	26.9	30.8	13.0	23.1	24.3
Clays, soils	5.4	3.7	7.7	4.3	23.1	5.4
Carbonates	3.2	1.3	0.0	0.0	0.0	5.4
Oxides, hydroxides, sulphides	7.2	3.8	0.0	13.0	3.8	13.5
Sulphates	1.4	0.0	0.0	0.0	3.8	0.0
Halides	0.5	0.0	7.7	0.0	0.0	2.7
Ores, fuels	14.0	23.1	15.4	13.1	19.2	13.5
Miscellaneous	39.5	38.1	38.4	43.6	27.0	29.8

Table 3. Analytical techniques used in the analysis of GRS samples

Instrumental technique	1960 - 1969 %	1970 - 1979 %	1980- 1 985 %
Neutron activation analysis	13.3	32.6	29.5
ICP emission spectrometry	-	1.1	18.9
X-ray fluorescence	12.1	13.8	15.2
Atomic absorption spectrometry	7.1	18.4	15.0
Mass spectrometry	2.7	3.0	2.1
Colorimetry	4.9	3.8	1.0
Other (incl.chemical)	59.9	27.3	18.3

Major fields of investigations

Some examples will illustrate the applications of different nuclear analytical techniques in the investigations of small and bulk geological samples.

1. Instrumental Thermal Neutron Activation Analysis (INAA) [3-5, 7-10]

   Fossil fuels and by-products, lunar samples, minerals and ores, rocks, soils, meteorites, ocean nodules, sediments, glacial till, cosmochemistry, borehole analysis.

2. Epithermal Neutron Activation Analysis (ENAA) [3-5, 10].

   F, Y, Er, Ba, Sn, W and Pt in rocks, As, Co, Eu, Fe, Ga, La, Na, Sb, Sc, Sm,U and W in phosphate rocks.

3. Fast Neutron Activation Analysis (FNAA) [3-5, 10-13].

   Major elements in drill cores, muds and rocks, Ti, Fe, Zr in sediments, B,F, O, Na, Al, Si, P in minerals, O in coals, C/O ratio in borehole analysis, U and Th by delayed neutrons.

4. Neutron Induced Prompt Gamma Analysis (PGNAA) [3-5, 10, 14-21].

   Borehole logging for coal, on-line analysis in the coal industry, B, Cd, Cl, Cr, Cu, Fe, Hg, Mn in sea-water, major elements in bulk concrete samples, in-situ multielemental analysis of coal beds and rocks, composition of cement, rare-earth elements in gelological samples and USGS standards, water in rocks, O, Mg, Al, Si, Fe in sands.

5. Radiochemical Neutron Activation Analysis (RNAA) and Preconcentration Methods [3-5, 20, 22, 23].

   Rare-earth elements in B-rich minerals by removing B from the matrices, multielemental analyis of rocks and minerals by using group separation techniques, isolation of Au, Cd, Cu, Mo, U, Zn, from sea-water, Cd, In, Tl, lanthanides in meteorites and lunar samples.

6. Isotope Dilution Analysis (IDA) [3-5].

   Cr, Ni, Zn, Sr, Mo, Cd, Sn, Sb, Tl, Pb, U in marine sediments, Pb, Tl and rare-earths in meteorite, U, Th in rocks.

7. Direct Counting of Natural and Long-Lived Radionuclides [3-5, 24-26].

   U, Ra, Th and K in rocks, soils and sediments by gamma spectrometry.

8. Particle Induced Gamma-ray Emission (PIGE) [3-5, 14, 30].

   Multielemental analysis of geological samples and standards, B, Li and F in minerals, Si, Al and Na in zeolites.

9. Particle Induced X-ray Emission (PIXE) [3-5, 30].

   Multielemental analysis of samples for $Z \geq 15$.

10. Transmission, Attenuation and Scattering Methods.

    - Gamma-ray Methods [3-5, 16, 31, 32, 57]. Thickness, atomic number and density of geological samples, level and mass flow gauges.

    - X-ray Fluorescence (XRF) and Total Reflection XRF (TXRF) Methods [3-5, 23, 33-37, 81]. Multielemental analysis of geological and GRS samples, determination of average atomic number of complex samples via the ratio of the elastic to inelastic scattering components in XRF, sediment, sea-water, manganese nodules and crusts by TXRF.

    - Beta-particle Method.[3-5, 82]. Concentration by beta-backscattering, K by beta counting.

    - Radon Method [38-42]. Exploration of subsurface U and Th.

    - Neutron Methods. [3-5, 10, 33, 43-47] Moisture, H, C/H ratio, B, Cl and Br in bulk samples by reflection and transmission methods.

11. Isotopic Dating Methods. [3-5,10,33,38,48,49]

    Ages of rocks and minerals by U-Pb, Rb-Sr, K-Ar and track-etch methods.Groundwater dating by using naturally produced radionuclides (^{3}H, ^{14}C, ^{39}Ar, ^{32}Si, ^{36}Cl, ^{81}Kr, ^{85}Kr, ^{129}I).

12. Nuclear Well Logging [3, 10, 50-53].

    Salinity, porosity, density, mineral content by (γ-γ), (n,γ), (n-n), γ-ray and (n,xγ) methods.

    - Spectral γ-γ logging [54-56, 67]. Lithological information, concentration of heavy elements, density of soils and rocks.

    - Neutron lifetime logging (prompt and delayed neutrons) [10, 58-60]. Oil, gas, water, rock matrix, U, Th, soil moisture and porosity.

    - Neutron scattering and reaction logging (prompt and delayed gammas) [10, 52, 61-69]. Multielemental analysis (H, C, O, Mg, Al, Si, P, S, Ca, Cl, Ti, Fe, Mn, Cu. Gd).

- Natural gamma-ray logging [3, 53, 55, 70-76]. U(+^{214}Bi), Th(+^{208}Tl), K, minerals exploration, bed correlation studies.

During the last decade large efforts have been made to improve the nuclear techniques and methods used in various fields of the earth sciences. Some examples, for the new developments in particle sources, measuring techniques and interpretation of analytical results can be given as follows:

- Use of cold neutron beam in INAA and PGNAA [77].
- Method for the determination of the average activating neutron flux without separation of highly absorbing elements in analyses of geological samples [78].
- Production of a thermal neutron flux of about 10^{10}n/cm^{2}s over 2-3 kg of crushed rock by a small electron Linac [7] (see Fig.1).
- Energy selective activation analysis using a proton Linac as an intense ($>10^{12}$n/s) fast neutron source [79].
- Use of nanosecond pulsed-beam with time-gating at the detector to reduce the background and interferences in PIGE analysis of small and bulk samples [19, 29, 52] (see Fig. 2).
- A combination of the sealed-tube neutron generator with the associated alpha-particle technique for field analysis of bulk mineral samples (see Fig.3) [80].
- Elastic scattering analysis by backscattered fast neutrons at the resonance energy [51].
- Utilization of TXRF method for multielement determination down to the ultratrace concentration level [81].
- High-intensity monoenergetic neutron production between 5 and 12 MeV by D-D and H-T sources [86, 87].
- Production of intense pulsed neutron beams (10^{16}n/s) in the 10^{-2}-10^4 eV range by the LAMPF facility [84].
- Small neutron generators for borehole application [61, 88, 89].
- Pulsed albedo method for measuring thermal diffusion parameters of non-moderating materials (see Fig. 4) [83, 106].
- A new logging method for determining formation porosity by observing the die-away of epithermal neutrons [66].
- A simple method for the determination of the volume averaged spectral flux density of neutrons [85].
- Radiation spectrometry in special conditions by using BaF$_2$, CdTe, CdWO$_4$, HgI$_2$, Bi$_4$Ge$_3$O$_{12}$ and Si-avalanche detectors [90-93, 101].
- Advanced signal processors and high temperature electronics [94,95].
- Development and adaption to the PC environment of computer codes for analysis of complex decay curves and high resolution gamma spectra [96-100].
- Advanced interpretation models in nuclear well logging [50, 53, 119].

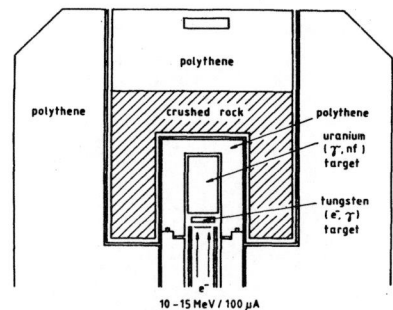

Fig. 1. An irradiation assembly for 2-3 kg rock sample around the W-U target of an electron Linac [7].

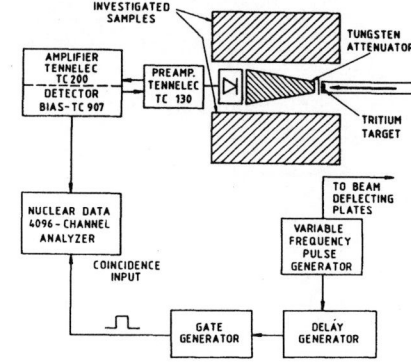

Fig. 2. A typical experimental arrangement used in PIGE analysis of bulk samples [52].

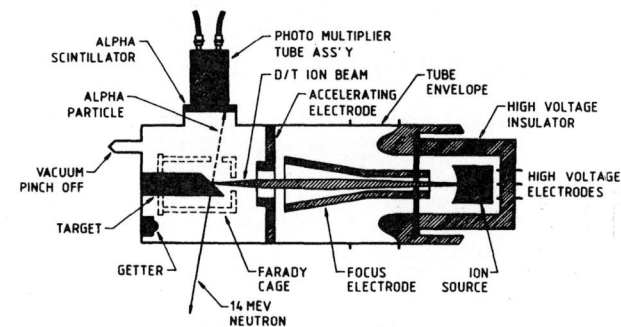

Fig. 3. A sealed-tube neutron generator combined with APM head for time correlated measurements [80].

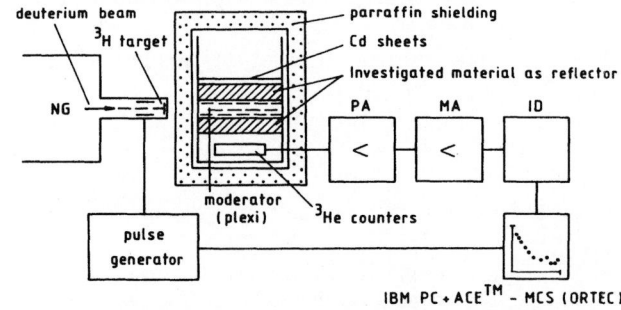

Fig. 4. Principle of pulsed neutron method used for the determination of diffusion parameters of non-moderating media [106].

Nuclear Data Requirements

The development and application of the wide variety of nuclear methods and associated technical devices for studying the samples and elements mentioned above require more accurate and complete sets of nuclear (and atomic) data.

Some conclusions regarding the micro cross section requirements in the earth sciences drawn by Clayton et al. [102] are as follows: a) Nearly all the elements may be encountered. b) The elemental analysis and the bulk media assay requires the gamma-ray production cross sections as a function of types and energies of the bombarding particles. c) Samples to be analysed vary from small dimensions (compared to the mean free path lengths of particles and photons) to very large in which the attenuation is high even for neutrons and gammas. d) Determinations of neutron and gamma-ray flux and energy distributions for realistic rock formations and measurement configurations require to increase the accuracy of a number of nuclear data. The required micro and macro cross sections as well as nuclear structure and decay data together with the nuclear properties of some geological formations are also given by Clayton et al. [102] in their comprehensive review.

Recently a number of new measured, evaluated and calculated nuclear (and atomic) data related to the earth sciences have been published. Some examples together with the requirements are as follows:

- Rock neutron parameters: Σ_a Σ_s, D, L_d, L_s, $\xi\Sigma_s/\rho$, β, t .

Measurements of Σ_a, Σ_s/ρ and $\xi\Sigma_s/\rho$ data for rock samples are in progress by neutron pulse methods [83, 103, 106, 107, 116] and steady state techniques [104, 108, 117].

Slowing down length, L_s, for rocks is deduced from the measured mass slowing down power. Neutron slowing down parameters (L_s, D_s, t_s) have been calculated by Czubek [109] for about 50 minerals and rocks. The determinations of the Σ_a, L_d and D parameters with high precision require the accurate knowledge of Σ_a for rock matrix, bound H content, rock porosity and pore saturation index [110].

The current and flux albedo, β, values for different media measured by neutron pulse method and steady state technique are scanty.

- σ_t, σ_s, σ_{act} at E_n < 10^{-2} eV.

Utilization of cold neutrons in elemental analysis needs the knowledge of micro cross sections as a function of energy below 10^{-2} eV.

- Activation cross-sections for dosimetry reactions.

Cross section curves of threshold reactions to unfold the spectral flux denisty of neutrons in different geological formations by multiple foil analysis have, mostly, not been measured with the required 3-5 % accuracy [111]. New data for ^{58}Ni(n,np) (^{235}U FS), ^{54,56}Fe(n,p) (9-15 MeV), ^{58}Ni(n,2n) (~ 14 MeV), ^{60}Ni(n,p) (5-12 MeV), ^{47}Ti(n,p) (5-10, 11-13 MeV), ^{58}Ni(n,p) (5-13 MeV), ^{197}Au(n,2n) (11-13 MeV), ^{115}In(n,n') (~ 14 MeV), ^{27}Al(n,α) (~ 14 MeV) dosimetry reactions were submitted to the present conference, in addition to the average cross sections measured in a continuous energy spectrum of neutrons ranging from a few 100 keV to 18 MeV for the ^{24}Mg, ^{46,48}Ti, ^{54}Fe, ^{58,60}Ni(n,p), ^{27}Al, ^{63}Cu(n,α) and ^{23}Na, ^{55}Mn, ^{58}Ni, ^{59}Co, ^{93}Nb, ^{197}Au(n,2n) reactions to control the validity of the recently evaluated dosimetry files.

- Nuclear data used in activation analysis with neutrons, charged particles and photons.

Cross section, nuclear structure and decay data are provided by the four International Data Centres. Data for the recommended reactions, sensitivity, determination limit, interfering reactions and gamma-lines are also available [10-13, 112-115].

Nuclear data for the estimation of the effects of secondary reactions on the accuracy of the elemental analysis of geological samples are very scanty and discrepant. In addition to the cross section curves of the energetic secondary charged particles, neutrons and gammas, data for the specific energy loss in rocks and minerals are also required. There is no reliable method for the determination of the average epithermal neutron flux in bulk samples of unknown compositions.

Prompt gammas produced in (a,bγ) reactions.

Prompt gammas emitted in charged particle or neutron induced reactions and inelastic scattering processes are widely used in elemental analysis. Data summarized in Table 4 demonstrate the lack of cross sections and gamma yields even for the charged particle resonance reactions.

Table 4. Nuclear data for PIGE resonance reactions

Reaction	Res. energy (keV)	Cross section (mb)	Gamma energy (MeV)	Emission per reaction (%)
^{7}Li(p,γ)	441	6	14.75 17.65	63
^{9}Be(p,γ)	991	-	7.5	100
^{11}B (p,γ)	163	0.16	4.43	96.5
^{12}C (p,γ)	459	0.13	2.363	100
^{13}C(p,γ)	1748	340	9.17 6.45 2.74	-
^{15}N(p,$\alpha\gamma$)	429	300	4.43	100
^{19}F(p,$\alpha\gamma$)	672 872 1375	57 661 300	6.131 6.131 6.131	81 68 87
^{23}Na(p,γ)	676.7 872.4 986.0	-	1.368 1.368 1.368	-
^{23}Na(p,p'γ)	1283.6	-	0.439	-
^{23}Na(p,$\alpha\gamma$)	1327.5	-	1.63	-
^{27}Al(p,γ)	632.6 991.8	-	1.778 1.778	-
^{31}P(p,γ)	541.5 811.3	-	2.237 2.237	-

Some recommended reactions for borehole and bulk-media assay based on PIGE technique are summarized in Refs. [14, 52].

The improvement of the PIGE technique requires a more precise knowledge of microscopic and macroscopic atomic and nuclear data (reaction and gamma-ray production cross section curves, decay data, scattering and absorption of gammas, slowing down and diffusion lengths of neutrons in samples having different physical and chemical properties). Measured and evaluated discrete γ-ray production cross sections required for geophysical applications are unsatisfactory [102, 113, 118, 120].

- For the application and improvement of the XRF and TXRF methods the absorption coefficient $\mu(<Z>,E)$

and fluorescence yields $\omega(<Z>,E)_{K,L}$ for various rocks and minerals should be measured as functions of energy and average atomic number, $<Z>$, of the sample.

- The fission track dating method requires the accurate knowledge of the range $R(<Z>)$ of fission fragments in various geological samples (the $R(<Z>)$ values can be determined e.g. by a ^{252}Cf fragment source), and the value of decay constant λ_{sf} of ^{238}U.

- The fast neutron backscattering analysis requires the cross section curves in the vicinity of the resonances especially for C, N and O [51].

The examples mentioned in this review indicate that the nuclear (and atomic) data required in geophysics cover at least such a large area as those needed in the nuclear energy program.

The author is indepted to Mrs. M. Juhasz for her kind assistance.

REFERENCES

1. T.Braun, Fresenius Z. Anal. Chem., **328** (1987) 1-9, Springer-Verlag
2. T.Braun and S.Zsindely, Trends in Analytical Chem. **9(5)** (1990) 144
3. W.D.Ehmann and S.W. Yates, Analytical Chemistry Fundamental Reviews, 1986, 58, 49R-65R
4. W.D.Ehmann and S.W.Yates, ibid., 1988, **60**, 42R-62R
5. W.D.Ehmann and S.W.Yates, ibid., 1990, **62**, 50R-70R
6. T.Braun and S.Zsindely, Trends in Analytical Chem. (1991) in press (private communication)
7. C.G.Clayton and R.Spackman, Nucl. Geophys. **2**, 25 (1988)
8. R.A.M. Rizk and A.Z. Hussein, Nucl. Geophys. **4**, 511 (1990)
9. H.K. Sharma, V.K. Mittal and H.S.Sahota, Nucl. Geophys. **3**, 141 (1989)
10. J.Csikai, Handbook of Fast Neutron Generators, CRC Press Inc., Boca Raton, Florida (1987) Vol.I. and II.
11. R.Pepelnik, Nucl. Geophys. **1**, 249 (1987)
12. R.Pepelnik, Fast Neutron Activation Analysis, IAEA-AGM on Low Energy Accelerators in Elemental Analysis, Chiang Mai, Thailand, 25-29 March 1991
13. S. S. Nargolwalla and E.P. Przybylowicz, Activation Analysis with Neutron Generators, John Wiley and Sons, New York, 1973
14. J.Csikai, Accelerator Based Prompt Gamma Analysis, IAEA-AGM on Low Energy Accelerators in Elemental Analysis, Chiang Mai, Thailand, 25-29 March 1991
15. D. Trubert, J.CH. Abbé and J. M. Paulus, J.Radioanal. Nucl. Chem., Articles **134** 405, (1989)
16. M.R. Wormald, Nucl. Geophys. **3**, 461 (1989)
17. M.R. Wormald, Nucl. Geophys. **3**, 373 (1989)
18. Z.Janout, J. Konicek, S.Pospisil, M.Vobecky and R.Drahonovsky, J.Radoianal. Nucl. Chem., Letters **136**, 423 (1989)
19. S. Hlavác and P. Oblozinsky, Nucl. Instrm, Meth. in Phys. Res. **B28**, 93 (1987)
20. L.G. Evans, J.I. Trombka, D.H.Jensen, W.A. Stephenson, R.A. Hoover, J.L.Mikesell, A.B.Tanner and F.E.Senftle, Nucl. Instrm. Meth. in Phys. Res. **219**, 233 (1984)
21. T.Gozani, Nucl. Geophys. **2**, 163 (1988)
22. J.C. Laul, At. Eng. Rev. **17**, 603 (1979)
23. W.Michaelis and A. Prange, Nucl. Geophys. **2**, 231 (1988)
24. R.J. de Meijer, H.M.E.Lesscher, R.D.Schuiling and M.E. Elburg, Nucl. Geophys. **4**, 455 (1990).
25. I.Pluta, I.Tomza and T.Pluta, Nucl. Geophys. **4**, 489 (1990)
26. J.S. Wykes, J.D.Hoddy, I. Adsley, G.M.Croke and G.J0. Haines, Nucl. Geophys. **3**, 203 (1989)
27. D.Gihwala and M.Peisach, J.Radioanal. Chem. **70**, 287 (1982)
28. B. Constantinescu, S. Dima, E. Ivanov and D.Plostinaru, Appl. Radiat. Isot. **37**, 53 (1986)
29. D.W.Mingay, V.M.Prozesky and P.B. Kotzé, Nucl. Instrm.Meth. in Phys. Res. **B35**, 339 (1988)
30. D.C. Cohen, Particle Induced X-ray Emission, IAEA-AGM on Low Energy Accelerators in Elemental Analysis, Chiang Mai, Tailand, 25-29 March 1991
31. M.B. Greenfield, R. J. de Meijer, L. W. Put, J.Wiersma and J.F.Donoghue, Nucl. Geophys. **3**, 231 (1989)
32. V. L. Gravitis, J. S. Watt, L. J. Muldoon and E.M.Cochrane, Nucl. Geophys. **1**, 111 (1987)
33. J.Csikai, in Proc. of the IAEA Consultants' Meeting on Nuclear Data for Borehole and Bulk-Media Assay using Nuclear Techniques, INDC(NDS)-151/L, IAEA Vienna (1984) p. 75
34. S.A. Raoof and N.A.R. Al-Sughayiar, Nucl. Geophys. **4**, 483 (1990)
35. K.J.S.Sawhney, G.S.Lodha and B.R.Singh, Nucl. Geophys. **3**, 125 (1989)
36. J.S. Wykes, J.D.Hoddy, G.M. Croke and A. Sandbank, Nucl.Geophys. **1**, (1987)
37. A. Starzec, M. Lankosz and L. Szostek, Nucl. Geophys. **4**, 365 (1990)
38. R.L. Fleischer, P.B. Price and R.M.Walker, Nuclear Tracks in Solids, Univ. of California Press, Berkeley, 1975
39. R.L. Fleischer and A. Mogro-Campero, General Electric Report No. 77CRD067 (1977)
40. L.-P. Cui, Nucl. Geophys. **4**, 353 (1990)
41. K. Wattananikorn, P. Asnachinda and S. Lamphunphong, Nucl. Geophys. **4**, 253 (1990)
42. R.C. Ramola, A.S.Sandhu, M.Singh, S. Singh and H.S.Virk, Nucl. Geophys. **3**, 57 (1989)
43. M.Buczkó, Z. Dezső and J. Csikai, J. Radioanal. Chem. **25**, 179 (1975)
44. J. Csikai, S.M. Al-Jobori, Cs. M. Buczkó and S.Szegedi, J.Radioanal. Chem. **71**, 215 (1982)
45. S.Szegedi and Z. Dezső, Radiochem. Radioanal. Lett. **52**, 343 (1982)
46. M. J. Millen, P. T. Rafter, B.D. Sowerby, M.T.Rainbow and L.Jelenich, Nucl. Geophys. **4**, 215 (1990)
47. M.J.Millen, P.T.Rafter and B.D. Sowerby, Nucl. Geophys. **2**, 207 (1988)
48. T.Florkowski, L. Morawska and K.Rozanski, Nucl. Geophys. **2**, 1 (1988)
49. J.L.Carroll and I.Lerche, Nucl. Geophys. **4**, 461 (1990)
50. Proceedings of the IAEA Consultants' Meeting on Nuclear Data for Borehole and Bulk-Media Assay using Nuclear Techniques, INDC(NDS)-151/L, Ed. K. Okamoto, IAEA, Vienna, 1984
51. S. W. Yates, Elemental Analysis with Accelerator-Produced Neutrons, IAEA-AGM on Low Energy Accelerators in Elemental Analysis, Chiang Mai, Thailand, 25-29 March 1991
52. K.Przewlocki, W.R.Mills and W.W.Givens, High Resolution Gamma-Ray Spectroscopy for Well Logging, INDC(NDS)-162/GM, IAEA, Vienna, 1984
53. Current Trends in Nuclear Borehole Logging Techniques for Elemental Analysis, IAEA-TECDOC-464, Vienna, 1988

54. P.G. Killeen and C.J. Mwenifumbo, in Ref. [53] p. 23
55. P. Dumesnil and K. Umiastowsky, in Ref. [50] p. 253
56. P. Eisleer, M. Borsaru, S. F. Youl and J. Charbucinski, Nucl. Geophys. 2, 43 (1988)
57. A.Z. Hussein, N.F. Soliman and L.S. Ashmawy, IAEA Research Coordination Meeting on Nuclear Techniques in Exploration and Exploitation of Natural Resources, 15-19 April 1991, Debrecen, Hungary
58. W.W.Givens and D.C.Stromswold, Nucl. Geophys. 3, 299 (1989)
59. Vo Dac Bang, Nguyen Phuc, Tran Dai Nghiep and Pham Duy Mien, in Ref. [50] p. 209
60. R.W. Barnard, W.A. Stephenson, J.H. Weinlein, D.H.Jensen and D.R.Humphreys, IEEE Trans. on Nucl. Sci. NS-30, 1664 (1983)
61. C. W. Tittle, C. Burger, G. Mathis and A.F.Veneruso, Nucl. Instrm. Meth. in Phys. Res. B10/11, 1038 (1985)
62. W. E. Schultz, H. D. Smith Jr., J. L. Verbout, J.R.Bridges and G.H.Garcia, Nucl. Instrm. Meth. in Phys. Res. B10/11, 1023 (1985)
63. M.Borsaru, P.L.Eisler and J.Charbucinski in Ref. [53] p. 31
64. F.E.Senftle and J.L.Mikesell, in Ref.[53] p.67
65. J.-L. Pinault, in Ref. [53] p. 93
66. W.R.Mills, D.C.Stromswold and L.S.Allen, Nucl. Geophys. 5, 13 (1991)
67. J.A.Grau and J.S.Schweitzer, Nucl. Geophys. 3, 1(1989)
68. E. Chrusciel, S. M. Kaczmarski, M. Kopec, J.Niewodniczanski, K.W.Palka and F.Wojda, Nucl. Geophys. 1, 345 (1987)
69. J.L. Mikesell, F.E.Sentftle, R.N.Anderson and M.Greenberg, Nucl. Geophys. 3, 501 (1989)
70. H.C.Sun, J.Q. Zhao, R.X. Zhang, C.L.Zhon, H.F.Ni, S.Y.Wu, G.S.Wuru, J.S. Wang, R.J. Han, B.H.Han, G.K.Ho, H.P.Liang, Q.X.Zhang. Z.C. Li, Z.H.Cheng and Z.Q. Xia, Nucl. Geophys. 5, 91 (1991)
71. P. Huppert, M. Borsaru, J. Charbucinski, C. Ceravolo and P.L.Eisler, Nucl. Geophys. 3, 381 (1989)
72. I.Wendt, K.-P. Sengpiel and H.Lenz, in Ref. [50] p. 280
73. D.M.Arnold, in Ref. [50] p. 291
74. The PGT Probe γ-Uranium Assa, Princeton Gamma-Tech
75. A.B. Tanner, R.M.Moxham and F.E.Senftle, Assay for uranium and determination of disequilibrium by means of in situ high-resolution gamma-ray spectrometry, U.S. Geological Survey, Open-File Report 77-751, June 1977
76. J. A. Aylmer, C. Ceravolo, J. Charbucinski, P. L. Eisler and S.F.Youl, in Ref. [50] p. 301
77. R.M. Lindstrom. R. Zeisler and M. Rossbach, J. Radioanal. Nucl. Chem. 112, 321 (1987)
78. Cs.M.Buczkó and A. Borbély, J. Radioanal. Chem. 42, 393 (1978)
79. B.G.Chidley, G.E.McMichael and S.O. Schriber, Nucl. Instrm. Meth. in Phys. Res. B10/11, 888 (1985)
80. C.M.Gordon and C.W. Peters, Nucl. Geophys. 2, 123 (1988)
81. A.Prange, Spectrochimica Acta, 44B, 437 (1989)
82. L.Botter-Jensen and V.Mejdahl, Nucl. Tracks 10, 663 (1985)
83. A.Demény, K.M.Dede and L. Vas, Nucl. Geophys. 2, 15 (1988)
84. C.D.Bowman, J.Phys. G: Nucl. Phys. 14S, 399 (1988)
85. J.Csikai, M.Váradi, Cs.M.Buczkó and S.Sudár, Nucl. Instrm. Meth. in Phys. Res. A269, 287 (1988)
86. S.M.Qaim, R.Wölfle, M.M.Rahman and H.Ollig, Nucl. Sci. Eng. 88, 143 (1984)
87. R.C.Haight and J.Garibaldi, Nucl. Sci. Eng. 106, 296 (1990)
88. R.S.Berg and E.L.Jacobs, IEEE Trans. Nucl. Sci. NS-30, 1459 (1983).
89. G.R.Morris, C.H.Bush and J.W.Reichardt, IEEE Trans. Nucl. Sci. NS-30, 1648 (1983)
90. E.Dafni, Nucl. Instrm. Meth. in Phys. Res. A254, 54 (1987)
91. C.L.Melcher, R.A.Manente and J.S.Schweitzer, IEEE Trans. Nucl. Sci. 36, 1188 (1988)
92. C.L.Melcher, Nucl. Instrm. Meth. in Phys. Res. B40/41, 1214 (1989)
93. D.Sueva, V.Spassov, N.Chikov and E.I.Vapirev, Nucl. Instrm.Meth. in Phys. Res. A288, 460 (1990)
94. T.Gozani, in Ref. (53) p.123
95. R.K.Traeger and P.C.Lysne, IEEE Trans. Nucl. Sci. 35, 852 (1988)
96. P. A. Aarnio, J. T. Routti and J. V. Sandberg, J.Radioanal.Nucl.Chem. 124, 457 (1988)
97. W.K.Hensley, E.A.Lepel, M.E.Yuli and K.H.Abel, J.Radioanal. Nucl. Chem. 124, 481 (1988)
98. S.Nagy, GAMANAL XT/AT User's Manual, Debrecen, 1989
99. J.Dolnicar, GANAAS, IAEA Vienna, 1990
100. J.O. Schmidt, Trace Microprobe Techn. 6, 371 (1988)
101. G.F.Knoll, Radiation Detection and Measurement, 2nd ed., John Wiley and Sons, New York, 1989
102. C.G. Clayton, B.H. Patrick, L.G. Sanders and M.G.Sowerby, in Ref. [50] p. 17
103. J. A. Czubek, K. Drozdowicz, E. Krynicka-Drozdowicz, A.Igielski and U.Woznicka in Ref. [50] p. 213
104. M. Ciechanowski, A. Kreft and K. Morstin, IAEA-SM-308/54, Vienna (1991)p. 279
105. IAEA Research Coordination Meeting on Nuclear Techniques in Exploration and Exploitation of Natural Resources, 15-19 April 1991, Debrecen, Hungary
106. A. Demény, J. Kuti Darai, K. M. Dede, S. Egri, A. I. Obi and L. Vas, see Ref. [105]
107. U. Woznicka, see Ref. [105]
108. Vo Dac Bang, Le Dai Dien, Nguyen Quang Hai, Vu The Ha, Nguyen Tien Nguyen, Hoang Anh Tuan, Nguyen Quang Trung, Vuong Huu Tan, Luong Ngoc Chau, Le Ba Phuong, Nguyen Van Loc, Nguyen Van Hai, see Ref. [105]
109. J.A.Czubek, Nucl. Geophys. 4, 143 (1990)
110. J.A.Czubek, Nucl. Geophys. 4, 293 (1990)
111. Nuclear Data for Fusion Technology, IAEA-TECDOC-457, Vienna (1988) p.9
112. Handbook on Nuclear Activation Data, Techn. Report Series No. 273. IAEA,. Vienna 1987
113. J.Csikai, Nucl. Instrm. Meth. in Phys. Res. A280, 233 (1989)
114. A.B.Pashchenko, Reaction cross-sections induced by 14.5 MeV and by Cf-252 and U-235 fission spectrum neutrons, INDC(CCP)-323/L, IAEA, Vienna, 1991
115. S.M.Qaim, Handbook of Spectroscopy, Vol. III, CRC Press Inc., Florida (1981) p.141
116. J. A. Czubek, K. Drozdowicz, B. Gabanska, A. Igielski, E. Krynicka-Drozdowicz and U. Woznicka, Nucl. Geophys. 5, 101 (1991)
117. A.Kreft, see Ref. [105]
118. P.Oblozinsky and S.Hlavác, Nucl. Geophys. 1, 263 (1987)
119. D. A. Kozhevinkov, IAEA Research Agreement No. 4201/R2/CF, IAEA, Vienna, 1991
120. P. F. Rose, T. Burrows and J. Tuli, Nucl. Geophys. 1, 277 (1987)

INTEGRAL EXPERIMENTS FOR REACTOR PRESSURE VESSEL NEUTRON EXPOSURE EVALUATION

B. Ošmera[+], M. Holman[++]

+) Nuclear Research Institute, 250 68 Řež, Czechoslovakia
++) Škoda, 316 00 Plzeň, Czechoslovakia

Abstract: Since the radiation embrittlement of the pressure vessel governs its service lifetime the radiation damage evaluation and prediction is an important part of the safety analysis of any power reactor of this type (PWR, VVER). The characterization of the neutron field at the pressure vessel is a main step in this analysis. To reach the target accuracy (20 % one sigma in fluence of neutrons impinging on VVER-440 pressure vessel and surveillance specimen at present) the calculations have to be verified by experiments (integral exp., benchmarks, relating to the computational model and, or input cross section data). Suitable geometrical conditions and technical arrangements of the LR-0 experimental reactor /1/ in N.R.I. Řež enable to construct full-scale physical models of a symmetry sector of the VVER type reactors in radial direction from the core to the biological shielding of the reactor. A set of experiments was realized for the VVER-440 reactor type and the data have been applied in the reactor dosimetry. The results of evaluations are discussed and generalized. The new series of experiments are devoted to the VVER-1000 type. After analysis of the first results it was concluded to examine also the benchmarks of general purpose. A spectrum of interest has been formed in a central zone of the LR-0 assembly replacing the fuel pins by the steel ones. This benchmark project is also supported by the IAEA (contract No 5964/RB). The preliminary analysis of this benchmark experiment is presented to dicsuss further activity in this programme.

(Integral experiments. VVER reactor dosimetry, pressure vessel exposure)

Introduction

The VVER-440 and VVER-1000 symmetry sector (60°) from the core periphery area to the inner region of the biological shield with the ionization chamber channel were modelled in the LR-0 facility. The core and radiation shielding models were located inside the aluminium tank of the LR-0, the pressure vessel (steel) and biological shielding simulators were outside the tank /2,3/.

The fuel composition in the core was chosen in such a way that the radial-azimuthal power distribution within its periphery area, was close as possible to the equilibrium power distribution in the core pheripheral area of the commercial VVER (standard loading).

The region of the maximum value of the fast neutron flux density incident on the pressure vessel is on the 0° and 8° direction to the mock-up symmetry axis in the VVER-440 and VVER-1000 mock-ups respectively.

For simulation of the water density reduction which takes place in the VVER operation conditions the steel displacing tanks is placed in the space between the barrel and the LR-0 tank.

The pressure vessel consists of the identical layers which can be successively shifted in radial direction in order to form an additional air gap layer for the spectrometric measurements "over the vessel thickness".

The neutron spectra were measured at the barrel simulator surface (surveillance capsule position), inner and outer pressure vessel (PV). Surfaces and over PV thickness (this measuring points are numbered in schemes of mock-ups in Fig. 1).

Neutron spectra calculations were performed by means of one- and two-dimensional transport codes ANISN and DOT-3.5 using data cross section libraries EURLIB-4, VITAMIN-C and CASK. The core calculations (source distribution) were performed in both cassette and pin-to-pin structures using different codes. The intercomparison of measured and computed data was done for differential spectra and integral quantities.

The physical model makes it possible to obtain directly
- basic data for neutron monitors (surveillance dosimetry, ex-vessel measurement) evaluation like integral cross sections and their uncertainties;
- relation of the neutron field parameters measured in the accessible locations to the critical ones of the power reactor (pressure vessel inner surface, 1/4 of the pressure vessel thickness, atc) including error propagation;
- experimental validation of the neutron calculations (including corresponding data) and their aplicability in the routine neutron field monitoring.
A review of the theoretical and experimental investigations is presented in /4/.

Mock-up experimental programme

The differential energy distribution of the fast neutron flux density measurements were carried out in the energy range 10 keV - 10 MeV and 100 keV - 10 Mev, respectively. The following proton recoil detectors were used:
a scintillation detector with a 10x10 mm stilbene crystal (pulse shape discrimination is used, gamma/neutron ratio about 100 is allowed),
a set of 40 mm diameter spherical proportional hydrogen-filled counters, gas pressure 981 kPa, 392.4 kPa, 98.1 kPa (with and without 3 He calibration admixture).

All the measurements were carried out at the central plane of the mock-up core. To perform the whole set of the spectra measurements the LR-0 power changed in the interval from 0.1 W to several kW. Thus the neutron flux density monitoring system playes an important role in the experiment and determines the uncertainty of the space distribution.

The ^{115}In(n,n') activation foils were used at the positions of the proton recoil measurements to prove the spatial distribution expecially.

The ^{58}Ni(n,p) detectors were used for azimuthal distributions measurements at the barrel simulator. The thermal neutron flux density distributions from the core boundary to the biological shielding were measured with Dy detector in the VVER-1000 mock-up. The low power of LR-0 limits the use of activation detectors generally. The azimuthal and axial distributions before and behind PV were performed by means of the proton recoil detectors in

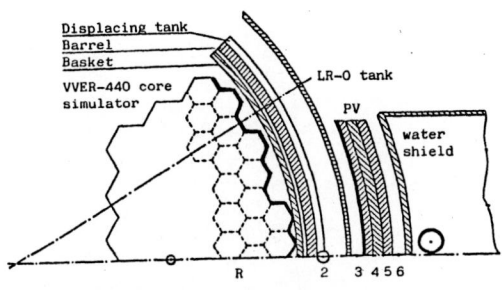

Schematic of VVER-440 mock-up.
(R2=1.625 m, R3=1.771 m, R6=1.92 m)

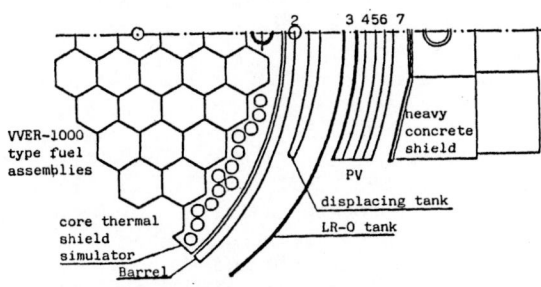

Schematic of VVER-1000 mock-up.
(R2=1.805 m, R3=1.830 m, R7=2.030 m)

Fig. 1

Tab. 1 - Statistical uncertainties (percent) of neutron flux density spectra

Measurement point	E energy Range, MeV				
	0.012-0.12	0.12-0.8	0.8-2.4	2.4-7.0	7.0-10.0
barrel surface	5-10	2-10	2-20	1-10	2-20
inner PV wall	10-25	3-20	2-15	2-10	15-30
outer PV wall	2-25	1-10	2-6	1-20	10-25

Tab. 2 - Proton recoil detectors energy resolutions

Ionization chambers				Stilben scintillator	
E /MeV/	resolution /%/	E /MeV/	resolution /%/	E /MeV/	resolution /%/
0.01	23.6	0.1	7.42	1	7.62
0.02	16.1	0.2	6.37	2	6.50
0.03	13.0	0.3	5.28	3	6.15
0.04	11.3	0.4	4.75	4	6.12
0.05	10.1	0.5	4.41	5	5.92
0.06	9.30	0.6	4.18	6	5.81
0.07	8.73	0.7	4.03	7	5.75
0.08	8.20	0.8	3.90	8	5.69
0.09	7.80	0.9	3.80	9	5.65

integral regime with gamma background discrimination.

The power distribution measurements were performed by means of the gamma activity measurement of about 600 fuel elements to determine the core radial power distribution. The axial power distribution was also obtained at selected points (about 20 points in mock-up core) by gamma scanning of the fuel elements.

All the mock-ups neutron spectra were mostly measured by two spectrometers (N.R.I., Škoda), the VVER-440 experiment was repeated in the same set-up to prove the results.

Discussion of results and experimental techniques incorporated

A special attention should be paid to the background due to albedo of core leakage neutrons at the measuring points and neutron field pertur-

bations due to neutron leakage at the boundary of the modeling sector. In experimental studies of neutron transport from the core to the biological shielding the 25 cm thick axial reflector suppress effectively the background in the distant measuring points.

The neutron spectrum measured in the water gap of the displacer corresponds to the required results in contrary to the hardly interpreted measurement in a dry cylindrical channel in the water layer (see comparison in /3/). For the same reason the mock-up was arranged in such a way that the PV simulator is placed out of the LR-0 tank.

To minimize the systematic uncertainties in neutron spectrometry the N.R.I. and Škoda spectrometers were intercompared in leakage spectra from simple iron, or iron-polyethylene assemblies with the 252 Cf source. The stochastic uncertainties of the measured spectrum depend on time of measurement, spectrum shape, gamma ray background (indirectly) and energy resolution. A typical stochastic uncertainties of the spectra measurements in the mock-up are listed in the Table 1. A review of representative proton recoil detectors energy resolutions is presented in Table 2. The listed resolutions include uncertainties in energy scale calibration and light anisotropy for the stilbene.

Passing through the water the neutron spectrum becomes harder, the gamma one gets softer and the experimental conditions are worsening especially in the energy range below 100 keV (proportional chamber area) and about 1 MeV (stilbene scintillator). Practical measureable water thicknes limit is about 20 cm and 15 cm only for energy region below 100 keV. The steel environment produces contradictory effect. The neutron spectrum gets softer and the gamma/neutron ratio becomes more favourable but the troubles rise in fast energy area above energy 3 - 5 MeV (depending on the steel layer thickness). The combination of steel and water layers in the reactor internals makes favourable experimental conditions in nearly all points where the neutron spectrum should be measured in the energy range 10 keV - 10 MeV.

The intercomparison of the experimental and calculated data were performed for relative distribution of the integral fluxes $\emptyset(E > E_i)$ (neutron flux density spectrum integrated over energy range E_i, 10 MeV) from the core to PV outer surface. The shape of the spectrum could be compared with the 33 group calculation. The estimated experimental uncertainties were 5 - 10 % including systematic of monitoring system.

Generally the discrepancies of the calculated and measured results in VVER-440 are less than 20 % for all desirable quantities in VVER-440 mock-up. The discrepacies were spread over this interval, without any system were found.

The source was well defined, the discrepancies of measured and calculated pin-to pin distribution were less than 10 %.

The VVER-440 mock-up results could be transfered to power reactors to evaluate surveillance capsule monitors ($^{54}Fe(n,p)$, $^{63}Cu(n,\alpha)$, $^{93}Nb(n,n')$) and measurements at the PV outer surface ($^{54}Fe(n,p)$, $^{63}Cu(n,\alpha)$, $^{93}Nb(n,n')$, $^{58}Ni(n,p)$, $^{46}Ti(n,p)$ detectors) /6/. The relative distributions at the points R_k and R_1, K/E_i, R_k/R_1) = $\emptyset(E > E_i, R_k) / \emptyset(E > E_i, R_1)$ and integral cross section of the monitors

$$\mathcal{G}(E \rangle E_i) = \int_0^{10} \mathcal{G}(E)\,\varphi(E)\,dE \;/\; \emptyset(E \rangle E_i),$$

where $\mathcal{G}(E)$ is detector cross section and $\varphi(E)$ mock-up spectrum, are listed in /4/, DOSCROSS cross-section library was used. The ex-vessel monitors are irradiated during the whole operating cycle ($\sim$ 1 year) the surveillance specimen from one to five years and this technical conditions limit the choise of the detectors. The results of ex-vessel measurements in VVER-440 power reactor with the core power distribution which was modelled in mock-up /4/ agreed with the calculations like in the mock-up. The evaluated $\emptyset(E \rangle 0.5$ MeV, PV inner surface) is 2.97 10^{15} m^{-2}sec^{-1} (10 %) in azimuthal maximum for nominal power. K (0.5, surveillance chain (PV inner surface) = 10.9; K (0.5, PV inner surface (PV outer surface) = 5.9. The shape of the neutron spectrum at the PV does not depend very much on the core distribution, it is determined by the total thickness of the water and steel layer between core and measuring point. It was also proved by ex-vessel distribution measurements /6/. The data of surveillance and ex-vessel monitors which are at our disposal show 20-25 % spread for different reactors and different fuel cycles. The high sensitivity of the PV fluence distribution to the core boundary power distribution was proved by ex-vessel measurements. This changes cannot be decribe by the routine core calculaltions.

It has been concluded to use the mock-up spectra and relative distributions /4/ for the evaluation of the surveillance and ex-vessel monitors. It is recommended to use ex-vessel monitors during all fuel cycles of VVER-440 to reduce the uncertainty in fluence estimation. The error propagation analysis should be completed by the evaluation of the integral cross section uncertainties (reevaluation with IRDF 90 cross section library).

After evaluation of the first VVER-1000 experiment the conclusion to repeat this measurement with some variations has been accepted. An important role the mock-up can play in the design of the surveillance programme. The surveillance specimen will be placed in the vicinity of PV in fluence 1.5 - 2 times greater than at PV inner surface. The fluence monitoring system combining surveillance and ex-vessel monitors should be designed and proved.

LR-O benchmark

The benchmark provides a possibility to compare the differential and integral methods of the neutron spectra measurement and test activation detectors in the spectrum of interest. The simple geometry benchmark core is composed of 19 VVER-1000 type cassettes (2 % enrichment in 235 U). In the central cassette the fuel pins were replaced by steel ones and the dry experimental channel was in the centre. The detailed description of geometry material compositions and nuclei densities are in /5/.

The neutron spectrum has been measured in the centre of the dry channel with proton recoil spectrometers N.R.I. and Škoda. The power was kept at a very low level during this measurement, the gamma ray flux density was also measured (Institute of Radiation Dosimetry, Prague) at this power level. Serious troubles with the gamma ray background were found at the energy above 1 MeV and below 60 keV and the results will have to be carefully evaluated.

Four multiple foil irradiations have been performed. The detectors set have been prepared by N.R.I., Škoda, ZfK Rossendorf (Germany). The following detectors, mostly the pure metals or certified alloys, were irradiated: ^{27}Al(n,α), ^{24}Mg(n,p), ^{58}Ni(n,p), ^{54}Fe(n,p), ^{47}Ti(n,p), ^{115}In(n,n'), ^{197}Au(n,γ), ^{55}Mn(n,γ), ^{23}Na(n,γ) (NaF pellet) and multicomponent detector (Ni, W, Au, Mn, Mo). The power and neutron flux density were several times less than the allowed ones (5 kW for one hour, 10^{13} m^{-2}s^{-1}, thermal). The induced activities of the used "heavy and great" foils were low. Nevertheless the possibility of the ex-vessel monitor testing there exists and detail error analysis before future measurement is desirable.

The gamma ray scanning of the fuel and stainless steel pins has been performed and the relative distributions have been obtained (fission products in fuel pins, (n,γ) reactions (^{58}Fe, ^{50}Cr) and (n,p) reactions (^{58}Ni, ^{54}Fe) in stainless steel pins). A separate radial distribution measurement in fission cassettes has been realized with Mn-Cu alloy activation detectors.

The first LR-O benchmark experimental programme has been realized in February, 1991. Some activation and distribution measurements finished in March, 1991. Additional calibration measurements are being performed. The detailed description of the LR-O benchmark is distributed to laboratories engaged in calculatlions, the experimental data will be available soon.

References

/1/ Sochor,Q., Starý,R., The experimental reactor LR-O after five years' operation, Nukleon, a special issue, p. 5, Nuclear Research Institute, Řež, 1988

/2/ Ošmera, B., Holman, M., Experimental and theoretical investigations for VVER pressure vessel neutron exposure evaluation, ibid., p. 25

/3/ Ošmera, B., Rataj,J., Racek, J., Turzík,Z., Černý,K.,Holman,M., Mařík,P., Vychytil,F., Brodkin,B.,Zaritskii,S.M., Moryakov,A.V., Khrustalev,A.V., Experimental validation of neutron calculaltions for VVER pressure vessel radiation load evaluation. Reactor dosimetry: Methods, applications and standardization. ASTM STP 1001, 1989, p. 130

/4/ Ošmera,B., Experimental and theoretical investigations for VVER pressure vessel neutron exposure evaluation in Czechoslovakia. IAE advisory group meeting on the status and requrirements of nuclear data for radiation damage and safety aspects, Vienna 19-22, September 1989, IAEA-TECDOC-572, p.9

/5/ Ošmera,B., Svoboda,Č., The benchmark experiment on the LR-O reactor, Nucleon No 2, 1991, p. 13, Nuclear Research Institute, Řež near Prague

/6/ Mehner,H.C., Böhmer,B.,Hagemann,U., Stephan,I., Meyer, N., Ošmera, B., Vychytil, F., Ex-vessel dosimetry at WWER 440, Jaderná energie, 36, 1990, p. 260

NEW NUCLEAR DATA REQUIREMENTS FOR RADIATION MONITORING AT HIGH INTENSITY PROTON ACCELERATORS

Yu.M.Nikolaev, S.V.Serezhnikov and V.E.Stepanov

Institute for Nuclear Research,
Academy of Science of the USSR,
117312 Moscow, USSR

Abstract: Some problems connected with new nuclear data requirements for radiation dosimetry at modern proton accelerators and in environment are discussed. Two important aspects of these problems are considered in the paper: measuring the fast neutron spectra (energy above ten MeV) for correct determination of dose equivalent; predictions and estimates of the environmental effect of airborne radioactivity releases from high-current proton accelerators.

(proton accelerators, neutron spectrum, plastic scintillator, differential efficiency, neutron-nuclear reaction, cross-section, airborne radioactivity emission, environment)

Introduction

Building high intensity proton accelerators of intermediate (400-800 MeV) energy (so called meson factories, such as LAMPF, TRIUMF, PSI) and designing 30-50 GeV proton accelerators with beam intensity about 100 μA (kaon factories) result in a number of significant radiation monitoring problems inside the accelerators hall and in environment, connected with a requirement for new nuclear data. We would like to discuss a few of them.

Measuring the fast neutron spectra and the integral characteristics of neutron fields

Experimental data analysis and our calculations show the significant part of dose equivalent to be caused by the hadrons with energy above a few tenths MeV at many points behind the accelerators shielding. By analyzing more than 130 neutron spectra behind the shielding of high and intermediate energy accelerators it was found that for 40% of them the contribution to dose equivalent from neutrons with energy above 10 MeV is more than 50% (see Fig. 1). It calls for creating suitable detectors for the efficient measurement of these components in mixed fields of radiation. The most perspective ones, we suppose, are plastic scintillation detectors operating in spectrometric mode [1,2]. For determining the integral characteristics of neutron (such as flux and dose equivalent) and for unfolding neutron spectra in mixed radiation fields behind the shielding it is enough to know the apparatus spectrum of these detectors in units of light output. It is also necessary to know the behavior of differential detection efficiency at total absorption

of initial particle energy in the detection volume. These events (if the initial particles energy is quite high) are a result of hadron-nuclear reaction on C-12 with a few fastemitted protons and/or fragments (clusters). The

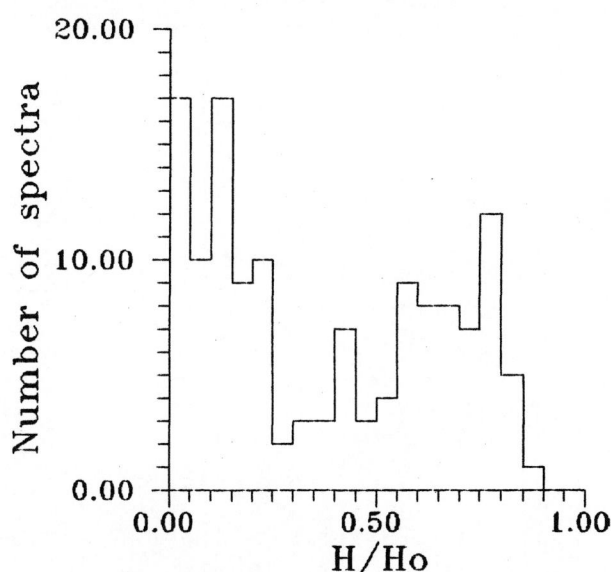

Fig. 1. The differential distribution of number of neutron spectra on the ratio neutron dose equivalent (En > 10 MeV) to total dose equivalent.

number of experimental data about cross section of these reactions are scant and a suitable calculation methods do not exist up to now. We are developing a method of calculation of the hadron-nuclear interaction on light nuclei with a few fastemitted nucleons and/or clusters in final state based on next scheme :
  - the quantum-mechanical calculation (in frame of DWBA) of the first

interaction of incident hadron and nucleons of nucleus with fast charged particles and clusters production in final state. Both pick-up and quasi-free knock-out mechanisms are taken into account in the calculation;

- Monte-Carlo calculation of intranuclear cascade initiated by these secondary particles.

However, the experimental confirmation for the calculations are also needed. Therefore the experimental data about cross section of nucleon-nuclei interaction for energy region 100-1000 MeV are very desirable especially for light nuclei.

There are also important for medical application.

Environmental effect of airborne radioactivity

One of the most major causes of public exposure attributable to proton accelerator operation are airborne radioactivity emissions. As our calculations show, under some conditions the significant part of effective dose equivalent is caused by the radionuclides Na-22, Na-24, Si-31, P-32, Cl-38, Cl-39 produced from interactions of high energy nucleons with Ar-40 contained in air. For the average time of air exchange within activation area of 10 hours the relative contribution of these radionuclides to the dose equivalent can exceed 50%. For example, results of the calculation [3] for contribution to effective dose equivalent from different radionuclides are presented in Table 1. Estimates were made for the point where the near-ground concentration is at a maximum. These data are obtained with following irradiation conditions: irradiation time – 10 hours; cooling time – 1 hour; thermal neutron flux 1.5 times lower than that of high energy neutrons; and

$$\int \Phi_n dV = 10^{17} \text{ cm/sec}$$

Table 1. Annual effective dose equivalent.

    1 – inhalation; 2,3 – external exposure from cloud by gamma and beta respectively; 4,5 –external exposure from ground by gamma and beta respectively; 6 – ingestion

Radio-nuclide	Concent. Bq/cm3	1 mSv/y	2 mSv/y	3 mSv/y	4 mSv/y	5 mSv/y	6 mSv/y	SUM mSv/y
^{3}H	7.90E-05	1.66E-05	0.00E+00	3.11E-10	0.00E+00	0.00E+00	4.20E-05	5.86E-05
^{7}Be	2.08E-03	1.09E-03	1.33E-04	0.00E+00	4.48E-02	0.00E+00	1.05E-01	1.51E-01
^{10}Be	2.05E-11	1.50E-08	0.00E+00	1.61E-14	0.00E+00	4.96E-11	3.75E-07	3.90E-07
^{11}C	6.79E-02	1.23E-03	8.99E-02	2.34E-04	8.29E-03	6.34E-05	0.00E+00	9.97E-02
^{14}C	5.46E-05	2.22E-04	0.00E+00	3.21E-09	0.00E+00	0.00E+00	8.68E-03	8.90E-03
^{13}N	4.97E-03	0.00E+00	6.59E-03	2.78E-05	2.97E-04	6.07E-06	0.00E+00	6.92E-03
^{14}O	1.35E-18	0.00E+00	5.79E-18	2.36E-20	2.67E-20	8.36E-22	0.00E+00	5.84E-18
^{15}O	1.64E-10	0.00E+00	2.17E-10	2.52E-12	2.00E-12	1.52E-13	0.00E+00	2.21E-10
^{17}F	2.70E-02	0.00E+00	3.58E-22	4.19E-24	1.74E-24	1.40E-24	0.00E+00	3.65E-22
^{18}F	7.58E-04	9.24E-05	1.00E-03	8.71E-07	4.98E-04	2.36E-07	0.00E+00	1.59E-03
^{24}Ne	1.36E-10	0.00E+00	9.51E-11	2.79E-12	1.44E-12	3.08E-13	0.00E+00	9.96E-11
^{22}Na	5.86E-07	1.07E-05	1.66E-06	3.41E-10	6.44E-03	3.27E-07	1.58E-01	1.65E-01
^{24}Na	4.75E-04	8.65E-04	2.53E-03	3.86E-06	8.36E-03	1.04E-04	1.86E-02	3.05E-02
^{25}Na	2.57E-23	0.00E+00	1.43E-23	3.80E-24	6.01E-26	9.10E-26	0.00E+00	1.83E-23
^{27}Mg	4.71E-06	0.00E+00	5.45E-06	6.83E-08	2.25E-07	2.12E-08	0.00E+00	5.76E-06
^{28}Mg	4.72E-05	3.45E-04	1.91E-04	3.10E-06	9.64E-04	1.12E-04	1.12E-03	2.73E-03
^{26}Al	4.25E-13	5.19E-11	1.45E-12	2.06E-15	1.56E-08	9.65E-11	6.98E-08	8.55E-08
^{28}Al	8.05E-12	0.00E+00	1.85E-11	5.26E-13	1.59E-13	3.41E-14	0.00E+00	1.92E-11
^{29}Al	1.01E-06	0.00E+00	1.81E-06	3.98E-08	4.82E-08	8.09E-09	0.00E+00	1.91E-06
^{31}Si	8.16E-04	2.98E-04	0.00E+00	7.83E-06	0.00E+00	3.84E-05	0.00E+00	3.44E-04
^{32}Si	1.02E-09	1.85E-06	0.00E+00	1.45E-11	0.00E+00	1.24E-06	7.28E-06	1.04E-05
^{30}P	2.08E-11	5.90E-13	2.76E-11	2.03E-12	3.11E-13	1.32E-13	0.00E+00	3.07E-11
^{32}P	9.68E-05	3.53E-03	0.00E+00	1.38E-06	0.00E+00	9.35E-04	3.11E-01	3.15E-01
^{33}P	2.41E-05	8.79E-05	0.00E+00	2.63E-09	0.00E+00	0.00E+00	2.41E-02	2.42E-02
^{35}S	1.46E-05	6.70E-05	0.00E+00	7.52E-10	0.00E+00	0.00E+00	8.43E-02	8.44E-02
^{37}S	6.84E-07	0.00E+00	2.58E-06	2.08E-08	4.35E-08	2.71E-09	0.00E+00	2.64E-06
^{38}S	2.73E-05	0.00E+00	1.10E-04	6.00E-06	7.03E-05	2.13E-05	0.00E+00	2.07E-04
^{34}Cl	7.40E-05	0.00E+00	2.37E-04	1.09E-05	3.04E-05	7.55E-06	0.00E+00	2.86E-04
^{36}Cl	1.22E-11	4.94E-10	0.00E+00	1.59E-14	0.00E+00	2.80E-10	1.21E-07	1.22E-07
^{38}Cl	8.93E-04	1.63E-04	1.72E-03	1.90E-04	2.40E-04	1.46E-04	0.00E+00	2.46E-03
^{39}Cl	3.52E-04	6.40E-05	6.50E-04	6.19E-06	1.46E-04	1.16E-05	0.00E+00	8.78E-04
^{40}Cl	1.56E-17	0.00E+00	7.65E-17	3.10E-18	3.54E-19	8.41E-20	0.00E+00	8.00E-17
^{37}Ar	1.59E-05	0.00E+00	4.53E-09	0.00E+00	9.67E-06	0.00E+00	0.00E+00	9.67E-06
^{39}Ar	4.07E-08	3.31E-18	3.31E-18	3.31E-18	0.00E+00	0.00E+00	0.00E+00	9.94E-18
^{41}Ar	5.86E-02	0.00E+00	9.72E-02	3.05E-04	4.39E-02	9.14E-04	0.00E+00	1.42E-01
SUM		8.09E-03	2.00E-01	7.98E-4	1.14E-01	2.35E-03	7.11E-01	1.04E+00

The annual effective dose equivalent was calculated with account of three main exposure pathways: inhalation, ingestion, and external exposure. As one can see from Table. 1, the summary contribution of the radionuclides, produced by fast neutron interaction with Ar-40 is more than 60%. The reliability of this estimation depends upon the radionuclide production cross sections in energy ranges 0.02 to 1 GeV. They have been chosen as follows:

(i) if there were experimental data, we preferred them over all other data;

(ii) if we had to decide between the data of Rindi [4,5] and the results obtained by SPALL program [6,7], we took the largest value;

(iii) if we had only the results calculated by SPALL, we took the value averaged over energy ranges.
Nevertheless data about cross sections of such an isotope production from Ar-40 are very scant which causes many difficulties for radiation monitoring and for the prediction of the environmental effect of airborn radioactivity releases from accelerators. As long as there are no reliable methods of calculating deep-inelastic nucleon-nuclear reactions available, the systematic measurements of yields of the above-mentioned nuclides are needed.

REFERENCES

1. G.N. Timoshenko and A.R. Krylov: Preprint JINR E-16-89-59, Dubna , (1989)
2. Yu.M. Nikolaev, V.B. Polushin and S.V. Serezhnikov: Nucl. Instr. Methods. (A), in press
3. I.N. Kopeikin, S.V. Serezhnikov and V.E. Stepanov: Proc. 23rd Intern. Symp. on Rad. Protec. Physics (1991)
4. A. Rindi and S. Charalambus: Nucl. Instr. Methods 47, 227 (1967)
5. A. Rindi: CERN Lab.11-RA/72-5 (1972)
6. R. Silberberg and C.H. Tcao: Astrophys. J. Suppl. 25, 315 (1973)
7. J.T. Routti and J.V. Sandberg: Comp. Phys. Commun. 23, 411 (1981)

CROSS-SECTIONS OF ^{113}In(n,n')^{113m}In AND ^{113}In(n,2n)^{112m,g}In REACTIONS AROUND 2 AND 14 MeV NEUTRON ENERGIES

Cs.M. Buczko, J. Csikai

T. Chimoye[*] and B.W. Jimba[**]

Institute of Experimental Physics, Kossuth University

Debrecen-4001, Pf. 105. Hungary

Abstract: Cross-section curves of (n,n') and (n,2n) reactions on ^{113}In leading to isomeric and ground states have been measured in the 2.1–2.9 and 13.5–14.8 MeV intervals using D–D and D–T neutrons, respectively. Results obtained for (n,n') and (n,2n) reactions in the 13.5–14.8 MeV interval are in good agreement with the recent recommended values. Precise cross-sections measured for the ^{113}In(n,n')^{113m}In reaction in the 2.1–2.9 and 13.5–14.8 MeV ranges are needed for standardization of the excitation function. Cross-sections of the ^{113}In(n,n')^{113g}In reaction were also deduced around 14 MeV as a difference of σ_{NE} and the sum of the measured partial data. A comparison of the isomeric cross-section ratios for (n,n') and (n,2n) reactions on ^{113}In and ^{113}In nuclei indicates the strong spin dependence of the n/γ branching ratio.

(cross-section, ^{113}In, (n,n') reaction, excitation function, isomeric and ground states)

Introduction

The accurate knowledge of excitation functions of fast neutron reactions are of interest from the point of view of nuclear reaction theory (spin distribution parameters, decay branching ratio), design of thermonuclear devices and for the applications of data in dosimetry, neutron flux standardization, elemental analysis, etc. The published experimental cross-sections for (n,n') and (n,2n) reactions [1-18] on ^{113}In around 14 MeV show large discrepancies, while the results for (n,n') process in the 2–3 MeV range are scanty. The large spread existing in the measured data can be attributed to the uncertainty in the bombarding neutron energy caused mainly by the thick target, furthermore, to the other sources of errors of measurements, such as the self-absorption, dead time, cascade-coincidences and flux variation in time. Contamination of the neutron energy in the 14 MeV region can be caused both by the room scattered background and the D–D neutrons from self-target. The possible sources of errors are also discussed. Investigations with ^{113}In and ^{115}In nuclei can give information on the dependence of the de-excitation process on the spin values of the final states.

Experimental procedure

The irradiation of high purity indium metal samples was performed at the home-made neutron generator of the Institute of Experimental Physics, Kossuth University, Debrecen. Neutrons were produced via the ^{3}H(d,n)^{4}He and ^{2}H(d,n)^{3}He reactions, using 200 keV analyzed deuteron ion beam. The neutron fluence in the position of sample has been measured by activation foils, while the flux-variation in time was monitored by a BF$_3$ long-counter. Samples and monitor foils were irradiated simultaneously with neutrons of different energies in a scattering-free arrangement [19]. To decrease the energy spread of neutrons a thin TiD target was cooled by an air-jet system. The neutron energies were changed by placing the samples on aluminium rings of 10 and 20 cm diams at different degrees to the direction of the incident deuteron beam. The irradiation time was choosen according to the half-lives of the residual nuclei. The shapes of indium samples and aluminium monitors of 0.25 and 1.0 mm thick, respectively, were rectangular with dimension of 1.5 x 1 cm^2. The activities of the indium samples and aluminium monitors have been measured by efficiency-calibrated Ge(Li) and NaI(Tl) gamma spectrometers, respectively, connected to an IBM compatible computer based analyzer. The decay characteristics [20] of the residual nuclei are listed in Table 1. The reaction cross-sections used as monitors were taken from Ref. 17.

Table 1. Nuclear data used in this experiment

Reaction	Half-life	E_γ (keV)	I_γ (%)
^{113}In(n,n')^{113m}In	1.66 h	391.7	64.2
^{113}In(n,2n)^{112m}In	20.9 m	155.5	12.8
^{113}In(n,2n)^{112g}In	14.4 m	511.0	43.6
Monitors:			
^{115}In(n,n')^{115m}In	4.49 h	336.2	45.8
^{27}Al(n,α)^{24}Na	15.02 h	1368.5	100

Permanent address: * Faculty of Science and Technology, Thammasat University, Bangkok, Thailand. ** Centre for Energy Research and Training, Ahmadu Bello University, Zaria, Nigeria.

This work was supported by the Hungarian Research Foundation (Contract No. 1734/91).

In the determination of cross-sections corrections were carried out for the following effects: cascade-coincidence, self-absorption, dead time and fluctuation of the neutron yield as a function of time. Other corrections, including the effect of low energy neutrons from the D–D drive-in target and the scattered fast neutrons were also taken into account. In the case of $^{113}In(n,n')^{113m}In$ reaction a simple experimental method was used for the estimation of the D–D contribution to the total activity. The total activity "A" of the irradiated sample produced both by the D–D and D–T neutrons can be expressed as $A=K(\phi_T \sigma_T + \phi_D \sigma_D)$. This relationship was used for two well defined neutron energies, namely at $0°$ and $90°$, where significant differences exist in the relative yield of D–D and D–T neutrons. The activity ratio for the two sample positions (1 and 2) is the following:

$$\frac{A_1}{A_2} = \frac{\phi_{1T}\sigma_{1T} + \phi_{1D}\sigma_{1D}}{\phi_{2T}\sigma_{2T} + \phi_{2D}\sigma_{2D}} \quad (1)$$

The relationships between the neutron fluxes at the indicated positions are $\phi_{1T}=a\phi_{2T}$ and $\phi_{1D}=b\phi_{2D}$. In the knowledge of "a" and "b" taking from calculations or measurements [19] the flux ratio can be determined by the following expression:

$$\frac{\phi_{2D}}{\phi_{2T}} = \frac{(A_1/A_2)\,\sigma_{2T} - a\,\sigma_{1T}}{b\,\sigma_{1D} - (A_1/A_2)\,\sigma_{2D}} \quad (2)$$

For the determination of the contribution of room scattered neutrons to the ^{113m}In activity, indium foils were placed at distances of 1, 2, 4, 8 and 16 cm from the collimated deuteron beam spot of 0.5 cm diameter on the target. The deviation of the activities from the inverse square law is caused by the contribution of scattered neutrons. This contribution is given by ϕ_r = Background + Yield/$4\pi r^2$, where ϕ_r is the measured neutron flux at distance r. A plot of ϕ_r vs. $1/r^2$ was used to determine the effect of the background, at a distance of 5 cm from the beam spot.

Results and discussion

The results obtained for the different reaction cross-sections are summarized in Tables 2, 3, 4 and 5. In general, the results obtained show a small spread.

Table 2. Activation cross-section of $^{113}In(n,n')^{113m}In$ reaction around 2.5 MeV

Neutron energy (MeV)	$\sigma^m_{n,n'}$ (m b)	
	Measured	Fitted
2.15	236.6 ± 10.9	228.2
2.21	228.7 ± 10.8	240.4
2.28	246.4 ± 12.0	252.8
2.38	279.3 ± 13.7	267.4
2.48	281.4 ± 15.7	278.5
2.58	286.5 ± 13.3	286.4
2.69	279.4 ± 12.1	292.0
2.80	301.4 ± 12.9	294.5
2.86	297.1 ± 12.6	294.8
2.94	290.4 ± 11.8	294.0

Table 3. Activation cross-section of $^{113}In(n,n')^{113m}In$ and $^{113}In(n,n')^{113g}In$ reaction around 14 MeV

E_n (MeV)	$\sigma^m_{n,n'}$ (mb)		$\sigma^g_{n,n'}$ (mb)
	Measured	Fitted	Deduced
13.43	70.4 ± 3.1	70.1	601.6
13.48	69.0 ± 2.6	68.9	592.5
13.64	63.4 ± 2.6	65.2	564.0
13.74	64.6 ± 2.3	63.2	547.1
13.79	61.8 ± 2.4	62.3	537.0
13.85	63.3 ± 2.5	61.2	529.3
13.97	56.7 ± 2.1	59.3	510.7
14.10	56.5 ± 2.3	57.4	491.8
14.22	58.6 ± 2.3	55.4	475.1
14.35	55.8 ± 2.2	54.6	458.4
14.46	51.6 ± 2.0	53.7	445.0
14.62	53.2 ± 2.1	52.5	426.8
14.74	50.3 ± 2.0	51.8	414.6
14.83	52.6 ± 2.0	51.4	405.4

Table 4. Activation cross-section of $^{113}In(n,2n)^{112m}In$ reaction around 14 MeV

E_n (MeV)	$\sigma^m_{n,2n}$ (mb)	
	Measured	Fitted
13.43	1005.4 ± 30.9	1011.4
13.48	1011.5 ± 29.1	1018.8
13.64	1025.0 ± 30.0	1041.5
13.74	1070.9 ± 31.6	1054.9
13.79	1083.7 ± 31.0	1061.4
13.85	1093.0 ± 31.3	1069.0
13.97	1075.6 ± 30.0	1083.5
14.16	1077.3 ± 32.0	1105.0
14.35	1124.3 ± 33.2	1124.6
14.54	1154.3 ± 31.2	1142.5
14.74	1139.4 ± 33.6	1159.6
14.83	1181.5 ± 34.4	1166.8

Table 5. Activation cross-section of $^{113}In(n,2n)^{112g}In$ reaction around 14 MeV

E_n (MeV)	$\sigma^g_{n,2n}$ (mb)	
	Measured	Fitted
13.43	223.4 ± 7.7	222.9
13.48	224.8 ± 7.3	225.4
13.64	227.8 ± 7.5	232.2
13.74	238.0 ± 7.8	235.6
13.79	240.8 ± 7.7	237.2
13.85	242.9 ± 7.8	238.9
13.97	239.0 ± 7.5	241.9
14.16	239.4 ± 8.0	245.7
14.35	249.8 ± 8.3	248.8
14.54	256.5 ± 7.8	251.8
14.74	253.2 ± 8.4	255.3

Fitted cross-sections were obtained from a least square fit. Figures 1, 2, 3, 4 and 5 show the measured cross-section curves.

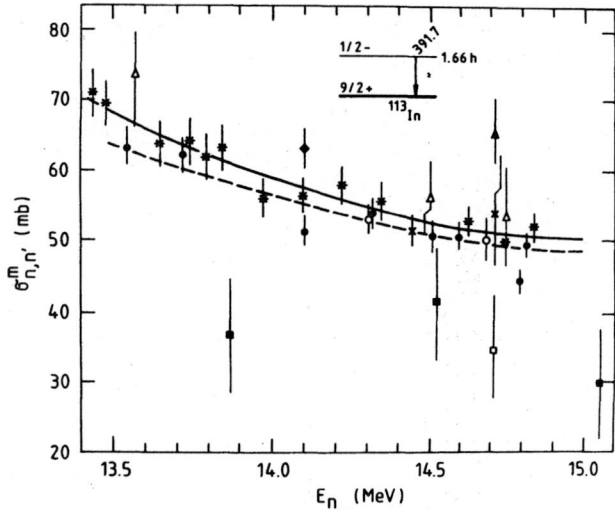

Fig.1. Cross-section curve of ^{113}In(n,n')^{113m}In reaction in 13.5-14.8 MeV incident neutron energy range (--- Lu Hanlin, —— Present work, □ Kozlowski, ♦ Temperley, ▲ Pazsit, ■ Decowski, △ Santry, ○ Ryves, × ANL/NBM, · Lu Hanlin, * Present work)

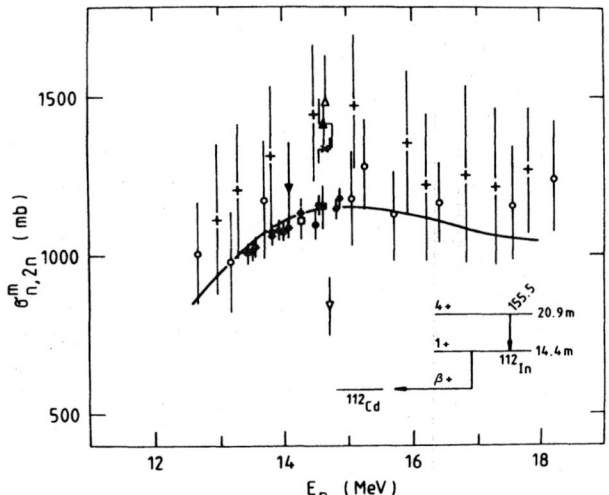

Fig.2. Cross-section curve of ^{113}In(n,2n)^{112m}In reaction in 13.5-14.8 MeV energy range (—— Ke Wei, ▽ Kozlowski, △ Rotzer, ▲ Minetti, + Decowski, × Reggoug, □ Ryves [83], ■ Ryves [81], ▽ Temperley, ○ Li Jianwei, · Ke Wei et al., * Present work)

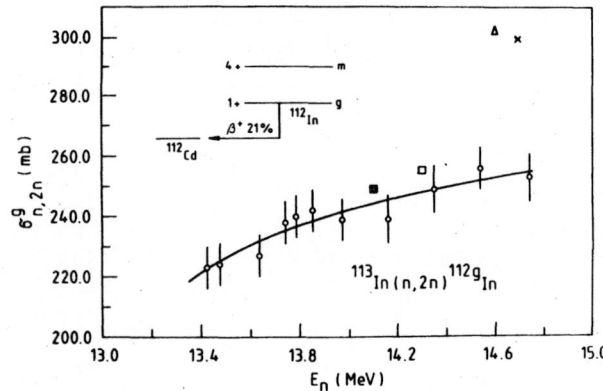

Fig.3. Cross-section curve of ^{113}In(n,2n)^{112g}In reaction in 13.5-14.8 MeV energy range (□ Ryves [83], ■ Temperley, △ Holub, × Rotzer, ○ Present work)

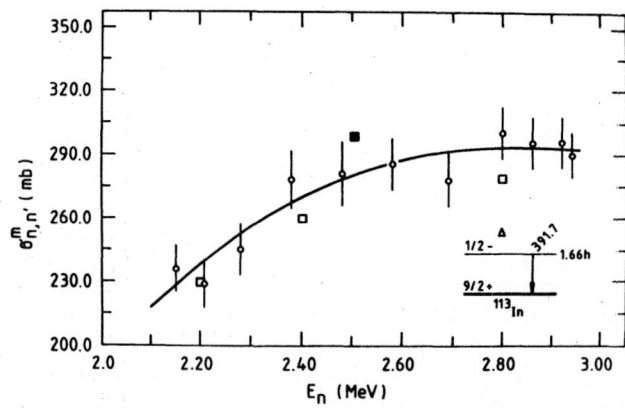

Fig.4. Cross-section curve of ^{113}In(n,n')^{113m}In reaction in 2.1-2.9 MeV energy range (□ Smith, ■ N.C.S.Vol.2., △ Pazsit, ○ Present work)

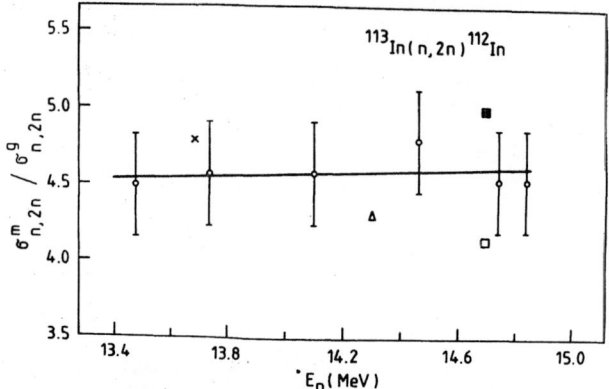

Fig.5. Isomeric cross-section ratio σ^m/σ^g for ^{113}In(n,2n)^{112m,g}In reaction (■ Rotzer, □ Minetti, △ Ryves [83], × Ke Wei, ○ Present work)

For a comparison the literature data are also indicated. As can be seen in Figs. 1 and 2 large discrepancies exist in the published data for $\sigma^m_{n,n'}$ and $\sigma^m_{n,2n}$ values. Data obtained in the present experiment for $\sigma^m_{n,n'}$ and $\sigma^m_{n,2n}$ are in agreement with The recommended curves of Lu Hanlin et al. [7] and Ke Wei et al. [12] respectively, around 14 MeV, as well as with the results of Ryves [6,9] for both cases. Very few results were reported for $\sigma^m_{n,n'}$ reaction in the 2–3 MeV range and for $\sigma^g_{n,2n}$ at around 14 MeV. The present results for $\sigma^g_{n,2n}$ shown in Fig.3 confirm the data given by Ryves [6] and Temperley [2], while the $\sigma^m_{n,n'}$ values in low energy range indicated in Fig.4 are in good agreement with the results of Smith [18].

Errors given in this work represent two sigma and are due to the following sources: neutron flux determination (2%), uncertainties in half-life and standard cross-section (1-1 %) detector efficiencies (1 % Ge(Li) and 5 % NaI). Correction for D–D contribution to the (n,n') reaction in the 14 MeV region has been determined with an error less than 2%. The contribution of room scattered neutrons at 5 cm from the beam spot was found to be about 0.5% of the total flux. In Fig.5 the measured σ^m/σ^g isomeric cross section ratios are given together with the previous results for the ^{113}In(n,2n)^{112m,g}In reactions in the 13.4 - 14.8 MeV range. The error in these data is about ±10%. The deduced cross-section values for the ^{113}In(n,n')^{113g}In reaction are presented in Table 3.

For the estimation of $\sigma_{n,n'}^g$, the following equations were used:

$$\sigma_{NE} = 31.128 \left(6.667 + \frac{4.59}{(E_n(MeV))^2} \right)^2 \text{ mb} \qquad (3)$$

$$\sigma_{NE} = \sigma_{n,n'}^m + \sigma_{n,n'}^g + \sigma_{n,2n}^m + \sigma_{n,2n}^g + \sigma_{n,ch} + \sigma_{n,\gamma} \qquad (4)$$

Eq. (3) was deduced [21] from the mass number dependence of the measured σ_{NE} cross section at 14 MeV. Because the sum of $\sigma_{n,ch}$ are a few mb and the $\sigma_{n,\gamma}$ is also small (^{252}Cf fission neutron spectrum averaged $\sigma_{n,\gamma}$ is 16.7 mb), therefore, to estimate the $\sigma_{n,n'}^g$ cross-section for ^{113}In the following expression can be used:

$$\sigma_{n,n'}^g = \sigma_{NE} - \left(\sigma_{n,n'}^m + \sigma_{n,2n}^m + \sigma_{n,2n}^g \right) \qquad (5)$$

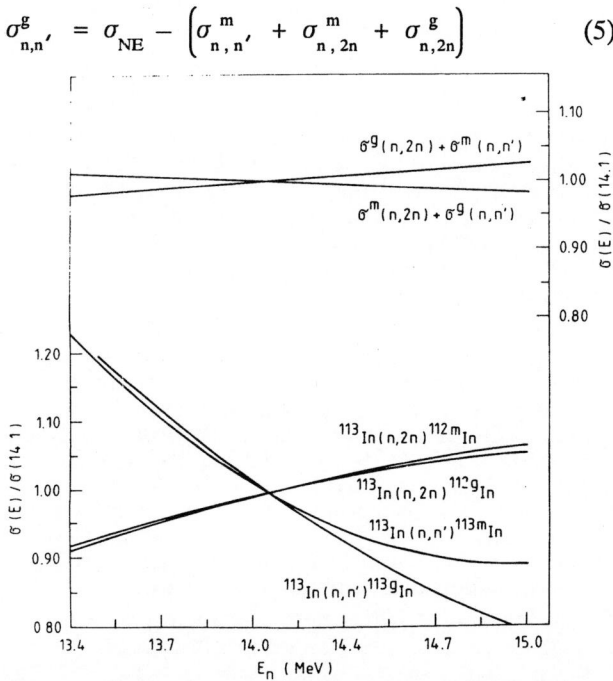

Fig .6.   Relative cross-section curves of (n,n') and (n,2n) reactions on ^{113}In around 14 MeV

In Fig.6 the fitted cross-section curves relative to the 14.1 MeV data are presented. As can be seen in the figure reverse trends are present in the (n,2n) and (n,n') reactions independently of the spin values of the metastable and ground states. In addition, the relative changes in the sums of $\sigma_{n,2n}^m + \sigma_{n,n'}^g$ (resulting high spin states) and $\sigma_{n,2n}^g + \sigma_{n,n'}^m$ (resulting low spin states) can be neglected. The values of the isomeric cross-section ratios σ^m/σ^g are 4.6 and 0.11 for (n,2n) and (n,n') reactions, respectively, at 14.1 MeV These ratios [22] for ^{115}In are as follows: $(\sigma^m/\sigma^g)_{n,n'} = 4.7$ and $(\sigma^m/\sigma^g)_{n,2n} = 0.2$. A comparison of the spin values of the target and residual nuclei indicates that the transition possibilities depend strongly on the spin values of the final states. The ^{113}In and ^{115}In nuclei are symmetrical in the spin values, i.e.

$I_g = 9/2+$ and $I_m = 1/2-$ for the ground and isomeric states, respectively. These data for the ^{112}In and ^{114}In nuclei produced in (n,2n) reactions are $I_g = 1+$ and $I_m = 4+$, $5+$, respectively. Cross-sections of (n,n') and (n,2n) reactions leading to the low spin states are almost the same for the two nuclei. However, significant difference exist in the cross-sections belonging to the high spin states. The $\sigma_{n,2n}^m$ values for ^{113}In(n,2n)^{112m}In and ^{115}In(n,2n)^{114m}In reactions are about 1100 and 1300 mb, respectively, around 14 MeV indicating the strong spin dependence of the (n,2n) cross-sections. The same behaviour of the σ values to the low spin states indicates the cascade saturation in the de-excitation processes.

The authors are indepted to Mrs. M Juhasz for her kind assistance.

REFERENCES

1. T. Kozlowski, Z. Moroz, E. Rurarz and J. Wojtkowska, Acta. Phys. Pol. 33, 409 (1968)
2. J.K.Temperley and D.E.Barnes, BRL-1491 (1970)
3. A. Pazsit and J.Csikai, Sov.J.Nucl.Phys., 15, 232 (1972)
4. P. Dekowski, W. Grochulski, J. Karolyi, A. Marcinkowski, J.Piotrowski, E.Saad and Z. Wilhelmi, Nucl. Phys. A204, 121 (1973)
5. D. G. Santry and J. P. Butler, Can. J. Phys. 54, 757 (1976)
6. T.B. Ryves, Ma Hongchang, S.Judge and P. Kolkowski, J. Phys. G. 9, 1594 (1983)
7. Lu Hanlin, Ke Wei Zhao Wenrong, Yu Weixiang and Yuan Xialin, Chin. J. of Nucl. Phys., 11, 53 (1989)
8. ANL/NBM (1989)
9. T.B. Ryves and P. Kolkowski, J. Phys. G: Nucl. Phys. 7, 115. (1981)
10. E. Holub and N. Cindro, J. Phys. G. Nucl Phys, 2, 405 (1976).
11. Li Jianwei et al. Chin. J. Nucl. Phys.,1,52(1988)
12. Ke Wei, Zhao Wenrong, Yu Weixiang, Yuan Xialin and Lu Hanlin, Chin. J.Nucl. Phys., 1(3), 11 (1989)
13. H.Rotzer, Sitz. Osterr. Acad. Wiss., 176, 289 (1968)
14. B. Minetti and A. Pasquarelli, Z.Physik, 217, 83 (1968)
15. A. Reggoug, G. Paic and A. Chiadli, Proc. of Int. Conf. on Nuclear Data for Science and Technology, Antwerp 6-10, Sept. 1982, (Edited by K.H.Böckhoff) (1983)
16. A. Rotzer, Nucl. Phys. A109, 694 (1968)
17. Neutron Cross Sections, Vol.2, Neutron Cross Section Curves, Eds., V. McLane, C.L.Dunford, P.F.Rose, Academic Press Inc. San Diego (1988).
18. L. D. Smith and J. W. Meadows, Nucl. Sci. Eng., 60, 319 (1976)
19. J. Csikai, Handbook of Fast Neutron Generators, CRC Press Inc., Boca Raton, Florida, (1987)
20. U. Reus and W Westmeier, Atomic Data and Nuclear Data Tables, Academic Press, Vol. 29, No. 1, (1983).
21. J.Csikai, Nucl. Instrum. Meth. Phys. Res., A280, 233 (1989)
22. J. Csikai, Zs. Lantos, Cs .M. Buczko and S. Sudar, Z. Phys A-Atomic Nuclei, 337, 39 (1990)

CROSS SECTION MEASUREMENTS OF NEUTRON INDUCED REACTIONS ON THE ZIRCONIUM ISOTOPES IN THE ENERGY RANGE OF 5.4 TO 12.3 MeV

P. Raics, S. Nagy, S. Szegedi,

Institute of Experimental Physics, Kosssuth L.University,
4001 Debrecen, P.O.B. 105, Hungary;

N.V. Kornilov, A.B. Kagalenko,
Institute of Physics and Power Engineering, Obninsk, USSR

Abstract: Cross sections were determined for the reactions ^{90}Zr(n,α)^{87m}Sr, ^{90}Zr(n,p)^{90m}Y, ^{91}Zr(n,p)^{91m}Y, ^{92}Zr(n,p)^{92}Y, ^{94}Zr(n,α)^{91}Sr, ^{94}Zr(n,p)^{94}Y and ^{96}Zr(n,2n)^{95}Zr using the foil activation technique. Neutrons have been produced via the ^{2}H(d,n)^{3}He reaction with a gas target on a variable energy cyclotron. The samples were irradiated with essentially monoenergetic neutrons at neutron energies of 5.4, 5.9, 6.4, 6.8, 8.2, 8.4, 8.9, 9.4, 10.2, 11.0, 11.6 and 12.3 MeV. Energy and energy spread of neutrons were calculated by the Monte-Carlo method and checked by scanning the neutron resonance absorption curve of ^{12}C around 6.3 Mev. Neutron flux density was determined by reactions ^{27}Al(n,α) and ^{56}Fe(n,p).

(cross section, excitation function, foil activation technique, D$_2$-gas target, monoenergetic fast neutrons, (n,α), (n,p), (n,2n) reactions, Zr isotopes)

Introduction

Excitation functions of several neutron induced nuclear reactions have been determined by our joint group using the EGP-10M tandem accelerator of the Institute of Physics and Power Engineering as neutron source [1,2,3,4]. As the first experiment on our new cyclotron-based variable energy monoenergetic neutron source, the determination of the excitation functions of neutron induced reactions on the isotopes of natural zirconium was chosen. Zirconium is an important constructional material for nuclear reactors, but the excitation function of its neutron induced reactions remained practically unmeasured below 12 MeV bombarding neutron energy until the last year [5].

Experimental

Irradiation circumstances were similar to those in our previous experiments, described e.g. in Ref. [2]. A similar D$_2$-gas target cell with 50 mm length was built and mounted on one of the analyzed beams of the MGC-20 compact cyclotron of the Nuclear Research Institute of the Hungarian Academy of Sciences (ATOMKI) [6]. The most important characteristics of the D$_2$-gas target are summarized in Table 1.

Table 1. Characteristics of the D$_2$-gas target.

Length:	50 mm
Diameter (outer):	10 mm
Gas pressure:	~1100 mbar
Entrance window: its thickness:	Mo 8.1±0.1 mg/cm^2
Wall: its thickness:	stainless steel 0.3 mm
Inner wall cover: its thickness:	Mo 14.1 mg/cm^2
Beam stopper: its thickness:	Pt 0.3 mm
Cooling:	pressed air

Neutrons have been produced via the ^{2}H(d,n)^{3}He reaction in the range from 5.4 MeV to 12.3 MeV at 12 different energies using approximately 3 μA beam current. Sample foils of 19 mm in diameter were positioned at a distance of 40 or 15 mm from the end of the gas target cell at forward direction compared to the bombarding deuteron beam. The energy distributions of neutrons activating the samples were calculated for all the irradiation arrangements by the Monte-Carlo method from the reaction kinematics on the basis of deuteron energy loss of deuterons in the entrance window and the D$_2$-gas of the target cell and the differential cross section. The energy of bombarding deuterons was measured by the calibrated analyzer magnet of the cyclotron. The calculated neutron energy distribution was checked by scanning the neutron resonance absorption curve of ^{12}C around 6.3 MeV. Irradiations with empty gas target cell were carried out to determine the contribution of background neutrons emerging from (d,n) reactions on the structural materials of the accelerator. The most important characteristics of each irradiation are summarized in Table 2.

Table 2. Characteristics of the irradiations

E_d	$<E>_n$	ΔE_n	Φ	l	T_{irr}	BG
MeV	MeV		10^6cm^{-2}s^{-1}	mm	h	
3.16	5.38±0.18		16.47	15	6.84	–
3.60	5.89±0.18		16.02	15	6.85	–
4.01	6.39±0.18		12.82	15	8.93	–
4.50	6.80±0.18		14.46	15	4.33	–
5.45	8.20±0.09		5.49	40	7.19	–
5.87	8.43±0.09		5.83	40	6.89	–
6.40	8.92±0.17		15.04	20	3.65	–
6.80	9.43±0.09		2.05	40	10.10	–
7.64	10.23±0.09		4.82	40	6.23	+
8.37	10.95±0.09		5.57	40	7.38	+
9.00	11.56±0.09		3.89	40	5.14	+
9.72	12.25±0.10		7.04	40	6.18	+

E_d deuteron beam energy,

$<E>_n$ average neutron energy,

ΔE_n spread of neutron energy,

Φ average neutron flux density,

l   distance of the samples from the D$_2$-gas target cell,

T_{irr} time of irradiation,

BG   background irradiation with empty gas target.

Neutron flux density was measured by the reactions of ^{27}Al(n,α) and ^{56}Fe(n,p) applying the foil activation technique (their cross sections were taken from IRDF-90 data file received from IAEA Nuclear Data Section). A flow-type fission chamber with depleted ^{238}U layer and a stilbene neutron spectrometer were used as monitors of the time variation of the neutron flux density. Activity contribution from the ^{2}H(d,np)^{2}H process at high energies were calculated on the basis of energy spectra from the literature. To check these contributions, the low threshold ^{115}In(n,n')^{115m}In and ^{238}U(n,f) processes are planned, but not yet evaluated (this means that the presented results have somewhat preliminary character). The irradiated foil stacks, in the order from the neutron source side, with the thicknesses of foils in brackets, were as follows: In (195 g/cm^2), Al (50 mg/cm^2), Fe (80 mg/cm^2), Ni (890 mg/cm^2), Zr (630 mg/cm^2), Al (50 mg/cm^2), U (on Al backing of 50 mg/cm^2, 0.19 mg/cm^2, in the fission chamber), each of them having the same diameter of 19 mm.

The activities of the flux monitor foils and the Zr samples were measured by a Ge(Li) and a HPGe γ-spectrometer. Several spectra were counted from each sample in the cooling time interval of 7 m to 100 h. Count losses caused by random pile up were determined using a known frequency pulse generator peak in the spectra. Absolute full-energy peak and total efficiencies of the detectors were determined experimentally for the 19 mm source diameter experimentally using standard point like and ^{226}Ra (19 mm in diameter) sources. An IBM XT/AT version of the program GAMANAL [7,8] was applied to the analysis of γ-spectra.

Table 3. Decay data of radionuclides used in the evaluation (A: Abundance of target nuclide)

Reaction	A %	$T_{1/2}$	E_γ keV	I_γ %
^{90}Zr(n,α)^{87m}Sr	51.12	2.805 h	388.3	81.8
^{90}Zr(n,p)^{90m}Y	51.12	3.19 h	202.5	96.5
			479.5	90.6
^{91}Zr(n,p)^{91m}Y	11.22	49.71 m	555.6	94.9
^{92}Zr(n,p)^{92}Y	17.4	3.54 h	934.5	13.9
^{94}Zr(n,α)^{91}Sr	17.57	9.48 h	555.6	61.3
			749.8	23.6
			1024.3	33.4
^{94}Zr(n,p)^{94}Y	17.57	18.70 m	918.8	73.3
^{96}Zr(n,2n)^{95}Zr	2.79	64.03 d	724.2	43.7
			756.7	55.4
^{27}Al(n,α)^{24}Na	100.00	14.97 h	1368.53	100.0
			2754.4	99.9
^{56}Fe(n,p)^{56}Mn	91.7	2.58 h	846.6	98.9
			1810.7	27.2

The decay curves of the different γ-lines were fitted by the method of least-squares. The extrapolated activities of the identified radionuclides were corrected for the cascade coincidence losses and for the shelf-absorption of the samples. The nuclear data for the absolute γ-branching, decay schemes and absorption coefficients were taken from Refs. [9], [10] and [11], respectively. The decay data of the radionuclides used in the evaluation are given in Table 3.

Results and discussion

The cross section data for the investigated eight nuclear reactions on various isotopes of zirconium are given in Table 4, and compared with results of Majah and Qaim [5] in Fig. 1.

Neutron flux densities determined by monitor reaction on Fe and Al are in agreement even at low energies where the cross section of ^{27}Al(n,α) reaction is very low; their average was used at the calculations. The cross section errors in Table 4. are the statistical uncertainties of the activity measurements only.

Our results seem to be in reasonable agreement with those measured by Majah and Qaim [5].

In addition to the six reactions presented in Ref [5], we could observe products of the ^{94}Zr(n,p)^{94}Y and ^{90}Zr(n,2n)^{89}Zr reactions. Since the short half-life of ^{94}Y (17 m), for the first reaction we could achieve good enough accuracy at the two highest energy irradiations only. For the second reaction our results are contradictory and need further investigations, not presented in this paper.

Acknowledgements

We thank Prof. J. Csikai for his active support of this research program, Dr. A. Valek and the crew of the Cyclotron MGC-20 for the conscientiously performed irradiations. The authors are grateful to V. Ja. Baryba for advices in the construction of the D$_2$-gas target cell, to students Sz. Márka, A. Némethy and D. Sohler students for their cooperation in the evaluation. This work was supported by the Hungarian National Science Foundation Contract No. OTKA 259/1986.

REFERENCES

1. N.V. Kornilov, B.V. Zhuravlev, O.A. Salnikov, P. Raics, S. Nagy, S. Daróczy, K. Sailer and J. Csikai: Izmerenie sechenija reakcii ^{238}U(n,2n)^{237}U v intervale energii nejtronov 6.5-10.5 MeV. Atomnaja Energija 49, 283 (1980)

2. P. Raics, S. Daróczy, J. Csikai, N.V. Kornilov, V.Ja. Baryba, and O.A. Salnikov: Measurement of the cross sections for the ^{232}Th(n,2n)^{231}Th reaction in the 6.745 to 10.450 MeV energy range. Phys. Rev. C32, 87 (1985)

3. N.V. Kornilov, V.Ja. Baryba, A.V Balickij, A.P. Rudenko, B.D. Kuzminov, O.A. Salnikov, E.A. Gromova, S.S. Kovalenko, L.D. Preobrazhenskaja, Ju.A. Selickiij, B.I. Tarler, V.B. Funstejn, B.A. Jakovlev, S. Daróczy, P. Raics and J. Csikai,: Izmerenie sechenij reakcij ^{237}Np(n,2n)^{236}Np(22.5 ch) v diapazone energii nejtronov 7-10 MeV. Atomnaja Energija 58, 117 (1985)
The measurement of ^{237}Np(n,2n)^{236}Np (22.5 h) reaction cross-sections in the energy range 7-10 MeV. Transactinium Isotope Nuclear Data, 1984 (IAEA-TECDOC-336),305, (Proc. 3. Advisory Group Meeting, Uppsala, 21-25 May 1984)

4. N.V. Kornilov, V.Ja. Baryba, A.V. Balickij, A.P. Rudenko, S. Daróczy, P. Raics and Z. Papp: Sechenie reakcii ^{58}Ni(n,p) dlja energii nejtronov 7-10 MeV. Atomnaja Energija, 58, 128 (1985)

5. M. Ibn Majah and S.M. Qaim, Nucl. Sci. Eng. 104, 271 (1990)
6. A. Valek, G. Bibok and A. Paál: Cyclotron laboratory in the Institute of Nuclear Research, Debrecen, International Symposium on in-beam nuclear spectroscopy, Debrecen, Hungary, May 14-18 1984 (Edited by Zs. Dombrádi and T.Fényes, Akadémiai Kiadó, Budapest)
7. R. Gunnink and J.B. Niday: Computerized quantitative analysis by gamma-ray spectrometry, Vol.I. Description, of the GAMANAL program, UCRL-51061, Vol.1, Lawrence Livermore Laboratory, 1 March 1972
8. S. Nagy: GAMANAL XT/AT user's manual, Institute of Experimental Physics, Kossuth University, Debrecen, Hungary, Version 1.1, December 1989 (unpublished)
9. U. Reus and W. Westmeier, Catalog of Gamma Rays from Radioactive Decay: Part I. and II, Atomic Data and Nucl. Data Tables 29, 1 and 194 (1983)
10. C.M. Lederer and V.S. Shirley (Editors), Table of Isotopes, Seventh Edition (John Wiley and Sons, Inc., 1978)
11. E. Storm and H.J. Israel, Nucl. Data Tables A7, 565 (1970)

Table 4. Measured cross sections of neutron induced reactions on some Zr isotopes

E_n	Nuclear reaction cross section (mb)						
	^{90}Zr(n,α)^{87m}Sr	^{90}Zr(n,p)^{90m}Y	^{91}Zr(n,p)^{91m}Y	^{92}Zr(n,p)^{92}Y	^{94}Zr(n,α)^{91}Sr	^{94}Zr(n,p)^{94}Y	^{96}Zr(n,2n)^{95}Zr
5.38	0.011 ±0.006	0.044 ±0.009	0.18 ±0.05				
5.89		0.145 ±0.008	0.42 ±0.04	0.46 ±0.16			
6.39	0.010 ±0.004	0.360 ±0.100	0.82 ±0.06	0.53 ±0.10			
6.80	0.028 ±0.004	0.580 ±0.140	0.95 ±0.06	1.25 ±0.73			
8.20	0.155 ±0.009	1.750 ±0.040	2.34 ±0.14	1.21 ±0.54	0.45 ±0.15		69.2 ±35
8.43	0.233 ±0.010	1.900 ±0.040	2.66 ±0.15	1.50 ±0.17	0.29 ±0.11		248.0 ±22
8.92	0.355 ±0.009	2.500 ±0.030	3.35 ±0.10	1.73 ±0.14	0.43 ±0.09		540.0 ±20
9.43	0.523 ±0.025	2.780 ±0.060	2.20 ±0.26	2.93 ±0.47	0.85 ±0.28		1103.0 ±43
10.23	0.869 ±0.020	3.860 ±0.050	3.51 ±0.17	4.37 ±0.33	1.06 ±0.18		1233.0 ±55
10.95	1.365 ±0.025	5.410 ±0.070	7.23 ±0.26	6.35 ±0.38	1.30 ±0.13		1152.0 ±37
11.56	1.355 ±0.037	5.620 ±0.080	6.75 ±0.25	6.07 ±0.56	1.46 ±0.55	1.36 ±0.65	1152.0 ±52
12.25	2.525 ±0.035	9.080 ±0.100	12.88 ±0.31	11.05 ±0.41	2.72 ±0.19	1.98 ±0.21	722.0 ±20

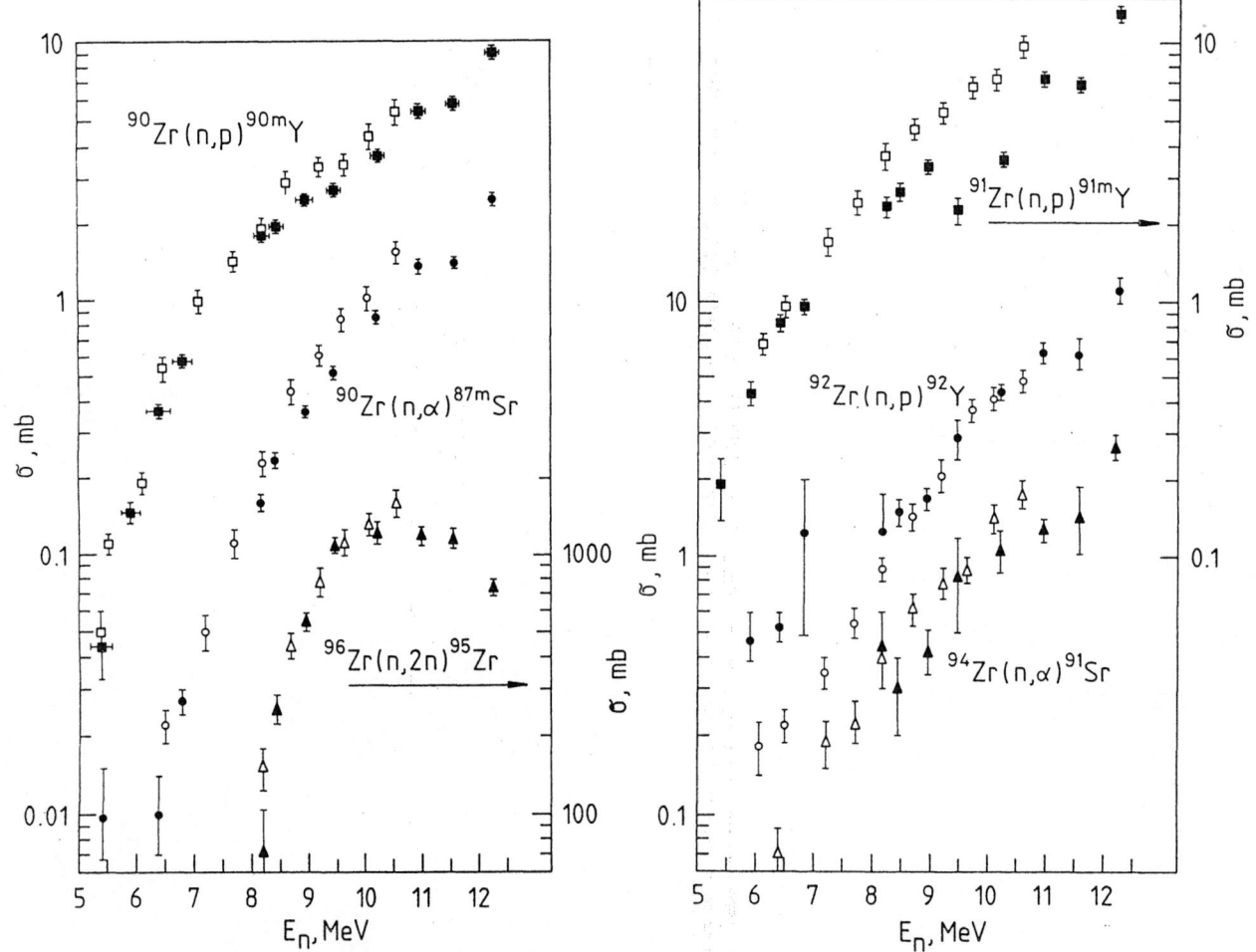

Fig 1. Measured cross sections of neutron induced reactions on some Zr isotopes
Full symbols: present experiment, open symbols: Majah and Qaim [5]

MEASUREMENT OF THE ^{23}Na(n,2n)^{22}Na CROSS SECTION AT E$_n$ = 19.45 MeV
AND MODEL CALCULATIONS OF CROSS SECTIONS FOR
NEUTRON INDUCED REACTIONS ON ^{23}Na

B. Strohmaier, M. Wagner and H. Vonach,
Institut für Radiumforschung und Kernphysik
der Universität Wien, A-1090 Vienna, Austria

Abstract: The ^{23}Na(n,2n)^{22}Na cross section was measured at an incident energy of 19.45 MeV using the activation method. In addition, model calculations were performed aiming at reproduction of the new experimental value together with existing cross-section data for the (n,2n) reaction in the first 2.5 MeV above threshold as well as for other neutron-induced reactions on ^{23}Na.

(^{23}Na(n,2n)^{22}Na, activation method, dosimetry reaction, cross section, excitation function, model calculations)

Introduction

The existing measurements of the ^{23}Na(n,2n)^{22}Na cross section /1-11/ exhibit serious discrepancies especially at neutron energies above 15 MeV. These inconsistencies of the experimental data have resulted in a large difference between the cross-section predictions of the two most recent evaluations for ENDF/B-V /12/ and for the International Reactor Dosimetry File /13/ of the IAEA. In order to improve this situation a new measurement of the ^{23}Na(n,2n)^{22}Na cross section was carried out at a neutron energy of 19.449 ± 0.025 MeV applying the activation method. Subsequently, a calculational investigation employing the coupled-channels optical model and the statistical model under consideration of preequilibrium emission was undertaken to inquire if the new experimental value could be described simultaneously with the existing data regarding other neutron-induced reactions on ^{23}Na.

Experiment

Quasi-monoenergetic neutrons were produced at the CN-type 7 MV Van-de-Graaff accelerator at the Central Bureau for Nuclear Measurements, Geel, by means of the T(d,n)^{4}He reaction by bombarding an air-cooled solid-state T-Ti target with a deuteron beam of 9.1 μA average current for $\approx$ 24 hours. The target consisted of a Ti layer (diameter = 3 cm, thickness = 4.42 mg cm^{-2}) which was backed by an Ag sheet of 0.25 mm thickness and glued in a cylindrical Al target holder with 1 mm wall thickness. The average energy of the deuterons in the target was 3.00 ± 0.02 MeV.

The sample, fine-grained NaCl powder of pro analysi purity (> 99.5%) which had been dried in a desiccator at 120°C immediately before weighing, was contained in a cylindrical Al box with a wall thickness of 1.0 mm, an inner diameter of 12.0 mm and an inner height of 10.0 mm. For the irradiation, the sample was sandwiched between two 0.125 mm thick, 12.0 mm Ø Ni foils and positioned at 12.3° relative to the incident d$^+$-beam; the distance of the midplane of the sample to the neutron source was 1.87 cm. The neutron energy distribution profile averaged over the area of the sample and the monitor foils was calculated by means of a Monte Carlo simulation providing both the average neutron energy and the energy spread (FWHM/2 = 0.350 MeV).

The ^{23}Na(n,2n)^{22}Na cross section was measured relative to the ^{58}Ni[(n,np) + (n,d)]^{57}Co cross section which had been accurately determined in previous experiments at the Institut für Radiumforschung und Kernphysik, Vienna, (IRK) /14/. Above a neutron energy of $\approx$ 12.5 MeV ^{57}Co is also produced via the reaction ^{58}Ni(n,2n)^{57}Ni → ^{57}Co. This contribution to the ^{57}Co activity was properly taken into account and corrected for, using the evaluated ^{58}Ni(n,2n)^{57}Ni cross section given in Ref. 15. The ^{57}Co measurements were not started before ^{57}Ni (t$_{1/2}$ = 36.08 ± 0.09 h) /16/ had completely decayed to ^{57}Co. The average neutron fluence incident on the NaCl sample was determined by aid of the ratio of the ^{57}Co activity induced in the two Ni foils. Assuming a fluence dependence like 1/r$^\alpha$ (r... distance sample to neutron source) the value of the exponent α was established to be 1.85 ± 0.01, which includes neutron absorption effects.

The induced ^{22}Na and ^{57}Co activities were measured at the IRK by standard γ-ray counting techniques employing an intrinsic Ge γ-ray detector. Due to the low ^{22}Na count rate ($\approx$ 4.87 cpm in the 1274.5 keV full-energy peak) four activity measurements, interspaced by background measurements, were carried out lasting 8.5×10^4 s each. In order to take into account the inhomogeneous activation due to the geometry and to neutron absorption in the 10.0 mm thick NaCl sample, the activity measurements were carried out with alternately the near end and the far end of the sample - as seen from the neutron source - facing the Ge γ-ray detector. The efficiency calibration for the γ-rays emitted in the decay of ^{22}Na and ^{57}Co was carried out by means of calibrated point-like ^{22}Na and ^{57}Co sources. These sources also served to study the variation of the efficiency with the distance from the detector and with the displacement from the axis of the detector set-up. The absolute efficiency for extended sources was then obtained by integration over the volume of the samples. In order to create similar conditions for the counting of the radiation emitted by ^{22}Na from the irradiated sample and from the standard source especially with respect to the annihilation of the positrons, the ^{22}Na point-like source was also put between two 1.0 mm thick Al disks. Thus, in assessing the attenuation of the ^{22}Na γ-rays in the irradiated sample, only the self-attenuation in the NaCl powder had to be taken into account; it amounted to (3.4 ± 0.3)% for the 1274.5 keV γ-ray. The self-attenuation of the ^{57}Co γ-rays in the Ni foils was also corrected for, the correction being (1.7 ± 0.2)%.

The activation enhancement due to elastic neutron scattering in the sample and the Ni monitor foils and to inscattering from the adjacent walls of the Al sample container was estimated. The correction amounted to $\approx$ 2.6% and $\approx$ 0.15% for the NaCl and the Ni samples, respectively.

The measured cross section for the ^{23}Na(n,2n)^{22}Na reaction is 114.5 ± 6.6 mb; this result is based on the evaluation of the 1274.5 keV full-energy peak and agrees very well with the result obtained by evaluating the 511 keV peak from the annihilation radiation. The half-life of ^{22}Na, 2.602 ± 0.002 a, was taken from Ref. 16.

The principal sources of uncertainty are the statistical uncertainties of the count rates (± 0.87% for ^{22}Na, ± 0.48% for ^{57}Co); the uncertainties of the efficiency calibration of the γ-ray detector (± 1.59% for ^{22}Na, ± 0.90% for ^{57}Co); the emission probability of the 122 keV + 136 keV γ-rays of ^{57}Co (± 0.21%), since the ^{57}Co reference source was not calibrated in activity but in γ-emissions per second; the number of ^{23}Na atoms (± 0.2% from the NaCl mass determination); the γ-ray self-attenuation in NaCl (± 0.3%) and in Ni (± 0.2%)); the activity enhancement due to elastic scattering of the neutrons in the sample (± 0.5%); the correction applied to the average of the neutron fluences hitting the Ni monitor foils, which accounts for the nonlinear decrease of the fluence in the NaCl sample (± 2.27%); and, finally, the ^{58}Ni(n,np)^{57}Co reference cross section including the correction for the contribution from the ^{58}Ni(n,2n)^{57}Ni reaction to the ^{57}Co activity (± 4.79%). The uncertainties of the ^{22}Na and ^{57}Co decay constants and of the number of ^{58}Ni atoms in the monitor foils were negligible. All uncertainty components are standard deviations or equivalent standard deviations; they are uncorrelated and were added in quadrature to give the total uncertainty.

In Fig. 1 the result of our measurement is compared with the experimental cross-section values published in the literature and with the ENDF/B-V and IRDF evaluations. The experimental cross sections were renormalized to recent decay data and monitor cross sections, if necessary. The ^{23}Na(n,2n)^{22}Na cross section found in this work lies between the predictions of the two evaluations mentioned above, but is much closer to the IRDF-85 than the ENDF/B- value. The result is lower by a factor of 0.6 than the value given in Ref. 2 at the corresponding incident energy, and is in good agreement with a measurement reported by Lu Hanlin /17/. Further support for these cross-section values in the energy range 15 to 18 MeV comes from a Japanese group /18/ whose measured excitation function, however, decreases even below the IRDF-85 prediction beyond 18 MeV incident energy. On the other hand, the experiment reported by Xu Zhijing /17,19/ reproduces the result given by Liskien and Paulsen /2/. Both groups measured the ^{22}Na activities by coincidence counting of the emitted annihilation radiation. The simultaneous discrepancy to recent results obtained by counting the 1274.5 keV γ-rays may point to a hitherto unrecognized systematic error in the 511 keV-ray coincidence counting.

Model calculations

For ENDF/B-V, Larson /12/ had performed comprehensive model calculations combining the DWBA with a statistical model description including preequilibrium emission. The choice of model parameters had been optimized to reproduce all relevant data and was completely unbiased with regard to the discrepant ^{23}Na(n,2n)^{22}Na measurements. The result of the (n,2n) calculation was in favor of the high-lying data sets /2/.

In view of the enhanced experimental evidence for the ^{23}Na(n,2n)^{22}Na cross section to lie well below the prediction of ENDF/B-V in the energy range 15 to 20 MeV, we decided to investigate to which extent a reduction of the calculated (n,2n) excitation function is compatible with the simultaneous reproduction of experimental data for other quantities, in particular the ^{23}Na(n,p)^{23}Ne and ^{23}Na(n,α)^{20}F activation cross sections.

Since ^{23}Na is a deformed nucleus, inelastic scattering to low-lying collective states needs to be considered. We did this in the frame of the coupled-channels optical model (CCOM) employing the computer code JUPITOR /20/. The CCOM calculations yielded the total cross section and differential elastic and inelastic cross sections for verification of the chosen optical potential through comparison of these results with measurements, and also served for generating the transmission coefficients for the statistical model (SM) calculations.

In the CCOM calculations, ^{23}Na was treated in the rotational model, coupling the ground state (3/2$^+$) and the excited states at 0.44 MeV (5/2$^+$) and 2.08 MeV (7/2$^+$). The choice of the CCOM parameters was guided by work by Whisnant /21/ and Schreder, Dagge and coworkers /22/ on scattering of neutrons on ^{27}Al, and by Walter et al. /23/ for scattering of nucleons on ^{28}Si. The deformation parameters were taken from these publications; necessary modifications of the potential depths may be ascribed to differences in the coupling schemes and resulted in an acceptable description of the total cross section and - under inclusion of the SM contribution - of the total elastic cross section and differential elastic and inelastic cross sections in the incident energy range between 5.44 and 14.1 MeV. The corresponding data are cited in Ref. 12.

For charged particles, the transmission coefficients were generated in the spherical optical model. A variety of proton and α-particle optical potentials from the literature was tested with regard to their ability to provide both considerable charged particle competition to neutron emission, particularly in the second emission step, and a good description of the (n,α) and (n,p) activation cross sections. We chose the optical potential from MacFadden and Satchler /24/ with an adjusted real-well depth for α-particles. For protons, a compromise between the potential used in Ref. 12 and the Becchetti and Greenlees prescription /25/ was constructed. Deuterons were also considered in the SM calculations, with the transmission coefficients generated with the optical potential given by Perey and Perey /26/.

The statistical model calculations were carried out with the code STAPRE /27/. Discrete levels were considered according to the information given in Refs. 28 and 29. The continuum level densities were described in the back-shifted Fermi-gas model with the moment of inertia chosen as 0.75 times the rigid body value. The parameters a and Δ were adjusted to fit the cumulated level numbers as a function of energy as far as possible. Slight deviations of these values were admitted later in the evaluation procedure to improve the description of the cross-section data. Also from Ref. 28, resonance data were taken which, however, do not pose stringent restrictions on the level-density parameters due to the large fractions of resonances with missing J^π assignment.

For γ-ray emission, the strength functions were calculated from the Brink-Axel model with a 30% decresed strength for electric dipole radiation, and from the Weisskopf model for all other radiation types.

In the exciton model, particle-hole state densities were calculated assuming the single-particle state densities to be in a constant relation to mass number A, and applying energy shifts of size 12 A$^{-1/2}$ for each nucleon type of even number to account for pairing. Charge conservation was considered in the nucleon emission rates. In the internal transition rates, the constant in the squared matrix element was equal to 500 MeV^3; the Pauli-principle correction was omitted. In order to achieve an increase at the high-energy end of particle-emission spectra, it was found favorable to exclude intrinsic transitions leading to an unchanged or decreased exciton number.

With the described options and parameters, we achieved a reduction of the (n,2n) excitation function by about 25% with respect to the ENDF/B-V values in the energy region 19 - 20 MeV, a much smaller reduction at lower incident energies. The results are displayed in Fig. 1. They are still about 20% high compared to the new measurements; also the reproduction of the shape is not satisfactory. The (n,p) and (n,α) excitation functions obtained under the same assumptions are shown in Figs. 2 and 3. Here, with regard to the measurements existing already in 1980 (for references, see the work by Larson /12/) the experimental information was completed mainly by the experiment by Weigmann et al. /30/. The (n,2n) cross-section values can be further diminished by using optical potentials which enhance low-energy charged-particle emission, at the price of overestimating the low-energy portions of the (n,p) and (n,α) activation cross sections. The production-cross section of the 1274.5 keV γ-line in ^{22}Ne reported by Zhou Hongyu et al. /31/ was compared to the calculation and found to agree roughly.

In addition to activation cross sections, we also checked particle emission data. For neutrons, measurements of the double-differential emission cross sections for Na at E$_n \approx 14$ MeV have been reported by Hermsdorf et al. /32/ and Takahashi et al. /33/. Both data sets, however, show considerable irregularities in the angular distributions of the low-energy part of the neutron spectrum. In fact, the Hermsdorf data have been changed with respect to the version used in Ref. 12 due to a re-evaluation /34/ of the original measurement /32/. In principle, the data of Ref. 33 were obtained under much more favorable experimental conditions and should, therefore, be more reliable, but the results indicate there may also be undetected systematic errors. Thus, both data sets cannot be used for a sensitive test of our evaluation. Comparison of the calculations with angle-integrated cross sections from these measurements shows better agreement with the results of Ref. 32. For charged particles, the production spectra were compared to the experimental data from Ref. 35. The agreement was found to be acceptable; the experimental data are unnormalized, however.

In summary, our goal to reproduce the most recent measurements of the ^{23}Na(n,2n)^{22}Na cross section together with the existing experimental data on other neutron-induced reactions on ^{23}Na has as yet been reached only to a limited extent: the (n,2n) cross section was reduced with respect to ENDF/B-V /12/ by a fraction varying over the energy range considered, but amounting to 25% at most. Subsequently, the calculations deviate from an eye-guide through the most recent measurements in magnitude as well as in shape. In order to resolve the remaining discrepancies, investigations will be continued.

We are indebted to Dr. H. Liskien and Mr. R. Widera for carrying out the irradiation of the NaCl sample at the CBNM, Geel, Belgium, and to Dr. O. Schwerer from the Nucl. Data Section, IAEA, Vienna, for supplying the cross-section data of the EXFOR library.

/1/ R.J. Prestwood, Phys. Rev. **98**, 47 (1955).
/2/ H. Liskien and A. Paulsen, Nucl. Phys. **63**, 393 (1965), and A. Paulsen, private communication to CCDN (1965).
/3/ J. Picard and C.F. Williamson, Nucl. Phys. **63**, 673 (1965).
/4/ H.O. Menlove, Phys. Rev. **163**, 1308 (1967).
/5/ R.C. Barrall, J.A. Holmes and M. Silbergeld, Rept. AFWL-TR-68-134, Air Force Weapons Lab., Kirtland, New Mexico (1968).
/6/ J. Araminowicz and J. Dresler, Rept. INR-1464/I/A, 14, Institute of Nuclear Research, Warsaw (1973).
/7/ G.N. Maslov, F. Nasyrov and N.F. Pashkin, Rept. YF-9, 50, FEI,

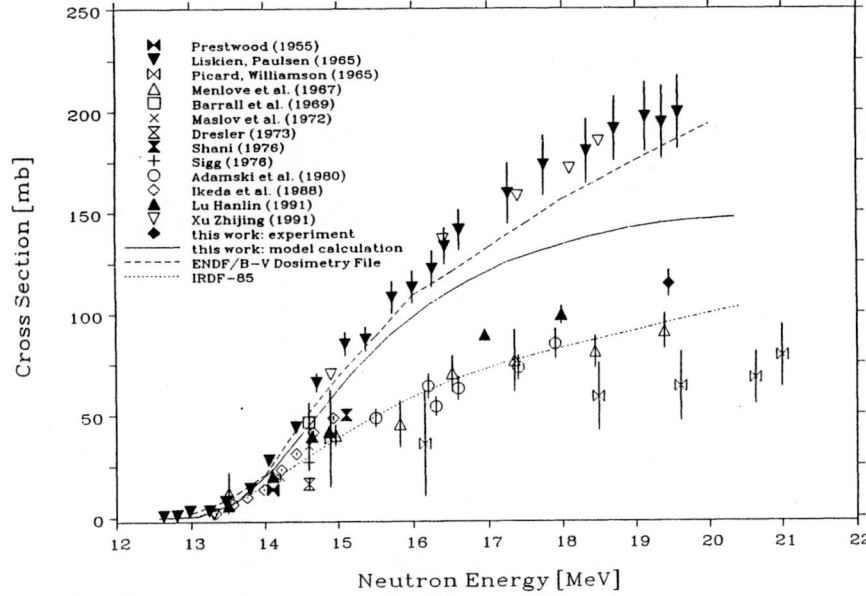

Fig. 1: Experimental results for the ^{23}Na(n,2n)^{22}Na cross section from the literature and the evaluations for ENDF/B-V and the International Reactor Dosimetry File (IRDF-85) as compared to the results of this work.

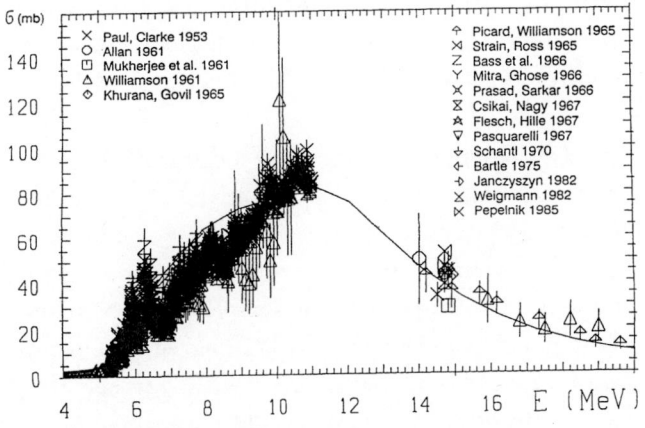

Fig. 2: Measured and calculated ^{23}Na(n,p)^{23}Ne cross sections.

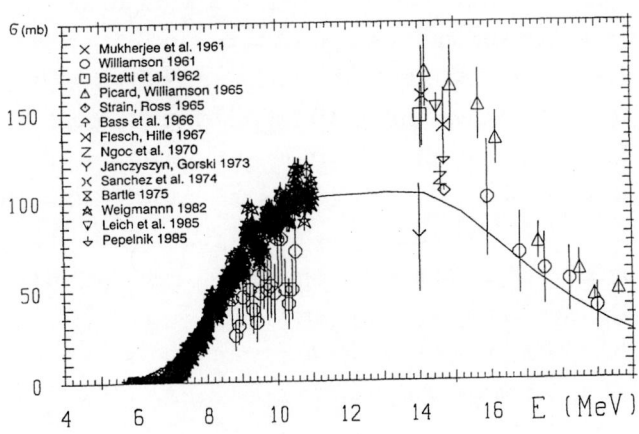

Fig. 3: Measured and calculated ^{23}Na(n,α)^{20}F cross sections.

Obninsk (1972) and Rept. INDC(CCP)-42, IAEA, Vienna (1974).

/8/ G. Shani, Rept. INIS-MF-3663, 83 (1976).

/9/ R.A. Sigg, Thesis, Univ. of Arkansas (1976), and Diss. Abstr. B 37, 2237 (1976).

/10/ L. Adamski, H. Herman and A. Marcinkowski, Ann. Nucl. Energy 7, 397 (1980).

/11/ Y. Ikeda, Ch. Konno, K. Oishi, T. Nakamura, H. Miyade, K. Kawade, H. Yamamoto and T. Katoh, Report JAERI 1312, Japan Atomic Research Institute, Tokai-mura (1988).

/12/ D. Larson, Report ORNL-5662, Oak Ridge Nat. Laboratory (1980), and Nucl. Sci. Eng. 78, 324 (1981).

/13/ L. Adamski, M. Herman and A. Marcinkowski, Prog. Rept. INR 1809/I/PL/A, Institute of Nuclear Research, Warsaw (1979), and Rept. INDC(POL)-10/G, IAEA, Vienna (1979).

/14/ A. Pavlik, G. Winkler, M. Uhl, A. Paulsen and H. Liskien, Nucl. Sci. Eng. 90, 186 (1985).

/15/ M. Wagner, H. Vonach, A. Pavlik, B. Strohmaier, S. Tagesen and J. Martinez-Rico, Physics Data 13-5, 75, Fachinformationszentrum Karlsruhe (1990).

/16/ J.K. Tuli in "Handbook on Nucl. Activation Data", IAEA Techn. Report Series 273, 3, IAEA, Vienna (1987).

/17/ Lu Hanlin, priv. comm. to H. Vonach (1991).

/18/ T. Nakamura, H. Sugita, Y. Kondo, Y. Uwamino and M. Imamura, Prog. Rept. JAERI July 1989 -June 1990, NEANDC(J)-155/U, 101 (1990).

/19/ Xu Zhijing et al., contribution to the present conference, to be published.

/20/ T. Tamura, Rev. Mod. Phys. 37, 679 (1965), and Rept. ORNL-4152, Oak Ridge Nat. Laboratory (1967).

/21/ C.R. Whisnant, J.H. Dave and C.R. Gould, Phys. Rev. C30, 1435 (1984).

/22/ G. Schreder et al., Nucl. Data for Science and Technology, Proc. Int. Conf. Mito, Japan, May 30 - June 3, 1988, 691, S. Igarasi, ed., Saikon Publ. comp., Tokyo (1988), and G. Dagge, W. Grum, J.W. Hammer, K.-W. Hoffmann and G. Schreder, Phys. Rev. C39, 1768 (1989).

/23/ R.L.Walter, C.R. Howell and W. Tornow, Nucl. Data for Science and Technology, Proc. Int. Conf. Mito, Japan, May 30 - June 3, Nucl. Data for Science and Technology, Proc. Int. Conf. Mito, Japan, May 30 - June 3, 1988, 675, S. Igarasi, ed., Saikon Publ. comp., Tokyo (1988), and C.R. Howell, R.S. Pedroni, G.M. Honoré, K. Murphy, R.C. Byrd, G. Tungate and R.L. Walter, Phys. Rev. C38, 1552 (1988).

/24/ L. MacFadden and G.R. Satchler, Nucl. Phys. 84, 177 (1966).

/25/ F.D. Becchetti and G.W. Greenlees, Phys. Rev. 182, 1190 (1969).

/26/ C.M. Perey and F.G. Perey, Phys. Rev. 132, 755 (1963).

/27/ B. Strohmaier and M. Uhl, Proc. IAEA Course Nuclear Theory for Applications, Jan. 17 - Feb. 10, 1978, Rept. IAEA-SMR-43, 313, IAEA, Vienna (1980).

/28/ P.M. Endt, Nucl. Phys. A521, 1 (1990).

/29/ F. Ajzenberg-Selove, Nucl. Phys. A473, 1 (1987).

/30/ H. Weigmann, G.F. Auchampaugh, P.W. Lisowski, M.S. Moore and G.L. Morgan, Nucl. Data for Science and Technology, Proc. Int. Conf. Antwerp, Sept. 6 - 10, 1982, 814, K.H. Böckhoff, ed., Reidel Publ. Comp., Dordrecht (1983).

/31/ Zhou Hongyu et al., Rept. CNDC-0003, No. 2, Chinese Nuclear Data Center (1989).

/32/ D. Hermsdorf, A. Meister, S. Sassonoff, D. Seeliger, K. Seidel and F. Shahin, Rept. ZfK-277/Ü, Zentralinstitut für Kernforschung, Rossendorf (1975).

/33/ A. Takahashi et al., OKTAVIAN Rept. A-87-01, Osaka University (1987).

/34/ D. Hermsdorf, priv. comm. to Nucl. Data Section, IAEA, Vienna (1982), see EXFOR Entry 30397.

/35/ B. Aldefeld, Nucl. Phys. A145, 569 (1970).

THE (n, 2n) CROSS SECTION MEASUREMENT FOR Na-23

Xu Zhi-zheng, Pan Li-min, Wu Zhi-hua

Dep. of Nuclear Science, Fudan University, Shanghai, China

Abstract: The cross section for Na-23(n,2n) reaction is important not only for practical use, but also for theoretical study. Only a few sets of experimental data exist for this reaction, and each differs from the other by approximately a factor of two. So we decided to remeasure it. Measurement was done by the activation method in the energy range of E_n=13.3 to 18.5 MeV. Monoenergetic neutrons were obtained from T(d,n)He-4 reaction at E_d=2.3 MeV. The induced specific activities were detected by a coincidence count setup. For comparison, the other measured, evaluated and calculated data from E_n=12 to 20 MeV are also given.

(Na-23, (n, 2n) reaction, natural target, fast neutron, activation cross section)

Introduction

The activation cross section for Na-23(n,2n) reaction is important not only for practical use but also for theoretical study. First, they are necessary for radiation protection calculations in fast reactor using Na as a coolant. From a theoretical point of view, the statistical theory can give good agreement with measured cross-section for medium and heavy mass nuclei. But neutron induced measured reactions on light elements show isolated resonance structure. This means breakdown of the statistical conditions. The nucleus Na-23 is just in the "Intermediate zone" between the region of validity of statistical theory and region of very light nuclei which requires detailed treatment. Only a few set of experimental data exist for the (n,2n) reaction on Na [1-3], and each differs from the others by approximately a factor of two. It causes difficulty in the use of these data. So we decided to remeasure it.

Experimental Procedure

The neutrons of En=13.3 to 18.5 MeV were obtained by bombarding a T-Ti target with Ed=2.3 MeV deuterons. The target was 2.65 mg/cm² in thickness. The deuteron beam was from 1 to 3 microampere. The samples were suspended on a ring of 70mm in radius.

The centre of the ring was geometricl centre of the beam. The size of the beam spot was defined by an aperture of 3mm in diameter. The tritium target was mounted on a thin copper tube. The thickness of the tube was 0.1mm. The neutron flux was determined by a calibrated long counter positioned in zero degree direction at a distance of 120mm from the target before and after the samples were irradiated. A BF₃ counter in 40 degree was used as monitor. The samples were irradiated for 60 hours.

The samples consisted of natural NaF powder, pressed firmly into cylindrical shape. The dimension of the cyclinder was 1cm in thickness and 2cm in diameter. Nine samples were irradiated simultaneously. The angle of the samples with respect to the beam were from zero to 121 degree. The samples were mounted at both sides with respect to the beam direction to counteract a possible shift of beam centre.

The induced specific activity of sample was low and determined by gamma-gamma coincidence spectrometer which consisted of two scintillation counters both with 10cm in thickness NaI(Tl) crystals and coupled to type GBD-100 photomultipliers. The high voltage of PM tube was 800 volts. The photopeak of 0.511 MeV were fed into the coincidence unit. The resolution time of this unit was 2.5 microsecond. The coincidence counting efficiency of the spectrometer was determined with a standard Na-22 source. The activity of the source was 25.8±0.1KBqs. The positron per decay 0.905[5] was used in this experiment. To eliminate difference in geometry and gamma self-absorbtion between the source and sample, the calibrated source was sandwiched in the block of sample material with different thickness and put in different position and then the average was adopted. The coincidence counting efficiency was 0.0735±0.0037. The spectrometer was checked several times per day and no significant drift was found during the period of measurement. When samples were measured they were mounted in a plastic box. The thickness of the box was 1mm.

Results

The cross-sections we measured are listed in Table 1.

Table 1. Na-23 (n, 2n) cross section

En(MeV)	±(MeV)	Sigma(mb)	±(mb)
18.5	0.27	185	33
18.1	0.23	171.2	20
17.5	0.4	157.7	20
16.4	0.5	136.5	20
14.9	0.5	70	19
13.3	0.4	1.9	2

In order to estimate the neutron attenuation within the sample, the Cu-63(n,2n) activation was used. The excitation curve of this reaction was rather similar to that of the Na(n,2n) reaction. Two copper sheets, each 0.2mm thick were used, one in front of the NaF sample and the other at the back. The diameter of these copper sheets was just the same as of the NaF sample. After ten minutes

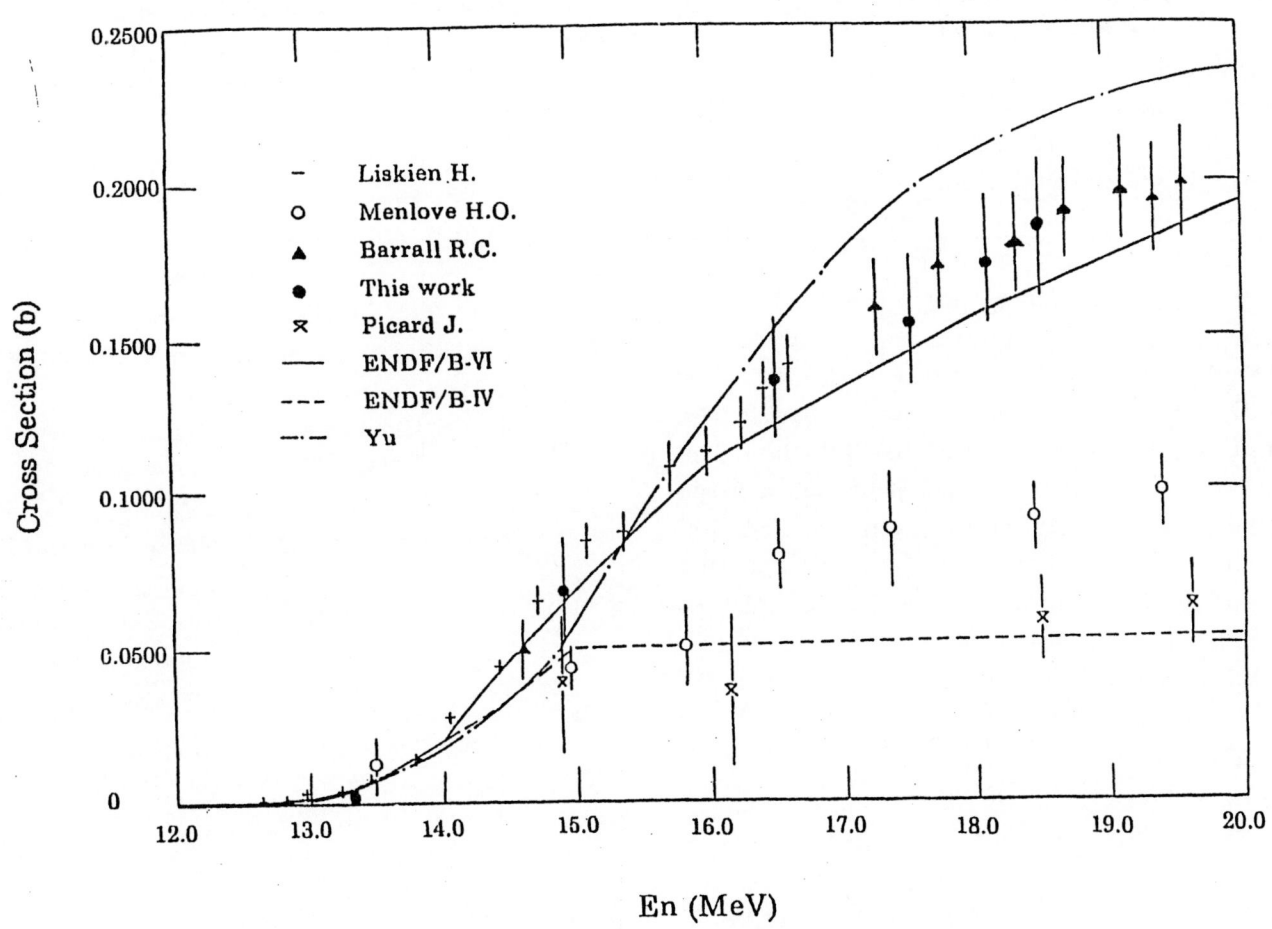

Fig. 1. Comparison of measured Na-23 (n, 2n) cross sections with available data.

668

irradiation the 0.511MeV photon coincidence was recorded. The count rate of the sheets was proportional to the average neutron flux in the NaF sample approximately. The ratio of average count rate and the front sheet count rate was used as neutron attenuation correction coefficient. It was 0.901 ± 0.005 in our case.

The experimental error in the resulted cross sections are aslo shown in Fig.1. They were calculated by compounding quadratically the uncertainties of the different factors which are listed in Table 2.

Table 2. Error of the cross section

1. Uncertainty of sample position	1.5%
2. Neutron flux measurement with long counter	6 %
3. Neutron attenuation correction in sample	5 %
4. Gamma coincidence counting efficiency	5 %
5. Neutron angular distribution calculation	2 %
6. Mass of sample	<0.01%
7. Satatistical errors of counting (according to the total counts)	>7 %
Total	~13%

Comparison

Fig.1 shows the result of this work. The other measured data are also shown in the figure. The dashed line is the evaluated ENDF/B-VI data. The solid line is the calculated result[4] which is based on the optical model, Haser Feshbach and pre-equlibrium theory. From this figure we can find that our result is very close to values from Liskien and Paulsen[1] but lower than Yu's calculation.

Y.N.Manokhin et al[6] have taken the IRDF data as their evaluation data but it was much lower than ENDF/B-VI data.

Takashi Nakamura presented their report recently[7]. They used a quasi-monoenergetic neutron source with energies from En=15 to 35 MeV produced by the Be(p,n) reaction. They gave the average cross-section about 120mb between En=14 to 19 MeV and about 210mb between En=19 to 24 MeV which is much closer to our result.

ACKNOWLEDGEMENT

The author wish to thank Mr. Wu Son-Mou, Wang Fa-Yang and Song Lin-Gang for many helpful discussions with us and the Van De Graff Laboratory of Fudan University for their skillful operation of the accelerator. The authors are grateful to "Chinese Nuclear Data Centre" for their financial support to this work.

References

1. H.Liskien, A.Paulsen: Nucl.Phys.63,393(1965)
2. H.O.Menlove, K.L.Coop: Phys.Rev. 163,1308(1967)
3. J.Picard, C.F.Williamson: Nucl.Phys. 63,673(1965)
4. Yu Ziqang: CNDC-85011,P.85(1985) (in Chinese)
5. C.M.Lederer: Table of Isotopes 7th Edition P.35(1978) Edited by C.M.Lederer and Virginia S. Shirley (Wiley New York, 1978)
6. V.N.Manokhin, A.B.Pashchenko, V.I.Plyaskin, V.M.Bychkov, V.G.Pronyaev: Handbook on Nuclear Activation Data, P.313, IAEA Vienna(1987)
7. T.Nakamura, Y.Kondo, H.Sugita Proc. Int. Conf. on Nuclear Data for Science and Technology, May/June, 1988, Mito, Japan (Edited by S. Igarasi), P.1017 Saikon, Tokyo (1988)

^{107,109}Ag(n,3/5n)^{105}Ag REACTION CROSS SECTION FOR 20 MeV < E_n < 70 MeV

U.J. Schrewe[1], H.J. Brede[1], M. Matzke[1], R. Nolte[1], J.P. Meulders[2], H. Schuhmacher[1] and I. Slypen[2]
[1]Physikalisch-Technische Bundesanstalt, W-3300 Braunschweig, Federal Republic of Germany
[2]Unité de Physique Nucléaire, Université Catholique de Louvain, B-1348 Louvain-la-Neuve, Belgium

Abstract: The activation of ^{105}Ag($T_{1/2}$ = 41 d) via the ^{107}Ag(n,3n) and the ^{109}Ag(n,5n) reactions has been investigated in quasi-monoenergetic neutron beams with mean energies of 27, 31, 38, 39, 60 and 64 MeV. The activity of ^{105}Ag was determined by using calibrated Ge detectors. The relative spectral fluences of the neutron beams were measured by applying time-of-flight techniques on scintillation detectors. The absolute fluences were determined with a proton recoil telescope. The fluence-weighted mean cross sections were 780(75), 820(50), 550(50), 530(50), 540(100) and 590(70) mb. The excitation function was derived with an unfolding procedure applied on the mean cross sections measured and on the spectral neutron fluences.

(^{107}Ag, ^{109}Ag, neutron activation cross section, 20 MeV < E_n < 70 MeV)

Introduction

There are scarcely any cross section data on activation for monoenergetic neutron beams of energies above 20 MeV, even though such data could be of interest for both basic and applied nuclear physics. There are various reasons for the lack of data in this energy region. First, the production of neutron beams is difficult due to the conflicting demands of monoenergetic and intense neutron sources, and secondly, the determination of total and spectral neutron fluences for E_n > 20 MeV contains large uncertainties, since precise reference cross section data are rare.

Neutron activation techniques could be applied in neutron spectrometry, since nuclear reactions with different reaction thresholds allow various energy components to be distinguished. The production of ^{105}Ag, for example, can be used as a high-energy neutron fluence monitor: (i) ^{105}Ag is efficiently produced in the entire energy range 20 MeV < E_n < 70 MeV, since two reactions, ^{107}Ag(n,3n) and ^{109}Ag(n,5n), with thresholds of 17.6 MeV and 34.6 MeV, respectively, contribute with comparable strength to the activation, (ii) the half-life of ^{105}Ag ($T_{1/2}$ = 41d) allows even low sample activities to be measured, (iii) due to the high reaction threshold energies, the activation is insensitive to low-energy neutrons.

In the course of an investigation of kerma factors [1], silver of a natural isotopic composition was activated in neutron beams at energies between 20 MeV and 70 MeV, simultaneously with the other investigations. The study has not been comprehensive enough to determine the excitation function with the required low uncertainties for the entire energy range. Here we report on the measurement of some of the fluence-weighted mean cross sections in various neutron fields and describe the procedure for unfolding the excitation function from the mean cross sections measured and from the spectral neutron fluences.

Measurements

The experiments were performed at the monoenergetic neutron beam facility of the Paul Scherrer Institute (PSI), Switzerland [2], and the Université Catholique de Louvain-la-Neuve (UCL), Belgium [3]. Intensive neutron beams were produced by bombarding a 2 mm thick beryllium target (Q-value of ^{9}Be(p,n) reaction: - 1.85 MeV) with protons of energies of 31.9, 43.3, 43.4, and 63.7 MeV at PSI, and a 3 mm thick lithium target (Q-value of ^{7}Li(p,n) reaction: -1.64 MeV) with protons of 35.0 and 65.0 MeV at UCL. Both facilities provided a collimator with a narrow neutron beam, e.g. at PSI the solid angle was $2.5 \cdot 10^{-5}$ sr and the fluence per unit proton beam charge was approximately 10 cm^{-2}nC^{-1} at a distance of 7.8 m from the neutron source.

The neutron beams were not ideally monoenergetic. Various effects influence the actual spectral distribution: (i) Proton energy loss in the target caused broadening of the neutron peaks; the energy losses range from 1.4 MeV for E_p = 65.0 MeV in Li to 5.9 MeV for E_p = 31.9 MeV in Be. (ii) Further energy broadening was due to neutron scattering in the surroundings of the target, the collimator shielding and the air. (iii) Break-up processes and other competing reactions that populate excited states in the residual nuclei produce significant low-energy tails.

Disks of natural silver, 25 mm in diameter and 1 mm thick, were activated during the time t_i at a fixed distance X_i from the neutron target. The disks were irradiated in the center of the neutron beams, perpendicular to the beam direction. After several days the γ-ray intensity emitted from the ^{105}Ag decay was measured with a calibrated HP-Ge detector.

Various scintillation detectors (NE213: 51 mm diameter and 51 mm length, NE213: 51 mm diam. and 102 mm length, Stilbene: 25.4 mm diam. and 25.4 mm length) applied with pulse-shape techniques for photon-neutron separation and time-of-flight techniques (TOF) for neutron energy determination were used.

In order to measure a wide neutron energy range with TOF techniques, the repetition rate must be adapted to the length of the flight path and the duration of the proton beam pulses must be short. In all experiments the duration of the pulses was approximately 1 ns. At PSI, the pulse repetition frequency produced by the cyclotron was reduced by means of a beam-deflector device which made possible repetition times of about 1 μs. At UCL, however, a beam-deflector device was not available and the pulse repetition times were about 70 ns, thus restricting the energy determination to high-neutron energies. The number of neutron-induced events was recorded as a function of TOF with thresholds between 0.4 and 17 MeV electron energies on the pulse heights. The detector efficiencies were calculated with the Monte Carlo codes CECIL [4] and SCINFUL [5]. In principle, each individual measurement also determines the absolute fluence. However, the agreement between the measured response functions of the scintillation detectors and the efficiency predicted by calculations was not satisfactory. Discrepancies up to 30 % in the efficiency as a function of the detector energy threshold have been found. Therefore only the relative spectral fluences, Φ_E, which were much less discrepant, were used for further analysis.

In addition, the absolute neutron fluence, Φ, was measured with a proton recoil telescope (PRT) of the Bame type [6] which was positioned at a distance X_r from the neutron target. The PRT measured the recoil protons emitted from a polyethylene (PE) radiator foil (mass per unit area: 144 mg cm^{-2}) mounted on a 1 mm thick Ta backing by means of four detectors arranged in series, two proportional counters P1 and P2 and two solid-state silicon detectors, S1 and S2, 1 mm and 5 mm thick, respectively [1]. At neutron energies above 40 MeV additional aluminium or iron absorbers were inserted between P2 and S1. The signals from the four detectors were analysed with a multi-parameter data acquisition system which allowed the protons emitted from the radiator foil to be distinguished from other charged particles. The proton background produced in the PE and outside the radiator foil was taken into account by replacing the PE radiator with a graphite disk.

The neutron fluence of the quasi-monoenergetic peak was determined from the number of recoil protons by using: (i) the geometry of the detector assembly [6], (ii) taking into account the influence of the PRT components such as scattering on apertures and the energy and angle-straggling of the protons in the radiator, the absorbers and the detector, the corrections being determined by Monte Carlo calculations [7], (iii) n-p cross sections of Arndt et al. [8], slightly modified on the basis of the recent work of Fink et al. [9], and (iv) further corrections for deadtime losses, inelastic proton scattering in the silicon detectors and neutron fluence attenuation in front of the radiator.

In order to normalize the fluences of the various measurements, the proton beam charge on the neutron target and the events produced by a thin NE102 scintillator behind the collimator exit both served as monitors.

Data Analysis and Results

The intensities of the γ-rays with energies of 0.281 MeV and 0.345 MeV, which have an emission probability of 0.302 and 0.414 per β-decay of ^{105}Ag [10], were measured. These values were then used to evaluate A_o, the activity of ^{105}Ag at the end of the activation process.

The fluence-weighted mean cross sections, $\bar{\sigma}$, defined by the relation:

$$\bar{\sigma} = (\int_{E_{th}}^{\infty} \sigma(E)\, \varphi_E\, dE\,) (\int_{E_{th}}^{\infty} \varphi_E\, dE\,)^{-1} \quad (1)$$

was evaluated from A_o by the relation:

$$\bar{\sigma} = A_o \cdot \Phi^{-1} \cdot Q \cdot \bar{I}^{-1} \cdot (1 - \exp(-\lambda t_i))^{-1} \cdot X_i^2 \cdot X_r^{-2} \cdot K_a \cdot N^{-1} \quad (2)$$

where φ_E denotes the spectral neutron fluence rate, E_{th} the reaction threshold, Φ/Q the absolute fluence per unit proton beam charge for all neutrons with energies $E_n > E_{th}$, $\bar{I}$ the mean proton beam current, λ the decay constant of ^{105}Ag [10], N the number of nuclei in the sample and K_a the correction factor for the attenuation of the neutron fluence in air between the positions X_i and X_r.

The spectral fluences obtained by combining the relative measurements of the scintillators with the absolute fluence measurements of the PRT are shown in Fig. 1 and the fluence-weighted mean cross sections are listed in Table 1.

It was intended to determine the excitation function $\sigma(E)$ on the basis of equation (1), but the quality of numerical unfolding techniques with only very few experimental data strongly depends on the reliability of the preinformation. The STAY'SL adjustment code [11] requires a preinformation (guess) function, $\sigma_g(E)$, and the adjusted excitation function, $\sigma_{ad}(E)$, is obtained by searching for the minimum of the χ^2 value which includes the two components: (i) the deviation of adjusted means and measured mean cross sections and (ii) the deviation between the adjusted and the guess function.

The guess function was obtained by combining experimental and theoretical cross section data as follows:
1. For neutron energies $E_n < 28$ MeV, experimental cross section data have been reported by Liskien [12] and Bayhurst et al. [13]. The guess function at neutron energies $E_n < 28$ MeV was taken from the evaluation of McLane et al. [14].
2. At neutron energies $E_n > 28$ MeV no experimental data were available, so calculated excitation functions were needed to continue the guess function.

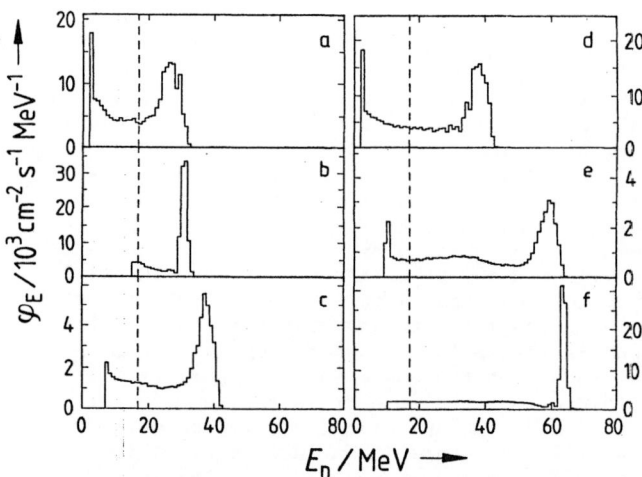

Fig. 1. The spectral fluence rates are shown as a function of neutron energy. The mean energy values of the quasi-monoenergetic peaks are: (a) 27±3 MeV; (b) 31±1 MeV; (c) 38.0±2.5 MeV; (d) 38.5±2.5 MeV; (e) 60±2 MeV; (f) 64±1 MeV. The dotted lines indicate the threshold energy of the ^{107}Ag(n,3n)^{105}Ag reaction.

The cross section values of the ^{107}Ag(n,3n) and the ^{109}Ag(n,5n) reactions were calculated with the ALICE nuclear evaporation code [15]. The shape of the theoretical ^{107}Ag(n,3n) excitation function fitted well to the experimental one [14] at lower energies, but the absolute values deviate by almost 50 %. The guess function was therefore extended for $E_n > 28$ MeV with the theoretical excitation functions which were adjusted to the experimental data in the energy range of 25 - 28 MeV. The uncertainties of the guess function needed by the STAY'SL code were taken from [12,13] below 28 MeV. The uncertainties of the data in the upper energy range were estimated to be 50 % to 100 %. The larger uncertainties were assumed for the decreasing part of the excitation functions, since it can be expected that the ALICE code is less precise in predicting the precompound reaction cross sections. Short-range Gaussian correlations were assumed for the guess function, which smoothed the results.

In Fig. 2 the experimental cross section data of other authors, the theoretical predictions of the ALICE code and the result obtained by applying the STAY'SL adjustment procedure are shown. The fluence-weighted mean cross sections were calculated using the guess function and the adjusted cross sections, and inserted in Table 1 for comparison with the experimental data.

Acknowledgement

We gratefully acknowledge the help of Dr. W. Mannhart and Mrs. G. Müller. The work was partly supported by the Commission of the European Communities, contract BI7-0030.

Table 1. Experimental results $\bar{\sigma}_{ex}$ and comparison with the mean values obtained with the preinformation, $\bar{\sigma}_g$, and the adjusted excitation function, $\bar{\sigma}_{ad}$.

$\bar{E}_n$ MeV	$\bar{\sigma}_{ex}$ mb	$\bar{\sigma}_g$ mb	$\bar{\sigma}_{ad}$ mb
27 ± 3	780 ± 75	737 ± 54	771 ± 40
31 ± 1	820 ± 50	781 ± 74	824 ± 40
38.0 ± 2.5	550 ± 50	411 ± 70	526 ± 31
38.5 ± 2.5	530 ± 50	396 ± 71	512 ± 32
60 ± 2	540 ± 100	421 ± 60	533 ± 45
64 ± 1	590 ± 70	323 ± 56	440 ± 40

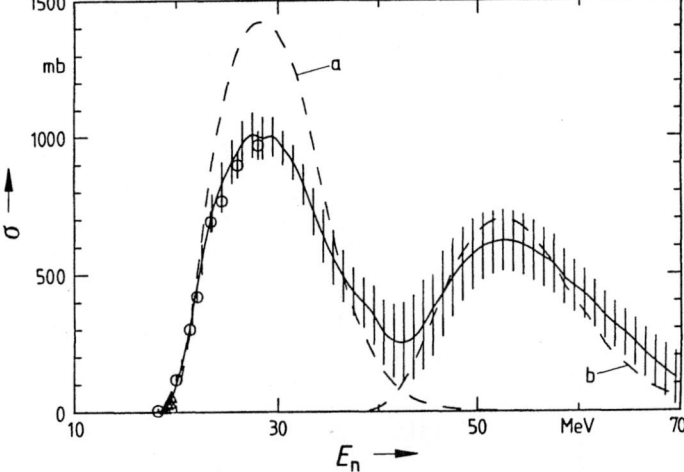

Fig. 2. Activation cross sections of ^{105}Ag produced via the ^{107}Ag(n,3n) and the ^{109}Ag(n,5n) reactions. The data points below energies of 28 MeV are experimental values of Liskien [12] (triangles) and Bayhurst et al. [13] (circles). The dotted lines are calculations of the ALICE code [15] for ^{107}Ag(n,3n) (a) and ^{109}Ag(n,5n) (b). The full line is the result of the adjustment procedure described in the text.

REFERENCES

1. H. Schuhmacher, H.J. Brede, R. Henneck, A. Kunz, J.P. Meulders, P. Pihet and U.J. Schrewe: submitted to Phys. Med. Biol., and U.J. Schrewe, H.J. Brede, R. Henneck, S. Gerdung, A. Kunz, H.G. Menzel, J.P. Meulders, P. Pihet, H. Schuhmacher, and I. Slypen: contribution to this conference
2. R. Henneck, C. Gysin, M. Hammans, J. Jourdan, W. Lorenzon, M.A. Pickar I. Sick, S. Burzynski and T. Stammbach: Nucl. Instr. and Meth. A259, 329 (1987)
3. M. Bosman, P. Leleux, P. Lipnik, P.Macq, J.P. Meulders, R. Petit, C. Pirart and G. Valenduc: Nucl. Instr. and Meth. A148, 363 (1978)
4. R.A. Cecil, B.D. Anderson and R. Madey: Nucl. Instrum. and Meth. 161, 439 (1979)
5. J.K. Dickens: Scinful, Oak Ridge National Laboratory (1988), Report: ORNL-6462
6. S.J. Bame, E. Haddard, J.E. Perry and R.K. Smith: Rev. Sci. Instr. 28, 997 (1957)
7. B.R.L. Siebert, H.J. Brede and H. Lesiecki: Nucl. Instr. and Meth. A235, 542 (1985), and B.R.L. Siebert: private communication

8. R.A. Arndt, J.S. Hyslop and L.D. Roper: Phys. Rev. D35, 128 (1987)
9. G. Fink, P. Doll, T.D. Ford, L. Garret, W. Heeringa, K. Hofmann, H.O. Klages and H. Krupp: Nucl. Phys. A518, 501 (1990)
10. D. De Frenne, E. Jacobs, M. Verboven and P. De Geldern: Nuclear Data Sheets 47 No. 2, 261 (1986)
11. F.G. Perey: Oak Ridge National Laboratory (1977), Report: ORNL/TM-6062, ENDF-254, and M. Matzke: Laboratory Report PTB 1991 (to be published)
12. H. Liskien: Nucl. Phys. A118, 379 (1968)
13. B.P. Bayhurst, J.S. Gilmore, R.J. Prestwood, J.B. Wilhelmy, Nelson Jarmie, B.H. Erkkila and R. A. Hardekopf: Phys. Rev. C12, 451 (1975)
14. V. McLane, C. L. Dunford and P. F. Rose: Neutron Cross Sections, Academic Press, Inc., San Diego, (1988)
15. M. Blann and J. Bisplingshoff: Lawrence Livermore National Laboratory (1983), LLNL Report: UCID 19614

GAMMA-INDUCED ^{115m}In PRODUCTION IN THE 0-50 MeV ENERGY REGION

L. Lakosi, J. Sáfár and Á. Veres

Institute of Isotopes, P.O.Box. 77, H-1525 Budapest, Hungary

T. Sekine, K. Yoshihara

Department of Chemistry, Faculty of Science, Tohoku University, Sendai 980,
Japan

Abstract: Integrated cross section data available in the literature were collected for ^{115m}In (half-life: 4.5 h) isomer production at gamma-energies corresponding to individual nuclear levels. The data determined from measured isomeric activities induced by intense isotopic gamma- and bremsstrahlung sources were compared with those calculated from nuclear spectroscopic data (level lifetimes, branching ratios) of the level scheme. A stepwise curve was yielded as a function of the exciting gamma-energy in the region of known individual levels, up to 2 MeV. Above this energy the cross section as a continuous curve was determined on the basis of statistical model calculations. Calculated values are experimentally supported at 4, 15, 20 and 50 MeV endpoint energies by means of bremsstrahlung irradiation from electron linear accelerators. The integrated cross section was found to be constant in the 15-50 MeV range within the experimental error, in accordance with the calculation.

(^{115m}In(γ,γ')^{115m}In, integrated cross section, statistical cascade model calculation)

Introduction

Photoexcitation of nuclear isomers involves transitions from the ground state of target nuclei to certain higher-lying (activation) levels which populate the metastable state via cascade decay. At higher energies, in the level continuum the cross section as a continuous curve has a steep rise with increasing energy, then drops sharply at the (γ,n) threshold and a second peak has long been expected to appear around the maximum of the giant resonance. However, experimental results are scarce and partly contradictory, and no consistent theoretical approach has been available for the amplitude of the second peak.

Among nuclides having a long-lived isomeric state ^{115}In has an outstanding significance in photoexcitation studies as it is relatively well-investigated due to the lack of disturbing reactions (abundance: 95.7 %) and the easy detection of the 336 keV (45.9 %) gamma-rays of the 4.5 h half-life isomer. In addition, since this isomer is produced also in (n,n') reaction, it has a practical role as a threshold detector in fast neutron dosimetry. The assessment of its gamma-response is therefore an important task when used in mixed neutron-gamma-radiation fields (e.g. reactor environment).

Photoactivation of ^{115m}In has also a practical importance in estimating induced radioactivity in products of industrial radiation processing.

In the present study we give a compilation of photoproduction cross sections of ^{115m}In available in the literature. In addition, we give results of recent experiments by means of bremsstrahlung irradiations at 4, 15, 20 and 50 MeV endpoint energies from electron linear accelerators[1,2]. A spin-dependent gamma-ray cascade calculation was also performed with open nucleon emission channels, taking into consideration the preequilibrium effects in the transition probabilities.

Low energy region (up to 2 MeV)

The level diagram of ^{115}In is given in Fig.1. Transitions and levels relevant to photoactivation are only indicated. The known major activation levels Ea_i can be excited by resonance energy gammas from continuous spectra of intense bremsstrahlung or isotopic gamma-sources, the line spectra of which undergo heavy degradation by Compton scattering in real geometries employed. Branching ratios for transitions leading to the metastable state Em are indicated in %. (Ground state branches of complementary value are not indicated.)

In Fig.2 a compilation of photoactivation cross sections are shown. In the low energy part cross sections integrated over single activation levels are given, which were determined from isomeric activity measurements according to the formula

$$\sigma_{int} = \frac{Y}{f(Ea)} \quad , \qquad (1)$$

where Y stands for the isomer production yield per target nucleus, f(Ea) for the flux of incident photons per unit energy interval usually determined by calculation at the energy of the activation level Ea.

At the energies where more than one activation levels are concerned, single-level cross sections summed up are displayed in the figure. Most of the data from irradiations by isotopic sources, related to the activation level at 1078 keV, are derived from ^{60}Co irradiations.

On the other hand, σ_{int} can be calculated for an individual level Ea from the level scheme as

$$\sigma_{int} = \frac{\lambda^2}{4} g \Gamma \frac{B_0}{\alpha+1} B_m \quad , \qquad (2)$$

if ground and metastable state branching ratios

B_o and B_m and level width Γ are known. Here λ denotes the photon wavelength at the level energy, α is the internal conversion coefficient for ground state transition and g is equal to $(2I+1)/(2Ig+1)$, where I and Ig are the corresponding spin values of excited and ground states. The level width Γ can be calculated from level lifetime. Since lifetimes are not known for the activation levels Ea_6 and Ea_7, calculation stopped at 1608 keV. The stepwise solid line in the low energy part of Fig.2 represents the build-up of calculated integrated cross sections with increasing energy as new activation levels enter.

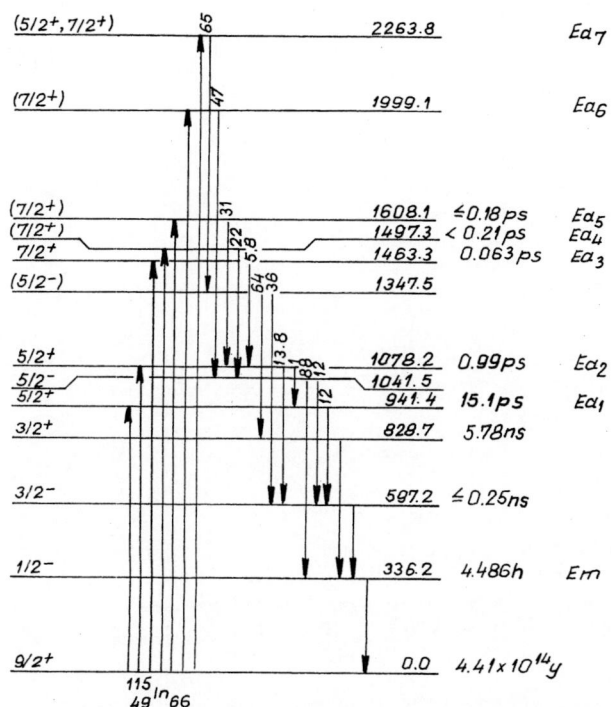

Fig. 1. A partial level diagram of ^{115m}In, showing only levels and transitions relevant to photoactivation. Transitions leading up to known strong activation levels Ea_i and down to the metastable level Em are only indicated. Level energies in keV, branching ratios in % are given. Data are taken from refs.[3,4].

Energy region beyond 2 MeV

Experiments

Electron linear accelerators of fixed 4 MeV endpoint energy 25 /uA beam intensity at the Institute of Isotopes, Budapest, and of 15-50 MeV endpoint energy, 100 /uA intensity at the Laboratory of Nuclear Science, Tohoku University, Sendai, were used to produce bremsstrahlung. Indium metal or oxide samples of natural isotopic composition were irradiated together will flux monitors and the 336 keV gamma-rays were detected by Ge(Li) and pure Ge spectrometers. Experimental details are described elsewhere [1,2].

Data analysis

The bremsstrahlung flux was calculated on the basis of the 1/E form of the bremsstrahlung energy spectrum of 4 MeV endpoint energy as well as by the Schiff's formula at 15, 20 and 50 MeV endpoint energies. Attenuation of the flux in the Pt converter was taken into account. The gamma-flux of the 4 MeV LINAC as a function of target-to-converter distance was determined by Al_2O_3 thermoluminescent and chlorbensol dosemeters as flux monitors, whereas metallic copper and gold foils were used to normalize the amplitude of the calculated spectral distribution in the range of 15-50 MeV.

Instead of unfolding the bremsstrahlung spectrum, the integrated cross section was evaluated by making a direct analysis of the data according to Eq.(1). However, instead of f (Ea) the weighted mean of the bremsstrahlung flux

$$\langle f \rangle = \frac{\int \sigma_o(E) f(E) dE}{\int \sigma_o(E) dE} \quad (3)$$

was determined, using a reference cross section $\sigma_o(E)$ and the actual flux density f(E) calculated and fitted to measured data of flux monitors as described above. The reference cross section was determined by model calculations, see below. The experimental results obtained in this way are represented at 4, 15, 20 and 50 MeV in Fig.2. No second increase was observed in the cross section after 9 MeV, the (γ,n) threshold, in contrast with earlier measurements [19].

Model calculation

Cascade gamma-ray calculations were performed to determine the cross section in the 2 to 23 MeV range. Branching ratios of each step were determined using the conventional optical statistical model. The preequilibrium correction was calculated in the framework of the fully spin dependent exciton model. E1 transitions were only considered. The results of the calculation are shown in Fig.2 (solid line). The curve displays a steep increase up to the (γ,n) threshold, then remains practically unchanged, in accordance with experimental results [2]. Similar calculations were performed and similar results were obtained for the $^{103}Rh(\gamma,\gamma')^{103m}Rh$ reaction recently [22].

The integrated cross section is expected to remain also further unchanged up to the meson production threshold.

674

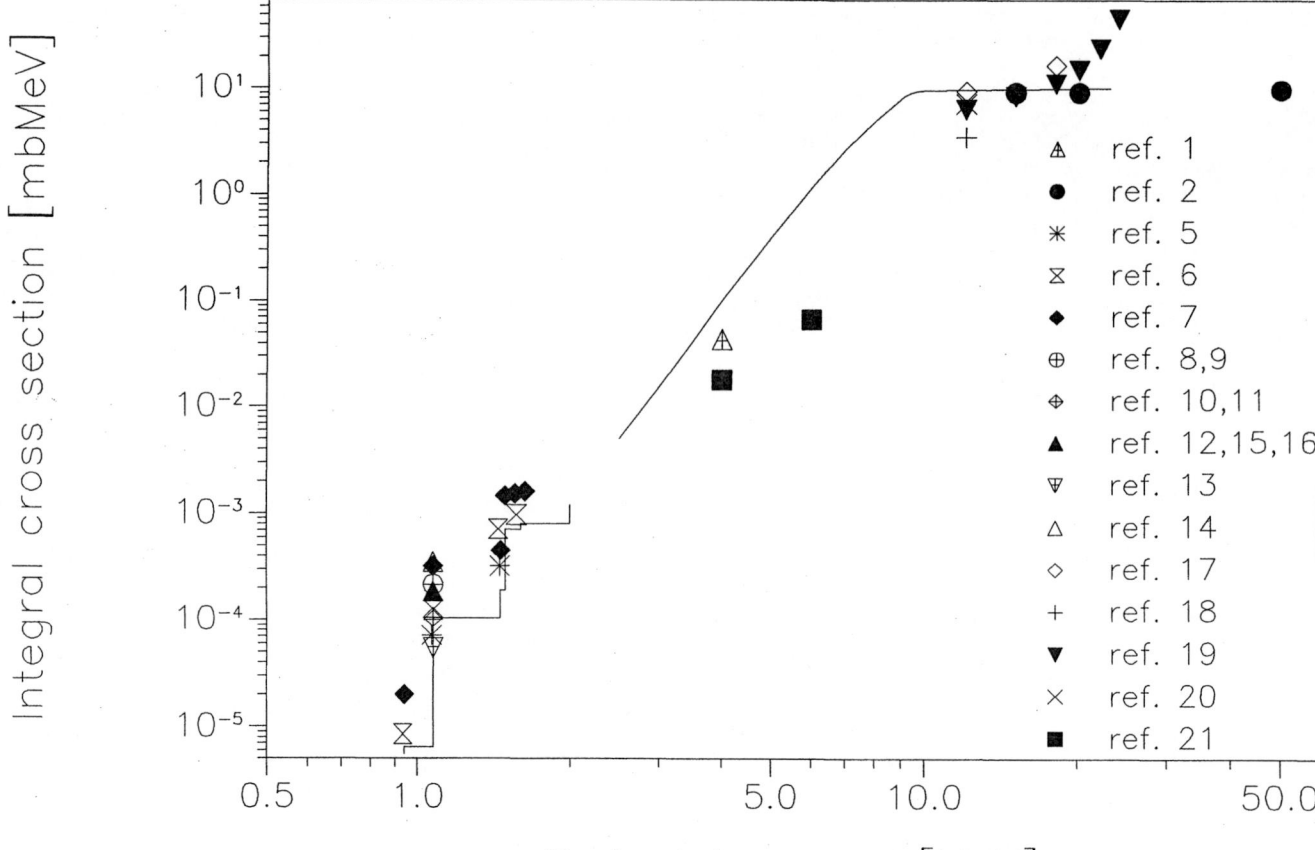

Fig. 2. The integrated cross section for ^{115m}In production by $(\gamma.\gamma')$ reaction as a function of the energy. Experimental results are marked by various symbols. The stepwise solid line in the region up to 2 MeV represents the increasing sum of integrated cross sections calculated for individual activation levels, whereas the solid line beyond 2 MeV represents theoretical values provided by model calculation.

Acknowledgement. The authors thank Dr. N.C. Tam for his assistance in the measurements and Mrs. Dr. M. Osvay for providing and evaluating thermoluminescent dosemeters. This work was partly supported by the Cooperative Research Project of the Hungarian Academy of Sciences and the Japan Society for Promotion of Science.

REFERENCES

1. L. Lakosi, I. Pavlicsek and Á. Veres: Acta Phys. Hung. 69, No.3-4 (1991), in press
2. T. Sekine, K. Yoshihara, L. Lakosi, Zs. Németh and Á. Veres: Appl. Radiat. Isotopes 42, 149 (1991)
3. J. Blachot and G. Marguier: Nucl. Data Sheets 52, 565 (1987)
4. A.B. Smith, P.T. Guenther, J.F. Whalen, I.J. van Heerden and W. R. McMurray: J. Phys. G: Nucl. Phys. 11, 125 (1985)
5. B.T. Chertok and E.C. Booth: Nucl. Phys. 66, 230 (1965)
6. E.C. Booth and J. Brownson: Nucl. Phys. A98, 529 (1967)
7. M. Boivin, Y. Cauchois and Y. Heno: Nucl. Phys. A176, 626 (1971)
8. N. Ikeda and K. Yoshihara: Radioisotopes 7, 11 (1958)
9. A. Veres: Appl.Radiat.Isotopes 14, 123 (1963)
10. E.A. Zaparov, Yu.N. Koblik, B.C. Mazitov and G.A. Tadyuk: in Direct Reactions and Isomer Transitions, FAN Uzb. SSR, Tashkent (1973) (in Russian)
11. L. Lakosi, M. Csürös and Á. Veres: Nucl. Instr. Meth. 114, 13 (1974)
12. Y. Watanabe and T. Mukoyama: Bull. Inst. Chem. Res., Kyoto Univ. 57, 72 (1979)
13. A. Ljubicic, K. Pisk and B.A. Logan: Phys. Rev. C23, 2238 (1981)
14. Y. Watanabe and T. Mukoyama: Nucl. Sci. Eng. 80, 92 (1982)
15. K. Yoshihara, Zs. Németh, L. Lakosi, I. Pavlicsek and Á. Veres: Phys. Rev. C33, 728 (1986)
16. C.B. Collins, J.A. Anderson, Y. Paiss, C.D. Eberhard, R.J. Peterson and W.L. Hodge: Phys. Rev. C38, 1852 (1988)
17. J. Goldemberg and L. Katz: Phys. Rev. 90, 308 (1953)
18. J.L. Burkhardt, E.J. Winhold and T.H. Dupree: Phys. Rev. 100, 199 (1955)
19. O.V. Bogdankevich, L.E. Lazareva and F.A. Nikolaev: J. Exp. Theor. Phys. (USSR) 31, 405 (1956)
20. W.J. Varhue and T.G. Williamson: Appl. Radiat. Isotopes 37, 155 (1986)
21. J.J. Carroll, M.J. Byrd, D.G. Richmond, T.W. Sinor, K.N. Taylor, W.L. Hodge, Y. Paiss, C.D. Eberhard, J.A. Anderson, C.B. Collins, E.C. Scarbrough, P.P. Antich, F.J. Agee, D. Davis, G.A. Huttlin, K.G. Kerris, M.S. Litz and D. A. Whittaker: Phys. Rev. C43, 1238 (1991)
22. J. Sáfár, H. Kaji, K. Yoshihara, L. Lakosi and Á. Veres: submitted to Phys.Rev.C

CHARGED PARTICLE REACTIONS FOR RADIATION PROTECTION OF A MEDIUM ENERGY CYCLOTRON

A. Sohn, M. Mattes, G. Hehn

Institut für Kernenergetik und Energiesysteme (IKE),
Universität Stuttgart, Pfaffenwaldring 31, D-7000 Stuttgart 80

Abstract: For calculating the radiation protection and optimizing the shielding of a negative ion cyclotron*, charged particle cross sections were needed to determine the production of neutron and gamma radiation. Double differential cross sections were calculated with the hybrid model of Blann using the program ALICE together with the systematics of continuum angular distributions by Kalbach (1988). Comparisons with published experimental data were made for ^{27}Al, ^{53}Cr, ^{56}Fe, ^{90}Zr, and ^{181}Ta at about 22 MeV which showed satisfying results with the preequilibrium-emission at higher energies so that the differential cross sections could be determined for the total energy range required. Finally the model calculations had to be extended to other target nuclides needed to predict all neutron and gamma sources. Integral measurements are planned, so that the final dose rates will be checked.

(charged particle, neutron emission, preequilibrium, hybrid model, differential cross sections, angular distribution, energy distribution)

Model description

Experiments showed that there are nuclear reactions which cannot be described by compound nucleus reactions or by direct reactions. At energies above 10 MeV these intermediate processes give a not negligible contribution to nearly all nuclear reactions.

With increasing projectile energy there are differences in the spectra of the emitted particles to the prediction of the compound nucleus evaporation model. Instead of a pure evaporation spectrum with a 90° symmetry in the center-of-mass system there are more emitted particles with higher energy preferably in the forward direction. The time for such a reaction between compound nucleus reactions and direct reactions is about 10^{-19} sec.

Because these reactions are explained as emission of particles before reaching the equilibrium state which is characteristic for a compound nucleus reaction, the emission is called preequilibrium emission.

For the calculation of single and double differential cross sections we used the program ALICE [1] which describes preequilibrium-emission according to the hybrid model of Blann. The advantage of ALICE comprises simple input preparation and fast calculation for different incident particles, including deuterons. Additionally we implemented the extended Kalbach systematics [2] in ALICE. It is a parametrization of many experimental angular distributions, where the double differential cross section can be written as

$$\frac{d^2\sigma}{d\Omega dE} = \frac{1}{4\pi}\frac{d\sigma}{dE}\frac{a}{\sinh a}[\cosh(a\cos\Theta) + f\sinh(a\cos\Theta)] ,$$

with a depending on the projectile energy, the emission energy and the separation energy. Originally Kalbach made a division of the cross section into multistep direct and multistep compound part, where f is the fraction of the multistep direct part. In ALICE no such division is made. Instead we used the fraction of the preequilibrium cross section, because this cross section determines nearly the whole multistep direct part [3].

Model verification for proton and deuteron reactions

To see how well the model predicts cross sections we compared results of the ALICE calculation with measurements [4] for the following reactions: 22.2 MeV protons on ^{27}Al, ^{56}Fe, ^{90}Zr, ^{181}Ta and 22.3 MeV deuterons on ^{27}Al, ^{53}Cr, ^{90}Zr, ^{181}Ta. Fig. 1 shows the neutron emission cross section for the proton and deuteron induced reactions. The experimental data were obtained from the double differential cross sections. The equilibrium part of the cross section underpredicts the emission of neutrons with higher energies as indicated by the dashed curves.

For gamma emission only a few data exist. Therefore we show in Fig. 2 the gamma emission cross section for 14.6 MeV incident neutrons on ^{93}Nb [5,6]. The ALICE results describe well the equilibrium part of the emission cross section but underestimate the preequilibrium part. For the calculation of the double differential cross section we used the extended Kalbach systematics. The results for neutron emission by the reactions p + ^{181}Ta and d + ^{53}Cr for several emission angles are given in Fig. 3. The neutron angular distribution for these reactions shows Fig. 4 for several neutron emission energies. Furthermore a comparison with the code EXIFON [7] was made for the reaction p + ^{181}Ta as an additional verification. The results of this calculation are indicated by the dotted curves in Fig. 1-4.

As can be seen especially from the neutron emission spectra direct processes are not taken into account by the ALICE calculation. But if one requires cross sections of various target nuclides and for several incident particles in a wide energy range, ALICE offers the opportunity of a fast calculation without great preparation to get a first reliable estimation. Such data are needed not only for particle accelerators but even more urgently for predicting the radiation exposure of man and instruments in space activities.

*Cyclotron MC 32 NI, Deutsches Krebsforschungszentrum Heidelberg

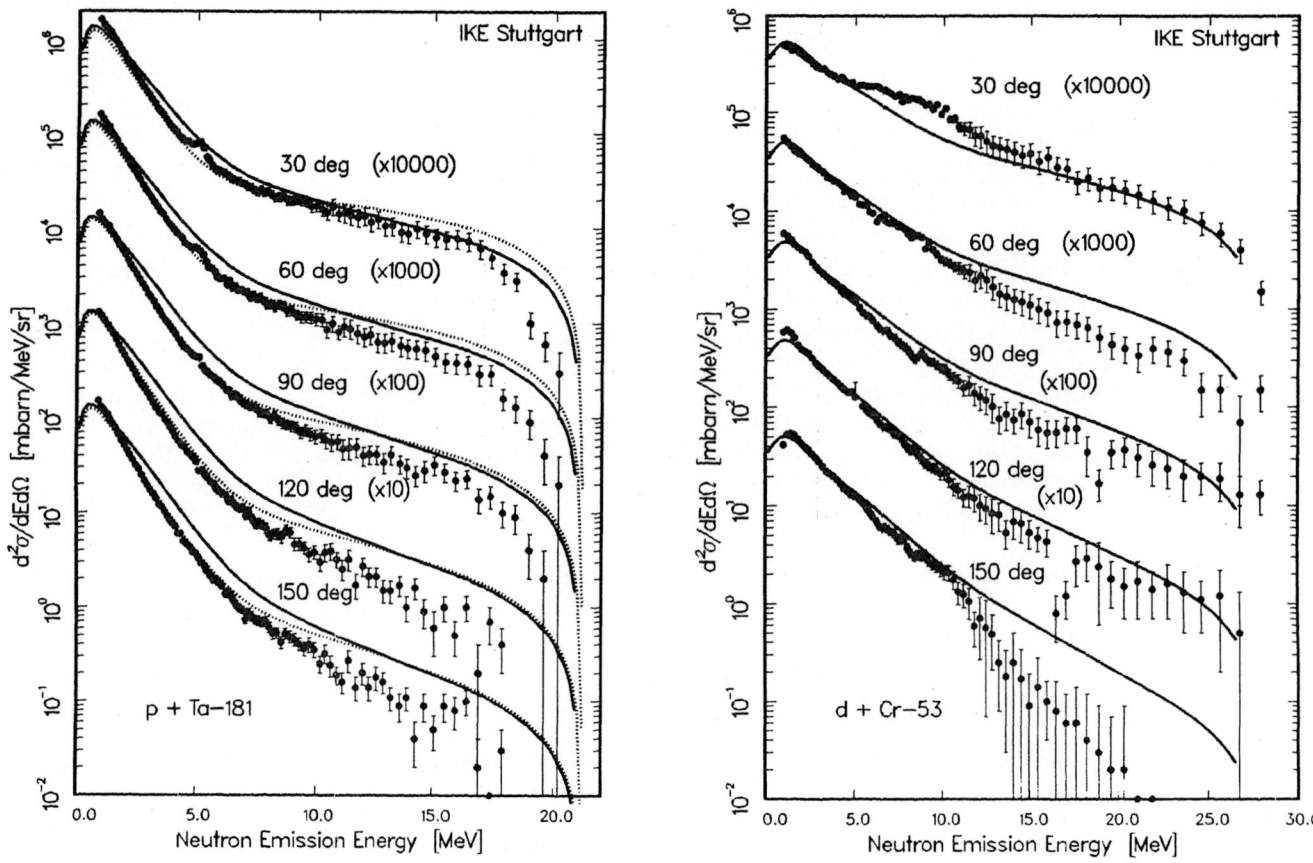

Fig. 3: Double differential cross section for incident protons of 22.2 MeV on ^{181}Ta (left) and for incident deuterons of 22.3 MeV on ^{53}Cr (right) for several neutron emission angles [4]. The solid curve shows the results of the ALICE calculation and the dotted curve (left) the results of a calculation with the code EXIFON [7].

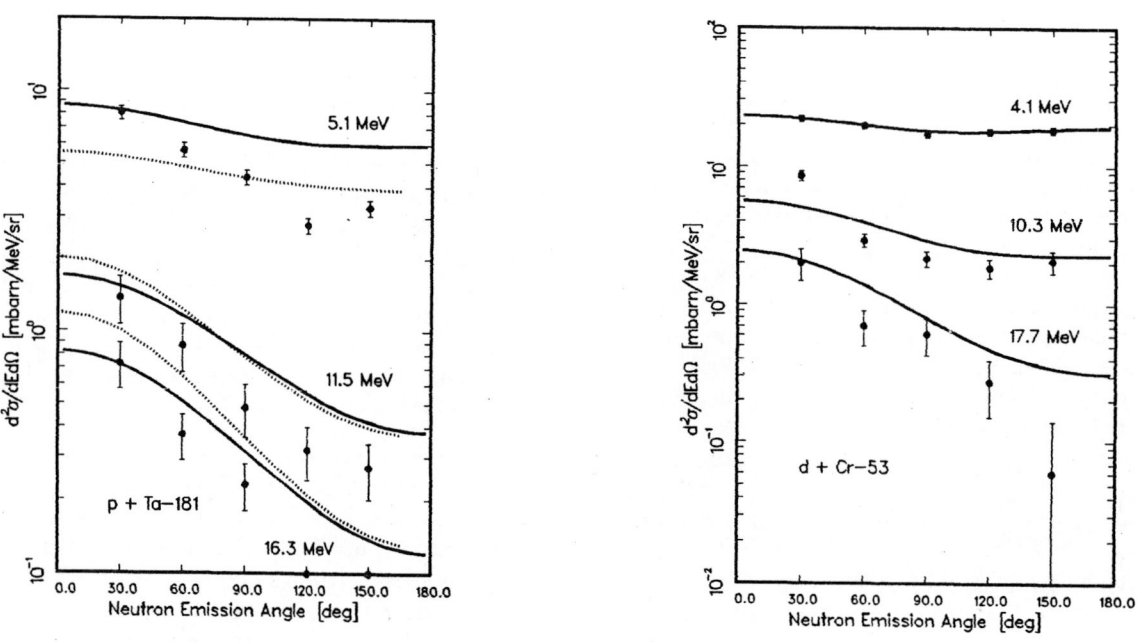

Fig. 4: Angular distribution for several secondary neutron energies for the same nuclear reactions as in Fig. 3.

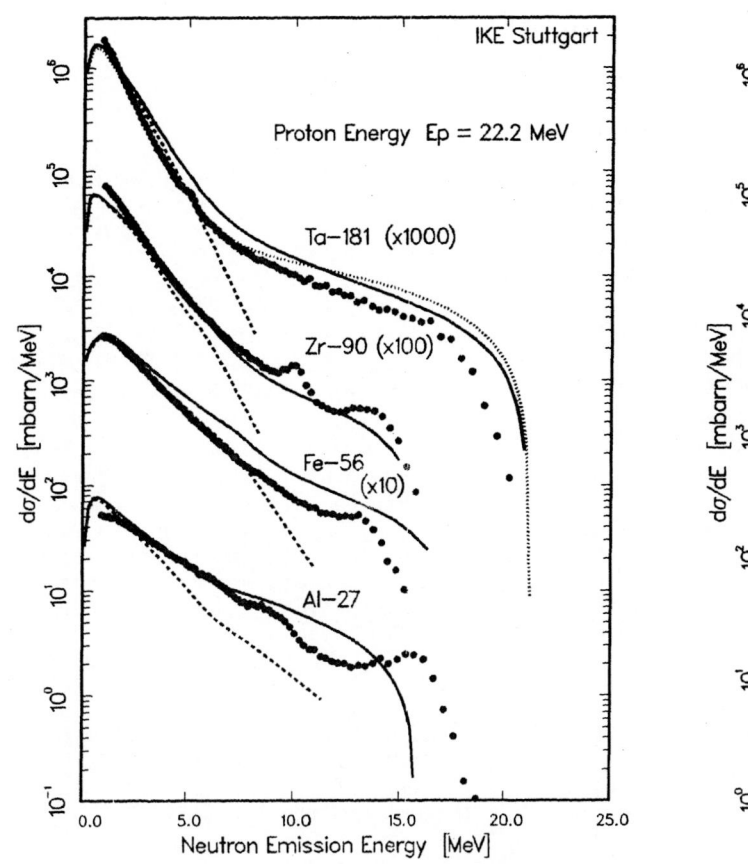
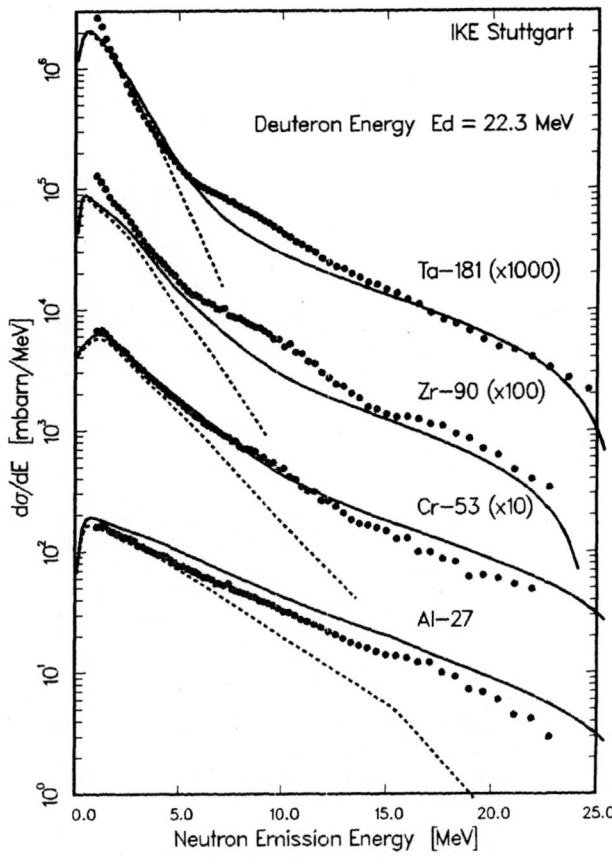

Fig. 1: Neutron emission cross section for 22.2 MeV incident protons (left) and 22.3 MeV incident deuterons (right) on several target nuclides [4]. The curves show the results of the ALICE calculation where the dashed curve represents the equilibrium part of the emission cross section. The dotted curve (left) gives the results of a calculation with the code EXIFON [7].

References

[1]   M. Blann, J. Bisplinghoff: Code ALICE/Livermore 82. UCID-19614 (1982) and Update UCID-20169 (1985)

[2]   C. Kalbach: Phys. Rev. C37, 2350 (1988)

[3]   A. Sohn: Berechnung von Wirkungsquerschnitten schneller Neutronen unter Berücksichtigung von Preequilibrium-Prozessen. Stuttgart: IKE 6-D-73 (1989)

[4]   N.S. Birjukov: (p,n)-Reactions at Proton Energy of 22.2 MeV, (1980). Data file EXFOR-50068. OECD/NEA Data Bank, Paris
N.S. Birjukov: Neutron from (d,n)-Reaction on ^{27}Al, ^{53}Cr, ^{90,94}Zr, ^{122}Sn, ^{181}Ta, (1986). Data file EXFOR-A0302. OECD/NEA Data Bank, Paris

[5]   D.M. Drake, E.D. Arthur and M.G. Silbert: Nucl. Sci. and Eng. 65, 49 (1978). Data file EXFOR-10684. OECD/NEA Data Bank, Paris

[6]   V.A. Plyuyko, G.A. Prokopets: Phys. Lett. 76B, 253 (1978)

[7]   H. Kalka: A model for statistical multistep reactions (code EXIFON). INDC(GDR)-60/L (1990)

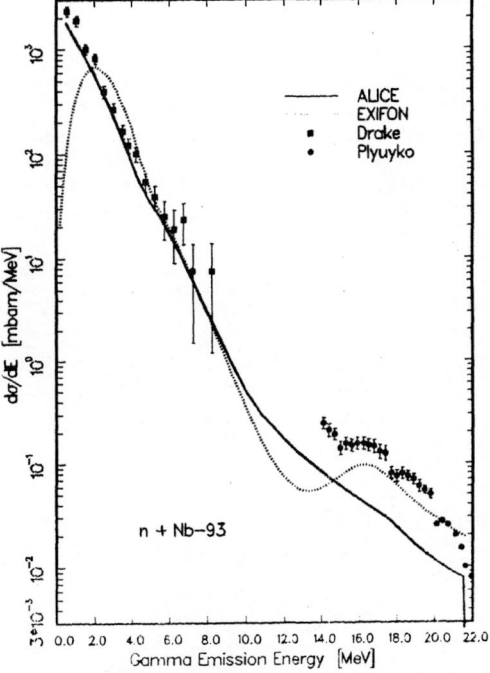

Fig. 2: Gamma emission cross section for 14.6 MeV incident neutrons on ^{93}Nb [5,6]. The dotted curve shows the results of a calculation with the code EXIFON [7].

NUCLEAR DATA NEEDED FOR INDUSTRIAL NEUTRON SOURCES

M. Aslam Lone

AECL Research
Physical Sciences
Chalk River Laboratories
Chalk River, Ontario, Canada K0J 1J0

ABSTRACT

Nuclear reactions suitable for compact accelerator-based thermal neutron sources are reviewed. Thick target total neutron production cross sections and point source peak thermal fluxes are evaluated.

Introduction

Applications of neutrons in industry can be grouped into three main categories: production of radioisotopes, materials analysis, and neutron radiography. Neutron irradiations for the production of radioisotopes are predominantly carried out at nuclear reactor facilities, but various types of neutron sources are used for materials analysis and neutron radiography. While facilities for these applications at multipurpose research reactors are of excellent quality these are not convenient for in situ inspection of large objects.

Much effort is presently devoted to the development of medium- to high-power accelerator-based neutron sources for in-house unrestrained use. A significant breakthrough would be the development of transportable or manoeuverable high-flux thermal neutron sources for radiography of large aerospace objects and for materials analysis. At present such sources rely on radioactive ^{252}Cf or sealed 14 MeV DT neutron generators that provide relatively modest thermal neutron fluxes.

In recent years several relatively compact accelerators have been developed for in-house production of radionuclides for nuclear medicine. These provide proton and deuteron beams of reasonably high currents and energies extending to 30 MeV that could be used to build high-flux thermal neutron facilities.

TABLE 1
SURVEY OF TOTAL NEUTRON YIELD DATA

Beam	Targets	Energy (MeV)	Method (Bath)	Accuracy (%)	Ref.
p	Li/Be	2-3	Vn	3	2
p	Li/Be	3-7	Mn		3
p	Li	3-10	Mn	3	4
p	Li/Be	5-10	Vn	3	5
p	Li/Be	14-23	TOF	20	6
p	Li/Be	18,32	Mn	7	7
p	Li	100	H$_2$O	7	8
d	Be	1-3	TOF		9
d	Li/Be	3-7	Mn		3
d	Be	10	Mn	7	10
d	Li/be	14-23	TOF	20	6
d	Be	15-40	Wax	20	11
d	Be	23	TOF		12
d	Be	0.4-0.8			13
d	Be	1.4-2.8	TOF	5	14

At energies below 30 MeV, proton and deuteron induced nuclear reactions in Li and Be are found to be the most efficient neutron sources. These reactions have been investigated for production of fast neutrons for materials testing and cancer therapy. The differential yields at forward angles needed for these applications were reviewed earlier [1]. For thermal neutron facilities however, data on total neutron yields and energy spectra integrated over all angles is required. These data have not been reviewed in the literature. This paper examines the data on total neutron yields and evaluates these source reactions for thermal neutron facilities.

Total neutron yields and cross sections

Total neutron yields have been determined by various techniques from proton and deuteron reactions in thick Li and Be targets at energies extending to 100 MeV. Table 1 gives a summary of the measurements reported in the literature. The vanadium (Vn) and the manganese (Mn) activation bath techniques [2-5,7,10] have provided the most accurate data at low energies. Another reliable method is mapping the thermal neutron flux in a large water moderator surrounding the source [8]. This technique also provides the thermalization factors for the source neutrons. The total yields have been reported [6,9,12-14] from integration of the neutron spectra measured with the time-of-flight technique. The primary uncertainty in this method is caused by the detector threshold that suppresses detection of lower energy neutrons. An upper limit on the total neutron yield can be calculated from total reaction cross sections measured with the beam attenuation technique [16]. In very thick ^{9}Be targets the (n,2n) reactions can lead to neutron multiplication and higher yields than would be predicted by the total reaction cross sections.

The data from protons on Li and Be targets are displayed in figures 1 and 2. In references [3-5] measurements were made at small energy intervals and only representative data are shown in these figures. The majority of the data lie on monotonically increasing yield curves. One exception is the data from reference [7], which fall below the systematic trend. The data from d + Be as displayed in figure 3 show a considerable scatter. From a global fit to the (d,n) thick-target yields from $7 \leq A \leq 27$ nuclei at energies from 1 to 40 MeV, Lucas et al.[11] proposed that a reasonable estimate of the yield can be obtained from

$$Y_n = 10^{15} \ (R/A) \ [0.82 \ \ln E_d + 0.63] \ (n/mC)$$

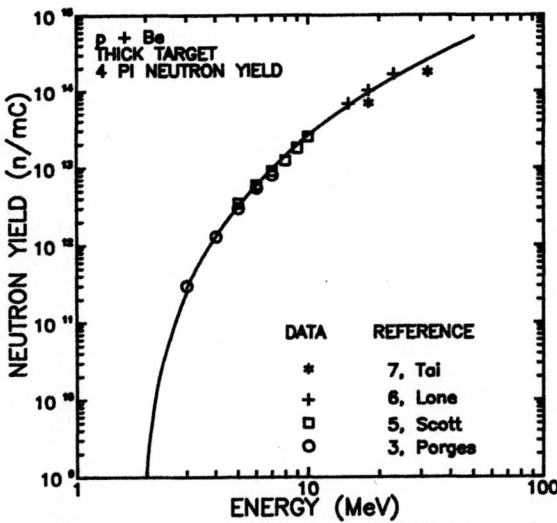

Figure 1. Thick target total neutron yield from p + Be

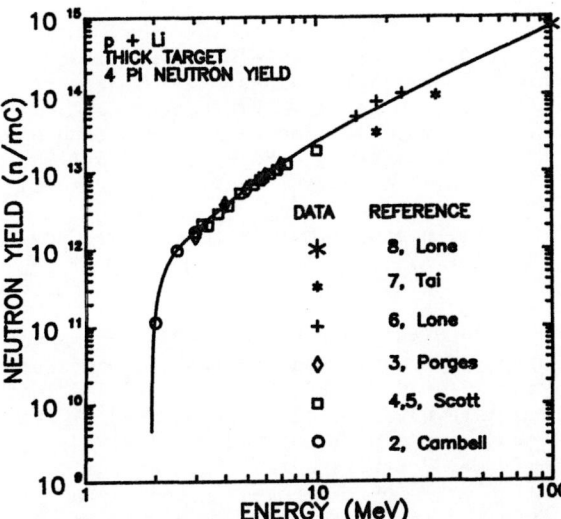

Figure 2. Thick target total neutron yield from p + Li

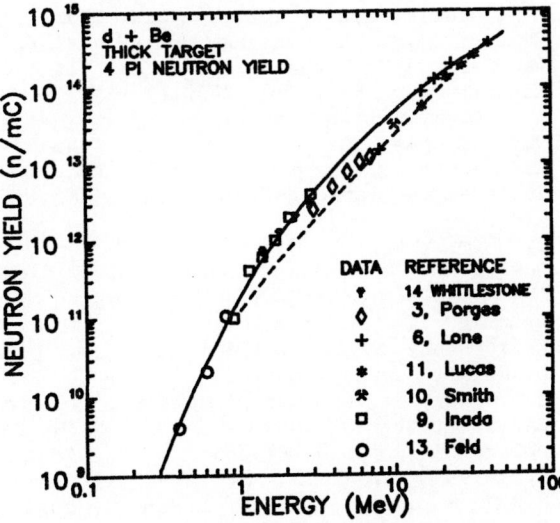

Figure 3. Thick target total neutron yield from d + Be

where R (g·cm^{-2}) is the range of deuterons of energy E_d (MeV). The dashed curve in figure 3 shows that the yields calculated from this global expression are somewhat lower than measured yields. This is not surprising since ^{9}Be is a loosely bound (2He+n) system that breaks apart very easily. For the d + Li reaction there are few data but the neutron yield is similar to that from the d + Be reaction.

For a thick target of mass number A the neutron yield per projectile particle is given by

$$Y_n(E_o) = \frac{0.6}{A} \int_o^{E_o} \frac{e^{-T(E)}}{S(E)} \Sigma_x [x \cdot \sigma_{xn}(E)] \, dE$$

where $\sigma_{xn}(E)$ is the cross section in barns for the (p,xn) channel and S(E) in (MeV·cm^2/g) is the stopping power for the projectile particle of energy E in the target medium. The attenuation of the beam in the thick target is determined by the exponent

$$T(E) = \int_E^{E_o} \frac{0.6}{A \cdot S(E)} \sigma_R(E) \, dE$$

where σ_R is the total reaction cross section that attenuates the projectile beam. At low and medium energies the dominant process for the beam degradation is the energy loss by ionizing collisions rather than the nuclear reactions.

In order to compare the neutron source reactions we define an effective thick target total neutron production cross section as

$$\sigma_{pn}(E_o) = 1.66 \cdot A \cdot Y_n(E_o) / R(E_o)$$

where R is the range of the projectile in the target material. The effective cross sections determined from the smooth yield curves shown in figures 1-3 are given in figure 4.

The neutron production cross sections are higher for the loosely bound Be nucleus. The energy dependence is consistent with the total reaction cross sections. It is, however, intriguing that above 20 MeV the thick target p + Be cross section is about 500 mb. This exceeds the thin target total reaction cross section [16] which decreases from 565 mb at 20.1 MeV to 353 mb at 46.2 MeV and could be due to the (p,2n) and (n,2n) channels.

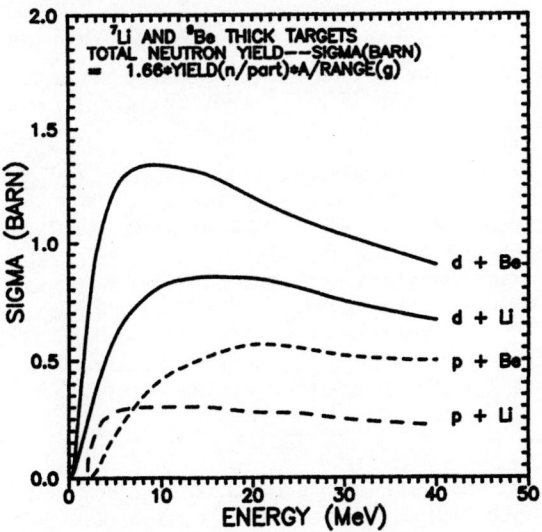

Figure 4. Thick target total neutron production cross sections.

680

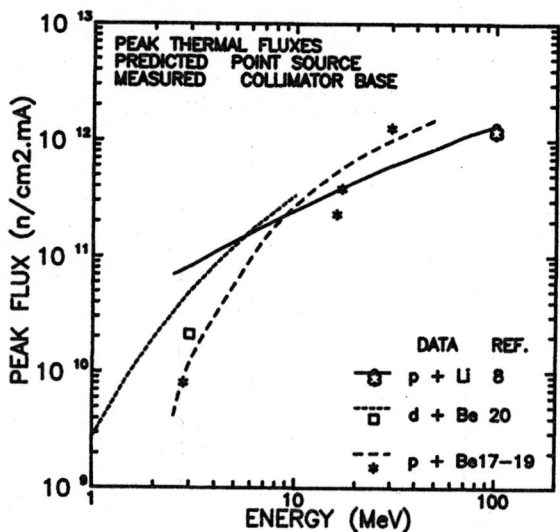

Figure 5. Calculated peak thermal fluxes and validation data.

Thermal neutron fluxes

The spatial distribution of thermal fluxes in a water moderator will depend on the source neutron energy and the angular distributions which dictate the geometrical coupling between the source and the moderator. The size of the moderator and the biological shielding need to be optimized for highest thermal flux and lowest leakage of primary and secondary radiation. Neutron absorption in hydrogen produces 2.2 MeV gamma-rays which requires shielding. Splitting of the moderator into two sections with ^{6}Li absorber in the outer section can reduce the requirement of gamma-ray shield. Such optimisation studies require a complete set of data on neutron energy and angular distributions. These data are not currently available.

The peak thermal flux in a hydrogen-based moderator surrounding the source depends primarily on the source energy distribution rather than the angular distributions [8]. For a point source with no void or neutron absorbing material in a moderator the thermalization efficiency is given by

$$\epsilon_{th} = 1.74 \ 10^{-2} \ E^{-0.715} \ (n_{th} \ /cm^2 \cdot n_{source})$$

where E is the source neutron energy in MeV. The efficiency is 1.74% at 1 MeV but drops to 0.26% at 14 MeV.

A reasonable estimate of thermal fluxes can thus be made by assuming an isotropic source with an effective one-group energy of $0.25(E_p-2)$ for protons and $0.25(E_d+4)$ for deuterons. Figure 5 shows a comparison of the calculated fluxes for these source reactions. The data show measured fluxes at collimator bases of external beam tubes. As expected the point source predicted values are about a factor of two higher.

Conclusions.

The nuclear data on neutron source reactions is adequate for preliminary estimates but is inadequate for engineering design studies. The most efficient reactions are: d + Be below 2.5

MeV, p + Li from 2.5 to 6 MeV, and p + Be from 6 to 30 MeV. Above 30 MeV, proton and deuteron reactions in heavy elements would be more efficient [15].

References.

1. M.A. Lone and C.B. Bigham, Neutron Sources For Basic Physics and Applications, (edited by S. Cierjacks),p.133, Pergaman Press (1983)
2. J. Campbell and M.C. Scott, in Proc. 4th Conf. on the Scientific & Industrial applications of Small Accelerators, Denton, Texas, p.517-521, IEEE, New York USA (1977)
3. K. Porges, J.L. Snelgrove, R. Gold, A. DeVolpi, R.J. Armani and C.E. Cohen, Argonne National Laboratory Progress Report, ANL-7910, p.361 (1971)
4. M.C. Scott: J. Nucl. Energy 25, 405 (1971)
5. M.A. Atta and M.C. Scott: J. Nucl. Energy 27, 875 (1973)
6. M.A. Lone, C.B. Bigham, J.S. Fraser, H.R. Schneider, T.K. Alexander, A.J. Ferguson and A.B. Mcdonald: Nucl. Inst. & Meth. 143, 331 (1977)
7. Y. K. Tai, P.G. Milbourn, G.P. Kaplan and B.J. Moyer: Phys. Rev. 109, 2086 (1958)
8. M.A. Lone, R.T. Jones, A. Okazaki, B.M. Townes, D.C. Santry, E.D. Earle, J.K.P. Lee, J.M. Robson, R.B. Moore, L. Nikkinen and V. Raut: Nucl. Inst. & Meth. 214, 333 (1983)
9. T. Inada, K. Kawachi and T. Hiramoto: J. Nucl. Sci. & Technology 5, 22 (1968)
10. L.V. Smith and P.G. Kruger: Phys. Rev. 83, 1137 (1951)
11. L.L. Lucas and J.W. Root: J. Appl. Phys. 43, 3886 (1972)
12. P.J. Persiani, W. Becker and J. Donahue, in proc. Symposium on Neutron Cross Sections from 10-40 MeV, (Edited by M.R. Bhat and S. Pearlstein), p.151, Brookhaven National Laboratory Report BNL-NCS-50681 (1977)
13. B.T. Feld, Experimental Nuclear Phsycis, Vol II, (Edited by E. Segre), p.388, John Wiley & Sons Inc. (1953)
14. S. Wittlestone: J. Phys. D: Appl. Phys. 10, 1715 (1977)
15. Summary Report of the NEANDC Working Group on Intense Neutron Sources for Nuclear Phyics & Nuclear Data Measurements, (Edited by S. Cierjacks), p.15, NEANDC-274-U (1990)
16. W.F. McGill, R.F. Carlson, T.H. Short, J.M. Cameron, J.R. Richardson, I. Slaus, W.T.H. van Oers, J.W. Verba, D.J. Margaziotis and P. Dohert: Phys. Rev. C10, 2237 (1974)
17. M.R.Hawkesworth and J. Walker, Neutron Radiography Using Birmingham Dynamitron Accelerator, in proc. Fourth Small Acc. Conference. Denton, Texas USA, p.508, IEEE (1976)
18. Y. Fukushima, T. Nakamara, E. Hiraoka, J. Sekita, H. Yokochi, T. Yamada and S. Yamaki, in proc. Neutron Radiography using Ultra Compact Cyclotron, in proc. Neutron Radiography Second World Conference, p.215, D. Reidel Company, Boston (1986)
19. S. Tazawa, M. Yano and T. Nakanii, Cyclotron-Based Neutron Radiography Facility, in proc. reference 18, p.18, (1986)
20. J. Stokes, L. Parks, C. Preskitt, J. John, H.R. Lukens Jr., J. Mackenzie, R. Vogt, N. Lurie and A.P. Yrippe, A new Accelerator-Based Neutron radiography System, IRT Corporation Report, IRT 4611-019 (1981)

PRODUCTION OF NEUTRONS FROM WATER, POLYETHYLENE, TISSUE EQUIVALENT MATERIAL AND CR-39 IRRADIATED WITH 2.5 - 30 MeV PHOTONS*

M. Anwar Chaudhri[1] and Peter D Allen[2]

[2]Australian Radiation Laboratory, Yallambie, Victoria 3085, Australia.

[1]Department of Medical Physics, Austin Hospital, Heidelberg, Victoria 3084,

Australia, and the University of Melbourne.

Abstract: Photoneutron yields from water, polyethylene, tissue equivalent material and CR-39 have been calculated for the photon energy range of 2 to 30 MeV, using a previously established method and photoneutron production data on hydrogen, carbon, nitrogen and oxygen. The rarer isotopes of the constituent elements of these compounds, namely ^{2}H, ^{13}C, ^{15}N, ^{17}O and ^{18}O, have been taken into account, and neutrons are shown to be produced for photon energies as low as 2.2 MeV, the (γ,n) threshold for ^{2}H. The data is useful for estimating neutron production in materials located in the vicinity of a megavoltage radiotherapy beam. Substances such as those considered here are often used as filtration, phantom or scattering material and as components of neutron dosimetry detectors. It is pointed out that photoneutrons produced in such materials may need to be taken into consideration when carrying out neutron dosimetry in the presence of even relatively low energy photons, especially when the neutron flux is several orders of magnitude less than that of the photons.

Keywords Radiation therapy, photonuclear reactions, neutron dosimetry, CR-39, Tissue equivalent material.

Introduction

There has been much recent interest in the problem of neutron contamination in the vicinity of high energy bremsstrahlung radiotherapy sources [1]. Neutrons are produced by photonuclear interactions in the accelerator target, in shielding and beam modification components, and in any materials in the treatment room which are traversed by the therapy beam. These neutrons contribute to background radiation levels in the treatment room and to patient dose.

Particular problems arise when an attempt is made to measure neutrons in the presence of photons with energy above a few MeV. Typically the ratio of neutron absorbed dose to photon absorbed dose within a high energy (>10 MeV) beam is of the order of 10^{-3} or less [2]. In these circumstances photonuclear interactions in the detector itself can interfere with the neutron measurements being made [1-3].

The aim of the present work is to calculate photoneutron production data, which are more accurate than previously available information, for certain low-Z elements and hydrogenous compounds. These types of compounds are often used as moderators or radiators in neutron detectors or as the neutron detector itself. They are also used as filtration, phantom or scattering materials in radiotherapy facilities.

In this paper we are presenting data on photoneutron production in CR-39, tissue, water and polyethylene. In order to obtain this information we have calculated the photoneutron yields for hydrogen, carbon, nitrogen and oxygen, using the best available photoneutron cross sections. The calculations include the effects of the low abundance isotopes of each element which have been shown to have a definite contribution at low energies and are therefore essential for accurate estimations at energies below 20 to 24 MeV, depending on the composition of the material involved.

Calculations

Absolute yields $y(k)$ (number of neutrons produced per incident photon) were calculated as a function of (monoenergetic) photon energy k for the elements hydrogen, carbon, nitrogen and oxygen, and for the compounds 'tissue equivalent material' $(C_5H_{40}O_{18})$[4,5], water (H_2O), CR-39 $(C_{12}H_{18}O_7)$ and polyethylene $((CH_2)_n)$, using published cross section measurements for each of the constituent isotopes (Table 1). These particular compounds were chosen as examples of materials used as phantom or scattering materials, or as components of neutron detectors. They also illustrate the range of effects expected for this type of hydrogenous compound which includes many plastics and organic substances.

*Presented at the ARPS/ACPSEM Conference, Adelaide, September 1990.

Table 1

Data on isotopes of oxygen, carbon, nitrogen and hydrogen

Isotope	Isotopic abundance(%)	Threshold(MeV) (γ,n)	(γ,2n)
^{1}H	99.99	-	-
^{2}H	0.015	2.2	-
^{12}C	98.89	18.7	31.8
^{13}C	1.11	4.9	23.7
^{14}N	99.63	10.6	30.6
^{15}N	0.37	10.8	21.4
^{16}O	99.76	15.7	28.9
^{17}O	0.04	4.1	19.8
^{18}O	0.20	8.0	12.2

For a target irradiated with a parallel beam of photons of energy k the yield y(k) is given by

$$y(k) = f_t(k) \, t \, \Sigma[\rho j \sigma j(k)] \qquad (1)$$

where ρj is the atomic density of the jth constituent isotope, $\sigma j(k)$ is the cross section for photoneutron production and t is the thickness of the irradiated material. The factor $f_t(k)$ corrects for photon absorption within the irradiated material:

$$f_t(k) = (1 - e^{-\mu t(k)t})/(\mu_t(k)t) \qquad (2)$$

where $\mu_t(k)$ is the appropriate attenuation coefficient for the material under consideration. The photoneutron yields for each compound were calculated from equation (1) at 0.1 MeV intervals using carefully selected representative recent cross section measurements for the relevant isotopes. The need to include the rarer isotopes of oxygen, carbon, nitrogen and hydrogen (^{17}O, ^{18}O, ^{13}C, ^{15}N and ^{2}H) in the calculations has been demonstrated in a previous paper [6]; the selection of the cross section measurements used is also discussed in that paper.

The yields of photoneutrons from the four compounds averaged over 1 MeV bins are shown in Table 2.

The data may be used to estimate photoneutron production in these materials for any incident photon spectrum simply by numerical integration of equation (1) over the assumed photon spectral shape. For general practical application the results should be correct to first order, but the yield is quite sensitive to the exact spectrum shape incident on the materials [6]. In a given situation therefore the most accurate available spectral information should be used.

Discussion

The present calculations illustrate the energy dependence of sources of neutrons that may arise from the interaction of a photon beam of 2 to 30 MeV energy with materials in a radiotherapy treatment room (Table 2). Some neutrons are produced at relatively low energies (below 16 MeV), a region usually ignored. These photoneutrons may be detected as an otherwise unexplained background in very sensitive neutron measurement systems.

For the compounds considered here Table 2 shows that the photoneutron yield can vary greatly at a given bremsstrahlung energy, particularly for energies less than about 24 MeV. In certain circumstances this could affect the choice of the most appropriate material for a particular application. For example, if internal photoneutron production in the moderator or converter of a neutron detector is a significant problem, then the optimum detector material for such components may depend on the bremsstrahlung energy at which measurements are being made.

The significance of internally produced photoneutrons or other photoeffects to the response of a neutron detector depends on experimental circumstances such as the ratio of neutrons to photons in the field being measured and the type of detector being used. When the photonuclear response of a neutron detector is substantial it is important that the evaluation of the appropriate correction is accurate. The data of Table 2 can be helpful for the accurate estimation of these effects.

Table 2

Photoneutron yield per incident photon for various compounds of mass thickness 1 g cm^{-2} irradiated by a monoenergetic beam of photons. No correction has been made for attenuation of the incident beam within the irradiated material. The yields presented are the average values over an energy bin of width 1 MeV centred on k_b.

Energy k_b (MeV)	'Tissue approximation'	Water	CR-39	Polyethylene
2.5	7.58×10^{-9}	8.46×10^{-9}	5.00×10^{-9}	1.09×10^{-8}
3.5	1.88×10^{-8}	2.10×10^{-8}	1.24×10^{-8}	2.70×10^{-8}
4.5	2.21×10^{-8}	2.47×10^{-8}	1.46×10^{-8}	3.18×10^{-8}
5.5	2.14×10^{-8}	2.39×10^{-8}	1.42×10^{-8}	3.07×10^{-8}
6.5	1.92×10^{-8}	2.14×10^{-8}	1.27×10^{-8}	2.75×10^{-8}
7.5	5.71×10^{-8}	1.88×10^{-8}	1.53×10^{-7}	2.55×10^{-7}
8.5	9.29×10^{-8}	5.41×10^{-8}	1.95×10^{-7}	2.95×10^{-7}
9.5	2.20×10^{-7}	1.88×10^{-7}	3.27×10^{-7}	4.07×10^{-7}
10.5	3.65×10^{-7}	2.96×10^{-7}	5.76×10^{-7}	7.32×10^{-7}
11.5	6.95×10^{-7}	4.46×10^{-7}	8.93×10^{-7}	1.13×10^{-6}
12.5	8.31×10^{-7}	2.44×10^{-7}	9.98×10^{-7}	1.46×10^{-6}
13.5	1.36×10^{-6}	5.26×10^{-7}	1.31×10^{-6}	1.76×10^{-6}
14.5	1.13×10^{-6}	6.50×10^{-7}	1.22×10^{-6}	1.51×10^{-6}
15.5	1.56×10^{-6}	6.24×10^{-7}	1.17×10^{-6}	1.45×10^{-6}
16.5	1.37×10^{-5}	1.57×10^{-5}	8.01×10^{-6}	1.32×10^{-6}
17.5	3.71×10^{-5}	4.45×10^{-5}	2.11×10^{-5}	1.09×10^{-6}
18.5	2.95×10^{-5}	3.20×10^{-5}	1.55×10^{-5}	1.27×10^{-6}
19.5	4.28×10^{-5}	4.14×10^{-5}	2.85×10^{-5}	1.55×10^{-5}
20.5	7.09×10^{-5}	6.09×10^{-5}	7.26×10^{-5}	7.28×10^{-5}
21.5	1.94×10^{-4}	1.78×10^{-4}	2.20×10^{-4}	2.25×10^{-4}
22.5	2.91×10^{-4}	2.64×10^{-4}	3.28×10^{-4}	3.37×10^{-4}
23.5	2.77×10^{-4}	2.42×10^{-4}	3.25×10^{-4}	3.48×10^{-4}
24.5	2.77×10^{-4}	2.71×10^{-4}	2.72×10^{-4}	2.41×10^{-4}
25.5	2.15×10^{-4}	2.01×10^{-4}	2.30×10^{-4}	2.24×10^{-4}
26.5	1.74×10^{-4}	1.63×10^{-4}	1.86×10^{-4}	1.80×10^{-4}
27.5	1.33×10^{-4}	1.27×10^{-4}	1.34×10^{-4}	1.23×10^{-4}
28.5	1.22×10^{-4}	1.22×10^{-4}	1.11×10^{-4}	8.92×10^{-5}
29.5	7.18×10^{-5}	5.46×10^{-5}	9.97×10^{-5}	1.22×10^{-4}

Conclusion

Data have been presented which enable estimation of the rate of photoneutron production from certain compounds irradiated with bremsstrahlung beams of energies up to 30 MeV. This information is necessary to assess the significance of internal photoneutron production in detectors used to determine the neutron contamination within a high energy photon field. The variation with energy of the photoneutron production in these materials demonstrates one factor which may affect the choice of the most appropriate detector in a given application. The date will also be useful for estimating the effects of any filtration, phantom or scattering materials of the type considered here through which a high energy photon beam passes; in this case the energy variation may help identify possible sources of neutron background flux in a radiotherapy treatment room.

References

1. Neutron Contamination from Medical Electron Accelerators, National Council on Radiation Protection and Measurement (NCRP) Report No. 79, (1984)

2. I. Gudwoska, Measurements of the Neutron Absorbed Dose from Medical Electron Accelerators, Department of Radiation Physics, Karolinska Institute, Internal Report RI 1984-04 (1984)

3. Neutron Measurements Around High Energy X-ray Radiotherapy Machines, American Association of Physicists in Medicine Report No. 19, (1986)

4. H.H. Rossi, and G. Failla, Tissue-equivalent ionization chambers, Nucleonics 14: 32-37 (1956)

5. Tissue Substitutes in Radiation Dosimetry and Measurements, International Commission on Radiation Units and Measurements (ICRP) Report 44, (1989)

6. P.D. Allen and M.A. Chaudhri, Photoneutron Production in Tissue During High Energy Bremsstrahlung Radiotherapy, Phys. Med. Biol. 33: 1017-1036, (1988)

Part IX

Medium Energy Data

NEW TECHNIQUES IN NEUTRON DATA MEASUREMENTS ABOVE 30 MeV

P.W. Lisowski and R. C. Haight

Los Alamos National Laboratory
Los Alamos, NM 87545 USA

ABSTRACT: Recent developments in experimental facilities have enabled new techniques for measurements of neutron interactions above 30 MeV. Foremost is the development of both monoenergetic and continuous neutron sources using accelerators in the medium energy region between 100 and 800 MeV. Measurements of the reaction products have been advanced by the continuous improvement in detector systems, electronics and computers. Corresponding developments in particle transport codes and in the theory of nuclear reactions at these energies have allowed more precise design of neutron sources, experimental shielding and detector response. As a result of these improvements, many new measurements are possible and the data base in this energy range is expanding quickly.

(Neutron sources, detectors, nuclear data, neutrons)

Introduction

Interest in neutron data above 30 MeV has increased in recent years due to data needs for applications and to the increased capability of sources and techniques in this region. Applications include neutron therapy in the treatment of cancer, space radiation effects, accelerator shielding, and materials irradiation test facilities. [1] In addition, many experiments in basic neutron nuclear physics are being undertaken to study nuclear structure and reaction mechanisms.

This range is above that traditionally studied for nuclear energy programs. Consequently, new approaches are often needed. It is the purpose of this paper to point out some of the techniques that are being used in the range 30-800 MeV. A complete review of the field is not possible, but we hope to capture a flavor of the challenges and excitement of this quickly developing field.

Neutron Sources

Two types of neutron sources are in common use above 30 MeV: the quasi-monoenergetic ^{7}Li(p,n) source and "white" neutron sources where a continuum of neutrons is produced. Just as at lower energies, these two types of sources are complementary. A quasi-monoenergetic permits the experiment to focus on results at one energy with more control of the neutron energy spectrum. The white source on the other hand allows a detailed mapping of the energy dependence of cross sections. Both span the energy range 30-800 MeV at various facilities.

The ^{7}Li(p,n) reaction with a thin ^{7}Li production target is generally preferred to other quasi-monoenergetic source reactions because the intensity is sufficient, tritium and other gaseous targets need not be involved, and the contamination by lower energy neutrons is less of a problem. This reaction has been compared with others. [2,3]. The proton beam is usually produced by a cyclotron. A typical source spectrum is shown in Fig.1. [3] Note that in addition to the strong peak there is also a continuum at lower energies. To separate reactions induced by peak and continuum neutrons, all such sources are pulsed and time-of-flight techniques must be used. A clearing magnet is installed downstream from the production target to remove the unreacted proton beam from neutrons produced at 0º, where the maximum intensity is obtained. For both of these reasons, the target to be irradiated is placed several meters from the source. This situation is qualitatively different from that at lower neutron energies where truly monoenergetic sources are available, time-of-flight is often unnecessary, and the investigated sample can be very close to the neutron source. A facility [4] for (n,p) measurements at Uppsala is illustrated in Fig. 2.

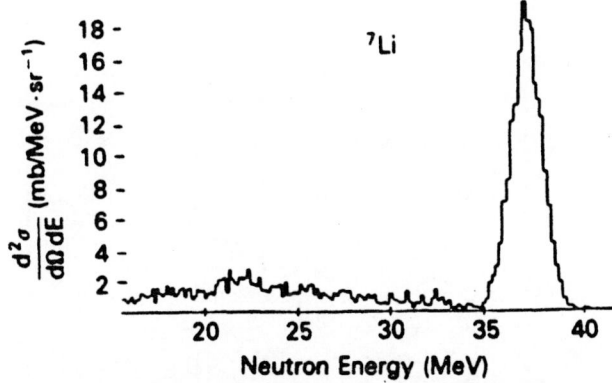

Fig. 1. Neutron source spectrum using the ^{7}Li(p,n) reaction at 39.3 MeV from the facility at the University of California at Davis. [3] Note the strong peak of this quasi-monoenergetic source and the continuum of neutrons at lower energy which, although small, is not insignificant.

White neutron sources are produced by the interaction of energetic ion or electron beams with suitable targets. For neutron energies in the range discussed here, proton or deuteron beams on thick targets offer the greatest neutron source strength. For many years a white source at Karlsruhe was used up to 50 MeV based on the ^{2}H(d,n) and U+d reactions. [5] If the ion beam has energy of several hundred MeV, neutrons are created efficiently by spallation. A facility using an 800 MeV proton beam incident on a tungsten target has recently become operational at LAMPF. [6] This source is pulsed with a time spread at the source of less than 1 ns so that time-of-flight techniques can be used to determine the neutron energy. The neutron spectra from this source are shown in Fig. 3 for the different neutron production angles available for experiments. Calculated spectra are also shown in the figure. The neutron flux from this source extends from below 0.1 MeV to well beyond 500 MeV for some production angles. An overview of the facility is shown in Fig. 4.

The types of neutron sources described above are suitable for measurements of nuclear data at well defined neutron energies. Other facilities that produce neutrons with a broad spectrum but without time-of-flight capability could be attractive for integral tests of nuclear data. We refer here, for example, to the significant number of new cyclotrons built for neutron therapy where neutrons up to 100 MeV are routinely produced at high intensity.

688

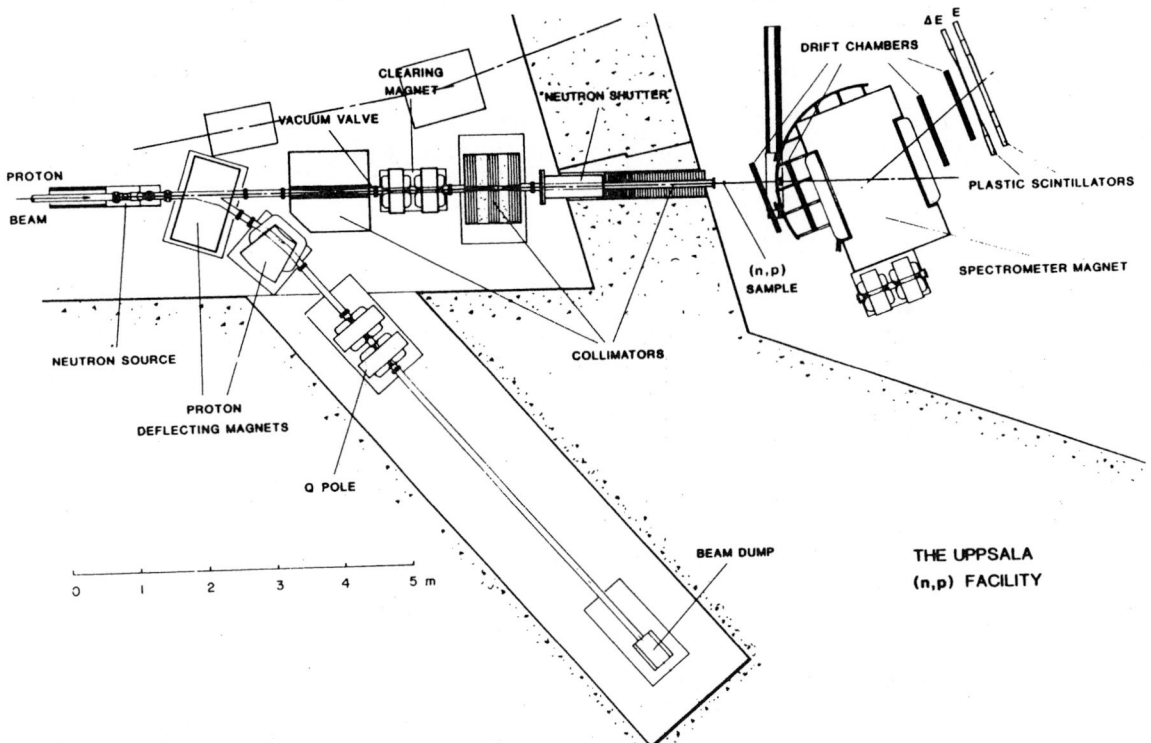

Fig. 2. Implementation of the quasi-monoenergetic neutron source at the The Svedberg Laboratory, Uppsala, Sweden. [4]

Background radiations provide new challenges for neutron sources in this energy range. Because of the higher neutron energy, the time-of-flight separation between neutrons and gamma-rays from the source is less and therefore care must be taken to separate the two. In practice this is not a big problem with ion-produced neutrons (as

opposed to sources based on electron accelerators) because the so-called gamma-flash is small. A more subtle background comes from charged particles generated by neutron interactions with collimators, beam pipes, shields, and residual gas in the flight paths. Magnets can sweep charged particles out of the neutron beam, but more will be generated by neutron interactions downstream. Collimators and shields must be much thicker at higher neutron energies because of the lower total cross section. [7]

Detectors

Detectors for neutron data measurements in the 30-800 MeV region have developed from those used in low and medium energy nuclear physics research. Detected radiations can be penetrating or not, so that a wide range of detector designs is necessary.

To detect neutrons in this energy range, moderately large plastic or liquid CH_x scintillators are preferred. The detection mechanism is not necessarily n-p elastic scattering as it is for lower energy neutrons, however. Reactions on carbon become more important as the neutron energy increases. Calculations of the efficiency of the detectors are then more complicated. [8,9] To enhance the efficiency of detectors to higher energy neutrons, a CH_2 radiator can be placed in front of the detector. These detectors are readily used in total cross section studies.[7]

Detection of scattered neutrons is more difficult because of the time-of-flight requirement usually imposed. For elastic scattering, quasi-monoenergetic beams can be used and the contaminant neutrons are at lower energies.[3] Beam swingers for the beams producing the neutrons allow an angular range to be covered.[10]

To investigate both elastic and inelastic scattering, a different approach must be used. One way is to convert the scattered neutrons to protons in a hydrogenous foil and then measure the energy and angle of the recoil proton.[11] This has been used at 65 MeV incident neutron energy at the University of California at Davis (Fig. 5). This technique or

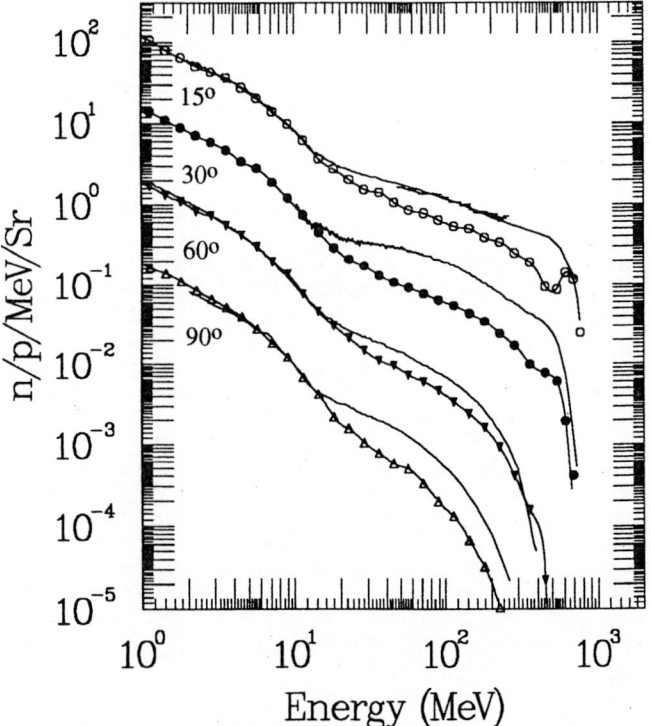

Fig. 3. Neutron source spectra for the different flight paths at the Target-4 spallation neutron source at LAMPF. The narrow curves are the measured spectra and the curves highlighted with symbols are calculations based on the LAHET code system for nuclear reactions and particle transport. [19]

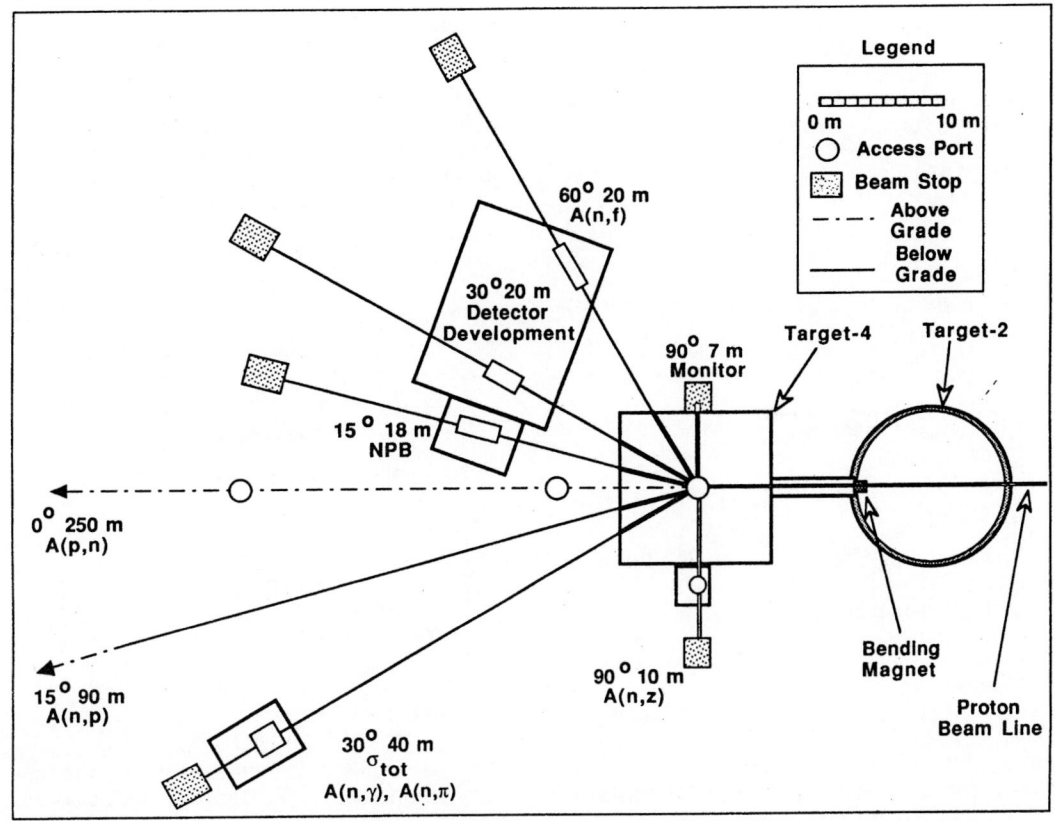

Fig. 4. Layout of the source and flight paths for the Target-4 white neutron source at LAMPF.

one similar to it needs to be exploited at higher energies where there are essentially no neutron elastic or inelastic scattering data.

Gamma-rays from neutron interactions above 30 MeV can range in energy all the way from very low energies to the energy of capture to the ground state. The low energy gamma rays, up to a few MeV, can serve as indicators of the partial reaction cross sections.[12-14] Fig. 6 shows the cross section for producing a characteristic gamma ray from the ^{56}Fe(n,2n)^{55}Fe reaction as a function of incident neutron energy.[12] The data were taken with a germanium gamma-ray detector. For neutron capture, much more energetic gamma rays are produced and different detectors, such as BGO, BaF$_2$, or NaI(Tl) are required. These detectors are

basically the same as those used in charged-particle experiments. They must however be shielded better against scattered and background neutrons. Often a layer of ^{6}LiH or ^{6}LiD is placed between the sample and the detector to reduce the number of scattered neutrons.

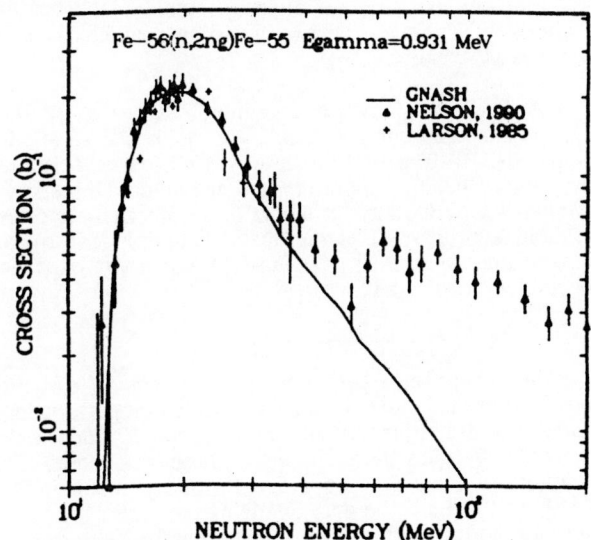

Fig. 6. Excitation function of the 0.931 MeV gamma-ray from the ^{56}Fe(n,2n)^{55}Fe reaction. The transition is in the residual nucleus. Data are from [12].

Charged-particle detectors are designed for different energy ranges. For particle energies up to 30 MeV or so, the detectors must be in an evacuated chamber with the target foil. [15] For higher outgoing energies, the charged particles have a reasonably long range in materials so that wire chambers and calorimeters are appropriate. Fig. 2 illustrates one arrangement of detectors.

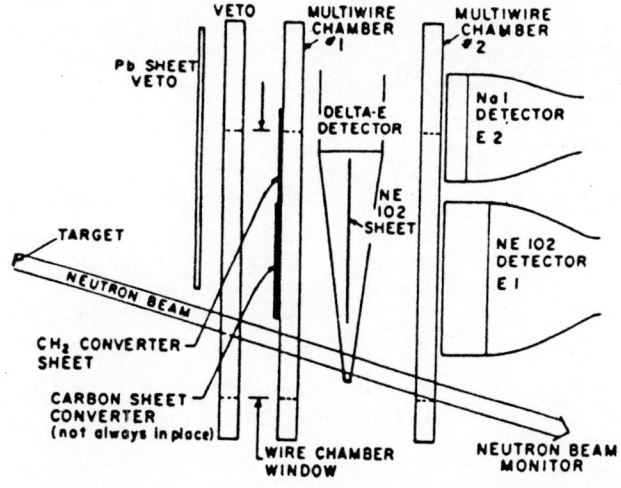

Fig. 5. Detectors for measuring elastic and inelastic neutron scattering in the 65-MeV region at the University of California at Davis. [11]

690

Fission induced by neutrons above 30 MeV can be detected in a manner similar to that used at lower energy. A difficulty at energies above a few 100 MeV is that spallation can occur in materials of the fission chamber and yield large pulses from the heavy fragments. At these high energies, background measurements need to be taken without the fissionable material.[16]

Electronics

Continuing advances in commercially available electronics makes many of these experiments possible. A partial list illustrates these advances:
CAMAC interfaces
FASTBUS interfaces
Multiple units per NIM module
High voltage supplies
Discriminators
Logic Units
Linear Gates
Fast amplifiers
Fast encoding and readout analog-to-digital converters
Multiple-stop time digitizers
Fiber-optic links
Improved oscilloscopes
Cables, connectors, etc.

Not only are new capabilities available, but the older capabilities are becoming much more reliable. This makes complex measurements possible where many components all need to be working for the experiment to succeed.

Computers and Software

For nuclear data measurements, the emergence of relatively inexpensive computers has meant that all of the data taking and most of the analysis can be accomplished with what are now considered small computers in the laboratory. This trend is certainly not finished now. Advanced workstations are beginning to appear in nuclear laboratories and will likely continue to change the way we handle data and control the experiments.

As important as the computer hardware are the software data collection programs. We use the XSYS and Q systems.[17,18] The first of these was written at Triangle Universities Nuclear Laboratory and at Indiana University. The latter was written at our laboratory. Other laboratories have their favorites. We predict that the sharing of software will continue to increase because of the great expense associated with writing it.

Tools to Design Experiments

In designing experiments, one needs to estimate counting rates and assess the probable radiation performance of shields, collimators, and other components. Codes exist now to permit calculation of many of these parameters. The predictions of one code are compared with experiment in Fig. 7. [19] In this case the only available data are for proton-induced reactions. It will be interesting and important to compare calculations and experiment for neutron-induced reactions.

In the future there will be data bases above 20 MeV that will provide data for neutron transport codes. At present, the beginnings of data bases are appearing. (See for example [20,21].) Integral benchmarks do not now exist to test the data and transport codes. A future program to provide such integral data will certainly be required.

Conclusion

Nuclear data measurements in the region above 30 MeV are progressing rapidly due to new techniques and new capabilities. Neutron sources, detectors, electronics, computers, software and tools to design experiments all have advanced significantly in the last few years. The next few years should show accelerated developments.

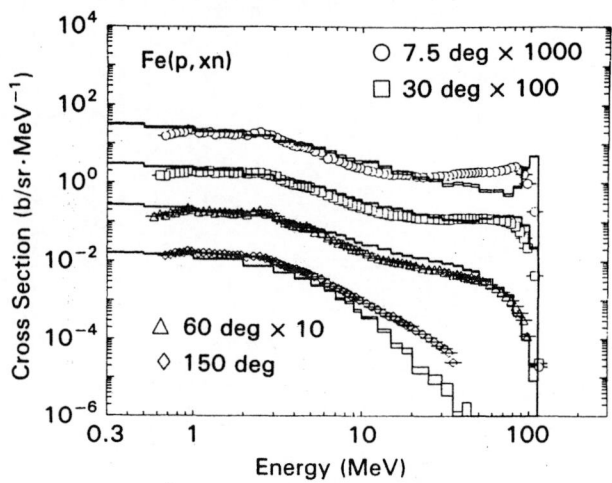

Fig. 7. Double differential cross-section data for 113 MeV protons incident on Fe compared with calculation.

References

[1] Proc. IAEA Specialists' Meeting on Intermediate Energy Nuclear Data for Applications, Vienna, 1990, ed. by N. Kocherov (in press)

[2]. J. Rapaport, F. P. Brady and J. L. Romero, Nucl. Sci. Eng. 106, 318 (1990)

[3] J. Rapaport, Proc Int. Conf. Nuclear Data for Basic and Applied Science, ed. P.G. Young et al. (Gordon and Breach Science Publishers, 1986) Vol. 2, P. 1229

[4] H. Conde, S. Crona, A. Hakansson, O. Jonsson, A. Lindholm, L. Nilsson, P.-U. Renberg, G. Tibell, I. Bergqvist, and P. Ekstrom, Can. J. Phys. 65, 643 (1987)

[5] H. O. Klages, H. Dobiasch, P. Doll, H. Krupp, M. Oexner, P. Lischke, B. Zeitnitz, F. P. Brady, and J. C. Hiebert, Nucl. Inst. Meth. A219, 269 (1984)

[6] P. W. Lisowski, C. D. Bowman, S. A. Wender, and G. J. Russell, Nucl. Sci. Eng. (1990)

[7] R. W. Finlay et al., these procedings

[8] W. Sailor, R. C. Byrd, and Y. Yariv, "Calculation of the Pulse-height Response of Organic Scintillators for Neutron Energies 28<E_n<492 MeV," Nucl. Inst. Meth. in Phys. Research A277, 599 (1977)

[9] J. K. Dickens, RSIC Peripheral Shielding Routine Collection SCINFUL, Oak Ridge National Laboratory report PSR-267 (1988)

[10] R. Bhownik et al., Nucl. Instr. Meth. 143, 63 (1977)

[11] F. P. Brady, T. D. Ford, G. A. Needham, J. L. Romero, C. M. Casteaneda, and M. L. Webb, Nucl. Instr. Meth. in Phys. Res. 228, 89 (1984)

[12] R. O. Nelson et al., <u>Proc. 7th Int. Symp. on Capture Camma-ray Spectroscopy and Related Topics</u>, Asilomar, California, (1990) (in press)

[13] D. C. Larson et al., <u>Proc. Int. Conf. on Nuclear Data for Basic and Applied Science</u>, ed. P. G. Young et al. (Gordon and Breach, 1986) vol. 1, p. 71

[14] R. C. Haight et al., Bull. Am. Phys. Soc. <u>35</u>, 1038 (1990)

[15] S. M. Grimes and R. C. Haight, these procedings

[16] P. W. Lisowski, these procedings

[17] IUCF XSYS Manual, Indiana University Cyclotron Facility report (1988)

[18] Q-User Information Manual, Clinton P. Anderson Meson Physics Facility Report MP-1-3401-4 (1989)

[19] M.M. Meier, D.A. Clark, C.A. Goulding, J.B. McClelland, G.L. Morgan, and W.B. Amian, Nuc. Sci. Eng., <u>102</u>, 210, (1989)

[20] P. G. Young et al., private communication (1991)

[21] S. Pearlstein, The Astrophysical Journal, <u>346</u>, 1049 (1989)

NUCLEAR DATA NEEDS FOR THE SPACE EXPLORATION INITIATIVE

Steven D. Howe and George Auchampaugh
Los Alamos National Laboratory
P. O. Box 1663
Los Alamos, NM 87545

ABSTRACT

On July 20, 1989, the President of the United States announced a new direction for the U.S.Space Program. The new Space Exploration Initiative (SEI) is intended to emplace a permanent base on the Lunar surface and a manned outpost on the Mars surface by 2019. In order to achieve this ambitious challenge, new, innovative and robust technologies will have to be developed to support crew operations. Nuclear power and propulsion have been recognized as technologies that are at least mission enhancing and, in some scenarios, mission enabling. Because of the extreme operating conditions present in a nuclear rocket core, accurate modeling of the rocket will require cross section data sets which do not currently exist.

In order to successfully achieve the goals of the SEI, major obstacles inherent in long duration space travel will have to be overcome. One of these obstacles is the radiation environment to which the astronauts will be exposed. In general, an unshielded crew will be exposed to roughly one REM per week in free space. For missions to Mars, the total dose could exceed more than one-half the total allowed lifetime level. Shielding of the crew may be possible, but accurate assessments of shield composition and thickness are critical if shield masses are to be kept at acceptable levels. In addition, the entire ship design may be altered by the differential neutron production by heavy ions (Galactic Cosmic Rays) incident on ship structures. The components of the radiation environment, current modeling capability and envisioned experiments will be discussed.

I. INTRODUCTION

Since the termination of the Apollo program in the early 1970's, NASA centers, DOE laboratories, and universities have continued to study the technological requirements to develop a lunar base or complete a manned mission to Mars. On July 20, 1989, President Bush validated such efforts by announcing [1] a new direction for the U.S. space program. The goals of the new Space Exploration Initiative were declared to 1) establish a permanent base on the Lunar surface and 2) develop a manned outpost on Mars by the year 2019. The President also tasked the new National Space Council with formulating the SEI program.

Following the announcement, NASA completed a "90-Day Study" [2] which assessed technologies needed for a Mars Mission. In addition, the National Academy of Sciences reviewed both the 90-Day Study and some alternative technology options [3]. Both of these reviews produced a list of critical technologies which needed further development to support SEI goals. Two of the critical technologies identified by these committees were advanced nuclear propulsion and protection against the space radiation environment. Both of these areas will require improvement of the nuclear data bases that currently exist.

II. SPACE RADIATION PROTECTION

A critical component of SEI is a prior knowledge of the radiation field inside of complex geometries, including the human body, that have been exposed to space radiation. This knowledge could be obtained from tests of actual space modules at an accelerator facility that simulates space radiation. But, currently there are no facilities that can produce the desired radiation field and the cost of such tests would be prohibitive. Therefore, validated models will play an important role in the design of spacecraft and human habitat modules and in providing an accurate representation of the radiation field inside these geometries. These models will encompass a wide range of nuclear physics disciplines, from medium energy to high energy and relativistic heavy

ion physics. They will use state-of-the-art representations of radiation interactions with a wide variety of structural and biological matter, and experimental validation will be used in every stage of the development to help establish confidence levels for the results. The interaction physics must treat e,γ, n, and H through, possibly, U ions and particle energies from MeV to tens of GeV/amu. This multidimensional parameter space is too large to presume that data will be obtained on all particle types and energies. Instead, extensive use will be made of the models to define the critical parameters for a wide range of mission scenarios that include, for example, space stations, interplanetary travel, and permanent habitation of the moon and planets. These parameters will then be used to guide the experimental efforts for development of the radiation transport code and the radiobiological response data base.

There are two sources of radiation that are major drivers in the SEI program; they are, Galactic Cosmic Radiation (GCR) and Solar Energetic Particle (SEP) events. A calculation of the dose received by an astronaut inside of a spacecraft for a three-year mission to Mars illustrates the importance of the GCR and SEP sources. For such a mission, the calculated dose for just the GCR component is as high as ~100 rem [4] which exceeds the recommended three year occupational dose by a factor of 6.7. If an intense SEP event should occur during the mission, the astronaut could receive a lethal dose from some of the more intense SEP events.

The GCR is a persistent source of radiation that originates outside of the solar system. The intensity of the GCR varies by as much as a factor of two or more with the eleven-year solar cycle, the maximum intensity occurs during solar minimum. The GCR is composed of ions from hydrogen to uranium with the most abundant ion being hydrogen. In Fig. 1, spectra are shown for classes of ions pertinent to SEI. Ions with Z > 28 are present in the GCR, but their relative biological importance is minimal. All ions have roughly the same energy distribution with the maximum in the distribution occurring at approximately 400 MeV/amu. Ion energies as high as 1000 GeV/amu have been measured, but their relative intensity is at least six orders of magnitude less than the peak intensity. The high-Z, high-energy (HZE) particles in the GCR are extremely penetrating and biologically destructive; for example, an iron ion with an energy near the peak of the distribution can effectively destroy every cell in its path in passing through the human body. Even though ions with $Z > 2$ represent less than two percent of the GCR fluence, they can contribute more than 50% to the dose. Nuclear data on high-Z projectile and target fragmentation processes and secondary particle production and transport are necessary to develop the radiation transport models, to build evaluated data libraries, and to validate the models.

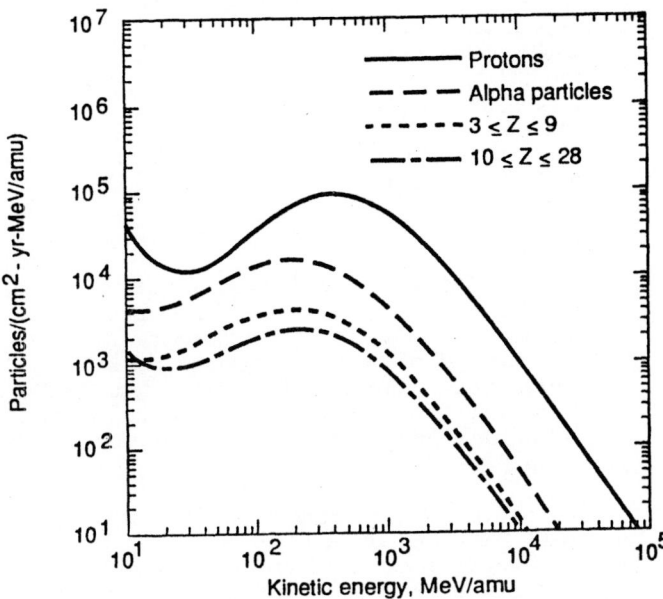

Fig. 1. Flux Versus Energy for GCR Ions During Solar Minimum [5]

The SEP events are sporadic and can last for hours to several days. These events are unpredictable but occur most frequently during solar maximum. They consist of mostly protons and alphas with energies as high as several GeV/amu. In Fig. 2, accumulative plots of proton intensity are shown for three characteristically different SEP events. The August 1972 event, which was the most intense SEP

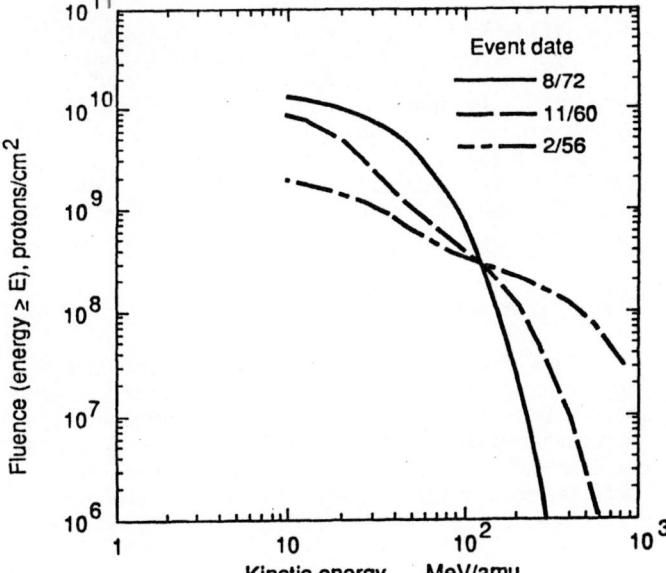

Fig. 2. Accumulative Proton Fluence Spectra for Three Large SEP Events [6]

event recorded prior to the October 1989 event, had a proton fluence of about 10^{10} protons/cm^2 and proton energies of several hundred MeV. On the other hand, the SEP event of February 1956 had a proton fluence of about an order of magnitude less, but it had protons with energies of several GeV. Shielding may be effective in protecting an astronaut from SEP events, but now secondary radiation, mainly high-energy neutrons, becomes important.

Secondary radiation occurs when GCR or SEP particles penetrate thick shields. This radiation manifests itself in the form of very high-energy neutrons and protons (hundreds of MeV). Calculations of the differential particle flux for particles at 100 g/cm^2 depth in carbon dioxide resulting from incident GCR at solar minimum are shown in Fig. 3. Similar spectra are obtained for other shielding materials, including lunar regolith. Nuclear data will be required for total, elastic, and inelastic neutron processes for energies up to several hundreds of MeV. Furthermore, to calculate a dose, (n,z) cross sections will be required for materials found in the human body.

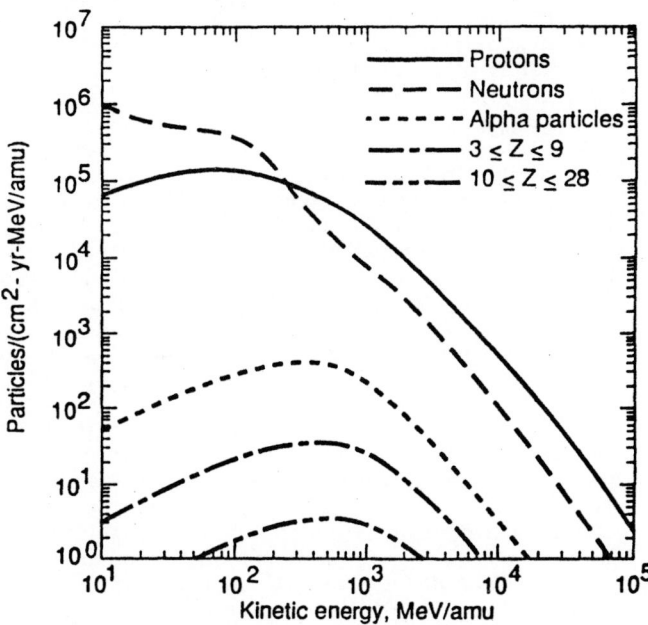

Fig. 3. Yearly Differential Flux Versus Energy for Particles at 100 g/cm^2 Depth in Carbon Dioxide Resulting from Incident GCR at Solar Minimum [7]

In this section we have ignored the man-made radiation from a nuclear reactor. If a nuclear propulsion system is used, then the total dose must include the effects from nuclear radiation. The nuclear data needs for calculating the radiation output from a nuclear propulsion system are addressed in the following sections.

III. SOLID CORE NUCLEAR PROPULSION

Although a nuclear rocket engine had been partially developed and tested during the ROVER/NERVA program [8,9] in the 1960's, substantial advances have occurred in several technological areas pertinent to nuclear propulsion systems. As a consequence, NASA and the DOE sponsored a pair of workshops in the summer of 1990 to examine all concepts for nuclear electric propulsion (NEP) and nuclear thermal propulsion (NTP). The purpose of the workshops was to allow a common panel of experts to evaluate several different reactor concepts, fuel forms, or engine systems. Approximately 60 individuals from NASA, DOE and the industrial community comprised 5 review panels, each reviewing all of the concepts. One of the primary findings of the panels was that NTP technologies for manned missions could probably be developed earlier than NEP systems but that higher operating temperatures would be required for the mission.

Subsequent to these findings, both NASA and the Department of Energy (DOE) have initiated a series of meetings to more completely examine the research needs in these areas. The more complete effort is a jointly NASA/DOE sponsored group of 6 Task Teams to examine nuclear propulsion from the perspectives of safety, mission benefit, nuclear thermal propulsion (NTP) technology, Nuclear Electric Propulsion (NEP) technology, fuels and materials, and test facilities. The NTP panel examines all technology needs including improvements in data bases necessary for high temperature nuclear reactor core operations.

The ROVER/NERVA program succeeded in performing 22 different reactor/engine tests between 1955 and 1971. The highest power engine tested was the Phoebus 2 at 4500 MW and a thrust of 1.1×10^6N. The hottest fuel temperature achieved during the tests was 2550K which corresponds to a specific impulse, Isp, (momentum delivered per unit mass) of about 850s. Because of the space radiation dose problem described in Section II, the manned flight to Mars must be made as quickly as possible to reduce GCR dose and potential solar flare coincidence. For example, in order to reduce the total round trip to around 1 year, an Isp of around 950 to 1000s must be achieved (for a reasonable mass of around 750 tons in Low Earth Orbit). Consequently, peak fuel operating temperatures of around 3200K would be required. To date, no institution has ever operated a critical system at steady state conditions at such temperatures.

During the ROVER program, computational modeling was used but with moderate success due to the complexity of the problem and the, then current, computer capability. Current constraints on the testing environment and the budget environment, however, will necessitate a much stronger dependence on computational simulation than before. Consequently, several improvements in cross section sets and computer code treatments will be required.

Although several different reactor concepts have been proposed, a generic geometry can be envisioned such that:

1) the reactor core will be a cylindrical assembly consisting predominantly (from the neutronics viewpoint) of UC_x ($x \geq 1$) with some ZrC intermixed with hydrogen coolant; and

2) the core is surrounded by insulator material (possibly aluminae) then reflector/moderator which is cooled by cold hydrogen gas.

Because of this geometry, moderated neutrons which are reflected back from the cold reflector will be upscattered by the insulator and fuel material. The upscattering kernels of aluminae, beryllium, and carbon need to be re-examined and probably improved. Accurate calculation of the neutron spectra will be necessary due to the strong dependence of the U^{235} fission cross section on neutron energy between 273K (σ_f =577b) and 3300K (σ_f =120b) temperatures. The spectral content of the neutron flux will determine the radial power profile in the core and thus engine life and performance.

Because of the need to achieve much higher fuel operating temperatures, new materials with higher melting points are being considered as flow channel coatings. The two most promising candidates are HfC and TaC which have melting points of 4120K and 4080K, respectively. These materials, however, also have significant neutron absorption cross sections. A review of the cross section data base between .01 to .3eV is needed. Furthermore, reactivity measurements of such materials in criticality assemblies will be necessary.

IV. ADVANCED PROPULSION CONCEPTS

As part of the effort of the Nuclear Thermal Propulsion Task Team, one of the authors (Howe) chaired a sub-panel to evaluate innovative propulsion concepts. The motivation to form such a panel was to investigate alternative methods of achieving much higher Isp's (> 2000s) in order to reduce Mars mission round trip times to 1-2 months. The goals of the sub-panel were to 1) evaluate a variety of concepts on a "level technological playing field", 2) identify critical research issues of each concept, and 3) identify proof-of-concept experiments for the leading concepts.

Because of substantial work performed in the 60's and 70's, the gas core nuclear rocket is the leading concept. Although several variations exist as to geometry, flow fields, and fuel forms, all of the concepts essentially involve the containment of a uranium gas or plasmoid surrounded by hydrogen coolant/propellant . Design chamber pressures range from 100 to 500 atmospheres. Uranium temperatures range from 5000K to 100,000K.

Neutronically, the upscatter problem is even more pronounced. Early findings indicate that the main fission generation region in a spherical uranium plasmoid will be a spherical shell at some radius inside the plasmoid radius. This effect is caused by the assumed upscattering of the reflected cold neutrons and may have a dramatic effect on the concept performance. Consequently, the upscatter cross sections for BeO,F ,and U need to be accurately determined.

In addition, some of the concepts rely on mechanical containment of the uranium fuel and on radiative coupling to the propellant through a fuzed silica window. The opacity of the window as a function of neutron fluence has not yet been thoroughly investigated.

The sub-panel also considered several concepts for long-term propulsion systems for solar system exploration. Of these, fusion and antimatter annihilation ranked the highest. While the fusion reaction has been well studied, no data exist for the antiproton annihilation cross sections below 2MeV, except for the totally stopped, thermal data point. Many of the proposed antiproton storage concepts are extremely dependent on the annihilation cross section in the 100eV to 100KeV regime. Determination of these cross sections will be necessary to pursue this propulsion concept.

V. SUMMATION

The President of the United States has initiated a new space initiative which will require technological advances in several areas. Two of these areas, space radiation protection and advanced propulsion will require improvements in the nuclear data base. Accurate knowledge of the secondary products of relativistic heavy ion collisions produced by Galactic Cosmic Rays will be necessary to protect the crew , the electronic controls, and to design the placement of the spaceship's main components such as fuel tanks. Determination of neutron upscatter cross sections and absorption cross sections for specific materials must be made in order to design advanced propulsion systems such as solid-core or gas-core nuclear rockets. In addition, the antiproton annihilation cross sections below 2MeV are unknown and will need to be measured to develop even higher performance propulsion systems. Such systems will be necessary to reduce trip times to planets for human crews. If mankind is to explore the solar system, significant extensions of our current nuclear reaction data base will be absolutely essential.

References

1. G. Bush: "Remarks by the President at the 20th Anniversary of the Apollo Moon Landing," Washington, D.C., July 20, 1989. Published in the Space Exploration Initiative Fact Sheet

2. A. Cohen: "Report of the 90-Day Study on Human Exploration of the Moon and Mars,"NASA, Nov. 1989

3. G. Stever: "Human Exploration of Space: A Review of NASA's 90-Day Study and Alternatives," National Research Council, National Academy Press. 1990

4. "Guidance on Radiation Received in Space Activities. NCRP Report No. 98, July 989

5. J. H. Adams, Jr., R. Silberberg, C. H. Tsao, "Cosmic Ray Effects on Microelectronics. Part 1: The Near-Earth Particle Environment." NRL-MR-4506-PT-1, Naval Research Laboratory (Aug. 1981)

6. J. W. Wilson, "Environmental Geophysics and SPS Shielding," Workshop on the Radiation Environment of the Satellite Power System," (Edited by W. Schimmerling and S. B. Curtis), LBL-8581, pp.33-116, Sept. 1978

7. L. C. Simonsen, J. E. Nealy, L. W. Townsend, and J. W. Wilson, "Radiation Exposure for Manned Mars Surface Missions." NASA Technical Paper 2979, March 1990

8. J. H. Altseimer, G. F. Mader, and J. J. Seinart: Journal of Spacecraft, Vol. 8, No. 7, July 1971

9. Koenig, D. R., Los Alamos National Laboratory Report, LA-1062-H, May 1986

NEUTRON SPECTRA MEASUREMENTS AND MONTE CARLO SIMULATIONS OF (p,xn) REACTIONS IN THE MEDIUM ENERGY RANGE

W. Amian, P. Cloth, P. Dragovitsch[*], V. Drüke, D. Filges, M.M. Meier[**]

Forschungszentrum Jülich, Institut für Kernphysik, Postfach 1913, D-5170 Jülich, Germany
[*]Supercomputer Computations Research Institute, University of Florida, USA
[**]Los Alamos National Laboratory, Los Alamos, NM 87545, USA

Abstract: Double differential neutron production cross sections were measured at the LAMPF-WNR facility at Los Alamos between 100 MeV and 800 MeV incident proton energy using time-of-flight techniques. Virtually systematic results were gained by a large number of different angles, energy points and target materials from low to high atomic masses. The measured spectra were compared to detailed Monte Carlo simulations using the KFA version of the intra-nuclear cascade code HETC. The results are also compared with earlier measurements and those of other authors. An overview of the status of neutron production data in the medium energy range is presented.

Introduction

The application potential of results of beam-material interaction investigations is large, especially in context with particle accelerators of high beam intensity, e.g. for neutron sources for solid state and nuclear physics, in material damage studies, medicine and radiobiology, and simulation of space irradiations. Nuclear data needed for these purposes include total, elastic, nonelastic, and production cross sections. Particle yields are needed for source calculations. Data for operation of advanced accelerators include cross sections for transport, dosimetry, damage energy, transmutation, and gas production. One important example of application is shielding of accelerators and neutron producing targets. An additional one may be the estimation of radiation levels in low orbit manned space stations.

Much of the needed data has to come from nuclear model calculations, mainly above 20 MeV. Such calculations require computer codes capable of multi-particle emission and pre-equilibrium treatment [1-3]. However, such calculations must be anchored and verified by experimental data. Unlike the situations at lower energies, there do not exist sufficient differential data at higher energies for important materials to provide complete evaluations.

Secondary neutron measurements at incident proton energies above 100 MeV are rather scarce and cover only few points in the double differential scale [4-8].

In this paper, systematic measurements of double differential neutron production cross sections, which were measured at the LAMPF-WNR facility at Los Alamos between 100 MeV and 800 MeV incident proton energy using time-of-flight techniques are summarized. These experiments were performed by the LANL-KFA collaboration in the period from 1981 to 1990. Virtually systematic results were gained by a large number of different angles, energy points and target materials from low to high atomic masses. Single measurements in this series and in one case comparison with Monte Carlo simulations have been published earlier [9-12]. The measured spectra were compared to detailed Monte Carlo simulations using the KFA version of the intra-nuclear cascade code HETC [1,13,14]. The results are also compared to earlier measurements.

Measurements

The neutron spectra measurements were performed in the time-of-flight technique utilizing the unique conditions for this type of experiments at the Los Alamos WNR facility. The flight path lengths were between 30 m and 60 m at several distinct angles. The time resolution was 200 ps, mainly due to the favourable pulse structure of the proton beam, and at a reasonable intensity of about 10^8 protons per micropulse. Other sources of time uncertainty could not be determined. But it is probable that they reduce the accuracy, which can be obtained with the above machine parameters. A list of the thin target (p,xn) measurements is given in Table 1.

Table 1. Thin Target (p,xn) Measurements

Energy (MeV)	Target Materials	Angle (degree)
113	Be,C,O,Al,Fe,W,Pb,U	7.5,30,60,150
256	Be,C,O,Al,Fe,Pb,U	7.5,30,60,150
318	C,Al,Ni,Ta,W,Pb,U	7.5,30
597	Be,B,C,N,O,Al,Fe,Pb,U	30,60,120,150
800	Be,B,C,N,O,Al,Cd,Fe,W,Pb,U	30,60,120,150
800	C,Al,Ni,Ta,W,Pb,U	7.5,30
160(a)	Al,Zr,Pb	0,11....145
597(b)	C,Al,Fe,Nb,In,Ta,Pb,U	30,90,150
800(c)	Al,Cu,In,Pb,U	0,30,45,112

(a) Scobel et al. [15], (b) Cierjacks et al. [16], Howe [5]

The experimental arrangement is shown in Fig. 1.

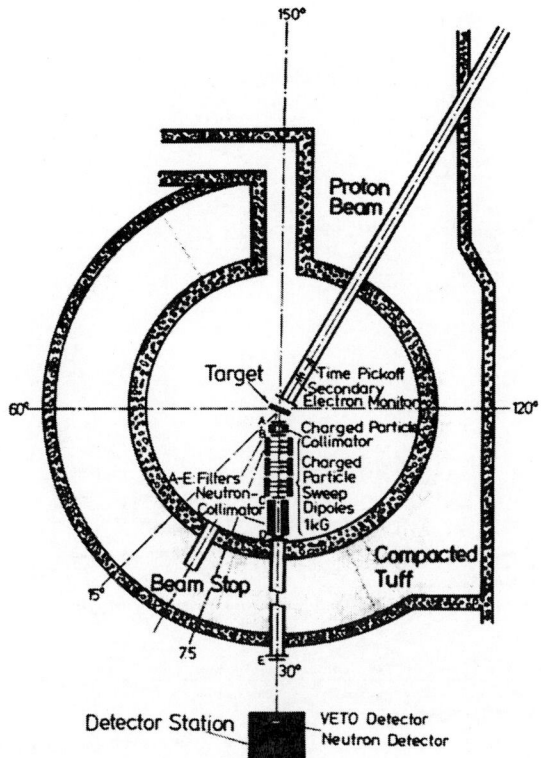

Fig. 1. The Time-of-Flight Facility

Another source of uncertainty, which has to be considered in neutron time-of-flight measurements, is the energy dependent efficiency of the neutron detector. Several efficiency curves for scintillator materials exist, which differ considerably at energies above 400 MeV. We used for our measurements the zero-degree - ^{7}Li curve, which was measured for our purposes. In Fig. 2 the used efficiency data are depicted together with earlier data.

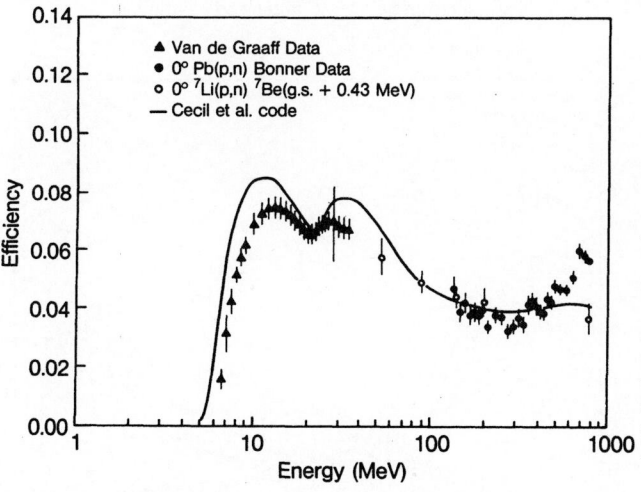

Fig. 2. Efficiency of one of the used time-of-flight detectors (2"x2" BC418) ("van de Graff", and Bonner data [8], Cecil et al. [17]

Monte Carlo Simulation

The intranuclear cascade evaporation model (INCE), as implemented in the high energy radiation transport code HETC of the HERMES system [1,13,14], is used in the calculations of double differential cross sections of proton induced neutron production. The predictions of the INCE model were compared to the experimental data of double differential cross sections. The calculations performed here were part of an experimental-theoretical program within the LANL-KFA collaboration concerning medium energy cross section measurements - mainly neutrons and state-of-the-art computer code validations of these measurements.

Main Features of the HETC Code
The Bertini Intranuclear Cascade Model and the
Excited Nucleus Evaporation Model

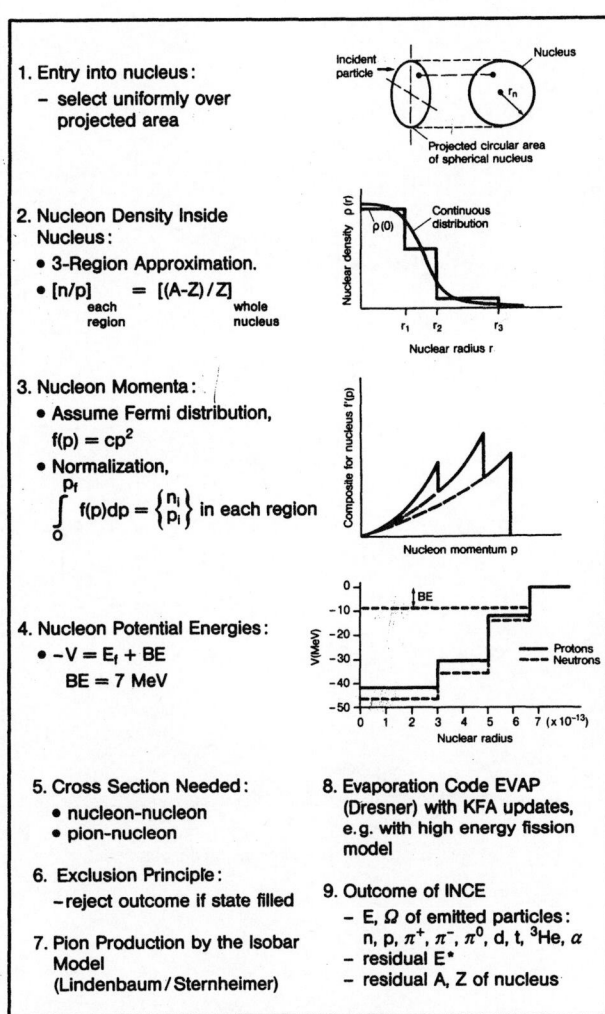

To achieve optimum comparison, identical energy intervals as used in the experimental analysis were provided in the HETC calculations. Depending on target material, emission angle and incident proton energy, the number of spallation events varied from 2.5×10^5 to 6.0×10^5 to achieve reasonable statistics in the calculations.

Experimental Results and Comparison with Simulation

Most of the measurements given in Table 1 were simulated and partly given in Figs. 3-16 as examples of the (p,xn) cross sections of the following materials. The kind of presentation (per units of lethargy) was chosen to pronounce discrepancies between experimental and simulated curves.

Carbon Data

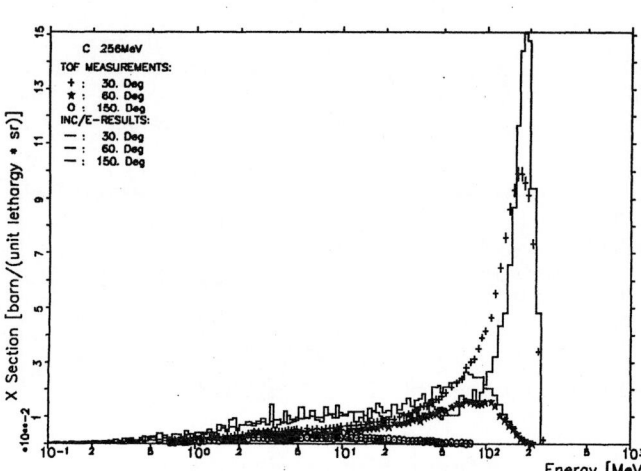

Fig. 3. Neutrons from 256 MeV protons on carbon

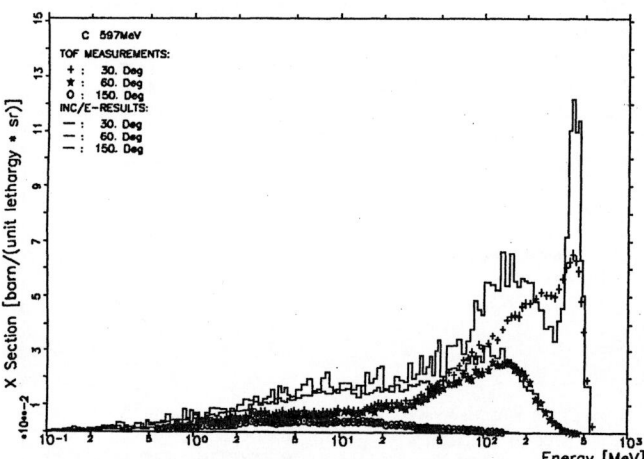

Fig. 4. Neutrons from 597 MeV protons on carbon

Aluminum Data

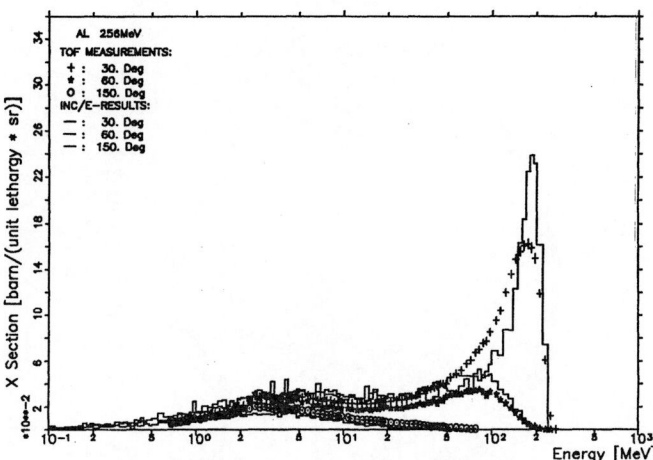

Fig. 5. Neutrons from 256 MeV protons on aluminum

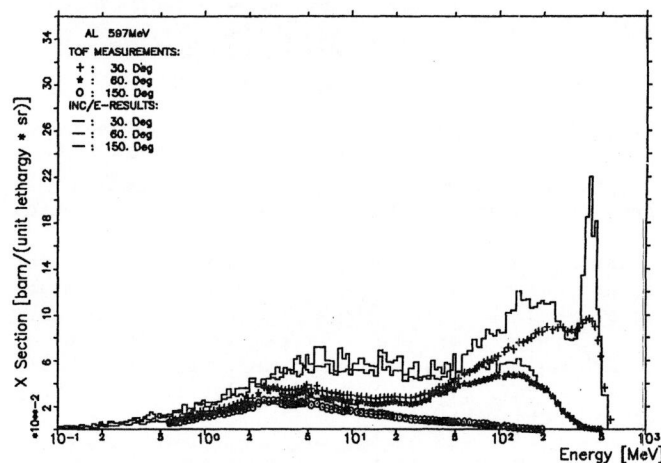

Fig. 6. Neutrons from 597 MeV protons on aluminum

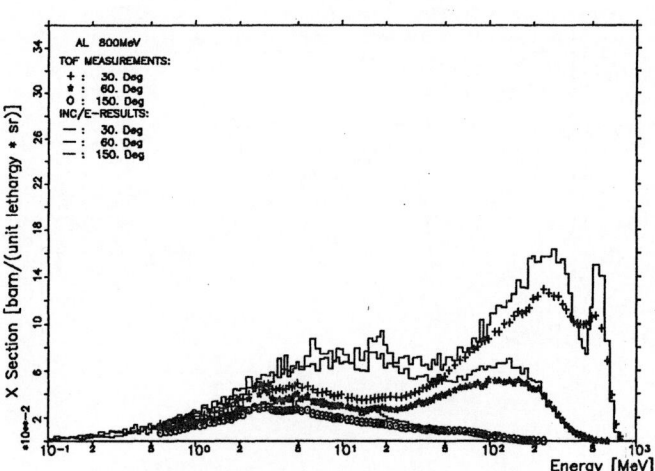

Fig. 7. Neutrons from 800 MeV protons on aluminum

Iron Data

Lead Data

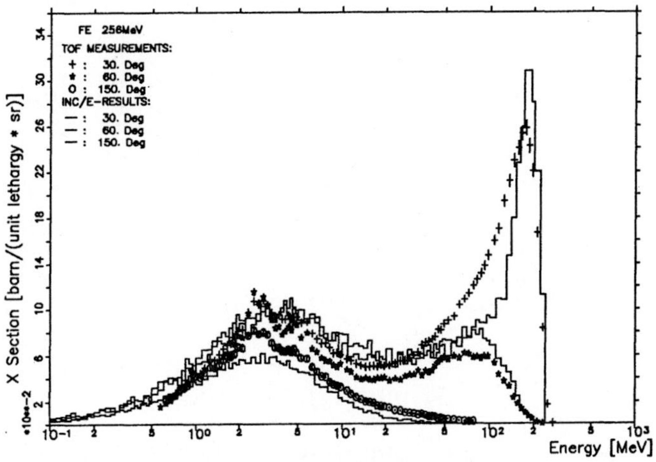

Fig. 8. Neutrons from 256 MeV protons on iron

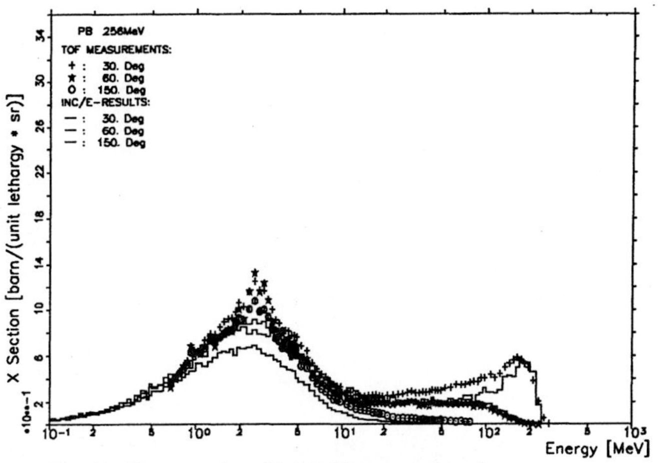

Fig. 11. Neutrons from 256 MeV protons on lead

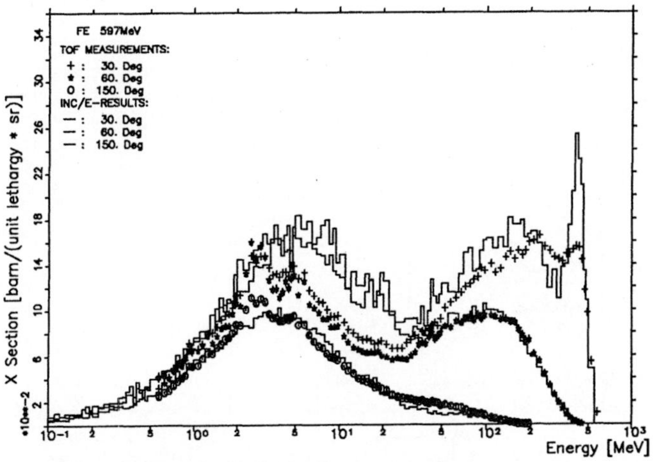

Fig. 9. Neutrons from 597 MeV protons on iron

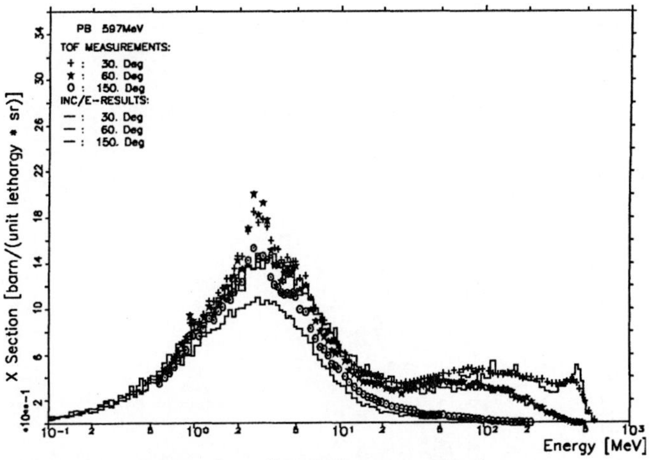

Fig. 12. Neutrons from 597 MeV protons on lead

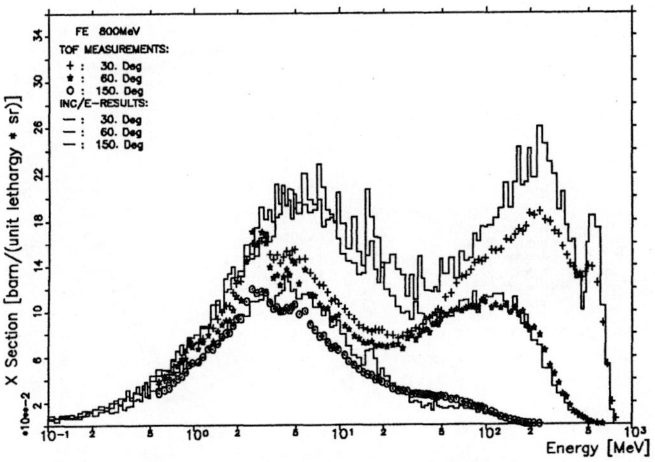

Fig. 10. Neutrons from 800 MeV protons on iron

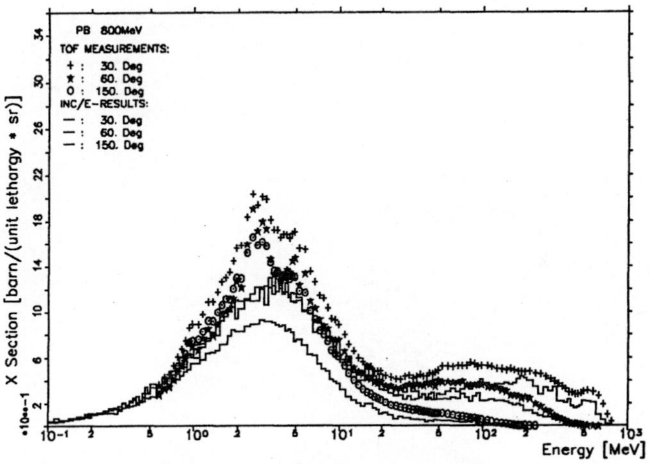

Fig. 13. Neutrons from 800 MeV protons on lead

Uranium Data

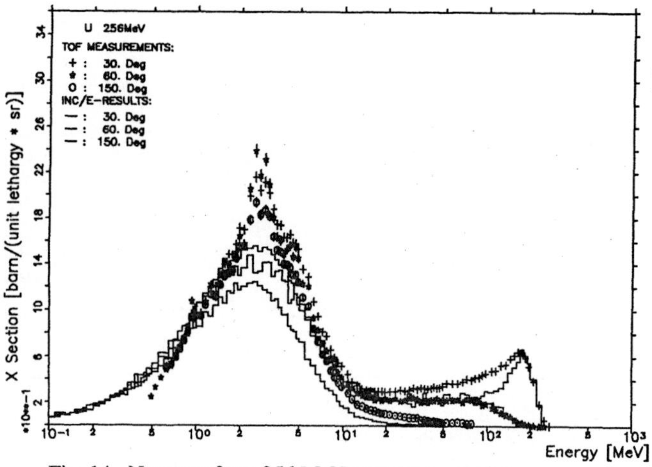

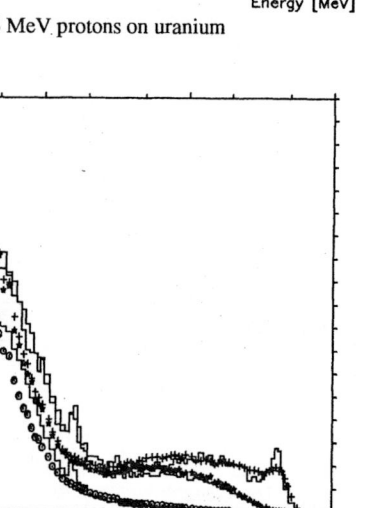

Fig. 14. Neutrons from 256 MeV protons on uranium

Fig. 15. Neutrons from 597 MeV protons on uranium

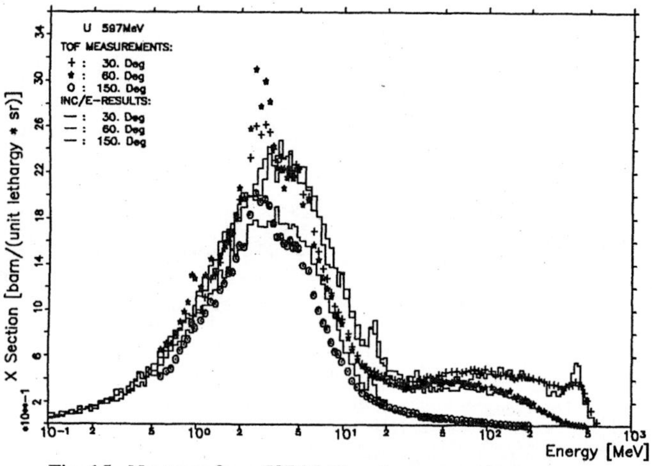

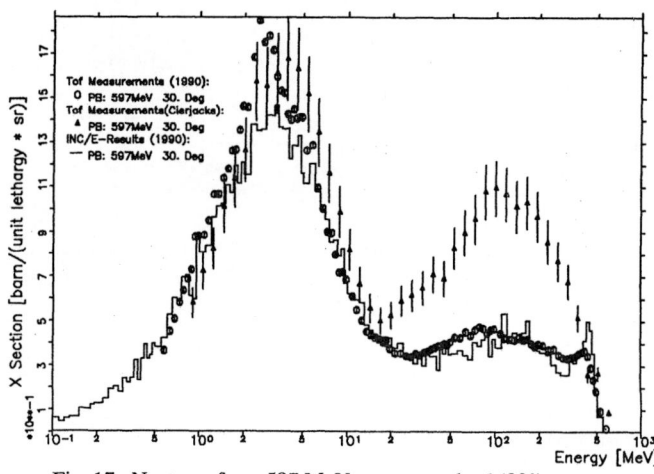

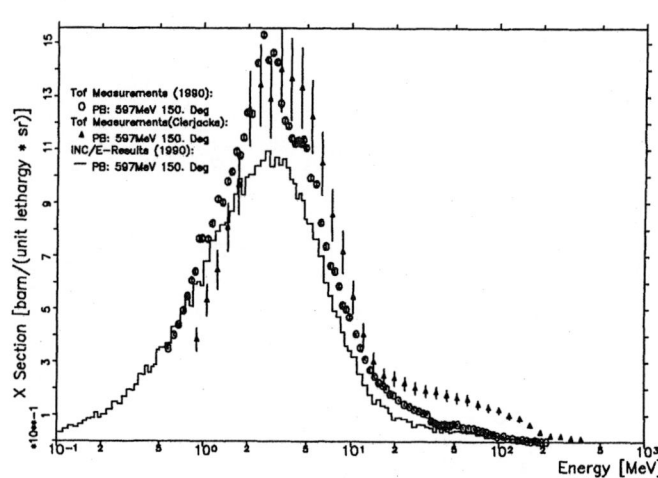

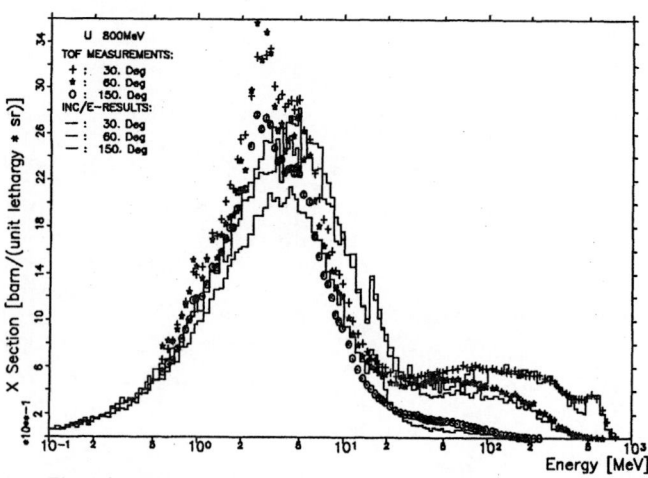

Fig. 16. Neutrons from 800 MeV protons on uranium

Comparison with Earlier Results

In Figs. 17-19 a comparison is made in some examples with older data of Cierjacks et al. [16] and Howe [5]. The disagreement between measurements and HETC-INCE simulation is clearly larger in the older measurements, than in the most recent ones. This is particularly true for the backward angles.

Fig. 17. Neutrons from 597 MeV protons on lead (30°)

Fig. 18. Neutrons from 597 MeV protons on lead (150°)

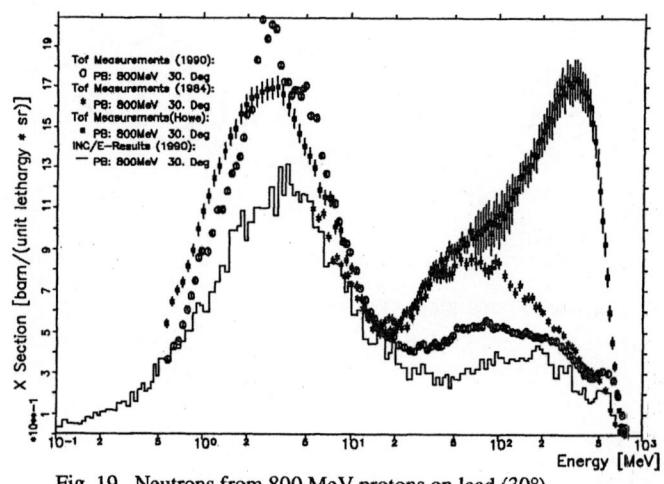

Fig. 19. Neutrons from 800 MeV protons on lead (30°)

Comparison of Mass Dependent (p,xn) Cross Sections for Neutrons above 50 MeV

The results of differential cross sections integrated over the high energy range from 50 MeV to incident proton energy versus the target mass is shown in Fig. 20-22.

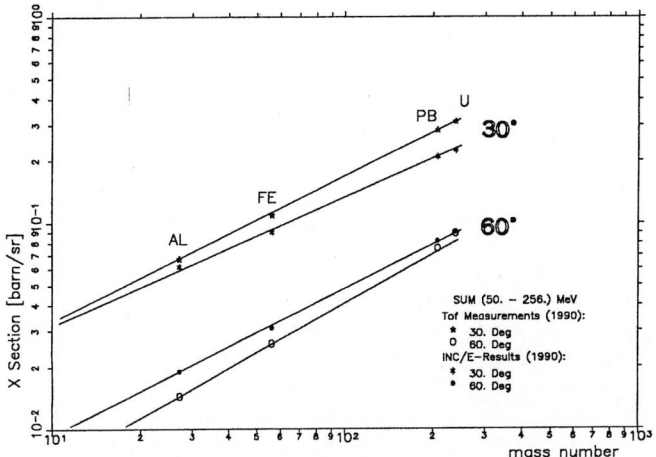

Fig. 20. Neutrons from 256 MeV protons (30° and 60°)

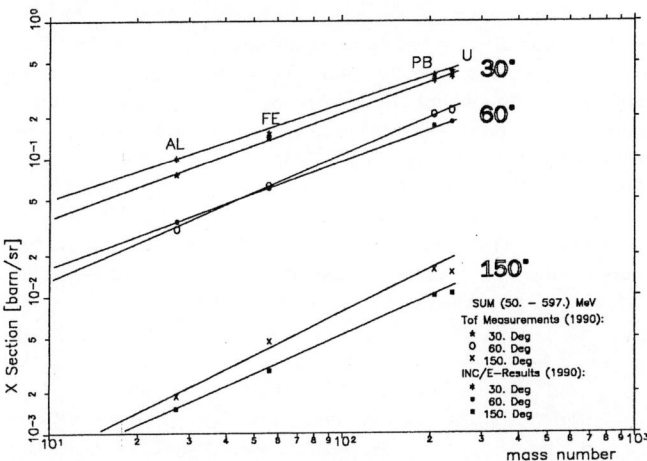

Fig. 21. Neutrons from 597 MeV protons (30°, 60° and 150°)

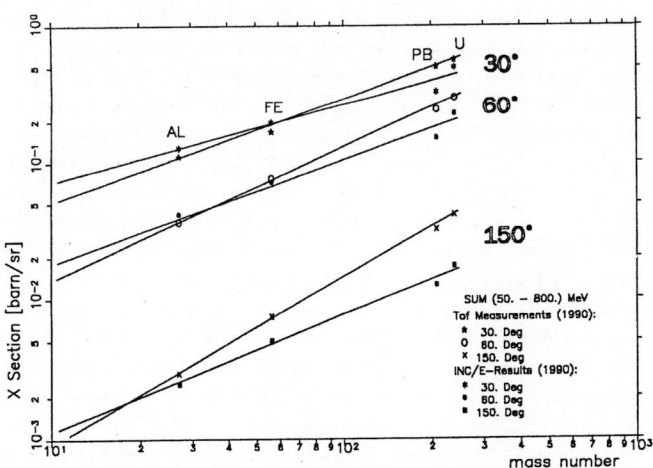

Fig. 22. Neutrons from 800 MeV protons (30°, 60° and 150°)

Summary

Measurements of proton induced neutron production cross sections are available now for a wide range of target masses, incident energies and neutron emission angles. Discrepancies between earlier measurements and INCE simulations are considerably smaller with the latest measurements.

References

1. P. Cloth, D. Filges, R.D. Neef, G. Sterzenbach, Ch. Reul, T.W. Armstrong, B.L. Colborn, B. Anders, H. Brückmann, KFA-Report Jül-2203, (1988)
2. V.S. Barashenkov, et al., Atomnaya Energiya 32, 217 (1972)
3. E.H. Auerbach, Computer Physics Comm. 15, 165 (1978)
4. J.W. Wachter, W.A. Gibson, W.R. Burrus, Phys. Rev. C6, 1496 (1972)
5. S.D. Howe, Ph.D. thesis, Kansas State University, 1980
6. R. Madey, F.M. Waterman, Phys. Rev. C8, 2414 (1973)
7. C.G. Cassapakis, H.C. Bryant, B.D. Dieterle et al., Phys. Lett. 64B, 35 (1976)
8. B.E. Bonner, J.E. Simmons, C.R. Newsom et al., Phys, Rev. C18, 1418 (1978)
9. M.M. Meier, D.B. Holtkamp, G.L. Morgan, H. Robinson, G.J. Russell, E.R. Whitaker, W. Amian, N. Paul, Radiations Effects 96, 73 (1986)
10. M.M.Meier, D.A. Clark, C.A. Goulding, J.B. McClelland, G.L. Morgan, C.E. Moss, Nucl. Sci. and Eng. 102, 310 (1989)
11. M.M. Meier, W.B. Amian, C.A. Goulding, G.L. Morgan, C.E. Moss, Report LA-11518-MS (1989)
12. P. Cloth, P. Dragovitsch, D. Filges, Ch. Reul, W.B. Amian, M.M. Meier, Report Jül-2295 (1989)
13. T.W. Armstrong, K.C. Chandler, Nucl. Sci. Eng. 49, 110 (1972)
14. P. Cloth, D. Filges, G. Sterzenbach, T.W. Armstrong, B.L. Colborn, KFA-Report Jül-Spez-196 (1983)
15. W. Scobel et al., Phys. Rev. C41, 2010 (1990)
16. S. Cierjacks et al., Phys. Rev. C36, 1976 (1987)
17. R.A. Cecil et al., NIM 161, 439 (1979)

PROTON-INDUCED SPALLATION BETWEEN 600 AND 2600 MeV

M. Lüpke[1], R. Michel[1], B. Dittrich[2,4], U. Herpers[2],
P. Dragovitsch[3], D. Filges[3], H.J. Hofmann[4], W. Wölfli[4]

1 Zentraleinrichtung für Strahlenschutz, Universität Hannover
Am Kleinen Felde 30, D 3000 Hannover 1, F.R.G.
2 Abteilung Nuklearchemie, Universität zu Köln
Zülpicher Str. 47, D 5000 Köln 1, F.R.G.
3 Institut für Kernphysik, KFA Jülich
Postfach 1913, D 5170 Jülich, F.R.G.
4 Institut für Mittelenergiephysik, ETH Hönggerberg HPK G7
CH 8093 Zürich, Switzerland

Abstract: In the course of a systematic investigation of proton-induced reactions fifteen target elements ranging from O to Au were irradiated with protons having energies of 600, 800, 1200, 1600 and 2600 MeV. Thin-target cross sections for the production of residual radionuclides were measured by X- and gamma-spectrometry and by accelerator mass spectrometry (AMS). In this work, we deal exemplarily with the production of ^{10}Be, ^{22}Na and ^{26}Al from target elements ($25 \leq Z \leq 28$). These data are essential for an understanding of the production of cosmogenic nuclides by the interaction of galactic cosmic ray protons with matter. The new cross sections are compared with literature data and discussed in the context of Monte Carlo calculations describing the production cross sections by an intranuclear cascade evaporation model.

Introduction

Thin-target cross sections for the production of residual nuclides by proton-induced reactions at intermediate energies (50 MeV – 10 GeV) are key quantities for a large number of applications, covering astro- and cosmophysics [1,2], space research and technology, radiation protection in space and at terrestrial accelerators as well as design and operation of accelerators, detector systems and spallation neutron sources. However, the existing experimental data base is neither comprehensive nor reliable, e.g. [2].

Therefore, integral production cross sections were measured in a systematic study of proton-induced reactions with energies between 600 and 2600 MeV, investigating 15 target elements ($8 \leq Z \leq 79$). This study extends our earlier work [3] and establishes a consistent data base of intermediate energy cross sections for a quantitative interpretation of the observed abundances of cosmogenic nuclides in terrestrial and extraterrestrial matter, which, moreover, allows for systematic tests of nuclear reaction models. In this work, we deal with a subset of experimental results from these investigations, namely production cross sections for ^{10}Be, ^{22}Na and ^{26}Al from target elements $25 \leq Z \leq 28$. For other published results see [4-7] and references therein.

Experimental

The elements O, Mg, Al, Si, Ti, V, Mn, Fe, Co, Ni, Cu, Zr, Rh, Ba, and Au were irradiated as high-purity foils or chemical compounds with protons of 600 MeV at CERN/Geneve, 800 MeV at LANL/Los Alamos and 1200, 1600 and 2600 MeV at LNS/Saclay. Small stacks of foils were used as targets which degraded the proton-energy by less than 15 MeV. Residual radioactive nuclides were measured by X- and gamma-spectrometry and by ac-celerator mass spectrometry. The proton-fluxes were determined via the reaction ^{27}Al(p,3p3n)^{22}Na adopting cross sections of 16.0 mb, 15.5 mb, 14.4 mb, 13.2 mb and 11.6 mb for 600, 800, 1200, 1600, and 2600 MeV, respectively, in accordance to the recommendations by Tobailem and de Lassus St. Genies [8]. Details of the experimental technique, of the chemical procedures and AMS measurements (including the description of standards used) and of the experimental uncertainties may be found elsewhere [3,4,7].

Results and Discussion

Up to now, cross sections have been determined for more than 200 reactions. Combining these results with those from earlier work of our group [3] and references therein, there now exists a consistent set of excitation functions from threshold energies up to 2600 MeV for more than 120 cosmophysically relevant reactions for the production of radionuclides from target elements ($Z \leq 29$). Further data from still ongoing measurements will be finalized in due course.

The cross sections for the production of ^{22}Na from Mn, Fe and Co (Fig. 1) are typical for the status of spallation cross sections in general. There is a number of investigations dealing with ^{22}Na from Fe [9-17], most data clustering around 600 MeV, but no earlier report on Mn and Co above 400 MeV. ^{22}Na from Fe is among the best investigated spallation reactions. Our data fit excellently to those by Raisbeck and Yiou [15] and describe the transition of the excitation function from a steep increase below 2 GeV into a constant plateau up to 30 GeV [15-17]. But there are still some discrepancies in this excitation function. The data at 585, 590 and 600 MeV [12, 13,4] deviate from each other by up to 40 %. The measurements at 660 MeV [11] and 730 MeV [14] are within errors identical to our measurement at 600

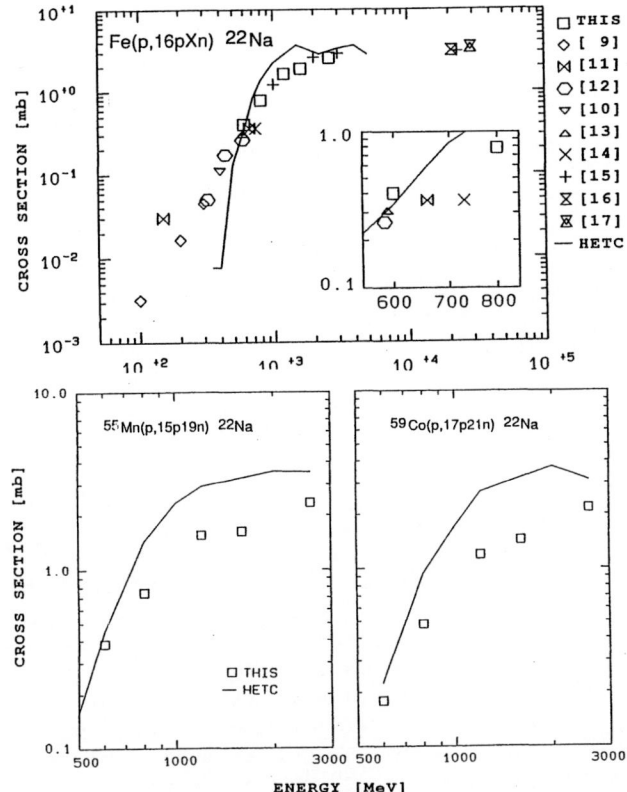

Fig. 1. Experimental cross sections for the reactions ^{55}Mn(p,15p19n)^{22}Na, Fe(p,16pXn)^{22}Na and ^{59}Co(p,17p21n)^{22}Na compared with HETC calculations. Errors are plotted only in this work if they exceed the symbol size.

MeV, though the excitation function increases by a factor of two between 600 and 800 MeV.

As discussed elsewhere in detail [2-4], such discrepancies are fairly common for spallation reactions and for many reactions different measurements deviate by up to an order of magnitude. Even among recent investigations inacceptable differences are found. As an example, we compare our data at 600 MeV with those reported in 1990 by Webber et al. [18]. These authors derived cross sections for the production of stable and radioactive residual nuclides by irradiation experiments in which targets of hydrogen and hydrocarbons were irradiated by heavy ions. Product nuclides were directly measured online using semiconductor telescopes. There are 25 reactions for which 600 MeV cross sections were determined in our work [4,5] and in that by Webber et al. [18], thus allowing for a direct comparison. The ratios of the cross sections from our work to those given by Webber et al. [18] range from 0.22 to 2.1 (Fig. 2), thus demonstrating severe discrepancies. There seems to be a tendency of the ratios to increase with decreasing relative mass loss, thus pointing to systematic problems.

Based on their measured cross sections Webber et al. [18] proposed a semiempirical formula, for which they claimed an accuracy "better than 10 % in most cases". From the above comparison we conclude that such a claim of accuracy surely cannot be made until the accuracy of the underlying data base has been proven. As discussed

elsewhere in detail [2], semiempirical formulas describe the measured data just within a factor of ten, in good cases within a factor of 2. Since, however, semiempirical formulas were derived in the past from highly unreliable data sets, one can hardly decide whether the insufficiencies of these formulas are due to the uncertainties of the underlying data bases or due to a general neglect of individual nuclear properties. In any case, they presently can scarcely be used if reliable cross sections for applications are required.

Our new and consistent data set is used to investigate the capabilities of an intranuclear cascade evaporation model to calculate cross sections for the production of residual nuclides. We have calculated production cross sections for the target elements O, Mg, Al, Si, Ca, Ti, V, Mn, Fe, and Co for energies between 200 MeV and 5 GeV by Monte Carlo techniques using the HETC module [19] of the HERMES code system [20]. For the production of ^{22}Na from Mn, Fe and Co the results are shown in Fig. 1. A general comparison of the measured and calculated data [2,4] demonstrates that the HETC calculations are capable to reproduce the measured ones within a factor of 3 on the average. The calculations are, however, not applicable to products predominantly produced by fragmentation as ^{7}Be and ^{10}Be, since this formation mode is not considered in HETC. Also in the case of ^{22}Na from Fe (Fig. 1) this problem shows up for energies below 600 MeV, where the theoretical data decrease much steeper with decreasing energy than the experimental ones. Considering, however, the discrepancies in the experimental data between 100 and 200 MeV, surely more measurements are necessary to confirm production of ^{22}Na at such low energies.

For most longlived and stable cosmogenic nuclides the availability of experimental cross sections is much worse than for ^{22}Na. This is demonstrated in Fig. 3 for the production of ^{10}Be from Fe and Ni and for ^{26}Al from Ni. For further data for these nuclides from other target elements see [6,7]. For ^{10}Be from Fe there is one

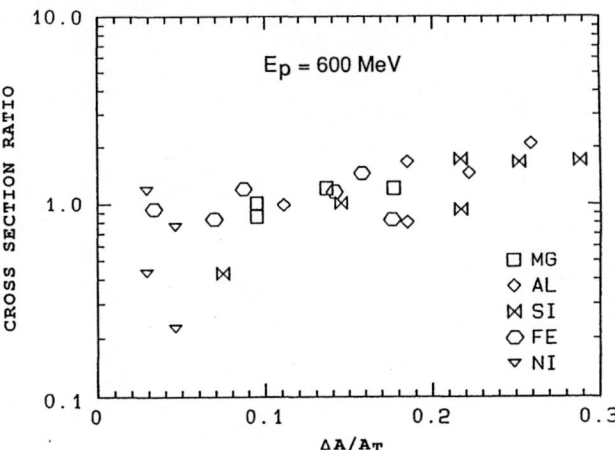

Fig. 2: Ratios of experimental production cross sections from this work and from the work of Webber et al. [18] for the target elements Al, Si, Fe and Ni at E_p = 600 MeV.

704

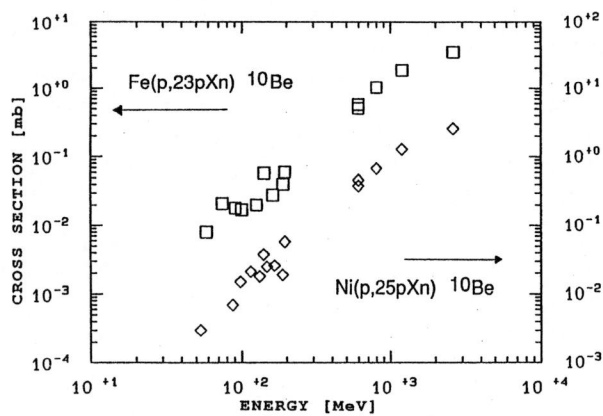

Fig. 3: Experimental cross sections for the reactions Fe(p,23pXn)^{10}Be and Ni(p,25pXn)^{10}Be.

earlier measurement at 730 MeV [14], which agrees well with our new data. Our data describe the excitation functions of ^{10}Be from Fe and Ni between 50 MeV and 2.6 GeV. For the low-energy data there is still some scatter in the experimental data, the cross sections being between 0.001 an 0.1 mb. It has to be investigated whether there are really structures in these excitation functions between 150 and 200 MeV. For ^{26}Al from Ni our cross sections are lower than those derived earlier from activity measurements by Rayudu [21,22] by factors between 1.5 to 3.

The cross sections measured in the course of our investigations will provide the largest consistent set of spallation cross sections measured up to now. It will allow for an improvement of model describing the production of cosmogenic nuclides in terrestrial and extraterrestrial matter, for detailed analyses of semiempirical formulas and of nuclear reaction theories. However, further investigations will be necessary to satisfy all needs for the various applications of intermediate energy nuclear data.

Acknowledgement The authors are grateful to the authorities of CERN/Geneve, LANL/Los Alamos, and LNS/Saclay for the beam-time and to the staffs of the accelerators for their good cooperation. The silicon targets were kindly provided by Wacker Chemitronic, Burghausen. This work was supported by the Deutsche Forschungsgemeinschaft and in part by the Swiss National Science Foundation.

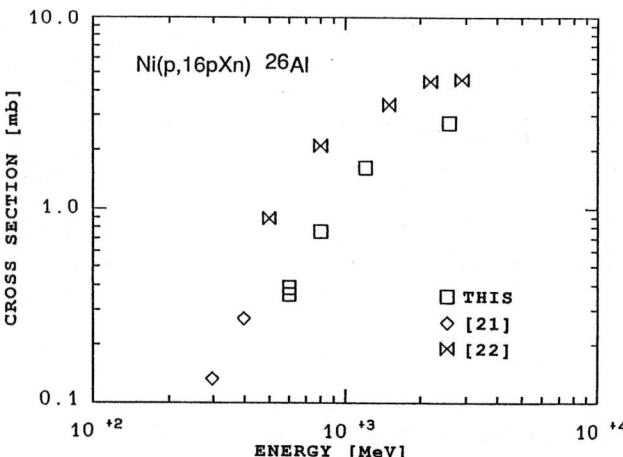

Fig. 4: Experimental cross sections for the reaction Ni(p,16pXn)^{26}Al.

References

1. S. Vogt, G.F. Herzog and R.C. Reedy, Rev. Geophys. 28, 253 (1990)
2. R. Michel, Medium Energy Nuclear Data to Understand the Interactions of Cosmic Ray Particles with Matter, in: N.P. Kocherov (ed.) IAEA Advisory Group Meeting on Intermediate Energy Nuclear Data for Applications, Oct. 9 – 12, 1990, Wien, in press
3. R. Michel and R. Stück, J. Geophys. Res. (Suppl.) 89 B, 673 (1984)
4. R. Michel, B. Dittrich, U. Herpers, T.Schiffmann, P. Cloth, P. Dragovitsch and D. Filges, Analyst 114, 287 (1989)
5. R. Michel, F. Peiffer, S. Theis, F. Begemann, H. Weber, P. Signer, R. Wieler, P. Cloth, P. Dragovitsch, D. Filges and P. Englert, Nucl. Instr. Meth. Phys. Res. B42, 76 (1989)
6. B. Dittrich, U. Herpers, M. Lüpke, R. Michel, H.J. Hofmann and W. Wölfli, Radiochimica Acta 50, 11 (1990)
7. B. Dittrich, U. Herpers, H.J. Hofmann, W. Woelfli, R. Bodemann, M. Luepke, R. Michel, P. Dragovitsch and D. Filges, Nucl. Instr. Meth. Phys. Res. B52, 588 (1990)
8. J. Tobailem and C.H. de Lassus St. Genies, Additif No. 2 a la CEA-N-1466(1) (1975), CEA-N-1466(5) (1981)
9. R.G. Korteling and A.A. Caretto, Phys. Rev. C1, 1960 (1970)
10. R. G. Korteling and A.A. Caretto, J. Inorg. Nucl. Chem. 29, 2863 (1967)
11. A.K. Lavrukhina, L.P. Moskaleva, V.V. Malyshev, L.M. Satarova, Sov. Phys. JETP 17, 960 (1963)
12. R.L. Brodzinski, L.A. Rancitelli, J.A. Cooper and N.A. Wogman, Phys. Rev. C4, 1257 (1971)
13. C.J. Orth, H.A. O'Brien, M.E. Schillaci, B.J. Dropesky, J.E. Cline, E.B. Nieschmidt and R.L. Brodzinski, J. Inorg. Nucl. Chem. 38, 13 (1976)
14. M. Honda, D. Lal and Phys. Rev. 118, 1618 (1960)
15. G.M. Raisbeck and F. Yiou, Proc. 14th Int. Cosmic Ray Conf., Munich, 2, 495 (1975)
16. C. Perron, Phys. Rev. C 14, 1108 (1976)
17. N.T. Porile and S. Tanaka, Phys. Rev. B135, 122 (1964)
18. W.R. Webber, J.C. Kish and D.A. Schrier, (1990) Phys. Rev. C 41, 547, 520, 533, and 566 (1990)
19. K.C. Chandler and T.W. Armstrong, ORNL-4744 (1972)
20. P. Cloth, D. Filges, R.D. Neef, G. Sterzenbach, Ch. Reul, T.W. Armstrong, B.L. Colborn, B. Anders and H. Brueckmann, Juel-2203 (1988)
21. G.V.S. Rayudu, Can. J. Chem. 42, 1149 (1964)
22. G.V.S. Rayudu, J. Inorg. Nucl. Chem. 30, 2311 (1968)

CROSS SECTIONS OF p-INDUCED REACTIONS UP TO 100 MeV
FOR THE INTERPRETATION OF SOLAR COSMIC RAY PRODUCED NUCLIDES

R. Bodemann[1], R. Michel[1], B. Dittrich[2,3], U. Herpers[2], R. Rösel[2],
H.J. Hofmann[3], W. Wölfli[3], B. Holmqvist[4], H. Condé[5], P. Malmborg[6]

1 Zentraleinrichtung für Strahlenschutz, Universität Hannover
Am Kleinen Felde 30, D 3000 Hannover 1, F.R.G.
2 Abteilung Nuklearchemie, Universität zu Köln
Zülpicher Str. 47, D 5000 Köln 1, F.R.G.
3 Institut für Mittelenergiephysik, ETH Hönggerberg HPK G7
CH 8093 Zürich, Switzerland
4 The Studsvik Neutron Research Laboratory, University of Uppsala
S-611 82 Studsvik, Sweden
5 Department of Neutron Research, University of Uppsala
S-751 21 Uppsala, P.O.B. 531, Sweden
6 Department of Physical Biology, University of Uppsala
S-751 21 Uppsala, P.O.B. 535, Sweden

Abstract: Integral excitation functions for the production of residual nuclides by proton-induced reactions are the basic data for an accurate modelling of the interactions of solar cosmic ray particles with terrestrial and extraterrestrial matter. In order to improve the still insufficient data base, cross sections were measured for for nuclear reactions of protons up to 100 MeV on 18 elements. Using the stacked foil technique, irradiation experiments were performed at an external beam of the cyclotron at the Svedberg Laboratory/Uppsala. Radioactive residual nuclide production was measured by X- and gamma-spectrometry and by accelerator mass spectrometry for more than 120 target/product combinations. Here, cross sections for the production of ^{7}Be from C, N, O, and Si and of ^{22}Na and ^{26}Al from Mg and Si are presented and discussed with respect to those from earlier work.

Introduction

Cosmogenic nuclides, produced by the interactions of solar cosmic ray (SCR) particles with terrestrial and extraterrestrial matter, are (stable and radioactive) natural tracers, which allow to investigate geo- and cosmochemical processes and to describe the cosmic ray exposure history of lunar samples, meteorites and cosmic dust. They provide the only tool by which SCR spectra and intensities can be determined over the entire history of the solar system.

Integral excitation functions for the production of residual nuclides by p-induced reactions are the basic data for an accurate modelling of SCR interactions with matter. Moreover, they are essential for accelerator technology as well as for design and optimization of radionuclide production and are useful to test nuclear reaction theories. The cross section data base for the production of geo- and cosmochemically relevant nuclides obtained in the past was neither comprehensive nor reliable.

Here, we report on experimental results from a systematic investigation of cosmochemically relevant nuclear reactions, which extends earlier work from our group [1,2]. The new cross sections later on will be used in model calculations describing the production of cosmogenic nuclides in terrestrial and extraterrestrial matter and for tests of nuclear reaction theories in the transition between preequilibrium and spallation.

Experimental

Eighteen elements (C, N as Si_3N_4, O as SiO_2, Mg, Al, Si, Ti, V, Mn as Mn/Ni-alloy, Fe, Co, Ni, Cu, Zr, Nb, Rh, Ba as a Ba-containing glass, and Au) were irradiated with 94.4 MeV and 98.9 MeV protons at an external beam of the cyclotron of the Svedberg Laboratory/Uppsala. Cross sections were determined using the stacked foil technique. The p-energies in the different target foils were calculated according to Andersen and Ziegler [3]. These calculations are in excellent agreement with those according to Williamson et al. [4] used in our earlier work [1,2]. The reaction ^{27}Al(p,3p3n)^{22}Na was used for beam monitoring, adopting cross sections of 19.3 mb and 18.9 mb at 94.4 and 98.9 MeV, respectively, according to Tobailem and de Lassus St. Genies [5]. Residual nuclides were measured by X- and gamma-spectrometry and by accelerator mass spectrometry (AMS). Details of the experimental technique, of chemical procedures and AMS measurements (including the standards used) and of the experimental uncertainties may be found elsewhere [1,6].

Results and Discussion

Up to now, cross sections were measured for more than 120 different reactions. Measurements of long-lived radionuclides by AMS are still going on. In order to check the quality of our experimental procedure some target elements ($22 \leq Z \leq 28$) were included in the new experiments,

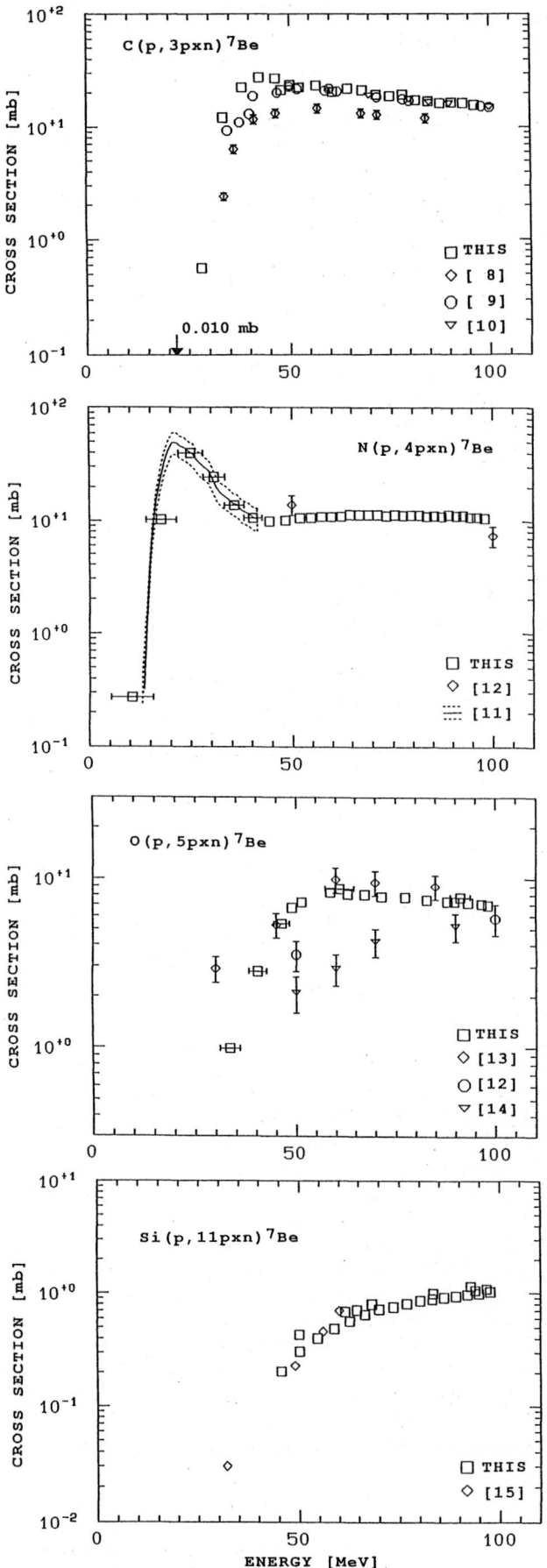

Fig. 1. Experimental cross sections for the production of ^{7}Be from C, N, O, and Si. Error bars are plotted only if they exceed symbol sizes.

which were formerly irradiated with protons having initial energies of 45 MeV at Jülich, 80 MeV at Louvain La Neuve and 168 and 201 MeV at IPN Orsay [1,2,7]. Excellent consistency was obtained between the new cross sections and our earlier measurements. In this work, we deal exemplarily with the status of experimental excitation functions for the production of some radionuclides relevant for SCR interactions with terrestrial and extraterrestrial matter: ^{7}Be from C, N, O, Si and ^{22}Na and ^{26}Al from Mg and Si.

For the production of ^{7}Be from carbon (Fig. 1) our measurements rule out earlier data by Dickson and Randle [8], while the measurements by Brun et al. [9] and by Gauvin et al. [10] are in excellent agreement with our data for energies above 50 MeV. Between threshold and maximum, however, our data are also significantly higher than those reported Brun et al. [9].

For ^{7}Be from nitrogen (Fig. 1) our data are in excellent agreement with those reported up to 42 MeV by Epherre and Seide [11]. For higher energies, there are two earlier measurements [12] at 50 MeV and 100 MeV, which are only in moderate agreement with our data.

In case of ^{7}Be from oxygen (Fig. 1) earlier data [12,13,14] show discrepancies of up to a factor of three. Our new measurements describe the excitation function in detail and resolve the existing discrepancies. Again some earlier data [12,14] must completely be ruled out according to our results. The data by Lafleur et al. [13] are in moderate agreement with our data with the exception of their cross section at 30 MeV, which is too high by more than a factor of 3.

For ^{7}Be from Si (Fig. 1) there is just one earlier report up to 60 MeV [15] which fits well to our measurements.

Our data for ^{22}Na from Mg and Si (Fig. 2) give fully consistent excitation functions from threshold up to 100 MeV together with those reported by Furukawa and coworkers [16,17]. In case of Mg, there are further data between 10 and 32 MeV by Batzel and Coleman [18] which show just small deviations from this excitation functions. The data by Bartell and Softky [19], however, have to be ruled out. At 100 MeV earlier measurements by Korteling and Caretto [20] and by Bimbot and Gauvin [12] agree with our new data, while one measurement at 50 MeV by Bimbot and Gauvin [12] deviates from ours by nearly a factor of two.

Also in case of Si there is a number of other measurements [12,15,21] which partially [12,21] are not or only in marginal agreement with our data.

The excitation functions for the production of ^{26}Al from Mg and Si (Fig. 2) were measured earlier up to 50 MeV by low-level counting after chemical separation [16]. For both reactions our cross sections are higher than the earlier ones and also the general shapes of the excitation functions are in disagreement. In view of the outstanding importance of cosmogenic ^{26}Al, these discrepancies have to by resolved by further investigations.

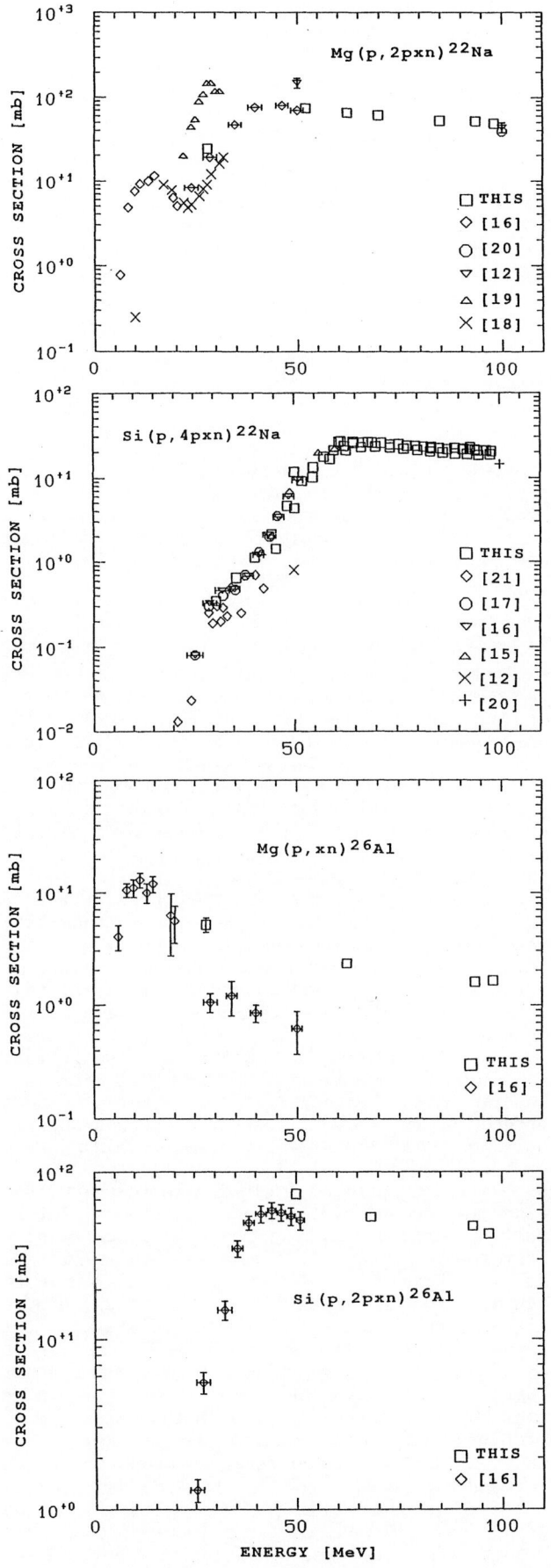

Fig. 2. Experimental cross sections for the production of [22]Na and [26]Al from Mg and Si. Error bars are plotted only if they exceed symbol sizes.

Acknowledgement: The authors are grateful to the authorities of the Svedberg Laboratory/University Uppsala for the beam-time and to the staff of the accelerator for the good cooperation. The silicon targets were kindly provided by Wacker Chemitronic, Burghausen. This work was supported by the Deutsche Forschungsgemeinschaft, the Swedish Natural Science Research Council and the Swiss National Science Foundation.

References

1. R. Michel and R. Stück, J. Geophys. Res. (Suppl.) 89 B, 673 (1984)

2. R. Michel, F. Peiffer and R. Stueck, Nucl. Phys. A 441 (1985) 617

3. H.H. Andersen and J.F. Ziegler, Hydrogen. Stopping Powers and Ranges in All Elements, Pergamon Press (1977)

4. C.F. Williamson, J.P. Boyot and J. Picard, CEA-R 3042 (1966)

5. J. Tobailem and C.H. de Lassus St. Genies, Additif No. 2 a la CEA-N-1466(1) (1975), CEA-N-1466(5) (1981)

6. B. Dittrich, U. Herpers, H.J. Hofmann, W. Woelfli, R. Bodemann, M. Luepke, R. Michel, P. Dragovitsch and D. Filges, Nucl. Instr. Meth. Phys. Res. B52, 588 (1990)

7. R. Michel, G. Brinkmann, H. Weigel and W. Herr, Nucl. Phys. A 322, 40 (1979)

8. J.M. Dickson and T.C. Randle, Proc. Phys. Soc. London A64, 902 (1951)

9. C. Brun, M. Lefort and X. Tarrago, J. Phys. et Radium 23, 371 (1962)

10. H. Gauvin, M. Lefort and X. Tarrago, Nucl. Phys. 39, 447 (1962)

11. M. Epherre and C. Seide, Phys. Rev. C3, 2167 (1971), errata ibid C4, 1494 (1971)

12. R. Bimbot and H. Gauvin, Compt. Rend. Acad. Sci. Paris B273, 1054 (1971)

13. M.S. Lafleur, N.T. Porile and L. Yaffe, Can. J. Chem. 44, 2749 (1966)

14. G. Albouy, J.P. Cohen, M. Gusakow, N. Poffe, H. Sergolle and L. Valentin, Phys. Letters 2, 306 (1962)

15. D.W. Sheffey, I.R. Williams and C.B. Fulmer, Phys. Rev. 172, 1094 (1968)

16. M. Furukawa, K. Shizuri, K.Komura, K. Sakamoto and S. Tanaka, Nucl. Phys. A174 (1971) 539

17. M. Furukawa, S. Kume and M. Ogawa, Nucl. Phys. 69, 362 (1965)

18. R.E. Batzel and G.H. Coleman, Phys. Rev. 93, 280 (1954)

19. F.O. Bartell and S. Softky, Phys. Rev. 84, 463 (1951)

20. R.G. Korteling and A.A. Caretto, Phys. Rev. C1, 1960 (1970)

21. J.R. Walton, D. Heymann, A. Yaniv, D. Edgerly and M.W. Rowe, J. Geophys. Res. 81, 5689 (1976)

MEASUREMENT AND ANALYSIS OF INTEGRAL EXCITATION FUNCTIONS FOR ^{nat}Cu+p UP TO 200 MeV: INFLUENCE OF THE $f_{7/2}$ SHELL CLOSURE

S.J. Mills, G.F. Steyn and F.M. Nortier

National Accelerator Centre, P.O. Box 72, Faure, 7131, South Africa

Abstract: Cross-sections were measured up to 200 MeV for the production of 24 radionuclides in the bombardment of natural copper with protons. The experimental results were compared with theoretical calculations performed within the framework of the geometry-dependent hybrid model in combination with the Weisskopf-Ewing evaporation theory. Surprisingly good overall agreement for such a broad range of nuclides in this mass region was obtained.

(Natural copper target, proton bombardment, 10-200 MeV, stacked-foil technique, cross-section, excitation function, geometry-dependent hybrid model, ALICE/85/300 code evaluation, shell structure effects)

Introduction

In recent years nuclear models which combine preequilibrium particle emission with equilibrium decay of the remaining compound nucleus have been increasingly successful in reproducing a broad range of experimental results. Among these, the hybrid/geometry-dependent-hybrid model in combination with the Weisskopf-Ewing evaporation theory, as embodied in successive versions of the computer code ALICE [1], has proved to be a particularly convenient and useful tool for predicting excitation functions of a wide variety of nuclear reactions [2,3]. On the other hand, some puzzling total failures of these models in certain "exceptional" cases still remain to be resolved (see for instance refs. [4-6]). One possible explanation for such failures - which would, of course, not affect the inherent validity of the relevant model, as such - is that they are due to the use of incorrect calculation parameters for certain nuclei involved in the decay process, especially those in the last stages of the evaporation cascade - i.e. those at relatively low excitation energies. They could, for instance, arise from the use of average (global) parameters or parameters which have been improperly adjusted to take into account properties peculiar to specific nuclei in the cascade, such as is well known to be the case for some discrepancies observed in the yields of nuclides with closed or nearly-closed nucleon shells.

In this work cross-sections were measured up to 200 MeV for the production of 24 radionuclides, ranging from ^{44m}Sc to ^{65}Zn, in the bombardment of ^{nat}Cu with protons. Although our primary aim was the measurement of suitable monitor excitation functions for proton beam-flux determination [7], the data obtained also enabled a stringent test of the above-mentioned model to be performed and the influence of the $f_{7/2}$ proton and/or neutron shell closures to be thoroughly investigated.

Experimental Procedure

The well-known stacked-foil technique was employed. High-purity Cu foils (99.9%, Koch Chemicals), 50 μm thick, were included in various composite foil stacks irradiated with the external proton beam of the separated-sector cyclotron facility of the National Accelerator Centre. Proton beam energies of 200 ± 0.5, 120 ± 0.4, 100 ± 0.3, 66 ± 0.3 and 40 ± 0.2 MeV were used. Further details of the experimental as well as the data-analysis procedures followed are indentical to those given in ref. [8].

Theoretical Calculations

A priori theoretical calculations of the excitation functions for the production of the various radionuclides of interest were performed by means of the computer code ALICE/85/300 [1]. All these calculations were done within the framework of the geometry-dependent hybrid model for the preequilibrium emission of neutrons and protons, in combination with the Weisskopf-Ewing evaporation theory for the subsequent equilibrium emission of neutrons, protons, deuterons and α-particles. Default initial exciton configurations were used for the preequilibrium calculations, and the intra-nuclear transition rates were based on nucleon mean free paths calculated from free nucleon-nucleon scattering cross-sections. Inverse reaction cross-sections were generated by the optical-model subroutine of the code.

In the calculation of the level densities, the default value of A/9 was used for the level-density parameter, and both methods offered by the code to take into account nucleon pairing and shell structure effects were explored:

Method 1: Experimental binding energies [9] are read in, when available; otherwise internally-generated binding energies, calculated from the Myers and Swiatecki mass formula, are used. Level densities are calculated with respect to an effective (fictitious) ground state which is shifted in energy relative to the actual ground state, the shift being the result of a pairing energy shift as well as a shell correction shift. These shifts are calculated from the corresponding terms in the mass formula of Myers and Swiatecki, and the pairing energy shift is taken as zero for even-even nuclei, $-\delta$ for odd-even nuclei and -2δ for odd-odd nuclei, where δ is the pairing energy term, given by $\delta = 11/\sqrt{A}$. The calculations incorporating this method were therefore basically identical to those performed in a number of previous evaluations of the code [7], in all of which reasonably good overall agreement - to within a factor of about 2 - was, in general, obtained between theory and experiment.

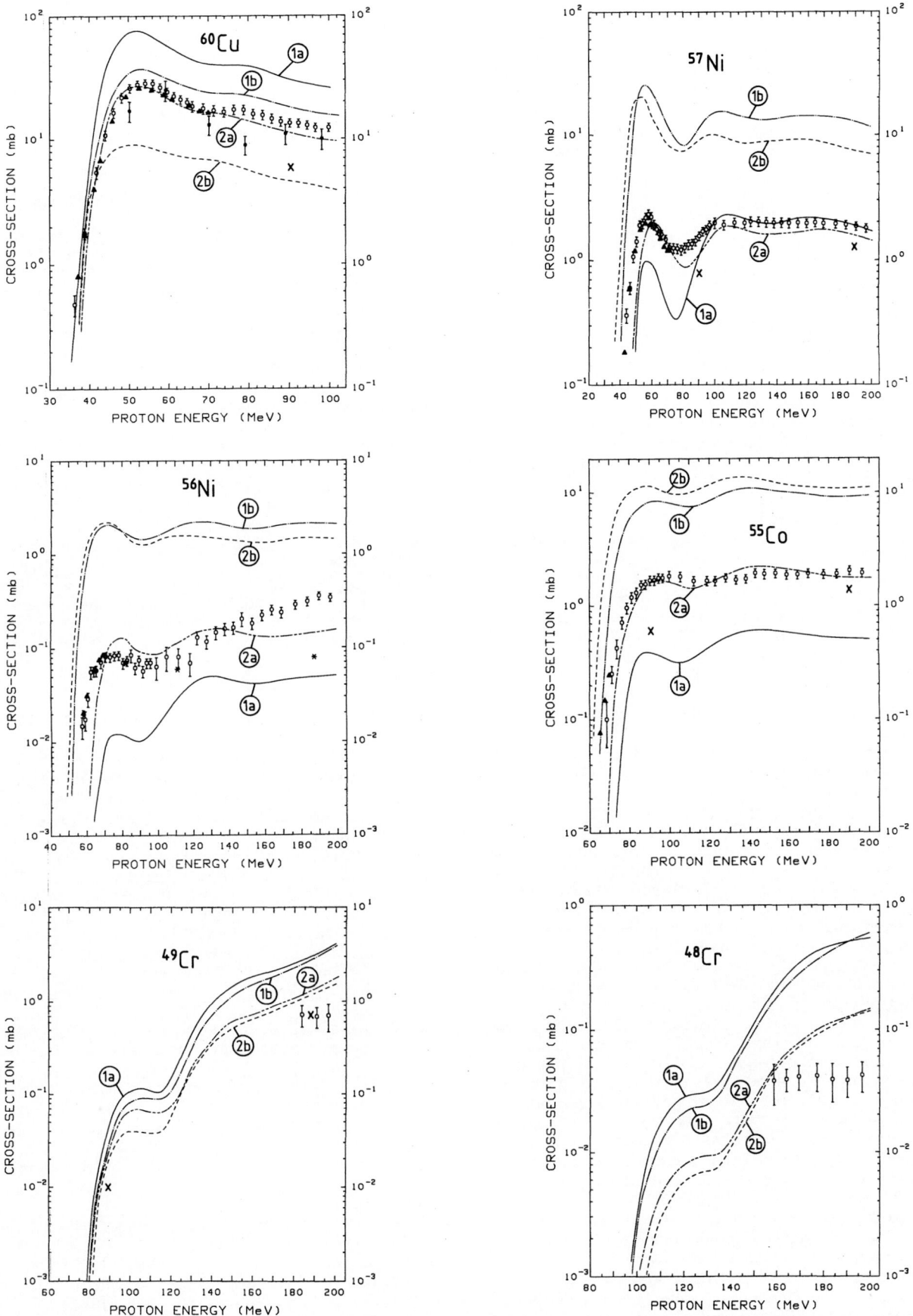

Fig. 1. Experimental and theoretical integral excitation functions for some radionuclides produced in the bombardment of ^{nat}Cu with protons. (Experimental points: O, this work; ▲, ref. [10]; ●, ref. [11]; *, ref. [12]; x, ref. [14]. Theoretical curves: 1(a), Method 1 with pairing plus shell corrections; 1(b), Method 1 with pairing corrections only; 2(a), Method 2 with pairing plus shell corrections; 2(b), Method 2 with pairing corrections only.)

710

Method 2: All masses, and thus binding energies, are calculated by means of the mass formula of Myers and Swiatecki, but with zero pairing-energy and shell-correction terms. This should amount to roughly the same effect, in priciple, as that of Method 1. In this case, however, the pairing energy shift is taken as zero for odd-even nuclei, δ for even-even nuclei and $-\delta$ for odd-odd nuclei.

Results and Discussion

Cross-section data in the energy range 10 to 200 MeV were obtained for the following 24 radionuclides: ^{62,63,65}Zn, ^{60,61,64}Cu, ^{56,57}Ni, ^{55,56,57,58,60}Co, ^{52g,59}Fe, ^{52g,54,56}Mn, ^{48,49,51}Cr, ^{48}V and ^{44m,46}Sc. In the majority of cases the excitation function for the production of the particular radionuclide was measured up to 200 MeV; for ^{63}Zn and ^{60,64}Cu, however, it only extends up to 100 MeV, while only partial excitation functions, in the high-energy region, were obtained for ^{52g}Fe, ^{48,49}Cr and ^{44m,46}Sc. The cross-sections measured for ^{62,63,65}Zn, ^{64}Cu, ^{58,60}Co, ^{54}Mn and ^{44m,46}Sc obviously are for their direct production only, while those for ^{56,57}Co and ^{48}V could be corrected for their precursor decay. In all the other cases, the measured cross-sections are cumulative cross-sections for the production of the particular radionuclide.

For many of the above-mentioned radionuclides extensive experimental data are available at the lower proton energies. In the most comprehensive previous investigation of ^{nat}Cu+p, Grütter [10] measured the excitation functions of 17 of these radionuclides up to 70 MeV, and compared his results with all relevant data published since 1970. Later, 8 of these were remeasured and extended up to 100 MeV by Aleksandrov et al. [11]. The present results were found to be in excellent agreement with those of Grütter, while the agreement with the data of Aleksandrov et al. is less satisfactory (see for instance Fig. 1). Especially in the cases of ^{63}Zn and ^{59}Fe large discrepancies between our results and those of the last-mentioned authors have been observed: The ^{63}Zn excitation function measured by them does not show any sign of the relatively deep minimum observed by us (and Grütter) at ~24 MeV, while no indication of the sharp peak at ~80 MeV in their ^{59}Fe excitation function could be found by us - in fact, at this energy a cross-section of nearly an order of magnitude lower was obtained, in very good agreement with the data point of Heydegger et al. [12] at 82 MeV. For comparison purposes above 100 MeV, only the data of Albouy et al. [13] for ^{65}Zn between 45 and 142 MeV, and the isolated data points measured for a number of radionuclides by Coleman and Tewes [14] at 90 and 190 MeV and by Heydegger et al. [12] at 82, 111 and 187 MeV could be found in the literature. Rather poor agreement with the measurements of Albouy et al. was obtained, their cross-sections being consistently higher than ours by a factor of more than two, while the agreement with the individual data points of both Coleman and Tewes and Heydegger et al. varies from excellent to very poor (see for instance Fig. 1).

As regards the comparison of the calculated and experimental excitation functions, exceptionally good overall agreement for such a large number of radionuclides over a very broad mass range in a region around the closures of both the $f_{7/2}$ proton and neutron shells was obtained with Method 2 - in most cases well within a factor of 2. The results obtained with Method 1 are less satisfactory. In fact, in 14 of the cases studied clearly better agreement was obtained with Method 2, while the opposite is true in only 3 cases. Especially striking is the improvement achieved by Method 2 over Method 1 in the cases of the doubly-magic nuclide ^{56}Ni and its immediate neighbours ^{57}Ni and ^{55}Co (see Fig. 1). In these cases the improvement obviously is primarily due to a better shell effect correction in Method 2. Remarkably, however, a significant improvement is also obtained in cases which would be classified as genuine "exceptional" cases on the basis of the results obtained with Method 1, such as ^{48}Cr and ^{49}Cr - i.e. cases in which the difference between the calculated and experimental excitation functions can not possibly be ascribed to an incorrect shell effect description (note that in ^{48}Cr both the $f_{7/2}$ proton and neutron shells are exactly half-filled) - and, to a lesser extent, ^{60}Cu (see Fig. 1). Also noteworthy is the fact that all three these nuclides have already previously been found in other types of reactions to be "exceptional" cases - ^{48}Cr in Mn+p [5,6] and Ni+p [6], ^{49}Cr in Mn+p [6] and Ni+p [6] and ^{60}Cu in Co+α [4] - indicating that the effect is related to specific nuclides rather than to specific reaction types.

REFERENCES

1. M. Blann, Code ALICE/85/300, UCID-20169, Lawrence Livermore National Laboratory (1984), and references therein
2. R. Nowotny and M. Uhl, in Handbook on Nuclear Activation Data (Technical Report Series No. 273), p. 441, IAEA, Vienna (1987)
3. M. Blann, in Proc. IAEA Consultants' Meeting on Data Requirements for Medical Radio-isotope Production, April 1987, Tokyo, Japan (Edited by K. Okamoto), INDC(NDS)-195/GZ, p. 115, IAEA, Vienna (1988)
4. E. Gadioli, E. Gadioli-Erba, J. Asher and D.J. Parker: Z. Phys. A 317, 155 (1984)
5. R. Michel, F. Peiffer and R. Stück: Nucl. Phys. A441, 617 (1985), and references therein
6. G.F. Steyn, S.J. Mills and F.M. Nortier, in Progress Report for South Africa (Edited by D.W. Mingay and W.R. McMurray), INDC(SAF)-010/G, p. 10, IAEA, Vienna (1989)
7. G.F. Steyn, S.J. Mills, F.M. Nortier and F.J. Haasbroek: Appl. Radiat. Isot. 42, 361 (1991), and references therein
8. G.F. Steyn, S.J. Mills, F.M. Nortier, B.R.S. Simpson and B.R. Meyer: Appl. Radiat. Isot. 41, 315 (1990)
9. A.H. Wapstra and G. Audi: Nucl. Phys. A432, 55 (1985)
10. A. Grütter: Nucl. Phys. A383, 98 (1982)
11. V.N. Aleksandrov, M.P. Semenova and V.G. Semenov: At. Énerg. 62, 411 (1987)
12. H.R. Heydegger, C.K. Garret and A. Van Ginneken: Phys. Rev. C 6, 1235 (1972)
13. P.G. Albouy, M. Gusakow, N. Poffé, H. Sergolle and L. Valentin: J. Physique Rad. 28, 1000 (1962)
14. G.H. Coleman and H.A. Tewes: Phys. Rev. 99, 288 (1955)

^{7}Be EMISSION IN PROTON AND NEUTRON INDUCED REACTIONS

B. Scholten, S.M. Qaim and G. Stöcklin

Institut für Chemie 1 (Nuklearchemie)
Forschungszentrum Jülich GmbH
5170 Jülich, Federal Republic of Germany

Abstract: Excitation functions were measured for (p,^{7}Be) reactions on natural V, Nb, Au and Bi over the proton energy range of 35 to 100 MeV using the stacked foil technique. Cross sections were measured for emission of ^{7}Be in 53 MeV d(Be) breakup neutron induced reactions on Al, Si, V, Fe, Co, As, Nb, Au and Bi. The (n,^{7}Be) cross sections lie between 0.3 and 5 μb. In all the cases ^{7}Be was separated radiochemically and measured by γ-ray spectroscopy. Some systematic trends in the data are discussed.

(^{7}Be emission, cross section, excitation function, 53 MeV d(Be) breakup neutrons, (p,^{7}Be), (n,^{7}Be))

Introduction

^{7}Be emission was studied so far in spallation reactions at proton energies of 300 MeV and up to a few GeV. At lower energies ($E_p \leq 100$ MeV), where compound and precompound reactions are possible, cross sections were measured only for the light mass elements C, O, Mg, Al and Si without chemical separation [cf. 1-3]. For other target nuclei only a few data exist [cf. 4]. We investigated the excitation functions of (p,^{7}Be) reactions on monoisotopic Nb, Au, and Bi and the quasi-monoisotopic V.

Regarding charged particle emission in interactions with 53 MeV d(Be) breakup neutrons, the reactions (n,t), (n,^{3}He) and (n,α) were measured at Jülich [cf. 5-7]. We investigated now for the first time the emission of the complex charged particle ^{7}Be by fast neutrons.

Experimental

Irradiations

Proton irradiations were done using the well known stacked-foil technique. One stack with six foil packages, each of highly pure V, Nb, Au and Bi, covered by Al- and Cu-foils for beam monitoring, was irradiated at the cyclotron in Uppsala (Sweden) for 5.5 h at a beam current of 100 nA. The primary proton energy was 99 MeV. It was degraded in the stack down to 40 MeV. At the Jülich Isochronous Cyclotron JULIC three stacks were irradiated with protons (Ep = 45-35 MeV) for about 10 h at beam currents of 0.5 to 2 μA.

For studies on neutron induced emission of ^{7}Be, irradiations of high purity Al, Si, V, Fe, Co, As, Nb, Au and Bi were carried out at the Jülich Isochronous Cyclotron JULIC with fast neutrons produced via breakup of 53 MeV deuterons on a Be target (E_n = 4-50 MeV; I_{max} at 22.5 MeV;

FWHM = 15.8 MeV). Monitor foils (Al, each 90 μm) were attached in front and at the back of the samples, and irradiations done in the 0° direction at a distance of 5 cm from the Be-target.

The neutron spectrum was characterized earlier by time-of-flight measurements [8,9]. The irradiation time was 20-40 hours at deuteron beam currents of about 2 μA.

Flux monitoring

The proton beam intensity was determined via the monitor reactions ^{27}Al(p,x)^{22}Na and Cu(p,x)^{62}Zn. The proton energies were also controlled via these monitor reactions. The monitor cross sections were taken from the literature. [cf. 10-13] The mean proton flux was obtained by taking a mean of the flux values from the front and back monitor foils of each target package. Proton flux values between 5 and 100 $\times 10^{11}$ s^{-1} were obtained.

The neutron flux density was determined via the ^{27}Al(n,α)^{24}Na monitor reaction. The neutron spectrum averaged cross section for this reaction is given as 20.7 mb [cf.7]. The mean flux densities were in the order of 5 $\times 10^{10}$ cm^{-2}s^{-1}.

Radiochemical separation

Due to very high matrix activity ^{7}Be could not be measured nondestructively. So Be-carrier added chemical separation was mandatory to remove the ^{7}Be from a number of radioactive isotopes and the target material. Solvent extraction, precipitation or ion exchange methods were carried out to remove the target material. The separation of the Be from other radioactive isotopes was done by successive anion and cation exchange chromatography and repeated precipitation of Be as hydroxide. A decontamination factor of 10^6 was estimated. To determine the chemical yield by weighing the precipitated Be(OH)$_2$ was heated at 850 °C to convert it to BeO. The chemical purity of the BeO

712

was checked by atomic emission spectroscopy with inductively coupled argon plasma (AES-ICP).

Measurement of radioactivity

The radioactivity of the radiochemically separated ^{7}Be ($T_{1/2}$ = 53.26d; E_γ = 477.6 keV; I_γ = 10.39%) was measured via conventional computer based γ-ray spectroscopy using a calibrated Ge(Li)- or Ge-detector. The peak area was determined from the spectrum by automatic peak search and calculation by the spectroscopy software in the case of clear peaks. For double peaks as well as for weak peaks on a high background, the peak area calculations were done manually. The spectra were carefully controlled for interfering γ-rays from other isotopes. The count rates were corrected for pile-up and coincidence effects.

Calculation of cross sections

Cross sections were calculated using the well known activation equation. The main sources of errors in studies on proton induced reactions were the low count rates of ^{7}Be (5 - 45 %) and the error in flux determination (5 - 20 %), which in the higher energy region is relatively high since the monitor excitation functions are not well known. The contribution of ^{7}Be due to impurities in the target material was estimated to < 2 %. The total error in the cross sections lies in the order of 20-60 %.

Regarding the neutron induced reactions the main error originates from the peak area calculation due to very low count rate on a high background (60 - 350 %). The other major error is in neutron flux density determination ($\approx$ 15 %)

Results and Discussion

The (p,^{7}Be) cross sections for the four investigated elements are shown in Fig.1 as a function of proton energy. For V and Nb the cross sections increase slightly with the proton energy and become flatter later. The cross sections for Au show only a very slight constant increase over this

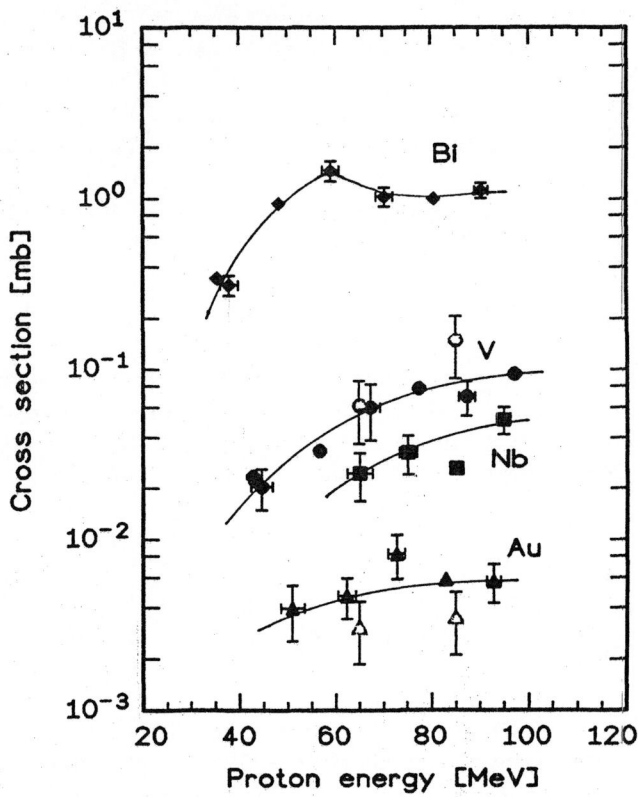

Fig. 1: (p,^{7}Be) cross section for V, Nb, Au and Bi as a function of proton energy. Closed symbols our data, open symbols Lafleur et al. [cf. 4]

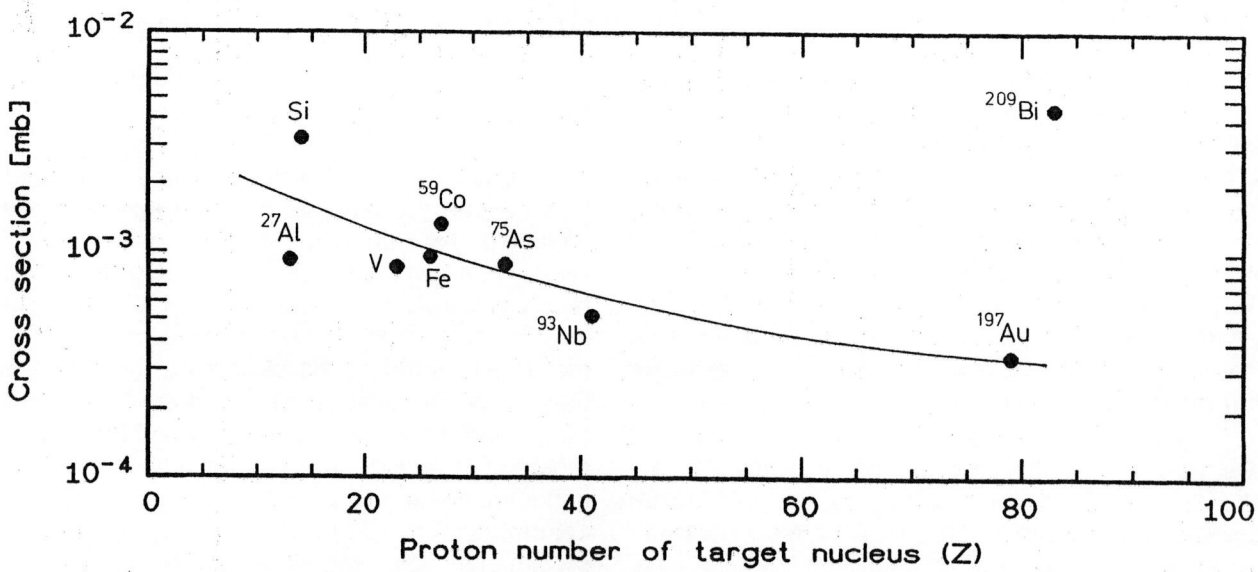

Fig. 2: 53 MeV d(Be)-breakup neutron induced (n,^{7}Be) cross sections versus proton number

energy region. Only the cross sections for Bi show a strong increase at lower energy, forming a plateau at 60 MeV to 90 MeV. Our data agree within the errors with the cross sections for V and Au at 65 and 85 MeV measured by Lafleur et al. [cf. 4].

The (p,^{7}Be) cross sections at a particular incident proton energy show a decrease with the increasing mass of the target nucleus. The Bi(p,^{7}Be) cross section on the other hand is as high as the Al (p,^{7}Be) cross section measured by us and other groups, and is much higher then the corresponding cross sections for U at 65 and 85 MeV [cf. 4]. This result indicates a different reaction mechanism for ^{7}Be emission from Bi, probably a fission reaction.

In Fig.2 the (n,^{7}Be) cross sections are given as a function of proton number of target nucleus. The cross sections lie between 0.3 and 5 μb and are more than three orders of magnitudes lower than the corresponding (n,t) and (n,^{3}He) cross sections [cf. 5-7]. The (n,^{7}Be) cross section decreases with increasing proton number of the target nucleus. Again the cross section for Bi is higher than expected; presumably a fission reaction is involved.

Acknowledgements

We thank the operating crews of the cyclotrons at Uppsala and Jülich for irradiations.
Acknowledgement is made to Dr. U. Herpers for arranging the irradiation in Uppsala.

References

1. J.B. Cumming: Ann. Rev. Nucl. Sci. 13, 261 (1963)
2. R. Bimbot and H. Gauvin: C.R. Acad. Sci. B273, 1054 (1971)
3. H.R. Heydegger, A.L. Turkevich, A. Van Ginneken and P.H. Walpole: Phys. Rev. C14, 1506 (1976)
4. M.S. Lafleur, N.T. Porile and L. Yaffe: Can. J. Chem. 44; 2749 (1966)
5. S.M. Qaim, R. Wölfle and G. Stöcklin: J. Inorg. Nucl.Chem. 36, 3639 (1974)
6. S.M. Qaim and R. Wölfle: Nucl. Phys. A295, 150 (1978)
7. S.M. Qaim, C.H. Wu and R. Wölfle: Nucl. Phys. A410, 421 (1983)
8. G.W. Schweimer: Nucl. Phys. A100, 537 (1967)
9. J.P. Meulders, P. Leleux, P.C. Macq and C. Pirart: Phys. Med. Biol. 20, 235 (1975)
10. A. Grütter: Nucl. Phys. A383, 98 (1982)
11. R. Michel, F. Peiffer and R. Stück: Nucl. Phys. A441, 617 (1985)
12. M.C. Lagunas-Solar, O.F. Carvacho and R.R. Cima: Appl. Radiat.Isot., Vol.39, 41 (1988)
13. G.F. Steyn, S.J. Mills, F.M. Nortier, B.R.S. Simpson and B.R. Meyer: Appl. Radiat. Isot., Vol. 41, 315 (1990)

MEASUREMENT OF LONG-LIVED ^{10}Be, ^{14}C AND ^{26}Al PRODUCTION CROSS SECTIONS FOR 10 - 40 MeV NEUTRONS BY ACCELERATOR MASS SPECTROMETRY

T. Nakamura and H. Sugita
Cyclotron and Radioisotope Center, Tohoku University
Aoba, Aramaki, Aoba-ku, Sendai 980, Japan

M. Imamura, Y. Uwamino and S. Shibata
Institute for Nuclear Study, University of Tokyo
Midori-cho 3-2-1, Tanashi, Tokyo 188, Japan

H. Nagai and M. Takabatake
College of Humanities and Sciences, Nihon University
Sakura-Johsui, Setagaya-ku, Tokyo 156, Japan

K. Kobayashi
Research Center for Nuclear Science and Technology, University of Tokyo
Yayoi 2-11-16, Bunkyo-ku, Tokyo 113, Japan

Abstract: We have measured the production cross sections of long-lived isotopes, ^{10}Be(1.5×10^6 y) by ^{14}N(n,pα)^{10}Be, ^{14}C(5.7×10^3 y) by ^{16}O(n,^{3}He)^{14}C and ^{26}Al(7.2×10^5 y) by ^{27}Al(n,2n)^{26}Al and ^{nat}Si(n,pxn)^{26}Al reactions in the energy range of 10 to 40 MeV. Vanadium nitride(VN) for ^{14}N, quartz plate(SiO_2) for ^{16}O and ^{nat}Si, silicon plate for natural Si and aluminum foils for ^{27}Al were irradiated with semi-monoenergetic p-^{9}Be and p-^{7}Li neutrons at several proton energies of 20 to 40 MeV using the AVF cyclotron. The activities of the irradiated samples have been measured by the accelerator mass spectrometry system using the internal beam monitor method which has been equipped at the tandem Van-de-Graaf accelerator. The excitation functions of these reactions were obtained from the neutron energy spectrum and the measured number of atoms.

(acclerator mass spectrometry(AMS), excitation function, unfolding, p-Be neutron, p-Li neutron, ^{27}Al(n,2n)^{26}Al, ^{nat}Si(n,pxn)^{26}Al, ^{16}O(n,^{3}He)^{14}C, ^{14}N(n,pα)^{10}Be)

Introduction

The production of long-lived isotopes of ^{10}Be (1.5×10^6 y), ^{14}C(5.7×10^3 y) and ^{26}Al(7.2×10^5 y) by neutrons is of special interest, since their occurrence in nature is closely related to the nuclear reactions due to cosmic-ray secondaries, mostly neutrons having energies of 10 to 100 MeV. It is also a great concern with the accumulation of neutron-induced activities of these isotopes in accelerator facilities during their long-term operation. We have measured the production cross sections of ^{10}Be by ^{14}N(n,p α)^{10}Be, ^{14}C by ^{16}O(n,^{3}He)^{14}C, ^{26}Al by ^{27}Al(n,2n)^{26}Al and ^{nat}Si(n,pxn)^{26}Al in the energy range of 10 to 40 MeV, by using the accelerator mass spectrometry(AMS) system of the tandem Van-de-Graaf accelerator at the Research Center for Nuclear Science and Technology, University of Tokyo. Semi-monoenergetic neutrons were produced from the ^{9}Be(p,n) and ^{7}Li(p,n) reactions using the AVF cyclotron at the Institute for Nuclear Study, University of Tokyo.

Experimental

Proton beams of energy at every 2.5 MeV interval between 20 and 40 MeV hit 1- and 2-mm thick ^{9}Be targets backed by water, and 2-mm thick ^{7}Li target backed by carbon, both of which are used as beam stoppers. The neutron energy spectra in the forward direction measured by the NE-213 scintillator showed a monoenergy peak, whose energy is 3 to 4 MeV lower than the proton energy, and a weak low energy continuum.

Vanadium nitride(VN) pellets and pure aluminum foils(50 μm thickness) were used as nitrogen and aluminum targets for the ^{14}N(n,p α)^{10}Be and ^{27}Al(n,2n)^{26}Al reactions, respectively. High-purity quartz plates(SiO_2, 1mm thickness) were used as oxygen and silicon targets and pure silicon plates also as silicon target for ^{16}O(n,^{3}He)^{14}C and ^{nat}Si(n,pxn)^{26}Al, respectively. These samples were irradiated at 5 or 10 cm distant from the target. The irradiated samples were chemically treated to get the chemical forms of BeO, C and Al_2O_3 appropriate to the AMS of ^{10}Be, ^{14}C and ^{26}Al.

The schematic diagram of the AMS system is illustrated in Fig. 1. The samples were set on a Cs sputtering negative ion source and analyzed by two analyzer magnets and the tandem Van-de-Graaf.

Isotopic ratios of ^{10}Be/ ^{9}Be, ^{14}C/^{12}C and ^{26}Al/^{27}Al were measured by the internal beam monitor method[1], where negative ions with the same atomic or molecular weight as the relevant radioactive ions are injected simultaneously. The isotopic ratio IR is given by

$$IR = (I_r^+/I_m^+) \, (I_m^-/I_s^-) \, f$$

where I_r^+ is the detector counts of radioisotope ions, I_m^+ the currents at the monitor Faraday cup of the positive monitor ion beam produced from the negative ion of the stable isotope, I_m^-/I_s^- the beam current ratio of the negative monitor/stable ions, and f is the correction factor for the difference of transmission efficiencies between radioactive and monitor ions.

The total number of nuclides, P_i , for ith experiment corresponding to ith proton energy E_i , is related to

$$P_i = N \int_0^{E_i} \sigma(E) \, \phi_i(E) \, dE$$

where N is the number of target nuclei, $\sigma(E)$ the production cross section and $\phi_i(E)$ the neutron flux. Since the neutron flux $\phi_i(E)$ is not purely monoenergetic but has a low energy continuum, the $\sigma(E)$ value can be obtained by unfolding this equation. We used a simple successive subtraction method to get the average cross section for ^{14}N(n,p α) ^{10}Be, ^{16}O(n, ^{3}He) ^{14}C and ^{nat}Si(n,pxn) ^{26}Al, while for ^{27}Al(n,2n) ^{26}Al, we used the SAND-II[2] and the NEUPAC[3] unfolding codes to get the excitation function with the use of the initial guess value calculated from the ALICE/LIVERMORE 82 code[4].

Results

Figures 2 to 5 show the obtained results of these four cross sections. For ^{14}N(n,pα)^{10}Be reaction in Fig. 2, the maximum cross section values are about 10 mb and our data are compared with other results of ^{14}N(p,X) ^{10}Be reaction[5,6]. Our data on neutron-induced reaction are much higher than those on proton-induced reaction. For ^{16}O(n, ^{3}He) ^{14}C and ^{nat}Si(n,pxn)^{26}Al reactions, the results obtained by p-Be and p-Li neutrons are shown in Figs. 3 and 4. Both results are close to each other. The maximum values are about 5 mb for ^{16}O(n, ^{3}He) ^{14}C and 20 mb for ^{nat}Si(n,pxn)^{26}Al. The shape of ^{nat}Si(n,pxn)^{26}Al reaction cross section is found to be very similar to that of ^{nat}Si(p,X)^{26}Al cross section[7], but the absolute value is about a half.

The unfolded results of ^{27}Al(n,2n)^{26}Al reaction cross section are shown in Fig. 5, together with other experimental data around 15 MeV[8]. Results given by SAND-II and NEUPAC unfolding show very good agreement each other. Our data near threshold energy are lower than the other data, because of large experimental errors.

We could first give the experimental data on the long-lived isotope, ^{10}Be, ^{14}C and ^{26}Al, production cross sections by neutrons having energy of 10 to 40 MeV. This work was supported in part by a Grant-in-Aid for Scientific Research of the Japanese Ministry of Education and Culture.

REFERENCES

1. M. Imamura, Y. Hashimoto, K. Yoshida, I. Yamane, H. Yamashita, T. Inoue, S. Tanaka, H. Nagai, M. Honda, K. Kobayashi, N. Takaoka and Y. Ohba: Nucl. Instr. and Meth. B5, 211(1984).
2. W. N. McElroy, S. Berg, T. Crockett and R. G. Hawkins: AFWL-TR 67-41, Air Force Weapons Laboratory (1967).
3. T. Taniguchi, N. Ueda, M. Nakazawa and A. Sekiguchi: NEUT Research Report 83-10, Department of Nuclear Engineering, University of Tokyo(1984).
4. M. Blann and J. Bisplingshoff: Code ALICE/LIVERMORE 82, Lawrence Livermore National Laboratory Report No. UCID 19614 (1983).
5. G. M. Raisbeck and F. Yiou: Phys. Rev C9, 1385(1974).
6. J. L. Reyss, Y. Yokoyama and F. Guichard: Earth Planet. Sci. Lett. 53, 203(1981).
7. R. M. Keyser, R. A. Blue and H. R. Weller: Nucl. Phys. A186, 528(1972).
8. S. Iwasaki, J. R. Dumais and K. Sugiyama: Proc. Int. Conf. on Nuclear Data for Science and Technology, May/June 1988, Mito, Japan (Edited by S. Igarashi), p.295 (1988).

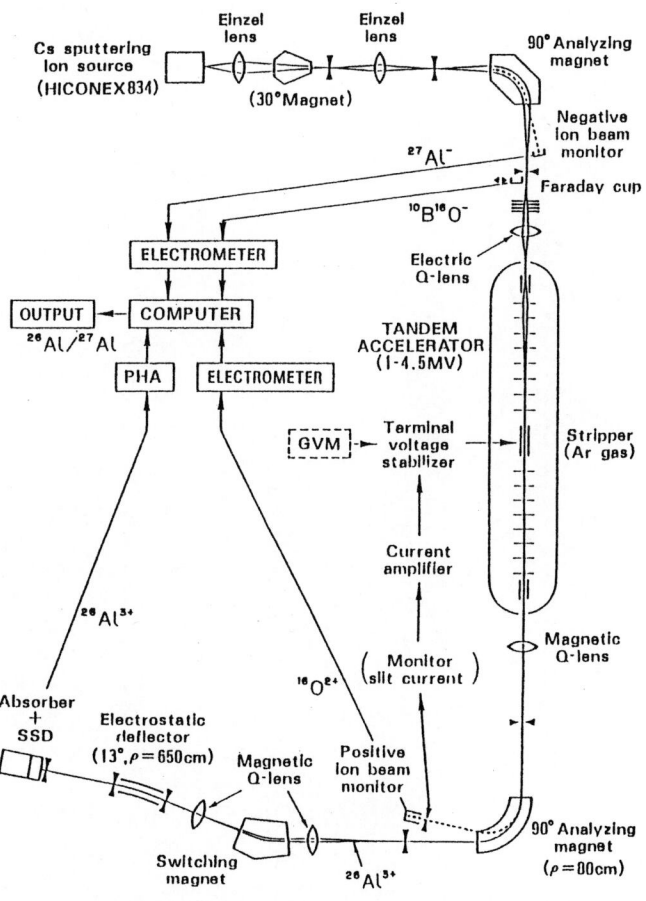

Fig. 1 Schematic diagram of the accelerator mass spectrometry system at the tandem Van-de-Graaf accelerator

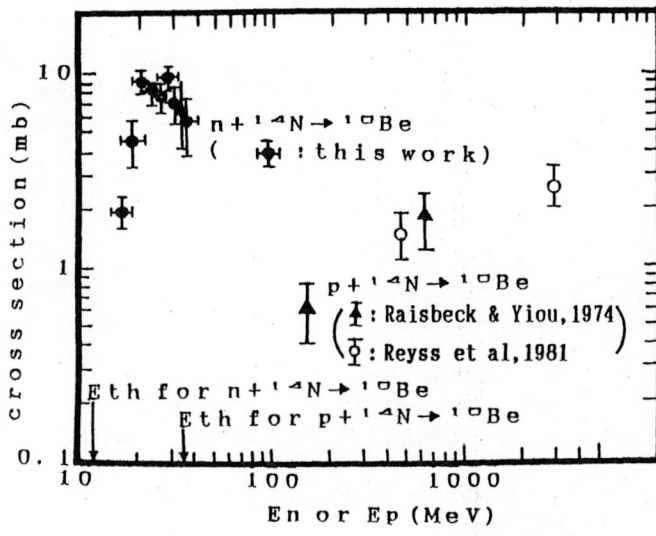

Fig. 2 Experimental excitation function of $^{14}N(n,p\alpha)^{10}Be$ reaction, compared with other results of $^{14}N(p,X)^{10}Be$

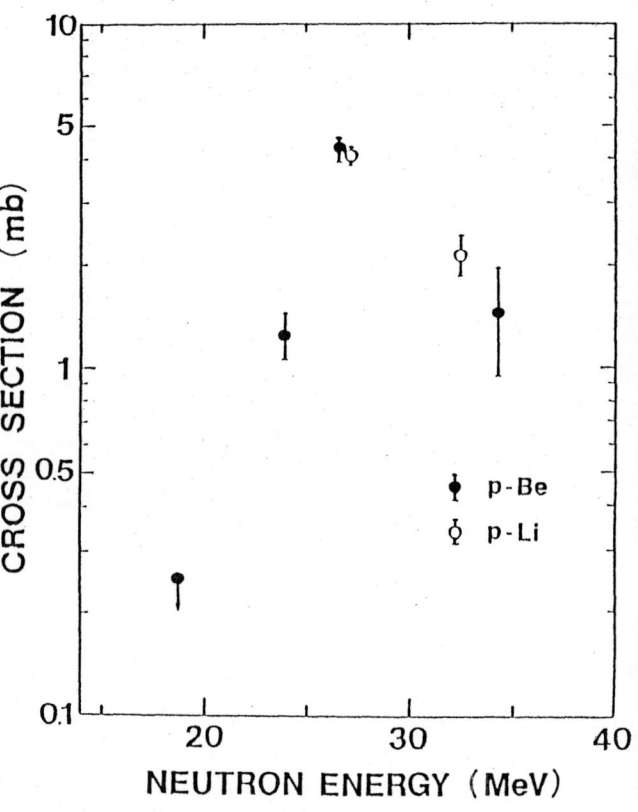

Fig. 3 Experimental excitation function of $^{16}O(n,^3He)^{14}C$ reaction

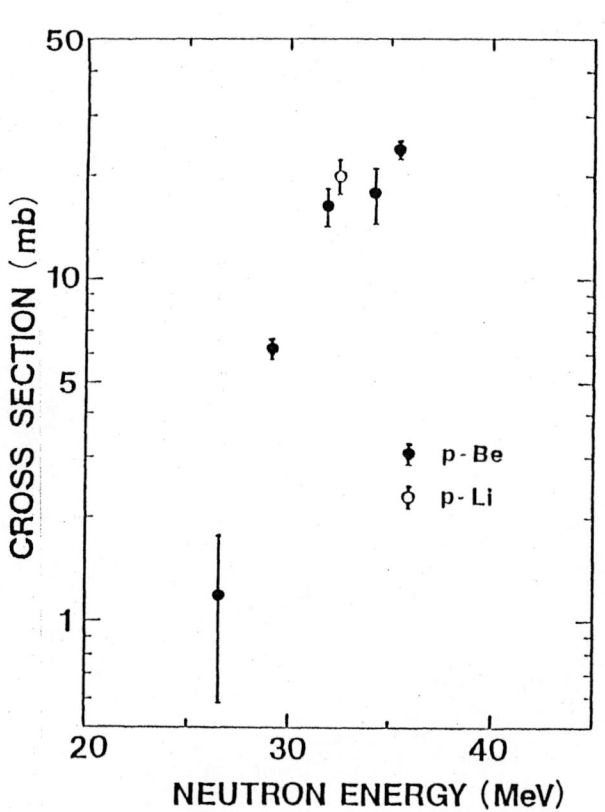

Fig. 4 Experimental excitation function of $^{nat}Si(n,pxn)^{26}Al$ reaction

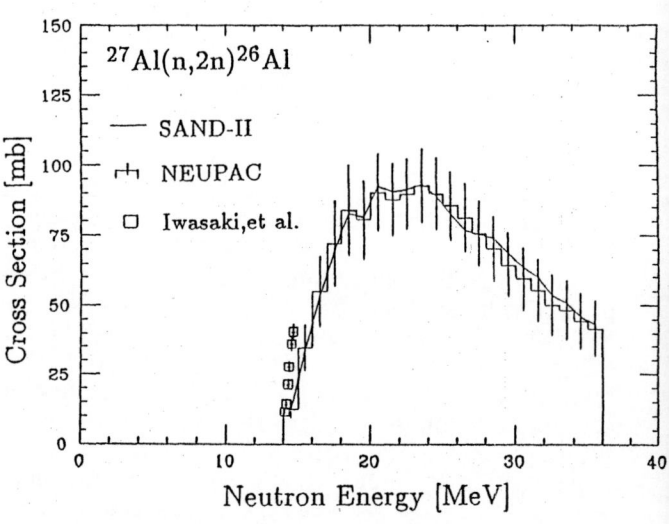

Fig. 5 Experimental excitation function of $^{27}Al(n,2n)^{26}Al$ reaction, compared with other results near threshold energy

SCATTERING OF 28.2 MeV NEUTRONS FROM ^{12}C AND 18.5 MeV NEUTRONS FROM ^{52}Cr AND ^{60}Ni

Y. Yamanouti, M. Sugimoto, S. Chiba, M. Mizumoto and K. Hasegawa

Japan Atomic Energy Research Institute
Tokai, Ibaraki 319-11, Japan

Y. Watanabe

Department of Nuclear Engineering, Kyushu University
Hakozaki, Fukuoka 812, Japan

Abstract: Differential cross sections for elastic and inelastic scattering of
28.2 MeV neutrons from ^{12}C, and 18.5 MeV neutrons from ^{52}Cr and ^{60}Ni were
measured by using the JAERI tandem accelerator. Scattered neutrons were
observed by a time-of-flight spectrometer with an array of four 20cmϕ x 35cm
NE213 liquid scintillator detectors. Neutron time-of-flight spectra were taken
at scattering angles from 20° to 140° in 10° steps. Inelastic scattering cross
sections of neutrons leading to the 4.439 MeV 2$^+$, and 9.641 MeV 3$^-$states of
^{12}C, 1.434 MeV 2$^+$ and 4.563 MeV 3$^-$ states of ^{52}Cr, and 1.332 MeV 2$^+$ and
4.045 MeV 3$^-$ states of ^{60}Ni were measured simultaneously with the elastic
scattering cross sections. The experimental cross sections were analyzed in
terms of phenomenological and microscopic optical model potentials and
coupled-channel calculations using the collective model.

(elastic and inelastic scattering, differential cross sections, neutron,
28.2 MeV, 18.5 MeV, ^{12}C, ^{52}Cr, ^{60}Ni, optical model potential,
coupled-channel calculations)

Introduction

Neutron scattering has been shown to give
direct information about the nucleon-nucleus optical potential and the mechanism for the excitation of low-lying nuclear states. In this work
differential cross sections for elastic and
inelastic scattering of 28.2 MeV neutrons from
^{12}C and 18.5 MeV neutrons from ^{52}Cr and ^{60}Ni were
measured in order to study the reaction mechanism
for the neutron scattering in the energy regions
around 20 MeV and 30 MeV. The experimental data
of neutrons in the energy region higher than
14 MeV are now required strongly for the research
fields such as nuclear engineering and neutron
radiotherapy. For the neutron scattering from ^{12}C
in the energy regions around 20 and 30 MeV, elastic and inelastic scatterng at incident energies
from 20.8 to 26 MeV /1/, and at 22.0 MeV /2/ has
been studied, and differential cross sections for
elastic scattering at 40 MeV have been reported
/3/. For ^{52}Cr in the energy range above 10 MeV
elastic and inelastic scattering cross sections
from natural chromium have been measured at 14MeV
/4/. For ^{60}Ni differential scattering cross
sections at neutron energies of 8-14 MeV /5/ and
24 MeV /6/ have been reported. Our experimental
data were analyzed by the phenomenological and
microscopic optical model potentials and the coupled-channel theory, and compared with proton
scattering in the framework of the Lane model.

Experimental method

The measurements were performed with pulsed
beam time-of-flight methods. Pulsed beams of
protons and deuterons with repetition rates of
4MHz and 2MHz, respectively, were provided by the
JAERI tandem electrostatic accelerator. Pulsed
beams with burst duration of 2nsec were used in
the experiments. Neutrons of 28.2 MeV and 18.5 MeV
were obtained by the ^{7}Li(p,n)^{7}Be reaction and the
^{2}H(d,n)^{3}He reaction, respectively. A cylindrical
graphite(3.5cmϕ x 3.2cm, 53.7g) was used as a
scattering sample of ^{12}C. The scattering sample of
^{52}Cr was pure metal disc of 46.5g with enrichment
of 99.84%, since ORNL kindly accepted our request
to borrow the separated isotope of ^{52}Cr converted
to metal from oxide. The ^{60}Ni scattering sample
of 38.67g with 99.79% enrichment was used on loan
from ORNL. The neutron detector is a 20cmϕ x 35cm
NE213 liquid scintillator viewed by RCA 8854 photomultiplier tubes at the front and rear scintillator faces. The neutron detector is similar in
principle to the very large detector developed at
Ohio University. Scattered neutrons were observed
by an array of these four neutron detectors for
efficient measurements. The neutron detectors are
placed in a detector shield tank. The neutron detector shields are mounted on a large turn table
which is rotatable around the scattering sample.
The flight path of the time-of-flight spectrometer is 8m.

718

Data reduction and experimental results

The experimental yields were calculated by fitting the peaks of interest with line shapes of a basic Gausian function with tailing components of the exponential shape multiplied by the complementary error function. The differential cross sections were normalized to the known n-p scattering cross sections by measuring neutrons scattered from a 1.2cmϕ x 4cm polyethylene scatterer. The relative efficiency of the neutron detector was determined by measuring the angular distribution of the n-p scattering. The resulting differential cross sections were corrected for dead time in the data acquisition system,and also for multiple scattering and flux attenuation in the scattering sample.Differential cross sections were determined for the elastic scattering and the inelastic scattering leading to the excited states at 4.439 MeV(2^+) and 9.641 MeV(3^-) of ^{12}C, at 1.434 MeV (2^+) and 4.563 MeV (3^-) of ^{52}Cr, and at 1.332 MeV (2^+) and 4.045 MeV (3^-) of ^{60}Ni,respectively. The differential cross sections were measured in the angular range from 20° to 140° in 10° steps at the incident energy of 28.2 MeV for ^{12}C, and at 18.5 MeV for ^{52}Cr and ^{60}Ni.The experimental cross sections are shown in figs.1,2 and 3 together with theoretical predictions.

Analysis

Optical model analysis

The elastic scattering data were analyzed by the phenomenological optical model with the standard Woods-Saxon form. The compound nuclear contribution estimated by the Hauser Feshbach formalism is small compared with the experimental cross sections for ^{12}C, ^{52}Cr and ^{60}Ni at these incident energies. The phenomenological global optical model potentials proposed by Walter/7/, Rapaport /8/ and Patterson/9/ were tested for the optical model calculations of ^{52}Cr and ^{60}Ni. The microscopic JLM potentials starting with the Reid's hard core interaction /10/ were examined to check the quality of the fit to the experimental data for ^{52}Cr and ^{60}Ni.

Coupled-channel calculations

The coupled-channel (CC) calculations based on the rotational model and the vibrational model were performed with the codes ECIS79 and JUPITOR1. In the rotational model calculations, CC calculations with the quadrupole deformation only, and with the quadrupole and hexadecapole deformation were carried out. In the CC calculations optical potential parameters and the deformation parameters were adjusted to get the best fit to the experimental cross sections for the elastic and inelastic scattering. The parameters for the spin orbit term were fixed in the calculations. The best fit was obtained in the rotational model calculation with the quadrupole and hexadecapole deformation for ^{12}C,whereas the vibrational model calculations well reproduced the experimental cross sections for ^{52}Cr and ^{60}Ni. The results of the CC calculations are shown in figs.1,2 and 3

for ^{12}C, ^{52}Cr and ^{60}Ni, respectively. The optical potential parameters and the deformation parameters used in the CC calculations are listed in table 1.

Comparison with proton scattering

Proton optical potentials obtained in the proton scattering for ^{52}Cr/11/ and ^{60}Ni/12/, respectively were transformed into neutron optical potentials in the framework of the Lane model, and the ability to predict the present experimental neutron cross sections was checked in the CC calculations. The same deformation parameters as those obtained in the proton scattering were used in the calculations. The values of the energy dependence and the Coulomb correction term were adopted from the global optical potential/7/.

Conclusions

Measurements of experimental neutron cross sections for elastic and inelastic scattering on ^{12}C, ^{52}Cr and ^{60}Ni have been made in the energy regions around 20 and 30MeV. It is to be noticed that neutron scattering data for ^{52}Cr have been measured in this energy range for the first time with the pure metal sample of 99.84% separated isotope. The phenomenological global optical potential by Walter reproduces very well the experimental data, and the microscopic JLM optical potential also well predicts the present differential cross sections on ^{52}Cr and ^{60}Ni. The CC calculations based on the collective model well described the experimental elastic and inelastic scattering cross sections of neutrons from ^{12}C, ^{52}Cr and ^{60}Ni in these energy regions. The optical potentials transformed from proton potentials give good fits to the experimental neutron cross sections.

The measured cross sections obtained in the present experiments are widely useful for the nuclear data for technical application.

References

1. A.Meigooni, R.W.Finlay and J.S.Petler: Nucl. Phys. A445, 304 (1985)
2. N.Olsson, B.Trostell and E.Ramstrom: Nucl. Phys. A496, 505 (1989)
3. P.R.Devito, Ph.D. thesis, Michigan State University (1979)
4. P.H.Stelson, R.L.Robinson, J.Rapaport and G.R.Satchler: Nucl.Phys. 68, 97 (1965)
5. P.P.Guss, R.C.Byrd, C.E.Floyd, C.R.Howell and R.L.Walter: Nucl.Phys. A438, 187 (1985)
6. Y.Yamanouti, J.Rapaport, S.Grimes, V.Kulkarni R.Finlay and D.Bainum, in Proc.Int.Conf. on Nuclear Cross Sections for Technology, October 1979, Knoxville,TN (Edited by J.L. Fowler), p.146, NBS SP 594 (1980)
7. R.L.Walter and P.P.Guss, in Proc.Int.Conf. on Nuclear Data for Basic and Applied Science, May 1985, Santa Fe, NM,(Edited by P.G.Young), p.1079, Gordon and Breach (1980)
8. J.Rapaport, V.Kulkarni and R.W.Finlay: Nucl.Phys. A330, 15 (1979)

9. D.M. Patterson, R.R. Doering and A.Galonsky:
    Nucl.Phys. A263, 261 (1976)
10. J.P.Jeukenne, A.Lejeune and C.Mahaux: Phys.
    Rev. C16, 80 (1976)
11. B.M.Preedom, C.R.Gruhn, T.Y.T.Kuo and
    C.J.Maggiore: Phys.Rev. C2, 166 (1970)
12. E.Fabrici, S.Micheletti, M.Pignanelli and
    F.G.Resmini: Phys.Rev. C21, 844 (1980)

Table 1   OP and deformation parameters

	^{12}C	^{52}Cr	^{60}Ni
V_R	45.45	45.19	45.43
r_R	1.196	1.219	1.219
a_R	0.663	0.587	0.688
W_I	4.04	1.499	1.499
r_I	1.196	1.452	1.442
a_I	0.663	0.493	0.497
W_D	8.18	5.67	6.52
r_D	1.187	1.444	1.282
a_D	0.160	0.454	0.512
V_{SO}	6.20	5.64	5.62
r_{SO}	1.05	1.103	1.103
a_{SO}	0.55	0.56	0.56
β_2	-0.601	0.223	0.213
β_3	0.247	0.180	0.156
β_4	0.05		

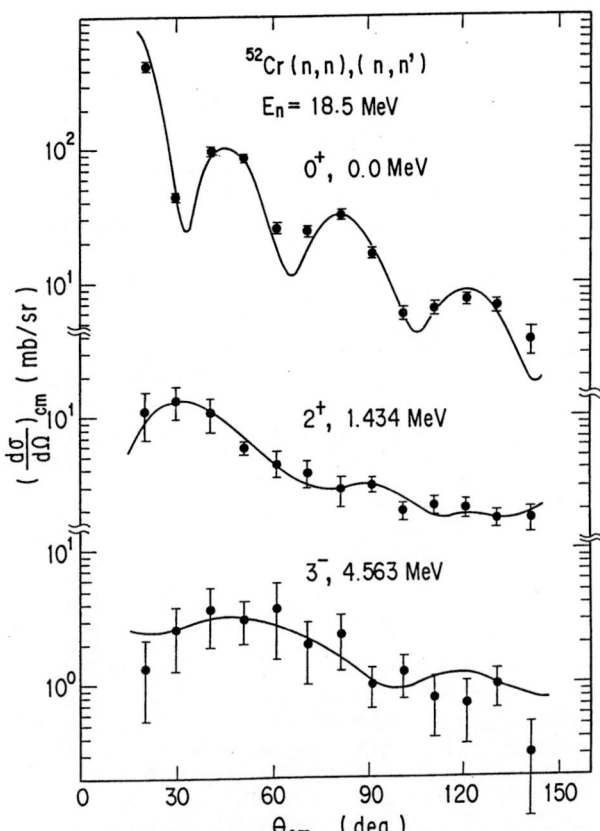

Fig.2   Experimental cross sections
and CC predictions for ^{52}Cr

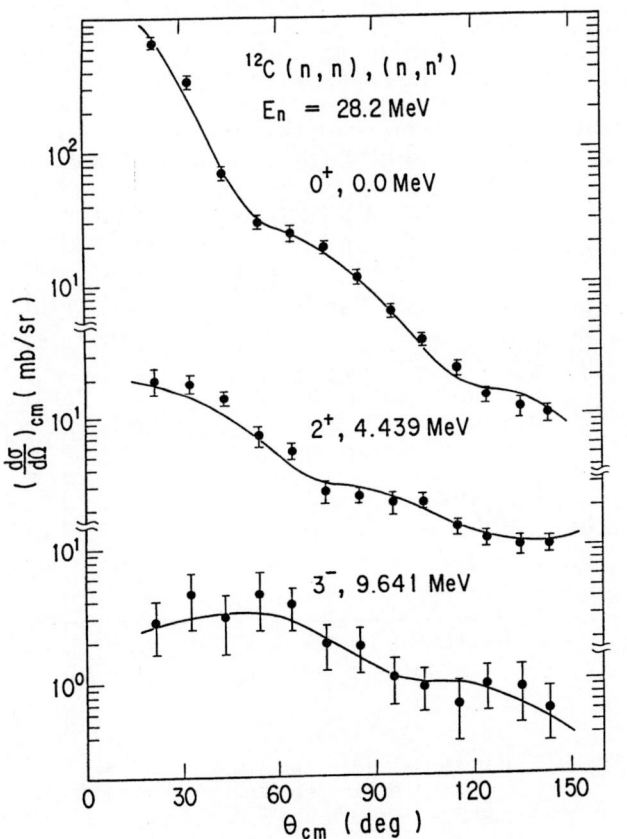

Fig.1   Experimental cross sections
and CC predictions for ^{12}C

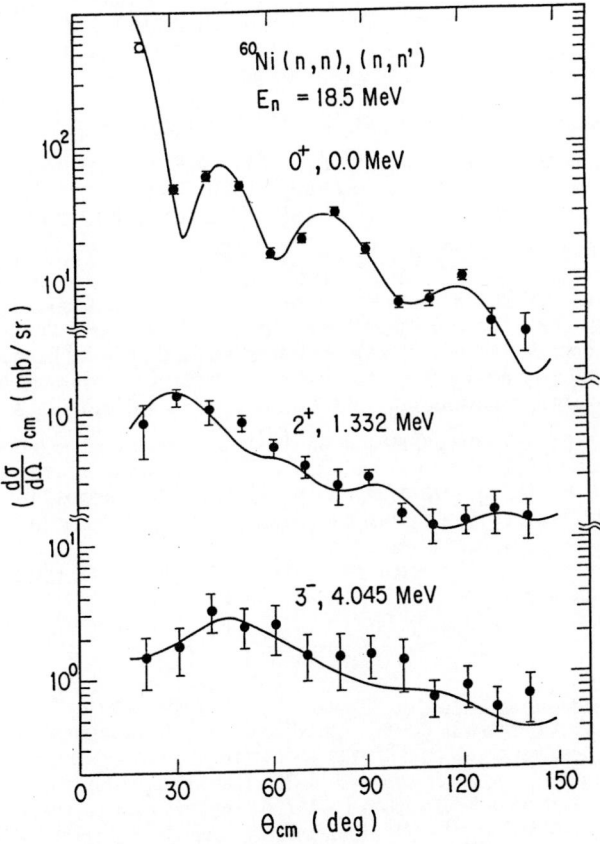

Fig.3   Experimental cross sections
and CC predictions for ^{60}Ni

Neutron Total Cross Section Measurements at Intermediate Energy

R.W. Finlay, G. Fink and W. Abfalterer
Department of Physics and Astronomy, Ohio University
Athens, OH 45701 USA

P. Lisowski, G.L. Morgan and R.C. Haight
P–Division, Los Alamos National Laboratory
Los Alamos, NM 87545 USA

Abstract: New measurements of neutron total cross sections have been performed as a function of neutron energy up to 600 MeV using the WNR facility at Los Alamos National Laboratory. Transmission measurements were performed for eighteen target nuclei with 9 < A < 209 including isotopically enriched samples of ^{40}Ca, ^{90}Zr and ^{208}Pb. While the goal of the experiment was to measure total cross sections above 100 MeV in small energy bins and with 1% statistical accuracy, much of the data at lower energy is significantly better than that. Rapid fluctuations in the cross sections of light nuclei below 20 MeV provide information on nuclear level density at high excitation energy. The overall dependence of the cross section on neutron energy can be interpreted in terms of optical models ranging in complexity from simple Ramsauer pictures to Dirac phenomenology. Preliminary results of some of these calculations will be presented.

Introduction

The 1988 edition of the Brookhaven "Barn Book" [1] displays the severe shortage of neutron cross section data in the intermediate energy region. Yet the value of such data has recently been demonstrated. Kozak and Madland [2,3] have attempted to construct a Dirac Optical Potential for nucleon–nucleus scattering focusing on the relatively rich data base for proton scattering and spin observables for ^{208}Pb in the energy range 50–1000 MeV. In order to establish the isovector terms of the potential, they simultaneously examined neutron data. Their analysis was ultimately restricted to the energy region 95–300 MeV because the available neutron total cross section data ended at about 240 MeV. Their model potential predicts strong differences in proton and neutron differential scattering cross sections and striking differences in spin-observables even in this limited energy interval.

Examination of ref. [1] reveals the following situation: Schimmerling et al. [4] measured $\sigma_T(E)$ at 15 energies between 379 and 1731 MeV for naturally abundant samples like Al, Fe, Ni, Cu, Zn, Ag and Sn. More recently, Franz et al. [5] measured $\sigma_T(E)$ for the nuclei Be, C, O, Al, V, Mn, Co, Cu, Ag, Ce, Ta, Pb, Bi and U at 22 energies from 160–575 MeV. Otherwise, no data set contains more than two points above 200 MeV for any nuclei between A = 6 and A = 208, except for Si which has been studied in detail. There is clearly a lack of total cross section measurements on separated isotope targets of interest in intermediate energy nuclear physics (e.g., ^{40,48}Ca, ^{90}Zr and ^{208}Pb). Differential cross section and analyzing power measurements are completely lacking. The purpose of the present work was to provide total cross section data for a large number of target nuclei in small energy bins and with good statistical and systematic accuracy.

Experiment

Experimental Arrangement. Total cross section data were taken in a classic "good geometry" transmission experiment. A continuous spectrum of neutrons up to about 600 MeV is emitted at 30° from the primary 800 MeV proton beam from LAMPF incident upon a thick tantalum target. Neutron energy was determined by time of flight with a 12 mm thick fast plastic scintillator viewed by two fast photomultipliers. The source–to–detector flight path was 38.14 m. A typical spectrum of neutrons transmitted through graphite is shown in fig. 1. The locations of the γ–flash and the narrow resonances in ^{12}C provide an automatic determination of the flight path.

The scattering samples were cylinders 2.54 cm in diameter and 10–20 cm long. Up to seven samples and one open beam position could be mounted on a 63.5 cm diameter sample wheel located 11.76 m from the source. A Geneva Drive rotary target changer could change the sample position in a few seconds.

The neutron flux was monitored by a 6 mm thick plastic scintillator located in front of the sample wheel. Sweeping magnets before and after the sample changer deflect charged particles out of the neutron flight path. Charged particles produced along the flight path were eliminated by a veto counter.

One other important feature of the experimental arrangement was the use of very tight collimation both before and after the scattering

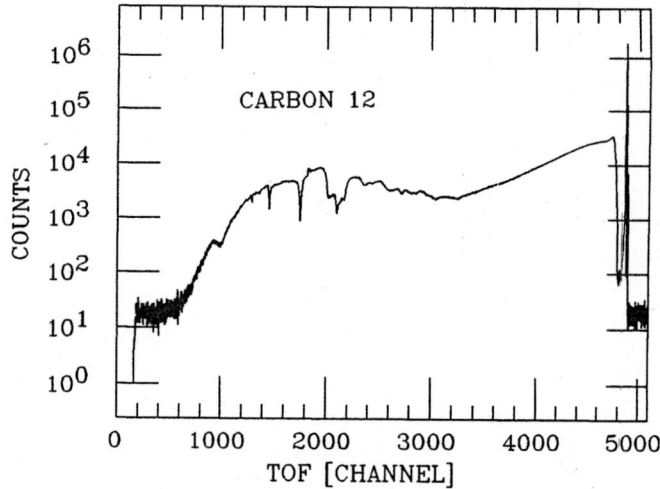

Fig. 1 Time–of–flight spectrum for neutrons transmitted through carbon. The peak near channel 4800 is the γ–flash.

sample. The solid angle subtended by the collimation system was 9.07×10^{-7} sr which provided manageable counting rates and reduced in—scattering to other experiments at the WNR facility.

Electronics. The time structure of the LAMPF—WNR beam provided good conditions for time—of—flight measurements. A micropulse spacing of 1.8 μs permitted observation from the γ—flash down to about 3—4 MeV. The pulse height threshold was set close to the frame overlap threshold in order to reduce the time—uncorrelated background. The start signal was derived by mean timing from the two phototubes viewing the neutron scintillator while a delayed stop signal was obtained from the WNR beam pulse pickoff. Time resolution of the γ—flash was about 0.75 ns.

The time—of—flight spectra were stored in a fast histogramming memory and transferred to disk at the end of a "run" which consisted of five complete rotations of the 8—position sample wheel. Counting time for the individual targets was set at 20 s.

Samples. Highest priority was given to the three separated isotope samples: ^{40}Ca, $^{90}ZrO_2$ and ^{208}Pb. The ^{208}Pb cross section was determined twice: once with a 325 g sample of highly enriched ^{208}Pb and then independently with two large samples of radiogenic lead of very different isotopic composition. The thicknesses of the radiogenic lead were chosen [6] (along with a small piece of natural Pb) to completely compensate for the A = 204, 206 and 207 in the "sample out" run thus yielding a true measure of the ^{208}Pb cross section.

A second class of samples consisted of readily available high—purity metals of normal isotopic abundance. The group consisted of Be, Al, Si, Cu, Nb, Sn, Ta and Bi.

The third group of samples consisted of ceramics and glasses: Al_2O_3, AlN, BeO, CaF_2, NaF, NaI, CsF and CsI. No attempt was made to make compensating pairs with these samples. Instead, the transmission of the compound was taken to be a product of the transmissions of the elements.

Typical samples were 2.54 cm diameter rods that varied in length from 6 to 20 cm. The $^{90}ZrO_2$ and the ^{40}Ca samples were encapsulated in metal cans, and matching cans were used for the sample—out runs. The amount of available ^{40}Ca was rather small (54 g) which resulted in somewhat larger errors in this cross section above 100 MeV.

Analysis

Data reduction and corrections were performed with the Fortran program WNRVAX.FOR which includes error propagation. Time of flight spectra were corrected for time—independent background charged—particle vetos and dead time. The time—independent background was determined to be negligible in a separate experiment using copper samples of very different thickness. Dead time corrections are large and must be treated carefully. Channel—independent correction factors ranged up to 1.8 but could be accurately determined from appro—priate scalar ratios. Channel—dependent corrections take into account the lower probability for counting a low—velocity neutron since the system is already busy with a high—velocity neutron or gamma ray from the same beam burst. These corrections, which were also as large as 1.7, were performed analytically. Finally a higher order correction to the channel—dependent correction [7] which accounts for variation in the beam intensity was found to be small (0.1%) but nonetheless included.

Corrected time—of—flight spectra were normalized to the number of counts in a monitor and then converted into energy spectra. Since the useful part of the time—of—flight spectrum extends over 3450 channels (corresponding to 5 MeV to 600 MeV), there is some flexibility in the choice of binning functions. The data shown here were all combined into 1 ns bins thus yielding about 1000 values for the cross section for each target. The energy bins increase from 7 keV at 5 MeV to 15 MeV at 600 MeV. Cross sections are then calculated for each energy bin from the familiar formula $\sigma(E) = \frac{1}{n\ell} \ell n\, T(E)$ where $T(E)$ is the tranmis—sion at energy E, n is the number of nuclei per unit volume and ℓ is the length of the sample.

Results

Typical results for several of the samples measured in the present experiment are shown in fig. 2. Data points are connected by a line and error bars are suppressed. Errors due to counting statistics were typically less than 1% for the metal targets and somewhat larger for elements that were measured in chemical compounds, especially those that depended on ^{40}Ca whose own counting error was 2—3% above 200 MeV.

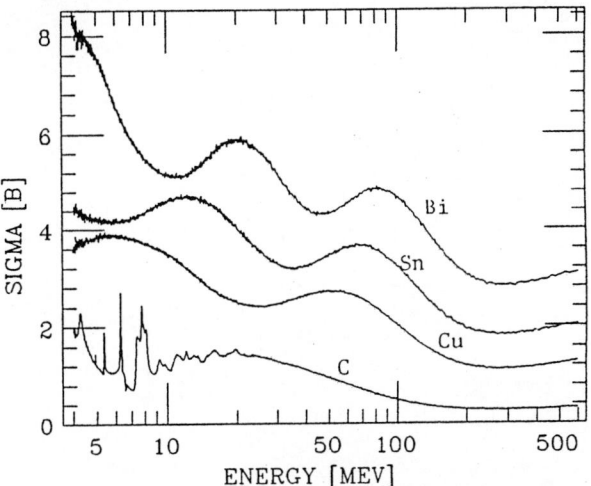

Fig. 2 Total cross sections for natural samples of (top to bottom) Bi, Sn, Cu and C.

A great many intercomparisons to earlier work were performed on line in order to verify our procedures. Details of these comparisons will be discussed elsewhere, but one typical case is shown in fig. 3. The solid line is the present work while the boxes are the National Bureau of Standards results from 1974 [8]. Overall agreement is excellent, and the present work seems to have somewhat better resolu—tion in this energy region.

Theory

While it is not possible to construct a reliable optical potential from total cross section data alone, these new measurements of total cross sections over a wide energy range are very useful for testing existing models and calculations. For example, the Global Dirac potentials of B.C. Clark and co—workers [9,10] do a superb job of describing proton scattering over a wide range of energy and mass. Their applicability to neutron scattering is demonstrated in fig. 4 for ^{16}O and ^{208}Pb [9,11]. In these calculations, the Coulomb interaction was set to zero but no other parameters were adjusted. The good agreement for ^{16}O (also for ^{12}C and ^{40}Ca, not shown) confirms the accuracy of the isoscalar part of the potential while the sharp

722

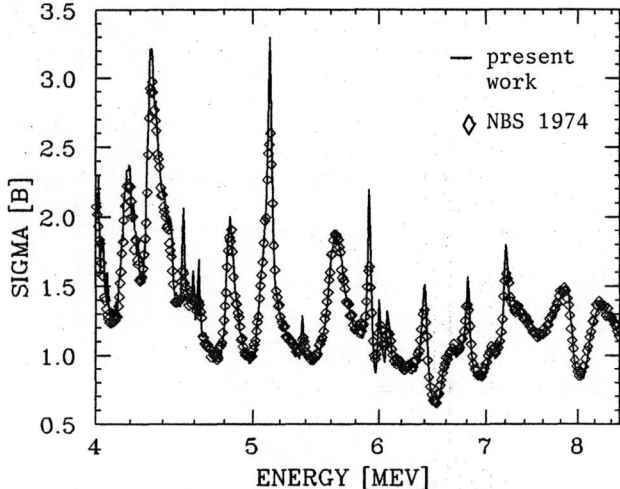

Fig. 3 Comparison of the present cross sections for Oxygen (line) with the work of Ref. 8.

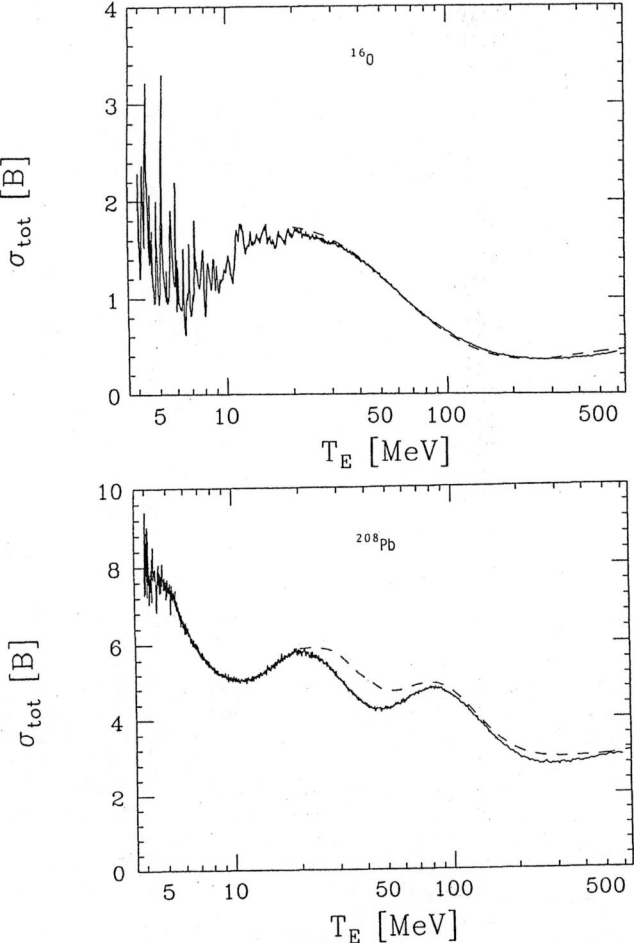

Fig. 4 Neutron total cross section data for ^{16}O (top) and ^{208}Pb (bottom) are shown by the solid lines. The dashed lines are calculations based on phenomenological Dirac proton potentials with the Coulomb term set to zero.

disagreement for ^{208}Pb indicates that, for neutron, the isovector potential is important but poorly understood.

Global potentials necessarily have a large number of parameters. However, at sufficiently high energy, the parameter–free relativistic impulse approximation (RIA) is quite successful at describing proton scattering. First application of the RIA to neutron scattering is shown in fig. 5. Here the present data for ^{208}Pb above 100 MeV are shown as points with error bars. The RIA calculations with the model by Horowitz et al. [12] are shown as solid circles. Here agreement with the data is very good indicating that isovector effects are given reasonably well in this approach. Unfortunately, the RIA is not useful below 100–200 MeV. The great bulk of the present data will need to be described phenomenologically. Work to this end is in progress.

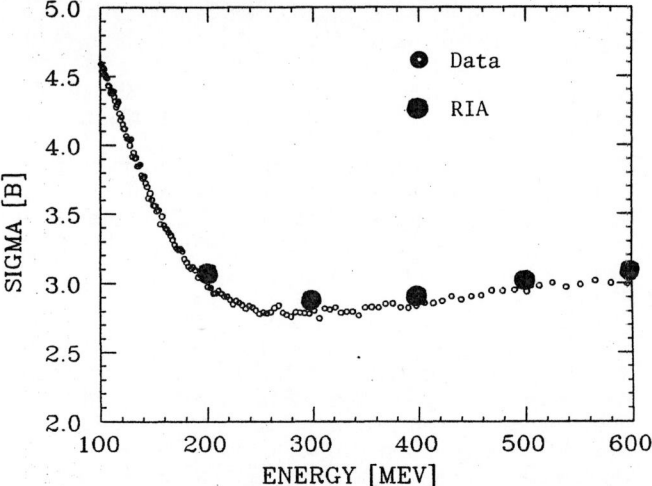

Fig. 5 Neutron total cross sections for ^{208}Pb are shown as points. Solid circles are the RIA calculations of ref. 12.

Acknowledgements

The authors are grateful to J.A. Harvey for suggesting the clever compensation technique with radiogenic lead ores and to J.A. Harvey, D.J. Horen and Oak Ridge National Laboratory for assistance with the separated isotope targets. We also wish to thank Eric Montei for his capable help with data reduction.

References

[1] V. McLane, C.L. Dunford and P.F. Rose, Neutron Cross Sections, Vol. 2. Academic Press, San Diego (1988)

[2] R. Kozak and D.G. Madland, Nucl. Phys. A509, 664 (1990)

[3] R. Kozak and D.G. Madland, Phys. Rev. C39, 1461 (1989)

[4] W. Schimmerling, T.J. Devlin, W.W. Johnson, K.G. Vosburg and R.E. Mischke, Phys. Rev. C7, 248 (1973)

[5] J. Franz, H.P. Grotz, L. Lehmann, E. Rössle, H. Schmitt and L. Schmitt, Nucl. Phys. A490, 667 (1988)

[6] D.J. Horen, C.H. Johnson, J.L. Fowler, A.D. MacKellar and B. Castel, Phys. Rev. C34, 429 (1986)

[7] M.S. Moore, Nucl. Inst. Meth. 169, 245 (1980)

[8] R.B. Schwartz, R.A. Schrack and H.T. Heaton, NBS Monograph 138 (1974)

[9] E.D. Cooper, B.C. Clark, S. Hama and R.L. Mercer, Phys. Rev. 206, 588 (1988)

[10] S. Hama, B.C. Clark, E.D. Cooper, H.S. Sherif and R.L. Mercer, Phys. Rev. C41, 2737 (1990)

[11] S. Hama, private communication (1991)

[12] C.J. Horowitz, D.P. Murdock and Brian D. Serot, Computational Nuclear Physics, Springer–Verlag (1991)

TRANSMISSION OF SEVERAL MEV NEUTRONS AND ASSOCIATED GAMMA RAYS THROUGH IRON, LEAD AND GRAPHITE SYSTEMS

Kazuo Shin, Yoshiaki Ishii, Kagetomo Miyahara
Department of Nuclear Engineering, Kyoto University
Yoshida, Sakyo-ku, Kyoto 606, Japan

Yoshitomo Uwamino
Institute for Nuclear Study, University of Tokyo
Midori-cho 3-2-1, Tanashi, Tokyo 188, Japan

Hideyuki Sakai
Department of Physics, University of Tokyo,
Hongo, Bunkyo-ku, Tokyo 113, Japan

Shigeo Numata
Institute of Technology,Shimizu Corporation,
Echujima, Koto-ku, Tokyo 135, Japan

This paper describes spectral measurements of several-tens-MeV neutrons and associated gamma rays transmitted through iron, lead and graphite shields. Monte Carlo calculations by MORSE with the DLC-87 HILO multigroup cross sections reproduced the measured data well for the graphite shield, but gave partly higher values at energies 15-25 MeV for the iron shield, and overpredicted the lead transmitted spectra above 15 MeV. Gamma-ray spectra obtained behind the graphite shield were well explained by the transmission of source gamma rays from the Cu target through the shield. However the measured photon spectra for the iron and lead shields were much higher than those of transmitted source photons. The discrepancies were not explained even by including neutron induced gamma rays caused by neutrons up to 15 MeV.
(benchmark experiment, neutron transport, intermediate energy, neutron, gamma ray, associated gamma ray, multigroup cross section, DLC87, Monte Carlo calculation)

Introduction

Neutron transport calculations in the energy range above 15 MeV are often made based on multigroup cross section data like DLC-87/HILO/1/. Since the multigroup cross sections were evaluated mostly depending on model calculations, it is necessary to test the data by benchmark experiments to assure the accuracy of design calculations made based on those tabulated cross sections. However, at this moment, there are very few neutron transmission experiments which can be utilized to benchmark those tables in the energy range above 15 MeV./2,3/ No experiment has been found for the secondary gamma ray production in and penetration through bulk materilas for source neutrons in the above energy range.

The objective of this work is to provide benchmark data for transmission calculation through several materials of neutrons in the energy range up to several tens of MeV and secondary gamma-rays associated with these neutrons.

Experimental Method

The experiment was made at the AVF cyclotron facility of Osaka University. A 1-cm thick (i.e., proton stopping range) Cu target was irradiated by 65-MeV protons and generated neutrons in the forward direction were passed to an experimental room through a 7.5-cm diameter iron-lined concrete collimator of 50-cm length. The experimental setup is depicted in Fig. 1. Shield systems of iron, lead and graphite were set at locations very close to the collimator exit. The size of the shields was about 40 cm by 40 cm in cross section and 10 cm to 90 cm in thickness.

Transmitted neutrons and gamma rays were measured just behind the shield system by a 3-inch (7.6 cm) diameter by 3-inch height NE-213 scintillator. The background data were obtained in the same geometrical condition except that the collimator was closed by an iron plug. The

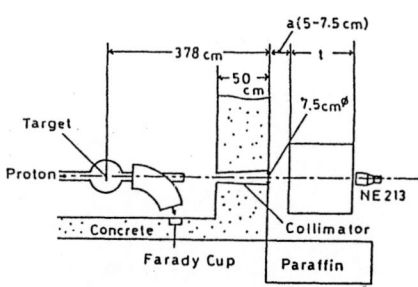

Fig. 1 Experimental setup

source spectrum measurement was made without the shield at a location 538 cm down from the Cu target.

Both of the obtained neutron and gamma-ray pulse-height spectra were unfolded to energy spectra using the FERDO-U code/4/ with neutron response fucntions constructed from measured response data,/5/ and those for gamma rays obtained by Monte Carlo calculations./6/

Results and Discussions

1.Source Spectrum and Transmission Calculation

The measured neutron source spectrum is showed in Fig. 2. The evaporation spectrum was utilized to extrapolate the spectrum to lower energy range beyond the energy accepted by the discrimination level of the measurement. This is shown by the solid line in Fig. 2.

The calculations of the transmitted neutron and photon spectra through the shields were made by the MORSE code/10/ with multigroup cross sections DLC-87,/1/ where source neutrons and gamma rays were emitted from the target within a sharp cone of 3.14×10^{-4} sr to account for the collimator effect.

2. Results for Neutrons

Spectra of neutrons transmitted through the iron shields are showed in Fig. 3. The measured

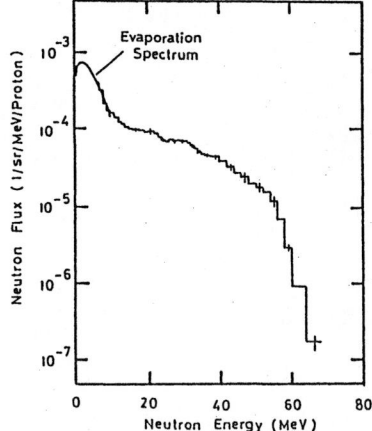

Fig. 2 Measured neutron source spectra

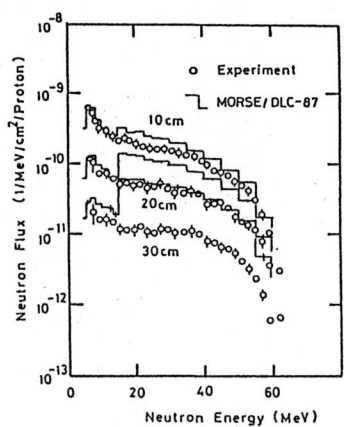

Fig. 4 Spectra of neutrons transmitted through the lead shields

spectra exhibit increasing fluctuation due to poorer statistics as the shield thickness becomes larger. The overall agreement between the calculation and the experiment is good except for the systematic disagreement seen at energies 15-30 MeV. By examination of the total cross sections of iron in the DLC-87, it was found that the total cross section was underestimated at energies 15-30 MeV in the DLC-87. This was the reason for the overprediction of the transmitted neutron fluxes by the MORSE/DLC87 calculation observed in the corresponding energy range. This caused about 50 % overestimation of the integrated neutron flux above 5 MeV by the calculation for 50-cm thickness.

Figure 4 shows comparison of the lead transmitted neutron spectra between the MORSE/DLC-87 calculation and the experiment. The MORSE calculation overestimates the spectra very much in the energy range above 15 MeV, the tendency becoming more pronounced as the lead thickness increases. Because of this overestimation, the fluxes in the lower energy range below 15 MeV are also slighly overpredicted for the 30-cm thick shield, probably through the down scattering of neutrons from higher energies.

The elastic scattering was ignored in evaluating the lead data above 15 MeV, and

nonelastic cross sections were treated as total cross sections in the DLC-87. Neglect of the elastic scattering is the main reason for the overprediction of the transmitted fluxes. The nonelastic cross sections seemed to be somewhat underestimated above 20 MeV. This may be another reason for the overprediction of the transmitted neutron fluxes.

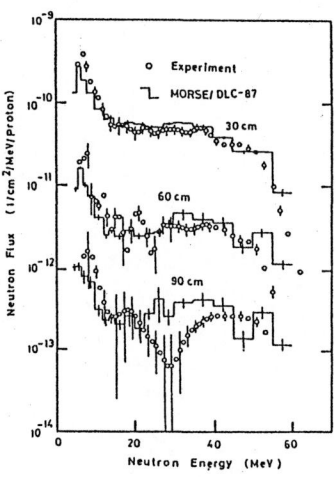

Fig. 5 Spectra of neutrons transmitted through the graphite shields

Spectra of neutrons transmitted through graphite shields are shown in Fig. 5. The overall agreement between the calculation and the experiment is good. A small underestimation by the MORSE/DLC-87 calculation is observed at the lowest energy part below 10 MeV. The reason for this is still not clear. The big depression around 30 MeV in the measured energy flux for 90-cm thickness is due to the subtraction of the fluctuating background spectrum and is not real.

3. Results for Gamma Rays

Gamma-ray spectra were analyzed only for thiner shield cases to avoid errors due to the relatively larger amount of the background.

The measured spectra were compared with transmitted spectra of source gamma rays. The gamma-ray Monte Carlo calculation was made by the MORSE/DLC-87 with the gamma-ray source spectrum. The calculated spectra agreed very well with the measured one for the graphite, as

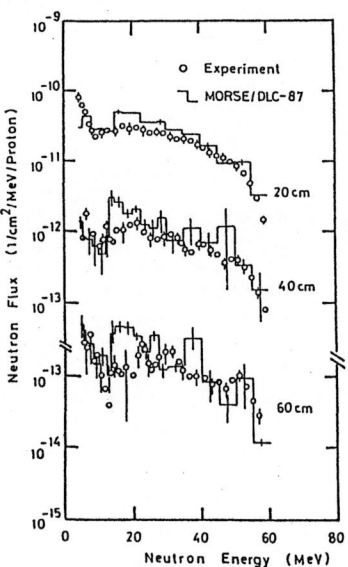

Fig. 3 Spectra of neutrons transmitted through the iron shields

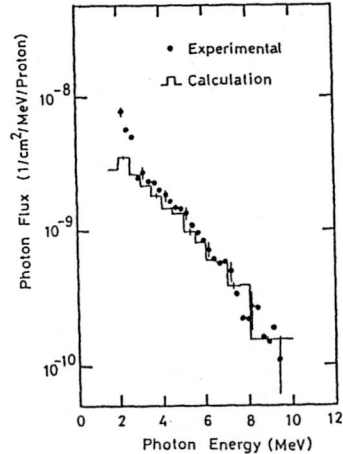

Fig.6 Spectra of gamma rays transmitted through
the 30-cm graphite shield

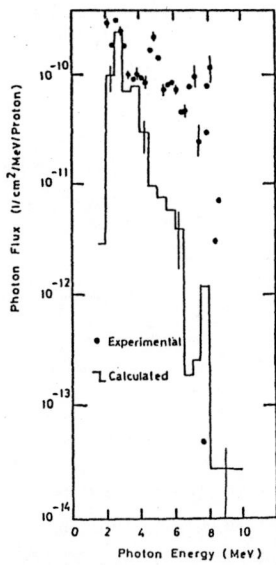

Fig.8 Spectra of neutron induced gamma rays
transmitted through the 10-cm lead shield

is showed in Fig. 6. Since the attenuation of gamma rays is not large in the light material, the gamma ray flux behind the graphite shield is dominated by the penetration of source gamma rays.

However this was not true for the heavier materials. The calculated values of penetrating gamma rays were only about one half of the measured values. The remaining part should be due to neutron induced secondary gamma rays produced in the shields.

The neutron-gamma ray coupled transport was traced with the aid of MORSE using the DLC-87 cross sections with the neutron source spectrum of Fig. 2. To yield experimental data for the neutron-induced gamma-ray spectrum, the penetrating fluxes of the source gamma rays were subtracted from the corresponding measured values.

The obtained results are compared in Figs. 7 and 8. The calculated values are very small as compared with the experimental values for both shields. Although the above calculation covered the neutron energy from the thermal group up to 65 MeV, the gamma-ray production cross-section data above 15 MeV were set zero in the DLC-87.

The discrepancies in Figs. 7 and 8 may be attributed to the contribution of reaction gamma rays by high energy neutrons in this energy range. The contribution from slow neutrons produced by multiple scattering of neutrons in the target room may be the other possible reason.

References

1) R. G. Alsmiller, Jr. and J. Barish, Nucl. Sci. Eng. 80, 448 (1982).
2) K. Shin et al., Nucl. Sci. Eng. 71, 294 (1979).
3) Y. Uwamino et al., Nucl. Sci. Eng. 80, 360 (1982).
4) K. Shin et al., Nucl. Technol. 53, 78 (1981).
5) K. Shin et al., to be published in Nucl. Instr. Methods.
6) K. Shin et al., J. Nucl. Sci. Technol. 16, 390 (1979).
7) T. Nakamura et al., Nucl. Instr. Methods 151, 493 (1978).
8) T. Nakamura et al., Nucl. Sci. Eng. 83, 444 (1983).
9) K. Shin et al., Phys. Rev. C29, 1307 (1984).
10) E. A. Straker et al., "The MORSE Code-A Multigroup Neutron and Gamma-Ray Monte Carlo Transport Code", ORNL-4585 (1970).

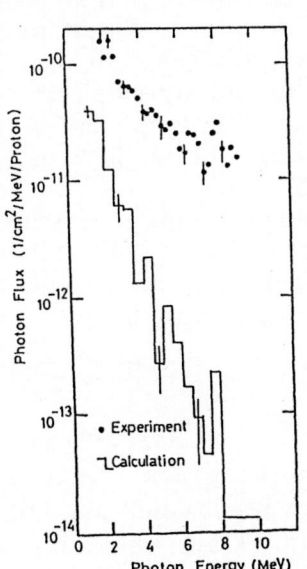

Fig.7 Spectra of neutron induced gamma rays
transmitted through the 20-cm iron shield

MEASUREMENT OF ACTIVATION CROSS SECTIONS BY USING p-Be NEUTRONS OF ENERGY UP TO 40 MeV

Yoshitomo Uwamino

Institute for Nuclear Study, University of Tokyo
3-2-1, Midori-cho, Tanashi, Tokyo 188 Japan

Hiroshi Sugita, Yuhri Kondo and Takashi Nakamura

Cyclotron and Radioisotope Center, Tohoku University
Aoba, Aramaki, Aoba-ku, Sendai 980 Japan

Abstract: Using a simple target system consisting of a Be disk and a water coolant, a semi-monoenergetic intense neutron field for an activation experiment was developed by using proton beams of 9 different energies between 20 and 40 MeV. Activation experiments were performed at this neutron field using Na, Mg, Al, Si, Ca, V, Cr, Mn, Cu, Zn and Au samples. Gamma rays from the induced radioactivities having half lives of between 2 minutes and 2.6 years were measured, and production rates of 32 radionuclides were obtained. The measured production rates were converted into excitation functions by using the least square method and the two unfolding codes of SAND-II and NEUPAC with an aid of quoted data and calculational results of the ALICE/LIVERMORE82 code.

Introduction

The situation of the activation cross section data compilation for the energy region above 20 MeV is still very poor, since preparing a monoenergetic neutron source for the measurement at this energy region presents some difficulties.

Using a proton beam and a simple target system consisting of a Be disk and a water coolant, an intense semi-monoenergetic neutron field for activation experiment of energy up to 40 MeV was developed [1] at the SF cyclotron of the Institute for Nuclear Study, University of Tokyo. The characteristics of the neutron field was studied, and activation experiments with 11 natural samples were performed.

Neutron Field and Activation Experiment

Proton beams of energy at every 2.5 MeV between 20 and 40 MeV were transported to 1-mm-thick (E_p=20 to 37.5 MeV) and 2-mm-thick (E_p=40 MeV) Be targets which were backed by a water coolant. Protons partly lose their energy at the Be target, and totally stop in the water coolant.

The neutrons produced by this target system were measured by an NE213 scintillation counter. The energy resolution of the unfolded spectrum is not very good due to the detector system and the unfolding process. The peak of a measured spectrum was replaced by a calculated one with the peak area conserved.[1]

The activation samples were irradiated by turns for one hour at 20 cm from the Be target. The Zn and Cu samples were also irradiated for 10 minutes, since these elements produce isotopes of short half lives. The irradiated samples were placed on a Ge detector for gamma-ray spectrometry, and these irradiations were repeated for 9 proton energies.

Evaluation of Cross Sections

The measured reaction rates were converted into excitation functions [2] by the SAND-II [3] and the NEUPAC [4] unfolding codes with an aid of the guess excitation functions which were mostly derived from calculational results of the ALICE/LIVERMORE82 code.[5] An estimated error of the initial guess is also required by the NEUPAC code. If experimental data are available, this error was estimated by comparing the initial guess with the experimental data. For the other case the error was set to be 100%.

The excitation functions were also evaluated by the least square fitting. (L.S.F.) A caluculated guess value was absolutely adjusted to the measured reaction rates as shown in the following,

$$R = \sum_i [A_r \sum_j \int_0^{E_{max}} \sigma_j(E)\Phi_i(E)dE]^2 \quad (1)$$

$$\sigma_j(E) = k_j g_j(E) \quad (2)$$

where A_i is an activation rate at the i-th neutron field, $g_j(E)$ is an calculated excitation function by ALICE for j-th reaction channel, and $\Phi_i(E)$ is the neutron flux. k_j is determined to make R minimum. If a fertile material consists of a single isotope and the reaction channel can be uniquely defined, the parameter j takes only 1.

Results

The reactions of which cross sections were measured are listed in table 1 with their threshold energies. Since the fertile element consists of a single isotope, the $^{23}Na(n,2n)^{22}Na$ reaction is definitely identified as a unique reaction. Natural V consists of ^{50}V and ^{51}V isotopes, but the ^{51}Ti production can be considered as a single reaction of $^{51}V(n,p)^{51}Ti$, since the production of ^{51}Ti from ^{50}V is impossible. Because of a high threshold energy of the $^{52}Cr(n,5n)^{48}Cr$ reaction, the $^{50}Cr(n,3n)^{48}Cr$ reaction is also clearly identified. From similar reasons, upper 17 reactions in table 1 could be identified.

The ^{24}Na can be produced by the $^{24}Mg(n,p)$, $^{25}Mg(n,d)$ and $^{26}Mg(n,t)$ reactions, and this production is noted as Mg(n,xnp) reaction. The lowest threshold energy among the possible reactions is listed with the corresponding reaction. The lower 15 reactions are treated in the similar way. Some Q values of the possible reactions are positive, for example $^{50}V(n,\alpha)^{47}Sc$. The cross sections for thermal neutrons, however, are negligibly small, and they are also treated as fast-neutron activation.

The obtained cross section curves are shown in the Figs. 1 through 6 for the reactions of $^{23}Na(n,2n)^{22}Na$, $^{27}Al(n,\alpha)^{24}Na$, $^{51}V(n,\alpha)^{48}Sc$, V(n,xn2p) ^{46}Sc, $^{197}Au(n,4n)^{194}Au$, and $^{197}Au(n,2n)$ ^{196}Au. The results obtained by the unfolding processes performed by the SAND-II and the NEUPAC codes are shown in a thick zigzag solid line and a histogram, respectively. The vertical bars attached on the histogram show errors estimated by the NEUPAC code. Results of the least square method are shown in a thick dashed line. If some reference data are available, they are also drawn in the same figure in a thin solid line for ENDF-B/V, in a thin dotted line for ENDF-B/IV, in a thin dashed line for IAEA Tech. Rep. 273, in a thin dash and dot line for the data of Greenwood, [6] and in circles for experimental values compiled by McLane et al., [7] which is the 4th edition of the BNL-325 and referred as BNL-325.

References

1. Y. Uwamino, T. Ohkubo, A. Torii, and T. Nakamura, Nucl. Instr. and Meth. **A271**, 546 (1988).
2. D. L. Smith, "A Least-Squares Method for Deriving Reaction Differential Cross-Section Information from Measurements Performed in Diverse Neutron Fields", ANL/NDM-77, Argonne National Laboratory (1982).
3. W. N. McElroy, S. Berg, T. Crockett and R. G. Hawkins, AFWL-TR-67-41, Air Force Weapons Laboratory (1967).
4. T. Taniguchi, N. Ueda, M. Nakazawa and A. Sekiguchi, NEUT Research Report 83-10, Department of Nuclear Engineering, University of Tokyo (1984).
5. M. Blann and J. Bisplinghoff, "Code ALICE/LIVERMORE 82", Lawrence Livermore National Laboratory Report No. UCID 19614 (1983).
6. L. R. Greenwood, " Extrapolated Neutron Activation Cross Sections for Dosimetry to 44 MeV", ANL/FPP/TM-115, Argonne National Laboratory (1978).
7. V. McLane, C. L. Dunford and P. F. Rose, "Neutron Cross Sections", Vol. 2, "Neutron Cross Section Curves", Academic Press, Inc. (1988).

Table 1. Reactions of which cross sections were measured and their threshold energies.

Reaction	Threshold of reaction energy	
$^{23}Na(n,2n)^{22}Na$	13.0 MeV	
$^{27}Al(n,\alpha)^{24}Na$	3.2	
$^{51}V(n,\alpha)^{48}Sc$	2.1	
$^{51}V(n,p)^{51}Ti$	1.7	
$^{50}Cr(n,3n)^{48}Cr$	24.1	
$^{50}Cr(n,2n)^{49}Cr$	13.3	
$^{55}Mn(n,p\alpha)^{51}Ti$	9.2	
$^{55}Mn(n,4n)^{52}Mn$	31.8	
$^{55}Mn(n,2n)^{54}Mn$	10.4	
$^{63}Cu(n,3n)^{61}Cu$	20.1	
$^{63}Cu(n,2n)^{62}Cu$	11.0	
$^{65}Cu(n,p)^{65}Ni$	1.4	
$^{64}Zn(n,t)^{62}Cu$	10.2	
$^{64}Zn(n,3n)^{62}Zn$	21.3	
$^{64}Zn(n,2n)^{63}Zn$	12.0	
$^{197}Au(n,4n)^{194}Au$	23.2	
$^{197}Au(n,2n)^{196}Au$	8.1	
Mg(n,xnp)^{24}Na	4.4	$^{24}Mg(n,p)$
Si(n,xnp)^{28}Al	3.5	$^{28}Si(n,p)$
Si(n,xnp)^{29}Al	2.5	$^{29}Si(n,p)$
Si(n,xn2p)^{27}Mg	4.3	$^{30}Si(n,\alpha)$
Ca(n,xnp)^{42}K	2.3	$^{42}Ca(n,p)$
Ca(n,xnp)^{43}K	5.4	$^{43}Ca(n,p)$
V(n,xn2p)^{46}Sc	10.1	$^{50}V(n,n\alpha)$
V(n,xn2p)^{47}Sc	-	$^{50}V(n,\alpha)$
Cr(n,xnp)^{52}V	2.7	$^{52}Cr(n,p)$
Cr(n,xnp)^{53}V	2.2	$^{53}Cr(n,p)$
Cu(n,xn2p)^{62m}Co	0.1	$^{65}Cu(n,\alpha)$
Zn(n,xn2p)^{65}Ni	-	$^{68}Zn(n,\alpha)$
Zn(n,xnp)^{64}Cu	-	$^{64}Zn(n,p)$
Zn(n,xnp)^{66}Cu	1.4	$^{66}Zn(n,p)$
Zn(n,xnp)^{68m}Cu	3.4	$^{68}Zn(n,p)$

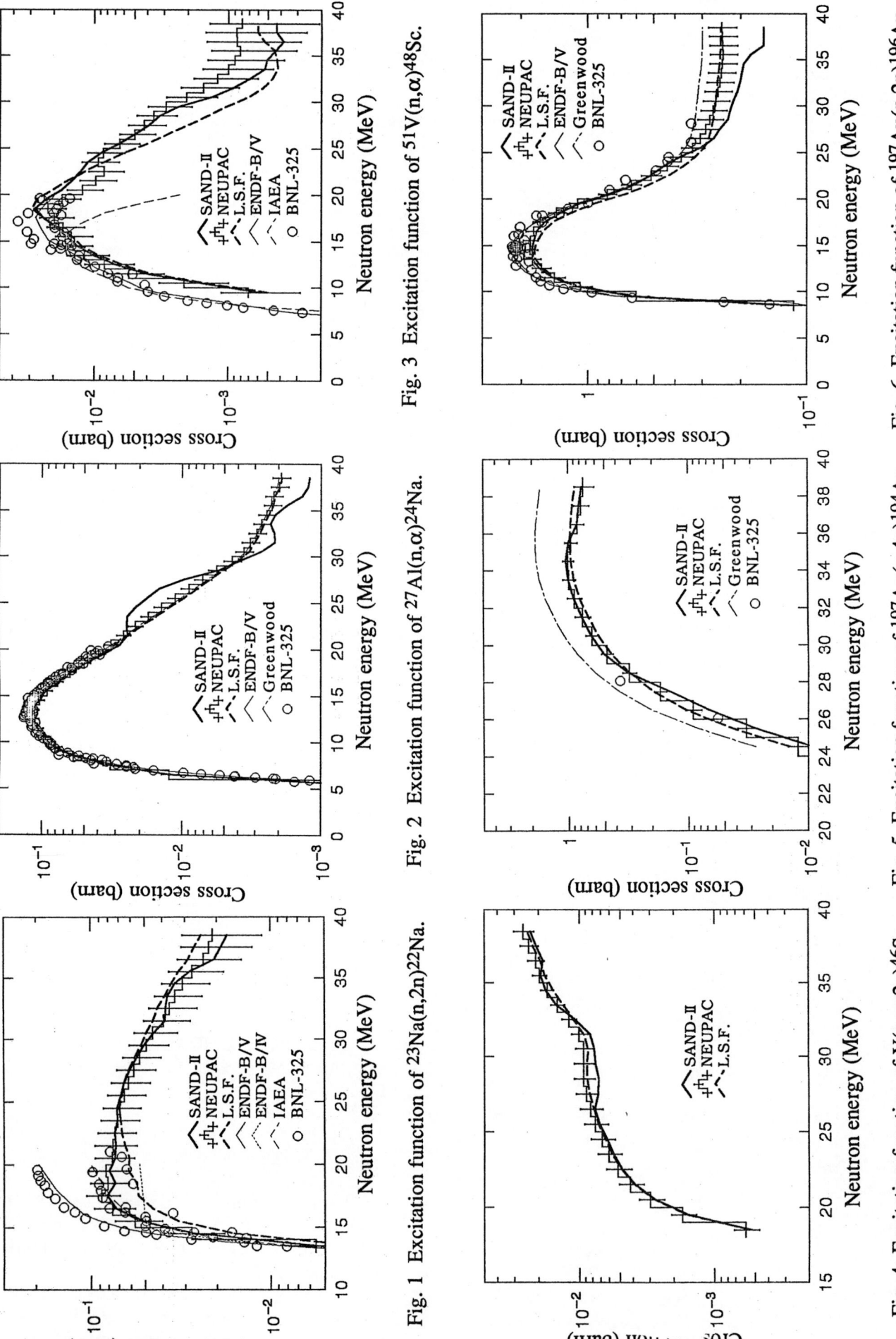

Fig. 1 Excitation function of ^{23}Na(n,2n)^{22}Na.

Fig. 2 Excitation function of ^{27}Al(n,α)^{24}Na.

Fig. 3 Excitation function of ^{51}V(n,α)^{48}Sc.

Fig. 4 Excitation function of V(n,xn2p)^{46}Sc.

Fig. 5 Excitation function of ^{197}Au(n,4n)^{194}Au.

Fig. 6 Excitation function of ^{197}Au(n,2n)^{196}Au.

MEASUREMENT OF THE NITROGEN TOTAL CROSS SECTION
FROM 0.5 eV to 50 MeV, AND ANALYSIS OF THE 433-keV RESONANCE

J. A. Harvey, N. W. Hill, N. M. Larson and D. C. Larson
Oak Ridge National Laboratory
Oak Ridge, Tennessee 37831-6356 USA

Abstract: High-resolution neutron transmission measurements have been made on several thicknesses of nitrogen gas samples from 0.5 eV to 50 MeV at the Oak Ridge Electron Linear Accelerator (ORELA). A preliminary R-matrix analysis has been done for resonances up to 800 keV. An R-matrix analysis of previous data was done by LANL for ENDF/B-VI, including the lowest energy resonance in ^{14}N at 433 keV. They found a spin of 3/2 (with $\ell = 1$) and a peak cross section of 7.0 b. Analysis of the present data yield a spin of 7/2 (requiring $\ell \geq 2$) and a peak cross section of 11.5 b for this resonance. These results are important for transport calculations of neutrons through air. Scattering measurements are planned to determine the parity of this resonance.

(^{14}N high-resolution transmission measurements, resonance parameters, R-matrix analysis)

Introduction

The recent (1990) ENDF/B-VI evaluation of the cross sections for nitrogen by Young, Hale and Chadwick [1] again emphasized the lack of reliable total cross-section data in the low energy region. Nitrogen total cross-section data are needed for investigating problems involving the transport of neutrons through air. The most recent measurement is thirteen years old, and the most recent measurement below 500 keV was in 1961. We have made a series of high-resolution neutron transmission measurements on nitrogen gas samples using several different detectors and flight path lengths at ORELA to provide accurate data from 0.5 eV to 50 MeV.

Experimental Measurements

High Resolution Measurements from 0.1 to 20 MeV

For the high-resolution measurements from 0.1 to 20 MeV we used unmoderated neutrons directly from the ORELA tantalum target produced by 5-nsec wide, 150-MeV electron pulses at a repetition rate of 800 pps. The detector was a 9-by-9 cm NE-110 scintillator, one-cm thick, epoxy-coupled to two 12.5-cm diameter RCA 8854 photomultipliers located 201.568 meters from the tantalum target. This bare (non-coated) scintillator is mounted in a 0.025-mm thick, 17.8-cm-diam mylar reflecting cylinder. A 0.30-g/cm^2 ^{10}B filter was used to eliminate overlap neutrons and a filter of 3.9 cm uranium plus 1.3 cm thorium was used to reduce the background due to gamma rays from the tantalum target. The phototube bases were gated off for 2 microsec to eliminate the gamma flash.

The nitrogen gas (99.995% purity) was contained in stainless steel cylinders [2] with approximately spherical end windows, an inside diameter of 16.14 cm, an inside length of 220.9 cm and a volume of 44,261 cm^3. An identical cylinder was used for the open beam measurement to cancel out the effects of the resonances due to the components of stainless steel. Careful inspection of the sample-in/sample-out ratio showed no sign of large resonances in iron, indicating acceptable compensation of the cylinder windows. Gas pressure gauges on each cylinder were calibrated to an accuracy of 1 psi.

Two transmission measurements were made with nitrogen in one cylinder with pressures of 488 and 490 psi and a vacuum in the second cylinder for the open beam spectrum. The two cylinders were cycled about 100 times. Three measurements were made with nitrogen in the second cylinder with pressures from 480 to 487 psi and a vacuum in the first cylinder. After the transmission measurements on each cylinder were completed the weight of nitrogen and the pressure were measured for each cylinder. The ratio of the weight of nitrogen to the pressure gauge reading for the first cylinder agreed with the similar quantity for the second cylinder to 0.34%. The total cross sections for the five runs were calculated as described below. Since the results were in good agreement, the five runs were added, resulting in a single data set with a weighted-average pressure of 487 psi, a weight of 1759 grams and a sample thickness of 0.3743 atoms/barn with an estimated uncertainty of 0.5%.

^{6}Li Measurements from 0.5 eV to 200 keV

For the transmission measurements in this energy region the collimation was changed so that only moderated neutrons from the water moderator of the ORELA target could reach the neutron detector, located at a flight path length of 80.117 meters. The detector was a ^{6}Li glass scintillator, 12.5 mm thick and 11.1 cm in diameter, mounted in a 0.025-mm-thick, 15-cm-diameter mylar reflecting cylinder between two RCA 8854 photomultiplier tubes. The accelerator was operated with burst widths of 30 nsec and a repetition rate of 80 pps. A 1.6-mm-thick cadmium filter was used to eliminate overlap neutrons and a 0.64-cm Pb filter to reduce the background due to gamma rays from the water moderator of the target.

Two transmission measurements were made, one with 993 g of nitrogen gas in one cylinder with a sample thickness of 0.2113 atoms/barn and a vacuum in the other cylinder as the compensator, and the second one with 979 g of nitrogen in the second cylinder with a sample thickness of 0.2092 atoms/barn and a vacuum in the first cylinder. Since the cross sections computed for these two runs agreed, the two data sets were averaged.

Measurements with the Be Block Target to 50 MeV

For measurements at high energies the ORELA Be block target was used and a thick NE-110 scintillation detector was located at a flight path length of 79.660 meters. The scintillator was 10.2 cm in diameter and 7.6 cm thick, and was mounted on a selected RCA 8854 photomultiplier. The ORELA was operated at a repetition rate of 800 pps with a pulse width of 24 ns. The flight path from 10 to 80 meters was filled with nitrogen gas to a pressure of one atmosphere for this high energy run.

Processing of the Transmission Data

The data for all three detectors have been corrected for the dead time (1104 ns) of the time digitizer, and several backgrounds which total less than 3% over most of the region for the ^{6}Li data and less than 1% and as small as 0.02% for the high resolution NE-110 data. The

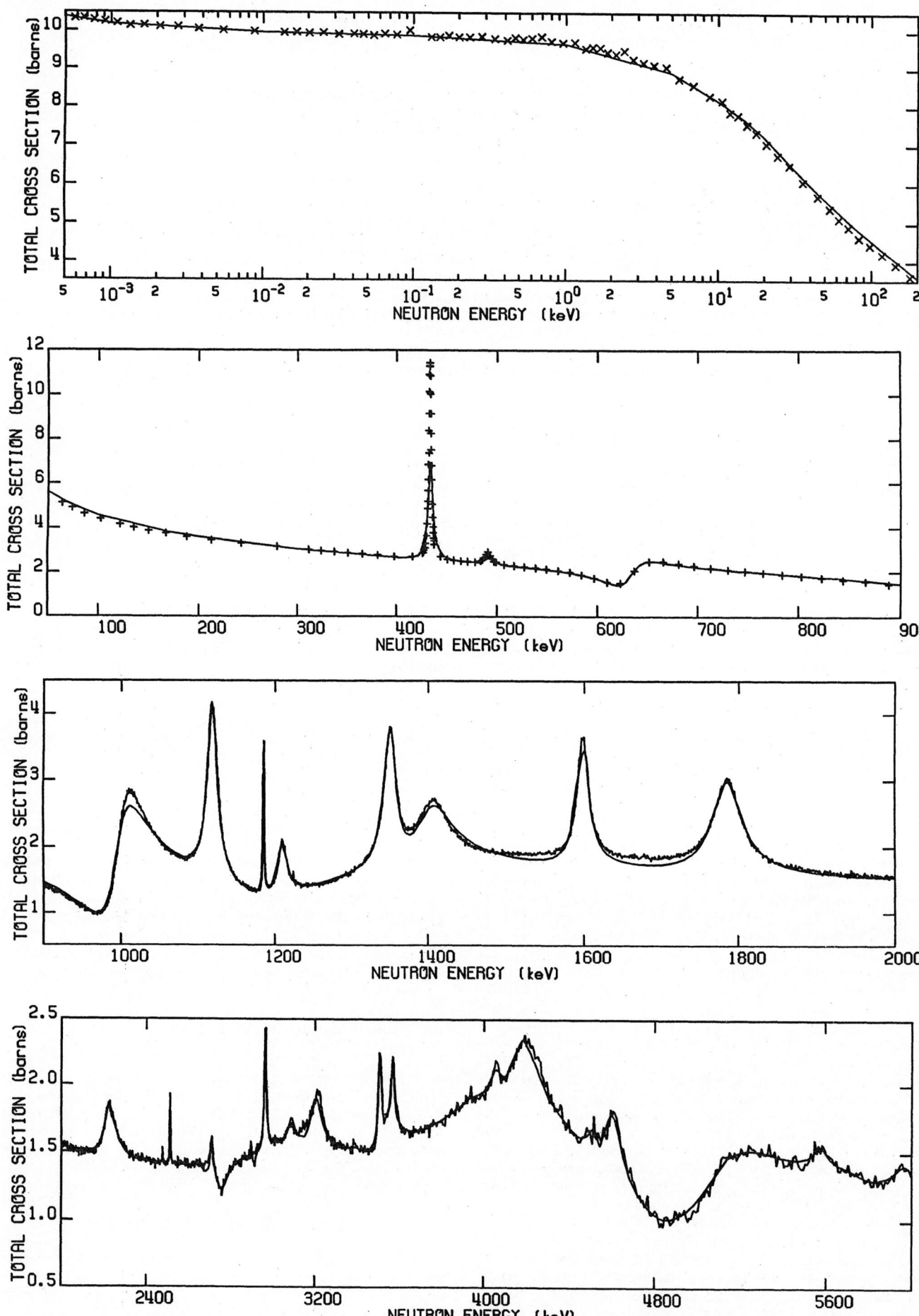

Fig. 1. Total cross-section data from the present measurements, compared with the ENDF/B-VI evaluation for ^{14}N.

backgrounds for the ^{6}Li data consist of a constant beam-independent background, a 17.6-μsec gamma-ray background from the capture of neutrons in the water moderator, overlap neutrons from the previous burst, and a time-dependent background arising from neutrons scattered from the ^{6}Li glass scintillator. For the high-resolution data, the backgrounds consist of only a constant and a very small 17.6-μsec gamma-ray background. Additional details on these backgrounds are discussed in a published paper and reports [3,4,5].

Total Cross-Section Results

The total cross sections for several energy regions are shown in Fig. 1, compared to the total cross section from the ^{14}N evaluation from ENDF/B-VI [1]. The upper plot is data from the ^{6}Li scintillation detector and the remainder are high-resolution 200-m data. Complete data from 0.5 eV to 50 MeV are available from the National Nuclear Data Center, BNL. The present results generally agree with the R-matrix analysis for ENDF/B-VI up to 2.5 MeV. In particular, our data confirm the evaluation in the region from 400 eV to 10 keV, where there was essentially no data. The primary difference is the 433-keV resonance.

R-Matrix Analysis Below 800 keV

As part of the ^{14}N evaluation [1] for ENDF/B-VI, the available total cross section data were included in an R-matrix analysis of the resonance region. For the resonance at 433 keV, $J = 3/2$ provided an acceptable fit to the total cross section data, with a peak cross section of 7.0 b. Except for $\ell = 0$ (which produces an interference minimum), the fits were insensitive to the ℓ-value; $\ell = 1$ was assumed.

A preliminary R-matrix analysis on our new data has been done below 800 keV using the computer code SAMMY. In this paper we present results for the two resonances below 500 keV, where the previous data base was poorest. The lowest energy resonance observed in ^{14}N is at 433 keV where we find the only acceptable fit to the data is with $J = 7/2$ which requires $\ell = 2$ or 3 (or up to 5). The parameters for this resonance are $E = 433.35 \pm 0.03$ keV, $\Gamma_n = 2.46 \pm 0.04$ keV, $\Gamma_\gamma = 40 \pm 70$ eV. Γ_p was held fixed at 0.0 since there is no evidence of the (n,p) reaction near this resonance. The large uncertainties on the capture width reflects the fact that the transmission data for this resonance are not sensitive to this parameter since it is so small compared to Γ_n. The fit for this resonance is shown in Fig. 2.

Attempts to fit this resonance with J less than 7/2 were not successful since the peak cross section of 11.5 barns could never be reached. For $J = 9/2$ a fit could be obtained which required the sum of Γ_γ and Γ_p to be approximately 500 eV. Since the Γ_p width must be very small, as noted above, this would require an unreasonable value for Γ_γ; hence J cannot be 9/2.

The other resonance below 500 keV is at 491.2 ± 0.1 keV which was assigned [6] as $J = 1/2$, $\ell = 1$ with a relatively large proton width. Our analysis for a fixed value of Γ_γ of 4.0 eV gives $\Gamma_n = 1.9 \pm 0.1$ keV and $\Gamma_p = 5.2 \pm 0.3$ keV which is in agreement with values listed in reference 6, namely $\Gamma_p = 6 \pm 2$ keV and a total width of 7.8 ± 0.5 keV.

The preliminary R-matrix analysis has been extended to 800 keV.

Conclusions

Accurate high-resolution transmission data have been obtained from 0.5 eV to 50 MeV using both unmoderated and moderated neutrons from the ORELA tantalum target and high-energy neutrons from the Be block target. The previous data below 500 keV, reported in 1961,

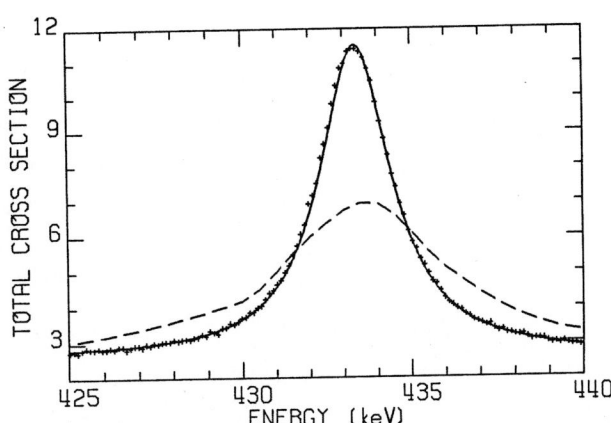

Fig. 2. Data for the 433-keV resonance, compared with our SAMMY fit (solid line). The dashed line is the ENDF/B-VI evaluation.

had inadequate resolution to define the peak cross section of the resonance at 433 keV. The present results significantly improve the data base below 500 keV, as well as provide the highest resolution data available above 500 keV. Using our new high resolution data, we have presented results for the two resonances below 500 keV, as analyzed with the R-matrix code SAMMY. The lowest lying resonance, at 433 keV, was found to have a much larger peak cross section (11.5 b rather than 7.0 b), and $\ell \geq 2$. These results will affect neutron transport calculations through air. Differential elastic-scattering measurements are planned for the resonance region to determine the ℓ values of the resonances.

Acknowledgments

The authors wish to thank Gerry Hale (LANL) for providing us with the resonance parameters used in the ^{14}N evaluation for ENDF/B-VI. This research was sponsored by the Office of Energy Research, Division of Nuclear Physics, U.S. Department of Energy, under Contract No. DE-AC05-84OR21400 with Martin Marietta Energy Systems, Inc., and under Interagency Agreement No. 0046-C083-A1 with the Defense Nuclear Agency.

References

1. P. Young, G. Hale and M. Chadwick, ENDF/B-VI evaluation for ^{14}N, MAT 725 (June 1990).
2. R. R. Winters, R. F. Carlton, C. H. Johnson, N. W. Hill and M. R. Lacerna, *Phys. Rev.* **C43**, 492 (February 1991).
3. M. S. Pandey, J. B. Garg and J. A. Harvey, *Phys. Rev.* **C15**, 600 (1977).
4. D. C. Larson, C. H. Johnson, J. A. Harvey and N. W. Hill, *Measurement of the Neutron Total Cross Section of Fluorine from 5 eV to 20 MeV*, ORNL/TM-5612 (October 1976).
5. D. C. Larson, N. M. Larson, J. A. Harvey, N. W. Hill and C. H. Johnson, *Application of New Techniques to ORELA Neutron Transmission Measurements and Their Uncertainty Analysis: The Case of Natural Nickel from 2 keV to 20 MeV*, ORNL/TM-8203 (October 1983).
6. S. F. Mughabghab, M. Divadeenam and N. E. Holden, *Neutron Cross Sections*, Vol. 1, Part-A, Academic Press, New York (1981).

FISSION CROSS SECTION RATIOS FOR ^{233,234,236}U RELATIVE TO ^{235}U FROM 0.5 to 400 MeV

P. W. Lisowski, A. Gavron, W. E. Parker,
S. J. Balestrini
Los Alamos National Laboratory, Los Alamos, NM, USA

A. D. Carlson, O. A. Wasson
National Institute for Science and Technology, Gaithersburg, MD, USA

N. W. Hill
Oak Ridge National Laboratory, Oak Ridge, TN, USA

ABSTRACT: Neutron-induced fission cross section ratios from 0.5 to 400 MeV for samples of ^{233,234,236}U relative to ^{235}U have been measured at the WNR neutron Source at Los Alamos. The fission reaction rate was determined using a fast parallel plate ionization chamber at a 20-m flight path. Cross sections over most of the energy range were also extracted using the neutron fluence determined with three different proton telescope arrangements. Those data provided the shape of the ^{235}U(n,f) cross section relative to the hydrogen scattering cross section. That shape was then normalized to the very accurately known value for ^{235}U(n,f) at 14.1 MeV which will allow us to obtain cross section section values from the ratio data and our values for ^{235}U(n,f).

(^{233,234,236,238}U cross section ratios to ^{235}U, proton-recoil telescope, neutron flux measurement techniques)

Introduction

Neutron induced fission is quite possibly the most important nuclear reaction used in modern technology. Especially for ratio measurements to ^{235}U, the experimental techniques involved are relatively simple compared to most nuclear physics experiments; yet, there are substantial disagreements among published data. In the region above about 20 MeV, there is virtually no information at all, except for the work of this group. Because of the availability of the Los Alamos WNR neutron source [1], it is now possible to perform fission cross section measurements for multiple samples in a single experiment covering a broad energy range. The measurements that we are reporting here are part of a series designed to help resolve discrepancies for all of the long-lived actinides and to provide information about fission cross sections at energies above 20 MeV.

Experimental Procedure

The Los Alamos National Laboratory high-intensity source [1] uses 800 MeV pulsed proton beam from the Clinton P. Anderson Meson Physics Facility (LAMPF) incident on a 7.5-cm long, 3-cm diam tungsten target to produce a 'white' source of neutrons extending to hundreds of MeV. This experiment was performed using a 20-m flight path which viewed the neutron source at a production angle of 60°. The proton beam consisted of 150 ps wide pulses separated by 3.6 ms with about 3×10^8 protons in each pulse. The macroscopic duty factor of LAMPF gave a rate of about 8000 of these proton pulses/second.

The neutron beam was contained in an evacuated flight tube and passed through a 2.54 cm thick polyethylene (CH$_2$) filter to reduce frame overlap; a permanent magnet was used to sweep out charged particles, and a system of three iron collimators as shown in Ref. 2, gave the beam diameter of 12.7 cm at the fission sample location. Monte Carlo calculations performed to determine neutron inscattering from the collimators showed the amount to be less than 0.01% at 10 MeV.

The fission reaction rate was measured in a fast parallel plate ionization chamber holding multiple foils of oxide fission material 10.2 cm in diameter. In contrast to earlier fission-foil deposits which were vacuum deposited, these were electroplated onto 50 mg/cm^2-aluminum covered 127 mm thick stainless steel backings. The foils used in this measurement were added to the chamber used earlier[2,3] and additional statistics for ^{232}Th, ^{237}Np, ^{238}U, and ^{239}Pu fission ratios to ^{235}U were also taken. A ^{252}Cf deposit was included in the chamber to gain match pulse height spectra and for diagnostic purposes. Flight paths used for these results were obtained using ^{12}C neutron transmission resonances.

After passing through the fission chamber the neutron flux was determined using two detector systems. Although the setup for these fission measurements is similar to that described in [3], there are some differences. In our previous experiments, we were limited in energy range by kinematic resolution and by the thickness of the Si(Li) detector used in the Annular Proton Telescope (APT) for flux measurements. More importantly, the APT design made adding a second instrument downstream to cover a higher energy region impossible. We therefore replaced the APT with a low-energy telescope (LET) which covered the region from about 3 MeV up to 30 MeV using a single 0.5-cm thick Si(Li) detector. The LET employed a 2.54×10^{-4} cm thick CH$_2$ radiator located in an evacuated cylinder with an arm positioned in the vertical plane at an angle of 16.4° to allow us to measure protons from H(n,p) scattering. We then installed a second flux measuring detector system, a medium-energy proton recoil telescope (MET), to cover the neutron energy range from about 26 MeV to 250 MeV. That device consisted of one 7.62-cm square x 0.16-cm thick plastic scintillator paddle, a second 7.62-cm square 0.64 cm thick paddle, and a $7.62 \times 7.62 \times 17.78$ cm^3 CsI(Tl) total energy detector aligned at a scattering angle of 15° relative to the incident neutron beam axis. Two thicknesses of carefully matched CH$_2$ and high-purity graphite targets were used in this system. The beam was then dumped in a concrete shield 15 meters from the MET. A measurement of the neutron fluence obtained using the ^{235}U(n,f) yield rate and the fission cross section results is shown in Ref 4, where the solid line is an intra-nuclear cascade calculation. These measurements agree with the fluence data up to 30 MeV previously obtained using the APT. Additional measurement details are available in another contribution to this conference [5].

An off-line analysis of the data was performed to subtract a small time-uncorrelated background, correct the fission rate spectra for the background resulting from high-energy (E$_n$ > 35 MeV) neutron interactions in the steel backing and oxide content of the fission deposits, and to correct for neutron transmission through any upstream material in the chamber. The chamber was also disassembled and the fission foils were alpha counted to determine their thicknesses. The results of a mass spectroscopic analysis provided by ORNL was used to identify contaminant mass contributions in cases where individual alpha particle peaks could not be resolved. We measured fission cross section ratio data for all of the major actinide contaminants during this experiment and those data were used to correct the results shown here.

Results and Conclusions

Values of our fission cross section ratios in 3% energy bins are shown in Fig. 1 for ^{233}Th,^{234}U, and ^{236}U relative to ^{235}U. This is more than half of the total data

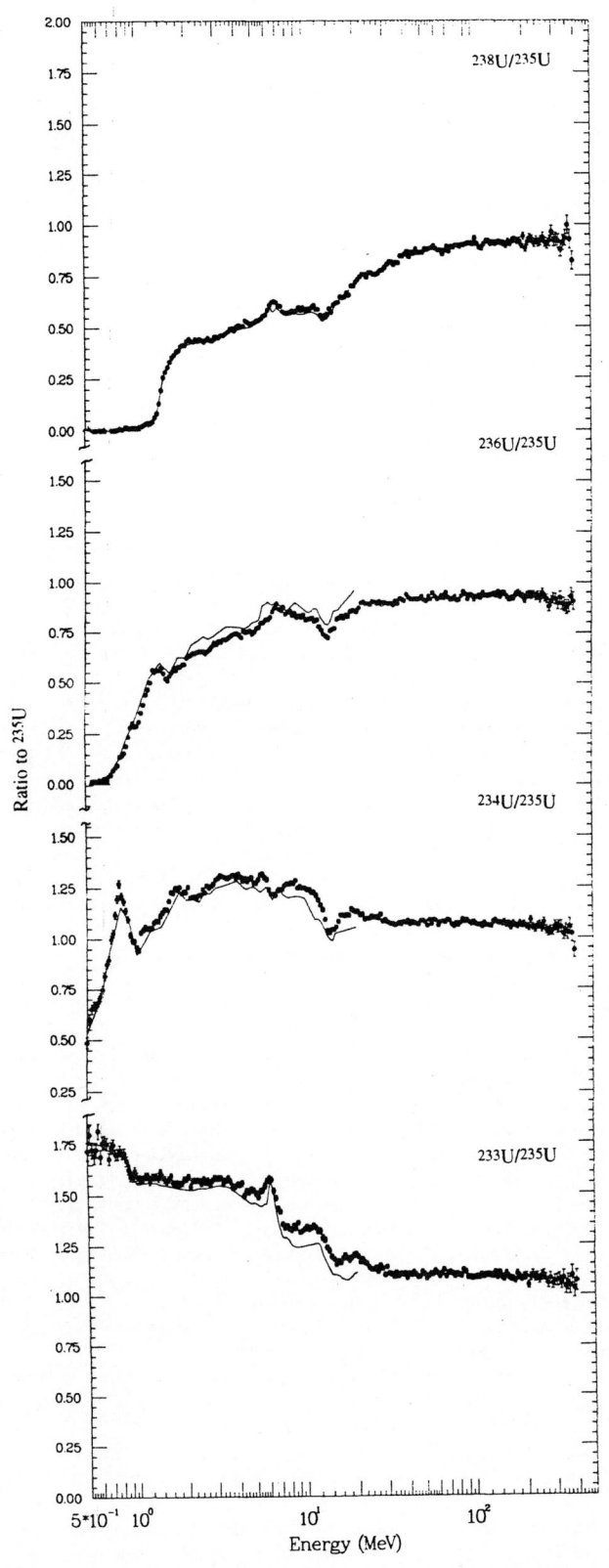

taken. The solid lines show ENDF/B-V or, in the case of ^{236}U and ^{238}U, ENDF/B-VI ratios, which for ^{234}U and ^{236}U have differences of about 2 - 4% from our new values. Also included in Fig. 1 is data for ^{238}U, obtained using this same apparatus and including data taken at the same time as these measurements which agrees with the ENDF/B-VI ratio within counting statistics. For ^{233}U, the ENDF/B-V ratio evaluation is about 4% lower than our results up to about 10 MeV and differs by as much as 15% above that. The data of Meadows [6] is about 2% lower than these results. In general, ^{234}U and ^{236}U show reasonable agreement with ENDF/B-V whereas ^{233}U has a larger than expected discrepancy.

Final analysis of all of the fission ratio data is nearly complete. Small corrections for fission fragment detection inefficiency still have to be applied to these data; those corrections are expected to have no effect in the ratio above about 30 MeV and be substantially below 1% at lower energies. We plan to use our $^{235}U(n,f)$ data to convert these ratio results to cross sections.

REFERENCES

1. P. W. Lisowski, C.D. Bowman, and S. A. Wender, Nucl. Sci and Eng., 106, 208 (1990)
2. P. W. Lisowski, J. L. Ullmann, S. J. Balestrini, A. D. Carlson, O. A. Wasson, and N. W. Hill, Proc. of Intl. Conf. on Nuclear Data for Science and Technology, Mito, Japan, May 30-June 3, 1988, p.145
3. A. D. Carlson, O. A. Wasson, P. W. Lisowski, J. L. Ullmann, and N. W. Hill, Proc. of Intl. Conf. on Nuclear Data for Science and Technology, Mito, Japan, May 30-June 3, p.1029, Saikon, Tokyo (1988)
4. P. W. Lisowski, "The Experimental Program at the WNR Neutron Source at LAMPF", Proc. Specialists Meeting on Neutron Cross Section Standards Above 20 MeV, May 21-23, 1991, Uppsala, Sweden, to be published
5. A. D. Carlson, O. A. Wasson, P. W. Lisowski, J. L. Ullmann, and N. W. Hill, "Measurements of the $^{235}U(n,f)$ Cross Section in the 3 to 30 MeV Neutron Energy Region", these proceedings
6. J. W. Meadows, "The Fission Cross Sections of Some Thorium, Uranium, Neptunium and Plutonium Isotopes Relative to ^{235}U", Argonne National Laboratory, ANL/NDM-33, Argonne National Laboratory (1983)

Fig. 1. Fission cross section ratios for $^{233,234,236,238}U$ from the present measurement. The sollid line extending to 20 MeV is from ENDF/B-V for ^{233}U and ^{234}U and from ENDF/B-VI for ^{236}U and ^{238}U.

MEASUREMENT OF NEUTRON INDUCED FISSION CROSS SECTION RATIOS FOR ^{235}U, ^{238}U, AND ^{232}Th FROM 1 TO 100 MeV

A. V. Fomichev, I. V. Tuboltseva, A. Yu. Donets
Khlopin Radium Institute, 194021 Leningrad, Shvernika Avenue, 28, U.S.S.R.

A. B. Laptev, O. A. Shcherbakov, G. A. Petrov
Konstantinov Nuclear Physics Institute, AS U.S.S.R., Leningrad, Gatchina

ABSTRACT: The method of the "shape" measurement was realized using an impulse spallation source of the synchrocyclotron at Leningrad Nuclear Physics Institute. The fission cross section ratios ^{238}U/^{235}U and ^{232}Th/^{235}U were measured as a function of neutron energy from 1 to 100 MeV. A multiplate methane gas ionization chamber located at a 50 m flight path was used to scan by time the fission count rates for all the three samples simultaneously. The data were compared with those obtained using linear accelerator.

(fission cross section, fast neutrons, uranium, thorium)

INTRODUCTION

Despite 30 years of extensive work the need for new neutron induced fission cross section measurements is still uncovered. This is a result of inconsistencies between the data of different researchers. Also grow the requirements for the accuracy and reliability of the results, arises interest in involving higher neutron energies and a greater number of the nuclei.

An electron linear accelerator as the "white" neutron source is most generally used for the shape measurements. It gives an adequate intensity for measuring fission cross section ratios below 30 MeV. Using the proton beam of the 1000 MeV synchrocyclotron allowed us to expand the measurements to higher neutron energies.

EXPERIMENTAL PROCEDURE

The measurements were performed using the GNEIS neutron time-of-flight spectrometer at Leningrad Nuclear Physics Institute [2]. The neutron source was an inner lead cyclotron target 40x 20x 5 cm^3 bombarded by the 1000 MeV proton bunches with the pulse width of 10 ns and the repetition rate of 50 Hz. The neutron detector was an ionization chamber located 50 m from the neutron producing target.

The multiplate ionization chamber was filled by methane under the pressure of 2.5 atmosphere. The electrode spacing was 1.3 cm, being sufficient for the fission fragments to be completely absorbed. The fissionable material was fabricated from the isotopes of mass separate purity. The fission samples were deposited onto aluminium foil 0.015 cm thick and 18 cm in diameter. The fissionable deposits were typically of ca 200 μg/cm^2 thick oxide material, which uniformity was verified.

The pulses from the ionization chamber were detected with a current preamplifier. The energy of neutrons was determined by measuring their TOF to the fission detector with a quartz time-digital converter. The width of the TOF channel was 10 ns. The response of the ionization chamber to the gamma-ray flash was our timing reference. The events from the ionization chamber were collected via CAMAC programmable controller into an array consisting of 1024 neutron channels. Simultaneously these events were accumulated in a two-dimensional array consisting of 128 TOF channels by 128 pulse height channels.

RESULT AND CONCLUSIONS

For each of the three nuclides 128 statistically reliable amplitude spectra were obtained corresponding to the particular TOF channels or to definite neutron energies. This enabled us not only to register the fission events but also to observe the transformation of the amplitude spectrum shape from two-humped into one-humped with increasing the neutron energy. So far as such transformation corresponds to a fission mechanism, all the 100 per cent of events were considered as fissions.

The data accumulated in the one-dimensional arrays were used to compute the cross section ratios ^{238}U/^{235}U and ^{232}Th/^{235}U Fig. 1, 2, 3, 4 show the comparison between our preliminary results and those obtained by other researchers using linear accelerators [3, 4]. It is found that for the overlapping energy range our results are in a good agreement with those of Behrens et al [3] and Difilippo et al [4]. This points to the absence of the background effects left out of account, particularly of those connected with room and beam transporting tube neutrons.

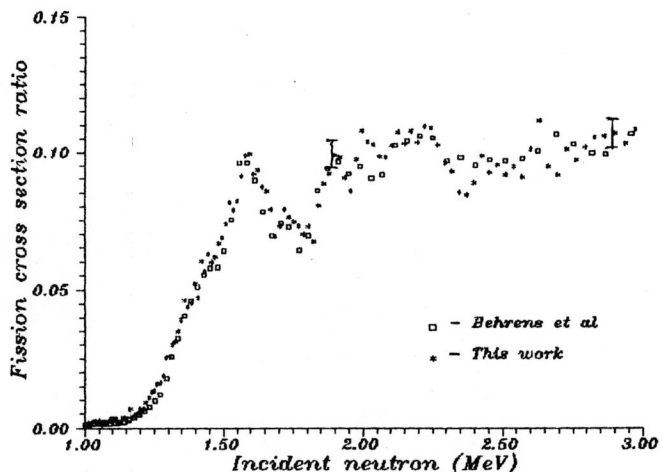

Fig. 1. Preliminary results of our measurements of the fission cross section ratio ^{232}Th to ^{235}U compared with the data of Behrens et al [3] over the energy range 1 to 3 MeV.

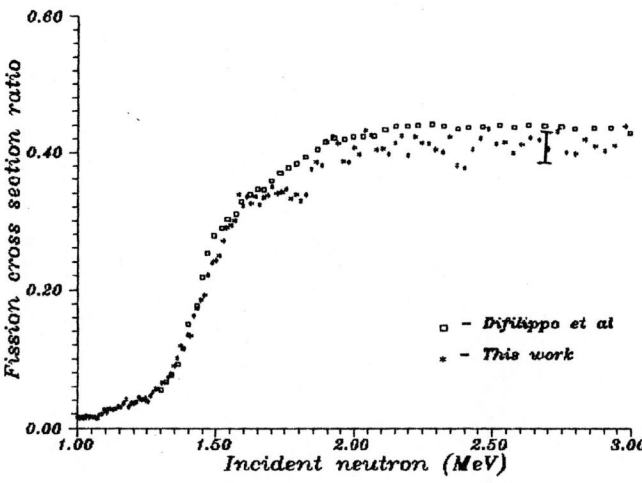

Fig. 2. Preliminary results of our measurements of the fission cross section ratio ^{238}U to ^{235}U compared with the data of Difilippo et al [4] over the energy range 1 to 3 MeV.

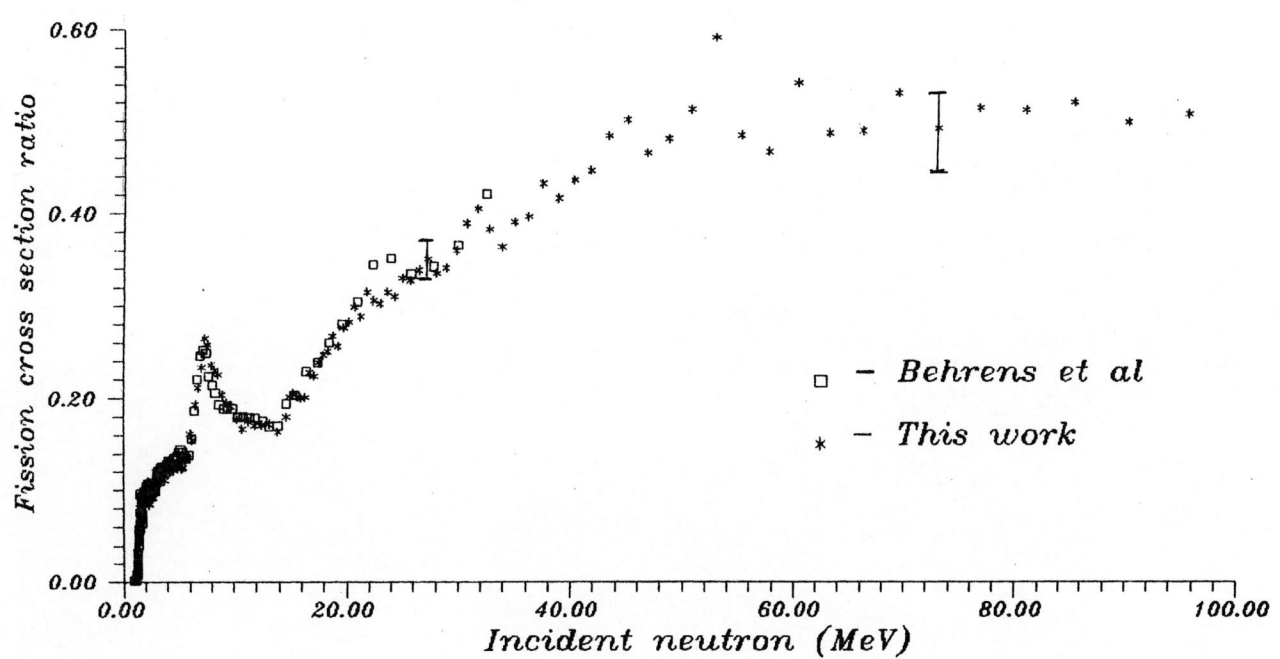

Fig. 3. Preliminary results of the present measurements of the fission cross section ratio ^{232}Th to ^{235}U compared with the data of Behrens et al [3] over the energy range 3 to 100 MeV.

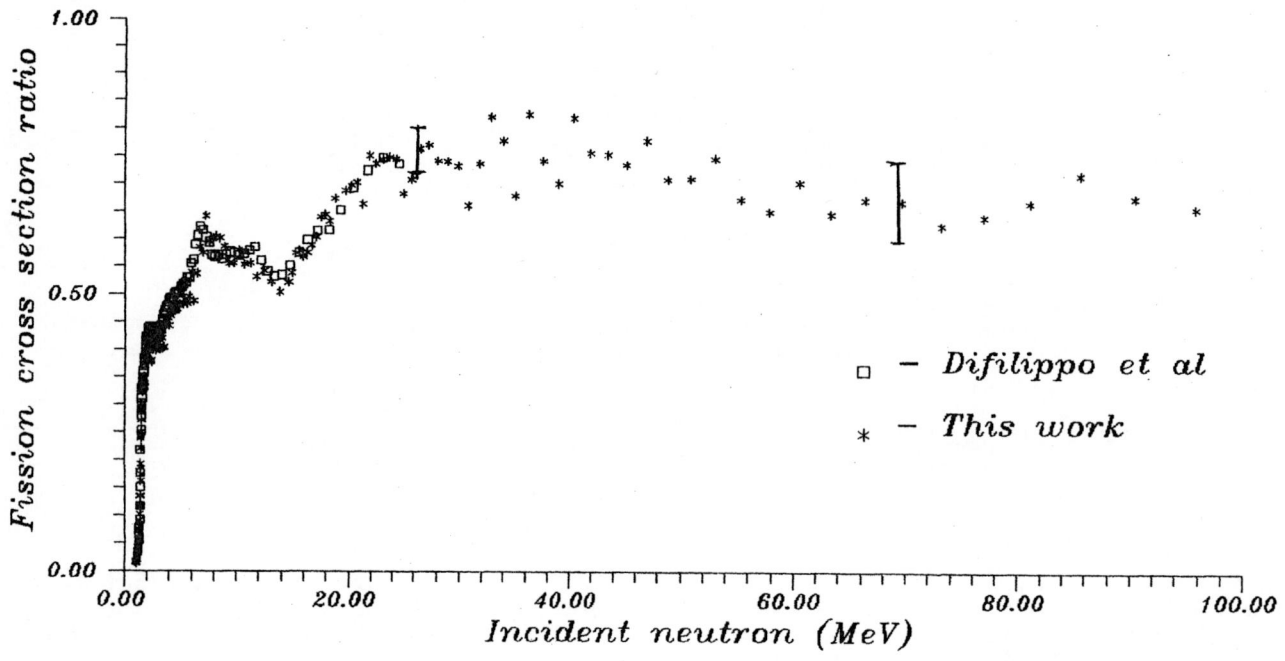

Fig. 4. Preliminary results of the present measurements of the fission cross section ratio ^{238}U to ^{235}U compared with the data of Difilippo et al [4] over the energy range 1 to 100 MeV.

REFERENCES

1. V.M. Pankratov, N.A. Vlasov, and B.V. Rybakov: Atomnaya Energiya, 14, 177 (1963)
2. N.K. Ambrosimov, G.Z. Borukhovich, A.B. Laptev, V.V Marchenkov, G.A. Petrov, O.A. Shcherbakov, Yu.V. Tuboltsev, V.I. Yurchenko: Nucl. Instr. Meth. A242, 121 (1985)
3. J.W. Behrens, J.C. Browne, and E. Ables: Nucl. Sci. Eng. 81, 515, (1982)
4. F.C. Difilippo, R.B. Perez, G. de Saussure, D.K. Olsen, and R.W. Ingl: Nucl. Sci. Eng. 68, 43, (1978)

CALCULATION OF GALACTIC COSMIC RAY INDUCED PLANETARY GAMMA-RAY SPECTRA AND ITS APPLICATION TO THE MARTIAN SURFACE

G. Dagge, D. Filges

Institut für Kernphysik, Forschungszentrum Jülich
5170 Jülich, Germany

and J. Brückner

Max-Planck-Institut für Chemie, Abt. Kosmochemie, 6500 Mainz, Germany

Abstract: The interaction of Galactic Cosmic Rays with planetary surfaces is simulated with the HERMES code system. The calculational procedure to obtain the planetary prompt gamma-ray spectrum by Monte Carlo methods is presented and applied to several chemical compositions which are of interest for the planet Mars. These calculations are necessary for the interpretation of experimental Martian gamma-ray spectra with respect to the surface composition. The analysis of gamma-ray spectra is discussed.

Monte Carlo Simulation, Galactic Cosmic Rays, Mars Observer

Introduction

Galactic cosmic rays (GCR) produce a shower of hadronic particles by spallation and evaporation processes in a planet's surface. The secondary neutrons are moderated in the material and are eventually captured or leave the planet's surface. Inelastic neutron scattering and neutron capture reactions produce gamma-ray lines which are characteristic for the target nucleus. The upcoming Mars Observer mission will provide gamma-ray measurements from the orbit of Mars. The elemental composition of the Martian surface can be deduced from an analysis of these data, which is of special interest for the investigation of the polar caps. An interpretation of experimental gamma-ray spectra can not be performed without theoretical simulations of the cosmic irradiation of the Martian surface, the production of gamma rays, and their transport to the orbiting spectrometer.

Method

The HERMES code system consists of four separate Monte Carlo Codes which can exchange data via so called submission files [1]. The HETC-KFA/2 code calculates the hadronic cascade resulting from the interaction with the planetrary surface. The code uses the INCE model (Intra-Nuclear Cascade Evaporation) for calculating the interaction of an average GCR spectrum (with energies up to 10 GeV per nucleon) with matter. Low energy neutron physics (<14.9 MeV) can not be performed properly by the HETC code, so the information on a neutron's location, direction and energy is taken over by the MORSE-CG code. A subsequent coupled n-γ transport calculation gives the thermal and epithermal neutron flux as well as the prompt gamma-ray spectrum as a function of depth and energy. A problem-independent multigroup library for 293.6K containing 118 neutron groups (19 of them thermal), 21 gamma-ray continuum groups, and 20 separate gamma-ray line groups was evaluated for this purpose. Prompt gamma rays of HETC-produced excited residual nuclei are generated by the NDEM module and transported again by MORSE.

The fourth module, EGS-4, is applied for the calculation of the prompt gamma-ray continuum produced by the decay of π^0-mesons. This method has been tested successfully for the lunar surface (2).

Results

The assumed chemical composition, which is based on Viking experiments [3], is varied with respect to its water concentration and macroscopic neutron capture cross section to study the basic parameters for the interpretation of experimental data. The influence of the thin Martian CO_2 atmosphere (about 16 g/cm^2) is investigated. The strong absorption of gamma-ray lines within the atmosphere is partly compensated by the reflection of neutrons back into the soil, which enhances the gamma-ray production. The solar modulation of the GCR spectrum is found to influence the overall normalization of particle and gamma-ray fluxes.

Possible water ice layers below the surface and polar caps of frozen CO_2 and H_2O are studied. Their influence on the emitted neutron and gamma-ray spectrum is presented in Figs. 1 and 2. These results can be used for the interpretation of experimental data. The distinction between frozen CO_2 and water ice, which is of interest for the polar caps, can be done easily [4,5]. The dust content of ice can also be measured.

If the experimental data indicate a homogeneous soil within the top 100 g/cm^2, the relative elemental abundances can be deduced. All elements of interest, except for oxygen and carbon, can be traced by their neutron-capture gamma-ray lines. Since the corresponding cross sections show a more or less perfect $1/v$ dependence, the analysis can be performed straightforward by using gamma-ray line ratios. An absolute normalization to a 50% oxygen concentration is applicable for relatively dry soils. Conversion factors k_j for transformation of gamma-ray signals S_j into mass fractions m_i have been evaluated applying the relation

$$m_i = \frac{S_j}{A\epsilon(E_j)} k_j x_{ij} \Sigma_{abs} \qquad (1)$$

738

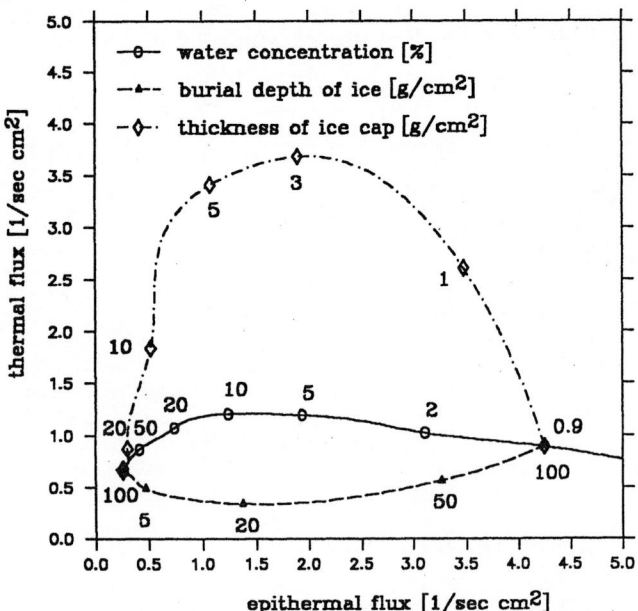

Fig. 1. Thermal versus epithermal neutron flux at the surface of Mars for 1) homogeneous soil with varying water content (solid line); 2) subsurface ice with varying burial depth (dashed line); 3) polar water icecap of variable thickness (dash-dotted line).

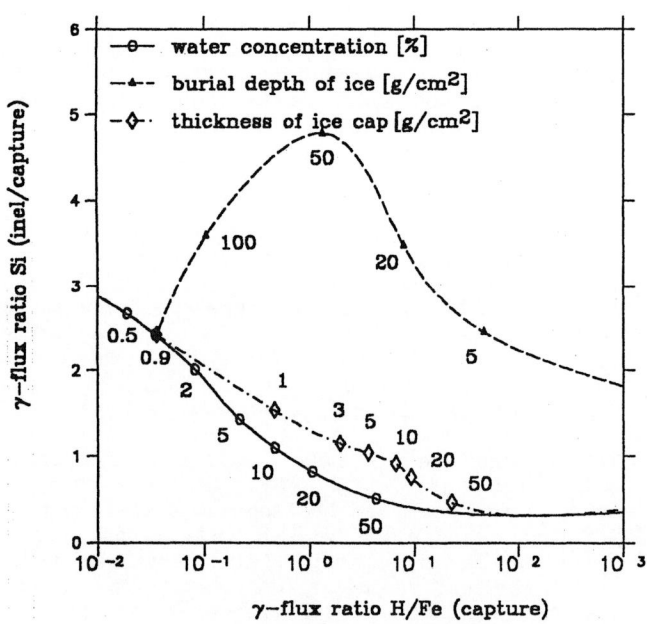

Fig. 2. The gamma-ray flux ratio of Si (inelastic/capture) is plotted against the gamma-ray flux ratio of H and Fe (capture). The flux is calculated for the Mars Observer orbital height of h=360 km. The following cases are presented: 1) homogeneous soil with varying water content (solid line); 2) subsurface ice with varying burial depth (dashed line); 3) polar water ice cap of variable thickness (dash-dotted line).

with the detector area A, the detector efficiency $\epsilon(E_i)$, the macroscopic capture cross section Σ_{abs} of the medium, and a constant x_{ij} depending on the element under consideration. The quantity k_i sums up the influence of parameters like orbiter height, GCR intensity and gamma ray attenuation within the atmosphere and soil. The energy dependence of conversion factors of $1/v$-capture lines is shown in Fig. 3 for a Viking soil composition and a composition with enhanced water concentration of 10%. This diagramm may be used for a consistency check and for normalizing purposes.

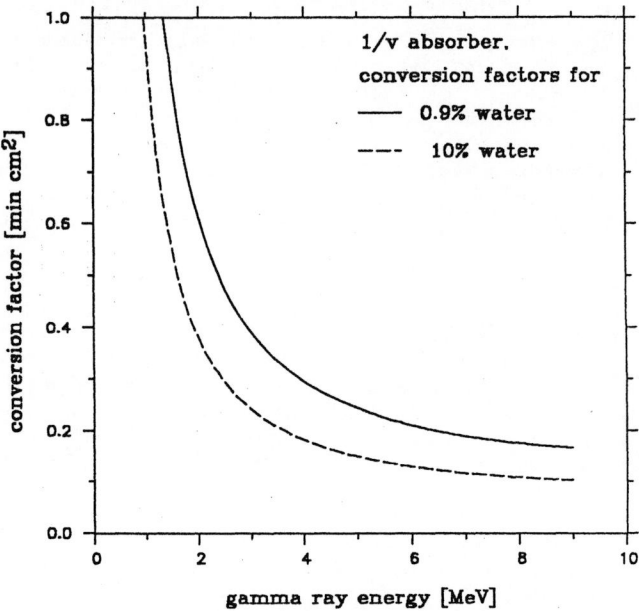

Fig. 3. Conversion factors as function of gamma-ray energy for a Viking soil composition (0.9%, solid line) and a Viking soil with 10% water (dashed).

Conclusion

A classification of Martian surface structures via orbital neutron and gamma ray spectroscopy can be performed with the aid of the Monte Carlo simulations of the GCR-irradiation. An elemental analysis for homogeneous compositions can be done easily using neutron capture lines. Detailed theoretical studies of the involved parameters are needed for layered surface structures. The presented parameter studies show that GCR-induced gamma-ray spectroscopy gives access to the first 100 g/cm² of Martian soil from the orbit, which by far exceeds the possibilities of other methods.

REFERENCES

1. Cloth P., Filges D., Neef R.D., Sterzenbach G., Reul Ch., Armstrong T.W., Colborn B.L., Anders B., and Brückmann H., HERMES User's Guide, Jül-2203 (1988).
2. Dagge G., Dragivitsch P. , Filges D. and Brückner J., in Proceedings of Lunar and Planetary Science, Vol. 21, pp. 425-435, Lunar and Planetary Institute, Houston (1991).
3. Clark B.C., Baird A.K., Weldon R.J., Tsusaki D.M., Schnabel L., Candelaria M.P., J. Geophys. Res. 87, 10059 (1982).
4. Evans L.G., and Squyres S.W., J. Geophys. Res. 92, 9153 (1987).
5. Drake D.M., Feldman W.C., and Jakosky B.M., J. Geophys. Res. 93, 6353 (1988).

RAD - SOFTWARE PACKAGE FOR CALCULATIONS ON RADIONUCLIDE PRODUCTION AND RADIATION DOSES

V.L.Matushko, S.V.Serezhnikov and Yu.M.Sheynov

Institute for Nuclear Research,
Academy of Science of the USSR,
117312 Moscow, USSR

Abstract: Software package RAD for calculations on radionuclide production and radiation doses induced by intermediate energy protons in a compound heterogeneous system is described. It is based on extensive experimental data combined in the following databases: radionuclide production cross section data, radionuclide decay chain data, and nuclide radioactive emission data. The evaluated excitation function database based on the existing experimental data and theoretical models for hadron-nucleus reactions is under development. The RAD application field concerns the following calculations: radionuclide production; airborn, soil and installation residual radioactivity; dose characteristics of induced radioactivity. The software package RAD is successfully used for designing experimental setups at the Moscow meson factory.

(software package, databases, hadron-nuclear reactions, excitation functions, cross sections, radionuclide production, induced radioactivity, radiation dose)

Introduction

Software package RAD created for solving the problems connected with radiation safety, radionuclide production, designing nuclear physical installations and experiments at high intensity proton accelerators (such as meson facilities) is described in the paper.

General principles of RAD organization

While designing the RAD we tried to follow the principles:
1. Possibility to carry out the whole continuous computations beginning from definition of nucleon fluxes in complex heterogeneous systems to calculation of dose characteristics of induced radioactivity.
2. Highest possible use of the experimental data systematized in the databases.
3. A modular organization. During the RAD development new modules can be added and some existent modules can be modified, but these innovations slightly affect another parts of the package.
4. Possibility to carry out both rough and more detailed calculations.
5. Simplicity for input data preparation procedure and clearly evident calculational result representation.

General scheme of computation

The package consists of the input data preparation procedure, program modules for calculation of the requisite physical processes, databases and a few independent codes running off-line. RAD block diagram is shown in Fig.1.

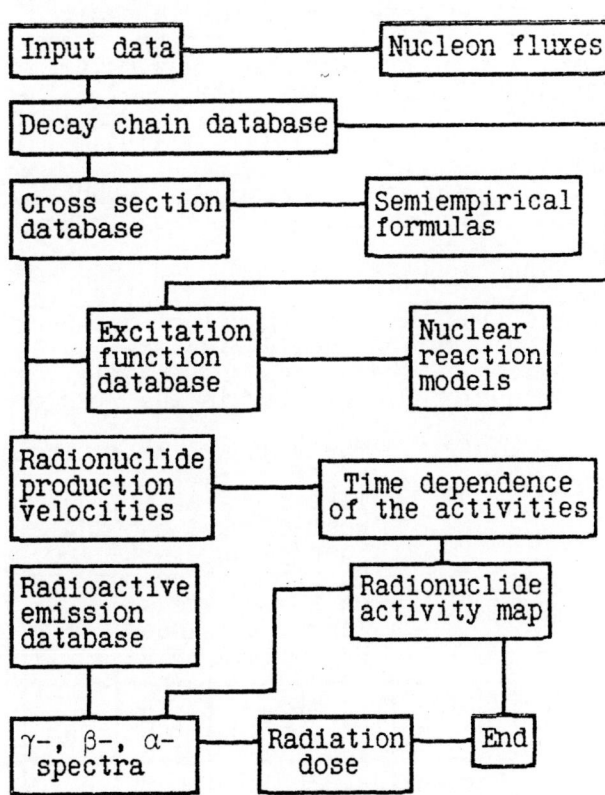

Fig. 1. RAD block diagram.

The input data preparation comprises six procedures for the target, the products, the fluxes, the time conditions, the geometric characteristics and some of the common characteristics of the problem. These procedures were realized in the interactive mode and provided with the detailed information about input data. The nucleon fluxes may be calculated a) by the parametric formulas representing the most typical forms of the energy dependencies in the medium energy range, b) by SUPER code for calculation of the internucleus cascade in the complex heterogeneous systems [1].

The first step of the calculation is the preparation of the radionuclide decay chain from the described product range. Then the cross sections for each reaction (isotope of the target + initial particle) -> (radionuclide of the chain + X) are calculated by the excitation functions database or by the semiempirical Rudstam's [2] and Silberberg-Tsao's [3] formulas normalized to the experimental data from the database of radionuclide production cross-sections in intermediate energy hadron-nucleus reactions [4].

The effective radionuclide production velocity obtained by the weighted means of the excitation function on the energy nucleon spectrum and the isotopic composition of the target. To define the time dependence of the radionuclide activities for all radionuclides of the chain the differential equation system is solved. The radionuclide activity map is formed during RAD processing. The analysis of this map defines the set of the radionuclides giving the greatest contribution to the total activity in each time moment.

The repeat run can be used to obtain the detailed data for these radionuclides.

The spectra of the γ-, β- and α-radiation can be calculated with the nuclide radioactive emission database and radionuclide activities map. The definition of the dose characteristics of residual radioactivity in the region around the target requires to take into account geometric factors: the size of target, the presence of the shielding shades etc. The package uses the approximate formulas for typical geometric configurations. It is sufficient for the most practical problems.

The RAD exploits three main databases comprising necessary information for calculations on radionuclide production and decays. The creation of database of radionuclide production cross-section in intermediate energy hadron-nucleus reactions was begun at INR AS USSR in 1981. By now this database contains about 20.000 data items in energy range from reaction thresholds up to 1 GeV.

Radionuclide decay chains database has been made specially for RAD. All radionuclides with mass number up to 211 and $T_{1/2} < 1$ sec were well-regulated to mutual independent chains. Each chain contains no more than 25 nuclides and its branching ratios.

The nuclide radioactive emission database consist of energy of γ-, β-, and α-particles and particle intensities per decay of the parent nucleus. The greater part of the data has been obtained by the transformation of ENSDF [5]

Besides that the evaluated excitation function database for important nuclear reactions is being created. The excitation functions in energy range from reaction thresholds to 200 MeV are described by parametric formulas of a few kinds. The kind and the parameters of the formula are fixed by an analysis of the existing experimental data and calculated data. Two computer codes for calculations running off-line are used: first one is ALICE code using preequilibrium exciton model [6] and second is CASCAD+EVAPOR code using cascade-evaporation model [1].

Table 1. The most active radionuclides.

Target - ^{87}Rb
Beam current - 50 μA
Energy range - 50-100 MeV
Duration of bombardment - 5 days
Wait - 0 days
Target thickness - 1 g/cm^2
Calculated total activity - 31.9 Ci

Radio-nuclide	^{87}Sr	^{83}Sr	^{84}Rb	^{82}Rb	^{82m}Rb	^{85}Sr	^{84}Rb	^{81}Rb	^{86}Rb	^{81m}Rb
Percent	44.82	14.22	10.39	6.40	5.18	3.68	2.08	2.02	1.94	1.75

Table 2. Same as Table 1 for:

Wait - 9 days
Calculated total activity - 2.16 Ci

Radio-nuclide	^{84}Rb	^{86}Rb	^{83}Rb	^{82}Rb	^{82}Sr	^{85}Sr	^{83}Sr	^{77}Br	^{79}Kr	^{75}Se
Percent	25.48	20.54	18.33	14.09	14.09	4.79	2.07	0.36	0.16	0.05

Results

A wide programme of radionuclide production is being designed at Moscow meson factory. First experiment for estimation of the Sr-82(Rb-82) production is being begun at 160 MeV proton beam. The RAD software package was used for optimization of the irradiation conditions and for calculation of the radionuclide activity map. The results of the calculations are shown in Tables 1,2.

Acknowledgements

The authors would like to express their gratitude to Professor M.V.Kazarnovsky for very useful and stimulating discussions and to A.I.Mayevskaya, A.K.Skasirskaya and L.Stepanova for help in the programming.

REFERENCES

1. A.S.Iljinov, M.V.Kazarnovsky, G.K.Matushko, V.L.Matushko, E.Ya.Paryev, S.V.Serezhnikov, N.M.Sobolevsky and B.E.Shtern: Software Library for High Energy Particle Interactions with Uniform Matter Calculations. User Manual (in Russian). Institute for Nuclear Research, Moscow (1985)

2. G.Rudstam: Z.f.Naturf. 21A, 1027 (1966)

3. R.Silberberg and C.H.Tsao: Astrophys. J. Suppp. 25, 315 (1973)

4. A.S.Iljinov, V.L.Matushko, V.G.Semenov, N.M.Sobolevsky and L.V.Udovenko : The Data File on Radionuclide Production Cross Sections in Intermediate Energy Hadron-Nucleus Reactions. Status and Perspectives (in Russian), P.N 0618 , Institute for Nuclear Research, Moscow (1989)

5. J.K.Tuli: Evaluated Nuclear Structure Data File. A Manual for Preparation of Data Sets. BNL-NCS-51655,Brookhaven National Laboratory (1983)

6. M.Blann and H.K.Vonach: Phys.Rev. 28C, 1475 (1983)

SYSTEMATIC EVALUATION OF NEUTRON-EMISSION CROSS SECTION FOR THE REACTIONS INDUCED BY PROTONS OF 80-800 MeV

K. Ishibashi, K. Higo, S. Sakaguchi, Y. Wakuta
H. Takada,* T. Nishida,* Y. Nakahara* and Y. Kaneko*

Department of Nuclear Engineering, Kyushu University
Hakozaki, Fukuoka 812, Japan
*Japan Atomic Energy Research Institute
Tokai-mura, Ibaraki 319-11, Japan

Abstract: A moving source model uses a relatively small number of parameters. The neutron data for the spallation reactions induced by protons of 600 to 800 MeV are well reproduced by the moving source model. At incident energies below 400 MeV, however, the use of this model leads to an unsuccessful result. A new idea of a double moving source model is introduced in the lower energy region. The combination of the two models enables us to parameterize the double differential cross sections of neutron emission for incident protons of 80 to 800 MeV.

(proton, spallation reaction, double differential cross section, neutron emission, systematics, moving source model, Watt distribution)

Introduction

The spallation reaction is caused by bombarding targets with particles having an energy above a few hundred MeV. The reaction produces a number of neutrons. This reaction then is usable for providing an intense spallation neutron source or transmuting long-lived radioactive wastes. In spite of the usefulness for the engineering purpose, the accumulation of neutron data is poor for the proton induced spallation reaction. Therefore, it is important to find the systematic behavior in the neutron data, and to parameterize [1] them to extend to different energy regions or target nuclei.

A moving source (MS) model [2,3] has been proposed for describing protons emitted by several-GeV proton-induced spallation reactions. The model expresses the angular distribution of emitted particles by the use of a single parameter. At present, the experimental neutron spectra in the spallation reaction are available at only a few incident energies and emission angles. Hence, we will apply this model to study the systematics of the neutron data in the incident proton energy region from 600 to 800 MeV. At incident energies below 400 MeV, we will introduce a double moving source (DMS) model in accordance with the change of the collision phenomena.

Moving Source (MS) Model

The situation of the MS model [2,3] is shown in Fig. 1, where a projectile interacts with a nucleus that is indicated by a circle. The illustrations on the left and the right hand sides show the collision phenomenon in the laboratory and the moving frames, respectively. A cascade collision is seen in the left hand side. When the velocity of the moving frame β is properly chosen, the collision is observed to be isotropically

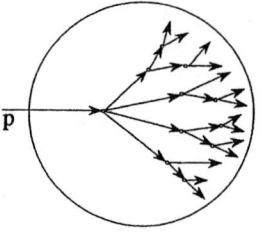
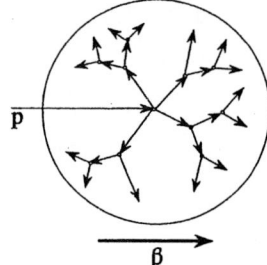

Laboratory Frame　　**Moving Frame**

Fig. 1. Illustration of the moving source model.

developed as shown in the right hand side. In this moving frame, neutrons are assumed to be emitted isotropically at a nuclear temperature T (MeV) with a Maxwellian energy distribution. The invariant cross section, therefore, is written as

$$E_{tot} (d^3\sigma/dp^3) = A \exp(-E_{kin}^*/T) , \qquad (1)$$

where E_{tot} (MeV) is the total neutron energy including rest mass, A (mb/sr(MeV/c)^2 c) is a variable parameter, and E_{kin}^* (MeV) the emitted neutron energy in the moving frame. The Lorentz transformation is written as

$$E_{kin}^*+m= (E_{kin}+m-p\beta \cos \theta)(1-\beta^2)^{-1/2}, (2)$$

where m is the neutron mass (MeV), E_{kin} (MeV) the kinetic energy , p (MeV/c) the momentum, and θ the emission angle. These are quantities in the laboratory frame. Then, eq. (1) is converted into the equation in the laboratory frame as

$$E_{tot} (d^3\sigma/dp^3) = p^{-1}(d^2\sigma/d\Omega dE_{kin})$$

$$=A \exp[-\{\frac{E_{kin}+m-p\beta \cos \theta}{(1-\beta^2)^{1/2}} - m \}/T]. (3)$$

The parameters A, T, and β are adjustable in fitting the equation with the experimental differential cross sections. We call A amplitude parameter, T temperature parameter, and β velocity parameter. The angular distribution in the laboratory frame is dominantly influenced by the value of β. To the spallation reaction, the MS model is applied in a form of summation of three components as

$$E_{tot}\,(d^3\sigma/dp^3)=p^{-1}(d^2\sigma/d\Omega\,dE_{kin})$$

$$=\sum_{i=1}^{3}A_i\exp\left[-\left\{\frac{E_{kin}+m-p\beta_i\cos\theta}{(1-\beta_i^2)^{1/2}}-m\right\}/T_i\right].\quad(4)$$

The subscripts i=1, 2 and 3 stand for the intranuclear-cascade, the preequilibrium and nuclear-evaporation (equilibrium) processes, respectively.

Figure 2 shows the results of the parameter fitting for the data of 585 MeV proton incident on a lead target. The experimental neutron spectra [4] are plotted by marks, and the fitting results

by solid curves. The agreement between the fitting and the experimental data is excellent. The solid curve at 150 deg is decomposed into the three processes, as plotted by the dashed lines. The parameters obtained for targets from carbon to uranium are listed in Table 1.

Double Moving Source (DMS) Model

For the measurement data of 113 MeV proton incidence on iron [5], the results of fitting by the MS model are shown by the dashed curves in Fig. 3. In the forward direction of 7.5 and 30 deg, the dashed curves are in a poor agreement with the data at neutron energies above several ten MeV. The neutrons in this energy range are emitted mainly from the intranuclear-cascade process. The MS

Table 1. Parameters obtained for the 585 MeV proton-induced reactions.

Amplitude mb/sr (MeV/c)^{2}c			Temperature MeV			Velocity		
A_1 10^{-4}	A_2 10^{-3}	A_3 10^{-2}	T_1	T_2	T_3	β_1 10^{-1}	β_2 10^{-2}	β_3 10^{-3}
C 6.77	2.20	8.20	82.0	16.7	2.23	2.89	8.83	15.8
Al 22.6	8.59	33.3	79.8	14.6	2.19	2.49	6.11	18.4
Fe 48.6	15.4	77.3	73.9	17.7	2.67	2.54	6.42	11.2
Nb 69.8	24.4	205.	72.8	19.3	2.61	2.39	6.07	7.24
In 89.5	35.7	325.	70.5	18.1	2.55	2.34	5.11	5.04
Ta 109.	54.3	705.	76.1	19.9	2.53	2.31	6.34	4.33
Pb 99.4	50.9	788.	75.9	22.9	2.57	2.26	6.02	3.70
U 127.	72.1	1370	78.8	20.4	2.40	2.17	6.95	3.49

Table 2. Parameters obtained for the 113 MeV proton-induced reactions.

Amplitude mb/sr (MeV/c)^{2}c			Temperature MeV				Velocity		
A_1 10^{-2}	A_2 10^{-3}	A_3 10^{-2}	T_d	E_d	T_2	T_3	β_1 10^{-1}	β_2 10^{-3}	β_3 10^{-3}
C 9.38	3.47	1.85	7.64	12.7	6.60	1.72	2.53	5.59	16.8
O 7.81	5.37	2.63	5.98	13.3	6.41	2.39	2.42	5.59	17.4
Al 13.3	9.02	14.7	7.11	13.5	7.78	2.12	2.26	2.50	5.44
Fe 27.0	17.1	62.9	7.25	11.7	7.05	1.89	2.04	5.48	4.92
W 101.	136.	1500	7.38	12.6	6.67	1.41	1.74	6.04	2.29
Pb 154.	70.8	1160	8.89	9.12	9.14	1.35	1.85	2.59	2.00
U 94.9	63.6	1530	8.34	11.2	9.64	1.65	1.81	10.5	2.25

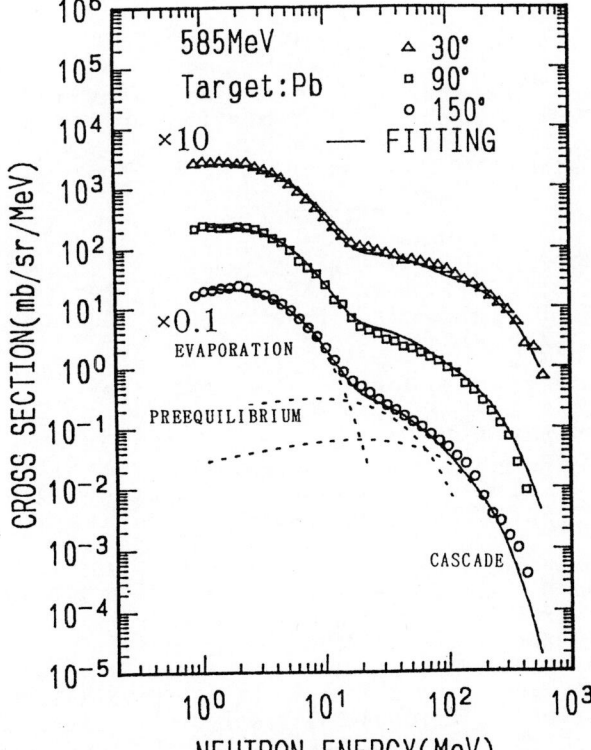

Fig. 2. Experimental neutron spectra and the results of fitting by the MS model for 585 MeV protons on lead.

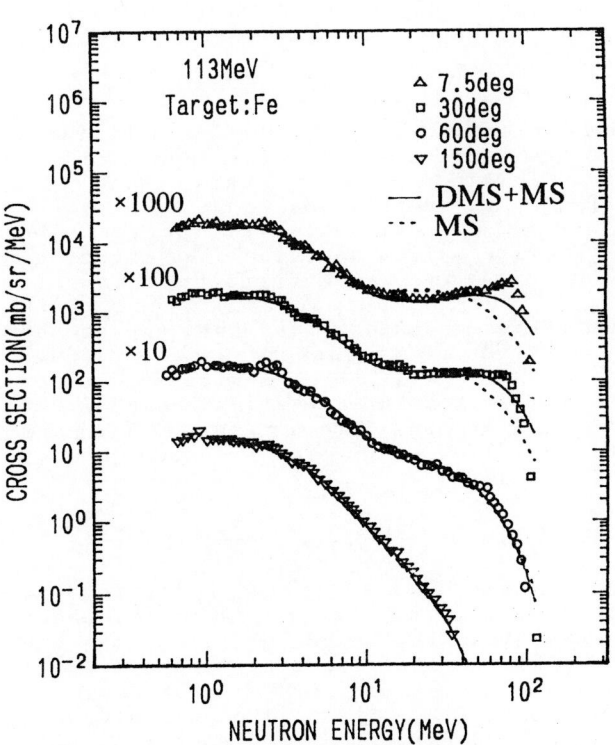

Fig. 3. Experimental neutron spectra and the results of fitting by the DMS model for 113 MeV protons on iron.

744

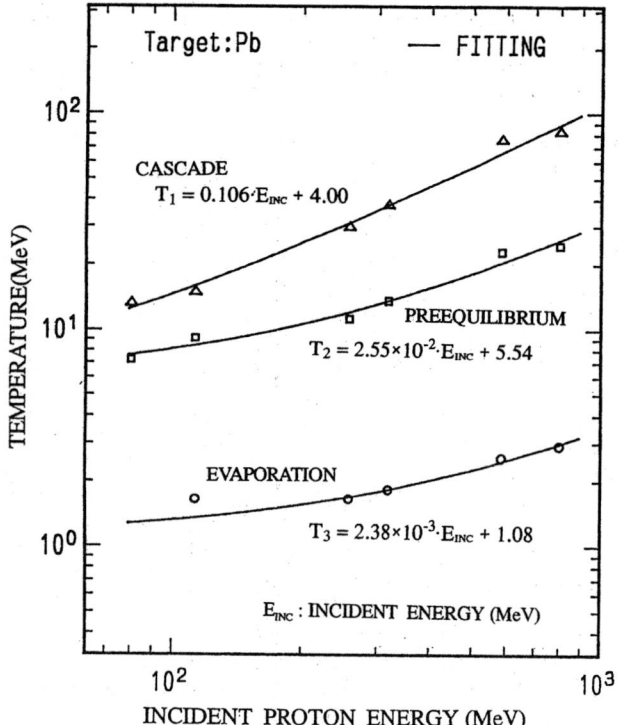

Target:Pb   — FITTING

CASCADE
$T_1 = 0.106 \cdot E_{INC} + 4.00$

PREEQUILIBRIUM
$T_2 = 2.55 \times 10^{-2} \cdot E_{INC} + 5.54$

EVAPORATION

$T_3 = 2.38 \times 10^{-3} \cdot E_{INC} + 1.08$

E_{INC} : INCIDENT ENERGY (MeV)

INCIDENT PROTON ENERGY (MeV)

Fig. 4. Temperature parameter T for 80-800 MeV proton incident on lead.

model reproduced the experimental data unsuccessfully in the incident proton energy region below a few hundred MeV.

The disagreement in Fig. 3 suggests that the energy spectra in the moving frame are different from the simple exponential function like the Maxwellian type, and may be reproduced by the Watt distribution [6] which produces a shoulder in the spectra. The Watt distribution has a form of

$$\exp(-E_{kin}^*/T_d)\sinh\{2(E_{kin}^* E_d)^{1/2}/T_d\} , \quad (5)$$

where the parameters T_d (MeV) and E_d (MeV) are the temperature and the energy of the Watt distribution, respectively. The value of E_d corresponds to the characteristic velocity of the distribution. The Watt distribution describes the situation where neutrons are evaporated from isotropically-moving sources. Since it is possible to consider that this phenomenon is observed in the original moving frame of the velocity β, we call this model the double moving source (DMS) model. For the DMS model, the cross section is written in the laboratory frame as

$$E_{tot} (d^3\sigma/dp^3) = p^{-1}(d^2\sigma/d\Omega\, dE_{kin})$$

$$= A_d p^{*-1}\exp\left(-\frac{E_{kin}^*}{T_d}\right)\sinh\frac{2(E_{kin}^* E_d)^{1/2}}{T_d} . \quad (6)$$

The superscript * indicates the values in the moving frame. The parameter A_d shows an amplitude in the DMS model. By the use of values in the laboratory frame, E_{kin}^* is given by Eq. (2) and p^* is written by $p^* = (E_{kin}^{*2} + 2mE_{kin}^*)^{1/2}$. The parameters T_d and E_d are designated as a double-moving-source temperature and energy, respectively. The Watt distribution produces an average kinetic energy $\langle E \rangle = 3/2T_d + E_d$ in

the first approximation, while the Maxwell distribution gives $\langle E \rangle = 3/2T$ in the moving frame. The two parameters in the DMS then are converted into T' as

$$T' = T_d + 2/3E_d . \quad (7)$$

The DMS model is applicable only to the initial process that is directly induced by the incident protons. The other remaining processes are treated by the MS model. The solid curves in Fig. 3 show the results of fitting by the combination of DMS and MS models. The experimental data are expressed better by the solid than the dashed curves. The parameters obtained are listed in Table 2 for 113 MeV proton incidence on targets of carbon to uranium.

Parameter sets were obtained for the neutron data [7,8,9,10] at incident proton energies of 80.5 to 800 MeV. Values of the temperature T are shown for lead in Fig. 4, where the subscripts 1, 2 and 3 indicate the intranuclear-cascade, pre-equilibrium, and nuclear-evaporation processes, respectively. For the cascade processes, the reduced values of T' are plotted in the proton energy region below 400 MeV. One can see that temperatures change smoothly in this figure.

Conclusion

The use of the moving source (MS) model with three components reproduced the experimental neutron spectra in the incident proton energy region of 600 to 800 MeV. At energies below 400 MeV, the introduction of the double moving source (DMS) model for the cascade process were found to lead to good agreement with the experimental spectra. The MS and DMS models were found to be applicable to the nuclear data evaluation for a wide range of target masses and incident energies.

Acknowledgments

The authors gratefully acknowledge Prof. A. Katase of Tohwa University and Prof. Y. Matsumoto of Teikyo Junior College for their useful discussions.

REFERENCES

1. S.Pearlstein: Nucl.Sci.Eng. 95,116(1987)
2. I.G.Bogatskaya: Phys.Rev. C22, 209(1980)
3. T.-A.Shibata, K.Nakai, H.En'yo, S.Sasaki, M.Sekimoto, I.Arai, K. Nakayama, K. Ichimaru, H.Nakamura-Yokota and R. Chiba: Nucl.Phys. A408, 525 (1983)
4. S.Cierjacks, Y.Hino, F.Raupp, L.Buth, D.Filges, P.Cloth and T.W.Armstrong: Phys.Rev. C36, 1976 (1987)
5. M.M.Meier, D.A.Clark, C.A.Goulding, J.B. McClelland, G.L.Morgan and C.E.Moss: Nucl.Sci.Eng. 102, 310 (1989)
6. J.Terrell: Phys.Rev. 113, 527 (1957)
7. M.Trabandt, W.Scobel, M.Blann, B.A.Pohl, R.C.Byrd, C.C.Foster, R.Bonetti: Phys. Rev. C39, 452 (1989)
8. M.M.Meier, W.B.Amian, C.A.Goulding, G.L. Morgan and C.E.Moss: LA-11656-MS (1989)
9. M.M.Meier, D.B.Holtkamp, G.L. Morgan, H. Robinson, G.J.Russel, E.R.Whitaker, W. Amian and N.Paul: Rad.Effect.96,73(1986)
10. S.D. Howe, P.W.Lisowsky, G.J. Russel: LA-UR-85-3660 (1985).

PRODUCTION OF ^{7}Be, ^{10}Be AND ^{26}Al BY ^{4}He-INDUCED REACTIONS UP TO 171 MeV

H.-J. Lange[1], T. Hahn[1], R. Michel[1], B. Dittrich[2,3], U. Herpers[2], R. Rösel[2],
H.J. Hofmann[3], W. Wölfli[3]

1 Zentraleinrichtung für Strahlenschutz, Universität Hannover
Am Kleinen Felde 30, D 3000 Hannover 1, F.R.G.
2 Abteilung Nuklearchemie, Universität zu Köln
Zülpicher Str. 47, D 5000 Köln 1, F.R.G.
3 Institut für Mittelenergiephysik, ETH Hönggerberg HPK G7
CH 8093 Zürich, Switzerland

Abstract: ^{4}He-induced reactions significantly contribute to the production of cosmogenic nuclides in terrestrial and extraterrestrial matter. The cross sections of these reactions are still poorly known. Therefore, irradiation experiments with ^{4}He were performed with energies of 119.7 MeV and 170.5 MeV in order to investigate the production of residual nuclei by ^{4}He-induced reactions on C, N, O, Mg, Al, and Si. Here, we report on the production of ^{7}Be, ^{10}Be and ^{26}Al. These nuclides were measured by gamma-spectrometry and by accelerator mass spectrometry. The new data are compared with some rare earlier measurements and analyzed by hybrid model calculations.

Introduction

^{4}He-induced reactions significantly contribute to the production of cosmogenic nuclides in extraterrestrial materials such as lunar samples, meteoroids and cosmic dust, as well as in the earth's atmosphere. The knowledge of thin-target cross sections of the underlying nuclear reactions is a basic requirement for any interpretation of the observed abundances of cosmogenic nuclides.

Up to now, most of the required excitation functions of ^{4}He-induced reactions are poorly known, if at all. Therefore, we extended our earlier studies of ^{4}He-induced reactions on Ti, V, Mn, Fe, Co and Ni [1-3] and investigated the target elements C, N, O, Mg, Al, and Si for ^{4}He-energies up to 170.5 MeV. Cross sections for the production of ^{7}Be, ^{10}Be, ^{22}Na, ^{24}Na, ^{28}Mg and ^{26}Al were determined. The new data are compared with earlier measurements and analyzed by hybrid model [4] calculations using the code ALICE LIVERMORE 87 [5]. The new excitation functions provide a basis to improve model calculations describing the depth-dependent production of cosmogenic nuclides by solar ^{4}He-particles.

Experimental

The stacked foil technique was used to investigate the production of residual nuclei by ^{4}He-induced reactions on C, N, O, Mg, Al, and Si. Except for N and O, the targets were pure element foils with thicknesses between 0.025 and 1 mm. For nitrogen and oxygen 1 mm thick foils made of Si_3N_4 and SiO_2, respectively, were used. The contributions to the residual nuclide production from Si in these targets were later corrected for. For each element individual stacks were irradiated. The ^{4}He-energies in the individual target foils were calculated according to Williamson et al. [6].

Irradiation experiments with ^{4}He were performed at the sector cyclotron at PSI/Villigen (E = 119.7 MeV) and at the isochronous cyclotron JULIC at KFA Jülich (E = 170.5 MeV). The beam currents were determined by Faraday cup measurements at the JULIC accelerator. At PSI the reaction ^{27}Al(^{4}He,4p5n)^{22}Na was used as monitor reaction adopting the recommended cross sections from the evaluation by Tobailem and de Lassus St. Genies [7]. The excitation functions determined from the irradiations at the two accelerators overlap between 120 and 80 MeV. Here, the cross sections are in excellent agreement, demonstrating the consistency of the flux determinations.

Residual nuclides were measured by gamma-spectrometry and by accelerator mass spectrometry (AMS). Details of the experimental technique, of the chemical procedures and AMS measurements (including the standards used) and of the experimental uncertainties may be found elsewhere [1-3,8,9].

Results and Discussion

The gamma-spectrometric measurements are finished both for the PSI and Jülich irradiations. They resulted in detailed excitation functions for the production of ^{7}Be, ^{22}Na, ^{24}Na and ^{28}Mg. The AMS measurements of ^{10}Be and ^{26}Al are still going on, in particular for the targets irradiated at Jülich. For ^{26}Al from Si also some gamma-spectrometric measurements after chemical separations were done. In this work, we exemplarily discuss production of ^{7}Be and ^{10}Be from C, N, O (Fig. 1) and of ^{10}Be and ^{26}Al from Mg, Al and Si (Fig. 2 and 3). These nuclides are of particular importance for geo- and cosmochemistry.

There are just very few earlier reports on cross section measurements for the nuclides dealt with in this work. For ^{7}Be from C there are two cross sections at 42.9 MeV and 90.0 MeV, measured by Vidal-Quadras and Ortega [10] and by Jung et al. [11], respectivly, which do not agree with shape and magnitude of our excitation function. The measurements between 100 and 140 MeV by Fontes and coworkers [12,13] agree within limits of errors with our results. Vidal-Quadras and Ortega [10] measured also cross sections for ^{7}Be from N

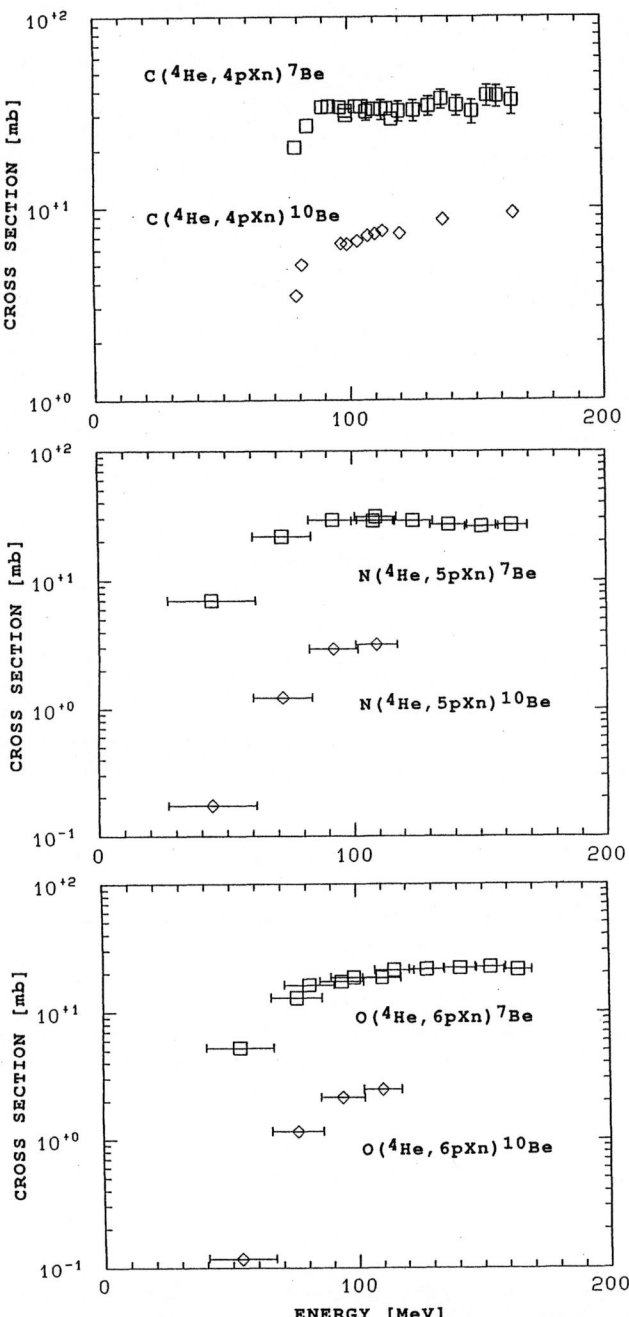

Fig. 1. Production of ^{7}Be and ^{10}Be by ^{4}He-induced
reactions on carbon, nitrogen and oxygen.
Errors are only plotted in this work if
they exceed symbol size.

For the production of ^{10}Be from Mg, Al and
Si (Fig. 2) no measurements existed up to now.
For the production of ^{26}Al from these elements
only some measurements below 45 MeV MeV existed
by Tanaka et al. [15] (Fig. 3). Generally, our
data tend to be somewhat higher, though they
agree within limits of errors for aluminum. For
magnesium, however, there are still some discrepancies which have to be further investigated.
For the target element Si our gamma-spectrometric
measurements confirm those done by AMS, thus
adding some certainty to the new measurements.

Considering the lack of cross sections for
^{4}He-induced reactions the question for nuclear
models capable of predicting reliably unknown excitation functions becomes crucial. For nucleon-
induced reactions a priori calculations of production cross sections using the hybrid model of
preequilibrium reactions [4] in the form of various versions of the code ALICE, e.g. [16], have
been very succesful [17,18]. Therefore, a hybrid
model analysis was performed for the reactions
measured in this work using the code ALICE LIVER-
MORE 87 [5]. For ^{4}He-induced reactions the initial exciton number is not uniquely defined as it
is for nucleon-induced reactions. Initial configurations with exciton numbers between 4 and 6
contribute to ^{4}He-induced reactions with various
combinations of particle hole combinations. In
this work, we distinguished calculations with
$n_0=4$ (p,n,h=2,2,0), $n_0=5$ (50% p,n,h=2,3,0; 50%
p,n,h =3,2,0) and $n_0=6$ (50% p,n,h=2,3,1; 50%
p,n,h= 3,2,1).

Our calculations do not adequately describe
the production of ^{7}Be and ^{10}Be from C, N and O.
However, this is not unexpected, since the nuclear reaction models of the code used do not
consider direct reactions, which most probably
are responsible for the observed underestimation
of the excitation functions by the hybrid model
calculations.

For the production of ^{22}Na, ^{24}Na, ^{26}Al and
^{28}Mg no objections against the applicability of
the calculational methods exist. This is revealed
by a comparison of the experimental and calculated excitation functions. However, the agreement is not as good as observed earlier [17] for

and O, which within limits of error are compatible with the new data. For ^{7}Be from O, there is
one more earlier report for energies up to 40 MeV
by Bouchard and Fairball [14]. Their low-energy
data fit well to the excitation function measured
in this work.

For the production of ^{10}Be from C, N and O
there only exist earlier measurements by Vidal-
Quadras and Ortega [10] at 42.9 and 50.6 MeV. All
their cross sections for ^{10}Be are higher than our
data by at least one order of magnitude. For ^{10}Be
from C their data even are in contradiction to
the general shape of our excitation function.

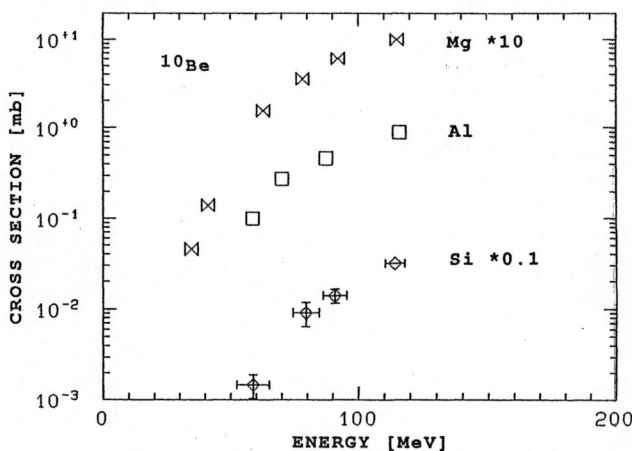

Fig. 2. Production of ^{10}Be by ^{4}He-induced reactions on magnesium, aluminum and silicon.

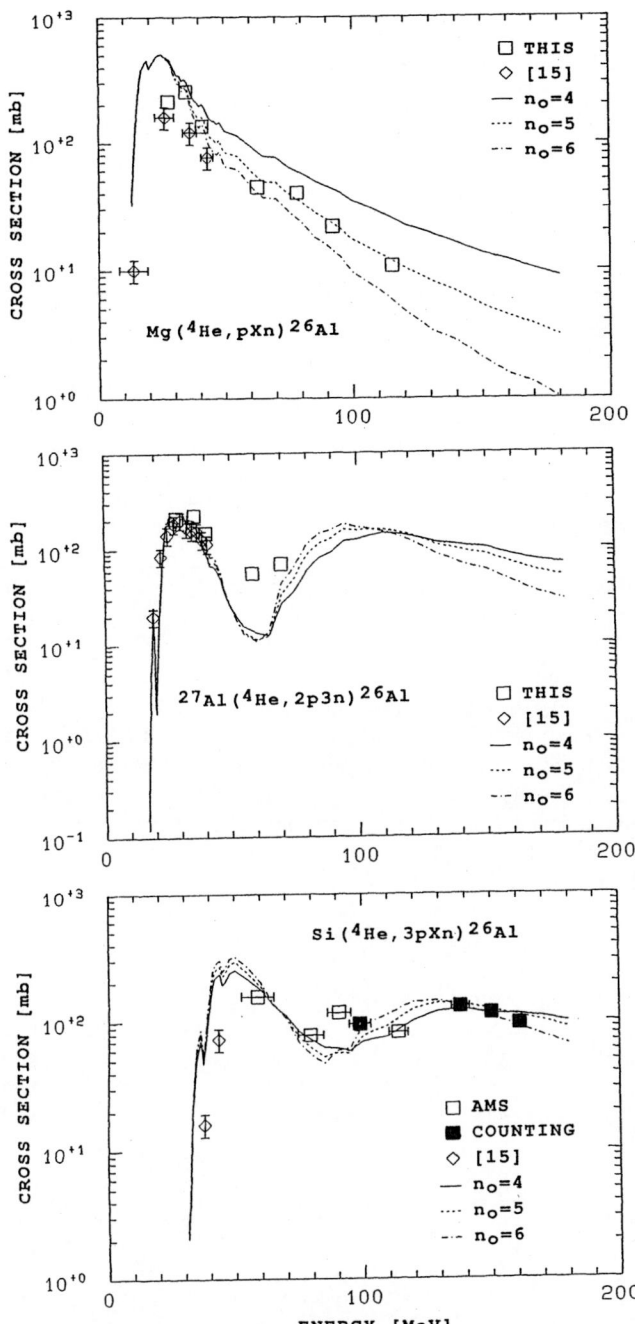

Fig. 3. Experimental cross sections for the production of ^{26}Al from magnesium, aluminum and silicon compared with hybrid model calculations. The discontinuities in the theoretical curves are due to too large energy-bins in the preequilibrium calculations.

nucleon-induced reactions, see e.g. Fig. 3. For Mg(^{4}He,pXn)^{26}Al the maximum of the excitation function is overestimated by more than a factor of 2. For ^{27}Al(^{4}He,2p3n)^{26}Al the discrepancies between 50 and 70 MeV are due to the neglect of preequilibrium ^{4}He-emission. There are general problems with the shape of the excitation function for Si(^{4}He,3pXn)^{26}Al, the reasons for which are not yet clear. Surely, the hybrid model analysis of all the reactions measured in this work deserves a more detailed discussion than possible

within the limited space available here. But, one can conclude already now, that these calculations still show uncertainties too large to allow for their applicability to predictions of integral excitation functions for applications. Consequently, systematic measurements are presently the only means to satisfy the data needs of cosmochemistry and astrophysics.

Acknowledgement: The authors are grateful to the authorities of PSI/Villigen and KFA/Jülich for the beam-time and to the staffs of the accelerators for their good cooperation. The silicon targets were kindly provided by Wacker Chemitronic, Burghausen. This work was suported by the Deutsche Forschungsgemeinschaft and in part by the Swiss National Science Foundation.

References

1. R. Michel and G. Brinkmann, Nucl. Phys. A338, 167 (1980)

2. R. Michel, G. Brinkmann and R. Stueck, Radiochim. Acta 32, 173 (1983)

3. R. Michel, G. Brinkmann and R. Stueck, in: K.H. Boeckhoff (ed.) Proc. Int. Conf. Nuclear Data for Science and Technology, Sept. 6-10, 1982, Antwerp, 599 (1983)

4. M. Blann, Phys. Rev. Lett. 27, 337 (1971)

5. M. Blann, priv. comm. (1987)

6. C.F. Williamson, J.P. Boyot and J. Picard, CEA-R3042 (1966)

7. J. Tobailem and C.H. de Lassus St. Genies, Additif No. 2 a la CEA-N-1466(1) (1975), CEA-N-1466(5) (1981)

8. B. Dittrich, U. Herpers, H.J. Hofmann, W. Woelfli, R. Bodemann, M. Luepke, R. Michel, P. Dragovitsch and D. Filges, Nucl. Instr. Meth. Phys. Res. B52, 588 (1990)

9. B. Dittrich, U. Herpers, M. Lüpke, R. Michel, H.J. Hofmann and W. Wölfli, Radiochimica Acta 50, 11 (1990)

10. A. Vidal-Quadras and M. Ortega, Nuovo Cim. 49a, 235 (1979)

11. M. Jung, C. Jacquot, C. Baixeras-Aiguabella, R. Schmitt, H. Braun and L. Girardin, Phys. Rev. 188, 1517 (1969)

12. P. Fontes, C. Perron, J. Lestringuez, F. Yiou and R. Bernas, Nucl. Phys. A165, 405 (1971)

13. P. Fontes, Phys. Rev. C15, 2159 (1977)

14. G.H. Bouchard and A.W. Fairhall, Phys. Rev. 116, 160 (1959)

15. S. Tanaka, K. Sakamoto, K. Komura and M. Furukawa, J. Inorg. Nucl. Chem. 37, 2002 (1975)

16. M. Blann and J. Bisplinghoff, UCID-19614 (1982)

17. R. Michel, F. Peiffer and R. Stück, Nucl. Phys. A441, 617 (1985)

18. M. Blann, in K. Okamoto (ed.) Proc. of the IAEA Consultant's Meeting on Data Requirements for Medical Radioisotope Production, INDC(NDS)-195/GZ, IAEA (1988) p. 115

THICK TARGET NEUTRON YIELD BY HEAVY IONS

Kazuo Shin and Kagetomo Miyahara
Department of Nuclear Engineering, Kyoto University
Yoshida, Sakyo-ku, Kyoto 606, Japan

Yoshitomo Uwamino
Institute for Nuclear Study, University of Tokyo
Midori-cho 3-2-1, Tanashi, Tokyo 188, Japan

Abstract: This paper describes the systematics study of the thick target neutron yield by light and heavy ions. Measurements of neutron yield data were made for combinations of thick targets (C, Al, Cu, Pb) and projectiles (40-MeV α, 120-MeV ^{12}C, 153-MeV ^{16}O). Obtained neutron angular spectra together with those in previous works (30- and 52-MeV p, 65-MeV α) were analyzed by the moving source model. Through the consideration on the nuclear temperature, it was found that the multiplicity of equilibrium neutrons was proportional to the target mass number. Using this fact, a simple expression for the systematics of equilibrium neutron yield was derived. For the nonequilibrium component, the neutron yield was proportional to the geometrical cross section. The threshold for the nonequilibrium neutron production affected the yield for low energy projectiles.

(heavy ion, neutron yield, thick target, neutron production, equilibrium neutrons, nonequilibrium neutrons, moving source model)

I. Introduction

Inclusive neutron production data are fundamental for the application of ion accelerators to the engineering purpose. There are very few inclusive neutron production data existing./1/ And it is important to study the systematics of the neutron yield to assess unknown data.

We measured the neutron production data from thick targets bombarded by light ions./2-4/ In this work the measurement was extended to heavy ions and the systematics of the inclusive neutron yield covering both light and heavy ions is studied.

II. Experimental Procedures

The measurement of secondary neutron-energy and angular distributions was made at the SF-cyclotron facility at the Institute for Nuclear Study, University of Tokyo. The experimental arrangement is shown in Fig. 1. The ion beams used in this work were 40-MeV α, 120-MeV ^{12}C and 153-MeV ^{16}O. Targets of C, Al, Cu and Pb which were thicker than the ion stopping range were bombarded by the ions and neutron spectra were detected by an NE-213 scintillator at angles 0°, 30°, 75°, 120° and 150°. The detector was located 2 m down stream

from the target. The spectra were analyzed by the unfolding method by FERDO-U code/5/ with a response matrix which was constructed from measured and calculated response functions./6/ The background was estimated by similar measurements with an iron and polyethylene shadow shield inserted between the target and the detector.

III. Results and Discussions

1. Moving Source Analysis

Measured neutron angular spectra of this work and the previous works (data of 30- and 52MeV p, and 65-MeV α) were fitted by the two-component moving source model of Eqs. (1) and (2):

$$\phi (E_n, \theta) = \sum_{i=1}^{2} M_i \frac{E_n}{2(\pi \tau_i)^{3/2}} \exp(-\frac{E_s}{\tau_i}), \quad (1)$$

$$E_s = E_n - 2 \varepsilon_i E_n \cos \theta + \varepsilon_i, \quad (2)$$

where M_i is the source intensity, τ_i the nuclear

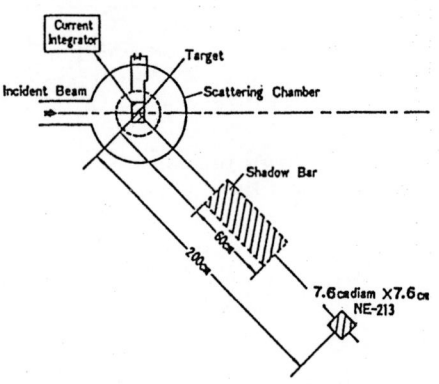

Fig.1 Experimental arrangement

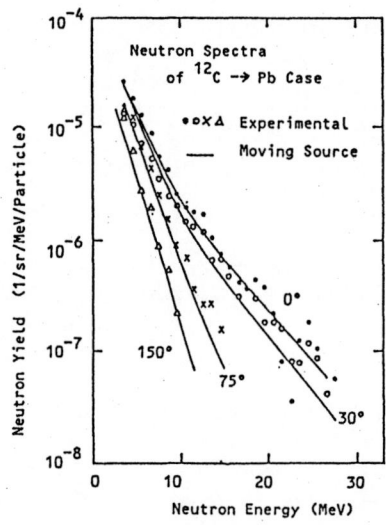

Fig.2 Fitting by the moving source model

temperature, ε ; the source moving energy in terms of the kinetic energy of a nucleon.

As is illustrated by Fig. 2, which is for the case of the ^{12}C ion on the Pb target, the neutron angular spectra are very well reproduced by two sources, one with smaller ε (equilibrium neutron: EN) and the other with higher source speed (nonequilibrium neutron: NEN). The total neutron yield is dominated by the EN component. The contribution by the NEN component is about one-tenth.

The systematics of the neutron yield was studied seperately for the EN and NEN components using M_i values obtained by data fitting.

2. Equilibrium Neutrons

Fig. 3 summarizes the nuclear temperature of the EN component, where the temperature τ ; is plotted vs the energy that projectile-like or target-like fragment shared. It shows that the nuclear temperature data for each target are closely on a unique curve and the shape of the curves for different targets resembles one another.

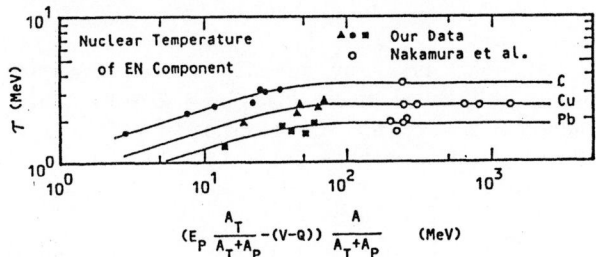

Fig.3 Nuclear temperature of EN neutrons

The temperature τ is proportional to $A_T^{-1/5}$, where A_T is the target mass. Then, we use the relation $E = a\tau^2$ between the excitation energy E and τ with the level density parameter a, and assume the proportionality of a to A_T, i.e. $a = A_T/8$, to obtain E is proportional to $A_T^{3/5}$. For this excitation energy, the neutron multiplicity from excited nuclei was estimated using calculated data of Dostrovsky et al./7/, with the assumption that the multiplicity was proportional to the excitation energy. The consequence was that the neutron multiplicity was approximately proportional to the target mass number A_T.

The neutron production cross section for the EN component from the target nucleus in the central collision would be expressed as the target geometrical cross section times the nucleon number in the projectile times the neutron multiplicity. The exchange of A_T with A_P would give the cross section for the projectile evaporation. So the total EN cross section would be,

$$\sigma_n = g(A_P, E_P)(A_P A_T^{5/3} + A_T A_P^{5/3})f , \quad (3)$$

where g is a function of the projectile mass A_P and the projectile energy E_P. The factor f is the correction for the Coulomb barrier,

$$f = 1 - \frac{A_T + A_P}{A_T}\frac{V-Q}{E_P} , \quad (4)$$

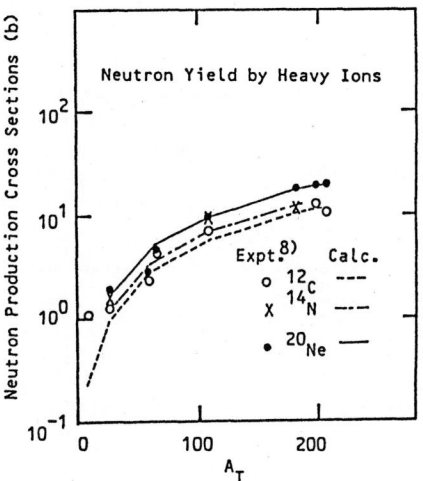

Fig.4 Neutron production cross sections by 10 MeV/u heavy ions

where V is the Coulomb barrier and Q the effective Q value.

In Fig. 4, the result of the equations (3) and (4) with the constant g is compared with measured data of the total neutron production cross section by heavy ions given by Hubbard et al./8/ The dependence of the neutron cross section on the target and projectile mass numbers is very well reproduced.

The thick target neutron yield is given by the following, neglecting the neutron production by secondary ions,

$$Y_n = \int_{E_m}^{E_P} N\sigma_n(E)(\frac{dE}{dx})^{-1} dE \quad (5)$$

where N is the target atomic density, E_m the minimum energy for $f \geqq 0$.

The level density parameter a exhibits oscillation with A_T at low excitation energies due to the shell effect./9/ This effect should be incorporated in Eq. (3) when E_P is small. To derive Eq. (3) we simply assumed that the neutron multiplicity was proportional to the nucleus excitation energy which in turn was

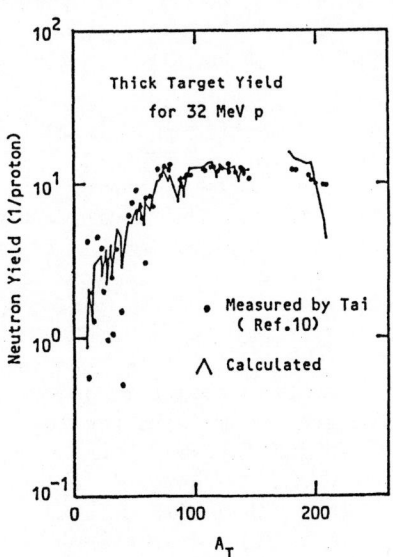

Fig.5 Neutron yield by 32-MeV protons

proportional to a. Then, for the correction for this effect, σ_n of Eq. (3) should be multiplied by the ratio of a to A_T.

Tai et al./10/ measured thick target neutron yield for 32-MeV p for many targets. The result of the thick target yield predicted by Eqs. (3)-(5) with the above level density parameter correction is compared in Fig. 5 with his data, where the level density parameter was taken from Baba's measurement./9/ The agreement is quite good.

The model was also applied to our measurements, and all the thick target neutron yield data were well fitted by the model. This is illustrated in Fig.6 for the case of 10-MeV/u ^{12}C ions.

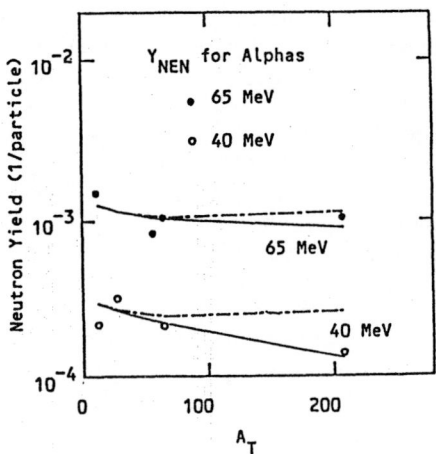

Fig.7 NEN neutron yield by 40- and 65-MeV alphas

nonequilibrium neutrons. Unless the velocity is larger tha E_Q in terms of nucleon kinetic energy, the NEN is not produced. Then the lower boundary E_m of the integration of Eq. (5) was given for the NEN component by,

$$E_m = A_P E_Q + (A_T + A_P)(V-Q)/A_T . \qquad (7)$$

All our data for the NEN yield were fitted well by the above model with E_Q = 2 MeV. This is demonstrated by the solid line in Fig. 7. The dashed line is the results for E_Q = 0 MeV.

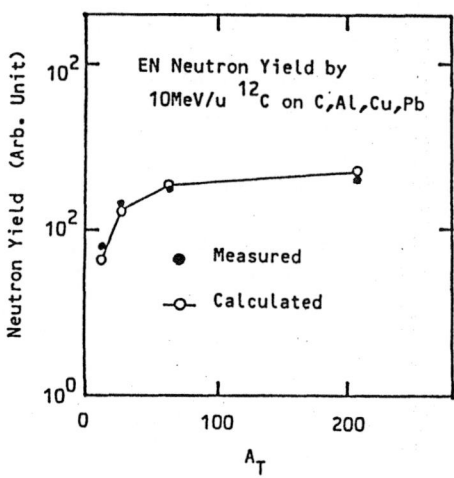

Fig.6 Neutron yield by 10-MeV/u ^{12}C ions

3. Nonequilibrium Neutrons

The temperature τ of nonequilibrium protons emitted from heavy ion reactions was reproduced well by the local hot spot model./11/ In the model, the nonequilibrium protons were emitted from a thin layer where the target and the projectile nuclei were contacting. The same model was assumed in this work for the NEN production. Then, the neutron production rate should be proportional to the probability of contact (geometrical cross section) times the probability of neutron existing in the region,

$$\sigma_n = s(A_P, E_P) f \frac{A_T - Z_T}{A_T} \frac{A_P - Z_P}{A_P} A_P^{2/3} A_T^{2/3} \Delta , \qquad (6)$$

where s is a function of A_P and E_P, and Z_T the Z number of the target. The factor Δ depends on the thickness of the hot spot. We assumed a constant value for Δ.

The thick target neutron yield of the NEN component was given by inserting Eq. (6) into Eq. (5), where we neglected the E_P dependence of s. This very simple description well reproduced the target dependence of our data for 30- and 52-MeV p and 65-MeV α but failed in the cases of 40-MeV α and heavy ions. The projectile energy in the latter cases are $\sim$ 10 MeV/u, while the former are 30, 52 and 16.3 MeV/u, respectively.

We considered that there was a threshold in the velocity of contacting nuclei to produce

References
1. T. Nakamura, Y. Uwamino, K. Sato, Y. Furuta, S. Tanaka, S. Ban, H. Hirayama, T. Kosako, K. Kawachi and Y. Nishihara, INS-TS-20, Institute for Nuclear Study, Univ. Tokyo (1981)
2. T. Nakamura, M. Yoshida and K. Shin, Nucl. Instr. Methods 151, 493 (1978)
3. T. Nakamura, M. Fujii and K. Shin, Nucl. Sci. Eng. 83, 444 (1983)
4. K. Shin, K. Hibi, M. Fujii, Y. Uwamino and T. Nakamura, Phys. Rev. C29, 1317 (1984)
5. K. Shin, Y. Uwamino and T. Hyodo, Nucl. Technol. 53, 78 . (1981)
6. K. Shin et al., to be published in Nucl. Instr. Methods
7. I. Dostrovsky, P. Rabinowitz and R. Bivins, Phys. Rev. 111, 1659 (1958)
8. E. L. Hubbard, R. Main and R. Pyle, Phys. Rev. 118, 507 (1960)
9. H. Baba, Nucl. Phys. A159, 625 (1970)
10. Y. K. Tai, G. Millburn, S. Kaplan and B. Moyer, Phys. Rev. 109, 2086 (1958)
11. T. Awes, G. Poggi, S. Saini C. Gelbke, R. Legrain and G. Westfall, Phys. Lett. 103B, 417 (1981)

RELATIVE CROSS SECTIONS
OF 25 TO 70 MeV PROTON-INDUCED FISSION
OF ^{232}Th, ^{233}U, ^{235}U, ^{237}Np AND ^{239}Pu

A. N. Smirnov, I. Yu. Gorshkov, A. V. Prokofyev, V. P. Eismont

V. G. Khlopin Radium Institute
Roentgen str., 1, 197022 Leningrad, USSR

Abstract: Measurements of relative cross sections of 25 to 70 MeV proton-induced fission of ^{232}Th, ^{233}U, ^{235}U, ^{237}Np and ^{239}Pu were performed at the V. G. Khlopin Radium Institute's proton synchrotron. To detect fission fragments the thin-film breakdown counters technique was used. The summed experimental errors were about 8-9% .

(MEDIUM-ENERGY PROTONS , relative fission cross sections,
^{232}Th, ^{233}U, ^{235}U, ^{237}Np, ^{239}Pu,
thin-film breakdown counters of fission fragments)

Introduction

The proton-induced fission cross sections of ^{232}Th, ^{233}U, ^{235}U, ^{237}Np and ^{239}Pu relative to that of ^{238}U have been measured for proton energies from 25 to 70 MeV to fill a gap in the experimental data of this kind. Absolute fission cross sections of ^{238}U have been measured rather well up to date, but there are serious discrepancies in the data for the above energy region. So we measured the energy dependence of ^{238}U proton-induced fission cross section earlier for proton energies from 10 to 100 MeV /1/ at the V. G. Khlopin Radium Institute's proton synchrotron with the use of the absorption foil technique to change the proton energy and thin-film breakdown counters (TFBC) /2/ to detect the fission fragments. The results obtained were in good agreement with the data of Baba et al. /3/ for the proton energies up to 55 MeV and Bychenkov, Shygaev et al. /4/ near 70 MeV.

Experimental Procedure and Results

The measurements reported in this paper have also been performed at the same synchrotron. Proton energies were varied by changing the accelerator regime. The beam from the evacuated exit tube was transported through a 0.2 mm titanium window and a short distance in air to the fission chamber. In the chamber were installed two pairs of "back-to-back" targets with the layers of ^{238}U and one of the nuclides under study. The targets were prepared by the technique of multiple smearing of the oxide layers onto a 100-μm thick aluminium foil backing with interim annealing. The thickness of the fissile layers was from 200 to 400 μg/cm^2 and was determined by measuring the

intrinsic activity (for ^{233}U, ^{237}Np and ^{239}Pu) or by the α-particle backscattering technique (for ^{232}Th, ^{235}U and ^{238}U). The targets were mounted in a circular holder moving by remote control. The fission fragment counters (TFBC) were placed on both sides of the target to detect the fragments of each nuclide separately. A schematic arrangement of the targets and detectors in the chamber and some dimensions are shown in fig.1. The use of TFBCs was required because of the hard background conditions caused by a number of reasons, the main being the pulsed proton beam (50 Hz, 100 μs, 10^8 proton/pulse) and the "thick" target backings. The threshold properties of TFBC permit selective detection of the fragments with an

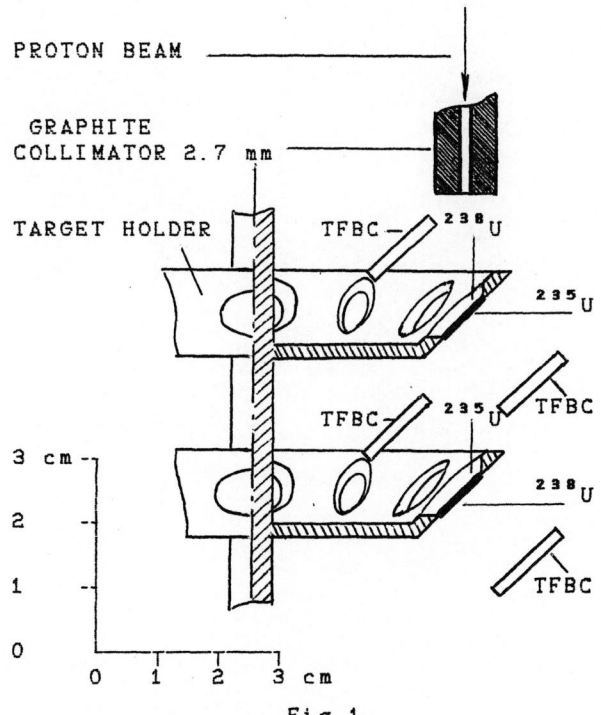

Fig.1.
Schematic arrangement
of the targets and detectors.

intense flux of accompanying particles and radiations. A significant advantage of TFBC is instantaneous detection of the events which makes it possible to work "on-line" with the use of various techniques to control the measurement run, to accumulate and process the preliminary results, to perform calibration measurements. The latter were most significant because of the gradual decrease of the TFBC detection efficiency with the increase of the number of the detected fragments /2/. To measure the relative detection efficiency two paired targets of ^{238}U were used which, furthermore, permitted to compensate the difference between the number of fragments leaving the target in the forward and backward directions relative to that of the proton beam, due to the linear momentum transfer from a projectile proton to a target nucleus. Calibration measurements were repeated after two or three runs with work targets. The computer program of data collection and processing allowed to obtain "on-line" the preliminary values of the cross sections ratios computed with the use of calibration data.

Experimental errors consisted of the uncertainty of the target weight (3%), the inhomogeneity of the layer thickness (3%) and the statistical errors (1-2%) of the main and the calibration measurements. Estimated corrections for the influence of the fission fragments angular anisotropy and its dependence on the proton energy, the dependence of the TFBC detection efficiency on the fragment

energy and angular distribution, as well as for the influence of the neutron background were less than 1%. The summed errors were about 8-9%. The relative character of the measurements allowed to eliminate the errors connected with proton beam monitoring.

The relative to ^{238}U cross section values versus the proton energy E_p are shown in the following table :

E_p MeV	^{232}Th	^{233}U	^{235}U	^{237}Np	^{239}Pu
24	0.75	1.16	1.08	1.06	1.13
27	0.74	1.14	1.06	1.12	1.13
29	0.77	1.14	1.00	1.08	1.10
34	0.84	1.08	1.03	1.05	1.12
37	0.84	1.09	--	1.06	1.17
42	0.83	1.10	0.95	0.98	1.14
49	0.85	1.09	1.05	1.05	1.16
54	0.84	1.08	0.97	1.04	1.15
59	0.81	1.11	1.04	1.09	1.16
69	0.85	1.06	1.01	0.99	1.13

References

1. I.Yu.Gorshkov, A.N.Smirnov, V.P.Eismont.Abstracts of International Conference "Fiftieth Anniversary of Nuclear Fission". Leningrad, USSR, October 16-20, 1989, p.82.
2. A.N.Smirnov, V.P.Eismont. Pribory i technika experimenta 6, 5 (1983) (in Russian).
3. S.Baba, H.Umezeva and H.Baba. Nucl. Phys., A175, 177 (1971).
4. V.S.Bychenkov, M.M.Lomanov, A.I.Obukhov, O.E.Shygaev. Preprint RI-17, Leningrad, 1973.

MEASUREMENT OF THE RATIO Γ_f/Γ_n AND THE FISSION BARRIER HEIGHT FOR p + ^{209}Bi

Bao Zongyu, Chen Jinhua, Meng Jiangchen and Huang Shengnian

Institute of Atomic Energy, P.O.Box 275, Beijing 102413, P.R.China

Abstract: The excitation curve of the reaction ^{209}Bi(p,f) is measured relatively to that of 209Bi(p,xn) for x=3,4 using a mica fission fragment detector and a Ge(Li) gamma-ray spectrometer at the 35.5 MeV proton linear accelerator. The energy dependence of the ratio Γ_f/Γ_n of the compound nucleus 210Po has been obtained according to the reaction excitation function ratio between its fission and neutron emission. Besides, the fission barrier height of p + 209Bi was deduced as 21.9 $\pm$ 1.5 MeV.

(^{209}Bi(p,f), ^{209}Bi(p,xn), Γ_f/Γ_n, excitation curve, fission barrier height)

Introduction

It is interesting to study the ratio of the de-excitation probabilities through fission and neutron emission for the heighly excited nucleus 209Bi, a heavy stable nucleus very close to the region of double magic number. From this quantity the ratio of its fission width Γ_f to the neutron width Γ_n can be acquired directly and some nuclear parameters can be deduced as well. Since the reaction (p,xn) for x=2,3,4 are predominant over all other ones in the range of Ep=10-50 MeV for p + ^{209}Bi, its ratio Γ_f/Γ_n can be expressed by:

$$\Gamma_f/\Gamma_n = \sigma_f/\sum_x \sigma_{p,xn} \quad x=2,3,4 \quad (1).$$

Here, σ_f and $\sigma_{p,xn}$ are the fission cross section and the neutron emission cross section, respectively. But as one knows, the fission cross section is extreemely difficult to measure due to the interference of the trace impurity from U and Th in the sample, around 3-8 orders of magnitude lower than that of them. Besides, its (p,xn) cross section is scarcely measured. Furthermore, there has not been any data of its ratio Γ_f/Γ_n from an absolute measurement up to now.

Khodai-Joopari measured the fission cross section of ^{209}Bi at Ep=15-50 MeV, using a mica fission fragment detector at the Berkely' 88 " cyclotron with energy spread 0.4 MeV /1/. The detection geometry was calibrated by use of the reaction ^{197}Au(α,f) and the integrated proton beam current was measured by an integrator. Instead of experimental neutron emission cross sections, a theoretical total reaction cross section σ_r was taken, calculated with the optical model, so that the ratio Γ_f/Γ_n was taken from the ratio of σ_f over σ_r.

Bell and Skarsgard measured the neutron emission cross sections of ^{209}Bi by the activation method at the Chalk River cyclotron at Ep=20-80 Mev /2/, relatively to the monitor reactions 12C(p,pn) etc. with proton energy uncertainty and spread around 2-3 Mev. The total experimental uncertainty was 16%. Besides, the measurement was carried out quite long ago, some basic data, such as branching ratios, have been improved a lot in the meantime.

It is the purpose of the present paper to measure the excitation functions of both reactions ^{209}Bi(p,f) and ^{209}Bi(p,xn) for x=3,4 under the equal experimental conditions at Ep=20-35 MeV, so that the ratio $\sigma_f/\sum_x \sigma_{p,xn}$ can be deduced directly without employing any other cross sections as nominal standard as follows:

$$\sigma_f/\sum_x N_{p,xn} = N_f/\sum_x N_{p,xn} \quad x=2,3,4 \quad (2),$$

where, N_f is the total number of fission events, produced in one piece of the samples and $N_{p,xn}$ is the absolute number of ^{209}Bi(p,xn) reaction products in the identical sample. The data of ^{209}Bi(p,f) below 30 MeV and the data of ^{209}Bi(p,2n) were taken from Ref.1 and Ref.2, respectively, to complement the present ones for final analyses.

Experiment

Proton beam, sample and absorber

The proton beam was acquired from the 35.5 MeV proton linear accelerator of the Institute of High Energy Physics in Beijing. Along the beam path 2 to 5 sandwiches were placed closely one after another. Each of them consisted of a Bi sample, a piece of mica fission detector and sheets of proton energy absorbers. The values of the degrated proton energy were enquired from the Proton Range-Energy Tables /3/. The energy resolution was 0.6 MeV. The irradition conditions were as the follows: peak of proton beam current 30 mA, beam pulse width 30 μs, frequency 2 or 5 Hz, beam spot diameter 11.6 $\pm$ 0.1 mm and the irradiation time 2-10 minutes. Three runs of irradiation were carried out. Each of them was normalized by the γ-activities of ^{206}Bi, one of the decay products of the reaction ^{209}Bi(p,4n).

The samples were made from ultrapure metal Bismuth powder of 99.9995% purity. By means of vacuum evaporation the purity has been improved further. As a results the content of U and Th of the samples were very low and were of the order 10^{-3} ppm, determined by the fission track method. Therefore, the background caused by them was negligible. The samples were around 1.5 mg/cm^2 thick and their backing were thin mica of 30 μm thickness. The absorbers were 0.1-0.5 mm Aluminium or stainless steel sheets. Their uniformity was better than 4%.

Measurement of the reaction ^{209}Bi(p,f)

The mica track detector was adopted to detect the fission fragments of ^{209}Bi(p,f).The impurity of U or Th in the mica was less than 10^{-3} ppm, determined by the method mentioned above, so that their interference could be ignored, too. The pre-etching of mica was carried out at 45-50 °C in 40% HF of analytical purity for 1 hour. the same condition was applied for etching after the irradiation, but only for 20 minutes. In this way the size of the signal track could be well distinguished from that of the background. The density of the tracks of each field of vision was counted by utilizing a microscope. Dozens to several hundreds fields of vision were even scanned over the whole irradiation area of a sample, so that the total absolute number of fission events can be deduced, considering the efficiency of mica detectors at 2π-geometry.

Measurement of the reactions ^{209}Bi(p,xn)

The excitation function of the reactions ^{209}Bi(p,xn) for x=3,4 was measured by the activation method. The gamma ray spectra of ^{209}Bi(p,xn) products were detected by use of a 130 cm^3 coaxial Ge(Li) detector with an energy resolution of 2 KeV at Eγ=1332 KeV and were recorded by a multichannel analyzer-computer system. The absolute efficiency of the Ge(Li) detector was determined by use of a series of standard γ-sources; while the energy calibration for the system was done with a ^{152}Eu source. A code of analyzing gamma ray spectra was used for searching peaks, energy calibration and calculation of the area under the peak.

According to the counts over the characteristic gamma ray peak area, the absolute number of the product nuclei of the reactions ^{209}Bi(p,xn) could be obtained, taking into account the efficiency of the Ge(Li) detector, the decay constant and the branching ratio of the reaction products, self-absorption of the sample, the dead time of the multichannel analyzer-system and the coincidence summing correction.

Results and Discussion

According to eq.(1) and (2), the energy dependence of Γ_f/Γ_n has been obtained directly from the excitation curves of ^{209}Bi(p,f) and ^{209}Bi(p,xn) for x=2,3,4 which is shown in Fig.1 as the curve 1. The experimental uncertainty is less than 13%, mainly from the uncertainties of the coincidence summing correction of γ-detection and the normalization factor.

The excitation energy dependence of Γ_f/Γ_n of Khodai-Joopari is also shown as curve 2 by full triangles /1/ and the result deduced from his σ_f over $\sum_x \sigma_{p,xn}$ (x=2,3,4) of Bell and Skarsgard /2/ is plotted there by crosses, too.

As can see, both triangles and crosses are about 3 times higher than the dots of the present work, or it could be said, there is an energy shift of about 3 MeV between our and their results. It is clear that the statistical uncertainty can not cover the gap between them.

This dicrepancy might be caused by the following:

1. The proton accelerators of Ref.1, Ref.2 and

the present one might differ by a few MeV in the energy absolute calibration.

2. The fission cross section of Ref.1 might be systematically larger because of the uncertainty of the integrated beam current. In addition, the thickness of the Bi sample was 15 mg/cm^2, which was too thick for detecting fission fragment and differed much from that of its monitor sample Au, 3 mg/cm^2.

3. The Γ_f/Γ_n of Ref.1 was from the theoretical total reaction cross section σ_r, but the one

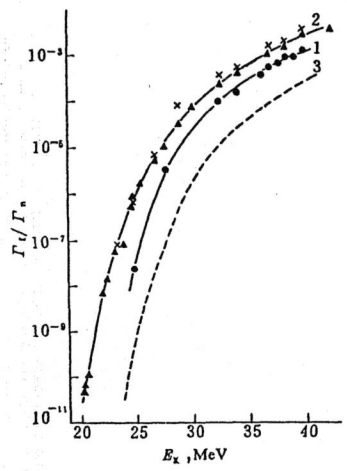

Fig.1. The Log Γ_f/Γ_n as a function of excitation energy Ex for p + ^{209}Bi.
- • Present work
- ▲ Khodai-Joopari /1/;
- x Deduced from σ_f /1/ and p,xn for x=2,3,4 /2/;
- -------- Theoertical curve /4/.

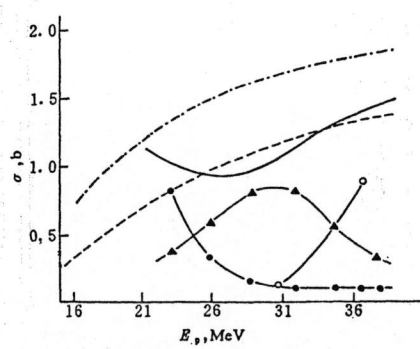

Fig.2. The experimental neutron emission cross sections of ^{209}Bi(p,xn) for x=2,3,4 and the calculated total reaction cross section σ_r.

(Exp.): • (p,2n); ▲ (p,3n); o (p,4n); – sum /2/
(Theo.): —·— σ_r /1/; ---- σ_r /2/.

of the present paper is from experimental neutron emission cross section. It seems that the sum of (p,xn) reaction cross sections for x=2,3,4 should be nearly equal to the calculated σ_r in this energy region. But in fact there is a deviation between them, as shown in Fig.2. Besides, the theoretical total reaction cross section obtained by various authors could be rather different (see Fig.2). Therefore, in order to get Γ_f/Γ_n, it seems to be more resonable to use the sum of (p,xn) experimental cross sections for x=2,3,4, rather than the calculated reaction cross section.

In Fig.1 a theoretical curve of Γ_f/Γ_n for ^{210}Po is also shown as curve 3 /4/, which was obtained by considering the shell influence and the dependence of single particle density on the deformation of nuclei, supposing that its fission barrier height was 22.9 MeV, as derived from the two-center model /5/. It is obvious that the points of the present work are much closer to this theoretical curve than the others.

Khodai-Joopari suggested that one can take the excitation energy where the value of Γ_f/Γ_n has dropped by a factor of about 30 or more per MeV as the fission barrier height with standard deviation about 1.5 MeV. Although this method was simple, the value of B_f obtained in this way was reliable and consistent with that deduced by fitting the experimental data of Γ_f/Γ_n with theoretical models /1,6/. Based on this definition, $B_f=21.9 \pm 1.5$ MeV is obtained using a polynomial fit to the present Γ_f/Γ_n curve by means of the least square method. Similarly, a value of 20.9 ± 1.5 MeV can be deduced from the data of σ_f of Ref.1 and $\sum_x \sigma_{p,xn}$ for x=2,3,4 of Ref.2.

In Table 1 the experimental fission barrier height B_f of p + ^{209}Bi are listed together with the theoretical ones of several authors /5-7/.

Table 1 The fission barrier height of p + ^{209}Bi--^{210}Po

Author	B_f, MeV
(Experimental)	
Present work	21.9 $\pm$ 1.5
Khodai-Joopari /1/	20.4 $\pm$ 0.5
Khodai-Joopari (σ_f) /1/ and Bell and Skarsgard ($\sum_x \sigma_{p,xn}$, x=2,3,4) /2/	20.9 $\pm$ 1.5
Moretto et al. /6/	21.4
(Theoretical)	
Mayers and Swiatecki /7/	
Liquid drop model	13.763
L.D.M. + shell and pairing effects	21.136
Mosel and Schmitt /5/	
Two-center model	22.9

It is obvious that the theoretical fission barrier height, derived with the liquid drop model, deviates quite a lot from the experimental ones of about 8 MeV. However, it is in good agreement with the experiments, if the shell and the pairing effects are taken into account.

Therefore, for the nuclei close to the shell region, these effects play an important role in the competition between fission and neutron emission and significantly influence the value of the fission barrier height.

References

1. A.Khodai-Joopari: UCRL-16489(1966).
2. R.E.Bell and H.M.Skarsgard: Can. J. Phys.,34, 745(1956).
3. J.F.Janni: Atomic Data and Nuclear Data Tables, 27, 147(1982).
4. R.Vandenbosch and U.Mosel: Phys. Rev. Lett., 28, 1726(1972).
5. U.Mosel and H.W.Schmitt: Phys. Rev. Lett., 37B, 335(1971).
6. L.G.Moretto, S.G.Thomposon, J.Routti and R.C.Gatti: Phys. Lett., 38B, 471(1972).
7. W.D.Myers and W.J.Swiatecki: Nucl. Phys., 81, 1(1966).

Part X

Nuclear Models and Evaluation Methodology

NUCLEAR MODELS RELEVANT TO EVALUATION

E. D. Arthur, M. B. Chadwick, G. M. Hale, and P. G. Young

Theoretical Division, MS B243
Los Alamos National Laboratory
Los Alamos, New Mexico 87545, USA

Abstract: The widespread use of nuclear models continues in the creation of data evaluations. The reasons include extension of data evaluations to higher energies, creation of data libraries for isotopic components of natural materials, and production of evaluations for radioactive target species. In these cases, experimental data are often sparse or nonexistent. As this trend continues, the nuclear models employed in evaluation work move towards more microscopically-based theoretical methods, prompted in part by the availability of increasingly powerful computational resources.

Advances in nuclear models applicable to evaluation will be reviewed. These include advances in optical model theory, microscopic and phenomenological state and level density theory, unified models that consistently describe both equilibrium and nonequilibrium reaction mechanisms, and improved methodologies for calculation of prompt radiation from fission.

(Keywords: Nuclear theory, nuclear models, R-matrix, optical model, statistical theory, fission theory, preequilibrium theory, nuclear data evaluation)

Introduction

Reviews by Uhl [1], Gardner [2], and Young [3] at earlier international nuclear data conferences have stressed the increased and important role that nuclear models play in the provision of data for a variety of applications. This trend continues with theoretical calculations representing a major component required for development of evaluated data files for ENDF-VI, JENDL, JEF, and the BROND libraries. These data systems have expanded their scope beyond the incorporation of neutron-induced reaction data for energies below 20 MeV. Extensions include provision of reaction data for incident charged-particles and higher energy data. In these extensions, nuclear models have played a significant and major role. Applications such as nuclear waste transmutation, radiation protection for manned space exploration, and radiotherapy involving energetic nucleon beams all require expanded nuclear data bases [4-6]. Nuclear models, will be used to produce major parts of the databases required for these new applications.

The application of nuclear models to data evaluation has largely involved use of a collection of models--optical, statistical, and preequilibrium--where the user compromises between the complexity of microscopically-based formulations and the simplicity (and hence calculational speed) of more phenomenological approaches. Much of the nuclear model effort directed towards evaluation uses results and insights from fundamental nuclear theory development to guide the choice of approximations required for applications. Determination of parameter systematics is a major effort area that aims at improving the interpolative or predictive capability of phenomenologically-based models. Finally, it is now a universal practice in model applications to simultaneously analyze data for a variety of reaction channels for the system of interest, as well as data that may be available for nearby nuclei or for other incident particles. Achieving an overall consistency in calculated results helps ensure the reliability of extrapolated results and guarantees that fundamental conservation principles are adhered to in the resulting evaluated data files.

This paper briefly reviews the main nuclear models used for applications and will indicate areas of major development and application activity. Discussion of more fundamental theoretical development is covered in other review papers at this conference [7-8]. Also the primary emphasis will be on model development and application during the period since the 1988 Mito Conference.

The areas to be covered include examples of models appropriate for the calculation of properties of reactions involving light nuclei as targets along with models describing reactions involving medium to heavy nuclei. This includes optical model development of interest for application along with models that make up components of statistical reaction theories (level density and gamma-ray strength formalisms). Nonequilibrium reaction theories (classical preequilibrium and quantum-mechanical models) will be discussed along with efforts aimed at creation of unified reaction models for applications. Finally, a brief description of recent developments in models used for calculation of prompt radiation emission from fission will be provided.

Methods for Light-Element Evaluations

The most successful and frequently used method for light-element evaluations is R-matrix theory. The theory is ideally suited for describing the resonances that are usually seen in light-element reactions, and at the same time it builds in the correct energy dependence of the transition (T) matrix elements at low energies by making explicit use of the solutions for the external (long-ranged) parts of the interaction (Coulomb and angular momentum barriers, etc.). Thus, the method gives reliable extrapolations to low energies for both neutron- and charged-particle-induced reactions.

Applications of the theory usually involve parametrizing the R matrix for a light system as the channel-space spectral expansion,

$$R^B_{c'c}(E) = \sum_\lambda \frac{\gamma_{c'\lambda}\gamma^T_{\lambda c}}{E_\lambda - E} , \qquad (1)$$

in terms of the reduced-width amplitudes $\gamma_{c\lambda}$ and eigenenergies E_λ. The $\gamma_{c\lambda}$ are channel-surface projections of eigenfunctions $|\lambda\rangle$ that are regular solutions of the Schrödinger equation governing the motion of the particles at small distances and satisfying at the channel radii $r_c = a_c$ real, energy-independent boundary values B_c. Using standard methods, it is convenient to transform Eq. (1) to the outgoing-wave R matrix, $R^L_{c'c}(E)$, having boundary values given by the logarithmic derivatives L_c of the external outgoing-wave solutions O_c at $r_c = a_c$. From the elements of the multichannel partial-wave transition matrix,

$$T_{c'c} = O^{-1}_{c'} R^L_{c'c} O^{-1}_c - O^{-1}_c \mathrm{Im}(O_c)\delta_{c'c} , \qquad (2)$$

can be calculated observables (cross sections, polarizations, etc.) for any two-body reaction in the system described by the R matrix in Eq. (1). Its parameters, $\gamma_{c\lambda}$, E_λ, and sometimes a_c, are usually determined by fitting experimental data for all the reactions in the system. Some evaluators consider only a subset of the possible reactions by using a variation of the method that involves a complex, reduced R matrix.

The full method was used in several of the light-element evaluations for ENDF/B-VI, including the standards evaluations for ^{1}H, ^{3}He, ^{6}Li, and ^{10}B. Other light-element ENDF evaluations that used R-matrix methods were those for ^{12}C [9], ^{14}N, and ^{16}O [10].

An example that illustrates the wide-ranging utility of this approach and the ability to incorporate fundamental theoretical constraints is the R-matrix description of the A=4 reactions being done at Los Alamos. All possible reactions involving the channels n+^{3}H, n+^{3}He, p+^{3}H, p+^{3}He, and d+d are described using a single set of Coulomb-corrected, charge-independent R-matrix parameters. Neutron cross sections from this analysis have been used in the ^{3}He evaluation and will be used in a future update of the ^{3}H evaluation where the previous low-energy total cross sections differ notably from the present calculations [11] and data [12].

A novel application of the analysis has been to compare its results [13] with measurements being done for the d+d muon-catalyzed fusion (μCF) reactions. Room temperature μCF measurements [14] give the surprising result that the branching ratio for the P-wave part of the d+d reactions favors the n+^{3}He branch over that for p+^{3}H by about 40%. At lower temperatures, where the molecular transitions allow increasing amounts of S-wave formation between the two deuterons, the branching ratio has recently been observed [15] to decrease toward unity. The P-wave branching ratio calculated from the analysis is 1.43, while that for the S-waves is 0.886, giving excellent agreement with the measured room-temperature number, and also accounting qualitatively for the decrease in the branching ratio toward unity as the S-wave admixture increases at lower temperatures. Thus, the analysis successfully accounts for all the new experimental information (including the surprises) about the d+d reactions that has come from this exciting new field, confirming *by individual partial waves* the reliability of R-matrix extrapolations to low energies.

One would like to connect the parameters obtained in R-matrix analyses with more microscopic models of nuclear reactions that start, *e.g.*, from the nucleon-nucleon interaction. An effort to make this connection with shell-model (SM) quantities was reported by Knox [16]; this work has been continued by Ressler [17]. An improved prescription that relates the SM energies to the correct boundary conditions and takes into account the renormalization and nonorthogonality of the SM states due to the finite extent of the nuclear region has been given by Hale [18]. Bolsterli [19] has developed a continuum SM approach that has many similarities to R-matrix theory, especially when the interactions are approximated by delta functions at the nuclear surface. By adjusting the levels and interactions phenomenologically, he is able to match closely with fewer parameters the S-wave T-matrix elements that Hale *et al.* [10] obtained from their R-matrix description of the n+^{14}N reactions.

Optical Model

Good optical model analyses are essential for thorough nuclear data evaluations. In the period since the Mito Conference, much of this activity has dealt with the use of dispersion relationships in optical model analyses. Because a review paper on the dispersive optical model is being given at this conference [8], we will limit our discussion to three examples of optical model application in current evaluations.

A number of optical model studies have been performed at Argonne National Laboratory (ANL) that utilize dispersion relationships. Isotopes studied include ^{51}V,

^{59}Co, ^{89}Y, ^{93}Nb, ^{115}In, and ^{209}Bi, and several of the analyses and accompanying data sets form the basis for ENDF/B-VI evaluations. In such analyses the real (V) and imaginary (W) optical model potentials are related through a dispersive correction term [20].

$$\Delta V(r,E) = \frac{1}{\pi} P \int_{\infty}^{\infty} dE' \frac{W(r,E')}{E' - E} , \qquad (3)$$

where P signifies the principal-value integral. The ANL studies usually involve fitting experimental neutron total and scattering data in the energy range $E_n \sim 0.5 - 10$ MeV, either with a spherical optical potential that includes a dispersion relation [21], or a standard spherical potential that is then interpreted using dispersion relationships and the method of moments [22].

Several of these analyses, when coupled with bound-state data, give clear evidence of Fermi surface anomalies, as illustrated by the n + ^{51}V [23] example in Fig. 1. Here the volume integral/nucleon of the real potential strength obtained using dispersion relations (curve) is compared to J_v values derived from experimental data (circles). While the data for $E_n > 0$ are consistent (approximately) with linear behavior, the values of J_v needed to reproduce the single-particle and hole bound-state energies shows characteristic anomalous behavior.

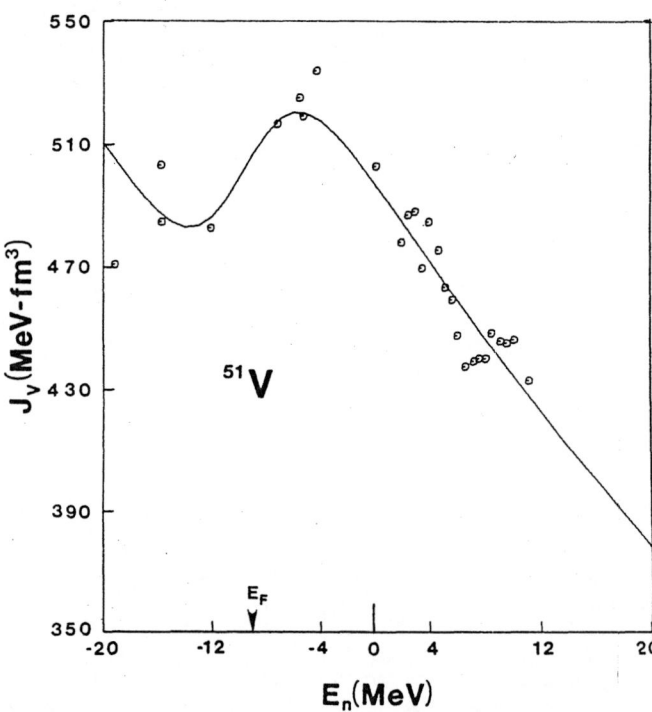

Fig. 1. The volume integral per nucleon of the real potential strength *vs* neutron energy for n + ^{51}V reactions.

The ANL optical-statistical analyses also highlight the difficulties associated with use of global and even regional optical model potentials at lower incident energies. The strength and/or diffuseness parameters associated with the imaginary potential are found to be strongly influenced by the level structure and properties of individual nuclei.

Coupled-channel optical model analyses for permanently deformed nuclei continue to be an essential tool in data evaluations. An example of coupled-channel calculations of ground state rotational bands is offered by the ENDF/B-VI analyses of neutron reactions on ^{235}U, ^{238}U, ^{237}Np, and ^{239}Pu [24]. For these calculations earlier "SPRT" analyses [25] were adjusted to improve agreement with σ_{tot} data up to 20 MeV and were then used to calculate neutron transmission coefficients for detailed Hauser-Feshbach statistical theory analyses as well as shape elastic and direct (n,n') cross sections. Very similar potentials were utilized for the four target nuclei. The ^{237}Np potential was obtained prior to release of new total cross section data [26] and is based on a simple interpolation of the neighboring potentials.

A third example of optical model phenomenology illustrates the movement to better characterize optical model potentials at higher energies. In this study Kozack and Madland [27] adapted and combined Dirac phenomenology [28] with a relativistic generalization [29] of the Lane model [30] to successfully fit ^{208}Pb neutron total cross sections and proton elastic scattering data between 95 and 300 MeV. This analysis used standard Lorentz potentials (scalar and vector, real and imaginary). Several different forms for the energy dependence of the potential strengths and two forms for the isoscalar and isovector potentials were tested in the analysis. Best results were obtained with strengths parameterized in the form

$$U_0(T) = A + \alpha \ln(T) \pm \eta B \qquad (4)$$

where T is the projectile kinetic energy, $\eta = (N-Z)A$, the plus (minus) sign refers to neutrons (protons), and the parameters A, B, and α were determined by the data fitting procedure. A total of 20 parameters were determined in the analysis, which is similar to the number of parameters involved in more usual Schrödinger analyses. Better fits were obtained to the higher energy data with the Dirac formalism. Kozack and Madland's results for the p + ^{208}Pb reaction cross section and for the n + ^{208}Pb total cross section compare well with experimental data. One outcome of this analysis is the observation that dramatic differences occur between the behavior of protons and neutrons elastically scattered from ^{208}Pb at incident energies in the neighborhood of 100 MeV, especially differential scattering cross sections and analyzing powers [31]. This prediction awaits experimental verification.

Nuclear Level Densities

Nuclear level densities are central to Hauser-Feshbach [32] and Weisskopf-Ewing [33] models used to determine equilibrium contributions to cross sections. The importance of realistic level density information in applied and basic calculations is reflected in the continuing body of literature aimed at both phenomenological and microscopic models. Application calculations continue to rely principally on simple phenomenological models, e.g., the Gilbert-Cameron (GC) [34] or the Back Shifted Fermi Gas (BSFG) [35]. These models feature unphysical assumptions and uncertainties in parametrization in areas pertaining to shell effects, equidistant spacing of particle levels, excitation energy dependence of the spin cutoff parameter and parity factors. For this reason much recent effort has focused on use of more sophisticated phenomenological models [36] or to use of microscopic Fermi gas models employing realistic single particle schemes [37].

Arthur and Guenther [38] explored the sensitivity of calculated particle emission spectra to level density models and associated parameters used. The A=60 region was chosen because of the existence of experimental spectral data for neutron, proton, and alpha-particle emission. BSFG model results were calculated with recent parameter improvements due to Ivascu et al. [39] and Zhuang et al. [40] while GC level densities were determined using systematics for the level density parameter [41] coupled with adjustment of the constant temperature potion of the GC model to fit low-lying discrete level data. Figure 2-a illustrates differences in ^{60}Ni level densities calculated with the models and parameters shown, while Fig. 2-b illustrates the rather negligible effect of these level density differences on calculated neutron emission spectra for ^{60}Ni(n,n'). More sensitivity was found for emission channels (proton and alpha particles) that are minor

in magnitude (as compared with neutron emission). This occurs because in Hauser Feshbach calculations the cross section is dominated by the ratio of the partial width for particle emission to the total width. Here the neutron emission channel dominates calculation of both partial and total widths; thus, large level density changes there produce relatively little impact on this ratio.

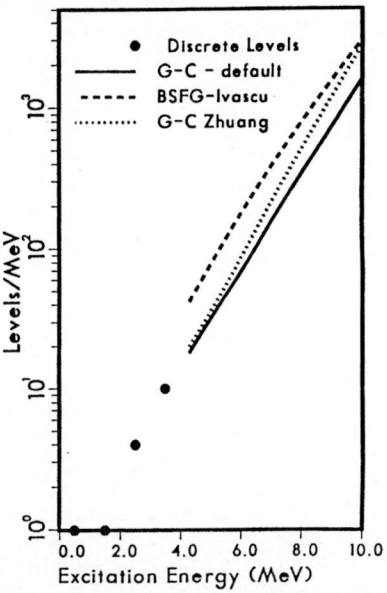

Fig. 2-a. Level densities for ^{60}Ni computed using the GC model with Cook parameters (solid curve), GC with Zhuang parameters (dotted curve), and the BSFG with Ivascu parameters (dashed curve).

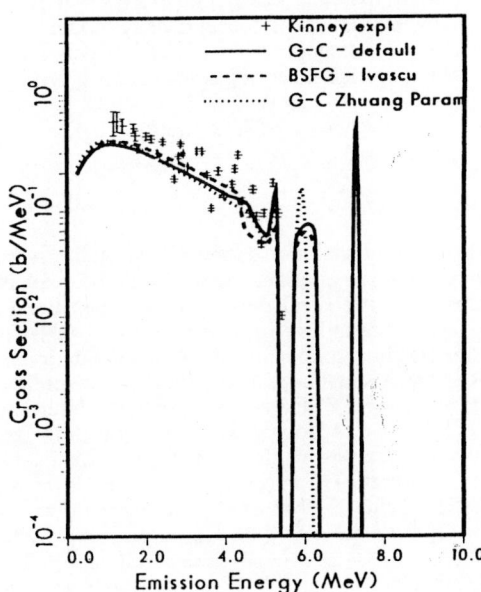

Fig. 2-b. Calculated ^{60}Ni(n,n') spectra using these level densities.

In many application calculations, emphasis is placed on the reliability of level density extrapolations to mass regions where little or no data exist. Analyses by Reffo [42] and Iijima [43], e.g., have produced sets of level density parameters optimized using relevant experimental data over a limited range of masses. These parameters exhibit significant differences when compared to straightforward applications of Gilbert Cameron systematics for the Fermi gas parameter a and U_x, the matching energy between constant temperature and Fermi gas forms used in this model. These parameters exhibit correlations with shell effects, not accounted for in more general parameter systematics. Such studies yield uncertainty estimates of the general magnitude $\Delta a = \pm 15\%$, $\Delta U_x = \pm 15\%$, and uncertainties in the spin cutoff parameter σ^2 of $\pm 40\%$. Calculated (n,γ) cross sections (the area of emphasis in Ref. [42]) are most sensitive to changes in the FG and matching energy. Uncertainties in these parameters translate into cross section variations about 2 to 3 times larger.

The GC and BSFG level density models commonly used in data application calculations do not properly account for the damping out of shell effects at higher excitation energies. Microscopic calculations predict such effects; they are included in more modern phenomenological level density expression, e.g., such as those due to Schmidt [37] and Ignatyuk [36]. The model derived by Schmidt is used in the STAPRE model code version described in Ref. [44] and results from microscopic calculations that employ realistic single particle levels and pairing corrections. These produce a simple relation between the entropy of a system, S, and the level density

$$\omega(E) = 1/(2\pi)^{3/2} (D)^{-1/2} \exp(S), \qquad (5)$$

where

$$S = 2[a(E)+\delta Uf(E)\delta Ph(E)]^{1/2} \text{ and } D = 18\pi^{-1/4} a^{1/2}E^{5/2}. \quad (6)$$

The high energy asymptotic level density parameter is determined from an expression for a Fermi gas with a diffuse surface. The difference between experimental and the liquid drop masses is used to determine the sum of δU plus δP, while δP is determined from standard pairing energy assumptions. The energy dependences of the shell effects and effective pairing corrections are described by the weighting functions $f(E) = (1-\exp(\gamma E))$ where $\gamma = 2.5aA^{-1/3}$. The function $h(E)$ describes the transition occurring in the pairing energy at the critical energy E_c at which at phase transition occurs from a superconducting state to a normal one.

The phenomenological level density model of Ignatyuk also accounts for disappearance of shell effects through use of an energy dependent Fermi gas parameter given by

$$a(E) = a_{LDM} (1+\delta W(1-\exp(-\gamma E)/E)), \qquad (7)$$

where the high energy asymptotic value

$$a_{LDM} = (0.154+6.3\times10^{-5}A) \text{ and } \gamma = 0.054. \qquad (8)$$

The experimental shell correction δW is again given by the difference between experimental and liquid drop masses.

The Ignatyuk level density model has been implemented into two codes, STAPRE [45] and GNASH [46], which are often used in data evaluation calculations. In the STAPRE implementation, the lower excitation energy region is described using a BSFG and a logarithmic interpolation is made to join results from this model with those from the Ignatyuk expressions at higher energies. In the GNASH implementation, the GC constant temperature form is used to describe low-lying regions of excitation energies. This expression is then joined to the Ignatyuk expression at a matching energy U_x in a manner described earlier.

Results calculated using the Ignatyuk and GC level density models have recently been compared [47] with $^{204-208}$Pb$(n,x\gamma)$ data. The data were measured for incident neutron energies up to 100 MeV by Haight [48]. The comparison, made for specific gamma rays from ^{208}Pb $(n,x\gamma)$ measurements, provides a unique test of shell effects and dissipation at higher excitation energies. Figure 3 compares the calculated excitation function for the ^{208}Pb$(n,3n\gamma)$ ^{206}Pb reaction using these two level density models. Because of the dissipation of shell effects, increasing the level density in the residual system of interest (here ^{206}Pb) and in competing residual systems (^{205}Pb), the Ignatyuk level density results peak earlier and drop more rapidly than the GC results. This behavior appears to be in better agreement with these data. For nuclei farther away from shell closures, calculated differences in such excitation functions are smaller.

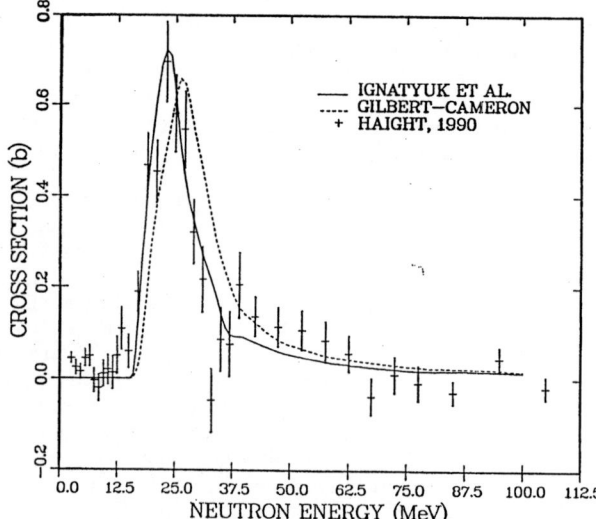

Fig. 3. Excitation functions for the ^{208}Pb$(n,3n\gamma)^{206}$Pb reactions using the Ignatyuk (solid curve) and GC (dashed curve) level density models.

Gamma Ray Strength Functions

The use of gamma-ray strength functions provides the preferred mechanism for determination of gamma-ray transmission coefficients in statistical model calculations. Recent studies of gamma-ray strength function systematics include the work of Gardner, Gardner, and Hoff [49], which extend earlier investigations centered primarily around rare earth nuclei to the actinide mass region. They modeled $f_{E1}(\varepsilon_\gamma)$ using an energy dependent Breit-Wigner line shape, while $f_{M1}(\varepsilon_\gamma)$ was represented by a constant value as suggested by the Weisskopf single-particle approximation. Comparisons were made with partial gamma-ray width data for ^{232}Th, ^{237}Np, and ^{238}U in statistical model calculations that employed large sets of modeled, discrete sets (> 100 levels) for the daughter nuclei. They found that the M1/E1 ratio was the same as earlier values they determined for the rare-earth mass region, but that for these actinide nuclei, the total transition strength was almost thirty percent greater.

Kopecky and Uhl [50] have also studied the impact of models for E1 and M1 strength functions on the results of statistical model calculations of average radiation widths, radiative neutron capture cross sections, and gamma ray spectra. Several models were compared that reproduced photoabsorption and other data but differed significantly at low gamma-ray energies. They found strong evidence to support an E1 strength function model, including an energy dependent spreading width in a Lorentzian form, as well as a non-zero, temperature-dependent limit for the strength function as ε_γ approaches zero. The specific form is given by

$$f_{E1} (\varepsilon_\gamma, T) = 8.68 \times 10^{-8} \text{ (mb}^{-1} \text{ MeV}^{-2})$$

$$\left[\frac{\varepsilon_\gamma \Gamma(\varepsilon_\gamma)}{(\varepsilon_\gamma^2 - E^2)^2 + \varepsilon_\gamma^2 \Gamma(\varepsilon_\gamma)^2} + \frac{0.7\Gamma 4\pi^2 T^2}{E^5} \right] \sigma_0\Gamma, \quad (9)$$

where

$$\Gamma(\varepsilon_\gamma) = \Gamma[(\varepsilon_\gamma^2 + 4\pi^2 T^2)/E^2] .$$

The rightmost portion of the expression is a term describing $f_{E1}(\varepsilon_\gamma=0, T)$, as proposed by Kadmenskij [51]. The overall expression is correct in the low-energy form and exhibits little difference with a Lorentzian near the neutron binding energy. It is, however, not valid near the giant resonance peak. Because of the temperature dependence of the expression, it partially violates the Brink-Axel model, [52] which calculates photoabsorption for excited states in the same fashion as for the ground state by assuming a giant resonance built on each excited state.

Data for ^{197}Au, ^{143}Nd, ^{105}Pd, and ^{93}Nb(n,γ) reactions were compared with predictions using this gamma-ray strength function form. Such data do not extend to low enough gamma-ray energies to unambiguously support use of this non-zero gamma-ray strength function in the limit of $\varepsilon_\gamma \to 0$. Comparisons with data for the total gamma ray width are sensitive to behavior of the gamma-ray strength function in the energy range between zero and a few MeV. In comparisons of gamma-ray strength models to these data, overprediction of total widths was found when the standard Lorentzian strength function was used. The best agreement resulted from use of the generalized Lorentzian form given above. Likewise use of this expression provided consistently better agreement with (n,γ) excitation functions as well as spectral data resulting from capture reactions.

These analyses indicate a preference for the generalized Lorentzian form for spherical nuclei. This form is significantly different than the standard Lorentzian often used in statistical model analyses. Likewise its use implies a partial breakdown of the Brink hypothesis--a principal assumption underlying most statistical model calculations. Because of the potentially important impact on applied calculations, similar analyses should be extended to other mass regions, particularly those involving deformed nuclei.

Preequilibrium Model Development

The study of preequilibrium particle and gamma ray emission continues to be an area of significant development for applied calculations. Here we choose several example areas to illustrate progress and development directions occurring during the interval since Mito. This period has seen the continued development and application of quantum mechanical theories such as those of Feshbach, Kerman, Koonin [53] (FKK model), Tamura, Udagawa, Lenske [54] (TUL model), and currently the multistep reaction model of Nishioka, Verbaarschot, Weidenmüller, and Yoshida [55] (NVW model). Recently Hodgson [56] reviewed the status of multistep compound analyses of (n,nx), (n,px), and (p,nx) reactions on medium weight nuclei for incident energies around 14 MeV. There, concurrent analyses of all open channels provided rather stringent tests of the theory as well as important parameters such as optical potential and the strength of the effective interaction used in the FKK model. This analysis illustrated that the FKK theory provides reliable spectral and angular distribution information for important reaction channels in this mass and energy range.

Comparisons of the multistep direct formalisms to higher energy reaction data (spectra and angular distributions) have been made. Examples include analyses of neutron and proton nonelastic scattering data by Marcinkowski [57], Kalka [58], Cowley [59], and recently Scobel [60]. Scobel found that the FKK theory was able to describe (p,n) angular distributions for incident energies up to 160 MeV, whereas the classical hybrid preequilibrium model failed to predict angular distributions correctly because of an overestimate of the first step cross section. An example of these results is presented in Fig. 4.

MSD analyses over extended incident energy ranges provide increased information concerning the energy dependence of the effective interaction strength, V_0. This parameter is fixed by fitting it to cross section data. Several models are used to describe the interaction potential, with the most realistic involving use of a Yukawa two-body interaction. A series of analyses using this Yukawa potential form has resulted in an energy dependent form given by $V = 31 - 0.15 E$ [56,60]. This form agrees with standard optical model results and applies to both MSC and MSD calculations within the FKK model framework.

Analyses using the other formalisms listed above have been performed. These efforts are often aimed at providing an integrated perspective on the similarities and differences of models for multistep emission processes. For example, in [Ref. 61] Herman and Reffo compare content and assumptions inherent in the FKK and exciton emission

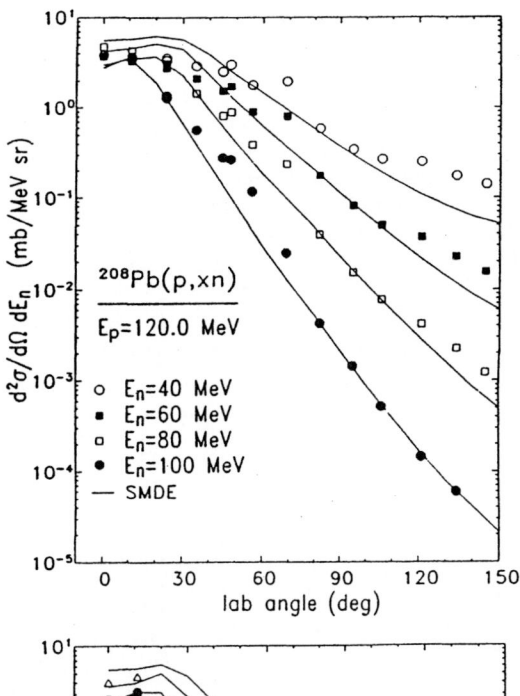

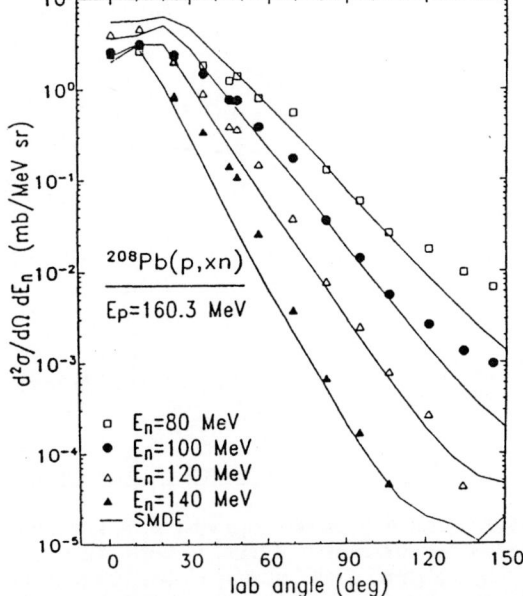

Fig. 4. Comparison of neutron angular distributions for 120- and 160-MeV ^{208}Pb(p,xn) measured by Scobel. Curves are obtained from statistical multistep direct emission calculations.

models. In Ref. [62], Köning and Akkermans explore different nonequilibrium reaction models (FKK, TUL, NVW) and classical exciton models, and they classify them in terms of the assumption of leading particle or residual system statistics. More information concerning quantum mechanical non-equilibrium reaction models can be found in the invited paper by Weidenmüller at this conference.

A second area of activity involves efforts to introduce parameter consistency between equilibrium and nonequilibrium portions of unified reaction models. A focus has been consistent treatments between exciton state densities used in preequilibrium models and total level densities used in equilibrium reaction calculations. Examples include work by Avrigeanu [63] to achieve consistent exciton state and level densities and a two fermion state density analysis by Herman, Reffo, and Costa [64] using realistic single-particle levels. Similarly, Fu [65] has extended his earlier analysis of unification and simplification of total and partial level densities performed in the framework of the one-fermion model to the case of two-fermion expressions. Pairing interaction effects and spin cutoff expressions were the main focus of the work. The effect of using the two-fermion model results in determination of preequilibrium state densities is illustrated in Fig. 5 where the ratios $\rho_1(n=2)/\rho_2(n=2)$ and $\rho_1(all\ n)/\rho_2(all\ n)$ are shown. Here the subscripts refer to the one-and two-fermion expressions and n refers to the exciton configuration. These ratios exhibit very different behavior as a function of the excitation energy U--a situation that is, quite difficult to approximate in preequilibrium Hauser-Feshbach calculations that attempt simple corrections for one-and two-fermion model results. Use of such formalisms achieves consistency in level density models required for compound and precompound reactions descripion.

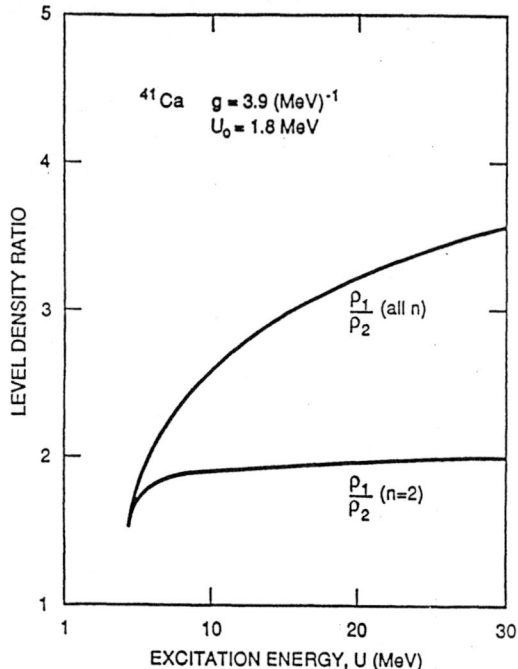

Fig. 5. Ratio of one-fermion and two-fermion level and state densities calculated for ^{41}Ca.

Non Equilibrium Gamma-Ray Emission

The emission of gamma-rays via preequilibrium mechanisms has been the subject of much recent work within both semi-classical and quantum-mechanical preequilibrium models. For gamma-rays that are emitted with energies below about 30 MeV, the dominant mechanism is a direct/semi-direct process via the excitation of the Giant Dipole Resonance (GDR). Akkermans and Gruppelaar [66] presented a description of such a process within the exciton model, based upon the application of detailed balance using the Brink-Axel hypothesis, which yields correct emission cross sections in the equilibrium limit. The exciton model description of gamma-ray emission in the GDR region is easy to implement and yields emission cross sections which generally describe the data well; for a full review see [67]. Oblozinsky has proposed [68] a hybrid model for GDR gamma-ray emission that is consistent with Akkerman's exciton model in its application of detailed balance The exciton model description of GDR gamma rays has also been applied with some success by Cvelbar et al [69] to calculate (n,γ) and (p,γ) activation cross sections.

There have been some recent theoretical developments into gamma-ray emission in the GDR region using quantum-mechanical reaction models. The FKK multistep compound theory has been developed by Oblozinsky and Chadwick [70] to include gamma-ray emission. This theory yields the correct Hauser-Feshbach spectrum in its equilibrium limit and is able to account for a considerable fraction of the high-energy gamma-ray cross section. Multistep compound and direct gamma-ray theories are also being developed within the NVW multistep formalism [55].

For photon energies above 30 MeV the two-body emission mechanism becomes important. Two nucleons collide, de-accelerate, and a bremsstrahlung photon is emitted. The two-body nature of this emission led Oblozinsky to present a preequilibrium model of hard photon emission using the quasi-deuteron concept [71]. As in all semi-classical preequilibrium models, this model uses the inverse photoabsorption cross section, applies detailed balance, and is easy to implement in a reaction model code. The success and simplicity of the quasi-deuteron approach to hard photon emission has led to an a collaboration between Los Alamos, Bratislava, Bologna, and Oxford to develop and implement this model in nucleon-induced and heavy-ion reaction codes. Since a thorough understanding of the photoabsorption mechanism is essential for a preequilibrium model of hard photon emission, a photoabsorption model has been developed [72] that describes the effects of Pauli-blocking theoretically and accounts for the quasi-deuteron photoabsorption data well. Only final states that can be reached via linear momentum, as well as energy conservation, were included in the accessible state densities. More refined hard-photon emission quasi-deuteron models are presently being developed by Chadwick and Oblozinsky. Hard photons from the first n-p collision (the n=1 preequilibrium stage) dominate the high-energy emission spectrum, and in one such model the n=1 theoretical emission rate is given by

$$\lambda_\gamma(\varepsilon,\varepsilon_\gamma) = \frac{\varepsilon_\gamma^2\ \sigma_{qd}(\varepsilon_\gamma)}{4\pi^3\hbar^3 c^2}\ \frac{\int\rho(2p1h,\varepsilon\text{-}\varepsilon_\gamma,k\text{-}k_\gamma)d\Omega_\gamma}{\rho(2p2h,\varepsilon_\gamma,k_\gamma)}\ , \quad (10)$$

where ε, k, ε_γ and k_γ are the nuclear excitation energy, the incident-nucleon momentum, the emitted photon energy and the emitted photon momentum respectively. The $2p1h$ and $2p2h$ state densities are determined within a Fermi-gas model and include linear momentum conservation, and the inverse quasi-deuteron cross section, $\sigma_{qd}(\varepsilon_\gamma)$, was taken from the photoabsorption model of Ref [72]. In the numerator we integrate the residual nucleus $2p1h$ state density over all directions of the $2p1h$ momentum, determined by the various photon emission directions Ω_γ. The $2p2h$ density in the denominator is independent of the direction of k_γ. In Fig. 6 the photon emission spectrum obtained in this model is compared with data for the reaction ^{197}Au (p,γ) induced by 72 MeV protons. It is evident that this model describes the data well except for the lowest energies where photons from more complex preequilibrium stages and the GDR-tail photons (which we have not included) will contribute. It should be noted that a calculation that used the full Fermi-gas state densities, and that ignored linear momentum effects, led to an overprediction of the data by a factor of 2-3 [73].

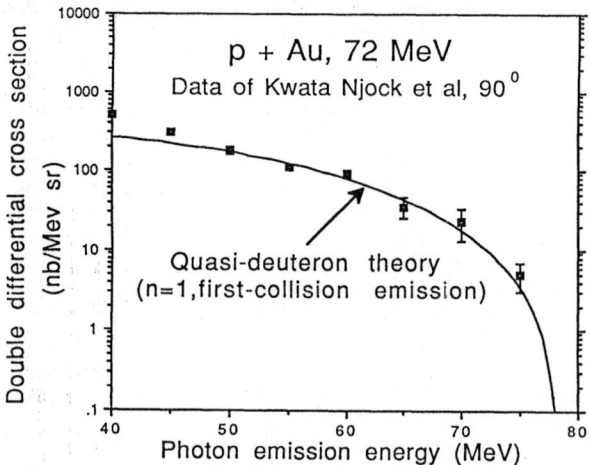

Fig. 6. Comparison of energetic gamma-ray emission from 72-MeV Au(p,γ) predictions of quasi-deuteron photon emission theory.

Prompt Radiations from Fission

Recent actinide data evaluations for libraries such as ENDF include descriptions of prompt radiation from fission (neutrons, specifically) based upon models such as those developed by Madland and Nix [74]. Likewise, recent higher-energy evaluated data libraries [75] have included prompt neutron emission contributions to total neutron emission spectra for neutron and proton interactions with ^{238}U. In this latter case, prompt neutron and gamma-ray emission were determined using cascade evaporation model calculations of the decay of fission fragments (distributions thereof) produced during nucleon-induced fission [76]. Figure 7 illustrates a reaction chain appropriate for the case proton-induced fission of ^{238}U. Here several points are important. Statistical model calculations of multi-chance fission components of the total fission cross section included both (p,xn) and (p,pxn) paths that populated both uranium and neptunium compound systems. For each compound system undergoing fission, separate model calculations were then performed for deexcitation of the binary distribution of fragments produced during fission of a specific compound system. This schematic representation indicates the complexity of such calculations and provides an introduction to models that can be used for determination of prompt fission radiation emission.

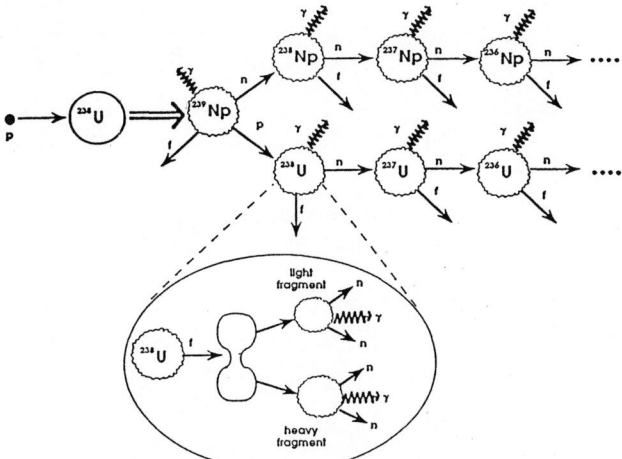

Fig. 7. Reaction chain used for calculation of $^{237}Np + p$ total fission cross sections up to 100 MeV.

Recent reviews by Madland [77] and Märten [78] summarize models used in calculation of prompt fission emissions. A short classification of these models is presented here. The first category is that of the temperature distribution model as typified by the Madland-Nix [74] and Dresden [79] formulations. These models address both neutron-induced and spontaneous fission and account for the distribution of fission fragment excitation energy, the energy dependence of the cross section for compound nucleus formation, the center-of-mass motion of the fission fragments, and multiple-chance fission at higher neutron energies. In these formulations, a triangular approximation to the corresponding fission fragment residual nuclear temperature distribution simulates the initial distribution of fission-fragment excitation energies and subsequent cooling as neutrons are emitted. A second approach, the cascade evaporation model (CEM [80]) is based upon the Weisskopf evaporation theory with factors applied to account for neutron- gamma-ray competition. This model, in addition to the features described with the temperature distribution model, includes the distribution of fission-fragment excitation energy in each step of the cascade process, the anisotropy of the center-of-mass neutron spectrum, complete fission-fragment mass and kinetic energy distributions, and fission-fragment nuclear level densities. The cascade-evaporation model has also been formulated to explicitly include effects of competition

between neutron and gamma-ray emission as described by Bozoian [81]. This later formulation provides a crude basis for calculation of prompt gamma-ray emission during deexcitation of fission fragment pairs. The final, and most sophisticated approach to calculation of prompt fission neutron, and particularly prompt gamma-ray, emission involves Hauser-Feshbach calculation of the deexcitation of the fission fragments. This model allows explicit competition between neutron and gamma-ray emission in the deexcitation of a given fission fragment, neutron transmission coefficients from optical model potentials, gamma-ray transmission coefficients, and angular momentum distributions for each fragment considered. Examples of this approach can be found in Ref. [82].

Important technical issues that impact modeling of prompt emissions from excited fission fragments include determination of the fission fragment mass distribution along with implementation of sampling procedures for fragment selection, determination of the total energy release in fission and its dependence on masses (experimental and theoretical) used to determine it, fission fragment kinetic energies, and, significantly, the partition of total fission fragment excitation energy between light and heavy fragments. This latter area is particularly important in calculations using the models described above. Several methods can be utilized in the energy partitioning process. The simplest assumption scales the total fragment excitation energy by the masses of fragments involved; $E_f^H = E^*/(1+A_L/A_H)$ and $E_f^L = E^*/(1+A_H/A_L)$. A similar approach partitions the excitation energy using the respective Fermi gas level density parameter a: $E_f^H = E^*/(1+a_L/a_H)$ and $E_f^L = E^*/(1+a_H/a_L)$. A standard approach for verifying choices for fragment excitation energy partitioning is by comparing calculations with data for neutron multiplicity as a function of fragment mass. Figure 8 illustrates such a comparison using partitioning based on the two mass and Fermi gas parameters. The example illustrates difficulties occurring when simple energy partition models are used. Better agreement can be achieved through adjustment of Fermi gas parameters [81] or through more detailed models for energy partitioning [83].

The question of better methods for partitioning of fission fragment energy at scission is a major uncertainty that restricts the applicability of fission neutron theory. Recent conclusions from specialists' workshops (e.g., Vienna, 1990 [84]) recommend that substantial effort be devoted towards improvement of models for energy partitioning.

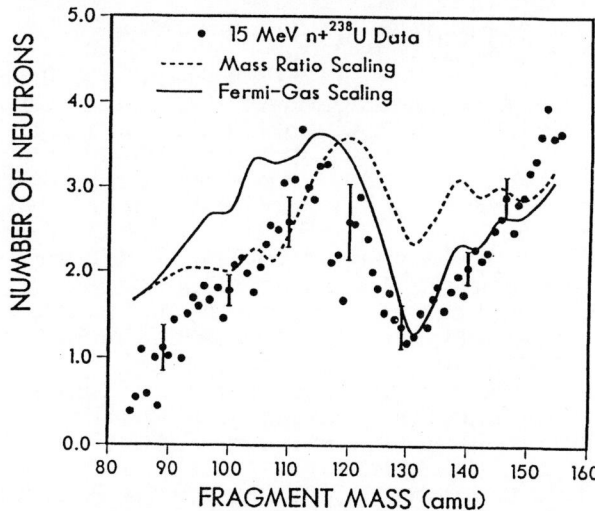

Fig. 8. Calculation of the average number of neutrons emitted per fragment using two methods for fragment excitation energy partitioning.

766

Summary

This paper has reviewed the application of nuclear models in current evaluation efforts. Selected examples were given for principal models and their components. Significant progress continues in model development that is directly translated into improved calculations for data libraries. This trend will continue into the future and is driven by factors such as increased computational capabilities and expanded requirements for evaluated data files. Model improvement will also occur with general trends being direct use of more microscopic models specifically in data evaluation and/or use of microscopic treatments to guide phenomenology developed for application uses.

References

1. M. Uhl, Proc. Int. Conf. Nuclear Data and Technology, May 1988, S. Igarasi, Ed, p. 435.
2. M. A. Gardner, Proc. Int. Conf. on Nuclear Data for Basic and Applied Science, Santa Fe, May 1985, Eds., P. Young, R. Brown, G. Auchampaugh, P. Lisowski, L. Stewart, p. 1481.
3. P. G. Young, Proc.Int. Conf. Nuclear Data and Technology, Sept. 1982, Antwerp, Ed., K. H. Bockhoff, p. 506.
4. P. W. Lisowski, C. D. Bowman, E. D. Arthur, and P. G. Young, Nuclear Physics Information Needed for Accelerator Driven Transmutation of Nuclear Waste, contributed paper, this conference.
5. S. Howe, Nuclear Data Needs for the Space Exploration Initiative, invited paper, this conference.
6. A. Wambersie, Nuclear Data Relevant to Cancer Therapy with Fast Neutrons and Charged-Particles, invited paper this conference.
7. H. A. Weidenmüller, Recent Progress in the Statistical Theory of Nuclear Reactions, invited paper, this conference.
8. P. E. Hodgson, Dispersion Relations in Optical Model Development, invited paper, this conference.
9. C. Y. Fu, Nucl. Sci Eng. 106, 489 (1990).
10. G. M. Hale, P. G. Young, M. Chadwick, and Z. P. Chen, New Evaluations for ^{14}N and ^{16}O, invited paper, this conference.
11. G. M. Hale, D. Dodder, J. D. Seagrave, B. Berman, and T. Phillips, Phys Rev C42, 438 (1990).
12. T. W. Phillips, B. L. Berman, and J. D. Seagrave, Phys. Rev. C 22 (1980).
13. G. M. Hale, Proc. Int. Conf on Muon Catalyzed Fusion, Vienna, May 1990.
14. D. V. Balin et al., Phys Rev Lett 141B, 173 (1984); JETP Lett. 40, 112 (1984).
15. D. V. Balin et al., Proc. Int. Conf on Muon Catalyzed Fusion, Vienna, 1990.
16. H. D. Knox, Proc. Int Conf on Nuclear Data for Basic and Applied Science, Santa Fe, NM, May 1985, Eds., P. G. Young, R. E. Brown, G. F. Auchampaugh, P. W. Lisowski, and L. Stewart, p. 1007.
17. D. A. Ressler, Advanced Modeling of Reaction Cross Sections for Light Nuclei, Proc. Wkshp. on Strong, Weak, and Electromagnetic Interactions in Nuclei and Astrophysics, Livermore, Calif., Jan. 1991; Lawrence Livermore Nat. Lab. Report, UCRL-JC106542.
18. G. M. Hale, Proc. Symp. in Honor of Akito Arima: Nuclear Physics in the 1990's, Santa Fe, N. M., May 1990, p. 8.
19. M. Bolsterli, A Scattering Theory for the Shell Model, Los Alamos Nat. Lab. report LA-UR-90-2506.
20. G. R. Satchler, *Direct Nuclear Reactions* (Clarendon Press, Oxford, 1983).
21. R. D. Lawson, P. T. Guenther, and A. B. Smith, Phys. Rev. C 34, 1599 (1986).
22. See. *e.g.*, S. Chiba, P. Guenther, R. Lawson, and A. B. Smith, Phys. Rev. C 42, 2487 (1990) and refs.
23. R.D. Lawson, P. T. Guenther and A. B. Smith, Nucl. Phys. A493, 267 (1989).
24. P. G. Young and E. D. Arthur, Proc. Int. Conf. *Nuclear Data For Science and Technology*, Jülich, Germany, 13-17 May 1991 (to be issued).
25. G. Haouat, et al., Nucl. Sci. Eng. 81, 491 (1982).
26. P. Lisowski, personal communication, Los Alamos National Laboratory (1989).
27. R. Kozack and D. G. Madland, Phys. Rev. C 39, 1461 (1990).
28. For example, B. C. Clark, R. L. Mercer, and P. Schwandt, Phys. Lett. 122B, 211 (1983).
29. B. C. Clark et al., Phys. Rev. C 30, 314 (1984).
30. A. M. Lane, Nucl. Phys. 35, 676 (1962).
31. R. Kozack and D. G. Madland, Nucl. Phys. A509, 664 (1990).
32. W. Hauser and H. Feshbach, Phys. Rev. 87, 366 (1952).
33. V. W. Weisskopf and D. Ewing, Phys. Rev. 57, 472 (1940).
34. A. Gilbert and A. G. W. Cameron, Can J. Phys 43, 1446 (1965).
35. W. Dilg et al., Nucl Phys A217, 269 (1973).
36. A. V. Ignatyuk et al., Sov. J. Nucl Phys 21, 450 (1979).
37. K. H. Schmidt et al., Zeits. für Physik A308, 215 (1982).
38. E. D. Arthur et al., Applied Uses of Level Density Models, Specialists' Meeting on Nuclear Level Densities, Bologna, 1989.
39. M. Ivascu et al., Rev Roum. Phys 32, 697 (1987).
40. Zhuang Youxiang et al., Chinese J. of Nucl Phys 8, 199 (1986).
41. J. L. Cook et al., Aust. J. Phys 20, 477 (1967).
42. G. Reffo et al., Nucl Ins Methods A267, 408 (1988).
43. S. Iijima et al., Nucl Science Technology 20, 77 (1983).
44. M. Avrigeanu et al., Zeits für Physik A335, 299 (1990).
45. M. Uhl and B. Strohmaier, Institut für Radiumforschung report, IRK 76/01, Vienna, 1976.
46. P. G. Young and E. D. Arthur, GNASH: A Preequilibrium Statistical Nuclear Model Code, Los Alamos Nat. Lab. Report LA-6947 (1977).
47. P. G. Young et al., Calculation of (n,xγ) Cross Sections Between Threshold and 100 MeV, Los Alamos Nat. Lab. report LA-UR-90-2129 (1990).
48. R. C. Haight et al., Bull. Am. Phys. Soc 35, 1038 (1990).
49. D. G. Gardner, M. A. Gardner, and R. W. Hoff, L. Livermore Nat. Lab. report UCRL-95966 (1987).
50. J. Kopecky and M. Uhl, Phys. Rev. C 41, 1941 (1990).
51. S. G. Kadmenskij et al., Sov. J. Nucl Phys. 37, 165 (1983).
52. D. M. Brink, Ph. D. Thesis, Oxford University, 1955.
53. H. Feshbach, A. Kerman, and S. Koonin, Ann of Phys. 125, 429 (1980).
54. T. Tamura, T. Udagawa, and H. Lenske, Phys Rev C 26, 379 (1982).
55. H. Nishioka, J.J.M. Verbaarschot, H. A. Weidenmüller, and S. Yoshida, Ann. of Phys. 172, 67 (1986).
56. P. E. Hodgson, Analysis of Multistep Reaction Cross Sections with the FKK Theory, Oxford Univ. (1990).
57. A. Marcinkowski et al., Nucl. Phys A501, 1, 1989.
58. H. Kalka, Proc. Specialists' Meeting on Preequilibrium Nuclear Reactions (sponsored by NEANDC/OECD), Semmering, Austria, 10 Feb. 1988, Ed., B. Strohmaier, p. 305.
59. W. Scobel et al., Phys. Rev. C 41, 2010 (1990).
60. A. A. Cowley et al., Phys Rev C 43, 678 (1991).
61. M. Herman and G. Reffo, Nuovo Cimento A103, 577 (1990).

62. A. J. Köning and J. M. Akkermans, Randomness in Multi-Step Direct Reactions, ECN Netherlands report, ECN-RX-90-065 (1990).

63. M. Avrigeanu and V. Avrigeanu, J. Phys. G15, L261, (1989).

64. M. Herman, G. Reffo, and C. Costa, Phys Rev C39, 1269 (1989).

65. C. Y. Fu, Pairing Corrections and Spin Cutoff Factors in Exciton Level Densities for Two Kinds of Fermion, Oak Ridge Nat. Lab. Preprint (1990).

66. J.M. Akkermans and H. Gruppelaar: Phys. Lett. 157B, 95 (1985).

67. P. Oblozinsky, Proc. Specialists' Meeting on Preequilibrium Nuclear Reactions, (sponsored by NEANDC/OECD), Semmering, Austria, Feb.10, 1988, Ed., B. Strohmaier, p. 157.

68. Oblozinsky: Phys. Lett. B215, 597 (1988).

69. F. Cvelbar, E. Betak, and J. Merhar: J. Phys. G17, 113 (1991).

70. P. Oblozinsky and M. B. Chadwick: Phys. Rev. C 42, 1652 (1990).

71. P. Oblozinsky: Phys. Rev. C 40, 1591 (1989)

72. M. B. Chadwick, P. Oblozinsky, P.E. Hodgson and G. Reffo, submitted to Phys. Rev. C.

73. M. B. Chadwick and G. Reffo, accepted for publication by Phys. Rev. C.

74. D. G. Madland and J. R. Nix, Nucl. Sci. Eng. 81, 213 (1982).

75. P. G. Young et al., Transport Data Libraries for Incident Proton and Neutron Energies to 100 MeV, Los Alamos Nat. Lab. report, LA-11753-MS (1990).

76. E. D. Arthur and P. G. Young, Proc. Conf 50 Years with Fission, Gaithersburg, 1989, Eds., J. W. Behrens and A. D. Carlson, p. 931

77. D. G. Madland, Theoretical Description of Neutron Emission in Fission, Los Alamos Nat. Lab. report, LA-UR-91-437 (1991).

78. H. Märten et al., Proc. Conf 50 Years with Fission, Gaithersburg, 1989, Eds., J. W. Behrens and A. D. Carlson, p. 743.

79. H. Märten and D. Seeliger, J. Phys. G10, 349 (1984).

80. K. Arnold et al., Nucl Phys A502, 325 (1989).

81. M. Bozoian et al., Proc. Conf. 50 Years with Fission, Gaithersburg, 1989, Eds., J. W. Behrens and A. D. Carlson, p. 624.

82. J. C. Browne and F. S. Dietrich, Phys Rev C10, 2545 (1974).

83. A. Ruben et al., Zeit. für Physik A338, 67 (1991)

84. IAEA Consutants' Meeting on Nuclear Data for Neutron Emission in the Fission Process, Vienna, October 1990.

THE DISPERSIVE OPTICAL MODEL

P.E. Hodgson

Department of Physics, University of Oxford, U.K.

Abstract:

The advantages of the dispersive optical model for the analysis of neutron interactions are described, and recent analyses reviewed.

1. Introduction

In recent years many studies have shown that the dispersive optical model is able to account for a wide range of data from negative energies (bound states) to positive energies (scattering states) in a unified way (Hodgson, 1989, 1990). It is able to fit the experimental data more accurately than the simple optical model, and thus provides a way of obtaining a more accurate global optical potential. For both these reasons it is potentially of use in nuclear data analyses.

In this review, the advantages of the dispersive optical model are summarised, and the areas where increased predictive power may be expected are discussed. The older work is summarised briefly and recent advances considered in more detail.

2. Historical Review

The simple optical model, with energy-independent form factors and energy-dependent potential depths, has proved able to give differential cross-sections, polarisations and total cross-sections in good accord with the experimental data for many nuclei and energies using parameters that are almost independent of the target nucleus (Rapaport, 1982; Hodgson, 1984). In the case of neutrons several successful global potentials have been proposed, and these have been extensively compared with experimental data (Young, 1986).

The predictions of even the best global potentials show definite divergences from the experimental data, and these can be attributed to nuclear structure effects. It is possible to improve the fits by parameter adjustment for each nucleus, but thereby the predictive advantages of the global potentials are lost. Such parametrisations often require notably different potentials below about 7 MeV, and in particular the radius parameter is greater in this region (Moldauer, 1963; Van Oers et al, 1974; Eck and Thompson, 1975; Gyarmati et al, 1981; Finlay et al, 1984, 1985ab, 1989; Annand et al, 1985; Smith, Guenther and Lawson, 1987,1988). More recently, a detailed analysis of the elastic scattering of neutrons and protons by ^{40}Ca and ^{208}Pb by Mahaux and Sartor (1986c, 1990ab) showed that the radial shape of the optical potential changes with energy. All these studies show that the extrapolation of optical potentials from high to low energies assuming that the radial shape does not depend on energy is unreliable. This is confirmed by analysis of the single-particle energies; the gap between the low-lying particle and hole states obtained by a linear extrapolation from higher energies with constant geometry is about twice the experimental value (Brown, Gunn and Gould,

1963; Mahaux and Sartor, 1990).

These difficulties can be overcome in a systematic and physically-realistic way by making use of the dispersion relation that connects the real and imaginary parts of the potential and thus removes the inconsistency in the simple optical model that treats them as independent. Microscopic derivations of the optical potential show that it is a complex non-local function $\mathcal{V}(k, E)$ of the momentum k and the energy E. It is an analytic function of the energy and therefore satisfies the dispersion relation

$$\mathcal{V}(k, E) = \frac{1}{2\pi i} \int_c \frac{\mathcal{V}(k, E')}{E' - E} dE' \qquad (2.1)$$

where the contour c encloses the singularity at $E' = E$. The physical basis of the dispersion relation is the requirement of causality. If we impose the condition that the scattered wave cannot start before the incident wave arrives, then the dispersion relation follows as a necessary consequence (Cornwall and Ruderman, 1962; Feshbach, 1958; Hagedorn, 1966; Mahaux et al, 1986).

If now we separate the potential into its real and imaginary parts

$$\mathcal{V}(k, E) = V(k, E) + iW(k, E) \qquad (2.2)$$

and insert in (2.1), then evaluation of the contour integral and separation into real and imaginary parts gives

$$V(k, E) = \frac{\mathcal{P}}{\pi} \int_{-\infty}^{\infty} \frac{W(k, E')}{E' - E} dE' \qquad (2.3)$$

and

$$W(k, E) = -\frac{\mathcal{P}}{\pi} \int_{-\infty}^{\infty} \frac{V(k, E')}{E' - E} dE' \qquad (2.4)$$

where $\mathcal{P}$ indicates principal value. To use these relations to analyse data it is assumed that they also hold for the equivalent local potentials.

The real part of the optical potential shows a linear fall with increasing energy, and this is attributed to the non-locality. Superposed on this linear variation is a modulation centred at the Fermi surface which is due to the imaginary part of the potential. The linear part may be identified with the Hartree-Fock potential $V_{HF}(E, r)$ and the modulation obtained from the dispersion relation

$$V(E, r) = V_{HF}(E, r) + \frac{\mathcal{P}}{\pi} \int_{-\infty}^{\infty} \frac{W(E', r)}{E' - E} dE'$$
$$= V_{HF}(E, r) + \Delta V(E, r) \qquad (2.5)$$

The great advantage of this relation is that it separates the smoothly varying Hartree-Fock potential from the dispersive term that contains the complicated nuclear structure effects. The Hartree-Fock potential is well described by a Saxon-Woods form factor with constant radius and diffuseness, and a depth that varies smoothly and monotonically with energy. The dispersive term depends on the imaginary potential that requires a more complicated parametrisation, as described in Section 4.

The imaginary potential has volume and surface components. The volume component adds a small term

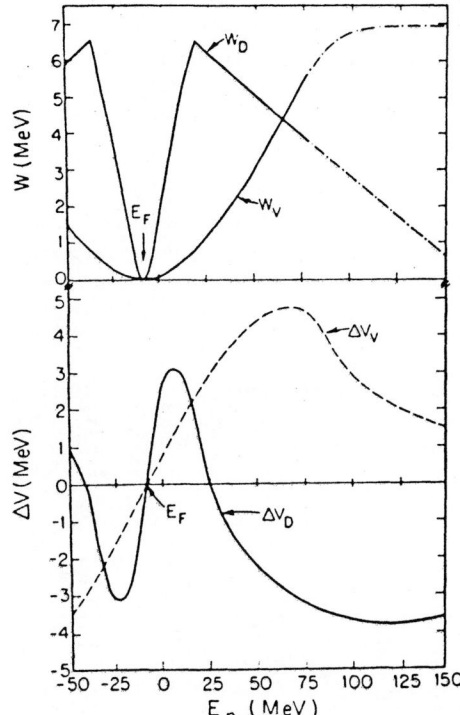

Fig.1. Energy variations of the volume (W_V) and surface (W_D) imaginary potentials for neutrons on ^{40}Ca, together with the corresponding dispersive additions to the real potential (Delaroche and Tornow, 1988).

of the same radial shape to the Hartree-Fock potential. The surface imaginary term generally has a larger radius than the volume term, and so has the effect of adding an energy-dependent surface-peaked term to the edge of the volume potential, thus increasing its radius. If the experimental data are analysed using a potential with a fixed radius the dispersion term produces a modulation of the potential depth as a function of energy that is known as the Fermi surface anomaly. The energy variations of the volume and surface imaginary potentials, and the corresponding dispersive additions to the real potential are shown in fig.1 for neutrons on ^{40}Ca.

Physically, the dispersion term gives the effect on the real part of the potential due to the coupling to the non-elastic channels, principally the inelastic channels associated with the excitation of low-lying collective states. This effect is greatest for incident energies comparable with the energies of these collective states, and thus the dispersion effects are greatest at low energies. The effect of the dispersion term is thus mainly to increase the radius of the real part of the potential at lower energies, in accord with phenomenological studies. It also reduces

the total cross-section for neutron scattering, bringing it into closer accord with the experimental data (Zong Di and Hodgson, 1988), and removes the need for an imaginary spin-orbit potential (Honoré et al, 1988; Delaroche and Tornow, 1988; Johnson and Mahaux, 1988; Tornow, Chen and Delaroche, 1990; Mahaux and Sartor, 1990b).

These and other deviations from the predictions of the simple optical model could be removed by the introduction of further parameters, or by defining a different potential for low energies (Moldauer, 1963) but this soon becomes unwieldy and unphysical. The dispersive optical model, however, enables the same data to be fitted to a higher accuracy with fewer parameters in a physically-realistic way. It also has improved predictive power for a particular nucleus from one energy to another. The accuracy with which a particular parametrisation of the dispersive term can represent the coupling to excited states has to be studied for each nucleus.

The effects of nuclear structure on the optical potential imply that a perfect global optical potential cannot be obtained. The dispersive optical model does, however, provide the means to define a global optical potential in which the nuclear structure effects are included by a single parameter, namely the strength of the imaginary potential. This possibility is discussed in Section 6.

Early attempts to apply the dispersion relations to the optical potential by Passatore (1968) encountered the difficulty of evaluating the dispersion integral over the whole energy range. A way of doing this was found by Ahmad and Haider (1976), making use of the different energy dependences of the volume and surface imaginary potentials. Subsequently many analyses have been made using the dispersion relation in the following ways:

(a) Since the dispersion correction is skew-symmetric about the Fermi surface, it is convenient to define it relative to the Hartree-Fock potential at the Fermi energy by the subtracted form

$$\Delta V(E,r) = V(E,r) - V(E_F,r) =$$
$$= \frac{1}{\pi} \int_{-\infty}^{\infty} \left(\frac{1}{E'-E} - \frac{1}{E'-E_F} \right) W(E',r)dE'$$
$$= \frac{E-E_F}{\pi} \int_{-\infty}^{\infty} \frac{W(E',r)}{(E'-E)(E'-E_F)}dE' \quad (2.6)$$

where E_F is the Fermi energy. This converges more rapidly than (2.1).

(b) If the imaginary potential has an energy dependence symmetric about the Fermi energy, (2.1) becomes

$$V(E,r) = V_{HF}(E,r) +$$
$$+ \frac{2}{\pi}(E-E_F) \int_{E_F}^{\infty} \frac{W(E')}{(E'-E_F)^2 + (E-E_F)^2}dE'$$
$$(2.7)$$

which also has improved convergence properties.

(c) Since phenomenological optical model analyses determine certain moments of the potential with greater accuracy than the individual parameters it is useful to work with moments of the radially integrated potentials defined by

$$J_V^{(q)}(E) = \frac{4\pi}{A} \int_0^{\infty} V(E,r)r^q dr \quad (2.8)$$

and a similar relation for the imaginary potential. These

radial moments also satisfy the dispersion relations.

The first detailed analyses were made by adopting various polynomial or exponential parametrisations of the energy dependence of the imaginary potential, assuming it to be symmetric about the Fermi surface. These were fitted to the potentials obtained by standard optical model analyses and the resulting potential used to calculate the dispersion correction to the real potential. The parameters of the Hartree-Fock potential were then adjusted to optimise the fit to the phenomenological real potentials. These calculations are more conveniently done using the volume integrals of the potentials. Many of these analyses have been made by Mahaux and Ngô (1981, 1982, 1983, 1984). This established the usefulness of the dispersion relation and gave a detailed understanding of the Fermi surface anomaly.

3. Dispersive Optical Model Analyses

Since the pioneer work of Ahmad and Haider many dispersive optical model analyses have been made. The aim is to find the parametrisation of the optical potential as a function of radial distance and also of energy that will give the best overall fit to the experimental data. The main difficulties are that at negative energies the only data are the bound state energies and sometimes the widths. At positive energies extensive total and differential cross-sections are available, but at the lower energies the interpretation is complicated by the presence of compound elastic scattering, and for protons the Coulomb barrier limits the range of available energies.

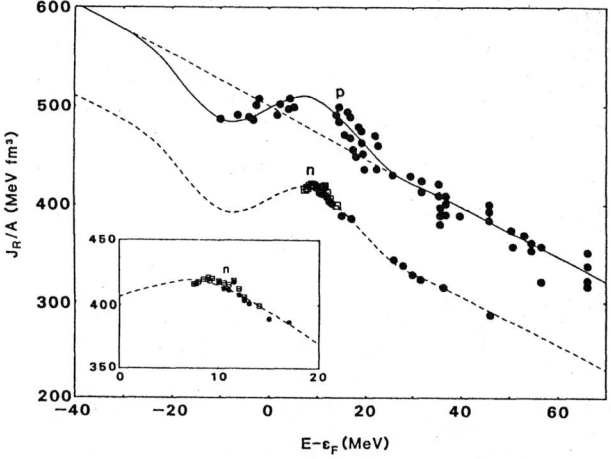

Fig.2. Energy dependence of the volume integral of the real potential for neutrons and protons on ^{208}Pb showing the Fermi surface anomaly (Finlay *et al*, 1985a).

There are two ways of carrying out a dispersive optical model analysis: the first parametrises the potential and seeks the optimum fit to selected data, while the second uses as data the phenomenological potentials obtained from standard optical model analyses. These will now be described.

In the first dispersive optical model method (DOMA), the Hartree-Fock potential is parametrised by a Saxon-Woods potential with a fixed radius and diffuseness and a depth that varies linearly or exponentially with energy. The imaginary potential is taken as the sum of volume and surface-peaked terms, usually parametrised

by Saxon-Woods and Saxon-Woods derivative radial form factors. The energy variations of these imaginary potentials are separately parametrised; some of the ways of doing this are described in the next section. The values of all the parameters of the potential are either fixed from previous work or adjusted to fit selected experimental data, usually total and differential elastic cross-sections for a range of energies. In these analyses, the dispersive term is calculated from the imaginary potential and added to the Hartree-Fock term to give the real potential as in (2.5). This ensures that, although all the radius parameters are fixed, the effective radius of the real potential varies in the way required by the experimental data. The energy dependence of the volume integral of the real potential then fits the experimental values, as shown in fig.2.

A detailed analysis of the interaction of neutrons with ^{208}Pb from -20 to $+165$ MeV by Johnson, Horen and Mahaux (1987) used straight-line segments (4.4) to represent the energy variation of the volume and surface-

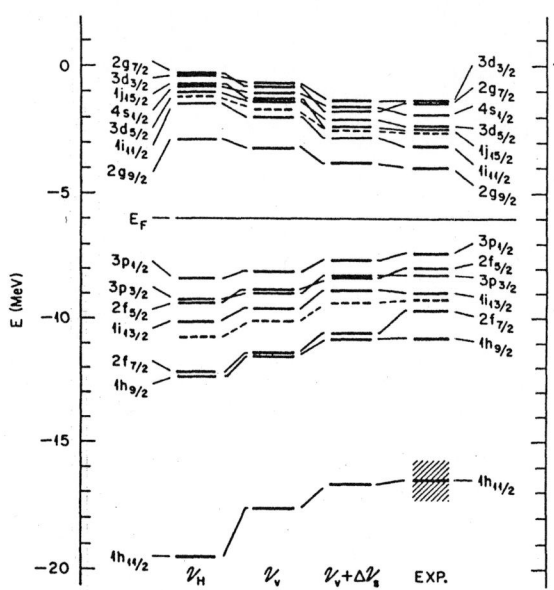

Fig.3. Energies of bound neutron states in ^{208}Pb, compared with Hartree-Fock calculations with and without the dispersion corrections (Johnson, Horen and Mahaux, 1987).

peaked imaginary potentials. This showed the usefulness of the dispersive optical model and gave excellent fits to the experimental data. The improved fit to the bound single-particle neutron states in ^{208}Pb due to the addition of the dispersion terms is shown in fig.3. At low energies ($0 < E < 10$ MeV), Jeukenne, Johnson and Mahaux (1988) showed that the fits can be improved by allowing the diffuseness of the surface imaginary potential to decrease with decreasing energies, and its depth to depend on the orbital angular momentum. Providing this is done, the radius of the surface absorption can be kept constant. Almost as good fits can be obtained with surface absorptions that depend on angular momentum but not on energy. The angular momentum dependence has one value of W_s for $L = 1$, 3 and 6 and another for the remaining partial waves. The first group corresponds to those with wave functions that peak in the nuclear surface and so couple to surface excitations, while those in the second group do not. Further evidence of angular momentum dependence has been found by Johnson and Winters (1988).

A similar analysis of the interactions of neutrons with ^{40}Ca from -80 to $+80$ MeV was made by Johnson and Mahaux (1988). The potentials were similar to those found for ^{208}Pb. A small difference between the calculated and experimental total cross-section at low energy could be fitted by allowing the surface diffuseness and radius parameters to depend on energy. The interactions of protons with ^{40}Ca from -60 to $+200$ MeV were analysed by Tornow, Chen and Delaroche (1990), and of neutrons with ^{86}Kr by Johnson, Carlton and Winters (1989). An analysis of the interactions of neutrons and protons with ^{208}Pb by Finlay et al (1989) showed dispersive effects for neutrons but not for protons. This is because they are prominent below 10 MeV where the proton interaction is inhibited by the Coulomb barrier. The interaction of neutrons with ^{208}Bi has been analysed by Das and Finlay (1990). The dispersive optical model has been extended to include analysing powers by Roberts et al (1988–9) and by Chen and Tornow (1988–9), and the Coulomb correction term has been studied by Tornow and Delaroche (1988) and by Mahaux and Sartor (1988b).

The dispersive model analyses that use potentials obtained by standard optical model fits can be made in several different ways. One of the simplest is to calculate the volume integrals given by (2.8) with $q = 2$, parametrise the energy variation of the imaginary part and then to use the dispersion relation to calculate volume integrals of the real potential at all energies. At negative energies this can be compared with the volume integrals of the potentials giving the experimental bound states. This method has been applied to the interactions of neutrons with ^{51}V, ^{59}Co, ^{89}Y, ^{93}Nb, In and ^{209}Bi by Smith, Guenther and Lawson (1988) by Lawson, Guenther and Smith (1986, 1987, 1989) and by Chiba et al (1990). In most cases there is good qualitative agreement, confirming the usefulness of the dispersive optical model analysis.

Several powerful methods using two or three radial moments of the phenomenological optical potentials have been developed by Mahaux and Sartor. The basic idea is that the moments of the potential suffice to define its three parameters V_u, r_u and a_u. The moments satisfy the dispersion relations and so can be used to fit data over a wide range of energies, treating the parameters of the Hartree-Fock field as variable. A few iterations suffice to give a good overall fit to both scattering and bound state data. The optimum moments are then used to obtain the energy variations of the optical model parameters.

In this method, the Hartree-Fock potential is parametrised by a Saxon-Woods form factor usually with fixed diffuseness $a = 0.7$fm. Its energy variation is taken to be linear or exponential, and the parameters describing this, together with the radius parameter, are adjustable. In the first of their studies, Mahaux and Sartor (1986a) calculated the moments (2.8) of the imaginary potential with $q = 0.8$, 2 and 4 for potentials fitted to the elastic scattering of neutrons by ^{208}Pb and then fitted their energy variations with the Brown-Rho parametrisation

(4.5a). The dispersion relation was then used to calculate the dispersion correction and this, together with the Hartree-Fock term, gave the moments of the real potential for all energies. These moments were compared with those of the potentials fitted to the bound and scattering state data, and the parameters of the Hartree-Fock potential adjusted to optimise the fit. From the three moments of the optimum real potential the parameters of the equivalent Saxon-Woods potential were obtained, and this confirmed that the main effect of the dispersive term is to modulate the radius of the potential around the Fermi energy.

This method, subsequently called the iterative moment approach (IMA), has been extended and applied to analyse the potentials fitted to data on neutron-^{208}Pb (Mahaux and Sartor, 1987ab, 1989a), proton-^{208}Pb (Mahaux and Sartor 1988a), neutron-^{40}Ca and proton-^{40}Ca (Mahaux and Sartor 1988c), neutron-^{89}Y (Jeukenne et al, 1977; Mahaux and Sartor 1987c), neutron-^{59}Co (Smith, Guenther and Lawson, 1988) and neutron-^{51}V interactions (Lawson, Guenther and Smith, 1989). There are several differences in the ways the analyses were carried out, in particular whether the radius and diffuseness parameters of the Hartree-Fock potential were fixed or allowed to vary, and whether its potential depth was constrained to fit the Fermi energy.

Detailed comparisons between the DOMA and the IMA show that they give almost identical radial moments, but rather different results for the central value of the effective mass. Furthermore, the IMA gives an unphysical energy dependence of the real potential around $E = 0$. These defects may be attributed to inadequacies in the Brown-Rho parametrisation of the radial moments (Mahaux and Sartor, 1989a). In addition, because it does not give the imaginary potentials themselves, the IMA cannot be used to calculate differential cross-sections, spectral functions and occupation probabilities. It was therefore extended so as to include an explicit parametrisation of the imaginary potential, but still using only the moments of the phenomenological potentials (Mahaux and Sartor, 1989a). This allowed the additional observables to be calculated, but other defects remained.

Since the IMA involves only three adjusted parameters compared with seven for the DOMA, it is worthwhile trying to remove its defects by a more sophisticated parametrisation, and this has been done by the variational moment approach (VMA) (Mahaux and Sartor, 1989a).

In the VMA, the total and volume imaginary potentials are parametrised separately, by the BR and JM forms respectively (see (4.5a) and (4.6)). The surface imaginary potential is then given by their difference, which falls to zero at higher energies as required. This is the BR-JM parametrisation (4.9). The parameters are fixed by requiring that their moments fit those of the phenomenological potentials. From the moments the dispersive correction is obtained, and subtraction from the phenomenological moments of the real potential gives those of the Hartree-Fock potential. This potential is taken to have the Saxon-Woods form with surface diffuseness 0.7 fm and depth depending exponentially on energy. The radius parameter is chosen to minimise the difference between the calculated and phenomenological moments. The optical potential is then specified for all energies and can be used to calculate all observable quantities for comparison with experimental data.

In another version of the VMA, the total and volume imaginary potentials are both parametrised by the BR form; this is the BR-BR parametrisation (4.8ab). It is found that while the calculated real potential is not very sensitive to the details of the $W(r, E)$ parametrisation, the total cross-section is better fitted by the BR-BR than by the BR-JM parametrisation (Mahaux and Sartor, 1990, 1991b).

The VMA is simpler than the extended IMA, and also ensures that the dispersion relation is satisfied at each value of the radius, and that the effective mass has physical values. It has been successfully applied to the interactions of neutrons and protons with ^{208}Pb (Mahaux and Sartor, 1989ab). The proton analysis encountered the difficulty that there is no scattering data at low energies due to the Coulomb repulsion, and this was overcome by requiring the extrapolated mean field to reproduce the energy gap between the valence shells. In both the neutron and proton analyses the overall quality of the fit is very similar to that of the DOMA, and it is achieved with fewer adjustable parameters. It is particularly notable that the VMA fixes the potential from the moments of the scattering data only, and its extrapolation to lower energies fits the bound state data. The method has also been successfully applied to the interactions of neutrons and protons with ^{40}Ca (Mahaux and Sartor, 1991b).

The dispersion relation method has been applied to a well-deformed nucleus by Merchant *et al* (1991). A coupled-channel analysis of the elastic, inelastic and total cross-sections for neutrons on ^{238}U gave a good overall fit to the experimental data, but this was appreciably improved by including the correction to the real potential obtained from the dispersion relation. To do this the imaginary potential was taken to have the form (4.5) and the parameters determined from the best fits to the data.

The energy dependence of both the surface and volume potentials have been taken to have the form (4.7a) by Niizeki *et al* (1989) and the parameters obtained by fitting to the experimental data for neutrons on ^{28}Si at six energies.

4. The Energy Variation of the Imaginary Potential

Evaluation of the dispersion integral requires a knowledge of the imaginary potential over the whole range of energies; for positive energies it can be obtained from optical model analyses of elastic scattering but for negative energies it is only obtainable with difficulty and with less accuracy from the widths of single-particle states. It is therefore usual in dispersive optical model analyses to assume that the imaginary potential is symmetric about the Fermi energy, and this enables the dispersion correction to be written in the form (2.6).

The reliability of this assumption may be studied by examining the energy dependence of the bound-state widths. Figs.4 and 5 show that the widths are given by (Mahaux and Sartor, 1990)

$$\Gamma_{nlj} = \alpha(E - E_F)^2 \qquad (4.1)$$

where $\alpha \simeq 0.04$ in the vicinity of the Fermi surface and $\alpha \approx 0.009$ for the deeper states.

The widths of the bound states in a complex potential are related to the imaginary part of the potential by

$$\Gamma = 2 \int_0^\infty W(r)|\psi(r)|^2 dr \qquad (4.2)$$

If the form factor of the imaginary potential is suitably parametrised, its depth can therefore be obtained

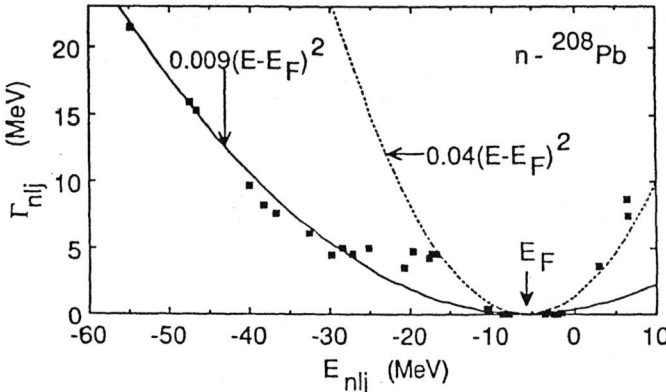

Fig.4. Calculated quasi-particle bound-state widths as a function of energy for neutrons on ^{208}Pb (Mahaux and Sartor, 1991a).

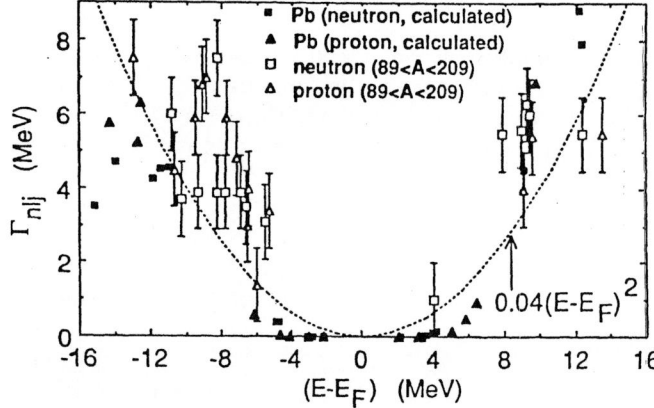

Fig.5. Empirical and calculated quasi-particle bound-state widths as a function of energy for neutrons and protons on ^{208}Pb (Mahaux and Sartor, 1991a).

using the experimental values of Γ. In the limit of infinite nuclear matter,

$$\Gamma = 2W \qquad (4.3)$$

If this relation is applied to finite nuclei, the symmetry assumption is approximately confirmed. Calculations with the more exact expression (4.2) are in progress.

Mahaux and Sartor (1991a) have investigated the consequences of a breakdown in the symmetry assumption, and find that it can be compensated by a change in the energy variation of the Hartree-Fock potential.

It is convenient to have an analytical form for the energy variation of the imaginary potential as this allows the dispersion integral to be evaluated analytically. Several forms have been used:

1. The straight-line segment potential of Johnson *et al* (1987) for ^{208}Pb

$$W_v(E) \begin{cases} = 0 & \text{for } E < 10 \text{ MeV} \\ = 0.17(E - 10) & \text{for } 10 < E < 50 \text{ MeV} \\ = 6.8 & \text{for } E > 50 \text{ MeV} \end{cases}$$

(4.4a)

and

$$W_s(E) \begin{cases} = 0.4(E - E_F) & \text{for } -6 < E < 10 \text{ MeV} \\ = -0.103(E - 27) & \text{for } 10 < E < 72 \text{ MeV} \\ = 0 & \text{for } E > 72 \text{ MeV} \end{cases}$$

(4.4b)

(and symmetric expressions for $E < E_F$).

A more complicated four-segment expression has been given by Mahaux *et al* (1986).

2. The Brown and Rho (1981) (BR) form

$$W_v(E) = \frac{W_1(E - E_F)^2}{(E - E_F)^2 + E_0^2}$$

(4.5a)

supplemented by

$$W_s(E) = \frac{W_1(E - E_F)^2}{((E - E_F)^2 + E_0^2)^2}$$

(4.5b)

3. The Jeukenne and Mahaux (1983) (JM) form

$$W(E) = W_1 \frac{(E - E_F)^4}{(E - E_F)^4 + E_0^4}$$

(4.6a)

This has been modified by Delaroche *et al* (1989) to give the exponential form

$$W(E) = W_1 \frac{(E - E_F)^4}{(E - E_F)^4 + E_0^4} \exp[-C(E - E_F)] \quad (4.6b)$$

and

$$W_s(E) = \frac{a_0 + a_2(E - E_F)^2}{\{E_0^2 + (E - E_F)^2\}^2}$$

(4.7b)

4. The potentials used by Hicks and McEllistrem (1988) for osmium and platinum

$$W_v(E) \begin{cases} = 0 & \text{for } E < 8 \text{ MeV} \\ = 2.33(E^{\frac{1}{2}} - 8^{\frac{1}{2}}) & 8 < E < 40 \text{ MeV} \\ = 2.33(40^{\frac{1}{2}} - 8^{\frac{1}{2}}) & \text{for } E > 40 \text{ MeV} \\ \equiv\approx 8.1 \text{ MeV} \end{cases}$$

(4.7a)

(and symmetric expressions for $E < E_f$).

5. The BR-BR parametrisation of Mahaux and Sartor (1991)

$$W_v(E) = \gamma_v \frac{(E - E_p)^2}{(E - E_p)^2 + \mu_v^2}$$

(4.8a)

$$W_s(E) = \beta_s \left[\frac{(E - E_p)^2}{(E - E_p)^2 + \rho_s^2} - \frac{(E - E_p)^2}{(E - E_p)^2 + \mu_v^2} \right]$$

(4.8b)

6. The BR-JM parametrisation of Mahaux and Sartor (1989a)

$$W_v(E) = \gamma_v \frac{(E - E_p)^4}{(E - E_p)^4 + \mu_v^4}$$

(4.9a)

$$W_s(E) = \beta_s \left[\frac{(E - E_p)^2}{(E - E_p)^2 + \rho_s^2} - \frac{(E - E_p)^4}{(E - E_p)^4 + \mu_v^4} \right]$$

(4.9b)

6. Conclusion

The analyses summarised above show that the dispersive optical model is able to give more accurate fits to the neutron data than the simple optical model, and also shows how a better global potential may be constructed. It is therefore desirable that standard optical model computer programs be modified to include the dispersion correction (2.5) to the real potential. This requires incorporation of a parametrised form for the energy dependence of the imaginary potential, and for this one of the analytical forms listed in the previous section can be used. With such a program, the optical model analysis can be carried out in the usual way. Alternatively, the optical model analysis can be made, followed by a variable moment analysis as described in section 3.

An improved global potential could be constructed by performing a dispersive optical model analysis for a large number of nuclei, fixing all the parameters except the depth of the imaginary potential to standard average values. This depth provides the normalisation of the assumed energy variation, and so is just one number for each nucleus that takes account in a global way of the coupling to excited states.

Acknowledgments

I am grateful to Professor C. Mahaux, Professor W. Tornow and Dr A.B. Smith for sending me details of their unpublished work and to Professor R. Keddy and Professor R.H. Lemmer for inviting me to visit the University of the Witwatersrand where much of this paper was written.

References

I. Ahmad and W. Haider, *J. Phys.* **G2**, L157 (1976).

J.R.M. Annand, R.W. Finlay and F.S. Dietrich, *Nucl. Phys.* **A443**, 249 (1985).

G.E. Brown, J.H. Gunn and P. Gould, *Nucl. Phys.* **46**, 598 (1963).

G.E. Brown and M. Rho, *Nucl. Phys.* **A372**, 397 (1981).

Z.P. Chen and W. Tornow, TUNL Report XXVIII 82 (1988-9).

S. Chiba, P.T. Guenther, R.D. Lawson and A.B. Smith, *Phys. Rev.* **C42**, 2487 (1990).

J.M. Cornwall and M.A. Ruderman, *Phys. Rev.* **128**, 1474 (1962).

R.K. Das and R.W. Finlay, *Phys. Rev.* **C42**, 1013 (1990).

J.P. Delaroche and W. Tornow, *Phys. Lett.* **B203**, 4 (1988).

J.P. Delaroche, Y. Wang and J. Rapaport, *Phys. Rev.* **C39**, 391 (1989).

J.S. Eck and W.J. Thompson, *Nucl. Phys.* **A237**, 83 (1975).

H. Feshbach, *Ann. Phys.* (New York) **5**, 357 (1958).

R.W. Finlay, J.R.M. Annand, J.R. Petler and F.S. Dietrich, *Phys. Lett.* **B155**, 313 (1985a); **B157**, 475 (1985b).

R.W. Finlay, J.R.M. Annand, T.S. Cheema, J. Rapaport and F.S. Dietrich, *Phys. Rev.* **C30**, 796 (1984).

R.W. Finlay, J. Wierzbicki, R.K. Das and F.S. Dietrich, *Phys. Rev.* **C39**, 804 (1989).

B.R. Fulton, D.W. Banes, J.S. Lilley, M.A. Nagarajan and I.J. Thompson, *Phys. Lett.* **B162**, 55 (1985).

B. Gyarmati, R.G. Lovas, T. Vertse and P.E. Hodgson, *J. Phys.* **G7**, L209 (1981).

R. Hagedorn, Preludes to Theoretical Physics. Ed. A. de Shalit, H. Feshbach and L. Van Hove, North-Holland (1966), 154.

P.E. Hodgson, *Rep. Prog. Phys.* **47**, 613 (1984); Proceedings of the International Conference on Nuclear Reaction Mechanism, Calcutta; World Scientific (1989) 39. *Cont. Phys.* **31**, 295 (1990).

S.E. Hicks and M.T. McEllistrem, *Phys. Rev.* **C37**, 1787 (1988).

G.M. Honoré, W. Tornow, C.R. Howell, R.S. Pedroni, R.C. Byrd, R.L. Walter and J.P. Delaroche, *Phys. Rev.* **C33**, 1129 (1986).

M. Jaminon, J.-P. Jeukenne and C. Mahaux, *Phys. Rev.* **C34**, 468 (1986).

J.-P. Jeukenne and C. Mahaux, *Nucl. Phys.* **A394**, 445 (1983).

J.-P. Jeukenne, C.H. Johnson and C. Mahaux, *Phys. Rev.* **C38**, 2573 (1988).

J.-P. Jeukenne, A. Lejeune and C. Mahaux, *Phys. Rev.* **C15**, 10 (1977).

C.H. Johnson, R.F. Carlton and R.R. Winters, *Phys. Rev.* **C39**, 415 (1989).

C.H. Johnson, D.J. Horen and C. Mahaux, *Phys. Rev.* **C36**, 2252 (1987).

C.H. Johnson and C. Mahaux, *Phys. Rev.* **C38**, 2589 (1988).

C.H. Johnson and R.R. Winters, *Phys. Rev.* **C37**, 2340 (1988).

R.D. Lawson, P.T. Guenther and A.B. Smith, *Phys. Rev.* **C34**, 1599 (1986); *Phys. Rev.* **C36**, 1298 (1987); *Nucl. Phys.* **A493**, 267 (1989).

C. Mahaux, P.F. Bortignon, R.A. Broglia and C.H. Dasso, *Phys. Rep.* **120**, 1 (1985).

C. Mahaux and H. Ngô, *Phys. Lett.* **B100**, 285 (1981); *Nucl. Phys.* **A378**, 205 (1982); *Nucl. Phys.* **410**, 271 (1983); *Nucl. Phys.* **A431**, 486 (1984).

C. Mahaux, H. Ngô and G.R. Satchler, *Nucl. Phys.* **A449**, 354 (1986).

C. Mahaux and R. Sartor, *Phys. Rev.* **C19**, 229 (1979); *Phys. Rev. Lett.* **57**, 3015 (1986a); *Nucl. Phys.* **A451**, 441 (1986b); *Nucl. Phys.* **A458**, 25 (1986c); *Nucl. Phys.* **A460**, 466 (1986d); *Phys. Rev.* **C34**, 2119 (1986e); *Nucl. Phys.* **A468**, 193 (1987a); *Phys. Rev.* **C36**, 1777 (1987b); *Nucl. Phys.* **A475**, 247 (1987c); *Nucl. Phys.* **A481**, 381 (1988a); *Nucl. Phys.* **A481**, 407 (1988b); *Nucl. Phys.* **A484**, 205 (1988c); *Nucl. Phys.* **A493**, 157 (1989a); *Nucl. Phys.* **A503**, 525 (1989b); *Nucl. Phys.* **A516**, 285 (1990); *Advances in Physics*, Ed. J.W. Negele and E. Vogt (1991a) (In press); *Nucl. Phys.* (1991b) (In press).

A.C. Merchant, P.E. Hodgson and H.R. Schelin. (In press) (1991).

P.A. Moldauer, *Nucl. Phys.* **47**, 65 (1963).

M.A. Nagarajan, C. Mahaux and G.R. Satchler, *Phys. Rev. Lett.* **54**, 1136 (1985).

T. Niizeki, H. Orihara, M. Ohura, M. Hosaka, G.C. Jon, K. Ishii, K. Miura and H. Ohnuma, CYRIC Annual Report, Tohoku University 10 (1989).

G. Passatore, *Nucl. Phys.* **A110**, 91 (1968).

J. Rapaport, *Phys. Rep.* **87**, 25 (1982).

M.L. Roberts, Z.M. Chen, P.D. Felsher, D.J. Horen, C.R. Howell, K. Murphy, H.G. Pfützer, W. Tornow and R.L. Walter, TUNL Report XXVIII 70 (1988-9).

A.B. Smith, P.T. Guenther and R.D. Lawson, *Nucl.Phys.* **A455**, 344 (1986); *Nucl. Phys.* **A483**, 50 (1988).

W. Tornow and J.P. Delaroche, *Phys. Lett.* **B210**, 26 (1988).

W. Tornow, Z.P. Chen and J.P. Delaroche, *Phys. Rev.* **C42**, 693 (1990).

W.T.H. van Oers, H. Haw, N.E. Davison, A. Ingemarsson, B. Fagerström and G. Tibell, *Phys. Rev.* **C10**, 307 (1974).

P.G. Young, Proceedings of the Specialists' Meeting on the Use of the Optical Model for the Calculation of Neutron Cross-Sections below 20 MeV., NEANDC-222"U", OECD, Paris (1986), 127.

Su Zong Di and P.E. Hodgson, *J. Phys.* **G14**, 1485 (1988).

Recent Progress in the Statistical Theory of Nuclear Reactions

H. A. Weidenmüller

Max-Planck-Institut für Kernphysik, 6900 Heidelberg, Germany

Recent progress in the theory of multistep nuclear reactions is reviewed. Emphasis is placed on applications of the formulas by Feshbach, Kerman and Koonin, on the implementation of the approach by Nishioka *et al.*, and on a comparison between both approaches and the one due to Tamura *et al.*.

1 Introduction. Basic Ideas

We are concerned with nuclear reactions induced by light particles of mass up to four and indident energy E. The energy domain E $\leq$ 10 MeV is governed by compound nucleus (CN) processes. In the domain above 150 or 200 MeV, semiclassical or classical approaches such as the Glauber model or the cascade model should become reliable, since the de Broglie wavelength of the incident particle becomes comparable with the internucleon distance. In this paper, attention is focussed on reactions in the domain 10 MeV $\leq$ E $\leq$ 100 $\cdots$ 200 MeV. This is the domain of precompound reactions, both in their multistep compound (MSC) and their multistep direct (MSD) forms.

Before entering this subject, we recall some pertinent features of CN theory. The CN is characterized by the inequality $\Gamma^\downarrow \gg \Gamma^\uparrow$: The internal spreading width $\Gamma^\downarrow$ (the inverse of the equilibration time) of the CN states is large compared to the decay width $\Gamma^\uparrow$ (the inverse of the decay time). Moreover, CN reactions have stochastic features. Therefore, a simple random matrix approach is sufficient to model all observable features of CN scattering. [This type of modelling is not restricted to CN reactions but applies universally to all chaotic quantum scattering processes in which all parts of phase space are equally accessible [1] from the channel(s)]. During the last decade, it has become possible to evaluate the above mentioned random-matrix model exactly [2]. For all values of $\Gamma^\uparrow/d$ (with d the average level spacing), the average CN cross-section is given as a three-fold integral involving the transmission coefficients in all open channels [2]. For $\Gamma^\uparrow \gg d$, the result yields the Hauser-Feshbach formula [3]. In the contribution C 29 to this meeting, Igarasi announces an application of this result to cross-section calculations, and a comparison with the formulas due to Hauser and Feshbach, and to Moldauer.

Precompound reactions are characterized by the approximate equality $\Gamma^\downarrow \approx \Gamma^\uparrow$. Since the early work by

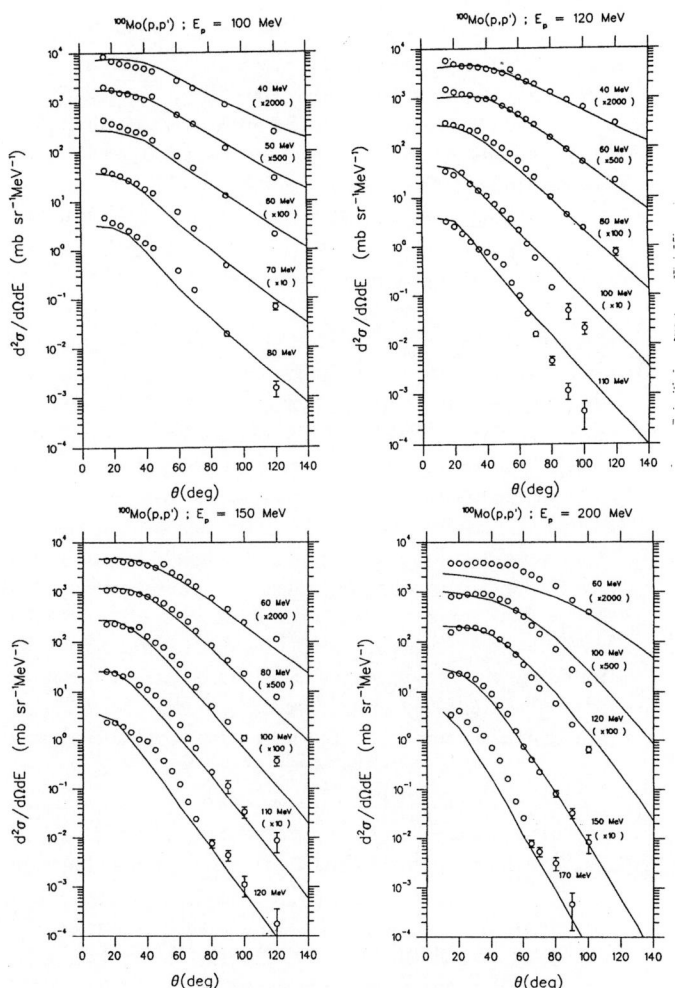

Figure 1: The reaction $^{100}Mo(p,p')$ at various incident energies. Circles: data. Lines: FKK results. Taken from [10].

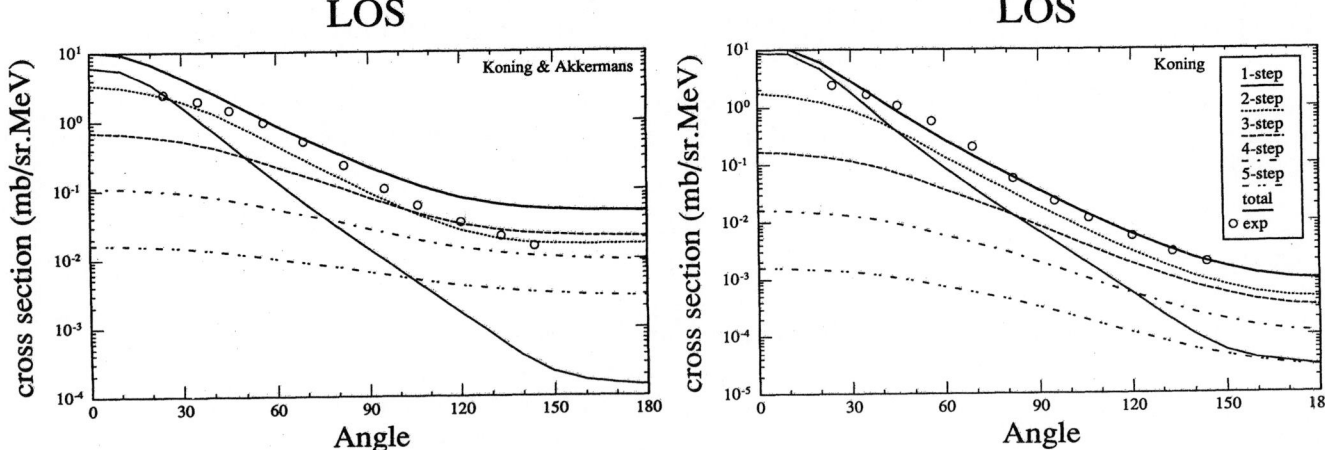

Figure 2: The reaction $^{90}Zr(p,p')$ at E = 120 MeV for a final proton energy E$_{p'}$ = 60 MeV. Circles: Data. Lines: Leading-particle-on-shell calculation. Taken from [11].

Figure 3: Same as Fig.2 for E$_{p'}$ = 80 MeV.

Griffin about 25 years ago, phenomenological models, most of them based on the concept of excitons (and therefore, on the shell model) have been available and useful. About 15 years ago, the advent of "leading particle" models permitted the incorporation of angular distributions in the phenomenology.

Theoretical developments in the last decade have focussed on a microscopic derivation of formulas which incorporate the essential aspects of the exciton models. I mention the work of Feshbach, Kerman and Koonin [4] (FKK), the multistep-direct reaction formalism by Tamura, Udagawa and Lenske [5] (TUL), and the work by Nishioka, Verbaarschot, Weidenmüller and Yoshida [6] (NWY). All these works use some statistical input and may be viewed as natural generalisations of the statistical CN theory to the situation where $\Gamma^\uparrow \approx \Gamma^\downarrow$. They implement the basic idea that in a series of two-body collisions, the incident particle populates CN states of increasing complexity ("Chaining hypothesis"). This same idea was, incidentally, used in a model by Kalka[7]. With the help of substantial simplifications, the author derives simple analytical expressions for angle-integrated reaction cross-section and achieves impressive agreement with data on several nuclei and for a set of incident energies.

FKK were the first authors to distinguish the multistep direct (MSD) and the multistep compound (MSC) reaction mechanism. This distinction is now part of all modern approaches. In the MSD process the reaction consists of a chain of states with one nucleon in a continuum orbital of the shell- model. In the MSC process, all nucleons occupy bound orbitals. The MSC process leads to a reaction cross-section which is symmetric about 90° c .m. while the MSD process is described by a sequence of DWBA matrix elements and yields a forward-peaked angular distribution. The MSC process, very important at E = 10 or 20 MeV becomes

relatively unimportant for E ≤ 60 MeV or so. Crossover between both mechanisms is possible in any stage of the reaction but has not been investigated much. Understanding this cross-over mechanism is one of the challenging open problems of the field.

The statistical modelling of precompound reactions encapsulates another generic case of quantum chaotic scattering: The case where not all parts of phase space are equally accessible from the channel(s). Examples are microwave scattering on a chain of irregularly shaped cavities or the passage of electrons through impurity devices of mesoscopic size.

2 Novel Results for FKK

The FKK approach uses statistical assumptions on matrix elements wherever they seem necessary and reasonable in the course of the argument. Therefore, the FKK form for the average cross-section is not based on a stringent derivation from a quantum-statistical input. In the hand of practitioners [8], the FKK formulas have become viable tools which unify the analysis of reactions on different nuclei, and at different incident energies, and which have predictive power. In a recent review talk, Hodgson [9] has summarized the situation. I limit myself to more recent works.

Cowley et al. [10] have extended their previous work on the (p,n) reaction to the (p,p') reaction. This provides a consistency check for the previous analysis. Fig. 1 shows the $^{100}Mo(p,p')$ reaction at various incident energies and the results of FKK MSD calculations. The overall agreement is good; for large E, the FKK angular distributions drop off too rapidly. It is surprising that the FKK MSD cross-section reproduces the data even at low E$_{p'}$, although here MSC processes are expected to be important. This point deserves clarification.

In a modified version of FKK, Akkermanns and Koning [11] calculate the MSD process for the reaction $^{90}Zr(p,p')$ at $E = 120 MeV$ using non-DWBA type

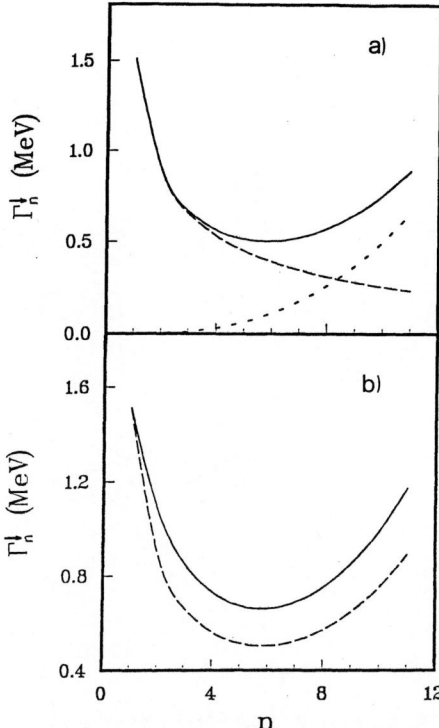

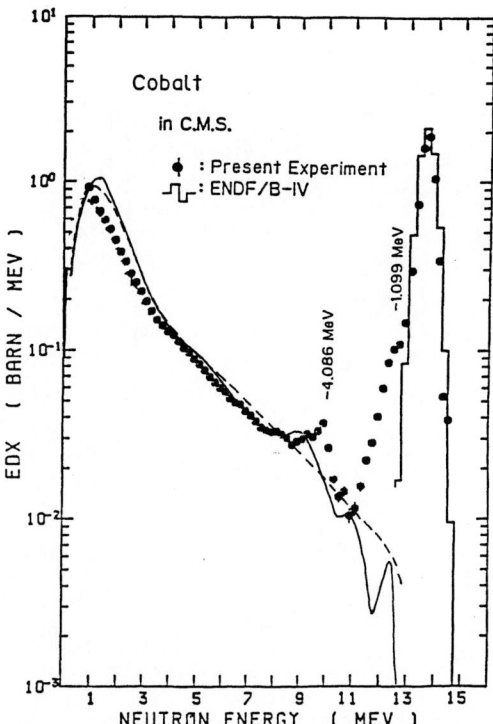

Figure 4: Upper part: Contributions to $\Gamma_n^\downarrow$ from transitions $n \rightarrow n+1$ (long dashes) and $n \rightarrow n-1$ (short dashes); the solid line is the sum. Lower part: Differences in implementing the theory to evaluate $\Gamma_n^\downarrow$ using two different prescriptions. Taken from [12].

Figure 6: Angle-integrated MSC emission spectra of Co using the Williams level-density formula (dashed curve) and the Nix-Moeller formula (full curve), together with the data. Taken from [13].

matrix elements. As done in FKK, the authors use "leading particle statistics" and the on-shell approximation but obtain a final formula which differs from the FKK expression. The calculated cross-sections are free of divergences and agree will with the data . With this modification, the authors have resolved a long-standing problem. Fitting the inelastic cross-section at $E_{p'} = 100$ MeV , the authors calculate in a parameter-free manner the inelastic cross-sections at $E_{p'} = 60$ and 80 MeV . Their results are compared with the data in Fig.2,3.

A recurrent theme - not taken up here - in FKK MSD calculations is the need to reduce the strength V_0 of the effective interaction as E increases.

3 Implementation of NWY and First Results

The NWY approach uses a well-defined quantum-statistical input and derives closed - form expressions for average cross-sections within controlled approximations. Recently, this approach has been implemented in practical calculations. First results are now available.

For the NWY MSD process [12], it was shown that ambiguities in implementing the theoretical expressions lead only to negligible differences in the calculated cross-sections. Such ambiguities arise in the calculations of the spreading widths $\Gamma_{n \rightarrow n+1}^\downarrow$ for the transition from exciton class n to exciton class $n+1$ (Fig.4), and in

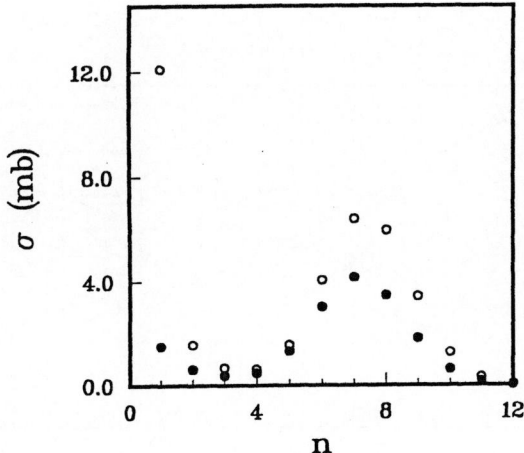

Figure 5: Contribution to the cross-section versus exciton number n for spin $\mathcal{J} = 12^+$ (open circles) and $\mathcal{J} = 2^+$ (full dots). Taken from [12].

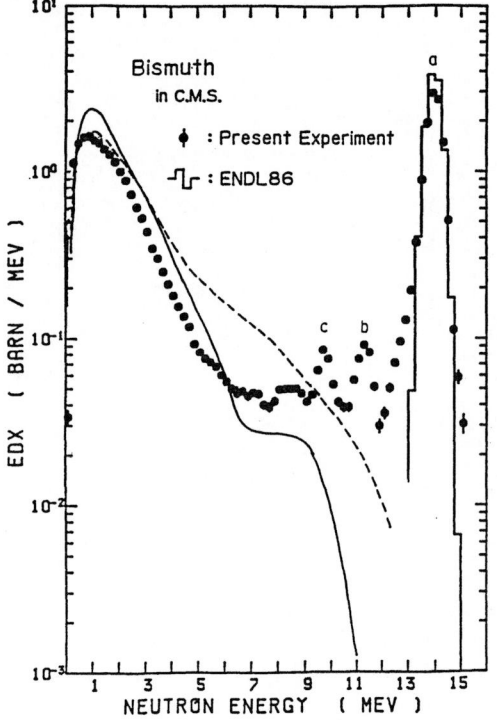

Figure 7: Same as Fig.6 for Bi.

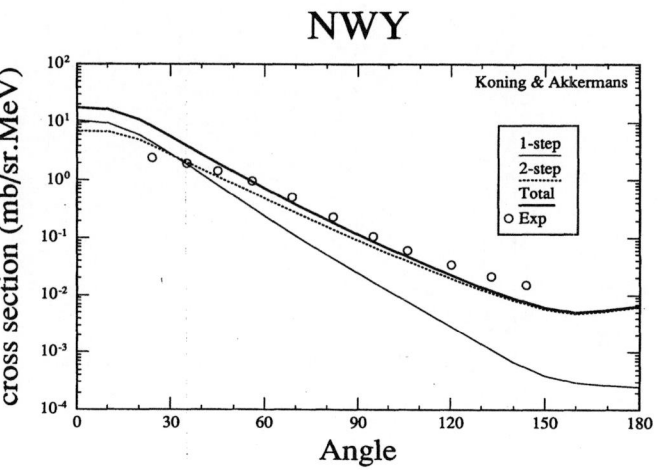

Figure 8: Same as Fig.2 for the NWY MSD process. Taken from [11].

the calculation of transmission coefficients for the decay of various classes. Both types of quantities are obtained from the optical-model potential in the entrance channel. For the partial level densities, the NWY formalism requires the use of expressions which take account of the residual interaction, i.e. which differ from the usual Fermi gas model. So far, the actual calculations have not complied with this requirement. I return to this point in section 5.

Figure 5 shows contributions to the $^{93}Nb(n,n')$ cross-section at E = 14.1 MeV due to states differing in exciton number n for two spin values. Precompound and CN contributions can be distinguished clearly. (In the NWY MSC calculations, the CN contribution is not grafted on to the results via the Hauser-Feshbach formula, but is calculated within the approach itself.) Figures 6 and 7 show preliminary results [13] for the angle-integrated emission spectrum for neutrons from the bombardment with 14.1 MeV neutron of Co and Bi, respectively, with details given in the captions.

The NWY MSD mechanism assumes that transitions between classes are so rapid as to justify the " sudden approximation". Then, only the final-state density enters into the cross-section formula. Both on - and off-shell DWBA contributions are taken into account. The formulars were simplified to be calculable with existing codes; this probably overestimates the cross-section. Results [11] are shown in figues 8 and 9 for the reaction $^{90}Zr(p,p')$ at $E = 120 MeV$. Only two steps were considered; the calculations were fitted to the data for $E_{p\prime} = 100 MeV$. The Williams level density was used (although NWY reqire inclusion of the residual inter-

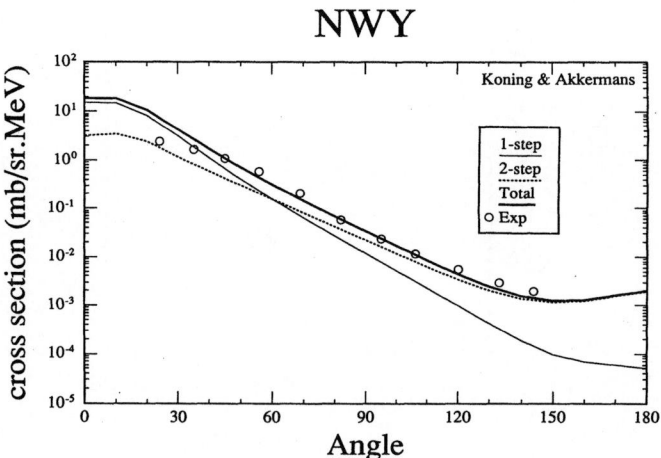

Figure 9: Same as Fig.8 for E $_{p\prime}$ = 80 MeV.

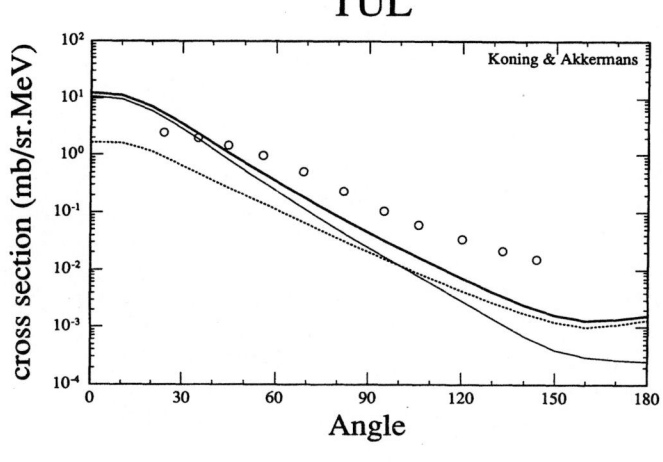

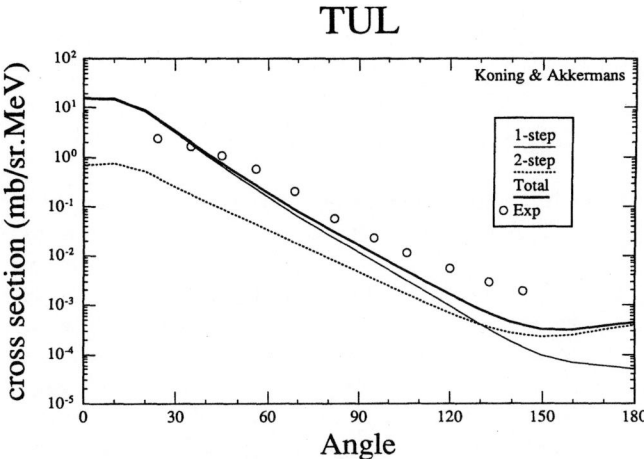

Figure 10: Same as Fig.2 for TUL. Taken from [11].

Figure 11: Same as Fig.10 for $E_{p'} = 80$ MeV.

action). Higher steps are "probably of marginal importance" [11].

4 Comparison FKK - NWY - TUL

For the MSD process, Akkermans and Koning [14] have carried out a detailed comparison between the structure of the three approaches. For the one-step process, all three agree. For the two-step process, the authors identified the following differences:

FKK use "leading particle statistics": They use statistical assumptions on the two-body matrix elements appearing in the DWBA expressions. And all processes are assumed to be on shell. This last assumption probably attributes higher weight to many-step MSD contributions than does the more complete quantal treatment of TUL or NWY.

NWY use the sudden approximation, take into account both on- shell and off- shell contributions, and use "residual system statistics": The levels populated at the end of the MSD chain in the residual nucleus are assumed to be statistical.

TUL use the adiabatic approximation (after each collision, the level density of the intermediate residual system is taken into account), also include both on- shell and off-shell contributions, and use "residual system statistics" to evaluate the average cross-section. Results for TUL obtained [11] under the same conditions as figures 2 and 3 for the modified FKK and figures 8 and 9 for NWY are shown in figures 10 and 11. Comparing these three sets of figures, we observe that NWY yields a particularly good agreement with the data. Future investigations will have to show whether this indication survives the test of time. We note that TUL typically yields smaller cross-sections than NWY, especially at backward angles. This may be a consequence of the above-mentioned simplification of the NWY expression.

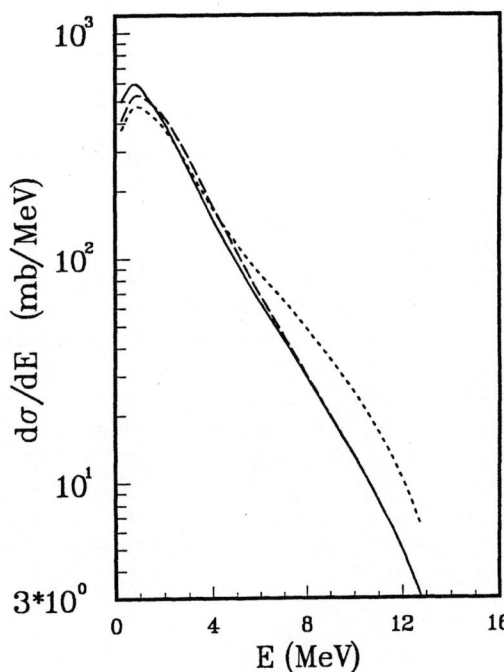

Figure 12: The NWY MSC process including all exciton classes (solid line), the same up to $n \sim 3$ plus the Hauser-Feshbach contribution (dashed line), and the FKK MSC process (dotted line). Taken from [12].

780

For the MSC process, some comparison between FKK and NWY is made in [12]. The main difference lies in the usage of transmission coefficients and of available state densities. The precise way in which this difference alters the calculations has not been pinned down. But the results do differ as shown in figure 12.

5   Open Problems

Aside from the problems of detail mentioned above, I see the following open problems: In the framework of NWY and TUL, it is necessary to use the level density of states with fixed $\mathcal{J}$, π and exciton number n including the effect of residual interaction. So far, the main work done on calculating such realistic partial level densities has come form the group of S. Yoshida [15]. Figure 13 shows for the nucleus Pb the difference between shell-model level densities and those including residual interactions. It is intuitively obvious that the large differences are relevant for precompound reactions. More work is needed.

Gamma decay in precompound reactions has - but for [16, 17] – not received the attention it deserves. Preequilibrium in heavy-ion induced reactions is an important topic [18] and should be related to the general framework reviewed above.

As E increases, we expect the semiclassical or classical approximation to become valid. When exactly does this happen?

6   Conclusions

During the last 10 years, great progress has been made in the conception, formulation and implementation of a quantum-statistical approach to nuclear reaction theory. During this time, FKK has emerged as a tool with predictive power. In the MSD domain, TUL is a viable competitor. Very recently, NWY has been implemented. To ascertain the validity of various approximations, to clarify substantial differences in detail, and establish their relative merits, further comparisons and applications of all three approaches are needed. Given the present pace of development, I am confident that in a few years' time, a statistical reaction theory based on first principles will be established. It is desirable to have an approach which, while based on reliable theory, is sufficiently simple to allow for speedy calculations.

7   Acknowledgement

The author is grateful to the authors of [10, 11, 13] for communications which form part of this review.

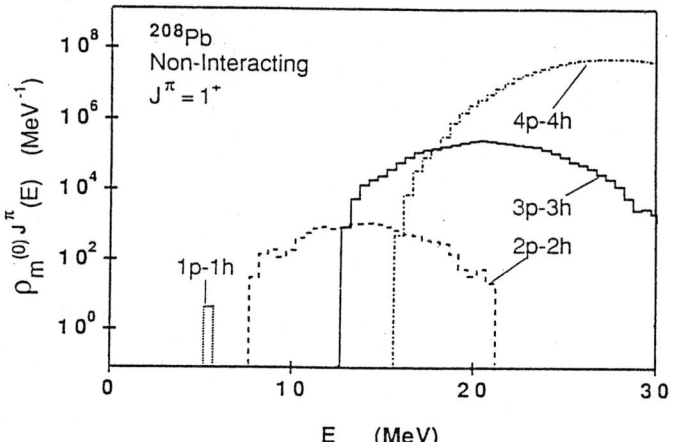

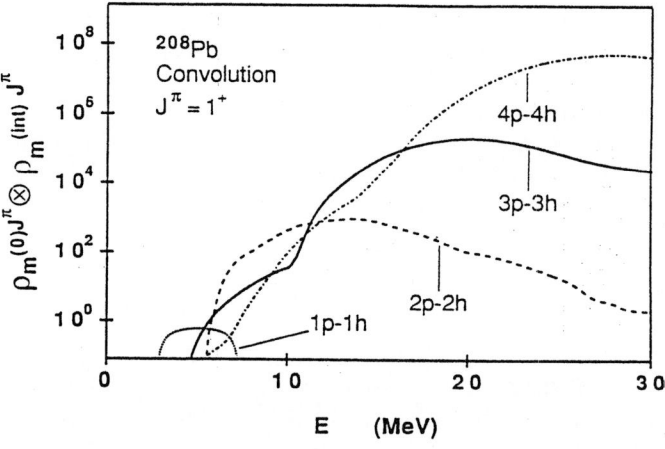

Figure 13: Comparison of shell-model level densities (upper part) and level densities taking into account the residual interaction (lower part). Taken from [15].

References

[1] C. H. Lewenkopf and H. A. Weidenmüller Ann. Phys. (N. Y.) (in press)

[2] J. J. M. Verbaarschot, H. A. Weidenmüller and M. R. Zirnbauer, Phys. Rep. 129 (1985) 367.

[3] J. J. M. Verbaarschot, Ann. Phys. (N. Y.) 168 (1986) 368.

[4] H. Feshbach, A. K. Kerman and S. Koonin, Ann. Phys. (N. Y.) 125 (1980) 429.

[5] T. Tamura, T. Udagawa and H. Lenske, Phys. Rev. C 26 (1982) 379.

[6] H. Nishioka, J. J. M. Verbaarschot, H. A. Weidenmüller, S. Yoshida, Ann. Phys. (N. Y.) 172 (1986) 67; H. Nishioka, H. A. Weidenmüller, S. Yoshida, Ann. Phys. (N. Y.) 183 (1988) 166; ibid 193 (1989) 195; Z. Phys. A 336 (1990) 197; Phys. Lett. B 203 (1988) 1

[7] H. Kalka, Z. Phys. A (in press) and Contribution O 72 to this Conference

[8] R. Bonetti, M. Camnasio, L. C. Milazzo and P. E. Hodgson, Phys. Rev. C 24 (1981) 71.

[9] P. E. Hodgson, invited talk at the 1990 IAEA meeting in Vienna.

[10] A. A. Cowley et al. Phys. Rev. C 43, (1991) 678 and A. A. Cowley, private communication.

[11] J. Akkermans and A. Koning, private communication, and A. Koning and J. Akkermans, Contribution O 69 to this Conference.

[12] M. Herman, G. Reffo and H. A. Weidenmüller submitted to Nucl. Phys. A.

[13] M. Herman, G. Reffo, private communication.

[14] A. Koning and M. Akkermans, Ann. Phys. (N. Y.) 208 (1991) 216

[15] K. Sato, Y. Takahashi and S. Yoshida, Z. Phys. A (in press).

[16] P. Ohlozinsky and M. B. Chatwick, Phys. Rev. C 42 (1990) 1652.

[17] A. Höring, private communication.

[18] J. P. Coffin et al. preprint 1991

STATUS OF THE JEF AND EFF PROJECTS

C. Nordborg
NEA Data Bank, 91191 GIF-SUR-YVETTE CEDEX, France

H. Gruppelaar
ECN, Postbus 1, 1755 ZG PETTEN, Holland

M. Salvatores
CEN Cadarache, 13108 ST. PAUL LEZ DURANCE, France

Abstract: Following the positive experience gained with the first version of the JEF (Joint Evaluated File) library, which was distributed to NEA Data Bank member countries in 1985, a significantly improved version of the library, JEF-2.0 (in ENDF-6 format), was distributed for testing at the end of February 1990. It is planned to compile, in the middle of 1991, following intensive benchmark testing, a version of the JEF-2 library for general use.

The EFF (European Fusion File) project and the associated EAF (European Activation File) project are sponsored by the European Community's Fusion Technology Programme. The programme of the EFF-project is in its second phase, after the successful completion of the EFF-1 data file. This phase has been defined from 1989 to 1991 and the emphasis is on the improvement of a shielding data base. The EFF-2 library will be ready at the end of 1991 and contain new evaluations for most of the important fusion reactor materials. A follow-up programme, including benchmark testing has been defined.

The creation of a joint JEF-EFF data file has been agreed. The file will be compiled during 1992.

Background

The Joint Evaluation File (JEF) project was started in October 1981 in order to coordinate the evaluation efforts within the NEA Data Bank member countries (Western Europe and Japan). A first version of the data library was distributed in 1985. The library was extensively benchmark tested [ref. 1], and the results showed very good agreement with experiments for different aspects of reactor physics calculations. Nevertheless, it was felt that significant improvements could be achieved, and it was decided to create an improved version in ENDF-6 format of the JEF-1 library. While starting the re-evaluation work for JEF-2, it was decided to release the JEF-1 data also to users outside the NEA Data Bank community. This decision affects also group cross section libraries derived from JEF-1.

The European Fusion File (EFF) project and the associated European Activation File Project (EAF) are sponsored by the European Community's Fusion Technology Programme. The first programme is directed to the short-term needs of the NET team, which designs the Next European Torus, whereas the second programme is directed to long-term needs in the development of a Fusion Demonstration Reactor, in particular in connection with the study of low activation materials. Various European laboratories participate in these projects.

The programme of the EFF-project is in its second phase, after the successful completion of the EFF-1 data file. The second phase programme has been defined from 1989 to 1991. The emphasis of the first phase was on the improvement of the tritium breeding and neutron multiplication cross-sections, whereas the second phase emphasizes the improvement of a shielding data base. In practice the EFF-1 and EFF-2 projects aim to supplement the JEF-1 and JEF-2 data files, respectively, with high-energy data relevant for fusion applications. Therefore the two projects are closely related.

Joint Evaluated File (JEF)

Content of the JEF-2 library

General Purpose Files

The JEF-2 general purpose tape contains 312 isotopes. A few of these isotopes have been taken over from the JEF-1 library, but the most important have been either updated or re-evaluated. Special attention has been given to the structural materials, the fission products and the actinide data. The stable isotopes of Cr, Fe and Ni have, for example, all been completely re-evaluated as well as the major actinides: ^{235}U, ^{238}U and ^{239}Pu. The unresolved data for ^{241}Pu and ^{242m}Am have been revised. Covariance data will be included for the most important isotopes.

The resolved resonance region of a large number of the fission products have been updated with recent information and improvements to the particle production reactions have been included. A few fission product isotopes (^{99}Tc, ^{129}I, ^{135}Cs and ^{107}Pd) have been selected for particular investigation in view of the interest in for example transmutation applications. An examination, covering both theory and experiments, of the inelastic cross section of the even-even fission products has been performed, but the application of the results to the JEF-2 data will be performed at a later stage.

Thermal Scattering Law Data

The JEF-2 thermal scattering law data (S(α,β)) tape will contain:

H (H_2O)   8 temperatures from 293.6 to 623.6 K
D (D_2O)   8 temperatures from 293.6 to 673.6 K
Graphite   11 temperatures from 293.6 to 3000. K
Be   8 temperatures from 293.6 to 1200. K
Polyethylene   2 temperatures 293.6 and 350. K

Pointwise data generated with NJOY-89.62+ will also be available for the following cold moderators:

para-Hydrogen   (H in para-H_2) T=20.38 K (liquid)
ortho-Hydrogen   (H in ortho-H_2) T=20.38 K (liquid)
para-Deuterium   (D in para-D_2) T=23.65 K (liquid)
ortho-Deuterium   (D in ortho-D_2) T=23.65 K (liquid)

The corresponding S(α,β) data for these cold moderators have 3-digit exponents. As this was not foreseen in the ENDF-6 format and as the processing can only be done on 64 bit computers, the pointwise files should be preferred.

Radioactive Decay and Fission Yield Data

The JEF-2 Radioactive Decay Data File will contain all necessary decay information, such as half-life, average decay energies, spectra etc., for about 2340 unstable isotopes. This is almost double as many isotopes as was present in JEF-1. The JEF-2 library has been updated and complemented with new information, mainly from the Evaluated Nuclear Structure Data File (ENSDF) compiled at the National Nuclear Data Center at Brookhaven, USA. This information has also been supplemented by recent experimental information, especially from G. Rudstam et.al., Studsvik, Sweden.

The Fission Yield Library for JEF-2 is completely new and is based on a recent evaluation by M. James et.al. [ref. 2]. It contains data for 39 different fissioning systems, and both independent and cumulative yields are given.

Photon Interaction Data

No special effort was devoted, within the JEF project, to the photon interaction data. The most recent data from D.E. Cullen et.al. [ref. 3] was adopted for the JEF-2 library. The tape contains evaluations of cross sections (MF=23) and form factors (MF=27) for 100 elements.

Status of the JEF-2 library

A preliminary version, called JEF-2.0, of the general purpose files was sent out at the end of February 1990 to laboratories involved in the benchmark testing of the library. Revised data for ^{55}Mn, ^{235}U, ^{239}Pu and the major structural materials were issued in October 1990. This library was named JEF-2.1. Following the initial processing of the data for benchmark testing, a few minor problems, mainly format problems, have been identified.

The Radioactive Decay Data and the Fission Yield Data were distributed for testing in the middle of June 1990. A final version of the Fission Yield library was issued in January 1991. The Radioactive Decay Data library would be ready in the early autumn of 1991, following the testing of an intermediate version of the library to be issued at the end of May 1991.

The work on the Thermal Scattering Law data (S(α,β) data) for the above mentioned moderators was terminated in the beginning of April 1991. Pointwise data have been processed using the computer code NJOY-89.62 with additional updates from IKE Stuttgart.

It is planned to have a version of the complete library, called JEF-2.2, available for general use in the Data Bank member countries in the autumn of 1991.

First results from the benchmark testing of JEF-2

The benchmark testing of the first version of the JEF-2 library has been started at several laboratories directly involved in the JEF project and the first preliminary result have already been discussed at a recent meeting of the JEF group.

Thermal reactor data

The integral validation for thermal neutron reactors of the JEF-2 major actinides (^{235}U, ^{238}U and ^{239}Pu) has been performed at Saclay, France, by H. Tellier et. al. [ref. 4]. The tendency research method was used and the comparison with calculations was performed for 66 different clean integral experiments.

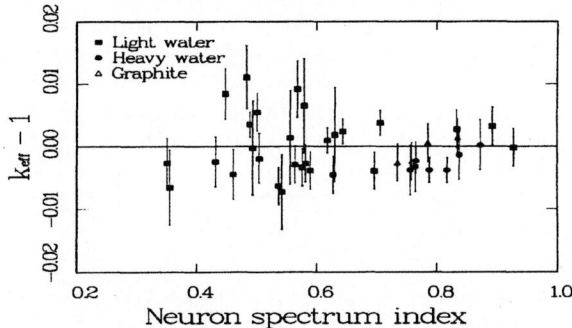

Figure 1 Residual deviations after tendency research for multiplying media which only contain uranium [ref. 4].

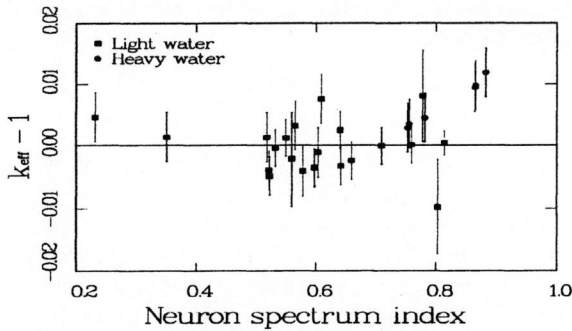

Figure 2 Residual deviations after tendency research for multiplying media which contain plutonium or an uranium and plutonium mixture [ref. 4].

The results could be summarised as follows:

- The moderators would need a small decrease of the light water migration area of about 1%. This modification affects only the leakage of small critical assemblies and will be of no importance for large size power reactors.

- No modifications are at present proposed for the fissile nuclei, ^{235}U and ^{239}Pu. The thermal data for these isotopes agree very well with the recommendations by Divadeenam [ref. 5] and Axton [ref. 6].

- The effective capture integral of ^{238}U may need to be decreased by about 0.25 barn. For a typical light water reactor this represents about 1.3 per-cent.

The above mentioned results has been fed back to the responsible evaluators for their consideration, and will be further checked on integral experiments..

Fast reactor data

Two independent set of results for the calculation of some fast critical assemblies were presented at the last meeting of the JEF working group in December 1990. E. Fort, CEN Cadarache, France and M. Caro, PSI, Switzerland, had calculated, using the preliminary JEF-2.1 data, the JEZEBEL, GODIVA, FLATTOP and BIG TEN cores and the results agreed well with each other and with the experimental results. The calculations recently performed at PSI will be presented at this conference [ref. 7].

Only limited new work has been devoted to the higher actinides of JEF-2. Most of the isotopes have been taken over from the JEF-1 library, as these isotopes, such as ^{237}Np, ^{241}Am and ^{243}Am, have shown excellent performance in both thermal and fast energy spectrum systems. Table 2 below shows the results of calculations using the ^{237}Np capture and (n,2n) data in the analysis of ^{237}Np sample irradiation in PHENIX [ref. 8].

Reaction	C/E
Capture	0.90 ± 0.05
(n,2n)	1.19 ± 0.15

Table 1 C/E values with JEF-2 data on ^{237}Np reaction rates, obtained from a ^{237}Np sample analysis after irradiation in the PHENIX core.

Fission products

The JEF-1 fission product data were thoroughly tested by H.J. Janssen et.al. [ref. 9]. This effort will not be repeated for the JEF-2 data, as most of the isotopes were taken over from the JEF-1 file with additional updates of the resonance region. A document containing a comparison between JEF-2 and other recent evaluated libraries for some simple integral data, such as thermal values, resonance integrals etc., will be issued.

A paper, presented at this conference, gives indications on the integral test of the fission product data in the thermal range [ref. 10].

Fission Yields and Decay Data

A provisional version of the JEF-2 Decay Data File was issued in June 1990 for testing purposes. A. Tobias, Nuclear Electric, UK and G. Rudstam, Studsvik, Sweden performed decay heat calculations using these data and concluded that the new data yielded improved agreement, relative to JEF-1, with measured results at most cooling times. The gamma heat part was particularly improved. A few isotopes in need of re-evaluation was identified. The delayed neutron yields have been examined by J. Blachot et.al. It is expected that the latest revisions to the JEF-2 Decay Data File will further improve the calculated results.

European Fusion File (EFF)

A break-down of the 1989-1991 EFF-2 programme, which also contains some benchmarking and test calculations to facilitate its use by the designers of NET, is as follows:

1. Completion of the working libraries derived from EFF-1, in particular with respect to problems with kerma, DDX, gas-production, MCNP library and benchmarking (Pb, ^{7}Li, Be).

2. Evaluate Fe, Cr, Ni cross-sections with emphasis on DDX, photon production and covariance data.

3. Create an EFF-2.0 starter file as an extension to EFF-1, with new evaluations for ^{7}Li, Be, Al, Fe, Cr, Ni, and Pb, process these data into the VITAMIN-J structure and perform benchmark calculations.

4. Create the EFF-2.1 data file with new or updated cross-sections for Si, Ti, V, Mn, Co, Cu, Zr, Nb, Mo, Ba, Ta, In and other materials.

5. Create the EFF-2.2 data file with cross sections for C, N, O, Mg, S, P, Ca, Sn, Re, and W.

6. Perform shielding and sensitivity calculations and estimate the effect of uncertainties in DDX calculations.

7. Perform benchmark and test calculations related to shielding, including the development of tools.

The EAF-programme was started in 1989, after an informal collaboration between Petten, Ispra and Harwell, resulting in UK and ECN data files with essentially the same contents. Early this year the first version of the European Activation File (EAF-2) has been released. This file is recommended for fusion reactor activation and transmutation calculations. The file contains all stable and unstable targets with half-lives longer than 0.5 day. Cross-sections leading to isomeric states are treated separately and further those isomeric states living longer than 0.5 day are included as targets. This version contains now about 11000 reactions on 667 target nuclides and is described in a paper at this conference [ref. 11].

Status of the EFF-2 project

Derived files from EFF-1, processing problems

Some errors were detected in the GEFF-1 multi-group library based upon EFF-1 in the VITAMIN-J structure (only in the photon production data). However, meanwhile there is a more extended library based upon JEF-1/EFF-1, which is PSI's MATXS library, in the same VITAMIN-J structure. Since this library was produced with an older version of NJOY, the Pb cross-sections were treated after a transformation of MF6 into MF4 and MF5, loosing the energy-angle coupling. Apart from this the library is very well applicable to calculate transport in fusion reactors blankets. A revision of the Pb cross-sections by calculations with the newest version of NJOY-89 is planned for this year. It is noted that the DDX calculations in GEFF-1 for lead were performed with the Petten GROUPXS code. Meanwhile an option to treat MF6 has been included in NJOY-89 as well. This option was tested for Pb of EFF-1 and ^{58}Ni of EFF-2 with very good results, reported at the NJOY workshop at the NEA Data Bank, 1989.

With respect to kerma calculations it was found that negative kerma values resulted from the expressions in NJOY, due to energy balance problems in some evaluations. A further problem was that users previously were adopting kerma values from MACKLIB-IV, which includes kermas due to radioactive decay as well, in contrast to the NJOY results. In fact the MACKLIB values overestimate the heating if one is interested in very short time intervals. This was pointed out both at PSI and at Bologna. In Bologna an EFF-1 based kerma library was made with the help of the MACK code with and without the inclusion of radioactive decay (up to 3 hours). This kerma library is recommended for NET calculations. The library of Bologna also contains other response functions such as gas-production data and DPA data (calculated with NJOY).

For EFF-2 evaluations the files contain revised photon production data and in many cases distributions of all emitted particles, including the recoil nuclei. In that case NJOY calculates the kerma very accurately, but still contributions from short-lived decay products should be added. Otherwise, the user needs to perform an additional calculation with a radioactive inventory code, such as FISPACT, recently developed at Harwell.

Monte Carlo libraries based upon EFF-1 have been produced at PSI and at ENEA-Frascati. Some problems were encountered for Be and Pb, due to format problems. For Be the problems were easily solved: here the LANL evaluation was adopted in EFF-1; for this evaluation the scattering matrices need to be multiplied by a factor of 2. For Pb the current version of NJOY cannot treat the MF6 format. It is understood that NJOY-89.62 has an option in the ACER module to convert the MF6 data into the format in which the Kalbach-Mann systematics is adopted. With this method it is possible to create MCNP libraries that include energy-angle coupling for continuum reactions.

Updates for Pb are forthcoming.

^{7}Li and ^{9}Be

The re-evaluation of ^{7}Li and ^{9}Be cross-sections is restricted to the energy-angle distributions of emitted neutrons. For the energy-angle integrated cross-sections almost exclusively the existing ENDF/B-VI evaluations are used. In fact there is an agreement with LANL to supply EFF-2 as an update for ENDF/B-VI.

This work is performed at Birmingham University by G.M. Field and D. Beynon, with experimental support from CBNM and formatting support from ECN-Petten. A new feature of this evaluation is the use of sophisticated kinematics with cm angular distributions fitted to experimental data where available. The neutron emission data are stored in MF6 of ENDF/B-VI, with an interesting application of the LIP parameter to distinguish between various physical processes.

The work for ^{7}Li and ^{9}Be has been completed and the files have been checked at ECN-Petten and have been combined with the energy-angle integrated ENDF/B-VI evaluated data. The comparison with experimental data is excellent. A problem is that there are no distributions of other particles than neutrons and that covariance data for the MF6 data are not given.

A review of the work for ^{7}Li and ^{9}Be was presented at Chengdu [ref. 12]. A paper on recent developments on ^{9}Be is given at this conference [ref. 13].

Fe, Cr, Ni

The work for these materials is supplementary to the JEF-2 evaluations for the stable isotopes of Fe, Cr, Ni. In fact, in the EFF-2 project the high-energy regions (above 1 MeV) of the JEF-2 evaluations for ^{56}Fe, ^{52}Cr, ^{58}Ni and ^{60}Ni are replaced by recent evaluations of Uhl et.al. [ref. 14]. These evaluations contain the lumped quantity MT10, describing all continuous particle and recoil emission. The EFF-2 evaluations also contain the individual reaction cross-sections, though without secondary distributions, as these are given in the lumped quantity MT10. This approach makes it very easy to create a group constant library for transport calculations. With respect to kerma all ingredients are on the file to perform these calculations. The present status is that all major structural isotopes (^{52}Cr, ^{56}Fe, ^{58}Ni and ^{60}Ni) are ready, including photon-production data. The other isotopes are taken from the JEF-2 data file.

Covariance data have been evaluated recently. The work was performed mainly at Vienna (see paper by Tagesen and Vonach [ref. 15]), but also at KfK and Petten. An international cooperation has been set up on the covariance problem for Fe and the results of this cooperation will be used to improve the quality of the covariance data gradually. The first priority was to evaluate uncertainties in the high-energy range, where the needs for these data are the highest. This has been performed initially by inspecting the dispersion of cross-sections in three new evaluations: ENDF/B-VI, JENDL-3 and EFF-2 [ref. 15] and by evaluating uncertainties for the strongest resonances

(KfK). The 14 MeV experimental data have also been re-evaluated [ref. 16], because these data are quite important in relatively thin coil shielding. Further methods, to determine uncertainties in secondary energy and angle distributions will match the capabilities of presently available user-oriented codes like SENSIT or SUSD.

Sensitivity and uncertainty calculations

At Petten, sensitivity calculations have been performed [ref. 17] to study the nuclear data sensitivity to shielding of super-conducting coils. Using the current NET design for the inboard shielding, mainly consisting of stainless steel and (borated) water, 1D neutron and photon transport calculations were performed using the EFF-1 data from the MATXS library. This library was connected to the SENSIT code to study the sensitivities to nuclear data. The nuclear heating in the super-conducting coils was taken as a representative quantity for the heat production.

It appeared that the sensitivity contributions of elastic and inelastic scattering cross-sections to the heating are about equal. The contribution from the 14 MeV peak is significant and there is very little contribution below a few hundred keV. This and other information was very useful to set the priorities in the calculation of covariance information. It appeared that due to the fact that the loss and gain terms of both elastic and inelastic scattering largely cancel, the sensitivities are rather small. With the available covariance information for EFF-1 the uncertainty in the heating was estimated. The uncertainty of the contribution of Fe was in the order of 10%. Similar calculations, recently performed at CEA-Cadarache [ref. 18] yield larger uncertainties. Further work will be necessary to solve this discrepancy. However, these numbers do not take into account uncertainties in the energy and angular distributions, nor the correlations between elastic and inelastic cross-sections.

A first estimate of the effect of angular distribution uncertainties was made using the uncertainties in the Legendre coefficients of elastic scattering, evaluated by Tagesen and Vonach [ref. 15]. The results indicate a contribution of 5 % due to elastic scattering above 0.8 MeV, which might increase if the uncertainties at low energies are added. The effect of inelastic scattering angular distributions can also be significant, more than that of secondary energy distributions [ref. 19].

Further work is performed to study uncertainties due to the neutronics modelling. At KfK rigorous 1D- and 2D-methods for studying the effect of large an-isotropic cross-sections has been developed. These methods will be used to study model errors from neutronics calculations. There is special emphasis on the study of the effect of coupled energy-angle calculations.

Furthermore, some realistic Fe shielding benchmarks are calculated to investigate possible other sources of uncertainties, e.g. due to the treatment of self shielding. Both KfK and CEA-Cadarache are involved in this activity.

Evaluations for other materials

New evaluations for Al and Si have been completed at ENEA-Bologna. An important step to store the data on a file in ENDF-6 format is the development of a code that reformats the output of the Bologna code system. This code has been used for Al and will also be applied for Si.

With respect to Pb some minor modifications are made by adding (n,p) and (n,α) reactions and by re-evaluating the (n,γ) cross-section. This work has been completed recently. The additions still have to be made in the file. Some small modifications in the (n,2n) cross-sections have already been applied.

A large part of the requirements for other materials is fulfilled by adopting evaluations from elsewhere, if the JEF-2 data were insufficient at high energy.

Benchmark and test calculations

Benchmark calculations for Pb and Be have been performed at KfK [ref. 20]. For Pb the newest measurements from Dresden, Osaka and Kurchatov have been analysed. The results are in agreement with the EFF-1 evaluations, if the (n,2n) cross-section is increased to about 2.2 barn at 14.5 MeV. This value is in agreement with a recent high precision evaluation of Vonach, based upon differential data. The EFF-2 evaluation for lead has been adjusted to this value.

Also for Be several integral experiments were used to check the EFF-1 evaluation. The Be evaluation in EFF-1 (obtained from LANL) gave a very good performance for the total multiplication and satisfactory results for the leakage spectra from these experiments. More work is scheduled at KfK, after the analysis of their KANT experiment. The forthcoming EFF-2 evaluation will be included.

In view of the emphasis of the NET project on shielding it has been decided to analyse some shielding benchmarks at KfK and CEA-Cadarache. At KfK the KANT facility will be used to study Fe as well. Finally, a new 14-MeV facility will be used at ENEA-Frascati for benchmark testing of the NET shielding.

Concluding remarks

The EFF-2 project is in good shape. The evaluations for ^{7}Li, ^{9}Be, Al, Si, Fe, Cr, Ni, and Pb have been completed, although some minor corrections may still be necessary. A detailed scheme for the completion of the EFF-2 project has been established and it is expected that the project will meet the dead line of December 31, 1991. By that time the EFF-2 data file will consist of new evaluations for the most important fusion reactor materials, partly from EFF evaluators and partly from elsewhere (selected from JEF-2 or other recent evaluations). Covariance data will be available for the most important materials only.

The close cooperation between users and evaluators in the framework of the NET project is quite important for both parties. Therefore a nuclear data activity, integrated in the neutronics programme of NET/ITER, is important also in the years after 1991. Currently a follow-up programme is defined. There are still many problems to be solved. First of all the EFF-2 file needs to be benchmark tested and the feedback of integral data experiments should be used to improve the data file. It is expected that still quite some work is needed to obtain reliable photon production data that are consistent with the neutron data. Also some work is needed to include distributions of all emitted particles and recoil nuclei on the files in order to enable accurate kerma assessments. A major activity will be the systematic update of covariance information, not only for smooth cross-sections, but also for resonance parameters and for energy-angle distributions of neutrons and photons. A covariance calculational benchmark will be defined to gain experience in the use of covariance data in shielding problems. Finally, there is a very large task to update the activation and transmutation library EAF with uncertainties. It should be stressed that the amount of work is quite large and therefore active involvement of users is required to set priorities.

Cooperation between the JEF and EFF projects

The present cooperation between the JEF and EFF projects has been discussed and it was decided to create a common library to be used both for fission and fusion purposes. A starter file of this common library, for which the acronym J/EFF is proposed, will be prepared during 1992, from an agreed choice of isotopes from both libraries. The two projects would still keep their identities and work in parallel and close cooperation on the improvement of the data. It has for example been decided that future releases of EFF-2 will only contain materials that are different from JEF-2 and that the meetings of the projects will be combined in the future.

References

1. M. Salvatores and C. Nordborg, "The JEF Evaluated Data File Validation of the Version 1 and Version 2 Development Status", Proceedings of the International Reactor Physics Conference, Jackson Hole, USA, Vol. 1, Page 257, September 1988.

2. M.F. James, R.W. Mills and D.R. Weaver," A New Evaluation of Fission Product Yields and the Production of a New Library (UKFY2) of Independent and Cumulative Yields", AEA-TRS-1015, 1990.

3. D.E. Cullen et.al., UCRL-50400, Vol. 6, Rev. 4, 1990.

4. H. Tellier, C. van der Gucht and J. Vanuxeem, "Integral Validation of the JEF-2 Major Actinides for Thermal Neutron Reactors", Contribution to the ANS International Topical Meeting on Advances in Mathematics, Computations and Reactor Physics in Pittsburgh, USA, 28th April- 2nd May 1991.

5. M. Divadeenam and J.R. Stehn, Annals of Nuclear Energy, 11, 375 (1984).

6. E.J. Axton, Geel Report GE/PH/01/86 (1986).

7. M. Caro and S. Pelloni, "Benchmark Test of JEF-2 and ENDF/B-VI Evaluations by Calculating some Criticals", Contribution to the this conference.

8. F. Cleri and A. D'Angelo, Private communication.

9. H.J. Janssen et.al., "Integral Data Test of JEF-1 Fission Product Cross Sections", ECN-175 (1985).

10. J. Mondot et.al., "Validation of fission product capture cross sections by the analysis of thermal and epithermal integral experiments", Contribution to this conference.

11. J. Kopecky, H. Gruppelaar and R. Forrest, "European Activation File for Fusion", Contribution to this conference.

12. H. Gruppelaar, "Double Differential Neutron Emission Cross-Sections for ^{7}Li, ^{9}Be and Pb", IAEA Adv. Gr. Mtg. Nuclear Data for Neutron Multiplication in Fusion Reactors, Chengdu, China, November 1990.

13. G.M. Field, T.D. Beynon and H. Gruppelaar, "Modelling inelastic emission cross sections for ^{9}Be(n,2n)", Contribution to this conference.

14. M. Uhl et.al., "The EFF-2 evaluation for the reaction systems n + ^{52}Cr, ^{56}Fe, ^{58}Ni and ^{60}Ni", Contribution to this conference.

15. S. Tagesen and H. Vonach, "Uncertainty estimates for the EFF-files for ^{52}Cr, ^{56}Fe, ^{58}Ni and ^{60}Ni", Contribution to this conference.

16. H. Vonach, S. Tagesen and M. Wagner, "Evaluation of 14 MeV cross sections for the main isotopes of the structural materials Cr, Fe and Ni", Contribution to this conference.

17. A. Hogenbirk and H. Gruppelaar, "Sensitivity and Uncertainty Analysis of the Nuclear Heating in the Coils of a Fusion Reactor", Proc. 16th SOFT Meeting, London, UK, September 1990, ECN-C--90-034, R--90-007 (1990).

18. A. Santamarina and H.J. Smith, CEA Cadarache, Private Communication (1990).

19. A. Hogenbirk, "Energy Self-Shielding and SED/SAD Effects in Sensitivity calculations of a NET shielding blanket", Sec. Int. Symp. on Fusion Nuclear Technology, Karlsruhe, Germany, June 1991.

20. U. Fischer, A. Schwenk-Ferrero, E. Wiegners, "Benchmark Analysis on the 14 MeV Neutron Transport in Beryllium", IAEA Adv. Gr. Mtg. Nuclear Data for Neutron Multiplication in Fusion Reactors", Chengdu, China, November 1990.

EVALUATED NUCLEAR DATA FILE
ENDF/B–VI

C.L. Dunford

National Nuclear Data Center
Brookhaven National Laboratory
Upton, NY 11973, USA

Abstract: For the past 25 years, the United States Department of Energy has sponsored a cooperative program among its laboratories, contractors and university research programs to produce an evaluated nuclear data library which would be application independent and universally accepted. The product of this cooperative activity is the ENDF/B evaluated nuclear data file. After approximately eight years of development, a new version of the data file, ENDF/B-VI has been released. The essential features of this evaluated data library are described in this paper.

(evaluated nuclear data, ENDF/B, nuclear reactions)

Introduction

In 1966, the United States Atomic Energy Commission, predecessor to the US Department of Energy, initiated a program to provide a nuclear data base which could be used in its research and development activities. This data base was intended to replace the many individual libraries then in use and provide a common reference point for evaluating proposals and research results. This data base was to be constructed from the results of nuclear data evaluation activities by contractors and coordinated through the Cross Section Evaluation Working Group (CSEWG) organized by the National Nuclear Data Center at BNL. The Evaluated Nuclear Data File data system which evolved in the intervening period consists of evaluated nuclear data in a format designed for computer processing along with the processing programs which can serve as a reference standard. The evaluated data added to the file are based on the best microscopic measurements and theoretical models available and validated against well tested benchmark experiments using the most advance calculational models available. The format and the cooperative approach to the production of an evaluated nuclear data library have served as models for similar activities throughout the world.

As a result of this coordinated effort, six versions of the ENDF/B data library have been completed. The most recent version, ENDF/B-V, was released in 1979 with an update in 1982. For the first time in its history, the distribution of significant parts of the ENDF/B data library was restricted to users from the United States and AECL, Chalk River, Canada by the sponsoring agency. Requests for release of specific evaluations were treated on a case by case basis. This decision to restrict ENDF/B distribution by the sponsoring agency did not have much support within CSEWG and caused much displeasure among the world wide community which used ENDF/B data. While the restriction on information flow generally has a strong negative effect on base technology development, in this case there was one positive result, namely the development of strong independent evaluation projects outside the United States.

The primary sponsorship of CSEWG and the ENDF evaluation activity changed in the early 1980's from reactor development projects to basic research programs. This change in sponsorship had two important manifestations, namely a drastic reduction in funding and a committment to remove restrictions on the distribution of the next version of the ENDF/B data file. The reduction in resources has led to increased interest on the part of CSEWG in developing non-CSEWG contributions to its activities.

In 1989 and 1990, a new version of the ENDF/B library, ENDF/B-VI, was released. Many of the evaluations for neutron interactions with important nuclides have been revised, most completely reevaluated. Separate Activation and Dosimetry libraries are no longer available. For consistency, they are derived from the full evaluations. The remaining portions of ENDF/B-VI,

The submitted manuscript has been authored under Contract No. DE-AC02-76CH00016 with the U.S. Department of Energy. Accordingly, the U.S. Government retains a nonexclusive, royalty-free license to publish or reproduce the published form of this contribution, or allow others to do so, for U.S. Government purposes.

namely the decay data and the neutron fission product yields, will be completed and released this year. We also expect to have a small charged particle library for light mass materials available in the near future. CSEWG laboratories supplying evaluations are

ANL	Argonne National Laboratory, Illinois
ANL-West	Argonne National Laboratory, Idaho
BNL	Brookhaven National Laboratory
HEDL	Hanford Engineering Development Laboratory
INEL	Idaho Nuclear Engineering Laboratory
LANL	Los Alamos National Laboratory
LLNL	Lawrence Livermore National Laboratory
ORNL	Oak Ridge National Laboratory

Goals for ENDF/B-VI

The initial planning for ENDF/B-VI began in 1980 with a conference on evaluation methodology[1], held at Brookhaven National Laboratory in September of that year. This conference attracted international participation with the aim of assessing the current state of the art of nuclear data evaluation and discussing developments which would affect the next version of ENDF/B and other evaluated nuclear data libraries.

Two major goals were adopted at the earliest planning stages of the evaluation effort. These were **the production of a self-consistent set of evaluated neutron standard cross sections and improved data at energies important to fusion reactor design**. In addition, CSEWG sought to improve the secondary energy balance in the important material evaluations, leading to an emphasis on isotopic evaluations in place of natural element evaluations where possible. It was recognized early in the evaluation cycle that the resonance region evaluations for the important fissile and fertile materials were out-of-date and in need of an intensive evaluation effort. However resources could not be identified immediately for the task. Only at a later stage were the resources found to accomplish this task using expertise from US and non-US evaluators.

It was recognized that the resources within CSEWG were insufficient to perform all the evaluations required. Work had to be deferred in some areas where we knew that significant improvements could be made. The most important of these areas were evaluations for fission product nuclides and for the trans-plutonium nuclides. Only limited improvements in these areas were made, in part by US evaluation efforts and in part by incorporating non-US evaluations in ENDF/B-VI. No separate evaluations of dosimetry and activation cross sections were made.

ENDF Formats

Traditionally, the ENDF data file format was the sole responsibility of the Methods and Formats Committee of CSEWG. However, increased use of the ENDF format by evaluators around the world in the early eighties led to its de-facto adoption as the international standard for evaluated nuclear data. In recognition of this fact, CSEWG has actively sought to incorporate needs from a wider community in its format determination process. In 1984, the IAEA sponsored a meeting[2] to discuss the international community's requirements for ENDF format improvements and revisions. Several suggestions which were proposed at this meeting were included in the latest ENDF format specifications[3], ENDF-6.

In order to represent charged-particle reactions in the ENDF format, information identifying not only the target material but also the incident particle is required. Such information was added to "File 1" and the concept of a sublibrary was developed. In general, a sublibrary represented all evaluations with a common incident particle. Special cases of this concept included a neutron thermal scattering law sublibrary, a decay data sublibrary and a neutron fission product yield sublibrary. Unique identification of a library, its version and its format were also added to the "File 1" contents.

Major new resonance region formats were adopted to permit use of R-function and Generalized R-matrix representations. The Reich-Moore format was approved for use in structural material evaluations. A flag was added to signal whether the unresolved resonance region parameters should be used to calculate the cross section in that energy region, or only for self-shielding calculations.

A new file, "File 6", has been added to the format to permit a complete description of secondary particle energy-angle distributions. This new format is important at energies above a few MeV for fusion applications. All outgoing particle distributions can be described, including those for recoil nuclei. In addition to improved transport calculations from the correlated energy-angle distributions, kerma calculations can be done directly with information given in file 6.

Beginning with ENDF/B-IV, the ENDF/B library included uncertainties and covariance matrices. The formats for this information have been constantly improved and broadened to cover more data types. The covariance formats for resonance parameter representation were improved for ENDF-6 and new formats for secondary neutron angular and energy distributions and for isomer production cross sections added. Very late in the format development came a proposal to create a more general format, "File 30", which could be used to include covariance data resulting from model parameterization. While approved, this format has not yet been implemented in ENDF/B-VI.

Standards Evaluation

Considerable resources were devoted to the simultaneous evaluation of the standard neutron reaction cross sections[4]. Several heavy element reaction cross sections were included in this evaluation because of their close link to the standards via ratio measurements. Also included in this activity were the thermal neutron cross sections for the important fissile and fertile nuclides. The evaluation process was a joint effort of ANL-West, LANL, ORNL and NIST under the auspices of the Standards subcommittee of CSEWG. Extensive non-overlapping experimental data bases were developed at LANL and ANL-West. LANL concentrated on R-matrix fits to the ^{6}Li and ^{10}B reactions, while ANL-West handled the remainder of the heavier standards plus some ^{6}Li and ^{10}B data with a generalized least squares simultaneous evaluation. The results from the two separate evaluations were merged at ORNL using both the data and covariance matrices in a comprehensive combination procedure.

The final results have been carefully reviewed by both CSEWG and the JEF Standards Committee. The results represent a considerable and perhaps definitive effort in this area. The major criticisms have been mainly due to a) the very small uncertainties which result from the analysis and b) the inconsistency in some of the results for ^{10}B, which probably result from the relatively poor experimental data base for that standard.

The cross sections for the neutron standard reactions resulting from this analysis were included in the ENDF/B-VI evaluations. However, representation of the covariance matrices can be best accomplished through use of the new generalized covariance format which has not yet been implemented. The cross sections for the thermal standards and for ^{238}U(n,γ) reaction were also incorporated in ENDF/B-VI. Results for the ^{239}Pu(n,f) cross section were also used after slight scaling.

Structural Materials

Complete new evaluated data files for the major structural materials were prepared at ORNL. New evaluations were made for the separate isotopes of Cr, Fe, Ni, Cu and Pb. The resonance parameters for the major Fe and Ni isotopes represent a comprehensive fit to transmission, differential scattering, and capture measurements using the Reich-Moore model. In these cases, the resonance energy region was extended to significantly higher energies. Extensive precompound/compound nuclear model calculations were performed above the resonance region to provide particle and gamma-ray production cross sections, and correlated energy-angle distributions for secondary neutrons in the "File 6" format. Careful attention was paid to secondary energy balance. Complete cross section uncertainty files were provided for each evaluation. An evaluation for ^{55}Mn was completed at ORNL by a visiting scientist from JAERI and is included also in the JENDL-3 evaluated data file.

Evaluations for natural V, ^{59}Co, ^{89}Y, ^{93}Nb, natural In, ^{209}Bi and ^{115}In isomer production were done as a cooperative effort between ANL and LLNL. The high energy part of the evaluation included recent experimental results for total and elastic scattering data measured at ANL. Optical model calculations based on fitting this experimental data were used in calculating charged particle production cross sections. Thermal data, resonance parameters and gamma-ray production cross sections were supplied by LLNL. Uncertainty files are included

Light Materials

The majority of the light material evaluations were the responsibility of LANL. R-matrix analysis for the standards materials, ^{1}H, ^{3}He, ^{6}Li, ^{10}B, ^{14}N and ^{16}O were included in the new evaluations. Charged particle production cross sections for ^{6}Li and ^{10}B were obtained from analysis of experimental data. Revised evaluations for ^{7}Li and ^{15}N were prepared as was a complete new evaluation for ^{11}B. The previous evaluation of ^{11}B had been done more than 20 years ago at AWRE and was badly out of date. No covariance files are supplied with these evaluations.

An extensive Monte-Carlo simulation to evaluate the (n,2n) reaction was included in a complete re-evaluation of ^{9}Be supplied by LLNL. In addition, a new evaluation of ^{19}F was performed at ORNL by a visiting scientist from P.R. China and included in ENDF/B-VI. A new evaluation for natural C was also supplied by ORNL. This evaluation extended up to 32 MeV and included all of the charged particle emission channels up to that energy. Covariance files were supplied for both C and ^{19}F.

Heavy Nuclides

The evaluation responsibility for the heavy nuclide evaluations for ^{235,238}U and ^{239,240}Pu were shared by LANL and ORNL. The data evaluation for energies above the unresolved resonance region was performed by LANL and is based on theoretical analysis of existing experimental data. The thermal and resonance region evaluations were the responsibility of ORNL. The evaluations for the standard and related cross sections were directly merged into the new evaluations. Complete evaluations for ^{241}Pu and ^{243}Am were performed at ORNL, for ^{237}Np at LANL and for ^{236}U at HEDL.

Significant new work for these nuclides included the reevaluation of the resonance parameters for ^{235,238}U and ^{239,241}Pu. Careful analysis of available high-quality experimental data was done using the latest resonance region fitting programs. The ^{238}U resolved resonance evaluation was done at Harwell and extended the resonance region to 10 keV. The unresolved resonance region evaluation was done at Karlsruhe. Included were 801 s-wave and 1112 p-wave resonances. The ^{235}U resonance region evaluation was performed at ORNL. The evaluation contains 3342 s-wave resonances in 10 energy regions up to 2.25 keV. The ^{239,241}Pu resonance evaluations were performed at ORNL by a visiting scientist from Caderache, France. These new evaluations extend the resolved resonance range up to 2 keV and 300 eV respectively, and are contained in both the ENDF/B-VI and JEF-2 data libraries, representing an example of significant data improvement resulting from international collaboration.

For the first time, complete evaluations performed outside the CSEWG organization have been included in a release of the ENDF data library. These are evaluations for ^{241}Am, ^{248}Bk and ^{248}Cf, provided by the Chinese Nuclear Data Center, Beijing.

Other Evaluations

Several fission product nuclide evaluations were modified at ORNL, primarily by updating the resonance parameters. Complete new evaluations for ^{151,153}Eu, ^{165}Ho and ^{197}Au were provided by LANL. The gold evaluation contained the recommended capture cross section from the standards analysis. Revised evaluations for ^{185,187}Re were provided by ORNL. The evaluations incorporated a new theoretical analysis of all reactions performed at LANL which took into account new capture measurements made at ORNL.

A significant improvement in the delayed fission neutron yields and spectra for all the important fissionable nuclides was made at LANL. The spectra were calculated from first principles using the latest decay data and neutron precursor yields and included in the ENDF/B-VI evaluations.

An updated decay data library has been prepared by INEL, LANL and HEDL. The starting point has been the ENSDF[5] data file supplemented by new data at INEL and theory at LANL. Final formatting and checking was done at HEDL. This data is being used in the evaluation of fission product yields at LANL. These parts of ENDF/B-VI have not yet been released.

For the first time, charged particle and high energy evaluations have been included in ENDF/B-VI. The p-p and p-^{3}He evaluations were supplied by LANL. Evaluations of neutron and proton induced reactions on ^{56}Fe up to 1 GeV were supplied by BNL.

Validation Procedures

The new and revised evaluations included in ENDF/B-VI have undergone extensive review. Completed versions of evaluations were sent to BNL for final processing by the ENDF checking codes maintained by BNL. For the first time these codes had been used by the evaluating laboratories as part of the final file preparation. Kits were prepared including the checking code results, listings and plots of the evaluated data. These review kits contained overlays of cross sections from the new evaluations, ENDF/B-V and experimental data which appeared in the neutron data atlas[6] produced by NNDC. These kits were reviewed in detail at regular and special CSEWG meetings with the evaluators present in most cases. A document summarizing the new neutron evaluations contained in ENDF/B-VI is ready for publication[7].

Testing of this library against integral experiments has begun. This work has been delayed by lack of funding and delays in making the processing systems operational

at some of the data testing laboratories. The results currently available are encouraging. Thermal reactor data testing results have been obtained only for ^{235}U assemblies. ENDF/B-VI results are similar to ENDF/B-V except the trend of increasing eigenvalue as a function of epithermal leakage and epithermal fission fractions has significantly improved. In the fast reactor area, results show a slight improvement, although there are still areas of concern. For example, k_{eff} for larger assemblies such as BIG-10, ZPR-6/6A and ZPR-6/7 appear to be about 1% high, although there is greater consistency between ^{235}U and ^{239}Pu assemblies. The k_{eff} for Pu-fueled assemblies has increased and is close to 1.0. The ^{238}U capture to ^{235}U fission ratio and to ^{239}Pu fission ratio have improved. The dosimetry reaction evaluations do not perform as well as in ENDF/B-V, in particular, the ^{58}Fe and ^{59}Co (n,γ) reactions.

Future Plans

There are no plans for a complete new release of the ENDF/B data library at this time. CSEWG will continue to operate in a maintenance mode while awaiting the results of integral data testing, the revitalization of reactor development programs and the growth of other nuclear applications. We expect to release an upgrade to ENDF/B-VI which will contain corrections to known errors, additional uncertainty files and a few new evaluations in 1992 or 1993. We will make maximum use of the NEACRP/NEANDC sponsored cooperation among the NEA countries on nuclear data evaluation and contribute where interest and resources permit. Where possible, we will make use of evaluations from other data files where they are better than the existing evaluation in ENDF/B.

Summary

A new version of the ENDF/B evaluated data library, ENDF/B-VI, has been completed and released for unrestricted distribution. It is available from the nuclear data centers at Brookhaven, Saclay, Vienna and Obninsk. Despite drastically reduced resources, significant improvements to the ENDF/B-V data library were made, in part due to increased international cooperation.

References

1. Proceedings of the Conference on Nuclear Data Evaluation Methods and Procedures, September 1980, Upton, NY, USA, BNL-NSC-51363, Brookhaven National Laboratory(1980)

2. IAEA Specialists' Meeting on the "Format for the Exchange of Evaluated Neutron Nuclear Data", April 1984, Vienna, Austria, INDC(NDS)-156, IAEA(1984)

3. ENDF-102, Data Formats and Procedures for the Evaluated Nuclear Data File, ENDF-6 (Edited by P.F.Rose and C.L.Dunford), BNL-NCS-44945, Brookhaven National Laboratory(1990)

792

4. ENDF-351, ENDF/B-VI Neutron Cross Section Standards, to be published US National Institute of Standards and Technology

5. Evaluated Nuclear Structure Data File, edited and maintained by the National Nuclear Data Center, Brookhaven National Laboratory on behalf of the International Network for Nuclear Structure Data Evaluation. Summary of file contents and published documentation appear on page ii of any issue of the Nuclear Data Sheets journal.

6. V.McLane, C.L.Dunford and P.F.Rose, Neutron Cross Sections, Volume 2, Neutron Cross Section Curves, Academic Press, San Diego(1988)

7. ENDF-201, ENDF/B-VI Summary Documentation (Edited by P.F.Rose), to be published by Brookhaven National Laboratory.

JAPANESE EVALUATED NUCLEAR DATA LIBRARY, VERSION-3
— JENDL-3 —

Yasuyuki KIKUCHI and Members of JNDC

Japan Atomic Energy Research Institute
Tokai-mura, Naka-gun, Ibaraki-ken 319-11, Japan

Abstract: The third version of Japanese Evaluated Nuclear Data Library (JENDL-3) has been developed aiming at really general applications such as fission, fusion and shielding calculations. The general purpose file of JENDL-3 contains neutron nuclear data for 324 nuclides in the ENDF-5 format. In the JENDL-3 evaluation, much effort has been devoted to improve reliability of high-energy data for fusion application, which were not satisfactory in JENDL-2, and to include gamma-ray production data. Some advanced nuclear theoretical models were adopted and recent experimental data of energy-angle double-differential cross sections (DDX) mainly measured in Japan were taken into account. Various benchmark tests have so far been made in order to verify the applicability of JENDL-3 to various fields. For fast reactor calculations, JENDL-3 gives satisfactory results for most of characteristics. Particularly, space dependences of reaction rates, sodium void coefficients and control rod worths, which were significant with JENDL-2, nearly disappear. This suggests that the JENDL-3 data are well balanced. Satisfactory applicability has been also proved for the other field such as thermal reactor calculations, fusion neutronics, shielding and dosimetry.

(JENDL-3, general purpose file, fission reactors, fusion neutronics, simulations evaluation, direct and pre-equilibrium processes, benchmark tests.)

Introduction

The Japanese Evaluated Nuclear Data Library (JENDL) has been developed by JAERI Nuclear Data Center in cooperation with Japanese Nuclear Data Committee since 1970's. Its first version was mainly aimed at fast reactor applications and was completed in 1977. The second version, which was completed in 1982, was applicable to all the fission reactor calculations, but was proved to be unsatisfactory for fusion neutronics. The third version (JENDL-3) has been compiled aiming at really general applications such as fission, fusion and shielding calculations. The general purpose file of JENDL-3 contains neutron nuclear data for 324 nuclides in the ENDF-5 format and was completed in 1989[1]. The specifications for JENDL-1, -2 and -3 are given in Table 1.

Table 1 Specification for each version of JENDL

	JENDL-1	JENDL-2	JENDL-3
Purpose	Fast Reactor	Fission Reactor	General
Completion	1977	1982	1989
Maximum Energy	15 MeV	20 MeV	20 MeV
Number of Nuclides*	72	181	324(53)
Light (Z=1~9)	4	8	14(10)
Medium light (Z=10~30)	23	33	56(23)
Fission Product (Z=31~69)	34	101	178(8)
Medium heavy (Z=70~87)	1	12	19(9)
Heavy (Z=88~94)	9	19	31(3)
Transplutonium (Z=95~100)	1	8	26(0)

( ) Number of nuclides with γ-ray production data

In the JENDL-3 evaluation, much effort has been devoted to improve reliability of high-energy data for fusion application, which were not satisfactory in JENDL-2, and to include gamma-ray production data. Some advanced nuclear theoretical models such as the coupled-channel model, DWBA, the pre-equilibrium model etc. have played an important role to achieve these purposes. Recent experimental data of energy-angle double-differential cross sections (DDX) mainly measured in Japan[2,3] were taken into account in the evaluation of emitted neutron spectra. The detailed evaluation will be given for each mass region.

The evaluation of JENDL-3 was completed in 1987, and a test version of JENDL-3, called JENDL-3T, was distributed for data validation. Since then various benchmark tests were made in order to verify the applicability of the data. The results of the benchmark tests were informed to the JENDL-3 compilation group in JNDC. The compilation group examined the results as a whole and informed the evaluators, if some modifications were to be made. The evaluators re-examined their evaluation by taking account of the comments from the compilation group. The final version of JENDL-3 was released in 1989, adopting the revised data.

Evaluation

Light Nuclides

The evaluation of light nuclides were performed mainly on the basis of experimental data, if they are available. A particular care was paid for DDX. The R-matrix theory was applied to the resonance structure.

Lithium: The tritium production cross sections of Li isotopes are important as a candidate material of the fusion blanket. The (n,t) cross section of ^{6}Li was evaluated with the R-matrix theory below 1 MeV and on the basis of the experimental data above 1 MeV. The (n,n't) cross section of ^{7}Li was evaluated on the basis of the newly measured data and is shown in Fig. 1. Energy distributions of both the isotopes were calculated with the phase-space model, and were given by using about 30 pseudo levels in the file.

Beryllium: The (n,2n) reaction cross section of ^{9}Be was evaluated on the basis of available experimental data. The 14 MeV value was based on the recent measurements in Japan, and is by 4% lower than that of JENDL-2 as shown in Fig. 2. The agreement with the experimental data is satisfactory.

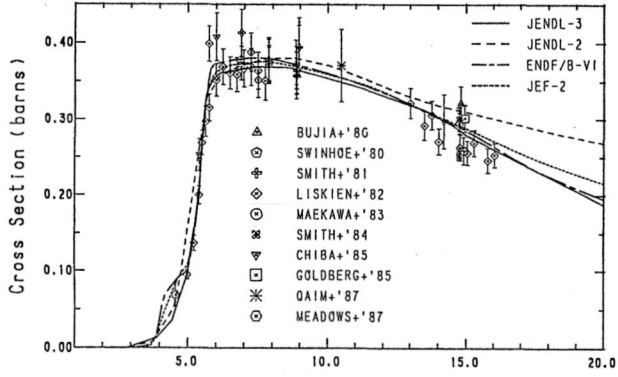

Fig. 1 ^{7}Li(n,n'α) T reaction cross section

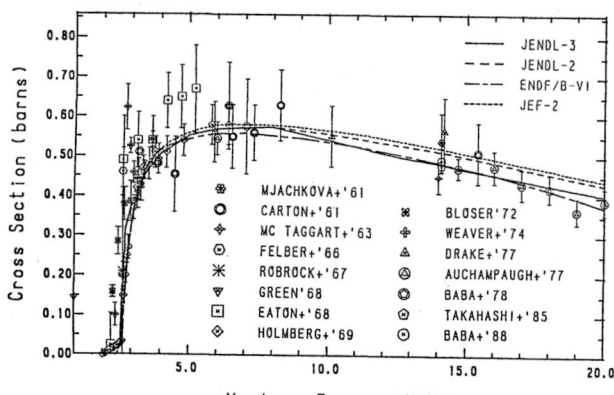

Fig. 2 ^{9}Be(n,2n) reaction cross section

Medium Nuclides

Among nuclides in the region of Z=10~87, fission products nuclides (Z=33~65) have been evaluated independently and systematically by a working group of JNDC, and the results are presented elsewhere in this proceedings. Hence the following are for the other nuclides.

Theoretical calculation: The spherical optical model and statistical model played an important role in the evaluation. In addition to these conventional models, more advanced models such as the coupled-channel optical model, DWBA and the pre-equilibrium model were applied particularly in the high energy region. Systematic trends of various parameters such as optical potentials and level densities were investigated. Figure 3 compares the DDXs of natural iron of JENDL-3 with the experimental data of Osaka and Tohoku Universities[2,3]. The JENDL-3 data reproduce the measured data well, while the JENDL-2 data, which were evaluated only with the conventional statistical model, give poor agreement.

Natural elements: In JENDL-3, the data of natural elements were evaluated on the basis of the experimental data of the natural elements independently of the evaluation of isotopic data, because there exist much more experimental data for the natural elements. If there exist some discrepancies between the natural element data thus evaluated and those composed of the isotopic data, we left them. Hence we recommend to use the natural element data for the natural elements. On the other hand, ENDF/B-VI and JEF-2 give only the isotopic data.

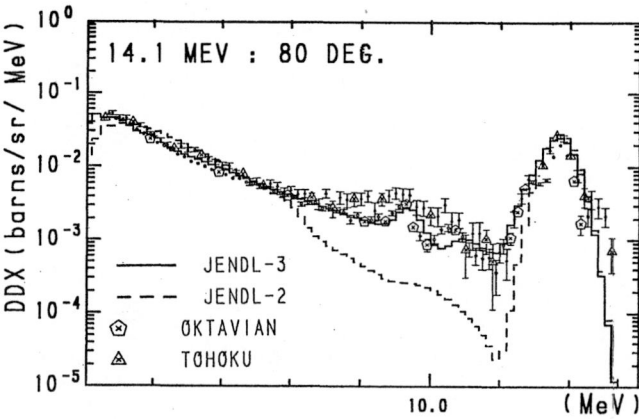

Fig. 3   DDX of Fe at 14 MeV

Threshold reaction cross sections are important for fusion and dosimetory applications. In most cases, they were calculated with the statistical model including the pre-equilibrium effects and were normalized to the experimental data.

However discrepancies still remain in some of the threshold reactions. Figure 4 compares the recently evaluated data of ^{56}Fe(n,α) reaction cross section. They agree at 14 MeV where the experimental data exist, but disagree in the other energy regions. To resolve these discrepancies, new measurements of the (n,α) cross sections are under planning at JAERI Tandem accelerator for main structural materials.

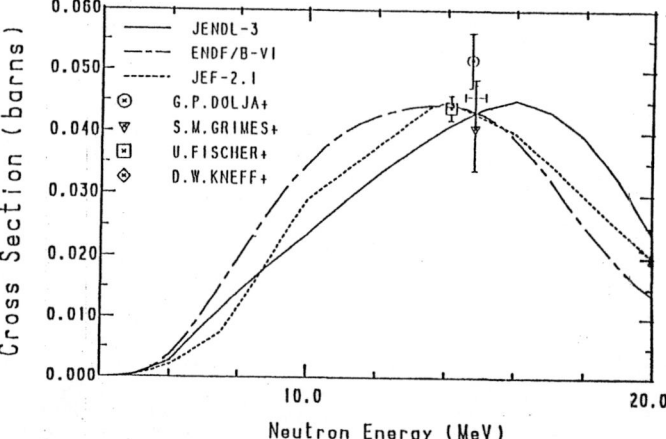

Fig. 4 ^{56}Fe(n,α) reaction cross section

Heavy Nuclides

Fifty-seven nuclides from ^{223}Ra and ^{225}Fm are contained as heavy nuclides. Among them a special care was taken to main fissile and fertile materials.

Simultaneous evaluation was applied[4] above 50 keV for the fission cross sections of ^{235}U, ^{238}U, ^{239}Pu, ^{240}Pu and ^{241}Pu and the capture cross sections of ^{238}U and ^{197}Au. All the absolute measurements and the ratio measurements such as $\sigma_f(^{239}$Pu$)/\sigma_f(^{235}$U$)$ were fitted by the generalized least squares method by a B-spline function. Covariance data required for this method were estimated from the experimental conditions. The evaluated results of ^{235}U and ^{239}Pu are shown in Fig. 5.

Capture cross section of ^{238}U has been carefully investigated. The values obtained with the simultaneous evaluation were found to be higher than the latest measurements of Kazakov et al.[5] in the energy range from 50 keV to 300 keV.

The results of the benchmark tests favor the smaller cross section, and Fröhner's theoretical calculation[6] also supports the lower values. Hence we adopted the lower values on the basis of the data of Kazakov et al. Thus the values of JENDL-3 are by 10% smaller than those of JENDL-2 in this energy region but agree with those of ENDF/B-VI and JEF-2 as seen in Fig. 6.

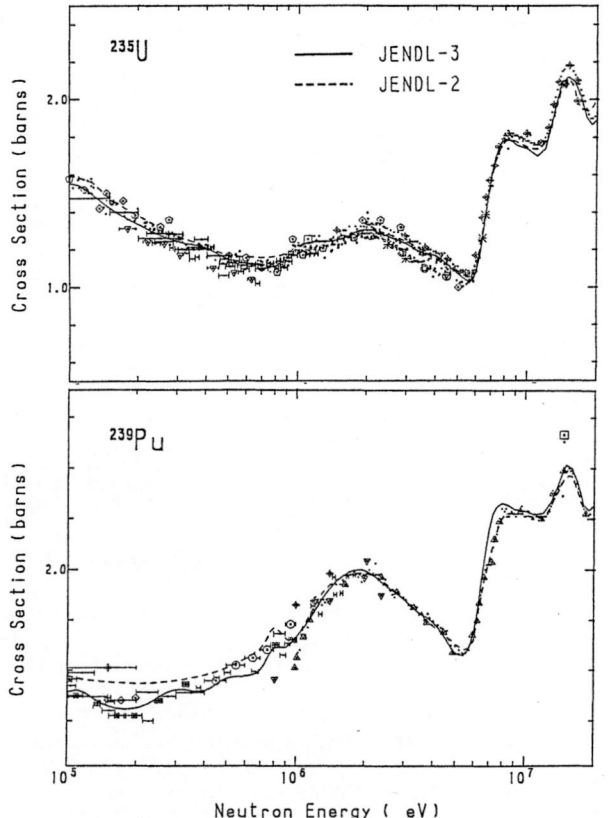

Fig. 5 Fission cross sections of ^{235}U and ^{239}Pu

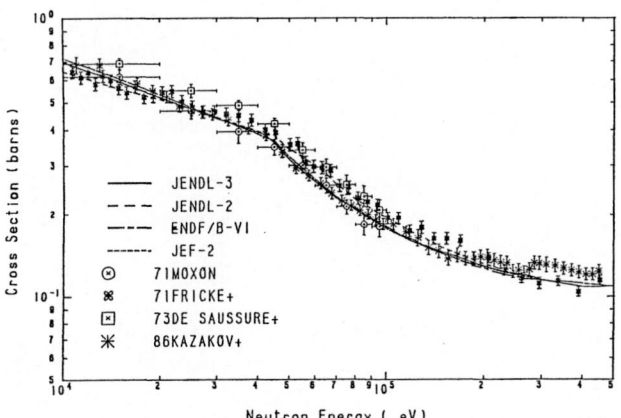

Fig. 6 Capture cross section of ^{238}U

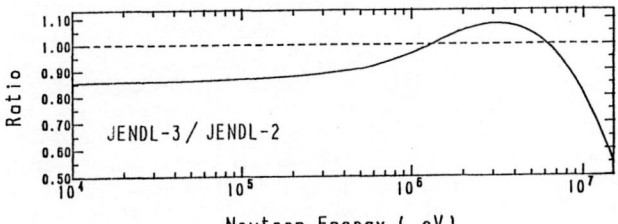

Fig. 7 Fission neutron spectrum of ^{239}Pu

Fission neutron spectrum: The prompt fission neutron spectrum formula of Madland and Nix[7] were adopted for ^{233}U, ^{234}U, ^{235}U, ^{238}U, ^{239}Pu and ^{240}Pu. This type of spectrum has larger average neutron energy than the Maxwellian or Watt types. The spectra for ^{239}Pu are shown in Fig. 7. The Maxwellian spectra were adopted for the other nuclides.

Gamma-ray Production Data

The gamma-ray production data were evaluated mainly on the basis of the statistical model with the pre-equilibrium effects. As will be described later, the gamma-ray spectra thus calculated was found to reproduce the measured data fairly well except the thermal neutron capture gamma-ray. Hence the gamma-ray data caused by the thermal neutron capture were re-evaluated on the basis of the available experimental data.

Benchmark Tests

Simple Fast Benchmark Assemblies[8]

Fast assemblies of simple geometry and simple components are useful to know the balance of data for main fissile nuclides. Eight assemblies such as JEZEBEL were chosen. The calculation was made with the ANISN code with S_{16} and P_5 approximation by using the group cross sections of 175 group structure. The C/E values are given in Table 2 for k-eff and central fission rate ratio of ^{238}U to ^{235}U. Very satisfactory results were obtained for the Pu and ^{235}U cores. For the ^{233}U cores, however, the k-eff values and the ^{238}U/^{235}U fission rate ratios are overestimated by 2 % and by 7 %, respectively. The sensitivity analysis revealed that the fission neutron spectrum of ^{233}U might be too hard and that the inelastic scattering cross sections might be too high. Some modification should be made on these quantities in future.

Table 2 C/E values of k-eff and fission rate ratio of ^{238}U to ^{235}U

Core		k-eff	$<^{28}\sigma_f> / <^{25}\sigma_f>$
Pu	JEZEBEL	1.0001	0.9944
	JEZEBEL-PU	0.9963	1.0063
	FLATTOP-PU	0.9974	1.0117
	THOR	0.9985	1.0034
^{235}U	GODIVA	1.0066	1.0006
	BIG TEN	1.0038	1.0195
	FLATTOP-25	1.0033	1.0697
^{233}U	JEZEBEL-23	1.0206	1.0619
	FLATTOP-23	1.0175	1.0696

Fast Critical Assemblies

For fast reactor benchmark tests of JENDL-1 and -2, 21 assemblies with one-dimensional model have been used, which consists of 18 international benchmark assemblies, two MOZART cores (MZA and MZB) and FCA-V-2. In the present tests, the JUPITOR reference core (ZPPR-9) and FCA-IX cores were added*. Most of the results were already published elsewhere[9].

---

* FCA-IX series consist of 7 EU cores. FCA-IX-1 ~ 3 are graphite moderated, IX-4 ~ 6 are SUS moderated and IX-7 is a 20 % EU metal core. Trends of prediction with JENDL-3 are very different between graphite moderated and SUS moderated cores. Hereafter FCA-IX-C represents the average of IX-1~3 and IX-S that of IX-4~6.

<u>Effective multiplication factor(k-eff)</u>:Table 3
The C/E values of JENDL-3 are satisfactory as seen
in Table 3. It is found, however, that the k-eff
values of U cores are lower by 0.7 % than those of
Pu cores. As to FCA IX cores, the C/E values are
much underestimated for IX-G cores, which are
graphite moderated, while satisfactory C/E values
are observed for SUS moderated cores.

Table 3   C/E values of k-eff

Cores	JENDL-2	JENDL-3
21 Benchmark cores	1.009	1.002
Pu cores	1.004	1.004
U cores	1.005	0.997
ZPPR-9	0.999	1.006
FCA-IX-C	0.992	0.984
FCA-IX-S	1.012	1.006
FCA-IX-7	1.009	1.007

<u>Central reaction rate ratios</u>: Table 4
The C/E values of important ratios are given in
Table 4. JENDL-3 gives better prediction for
fission rate ratios of ^{239}Pu and ^{235}U except for
FCA-IX-C. This good balance may come from our
adoption of simultaneous evaluation. As for the
ratio of ^{238}U fission to ^{235}U fission, JENDL-3
looks to overestimate the ratio for 21 benchmark
cores and FCA cores. This ratio is, however,
sensitive to the measurements, i.e., perturbation
caused by the detectors. Most of these
measurements were made with micro fission chambers
and the perturbation must be serious. On the
other hand, the recent measurements in ZPPR-9 were
made with foil techniques and very precise
analyses were made. As the C/E value of ZPPR-9 is
near unity, we believe that this quantity is well
predicted. The ratio of ^{238}U capture to ^{239}Pu
fission is well predicted for the benchmark cores,
but is overestimated by 4 % for ZPPR-9. As this
ratio is an important parameters for breeding
ratio, further study will be required.

Table 4 C/E values of central reaction rate ratios

Quantities*	Assemblies	JENDL-2	JENDL-3
$\dfrac{F(^{239}Pu)}{F(^{235}U)}$	21 Benchmark		
	Pu cores	0.97	0.99
	U cores	0.99	0.99
	All cores	0.97	0.99
	ZPPR-9	0.98	1.00
	FCA-IX-C	1.03	1.05
	FCA-IX-S	0.97	0.98
	FCA-IX-7	1.02	1.02
$\dfrac{F(^{238}U)}{F(^{235}U)}$	21 Benchmark		
	Pu cores	1.05	1.12
	U cores	0.98	1.04
	All cores	1.03	1.10
	ZPPR-9	0.94	1.00
	FCA-IX-C	1.22	1.30
	FCA-IX-S	1.12	1.19
	FCA-IX-7	1.05	1.11
$\dfrac{C(^{238}U)}{F(^{239}Pu)}$	21 Benchmark		
	Pu cores	1.02	1.00
	U cores	0.96	0.94
	All cores	0.99	0.98
	ZPPR-9	1.05	1.04

* F means fission and C means capture

<u>Doppler coefficients</u>: The C/E values of
Doppler coefficients of natural UO_2 samples in
ZPPR-9 were calculated. The average C/E value
with JENDL-3 is 0.94 and better than that of 0.91
with JENDL-2.
<u>Sodium void coefficients</u>: Fig. 8
The sodium void coefficients were much
overestimated (by 20 %) with JENDL-2, and the
overestimation becomes larger when the voided
volume increases as seen in Fig. 8. With JENDL-3,
the overestimation is much reduced and the C/E
values show little dependence on the voided
volume.

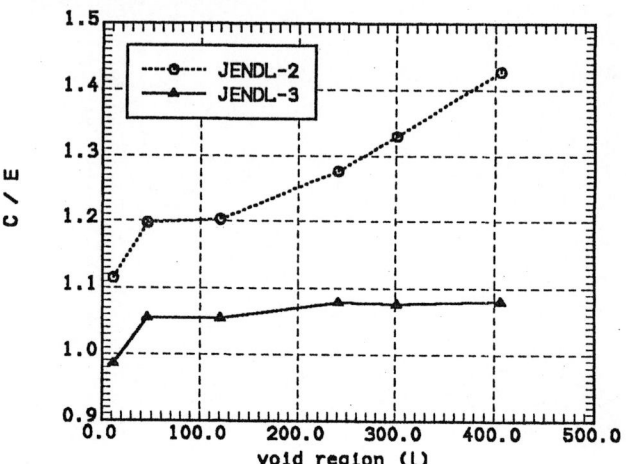

Fig. 8   Sodium void coefficients in ZPPR-9

<u>Space dependence</u>: With JENDL-2, considerable
space dependences were observed for the C/E values
of reaction rates, control rod worths and sodium
void coefficients. These space dependences are
much reduced with JENDL-3. For the ^{239}Pu fission
rate distribution in ZPPR-9, the C/E values remain
± 2% even in the outer core, while they reach 6%
in the outer core with JENDL-2. As to the control
rod worths in ZPPR-9, the maximum C/E deviation is
2% with JENDL-3, while it is 4.5% with JENDL-2 as
is seen in Fig. 9.

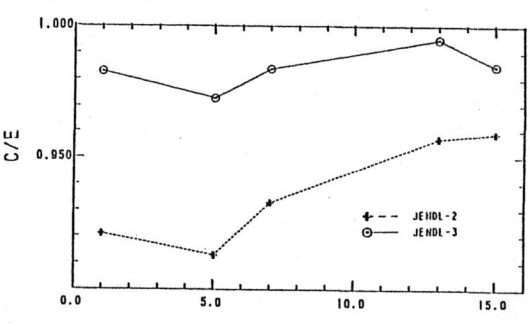

Fig. 9   Control rod worths in ZPPR-9

<u>Thermal Benchmark Tests</u>
Various thermal critical assemblies have been
analyzed. The benchmark calculations were
performed with SRAC, a thermal reactor standard
code for reactor design and analyses. Some
results were already published [8].
Here the results on TRX-1, -2, BAPL-1, -2 and
-3 are shown, which are U- cores selected for
'WIMS Library Update' project proposed by IAEA.
Table 5 shows the C/E values of k-eff and lattice
parameters calculated with JENDL-3 and JENDL-2.
Both JENDL-2 and -3 slightly underestimate the
k-eff values but predict well the lattice
parameters.

Table 5 The C/E values of k-eff and lattice parameter for thermal assemblies

Core	Quantities+	JENDL-2	JENDL-3
TRX-1	k-eff	0.9939	0.9956
	ρ-28	1.02	1.02
	δ-25	1.00	0.98
	C*	1.01	1.00
TRX-2	k-eff	0.9963	0.9979
	ρ-28	1.00	1.00
	δ-25	0.99	0.97
	C*	1.00	0.99
BAPL-1	k-eff	0.9956	0.9953
	ρ-28	1.01	1.01
	δ-25	1.00	0.98
BAPL-2	k-eff	0.9966	0.9960
	ρ-28	1.04	1.04
	δ-25	1.01	0.99
BAPL-3	k-eff	0.9979	0.9977
	ρ-28	1.01	1.01
	δ-25	1.01	0.99

ρ-28: Ratio of epithermal to thermal ^{238}U capture.
δ-25: Ratio of epithermal to thermal ^{235}U fission.
C* : Ratio of ^{238}U capture to ^{235}U fission.

Fusion Neutronics

As one of the main aims of JENDL-3 is its fusion application, intensive testing has been made for fusion neutronics benchmarking. Most of benchmark experiments were selected from those made in Japan such as TOF neutron spectrum measurements, tritium breeding and neutron multiplication experiments and gamma-ray leakage spectrum measurements performed in FNS and OKTAVIAN facilities. The precise discussion was published in Ref. [10].

TOF Spectrum Experiments are useful for direct validation of inelastic scattering and neutron emission reaction data. Very satisfactory agreement has been observed for most of nuclides. Figure 10 shows the spectrum of Oxygen.

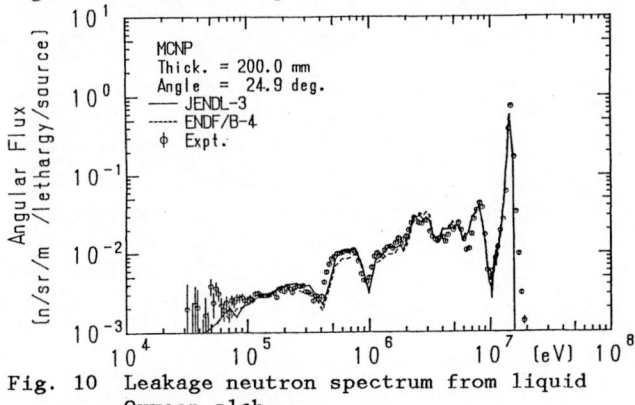

Fig. 10 Leakage neutron spectrum from liquid Oxygen slab

Integral Experiments on tritium breeding rate or neutron multiplication by Be or Pb were performed in FNS and OKTAVIAN facilities and analyzed with JENDL-3. It was found that the tritium breeding rate could be predicted within error of 5 ~ 10 % for a blanket consists of Li and Be.

Gamma-ray benchmarking: The gamma-ray leakage spectrum and gamma-ray heating experiments were analyzed. The results were satisfactory for Li_2O, C, Al and Si but some problems were pointed out for Be and W. Figure 11 shows the results of gamma-ray heating in a Be assembly.

Conclusion: As to the neutron data, JENDL-3 is satisfactory for most of cases. For some important nuclides such as Be, it was pointed out that the angle and energy distribution of inelastic scattering and (n,2n) reaction should be further modified. The gamma-ray benchmark tests are still going on.

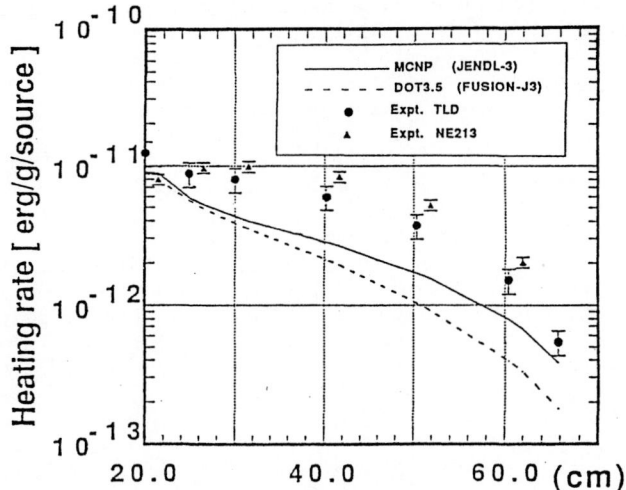

Fig. 11 Gamma-ray heating rate in Be slab

Shielding

As for neutron shielding, typical 9 benchmark experiments have been analyzed, such as Broomstick, ASPIS, ORNL, KFK etc. The results were satisfactory as a whole. Figure 12 shows neutron fluxes in iron of the ASPIS experiment.

For secondary gamma-ray shielding problems, studies are going on in cooperation with the fusion neutronics benchmarking group.

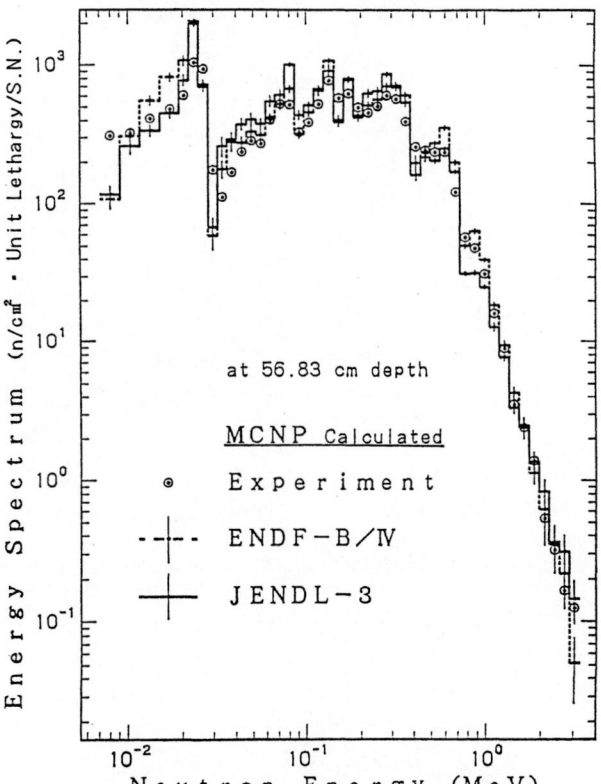

Fig. 12 Neutron energy spectra at 56.83-cm-depth in the iron shield of the ASPIS experiment

Gamma-ray Production Data

Emitted gamma-ray spectra arising from thermal neutron capture and fast neutron reactions were calculated and compared [11] with the experimental data measured at ORNL Tower Shielding Facility. The agreement was satisfactory for gamma-rays arising from fast neutron reactions. It was found, however, that the gamma-ray spectra caused by thermal neutron capture could not be well reproduced with JENDL-3T. Hence re-evaluation works were made and the modified data were adopted in JENDL-3.

Dosimetry

As one of JENDL special purpose files, a dosimetry file was provided and its applicability has been tested. The cross section data were mainly taken from JENDL-3 general purpose file and the covaricance data were taken from IRDF-85. Total of 61 reactions were adopted.

The data were tested with the activation data measured in various standard spectra. Figure 13 shows the C/E values in ^{252}Cf spontaneous fission spectrum (NBS standard) and CFRMF spectrum. Most of C/E values fall on unity within the experimental errors. Disagreement is remarkable, however, for ^{58}Ni(n,2n), ^{63}Cu(n,2n) etc. Both the cross sections and the integral experiments are being studied for these cases.

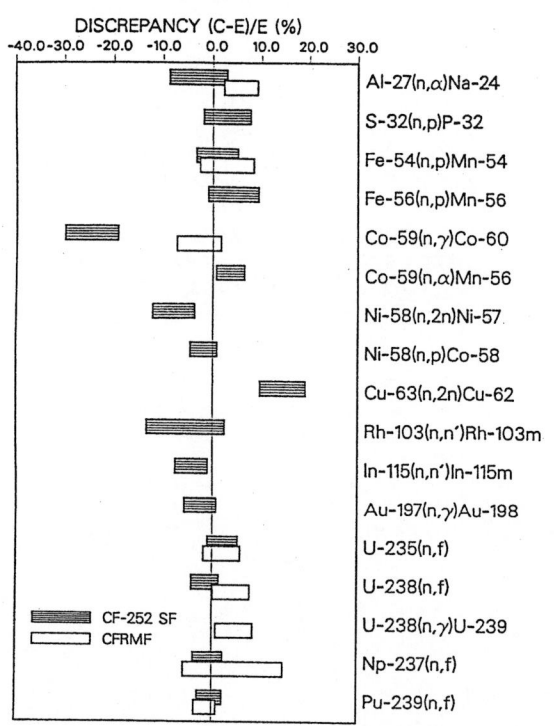

Fig. 13 C/E values of dosimetry reaction rates in ^{252}Cf fission and CFRMF spectra

Table 6 JENDL Special Purpose Files

File	Contents*	Completion
Dosimetry	61 R	1991
Gas production	23 N	1991
Activation	1000 R	1991
(α,n)	11 E	1991
KERMA/DPA	46 E	1992
Actinide	73 N	?

* R: Reaction, N: Nuclides, E: Elements

Future Scope

Revision of JENDL-3

The results of the benchmark tests have been satisfactory as a whole. JENDL-3 will be used widely for general purposes. For some nuclides, however, drawbacks have been already pointed out. More problems will be revealed in near future. The revision works have already started. The revised version of JENDL-3 (JENDL-3 Revision 2*) will be released at the end of 1993.

Special Purpose File

Aside from JENDL general purpose file, various special purpose files will be provided. They are tabulated in Table 6.

High Energy Nuclear Data Files

Recently high energy accelerator techniques become to attract much attention particularly from the viewpoint of transactinide burning, accelerator breeding, high intensity neutron sources etc. These techniques need nuclear data for high energy incident particles, which are beyond the scope of JENDL-3. These high energy nuclear data will be one of the main challenging targets in future.

Evaluation of neutron data up to 50 MeV was already started for the conceptional design of ESNIT project (d-Li neutron source energy selective irradiation test facility). Evaluation of high energy neutron and charged particle (up to a few hundred MeV) induced nuclear data is now under planning. The photo reaction cross section data up to 100 MeV is now going on.

---

* Revision 1 was already released in 1990 after correcting trivial compilation errors.

REFERNCES

1. K. Shibata, T. Nakagawa, T. Asami, T. Fukahori, T. Narita, S. Chiba, M. Mizumoto, A. Hasegawa, Y. Kikuchi, Y. Nakajima and S. Igarasi, Japanese Evaluated Nuclear Data Library, Version-3, JAERI 1319, JAERI (1990).

2. A. Takahashi, J. Nucl. Sci. Technol., 26, 15 (1989).

3. M. Baba, M. Ishikawa, T. Kikuchi, H. Wakabayashi and N. Hirakawa, in Proc. Int. Conf. Nuclear Data for Science and Technol., May/June 1988, Mito, Japan, p.209 and 291, Saikon (1988).

4. Y. Kanda, Y. Uenohara, T. Murata, M. Kawai, H. Matsunobu, T. Nakagawa, Y. Kikuchi and Y. Nakajima, in Proc. Int. Conf. Nuclear Data for Basic and Applied Science, May 1985, Santa Fe, USA p.1567, Gordon and Breach (1986).

5. L.E. Kazakov, V.N. Kononov, G.N. Manlurov, E.D. Paletaev, M.V. Bokhovko, V.M. Timokhov and A.A. Voevodskij, Yad. Konst., 3, (1986).

6. F.H. Fröhner, Nucl. Sci. Eng., 103, 119 (1989).

7. D.G. Madland and J.R. Nix, Nucl. Sci. Eng., 81, 213 (1982).

8. D. Tian, A. Hasegawa, T. Nakagawa and Y. Kikuchi, in Proc. 1990 Symp. Nuclear Data, Nov. 1990, JAERI, Japan, p.148, JAERI-M 91-032, JAERI (1991).

9. H. Takano, K. Kaneko and T. Nakagawa, in Proc. Int. Conf. Phys. Reactors, April 1990, Marseille, France, Vol.3, PI-21 (1990).

10. Y. Nakajima and H. Maekawa (Edt.), Proc. 2nd
    Specialists' Meeting Nucl. Data for Fusion
    Reactors, Dec. 1990, JAERI-M 91-062, JAERI
    (1991).
11. S. Cai, A. Hasegawa, T. Nakagawa and
    Y. Kikuchi, J. Nucl. Sci. Technol., $\underline{27}$, 844
    (1990).

LIBRARY OF EVALUATED NEUTRON DATA FILES

Blokhin A.I., Ignatyuk A.V., Kuzminov B.D., Koshcheev V.N.,
Manokhin V.N., Manturov G.N., Nikolaev M.N.

Institute of Physics and Power Engineering,
Obninsk, USSR

Abstract: The second version of the recommended evaluated neutron data library (BROND) has been formed in IPPE. The library contains the data in ENDF/B -format for the main materials used in reactor technology and radiation shielding. Most of the files were compiled from the Soviet original evaluations. The evaluated cross sections were converted into group constants, which were tested against the macroscopic experiments on critical assemblies and in reactor calculations. The results displayed good accuracy and general consistency of the evaluated microscopic data.

(BROND-2, nuclear data evaluation, neutron cross - sections, neutron - production, gamma - production, benchmark tests, Keff)

During the last years the problems of nuclear safety, ecological cleanness and economic effectiveness became of great importance in nuclear power engineering. Wide discussions were conducted on the design of inherent passive safety reactors, environmental aspects of the storage and interment of radioactive wastes and its transmutation, reactor decommissioning, new fuel cycles, the international ITER project and so on.

In the process of elaboration of the second version of the recommended evaluated data file library (BROND-2) we have tried to take into account all these new trends and to meet the relevant requirements in nuclear data. The selection of evaluated data for the BROND-2 library, its analysis and handling have been completed in 1990. For principal reactor materials the evaluated data developed by Soviet specialists have been included into the library. In the case of materials used as neutron standards the data recommended by the IAEA were accepted. For remaining materials, the files which Soviet specialists considered as the most reliable were taken from existing foreign libraries and used as a basis for thorough revision. As a result of such revision the evaluated data were changed or completed in correspondence with recent experimental data and theoretical calculations (replacement of resonance parameter file, inclusion of the data for the (n, 2n) and (n, 3n) reactions and proper modification of spectra of neutron inelastic scattering, reexamination of photon production data in neutron reactions and so on).

The number of files included in the BROND-2 has increased by more then twice in comparison with the BROND-1 [1]. The files are written in the ENDF-format [2]. The content of the library is given in the Table. It should be noted that the whole number of the files used in practical application exceeds that included in Table. The BROND-2 files together with additional files (as a rule of foreign origin), which are not completely expertised yet, are named as the FUND - library. All materials were checked by the ENDF Utility codes [2] and converted into the 28 - group constant library ABBN-90 [3] by means of the GRUCON program system [4]. Using ABBN-90 the calculations of a wide set of benchmark experiments on fast critical assemblies were made and the estimation of accuracy of the main reactor characteristics was obtained [5]. As a result, uncertainty in Keff calculation was equal to (1-1.5)%.

During the process of the library formation we aimed to obtain the data sets for the natural mixture of isotopes. The data for each isotope are given as far as it seems to be necessary. We also aimed to give all the data concerning radionuclide production.

The data on total neutron cross sections, energy-angular distributions of secondary neutrons, photon production in neutron reactions were given only for the natural mixture of isotopes in those cases when there is no risk to lose an information. This approach allows easily to take into account experimental information available often only for natural mixture of isotopes (total cross section, photon production data and so on). There are no difficulties for users because of necessity to take into account overlapping resonances of different kinds at the formation of cross sections for natural mixture of isotopes. The amount of information is decreased. In this connection we intend to include into the library the files for separate isotopes only when it is necessary (isotopes of boron, lithium, fission products, actinides). In the case of data taken from other libraries and presented isotopically we produced data sets for natural mixture in resonance region only.

Now one should underline some results of comparison of the nuclear data for important materials from the BROND-2 and ENDF/B-VI libraries.

Figure 1 shows the group neutron capture cross sections for the natural iron, chromium, nickel calculated from the data of the libraries BROND-2 and ENDF/B-VI. For the iron the ENDF/B-VI data are systematically below the BROND-2 data by about 3%. In case of chromium the discrepancy amounts to 10%. The principal discrepancies of the data for nickel are due to missing low lying resonances in the ENDF/B-VI. The inclusion of these resonances removes the discrepancies mentioned.

Figure 2 gives neutron capture cross section for U-238 from BROND-2 and ENDF/B-VI. The differences are insignificant.

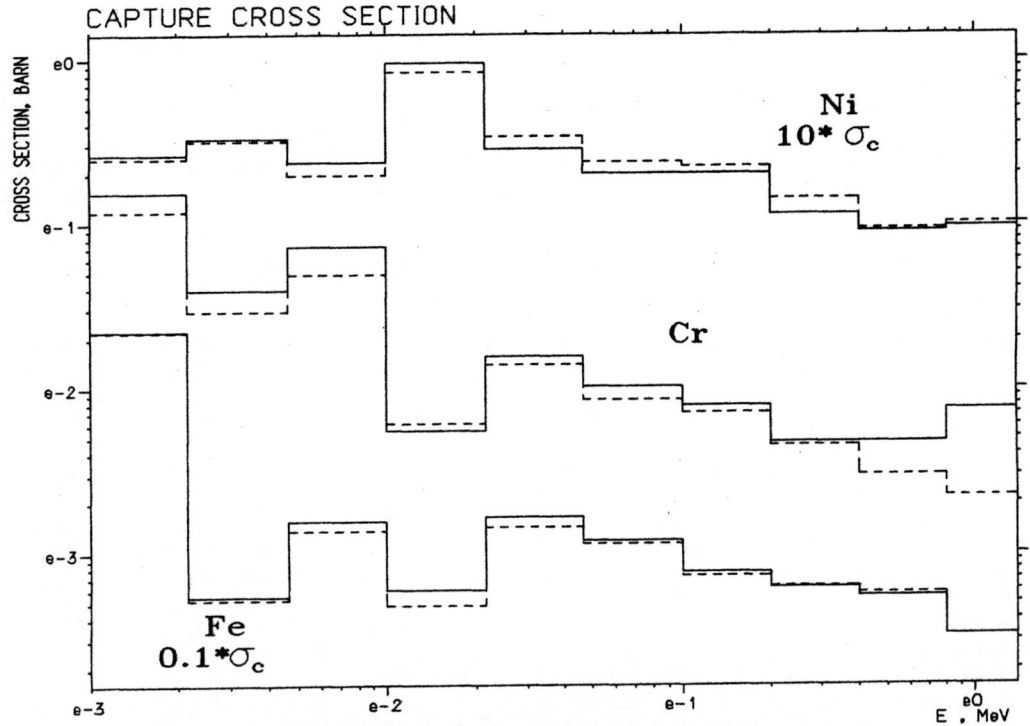

Fig.1. Different evaluations of capture
cross sections for Cr,Fe and Ni :
solid— BROND-2, dash— ENDF/B-VI

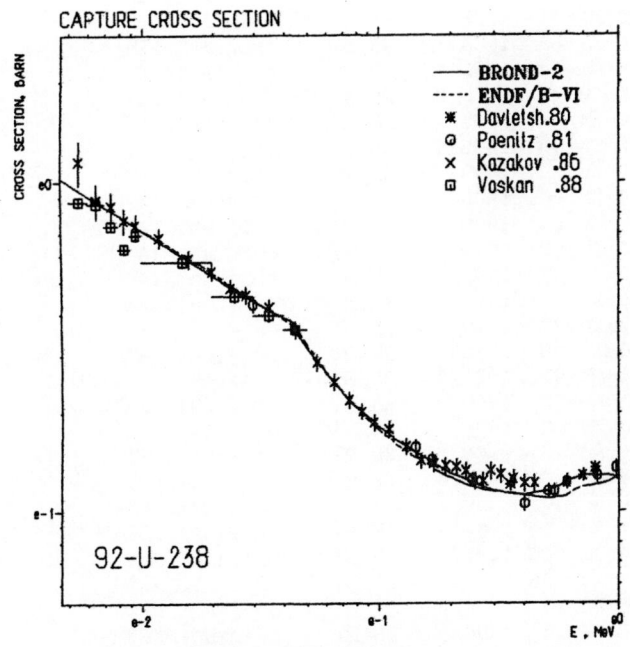

Fig.2. U-238 capture cross section in energy
range 4.65–1000 keV

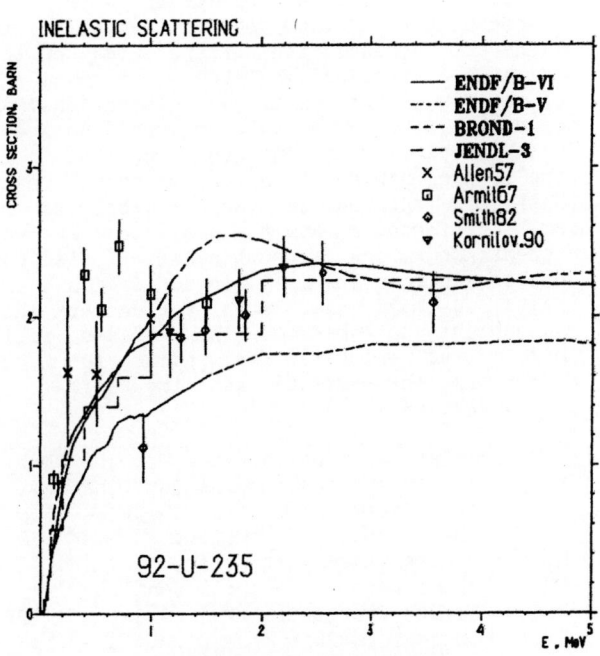

Fig.3. U-235 inelastic cross section
from different evaluations

The inelastic cross sections for U-235
from ENDF/B-V, ENDF/B-VI and BROND-2 are given
in Figure 3. The recent experimental data are
close to the ENDF/B-VI data.

There are uncertainties in the choice of
the prompt neutron spectra of the U-235 fission
induced by thermal neutrons. The averaging of
the most experimental results gives the mean

neutron energy E = 1.98 MeV. Only the unique work [6] gives the value E = 2.03 MeV and this high magnitude essentially better describes spectral indexes of such assemblies as GODIVA.

For further development of BROND-2 it is necessary to expand the radionuclide production cross sections and improve the quality of the data on photon production in neutron reactions.

An expansion of the library by evaluated data on photon production encounters considerable difficulties. For conservation of energy balance it is needed to evaluate photon production data simultaneously with an evaluation of neutron data for each neutron reaction. Such an evaluation is based as a rule on theoretical calculations. There is often a rich information on photon production-spectra and multiplicities of photons, measured for several energies of incident neutrons, but as a sum of all reactions on all isotopes of natural mixture. A search of agreement between theoretical investigations and experimental information is a very complicated process. Often the available evaluated data can be represented in the existing ENDF-format only with great complexity. For example, to describe the energy dependence of radiative capture spectra we need to represent in explicit form the energy dependence of excitation of discrete levels (and in this case the file MF=12 becomes too complex), or represent photon spectra in group form, but in this case we lose the information about discrete part of spectrum.

It was noted in a memorandum by J.Schmidt [7] to the XVIII INDC meeting that the development of improved nuclear data libraries (ENDF/B-VI, BROND, JEF-2, JENDL-3) and a free access to them created favorable situation for organization of the work on their comparison, testing, consistency and development on their base for a unified international nuclear data library.We share this opinion and are ready to participate in such a work.To our mind the sets of neutron data for principal reactor materials accepted in various libraries have small differences and the evaluators from different countries without great efforts could choose and form the best version from available files. In those cases when one can not manage to achieve an agreement, the problems will be highlighted which require primary solution by additional experiments and theoretical investigations.

Some explanations for the Table:

RR = in the resonance region the neutron cross-sections must be computed from the resonance parameters
NT = neutron transport data
GP = gamma-production cross-sections
FEI = Fiziko-Energeticheskij Institut (IPPE), Obninsk, USSR
IATE = Inst. Atomnoj Energetiki, Obninsk, USSR
IJE = Inst. Jadern. Energetiki, Acad. Nauk BSSR, Minsk
TUD = Technische Univ. Dresden, Germany

Table. Content of the BROND-2 library.

Nuclide	Date eval/rev	Contents	Origin
1-H-1	1989/90	NT, GP	modified ENDF/B-V
1-H-2	1975/89	NT	original, FEI
1-H-3	1974/88	NT	modified ENDF/B-IV
2-He-3	1976/89	NT	original, FEI
2-He-4	1976	NT	original, FEI
3-Li-6	1984/89	NT	original, FEI
3-Li-7	1985/89	NT, GP	original, FEI
6-C	1973	NT, GP	ENDF/B-V
7-N-14	1985/90	NT, GP	original, FEI
7-N-15	1989	NT, GP	original, FEI
8-O	1975	NT, GP	original, FEI
9-F-19	1990	NT, GP	original, FEI
11-Na-23	1978	RR, NT	original, FEI
13-Al-27	1973/90	NT, GP	ENDF-VI
14-Si	1985/89	RR, NT, GP	original, FEI+TUD
15-P-31	1978/89	RR, NT	original, FEI
16-S	1979/89	RR, NT, GP	original, FEI
17-Cl	1975/89	NT, GP	original, FEI
24-Cr	1987/90	RR, NT, GP	original, FEI
24-Cr-50	1987/89	RR, NT	original, FEI
24-Cr-52	1987/89	RR, NT	original, FEI
24-Cr-53	1987/89	RR, NT	original, FEI
24-Cr-54	1987/89	RR, NT	original, FEI
25-Mn-55	1988	RR, NT, GP	ENDF-VI
26-Fe	1985/89	RR, NT, GP	original, FEI
26-Fe-54	1985/90	RR, NT	original, FEI
26-Fe-56	1985/90	RR, NT	original, FEI
26-Fe-57	1985/90	RR, NT	original, FEI
26-Fe-58	1985/90	RR, NT	original, FEI
27-Co-59	1989	RR, NT, GP	ENDF-VI
28-Ni	1983/89	RR, NT, GP	ENDF-VI
28-Ni-58	1983	RR, NT	original, FEI
28-Ni-60	1983	RR, NT	original, FEI
28-Ni-61	1983	RR, NT	original, FEI
28-Ni-62	1983	RR, NT	original, FEI
28-Ni-64	1983	RR, NT	original, FEI
29-Cu	1973/81	RR, NT, GP	original, FEI
30-Zn	1985/89	RR, NT, GP	original, FEI
38-Sr-90	1985/89	RR, NT	original, FEI
40-Zr	1990	RR, NT, GP	original, IATE
40-Zr-90	1990	RR, NT	original, IATE
40-Zr-91	1990	RR, NT	original, IATE
40-Zr-92	1990	RR, NT	original, IATE
40-Zr-93	1990	RR, NT	original, IATE
40-Zr-94	1990	RR, NT	original, IATE
40-Zr-95	1990	RR, NT	original, IATE
40-Zr-96	1990	RR, NT	original, IATE
41-Nb-93	1981/89	RR, NT, GP	original, FEI+TUD
41-Nb-95	1990	RR, NT	original, IATE
43-Tc-99	1984	RR, NT	original, FEI
44-Ru-101	1984/85	RR, NT	original, FEI
44-Ru-102	1984/85	RR, NT	original, FEI
44-Ru-103	1990		original, FEI
44-Ru-104	1984	RR, NT	original, FEI
44-Ru-105	1985	RR, NT	original, FEI
44-Ru-106	1985	RR, NT	original, FEI
45-Rh-103	1984/89	RR, NT	original, FEI
46-Pd-105	1984/90	RR, NT	original, FEI
46-Pd-106	1987	RR, NT	original, FEI
46-Pd-107	1985	RR, NT	original, FEI
46-Pd-108	1987	RR, NT	original, FEI
46-Pd-110	1987	RR, NT	original, FEI
47-Ag-107	1976/89	RR, NT	revised ENDF/B-V
47-Ag-109	1984/89	RR, NT	original, FEI

Table.   Content of the BROND-2 library (cont.)

Table.   Content of the BROND-2 library (cont.)

Nuclide	Date eval/rev	Contents	Origin
50-Sn	1985/89	RR, NT	original, FEI
53-I- 127	1977/84	RR, NT	ENDF/B-V
53-I- 129	1985/89	RR, NT	original, FEI
54-Xe-131	1985	RR, NT	original, FEI
54-Xe-135	1977/85	RR, NT	ENDF/B-V, min.cor.
55-Cs-133	1981/84	RR, NT	revised JENDL-1
55-Cs-135	1981/85	RR, NT	modified JENDL-1
57-La-139	1977/84	RR, NT	ENDF/B-V, min.cor.
58-Ce-140	1980/90	RR, NT	modified JENDL-1
58-Ce-142	1980/90	RR, NT	the same
58-Ce-144	1979/85	RR, NT	the same
59-Pr-141	1977/84	RR, NT	ENDF/B-V, min.cor.
60-Nd-143	1985	RR, NT	original, FEI
60-Nd-145	1985	RR, NT	original, FEI
61-Pm-147	1984	RR, NT	original, FEI
62-Sm	1990	RR, NT	original, FEI
62-Sm-144	1990	RR, NT	original, FEI
62-Sm-147	1984	RR, NT	original, FEI
62-Sm-148	1989	RR, NT	original, FEI
62-Sm-149	1984	RR, NT	original, FEI
62-Sm-150	1989	RR, NT	original, FEI
62-Sm-151	1985	RR, NT	original, FEI
62-Sm-152	1985	RR, NT	original, FEI
62-Sm-154	1989	RR, NT	original, FEI
63-Eu-151	1977/85	RR, NT, GP	ENDF/B-V, min.cor
63-Eu-153	1985/89	RR, NT, GP	original, FEI
63-Eu-155	1979/85	NT	ENDF/B-V, min.cor.
64-Gd	1990	RR, NT, GP	original, FEI
64-Gd-152	1977/90	RR, NT	re-eval. of FEI
64-Gd-154	1977/90	RR, NT	the same
64-Gd-155	1977/90	RR, NT	the same
64-Gd-156	1974/90	RR, NT	the same
64-Gd-157	1977/90	RR, NT	the same
64-Gd-158	1977/90	RR, NT	the same
64-Gd-160	1977/90	RR, NT	the same
68-Er-162	1976/89	RR, NT	original, FEI
68-Er-164	1976/89	RR, NT	original, FEI
68-Er-166	1976/89	RR, NT	original, FEI
68-Er-167	1976/89	RR, NT	original, FEI
68-Er-168	1976/89	RR, NT	original, FEI
68-Er-170	1976/89	RR, NT	original, FEI
73-Ta	1974/89	RR, NT, GP	original, FEI
74-W- 182	1973/83	RR, NT, GP	original, FEI
74-W- 183	1973/83	RR, NT, GP	original, FEI
74-W- 184	1973/83	RR, NT, GP	original, FEI
74-W- 186	1973/83	RR, NT, GP	original, FEI
75-Re	1988	RR, NT	original, FEI
75-Re-185	1968/88	RR, NT	modified ENDF/B-IV
75-Re-187	1968/88	RR, NT	the same
76-Os	1990	RR, NT, GP	original, FEI
77-Ir	1990	RR, NT, GP	original, FEI
79-Au-197	1987	RR, NT, GP	ENDF-VI
82-Pb	1984/90	RR, NT, GP	original, FEI+TUD
82-Pb-204	1990	RR, NT	original, FEI
82-Pb-206	1990	RR, NT	original, FEI
82-Pb-207	1990	RR, NT	original, FEI
82-Pb-208	1990	RR, NT	original, FEI
83-Bi-209	1990	RR, NT, GP	original, FEI
90-Th-232	1978/83	RR, NT	original, FEI
92-U- 233	1990	RR, NT	original, FEI
92-U- 234	1990	RR, NT	original, IJE
92-U- 235	1985	RR, NT	original, IJE

Nuclide	Date eval/rev	Contents	Origin
92-U- 236	1986	RR, NT	original, IJE
92-U- 238	1978/89	RR, NT	original, FEI
93-Np-237	1981	RR, NT	revised INDL/A
94-Pu-238	1989	RR, NT	original, IJE
94-Pu-239	1980	RR, NT, GP	original, IJE
94-Pu-240	1980	RR, NT, GP	original, IJE
94-Pu-241	1979	RR, NT, GP	original, IJE
94-Pu-242	1980	RR, NT, GP	original, IJE
95-Am-241	1990	RR, NT	original, FEI+IJE
95-Am-242	1990	RR, NT	original, FEI+IJE
95-Am-242m	1990	RR, NT	original, FEI+IJE
95-Am-243	1990	RR, NT	original, FEI+IJE
96-Cm-242	1989	RR, NT	original, IJE
96-Cm-243	1982	RR, NT	revised INDL/A
96-Cm-244	1989	RR, NT	original, IJE

REFERENCES

1. A.I.Blokhin,   A.V.Ignatyuk, V.N.Koshcheev, B.D.Kuz'minov,   V.N.Manokhin,   G.N.Manturov, M.N.Nikolaev. The Proc.Int. Conf. on Nuclear Data for Science and Technology, Japan, Mito 1988, Editor S.Igarasi, p.611

2. ENDF-102. Data formats and procedures for the evaluated nuclear data file ENDF-VI. Report BNL-NCS-44945, July 1990. Edited by P.F.Rose, C.L.Dunford

3. M.N.Nikolaev, A.M.Tsibulya.   New nuclear data set ABBN-90 and its neutron data for some structural materials on the basis of macroscopic experiments. Int.Conf. on Nuclear Data , for Science and Technology, May, 1991, Julich, FRG, Abstracts, p.154.

4. V.V.Sinitsa, A.A.Rineiskij. Applied program package GRUCON-6 for neutron group constants calculation. Report on the 1st Inter. Neutron Physics Conf., 14-18 sept. 1987, Kiev, USSR, vol.1, p.439

5. S.M.Bednyakov,   G.N.Manturov,   K.Dietze. J., Atomnaya Energiya, 1990, v.69, p.31 (in Russ)

6. P.I.Johansson, B.Holmqvist, T.Wiedling. NBS Spec.Publ. 425, (1975), p.572

7. J.J.Schmidt. Report INDC/P(90)-5, IAEA, Vienna, 1990

CENDL-2, CHINESE EVALUATED NUCLEAR DATA LIBRARY, VERSION-2

Liu Tingjin,   Liang Qichang,   Su Zongdi

Chinese Nuclear Data Center
Institute of Atomic Energy
P. O. Box 275 (41), Beijing, P. R. China

Abstract: CENDL-2 was developed at the Chinese Nuclear Data Center (CNDC), Beijing, China. It contains cross sections of all reactions, angular distributions and spectra of secondary neutrons from 54 nuclides in the energy rang from 10^{-5} eV to 20 MeV. To complete CENDL-2, some evaluation and theoretical calculation methods as well as relative programs have been developed and used. Comparisons with ENDF / B-VI and JENDL-3 show that most of the data agree with each other within experimental error but some are different. To improve the evaluation further, advanced evaluation and theoretical calculation methods and programs, such as simultaneous evaluation, covariance matrix, MUP-3, UNIFY-2 have been or are being developed.

(CENDL, evaluated nuclear data library, evaluated neutron data, nuclear data evaluation, nuclear data calculation, data processing)

1   Introduction to CENDL-2[1]

The second version of the Chinese Evaluated Nuclear Data Library (CENDL-2) has been prepared based on the CENDL-1, it contains 54 nuclides, as shown in Table 1, among them 36 nuclides are from CENDL-1, but have been re-evaluated or extensively revised around the years 1989 / 1990, the rest are newly evaluated. Most of the evaluations were performed at the CNDC and under Chinese Nuclear Data Coordination Network (CNDCN), some evaluations were completed by Chinese evaluators abroad under the international cooperation, and some evaluations are based on foreign libraries, e.g. ENDF / B-VI, JENDL-3 and BROND, with partial updates and revisions performed by Chinese evaluators.

Table 1   Nuclides contained in CENDL-2

^{1}H, ^{2}H, ^{3}H, ^{3}He, ^{4}He, ^{6}Li, ^{7}Li, ^{9}Be, ^{10}B, ^{11}B, ^{14}N, ^{16}O, ^{19}F, ^{23}Na, Mg, ^{27}Al, Si, ^{31}P, S, K, Ca, Zn, Ti, ^{51}V, Cr, ^{55}Mn, Fe, ^{59}Co, Ni, Cu, Zr, ^{93}Nb, Mo, Ag, ^{107}Ag, ^{109}Ag, Cd, In, Sn, Sb, Hf, Ta, W, ^{197}Au, Pb, ^{232}Th, ^{235}U, ^{238}U, ^{237}Np, ^{239}Pu, ^{240}Pu, ^{241}Am, ^{249}Bk, ^{249}Cf

The library contains full sets of neutron data, i.e. cross sections of all reactions and energy and angular distributions of secondary neutrons, in the energy range from 10^{-5} eV to 20 MeV. Most of the evaluations are presented in the ENDF / B-V format, some in ENDF / B-VI which contains double differential cross sections or Reich-Moore resonance parameters in re-solved resonance region, which are not permitted in ENDF / B-V format. CENDL-2 contains data for nuclear engineering calculations, including light nuclides, structure materials and fuel in nuclear reactors, radiation shielding materials and so on.

All the evaluations were carefully checked and reviewed by Chinese specialists; the Chinese Nuclear Data Center also organized a collective examination carried out by specialists from various institutes and universities.

Each evaluation in CENDL-2 contains a short description giving:

— Information about the authors and dates of the evaluation;

— Brief information on the sources of the experimental data, the evaluation method, and the theoretical calculations;

— Contents of the evaluation and related references.

In order to manage the library, the data base system developed at NNDC, U.S.A., has been installed in Micro-VAX-II computer at CNDC. The ENDF utility programs from NNDC, version 6.6, have also been transplanted on Micro-VAX-II computer and used for checking and processing evaluated neutron data for CENDL-2.

2   Evaluation Method

The experimental data up to 1987～1989 (depending on the completion time of the evaluation) were collected, mainly taken from EXFOR experimental neutron data library of NDS, but also supplemented with some newly measured data from publications or private communications. The data were compared, carefully analyzed and, if necessary, were corrected (renormalized, changed error etc.).

The EXFOR data were processed using Nuclear Cross Section Evaluation System NCSES[2]. Through the system, the data are first retrieved (NEXFOR), then changed in format and primarily processed (energy point selection, data renormalization, error correction etc.) (FORMAT), then plotted, compared, identified (SIG), and then fitted with spline function (SPF); finally, with theoretically calculated data together, they are comprehensively adjusted to make the data consistent in physics (CCSC3).

Curve fit program SPF[3] can be used for multi-set of data. The knots can be optimized, this not only makes the fitting easier, but also makes the fit result less artificial. Also the order of spline function can be selected according to the requirement, making the code use convenient for different situations.

If necessary and possible, the resolved resonance parameters are given in CENDL-2. The parameters were evaluated by ourselves or taken from BNL-325. In any case, the parameters were adjusted to insure the continual connection with the smooth region at the bound energy point, and also the negative level parameters were adjusted to make the calculated cross section consistent with experimental data at thermal energy region. It was found that in some cases (e.g. for Fe), the resonances above the resonance energy region must be given, because there is a strong interference effect of these resonances with those in the resonance region.

To insure the correctness of the data in physics and format, before they were accepted and stored in CENDL, they all had been reviewed by our Evaluation Working Group and checked by using programs CHECKER, PHYCHE, FIZCON, which were transplanted from NNDC of America. For some nuclides, the inconsistency of the cross sections at some energy points, discomtinuity at the resonance bound point, the negative values of angular distribution, unnormalization of neutron spectrum were found. In this case, the adjusting of the cross sections, Legendre coefficients and renormalization of the spectrum were done correspondingly.

3 Theoretical Calculational Method

1) Light nuclides

By using phase shift analysis, the calculations of n-H, n-D and n-T scattering cross sections have been carried out[4]. The results are much better than those from the earlier similar works and resonance group theory, and agree with experimental data within relative error of 1%.

The double differential cross sections for n+D have been calculated with code TSD, which is based on solving Faddeev equation[5]. A set of low rank separable nucleon-nucleon potentials has been constructed, which includes different partial waves and tensor force. The calculated results are in good agreement with experimental data. The results also show that the contribution from off-shell effect is more significant than those from three body force and relativistic correlation.

To calculate the data of nuclei such as ^{6,7}Li, programs DRM, ROP and CRAP were developed[6]. DRM is a unified code of direct reaction mechanism and optical model for calculating the inelastic scattering. ROP and CRAP are two codes of R-matrix analysis. The former is a unified code of optical model and R-matrix theory. The results obtained by means of the optical model are used as background of R-matrix analysis in the resonance region. The latter is able to fit simultaneously the experimental data of both a reaction and the corresponding inverse reaction in a system.

2) Medium-heavy and fission nuclides

To calculate neutron nuclear data for medium-heavy and fission nuclides, some theoretical methods have been studied and relative codes have been developed.

In pre-equilibrium statistical theory, the pick-up mechanism for composite particle (d,t,^{3,4}He) emission was adopted[7]. The calculated results of (n,α), (n,d) ··· reaction cross sections were improved obviously in the high energy region. In addition, the Pauli exclusion effect in the exciton state density[8] and the effect of the Fermi motion, Pauli principle in the exciton model were taken into account[8] to improve the agreement between the calculated results and experimental data.

In the compound nucleus theory, width fluctuation correction, level-level correlation and discrete level effect etc. have been studied[9]. Width fluctuation correction for all kinds of reaction channels in a compound nucleus reaction was unitedly treated and successfully applied to the calculations. In order to calculate the spectra of the secondary emission particle in the continuum region at given outgoing angle by the width fluctuation corrected Hauser-Feshbach theory, a discretization method and a statistical method for the continuum levels of the residual nucleus have been used to deal with the width fluctuation correction factor[9].

For optical model, with certain versions of the Skyrme interactions we have obtained a microscopic optical potential without any free parameters[10], which shows that for certain energy regions the potential depth, shape, relative contributions of the surface and volume parts, as well as the energy dependences are in reasonable agreement with the results obtained from the phenomenological optical potential and those based on realistic nucleon-nucleon interaction. The calculated results, as a whole, are in agreement with experimental data. For Z, A region, where there are lack of experimental data, the microsopic optical potential may also be adopted to calculate the nuclear data, besides the phenomenological global optical potential.

In the research on the radiative neutron capture, three nonstatistical effects, the capture in shape elastic channels, compound elastic channels and compound inelastic channels, have been studied[11]. The research shows that the contributions of some nonstatistical effects must be taken into account carefully even if the incident neutron energy is smaller than 3 MeV. Nonstatistical effects lead to an improvement of the gamma emission spectrum in high energy region.

On the basis of research on theory mentioned above, some codes for data calculations have been developed.

MUP2[12] is the second version of a unified program for calculation of fast neutron data of medium-heavy nuclei by using optical model, Hauser-Feshbach theory with width fluctuation correction and pre-equilibrium statistical theory based on the exciton model and evaporation model. AUJP[13] is an associated code of MUP2, which is used for automatically searching for a set of optimal optical potential parameters. We have completed the intercomparison between MUP2 and some international nuclear model progams[14]. It was found that the accuracy of MUP2 is comparable with that of similar international codes and the results calculated with MUP2 are reliable.

For the calculation of fissile nuclides, FUP1[15] code has been developed based on MUP2, and ASFP[15] is its associated code used to automatically search for group of optimal fission parameters for a group of successive fissile nuclides.

CCOM[16] code is one of coupling channel optical model for deformed nucleus to calculate the direct reaction components of inelastic scattering. These direct components are added to the calculated compound nucleus cross section and angular distribution.

To calculate better, in addition to developing theoretical method and codes, the level density parameters also have been studied. Based on more recent reliable experimental data, two sets of parameters for Gilbert Cameron and back-shifted have been recommended[17]. Comparison with others shows that ours could give good argeement with experimental data. They have been used in the calculations for many nuclides and good results were obtained.

4　Comparison of CENDL-2 with ENDF／B-VI and JENDL-3

Compared with newest evaluated data ENDF／B-VI and JENDL-3, as there are only 54 nuclides and only files 1～5 for most nuclides in CENDL-2, the nuclides and files of CENDL-2 are much less than ENDF／B-VI and JENDL-3. Comparisons show that the data for most nuclides, which are included in all CENDL-2, ENDF／B-VI and／or JENDL-3, agree with each other within experimental er-

ror, but some are considerably different. For example, the double differential cross sections of D(n,2n) neutrons are given systematically in CENDL-2, which were calculated with Faddeev equation[5] and the results are consistent with experimental data quite well (Fig. 1). Compared to ENDF／B-VI, the cross section of inelastic scattering to 4.63 MeV level for ^{7}Li is considerably lower[18], this is based on newly measured data, in which the correction to D-D break-up neutron was made. For structural materials, some examples are the cross section of Ni(n,α)[19], Cr(n,2n) and (n,α)[20], Ta(n,2n) and (n,3n)[21], Nb(n,α)[22], they are shown in Figs. 2 to 6. For fissile nuclides, the comparison[23] with ENDF／B-VI of ^{239}Pu fission cross section is shown in Fig. 7, they are consistent with each other within 4%. The examples of the differences are the cross sections of ^{238}U(n,n$'$)[24], and ^{239}Pu(n,2n)[23] (Figs. 8, 9).

5　Development of Evaluation and Theoretical Calculational Method for Future

Although we have made great efforts to complete CENDL-2, it still is a primary one. It will be developed, including addition of γ-production data, double differential cross section and covariance file. For this purpose some evaluation and theoretical calculational methods and some programs concerned have been or are being developed. They will be used to develop CENDL further.

1)　Simultaneous evaluation method

It is based on spline fitting for multi-curves[25]. Several relative reaction cross sections and their ratios for different nuclides are fitted simultaneously. As a result, they are made consistent with each other, and the correlation between different reactions and nuclides can be calculated. Using this method, the fission cross sections of ^{235}U, ^{238}U, ^{239}Pu and capture cross section of ^{238}U have been evaluated[26].

2)　Covariance problem and correlative data processing

A method, called parameter analysis method, and a program have been developed to calculate the covariance matrix of evaluated data. Using the program, putting in the error information, the experimenters and evaluators can conveniently calculate the covariance matrix of the data. Correlative data processing, including average of single energy point data, curve fitting, and covariance production and propagation in data evaluation processing have been studied[27]. The spline fit and simultaneous evaluation programs mentioned above have been developed to be used for correlative data. Using these methods and programs, some covariance matrices for experimental data and covariance files for evaluated data have been tried to calculate[28], and the covariance files for more nuclides will be given in future.

3) Theoretical calculational codes MUP−3 and UNIFY−2

Recently theoretical calculational programs MUP−3[29] and UNIFY−2[30] have been developed. MUP−3 is to calculate double differential cross section of the secondary particles and to deal with composite particle emission. UNIFY−2 code is a unified treatment for the pre−equilibrium and equilibrium reaction processes with angular momentum and parity conservation. In addition to neutron data, this code can also calculate γ−production data, double differential cross sections for all kind of emitted particles and recoil nucleon. The contributions of some nonstatistical effects, such as, the capture in elastic and inelastic channels of the compound nucleus and direct−semi−direct capture etc. are carefully taken into account in the calculation of gamma production data.

Acknowledgments

The authors express their appreciation to CNNC / China for their support and organization to this work. Appreciation also to IAEA / NDS for their support (CRP, RCP, expert service etc.) and supply of EXFOR experimental data and some evaluated data, to BNL / NNDC for their technical help and supply of ENDF / B−VI data, format manual, and program system, to JAERI / NDC for their technical help and supply of JENDL−3 data and some programs.

REFERENCES

1. Liang Qichang, Ma Lizhen, Liu Tingjin, Sun Naihong, Zhang Limin: CNDP 1, 75(1989); 3, 56(1990); 4(1991)

2. Liu Tingjin, Shen Linxing, Liang Qichang, Liu Renqiu, Ma Lizhen: Chinese J. of Nucl. Sci. & Eng. 1, 79(1988) (in Chinese)

3. Liu Tingjin, Zhou Hongmo, Liu Renqiu: CNDP 2, 58(1989)

4. Wang Yansen, Chen Jianxin, Qiu Zhihong, Yuan Zhushu, Chen Chiqing: CNDP 3,24(1990)

5. Chu Lianyuan, Wang cuilan, Zhuang Youxiang: CNDP 6(to be published)

6. Zhu Yaoyin, Liu Manfen, Zhang Yujun, Hui Ruifa, Li Zhiwen, Piao Zhenlie: CNDP 1, 29(1989); Chen Zhenpeng, Wang Wanhong, Chen Qiankun, Chen Guojie: CNDP 1, 39(1989)

7. Zhang Jingshang: Commun. in Theor. Phys. 10, 33(1988)

8. J. S. Zhang, X. J. Yang: : Z. Phys. A 329, 69(1988); Sun Ziyang, Wang Shunuan, Zhang Jingshang, Zhuo Yizhong: Z. Phys. A 305, 61(1982)

9. Su Zongdi, Shi Xiangjun: Chin. Phys. 1, 952(1981); INDC (CPR)−2, 1986

10. Shen Qingbiao, Li Zhuxia, Shi Xiangjun: Proc. Inter. Summer School on N−N Inter. and Nucl. Many−Body Problem,Chang Chun, China, July 25~31,p 618(1983)

11. Liu Jianfeng, Zhang Xizi, Lu Zuhui: CNDP 1, 36(1989); 2,19(1989)

12. Cai Chonghai, Yu Ziqiang, Su Zongdi, Shi Xiangjun, Wang Shunuan: Chin. J. Sci. and Tech. of Atomic Energy 23, 2,7(1989) (in Chinese)

13. Zhou Hongmo, Yu Ziqiang, Zhang Xiaocheng, Zuo Yixin: (CNDC−85011), P138

14. Su Zongdi, P. E. Hodgson: Chin. J. Nucl. Phys. 11, 441(1989)

15. Cai Chonghai, Zuo Yixin, Wang Zhengxing: CNDP 3, 26,29(1990)

16. Yang Zesen, Zhou Zhiming: High Energy Phys. and Nucl. Phys. 4, 374(1980) (in Chinese)

17. Su Zongdi, Huang Zhongfu, He Ping, Zhou Chunmei: Chin. J. Nucl. Phys. 13, 2, 147(1991)

18. Yu Baosheng, Cai Dunjiu: The Evaluation of ^{7}Li Neutron Data (to be published)

19. Ma Gonggui, Zou Yiming, Wang Shiming: CNDP 6(to be published, 1991)

20. Ma Gonggui, Zou Yiming, Wang Shiming: CNDP 6(to be published, 1991)

21. Yao Lishan: CNDP 6(to be published, 1991)

22. Ma Gonggui, Zou Yiming, Wang Shiming: CNDP 6(to be published, 1991).

23. Liang Qichang: CNDP 6(to be published, 1991)

24. Tang Guoyou: CNDP 6(to be published, 1991)

25. Chen Baoqian, Liu Tingjin: Simultaneous Evaluation Method and Program (to be published)

26. Liu Tingjin, Deng Jingshan: CNDP 4, 49(1991); proceedings of this conference

27. Liu Tingjin, Chen Baoqian, Zhou Hongmo: Chin. J. of Atomic Ener. and Tech. 24, 1, 15(1990) (in Chinese)

28. Liu Tingjin: CNDP 6(to be published, 1991)

29. Yu Ziqiang, Zhang Xiaocheng, Zhou Hongmo: CNDP 1, 41(1989)

30. Zhang Jingshang: (to be published)

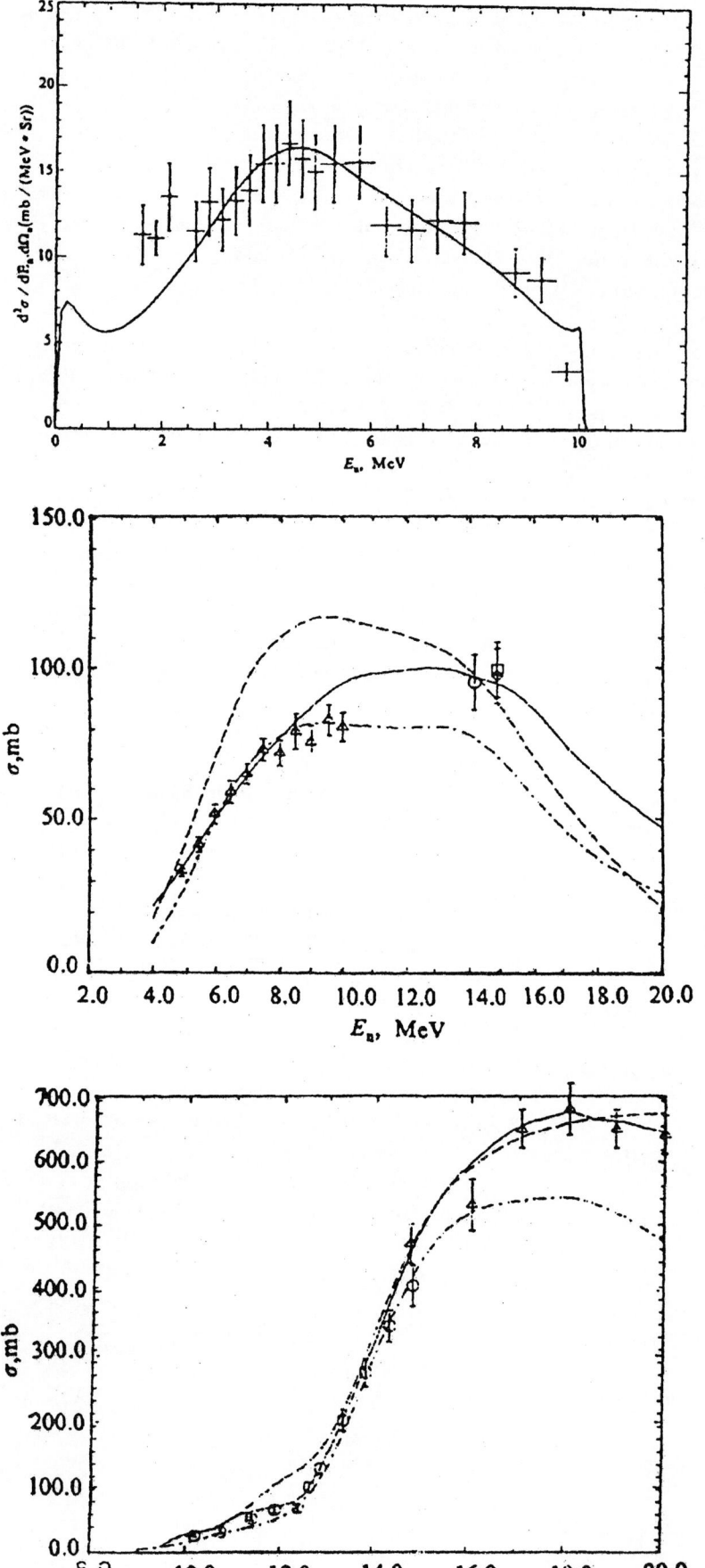

**Fig. 1 D(n,2n) Double
Differential Cross Section
at θ = 25°,
E_n = 13.6 MeV**
—— Present work,
+ Experimental data (China)

Fig. 2   Ni(n,α) Cross Section
—— Present work
— — ENDF / B–VI
— · — JENDL–3
⊙ GEL Wattecamps (83)
△ GEL Paulsen (81)
⊡ AI Kneff (86)
◇ LRL Grimes (79)

Fig. 3   Cr(n,2n) Cross Section
—— Present work
— — ENDF / B–VI
— · — JENDL–3
⊙ BRC Frchuat (80)
△ LAS Auchamp (77)

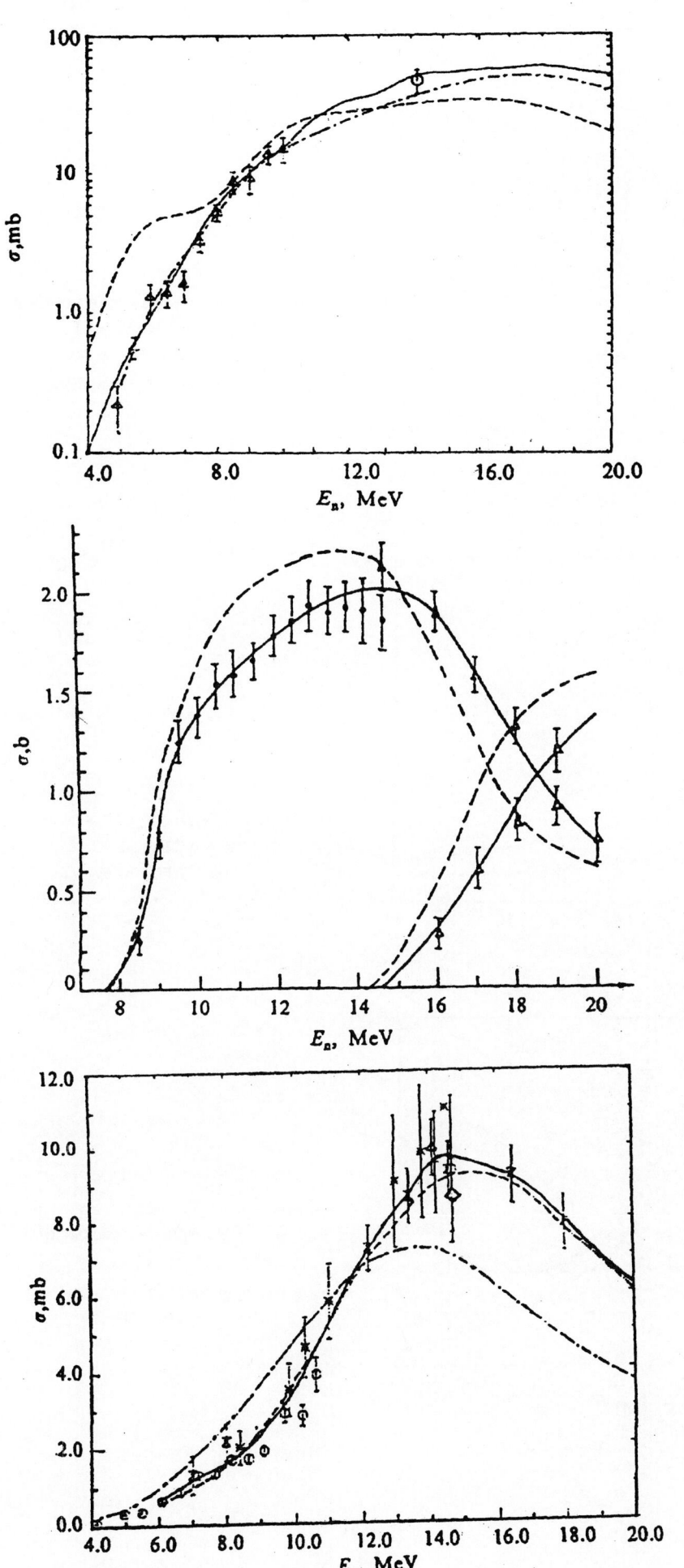

Fig. 4 Cr(n,α) Cross Section
——— Present work
– – – ENDF / B–VI
–·– JENDL–3
⊙ GEL Wattecamps (83)
△ GEL Paulsen (81)

Fig. 5 ^{181}Ta(n,2n) Cross Section
——CENDL–2,
– – – ENDF / B–VI

Fig. 6 Nb(n,α) Cross Section
——CENDL–2,
– – – ENDF / B–VI,
–·– JENDL–3
⊙ KFA Mannann (88),
△ IRK Fischer (83)
+ HAM Bormann (72),
⊡ RBZ Kulisic (64)
◇ ARK Bramlitt (63),
X LAS Bayhurst (61)
∗ LRL Tewes (60)

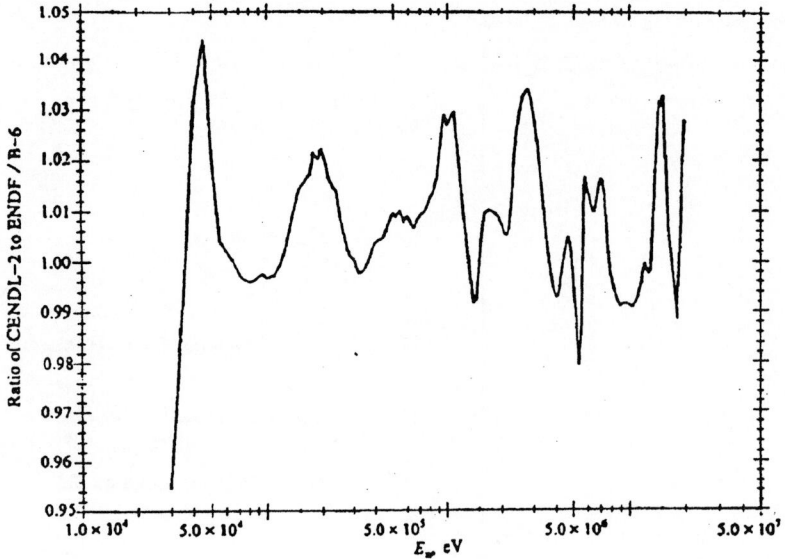

Fig. 7 The Comparison between
ENDF / B–VI and
CENDL–2 for
^{239}Pu(n,f) Cross Section

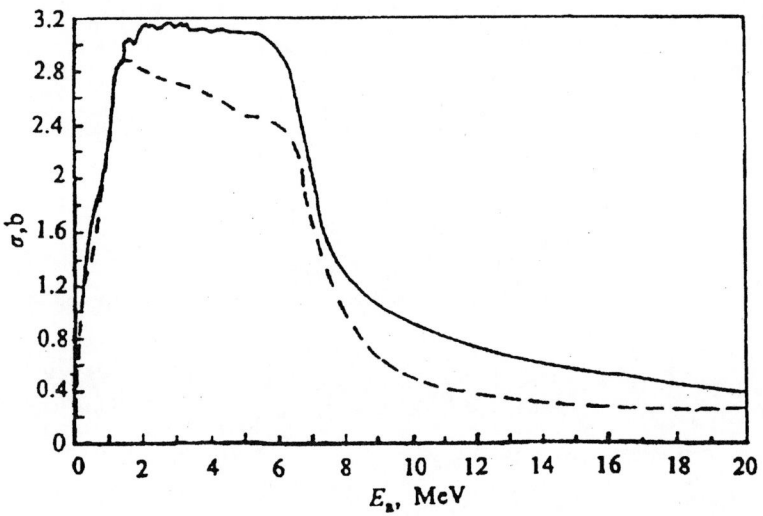

Fig. 8 ^{238}U(n,n′) Cross Section
– – – CENDL–2,
——— ENDF / B–VI

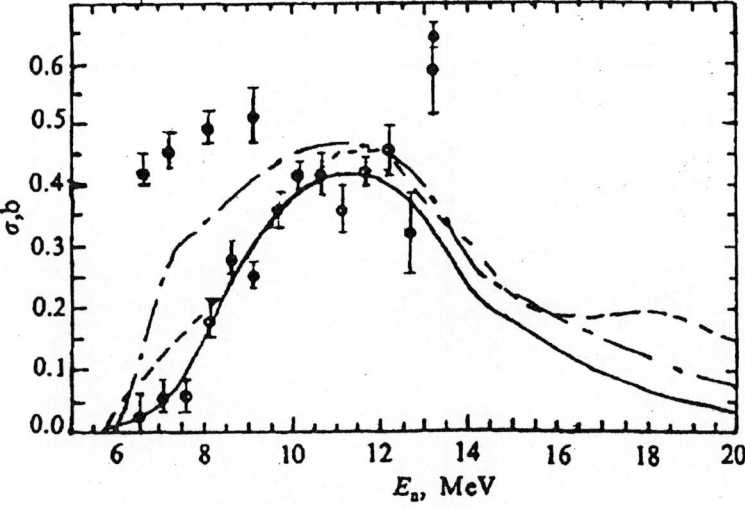

Fig. 9 ^{239}Pu(n,2n) Cross Section
——— CENDL–2,
— · — ENDF / B–VI,
– – – JENDL–3

THE NETWORK OF THE NUCLEAR REACTION DATA CENTRES

H.D. Lemmel
Nuclear Data Section, International Atomic Energy Agency
Vienna, Austria

V.N. Manokhin
USSR Nuclear Data Centre, Physics Energetics Institute
Obninsk, USSR

V. McLane
US National Nuclear Data Center, Brookhaven National Laboratory
Upton, New York, USA

S. Webster
Nuclear Energy Agency Data Bank
Gif-sur-Yvette, France

Abstract: Nine Nuclear Reaction Data Centers co-operate in the maintenance of international databases of experimental and evaluated nuclear reaction data for incident neutrons, charged particles and photons in order to satisfy the nuclear data needs in various fields of science and technology. The status of these databases is reviewed, and the services that scientists may obtain from the data centers are described. Emphasis is given to the current problems confronted by the data centers.

For more than 20 years national and regional nuclear data centers co-operate in the compilation of internationally available nuclear data libraries, to the effect that

- data obtained in a given country are made available speedily to data users in all countries;

- data are available world-wide in uniform computerized formats for which data-processing computer codes are also provided;

- data handbooks are produced from the international data files for the convenience of data users;

- for some more important nuclear data, international standard reference values have been established with high precision.

This has been achieved by a joint effort of nine nuclear reaction data centers co-ordinated by the IAEA Nuclear Data Section under the guidance of the International Nuclear Data Committee.

For the existing nuclear data centers and their acronyms see table 1.

The nuclear data centers provide an essential service required for pure and applied nuclear science and technology. The co-operation of the nuclear data centers can be considered as a model for various other scientific disciplines (such as materials properties data) for which similar international databases do not yet exist.

Unfortunately, the continuing high quality of the data center services is now seriously endangered due to repeated budget cuts or changing work priorities of the data centers involved. This has the consequence that

- the updating of some of the databases will be less complete and less speedy than it used to be;

- there are new data requirements which cannot be taken care of by the staff available at the existing data centers (such as medium energy nuclear reaction data for research on actinide burning in accelerators and for medical applications and radiation damage in space technology; or nuclear activation reactions for a large number of isotopes occurring in nuclear waste or in fusion technology; or charged particle reaction data for various scientific and industrial applications);

- database software maintenance and development will become increasingly inadequate, thus reducing the speed and efficiency of retrievals; notably there is only slow progress at most centres in the development of an effective system of on-line access.

Complete, speedy and reliable nuclear data center services can be continued only,

- if the gradual decrease in data center support is reversed,

- and if additional data center manpower is granted either by additional staff at the existing centers or by the creation of new specialized data centers within the co-ordinated data center network.

Table 1. List of Cooperating Nuclear Reaction Data Centers

Code	Address	Type of Data Compiled
CJD	USSR Nuclear Data Center, Fiziko Energeticheskij Institut, Obninsk, Kaluga Region, USSR	Neutron reaction data
NDS	IAEA Nuclear Data Section P.O. Box 100, A-1400 Vienna, Austria DATEX-P 2322 Host ID 6221047 TYMNET 2329 Host ID 11507701 EARN/BITNET: "RNDS@IAEA1"	Neutron reaction data and others
NEADB	NEA Data Bank, F-91191 Gif-Sur-Yvette, France EARN/BITNET: "DBMAIL@FRNEAB51" TRANSPAC (PSN): (0)208 – (0)or(1)91040946 (Username: NEADB) PHYSNET/HEPNET: Decnet node address 32.9	Neutron reaction data
NNDC	US National Nuclear Data Center, Brookhaven National Laboratory, Upton, N.Y., U.S.A. 11973 BITNET "NNDC@BNL"	Neutron and charged particle reaction data
CAJaD	Center for Nuclear Structure and Reaction Data of the USSR State Committee on the Utilization of Atomic Energy, I.V. Kurchatov Institute of Atomic Energy Moscow, USSR	Charged particle reaction data
CDFE	Center for Photonuclear Experimental Data Nauchno-Iss. Inst.Yad.Fiz., Moskovskiy Gos. Unviersitet Leninskiye Gory, Moscow, USSR	Photonuclear data
CNDC	Chinese Nuclear Data Center, Institute of Atomic Energy P.O. Box 275, Beijing, People's Republic of China	Neutron and charged particle reaction data
RIKEN	Nuclear Data Group, RIKEN Institute of Physics and Chemistry Research, Wako-Shi, Saitama, Japan 351-01	Charged particle reaction data for medical radioisotope production
Study Group	Japanese Nuclear Data Study Group, c/o Dr. M. Chiba, Hokkaido University, Computing Center Kita-11 Nishi-5, Kita-Ku Sapporo 060, Japan	Differential charged particle reaction data

The following center has made significant contributions in the past but was discontinued:

KaChaPaG	Karlsruhe Charged Particle Group Federal Republic of Germany	Charged particle reaction data

The following centers are not formal members of the network but contributed important comprehensive evaluated data files:

JAERI	Nuclear Data Center, Japan Atomic Energy Research Institute, Tokai-Mura, Naka-Gun, Ibaraki-Ken 319-11 Japan	Evaluated neutron reaction data
LLNL	Nuclear Data Group, Lawrence Livermore National Laboratory, P.O. Box 808, Livermore, California 94550 U.S.A.	Evaluated neutron, photon and charged particle reaction data

and
others

1. Data center services:
   The following centers should be contacted for data retrievals, copies of data libraries, related documents and data processing computer codes, depending on the geographical location:

   CJD - USSR; (resp. CAJaD in case of charged particle and photonuclear data);
   NEADB - OECD countries in Europe and Japan;
   NNDC - USA and Canada;
   CNDC - the People's Republic of China (services being developed);
   NDS - all other countries.

   Of these, NEADB and NNDC offer on-line computer services via telephone and computer networks such as HEPNET or INTERNET. Included are the CINDA, EXFOR, and evaluated nuclear data files. Additionally, the two centers also offer on-line access to the Nuclear Structure References and ENDSF data files. Contact the appropriate center for more information (see Table 1). At the other centers on-line services are so far only in a preparatory stage. The services of the centers are advertised by means of newsletters [1].

   Whereas the dissemination of data to customers upon their request is concentrated at the above centers, the data compilation and the production of evaluated data files involve contributions from various additional centers and research groups.

2. Neutron reaction data:
   CJD, NEADB, NNDC and NDS co-operate in the following projects.

2.1 WRENDA [2]
   This is a computer produced periodical document which lists (mostly) neutron nuclear reactions for which adequate data are missing or where existing data have insufficient accuracy to meet the requirements in the physics of fission and fusion reactors and nuclear material safeguards. The requests included have been reviewed by official bodies such as national or international nuclear data committees.

2.2 CINDA
   This is a bibliographic index to the literature and numerical databases on microscopic neutron data. The centers maintain a computer file from which selective retrievals are available. CINDA handbooks are published annually by IAEA [3]. The present file has 240 000 entries. The unique feature of CINDA compared to other bibliographies is that references referring to the same data set are blocked together, and that superseded references are either labelled as such or omitted from the published book, thus avoiding "noise" in retrievals.

2.3 EXFOR [4]
   This is a world-wide comprehensive compilation of experimental neutron reaction data, containing about 50 000 data sets for 600 nuclides in 4 million records (=320 Megabytes). It includes not only the numerical values but also detailed information on the measurement method and data uncertainty analysis. In particular, EXFOR includes data
   - that have not been published in numerical form,
   - and data that supersede published values.

   Reliability and accuracy are ensured as far as possible through a combination of measures: the soliciting of large files of experimental data either on tape or by electronic mail directly from the experimentalist, the use of rigorous checking programs, and where possible the approval by the author or experimentalist concerned of the EXFOR proof copy.

   For a set of published handbooks on experimental neutron reaction data see [5].

   The EXFOR database of experimental data is the basis for the development of evaluated data libraries with "best" values for all neutron nuclear reactions.

2.4 ENDF - evaluated data [6]
   Except for selected standard reference data, there is not yet one universal file with recommended best neutron cross-section values. Different data libraries with often disagreeing numerical values are used in different countries. However, all these data libraries are now in the internationally recognized ENDF format [7]. The following comprehensive evaluated neutron data libraries exist at present, containing data for 411 elements or isotopes [8]:
   - BROND (USSR), 94 elements/isotopes [9]; BROND-2, 163 elements/isotopes, being presented at this conference;
   - ENDF/B-VI (USA), 320 elements/ isotopes [10];
   - JEF-1 (NEA), 303 elements/isotopes [11]; JEF-2 has been prepared but its availability at present is restricted to users within the NEA Data Bank community;
   - JENDL-3.1 (Japan), 324 elements/ isotopes [12].

   Partly, these data libraries complement each other: an isotope not contained in the one library may be included in the other. In general, these data libraries satisfy most of the data needs. But continuing work is required in many areas:
   - Discrepancies among the data libraries must be analyzed in order to decide, which data of which library may be the best choice for a given nuclear reaction.
   - A large number of reactions, in particular neutron activation cross-sections required for applications in nuclear fusion, nuclear waste disposal, reactor decomissioning, etc., are not yet contained in any of the four major evaluated data libraries.

   Data evaluation activities are reported in an international newsletter [13], and the evaluation efforts in OECD countries are coordinated by a newly established working group on evaluation cooperation.

2.5 ENDF data processing codes

The ENDF-formatted data files are accompanied by data processing computer codes [14] by which the original ENDF formatted data files can be transformed to other formats, in particular multigroup formats required as input to application codes. This includes processes like interpolation, avaraging, graphical plotting, and specifically the reconstruction of cross-section data from the resonance-parameter data.

In some respects the complexity of the new data files presents problems, especially for the user with more modest computer resources at his disposal. For example, pointwise cross-section generation from parametric files containing many hundreds of resolved resonances will inevitably incur long run times and produce voluminous output files [15]. Clearly there is a need for improved and more user-friendly data handling codes, capable of sifting the vast amounts of data present in these original libraries and producing optimized files tailored to specific applications.

2.6 Standard reference data

So far, there are only a few neutron data areas where internationally recommended values exist. These are
- Standard cross-sections for nuclear measurements, i.e. the ENDF/B-VI standards file [16];
- IRDF-90, the international reactor dosimetry file containing recommended values of neutron activation cross-sections for selected materials used for reactor neutron dosimetry by foil activation [17];
- Actinide decay data [18];
- X-Ray and Gamma-ray standards for detector efficiency calibration [19].

2.7 Specialized neutron reaction data libraries

In addition to the ENDF-formatted evaluated data libraries for general application (see item 2.4 above), there are many specialized evaluated neutron data libraries [20] available from the nuclear data centers. Space limitations permit only a few examples to be quoted here:
- JEF-1 Thermal Neutron Scattering Law Library [21];
- Chinese Evaluated Fission Product Yield Library [22];
- UKFY2, new evaluated library of fission product yields from UK [23];
- JNDC-FP2, the fission product nuclear data library of the Japanese Nuclear Data Committee, including data for yields, decay data and neutron cross-sections of fission product nuclei [24];
- USSR Evaluated Nuclear Data Library for Neutron Activation Analysis [25].
- USSR Fast Neutron Activation Cross-section file ACTIV87 [26].

Data compilations for special applications are also issued as handbooks [27, 28, 29].

3. Charged-particle reaction data
Some of the data centers (see Table 1) are specialized in the compilation of charged-particle and heavy-ion reaction cross-sections or in photonuclear data in the international EXFOR format. For an index to the data already compiled see [30].

Due to lack of manpower, a systematic compilation to a similar degree of completeness as for neutron data has not yet been possible. Special emphasis is given to
- charged-particle induced neutron source reactions [31];
- cross-sections used as proton beam monitors [32], specifically for radioisotope production [33];
- cross-sections for the production of radioisotopes for medical applications [34];
- (α,n) reactions for selected materials [35];
and others. The existing data compilations are still incomplete and evaluated data files with recommended values are seriously lacking for many applications. The creation of additional specialized data centers embedded in the international data center network is required in order to be able to provide the scientific community with reliable nuclear data for applications. This is particularly needed for intermediate energy data and for medical application data.

The ENDF format originally developed for neutron reaction data has been widened to include also evaluated data of charged-particle induced reactions. A first demonstration file has been produced for neutron and proton reaction data up to 1 GeV for Fe-56 [36].

4. Photonuclear and photo-atomic reaction data
For photonuclear data a report series [37] is issued by the USSR Photonuclear Data Center in Russian and English, containing partly bibliographic compilations, partly data compilations on selected topics such as photofission.

Numerical photoneutron data and selected other photonuclear reaction data have been compiled in EXFOR [4, 30]. For photo-atomic interaction data see
- the ENDF/B-VI Photon Interaction Data Library [38],
- XCOM, a PC database for various photon cross-sections in any element or chemical compound [39].

References:

[1] NNDC Newsletter, issue 103, No. 91-1 (1991). - News from the NEA Data Bank No. 12 (1991). - IAEA Nuclear Data Newsletter, issue 15 (1991). US National Nuclear Data Center, Fast Neutron Cross-Section Newsletter 14 (1989), 15 to be publ. 1991. - M. Lammer, Progress in Fission Product Nuclear Data No. 13, report INDC(NDS)-222 (1990).

[2] WRENDA 87/88. Wang Dahai (ed.), IAEA report INDC(SEC)-95 (1989). The preparation of a new issue is underway.

[3] CINDA-A, covering the literature 1935-1987, 5 volumes, IAEA, Vienna, 1990. - CINDA 90, covering the literature 1980-1990, IAEA, Vienna, 1990. - CINDA 91 in preparation. - S. Webster, NEA-DB, CINDA Manual, report IAEA-NDS-109 Rev. 90/2.

[4] H.D. Lemmel, Short Guide to EXFOR, report IAEA-NDS-1 Rev. 5 (1986). - V. McLane, CSISRS Users' Manual, report BNL-NCS-41300 (1988). - NEA Data Bank, User Guide to numerical neutron data retrievals, and Data Bank guides to on-line and neutron data services; numbered reports available from the NEA Data Bank. - V. McLane, EXFOR Manual, report IAEA-NDS-103 Rev. 88-1 (1988).

[5] Neutron Cross-Sections, Academic Press. Vol. 1A: S.F. Mughabhab, M. Divadeenam, N.E. Holden, Neutron Resonance Parameters and Thermal Cross-Sections, Z = 1-60 (1981). Vol. 1B: S. F. Mughabhab, Neutron Resonance Parameters and Thermal Cross-Sections, Z = 61-100 (1984). Vol. 2: V. McLane, C.L. Dunford, P.F. Rose, Neutron Cross-Section Curves (1988).

[6] For the concept of "evaluated data" versus "experimental data" see, e.g., H.D. Lemmel, D.E. Cullen, J.J. Schmidt: Nuclear data files for reactor calculations and other applications - experimental data/evaluated data. Computer Physics Communications 33 (1984) 161-171.

[7] ENDF format manuals "ENDF-102" for different ENDF versions:
ENDF-6, P.F. Rose, C.L. Dunford, report BNL-NCS-44945 (July 1990). (Also report IAEA-NDS-76 Rev. 3.)
ENDF-5, R. Kinsey, report BNL-NCS-50496 2nd Ed. (Oct. 1979) updated with B.A. Magusno, report BNL-NCS-50496 2nd Ed. Rev. (Nov. 1983). (Also report IAEA-NDS-75 Rev. 1.)

[8] H.D. Lemmel, Index to BROND, ENDF/B-6, JEF-1, JENDL-3, report IAEA-NDS-107 Rev. 5 (1991).

[9] BROND-USSR Evaluated Neutron Data Library, by V.N. Manokhin et al. Documentation by H.D. Lemmel, P.K. McLaughlin, report IAEA-NDS-90 Rev. 4 (1991).

[10] ENDF/B-VI. The U.S. Evaluated Nuclear Data Library for Neutron Reaction Data. Summary of contents by H.D. Lemmel, report IAEA-NDS-100 Rev. 3 (Sept. 1990).

[11] JEF-1, the Joint Evaluated File version 1. Comprehensive index in JEF Report 1, 2 vols. (1985), available from the NEA Data Bank. - Y. Nakajima, Graphical comparison of JEF-1 with EXFOR cross-section data, JEF Report 8 (1986). - Summary of contents by H.D. Lemmel, report IAEA-NDS-120 (Sept. 1990).

[12] K. Shibata et al., Japanese Evaluated Nuclear Data Library, Version 3 - JENDL-3, report JAERI-1319 (1990). Supplemented by JENDL-3-FP, a special data library for the neutron reaction data of fission product nuclei. JENDL-3.1, Summary of contents by H.D. Lemmel, report IAEA-NDS-110 Rev. 3 (1991).

[13] NEA Data Bank, Neutron Nuclear Data Evaluation Newsletter issued annually, latest issue NNDEN-44 (1990).

[14] D.E. Cullen, P.K. McLaughlin, S. Ganesan, The 1989 ENDF Pre-processing Codes, report IAEA-NDS-39 Rev. 6 (1991). - Same, PC version, report IAEA-NDS-69. - US National Nuclear Data Center, ENDF Utility Codes Version 6.6; documentation by P.K. McLaughlin, report IAEA-NDS-29. - NJOY, a code system to be requested from the Reactor Shielding Information Center (RSIC), Oak Ridge National Laboratory; introduction see, e.g., D.W. Muir, report IAEA-NDS-119 (1990).

[15] S. Ganesan, D.W. Muir, P.K. McLaughlin, Cross-section data files generated from ENDF/B-VI resonance parameter data files, report IAEA-NDS-130 (1991). The ENDF/B-VI file for U-238 has now 1913 resolved resonances in multilevel Reich-Moore formalism. To compute the cross-section data from such a large number of resonances with the codes LINEAR/RECENT at 0.1% accuracy requires 19.5 hours of CPU time on an IBM-3081 computer, and the resulting data file containing the U-238 cross-sections as a function of energy has a size of 937252 records, that is 75.2 Megabytes. When the same problem was solved with the reduced accuracy of 0.5% and the code RECONR/NJOY8931, 156 minutes CPU time were required, and the resulting cross-section data file had a size of 150 000 records = 12 Megabytes.

[16] US National Nuclear Data Center, ENDF/B-VI standards file. Summary documentation see H.D. Lemmel, P.K. McLaughlin, ENDF/B-VI Standards Library, report IAEA-NDS-88 Rev. 2 (1990).

[17] N.P. Kocherov, P.K. McLaughlin, IRDF-90, the International Reactor Dosimetry File, report IAEA-NDS-141 (1990).

[18] A. Lorenz (ed.), Decay Data of the Transactinium Nuclides, IAEA Technical Report No. 261, Vienna 1986. (To be updated).

[19] International Atomic Energy Agency, X-Ray and Gamma-Ray Standards for Detector Calibration, A. Lorenz and H.D. Lemmel (ed.), IAEA Technical Document, to be published 1991.

816

[20] H.D. Lemmel, Index of data libraries available on magnetic tape from the IAEA Nuclear Data Section, report IAEA-NDS-7 Rev. 91/4.

[21] J. Keinert, M. Mattes, JEF-1 Scattering Law Data, JEF Report 2, JEF/DOC 41.2, IKE 6-147 (1984), available from NEA Data Bank. - Compare summary documentation in report IAEA-NDS-121.

[22] Wang Dao, Zhang Dongming, Chinese Evaluated Fission-Product Yield Library 1987. Report IAEA-NDS-91.

[23] M.F. James, R.W. Mills, D.R. Weaver. A new evaluation of fission product yields and the production of a new library (UKFY2) of independent and cumulative yields. Parts I - III. Reports AEA-TRS-1015, -1018 and -1019. Summary documentation see M. Lammer, report IAEA-NDS-124.

[24] The JNDC Nuclear Data Library of Fission Products, second version. Summary documentation by O. Schwerer, report IAEA-NDS-51 Rev. 3 (1991). Detailed documentation: H. Ihara (ed.), report JAERI-M-89-204 (1989) and K. Tasaka et al., report JAERI-1320 (1990).

[25] V.M. Bychkov et al., USSR Evaluated Nuclear Data Library for Neutron Activation Analysis 1988. Documented by A.B. Pashchenko in report IAEA-NDS-125.

[26] V.N. Manokhin, A.B. Pashchenko, V.I. Pljaskin, M.V. Bychkov, V.G. Pronjaev, Fast Neutron Activation Cross-Section File, see ref. [27]. Summary documentation see O. Schwerer, report IAEA-NDS-96.

[27] K. Okamoto (ed.), Handbook on Nuclear Activation Data, IAEA Technical Report 273 (1987).

[28] N. Kocherov, Handbook on nuclear data for applications in nuclear geophysics, IAEA Technical Document, to be published 1991.

[29] M. Lammer, O. Schwerer, Draft for a handbook on nuclear data for safeguards, report INDC(NDS)-..., to be published 1991.

[30] V.V. Varlamov, G.M. Zhuravleva, V.V. Surgutanov, F.E. Chukreev, Jadernye Reakcii pod Dejstviem Zarjazhennykh Chastic i Fotonov v Sisteme EKSFOR (Nuclear Reactions of Charged Particles and Photons in the EXFOR System), a data handbook, Moskva, CNII Atominform 1987.

[31] M. Drosg, Neutron source reactions, data library with computer code, 1990. Documentation: O. Schwerer, report IAEA-NDS-87 Rev. 2 (1990).

[32] V.A. Vukolov, F.E. Chukreev, Evaluated cross-sections used as proton beam monitors, English translation by A. Lorenz, report INDC(CCP)-330 (1991).

[33] O. Schwerer, K. Okamoto, Status report on cross-sections of monitor reactions for radioisotope production, report INDC(NDS)-218 (1989).

[34] A. Hashizume, Y. Tendow, K. Kitao, Compilation of excitation functions for the production of the radionuclides I-123, Xe-123 and Cs-123 by charged-particle induced reactions, report INDC(JPN)-144 (1990).

[35] V.A. Vukolov, F.E. Chukreev, Tables of recommended standard data: neutron yields from alpha particle induced reactions on Li, Be, B, C, O and F for energies up to 10 MeV, English translation by A. Lorenz, report INDC(CCP)-331 (1991).

[36] S. Pearlstein, ENDF/B-6 High-Energy Data Library, see H.D. Lemmel, report IAEA-NDS-113 (1990).

[37] I.N. Boboshin, V.V. Varlamov, V.V. Sapunenko, M.E. Stepanov, V.V. Surgutanov, Fotojadernye Dannye - Photonuclear Data No. 13 1989 (1990).

[38] US National Nuclear Data Center, ENDF/B-VI Photon Atomic Interaction Data Library (1990). Summary documentation H.D. Lemmel, report IAEA-NDS-58 Rev. 2 (1990).

[39] M.J. Berger, J.H. Hubbell, XCOM: Photon cross-sections on a PC, version 1.2 (1987). Summary documentation H.D. Lemmel, report IAEA-NDS-89 (1987).

EVALUATED NUCLEAR STRUCTURE DATA FILE (ENSDF)

M. R. Bhat

National Nuclear Data Center,
Brookhaven National Laboratory,
Upton, NY 11973 USA

Abstract: The Evaluated Nuclear Structure Data File (ENSDF), is maintained by the National Nuclear Data Center (NNDC) on behalf of the international Nuclear Structure and Decay Data (NSDD) network organized under the auspices of the International Atomic Energy Agency. ENSDF provides evaluated experimental nuclear structure and decay data for basic and applied research. The activities of the NSDD network, the publication of the evaluations, and their use in different applications are described. Since 1986, the ENSDF and related numeric and bibliographic data bases have been made available for on-line access. The current status of these data bases, and future plans to improve the on-line access to their contents are discussed.

(evaluated nuclear structure data, nuclear structure references, evaluated nuclear reaction data, on-line access)

Introduction

The data on discrete bound states of nuclei are of interest to basic and applied research. This paper describes the international effort organized to evaluate these data, the status of this work and the availability of the evaluations for various applications.

Nuclear Structure & Decay Data (NSDD) Network

The international Nuclear Structure & Decay Data (NSDD) network is organized under the auspices of the International Atomic Energy Agency, and co-ordinated by the National Nuclear Data Center (NNDC) at the Brookhaven National Laboratory. The NSDD network is made up of evaluators in 10 countries and 15 institutions shown in Table 1. The activities of the Chinese Nuclear Data Center, Beijing, include evaluations done at Jilin University, Changchun, and the Institute of Nuclear Research, Shanghai. Evaluations are also performed at the National Tsing Hua University, Taiwan. The network is responsible for the evaluation of all the mass-chains on a continual basis. A mass-chain is formed by all the nuclides of the same mass and different chemical elements. Each of the members of the network has a permanent assignment of mass-chains depending on the number of evaluators, their research interests, or special expertise in a particular mass-region. Temporary assignments of mass-chains are also possible depending on the current interests of the evaluators and also to help out other members of the network in updating the mass-chains that are old.

Table 1 The NSDD Network	
Country	Institution
Belgium	Ghent University
Canada	McMaster University
France	Centre d'Etudes Nucleaires de Grenoble
Japan	Japan Atomic Energy Research Institute
Kuwait	Kuwait Institute for Scientific Research
The Netherlands	University of Utrecht
Peoples' Republic of China	Chinese Nuclear Data Center
Sweden	University of Lund
Union of Soviet Socialist Republics	Institut Atomnoi Energii I.V.Kurchatova Data Center, Leningrad Nuclear Physics Institute
United States of America	Brookhaven National Laboratory Idaho National Engineering Lab Lawrence Berkeley Laboratory Oak Ridge National Laboratory Triangle Universities Nuclear Lab

The submitted manuscript has been authored under Contract No. DE-AC02-76CH00016 with the U.S. Department of Energy. Accordingly, the U.S. Government retains a nonexclusive, royalty-free license to publish or reproduce the published form of this contribution, or allow others to do so, for U.S. Government purposes.

Evaluated Nuclear Structure Data File (ENSDF)

The Evaluated Nuclear Structure Data File[1] (ENSDF), is made up of evaluated experimental data for A=1–266. ENSDF is maintained by the NNDC on behalf of the international NSDD network. For A>44, the evaluations are coded in the ENSDF format for entry into the data file and are published in the Nuclear Data Sheets. For 3<A<45, the evaluations are published in the journal Nuclear Physics and a subset of the published data is extracted and entered into the ENSDF. It is expected that future evaluations of A=3–20 by the Triangle Universities group will be coded into the ENSDF. The publication status of the evaluations for the light nuclei A<45 is shown in Table 2. For the remaining nuclei in the ENSDF, the status is shown in Fig. 1. The ENSDF is updated when new evaluations are published in the Nuclear Data Sheets.

Table 2 Publication Status of Evaluations for A=1–44		
A–range	Date*	Reference
1–2		@
3	6/87	Nuc. Phys. **A474**, 1 (1987)@
4#	12/72	Nuc. Phys. **A206**, 1 (1973)@
5–10	6/88	Nuc. Phys. **A490**, 1 (1988)
11–12	6/89	Nuc. Phys. **A506**, 1 (1990)
13–15	7/90	Nuc. Phys. **A523**, 1 (1991)
16–17	6/86	Nuc. Phys. **A460**, 1 (1986)
18–20	6/87	Nuc. Phys. **A475**, 1 (1987)
21–44	6/90	Nuc. Phys. **A521**, 1 (1990)
* Literature cut-off dates for published evaluations @ Unpublished partial evaluations are included in ENSDF # A new evaluation by the Triangle Universities group is in progress		

The Contents of ENSDF

ENSDF contains evaluated experimental data summarizing the present knowledge on the structure and decay of nuclei. If there are gaps in experimental data, they are not filled in with theoretical or nuclear model calculations. While the emphasis is on experimental data, the evaluations do use well-founded systematics or theory. Thus the ENSDF can serve as a guide for planning future experimental work, and to test or develop new theories. It can also be used as input for nuclear model code calculations of reaction data or other applications such as decay heat calculations in reactors or nuclear medicine.

The ENSDF is organized as a collection of data sets for each nuclear species. The data sets are of the following types: adopted properties of the nucleus, and source data sets each of which gives the evaluated data of a single type of measurement such as radioactive decay or reaction experiments. The contents of the source data are combined together by an evaluator to arrive at the adopted properties which may be considered as the "best" or the recommended properties of the nuclear levels or radiations.

In the adopted data sets, β^- and α decay energies of the ground state, and neutron and proton separation energies are given for each nuclide. For each level the following information is given: its excitation energy, J^π with arguments supporting the assignment, half-life or total width, decay branching for ground state and isomers, static electric and magnetic moments, flags indicating in which decay and reaction data sets the level is seen, the configuration assignments (Nilsson orbitals or shell model), band parameters, isomer and isotope shifts, references to data on charge distribution of ground states, deformation parameters and electric and magnetic excitation probabilities. For the γ-rays, their placement in the level scheme, measured γ-ray or E0 transition energy, relative photon intensity from each level, electric or magnetic multipole character, the mixing ratio, and nuclear penetration parameter, total internal-conversion coefficient when appreciable and reduced transition probabilities are given. In the decay or reaction data sets, nuclear structure data extracted from β decay, α decay, isomeric decay, Coulomb excitation, particle transfer reactions such as (d,p), (t,p) etc., heavy ion reactions, or mesic atom studies are presented. The contents of the ENSDF at this time are shown in Table 3.

Table 3 Current Contents of ENSDF	
Card Images:	$\approx 990,000$
Data Sets:	11,002
Nuclides:	2,325
Data Sets:	
Adopted Levels, Gammas*	2,321
Decay Data	2,917
(including spontaneous fission)	
Reactions	5,390
Muonic Atom	28
Mossbauer	18
Comments	223
References	252
*Includes decay and reaction data sets for nuclei which have no adopted data sets	

The evaluations for A>44 are sent as a computer file by the evaluator to the NNDC for further processing. The NNDC maintains and distributes to the network members a number of format and physics checking codes to aid in evaluating the data and in assembling the data files. These data files are processed and checked at the NNDC to prepare a pre-review copy which is sent back to the evaluator. The evaluator proofreads this copy, checks the evaluation and sends it back with any corrections. These corrections are implemented, and a review copy prepared and sent for refereeing which checks the evaluation for completeness and correctness of data, physics content, documentation and style of presentation. The referee's comments are sent to the evaluator along with any corrections or suggestions from the Editor-in-chief. These corrections, changes and any updates to the data files by the evaluator are included in the post-review copy

A-Chain Status in ENSDF (A>44)
Center - ALL
25-APR-91

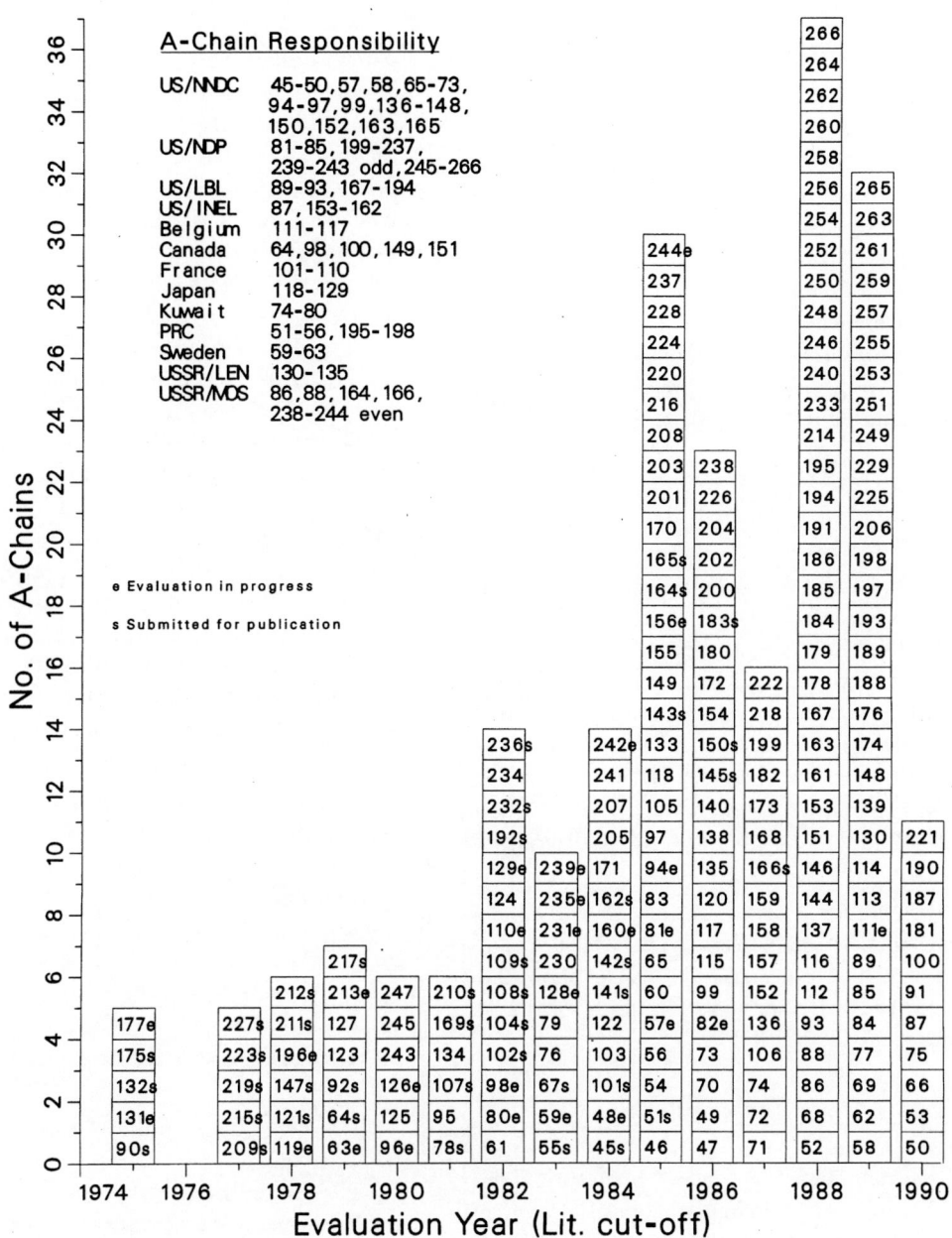

Fig. 1. The status of A-chains in ENSDF for A>44.

which is sent to the reviewer for approval of changes. If there are no problems, the manuscript is then prepared in publication format. This involves suppressing several computer-generated numbers, tables and drawings and generating a new lay-out without in any way sacrificing the essential physics contents of the publication. The final camera-ready publication copy is then sent to the publisher and after an evaluation has been published in the Nuclear Data Sheets, it is merged into the ENSDF.

New mass-chain evaluations are published in 9 issues of the Nuclear Data Sheets per year. In this publication, the format is designed to convey to the user the necessary physics information in the evaluation in a direct transparent manner. Drawings of radioactive decays and reactions with gammas are given showing band structure where needed. Detailed results that could not be adequately shown in the drawings are given in Tables. An Index to the Drawings and Tables precedes these; a list of References to the literature used in the evaluation is also given.

Applied Uses of ENSDF

The data in the ENSDF can be selectively extracted, processed and re-formatted to suit special applications. A program MEDLIST [2,3] was written in 1975 to use ENSDF data for nuclear medicine. RADLST [4] is a much enhanced and improved version of this program. It is designed to calculate the nuclear and atomic radiations associated with radioactive decay of nuclei for a variety of applications such as nuclear medicine, health physics, industry, nuclear power, geophysics, and environmental studies.

The ENSDF is also a source from which a number of publications such as the Nuclear Wallet Cards[5] published by the NNDC are derived. It is also planned to extract data for the projected new 8th edition of the Table of Isotopes from the ENSDF.

Nuclear Structure References (NSR) File

The NNDC maintains the Nuclear Structure References (NSR) file which is a computer file of indexed references to low and intermediate energy nuclear physics. This file was initially begun in support of the nuclear structure data evaluations for the ENSDF. However, this bibliographic file has found a much larger class of basic and applied physics users in addition to the network evaluators. The scope of literature coverage in the NSR has also expanded over the years to include new areas of active research published in the leading publications in the field such as Physical Review C or Nuclear Physics A.

The updates to the NSR are published every four months in three issues of the Nuclear Data Sheets as Recent References; the file however, is updated weekly. This publication is divided into four sections: (i) key numbers and keywords: each key number is a unique index for each entry followed by a keyword abstract containing information on specific nuclei; (ii) reaction index: an index to experimental papers on nuclear reactions; (iii) references: a bibliography of experimental nuclear reaction and structure articles ordered by year of publication and last name of the first author and (iv) secondary sources: contain an index to non-journal literature. Though the emphasis is on experimental papers in literature coverage, theoretical papers dealing with specific nuclei and reactions are also scanned and assigned keywords. The NNDC regularly scans 80 journals; these are listed in the introductory section of the Recent References. In addition to these primary references, conference proceedings published in Nuclear Physics A, Nuclear Instruments & Methods in Physics Research and other journals are also coded as primary references. Secondary references published in progress reports of leading laboratories around the world and important conference proceedings are scanned regularly. Other secondary sources are covered on a time available basis or on request from the network evaluators for mass-chain evaluations. Over the past few years, the NNDC has received continued cooperation from the compilers at the Gatchina data center in the USSR in coding for the NSR conference proceedings and obscure sources from the USSR. Periodically, the NNDC has also received entries prepared from Japanese laboratory reports by the compilers at the RIKEN data center in Japan. At present, there are $\approx 120,000$ references in the NSR file; and $\approx 2,600$ primary references and $\approx 1,100$ secondary references are added every year.

In the NSR, each reference is uniquely identified by a keynumber; this is used in identifying the reference in the ENSDF. Most NSR entries made since 1968 have a keyword abstract describing the contents of the article. This abstract consists of a complete list of nuclides, reactions, decays and nuclear properties measured or deduced in it. It is possible to make retrievals from NSR using a group of "selectors" which form a subset of keywords; further narrowing down of selected references is possible by imposing boolean restrictions on the selected groups.

The On-line Access System

The data in the ENSDF may also be accessed by electronic means using computer networks as an alternative to the hard copy publication. The advantages of such access are many: most recent data are obtained from one source, quick computer search and retrievals of data are possible, users can tailor retrievals for special applications and the retrieved data can be transferred to the user's computer for further processing or printed out to obtain a hard copy. Since 1986, on-line access has been provided for the computerized numeric and bibliographic nuclear physics information available at the NNDC to users in the United States and Canada. Similar services may also be available from other members of the NSDD network. The service is available on the NNDC VAX cluster and may be accessed by computer networks such as INTERNET or ESNET or by telephone. No special authorization is required to access the NNDC on-line newsletter, mail facility, and HELP files to become acquainted with the system. Users have to contact the NNDC for authorization to access other data bases such as the ENSDF and the NSR discussed earlier and the following.

NUDAT–NUclear DAta–evaluated numeric data extracted or derived from the ENSDF for nuclear level properties and radiations, from the Nuclear Wallet Cards[5] for nuclear ground and metastable state properties and evaluated thermal neutron cross sections and resonance integrals from Neutron Cross Sections, Vol. 1[6,7]

CINDA – Computer Index of Neutron Data – bibliographic references to neutron reaction data.

CSISRS – Cross Section Information Storage and Retrieval System – experimental data on neutron, photon, and charged particle reactions.

ENDF – Evaluated Nuclear Data File – evaluated neutron-induced reaction and decay data. Currently the data base contains the US evaluated file, ENDF/B-VI.

MIRD–Medical Internal Radiation Dose–plots and tables of nuclide decay radiations derived from the ENSDF and in the format of the ICRP Publication 38[8].

PHYSCO–PHYSics COdes–codes to calculate physics quantities e.g., internal conversion coefficients, logft values and related quantities.

XRAY–Photo Atomic Data–evaluated photo-atomic cross sections and attenuation coefficients for elements, compounds and mixtures and polarized scattering cross sections including Compton and Rayleigh scattering and Rayleigh scattering with anomalous corrections.

In the above data bases, at the present time, there is a terminal plotting capability for ENSDF, CSISRS, ENDF and XRAY and the disk output may be used in a PLOT utility program to generate a graphics file for ENSDF and CSISRS.

The use of the NNDC on-line system has grown over the past few years and the user statistics for 1986–1991 are shown in Table 4.

The NNDC assisted in installing in 1987 its nuclear structure related data bases in the on-line system of the Nuclear Energy Agency Data Bank, Saclay, which provides similar services locally in Europe. A data base

system for radioactivity gamma rays and nuclear structure references from the NSR file have been made available by the evaluation group at Lund University. These are available for on-line access via the NORDic University NETwork (NORDUNET) and the Swedish University computer NETwork (SUNET). The Fachinformationszentrum (FIZ), Karlsruhe, a former member of the NSDD network, offers ENSDF, NSR and MEDLIST data bases for on-line access via telecommunication networks. MEDLIST is a nuclear medicine data base derived from ENSDF using the program MEDLIST [2,3].

References

1. J. K. Tuli, Evaluated Nuclear Structure Data File– A Manual for Preparation of Data Sets, National Nuclear Data Center, Brookhaven National Laboratory (1987) Brookhaven National Laboratory Report BNL-NCS-51655-Rev.87

2. W. B. Ewbank and M. J. Kowalski, documented in reference [3]

3. M. J. Martin (Ed), Nuclear Decay Data for Selected Radionuclides, Nuclear Data Project, Oak Ridge National Laboratory (1976) Oak Ridge National Laboratory Report ORNL-5114

4. T. W. Burrows, The Program RADLST, National Nuclear Data Center, Brookhaven National Laboratory (1988) Brookhaven National Laboratory Report BNL-NCS-52142

5. J. K. Tuli, Nuclear Wallet Cards, National Nuclear Data Center, Brookhaven National Laboratory (1990)

6. S. F. Mughabghab, M. Divadeenam & N. E. Holden, Neutron Cross Sections, Vol. 1, Part A, Z=1-60, Academic Press (1981)

7. S. F. Mughabghab, Neutron Cross Sections, Vol. 1, Part B, Z=61-100, Academic Press (1984)

8. ICRP Publication 38 - Radionuclide Transformations - Report of a Task Group of the International Commission on Radiological Protection, Pergamon Press (1983)

Table 4 On-line Access Statistics												
Year	Runs	Retrievals*	NSR	ENSDF	NUDAT	CINDA	CSISRS	ENDF	MIRD	PLOT	PHYSCO	XRAY
1986	648	1621	814	142	536	129						
1987	1275	4263	2521	863	815	60						
1988	2264	8748	5022	1303	1492	285	459	187				
1989	3374	8406	3253	850	1841	522	1649	150	121	11	9	
1990	5436	12067	5613	1256	2204	187	1623	1019	53	39	65	8
1991@	3185	6452	2983	690	1806	78	278	395	5	23	107	87

@ January to April
* The number of pieces of information in each retrieval depends on the complexity of the retrieval

FISSION DATA SYSTEMATICS

H. Märten, A. Ruben, D. Seeliger, I. Düring, and D. Polster

TU Dresden, Mommsenstr. 13, D-O-8027 Dresden, Germany

Abstract: Fission fragment characteristics as well as multiplicities $\bar{\nu}$, energy spectra N(E) and angular distributions of fission neutrons were analyzed in the framework of a phenomenological scission point model including semi-empirical, temperature- dependent shell correction energies for describing energy partition in fission as function of mass asymmetry and a statistical-model approach to post-fission neutron emission, respectively. In the case of multiple-chance fission reactions, statistical multi-step reaction theory including fission channel in the multi-step compound part was used to calculate partial fission cross sections together with pre-fission neutron spectra. The physical consistency achieved in the calculations concern exact energy conservation within the fission model applied as well as reaction theory with account for multi-step direct (particle-hole and phonon excitations) and multi-step compound processes including fission channel (master equation approach). Calculations were performed for a large number of spontaneously fissioning nuclei as well as neutron induced fission of actinides important for nuclear technology in the incidence energy range 0-20 MeV. New $\bar{E}(\bar{\nu})$ relations are presented.

(fission, prompt neutron emission, neutron multiplicity, neutron spectra)

Introduction

Fission neutron spectra and average numbers of fission neutrons are essential data for nuclear technology. They were measured, calculated, compiled, and evaluated in many works. However, fission neutron data files do not correspond neither to the progress achieved in the field of experimental techniques and the theoretical understanding nor to various data requests. Sparse (and sometimes contradictory) experimental fission neutron data are not an adequate basis for evaluations at neutron incidence energies up to 20 MeV and beyond. The present status can be summarized as follows:

- Experimental data are available for thermal-neutron induced fission of the major and some of the minor actinides, for fast-neutron induced fission at some incidence energy (E_i) points (MeV region, around 14 MeV), and for several spontaneous fission reactions (cf. CINDA). Most of the data were measured a long time ago on the basis of simple experimental techniques. The data are often contradictory.

- Theoretical approaches to fission neutron emission have not yet been applied in a systematic manner for calculations of fission neutron data and for their check in regard of physical consistency (with the exception of the Madland-Nix model (MNM) [1] applied to ^{235}U [2]).

- Data evaluations were directly based on the few experimental data in combination with semi-empirical or empirical relations.

- Nuclear data libraries do not correspond to the present status of experiment and theory. They do not include fission neutron data in a reliable complexity, e.g. average parameters ($\bar{E}$ and $\bar{\nu}$) are not considered in their dependence on E_i in most cases. Spectral shapes assumed are either Watt or Maxwellian distributions, whose parameters are taken as identical and E_i-independent for most of the actinides.

Considerable effort has been devoted to the precise determination of the ^{252}Cf(sf) standard in the last years [3-5].

The requirements to be met in fission data measurements and analysis were specified in detail. Remarkable progress was achieved in the theoretical understanding of fission and fission neutron emission [6]. The ^{252}Cf example shows the present possibilities in experiment and theory. Since a general improvement of the experimental fission neutron data basis is infeasible, further activities to improve the precision and the complexity of fission neutron data files should be based on adequate (i.e. physically consistent) theoretical approaches and a few precision experiments in order to check (or to adjust) the theory at typical points.

The present work relies on recent theoretical approaches developed at TU Dresden involving fission theory, complex fission neutron (evaporation) theory, and reaction theory with fission channel (applications to multiple chance fission).

Theoretical Basis

The yield and spectral distribution of fission neutrons are strongly dependent on fragment mass number. Accordingly, fission neutron data calculations should be performed in an adequately complex manner in order to provide physical consistency. Necessary preconditions are:

- the application of the statistical theory of neutron emission from highly excited, rapidly moving fragments (evaporation theory) to the fragment diversity represented by a complex occurrence probability in mass number A, excitation energy E^*, and kinetic energy E_k (or total kinetic energy TKE), i.e. P(A,E^*,TKE),

- the knowledge of the fragment distribution P(A,E^*,TKE), i.e. application of fission theory to deduce necessary informations,

- the application of reaction theory with fission channel to multiple chance fission reactions (n,jnf) in order to calculate the partial fission cross sections $\sigma_{f,j}(E_i)$ (i.e. the weight of the chances j) as well as the spectral distribution of pre-fission neutrons.

After calculating the yield $\bar{\nu}(A)$ and the LS distribution $N(E,\Theta:A)$ (with norm 1.) for given mass number (Θ - angle of neutron emission with reference to fission axis), the total neutron yield and spectrum are given by

$$\bar{\nu} = \sum Y(A)\bar{\nu}(A),$$

$$N(E) = \sum Y(A) \int d\Omega \frac{\bar{\nu}(A)}{\bar{\nu}} N(E,\Theta:A).$$

In the case of multiple chance fission of the type (n,jnf), j=0,1,2,... , Eqs. (1) and (2) are separately solved for each fission chance j, i.e. we obtain $\bar{\nu}_j$ and $N_j(E)$, with the weight

$$w_j = f(E_i) = \frac{\sigma_{f,j}}{\sigma_f}, \qquad \sigma_f = \sum_j \sigma_{f,j}$$

The total value $\bar{\nu}$ includes post-fission neutrons (number $\bar{\nu}_j$) as well as pre-fission neutrons (number j):

$$\bar{\nu} = \sum_j w_j(\bar{\nu}_j + j)$$

The total emission cross section of fission neutrons (which are measured in coincidence with fission events, i.e. including pre-fission neutrons with the spectral distribution $S_j(E) = d\sigma_j/dE$) is given by

$$\frac{d\sigma}{dE} = \sigma_{f,0}\bar{\nu}_0 N_0(E) + \sum_{j\geq1} \sigma_{f,j}[\bar{\nu}_j N_j(E) + S_j(E)],$$

where $S_j(E)$ is normalized to 1. in the energy interval allowed because of energy conservation restrictions.

The following nuclear models were applied:

Scission point model (two-spheroid model TSM [7] for solving the energy partition problem : It is based on potential energy minimization at scission point. In connection with a detailed energy balance equation at scission with reference to saddle B conditions, the model is suitable to predict the partition of total available energy on both complementary fragments (E_k and E^*) as function of mass asymmetry. Microscopic effects (influencing the stiffness of the fragments at scission point strongly) are considered in a phenomenological way, i.e. shell correction energies as function of A were deduced within the model on the basis of well-known experimental data. Their dependence on nuclear temperature is taken into account.

5 - *Gaussian approach* for calculating mass yield curves $Y(A:E_i)$: As found phenomenologically [8] as well as theoretically [9], mass yield curves can be well represented by a 5-Gaussian-approach corresponding to two asymmetric and one symmetric fission mode. On the basis of experimental data the parameters average mass number, width, and weight of the five Gaussians were deduced as function of E_i (and separately for all possible fission chances at $E_i \geq 6$ MeV) for fission reactions in the Th-Cf region. The dependence on E_i was represented by data fits to several functions.

Temperature distribution model (code FINESSE) [10] for describing the neutron yield and the spectral distribution of fission neutrons as function of A and in total form: The model relies on basic ideas of the MNM, but accounts for:

- the explicit dependence of fission neutron characteristics on A, e.g. a realistic distribution in rest-nucleus temperature deduced from Gaussian distribution in E^*,
- model parameter averages over the charge distribution for given A,
- a modified evaporation ansatz including higher-order terms of entropy expansion in powers of excitation energy,
- emission anisotropy in the centre-of-mass system due to fragment angular momentum,
- competition of neutron and γ-ray emission (simulation),
- angular distribution of fission neutrons in the laboratory frame (with reference to fission axis and incidence beam direction, etc.

Statistical multistep reaction theory for predicting fission cross sections and spectra of pre-fission neutrons: Whereas in earlier calculations the Hauser-Feshbach theory with account for pre-equilibrium effects and fission (code STAPRE [11]) was applied, the statistical multistep (direct/compound) reaction theory SMD/SMC [12] with the direct incorporation of the fission channel in the SMC part (code EXIFON(fis) [13]) has recently been proposed. The fission escape width depending on exciton number (n) is given by statistical arguments on the basis of state densities ρ_n:

$$\Gamma_{nf}^\uparrow(E_{FN}^*, E') = \frac{\int_0^{E_{FN}^* - B_f} \rho_n(E_{FN}^* - B_f - E')dE'}{2\pi\rho_n(E_{FN}^*)},$$

where E_{FN}^* is the excitation energy of the fissioning nucleus decaying either by particle (neutron) emission, γ-ray emission, or fission. B_f is the height of the fission barrier hump A or B, E' is the kinetic energy of the fission degree of freedom. Considering a double-humped fission barrier, the total fission width reads

$$\Gamma_{nf}^\uparrow = \frac{\Gamma_{nf}^\uparrow(A)\Gamma_{nf}^\uparrow(B)}{\Gamma_{nf}^\uparrow(A) + \Gamma_{nf}^\uparrow(B)}$$

The calculated cross section of ^{238}U fission induced by neutrons with energies up to 20 MeV is represented in Fig. 1 as an example. According to the SMD/SMC model, pre-fission neutrons are emitted due to different reaction mechanisms. Besides direct (exciton and collective excitation) processes they are mainly emitted from compound (pre-eqilibrium as well as eqilibrium) states. The SMC concept is based on

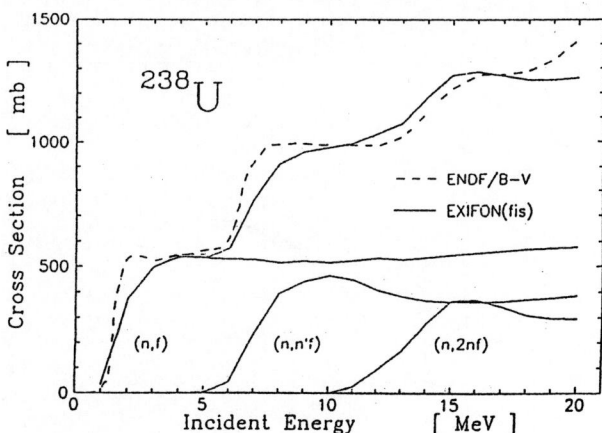

Fig. 1. Calculated fission cross section for ^{238}U in comparison with ENDF/B-V data.

a full-scale calculation within the master equation approach also applied to multiple emission.

The theoretical scheme outlined above was applied to calculate fission neutron data for important actinide nuclei in the energy range from thermal energy (or threshold) to 20 MeV. The results are presented and discussed in the following sections.

Fission Neutron Multiplicities

Average numbers of fission neutrons calculated within the above theoretical complex without any parameter fit are shown in the Figs. 2-4. Whereas the $\bar{\nu}$ results for spontaneous fission and pure (n,f) reactions are only defined by the predictive power of the scission point model and the accuracy

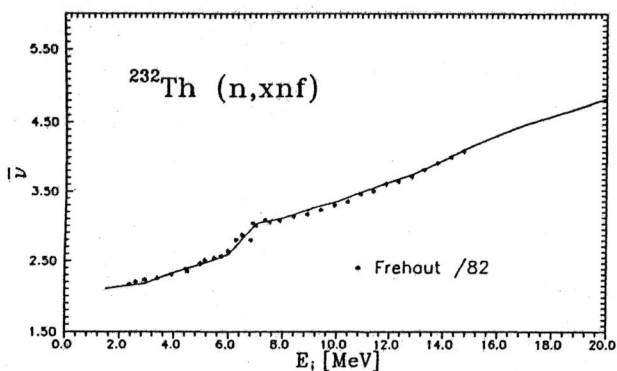

Fig. 2. Calculated $\bar{\nu}$ data for ^{232}Th in comparison with experimental data [14].

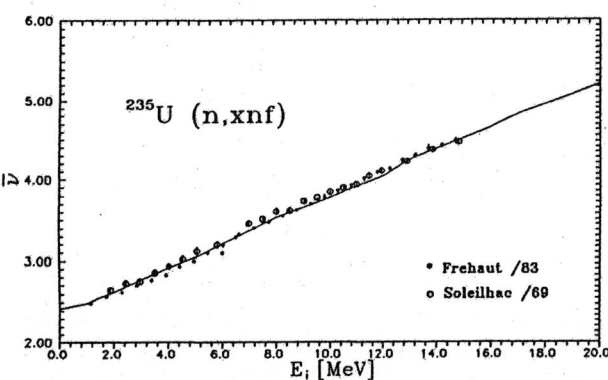

Fig. 3. Calculated $\bar{\nu}$ data for ^{235}U in comparison with experimental data [14,15].

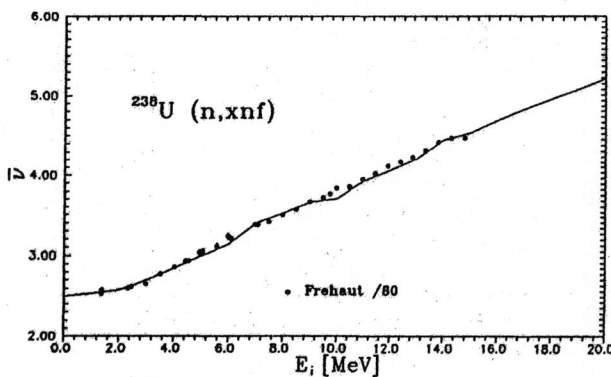

Fig. 4. Calculated $\bar{\nu}$ data for ^{238}U in comparison with experimental data [16].

of the de-excitation balance equation within FINESSE, the data for multiple chance fission are influenced by the calculated partial fission cross sections, i.e. the weight of the fission chances. This is illustrated in Fig. 2 showing the results for ^{232}Th. There is a remarkable step-like increase of $\bar{\nu}$ at the threshold of the second chance (note that $\sigma_{f,1} > \sigma_{f,0}$).

Fission Neutron Spectra

Post-fission neutron spectra are neither Maxwellian nor Watt distributions [1,10]. Nevertheless, most of the experimental spectra were approximated by at least one of both for data reduction. The spectral shape of the ^{252}Cf(sf) neutron spectrum was carefully investigated experimentally as well as theoretically [3-5]. This nuclear standard can now be considered as well- established [17].

As shown in Ref. [10], FINESSE calculations reproduce the ^{252}Cf(sf) standards N(E) and $\bar{\nu}$ within experimental uncertainties. Fig. 5 shows calculated ^{252}Cf(sf) neutron data represented by Legendre polynome coefficients (angular distribution with reference to fission axis).

In the present work, some of the calculational results are represented with reference to the standard spectrum $N_{Cf}(E)$, i.e. the spectral ratio $R(E,E_i) = N(E,E_i)/N_{Cf}(E)$ is analyzed. Note that both spectra are normalized to 1, so that neutron yield differences are eliminated (see definitions above). The matrix $R(E,E_i)$ calculated for ^{238}U fission induced by neutrons with energies up to 20 MeV is represented in Fig. 6. Note the remarkable influence of the multiple-chance fission at the thresholds 6.5 (j=1) and 12 MeV (j=2). Here, the emission of pre-fission neutrons reduces the total excitation energy of the fragments from higher-order fission chances and, consequently, the average emission energy for $j > 1$.

In the following, FINESSE results obtained without any parameter fit are shown in comparison with recent experimental data (Figs. 7-10). The Figs. 9-10 show the spectra of pre-fission neutrons explicitly. As already discussed, equilibrium, pre-equilibrium as well as direct processes are the mechanisms of their emission [6,13].

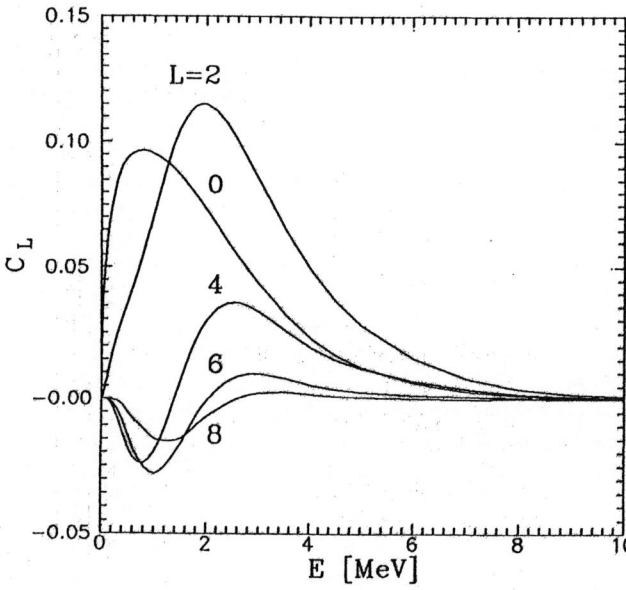

Fig. 5. The energy and angular distribution of ^{252}Cf neutrons represented by Legendre polynome coefficients. $C_0(E)$ is the energy spectrum.

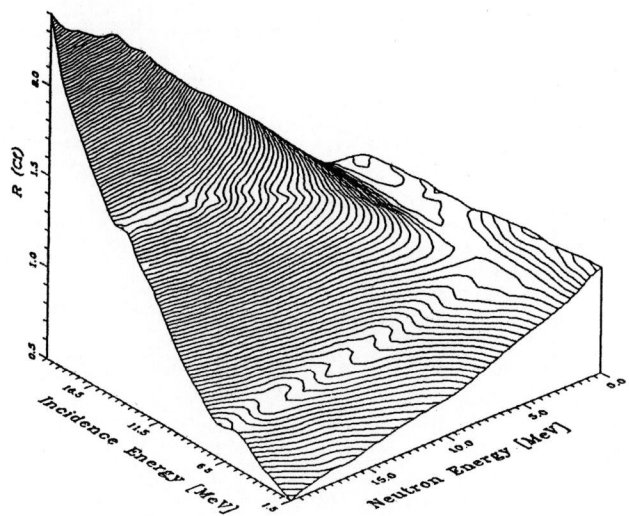

Fig. 6. The spectral matrix of ^{238}U(n,jnf) neutrons (here, only post-fission neutrons) represented as ratio to the ^{252}Cf(sf) spectrum.

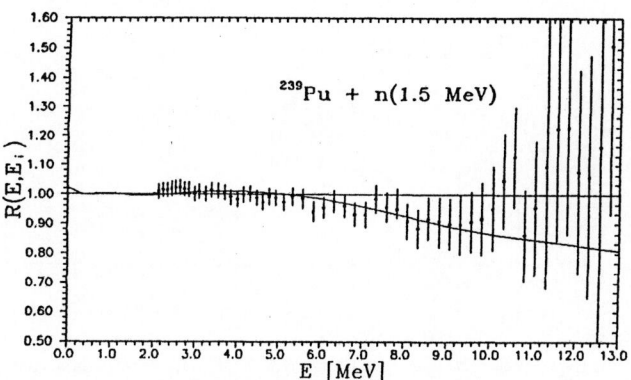

Fig. 7. Spectral ratio to the ^{252}Cf(sf) standard for ^{239}Pu at 1.5 MeV incidence energy (experimental data - [18])

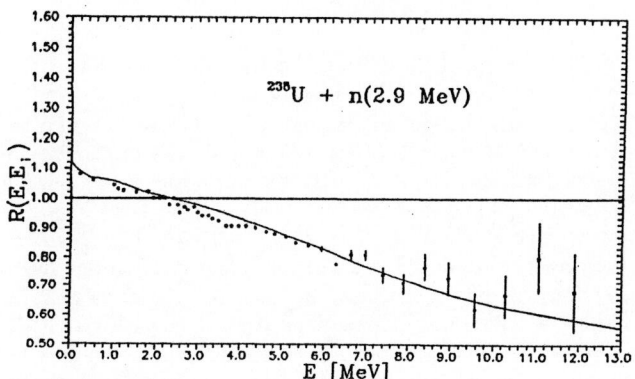

Fig. 8. Spectral ratio to the ^{252}Cf(sf) standard for ^{238}U at 2.9 MeV incidence energy (experimental data - [19])

Group data were deduced for some thermal-neutron induced fission reactions. Examples are shown in the Figs. 11-12 in comparison with an evaluation by Starostov et al. [21]. From the calculations, the $\bar{E}$ values 1.969 and 2.078 for ^{235}U and ^{239}Pu, respectively, were deduced (cf. evaluated data [21]: 1.970 ± 0.015, 2.087 ± 0.015 MeV, respectively).

Besides $\bar{\nu}$, the average emission energy $\bar{E}$ is an essential parameter characterizing fission neutron emission. Since the calculations are performed for the full energy region, it can

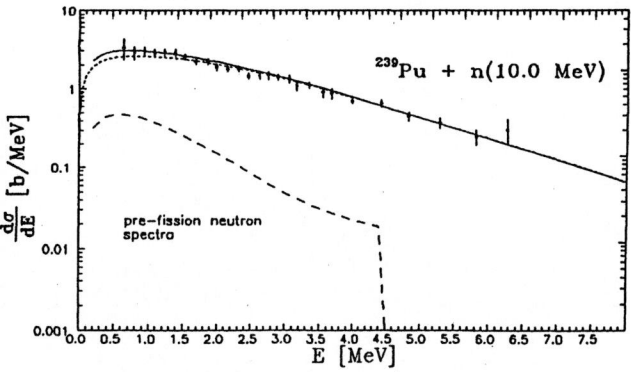

Fig. 9. Emission cross section of neutrons from ^{239}Pu fission at 10.0 MeV incidence energy (experimental data - [20]). The long-dashed line represents the pre-fission neutron spectrum.

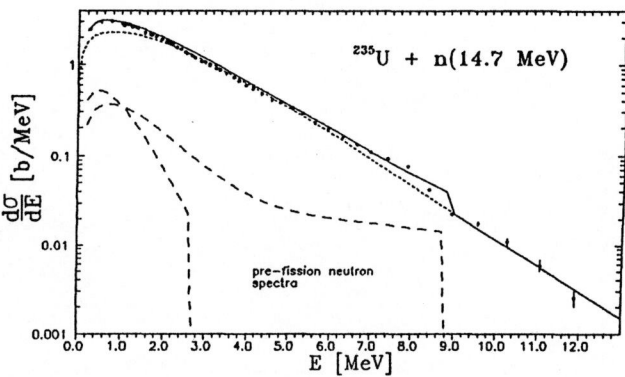

Fig. 10. Emission cross section of neutrons from ^{235}U fission at 14.7 MeV incidence energy (experimental data - [19]). The long-dashed lines represent the pre-fission neutron spectra.

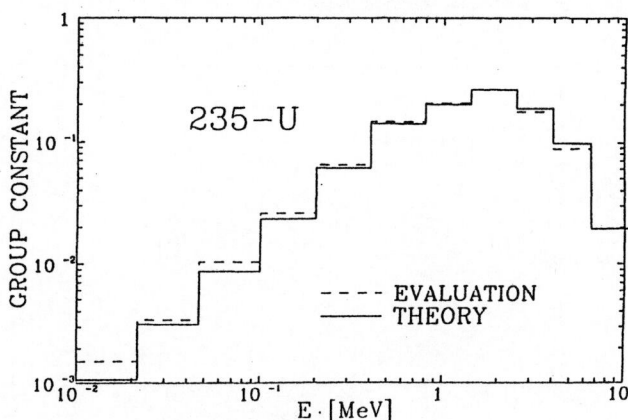

Fig. 11. Calculated group data for ^{235}U(n_{th},f) in comparison with evaluated data [21].

be obtained by direct averaging, i.e. $\bar{E} = \int E\ N(E)$. In contrast to this theoretical treatment, experimental fission neutron spectra are commonly fitted to either a Maxwellian or a Watt distribution yielding $\bar{E}$ from the spectrum parameters. However, the values deduced depend on the energy range covered in the experiment. Therefore, a direct comparison between experimental and theoretical spectra is inevitable. Experimental $\bar{E}$ data are systematically to high (slow) if the low-energy (high-energy) range is preferentially covered in the experiment. This is a consequence of the typical spectral shape of fission neutron spectra [1,10]. The influ-

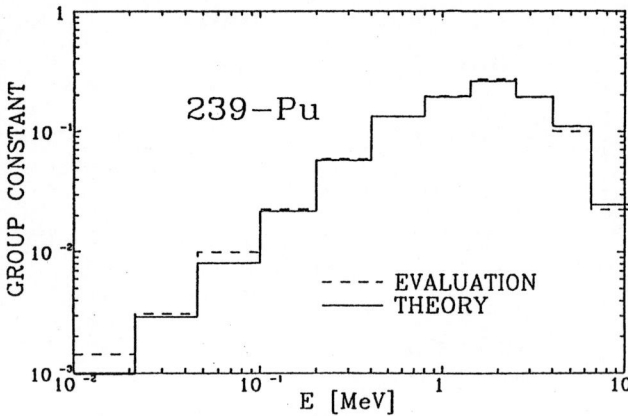

Fig. 12. Calculated group data for ^{239}Pu(n_{th},f) in comparison with evaluated data [21].

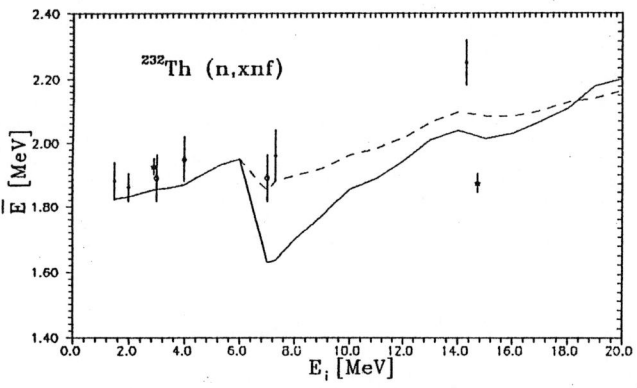

Fig. 13. $\bar{E}$ data calculated for ^{232}Th in comparison with experimental data (cf. CINDA, [18,19]). The dashed line represents the pure post-fission neutron data.

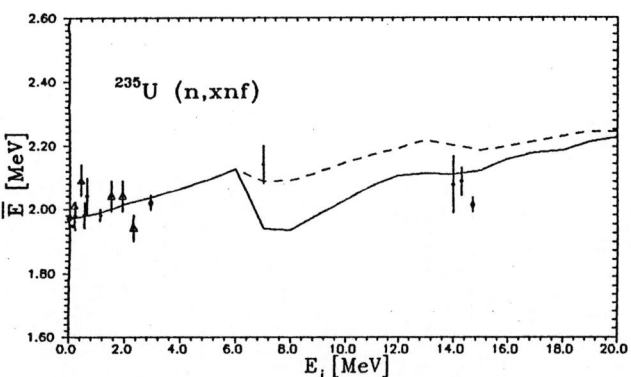

Fig. 14. $\bar{E}$ data calculated for ^{235}U in comparison with experimental data (cf. CINDA, [18,19]).

ence of the experimental energy range is much more crucial in the case of multiple-chance fission reactions, e.g. just above the chance thresholds, where the spectrum of the pre-fission neutrons is limited to low energy. At $E_i < 6$ MeV, $\bar{E} = f(E_i)$ is a (approximately) linearly increasing function. At $E_i > 6$ MeV, $\bar{E}$ is drastically reduced due to the influence of the pre-fission neutrons yielding a step-like behaviour of $\bar{E} = f(E_i)$. Average emission energies of fission neutrons were analyzed on the basis of the present calculations. The Figs. 13-16 represent calculational results in comparison with experimental data (cf. CINDA, [18-21]). For comparison, the $\bar{E}$ values for pure post-fission neutrons at $E_i > 6$ MeV are included.

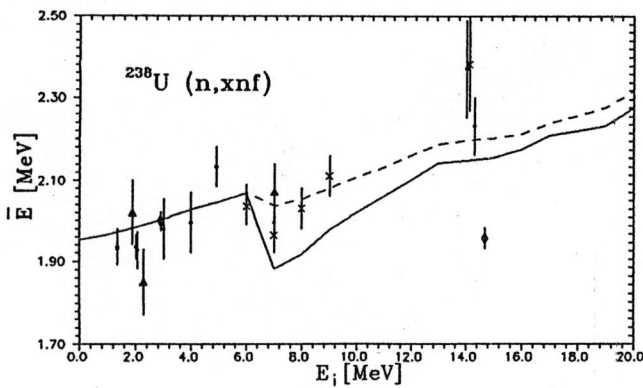

Fig. 15. $\bar{E}$ data calculated for ^{238}U in comparison with experimental data (cf. CINDA, [18,19]).

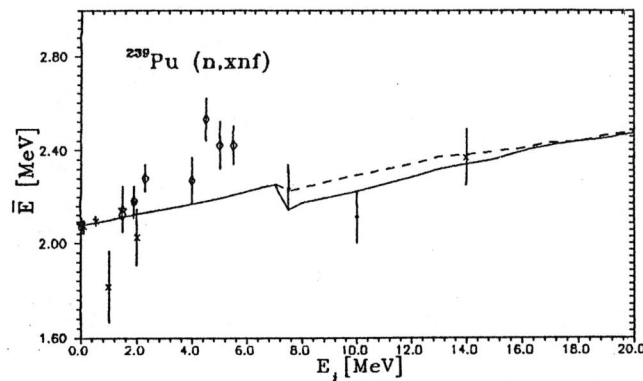

Fig. 16. $\bar{E}$ data calculated for ^{239}Pu in comparison with experimental data (cf. CINDA, [18-20]).

$\bar{E}(\bar{\nu})$ Relations

The average number of neutrons is a direct measure of the total excitation energy of the fragments. As higher E_{tot}^* as higher $\bar{\nu}$ and, consequently, as higher $\bar{E}$. The correlation function $\bar{E} = f(\bar{\nu})$, which is of essential importance in the field of fission neutron data, was postulated as an universal function by Terrell [22], $\bar{E} = 0.74 + 0.645\sqrt{\bar{\nu} + 1}$, and later modified by Knitter et al. [23], $\bar{E} = 0.74 + 0.35(\bar{\nu} + 1)$. However, the present systematic calculations do not confirm the above relations. As firstly discussed in Ref. [24], the $\bar{E}(\bar{\nu})$ relation is different for the (n,f)-reactions studied. The Figs. 17-18 show calculational results for various reactions. The following parameterization, which reproduces the calculational results within (1-2) per cent accuracy, includes the dependence on the fissility parameter x = Z^2/A:

$$\bar{E} = (0.0698x - 0.8825) + (0.641 - 0.0133x)\bar{\nu}$$

(valid for (n,f)-reactions). Spontaneous fission neutron data do not follow this relation (Fig. 18), but can be well reproduced by

$$\bar{E} = 0.1181x - 2.35907,$$

i.e. their is no explicit dependence on $\bar{\nu}$ (cf. Fig. 18). It is emphasized that the fission neutron data at $E_i > 6$ MeV

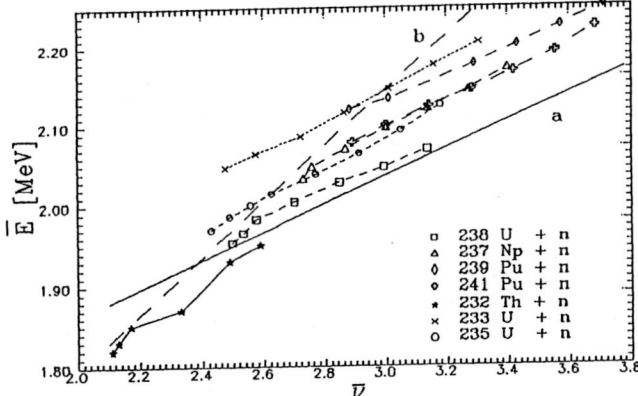

Fig. 17. The $\bar{E}(\bar{\nu})$ correlation: Calculational results for various fission reactions in comparison with previous parameterizations (a - [22], b - [23]).

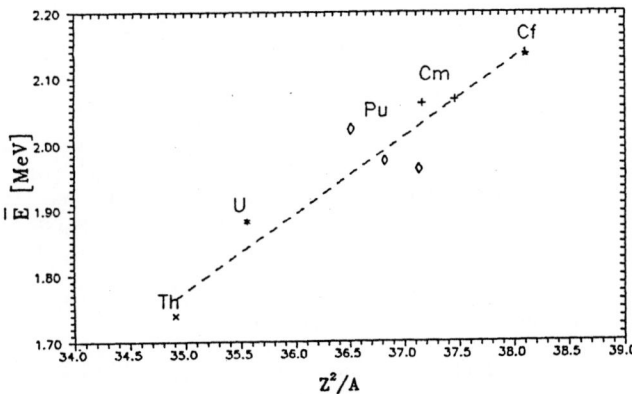

Fig. 18. The $\bar{E}(Z^2/A)$ correlation for spontaneously fissioning nuclei.

cannot be parameterized in a simplifying manner. Here, the spectral shape as well as $\bar{E}(\bar{\nu})$ relation are strongly influenced by pre-fission neutron emission. New evaluations of neutron data for actinide nuclei should account for this fact adequately, i.e. inclusion of post-fission and pre-fission neutron data in complex form.

Summary

The present work shows the predictive power of recent nuclear model approaches to fission neutron emission. All calculations were performed in the framework of a model complex which has successfully been tested in the case of well-investigated fission reactions. The calculations are consistent in regard of energy conservation. As already described in [7], the scission point model reproduces average TKE data as function of E_i for major actinides. Exact consideration of energy conservation means that the total excitation energy of the fragments is fixed correctly. The scission point model describes the partition of E^*_{tot} on both complementary fragments including the temperature dependence of shell effects so that the E^* data for individual fragments are quite reliable. Note that the solution of the energy partition problem is given as function of mass asymmetry. Accordingly, the evaporation model approach (temperature distribution model) includes the full dependence of fission neutron observables on fragment mass number. A further point is that the calculations of neutron multiplicity as function of E_i are in good agreement with experimental/evaluated data. This indicates

again the reliable description of the energetic conditions in nuclear fission. All calculations performed reproduce recent experimental fission neutron spectra without parameter fit. Based on the calculations a new systematics of average fission neutron energies in correlation with the average number of fission neutrons was presented. It includes the dependence on fissility parameter $x = Z^2/A$. In the case of spontaneous fission, $\bar{E}$ data are best reproduced if considering only the dependence on x.

References

[1] D.G. Madland and J.R. Nix, Nucl. Sci. Eng. **81**,213 (1982)

[2] D.G. Madland et al., Proc. IAEA Consultants' Meeting on Physics of Neutron Emission in Fission, Mito City, 24-27 May, 1988, ed. by H.D. Lemmel (IAEA, Vienna, 1989) INDC(NDS)-**220**, 259

[3] Proc. IAEA Consultants' Meeting on the U-235 Fast-Neutron Fission Cross-Section, and the Cf-252 Fission Neutron Spectrum, Smolenice, 28 March - 1 April, 1983, ed. by H.D. Lemmel and D.E. Cullen (IAEA, Vienna, 1983) INDC(NDS-**146**/L

[4] Proc. IAEA Advisory Group Meeting on Properties of Neutron Sources, Leningrad, 9-13 June, 1986 (IAEA, Vienna, 1987) IAEA-TECDOC-**410**

[5] Proc. IAEA Consultants' Meeting on Physics of Neutron Emission in Fission, Mito City, 24-27 May, 1988, ed. by H.D. Lemmel (IAEA, Vienna, 1989) INDC(NDS)-**220**

[6] H. Märten, Proc. IAEA Consultants' Meeting on Nuclear Data for Neutron Emission in the Fission Process, Vienna, 22 - 24 October, 1990, in print (cf. references therein)

[7] A. Ruben et al., Z. Phys. **A338**,67 (1991)

[8] S. Nagy et al., Nucl. Sci. Eng. **88**,154 (1984)

[9] U. Brosa et al., Z. Naturforschung **41a**,1341 (1986)

[10] H. Märten et al., Proc. Int. Conf. on 50 Years with Nuclear Fission, 25-28 April, 1989, Gaithersburg, ed. by J.W. Behrens and A.D. Carlson (American Nuclear Society, 1989) Vol. II, p. 743

[11] M. Uhl and B. Strohmaier, Report IRK-**76/01**, IRK Vienna (1976) and Program Manual for STAPRE code, NEA Data Bank (1981)

[12] H. Kalka et al., Z. Phys. **A329**, 331 (1988)

[13] D. Polster, Proc. Int. Conf. on Nucl. Phys., Gaussig, 12-16 Nov, 1990, in print, and D. Polster and H. Kalka, submitted to Z. Phys. A

[14] J. Frehaut et al., Proc. Int. Conf. on Nuclear Data for Science and Technology, Antwerp, 6 - 10 September, 1982, D. Reidel Publ. Comp. (Eindhoven, 1983) 78

[15] M. Soleilhac et al., Nucl. Energy **23**, 257 (1969)

[16] J. Frehaut et al., Report "Recent results on ν-prompt measurements between 1.5 and 15 MeV", Paris (1980)

[17] W. Mannhart, Handbook of Nuclear Activation Data, Technical Report Ser. **273** (IAEA, Vienna, 1987), 163

[18] G.N. Lovchikova et al., Proc. Int. Conf. on 50 Anniversary of Nuclear Fission, Leningrad, 16 - 20 October, 1989, in print

[19] G.S. Boykov et al., ibid.

[20] I. Düring, Diploma Thesis, TU Dresden, 1990, and to be published

[21] B.I. Starostov et al., Yad. Const. 3 (1985) 16 (identical to INDC(CCR)-252/G)

[22] J. Terrell, Proc. IAEA Symp. on Physics and Chemistry of Fission, Salzburg, 22 - 26 March, 1965 (IAEA, Vienna, 1965) Vol. II, p. 3

[23] H.H. Knitter et al., Proc. IAEA Consultants' Meeting on Prompt Fission Neutron Spectra, Vienna, 25 - 27 August, 1971 (IAEA, Vienna, 1972) 41

[24] H. Märten et al., Proc. Int. Conf. on Nuclear Data for Science and Technology, Mito, 1988, ed. by S. Igarasi (JAERI, Saikon Publ. Co., Ltd., 683 (1988)

EUROPEAN ACTIVATION FILE FOR FUSION

J. Kopecky[1], H. Gruppelaar[1] and R.A. Forrest[2]

[1] Netherlands Energy Research Foundation ECN,
P.O.Box 1, 1755 ZG Petten, The Netherlands
[2] AEA Industrial Technology, Harwell Laboratory,
Oxfordshire OX11 0RA, United Kingdom

Abstract: This paper describes the work performed to revise and extend the European Activation File (EAF). The extensions, revisions and improvements made in the first two versions of the file (EAF-1, EAF-2) are discussed as well as the work which is planned for EAF-3. The EAF-2 file contains cross-sections for all stable and unstable nuclides with half-lifes exceeding 0.5 day. Cross-sections leading to isomeric states are given separately and those isomeric states with $T_{1/2} > 0.5$ d are included as targets. The EAF-2 version contains about 11000 reactions on 667 targets. The EAF-2 data file will be further extended and improved and will be supplied with uncertainty information.

(Neutron activation cross sections, incident energies up to 20 MeV).

Introduction

The aim of this work was to provide a complete system for a European activation data library (EAF) which together with the inventory code FISPACT [1] and its associated data libraries will be used for activation and transmutation calculations in fusion reactor technology problems. The importance of studying low activation materials to ensure that future reactors aer as safe as possible and environmentally benign has accelerated the need for a complete and yet high-quality activation file. Another emphasis has been on improvement of the (n,γ) data in view of the importance of the low-energy component in neutron spectra in fusion reactors. The strategy and goals for the creation of activation files in the framework of the European Technology Programme are as follows:

1. Address the completeness of the library using the well tested mass-production means of the EAF files such as simple model codes and systematics. Extend further the library with actinide targets (EAF-3).

2. Provide the library with a simple but complete uncertainty file which will enable routine uncertainty estimates and sensitivity studies to be carried out (EAF-3).

3. Based on sensitivity studies, the feedback of information from users and benchmark testing, will improve the quality of selected important reaction cross sections with full evaluations or data adopted from other libraries (e.g. JEF-2, EFF-2, ENDF/B-6, JENDL-3). This will be a continuing activity for all future versions of EAF. As a starter file for this exercise the last version from the series of REAC-ECN files [2,3] – REAC-ECN-5 – has been used, together with some information from the UKACT1 file of AEA Harwell. The basis of the revision into EAF-1 was an update of renormalizations to 14.5 Mev experimental data or to data from systematics. In particular data for the neutron emmission channels (n,n'), (n,2n) and (n,3n) have been improved. Another aspect was to extend the library so as to achieve a library which contains all stable and unstable targes with half-lifes longer than 0.5 day. Cross sections leading to isomeric states are treated separately and further those isomeric states with half-lifes longer than 0.5 day are included

as targets. To ensure that gas production (e.g. ^{3}H and ^{4}He) is calculated correctly, different reactions leading to the same daughter nuclide are given separately. The present version of EAF-2 contains now about 11000 reactions on 667 targets. The major improvement, however, was a complete revision and re-evaluation of radiative neutron capture cross sections.

The point wise file EAF-1 was released in January 1990. The present version EAF-2 which was released in May 1991 is described in this paper and in more detail in the laboratory report [5]. Both data files have been processed into the multigroup library GEAF with either 100- (GAM-II) or 175 group (VITAMIN-J) structures.

Evaluation methods

Techniques which have been applied in the evaluation of the EAF files have been documented in refs. [2-4]. They consisted mainly of two codes to calculate excitation funtions (THRES and FISPRO-ECN), the application of cross-section systematics of Forrest [6], Vonach [7] and Kopecky [2] and the use of Kopecky et al. systematics of isomer ratios [3,8]. Recent extensions and improvements are discussed below.

1. Nuclear models for excitation functions

In general we have still adopted the code THRES [9] as a main source for calculation of excitation curves (except for radiative capture) if no other evaluation was available for the target nuclide. Because of severe shortcomings of THRES, we have applied several corrections to the calculated data, based on comparisons with "horizontal" evaluations and/or experimental data.

The neutron emission cross sections as modelled in THRES have been improved in the following way:

(i) The (n,n'γ) cross sections to metastable states were derived in two steps. First the branching ratio systematics for one-step reactions [8] was applied to the THRES results of the total inelastic cross section. Then a constant component was added, modelling in a crude way the pre-equilibrium part, to reach the value of the systematics of Vonach [7] at 14.5 MeV.

(ii) In order to correct for (n,3n) cross sections, set too high by THRES, all (n,3n) cross sections were decreased by 20% so that neutron emmission cross sections were in agreement with systematics. This factor has been derived from available experimental information and seems to be a rather reasonable approximation [5].

(iii) For target nuclides with the (n,3n) threshold below 14 MeV, and thus the (n,2n) excitation function around 14 MeV possesing a steep fall, no renormaliza- tions have been applied.

The effective threshold energies for all reactions leading to isomeric states have been implemented. Cross sections for isomeric targets are so far assumed to be identical to those for the ground-state.

2. Evaluation of radiative capture data

Completeness and quality of the radiative capture data were the main objectives of the recent version of the EAF. The (n,γ) reaction is important at relatively low energies and therefore the thermal and resonance ranges have been fully reconsidered.

In the re-evaluation of the radiative cross sections the targets were categori- zed in three groups:

1. complete evaluation available, including resolved resonance parameters, in existing general purpose files;
2. no evaluation is available but experimental information exists;
3. no experimental information exists.

In the first group all existing evaluations (JEF-1, JENDL-3, ENDF/B-6) have been adopted to replace the old data. These evaluations are tested against the experimental data and include a resolved resonance region. Thus they belong to the category with the highest accuracy, however, some do not have a proper direct/semidirect component. This omission results in a 14.5 MeV value well below the systematic value. For the re maining targets (groups 2 and 3) new calculations have been performed with two ECN codes, FISGIN and MASGAM. In the case of group 2 a special evaluation in the resolved and thermal energy range was made with the help of the code SIGECN and included in the file. The contents of the EAF-2 (n,γ) subfile is summarized in Table 1.

Table 1. List of sources for (n,γ) data in EAF-2

Data source	Number of targets
JEF-1	246
ENDF/B-5,6	18
JENDL-3	32
Other evaluations	5
FISGIN	6
MASGAM + SIG-ECN	57
MASGAM	303
Total	667

2.1 Applied nuclear models for calculation of (n,γ) excitation functions

FISGIN [10] is an improved version of the code FISPRO (originating from ENEA Bologna), a Hauser-Feshbach statistical- model code with a direct-semidirect component of Lane-Lynn-Brown. It also allows the calculation of smooth backround cross sections due to missed resonances. Available information on the resolved resonance region is generated with the code SIG-ECN and included in a point-wise format. The calculation is checked against experimental data and these evaluations can be considered to reach the same quality as the first category. This approach has been applied for reactions from group 2 having the status of an important reaction.

MASGAM [11] is a simplified version of FISGIN provided with 1/v-, statistical- and pre-equilibrium components. Recognizing difficulties of the direct-semidirect approach to reproduce the experimental data of excitation curves above 6 MeV, the systematics of Ref.[12] was implemented instead. This rather empirical systematics, despite several simple assumtions, fits the experimental data reasonably well. The final formula reads as

$$\sigma_{n\gamma}(DSD) = C \frac{\int_0^{S_n+E_n} E_R^4 (E_\gamma^2 - E_R^2)^2 + E_\gamma^2 \Gamma^2}{E_n^3 [1 + 0.035A(1 + S_n/E_n)^3]} dE,$$

where E_γ, E_n and E_R are gamma-ray-, neutron and the giant-resonance energies, Γ is the spreading width, S_n is the neutron threshold (all in MeV) and C is a constant independent of energy. Contrary to the original approach, where this constant was globaly adjusted to experimental data to reproduce absolute cross-section values, we applied the systematics at 14.5 MeV [2] instead. The input of data to MASGAM is fully automated. It uses a minimal input information such as Z, A and the level scheme of the target nuclide. It assumes only E1 radiation and all the other relevant parameters, such as the optical-model parameters, the level density- and the E1-giant resonance parameters are derived from global systematics. An example of the calculation for a target nuclide with A≈150 is given in Fig. 1. This procedure is discussed in detail in Ref. [11]. The smooth statistical region is coupled to a 1/v-component at the energy $E_L = 0.5 D_0$. The calculated values can be checked and renormalized at three neutron energies, namely 0.0253 eV, 30 keV and 14.5 MeV. A comparison of MASGAM results (before renormalization) for the ^{142}Nd target with the JEF-1 evaluation is displayed in Fig. 2.

If experimental information on the resolved resonance region exists [13], the code SIG-ECN has been used to generate a point-wise file based on the resonance parameters. Merging these data with the MASGAM smooth results have been carried out within the processing code SYMPAL. The energy E_H, at which both data are joint, is determined in the following way. Both files have been processed into the Vitamin-J group structure and were graphically compared to identify the energy at which it seems resonances start missing, as compared to the calculated smooth statistical part. The group boundary

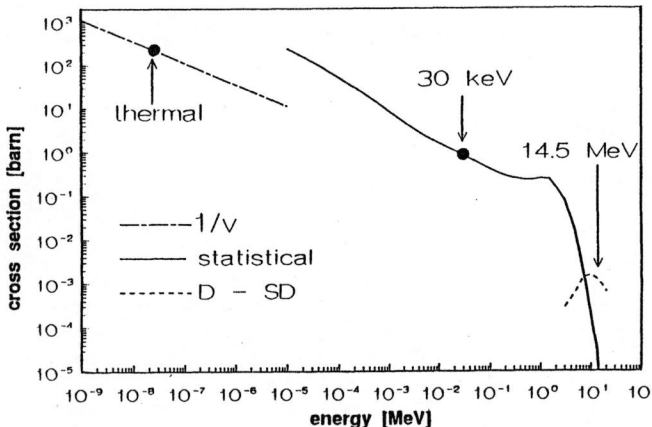

Fig.1. The (n,γ) excitation curve as calculated with the code MASGAM for an even target with A ≈ 150. Three components are dispayed separately with their normalization points.

prior to this energy has been chosen as E_H. This rather simple and quick procedure has been tested against some existing earlier evaluations and turned out to be reasonably good. These data are denoted in Table 1 as MASGAM + SIG-ECN.

2.2 Applied renormalizations

If experimental information on cross sections is available, it is used to renormalize the calculated component of the excitation curve separately. The 1/v-component is adjusted to agree at 0.0253 eV with the thermal cross-section value [13] and the smooth statistical component is adjusted at 30 keV, where a large amount of cross section data exist from astrophysics studies [14]. The compilation of Wagner and Warhanek [15] was used for the direct/semidirect (DSD) component at 14.5 MeV. The calculated excitation curve, renormalized in this way, has a large uncertainty mainly in the resonance region, if the resolved resonance data are unavailable (see table 2 for details).

In order to provide additional renormalization systematics, if no experimental information is available, e.g. for radioactive targets, an attempt has been made to develope systematics both for thermal data and for the 30 keV data. The prediction of thermal cross sections is almost impossible, due to the fact that only a limited number of resonances determine these values and thus no statistical assumptions can be applied. However, in spite of the expected very large uncertainty in these "predictions" some attempt is made to account for at least the global trend.

Starting from the expression for the average capture cross section, after several simplifications, the parameterized formula

$$\sigma_{n\gamma} = C\,(aU)^X$$

can be used to fit the constants C and X to the measured data. U is the effective excitation energy, defined as $U = S_n$ - pairing energy, and a is the level density parameter.

First a least-squares fit was applied to the thermal cross-section data (compliled in [13]), with a and U values derived from systematics appled in the code MASGAM. The results are displayed in Table 2 and fig.3 and show that the pairing correction smoothed to some extent the

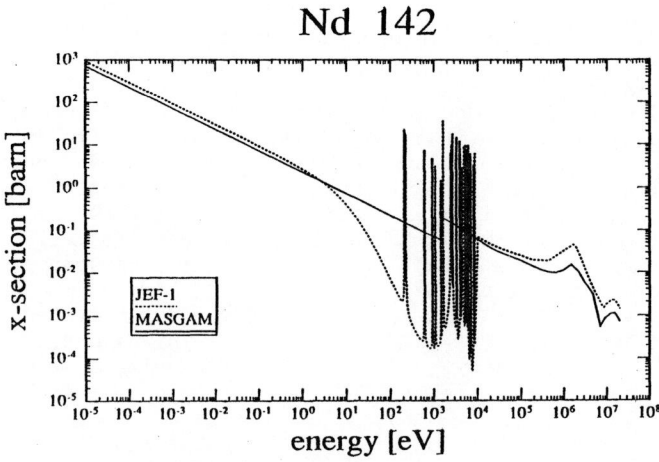

Fig.2. A comparison of MASCAM and JEF-1 data for the ^{142}Nd(n,γ) reaction before renormalization at 0.0253 eV and 30 keV. Mind the D-SD component in NEF-1, which is too large compared to the systematics.

raw data. However, as expected the scatter of the data around the fitted curve remains large with deviations of a factor f=50 or more.

A similar procedure has been applied to the 30 keV data (compiled in [14]) separately for odd- and even-Z targets. The results are displayed in figs. 4 and 5 and in Table 2 and show an acceptable scatter of data arround the fitted curves with deviations of a factor 1.5 - 2.0.

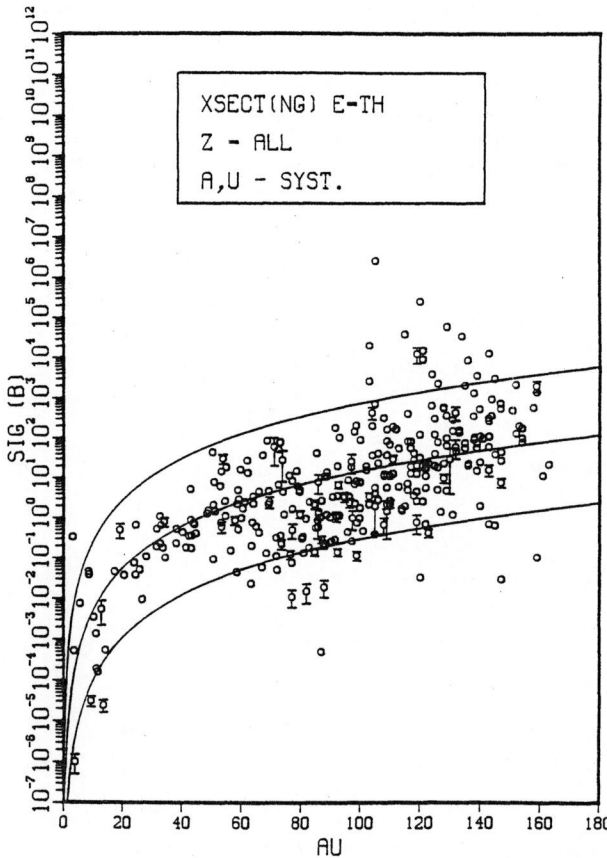

Fig.3. Thermal capture cross sections (taken from ref. [13]) plotted against aU together with the fitted curve. The uncertainty band with f=50 is indicated.

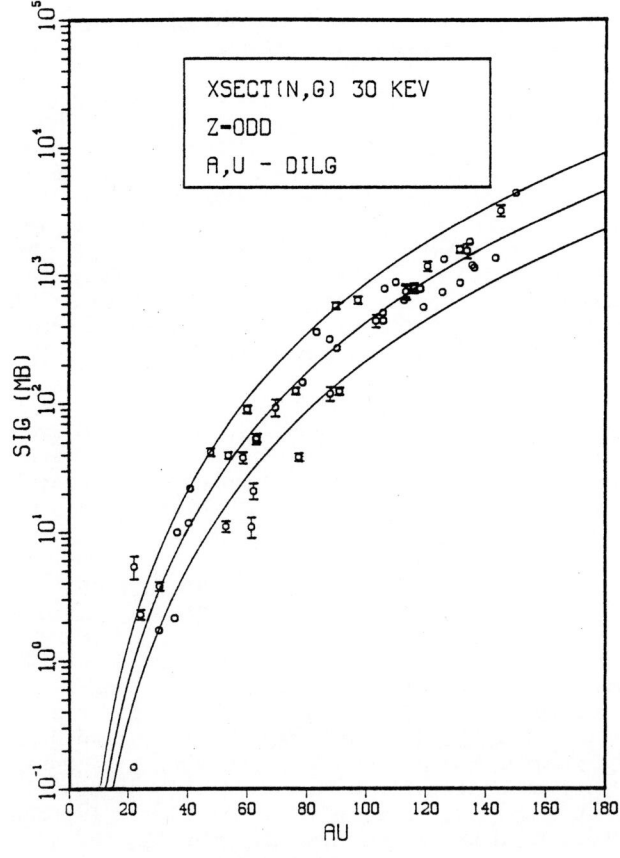

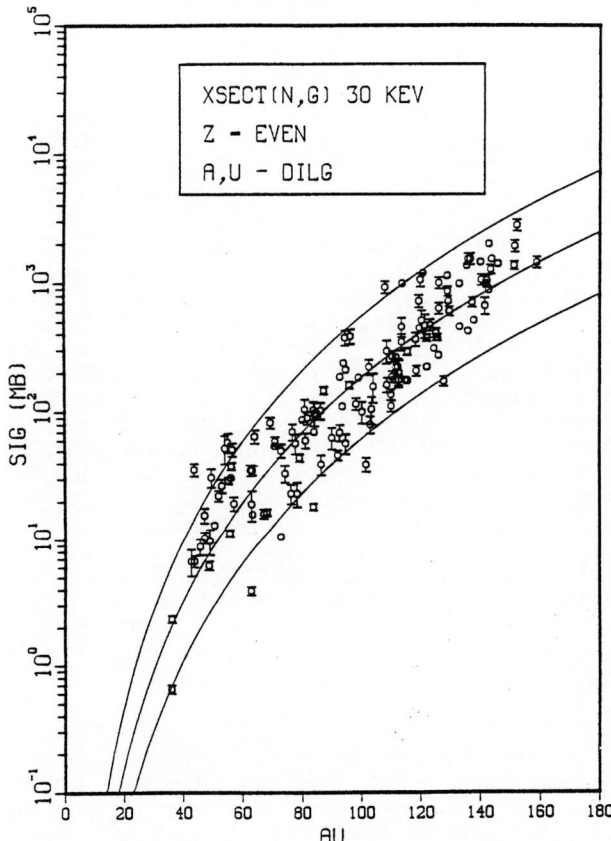

Fig.4 and 5. Average capture cross sections at 30 keV (adopted from [14]) or odd-(even)-Z targets, respectively, plotted against aU together with their uncertainties of f=1.5 and f=2, respectively.

Table 2. Derived systematics to fit the MASGAM calculations

$\sigma_{n\gamma}(E_n)$	LSQ-fit
$\sigma_{n\gamma}$(th)	1.5E-6 $(aU)^{3.5}$
$\sigma_{n\gamma}$(30 keV)	
Odd-Z	3.346E-6 $(aU)^{4.025}$
Even-Z	2.461E-7 $(aU)^{4.41}$
$\sigma_{n\gamma}$(14.5 MeV)	1.18-1.13 $e^{-0.01338A}$

The full procedure of MASGAM calulations and renormalizations has been tested against several JEF-1 evaluations across the entire mass range and showed a very good agreement, considering all the approximations applied (for results see [11]). The adopted uncertainties, both due to the applied model and the renormalizations, are given in table 3.

Table 3. The estimated uncertainties of the MASGAM calculations

MASGAM renormalization	Energy component		
	1/v	Stat.	D-SD
experiment	$\Delta\sigma_{n\gamma}$	$1.5^{a,b}$	1.5^b
systematics	100	3.0^b	1.5^b

[a] Except the energy region between $D_0/2$ and several tens of D_0, where the absent resonance cross sections introduce a large uncertainty.

[b] Includes also the guessed uncertainty of the statistical and DSD model in the energy ranges considered.

3. Renormalizations at 14.5 MeV to systematics

Several new systematics have been derived in recent few years. They estimate mainly the cross-section values at 14.5 MeV. However, some derive also the shape of the excitation curve. The list concerning only the 14.5 MeV values can be summarized as follows: (n,2n) [15,16]; (n,p) [15,17,18]; (n,d+n,np) [15,19]; (n,α) [15]. The references on the systematics of the excitation curves are: (n,charged particle) [20]; (n,np),(n,nα) [21] and (n,γ) [22]. These systematics have been compared with the set of those adopted earlier during the development of REAC-ECN files, based essentially on three authors: Forrest [6], Vonach [7] and Kopecky [2]. The comparison was done only graphically, the systematics against each other together with the available experimental data. We did not find any significant improvement with the new proposals and therefore we decided to keep the original set also for the EAF-1 and 2 files.

Table 4. Recommended systematics with their uncertainties.

Reaction	Systematics	Uncertainty[a]
(n,γ)	Kopecky [2]	1.5
(n,p)	Forrest [6]	1.5
(n,h)	Forrest [6]	1.9
(n,α)	Forrest [6]	1.6
(n,n')	Vonach [7]	1.5
(n,2n)	THRES	1.2
(n,3n)	0.8 x THRES	3.0
(n,nα)	0.125 x (n,a)	3.0
(n,d+np)	Forrest [6]	2.0
(n,t+nd)	Forrest [6]	1.6
(n,nt)	-	
(n,nh)	-	
(n,2p)	0.75 x (n,t)	3.0

[a] The uncertainty factors have been de- rived either from a rigorous error treatment of experimental data and fitted curves [6] or from a simplified graphical treatment (in all remaining cases). The uncertainty (standard deviation) in σ can vary from $-\sigma/f$ to $+f\sigma$.

A remark can be made concerning (n,2n) systematics. By a carefull comparison of all formulae [2,15,16] for new (n,2n) systematics, both against each other and against experimental data, we have realised that the results differ depending on the data set used for fitting (see ref.[5]). Thus we have decided at this moment to rely on the original systematics within the THRES code.

4. Renormalizations at 14.5 MeV to experimental data

The main source of 14.5 MeV data in the EAF-1 file has been the extensive compilation of Qaim [23]. In addition some more recent references were used in connection with the special treatment for 102 important reactions [26]. We felt that a large-scale update of the experimental renormalizations is needed. The literature was scanned to find new experimental data at 14 to 15 Mev in the period up to 1990. These data were used to update experimental renormalizations in the EAF-2 file and thus this file reflects the present experimental situation rather well. A listing of all renormalizations is included in the final EAF-2 document [5].

In all new renormalizations, both to systematics and to experimental values, the shape of the excitation curve has been inspected in order to avoid an erroneous application of this procedure for curves with either a steep rise (e.g. (n,3n)) or fall (e.g. (n,2n)) around the 14.5 MeV energy point. The checking procedure was built into our processing code SYMPAL and applied to all renormalizations.

5. Treatment of isomer ratios

Many nuclides have isomer states that are either long lived or which have a different decay mode from the ground state. It is therefore important that in these cases the total reaction cross section for a channel be split by use of a branching ratio so that the cross sections to the ground state and isomer are given separately.

The recommended branching ratios, based on a simplified model calculation with the code GNASH [8] and tabulated in refs. [2,5] have been applied without change for all reactions, this time also including the (n,γ) reaction.

The branching ratio for the capture data was set to a value taken from the systematics for the single-particle emission [8]. In cases where the thermal and resonance regions are included and the thermal experimental isomer ratio exists, this ratio was applied for energies upto E_H.

The branching ratios, predicted from the systematics [8], were recently supported by a theoretical study of Chadwick [24] and the recent, still in progress, irradiation experiment at AEA Harwell on branchings for several long-lived Hf isotopes with high spins of isomeric states [25].

Check and improvement of cross-sections for important reactions

A part of the EAF-1 file was a set of 102 important reactions leading to long-lived products which were evaluated or tested individually [26]. These data are included in the present EAF-2 version, however, no other reactions received special treatment separately in this version. Such action is planned for the future and be based on the results of the sensitivity study (see next paragraph).

Future plans

Further improvements of the EAF-2 file are foreseen. They will be included in the next version, EAF-3, which will be released during the spring of 1992. They can be categorized in five major steps:

1. The issue of completeness needs to be further adressed. The effect of the presence of U and Th impurities in structural materials for fusion reactors has been ignored untill now . However, these actinides may be present at level of a few ppm and should be considered [27] in activation and transmutations calculations for low activation materials. To do this the cross section data for 52 actinide targets will be included into the EEF-3 version. The data for these will be generated using THRES, but with (n,γ), (n,f), (n,2n) data taken, wherever available from evaluated files. The remaining (n,γ) reactions will be calculated by the methods used for EAF-2. The missing (n,f) data will either use available data to renormalize data from neighbouring nuclide or will use a simple model calculation.

2. The second step will be the production of the uncertainty file for all cross section data. In the first phase the EAF-3 file will be provided with one-group error file in the data for the (n,x) reactions and with two-group file for the radiative capture data. The latter will cover the thermal and resonance range for capture reactions upto the energy E_H and the remaining part to 20 MeV. To generate the error estimates, both the experimental information and the uncertainties from systematics will be used. Such a data base will enable us to carry out sensitivity studies with the inventory code FISPACT. Also using the dominant nuclides at a

particular time and calculating the sensitivity coefficients of these in conjuction with the uncertainty file, it will be possible to give the uncertainty in the total quantities such as activity.

3. A new code will be selected to replace THRES. This code should be fast and yet include a reliable nuclear model with relatively simple input.

4. An important task will be the improvement of the accuracy of the cross section data base. It may be possible to replace all THRES generated cross sections by a more physically based model, but this activity will be concentrated on the selected amount of important reactions identified in previous studies, such as in ref. [28,29], or from new calculations with EAF-2 data base. The data for these materials will be replaced from the available high-quality libraries as general purpose (EFF/JEF-2, ENDF/B-6 and JENDL-3) files or special files as JENDL-3 activation file or the dosimetry file IRDF-90. For some special cases our own evaluation is not excluded.

5. Finally new reactions, e.g. the sequential reactions discussed by Cierjacks et al. [30], will be included in EAF-3, while gas production cross sections and kerma factors are considered for inclusion.

Acknowledgment

The European Activation File is sponsored by the Fusion Technology Programme of the European Community (Task LAM-2).

References

1. R.A. Forrest, D.A.J Endacott and A. Khurseed: AERE M 3655 (1988).
2. H. Gruppelaar, H.A.J. van der Kamp, J. Kopecky and D. Nierop: ECN-207 (1988).
3. J. Kopecky and H. Gruppelaar: Proc. Int. Conf. on Nuclear Data for Science and Technology, Mito (Saikon Publ. Comp. Ltd., Tokyo), 245 (1988).
4. H. Gruppelaar, H.A.J. van der Kamp, J. Kopecky and D. Nierop: ECN-87 (1988).
5. J. Kopecky et al.: ECN Report to be published.
6. R.A. Forrest: AERE-R-12419 (1986).
7. H. Vonach: IAEA-TECDOC-457, 127 (1986).
8. J. Kopecky and H. Gruppelaar: ECN-200 (1987)
9. S. Pearlstein: J. of Nucl. Energy 27, 81 (1973).
10. M.G. Delfini and H. Gruppelaar: ECN-FYS-STEK-144 (1988)
11. J. Kopecky and M.G. Delfini: ECN Report, to be published.
12. Zhao Zhixiang and Zhou Delin: CNDC-0004 (1990).
13. S.F. Mughabghab et al.: Neutron Cross Sections; Neutron Resonance Parameters and Thermal Cross Sections Vo.1, (Academic Press New York), Part A (1981) and Part B (1984).
14. Z.Y. Bao and F. Kappeler: Atomic Data and Nuclear Data Tables 36, 447 (1987).
15. Y. Ikeda et al.: Proc. Int. Conf. on Nuclear Data for Science and Technology, Mito (Saikon Publ. Comp. Ltd., Tokyo) 257 (1988).
16. Zhang Jin, Zhou Delin and Cai Dunjiu: CNDC-0002 (1988).
17. S. Selvi and H.H. Erbil: INDC(TUR) - 002/4 (1989).
18. S.A. Bedikov and A.B. Pashchenko: INDC(CCP)-325 (1991).
19. Zhou Yongi, Han Min, Wang Guanghou: CNDC-0003 (1989).
20. Zhao Zhixiang and Zhou Delin: CNCD priv. communication.
21. Zhao Zhixiang and Zhou Delin: CNCD priv. communication.
22. Zhao Zhixiang, Zhou Delin and Cai Diunjiu: CNCD priv. communication.
23. S.M. Qaim: 14 MeV Neutron Activation Cross-sections, Handbook of Spectroscopy, (CRC Press Inc., Boca Raton, Florida), Vo.1 (1981).
24. M. B. Chadwick and P.G. Young: Proc. of a Specialists' Meeting on Neutron Activation Cross Sections for Fusion and Fusion Energy Applications, NEANDC-259 'U', 241 (1990) and Nucl.Sci.Eng., (1991) to be published.
25. B. Patrick, AEA Harwell (private communication.
26. J. Kopecky and H.A.J. van der Kamp: ECN-89-181 (1989).
27. M.G. Sowerby: private communication.
28. R.A. Forrest and D.A.J. Endacott: AERE R 13402 (1989).
29. M.G. Sowerby and R.A. Forrest: AERE R 13708 (1990).
30. S. Cierjacks, P. Oblozinsky and B. Rzehorz: KfK-4867 (1991).

EVALUATION OF CHARGED-PARTICLE REACTIONS FOR FUSION APPLICATIONS

Roger M. White, David A. Resler, and Stephen I. Warshaw

University of California
Lawrence Livermore National Laboratory
Livermore, CA 94550 U.S.A.

Abstract: New evaluations of the total reaction cross sections for ^{2}H(d,n)^{3}He, ^{2}H(d,p)^{3}H, ^{3}H(t,2n)^{4}He, ^{3}H(d,n)^{4}He, and ^{3}He(d,p)^{4}He have been completed. These evaluations are based on all known published data from 1946 to 1990 and include over 1150 measured data points from 67 references. The purpose of this work is to provide a consistent and well-documented set of cross sections for use in calculations relating to fusion energy research. A new thermonuclear data file, TDF, and a library of FORTRAN subprograms to read the file have been developed. Calculated from the new evaluations, the TDF file contains information on the Maxwellian-averaged reaction rates as a function of reaction and plasma temperature and the Maxwellian-averaged average energy of the interacting particles and reaction products. Routines are included that provide thermally-broadened spectral information for the secondary reaction products.

[^{2}H(d,n)^{3}He, ^{2}H(d,p)^{3}H, ^{3}H(t,2n)^{4}He, ^{3}H(d,n)^{4}He, ^{3}He(d,p)^{4}He reactions, charged-particle evaluations, fusion reactions, fusion reactivities, Maxwellian-averaged reaction rates, calculated emission spectra, astrophysical S-factors, R-matrix analyses]

Introduction

This work has been done to provide a consistent and well-documented set of evaluated cross sections from which processed information can be generated for use in fusion applications. An important part of this is to provide the user with a realistic assessment of the uncertainties remaining in these reaction cross sections. We have developed a processed data file which accurately represents the information needed in practical calculations. *This paper contains only the most brief summary of an extensive final report to be produced later this year* and is referenced here as LLNL evaluation [91].

We have assessed and extracted total reaction cross sections as a function of energy. The sources of experimental data were in the form of integrated cross sections, angular distributions of secondary particles, measurements at one angle multiplied by 4π where the angular distribution is assumed or known to be isotropic, or, astrophysical S-factors—basically a quantity which is the cross section divided by the Coulomb penetrability and incident energy. It is usually smoothly varying with energy and represents the nuclear part of the cross section. Some data sets were available only in graphical form and were scanned with a digitizing program written for this application to insure that no additional error (beyond that introduced by the draftsman) resulted. Integrated cross sections were obtained from measured angular distributions by converting them to center-of-mass values and using a least-squares fitting procedure to obtain Legendre polynomial coefficients according to a consistent prescription.

Cross section evaluations

References and graphical symbols for the experimental data bases used in the five evaluated reaction cross sections are given in Figs. 1, 4, 8, 10, and 14 and in the references section. Also included in these figures are the uncertainties (at the 95% confidence level) we place on the evaluations *over the energy range of importance to fusion applications*.

In Fig. 2, we show the two most recent measurements of the ^{2}H(d,n)^{3}He reaction by Brown[90] and Krauss[87] along with the older measurement of Arnold[54] plotted in terms of the astrophysical S-factor from E_d=0 to 0.12 MeV (in all figures the scale is the kinetic energy of the incident particle in the laboratory frame). The measurements of Krauss[87] are separated into (b) and (m) because they were carried out at two different facilities. The measurements of Brown[90] are approximately 8% higher than the previous measurements. In Fig. 3, an extrapolation of the S-factor from higher energies clearly favors the measurements of Brown[90]. Current knowledge of the structure of ^{4}He indicates that the S-factor is smooth over the 400 keV range plotted in Fig. 3. Not shown in Fig. 3 are the data of Davidenko[57] because of a shape difference and large error bars and the data of Chagnon[56] which are considerably different from the other measurements. As with the other reactions discussed below, data of some authors listed in Fig. 1 for the ^{2}H(d,n)^{3}He reaction are not shown because they are not in the energy range plotted in the figure.

In Fig. 5, we show all the data for the ^{2}H(d,p)^{3}H reaction from E_d=0 to 0.2 MeV. The data are somewhat discrepant but an extrapolation of the S-factor from higher energies together with the data of Brown[90] and Arnold[54] and knowledge of the structure of ^{4}He give us confidence in the evaluation to ±3%. In Fig. 6 we show the high energy evaluation of the ^{2}H(d,p)^{3}H reaction as the S-factor vs. E_d. Plotted in this way, an extrapolation to 30 MeV is less difficult to make. Figure 7 shows the same data plotted in terms of cross section vs. E_d.

The data base for the ^{3}H(t,2n)^{4}He reaction is more sparse as can be seen in Figs. 8 and 9. The S-factor vs. E_t shows a clear indication of changing slope from E_t=0 to 0.4 MeV. The evaluation above 1 MeV is heavily weighted in favor of the measurements of Govorov[62] and Jarmie[58]. Further details describing this evaluation will be given in the final report.

The ^{3}H(d,n)^{4}He reaction evaluation shown in Fig. 11 is based on a single-level R-matrix fit to all but three measurements. The three data sets not used are shown in

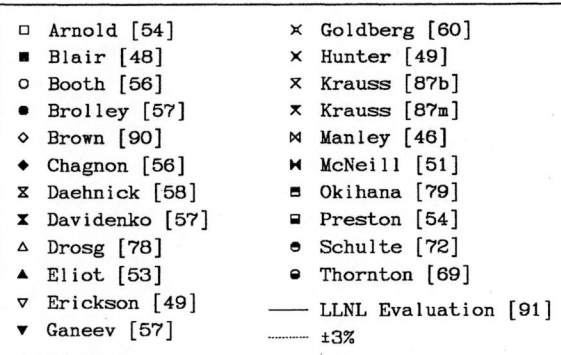

Fig. 1. References and plotting symbols used for the ²H(d,n)³He reaction data as shown in Figs. 2 and 3.

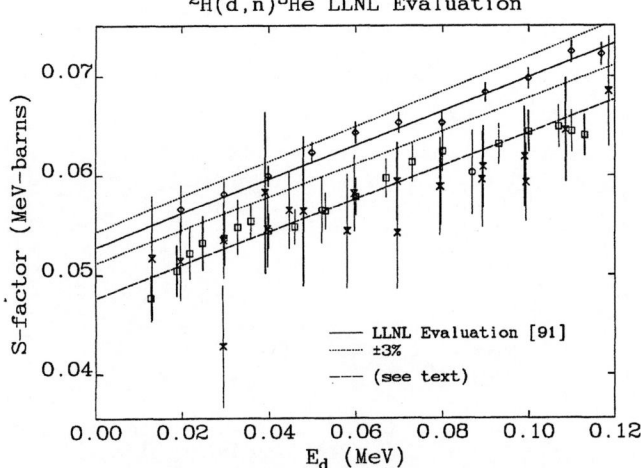

Fig. 2. The recent measurements of Brown[90] and Krauss[87] are plotted with the work of Arnold[54] as the astrophysical S-factor vs. E_d in the laboratory frame. The data of Brown[90] are approximately 8% higher than the previous measurements.

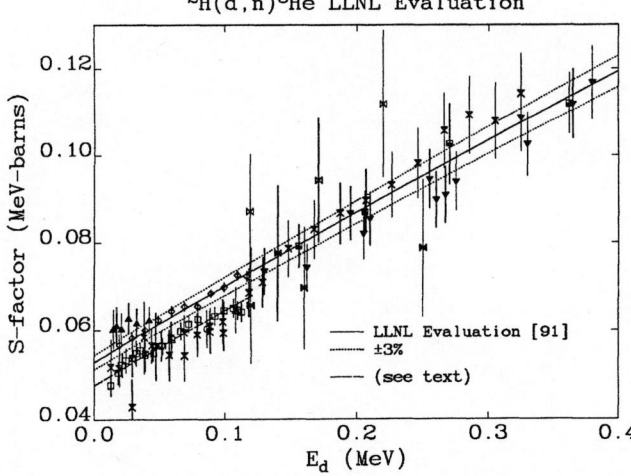

Fig. 3. The dashed line here and in Fig. 2 represents the most likely evaluation in the energy range E_d=0 to 0.12 MeV without the Brown[90] measurement and if no data had existed above 0.12 MeV. Not all data are plotted in this figure (see text).

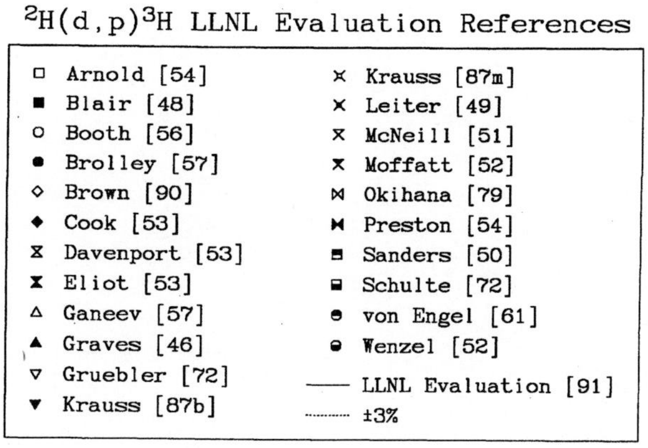

²H(d,p)³H LLNL Evaluation References

Fig. 4. References and plotting symbols used for the ²H(d,p)³H reaction data as shown in Figs. 5, 6, and 7.

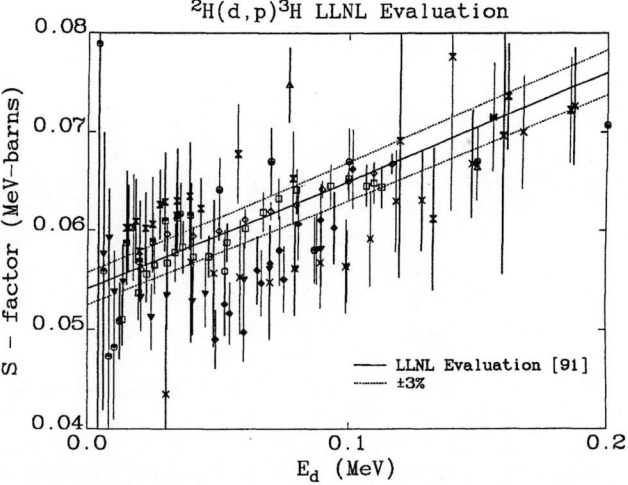

Fig. 5. All ²H(d,p)³H experimental data and the evaluation plotted as the S-factor vs. E_d from 0 to 0.2 MeV.

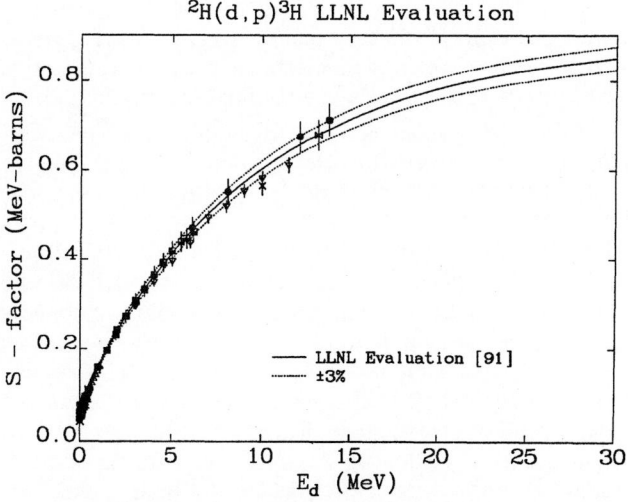

Fig. 6. Data and the evaluation of the high-energy portion of the ²H(d,p)³H reaction plotted in terms of the S-factor vs. E_d to show how the evaluation was carried out to 30 MeV.

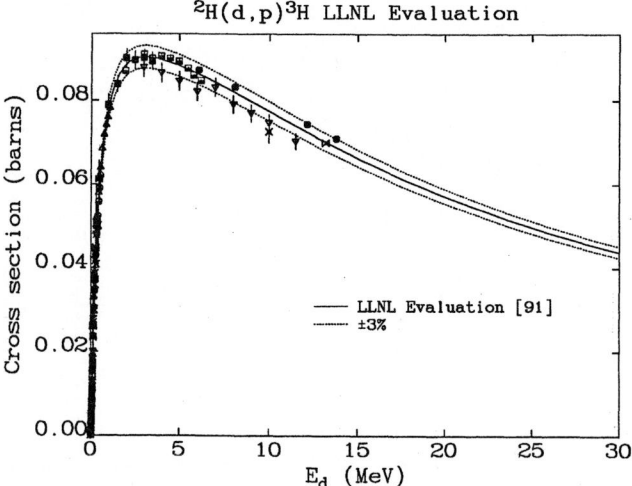

Fig. 7. Same as in Fig. 6 except plotted as cross section vs. E_d to show how the cross section is extrapolated to 30 MeV.

Fig. 12 as ratios to the evaluation with the dashed lines serving to guide the eye. Numerous R-matrix calculations were performed using many energy ranges and using various subsets of all the data. All of the results fall within the ±2% indicated. The final evaluation is based upon allowing the normalization of each measurement to vary while simultaneously performing a least-squares fit with a single-level R-matrix calculation. Shown in Fig. 11 are the measurements used in the R-matrix fit, the average percent error for each data set, and the percent change in the overall normalization for arriving at the best fit. We assumed that the overall normalization of the entire data base is roughly correct and therefore the normalizations of the individual data sets were allowed to vary subject to the constraint that the average normalization was unity. The data plotted in Figs. 11 and 12 have been renormalized by the amount indicated in Fig. 11. In Fig. 13, we show the R-matrix analyses by Jarmie[84] and Brown[87] as ratios to the LLNL evaluation [91]. We also show the effect of not allowing the individual data sets to change in normalization. All of these results fall within ±2% and indicate to us the uncertainty limits of this data base. More specific details of the R-matrix calculations and renormalization procedures will be presented in the final report.

Of the five reactions evaluated in this work, the measurements for the ^{3}He(d,p)^{4}He reaction (Fig. 15) are the most discrepant. The absolute values differ by more than the experimenters' quoted errors. As shown in Fig. 16, except for the data sets of Bonner[52] and Jarvis[53] and the low-energy portion of the data sets of Carlton[70] and Kliucharev[56], the shapes are in good agreement. For E_d=0 to 800 keV, the evaluation is based on a single-level R-matrix fit to all the available data except for those discussed above. Many R-matrix calculations were performed under a variety of conditions and the best fit was obtained by simultaneously allowing the individual data set normalizations to vary. The normalizations of many data sets differ with one another by more than the quoted errors and it is not obvious which measurement is correct. Therefore, we assumed that, on the average, the overall normalization of all the measurements is correct and the individual data set normalizations should average to unity. In Fig. 17, we show the average percent error for each data set and the

percent change in normalization for arriving at the best fit. The data plotted are only those points included in the fitting process and have been renormalized by the indicated amounts. We believe the evaluation to be good to no better than ±8% in normalization even though we are certain that the shape of the ^{3}He(d,p)^{4}He reaction is known much better.

Applications File—TDF

We have developed a thermonuclear data file, TDF, which is an ASCII file that contains thermonuclear reaction rates and spectral information on the outgoing particles as a function of plasma temperature. This file contains interpolatable data for all plasma temperatures from 100 eV to 1 MeV. It is assumed that the distribution of the reacting particles in the plasma is Maxwellian. Currently, TDF contains reaction rates and spectral information calculated

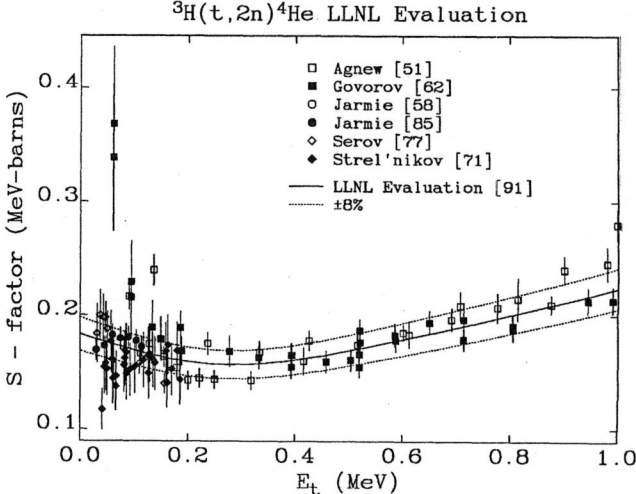

Fig. 8. Plot of all experimental data and evaluation of the ^{3}H(t,2n)^{4}He reaction from E_t=0 to 1.0 MeV. The change in slope of the evaluation between 0 and 0.5 MeV is independent of any particular data set or of the evaluation techniques we used on this data base.

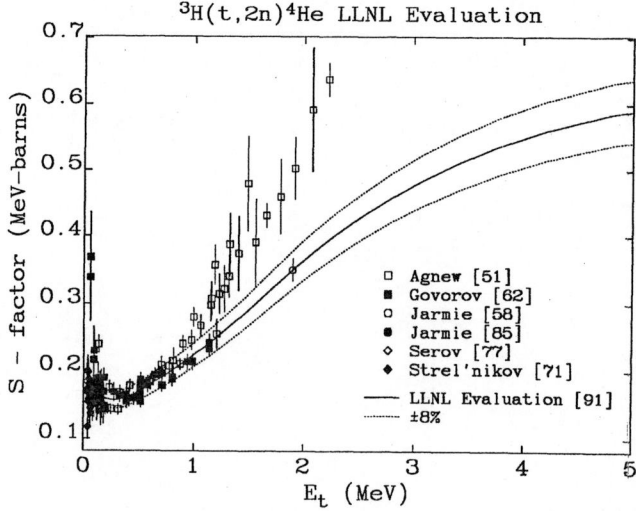

Fig. 9. Plot of the ^{3}H(t,2n)^{4}He data and extrapolation of the evaluation to higher energies in terms of the S-factor vs. E_t. Further details describing this evaluation will be given in the final report described in the text.

^{3}H(d,n)^{4}He LLNL Evaluation References

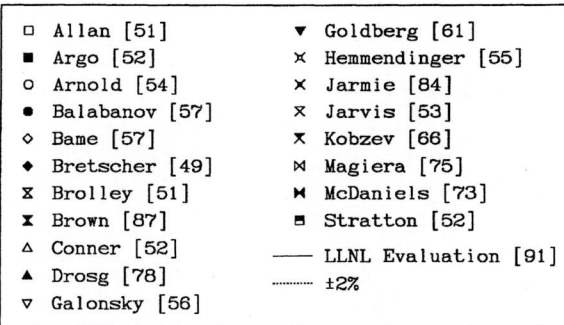

□	Allan [51]	▼	Goldberg [61]
■	Argo [52]	✕	Hemmendinger [55]
○	Arnold [54]	✕	Jarmie [84]
●	Balabanov [57]	✕	Jarvis [53]
◇	Bame [57]	✕	Kobzev [66]
◆	Bretscher [49]	⋈	Magiera [75]
⋉	Brolley [51]	⋈	McDaniels [73]
⋈	Brown [87]	▣	Stratton [52]
△	Conner [52]	——	LLNL Evaluation [91]
▲	Drosg [78]	······	±2%
▽	Galonsky [56]		

Fig. 10. References and plotting symbols used for the ^{3}H(d,n)^{4}He reaction data as shown in Figs. 11, 12, and 13.

^{3}H(d,n)^{4}He LLNL Evaluation

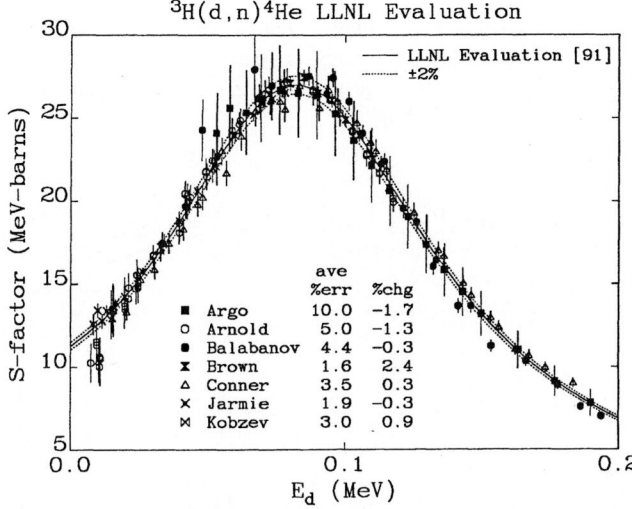

Fig. 11. Experimental data and evaluation for the ^{3}H(d,n)^{4}He reaction in terms of the S-factor vs. E_d. The *ave %err* is the average total error of each data set. The *%chg* is the renormalization of each data set for the best R-matrix fit (see text).

^{3}H(d,n)^{4}He Reaction

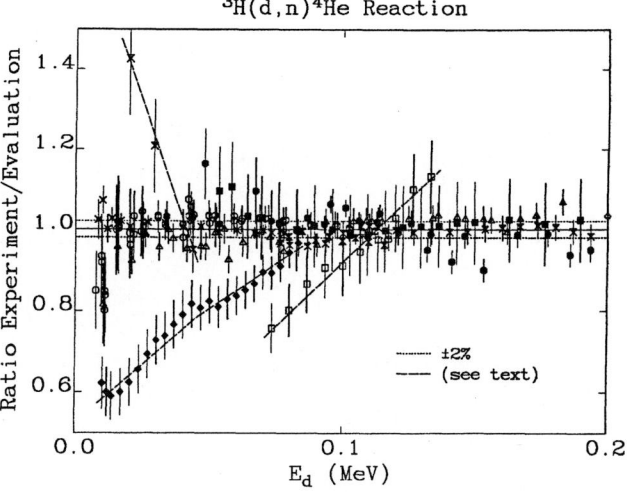

Fig. 12. Ratio of experimental data to evaluation for the ^{3}H(d,n)^{4}He reaction showing which data sets were excluded from the R-matrix fitting procedure.

^{3}H(d,n)^{4}He Reaction

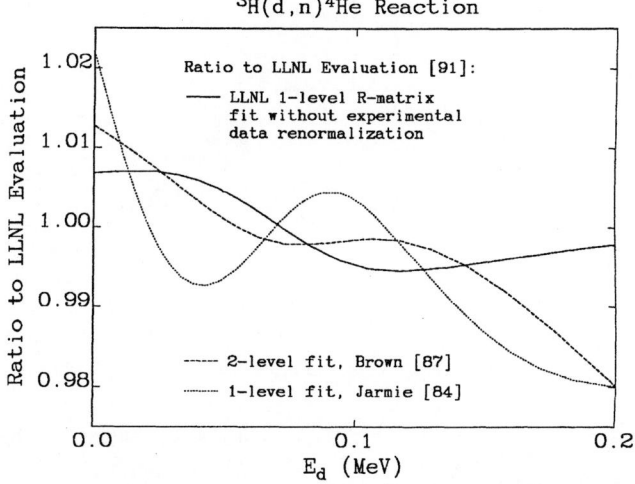

Fig. 13. Ratio comparisons of our R-matrix fit without data renormalization and the R-matrix fits of Jarmie[84] and Brown[87] to our evaluation of the ^{3}H(d,n)^{4}He cross section.

from the five evaluated reactions discussed in the previous section in units selectable by the user.

We have also developed a library of four subprograms written in FORTRAN77 for accessing TDF. The first subprogram is simply a reader and is called only once in an applications program prior to any call of the other three subprograms. The second subprogram returns the value of the reactivity (Maxwellian-averaged reaction rate) given a reaction number and a plasma temperature. The third subprogram returns the reactivity and the Maxwellian-averaged average energies of the two interacting particles and the reaction products given a reaction number and a plasma temperature. The fourth subprogram is a spectral lookup routine (SPECLU) which is slightly more complicated. This subprogram will return the energy of a secondary reaction particle and the value of the corresponding spectral shape function in the frame of a laboratory detector as a function of a real number, RN, between 0 and 1. We show in Fig. 18 plots of the normalized spectral distributions vs. neutron emission energy for the ^{3}H(d,n)^{4}He reaction at temperatures of 5, 10, 20, and 50 keV. These plots were generated by repeated calls to SPECLU for 1000 input RN's equally spaced between 0 and 1. The filled

diamonds are the result of a CALL SPECLU as described above but for 2500 random numbers, RN, where the neutron energies are grouped into 50 bins. This gives one example of how this routine and TDF could be used in a Monte Carlo type calculation.

The calculations for the spectral information are carried out during the production of the TDF file so the file itself contains spectral data which lend themselves to fast look-up techniques. Because this information is obtained from complex calculations using the more fundamental cross sections, considerable effort was put into the computational algorithms to insure accuracy first and then speed. The processing code which produces the TDF file and the subprograms which provide the interpolations and look-ups introduce errors at a level of not more than 0.1%. The TDF file and the library of subprograms have been successfully tested on computers using three different operating systems and we consider them to be machine-independent. They will be available for public distribution later this year.

^{3}He(d,p)^{4}He LLNL Evaluation References

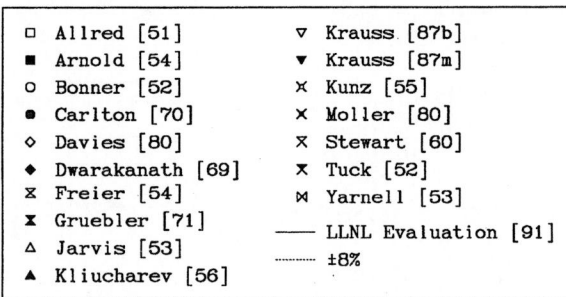

Fig. 14. References and plotting symbols used for the ^{3}He(d,p)^{4}He reaction data as shown in Figs. 15, 16, and 17.

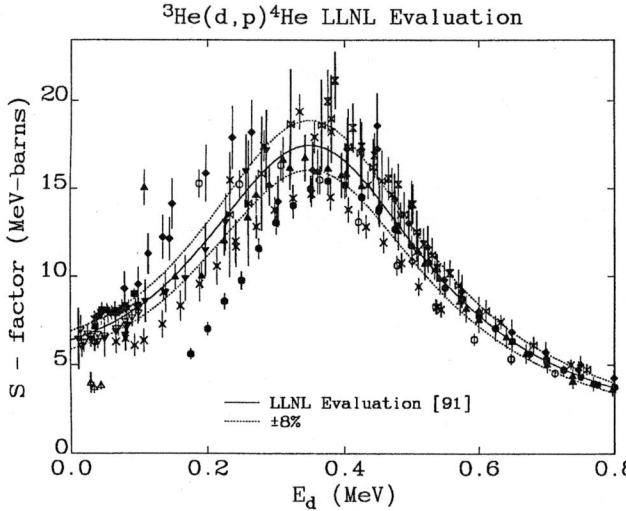

Fig. 15. Plot of all experimental data and evaluation of the ^{3}He(d,p)^{4}He reaction in terms of the S-factor vs. E_d from 0 to 0.8 MeV.

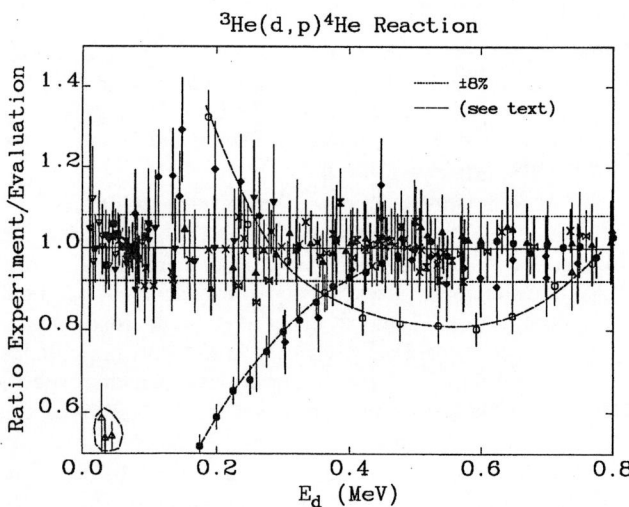

Fig. 16. Ratio of experimental data to evaluation for the ^{3}He(d,p)^{4}He reaction showing which data sets were excluded from the R-matrix fitting procedure.

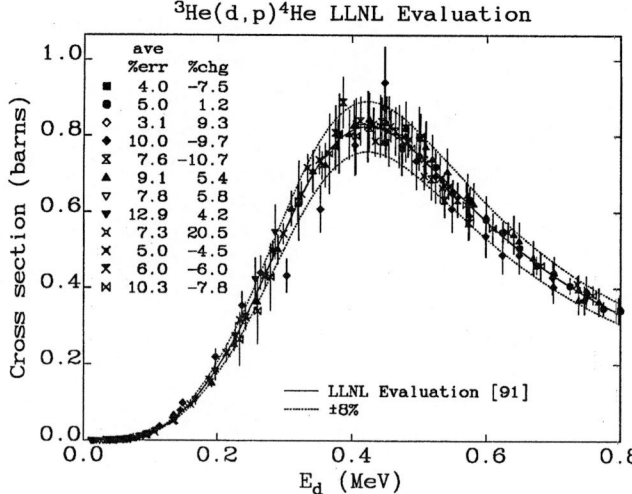

Fig. 17. Experimental data and evaluation for the ^{3}He(d,p)^{4}He reaction in terms of cross section vs. E_d. The numbers have the same meaning as described in Fig. 11.

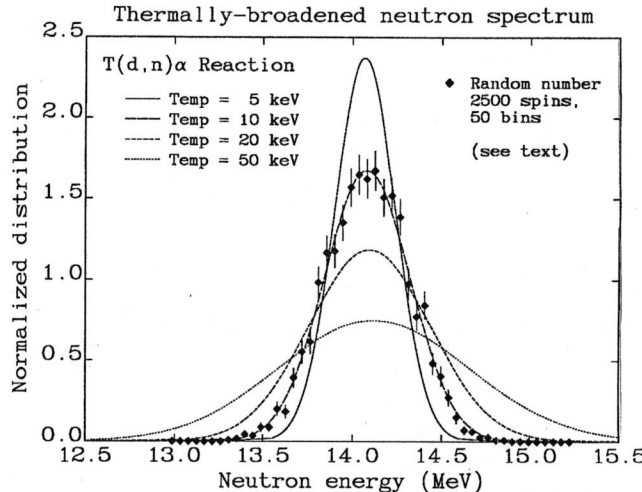

Fig. 18. Plot of the thermally-broadened neutron spectrum for the ^{3}H(d,n)^{4}He reaction for several plasma temperatures. The filled diamonds are described in the text. This information is obtained directly from TDF using the subprograms library.

Conclusions

We have given a brief overview of new evaluations of the most important charged-particle reaction cross sections needed for current fusion applications. We have also discussed a new applications file, based on these evaluations, which is machine-independent and provides the most important kinds of information necessary for fusion applications. This paper is intended to provide only a brief summary of a more extensive report to be produced later this year.

Work performed under the auspices of the U.S. Department of Energy by the Lawrence Livermore National Laboratory under contract number W-7405-ENG-48.

References

Agnew[51]	H.M. Agnew, W.T. Leland, H.V. Argo, R.W. Crews, A.H. Hemmendinger, W.E. Scott, and R.F. Taschek, Phys. Rev. **84**, 862 (1951).
Allan[51]	D.L. Allan and M.J. Poole, Proc. R. Soc. London, Ser. A **204**, 488 (1951).
Allen[58]	R.C. Allen and N. Jarmie, Phys. Rev. **111**, 1129 (1958).
Allred[51]	J.C. Allred, Phys. Rev. **84**, 695 (1951).
Argo[52]	H.V. Argo, R.F. Taschek, H.M. Agnew, A. Hemmendinger, and W.T. Leland, Phys. Rev. **87**, 612 (1952).
Arnold[54]	W.R. Arnold, J.A. Phillips, G.A. Sawyer, E.J. Stovall, Jr., and J.L. Tuck, Phys. Rev. **93**, 483 (1954); Los Alamos Scientific Laboratory Reports LA-1479 and LA-1481 (unpublished).
Balabanov[57]	E.M. Balabanov, I.Ia. Barit, L.N. Katsaurov, I.M. Frank, and I.V. Shtranikh, Sov. J. Atomic Energy, Atomnaya Energiya, Supp. No. 5, 43 (1957); L.N. Katsaurov, Akad. Nauk. USSR, Trudy Fiz. Inst. **14**, 224 (1962).
Bame[57]	S.J. Bame, Jr. and J.E. Perry Jr., Phys. Rev. **107**, 1616 (1957).
Blair[48]	J.M. Blair, G. Freier, E. Lampi, W. Sleator, Jr., and J.H. Williams, Phys. Rev. **74**, 1599 (1948).
Bonner[52]	T.W. Bonner, J.P. Conner, and A.B. Lillie, Phys. Rev. **88**, 473 (1952).
Booth[56]	D.L. Booth, G. Preston, and P.F.D. Shaw, Proc. R. Soc. London, Ser. A **69**, 265 (1956).
Bretscher[49]	E. Bretscher and A.P. French, Phys. Rev. **75**, 1154 (1949).
Brolley[51]	J.E. Brolley, Jr., J.L. Fowler, and E.J. Stovall, Jr., Phys. Rev. **82**, 502 (1951).
Brolley[57]	J.E. Brolley, Jr., T.M. Putnam, and L. Rosen, Phys. Rev. **107**, 820 (1957).
Brown[87]	R.E. Brown, N. Jarmie, and G.M. Hale, Phys. Rev. C **35**, 1999 (1987).
Brown[90]	R.E. Brown and N. Jarmie, Los Alamos National Laboratory Report LA-UR-89-953; Phys. Rev. C **41**, 1391 (1990).
Carlton[70]	R.F. Carlton, Ph. D. Thesis, University of Georgia (1970); private communication to N. Jarmie (1978).
Chagnon[56]	P.R. Chagnon and G.E. Owen, Phys. Rev. **101**, 1798 (1956).
Conner[52]	J.P. Conner, T.W. Bonner, and J.R. Smith, Phys. Rev. **88**, 468 (1952).
Cook[53]	C.F. Cook and J.R. Smith, Phys. Rev. **89**, 785 (1953).
Daehnick[58]	W.W. Daehnick and J.M. Fowler, Phys. Rev. **111**, 1309 (1958).
Davenport[53]	P.A. Davenport, T.O. Jeffries, M.E. Owen, F.V. Price, and D. Roaf, Proc. R. Soc. London, Ser. A **216**, 66 (1953).
Davidenko[57]	V.A. Davidenko, A.M. Kucher, I.S. Pogrebov, and Iu.F. Tuturov, Sov. J. At. Energy, Suppl. 5, 7 (1957).
Davies[80]	J.A. Davies and P.R. Norton, Nucl. Instrum. Methods **168**, 611 (1980).
Drosg[78]	M. Drosg, Nuc. Sci. Eng. **67**, 190 (1978); Z. Phys. A **300**, 315 (1981).
Dwarakanath[69]	M.R. Dwarakanath, Ph. D. Thesis, California Institute Technology (1969).
Eliot[53]	E.A. Eliot, D. Roaf, and P.F.D. Shaw, Proc. R. Soc. London, Ser. A **216**, 57 (1953).
Erickson[49]	K.W. Erickson, J.L. Fowler, and E.J. Stovall, Jr., Phys. Rev. **75**, 894 (1949); Phys. Rev. **76** 1141 (1949).
Freier[54]	G. Freier and H. Holmgren, Phys. Rev. **93**, 825 (1954).
Galonsky[56]	A. Galonsky and C.H. Johnson, Phys. Rev. **104**, 421 (1956).
Ganeev[57]	A.S. Ganeev, A.M. Govorov, G.M. Osetinskii, A.N. Rakivnenko, I.V. Sizov, and V.S. Siksin, Sov. J. At. Energy, Suppl. 5, 21 (1957).
Goldberg[60]	M.D. Goldberg and J.M. Le Blanc, Phys. Rev. **119**, 1992 (1960).
Goldberg[61]	M.D. Goldberg and J.M. Le Blanc, Phys. Rev. **122**, 164 (1961).
Govorov[62]	A.M. Govorov, Li Ka-Yeng, G.M. Osetinskii, V.I. Salatskii, and I.V. Sizov, Sov. Phys. JETP **15**, 266 (1962).
Graves[46]	A.C. Graves, E.R. Graves, J.H. Coon, and J.H. Manley, Phys. Rev. **70**, 101 (1946).
Grüebler[71]	W. Grüebler, V. König, A. Ruh, P.A. Schmelzbach, R.E. White, and P. Marmier, Nucl. Phys. **A176**, 631 (1971).
Grüebler[72]	W. Grüebler, V. König, P.A. Schmelzbach, R. Risler, R.E. White, and P. Marmier, Nucl. Phys. **A193**, 129 (1972).
Hemmendinger[55]	A. Hemmendinger and H.V. Argo, Phys. Rev. **98**, 70 (1955).
Hunter[49]	G.T. Hunter and H.T. Richards, Phys. Rev. **76**, 1445 (1949).
Jarmie[58]	N. Jarmie and R.C. Allen, Phys. Rev. **111**, 1121 (1958).
Jarmie[84]	N. Jarmie, R.E. Brown, and R.A. Hardekopf, Phys. Rev. C **29**, 2031 (1984).
Jarmie[85]	N. Jarmie and R.E. Brown, Nucl. Instrum. Methods B **10/11**, 405 (1985); N. Jarmie, private communication (1989).
Jarvis[53]	R.G. Jarvis and D. Roaf, Proc. R. Soc. London, Ser. A **218**, 432 (1953).
Kliucharev[56]	A.P. Kliucharev, B.N. Esel'son, and A.K. Val'ter, Sov. Phys. Doklady **1**, 475 (1956).
Kobez[66]	A.P. Kobzev, V.I. Salatskii, and S.A. Telezhnikov, Sov. J. Nucl. Phys. **3**, 774 (1966).
Krauss[87]	A. Krauss, H.W. Becker, H.P. Trautvetter, C. Rolfs, and K. Brand, Nucl. Phys. **A465**, 150 (1987).
Kunz[55]	W.E. Kunz, Phys. Rev. **97**, 456 (1955), and Ph. D. Thesis, University of Tennessee, (1954).
Leiter[49]	H.A. Leiter, R.E. Meagher, F.A. Rodgers, and P.G. Kruger, Phys. Rev. **76**, 167 (1949).
Magiera[75]	E. Magiera, M. Bormann, W. Scobel, and P. Heiss, Nucl. Phys. **A246**, 413 (1975).
Manley[46]	J.H. Manley, J.H. Coon, and E.R. Graves, Phys. Rev. **70**, 101 (1946).
McDaniels[73]	D.K. McDaniels, M. Drosg, J.C. Hopkins, and J.D. Seagrave, Phys. Rev. C **7**, 882 (1973).
McNeill[51]	K.G. McNeill and G.M. Keyser, Phys. Rev. **81**, 602 (1951).
Moffatt[52]	J. Moffatt, D. Roaf, and J.H. Sanders, Proc. R. Soc. London, Ser. A **212**, 220 (1952).
Möller[80]	W. Möller and F. Besenbacher, Nucl. Instrum. Methods **168**, 111 (1980).
Okihana[79]	A. Okihana, N. Fujiwara, H. Nakamura-Yokota, T. Yanabu, K. Fukunaga, T. Ohsawa, and S. Tanaka, J. of Phys. Soc. Japan **46**, 707 (1979).
Preston[54]	G. Preston, P.F.D. Shaw, and S.A. Young, Proc. R. Soc. London, Ser. A **226**, 206 (1954).
Sanders[50]	J.H. Sanders, J. Moffatt, and D. Roaf, Phys. Rev. **77**, 754 (1950).
Schulte[72]	R.L. Schulte, M. Cosack, A.W. Obst, and J.L. Weil, Nucl. Phys. **A192**, 609 (1972).
Serov[77]	V.I. Serov, S.N. Abramovich, and L.A. Morkin, Sov. J. At. Energy **42**, 66 (1977).
Stewart[60]	L. Stewart, J.E. Brolley, Jr., and L. Rosen, Phys. Rev. **119**, 1649 (1960).
Stratton[52]	T.F. Stratton and G.D. Freier, Phys. Rev. **88**, 261 (1952).
Strel'nikov[71]	Yu.V. Strel'nikov, S.N. Abramovich, L.A. Morkin, and N.D. Yur'eva, Bull. Acad. Sci. USSR Phys. Sci. (Isv.) **35**, 149 (1971).
Thorton[69]	S.T. Thornton, Nucl. Phys. **A136**, 25 (1969).
Tuck[52]	J.L. Tuck, W.R. Arnold, J.A. Phillips, G.A. Sawyer, and E.J. Stovall, Jr., Phys. Rev. **88**, 159A (1952).
vonEngle[61]	A. von Engel and C.C. Goodyear, Proc. R. Soc. London, Ser. A **264**, 445 (1961).
Wenzel[52]	W.A. Wenzel and W. Whaling, Phys. Rev. **88**, 1149 (1952).
Yarnell[53]	J.L. Yarnell, R.H. Lovberg, and W.R. Stratton, Phys. Rev. **90**, 292 (1953).

INTERNATIONAL COOPERATION FOR COMPILATION AND EVALUATION OF THE DATA ON CHARGED PARTICLE NUCLEAR REACTION

F.E.Chukreev
I.V.Kurchatov Institute of Atomic Energy,CAJAD.
123182, Moscow, USSR

Abstract: The report presents brief information on the activities of the network of Centers in the field of nuclear reaction induced by charged particles and photons.

During 1973-75 a cooperation of Nuclear Data Centers on compilation and evaluation of data on nuclear reactions induced by charged particles and photons was organized. The IAEA Nuclear Data Section is the initiator of this network.

The necessity to perform such a work and international cooperation is evident - practical applications of nuclear physics methods should be provided with nuclear data of the known accuracy.

This network of cooperating centers consists now of:

1. The USSR Center for Nuclear Structure and Reactions Data (CAJAD);

2. The IAEA Nuclear Data Section ;

3. The USA National Nuclear Data Center (NNDC);

4. The Computer Center of the University of Hokkaido (Japan);

5. Nuclear Data Groups of the Institute of Physical and Chemical Research (Japan);

6. Center for Photonuclear Experimental Data of the Moscow State University (CDFE);

7. Charged-Particle Nuclear Data Group, Atomic Energy Institute of the Peoples Republic of China.

Till 1981 the group headed by H.Münzel (KFK,FRG) took a very active part in the data compilation activity.

In the network of these centers there is no limitation on their activity on the compilation and evaluation of nuclear reactions data except for one : an intercenter exchange should be accomplished in accordance with the "EXFOR" system rules.

This system requires, that information prepared for an exchange be presented in a coded form permitting reading practically by any computer system.

Coding of information itself is performed by a specialist - physicist and its result is a maximum complete transfer of experimental material together with information on specific features of an experimental facility and data analysis.

Information on a theoretical interpretation of experimental data by authors of a paper is, as a rule, seldom reflected.

Of course, there is no point to affirm that reading of the EXFOR-text can replace reading of an original article. The goal of the system is to present experimental data in a computer-readable form and to simplify selection of information sources necessary for applied and basic research activities. The significant difference between the EXFOR-text and an original article is it conveys results rather than the authors thoughts. The experience shows that one reads easily EXFOR-texts and a computer helps him to choose them.

Undoubtedly, a selection of material to include them into the EXFOR system is of a subjective character, to a certain degree: a compiler selects original articles on a subject, publication language, confidence in author group data. It is also important that EXFOR-texts are not "frozen" collection - information relative to each publication can be corrected in due time by means of:

1. Correction of annoying but unavoidable misprints.

2. Information on the reaction of scientific community to the quality of data.

3. Personal contacts with authors of original papers which promote much an understanding of existing discrepancies in experimental data.

There is one limitation for EXFOR compilation activity. In order to avoid repetition, every Center much inform CAJAD about their compilation plans.

Although every Center selects compilation sujects and degree of plentitude independently, the largest part of compilations take account of the fields, which are recommended as first priority data by IAEA consultants, meetings.

These data are:

1.Cross-sections for medical radio-nuclide production.

2.Cross-sections for thermonuclear light ion reactions.

3.The data about neutron source reactions.

4. Data on (α,n) reactions at energies peculiar to actinide group nuclei.

5. Data on monitor-reactions.

It is seen that primary attention is given by the Centers to reactions important to nuclear physics applications. But while in this list, for instance, the astrophysical problems are lacking, nevertheless a great deal of such data are available in the collections of Centers. The data of points 2, 3, as a rule, are interesting to nuclear astrophysics specialists. This list of priorities does also not include data aimed at an analysis of a substance but compilations on points 1, 3, 4 can be very much helpful for specialists or the activation analysis.

As of April, 1991 the non-neutron data library in the "EXFOR" system contained data from 1363 publications with total volume of approximately 44 Mbytes.

Activity on data compilation from publications is accompanied by their dissemination among scientists working in the nuclear physics applications. Dissemination of data is accomplished by:

CENTER	REGION
NNDC	USA and Canada
CAJAD	USSR and (?) East European countries
NEA-Data bank	Developed countries except USSR, USA and Canada
IAEA Nuclear Data Section	All countries except those mentioned above
CDFE	All countries. (photonuclear data only)

Data Publications

Cooperation between these Centers results, in addition to data collection, compilation, evaluation and replies to inquiries from specialists, in publication of a number of useful issues.

Among them of particular interest should be the annual bibliography publications NNDC [1] ("The studies of integral nuclear reaction cross-sections induced by charged particles" and CDFE [2] ("Photonuclear reactions").

The trial issue of the "EXFOR" library content [3] turned out to be very popular and, evidently, will be published regularly.

Interesting to specialists are also two publications of the Nuclear Data Section - on nuclear monitor reactions

[4] and cross-sections for medical radionuclide productions [5].

The development by NNDC of a new format for evaluated data (ENDF/B-VI) suitable for data on nuclear reactions induced by charged particles enables us to expect that the evaluated data libraries will be also added through the mutual efforts of the Centers.

REFERENCES

1. N.E.Holden, S.Ramavataram. "Integral charged particle nuclear data bibliography", BNL-NCS-51771,1989. (annually).

2. "Photonuclear data ", Ed. by MGU Moscow. (annual)

3. V.V.Varlamov, G.M.Zhuravleva, V.V.Surgutanov, F.E.Chukreev. "Nuclear reactions induced by charged partiles and photons in the "EXFOR" system (reference data)". Moscow, 1987.

4. O.Schwerer, K.Okamoto. "Status report on cross-sections of monitor reactions for radioisotope prodution". INDC(NDS)-218/GZ+, Vienna, 1989.

5. D.Gondarias-Cruz, K.Okamoto. "Status on the compilation of nuclear data for medical radioisotopes produced by accelarators". INDC(NDS)-209/GZ, Vienna,1988.

CALCULATIONS OF PHOTOPRODUCTION REACTIONS

P. Obložinský

Kernforschungszentrum Karlsruhe, Institut für Materialforschung
7500 Karlsruhe, Postfach 3640, Federal Republic of Germany
and
Institute of Physics, Slovak Academy of Sciences
842 28 Bratislava, Czechoslovakia

M.B. Chadwick

Theoretical Division, Los Alamos National Laboratory
Los Alamos, New Mexico 87545, U.S.A.

Abstract: We discuss current theoretical methods used in calculations of photoproduction in reactions induced by fast neutrons (protons). A basic tool in these calculations still remains the well established statistical model of nuclear reactions that includes γ emission down to realistic decay schemes of low lying levels. The statistical model fails in describing the giant dipole resonance range of γ spectra and things are even worse for hard photons; also discrete γ ray production cross sections may be underpredicted by the statistical model starting from incident energies of a few tens of MeV. A considerable fraction of these production cross sections are well accounted for by preequilibrium models of γ ray emission, and this is discussed in some depth. An overview of available codes is given and illustrative examples characterizing the state-of-the-art in the field are provided.

(Reactions $(n, x\gamma)$, fast neutrons, theoretical methods, statistical model, preequilibrium γ emission, codes, photoproduction calculations, γ ray spectra, discrete γ rays)

Introduction

It is the purpose of the present paper do discuss the current status of theoretical methods used to calculate photoproduction cross sections in reactions such as $(n, x\gamma)$ induced by fast neutrons. We are not concerned with slow neutron capture processes, which are discussed elsewhere in this conference [1].

Photoproduction cross sections are of practical interest for several reasons. One is interested in bulk γ ray output, total γ ray spectra, discrete γ production cross sections, γ ray multiplicities, and, to lesser extent, angular distributions of γ rays. Probably the main argument for having these data in the evaluation form is due to calculations of γ ray heating and shielding within fusion reactors. A more detailed discussion of this aspect can be found e.g. in Ref. [2]. Discrete γ ray production cross sections are also of interest for nuclear geophysics. There is generally a growing demand for data at higher incident energies.

The status in measurement, calculation and evaluation of photon production cross sections has been reviewed recently at the IAEA Specialists' Meeting in the Smolenice castle [3]. Probably the most significant progress in the last few years has been in the understanding of preequilibrium γ emission and this was most recently reviewed at the 7th International Symposium on Capture γ-Ray Spectroscopy at Asilomar [4].

Theoretical methods

The essential theoretical tool for calculating γ production data is the statistical model of nuclear reactions, with the preequilibrium emission component added whenever necessary. It is generally important to include preequilibrium processes when the incident neutron (proton) energy exceeds about 10 MeV. Its influence becomes dominant in the giant dipole spectral energy range as well as in the even higher spectral energy range governed by the quasi-deuteron radiative mechanism. Below we give a brief account of recent theoretical developments, as well as the present status of nuclear model codes used for calculations of photoproduction reactions.

Statistical model

In a simple evaporation formulation of the statistical model one neglects angular-momentum effects and a book-keeping of spin and parity of states is not included. In this case only γ transitions of the continuum-continuum type are considered. In the Hauser-Feshbach quantum-mechanical formulation such a book-keeping is essential, and continuum-continuum, continuum-discrete and discrete-discrete γ ray transitions are included.

Key quantities for calculating γ production are the γ ray strength functions, $f_{XL}(\epsilon_\gamma)$, since they determine the γ ray transmission coefficients

$T_{XL}(\epsilon_\gamma)$ through the relation

$$T_{\gamma XL}(\epsilon_\gamma) = 2\pi\epsilon_\gamma^{2L+1} f_{\gamma XL}(\epsilon_\gamma), \qquad (1)$$

XL being the multipole type and ϵ_γ the transition energy. The E1 strength function is the most important term, and the usual approach is to apply the giant dipole resonance (GDR) prescription based on the Lorentzian functional form as fitted to photoabsorption data [5], and extended to higher excitation energies by applying the Brink-Axel hypothesis. A number of alternative functional forms can also be used [6].

We would like to call to attention an interesting development advocated so far in low energy neutron capture studies that suggests a violation of the Brink-Axel hypothesis. This shows up in the strength function as a temperature-dependent GDR spreading width, and there is a non-zero limit for the strength function as $\epsilon_\gamma \to 0$. As pointed out recently by Kopecký and Uhl [7] such a γ ray strength function is in accordance with theoretical proposals [8] and for E1 it reads

$$f_{\gamma E1}(\epsilon_\gamma, T) = \frac{8.68 \times 10^{-8}}{\mathrm{mb\,MeV}^2}\sigma_0\Gamma \times$$
$$\left(\frac{\epsilon_\gamma \Gamma(\epsilon_\gamma)}{(\epsilon_\gamma^2 - E^2)^2 + \epsilon_\gamma^2\Gamma(\epsilon_\gamma)^2} + \frac{0.7\Gamma 4\pi^2 T^2}{E^5} \right)(2)$$

σ_0, Γ and E being resonance parameters. The spreading width depends on the temperature T as

$$\Gamma(\epsilon_\gamma) = \Gamma\frac{\epsilon_\gamma^2 + 4\pi^2 T^2}{E^2}, \qquad (3)$$

for a more detail discussion see Ref.[9]. The useful role of this strength function has already been demonstrated in calculations of slow neutron capture [1]. It would, therefore, be of interest also to apply it consistently in photoproduction calculations at higher incident energies, and such work is presently in progress.

Some other developments of interest are of a more general nature, such as improved knowledge of level densities [10]. If one wants to calculate γ cascades down to low lying discrete levels, it is generally agreed that one should use realistic decay schemes and branching ratios rather than theoretical models. Therefore the improved spectroscopic knowledge that has been achieved in recent years is welcome.

Preequilibrium γ emission

Preequilibrium reaction models for nucleon emission are well established (for a review see [11]). In order to calculate γ emission cross sections, one must determine γ emission rates as functions of particle-hole number. These rates have so far been derived within the exciton and hybrid models as well as the Feshbach-Kerman-Koonin (FKK) statistical multistep compound theory.

Preequilibrium models are statistical in nature, and consequently γ emission rates should

be evaluated by applying microscopic reversibility (detailed balance), utilizing the photo-absorption cross section. As discussed in Ref.[4], two mechanisms, valid at different energies, should be distinguished. In the spectral energy range of the giant dipole resonance, photo-absorption proceeds via single-particle transitions that predominantly lead to 1p-1h states, and the number of excitons can change by $\Delta n = 0, +2$. In the spectral energy range of hard photons ($\epsilon_\gamma \approx 30 - 140\,\mathrm{MeV}$), the two-particle neutron-proton (quasi-deuteron) process is dominant and this gives rise to 2p-2h states, $\Delta n = -2, 0, +2, +4$. An important step consists of introducing branching ratios for photo-absorption that allows one to distinguish between the various Δn processes, enabling an application of detailed balance in a straightforward and clear manner.

As regards single-particle γ rays, the exciton model emission rate is given by [12]

$$\lambda_\gamma^{exc}(\epsilon_\gamma) = \frac{\epsilon_\gamma^2 \sigma_{GDR}(\epsilon_\gamma)}{\pi^2\hbar^3 c^2} \times$$
$$\frac{\omega_{n-2}(E - \epsilon_\gamma)b_{n-2}^n + \omega_n(E)b_n^n}{\omega_n(E)}, \qquad (4)$$

where b_{n-2}^n and b_n^n are branching ratios that sum up to unity, the excitation energy is denoted as E, ω stands for the state density and σ_{GDR} is the GDR photo-absorption cross section. A full account of angular-momentum conservation makes the formalism considerably more complicated but yields approximately the same γ ray spectra [13]. This concept seems to be able to explain radiative capture cross sections at excitation energies up to a few tens of MeV, as can be seen in Fig.1.

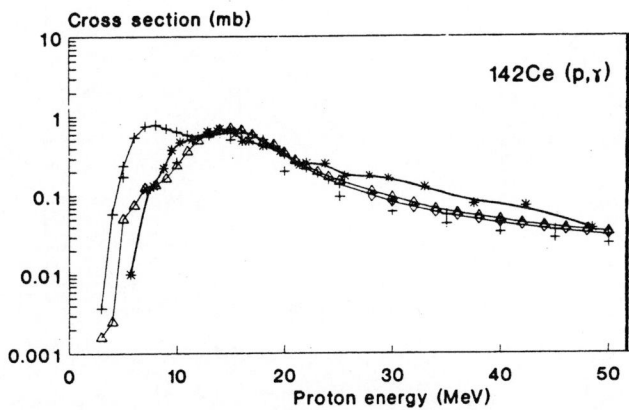

Fig. 1. Excitation function for ^{142}Ce(p, γ). Experimental data (denoted by *) are compared with calculated cross sections using the concept of single-particle preequilibrium γ emission. Calculated values refer to different sets of input parameters applied in the code PEQAG. (Taken from Ref.[14].)

In the hybrid model one needs the γ emission rate for selected particles or holes with the excitation energy ϵ. The rate is given as [15, 16]

$$\lambda_\gamma^{hyb}(\epsilon,\epsilon_\gamma) = \frac{\epsilon_\gamma^2 \sigma_{GDR}(\epsilon_\gamma)}{\pi^2 \hbar^3 c^2} \frac{g}{gn + g^2 \epsilon_\gamma}, \quad (5)$$

where g is the single-particle state density.

The γ emission rate within the FKK multistep compound theory has been determined in Ref.[19]. A γ ray transition between two bound states characterized by the exciton number, energy and spin proceeds with the rate

$$< d\Gamma_{nJ}/d\epsilon_\gamma > = \frac{\epsilon_\gamma^2 \sigma_{GDR}(\epsilon_\gamma)}{\pi^2 \hbar^3 c^2} \frac{\rho_{n'}(U,S)}{\rho_n(E,J)} \beta_{n'S}^{nJ}, \quad (6)$$

where $n' = n, n-2$, level density is denoted by ρ, and the branching ratios β refer to bound states and sum up to a quantity smaller than unity. Shown in Fig.2 are the multistep compound γ emission spectra for ^{181}Ta(n,γ) at a neutron incident energy of 14 MeV. The theory underpredicts the data since the multistep-direct contribution has not yet been calculated in the FKK theory. Attempts to account for multistep compound decay together with semi-direct decay of giant E1 resonances are also being undertaken, see Refs.[17, 18].

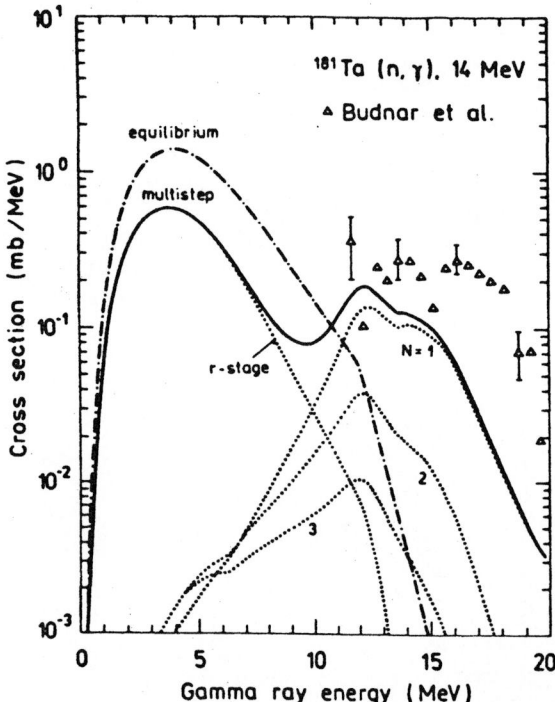

Fig. 2. γ ray spectra from ^{181}Ta(n,γ) at neutron incident energy of 14 MeV. Experimental data are compared with calculations performed using the statistical multistep compound theory. Shown are contributions from the first three stages of the reaction and from the r-stage. Shown for comparison is the equilibrium spectrum as obtained from statistical model calculations. (Taken from Ref. [19].)

Considerable progress has been achieved in understanding hard photon emission [20], [21]. Using the concept of the quasi-deuteron model of photo-absorption the emission rate was derived within the hybrid model to be [21]

$$\lambda_\gamma(\epsilon,\epsilon_\gamma) = \frac{\epsilon_\gamma^2 \sigma_{qd}(\epsilon_\gamma)}{\pi^2 \hbar^3 c^2} \frac{\omega_3(\epsilon - \epsilon_\gamma)}{\omega_4(\epsilon_\gamma)}, \quad (7)$$

where σ_{qd} refers to the quasi-deuteron component of the photo-absorption cross section. It has been pointed out that in the above state densities it is important to reproduce the correct number of neutrons and protons in the nucleus, and that the quasi-deuteron model yields photon emission rates that are analytically almost identical to rates obtained using bremsstrahlung theory [22]. The photo-absorption cross section σ_{qd} has been studied in depth [23], and it was shown that it can be expressed in the form

$$\sigma_{qd}(\epsilon_\gamma) = 6.5 \frac{NZ}{A} \sigma_d(\epsilon_\gamma) f(\epsilon_\gamma), \quad (8)$$

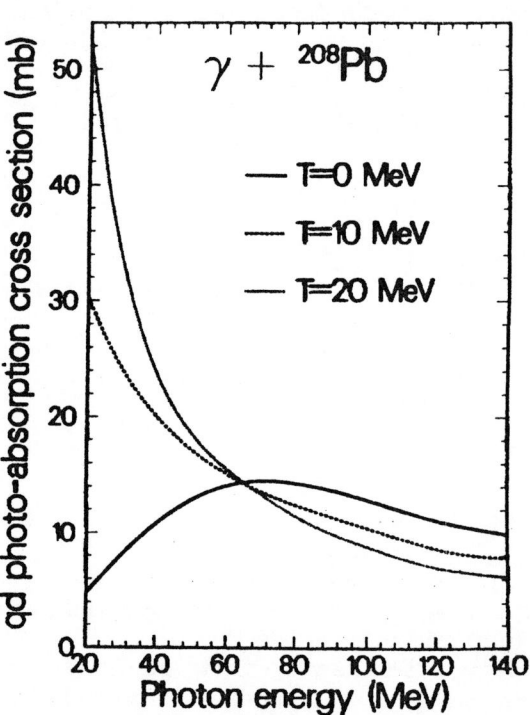

Fig. 3. Calculated quasi-deuteron photo-absorption cross sections σ_{qd} for ^{208}Pb as functions of the photon energy for three different values of the nuclear temperature $T = 0, 10, 20$ MeV. The finite-temperature values of σ_{qd} should be used in calculations of hard photon emission from hot nuclei. (Taken from Ref.[23].)

where σ_d stands for the free deuteron photo - disintegration. The Pauli-blocking function f has been calculated for the first time and fitted by the polynomial [23]

$$f(\epsilon_\gamma) = \sum_{i=0}^{4} a_i \epsilon_\gamma^i, \quad (9)$$

where the coefficients are 8.3714×10^{-2}, -9.8343×10^{-3}, 4.1222×10^{-4}, -3.4762×10^{-6}

and 9.3537×10^{-9}, respectively. Furthermore, the temperature dependence of qd photo-absorption cross sections was also determined, and this is shown in Fig.3. This is of importance when calculating photon emission rates from hot nuclei. With these results an improved formulation of the hard photon emission model can be achieved [24], an example of calculated spectrum being shown in Fig.4.

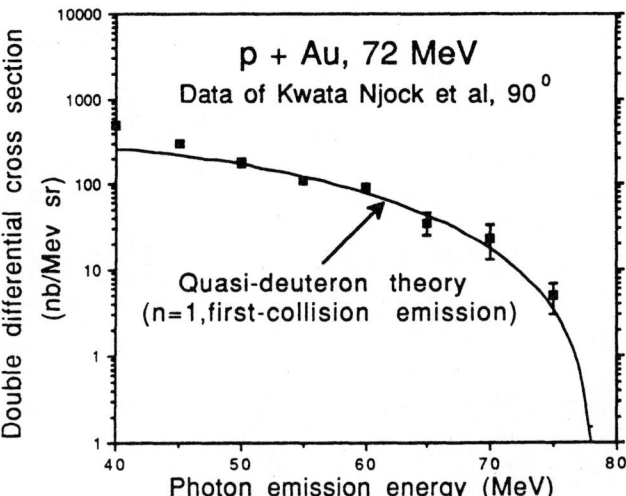

p + Au, 72 MeV
Data of Kwata Njock et al, 90°

Quasi-deuteron theory
(n=1,first-collision emission)

Fig. 4. Hard photon spectra from ^{197}Au bombarded with 72 MeV protons at a $90°$ emission angle in the laboratory frame. Experimental data are compared with the calculation performed within the improved preequilibrium quasideuteron radiative concept. (Taken from Ref. [24].)

Codes

There are several codes of interest for photoproduction calculations. In Tab.1 we list 9 codes that appear to be well suited for evaluation purposes. Each of them is based on the statistical model plus the exciton or hybrid preequilibrium model, or on some kind of unified approach. The codes differ to various extents regarding their sophistication, treatmet of spin effects, preequilibrium γ emission, and γ cascades.

ALICE and PEQAG do not consider spins and so they can only provide continuum γ rays. This also applies to EXIFON which is based on the statistical multistep approach [34, 35]. The other codes, including widely used GNASH, TNG and STAPRE, consider angular-momenta and can calculate discrete γ rays as well (denoted in table by 'all γ'). The treatment of spin effects in these latter codes is done rigorously in the statistical model part, though various approximations are made in the preequilibrium calculations. In older codes, such as STAPRE, spin distribution is considered to be governed fully by the statistical model. In the fairly sophisticated and more recent MAURINA, angular momentum conservation in the preequilibrium stage is treated in the frame of the 'mean-lifetime Ansatz' of Ref.[36] which is much better though still approximate. It seems worthwhile to incorporate a full angular-momentum treatment since the formalism is available [13, 19].

Preequilibrium γ emission is now included in most of these codes but they consider the single-particle mechanism only. At least two groups (GNASH, PEQAG) are also presently incorporating the two-particle quasi-deuteron radiative mechanism to describe hard photon emission.

The above codes also differ in the complexity of calculations that they can handle. The most sophisticated code seems to be, at the moment, MAURINA. It has a built in DWBA routine (essentially DWUCK-4) and can also calculate γ ray multiplicities M_γ and some other γ ray data that can be measured in coincidence experiments. Effective algorithms have been developed for PEQAG and EXIFON that are based on unified approaches (exciton master equation and statistical multistep, respectively). This allows these codes to be effectively used up to a few tens of MeV on personal computers. On the other hand, some other codes (PENELOPE, MAURINA) are available in their home institutes only.

Tab.1. Codes for Photoproduction Calculations. By the 'all γ' in the last column we mean continuum plus discrete γ rays, M_γ denotes the γ multiplicity.

Name	Author	Affiliation	Comment
GNASH	Arthur [25]	Los Alamos	all γ
TNG	Fu [26]	Oak Ridge	all γ
STAPRE	Uhl [27]	Vienna	all γ
PENELOPE	Fabbri [28]	Bologna	also M_γ
EMPIRE	Herman [29]	Warsaw	all γ
MAURINA	Uhl [30]	Vienna	also M_γ
ALICE	Blann [31]	Livermore	no discrete γ
PEQAG	Běták [32]	Bratislava	no discrete γ
EXIFON	Kalka [33]	Dresden	no discrete γ

Calculations of photoproduction

A number of photoproduction calculations has been made in the last 2-3 years. They were mostly motivated by evaluation purposes and by the physics of γ emission, see e.g. [37, 38, 39, 40, 42, 43]. The quantities calculated are most often overall γ production spectra, to a lesser extent production cross sections of specific discrete γ rays and in some cases also average γ ray multiplicities.

Spectra of γ rays. Spectra of γ rays are of most interest for evaluation data files. Advanced calculations also include the preequilibrium γ ray component that seems to account quite reasonably for the hardest part of spectra. As an example we show in Fig.5 the case of $^{56}Fe(n, x\gamma)$ at 14.6 MeV neutron incident energy. Experimental data are compared with calculations performed with the code STAPRE, preequilibrium (n, γ) component was calculated with the exciton model as well as the hybrid model [4]. We note that the hybrid model spectrum should be generally multiplied by about a factor of 4-6 to reach agreement with data [15].

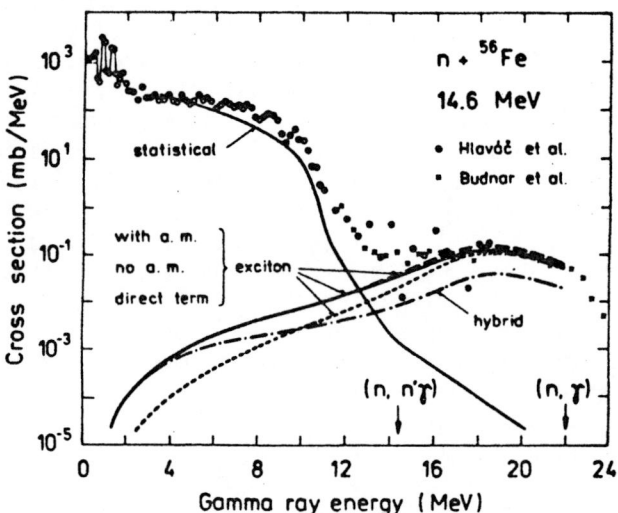

Fig. 5. Calculated γ ray spectra are compared with experimental data for the reaction ^{56}Fe$(n, x\gamma)$ at 14.6 MeV neutron incident energy. Shown are spectra calculated with the statistical model using the code STAPRE, the preequilibrium exciton model (with angular momentum, without angular momentum and the most important n=1 term) as well as by the preequilibrium hybrid model. (Taken from Ref.[4].)

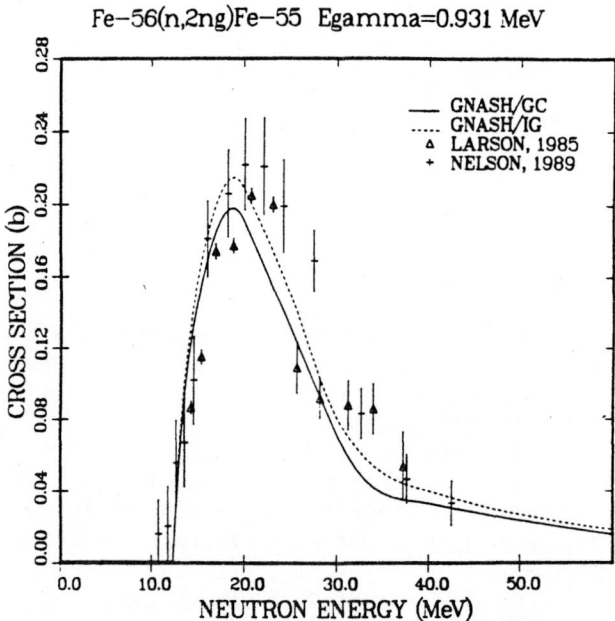

Fig.6. Comparison of calculated (code GNASH) cross sections from ^{56}Fe$(n, 2n\gamma)$ for the discrete γ ray transition $E_\gamma = 0.931\,MeV$. (Taken from Ref.[39].)

Discrete γ rays. In some cases discrete γ ray production cross sections are required. One should then use a code that considers angular momentum conservation (most often GNASH and TNG). We show in Fig.6 the case of ^{56}Fe$(n, 2n\gamma)$

studied recently at Los Alamos for incident neutron energy up to a few tens of MeV [39]. The calculations were performed with the code GNASH and the agreement with data seems to be good. We note, however, that the ^{56}Fe$(n, n'\gamma)$ channel was reproduced substantially worse at higher incident energies [41]. This may indicate that one should also include preequilibrium γ emission after neutron inelastic scattering.

A rather detailed study of the ^{52}Cr$(n, x\gamma)$ case at 14.6 MeV made in Ref.[40] suggests that theoretical predictions of discrete γ ray production cross sections may generally fail to reproduce data by as much as a factor of two.

γ multiplicities. Average γ ray multiplicities are of interest since they provide information about details of the reaction, and they can be measured in coincidence experiments. We show in Tab.2 a case for average γ ray multiplicities related to specific discrete γ rays from ^{52}Cr$(n, n'\gamma)$ reactions at 14.6 MeV as studied in Ref.[40]. The calculations were performed with the code PENELOPE and the agreement with data is very good.

Tab.2. γ Multiplicities for ^{52}Cr$(n, n'\gamma)$ at 14.6 MeV. Theoretical values were calculated with the code PENELOPE. (Taken from Ref.40.)

Transition			Multiplicity	
J_i^π	J_f^π	$E_\gamma(MeV)$	Exper.	Theor.
2_2^+	0_1^+	1.434	3.7 ± 0.2	3.56
4_2^+	2_1^+	1.334	3.7 ± 0.4	3.85
5_1^+	4_1^+	1.246	4.4 ± 1.4	4.49
4_1^+	2_1^+	0.936	3.7 ± 0.3	4.01
5_1^+	4_1^+	0.848	4.8 ± 1.5	4.55
6_1^+	4_1^+	0.744	5.8 ± 0.6	5.94
3_1^+	4_2^+	0.704	6.8 ± 0.7	5.48
4_3^+	4_2^+	0.647	5.6 ± 0.6	5.49

Evaluated libraries. All new versions of major evaluated nuclear data libraries do contain, at least to some extent, γ production data that refer to $(n, x\gamma)$ processes. New evaluations are generally backed by model calculations. A brief overview is shown in Tab.3 where we indicate the number of photoproduction evaluations included within general purpose files of 4 major libraries.

Traditionally the most thorough evaluations are contained within the US library ENDF. The recently released ENDF/B-6 provides photoproduction data in its general purpose file for as many as 85 isotopes. It should be pointed out that there are still a number of (older) evaluations based on very simple models or estimates. The $(n, x\gamma)$ data are to a much lesser extent included in the Soviet BROND, probably the most thorough case still being silicon [48]. Evaluation of photon production nuclear data was one of the important targets for the Japanese file JENDL-3 [49]. It is worth to note that the earlier file JENDL-2 did not contain any γ production data.

Tab.3. γ Production in Evaluated Nuclear Data Libraries. Given are numbers of isotopes/elements having photoproduction data in general purpose files.

Name	Country	Year	$(n, x\gamma)$	Ref.
ENDF/B-6	USA	1990	85	[44]
BROND-2	USSR	1991	38	[45]
JENDL-3	Japan	1990	59	[46]
JEF-1	Europe	1989	49	[47]

REFERENCES

1. M. Uhl and J. Kopecký: Contribution to this Symposium

2. P. Obložinský: in Nuclear Data dor Fusion Reactor Technology, IAEA-TECDOC-457, p.97, Vienna (1988)

3. Ed. N.P. Kocherov: Proc. IAEA Specialists' Meeting on Measurement, Calculation and Evaluation of Photon Production Cross Sections, February 1990, Smolenice, Czechoslovakia, Report IAEA INDC(NDS)-238/L, Vienna (1990)

4. P. Obložinský, in Proc. 7th Int. Symp. on Capture γ Ray Spectroscopy and Related Topics, October 1990, Asilomar, USA (Edited by R.W. Hoff), AIP, New York (1991), in press

5. S.S. Dietrich and B.L. Berman: At. Nucl. Data Tables 38, 199 (1989)

6. D.G. Gardner, in Neutron Radiative Capture (Edited by R.E. Chrien), p.62, Pergamon, Oxford (1984)

7. J. Kopecký and M. Uhl: Phys. Rev. C41, 1941 (1990)

8. S.G. Kadmenskij, V.P. Markushev, and V.I. Furman, Sov. J. Nucl. Phys. 37, 165 (1983)

9. M. Uhl and J. Kopecký: As Ref.[4], in press

10. Eds. G. Reffo and M. Herman, Proc. of the Specialists' Meeting on Nuclear Level Densities, November 1989, Bologna, Italy, in press

11. H. Gruppelaar, P. Nagel and P.E. Hodgson: Riv. Nuovo Cim. 9 no.7, 1 (1986)

12. J.M. Akkermans and H. Gruppelaar: Phys. Lett. B157, 95 (1985)

13. P. Obložinský: Phys. Rev. C35, 407 (1987)

14. F. Cvelbar, E. Běták and J. Merhar: J. Phys. G17, 113 (1991)

15. G. Reffo, M. Blann and B.A. Remington: Phys. Rev. C38, 1190 (1988); Phys. Rev. C41, 403 (1990)

16. P. Obložinský: Phys. Rev. C41, 401 (1990)

17. A. Höring, H.A. Weidenmüller, F.S. Dietrich, M. Herman and G. Reffo: As Ref.[4], in press

18. H. Kalka: Private communication (April 1991)

19. P. Obložinský and M.B. Chadwick: Phys. Rev. C42, 1652 (1990)

20. B.A. Remington, M. Blann and G. Bertsch: Phys. Rev. C35, 1720 (1987)

21. P. Obložinský: Phys. Rev. C40, 1591 (1989)

22. M.B. Chadwick and G. Reffo: Report LA-UR 91-1265, Los Alamos (1991); Phys. Rev. C, in press

23. M.B. Chadwick, P. Obložinský, P.E. Hodgson and G.Reffo: Report LA-UR 91-1266, Los Alamos (1991); Phys. Rev. C, submitted

24. M.B. Chadwick and P. Obložinský: under preparation

25. E.D. Arthur: The GNASH Preequilibrium-Statistical Nuclear Model Code, LA-UR 88-1753, Los Alamos (1988)

26. K. Shibata and C.Y. Fu: Recent Improvements of the TNG Statistical Model Code, ORNL/TM-10093, Oak Ridge (1986)

27. M. Uhl and B. Strohmaier: STAPRE, Statistical Model Code with Consideration of Preequilibrium Decay, p.313, IEAE-SMR-43, Vienna (1980)

28. F. Fabbri and G. Reffo: PENELOPE, a Fortran-IV Code for Multiple Particle and γ-Ray Cascades up to 50 MeV, ENEA, Bologna (1980), unpublished

29. M.Herman, A.Marcinkowski and K.Stankiewicz: Comp. Phys. Com. 33, 373 (1984)

30. M. Uhl: Code MAURINA, IRK Vienna, unpublished

31. M. Blann and J. Bisplinghoff: Code ALICE/Livermore 82, UCID 19614, Livermore (1982); for γ rays see recent modifications, unpublished

32. E. Běták: A PC Version of Fully Pre-equilibrium Computer Code with γ Emission, INDC(CSR)-016/LJ, Vienna (1989)

33. H. Kalka: A Model for Statistical Multistep Reactions (Code EXIFON), INDC(GDR)-60, Vienna (1990)

34. H. Kalka, M. Tojrman and D. Seeliger: Phys. Rev. C40, 1619 (1989)

35. H. Kalka, M. Tojrman, H.N. Lien, R. Lopez and D. Seeliger: Z. Phys. A335, 163 (1990)

36. Shi Xiangjun, J.M. Akkermans and H. Gruppelaar: Nucl. Phys. A466, 333 (1987)

37. K. Shibata: J. Nucl. Sci. Tech. 26, 955 (1989)

38. T. Liu, K. Shibata and T. Nakagawa: J. Nucl. Sci. Tech. 27, 756 (1990)

39. P.G. Young et al: Calculation of $(n, x\gamma)$ Cross Sections Between Threshold and 100 MeV for Fe and Pb Isotopes, LA-UR 90-2129, Los Alamos (1990)

40. G. Maino, A. Mengoni and P. Obložinský: As Ref.[3], p.91

41. R.O. Nelson et al: As Ref.[4], in press

42. D.C. Larson and P.G. Young: Trans. Am. Nucl. Soc. 60, 606 (1989)

43. D.C. Larson, D.M. Hetrick, C.Y. Fu, and S.J. Epperson: Improvements in ENDF/B-6 Silicon, Chromium, Iron and Nickel Evaluations for Radiation Damage Studies, in 7th ASTM-EURATOM Symp. on Reactor Dosimetry, Strasbourg, France, August 27-31, 1990

44. ENDF/B-6: The U.S. Evaluated Nuclear Data Library for Neutron Reaction Data (Summary of Contents by H.D. Lemmel), IAEA-NDS-100, Vienna (1990)

45. BROND: USSR Evaluated Neutron Data Library, ed. V.N. Manokhin, Report IAEA-NDS-90, Rev.4, Vienna (1991)

46. JENDL-3: Japanese Evaluated Nuclear Data Library, Version-3 (Summary by K. Shibata et al), JAERI-1319, Japan Atomic Energy Research Institute (1990)

47. JEF-1: Joint Evaluated File, Version-1. (Tables of Available Evaluated Photon Production Data, C. Nordborg, NEA Data Bank (1989))

48. D. Hermsdorf: Description of File 1402 for Silicon for the Library BROND, INDC(GDR)-038/L, Vienna (1986)

49. S. Igarasi: J. Nucl. Sci. Tech. 26, 5 (1989)

Cooperation in Nuclear Data Evaluation Among the OECD Countries

C.L. Dunford

Brookhaven National Laboratory, Upton, NY 11973, USA

Y. Kikuchi

Japan Atomic Energy Research Institute, Tokai-Mura, Japan

M. Salvatores

C.E.N. Caderache, Saint-Paul-Lez-Durance, France

Abstract: In the fall of 1988, agreement was reached on a collaborative effort between the four nuclear data evaluation projects which exist within the OECD countries. Those projects participating in this effort are the ENDF/B project in the United States, the JENDL project in Japan and the JEF and EFF projects in Western Europe. The cooperation among these projects has been proceeding under the sponsorship of the NEA Committee on Reactor Physics and the NEA Nuclear Data Committee since 1989. The goals and accomplishments of the Working Group on Evaluation Cooperation and the work of its seven ongoing projects are briefly described.

(evaluated nuclear data, ENDF/B, JENDL, JEF, EFF, nuclear reactions)

Introduction

The production of evaluated nuclear data libraries covering materials of interest for applied research is a very expensive activity. In the 1960's there were many different evaluation projects and data libraries within the OECD countries including several in the United States. An AEC-ENEA sponsored seminar, held at Brookhaven National Laboratory in May 1965 reviewed the status of nuclear data evaluation. The participants recommended that European-American cooperation in the field be established[1].

The increasing scope of materials, data types and energy range needed to satisfy demands of a growing user community resulted in the merger of some projects and termination of others. In the United States, nearly all US government sponsored nuclear data activity was merged to form the Cross Section Evaluation Working Group. This resulted in the development of the ENDF system and the ENDF/B data file, a product only of United States and Canadian efforts. Because of the comprehensiveness of ENDF, its associated processing programs and its free availability, the ENDF/B data file became the major source of evaluated nuclear data used in the OECD countries (except for Japan) in the 1970's. The JENDL project was started in the early 1970's in Japan to provide evaluated data for Japanese reactor development programs. A comprehensive summary of the international evaluation activity at that time was made by the IAEA Panel on Nuclear Data Evaluation held in Vienna,

Austria in 1971[2]. During this period, evaluated nuclear data files were generally available without restriction.

The US decision to restrict distribution of ENDF/B-V spurred formation of other evaluation projects within the European countries, namely JEF and EFF in the 1980's, while Japan continued with the JENDL project. As the end of the decade approached, each project had matured and completed release of an evaluated data file which was considered adequate for use by their user community. Along with this maturity and success came serious funding reductions along with the realization that concentrated and specialized efforts would be required to make future extensions and improvements.

As a result of NEANDC interest in the matter, the NEANDC chairman, Alan B. Smith, Argonne National Laboratory, wrote a letter in January 1988 to chairmen of the JENDL, ENDF and JEF projects stressing the importance which the NEANDC placed on future cooperation between the different evaluation projects and soliciting their suggestions. The positive responses from the projects led to a small exploratory meeting during the Mito Conference to exchange views.

At the NEANDC meeting in Los Alamos, USA, in September 1988, a NEACRP/NEANDC Task Force on Evaluation Cooperation was formed under the chairmanship of John Rowlands, Winfrith, UK. The subcommittee included the chairmen of the the three projects and NEANDC members. During the meeting, agreement in principle was reached on the following items, 1) evaluated

This work has been supported in part by the U.S. Department of Energy under Contract No. DE-AC02-76CH00016.

nuclear data files and associated documentation would be freely exchanged between the different projects, 2) the different evaluated nuclear data files would remain separate entities for the forseeable future, but through increased cooperation, the contents of the different files would tend to merge, 3) cooperative evaluation activities should be promoted in order to prevent unnecessary duplication, 4) personnel exchanges would be encouraged, and 5) exchange of benchmark testing results would be considered.

Following formal agreement from the participating projects, the Task Force met in Paris, France, in May 1989 to drafted the working arrangements of the cooperation which established a NEACRP/NEANDC Working Group on Evaluation Cooperation. The Working Group would consist of up to four members from each of the projects and two representatives from the parent NEA committees. The NEA would serve as secretariat to the Working Group. Technical activities of the Working Group are carried out by Subgroups formed to solve specific evaluation tasks. Six high priority tasks were identified as the first cooperative evaluation activities to be supported by the Working Group. The Task Force held its final meeting at Brookhaven National Laboratory in October 1989 to prepare its final report.

Following acceptance of this report by the NEACRP and the NEANDC, the Task Force was dissolved and the Working Group established. The Working Group met briefly at Argonne National Laboratory following the NEACRP meeting to organize and establish the Subgroups. One member of the Working Group was assigned to the monitor work of each subgroup. The Working Group held its first full meeting at Marseille, France, in April 1990 and will hold its next meeting in Petten, the Netherlands, following this conference.

Active Subgroups

SG1
Intercomparison of files for ^{52}Cr, ^{56}Fe and ^{58}Ni
— C.Y. Fu, ORNL, Oak Ridge, coordinator

The subgroup was organized to intercompare the evaluations for these three nuclides available from the cooperating projects. The nuclides are major components of stainless steel and have recently been evaluated by all the projects. Graphical intercomparisons have been nearly completed. An effort to understand reasons for the observed discrepancies is in progress. Many suggested improvements to the existing evaluations have been made. The status of this task is described in a paper[3] at this conference.

SG2
Generation of Covariance Files for ^{56}Fe and ^{nat}Fe
— H. Vonach, IRK, Vienna, coordinator

Covariance data are required to assess uncertainties in design parameters and to refine use of nuclear data in reactor applications. Technology for generating evaluated covariance data is still primitive compared to that

employed in evaluating cross sections and other nuclear data. This subgroup has been assigned the task of reviewing the existing technology, particularly with respect to these two important reactor materials and recommend methods for providing improved evaluations.

SG3
Actinide Data in the Thermal Energy Range
— H. Tellier, CEN, Saclay, coordinator

This subgroup was formed to investigate apparent discrepancies between evaluated thermal nuclear constants and those values implied by thermal reactor benchmark studies. It is believed that the discrepancies are related to the temperature coefficient calculation problem. The nuclear data for the major heavy nuclides were reviewed. The only remaining problem concerns ^{235}U η. The Geel and Grenoble measurements do not agree with the ones from Harwell and Oak Ridge. The origin of this discrepancy, which could be due to analysis of the raw data, is under investigation.

SG4
^{238}U Capture and Inelastic Scattering Cross Sections
— Y. Kanda, Kyushu U., Fukuoka, coordinator

These cross sections are of primary importance for fast reactors. Recently evaluated data for the capture cross section are significantly lower than the average of measured data in the unresolved resonance region. The normalization of measured data in the resolved resonance region has been studied and it was concluded that older measured values should be renormalized downward. The improved agreement seems to resolve this discrepancy. Inelastic data evaluations in JENDL-3 and ENDF/B-VI agree fairly well but are higher than the measured data. New measurements from ANL are expected. The status of this task is described in a paper[4] this conference.

SG5
^{239}Pu 1–100 keV Fission Cross Section
— E. Fort, CEN, Caderache, coordinator

The ^{239}Pu fission cross section is important for calculating performance of fast reactors. The high accuracy values obtained from simultaneous evaluation of the standards cross sections for ENDF/B-VI give significantly higher values at the top of the resolved resonance region than does the resonance region evaluation by Derrien and the latest experimental data of Weston.

Experimental programs are being prepared at CBNM and ORNL to check the normalization procedure in Weston's experiments. A re-evaluation of the cross section in now underway at both Los Alamos and Cadarache, taking into account new microscopic and integral information. The work of this subgroup is described in a paper[5] at this conference.

SG6
Delayed Neutron Data Benchmarking
— A. Filip, CEN Caderache,
— G. Rudstam, Studsvik Energiteknik, coordinators

The discrepancy of up to 10% in the calculation-to-experiment ratio for integral measurements of β_{eff} result in undesirable conservatism in design and operation of reactor control systems. This subgroup is investigating both experimental and theoretical avenues to reduce the uncertainty.

A careful review of present evaluated data uncertainties indicates that they are probably underestimated. It has been suggested that ν_d for ^{238}U may be underestimated. A series of benchmark integral experiments will be performed with the MASURCA reactor in 1992–1993. The accuracy and analysis method for these integral experiments should produce significant new information on ν_d of ^{235}U, ^{238}U and ^{239}Pu.

SG7
Multigroup Cross Section Processing
— R. Roussin, ORNL, Oak Ridge, coordinator

This is a new subgroup formed to coordinate production of a comprehensive library from the ENDF/B–VI, JENDL–3 and JEF–2 libraries using VITAMIN–J specifications . Mainly, the NJOY processing system will be used. Output will be available in the MATXS and AMPX formats. The status of this task is described in a paper[6] at this conference.

Results of Cooperation to Date

The cooperating projects have agreed that reports, conclusions and any evaluated data resulting from the cooperation will be available world-wide without restriction. Those wishing to obtain specific information should contact the secretariat at the Nuclear Energy Agency Data Bank.

The evaluated data files, ENDF/B–VI, JENDL–3 and JEF–1, have been released for distribution world-wide. Those wishing to receive information from one of the files should contact the neutron data center responsible for servicing your country. It is expected that the JEF–2 file will be released for unrestricted distribution after integral benchmark testing is completed.

Seven cooperative projects have been initiated under the sponsorship of this Working Group. Significant progress has already been made in several of these tasks. While there have been important contributions to the ENDF/B–VI, JEF–2 and JENDL–3 libraries from in-

formal cooperative activities, formation of this Working Group has provided a mechanism for exchange of information and for expansion in a more formal way of the sharing of evaluation tasks within available resources.

The Future

This cooperative effort involves only the evaluation activities of OECD countries. However experts from non-OECD countries are permitted to participate in the work of the subgroups when their special expertise is needed. We shall be exploring the possibility of extending this cooperative effort to projects outside the OECD in the near future.

While we expect existing data files to continue as separate entities for the forseeable future, contents of the files should converge with time. The work of the subgroups will tend to resolve differences in approach and conflicts in conclusions of the different evaluators by involving them together in the search for a solution to common problems. The revised evaluated data coming from these joint studies will be reflected in future releases of the data files. In deciding on future tasks for investigation, the Working Group will attempt to assess the priority requirements for nuclear data and to anticipate data needs for innovative reactor concepts.

References

1. I.Zartman, R.P.Perret, Conclusions of AEC–ENEA Seminar on the Evaluation of Neutron Cross-Section Data, Brookhaven National Laboratory, May 1965

2. Neutron Nuclear Data Evaluation, STI/DOC/10/146, IAEA, Vienna, 1973

3. C.Y.Fu et.al., "Intercomparision of Evaluated Files for ^{52}Cr, ^{56}Fe, and ^{58}Ni", these proceedings

4. Y.Kanda et.al., "A Report on Evaluated ^{238}U(n,γ) Cross Section", these proceedings

5. E.Fort et.al., "Progress Report on the Subgroup on Pu-239 Fission Cross Section between 1 keV and 100 keV", these proceedings

6. R.Roussin et.al., "NEACRP/NEANDC Working Group on Evaluation Cooperation: Progress of the Subgroup on Multigroup Cross Section Processing", these proceedings

A REPORT ON EVALUATED ^{238}U(n, γ) CROSS SECTION

Y. Kanda

Kyushu University, Kasuga, Fukuoka 816, Japan

Y. Kikuchi and Y. Nakajima

Japan Atomic Energy Research Institute, Tokai, Ibaraki 319-11, Japan

M. G. Sowerby and M. C. Moxon

Atomic Energy Research Establishment, Harwell, Oxon, OX11 ORA, U.K.

F. H. Fröhner

Kernforschungzentrum Karlsruhe, Karlsruhe, F. R. Germany

W. P. Poenitz

Argonne National Laboratory-West, Idaho Falls, ID 83403-2528, U.S.A.

L. W. Weston

Oak Ridge National Laboratory, OakRidge, TN 37830, U.S.A.

Abstract : A longstanding and difficult problem in nuclear data evaluation has been that ^{238}U capture cross sections in the unresolved resonance region evaluated on the basis of available experiments are larger than the ones expected from the reactor physics analysis. However, the new versions of the major files, ENDF/B-VI, JEF-2 and JENDL-3 have now adopted lower values than the respective previous versions. To be convinced that these new evaluations are correct, a subgroup in the NEACRP/NEANDC Working Group on Internaional Evaluation Cooperation has studied the problem.

New resonance parameters deduced from capture and transmission data by the shape analysis method can be used to renormalise early capture cross section experiments. The average capture cross sections calculated from these then has high weight in a simultaneous evaluations which leads to lower evaluated capture values. Theoretical model fitting to available experiments gives similar values. Two recent measurement agree with the lower ones. Therefore, the subgroup concludes that the lower capture cross sections of ^{238}U recomended in ENDF/B-VI, JEF-2 and JENDL-3 are reasonable.

(^{238}U, capture cross section, shape analysis, neutron width, gamma width, simultaneous evaluation, ENDF/B-VI, JEF-2, JENDL-3)

Introduction

Neutron capture cross section of ^{238}U in unresolved resonance region is an important quantity for reactor calculation. There was, however, a longstanding difficulty: Earlier evaluations resulted in a higher capture cross sections than expected from reactor physics analysis. This was common in three major evaluated data files i.e. ENDF/B-V, JEF-1 and JENDL-2. The available differential measurements on which they were scattered by more than 15 % depending on the neutron energy region. As shown in Fig. 1 there were substantial discrepancies even in the careful experiments e.g. Moxon[1] and de Saussure et al. [2] which had been undertaken to solve the problem. Evaluators recommended mean values of experiments as the best cross section values as there was no reason to modify the experimental results. Nevertheless, the new versions of the major files, ENDF/B-VI, JEF-2 and JENDL-3 have now adopted smaller capture cross sections in the unresolved region than the previous versions. In Fig. 1, as a typical example, JENDL-3 is shown in comparison with the experiments and the previous version JENDL-2 as a typical example. As can be seen in this figure JENDL-3 follows the lowest values of experimental data, while JENDL-2 follows the average values.

To have a full understanding of the smaller capture cross sections adopted in the new major files, the subgroup on ^{238}U capture and inelastic scattering cross sections was established by the NEACRP/NEANDC Working Group on International Evaluation Cooperation. The problems concerning on the capture cross sections have been intensively studied and resolved. The results are reported here. The project on the inelastic scattering cross section is in progress.

Validity of the New Versions of three Major Files

The Resolved Resonance Range

It is believed that measurements of the ^{238}U capture cross sections made with white neutron sources (e.g. Linacs) can be accurate. Normally these are normalised at very low energies using resonances where the neutron width Γ_n is much smaller than the capture width Γ_γ and the sample is "black" so that at the resonance peak the capture yield gives directly the normalisation constant. Therefore, the measurements with white neutron sources must be renormalised when it is found that incorrect values of Γ_n and Γ_γ were used in their original data procedures.

Sowerby et al. [3,4] and Moxon et al. [5,6]

The NEANDC Task Force on ^{238}U was set up to deal with two problems; the neutron widths of the resolved resonances above 1.4 keV and the capture cross section in the resolved and unresolved resonance regions. In the Task Force, the new evaluation was carried out over the whole energy range below 10 keV using the shape analysis code REFIT[7].This code can simultaneously analyse both capture and transmission data. The shape analysis is able to identify errors in background and normalisation in both the measurements. It should be noted that previous evaluations are mainly based on area analysis.

The normalisation of capture data is no longer necessary to consider only very low energy resonances as in principle the experiments can be normalised at any resonance where both Γ_n and Γ_γ are well known. The peak height in a capture measurement and its capture area can be derived from the resonance parameters obtained from transmission measurements. For resonances which are isolated and do not overlap with others the derived quantity is accurate and hence can be used to normalise the experiment. As a result of this ability, it was found that the capture cross section of de Saussure et al. [2] were inconsistent with the parameters obtained from the transmission data and an assumed values of Γ_γ (23~23.5 meV) unless the capture data were renormalised by a factor of ~0.9 near 1.8 keV neutron energy. This renormalisation is correct for a wide range of resonances neutron widths from $\Gamma_n \ll \Gamma_\gamma$ to $\Gamma_n \gg \Gamma_\gamma$. The orignal normalisation is correct at the first resonance at 6.67 eV, where the measured data were normalised by the authors, but at higher resonance energies the capture yields calculated from transmission data increasingly deviate from the experimental values. The average capture cross sections published by de Saussure et al. [2] must be corrected by multiplying by the following correction factor F:

$$F = 0.845 \exp\left(0.38421/\sqrt{E}\right),$$

where E is in eV[8]. It is probable that this tendency will continue in the unresolved resonance region. This correction bring the data of de Saussure et al. into good broad agreement with the data of Moxon[1,9] as seen in Fig. 2. The reason for the above correction factor is uncertain. It is worth noting that the normalisation of the Moxon data[1] has also been checked by the same method and it was found that the normalisation was correct.

The Unresolved Resonance Range

Evaluation by Fröhner[10,11,12] The evaluation in the unresolved resonance region (~10 to 300 keV) is based on simultaneous fits with the FITACS code[13], which employs Hauser-Feshbach theory with width fluctuation and the generalized (Bayesian) least squares technique, to a large body of total (5 sets), inelastic scattering (4 sets) and capture cross section data (27 sets), and on rigorous (Bayesian) inclusion of prior knowledge from resolved resonances and from optical model fits at higher energies. Multiple scattering corrections applied to capture yield data produced lower capture cross sections, consistent with the resolved-resonance analysis.

Utilisation of theory permits simultaneous description of all the observations in terms of average resonance parameters. Since this theory relates averaged cross sections for all reaction channels, a coherent evaluation of all information provides powerful physical constraints and reduces uncertainties.

Evaluation by Poenitz[14,15,16] The evaluation of the ^{238}U capture cross section was part of a simultaneous evaluation of ten cross sections and later combination with a R-matrix analysis of additional data for five of these cross sections[16]. For ^{238}U the average capture cross sections in the resolved resonance region as well as data above the unresolved resonance range were used in addition to the cross section for the unresolved region. The resulting evaluated cross section lies on the lower side of the bulk of the available measurements and is accurate to ~ ±2-5 % or better over most of the energy range between 10 keV and 500 keV as shown in Fig. 1.

Contents of Three Major Libraries

ENDF/B-VI and JEF-2. These two files adopted the same evaluation in the resolved and unresolved resonance region[17]. For the resolved resonance region from 10^{-5} eV to 10 keV the evaluation by Moxon and Sowerby[3,4,5,6] was used. Fröhner's evaluation[10,11] was adopted in the unresolved resonance region from 10 to 149 keV. The evaluation from 149 keV to 20 MeV is taken directly from the simultaneous standards evaluation[14,16].

JENDL-3 The resolved resonance region from 10^{-5} eV to 4 keV. For the unresolved resonance region, the measurement by Kazakov et al. [18] was adopted because it was the recent experiment with good resolution and in addition it was consistent with reactor physics analysis. Although the ^{238}U capture cross section was included in the original simultaneous evaluation for JENDL-3 by Kanda et al. [19], the renormalisation of the average cross section in the resolved resonance region could not be utilized since the simultaneous evaluation was made only on the basis of the data above 50 keV. The recent measurement of Quang and Knoll [20] agrees also with the new evaluations.

Impact on Integral Parameters

The lower capture cross section of ^{238}U has been required from the reactor analysis. Hence it is interesting to know how the present change to lower values impacts on various integral characteristics of reactors. Some results of sensitivity analysis on JENDL-3 are presented here as an example. The sensitivity coefficients were calculated for ZPPR-9 critical assembly, which is a clean homogeneous physical mock-up core of large demonstration fast breeder reactor.

The sensitivity coefficients of ^{238}U capture cross section in the energy range from 1 keV to 500 keV are particularly significant to k-eff and the reaction rate of ^{238}U capture to ^{239}Pu fission (or ^{235}U fission). The deferences of ^{238}U capture cross section from JENDL-2 to JENDL-3 causes increase of k-eff by 0.4 % and decrease of ^{238}U capture to ^{239}Pu fission rate ratio by 1.3 %. Contributions from the other energy regions are negligible.

Conclusion

The subgroup has studied the reason why the recent evaluated data of the ^{238}U capture cross section are lower than the average of the available measured data. Sowerby and Moxon found in their shape analysis of resolved resonances that some of the measured capture data should be renormalised. After the renormalisation, the measured data converge to the lower values, with which the recent evaluated data agree. The lower values could be reproduced with fitting the theoretical model to available experiments by Fröhner. From the multiple-scattering effect it is understood why many of the old capture data sets were too high: This confirms and explaines Poenitz' renormalisation based on the resolved-resonance analysis. Furthrmore, the lower capture cross section of ^{238}U has significant influence on the integral parameters of fast reactors, whose direction has been predicted by reactor physicists. Thus the subgroup concludes that the lower capture cross section of ^{238}U in the unresolved recorance region, which was adopted in three of the recent major files, i.e. ENDF/B-VI, JEF-2 and JENDL-3, are reasonable and that the earlier evaluations should be superceded.

The authors are indebted to Mr.Y.Nakagawa of JAERI for preparing the figures.

REFERENCES

1. M.C. Moxon, The Neutron Capture Cross Section of ^{238}U in the Energy Region 0.5 to 100 keV, AERE-R 6074, Atomic Energy Research Establishment (1967)
2. G. de Saussure *et al.* ,: Nucl. Sci. Eng.51, 385 (1973)
3. M. Sowerby *et al.* , in Proc. Int. Conf on Nuclear Data for Basic and Applied Science, May 1985, Santa Fe, USA (Edited by P.G. Young *et al.*), Vol. 2, p.1511, Gordon and Breach, New York (1986)
4. M. Sowerby and F. Corvi, in Proc. Int. Conf. on Nuclear Data for Science and Technology, may/June 1988, Mito, Japan (Edited by S. Igarasi), p.37, Saikon, Tokyo (1988)
5. M.C. Moxon *et al.* , in Proc. 1988 Int. Reactor Physics Conf., Sept., Jackson Hole, USA, p. I-281 (1988)
6. M.C. Moxon *et al.* , in Proc. 1990 Int. Reactor Physics Conf., PHYSOR'90, April, Marseille, France, p.III-41 (1990)
7. M.C. Moxon, in Neutron Data of Structural Materials for Fast Reactors, Pergamon Press, Oxford, p.644 (1979)
8. M.G. Sowerby and M.C. Moxon, Private Communication (1990)
9. M.C. Moxon, Private Communication (1991)
10. F.H. Fröhner, in Ref. 5, p.III-171 (1988)
11. F.H. Fröhner, Nucl. Sci. Eng., 103, p. 119 (1989)
12. F.H. Fröhner, Private Communication (1991)
13. F.H. Fröhner *et al.* , in Proc. Mtg. Fast Neutron Capture Cross Sections, Sept. 1982, Argonne, USA, ANL-83-4, p.116, Argonne National Laboratory (1983)
14. W.P. Poenitz, Private Communication to S. Igarasi (1989)
15. W.P. Poenitz, in Proc.Conf. on Nuclear Data Evaluation Methods and procedures, Sept. 1980, Brookhaven, USA, BNL-NCS-51363, Vol.1, p.249, Brookhaven National Laboratory(1981)
16. A.D. Carlson *et al.* , in Ref. 3, Vol. 2, p.1429, Gordon and Breach, New York (1986)
17. L.W. Weston, Private Communication (1991)
18. L.E. Kazakov *et al.* , Yad. Const., 3, p. 37, (1986) (in Russian) and Report, INDC(CCP)-319/L, p.45 (1990) (in English)
19. Y. Kanda *et al.* , in Ref. 3, Vol. 2, p.1567

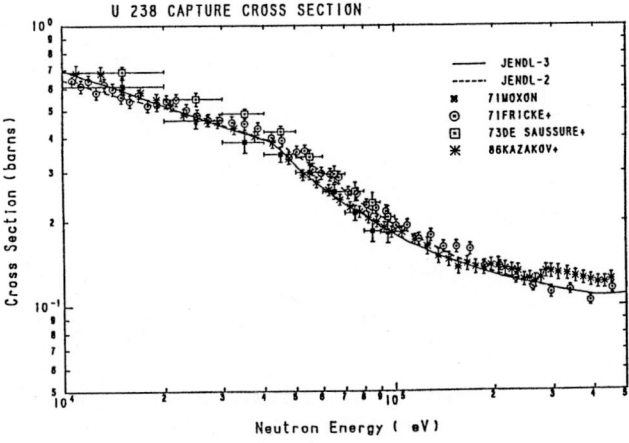

Fig.1 Comparison of evaluated data with experiments for ^{238}U capture cross sections in unresolved resonance region. JENDL-3 and JENDL-2 are shown as the typical examples of a new version and previous one of the major files, respectively. There are also typical experiments in available ones.

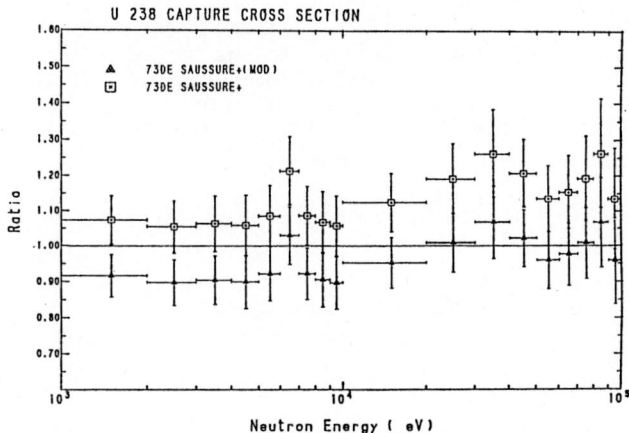

Fig.2 Comparison of ratios of the original de Saussure *et al.* data[2] and the de Saussure *et al.* ones renormalised by Moxon[9] to the Moxon measurement[1].

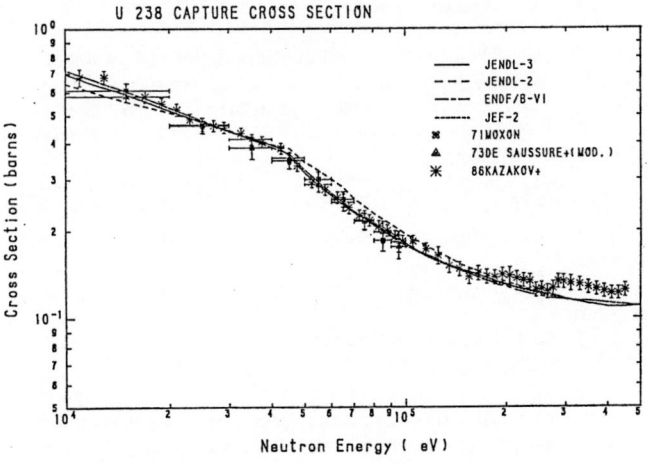

Fig.3 Comparison of the new versions of the major files with the experiments of Moxon, renormalised de Saussure *et al.* and Kazakov *et al.* .

INTERNATIONAL EVALUATION COOPERATION
PROGRESS REPORT OF THE SUBGROUP ON
"239Pu FISSION CROSS SECTION BETWEEN 1 KeV AND 100 KeV"

E. FORT, H. DERRIEN, M. KAWAï, C. LAGRANGE, NAKAGAWA, C. WAGEMANS, L. WESTON,
P.G. YOUNG

Abstract :

We present the current status of our efforts to try to solve the problem of the discrepancy in measurements of the ^{239}Pu fission cross section in an energy range of particular importance in FBR calculations.
The researches are progressing along two axes :
- Critical examination of WESTON's measurements with the objective of performing additional experiments to confirm the conclusions of the subgroup.
- Examination of all other sources of information relating to the fission cross section including the competitive cross section data (when existing) and integral data.
To finalize the investigation, it is envisaged that a new ^{239}Pu fission cross section measurement will be carried out in conditions to be defined in the course of the present study.

In 1984 WESTON and TODD published [1] the results of a ^{239}Pu fission cross section measurement in the range 20 eV to 100 KeV performed in excellent energy resolution conditions. While analysed as being excellent in the resolved range [2], WESTON's data in the unresolved range appear to be significantly lower (5 %) than almost all measurements and all the recent major evaluations (JENDL3, ENDFB6, JEF2) see fig. 1.

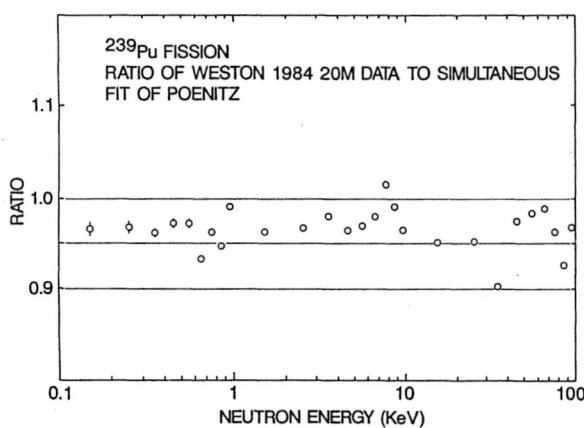

Fig. 1 :   Ratio of WESTON's data (1984) to simultaneous fit of POENITZ

If confirmed this difference would induce significant discrepancies in Keff, control rod worth or void coefficient calculations for FBRs for which the range 1 KeV - 100 KeV is of important sensitivity.

This explains the interest of the NEANDC/NEACRP task force which set up a subgroup to try to understand and possibly solve this problem.

The following scheme of work has been adopted :

1 Critical examination of the WESTON and TODD experiment.

2 Reevaluating the ^{239}Pu data taking into account new information not used in the above cited evaluations such as microscopic data recently produced and chosen selected integral information.

3 Global analysis of phases 1 and 2 so as to be able to produce a qualitative explanation and, possibly, a set of recommended data.

1 - CRITICAL EXAMINATION OF THE WESTON AND TODD EXPERIMENT

This is a T.O.F. experiment performed with ORELA, using a multiparallel plate fission chamber ; the shape of the neutron flux was measured relative to a ^{10}BF$_3$ chamber up to a neutron energy of 1 KeV and to a ^{6}Li glass scintillator at higher energies. The flux internormalization was made in the energy interval 100 eV - 1000 eV with a statistical uncertainty less than 0.1 %. Finally the σ_f curve was normalized to the thermal value. For this experiment conceived for a measurement at high energy, the following aspects were considered to detract from such way of normalization : sample thickness self absorption, dead time. On the other hand, the fission integral for the interval 100 eV - 1000 eV (I_f = 8996 b.eV) is known only with a fairly rather significant uncertainty (1.9 % for normalization, 1 % systematic, 0.15 % statistical). The comparison with the I_f values obtained from GWIN's measurements (I_f = 9268 b.eV) [3] and (I_f = 9365 b.eV) [4] reveals a 4 % difference not in contradiction with the above quoted errors. But since GWIN's measurements are affected by a large uncertainty (11 %) in the ^{10}B content of the neutron flux counter, they have not been considered as absolutely reliable references. Therefore it has been decided to plan experimental programs both at OAK-RIDGE and GEEL to check this normalization point.

Nevertheless the possibility of energy dependent sources of discrepancy is not totally to be excluded, although the fission cross section of ^{235}U measured in the same experiment is consistent with measurements and evaluations.

2 - 239Pu REEVALUATION BASED ON NEW OR ADDITIONAL INFORMATION

2.a. New microscopic information

Transmission data have been obtained at ORELA by J. HARVEY et al [5] with good energy resolution (1.6 ns/m to 0.1 ns/m from 0.5 eV to several hundred KeV), three samples cooled to liquid nitrogen temperature whose thicknesses were chosen to be a good compromise between opposite conditions to get on the one hand accurate experimental data and at the same time only moderate self screening effects. They have been analysed to derive first a resonance parameter set up to 2 KeV recognized as being of excellent quality (no need of a MT-3 cross section in the resolved region, validation on integral experiment) and second a total cross section up to 500 KeV [6], referred to in the following as DERRIEN's σ_t.

This σ_t curve appears to be in agreement with the POENITZ et al experimental data [7], but significantly lower (3 % - 4 %) than ENDFB6 and JEF2. The difference results, in our judgement, from the self screening correction and also from the better quality of the raw transmission data. That's the reason why we have considered the data-base formed by POENITZ's and DERRIEN's data as a reference in the range 1 KeV - 500 KeV in the procedure to produce a new Optical Model Parametrization (OMP).

This parametrisation, which is to be used in coupled channel calculations, has been derived in the same way as the one used in the JEF2 evaluation [8]. In what follows both parametrizations will be referred to respectively as OMP90 and OMP86.

All other reference data being identical (σ_T above 500 KeV, angular distributions) the consideration of the new σ_T data-base led to the renormalization of the real and the imaginary parts of the potential, the ranges of the effective interactions and the spin-orbit potential being unchanged.

The quality of the fit is improved over the whole energy range compared to the previous situation.

It must be stressed that the new potential is totally consistent with the parameters in the resolved range thanks to the so called "SPRT" method.

With this new OMP, the fit is globally improved : the neutron scattering angular distributions (see fig. 2 below) but essentially the total cross section above 0.5 MeV in an energy range rich in data ; see fig. 3.

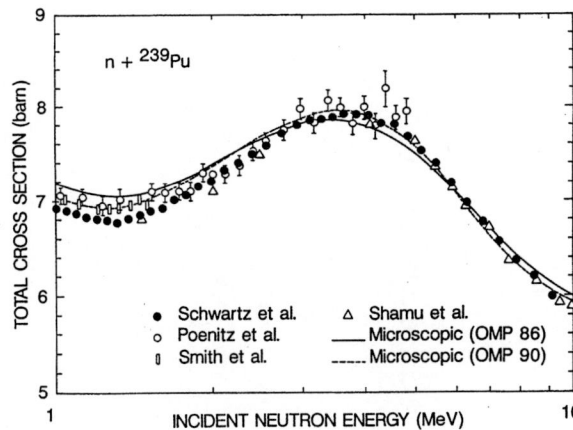

Fig. 3 : Experimental total cross sections compared with data calculated using OMP86 and OMP90

With respect to above cited evaluations, changes appear essentially below 3 MeV but they are significant : a 5 % lower σ_T between 0.5 MeV and 1.5 MeV, a 10 % - 15 % lower compound nucleus formation cross section as a result of a lower σ_T and the adoption of a greater scattering radius (9.45 fm against 9.15 fm) :

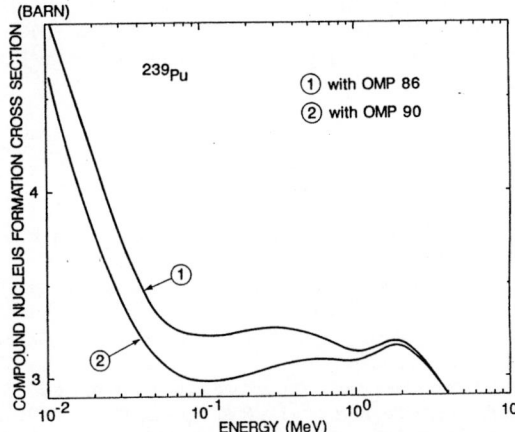

Fig. 4 : Comparison of the compound nucleus formation cross section obtained by using OMP86 (curve 1) and OMP90 (curve 2)

Using it to describe the neutron channel two model evaluations are underway using the POENITZ evaluation (ENDFB6) on the one hand and WESTON's data on the other as fission data bases. The difficulty in the last case lies in that :

<u>1</u> The fission data base is formally truncated above 100 KeV.

<u>2</u> There are no reliable experimental data for the competitive cross section (σ_c, $\sigma_{n,n'}$, σ_{ne}) or when existing (alpha) they are not very helpful in solving the present dilemma. The same conclusion applies more or less at higher energies so that it is difficult to draw conclusions for the range of interest from conclusions in the higher energy range. Fortunately it has been possible to reproduce WESTON's data without any change in the fission channel distribution above the ground state. Approximate agreement with JEF2, is obtained at 1 MeV an energy where the old and new OMP practically coincide.

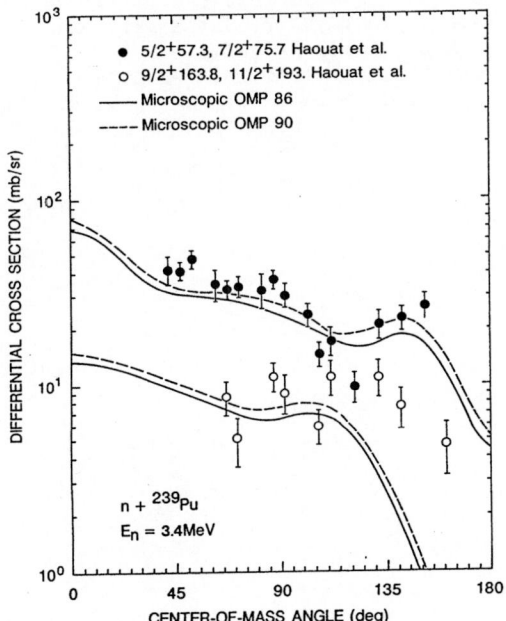

Fig. 2 : Comparison of experimental inelastic scattering cross sections at 3.4 MeV with data calculated with OMP86 and OMP90

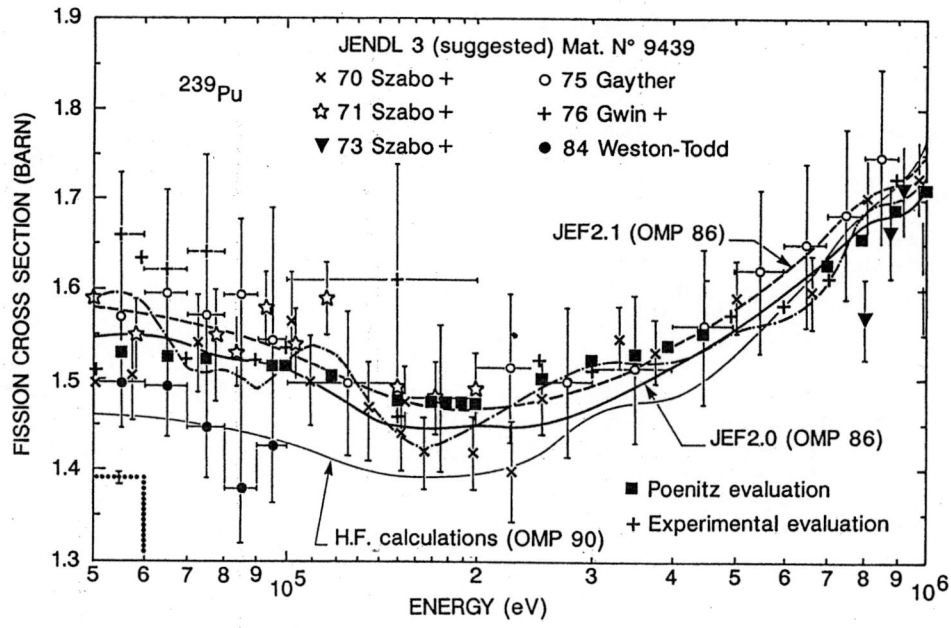

Fig. 5 :   Experimental fission cross section data compared with the model calculations (OMP90) based on WESTON's data

This means that the fission data-base is the same above 1 MeV for both model evaluations. It is also concluded that if the true fission cross section is WESTON's it should also be lower than the present evaluations over a larger energy range and this should be confirmed by an additional experiment.

In the future, when both evaluations are complete and available in ENDFB6 format they will be validated against clean integral data obtained from :

- JEZEBEL, JEZEBEL-Pu, FLATTOP-Pu (sensitivity range : 50 KeV < E < 10 MeV).

The microscopic data will be processed in 33 groups and SN transport calculations performed in P_3S_{16} (with a correction to account for the S_N effect).

- The set of experiments performed on Pu fuelled criticals, MASURCA, SNEAK, ZEBRA, considered in the JEF2 benchmarking (sensitivity range : 1 KeV < E < 1 MeV).

The microscopic data will be processed in 1968 groups and resonance structure within these groups represented by probability tables. Since the selected experiments are asymptotic flux (fundamental mode conditions) the neutronics calculations will be cell calculations using the new European cell code ECCO.

Sensitivity calculations and a statistical adjustments procedure will locate the energy regions where microscopic and integral data disagree and will quantify the amplitudes of the disagreement.

3 - GLOBAL ANALYSIS OF PHASE 1 AND 2

Phases 1 and 2 will demonstrate the degree of consistency of microscopic and integral information in the present state of art but a final conclusion is dependent on the availability of the results of the experiments planned at GEEL and OAK-RIDGE.

The production of recommended data also depends on the possibility of carrying out a new measurement in experimental conditions to be defined in the frame of this task force.

REFERENCES

1. L.W. WESTON and J.H. TODD : N.S.E 88, 567 (1984).

2. H. DERRIEN,   G. de SAUSSURE   and R.B. PEREZ : N.S.E. 106, 4, 434 (1990).

3. R. GWIN,   L.W. WESTON,   G. de SAUSSURE, R.W. INGLE,   J.H. TODD,   F.E. GILLESPIE, R.W. HOCKENBURY and R.C. BLOCK : N.S.E., 45, 25 (1971).

4. R. GWIN,   E.G. SILVER,   R.W. INGLE   and H. WEAVER : N.S.E. 59, 79 (1976).

5. J.A. HARVEY,   N.W. HILL,   F.G. PEREY, G.L. TWEED and L. LEAL : In Proc. Int. Conf. on Nuclear Data for Science and Technology, May/June 1988, Mito, JAPAN (edited by S. Igarasi) p. 115.

6. H. DERRIEN : JEF-DOC report 291.

7. W.P. POENITZ, J.F. WHALEN and A.B. SMITH : N.S.E. 78, 333 (1981).

8. Ch. LAGRANGE and D.G. MADLAND : Physical Review C, Volume 33, number 5, p. 1616.

INTERNATIONAL EVALUATION COOPERATION TASK 1.1: INTERCOMPARISON OF EVALUATED FILES FOR ^{52}Cr, ^{56}Fe, AND ^{58}Ni

C. Y. Fu, D. C. Larson, D. M. Hetrick
Oak Ridge National Laboratory, Oak Ridge, TN 37831-6356, USA

H. Vonach
Vienna University, Vienna, Austria

J. Kopecky
Netherlands Energy Research Foundation, Petten, Netherlands

S. Iijima
Nuclear Engineering Laboratory, Kawasaki, Kanagawa, Japan

N. Yamamuro
Data Engineering, Inc., Yokohama, Kanagawa, Japan

G. Maino
ENEA, Bologna, Italy

Abstract: The cross sections and energy and angular distributions in the JENDL-3, JEF-2/EFF-2, and ENDF/B-VI evaluations for ^{52}Cr, ^{56}Fe, and ^{58}Ni are compared graphically. The purpose is to understand the reasons for observed discrepancies among the evaluations, and to suggest recommendations for improvements. The accomplishments to date are presented.

(^{52}Cr, ^{56}Fe, ^{58}Ni, cross section, evaluation)

1. Introduction

Task 1.1 is one of seven in a NEACRP/NEANDC Task Force on Evaluation Cooperation organized to promote international evaluation collaboration. The purpose of Task 1.1 is to graphically compare selected evaluations for structural materials from ENDF/B-VI (USA), JEF-2/EFF-2 (Europe), and JENDL-3 (Japan). The isotopes ^{52}Cr, ^{56}Fe, and ^{58}Ni are especially interesting because they are important components of steel and because new evaluations for them have recently been completed for each library. In Section 2, a summary of the completed comparisons is presented. An attempt to resolve major discrepancies is described in Section 3. Section 4 contains recommendations for remaining work and for improvements.

2. Summary of the Comparisons

The following comments are based on the consensus of a meeting of the subgroup held on Dec. 3, 1990, at the NEA Data Bank, Saclay, France. Detailed recommendations for improvements are available upon request. Only results of general interest or having the largest discrepancy are presented here.

The spread for the **nonelastic cross sections** is less than 3% for ^{52}Cr and ^{56}Fe and less than 15% for ^{58}Ni. This deviation is correlated with the partial reaction cross sections discussed below. For example, a 3% difference in the ^{56}Fe nonelastic cross section at 14 MeV could result from a 10% deviation in the ^{56}Fe$(n, 2n)$ cross section and result in a 10% discrepancy in the total gamma-ray production cross section.

The **inelastic scattering cross sections** behave much like the nonelastic. Because these cross sections are important for shielding (pressure vessel surveillance, fusion blanket) calculations, perhaps 2% accuracy up to 14 MeV is needed. The intercomparison shows differences as much as 20% in the 2 to 4 MeV region.

The $(n, 2n)$ **cross sections** seem satisfactory for chromium and nickel due to adequate data bases. However, the 20% spread in ^{56}Fe at 14 MeV could result in large discrepancies in calculated neutron transmission through thick iron spheres with a central 14-MeV source.

The (n, α) **cross sections** have large spreads among the evaluations. In general, there are two different shapes. Below 14 MeV, EFF-2 is relatively low and ENDF/B-VI is relatively high. The worst case is for ^{58}Ni which is shown in Fig. 1. The data shown are from Qaim et al. [1,2]. The possible explanation for this discrepancy is explored on the basis of further information furnished by the evaluators and is described in Section 3. It is recommended that more experimental data be taken at energies between 8 and 10 MeV for all three isotopes in view of the importance of this cross section in radiation damage studies. An IAEA CRP has been proposed to study this problem.

The **energy distributions of outgoing neutrons** show large differences, in particular for ^{58}Ni at 11 MeV, shown in Fig. 2. It was decided that more information was needed from the evaluators to understand these discrepancies. This follow-through is discussed in Section 3. Furthermore, experimental double differential measurements for iron at energies below 14 MeV are strongly encouraged.

The **cross sections for photon production** as a function of incident neutron energy were found to be in substantial disagreement: from 4 to 10 MeV for ^{52}Cr, above 8 MeV for ^{56}Fe, and above 4 MeV for ^{58}Ni. The discussion was centered on ^{56}Fe, shown in Fig. 3. Two suggested reasons for the discrepancies are given in Section 3.

3. The Larger Discrepancies

The following discussions are based on data supplied by the evaluators after the Dec. 3, 1990, meeting.

Possible reasons suggested for the discrepancies seen in Fig. 1 for the (n, α) cross sections were: competition of other channels, alpha-particle optical model, level density in the residual nuclides, and preformation factors in pre-equilibrium models. It was concluded during the meeting that the last point should be studied first. The findings are: JENDL-3 used the Iwamoto-Harada model [3] for the preformation factors, EFF-2 used the Milazzo-Colli model [4], and ENDF/B-VI used the Kalbach model

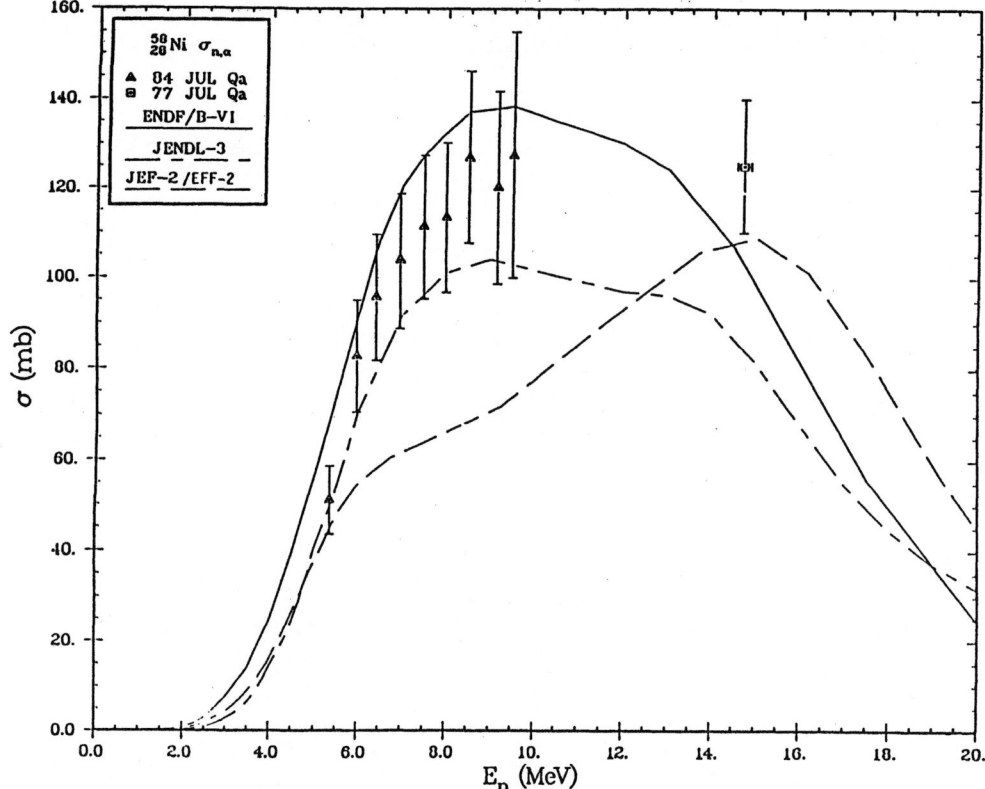

Fig. 1. Comparison of the ^{58}Ni(n, α) cross sections from the three evaluations.

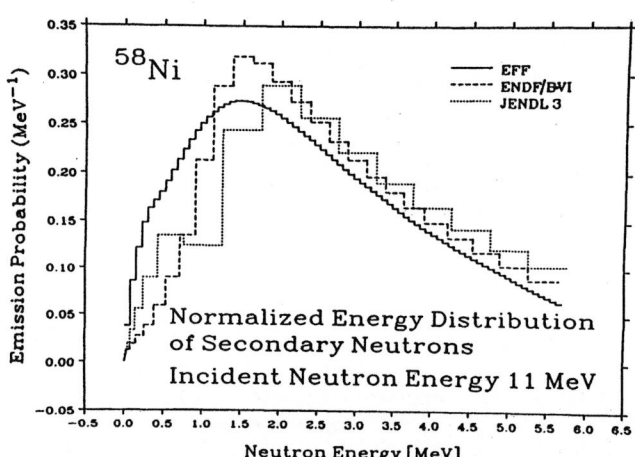

Fig. 2. Comparison of the ^{58}Ni secondary neutron distributions from the three evaluations.

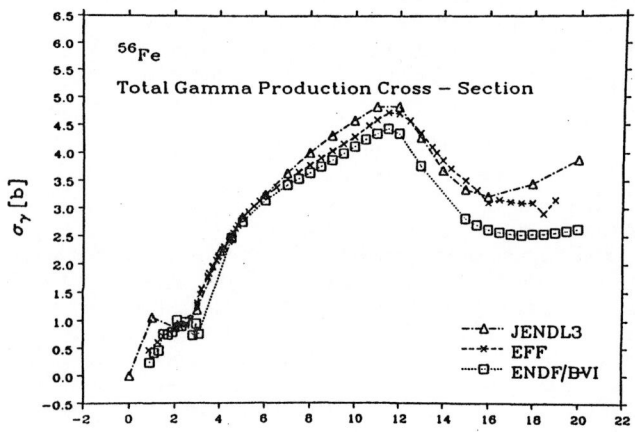

[5] modified to include spin dependence [6]. The differences in these formulas cannot be easily sorted out because the precompound models in which these preformation factors reside are also different. This difficulty could be an interesting topic for further international cooperation work.

The evaluators for JENDL-3 and ENDF/B-VI have also furnished the **level densities for** ^{55}**Fe**, the residual nuclide in the ^{58}Ni(n, α) reaction. The JENDL-3 values are substantially larger. However, trends in level densities and shape of the ^{58}Ni(n, α) cross sections are not correlated. So the problem remains unresolved.

The **level densities for** ^{58}**Ni** used in JENDL-3 for calculating neutron emission spectra have the lowest temperature (steepest rise) and those for ENDF/B-VI the highest temperature. This trend is inconsistent with observed differences in the neutron emission spectra for E_n = 11 MeV shown in Fig. 2. The largest discrepancy seen in Fig. 2 is in the low $E_{n'}$ side where the calculated neutron emission spectra are the most sensitive to the level-density variations. Other explanations have to be sought.

The **cross sections for photon production in** ^{56}**Fe**, shown in Fig. 3, could be improved by consideration of two possibilities. The first is that the two major measurements for iron, Dickens et al. [7] and Chapman et al. [8], disagree. The former data are lower than the latter by nearly a factor of 2 around 8 MeV and 14 MeV. This may have caused some confusion. The preliminary conclusion is that Chapman's data are better at 8 MeV while Dickens' are better at 14 MeV. The second possibility has to do with a 3.5% spread in the evaluated nonelastic cross sections near 14 MeV for ^{56}Fe. Due to a photon multiplicity of 3, the spread in the nonelastic cross section at 14 MeV can produce a 10% discrepancy in photon production.

4. Recommendations

Due to the importance of the **resolved resonance range** in fission reactor applications it should be studied in detail in the near future. In addition, it was suggested to compare group constants (3 per decade) in order to find important differences.

The **double differential neutron emission** cross sections can be correctly represented only by the File-6 format. Since JENDL-3 does not use File-6 for this purpose, some problems in the comparison can be anticipated. A revision in JENDL-3 for at least one of the three isotopes to make use of the File-6 format is needed.

Vonach et al. [9] have provided an independent evaluation for the **14-MeV cross sections** of ^{52}Cr, ^{56}Fe, and ^{58,60}Ni based solely on experimental data. The initially evaluated nonelastic cross section and the sum of the partial cross sections were inconsistent and have been resolved by least-squares adjustment. It is recommended that all evaluators consider this evaluation in their future revisions.

Acknowledgments

The contributions of graphs, documents, or advice by C. L. Dunford, H. Gruppelaar, S. Tagesen, and T. Asami to this working group are gratefully acknowledged. Part of this work was sponsored by the Office of Energy Research, Division of Nuclear Physics, U. S. Department of Energy, under contract DE-AC05-84OR21400 with Martin Marietta Energy Systems, Inc.

References

1. S. M. Qaim and N. I. Molla, *Nucl. Phys.* **A283**, 269 (1977).
2. S. M. Qaim, R. Wolfe, N. M. Rahman, *Nucl. Sci. Eng.* **88**, 143 (1984).
3. A. Iwamoto and K. Harada, *Phys. Rev.* **C26**, 1821 (1982).
4. L. Milazzo-Colli and G. M. Braga-Marcazzan, *Nucl. Phys.* **A210**, 297 (1973).
5. C. Kalbach, *Z. Phys.* **A283**, 401 (1977).
6. C. Y. Fu, *Nucl. Sci. Eng.* **100**, 61 (1988).
7. J. K. Dickens, G. L. Morgan, and F. G. Perey, *Nucl. Sci. Eng.* **50**, 311 (1973).
8. G. T. Chapman, G. L. Morgan, and F. G. Perey, *A Re-Measurement of the Neutron-Induced Gamma-Ray Production Cross Sections for Iron in the Energy Range 850 keV < E_n < 20 MeV*, ORNL/TM-5416, Oak Ridge National Laboratory (1976).
9. H. Vonach, S. Tagesen, M. Wagner and A. Pavlik, *Uncertainty Estimates for the Fast Neutron Cross Sections of the European Fusion File EFF for ^{52}Cr, ^{56}Fe, and ^{58,60}Ni and Evaluation of the 14-MeV Cross Sections of these Isotopes from the Existing Experimental Data Base*, EFF-DOC-85, Institut fur Radiumforschung und Kernphysik der Unversitat Wien (1991).

INTERNATIONAL EVALUATION COOPERATION SUBGROUP 7: MULTIGROUP CROSS SECTION PROCESSING

R. W. Roussin and J. E. White
RSIC, Oak Ridge National Laboratory,*
Oak Ridge, TN, USA 37831

D. Muir
IAEA, Nuclear Data Section,
Vienna, Austria

E. Sartori
Nuclear Energy Agency Data Bank, Saclay, France

M. Mattes
IKE, Stuttgart Technical University, Germany

G. Panini
ENEA, Bologna, Italy

A. Hasegawa and H. Takano
Japan Atomic Energy Research Inst., Tokai-mura, Japan

R. MacFarlane
Los Alamos National Laboratory,
Los Alamos, NM, USA

F. Mann
Hanford Engineering Development Laboratory,
Richland, WA, USA

ABSTRACT: The chairmen of the ENDF/B, JEF, EFF, and JENDL evaluated data files adopted a proposal to develop a fine-group processed cross section library based on the "VITAMIN" concept. The authors listed above, with support from others, are participating in this project. The end result will be a pseudo-problem-independent fine-group cross section library generated from the latest evaluated data in ENDF/B-VI, JEF-2, EFF-2, and JENDL-3. Initial applications of the library will be for shielding, fast reactor physics, and fusion neutronics. Progress made to date will be discussed. (Cross sections, EFF, ENDF, JEF, JENDL, multigroup, radiation transport, shielding)

Introduction

A desirable outcome of international cooperation on evaluation of cross section data is the development of the corresponding processed libraries that can be used in radiation transport applications. The goal of Subgroup 7 is to produce processed multigroup libraries based on four major evaluated cross-section libraries. These include ENDF/B, developed in the United States, JEF, developed in Europe under NEA sponsorship, EFF, developed in Europe under ECC sponsorship, and JENDL, developed in Japan. A common set of specifications for processed data is suggested so that differences in preparation of libraries can be minimized. Data derived from the three evaluation sources will be available in common formats so that they can be incorporated into radiation transport analysis schemes with a minimum of difficulty. This will allow the assessment of the impact of various data sets on results and, if major discrepancies are encountered, provide additional guidance to the evaluation process.

The initial project will focus on producing a library based on the "VITAMIN" concept that has been used for libraries based on ENDF/B-IV (VITAMIN-C [1]), ENDF/B-V (VITAMIN-E [2]), and JEF-1 (VITAMIN-J [3]). This concept involves the generation of a fine-group pseudo-problem-independent cross-section library using the shielding factor method [4] to account for resonance self-shielding and temperature effects. The successful application of the libraries to a variety of problems, including shielding, fast reactor physics, and fusion neutronics provides the impetus to select a similar approach for the first task undertaken by Subgroup 7.

Specifications

The suggested specifications for processed data are those used for VITAMIN-J, which were based on those developed earlier for VITAMIN-C and VITAMIN-E. Some details are provided in the sections that follow.

Energy Group Structures

The suggested neutron energy group structure includes the 175 groups in VITAMIN-J which span the energy range from thermal to 20 MeV. For photons, a 42 group structure extending up to 50 MeV is suggested. Both these energy structures contain as subsets the group structures of many broad group libraries in common use in the international community.

Weighting Spectrum

The suggested weighting spectrum for neutron groups is Maxwellian in the thermal region, and also includes a fission spectrum and a fusion peak. The three features are joined by "1/E" slowing-down regions. For photon groups the weighting spectrum is "constant."

Resonance Reconstruction Tolerance

The suggested tolerance for reconstructing 0 Kelvin resonance data or linearizing is 0.1%.

Temperature Grid

The range of temperatures at which cross sections will be generated is between about 300 and 2100 Kelvin on a grid of 4 to 10 values.

*Managed by Martin Marietta Energy Systems, Inc., under contract DE-AC05-84OR21400 with the U.S. Department of Energy.

Background Cross Sections

Ten or fewer background cross sections (including infinity and zero, essentially) will be used, representing a range of values expected in actual calculations.

Legendre Polynomial Expansion Order

The expansion order for neutron and photon scattering will be P_5 or higher.

Output Formats

Libraries will be available in a variety of standard formats from which problem-dependent libraries can be derived, primarily MATXS and AMPX.

Processing Codes

It is expected that the NJOY91 [5] processing system will be used for generating cross sections from ENDF/B-VI and JEF-2 and the PROF-GR system for JENDL-3. The NJOY processing system handles the ENDF-6 format which is used by ENDF/B-VI and JEF-2/EFF-2. Los Alamos developed and maintains NJOY with feedback from various users that rely on the Los Alamos National Laboratory (LANL) for help in supporting their processing efforts. ENEA/Bologna, ECN Petten, and Oak Ridge National Laboratory (ORNL) are working on GENDF to AMPX form translation.

Progress to Date

Processing tasks have already begun on evaluations from the major evaluated libraries. Specifications have been developed and procedures outlined for producing initial versions that will feed into the data testing activities supporting the ENDF/B, JEF/EFF, and JENDL evaluation projects.

JENDL

At the Japan Atomic Energy Research Institute, a JENDL Shielding Standard Library (JSSTDL) has been produced [6] from JENDL-3 with 295 neutron and 104 photon groups. The neutron and photon groups from VITAMIN-J are subsets of JSSTDL with the exception of one neutron group boundary at 70 keV. The neutron weighting function is Maxwellian to 0.3224 eV and "1/E" at higher energies (because of the increased number of groups, weighting function energy dependence may be less important for JSSTDL). The Legendre order is P_5.

The temperature grid is 300, 600, 900, and 2100 K and the background cross-section grid is 0, 0.1778, and 1 to 10^6 in decades. Some 60 elements and isotopes have been completed using the PROF-GR system into an output format that can be processed by the MICROJ system to account for self-shielding and temperature effects.

JEF/EFF

The OECD/NEA Data Bank has issued a draft of Revision 3 of JEF/DOC-315 [7] on standard energy group structures for benchmarking the newly released major evaluated libraries. Some 38 data sets had already been processed at 300 K and infinite background cross

section at IKE, Stuttgart, Germany, and PSI, Villigen, Switzerland, at the time of the Draft (Dec. 1990). The Draft also indicates a temperature range from 293.4 to 2100 K on a grid of 10 values and also a grid of 10 background cross sections from 10^{-3} to 10^{10}. Neutron data are to be produced at P_5, photon data at P_8.

The recommended processing codes are NJOY89.62 from Los Alamos, THEMIS from CEA/CEN Saclay, France (the physics modules in NJOY and THEMIS are identical) and MILER from ENEA, Bologna, Italy [8]. Recommended output formats include GENDF, MATXS, AMPX (produced from GENDF with MILER), and ANISN, all formats currently in wide use.

ENDF/B-VI

At ORNL plans call for an ENDF/B-VI based library using the NJOY processing system. Some 37 data sets have already been processed with the VITAMIN-J group structures and weighting functions. Both neutron and photon data were generated at P_5. The final version will contain a temperature grid of 300, 900, 1500, and 2100 K and a range of background cross sections from 0 to ∞. The output format will be AMPX, which will be produced with MILER or another module now under development at ORNL. Data from this initial set are to be used in CSEWG data testing of ENDF/B-VI.

Hanford Engineering Development Laboratory (HEDL) has agreed to provide ORNL with ENDF/B-VI pointwise tapes for over 200 nuclides at several temperatures.

The IAEA Nuclear Data Section plans to process the international Fusion Evaluated Nuclear Data File (FENDL) with NJOY according to VITAMIN-J specifications into GENDF output format. Since much of FENDL is based on ENDF/B-VI, discussions are underway with ORNL on sharing the processing chore.

Conclusions

Three major evaluated cross section data libraries have been released in recent months. After satisfactory review and testing it is anticipated that they will be released with no restrictions on their distribution. International cooperation in the evaluation arena is being actively pursued to help enhance the overall quality of these major evaluated data libraries. This era of cooperation extends into the cross-section processing field and plans are well underway to develop fine-group libraries that can be shared and used for benchmark testing of the evaluated data. The resulting data libraries will be distributed in standard formats via the NEA Data Bank, Saclay, France, and the Radiation Shielding Information Center (RSIC), Oak Ridge, Tenn., USA.

Acknowledgements

Appreciation is extended to C. L. Dunford, M. Salvatores, H. Gruppelaar, and Y. Kikuchi, chairmen of the ENDF/B, JEF, EFF, and JENDL activities, respectively for establishing Subgroup 7 on Multi-

862

group Cross Section Processing and to E. Menapace for providing the liaison with the NEACRP/NEANDC Working Group on Evaluation Cooperation.

References

[1]. R. W. Roussin, *et al*, ORNL/RSIC-37 (1979).

[2]. C. R. Weisbin *et al*, ORNL-5505 (February 1979).

[3]. E. Sartori, JEF/DOC-100 (1985).

[4]. I. I. Bondarenko, Ed., Consultant's Bureau (1964).

[5]. R. E. MacFarlane, LA-UR 89-2057 (June 1989).

[6]. A. Hasegawa, JAERI-M 88-221 (1988). (See also the A. Hasegawa paper at this conference.)

[7]. E. Sartori, JEF/DOC-315, Rev. 3 (Draft) (1990).

[8]. G. C. Panini and M. Pescarini, ENEA Informal Notes (1988).

IAEA Coordinated Research Programme on
Methods for Calculation of Fast Neutron Data
for Structural Materials

A. Marcinkowski
Soltan Institute for Nuclear Studies, 00–681 Warsaw, Hoza 69, Poland

D. Muir
International Atomic Energy Agency, RIPC/Nuclear Data Section
P.O. Box 100, A–1400 Vienna, Austria

and

D. Seeliger
Technische Universität Dresden, Sektion Physik,
Mommsenstrasse 13, D–0 8027 Dresden, Germany

Abstract: The calculational methods and the associated computer codes currently used within the IAEA coordinated research programme (CRP) on the calculation of neutron cross sections for structural materials are reviewed. This project coordinates activities of neutron laboratories from 13 countries: ANL (USA), BARC (India), CTA (Brazil), EPRC (Czechoslovakia), ENEA (Italy), ECN (Netherlands), FEI (USSR), IAE (China), INRK (Germany), INPR (Romania), JAERI (Japan), LANL (USA), LLNL (USA), SINS (Poland), TUD (Germany) and University of Oxford (UK).

Introduction

Structural materials and blankets for fusion and fission energy development contain a number of components and associated multipliers, which are confined to the mass ranges A = 50–60, 90–100 and 204–209 mainly. Although many of the requested cross sections for these elements have been measured, many others are required and it would be impractical to measure them all. We, therefore, resort to theoretical calculations.

Neutrons from the fission or fusion sources span over the energy range from 10^{-5} eV to 30 MeV. In this energy range the mechanism of neutron–induced reactions changes, and this has an important implication that different reaction theories and models have to be used in order to describe the required cross sections. Usually the interaction appears predominantly compound just above the reaction threshold and changes with increasing energy to predominantly precompound or direct.

The cross sections for the direct processes are provided by the DWBA, CCBA or CRC methods, whereas the evaporation from the compound nucleus is described by the Weisshopf–Ewing and Hauser–Feshbach theories. The first models of the preequilibrium processes were semiclassical in nature and provided expressions for calculating the energy spectra of the emitted particles, using parameterized interaction matrix element. Subsequently these models were extended to give angular distributions. More recently fully quantum–mechanical theories [1–3] were derived. The applied versions of these theories make approximations, which require thorough test before they will become trustworthy predictive tools for nuclear data evaluation. Nonetheless, as the best future hopes of a physical solution to the modeling of differential cross sections lies in the quantal approaches, a considerable effort was devoted within the CRP to develop computer codes that make them applicable.

Still the optical model provides the important neutron strength functions, scattering and total cross sections, and its latest developments are discussed altogether with the statistical methods in the following.

Optical Models for Neutron Nuclear Data Evaluation

The neutron optical potential is an essential tool for evaluation of neutron scattering and reaction data, and efforts devoted to determining it with higher accuracy were continued within this CRP. The emphasis was put on selected nuclei and energies below 10 MeV, which are most important for practical calculations. Extensive optical model analyses have been completed at ANL, by Smith et al. [4–6] for vanadium, cobalt, nickel, yttrium, zirconium, niobium and bismuth and now are in progress for chromium, iron and strontium. They involve differential elastic and inelastic cross sections and total neutron cross sections measured with high precision. The calculations are done with use of the ABAREX code, which allows for the Hauser–Feshbach statistical model with the width–fluctuations corrections by Moldauer beside the optical model fits. A standard, spherical optical model was used. These analyses have shown that a simple optical model parameterization is no longer able to give satisfactory fits to experimental data.

Most striking are the results obtained from fitting the high precision data for ^{59}Co, which range from 0.36 MeV to 10 MeV. Strong energy dependence of the potential's real and imaginary depths together with strong energy variation of the geometries were found at low energies, which change for energies higher than 7.5 MeV to those commonly encountered in global potentials. Furthermore, the imaginary volume–integral J_W decreases with increasing energy, which can be attributed to the possible vibrational features of ^{59}Co. For ^{51}V the change from the rapid energy dependence of the parameters to the one known from global optical models occurs at about 6 MeV and J_W increases with neutron energy below this point in accordance with expectations. On the other hand, the analysis of the data on ^{89}Y provide a spherical optical model with energy–independent real geometry and the real and imaginary volume integrals varying with energy as encountered in the global potentials.

It was, of course, possible to accommodate all the departures from the standard optical model by more complicated parameterizations. However, these parameters are restricted to the nuclei for which the neutron data were available. They differ from nucleus to nucleus and render the concept of a global optical potential that spans a mass range wide enough to include all the structural materials not useful, at least if one wishes to reproduce the high precision data. However, it is possible that regional optical models confined to limited mass ranges can be used. Such hopes are held out by results of the application of dispersion relations that connect the real and imaginary parts of the optical potential. When applied to precise neutron data, they are able to account in some detail for the anomalous behavior, referred to as the Fermi surface anomaly, of some optical model parameters. They result in a surface peaked real term in the potential. Analyses including the real dispersion correction term were reported to the CRP for ^{59}Co and ^{89}Y by Lawson et al. [4]. The parameterization of this potential appeared much simpler in comparison to that obtained for the standard optical model.

Semiclassical Preequilibrium Models

A variety of semiclassical models is in use that characterize a nuclear state by its excitation energy E and the total number n of particles (p) above and holes (h) below the Fermi surface, and the cross section is obtained as a sum of contributions from states of all possible n values,

$$\frac{d\sigma}{d\epsilon}(a,b) = \sigma_a \sum_{\substack{n \\ \Delta n=2}} \frac{W_b(n,\epsilon)}{\lambda + \lambda_+}, \qquad (1)$$

where σ_a is the projectile reaction cross section. The partial $W_b(n,\epsilon)$ and total λ nucleon emission rates are expressed by the inverse reaction cross section and λ_+ is the rate of internal transitions creating a particle-hole pair $\Delta n = +2$, which can be obtained from Fermi's golden rule.

The exciton model refers to the nuclear system as a whole and uses the internal transition rate $\lambda_+(n,E)$ depending on the number of excitons and the excitation energy. On the other hand, by referring to the leading particle to be emitted with continuum energy ϵ, the hybrid model was formulated, in which $\lambda_+(\epsilon)$ gets related to the mean free path of the particle in nuclear matter that depends upon the outgoing energy ϵ.

Note that $\tau = (\lambda + \lambda_+)^{-1} \approx \lambda_+^{-1}$ are the lifetimes, which govern the emission in either of the two approaches. These lifetimes are subject to different approximations. One way is to calculate them by solving the time–integrated master equation. This procedure is used for calculating the preequilibrium cross sections in the GNASH code [7]. However, when calculated from a random–walk equation, they provide the basis for formulation of the exciton model used in the STAPRE code [7]. On the other hand, the mean free path used in the hybrid model may be calculated from the imaginary optical potential, as in the EMPIRE code [7], or from the Pauli corrected nucleon–nucleon scattering cross sections in nuclear matter, which is the most commonly used option in the ALICE code [7]. The hybrid model was extended by allowing for a longer mean free path in the diffuse nuclear surface. This geometry–dependent model has proved better in describing the very high energy end of nucleon emission spectra [7].

The above listed codes include the decay of the compound nucleus, which is calculated according to the Hauser–Feshbach theory in all but the ALICE code where the angular momentum independent approach of Weisshopf–Ewing is utilized. The last code has been designed and maintained to emphasize ease of use, high speed and versality. However, other codes will be preferable when accuracy better than 50% is more important than speedy calculation. Examples of such codes may be the GNASH or STAPRE codes.

The first of these codes specialized for the calculation of 20–100 MeV neutron cross sections is used in a program that is currently in progress at the LANL's LAMPF facility to perform high resolution measurements of (n,xγ) cross sections for individual gamma transitions up to incident neutron energies in the medium energy range for a variety of target materials [8].

The GNASH code has been extended at JAERI for the purpose of accurate calculations of neutron capture cross sections, with width fluctuations corrections and with account for the direct component of neutron inelastic scattering as obtained with use of the DWUCK–4 code. It was used by Nakajima and Yamamuro for an extensive evaluation of fast neutron activation cross sections on Ni, Zr, Mo and Nb isotopes [9]. The evaluation of more than 30 excitation curves for (n,p), (n,α) and (n,2n) reactions, which extend from thresholds up to 20 MeV, were reported for these target nuclei.

Konshin and Konshin [10] adapted GNASH for use with a PC computer and investigated the impact of uncertainties of the input level densities and optical model transmission coefficients on the evaluated cross sections. Commonly–used global optical potentials and level density parameterizations were tested for ^{56}Fe and ^{93}Nb. It was found that the assumed optical models provide $\sigma_{n,2n}$ values that deviate by up to 25%. The level density model of Blokhin and Ignatyuk [11] was favored, though the deviations produced by different optical models were most important.

The STAPRE code was used in a series of evaluations of neutron threshold reactions for components of structural materials completed by Ivascu et al. [12]. For this purpose the code has been equipped with a subroutine using the formalism of the geometry–dependent hybrid model for calculating angular momentum–dependent preequilibrium cross sections. Special attention has been paid to a consistent use of the density of particle–hole states, describing the first stages of the reaction and the level density of the compound nucleus. The two densities have been matched by using an energy–dependent level density parameter. This approach utilizing regional optical potentials was used for calculating isotopic cross sections for (n,p), (n,α) and multi-particle emission reactions on Cr, Fe, Ni [12].

Unified Precompound/Compound Models and Angular Distributions

A considerable interest in the unification of the Hauser–Feshbach and the exciton model was reflected in the contributions of Xiangjung et al. [13] and Zingshang et al. [14]. In these reports the master-equation theory was generalized to account for the conservation of angular momentum or to include the light composite particle emissions by simplifying the Iwamoto cluster model for practical use. The constructed models retain the Hauser–Feshbach and standard exciton model as limiting cases.

Computationally these models have been

optimized such that the related numerical effort is comparable to standard Hauser–Feshbach calculations. A comparison of the more detailed description provided by the GNASH code with the unified exciton model as used in the GRAPE code [15] and with ALICE code, which includes the geometry–dependent hybrid model was reported by Garg [7]. Special reference was made to the neutron–induced primary and binary reaction cross sections on ^{58}Ni and ^{60}Ni. In particular the total production cross sections for neutrons, hydrogen, helium and γ–rays were evaluated in the incident energy interval from 1 to 26 MeV. This evaluation served also for the generation of previously unmeasured neutron cross sections on the Ni isotopes. The cross sections predicted by the more advanced methods of GNASH and GRAPE deviated by only 1.5% to 15% for neutrons.

The main conclusion from numerical calculations is that though the unified methods do not provide better cross sections, they may be easier extended to predict angular distributions. This is usually done by assuming that the most important contribution to the preequilibrium angular distribution comes from the single collision of the projectile with a target nucleon. This approach allows to calculate the angular distribution of particles emitted after just one collision with nucleus having a Fermi gas momentum distribution from the nucleon–nucleon scattering cross sections. Blann et al. [17] have extended this method to obtain the angular distributions of particles following the first, second and higher order scattering by folding the single scattering result within the geometry–dependent hybrid model. This model has been compared with different approaches, e.g., with the one of Akkermans et al. [18], with phenomenological systematics of Kalbach [19], or with predictions of the quantum mechanical theory of Feshbach, Kerman and Koonin (FKK) [2] for the (n,xn) reaction data at 14 MeV incident energy. The results of this analysis indicate that the semiclassical exciton approaches to angular distributions may not be extrapolated much beyond the 14 MeV region where they were designed for. They lack an adequate treatment of physics, which is provided by the more rigorous quantal theory of FKK.

Quantum Mechanical Theories

The treatment of the multistep direct (MSD) reactions by Tamura et al. (TUL) [1] using the distorted–wave Born approximation has been used to analyze (n,n′) and (n,p) data and the statistical multistep reaction theory formulated by FKK [2] has been used in studying some (n,p), (n,n′) and (n,2n) reaction cross sections at lower incident energies [20]. The FKK theory introduced the multistep compound (MSC) and the MSD mechanisms in parallel as complementary contributions to preequilibrium decay. The MSC emission results in symmetric angular distributions of particles, whereas the MSD emission gives rise to forward–peaked angular distributions. Recently, Nishioka et al. (NWY) [3] treated the MSC reactions in the presence of direct processes and also derived the cross sections for the MSD processes as a sum of one and two–step DWBA terms.

The MSD theory of FKK assumes a continuum or leading particle statistics, which results in the simple, convolution–like MSD cross section formula, which contains only onestep terms and is easy to apply,

$$\frac{d^2\sigma_N}{dU d\Omega} = \sum_N W_{N,N-1} \cdots W_{2,1} \frac{d^2\sigma_{1,i}}{dU d\Omega}. \quad (2)$$

The FKK theory provides basis for expressing the transition probabilities $W_{N,N-1}$ and the onestep cross section via the DWBA angle–differential cross sections.

As far as angle integrated MSC cross sections are considered, the FKK theory gives a formula similar to eq. (1) expressed in terms of energy averaged entrance, spreading, emission and total widths, which are related to the transition rates λ by $\Gamma = \hbar\lambda$.

The MSC theory of FKK has been incorporated into the EMPIRE code [7] and used for description of neutron emission spectra and cross sections for the (n,2n) reaction on Mo isotopes [21]. In conjunction with an MSD subroutine, which offers optionally the formulation of FKK or of TUL, the EMPIRE code was also used by Marcinkowski and Kielan [21] in evaluation of the excitation curves for the (n,p) reaction on Zr isotopes as well as the emission spectra and angular distributions from the (n,xn) reaction at 20 MeV incident energy on ^{93}Nb and ^{209}Bi. The EMPIRE code was extended at Bologna to allow for the MSC emission according to the formalism of NWY. This theory provides the average cross section connecting channel a and b via the MSC mechanism in the form,

$$\sigma_{ab} = (1 + \delta_{ab}) \sum_{m,n} T_m^a \, \Pi_{mn} \, T_n^b. \quad (3)$$

The sums run over classes (m or n) of particle–hole excitations and the partial transmission coefficients T_m^a, which couple channel a with class m, when summed over all m–states give the usual optical model transmission coefficient in channel a. The probability matrix Π_{mn} is defined via the mean squared matrix element and the emission and spreading widths. Herman et al. [22] calculated the spreading width from the imaginary optical potential and used this MSC model in conjunction with the MSD theory of TUL to describe neutron scattering on ^{93}Nb at 14 MeV incident energy. Agreement with experiment was possible only after taking into account of the low energy collective excitations in the transition density.

The recent calculations of the MSC and/or the MSD cross sections, including those for the (n,γ), (n,xn) and (p,xn) reactions done at Oxford, which make use of the FKK theory were reported by Hodgson [23]. A generally satisfactory agreement with experimental data was found with consistent values of the interaction strength V_0. Further work, however, is required to extend these analyses over a wider range of nuclei and to determine the parameters with better accuracy.

Most recently Kalka et al. [24] developed quite simple expressions for calculating both MSC and MSD cross sections using a random matrix and Green's function formalism. Their theory provides consistently cross sections for collective multi–phonon excitations. However, it does not provide angular distributions. This theory has been used extensively in evaluations of neutron emission spectra on many nuclei ranging from Al to Bi and at incident energies from 5 to 26 MeV. The same authors analyzed neutron and proton emission spectra following interaction of 14 MeV neutrons with Ti, Cr, Fe, Co, Ni and Cu targets along with excitation functions for the (n,p) and (n,2n) reactions obtained in activation experiments [24]. A very good agreement between theory and experiment was obtained by using only global parameters for all but the magic ^{58}Ni, ^{208}Pb and ^{209}Bi nuclei. The associated EXIFON code has been adapted for an AT personal computer [24].

Level Densities for Statistical Models and Theories

The decay of the compound nucleus has been for fifty years a level density problem. The last two decades have added to it the partial, model particle–hole state densities used in semiclassical preequilibrium calculations and the last two years brought calls for true partial level densities [3] needed for calculations using some of the quantum–mechanical theories. Since the evaluation of neutron nuclear data for applications requires reasonable accuracy, the majority of calculations employ phenomenological level density models, which have proved more reliable.

The two most popular compound nucleus level density models are the Gilbert and Cameron two–component Fermi gas formula and the back–shifted Fermi–gas model. Very extensive parameterizations of both models are available. The first of these models has been used in the GNASH, EMPIRE and the original version of the STAPRE codes. The back–shifted Fermi gas level density has been used for the purpose of fast neutron data evaluation in the PC version of the GNASH code by Konshin [10]. The models described above use an energy–independent level density parameter a. The effects of shell closures on this parameter and of their propagation to higher energies were addressed in the approach developed by Blokhin and Ignatyuk [11]. These authors employ the continuous spectrum approximation in order to work out a practical method of calculating level density. The shell effects are allowed for by assuming an energy–dependent level density parameter $a(U) = a[1+f(U) \delta W/U]$, with a being the asymptotic value of a occurring at high excitation energies, and δW determined via nuclear masses $\delta W = M_{exp}(Z,A) - M_{ld}(Z,A)$. The term $f(U) = 1 - \exp(-\gamma U)$, with $\gamma = 0.05$ MeV. The model of Ignatyuk was implemented by Young et al. [8] in the GNASH code used in the analysis of the $(n,xn\gamma)$ reactions. Also Kalka et al. use this model in their EXIFON code [25]. This model accounts in addition for the increase of level density due to vibrational excitations [11].

The effect of the different level densities including these described above, on the calculated neutron reaction cross sections was investigated by Garg [16]. This author found that the Gilbert and Cameron model and the Ignatyuk model yield most identical results for many reactions investigated in his study.

In the semiclassical preequilibrium models and some of the rigorous quantal theories, e.g., in the FKK model, the state density for a system of excited particles p and holes h is taken from the Ericson formula, which, when summed over all particle–hole pairs, is supposed to be consistent with the Fermi gas state density. The consistency conditions for the partial level densities have been investigated by Ivascu et al. [12].

Another kind of partial particle–hole state density is used in the quantal theories of the MSC process. There the decaying states are supposed to include only bound–particle–hole orbitals [2]. Numerical methods that truncate states involving continuum orbitals have been developed as well as analytic expressions have been derived for the quasibound particle–hole state density and used in the EMPIRE code.

Microscopic approaches were developed for the calculation of exciton level densities based on combinatorial method for the determination of the various configurations generated according to a given shell model spectrum of single particle levels. In these calculations, total configuration energies are determined in the frame of the BCS theory. This method allows for both spin and parity distributions and works for spherical and deformed nuclei within the ALICE code. Gardina et al. [25] have analyzed the extensive set of single particle levels determined by Nix and Möller, which contains results for more than 4,000 isotopes ranging from A = 12 to A = 269. As a result, the dependence of the shell–model spacings for excited nucleons and for nucleon holes on energy, neutron number N and atomic number Z has been systematized.

Summary

The standard optical model analysis of high quality scattering and total cross sections have revealed structural effects, which differentiate the parameterization of the optical potential for the individual nuclei and render a global approach difficult. New methods such as the inclusion of dispersion relations hold out the hope that it will prove possible to obtain global parameterization.

The physical solution to the modeling of reaction cross sections lies in the quantum–mechanical theories. On the other hand, the semiclassical preequilibrium models provide good angle–integrated cross sections and are computationally fast and easily used tools.

References

1. T. Tamura, T. Udagawa and H. Lenske, Phys. Rev. C26, 379 (1982)
2. H. Feshbach, A. Kerman and S. Koonin, Ann. Phys. 125, 429 (1980)
3. H. Nishioka, H.A. Weidenmüller and S. Yoshida, Ann. Phys. 193, 195 (1989); Ann. Phys. 183, 166 (1988)
4. R.D. Lawson and A.B. Smith, IAEA Report, INDC (NDS)–193/L, 137 (1988)
5. A.B. Smith and R.D. Lawson, IAEA Report, INDC (NDS)–214/LJ, 7 and 25 (1989)
6. S. Chiba, P.T. Guenther, R.D. Lawson and A.B. Smith, in IAEA Report, INDC (NDS)–247/L, (1991)
7. H. Gruppelaar and P. Nagel, Newsletter of NEA Data Bank 32, NEANDC–204U (1985)
8. P. Young, R. Haight, R. Nelson, S. Wender, C. Laymon, G. Morgan, D. Drake, M. Drosg, H. Vonach, A. Pavlik, S. Tagesen, D. Larson and D. Dale, in IAEA Report, INDC (NDS)–247/L, (1991)
9. Y. Nakajima and N. Yamamuro, IAEA Report, INDC (NDS)–247/L, (1991)
10. V.A. Konshin and O.V. Konshin, in IAEA Report, INDC (NDS)–247/L, (1991)
11. A.I. Blokhin and A.V. Ignatyuk, in IAEA Report, INDC (NDS)–193/L, 113 (1988)
12. M. Ivascu, M. Avrigeanu and V.A. Avrigeanu, IAEA Report, INDC (NDS)–193/L, 127 (1988)
13. Shi Xiangjun, H. Gruppelaar and J.M. Akkermans, IAEA Report, INDC (NDS)–193/L, 61 (1988)
14. Zhang Jingshan, Wen Yuanqi, Wang Shunuan and Shi Xiangjun, IAEA Report, INDC (NDS)–193/L, 21 (1988).
15. H. Gruppelaar and J.M. Akkermans, Report ECN–164, Petten (1985)
16. S.B. Garg, in IAEA Report, INDC (NDS)–247/L, (1991)

17. M. Blann, B. Pohl, B. Remington, W. Scobel, M. Trabant, R. Byrd, C. Foster, R. Bonetti, C. Chiesa and S. Grimes, in IAEA Report, INDC (NDS)–247/L, (1991)
18. J.M. Akkermans, H. Gruppelaar and G. Reffo, Phys. Rev. C22, 73 (1980)
19. C. Kalbach, Phys. Rev. C37, 2350 (1988)
20. A. Marcinkowski, in Proc. Spec. Meeting on Preequilibrium Reactions, Feb. 1988, Semmering, Austria (edited by B. Strohmaier), p. 209, OECD, Paris (1988)
21. A. Marcinkowski and D. Kielan, in IAEA Report, INDC(NDS)–247/L, (1991)
22. M. Herman, G. Reffo and H.A. Weidenmüller, in IAEA Report, INDC (NDS)–247/L, (1991)
23. P.E. Hodgson, in IAEA Report, INDC (NDS)–247/L, (1991)
24. H. Kalka, M. Torjman and D. Seeliger, Phys. Rev. C40, 1619 (1989); Z. Phys. A335, 163 (1990); IAEA Report, INDC (GDR)–59/L, 1 (1990)
25. G. Gardina, M. Herman, A. Italiano, G. Reffo and M. Rosetti, in IAEA Report, INDC (NDS)–247/L, (1991)

IAEA CO-ORDINATED RESEARCH PROGRAMME ON
MEASUREMENT AND ANALYSIS OF DOUBLE DIFFERENTIAL NEUTRON
EMISSION SPECTRA IN (p,n) AND (alpha,n) REACTIONS

M.K. Mehta* and H.K. Vonach+

* Vikram A. Sarabhai Community Science Centre, Navrangpura,
  Ahmedabad 380 009, India
+ Institut für Radiumforschung und Kernphysik, Boltzmanngasse 3,
  A-1090 Vienna, Austria

Abstract: A summary report on the Co-ordinated Research Programme on Measurement and Analysis of Double Differential Neutron Emission Spectra from (p,n) and (α,n) reaction is presented. Eight laboratories participated in the programme under which measurements were performed on 18 nuclides in the mass region 45-65, 90-100 and 180-210 at bombarding energy ranges of 9 to 12 MeV for alpha particles. 5-25 MeV from protons and 5 to 8 MeV for neutrons. Extraction of level density information through systematic and constant analysis of all the data is being done. General conclusions arising out of this work are discussed.

Nuclear level density, neutron emission spectra, (p,n), (α,n) reactions, statistical model analysis.

Introduction

Information on nuclear level density – specifically nuclear level density parameter – "a" – as a function of nuclear excitation energy (temperature) and as function of nuclear shape configuration – is one of the most important inputs for the calculation of nuclear reaction cross sections needed for technology applications. Nuclear level densities also play a very significant role in basic nuclear structure physics and nuclear physics applications. Thus a considerable effort, both theoretical and experimental, has been expanded in investigations specifically aimed at obtaining nuclear level density information.

The International Atomic Energy Agency organised an Advisory Group Meeting (AGM) on Basic and Applied Problems of Nuclear Level Densities at Brookhaven National Laboratory in 1983 to review the situation and make recommendations for actions necessary to satisfy the requirements in nuclear technology applications. It was realised that very good progress has been made in the fully microscopic fundamental approaches to the calculation of nuclear level densities. However it was not yet possible to use these developments directly for detailed calculations for practical applications. On the other hand a number of simpler prescriptions to calculate nuclear level densities based on semi-microscopic phenomenological models have proven to be very promising. Such phenomenological models depend heavily on experimentally measured systematics to determine the dependence of the various semiempirical parameters (level spins, shell corrections, pairing correction etc.) on microscopic nuclear structure systematics. However the experimental knowledge of nuclear level densities is largely restricted to the region of resolved nuclear levels at low excitation energy and to "one energy point" – the neutron binding energy, where the density of s-wave neutron resonances have been measured for many nuclides. The group recommended that it would be highly desirable to have more extensive experimental knowledge of the level densities over a large range of nuclides and excitation energy. It was recognised that an efficient approach to this objective would be the study of neutron spectra (energy angle double differential neutron emission spectra from (p,n) and (α,n) reactions [1]. It was suggested at that meeting that the Nuclear Data Section of IAEA may organise a Co-ordinated Research Programme (CRP) to obtain such information.

Objectives, Participants and Programme

The CRP was initiated by IAEA in 1985-86 with the following objectives:
(1) to extract systematic information about nuclear level densities as function of excitation energy by analysing the neutron emission spectra from (p,n) and (α,n) reactions on properly selected targets and bombarding energy range.
(2) to parameterise this information into appropriate phenomenological models to enable reliable extrapolation of level density information for general use in basic and applied nuclear physics related problems.

Eight laboratories participated in this programme, the names of the participant laboratories and of the concerned investigators are given in Table 1.

Table 1: List of Participants and Laboratories

Sr. No.	Name of the Investigator	Institution/Organisation
1.	S.M. Grimes	Ohio University, Athens, Ohio, USA
2.	P.T. Guenther	Argonne National Lab., Argonne, USA
3.	S.K.Kataria/ V.S. Ramamurthy	Bhabha Atomic Research Centre, Bombay, India
4.	T. Hongking	Institute of Atomic Energy Beijing, P.R. of China
5.	W. Scobel	University of Hamburg, Hamburg, Germany
6.	H. Maerten	Technical University of Dresden, Dresden, Germany
7.	H.K. Vonach	Institut für Radium-forschung und Kern-physik, University of Vienna, Vienna, Austria
8.	B.V. Zhuravlev	Institute of Physics and Power Engineering, Obninsk, USSR

At the first Research Co-ordination Meeting (RCM) in June 1986 the participants agreed that measurements needed to be carried out on targets with mass numbers around 60, 100 and 208 and the bombarding energies should be chosen within three ranges namely 5-10 MeV, 11-15 MeV and around 25 MeV. Five laboratories (at Sr.Nos. 1,2,4,5 and 8 in Table 1) carried out the measurements and three laboratories (at Sr.Nos. 3,6 and 7 in Table 1) agreed to carry out systematic and consistent analysis of all the data and to incorporate the results into an appropriate phenomenological model. The progress of work was reviewed at two more RCMs held in February 1988 and November 1989 respectively when measured data and developments in theoretical techniques were reported. Most of the final measurements and results of analysis carried out on individual sets of measured data as well as theoretical work carried out under this programme have been published as an INDC report [2], and/or in papers reported at this Conference or will be published in journals by the individual laboratories.

The reactions investigated experimentally under this CRP and names of the laboratories which carried out the respective experiments are listed in Table 2. All the numerical data resulting from the measurements listed in Table 2 are available from individual authors or from the Nuclear Data Section of IAEA.

Amongst the theoretical investigators, the Bombay group has incorporated their own phenomenological prescription for calculating level densities into the code Alice and tested it on Mo data of Hamburg. They find that a simultaneous fit to neutron, proton and alpha particle exit channels data, including the respective excitation functions, would be necessary to obtain unambiguous level density information. The Dresden group has developed a full scale SMD/SMC code, including

Table 2: Reactions Investigated in this CRP

Sr. No.	Reaction	Incident Energy (MeV)		Laboratory
1.	$^{56}Fe(\alpha,n)^{59}Ni$	12.40,	8.95	Ohio
		10.92		"
2.	$^{48}Ti(\alpha,n)^{51}Cr$	8.96,	10.93	"
		12.40		"
3.	$^{51}V(p,n)^{51}Cr$	5.00,	6.00	"
		7.00,	8.00	"
		8.49		"
4.	$^{59}Co(p,n)^{59}Ni$	5.00,	6.00	"
		7.00,	8.00	"
		8.93,	10.93	Beijing
		12.93,	14.93	"
5.	$^{95}Mo(p,n)^{95}Tc$	8.95,		"
		15.00,	25.60	Hamburg
6.	$^{94}Mo(p,n)^{94}Tc$	13.10,	25.60	"
7.	$^{96}Mo(p,n)^{96}Tc$	8.95		Beijing
		13.10,	25.60	Hamburg
8.	$^{97}Mo(p,n)^{97}Tc$	8.95		Beijing
		13.10,	25.60	Hamburg
9.	$^{98}Mo(p,n)^{98}TC$	8.95		Beijing
		13.10,	25.60	Hamburg
10.	$^{100}Mo(p,n)^{100}TC$	8.95		Beijing
		13.10,	25.60	hamburg
11.	$^{181}Ta(p,n)^{181}W$	11.20,	7.00	Obninsk
		8.00,	9.00	"
		10.00		"
12.	$^{165}Ho(p,n)^{165}Er$	11.20,	6.95	"
13.	$^{197}Au(p,n)^{197}Hg$	11.20		"
14.	$^{204}Pb(p,n)^{204}Bi$	11.20		"
15.	$^{206}Pb(p,n)^{206}Bi$	11.20		"
16.	$^{207}Pb(p,n)^{207}Bi$	11.20,	6.95	"
17.	$^{208}Pb(p,n)^{208}Bi$	11.20,	6.95	"
18.	$^{209}Bi(p,n)^{209}Po$	11.20,	6.95	"
19.	(n,xn) on Nb, Rh, In, Ho, Ta, Au and Bi	5 to 8		Argonne

non-equilibrium contribution calculations and tested it on most of the data produced under this CRP. They find that the model works very well for prediction of a wide variety of data without any parameter adjustment. Some of the experimental groups also fitted their own data with one of the statistical model codes and extracted information about level density parameters for respective residual nuclides.

The main objective of the CRP is a systematic, consistent analysis of all the data measured under the CRP to extract the level density information. Each of the three theoretical groups (Vienna, Bombay and Dresden) would carry out such analysis using their own techniques. This work is underway. The nuclides for which level density information can be extracted from the CRP data are listed in Table 3, which also contains the highest excitation energy to which this information can be obtained, the reactions investigated under the CRP to be utilized for this purpose and the available information on resolved levels for each of these nuclides.

Table 3. **The nuclides and excitation energies**
upto which LD information can be
extracted from the CRP data

Nuclide	Excitation Energy Range MeV	Reactions (Sr.No. in Table 2) from which this information can be obtained	No. of resolved levels known	
			No. of levels	Excit. Energy range (MeV) upto ...
^{51}Cr	0 – 7.7	2,3	43	3.5
^{59}Ni	0 – 10.5	1,4	13	2.0
^{94}Tc	0 – 7.0	6	26	1.5
^{95}Tc	0 – 9.0	5	22	1.5
^{96}Tc	0 – 8.0	7	21	0.5
^{97}Tc	0 – 10.5	8	16	1.0
^{98}Tc	0 – 9.0	9	24	0.5
^{100}Tc	0 – 10.5	10	–	–
^{165}Er	0 – 9.0	12	55	1.0
^{181}W	0 – 9.0	11	55	1.0
^{204}Bi	0 – 4.3	14	23	1.5
^{206}Bi	0 – 5.7	15	–	–
^{207}Bi	0 – 6.9	16	31	2.0
^{208}Bi	0 – 6.5	17	28	1.7
^{209}Po	0 – 7.4	18	–	–

Results and Conclusions

Although the results of the systematic consisent analysis of all the data measured under the CRP are not yet available, critical examination and assessment of all the work done under the CRP led the participants to a number of general conclusions which are summarised below:

1. Extensive new measurements have been carried out under the CRP spanning 18 nuclides in the mass ranges 45–65, 70–100, and 180–210 with bombarding energies ranging from 9 to 12 MeV for alpha particles, 5 to 25 MeV for protons and 5 – 8 MeV for neutrons as summarised in Table 2. The nuclides and excitation energy range for which the level density information can be extracted from these data are listed in Table 3.

The errors reported for all measurements were within limits required for the proposed level density analysis. It is also necessary to recognise that the errors in the neutron energy at high energy end of the spectrum are very critical for matching the extracted level density information with the experimental values based on discrete level information especially for those odd-odd nuclides where level density was already as high as 50 MeV. The instrumental error in timing for the time of flight technique were the major contributions to this error which was about 200 keV for most of the reported data. The participants were of the opinion that extending the work of determination of level densities based on discrete level counting to higher energies should be encouraged. Reported analysis of some of the individual sets of data brought out the importance of contributions from non-equilibrium processes in (p,n) reaction at bombarding energies higher than 8 MeV.

2. Sufficient data were measured under the CRP to carry out the work of extraction of level density information using consistent analytical procedures. However measurements on Pb isotopes at around 14 MeV and ^{197}Au at 9 MeV would add useful information towards extending the range of nuclear level density information for these nuclides.

3. For the purpose of extracting meaningful information, "nuclide specific" or "regional" optical model parameters would be preferable to the use of "global" parameters. It is also important that such analysis should explicitly report on the non-compound contributions, if any, and the way in which they have been accounted for.

The NDS will publish a final report containing the systematic analysis and the level density results extracted from all the data. According to a conclusion of the participants of the Second AGM on Level Densities organised by IAEA at Bologna, Italy in November 1989, the results will contribute to substantial improvement in our knowledge of the level density for a number of nuclei. The Second AGM also stressed the international collaborations and cooperation between theorists and experimentalists as two very significant aspects of this CRP.

References

1. H. Vonach – Proc. of IAEA Advisory Group Meeting on Basic and Applied Problems of Nuclear Level Densities, April 1989 at Brookhaven National Laboratory, USA BNL-NCS-51694 (1983) 247.

2. Measurement and Analysis of Double Differential Neutron Emission Spectra in (p,n) and (α,n) Reactions. Proc. of the Final Meeting of CRP organised by IAEA, November 1989 at Bologna, Italy (Compiled by N.P. Kocherov) INDC(NDS)-234 (1990).

UNCERTAINTY ESTIMATES FOR THE EFF-FILES FOR
^{52}Cr, ^{56}Fe, ^{58}Ni AND ^{60}Ni

S. Tagesen and H. Vonach

Institut für Radiumforschung und Kernphysik
Universität Wien, Boltzmanngasse 3, A - 1090 Wien, Austria

Work performed under contract Nr. 395-89-8/FU-D/NET,
available as report EFF-DOC-85.

Abstract: Uncertainty estimates and estimates of correlation between cross sections at different energies of the respective excitation functions were derived for the European Fusion File (EFF) for the main isotopes of the structural materials Cr, Fe and Ni. The estimates are based on a comparison of the EFF results with the corresponding results of other recently evaluated neutron cross section files, especially ENDF/B-VI, JENDL-3 and BROND, and a comparison with new high-quality measurements published after the literature cutoff of the EFF file. A test of the results has been possible by comparison to our experimental evaluation of 14 MeV cross sections also reported at this conference.

Uncertainty estimates (standard deviations) were derived for all important neutron cross sections, for the total gamma production cross section and for the reduced Legendre coefficients of the P_1, P_2 and P_3 components of elastic scattering angular distributions. In addition some very rough estimates are given for the uncertainty of the shapes of the secondary neutron and gamma emission spectra.

The information on cross section uncertainties and correlations was put into ENDF/B-VI format (file 33), the uncertainty estimates for the Legendre coefficients are given as file 34.

(^{52}Cr, ^{56}Fe, ^{58}Ni, ^{60}Ni, uncertainties, correlations)

Introduction.

The NET - team has established a data file, EFF (European Fusion File) [1], specifically assembled for application in neutronics calculations for fusion reactor design. The evaluations for the EFF files were done by nuclear model calculations with parameters adjusted to fit the existing experimental data base in the best possible way. This method, however, does not provide any information on the uncertainty of the calculated cross sections.

It is evident, that the reliability of shielding design studies depend, amongst other contributions, to a large extent on the reliability of the involved neutron nuclear reaction data, entering into the calculations. The EFF data therefore had to be amended with sufficiently reliable uncertainty and correlation information at least for the 4 main isotopes of the structural elements, Cr-52, Fe-56, Ni-58 and Ni-60.

Method for deduction of uncertainty and correlations:

In an evaluation of experimental data with related uncertainties this information is automatically generated, in principle by adequate propagation of uncertainties of single measurements, but this method is applicable only for those excitation functions, where a sufficiently extensive set of data is available. For an evaluation, based on theoretical calculations, a suitable method would be variation of the nuclear model parameters within theoretically justified limits and observing the resulting cross section variations. This is, however, not reasonably applicable either, because of computer time and manpower requirements. The authors therefore chose the method of comparison of different evaluations, suggested by Vonach 1987 [2].

For all four isotopes three evaluations have been performed recently in Japan (JENDL-3) [3], U.S. (ENDF/B-VI) [4] and U.S.S.R. (BROND) [5]. All these evaluations are based on about the same experimental data base and the same knowledge of nuclear reaction theory. Thus their results and the results in the EFF files represent the cross section predictions of four independent experienced experts based on our present experimental and theoretical knowledge. It can be expected that all important uncertainties in the evaluation process, such as different judgement of the quality of different experimental data sets, use of different parametrizations in the nuclear reaction models, uncertainties in the preferred input parameters e.g. level densities and use of slightly different approaches concerning the description of direct and preequilibrium processes will manifest themselves as differences between the four evaluations. Fig. 1 may serve to illustrate this situation.

COMPARISON OF CR-52 DATA
INELASTIC CROSS SECTION

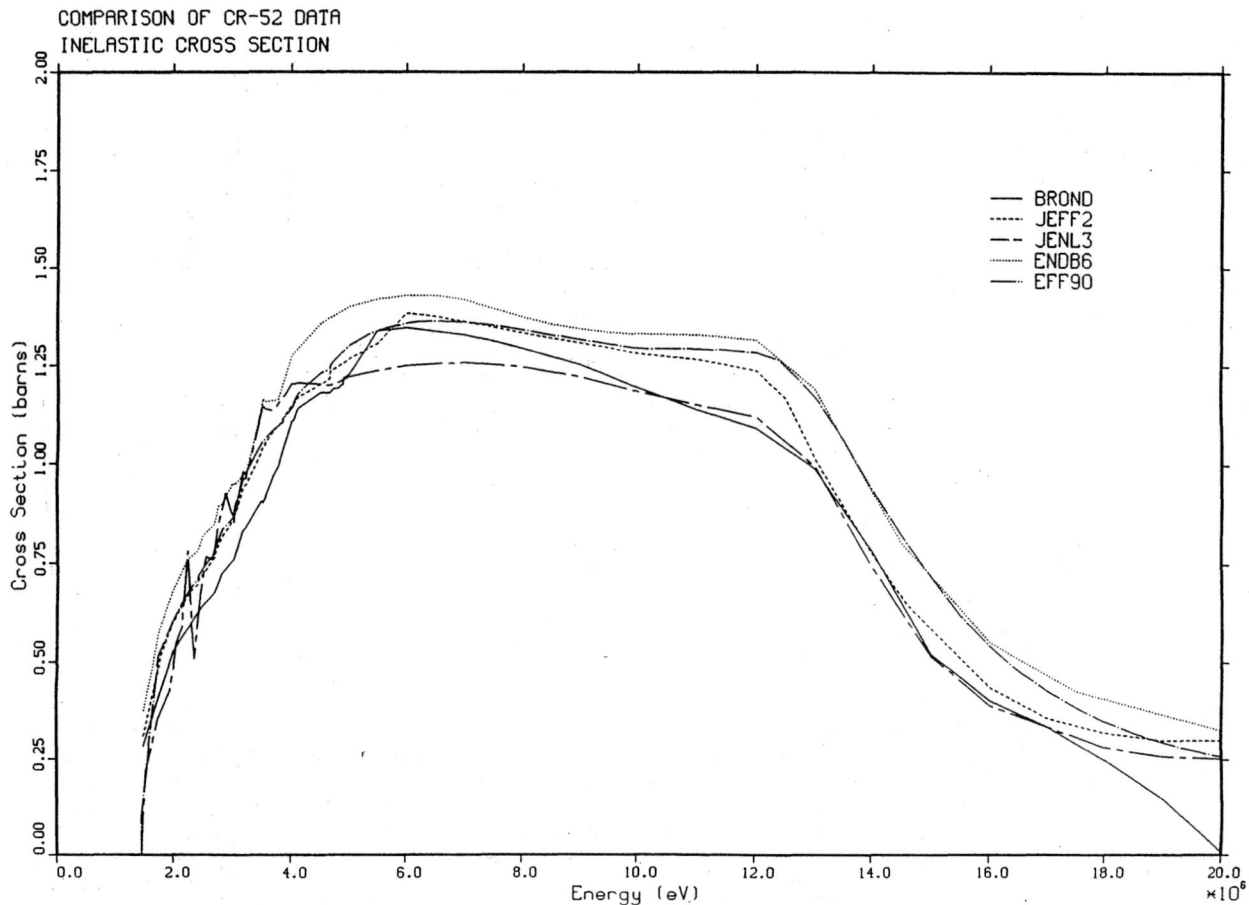

Fig. 1: Comparison of ^{52}Cr inelastic cross sections in various evaluated data files.

As we judge the different evaluations as equal in quality, we adopted as our basic working hypothesis, that the uncertainty of any EFF cross section at any energy is approximately given by the maximum difference to any of the results of the remaining three evaluations. In some cases where these uncertainties appeared too small because of accidental coincidence of all four evaluations we took the uncertainty level of the underlying (usually common single) database as uncertainty estimate for the evaluation.

As we define our uncertainties in the usual way as effective standard deviations, this is a quite conservative assumption, since we admit a chance of at least 35%, that the "true cross section value" may be found outside the "band" spanned by the four evaluations. In absence of more precise methods for the uncertainty estimates we decided however, that this safety margin was necessary.

For a relatively small part of the data high precision measurements have been performed in the last years. Those results were not yet available to the evaluators at the time of their decision. In these cases we used the differences of the EFF predictions to the measurements as our best estimate for the EFF uncertainties, which allowed to reduce these uncertainties considerably for a number of cross sections.

In order to actually implement the method plots were made comparing the four evaluations for each of the four isotopes and the rections characterized by the following ENDF/B "MT" numbers:

MT1(total), MT3(non), MT4(inel), MT16(n,2n), MT28(n,np), MT51(n,n1), MT102(n,γ), MT103(n,p) and MT107(n,α).

In a similar manner the uncertainties for the three lowest order coefficients of the Legendre representation of the elastic angular distributions as given in ENDF/B file 4 have been determined. Higher order components are usually not used in neutron transport calculations since their influence seems to be negligible.

By looking at fig. 1, we can also state, that there is certainly some correlation, but generally not extending over energy ranges of more than 5 MeV. After close inspection of a total of 45 sets of excitation functions we decided to describe the correlations between the cross section uncertainties by a gaussian distribution function with a half-width of 4 MeV.

The choice of the correlation function determined in turn to a large extent the energy grid for the representation of the covariance matrices. Adjacent energy bins of 1 MeV width are 84% correlated, thus intervals smaller than .5 MeV are certainly of no practical use. For future improvement by adding new information it appeared, however, favorable to use not too crude a structure. We therefore decided to use an energy grid of 28 bins, identical for all excitation functions, with .5 MeV bins from the resonance region up to 6 MeV, then 1 MeV groups up to 13 MeV, again .5 MeV groups for the important energy range up to 15 MeV and 1 MeV bins beyond.

Results of the uncertainty analysis:

Results of the outlined procedures have been put into ENDF/B-VI format and added to the EFF file, using MF33 for the cross section uncertainties and correlations and MF34 for the respective elastic scattering angular distribution data. Fig. 2 displays relative uncertainties of some reactions for ^{52}Cr.

A rough estimate of the uncertainties of secondary neutron energy distributions has been calculated in terms of "spectral shape uncertainty parameters [6], defined as relative uncertainty of the area below or above the medium of the respective energy distribution.

Uncertainties of total gamma ray production cross sections have been estimated according to our general procedures in a structure of 7 incident neutron energy groups only.

Uncertainties of gamma energy distributions have been estimated by means of the uncertainty of a "spectral index" giving a fraction of photons emitted with energies above 2 MeV ("hardness" of the spectrum).

References:

1   C. Nordborg and H. Gruppelaar, IP11 on JEF/EFF data files, this conference.

2   H. Vonach, Proc. Advisory Group Meeting, Beijing, Oct. 12-16, 1987, IAEA TECDOC-483, 37, IAEA Vienna (1988).

3   K. Shibata et al., Report JAERI 1319, Japan Atomic Energy Research Institut, Tokai-mura (1990).

4   H.D. Lemmel, Report IAEA-NDS-100, Rev.3, IAEA, Vienna (1990).

5   V.N. Manokhin ed., Report INDC (CCP)-283, IAEA, Vienna (1988).

6   S.A.W. Gerstl, Report ORNL/RSIC-42, p 219, Oak Ridge (1979).

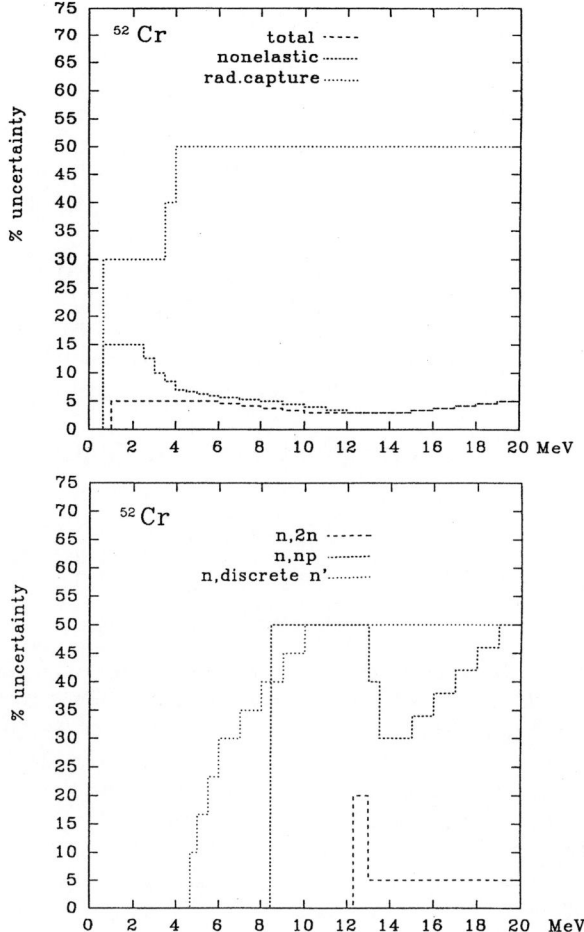

Fig. 2: Relative uncertainties of various excitation functions for ^{52}Cr.

SCATTERING LAWS FOR MODERATORS, REFLECTORS AND FILTERS FOR APPLICATION IN DESIGN CALCULATIONS OF COLD AND SUPERTHERMAL NEUTRON SOURCES [1]

W. Bernnat, J. Keinert, M. Mattes

Institut für Kernenergetik und Energiesysteme (IKE),
Universität Stuttgart, Pfaffenwaldring 31, D-7000 Stuttgart 80

Abstract: Scattering laws $S(\alpha,\beta)$ were derived in ENDF-6 format for the most important materials as liquid H_2, D_2, ^4He and solid Be, Bi used in cold and superthermal neutron sources by considering the incoherent as well as the coherent scattering dynamics at working temperature. For cold and superthermal neutron source design calculations including the determination of heat production coupled n-γ cross-section matrices and Kerma factors were generated.

(scattering law, total neutron cross sections, cold neutrons, neutron filters, cold moderators)

Introduction

For the basic research with cold and ultracold neutrons cold (moderators : liquid H_2 at 20.38K or liquid D_2 at 23.65K) and superthermal (H_2 source as input, Bi-filter at 77K and ^4He at 1K as moderator) neutron sources are in use. The optimisation of these sources is an important task for an efficient generation of cold neutrons and minimising the radiation heat production in the cold moderator zone. For a realistic design of such sources adequate cross-section data are needed for the different materials at working temperature considering the incoherent as well as the coherent neutron scattering, especially at the energy range of cold neutrons.

Improved cross-section data for important materials

For the moderators and neutron filters scattering law data $S(\alpha,\beta)$ were generated in ENDF-6 format in an adequate α and β grid to minimise interpolation errors. To validate the cross-section libraries, differential and integral cross-sections are calculated and compared with measurements.

- Liquid hydrogen at working temperature

  For liquid H_2 at 20.38K and liquid D_2 at 23.65K we have derived scattering law data which consider the incoherent scattering as well as both the intramolecular and intermolecular interference scattering [1]. The quality of calculated integral cross-sections may be demonstrated in Fig. 1.

- Liquid ^4He at T=1K (ultracold neutrons)

  The neutron scattering dynamics in the supra-fluid helium is pronounced by the purely coherent scattering. The scattering law data are derived by first

assuming diffusive motion or free translations in incoherent approximation for the atomic scattering and then connecting the interatomic interference scattering by a realistic static structure factor $S(\kappa)$ (see Fig. 2) according to the Sköld-approximation [2]:

$$S_{coh}(\alpha,\beta) = S_{inc}(\alpha',\beta) * S(\kappa)$$

with

$$\alpha' = \alpha / S(\kappa)$$

As is seen from Fig. 3, the commonly used Vineyard-approximation for calculating the coherent scattering is not adequate for liquid He-4.

- Bismuth (neutron filter at 77K)

  For Bi the incoherent neutron scattering is generated with a realistic frequency distribution derived from Kress [3,7]. The coherent elastic scattering is formulated by a transformation of the rhomboedric crystal lattice structure into an equivalent hexagonal one and deriving the corresponding form function. The coherent cross-section then may be calculated by implementing the formulas in HEXSCAT or NJOY. The quality of the derived total neutron cross-section is demonstrated in Fig. 4.

- Beryllium (He-tank at T= 2K to 25K)

  The cold Be serves as a total reflexion medium for the ultracold neutrons. For the design of the superthermal neutron source low temperature neutron cross-sections must be supported. For the hexagonal close packed Be the approved phonon

[1]funded by BMFT, Project 'Erforschung kondensierter Materie und Atomphysik'

spectrum of Koppel [4] is used for the incoherent scattering which is then completed by the coherent scattering with HEXSCAT or NJOY. In Fig.5 the total neutron cross-section is represented.

- D_2O (reflector material in research reactor)

For D bound in D_2O the incoherent neutron scattering is generated with a modified version of the IKE phonon spectrum [5] which essentially is a correction of the formerly free translational mode. In addition, now the intramolecular interference scattering is considered and the intermolecular interaction is calculated by using a realistic static structure factor for the motion of the molecular centre according to the Sköld-approximation. The derived total cross-section at room temperature is represented in Fig. 6.

- Structure materials (aluminium and lead)

To take care of a realistic neutron transmission, the commonly calculated scattering cross-sections for Al or Pb (free gas approximation) are corrected. We plan to consider the real crystal lattice structure for these fcc-materials similar to Be or Bi.

Neutron transport calculations and radiation heat production

For an advanced design of cold and especially superthermal neutron sources adequate radiation transport methods for the calculation of neutron flux spectra and heat production must be used. Especially for the helium zone the calculation must be done very careful because from technical aspects the heat production may not increase 200 mW [6]. Therefore transport calculations with coupled neutron/gamma cross-section matrices with S_N (ANISN) and Monte Carlo codes (MORSE, MCNP) for the determination of neutron flux spectra in the different zones and heat production due to neutron and gamma interaction with moderators and structure materials have been carried out for example for the planned research reactor FRM-II in Munich (Fig. 7). Results of the calculations are shown in Fig. 8.

References

[1] Bernnat, W.; Emendörfer, D.; Keinert, J.; Mattes, M.: Flüssigkeitsmodelle für die Neutronenstreuung an Wasserstoff und Deuterium unter Berücksichtigung der innermolekularen und zwischenmolekularen Interferenz. Stuttgart: IKE, 1989 (IKE 6-175)

[2] Methods of Experimental Physics, Vol. 23, Neutron Scattering Part A, 45 (1986). Academic Press, Inc.

[3] Kress, W.: Phonon Dispersion Curves, One-Phonon Densities of States and Impurity Vibrations of Metallic Systems. Physik Daten 26-1 (1987)

[4] Koppel, J.U.; Houston, D.H.: Reference Manual for ENDF Thermal Neutron Scattering Data. GA-8774 (ENDF-269) (1968)

[5] Keinert, J.: Re-evaluation of the Neutron Scattering Dynamics in Heavy Water, Generation of Multigroup Cross Sections for THERM-126. Stuttgart: IKE, 1982 (IKE 6-138)

[6] Golub, R.: UCN at the FRM II (1988), private communication

[7] Käfer, S.; Keinert, J.; Mattes, M.: Berechnung von inkohärenten und kohärenten Neutronen-Wirkungsquerschnitten in kristallinem Wismut bei verschiedenen Temperaturen. IKE 6-181 (1991). Report in preparation

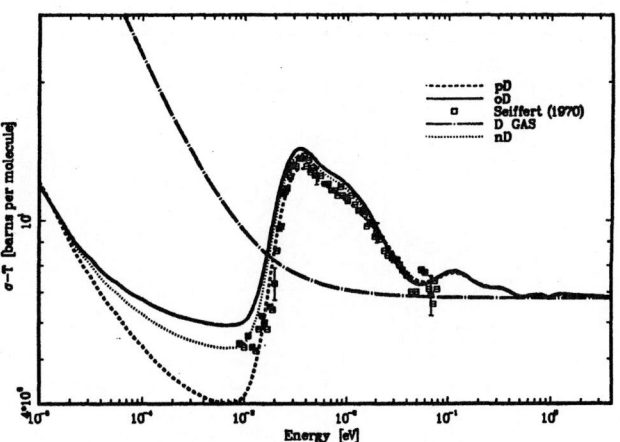

Fig. 1: Total neutron cross-sections of deuterium at T = 19 K

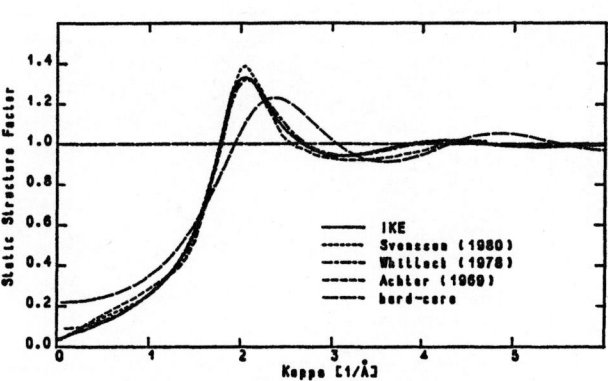

Fig. 2: Static structure factors for liquid ^{4}He at T = 1 K

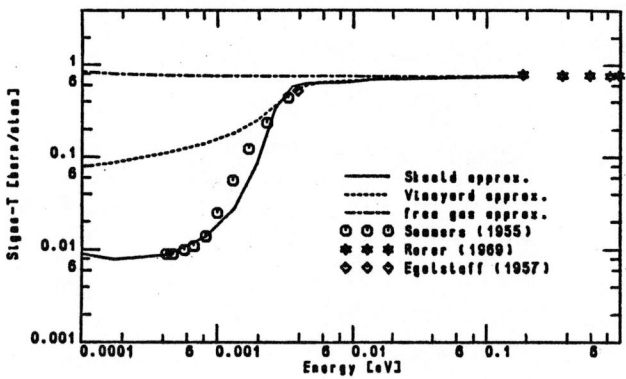

Fig. 3: Total neutron cross-section of liquid ^{4}He around T = 1 K

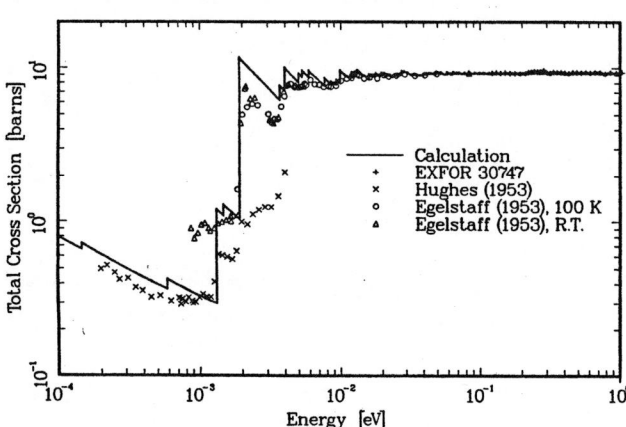

Fig. 4: Total neutron cross-section of bismuth at T = 77 K

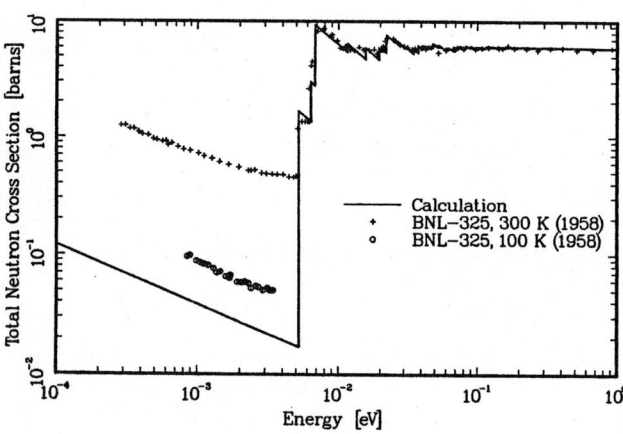

Fig. 5: Total neutron cross-section of beryllium at T = 2.5 K

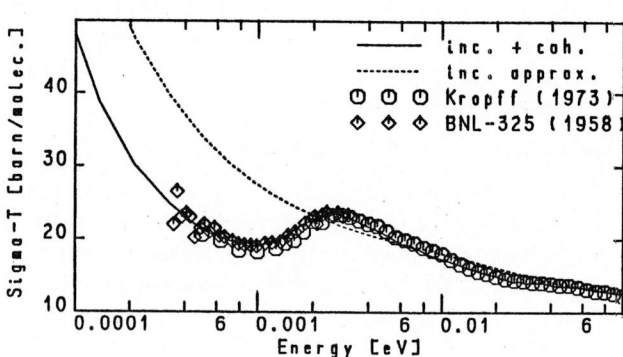

Fig. 6: Total neutron cross-section of D$_2$O at room temperature

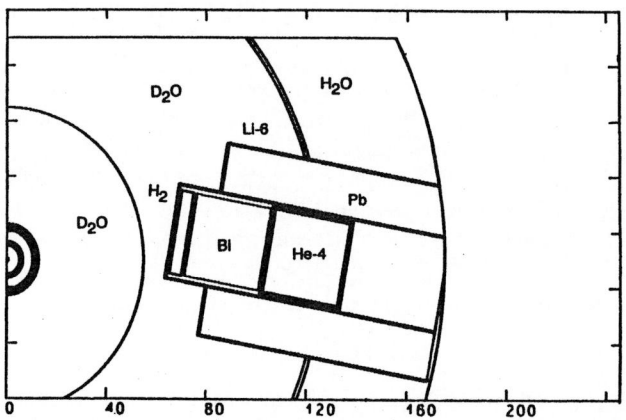

Fig. 7: Geometrical configuration of a superthermal neutron source

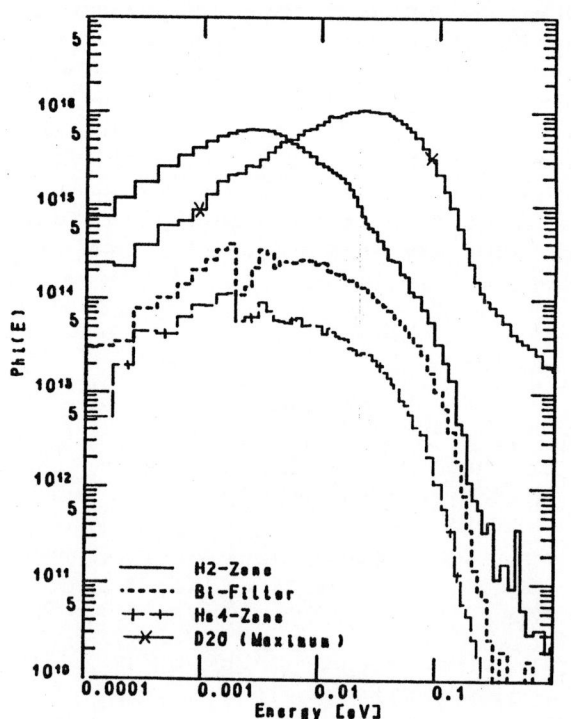

Fig. 8: Flux spectra in a superthermal neutron source

ANALYSIS OF POSSIBLE SOURCES OF UNCERTAINTIES
IN NEUTRON DATA CALCULATIONS

V.A. Konshin and O.V. Konshin*

International Atomic Energy Agency, Vienna, Austria
* Byelorussian State University, Minsk, USSR

Abstract. Possible sources of uncertainties in calculations of neutron cross sections and spectra of secondary particles, in particular OM parameters, level density models and model parameters, were analysed for iron and ^{93}Nb taken as examples in the energy range up to 90 MeV. It was found that the parametrization by Put and Paans with the respective modification of the parameter r_v and taking into account the energy dependence a(u), allows a better description of experimental data for (n,α)-cross sections and alpha spectra than other parametrizations. The importance of using correct neutron and proton OM parametrizations is emphasized and analysed for different neutron reaction cross sections.

(neutron cross section, optical model parameters, computer codes, level density model, alpha-particle emission spectrum, proton spectrum)

Introduction

With the increasing significance of reliable theoretical prediction of neutron data for different applications, the demand for accurate computer codes and correct parametrizations for different reaction channels has been increasing in recent years. Development of theoretical methods led to the fact that neutron and proton cross sections can be predicted more accurately than α-particle cross sections. The parametrizations available for the alpha-channel [1–5] do not provide an adequate description of experimental data without the essential changing of parameters often individually for each nucleus, so the (n,α)-cross sections cannot be adequately predicted. The search for respective OM parameters for the alpha-channel is therefore an important task. The most known existing codes GNASH [6], ALICE [7], TNG [8] and STAPRE-H [9, 10] were used in the present work for intercomparison purposes to investigate the difference in model approaches used and then the code chosen (GNASH) was used for calculation of neutron cross sections with emission of α, p and n in the energy region up to 90 MeV.

Results of calculations and discussion

The (n,α)-cross section calculations with OM parameters for α-particles from [1, 3, 4] lead to too low values (Fig. 1, GNASH calculations — curves 7, 6, 10, ALICE — curve 11, TNG — curve 12). Wilmore et al. [11] OM parameters for neutrons and level density parameters by Gilbert et al. [12] were used in these calculations. The curve 10 represents the results obtained using alpha OM parameters of Put et al. [5] and Nolte et al. [4] for elastic scattering of α-particles from ^{90}Zr at energies between 40 and 120 MeV:

$$V = 101,1 + 6,051 \, (Z/A^{1/3}) - 0,248E,$$
$$r_v = 1,245fm, \ a_v = 0,817 - 0,0085A^{1/3},$$
$$W = 26,82 - 1,706A^{1/3} + 0,006E, \quad (1)$$
$$r_w = 1,570fm, \ a_w = 0,692 - 0,020A^{1/3}.$$

The (n,α)-cross section can be increased by increasing r_v (1.57) but it is not sufficient (curve 5). An additional large increase of V to the value 185-0,243E does not solve the problem (curve 3).

It appeared that neutron OM parameters used for (n,α) calculations can be very essential. As it was shown in [13], a deep minimum in the total cross sections at energies below 3 MeV as well as level excitation functions can be well reproduced by using OM parametrization by Rapaport et al. [14] with the following modification of the radius of the real part of the potential:

$$r_v(E) = 1,315 - 0,0167E \quad (2)$$

The use of this neutron OM parameters with $r_v(E)$ (2) and alpha OM parameters (1) with modified r_v, see Eq. (3) (for ^{56}Fe $r_v = r_w = 1,570fm$),

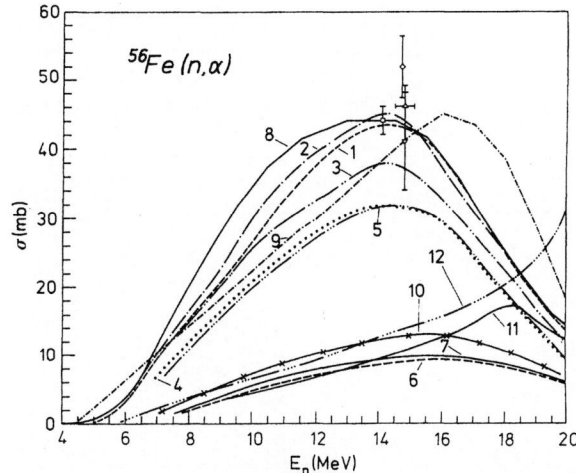

FIG. 1. ^{56}Fe (n,α)-cross section calculations. Curves: 6 — GNASH calculations, OMα-[3], OMn-[11], LDM-[12]; 7 — same as 6, but OMα-[1]; 10 — same as 6, but OMα-[4]; 11 — ALICE calculations; 12 — TNG with OMn-[11], OMα-[1], LDM-[12]; 5 — GNASH, OMα-[4] with $r_v = r_w = 1,57fm$, OMn-[11], LDM-[12]; 4 — same as 5, but LDM-a(u) [15]; 3 — OMα-[4] with $V = 185-0,248E$, $r_v = r_w = 1,57fm$, $a_v = a_w = 0,785f$; 1 — OMα from [4] with r_v modified Eq. (3), OMn-[14] with modified r(E) [13] Eq.(2), LDM-[12]; 2 — same as 1, but LDM-a(u) from [15]; 8 — ENDF/B-VI; 9 — BROND Library.

allow to describe satisfactorily the (n,α)-cross section (curve 1). The use of $r_v(E)$ dependence increases σ_{na} by 40% compared to OM [11]. The energy dependence of the level density parameter a(u) in the form proposed by Ignatyuk et al. [15] slightly increases the (n,α)-cross section (by 5%) (curve 2).

The use of such parameters allows to describe better the experimental data for ^{56}Fe(α,n) at 9, 11 and 13 MeV [16], as well as the α-spectrum at 14,6 MeV (Fig. 2.) Curves 3, 4 and 5 represent the results of calculations by the codes GNASH, ALICE and TNG, respectively, using for GNASH and TNG neutron OM — [11], α OM — [1], LDM — [12]. The difference between curves 6 and 1 is due to the neutron OM used — [11] or [14] with $r_v(E)$. A very weak impact of a(u) on the α-spectrum is seen.

Much stronger impact of a(u) can be seen for binary reaction cross sections, such as (n,2α), (n,pα), (n,2nα) (Figs 3 and 4) at higher energies. At energies higher than 30 MeV neutron OM by Schwandt et al. [17] was used as it was recommended in [18]. The use of a(u) leads to a sharp decrease of the (n,2α) and (n,pα) cross sections (curves 2 and 3, 5 and 6, respectively, in Fig. 3). ALICE calculations underevaluate (n,α) and (n,2α)-cross sections due to the method used for OM cross section calculations. The advantages of the ALICE code could be increased if the use of more correct OM could be made.

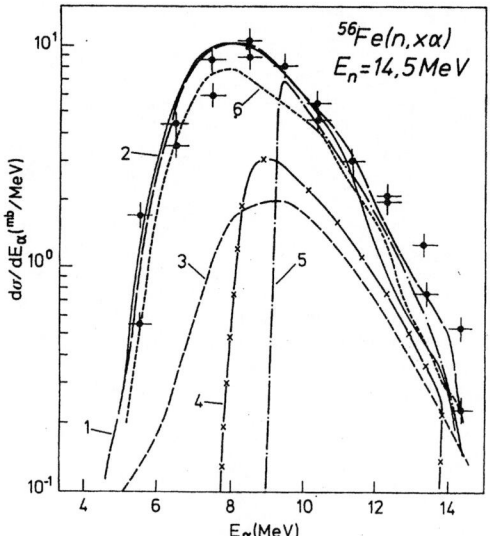

FIG. 2. ^{56}Fe alpha emission spectra at 14,5 MeV. Curves: 1 — GNASH calculations, OMn-[14] with r(E) [13], OMα — Eq. (1) with r_v from Eq. (3), LDM-a(u) from [15]; 2 — same as 1, but LDM-[12]; 6 — same as 1, but OMn-[11]; 3 — OMn-[11], OMα-[1], LDM-[12]; 4 — ALICE calculations; 5 — TNG calculations, with the same parameters as 3.

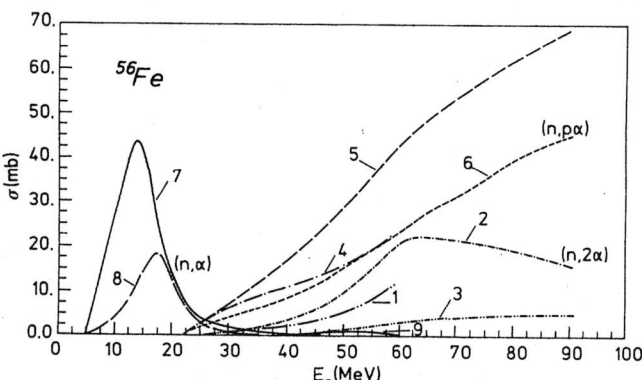

FIG. 3. ^{56}Fe (n,α), (n,pα) and (n,2α)-cross sections. (n,2α)-cross section curves: 1 — OMα-Eq. (1) with r_v from Eq. (3), OMn-[14] with modified r(E) [13] up to 30 MeV, above-[11], LDM-[12]; 2 — same as 1, but above 30 MeV OMn — Schwandt [17]; 3 — same as 2, but LDM-a(u) [15]; (n,pα)-cross section curves: 4, 5, 6 — same as 1, 2, 3; (n,α)-cross section curves: 7 — same as 3; 8 — ALICE calculations; 9 — ALICE results for (n,2α).

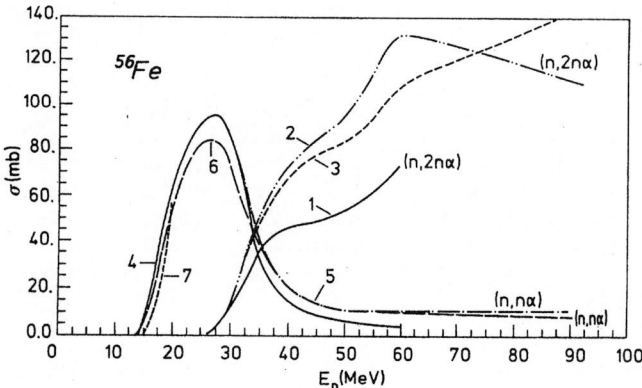

FIG. 4. ^{56}Fe (n,nα) and (n,2nα)-cross sections. (n,2nα)-cross section curves: 1 — OMα — Eq. (1) with r_v from Eq. (3), OMn-[14] with modified r(E) [13] up to 30 MeV, and above [11], LDM-[12]; 2—same as 1, but above 30 MeV OMn and OMp — Schwandt [17]; 3 — same as 2, but LDM-a(u) [15]; (n,nα) curves: 4, 5, 6 — same as 1, 2, 3 respectively; (n,nα) means (n,nα)+(n,αn)-reactions; 7 — BROND Library.

The calculations of the cross sections of the **proton channel** showed that the (n,p)-cross section for ^{56}Fe can be well reproduced using neutron OM [11] and proton OM [19]; a(u) has no impact on σ_{np}. ALICE underestimates σ_{np} in the peak (~ 70 mb) giving a correct shape of the curve, TNG strongly overestimates σ_{np} in the region above 16 MeV. STAPRE-H calculations showed that the use of LDM by Schmidt et al. [20] leads to a very low σ_{np} (36,1 mb) at 14,1 MeV and therefore its application at low energies looks unreasonable. Due to the fact that in STAPRE-H a rather artificial region from 12 to 50 MeV for joining LDM was used [10] and because of difficulties to use input data others than default, STAPRE-H was not extensively used.

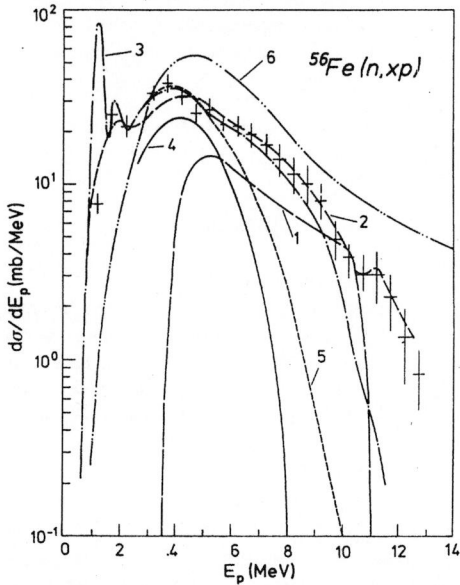

FIG. 5. ^{56}Fe proton emission spectra. Curves: 2 — GNASH calculations at $E_n = 14,8$ MeV, OMn — Wilmore [11], OMp — Perey [19], LDM — Gilbert [12]; 6 — same as 2, at 20 MeV; 3 — TNG calculations, with the same parameters as 2, at 14,8 MeV; 4 — same as 3, at 18 MeV; 5 — same as 3, at 20 MeV; 1 — ALICE calculations at 14,8 MeV.

A comparison of calculated proton emission spectra with experimental data (Fig. 5) for ^{56}Fe shows that a good agreement is observed for GNASH calculations using neutron OM [11], proton OM [19] and LDM — [12] (curve 2 at 14,8 MeV and 6 at 20 MeV). ALICE calculations underestimate the equilibrium component of the spectrum (curve 1) and hence the application of this code at low energies may be questionable. TNG describes quite well the experimental data at 14,8 MeV (curve 3), but with the incoming neutron energy increase the results become inconsistent (curves 4 and 5) — proton spectra are softer at 18 and 20 MeV than at 14,8 MeV, which can be attributed to the version or parameters used of the preequilibrium model.

A good agreement with experimental data for the (n,2n) cross sections for ^{56}Fe can be obtained by using GNASH and neutron OM [11], or OM [14] with r_v(E), and LDM [12]. Both neutron OM potentials give similar results (the difference is 5%), although in the region below 20 MeV OM [14] with r_v(E) describes better the experimental data than [11]. The use of a(u) leads to the non-significant increasing of (n,2n) at the peak ($\sim 5\%$) (Fig. 6, curves 1 and 2) and to the essential increasing of (n,3n) ($\sim 30\%$). ALICE calculations are satisfactory (curves 5 and 6) and TNG results are too high above 18 MeV. Any experimental data for (n,3n) would be desirable for LDM testing.

The (n,α) cross section calculations made for ^{54,56,57,58}Fe, ^{50,52,53}Cr, ^{62,63}Ni, ^{63}Cu, ^{48}Ti, ^{93}Nb, as well as (p,α) and (p,αn)-cross sections for $^{54-58}$Fe, ^{93}Nb, ^{90}Zr showed that a satisfactory agreement with experimental data can be achieved using alpha OM parameters (1) with

$$r_v = 2,611 - 2,45 \cdot 10^{-2}A + 1,055 \cdot 10^{-4}A^2 \qquad (3)$$

valid for $A \leq 100$ and $E_\alpha \leq 100$ MeV.

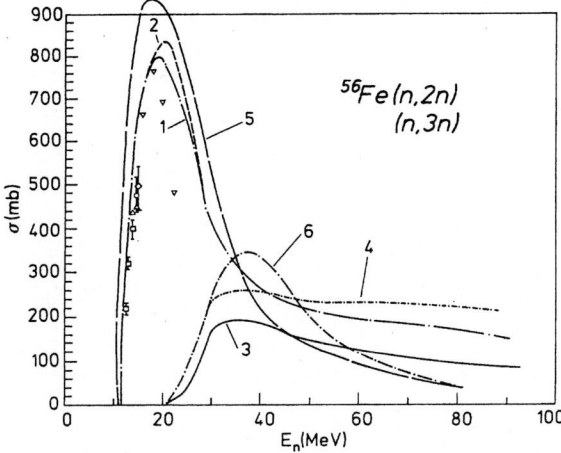

FIG. 6. ^{56}Fe (n,2n) and (n,3n)-cross sections. (n,2n) curves: GNASH, 1 — OMn-[14] with modified r(E) [13] up to 30 MeV, and above [17], OMp-[17], OMα — Eq. (1) with r_v from Eq. (3), LDM-[12]; 2 — same as 1, but LDM-a(u) [15]; 5 — ALICE calculations; (n,3n) curves: GNASH, 3 — same as 1; 4 — same as 2; 6 — ALICE calculations.

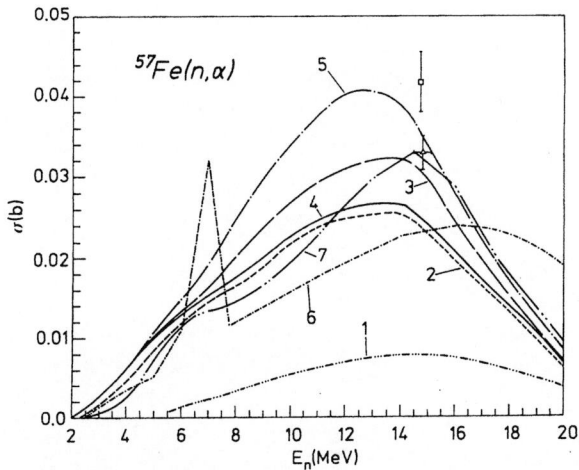

FIG. 7. ^{57}Fe (n,a)-cross section calculations. Curves: 1 — OMα-[1], OMn-[11], LDM-[12]; 2 — OMα Eq. (1) with r_v from Eq. (3), OMn-[11], LDM-[12]; 4 — same as 2, but $a_v = a_w = 0,785$ fm; 3 — OMα Eq. (1) with r_v from Eq. (3), OMn-[14] with modified r(E), LDM-[12]; 5 — same as 3, but LDM-a(u); 6 — BROND Library; 7 — ENDF/B-VI.

Neutron OM parameters were those of [14] with modified r_v(E) and a(u) should be used. Figures 7 and 8 illustrate such an approach for (n,α) and (p,αn) cross section calculations. There might still be a need for the alpha OM parameter fitting for individual nuclei, as the occasional failure of theory for some nuclei is possible.

The calculations made for ^{93}Nb showed that good agreement with experimental data on (n,α) and α-spectra can be achieved by using alpha OM of (1) with r_v from (3), neutron OM of [14] with modified r_v(E) and LDM — a(u). The difference in using a(u) and LDM [12] is not very essential for the alpha channel (Fig. 9) (about 20% in the peak). The use of the energy dependence of a is very essential for (n,p) and (n,np) cross sections (Fig. 10). The difference is a factor of about two at the peaks (14 and 25 MeV). Good agreement with experimental data for (n,p) is observed when using a(u) — curve 1 (Fig. 10). For the ^{93}Nb proton emission spectrum, contrary to ^{56}Fe, using a(u) leads to decreasing calculational results by a factor of three at the peak of the spectrum (from 23 to 7 mb/MeV) and to good agreement with experimental data.

For the ^{93}Nb neutron channel the use of neutron OM [14] with r_v(E) allows to describe the experimental data [22] for (n,2n) in the low energy region (up to 15 MeV), and the use of [11], as was correctly shown in [23], gives the results which are higher than [22].

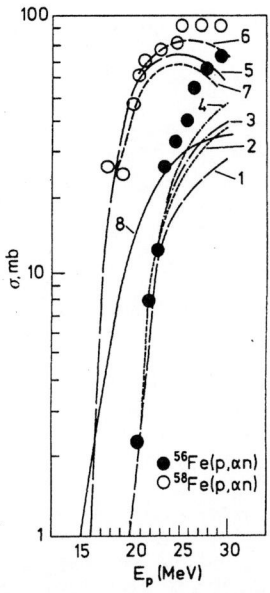

FIG. 8. ^{56}Fe and ^{58}Fe (p,αn), (p,αp)-cross section calculations. Experimental data for (p,αn) are taken from [24]. For ^{56}Fe (p,αn) curves: 1 — OMα Eq. (1) with r_v from Eq. (3), OMn-[14] with modified r(E), OMp-[19], LDM-[12]; 2 — same as 1, but above 14 MeV for OMp and above 20 MeV for OMn-[17]; 4 — same as 2, but LDM-a(u); 3 — same as 2, but OMn-[11] up to 20 MeV; 8 — (p,αp), same as 2; ^{58}Fe (p,αn) curves: 5 — same as 2; 7 — same as 2, but OMn-[11]; 6 — same as 2, but LDM-a(u).

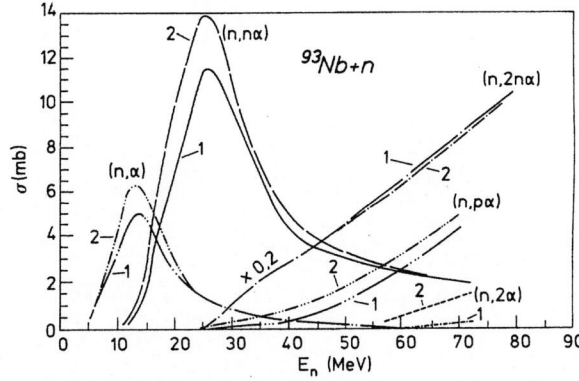

FIG. 9. ^{93}Nb alpha channel reaction cross sections. Curves: 1 — OMn-[11] up to 20 MeV and above-[17], OMp-[19] up to 14 MeV and above-[17], OMα — Eq. (1) with r_v from Eq. (3), LDM-a(u) [15]; 2 — same as 1, but LDM-[12].

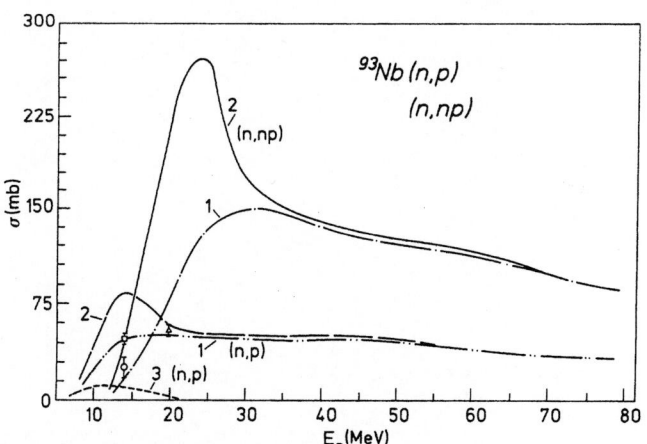

FIG. 10. ^{93}Nb (n,p) and (n,np)-cross section calculations. Curves: 1 — OMn-[11] up to 20 MeV and above-[17], OMp-[19] up to 14 MeV and above-[17], OMα-Eq. (1) with r_v from Eq. (3), LDM-a(u) [15]; 2 — same as 1, but LDM — [12]; (n,np) means (n,np)+(n,pn)-reactions; 3 — calculations [21].

REFERENCES

[1] McFADDEN, L., SATCHLER, G.R., Nucl. Phys. **84** 177 (1966).

[2] HUIZENGA, J.R., IGO, G., Nucl. Phys. **29** 462 (1962).

[3] LEMOS, O.F., Diffusion Elastique de Particules Alpha de 21 à 29.6 MeV sur des Noyaux de la Région Ti-Zn, Orsay report, Series A, No.136 (1972).

[4] NOLTE, M., MACHNER, H., BOJOWALD, J., Phys. Rev. C, **36** 1312 (1987).

[5] PUT, L.W., PAANS, A.M.J., Nucl. Phys. **A291**, 93 (1977).

[6] YOUNG, P.G., ARTHUR, E.D., A Preequilibrium, Statistical Nuclear Model Code for Calculation of Cross Sections and Emission Spectra, LA-6947, Los Alamos Nat. Lab. (1977).

[7] BLANN, M., Code ALICE/85/300, VCID-20169, Livermore Nat. Lab. (1984).

[8] FU, C.Y., A Consistent Nuclear Model for Compound and Precompound Reactions with Conservation of Angular Momentum, ORNL/TM-7042 (1980).

[9] UHL, M., STROHMAIER, B., Program Manual for STAPRE code, IRK-76/01 (1976).

[10] AVRIGEANU, M., IVASCU, M., AVRIGEANU, V., Z. Phys., A **329** 177 (1988).

[11] WILMORE, D., HODGSON, P.E., Nucl. Phys. **55** 673 (1964).

[12] GILBERT, A., CAMERON, A.G.W., Can. J. Phys. **43** 1466 (1965).

[13] KONSHIN, O.V., in Proc. of Advisory Group Meeting on Nuclear Theory for Fast Neutron Nuclear Data Evaluation, Oct. 1987, Beijing, China, p. 85, IAEA, Vienna (1988).

[14] RAPAPORT, J., KULKARNI, V., FINLAY, R.W., Nucl. Phys. **A330**, 15 (1979).

[15] IGNATYUK, A.V., SMIRENKIN, G.N., TISHIN, A.S., Sov. J. Nucl. Phys. **21** 255 (1975).

[16] GRIMES, S.M., PEDRONI, R.S., BRIENT, C.E., MISHRA, V., BOUKAROUBA, N., in Proc. on Measurements and Analysis of Double Differential Neutron Spectra in (p,n) and (α,n)-reactions, Nov. 1989, Bologna, Italy, p. 7, IAEA, Vienna (1990).

[17] SCHWANDT, P., MEYER, H.O., JACOBS, W.W., BACHER, A.D., VIGDOR, S.E., KAITCHUK, M.D., Phys. Rev. **26** 55 (1982).

[18] MADLAND, D.G., in Proc. Advisory Group Meeting on Nuclear Theory for Fast Neutron Nuclear Data Evaluation, Oct. 1987, Beijing, China, p. 80, IAEA, Vienna (1988).

[19] PEREY, F.G., Phys. Rev. **131** 745 (1962).

[20] K.H. SCHMIDT, DELAGRANGE, H., DUFOUR, J.P., CARJAN, N., FLEURY, A., Z. Phys. A — Atoms and Nuclei **308** 215 (1982).

[21] HODGSON, P.E., in Proc. Intern. Conf. on Neutron Physics, Sept. 1987, Kiev, USSR, p. 76 (1988).

[22] FREHAUT, J., BERIN, A., BOIS, R., JARY, J., in Proc. Symp. Neutron Cross Sections from 10 to 50 MeV, 1980, Upton, New York, V. 1, p. 399, BNL (1980).

[23] STROHMAIER, B., Ann. Nucl. Energy **16** 461 (1989).

[24] LEVKOVSKIJ, V.N., REUTOV, V.F., BOTVIN, K.V., At. Ehnerg. **69** 180 (1990).

FAST NEUTRON NUCLEAR DATA : Pu-239 REVISION AND Am STATUS

A.B. Klepatskij and V.M. Maslov

Institute of Nuclear Engineering, Minsk, USSR

Abstract: Neutron cross sections for Pu-239 and Am-241,-242m,-243 have been analyzed with the aid of theoretical models. A deformed optical potential that fits total, elastic and inelastic differential cross section data and neutron strength functions for Pu-239 and Am-241 have been used. In case of Pu-239 the consistency of absolute fission data and (n,2n) cross section is investigated. Because of the strong discrepancies in Am fission cross section data a consistent set of calculated cross section values for the chain of Am nuclei is proposed. The present state of knowledge concerning first chance fission cross section allows to analyze fission cross section data of Am-241,-242m,-243, up to 20 MeV. The results thus obtained are compared with ENDF/B-V and JENDL-3 libraries.

(Pu-239, Am-241,-242m,-243, neutron data evaluation, fission cross sections, deformed optical potential, (n,2n) reactions,(n,3n) reactions)

Introduction

Recent measurements of Pu-239 absolute fission cross section /1/ show the need to revise current evaluated data files even for that thoroughly studied nucleus. The analysis of current JENDL-3 and ENDF/B-V files shows rather large discrepancies in (n,f) and (n,xn) cross sections.Due to inherent experimental difficulties of measuring neutron cross sections for active targets the role of theoretical analysis increases. That is a way to improve the quality of evaluated data files for Am nuclei as that achieved for U-235 and U-238. In case of Am-241,-242m, -243 it is especially important because various fission data sets differ greatly.

Theoretical Analysis of Neutron Data

In the fast neutron region traditional analysis tools are coupled-channel optical method and Hauser-Feshbach statistical theory. The coupled-channel calculations were performed with the computer code COUPLE /2/.In the case of n+Pu-239 reaction first six levels of ground state band were coupled. The coupling potential was the same for all channels, coupling form factors used to expand optical parameters were assumed complex.Potential parameters were obtained by fitting s- and p-wave strength functions, total cross section /3/ and angular distributions of inelastically scattered neutrons /4/. The potential parameters are given in Table 1. Calculated angular distributions at 14 MeV are compared to data /5/ in Fig. 1. Since the experimental resolution is ≈250 keV, the measured distribution is compared with elastic distribution and with the successive sum of differential cross sections up to 9/2+ level.

In the case of n+Am-241 reaction the optical potential given in Table 1 fits strength functions and total cross section data /6/ quite well when β_2 =0.21 and β_4 =0.065 with 3 ground state band levels coupled. However, the calculated curve

lies higher than data /6/ for E > 12 MeV and E=1-4 MeV, but the authors /6/ claim there is a systematic error ≈0.5 b in the total data.

Hauser-Feshbach statistical theory calculations have been performed with a modified version of STAPRE code /7/. Neutron transmission coefficients were obtained with the potential, given in Table 1. Transmission coefficients for competing decay channels were calculated with the level density model, taking into account shell, pairing and collective effects /8/. In an adiabatic approximation level density is defined as

$$\rho(U,J,\pi)=Krot(U,J)Kvib(U) \rho_{qp}(U,J,\pi).$$

Here, $\rho_{qp}(U,J,\pi)$ is the quasi-particle level density, Krot and Kvib - factors of rotational and vibrational enchancement. Shell effects were modelled as in /8/. At low excitation energies $\rho(U,J,\pi)$ was approximated by temperature dependence /9/ as in Gilbert-Cameron approach. Fission transmission coefficients were calculated within a simple version of double-humped barrier. Asymptotic value of level density parameter, shell corrections, barrier heights and correlation function for fissioning nuclei were considered model parameters. In case of Pu and Am nuclei of interest the inner saddle A possesses axial asymmetry, while the outer saddle B is mass-asymmetric. That is, $Krot^A =2\sqrt{2\pi}\sigma_{\perp}^2\sigma_{\parallel}$, and $Krot^B = 2\sigma_{\perp}^2$. The resulted fission barrier parameters are generally consistent with theoretical values /10/.

At neutron energies higher than (n,nf) reaction threshold, if fission parameters of residual nuclei, formed after emission of 1, 2... neutrons are fixed, the approach adopted in /9/ to calculate fission probability successfully describes the experimental data on (n,f) and (n,xn) reaction data for U-235 and U-238 targets. The important point here is the correct parametrization of the secondary neutron spe-

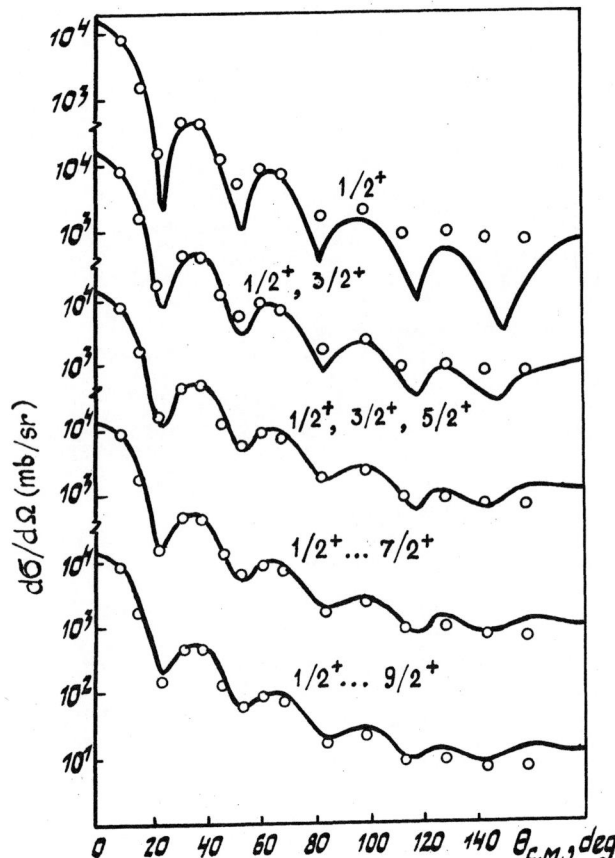

Fig. 1 Measured /5/ and calculated neutron scattering cross section of 239-Pu at E=14.1 MeV

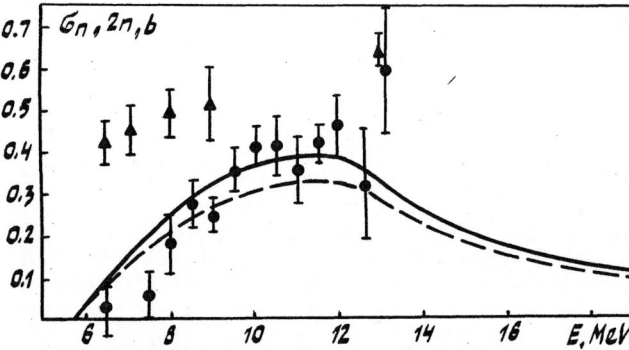

Fig. 3 Calculated 239-P(n,2n) cross section (solid curve) compared with measured data: ●-/13/; ▲-/14/; dashed curve -for renormalized fission cross section

Table 1. Deformed optical potential

Potential parameters

V_R =46.1-0.3E r_R =1.256 a_R =0.626
W_{SD} =3.1+0.4E,E<10 MeV r_W =1.260 a_W =0.555
W_{SD} = 7.1   E>10 MeV
V_{SO} = 6.2 r_{SO} =1.120 a_{SO} =0.470

β_2 = 0.216 β_4 =0.080

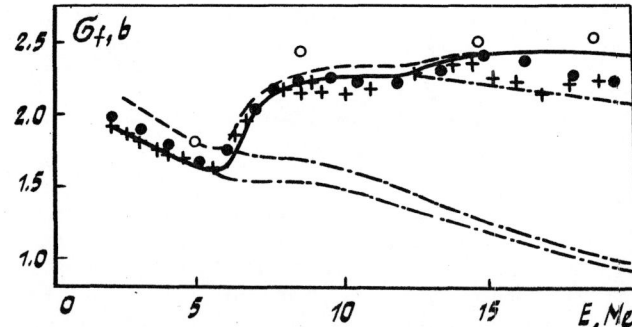

Fig.2 Calculated 239-Pu(n,f) cross section (solid curve) compared with measured data: o-/1/, ●-/11/,+-/12/; dashed curve- tried renormalization, dot-dashed- first chance fission cross section

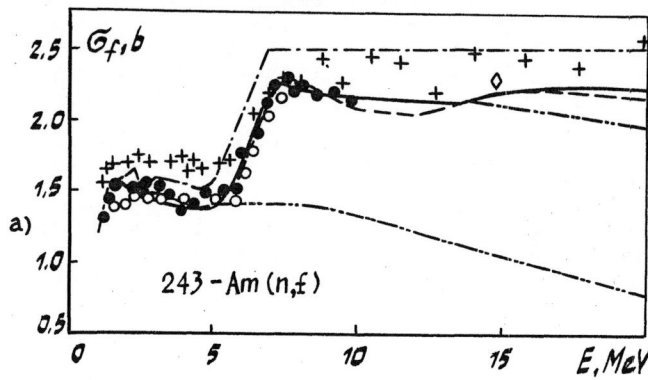

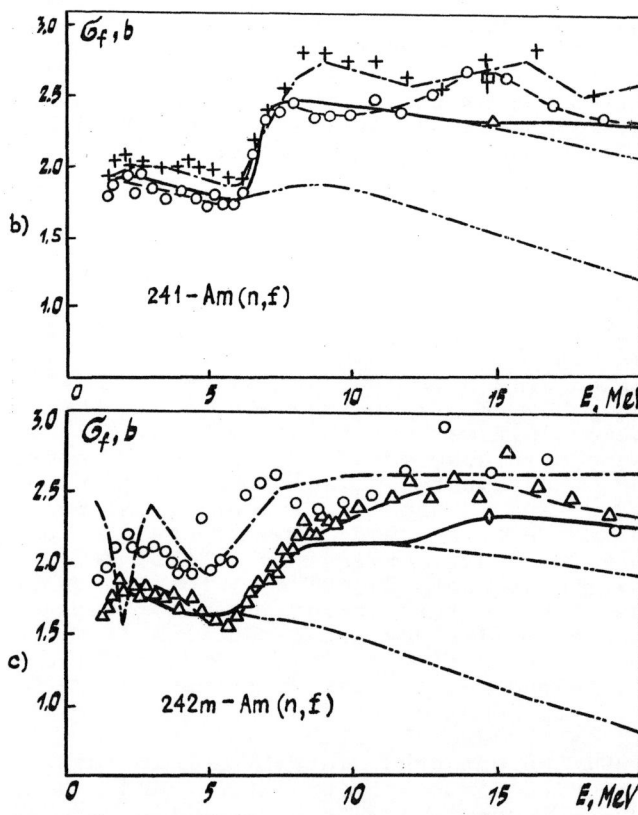

Fig. 4 Calculated Am neutron induced fission cross section (solid curve), first chance fission-dashed curve, dot-dashed curve-JENDL-3,double-dot-dashed curve-ENDF/B-V;data points:a):● -/15/,o -/16/, + -/17/,◊- /18/;b):+-/17/,o-/20/,□-/22/, Δ -/23/;c):o-/19/, Δ -/24/, ◊ -/25/

ctra within a preequilibrium model, giving a good fit of high energy tail of 238-U(n,2n) cross section. The same approach gives a fair description of data on reactions 239-Pu(n,f) and 239-Pu(n,2n). The noticeable spread in relative and absolute fission cross sections around 14 MeV is well known. But when data /1/ became available the discrepancy got a systematic character. So, it is of interest, whether it is possible to reproduce consistently fission /1/ and (n,2n) data within a statistical model framework. If the calculated curve is shifted before the onset of second chance fission by 0.100 b up, after the onset of (n,nf) reaction the shift amounts to 0.050 b, that is 5 times less than needed. After the onset of (n,2nf) reaction the shift you are looking for, disappears (Fig.2).
There is some scatter in (n,2n) data, but with fission cross section thus changed the calculated curve slightly underpredicts (n,2n) cross section (Fig.3). Thus, it seems impossible to renormalize fission cross section with the current state of theoretical models.

In case of Am nuclei the various data sets are at odds with each other both in shape and absolute values. For reaction 243-Am(n,f) data /15,16/ could be fitted only if residual nucleus 243-Am fission parameters are inferred from data on 242m-Am(n,f) reaction /24/. After the onset of (n,2nf) reaction the calculated curve agrees with data /18/ (Fig. 4a).

For 241-Am target only data /20/ up to the onset of (n,2nf) fission could be reproduced /21/. After the onset of reaction (n,2nf) there is a steep rise in data /20/,which is irreproducible. We suppose that due to a high fissility of target nucleus the (n,2nf) fission plateau must be rather flat (see Fig. 4b).

The most severe is the state of data on 242m-Am(n,f), where two data sets differ completely. The calculated curve consistent with 243-Am(n,f) data /15,16/ and 241-Am(n,f) data /20/ exhibits shape difference with data /24/. However the shape of data /19/ agrees quite well with our curve. The data /25/ support our prediction (see Fig. 4c).

There is a drastic disagreement between calculated (n,2n) and (n,3n) cross sections and evaluated data of ENDF/B-V and JENDL-3.They greatly diverge in shape and absolute values. We suppose it is due to approximations in description of fission competition and level density and neglect of preequilibrium effects.

Summary

The present analysis shows that existing data base on 239-Pu neutron reactions is incompatible with absolute fission cross section data /1/.
The analysis of fission cross section data for Am nuclei offers a possibility to define a consistent data set for 241-Am(n,f), 243-Am(n,f) and predict a fission cross section 242m-Am(n,f) above the onset of (n,nf) reaction. There is an urgent need to revise Am neutron data files.

REFERENCES
1. C.M. Herbach, K. Merla, G. Musiol, H.-G. Ortlepp, G. Pausch, W. Wagner, I.D. Alchazov, L.V. Drapchinsky, E.A. Ganza, S.S. Kovalenko, O.I. Kostochkin, S.M. Soloviev, V.I. Shpakov, in Proc. Int. Conf.on Neutron Physics Sept. 1987, Kiev, USSR, Vol.3, p.260, Moscow (1988)
2. A.B. Klepatskij, V.A. Konshin, E.Sh. Sukhovitskij, INDC(CCP)-161/L,p.9, Vienna (1981)
3. W.P. Poenitz, J.F. Whalen, ANL/NDM-80 (1983)
4. G. Haouat, J. Lachkar, Ch. Lagrange, J. Jary, J. Sigaud, Y. Patin.: Nucl. Sci. Eng. 81, 491 (1982)
5. L.F. Hansen, B.A. Pohl, C. Wong, R.C. Haight, Ch. Lagrange.: Phys. Rev. C34, 2075 (1986)
6. T.W. Phillips, R.E. Howe.: Nucl. Sci. Eng. 69, 375 (1979)
7. M. Uhl, B. Strohmaier IRK-76/01(1976)
8. A.V. Ignatjuk, K.K. Istekov,G.N. Smirenkin: Sov.J.Nucl.Phys. 29,875(1979)
9. A.V. Ignatjuk, V.M. Maslov, A.B. Paschenko:Sov.J.Nucl.Phys. 47,224 (1988)
10. W.M. Howard, P. Moller: Atom.Data and Nucl.Data Tables 15, 219 (1980)
11. K. Kari, S. Cierjacks,KFK-2673 (1978)
12. G.W. Carlson, J.W. Behrens: Nucl. Sci. Eng. 66, 205 (1978)
13. J. Frehaut, A. Bertin, R. Bois, E. Gryntakis, C.A. Philis: Rad. Effects 95, 215 (1986)
14. D.C. Mather, P.F. Bampton, R. Coles AWRE 072/72 (1972)
15. H.H. Knitter, C. Budtz-Jorgensen: Nucl. Sci. Eng. 99, 1 (1988)
16. B.I. Fursov, E.Y. Baranov, M.P. Klemishev, B.F. Samylin, G.N. Smirenkin, Y.M.Turchin: Rad.Effects,93,305(1986)
17. J.W. Behrens, J.C. Browne, E. Ables: Nucl. Sci. Eng. 77, 44 (1981)
18. E.F. Fomushkin,G.F. Novoselov,Y.I.Vinogradov:Jadern.Konstanty,3,17 (1984)
19. J.W. Dabbs, C.E. Bemis,Jr., S. Raman, R.J. Dougan, R.W. Hoff: Nucl. Sci. Eng. 84, 1 (1983)
20. J.W. Dabbs, C.H. Johnson, C.E. Bemis, Jr.: Nucl. Sci. Eng. 83, 22 (1983)
21. V.M. Maslov: Sov. J. At. Energy, 64, 478 (1988)
22. M. Cance, G. Grenier,CEA-N-2194 (1981)
23. A. Prindle D. Sisson, D. Nethaway: Phys. Rev. C20, 1984 (1979)
24. J.C. Browne, R.M. White, R.E. Howe, J.H. Landrum, R.J. Dougan,R.J. Dupzyk: Phys. Rev. C29, 2188 (1984)
25. E.F. Fomushkin, G.F. Novoselov, J.I. Vinogradov, V.V. Gavrilov, V.I. Inkov, B.K. Maslennikov, V.N. Polynov, V.M. Surin, A.M. Shvetsov: Sov. J. Nucl. Phys. 33, 324 (1981)

STUDY OF A FUNDAMENTAL NEAREST LEVEL SPACING DISTRIBUTION OF NEUTRON RESONANCES

G. Rohr

Commission of the European Communities,Joint Research Centre
Central Bureau for Nuclear Measurements, 2440 Geel, Belgium.

Abstract: Regular spacings of s-wave neutron resonances for the target nuclides ^{40}Ca, ^{54}Fe, ^{58}Fe, ^{96}Zr and ^{238}U are discussed. An attempt is made to interpret these spacings with the phonon representation. The maximum and the width of the obtained Gaussian-like distribution for the nearest level spacings determine the phonon energy and their lifetime respectively. There is good agreement between the lifetime of phonons and that of neutron resonances based on the average neutron width. The long lifetimes of phonons for ^{96}Zr and ^{238}U allow to distinguish subharmonic resonances. The finite lifetime of phonons and the observation of subharmonics characterize the neutron resonance spectrum as a spectrum of an anharmonic oscillator. These spectra are observable at neutron separation energy for neutron closed shell and neutron excess nuclei, which have a small neutron separation energy. Therefore the study of neutron resonances provides us with an experimental signature that the nucleus has properties of a non-linear system.

(regular spacings of s-wave resonances, phonons, subharmonics, anharmonic oscillator)

Introduction

In contrast to the Bohr assumption of statistical equilibrium there is evidence in the literature that the compound nuclear reaction process can be very simple. Calculations of the level density of doorway states (2p-1h) at neutron separation energy, based on the independent particle model including residual interaction, result in spacings which agree reasonably well with those observed in medium light (A<38) and closed shell nuclei [1,2,3]. In this case the resonances can be considered as created in a single two-body collision process which is the lowest mode of nucleon-nucleon interaction and the spacing distribution belonging to it is fundamental. Beyond A = 38 more complicated states (3p-2h,4p-3h etc.) become energetically possible, indicated by steps in the level density systematics [3] and the additional states may obscure this fundamental spacing distribution. There is no possibility to reconstruct it since in general there is no possibility to distinguish experimentally between resonances of different seniority. Therefore only a level spacing distribution based on doorway resonances characterized by a defined spin,parity and s(seniority) = 3 will reflect the properties of nucleon-nucleon interaction in a nucleus.
In this contribution the regular spacings in the neutron spectra of ^{40}Ca, ^{54}Fe, ^{58}Fe, ^{96}Zr and ^{238}U are discussed. It is attempted to interpret the properties of the spectra with the well known phonon representation.

Regular spacings for simple resonances

The limitation to resonances for A < 38 and closed shell nuclides considerably reduces the number of states available for the adjacent level spacing fluctuation studies. Therefore only a narrow distribution of these values is expected to become detectable. In this section examples for a Gaussian-like distribution using s-wave neutron resonance data of ^{40}Ca, ^{54}Fe, ^{58}Fe and ^{96}Zr are presented [4,5]. The data have been taken from the literature and are used up to a maximum energy that prevents inclusion of 3p-2h states. It is also assumed that the resonance data up to this energy are complete. In this small energy range no correction for the energy dependence of the level spacings has been applied since this effect is assumed to be small compared with the inherent spacings fluctuation.
In fig. 1 the s-wave resonance parameters of $^{40}Ca + n$ up to 740 keV are plotted versus their resonance energy. Resonance energies and their corresponding reduced neutron widths have been taken from [6]. The plot contains no small spacings, indicating a strong mutual level repulsion effect. This property is seen even more clearly in fig. 2, which represents a plot of the 15 adjacent level spacings in a histogram. For comparison fig. 2 includes the Wigner distribution.

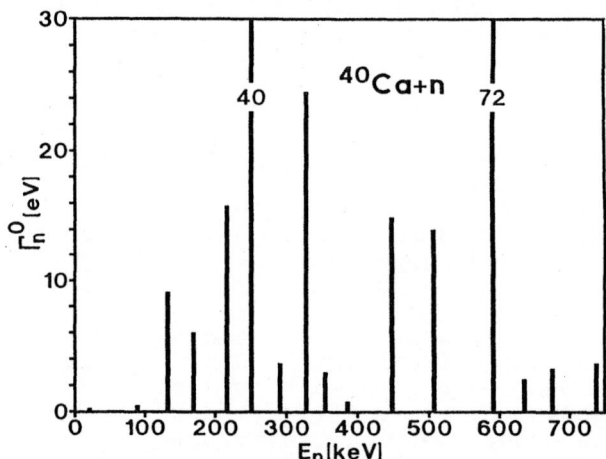

Fig. 1: s-wave resonance energies and reduced neutron with for $^{40}Ca + n$.

The experimental spacing distribution is narrower and is represented by the Gaussian distribution using a standard deviation calculated from the experimental spacings. The level spacings are distributed in an interval $0.56 \leq D/<D> \leq 1.76$; the probability that a spacing inside this interval is distributed according to a Wigner distribution is 0.693 and consequently the probability for fifteen spacings is $P_{wig} = (0.693)^{15} = 0.41\%$. ^{40}Ca is the largest natural nucleus which has the same neutron and proton number. Since it has a double closed shell (N = P = 20) the resonances at neutron separation energy are doorway resonances.

Although the truncation of the sequence of levels for ^{40}Ca at 740 keV cannot be justified experimentally for an onset of 3p-2h states at this energy the remaining number of spacings clearly indicates a deviation from the Wigner nearest level spacing distribution.

Fig. 3 contains the s-wave resonance parameters of ten resonances in ^{54}Fe up to an energy of 250 keV.

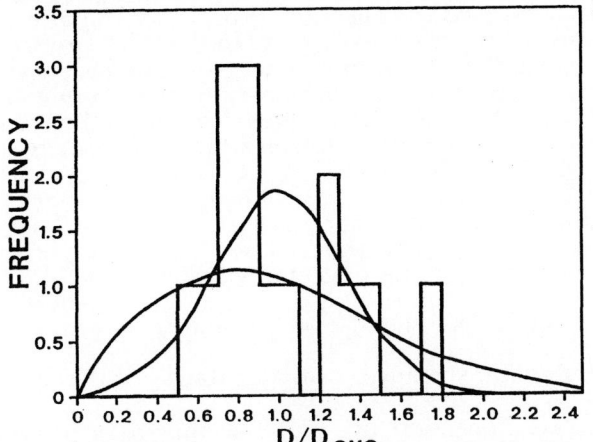

Fig. 2: Histogram of the level spacing distribution for ^{40}Ca + n resonances compared with Wigner and Gaussian functions.

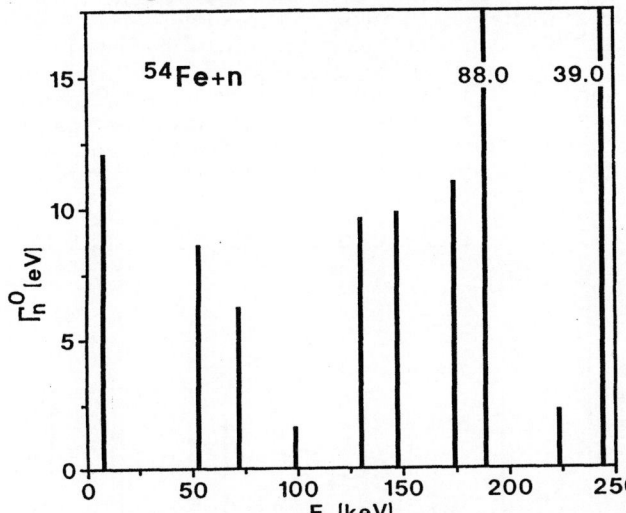

Fig. 3   s-wave resonance energies and reduced neutron widths for ^{54}Fe + n.

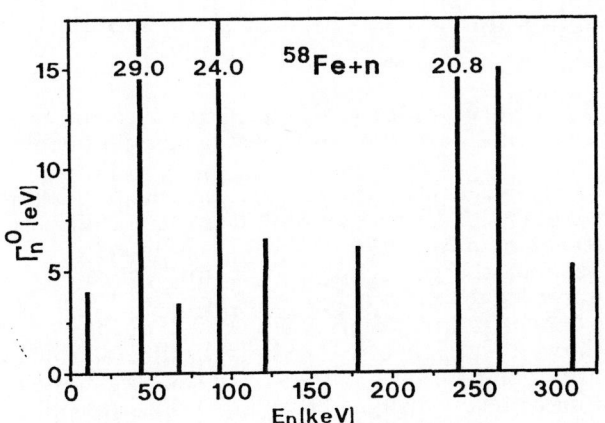

Fig. 4: s-wave resonance energies and reduced neutron widths for ^{58}Fe + n

The resonance data are taken from ref. [6], except the 231.0 keV resonance which is not assigned as a s-wave resonance in experimental papers [7]. The adjacent level spacings are distributed in the interval $0.55 \leq D/<D> \leq 1.71$; the probability that the resonances are distributed according to a Wigner distribution is 3.4.%. Also ^{54}Fe has a closed neutron (N = 28) shell and the resonances are expected to be of 2p-1h type.

A similar spectrum to ^{54}Fe is observed for ^{58}Fe and the resonance parameters of nine resonances up to 310 keV [6,8,9] are represented in fig 4. The spacings are distributed in the interval of $0.62 \leq D/<D> \leq 1.69$ and the probability that they obey a Wigner distribution is $P_{wig} = 1.5\%$. ^{58}Fe has no shell closure but has a neutron separation energy which is 2.72 Mev lower than that of ^{54}Fe. The low excitation energy results in an average level spacing which is even larger than that for ^{54}Fe. Therefore both Fe isotopes should have resonances of the same type at neutron separation energy.

A very regular neutron spectrum is seen in fig. 5 which represents the eight ^{96}Zr resonances up to an energy of 100 keV taken from the literature [6,10].

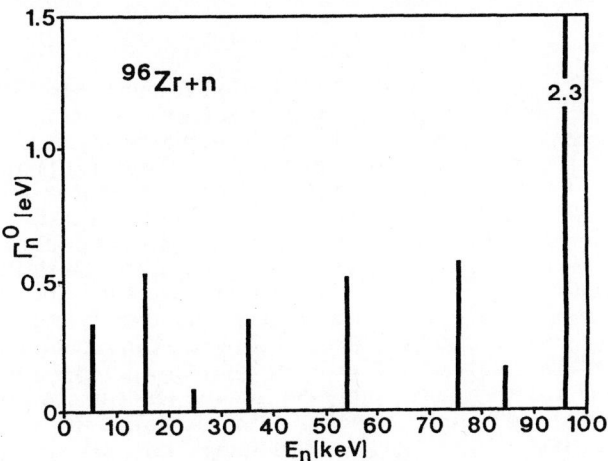

Fig. 5: s-wave resonance energies and reduced neutron widths for ^{96}Zr + n

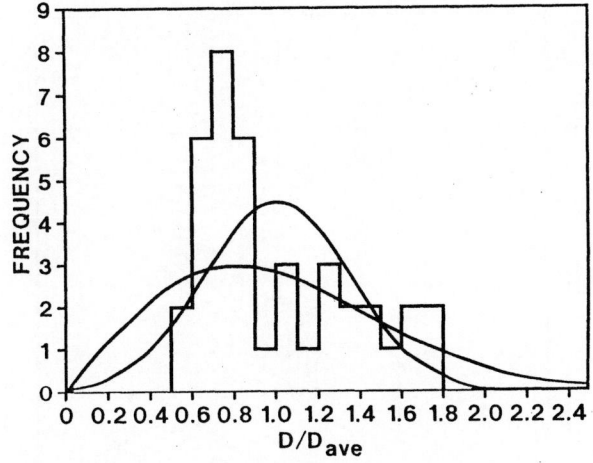

Fig. 6: Distribution of the nearest level spacings for ^{41}Ca, ^{55}Fe, ^{59}Fe and ^{97}Zr resonances (39 spacings)

Seven spacings are distributed in the interval $0.62 \leq D/<D> \leq 1.69$ and the distribution has a pro-

886

bability of being represented by a Wigner distribution of $P_{wig} = 2\%$. ^{96}Zr is a closed sub-shell (N = 40) nucleus and has a neutron separation energy which is 0.9 Mev lower than that of ^{94}Zr. Both effects suggest that the ^{96}Zr resonances are of the 2p-1h type.

The plots of figs. 1,3,4,5 contain no resonances of small spacing indicating a strong mutual level repulsion effect. This property is seen even more clearly in fig. 6, which represents a plot of all 39 adjacent level spacings from the four above mentioned nuclides in one histogram. For comparison this figure includes the Wigner distribution. The level spacings are distributed in an interval $0.55 \leq D/\langle D \rangle \leq 1.76$; the probability that a spacing inside this interval is distributed according to a Wigner distribution is 0.705 and the probability for thirty nine spacings can be estimated to be $P_{wig} = (0.705)^{39} = 1.2 \ 10^{-4} \%$. The result excludes the Wigner distribution and supports a Gaussian-like distribution. However the histogram is not symmetric around the average level spacing and may indicate larger spacings (gaps) compared with the spacing defined at the peak of the distribution.

The lifetime of neutron resonances

The Gaussian-like distribution discussed in the last section represents a spectrum of almost equidistant levels, which may be explained with the lattice vibrational model [11]. The neutron resonance in this model can be interpreted as consisting of a large number of vibration quanta, the so-called phonons. However the observed resonance spectra are not continuous phonon spectra, they are mixed with phonons of about double the energy, as can be seen in figs. 5 and 6. This fact, and the small number of relevant spacings (in total of about 70), prevent a proof of phonons in nuclei with a reasonable confidence level based on the nearest level spacing distribution. But the Gaussian-like distribution contains a further property of phonons. The energies of the phonons are not constant and therefore have finite lifetime. Based on the standard deviation of the Gaussian-like distribution $\Delta\varepsilon_q$ [11] the lifetime of the phonons

can be calculated by $\tau_q = \hbar / \Delta\varepsilon_q$. These values (full dots) of six nuclides are shown in fig. 7 together with the lifetime of neutron resonances calculated from the average neutron width $\tau_{p-h} = \hbar / \langle \Gamma_n \rangle$. A reasonable agreement is observed, especially for nuclides with ten and more resonances and the lifetime is seen to increase with the atomic number (dotted eyeguided line). This dependence explains the high regularity of the Zr spectrum (fig.5) compared to the other examples. In the ^{96}Zr spectrum phonons of two different energies can be clearly distinguished, where the larger one is double the smaller. These phonon frequencies are smaller than the frequency of normal modes corresponding to an oscillator energy of $\hbar\omega = 41 \ A^{-1/3}$ and these are called subharmonics [12]. They characterize an anharmonic oscillator spectrum. If the larger phonons in ^{96}Zr are replaced by two phonons of the smaller energy, the lifetime increases from 14.3 to 72.3 10^{-20} s. This value is in better agreement with the lifetime calculated from the neutron width.

Regular spacings in complicated resonances

In this section it will be shown that the regular spacings are not limited to simple excited resonances. They are also observable for nuclei with a low neutron separation energy which is observed for closed neutron shell nuclei and or for neutron excess nuclei. Examples are the nuclei ^{140}Ce, ^{206}Pb and ^{238}U. The lifetimes of phonons for these nuclei are large enough to distinguish subharmonic resonances. In fig. 8 the spacings of ^{238}U s-wave resonances are plotted as a function of the spacing sequence (number) as they are observed with increasing neutron energy. Sixteen spacings are detected up to 380eV.

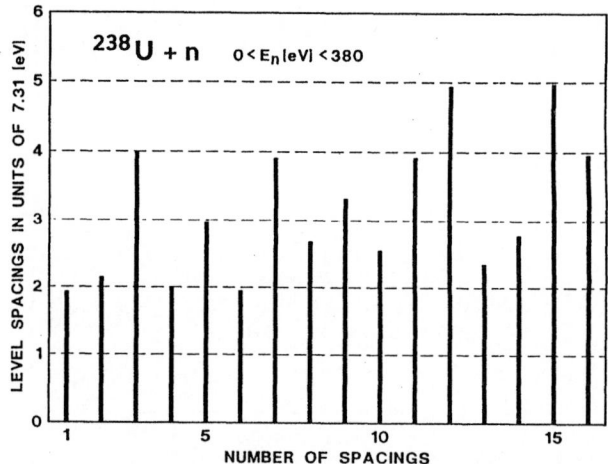

Fig. 8: Spacings of ^{238}U + n s-wave resonances plotted as the function of spacing number.

The first seven spacings are part of a clean phonon spectrum with three different energies, which are different by a multiple of the 7.5 eV energy. With increasing energy, distortions in this property can be observed for the spacing numbers 8, 9, 10 and 13. The sum of spacings for 8 and 9 as well as for 10 and 13 have a multiple of 7.5 and may indicate that the smaller phonons are obtained from the larger one by a bifurcation due to a phonon-phonon interaction. However, it is surprising that bifurcation processes can create regular phonons as observed in the level spacing of ^{238}U resonances starting from phonon energies of the normal modes. Because of the large ratio of the observed level spacing (21.5eV [3]) and

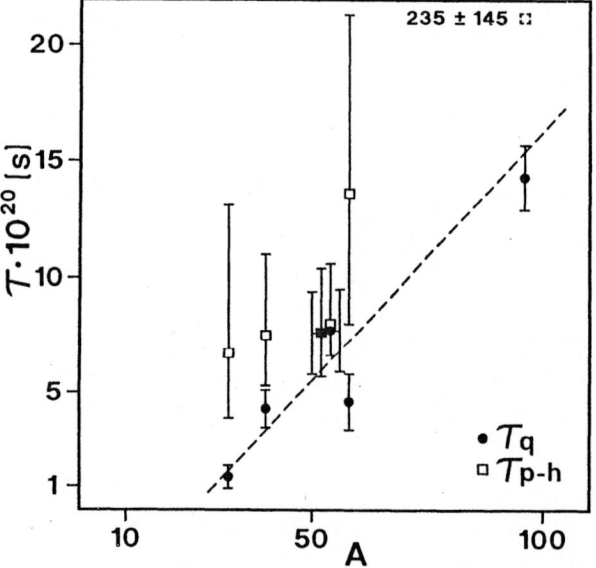

Fig. 7: The lifetime of phonons (full dots) and the average lifetime of resonances (open squares) for six nuclides discussed in [5].

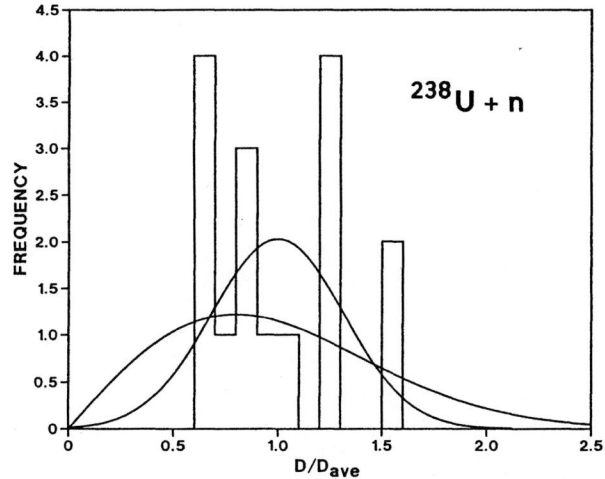

Fig. 9:   Histogram of the level spacing distribution for $^{238}U + n$ resonances compared with Wigner and Gaussian functions.

the calculated level density of doorway states (18.keV [3]), the ^{238}U resonances are certainly not simple resonances. The property of subharmonics in the ^{238}U spectrum disappears with increasing energy; but can still be clearly seen in resonances up to 855 eV neutron energy.

The attributed nearest level spacing distibution is shown in fig. 9, which includes three bumps indicating subharmonics. The deviation from a Wigner distribution in this case can, in contrast to the examples in the second section, be shown from a chi-squared test. The observed value of $\chi^2 = 30.2$, calculated in the interval $0 \leq D/<D> \leq 2.0$, rejects the hypothesis of a Wigner distribution with a 5% level of significance.

Indications for such distributions are observed also in ^{56}Fe [13] and ^{60}Ni resonance data (using data of A. Brusegan et al. a contribution of this conference). Both nuclei have no closed neutron shell and also a high neutron separation energy.

Summary and conclusions

Simple excited neutron resonances of the target nuclides ^{40}Ca, ^{54}Fe, ^{58}Fe and ^{96}Zr have been studied, where neutron resonances are created in a one collision process, and where the nearest level spacing distribution is expected to give a fingerprint for the dynamics of nucleons in a nucleus. A Gaussian-like distribution is observed, which represents a spectrum of almost equidistant levels and as such can be interpreted as phonons. Although the spectrum is not a continuous phonon spectrum, the maximum and the width of the distribution allow the determination of energy and lifetime of the phonons respectively. A comparison of the lifetime of phonons with that of neutron resonances based on the average neutron width, gives a reasonable agreement for ^{40}Ca, ^{54}Fe and ^{58}Fe. In general the lifetime of both sets increases with the atomic number. The long lifetime of ^{96}Zr allows to distinguish two different phonons where one has twice the energy of the other. Using this property of the so-called subharmonics for the

calculation of the lifetimes of phonons, the lifetime increases and there is now also, for ^{96}Zr, a reasonable agreement with the value based on the neutron width. With the subharmonic spectrum of ^{238}U it is shown that regular spacings are not only limited to simple excited resonances. These spectra can be observed at neutron separation energy for closed shell and neutron excess nuclei, which have a low neutron separation energy. The subharmonics introduce bumps in the nearest level spacings distribution, which are smoothed out due to the increased number of phonons created by phonon-phonon interaction with increasing energy. This process may result in a Wigner distribution for the nearest level spacings where the remaining repulsion effect is due to the vibrational properties of the nucleons in the nucleus. One can conclude that the finite lifetime of phonons and the existence of subharmonics in the neutron resonance spectra support an oscillator model of a nucleus and characterize the neutron resonance spectrum as one obtained from an anharmonic oscillator. On the other hand the anharmonic spectrum of neutron resonances can be considered to reflect the fine structure of a liquid drop and in doing so a challenging step is made in the direction of a single nuclear model.

References

1.   G.L. Payne, Phys. Rev. 174, 1227 (1968)
2.   K.N. Müller and G. Rohr, Nucl.Phys.A164, 97 (1971)
3.   G. Rohr, Z.Phys. A318, 299 (1984)
4.   G. Rohr, Nuclear Data for Science and Technology (1988 Mito), Editor S. Igarasi, Saikon Publishing Co. Ltd., 707 (1988)
5.   G. Rohr, Proceedings of the 7th International Conference on Capture Gamma-Ray Spectroscopy, Asilomar, California 1990, to be published
6.   S.F. Mughabghab, M. Divadeenam and N.E. Holden, Neutron Cross Sections Vol. 1 Academic Press (Part A 1981, Part B 1984)
7.   E.M.R. Cornelis, C.R. Jungmann, L. Mewissen and F. Poortmans, Neutron Cross Sections for Technology (1979 Knoxville) Editors J.L. Fowler, C.H. Johnson and C.D. Bowman U.S. Government Printing Office:Washington:1980. p. 159 and refs. cited therein.
8.   E. Beer, Ly Di Hong, and F. Käppeler, Nucl. Sci. Eng. 67, 184 (1978)
9.   J.B. Garg, S. Jain and J.A. Harvey, Phys. Rev. C18, 1141 (1978)
10.   C. Coceva, P. Giacobbe and M. Magnani, Neutron Cross Sections for Technology (1979 Knoxville) Editors J.L. Fowler, C.H. Johnson and C.D. Bowman, U.S.Government, Printing Office: Washington: p. 319 (1980)
11.   D. Mattuck, A Guide to Feyman Diagrams in the Many-Body Problem, Mc. Graw-Hill Publishing Company Limited (1967)
12.   H. Margenau and G.M. Murphy, The Mathematics of Physics and Chemistry, D.Van Nostrand Company, INC. Princeton. Vol. II p. 376 (1964)
13.   C.M. Perey, F.G. Perey, J.A. Harvey, N.W. Hill and N.M. Larson, Oak Ridge National Report ORNL/TM-11742, (1990).

THE PRE-EQUILIBRIUM PICTURE OF NUCLEON RADIATIVE CAPTURE
AND THE NEUTRON-TO-GAMMA COMPETITION

E. Běták [1,2], F. Cvelbar [3] and J. Kopecky [2]

[1] Institute of Physics, Slovak Acad. Sci., CS-84228 Bratislava, Czechoslovakia*
[2] ECN - Netherlands Energy Res. Found., NL-1755 ZG Petten, the Netherlands
[3] Institute J. Stefan, E. Kardelj University, YU-61111 Ljubljana, Yugoslavia

Abstract: The pre-equilibrium (exciton) model is used to describe the continuous γ spectra as well as both the activation (σ_{act}) and the integrated (σ_{int}) cross sections (and their excitation functions) of the radiative capture of nucleons with energies ranging from 0 up to 80 MeV. The γ cascades are taken fully into account to evaluate properly σ_{act} (and its distinction with respect to σ_{int}). Our calculations do not contain any free parameters for the γ emission. Comparison to the available experimantal data demonstrates clearly the necessity to include properly the level density parameters for individual nuclei. The low-energy behaviour of the proton radiative capture excitation functions supports the use of energy-dependent GDR width, which we have employed in the generalized Lorentzian.

[Pre-equilibrium model, neutron-to-γ competition, nucleon radiative capture, activation cross section, integrated cross section, γ spectra, excitation function, level density parameters, giant dipole resonance (GDR), energy-dependent damping width, generalized Lorentzian]

Introduction

Different models are presently used to describe the continuum γ emission in nuclear reactions. For long, the direct-semidirect capture model [1] has been clearly superior to the other ones. The appearance and the following expasion of pre-equilibrium models, originally developed for the particle emission, has reached the domain of the γ emission in the eightieths. The pre-equilibrium models succeeded to describe the γ spectra and corresponding cross sections at energies above 10 MeV better than it was possible within the other approaches.

The model

In the following, we shall restrict ourselves to the exciton model, and use the single-particle emission mechanism therein. Thus, the γ emission is associated with the change of the energy of a single nucleon (which eventually may fill in the corresponding hole, decreasing thus the exciton number by -2). The γ emission rate λ_γ^c from an n-exciton state is [2, 3]

$$\lambda_\gamma^c(n, E, \epsilon_\gamma) = \frac{\epsilon_\gamma^2 \sigma_a(\epsilon_\gamma)}{\pi^2 \hbar^3 c^2} \frac{\sum_{m=n,n-2} b(m, \epsilon_\gamma)\omega(m, E - \epsilon_\gamma)}{\omega(n, E)}. \quad (1)$$

Here, $\omega(n, E)$ is the density of the n-exciton states at excitation energy E, $\sigma_a(\epsilon_\gamma)$ is the photo-absorption cross section (the giant dipole resonance), and the corresponding branching ratios b's are given by ref. [3].

The γ emission energy spectrum (from the primary composite nucleus) is given by

$$\frac{d\sigma_\gamma}{d\epsilon_\gamma} = \sigma_R \sum_n \int \tau(n, E') \lambda_\gamma^c(n, E', \epsilon_\gamma) dE', \quad (2)$$

where σ_R is the reaction cross section and $\tau(n, E)$ stands for the total time spent by a nucleus in an n-exciton state,

$$\tau(n, E) = \int_0^\infty P(n, t, E) dt, \quad (3)$$

*EB permanent and present address

and P is the probability of finding the nucleus at time t with the excitation energy E in the state of n excitons, which is a solution of the corresponding set of master equations [4]

$$
\begin{aligned}
\frac{dP(n, t, E, i)}{dt} =& \\
& P(n - 2, t, E, i)\lambda^+(n - 2, E, i) \\
+& P(n + 2, t, E, i)\lambda^-(n + 2, E, i) \\
-& P(n, t, E, i)[\lambda^+(n, E, i) + \lambda^-(n, E, i) + L(n, E, i)] \\
+& \sum_{j,m,x} \int_\epsilon P(m, t, E', j)\lambda_x^c(m, E', j, \epsilon) d\epsilon. \quad (4)
\end{aligned}
$$

Here, the coupling of both different excitation energies (E) and different nuclei in the reaction chain (i, j) is ensured by the last term. Obviously, the coupling of nuclei may be omitted just for the purpose of the radiative capture calculations. The symbols $\lambda^\pm$ denote the intranuclear transition rates, and L is the total emission rate, i.e. the energy integral of the emission rates to continuum summed over all ejectiles (including γ).

Though the exciton model contains usually a parameter $|M|^2$, the square of the average matrix element of intranuclear transitions, its value is taken fixed from nucleon emission systematics. For the γ emission itself, there are no free parameters in our calculations. The description of spectra from (n, γ) reactions is reasonable over a wide range of nuclei: good agreement in shape; and the absolute value typically within a factor of 2 (see [5]).

Int. and act. radiative capture c. s.

The integrated cross section of the radiative capture is simply the integral of (2) over the γ energies leading directly to the bound states (given usually by the neutron binding energy). In practice, however, the result is identical to that we obtain if we consider just the initial excitation energy E_0 of the composite nucleus, as the cascade contribution is vanishibly small to that process. The difference of the activation cross section with respect to the integrated one is based just on the presence of the γ cascades. The corresponding full expression is [6]

$$\sigma_{act} = \sigma_R \sum_n \int_{unb} \int_{\epsilon_\gamma} \quad \tau(n, E') \lambda_\gamma^c(n, E', \epsilon_\gamma) \cdot$$
$$\cdot [1 - \Theta(E' - B - \epsilon_\gamma)] d\epsilon_\gamma dE', \quad (5)$$

where the Heaviside funcion Θ ensures that we reach the region of the bound states (B being the particle binding energy) by the γ quantum of the energy ϵ_γ emitted from an unbound state of the excitation energy E'.

Though the level densities of the exciton model are commonly expressed using the global parameters (e.g. $g = A/13$ and no pairing), the recent study of nucleon radiative capture [6] demonstrated the significance of choosing proper level density parameters for individual nuclei.

The relative comparison of σ_{act} and σ_{int} was done in [5] -for 14 MeV neutron-induced reactions, and more generally in [6]. Fig. 1 brings such an illustrative comparison. At incident energies below 30 MeV, the ratio $\sigma_{int}/\sigma_{act}$ varies strongly with the energy. Generally, it is higher in the neutron-induced reactions than in those by protons. At the incident energies (10–15) MeV, its value is between 0.9 and 1 in for the incoming neutron. At higher energies, the value remains practically constant with only a weak reflexion of new open channels.

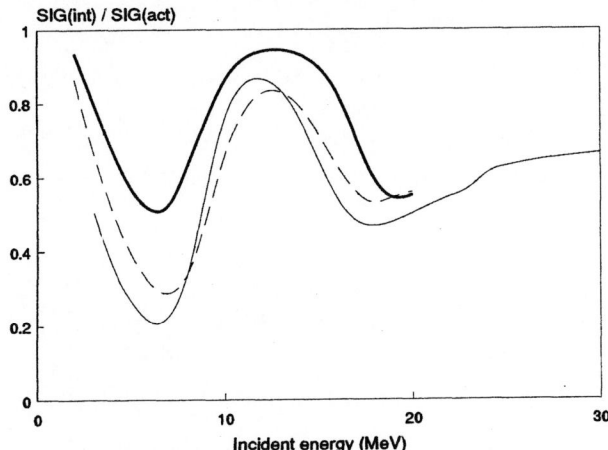

SIG(int) / SIG(act)

Incident energy (MeV)

Fig. 1. Calculated ratio $\sigma_{int}/\sigma_{act}$ of the radiative capture in reactions ^{89}Y+n (heavy line), ^{142}Ce+p (thin line) and ^{130}Te+p (dashed). The level density parameters taken in accord with [7].

GDR form

The γ emission depends on the form of the giant dipole resonance. That is often taken in the Lorentzian form (or a double Lorentzian for the deformed nuclei) with parameters in accord with the tables (e.g. [8]). However, as was shown in [6], the calculated radiative cross sections manifest a slight excess with respect to the experimental data in the low-energy region. This cannot be removed by standard approaches, and a possible way is to try to use some modified form of the GDR (this possibility was mentioned already in [6]). Probably, a simple *ad hoc* modification of the low-energy GDR tail, as used by [9], would yield reasonable results, but we preferred to apply the approach of [10], which is — at least partially — theoretically backed [11]. Therein, the Lorentzian shape, which in its standard form is

$$\sigma_a(\epsilon_\gamma) \propto \frac{\Gamma^2}{(\epsilon_\gamma^2 - E_{GDR}^2)^2 + \epsilon_\gamma^2 \Gamma^2}, \quad (6)$$

is replaced by

$$\sigma_a(\epsilon_\gamma) \propto$$
$$\left(\frac{\Gamma(\epsilon_\gamma)}{(\epsilon_\gamma^2 - E_{GDR}^2)^2 + \epsilon_\gamma^2 \Gamma(\epsilon_\gamma)^2} + \frac{2.8\pi^2 \Gamma T^2}{E_{GDR}^5}\right) \cdot \Gamma. \quad (7)$$

In (7), $\Gamma(\epsilon_\gamma)$ is the width, which depends both on ϵ_γ, the energy of the emitted γ, and the excitation energy of the nucleus via its temperature T,

$$\Gamma(\epsilon_\gamma) \propto \frac{\epsilon_\gamma^2 + 4\pi^2 T^2}{E_{GDR}^2}. \quad (8)$$

The last term in (7) removes troubles which appear at $\epsilon_\gamma \to \infty$ and $T \to \infty$ simultaneously. Even with the extra term added in (7), the low-energy part of GDR is *below* that of (6) for greater part of the energy interval considered, because of lower $\Gamma(\epsilon_\gamma)$.

Comparison to the data

We have calculated excitation functions of (p,γ) and (n,γ) reactions, continuing thus the analysis performed in [6]. Therein, typically a small excess of calculated cross section (when compared to the experiment) was reported at proton incident energies between 5 and 10 MeV (neutron experimental data did not cover sufficiently this energy region). The results are depicted in Figs. 2. With the generalized Lorentzian (7) we have arrived to a significantly better reproduction of the data for all calculated nuclei. At higher energies, however, the conclusion is not so straightforward: some excitation curves became improved, whereas other worsened by this attempt. Anyway, the Lorentzian modification [11] has been devised for excitation energy not exceeding the neutron binding energy, so that we are above the limits of validity already.

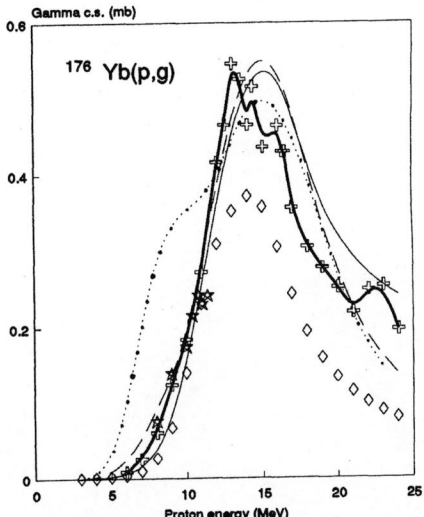

Gamma c.s. (mb)

^{176}Yb(p,g)

Proton energy (MeV)

Fig. 2a. Excitation function of proton radiative capture on ^{176}Yb. The experimental data [12] of the activation cross section are drawn in a heavy line with crosses, those of the integrated cross section by isolated stars. The calculated activation cross sections are drawn as: dotted line ($g =$A/13, no pairing), dashed line (Gilbert and Cameron [7] level density parameters) and a full thin line (generalized Lorentzian); the integrated cross sections with Gilbert and Cameron parameters are as isolated diamonds.

Conclusions

We present a method to calculate both the integrated and the activation cross sections of the nucleon radiative capture. Analyses of available excitation curves strongly suggest for the necessity of inclusion of the individual level density parameters for individual nuclei, and the overall agreement to the data, extended to the energies up to about 50 MeV, supports our belief in the validity of the model used. A small excess in the calculated cross section in the energy region below 10 MeV, which was a common feature in our previous study, can be removed by using the generalized Lorentzian shape of the GDR (with the spreading width depending both on the γ and the excitation energy) instead of its standard form. This is a new indication of this possible phenomenon.

One of us (EB) is grateful to ECN for its hospitality. This work has been supported in part by the International Atomic Energy Agency contract No. 5148/RB. We acknowledge also the technical assistance of A.J. Koning.

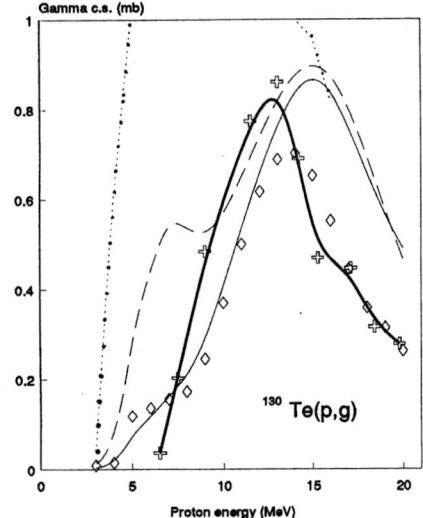

Fig. 2b. The same as in Fig. 2a, but for ^{130}Te+p. Only the activation data are available from the experiment [13]. The curves have the same meaning as in Fig. 2a.

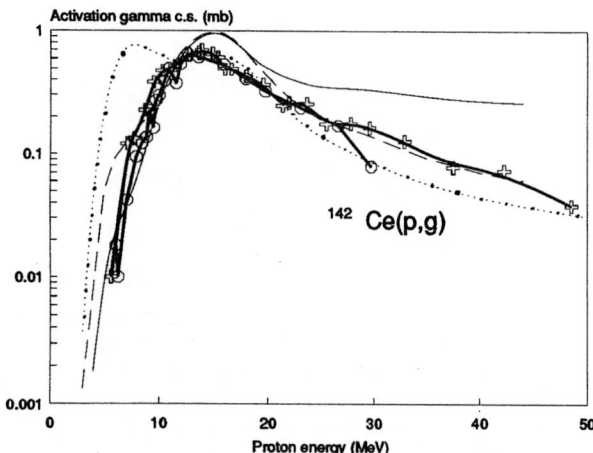

Fig. 2c. The same as in Fig. 2b, but for ^{142}Ce+p. Two experimental sets are available, those by Dally et al. [13] are marked with crosses, that of Verdieck et al. [14] as circles. Only the activation excitation functions are drawn; the meaning of lines is the same as in preceding Figures.

References

[1] C.L. Clement, A.M. Lane, J.A. Rook: Nucl. Phys. <u>66</u>, 273 *and ibid.*, 293 (1965);
M. Potokar: Phys. Lett. <u>46B</u>, 346 (1973)

[2] E. Běták and J. Dobeš: Phys. Lett. <u>84B</u>, 368 (1979)

[3] J.M. Akkermans and H. Gruppelaar: Phys. Lett. <u>157B</u>, 95 (1985)

[4] E. Běták and J. Dobeš: Report IP EPRC SAS 43/1983, Bratislava (1983);
E. Běták, I. Brezník, S. Hlaváč and P. Obložinský: Izv. AN SSSR, ser. fiz., <u>49</u>, 1023 (1985)

[5] F. Cvelbar and E. Běták: Z. Phys. <u>A332</u>, 163 (1989)

[6] F. Cvelbar, E. Běták and J. Merhar: J. Phys. <u>G17</u>, 113 (1991)

[7] A. Gilbert and A.G.W. Cameron: Can. J. Phys. <u>43</u>, 1446 (1965)

[8] B.L. Berman: At. Data Nucl. Data Tbl. <u>15</u>, 319 (1975)

[9] S. Joly, D.M. Drake and L. Nilsson: Phys. Rev. <u>C20</u>, 2072 (1979)

[10] J. Kopecky and R.E. Chrien: Nucl. Phys. <u>A468</u>, 285 (1987);
J. Kopecky and M. Uhl: Phys. Rev. <u>C41</u>, 1941 (1990)

[11] S.G. Kadmenskii, V.P. Markushev and V.I. Furman: Yad. Fiz. <u>37</u>, 277 (1983)

[12] B. Palsson, J. Krumlinde, I. Bergqvist, L. Nilsson, A. Lindholm, D.C. Santry and E.D. Earle: Nucl. Phys.

[13] B. Dally and S. Shaw: Nucl. Phys. <u>56</u>, 322 (1964) <u>A345</u>, 221 (1980)

[14] E.V. Verdieck and J.M. Miller: Phys. Rev. <u>153</u>, 1253 (1967)

Intercomparison of Multi-Step Direct Reaction Models

A.J. Koning* and J.M. Akkermans[‡]

Netherlands Energy Research Foundation ECN
*Nuclear Analysis Department
[‡]Software Engineering & Research Department
P.O. Box 1, NL-1755 ZG Petten, The Netherlands

Abstract: We propose a quantum-statistical framework that provides an integrated perspective on the differences and similarities between the many current models for multi-step direct reactions in the continuum. In a statistical MSD theory two physically different approaches are conceivable to postulate randomness, respectively called *leading-particle statistics* and *residual-system statistics*. We have newly implemented all MSD theories on a consistent basis, and performed a computational comparison for the reaction $^{90}Zr(p, p')$ at 120 MeV.

[MSD reactions, pre-equilibrium, ECIS, non-DWBA matrix elements, ^{90}Zr proton inelastic scattering, model comparison]

Introduction

The various Multi-Step Direct (MSD) reaction models that have been developed in the past years are now being recognized as an adequate quantum-mechanical tool for the description of the intermediate region between direct and compound nuclear reactions. They are all based on the extrapolation of the (discrete) direct reaction concepts to the continuum. MSD reactions are characterized by high energy tails in the emission spectra and forward-peaked, but smooth, angular distributions. The high level density that is involved in these reactions necessitates the introduction of statistical assumptions in order to obtain a manageable expression for the cross section. We give a short review of the quantum-mechanical framework that has been proposed in [1] and we present the results of our computational model comparison.

Theoretical comparison

In [1], we have argued that a whole spectrum of MSD-models can be generated on the basis of two different statistical postulates. The first, leading-particle statistics (LPS), is based on the assumption that many states are accessible to the leading particle and that the associated matrix elements are randomly distributed. It is a necessary condition for the – relatively simple – convolution-type models and also the intuitive picture behind the FKK model and most of the semi-classical pre-equilibrium models [2]. The alternative postulate concerns residual-system statistics (RSS), which is represented by random configuration mixing due to the residual interactions within the residual nucleus. RSS provides the foundation for the TUL [3] and NWY [4] theories. It does not make any statistical assumptions about the interactions of the leading particle. Applying these physically different random postulates on first and higher order distorted wave theory leads to two distinct classes of statistical MSD theories (see fig.1). One salient feature is that the FKK model can be derived using *only* leading-particle statistics and the on-shell approximation. There is no need at all to invoke residual-system statistics, although this was explicitly mentioned in [5].

Computational comparison

We have constructed a new MSD code system which calculates continuum cross sections according to all aforementioned models (see fig.2). The core of this code system consists of the coupled channels code ECIS. This choice was motivated by ECIS' overall success in discrete direct reaction calculations and its capability of handling a large variety of reaction types over a wide energy range. Moreover, ECIS can easily switch from coupled channels to pure (first and higher order) DWBA calculations (which is sufficient for our purposes). For each MSD model, our system contains a program that generates an energy grid, the corresponding optical potentials and the appropriate input files for ECIS. Subsequently, the calculated DWBA cross sections are digested by the corresponding MSD programs which deliver the continuum cross sections. For the LPS models, only one-step DWBA cross sections are required (the multi-step continuum cross sections are calculated by convolution), whereas we have performed the calculations for the RSS-models with first and second order DWBA (inclusion of higher orders is too time consuming and probably of marginal importance). In this way, we are able to calculate the continuum cross sections according to the

- LPS model [2]

- LOS (LPS on-shell) model, which is the on-shell limit of the above LPS model

- FKK model [5]

- TUL model [3]

892

- NWY model [4]. Since this theory also uses microscopic information, our implementation is approximate and represents an upper estimate.

In figs.3 and 4 we present a comparison between the calculated angular distributions, for inelastic proton scattering on 90 Zr at 120 MeV of incident energy and 80 and 60 MeV of outgoing energy (the cross sections were fitted at 100 MeV of outgoing energy). The experimental data were taken from [6]. We used the Willams level density in all models (also in that of TUL and NWY, although the associated theories rely on true partial level densities), so that resulting differences can be said to be of a purely statistical nature.

Conclusions

We have implemented the various quantum MSD models in a single computer system. All models employ the same set of parameters (optical models, DWBA cross sections, etc.). This enables for the first time, to carry out a comparison between different MSD theories on a consistent basis. The results for the reaction $^{90}Zr(p,p')$ at 120 MeV show that:

- All theories provide a good *overall* fit to the MSD spectra and angular distributions. The importance of individual higher-step contributions may however differ.

- The LPS theories are clearly computationally advantageous compared to the RSS ones, but also employ stronger physical postulates.

- The on-shell approximation gives fairly reasonable results but tends to overestimate the cross sections.

- The LPS and LOS models use non-DWBA matrix elements (as required by the theory) with good results. This resolves a long-standing debate. It appears to be unnecessary to introduce the DWBA approximation as proposed by Feshbach [5,7] and criticised by Tamura [8].

A much more comprehensive MSD model comparison is currently in progress.

We would like to thank Dr. A. Hogenbirk for some useful discussions concerning ECIS.

References

1. A.J. Koning & J.M. Akkermans, in press (*Ann. of Phys.*)(1991).

2. J.M. Akkermans & A.J. Koning, *Phys. Lett.* 234B(1990), 417.

3. T. Tamura, T. Udagawa & H. Lenske, *Phys. Rev.* C26(1982), 379.

4. H. Nishioka, H.A. Weidenmüller & S. Yoshida, *Ann. Phys.* 183(1988), 166.

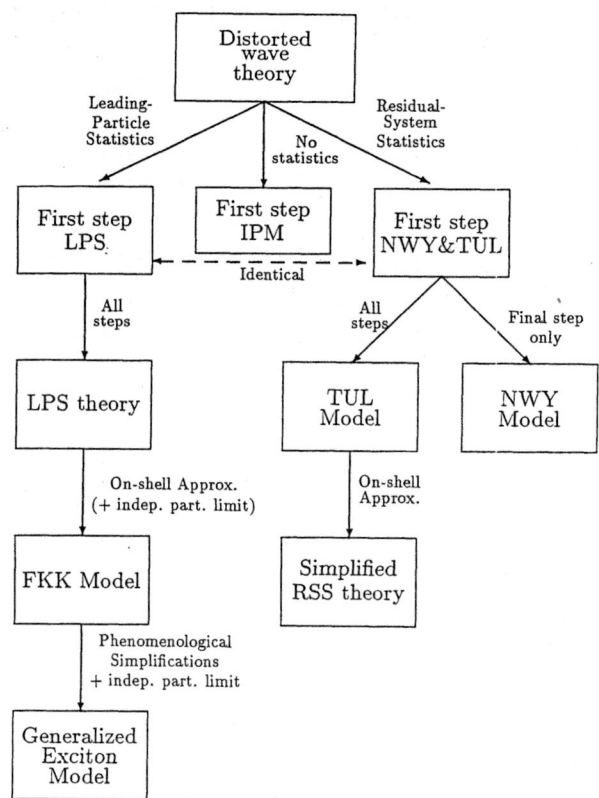

Fig. 1: Family tree of statistical MSD-theories

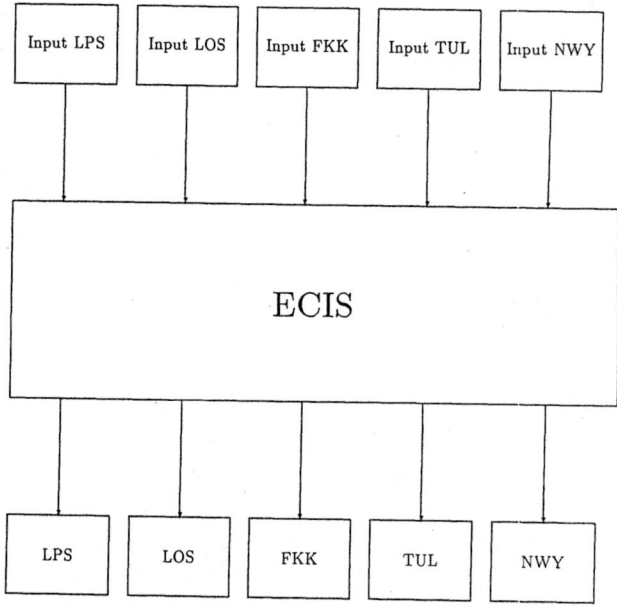

Fig. 2: MSD Code System

5. H. Feshbach, A. Kerman & S. Koonin, *Ann. Phys.* 125(1980), 429.

6. A.A. Cowley et al., *Phys. Rev.* C43(1991), 678.

7. H.Feshbach, *Ann. Phys.* 159(1985), 150.

8. T. Udagawa, K.S. Low & T. Tamura, *Phys. Rev.* C28(1983), 1033.

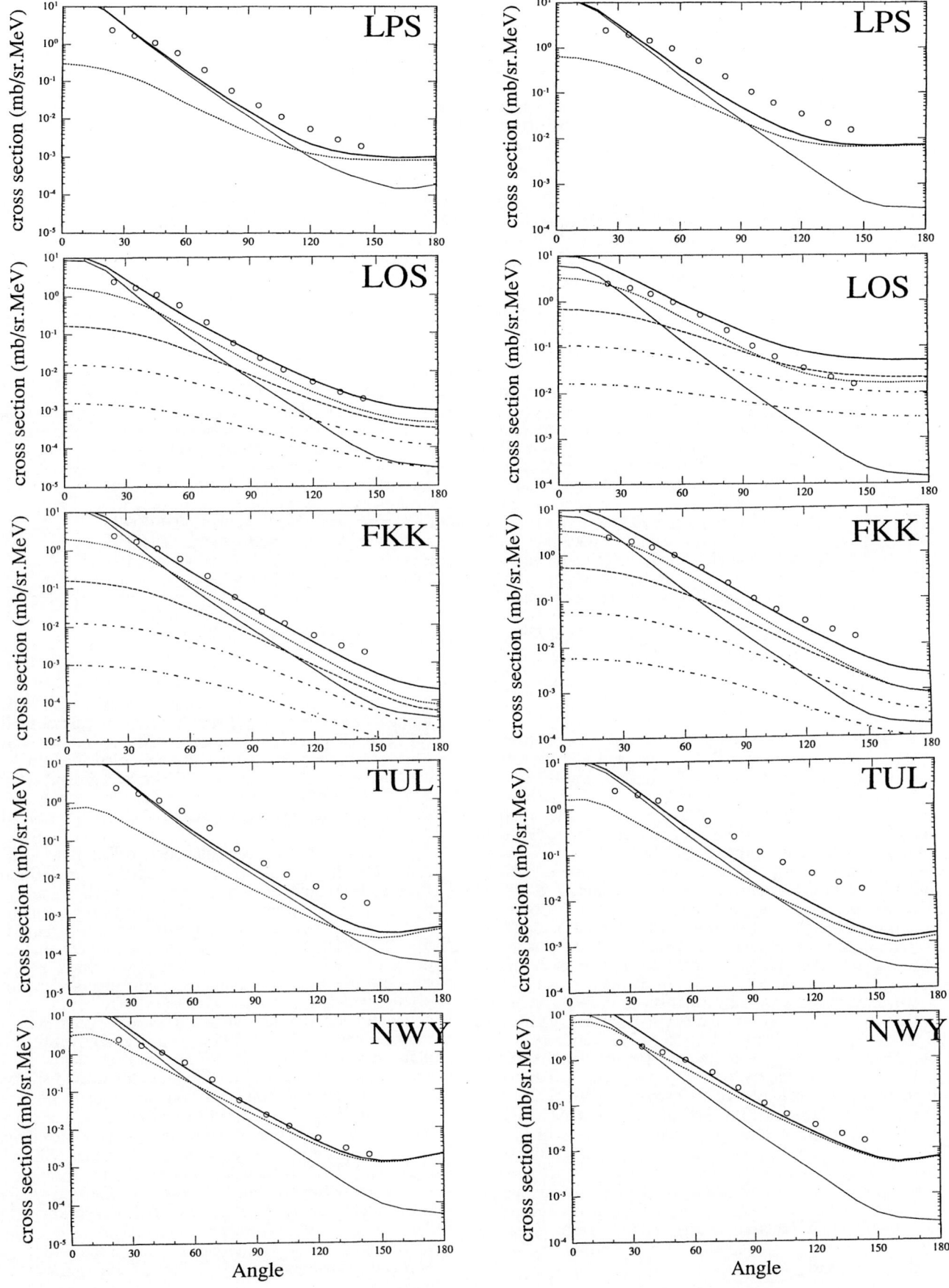

Figs.3 and 4: Model comparison for $^{90}Zr(p,p')$ with $E_{in} = 120$ MeV and $E_{out} = 80$ MeV (left column) and $E_{out} = 60$ MeV (right column). Thin solid line: 1-step, dotted line: 2-step, dashed line: 3-step, dashed-dotted line: 4-step, dashed-dotted-dotted line: 5-step, Fat solid line: Total MSD cross section.

THEORETICAL ANALYSES OF (n,xn) REACTIONS ON ^{235}U, ^{238}U, ^{237}Np, AND ^{239}Pu FOR ENDF/B-VI

P. G. Young and E. D. Arthur

Theoretical Division, MS B243
Los Alamos National Laboratory
Los Alamos, New Mexico 87545, USA

Abstract: Theoretical analyses were performed of neutron-induced reactions on ^{235}U, ^{238}U, ^{237}Np, and ^{239}Pu between 0.01 and 20 MeV in order to calculate neutron emission cross sections and spectra for ENDF/B-VI evaluations. Coupled-channel optical model potentials were obtained for each target nucleus by fitting total, elastic, and inelastic scattering cross section data, as well as low-energy average resonance data. The resulting deformed optical model potentials were used to calculate direct (n,n') cross sections and transmission coefficients for use in Hauser-Feshbach statistical theory analyses. A fission model with multiple barrier representation, width fluctuation corrections, and preequilibrium corrections were included in the analyses. Direct cross sections for higher-lying vibrational states were calculated using DWBA theory, normalized using B(Eℓ) values determined from (d,d') and Coulomb excitation data, where available, and from systematics otherwise. Initial fission barrier parameters and transition state density enhancements appropriate to the compound systems involved were obtained from previous analyses, especially fits to charged-particle fission probability data. The parameters for the fission model were adjusted for each target system to obtain optimum agreement with direct (n,f) cross section measurements, taking account of the various multichance fission channels, that is, the different compound systems involved. The results from these analyses were used to calculate most of the neutron (n,n), (n,n'), and (n,xn) cross section data in the ENDF/B/VI evaluations for the above nuclei, and all of the energy-angle correlated spectra. The deformed optical model and fission model parameterizations are described. Comparisons are given between the results of these analyses and the previous ENDF/B-V evaluations as well as with the available experimental data.

(Keywords: ^{235}U, ^{238}U, ^{237}Np, ^{239}Pu, neutron reactions, data evaluation, nuclear models, coupled-channel optical model, fission theory)

Introduction

We have completed theoretical analyses of neutron-induced reactions on ^{235}U, ^{238}U, ^{237}Np, and ^{239}Pu over the incident energy range 0.01-20 MeV in support of the ENDF/B-VI evaluation effort. Preliminary results from the ^{235}U analysis were reported at the Mito Conference. [1] The primary purpose for performing the analyses is to provide data on the reactions and energy ranges where little or no experimental data exist, especially for neutron emission reactions and with particular emphasis on odd-A actinides. For most of these nuclei, neutron total and fission cross section measurements exist, [2] so that parameters in the calculations can be optimized to those data. Additionally, limited elastic and inelastic angular distribution data were available for the analyses of ^{235,238}U and ^{239}Pu. [3] In the case of n+^{237}Np, however, there were virtually no elastic or inelastic scattering data, only fragmentary information on (n,γ) and (n,2n) reactions, and no experimental data on (n,3n) reactions or secondary neutron energy distributions. The reaction that is best described experimentally for ^{237}Np is fission, as new fission ratio measurements have recently been completed at LAMPF/WNR [4] and Argonne. [5] For all these actinides, prompt nubar measurements have been made over much of the energy range of interest, but almost no data are available on neutron energy and angular distributions at energies above a few MeV. Therefore, depending on the specific nuclide involved, the main function of the theoretical analyses is to provide total, elastic, inelastic, (n,2n), and (n,3n) cross sections, and in all cases, the angular and energy distributions of secondary neutrons.

Theoretical Analyses and Results

To summarize the analyses briefly, coupled-channel deformed optical model calculations were performed with the ECIS code [6] over the incident neutron energy range from approximately 0.001 to 20 MeV. The starting point for our optical model analyses were usually extensions [7] of the potentials of Lagrange, [3,8] which were then further modified for the present analysis to improve the calculations above 10 MeV. The role of the coupled-channel calculations in the present analysis are to obtain total, elastic, and ground-state rotational-band (n,n') cross sections, and to provide neutron transmission coefficients for Hauser-Feshbach statistical theory calculations.

The Hauser-Feshbach statistical calculations were performed with the COMNUC [9] and GNASH [10] codes. Both codes include a double-humped fission barrier model, using uncoupled oscillators for the barrier representation in GNASH and coupled or uncoupled oscillators in COMNUC. The COMNUC calculations include width-fluctuation corrections, which are needed at lower energies, whereas GNASH provides the preequilibrium corrections that are required at higher energies. Accordingly, COMNUC was used in the calculations below the threshold for second chance fission (approximately 5 MeV), utilizing fairly strongly damped coupled oscillators. The GNASH code was employed at higher energies, using uncoupled oscillators for second and higher chance fission. Fission transition state spectra were calculated from inputted bandhead parameters or were constructed by taking known (or calculated) energy levels and compressing their spacing by a factor of 2. As usual, Gilbert and Cameron [11] phenomenological level density functions were used to represent continuum levels at ground-state deformations, appropriately matched to available experimental level data. Multiplicative factors were applied to the level density functions to account for enhancements in the fission transition-state densities at barriers due to increased asymmetry conditions.

The fission barrier parameters for the ^{238}U, ^{237}Np, and ^{239}Pu calculations are given in Table 1. The fission cross sections from the calculations for these actinides are compared with experimental data in Fig. 1. The dashed curves given in Fig. 1 for ^{237}Np illustrate the contributions from first-, second-, and third-chance fission.

Table 1. Barrier Parameters Used in the Fission Calculations for ^{238}U, ^{237}Np, and ^{239}Pu.

| | n + ^{237}Np Compound Systems | | | |
	^{238}Np	^{237}Np	^{236}Np	^{235}Np
E_A (MeV)	5.87	6.20	5.70	6.40
$\hbar\omega_A$ (MeV)	0.31	0.85	0.50	0.85
E_B (MeV)	5.40	5.50	5.40	5.90
$\hbar\omega_B$ (MeV)	0.36	0.55	0.40	0.55
Density Enhancements:				
Barrier A	4.5	4.5	1.0	1.0
Barrier B	4.5	4.5	1.0	1.0

| | n + ^{238}U Compound Systems | | | |
	^{239}U	^{238}U	^{237}U	^{236}U
E_A (MeV)	6.25	5.83	6.03	6.10
$\hbar\omega_A$ (MeV)	0.75	0.50	0.50	0.50
E_B (MeV)	6.00	5.33	5.63	5.90
$\hbar\omega_B$ (MeV)	0.50	0.50	0.50	0.50
Density Enhancements:				
Barrier A	15.	3.8	1.8	1.0
Barrier B	2.	2.0	1.8	1.0

| | n + ^{239}Pu Compound Systems | | | |
	^{240}Pu	^{239}Pu	^{238}Pu	^{237}Pu
E_A (MeV)	5.78	5.75	5.65	5.65
$\hbar\omega_A$ (MeV)	0.80	0.63	0.90	1.00
E_B (MeV)	5.46	5.10	5.10	5.10
$\hbar\omega_B$ (MeV)	0.60	0.52	0.85	0.55
Density Enhancements:				
Barrier A	16.	1.1	1.0	2.5
Barrier B	2.	1.1	1.0	2.5

Although most of the direct reaction contribution to inelastic scattering is provided by the coupled-channel calculations of the ground state rotational bands (usually the first three members), additional direct contributions come from vibrational states, generally lying at higher excitation energies. Because of the close spacing of levels in ^{235}U, ^{237}Np, and ^{239}Pu, experimental information on (n,n') reactions to such vibrational states is essentially nonexistent. Therefore, to account for such contributions, we performed distorted-wave Born approximation (DWBA) calculations on nearby even-even nuclei (^{234,238}U, ^{238,240}Pu), using reduced transition probabilities B(Eℓ) from (d,d') and Coulomb excitation measurements [12] to obtain absolute (n,n') cross sections, and a weak coupling model [13] to apply the results to states in ^{235}U, ^{237}Np, and ^{239}Pu. The strongest transitions observed in the (d,d') measurements involve population of 3⁻ and 2⁺ vibrational states, corresponding to angular momentum transfers of ℓ=3 and ℓ=2, respectively. For ^{237}Np, the deformation parameters needed for normalizing the DWUCK calculations were estimated from systematics, that is, the required B(E2) and B(E3) values were estimated from those determined in the analyses for ^{235,238}U and ^{239}Pu, and all the ℓ=2 and ℓ=3 vibrational strength was placed into two fictitious states near E_x = 1 MeV. In the case of ^{235}U, the sum of (n,n') cross sections calculated from the dominant ℓ=3 and ℓ=2 transitions amounted to approximately 10% of the coupled-channel direct reactions at a neutron energy of 3 MeV, 30% at 8 MeV, and 23% at 20 MeV. While these direct contributions are not large, they do lead to a hardening of the inelastic neutron spectrum that should be included in the ENDF/B-VI evaluations. All the DWBA calculations were performed with the DWUCK code.[14]

As stated above, very limited experimental data exist for elastic scattering, inelastic scattering, and (n,xn) cross sections for the odd-A actinides, and some of the data that do exist are discrepant. In Fig. 2 elastic scattering angular distributions from the ^{239}Pu analysis are compared to experimental data and to earlier ENDF/B evaluations at a few incident energies.

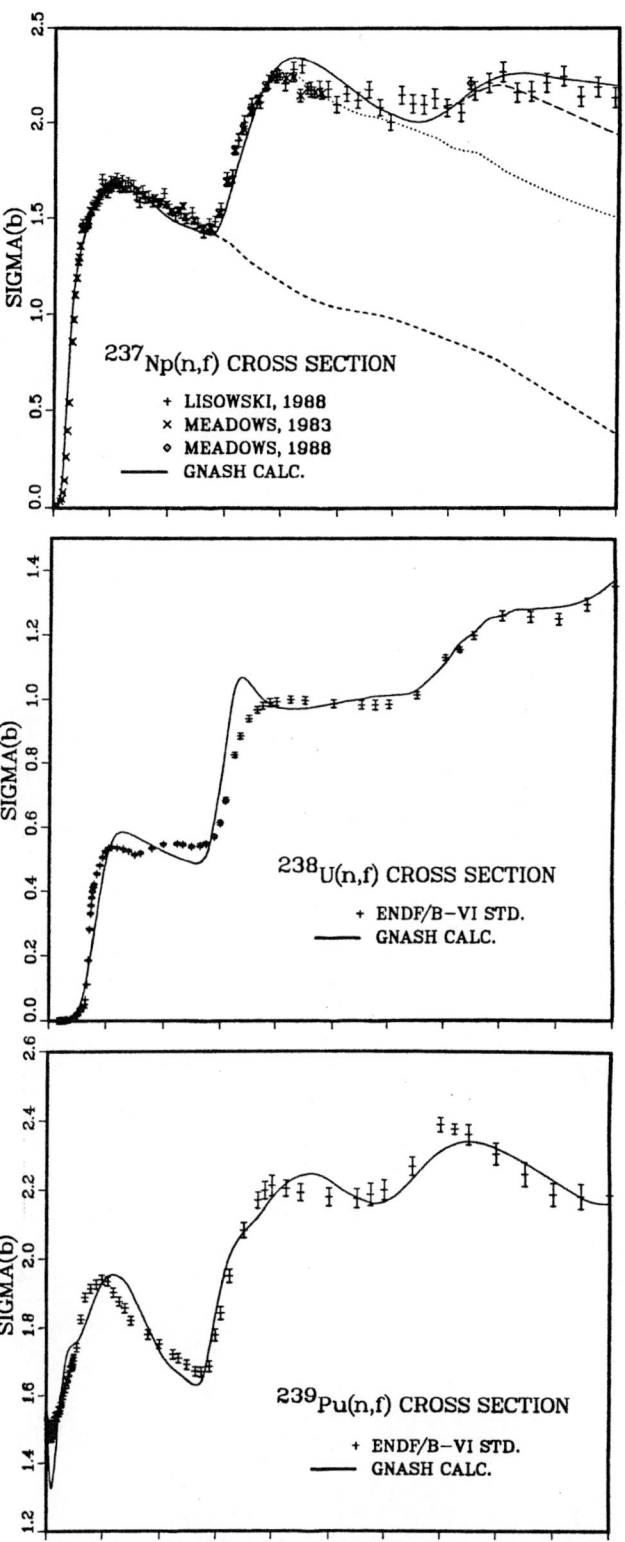

Fig. 1. Calculated and measured neutron-induced fission cross sections for ^{237}Np, ^{238}U, and ^{239}Pu from approximately 50 keV to 20 MeV. The points represent experimental data[2] and the curves are the results of our theoretical analysis. The dashed and dotted curves shown for ^{237}Np represent the calculated contributions from first-, second-, and third-chance fission.

896

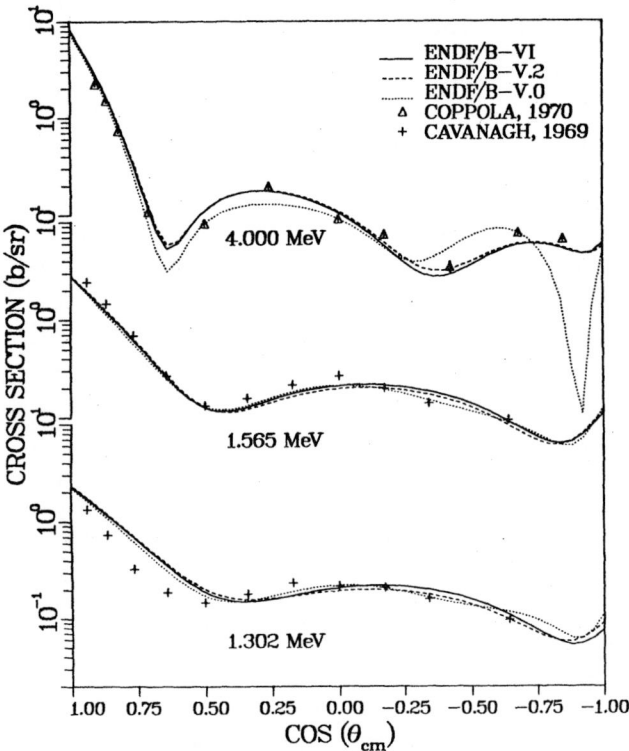

Fig. 2. Experimental and calculated elastic scattering angular distributions for n + ^{239}Pu at incident neutron energies of 1.302, 1.565, and 4.0 MeV. The dashed and dotted curves represent previous ENDF/B-V.2 and ENDF/B-V.0 evaluations, respectively.

Comparisons between experimental data and results from the present calculations of the ^{239}Pu(n,n') and ^{239}Pu(n,2n) cross sections are given in Fig. 3. Also shown

^{239}Pu(n,2n) Cross Section

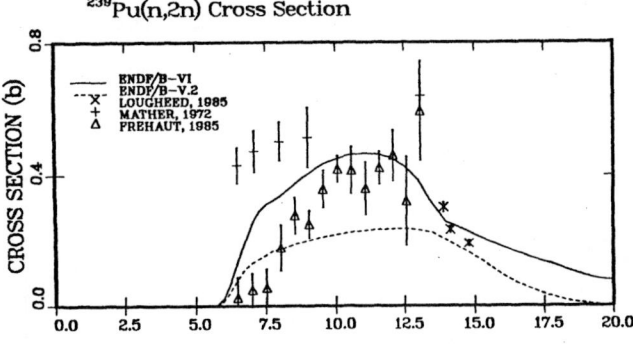

^{239}Pu(n,n') Cross Section

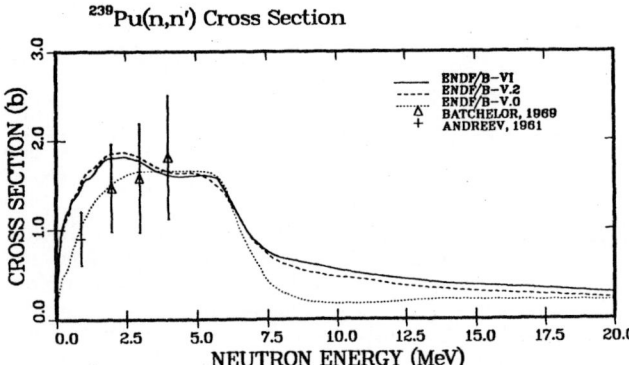

Fig. 3. Calculated and measured n + ^{239}Pu inelastic scattering and ^{239}Pu(n,2n)^{238}Pu cross sections. The dashed and dotted curves represent previous ENDF/B-V.2 and ENDF/B-V.0 evaluations, respectively.

are the previous ENDF/B-V.0 and ENDF/B-V.2 evaluations of these reactions. The present analysis leads to significant improvement in the (n,2n) cross section in the 8-15 MeV region, although discrepancies in the data are apparent. The influence of better accounting for direct reactions results in a somewhat higher (n,n') cross section than the previous ENDF/B-V.0 evaluation, particularly at higher energies.

Conclusions

In conclusion, the present analyses result in substantially improved agreement with the available data for the major odd-A actinides. Except for the fission and total cross sections, which are accurately determined from experiments, results from the present analyses are being used in the MeV region for all major cross sections, angular and energy distributions in the ENDF/B-VI evaluations for ^{235}U, ^{237}Np, and ^{239}Pu. Because more experimental data are available for n + ^{238}U reactions, we have used covariance analyses of the experimental data as well as previous ENDF/B-V analyses of data to represent the cross sections for several of the ^{238}U reactions, mainly utilizing our theoretical results for continuum neutron energy and angular distributions.

References

1. P. G. Young and E. D. Arthur, in Proc. Int. Conf. on *Nuclear Data for Science and Technology*, Mito, Japan, May 30 - June 3, 1988 (Ed. S. Igarasi, Saikon Publ. Co., Ltd., 1988), p. 603.

2. Experimental data available from the CSISRS compilation by the National Nuclear Data Center, Brookhaven National Laboratory.

3. G. Haouat, J. Lachkar, Ch. Lagrange, J. Jary, J. Sigaud, and Y. Patin, *Nucl. Sci. Eng.* 81, 491 (1982).

4. P. W. Lisowski, J. L. Ullmann, S. J. Balestrini, A. D. Carlson, O. A. Wasson, and N. W. Hill, in Proc. Int. Conf. on *Nuclear Data for Science and Technology*, Mito, Japan, May 30 - June 3, 1988 (Ed. S. Igarasi, Saikon Publ. Co., Ltd., 1988) p. 97.

5. J. W. Meadows, *Nucl. Sci. Eng.* 85, 271 (1983); J. W. Meadows, *Ann. Nucl. En.* 15, 421 (1988).

6. J. Raynal, "Optical-Model and Coupled-Channel Calculations in Nuclear Physics," IAEA SMR-9/8, Int. Atomic Energy Agency (1970).

7. P. G. Young, "Rare Earth-Actinide Potentials: ^{165}Ho, ^{238}U, ^{242}Pu," in Applied Nuclear Science Research and Development Semiannual Progress Report (Cps. E. D. Arthur, A. D. Mutschlecner) LA-10689-PR (1986) p. 48.

8. G. Haouat, Ch. Lagrange, J. Lachkar, J. Jary, Y. Patin, and J. Sigaud, in Proc. Int. Conf. on *Nuclear Cross Sections for Technology*, Knoxville, Tennessee (Oct. 22-26, 1979) p. 672.

9. C. L. Dunford, "A Unified Model for Analysis of Compound Nucleus Reactions," AI-AEC-12931, Atomics International (1970).

10. P. G. Young and E. D. Arthur, "GNASH: A Preequilibrium Statistical Nuclear-Model Code for Calculation of Cross Sections and Emission Spectra," Los Alamos Scientific Laboratory report LA-6947 (Nov. 1977); E. D. Arthur, "The GNASH Preequilibrium-Statistical Model Code," LA-UR-88-382 (1988).

11. A. Gilbert and A. G. W. Cameron, *Can. J. Phys.* 43, 1446 (1965).

12. Y. A. Ellis-Akovali, *Nucl. Data Sheets* 40, 523 (1983); E. N. Shurshikov, *Nucl. Data Sheets* 38, 277 (1983).

13. D. M. Brink, Nucl. Phys. 4, 215 (1957); P. Axel, Phys. Rev. 126, 671 (1962).

14. P. D. Kunz, "DWUCK - A Distorted-Wave Born Approximation Program," unpublished.

STATISTICAL MULTISTEP REACTION MODEL FOR NUCLEAR DATA

H. Kalka

TU Dresden, Mommsenstr.13, O-8027 Dresden, Germany

Abstract: A unique description of emission spectra, angular distributions and activation cross sections is proposed within a pure statistical multistep approach. The analytical model is based on both many-body theory (by Green's function formalism) and random matrix physics (EGOE). Calculations were performed for about 120 nuclei with code EXIFON.

(statistical multistep direct, statistical multistep compound reactions)

Introduction

The step from simple compound-nucleus models and single-step direct reaction models towards statistical multistep theory was mainly influenced by three basic ideas:

(i) Griffin [1]: classification of nuclear states by their complexity (exciton number);

(ii) FKK [2]: distinction between bound and unbound configurations;

(iii) AWM [3]: the (chaotic) nuclear Hamiltonian treated as a random matrix.

In many-body theory, these ideas are realized by (Born-series) expansion of the (retarded part of the) mass operator in powers of the residual interaction. The latter are replaced by random matrices from EGOE. After energy (ensemble) averaging analytical expressions for the (differential) cross section were obtained [4],

$$(a, xb) = (SMD) + (SMC) + (MPE)$$

The first term denotes the statistical multistep direct part. The second term symbolizes the multistep compound emission. Both terms together (SMD+SMC) represent the *first-*chance emission process. In the last term multiple particle emissions (MPE) are considered which are calculated in a pure SMC concept. We account for reactions with neutrons, protons, α-particles and photons.

Fundamental Processes

The cross section can be constructed from few fundamental graphs. We distinguish between 4 types of mean squared matrix elements between bound and/or unbound configurations ($\overline{I^2_{BB}}\ \overline{I^2_{BU}}\ \overline{I^2_{UB}}\ \overline{I^2_{UU}}$):

They are defined by the same (surface-delta) residual interaction of strength $F_0 = 27.5$ MeV [5]. In addition, collective amplitudes for phonon excitations and γ-emissions (via E1) are considered:

$$E1$$

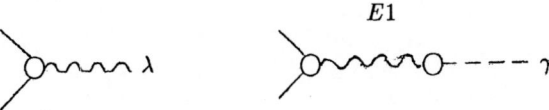

The first is obtained from deformation β_λ, the second from E1-sum rule.

Bound particles and holes (solid line) and unbound particles (double line) are distinguished by the s.p. state density $\rho(E) = 4\pi V m\,(2mE)^{1/2}\,/\,(2\pi\hbar) \propto r_0^3 AE^{1/2}$ in the nuclear volume V. Bound states are defined by $\rho(E_F)$ at Fermi energy E_F and unbound states by $\rho(E_c)$ at the kinetic energy E_c. An extra introduction of state density g is not necessary.

SMD Cross Section

The SMD process is based on the unbound-unbound matrix elements $\overline{I^2_{UU}}$. It is a sum over s-step direct contributions,

$$(SMD) = (1) + (2) + (3) + \cdots$$

Considering also collective phonon excitations, each contribution splits into the following processes: $(1) = (ex) + (vib)$, $(2) = (2ex) + (ex, vib) + (vib, ex) + (2vib)$, etc. They are denoted according to the sequence of exciton and phonon excitations.

SMC Cross Section

The graphical representation of the SMC process is

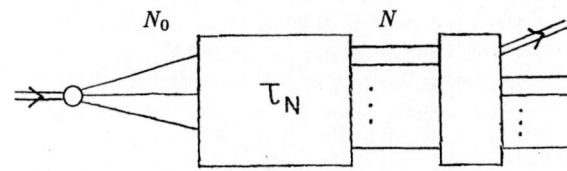

The open circle represents the unbound-bound matrix element $\overline{I^2_{UB}}$ which defines the SMC-formation cross section. In this way, there is no reference to optical model calculations.

The left box describes the transition from the exciton state $N_0 = 3$ up to N. This is organized by the master equation which predicts the mean life time τ_N. The main contributions to this box come from bound-bound graphs ($\overline{I^2_{BB}}$), which define the damping widths $\Gamma_N \downarrow$.

The right box contain bound-unbound diagrams ($\overline{I^2_{BU}}$) which define the escape widths $\Gamma_N \uparrow$.

For *proton*-induced reactions isospin conservation is taken into account [6].

Gamma Emission

The direct part of the γ-spectrum is described by the direct-semidirect process DSD (formation and decay of GDR),

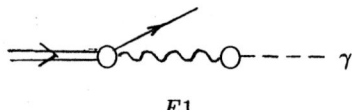

$E1$

The SMC γ-emission is described by inclusion into the right box of the above SMC diagram two modes ($\Delta N = 0, -2$) of γ-escape widths.

Results

Calculations for about 120 nuclei (below 100 MeV incident energy) were performed with code EXIFON [7]. This was done with *one* parameter set (F_0, r_0, E_F). Binding energies and phonon parameters were taken from nuclear tables. The only adjustable quantity is the pairing shift (standard value $\Delta = 12.8 A^{-1/2}$). This value is diminished near closed shell nuclei.

The formation and decay of the compound system as well as the direct processes were calculated by the same interaction strength F_0 taken from nuclear structure considerations.

Some results of (standard-value) calculations of emission spectra and activation cross sections are presented in Figs. 1 to 9. The agreement with experimental data is very good. (The calculation time is about 20 sec per incident energy at AT/386.)

For energies above about 80 MeV the strength of the residual interaction F_0 should be reduced.

The output of code EXIFON can also be arranged into ENDF-6 format [8].

References

[1] J.J. Griffin, Phys. Lett. **24B**, 5 (1967)

[2] H. Feshbach, A. Kerman, and S.E. Koonin, Ann. Phys. (N.Y.) **125**, 429 (1980)

[3] D. Agassi, H.A. Weidenmüller, and G. Mantzouranis, Phys. Rep. **22**, 145 (1975)

[4] H. Kalka, Z. Phys. A, in press (1991)

[5] A. Faessler, Fortschr. Phys. **16**, 309 (1968)

[6] C. Kalbach, S.M. Grimes, C. Wong, Z. Phys. **A275**, 175 (1975)

[7] H. Kalka, code EXIFON (version 2.0), (1991)

[8] M. Toepfer, and E. Schubert, code MAKE6 (1991)

[9] M.Baba, M.Ishikawa, N.Yabuta, T.Kikuchi, H.Wakabayashi, N.Mirakawa, Int. Conf. on Nucl. Data, Mito, 291 (1988)

[10] S.M. Grimes, R.C. Haight, K.R. Alvar, H.H. Barschall, and R.R. Borchers, Phys. Rev. **C1**, 2127 (1979)

[11] R. Fischer, C. Derndorfer, B. Strohmaier, M. Uhl, and H. Vonach, Ann. Nucl. Eng. **90**, 174 (1982)

[12] D.M. Drake, E.D. Arthur, and M.G. Silbert, Nucl. Sc. Eng. **65**, 48 (1978)

[13] M. Budnar et al., *Report INDC(YUG)-6/L* (1979)

[14] D.C. Santry and J.P. Butler, Can. J. Phys. **42**, 1030 (1964)

[15] A.Marcinkowski, R.Finlay, G.Randeva, C.Brient, R.Kunup, S.Hellema, S.Meigaoni, R.Tailor, Nucl.Sci.Eng. **13**, 13 (1983)

[16] E. Mordhorst, M. Trabandt, A. Kaminsky, H. Krause, W. Scobel, R. Bonetti, and F. Crepti, Phys. Rev. **C34**, 103 (1986)

[17] Y. Watanabe, I. Kumabe, M. Hyakutake, N. Koori, K. Ogawa, K. Orito, K. Akagi, and N. Oda, Phys. Rev. **C36**, 1325 (1987)

[18] M. Trabandt, W. Scobel, H.M. Blann, R. Pohl, C. Byrd, C.C. Foster, and R. Bonetti, Phys. Rev. **C39**, 452 (1989)

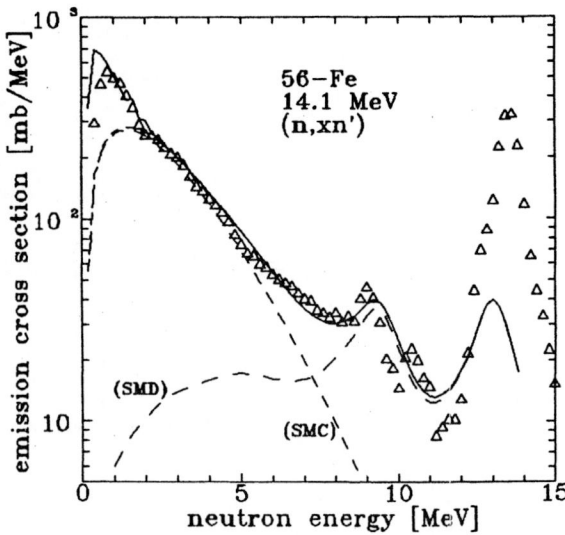

Fig. 1. (n, xn')-Emission spectrum at 14.1 MeV incident-energy. Experimental data from [9]. For abbreviations see text.

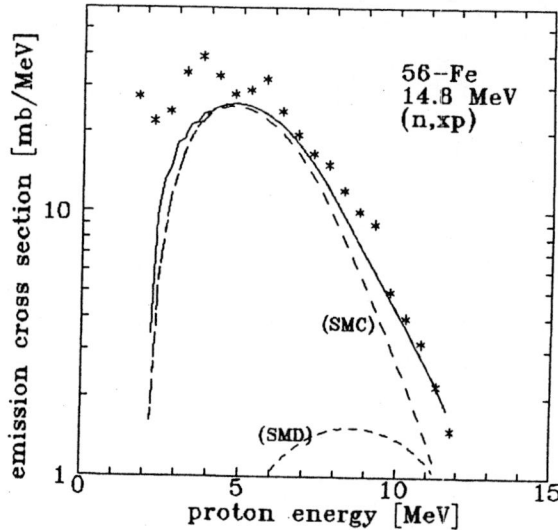

Fig. 2. (n, xp)-Emission spectrum at 14.8 MeV incident energy. Experimental data from [10].

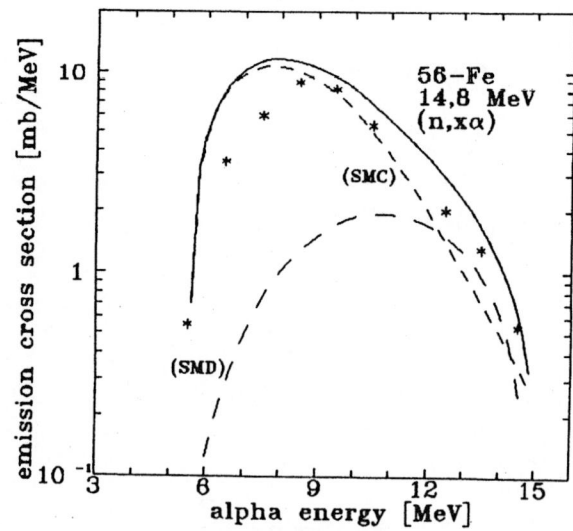

Fig. 3. $(n, x\alpha)$-Emission spectrum at 14.8 MeV incident energy. Experimental data from [11].

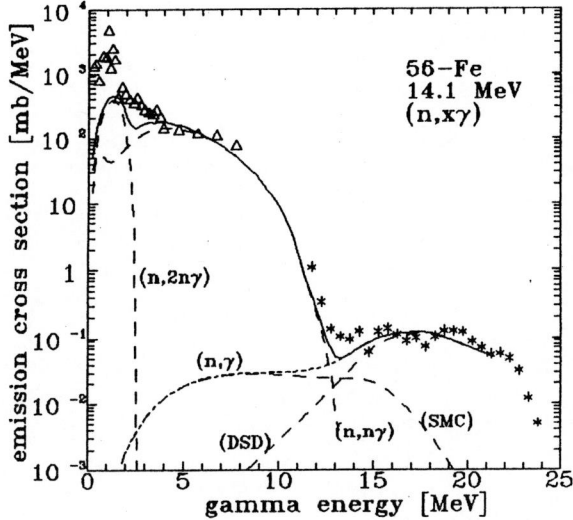

Fig. 4. $(n, x\gamma)$ emission spectrum at 14.1 MeV incident energy. Experimental data from [12,13].

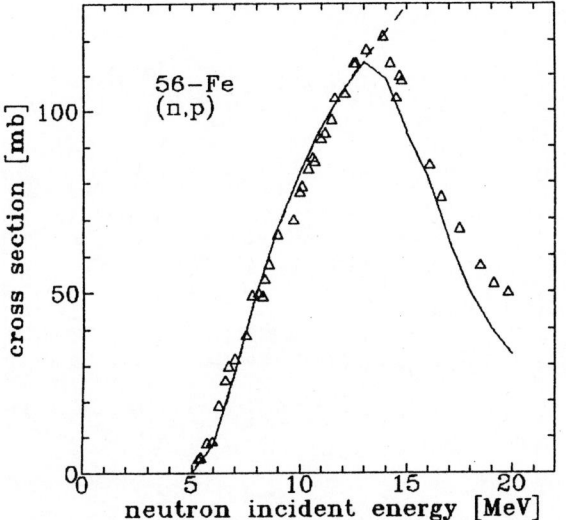

Fig. 5. (n, p)-Activation cross section for 56-Fe. Experimental data from [14].

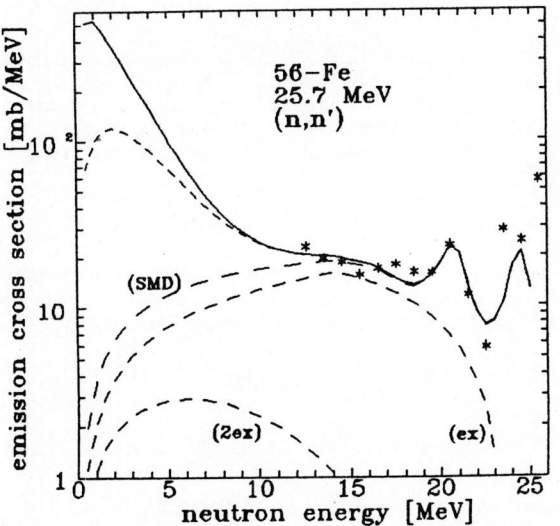

Fig. 6. (n, xn')-Emission spectrum at 25.7 MeV incident energy. Experimental data from [15]. (ex) and (2ex) denote the non-collective one- and two-step processes.

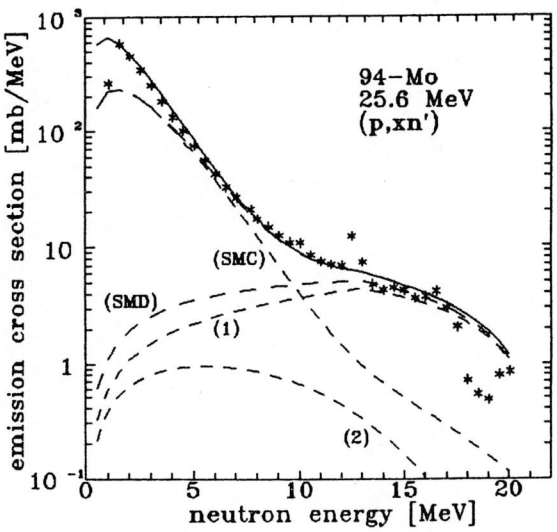

Fig. 7. (p, xn)-Emission spectrum at 25.6 MeV incident energy. Experimental data from [16].

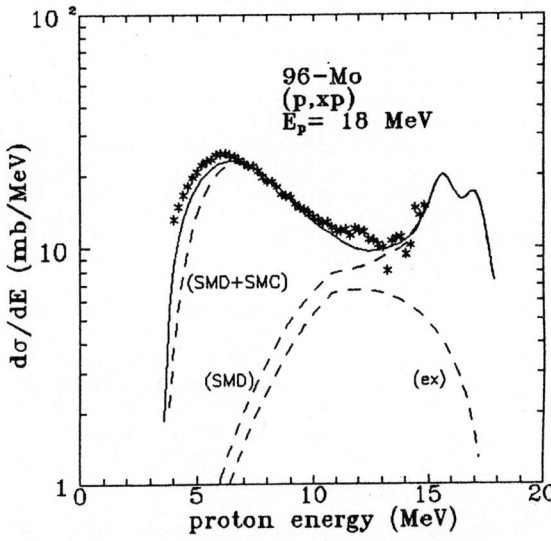

Fig. 8. (p, xp')-Emission spectrum at 18 MeV incident energy. In SMD pairing effects are neglected. Experimental data from [17].

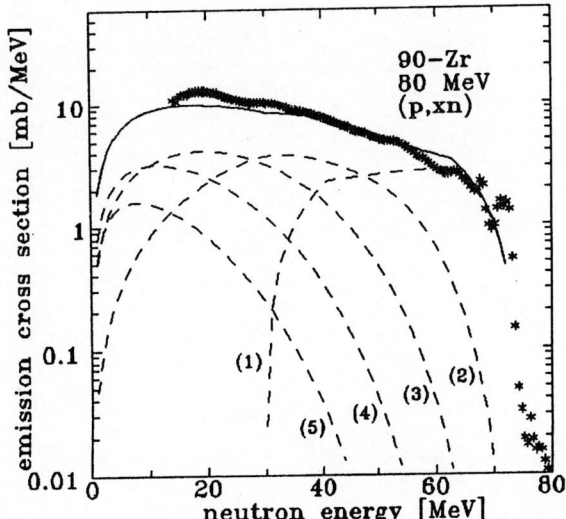

Fig. 9. (p, n)-SMD spectrum at 80 MeV incident energy. Experimental data from [18].

NEUTRON OPTICAL POTENTIAL OF ^{28}Si DERIVED FROM THE DISPERSION RELATION

H. Kitazawa, S. Igarasi[†], D. Katsuragi and Y. Harima

Tokyo Institute of Technology
2-12-1 O-okayama, Meguro-ku, Tokyo 152, Japan

† Nuclear Energy Data Center
2-4 Shirakata, Tokai-mura, Naka-gun
Ibaraki-ken 319-11, Japan

Abstract: Based upon the dispersion theory, an optical potential of ^{28}Si was determined at the neutron energies from the Fermi energy to 20 MeV. In particular, discussion was given on a characteristic behavior of the optical potential for low-energy neutrons. Moreover, the validity of the dispersion theory was investigated for neutron single-particle bound states in ^{29}Si.

(^{28}Si, neutron optical potential, dispersion theory)

Introduction

A lot of analyses of experimental data on neutron scattering have provided several systematics of the complex optical potential, $U(r,E) = V(r,E) + iW(r,E)$, which describes an averaged interaction between neutrons and nuclei in the large domain of neutron energy and nuclear mass. However, continuing efforts are still being devoted to a more precise determination of the neutron optical potential. On the other hand, recent studies [1,2] have shown that an optical potential derived from the dispersion theory satisfactorily represents both the neutron scattering states and bound states in the wide neutron-energy range.

Motivated by those results, we planned to determine the optical potential of ^{28}Si, using the dispersion theory, because reliable sufficient data on neutron scattering on ^{28}Si are available and neutron single-particle bound states in ^{29}Si have been observed well.

Dispersion Relation

According to Feshbach's generalized optical model[3], the nucleon mean field is described as the sum of a Hartree-Fock component and a dispersive component. The Hartree-Fock component is theoretically approximated by a local potential, $V_H(r,E)$, whose strength decreases linearly with increasing incident particle energy. While, the dispersive component, $\Delta V(r,E)$, is represented by a dispersion relation that connects the real part of an optical potential with the imaginary part consisting of the surface absorption potential, $W_S(r,E)$, and the volume absorption potential, $W_V(r,E)$:

$$V(r,E) = V_H(r,E) + \frac{P}{\pi}\int_{-\infty}^{\infty} \frac{W(r,E')}{E'-E} dE'$$
$$= V_H(r,E) + \Delta V(r,E)$$
$$= V_H(r,E) + \Delta V_S(r,E) + \Delta V_V(r,E),$$

where P indicates a principal value, and $\Delta V_S(r,E)$ and $\Delta V_V(r,E)$ are the dispersion potentials due to surface and volume absorption, respectively.

For a nucleon potential, the absorption potential, $W(r,E)$, is usually assumed to be symmetric about the Fermi energy ε_F:

$$V(r,E) = V_H(r,E) + \frac{2}{\pi}(E-\varepsilon_F) P \int_{\varepsilon_F}^{\infty} \frac{W(r,E')}{(E'-\varepsilon_F)^2-(E-\varepsilon_F)^2} dE'.$$

From the above equation, it is found that the dispersive component is zero at the Fermi energy. When the nucleon energy is measured from the Fermi-energy level, the real term of the optical potential is written as

$$V(r,E) = V_H(r,E) + \frac{2E}{\pi} P \int_{0}^{\infty} \frac{W(r,E')}{(E'^2-E^2)} dE'.$$

Optical Potential

In order to calculate the dispersive component of the optical potential, we assumed the neutron-energy dependence of the absorption potentials, $W_S(r,E)$ and $W_V(r,E)$, as shown by the solid lines in fig. 1[4,5]:

$$-W_S(E_n) = 0.30E_n + 3.84 \text{ MeV}, \quad E_n \leq 11.0 \text{ MeV}$$
$$= -0.084E_n + 8.06 \text{ MeV}, \quad 11.0 \leq E_n \leq 96.0 \text{ MeV}$$
$$= 0.0, \quad E_n \geq 96.0 \text{ MeV},$$

and

$$-W_V(E_n) = 0.0, \quad E_n \leq 13.0 \text{ MeV}$$
$$= 0.15(E_n-13.0) \text{ MeV}, \quad 13.0 \leq E_n \leq 60.6 \text{ MeV}$$
$$= 7.14 \text{ MeV}, \quad E_n \geq 60.6 \text{ MeV}.$$

The Hartree-Fock potential was determined so that the real Woods-Saxon potential strength in the full optical potential equalized to that of the phenomenological optical potential of Whisnant et al.[5] at the neutron energy for $V_S(r,E)=0.0$:

$$V_H(E_n) = 48.54 - 0.31E_n \text{ MeV}.$$

The geometrical parameters were taken from the work of Whisnant et al. except the radius parameter r_0 of the real Woods-Saxon potential:

$$r_W = 1.26 \text{ fm}, \quad a_W = 0.58 \text{ fm}, \quad a_0 = 0.64 \text{ fm},$$
$$r_{SO} = 1.01 \text{ fm}, \quad a_{SO} = 0.50 \text{ fm}, \quad V_{SO} = 6.0 \text{ fm}.$$

The Fermi energy ε_F was defined as an average of the minimum energy ε_F^+ of particle states and the maximum energy ε_F^- of hole states in ^{28}Si:

$$\varepsilon_F = \frac{\varepsilon_F^+ + \varepsilon_F^-}{2} = -12.8 \text{ MeV}.$$

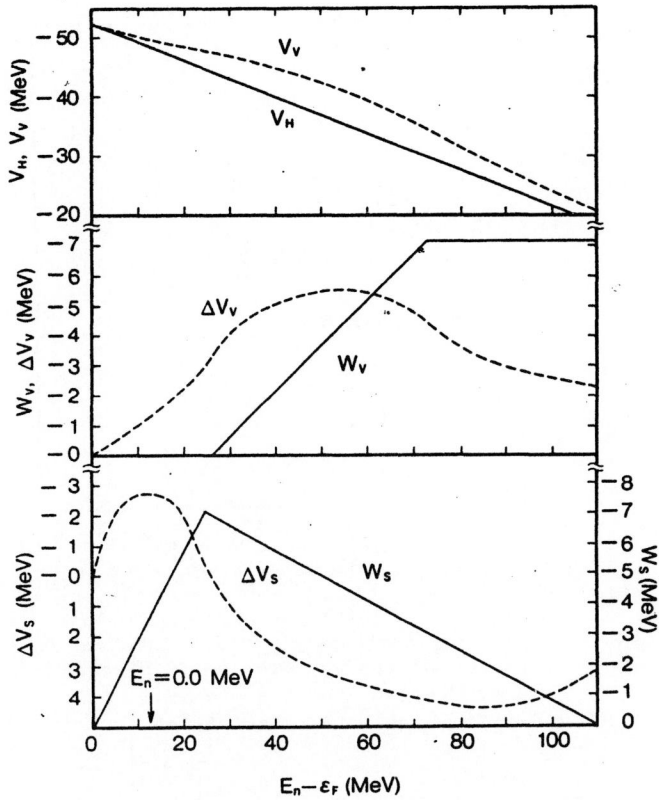

Fig. 1. Hartree-Fock potential and dispersion potential.

Calculations and Discussion

Fig. 2 shows a comparison between the measured and calculated total cross sections of ^{28}Si. The dashed curve was calculated using the optical potential predicted by the dispersion theory. In the calculation, the radius parameter r_0 was increased with decreasing neutron energy below 11.0 MeV to reproduce the average total cross sections, as shown in fig. 5. The solid curve was calculated without the surface dispersion potential. The calculations produce a considerable overestimation of the average total cross sections below 2 MeV.

In order to investigate effects of the surface dispersion potential, we obtained the dependence of total cross sections on the radius parameter r_0. As seen from the solid curves in fig. 3, the total cross sections above 2 MeV increase with the increasing radius parameter, which may be predicted by a classical picture. However, the total cross sections below 2 MeV decrease with the increasing radius parameter and reach a minimum value at $r_0 = 1.35 - 1.40$ fm. The behavior is more remarkable at lower neutron energy. In the figure, the dashed curve for $E_n = 0.7$ MeV was calculated without the surface dispersion potential. From the comparison between the solid and dashed curves for $E_n = 0.7$ MeV, it is easily found that an addition of $\Delta V_s(r,E)$ is equivalent to increasing the radius parameter r_0 in the calculation which uses only a Woods-Saxon potential.

A physical explanation to the characteristic behavior of total cross sections is provided by fig. 4. That is to say, in the low energy region where there is no essential contribution of partial waves with angular momentum $l>0$, it originates from the Ramsauer-like interference[6] produced by a gradual approach to 180° of the s-wave nuclear phase shift with an increase of the radius parameter r_0.

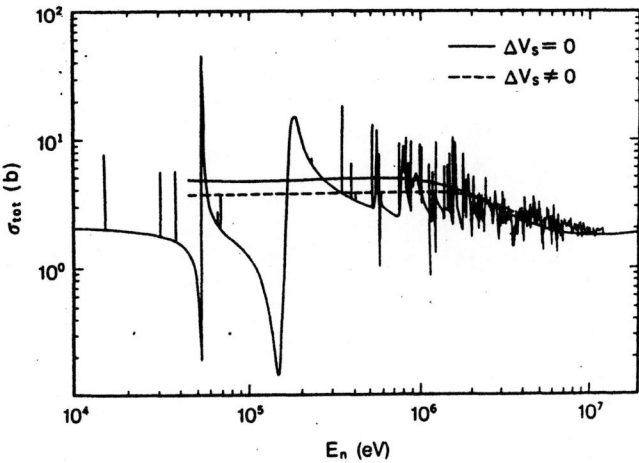

Fig. 2. Neutron total cross sections of ^{28}Si.

Using the full optical potential obtained above, moreover, we calculated the energy E_n of neutron single-particle bound states in ^{29}Si. Those results are listed in table 1. In the calculation, the radius parameter r_0 was adjusted to reproduce the single-particle energy, which is the center-of-gravity of single-particle strengths observed by El-Naiem et al.[7] and Mermaz et al.[8] in the ^{28}Si(d,p) reactions. For the $2s_{1/2}$, $1d_{3/2}$ and $1f_{7/2}$ states, 90-100% of the sum-rule strength has been observed. However, the 2p states are more fragmented, and only 50-60% of the strength was found. As seen from the table, the surface dispersion potential produces a large effect on the single-particle energies, and those energies approach toward the Fermi-energy level in addition of the potential.

Figs. 5 and 6 show the neutron-energy dependence of the radius parameter r_0 and the potential integral, respectively. Both of those quantities reach a peak in the vicinity of $E_n=0$.

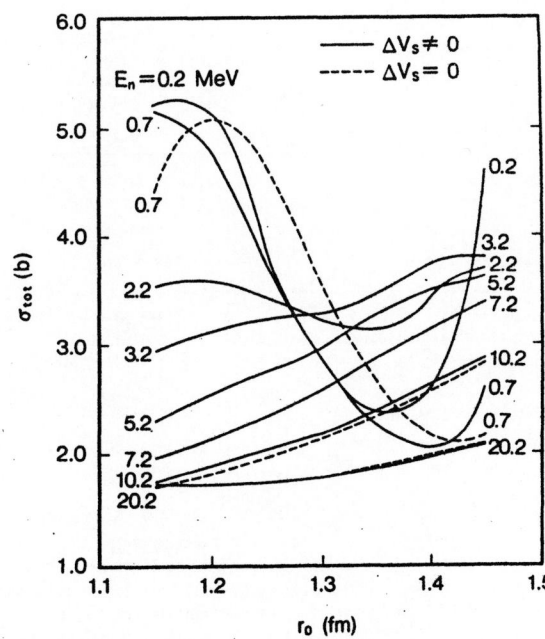

Fig. 3. Dependence of the total cross section on the radius parameter r_0.

902

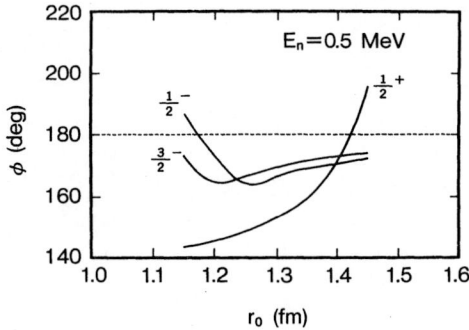

Fig. 4. Dependence of the nuclear phase shift φ on the radius parameter r_0.

Table 1. Neutron single-particle energies in ^{29}Si

nlj	$-E_n$ (MeV)		
	V_H	$V_H + \Delta V_V$	$V_H + \Delta V_V + \Delta V_S$
$2s_{1/2}$	7.31	7.55	8.52
$1d_{3/2}$	5.64	5.97	7.21
$1f_{7/2}$	1.73	2.19	3.88
$2p_{3/2}$	1.90	2.26	3.30
$2p_{1/2}$	0.71	1.08	2.06

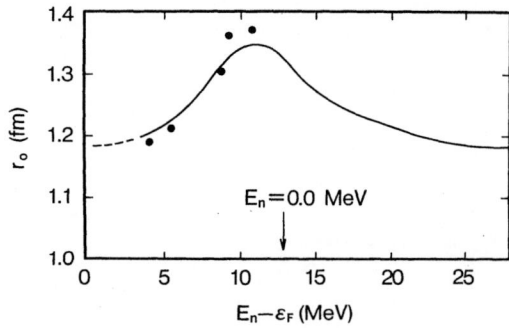

Fig. 5. Dependence of the radius parameter r_0 on neutron energy E_n. The solid circles were obtained from the analysis of neutron single-particle states in ^{29}Si.

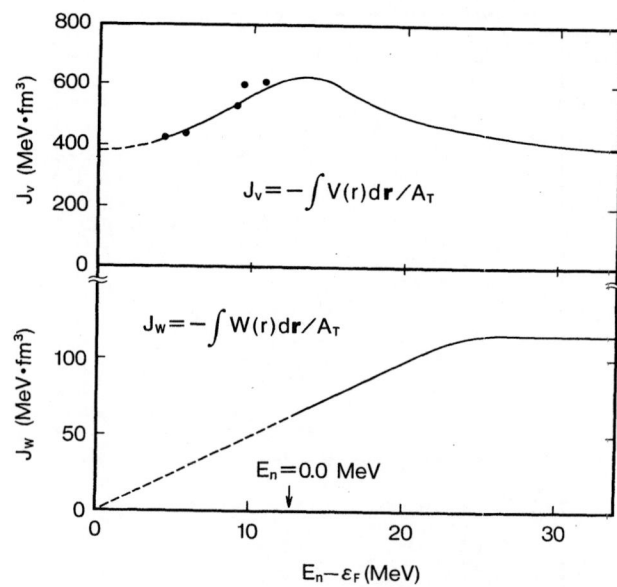

Fig. 6. Dependence of the potential integral on neutron energy E_n. The solid circles were obtained from the analysis of neutron single-particle states in ^{29}Si.

REFERENCES

1. C.H. Johnson, D.J. Horen and C. Mahaux : Phys. Rev. C36, 2252 (1987)
2. R.K. Das and R.W. Finlay : Phys. Rev. C42, 1013 (1990)
3. H. Feshbach : Ann. Phys. 5, 357 (1958)
4. R.L. Walter, C.R. Howell and W. Tornow : Proc. Int. Conf. on Nuclear Data for Science and Technology, Mito, 1988 (Saikon, Tokyo, 1988), p.675
5. C.S. Whisnant, J.H. Dave and C.R. Gould : Phys. Rev. C30, 1435 (1984)
6. P.E. Hodgson : Proc. Int. Conf. on Nuclear Reactions, Calcutta (1989), p.39
7. A. El-Naiem and R. Reif : Nucl. Phys. A189, 305 (1972)
8. M.C. Mermaz, C.A. Whitten, Jr., J.W. Champlin, A.J. Howard and D.A. Bromley : Phys. Rev. C4, 1778 (1971)

Summary

A considerable overestimation of the total cross sections of ^{28}Si for low-energy neutrons, which is produced by a global optical potential, was removed by using an optical potential predicted by the dispersion theory. Moreover, we have shown that the potential successfully reproduces the energies of observed single-particle bound states in ^{29}Si. Those results were chiefly obtained by adding the surface dispersion potential to the real Woods-Saxon potential, rather than by adding the volume dispersion potential.

ON APPLICATION OF THE S-MATRIX TWO-POINT FUNCTION TO NUCLEAR DATA EVALUATION

S. Igarasi

Nuclear Energy Data Center
Tokai-mura, Naka-gun, Ibaraki-ken 319-11, Japan

Abstract: Statistical model calculation using S-matrix two-point function (STF) was tried. The results were compared with those calculated with the Hauser-Feshbach formula (HF) with and without resonance level-width fluctuation corrections (WFC). The STF gave almost the same cross sections as calculated using Moldauer's degrees of freedom for the χ^2-distributions (MCD). The effect of the WFC to the final states in continuum was also studied using the HF with WFC of the MCD and of Porter-Thomas distribution (PTD). The HF with the MCD is recommended for practical calculation of the cross sections.

(statistical models, S-matrix two-point function, Hauser-Feshbach, level-width fluctuation correction, cross sections, χ^2-distribution)

Introduction

Evaluation work for a nuclear data library requires a huge amount of numerical data for many nuclides in wide energy ranges. Data for minor isotopes, for radioactive species and for higher energy regions are often requested. Frequently, experimental data are not available for such requests, because available data are restricted to limited energies, stable isotopes and measurable quantities. Hence, nuclear model calculations are indispensable for production of data for these nuclides and energies. The model calculations are useful also for supplementing deficiencies of experimental data, compensating for the lack of data and polishing the quality of the evaluated library.

It is fortunate for nuclear data evaluation that many rigorous, modern, sophisticated and utilizable theoretical models exist. The statistical model is an example. Together with the others, such as the optical model, it has been widely used to perform nuclear model calculations for evaluation. In particular, the Hauser-Feshbach formula (HF) /1/ with and without resonance level-width fluctuation correction (WFC) /2,3,4/ has been generally applied to nuclear data production.

This correction enhances the compound elastic scattering cross section (σ_{ce}) and decreases the inelastic scattering cross sections (σ_{in}), in general. Therefore, the spectra of scattered neutrons are changed by taking this correction into account. Especially, the degrees of freedom (DOF) for χ^2-distribution of the WFC are artificial parameters which affect increase and decrease of the cross sections. In this sense, arbitrary selection of the DOF for the WFC produces ambiguity of the evaluated nuclear data.

Recently, an excellent and rigorous formula of the S-matrix two-point function (STF) for the compound nuclear reaction has been derived /5,6/. This is based on the Gaussian Orthogonal Ensemble (GOE) of matrix elements for the nuclear Hamiltonian. It is composed of a triple integral without any optional parameters like the DOF. Although it seems too complicated to apply it to the nuclear data evaluation, it should be tried to examine

whether it is applicable or not, and what relation to the existing formulas appears.

In this paper, σ_{ce} and σ_{in} of ^{238}U obtained with the STF are presented for several energy points, and are compared with the results calculated using the HF with the MCD and PTD. Effect to the level excitation cross sections which are proportional to the neutron spectra are also examined. The STF gave almost the same results as those obtained with the MCD for these cross sections.

The effect of the WFC to the final states in continuum (overlapping levels) is studied. The STF is not available for the study on this problem, because it is only applicable to discrete levels in the final states. Hence, the HF with the WFC of MCD and PTD are used for this purpose. It is noticed from this study that the effect of the WFC still remains in the continuum region. In practical calculation of the cross section, in this sense, the HF with the MCD is recommended, not only for the discrete levels but for the states in continuum.

Table 1. Comparison between the cross sections obtained by the STF and those obtained by the HF with and without the WFC. The STF gives almost the same results as the HF with the MCD.

MeV	Barns	STF	MCD	PTD	HF
0.1	σ_{ce}	3.21	3.21	3.25	2.61
	σ_{in}	0.878	0.883	0.836	1.48
0.3	σ_{ce}	2.63	2.62	2.86	1.85
	σ_{in}	1.67	1.68	1.45	2.45
0.5	σ_{ce}	2.02	2.02	2.27	1.39
	σ_{in}	1.78	1.78	1.54	2.42
0.8	σ_{ce}	1.33	1.33	1.55	0.836
	σ_{in}	2.24	2.24	2.02	2.74
1.0	σ_{ce}	1.03	1.04	1.26	0.610
	σ_{in}	2.68	2.67	2.45	3.10

Cross Section Formula and WFC

Differential cross section for the reaction between entrance channel α and exit channel α' is given as follows, using the Legendre expansion:

$$\frac{d\sigma(\alpha \to \alpha')}{d\Omega_{\alpha'}} = \frac{1}{k_\alpha^2} \frac{1}{(2I_1+1)(2I_2+1)} \sum_L B_L(\alpha \to \alpha') P_L(\cos\theta_\alpha) \qquad (1)$$

where I_1 and I_2 are the spins of the projectile and the target nucleus, respectively, and k_α is the wave number of the relative motion.

The Legendre coefficient B_L is written using the STF which is composed of the triple integrals;

$$B_L(\alpha \to \alpha') = \frac{1}{16} \int_0^1 dx_0 \int_0^1 dx_1 \int_0^{x_1} dx_2 \frac{x_0(1-x_0)(x_1-x_2)}{(x_0+(1-x_0)x_1)^2(x_0+(1-x_0)x_2)^2(x_1x_2)^{1/2}} \sum_J (2J+1)^2 \sum_{\ell j} \prod_{\ell' j'}^J (1-T_c x_0) \left\{ \frac{(1-x_1)(1-x_2)}{[1-(1-T_c)x_1][1-(1-T_c)x_2]} \right\}^{1/2} \left\{ \delta_{\alpha\alpha'}(Z(\ell j \ell' j'; I_1 L)W(Jj Jj'; I_2 L))^2 \right.$$

$$\times (X^J_{(\alpha j \ell),(\alpha j \ell')} + Y^J_{(\alpha j \ell),(\alpha j \ell')}) + (-1)^{I_2-I'_2+I'_1-I_1} Z(\ell j \ell j; I_1 L)Z(\ell' j' \ell' j'; I'_1 L) W(Jj Jj; I_2 L)W(Jj' Jj' : I'_2 L) Y^J_{(\alpha j \ell)(\alpha' j' \ell')} \right\} \qquad (2)$$

where X_{ab} and Y_{ab} correspond to the elastic and inelastic components of the STF, respectively. They are written as

$$X_{ab} = T_a \sqrt{1-T_a}\, T_b \sqrt{1-T_b} \left(\frac{X_1}{1-(1-T_a)X_1} + \frac{X_2}{1-(1-T_a)X_2} \right.$$

$$\left. + \frac{2X_0}{1-T_a X_0} \right) \left(\frac{X_1}{1-(1-T_b)X_1} + \frac{X_2}{1-(1-T_b)X_2} + \frac{2X_0}{1-T_b X_0} \right). \qquad (3)$$

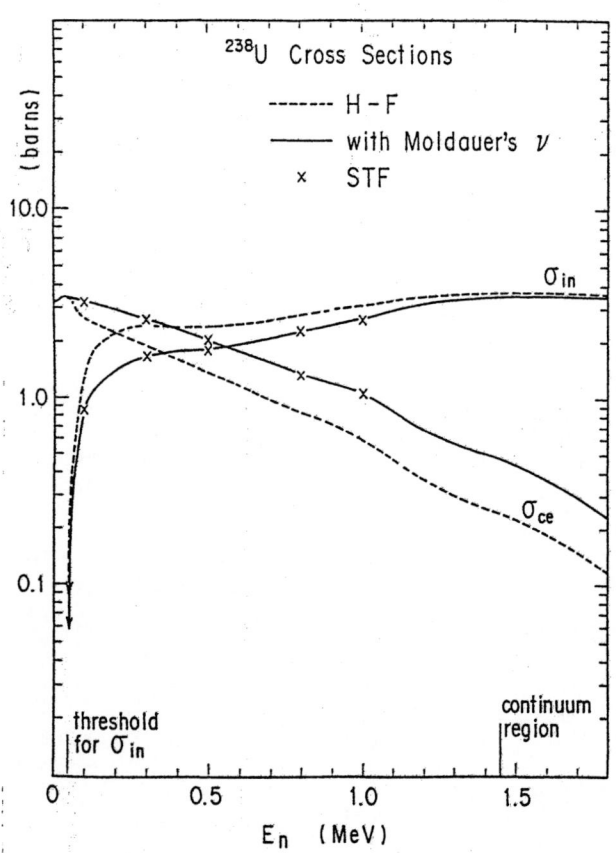

Fig. 1. Calculated cross sections of σ_{ce} and σ_{in}. The STF shows very good agreement with the results of the HF with the MCD. Significant differences exist between the HF with and without the WFC in continuum state.

Table 2. Level excitation cross sections (barns) calculated at 800 keV. The STF and the MCD give almost the same values, but there are some subtle differences, depending on the spin values.

keV	spin parity	STF	MCD	PTD	HF
0.0	0+	1.327	1.333	1.555	0.8359
44.89	2+	1.417	1.408	1.229	1.792
148.4	4+	0.5082	0.5109	0.4808	0.5639
307.2	6+	0.01707	0.01679	0.01821	0.01493
680.1	1-	0.2277	0.2287	0.2137	0.2892
731.9	3-	0.07379	0.07334	0.07452	0.07520

and

$$Y_{ab} = T_a T_b \left[\frac{X_1}{[1-(1-T_a)X_1][1-(1-T_b)X_1]} \right.$$

$$\left. + \frac{X_2}{[1-(1-T_a)X_2][1-(1-T_b)X_2]} + \frac{2X_0(1-X_0)}{(1-T_a X_0)(1-T_b X_0)} \right] \qquad (4)$$

The cross sections are proportional to the coefficient B_0.

In the expression of Eq.(2), variable transformation was made concerning two integrals in the original formula /5,6/. Geometrical factors W and Z are usual ones.

In this paper, relation between the STF and the HF with and without the WFC was examined. The WFC used here are

$$W_{ab} = \int_0^\infty dt \frac{(1+\frac{2}{\nu_b}\delta_{ab}) e^{-\langle\Gamma_c\rangle t/\langle\Gamma\rangle}}{(1+\frac{2}{\nu_a}\frac{\langle\Gamma_a\rangle}{\langle\Gamma\rangle}t)(1+\frac{2}{\nu_b}\frac{\langle\Gamma_b\rangle}{\langle\Gamma\rangle}t)\prod_{c'\neq c}(1+\frac{2}{\nu_{c'}}\frac{\langle\Gamma_{c'}\rangle}{\langle\Gamma\rangle}t)^{\nu_{c'}/2}} \qquad (5)$$

for the final states in discrete levels, and

$$W_{ac} = \int_0^\infty dt \frac{e^{-\frac{\langle\Gamma_c\rangle}{\langle\Gamma\rangle}t}}{(1+\frac{2}{\nu_a}\frac{\langle\Gamma_a\rangle}{\langle\Gamma\rangle}t)\prod_{c'\neq c}(1+\frac{2}{\nu_{c'}}\frac{\langle\Gamma_{c'}\rangle}{\langle\Gamma\rangle}t)^{\nu_{c'}/2}} \qquad (6)$$

for the states in continuum, respectively. Here, symbol a denotes the entrance channel, b the final states in discrete levels, and c the states in continuum. Symbol c' in the denominators in Eqs.(5) and (6) stands for the discrete levels only. The denominators $\langle\Gamma\rangle$ of the branching ratios, on the other hand, include both the states in continuum and the discrete levels.

For the DOF of the WFC, Moldauer /7/ gave the following formula,

$$\nu_c = 1.78 + (T_c^{1.212} - 0.78) e^{-0.228T} \qquad (7)$$

where

$$T = \sum T_c + \int T_{c'} \rho_{c'} d\mathcal{E}_{c'} \qquad (8)$$

In this paper, Eq.(7) for the DOF and the PTD ($\nu=1$) were used for comparison.

Numerical Calculations

Calculations were performed for σ_{ce} and σ_{in} of ^{238}U. Optical potential, level parameters and level density parameters used in the present calculations were those taken for JENDL-3 evaluation /8/. For the STF calculation, five energy points were taken. Numerical integrations were performed by applying the Gauss-Legendre integration method. The second integrand was mild, but the first and third ones changed very steeply. Hence, the range for the former integral was divided into only two intervals, whereas for the latter two, several intervals were taken so that the sum of the σ_{ce} and σ_{in} could be equal to the total reaction cross section, σ_R. In order to keep this relation, the calculation had to be done rather by brute force, and it consumed very much computer time: about one hundred times more than that of the HF calculations.

Calculations were also carried out for the HF with and without the WFC. The results are shown in Table 1, and Figs. 1 and 2. It is easily seen that the STF gives almost the same results as the HF with the MCD does. Since the STF has no optional parameters and is formulated rigorously, it seems to give correct answers. In this context, the HF with the MCD is considered a very good approximation for the statistical model calculations.

In Table 2, the level excitation cross sections calculated at 800 keV are shown. The STF and the MCD give almost the same values, but it can be seen that there are some subtle differences between them, depending on the spin values. It is noticed also that the HF gives the smallest value of σ_{in} for high spin state, while the PTD gives the largest one. This is a reverse phenomenon to that in the low spin state. The WFC has a character of raising the contribution of smaller T_b and reducing that of larger T_b.

The effect of the WFC to the final states in continuum was studied. As the energy increases, this effect gradually decreases, but in the lower energy region in continuum, it still remains obviously. This may reflect existence of the level-width fluctuation in the overlapping levels. Hence, the WFC cannot be always neglected for the state in continuum.

Concluding Remarks

The statistical models with the STF as well as the HF with and without the WFC were studied from the viewpoint of the nuclear data evaluation. Although the former is rigorous and gives precise results, it consumes the computer time more than hundred times of the latter. In order to use it for the nuclear data evaluation, much shorter computer time should be attained. In this sense, further efforts should be done to develop more appropriate numerical calculation method for use of the STF. However, the STF is useful to make an accurate estimate of the cross sections for a few energy points, and to make use of them as stepping stones for the evaluation.

For practical purposes, the HF with the MCD seems good approximation to the STF.

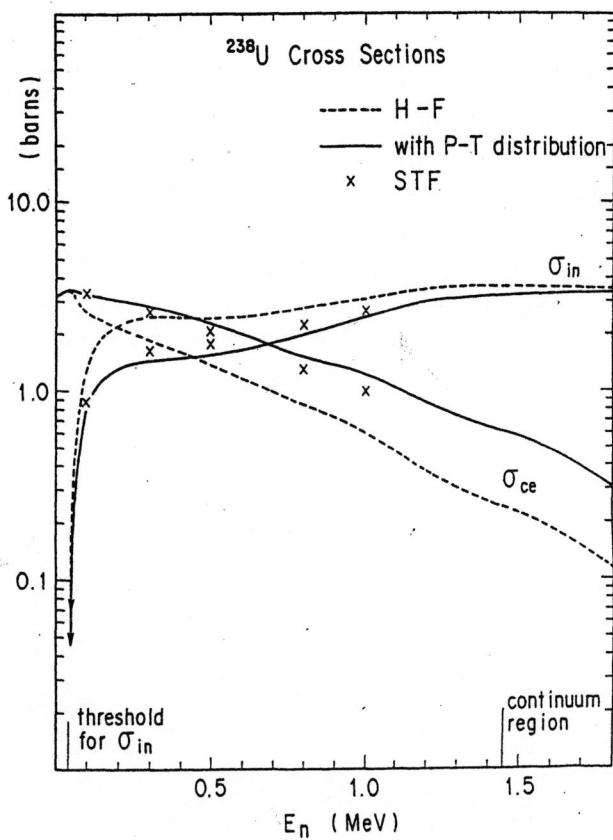

Fig. 2. Calculated cross sections of σ_{ce} and σ_{in}. The STF shows some differences from the results of the HF with the PTD and the HF without the WFC. Significant differences exist between the HF with and without the WFC in continuum state.

Hence, this scheme of the statistical model calculation is recommended for the nuclear data evaluation.

REFERENCES

1. W. Hauser and H. Feshbach: Phys. Rev. 87, 366 (1952)
2. L. Dresner: Proc. Int. Conf. on Neutron Interaction with the Nucleus: Columbia Univ. 1957, CU-175, p.71 (1957)
3. J.E. Lynn: Theory of Neutron Resonance Reaction, Clarendon Oxford (1968) p.230
4. H. Gruppelaar and G. Reffo: Nucl. Sci. Eng. 62, 756 (1977)
5. J.J.M. Verbaarschot, H.A. Weidenmuller and M.R. Zirnbauer: Phys. Rep. (Phys. Lett. C) 129, 367 (1985)
6. J.J.M. Verbaarschot: Ann. Phys. 168, 368 (1986)
7. P.A. Moldauer: Nucl. Phys. A344, 185 (1980)
8. K. Shibata, T. Nakagawa, T. Asami, T. Fukahori, T. Narita, S. Chiba, M. Mizumoto, A. Hasegawa, Y. Kikuchi, Y. Nakajima and S. Igarasi: JAERI 1319, p.417 (1990)

EVALUATION OF 14 MeV CROSS SECTIONS FOR THE MAIN ISOTOPES OF THE STRUCTURAL MATERIALS Cr, Fe AND Ni

H. Vonach, A. Pavlik, S. Tagesen and M. Wagner
Institut für Radiumforschung und Kernphysik der Universität Wien
Boltzmanngasse 3,
A-1090 Wien, Austria

Abstract: Evaluated data including uncertainty estimates were derived for the following quantities:

$$\sigma_{tot}, \ \sigma_{non}, \ \sigma_{el}, \ \sigma_{inel}, \ \sigma_{n,p}, \ \sigma_{n,np}, \ \sigma_{n,\alpha}, \ \sigma_{n,n1}, \ \sigma_{n,2n}, \ \sigma_{n-em}, \ \sigma_{p-em} \ \sigma_{\alpha-em}$$

and the reduced Legendre coefficients of the P_1, P_2 and P_3 components of the elastic angular distribution at $E_n = 14.0$ MeV. For the purpose of evaluation all measurements in the neutron energy range 13.5 - 14.8 MeV were critically reviewed, reduced to the nominal energy of 14.0 MeV and the evaluation results derived as weighted averages from the accepted and - if necessary - properly renormalized data. The results of this evaluation (with few exceptions) lie within the range of predictions of the new evaluated data files BROND, EFF-2, ENDF/B-VI and JENDL-3 and thus confirm the assumption that comparison of different evaluated files may be used for uncertainty estimates.

1. Introduction

The interaction of 14 MeV neutrons with the structural materials Cr, Fe, Ni and in particular their main isotopes ^{52}Cr, ^{56}Fe, ^{58}Ni and ^{60}Ni has been thoroughly investigated in many experiments. However, no systematic evaluation of this data base has been performed so far, and thus the information contained in this data base has not yet been fully utilized. Therefore, we have started such an evaluation especially for the following purposes:

1. The evaluation allows a very sensitive test of the accuracy of the new evaluated data files EFF-2 [1,2], ENDF/B-VI [3], JENDL-3 [4] and BROND [5] which were derived from nuclear reaction model calculations.

2. In addition, the evaluation can be used to improve the accuracy of these files around $E_n = 14$ MeV. This is especially important for fusion neutronics calculations which are especially sensitive to the 14 MeV cross section.

According to this purpose we selected for the evaluation all quantities needed in neutron-transport calculations, and those needed for the calculation of activation and gas-production.

2. Evaluation Method

2.1 Separate evaluation of the individual cross sections

The experimental results in the literature were reviewed with respect to the nuclear data and the reference cross sections upon which the reported data are based. If necessary, the results were renormalized using recent nuclear data and evaluated cross-section standards [6,7]. Superseded and obviously wrong data sets were disregarded. The error estimates given by the authors were checked for inclusion of all relevant sources of systematic errors. If necessary, the uncertainties were increased according to the best estimate of the evaluators in order to obtain a realistic total uncertainty of one standard deviation for each accepted data point. By applying a linear least-squares fit method, gradients $d\sigma/dE_n$ of the respective excitation functions in the considered 14 MeV region were extracted from those works which report measurements for several neutron energies. From these gradients an averaged gradient was calculated, which then served to normalize the results of the single energy measurements to

a nominal neutron energy of 14.0 MeV. Finally, the evaluated cross section at this energy was obtained as the weighted average of the normalized cross sections. Both the internal and the external errors of these averages were calculated and the larger of the two was taken as the uncertainty of the evaluated cross section.

For some of the cross sections in addition to this general scheme the following special procedures and/or corrections had to be used:

a) For a considerable part of the cross sections (e.g. σ_{tot}, σ_{non}, σ_{n-em}) very accurate data are available for the natural elements, whereas isotopic data are either lacking completely or are of much poorer quality. In these cases which are indicated on the corresponding tables the evaluated cross sections for the natural elements are given if the cross section differences between the natural elements and the corresponding isotopes are negligible compared to the corresponding cross section uncertainties. In two cases where the difference between the elemental and the isotopic cross sections are small but not completely negligible, the elemental cross sections were converted into isotopic cross sections using the ENDF/B-VI values for the cross sections of the minor isotopes of the corresponding elements.

b) For some cross sections recent high-quality evaluations are available; in these cases [6-8] duplication of work was avoided and the result incorporated in this evaluation. This applies to $\sigma_{n,p}$ for ^{56}Fe [6], $\sigma_{n,2n}$ for ^{52}Cr and ^{58}Ni [7] and σ_{non} for all 4 isotopes and σ_{n-em} for ^{52}Cr and ^{56}Fe [8]. In the latter cases the cross sections $\sigma_{n,n1}$ for inelastic excitation of the first excited levels were added to the results of ref. [8] as this evaluation is restricted to processes with an energy loss of at least 2 MeV.

c) Total elastic cross sections as well as the reduced Legendre coefficients of elastic scattering were derived from a Legendre fit to all existing differential elastic scattering cross sections.

d) For some cross section types only one measurement exists which was accepted as evaluation result. This unfavorable situation exists for the total deuteron emission cross sections and for the total proton emission cross sections for ^{52}Cr, ^{56}Fe and ^{60}Ni [9].

e) (n,np) cross sections for ^{52}Cr, ^{56}Fe and ^{60}Ni were derived as $\sigma_{n,np} = \sigma_{p-em} - \sigma_{n,p}$ as there are no direct measurements of these quantities. For ^{58}Ni, where very accurate values of both $\sigma_{n,p}$ and $\sigma_{n,np}$ could be derived from activation cross section measurements, the sum of these two was used as evaluated value of σ_{p-em}.

f) Total α-emission cross sections could be derived by the standard evaluation procedure for all isotopes. These were subsequently used to also derive (n,α) cross sections by substracting a (small) theoretically estimated contribution for the (n,n'α) process.

g) Inelastic cross sections for excitation of the first excited states $\sigma_{n,n1}$ were derived by numerical integration of the existing differential cross sections over the angular range covered by the data and using theoretical estimates for the small angular ranges not covered by data.

h) Total inelastic cross sections σ_{inel} were derived from the measured production cross sections of the γ-ray from the decay of the first excited (2$^+$) level of the considered nuclei. As the γ-decay following any inelastic scattering proceeds via this level with a very high probability (85 - 95%) the γ-ray production cross sections, corrected theoretically for the small fraction of transitions not populating the first 2$^+$ states give rather accurate values of σ_{inel}.

i) For the ^{60}Ni(n,2n)^{59}Ni reaction two measurements exist: $\sigma_{n,2n}$ = 104 ± 25 mb [10] and $\sigma_{n,2n}$ = 407 ± 102 mb [11]. The latter value was determined by accelerator mass-spectrometry and agrees very well with our theoretical expectations; the low value of [10] cannot be understood in any nuclear reaction model. Thus it was decided to use the value of [11].

2.2 Least-square adjustment of the evaluated cross-section sets

Three out of the thirteen evaluated cross sections are redundant, that is, they may be expressed as linear functions of the remaining ten basic cross sections. These relations are the following

$$\sigma_{tot} = \sigma_{non} + \sigma_{el} \tag{1}$$
$$\sigma_{non} = \sigma_{inel} + \sigma_{n,2n} + \sigma_{n,p} + \sigma_{n,np} + \sigma_{\alpha-em} \tag{2}$$
$$\sigma_{n-em} = \sigma_{inel} + 2\,\sigma_{n,2n} + \sigma_{n,np} + \sigma_{n,n\alpha} \tag{3}$$

These constraints were used to further improve the evaluation by least square adjustment for the following reasons:

a) As a result of this adjustment we obtain final cross-section sets which fulfill equ. 1-3 exactly and thus are completely internally consistent as required for practical use in evaluated data files.

b) The uncertainties are considerably reduced for a number of important cross sections.

3. Results and comparison to recent evaluated data files

The results of this evaluation are given in tables 1-3. As a consequence of the described least-squares adjustment procedure considerable correlations do exist between different cross-sections. As an example table 3 gives the correlation matrix for ^{56}Fe, the other matrices are available on request from the authors.

The results of this cross section evaluation for σ_{tot}, σ_{el}, σ_{non} and σ_{inel} are compared to the recent evaluated nuclear data

files EFF-2, ENDF/B-VI, JENDL-3 and BROND in fig. 1. The figures show that all of the files need some adjustment at E_n = 14 MeV, e.g. all files overestimate the total elastic cross sections by about 5%. In general EFF-2 and ENDF/B-VI seem to describe the experimental data somewhat better than JENDL-3 and BROND. Especially the total nonelastic cross sections, which are the most important quantities for shielding calculations, are predicted much better by EFF-2 and ENDF/B-VI than by the other files.

Finally it can be seen from fig. 1 that most of the experimental cross sections lie within the range of predictions of the four discussed evaluations and all except one are compatible with this assumption within their

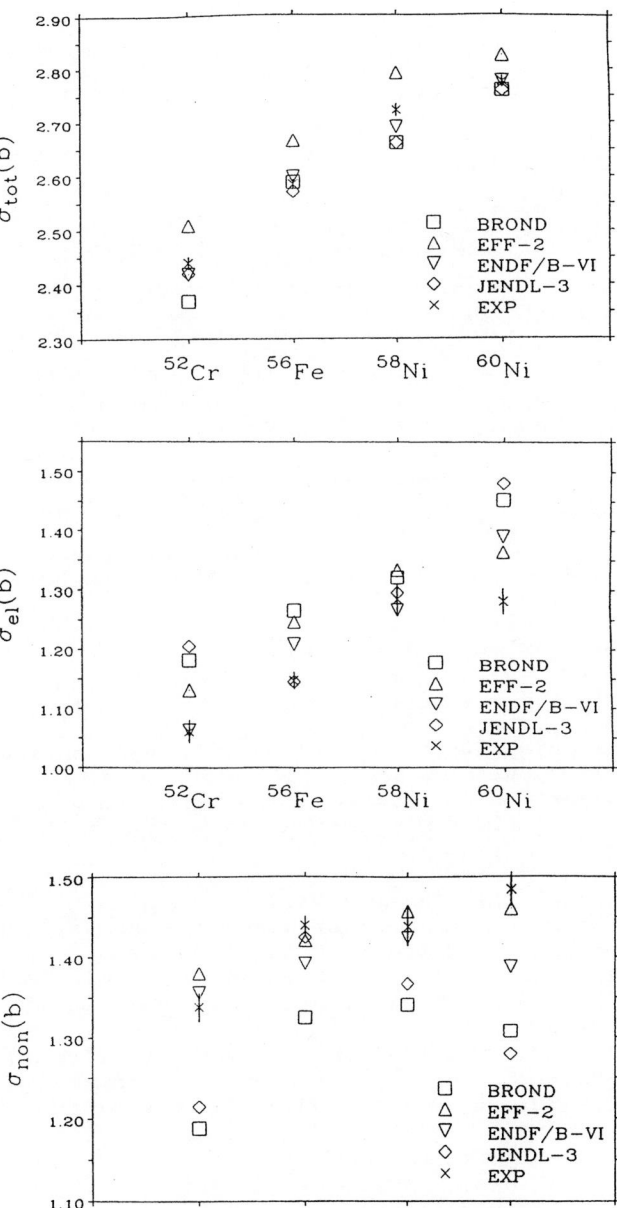

Fig. 1: Comparison of experimental 14.0 MeV evaluation with new evaluated data files

Table 2: Evaluated cross sections for ^{52}Cr, ^{56}Fe, ^{58}Ni and ^{60}Ni at E_n = 14.0 MeV after least-squares adjustment

	reac code	^{52}Cr	^{56}Fe	^{58}Ni	^{60}Ni
σ_t(b)	1	2.438 ± 0.013[a]	2.586 ± 0.010[a]	2.723 ± 0.012	2.766 ± 0.010
σ_{el}(b)[a]	2	1.061 ± 0.020	1.147 ± 0.015	1.285 ± 0.026	1.292 ± 0.022
σ_{non}(b)	3	1.377 ± 0.019[a]	1.439 ± 0.012[a]	1.438 ± 0.025[c]	1.484 ± 0.021[c]
σ_{inel}(b)	4	0.924 ± 0.021	0.8075 ± 0.020	0.358 ± 0.026	0.784 ± 0.050
$\sigma_{n,n1}$(mb)		52 ± 4	72 ± 4	45 ± 6	61 ± 5
σ_{n-em}(b)	5	1.510 ± 0.020[b]	1.681 ± 0.019[b]		
σ_{p-em}(mb)	6	155 ± 13	173 ± 13	923 ± 19	311 ± 55
$\sigma_{\alpha-em}$(mb)	7	34 ± 6	43.5 ± 1.6	120 ± 11	71 ± 3
σ_{d-em}(mb)	12	8 ± 3	8 ± 3	15 ± 6	14 ± 6
$\sigma_{n,2n}$(mb)	8	256 ± 5	407 ± 14	22 ± 0.4	304 ± 34
$\sigma_{n,p}$(mb)	9	82 ± 2	115.2 ± 1.2	368 ± 10	145 ± 3.4
$\sigma_{n,np}$(mb)	10	73 ± 13	57.8 ± 13	555 ± 16	159 ± 34
$\sigma_{n,\alpha}$(mb)	11	34 ± 6	42.4 ± 1.7	102 ± 19	61 ± 5

[a] Cross section for natural element assigned to isotope as difference is known to be small
[b] Cross section derived from elemental cross sections using ENDF/B-VI cross sections for minor isotopes
[c] Cross section for natural element assigned to isotope, possible difference taken into account by an additional uncertainty component

Table 1: Evaluated reduced Legendre coefficients a_1-a_3 * for the elastic scattering of 14 MeV neutrons on Cr, Fe and Ni

	Cr	Fe	Ni
a_1	0.819 ± 0.008	0.835 ± 0.008	0.854 ± 0.011
a_2	0.685 ± 0.007	0.727 ± 0.007	0.746 ± 0.007
a_3	0.584 ± 0.007	0.617 ± 0.006	0.637 ± 0.006

* The coefficients are defined by

$$\frac{d\sigma}{d\Omega}(\cos\theta) = \frac{\sigma_{el}}{4\pi}\left[1 + \sum_{l=1}^{\infty}(2l+1)\,a_l P_l(\cos\theta)\right]$$

Table 3: Correlation matrix for adjusted cross sections of ^{56}Fe

reac code	Correlation [%]
1	100
2	59 100
3	13 -71 100
4	6 -33 47 100
5	8 -47 66 -8 100
6	1 -7 10 -50 -2 100
7	0 0 1 -5 0 -1 100
8	0 -3 4 -55 75 -12 -1 100
9	0 0 0 -2 -5 3 0 0 100
10	1 -7 10 -50 -1 99 -1 -12 -4 100
11	0 0 0 0 -7 0 0 0 0 0 100
12	0 -3 4 -8 -10 -2 0 0 0 -2 0 100

uncertainties. This confirms the assumption made by Tagesen and Vonach [12] that the spread of values between the 4 new data files can be used as a basis for uncertainty estimates for these files.

Furthermore it is apparent from table 1 and fig. 1 that for most cross sections only precision measurements of very high quality can noticably improve the present data base. In a few cases, however, new measurements are highly desirable. There exists only one reliable measurement of total proton emission for all isotopes discussed in this work and also the ^{58}Ni(n,α) and ^{60}Ni(n,2n) cross sections for formation of the long-lived radio-nuclides ^{55}Fe and ^{59}Ni can probably be improved considerably by further measurements.

A more detailed description of this evaluation including full documentation of the data base will appear in Physics Data, Fachinformationszentrum Karlsruhe.

This work was supported by the NET TEAM under contract 395-89-8/FU-D/NET.

References

1. B. Strohmaier and M. Uhl, Proc. Int. Conf. on Nucl. Data for Science and Technology, Antwerp, Sept. 6-10, 1982, 522, Reidel Publ. Comp., Dordrecht (1983).
2. M. Uhl et al., Contribution to this conference.
3. P. Rose, ed. ENDF/B-VI Summary Documentation, Report ENDF-201, 4th edn., BNL, in press.
4. K. Shibata et al., Report JAERI-1319 (1990).
5. V.N. Manokhin, ed., Report INDC(CCP)-283, IAEA (1988).
6. T.B. Ryves, Report EUR-11912-EN (1989), and Ann. Nucl. Energy 16, 307 (1989).
7. M. Wagner et al., Physics Data 13-5 (1990), Fachinformationszentrum Karlsruhe.
8. A. Pavlik and H. Vonach, Physics Data 13-4 (1988), Fachinformationszentrum Karlsruhe
9. S.M. Grimes et al., Phys. Rev. C 19, 2127 (1979).
10. L.R. Greenwood and D.L. Bowers, J.Nucl.Mat. 155, 588 (1988).
11. D. Weselka et al., Contribution to this conference.
12. S. Tagesen and H. Vonach, Contribution to this conference.

EVALUATION OF THE SILICON ISOTOPES FOR ENDF/B-VI

D. M. Hetrick, D. C. Larson, N. M. Larson, C. Y. Fu, and S. J. Epperson

Oak Ridge National Laboratory
P. O. Box 2008
Oak Ridge, TN 37831-6356 USA

Abstract: Isotopic evaluations for ^{28,29,30}Si performed for ENDF/B-VI are briefly reviewed. The evaluations are based on analysis of experimental data and results of model calculations. Evaluated data are given for neutron induced reaction cross sections, angular and energy distributions, and gamma-ray production cross sections. All necessary data are given to allow KERMA (Kinetic Energy Released in MAterials) and displacement cross sections to be calculated directly from information available in the evaluations. These quantities are fundamental to studies of neutron heating and radiation damage.

(^{28,29,30}Si for ENDF/B-VI, cross sections, neutron heating, radiation damage)

Introduction

Silicon is important for radiation damage studies and neutron and gamma-ray transport calculations. For ENDF/B-VI, separate evaluations were done for the three stable isotopes of silicon. The evaluations are based on analysis of experimental data, supplemented by results of nuclear model calculations. This paper summarizes the evaluations and notes the measured data considered and the model codes used. Examples of evaluations are compared to data.

Computational Methods and Procedures

The primary code used for this evaluation work was TNG [1], an advanced multistep Hauser-Feshbach model. TNG includes precompound and compound contributions to cross sections in a self-consistent manner, provides correlated angular and energy distributions, calculates gamma-ray production, and conserves angular momentum in all steps.

Calculations for ^{28,29,30}Si at a number of incident energies from 0.01 to 20.0 MeV were performed. Parameters required as input to TNG for the silicon isotopes are discussed in detail in [2]. The results from TNG are found to agree reasonably well with available data.

The resonance parameters are given in File 2 ENDF format for the evaluations. For File 3, when sufficient cross-section data were available, they were evaluated and used; otherwise calculations are used. Energy spectra for all outgoing particles, including photons, are given in File 6. Angular distributions are given for the neutron emission spectra for ^{28}Si only and isotropy is assumed for the other isotopes and particles. Branching ratios for the discrete levels are given directly in File 12; i.e., the continuum and discrete gammas are given separately. For the first time, all necessary nuclear data are given in the evaluations to allow direct computation of KERMA and displacement cross sections. These quantities are fundamental in studies of neutron heating and radiation damage.

Resonance Parameters

In the resonance region (thermal to 4 MeV), the available total [3,4,5] and inelastic [6,7] data for natural silicon, and total cross section data for ^{29,30}Si dioxide [6,7], were analyzed using the multilevel R-matrix code SAMMY [8]. Included in the analysis were thermal values of total [9], elastic [9], and capture [10] cross sections for all three silicon isotopes, and for oxygen. Resonance parameters used in ENDF/B-VI were derived from and are consistent with all these data.

Cross Sections

The total cross section in the resonance region is represented by resonance parameters for each isotope, taken from a SAMMY analysis of measured data. Above the resonance region there are no new isotopic data, so extensive checks were done comparing the ENDF/B-V total cross section with recent data for ^{nat}Si. Based on these comparisons, the ENDF/B-V results were retained. Details are given in [2].

The nonelastic cross section is the sum of all reaction processes except elastic scattering; the elastic cross section is derived by subtracting the nonelastic from the total.

Inelastic scattering data to the 1.779- and 4.617-MeV levels were taken from ENDF/B-V and were based on measurements [2]. Inelastic scattering data to higher states in ^{28}Si are sparse, and evaluated cross sections for these levels (up to 8 MeV) are from the TNG calculation. A direct interaction component was included for the 6.276-, 6.879-, and 6.889-MeV levels.

Discrete inelastic levels up to 6.0 MeV in ^{29}Si and 5.0 MeV in ^{30}Si were included in the respective evaluations. The TNG analyses were used exclusively since there are no available cross-section data.

The (n, p) cross section for ^{28}Si is based on measurements from threshold to 9 MeV; TNG results are used from 9 to 20 MeV. Cross sections are given for individual levels up to 3 MeV in the residual nucleus ^{28}Al, again based on calculation and using available data as a guide. Results for ^{28}Si$(n, p)^{28}$Al are shown in Fig. 1; the TNG calculations (for $E_n > 9.0$ MeV) are a good compromise.

From threshold to 8.4 MeV, the evaluation of Drake [11] is adopted for the ^{28}Si$(n, \alpha)^{25}$Mg reaction. Above 8.4 MeV, TNG is used as other alpha particle groups are contributing to the cross section but can not be resolved. Cross sections are given for levels up to 4.8 MeV in the residual nucleus ^{25}Mg, based on TNG and using data as a guide. TNG calculations are used for ^{29}Si(n, p), ^{29}Si(n, α), ^{30}Si(n, p) and ^{30}Si(n, α) as the measurements are quite discrepant and TNG offers a good compromise.

There have been no new measurements of the (n, d) reaction since the evaluation of Drake [11], so that work is adopted. Few data exist for tertiary reaction cross sections of the silicon isotopes and thus TNG results are used.

Angular Distributions

Angular distributions of secondary neutrons for elastic scattering are taken from ENDF/B-V [2]. For the discrete levels at 1.779, 4.617, 6.276, 6.879, and 6.8899 MeV in ^{28}Si, the angular distributions in ENDF/B-VI are a weighted sum of the Legendre coefficients from the TNG and DWUCK calculations. The angular distributions for all other levels of ^{28,29,30}Si were taken from TNG results.

Energy-Angle Correlated Distributions

The computed angular distributions of neutron production cross sections for silicon at $E_n = 14.5$ MeV and for three outgoing energy bins are compared with experiments in Fig. 2. The results for $E'_n = 7.0 - 8.0$ MeV are in the discrete region and include the sum of TNG and DWUCK calculations. For $E'_n = 6.0 - 7.0$ MeV, the results are also in the discrete region but are symmetric since there was no direct interaction contribution.

Neutron emission measurements are available only for incident energies from 14.0 to 14.6 MeV and there are no charged particle production measurements. Particle emission spectra were computed via TNG for 35 incident energies for each silicon isotope. The computed neutron spectra at $E_n = 14.5$ MeV (weighted sum of ^{28,29,30}Si) are compared with the natural silicon measured data in Fig. 3.

It appears that the calculated neutron emission is too small at low outgoing energies. Experimentally, low energy neutrons form the largest part of the background, so background subtraction is quite difficult in this region of the spectrum. It has been observed that the data of Clayeux and Voignier [12] and Takahashi et al. [13] for several elements differ significantly at low energy from the data of Hermsdorf et al. [14] and others [15],[16]. There is good overall agreement between experiment and calculation in the nonelastic cross section, the various partial reaction cross sections, and gamma-ray production spectra (see below). Since energy conservation must be satisfied, the computed neutron emission as shown in Fig. 3 is judged to be acceptable.

Prior to incorporation in File 6, the neutron and charged particle energy distributions from TNG were input to the RECOIL code [17], which converts the distributions from the center of mass to the laboratory frame, and computes the energy spectra of the heavy recoil nucleus. The recoil nucleus distributions are also entered in File 6; isotropy is assumed.

The calculated gamma-ray emission spectrum at 14.5 MeV is compared to measured data [18] in Fig. 4. The calculation provides a good reproduction of the data, and in addition provides information on the cross sections for $E_\gamma < 0.7$ MeV. The computed energy-dependent yield and the TNG normalized distributions for the gamma rays at several incident energies are given for each reaction using File 6 in ENDF/B-VI.

Uncertainty Information

Uncertainty files are given for all cross sections in File 3, but not for the resonance parameters, energy distributions, or angular distributions. Fractional and absolute components, correlated only within a given energy interval, are based on data and estimates of uncertainties associated with the model calculations (see Hetrick et al. [19]).

KERMA and Damage Calculations

Improvements to ENDF/B-VI for KERMA and damage calculations are discussed in Ref. [20]. Briefly, prior to ENDF/B-VI, evaluations did not contain spectral distributions for outgoing charged particles, so only approximations could be made for heating and KERMA. Now, the nuclear data are given to compute these quantities directly.

Summary and Conclusions

Advanced nuclear model codes, an improved experimental data base, more flexible ENDF formats, and isotopic evaluations were used for the evaluation of silicon in ENDF/B-VI. Cross sections for all important reactions are included. Measurements are used to benchmark the calculations. Charged particle spectra are provided. Energy conservation is achieved to less than 1% for all reactions at all energies. The required data are given to allow KERMA and heating to be calculated directly from the evaluation.

This evaluation is much improved over ENDF/B-V. However, the evaluations would benefit from improving the data base further. For example, isotopic total cross section data need to be made available, particularly in the resonance region. Proton and alpha emission spectra are nonexistent and are needed to verify the model calculations, as well as neutron emission spectra at energies other than 14.5 MeV. Little or no data exist for the tertiary reactions with which to benchmark model calculations. Uncertainties should be given for important resonance parameters, and angular and energy distributions.

Acknowledgements

This research was sponsored by the office of Energy Research, Division of Nuclear Physics, U.S. Department of Energy, under contract DE-AC05-84OR21400 with Martin Marietta Energy Systems, Inc., and the Defense Nuclear Agency under Interagency Agreement Number 0046-C084-A1. Also, sincere thanks goes to Sue Damewood for typing this report and to Larry Weston and Bob Roussin for their helpful comments upon reviewing this paper.

References

1. C. Y. Fu, *Nucl. Sci. Eng.* **100**, 61 (1990).
2. D. C. Larson, D. M. Hetrick, N. M. Larson, C. Y. Fu, and S. J. Epperson, *Evaluation of* ^{28,29,30}*Si for ENDF/B-VI*, ORNL/TM-11825 (1991).
3. R. B. Schwartz, R. A. Schrack, and H. T. Heaton, *Bull. Amer. Phys. Soc.* **16**, 495 (1971).
4. H. Weigmann, P. W. Martin, R. Kohler, I. van Parijs, F. Poortmans, and J. A. Wartena, *Phys. Rev.* **C36**, 585 (1987).
5. D. C. Larson, C. H. Johnson, J. A. Harvey, N. W. Hill, *Measurement of the Neutron Total Cross of Silicon from 5 eV to 730 keV*, ORNL/TM-5618 (1976).
6. J. A. Harvey, private communication, ORNL, (1991)
7. J. A. Harvey, W. M. Good, R. F. Carlton, B. Castel, J. B. McGrory, and S. F. Mughabghab, *Phys. Rev.* **C28**, 24 (1983).
8. N. M. Larson, *Updated Users' Guide for SAMMY: Multilevel R-Matrix Fits to Neutron Data Using Bayes' Equations*, ORNL/TM-9179/R2 (1989).
9. S. F. Mughabghab, *Neutron Cross Sections*, Academic Press, Inc. (1984).
10. M. A Islam, T. J. Kennett, and W. V. Prestwich, *Phys. Rev.* **C41**, 1272 (1990).
11. M. K. Drake, Gulf General Atomic Report GA-8628 (1968).
12. G. Clayeux and J. Voignier, CEA Report CEA-R-4279 (1972).
13. A. Takahashi, J. Yamamoto, T. Murakami, K. Oshima, H. Oda, K. Fujimoto, M. Ueda, M. Fukazawa, Y. Yanagi, J. Miyaguchi, and K. Sumita, Octavian Rpt. A-83-01, Osaka Univ., Japan (1983).
14. D. Hermsdorf, A. Meister, S. Sassonoff, D. Seeliger, K. Seidel, and F. Shahin, Zentralinstitut fur Kernforschung Rossendorf Bei Dresden, ZfK-277 (U) (1975).
15. D. M. Hetrick, D. C. Larson, and C. Y. Fu, *Status of ENDF/B-V Neutron Emission Spectra Induced by 14-MeV Neutrons*, ORNL/TM-6637, ENDF-280 (1979).
16. C. Y. Fu and D. M. Hetrick, *Trans. Am. Nucl. Society* **53**, 409 (1986).

17. C. Y. Fu and D. M. Hetrick, Computer Code RE-COIL, ORNL, Unpublished (1985).
18. J. K. Dickens, T. A. Love, and G. L. Morgan, *Gamma-Ray Production from Neutron Interactions with Silicon for Incident Neutron Energies Between 1.0 and 20 MeV: Tabulated Differential Cross Sections*, ORNL-TM-4389 (1973).
19. D. M. Hetrick, D. C. Larson, and C. Y. Fu, *Generation of Covariance Files for Isotopes of Cr, Fe, Ni, Cu, and Pb in ENDF/B-VI*, ORNL/TM-11763 (1991).
20. D. C. Larson, D. M. Hetrick, C. Y. Fu, and S. J. Epperson, in *Proc. Seventh ASTM-EURATOM Symposium on Reactor Dosimetry, August 1990, Strasbourg, France* (Edited by G. Tsotridis), (1991).

ORNL-DWG 91-2268

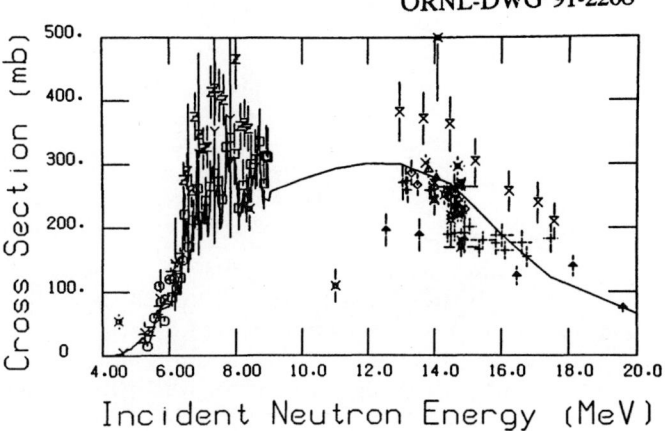

Fig. 1. $^{28}Si(n,p)$ in ENDF/B-VI compared to data (see Ref. 2.)

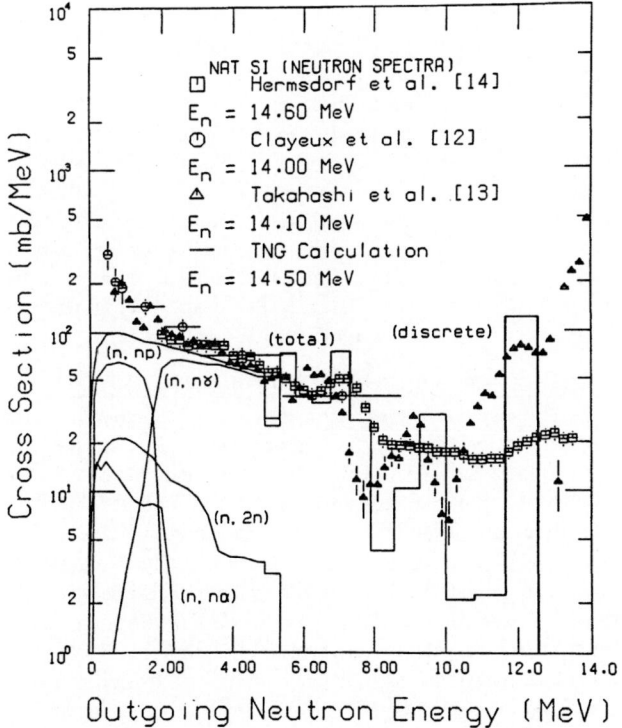

Fig. 3. Calculated neutron emission spectra compared to data.

ORNL-DWG 91M-8329

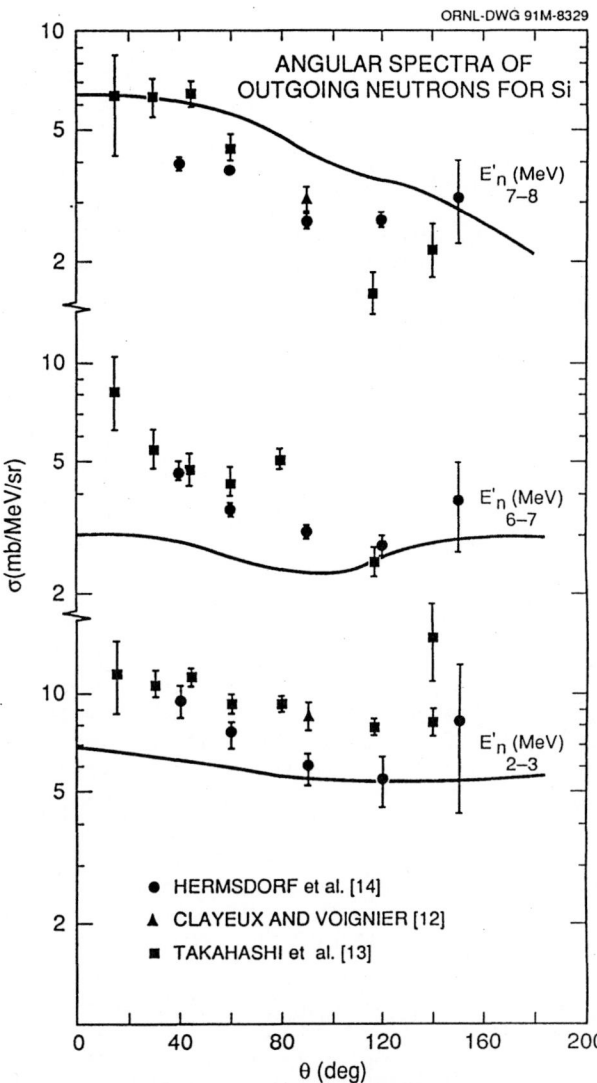

Fig. 2. Comparison of computed and measured neutron production cross sections.

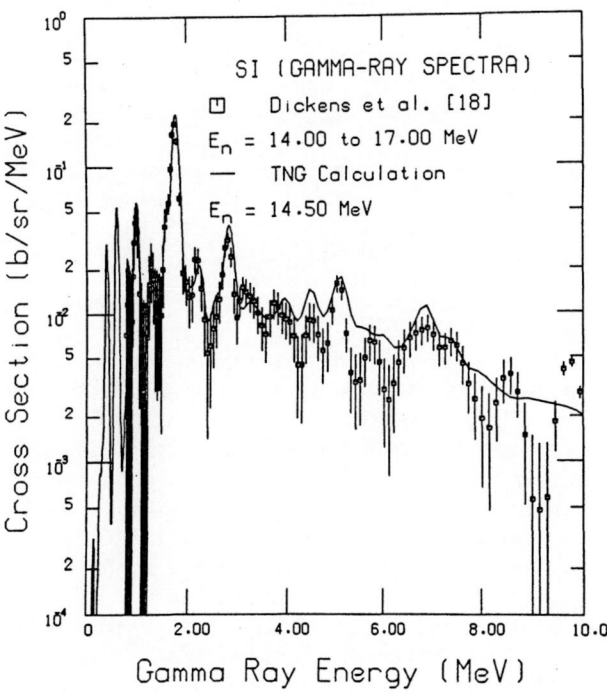

Fig. 4. Calculated gamma-ray spectra compared to data.

EVALUATION OF CROSS SECTIONS FOR THE DOSIMETRY REACTIONS OF NIOBIUM

N. Odano, S. Iwasaki and K. Sugiyama

Department of Nuclear Engineering, Tohoku University,
Aramaki-Aza-Aoba, Aoba-ku, Sendai 980, JAPAN

Abstract: Cross sections for ^{93}Nb$(n,n')^{93m}$Nb and ^{93}Nb$(n,2n)^{92m}$Nb reactions were evaluated using nuclear model calculation and experimental data. The ELIESE-GNASH joint program, the Hauser-Feshbach code with preequilibrium correction, was used for calculation of cross sections to obtain consistent data set of all the cross sections and particle emission spectra. To establish reliable excitation functions for the dosimetry application, the cross sections and their covariances were evaluated from the results of the nuclear model calculation and recent experimental data using a generalized least squares program, GMA. The fission spectrum and Li(d,n) neutron spectrum averaged cross sections calculated from the evaluated cross sections were compared with the recent integral experiments. The degree of agreement for both the reactions was good.

(dosimetry reaction, evaluation, ^{93}Nb, nuclear model calculation, experimental data, least squares method, integral test)

Introduction

The ^{93}Nb$(n,n')^{93m}$Nb and ^{93}Nb$(n,2n)^{92m}$Nb reactions are of importance for dosimetry applications. Specifically, the (n,n') reaction is considered as a reaction for the measurement of fast neutron fluence by activation in the pressure vessel of a light water reactor. Its threshold energy of 30 keV and half life of ^{93m}Nb (13.6 year) are suitable for the fluence monitor. On the other hand, the $(n,2n)$ reaction cross section is interesting as a monitor for the fusion reactor dosimetry as well as a high energy neutron monitor in the fission reactor because of the half life of the produced ^{92m}Nb (10.15 day) and its high threshold energy, about 9 MeV.

In spite of the importance of the (n,n') reaction, experimental data of the cross section is sparse and the uncertainties of the data are large due to the difficulty of the activity measurement. Recently, evaluation works of (n,n') reaction cross section were performed [1-3] for ENDF/B-VI [4], JENDL Dosimetry File [2] and IRDF-90 [5]. The evaluated data for the ENDF/B-VI and the IRDF-90 contain theoretical calculation by Strohmaier [6]. We also intended to improve the present status of the reaction cross section using theoretical calculation [7]. For the $(n,2n)$ reaction cross section, there is a discrepancy between the existing evaluations in the energy range except for around 14 MeV. More accurate evaluation of excitation functions of these reactions is necessary for dosimetry application.

Evaluation Procedure

For the (n,n') reaction, it is difficult to evaluate the reaction cross section from threshold energy to 20 MeV based only on the experimental data because there are a few sets of experimental data [8-10] which do not cover the entire energy range. One of the methods to evaluate the cross sections in such a case is to employ nuclear model calculations. Though the result of the nuclear model calculation must be a consistent data set in all kinds of cross sections for a particular nuclide, a more accurate evaluation taking into account the experimental data is desirable for the dosimetry cross sections. In the present study, we adopted a method based on a statistical least squares method, where we incorporated not only the experimental data but also the nuclear model calculation. With the least squares method, we can obtain the covariances of the evaluated cross sections which play an important role in the reactor dosimetry.

Nuclear Model Calculation

Nuclear Model Codes

A joint program of ELIESE [11] and GNASH [12] (EGNASH [13]), and DWUCK [14] code were used for the calculation of the neutron induced reaction cross sections and particle emission spectra. The contribution of the direct inelastic scattering cross sections was calculated by the DWUCK code and taken into account in the calculation of EGNASH code.

Parameters for the Nuclear Model Codes

To determine optical model potential parameters, so-called "SPRT" data, i.e., the s- and p-wave neutron strength functions, scattering radius and total cross section, were calculated using several parameter sets and compared with the experimental data. According to checking of the "SPRT" data, finally, an optical model potential parameter set by Smith (ANL) [15] was adopted for the calculation.

Energy dependence of level density parameters by Ignatyuk [16], which is effective for the calculation for nucleus in the neighborhood of the closed shell was implemented [7] in place of the Gilbert-Cameron formula [17]

The discrete level data (spin and parity of discrete levels, and γ-ray decay scheme) was found to be important input data in the present calculation. The data for ^{93}Nb were taken from the evaluation by Demanins [18]. The data of ref.19 were used for the rest of the nuclei related to the calculations.

Discussion of Result of Nuclear Model Calculation

The calculated excitation function of the (n,n') reaction was slightly lower than the experimental data in the energy range from 2 MeV to 4 MeV. The present calculated result was much lower than the experimental data of Wagner [8] at 8 MeV and about 20 % higher than the experimental data of Ryves [9] at 14 MeV. The excitation function of the isomeric state production cross section via the $(n,2n)$ reaction was higher than the experimental data above 16 MeV while the excitation function agreed with the experimental data below 16 MeV. Total $(n,2n)$ reaction cross section was somewhat higher from the threshold energy to 12 MeV. For other reactions, for example, the (n,α) and the (n,γ) reaction were in good agreement with experimental data. The particle emission spectra calculated around 14 MeV reproduced well the experimental data.

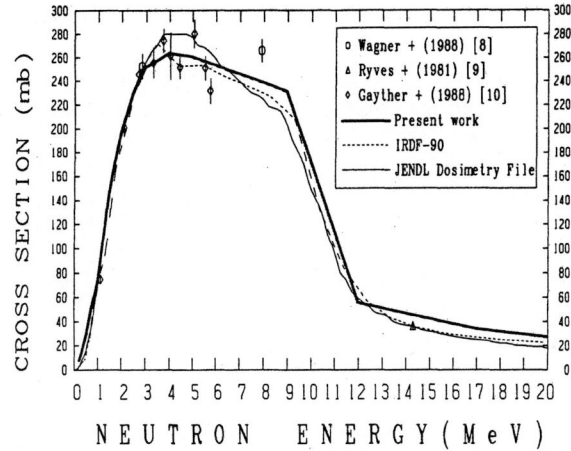

Figure 1: The experimental data base used in the present evaluation and the evaluated cross sections for the ^{93}Nb$(n, n')^{93m}$Nb reaction.

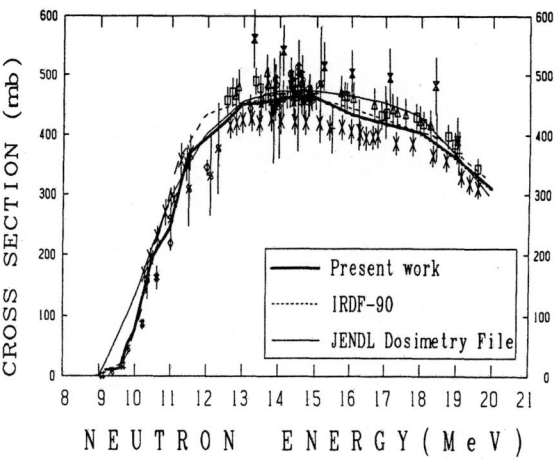

Figure 2: The experimental data base (except for 14 MeV) used in the present evaluation and evaluated cross sections for the ^{93}Nb$(n, 2n)^{92m}$Nb reaction .

Evaluation of Cross Sections

Evaluation by the GMA code

A FORTRAN version of the GMA code [20], Gauss-Markov-Aitken least squares nuclear data evaluation program, was used for evaluating the cross sections with their uncertainties and covariance matrices.

As the first step of the evaluation, the experimental data were surveyed. Some experimental data were rejected because they were apparently inconsistent with other experimental data. The experimental data shown in Fig.1 were used for the evaluation of the (n, n') reaction cross section. For the $(n, 2n)$ reaction cross section, 33 experimental data sets (181 energy points) were used. New experimental data from 11 MeV to 13 MeV by Ikeda [21], JAERI, were also taken into account. The uncertainties of the experimental data were taken from the EXFOR [22] and literature. Correlations of the experimental data were also estimated from the description in the references. The result of the present nuclear model calculation prepared at every 0.5 or 1 MeV energy interval was used for the input data as well as the experimental data. The uncertainty of result of the theoretical calculation was assumed to be 20 % on the basis of the degree of discrepancy between the theoretical calculation and experimental data. In the uncertainty of the theoretical calculation, systematic error was assumed to be 10 %.

After the assignment of the uncertainties, all input data used in the evaluation were converted at an energy grid which was determined in advance. We chose the following energy grid for the (n, n') reaction considering the distribution of the experimental data, i.e., 0.5 ~ 1 MeV step below 6 MeV, whereas 3 MeV step above 6 MeV.

^{93}Nb$(n, n')^{93m}$Nb Reaction

The evaluated results are also shown in Fig.1. Variance of the present evaluated result was 3 % in the energy range from 2 to 6 MeV. In the other energy ranges, the variance was 7 to 18 %. The present evaluation reasonably follows the experimental data below 6 MeV. Compared with other evaluations, the JENDL Dosimetry File seemed to be overestimation in the energy range between 4 and 6 MeV, while the present result agreed with the IRDF-90. There was a discrepancy between the present evaluation and the other libraries above 6 MeV due to the different methods of evaluations.

^{93}Nb$(n, 2n)^{92m}$Nb Reaction

The evaluated results are shown in Fig.2 with experimental data. Variance of the present evaluated results was 0.7 % around 14 MeV and 2 to 20 % in the other energy ranges. Around 12 MeV, the present evaluated result is lower than the evaluation of the IRDF-90.

Integral Test of Present Evaluation

To validate the present results, the spectrum averaged cross sections were calculated. As the reference neutron field, spontaneous fission neutron spectrum of ^{252}Cf, evaluated by Mannhart [23], was adopted for the calculation of the average cross section of the (n, n') reaction. For the $(n, 2n)$ reaction, the neutron field from Li(d,n) reaction [24] was used as well as the thermal fission neutron spectrum of ^{235}U of the ENDF/B-VI.

Table 1 shows the comparison of the average cross sections of the (n, n') reaction calculated from the present evaluation and the other evaluated cross sections, and experimental data [25-27]. The average cross section calculated using the present result agrees with the recent integral experiments. Since the (n, n') reaction cross section has high sensitivity for the ^{252}Cf fission spectrum near 2 MeV, it is difficult to deduce the performance of the present evaluation in the energy range above 5 MeV. Thus, the neutron field whose average neutron energy is higher than that of ^{252}Cf fission neutrons should be used to check the cross section above 5 MeV.

The calculated average cross section of the $(n, 2n)$ reaction from the present evaluation is shown in Table 2 with experimental data [28-30] and the calculated ones using the other evaluated data. The present average cross section agrees with that of the IRDF-90 and is much lower than that of the JENDL Dosimetry File. This can be attributed to the low cross section of the present evaluation from the threshold energy to 12 MeV. Compared with the recent experimental data of the average cross section by Williamson [30], the present result was lower than the experimental data by 5 %.

The average cross section in Li(d,n) neutron field calculated by the present evaluation was 142±12.9 mb and in good agreement with the experimental data of 138±11.7 mb [31].

914

Table 1: The ^{252}Cf fission spectrum averaged cross sections of the ^{93}Nb$(n, n')^{93m}$Nb reaction.

Exp.	Alberts (1983)	[25]	149 ±10	mb	
	Williams (1989)	[26]	150.6± 3.9		
	Alberts (1989)	[27]	145 ± 5		
Eval.	Strohmaier (1989)	[6]	155.4±		
	IRDF-90 (1990)		142.7± 2.4	*	
	JENDL Dosimetry File		149.5± 1.4	*	
	Present Work		151.6± 9.7	*	

*)calculated by the present authors.

Table 2: The ^{235}U fission spectrum averaged cross section of the ^{93}Nb$(n, 2n)^{92m}$Nb reaction.

Exp.	Kimura (1971) (a)	[28]	0.432±0.033	mb	
	Kimura (1971) (b)	[28]	0.402±0.034		
	Fabry (1972) (c)	[29]	0.47 ±0.03	**	
	Williamson (1990) (c)	[30]	0.433±0.011		
Eval.	Strohmaier (1989) (d)	[6]	0.44 ±	**	
	Strohmaier (1989) (e)	[6]	0.43 ±	**	
	IRDF-90 (1990) (f)		0.412±0.008	*	
	JENDL Dosimetry File (f)		0.478±0.006	*	
	Present Work (f)		0.410±0.018	*	

(a)reactor spectrum, (b)fission plate, (c)^{235}U fiss. spec., (d)NIST fiss. spec., (e)ENDF/B-V fiss. spec., (f)ENDF/B-VI fiss. spec. *)calculated by the present authors. **)reported in ref.30.

Conclusion

The cross sections of the dosimetry reaction of ^{93}Nb, the ^{93}Nb$(n, n')^{93m}$Nb and the ^{93}Nb$(n, 2n)^{92m}$Nb reactions were evaluated by the statistical method using the GMA based on the nuclear model calculation and experimental data.

For the (n, n') reaction, the evaluated result reasonably followed the experimental data. The fission spectrum averaged cross section was in good agreement with experimental data. The evaluated $(n, 2n)$ reaction cross section was slightly lower than that of the recently compiled libraries from the threshold energy to 12 MeV. The fission spectrum averaged cross section was lower than experimental data while the average cross section in the Li(d,n) neutron field agreed with experimental data.

Differential measurement is required to evaluate the cross sections more precisely in the energy range above 6 MeV for the (n, n') reaction, and around 12 MeV for the $(n, 2n)$ reaction. Improvement of the nuclear model calculations is also required.

References

1. D.L.Smith and L.P.Geraldo, ANL/NDM-117 (1990)
2. S. Iijima, M .Nakazawa, K.Kobayashi, S.Iwasaki, T. Iguchi, Y.Ikeda, K.Sakurai and Y.Nakagawa, to be published in JAERI-M (1991).
3. M.Wagner, H.Vonach, A.Pavlik, B.Strohmaier, S.Tagesen and J.Martinez-Rico, Physics Data, 13-5 (1990).
4. BNL/ National Nuclear Data Center, ENDF/B-VI (1990).
5. N.P. Kocherov and H. Vonach, The Seventh ASTM-EURATOM Symposium on Reactor Dosimetry, Strasbourg, France, 27-31 August 1990 (to be published).
6. B.Strohmaier, Ann. Nucl. Energy, 16, p.461 (1989).
7. N.Odano, S.Iwasaki and K.Sugiyama, The Seventh ASTM-EURATOM Symposium on Reactor Dosimetry, Strasbourg, France, 27-31 August 1990 (to be published).
8. M.Wagner, G.Winkler, H.Vonach and H.Liskien, Proc. Int. Conf. Nucl. Data for Science and Technology, Mito, Japan, May 30 - June 3, 1988, p.1049 (1988).
9. T.B.Ryves and P.Kolkowski, J. Phys. G., 7, p.529 (1981).
10. D.B.Gayther, C.A.Uttley, M.F.Murphy, W.H.Taylor and K.Randle, Proc. Int. Conf. Nucl. Data for Science and Technology, 1988, Mito, Japan, p.1057 (1988).
11. S.Igarashi, JAERI 1224 (1972).
12. P.G.Young and E.D.Arthur, LA-6947 (1977).
13. N.Yamamuro, JAERI-M 88-140 (1988).
14. P.D.Kunz, "DWUCK", University of Colorado Report (1974).
15. A.B.Smith, P.T.Guenther and R.D.Lawson, ANL/NDM-91 (1985).
16. A.V.Ignatyuk, INDC(CCP)-233/L (1985).
17. A.Gilbert and A.G.W.Cameron, Can. J. Phys., 43, 1446 (1965).
18. F.Demanins, U.Abbondano and F.Raicich, Report INFN(REP)-014/88, Istituto Nazionale de Legnaro, Annual Report 1987 (1988).
19. C.M.Lederer and V.S.Shirley, "Table of Isotopes, Seventh Edition", John Wiley & Sons, Inc., (1978).
20. W.P.Poenitz, BNL-NCS-51363, p.249 (1981). (modified by M.Sugimoto of JAERI for PC (Pascal) version, and modified by S.Chiba of JAERI for FORTRAN version.)
21. Y.Ikeda, C.Konno, M.Mizumoto, K.Hasegawa, S.Chiba, Y.Yamanouchi and M.Sugimoto, JAERI-M 91-032, p.281 (1991).
22. "EXFOR International Data File for Exchange of Experimental Data", available from the Data Compilation Centers.
23. W.Mannhart, IAEA-TECDOC-410, p.158 (1987).
24. J.R.Dumais, S.Iwasaki, S.Tanaka, N.Odano and K.Sugiyama, The Seventh ASTM-EURATOM Symposium on Reactor Dosimetry, Strasbourg, France, 27-31 August 1990 (to be published).
25. W.G.Alberts, R.Hollnagel, K.Knauf, M.Matzke, W.Pessara, Proc. of the 4th ASTM-EURATOM Symposium on Reactor Dosimetry, NBS, p.433 (1983).
26. J.G.Williams, C.O.Cogburn, L.M.Hodgson, S.C.Apple, E.D.McGarry, G.P.Lamaze, F.J.Schima, J.W Rogers, R.J.Gehrke, J.D.Baker and F.J.Wheeler, Proc. of the 6th ASTM- EURATOM Symposium on Reactor Dosimetry, ASTM STP 1001, p.235 (1989).
27. W.G.Alberts, U.Schotzig and H.Siegert, ibid., p.223 (1989).
28. I.Kimura, K.Kobayashi and T.Shibata, J. Nucl. Sci. Tech., 8, p.59 (1971).
29. A.Fabry, Rep. BLG-465 (1972).
30. T.G.Williamson and G.P.Lamaze, The Seventh ASTM-EURATOM Symposium on Reactor Dosimetry, Strasbourg, France, 27-31 August 1990 (to be published).
31. J.R.Dumais, S.Iwasaki, N.Odano, M.Sakuma and K.Sugiyama, these proceedings.

NUCLEAR DATA EVALUATION FOR U-233

L.A.Bakhanovich, A.B.Klepatskij, V.M.Maslov,
G.B.Morogovskij, Yu.V.Porodzinskij, E.Sh.Sukhovitskij

Institute of Nuclear Engineering, Minsk, USSR

Abstract:Cross section of neutron-induced reactions on U-233 have been evaluated between 10^{-5} eV and 20 MeV for the BROND data library. In the thermal and resonance regions evaluation relies on experimental data.Fission and total cross sections have been evaluated on the basis of recent experimental data.A deformed optical potential that fits total, elastic and inelastic cross sections, s- and p-wave strength functions was used to calculate direct inelastic and other cross sections. The available data on total, fission,capture and inelastic scattering are reproduced quite well. Above (n,n'f)reaction threshold preequilibrium emission has been taken into account. The calculated (n,2n) reaction cross section disagrees with available fission spectrum-averaged data and other evaluations. Partial energy distributions of secondary neutrons are given in tabulated form. The evaluated data are compared with ENDL-78 and JENDL evaluations.

(U-233, neutron data evaluation, deformed optical potential, fission cross section, capture cross section, (n,2n) cross section)

Introduction

In recent time U-Th fuel cycle has been acknowledged as an alternative to traditional U-Pu one. The available U-233 evaluated data files are discrepant in numerous ways.The existing experimental data base for U-233 alongside with modern theoretical tools provides rather strong grounds for consistent evaluation of neutron data.

Thermal cross sections

The available experimental data on total, fission, capture, elastic scattering and η-value /1-11/ between 10^{-5} eV and 1 eV are normalized to thermal cross section which we consider the most reliable /12/. At first the partial cross sections were evaluated independently. Then evaluated total and fission cross sections were fixed,other cross sections were derived self-consistently. Scattering cross section was calculated with resolved resonance parameters.The agreement of evaluated data with η and α values appears reasonable.

Resonance parameters

Resolved resonances.

Between 1 eV and 100 eV resonance parameter sets for Breit-Wigner (single level) and Adler-Adler formalisms have been derived. Both parameter sets describe experimental data on total, fission and capture cross sections /1,8,10,13/.However, we have adopted Adler-Adler parameters set since they give a better least squares fit.Inconsistencies between experimental and calculated data have been compensated by adding smooth background to every partial cross section.

Unresolved resonances

Unresolved resonance region lies between 0.1 -40.35 keV. We aimed at getting a physically justified parameter set, describing average evaluated data. Evaluated total data were obtained for chosen intervals using measured data from /1,7, 14/. In case of fission cross section we averaged data /13,15,16,17/,normalized to fission integral /8/ or thermal point cross section.Developed in /18/ method of statistical resonance parameter analysis was used to get average resonance parameter values: $\langle D \rangle$=0.42 eV,S_0=0.97, R =9.6 fm. Our

$\langle D \rangle$ value is appreciably lower than other values, but we suppose it to be more justified, since our approach takes into account missing of resonances due to clustering and low amplitude. Moreover, it ensures consistency between experimental and theoretical distributions of resonance spacings and neutron widths. S_0 and R' values given above are in reasonable agreement with R'=9.6 fm and S_0= 0.99, which result from total data fit in the interval 0.1-100 keV, when S_1=1.54. This S_0 value is about 10% higher than S_0 in JENDL-3, whereas S_1 value is about 15% higher. Fission width fits average resonance data on $\langle \sigma f \rangle$. Radiative capture width was calculated using giant dipole resonance splitting and was normalized to S_γ=0.928. Total radiative width equals 39 meV, while (n,γf)-process width equals 7 meV. These energy dependent widths give a reasonable fit of experimental data on total, fission and capture cross section.To get a perfect fit to average cross sections S_0 and $\langle \Gamma f \rangle$ values were slightly varied.

Smooth cross section

At energy higher than $7/2^+$-level excitation threshold total and fission cross section evaluation relies on experimental data. Other cross sections are provided by the coupled-channel and statistical model calculations.

Total cross section

Total data evaluation is performed by the least squares method up to 12 MeV neutron energy. Thus derived energy dependence curve is fitted with the deformed optical code COUPLE /19/ within 2%. That is why above 12 MeV the evaluation relies on calculated curve and not on the data of /20/,which cover the whole energy range, but deviate from our evaluation between 4 and 12 MeV. The first five members of ground state band were coupled. The potential parameters are given in Table 1. The deformation parameters were obtained by fitting S_0 and R'. We have got S_0=0.99 and R'=9.51 fm.

Fission cross section

Among the 233-U (n,f)/235-U (n,f) ratio results two data groups could be distinguished.The first data group generally agree with data /21/. The second data group /22,23/,though less numerous,are lying significantly higher (up to 10%). The absolute measurements /24/ have rather large

916

scatter.but as well as data /25/ they are in
reasonable agreement with the first group data,
unlike the /26/ data. The evaluated curve fits
the first group data since they agree with each
other both in shape and absolute values quite
reasonably.Between 40 and 100 keV it fits data
/15/.Generally smooth curve has three step-like
irregularities: around 0.2, 0.55 and 1.2 MeV.

Statistical theory calculations

For neutron energies less than (n,n'f) reaction threshold we have used computer code /27/.
Above the threshold calculations were performed
with a modified version of STAPRE code /28/. The
level density calculation is described elsewhere
/29/. Here we will comment only the results obtained in our calculations.

The fission barrier parameters for U-234
agree with theoretical values /30/. The inner
and outer saddle heights are given in parentheses after the mass number: 234(4.8;5.6),233(5.3;
5.9),232(5.1;5.8). The 233-U and 232-U barriers
have been fixed after 232-U(n,f) /31/ data analysis. The correlation function for outer saddle
is derived from energy slope of fission cross
section. For U-234 nucleus the collective bandheads were defined according to axial symmetry
of both saddles and mass- asymmetry of outer
saddle. The first chance fission behaviour follows from /32/.

The approach used allows to fit measured
data up to the (n,n'f) onset better than 50 mb.
The contribution of (n,↑f)-reaction into fission
cross section weakly increases from 3% at 0.1 keV
up to 7% at 1.5 MeV and then decreases to 1% at
5 MeV. The decrease is due to (n,↑n') competition. It appears impossible to describe all step-
like irregularities. However the step at E=0.55
could be interpreted as a manifestation of two-
quasi-particle excitation of e-e fissioning nucleus at outer saddle. The calculated and evaluated curves are compared in Fig.1 with JENDL-3
evaluation. The most significant discrepancy
between our curves occurs above the (n,2nf) reaction onset.

The Fig.2 displays the comparison of measured and calculated data on elastic and inelastic
angular distributions at 1.5 MeV.The evaluated
total inelastic scattering is rather discrepant
with JENDL and ENDL evaluations. The "experimental" data shown are derived from data/14,34/on
total and total scattering cross sections and
fission, capture and low-lying levels excitation
cross sections /see Fig.3/.

The capture cross section is compared to measured data, derived using capture-to-fission ratio in Fig.4. The radiative strength function is
defined above. The sharp descent occurs in calculated curve above 1 MeV due to (n,↑n') competition. Above 10 MeV the direct capture cross section equals 1 mb.

The (n,2n) and (n,3n) cross sections are compared with JENDL and ENDL libraries in Fig.5.
The most important is the discrepancy in (n,2n)
reaction cross section,leading to U-232 formation. The fission spectrum average cross section
have been measured in /37/:<⊖n,2n>=4.08 0.3 mb.
Assuming a fission spectrum of the type /38/ for
JENDL-2,JENDL-3 and our curve we will get 3.8 mb,
1.6 mb and 2.9 mb,respectively. However, the model we have used, successfully describes data on
238-U(n,xn), 235-(n,xn), 237-Np(n,2n), 232-Th(n,
2n) and 239-Pu(n,2n) cross sections.

Summary

The present evaluation successfully describes much of the measured data on n+233-U reacti-
ons.In some respects it deviates strongly from
the available files. In current state U-233 file
have been incorporated into BROND library.

References

1. W. Kolar, G. Carraro, G. Nastri, in Proc. Int.
   Conf. on Nucl. Data for Reactors, 1970,Helsin-
   ki,1,387 (1970)
2. V.A. Pshenichny, A.I. Blankovsky, M.L. Gnidak,
   E.A. Pavlenko, INDC(CCP)-111(U), 23 (1978)
3. V.L. Sailor: Phys.Rev. 100, 1249 (1955)
4. M.S. Moore, L.G. Miller, O.D. Simpson: Phys.
   Rev. 118, 714 (1960)
5. J.A. Harvey, C.L. Moore, N.W. Hill, in Proc.
   Int. Conf. on Nucl. Cross-Sect. and Technology, Knoxville, USA, p.690(1979)
6. R.C. Block, G.G. Sloughter, J.A. Harvey: Nucl.
   Sci. Eng. 17, 404 (1963)
7. N.J. Pattenden, J.A. Harvey: Nucl. Sci. Eng.
   8, 112 (1960)
8. A.J. Deruytter, C. Wagemans: Nucl. Sci. Eng.
   54, 423 (1974)
9. L.W. Weston, R. Gwin, G.De Saussure: Nucl.
   Sci. Eng. 42, 143 (1970)
10.L.W. Weston, R. Gwin, G.De Saussure: Nucl.
   Sci.Sng. 34, 1 (1968)
11.J.R. Smith, S.D. Reeder, in Proc. Int. Conf.
   on Neutron Cross Section and Technol.,1968,
   Washington, 1, p.589 (1968)
12.M. Divadeenam, J.R. Stehn. IAEA-TEC-DOC-335,
   p.238, Vienna (1985)
13.J. Blons: Nucl. Sci. Eng. 51, 130 (1973)
14.W.P. Poenitz, J.F. Whallen, P. Guenter, A.B.
   Smith: Nucl. Sci. Eng. 68, 358 (1978)
15.R. Gwin, E. Silver, R.W. Ingle, H. Weaver:
   Nucl. Sci. Eng. 59, 79 (1976)
16.T.A. Mostovaya, B.I. Mostovoj, C.A. Berukov,
   in Proc. Int. Conf. on Neutron Physics, 1980,
   Kiev, 3, p.30 (1980)
17.A.A. Bergman, A.G. Kolosovskij, A.N. Medvedev,
   in Proc. Int. Conf. on Neutron Physics, 1980,
   Kiev, 3, p.54 (1980)
18.Ju.V. Porodzinskij, E.Sh. Sukhovitskij: Izv.
   Akad. Nauk BSSR, Ser. Fiz.-En. 3, 19 (1986)
19.A.B. Klepatskij, V.A. Konshin, E.Sh. Sukhovitskij, INDC(CCP)-161/L, p.2 (1981)
20.W.P. Poenitz, J.F. Whalen, ANL/NDM-80,Argonne
   National Laboratory (1983)
21.G.W. Carlson, J.W. Behrens: Nucl. Sci. Eng.
   66, 205 (1978)
22.J.W. Meadows:Nucl. Sci. Eng. 54, 317 (1974)
23.K. Kanda, H. Imaruoka, K. Yoshida, O. Sato, N.
   Hirakawa: Rad. Effects 93, 233 (1986)
24.W.P. Poenitz: ANL/NDM-36 (1978)
25.V.N. Dushin, A.V. Fomichev, S.S. Kovalenko,
   K.A. Petrzhak, V.I. Shpakov, R. Arlt, M. Josch
   G. Musiol, H.-G. Ortlepp, W. Wagner:Sov.J.At.
   Energy, 55, 218 (1983)
26.K.R. Zasadny, H.M. Agrawal, M. Mandavi, G.F.
   Knoll: Trans. Amer. Nucl. Soc. 47, 425 (1984)
27.A.V. Ignatjuk, A.B. Klepatskij, V.M. Maslov,
   E.Sh. Sukhovitskij: Sov.J.Nucl.Phys. 42, 569
   (1985)
28.M. Uhl, B. Strohmaier IRK-76/01 (1976)
29.A.V. Ignatjuk, V.M. Maslov: Sov.J.Nucl.Phys.
   51, 1227 (1990)
30.W.M. Howard, P.Moller: Atom.Data and Nucl.
   Data Tables 15, 219 (1980)

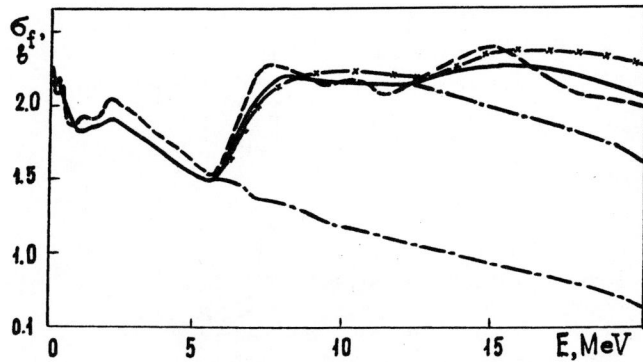

Fig. 1 Calculated 233-U(n,f) cross section(--x--) compared with evaluated (----) and JENDL-3(- - -) data; dot-dashed curve -first and second chance fission contribution.

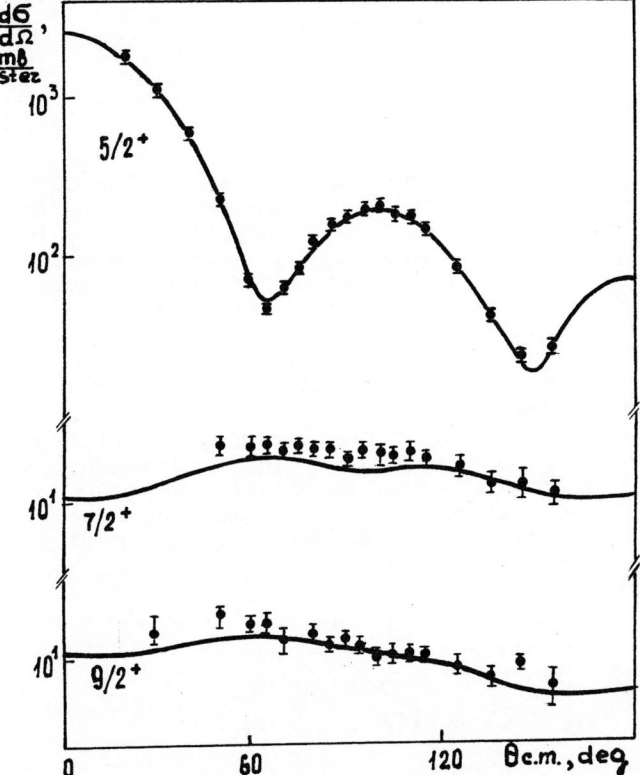

Fig. 2 Differential elastic and inelastic scattering cross sections at 1.5 MeV; data-/33/.

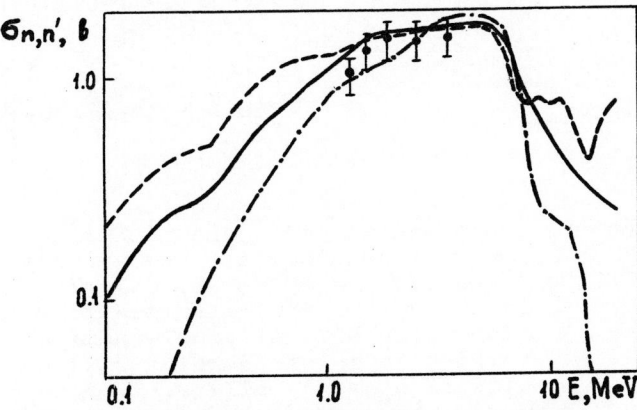

Fig. 3 Total inelastic scattering (solid curve) compared with JENDL-3 (dashed curve) and ENDL (dot-dashed curve) evaluations. Data points see text.

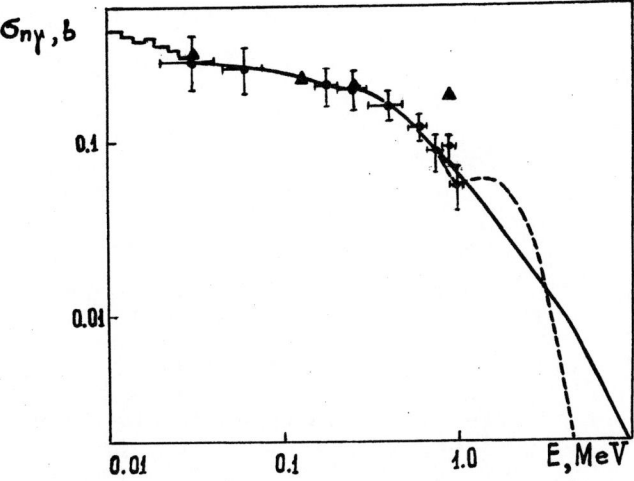

Fig. 4 Capture cross section (solid curve) compared with measured data : ● -/35/, ▲ -/36/; dashed curve -JENDL-3.

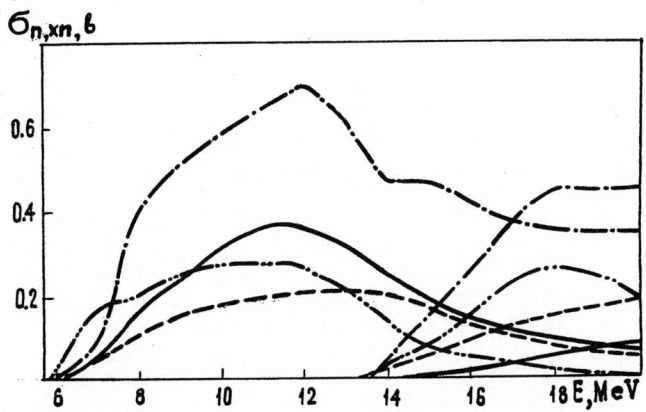

Fig. 5 (n,2n) and (n,3n) cross sections (solid curves) compared with JENDL-2 (double-dot-dashed curve), JENDL-3(dashed curve) and ENDL (dot-dashed curve) evaluations.

Table 1. Deformed potential parameters.

$V_R = 46.14 - 0.3E$ MeV　　$r_R = 1.256$ fm $a_R = 0.626$ fm
$W_S = 3.12 + 0.4E$ MeV, $E < 10$ MeV　$r_S = 1.260$ fm $a_S = 0.555$ fm
$W_S = 7.12$ MeV　　, $E > 10$ MeV
$V_{SO} = 6.2$ MeV　　　$r_{SO} = 1.12$ fm $a_{SO} = 0.47$ fm
$\beta_2 = 0.186$　　$\beta_4 = 0.090$

31. B.N. Fursov, E.Yu. Baranov, M.P. Klemyshev, B.F. Samylin, G.N. Smirenkin:Sov.J.At.Energy 61, 382 (1986)
32. A.V. Ignatjuk, A.B. Paschenko, V.M. Maslov: Sov.J.Nucl.Phys. 47, 224 (1988)
33. G. Haouat, J. Lachkar, Ch. Lagrange, J. Jary, J.Sigaud, V. Patin: Nucl. Sci. Eng. 81, 491 (1982)
34. A.B. Smith, P.T. Guenter,in Proc. Int. Conf. on Nucl. Data for Sci.& Technol., 1982, Antwerpen,p.39 (1982)
35. J.C. Hopkins, B.C. Diven: Nucl. Sci. Eng. 12, 169 (1962)
36. P.E. Spivak, B.G. Erozolimsky, G.A. Dorofeev, V.N. Lavrenchik, I.E. Kutikov, Ya.P. Dobrynin: Sov.J.At.Energy 1, 21 (1956)
37. K. Kobayashi, T. Hashimoto, I. Kimura: J.Nucl. Sci.Technol. 10, 668 (1973)
38. L. Cranberg, G. Frye, N. Nereson, L. Rosen: Phys.Rev. 103, 662 (1956)

MEASUREMENT AND EVALUATION OF (n,γf)-REACTION EFFECTS IN THE RESONANCES OF U-235 AND Pu-239

O.A. Shcherbakov

Leningrad Nuclear Physics Institute
Academy of Sciences of the USSR
188 350 Gatchina, Leningrad district, USSR

Abstract: A review of the recent results of (n,γf)-reaction investigations for U-235 and Pu-239 in the resonance energy range is given. The analysis of the present data reveals very large dispersion of the experimental values of $\overline{\Gamma}_{\gamma f}$ and shows the necessity of further measurements. Moreover, this analysis has provided the evidence that M1-radiation dominates in the prefission γ-rays of U-236 while E1-transitions prevail in Pu-240. The best agreement between the experimental and calculated values of $\overline{\Gamma}_{\gamma f}$ can be attained using the assumption of intermediate damping of the vibrational states in the second well.

(U-235, Pu-239, (n,γf)-reaction, fission, time-of-flight, γ-rays, neutron, multiplicity, two-humped barrier, prefission transition)

Introduction

The two-step (n,γf)-reaction first predicted and estimated by Stavinsky and Shaker [1], Lynn [2] gives the unique opportunity to study not only the fission process itself, but also the structure of highly excited states in heavy nuclei and radiative transitions between them. Now, the methods employed for studies of this reaction are based on the registration of fission neutrons and γ-rays in the resonance energy region. The detailed survey of these methods can be found in a review article [3]. Usually, the following quantities are measured: fission γ-rays multiplicity ν_γ, fission neutrons multiplicity ν_n, average total energy of fission γ-rays $\overline{E}_\gamma$, characteristic X-rays yield Y_γ. The measurements performed in 1971-1975 led to experimental evidence of the (n,γf)-reaction, and also the evaluations of $\overline{\Gamma}_{\gamma f}$-widths for U-235, Pu-239 and Pu-241 were made. The results of the first decade of (n,γf)-investigations were summarized by Trochon [4].

During the next few years some new experimental data have been obtained for U-235 and Pu-239 which are discussed in this article.

Experimental data

The measurements of the fission γ-ray multiplicities for U-235 in the energy range from 4 eV to 130 eV and for Pu-239 in the range from 14 eV to 205 eV have been carried out at Gatchina using the neutron TOF-spectrometer GNEIS [5]. Fission fragments were detected with a fast multiplate ionization chamber while two large NaI(Tl) - scintillators were used as a γ-ray detector (spectrometer). The $\overline{\Gamma}_{\gamma f}$-widths were obtained from the experimental data after the manner of Shackleton [6]. Fig.1 shows $\overline{\Gamma}_{\gamma f}$-widths as

a function of the ratio of E1- to M1-transition intensities in the prefission γ-rays. The best agreement between the experimental results and model calculations is reached using the giant dipole resonance model and the modified doorway-state model [7] with intermediate damping of the class-II states. The scheme of transition states obtained by Goldstone [8] from the (d,pf)-reaction was used. The comparison of the experiment with the calculations also shows the predominance of M1-transitions in U-236 and that of E1-transitions in Pu-240 in the spectra of γ-transitions between highly excited states ~1 MeV below the neutron binding energy B_n.

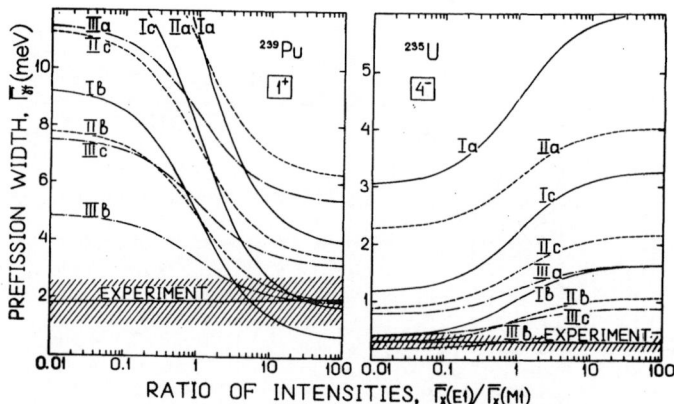

Fig. 1. Experimental and calculated $\overline{\Gamma}_{\gamma f}$ for 1^+-resonances of Pu-239 and 4^--resonances of U-235. Calculation model: I-single-humped fission barrier, II-double-humped barrier (complete damping of the clas-II states), III-double-humped barrier (intermediate damping); a-single-particle model (Weisskopf) of γ-transitions; b,c-giant dipole resonance model (Lorentz line shape).

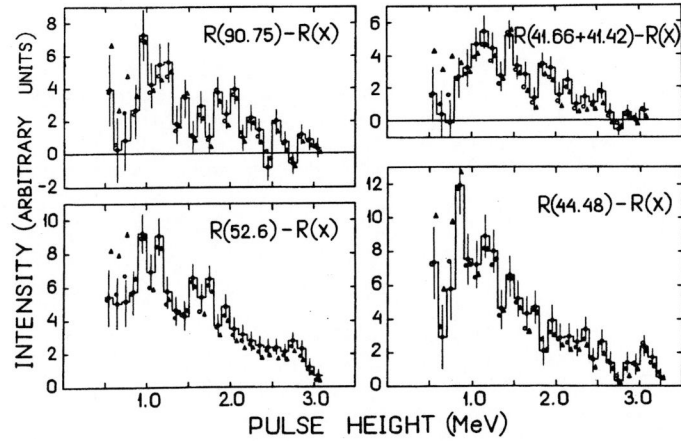

Fig. 2. Difference pulse-height spectra of fission γ-rays for 1⁺-resonances of Pu-239. Resonance energy:
● R(x)=R(10.93 eV), ○ R(x)= =R(17.66 eV), Δ R(x)=R(22.93 eV)

The analogous conclusion about the nature of prefission γ-rays in U-236 was made by Dlouhy et al [9] at Dubna, where the measurements of characteristic X-ray yield were carried out for resonances of U-235 in the neutron energy range from 0.5 eV to 35 eV. The opposite result was obtained by Petrov et al [10] at Gatchina where the measurements of electron conversion spectra in Pu-239 fission were performed with resonance (0.298 eV) and thermal neutrons. From analysis of the experimental data it was concluded that M1-transitions should dominate in prefission γ-rays for Pu-240.

In the another experiment at the GNEIS facility [3,5], the pulse-height spectra of fission γ-rays have been measured for isolated resonances of Pu-239 in the energy range from 10 eV to 91 eV including 1⁺-resonances with very narrow fission widths. Fig.2 shows the difference pulse-height spectra for weak (Γ_f<10 meV) and strong (Γ_f>40 meV) resonances of Pu-239. Besides the known maximum at E_γ = 2 MeV first observed by Trochon et al [11], these spectra display some more structures which could be interpreted as the prefission γ-transitions to the levels at excitation energies 1 MeV-3 MeV below B_n. The following decay from these levels into the exit fission channel could go via the intermediately damped states in the second well of Pu-240 fission barrier.

Discussion

The collected $\overline{\Gamma}_{\gamma f}$-widths for U-235 and Pu-239 are given in Table 1. Most of these values were deduced from the experimental data using a linear relationship between measured quantities ($\nu_\gamma, \nu_n, \overline{E}_\gamma, Y_\gamma$) and $1/\Gamma_f$-the reciprocal of the resonance fission width. The $\overline{\Gamma}_{\gamma f}$-widths ascribed to Howe et al.[12], Weston and Todd [14] were deduced by the author of present work.

It is evident that dispersion of the data on $\overline{\Gamma}_{\gamma f}$ is too large not only for pra-

Table 1. Resonance values of $\overline{\Gamma}_{\gamma f}$.

Nucleus J^π	$\overline{\Gamma}_{\gamma f}$ meV	Comments
235U 4⁻	0.83±0.13	Shackleton [6], $\overline{E}_\gamma$ a)
	2.1±0.7	Trochon [4], $\nu_n, \overline{E}_\gamma$ b)
	0.88±0.44	Howe [12], ν_n c)
	$2.1^{+1.5}_{-1.7}$	Dlouhy [9], Y_γ d,e)
	0.32±0.13	Shcherbakov [3], ν_γ
	0.33±0.13	Gohs [17], evaluation
235U 3⁻	6.6±3.0	Shackleton [6], ν_n a)
	4.7±2.3	Trochon [4], $\nu_n, \overline{E}_\gamma$ b)
	0.13±0.86	Howe [12], ν_n c)
	0.87±0.89	Shcherbakov [3], ν_γ
239Pu 1⁺	3.4±0.3	Shackleton [6], $\nu_n, \overline{E}_\gamma$
	4.2±0.4	Trochon [4], $\nu_n, \overline{E}_\gamma$ f)
	2.2±0.8	Weston [14], ν_n c)
	1.9±0.8	Shcherbakov [3], ν_γ
	~1.6	Petrov [11], $R_{\gamma e}$ e)
239Pu 0⁺	5.8±1.4	Shackleton [6], $\overline{E}_\gamma$ a)
	7.3±1.8	Trochon [4], $\nu_n, \overline{E}_\gamma$ f)
	7.4±6.1	Weston [14], ν_n c)
	2.8±9.2	Shcherbakov [3], ν_γ

a) derived from $(\overline{\Gamma}_{\gamma f} \cdot \overline{E}_{\gamma f})_{exp}$ using $\overline{E}_{\gamma f}$ = 1.08 MeV(4⁻), 1.10 MeV(3⁻), 1.36 MeV(1⁺) 1.38 MeV(0⁺).

b) averaged $(\overline{\Gamma}_{\gamma f} \cdot \overline{E}_{\gamma f})_{exp}$ from [12,13]

c) $d\nu_n/dE^* = 0.135$ n/MeV, $\overline{E}_{\gamma f} = 1$ MeV, $\langle \nu_n \rangle$ = 2.409 for ²³⁵U.

d) deduced regardless of J^π.

e) supposed pure M1-transitions.

f) revized data from [6].

ctical use in the reactor applications but also for the correct comparison with the theory. Moreover, the $\overline{\Gamma}_{\gamma f}$-widths obtained at Saclay [4,6,13] for Pu-239 are at least twice as large as those obtained at Gatchina, whilst for U-235 a discrepancy is still larger. It ought to discuss the origins of this discrepancy.

Firstly, it may be shown, that data of Shackleton [6] for U-235 could be reanalysed using the resonance parameters published after 1973 (e.g. [16]). Then, one can get $\overline{\Gamma}_{\gamma f}$=(0.63±0.54) meV for 4⁻-resonances using the data on $\overline{E}_\gamma$ and (3.5 ± 3.0) meV for 3⁻-resonances (data on ν_n).

Furthermore, it is common practice to analyse the data on ν_n using parameter $d\nu_n/dE^*$ derived from the fast neutron measurements. The analysis [15] of the experimental data on TKE shows that excitation energy region $4.5 < E^* < 5.5$ MeV (where the maximum of the prefission γ-ray spectrum is localized) is a transitional one lying between two regions connected with different types of fission. The superfluid region ($E^* < 4.5$ MeV) is characterized by very weak damping when all increase in excitation energy transforms almost entirely to an increase in $\overline{TKE}$. In the moderate and strong damping region ($E^* > 5.5$ MeV) $\overline{TKE}$ decreases when E^* is increased, while ν_n is increased too. So far, nearly nothing is known about the behavior of $\nu_n(E^*)$ in the range below B_n, including the transitional region. That is why the linear extrapolation of $\nu_n(E^*)$ to the excitation energy region below B_n is incorrect and it should be considered as the probable origin of the systematic error in $\overline{\Gamma}_{\gamma f}$ deduced from the data on ν_n.

The thorough analysis of the techniques used by Dlouhy et al and Petrov et al is beyond the scope of this paper. It is worth noting here, that due to the properties of the γ-ray conversion, $\overline{\Gamma}_{\gamma f}$ obtained by these authors are naturally the lower limits, on condition that the false background effects were negligibly small. The value obtained by Dlouhy et al for U-236 using an assumption about the predominance of M1-transitions before fission is not inconsistent with the other data. An analogous assumption for Pu-240 made by Petrov et al is scarcely correct because the calculations show that for pure M1-transitions before fission the $\overline{\Gamma}_{\gamma f}$-width for 0^+-resonances of Pu-239 does not exceeds ~ 1 meV. It can be seen from the Table 1 that it is not the case.

Unlike to the other authors, who obtained $\overline{\Gamma}_{\gamma f}$ using the "traditional" $1/\Gamma_f$-dependence, Gohs [17] made an evaluation of the fission characteristics for 4^--resonances of U-235 on the basis of a multimode model of Brosa [18] and a joint fission mode/fission channel representation [19]. Apparently, it is the first successful attempt to describe in a self-consistent manner the resonance fluctuations of $\overline{TKE}$ [20], ν_n [12] and ν_γ [3] caused both by fluctuations in the yield of fission modes and by the $(n, \gamma f)$-reaction. The value of $\overline{\Gamma}_{\gamma f}(4^-)$ obtained by Gohs is in a good agreement with that obtained at GNEIS [3].

Though effects of the $(n, \gamma f)$-reaction in resonances of U-235 and Pu-239 on the average seem to be small, the data for this reaction were recommended to be included in data files. A significance of the $(n, \gamma f)$-effects in resonance energy range for reactor applications was demonstrated in a recent evaluation of ν_n for Pu-239 made by Fort et al [22]. Formerly, an importance of this reaction for Pu-239 in the fast neutron range was shown by Konshin [23].

Conclusion

One can admit that summarizing results of the $(n, \gamma f)$-reaction investigations we should be modest about our achievements. Up to now, there is a need for new high quality experimental data on this reaction for U-235 and Pu-239 in the resonance energy range. Apart from the important problems of fission physics, more background to this statement is given by a constant growth of an accuracy of the nuclear constants needed for applications.

References

1. V. Stavinsky and M.O. Shaker: Nucl. Phys. **62**, 667 (1965)
2. J.E. Lynn: Phys. Lett. **18**, 31 (1965)
3. O.A. Shcherbakov: Phys. Elem. Part. and Atom. Nucl.: **21**, 419 (1990)
4. J. Trochon, in Proc. Int. Symp. on Phys. and Chem. of Fission, May 1979, Julich, FRG. IAEA, p.87, Vienna (1980)
5. N.K. Abrosimov, G.Z. Borukhovich, A.B. Laptev, V.V. Marchenkov, G.A. Petrov, O.A. Shcherbakov, Yu.V. Tuboltzev and V.I. Yurchenko: Nucl. Instr. Meth. **A242**, 121 (1985)
6. D. Shackleton, These de Doctorat, Paris (1974)
7. P.D. Goldstone and P. Paul: Phys. Rev. **C18**, 1733 (1978)
8. P.D. Goldstone, F. Hopkins, R.E. Malmin and P. Paul: Phys. Rev. **C18**, 1706 (1978)
9. Z. Dlouhy, A. Duka-Zolyomi, J. Kristiak and Ts. Panteleev: Czech. J. Phys. **B30**, 1101 (1980)
10. G.A. Petrov, L.A. Popeko and Yu.P. Rudnev: Yadern. Fiz. **49**, 1261 (1989)
11. J. Trochon, Y. Pranal, G. Simon and C. Sucosd, in Proc. 3 All Union Conf. on Neutron Physics, June 1975, Kiev, USSR, v.5, p.323, Moscow (1976)
12. R.E. Howe, T.W. Phillips and C.D. Bowman: Phys. Rev. **C13**, 195 (1976)
13. G. Simon, These de Doctorat, Paris (1975)
14. L.W. Weston and J.H. Todd: Phys. Rev. **C10**, 1402 (1974)
15. Y. Patin, J. Lachkar and J. Sigaud, in Proc. 3 All Union Conf. on Neutron Physics, June 1975, Kiev, USSR, v.5, p.300, Moscow (1976)
16. S.F. Mughabghab, Neutron Cross Sections: Vol.1, Part B, Academic Press, London (1984)
17. U. Gohs, contribution to this Conference
18. U. Brosa, S. Grossmann and A. Muller: Z. Naturforsch. **41a**, 1341 (1986)
19. V.I. Furman and J. Kliman, in Proc. XVII Int. Symp. on Nuclear Physics, Nov. 1987, Gaussig, p.142 (1988)
20. F.J. Hambsch, H.H. Knitter, C. Budtz-Jorgensen and J.P. Theobald: Nucl. Phys. **A491**, 56 (1989)
21. Summary and Recommendations, in Proc. of a Consulting Meeting on Physics of Neutron Emission in Fission, May 1988 Mito, Japan (Edited by H.D. Lemmel), INDC-220, IAEA, Vienna (1988), p.9
22. E. Fort, J. Frehaut, H. Tellier and P. Long: Nucl. Sci. Eng. **99**, 375 (1988)
23. V.A. Konshin: Izv. Acad. Sci. BSSR, Fiz.-Energ. Sci. Series **1**, 16 (1979)

NEW EVALUATIONS OF NEUTRON CROSS SECTIONS FOR ^{14}N AND ^{16}O

G. M. Hale, P. G. Young, M. Chadwick

Theoretical Division, Los Alamos National Laboratory
Los Alamos, New Mexico 87545 USA

and

Z.-P. Chen

Qinghua University, Beijing, PRC

Abstract: New evaluations of the neutron cross sections for ^{14}N and ^{16}O have been made for ENDF/B-VI. The evaluations are based at low energies on R-matrix analyses of reactions in the ^{15}N and ^{17}O systems, and at higher energies on GNASH calculations and experimental data evaluations, including covariance analyses. The ^{15}N system R-matrix analysis includes data from reactions among the channels n+^{14}N, p+^{14}C, and α+^{11}B at energies corresponding to excitations in ^{15}N below E_x=13 MeV. The resonance structure of all cross sections in this energy range is fairly well reproduced. New data indicate a different J-value for the first resonance, however. Sub-threshold S-wave levels required to explain the large n+^{14}N total and elastic cross sections near zero energy give scattering lengths that differ significantly from the previous values. The R-matrix analysis of the ^{17}O system includes many new measurements of the n+^{16}O total cross section, done primarily at Oak Ridge and at Karlsruhe. The resonance structure of all the cross sections [total, (n,n), (n,α), and (α,α)] is well represented by the fit in the region below E_n = 6.5 MeV. The new total cross section information gives different positions for some of the resonances and implies a different normalization for the (n,α) cross sections than that obtained in the ENDF/B-IV analysis. The evaluations at energies above the ranges of the R-matrix analyses incorporate results from a number of experiments performed since the previous ENDF/B evaluations. Especially important are new measurements of the total cross sections and differential elastic, inelastic, and gamma-ray production cross sections.

(^{14}N, ^{16}O, neutron-induced reactions, R-matrix analysis, data evaluation)

Introduction

Because of renewed interest in the transport of neutrons in air, encouraged by a National Academy of Sciences study of radiation effects in the early above-ground nuclear explosions, we have re-evaluated the neutron cross sections for ^{14}N and ^{16}O for ENDF/B-VI. The evaluations are based at low energies on R-matrix analyses of reactions in the ^{15}N and ^{17}O systems, and at higher energies on GNASH calculations and variance-covariance fits to the experimental data. We will first discuss the R-matrix analyses that were used to provide the low-energy cross sections, and then briefly summarize the experimental data evaluations that were used at the higher energies.

R-Matrix Analysis for the ^{15}N System

The ^{15}N system R-matrix analysis includes data from reactions among the channels n+^{14}N, p+^{14}C, and α+^{11}B at energies corresponding to excitations in ^{15}N below E_x = 13 MeV. The channel configuration for the analysis and a summary of the data included for each reaction are given in Table 1. The n+^{14}N total cross section fitted was a smoothed composite [1] of experimental measurements made in the years 1950-1970, and the elastic angular distributions were those of Fowler et al. [2]. ^{14}C(p,n)^{14}N angular distributions [3] provided important information about the values of the dominant n-^{14}N channel spins for the resonances, since the doublet-doublet and quartet-doublet transitions give shapes that have opposite sign.

The resonance structure of all the data in this energy region is fairly well reproduced by the analysis, which was started with resonance parameters from the tabulation of Ajzenberg-Selove [4]. Some changes in the level assignments were required in the range E_x = 11.9 - 12.3 MeV, and an additional 1/2$^+$ level was found at E_x = 11.96 MeV in order to improve the fit to the total cross section, as is shown in Fig. 1. The first resonance visible in the total cross section at E_n = 0.43 MeV was assumed to have J^{π}=3/2$^-$. High-resolution measurements of σ_T just completed by Harvey and Larson [5] indicate that it is consistent with J=7/2,

however. These new data, along with elastic scattering angular distributions that will be measured at ORELA this summer, undoubtedly will permit further refinement of the ^{15}N resonance parameters. Some results of the ^{15}N R-matrix analysis, and the predicted n+^{14}N cross sections at low energies are therefore likely to change in the near future.

The region below the first resonance is also interesting because of the rapid rise of the elastic and total cross sections with decreasing energy. This behavior could only be explained (and not entirely satisfactorily) by the presence of levels below the n+^{14}N threshold in both the J=1/2 and J=3/2 S-waves. The resulting fit gives nearly equal S-wave scattering lengths, so that the low-energy scattering cross section is only slightly larger than the coherent cross section. The J=1/2 scattering length is quite different from the one obtained by Mughabghab et al. [6], having been at one point in the analysis larger than the J=3/2 scattering length.

Table 1. Channel configuration and data summary for ^{15}N system analysis.

Channel	l_{max}	a_c (fm)
n-^{14}N	2	2.6
p-^{14}C	3	4.3
α-^{11}B	2	5.1

Reaction	Energy Range	Observable Types	# Data Points
^{14}N(n,n)^{14}N	E_n=0-2.3 MeV	σ_T, $\sigma_{nn}(\theta)$	793
^{14}N(n,p)^{14}C + inv.	E_n=0-2.3 MeV	σ_{np}, $\sigma_{pn}(\theta)$, $A_p(\theta)$	714
^{14}N(n,α)^{11}B	E_n=1.3-2.3 MeV	$\sigma_{n\alpha}$	112
^{11}B(α,p)^{14}C	E_α=1.4-2.6 MeV	$\sigma_{\alpha p}$, $\sigma_{\alpha p}(\theta)$	110
Totals:		8 obs.	1729

R-Matrix Analysis for the ^{17}O System

The R-matrix analysis of the ^{17}O system is an extensive update of the one used to provide cross sections at energies up to 6 MeV for the ENDF/B-IV evaluation. The channel configuration for the analysis and a summary of the data included for each reaction are given in Table 2. Many new measurements of the n+^{16}O total cross section, done primarily at Oak Ridge [7] and at Karlsruhe [8], were included, as well as new measurements [9] of the differential

922

elastic cross section and polarization at neutron energies between 2 and 4 MeV. The resonance structure of all the cross sections is well represented by the 45 R-matrix levels included in the fit at energies below $E_n = 6.5$ MeV, as is illustrated in Fig. 2. The level structure found agrees for the most part with the recommended data, [10] but with different parity assignments for some of the resonances and minor differences in positions and widths for the others. The new total cross section information gives a different position, especially, for the first resonance at $E_n = 435$ keV, and implies a different normalization for the (n,α) cross sections than that obtained in the ENDF/B-IV analysis.

Table 2. Channel configuration and data summary for ^{17}O system analysis.

Channel	l_{max}	a_c (fm)
n-^{16}O	4	4.44
α-^{13}C	4	5.69

Reaction	Energy Range	Observable Types	# Data Points
$^{16}O(n,n)^{16}O$	E_n=0-6.5 MeV	σ_T, $\sigma_{nn}(\theta)$, $A_n(\theta)$	2421
$^{16}O(n,\alpha)^{13}C$	E_n=0-6.0 MeV	$\sigma_{n\alpha}$, $\sigma_{n\alpha}(\theta)$, $A_n(\theta)$	904
$^{13}C(\alpha,\alpha)^{13}C$	E_α=0-4.6 MeV	$\sigma_{\alpha\alpha}(\theta)$	207
Totals:		7 obs.	3532

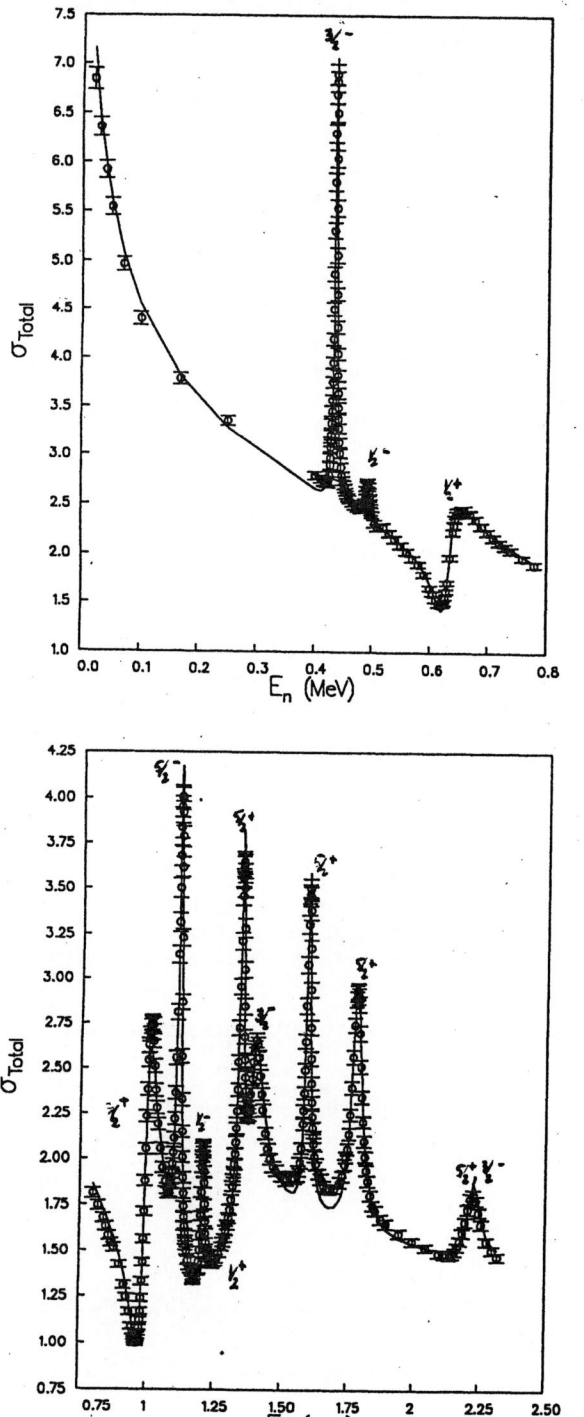

Fig. 1 Neutron elastic scattering cross section for ^{14}N at energies up to 0.8 MeV (top) and at energies between 0.8 and 2.3 MeV (bottom). The solid curve is tthe R-matrix fit, and the points are experimental data. [1]

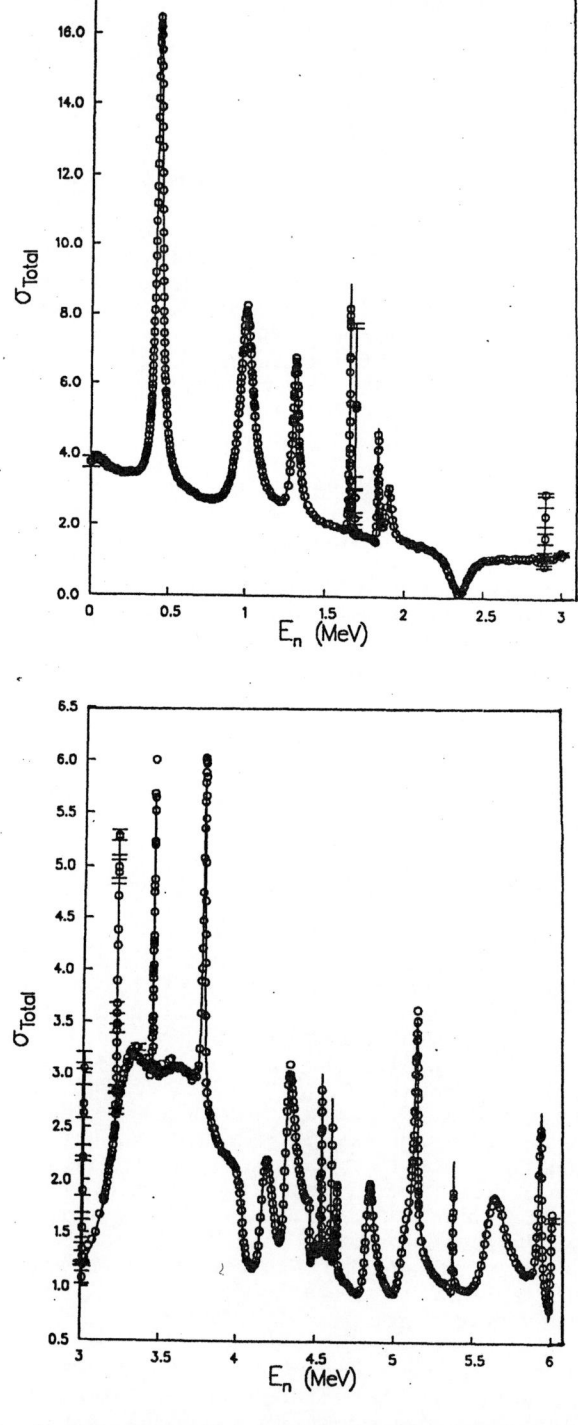

Fig. 2. Neutron total cross section for ^{16}O at energies up to 3 MeV (top) and at energies between 3 and 6 MeV (bottom). The solid curve is the R-matrix fit, and the points are experimental data. [12]

Experimental Data Evaluations

Above the thresholds for inelastic scattering, the ^{14}N ($E_n \geq 2.30$ MeV) and ^{16}O ($E_n \geq 6.25$ MeV) evaluations are based on analyses of the available experimental data, supplemented in regions where data are unavailable by Hauser-Feshbach statistical theory calculations. For both isotopes important new differential elastic and inelastic scattering data as well as gamma-ray production measurements were included in the evaluations. Hauser-Feshbach statistical theory (GNASH) calculations were used to interpolate and extrapolate the data at higher energies.

In the case of n+^{14}N, major experiments completed since 1973 include measurements by Chardine *et al.* [11] of neutron elastic and inelastic scattering angular distributions to the lowest 5 excited states of ^{14}N for incident neutron energies between 7.7 and 13.5 MeV, plus several other measurements spanning the energy range from 4.3 to 25 MeV.[12] New information on neutron inelastic scattering up to $E_n = 20$ MeV is provided by the measurements of Nelson *et al.* [13] and Auchampaugh and Wender, [14] as well as extensive double-differential neutron emission measurements for $E_n \cong 14$ MeV. [12] A comprehensive new measurement of $^{14}N(n,p)$ and $^{14}N(n,\alpha)$ cross sections to discrete states by Morgan *et al.* [15] up to $E_n = 14$ MeV is useful for the experimental data evaluation as well as the R-matrix studies.

The only major new total cross section measurement, [16] covering the energy range $E_n = 0.97$ to 5.3 MeV, is in substantial disagreement (~14%) with older, precision measurements and was not included in the ENDF/B-VI evaluation. Consequently, only minor changes were made in the evaluated total cross section above 2.5 MeV. However, the new ORELA measurement of the ^{14}N total cross section [5] will be incorporated into the present analysis as an update to the ENDF/B-VI evaluation.

The n+^{14}N elastic scattering cross section between 2 and 20 MeV is illustrated in Fig. 3, where a comparison is given between the experimental data base and the ENDF/B-V and ENDF/B-VI evaluations. Note that the cross section has increased by some 9% near $E_n = 14$ MeV.

Several new neutron total cross section measurements for ^{16}O [12] have become available since the previous ENDF/B evaluation. Most important for the present experimental (and R-matrix) evaluations is the measurement of Cierjacks *et al.* [8] To obtain average total cross sections, we performed a variance-covariance analysis of all the available data, and then relied on the higher resolution measurements (*e.g.*, Ref. 9) to define local structure. Important new differential elastic and inelastic scattering measurements have also become available, [12] especially the precision data of Börker *et al.* [17] Finally, several n + ^{16}O gamma-ray production measurements have been completed since the last ENDF/B evaluation, and one very recent one [13] contributed significantly to the present work.

Conclusion

The new ENDF/B-VI evaluations of neutron-induced reactions on ^{14}N and ^{16}O at energies between 10^{-11} and 20 MeV are overall improvements on the previous evaluations. We believe the results for ^{16}O to be significantly better over the whole energy range, where the new total cross-section information has given firmer positions and shapes for the resonances and improved consistency with the reaction data. The shift in position of the first oxygen resonance and revised normalization scale of the (n,α) reaction cross section may have important bearing in nuclear data applications.

The results for the low-energy (≤ 2.5 MeV) nitrogen cross sections are relatively more uncertain in the light of the new total cross-section measurement [5] at Oak Ridge. We know already that the J-value of the first resonance differs from that used in the evaluation. Refinements of the n+^{14}N cross sections based on these data and new measurements of the differential elastic scattering cross section will be incorporated in an updated evaluation in the near future.

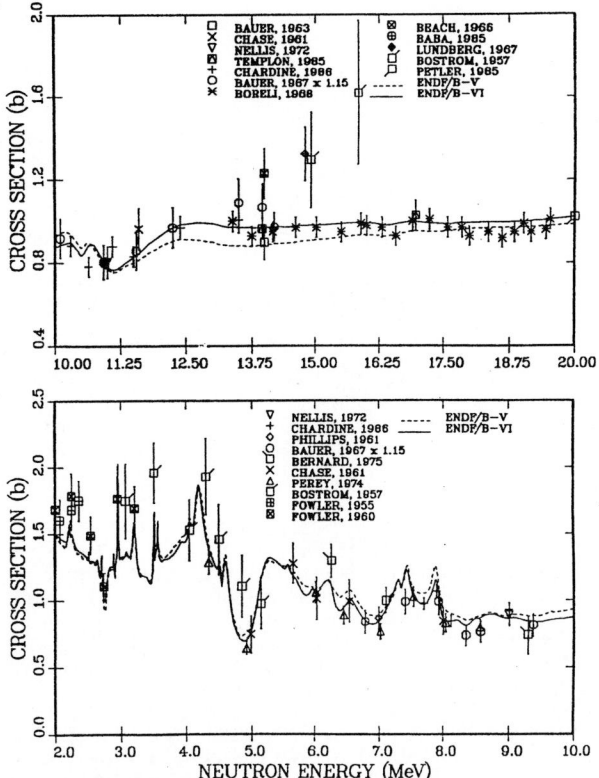

Fig. 3. Neutron elastic scattering cross section for ^{14}N between 2 and 20 MeV.

References

1. P. G. Young and D. G. Foster, An Evaluation of the Neutron and Gamma-Ray Production Cross Sections for Nitrogen, LA-4725, Los Alamos (1972).
2. J. L. Fowler and C. H. Johnson: *Phys. Rev.* **98**, 728 (1955); J. L. Fowler, C.H. Johnson, and R.L. Kernell, in Proc. Conf. on Neutron Cross Secs.& Tech. Washington, DC (March 1968), p. 653, NBS (1968).
3. G. Bartholomew: *Can. J. Nucl. Phys.* **33**, 441 (1955).
4. F. Ajzenberg-Selove: *Nucl. Phys.* **A449**, 113 (1986).
5. J. Harvey and D. Larson, Oak Ridge Nat. Lab., pers. comm. (April 1991); this conference (May 1991).
6. S. F. Mughabghab, M. Divadeenam and N. Holden, Neutron Cross Sections, Vol.1, Part A, Academic Press, New York (1981).
7. D. C. Larson, "ORELA Measurements to Meet Fusion Energy Neutron Cross Section Needs, in Proc. Symp. on Neutron Cross Sections from 10 to 50 MeV (BNL-NCS-51245), p.277, BNL, New York (1980).
8. S. Cierjacks, F. Hinterberger, G. Schalz, and D. Erbe: *Nucl. Inst. Meth.* **169**, 185 (1980).
9. L. Drigo, G. Tornielli, and G. Zannoni: *Nuov. Cim.* **31**, 1 (1976).
10. F. Ajzenberg-Selove: *Nucl. Phys.* **A460**, 77 (1986).
11. J. Chardine, G. Haouat, S. Seguin, and C. Humeau, Diffusion Elastique et Inelastique de Neutrons sur ^{14}N Entre 7.7 et 13.5 MeV, Commissariat à l'Energie Atomique report CEA-N-2506 (1986).
12. Exp. data from CSISRS compilation, NNDC, BNL.
13. R Nelson, S. Wender, C. Laymon, H. O'Brien, R. Haight, G. Morgan, D. Drake, P. Young, M. Drosg, H. Vonach, A. Pavlik, D. Larson, P. Englert and J. Brueckner: *Bull. Am. Phys. Soc.* **35**, 1038 (1990).
14. G. Auchampaugh and S. Wender, Los Alamos National Laboratory, personal communication (1986).
15. G. L. Morgan: *Nucl. Sci. Eng.* **70**, 163 (1979).
16. J. Bommer: *Nucl. Phys.* **A263**, 86 (1976).
17. G. Börker, R. Bottger, H. J. Brede, H. Klein, W. Mannhart and B. R. L. Siebert, Elastic and Inelastic Differential Neutron Scattering Cross Sections of Oxygen Between 6 and 15 MeV, PTB-N-1 (1989).

THE EFF-2 EVALUATION FOR THE REACTION SYSTEMS
n + ^{52}Cr, ^{56}Fe, ^{58}Ni AND ^{60}Ni

M. Uhl
Institut für Radiumforschung und Kernphysik,
University of Vienna, Boltzmanngasse 3, A1090 Vienna, Austria
and
H. Gruppelaar, H. A. J. van der Kamp, J. Kopecky, and D. Nierop,
Netherlands Energy Research Foundation ECN, P.O. Box 1, 1755 ZG Petten, The Netherlands

Abstract: The evaluated cross sections for incident energies between 1 and 20 MeV are primarily based on nuclear reaction model calculations. The model parameters were chosen so as to simultaneously reproduce many available experimental data. At lower energies the cross sections were taken from experiment. The calculations consider the spherical optical model, the statistical compound nucleus model under consideration of angular momentum and parity, the exciton model for first-chance preequilibrium emission of particles and photons and a direct reaction model for inelastic scattering. The exciton model employed accounted for angular momentum conservation.

For each reaction system the evaluation comprises the following cross sections: total, elastic, inelastic, production of photons, neutrons, protons and alphas and production of about 10 residual nuclei for each target. Furthermore included are the energy-angle distributions of the light and of the heavy reaction products.

(Evaluation for incident energies E_n up to 20 MeV, model calculations for $E_n \geq 1$ MeV)

Introduction

The evaluation comprises the most abundant isotopes of the structural materials Cr, Fe and Ni for the European Fusion File EFF-2. Its main objectives are technological applications related to energy production. All important cross sections required for this purpose are given for incident neutron energies up to 20 MeV. Different procedures have been adopted in two regions of incident energy. Below one MeV the evaluation is primarily based on experimental data. For incident energies from 1 to 20 MeV all evaluated reaction cross sections result from model calculations with parameters adjusted to simultaneously reproduce many available experimental data. Most transmutation cross sections vanish in the first region. Those cross sections which are non-zero in both regions smoothly join at $E_n = 1$ MeV. The advantage of evaluated cross sections from model calculations is the guarantee of flux- and energy conservation; some disadvantages will be discussed below. In the adaptation for EFF-2 the total cross section was replaced by an evaluation of experimental data representing nuclear structure effects; the total elastic cross section was correspondingly adjusted and all other data were left unchanged.

The model calculations were carried out with the code MAURINA /1/; those for ^{58,60}Ni, performed in 1987, employed an older version than the computations for ^{52}Cr and ^{56}Fe done in 1988-1989. For this reason the treatment of preequilibrium emission slightly differs for the two Ni isotopes and for ^{52}Cr and ^{56}Fe. Extensive comparisons of our results with experimental data will be presented in a forthcoming paper /2/.

Incident energies below 1 MeV

Resonance parameters were used exclusively to provide the total, elastic, inelastic and capture cross sections in the resolved resonance regions below 1 MeV. The smooth parts of these cross sections were adjusted to reproduce the evaluated experimental data, cf. Ref. 2.

Incident energies from 1 MeV to 20 MeV

The following models were considered: the optical model, the statistical compound nucleus model (CNM), the exciton model (EM) for first chance particle and photon emission in the preequilibrium stage and a simple model for direct inelastic scattering. In the CNM and EM we accounted for the (competing) emission of neutrons, protons, alpha-particles and photons.

The **optical model** and the underlying optical model potential (OMP) are required in the other models, too. As OMP's are needed for many nuclei with no relevant experimental data to determine them, we chose regional or global potentials. These were then used for **all** nuclei, i.e. even for those with experimental data.

Of special importance for this evaluation is the **neutron OMP**. We assumed a conventional potential which in standard notation reads

$$-U(r,E) = V_R(E)f_R(r) + i\left[W_V(E)f_V(r) - 4a_S W_S(E)\frac{df_S(r)}{dr} \right] \\ - V_{so}(E)(\vec{\ell}\cdot\vec{\sigma})\lambda_\pi^2 \frac{1}{r}\frac{df_{so}(r)}{dr} . \tag{1}$$

The radial form factors $f_x(r) = 1/\{1 + \exp[(r - r_x A^{1/3})/a_x]\}$, where A is the mass number, are of Woods-Saxon type. The potential parameters were determined so as to simultaneously reproduce for nuclei with mass numbers between 46 and 60 and incident energies between 10 keV and 30 MeV the available experimental s- and p-wave strength functions, scattering radii, total cross sections and differential elastic cross sections. We started with the global potential proposed by Rapaport et al. /3/. Within the range of its validity (7 MeV $\leq$ E $\leq$ 30 MeV) this OMP reasonably describes the experimental data; it fails, however, in in the low energy region. In order to reproduce at least approximately the minimum of the total cross section around 1 MeV and the average resonance data we had to introduce below E = 5 MeV energy dependent geometry parameters a_R, r_S and a_s. Furthermore the depth of the surface absorptive potential was changed below 10 MeV. The resulting OMP parameters are:

$$V_R(E) = 54.19 - 0.33E - (22.7 - 0.19E)(N - Z)/A \text{ MeV}$$
$$r_R = 1.198 \text{ fm}$$
$$a_R(E) = 0.63 + 0.00714(1 - E/5)\Theta(5 - E)(\max(A,46) - 46) \text{ fm}$$

$$W_V(E) = (-4.3 + 0.38E)\Theta(E - 15) \text{ MeV}$$
$$r_V = 1.295 \text{ fm} \quad a_V = 0.53 \text{ fm}$$

$$\begin{aligned} W_S(E) &= 8.28 - 12.8(N - Z)/A \text{ MeV} & E < 10 \text{ MeV} \\ &= 4.28 - 12.8(N - Z)/A + 0.40E \text{ MeV} & 10 \leq E < 15 \text{ MeV} \\ &= 14.0 - 10.4(N - Z)/A - 0.39E \text{ MeV} & E \geq 15 \text{ MeV} \end{aligned}$$
$$r_S(E) = 1.295 + 0.096(1. - E/5)\Theta(5 - E) \text{ fm}$$
$$a_S(E) = 0.530 + 0.199(E/5. - 1)\Theta(5 - E) \text{ fm}$$

$$V_{so} = 6.20 \text{ MeV}$$
$$r_{so} = 1.010 \text{ fm} \quad a_{so} = 0.750 \text{ fm}$$

The energy E (LAB-system) is given in MeV and $\Theta(x)$ is 1 for positive arguments and zero otherwise. Following Strohmaier /4/ the high energy values of the diffuseness parameters a_R, a_V and a_S were slightly reduced compared to the original values of Ref. /3/. Our neutron OMP exhibits a complicated energy dependence. Difficulties to reproduce experimental data in this mass region with OMP's with energy independent geometry parameters were also reported by the Argonne group /5/.

The **proton OMP** was taken from Mani et al. /6/. For the **alpha-particle OMP** we slightly modified the potential proposed by Mc Fadden and Satchler /7/. The respective parameters for Eq. (1) read:

$$V_R(E) = 173.00 - 0.30E \text{ MeV}$$
$$r_R = 1.445 \text{fm} \quad a_R = 0.510 \text{ fm}$$

$$W_V(E) = 20.5 + 0.10E \text{ MeV}$$
$$r_V = 1.445 \text{fm} \quad a_V = 0.510 \text{ fm}$$

The Coulomb potential was that of a uniformly charged sphere with radius 1.3A$^{1/3}$ fm. As described on the occasion of a previous version of this evaluation /8/, the charged particle OMP's were tested by comparing (p,n)- and (α,n) cross sections to experimental data.

The **statistical compound nucleus model** was employed under consideration of angular momentum and parity conservation.

The required **level schemes** of the residual nuclei were taken from recent issues of the journal "Nuclear Data Sheets" /9/. For higher excitation energy the excited states were described by a **level density** formula. We chose the backshifted Fermi-gas model /10/ with a spin cut-off factor based on the rigid body moment of inertia. The level density parameter a and the backshift Δ were determined from the number of low lying levels and from the spacing of s-wave resonances (Refs. 11-14). For nuclei with no resonance data we had to resort to systematics. Occasionally we varied the level density parameters a and Δ within their uncertainties in order to improve the reproduction of well known cross sections.

The **particle transmission coefficients** were calculated from the OMP's described before. The **gamma-ray transmission coefficients** $T_{XL}(ε_γ) = 2πε_γ^{2L+1}f_{XL}(ε_γ)$ of multipole type XL and for energy $ε_γ$ are related to the corresponding strength function f_{XL}. The E1 strength f_{E1} was deduced from a giant resonance of Lorentzian shape with energy $E_r = 75/A^{1/3}$ MeV, width $Γ_r = 5.5$ MeV and peak cross section $σ_r = 60x(NxZ/A)x(2/πΓ_r)$ mb corresponding to the TRK sum rule. For the strength of M1-, E2- ...M3 radiation we used the single particle model /15/ ; the absolute values were chosen according to the systematics of McCullagh /16/ in case of M1- and as one Weisskopf unit per MeV for E2-, M2-, E3- and M3 radiation. An exception is ^{57}Fe where we used the results of a recent study of gamma-ray strength functions /17/: two Lorentzians for f_{E1} as suggested by fits to the photoabsorption cross section /18/ and a single Lorentzian approximating a spin-flip giant resonance for f_{M1}. In order to improve the reproduction of gamma-ray production spectra the E1 strength for all nuclei was reduced by 0.55 below 6.0 MeV. An overall normalization factor for all strength functions was determined by reproducing the low energy capture cross sections in this mass region; the average of these factors was used for all nuclei with no capture data.

For the **angular distributions** of first-chance emitted particles the exact formulas of the CNM were applied. Isotropy was assumed for the emission of photons and higher chance particles.

The essential ingredients of the **exciton model** calculations were: i) dependence of the matrix element on exciton number and energy according to Kalbach /19/ and its normalization constant K' chosen to reproduce experimental neutron and proton production spectra, ii) particle hole state densities according to Williams /20/ with the single particle state density $g = (6/π^2)A/8$ MeV^{-1} and a conventional pairing shift $Δ_P = 12A^{-1/2}$ MeV, iii) emission rates for nucleons according to Gadioli et al. /21/, for alphas according to Milazzo-Colli and Braga-Marcazzan /22/ with an α- preformation factor of 0.2 and for photons according to Akkermans and Gruppelaar /23/, iv) spin distribution parameters $σ^2(n) = 0.24A^{2/3}n$ depending on the exciton number n, as suggested by Feshbach et al. /24/.

Angular momentum conservation was considered in two different ways. For the two Ni isotopes the distribution of the angle integrated emission spectrum $dσ_{ab}/dε$ (itself calculated using the angular momentum independent model) among the individual spins of the residual nucleus is obtained under consideration of angular momentum coupling. The more recent results for ^{52}Cr and ^{56}Fe are based on the "mean-lifetime Ansatz" proposed by Shi Xiangjun et al. /25/ and thus also the emission spectra (weakly) depend on angular momentum. In view of the relatively small energies and angular momenta involved both approaches give very similar results.

Preequilibrium **photon emission** was considered only for ^{52}Cr and ^{56}Fe. The result was normalized so as to reproduce the capture cross section around 14 MeV. For the two Ni isotopes the capture cross section for incident energies beyond some MeV was calculated by means of simple models by Lane and Lynn /26/ and Brown /27/ for direct and semidirect capture.

The **angular distributions** of particles emitted in the preequilibrium stage were obtained by means of the systematics by Kalbach and Mann /28/. For simplicity we considered the total EM contributions as "multistep direct". In Ref. 29 it is shown that the combination of such an angular distribution for the PE contribution with the exact result for the CNM portion provides a good description of angular distributions of protons from (n,p) reactions at neutron energies around 14 MeV. Isotropy was assumed for photons emitted in the preequilibrium stage.

Direct Reaction contributions were accounted for in case of inelastic neutron scattering exciting final states with predominantly collective character. We employed the distorted wave approximation (DWBA) with macroscopic form factors corresponding to vibrations with multipolarity $ℓ$ and a deformation parameter $β_ℓ$. The deformation parameters were taken from the journal Nuclear Data Sheets /9/ and recent references quoted therein; they mainly stem from (n,n') experiments but for some higher excited states we used also electromagnetic and charged particle data.

The **combination of the contributions of the various models** is given by the incoherent sum; however, to guarantee flux conservation the EM- and the CNM contributions are corrected for depletion by direct and by direct and preequilibrium reactions, respectively. The production cross sections of the various reaction products and the corresponding spectra are in general obtained by summing the contributions of many reaction paths. The calculation of the **recoil spectra** is described in detail in Ref. 30.

Evaluated cross sections

The evaluation comprises the following results:

- The total, elastic, inelastic and differential elastic and inelastic cross sections.
- Production cross sections of the light reaction products: neutrons, protons and alpha-particles.
- Production cross sections of the heavy reaction products. These include cross sections as (n,pγ), (n, αγ), (n, 2nγ)....
- Energy-angle distributions of the light reaction products. The contributions resulting from first-chance particle emission exciting discrete levels of the residual nucleus are given separately.
- Energy-angle distributions of the heavy reaction products (recoil spectra).
- Photon production spectra and (isotropic) angular distributions. They are given for lumped reaction channels except for (n, γ); in case of ^{52}Cr and ^{56}Fe also the gamma-rays resulting from first-chance particle emission exciting discrete levels of the residual nucleus are given separately.

A detailed description of the data file is given in Ref. 2.

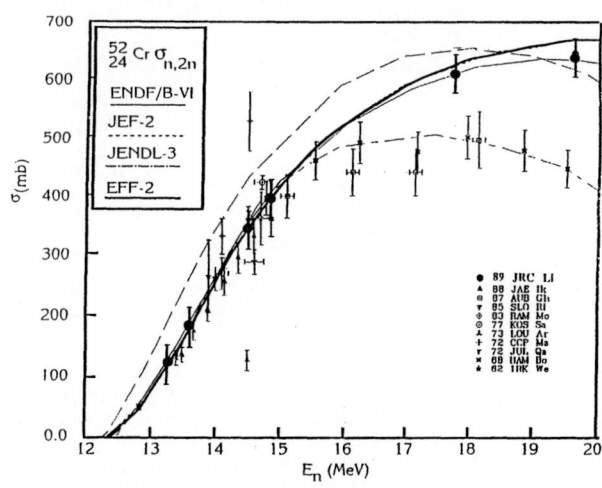

Fig. 1 Evaluated ^{52}Cr(n,2n) cross section compared to experimental data and other recent evaluations.

Discussion

Our model calculations are based on a consistent set of model parameters which aims at a simultaneous description of many experimental data. Of course, such a requirement often demands compromises: a particular cross section may not reproduce the experimental data as well as if model parameters were adjusted to fit this very cross section. On the other hand, the use of such a consistent set of model parameters hopefully increases the predictive power of unknown cross sections. We recall here that in order to reproduce structure effects in the total cross section this cross section and the total elastic one were adjusted to experimental data.

In nuclear data evaluation model calculations are mainly applied in case that experimental data are lacking. In the present evaluation we use the results of model calculations also in energy regions where many good experimental data are available, as e.g. for incident energies around 14 MeV. Therefore, compared to an evaluation which in such a case relies on experimental data, our results may be less accurate. The advantage of our method is the guarantee of flux and energy conservation.

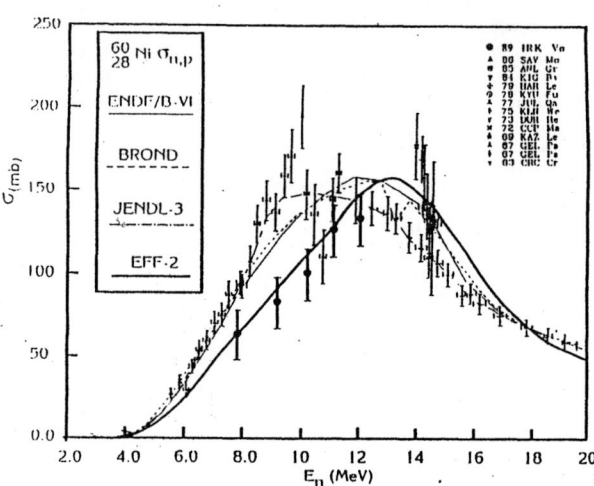

Fig. 2 Evaluated ^{60}Ni(n,p) cross section compared to experimental data and other recent evaluations.

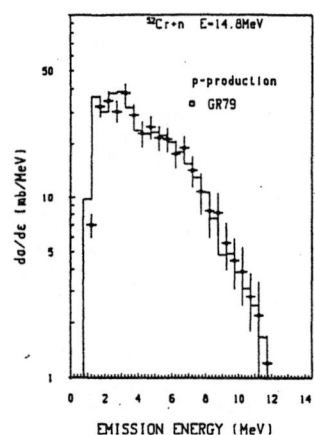

Fig. 3 Evaluated proton production spectrum for ^{52}Cr + n compared to experimental data; the incident energy is 14.8 MeV.

Where experimental data are available the accuracy of the evaluated cross sections can be assessed by inspection. Comparisons of many excitation functions to experimental data showed that the energy dependence of transmutation cross sections is quite well described by the calculations. Two interesting examples, namely the ^{52}Cr(n,2n)- and the ^{60}Ni(n,p) cross sections, are displayed in Figs. 1 and 2, respectively. The calculations disagree with older experimental data but reasonably well reproduce the most recent results by Liskien et al. /31/ for ^{52}Cr(n,2n) and by Vonach et al. /32/ for ^{60}Ni(n,p); both

measurements were performed after the calculations. As an example for the simultaneous reproduction of related cross sections we show in Fig. 3 also the proton production spectrum for ^{52}Cr + n compared to the data by Grimes et al. /33/.

For energy regions with no experimental data the accuracy of the calculation depends on the quality of the model parameters and the adequacy of the models. While the effect of inaccuracies of model parameters can in principle be assessed by (tedious) sensitivity studies a *meaningful* prediction of the consequences of inadequacies of the models seems very difficult. Because of the use of a consistent parameter set and the reasonable overall agreement between calculations and available experimental data we hope that the predictive power of this evaluation in regions with no data is in general not much worse than that for experimentally known cross sections.

ACKNOWLEDGMENT: We thank Dr. S. Wiboolsak for her help with the neutron OMP determination.

References

1. M. Uhl: (unpublished)
2. M. Uhl, H. Gruppelaar, H. A. J. van der Kamp, J. Kopecky and D. Nierop: to be published
3. J. Rapaport, V. Kulkarni and R. W. Finlay: Nucl. Phys. **A330**, 15 (1979)
4. B. Strohmaier: Report IRK 84/01 (1984)
5. A. B. Smith, R. D. Lawson and P. T. Guenther: Proc. Specialists' Meeting on the Use of the Optical Model for the Calculation of Neutron Cross Sections Below 20 MeV, Paris 13th-15th November 1985, OECD, p. 127
6. G. S. Mani, M. A. Melkanoff, and I. Iori: Report CEA-2379, Centre d'Etudes Nucleaires de Saclais (1963)
7. L. M. McFadden and G. R. Satchler: Nucl. Phys. **84**, 177 (1966)
8. B. Strohmaier and M. Uhl: Proc. of the Int. Conf. on Nuclear Data for Science and Technology, Antwerp, 1982 (Reidl, Dordrecht, 1983), p. 552
9. Nuclear Data Sheets, Academic Press, Inc..
10. W. Dilg, W. Schantl, H. Vonach, and M. Uhl: Nucl. Phys. **A217**, 269 (1973)
11. G. Rohr: Proc. Specialists Meeting on Neutron Data of Structural Materials for fast Reactors, Geel, 5-8 Dec. 1977, Pergamon Press (1979), p. 614
12. F. H. Fröhner: Proc, Int. Conf. on Neutron Physics and Nucl. Data for Reactors and other Applied Purposes, Harwell, 25-29 Sept. 1978, p. 268
13. S. F. Mughabhab, M. Divadeenam, and N. E. Holden: Neutron Cross Sections (Academic, New York, 1981), Vol. 1, Part A
14. H. Vonach et al.: Phys. Rev. **C38**, 2541 (1988)
15. M. Blatt and V. Weisskopf: Theoretical Nuclear Physics (Wiley, New York, 1952)
16. Carol M. McCullagh, M. L. Stelts, and R. E. Chrien: Phys. Rev. **C23**, 1394 (1981)
17. M. Uhl and J. Kopecky: Proc. IAEA Specialists' Meeting on the Measurement, Calculation and Evaluation of Photon Production Cross Sections, Febr. 5-7, 1990, Smolenice, CSFR, INDC(NDS)-238 (1990)
18. S. S. Dietrich and B. L. Bermann: At. Nucl. Data Tables **38**, 199 (1989)
19. C. Kalbach: Z. Phys. **A287**, 319 (1978)
20. F. C. Williams, Jr.: Nucl. Phys. **A166**, 231 (1971)
21. E. Gadioli, E. Gadioli-Erba, and P. G. Sona: Nucl. Phys. **A217**, 589 (1973)
22. L. Milazzo-Colli and G. M. Braga Marcazzan: Nucl. Phys. **A210**, 297 (1973)
23. J. M. Akkermans and H. Gruppelaar: Phys. Lett. **157B**, 95 (1985)
24. H. Feshbach, A. Kerman, and S. Koonin: Ann. Phys. (NY) **125**, 429 (1980)
25. Shi Xiangjun, H. Gruppelaar and J. Akkermans: Nucl.Phys. **A466**, 333 (1987)
26. A. M. Lane and J. M. Lynn: Nucl. Phys. **11**, 625 (1959)
27. G. E. Brown: Nucl. Phys. **57**, 339 (1964)
28. C. Kalbach and F. M. Mann, Phys. Rev. **C23**, 112 (1981)
29. R. Fischer, M. Uhl, and H. Vonach: Phys. Rev. **37**, 578 (1988)
30. M. Uhl: Nucl. Sci. Eng. **100**, 77 (1988)
31. H. Liskien, M. Uhl, M. Wagner, and G. Winkler: Ann. Nucl. Energy **16**, 563 (1989)
32. H. Vonach, M. Wagner, and R. C. Haight: Proc. Specialists' Meeting on Neutron Activation Cross Sections for Fission and Fusion Energy Applications, Argonne, USA, Sept. 13-15, 1989, p. 165
33. S. M. Grimes et al.: Phys. Rev. **C19**, 2127 (1979)

PHOTON PRODUCTION CROSS SECTIONS AND SPECTRA FOR (n,xγ) ON ^{56}Fe

F. Fabbri, G. Maino, E. Menapace and A. Mengoni

ENEA, INN.SVIL, Divisione Calcolo, Laboratorio Dati Nucleari e Codici
Viale G.B. Ercolani 8, 40138 Bologna, Italy

Abstract: Two commonly adopted techniques for the calculation of nuclear level densities, employed in neutron-induced reaction theories, have been investigated and tested in cross section calculations on Fe-56 in the MeV incident-neutron energy region. The effect on various reaction channels and particularly on some gamma-ray production cross section in the (n,n'γ) and (n,2nγ) reactions have been investigated and compared for the two different approaches.

Introduction

The nuclear level densities play a fundamental role in neutron cross section calculations. Ab-initio techniques can be reliable in a number of cases and for specific excitation energy-ranges but they are seldom, if not at all, included in the nuclear reaction codes. Phenomenological approaches are usually preferred for the massive calculations required, for instance, in codes based on Hauser-Feshbach statistical model theory. Phenomenological models require a careful determination of the parameters which are usually derived directly from experimental quantities or from systematics constructed from experimental quantities available for neighboring nuclei in the same mass region. This procedure is acceptable when the parameters possess a clear physical meaning. For instance, in the Fermi-gas model at a given excitation energy, the level density parameter a, related to the single particle level spacing, follows a systematic trend as a function of the nuclear mass and the systematics is "smooth" as long as effects such as pairing correlations, shell inhomogeneities and others are taken into account. It is intended here to introduce two of the most commonly used model prescriptions for the calculation of nuclear level densities, determine their parameters in the mass region of interest, A $\simeq$ 60, and compare the results obtained for the cross section calculations. The calculations of discrete gamma-ray production cross sections in the inelastic neutron channels (n,n'γ) and (n,2nγ) on the n + Fe-56 induced reactions will be presented.

Two phenomenological level density models

The Gilbert-Cameron prescription (here referred to as GC) for the calculation of nuclear level densities is widely adopted in nuclear reaction codes. Different techniques used to determine parameters for the low and for the high excitation energy parts can differ considerably from the way the mentioned corrections are included into the model. However, the basic assumptions made in the calculations of level densities are: i) the constant temperature assumption for low excitation energies, ii) the Fermi-gas model with pairing corrections for the high excitation energy part and iii) equal probability for the two different parity distributions.

We have determined the parameters of the GC level density according to Ref. 1. Essentially, the level density formulae depend on the the parameter a, the nuclear temperature T, the odd-even correction energy Δ, the matching energy Ux and the low excitation energy parameter E_0. In addition, a spin cut-off factor $0.146 \ A^{2/3} (aU')^{1/2}$ has been used as suggested in Ref. 1. This parametrization usually assumes a 1/2 probability for the parity distribution of the levels starting from the energy Ecut, where the level density formulae are introduced.

The formulation of Ignatyuk et al. (2), here referred to as IG level density, starts from the inclusion of shell inhomogeneities on the single particle level spacing of the Fermi-gas. From this fact it follows that the level density parameter a depends on the excitation energy in the form

$$a(U')=a(*)\{1+E_{sh}/U' \ (1-\exp - \gamma U')\}$$

where the shell correction energy E_{sh} is determined from nuclear masses and the liquid drop model (see Ref. 2), γ =0.054 is a parameter determined from systematics (2), and a(*) is the asymptotic value of the level density parameter a. The excitation energy is redefined as usual by $U' = U - \Delta$ with Δ= 12/A for even-even and Δ =0 for even-odd nuclei.

In order to take into account the parity distribution of the levels which tends to 1/2 at high excitation energy, we adopt a correction term to the total level density which approximates microscopic Nilsson-BCS calculations (3)

$$F \ (U,\pi) = 1 - 1/2 \ \tanh(\alpha \ U /2)$$

for states of the same parity as the ground state. Using this approach we have realized that it was possible to fit the cumulative number of levels at low excitation energy as well as the level spacing at the neutron binding if, in addition to the parameter a(*) we also fit a parameter δ to be added to the effective excitation energy U'.

Table 1: parameters of the GC and IG level density models. Those parameters not given in the text are defined in Refs. 1-4.

	Fe-55	Fe-56	Fe-57	
a (MeV^{-1})	7.169	7.665	7.970	
Δ (MeV)	1.540	2.810	1.540	
T (MeV)	1.311	1.283	1.228	
Ux (MeV)	7.90	8.317	7.914	
E_0 (MeV)	-1.09	-0.17	-1.27	
σ^2_{LEV}	8.675	5.640	4.687	GC set
a(*) (MeV^{-1})	7.169	6.75	6.691	
Δ (MeV)	0.0	1.470	0.0	
γ (MeV^{-1})	0.054	0.054	0.054	
α (MeV^{-1})	0.20	0.40	0.20	
δ (MeV)	0.283	0.418	0.370	
$C(\sigma^2)$	0.014	0.014	0.014	IG set
En = 0 < D > (Kev)	15.4	1.7	28.9	
En = 14 MeV < D > (eV)	5.76	0.52	4.08	GC
	7.73	1.74	13.35	IG

In the IG formulation we have fitted only two parameters for each nucleus where low-lying levels and level spacings were known and only one parameter, ζ , for nuclei of unknown level spacings at the neutron binding. For the latter case, the parameter a(*) was taken from the systematics which is much more reliable once that the corrections described above are applied.

The parameters for the GC and for the IG level densities of the three iron isotopes needed for the calculation of neutron induced cross sections of Fe-56 are given in Table 1. The parameters for the other nuclei involved in the calculations can be found in Ref. (4). The parameters for Fe-55 and Fe-57 have been derived from the experimental level spacings of s-waves neutron resonances (5) and from the low excitation energy spectra. The parameter a(*) for Fe-56 has been derived from the systematics of the A≅60 mass region and then, on the base of the the calculated s-wave level spacing with the IG level density, the GC parameters has been derived. In this way, the level densities in the IG or GC approach are the same at the neutron binding for the three iron isotopes.

Nuclear reaction models

The cross section calculations have been performed within the Hauser-Feshbach statistical model theory. Particle transmission coefficients have been determined within the spherical optical model approximation. The optical model parameters (OMP) are based on the values given in Ref. (6) for the neutron channel. For the proton- and alpha-channels, the OMP of Becchetti and Greenlees (7) and of Igo and Huizenga (8) respectively has been adopted. Gamma-ray transmission coefficients are calculated making use of the Brink-Axel hypothesis. A two-humped Lorentzian giant resonance for E1 transitions has been assumed because of isospin splitting of the dipole strength whereas a single-humped giant quadrupole resonance has been adopted, whose parameters were derived from a semiclassical sum-rule estimate. In addition a M1 resonance, fragmented and partially suppressed, has been included for some nuclei involved in the calculations. Experimental branching-ratios have been included for the transitions between low-lying levels.

Preequilibrium emission has been included via the simplified treatment of the exciton model with the exciton state densities of Kalbach (9). We have considered up to two particles and up to seven gamma rays emission chances in the calculations. The code used for the calculation is a modified version of the Penelope code developed at ENEA, Bologna (1).

Results

The calculations have been made in the incoming neutron energy range from 2.0 to 20.0 MeV. In our two sets of calculations, all the parameters have been kept the same, except for the level density parameters of the three iron isotopes. The calculated level spacings are given in Table 1 for excitation energies corresponding to an incoming neutron energy of 14 MeV. The values obtained using the two level density models differ up to a factor of about 3 for Fe-57. This difference is due to the combination of several effects. One is the energy dependence of the level density parameter, a(U), in the IG formalism. Another arises from the parity dependent level density adopted in the IG case (see the equation above). Finally, there is a different prescription given for the calculation of the spin cut-off factor in the two models (see Refs. 3 and 4).

The comparison of the calculated cross sections with experimental values shows an overall agreement. A partial set of cross sections is

shown in Table 2 for comparison' sake of the different techniques used. Here, we will concentrate on the effect of using the GC or IG level densities on the inelastic and on the gamma-ray production cross sections. The result of our comparison shows that the two level density parameter sets produce different result for energies of the incoming neutron around and above 10 MeV.

Table 2: cross sections (mb) at En=14 MeV.

reaction	GC	IG
n,n'γ	600.3	695.5
ε_γ =847 KeV	449.3	521.5
n,2n	502.5	426.6
ε_γ =931 KeV	55.6	44.4
n,np	95.6	42.1
n,pγ	110.1	109.6
n,αγ	18.3	14.4

In fact, the production cross sections for the gamma-ray of 847 KeV (Fig. 1), are within 5% each other up to 10-12 MeV. Above this energies the IG level density gives a up to 50% higher cross section for the production of this gamma ray.

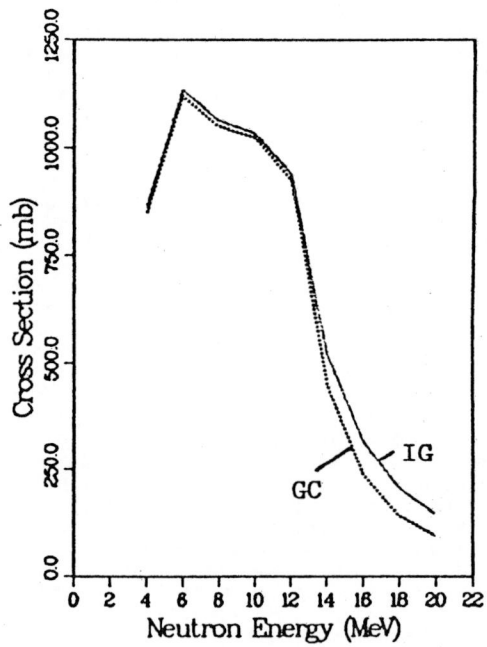

Fig. 1. Gamma-ray production cross section for the = 847 KeV in the Fe-56(n,n'γ) reaction, corresponding to the transition $2_1^+ \to 0_1^+$ in Fe-56.

The (n,2n) channel opens at about 11 MeV and in this case the production cross section for the gamma of 931 KeV (Fig. 2) shows an over production for the GC level density above 15 MeV. As for the comparison of the results of the calculations with the experimental values we should mention that the

experimental data are fragmentary and with
discrepancies up to a factor of 2 for the
gamma-rays production cross sections so that it
is, at the present, difficult to decide which of
the two level densities produces better results.
It has to be pointed out, however, that the IG
level density produces a cross section for the
(n,2n) reaction in better agreement with the
available measured values on the whole energy
range.

Holden, Neutron Cross Sections, Vol. 1, part A,
Academic, New York (1981).

6) E. D. Arthur and P. G. Young, LANL Report,
LA-8626-MS(ENDF-304), (1980).

7) F. D. Becchetti and G. W. Greenless, Phys. Rev.
182, 1190, (1969).

8) G. Igo and J. R. Huizenga, Nucl. Phys. 29,
462, (1962).

9) C. Kalbach, BNL Report 1983, BNL-NCS-51594,
pag. 113.

10)V. J. Orphan, C. G. Hot and V. C. Rogers, Nucl.
Sci. Eng. 57, 309 (1975).

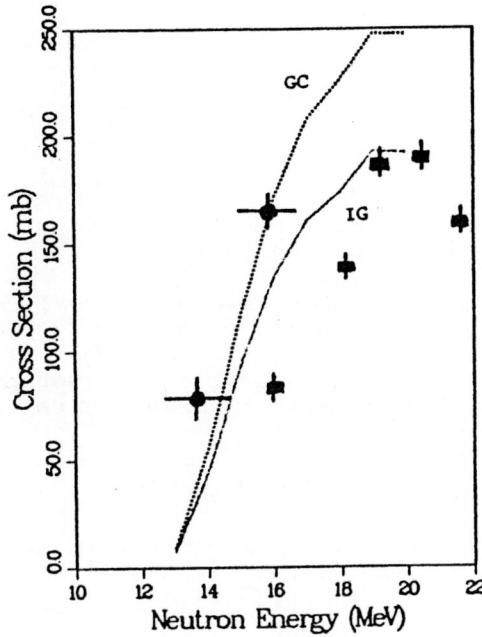

Fig. 2. Gamma-ray production cross section for
the $\mathcal{E}_\gamma$=931 KeV in the Fe-56(n,2nγ) reaction,
corresponding to the transition $5/2^+_1$ ->
$1/2^+_1$ in Fe-55. The experimental values are
as from Ref. 6 (■) and from Ref. 10 (●).

Conclusions

The comparison made for the cross sections
calculations of neutron induced reactions on Fe-56
between two commonly adopted models for level
density gives a general common results for
particles channels and for gamma-rays production
cross sections for energies of the incoming
neutron up to 10 MeV. Above this, the two level
densities give differences up to 50% in the cross
sections. The IG set has to be preferred for the
better estimate of various quantities in the upper
part of the energy range considered here.

REFERENCES

1) G. Reffo, IAEA-SMR-43, 205, Vienna (1980).

2) A. V. Ignatyuk, G. N. Smirenkin and A. S.
Tishin, Sov. J. Nucl. Phys. 21, 255, (1975).

3) A. Mengoni, G. Maino and F. Fabbri, Nuovo Cim.
A 94, 207, (1986).

4) G. Maino, A. Mengoni and P. Oblozinsky,
INDC(NDS)-238, IAEA Vienna (1990).

5) S. F. Mughabghab, M. Devadeenam and N. E.

CROSS-SECTION EVALUATIONS FOR ACTIVATION DATA LIBRARY

S.A.Badikov, O.T.Grudzevich, A.V.Ignatyuk, A.B.Pashchenko[*],
A.V.Zelenetsky, K.I.Zolotarev

Institute of Physics and Power Engineering, Obninsk, USSR
[*]International Atomic Energy Agency, Vienna, Austria

Abstract: This paper describes the work performed in the Nuclear Data Center (Obninsk) to create the Activation Data Library (ADL) for use in calculations of activation and transmutation of materials. The ADL-90 library contains cross-section data between 0 and 20 MeV for more than 5000 neutron induced reactions.The library covers almost all stable and unstable nuclides with half-lifes more than 1 day and with atomic numbers less than 85.The evaluation of cross-sections was performed in terms of the simplified theoretical models with the normalizations of resulting excitation functions to the experimental data at 14.5 MeV or to predictions of revised 14.5 MeV systematics.

(Activation and transmutation of materials, neutron induced reactions,evaluation of cross-sections,stable and unstable nuclides,theoretical models)

The safety and technological problems of fusion reactors require the activation cross-section data in wide mass and energy regions [1].It is obvious that existing needs can not be satisfied by experimental data only.Theoretical calculations on the basis of simplified models are necessary for preparation of large scale libraries.

Calculations with theoretical models

Nowadays calculations of neutron cross sections are usually carried out in framework of the optical and statistical models with some addition of nonequilibrium processes contribution.The main difficulties of such approach are connected with i) uncertainties of the level density representation [2], ii) the choice of inverse reaction cross-sections particularly for charged particles [3], iii) discrepancies in nonstatistical process interpretations [4].In view of these difficulties the detailed fitting of the model parameters is needed for a good description of the observed cross-sections.

To meet current needs in activation data many thousands of excitation functions are required.Keeping it in mind the use of rigorous theoretical models seems questionable. The most economical way of proceeding is to develop essentially simplified versions of theoretical models. The use of Weiskopf-Ewing expressions for the equilibrium stage of reactions [5],the exciton model for preequilibrium decay of the composite systems [6] and a semi-empirical estimation of the direct process contribution [7] appears to be the most suitable.Nevertheless the problem of setting of model parameters still remains.We prefered the existing global systematics of parameters - the level density description based on the generalized superfluid model [2],the optical potential parameters of the Wilmore-Hodgson for neutrons, the Perey's parameters for protons and McFadden-Satchler's ones for alpha-particles. The matrix elements of the preequilibrium model were taken in the form $\bar{M}^2 = K A^{-1} E^{-3}$

with $K=500$ MeV^3 for the nucleon channels and the Iwamoto-Harata model [8] was used for the alpha-particles and other clusters.The contributions of the direct processes were evaluated by means of the empirical expression [7]:

$$\sigma_{dir}(E)=[d\sigma/dE]_{obs}(U-E_o),$$

where U is the maximum excitation energy of residual nucleus, E_o is the effective boundary energy and $d\sigma/dE$ is the averaged observed cross-section in the hard part of particle emission spectra.The analysis of experimental emission spectra in wide mass region for 14.5 MeV neutrons testifies [9] that the values of $[d\sigma/dE]_{obs}$ can be estimated as 20 mb/MeV for neutrons,2-3 mb/MeV for protons and 0.5-1.5 mb/MeV for alpha-particles.

To test the proposed approach the calculations of excitation functions of the (n,p),(n,α) and (n,2n) reactions were performed for three hundred targets where experimental data are available. The typical example is shown in Fig.1.The results of calculations point out that the proposed approach describes the shape of excitation functions rather reliably while the normalization of the calculated curves to experimental data and 14.5 MeV systematics is necessary for the fitting of the absolute values of cross sections.

Systematics of cross-sections

The contribution of the (n,2n) and (n,p) reactions to the nonelastic interactions is the largest for over threshold neutron energies.Due to the importance of these reactions we revised the parameters of existing 14.5 Mev systematics.The (n,p) and (n,2n) cross-section libraries at $E_n=14.5$ MeV containing information on the experimental data published up to 1990 were prepared [10,11].The libraries cover 156 and 126 nuclides respectively.Statistical analysis of the cross section data

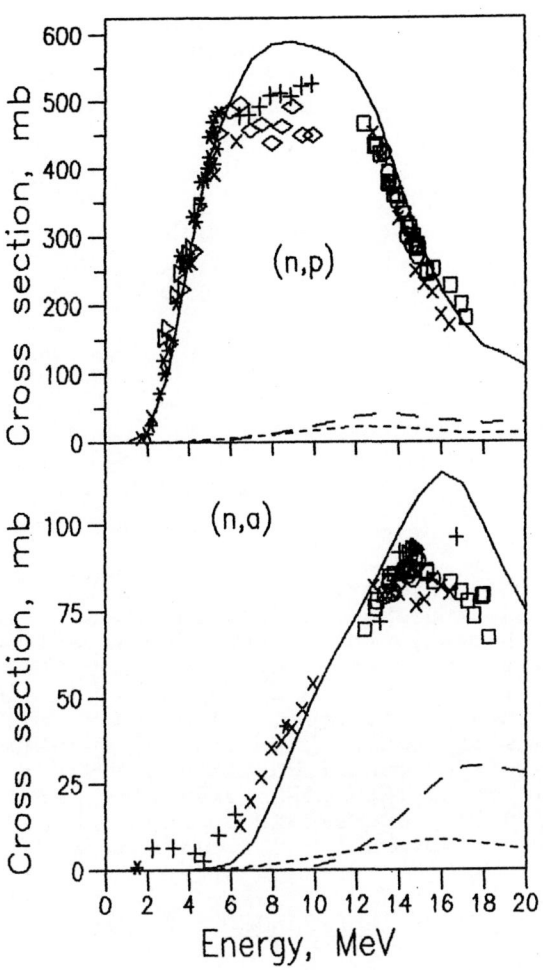

let's consider the distribution of ratios over the mass number for (n,p) reaction. The number of points above and below unit for nuclei with A≤100 are 29 and 37 while for A>100 they are 31 and 59. The difference between ratios greater and lesser than unit increases with A. Analysis of the Lu-Fink formula leads to the same inferences – the Forrest and Lu-Fink formulae predict on the average values lower than experimental data mainly for the heavier nuclei. In our opinion this difference may be attributed to the nonequilibrium part of cross-section not taken into account by the systematics formulae.

For the (n,α) reaction the Forrest's formula [13] with original parameters has been used.

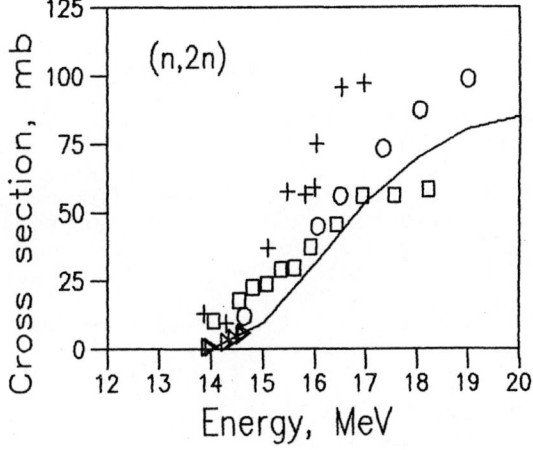

Fig.1. Excitation functions of threshold reactions on the ^{54}Fe. The solid curves are calculated for global sets of model parameters, the dot and dash curves show the contributions of direct and preequilibrium processes.

was performed within the framework of the nonlinear regression model [10]. For different systematics χ^2 values are in the interval from 6.14 (the Levkovskii formula) to 3.12 (Forrest) for (n,p) reaction and from 9.02 (Kopecky-Gruppelaar) to 5.7 (Lu-Fink) for (n,2n) reaction [11].

We recommend the Forrest and the Lu-Fink dependences to be used for the (n,p) and (n,2n) reactions with corrected parameters (cross-section in mb, A and s – the mass number and asymmetry parameter of the target nucleus):

$$\sigma_{n,p}=5.2093(A^{1/3}+1)^2 \times$$
$$\exp\{-23.486s-85.044s^2+0.25406A^{1/2}\}$$

$$\sigma_{n,2n}=47.015(A^{1/3}+1)^2 \times$$
$$\{1-3.9777\exp(-24.116s)\}$$

Fig.2 shows the ratio $\sigma_{calc}/\sigma_{exp}$ plotted against A for the recalculated Lu-Fink and Forrest formulae. The considerable disbalance of ratios greater and lesser than unit (60 and 96 for the (n,p) reaction and 53 and 73 for the (n,2n) reaction respectively) attracts attention. As the Wald-Wolfowits criterion [12] points out this effect has a regular nature. To understand it

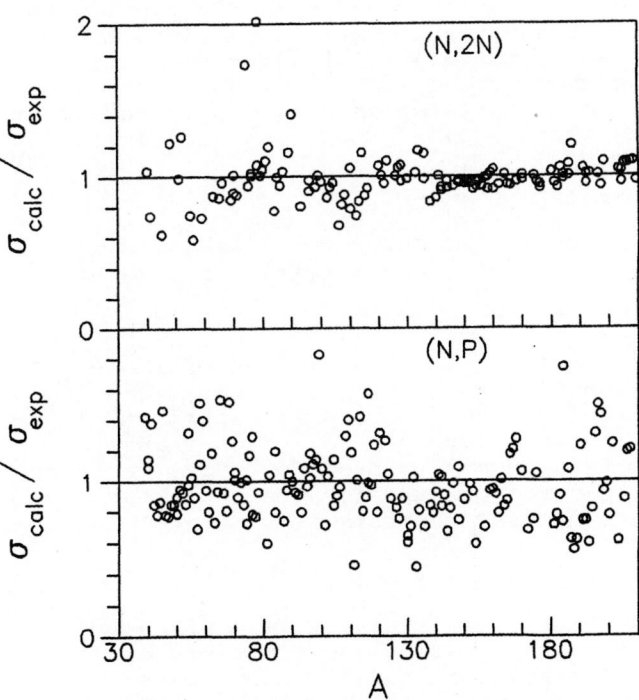

Fig.2. Ratios of cross-sections predicted by the systematics to experimental data at 14.5 MeV neutrons.

932

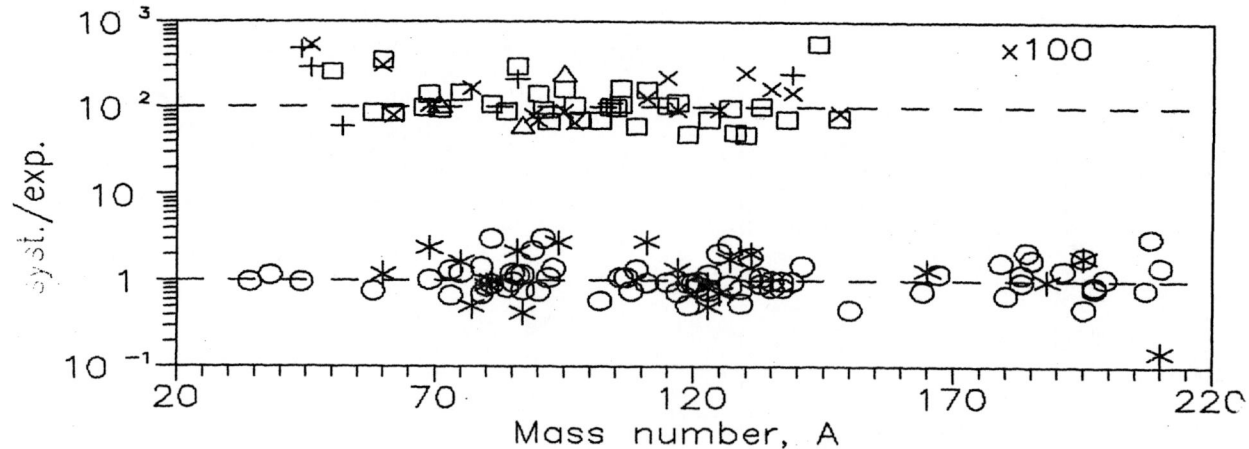

Fig.3. Ratios of the R values predicted by the systematics to experimental data for different reactions: □ - (n,p), Δ - (n,d), + - (n,t), × - (n,α), o - (n,2n) for 14.5 MeV neutrons, * - (n,γ) for fast neutrons.

Systematics of isomeric ratios

Excitation functions of isomeric levels in neutron induced reactions cover nearly 30% of the list of practical requests [1]. To meet these needs we analysed the available experimental data and theoretical calculations performed within the framework of Hauser-Feshbach's theory of particle and gamma cascades.

The comparison of calculations with the experimental data shows:

1) the isomeric ratio $R(E_n)=\sigma_m/(\sigma_m+\sigma_g)$ changes slightly in the energy region from threshold up to 20 MeV;

2) a good agreement between the theoretical calculations and the experimental values of isomeric ratios can be achieved in many cases without parameters fitting;

3) the description of the excitation functions can be improved considerably if the calculated results are normalized to the experimental values of isomeric ratio at any energy.

Some systematics of isomeric ratios on the isomer spin J_m were proposed in [14].

From our analysis we concluded that the formula

$$R = \frac{4(2J_m+1)}{(J_m-J_g)^2} \exp\{-(J_m+1/2)^2/2(J_m-J_g)^2\}$$

ensures a better description of the experimental data. The Fig.3 shows the ratios of calculated R values to experimental data. It should be emphasized that proposed systematic may be used both for the threshold reactions and for the radiative capture of fast neutrons.

Conclusions

On the basis of the methods described above the excitation functions of about 5000 reactions have been evaluated and compiled as the ADL-90 library [15]. Of course, these evaluations cannot compete with analogous data for nuclei where good experimental data are available and the calculations can be performed by means of the direct fitting of parameters. We tried to predict cross-sections for nuclei far from stability valley where no experimental data are available. As follows from our calculations the use of the simplified theoretical models with the normalizations of resulting excitation functions to the experimental data at 14.5 MeV or to values of 14.5 MeV systematics permits to predict cross-sections with accuracy required for practical applications.

REFERENCES

1. Summary Report of Consultants Meeting on FENDL, INDC(NDS)-241, IAEA(1990).
2. A.V.Ignatyuk:Stat. properties of excited nuclei,Energoatomizdat,Moscow(1983)
3. V.M.Bychkov et al,in Proc.Int.Conf.on Nucl.Data for Science and Techn.,1988, Mito,Japan,p.1221,Saikon,Tokyo(1988)
4. A.V.Ignatyuk,V.P.Lunev,in Proc.Meeting on Preequilibrium Nucl.Reactions,p.233, OECD,Paris(1988)
5. V.F.Weisskopf,D.H.Ewing:Phys.Rev.57, 472(1940)
6. J.J.Griffin:Phys.Rev.Lett.17,478(1966), C.Cline,M.Blann:Nucl.Phys.A172,225(1971)
7. O.T.Grudzevich et al,in Proc.IAEA RCM on Methods for Calc.of Neut.Nucl.Data for Struct.Materials,p.81,Bologna(1986)
8. A.Iwamoto,K.Harada:Phys.Rev.C26,1821 (1982)
9. S.Grimes et al:Phys.Rev.C19,2127(1979), R.C.Haight et al:Phys.Rev.C23,700(1981)
10. S.A.Badikov,A.B.Pashchenko: Preprint FEI-2055(1989)
11. S.A.Badikov,A.B.Pashchenko:Vopr.Atom. Nauki i Tekhniki.Ser.Yad.Konstanty (1991),to be published
12. M.G.Kendall,A.Stuart,The Advanced Theory of Statistics,v.2,Hafner,New York (1961)
13. R.A.Forrest,Systematics of Neutron Induced Threshold Reactions with Charged Products at about 14.5 MeV,AERE-R12149, Harwell(1986)
14. J.Kopecky,H.Gruppelaar:Report ECN-200 (1987)
15. O.T.Grudzevich,A.V.Zelenetsky,A.B.Pashchenko:Preprint FEI-2058(1989)

FENDL: A REFERENCE NUCLEAR DATA LIBRARY FOR FUSION APPLICATIONS

D.W. Muir, S. Ganesan and A.B. Pashchenko

International Atomic Energy Agency (IAEA)
RIPC – Nuclear Data Section
P.O. Box 100, A-1400 Vienna, Austria

Abstract: The IAEA, in cooperation with several national nuclear data centres and research laboratories, is creating a modern and internationally available Fusion Evaluated Nuclear Data Library (FENDL), which will serve as a comprehensive source of processed and tested nuclear data for fusion applications.

(FENDL, fusion, multigroup cross sections, processing codes, NJOY, RECENT, code verification, activation cross sections, decay data, EXFOR).

Background

A series of international meetings, having taken place in 1986, 1987, 1989 and 1990 and projected to continue with about the same frequency in the next few years, has been organized by the International Atomic Energy Agency (IAEA), with goal of assembling, processing and testing a comprehensive, fusion-relevant Fusion Evaluated Nuclear Data Library (FENDL) with unrestricted international distribution [1,2].

FENDL will be composed of "sublibraries" containing different data types, describing the transport of both the plasma-source neutrons and secondary gamma rays through reactor components, as well as the resulting radiation effects, such as nuclear heating, tritium breeding, activation and material damage. Also included will be cross sections for fusion and other important charged-particle nuclear reactions of the plasma constituents, as well as data for fusion-relevant neutron dosimetry.

With the welcomed international release of ENDF/B-VI, JENDL-3 and BROND in early 1990, the FENDL project has turned from its earlier emphasis on the review and selection of candidate neutron-interaction evaluations to (a) processing and testing of cross sections for neutron and gamma-ray transport and (b) assembling, processing and testing of large special-application files needed to supplement the available files, especially in the field of activation and decay data.

Processed data for neutron and gamma-ray transport

A major focus of the present FENDL activity in the IAEA Nuclear Data Section (NDS) is the production of a sublibrary of multigrouped data for input to the neutron/gamma-ray transport codes used in fusion design. In the first stage of this work, all materials of significance in fusion design that are represented in the ENDF/B-VI library are being processed at the NDS into point-data and multigroup form, using the IBM-3081 mainframe computer at the IAEA. The bulk of the data processing is being performed with the widely used NJOY nuclear data processing system [3]. The first version of the multigroup sublibrary is expected to be completed in the Fall of 1991. At a later stage, this will be supplemented with companion sublibraries of input data for continuous-energy Monte Carlo codes and multigroup covariance matrices for sensitivity studies.

After detailed discussions with fusion nuclear-data and neutronics specialists [1,2], the materials in Table I were selected for inclusion in the multigroup sublibrary. Similarly, Table II lists the detailed specifications being used to prepare the multigroup sets.

With the assistance of R.E. MacFarlane, version 89.31 of the NJOY code system was installed on the IAEA IBM-3081 mainframe computer in late 1989. Also installed was the related source-code maintenance program UPEML [5]. Since the original installation of NJOY 89.31, we have employed UPEML to incorporate and document a number of local NDS improvements to the NJOY code system. The multigroup processing of FENDL is a large task, involving as it does many materials, a large number of energy groups, new (ENDF-6) formats, many new evaluations, and machine-precision concerns.

As the first step in the processing task, NJOY 89.31 was used to produce zero-Kelvin resonance-reconstructed data and 300-Kelvin Doppler-broadened data for all of the isotopes and elements listed in Table I.

In order to validate the processing of ENDF/B-VI for all FENDL materials, the IAEA "pre-processing" codes LINEAR, RECENT and SIGMA1 [6] were also run to generate zero- and 300-K pointwise cross section for all of the FENDL materials. A reconstruction tolerance of 0.1% was used in LINEAR, RECENT, SIGMA1 and in the corresponding modules, RECONR and BROADR, of NJOY. The point-data results obtained from the two code systems were then compared in considerable detail, using the COMPLOT code [6].

For most isotopes, the two processing code systems were found to produce essentially identical results. In some cases, however, there were differences in some energy regions that far exceeded the error tolerances specified. We list below some of the more important results of this intercomparison.

1) The IBM-3081 employs 32-bit words, which has the consequence that single-precision neutron energies cannot be reliably represented with more than 6 digits. At high energies, many differences were observed between the zero-Kelvin results from the single-precision NJOY and those from RECENT. For example, the NJOY and RECENT cross sections for capture in ^{60}Ni differ by

Table I. Contents of Multigroup Sublibrary

^{1}H, ^{2}H, ^{3}H, ^{6}Li, ^{7}Li, ^{9}Be, ^{10}B, ^{11}B, C, ^{14}N, ^{15}N, ^{16}O, ^{19}F, ^{27}Al, Si, Ti, ^{51}V, $^{50,52-54}$Cr, ^{55}Mn, $^{54,56-58}$Fe, ^{59}Co, $^{58,60-62,64}$Ni, ^{63,65}Cu, $^{90-92,94,96}$Zr, ^{93}Nb, Mo, Sn, $^{134-137}$Ba, $^{182-184,186}$W, $^{206-208}$Pb, ^{209}Bi, ^{232}Th and ^{238}U.

Table II. Specifications for Multigroup Processing

Neutron groups: 175 (Vitamin-J structure [4])
Gamma groups: 42 (Vitamin-J structure [4])
Neutron weight function: Thermal + 1/E + Fission + Fusion (IWT = 6 in NJOY)
Gamma weight function: 1/E with rolloffs (IWT = 3 in NJOY)
Legendre order for neutrons:
P_6 for transport correction to P_5
Legendre order for gammas:
P_8 for transport correction to P_7
Temperatures: 300, 900 and 1500 Kelvin
Dilution factors: 10^0, 10^1, 10^2, 10^3, 10^4, and 10^{10} barns
Reconstruction, linearization and thinning tolerances: 0.1%
Thermal scattering-law data included for Be in Be metal, C in graphite and H in water

Source of basic data: ENDF/B-VI, supplemented with JENDL-3.

up to a factor of 4 in the vicinity of narrow, p-wave resonances around 200 keV. Most of the differences can be attributed to the fact that, unlike NJOY, RECENT uses a double-precision representation for neutron energies and outputs the energies with up to 9 digits of precision. Using the AUTODBL automatic double-precisioning option in IBM VS-Fortran, double-precision versions of the NJOY driver and the RECONR and BROADR modules were successfully created at the IAEA and tested on ^{60}Ni. Differences near the p-wave resonances were found essentially to disappear when double precision was employed. Unfortunately, the double-precision version of NJOY is about a factor of 5 slower than the single-precision one, so we will continue to use single precision to produce the multigroup sublibrary of FENDL and will use the double-precision version to spot check the results to guarantee that no unacceptable loss of accuracy occurs.

2) The 300 K data for ^{16}O and ^{28}Si showed unphysical line shapes in the region around 1 MeV when the data were Doppler-broadened from zero K using NJOY. This effect was ultimately traced by R.E. MacFarlane to a small error in the BROADR module. All FENDL production runs will be made with the corrected version.

3) For ^{207}Pb, we found that the potential scattering cross section calculated by NJOY and RECENT differed substantially from each other. The 35% difference that was found in the elastic cross section for ^{207}Pb is due to the presence of j-multiplicity for both L = 1 and L = 2. The U.S. Cross Section Evaluation Working Group is presently addressing the problem of defining the issues here and recommending necessary corrective actions in both the data files and the processing codes.

4) There were some instances where the cross sections in the unresolved-resonance region showed noticeable discrepancies. For ^{nat}W, ^{nat}Mo and ^{238}U, differences of a few percent were seen in the cross-section shapes. These differences are not yet fully explained, but it is expected that they will be fairly easy to correct. A document giving additional details of all of our code comparisons for the FENDL materials is available separately from the NDS. Work is in progress to complete the few remaining NJOY code updates needed prior to submission of the FENDL multigroup production runs.

FENDL activation sublibrary

Recent interest in low-activation materials for fusion reactor technology has accelerated the development of national activation cross-section libraries with a large number of target nuclei and reaction types. There are no experimental data for most of the important activation reactions, and highly simplified model calculations are often used to supply the needed data.

At the May 1989 FENDL meeting [1], the Working Group on Neutron Activation Data initiated an intercomparison of activation cross sections important for fusion reactor technology. It was agreed that national nuclear data centers will send to the NDS their contributions, according to a list of reactions selected on the basis of inventory calculations.

A list of 256 important reactions that are significant in producing activation both at short and long cooling times has been compiled by R. Forrest at Harwell Laboratory, UKAEA. The list was distributed to all interested parties and now many activation data files have been received at the NDS from institutes participating in this exercise. Very detailed graphical intercomparisons have been prepared at the NDS, plotting, for each reaction, overlays of the various submitted evaluated data sets and experimental data from EXFOR.

Groups that are participating in the intercomparison are listed below, along with their libraries:

REAC-2 (F.M. Mann, Hanford Engineering Development Laboratory) [7]

REAC-ECN-5 (J. Kopecky and H. Gruppelaar, Netherlands Energy Research Foundation) [8]

ENDF/B-VI
All extracted individual reaction cross sections of the ENDF/B-6 general purpose files that are contained in the list of 256 high-priority reactions.

BOSPOR-86 (V. Bychkov et. al., Institute of Physics and Power Engineering) [9,10]
BOSPOR, which means "library of evaluated threshold activation reaction cross sections", was organized in 1980 in Obninsk Nuclear Data Centre. In 1986 the BOSPOR-80 library was

revised and its new, extended version BOSPOR-86 was created in ENDF/B-5 File-3 format. Energy range is from threshold to 20 MeV.

SINCROSACT (N. Yamamuro, Data Engineering, Inc.) [11]

These activation data are almost all the cross-sections in ENDF-5 File-10 format for ^{59}Co, $^{64,66-68,70}Zn$, $^{58,60-62,64}Ni$, ^{63}Cu, $^{90-92,94,96}Zr$, ^{93}Nb, $^{92,94-98,100}Mo$, $^{107,109}Ag$, $^{121,123}Sb$, $^{185-187}Re$, ^{197}Au and $^{196,198-200,204}Hg$

Ground state production and isomeric state production cross-sections are given separately. The energy region covered is from threshold to 20 MeV.

ADL-90 (O. Grudzevich et. al. Institute of Physics and Power Engineering + Institute of Atomic Energetics, Obninsk) [12]

The Activation Data Library (ADL-90), for calculation of activation and transmutation in materials used in fusion reactor technology, was created to contribute to the development of FENDL activation sublibrary. This version of library now contains cross sections for about 5000 (n,p), (n,α), (n,γ), (n,2n) and (n,3n) reaction, for all stable targets.

All data contributed by the participants were kept unchanged, except that the data format was in all cases changed to ENDF-6 at the NDS.

This review and intercomparison activity has the practical goal of creating activation (and later) decay data sublibraries of FENDL. Selection of the 256 specific evaluations which will constitute the first version of the activation cross-section sublibrary was made at the June 1990 FENDL meeting [2]. This selection was considerably aided by the availability of the large number of overlay plots mentioned above. This list of evaluations, with a few subsequent modifications, will be the basis of the first version of the sublibrary, to be issued in the Fall of 1991. At the next FENDL meeting, planned for 18-22 November 1991 in Vienna, the review and selection process will be renewed, and this will lead to the creation of a second version early in 1992.

The second outcome of the June 1990 FENDL meeting was the initiation of the Second International Fusion Activation Calculation Comparison Study. This benchmark exercise, organized by E.T. Cheng, compares inventory codes and assesses the accuracy of the prediction of radionuclide production in materials at the fusion-reactor first-wall location. R. Forrest provided an input parameter set for this benchmark calculation, namely a reduced cross section library (in 100 energy groups), the corresponding decay data library, and irradiation conditions. The participants, including groups from the U.S.A., Western Europe, the U.S.S.R., China and Japan, were requested to calculate the number of atoms a various nuclides present after an irradiation of 1 year under typical fusion-reactor first-wall conditions. E.T. Cheng has kindly agreed to coordinate and evaluate the results of this benchmark exercise. The report on this international benchmark should be available for discussion at the next FENDL Meeting.

Following the recommendation of the June 1990 FENDL Meeting [2] future work will include the development of FENDL-2 activation sublibrary. Among other things, this sublibrary should include:

1) a complete cross-section library (i.e., containing at least all targets with $t_{1/2} > 10$ days and all reactions energetically possible for $E_n < 20$ MeV); and

2) a decay data library.

References

1. Proceedings of the IAEA Specialists' Meeting on Fusion Evaluated Nuclear Data Library (FENDL), Vienna, 8-11 May 1989, Report INDC(NDS)-223/GF, August 1989.

2. Summary Report of the Consultants' Meeting on First Results of FENDL-1 Testing and Start of FENDL-2, Vienna, 25-28 June 1990, Report INDC(NDS)-241/LF, November 1990.

3. R.E. MacFarlane: NJOY 89, p. 21-49 in Los Alamos National Laboratory report LA-11972-PR (December 1990). Also see other NJOY documentation referenced therein.

4. E. Sartori: Standard Energy Group Structures of Cross Section Libraries for Reactor Shielding, Reactor Cell and Fusion Neutronics Applications: VITAMIN-J, ECCO-33, ECCO-2000 and XMASS, OECD/NEA Data Bank report JEF/DOC-315 (December 1990).

5. T.A. Mehlborn: UPEML 2.1: A Machine Portable CDC Update Emulator, Radiation Shielding Information Center report PSR-245 (1988).

6. D.E. Cullen, P.K. McLaughlin and S. Ganesan: The 1989 ENDF Pre-Processing Codes: IAEA-NDS-39 (Rev. 6) (April 1991).

7. F.M. Mann: Status of Dosimetry and Activation Data, Proceedings of the International Conference, May 30 - June 3, 1988, Mito, Japan, p. 1013-1015.

8. J. Kopecky and H. Gruppelaar: Proceedings of the International Conference, May 30 - June 3, 1988, Mito, Japan, p. 245-248.

9. V.M. Bychkov, K.I. Zolotarev, A.B. Pashchenko: Creation of Computerized Library of Evaluated Threshold Reaction Cross-sections (BOSPOR-80) and Testing BOSPOR-80 by Means of Integral Experimental Data, Problems of Atomic Science and Engineering. Ser.: Nucl. Const., 1981, No. 3(42), p. 60.

10. V.N. Manokhin, A.B. Pashchenko, V.I. Plyaskin: et al, Activation Cross-Sections Induced by Fast Neutrons: Handbook on Nuclear Activation Data, Vienna, 1987, p. 305-411.

11. N. Yamamuro: Private communication.

12. O.T. Grudzevich, A.V. Zelenetskij, A.B. Pashchenko: Preprint FEI-2043, Obninsk, 1989.

REAC*3 NUCLEAR DATA LIBRARIES

F. M. Mann and D. E. Lessor

Westinghouse Hanford Company
HO-36, P. O. Box 1970
Richland, WA 99352 USA

Abstract: The libraries (transport cross section, transmutation cross section, and decay data) associated with the REAC*3 nuclear transmutation code system are described.

[REAC*3, ENDF/B-VI, activation library, decay library]

Introduction

In order to calculate neutron activation and the resultant dose, a series of libraries are needed. The neutrons must be transported from their source to the point where transmutation occurs (which requires a neutron transport cross section library). The probability for transmutation must be calculated (which requires a transmutation cross section library). The decay of the transmutation product must be followed, including the production of photons (which requires a decay data library). Finally the photons must be followed to the point of interest (which requires a photon transport cross section library).

The REAC*3 nuclear transmutation code system [1] contains modern versions of each of these data.

Transport Cross Section Library

The NJOY nuclear data processing system (Version 89) [2] was used to process about 300 materials from the latest versions of the Evaluated Nuclear Data File (ENDF/B-V and ENDF/B-VI) [3]. Except for a few isotopes in the fission product region not reached from fission, all ENDF/B-VI evaluations were processed. Pointwise ENDF (PENDF) and groupwise ENDF (GENDF) tapes were created. The specifications for the PENDF tapes were reached after consultation with Bob MacFarlane of the Los Alamos National Laboratory (LANL). Resonance reconstruction was done to 0.2% and then Dopler broadened to 300, 400, 600, 900, and 1200 K. Cross sections for actinide isotopes were further broadened to 1600, 2000, 3000, and 4000 K. Bound effects were included. Also calculated were KERMA and damage energy parameters. The GENDF tapes were generated in an 89 neutron group structure and a 20 photon group structure. The neutron group structure is the same below 4 eV as that of the standard 69 group WIMS structure [4]. Two flux weightings were used, one typical of thermal reactors and the other typical of fast reactor cores. A minimum of 6 sigma-o's were used for each material with the most important actinides having 8 sigma-o's. Cross sections for the MCNP [5] ACE tapes were created for all evaluations except for the ENDF/B-VI evaluations from the Oak Ridge National Laboratory (ORNL), which used the R-matrix resonance formalism. Plots of the total, capture, fission, (n,2n), (n,p), (n,alpha) cross sections were created from PENDF tapes whenever the reactions were present.

Transmutation Cross Section Library

A new transmutation cross section library is being created in ENDF/B-VI format [6]. About 500 non-actinide isotopes will be present with evaluations extended in most cases to 40 MeV. Cross sections to the first and second isomeric states are included whenever these states are present in the product isotopes. All transmuting reactions in ENDF/B-VI have been placed in this library. In particular, (n,gamma) evaluations were processed into a fine group structure (399 groups). The REAC-ECN-5 [7] activation cross section library which is also known as EAF-1 is being used for target isotopes having short half-lives. Also experimental and systematic data in the REAC-ECN-5 library are being used for normalization.

After creation as a pointwise library, the library will be processed into various group structures.

Decay Data Library

A new decay data library has also been formed in ENDF/B-VI format. This library consists of evaluations created by Charles Reich of the Idaho National Engineering Development Laboratory (INEL) for ENDF/B-VI and data from the Evaluated Nuclear Structure Data File (ENSDF) (May 1990) [8]. Once Tal England of LANL completes the ENDF/B-VI decay data file of the fission products, his evaluations will replace those currently in the library. Table 1 shows the contents of the current library and of that expected this fall. As the libraries are designed to have data for all isotopes between the neutron-drip and proton-drip lines, some data will have only half-life and branching information from the Table of Isotopes [9].

Table 1. Source of Data for Decay Data Libraries

Source	Number of Isotopes	
	Current status	Fall 1991
Stable isotopes	262	262
C. Reich (INEL)	724	---
ENDF/B-VI	---	977
ENSDF (90)	1316	1279
Table of Isotopes	459	370
Total	2761	2888

Summary

Three types of nuclear data libraries will soon be complete for use with the REAC transmutation code system: a transport cross section library, a transmutation cross section library, and a decay data library. Extensive documentation for each of the libraries is planned.

References

1. F. M. Mann, "REAC*2: User's Manual and Code Description", WHC-EP-0282, Westinghouse Hanford Company, Richland, WA, December, 1989.

2. R. E. MacFarlane, D. W. Muir, and R. M. Boicourt, "The NJOY Nuclear Data Processing System, Volume I: User's Manual," "LA-9309-M, Vol. I," Los Alamos National Laboratory, Los Alamos, NM, May 1982.

3. The ENDF/B libraries are produced by the Cross Section Evaluation Working Group and are maintained by the National Nuclear Data Center, Brookhaven National Laboratory, Upton, NY.

4. R. E. MacFarlane and D. W. Muir, "The NJOY Nuclear Data Processing System, Volume III: The GROUPR, GAMINR, and MODER Modules," "LA-9303-M, Vol. III," Los Alamos National Laboratory, Los Alamos, NM, October, 1987.

5. J. F. Briesmeister, Editor, "MCNP- A General Monte Carlo Code for Neutron and Photon Transport," Version 3A, LA-7936-M, Rev. 2, Los Alamos National Laboratory, Los Alamos, NM, 1986.

938

6.   P. F. Rose and C. L. Dunford, Editors, "Data Formats and Procedures for the Evaluated Nuclear Data File, ENDF-6", ENDF, Version 6, ENDF-102, National Nuclear Data Center, Brookhaven National Laboratory, Upton, NY, 1990.

7.   J. Kopecky and H. Gruppelaar, "The REAC-ECN-3 Data Library with Activation and Transmutation Cross Sections for Use in Fusion Reactor Technology," Proceedings of the International Conference on Nuclear Data for Science and Technology, Mito, Japan, 1988, p. 245. Published by Saikon Publishing Co., Ltd., Tokyo, Japan.

8.   The ENSDF data are produced by the International Nuclear Data Project and are maintained by the National Nuclear Data Center, Brookhaven National Laboratory, Upton, NY.

9.   C. M. Lederer and V. S. Shirley, Editors, "Table of Isotopes," Seventh Edition, Wiley-Interscience, New York, NY, 1978.

JENDL-3 FP NUCLEAR DATA LIBRARY

T. Nakagawa[1], M. Kawai[2], S. Iijima[2], H. Matsunobu[3], T. Watanabe[4]
Y. Nakajima[1], T. Sugi[1], M. Sasaki[5], A. Zukeran[6]
K. Kaneko[7], and H. Takano[1]

Japanese Nuclear Data Committee FP Nuclear Data Working Group
Japan Atomic Energy Research Institute
Tokai-mura, Naka-gun, Ibaraki-ken 319-11, Japan

1) Japan Atomic Energy Research Institute
2) Toshiba Corporation, Kawasaki-ku, Kawasaki 210
3) Sumitomo Atomic Energy Industries, Ltd., Chiyoda-ku, Tokyo 101
4) Kawasaki Heavy Industries, Ltd., Koto-ku, Tokyo 136
5) Mitsubishi Atomic Power Industries, Inc., Minato-ku, Tokyo 105
6) Hitachi, Ltd., Hitachi-shi, Ibaraki-ken 316
7) The Japan Research Institute, Ltd., Minato-ku, Tokyo 107

Abstract: Evaluation of neutron nuclear data for 172 nuclides from As to Tb has been made for JENDL-3. Evaluated quantities are the total, elastic and inelastic scattering, capture and threshold reaction cross sections, the angular and energy distributions of secondary neutrons in the incident neutron energy range from 10^{-5} eV to 20 MeV. The evaluation was made on the basis of theoretical calculations with optical, statistical, preequilibrium and multi-step evaporation models as well as available experimental data. Below 100 keV, resolved and unresolved resonance parameters were evaluated. The benchmark tests performed by comparing with experimental data at STEK, EBR-II and CFRMF have confirmed the reliability of the present evaluation.

(nuclear data, cross sections, resonance parameters, angular distributions, energy distributions, evaluation, fission products, benchmark tests, JENDL-3)

Introduction

After the completion of evaluation work/1/ for the JENDL-2 FP library, new evaluation work for JENDL-3 started in 1985 in order to make a data library applicable to much wider fields. In the present work, the number of nuclides was increased to 172 from 100 of the JENDL-2 FP library, and threshold reaction cross sections were newly evaluated for all the nuclides. The results were compiled in the ENDF-5 format as a part of JENDL-3 and released in December 1990.

The present work covers neutron induced reaction cross sections, angular and energy distributions of secondary neutrons in the energy range from 10^{-5} eV to 20 MeV for nuclides listed in Table 1. The evaluation method is described in the next section.

Benchmark tests for the present evaluation were performed by using integral experimental data. Their results are given in the third section.

Evaluation

Below 100 keV, cross sections were represented with resolved and unresolved resonance parameters. Above 100 keV, optical and statistical model calculations were performed for the total, elastic and inelastic scattering and capture cross sections, by taking account of threshold reaction cross sections as competing processes. Threshold reactions were evaluated for all the nuclides by means of the preequilibrium and multi-step evaporation models. The evaluation was carried out to reproduce reliable experimental data: By using the experimental data, the parameters of theoretical calculation were determined, and the calculated results were modified.

Resonance Parameters

Resolved resonance parameters for the MLBW formula for 141 nuclides were taken from recommended sets in JENDL-2 and Ref.2, or were evaluated on the basis of recent data measured at JAERI, ORNL and NIR. The parameters of negative and low-lying positive resonances were adjusted so as to well reproduce measured thermal cross sections and resonance integrals.

For nuclides without measured resonance parameters, constant elastic scattering and $1/v$ shape capture cross sections were assumed in the thermal energy region. The thermal capture cross sections of 10 nuclides having no measured

Table 1 Nuclides in JENDL-3 FP Nuclear Data Library

As-75, Se-74, 76, 77, 78, 79, 80, 82, Br-79, 81, Kr-78, 80, 82, 83, 84, 85, 86, Rb-85, 87, Sr-86, 87, 88, 89, 90, Y -89, 91, Zr-90, 91, 92, 93, 94, 95, 96, Nb-93, 94, 95, Mo-92, 94, 95, 96, 97, 98, 99, 100, Tc-99, Ru-96, 98, 99, 100, 101, 102, 103, 104, 106, Rh-103, 105, Pd-102, 104, 105, 106, 107, 108, 110, Ag-107, 109, 110m, Cd-106, 108, 110, 111, 112, 113, 114, 116, In-113, 115, Sn-112, 114, 115, 116, 117, 118, 119, 120, 122, 123, 124, 126, Sb-121, 123, 124, 125, Te-120, 122, 123, 124, 125, 126, 127m, 128, 129m, 130, I -127, 129, 131, Xe-124, 126, 128, 129, 130, 131, 132, 133, 134, 135, 136, Cs-133, 134, 135, 136, 137, Ba-130, 132, 134, 135, 136, 137, 138, 140, La-138, 139, Ce-140, 141, 142, 144, Pr-141, 143, Nd-142, 143, 144, 145, 146, 147, 148, 150, Pm-147, 148, 148m, 149, Sm-144, 147, 148, 149, 150, 151, 152, 153, 154, Eu-151, 152, 153, 154, 155, 156, Gd-152, 154, 155, 156, 157, 158, 160, Tb-159.

values were determined from systematics of the capture cross section vs. level spacing.

Unresolved resonance parameters were given in the energy range up to 100 keV. The neutron strength functions were taken from the compilation of Mughabghab et al./2/ or from the calculation with an optical and statistical model code CASTHY/3/. The radiative widths were adopted from Ref. 2 or its systematics. The level spacing was adjusted so as to reproduce the evaluated capture cross section, and the effective scattering radius to reproduce the total cross section at 100 keV.

Optical Model and Level Density Parameters

Spherical optical potential parameters for various mass intervals were determined on the basis of typical measured total cross sections, neutron strength functions and scattering radii recommended by Mughabghab et al./2/ For calculation of inverse-cross sections, global sets of OMP were used for proton/4/, α/5/, deuteron/6/, and triton and He-3/7/.

Level density parameters of a and T for the composite formula of Gilbert and Cameron were determined from numbers of low-lying excited levels adopted in ENSDF/8/ and resonance level spacings recommended in Ref. 2 for about 320 nuclides including residual nuclides of the (n,2n), (n,p) and (n,α) reactions. The systematics of a and T were deduced in order to be applied to the other residual nuclides of the threshold reactions.

Total and Capture Cross Sections

Those were calculated with CASTHY/3/. Gamma-ray strength functions $S\gamma$ were adjusted to reproduce the available capture cross sections measured in the keV region. If no experimental data were available, they were determined from $\langle\Gamma\gamma\rangle$ and the resonance level spacing, or systematics of $S\gamma$. For Xe-132, Xe-134, Eu-152 and Eu-154, the $S\gamma$'s were selected by considering the results of cross section adjustment based on JENDL-2/9/. Above 1 MeV, the direct and semi-direct capture cross sections were calculated from a simple formula given by Benzi and Reffo/10/ by normalizing to 1 mb at 14 MeV, and added to the CASTHY calculation. Figure 1 shows the capture cross section of Br-81.

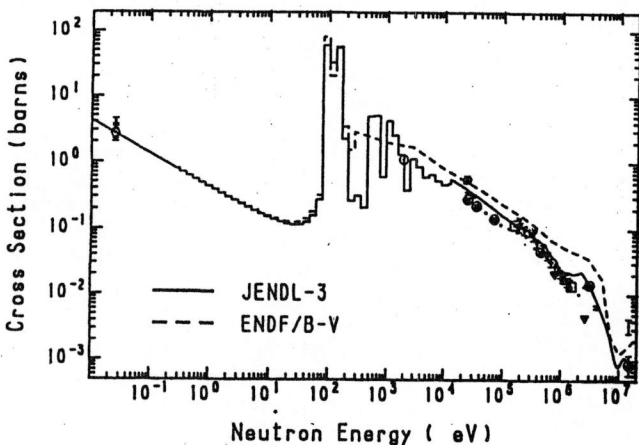

Fig. 1 Capture cross section of Br-81

Inelastic Scattering Cross Sections

The inelastic scattering cross sections were also calculated with CASTHY. For nuclides contained in JENDL-2, the same level schemes were

adopted. For the other nuclides, they were evaluated on the basis of ENSDF and the Nuclear Data Sheets up to 1987. For the even mass isotopes of Ru, Pd, Cd, Ba, Ce, Nd and Sm-144, the direct inelastic scattering cross section was calculated with a DWBA code DWUCK-4/11/, because Gruppelaar and van der Kamp/12/ pointed out the direct inelastic scattering played an important role in prediction of STEK sample worth experiments/13/. Deformation parameters were mainly taken from Refs. 14 and 15. The results were added to the cross sections and angular distributions calculated with CASTHY. The energy distributions of secondary neutrons in the continuum were calculated with a preequilibrium and multi-step evaporation model code PEGASUS/16/.

Threshold Reaction Cross Sections

The cross sections of the (n,2n), (n,p), (n,α), (n,d), (n,t), (n,He-3), (n,np), (n,nα), (n,nd), (n,nt), (n,2p) and (n,3n) reactions were evaluated for all the nuclides by using PEGASUS/16/. The Kalbach's constant K was estimated as $K=0.1/(g_c/A_c)^3$ where g_c and A_c are the single particle level density and the mass number of the compound nucleus. For Ce, Nd and Sm isotopes, the values of K were adjusted to reproduce the measured (n,2n) cross sections.

The systematics/17/ and/or recommended values/18/ around 14 MeV were used to renormalize the calculated (n,2n) cross sections of about 30 nuclides. For several nuclides, the (n,2n) cross section was determined by smoothing out the experimental data. The calculated (n,p) and (n,α) cross sections of most of nuclides were renormalized to the systematics or recommendation at 14.5 MeV by Forrest/19/. An example of the (n,p) cross section is given in Fig. 2.

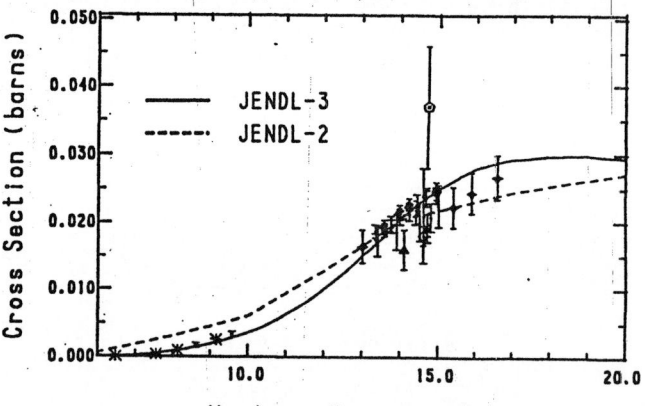

Fig. 2 (n,p) cross section of Mo-96

For nuclides with the measured (n,α) cross section at thermal energy, its cross section in the resolved resonance region was calculated from the resonance parameters by adjusting an average α width so as to reproduce the thermal cross section, and was averaged in quarter lethargy intervals.

The angular distributions of secondary neutrons were assumed to be isotropic in the laboratory system. The energy distributions were also calculated with PEGASUS.

Elastic Scattering Cross Section

The elastic scattering cross section was calculated as (total cross section) - (sum of partial cross sections). The angular distributions of elastically scattered neutrons were taken from the CASTHY calculation.

Benchmark Tests

Group constants were produced with TIMS-PGG/20/. Benchmark tests were performed by comparing calculations with sample worth measured in STEK/13/ and reaction rates in CFRMF/21/ and EBR-II/22/. Figure 3 shows examples for Ru-102 (weak absorber) and I-129 (strong absorber). As a whole, the results were almost satisfactory except for Sb-121, 123, Sm-147 and 149, whose capture cross sections were evaluated on the basis of the experimental data.

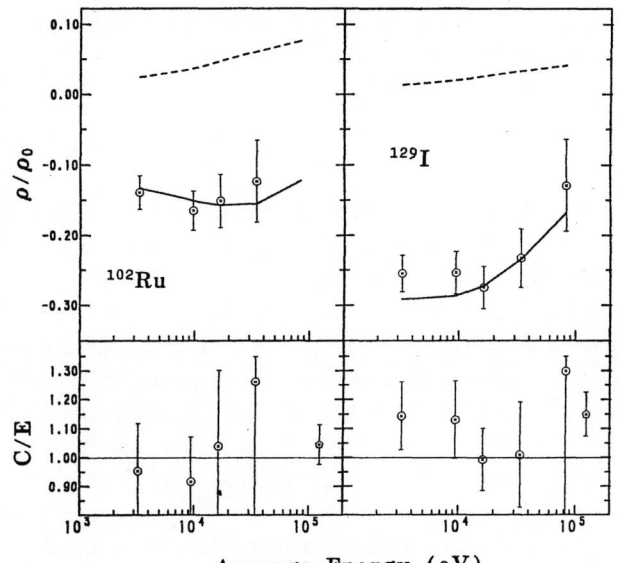

Average Energy (eV)

——— Calculation of sample worth   ⊙   Sample worth in STEK

- - - Calculation of scattering component   ⬠   Activation data in CFRMF

Fig. 3 Typical results of benchmark tests Upper graphs show sample worth and lower ones the C/E values as a function of average neutron energy.

STEK sample worth experiments

For important FP nuclides, calculated values are in agreement with the experimental data within 15 %. The C/E values for STEK 4000 with a soft spectrum were improved from JENDL-2 for Zr-91, 93, Ru-102, Ce-140, 142 and Sm-148 whose resolved resonance regions were extended to much higher energies.

The effect of direct inelastic scattering was investigated. It was found that the contribution from the direct inelastic scattering was less than 10 % of the scattering component. This amount cannot be ignored for the weak absorbing nuclides, since scattering and absorbing components almost cancel out each other.

CFRMF and EBR-II reaction rate measurements

The present results give predictions better than or equal to JENDL-2. In the case of Xe-132, 134, Eu-152 and Eu-154 whose capture cross sections were determined on the basis of the JENDL-2 benchmark tests/9/ mentioned above, the C/E values became better than JENDL-2, but still deviate (about 15 %) from unity.

Concluding Remarks

The present evaluation work for the 172

nuclides in the FP region was made on the basis of the theoretical calculations and recent experimental data. By increasing the number of nuclides and including the threshold reaction cross sections, the quality of the evaluated data library has been much improved. The benchmark tests showed, as a whole, that the present results were reasonable except for some nuclides. Major part of the discrepancy can be attributed to an inconsistency between the differential and integral experimental data. The present evaluation relied more on the differential experimental data.

References

1. T.Aoki, S.Iijima, M.Kawai, Y.Kikuchi, H.Matsunobu, T.Nakagawa, Y.Nakajima, T.Nishigori, M.Sasaki, T.Watanabe, T.Yoshida, and A.Zukeran: Proc. Int. Conf. on Nuclear Data for Basic and Applied Science, 13-17 May 1985, Santa Fe, USA, vol.II, p.1627 (1986)
2. S.F.Mughabghab, M.Divadeenam and N.E.Holden: "Neutron Cross Sections, Vol.I, Part A", Academic Press (1981), and S.F.Mughabghab: "Neutron Cross Sections, Vol.I. Part B", Academic Press (1984).
3. S.Igarasi and T.Fukahori: JAERI 1321 (1991).
4. F.G.Perey: Phys. Rev., 131, 745 (1963).
5. J.R.Huizenga and G.Igo: Nucl. Phys., 29, 462 (1962).
6. J.M.Lohr and W.Haeberli: Nucl. Phys., A232, 381 (1974).
7. F.D.Bechetti Jr. and G.W.Greenlees: Polarization Phenomena in Nuclear Reactions, p.682, The University of Wisconsin Press (1971)
8. BNL/NNDC: Evaluated Nuclear Structure Data File (1987)
9. T.Watanabe, T.Nishigori, S.Iijima, M.Kawai, M.Sasaki, T.Nakagawa, H.Matsunobu and A.Zukeran: JAERI-M 88-065, p.148 (1988)
10. V.Benzi and G.Reffo: CCDN-NW/10 (1969)
11. P.D.Kunz: private communication.
12. H.Gruppelaar and H.A.J.van der Kamp: Proc. Specialists' Meeting on the Use of the Optical Model for the Calculation of Neutron Cross Sections below 20 MeV, 13 - 15 Nov. 1985, Paris, NEANDC-222"U", p.221 (1986)
13. J.J.Veenema and A.J.Janssen: ECN-10 (1976)
14. S.Raman, C.H.Malarkey, W.T.Milner, C.W.Nestor, Jr. and P.H.Stelson: Atomic Data and Nucl. Data Tables, 36, 1 (1987)
15. R.H.Spear: ibid., 42, 55 (1989)
16. S.Iijima, T.Sugi, T.Nakagawa and T.Nishigori: JAERI-M 87-025, p.337 (1987)
17. Wen Den Lu and R.W.Fink: Phys. Rev., C4, 1173 (1971)
18. V.M.Bychkov, V.N.Manokhin, A.B.Pashchenko and V.I.Plyaskin: INDC(CCP)-146/LJ (1980)
19. R.A.Forrest: AERE-R 12419 (1986)
20. H.Takano, A.Hasegawa and K.Kaneko: JAERI-M 82-072 (1982).
21. R.A.Anderl, F.Schmittroth and Y.D.Harker: EGG-PHYS-5182 (1981)
22. R.A.Anderl, Y.D.Harker and F.Schmittroth: Proc. Specialists' Meeting on Neutron Cross Sections of Fission Product Nuclei, 12 - 14 Dec. 1979, Bologna, NEANDC(E)-209"L", p.363 (1979)

Evaluations of Secondary Neutron Energy Spectra for Cr, Fe, Ni and Pb

Zhou Delin,   Zhang Jin and Liu Tong

China Institute of Atomic Energy , P. O. Box 275 (41), Beijing, PR China

Abstract :The angle—integrated neutron emission spectra have been evaluated for Cr, Fe, Ni and Pb based on measured data. Two simplified methods for combination of correlated data are advanced and adopted in the present work with reasonable combined results obtained by least—squares or Bayesian methods. The evaluated results have been compared with Pavlik and Vonach' s evaluations and with those deduced from BROND, ENDF / B—VI and JENDL—3.

(Cr, Fe, Ni, Pb; evaluation, neutron emission spectrum, combination of correlated data)

I. Introduction

The evaluations of secondary neutron energy spectrum for a series nuclides have been carried out by Pavlik and Vonach on comprehensive experimental data base. [1] In view of the most recent high resolution, high statistics measurements performed by Takahashi et al. [2] and Baba et al. [3] , the new evaluations of the angle—integrated neutron emission spectrum (EDX) for Cr, Fe, Ni and Pb at 14.1 MeV are carried out. The evaluated results are compared with the evaluations performed by Pavlik and Vonach and also compared with the calculations (recommendations) for BROND, ENDF / B—VI and JENDL—3 respectively.

During these evaluations, attention has been paid to the correlated data combination. For getting reasonable combinations, two simplified ways have been advanced and used in the present evaluations to resolve the problem so called PPP.

A code system has been developed for deducing EDX and DDX from various libraries in ENDF / B format and plotting. [4]

II. Information Used in the Evaluations

Except for the most recent measurements performed by Takahashi et al. and Baba et al., all other available EDX data for Cr, Fe, Ni and Pb [5—12] have been taken into account in the Pavlik and Vonach's evaluations. These measurements including the new ones are generally in agreement with each other except the low and high energy parts of the spectra where the data are scattered. [1,13] In the energy region below 1.5 MeV for Fe for instanse some measurements are much higher than other measurements. Since the neutrons in this eneregy region mainly come from (n,2n) reaction, so for checking the spectrum and the total emission cross section evaluations the (n,2n) cross sections of Cr, Fe, Ni and Pb at 14.1 MeV are evaluated based on the available measured data including those extracted from secondary neutron energy spectra.As shown in table 1, the results are very good in agreement with Pavlik and Vonach's evaluations since no more newly measured data are available. It has been pointed out that after appropriate corrections the measured (n,2n) cross sections by liquid scientilation tank by

Frehaut et al., [14] are in agreement with those measured by other methods and with those deduced from neutron emission spectrum measurements. [1,15]

The high energy part of the spectrum mainly consists of neutron from (n,n') reaction. The total inelastic scattering cross section and other integral cross sections of Fe—56 have been evaluated individually and comprehensively as showing in table 2. [15]

The results of the comprehensive evaluation including the neutron emission cross section for Fe—56 are consistent very well with the values deduced from the evaluations of nonelastic, (n,2n), (n,p), and (n,a) reaction cross sections for Fe. Actually the way provided by Pavlik and Vonach for checking the evaluations of neutron emission cross section is more general. So their evaluations on these cross sections for Cr, Fe, Ni and Pb are adopted to check the present total neutron emission cross sections as well as the EDX evaluations since no more information significantly is available.

III. Combination of Correlated Data.

In the present work, the Pavlik and Vonach's evaluations for Cr, Fe, Ni and Pb are adopted as a priori ones and taking the Takahashi et al. and Baba et al.'s measurements into account, then to combine them by Bayesian or least—squares method to obtain the updated evaluations respectively. Since the long range correlations existed or are assumed for these data sets,and in general case the correlations may exist between the different measurements . So the combination of such data sets by usual way results in wrong estimations. This is just the problem called PPP (Peele' Pertinent Puzzle) which may be solved in a way advanced by Zhao Zhixiang while he visited at ORNL. [16] But as many practical cases presently no detailed information on the directely measured quantities are available. It is difficult to combine the data sets by using the method provided by Zhao in 'data space' or in 'parameter space (derived quantities)'. So we have to find alternative way for practical use. Two simplified methods have been advanced and used in the present work .

Method 1:

Based on the original uncertainties for measured data or parameters, the covariance of the parameters may be constructed. Then the elements of the covariance $v_{kl,kj}$ ($k = 1,...,L$, $L-$ the number of measurements) for the k−th measurement (or the a priori evaluation) are reconstructed as:

$$v_{kl,kj}^{0} = \frac{v_{kl,kj}}{y_{kl} y_{kj}} y_i^0 y_j^0 = d_{kl,kj} y_i^0 y_j^0$$

y_{kl}, y_{kj}($i,j = 1,2,....$) are the measured values or the a priori evaluated values. y_i^n, y_j^n are the combination results of n−th iteration of the 'parameters (derived quantites)'. Obviously, in this method the relative errors $d_{kl,kj}$ are kept unchanged and the relative weights of the measurements to be combined are dominated by them. This is just the point of approximation involved in this method. But it is acceptable in most practical cases.

Method 2

Suppose all final results y_{kl} of k−th measurement are expressed as :

$y_{kl} = c_k c_{kml} a_{kl}$

and

$c_k = 1 \pm \Delta c_k$

$c_{kml} = 1 \pm \Delta c_{kml}$

So numerically

$a_{kl} = y_{kl} \pm \Delta a_{kl}$

and only uncorrelated errors exist in a_{kl} for all measurements . Then the elements of the covariance of the final results are reconstructed after n−th iteration :

$$var(y_{kl}) = var(a_{kl}) + y_i^0 y_i^0 var(c_k) + y_i^0 y_i^0 var(c_{kml})$$
$$cov(y_{kl}, y_{kj}) = y_i^0 y_j^0 var(c_k)$$
$$cov(y_{kl}, y_{ml}) = y_i^0 y_j^0 var(c_{kml}) \quad \text{for } k \neq m, \, i = j$$
$$= 0 \quad \text{for } k \neq m, \, i \neq j$$

Obviously the significant approximation involved in method 2 are that the derived quantities are simplified as the expressions mentioned above and the constant relative correlated errors for all measurements. It is also supposed that the correlated−error−related quantities are measured with uncertainties but without discrepencies, so the measured values for these quantities are do the expectations. In many practical cases in neutron nuclear data evaluation these approximations are also reasonable and acceptable.

IV. Results and Comparisons

The results of the present evaluations of EDX for Cr, Fe, Ni and Pb are shown in figs. 1−4. The corresponding total neutron emission cross sections and related reaction cross sections are shown in table 1. For comparison, the evaluations of EDX, total neutron emission cross sections and the related reaction cross sections performed by Pavlik and Vonach, and all these data for these nuclides deduced from BROND, ENDF / B−VI and JENDL−3 respectively are also shown in table 1 and figs. 1−4.

Acknowlegement

We would like to express our thanks to H. Vonach, A. Takahashi and M. Baba for providing us their evaluated and measured data respectively and Zhao Zhixiang for the beneficial discussion on the data combination. This work was supported by a grant from the IAEA .

Note added in proof :

After this report has been finished , we have noticed that the method similar to the method 1 of this work has been used by A.B. Smith et al., in their evaluation work (for example, ANL / NDM−115).

REFERENCE

〔1〕 A. Pavlik, H. Vonach, ISSN 0344−8401, Fachinformationszentrum Karlsruhe No. 13−4 (1988)

〔2〕 Á. Takahashi, Y. Sasaki, H. Sugimoto, JAERI−M 88−065 p.279

A. Takahashi, private communication (1990)

〔3〕 M. Baba, M. Ishikawa, N. Yabuta, T. Kikuchi, H. Wakabayashi, N.Hirakawa, Proc. Conf. on Nuclear Data for Science and Technology, May 30−June 3, 1988, Mito, Editor S. Lgarasi. Japan Atomic Energy Research Institute . p.291

M. Baba, private communication (1990)

〔4〕 Zhang Jin, Liu Tong, " A Code System for Deducing EDX in C.M. and DDX in Lab. from File 4 and 5 or from File 6 and Plot ting "

〔5〕 G. Stengl, M. Uhl, H. Vonach, Nucl. Phys., A290 109 (1979)

H. Vonach, A. Chalupka, F. Wenninger, G. Staffel, Proc. Conf. on Symposium on Neutron Cross Section from 10−50 MeV, BNL, Upton, N,Y,12−14 May 1980. p.343

〔6〕 O. Salikov, G. Lovchikov, G. Kotelnikova, V. Nesterenko, N. Fetisov, A. Trufanov, Y. F. 12, 1132 (1970)

〔7〕 D. Hermsdorf, A. Meister, S. Sassonoff, D. seeliger, K. Seidel, N. Fetisov, N. Shahin, ZfK−277 (1975)

〔8〕 R. Schectman, J. Anderson, Nucl. Phys. 77, 241 (1966)

〔9〕 A. Lychageen, V. Lunev, O. Salnikov, V. Vinogradov, B. Devkin, S. Sukhikh, Y. F., 45 1226 (1987)

〔10〕 Yu. Kozyr, G. Prokopets, Y.F., 26, 927 (1977)

〔11〕 S. Iwasaki, N. Odano, S. Tanaka, J. Dumais, K. Sugiyama, Proc. Conf. on Nuclear Data for Science and Technology, May 30−June 3, 1988, Mito, Editor S. Lgarasi. Japan Atomic Energy Research Institute . p.229

〔12〕 G. Clayeux, J. Voignier, CEA−R−4279 (1972).

〔13〕 Zhou Delin, Zhang Jin, Liu Tong, "Intercomparison of Evaluations of Neutron Emission Spectra for Fe, Cr, Ni and Pb for Various Libraries", report submitted to IAEA, (1990)

〔14〕 J. Frehaut, A. Bertin, R. Bois, J. Jary, BNL−NCS−51245 p. 399 (1980).

〔15〕 Zhou Delin, " Evaluation of Pb(n,2n) Cross Section at 14.1 MeV " (to be published)

Zhou Delin, Zhang Jin, Liu Tong, " Evaluations of

neutron emission and related cross sections for Cr, Fe, Ni and Pb at 14.1 MeV " (to be published)

[16] Zhao Zhixiang, " The Covariance Matrix of Derived Quantities and Their Combination " (to be published)

Table 1 Comparison of EDX and Related Cross Sections

	σ_t	σ_{el}	σ_{non}	$\sigma_{n,2n}$	$\sigma_{n,p}$	$\sigma_{n\alpha}$	σ_{nM}^{cal+}	$\sigma_{nM}^{eval}(E'max)$	R_1^*	R_2^{**}
Cr										
P.V.			1320±25	350±27	87±5	35±7	1548	1582(12.0)	0.98	1.02
Z.Z.L.++			1320±25	355±19	87±5	35±7	1553	1509(13.0)	0.98	1.03
BROND	2413	1098	1315	328.6	86.7	39.9	1517		0.96	1.00
B-VI	2556	1060	1496	338.5	83.9	29.7	1721		1.09	1.14
JENDL-3	2450	1229	1221	317.4	87.5	36.0	1415		0.89	0.94
Fe										
P.V.			1369±13	411±35	119±3	48±4	1613	1648(12.0)	0.98	1.00
Z.Z.L.++			1369±13	417±16	119±3	48±4	1619	1610(12.5)	0.98	1.00
BROND	2580	1256	1324	433	113.2	38.4	1605		0.97	1.00
B-VI	2588	1199	1389	415	114.1	44.3	1646		1.00	1.02
JENDL-3	2565	1138	1427	453	113.0	41.2	1726		1.05	1.07
Ni										
P.V.			1400±25	163±30	278±5	101±10	1184	1180(12.0)	1.00	1.03
Z.Z.L.++			1400±25	162±10	278±5	101±10	1183	1146(12.5)	1.00	1.03
BROND	2707	1330	1377	145.4	315.4	98.4	1109		0.94	0.97
B-VI	2700	1288	1412	120.9	303.3	96.8	1133		0.96	0.99
JENDL-3	2680	1337	1343	130.4	289.2	80.3	1104		0.94	0.96
Pb										
P.V.			2515±16	2150±160			4665	4871(12.0)	0.96	0.99
Z.Z.L.++			2515±16	2252±40			4767	4720(13.0)	0.98	1.01
BROND	5418	2743	2675	2263			4938		1.01	1.05
B-VI	5438	2963	2475	2078			4553		0.93	0.96
JENDL-3	5433	2740	2693	2191			4884		1.00	1.03

+ $\sigma_{nM}^{cal} = \sigma_{non} + \sigma_{2n} - \sigma_{np} - \sigma_{n\alpha}$.   * R_1, the ratio of the calculated neutron emission cross section to the evaluated one by pavlik and Vonach.   ** R_2, the ratio of the calculated neutron emission cross section to the evaluated one by Zhou et al..   ++ The evaluated values of σ_{non}, σ_{np} and $\sigma_{n\alpha}$ by Pavlik and Vonach are adopted.

Table 2. Evaluation of integral cross sections for Fe-56 at 14.1 MeV individually and comprehensively

	(n,non)	(n-em.)	(p-em)	(α-em.)	(d-em.)
Individually	1370±13 *	1610±22+	155±10	41.5±1.5	8±3
Comprehensively	1377±13	1633	155±10	41.5±1.5	8±3
BROND		1616	166	38.8	
ENDF/B-VI		1643	165	45.8	
JENDL-3		1739	173	42.8	

* Pavlik and Vonach, + Evaluated value for Fe

Continue of Table 2 .

	(n,n′)	(n,2n)	(n,p)	(n,np)	(n,a)	(n,na)	(n,d)
Individually	788 ± 40	417 ± 17	114.5 ± 3	40.5 ± 10	40 ± 1.5	1.5 ± 1	8 ± 3
Comprehensively	757 ± 40	417 ± 17	114.5 ± 3	40.5 ± 10	40 ± 1.5	1.5 ± 1	8 ± 3
BROND	659	447	112.4	63.1	38.4		
ENDF / B–VI	756	413	114.5	50.2	44.2	1.5	6.5
JENDL–3	743	467	113.0	60.4	41.5	1.3	

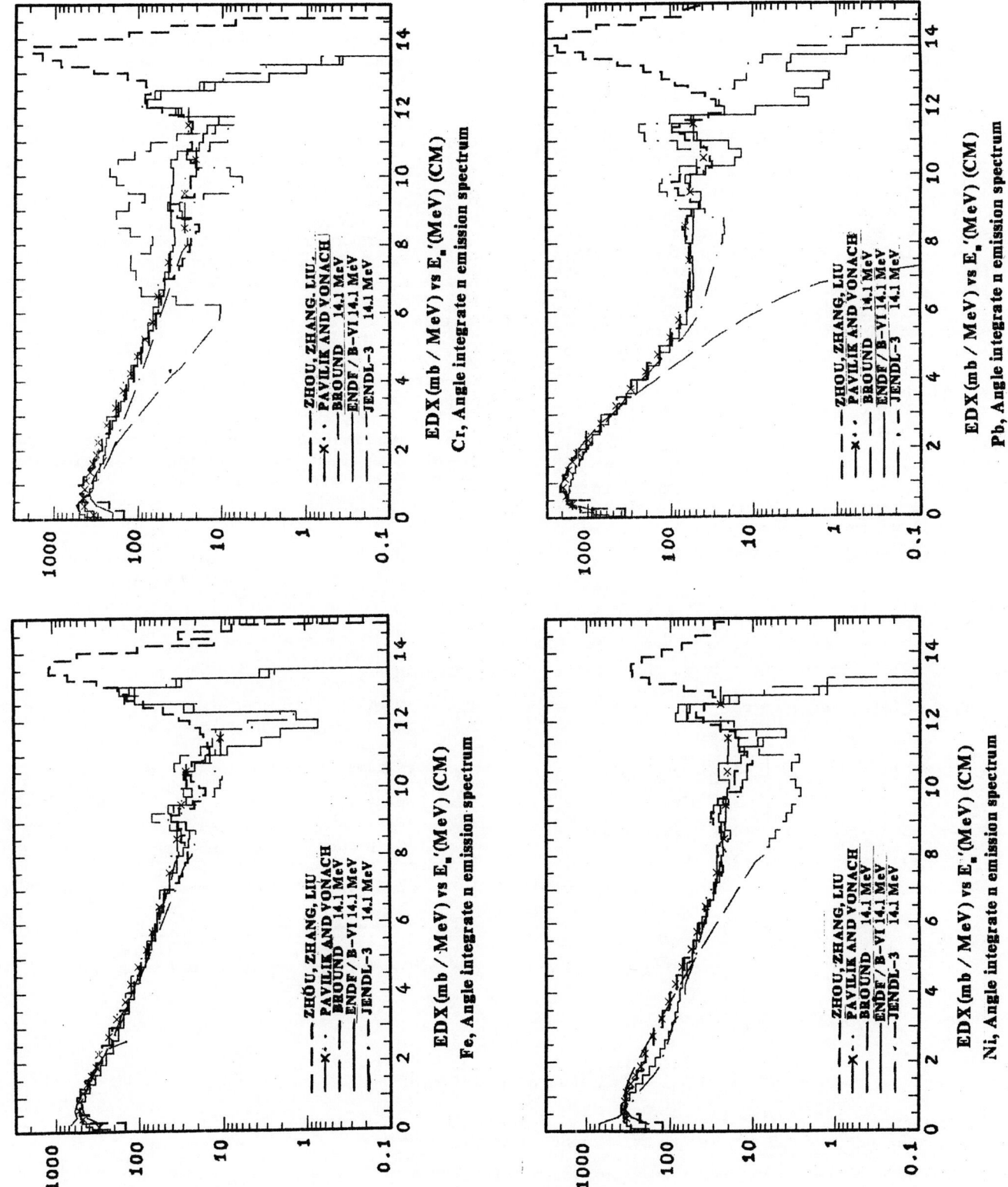

DELAYED NEUTRON DATA AND FISSION REACTOR REACTIVITY SCALE

A. FILIP (*) and A. D'ANGELO (**)

(*) CEA, Centre d'Etudes Nucléaire de Cadarache, 13108, St Paul-Lez-Durance, FRANCE
(**) ENEA-CRE Casaccia, S.P. Anguillarese 301, I-00100 Roma A.D. ITALY

Abstract : Delayed neutron data play a key role in nuclear energy technology, in particular in the estimation of reactor physics safety and economy related parameters. The data are particularly important in the analysis of reactivity measurements where the effective delayed neutron fraction, β_{eff}, is often used to define the reactivity scale. The uncertainty is estimated to be ± 5 %[*]. Improved safety margins and economical performances can be achieved by reducing drastically (to half) this number. We present in this paper the work, now in progress, to achieve this goal, specifically DN data evaluation and integral adjustment.

Delayed Neutrons (DN) data, DN data evaluation, reactivity and reactivity scale, β_{eff} calculus/measure, Perturbation/sensitivity methods, reactor noise, β_{eff} experiment.

Introduction

The goal to reduce the DN data and the reactivity scale uncertainties constitute the objective of an international (OCDE/NEANDC/NEANCRP) cooperation in two domains :
1) The DN data evaluation and,
2) the β_{eff} experimental benchmarking.
By the correlated treatment of these two sets of information [1,2,3] the overall precision of DN data and β_{eff} parameter can be significantly improved. In the first section we present the results of the most recent evaluation of DN data [1] focused on the fissile nuclides of primary importance for Nuclear Reactor Technology (NRT), U^{235}, U^{238}, Pu^{239}. In the second section, the concepts related to the reactivity scale and the current methods for β_{eff} measurement and DN data adjustment are shortly reviewed. A new strategy (now in progress) is presented in section III. The main information is assembled in the fourth (concluding) section together with some recommendations for further work.

1. INDIVIDUAL LEVEL (precursor)

I - DELAYED NEUTRON DATA

In a recent study [3], the overall consistency of the Measurement, Modelling and Evaluation (MM&E) process for the DN data and for the induced reactor kinetic parameters (β_{eff}) was defined through 3 (connected) levels of analysis (see Fig. 1) :
1) The individual (precursor) level ; this is the most extensive and physically based level. All the parameters are issued from the MM&E process (except ν_d, which results by summation).
2) The global level, resulting from MM&E of aggregate of precursor is currently mathematically split into 6 synthetic temporal groups. A stringent test of the overall consistency of DN parameters results from a direct comparison of the two values of the same ν_d : obtained from the summation at level 1 and from the direct MM&E at level 2.

$$\bar{\nu}_d(E_f) = \sum_{A,Z} \lambda(A,Z)Y_c(E_f,A,Z)P_n(A,Z) \int_0^\infty \chi_d(E,A,Z)dE \int_0^\infty e^{-\lambda(A,Z)t} dt$$

(uncertainties) : ±2 to 100% ; ±2 to 20% ; ±10 to 100%

2. GLOBAL LEVEL (aggregate)

$$\bar{\nu}_d(E_f) = \sum_{k=1}^6 \lambda_k(E_f)a_k(E_f) \int_0^\infty \chi_k(E)dE \int_0^\infty e^{-\lambda_k t} dt$$

±3 to 8% ; ±3 to 10% ; ±4 to 20% ; ±5 to 100%

3. INTEGRAL LEVEL (reactors)

$$\frac{\bar{\nu}_d}{\nu} \Rightarrow \beta_{eff} = I_o(\chi_d/\chi_t) \sum_i \frac{\nu_{d,i}}{\bar{\nu}} \frac{\overline{(\nu\Sigma_f)}_i}{\sum_i \overline{(\nu\Sigma_f)}_i}$$

±4 to 5% ; ±2% ; to be tested ; ±2%

Notations

ν_d	DN yield
E_f	Incident neutron energy
E	DN energy
Y_c	DN cumulative fission yield
P_n	DN emission probability
χ_d	DN spectra
M,Z	Masse, charge of precursors
λ	Decay constant of precursors
k indexes :	K temporal group
a_k	DN yield in temporal group

$$a_k/\nu_d = \bar{a}_k \Longrightarrow \sum_{k=1}^6 \bar{a}_k = 1$$

For Integral level formula, see text, Sect. III

Figure 1 - Delayed Neutron Parameters

N.B. : The quoted uncertainties on levels 1 and 2 are general : for reactor application the rather low part of the ranges are to be considered].

*) The uncertainties are given throughout as one standard deviation.

3) The integral level is related to applications, one of the most important being the reactor kinetic parameter calculation, namely β_{eff}. The direct MM&E of β_{eff} is another stringent test of cited consistency.

The more recently evaluated data for levels 1 and 2 are abundantly published ; see exhaustive bibliography cited in [1]. Here we just point out some unsolved problem (in decreasing importance for reactor technology) :

1) The great uncertainties in the measured/calculated β_{eff} (~ 5 %).

2) The probable underestimation of ν_d for U^{238} : recent measured values are greater by 3 to 4 % with respect to the earlier, generally accepted Tuttle evaluations (see Fig. 2).

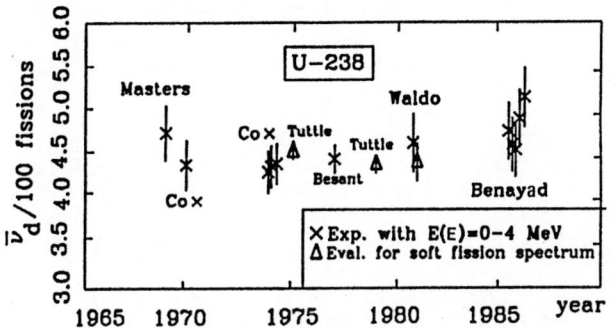

Figure 2 : ν_d experiments after 1965 with uncertainty < 10 %

3) The underestimation of the evaluated uncertainties associated with generally accepted evaluated values of ν_d for U^{235}, U^{238}, Pu^{239} (the quoted more probably uncertainties are : 3 %, 7 to 8 %, 4 to 6 % respectively [1]).

4) Ambiguity in the incident neutron dependence of ν_d (in the range 0 to 4 MeV).

The recommended values of the most important (for NRT) DN parameters are the ν_d for : U^{235} (0.0166 ± 3 %), Pu^{239} (0.0065 ± 5 %), U^{238} (0.043 to 0.047) [see ref. 1].

II - REACTIVITY AND REACTIVITY SCALE (β_{eff})

Two models of the Boltzmann (B) equation, calculating the neutron flux in a reactor, are currently used for <u>near</u> criticality analysis in the NRT field :

- The Static Model (SM) :

$$(1) \quad B\phi_\lambda - (1 - K_{eff})^{-1} F\phi_\lambda = \rho F\phi_\lambda \; ; \; \phi_\lambda = \phi_\lambda \, (\vec{r}, \vec{\Omega}, E)$$
(NB : The reactivity, ρ, caracterizes the "Out of Balance" (OoB) state).

- The Dynamic Model (DM) :

$$(2) \quad B\phi_\alpha = - \frac{1}{v} \frac{\partial\phi_\alpha}{\partial t} \; ; \; \phi_\alpha \propto \phi(\vec{r}, \vec{\Omega}, E) \, e^{\alpha t}$$

The notation is the usual one (see e.g. [3, 7]).

At equilibrium ("criticality"), the two models are identical $(\rho = \alpha = 0 \Rightarrow \phi_\lambda = \phi_\alpha)$ In the OoB state $(\rho \# \alpha \# 0 \Rightarrow \phi_\alpha \# \phi_\lambda)$, the consistency of the two models is imperative. This may be achieved by directly relating the weighted integrals of eqs. (1) and (2) (ref. 3). There results consistently the Point Kinetic (3) and the Inhour (4) equations :

$$(3) \quad p(t) \frac{\Lambda_{eff}}{\beta_{eff}} = (\$-1)p(t) + \sum_{k=1}^{6} \bar{a}_k \lambda_k \int_{-\alpha}^{t} e^{-\lambda_k(t-t')} p(t') dt$$

$$(4) \quad \$ = \rho/\beta_{eff} = \alpha(\Lambda/\beta)_{eff} + \alpha \sum_{k=1}^{6} \bar{a}_k/(\alpha+\lambda_k),$$

together with the associated Kinetic Parameters (KP).

The full expression of β_{eff} [3, 7], is :

$$(5) \quad \beta_{eff} = \sum_{ik} \frac{\int dr \int dE (\nu_{dki}\Sigma_{fi}\phi_\lambda)(E,r) \int dE' \chi_{dki}(E')\phi^+_\lambda(E',r)}{\int dr \int dE (\nu\Sigma_{fi}\phi_\lambda)(E,r) \int dE' \chi_{t,i}(E')\phi^+_\lambda(E',r)}$$

The notation is the usual one [3, 7]) ; ϕ^+_λ is the solution of the adjoint of eq. (1).

Eqs. (3) and (4) are modelling the measurable KP, namely $\$=\rho/\beta_{eff}$. Then β_{eff} is the "scale" of the reactivity, ρ, and their independent measure is fundamental. Such measurements, amongst the more complexes in nuclear reactor physics are carried out by means of essentially two classes of techniques. The "<u>deterministic</u>" one is based on the measurement of the (mean) level of neutron flux in "OoB" states induced by an external perturbation (namely a neutron source [5, 6]) combined with an independent absolute reactivity determination [7] . The "<u>noise</u>" methods consider the random "OoB" states induced by the (natural) fluctuations in nuclear reactions, namely neutron emission. The value of β^2_{eff} is directly deduced from the "treatment of (random) signal" in detectors (Fourier analysis, coincidence) (ref. 8, 9).

The current uncertainties for such measurements, carried out on Fast Reactor Mock-up (and using DN data with associated, quoted errors (see Fig. 1)) are ~ 5 % [10].

This state suggests to adopt the strategy, usual in similar situations i.e. to develop methods to improve, simultaneously, the measured analysis of β_{eff} and the (integral) adjustment of DN data (namely of ν_d). One can consider two methods of doing this :

1) The first one is <u>the statistical treatment of the set of existing</u> β_{eff} measurements, through the Generalised Perturbations & Sensitivity Analysis [7] in correlation with DN data on level 2 of Fig. 1. Such a method was developed [2] for a set of β_{eff} measurement realised on 10 fast critical facilities (ZPR, in Argonne Lab., and SNEAK in Karlsruhe Lab.). The results obtained in this way cannot be considered as conclusive [2]. In particular some important irreducible discrepancies between the calculated and the experimental β_{eff} values hinder any definitive integral testing of DN yields (ν_d) with satisfactory precision.

2) The second, a new approach, is developed in the followings.

III - AN IMPROVED, SEMI-ANALYTICAL METHOD FOR ν_d ADJUSTMENT FROM "CLEAN" (INTEGRAL) β_{eff} MEASUREMENTS

The main difficulties in extracting unambiguous informations on DN data from β_{eff} measurement are evident from eq (5).

1) Firstly we note the problem of <u>separating the contribution of fissile nuclides</u> in mixtures : U238 & U235 (Uranium cores) or U238 & Pu239 (Plutonium cores). A simple

solution consists of making measurements in a set of cores with different ratios of these nuclides. In Cadarache, two sets of three such cores are to be realised : one set, in the fast critical assembly, Masurca [10] and one set in the thermal facility, Eole [11]. For each set, one core is U235 enriched, the others two, Pu239 enriched. The fractional β_{eff} (Pu239/Pu238) are ~ 1.2 and 0.5 for the fast, respectively 66 and 77 for the thermal assemblies. So the measurements results on these cores transform eq. (5) into a linear system in ν_d (see next point).

2) The second class of difficulties is related to the <u>interpretations of the experiments by means of the basic eq.(5)</u>. The measured values, in addition to those of β_{eff}, are linear functional ($\iint\Sigma_f\phi$, $\iint\chi\phi^+$), while the related quantities in eq. (5) are bilinear functional ($\iint\Sigma_f\phi\phi^+\chi..$). Then, the necessary calculation of the transposition factors may introduce significant errors for usually complex assemblies. On the other hand the DN parameters, $\nu_{d,i}$ and $\chi_{d,i}$, are not "factorised" in eq.(5), so separate information about each of them is difficult to obtain. All these problems are avoided for "clean" cores (one zone, low heterogeneity and reflector and/or blanket contributions), for which the eq. (5) may be fully factorisable by means of an adapted "<u>Perturbed Fundamental Model</u>" (PFM) [3].

This model results from splitting the neutron flux and the adjoint (source importance) into a fundamental, factorisable term and a (weak) perturbation term :

(6) $\vec{\Phi}(r,E) = \phi(r)\varphi(E) + \Delta\Phi(E,r)$, (idem for ϕ^+)

Introducing (6) in (5), one obtains :

(7) $\beta_{eff} = I^0\beta^0\eta$ (upper 0 index relates to FM)

where :

(8) $\beta_{eff}^0 = \sum_i \dfrac{\nu_{d,i}}{\nu_0} f_i^0$; $f_i^0 = \dfrac{(\overline{\nu\Sigma_f})_i^0}{\sum_i (\overline{\nu\Sigma_f})_i^0}$

F.M., space independent fission rate ratio,

(9) $\eta = \sum_i f_i^0\alpha_i$; α_i: space dependent perturbation

(10) $I^\circ = \int\chi_d\,\varphi^+ / \int\chi_t\,\varphi^+$

FM, delayed/total, importance ratio.

Eqs. (7) to (10) permit one : to relate, directly measured to calculated quantities (f_i^0, I_d^0) ; to separate ν_d from χ_d ; to calculate (with experimental testing) the (spatial/) perturbation, η. The FM accurate interpretation (representative of cores'centre measurements) together with the very weak, higher order (η) correction allow to extract the ν_d parameters for U^{235}, U^{238}, Pu 239 with only weak amplification of the measurement errors.

These last, on the other hand, may be significantly reduced by using the two independent kinds of measurements cited : "deterministic" and "noise".

Preliminary analysis of planed β_{eff} experiments, by the exposed method, shows that precisions better than 2 %, 3 % and 4 % for ν_d of U235, Pu 239 and U238, respectively, are expected. More specifically, precision of ≤ 2 % for β_{eff} measurement and of ≤ 1 % for overall correction are expected (note that for Masurca cores the factor, η, is, only, a few percent).

IV - CONCLUSIONS

In order to improve the precision in DN data to the level required in the NRT Kinetic parameters a major effort is being made in the framework of an international cooperation [1]. This work involves a new detailed, evaluation of data and a integral benchmarking, via the reactivity scale (β_{eff}) measurement.

We propose, in this paper, a synthetic "three level", schema for consistent analysis of the bulk of DN and related kinetic parameters. A better transparency in the transfer of information between different levels of treatment (DN precursor/DN aggregate/Integral KP) is than assured. Several unsolved problems are pointed out, namely the probable underestimation of ν_d for U238. A new careful measurement of this parameter (at level 2) is therefore highly desirable.

On the other hand, the cited β_{eff} benchmarking will allow to improve significantly the precision of ν_d for U235, U238, Pu 239.

For achieving this goal a new elaborate strategy is adopted, namely in the experiment conception (several independent measurement techniques on several "parametric, clean" cores) and in the analysis (a new semi-analytical model). A large cooperation (open for all interested specialists) is hoped for.

REFERENCES

1. J. BLACHOT, M.C. BRADY, A. FILIP, R.W. MILLS and D.R. WEAVER : Status of DN data - 1990. Rapport NEACRP-L-323, NEANDC-299 "U", Dec. 1990.
2. A. D'ANGELO : Proc. ANS, PHYSOR, Conf. Marseille, France, Apr. 1990.
3. A. FILIP : Seminar in CEN Cadarache, June 1990 (to be published).
4. A. GANDINI: RT/FI (73) 5 and 22, ENEA 1973.
5. S.G. CARPENTER, J.M. GASILDO and J.M. STEVENSON : NSE, 49, 2, 236 (1972).
6. E.A. FISCHER : NSE, 62, 105 (1977).
7. G.R. KEEPIN : Physics of nucl. kinetics, Addison 1965.
8. C.E. COHN : NSE, 7, 472 (1960).
9. E.F. BENNETT : Rap. ANL-81-72 (09/1981).
10. M. MARTINI, J.C. GAUTHIER, G. GRANGET, R. SOULE, NEACRP-A-1064, meet. Oct. 1990.
11. J. MONDOT, J.P. CHAUVIN, J.C. LEFEBVRE, A. VALLEE. Proc. ANS, PHYSOR conf. Marseille, France, Apr. 1990.

CALCULATIONS ON NUCLEAR REACTIONS OF NEUTRONS WITH ^{138}Ba IN THE ENERGY RANGE OF 0.1 TO 20 MeV

Liu Yanping, Li Shiqing

Dept. of Phys., Wuhan Univ., P.R.China

Abstract: The phenomenological optical potential was used in the calculations on nuclear reactions of neutrons with ^{138}Ba in energy range of 0.1 to 20 MeV. For lack of experimental total cross sections in energy range of 15 to 20 MeV, the microscopic optical model based on Skyme force was used to extend the experimental total cross sections to the energy range. The optimal phenomenological optical potential parameters were incorporated in the calculations of statistical theory.   The calculated results   on many kinds of cross sections, the angular and energy distributions of secondary neutrons have been completely written on tape to store in CENDL-2.

(^{138}Ba, phenomenological optical potential,   extension of experimental data, microscopic optical model, nuclear data calculation)

Introduction

^{138}Ba was regarded as a kind of fission product (FP) in our FP Neutron Data Library, since stable ^{138}Ba can be produced from β-decay of fissionfragment ^{138}I.

In the calculations of neutron data of ^{138}Ba, the optical model and statistical theory were used. The latter includes Hauser-Feshbach theory with the width-fluctuation correction and statistical theory with pre-equilibrium emission.

For calculations of optical model, we made use of the global phenomenological optical potential which includes central potential and spin-orbit coupling potential. The central potential was divided into real part and imaginary part, but the spin-orbit coupling potential took only real part. Woods-Saxon form was used for these potential functions.

For the central potential we assume that the real part depth V and the imaginary part depths U and W (corresponding to body absorption and surface absorption) depend on incident neutron energies as below:

$$V=V_0+V_1E+V_2E^2+V_3(N-Z)/A+V_4Z/A^{1/3}$$

$$U=U_0+U_1E$$

$$W=W_0+W_1E+W_2(N-Z)/A,$$

where V_0, V_1, V_2, U_0, U_1, W_0 and W_1 are 7 variable depth parameters, but V_3, V_4 and W_2 were all taken as constants (by universal constants).

The 7 variable parameters mentioned above, and the radius and diffuseness parameters (r_1 and a_1) of real part potential function as well as the radius and diffuseness parameters (r_2 and a_2) of imaginary part potential function were taken as adjustable parameters /1/. The code AUJP program /2/ was used to adjust the 11 adjustable parameters. The optimal parameters were imparted to the code MUP2 program /3/ to calculate various kinds of neutron data for n-^{138}Ba reactions.

Selection of First-hand Data

Experimental Data: In AUJP program, three kinds of experimental data—total cross section σ_t, nonelastic scattering cross section $\sigma_{n,on}$ and elastic scattering angular distribution $\sigma_{el}(\theta)$ are needed. For this purpose, we collected the experimental data from EXFOR (till the end of 1987). Because of no experimental data on total cross sections of ^{138}Ba in some energy ranges and due to high abundance of ^{138}Ba in natural Ba (~71.7%), such total cross sections of ^{138}Ba were taken from the experimental data of natural Ba.

For σ_t, we selected 9 groups of data from EXFOR, the data are in the energy range of 0.1 to 15 MeV, and measuring accuracies are 2 to 5% . The spline fiting of above-mentioned total cross sections was done, the fitted results (expressed as σ_t^E) are given in Table 1.

Table 1.   Experimental Data

Energy (Mev)	σ_t^E (b)	$\sigma_{n,on}$ (b)	Number of θ for $\sigma_{el}(\theta)$
0.1	5.928		0
0.5	5.942		8
1.0	7.072	0.2	9
1.5	7.265		0
2.0	6.867	0.8	0
2.5	6.413		0
3.2	5.785		13
4.1	5.100	2.2	15
5.0	4.636		10
6.0	4.375		0
8.0	4.475		0
10.0	4.753		0
12.0	5.000		0
14.0	5.007	1.9	0
15.0	5.034		0

For $\sigma_{n,on}$ and $\sigma_{el}(\theta)$, since in EXPOR there are not many the experimental data, we took all of them and gave also in Table 1.

Initial Values of Potential Parameters: The potential depth, radius and difuseness parameters of n, p, t, ^{3}He, d and α reaction channels are taken from /4/. For parameters lacking in that reference, we took universal parameters obtained by systematics analyses.

Calculation

Extension of σ_t^E Data: For lack of experimental total cross sections in the energy range of 15 to 20 MeV, we resorted to the calculations of the microscopic optical model (MOM) without any free parameters /5/, the calculated results (σ_t^M) are given in Fig.1. Because of the similarity in the tendency of change in σ_t^E and σ_t^M with incident neutron energies, we used MOM to extend the total cross sections to energy range without experimental data. A comparison between experimental data σ_t^E and MOM calculation data σ_t^M is given in Table 2.

Table 2. Comparison Between σ_t^E and σ_t^M

E (MeV)	12	14	15
σ_t^E (b)	5.00	5.01	5.03
σ_t^M (b)	4.15	4.26	4.30

On the basis of Table 2., a formula based on experience was obtained as follows:

$$\sigma_t^{EX} = \sigma_t^M + 0.85 - 0.04 \ (\ E - 12.0 \ )$$

This extended values to energy range of 15 to 20 MeV (Table 3).

Table 3. Extended Values σ_t^{EX}

E (MeV)	16	17	18	19	20
σ_t^M (b)	4.32	4.31	4.28	4.23	4.17
σ_t^{EX} (b)	5.01	4.96	4.90	4.80	4.70

Influence of Extended Data σ_t^{EX}: The code AUJP program was used to adjust 11 adjustable parameters under two kinds of circumstances whether there are extended data or not.

For first kind of circumstance (there are extended data), the optimal phenomenological optical potential parameters were called "1st group" in this paper ; and for the second circumstance (there are no extended data), the parameters were called "2nd group"; both are given in Table 4.

The " 1st group " parameters were imparted to MUP2 program, the calculated total cross sections σ_t^T agree with σ_t^{EX} within 3%. But the greatest difference between the calculated total cross sections $\sigma_t^{T'}$ using " 2nd group " parameters and σ_t^{EX} in the energy range of 15 to 20 Mev (see Fig.1) was >8%.

Table 4. Optimal Parameters

	1st group	2nd group
V_0	46.7411	47.0921
V_1	-0.3216	-0.3604
V_2	0.0266	0.0211
U_0	-1.5937	-1.5433
U_1	0.2527	0.2119
W_0	10.7577	11.0218
W_1	-0.2407	-0.2872
r_1	1.3647	1.3538
r_2	1.3216	1.3925
a_1	0.5492	0.5388
a_2	0.3950	0.3533

Calculated Results

The "1st group" calculated results have been completely written into tape as consistent with ENDF-IV format, and stored in CENDL-2-FP.

The calculated data include cross sections of total, elastic scattering, nonelastic scattering, inelastic scattering, (n,2n), (n,3n), (n,γ), (n,p), (n,d), (n,t), (n,^{3}He) and (n,α) etc. The angular and energy distributions of secondary neutrons were also obtained, the contents are given in Table 5.

Table 5. Contents of Calculated Data

MF-number	MT-number
3	1,2,3,4, 16,17,22,28,32,33,34, 51--70, 91, 102--107, 251--253
4	2,16,17,22,28,32--34,51--70,91
5	16, 17, 22, 28, 32, 33, 34, 91

The "1st group" calculated results were compared with experimental data, it was found that calculated results can fit the experimental data very well. The details for σ_t, $\sigma_{n,on}$, $\sigma_{n,2n}$ and $\sigma_{el}(\theta)$, for example, are shown in Fig.1. and Fig.2. The calculated $\sigma_{n,2n}$ has been notably improved over that of ENDF/B-VI.

Discussion

As is well known, optical model is one of the most important theoretical approaches in nuclear data calculations. However, it is not very good to use the phenomenological optical potential parameters to predict the nuclear data in energy ranges without experimental data. This work shows that the method to extend experimental data by microscopic optical model has great use in nuclear data calculations for nuclides which lack experimental data.

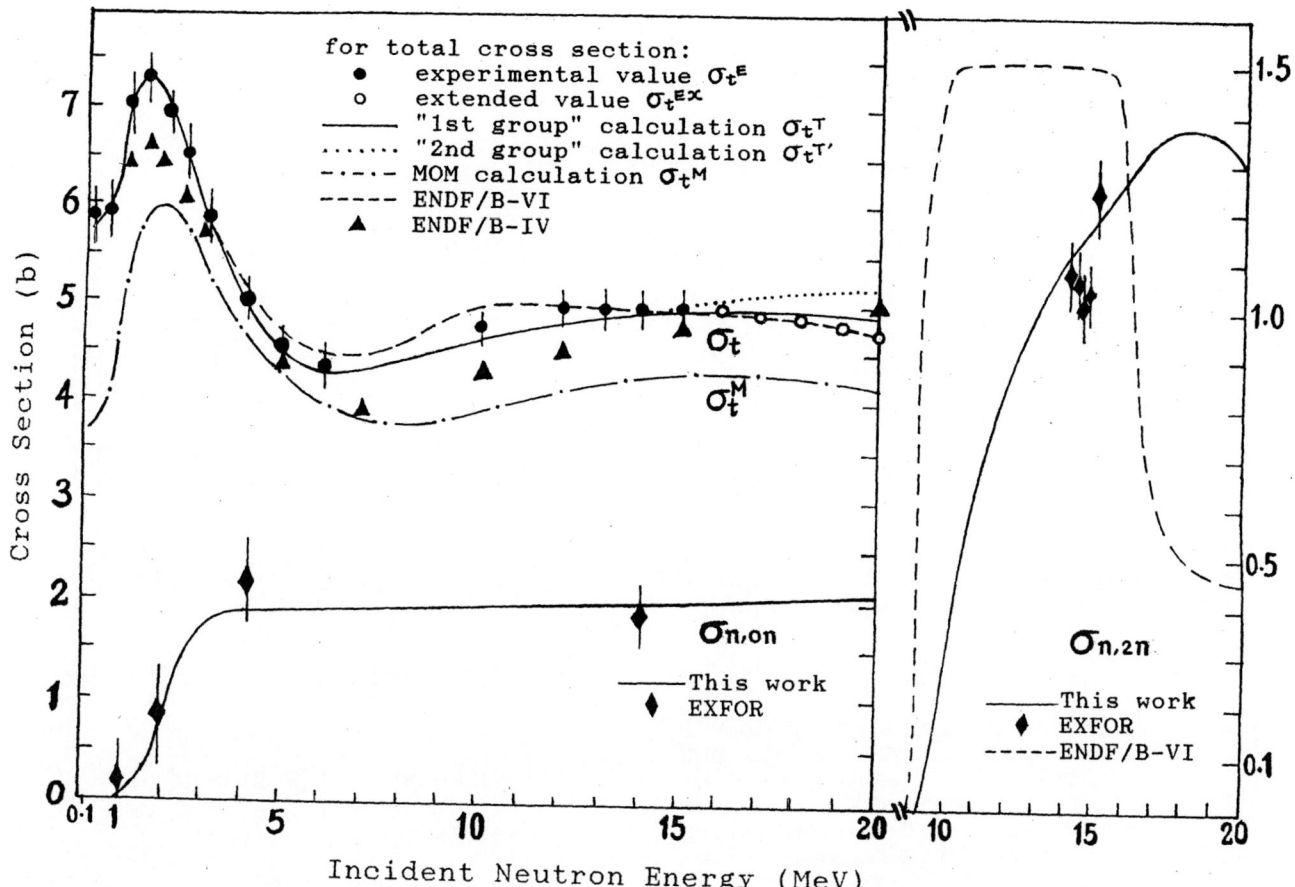

Fig. 1. Comparison between calculated and experimental cross sections for n - ^{138}Ba reactions.

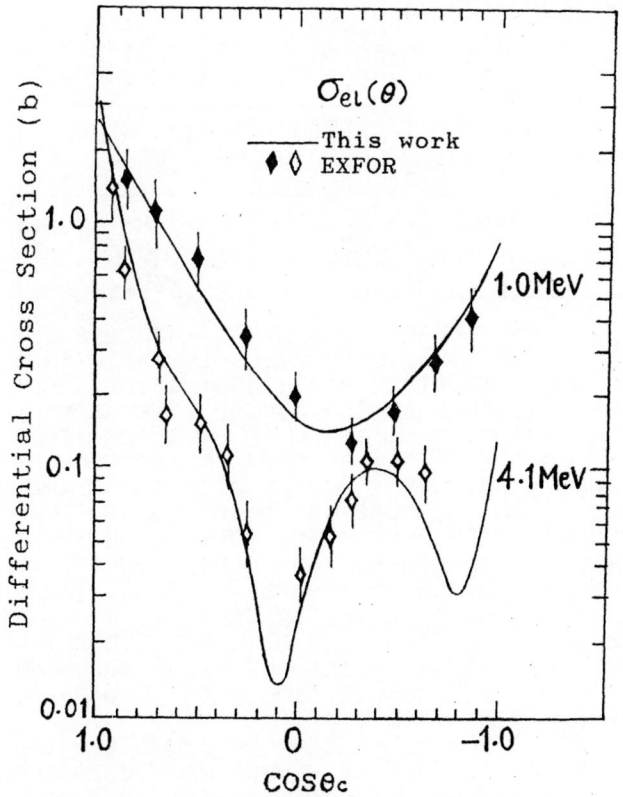

Fig.2. Elastic scattering angular distributions for En = 1.0 and 4.1 MeV

REFERENCES

1. Liu Yanping, The Calculations on Fast Neutron-induced Ni Reactions, CNDC-85011, 99 (1986)
2. Zhou Hongmo, Cai Chonghi, Zuo Yixing, The Program AUJP of Automatic Adjusting Parameters of Optical Model, CNDC0011 (1989.11.)
3. Cai Chonghi, Zhou Hongmo, Zhang Xiaocheng, The Program MUP2 for Calculation of the Fast Neutron Data of Medium and Heavy Mass Nuclei, CNDC0013 (1989.11.)
4. C.M. Perey, Atomic and Nuclear Data Tables, 17 (1976)
5. Tian Ye, Shen Qingbiao, Chinese Journal of Nuclear Physics, 7, 154 (1985) and 8, 28 (1986)

ACKNOWLEDGMENTS

The authors gratefully acknowledge the support of K. C. Wong Education Foundation, Hong Kong.

The authors would like to thank the Chinese Nuclear Data Center for all manner of help.

APPLICATION OF CROSS SECTION DISTRIBUTION FUNCTION TO THE ESTIMATION OF FLUX AT A POINT FOR RADIATION TRANSPORT PROBLEM

N.A.Solovjev and V.E.Kolesov, Institute of Physics and
Power Engineering, Obninsk, Kaluga region, USSR
and
V.V.Kolesov and V.F.Ukraintsev, Obninsk Institute of
Nuclear Power Engineering, Obninsk, Kaluga region, USSR

Abstract: Results of Monte Carlo test calculations on the influence of energy structure cross sections on the estimation of flux at a point is demonstrated. The problem of deep penetration of neutrons through iron slabs with different thicknesses is solved. Calculations were performed with a special MORSE multigroup Monte Carlo code,which uses group constants of ABBN library and the distribution functions of the cross sections in the subgroup approach. The calculation results show that subgroup approach application to neutron data permits an increase in the accuracy of the estimation of flux at a point.

(Monte Carlo method, estimation of flux at a point, integral transport equation, cross-sections energy structure, group-averaged cross-sections, subgroup approach parameters, cross-section distribution function)

Introduction

In the radiation transport problems a necessity frequently arises of flux and flux function determination at the relatively little volumes or at the separate fixed space points far from radiation source. For such a purpose analytical evaluation of expected values is usually applied. The procedure of point flux estimation by Monte Carlo method is widespread [1, 2].

For receiving correct result it is necessary to account correctly the energy cross section structure and scattering anisotropy. The influence of energy cross section structure on point flux estimation was investigated in [3]. Conclusion was made about possibility to get transport equation decision by Monte Carlo method in group approach without great harm.On the contrary,the use of the group neutron data for the next flight estimate in statistical estimation technique gives a large error. Here it needs to account the energy dependence of cross section more correctly. It was pointed out that neutron resonance cross section structure data can be represented by cross section distribution function.

A description is presented of a numerical experiment, which shows influence of energy neutron cross section structure on the "next event" estimation in Monte Carlo method.

Numerical Experiment

Exact contribution to detector can be calculated for only some particular cases (for example,contribution of uncollided flux from point source), so dependence of contribution on cross section structure can be estimated by using Monte Carlo procedure only. As a model task,the neutron transmission through iron slabs was investigated (density 7.79g/cm^3, diameter 152 cm,element composition is given in Table 1).

Table 1. Composition of the Slabs in [at./(b cm)]

	Density
Carbon	$9.815*10^{-4}$
Manganese	$5.150*10^{-4}$
Iron	$8.372*10^{-2}$

Calculations were performed for thicknesses 15.39, 30.81 and 62 cm. Point detectors were placed behind the slab on the center line, and separate shift from this center line was used. Variation of detector placing is given in Table 2.

Table 2. Position Configuration for Detectors

Detector	Center Line Distance behind Slab (cm)	Radial Distance from Center Line (cm)
1	427.05	0
2	411.81	102.87
3	310.21	279.40

Three point radiation sources for 10th ABBN [4] group were investigated: the one directed along the center line; isotropic one on the center line and on the slab surface, and only neutrons falling on the slab take part in the simulations; isotropic source with vector angles into the cone with the angles between its axis and cosine equal to 0.9989. The last was placed at cone top 173 cm from slab plane on the center line. Lighting spot on the slab surface from this source would be a circle 16.5cm diameter. From principal point of view the same geometry and composition are close to the experiment [5].

The 28-group ABBN-type cross section library was used as a neutron constant base and ARAMACO[6] code system as a macroconstant code supply. P5-approximation for elastic scattering angle distribution was used. Microconstants for 10th group of iron, carbon, manganese and the sample macroconstants are presented in Table 3 and Table 4, respectively.

Table 3. Microscopic Group Constants for 10th Group from ABBN

	Subgroup Constants			Group Averaged Constants	
	a	σ_t(b)	σ_e(b)	σ_t(b)	σ_e(b)
Iron	.0915 .6696 .2389	69.55 9.44 0.40	69.4716 9.4260 0.3910	12.7787	12.76
Carbon				4.62	4.62
Manganese				13.40	13.35

Table 4. Macroscopic Group Constants for 10th Group Calculated by Using ARAMACO Code

Subgroup Constants			Group Averaged Constants	
a	Σ_t (cm^{-1})	Σ_e (cm^{-1})	Σ_t (cm^{-1})	Σ_e (cm^{-1})
.0915 .6696 .2389	.8805 .1195 .0051	.8782 .1191 .0049	.1621	.1613

The MORSE code [7] was used as a basis for Monte Carlo calculations. Some subroutines of the complex were changed for subgroup formalism using our calculations. Uncollided flux of source -to- detector, placed on center line was calculated theoretically. Subgroup (group) approach contribution to the detector accounts from uncollided component C_{un}

$$C_{un} = W \sum_{i=1}^{N} a_i \exp(-\Sigma_{ti} x_j),$$

where

W– probability for source neutron to take flight to detector direction per steradian,

$$W = \begin{cases} 1 & \text{for point monodirected source,} \\ 8.73*10^{-7} & \text{for isotropic point source for a one half-plane,} \\ 6.027*10^{-4} & \text{for point source in the cone;} \end{cases}$$

a_i, Σ_{ti} –subgroup parameters,for group calculations $a_i=1$ and $\Sigma_{ti}=\Sigma_t$, all the rest equal zero;

x_j–j scatter thickness.

Three calculations by MORSE code were performed for each sample thickness: the group data were used in the random walk procedure and the next flight estimator also; the group data were used in the random walk procedure and the subgroup data were used in the next flight estimator; subgroup data were used in the random procedure and in the next flight estimator. 15000 particles were allowed for each calculation. Air scattering was not taken into account.

Table 5. The Ratio of Uncollided Flux to Total Flux at a Point for Isotropic Source at a Point on Centre Line behind Slab

Type Calculation*	Thicknesses		
	15.39	30.81	62.00
Group+Group	0.152	0.03	0.0013
Group+Subgroup	0.275	0.21	0.20
Subgroup+Subgroup	0.441	0.38	0.36

*In this table and in the next tables: Group+Group means, that group data were used in the random walk procedure and the next flight estimator; Group+Subgroup means, that group constants were used in the random walk procedure and subgroup constants were used in the next flight estimator; Subgroup+Subgroup means, that subgroup constants were in the random walk procedure and the next flight estimator.

Table 6. Uncollided Flux from Monodirectional Source on the Center Line behind Slab

Calculation	Neutron Flux n*(cm^{-2})*(n source)$^{-1}$		
	Thicknesses		
	15.39(cm)	30.81(cm)	62.00(cm)
Using Groups	$8.52*10^{-2}$	$6.77*10^{-3}$	$4.32*10^{-5}$
Using Subgroups	0.33	0.22	0.17

Table 7. Total Flux at Point Detectors for the Point Cone Source behind 30.81 cm

Detector	Neutron Flux n*(cm^{-2})*(n source)$^{-1}$		
	Group+Group	Group+Subgroup	Subgroup+Subgroup
1	$2.32*10^{-6}$	$1.06*10^{-4}$	$1.06*10^{-4}$
2	$4.56*10^{-7}$	$8.28*10^{-7}$	$3.19*10^{-7}$
3	$2.56*10^{-8}$	$7.69*10^{-7}$	$2.98*10^{-7}$

Discussion

Uncollided flux for detector on the center line (see Table 5) in group constant calculations strongly decreases with thickness. For another detector situation is much the same. It is related with the great absolute value (–0.1621*x) exponent function argument even for low thickness and the slope of function is exponential,as in Fig.1(curve 1). At the same time the number of neutrons reaching the neighbourhood of detector slice is great enough (probability of unabsorption equals 0.995 and the probability of remaining at the same group equals 0.952). While evaluating uncollided component, the subgroup cross section is included to the absolute exponent argument, which can be very small and so the exponent may be

954

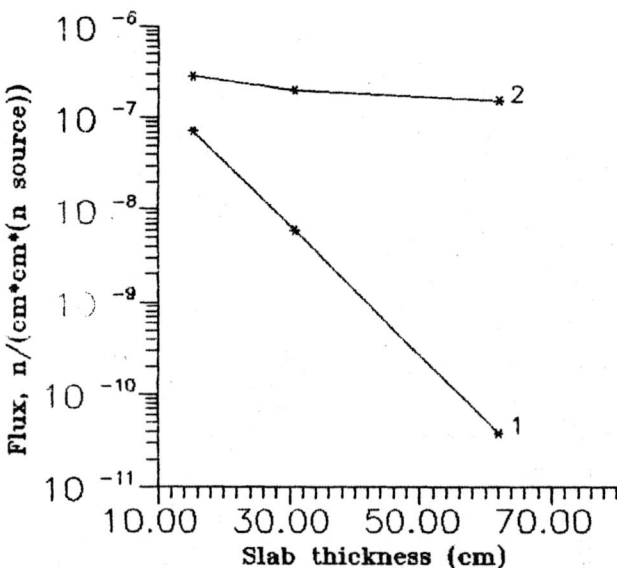

Fig.1. Uncollided flux at a center line point for the isotropical point source behind 15.39, 30.81 and 62 cm iron slabs.

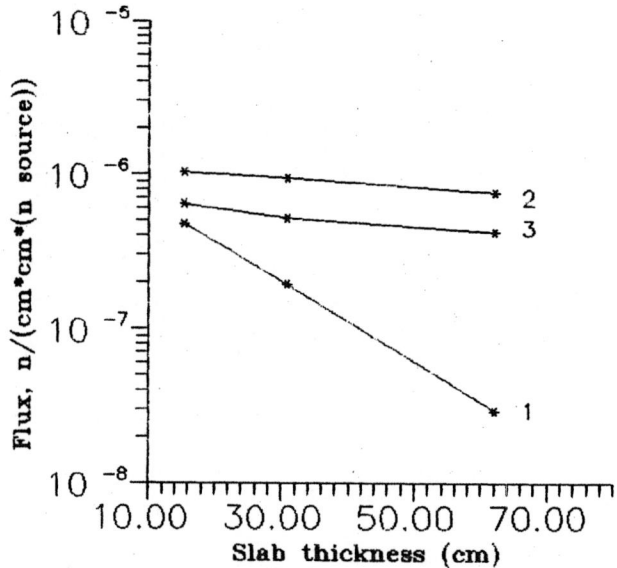

Fig.2. Total flux at a center line point for the isotropical point source behind 15.39, 30.81 and 62 cm Iron Slabs.

approximated even for thickness 62 cm as $\exp(-0.005*x) \sim (1-0.005*x)$, the $(1-0.005*x)$ function changing weakly till 62 cm. It is well illustrated by Fig.1(curve 2) and by Table 6.

Uncollided component contribution is a basic constituent part for total flux for detector on the axis from beam of monodirected or few angle divergent point source.It can be seen from Table 7, that the first detector flux is by an order greater than the second one where unscattered component is present.

The flux to point detector from isotropic point source is shown in the Fig.2. It must be noted, that the calculation, using group cross section (curve 1), gives essentially lower values, compared with another calculations for detector and the difference increases with scatterer thickness.The calculations using group constants in the random walk procedure and subgroup constants in the next flight estimator (curve 2), give greater flux compared with the result of all subgroup parameters (curve 3). Moreover this ratio weakly varies with thickness and equals 1.7 for this variant.Such increase of component is connected with the fact, that the number of collisions for neutron,calculated by using only subgroup parameters is about 2 times lower than that calculated by group parameters.

Conclusion

Local estimations obtained by Monte Carlo method should be performed with resonance structure accounting.It is extremely important for calculation with different filters with nearly monodirected beam. The number of the source particles in this case must be increased, compared with group calculations.

When considering local evaluation from isotropic source, the random walk procedure can be made by group method,but it is necessary to calculate the contribution of the next flight estimator with resonance structure accounting. Specially it can be performed with the help of the distribution cross section function. In this case the evaluations will be overestimations. The overestimation can be calculated by performing calculations for some thicknesses using cross section structure influence.

REFERENCES

1. V.G.Zolotuhin and S.M.Ermakov, in Voprosy Phisiki Zashchity Reactorov (Edited by D.L.Broder),p.171, Gosatomizdat, Moscow (1963) (in Russian)
2. M.H.Kalos: Nucl.Sci.Eng.16,111(1963)
3. V.E.Kolesov and N.A.Solovjev: VANT Yadernye Constanty 2,100 (1990) (in Russian)
4. L.P.Abagjan, N.O.Bazazjantz, M.N.Nikolaev and A.M.Tsyboolja, Gruppovye Constanty dlja Rashcheta Reactorov i Zashchity, Energoizdat, Moscow (1981)(in Russian)
5. R.E.Maerker and F.J.Muckenthaler: Nucl.Sci.Eng.52,227(1973)
6. M.N.Nikolaev and M.M.Savoskin: VANT Yadernye Constanty 5,24(1984) (in Russian)
7. E.A.Straker, P.N.Stevens, D.C.Irving and V.R.Cain, The Morse Code - A Multigroup Neutron and Gamma-Ray Monte Carlo Transport Code, ORNL-4585, Oac Ridge National Laboratory (1970)

ANALYSIS OF O-16 PHOTO-REACTION CROSS SECTIONS

Toru Murata
Nuclear Engineering Laboratory
Toshiba Corporation
4-1 Ukishima-cho,Kawasaki-ku,Kawasaki,
210,Japan

Abstract: O-16 photo-reaction cross sections were analysed and evaluated in the photon energy region below 150 MeV. In the low energy region below 30 MeV, resonance analyses were made for experimental photo-absorption cross section and photo-neutron emission cross section. Resonance parameters of 22 resonance levels were determined. In the energy region above about 30 MeV, quasi-deuteron model was adopted to reproduce the photo-absorption cross section and neutron emission cross section. Based on these theoretical models, other particle emission cross sections were estimated. For quasi-deuteron model, a new modification were made to obtain energy spectrum and angular distribution of neutron or proton emitted by photo-reaction of light elements.

(Oxygen-16, photo-reaction, resonance analysis, quasi-deuteron model, photo-absorption cross section, emission cross section of neutron, proton,deuteron,triton,helium-3 and alpha-particle)

Introduction

As a part of photo-reaction nuclear data evaluation program of Japanese Nuclear Data Committee, O-16 photo-reaction cross sections were analysed and evaluated in the incident photon energy region below 150 MeV. The nuclear data of elements contained in air and human tissue would be important for the estimation of radiation dose rate of working personnel of photon production accelerator and patients under radiation therapy. No data files are available presently for the purpose.

For oxygen, experimental data of photo-absorption and photo-neutron emission cross sections are available in the photon energy region for the present work. Other photo-reaction data are not so plenty to evaluate nuclear data on the experimental data. So, the photo-absorption and photo-neutron data were analysed with theoretical models, and estimation of other quantities were made with the theoretical models.

In the energy region below about 30 MeV, photo-absorption and photo-neutron cross section data show resonance structure. Resonance analyses were made in the energy region. In the higher energy region,experimental data show rather smooth structure, and the quasi-deuteron model was applied.

Analysis Methods and Results

Resonance Region

Photo-absorption cross section of Ahrens et al./1/ and photo-neutron cross sections/2/ were reproduced with the superposition of single-level formula. All resonance levels were assumed to be excited with E1 absorption. Comparison of the calculated results with the experimental cross sections is shown in Fig.1.

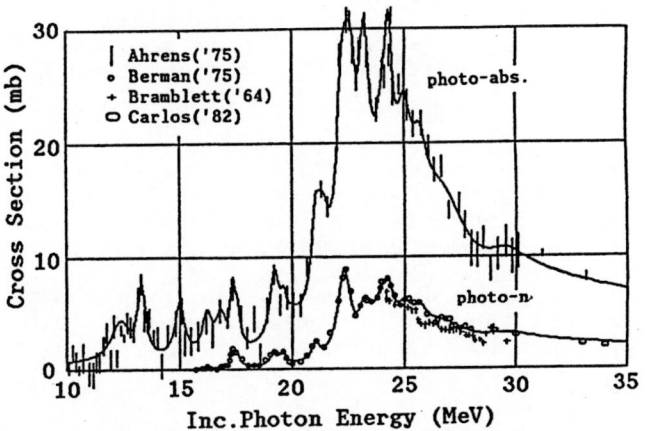

Fig.1 Resonance analysis of O-16 photo-absorption and photo-neutron cross section

Detailed resonance structures were determined also by consulting with photo-proton excitation function data/3/ and the level data compiled by Ajzenberg-Selove/4/. Fitting parameters of resonances are given in Table 1.

For other particle emission cross sections, only single particle emission was considered at present,and these cross sections were obtained, as ordinary compound nucleus decay process, by multiplying the ratio of transmission coefficient of the inverse process to the difference of absorption cross section and neutron emission cross section. The transmission coefficients were calculated for the final nucleus discrete levels up to forth and continuum levels over the fifth level, with the potential of square-well plus centrifugal and Coulomb barrier. The continuum levels were assumed to be given by the constant temperature, spin-dependent level density formula,of which parameters were determined to reproduce the observed level scheme on the average. Calculated cross sections are shown in Fig.2.

Table 1 Resonance parameters to reproduce
O-16 photo-reaction in Fig.1

Res.E. (MeV)	Tot.Width (MeV)	Gamma Ratio	Neutron Ratio
12.40	1.0	7.0E-5	0.0
13.30	0.4	1.5E-4	0.0
15.00	0.5	1.5E-4	0.0
16.22	0.5	1.2E-4	0.05
16.80	0.5	1.2E-4	0.0
17.30	0.2	2.0E-4	0.24
17.50	0.3	1.5E-4	0.32
19.20	0.5	3.0E-4	0.18
19.60	0.3	1.2E-4	0.35
20.00	0.8	8.0E-5	0.0
21.00	0.4	3.0E-4	0.26
21.30	0.8	5.0E-4	0.0
22.25	0.7	1.1E-3	0.45
22.60	0.6	8.0E-4	0.0
23.20	0.6	1.2E-3	0.18
23.50	0.7	1.0E-4	0.60
24.00	0.7	5.0E-4	0.50
24.30	0.4	9.0E-4	0.20
25.00	3.0	1.0E-3	0.23
25.00	0.6	4.0E-4	0.30
25.70	0.6	3.0E-4	0.33
26.80	3.0	8.0E-4	0.26

Res.E.: Resonance energy
Tot.Width : Total resonance width
Gamma Ratio : Ratio of ground state
gamma transition width to total width

Partial cross sections to each level
were also calculated with the same method,
including the continuum states. As an example of the calculation, partial cross
section of neutron emission is shown in
Fig.3.

Angular distributions were calculated with the single level resonance theory given by Simon/5/.

High Energy Region

Photo-absorption cross section/1/
and photo-neutron cross section/6/ data
show rather smooth trend in high energy
region above 30 MeV. The quasi-deuteron
model proposed by Levinger/7/ was adopted
to reproduce the cross sections in the
high energy region. Theoretical cross
section of deuteron disintegration/8/ was
normalized to experimental data of Ahrens
et al./1/ above 35 MeV. Below 35 MeV,
tail of the resonance region was added.
The quasi-deuteron cross section was multiplied by Lorentz type function to decrease smoothly, below 30 MeV, to zero at
about 22 MeV.

Photo-neutron cross section, in high
energy region, were reproduced well by
multiplying factor 0.3 to the photo-absorption cross section. Figure 4 shows
comparison between experimental cross
sections and results of present calculation.

As it is inferred from the nature
of quasi-deuteron model, photo-proton
cross section is assumed to be the same
value as that of photo-neutron. Other
particle emission cross sections were
obtained by sharing the rest of cross
section according to the ratios of transmission coefficients.

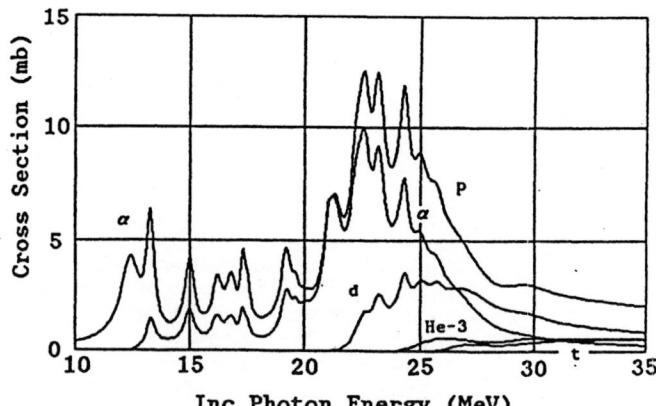

Fig.2 Calculated charged particle
emission cross sections from
O-16 photo-reaction (Each cross
section includes multi-particle
emission process,if any.)

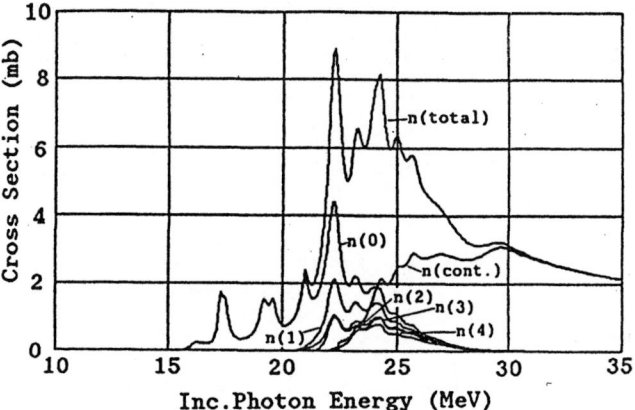

Fig.3 Calculated photo-neutron partial
cross sections of O-16

Analytical model of angular distribution and energy spectrum of the quasi-
deuteron model are given by Levinger/7/
originally and modified by Dedrick/9/.
Present calculation were made with the
same method as that of Dedrick,essential-
ly. The model describes that quasi-deuteron in the nucleus is moving and absorbs
photon, then dissociate itself into neutron and proton. So, angular distribution
and energy spectrum of emitted particles
are given by modulation of those of deuteron photo-reaction with quasi-deuteron
motion. For the motion of neutron or pro
ton in the quasi-deuteron, Levinger assumed Fermi distribution with a temperature of 8 MeV, and Dedrick adopted Gaussian distribution with width of 16 MeV.
Dedrick included the effect of potential
barrier penetration of emitted particles.

Kaushal et al./10/ measured spectra
of photo-neutrons emitted from 18 ele-
ments(Li to U,including O) at observation
angle of 67.5°,by photon differnce method
of 55 MeV and 85 MeV bremsstrahlung. The
results show that the spectra decrease
monotonously with neutron energy, for
medium-weight and heavy nuclei, as it is
predicted by the theory. However,for
light elements, such as Oxygen, spectra
show the broad peak at about neutron

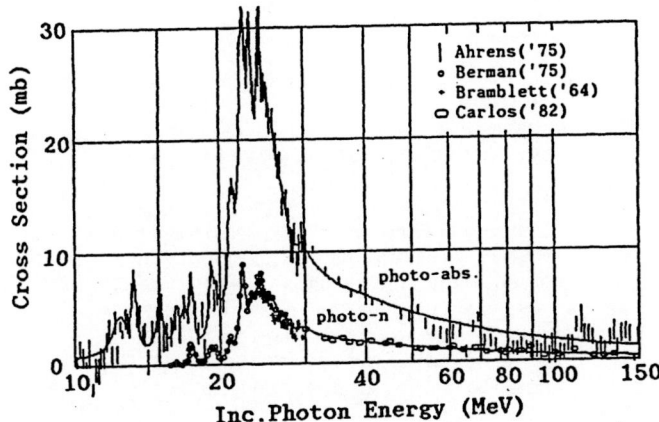

Fig.4 O-16 photo-absorption and photoneutron cross sections reproduced
with resonance and quasi-deuteron
model (solid lines).

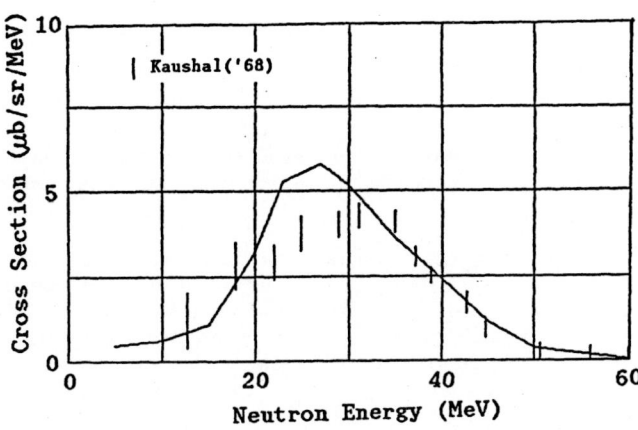

Fig.5 O-16 photo-neutron energy spectrum,
(solid line:present quasi-deuteron
model calculation)

energy of 30 MeV,and show somewhat different shape from nucleus to nucleus,
which can not be explained by the former
theory.

Present calculations of energy spectrum and angular distribution were made
by assuming that the quasi-deuterons are
excited by absorption of photon and then
disintegrate. The energy distribution of
excited quasi-deuterons determines, mainly, energy spectrum and angular distribution of emitted particles. For heavy
nuclei many quasi-deuterons are excited
and the distribution is statistical. But,
for light nuclei, number of quasi-deuteron is small and the distribution of
excitation differs from nucleus to nucleus. Calculated energy spectrum is shown
in Fig.5, comparing with the experimental
data of Kaushal et al. The excitation
distribution were assumed to be Lorentzian with width of 8 MeV and centered at
about a half of incident photon energy.
Figure 6 shows, as an example, calculated
energy and angular distribution of neutrons emitted by photo-reaction of O-16
with mono-energitic photons of energy 100
MeV.

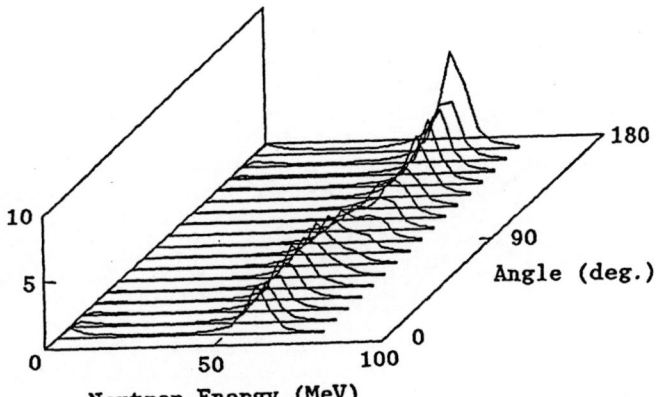

Fig.6 O-16 photo-neutron energy and
angular distribution calculated
with present quasi-deuteron model
for 100 MeV monoenergitic photons.

Conclusion

To evaluate nuclear data of O-16
photo-reactions, resonance analysis and
quasi-deuteron model analysis were made
for photo-absorption and photo-neutron
cross sections. Other photo-reaction data
were calculated with the theoretical models. At present,though the multi-particle
emission process was not considered,cross
sections of the process will be calculated using the present particle emission
cross section and branching ratio obtained with transmission coefficients.

For quasi-deuteron model, a new concept is presented to explain spectrum of
emitted particles from photo-reaction of
light nucleus. Further theoretical study
should be made on the present model.

REFERENCES

/1/J.Ahrens,H.Borchert,K.H.Czock,
H.B.Eppler,H.Gimm,H.Gundrum,M.Kroning,
P.Riehn,G.SitaRam,A.Zieger and
B.Ziegler:Nucl.Physics A251,479(1975)
/2/B.L.Berman:A.D.&N.D.Tables 15,No.4
(1975),R.L.Bramblett,J.D.Cadwell,R.R.
Harvey and S.C.Fultz:Phys.Rev.133,B869
(1964),J.W.Jury,B.L.Berman,D.D.Faul
and J.D.Woodworth:Phys.Rev.C21,503
(1980),B.L.Berman,J.W.Jury,R.E.Pywell,
K.G.McNeill,M.N.Thompson and
J.G.Woodworth:Phys.Rev.C27,1(1983),
A.Veyssiere,H.Beil,R.Bergere,P.Carlos,
A.Lepretre and A.DeMiniac:Nucl.Physics
A227,513(1974)
/3/V.P.Denisov,A.P.Komar and
L.A.Kulchitsky:Nucl.Physics A113,289
(1968)
/4/F.Ajzenberg-Selove:Nucl.Physics A460
(1988)
/5/A.Simon:Phys.Rev.92,1050(1953)
/6/P.Carlos,H.Beil,R.Bergere,B.L.Berman,
A.Lepretre and A.Veyssiere:
Nucl.Physics A378,317(1982)
/7/J.S.Levinger:Phys.Rev.84,43(1951)
/8/J.F.Marshall and E.Guth:
Phys.Rev.78,738(1950)
/9/K.G.Dedrick:Phys.Rev.100,58(1955)
/10/N.N.Kaushal,E.J.Winhold,P.F.Yergin,
H.A.Medicus and R.H.Augustson:
Phys.Rev.175,1330(1968)

PHENOMENOLOGICAL NUCLEON-NUCLEON K-MATRIX PARAMETRIZATIONS UP TO 1 GEV

S. Morioka, T. Ueda*, and H. Kadotani

Century Research Center Corporation, 1-3-D17, Nakase, Chiba 261-01, JAPAN
*Faculty of Engineering Science, Osaka University, Toyonaka, Osaka, JAPAN

Abstract: Nucleon-nucleon amplitudes are parametrized to reproduce the empirical inelasticity parameters by K-matrix method up to 1 GeV. Below the π production threshold energy the one-boson-exchange model is employed, and above this energy the phenomenological amplitudes for each partial waves are introduced to describe the nucleon-nucleon inelasticities. We obtained a set of parameters for the phenomenological amplitudes with which the empirical inelasticity parameters are well reproduced. The present parametrized amplitudes lead to a phenomenological nucleon-nucleon optical potential up to 1 GeV except the low partial waves.

(nucleon-nucleon scattering, K-matrix, π production, inelasticity)

Introduction

The field of nuclear data evaluation in the medium energy region has been increasing in view of need for accelerator development and spallation research [1]. Detailed study of nucleon-nucleon (NN) processes is thus essential because one needs knowledges of the NN processes in many cases for evaluation of the nuclear reactions.

Below the π production threshold energy (hereafter referred to as Region I) several realistic NN potentials [2, 3] which accurately reproduce the available NN data have been nearly established. On the other hand, above the π production threshold energy (Region II) no such a practical NN potential is obtained since the NN processes in Region II involve few-body problem which is very complicated even in theoretical approaches [4].

Here we would like to present a method to obtain a phenomenological NN potential which is capable of describing the NN processes up to 1 GeV. In order to do so, we work on K-matrix method and parametrize the NN amplitudes for each partial waves by assuming that mainly the NΔ intermediate state contributes to the NN processes in Region II [4, 5].

Method

We employ K-matrix method,

$$S_{\ell J} = \frac{1 + i a_{\ell J}/2}{1 - i a_{\ell J}/2} \qquad (1)$$

for uncoupled states (otherwise see [6]), where ℓ and J respectively indicate angular momentum and the total angular momentum for a certain channel, and $a_{\ell J}$ are the NN amplitudes for this channel. In this report we employ the NN amplitudes $a_{\ell J}^{B}$ of the one-boson-exchange model (OBEM) [6], so that these amplitudes are real numbers. In Region II however we need imaginary amplitudes to describe the inelasticities associated mainly with π productions. Let us take,

$$a_{\ell J} = a_{\ell J}^{B} + a_{\ell J}^{Ph} \qquad (2)$$

Where $a_{\ell J}^{Ph}$ are our phenomenological NN amplitudes (complex numbers), and we assume them as the following form,

$$a_{\ell J}^{Ph} = N(E) \sum_{\ell'} \left(g_{\ell J} \frac{E - E_\pi}{E + E_\pi} \right)^{\ell' + 1/2} \frac{\xi_{\ell J}(E - E_\Delta) + i\Gamma_{\ell J}/2}{(E - E_\Delta)^2 + \Gamma_{\ell J}^2/4} Q_\ell(x) \qquad (3)$$

$$x = 1 + \mu^2/2p^2 \qquad (4)$$

where N(E) is a trivial kinematical factor $(E+M)^2/2pE$, E and p are the incident nucleon energy and momentum respectively, E_π and E_Δ the threshold energies of π and Δ productions respectively. Q_ℓ is the Legendre function of the second kind. $g_{\ell J}$, $\xi_{\ell J}$ and $\Gamma_{\ell J}$ are our parameters to seek by χ^2 fitting to the empirical phase and inelasticity parameters [7]. ℓ' are possible angular momenta [8] of the NΔ system which is described by 2π-exchange box diagram [5]. We take thus $\mu = 2m_\pi$ where m_π is the π mass. The threshold behavior of eq. (3) is required of a quasi two-body channel. When $\xi_{\ell J}=1$, the energy dependence of eq. (3) except for the threshold behavior is a simple form of the Breit-Wigner one-level formula.

Under the conservations of parity and the total angular momentum for the incident channel and the NΔ system, we get several possible ℓ' for the incident ℓ. Among the possible ℓ' we take the lowest one, assuming that the state with the lowest angular momentum dominates in the NΔ system.

Since the real part of the amplitudes except for the low partial waves (S and P) do not sensitively contribute to the inelasticities, we employ a simple OBEM where we consider only four bosons π, ρ, ω and a fictitious scalar boson, and also we employ the damping factor [2] for the vector bosons ρ and ω. The damping factor remedies the divergent behaviors of the amplitudes in the high energy limit. We seek coupling constants of these bosons, mass of the scalor boson, and the damping factor (OBEM parameters) by χ^2 fitting to the empirical phase parameters in Region I. With these obtained OBEM parameters we then seek $g_{\ell J}$ and $\Gamma_{\ell J}$ by χ^2 fitting to the empirical inelasticity parameters in Region II. The rest parameter $\xi_{\ell J}$ is finally sought by χ^2 fitting to the phase parameters in Region II.

Results

Figs. 1~3 show our results with the parameters given in Table 1 and 2. For P waves we set $\xi_{\ell J}=0$, and we do not consider S waves because K-matrix method is not reliable for these low partial waves. As for $^1D_2$, $^3F_3$, $^3P_0$ and $^3P_2$, we took $\ell'=1$ (0), 2 (1), 3 (1) and 2 (1) respectively where the numbers in the parentheses are the expected values (see the previous section). These partial waves involve either resonances [4] or large phase shifts and inelasticities, so that the form (3) may be too simple to possess the theoretical features.

The resonance structure in the phase parameters of $^1D_2$ is clearly seen from our result (see Fig. 2). No such structure is observed without the real part of the phenomenological amplitudes.

Summing up the partial waves J=1~6, we obtain the total inelastic cross sections. They are shown in Fig. 3 where we can see that our results agree with the empirical data very well. The declining tendency of the total inelastic cross sections above 1 GeV is due to our form (3) where only the NΔ system is considered. To describe the NN processes above 1 GeV one needs to consider the $\Delta\Delta$ system [5] which involves the partial waves of the total isospin I=0 in addition to the present cases of I=1.

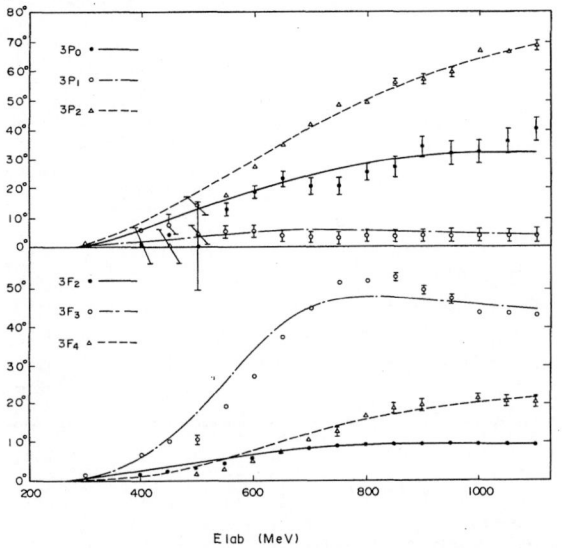
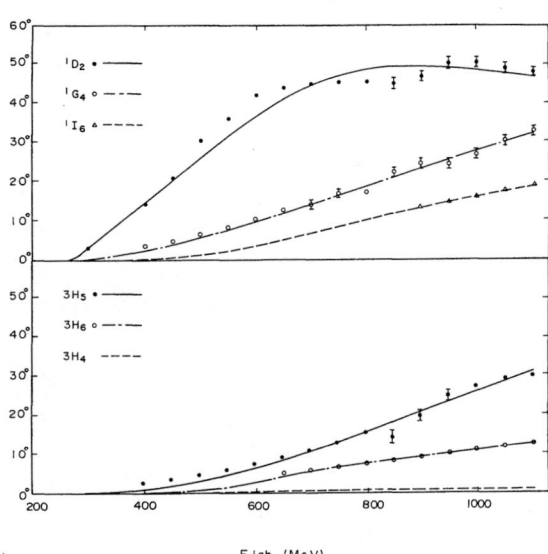

Fig. 1   Inelasticity Parameters. Circles indicate the empirical ones [7]. The curves correspond to the present results. The data for $^3P_1$ and $^3P_2$ are overlapping at 400 MeV. The error bar at this energy is for $^3P_2$. In spite of no imaginary part of our phenomenological amplitude for $^3H_4$, the inelasticity parameters for this channel are not completely zero because of the coupling to the state of $^3H_6$.

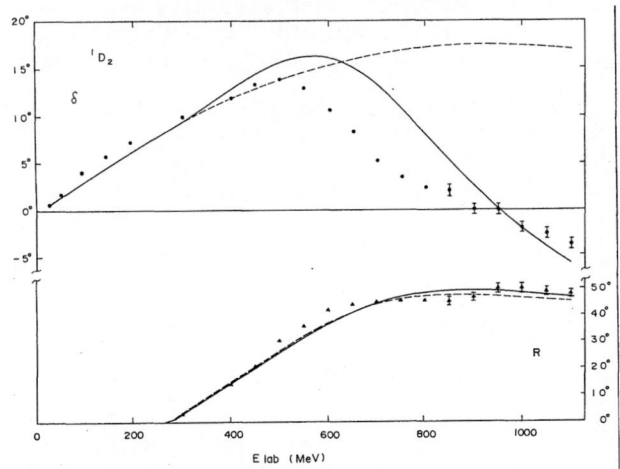
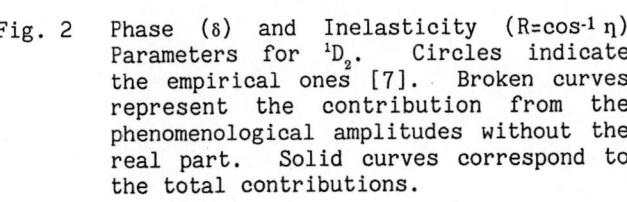
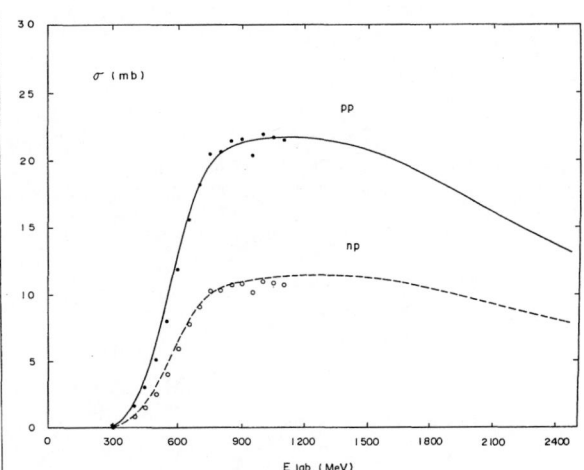

Fig. 2   Phase (δ) and Inelasticity (R=cos^{-1} η) Parameters for $^1D_2$. Circles indicate the empirical ones [7]. Broken curves represent the contribution from the phenomenological amplitudes without the real part. Solid curves correspond to the total contributions.

Fig. 3   The Inelastic Total Cross Sections. Circles indicate the empirical ones [7]. Solid curves correspond to the present results.

Table 1 OBEM Parameters. The asterisk indicates fixed parameters.

Boson	Mass (MeV)		Coupling constant	Damping factor Ω(MeV) Ref. 2
π	138.7*		13.8	x
ρ	759.1*		0.82	3250
		(f/g)	−2.40	
ω	782.8*		9.00	3250
		(f/g)	0.11	
s	591.0		9.80	x

Table 2 Parameters for Phenomenological Amplitudes.

	I = 1		
State	ξ	g	Γ
$^3P_0$	0	17.7	722.1
$^3P_1$	0	0.12	92.6
$^3P_2$	0	6.60	276.1
$^1D_2$	−1.60	3.38	150.5
$^1G_4$	1.85	7.96	691.7
$^1I_6$	0.75	8.14	131.1
$^3F_2$	−9.60	0.51	174.0
$^3F_3$	−0.45	7.61	87.4
$^3F_4$	−0.43	5.25	128.3
$^3H_4$	0	0	x
$^3H_5$	1.51	8.78	334.1
$^3H_6$	0.30	7.55	69.2

Discussion

Our phenomenological amplitudes can be used as a phenomenological NN potential. By solving Schrödinger equation with this potential, for high partial waves one would obtain the similar results to the present ones, whereas for low partial waves such as P waves the results may not be as good as the present ones. Thus one must reseek parameters for these low partial waves. In addition to that, as real part of the amplitudes (or potential) one requires to use the best available NN potentials [2, 3] rather than the present simple OBEM in order to refine the present NN phenomenological potential.

Acknowledgement

One of the authors (S.M.) would like to express his gratitude to the Paris potential group of l'Institut de Physique Nucléair at Orsay. The present work was essentially initiated while he joined in this group.

References

[1] S. Pearlstein, in Proc. Int. Conf. on Nuclear Data for Science and Technology, May/June 1988, Mito, Japan (Edited by S. Igarasi), p. 1115, Saikon, Tokyo (1988).

[2] T. Ueda and A.E.S. Green, Phys. Rev. C18, 337 (1978), C8, 2061 (1973).

[3] M. Lacombe et al., Phys. Rev. C21, 861 (1980).
R. Machleidt et al., Phys. Rep. 149, 1 (1987).

[4] T. Ueda, Nucl. Phys. A463, 69c (1987), and references cited therein.

[5] M. Lacombe et al., in Book of Abstracts Vol. 1 of Int. Conf. on Particles and Nuclei, 1984, Heidelberg, Germany (Edited by F. Güttner et al.), p. C10, Springer-Verlag.

[6] S. Ogawa et al., Prog. Theor. Phys. Suppl. 39, 140 (1967).

[7] R.A. Arndt et al., Phys. Rev. D35, 128 (1987).

[8] S. Mandelstam, Proc. Roy. Soc. (London) A244, 491 (1958).

RATES OF THE MAIN THERMONUCLEAR REACTIONS

S.N.Abramovich, B.Ja.Guzhovskii, S.A.Dunaeva
E.F.Fomushkin

All-Union Scientific Research Institute Of Experimental Physics
607200, Arzamas-16, Nizhnij Novgorod region, USSR

Abstract:The data on the cross sections of main thermonuclear reactions have been estimated with an account of the latest experimental results in a form of S-factor spline presentation. Based on this estimation, the rates of these reactions in 0.0001-1 MeV temperature range in the supposition of Maxwell distribution of relative velocities have been computed. The Maxwell-Boltzmann averaged <S>-factors were calculated according to the table values of the reaction rates. Then the <S>-factors were approximated with the 3 order spline-function. The necessity of the account of electron shielding and intramolecular movement at low temperatures is discussed.

[Rate of reaction, numerical integration, Maxwellian S-factor, spline-function]

Introduction

Reaction rate, i.e., number of reaction events per unit time in thermonuclear plasma volume unit at unit volume densities of interacting particles is determined by:

$$\langle \sigma V \rangle = \int_0^{\infty} \sigma \cdot V \cdot f(V,T)\, dV, \qquad (1)$$

where σ is reaction cross section, V - relative velocity of interacting particles, T - plasma temperature and $f(V,T)$ is normalized distribution function over relative rates.

For a thermodynamic equilibrium system the relative rate distribution has the form of Maxwellian distribution:

$$f(V,T) = 4\pi \left(\frac{M}{2kT}\right)^{3/2} V \cdot \exp\left(-\frac{MV^2}{2kT}\right), \qquad (2)$$

where M is the reduced mass of interacting particles and k is the Boltzmann constant.

Then, Eqn. (1) may be written in the form:

$$\langle \sigma V \rangle = \frac{(8/\pi)^{1/2}}{M^{1/2}(kT)^{3/2}} \int_0^{\infty} \sigma \cdot E \cdot \exp\left(-\frac{E}{kT}\right) dE \qquad (3)$$

where E is the energy of the interacting particles relative motion.

Calculations

To calculate $\langle \sigma V \rangle$ from expression (3) we used estimated data on reaction cross-section from ref. [1], taking into account the latest experimental data. Error due to numerical integration didn't exceed 0.05%. The main contribution to the error of the obtained $\langle \sigma V \rangle$ values is due to the errors of reaction cross-section absolute values. Tabulated calculational results are given in ref.[2]. The representation of reaction rates through analytic or piecewise analytic functions is of considerable interest. Such attempts have been made repeatedly [3-7], however, description accuracy in a wide temperature range (0.0001-1 MeV) even for the most successful work [4] is worse, than the accuracy of the obtained $\langle \sigma V \rangle$ values. In this connection we searched for an approximation, easy to use in computed calculations and which provides the tabulated representation accuracy. For that reason, we have used spline-approximation of the Maxwellian S-factor, associated with $\langle \sigma V \rangle$ by the ratio (see ref.[8]):

$$\langle \sigma V \rangle = \frac{8\pi \langle S \rangle}{3^{5/2}\pi Z_1 Z_2 e^2 M}\left(1+\frac{5}{12\tau}\right)\tau^2 \exp(-\tau),$$

$$\tau = \frac{3(\pi e^2 Z_1 Z_2 M^{1/2})^{2/3}}{(2kT\hbar^2)^{1/3}}, \qquad (4)$$

where Z_1, Z_2 are atomic numbers of the colliding particles and $\langle S \rangle$ is Maxwellian S-factor.

Calculated from this ratio values $\langle S \rangle$ for the considered reactions were approximated by the 3rd order spline-functions. First, we have searched for knots location for each reaction using maximum likelihood method starting from their minimum number at specified maximum deviation of spline-description from tabulated representation, which doesn't exceed 1% . This search technique has been treated in details elswhere [1]. The results are given in Table 1.

TABLE 1. SPLINE-FUNCTION COEFFICIENTS FOR <S>

Knot №	Knot temp. MeV	Spline-function coefficient			
		A_0	A_1	A_2	A_3
#	reaction ^{3}H(d,n)^{4}He				
1	.0001	12.119	.46066	-.47387	.20145
2	.00522	18.994	6.1672	1.9166	-5.0665
3	.01158	22.562	-.41778	-10.186	3.1819
4	.05001	10.111	-9.7910	3.7801	-.54457
5	.31210	1.5133	-1.4253	.78847	-.18110
6	1	.63680	0	0	0
	reaction ^{2}H(d,n)^{3}He				
1	.0001	.05056	.00187	-.00107	.00023
2	.00658	.05630	.00486	.00179	.00191
3	.02822	.07305	.02218	.01012	.00083
4	1	.31957	0	0	0
	reaction ^{2}H(d,p)^{3}H				
1	.0001	.05235	.00151	-.00073	.00012
2	.00276	.05361	.00052	.00043	.00076
3	.05405	.07898	.02327	.07202	.00295
4	1	.28142	0	0	0

TABLE 1. Continued.

K n o t #	Knot temp. MeV	Spline-function coefficient			
		A_0	A_1	A_2	A_3
		reaction $^3H(t,2n)^3He$			
1	.0001	.18241	-.00088	.00050	-.00016
2	.02192	.16762	-.00918	-.00204	.00822
3	.17825	.21503	.09051	.04961	-.02444
4	1	.03949	0	0	0
		reaction $^3He(d,p)^4He$			
1	.0001	6.2408	.04226	-.05089	.01971
2	.00329	6.6079	.40847	.15572	.34387
3	.01658	9.1318	3.6118	3.9326	-3.2796
4	.05163	13.502	-.14666	-7.2418	2.4997
5	.19031	6.5365	-6.2795	2.5423	-.37841
6	1	1.3880	0	0	0
		reaction $^3He(^3He,2p)^4He$			
1	.0001	5.2637	-.00245	.00715	-.00296
2	.00504	5.1860	-.08263	-.02761	-.00937
3	.06412	4.6434	-.40481	-.09911	.09840
4	.50227	4.2484	.43800	.50855	-.33303
5	1	4.6824	0	0	0

<S> values can be calculated from the following expression:

$$\langle S \rangle = \sum_{k=0}^{3} A_k [\ln(\frac{kT}{kT_j})]^k, \qquad (5)$$

where A_j is spline-function coefficients, corresponding to j knot, kT_j is j knot temperature, MeV; and kT is temperature running value, corresponding to the condition $kT_{j-1} < kT < kT_j$.

After that, we seached for the common knot set for all reactions with the requirement, that the maximum deviation from tabulated values should not be in excess of 1%. We obtained good results at knot number of 10. The results obtained are given in Table 2.

<S>-values can be calculated by formula (5), and <σV> by formula (4). Relative deviations δ_i of our <σV> data from the results of other known works $\langle \sigma V \rangle_i$

$$\delta_i = \frac{\langle \sigma V \rangle - \langle \sigma V \rangle_i}{\langle \sigma V \rangle} \cdot 100\%$$

are given for comparison in Figs. 1-6.

TABLE 2   SPLINE-FUNCTION COEFFICIENTS FOR S-FACTORS WITH CONSTANT NET

K n o t #	Knot temp. MeV	Spline-function coefficient			
		A_0	A_1	A_2	A_3
		reaction $^3H(d,n)^4He$			
1	.0001	12.247	-.13220	.09689	.04996
2	.0005	12.494	.56788	.33809	.25758
3	.005	18.739	6.2218	2.1174	-5.0533
4	.01	22.386	1.8735	-8.3907	.47143
5	.016	21.462	-5.7014	-7.7260	3.4843
6	.05	10.089	-9.7367	4.1845	-1.3732
7	.066	7.6789	-7.7307	3.0407	-.37066
8	.16	2.9602	-3.2175	2.0560	-.70315
9	.3	1.5754	-1.4661	.73002	-.13010
10	1	.64140	0	0	0

TABLE 2. Continued.

K n o t #	Knot temp. MeV	Spline-function coefficient			
		A_0	A_1	A_2	A_3
		reaction $^2H(d,n)^3He$			
1	.0001	.05122	-.00084	.00130	-.00035
2	.0005	.05176	.00059	-.00042	.00033
3	.005	.05496	.00394	.00187	.00107
4	.01	.05895	.00807	.00409	.00100
5	.016	.06375	.01258	.00551	.00242
6	.05	.08882	.03457	.01379	-.00036
7	.066	.09947	.04214	.01348	-.00032
8	.16	.14714	.06526	.01263	-.00084
9	.3	.19336	.08214	.01422	.00385
10	1	.31958	0	0	0
		reaction $^2H(d,p)^3H$			
1	.0001	.05281	-.00065	.00139	-.00045
2	.0005	.05347	.00031	-.00079	.00034
3	.005	.05421	.00215	.00159	.00080
4	.01	.05673	.00550	.00324	.00054
5	.016	.06008	.00891	.00409	.00113
6	.05	.07711	.02244	.00787	.00041
7	.066	.08395	.02690	.00821	.00326
8	.16	.11647	.04909	.01686	.00311
9	.3	.15476	.07397	.02272	.00258
10	1	.28126	0	0	0
		reaction $^3H(t,2n)^4He$			
1	.0001	.18182	.00216	-.00243	.00060
2	.0005	.18152	-.00097	.00048	-.00035
3	.005	.17755	-.00433	-.00194	-.00052
4	.01	.17345	-.00777	-.00302	.00203
5	.016	.16934	-.00926	-.00016	.00366
6	.05	.16400	.00465	.01236	.01076
7	.066	.16648	.01400	.02132	.01348
8	.16	.20495	.08346	.05713	-.02789
9	.3	.27306	.12222	.00454	-.01861
10	1	.39430	0	0	0
		reaction $^3He(d,p)^4He$			
1	.0001	6.2703	-.13526	.14669	-.38232
2	.0005	6.2732	.03982	-.03791	.05695
3	.005	6.8592	.77115	.35552	.38108
4	.01	7.6915	1.8133	1.1479	1.5223
5	.016	8.9554	3.9012	3.2944	-2.8445
6	.05	13.470	.32946	-6.4290	.47659
7	.066	13.076	-3.1301	-6.0321	2.9924
8	.016	7.6518	-6.7738	1.9174	.09723
9	.3	4.1755	-4.2480	2.1007	-.41184
10	1	1.3874	0	0	0
		reaction $^3He(^3He,2p)^4He$			
1	.0001	5.2665	-.01324	.01624	-.00510
2	.0005	5.2660	-.00059	-.00838	-.00280
3	.005	5.1860	-.08374	-.02773	-.00426
4	.01	5.1132	-.12833	-.03660	-.02197
5	.016	5.0426	-.17730	-.06758	-.01129
6	.05	4.7361	-.37529	-.10618	.11551
7	.066	4.6262	-.40754	-.00998	.02850
8	.16	4.2773	-.35817	.06573	.23116
9	.3	4.1355	-.00150	.50166	-.09779
10	1	4.6902	0	0	0

Subscript i, corresponds to the ordinal number of work, given in references.

Results

In Fig.1 One can see the data for reaction $^3H(d,n)^4He$. It should be noted that among analytic representations [3-6] of the $^3H(d,n)^4He$ reaction rate, Kozlov formula [4] describes our data more exactly; the deviations from our data do not exceed 3% for the most important

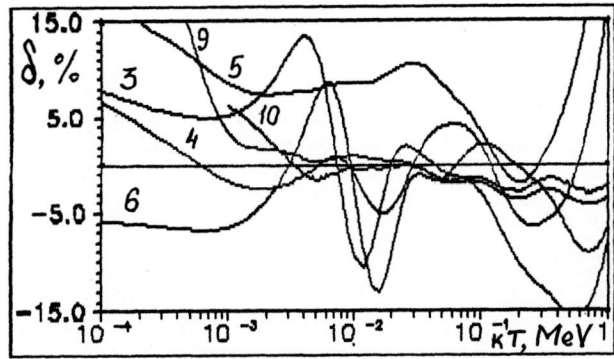

Fig. 1. Reaction rate deviations of ^{3}H(d,n)^{4}He

temperature range 1-100 keV for calculation of thermonuclear processes and devices. The formula obtained in work [6] gives rather poor description of the numerical integration results, in spite of the numerous fitting parameters. All the more, the sense of seach for the new analytic representations [5] is incomprehensible, since they have led to poor results, as compared to the Kozlov formula. It should be noted, that the discrepancy between our results and tabulated data from ECPL library of Livermore National Laboratory [9] and Oak Ridge National Laboratory data [10] does not exceed 5% for temperatures above 1 keV. However, at lower temperatures the discrepancy with ECPL data increases 8 times at temperature of 0.1 keV. The values of ^{3}H(d,n)^{4}He reaction cross section, contained in ECPL, are in good agreement with our estimate. Analysis of all conditions made it possible to conclude, that this discrepancy was the result of inadequate interpolation procedure, used for numerical integration in the calculation of the reaction rates, given in ECPL [9]. Evidently the work [9] used interpolation cross section instead of S-factor, that can lead to serious errors at temperature below 1 keV due to strong energy dependence.

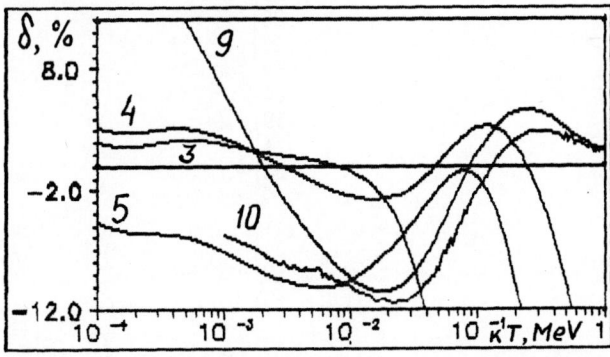

Fig. 2. Reaction rate deviations of ^{2}H(d,n)^{3}He

Data for ^{2}H(d,n)^{3}He reaction are given in Fig.2. It should be pointed out, that Kozlov analytic representation [4] has the best agreement with our data, and as for Putvinski formula [5], it has considerably worse description in the most important temperature range 1-100 keV for calculation of thermonuclear systems. One can see a great difference between our results and Livermore and Oak Ridge National Laboratories

data [9,10]. In the temperature range of 1-100 keV this difference is due to the estimate of the reaction cross section, which we have made, using data of the recent publication [11], that were not available for the authors [9,10], and essentially differ (up to 10 %) from the reaction cross section data, used in ref. [9,10]. Besides, the remarks, made in the previous section, concerning the errors, caused by inadequate interpolation procedure and extrapolation of cross section in the temperature range below 1 keV, remain valid.

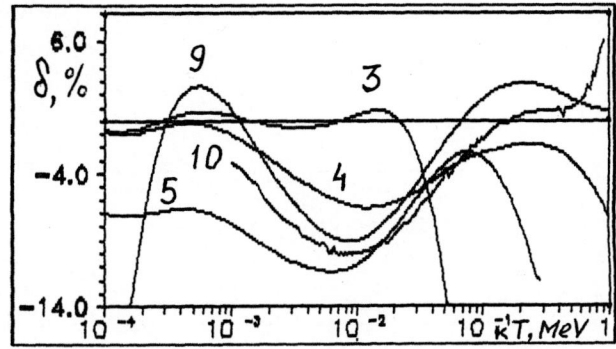

Fig. 3. Reaction rate deviations of ^{2}H(d,p)^{3}H

The data for ^{2}H(d,p)^{3}H reaction are given in Fig.3. The systematic shift of curves, observed in Fig.3 taken from the works [4,5,9,10], as in the case of ^{2}H(d,n)^{3}He reaction, is due to the fact, that in the reaction cross section estimation we have used the data from work [11], which were not available for the authors [4,5,9,10]. These data are taken into consideration in work [3], that's why one can observe a good agreement between our results and the results [3] at temperatures below 50 keV. The discrepancy at higher temperatures is due to the fact, that approximations, used in the work [3], become in valid.

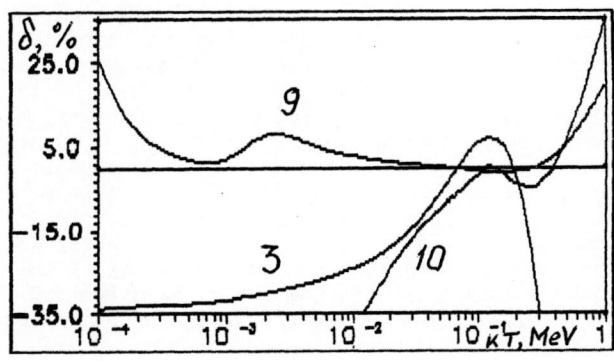

Fig. 4. Reaction rate deviations of ^{3}H(t,2n)^{4}He

The ^{3}H(t,2n)^{4}He reaction data are shown in Fig.4. Our results are compared with that of works [3,9,10].

The data for ^{3}He(d,p)^{4}He reaction are given in Fig.5. One can observe considerable data discrepancy [9], which is, first of all, due to the data for ^{3}He(d,p)^{4}He reaction cross section, used in work [9], that are significantly different from well-known and recent measurements [11].

964

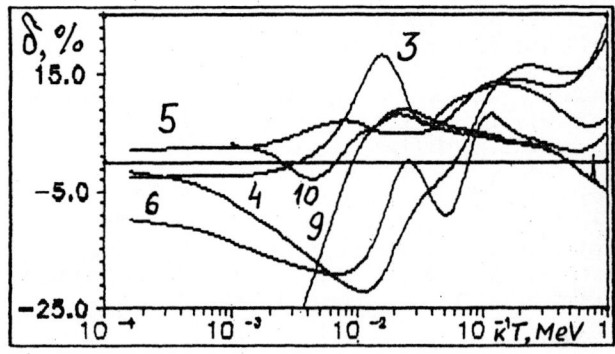

Fig. 5. Reaction rate deviations of ^{3}He(d,p)^{4}He

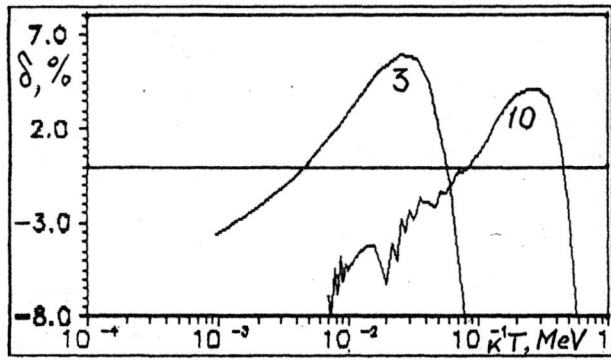

Fig. 6. Reaction rate deviations
of ^{3}He(^{3}He,2p)^{4}He

The data for ^{3}He(^{3}He,2p)^{4}He reaction presented in Fig.6.

CONCLUSIONS

The reaction rates, considered in this paper, were calculated under assumption, that plasma is "hot",i.e., perfect decomposition of molecules and total atomic ionization are believed to have occurred. Then, we can use a cross-section for bare nuclei (i.e., for atoms entirely free of electrons) as the reaction cross section in expression (3). However, at temperatures below 1 keV, even for systems consisting of nuclei with Z=1, this assumption is not satisfied. Hence, calculating reaction rates at so low temperatures we must make allowance for intramolecular motion and electronic shielding. The errors, being the result of neglect of these considerations can be too large, as evidenced by the estimates and experimental investigations [14,15].

REFERENCES

1.Abramovich S., Guzhovskii B., Zherebcov V., Zvenigorodskii A.: Jaderno-fizicheskie konstanty termojadernogo sinteza, spravochnoe posobie, Moskva, CNIIAtominform, 1989.
2.Abramovich S., Guzhovskii B., Dunaeva S.: Izvestija AN SSSR, ser. fiz., 55, 1050, (1991).
3.Caughlan G., Fowler W.: At. Data and Nucl. Data Tables 40, 283,(1988).
4.Kozlov B.: Atomnaja energija, 12, 238, (1962).
5.Putvinskij S.: VANiT, ser.Termojadernyj sintez, 2,3,(1988).
6.Gus'kov S., Il'in D., Levkovskij A., Rozanov

B.: Atomnaja energija, 63, 252, (1987).
7.Hively L., Nucl.Fus.: 17, 873, (1977).
8.Jarmie N., Brown R.: Phys. Rev. C29,2031,(1984).
9.Howerton R., Dye R., McGregor M., Perkins S.: LLNL Evaluated Charged-Particle Library (ECPL), Report UCRL-50400, v. 28, (1986).
10.McNally J., Rothe K., Sharp R.: Fusion Reactivity Graphs and Tables for Charged Particle Reactions, Report ORNL TM-6914, (1979).
11.Krauss A., Becker H., Trautvetter H. Rolfs C.: Nucl. Phys., A465, 150, (1987).
12.Jarmie N., Brown R.: Nucl. Inst. and Meth. in Phys. Res., B10/11, 405, (1985).
13.Krauss A., Becker H., Trautvetter H.: Nucl.Phys., A467, 273, (1987).
14.Assenbaum H., Langanke K., Rolfs C.: Z. Phys. A327, 461, (1987).
15.Engstler S., Krauss A., Neldner K., Rolfs C., Schroder V., Langanke K.: Phys. Lett. B202, 179, (1988).

ANALYSIS OF FISSION NEUTRON SPECTRA
BY NON–EQUITEMPERATURE MADLAND–NIX MODEL

Takaaki Ohsawa and Toshikazu Shibata

Atomic Energy Research Institute, Kinki University
3–4–1 Kowakae, Higashi–osaka, Japan

Abstract: An attempt was made to improve the Madland–Nix model for the calculation of the fission neutron spectrum. In the original Madland–Nix model, statistical equilibrium was assumed to hold between the two fission fragments at the scission point. This was the ground for using a single value of the maximum nuclear temperature for both fragments. However, this assumption seems questionable, since the deformation energies of the fragments at the scission point, which are generally different for the nascent fragments, eventually converts into the internal excitation energy. The author tried to take into account the difference in the nuclear temperature of the light and heavy fragments in a empirical manner. This non–equitemperature model was applied to analyze the fission neutron spectra for neutron–induced fission of Th–232, U–235 and Pu–239 and spontaneous fission of Cf–252. It was found that (i) consideration of the temperature difference had greater effects on the spectral shape than the previous attempts, and (ii) this modification of the model gave better account of the experimental spectra for these nuclides.

(fission neutron spectra, model calculation, nuclear temperature, Th–232, U–235, Pu–239, Cf–252)

Introduction

Recent sensitivity analysis[1] shows that the fission neutron spectrum is one of the important quantities that affect the nuclear characteristics of fast reactors. Recent interests on extended burnup of LWR–fuels and of nuclear incineration add further importance to the fission neutron spectrum data for many actinides at higher as well as at lower incident energies.

The Madland–Nix (MN) model[2] for fission neutron spectrum calculation has been successfully applied to generate evaluated data. However it has been recognized that the MN model underestimates the spectrum in the regions below about 0.5MeV and above 7MeV. Several attempts have been made to improve the model. Walsh et al.[3] took into account the anisotropy of neutron emission in the CM–system. Madland et al.[4] replaced the average values of the fragment mass, charge, and kinetic energy with the distributions themselves. Märten et al.[5] extended the model by considering the mass dependence of the fission quantities in the model.

One of the important assumptions of the MN model is the triangular distribution of the nuclear temperature. This assumption is equivalent to assuming that the excitation energy distribution is uniform, which is appropriate at high excitation energies but become less adequate at low excitation energies. So there may be some room for improvement in this respect.

As an attempt in this direction, we tried to take into consideration the difference in the nuclear temperatures of the two fragments.

The Non–equitemperature Assumption

The original MN model assumes that the same temperature distribution $P(T)=2T/Tm^2$ ($T \leq Tm$) applies to both of the fragments. This would be the case, if the nuclear system were in statistical equilibrium at the scission point, with the excitation energy and level density parameter of each fragment proportional to its mass number ($a=A/C$, $C=$const.). However, it is questionable if this assumption should be valid. Even if equilibrium were established at the scission point, the total excitation energy available for neutron emission is a sum of internal excitation energy $a_i T_{oi}^2$ and the deformation energy D_i at the scission point. Thus,

$$\langle E^*_i \rangle = a_i T_{oi}^2 + D_i = a_i T_{mi}^2 \qquad (1)$$

where T_{mi} is the maximum temperature for fragment i, i=L and H standing for light and heavy fragments, respectively. The deformation energy D_i at the scission point is strongly affected by the nuclear structure of the fragments so that the temperatures T_{mi} for the two fragments are generally not equal. In this respect, it is interesting to note that Wilkins et al.[7] have found that the fragment deformation β (A) at the scission point show a saw–tooth behavior very similar to the neutron multiplicity $\nu(A)$. This suggests that D_i is the major contributor to $\langle E^* \rangle$. This fact accounts for the non-uniform distribution of the nuclear temperature vs mass number A, as was observed in the Geel data[6].

Calculation with the Non-equitemperature Model

The present calculation is essentially based on the formalism of Madland and Nix [2]. The constant compound-formation cross section model was used for the sake of simplicity. The maximum nuclear temperature T_m is approximately related to the average total excitation energy $\langle E^* \rangle$ by

$$\langle E^* \rangle = \langle E_r \rangle + B_n + E_n - \langle E_k \rangle = aT_m^2 \quad (2)$$

where B_n is the neutron separation energy, E_n the incident neutron energy. The total energy release $\langle E_r \rangle$ was calculated according to the seven-fragment approximation[2] using the TUYY mass formula[8]. The total kintic energy(TKE) $\langle E_k \rangle$ of the fragments was taken from measurements[9,17].

Since the nuclear system is not in statistical equilibrium and the excitation energy is not proportional to the fragment mass number, we can write as follows:

$$\langle E_i^* \rangle = (A_i/C)T_{mi}^2, \quad (i=L \text{ or } H) \quad (3a)$$

$$\langle E^* \rangle = \langle E_L^* \rangle + \langle E_H^* \rangle = (A/C)T_m^2. \quad (3b)$$

Then we have

$$A_L T_{mL}^2 + A_H T_{mH}^2 = AT_m^2. \quad (4)$$

Defining the ratio of the temperatures for the light and heavy fragments as $R_T = T_{mL}/T_{mH}$, we obtain

$$T_{mL} = [AR_T^2/(A_L R_T^2 + A_H)]^{1/2}T_m, \quad (5a)$$

$$T_{mH} = [A/(A_L R_T^2 + A_H)]^{1/2}T_m. \quad (5b)$$

Results and Discussion

The non-equitemperature model was applied to analyze the data of the fission neutron spectra for Th-232(n,f), U-235 (n,f), Pu-239(n,f) and Cf-252(sf). The quantities used in the present calculation are summarized in Table 1. Figure 1 compares the spectra of U-235(n,f) for En=0.53 MeV for different ratios R_T of the two maximum temperatures. It was observed that if R_T was taken greater than unity, then both the low- and high-energy components were increased, and the spectrum fits better with the experimental data. The value R_T=1.13 was taken from the Geel data [6]. Though this value was obtained for Cf-252(sf), for lack data for U-235 (n,f), we tentatively used this value. This value was found to give a spectrum in better agreement with the measured data of Johansson[10]. The value R_T=1.34, suggested by Kapoor[11], seems to be too large.

Also for Cf-252(sf), better agreement with experiments[12,13] was obtained by assuming non-equality of nuclear temperatures, although there still remain some discrepancies in the high and low energy ends of the spectrum. The case for Pu-239 (n,f) was rather uncertain, because the two sets of measured data[14,15] showed different behavior in the region above 5 MeV.

For Th-232(n,f), recent experimental data of Baba et al.[16] for En=2MeV were analyzed. The data for TKE and mass distribution of the fragments were taken from Trochon et al.[17]. It was found that the spectral shape was reproduced with a=A/12 and R_T=1.2 (Fig.2). The lower level density parameter compared to other nuclides can be interpreted as follows: since the light fragment peak of Th-232 is shifted toward a shell region close to A≅80, the corresponding level density parameter is less, as can be seen from Fig.3. However, for lack of spectral data below 2MeV, the present analysis should be considered as a preliminary one.

Table 1. Parameter values used in the present calculation.

Quantity	Th-232	U-235	Pu-239	Cf-252
$\langle E_r \rangle$	170.733 MeV	185.896 MeV	198.088 MeV	218.886 MeV*
$\langle E_k \rangle$	163.49 MeV	171.8 MeV	177.1 MeV	185.9 MeV
a	A/12 MeV^{-1}	A/9.6 MeV^{-1}	A/8.5 MeV^{-1}	A/8.0 MeV^{-1}
A_H	141	140	140	144
A_L	92	96	100	108

Note: The value marked with * is that calculated with the Möller-Nix mass formula, as used by Madland et al.[18]. This value was chosen just for comparison purpose. The TUYY mass formula yielded 215.998MeV.

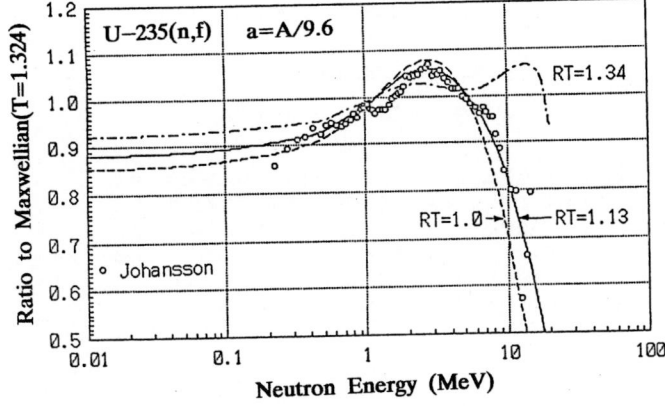

Fig.1 Fission neutron specra for U-235(n,f) for En= 0.53 MeV. The ratios to Maxwellian spectrum with T_M= 1.324 MeV (the value adopted in JENDL-2) are plotted.

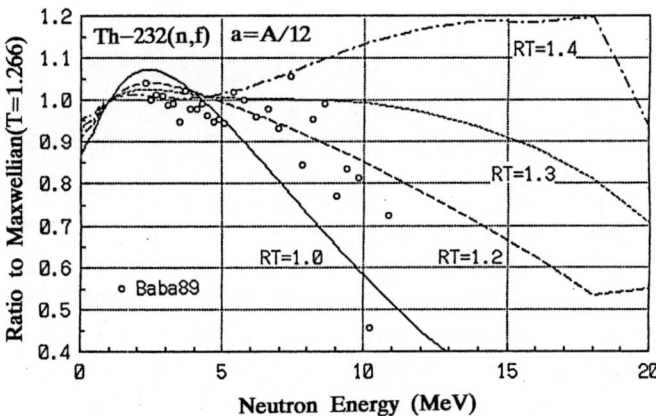

Fig.2 Fission neutron spectrum for Th-232 (n,f) for En=2 MeV.

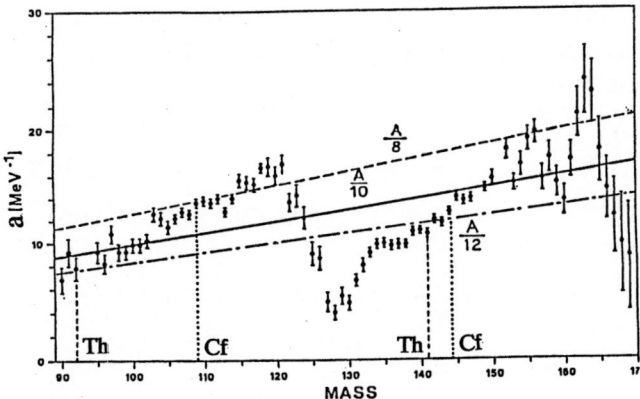

Fig.3 Nuclear level density parameters for fission fragments[6]. The dotted vertical lines indicate apppproximate position of light and heavy fragment peaks for Th-232 and Cf-252.

Concluding Remarks

Based on the present calculations, the following observations can be made: (a) Taking into account the non-equality of the nuclear temperature for the two fragments has considerable effects in improving the calculated fission spectrum in the frame-work of the MN formalism. (b) The best-fit level density parameter " a " tends to decrease as the mass number A of the fissioning nucleus decreases. This is interpreted as due to a shift of the light-fragment peak toward a closed-shell region.

The non-equitemperature MN model should further be tested on other nuclides and at higher incident energies. It would be interesting to know how the temperature ratio changes when the excitation energy of the fissioning system is increased.

References

1. T.Takeda, H.Matsumoto and Y.Kikuchi, J. Nucl. Sci. Technol.27, 581 (1990)
2. D.G.Madland and R.J.Nix, Nucl. Sci. Eng. 81, 213 (1982)
3. R.L.Walsh, Nucl. Sci. Eng. 102, 119 (1989)
4. D.G.Madland, R.J.LaBauve and J.R.Nix, Physics of Neutron Emission in Fission, Proc. of a Consultants' Meeting, Mito, 1988, INDC(NDS)-220, p.259 (1989)
5. H.Märten and D.Seeliger, Nucl. Sci. Eng. 93, 370 (1986)
6. C.Budtz-Jørgensen and H.H.Knitter, Physics of Neutron Emission in Fission, Proc. of a Consultants' Meeting, Mito, 1988, INDC(NDS)-220, p.181 (1989); Nucl. Phys. A490, 307 (1988)
7. B.D.Wilkins et al., Phys. Rev. C14, 1832 (1976)
8. T.Tachibana, M.Uno, M.Yamada and S.Yamada, Atomic Data and Nuclear Data Tables 39, 251 (1988)
9. J.P.Unik, J.E.Gindler, L.E.Glendenin, K.F.Flynn, A.Gorski, and R.K. Sjoblom, Physics and Chemistry of Fission, Int. Symposium, Rochester, Vol.II, p. 15 (1973)
10. P.I.Johansen and B.Holmqvist, Nucl. Sci. Eng. 62, 695 (1977)
11. S.S.Kapoor, R.Ramanna and P.N.Rama Rao, Phys. Rev. 131, 283 (1963)
12. W.P.Poenitz and T.Tamura, Proc. Int. Conf. on Nuclear Data for Science and Technology, p.473 (1983)
13. O.I.Batenkov, M.V.Blinov, G.S.Boykov, V.A.Vitenko, V.A.Rubchenya, Proc. IAEA Consultants' Meeting on the U-235 Fission Cross-Sections and the Cf-252 Fission Neutron Spectrum, Vienna 1983, p.161 (1983)
14. P.I.Johansson, B.Holmqvist, T.Wielding and L.Jeki, Proc. Conf. on Nuclear Cross Sections and Technology, Washington, Paper GB-8 (1975)
15. H.H.Knitter, Atomkernenergie 26, 76 (1975)
16. M.Baba et al., JAERI-M 89-143 (1989)
17. J.Trochon, H.Abou Yehia, F.Brisad and Y.Pranal, Nucl. Phys. A318, 63 (1979)
18. D.G.Madland, R.J.LaBauve and J.R.Nix, Nuclear Standard Reference Data, Proc. of IAEA/AG Meeting, Geel, 1984, IAEA-TECDOC-335 p.267 (1985)

MODELING OF CROSS SECTIONS IN THE UNRESOLVED RESONANCE REGION *

M. Alami, N. Koyumdjieva, A. Lukyanov, N. Janeva

Inst. Nucl. Res. Nucl. Energy, Blvd. Lenin 72, 1784 Sofia, Bulgaria

Abstract: In the unresolved resonance region the statistical modeling of the R-matrix elements is used. We solve this problem by determination of the distribution functions of the R-matrix parameters. We determine the characteristic function, which is common for all functionals of the cross sections. For any modeling of cross sections in this region the average functional $\langle F \rangle$ will be calculated with the same accuracy as the corresponding characteristic function.

(statistical R-matrix, a characteristic function of the R-matrix elements)

Following the resonance theory of nuclear reactions the collision matrix and the cross sections are to be constructed as appropriate combinations of R-matrix elements:

$$R_{cc'} = \frac{1}{2} \Sigma \Gamma_{\lambda c}^{1/2} \Gamma_{\lambda c'}^{1/2} / (E_{\lambda} - E - i\Gamma_{\gamma}/2) \qquad (1)$$

where $\Gamma_{\lambda c}$ are the resonance widths in the reaction channels c, E_{λ} -resonance energies, Γ_{γ} radiative capture width, supposed to be the same for all the resonance levels λ [1]. In the region of experimentally resolved resonances it is possible to find all resonance parameters in the sum (1). In the unresolved resonance region only mean values of the resonance parameters may be evaluated from analyses of the cross sections. For modeling of the resonance structure in this region the statistical R-matrix (1) is used, where the resonance parameters are distributed with appropriate statistical laws (Porter-Thomas distribution for $\Gamma_{\lambda c}$ and Wigner distrubution for E_{λ} [1]).

Meanwhile it becomes possible to establish also the distribution of R-matrix elements directly. Using the information theory the probability distribution of (nonreduced) R-matrix elements has been already established [2].

We have found [3] the characteristic function for the distribution of the diagonal elements R_{cc} in the form:

$$P = \langle \exp i(R_{cc}t - R_{cc}^*t') \rangle = \qquad (2)$$

$$\frac{1}{2} \int_{-\pi/2}^{\pi/2} dx \left[\frac{\sin(x-iy) \sin(x+iy)}{\sin(x-if-i\beta) \sin(x+if-i\beta)} \right]$$

where $y = \pi\Gamma_{\gamma}/2D$ (D-mean resonance spacing), $f^2 = \beta^2 + 2\gamma y + y^2$, $\beta = S_c(t-t')$, $\gamma = S_c(t+t')$, $S_c = \pi\Gamma_c/2D$ (Γ_c-average width). This is a result of averaging over the resonance parameter distributions and the energy in the R_{cc} (1).

The real function P depends on three parameters - y, β and f; $(0 < y, \beta < \infty, (\beta+y) < f < \infty)$.

For $f = \beta + y$ (t'=0) $P = \exp(-\beta)$ (3)

For $f \to \infty$ (4)

$P \to 2/\pi \exp(-(f-y)(1+\exp(-2y))) E(1/chy)$ where E is the complete elliptical integral of the second kind [3].

In order to express P in terms of special functions in the general case we reduce the integral (2) to the form:

$$P = \frac{2}{\pi} \int_0^{\pi/2} d\varphi \frac{[A \cdot B - sh^2\beta \cos^2\varphi \sin^2\varphi]^{1/2}}{ch^2 f - \cos^2\varphi} \qquad (5)$$

$A = (ch^2 f - \cos^2\varphi)$; $B = (ch^2 y - \cos^2\varphi)$

Then changing the variables as:

* Research sponsored by IAEA under contract No. 6452/RB.

$$thf=\sin\psi_1/\sin\psi_2; \quad thy=\cos\alpha\sin\psi_1\sin\psi_2$$

$$th\beta\cos\psi_1\sin\psi_2(1-\sin^2\psi_1\cos^2\alpha)^{1/2}=$$

$$\cos\psi_1\sin\psi_1(1-\sin^2\psi_2\cos^2\alpha)^{1/2} \qquad (6)$$

we obtain the difference of elliptical integrals:

$$P=ch\beta\, \Lambda_{(\psi_1,\alpha)} - sh\beta\, \Lambda_{(\tilde{\psi}_2,\alpha)} \qquad (7)$$

Λ is the complete elliptical integral of the third kind, depending on ψ, which is related to $\tilde{\psi}$ by $tg\psi\ tg\tilde{\psi}\cdot\sin\alpha=1$. The parameters in (7) can be expressed by y, β and f:

$$\sin\psi_1 =ch\beta(1+K)/[ch(f-y)+Kch(f+y)]$$
$$\sin\psi_2 =sh\beta(1+k)/[sh(f-y)+Ksh(f+y)]$$

$$K=\frac{1-\cos\alpha}{1+\cos\alpha}=\left[ch^2(f+y)-ch^2\beta\right]^{-1/2}\cdot$$

$$\left[ch^2(f-y)-ch^2\beta\right]^{1/2} \qquad (8)$$

On the other hand P is related to an integration of the function Φ [3]:

$$\Phi= -\frac{2}{sh2f}\,\frac{\partial F}{\partial f} =2\theta(K) \cdot$$

$$\left[\sqrt{sh(f+\beta+y)sh(f-\beta+y)}\,/sh^2 2f\right] \qquad (9)$$

where $\theta(K)=\frac{2}{\pi}(1+K)\,E\left(\frac{2\sqrt{K}}{1+K}\right)$

Using the asymptotic behaviour of P (4) and the parameter $z=f-\beta-y$ we have calculated the product

$$N=\exp(z+\beta)P=e^z\int_z^\infty e^{-z}M(z')dz' \qquad (10)$$

where $M(z')=\left[1-e^{-2(z+2\beta+2y)}\right]^{1/2}\cdot$

$$\left[1-e^{-2(z+2y)}\right]^{1/2}\left[1-e^{-4(z+\beta+y)}\right]^{-1}\theta(K^2)$$

This method demonstrates the transition to the asymptotic form (4) at $z\gtrless2$, where $k\to\exp(-2y)$. The results from numerical calculations by different methods are identical. Nevertheless we prefer the method with elliptical integrals (7), and a rapidly converging scheme of arithmetical-geometric means calculations has been found for it [4].

Knowledge of the characteristic function P gives the possibility to find any averaged physical functional F. Let's suppose that F may be written as a Laplace transformation of two variables:

$$F(1-iR, 1+iR^*)=\int_0^\infty dt\int_0^\infty dt'\varphi(t,t')e^{-(t+t')}\cdot$$

$$\exp(iRt-iR^*t') \qquad (11)$$

and average over energy and the resonance parameter distributions

$$\langle F\rangle=\int_0^\infty dt\int_0^\infty dt'\varphi(t,t')e^{-(t+t')}P(t,t') \qquad (12)$$

Then the problem for calculation of $\langle F\rangle$ is divided into: i/ calculation of the source function $\varphi(t,t')$ for a given functional F and ii/ calculation of the characteristic function P, which is unique for all functionals F. Thus for any R-matrix modeling of the resonance cross sections in the unresolved energy region the averaged functional $\langle F\rangle$ will be calculated with the same accuracy, as the corresponding characteristic function $\langle\exp(iRt-iR*t')\rangle$ for each t and t' coincides with the function P, eq. (2).

REFERENCES

1. A. Lane, R. Thomas, Rev. Mod. Phys., 30, 257 (1958)

2. F. H. Frohner, Proc. Santa Fe Conf., 1985, vol. 2, pp1541

3. N. Koyumdjieva, N. Savova, N. Janeva, A. Lukyanov, Bulg. J. Phys., No 16, 13 (1989)

4. A. Lukyanov, N. Koyumdjieva, M. Alami, N. Janeva, Yadernye Konstanty, No 3 7 (1990) /in Russian/

Table 1. Analytical calculations of the product $N=\exp(z+\beta)P$

y=0.001

β \ z	0.1	0.2	0.5	1.0	1.5	2.0
0.01	1.0373	1.0691	1.1457	1.2199	1.2519	1.2645
0.1	1.0518	1.0863	1.1602	1.2270	1.2548	1.2656
0.5	1.0643	1.1068	1.1838	1.2396	1.2600	1.2676
1.0	1.0660	1.1102	1.1884	1.2423	1.2612	1.2680

y=0.01

	0.1	0.2	0.5	1.0	1.5	2.0
0.01	1.0322	1.0620	1.1376	1.2101	1.2413	1.2535
0.1	1.0465	1.0798	1.1518	1.2170	1.2441	1.2546
0.5	1.0596	1.1005	1.1750	1.2293	1.2492	1.2565
1.0	1.0615	1.1039	1.1796	1.2320	1.2503	1.2570

y=0.1

	0.1	0.2	0.5	1.0	1.5	2.0
0.01	1.0202	1.0315	1.0845	1.1383	1.1614	1.1705
0.1	1.0225	1.0438	1.0952	1.1434	1.1635	1.1713
0.5	1.0345	1.0612	1.1133	1.1527	1.1679	1.1727
1.0	1.0365	1.0643	1.1170	1.1547	1.1681	1.1730

INTERNATIONAL NUCLEAR MODEL CODE COMPARISON STUDY OF HAUSER-FESHBACH CALCULATIONS

P.E. Hodgson
Nuclear Physics Laboratory, Oxford, U.K.

E. Sartori and K. Shibata
OECD/NEA Data Bank, Gif-sur-Yvette, France

Abstract: Participants were invited to calculate the elastic and inelastic scattering of neutrons from a fictitious nucleus ^{60}Co (Z=27, N=33) at incident laboratory energies of 0.2, 0.5, 1 and 2 MeV. The optical potential was specified. The differential shape elastic, compound elastic and inelastic cross-sections were tabulated. Among the twenty-five sets of results, twelve were sufficiently consistent with each other to be accepted as benchmark values. These fell into two sets, corresponding to calculations with and without the width fluctuation correction. The differences between the results corresponding to different forms of the width fluctuation correction were less than 2 percent.
(nuclear model, code comparison, Hauser-Feshbach theory, MeV range 0.2 - 2.0, differential cross sections, elastic scattering, inelastic scattering, codes)

Introduction

Several intercomparisons [1,2] of computer codes used to calculate nuclear cross sections show general agreement but significant differences in detail. Since these codes embody the same mathematical functions and the intercomparisons use the same input parameters one would expect, if the codes are indeed identical, that the results would be the same to a high degree of accuracy, limited only by the rounding errors of the computer used. The motivation for the present study is to remove these uncertainties by providing reliable benchmarks for such calculations.

A previous report [3] contained the results of an intercomparison of nuclear optical model codes for charged and uncharged particles. The present study is devoted to Hauser-Feshbach calculations. In the next section the specifications of the test calculations are given, and the last section gives and discusses the calculated results.

Specification of the Calculations

The Hauser-Feshbach theory enables the cross-sections of compound nucleus reactions to discrete final states to be calculated from a knowledge of the transmission coefficients in all channels and the energies and total angular momenta of the states of the residual nuclei.

Since the purpose of this intercomparison is to compare the computer codes, we calculate only inelastic neutron scattering from a fictitious nucleus ^{60}Co (Z=27, N=33) with states at the following energies (MeV) : 0+, 0; 2+, 0.1; 4+, 0.3; 0+, 1.0. The calculations were made at the following incident (lab) energies : 0.2, 0.5, 1, 2 MeV and the differential shape elastic, compound elastic and inelastic scattering cross-sections were tabulated at 10° intervals in the CM system.

The transmission coefficients were calculated from optical potentials specified for all channels by

$$V(r) = U f_u(r) + i W g(r)$$

where

$$f_u(r) = \left[1 + exp\left\{ \frac{r - r_u A^{1/3}}{a_u} \right\} \right]^{-1}$$

$$g(r) = -4 a_w \frac{df_w(r)}{dr}$$

with U = 50 MeV, W = 10 MeV, r_u = r_w = 1.2 fm; a_u = a_w = 0.6 fm. As in the optical model intercomparison, the wave number parameter was 0.218732. For consistency reasons, integer masses were used.

Hauser-Feshbach calculations may differ in several respects, in particular concerning the method used to take account of the correlation between the incident channel and the outgoing compound elastic channel. This correlation affects the compound elastic cross-section at low energies, and consequently the non-elastic channels open at these energies. Its magnitude can be calculated in two ways: (i) by the Moldauer width fluctuation correction (WFC) [4,5] and (ii) by the HRTW formalism [6,7,8]. The Moldauer WFC requires the evaluation of an integral and the HRTW formalism in its simpler form requires an iteration calculation. Since earlier codes have the Moldauer WFC and later codes the HRTW method, calculations were made with whatever options were available. The specific WFC options used are as follows:

The Width Fluctuation Correction (WFC)

There are several ways of making the WFC:

1. Moldauer [4]

    W as integral

2. Moldauer (M) [5]

    $$\nu_a = 1.78 + (T_a^{1.212} - 0.78)\, e^{-0.228\, \Sigma_c T_c}$$

972

Then $\quad W_a = 1 + \frac{2}{\nu_a}$

3. Tepel-Hofmann-Weidenmüller (THW) [6]

$$W_a = 1 + \frac{2}{1+\sqrt{T_a}}$$

4. Hofmann-Richert-Tepel-Weidenmüller (HRTW) [7,8]

$$W_a = 1 + \frac{2}{1+T_a^F} + 87 \left(\frac{T_a-\bar{T}}{\Sigma_c T_c}\right)^2 \left(\frac{T_a}{\Sigma_c T_c}\right)^5$$

$$\bar{T} = \frac{\Sigma_c T_c^2}{\Sigma_c T_c}$$

$$F = \frac{\frac{4\bar{T}}{\Sigma_c T_c}\left(1+\frac{T_a}{\Sigma_c T_c}\right)}{\left(1+\frac{3\bar{T}}{\Sigma_c T_c}\right)}$$

Iterative method.

It is worth noting that the previous statistical model intercomparison exercise did not include differential cross-sections, and in some of the outgoing channels the cross-sections differed by quite large factors. It therefore remained desirable to make a precise intercomparison of statistical model codes so as to establish the correct values, and this is the principal motivation of the present study.

Results and Comments

Ten participants presented twenty-five solutions to this benchmark obtained with thirteen different codes (Annex). Most of the results received showed consistency within each of the two classes, without the width fluctuation correction (WFC), including the WFC in either the Moldauer [4,5] or the HRTW [6,7,8] form. The results are tabulated in a report [9], and we recommend them as benchmark values. Figures 1 and 2 compare the recommended angular distributions for two different reactions.

Some other calculations, however, showed marked differences from these standards, and are excluded from the report [9].

All the calculations summarised have total reaction cross-sections within 1 percent of each other, indicating that the optical model calculations are consistent.

ACKNOWLEDGMENTS

This comparison study would have been impossible without the effort and dedication of the participants. We wish to thank all of them: M. Avrigeanu, B.V. Carlson, M. Chadwick, R.W. Finlay, D.G. Gardner, M. Gardner, D. Hirata, A. Merchant, E. Sheldon, M. Uhl, J.L. Weil and P.G. Young.

REFERENCES

1. E. Sartori, Report on the International Nuclear Model Code Intercomparison, Coupled Channel Model Study, NEANDC-182A, INDC(NEA)3, 1984.

2. A. Prince, G. Reffo and E. Sartori, Report on the International Nuclear Model Code Intercomparison Spherical Optical and Statistical Model Study, NEANDC-152A, INDC(NEA)4, 1983.

3. P.E. Hodgson and E. Sartori, International Nuclear Model Code Comparison Study of Spherical Optical Model for Charged Particles, NEANDC-198-U, INDC(NEA)5, 1985.

4. P.A. Moldauer, Phys. Rev. 135, B642 (1964) and Phys. Rev. 123, 968 (1961).

5. P.A. Moldauer, Nucl. Phys. A344, 185 (1980).

6. J.W. Tepel, H.M. Hofmann and H.A. Weidenmüller, Phys. Lett. 49B, 1 (1974).

7. H.M. Hofmann, J. Richert, J.W. Tepel and H.A. Weidenmüller, Ann. Phys. (N.Y.), 90, 403 (1975).

8. H.M. Hofmann, T. Mertelmeier, M. Herman and J.W. Tepel, Z. Physik A297, 153 (1980).

9. P.E. Hodgson, E. Sartori and K. Shibata, International Nuclear Model Comparison Study of Hauser Feshbach Calculations, NEANDC-298-U, INDC(NEA)8, March 1991.

ANNEX

Statistical Model Participants and Contributions

Participant	Computer Code	Comments
Avrigeanu, Rumania	STAPRE	M-no ang. dist.
Chadwick, UK	WILMORE6	HF, M, HRTW
Gardner, USA	COMNUC	M
Finlay, USA	OPSTAT	HF, M, THW, HRTW
Finlay, USA	HELENE	HF, M
Hirata, Brazil	HAUSER5	HF, HRTW
Merchant, Brazil	POLIFEMO	HF, M
Sheldon, USA	CINDY	HF, M
Sheldon, USA	JACQUI	HF, M-no ang. dist.
Sheldon, USA	NANCY	HRTW
Uhl, Austria	MAURINA	M
Weil, USA	HELGA	THW, HRTW
Weil, USA	HFCODE	THW-no ang. dist.
Young, USA	COMNUC	M

HF:	Hauser-Feshbach calculations without WFC
M:	Moldauer WFC
THW:	Tepel-Hofmann-Weidenmüller WFC
HRTW:	Hofmann-Richert-Tepel-Weidenmüller WFC

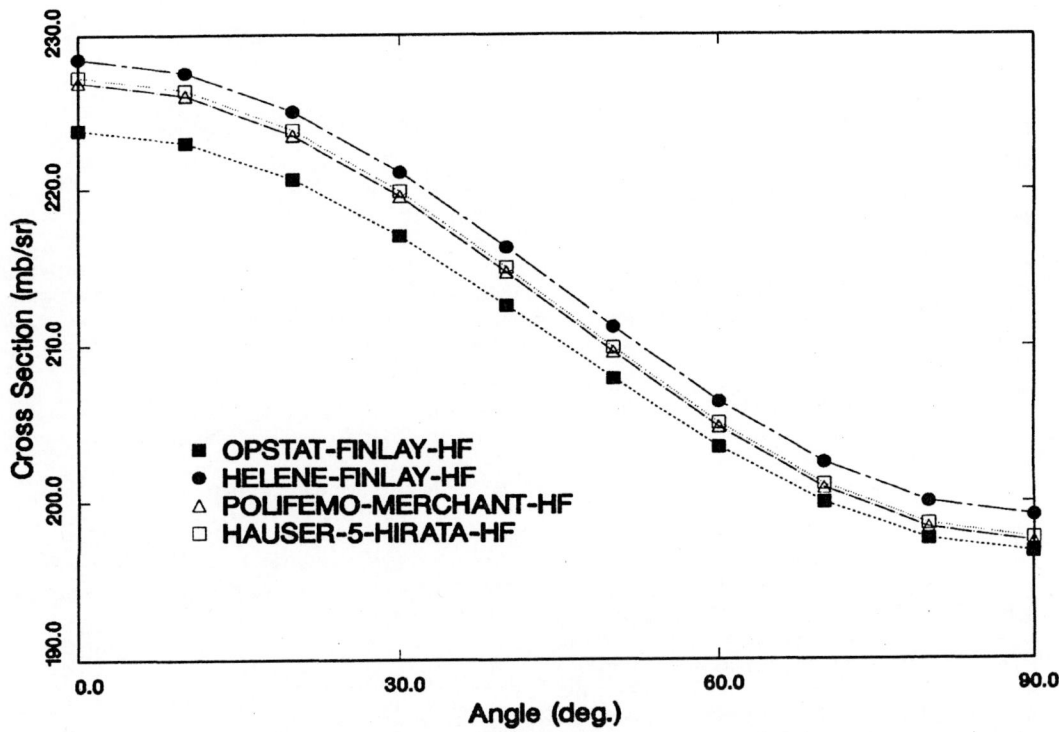

FIGURE 1: Compound elastic scattering cross sections without WFC at 0.2 MeV.

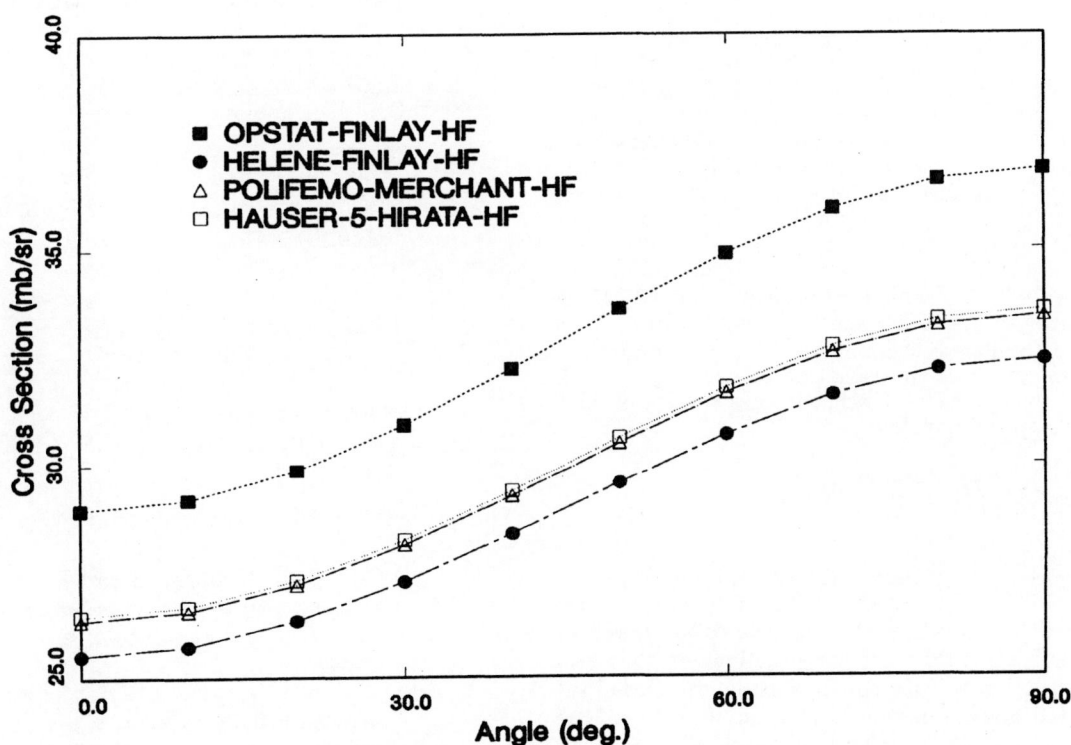

FIGURE 2: Compound inelastic scattering cross sections for the 1st level (0.1 MeV, 2+) without WFC at 0.2 MeV.

Estimation of Nuclear Reaction Model Parameters for ^{59}Co, ^{58}Ni, and ^{60}Ni

Toshihiko Kawano, Hiroya Tanaka, Kohta Kamitsubo, and Yukinori Kanda,

Department of Energy Conversion Engineering,
Kyushu University, Kasuga, Fukuoka 816, Japan

Abstract : Optical model parameters for ^{59}Co, ^{58}Ni, and ^{60}Ni have been estimated by a statistical method based on the Bayes' theorem, and energy dependencies of the parameters have been investigated in the few-MeV energy region. A successful reproduction of differential elastic scattering cross section and total cross section has been achieved by energy dependent potential. The tendencies of the estimated parameters are similar to those reported for the global parameters in the high energy region, however, the parameters show large energy dependencies in the low energy region.

The estimated optical model parameters were used in Hauser-Feshbach calculations, and level density parameters were estimated by the statistical method in order to inspect consistency among the reaction model, the parameters, and the measurements.

(^{59}Co, ^{58}Ni, ^{60}Ni, optical model, energy dependency, level density parameter)

Introduction

Hauser-Feshbach nuclear reaction model parameters that yield consistent cross sections with experimental data are estimated simultaneously by a statistical method which is based on the Bayes' theorem and has been developed and applied to the nuclear data evaluation in this laboratory. Level density parameters are especially noticed because they act effectively on partition of a compound nucleus formation cross section into individual reaction channels, and calculated reaction cross sections are adjustable by those parameters. To achieve this aim, transmission coefficients calculated from an optical model (OM) should be reliable sufficiently. In the previous study[1], neutron transmission coefficients obtained by the optical model were calculated by the Becchetti-Greenlees' global parameter[2]. This parameter set, however, tends to yield larger values than measurements in the low energy region, and the reaction cross sections by the Hauser-Feshbach model calculation are sometimes inconsistent with the experimental data. Walter-Guss' global parameter set[3] yields preferable results, however, it has energy limits and it is not available in the low energy region. Therefore, the optical model potential parameter (OMP) in the low energy region is required.

In this study, the statistical method is also applied to estimate the OMP. The spherical optical model (SOM) is adopted to reproduce differential elastic scattering cross sections and total cross sections of ^{59}Co, ^{58}Ni, and ^{60}Ni. The method can treat the variation of parameters with energy, and the estimated OMP becomes special parameter set for a specific nucleus.

OMP Estimation

Calculation Method

Procedure of OMP estimation is described in Ref. 5. OMPs are estimated so as to reproduce the experimental elastic scattering cross sections and the total cross section. The experimental total cross sections of ^{59}Co and ^{60}Ni are smoothed by a first order B-spline function that is a truncated linear function. Fluctuations are reduced by smoothing with 500 keV node intervals. The optical-statistical model calculation is carried out using the computer program ELIESE-3[6]. Below ≈ 3.5 MeV, compound elastic scattering cross sections are calculated with the Hauser-Feshbach-Moldauer theory. From ≈ 3.5 MeV to ≈ 6.5 MeV, the compound nuclear process is calculated with the Hauser-Feshbach theory with the level density formula by Gilbert and Cameron[7], and the cross sections are slightly adjusted so that the calculated values do not exceed the experimental ones at the minima of angular distributions.

A volume type imaginary potential is negligible in the low energy region, then the OMPs to be estimated are potential depth (V, W_D), radius (r_v, r_{wD}), and diffuseness (a_v, a_{wD}).

Results and Discussion

The estimated OMPs show large energy dependencies in the low energy region, and the tendencies are similar to those reported for the global parameters in the high energy region. The estimated real and imaginary potential parameters are expressed as volume integrals per nucleon (J_v, J_w), and J_v, J_w for ^{58,60}Ni are shown in Figs. 1-(a), and (b), respectively. In these figures, OMPs in Ref. 8, 9, and 10 are compared with the present results. The estimated OMPs are smoothed by the least square fitting as follows:

^{60}Ni

$$J_v = \begin{cases} 506.4 - 16.84E & E < 3.5 \\ 465.1 - 3.602E & E \geq 3.5 \end{cases}$$

$$J_w = \begin{cases} 50.43 + 17.03E & E \leq 2.5 \\ 87.06 + 0.5013E & 2.5 < E \leq 10.0 \\ 110.1 - 1.754E & E > 10.0 \end{cases}$$

^{58}Ni

$$J_v = \begin{cases} 513.1 - 16.88E & E < 3.5 \\ 471.8 - 3.638E & E \geq 3.5 \end{cases}$$

$$J_w = \begin{cases} 57.20 + 17.05E & E \leq 2.5 \\ 93.83 + 0.5225E & 2.5 < E \leq 10.0 \\ 116.9 - 1.775E & E > 10.0 \end{cases}$$

The geometrical parameters are common for ^{58,60}Ni:

$$r_v = \begin{cases} 1.3754 - 0.02932E & E < 5.0 \\ 1.2294 & E \geq 5.0 \end{cases}$$

$$r_{wD} = 1.1927$$

$$a_v = \begin{cases} 0.18484 + 0.09939E & E < 4.5 \\ 0.63341 & E \geq 4.5 \end{cases}$$

$$a_{wD} = \begin{cases} 0.18595 + 0.07811E & E < 6.0 \\ 0.63000 & E \geq 6.0 \end{cases}$$

where potentials and energies are in MeV(Lab), geometrical parameters in fm. The volume type imaginary potential parameters are $W_V = 0.1(E - 11.0)$, $r_{wV} = 1.165$, $a_{wV} = 0.656$, and the spin-orbit potential parameters are $V_{S.O.} = 6.50$, $r_{S.O.} = 1.017$, $a_{S.O.} = 0.6$, and they are identical to the value of Ref. 8.

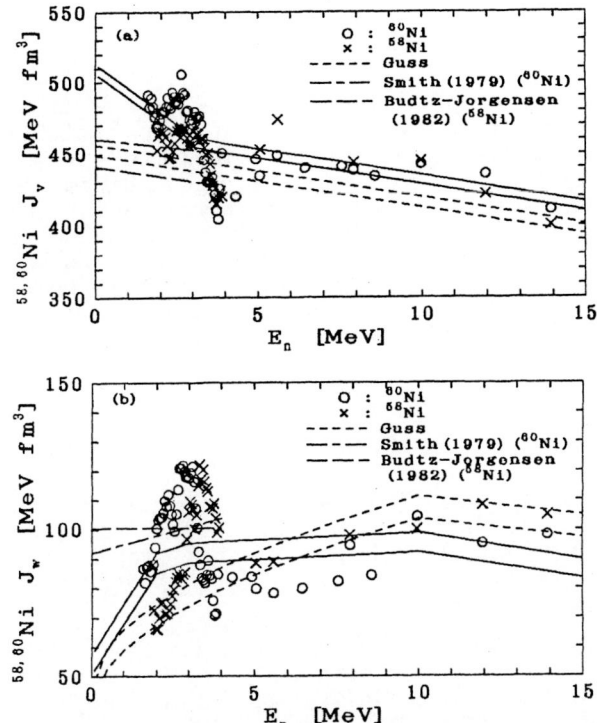

Fig. 1 : Volume integrals of (a) : real, and (b) : imaginary potentials as functions of incident neutron energy for n-^{58,60}Ni. The symbols are posterior parameters of the estimates. For the solid line and the dotted line, upper lines are for ^{58}Ni, and lower lines are for ^{60}Ni.

^{59}Co : An expression of OMP for ^{59}Co is an original Woods-Saxon form, as follows;

$$V = \begin{cases} 45.925 + 0.64777E & E \leq 3.5 \\ 49.511 - 0.38272E & 3.5 < E \leq 10.0 \\ 48.460 - 0.28711E & E > 10.0 \end{cases}$$

$$W_D = \begin{cases} 9.2865 - 0.21851E & E \leq 6.0 \\ 7.3011 + 0.15910E & 6.0 < E \leq 11.0 \\ 11.245 - 0.18422E & E > 11.0 \end{cases}$$

$$r_v = \begin{cases} 1.3886 - 0.04281E & E \leq 3.0 \\ 1.2566 & E > 3.0 \end{cases}$$

$$r_{wD} = 1.2118$$

$$a_v = \begin{cases} 0.28675 + 0.07759E & E \leq 4.5 \\ 0.62816 & E > 4.5 \end{cases}$$

$$a_{wD} = 0.48812$$

The spin-orbit parameters are equal to the values in Ref. 9, and they are $V_{S.O.} = 5.50$, $r_{S.O.} = 1.005$, $a_{S.O.} = 0.65$. The volume type imaginary potential is identical to the Walter-Guss' global parameter, and they are $W_V = -0.963 + 0.153E$, $r_{wV} = 1.444$, $a_{wV} = 0.497$.

Figure 2 gives a comparison of calculated total cross sections of ^{59}Co with the evaluated results. In this figure, the dotted line and symbols indicate the evaluated total cross sections by the B-spline function, the solid line is the OM calculation with the estimated OMP enumerated above, and the dot-dashed line and the dashed line are calculated with Walter-Guss' and Becchetti-Greenlees' global OMP, respectively. Differential elastic scattering cross sections at E_n =2.0 and 7.9 MeV for ^{60}Ni are shown in Fig. 3. As seen in these figures, the estimated OMP yields successful reproductions of the experimental total cross sections up to 30 MeV, and the differential elastic scattering cross sections in the low energy region.

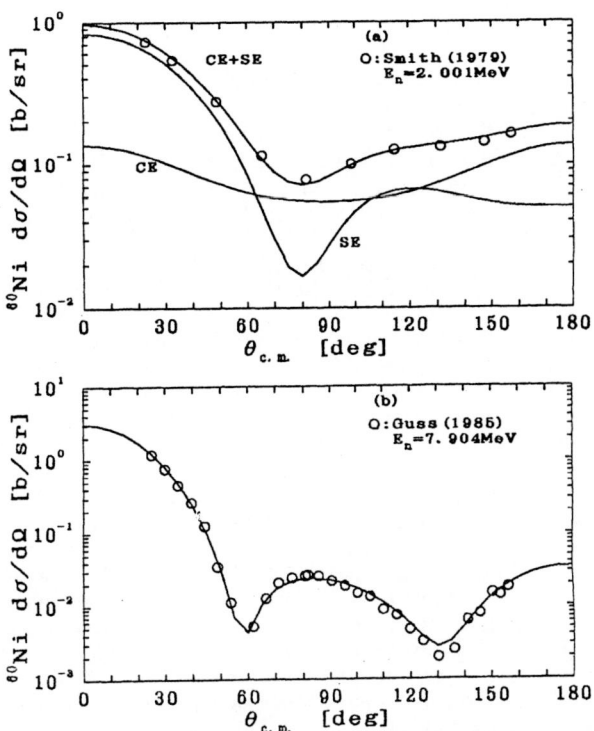

Fig. 2 : Comparison of the differential elastic scattering cross sections at (a)E_n =2.0MeV, and (b)E_n =7.9MeV. The solid line is calculated using the estimated OMP.

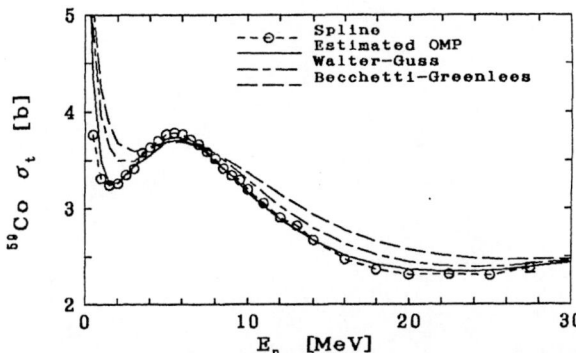

Fig. 3 : Comparison of the total cross sections for ^{59}Co. The symbols and dotted line are evaluated data by the spline function. The solid line is calculated using the estimated OMP, the dot-dashed line is Walter-Guss' global parameter, and the dashed line is Becchetti-Greenlees' global parmeter.

Level density parameter estimation

(n, p), (n, α), and $(n, 2n)$ reaction cross sections of ^{58,60}Ni and ^{59}Co are calculated by the Hauser-Feshbach model with the estimated OMP. Subsequently, the level density parameters are estimated so as to reproduce available experimental data. In order to compare with the other OMPs, some global parameters are picked up in this section. They are Becchetti-Greenlees' and Walter-Guss' OMP for neutron, Menet et al. [11] and Arthur-Young[12] for proton. Walter-Guss' OMP is not available at E <10.0 MeV, then the OMP modified by Yamamuro[13] is used. OMP for α-particle is fixed to Lemos' OMP[14]. The combinations of neutron and proton OMPs are; (1) estimated OMP – Menet et al. , (2) Becchetti-Greenlees – Menet et al. , (3) modified Walter-Guss – Arthur-Young.

976

The procedure of level density parameter estimation is discussed in Ref. 1. Gilbert-Cameron's level density parameters are employed as the prior values. The calculated ^{58}Ni(n,p) cross section is displayed in Fig. 4. In this figure, the solid line corresponds to case (1), the dotted line is the case (2), and the dot-dashed line is case (3). The estimated level density perameters are shown in Fig. 5. For the cases (1) and (2), proton emission process is overestimated, and this is attributed to the proton OMP. Generally, proton OMP by Menet *et al.* brings large penetrability. When the Becchetti-Greenlees' OMP is adopted as the neutron OMP, the proton OMP by Menet *et al.* is competent, and the cross sections concerning the proton can be adjusted by means of level density parameters only. For the case (3), it also shows good balance between neutron and proton penetrability. Then the posterior parameters seems to be proper values. On the other hand, the estimated OMP yields relative small penetrability, and Menet *et al.* 's parameter is not suitable. As shown in Fig. 5, some of the parameters are estimated to impractical values, and it seems that reproduction of the experimental data by this OMP combination is impossible. Therefore it is concluded that when strengths of both neutron and proton OMPs are competent, the calculated reaction cross sections are adjustable by level density parameters only, though the OMP are not always suitable for the reproductions of the elastic scattering and total cross section data.

Conclusion

Spherical optical model potential parameters for ^{59}Co, ^{58}Ni, and ^{60}Ni were estimated by the statistical method, and energy dependencies of the parameters were investigated in the few-MeV energy region. Experimental differential elastic scattering cross sections and total cross sections were reproduced by the estimated OMP.

When the estimated OMP was adopted to level density parameter estimation, the parameters were estimated to impractical values, and this is attributed to proton OMP. Therefore reliable proton OMP is necessary.

References

[1] Y.Uenohara, T.Tsuji, and Y.Kanda : Proc. Int. Conf. on Nuclear Data for Science and Technology, 1988, Mito, (Ed. S.Igarasi, Saikon, Tokyo), p.481.

[2] F.D.Becchetti,Jr. and G.W.Greenlees : Phys. Rev., **182**, 1190(1969).

[3] R.L.Walter and P.P.Guss : Proc. Int. Conf. Nucl. Data for Basic and Applied Science, 1985, Santa Fe, (Eds. P.G.Young, R.E.Brown, G.F.Auchampaugh, P.W.Lisowski, and L.Stewart, Gordon and Breach, New York), p.1079.

[5] T.Kawano, H.Tanaka, K.Kamitsubo, and Y.Kanda : Proc. of the 1990 Symposium on Nuclear Data, JAERI-M 91-032, NEANDC(J)-160/U, (Eds. M.Igashira and T.Nakagawa), p.355(1991).

[6] S.Igarasi : "Program ELIESE-3, Program for Calculation of the Nuclear Cross Sections by Using Local and Non-Local Optical Models and Statistical Model", JAERI-1224, (1972).

[7] A.Gilbert and A.G.W.Cameron : Can. J. Phys., **43**, 1446 (1965).

[8] P.P.Guss, R.C.Byrd, C.E.Floyd, C.R.Howell, K.Murphy, G.Tungate, R.S.Pedroni, R.L.Walter, J.P.Delaroche, and T.B.Clegg : Nucl. Phys., **A438**, 187(1985).

[9] A.B.Smith, P.T.Guenther, and R.D.Lawson : Nucl. Phys., **A483**, 50(1988).

[10] C.Budtz-Jørgensen, P.T.Guenther, A.B.Smith, and J.F.Whalen : Z. Phys., **A306**, 265(1982).

[11] J.J.H.Menet, E.E.Gross, J.J.Malanify, and A.Zucker : Phys. Rev., **C4**, 1114 (1971).

[12] E.D.Arthur and P.G.Young : Proc. of Sym. on Neutron Cross Sections from 10 to 50 MeV, 1980, BNL, (Eds. M.R.Bhat and S.Pearlstein) p.731.

[13] N.Yamamuro : JAERI-M 88-140(1988).

[14] O.F.Lemos : "Diffusion Elastique de Particules Alpha de 21 a 29.6 MeV sur des Noyaux de la Region Ti-Zn", Orsay report, Series A, No.136(1972).

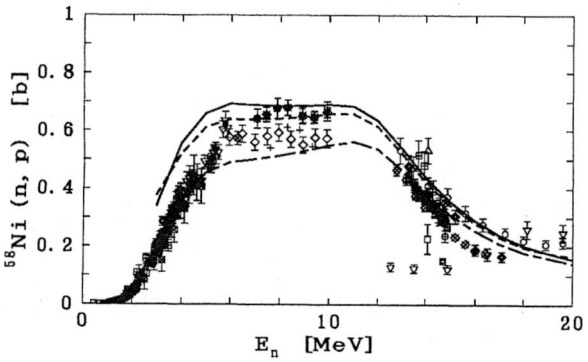

Fig. 4 : Calculated ^{58}Ni(n,p) reaction cross sections. The solid line corresponds to the case (1) explained in the text, the dotted line is the case (2), and the dot-dashed line is the case (3).

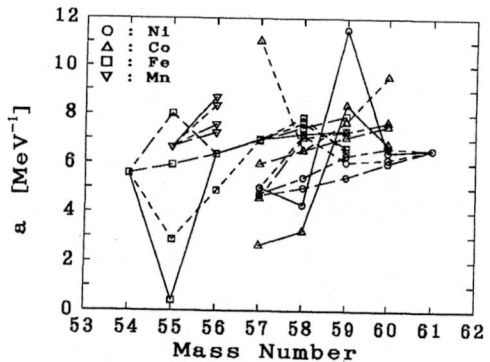

Fig. 5 : Comparison of the estimated level density parameters when neutron and proton OMPs are changed. The solid line is corresponding to the case (1) explained in the text, the dotted line is the case (2), and the dot-dashed line is the case (3).

THE SENSITIVITY OF STATISTICAL MODEL CAPTURE CALCULATIONS TO MODEL ASSUMPTIONS

M. Uhl

Institut für Radiumforschung und Kernphysik, Universität Wien
A1090 Wien, Boltzmanngasse 3, Austria

J.Kopecky

Netherlands Energy Research Foundation ECN
P.O. Box 1, 1755 ZG Petten, The Netherlands

Abstract: The most important ingredients of statistical model capture calculations are gamma-ray strength functions f_{XL} for E1 and M1 radiation, level densities and optical potentials. We studied the influence of uncertainties of these quantities with the aim to improve the predictive power of such calculations. Spherical and deformed targets in the mass range A = 55 - 197 were considered. Among different strength function models compatible with idependent experimental data the best results were obtained with f_{E1} derived from a "generalized" Lorentzian and f_{M1} from a standard one. The former is characterized by an energy dependent width and results in a non-zero limit of f_{E1} as the energy tends to zero. A generalization of the energy dependence of the width was required for deformed nuclei with masses between 151 and 175. We observed a marked sensitivity on different level density models with parameters based on the same experimental data. The influence of the neutron optical potential is in general rather weak.

(statistical model calculations of neutron capture, sensitivity to gamma ray-strength functions, level densities and neutron transmission coefficients)

Introduction

Many scientific and energy related applications require capture cross sections for incident energies ranging from overlapping resonances up to a few MeV. Often this information must be supplied by model calculations. In general the compound nucleus mechanism by far dominates the capture process in this energy range; exceptions occur in mass regions with strong valence capture as e.g. for the nuclei around A = 60. Therefore the simple statistical (compound nucleus) model is very useful to assess capture cross sections and gamma-ray production spectra. The results of this model depend on several additional quantities as gamma-ray strength functions, level densities of the residual nuclei and transmission coefficients of projectile and emitted particles. These are auxiliary means from the viewpoint of cross section calculations but on the other hand they represent by themselves objects of intensive investigations and are described by corresponding models with parameters relying on on specific experimental data. The auxiliary quantities are afflicted with characteristic uncertainties which limit the accuracy of the capture calculations, even if the reaction model is quite adequate. One can distinguish two types of these uncertainties. The first ones stem from the underlying experimental data as e.g. primary gamma-ray resonance transitions in case of the gamma-ray strength functions or average resonance spacings in case of level densities. The others result from the models themselves and therefore cannot be reduced by improved experimental data. In the following we mainly concentrate on this more fundamental type of uncertainties.

In this contribution we studied the effects of uncertainties of the auxiliary quantities on capture cross sections, gamma-ray spectra and average radiation widths. We considered targets with mass numbers between 55 and 197 and neutrons as projectiles. By comparison with appropriate experimental data we tried to find the best models and to gain some insight into the accuracy of the calculations.

Cross sections and auxiliary quantities

In spite of recent progress in the statistical theory of nuclear reactions /1/ we employ the simple Hauser-Feshbach theory in the formulation of Moldauer /2/. To recall the dependence of the cross sections on auxiliary quantities we consider the expression for the capture of a neutron (energy ε_n) with the emission of one (primary) gamma-ray (energy ε_γ) leaving the residual nucleus in states with spin J', parity Π' around excitation energy E'

$$\frac{d\sigma_{n\gamma}^{(1)}(\varepsilon_n, E'J'\Pi')}{d\varepsilon_\gamma} = \rho_\gamma(E'J'\Pi') \frac{\pi}{k_n^2} \sum_{J\Pi} g_J$$
$$\times \sum_{j\pi} \sum_{XL} \frac{T_{nj\pi}(\varepsilon_n) T_{\gamma XL}(E - E')}{T(EJ\Pi)} W_{j\pi, XL}^{J\Pi} \quad (1)$$

in terms of the transmission coefficients $T_{nj\pi}$ and and $T_{\gamma XL}$ for neutrons and gammas, respectively, and the level density ρ_γ. The quantity $T(E\,J\Pi)$ represents the sum of the transmission coefficients of all open channels compatible with the conservation of the excitation energy E, spin J and parity Π of the compound nucleus. The width fluctuation correction factor $W_{j\pi, XL}^{J\Pi}$ depends on $T_{nj\pi}$, $T_{\gamma XL}$ and $T(EJ\Pi)$. The symbols g_J and k_n stand for the statistical factor and the wave number of the incoming neutron. For low excitation energies the level densities are replaced by the known levels. For the actual calculations Eq.(1) must by supplemented by an appropriate treatment of gamma-ray cascades the results of which depend essentially on the same quantities. The computations were performed with the code MAURINA /3/.

The **gamma-ray transmission coefficients** $T_{\gamma XL}$ of multipole type XL are related to the corresponding gamma-ray strength functions f_{XL} by $T_{\gamma XL}(\varepsilon_\gamma) = 2\pi\varepsilon_\gamma^{2L+1} f_{XL}(\varepsilon_\gamma)$. The most important multipoles are E1 and M1. According to Brink /4/ gamma-ray strength functions may be related to the photoabsorption cross section. By assuming that the latter is dominated by a giant resonance of Lorentzian shape (or two of them, in case of deformed nuclei) one obtains a representation of $f_{XL}(\varepsilon_\gamma)$ by a **standard Lorentzian (SLO)**. This representation we used for E1, M1 and E2 with parameters based on experiment and/or on systematics; more details can be found in Ref. 5. For f_{E1} we considered also a **generalized Lorentzian (GLO)** as proposed by Kopecky and Chrien /6/

$$f_{E1}(\varepsilon_\gamma, T) = 8.68 \times 10^{-8} \, \text{mb}^{-1}\text{MeV}^{-2} \sum_{i=1}^{N} \sigma_{0i}\Gamma_{0i}$$
$$\times \left[\frac{\varepsilon_\gamma \Gamma_i(\varepsilon_\gamma, T)}{(\varepsilon_\gamma^2 - E_{0i}^2)^2 + \varepsilon_\gamma^2 \Gamma_i(\varepsilon_\gamma, T)^2} + \frac{0.7\Gamma_{0i}4\pi^2 T^2}{E_{0i}^5} \right] \quad (2)$$

Here N = 1 or 2, mainly depending on the mass region, and T represents the temperature of the final state. This formula exhibits the main features of microscopic models of the E1 strength developed by Kadmenskij et al. /7/ and by Sirotkin /8/: an energy dependence of the width(s) as prescribed by the theory of Fermi liquids /9/

$$\Gamma_i(\varepsilon_\gamma, T) = \beta_i(\varepsilon_\gamma^2 + 4\pi^2 T^2), \quad (3)$$

and a non-zero value of $f_{E1}(\varepsilon_\gamma, T)$. For the normalization(s) β_i the value(s) $\beta_i = \Gamma_{0i}/E_{0i}^2$ proposed by Kadmenskij et al. /7/ is (are) assumed. If, in Eq. (2), the second term in the brackets is omitted $f_{E1}(\varepsilon_\gamma)$ is derived from a **Lorentzian with energy dependent width (ELO)**; this representation has similar properties as a "depressed Lorentzian" or other empirical prescriptions invented to improve the agreement with experiment. The parameters $(\sigma_{0i}, E_{0i}, \Gamma_{0i})$ were taken from fits to photonuclear data /10/.

The **single particle model** /11/ was used for M2-, E3- and M3 radiation and occasionally for M1 radiation, too. While in the former case f_{XL} was chosen as 1 Weisskopf unit/MeV we adjusted the strength of M1 to primary resonance data (**adjusted single particle model (ASP)**).

For ^{93}Nb and the targets with $A \geq 181$ the contribution of a Lorentzian **pygmy resonance** was added to f_{E1}; the parameters were chosen so as to improve the reproduction of experimental data (cf. Ref. 5).

For the **level density** we selected two semi-empirical models with the most important parameters determined by the number of low lying levels and the average s-wave resonance spacing D_0 and a spin distribution parameter σ based on the rigid body moment of inertia. The first one is the **backshifted Fermi gas model (BSFG)** /12/ with the level density parameter a and the backshift Δ_B. These parameters are affected by shell effects and therefore difficult to extrapolate to nuclei without experimental data. This drawback is strongly reduced in the class of models which parameterize shell effects in terms of the shell correction to the binding energy. We chose the **model by Kataria, Ramamurthy and Kapoor (KRK)** /13/ with the fundamental frequency $\omega_0 = 0.185A^{1/3}$ and a conventional pairing shift Δ_P according to Gilbert and Cameron /14/. The asymptotic level density parameter $\tilde{a}$ is found by reproducing D_0. For lower excitation energies $E \leq E_x$ the KRK model was supplemented by a constant temperature portion parameterized as in Ref. 14; in this energy region we linearly interpolated σ^2 between the value σ^2_{lev} deduced from low lying levels and the value prescribed at E_x. Both models neglect parity effects.

Particle transmission coefficients were calculated with optical model potentials (OMP) taken from the literature; coupled channels calculations were performed for deformed nuclei.

E1 and M1 gamma-ray strength functions

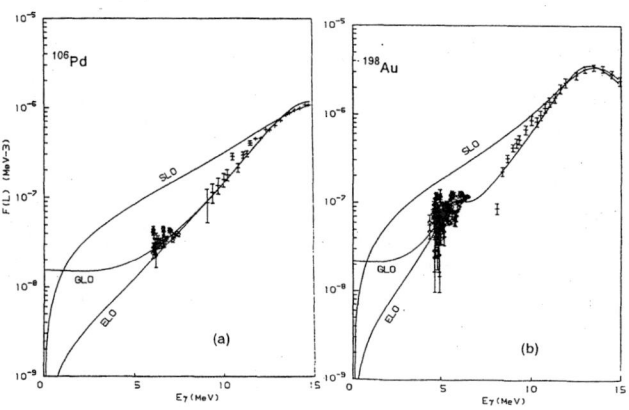

Fig. 1 E1-strength functions for ^{106}Pd (a) and ^{198}Au (b); see text.

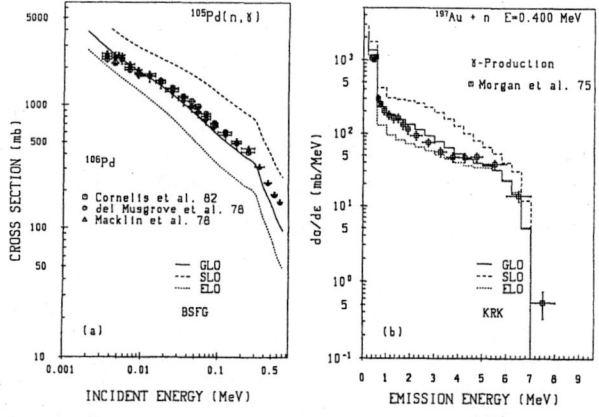

Fig. 2 Capture cross section for ^{105}Pd (a) and gamma-ray production spectrum for ^{197}Au (b) calculated with three different models for the E1-strength. The level density model employed is indicated.

E1 strength: For many nuclei E1-strength functions derived from a SLO overpredict primary resonance data. This is shown in Fig. 1 for ^{105}Pd and ^{197}Au , where we compare f_{E1} derived from a SLO, a ELO and a GLO to

photonuclear and to average resonance capture (ARC) data. Both, the GLO and the ELO, are compatible with experimental data. Fig. 2, however, shows that the resulting cross sections and spectra differ considerably. These differences can easily be traced back to the products $\rho_\gamma(E'J'\Pi')T_{\gamma XL}(E - E')$ in Eq. (1) and in similar expressions governing the gamma-ray cascades. The experimental data, in particular the slope of the spectrum for emission energies between 1 and 4 MeV favour the GLO. This is gratifying, as the non-zero $\varepsilon_\gamma \to 0$ limit exhibited by this model is founded by microscopic theory. A similar sensitivity of model calculations to E1 strength function models was found for all investigated nuclei. For the essentially spherical nuclei ^{93}Nb, ^{105}Pd, ^{143}Nd and ^{197}Au and the deformed nuclei ^{181}Ta, ^{182,183,184,186}W and ^{185,187}Re the experimental data support the GLO Eq.(2). Also the experimental data for nuclei with mass numbers between 50 and 60 can be reproduced with f_{E1} derived from a GLO. However, definite conclusions are complicated by the presence of strong nonstatistical effects (cf. Ref. 15). Several comparisons showed that Eq.(2) and the original formula by Kadmenskij et al. /7/ result in very similar cross sections and spectra.

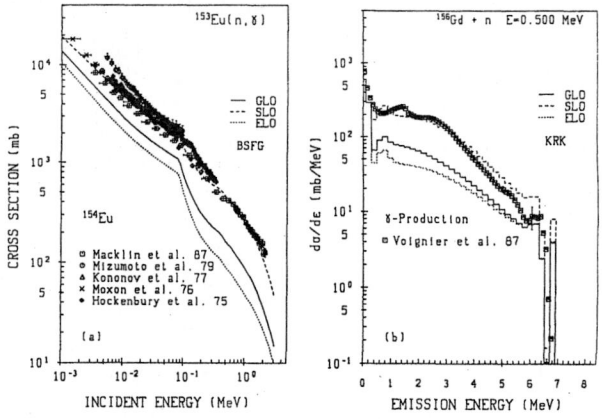

Fig. 3 Capture cross section for ^{153}Eu (a) and gamma-ray production spectrum for ^{156}Gd (b); see Fig. 2.

For the lighter rare-earth nuclei investigated the experimental data are best reproduced with an E1 strength derived from a SLO; this holds in particular for ^{151,153}Eu, ^{155,156}Gd, ^{169}Tm and ^{175}Lu. Characteristic examples we show in Fig. 3. This failure of the GLO shows up also in comparisons with ARC data. The GLO can be reconciled with **all** experimental data of these nuclei by increasing the normalization β_i in Eq. (3) (see Ref. 15). In order to guarantee also the reproduction of photonuclear data we assumed the following linear energy dependence:

$$\beta_i(\varepsilon_\gamma) = \frac{\Gamma_{0i}}{E_{0i}^2}\left(k_0 + \frac{\varepsilon_\gamma - \varepsilon_\gamma^0}{E_{0i} - \varepsilon_\gamma^0}(1 - k_0)\right). \qquad (4)$$

The constants k_0 and ε_γ^0 can be used to reproduce ARC data; $k_0 = 1$ yields the original value of β_i. The data of the Eu, Gd and Lu isotopes considered here can reasonably well be fitted with values of k_0 between 2 and 4 and ε_γ^0 around 5 MeV, while for the aforementioned spherical and heavier rare-earth nuclei k_0 is close to 1.

M1 strength: So far the M1 strength functions were derived from a SLO representing a spin-flip giant resonance /17/ with $E_0 = 41A^{-1/3}$ MeV, $\Gamma_0 = 4$ MeV and σ_0 adjusted to reproduce ARC data or systematics (see Ref. 5). As many authors employ for f_{M1} an energy independent strength we performed also calculations with the ASP model adjusted to the same strength as the SLO. The results showed that the preceding conclusions on the validity of models for the E1 strength depend on the underlying prescription for f_{M1}.

A combination of the ASP model for f_{M1} and a ELO for f_{E1} gives similar results as a SLO and a GLO, respectively. Therefore, the former combination is also appropriate for the spherical nuclei considered and the Ta, W and Re isotopes. On the other hand, for the lighter rare-earth nuclei where the combination SLO (M1) and GLO (E1) failed, a reasonable description of the data is achieved by assuming the ASP model for f_{M1} and a GLO for f_{E1}.

These ambiguities in the models for dipole radiation are not satisfying. They reflect the oversimplified nature of models for the M1 strength. As explained in Ref. 5 we prefer for f_{M1} a SLO as the ASP model is at variance with a finite energy-weighted sum rule and fails to reproduce information on transitions between low lying levels.

Level density

Both level density models, though relying on the same experimental data, result in appreciably different cross sections, spectra and average radiation widths; a typical example is shown in Fig. 4 for ^{105}Pd. The effect is caused by the different level densities a few Mev below the excitation energy E where $\rho_\gamma(E'J'\Pi')T_{\gamma XL}(E - E')$ in Eq. (1) has a maximum. In the energy range of interest here and in most of our examples the relevant part of the KRK level density is the constant temperature portion and not the region of the genuine KRK formula. Effects of similar magnitude as in Fig. 4 were found also for the other nuclei. The actual value of the difference depends on the neutron binding energy and the target spin. For the spectra magnitude and slope differ for the KRK and the BSFG model.

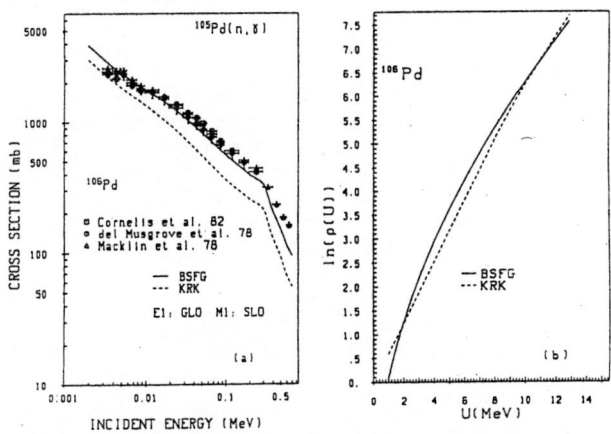

Fig. 4 (a) Capture cross section of ^{105}Pd calculated with the KRK and the BSFG model for the level density; (b) The corresponding total level density of ^{106}Pd.

This unpleasant uncertainty is the prize for using simple semi-empirical level density models. By comparing the results of model calculations, performed with our preferred strength function prescriptions, to experimental data in the mass region A = 93-197 we found slightly better results with the BSFG model. However, we emphasize that this holds for capture calculations at moderate incident energies and depends on the underlying strength functions. Recently Krusche /18/ studied the average s-wave radiation width for a large number of nuclei. The best agreement with experiment was obtained with a GLO for E1 and the ASP model for M1 radiation and level densities according to the constant temperature formula for a < 15 MeV^{-1} and the BSFG model for a $\geq$ 15 MeV^{-1} (a is the BSFG level density parameter). With our different gamma-ray strength functions we found also for some nuclei with a > 15 MeV^{-1}, as e.g. ^{185}Re and ^{197}Au a better results with the KRK model (essentially with its constant temperature portion). Test calculations showed that the consequence of parity effects and of uncertainties in the effective moment of inertia are in general smaller than the difference between the BSFG and the KRK model.

With increasing incident energy also the level densities of the nuclei populated by competing particle emission come into play and increase the uncertainties.

Neutron optical potential

The most important particle transmission coefficients are those for neutrons. The results of capture calculations obtained with different neutron OMP's showed, that in general the dependence on the OMP is much weaker than

that on gamma-ray strength functions and level densities. For incident energies below the inelastic scattering threshold this follows immediately from the structure of the Hauser-Feshbach formula. As long as the average radiation width is much smaller than the average neutron width the capture cross section is very small compared to the absorption cross section and nearly independent of the neutron transmission coefficients. In the other extreme, if the capture cross section represents a large fraction of the absorption cross section, the dependence on the neutron OMP is stronger. This may occur for targets with large neutron separation energies. In any case, accurate capture calculations should use specific neutron OMP's, which reproduce neutron strength functions as well as elastic and total cross sections.

Conclusion

We recommend to use for capture calculations in the mass region A = 50-210 for f_{E1} a GLO according to Eq. (2) and for f_{M1} a SLO; for f_{E1} of nuclei with A = 150-175 either the modification given by Eq. (4) with k_0 around 3 or a SLO should be applied. The resulting strength functions rely on independent data or well established systematics and are free of normalizations.

In case of the level density definite recommendations are more difficult. Though for the nuclei considered here the BSFG model was slightly superior, the KRK model is better suited for extrapolations; further investigations are required. So far we suggest to apply both models and to use the difference to assess the uncertainties.

By comparing model calculations performed in this way to experimental data we roughly assess a predictive power around of 30% for unknown capture cross sections. As the shape of the capture excitation function is rather insensitive to model assumptions a better accuracy may eventually be achieved by normalizing the calculations to a good experimental datum at one energy. In all these conclusions we exempt the actinide nuclei which we have not studied so far.

References

1. J. J. M. Verbaarschot, H. A. Weidenmüller, and M. R. Zirnbauer: Phys. Rep. *129*, 367 (1985)
2. P. A. Moldauer: Nucl. Phys. *A344*, 185 (1980)
3. M. Uhl: unpublished
4. D. M. Brink: Ph. D. thesis, Oxford University, 1955
5. J. Kopecky and M. Uhl: Phys. Rev. *C41*, 1941 (1990)
6. J. Kopecky and R. E. Chrien: Nucl. Phys. *A468*, 285 (1987)
7. S. G. Kadmenskij, V. P. Markushev, and V. I. Furmann: Sov. J. Nucl. Phys. *37*, 165 (1983)
8. V. K. Sirotkin: Sov. J. Nucl. Phys. *43*, 362 (1986)
9. D. Pines and P. Noziers: The Theory of Quantum Liquids, (Benjamin, N. Y. 1966)
10. S. S. Dietrich and B. L. Bermann: At. Nucl. Data Tables 38, 199 (1989)
11. M. Blatt and V. F. Weisskopf: Theoretical Nuclear Physics (Wiley, N. Y. 1952)
12. W. Dilg, W. Schantl, H. Vonach and M. Uhl: Nucl. Phys. *A217*, 269 (1973)
13. S. K. Kataria, V. S. Ramamurthy, and S. S. Kapoor: Phys. Rev. *C18*, 549 (1978)
14. A. Gilbert and A. G. W. Cameron: Can. J. Phys. *43*, 1446 (1985)
15. M. Uhl and J. Kopecky: Proc. IAEA Specialists' Meeting on the Measurement, Calculation and Evaluation of Photon production Cross sections, Febr. 5-7, 1990, Smolenice, CSFR, INDC(NDS)-238(1990), p. 113
16. J. Kopecky and M. Uhl: Proc. Seventh Intern. Symposium on Capture Gamma-ray Spectroscopy and Related Topics, Oct. 14-19, 1990, Asilomar, USA, in press
17. A. G. Bohr and B. R. Mottelson: Nuclear Structure (Benjamin, London, 1975), Vol. II, p.636
18. B. Krusche: Ref. 15

MODIFIED STATISTICAL CALCULATIONS OF AVERAGE DIPOLE RADIATIVE STRENGTH FUNCTIONS AND CROSS SECTIONS OF NEUTRON RADIATIVE CAPTURE IN THE ENERGY INTERVAL 10^{-3} - 2 MEV

Vyong Hyu Tan, G.A.Prokopets

Kiev State University, USSR

Abstract: The energy dependence of averaged total electric dipole radiative width is connected to the relative variance of the exciton number $D_n/\bar{n}$. If it is used together with experimental neutron strength functions for s- and p-waves to estimate transmission coefficients, one has a useful version of the compound-nucleus model. The calculations have been carried out for a number of nuclei in the mass region $A = 50 \div 250$ and for neutron energies up to 2 MeV. The results of comparison with experimental data are discussed.

(compound-nucleus model calculations, exciton number variance, dipole radiative strength function, neutron radiative capture cross sections).

Introduction.

The precision of neutron capture cross section evaluations for the keV-MeV region is still not enough to satisfy the required accuracy of $(10 \div 30)\%$. It seems likely that the uncertainty might partly come from the gamma ray strength functions and partly from the neutron transmission coefficients commonly used in compound-nucleus model calculations. In this paper a modification of the statistical approach to the neutron capture problem is presented. The energy dependences of averaged total electric dipole radiative widths of nuclei are evaluated on the basis of phenomenological exciton model. Next step was to calculate compound-nucleus radiative capture cross sections to test the fruitfulness of the approach.

Nuclear electric dipole radiative strength function.

Let us suppose that highly excited compound nucleus states may be treated as a superposition of particle-hole configurations. One may then differentiate these compound nucleus states according to mean exciton numbers $\bar{n} = \bar{n}_p + \bar{n}_h$ and to variances of their distribution $D_n = D_{np} + D_{nh} + 2 \cdot (D_{np} \cdot D_{nh})^{1/2} \cdot \rho_{ph}$. In Fermi-gas approximation the particles and holes are independent (the correlation coefficient $\rho_{ph} = 0$) and the following relations take place [1]:

$$\bar{n}_p = \bar{n}_h; \qquad D_{np} = D_{nh}$$
$$\bar{n} = g_F t \cdot \ln 4 ; \quad D_n = g_F t \qquad (1)$$

Here g_F is the one-particle level density on the Fermi surface and t- the nuclear temperature.
Therefore, the relative variance of the Fermi-gas exciton distribution is the same value for all nuclei and excitation energies,

$$(D_n/\bar{n})_F = 1/\ln 4 = 0.721 \qquad (2)$$

Nuclear effective forces would be introduced in the statistical consideration by means of a particle-hole correlation coefficient, so that

$$D_n/\bar{n} = (D_n/\bar{n})_F \cdot [1 + \rho_{ph}] \qquad (3)$$

The probability of radiative transitions may be found if it originates when one particle from an n-particle configuration with average exciton number $\bar{n}$ changes its state. Accordingly, the reduction of one-particle strength is determined as probability:

$$R(n) = (2\pi \cdot D_n)^{-1/2} \exp[-(n-\bar{n})^2/2D_n] \qquad (4)$$

The next step is to introduce the limitation on energies E_γ of γ-rays that may be emitted from the n-exciton configuration with excitation U:

$$n(U, E_\gamma) = \bar{n}(U) F(\Gamma_R, E_R, E_\gamma). \qquad (5)$$

The function $F(\Gamma_R, E_R, E_\gamma)$ connects the phenomenon of the giant electric dipole resonance (GDR) with the transition energy E_R in a spherical system of nucleons, which is associated with the most probable exciton configuration $n = \bar{n}$. Then:

$$F = \frac{[(\sigma\Gamma)_{R1} \cdot F_1 + (\sigma\Gamma)_{R2} \cdot F_2]}{[(\sigma\Gamma)_{R1} + (\sigma\Gamma)_{R2}]} \qquad (6)$$

$$F_1 = \{E_\gamma^2 \Gamma_{Ri}^2 / [(E_\gamma^2 - E_{Ri}^2)^2 + E_\gamma^2 \Gamma_{Ri}^2]\}^k \qquad (7)$$

Formula (6) reflects the splitting of the GDR excitation curve for deformed nuclei in two components with maximum $R1$ at energy E_{R1} and width Γ_{R1}, i=1,2. Then we obtain for the average electric dipole radiative width of a compound nucleus ^AX_Z with spin J and excitation energy U:

$$\bar{\Gamma}_\gamma(U,J) = \frac{10^2[(Z/A)^2 + (1-Z/A)^2] \cdot A^{2/3}}{d_0 \cdot \rho(U,J)} \left[\sum_{I=J-1}^{J+1} \int_0^{U-E_N} R(E_\gamma) E_\gamma^3 \cdot \rho(U-E_\gamma, I) dE_\gamma + \right.$$

$$\left. + \sum_{i=0}^{N-1} R(U-E_i) \cdot (U-E_i)^3 \right] = \frac{1}{\rho(U,J)} \sum_{I=J-1}^{J+1} \int_0^U S_{E1}(E_\gamma) E_\gamma^3 \cdot \rho(U-E_\gamma, I) dE_\gamma \qquad (8)$$

The electric dipole strength function is denoted here as $S_{E1}(E_\gamma)$ and $d_0 = E_N/(N-1)$ is the mean level spacing for discrete states taking part in E1 transitions. The constant 10^2 appears if $[\rho^{-1}] = [E_\gamma] = [d_0] = MeV$ and $[\bar\Gamma_\gamma] = meV$.

The expressions above were used to calculate average radiative widths $\bar\Gamma_{\gamma(calc)}$ of neutron s-resonances for 110 nuclei with mass numbers $A = 50 \div 250$ to extract from the comparison with experimental widths $\bar\Gamma_\gamma(exp)$ the only unknown parameter, the particle-hole correlation coefficient $-1 \le \rho_{ph} \le 1$ (or $D_n/\bar n$). All the other necessary values were fixed by independent experimental data and systematics based on them. In particular, for level density $\rho(U,J)$ we used systematics [2] with level density parameter a and pairing parameter Δ ("back shifted" Fermi-gas model). The exponent k in (7) was fixed by the condition $\langle D_n/\bar n\rangle_{z,A} \approx (D_n/\bar n)_F = 0.721$ and appeared to be equal to $k=1/4$. As a result it was established that the statistical dependence between $D_n/\bar n$ values and parameters Δ/U is close to linear, so that the series

$$D_n/\bar n = \sum_m B_m(\Delta/U)_m \qquad (9)$$

may be limited to a few terms. The correlation coefficient $r = 0.9$ and a mean relative deviation $\delta \approx 10\%$ is reached for the parabolic approximation with $B_0 = 0.767$, $B_1 = 1.662$, $B_2 = 0.975$.
The underlying physics implies the existence of a connection between the empirical "back-shift" parameter Δ and the particle-hole correlation coefficient:

$$\rho_{ph} = 1.352(\Delta/U)^2 + 2.305(\Delta/U) + 0.064 \qquad (10)$$

Accordingly one may see, that when the excitation energy U increases, ρ_{ph} tends to zero as the Fermi-gas model predicts. If the magic nuclei are not taken into consideration, the systematics reproduce average radiative widths of neutron resonances with an accuracy of $\approx 30\%$ in the mass region $50 < A < 250$. Under this condition a small variation of the level density parameter a is enough to come to the equality $\bar\Gamma_\gamma(calc) = \bar\Gamma_\gamma(exp)$. Only for individual magic nuclei it needs $\approx 20\%$ a-variation that is comparable with the precision of a-determination from the experimental data.

The calculations of the averaged neutron radiative capture cross sections in the energy interval 10^{-3} - 2 MeV

Now let us make use of the formerly developed approach to the problem of electric dipole radiative strength functions to calculate neutron radiative capture cross sections and to reproduce their energy dependence. For nuclei that are heavy enough $(A \gtrsim 50)$ and for energies from some keV to some MeV the compound nucleus formation and its subsequent decay dominate the neutron radiative capture mechanism [3]. The underlying procedure of statistical calculations for energy averaged cross sections is well known [4]. The key values in such calculations are the particle and γ-ray transmission coefficients $T_\gamma(E,l,J)$ as well as nuclear level schemes and level densities. These last were discussed previously. The transmission coefficient for a specific channel

$$T_\gamma(E,l,J) = 1 - \exp[-2\pi(\bar\Gamma_\gamma/D)_{E,l,J}] \qquad (11)$$

is connected with a corresponding strength function. Thus, for the channel of electric dipole emission one has:

$$(\bar\Gamma_\gamma/D)_{E_\gamma,E1} = E_\gamma^3 \cdot S_{E1}(E_\gamma) \qquad (12)$$

where $S_{E1}(E_\gamma)$ is determined by formula (8). The most important particle-channels in the case under investigation are neutron emission channels, and the main contribution to the low energy cross sections originates from angular momenta $l=0$ (s-wave) and $l=1$ (p-wave). In our calculations we make use of experimental data [3] on neutron s- and p-strength functions:

$$(\bar\Gamma_n/D)_{E_n,l} = P_l(E_n) \cdot \sqrt{E_n} \cdot S_{n,l} \cdot g^{-1} \qquad (13)$$

where in usual notations, g is the spin-statistical factor, $P_l(E)$ - centrifugal barrier penetrability and E_n - neutron energy in eV. For all others channels, that are of minor importance, the optical model values of $T_\gamma(E,l,J)$ have been taken. It must be noted too that width fluctuation corrections have been made following the recommendation given in [5].

Some results of the calculations are shown in the table 1. The agreement with experimental data up to 2 MeV neutron energy is sufficiently good, especially for $A = 100 \div 200$. The improvement in the low energy part of excitation curves is achieved mainly on account of the usage of experimental s- and p- neutron strength functions instead of optical model prescriptions. In the energy region $E_n > 0.5$ MeV it is important to use a correct form of the γ-ray strength function for better agreement with experiment and in particular it is necessary to take into account the energy dependence of the particle-hole correlation coefficient, in accordance with the expression (10).

The presented systematics has a clear physical meaning, ensures better approximation of the experimental data on radiative widths and neutron radiative cross sections than earlier ones and may be fruitfully used for prediction of the unknown values.

Table 1. The calculated average neutron radiative capture cross sections.

Target nuclei	$\sigma_{n\gamma}$, mb, at neutron energy, MeV							
	0.001	0.005	0.01	0.05	0.1	0.5	1.0	2.0
$^{51}V_{23}$	755	159	81	17.5	9.3	3.9	3.1	2.1
$^{55}Mn_{25}$	1310	288	149	33	18	5.8	4.9	6.3
$^{65}Cu_{29}$	821	189	106	31	19	14	7.9	1.2
$^{93}Nb_{41}$	2230	876	616	255	160	60	44	20
$^{98}Mo_{42}$	562	237	165	54	32	12	6.6	4.8
$^{100}Mo_{42}$	555	230	155	47	27	10	2.2	4.1
$^{103}Rh_{45}$	4150	1780	1290	753	544	172	95	84
$^{107}Ag_{47}$	6070	2310	1570	720	566	166	96	72
$^{109}Ag_{47}$	3480	1540	1130	648	579	199	100	74
$^{115}In_{49}$	2610	1200	934	549	405	194	169	71
$^{127}I_{53}$	5220	1940	1310	614	346	107	59	54
$^{133}Cs_{55}$	4370	1550	991	389	252	155	60	50
$^{139}La_{57}$	394	131	85	34	23	9.3	11	6.5
$^{151}Eu_{63}$	40200	15800	10400	3440	2240	593	445	270
$^{153}Eu_{63}$	25300	9430	6020	2090	1150	229	183	117
$^{159}Tb_{65}$	14000	5030	3260	1340	743	228	164	128
$^{165}Ho_{67}$	13400	4440	2730	922	546	152	99	83
$^{185}Re_{75}$	15300	5100	3190	1180	814	232	160	89
$^{197}Au_{79}$	8670	2810	1740	632	366	96	54	52
$^{205}Tl_{81}$	1170	276	150	39	11	1.8	1.5	0.9
$^{232}Th_{90}$	4090	1380	943	413	203	88	57	31

REFERENCES:

1. A.Bohr, B.Mottelson, Nuclear Structure 1, W.A.Benjamin Inc., N.Y., 1969.
2. W.Dilg, W.Schantt, H.Vonach, H.Uhl: Hucl.Phys. A217, 269 (1973).
3. T.S.Belanova, A.V.Ignatjuk, A.B.Pashenko, V.I.Plyaskin, Radiative Neutron Capture, Atomizdat, Moscow, 1986.
4. Neutron Physics and Nuclear Data in Science and Technology (edited by R.E.Chrien), 3, 270 (1984).
5. A.A.Lukyanov, Neutron Cross Sections Structure, Atomizdat, Moscow, 1978.

THE INELASTIC SCATTERING IN THE PRESENCE OF DIRECT REACTIONS: DESCRIPTION OF THE HIGH-LYING STATES EXCITATION ALLOWING FOR UNITARITY

S.N.Ezhov[1], V.A.Plujko[2], G.A.Prokopets[1]

[1]Kiev State University, Kiev, USSR
[2]Institute for Nuclear Research, Kiev, USSR

Abstract: A method is formulated and studied which allows to calculate the fluctuation component of the average cross section for high incident energy. The method takes into account the condition of the unitarity as in the HRTW method but with including of the highly excitated states of the residual nucleus. The comparing of this approach with standard one had been carried out for the elastic and inelastic scattering neutrons by the tungsten.

(Random matrix method, optical model, coupled channels, level density, $^{182}W(n,n')$, cross section, angular distribution)

Introduction

There is a considerable variety of nuclear models and nuclear reaction mechanisms that are used for nuclear data evaluation. But the problem of an adequate and physically meaningful formalism still exists.

It is known that Weidenmüller et al. [1-3] and Moldauer [4,5] developed the methods for calculation of the average fluctuation cross sections that take into account correlations between the partial width amplitudes caused by direct processes. Unfortunately the applicability of these methods is restricted by quite low incident particle energy when the spectrum of the residual nucleus can be taken into account explicitly.

The number of the open channels grows exponentially with incident energy and in this case the approach mentioned above becomes impracticable. As a rule the highly excited states are included in consideration on the phenomenological ground (e.g.[6]) by the addition of special term to the escape width that is calculated independently of the transmission coefficients for the low-lying states.

We should like to point out that the procedure of such kind may cause the violation of probability current conservation. In this paper we formulate the method which avoids the above disagreement and compare it with standard one.

The description of the procedure

According to [1,2] the second moments of the S-matrix (for fixed spin and parity) are

$$<S^{fl}_{ab}S^{fl*}_{cd}> =$$

$$\sum_{efgh} U^*_{ea}U^*_{fb}U_{gc}U_{hd}<\widetilde{S}^{fl}_{ef}\widetilde{S}^{fl*}_{gh}> , \quad (1)$$

where $S^{fl} = S - <S>$, U is Engelbrecht-Weidenmüller transformation and $\widetilde{S} = USU^T$ describes the scattering in the absence of the direct inelastic transitions.

The second moments $<\widetilde{S}^{fl}\widetilde{S}^{fl*}>$ are determined from equations [2,7]

$$<\widetilde{S}^{fl}_{ab}\widetilde{S}^{fl*}_{cd}> = (\delta_{ac}\delta_{bd} + \delta_{ad}\delta_{bc})*$$

$$V_aV_b/SpV + \delta_{ab}\delta_{cd} ((W_a-2)(W_c-2))^{1/2}* \cdot$$

$$V_aV_c/SpV , \quad (2)$$

where the unitary condition restricts the V_a's and the elastic enhancement factors W_a by the relations

$$V_a[1+V_a(W_a-1)/SpV]= r_a =1-|<\widetilde{S}_{aa}>|^2 . \quad (3)$$

For the incident particle with the energy of a few MeV the elastic channel is coupled strongly with some inelastic channels only. As consequence we may separate the set of the open channels on two groups: "strongly" coupled channels (i), for which the average S-matrix is calculated in the frame of the coupled channels method, and "weakly" coupled channels (α), for which the perturbation theory is valid [8].

In the first order the coefficients r_α for "weak" channels coincide with the diagonal elements of the Satchler's penetration matrix [9]: $r_\alpha = P_{\alpha\alpha}= T_\alpha$.
As result the Eqs. (3) become

$$V_i[1+V_i(W_i-1)/SpV]= r_i =1-|<\widetilde{S}_{ii}>|^2 , \quad (4)$$

$$V_\alpha[1+V_\alpha(W-1)/SpV]= T_\alpha, \quad (5)$$

$$SpV = \sum_{a=(i,\alpha)} \int_0^{E_0} \rho_a(E)V_a(E_0-E)dE . \quad (6)$$

Here E_0 is the energy of incident particle; $\rho_a(E)$ - the level density of the residual nucleus:

$$\rho_a(E)= \begin{cases} \sum_i \delta(E-E_i), & E < E^*, \\ \rho(E,I_\alpha), & E > E^*, \end{cases} \qquad (7)$$

E^* - the energy of the highest level of the residual nucleus for which all quantum characteristics are known and that had taken into account in the coupled channel scheme explicitly. The function $\rho(E,I_\alpha)$ is the average density for levels with spin I_α and excitation energy E.

The index of any channel "a" denotes the set of the fragment's quantum numbers (E_a,I_a,l_a,j_a) and as sequence we can write

$$SpV = \sum_i V_i + \sum_{Ilj}{}' \int_{E^*}^{E_0} \rho(E,I) V_{1j}(E_0-E)dE, \qquad (8)$$

where $I + j = J$.

To evaluate of the integral in the (8) one can use any numerical quadrature with $\mathcal{E}_\gamma$ as nodes and ω_γ as weights. As a result one can rewrite Eqs. (5) and (8) in the form

$$V_{1j}(\mathcal{E}_\gamma')[1 + \frac{V_{1j}(\mathcal{E}_\gamma')(W_{1j}(\mathcal{E}_\gamma')-1)}{SpV}]=T_{1j}(\mathcal{E}_\gamma') \qquad (9)$$

$$SpV = \sum_i V_i + \sum_{\gamma Ilj} \rho(\mathcal{E}_\gamma,I)V_{1j}(\mathcal{E}_\gamma')\omega_\gamma, \qquad (10)$$

where $\mathcal{E}_\gamma' = E_0 - \mathcal{E}_\gamma$.

Starting from the parametrization [2] for $<\widetilde{S}^{f1}{}_{aa}\widetilde{S}^{f1*}{}_{bb}>$

$$<\widetilde{S}^{f1}{}_{aa}\widetilde{S}^{f1*}{}_{bb}> = \chi_a \chi_b \qquad (11)$$

we find [7] the following expression for the elastic enhancement coefficient in the channel a:

$$W_a=2+(1-r_a)(1-y_a+4y_a^2)^2/(1+0.15SpP)^2, \qquad (12)$$

$$y_a=V_a(1-r_a)^{1/2}/((1+0.15SpP)(SpV)^{1/2}),$$

where index "a" may denote "strong" or "weak" channels.

Now we can solve the system of the coupled equations (4),(9), (10) and (12) by means of an iteration procedure to determine coefficients V_a and W_a and then we can use the formulas (1) and (2) for obtaining of the average cross sections and angular distributions.

Comparison with the standard approach and the experimental data

The comparison with experimental data on fast neutrons interaction with nucleus ^{182}W, some first levels wich belong to the ground state rotational band, has been chosen to test the model.

The optical-potential parameters were described in [10] and we used $0^+-2^+-4^+$ coupled scheme [11]. Only the partial waves with l 6 were taken into account.

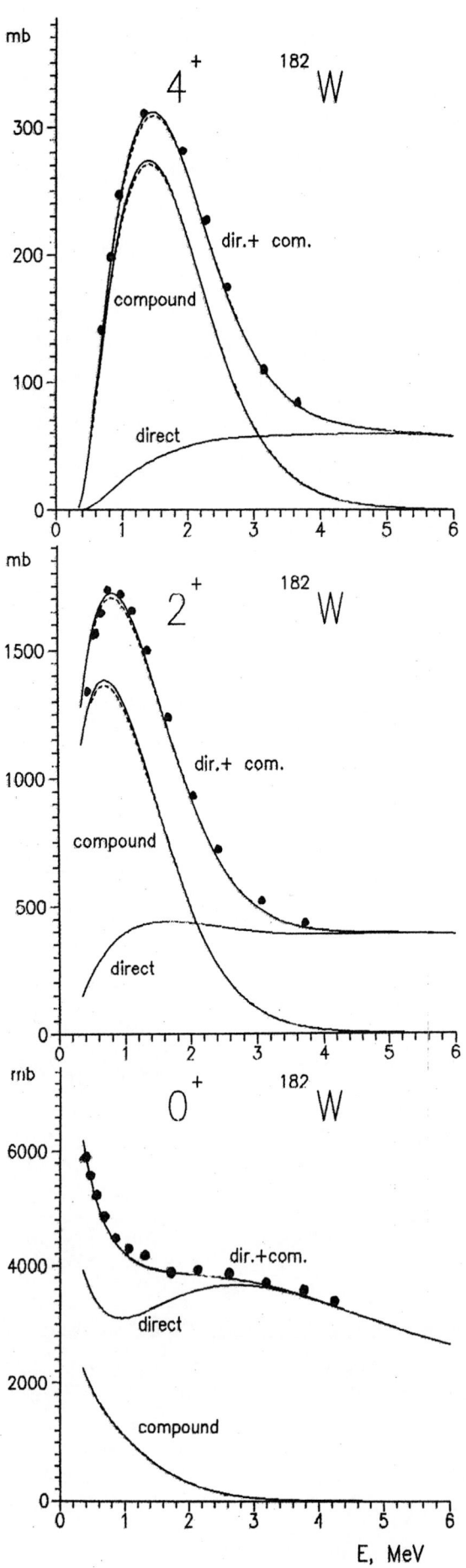

Fig. 1. The excitation functions

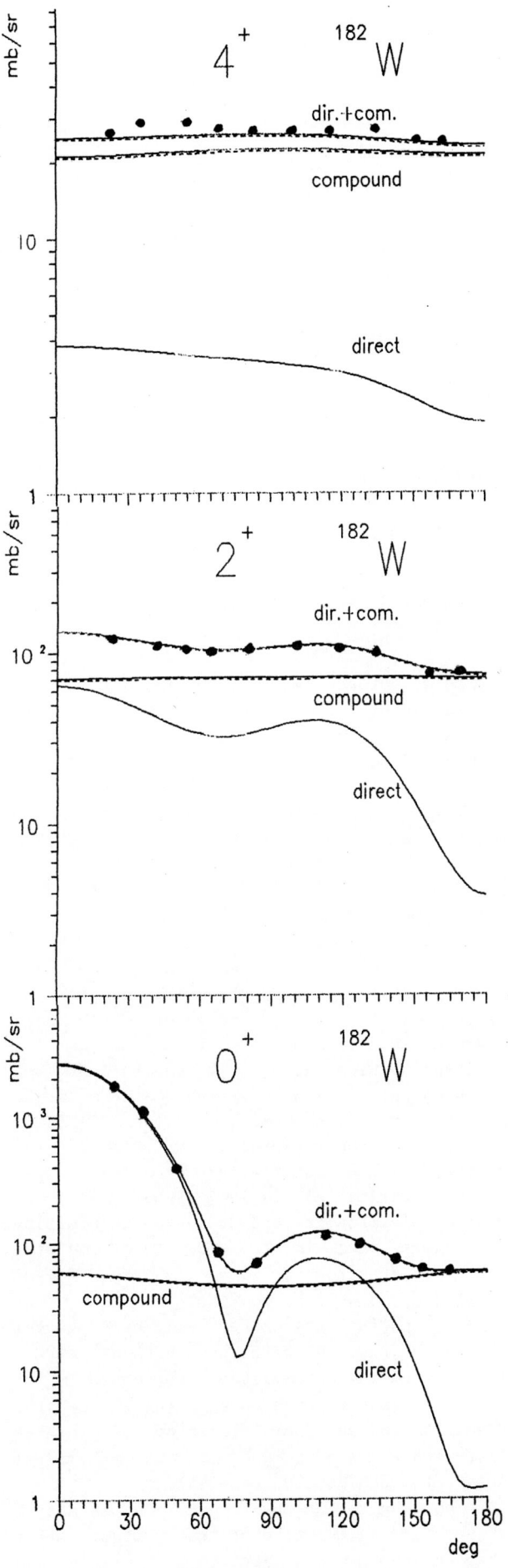

Fig. 2. Angular distribution, E(lab)=1.5 MeV

In Figs. 1,2 we compare the excitation functions and the angular distributions for three first levels of [182]W with experimental data [12]: solid curves - our approach, dashed curves - "standard" approach in which coefficients V_α are determined as

$$V_\alpha = r_\alpha = T_\alpha, \qquad (13)$$

where transmission coefficients T_α may be calculated in the frame of the usual optical model or with the help of any other parametrization. In particular we used the parametrization [13].

Formulas of Gilbert and Cameron [14] were used for calculating of the level density $\rho(E, I_\alpha)$.

The difference between the results of two calculation methods (refined and standard) does not exceed 5% and falls down when the incident particle energy increases. Therefore it is important take into account of this relatively small effect if one is forced to use evaluating code of high precision.

REFERENCES

1. C.A.Engelbrecht, H.A.Weidenmuller : Phys.Rev. C8, 859 (1973).
2. H.M.Hofmann, J.Richert, W.Tepel, H.A. Weidenmuller: Ann.Phys. 90,403(1975).
3. J.J.M.Verbaarshot, H.A.Weidenmuller, M.R.Zirnbauer: Phys.Rep. C129, 367 (1985).
4. P.A.Moldauer:Phys.Rev. C12,744(1975).
5. P.A.Moldauer:Phys.Rev. C11,426(1975).
6. A.E.Saveljev,I.K.Averjanov, B.M.Dzjuba: Jadernaja Fizika 12, 704 (1970).
7. S.N.Ezhov, N.E.Kabakova, V.A.Plujko : Sov.J. Nucl. Phys. 42, 97 (1985).
8. S.N.Ezhov, V.A.Plujko: Sov. J. Nucl. Phys. 39, 248 (1984).
9. G.R.Satchler: Phys.Lett. 7, 55(1963).
10. J.P.Delaroche, G.Haouat, J.Lachkar, Y.Patin, J.Sigaud, J.Chardine: Phys. Rev. C23, 136 (1981).
11. T.Tamura: Rev.Mod.Phys. 37,679(1965).
12. P.T.Guenther, A.B.Smith, J.F.Whalen : Phys. Rev. C26, 2433 (1982).
13. K.H.M.Murthy, S.K.Gupta, A.Chatterjee: Z. Phys. A305, 73 (1982).
14. A.Gilbert, A.G.W.Cameron: Can.J.Phys. 43, 1446 (1965).

DOUBLE DIFFERENTIAL NEUTRON EMISSION CROSS SECTIONS
OF (p,n) REACTIONS ON ^{59}Co and Mo ISOTOPES

Zuying Zhou, Hongqing Tang, Qingchang Sui,
Jun Sa, Bujia Qi and Guanren Shen

China Institute of Atomic Energy (CIAE)
P. O. Box 275 (46), Beijing 102413, P. R. China

Abstract : Double differential neutron emission cross sections for the (p,n) reactions on ^{59}Co and Mo isotopes ^{95}Mo, ^{96}Mo, ^{97}Mo, ^{98}Mo and ^{100}Mo at proton energy of 9 MeV, on ^{95}Mo at 13 and 15 MeV and on ^{59}Co at 11, 13, and 15 MeV have been measured. The presented level density parameters were derived from comparing the calculations using conventional evaporation model with the experimental data.

(level density, time of flight, evaporation model, Co, Mo)

Introduction

At low excitation energy the nuclear level densities can be determined by direct counting of the number of levels and at about neutron binding energy from resonance data, but level missing could be caused due to the limitation of experimental energy resolution. Alternatively, nuclear level densities can be derived from energy differential particle emission cross sections in compound nucleus reactions. For this purpose (p,n) and (α,n) reactions are recommended since they have smaller non-compound contributions at excitation energy below neutron binding energy.

Double differential neutron emission cross sections for (p,n) reactions on ^{59}Co and Mo isotopes ^{95}Mo, ^{96}Mo, ^{97}Mo ^{98}Mo and ^{100}Mo at proton energy of 9 MeV, on ^{95}Mo at 13 and 15 MeV and on ^{59}Co at 11, 13, 15 MeV have been measured. The presented level density parameters were derived from comparing the calculations using conventional evaporation model with the experimental data.

Experiment and Data Reduction

The experiments were carried out at the HI-13 tandem Van de Graaff accelerator in CIAE. A schematic layout of the experimental arrangement including the neutron time-of-flight setup, target chamber and the Faraday cup is shown in Fig. 1.

The neutron time-of-flight setup consists of three detectors. Each of them is composed of a liquid scintillator (10.5 cm diam *5 cm, filled with liquid scintillator ST-451) and a photomultiplier tube XP-2041. The detectors are massively shielded and well collimated to the target. The flight path is 5-7 m. The target chamber is an aluminium cylinder with a diameter of 44 cm, a height of 45 cm and a thin wall of 3 mm. The pulsed proton beam impinged on a self supporting target foil.

A beam burst of about 1 ns width (FWHM) with duration of 500 ns was used. The beam pick-off signal was taken out right in the upstream of the target chamber. The beam average current is about 100 nA.

Detector bias was set at about 0.85 MeV neutron energy. To reduce γ-ray background an n-γ pulse shape discriminator modular (CANBERRA 2160A) was used. Detector efficiency was calibrated with n-p scattering experiments [1] and Monte-Carlo calculation. Three sets of standard electronics and a data acquisition system were used in the TOF spectrometers. Another TOF spectrometer with a shorter flight path of 3 m was used as monitor.

For each incident energy and each target both neutron spectra with and without target were measured for background subtraction. The correction has been done considering the neutron attenuation through 3 mm aluminium of chamber wall for each angle (< 2%) and through target and target frame for 90° (5%-11%).

According to the position of the target γ peak and time calibration of the system, after background subtraction, detector efficiency and neutron attenuation correction the TOF spectra

reaction were calculated using the conventional evaporation model. The optical potential

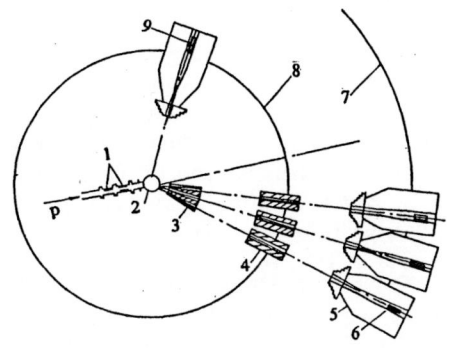

Fig. 1 Three-detector TOF setup

1.Beam pick off 2.Target chamber 3.Collimator-1
4. Collimator-2 5. Shielding 6. Neutron detector
7. R=6m rail 8. R=3m rail 9. Neutron monitor

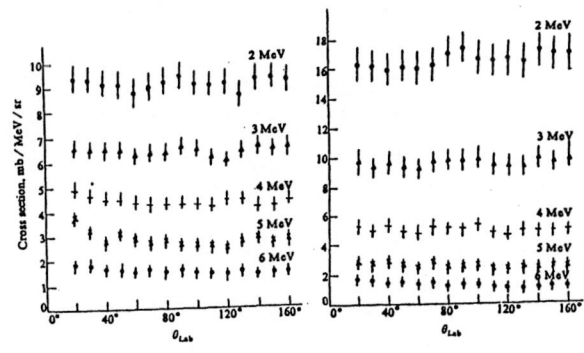

Fig.3 Angular distribution

Left: ^{59}Co(p,n)^{59}Ni, Ep=15 MeV

Right: ^{95}Mo(p,n)^{95}Tc, Ep=15 MeV

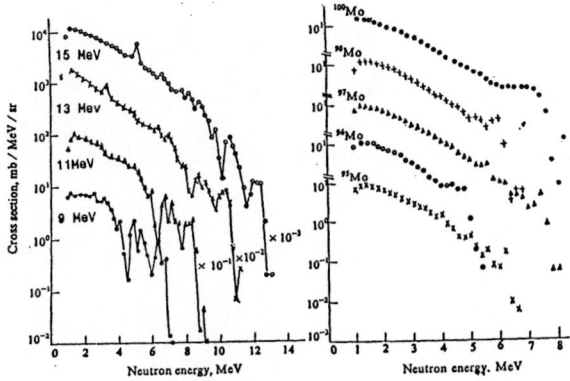

Fig.2 Neutron emission spectra at 50 deg..

Left: ^{59}Co(p,n)^{59}Ni

Right: Mo isotopes (p,n) reactions Ep=9 MeV

were converted into the energy spectra with bin size 200 kev in the centre of mass system. The neutron energy spectra are shown in Fig. 2 and some angular distributions are given in Fig. 3. Angle integrated neutron energy spectra were obtained using Legendre polynomial fitting to the data.

Total uncertainties of 7 - 8.5 % for double differential cross sections come from the error sources of statistics, detector efficiency, beam current integration and target thickness, etc..

Neutrons with energies less than 1.4 MeV were not included in the spectra since the detector efficiency changed rapidly with energy near the bias.

Discussion

The isotropic angular distributions for neutrons with energies less than 6 MeV show that compound nucleus reactions predominate in that region. Neutron emission spectra for (p,n)

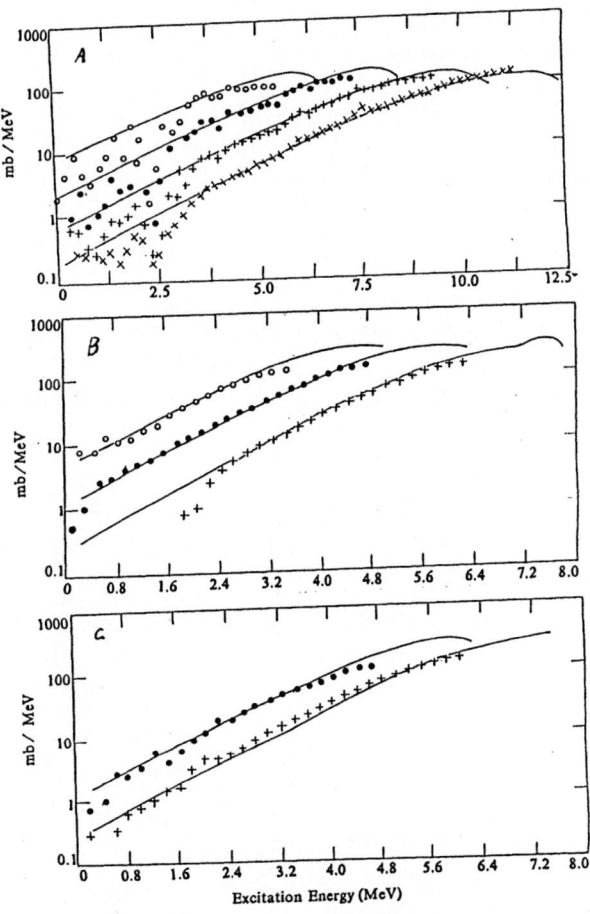

Fig.4 Comparison of experimental neutron spectra and calculation, solid lines represent calculation results.

A: For ^{59}Co(p,n)^{59}Ni, ○: Ep=9 MeV, •: Ep=11 MeV
+: Ep=13 MeV, x: Ep=15 MeV

B: Points labeled ○ for ^{96}Mo(p,n)^{96}Tc,
• for ^{98}Mo(p,n)^{98}Tc and + for ^{100}Mo(p,n)^{100}Tc

C: • For ^{95}Mo(p,n)^{95}Tc and + for ^{97}Mo(p,n)^{97}Tc

parameters of Perey [2] for proton and of Wilmore and Hodgson [3] for neutron were applied. The level density formula used in the calculation is in Gilbert-Cameron form with the new statistics of Su Zongdi et al. [4] :

$$Ux=1.4+263A$$
$$a =[0.00880(S(z)+S(n))+QB]A$$

where the QB are taken as :

$$QB= \begin{cases} 0.142, & 29 \leq z \leq 62 \text{ and } n \leq 89; \ 79 \leq z \leq 85 \\ 0.120, & z<28; \ 62(\text{and } n>89) \leq z \leq 78; \ z \geq 86. \end{cases}$$

and their pairing and shell correction values based on 667 sets of experimental data. The comparison of the calculations with the experimental data is shown in Fig.4. The level density parameters "a" were determined from the optimal description for experimental spectra and are given in the following table. To make the experimental spectra contain as low non-compound component as possible the data shown in the Fig. 4 are angle integrated based on those in backward hemisphere.

The cross sections for the competition channels from proton induced reaction on Mo isotopes are much lower as compared with those of (p,n) reactions. If changes in transmission coefficients can be ignored , the cross sections of (p,n) reactions on Mo isotopes will be determined by level densities of the residual nuclei. However in the case of ^{59}Co, the cross sections of (p,p') process are comparable with those of (p,n) reaction and there may be some ambiguity in the analysis of level density of ^{59}Ni.

Acknowledgement

The authors are grateful to Professor W. Scoble for his help with Mo isotopic targets.

Level density parameters a and δ

Nucleus	^{59}Ni	^{95}Tc	^{96}Tc	^{97}Tc	^{98}Tc	^{100}Tc
a (MeV^{-1})	6.3	10.7	11.0	11.2	11.2	11.8
δ (MeV)	1.00	0.60	-0.45	0.97	-0.15	-0.15

References

1. Shen Guanren, Sa Jun, Huang Tangzi, Yu Chunying, Tang Hongqing and Sui Qingchang: Chinese Atomic Energy Science and Technology, 21, 9 (1987)

2. F. G. Perey and C. M. Perey: Phys. Rev., 131, 745 (1963)

3. D. Wilmore and P. E. Hodgson: Nucl. Phys., 55, 673 (1964)

4. Su Zongdi, Wang Cuilan, Zhuang Youxiang and Zhou Chunmei: INDC(CPR)-2 (1985)

FAST-NEUTRON INELASTIC SCATTERING CROSS SECTIONS
FROM 2.3 TO 3.0 MeV FOR THE ACTINIDE NUCLEI 232Th AND 238U

E. Sheldon, A. Aliyar, L.E. Beghian, J.J. Egan, G.H.R. Kegel, A. Mittler and E.D. Arthur*

Department of Physics and Applied Physics, University of Lowell, 1 University Avenue, Lowell, MA 01854-2881, U.S.A. and
* *T-DO, Mail Stop B-210, Los Alamos Scientific Laboratory, Post Office Box 1663, Los Alamos, NM 87545, U.S.A.*

Abstract: Experimental and theoretical (CN + DI) excitation functions to 3.5 MeV and angular distributions at E_n = 2.4 and 2.8 MeV are presented for inelastic neutron scattering on 232Th and 238U to individual or grouped rotational/vibrational levels up to $E^* \cong 1,000$ keV.

[232Th, 238U(n,n'): inelastic neutron scattering, actinides, cross sections, differential cross sections, excitation functions, angular distributions, collective-model rot./vibrational levels.]

Introduction

Herein we update from 1988 the Lowell group's results for neutron scattering on the principal actinide nuclei, 232Th and 238U, concentrating on elevated incident energies in the 2 - 3 MeV range. At the 1988 Mito Conference [1], measured and theoretical *elastic* (n,n) data were presented for these (and other actinide) nuclides (*e.g.*, excitation functions [EF's] to 2.5 MeV and angular distributions [AD's] at E_n = 0.185, 0.55, 2.4 and 2.8 MeV) to supplement earlier [2] *inelastic* data [*e.g.*, (n,n') AD's from (n,n'γ) at 1.1, 1.9, 2.5 and 3.1 MeV, and from (n,n') at 0.185, 0.55, 1.2, 1.5, and 2.0 MeV]. The Mito report [1] detailed the experimental and theoretical procedures adopted by the Lowell group for such data acquisition, together with the relevant references (which, for reasons of brevity, will not be repeated here).

Experimental and Theoretical Techniques

The experimental techniques utilizing subnanosecond-resolution time-of-flight neutron spectroscopy have been described, including the latest corrections and refinements, by Goswami *et al.* [3], Aliyar [4], Kegel *et al.* [5], and Egan *et al.* [6]. Even with premium resolution, the measurements at these elevated (n,n') energies cannot in some instances resolve separate neutron groups going to the closely-spaced high-lying vibrational levels in these nuclei, and consequently the data in such cases comprise a composite for a *group* of excited states. Specifically, the data for 232Th(n,n') constitute EF's and AD's for the following five groups:

(i) the 333.1-keV (6+) [K = 0+ ground-state rotational] niveau;
(ii) the pair of states at 714.25 keV (1-) [K=0- octupole vib.] and 730.10 keV (0+) [K = 0+ β-vibrational];
(iii) a composite of five levels in the E^* = 774 - 830 keV range comprising the 774.1-keV (2+) [K=0+ β-vib.], 774.4-keV (3-) [K=0- oct. vib.], 785.2-keV (2+) [K=2+ γ-vib.], 828.1-keV (10+) [K=0+ g.s. rot.] and 829.6-keV (3+) [K=2+ γ-vibration] states;
(iv) a combination of three levels in the E^* = 873-890 keV range consisting of the 873-keV (4+) [K=2+ β-vib.], 883.3-keV (5-) [K=0- oct. vib.] and 890.1-keV (4+) [K=2+ γ-vibration] states;
and (v) the 960.2-keV (5+) [K = 2+ γ-vibrational] level.

For 238U(n,n'), the somewhat wider level spacing permitted better individualization of states into the following five groups:

(i) the 307.2-keV (6+) [K = 0+ ground-state rotational] level;
(ii) the 680.1-keV (1-) [K = 0- octupole vibrational] state;
(iii) the next 731.9-keV (3-) [K = 0- octupole vibrational] level;
(iv) the next 827.2-keV (5-) [K = 0- octupole vibrational] level;
and (v) a composite of eight levels in the E^* =927 - 998 keV range consisting of the 927.2-keV (0+) [K =0+ 2-phonon γ-vib.], the 930.8-keV (1-) [K=1- oct. vib.], the next 950.0-keV (2-) [K =1- oct. vib.], the 966-keV (7-) [K = 0- oct. vib.], 966.3-keV (2+) [K = 0+ 2-phonon γ-vib.], 993-keV (0+) [K = 0+ β-vib.] and 997.5-keV (3-) [K = 1- octupole vibrational] states.

In the theoretical analyses of EF's and AD's it was necessary to compute the requisite total (angle-integrated) and differential cross-sections as a function of energy and angle for each of the above states (and to sum the respective contributions whenever levels were combined in a composite group). To mitigate against overly long computation times, the calculations were restricted to the standard (compound-nucleus [CN] + direct-interaction [DI]) formalism and were *not* performed in the intricate HRTW mode. For consistency, the optical -potential parameters and deformations proposed by the Bruyères-le-Châtel group [7,8] were adopted throughout. In addition to the Lowell computations, performed with the CN program CINDY (with provision for the effects of Moldauer level-width fluctua-

tions and competing exit channels, including radiative capture, fission, and continuum channels) and the coupled-channels DI program KARJUP, some earlier theoretical Los Alamos results [9,10] were incorporated (in Fig. 1 overleaf) for comparison. In the latter, the CN cross sections were generated by the Hauser-Feshbach program COMNUC while the DI cross sections for the vibrational states were furnished by DWUCK (a simpler and quicker DWDI code than the above, but devoid of interband coupling).

Results

The excitation functions for 232Th and 238U(n,n') to 3.5 MeV are shown in Figs. 1(a,b) respectively. The points with error bars represent the measurements by Aliyar *et al.* [4,5], while the solid curves depict the theoretical Lowell results and the broken curves denote the Los Alamos [9,10] findings. A similar scheme is adopted for the 232Th(n,n') angular distributions at 2.4 and 2.8 MeV shown in Figs. 2(a,b) and for the 238U(n,n') angular distributions at 2.4 and 2.8 MeV displayed in Figs. 3(a,b). Pursuant to Arthur's suggestion [10] that previous coupling strengths had been rather too high, they were reduced to 0.01 for the β-band and 0.05 for the other vibrational bands, which advantageously lowered the magnitude of the DI contribution. No attempt was made to "fine-tune" these sole adjustable parameters further [B(Eλ)-values substantiate the above choices] in the quest for still further improving the fits [11] to the measured data.

REFERENCES AND ACKNOWLEDGEMENTS

1. Eric Sheldon, in *Proc. Int. Conf. on Nuclear Data for Science and Technology, May/June 1988, Mito, Japan* (Edited by S. Igarasi), pp. 105 - 110, Saikon, Tokyo (1988).
2. E. Sheldon, L. E. Beghian, D. W. S. Chan, A. Chang, C. A. Ciarcia, G. P. Couchell, J. H. Dave, J. J. Egan, G. Goswami, G.H.R. Kegel, S.Q. Li, A. Mittler, D.J. Pullen, W.A. Schier, J.Q.Shao and A. Wang: J. Phys. G.: Nucl. Phys. 12, 237 - 255 & 443 - 463 (1986).
3. G.C. Goswami, J.J. Egan, G.H.R.Kegel, A. Mittler and E. Sheldon: Nucl. Sci. Eng. 100, 48 - 60 (1988).
4. Abobakr Aliyar, *Inelastic Neutron Scattering Studies of 238U and 232Th on States above 300 keV for Incident Energies above 2.2 MeV.* Ph.D. Dissertation, University of Lowell (1988).
5. G.H.R. Kegel, A. Aliyar, J.J. Egan, C.A. Horton and A. Mittler: Bull.Am. Phys. Soc., Ser. II, 33(8), 1568, Paper AF 13 (1988).
6. J.J. Egan, G. H. R. Kegel, G. Yue, A. Mittler, P.A. Staples, D.J. DeSimone and M.L. Woodring (Contributed Paper, this Conference).
7. G. Haouat, Ch. Lagrange, J.Lachkar, J.Jary, Y. Patin and J. Sigaud in *Proc. Int. Conf. on Nuclear Cross Sections for Technology, Oct. 1979, Knoxville, USA* (Edited by J.L. Fowler, G.H.Johnson and C.D. Bowman), Report CP 791022, pp. 672 - 676, U.S. Government Printing Office, Washington (1980).
8. G. Haouat, J. Lachkar, Ch. Lagrange, J.Jary, J.Sigaud and Y.Patin: Nucl. Sci. Eng. 81, 491 - 511 (1982).
9. E.D. Arthur: Private communication (1984); J.J. Egan, E.D. Arthur, G.Kegel, A. Mittler and J.Q. Shao, in *Proc. Int. Conf. on Nuclear Data for Basic and Applied Science, May 1985, Santa Fe, USA,* Vol. 2, pp. 1209 - 1212, Gordon and Breach, New York (1986).
10. E.D. Arthur: Bull. Am. Phys. Soc., 31(8), 1238, Paper ED10 (1986).
11. E. Sheldon, J. J. Egan, G. H. R. Kegel, A. Mittler and A. Aliyar: Bull. Am. Phys. Soc. (1991) (in publication).

This work was supported in part by a research grant from the U.S. Department of Energy. The extensive use of experimental and computational facilities at the University of Lowell, as well as the valued assistance of accelerator personnel and computer staff, is gratefully acknowledged.

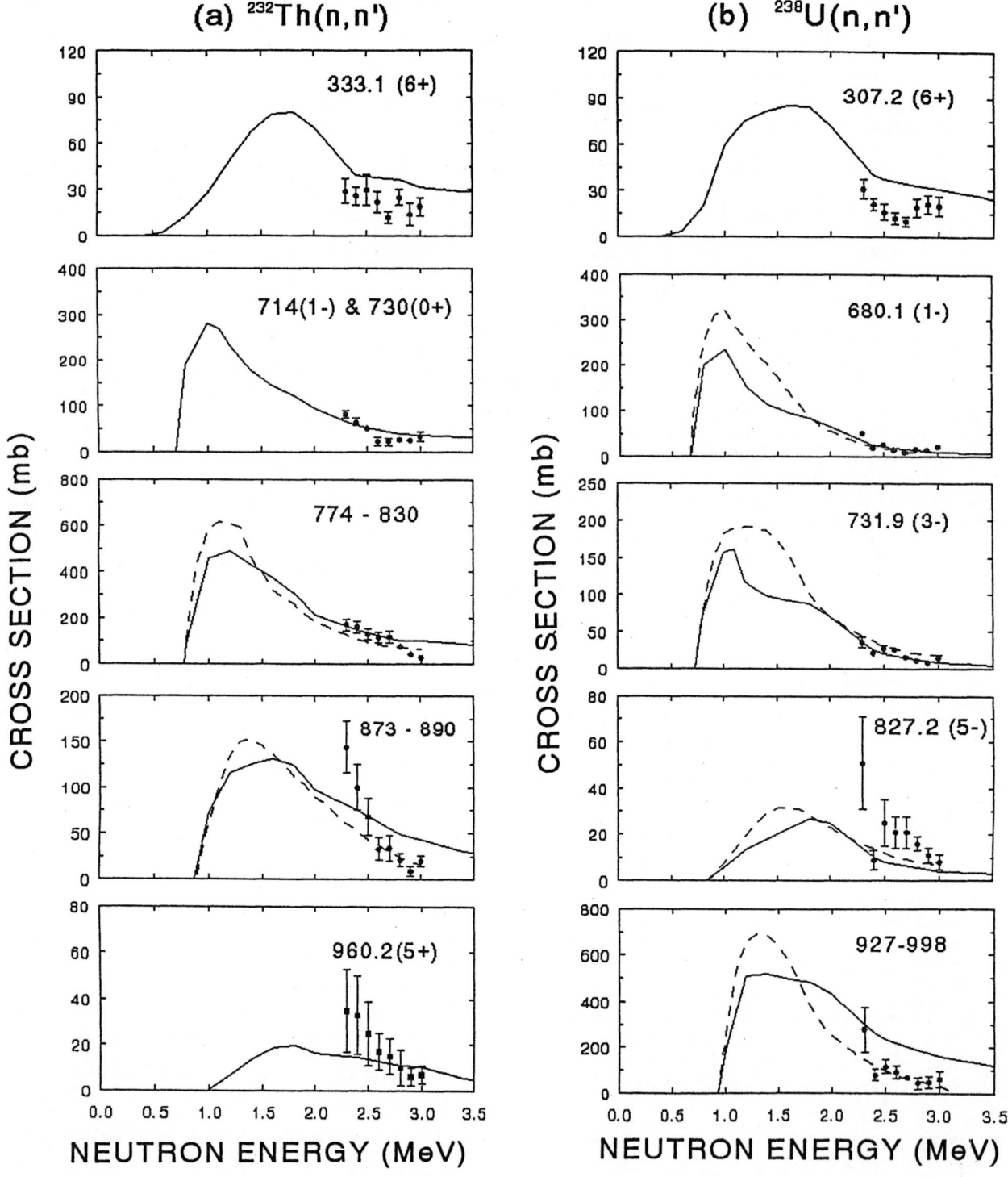

(a) ^{232}Th(n,n')   **(b)** ^{238}U(n,n')

CROSS SECTION (mb)

NEUTRON ENERGY (MeV)

Fig. 1. Excitation functions for inelastic neutron scattering on (a) ^{232}Th and (b) ^{238}U to individual or grouped (unresolvable) levels in rotational (uppermost plots) and vibrational bands up to E*≅1,00 keV, comparing the Lowell group's measured data (points, with error bars) from 2.3 to 3.0 MeV in 0.1-MeV steps [4,5] with the theoretical predictions from the standard (CN + DI) formalism, as evaluated at Lowell (solid curves) and Los Alamos [9,10] (broken curves) to 3.5 MeV. The Bruyères-le-Châtel [6,7] optical-potential parameters and deformations were used throughout. The (Lowell/LosAlamos) CN computations (with CINDY/COMNUC, respectively) allowed for the effects of Moldauer level-width fluctuations, (48/39, respectively) competing n' exit channels, continuum competition, as well as (n,γ) and (n,f) competition; the DI computations (with KARJUP/DWUCK, respectively) included/excluded interband coupling. In the uppermost (6+) plots, the theoretical cross sections at high energies tend to be elevated by the onset of substantial direct-interaction (DI) contributions (as compound-nucleus [CN] admixtures become vanishingly small).

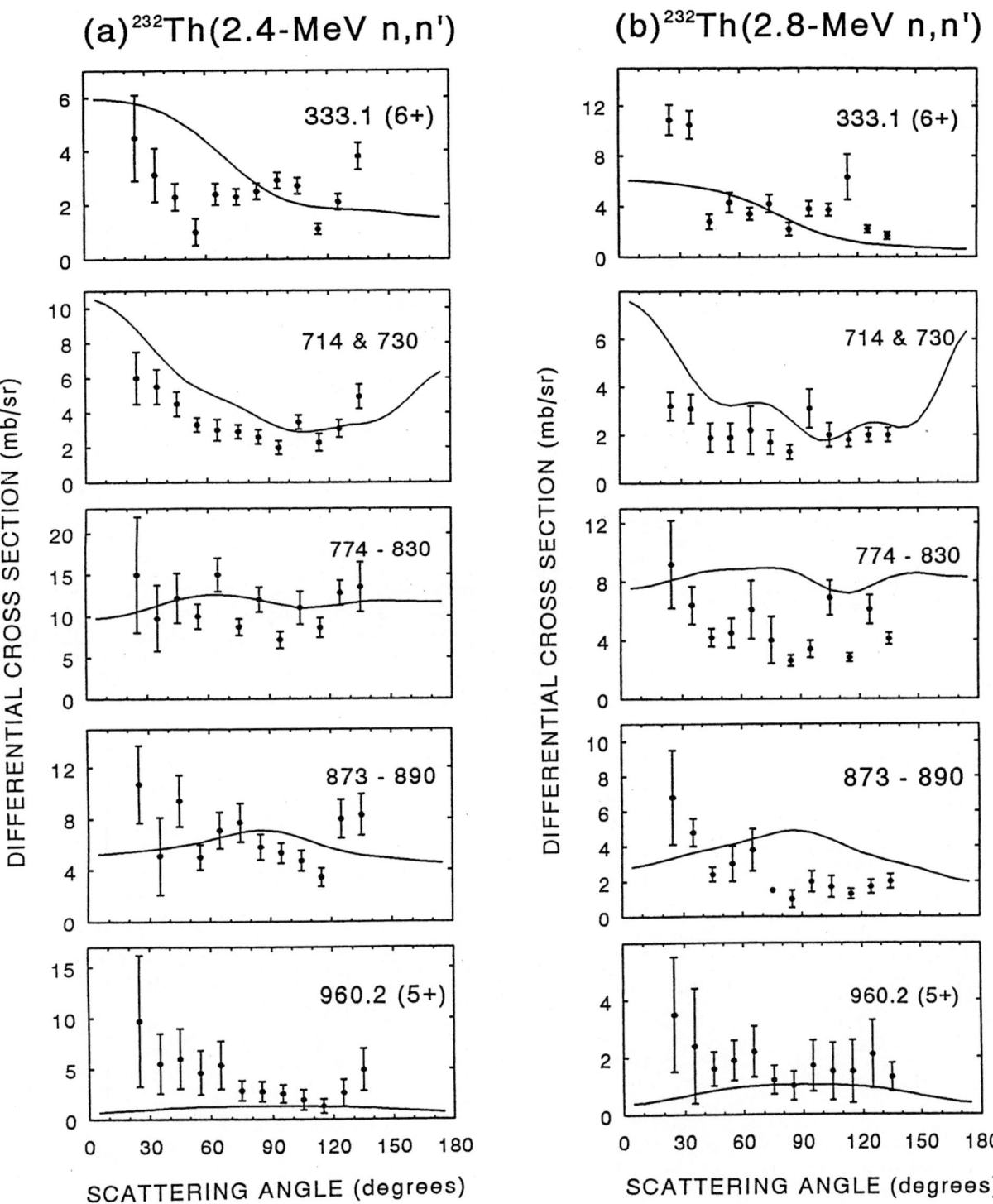

Fig. 2. Angular distributions for ^{232}Th(n,n') scattering to individual or grouped (unresolvable) levels up to E* = 960.2 keV at (a) E_n = 2.4 MeV and (b) E_n = 2.8 MeV, comparing the Lowell group's measured data [4,5] (points, with error bars) with theoretical (CN + DI) differential cross sections [11] computed with the programmes CINDY and KARJUP. The Bruyères optical-potential parameters for ^{232}Th were: V = 46.4 - 0.3E_n MeV, W = 3.6 + 0.4E_n MeV, r_0 = r_0' = 1.26 fm, a=0.63 fm, a' = 0.52 fm, V_{so} = 6.2 MeV, $(r_0)_{so}$ = 1.12 fm, a_{so} = 0.47 fm, and the deformations were: β_2 = 0.190, β_4 = 0.071. The scatter of the measured data-points renders a shape-comparison difficult, but the magnitudes are in reasonably close agreement with theory {with the possible exception of the data for the 960.2-keV (5+) [K = 2+ γ-vibrational] state, for which larger DI band-coupling strengths would be beneficial; however, the experimental error limits are rather large in these cases and "fine-tuning" of the theoretical parameters would appear to be unjustified}.

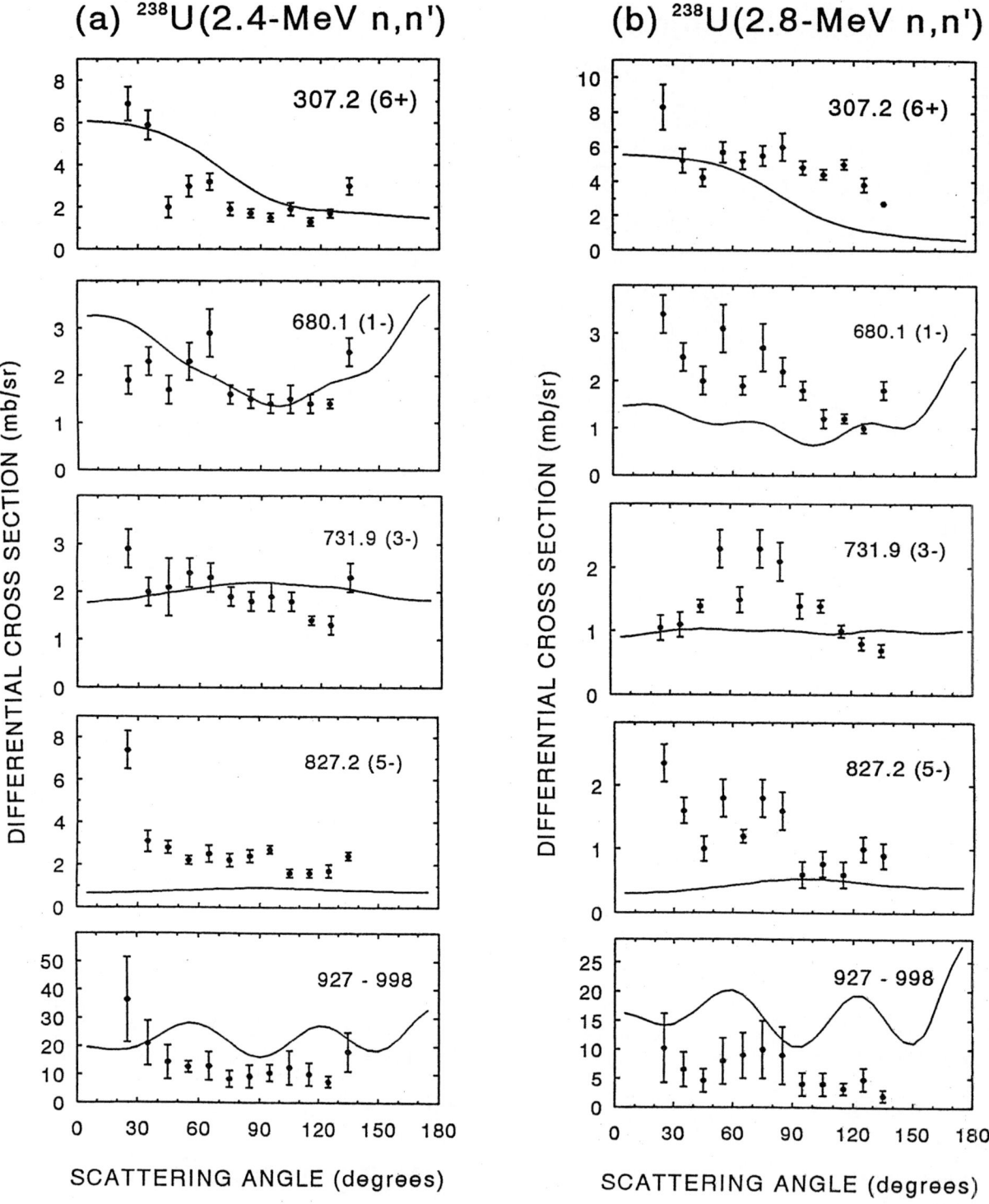

Fig. 3. Angular distributions for ^{238}U(n,n') to individual or composite (unresolvable) levels up to E*=997.5 keV at (a) E_n=2.4 MeV and (b) E_n = 2.8 MeV, comparing Lowell experimental data [4,5] with theoretical (CN + DI) differential cross sections [11], computed using the programmes CINDY and KARJUP with weakened coupling strengths (0.01 for β-bands and 0.05 for others). The Bruyères optical-potential parameters for ^{238}U are as for ^{232}Th (specified in the caption to Fig.2), except that V = 46.2 - 0.3E_n MeV and the deformations are: β_2 = 0.198, β_4 = 0.057. Again, the theoretical magnitudes match the measured differential cross sections quite closely, but a structural fit is unattainable. The rather poor fits in the case of the 827.2-keV (5-) [K = 0- octupole vibrational] state may be attributed to use of too small a value for the band-coupling parameter, but a larger value would have destroyed the fit to the other members of this octupole band, shown immediately above these plots.

Q_{gg} - SYSTEMATICS OF FAST NEUTRON REACTIONS IN THE MASS RANGE A=50 AND NUCLEAR STRUCTURE EFFECTS ON CALCULATED CROSS SECTIONS

M. Avrigeanu and V. Avrigeanu

Institute of Physics and Nuclear Engineering
P.O.Box MG-6, 76900 Bucharest, Romania

Abstract: Correlations of the isotope effect for the (n,p) and (n,α) reaction cross sections at 14.7 MeV, and significant quantities of both the statistical and preequilibrium emission models are made by means of Hauser-Feshbach and Geometry-Dependent Hybrid calculated cross sections. The accuracy of the nuclear model predictions has been proved by extensive comparison to experimental fast neutron reactions data for the stable isotopes of Ti, V, Cr, Mn, Fe, Co and Ni. The calculated cross sections have been furthermore used to discuss the applicability of the Q_{gg} - systematics of the deep inelastic collisions between complex nuclei to fast neutron reactions. Consequences of accurate low-lying discrete levels on confident statistical model calculations are analyzed and necessity of microscopical level schemes is pointed out.

(cross section, preequilibrium, statistical models, isotope effects, Q_{gg} - systematics, low-lying level scheme effects).

Introduction

Test of nuclear model calculations across the valley of stability is of considerable significance for further improvements of the models and their parametrization. Following a unitary account of a whole body of experimental fast reaction data over a large incident energy range for the stable isotopes of Ti, V, Cr, Mn, Fe, Co and Ni [1], in this work both the absolute (n,p), (n,α) and (n,2n) reaction cross section values at 14.7 MeV and steepness of the isotope trend have been found in good agreement with the experimental data for the isotope chains of Ti, Cr, Fe, and Ni (Fig. 1). The calculated cross sections

have been furthermore used to correlate the isotope effects with significant quantities of the statistical as well as preequilibrium emission models.

Generalized Geometry-Dependent Hybrid (GDH) preequilibrium emission model, including alpha-particle emission and angular momentum and parity conservation, and Hauser-Feshbach model calculations were performed with the STAPRE-H code [2], GDH version of STAPRE [3]. The DWBA method has been used to describe the neutron direct inelastic scattering on discrete excited nuclear states by means of the code DWUCK4 [4], while the GDH model has been proved able to account for the same process in the continuum. The same optical model potentials and level density parameters were employed in the three reaction mechanism models. A realistic nuclear level density approach 1 has been used. No free internal parameter for the preequilibrium emission model was involved.

Isotope Effects and Q_{gg} - Systematics

The isotopee effect for the (n,p) reaction cross sections of 14 MeV neutrons, pointed out

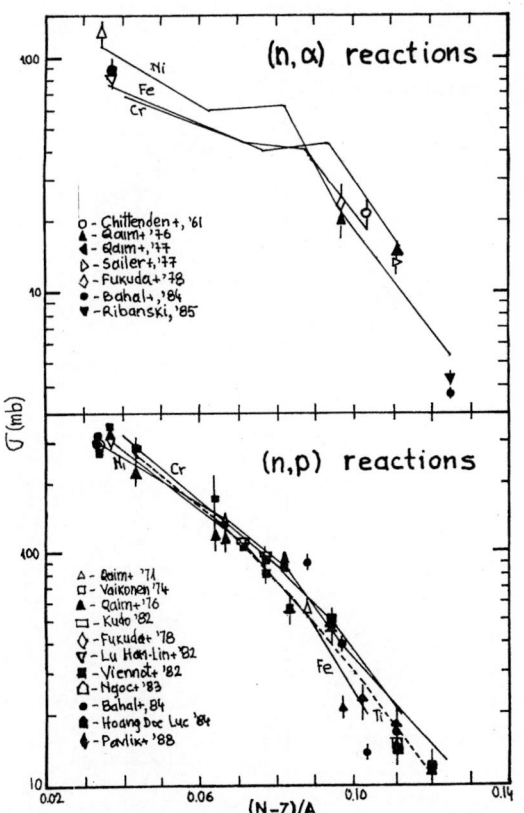

Fig. 1. Calculated and experimental cross-sections at 14.7 MeV.

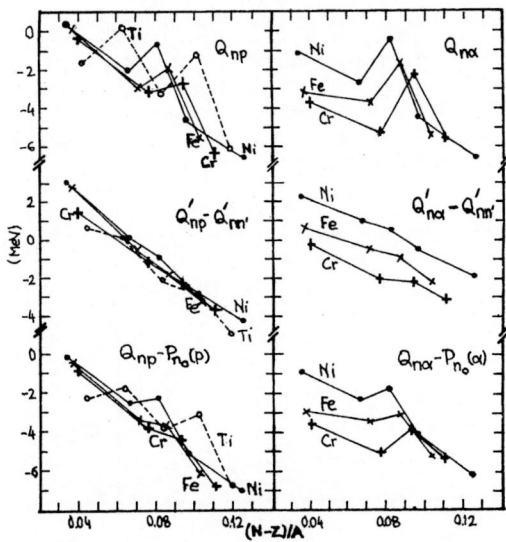

Fig. 2. Q-values and effective Q-value differences for stable isotopes of Ti, Cr, Fe and Ni.

by Gardner [5] as a Q-value effect, was interpreted by Molla and Qaim [6] in terms of the proton binding energy as a function of the asymmetry parameter $(N-Z)/A$. Pai [7] showed by means of statistical model calculations that the large changes in the total proton emission cross sections σ_{np} between neighbouring nuclei are accounted for by an effective Q-value $Q'_{np} = Q_{np} + \delta_n - \delta_p$, where δ_n and δ_p are depressions of the ground state energies of the related nuclei produced by pairing and shell correlations; this quantity has also been proved responsible for the exponential decrease of the cross sections ratio $\sigma'_{np}/\sigma'_{nn}$.

Following the generalized Q_{gg}-systematics of the deep inelastic collisions between complex nuclei [8], it seems worthwhile to introduce also for fast neutron reactions $(n, x_1 x_2)$ an empirical effective Q-value given by

$$Q'_{nx_1x_2} = Q_{nx_1x_2} - \sum_i \Delta_{x_i} \qquad (1)$$

where Δ_{x_i} are the ground state backshift parameters of the BSFG level density for the nuclei along the reaction path. The effective Q-value difference $Q'_{nx} - Q'_{nn}$ thus represents the difference between excitation energy regions of rather equal level densities in the neutron and particle x channels, respectively. A correlation seems to exist even betweens these differences (Fig. 2) and the exponential decreasing trend of the (n,p) and (n,α) reaction cross sections (Fig.1), which are only parts of the respective total particle emission cross sections. Still evident is the dependence of the calculated ratios $\sigma'_{np}/\sigma'_{nn}$ and $\sigma'_{n\alpha}/\sigma'_{nn}$ on the related effective Q-value differences, the straight lines in Fig. 3 corresponding to different emitted particles or compound nuclei at comparable excitation energies (the similar slopes being related to a temperature parameter characterizing the partial statistical equilibrium of the compound nuclei [7]).

On the other hand, the preequilibrium emission contribution depends on the maximum energy of the emitted particles, $Q_{nx} - P_n$, where P_n is the pairing correction [9] for the exciton state density of initial exciton configuration (giving by far the main contribution to this process). By comparing the dependence of this quantity by the $(N-Z)/A$ parameter (Fig. 2) to that of the (n,p) and (n,α) reaction cross-sections (Fig. 1), there results that the isotope effect for the (n,α) reactions is mainly determined by the preequilibrium emission.

(n, n'p) + (n, pn) Reaction Isotope Effect

Calculated cross sections for $(n,n'p) + (n,pn)$ processes are compared to both experimental and systematic values, the activation ones including the (n,d) reaction products (Fig. 4a). Actually, the (n,d) reaction cross section - of the order of 10 mb along the valley of stability [10] - is not significant for the lightest isotopes (daughter nuclei with binding energies $S_n > S_p$) but seems to be most important one for neutron-rich isotopes $(S_n < S_p)$. It could be considered as responsible for the systematic curve B derived by Qaim [10] for $[(n,d)+(n,n'p)-(n,pn)]$ reaction cross sections for nuclei with $S_n < S_p$.

The calculated $(n,n'p)$ and (n,pn) reaction cross sections (Fig. 4a) are depicting their contribution to the activation process. The (n,pn) cross section decreasing with the increase of the asymmetry is mainly a consequence of the above discussed behaviour of first proton emission cross section σ'_{np}. Additionally, it is reduced by the related increase of S_p. On the other hand, the $(n,n'p)$ reaction plays indeed a dominant role for the daughter nuclei with $S_n > S_p$ [10]. Once the difference $Q_{n,n'p} - Q_{n,2n}$ becomes lower than the effective V_c, $(n,n'p)$ decrease faster than $\sigma_{n,pn}$. This trend could be correlated with the constant diminution of the effective Q-value difference $Q'_{n,n'p} - Q'_{n,2n}$.

Low - Lying Level Scheme Effects

Accurate low-lying level schemes are also of considerable significance for the correct evaluation of the nuclear level density parameters. Confident statistical model results, highly determined by these parameters, are even directly

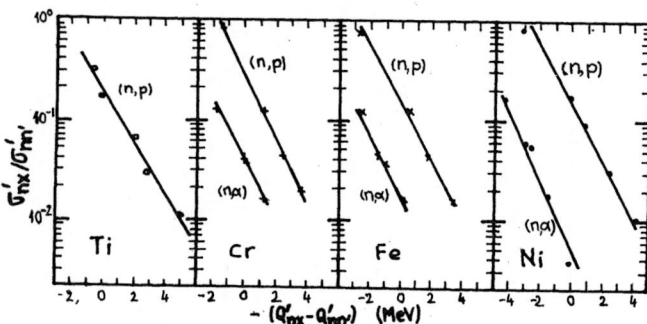

Fig. 3. Dependence of ratio $\sigma'_{nx}/\sigma'_{nn}$ at 14.7 MeV on effective Q-value differences.

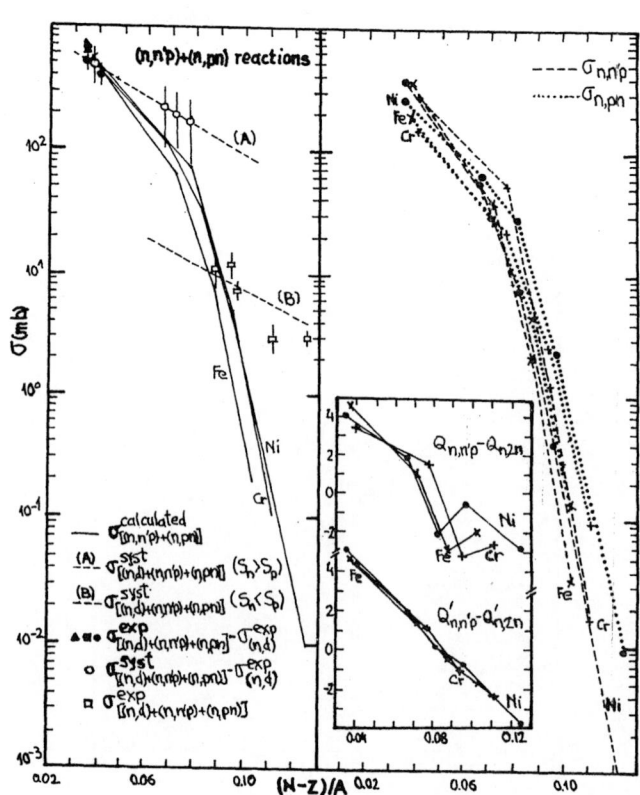

Fig. 4. Calculated $(n,n'p)$ and (n,pn) reaction cross sections at 14.7 MeV

dependent by the correctness of the discrete levels taken into account when the continuum population is less important. The usefulness of theoretical low-lying level schemes is evident both to provide these data when they are scarce or even missing, and to suggest the energy limit of the levels considered in level density parameter evaluation (e.g. for calculation of the $^{52}Cr(n,p)^{52}V$ reaction excitation function, Fig. 5). An example of problems arising probably

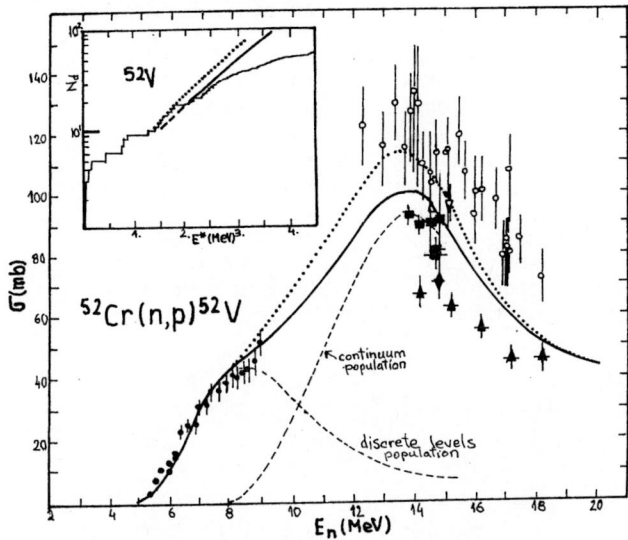

Fig. 5. Excitation function of $^{52}Cr(n,p)^{52}V$ reaction.

due to inaccurate discrete level scheme could be the low-energy side of the calculated $^{53}Cr(n,p)^{53}V$ reaction excitation function. It seems that the level scheme taken into account in the Hauser-Feshbach model calculation (Fig. 6a), as well as in the level density parameter analysis (Fig. 6b) are still incomplete.

Summary Remarks

Isotope effect of fast neutron reaction cross-sections in the mass range A = 50 are analyzed by using predictions of nuclear models previously tested over a large incident energy range. No free internal parameter for the GDH preequilibrium emission model has been involved to get the agreement with the available experimental data. Both the statistical and preequilibrium processes appeared to be significant for the (n,p) and (n, α) reaction isotope effects. The applicability of the deep inelastic collisions between complex nuclei to fast neutron reactions is discussed.

Work performed under IAEA research Contract No.3802/R4/RB.

References

1. M. Avrigeanu, M.Ivaşcu and V. Avrigeanu: Z.Phys. A335, 299 (1990), and to be published
2. M. Avrigeanu, M. Ivaşcu and V. Avrigeanu, STAPRE-H, NP-63-1987/Rev. 1, Institute of Atomic Physics, Bucharest (1988)
3. M. Uhl and B. Strohmaier, STAPRE, IRK-76/01. Institut für Radiumforschung und Kernphysik, Vienna (1976).
4. P.D. Kunz, DWUCK4 User Manual, NEA Data Bank (1984)
5. D.G. Gardner: Nucl. Phys. 29, 373 (1962)
6. N.I. Molla and S.M. Qaim: Nucl. Phys. A283 269 (1977)
7. H.L. Pai: Can.J.Phys. 54, 1421 (1976)
8. Y.Y. Yolkov: Nukleonika 21, 53 (1976)
9. C.Y.Fu: Nucl.Sci.Eng. 86 344 (1984)
10. S.M. Qaim: Nucl.Phys. A382, 255 (1982).

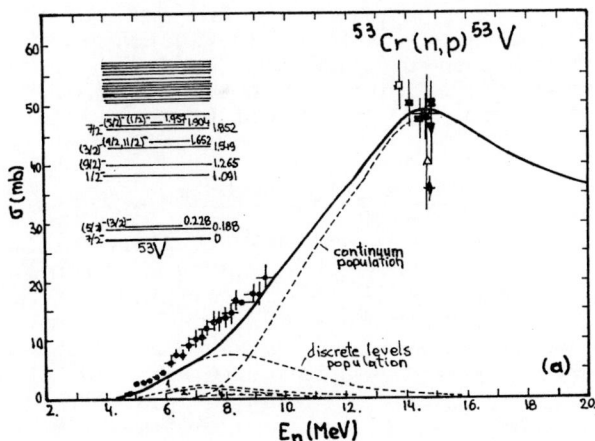

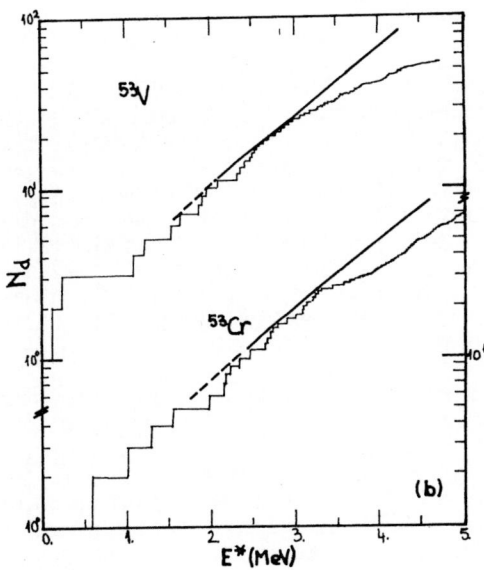

Fig. 6(a). Excitation function of $^{53}Cr(n,p)^{53}V$ reaction and the low-lying levels of ^{53}V taken into account in Hauser-Feshbach calculation. (b). Total numbers of discrete levels and BSFG level density model predictions (full curves for densities used in calculations).

ENERGY DEPENDENCE OF THE ISOTOPIC EFFECT
IN THE (n,p) REACTION ON MEDIUM WEIGHT NUCLEI

H.M. Hoang, U. Garuska, D. Kielan, A. Marcinkowski and B. Zwieglinski

Department of Nuclear Reactions
A.Soltan Institute for Nuclear Studies
00-681 Warsaw, Hoża 69, Poland

Abstract: Excitation functions for the (n,p) reaction on the ^{70,72,73,74}Ge, ^{74,76,77,78}Se, ^{90,91,92,94}Zr and ^{106,108}Pd nuclei were measured in the neutron energy range from 13 to 16.6 MeV with the aid of the activation method. Neutrons were produced with the T(d,n)^{4}He reaction in the Ti-T target. The cross sections decrease exponentially with increasing target asymmetry parameter $(N_t-Z_t)/A_t$ (isotopic effect) at fixed neutron energies for each of the studied isotopes. The slope coefficients of the exponent are found decreasing functions of energy for the Ge and Se series while for Zr they stay approximately constant in the investigated energy range. The data are interpreted in terms of the reaction model incorporating preequilibrium and equilibrium emission processes. The differences in energy dependence of the isotopic effect are ascribed by the assumed reaction model to the compound component which slopes more steeply with $(N_t-Z_t)/A_t$ than the precompound one and whose relative contribution decreases with increasing Z_t.

((n,p) reaction mechanism, activation method, cross section, excitation function, T(d,n)^{4}He reaction, Ti-T target, samples with natural isotopic abundance, preequilibrium and equilibrium reaction mechanisms).

Introduction

Gardner [1] and independently Levkovskii [2] in an attempt to systematize the (n,p) cross sections at 14.5 MeV available at the beginning of the sixties discovered that within the series of stable isotopes of a given element, Z_t, the cross sections reveal a simple exponential dependence on target neutron number, N_t, or on a variable equivalent to N_t - the asymmetry parameter, $(N_t-Z_t)/A_t$:

$$\sigma(n,p) = \sigma_0 \exp\left[-K(N_t-Z_t)/A_t\right], \qquad (1)$$

where σ_0 and K are fitting parameters. This dependence was subsequently termed isotopic effect in the literature. The data accumulated throughout the late sixties and the seventies (see notably an extensive work of the Julich group [3]) have confirmed this trend and called for an interpretation in terms of the compound nucleus mechanism [4]. However, energy dependence of the isotopic effect has not been adequately studied in the past. An obvious question as to whether eq. (1) gives a valid description of the $\sigma(n,p)$ vs N_t dependence also beyond the immediate vicinity of 14.5 MeV has not yet received a decisive answer. This prompted us to undertake the (n,p) excitation function measurements in the neutron energy range from 13.0 to 16.6 MeV on the Ge, Se, Zr and Pd isotopic sequences. The results are confronted with predictions of the reaction model incorporating preequilibrium and equilibrium reaction mechanisms.

Experimental method

Neutrons have been produced in the T(d,n)^{4}He reaction with deuterons of 440 and 990 keV accelerated in the Van de Graaff accelerator. Up to five samples of the studied element set at different angles relative to the deuteron beam direction were simultaneously irradiated with neutrons. The excitation curve measurement thus relies upon the E_n vs θ_n dependence determined by the T(d,n) reaction kinematics. Samples had natural isotopic content. Beam currents of 30-50 μA bombarded a thin Ti-T target on the water - cooled Cu backing during 3-4 half-lives to reach the saturation activity. For very long half-lives the activation time was limited to about 50 hours. The absolute cross section scale was established by referring the initial activities to ^{56}Mn from the reference reaction ^{56}Fe(n,p)^{56}Mn irradiated in the identical geometry but in a separate accelerator run.

The (n,p) reaction products were identified by their characteristic γ-ray energies and decay half-lives with an 80 cm^3 Ge(Li) detector coupled to the ND-4420 analysing system, which allowed for an automatic timing of the measurement of decay curves. The discussed data are corrected for (n,np)+(n,pn)+(n,d) and (n,α) contributions in those cases in which these reactions on the adjacent stable isotopes gave the same final activity as the (n,p) reaction of interest.

Theoretical formalism

The measured excitation functions were interpreted in terms of the reaction model incorporating precompound (PE) and compound emission (CE) processes using the code EMPIRE [5]. This employs the Hauser-Feshbach (H-F) theory for CE and the geometry-dependent hybrid (GDH) model of Blann for PE calculations. Emission of neutrons, protons, alphas and γ-rays has been taken into account in the H-F theory. The optical model parameters of Moldauer [6] served for the transmission coefficients calculation for neutrons below E_n = 4 MeV and those of Bjorklund and Fernbach [7] above this energy. For protons the parameters of the latter authors were used in the whole energy range. For the level densities of the final nuclei H-F theory employs the composite formula of Gilbert and Cameron [8]. The GDH model applied the Ericson [9] formula to calculate the exciton state densities. However with the standard choice of parameters a systematic overprediction of the theory over experiment with increasing N_t was observed for some of the studied isotopic series. Therefore the parameter a_n describing an exponential growth of the level density in the former and the single-particle state density parameter, g, in the latter theory were treated as adjustable and were fitted for best agreement with σ(n,p) at 15 MeV.

Discussion of excitation curves

The density parameters adjusted at 15 MeV were used to calculate the excitation curves in the 4-20 MeV range. Fig. 1 intercompares the results for the Ge(n,p)Ga reactions. Solid lines incorporate both PE and CE, while the dashed ones describe the pure PE process. The reaction mechanism evolves with the increasing N_t from CE dominated for the lightest ^{70}Ge in nearly the entire energy range to PE dominated for the heaviest ^{74}Ge. There is an overall tendency towards increasing of the relative contribution of PE with increasing E_n resulting in its dominance around E_n = 20 MeV for all the studied Ge targets. The dependence of σ(n,p) on $(N_t-Z_t)/A_t$ at the three indicated energies (Fig. 2) demonstrates a decreasing slope of the isotopic effect (eq. (1)) in between 13.0 and 14.5 MeV, correlated with the strongly decreasing contribution of CE in Fig.1, and a rather constant value of the K-coefficient with the further increase of E_n to 16.6 MeV. This observation is further corroborated by the Se isotopes (Fig.3) which show slope fall-off in the whole energy range. One might thus conclude that CE is typified by a more steep isotopic dependence than PE. Fig.4, containing the excitation curves for the Zr(n,p)Y reactions, reveals the pattern evolution with increasing Z_t. Comparing the lightest isotopes in Fig.1 and Fig.4 one notes a strong reduction of CE following an increase of the Coulomb barrier height for protons when going from Z_t = 32 to Z_t = 40. Fig.5 shows that the slope coefficients for Zr do not change with energy. This supports the above conclusion that significant variation of $K(E_n)$ is associated with the CE component, which is weak for the zirconiums. For the ^{106,108}Pd(n,p)Rh reactions (not shown) the excitation curves increase monotonically with E_n, which is interpreted as due to negligible contribution of CE.

In summary, eq.(1) with the energy dependent slope coefficient $K(E_n)$, gives a valid description of the (n,p) reaction excitation curves in the 13-16.3 MeV energy range for the medium mass domain. The assumed reaction model interpretes the observed energy dependence of $K(E_n)$ as a consequence of the rapid decrase of CE with E_n at these energies.

Shell effects in single-particle density g

Fig.6 presents the level density parameters in the neutron channels, a_n (open points), and the single particle density parameters, $\pi^2 g/6$ (filled points), obtained in the fitting procedure at 15 MeV as a function of N. Solid lines denote the trend of empirical a_n parameters determined in [10]. The open points do not deviate from the solid line by more than ±15%, which is the error of a_n estimated in [10]. On the other hand, the solid points reveal an important modulation associated with the neutron shell closure at N=50. This contradicts the monotonic N-dependence predicted for g by the infinite nuclear matter assumed in GDH

(dash-and-dot lines). Our analysis indicates therefore that the Saxon-Woods or Nilsson potentials, which predict shell-effects in g, are the preferred models for PE in the (n,p) reaction in this mass domain.

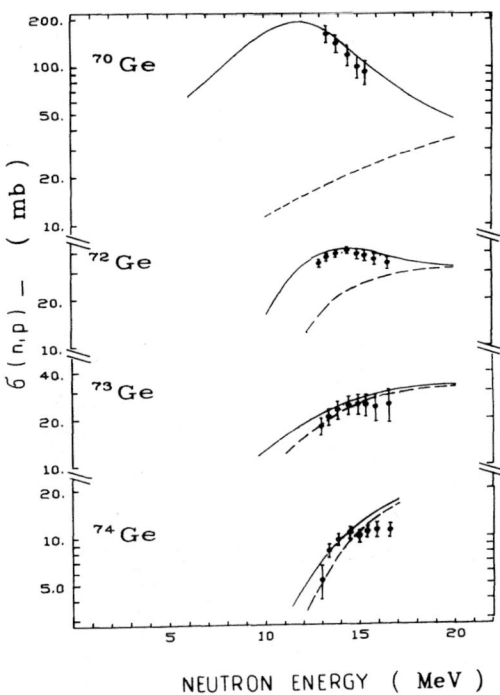

Fig.1. Comparison of the experimental and calculated excitation curves for the Ge(n,p)Ga reactions. Broken lines indicate the PE contribution, solid lines the sum of PE and CE components.

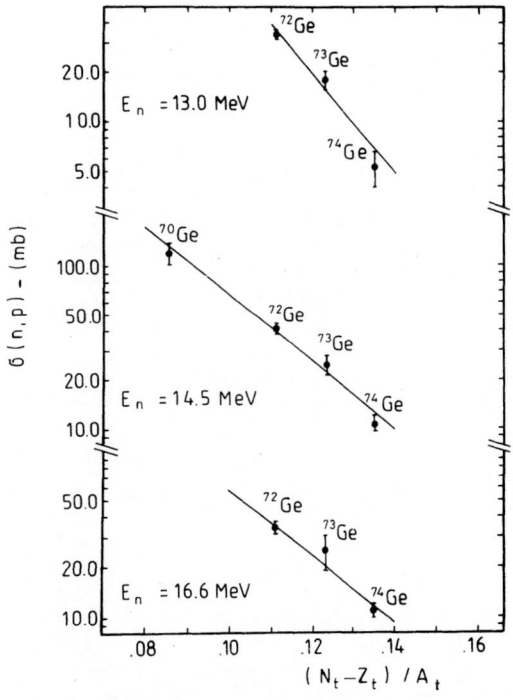

Fig.2. Dependence of the Ge(n,p)Ga cross sections on the asymmetry parameter $(N_t-Z_t)/A_t$ at the indicated neutron energies. Solid lines are the straight-line fits to the data points.

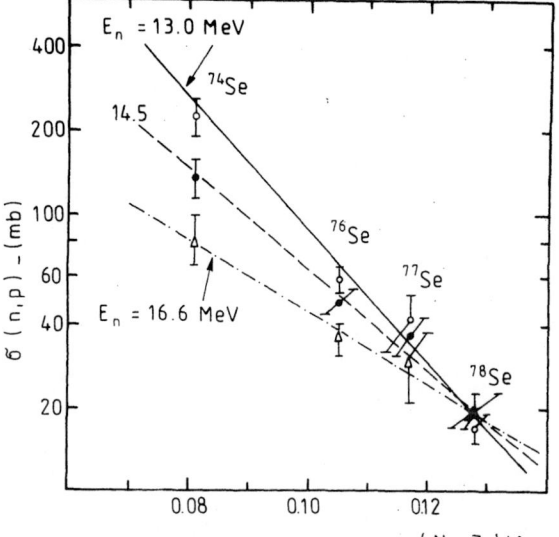

Fig.3. Dependence of the Se(n,p)As cross sections on the asymmetry parameter $(N_t-Z_t)/A_t$ at the indicated neutron energies.

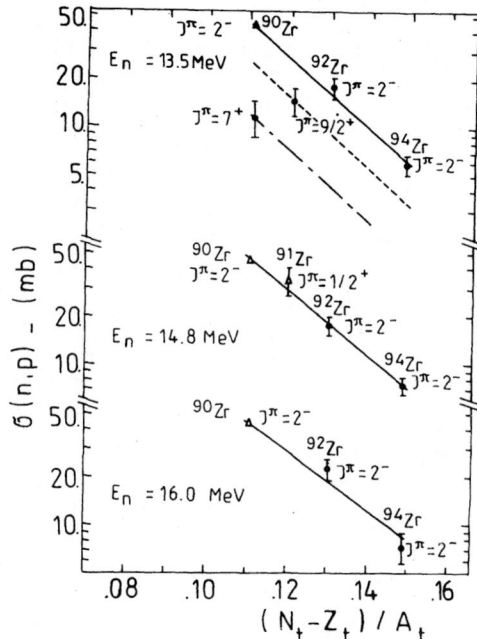

Fig.5. Dependence of the Zr(n,p) cross sections on the asymmetry parameter $(N_t-Z_t)/A_t$ at the indicated neutron energies. Solid lines are the straight-line fits to the data points. Dashed and dash-and-dot lines denote the postulated dependence for the isomeric states.

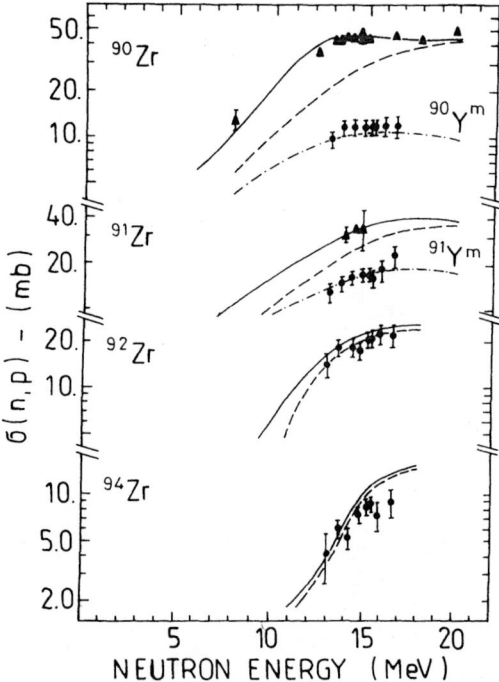

Fig.4. Comparison of the experimental and calculated excitation curves for the Zr(n,p)Y reactions. Triangles mark the results taken from [11]. Broken lines are for the PE contribution, solid lines for the sum of PE and CE components. For the isomeric states in ^{90}Y and ^{91}Y only the sum (dash-and-dot line) is indicated.

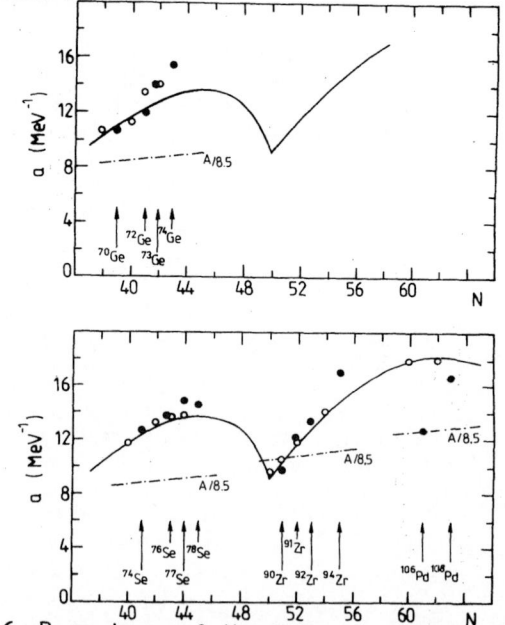

Fig.6. Dependence of the density parameters on neutron number. See the text for further details.

REFERENCES

1. D.G.Gardner: Nucl. Phys. <u>29</u>, 373(1962)
2. V.N.Levkovskii: Zhur. Exp. and Theor. Phys. <u>45</u>, 305 (1963)
3. N.I.Molla, S.M.Qaim: Nucl. Phys. <u>A283</u>,269(1977)
4. H.L.Pai, D.G.Andrews: Can.J.Phys. <u>56</u>,944(1978).
5. M.Herman, A.Marcinkowski, K.Stankiewicz: Comp. Phys. Comm. <u>33</u>, 373 (1984).

6. P.Moldauer: Nucl. Phys. <u>47</u>, 65 (1963)
7. F.Bjorklund, S.Fernbach: Phys. Rev. <u>109</u>, 295 (1958)
8. A.Gilbert, A.G.W.Cameron: Can.J.Phys. <u>43</u>, 1446 (1965)
9. T.Ericson: Adv. Phys. <u>9</u>, 425 (1960)
10. G.Reffo: CNEN Report RT/FI/78/11, Bologna (1978)
11. B.P.Bayhurst, J.R.Prestwood: J. Inorg. and Nucl. Chem. <u>23</u>, 173 (1961)

MODELLING INELASTIC EMISSION CROSS SECTIONS FOR ^{9}Be(n,2n)

G.M. Field[†], T.D. Beynon[†] and H. Gruppelaar[*]

† School of Physics and Space Research, The University of Birmingham, Edgbaston, Birmingham, B15 2TT, UK

* The Netherlands Energy Research Foundation ECN, P.O. Box 1, 1755ZG Petten, The Netherlands

Abstract: Detailed comparisons are made with the double differential ^{9}Be(n,2n) cross sections in the EFF-2 File, based on exact analytical modelling and the evaluation by Drake et al, with some of the recent measurements by Takahashi et al and Geel. Preliminary analysis indicates that the agreement, although good, could probably be improved by a re-evaluation of the angular distributions of the ^{9}Be inelastic contributions and the ^{8}Be contributions.

(^{9}Be(n,2n), neutron cross sections, modelling, double differential emission, EFF-2, comparison with new measurements)

Introduction

Beryllium remains an important candidate material for the breeder blanket of controlled thermonuclear reactors. Of particular interest is the reaction ^{9}Be(n,2n) which allows neutron multiplication and minimum lithium inventories in ^{6}Li enriched blankets. A recent analysis [1] of double differential neutron emission cross sections, using the evaluation of double differential spectra made by Drake et al [2], now forms the basis of the appropriate EFF-2 File. This has been based on the analytical modelling of Beynon and Sim [3] and is comparable to the Monte Carlo simulations of the sequential breakup channels made by Perkins et al [4] for the same measurements. The present paper uses the analytical model of ref. [3] to examine the recent measurements of double differential spectra made by Takahashi et al [5] and Geel [6] and includes some relevant comparisons with the earlier measurements of Drake et al. In essence a comparison is presented of the ^{9}Be(n,2n) EFF-2 File with later measurements than those used to produce the file, with a view to further data evaluation and a subsequent improvement of the file.

Methodology and Data Sources

The principal reaction channels are listed in Table 1 in which neutron emission is seen to proceed via compound nucleus formation and directly through levels in ^{9}Be, ^{8}Be, ^{6}He and ^{5}He. Following Perkins et al [4] no contribution from three-body breakup (cf. channel (v) in Table 1) has been included. The detailed analysis necessary for channels (i)–(iv) is described in ref. [3] and the values of relevant data, such as partial cross sections, branching ratios and intermediate state centre-of-mass angular distributions, are set out in both [1] and [3]. It is this data which as an input requirement to the analytical modelling–which preserves exact emitted energy-angle correlation–can therefore be adjusted to improve agreement with measured emission spectra.

Comparisons with Measurements

Takahashi et al

A recent set of measurements by Takahashi et al [5] covers sixteen laboratory angles θ_ℓ for the emitted neutron for an incident neutron energy E_n of 14.1 MeV. Figures 1 and 2 show the comparisons of these measurements with the EFF-2 File at θ_ℓ=60° and 120° respectively. In each case the elastic peak is included in both the measured data and the file data comparisons and is used to gaussian broaden the δ-function representation of the elastic double differential cross section in the file to agree with the width of the measured elastic peak. The broadening function thus obtained is then used to broaden all the file data used in the comparison. Whilst the overall agreement is good, a preliminary analysis indicates that further evaluation of the ^{9}Be inelastic angular distributions, together with a possible adjustment of the branching ratios for the ground and first excited states in ^{8}Be, would

Table 1. Reaction Mechanisms leading to the ^{9}Be(n,2n) reaction

	Q-value (MeV)	
^{9}Be+n $\rightarrow$ (i) ^{9}Be*+ n_1		inelastic followed by decay
^{9}Be* $\rightarrow$ ^{8}Be*+ n_2	-1.6655	through levels in ^{8}Be
^{8}Be* $\rightarrow$ 2α	$+0.09$	
(ii) ^{9}Be*+ n_1		inelastic followed by decay
^{9}Be* $\rightarrow$ ^{5}He* + α_1	-2.47	through levels in ^{5}He
^{5}He* $\rightarrow$ $n_2 + \alpha_2$	$+0.89$	
(iii) ^{5}He* + ^{5}He*	-3.36	2 ^{5}He breakup
^{5}He$^*_{1,2}$ $\rightarrow$ $n_{1,2} + \alpha_{1,2}$	$+0.89$	
(iv) ^{6}He* + α_1	-0.598	3–stage 2–body sequential process
^{6}He* $\rightarrow$ ^{5}He*+ n_1	-1.87	through levels in ^{5}He
^{5}He* $\rightarrow$ $n_2 + \alpha_2$	$+0.89$	
(v) ^{8}Be*+ n_1+ n_2	-1.6655	3–body breakup

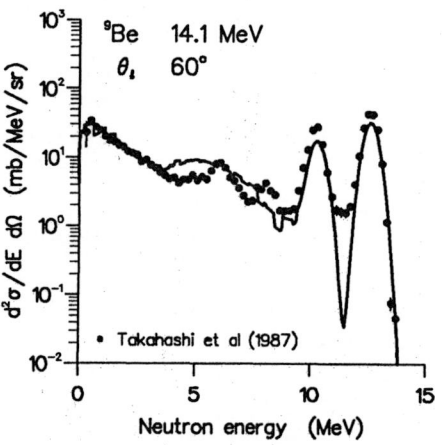

Fig. 1

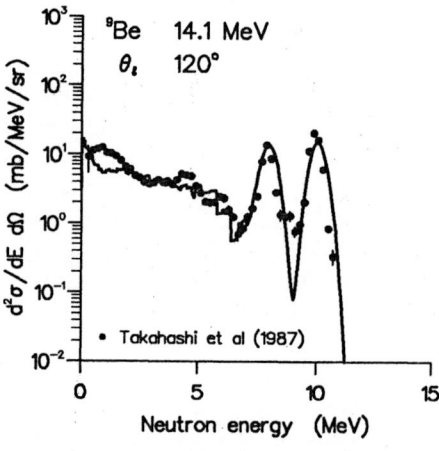

Fig. 2

lead to improved agreement over the complete range of θ_ℓ-values in the Takahashi measurements. For comparison's sake the emission data of Drake et al [2] at θ_ℓ=60° for E_n=14.2 MeV is shown in Fig. 3, where it should be noted that the measurements exclude the elastic n_o contribution and the n_1 contribution from the 2.43 MeV level in ^{9}Be.

Geel Measurements

Figs. 4 and 5 show a selection of similar comparisons with some of the Geel emission spectra measurements [6] at θ_ℓ=24° and 150° respectively, with the broadening procedure outlined above. The measurements are given at eight θ_ℓ-values, from θ_ℓ= 24° to θ_ℓ=150° and have a specified incident energy range of 8.37–8.75 MeV. Examination of the position of the measured elastic peaks show this is consistent with an incident energy of 8.75 MeV at θ_ℓ=24° and 8.37 MeV at θ_ℓ=150° and the comparisons therefore use EFF-2 File data at E_n=8.75 MeV, θ_ℓ=24° in Fig. 4 and E_n=8.37 MeV, θ_ℓ=150° in Fig. 5. As with the Takahashi data, a preliminary analysis indicates that for the complete set of Geel data (not shown) a re-evaluation of the ^{9}Be inelastic distributions and ^{8}Be contributions could

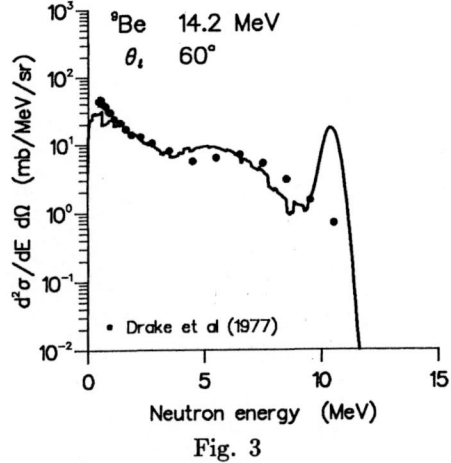

Fig. 3

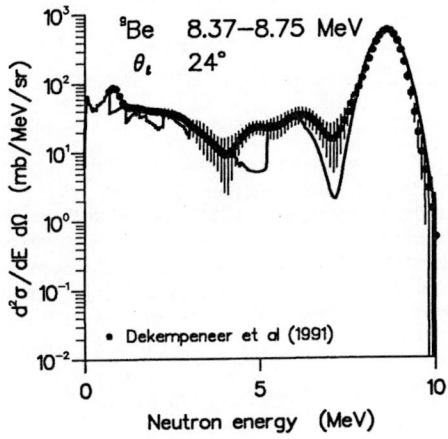

Fig. 4

lead to an improved agreement.

Conclusions

Comparisons of the EFF-2 File data with the measured emission spectra of Takahashi et al [5] and Geel [6] indicate that there is an a priori case for a re-evaluation of the angular distributions of the ^{9}Be inelastic contributions and the ^{8}Be contributions. As in our earlier analysis [1] there does not appear to be a strong case for including a significant contribution from the direct three-body reaction.

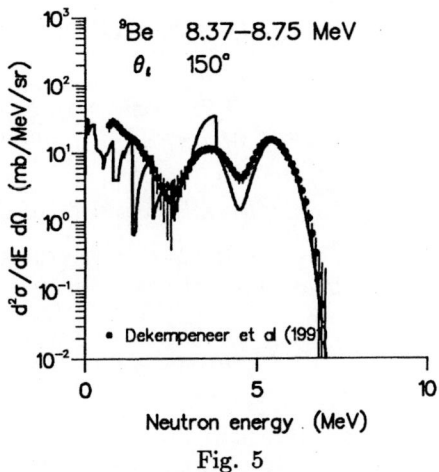

Fig. 5

References

1. T.D. Beynon and G.M. Field: Final Report for NET Contract 311/88-7/FU-UK (1990)
2. D.M. Drake, G.F. Auchampaugh, E.D. Arthur, C.E. Ragan and P.G. Young: Nucl. Sci. and Eng. 63, 401 (1977)
3. T.D. Beynon and B.S. Sim: Ann. Nucl. Energy 15, 27 (1988)
4. S.T. Perkins, E.F. Plechaty and R.J. Howerton: Nucl. Sci. and Eng. 90, 83 (1985)
5. A. Takahashi, E. Ichimura, Y. Sasaki and H. Sugimoto: OKTAVIAN Report, A-87-03, Osaka University (1987)
6. E. Dekempeneer, F. Poortmans, H. Weigmann, J.A. Wartena and C. Burkholz: See this conference; H. Weigmann: Private communication.

POLARIZED PROTON INDUCED BREAKUP OF ^{12}C AT 14 AND 16 MEV

Y. Watanabe, H. Kashimoto, H. Hane, and A. Aoto
Department of Energy Conversion Engineering, Kyushu University, Kasuga, Fukuoka 816, Japan

N. Koori
College of General Education, The University of Tokushima, Tokushima 770, Japan

A. Nohtomi, Susilo Widodo, O. Iwamoto, and R. Yamaguchi
Department of Nuclear Engineering, Kyushu University, Fukuoka 812, Japan

K. Sagara, H. Nakamura, K. Maeda, and T. Nakashima
Department of Physics, Kyushu University, Fukuoka 812, Japan

Abstract: Double differential cross sections and analyzing powers are measured of protons and α particles emitted from the bombardment of ^{12}C with 14 and 16 MeV polarized protons. The measured energy spectra of protons and α particles are analyzed on the basis of the reaction model in which three or four-body simultaneous breakup process is taken into account. The calculated proton and α spectra show good agreement with the continuous spectra observed in the low outgoing energy range.

(^{12}C(p,p')3α reaction, polarized proton experiment, simultaneous breakup process, phase space model)

Introduction

In interactions of fast neutrons with carbon, the contribution of the ^{12}C(n,n')3α reaction to nonelastic cross sections becomes dominant at neutron energies above 10 MeV. A more detailed study of the four-body breakup reaction, therefore, is important for several applications such as estimation of radiation damage and neutron shielding design in fusion energy development and evaluation of kerma factors needed for high energy neutron radiotherapy. Double differential cross sections (DDXs) for the reaction are required in more accurate calculations of neutron transport and kerma factors as well as in an understanding of the reaction mechanism. Several measurements of the ^{12}C(n,n')3α cross section have been made in the high incident energy range between 14 and 60 MeV [1,2,3]. However, there are only a few direct measurements of DDXs of both emitted neutrons and α particles at the same incident energy [4,5].

So far, some of the authors have studied neutron-induced reactions for lithium isotopes [6] and several medium heavy nuclei [7] through experimental investigations of proton-induced reactions analogous to neutron-induced reactions. In the present work, this approach is also applied to investigation of the ^{12}C(n,n')3α reaction. DDXs and analyzing powers have been measured of protons and α particles emitted from the bombardment of ^{12}C with 14 and 16 MeV polarized protons. The measured spectra of protons and α particles are analyzed on the basis of the reaction model in which simultaneous breakup process is taken into account. The aim of the present work is to establish a reliable model to evaluate the ^{12}C(n,n')3α or ^{12}C(p,p')3α breakup cross section.

Experimental procedure and data processing

The experiment was performed using 14 and 16 MeV polarized proton beams from the tandem Van de Graaff accelerator at Kyushu University. Before this experiment, double differential proton emission cross sections for the ^{12}C(p,p') reaction had been measured with a 16 MeV unpolarized proton beam as a preliminary experiment. These experimental procedures are almost the same as those reported elsewhere [6,7]. A ΔE-E counter telescope consisting of three silicon surface barrier detectors(E_1: 20μm, E_2: 75μm, E_3: 2000μm) was employed to detect protons and α particles emitted with as low energies as possible ($E_{p'}$>1 MeV and E_α> 1.5 MeV). A target was a self-supporting foil of natural carbon: its thickness was 0.116 mg/cm² for the polarized beam experiment and 0.524 mg/cm² for the unpolarized beam one. Beam polarization was monitored using a polarimeter

consisting of ^{4}He gas target and two ΔE-E silicon detectors [8] at the down stream of a scattering chamber.

To detect α particles with low outgoing energy, the signals from E_1 detector were separately stored in anticoincidence with the signals from E_2 and E_3 detectors. It is possible to identify two kinds of particle (proton and α particle) by making use of the difference of maximum energy loss in the E_1 detector; the energy to be deposited in the E_1 detector is below about 1.1 MeV for protons and about 4.3 MeV for α particles, if the energy loss straggling is neglected. At the forward angle, however, there appears some contribution from the nucleus ^{12}C recoiled by the elastic and inelastic scatterings. The lower and upper limits in low α particle energy region were determined under consideration of such effects.

Experimental results

Differential cross sections and analyzing powers for elastic scatterings at 16 MeV are shown together with the other experimental data [9] and those calculated using the optical potential parameter by Nodvik et al.[10] in Fig.1. The present experimental data are consistent with the other experimental data as can be seen in Fig.1.

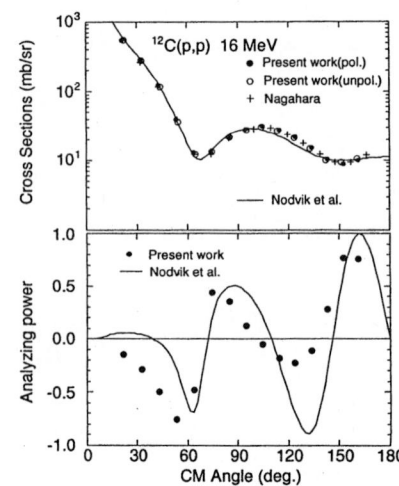

Fig.1 Differential cross sections and analyzing powers of ^{12}C(p,p) scattering at 16 MeV. Solid lines are for calculations with the optical potential parameter by Nodvik et al.(ref.[10]).

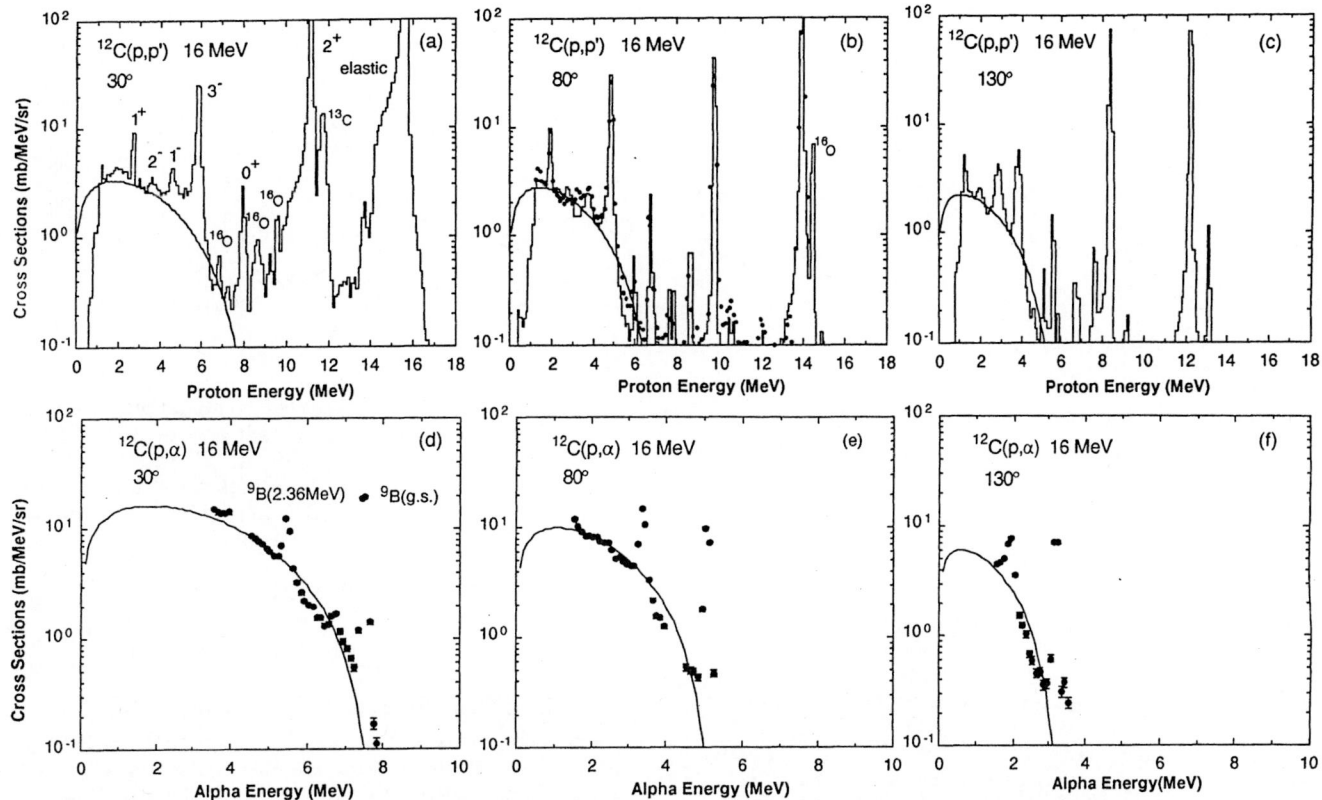

Fig.2 Double differential proton and α particle emission cross sections of the ^{12}C(p,p')3α reaction at 16 MeV. The measured angles are 30°, 80°, and 130°. Histograms and solid circles are the experimental data and solid lines show the calculated 4BSB components.

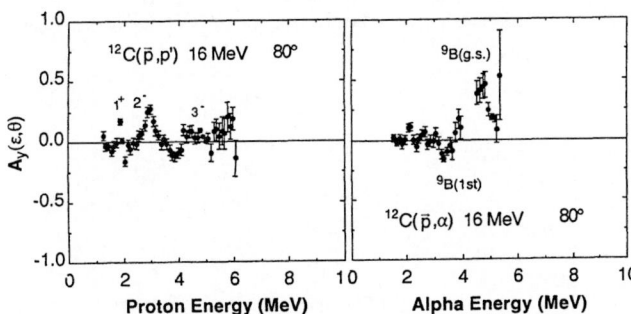

Fig.3 Experimental analyzing powers of protons and α particles emitted at 80° for the $\vec{p}$ + ^{12}C reaction at 16 MeV.

Figure 2 shows measured double differential protons and α particles emission cross sections for 16 MeV. In Fig.2(b), the experimental data for unpolarized (histograms) and polarized beams (solid circles) are compared and the consistency between both the data is confirmed. The observed proton spectra exhibit a distinct continuum underlying the peak structure corresponding to excited states of ^{12}C. Two peaks observed in Fig.2(d)-(f) correspond to the transitions to the ^{9}B(g.s.) and ^{9}B(2.36MeV) via the ^{12}C(p,α)^{9}B reaction. Under those peaks, there is a continuous component that is due to three α breakup processes of our interest.

Experimental analyzing powers of emitted protons and α particles at 16 MeV are shown in Fig.3. There is a tendency that the analyzing power of α particles at low energies becomes small and approaches to zero. Here it should be noted that the following analysis will be restricted to only the energy spectra of proton and α particles and the analyzing power will not be mentioned.

Theoretical analysis and discussion

Multiparticle breakup processes of ^{12}C induced by proton or neutron are mainly classified into the sequential breakup process via some intermediate states and the simultaneous breakup process [3]. Here we consider the following simultaneous breakup processes in analyzing the observed continuous spectra:

(i) ^{12}C + p → p + α + ^{8}Be
　　　　　　　　　　└→ α + α

(ii) ^{12}C + p → p + α + α + α

The processes (i) and (ii) are referred to as three-body simultaneous breakup (3BSB) process and four-body simultaneous breakup (4BSB) process, respectively. In the 3BSB process, the sequential decay of ^{8}Be into two α particles occurs and finally three α particles are generated as the reaction products.

The energy spectra of particles emitted in the simultaneous breakup process can be calculated by using the phase space distribution $\rho_f(E_1)$ derived on the basis of the reaction kinematics [11] as follows:

$$\frac{d^2\sigma}{dE_1 d\Omega_1} = \frac{2\pi}{\hbar^2} \frac{m_p}{k_p} |M|^2 \rho_f(E_1) \tag{1}$$

where m_p and k_p are the mass and the wave number of incident proton, respectively. $|M|^2$ is the square of the transition matrix element. For simplicity, we assume that $|M|^2$ is an adjustable parameter independent of the angle and outgoing energy. Namely, the shape of energy spectrum is provided by the phase space distribution and the absolute value is determined by normalization to the experimental value under the assumption that there is no other competing

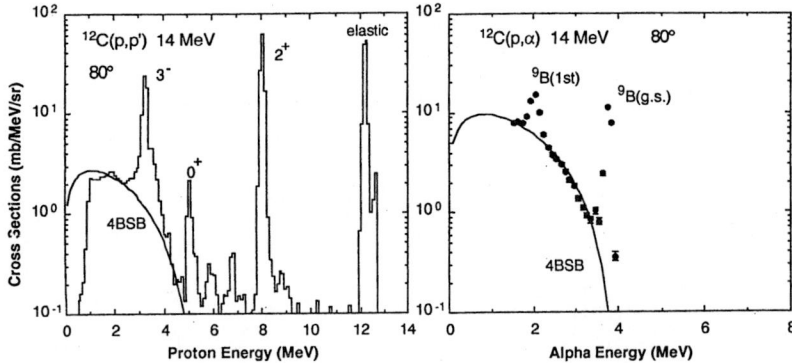

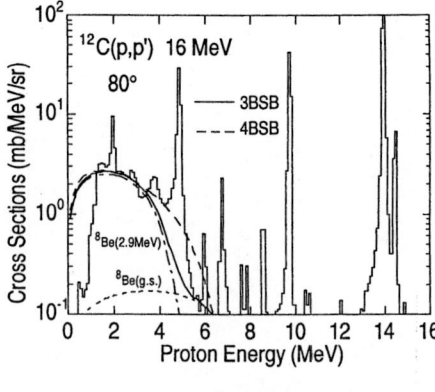

Fig.4 Double differential proton and α particle emission cross sections for the ^{12}C($\vec{p}$,p')3α reaction at 14 MeV. Solid lines show the calculated 4BSB components.

Fig.5 Double differential proton emission cross sections of the ^{12}C(p,p') reaction at 16 MeV. Solid and dashed lines show the calculated 3BSB and 4BSB components, respectively.

processes. In the present analysis, this adjustable parameter is determined so as to reproduce reasonably the continuum component of proton spectra for 16 MeV. The same value is used as the normalization parameter in the calculation of α spectra.

The calculated 4BSB spectra for protons and α particles are shown by solid lines in Fig.2. Since three α particles emitted in the process (ii) can not be distinguished by the measurement using the present detector system, the right hand side of eq.(1) is multiplied by a factor of three and the obtained α spectra are compared with the experimental data. The calculated proton and α spectra show good agreement with the experimental ones in shape and magnitude as shown in Fig.2. Therefore, the 4BSB process seems to be responsible for emissions of protons and α particles with low outgoing energies. However, there is somewhat overestimation in the region around the threshold energy of three α breakup in both those proton and α spectra. Furthermore, we have applied the similar 4BSB calculations to the ^{12}C(p,p')3α reaction at 14 MeV. The result is shown in Fig.4. In the calculation, the square of the matrix element $|M|^2$ is multiplied by a factor of 1.9 to reproduce the experimental spectra. This indicates that there is the incident energy dependence of $|M|^2$. Antolkovic' et al. [12] have also reported the strong incident energy dependence of $|M|^2$ (4BSB) in their analysis of the ^{12}C(n,n')3α reaction at 16-27 MeV.

According to the analysis by Antolkovic' et al.[12]· the sequential decay process involving the n + α + ^{8}Be$_{g.s.}$(or ^{8}Be$_{2.9}$) system is the dominant reaction mechanism and the contribution of the 4BSB process is appreciably smaller than that of the 3BSB process. To estimate the contribution of 3BSB in the ^{12}C(p,p')3α reaction, therefore, only the proton spectra for 16 MeV were calculated using the reaction model in which the α-α final state interaction is taken into account in the sequential decay from the ^{8}Be ground state and the 2.9 MeV state. The absolute values for each component are normalized so as to reproduce the experimental spectra. As shown in Fig.5, the difference between the calculated 3BSB and 4BSB spectra appears obviously near the threshold energy (5-6MeV), while both the spectra have similar shape in the low outgoing energy region. From the fitting of spectra near the threshold energy, the transition to ^{8}Be ground state (dotted line) in 3BSB is found to be much smaller than that to the 2.9 MeV excited state (dot-dashed line). Further calculation of 3BSB spectra for α emissions will be necessary to identify which process is dominant in the continuum underlying the peak structure, 3BSB or 4BSB process.

Summary

The double differential proton and α particle emission cross sections were measured for ^{12}C($\vec{p}$,p')3α reaction at 14 and 16 MeV, with better resolution than several neutron induced experiments. For the three or four-body simultaneous breakup process, the measured energy spectra of protons and α particles were analyzed in terms of the phase space model with the transition matrix element independent of the angle and outgoing energy. The calculated 4BSB spectra of both protons and α particles reproduced well the continuum observed in low outgoing energy region. Further analysis of the experimental data including the analyzing power for 14 and 16 MeV will be required to enhance an understanding of the p + ^{12}C breakup reaction mechanism.

References

1 D.J. Brenner, M. Zaider, J.J. Coyne, H.G. Menzel and R.E. Prael: *Nucl. Scie. Eng.*, **95**, 311 (1987)
2 D.J. Brenner and R.E. Prael: *Atomic Data and Nuclear Data Tables*, **41**, 71 (1989)
3 B. Antolkovic', G. Dietze and H. Klein, *Nucl. Scie. Eng.*, **107**, 1 (1991)
4 R.C. Haight , S.M. Grimes, R.G. Johnson and H.H. Barschall: *Nucl. Scie. and Eng.*, **87**, 41 (1984)
5 A. Takahashi, M. Ichimura, Y. Sasaki and H. Sugumoto: *J. Nucl. Sci. Technol.* **25**, 215 (1988); M. Baba, M. Ishikawa, T. Kikuchi, H. Wakabayashi and N. Hirakawa: in *Proc. Int. Conf. on Nuclear Data for Science and Technology,* May/June 1988, Mito, Japan (Edited by S. Igarashi), p. 209, Saikon, Tokyo (1988)
6 N. Koori, I. Kumabe, M. Hyakutake, K. Orito, K. Akagi, A. Iida, Y. Watanabe, K. Sagara, H. Nakamura, K. Maeda, T. Nakashima, M. Kamimura and Y. Sakuragi: *JAERI-M* 89-167, Japan Atomic Enegy Research Institute (1989)
7 Y.Watanabe, N. Koori, M. Hyakutake and I. Kumabe: *JAERI-M* 90-025, 216 (1990)
8 K. Sagara, K. Maeda, H. Nakamura, M. Izumi, T. Yamaoka, Y. Nishida, M. Nakashima and T. Nakashima, *Nucl. Inst. and Meth.* A270, 444 (1988)
9 Y. Nagahara, *J. Phys. Soc. Jpn*, **16**, 133 (1961)
10 J.S. Nodvik, C.B. Duke and M.A. Melkanoff, *Phys. Rev.* **125**, 975 (1962)
11 G.G. Ohlsen: *Nucl. Inst. and Meth.* **37**, 240 (1965).
12 B. Antolkovic', I. Šlaus and D. Plenkovic': *Nucl. Phys.* **A394**, 87 (1983).

Exciton Model Analysis of Double Differential Neutron Production Cross Sections for 113–MeV Protons

Norio Kishida and Hiroyuki Kadotani

Advanced Technology Department
Century Research Center Corporation
1–3–D17, Nakase, Chiba 261–01, Japan

Abstract: Differential neutron production cross sections at 113-MeV protons for Al, Fe, W and Pb target nuclides were analyzed by the MCEXCITON code based on an exciton model. The exciton model was able to reproduce all the measured DDX. Agreement between the measured and the calculated DDX is satisfactory at 60 deg. and 150 deg. However, at 150 deg., contrary to the HETC code, the MCEXCITTON code slightly overestimates the measured DDX for all targets. A slightly modified calculation of the exciton model was carried out by taking the direct knock-on process and the quasi-free nucleon-nucleon scattering such as (p,n) and (p,pn) reactions into account. However, no remarkable improvement was obtained for the disagreement between the measured and the calculated DDX.

(Al(p,xn), Fe(p,xn), W(p,xn), Pb(p,xn), E=113 MeV, DDX, exciton model, analysis)

Introduction

Double differential neutron production cross sections (DDX) for 80- to 800-MeV incident protons have been measured for various kind of nuclei and reported in the literatures[1,2,3,4,5]. These cross sections have been usually analyzed using the intranuclear cascade and evaporation model (INCM)[2,3,4,5] with the two exceptions of exciton model analyses[1,6] and an analysis by a quantum statistical multistep model[1].

These analyses clarified that INCM underpredicts the back angle DDX by a few orders of magnitude and the exciton model reproduces the DDX to the same extent as INCM. Although the quantum statistical multistep model reproduced the DDX very well, the analysis was carried out only for 80-MeV data. Hence, it is still unknown whether this model can provide the correct DDX at higher incident energies. Under these circumstances, new hybrid model calculations[7,8], in which exciton models are incorporated between the intranuclear cascade process and the evaporation process, have been performed as one of methods to reduce the disagreement between the measured cross sections and the calculated ones. The disagreement reduced to a great extent.

However, we feel that this hybrid model is not correct from the theoretical nuclear reaction model point of view. First, this is because both the exciton model and the cascade model descrive the same pre-equilibrium reaction. Second, this is because it is difficult to determine the distribution of exciton numbers when the cascade process has finished. Thus this hybrid model always involves large theoretical ambiguity. In fact, Ishibashi et al.[7] and Bozoian and Prael[8] did not present which exciton numbers they employed. It is, therefore, important to verify to what extent the exciton model can reproduce in itself the measured DDX. Thus we decided to analyze the 113-MeV data[2] from the following three reasons:

1. at this energy, the intranuclear cascade and exciton hybrid model most acceptably reproduces the measured DDX[8];

2. in incident energy of nucleons below ∼100 MeV, the exciton model generally achieves a great success in analyses for various pre-equilibrium reactions;
3. the experimental DDX were measured for many target nuclides in comparison with the 80-MeV experiment[1].

Analysis and Discussion

We will present only the analysis of DDX for Al, Fe, W and Pb targets on account of our restricted space. However, our results will not probably lose generality.

MCEXCITON code, which was developed by Kishida and Kadotani[6], was used to perform exciton model calculation of theoretical DDX. This code adopts the random walk exciton model[9] as a pre-equilibrium process model and the statistical evaporation model as an equilibrium one. Since detailed calculational methods about energy spectra and angular distributions are provided in Ref. [6], we would like to pass those accounts in this paper.

The measurement[2] was performed using natural isotope targets, but on the other hand the calculation was carried out for ^{27}Al, ^{56}Fe, ^{184}W and ^{208}Pb nuclides. Total reaction cross sections (σ_{tot}) were calculated from a relation $\sigma_{tot} = \pi(1.26A^{1/3})^2 - 50$ fm^2 which was empirically obtained by Pollock and Schrank[10].

Figure 1 shows comparisons between the experimental (symbols) and the calculated (solid lines) DDX. For all the nuclides, overall agreement between the calculated DDX and measured ones seems to be fairly well. However, strictly speaking, the calculation exceeds the measured DDX in the neutron energies between ∼5 MeV and ∼25 MeV at both the 7.5 deg. and 30 deg. The MCEXCITON code reproduces very well all the 60-deg. data except at low neutron energy (below ∼5 MeV) of the W target. It slightly overestimates all the 150-deg. DDX and the disagreement grows larger as the target mass number becomes smaller. This 150 deg. result is completely contrary to the HETC[11] calculations by Meier et al.[2], which underestimated the data. We conclude that the exciton model can

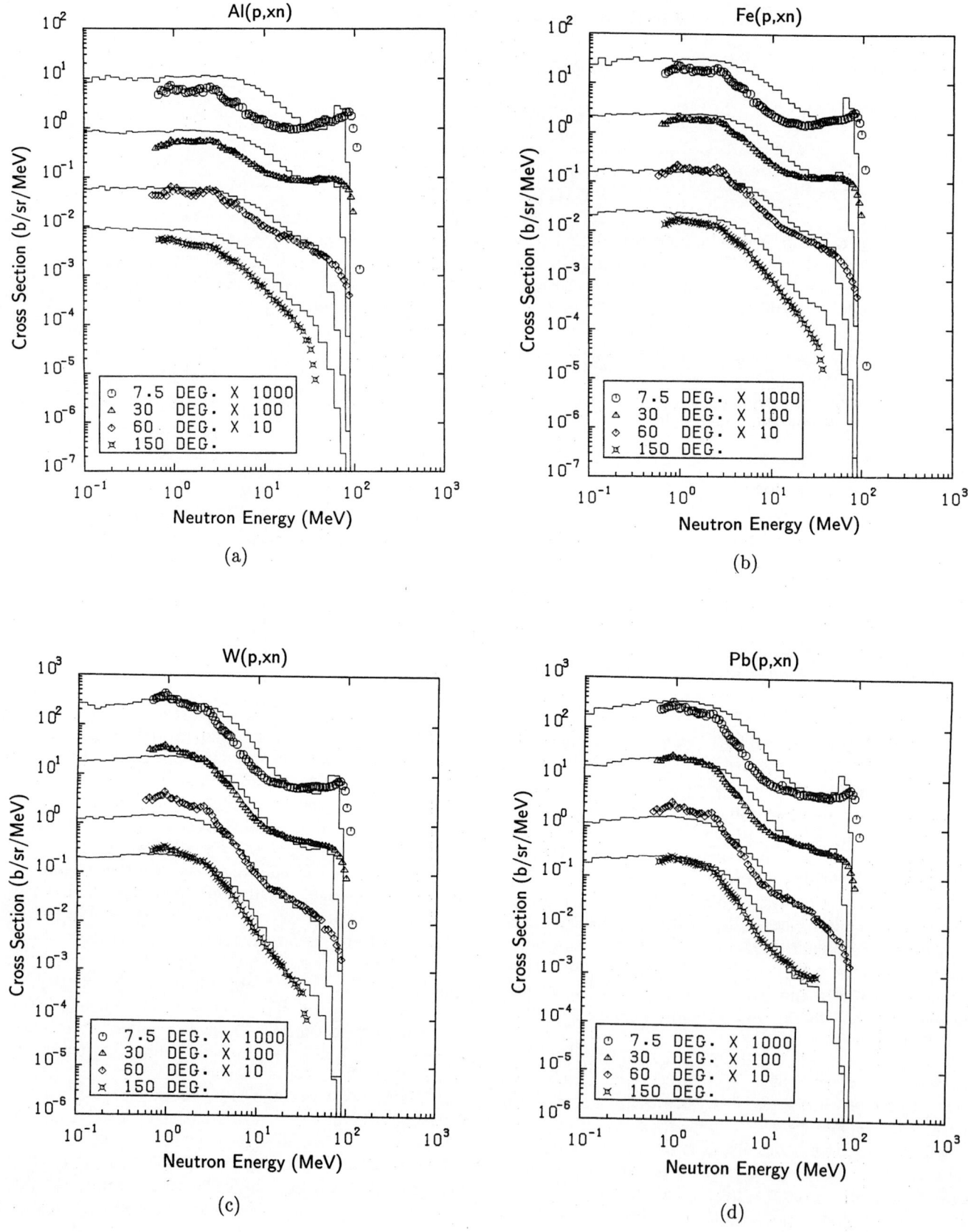

Fig. 1 Comparison between experimental differential cross sections at 113 MeV (symbols) and MCEXCITON calculations (solid lines) for (a) Al, (b) Fe, (c) W and (d) Pb targets.

reproduce the measured DDX to the same extent as the intranuclear cascade model.

The exciton model[9] does not consider the one-step process like the quasi free scattering and the knock-on reaction, because in this model an emitted particle is produced only through two-step or more complex processes such as disintegration of a particle-hole excitation state, which is formed by interaction between an incident proton and a nucleon inside the target nucleus. Cross sections of this process is large for light nuclei rather than for heavy nuclei, because probability that the neutron struck by the incident proton leaves away from a nucleus without the second collision becomes large as numbers of nucleons in the nucleus is small. Hence, in order to estimate the effect of this one-step process, we performed the calculation for the Al target.

First, we calculated DDX for the one-step process, namely, (p,n), (p,p), (p,pn) and (p,2p) reactions using the HETC code, but in this calculation we excluded the (p,pn) and (p,2p) reactions including an evaporation process. Second, we added the (p,n) and (p,pn) neutron production cross sections to the exciton model ones. In this operation, the total reaction cross section (σ_{tot}) was not altered, but it was re-distributed to the one-step process (σ_{QF}) and the exciton process (σ_{PE}) in such a way to equalize the sum of σ_{QF} and σ_{PE} with σ_{tot}. The reaction cross section of the one-step process amounts 35 percent of the total reaction cross section. Figure 2 shows comparison between this calculation and the measured DDX for the Al target. As was expected, the cross section near the incident energy somewhat increases at 7.5 deg. and 30 deg., and the disagreement decreases, especially at 30 deg. However, at 60 deg. and 150 deg., no changes in energy spectra is seen at all. It is natural since the one-step processes having strongly forward peaking angular distributions.

Summary and Conclusions

We analyzed the neutron production differential cross sections for incident protons at 113-MeV using the exciton model. The exciton model can reproduce fairly well the measured DDX for all the nuclides. Especially, at 60 deg. and 150 deg., the agreement between the measured and the calculated DDX is satisfactory. At 150 deg., contrary to the HETC code, the MCEXCITON code slightly overestimates the measured DDX.

We also performed the slightly modified exciton model calculation in such a way that the one-step process like (p,n) and (p,pn) reactions is taken into account. However, no remarkable improvement was obtained by this modification. We believe that the exciton model has the ability of reproducing the measured DDX around this incident energy to the same extent as the intranuclear cascade model.

Acknowledgment

One of us(N.K.) is greatly indebted to Dr. M. Meier and his collaborators for providing the numerical values of their experimental double differential cross sections.

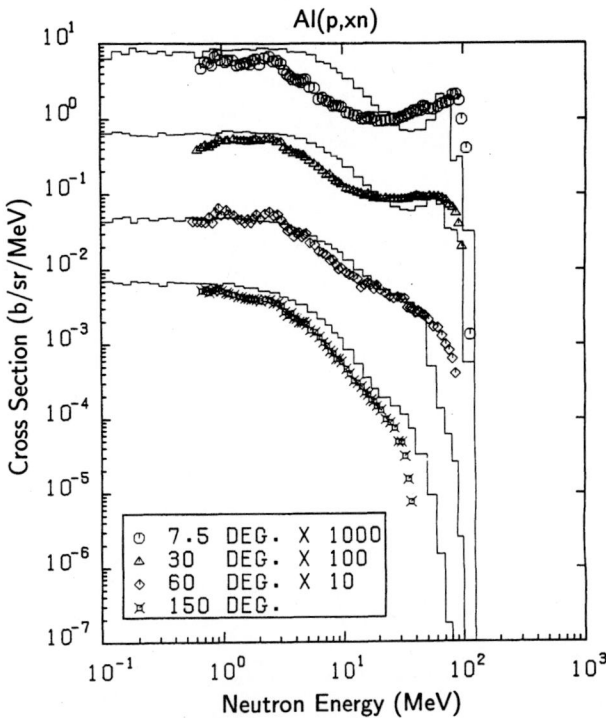

Fig. 2 Comparison between experimental differential cross sections at 113 MeV (symbols) and MCEXCITON calculations (solid lines) with direct (p,n) and (p,pn) reactions for the Al target.

REFERENCES

1. M. Trabandt, W. Scobel, M. Blann, A. Pohl, R.C. Byrd, C.C. Foster and R. Bonetti: Phys. Rev. **C39**, 452(1989)

2. M.M. Meier, D.A. Clark, C.A. Goulding, J.B. McClelland, G.L. Morgan and C.E. Moss: Nucl. Sci. and Eng. **102**, 310(1989)

3. M.M. Meier, C.A. Goulding, G.L. Morgan and J.L. Ullmann: Nucl. Sci. and Eng. **104**, 339(1990)

4. M.M. Meier, D.B. Holtkamp, G.L. Morgan, H. Robinson, G.J. Russell, E.R. Whitaker, W. Amian and N. Paul: Rad. Eff. **96**, 1415(1986)

5. S. Cierjacs, Y. Hino, F. Raupp, L. Buth, D. Filges, P. Cloth and T.W. Armstrong: Phy. Rev. **C36**, 1976(1987)

6. N. Kishida and H. Kadotani, in Proc. Int. Conf. on Nuclear Data for Science and Technology, May/June 1988, Mito, Japan(Edited by S. Igarasi), p. 1209, Saikon, Tokyo (1988)

7. K. Ishibashi, H. Takada, T. Sakae, Y. Nakahara and T. Nishida, in Proc. Int. Conf. on Nuclear Data for Science and Technology, May/June 1988, Mito, Japan(Edited by S. Igarasi), p. 1143, Saikon, Tokyo (1988)

8. M. Bozoian and R.E. Prael: Trans. of American Nuclear Society, **60**, 261(1989)

9. K.K. Gudima, G.A. Osokov and V.D. Toneev: Yad. Fiz. **21**, 260(1975); Engl. transl. in: Sov. J. Nucl. Phys. **21**, 138(1975)

10. R.E. Pollock and G. Schrank: Phys. Rev. **140**, B575(1965)

11. T.W. Armstrong and K.C. Chandler: Nucl. Sci. and Eng. **49**, 110(1972)

Closing Session

PROGRESS AND PROBLEMS IN ENERGY-RELATED NUCLEAR DATA[*]

Robert W. Peelle

Oak Ridge National Laboratory, P. O. Box 2008
Oak Ridge, Tennessee 37831-6354 USA

Abstract: The International Conference is reviewed relative to needs for measured data in nuclear energy applications. Notable achievements were recorded in several research areas. Elsewhere, obstacles have inhibited development of adequate data. Major data areas are identified with various levels of progress during recent years. The present energy-related nuclear data base is characterized as inadequate relative to a rising standard of quality. Ongoing efforts can achieve the continual improvement required for future technology.

Introduction

At this conference a wealth of information has been presented that will facilitate the development of options for nuclear energy. The goal of this paper is to identify large-scale trends in the progress achieved. The focus is on neutron reactions.

The summary presentations offer three views of but one discipline. There is great overlap among the topics, so what turns out to be the most important development for energy applications may well be mentioned this afternoon under another umbrella topic. For example, data _evaluations_ crucial for energy applications are covered elsewhere.

Our _nuclear data_ subdiscipline forms the strongest interface to the larger society from the body of nuclear physics. It is our task to develop this interface, and our privilege to be concerned with the difficult and fascinating puzzles it presents.

This week we have seen much evidence of skillful application of clever techniques to help solve these puzzles. Let us take a few moments this afternoon to celebrate our successes and to ponder some of the obstacles.

Areas of Great Progress

First, we list some of the areas of greatest progress since the previous conference in this series. Continued work in these areas will meet stated nuclear data needs. Surely additional topics could have been included if the author enjoyed a broader perspective.

1. "Dosimetry" cross sections are greatly improved. These energy-dependent reaction cross sections are the flux weighting functions for integral reaction rates measured in nuclear facilities. Many of us can recall an earlier era when there was little expectation that "effective cross sections" developed for dosimetry applications would have the same values as those for the same reactions used for other applications. Now all that has changed, and we are concerned about a 2-3% difference between a ^{238}U fission cross section measured against dosimetry standards

and the value from ENDF/B-VI. However, good success in "unfolding" neutron spectra using dosimetry data will require even more detailed measurements.

2. There is continuing improvement in experimental data such as total cross sections and elastic scattering angular distributions, data that are fundamental to comprehension of the neutron cross sections of a nuclide. These improvements have come from more sensitive fast-neutron detectors, improved electronics, and assiduous attention to detail.

3. Relatively complete and reliable resonance data are becoming available for more and more nuclides. The untangling of the weighting-function puzzle for "total energy" detectors as the result of the labors of an NEANDC working group now makes possible reliable resonant capture data for medium-A nuclides, for which big changes in capture gamma ray spectra from resonance to resonance are exhibited.[1] If this trend toward reliable data continues, we shall one day have available the resonance parameters required to calculate reaction rates (and thereby temperature coefficients) accurately when resonance self-shielding is significant.

4. Many good data are now available for gas production in reactions at neutron energies important to fusion energy. The accumulated gas is known to cause changes in material properties. Data for various reactions are available from activation studies and from direct observation of secondary charged particles from thin samples.

5. The total capabilities for measurements with "white" neutron sources have been greatly enlarged as the WNR and LANSCE portions of the Los Alamos Meson Physics Facility have become increasingly useful. The first facility is particularly valuable for neutron energies above several MeV, the second for energies below about 20 eV. Improved facilities in the USSR were also described.

6. There have been marked advances in availability of fission product data useful for estimation of delayed neutron effects and after-shutdown decay heat. Many of these new data were obtained using isotope separators. In use, these are combined with analyses of the β-decay process in the case for which many states

[*]Research sponsored by the U. S. Department of Energy, Office of Nuclear Physics, under contract DE-AC05-84OR21400 with Martin Marietta Energy Systems, Inc.

are available in the daughter nucleus. Results from the various modeling techniques tend to agree when averaged over the fission product ensemble, but often differ widely when data for particular nuclides are compared.

Areas of Good Progress But Great Need

Other energy-related areas where progress is promising have not yet reached the stage for a major celebration. Some of these topics are noted below.

1. For ^{235}U the quantity $\eta(E)$, the number of fission neutrons emitted per neutron absorbed, has now been shown to decrease for incident neutrons below thermal energies. Integral temperature coefficient data observed for thermal critical systems motivated the differential experiments that demonstrated an increase in the ratio of neutron capture to fission at subthermal energies. The exact magnitude of the effect is still being debated, and must be established because the effect impacts analyses of the temperature coefficient of reactivity for light water power reactors.

2. At this conference there was little mention of measurements on resonant neutron capture in ^{238}U, though the correct analysis of resonance absorption in this isotope has been a central problem in reactor physics. An interim set of resonance parameter results, far more complete than those of the past, was made available for the evaluated data sets now being completed.[2] These results were based to a great extent on new experimental data. Some work has continued in the pertinent laboratories to refine the resonance analysis and include the very latest flight-time spectra. It is important that this task be brought to a logical conclusion.

3. The NEANDC established a working group on the ^{10}B cross sections because evaluation efforts for ENDF/B-VI illustrated significant uncertainties and inconsistencies in the underlying data. New experimental results are starting to appear, and discussions during this meeting suggest that additional data will become available.[3]

Problems That Require Greater Effort

Some types of measurements have proven difficult in that significant careful efforts have left us with inadequate data. In some cases it is not clear how to obtain the needed experimental results.

1. In some fission reactor systems, inelastic scattering in fissile materials is an important mechanism of energy loss that therefore impacts the neutron spectrum. Experiments for these nuclides suffer from interference from prompt fission neutrons in addition to all the other difficulties of neutron inelastic scattering work. Uncertainty in neutron spectral shape is also caused by inadequate experimental data on prompt fission neutron spectra for other than ^{252}Cf and perhaps thermal fission in ^{235}U. Model calculations are proving to be very helpful for both inelastic scattering and fission neutron spectrum calculations, but some really clean experiments are needed if they are possible.

2. Strong efforts have been made to measure doubly differential neutron emission spectra for materials important in fusion systems, particularly near 14 MeV. Even in this range where there has been increasing experimental success over the years, the data obtained are marginally adequate for benchmarking the ever-more thorough pre-equilibrium models used to help produce evaluated cross sections and angle-energy distributions. Outside the 14-MeV region the data are less adequate. Experimental problems include interference from elastically scattered neutrons. There are too many theoretical difficulties for us to expect model calculations to suffice until some precise data can be obtained at least for the most important nuclides.

3. Energy-dependent neutron capture in fissile nuclides is very difficult to measure. De-excitation gamma rays are used to identify capture events, and the detectors are sensitive to the more-numerous prompt fission gamma rays. Neutron-induced backgrounds tend to be very large in most experiments. Because of a variety of constraints, promising results on ^{235}U obtained a decade ago by Muradyan[4] using a photon multiplicity detector have not yet been extended to the other important fissile nuclides.

4. Much has been learned in the last two decades about how to understand and quantitatively express the uncertainties in nuclear data experiments. Nevertheless, published experimental results often do not contain all the needed information available to the experimenter. Lacking such information, the user cannot know the value of the data either for direct use or for combination with other information. Comparisons among independent data provide a check, but the information content of such comparisons is often elusive. Overall, the trend toward better documentation of uncertainties and covariances needs to be accelerated. Considerable effort is involved.

Looking Forward

Beyond the specific suggestions above, there are numerous important goals for nuclear data experiments. Relative to a rising standard of quality, the improving evaluated data base is still inadequate for applications and combinations of theory with existing experimental data can meet only a portion of the needs. The standards rise because of demands that any new projects meet advanced design goals with minimum engineering margins. (For example, consider design and operation of a power reactor with inherent safety features and long core lifetime. The first criterion depends in part on low excess reactivity that makes the second criterion very hard to meet.) Highly optimized designs cannot be achieved without refined data and analysis methods.

Data request lists are a source of measurement priorities. More precisely, the needs that show up on published lists have usually been communicated directly to those who may be able to assist. In particular, cross section evaluators often identify measurement needs.

As designs are improved, a variety of competing materials are considered for use in fusion and fission systems, particularly for structural components. Often we have found our data base for such "new" materials to be obsolete or non-existent, and have hurried to meet the minimum need. Since we cannot know which nuclear applications will receive future emphasis, we must be ready with resilient data and methods.

If nuclear technology is to be efficient and safe, there must be continual improvement in the base of experimental nuclear data. The work can be accomplished stepwise by the logical continuation of programs of the present and recent past. We must do our best to inform our leaders of this need.

Final Remarks

During this week of immersion in nuclear data progress we have renewed our professional contacts across the world. Let us keep them fresh and beneficial.

I give my own best feelings to you all, with wishes for good health and for success in your research in the years ahead. We hope to see you at the 1994 nuclear data conference in Oak Ridge.

References

1. M. G. Sowerby and F. Corvi, *in Proc. Int. Conf. on Nuclear Data for Science and Technology,* May/June 1988, Mito, Japan (Edited by S. Igarasi), p 37.

2. Y. Kanda, Y. Kikuchi, Y. Nakajima, M. G. Sowerby, M. C. Moxon, F. H. Frohner, W. P. Poenitz, and L. W. Weston, "A Report on Evaluated ^{238}U(n,γ) Cross Sections," Paper O51 at this conference, includes a discussion of the new evaluated results for the ^{238}U resonance region.

3. A. Carlson, *Minutes of the Second Working Group Meeting on the ^{10}B(n,α) Standard,* held May 15, 1991 at Julich, FRG. Private communication, July 1991.

4. G. V. Muradyan, Y. G. Schepkin, Y. V. Adamchuk, M. A. Voskanyan, in *Proc. of Nuclear Cross Sections for Technology,* Oct. 22, 1979, Knoxville, Tennessee, National Bureau of Standards report NBS-594, p 488, (1980).

Summary on Non–Energy Related Nuclear Data

O.W.B. Schult

Institut für Kernphysik, Forschungszentrum Jülich, D–5170 Jülich

I will try to briefly summarize the contributions dealing with non–energy related nuclear data. Nuclear data are not only needed for fission– or fusion reactor technology. They are also essential for ensuring progress in a variety of fields which are often strongly connected. Dividing the summary, therefore, requires drawing boarders between regions of strong overlap.

The interest in the conference at the Forschungszentrum Jülich has been very strong, and we were very glad that more scientists from the Soviet Union could participate than in previous conferences. We hope very much that the conditions in all parts of the world do improve so that more scientists from other countries will be able to attend and contribute to the next conferences.

Very good for young people were the tutorial sessions on Monday afternoon.

I can only mention or remind the participants of part of the contributions to this conference. But I did benefit a lot from listening or reading abstracts about many more, although I cannot discuss them. The excellent talks by Prof. Wolf Häfele, Prof. Peter Hodgeson and Prof. Hans Weidenmüller are examples.

Neutrons and n–Resonances

Data on the neutron itself belong to the fundamental data on elementary particles. Shcherbakov and coworkers have measured the total neutron cross sections of ^{208}Pb and Bi with high accuracy in the 1 eV–keV region. They find $\alpha_n=(25\pm11)\cdot10^{-3}$fm^3 from the Bi experiment and $\alpha_n=(2.6\pm1.4)\cdot10^{-3}$ fm^3 from the ^{208}Pb data. The latter value is consistent with the value obtained through a neutron transmission measurement on ^{208}Pb carried out by Riehs and his colleagues for $0.1 < E_n < 500$ keV. In a previous experiment the group had obtained $\alpha_n=(1.2\pm1.0)\cdot10^{-3}$fm^3. The new value is the same, but the total uncertainty is only around 0.3. It would be interesting to compare this number with a QCD prediction.

Neutron resonances of ^{155,157}Gd were investigated and correlations between reduced neutron widths and the intensities of selected γ–transitions were studied by Muradyan and his coworkers. For ^{157}Gd the strength function was found to depend on spin. In a study of resonance neutron scattering and absorption on ^{238}U the group observed an anomalously small total γ–width for the 721 eV resonance. This might be evidence for a state in the second minimum. At the pulsed reactor in Dubna the neutron capture– and scattering cross sections and the γ–ray multiplicity M of ^{147}Sm were measured for $E_n = 15 - 400$ eV by Georgiev et al., who also studied the correlation between resonance spin and M. They found M ~ 3.4 for $I^\pi = 3^-$ and M ~ 3.7 for $I^\pi = 4^-$.

Also Coceva, Rohr and others have carried out interesting work in the neutron resonance region.

Fission

Cold fragmentation

A larger number of contributions deals with fission of various nuclear systems, with cold fragmentation, thermal– and resonance neutron induced fission, fragment yields, proton and neutron even–odd effects, fission dynamics and fission–neutron spectra. Hambsch and his coworkers have investigated cold fragmentation of ^{252}Cf and correlations between prompt γ–rays and fission fragments. They found 59 very neutron–rich nuclides which were not on the Karlsruhe chart. The conclusion from the very interesting and detailed study is

that the proton– and neutron odd–even effects δ_p, δ_n cannot be interpreted as indicators of the intrinsic excitation energy at scission, and that the fission yields are distributed according to the available phase space at scission.

An interesting study of cold fragmentation of ^{235}U + n has also been performed by V. Khrjachkov and his coworkers who did not only measure all principal fragment pairs for thermal but also for 1 MeV neutrons. As in the latter case the fragment mass spectra do not change as function of $E_{kin\ light}$ in contrast to the $n_{thermal}$ induced fission, the conclusion is that the excitation energy of the fissioning system is conserved on the way from saddle to scission. Thus it seems that true cold fragmentation is not obtained in ^{235}U + $n_{thermal}$ fission.

Fission induced by thermal neutrons

Thermal neutron induced fission of ^{249}Cf has been studied at LOHENGRIN by R. Hentzschel and coworkers, who observed a very strong enhancement of the yields compared with expectation, especially for the lightest fragments. Mass and isotope yields were obtained for very asymmetric fission. It was found that δ_p, δ_n increase with mass asymmetry which suggests colder fission.

Theobald et al. have used the mass separator LOHENGRIN at the Institute Laue–Langevin (ILL) for measuring the light fission fragments from neutron induced fission of the odd Z ^{243}Am which was formed by consecutive neutron capture. The motivation of the nice study was the hope to learn about the influences of pairing and of fragment proton shell structure on the fragmentation. Also light charged particles from ternary fission were measured with a sensitivity of $10^{-8} - 10^{-9}$ per fission.

Neutrons of 4.5 MeV were used by Makarenko and coworkers to populate in (n,n') the 0.27 μs fission isomer in ^{238}U and to measure the energy spectrum of long–range particles in ternary fission. The data were found to agree with those measured in prompt ternary fission.

Investigating ^{241}Am (2n,f) Stumpf et al. found that light fragments are preferentially even and heavy fragments mainly odd, and explain this assuming statistical distribution of the odd proton among the fragments. They conclude that in odd fissioning nuclei superfluidity is preserved like in even systems.

A very fine system for coincidence measurements of fission fragments, Cosi Fan Tutte, has been employed at the ILL in order to study fragment yields of neutron induced fission of ^{239}Pu and ^{232}U. Whereas global data about ^{239}Pu(n,f) and the odd–even charge effect δ_p agree with the old LOHENGRIN data, Kaufmann et al. find the neutron odd–even effect to be a factor of 2 smaller. The ^{232}U(n,f) information is much more detailed than previous data. A very pronounced δ_p effect was observed.

Djebara and Asghar have studied the neutron odd–even effect δ_n in various nuclei as function of the neutron number N between 50 and 70. They observe peaks at N = 60 and 66 which behave differently as function of fragment kinetic energy. For testing the relation between primary distributions and distributions after prompt neutron evaporation, Monte Carlo simulations were carried out. These indicate that neutron evaporation destroys structures due to pairing and may create $\delta_n \approx 15\%$ effects when one starts with a smooth distribution. It seems that this needs further careful investigation.

Gundorin et al. have compared the results for the independent fragment yields $^{239}Pu(n_{res},f)$ with $(n_{thermal},f)$ data and model calculations. El–Bakly and El–Mekkawi have investigated angular momentum effects in fission. In order to explain the very short half–lives of the spontaneously fissioning ^{258}Fm and ^{258}No Bhandari and Bendardaf used the concept of resonant tunneling and succeed to also explain the nearly equal probabilities for the two modes of fission.

A. Grashin et al. have carried on their investigation of the problem of neutron emission from a fissioning system on its way from scission to full acceleration of the fragments. Ignatjuk and Maslov have been studying pairing correlation effects, few–quasiparticle excitations and step–like structures in fission cross sections, for gaining information on the gap parameter and the higher barrier. B. Gerasimenko and V. Rubchenya have generalized their model and calculated fission neutron spectra of ^{232}Th and ^{235}U for $E_n < 20$ MeV.

Batenkov and Blinov have compared neutron spectra emitted from fragments of spontaneous fission and from nuclei emitting neutrons in (projectile, xn) reactions. They find very similar equilibrium spectra if the masses and excitation energies are nearly the same, and for A far from 132.

Medium Energy

In recent years nuclear physics interest has shifted more into the field of medium energy, forming a bridge between nuclear and elementary particle physics. Thus there is need for a lot of medium energy nuclear data both for groups planning experiments, where "the physics of today is the background of tomorrow", and for people who build machines and take care of proper shielding. I am sure, also here other groups will follow and apply the techniques which the nuclear physicists have then developed. Muon spin rotation[1] is just one example.

Paul Lisowski has given a nice and comprehensive overview on accelerator based sources for neutrons with energies up to 800 MeV, about the n–spectra, the advantages of using monoenergetic or white sources, depending on the problem, improvements of detection systems with beautiful examples for medium energy reaction studies and about recent progress in related fields.

Fission

Lisowski and collaborators have used the powerful WNR at Los Alamos for measuring the cross sections for fast neutron induced fission of $^{233\ 234\ 236}U$ relative to ^{235}U for 0.5 MeV $< E_n <$ 400 MeV. For neutron energies from 0.7–45 MeV fission cross sections have also been measured of ^{232}Th and ^{238}U relative to ^{235}U by Fomichev et al. making use of the 1 GeV proton synchro cyclotron at Gatchina.

Smirnov and his coworkers have measured relative cross sections for fission of ^{232}Th, $^{233\ 235}U$, ^{237}Np and ^{239}Pu bombarded with $25 - 70$ MeV protons for the study of the energy dependence of Γ_n/Γ_f at excitation energies of several tens of MeV. With the same goal Bao Zongyu et al. have measured the excitation functions during bombardment of ^{209}Bi with $20 - 35$ MeV protons and obtained a fission barrier height of 21.9 ± 1.5 MeV.

Radiation Protection

I enjoyed listening to R.O. Nelson's talk about nuclear data needs for the permanent base on the moon and the manned outpost on Mars as planned in the US. The idea to get nearer to the stars has always been fascinating and the much improved possibilities not only for stellar and space studies from outside the absorbing atmosphere justify orbiting bases. It is not clear to me that a manned outpost or a base on the moon is the optimum in view of robots and their expected development within the next 25 years. Radiation protection of astronauts is only one of the problems man has to cope with in space. Manned outposts are probably a matter of prestige rather than a necessity for measuring desired data.

Proper radiation protection and monitoring at high intensity medium high energy proton accelerators is by no means trivial, as has been explained well by Yu. Nikolaev. He and his coworkers show that energetic hadrons and airborn radioactivity contribute significantly to the dose. Cross sections for nuclear fragment production were mostly unknown. Matushko et al. have written a computer code which should allow the calculation of the production and dose of radioactivity when heterogeneous materials are irradiated with medium energy nucleons. Kazuo Shin et al. studied the transmission of neutrons and associated γ–rays, obtained from bombardment of copper with 65 MeV protons, through shields of C, Fe and Pb. Calculations using the DLC–87 data reproduced the graphite data well, but revealed discrepancies for Fe and Pb also for the photons. Further work is needed to clarify this.

Spallation

For various applications more information is needed about spallation.

It was a pity that no member of A.V. Daniel's group from Leningrad was present to talk about neutron fields of a thick lead target bombarded with 1–3.7 GeV protons and deuterons and the comparison with calculations.

Ishibashi et al. have carried out studies of neutron emission cross sections and describe data on C–U targets for 80–800 MeV protons successfully by their model. Such data are essential for spallation–neutron sources.

Filges and his coworkers have measured at the LAMPF White Neutron Research Facility double differential (p,xn) cross sections for proton energies between 100 and 800 MeV, for a large number of angles and for numerous targets by time of flight spectroscopy. The obtained spectra were compared with Monte Carlo simulations using the HETC code. This combination of extensive data and simulations provides a very good basis for description of A(p,xn) reactions even at large neutron emission angles.

Spallation is an efficient method for radioisotope production[2].

Radioisotope Production and Excitation Functions

Michel and collaborators have in view of the wide field of applicability measured production cross sections of numerous radionuclei in proton bombardment of a variety of elements for $0.6 < E_p < 2.6$ GeV. Combining these data with previous results the group has obtained a consistent set of excitation functions from threshold up to 2.6 GeV. The experimental results are compared with model calculations.

The production of ^7Be by $35 - 100$ MeV proton bombardment of medium and heavy mass nuclei and also by break up neutrons from 53 MeV deuterons was measured by Scholten and his colleagues, who found cross sections from 10 $\mu b - 1$ mb and in the μb region, respectively, but surprisingly large cross sections for the production of ^7Be on Bi.

Yoshi Uwamino and his coworkers at the INS have produced Be(p,n) neutrons with energies up to 40 MeV and determined activation cross sections for ~ 32 reactions on targets from Na $-$ Au. The obtained excitation functions were compared with previous data, and for $^{23}Na(n,2n)$ at 20 MeV the ENDF/B–V value was found to be 3 x the measured value. Ismail and Divatia measured excitation functions for the production of various nuclei during the bombardment of Co, Ga, Ag, Sb and Ta with $10 - 65$ MeV α beams and compared their results with Blann's model.

Reactions and n–Yields

For the study of angular momentum effects in α induced reactions Nassiff and his coworkers have measured the isomer ratios of [184]Re.

Mills and coworkers have measured production cross sections for 24 radioisotopes in the bombardment of Cu with $10 - 200$ MeV protons in order to investigate the influence of the $f_{7/2}$ shell closures and found reasonable reproduction of the results when using the ALICE code and zero shell correction and pairing energy terms in the mass formula.

The total cross section of nitrogen for neutrons from 0.5 eV up to 50 MeV was measured by Jack Harvey and his coworkers at ORELA using a gas sample, 80 and 200 m flight paths and advanced neutron detection. The results were found to differ by up to 10% from earlier data but in support of the R matrix analysis results used in the ENDF/B–VI evaluation. For studying nuclear reaction mechanisms Yamanouti et al. have measured the elastic and inelastic differential cross sections of 28.2 MeV neutrons scattered from ^{12}C and of 18.5 MeV neutrons scattered from [52]Cr and [60]Ni for angles from 20 to 140[0] and compared the results with various models.

Finlay has given a very interesting report on the work which he and his collaborators have carried out at LAMPF. They have measured with high accuracy total cross sections for $4.5 - 600$ MeV neutrons impinging on ~ 20 targets from Be to Bi. The results were used to test nuclear models, in particular the optical model and the relativistic impulse approximation.

A systematic study of thick C–Pb target neutron yield resulting from $30 - 50$ MeV p, 65 MeV d, and ~ 10 MeV/u C and O bombardment was performed by Shin and coworkers. The data measured from $0 - 135^0$ were analyzed in terms of the moving source model and e.g. a systematic expression was derived for the equilibrium component. Daniel and Perov have compared experimental data on high energy neutron yields with expectations based on thin target data, draw attention to secondaries and discussed limitations and improvements of calculations.

Cosmic Radiation Effects

Cosmic Radiation effects are an important issue connecting medium energy reactions with nuclear astrophysics.

In order to improve the cross section data base and to test models in the region between preequilibrium and spallation, Herpers et al. have measured production cross sections of radioisotopes during irradiation of numerous targets with 100 MeV protons. These data are also basic for solar cosmic ray production of nuclides in meteorites and lunar surface materials. Herpers has correctly stressed the importance of reliable data and meaningful errors.Synthetic Stony meteorite samples were bombarded with 200 MeV protons at the BNL linac by Divadeenam et al. for simulation of cosmic ray exposures. Long lived radioisotopes were measured and their production was found to be well reproduced by the HETC program.

Dagge and coworkers have applied the Hermes code for a Monte Carlo simulation of the cosmic radiation induced production of $\gamma-$ radiation emitted from a planet in view of the Mars Observer project.

As the occurrence of the long lived isotopes [10]Be, ^{14}C and [26]Al is closely related to reactions induced by secondary cosmic rays, their production has been measured by Nakamura and coworkers. They bombarded suitable targets with $10 - 40$ MeV neutrons produced in Be, Li (p,n) knock out and measured the long lived isotopes by AMS.

Dittrich et al. measured the production of [7] [10]Be, [22]Na, [26]Al et al. in reactions induced by [4]He with energies up to 172 MeV at the PSI and the KFA to also contribute to the data base of cosmogenic reactions. They obtained their results through $\gamma-$spectrometry and AMS and compared them with ALICE LIVERMORE 87 calculations.

Nuclear Astrophysics

Many of the astrophysical processes occur under conditions that cannot be realized in our laboratories. Only with most advanced techniques one can measure reactions in an energy regime which the astrophysicist is pushing further and further into the region of the real world, where short lived nuclei far from the valley of stability play the crucial role in nucleosynthesis. Claus Rolfs has in his very lively lecture explained why we like nuclear astrophysics, how one has to work in this field and which nuclear data are needed.

Wisshak and his colleagues utilize a pulsed [7]Li(p,n) source, neutron time of flight and their 4π BaF$_2$ detector for measurements of capture cross sections of various materials for $3 - 200$ keV neutrons. A significant improvement of accuracy has been obtained and also cross sections for Te isotopes were measured because these are crucial for precise tests of s–process models. The [147]Pm capture cross section has been measured by Gerstenhöfer et al. for keV neutrons applying 2 independent methods. [147]Pm, with a half life of ~ 2.6y, has a decay rate comparable to the n– capture rates in the s–process so that information can be obtained about the stellar n density. Activation in [7]Li(p,n) neutron fields was employed by Käppeler and his colleagues for measurements of stellar neutron capture cross sections at ~ 25 keV. They resolved a discrepancy for [87]Rb and carried out a first measurement of the [192]Pt cross section. This is relevant for obtaining information on the neutron density in the s–process during He shell burning in red giants. Druyts et al. have been studying (n,p) and (n,α) reactions on $26 \leq A \leq 50$ nuclei in the $E_n \sim 35$ keV region, relevant for red giants. They made use of the reactor at the ILL and its thermal neutrons for normalization and of GELINA, where keV neutrons are available.

Careful investigations of the r–process path were carried out by Kratz and collaborators, who emphasize the importance of reliable data on the ß–delayed n–emission and of improved quasiparticle RPA calculations. Also Klapdor has performed detailed studies in this field.

I am convinced that most of the work in the field of nuclear astrophyiscs, where we need very good and specific nuclear data, still has to be carried out. It is very difficult work, but definitely worth doing.

Decay Data

In the field of low energy nuclear physics most of the data on decay schemes relevant for application exist. We thus had only few such contributions at this conference.

Selinov and Chechev proposed to form an international group of scientists who prepare a Table of Nuclides free of discrepancies.

Woods has been employing at the NPL fine instruments and techniques for determination of particular nuclear decay data, relevant mainly for applied research. It is important that the availability of high quality instruments is guaranteed and that we do not loose the know–how and the advanced techniques for determining nuclear data. Schötzig and his colleagues have carried on their very precise measurements of nuclear half–lives and studied $\gamma-$ and x–ray emission probabilities with an accuracy of $\sim 1\%$. They present a list of such data on ~ 200 nuclei for use in activity measurements. The Manchester–NPL–ANL collaborators, Daniels et al., have begun to investigate the conversion electron spectrum from ^{245}Cm for improving its decay scheme. J. Stander and coworkers have performed $(\alpha,p\gamma)$ level scheme studies of ^{22}Ne and ^{29}Al and compared the measured cross sections with Hauser Feshbach predictions.

Nuclear Data for Application in Medicine

The possibility to use radionuclidic probes has opened a wide field for application in medicine, in order to study metabolisms, for imaging and for therapy, for which also the use of particle beams is essential.

Gerhard Stöcklin has in a very nice talk made clear that physiological substances labeled with suitable radioisotopes are crucial for good Positron Emission— or Single Photon Emission Computer Tomography. He has well reviewed which isotopes can be produced at accelerators with < 20 MeV and more energetic beams, drawn attention to processes where further work is needed, illustrated the limitations of relying on model calculations and emphasized radionuclidic purity for medical application.

Cancer therapy with fast neutrons or charged particles, protons, helium or heavier ions is one of the promising but also difficult approaches to fight one of the worst diseases of mankind. André Wambersie has very well informed also the less specialized scientists about the status of the research in this field, about the need to apply low and high linear energy transfer radiation to induce different biological effects, about the prospects to use heavy ions, and about the research needs on nuclear data for key substances. I was very fond of learning that in 45% of the cases cancer can be treated successfully.

At the Paul Scherrer Institute U. Schrewe and coworkers have determined the kinetic energy release for neutrons with 26 and 38 MeV in matter and neutron fluence for the tissue substitute A – 150 plastic and of carbon, in view of therapy facilities that use n–beams up to 65 MeV. In the 18–27 MeV neutron–energy region Hartmann and coworkers have measured the ^{19}F$(n,2n)$ cross section and the Kerma factors in C, Mg and Fe allowing for contributions from secondary neutrons, which is necessary in a careful analysis.

Mirzadeh and colleagues at the ORNL have been investigating the possibility to use double neutron capture in order to produce ^{188}Re and ^{194}Ir with ~ 1 m Ci/mg target material for therapeutic application and imaging.

Saed Mirzadeh and Yung Yee Chu have reported their results on the production of ^{66}Ga, an attractive candidate for therapeutic applications. They studied the ^{64}Zn$(\alpha,2n)$ and (α,pn) reactions and find especially the latter suitable to produce high purity material. Kovacz has together with Tarkanyi, Qaim and Stöcklin studied the production of ^{81}Rb and ^{82m}Rb in proton bombardment of Kr gas, because ^{81m}Kr from ^{81}Rb decay is well suited for studies of lung ventilation and 6.5 h ^{82m}Rb might be a good substitute for ^{82}Rb used for myocardial blood flow studies. The optimum beam energies were determined.

Szelecsenyi et al. have extended the previous work aiming at large scale production of ^{111}In. Zaitseva et al. have determined for proton energies up to 100 MeV cross section and yields for several radioisotopes which are frequently used in nuclear medicine. Zaitseva and colleagues have measured the excitation function of the ^{99}Tc$(p,3n)$ ^{97}Ru reaction, found a peak cross section of 460 mb at 32 MeV and determined the cumulative yield between 72 and 22 MeV.

Zaidi and coworkers have together with S. Qaim and G. Stöcklin measured ^{95}Mo(n,p) and ^{98}Mo(n,α) production of long lived contaminants of the ^{99}Mo/^{99m}Tc generator for the averaged fission neutron spectrum. Although the impurities are insignificant in eluted ^{99m}Tc they should be considered when disposing waste. Investi-

gation of the production of ^{57}Co from ^{58}Ni in a reactor showed that separation of ^{57}Ni allows ^{57}Co production in 10^5 Bq quantities.

Antolkovic has bombarded nuclear emulsions with 22.5 and 25.4 MeV neutrons in order to measure neutron induced charged particle production cross sections on tissue elements. She determined the ^{12}C$(n,n3\alpha)$ and ^{16}O$(n,n4\alpha)$ break up in a kinematically complete way through an elegant, basically simple experiment. Data have been obtained in a region, where little was known before. Intermediate–state structure has been observed.

Lunev and Shubin have compared calculated and measured production cross sections of radioisotopes for medical application and recommend to take into account the radiative channel near threshold. Hermanne et al. have in a study to optimize yields and contamination ratios investigated the ^{203}Tl$(p,3n)$ and ^{68}Zn$(p,2n)$ reactions and find generally good agreement with previous data. Afarideh and coworkers have measured very low charged particle induced reaction cross sections at their 3 MeV van de Graaff accelerator. I support such work because it is also a prerequisite for training people.

There is still room to optimize production of radionuclei for medical application. In particular, one has to allow also for half–lives, transport and the availability of accelerators. There may also be important technical aspects — waste disposal for example — which are crucial for practical isotope production.

Methods, Facilities, Applications

Nuclear Physics methods have found wide application in many fields of science and technology, mainly due to their extreme sensitivity. On the other hand, input is needed from other disciplines, in order to develop the methods to the highest level or to optimize applicability.

M. Suter has well described accelerator mass spectrometry, AMS, and which improvements could be achieved in sensitivity and selectivity. This analytical method allows detection of long lived radioisotopes that are produced cosmogenically in very small quantities. The main application is in geophysical studies of transport and exchange processes or for dating. Generally, half–lives and other data must also be known. A large discrepancy between the previously derived and theoretically predicted ^{60}Ni$(n,2n)$ cross sections led Weselka et al. to directly determine the number of produced ^{59}Ni atoms through AMS. With the radioactivity data they found $t_{1/2}$ to be four times the previous value and the cross section to well agree with theory.

H. Yamamoto, K. Kawade and coworkers have nicely shown how systematic errors can be reduced to < 0.1% during half–life measurements of short lived radioisotopes. For studies of short lived nuclei or of chemical properties of very heavy elements one must rapidly remove disturbing contaminants. My recommendation is to consult Trautmann and Herrmann for high yield, very selective and fast separation. In their contribution ARCA and SISAK are described and examples are given.

Allen and Chaudhri have drawn attention to the fact that in the presence of photons with E even less than 15 MeV (γ,n) reactions on rarer isotopes of constituents of the materials used in neutron dosimetry should be taken into account.

Scientists, who are interested in reactions induced by neutrons with $E_n \geq 10$ MeV and who do not have a suitable facility at home, should contact Conde, Klein, Haight or Lisowski who joined their efforts — which I would always recommend — to describe their machines that are able to produce "monochromatic" neutrons up to 200 MeV and white neutron spectra up to 800 MeV. I am convinced that they will also help interested users.

1018

In view of the possiblity to use compact accelerators as sources for neutron radiography Aslam Lone has reviewed suitable reactions for production of intense neutron fluxes. For the favorite Be(p,n) reaction he showed where data improvements are necessary for optimization of a facility.

Csikai has well presented a very detailed review on the status of nuclear techniques for geology and mining and he has discussed extensively which data are needed to improve and extend the applicability of these methods.

These are very good examples of what can and should be done.

A variety of widely differing research activities cannot be summarized in a few sentences. I could only touch upon part of the contributions. Most of the work has been accomplished by combining efforts in national and international cooperations. I want to remind you of Prof. Treusch, who showed in his very comprehensive talk about the Research Center Jülich how important interdisciplinary cooperation is so that energy—, materials—, information— and environmental research merge, in order to serve mankind. From my experience during the last 30 years I would like to recommend everybody: use possibilities for cooperation even more!

I have enjoyed very much learning from the excellent talks and well prepared posters of my old friends, and of colleagues which I met during this conference for the first time, how well their work is progressing. This was scientifically and personally a great pleasure for me. I wish all of you the best luck and success not only for your research work!

References:
1. Yamada Conference on Muon Spin Rotation and Associated Problems, April 18–22, 1983, Shimoda, Japan, Abstracts of Papers, Editors T. Yamazaki and K. Nagamine, Institute for Nuclear Study, Tokyo, Japan
2. H.L. Ravn, Nucl. Instr. & Meth. B26 (1987) 72

NUCLEAR DATA EVALUATIONS

S. Igarasi

Nuclear Energy Data Center
Tokai-mura, Naka-gun, Ibaraki-ken 319-11, Japan

Mr. Chairman, ladies and gentlemen.

It is my great pleasure to be given the role of the final speaker in this splendid Conference. I have already retired from JAERI and lived leisurely doing small job, such as study of application of S-matrix two-point function to nuclear data evaluation. So, little did I dream of standing at this place, before I was invited by the Program Committee of this Conference. I was excited and stimulated receiving the invitation letter, and probably I was rejuvenated for ten years by that stimulant. With the recovered youth, I eagerly listened to lecturers and contributors with great interest.

Introduction

Papers assigned to me by the Program Committee are those dealing with the nuclear data evaluations, nuclear models and systematics, including those about evaluated data files as well as activities on international cooperation for nuclear data evaluations. I was pleased to know that the sizeable progress was made in these fields, and a variety of high quality papers were contributed. I wished to touch as many papers as possible, but I could not have enough time to chew over and digest them, because there were too many papers for me to understand during limited time. So, I cannot help doing my talk with my prejudice, misunderstanding and ignorance. I apologize to you in advance for these undesirable points.

Contents of my talk are (i) Evaluated Data Files, (ii) International Cooperation for Nuclear Data Evaluation, and (iii) Nuclear Models, Systematics and Nuclear Data Evaluations.

Evaluated Data Files

One of the impressive events in this Conference was the presentation of the various evaluated data files and libraries; especially, big general purpose data files. They are BROND-2, CENDL-2, ENDF/B-VI, JEF-2 AND JENDL-3. Most of them have already been released as the final version or released for a tentative use with various purposes. I could see many papers in which the authors took the data of these files for their studies.

Besides the big files, the files of the nuclear structure data, the activation cross section data, and the data for fusion application were also introduced. Some of them were made under international cooperation or inter-laboratory collaboration. I should like to show them in Table 1, together with the general purpose files, with short descriptions.

I have seen many papers concerning the nuclear data evaluations for these data files and libraries. I was very impressed that a variety of skilled and comprehensive techniques were developed for the evaluations of the nuclear data for the files. In fact, full

Table 1. Evaluated Data Files and Libraries presented in the Conference.

File Name	Descriptions	Paper No.
ADL-90	activation data file, more than 5,000 reactions included (USSR)	C39
BROND-2	general purpose data file produced in USSR	IP14
CENDL-2	general purpose data file, 49 nuclides included, produced in P.R.China	IP15
EAF-2	activation data file, about 11,000 reactions for 667 nuclides included (EC)	IP19
ENDF/B-VI	general purpose data file, charged particle and medium energy data included (USA)	IP12
ENSDF	nuclear structure data file, promoted under international cooperation	IP17
FENDL	collective data file for fusion, promoted by IAEA	C40
JEF-2/EFF	general purpose data file produced by the cooperative work of the NEA Data Bank member countries	IP11
JENDL-3	general purpose data file, 171 nuclides included, produced in Japan	IP13
JENDL-3 FP Data File	data for 172 nuclides from ^{75}As to ^{159}Tb included, produced in Japan	C42
REAC*3 Data File	data library for REAC*3 code, about 500 nuclides included (USA)	C41
TDF	thermonuclear data file produced in USA	IP20

evaluation for the general purpose data files needs various kinds of nuclear reaction models, model parameters and systematics as well as the reliable experimental data. From the Dr. Young's excellent invited talk (IT15), I felt that the models and their parameters were mutually related with each other, and the data evaluation requested a lot of expert knowledge and experience to utilize them. These matters could be seen also in the papers about the evaluations of the data not only for ENDF/B-VI (O71, C31, C36) but for BROND-2 (C34) and JEF-2 (C37, D43).

I was interested also in the papers which made use of the data of the general purpose files for checking their own evaluation works (O55, O58, C30). They critically checked and reviewed the data of the files, and estimated a measure of permissible uncertainties of their planned evaluations using the data of the existing files. Such utilization of the data files is useful not only for the new evaluations but also for polishing the old data.

International Cooperation for Nuclear Data Evaluation

In order to compare and polish the data in the JENDL-3, JEF-2 and ENDF/B-VI, NEANDC/NEACRP organized six working groups. Status reports on these working groups were presented in this Conference. Although this cooperative activity has started in the frame work of the OECD/NEA member countries at present, it seems to stimulate other cooperative activities as well as individual researches. These trends may supplement funding and manpower decrease in the nuclear data community.

Status reports on the cooperative activities under IAEA, activities of the Reaction Data Center Network, and Nuclear Structure and Decay Data Network were also presented in the Conference. Nuclear data obtained, or to be obtained by these activities are very valuable and irreplaceable properties for human being. Hence, I hope that the International Organizations would continue holding the data, at least, and supporting these activities as successively as possible.

Nuclear Models, Systematics and Evaluations of Nuclear Data

Since my main task assigned by the Program Committee is to summarize the papers dealing with the nuclear models, systematics and nuclear data evaluations, as mentioned above, I would like to go about my business.

Papers about Fission

First of all, I would like to touch on the papers about fission. I saw four papers dealing with the fission. Two of them treated the mechanism of fission neutron emission (IP18, D29), and the other two (C35, D50) did resonance analyses. Since the fission neutron spectra affect greatly the results of the reactor calculations, careful evaluation has been carried out in the compilation of the evaluated data files. But, a little question still remains on this quantity. In this sense, I was interested in a non-equi-temperature fragment model for fission (D29) as a modifi-

cation of the Madland-Nix model.

Dr. Maerten (IP18) presented his nice systematics of the quantities on the fission fragment characteristics and fission neutrons based on the theoretical calculations. I hope that his systematics would be helpful to the nuclear data evaluation as well as to the analyses of the fission mechanism.

Papers C35 and D50 took an averaged width of photo-fission and identification of the p-wave resonance, respectively, for ^{235}U. These must be useful information to the evaluators, because very scarce resonance levels are assigned their spins and angular momenta.

New Sprouts

Many papers dealing with the cross-section calculations were submitted in the present Conference. They were reports on the cross-section evaluations for the above-mentioned data files and libraries, papers on the evaluations of the activation cross sections, and on the calculations of the elastic, inelastic and capture cross sections. They have played a leading part of the Conferences in the past, too.

Mingling with these, I saw a paper which calculated the photo-reaction cross sections (C50). The photo-reaction data may be one of the candidates for the important future subjects of the nuclear data evaluations.

Medium and high energy nuclear data evaluation is also a forthcoming important subject. In this Conference, some papers (C43, C51, C53, D52) took this new sprout. However, I felt that ambitious and challenging papers to the new subjects were scarce. I hope many new sprouts will be presented in the next Conference.

Nuclear Reaction Models

Needless to say, the multi-step process is the main part of the nuclear reactions in the medium and high energy region. Various models concerning the multi-step direct and multi-step compound processes have been investigated in order to calculate the cross sections for the high energy nuclear reactions.

In the traditional nuclear data evaluation below 20 MeV also, some model codes including the multi-step compound nuclear process as well as the pre-compound nuclear process have been used. Dr. Young (IT15) stressed the importance of the role of the pre-compound processes in the nuclear data evaluation around 14 MeV neutron energy. In fact, there were many papers in this Conference which used the model codes based on these processes, in particular, papers on the neutron cross-section evaluation for fusion.

Professor Weidenmueller reviewed the recent progress of the statistical theory of the nuclear reactions in his impressive talk (IT18), focusing on the multi-step reaction processes. He took three main theoretical models, and compared their characteristics. They were FKK (Feshbach, Kerman, Koonin), TUL (Tamura, Udagawa, Lenske) and NWY (Nishioka, Weidenmueller, Yoshida). Although they are sophisticated quantum mechanical models each, they require time consuming calculations. From a standpoint of the nuclear data evaluation, it is favorable that more results of the model calculations are obtained with shorter computer time, because a huge amount of the calculations must be done for many nuclides in wide energy ranges during a limited computer time. In this sense, it is laborious but valuable tasks to make utilizable nuclear model codes for the nuclear data evaluations using the rigorous, modern and sophisticated theory.

Dr. Koning performed an inter-comparison of the existing multi-step direct reaction models from a quantum mechanical standpoint, and proposed his own new theory with a leading-particle statistics for the multi-step direct reaction (069). Dr. Kalka showed his model code EXIFON which was based on the statistical multi-step reactions (072). In their model calculations, important quantities were interaction strength and level density. I hope such models as they have presented would be widely used for the calculations of the nuclear data evaluations.

Level Density

In general, the level density and the optical potential play principal role for the nuclear model calculations. In the present Conference, there were many papers such as 068, D34, D38, D39, D42, D45 and D46 dealing with the level density with which statistical model calculations for the reaction cross sections were performed. Various models of the level density such as an improved Gilbert-Cameron, Ignatyuk and generalized super-fluid models have been used in these papers. They studied also the shell effects, temperature dependence, effects of the collective motion, etc. Some papers took the back-shifted Fermi gas model.

Dr. Ignatyuk (IT17) reviewed the status of the level density, and stressed the importance of the collective enhancement effects. He showed the effects of the temperature change and quantum fluctuation to the equilibrium deformation and the nuclear shape. These must be much important in the highly excited states.

Although many investigations about the level density have been made, it seems that some vague problems still remain; such as those about the shell effects, the temperature dependent level density parameters, and the effects of the collective motion. Including these problems, I hope that the relations between corresponding parameters in the different models should be more clear in the next Conference.

Optical Potential

For the nuclear data evaluation, the optical potential model is the most fundamental tool for the cross-section calculations. There have been many studies on the systematics of the potential parameters and on the search for the best set of the parameters of the regional and global potentials for both spherical and deformed nuclei. These potentials were used in the evaluation work for the above-mentioned evaluated files and in some contributions (073, C32, C51, D40) presented in this Conference.

As Professor Hodgson mentioned in his nice invited talk (IT16), it is difficult, rather impossible in a sense, to get any reliable and satisfactory set of the potential parameters by the traditional way of the parameter search. One reason behind this difficulty is that the real and imaginary parts of the optical potential are determined indirectly with each other. This defect has been overcome partly applying the dispersion relation of the optical potential (073).

Professor Hodgson presented the recent development on the dispersion relation between the real and imaginary parts of the optical potential, and stressed that further improvement of the potential parameters would be attained using this relation together with certain moments of the integrated potential per nucleon. It is desirable that new systematic sets of the potential parameters would be looked for by this advanced way.

Sensitivity and Uncertainty Tests

Although nuclear models and parameters used in the nuclear data evaluations are those given as precisely and rigorously as possible, they may still include their limitation, some unexpected uncertainties and hidden faults. Nuclear model codes have also inheritances of these drawbacks, in general, depending on the approximations of the calculation method, default values of the parameters included, etc. Hence, inter-comparison of the model codes (D32), exploration of their uncertainty sources (057) as well as sensitivity tests of the model parameters (D33, D34) were useful to make these points clear.

I have roughly summarized the papers assigned me. All the papers were wonderful and valuable. I wanted to touch all of them, but it was beyond my ability to practice such a great plan.

Before closing my summary, I would like to express my sincere thanks to the members of the Organizing Committee and the Conference Secretariate for their outstanding Conference arrangement and warmhearted nice hospitality. In particular, I would like to thank Dr. Qaim for his ceaseless efforts to make this Conference successful. I also thank you all the contributors who took part in making the Conference vital and fruitful.

CLOSING REMARKS

A.B. Smith (ANL, Argonne, USA)

Ladies and Gentlemen,

It is my outstanding privilege and distinct honor to chair the closing session of this Conference. Over well more than a quarter of a century it has been my opportunity and my education to participate in a number of conferences of this type. It is at such conferences that I have established many lifelong professional and personal friendships. I recall attending the first of these conferences, in the late '50s or early '60s, at Columbia University in the lecture room of the Pupin Laboratory. Even then, as now, there was good foreign attendance from what later became known as "east" and "west" blocks. At that time neutrons were on the "cutting edge" of science, and I seem to recall that several Nobel Laureates were present (Wigner, Rabbi and perhaps Rainwater), as well as other notable scientists who played important roles in the historical development of nuclear energy (such as Dunning, Newson and Weinberg). In subsequent years, members of the research group I remain responsible for included individuals who were on the squash court under Stagg Field that memorable December 2, 1942. Since then, Nuclear Data (and I) have "matured", and both of us are facing some transitions. Our success in negotiating them will very likely determine our respective futures, and both of us are being questioned. Of little note is whether or not this aging man can gracefully fade from the scene. It may well be critical to the future of mankind that public confidence be restored to the application of nuclear processes to energy and other benefits, with all the consequent connotations of safety, energy density and benign environmental impact. Success will require continued and comprehensive understanding of the underlying physical properties -- "nuclear data" if you will. It is this understanding that has been so well addressed at this conference. The issue may remain in doubt, but it is encouraging to note the younger faces evident here (among them may again be a Nobel Laureate), and there is apparent a new and encouraging international breath of the endeavor with wide representation from around the world.

In this final session of the Conference, we have distinguished speakers for summary and guidance. Dr. Robert Peelle summarizes the energy-related aspects of the Conference. It is that area that has long sustained nuclear data, and where there remain critical outstanding issues, particularly relevant to the public acceptance of nuclear energy. Non-energy related nuclear data is relevant to a diversity of applications including our health, environment and technological industries. In my view, this aspect of nuclear data should rapidly increase. Professor Otto Schult will ably summarize the non-energy aspects of this Conference. A major portion of this Conference dealt with the calculational and evaluational aspects of nuclear data. These activities are of increasing importance and form an essential interface between the fundamental studies and the applications usage. Dr. Sin-iti Igarasi will give a good summary of the calculational and evaluational aspects of this Conference.

The final remarks from this Chair must be directed to Dr. Syed Qaim, Conference Chairman. I am certain that you all join me in thanking Dr. Qaim for the very impressive effort that has made this Conference so successful. He has set a standard for technical excellence and hospitality that will be very difficult to match. Able though he is, Dr. Qaim could not have succeeded without the support of this institution. Therefore, I would like to ask him to convey our appreciation to the senior management of the Jülich Laboratory, and also to the organisational support staff (co-workers of the Institute of Nuclear Chemistry and Mr. Krahl-Urban and his lovely ladies) who gave such careful attention to our every day needs.

Finally, there are the speakers of this session, all sessions, and the attendees. They too are essential ingredients for success. To each of you I offer my best personal wishes. May you have a safe return to your homes, and a challenging and productive future.

S.M. Qaim (KFA Jülich, Germany)

Ladies and Gentlemen,

I would like to thank you heartily for the overwhelming and warm expression of appreciation shown to me.

We are reaching the end of the conference and for me now remain conjectural impressions of this event. Time will show whether this conference was worthwhile. Some of the projected newer ideas, themes and trends, both in format and content of the conference, may find acceptance or may fall altogether.

The conference was organized through the help and support of many. I would first of all like to thank the International Programme Committee and the International Advisers through whose guidance, cooperation, constructive criticism and useful advice the scientific programme of the conference was established. I am grateful to the OECD-Nuclear Energy Agency (Paris) and the International Atomic Energy Agency (Vienna) for good cooperation. I also thank the invited speakers and session chairmen for their efforts. My special appreciation to the summary speakers who accepted the arduous task and performed it with great skill.

Turning to the home institution I would like to express my gratitude to the Board of Directors of the Research Centre Jülich for providing financial support and the necessary conference infrastructure to hold the meeting. My special thanks are due to the Head of the Institute of Nuclear Chemistry for constant encouragement, scientific counsel and technical advice. The secretariat and staff of the Institute as well as several young students worked hard and helped considerably at various stages of organization.

The Organizing Committee deserves special thanks. In particular I would like to thank the Conference Service for handling all the organizational matters. The Conference Secretary, Mrs. R. Mengels, worked with great efficiency, diligence, and, in spite of stress, with extreme courtesy. We should express our appreciation to her.

Finally, ladies and gentlemen, I would like to make an announcement as Chairman of NEANDC. The US Nuclear Data Committee made a proposal about the next conference. This proposal was approved by the NEANDC and the next conference will be held in the Oak Ridge area in the spring of 1994.

I wish you now a safe journey home and all the best for the future.

Index of Authors